# *the* ANTS

# *the* ANTS

*Bert Hölldobler*
*and*
*Edward O. Wilson*

*The Belknap Press of*
*Harvard University Press*
*Cambridge, Massachusetts*

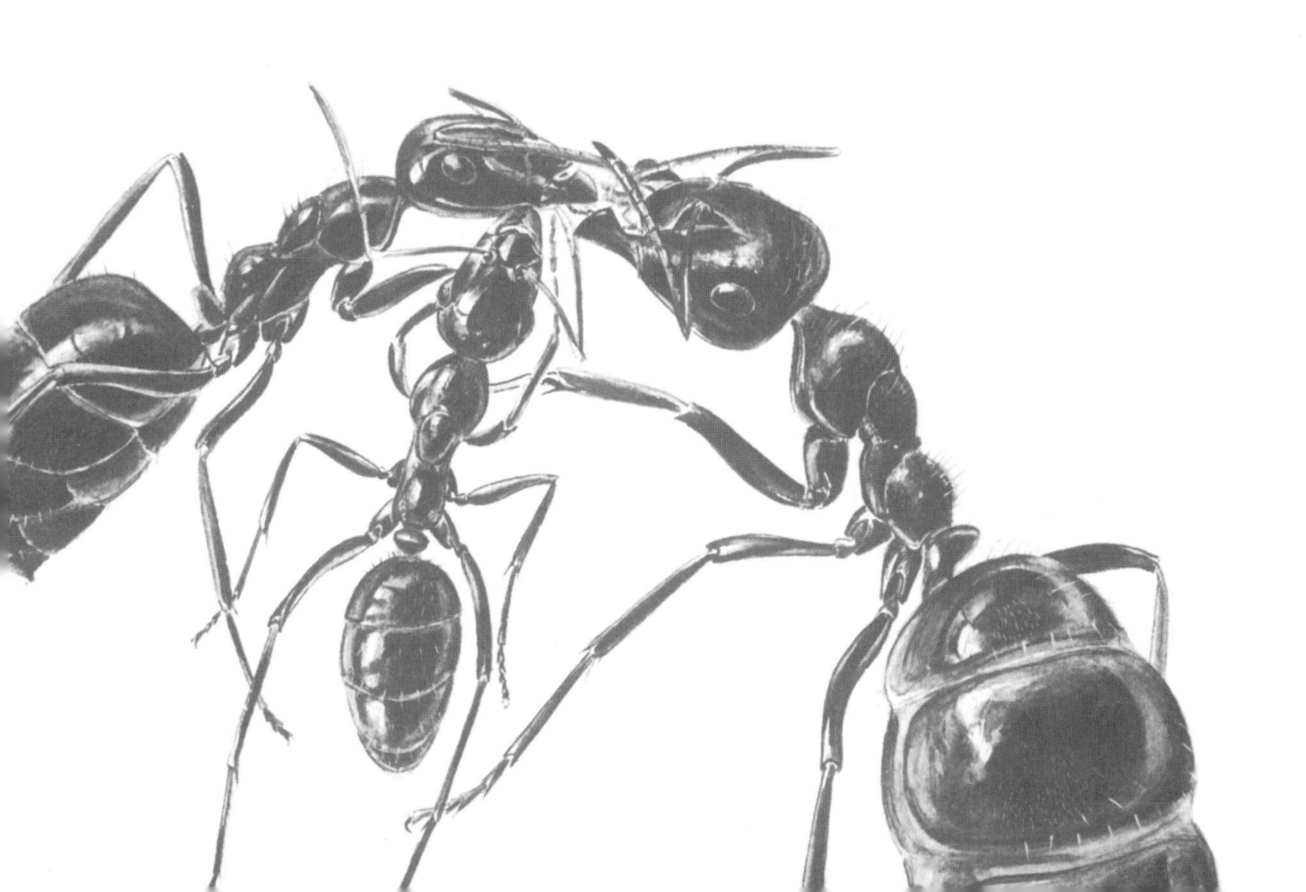

Printed in the United States of America

This book is printed on acid-free paper, and its binding materials have been chosen for strength and durability.

Designed by Marianne Perlak in Linotron Palatino.

Unless otherwise indicated, all artwork is by the authors.

Library of Congress cataloging information is on page 733.

*For the next generation of myrmecologists*

# Contents



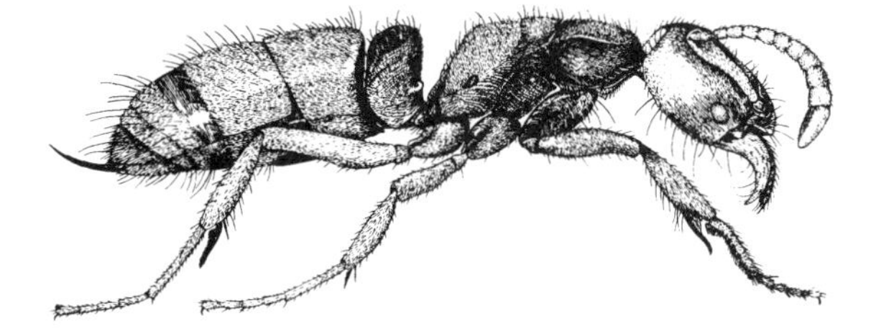

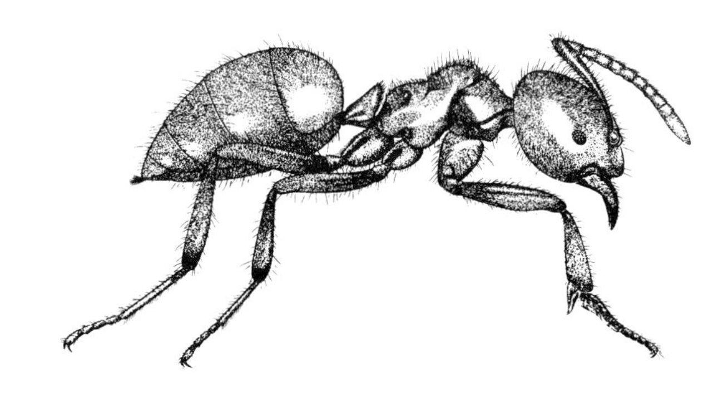

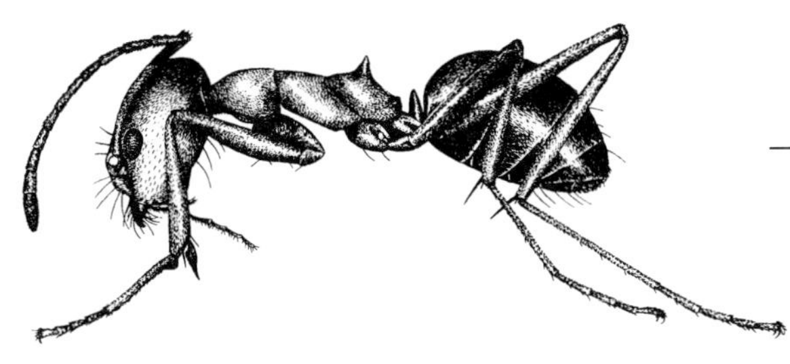

## 11. *The Organization of Species Communities* 419



## 12. *Symbioses among Ant Species* 436



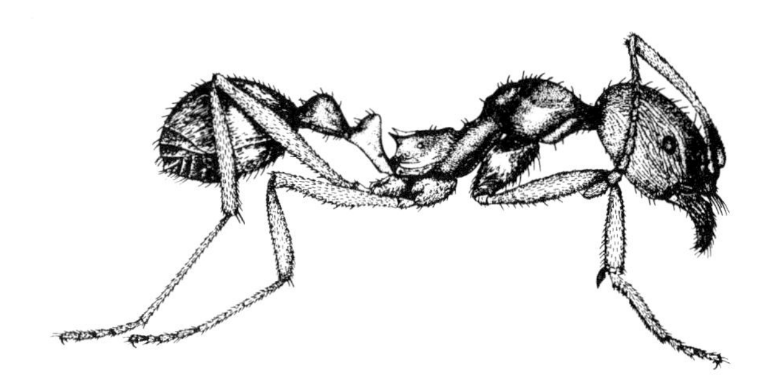

## 13. *Symbioses with Other Arthropods* 471



## 14. *Symbioses between Ants and Plants* 530

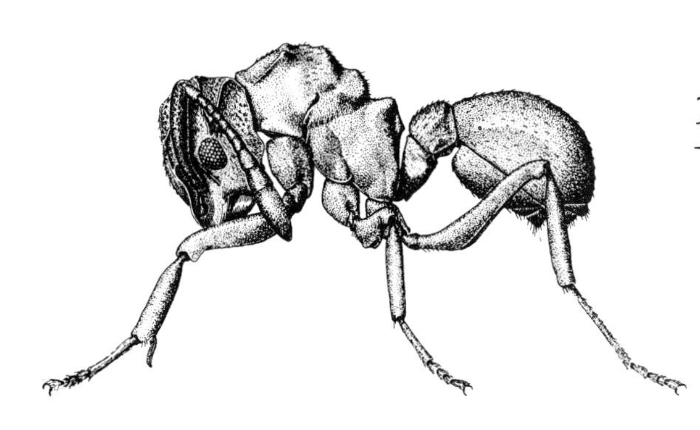

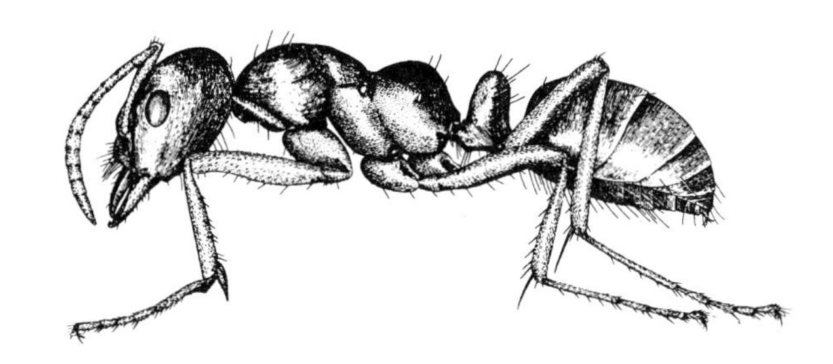

# *the* ANTS

# CHAPTER 1

# The Importance of Ants

Ants are everywhere, but only occasionally noticed. They run much of the terrestrial world as the premier soil turners, channelers of energy, dominatrices of the insect fauna—yet receive only passing mention in textbooks on ecology. They employ the most complex forms of chemical communication of any animals and their social organization provides an illuminating contrast to that of human beings, but not one biologist in a hundred can describe the life cycle of any species. The neglect of ants in science and natural history is a shortcoming that should be remedied, for they represent the culmination of insect evolution, in the same sense that human beings represent the summit of vertebrate evolution.

During the second half of the Paleozoic Era a sequence of three events occurred that predetermined the character of the modern insect fauna and set the stage for the origin of ants (F. M. Carpenter, 1989). The first was the invention of flight, never before achieved by any group of organisms in the history of life. One insect line out of the many existing in the Paleozoic somehow evolved wings in the adult stage, and it soon afterward enjoyed extensive species formation and diversification. Then a single line within this expanding "paleopterous" group attained the ability to fold the wings back over the body when not in use, giving the adults greater mobility on the ground after alighting. From this "neopterous" group, which also prospered, came still another line that attained complete metamorphosis. Now highly specialized larvae occupying one ecological niche could be transfigured into radically different adult forms occupying another ecological niche. This third, holometabolous group of insects exceeded its predecessors in evolutionary attainment. It radiated extensively to produce the most diverse and abundant insect orders now in existence, namely the beetles (Coleoptera), the flies (Diptera), and the bees, wasps, and their relatives (Hymenoptera).

The story did not quite end at this point. During Cretaceous times, spanning 140 million to 65 million years before the present, a fourth and wholly different kind of event occurred, the origin of advanced social life. All of the insects in the fourth category, comprising the termites, the ants, and some of the bees and wasps, are colonial. More precisely, they are eusocial in their habits. This means that two or more generations overlap in the society, adults take care of the young, and, most importantly, adults are divided into reproductive and nonreproductive castes, in other words queens and kings versus workers (Michener, 1969, 1974; Wilson, 1971).

To a degree seldom grasped even by entomologists, the modern insect fauna has become predominantly social. Recent measurements suggest that about one-third of the entire animal biomass of the Amazonian *terra firme* rain forest is composed of ants and termites, with each hectare of soil containing in excess of 8 million ants and 1 million termites. These two kinds of insects, along with bees and wasps, make up somewhat more than 75 percent of the total insect biomass (Beck, 1971; Fittkau and Klinge, 1973). Ants and termites similarly dominate the forests and savannas of Zaïre (Dejean et al., 1986). Although comparable biomass measurements have not yet been made elsewhere, it is our subjective impression that the eusocial insects, ants foremost among them, are comparably abundant in most other principal habitats around the world.

For example, on the Ivory Coast savanna the density of ants is 7,000 colonies and 20 million individuals per hectare, with one species alone, *Camponotus acvapimensis,* accounting for 2 million of the individuals (Lévieux, 1966, 1982). Such African habitats are often visited by driver ants (*Dorylus* spp.), single colonies of which occasionally contain more than 20 million workers (Raignier and van Boven, 1955). And the driver ant case is far from the ultimate. A "supercolony" of the ant *Formica yessensis* on the Ishikari Coast of Hokkaido was reported to be composed of 306 million workers and 1,080,000 queens living in 45,000 interconnected nests across a territory of 2.7 square kilometers (Higashi and Yamauchi, 1979).

The local diversity of ants is substantial, far exceeding that of other social insects and reflecting the manner in which ant species have evolved to saturate a wide range of feeding niches in the soil and vegetation. In lowland rain forest at the Busu River, northeastern Papua New Guinea, Wilson (1959c) collected 172 species of ants belonging to 59 genera in an area of about 1 square mile (2.6 $km^2$). Barry Bolton (in Room, 1971) recorded 219 species in 63 genera in a square mile of cocoa plantation and forest at Tafo, Ghana, while Kempf (1964a and personal communication) found 272 species belonging to 71 genera in a comparable area at Agudos, São Paulo State, Brazil. During two years of fieldwork, Manfred Verhaagh (personal communication) collected at least 350 species belonging to 71 genera at the Rio Yuyapichis, in the larger valley of the Rio Pachitea, Peru; the western Amazon Basin, in which this watershed is located, may have the richest ant fauna in the world. Moving to more limited sample spaces, Room (1971) recorded 48 genera and 128 species from only 250 square meters in a cocoa farm in Ghana. Wilson (1987c) identified 43 species in 26 genera in a single *tree* in the Tambopata Reserve in the Peruvian Amazon. If the terrestrial fauna had been assayed around the tree as well, the total point diversity would probably have rivaled that of Room's Ghanaian sample. Temperate faunas are less rich but often impressive nonetheless: 23 genera and 87 species in 5.6 square kilometers at the E. S. George Reserve in Michigan (Talbot, 1975), and 30 genera and 76 species in 8 square kilometers of the Welaka Reserve, Florida (Van Pelt, 1956).

The impact of ants on the terrestrial environment is correspondingly great. In most terrestrial habitats they are among the leading predators of other insects and small invertebrates (Wilson, 1971; Jeanne, 1979; Lévieux, 1982; Sörensen and Schmidt, 1987). Leafcutter ants, in other words members of the genera *Acromyrmex* and *Atta*, are species for species the principal herbivores and the most destructive insect pests of Central and South America (Weber, 1972; Cherrett, 1982). *Pogonomyrmex* and other harvester ants rank among the principal granivores, competing effectively with mammals for seeds in deserts of the southwestern United States (Davidson et al., 1980). In another adaptive zone, ants are sufficiently dense to reduce the abundance of ground-dwelling spiders and carabid beetles, especially when these arthropods are specialized to live in the soil and rotting vegetation (Darlington, 1971; Cherix, 1980; Wilson, 1987b). Where montane habitats are high enough to be mostly free of ants, such as the summit of Mt. Mitchell in North Carolina and the Sarawaget Mountains of Papua New Guinea above 2,500 meters, carabids and spiders increase markedly in numbers.

It is not surprising to find that ants also alter their physical environment profoundly. In the woodlands of New England, they move approximately the same amount of soil as earthworms, and they surpass them in tropical forests (Lyford, 1963; Abe, 1982). In the temperate forests of New York, they are responsible for the dispersal of nearly one-third of the herbaceous plant species, which in turn constitute 40 percent of the aboveground biomass (Handel et al., 1981). They aid in the spread of forest vegetation onto bare rocks in Finland (Oinonen, 1956) and foredune vegetation onto salt lakes in the USSR (Pavlova, 1977). Because ants transport plant and animal remains into their nest chambers, mixing these materials with excavated earth, the nest area is often charged with high levels of carbon, nitrogen, and phosphorus. The soil surface is consequently broken into a mosaic of nutrient concentrations, and this in turn creates patchy distributions of plant growth, especially during the early stages of succession (Beattie and Culver, 1977; Pętal, 1978; Briese, 1982b). The great earthen nests of some of the leafcutter ants belonging to the genus *Atta* have a particularly strong impact on local environments. In tropical rain forests, where less than 0.1 percent of nutrients normally filter deeper than 5 centimeters beneath the soil (Savage, 1982), the leafcutter workers carry large quantities of freshly cut vegetation into nest chambers as deep as 6 meters. Haines (1978) found that the flow of 13 elements through the underground refuse dumps of *Atta colombica* was 16 to 98 times the flow in undisturbed leaf litter beneath equivalent sample areas. The enrichment of materials resulted in a fourfold increase in the quantity of fine tree roots in the dump. Energy flow through the *A. colombica* nests was about ten times greater on a per-square-meter basis than in forest areas away from the nests. Other ants turn and modify the soil in ways that have just begun to be assessed, and to an especially important degree in deserts, savannas, and tropical forests (Lévieux, 1976d; Pętal, 1978; Whitford et al., 1986). According to Graedel and Eisner (1988; see also Monastersky, 1987), formicine ants may be responsible for much of the formic acid found in previously unexplained quantities in the atmosphere above the Amazon forest and other habitats rich in these insects. Graedel and Eisner estimate, very roughly, that formicine ants may release $10^{12}$ grams of formic acid globally each year.

The abundance and ecological dominance of ants are matched by their extraordinary geographic range. Various of the approximately 8,800 known species are found from the arctic circle to the southernmost reaches of Tasmania, Tierra del Fuego, and southern Africa. The only places free of native species are Antarctica, Iceland, Greenland, Polynesia east of Tonga, and a few of the most remote islands in the Atlantic and Indian oceans (Wilson and Taylor, 1967b). Four genera (*Camponotus, Crematogaster, Hypoponera,* and *Pheidole*) extend individually over most of this vast range (Wilson, 1976e).

Some species of ants have adapted very well to even the most disturbed habitats. Most cities in the tropics are homes to "tramp species," forms that have been carried worldwide by human commerce. The little myrmicine *Tetramorium simillimum* is equally likely to turn up in an alley in Alexandria or on a beach in Tahiti. "Crazy ants" (*Paratrechina longicornis*) swarm under debris in vacant lots; colonies of the tiny dolichoderine *Tapinoma melanocephalum* nest in abandoned plumbing, dead plant stems, and even soiled clothing. Pharaoh's ants (*Monomorium pharaonis*) are worldwide household pests. Their vast, multi-queened colonies thrive in wall spaces and detritus. In hospitals they often visit soiled bandages and track pathogenic microbes onto clean dressings and food. A notorious colony occupied the entire Biological Laboratories of Harvard University during the 1960s and 1970s. An extermination campaign was finally undertaken when workers were discovered carrying radioactive chemicals from culture dishes into the surrounding walls. (The incident was made the basis of the melodramatic scientific novel *Spirals,* by William Patrick, Houghton Mifflin, Boston, 1983.)

Ants are resistant to hard radiation. Colonies exposed to intense cesium-based irradiation in a French forest suffered no evident decline or change in behavior during 11 months, even when some of the surrounding plants were dying or losing their leaves (Le Masne and Bonavita-Cougourdan, 1972). At least some ant species are also highly resistant to industrial pollution. Near a nitrogen plant in Poland, populations of *Myrmica ruginodis* and *Lasius niger* remained robust after other invertebrates became scarce. They actually reduced the concentration of the nitrate, apparently by stimulation of microorganisms that bind the pollutant (Pętal, 1978).

Surprisingly, some species are even able to survive under water. Queens and workers of *Formica* species can live for up to 14 days or longer while submerged, during which time they are in an anesthetized condition and their oxygen consumption falls to between 5 and 20 percent of the usual resting rate. Oxygen consumption under water is highest in *Formica uralensis,* which lives in bogs and is most likely to suffer periodic flooding of the nest (Gryllenberg and Rosengren, 1984). In preliminary experiments, E. O. Wilson (unpublished) found that *Cardiocondyla venustula* colonies living close to water on islands in the Florida Keys can withstand submergence in salt water for at least several hours.

Ants offer special advantages for some important kinds of basic biological research. The colony is a superorganism. It can be analyzed as a coherent unit and compared with the organism in the design of experiments, with individuals treated as the rough analogues of cells. The aims of much of contemporary research on ants, as well as that on other kinds of social insects, are first to identify more fully the mechanisms by which colony members differentiate into castes and divide labor, and second to understand why certain combinations of these mechanisms have generated more successful products than others. A larger hope is that more general and exact principles of biological organization will be revealed by the meshing of information from insect sociobiology with equivalent information from developmental biology. The definitive process at the level of the organism is morphogenesis, the set of procedures by which individual cells or cell populations undergo changes in shape or position incident to organismic development. The definitive process at the level of the colony is sociogenesis, the procedures by which individuals undergo changes in caste, behavior, and physical location incident to colonial development. The question of interest for general biology is the nature of the similarities between morphogenesis and sociogenesis.

The study of ant social organization is by necessity both a reductionistic and a holistic enterprise. The behavior of the colony as a whole can be understood only if the programs and positional effects of the individual members are both specified and explained more deeply at the physiological level. But such accounts are still far from complete. The information makes full sense only when the colonial pattern of each species is examined as an idiosyncratic adaptation to the natural environment in which the species lives (Wilson, 1971; Sudd and Franks, 1987).

At both the individual and the colonial levels, social insects offer great advantages over ordinary organisms for the study of biological organization. Although it is virtually impossible to dissect a higher organism into its constituent parts for study and then put it back together again, alive and whole, this can easily be done with an insect colony. The colony, in other words the superorganism, can be subdivided into any conceivable combination of sets of its members. It can then be manipulated experimentally and reconstituted at the end of the day, unharmed and ready for replicate treatment at a later time. One technique used successfully for the analysis of optimization in social organization is the following. The colony is modified by changing caste ratios, as though it were a mutant. The performance of this "pseudomutant" is compared with that of the untransformed colony as well as with that of other modified versions. The same colony can be turned repetitively into pseudomutants in random sequences on different days, eliminating the variance that would otherwise arise from between-colony differences (Wilson, 1980b). Fragments of colonies can be separated for more intensive short-term studies of many kinds and then either discarded or rejoined. They can be shifted about in various geometric configurations to study position effects, rather like moving the brain or the liver to novel locations in order to study the effects on other parts of the body.

At the highest level of explanation, that of the ecosystem, the large numbers of kinds of ants (including more than 1,000 species in the ant genus *Pheidole* alone) give a panoramic view of the evolution of colonial patterns. The very exuberance of diversity makes the correlative analysis of adaptation easier and more rigorous. In a subsequent phase of research, hypotheses concerning the functions of body forms, caste systems, and other biological traits can be subjected to experimental tests in the field and in the laboratory. The small size of these insects, as well as the ease with which they can be cultured in the laboratory, facilitates this research. Finally, ants are extraordinary among social animals in the swiftness with which they adapt to radically altered environments in the laboratory and resume normal behavior under the gaze of the investigator.

Ants are premier organisms for research in behavioral ecology and sociobiology. They exemplify principles in these relatively new disciplines and offer exceptional opportunities for the testing and extension of theory. They provide, for example, some of the best documentation of the following phenomena:

- Kin selection and selection at the level of the colony.
- Competition at each of the three levels of organization: among individuals of the same colony, among colonies of the same species, and among species.
- The effects of competition on community structure.
- The shaping of the organization and the development of societies by natural selection.
- The shaping of castes by natural selection to create an "adaptive demography," in which the size and age distributions of colony members contribute to the genetic fitness of the colony as a whole.
- The nature of physiological and behavioral regulatory processes in social organization.
- Hierarchy in control processes.

Ants and other social insects have been underutilized in textbooks and the review literature on behavior and ecology, which tend to favor vertebrate examples. This has been true even when, as in the case of pheromones and kin selection, the basic concepts were pioneered during studies on insects. The disproportion has two causes. First, vertebrates attract more enthusiasts among scientists and naturalists simply because they are more nearly human in size. This bias is understandable but is not always a wise research strategy. The strongest efforts by gifted investigators often yield results that could have been achieved in less time and more convincingly with social insects. The second reason for the relative neglect of social insects is that their study seems more "technical." That is, it requires a specialized knowledge of anatomy, physiology, and other topics of entomology not ordinarily acquired by biologists during their basic university education.

We confess to having written this book mainly to celebrate a personal muse, perhaps the least explicable yet most sympathetic reason for writing any book. But in so doing we have tried to reorganize myrmecology into a form that addresses biological principles even as it makes ants, the paragons of the insect world, more accessible for future study by others.

# CHAPTER 2

# Classification and Origins

The ants are classified as a single family, the Formicidae, within the order Hymenoptera, which also includes the bees, wasps, sawflies, ichneumons, and similar forms. The known living ants comprise 11 subfamilies, 297 genera, and approximately 8,800 species (see Table 2–1). The rate of discovery of genera since 1940 and the alacrity with which incontestable novelties are publicized suggest that fewer than 100 genera are likely to remain unrecognized. In contrast, the number of undescribed species is immense. Many parts of the tropics still remain poorly collected, and this is especially true of the moist continental forests. "Sibling" species, that is, populations that are reproductively isolated but difficult to distinguish by means of ordinary anatomical traits, are notoriously rife among the ants. They often come to light during biometrical studies of large samples. A few have even been identified primarily or solely by differences in chromosomes or electrophoretically separated allozymes. Examples of complexes that have been "broken open" by these techniques occur in *Amblyopone, Aphaenogaster, Camponotus, Conomyrma, Myrmica,* and *Rhytidoponera,* and many more are thought to exist (Ward, 1980; Crozier, 1981). An extreme case of imperfect classification is provided by *Pheidole,* the most speciose of all ant genera. There are 524 currently unchallenged names in the New World alone, of which as many as 300 may represent distinct species, with several hundred additional, undescribed species present in collections. Similar "problem" genera, any one of which could be called the *crux myrmecologorum,* include *Camponotus, Crematogaster, Iridomyrmex,* and *Solenopsis.* Turning to an entire fauna, R. W. Taylor (personal communication) estimates that only one-fourth to one-third of the ants of Australia have been described. Overall, it is quite possible that 20,000 or more species of ants, constituting as many as 350 genera, exist in the world.

**TABLE 2–1** The number of living ant species described to the present time. The worldwide total must be regarded as only a rough estimate, because of uncertainties concerning the African fauna and the existence of a small percentage of species that occur in more than one region.

| Region | Number of species | Authority |
|---|---|---|
| Neotropical (West Indies, all of Mexico, Central and South America) | 2,162 | Kempf (1972b) |
| Nearctic (North America north of Mexico) | 580 | Smith (1979) |
| Europe | 180 | Bernard (1968) |
| Africa (sub-Saharan) | 2,500 | B. Bolton (rough estimate, personal communication, 1986) |
| Asia (temperate and tropical, from Mongolia, Japan, and Afghanistan to New Guinea, New Britain, and New Ireland) | 2,080 | Chapman and Capco (1951) |
| Melanesia exclusive of New Guinea, New Britain, and New Ireland | 275 | R. W. Taylor (personal communication, 1986) |
| Australia | 985 | R. W. Taylor (personal communication, 1986) |
| Polynesia (native only) | 42 | Wilson and Taylor (1967b) |
| Total | 8,804 | |

## THE TAXONOMY OF ANTS

Diagrams of external anatomy, basic to classification, are provided in Figures 2–1 through 2–11. (Internal anatomy, especially that of the exocrine glands used in communication and combat, is reviewed in Chapter 7.) A concise introduction to the higher classification of ants is presented in a list of the subfamilies, tribes, and genera in Table 2–2, and a catalog of regional faunistic studies and checklists in Table 2–3. At the end of the chapter, a glossary of special taxonomic terms is given in Table 2–5, and keys to all of the living subfamilies and genera of the world in Tables 2–6 through 2–13. In addition, we have illustrated at the end of the chapter representatives of virtually every one of the subfamilies and genera known from the worker or queen castes; the relatively few not included there are illustrated in later chapters.

The taxonomy of the world ant fauna is still very incomplete, with much more research needed in every taxonomic category from species to subfamily. There are few serviceable regional monographs.

*text continues on page 21*

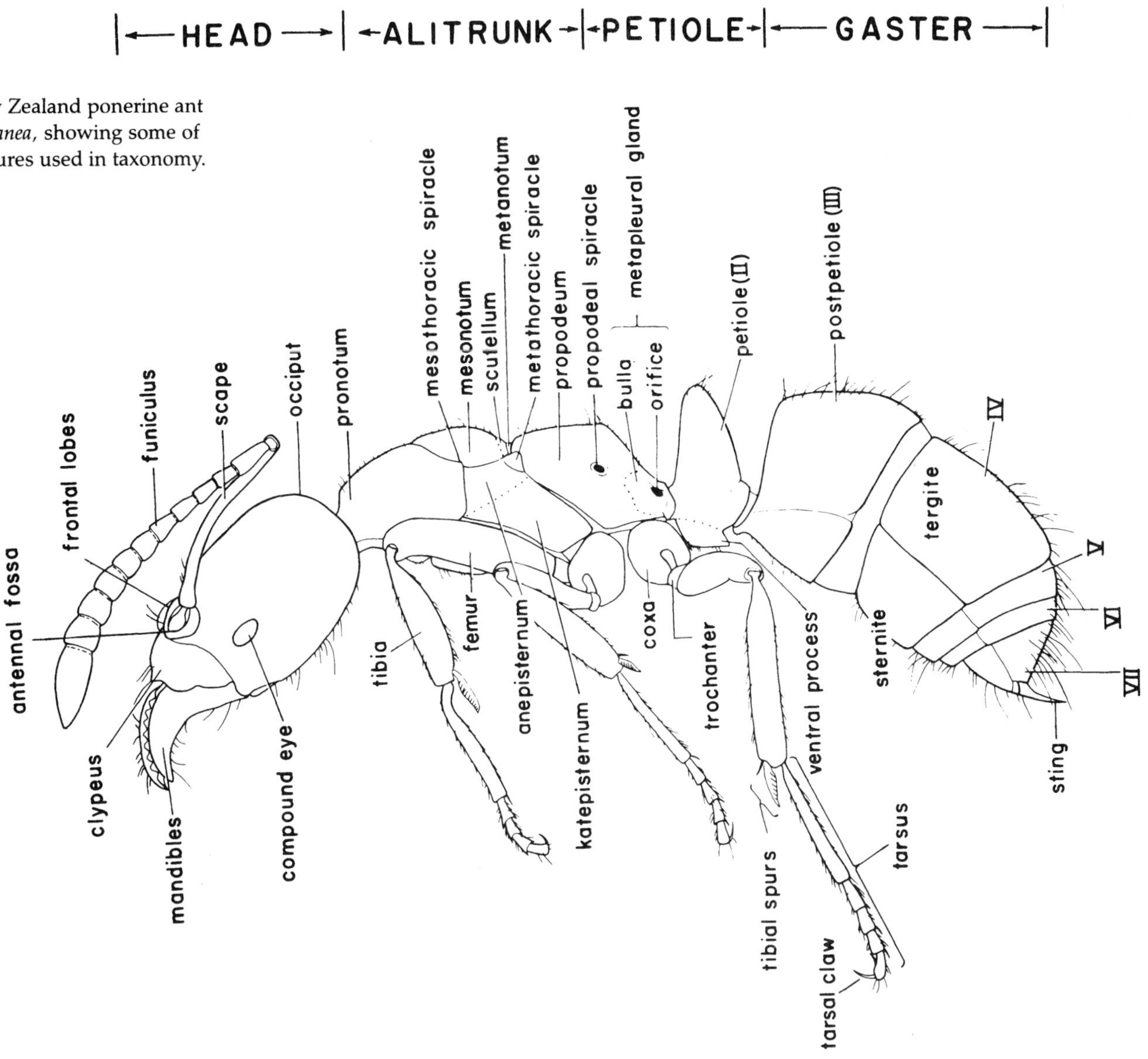

**FIGURE 2–1** Worker of the New Zealand ponerine ant *Pachycondyla* (= *Mesoponera*) *castanea*, showing some of the principal morphological features used in taxonomy. (From Brown, 1958a.)

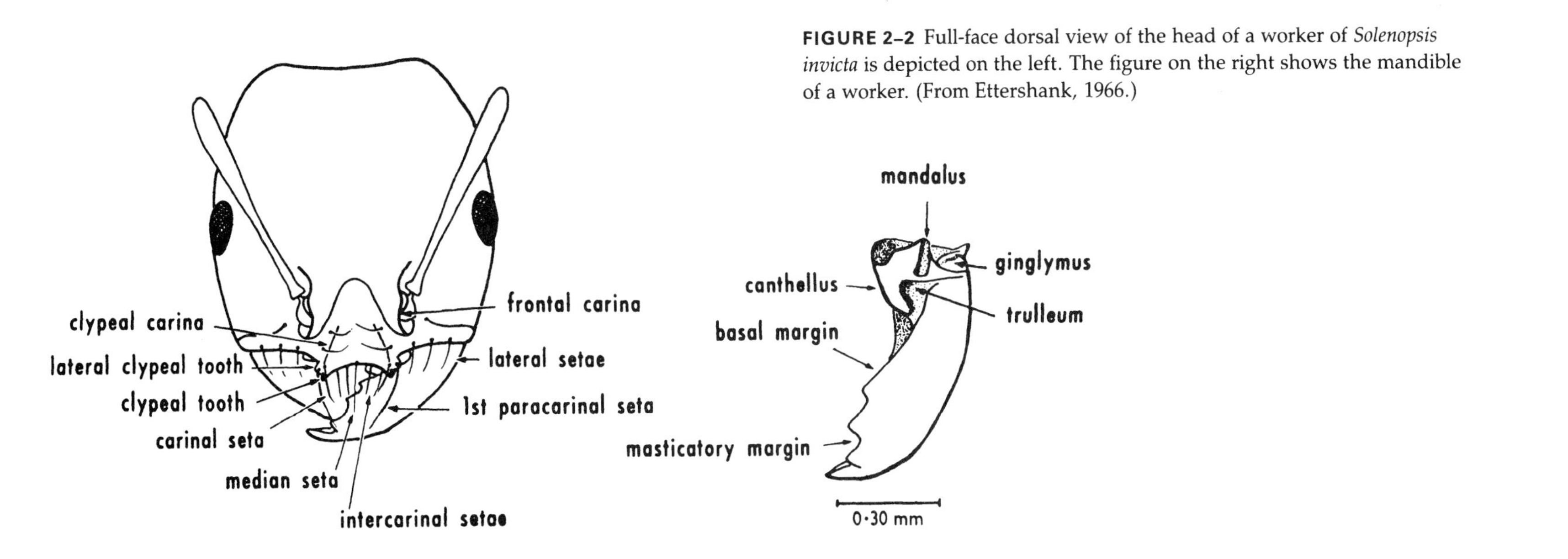


**FIGURE 2–2** Full-face dorsal view of the head of a worker of *Solenopsis invicta* is depicted on the left. The figure on the right shows the mandible of a worker. (From Ettershank, 1966.)

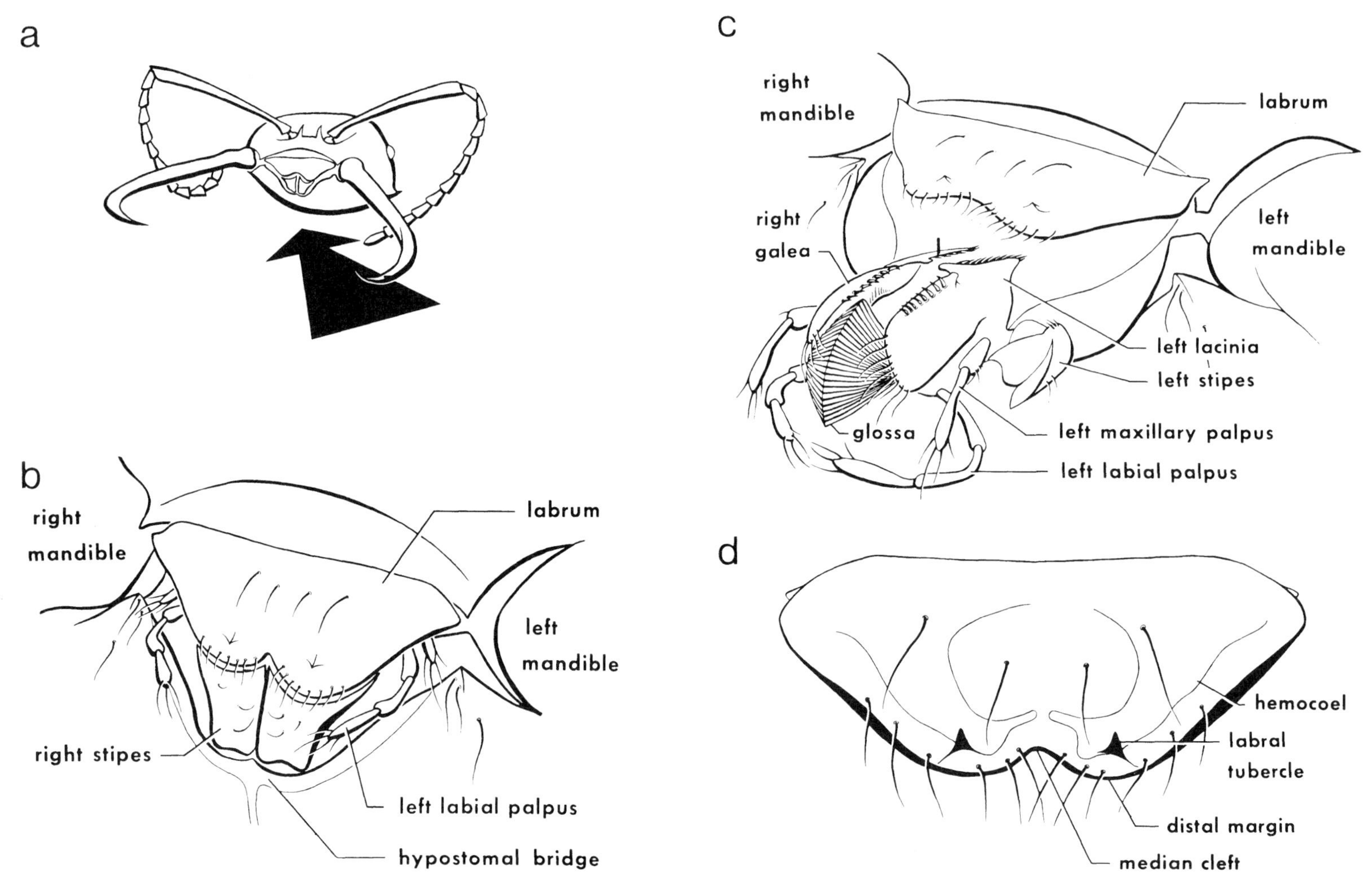


**FIGURE 2–3** Mouthparts of a soldier of *Eciton mexicanum*. (*a*) Anterior view of the head with its maxillo-labial apparatus retracted. The arrow indicates the direction of the view in *b* and *c*. (*b*) Diagrammatic representation of mouthparts in situ, with the maxillo-labial apparatus retracted. (*c*) Mouthparts in situ, with the maxillo-labial apparatus extended. (*d*) External surface of the labrum. (From Gotwald, 1969.)

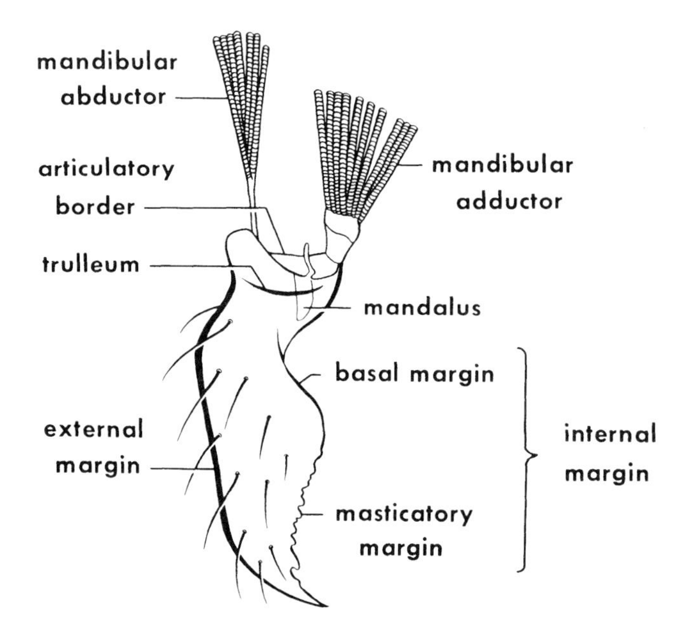


**FIGURE 2–4** Dorsal view of the right mandible of an *Eciton mexicanum* worker. (From Gotwald, 1969.)

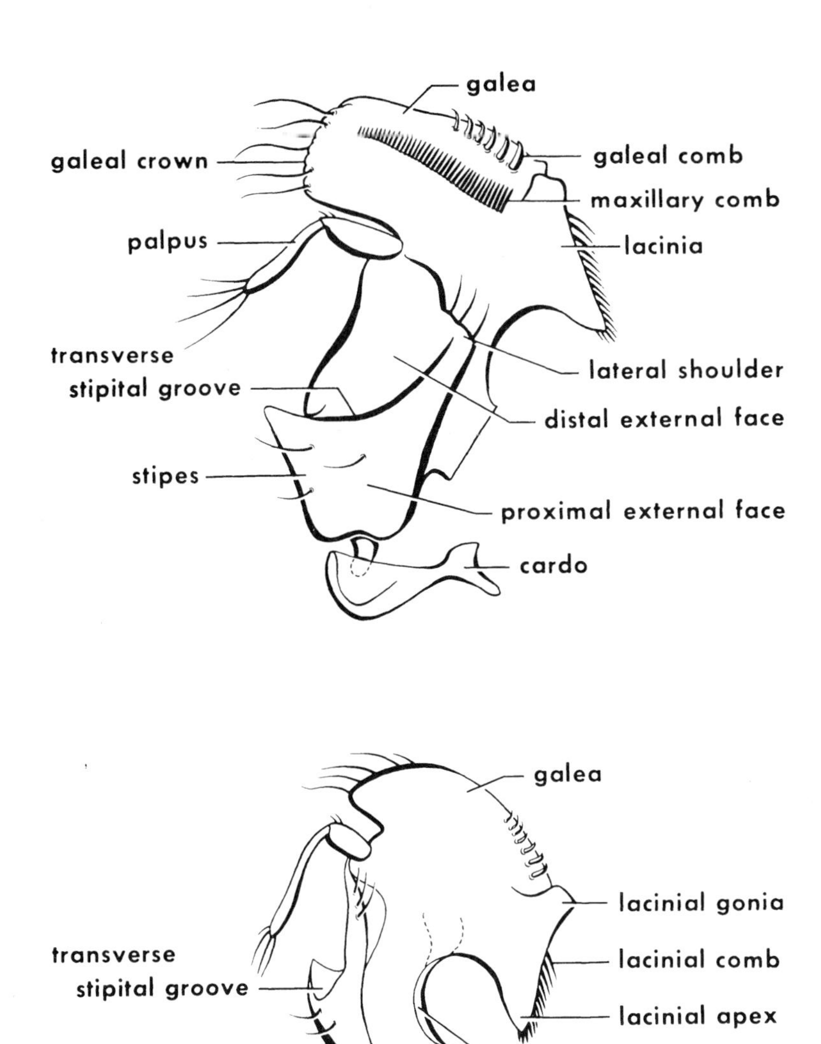


**FIGURE 2–5** *(left)* Left maxilla of a soldier of *Eciton mexicanum*. The diagram above depicts the maxilla somewhat flattened on a microscope slide, showing in greater detail the anatomical structures. The one below illustrates the maxilla (without cardo), showing the natural relationship of the components. (From Gotwald, 1969.)

**FIGURE 2–6** *(below)* Lateral view of the labium of a soldier of *Eciton quadriglume*. On the lower left is a ventral view of the postmentum. (From Gotwald, 1969.)

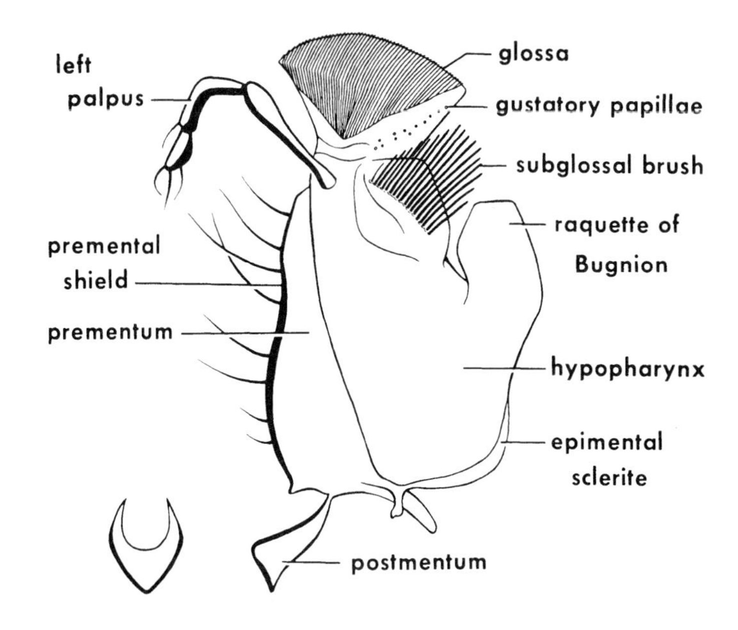


**FIGURE 2–7** Extended mouthparts of a soldier of *Eciton mexicanum*. The diagram on the left presents the dorsal view, that on the right the ventral view. (From Gotwald, 1969.)

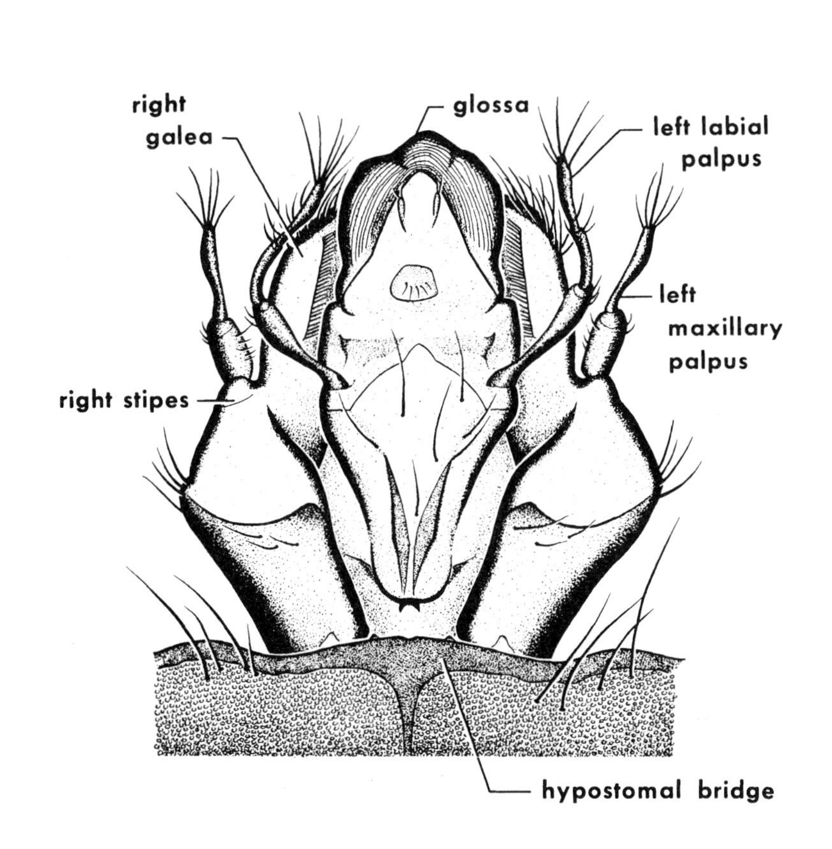


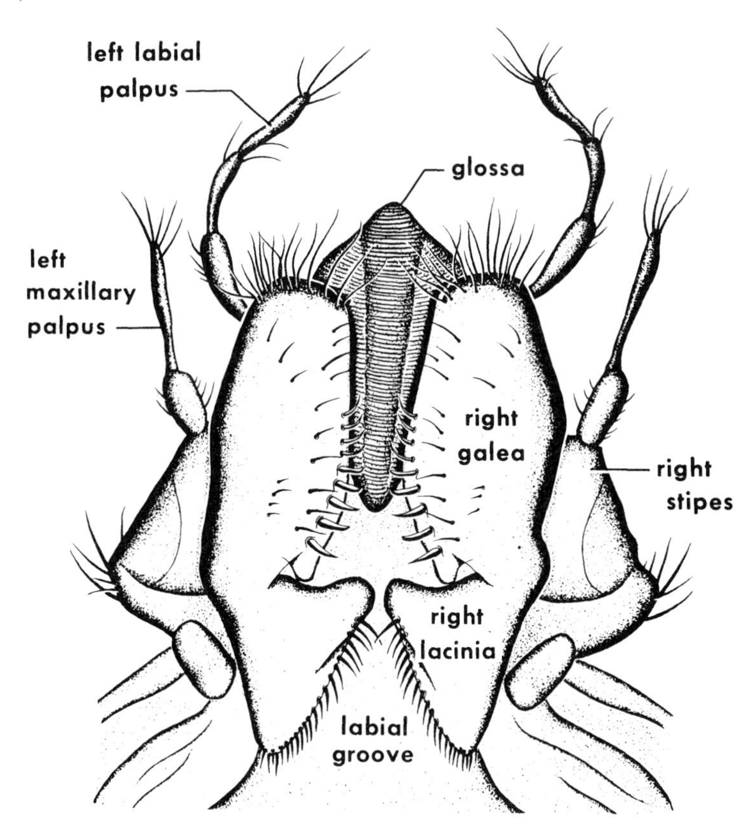

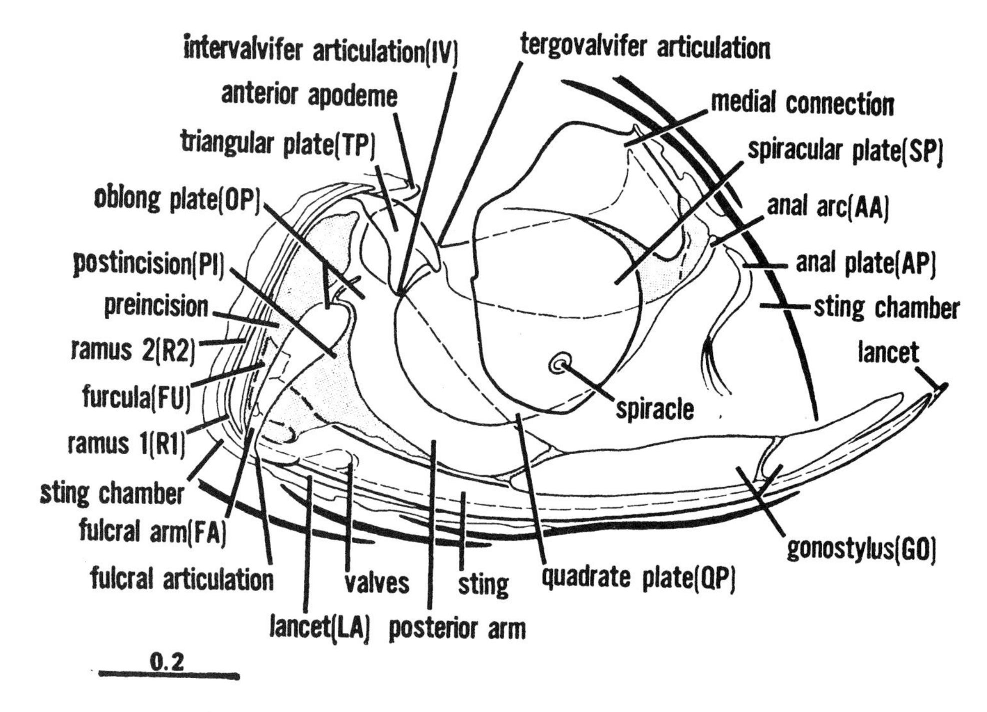


FIGURE 2–9 Sclerites of the sting apparatus of *Amblyopone pallipes*. The scale is in millimeters. (From Kugler, 1978a.)

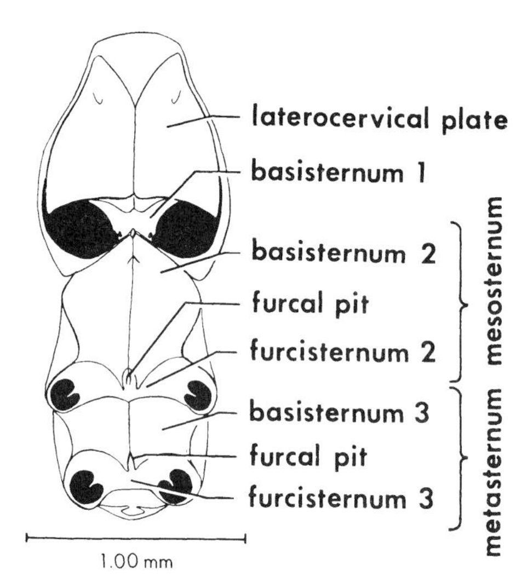


FIGURE 2–8 The ventral view of the alitrunk of *Cheliomyrmex morosus*. (From Gotwald and Kupiec, 1975.)

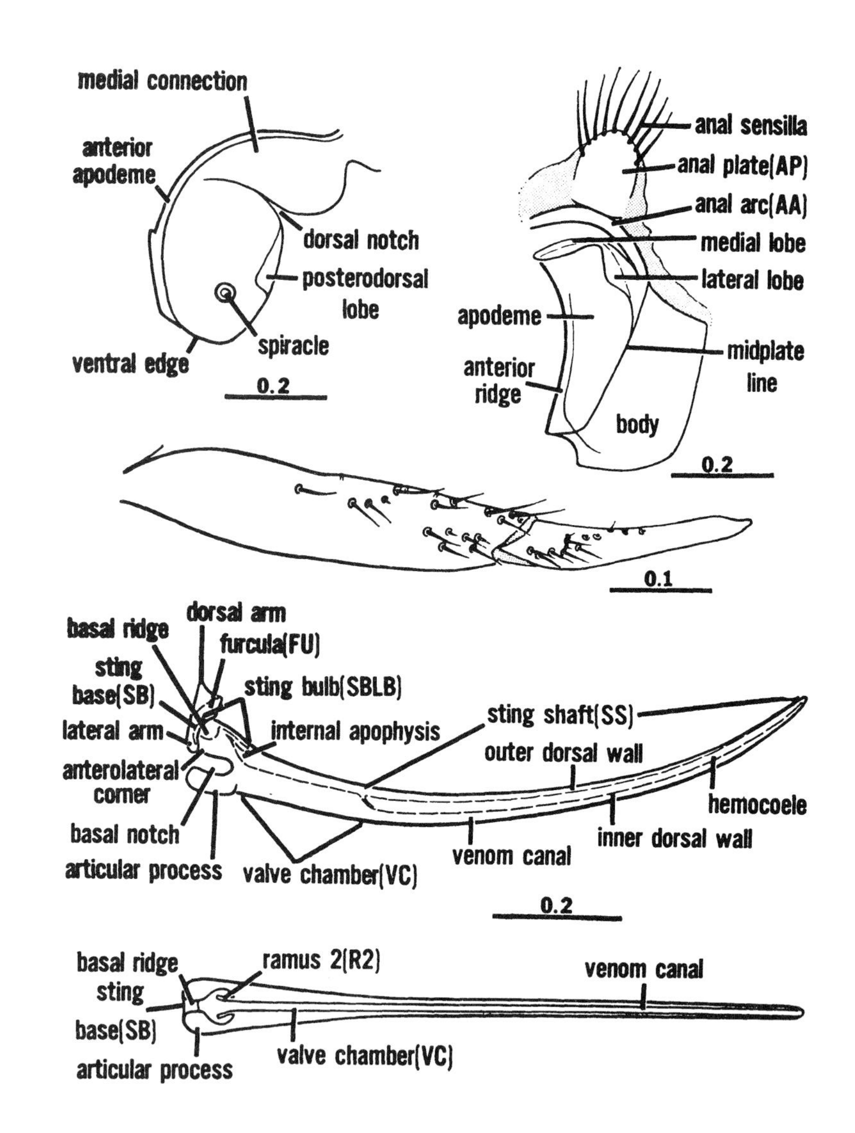


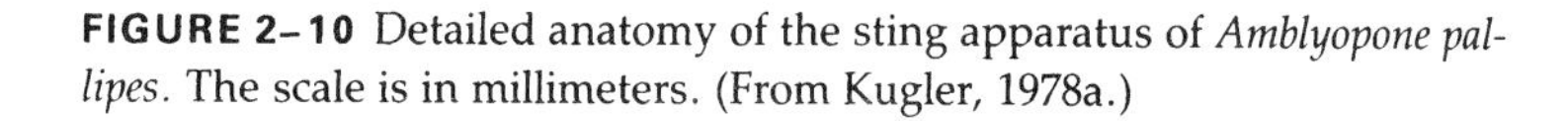
FIGURE 2–10 Detailed anatomy of the sting apparatus of *Amblyopone pallipes*. The scale is in millimeters. (From Kugler, 1978a.)

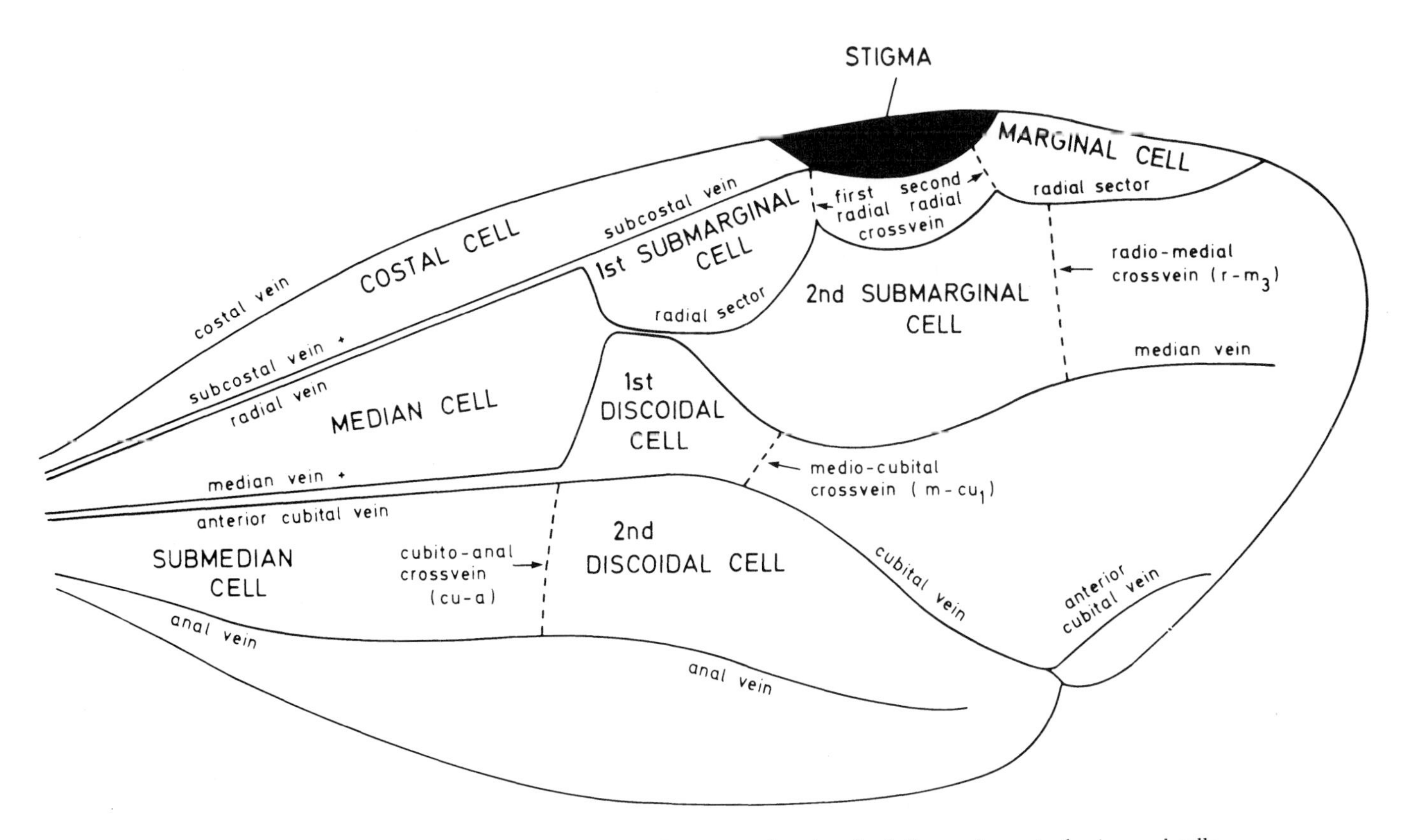


**FIGURE 2–11** Idealized representation of an ant forewing, showing the full complement of veins and cells.

**TABLE 2–2** The subfamilies and genera of living and fossil ants. †, extinct. Zoogeographic regions in which particular living genera occur: AUS, Australian; ETH, Ethiopian; MAL, Malagasy; NEA, Nearctic; NEO, Neotropical; OR, Oriental; PAL, Palaearctic. Extinct genera: ARK, Arkansas, USA, amber (mid-Eocene); BAL, Baltic amber, northern Europe (early Oligocene); BRI, Britain, including Isle of Wight (Oligocene); DOM, Dominican amber, Dominican Republic, West Indies (late Miocene); FLO, Florissant shales, Colorado, USA (Oligocene); SHA, Shanwang shales, China (Miocene); SIC, Sicilian amber, Sicily (Miocene). (Compiled chiefly from Emery, 1891; Wheeler, 1914b; Donisthorpe, 1920; Carpenter, 1930; Viana and Haedo Rossi, 1957; Brown, 1973a; Hong et al., 1974; Snelling, 1981; Wilson, 1985c–f,h; Barry Bolton, personal communication; André Francoeur, personal communication; and R. W. Taylor, personal communication.)

| Genera | Monographs |
|---|---|
| **†SUBFAMILY SPHECOMYRMINAE** | |
| †*Cretomyrma* Dlussky 1975 (Taymyr Peninsula, USSR, Upper Cretaceous) | *Cretomyrma:* Wilson (1987c) |
| †*Sphecomyrma* Wilson and Brown 1967 (New Jersey, United States; Alberta, Canada; and central and eastern Siberia; Upper Cretaceous) SYNONYMS: *Archaeopone* Dlussky 1975; *Armania* Dlussky 1983; *Armaniella* Dlussky 1983; ? *Cretopone* Dlussky 1975; *Dolichomyrma* Dlussky 1975; *Paleomyrmex* Dlussky 1975; ? *Petropone* Dlussky 1975; *Poneropterus* Dlussky 1983; *Pseudarmania* Dlussky 1983 | *Sphecomyrma:* Wilson (1985f, 1987c) |
| **SUBFAMILY PONERINAE** | |
| **Tribe Amblyoponini** | Amblyoponini: Wilson (1958c), Brown (1960b), Baroni Urbani (1978b), Taylor (1978b) |
| *Amblyopone* Erichson 1842 (AUS, ETH, MAL, NEA, NEO, PAL, OR) SYNONYMS: *Amblyopopone* Dalla Torre 1893 [emendation]; *Arotropus* Provancher 1881; *Ericapelta* Kusnezov 1955; *Fulakora* Mann 1919; *Lithomyrmex* Clark 1928; *Neoamblyopone* Clark 1927; ? *Paraprionopelta* Kusnezov 1955; *Protamblyopone* Clark 1927; *Stigmatomma* Roger 1859; *Xymmer* Santschi 1914 | |

*continued*

**TABLE 2–2** *(continued)*

| Genera | Monographs |
|---|---|
| *Apomyrma* Brown, Gotwald and Lévieux 1970b (ETH) | |
| *Concoctio* Brown 1974 (ETH) | |
| *Myopopone* Roger 1861 (AUS, OR) | |
| *Mystrium* Roger 1862 (AUS, ETH, MAL, OR) | |
| *Onychomyrmex* Emery 1895 (AUS) | *Onychomyrmex:* Wheeler (1916a) |
| *Prionopelta* Mayr 1866 (AUS, DOM, ETH, MAL, NEA, NEO, OR)<br>SYNONYMS: *Examblyopone* Donisthorpe 1949; *Renea* Donisthorpe 1974 | *Prionopelta:* Hölldobler and Wilson (1986a) |
| †*Protamblyopone* Dlussky 1981 (Kirgizia, USSR, Miocene) | |
| **Tribe Ectatommini** | Ectatommini: Brown (1958b) |
| *Acanthoponera* Mayr 1862 (NEO) | *Acanthoponera:* Brown (1958b) |
| *Aulacopone* Arnoldi 1930 (PAL) | *Aulacopone:* Taylor (1979) |
| †*Bradoponera* Mayr 1868 (BAL) | |
| *Discothyrea* Roger 1863 (AUS, ETH, NEA, NEO, OR)<br>SYNONYMS: *Prodiscothyrea* Wheeler 1916; *Pseudosphincta* Wheeler 1922 [variant spelling]; *Pseudosysphincta* Arnold 1916 | |
| *Ectatomma* F. Smith 1858 (NEO) | *Ectatomma:* Brown (1958b), Kugler and Brown (1982) |
| †*Electroponera* Wheeler 1914 (BAL) | |
| *Gnamptogenys* Roger 1863 (AUS, DOM, NEA, NEO, OR)<br>SYNONYMS: *Alfaria* Emery 1893; *Barbourella* Wheeler 1930; *Commateta* Santschi 1929; *Emeryella* Forel 1901; *Holcoponera* Mayr 1887; *Mictoponera* Forel 1901; *Opisthoscyphus* Mann 1922; *Parectatomma* Emery 1911; *Poneracantha* Emery 1897; *Rhopalopone* Emery 1897; *Spaniopone* Wheeler and Mann 1914; *Stictoponera* Mayr 1887; *Tammoteca* Santschi 1929; *Wheeleripone* Mann 1919 | *Gnamptogenys:* Brown (1958b) |
| *Heteroponera* Mayr 1887 (AUS, NEO)<br>SYNONYMS: *Anacanthoponera* Wheeler 1923; *Paranomopone* Wheeler 1915 | *Heteroponera:* Brown (1958b), Kempf (1962) |
| *Paraponera* F. Smith 1859 (DOM, NEO) | |
| *Proceratium* Roger 1863 (AUS, ETH, MAL, NEA, NEO, OR, PAL)<br>SYNONYMS: *Sysphincta* Mayr 1865 [emendation]; *Sysphingta* Roger 1863 | *Proceratium:* Brown (1979), Ward (1988) |
| *Rhytidoponera* Mayr 1862 (AUS, BAL, SIC, OR)<br>SYNONYM: *Chalcoponera* Emery 1897 | *Rhytidoponera:* Clark (1936), Brown (1958b), Wilson (1958d), Ward (1980, 1984), Crozier et al. (1986) |
| †*Syntaphus* Donisthorpe 1920 (BRI) | |
| **Tribe Typhlomyrmecini** | Typhlomyrmecini: Brown (1965) |
| *Typhlomyrmex* Mayr 1862 (NEO) | |

**TABLE 2–2** *(continued)*

| Genera | Monographs |
|---|---|
| **Tribe Platythyreini** | Platythyreini: Wilson (1958c), Brown (1975) |
| *Platythyrea* Roger 1863 (AUS, BAL, DOM, ETH, NEA, NEO, OR)<br>SYNONYM: *Eubothroponera* Clark 1930 | |
| *Probolomyrmex* Mayr 1901 (AUS, ETH, NEO, OR)<br>SYNONYM: *Escherichia* Forel 1910 | *Probolomyrmex:* Taylor (1965a) |
| **Tribe Cerapachyini** | Cerapachyini: Wilson (1959f), Brown (1975) |
| *Cerapachys* F. Smith 1857 (AUS, BAL, ETH, MAL, NEA, NEO, OR, PAL)<br>SYNONYMS: *Ceratopachys* Schulz 1906 [emendation]; *Chrysapace* Crawley 1924; *Cysias* Emery 1902; *Lioponera* Mayr 1878; *Neophyracaces* Clark 1941; *Ooceraea* Roger 1862; *Parasyscia* Emery 1882; *Phyracaces* Emery 1902; †*Procerapachys* Wheeler 1915; *Syscia* Roger 1861 | *Cerapachys:* Brown (1975) |
| *Leptanilloides* Mann 1923 (NEO) | *Leptanilloides:* Brown (1975) |
| *Simopone* Forel 1891 (AUS, ETH, MAL, OR) | *Simopone:* Taylor (1965d), Brown (1975) |
| *Sphinctomyrmex* Mayr 1866 (AUS, ETH, NEO, OR)<br>SYNONYMS: *Aethiopopone* Santschi 1930; *Eusphinctus* Emery 1893; *Nothosphinctus* Wheeler 1918; *Zasphinctus* Wheeler 1918 | *Sphinctomyrmex:* Brown (1975) |
| **Tribe Cylindromyrmecini** | Cylindromyrmecini: Brown (1975) |
| *Cylindromyrmex* Mayr 1870 (DOM, NEO)<br>SYNONYMS: *Holcoponera* Cameron 1891 (not Mayr 1887); *Hypocylindromyrmex* Wheeler 1924; *Metacylindromyrmex* Wheeler 1924 | |
| **Tribe Acanthostichini** | Acanthostichini: Kusnezov (1962), Brown (1975) |
| *Acanthostichus* Mayr 1870 (NEO) | |
| *Ctenopyga* Ashmead 1905 (NEA) | |
| **Tribe Thaumatomyrmecini** | |
| *Thaumatomyrmex* Mayr 1887 (NEO) | *Thaumatomyrmex:* Kempf (1975), Longino (1988) |
| **Tribe Ponerini** | Ponerini: Wilson (1958c) |
| †*Archiponera* Carpenter 1930 (FLO) | |
| *Asphinctopone* Santschi 1914 (ETH)<br>SYNONYM: *Lepidopone* Bernard 1952 | |
| *Belonopelta* Mayr 1870 (NEO, OR)<br>SYNONYMS: *Emeryopone* Forel 1912; *Leiopelta* Baroni Urbani 1975a | *Belonopelta:* Wilson (1955b), Baroni Urbani (1975a) |
| *Brachyponera* Emery 1901 (AUS, ETH, OR) | |
| *Centromyrmex* Mayr 1866 (ETH, NEO, OR) | |

*continued*

**TABLE 2–2** *(continued)*

| Genera | Monographs |
|---|---|
| SYNONYMS: *Glyphopone* Forel 1913; *Leptopone* Arnold 1916; *Promyopias* Santschi 1914; *Spalacomyrmex* Emery 1889; *Typhloteras* Karavaiev 1925 | |
| *Cryptopone* Emery 1892 (AUS, ETH, NEA, OR, PAL) | |
| *Diacamma* Mayr 1862 (AUS, OR) | |
| *Dinoponera* Roger 1861 (NEO) | *Dinoponera:* Kempf (1971) |
| *Dolioponera* Brown 1974 (ETH) | |
| †*Emplastus* Donisthorpe 1920 (BRI) | |
| *Euponera* Forel 1891 (MAL) | |
| *Hagensia* Forel 1901 (ETH) | |
| *Harpegnathos* Jerdon 1851 (OR)<br>SYNONYM: *Drepanognathus* F. Smith 1858 | |
| *Hypoponera* Santschi 1938 (AUS, DOM, ETH, MAL, NEA, NEO, OR, PAL) | |
| *Leptogenys* Roger 1861 (AUS, ETH, MAL, NEA, NEO, OR, PAL)<br>SYNONYMS: *Dorylozelus* Forel 1915; *Lobopelta* Mayr 1862; *Machaerogenys* Emery 1911; *Microbolbos* Donisthorpe 1948; *Odontopelta* Emery 1911; *Prionogenys* Emery 1895 | *Leptogenys:* Taylor (1969), Bolton (1975a) |
| *Megaponera* Mayr 1862 (ETH)<br>SYNONYM: *Megaloponera* Emery 1877 | |
| *Myopias* Roger 1861 (AUS, OR)<br>SYNONYMS: *Bradyponera* Mayr 1886; *Trapeziopelta* Mayr 1862 | *Myopias:* Willey and Brown (1983) |
| *Odontoponera* Mayr 1862 (AUS) | |
| *Ophthalmopone* Forel 1890 (ETH) | |
| *Pachycondyla* F. Smith 1858 (AUS, BAL, BRI, DOM, ETH, MAL, NEA, NEO, OR)<br>SYNONYMS: *Bothroponera* Mayr 1862; *Ectomomyrmex* Mayr 1867; *Eumecopone* Forel 1901; *Hiphopelta* Forel 1913; *Mesoponera* Emery 1901; *Neoponera* Emery 1901; *Pseudoneoponera* Donisthorpe 1943; *Pseudoponera* Emery 1901; *Syntermitopone* Wheeler 1936; *Termitopone* Wheeler 1936; *Trachymesopus* Emery 1911; *Trachyponera* Santschi 1928 [lapsus for *Trachymesopus*]; *Wadeura* Weber 1939; *Xiphopelta* Forel 1913 | |
| *Paltothyreus* Mayr 1862 (ETH) | |
| *Phrynoponera* Wheeler 1920 (ETH) | |
| *Plectroctena* F. Smith 1858 (ETH)<br>SYNONYM: *Cacopone* Santschi 1914 | *Plectroctena:* Bolton (1974b) |
| *Ponera* Latreille 1804 (AUS, BAL, BRI, MAL, NEA, NEO, OR, PAL, SIC)<br>SYNONYMS: *Pseudocryptopone* Wheeler 1933; *Pteroponera* Bernard 1949; *Selenopone* Wheeler 1933 | *Ponera:* Taylor (1967a) |
| *Psalidomyrmex* André 1890 (ETH) | *Psalidomyrmex:* Bolton (1975b) |
| *Simopelta* Mann 1922 (NEO) | *Simopelta:* Gotwald and Brown (1966) |
| *Streblognathus* Mayr 1862 (ETH) | |

**TABLE 2–2** *(continued)*

| Genera | Monographs |
|---|---|
| **Tribe Odontomachini** | Odontomachini: Wilson (1959e), Brown (1976b, 1977b, 1978) |
| *Anochetus* Mayr 1861 (AUS, DOM, ETH, MAL, NEO, OR)<br>SYNONYMS: *Myrmapatetes* Wheeler 1929; *Stenomyrmex* Mayr 1862 | |
| *Odontomachus* Latreille 1804 (AUS, DOM, ETH, MAL, NEA, NEO, OR, PAL)<br>SYNONYMS: *Champsomyrmex* Emery 1891; *Myrtoteras* Matsumura 1912; *Pedetes* Bernstein 1861 | |
| **Tribe Aenictogitini** | Aenictogitini: Brown (1975) |
| *Aenictogiton* Emery 1901 (ETH) | |
| **Incertae Sedis** | |
| †*Eomyrmex* Hong et al. 1974 (Manchuria, early Eocene) | |
| **SUBFAMILY NOTHOMYRMECIINAE** | |
| *Nothomyrmecia* Clark 1934 (AUS) | *Nothomyrmecia*: Taylor (1978c) |
| **SUBFAMILY MYRMECIINAE** | |
| †*Ameghinoia* Viana and Haedo Rossi 1957 (Argentina, early Tertiary) | *Ameghinoia*: Viana and Haedo Rossi (1957) |
| *Myrmecia* Fabricius 1804 (AUS)<br>SYNONYMS: *Halmamyrmecia* Wheeler 1922; *Pristomyrmecia* Emery 1911; *Promyrmecia* Emery 1911 | *Myrmecia*: Clark (1951), Brown (1953c) |
| †*Prionomyrmex* Mayr 1868 (BAL) | |
| **SUBFAMILY DORYLINAE** | Dorylinae: Schneirla (1971), Gotwald (1985) |
| **Tribe Aenictini** | |
| *Aenictus* Shuckard 1840 (AUS, ETH, OR)<br>SYNONYMS: *Paraenictus* Wheeler 1929; *Typhlatta* F. Smith 1857 | *Aenictus*: Wilson (1964), Schneirla (1971) |
| **Tribe Dorylini** | |
| *Dorylus* Fabricius 1793 (ETH, OR, PAL)<br>SYNONYMS: *Alaopone* Emery 1881; *Anomma* Shuckard 1840; *Cosmaecetes* Spinola 1853; *Cosmaegetes* Dalla Torre 1893 [variant spelling]; *Dichthadia* Gerstaecker 1863; *Rhogmus* Schuckard 1840; *Shuckardia* Emery 1895; *Sphegomyrmex* Imhoff 1852; *Typhlopone* Westwood 1839 | *Dorylus*: Wilson (1964), Barr and Gotwald (1982), Gotwald and Schaefer (1982), Barr et al. (1985), Gotwald (1985) |
| **SUBFAMILY ECITONINAE** | Ecitoninae: Smith (1942b), Borgmeier (1955), Schneirla (1971), Gotwald (1971, 1982, 1985), Watkins (1976, 1982) |
| **Tribe Cheliomyrmecini** | |
| *Cheliomyrmex* Mayr 1870 (NEO) | |

*continued*

**TABLE 2–2** *(continued)*

| Genera | Monographs |
|---|---|
| **Tribe Ecitonini** | |
| *Eciton* Latreille 1804 (NEO) | |
| SYNONYMS: *Ancylognathus* Lund 1831; *Camptognatha* Gray 1832; *Holopone* Santschi 1924; *Mayromyrmex* Ashmead 1905 | |
| *Labidus* Jurine 1807 (NEA, NEO) | |
| SYNONYMS: *Nycteresia* Roger 1861; *Pseudodichthadia* André 1885 | |
| *Neivamyrmex* Borgmeier 1940 (DOM, NEA, NEO) | |
| SYNONYMS: *Acamatus* Emery 1894 [preoccupied]; *Woitkowskia* Enzmann 1952 | |
| *Nomamyrmex* Borgmeier 1936 (NEA, NEO) | |
| **SUBFAMILY LEPTANILLINAE** | |
| **Tribe Leptanillini** | Leptanillini: Baroni Urbani (1977), Kugler (1986) |
| *Leptanilla* Emery 1870 (AUS, ETH, OR, PAL) | |
| SYNONYM: *Leptomesites* Kutter 1948 | |
| *Noonilla* Petersen 1968 (OR) | |
| *Phaulomyrma* G. C. and E. W. Wheeler 1930 (OR) | |
| *Scyphodon* Brues 1925 (OR) | |
| *Yavnella* Kugler 1986 (OR, PAL) | |
| **Tribe Anomalomyrmini** | Anomalomyrmini: manuscript name by R. W. Taylor, not formally published here; see Chapter 16 |
| *Anomalomyrma* Taylor ms (OR, PAL) | |
| *Protanilla* Taylor ms (OR) | |
| **SUBFAMILY PSEUDOMYRMECINAE** | |
| **Tribe Pseudomyrmecini** | |
| *Pseudomyrmex* Lund 1831 (DOM, FLO, NEA, NEO) | *Pseudomyrmex*: Creighton (1955), Kempf (1958b, 1960a, 1961a, 1967b), Ward (1985) |
| SYNONYMS: *Condylodon* Lund 1831; *Leptalaea* Spinola 1851 [variant spelling]; *Leptalea* Erichson 1839; *Myrmex* Guérin 1845; *Pseudomyrma* Guérin 1844 | |
| *Tetraponera* F. Smith 1852 (AUS, BAL, ETH, MAL, OR, PAL) | *Tetraponera*: Terron (1967, 1970) |
| SYNONYMS: *Pachysima* Emery 1912; *Parasima* Donisthorpe 1948; *Sima* Roger 1863; *Viticicola* Wheeler 1920 | |
| **SUBFAMILY MYRMICINAE** | |
| (Genera are listed alphabetically, and some are tentatively grouped into tribes at the end) | |
| *Acanthognathus* Mayr 1887 (NEO) | *Acanthognathus*: Brown and Kempf (1969) |
| *Acanthomyrmex* Emery 1892 (AUS, OR) | *Acanthomyrmex*: Moffett (1986b) |
| *Acromyrmex* Mayr 1865 (NEA, NEO) | *Acromyrmex*: Gonçalves (1961), |

**TABLE 2–2** *(continued)*

| Genera | Monographs |
|---|---|
| SYNONYMS: *Moellerius* Forel 1893; ? *Pseudoatta* Gallardo 1916 | Weber (1972), Wilson (1986b) |
| *Adelomyrmex* Emery 1897 (AUS, NEO, OR) | |
| SYNONYMS: *Apsychomyrmex* Wheeler 1910; *Arctomyrmex* Mann 1921 | |
| *Adlerzia* Forel 1902 (AUS) | |
| SYNONYM: *Stenothorax* McAreavey 1949 | |
| †*Agroecomyrmex* Wheeler 1914 (BAL) | |
| *Allomerus* Mayr 1877 (NEO) | |
| *Ancyridris* Wheeler 1935 (OR) | |
| *Anergates* Forel 1874 (NEA introduced, PAL) | |
| *Anillomyrma* Emery 1913 (OR) | *Anillomyrma*: Bolton (1987) |
| *Anisopheidole* Forel 1914 (OR) | |
| *Ankylomyrma* Bolton 1973 (ETH) | *Ankylomyrma*: Bolton (1973c, 1981b) |
| *Antichthonidris* Snelling 1975 (NEO) | *Antichthonidris*: Bolton (1987) |
| *Aphaenogaster* Mayr 1853 (AUS, BAL, DOM, FLO, MAL, NEA, NEO, OR, PAL) | *Aphaenogaster*: Brown (1974e) |
| SYNONYMS: *Attomyrma* Emery 1915; *Brunella* Forel 1917; *Deromyrma* Forel 1913; *Monomarium* F. Smith 1859; *Novomessor* Emery 1915; *Nystalomyrma* Wheeler 1916; *Planimyrma* Viehmeyer 1914 | |
| *Apterostigma* Mayr 1865 (NEO) | *Apterostigma*: Weber (1972) |
| *Asketogenys* Brown 1972 (OR) | |
| *Atopomyrmex* André 1889 (ETH) | *Atopomyrmex*: Bolton (1981b) |
| *Atta* Fabricius 1804 (NEA, NEO) | *Atta*: Borgmeier (1959), Weber (1972), Wilson (1986b) |
| SYNONYMS: *Archaeatta* Gonçalves 1942; *Epiatta* Borgmeier 1950; *Myrmegis* Rafinesque 1815; *Neoatta* Gonçalves 1942; *Oecodoma* Latreille 1818; *Palaeatta* Borgmeier 1950 | |
| †*Attaichnus* Laza 1982 (Miocene, Argentina; based on fossilized nest only) | |
| *Baracidris* Bolton 1981 (ETH) | *Baracidris*: Bolton (1981b) |
| *Basiceros* Schulz 1906 (NEO) | *Basiceros* and other Basicerotini: Brown and Kempf (1960), Taylor (1968b, 1970a), Wilson and Hölldobler (1986) |
| SYNONYMS: *Aspididris* Weber 1950; *Ceratobasis* F. Smith 1860 [preoccupied] | |
| *Blepharidatta* Wheeler 1915 (NEO) | |
| *Bondroitia* Forel 1911 (ETH) | *Bondroitia*: Bolton (1987) |
| *Calyptomyrmex* Emery 1887 (AUS, ETH, OR) | *Calyptomyrmex*: Baroni Urbani (1975b), Bolton (1981a) |
| SYNONYM: *Weberidris* Donisthorpe 1948 | |
| *Cardiocondyla* Emery 1869 (AUS, ETH, MAL, NEA introduced, NEO introduced, OR, PAL) | *Cardiocondyla*: Bernard (1956), Reiskind (1965) |
| SYNONYMS: *Dyclona* Santschi 1930; *Emeryia* Forel 1890; *Loncyda* Santschi 1930; *Prosopidris* Wheeler 1935; *Xenometra* Emery 1917 | |

*continued*

**TABLE 2–2** *(continued)*

| Genera | Monographs |
| --- | --- |
| *Carebara* Westwood 1849 (ETH, NEO, OR) | |
| *Carebarella* Emery 1905 (NEO)<br>SYNONYM: *Carebarelloides* Borgmeier 1947 | |
| *Cataulacus* F. Smith 1853 (ETH, MAL, OR, SIC)<br>SYNONYM: *Otomyrmex* Forel 1891 | *Cataulacus*: Bolton (1974a, 1982) |
| *Cephalotes* Latreille 1802 (NEO)<br>SYNONYMS: *Cryptocerus* Latreille 1804; ? *Eucryptocerus* Kempf 1951 | *Cephalotes*: Kempf (1951), Corn (1976) |
| *Chalepoxenus* Menozzi 1923 (PAL)<br>SYNONYMS: *Icothorax* Hamann and Klemm 1967; *Leonomyrma* Arnoldi 1968 | *Chalepoxenus*: Buschinger (1981, 1987a), Ehrhardt (1982), Buschinger et al. (1988) |
| *Chimaeridris* Wilson 1989 (OR) | |
| *Cladarogenys* Brown 1976 (ETH) | |
| *Codiomyrmex* Wheeler 1916 (NEO) | *Codiomyrmex*: Brown (1948) |
| *Codioxenus* Santschi 1931 (NEO) | *Codioxenus*: Brown (1948) |
| *Colobostruma* Wheeler 1927 (AUS)<br>SYNONYMS: *Alistruma* Brown 1948; *Clarkistruma* Brown 1948 | |
| *Creightonidris* Brown 1949 (NEO) | |
| *Crematogaster* Lund 1831 (AUS, DOM, ETH, MAL, NEA, NEO, OR, PAL, SIC)<br>SYNONYMS: *Acrocoelia* Mayr 1852; *Apterocrema* Wheeler 1936; *Atopogyne* Forel 1911; *Colobocrema* Wheeler 1927; *Cremastogaster* Mayr 1861 [emendation]; *Decacrema* Forel 1910; *Eucrema* Santschi 1918; *Mesocrema* Santschi 1928; *Nematocrema* Santschi 1918; *Neocrema* Santschi 1918; *Orthocrema* Santschi 1918; *Oxygyne* Forel 1901; *Paracrema* Santschi 1918; *Physocrema* Forel 1912; *Rhachiocrema* Mann 1919; *Sphaerocrema* Santschi 1918; *Tranopeltoides* Wheeler 1922; *Xiphocrema* Forel 1913 | *Crematogaster*: Buren (1958, 1968b) |
| *Cyphoidris* Weber 1952 (ETH) | *Cyphoidris*: Bolton (1981b) |
| *Cyphomyrmex* Mayr 1862 (DOM, NEA, NEO)<br>SYNONYM: *Cyphomannia* Weber 1938 | *Cyphomyrmex*: Kempf (1964a, 1965) |
| *Dacetinops* Brown and Wilson 1957 (AUS, OR) | *Dacetinops*: Taylor (1985) |
| *Daceton* Perty 1833 (NEO)<br>SYNONYM: *Dacetum* Agassiz 1846 [emendation] | *Daceton* and other Dacetini: Brown (1948, 1950, 1952a,b, 1953a,d, 1954b, 1959a,b, 1964b), Kempf (1958c), Brown and Wilson (1959), Brown and Kempf (1969), Taylor (1973, 1977, 1978a), Bolton (1983) |
| *Decamorium* Forel 1913 (ETH) | *Decamorium*: Bolton (1976) |
| *Dicroaspis* Emery 1908 (ETH)<br>SYNONYM: *Geognomicus* Menozzi 1924 | *Dicroaspis*: Bolton (1981a) |
| *Dilobocondyla* Santschi 1910 (AUS, OR)<br>SYNONYM: *Mesomyrma* Stitz 1911 | |

**TABLE 2–2** *(continued)*

| Genera | Monographs |
| --- | --- |
| *Diplomorium* Mayr 1901 (ETH) | *Diplomorium*: Bolton (1987) |
| *Dorisidris* Brown 1948 (NEO) | |
| *Doronomyrmex* Kutter 1945 (NEA, PAL) | *Doronomyrmex*: Buschinger (1981, 1987a) |
| *Dysedrognathus* Taylor 1968 (OR) | |
| †*Electromyrmex* Wheeler 1914 (BAL) | |
| †*Enneamerus* Mayr 1868 (BAL) | |
| †*Eocenidris* Wilson 1985 (ARK) | |
| *Epelysidris* Bolton 1987 (OR) | |
| *Ephebomyrmex* Wheeler 1902 (NEA, NEO) | |
| *Epimyrma* Emery 1915 (PAL)<br>SYNONYM: ? *Gonepimyrma* Bernard 1948 | *Epimyrma*: Buschinger (1981, 1985, 1987a), Buschinger et al. (1986) |
| *Epitritus* Emery 1869 (ETH, OR, PAL) | *Epitritus*: Bolton (1983) |
| *Epopostruma* Forel 1895 (AUS)<br>SYNONYM: *Hexadaceton* Brown 1948 | |
| *Erebomyrma* Wheeler 1903 (DOM, NEA, NEO) | *Erebomyrma*: Wilson (1986a) |
| *Eurhopalothrix* Brown and Kempf 1960 (AUS, NEA, NEO, OR) | *Eurhopalothrix*: Brown and Kempf (1960), Taylor (1980) |
| *Eutetramorium* Emery 1890 (MAL) | |
| *Formicoxenus* Mayr 1855 (NEA, PAL)<br>SYNONYM: *Symmyrmica* Wheeler 1904 | *Formicoxenus*: Francoeur et al. (1985), Buschinger (1987a) |
| *Glamyromyrmex* Wheeler 1915 (AUS, ETH, MAL, NEO, OR)<br>SYNONYMS: *Borgmeierita* Brown 1953; ? *Chelystruma* Brown 1950 | *Glamyromyrmex*: Bolton (1983), Kempf (1960d) |
| *Goniomma* Emery 1895 (PAL) | |
| *Gymnomyrmex* Borgmeier 1954 (NEO) | *Gymnomyrmex*: Kempf (1960d) |
| *Harpagoxenus* Forel 1893 (NEA, PAL)<br>SYNONYM: *Tomognathus* Mayr 1861 [preoccupied] | *Harpagoxenus*: Buschinger (1981) |
| *Huberia* Forel 1890 (AUS) | |
| *Hylomyrma* Forel 1912 (NEO)<br>SYNONYM: *Lundella* Emery 1915 | *Hylomyrma*: Kempf (1973) |
| *Hypocryptocerus* Wheeler 1920 (NEO) | *Hypocryptocerus*: Wheeler (1936b), Kempf (1951) |
| †*Hypopomyrmex* Emery 1891 (SIC) | |
| †*Ilemomyrmex* Wilson 1985 (DOM) | |
| *Indomyrma* Brown 1985 (OR) | |
| *Ireneopone* Donisthorpe 1946 (MAL) | |
| *Ishakidris* Bolton 1984 (OR) | |
| *Kyidris* Brown 1949 (AUS, MAL, OR)<br>SYNONYM: *Polyhomoa* Azuma 1950 | |
| *Lachnomyrmex* Wheeler 1910 (NEO) | |
| *Leptothorax* Mayr 1855 (AUS, BAL, DOM, ETH, MAL, NEA, NEO, OR, PAL)<br>SYNONYMS: *Caulomyrma* Forel 1915; *Goniothorax* Emery 1896 [preoccupied]; *Limnomyrmex* Arnold 1948; *Mycothorax* Ruzsky 1904; *Myrafant* M. Smith 1950; *Temnothorax* Mayr 1861; *Tetramyrma* Forel 1912 | *Leptothorax*: Kempf (1958d), Bolton (1982), Buschinger (1987a)<br>Parasitic *Leptothorax* and satellite genera: Buschinger (1981, 1987a) |
| *Liomyrmex* Mayr 1865 (AUS, OR)<br>SYNONYMS: *Aratromyrmex* Stitz 1938; *Laparomyrmex* Emery 1887; *Promyrma* Forel 1912 | |

*continued*

**TABLE 2–2** *(continued)*

| Genera | Monographs |
|---|---|
| †*Lithomyrmex* Carpenter 1930 (FLO) | |
| *Lophomyrmex* Emery 1892 (OR) | *Lophomyrmex*: Kugler (1986), Moffett (1986e) |
| *Lordomyrma* Emery 1897 (AUS, OR)<br>SYNONYMS: *Prodicroaspis* Emery 1914; *Promeranoplus* Emery 1914 | |
| *Machomyrma* Forel 1895 (AUS) | |
| *Macromischa* Roger 1863 (NEA, NEO)<br>SYNONYMS: *Antillaemyrmex* Mann 1920; *Croesomyrmex* Mann 1920 | *Macromischa*: Baroni Urbani (1978c) |
| *Manica* Jurine 1807 (NEA, PAL)<br>SYNONYMS: *Neomyrma* Forel 1914; *Oreomyrma* Wheeler 1914 | |
| *Mayriella* Forel 1902 (AUS, OR) | |
| *Megalomyrmex* Forel 1885 (NEO)<br>SYNONYMS: *Ceprobroticus* Wheeler 1925; *Wheelerimyrmex* Mann 1922 | *Megalomyrmex*: Kempf (1970b), Brandão (1987) |
| *Melissotarsus* Emery 1877 (ETH, MAL) | *Melissotarsus*: Bolton (1982) |
| *Meranoplus* F. Smith 1853 (AUS, ETH, MAL, OR)<br>SYNONYM: *Cryptocephalus* Lowne 1865 | *Meranoplus*: Bolton (1981b) |
| *Mesostruma* Brown 1948 (AUS) | *Mesostruma*: Taylor (1973) |
| *Messor* Forel 1890 (ETH, FLO, NEA, OR, PAL)<br>SYNONYMS: *Cratomyrmex* Emery 1892; *Lobognathus* Enzmann 1947; *Veromessor* Forel 1917 | *Messor*: Bernard (1981), Bolton (1982) |
| *Metapone* Forel 1900 (AUS, MAL, OR) | |
| *Microdaceton* Santschi 1913 (ETH) | *Microdaceton*: Bolton (1983) |
| *Monomorium* Mayr 1855 (AUS, BAL, ETH, MAL, NEA, NEO, OR, PAL)<br>SYNONYMS: *Chelaner* Emery 1914; *Corynomyrmex* Viehmeyer 1916; *Epixenus* Emery 1908; *Epoecus* Emery 1892; *Equessimessor* Santschi 1935 [emendation]; *Equestrimessor* Santschi 1919; *Holcomyrmex* Mayr 1878; *Irenidris* Donisthorpe 1943; *Isholcomyrmex* Santschi 1936 [emendation]; *Isolcomyrmex* Santschi 1917; *Lampromyrmex* Mayr 1868; *Mitara* Emery 1913; *Notomyrmex* Emery 1915; *Paraholcomyrmex* Emery 1915; *Paraphacota* Santschi 1919; *Parholcomyrmex* Emery 1915 [emendation]; *Pharaophanes* Bernard 1952 [nomen nudum]; *Protholcomyrmex* Wheeler 1912; *Schizopelta* McAreavey 1949; *Syllophopsis* Santschi 1915; *Trichomyrmex* Mayr 1865; *Wheeleria* Forel 1905; *Wheeleriella* Forel 1907; *Xenhyboma* Santschi 1919; *Xeromyrmex* Emery 1915 | *Monomorium* and related genera: Ettershank (1966), DuBois (1986), Berndt and Eichler (1987), Bolton (1987) |
| *Mycetarotes* Emery 1913 (NEO) | *Mycetarotes*: Kempf (1963a), Weber (1972) |
| *Mycetophylax* Emery 1913 (NEO)<br>SYNONYM: *Paramycetophylax* Kusnezov 1956 | *Mycetophylax*: Weber (1972) |

**TABLE 2–2** *(continued)*

| Genera | Monographs |
|---|---|
| *Mycetosoritis* Wheeler 1907 (NEA, NEO) | *Mycetosoritis*: Weber (1972) |
| *Mycocepurus* Forel 1893 (NEO)<br>SYNONYM: *Descolemyrma* Kusnezov 1951 | *Mycocepurus*: Kempf (1963a), Weber (1972) |
| *Myrmecina* Curtis 1829 (AUS, NEA, NEO, OR, PAL)<br>SYNONYM: *Archaeomyrmex* Mann 1921 | |
| *Myrmica* Latreille 1804 (BAL, NEA, PAL)<br>SYNONYMS: *Dodecamyrmica* Arnoldi 1968; *Paramyrmica* Cole 1957; *Sifolinia* Emery 1907; *Sommimyrma* Menozzi 1925; *Symbiomyrma* Arnoldi 1930 | *Myrmica*: Weber (1947a, 1948), Collingwood (1958a), Bolton (1988a) |
| *Myrmicaria* Saunders 1841 (ETH, OR)<br>SYNONYMS: *Heptacondylus* F. Smith 1857; *Physatta* F. Smith 1857 | |
| *Myrmicocrypta* F. Smith 1860 (NEO)<br>SYNONYM: *Glyptomyrmex* Forel 1885 | *Myrmicocrypta*: Weber (1972) |
| *Myrmoxenus* Ruzsky 1902 (PAL)<br>SYNONYM: *Myrmetaerus* Soudek 1925 | |
| *Neostruma* Brown 1948 (NEO) | *Neostruma*: Brown (1959a) |
| *Nesomyrmex* Wheeler 1910 (NEO) | *Nesomyrmex*: Kempf (1959b) |
| *Nothidris* Ettershank 1966 (NEO) | *Nothidris*: Snelling (1975), Bolton (1987) |
| †*Nothomyrmica* Wheeler 1914 (BAL) | |
| *Ochetomyrmex* Mayr 1877 (NEO)<br>SYNONYM: *Brownidris* Kusnezov 1957 | |
| *Octostruma* Forel 1912 (DOM, NEO) | *Octostruma*: Brown and Kempf (1960) |
| *Ocymyrmex* Emery 1886 (ETH) | *Ocymyrmex*: Bolton (1981b) |
| *Oligomyrmex* Mayr 1867 (AUS, BAL, ETH, MAL, OR, PAL, SIC)<br>SYNONYMS: *Aeromyrma* Forel 1891; *Aneleus* Emery 1900; ? *Crateropsis* Patrizi 1948; *Hendecatella* Wheeler 1927; *Lecanomyrma* Forel 1913; ? *Nimbamyrma* Bernard 1952; *Octella* Forel 1915; *Solenops* Karavaiev 1930; *Spelaeomyrmex* Wheeler 1922; *Sporocleptes* Arnold 1948 | *Oligomyrmex*: Kugler (1986), Wilson (1986a) |
| *Orectognathus* F. Smith 1853 (AUS)<br>SYNONYM: *Arnoldidris* Brown 1950 | *Orectognathus*: Taylor (1977, 1978a) |
| *Oxyepoecus* Santschi 1926 (NEO)<br>SYNONYMS: *Forelifidis* M. Smith 1954 [new name for *Martia* Forel]; *Martia* Forel 1907 [preoccupied] | *Oxyepoecus*: Kempf (1974), Bolton (1987) |
| †*Oxyidris* Wilson 1985 (DOM) | |
| *Oxyopomyrmex* André 1881 (PAL) | |
| *Paedalgus* Forel 1911 (ETH, OR) | |
| †*Parameranoplus* Wheeler 1914 (BAL) | |
| †*Paraphaenogaster* Dlussky 1981 (Kirgizia, USSR, Miocene; SHA) | *Paraphaenogaster*: Hong (1983) |
| *Paratopula* Wheeler 1919 (OR) | *Paratopula*: Bolton (1988b) |
| *Pentastruma* Forel 1912 (OR) | *Pentastruma*: Brown and Boisvert (1978) |
| *Peronomyrmex* Viehmeyer 1922 (AUS) | *Peronomyrmex*: Taylor (1970b) |
| *Perissomyrmex* M. Smith 1947 (NEO, OR, probably of OR origin) | |
| *Phacota* Roger 1862 (PAL) | *Phacota*: Bolton (1987) |
| *Phalacromyrmex* Kempf 1960 (NEO) | *Phalacromyrmex*: Bolton (1984) |

*continued*

**TABLE 2–2** *(continued)*

| Genera | Monographs |
|---|---|
| *Pheidole* Westwood 1840 (AUS, DOM, ETH, FLO, MAL, NEA, NEO, OR, PAL) SYNONYMS: *Allopheidole* Forel 1912; *Anergatides* Wasmann 1915; *Bruchomyrma* Santschi 1922; *Cardiopheidole* Wheeler 1914; *Cephalomorium* Forel 1922; *Ceratopheidole* Pergande 1895; *Conothoracoides* Strand 1935; *Conothorax* Karavaiev 1935; *Decapheidole* Forel 1912; *Elasmopheidole* Forel 1913; *Electropheidole* Mann 1921; *Epipheidole* Wheeler 1904; *Eriopheidole* Kusnezov 1952; *Gallardomyrma* Bruch 1932; *Hendecapheidole* Wheeler 1922; *Ischnomyrmex* Mayr 1862; *Isopheidole* Forel 1912; *Leptomyrma* Motschulsky 1863; *Macropheidole* Emery 1915; *Oecophthora* Heer 1852; ? *Parapheidole* Emery 1915; *Pheidolacanthinus* F. Smith 1864; *Phidole* Bingham 1903 (variant spelling); *Scrobopheidole* Emery 1915; *Stegopheidole* Emery 1915; *Sympheidole* Wheeler 1904; *Trachypheidole* Emery 1915; *Xenoaphaenogaster* Baroni Urbani 1964 | *Pheidole*: Gregg (1958a), Kempf (1972c), Naves (1985) Parasitic *Pheidole* and satellite genera: Wilson (1984c) |
| *Pheidologeton* Mayr 1862 (AUS, ETH, OR) SYNONYMS: *Amauromyrmex* Wheeler 1929; *Idrisella* Santschi 1937 | *Pheidologeton* and related genera: Ettershank (1966), Kugler (1986), Moffett (1987b,c) |
| *Pilotrochus* Brown 1977 (MAL) | *Pilotrochus*: Bolton (1984) |
| *Podomyrma* F. Smith 1859 (AUS) SYNONYMS: *Acrostigma* Emery 1891; ? *Dacyron* Forel 1895; ? *Pseudopodomyrma* Crawley 1925 | |
| *Poecilomyrma* Mann 1921 (OR: Fijian endemic) | |
| *Pogonomyrmex* Mayr 1868 (FLO, NEA, NEO) SYNONYMS: *Forelomyrmex* Wheeler 1913; *Janetia* Forel 1899 | *Pogonomyrmex*: Kusnezov (1951b), Cole (1968), MacKay (1981) |
| *Pristomyrmex* Mayr 1866 (AUS, ETH, MAL, OR) SYNONYMS: *Dodous* Donisthorpe 1946; *Hylidris* Weber 1941; *Odontomyrmex* André 1905 | *Pristomyrmex*: Taylor (1965b), Bolton (1981b) |
| *Proatta* Forel 1912 (OR) | *Proatta*: Wheeler and Wheeler (1985b) |
| *Procryptocerus* Emery 1887 (NEO) | *Procryptocerus* and other Cephalotini: Kempf (1951, 1958a, 1967a) |
| *Protalaridris* Brown 1980 (NEO) | |
| *Protomognathus* Wheeler 1905 (NEA) | *Protomognathus:* Buschinger (1981, 1987b); recognized for *Harpagoxenus* (partim) *americanus* |
| *Quadristruma* Brown 1949 (AUS, ETH) | |
| *Rhopalomastix* Forel 1900 (AUS, OR) | |

**TABLE 2–2** *(continued)*

| Genera | Monographs |
|---|---|
| *Rhopalothrix* Mayr 1870 (AUS, NEO) SYNONYMS: *Acanthidris* Weber 1941; *Heptastruma* Weber 1934 | *Rhopalothrix*: Brown and Kempf (1960) |
| *Rhoptromyrmex* Mayr 1901 (AUS, ETH, OR) SYNONYMS: *Acidomyrmex* Emery 1915; *Hagioxenus* Forel 1910; *Ireneella* Donisthorpe 1941 | *Rhoptromyrmex*: Brown (1964a), Bolton (1986a) |
| *Rogeria* Emery 1894 (AUS, NEA, NEO) SYNONYM: *Irogera* Emery 1915 | *Rogeria*: Kempf (1963b) |
| *Romblonella* Wheeler 1935 (AUS, OR) | |
| *Secostruma* Bolton 1988 (OR) | *Secostruma*: Bolton (1988c) |
| *Sericomyrmex* Mayr 1865 (NEO) | *Sericomyrmex*: Weber (1972) |
| *Serrastruma* Brown 1948 (ETH, MAL) | *Serrastruma*: Brown (1952a), Bolton (1983) |
| *Smithistruma* Brown 1948 (DOM, ETH, MAL, NEA, NEO, OR, PAL) SYNONYMS: *Cephaloxys* F. Smith 1864 [preoccupied]; *Miccostruma* Brown 1948; *Platystruma* Brown 1953; *Weberistruma* Brown 1948; *Wessonistruma* Brown 1948 | *Smithistruma*: Brown (1953d), Bolton (1983) |
| *Solenopsis* Westwood 1840 (AUS, DOM, ETH, NEA, NEO, OR, PAL) SYNONYMS: *Diagyne* Santschi 1923; ? *Diplorhoptrum* Mayr 1855; *Disolenopsis* Kusnezov 1953; *Euophthalma* Creighton 1930; *Granisolenopsis* Kusnezov 1957; *Labauchena* Santschi 1930; ? *Lilidris* Kusnezov 1957; *Oedaleocerus* Creighton 1930; *Paranamyrma* Kusnezov 1954; *Synsolenopsis* Forel 1918 | *Solenopsis* and related genera: Creighton (1930), Kusnezov (1957a), Snelling (1963), Ettershank (1966), Buren (1972), Bolton (1987) |
| *Stegomyrmex* Emery 1912 (NEO) | |
| *Stenamma* Westwood 1840 (BAL, NEA, NEO, OR, PAL) SYNONYMS: *Asemorphoptrum* Mayr 1861; *Theryella* Santschi 1921 | *Stenamma*: Smith (1957), Snelling (1973c) |
| *Stereomyrmex* Emery 1901 (OR) | |
| †*Stigmomyrmex* Mayr 1868 (BAL) | |
| †*Stiphromyrmex* Wheeler 1914 (BAL) | |
| *Strongylognathus* Mayr 1853 (PAL) SYNONYM: *Myrmus* Schenck 1853 | *Strongylognathus*: Bolton (1976) |
| *Strumigenys* F. Smith 1860 (AUS, ETH, MAL, NEA, NEO, OR) SYNONYMS: *Eneria* Donisthorpe 1948; *Labidogenys* Roger 1863; *Proscopomyrmex* Patrizi 1946; *Pyramica* Roger 1862 | *Strumigenys*: Brown (1962), Bolton (1983) |
| *Talaridris* Weber 1941 (NEO) | |
| *Tatuidris* Brown and Kempf 1968 (NEO) | |
| *Teleutomyrmex* Kutter 1950 (PAL) | |
| *Terataner* Emery 1912 (ETH, MAL) SYNONYM: *Tranetera* Arnold 1952 | *Teratаner*: Bolton (1981b) |
| *Tetramorium* Mayr 1855 (AUS, ETH, MAL, NEA, OR, PAL) SYNONYMS: *Atopula* Emery 1912; *Lobomyrmex* Kratochvil 1941; | *Tetramorium* and related genera: Bolton (1976, 1977, 1979, 1980) |

*continued*

**TABLE 2–2** *(continued)*

| Genera | Monographs |
|---|---|
| *Macromischoides* Wheeler 1920; *Sulcomyrmex* Kratochvil 1941; *Tetrogmus* Roger 1863; *Triglyphothrix* Forel 1890; *Xiphomyrmex* Forel 1887 | |
| *Tingimyrmex* Mann 1926 (NEO) | |
| *Trachymyrmex* Forel 1893 (DOM, NEA, NEO) | |
| *Tranopelta* Mayr 1866 (NEO) | |
| *Trichoscapa* Emery 1869 (ETH, NEA introduced, NEO introduced, OR, PAL) | |
| *Trigonogaster* Forel 1890 (OR) | *Trigonogaster*: Kugler (1986) |
| *Vollenhovia* Mayr 1865 (AUS, BAL, NEA introduced, OR)<br>SYNONYMS: *Acalama* M. Smith 1948; *Dorothea* Donisthorpe 1948; *Dyomorium* Donisthorpe 1947; *Gauromyrmex* Menozzi 1933; *Heteromyrmex* Wheeler 1920; *Propodomyrma* Wheeler 1910; *Solenomyrma* Karavaiev 1935; *Vollenhovenia* Dalla Torre 1893 [emendation] | |
| *Wasmannia* Forel 1893 (AUS, ETH and NEA introduced, NEO)<br>SYNONYM: *Hercynia* Enzmann 1947 | |
| *Willowsiella* Wheeler 1934 (OR: Solomon Islands) | |
| *Xenomyrmex* Forel 1884 (NEA, NEO)<br>SYNONYM: *Myrmecinella* Wheeler 1922 | |
| *Zacryptocerus* Ashmead 1905 (DOM, NEA, NEO)<br>SYNONYMS: *Cyathocephalus* Emery 1915; *Cyathomyrmex* Creighton 1933; *Harnedia* M. Smith 1949; *Paracryptocerus* Emery 1915 | *Zacryptocerus*: Kempf (1951, 1958a) |

**Myrmicinae Incertae Sedis**

†*Cephalomyrmex* Carpenter 1930 (FLO)
*Tricytarus* Donisthorpe 1947 (OR)
*Leptoxenus* Forel 1917 *nomen nudum* (attributed to Santschi)

The tribal classification of the Myrmicinae is still in a state of severe disarray. The following assignments of some of the living genera can be tentatively made, with the tribes Myrmicariini, Crematogastrini, Tetramoriini, Cephalotini, Basicerotini, Dacetini, and Attini being among the most discrete and hence easily identified:

Myrmicini: *Ephebomyrmex, Eutetramorium, Huberia, Hylomyrma, Manica, Myrmica, Pogonomyrmex*
Pheidolini: *Acanthomyrmex, Adlerzia, Ancyridris, Aphaenogaster, Chimaeridris, Cyphoidris, Goniomma, Lachnomyrmex, Lordomyrma, Machomyrma, Messor, Ocymyrmex, Oxyopomyrmex, Perissomyrmex, Pheidole, Pristomyrmex, Proatta, Rogeria, Stenamma*
Melissotarsini: *Melissotarsus, Rhopalomastix*

**TABLE 2–2** *(continued)*

| Genera | Monographs |
|---|---|

Metaponini: *Liomyrmex, Metapone, Vollenhovia, Xenomyrmex*
Leptothoracini: *Cardiocondyla, Chalepoxenus, Doronomyrmex, Epimyrma, Formicoxenus, Harpagoxenus, Leptothorax, Myrmoxenus, Nesomyrmex, Protomognathus*
Myrmicariini: *Myrmicaria*
Crematogastrini: *Crematogaster*
Solenopsidini: *Allomerus, Anillomyrma, Antichthonidris, Bondroitia, Carebarella, Diplomorium, Epelysidris, Megalomyrmex, Monomorium, Nothidris, Oxyepoecus, Phacota, Solenopsis*
Pheidologetini: *Anisopheidole, Carebara, Erebomyrma, Lophomyrmex, Oligomyrmex, Paedalgus, Pheidologeton, Trigonogaster*
Meranoplini: *Meranoplus, Romblonella, Willowsiella*
Tetramoriini: *Anergates, Decamorium, Rhoptromyrmex, Secostruma, Strongylognathus, Teleutomyrmex, Tetramorium*
Ochetomyrmecini: *Ochetomyrmex, Tranopelta*
Cephalotini: *Cephalotes, Procryptocerus, Zacryptocerus*
Basicerotini: *Basiceros, Creightonidris, Eurhopalothrix, Octostruma, Protalaridris, Rhopalothrix, Talaridris*
Dacetini: *Acanthognathus, Asketogenys, Cladarogenys, Codiomyrmex, Codioxenus, Colobostruma, Daceton, Dorisidris, Dysedrognathus, Epitritus, Epopostruma, Glamyromyrmex, Gymnomyrmex, Kyidris, Mesostruma, Microdaceton, Neostruma, Orectognathus, Pentastruma, Quadristruma, Serrastruma, Smithistruma, Strumigenys, Tingimyrmex, Trichoscapa*
Agroecomyrmecini: *Tatuidris*
Phalacromyrmecini: *Ishakidris, Phalacromyrmex, Pilotrochus*
Attini: *Acromyrmex, Apterostigma, Atta, Cyphomyrmex, Mycetarotes, Mycetophylax, Mycetosoritis, Mycocepurus, Myrmicocrypta, Sericomyrmex, Trachymyrmex*

**SUBFAMILY ANEURETINAE**

**Tribe Aneuretini**

| Genera | Monographs |
|---|---|
| *Aneuretus* Emery 1892 (OR) | *Aneuretus*: Wilson et al. (1956), Traniello and Jayasuriya (1981a,b, 1985), Jayasuriya and Traniello (1985) |

*continued*

**TABLE 2–2** *(continued)*

| Genera | Monographs |
| --- | --- |
| †*Mianeuretus* Carpenter 1930 (FLO) | |
| †*Paraneuretus* Wheeler 1914 (BAL) | |
| †*Protaneuretus* Wheeler 1914 (BAL) | |
| **SUBFAMILY DOLICHODERINAE** | |
| **Tribe Leptomyrmecini** | |
| *Leptomyrmex* Mayr 1862 (AUS, DOM) | *Leptomyrmex*: Wheeler (1934a), Baroni Urbani and Wilson (1987) |
| †*Leptomyrmula* Emery 1912 (SIC) | |
| **Tribe Dolichoderini** | |
| *Dolichoderus* Lund 1831 (DOM, NEO) | |
| *Hypoclinea* Mayr 1855 (AUS, BAL, BRI, DOM, FLO, NEA, NEO, OR, PAL) SYNONYMS: ? *Acanthoclinea* Wheeler 1935; *Diabolus* Karavaiev 1925 [preoccupied]; ? *Diceratoclinea* Wheeler 1935; ? *Karawajewella* Donisthorpe 1944 | *Hypoclinea*: Lattke (1986) |
| *Monacis* Roger 1862 (NEO) | *Monacis*: Kempf (1959a, 1972a), Swain (1977) |
| *Monoceratoclinea* Wheeler 1935 (OR: New Guinea) | |
| **Tribe Tapinomini** | |
| *Amyrmex* Kusnezov 1953 (NEO) | |
| *Anillidris* Santschi 1936 (NEO) | |
| †*Asymphylomyrmex* Wheeler 1914 (BAL) | |
| *Axinidris* Weber 1941 (ETH) | |
| *Azteca* Forel 1878 (DOM, NEO) SYNONYM: *Aztecum* Bertkau 1879 [emendation] | *Azteca:* Emery (1893) |
| *Bothriomyrmex* Emery 1869 (AUS, OR, PAL) SYNONYM: ? *Chronoxenus* Santschi 1920 | *Bothriomyrmex*: Santschi (1920) |
| *Conomyrma* Forel 1913 (NEA, NEO) SYNONYMS: *Ammomyrma* Santschi 1922; *Araucomyrmex* Gallardo 1919; *Biconomyrma* Kusnezov 1952 | *Conomyrma*: Snelling (1973b), Trager (1988) |
| *Dorymyrmex* Mayr 1866 (NEO) SYNONYMS: *Psammomyrma* Forel 1912; *Spinomyrma* Kusnezov 1952 | |
| *Ecphorella* Forel 1909 (ETH) | |
| †*Elaeomyrmex* Carpenter 1930 (FLO) | |
| *Engramma* Forel 1905 (ETH) | |
| *Forelius* Emery 1888 (NEA, NEO) | |
| *Froggattella* Forel 1902 (AUS) | |
| *Iridomyrmex* Mayr 1862 (ARK, AUS, BAL, DOM, NEA, NEO, OR) SYNONYMS: ? *Anonychomyrma* Donisthorpe 1947; *Ctenobethylus* Brues 1939; *Doleromyrma* Forel 1907 | |
| *Linepithema* Mayr 1866 (NEO) | |
| *Liometopum* Mayr 1861 (BAL, FLO, NEA, OR, PAL, SHA) | |
| †*Miomyrmex* Carpenter 1930 (FLO) | |

**TABLE 2–2** *(continued)*

| Genera | Monographs |
| --- | --- |
| *Neoforelius* Kusnezov 1953 (NEO) | |
| †*Petraeomyrmex* Carpenter 1930 (FLO) | |
| †*Protazteca* Carpenter 1930 (FLO) | |
| *Semonius* Forel 1910 (ETH, OR) | |
| *Tapinoma* Foerster 1855 (AUS, DOM, ETH, MAL, OR, NEA, NEO, PAL, SIC) SYNONYMS: *Micromyrma* Dufour 1857; *Neoclystopsenella* Kurian 1955; *Tapinoptera* Santschi 1925 | *Tapinoma:* Brown (1987) |
| *Technomyrmex* Mayr 1872 (AUS, ETH, MAL, OR, SIC) SYNONYM: *Aphanotolepis* Wheeler 1930 | |
| *Turneria* Forel 1895 (AUS) | |
| *Zatapinoma* Wheeler 1928 (AUS, OR) | |
| **Dolichoderinae Incertae Sedis** | |
| †*Kotshkorkia* Dlussky 1981 (USSR, Miocene) | |
| **†SUBFAMILY FORMICIINAE** | |
| †*Formicium* Westwood 1854 (Lower Eocene, Tennessee; Middle Eocene, England and Germany) SYNONYMS: *Eoponera* Carpenter 1929; *Pseudosirex* Handlirsch 1908 | *Formicium*: Lutz (1986) |
| **SUBFAMILY FORMICINAE** | |
| **Tribe Myrmoteratini** | Myrmoteratini: Moffett (1985c) |
| *Myrmoteras* Forel 1893 (OR) | |
| **Tribe Oecophyllini** | |
| *Oecophylla* F. Smith 1860 (AUS, BAL, BRI, ETH, OR, SIC) | |
| **Tribe Gesomyrmecini** | |
| *Gesomyrmex* Mayr 1868 (BAL, OR, SIC) SYNONYM: *Dimorphomyrmex* André 1892 | |
| †*Prodimorphomyrmex* Wheeler 1914 (BAL) | |
| †*Sicilomyrmex* Wheeler 1914 (SIC) SYNONYM: *Sicelomyrmex* Wheeler 1914 | *Sicilomyrmex*: Brown and Carpenter (1978) |
| **Tribe Myrmecorhynchini** | |
| *Myrmecorhynchus* André 1896 (AUS) | |
| **Tribe Melophorini** | |
| *Lasiophanes* Emery 1895 (NEO) | |
| *Melophorus* Lubbock 1883 (AUS) SYNONYMS: *Erimelophorus* Wheeler 1935; *Trichomelophorus* Wheeler 1935 | |
| *Notoncus* Emery 1895 (AUS) SYNONYM: *Diodontolepis* Wheeler 1920 | *Notoncus*: Brown (1955a) |
| *Prolasius* Forel 1892 (AUS) | |
| *Pseudonotoncus* Clark 1934 (AUS) | |

*continued*

**TABLE 2-2** *(continued)*

| Genera | Monographs |
|---|---|
| **Tribe Plagiolepidini** | |
| *Acantholepis* Mayr 1861 (AUS, ETH, OR, PAL)<br>SYNONYM: *Lepisiota* Santschi 1926 | |
| *Acropyga* Roger 1862 (AUS, ETH, NEA, NEO, OR, PAL)<br>SYNONYMS: *Atopodon* Forel 1912; *Malacomyrma* Emery 1922; *Rhizomyrma* Forel 1893 | |
| *Agraulomyrmex* Prins 1983 (ETH) | |
| *Anoplolepis* Santschi 1914 (AUS, ETH, OR)<br>SYNONYMS: ? *Mesanoplolepis* Santschi 1926; *Tapinolepis* Emery 1925; *Zealleyella* Arnold 1922 | |
| †*Dryomyrmex* Wheeler 1914 (BAL) | |
| *Plagiolepis* Mayr 1861 (AUS, BAL, ETH, MAL, OR, PAL, SIC)<br>SYNONYMS: *Anacantholepis* Santschi 1914; *Aporomyrmex* Faber 1969; *Paraplagiolepis* Faber 1969 | |
| †*Rhopalomyrmex* Mayr 1868 (BAL) | |
| **Tribe Myrmelachistini** | |
| *Aphomomyrmex* Emery 1899 (ETH, MAL) | *Aphomomyrmex*: Snelling (1979) |
| *Brachymyrmex* Mayr 1868 (MAL, NEA, NEO)<br>SYNONYM: *Bryscha* Santschi 1925 | |
| *Cladomyrma* Wheeler 1920 (OR) | |
| *Myrmelachista* Roger 1863 (NEO)<br>SYNONYMS: *Decamera* Roger 1863; *Hincksidris* Donisthorpe 1944; *Neaphomus* Menozzi 1935 | |
| *Petalomyrmex* Snelling 1979 (ETH) | *Petalomyrmex*: Snelling (1979) |
| *Pseudaphomomyrmex* Wheeler 1920 (OR)<br>SYNONYM: *Aphomyrmex* Ashmead 1905 [preoccupied] | |
| **Tribe Prenolepidini** | |
| *Euprenolepis* Emery 1906 (AUS, OR)<br>SYNONYM: *Chapmanella* Wheeler 1930 | |
| *Paratrechina* Motschulsky 1863 (AUS, DOM, ETH, MAL, NEA, NEO, OR, PAL)<br>SYNONYMS: *Nylanderia* Emery 1906; *Paraparatrechina* Donisthorpe 1947 | *Paratrechina:* Trager (1984) |
| *Prenolepis* Mayr 1861 (BAL, DOM, NEA, NEO, OR, PAL) | |
| †*Protrechina* Wilson 1985 (ARK) | |
| *Stigmacros* Forel 1905 (AUS)<br>SYNONYMS: *Acrostigma* Forel 1902 [preoccupied]; *Campostigmacros* McAreavey 1957; *Chariostigmacros* McAreavey 1957; *Cyrtostigmacros* McAreavey 1957; *Hagiostigmacros* McAreavey 1957; *Pseudostigmacros* McAreavey 1957 | |

**TABLE 2-2** *(continued)*

| Genera | Monographs |
|---|---|
| **Tribes Formicini and Lasiini** | |
| *Acanthomyops* Mayr 1862 (NEA) | *Acanthomyops*: Wing (1968) |
| *Andragnathus* Emery 1922 (OR) | |
| *Bregmatomyrma* Wheeler 1929 (OR) | |
| *Cataglyphis* Foerster 1850 (ETH, PAL)<br>SYNONYMS: *Eomonocombus* Arnoldi 1968; *Machaeromyrma* Forel 1916; *Monocombus* Mayr 1855; *Paraformica* Forel 1915 | |
| *Formica* Linné 1758 (BAL, FLO, NEA, PAL)<br>SYNONYMS: *Adformica* Lomnicki 1925; *Coptoformica* Mueller 1933; ? *Hypochira* Buckley 1866; *Neoformica* Wheeler 1913; *Raptiformica* Forel 1913; *Serviformica* Forel 1913 | *Formica*: Yarrow (1955), Lange (1958), Betrem (1960), Dlussky (1967), Dlussky and Pisarski (1971), Letendre and Huot (1972), Francoeur (1973, 1974), Francoeur and Snelling (1979), Vepsäläinen and Pisarski (1981) |
| †*Glaphyromyrmex* Wheeler 1914 (BAL) | |
| *Lasius* Fabricius 1804 (BAL, FLO, NEA, OR, PAL, SHA)<br>SYNONYMS: *Austrolasius* Faber 1969; *Cautolasius* Wilson 1955; *Chthonolasius* Ruzsky 1912; *Dendrolasius* Ruzsky 1912; *Donisthorpea* Morice and Durrant 1914; *Formicina* Shuckard 1840 | *Lasius*: Wilson (1955a), Bourne (1973) |
| †*Leucotaphus* Donisthorpe 1920 (BRI) | |
| *Myrmecocystus* Wesmael 1838 (NEA)<br>SYNONYMS: *Endiodioctes* Snelling 1976; *Eremnocystus* Snelling 1976 | *Myrmecocystus*: Snelling (1976, 1982) |
| *Polyergus* Latreille 1804 (NEA, PAL) | |
| *Proformica* Ruzsky 1903 (PAL)<br>SYNONYM: *Alloformica* Dlussky 1969 | |
| *Pseudolasius* Emery 1886 (AUS, BAL, ETH, OR)<br>SYNONYM: *Nesolasius* Wheeler 1935 | |
| *Rossomyrmex* Arnoldi 1928 (PAL) | |
| *Teratomyrmex* McAreavey 1957 (AUS) | |
| **Tribe Santschiellini** | |
| *Santschiella* Forel 1916 (ETH) | |
| **Tribe Gigantiopini** | |
| *Gigantiops* Roger 1862 (NEO) | |
| **Tribe Camponotini** | |
| *Calomyrmex* Emery 1895 (AUS) | |
| *Camponotus* Mayr 1861 (AUS, BAL, BRI, DOM, ETH, FLO, MAL, NEA, NEO, OR, PAL, SHA)<br>SYNONYMS: *Condylomyrma* Santschi 1928; *Dinomyrmex* Ashmead 1905; *Hypercolobopsis* Emery 1920; *Karavaievia* Emery 1925; *Manniella* Wheeler 1921; *Mayria* Forel 1878; *Myrmacrhaphe* Santschi 1926; | *Camponotus*: Creighton (1950), Bernard (1968), Hashmi (1973), Dumpert (1985), Snelling (1988) |

*continued*

**TABLE 2–2** *(continued)*

| Genera | Monographs |
|---|---|
| *Myrmamblys* Forel 1912; *Myrmaphaenus* Emery 1920; *Myrmentoma* Forel 1912; *Myrmepinotus* Santschi 1921; *Myrmepomis* Forel 1912; *Myrmespera* Santschi 1926; *Myrmetaerus* Soudek 1925; *Myrmeurynota* Forel 1912; *Myrmisolepis* Santschi 1921; *Myrmobrachys* Forel 1912; *Myrmocamelus* Forel 1914; *Myrmocladoecus* Wheeler 1921; *Myrmodirhachis* Emery 1925; *Myrmogigas* Forel 1912; *Myrmogonia* Forel 1912; *Myrmomalis* Forel 1914; *Myrmonesites* Emery 1920; *Myrmopalpella* Staercke 1934; *Myrmopelta* Santschi 1921; *Myrmoplatys* Forel 1916; *Myrmopsamma* Forel 1914; *Myrmopytia* Emery 1920; *Myrmosaga* Forel 1912; *Myrmostenus* Emery 1920; *Myrmotarsus* Forel 1912; *Myrmotemnus* Emery 1920; *Myrmothrix* Forel 1912; *Myrmotrema* Forel 1912; *Myrmoturba* Forel 1912; *Myrmoxygenys* Emery 1925; *Neocolobopsis* Borgmeier 1928; *Neomyrmamblys* Wheeler 1921; *Orthonotomyrmex* Ashmead 1906; *Orthonotus* Ashmead 1905 [preoccupied]; *Paramyrmamblys* Santschi 1926; *Pseudocolobopsis* Emery 1920; *Rhinomyrmex* Forel 1886; *Thlipsepinotus* Santschi 1928 | |
| *Colobopsis* Mayr 1861 (NEA, PAL) | |
| *Dendromyrmex* Emery 1895 (NEO) | |
| *Echinopla* F. Smith 1857 (AUS, OR)<br>SYNONYM: *Mesoxena* F. Smith 1860 | |
| *Forelophilus* Kutter 1931 (OR) | |
| *Notostigma* Emery 1920 (AUS) | |
| *Opisthopsis* Emery 1893 (AUS) | *Opisthopsis*: Wheeler (1918b) |
| *Overbeckia* Viehmeyer 1915 (OR) | |
| *Phasmomyrmex* Stitz 1910 (ETH)<br>SYNONYMS: *Myrmacantha* Emery 1920; *Myrmorhachis* Forel 1912 | |
| *Polyrhachis* F. Smith 1857 (AUS, ETH, OR, PAL)<br>SYNONYMS: *Anoplomyrma* Chapman 1963; *Aulacomyrma* Emery 1921; *Campomyrma* Wheeler 1911; *Cephalomyrma* Karavaiev 1935; *Chariomyrma* Forel 1915; *Cyrtomyrma* Forel 1915; *Dolichorhachis* Mann 1919; *Evelyna* Donisthorpe 1937; *Florencea* Donisthorpe 1937; *Hagiomyrma* Wheeler 1911; *Hedomyrma* Forel 1912; *Hemioptica* Roger 1962; *Hoplomyrmus* Gerstaecker 1858; *Irenea* Donisthorpe 1938; *Johnia* Karavaiev 1927; *Morleyidris* Donisthorpe 1944; *Myrma* | *Polyrhachis*: Bolton (1973b, 1975c) |

**TABLE 2–2** *(continued)*

| Genera | Monographs |
|---|---|
| Billberg 1820; *Myrmatopa* Forel 1915; *Myrmhopla* Forel 1915; *Myrmothrinax* Forel 1915; *Polyrhachis* Shuckard 1840 [*nomen nudum*]; *Pseudocyrtomyrma* Emery 1921 | |
| †*Pseudocamponotus* Carpenter 1930 (Elko, Nevada, Oligocene or Miocene) | |
| **Formicinae Incertae Sedis** | |
| †*Paleosminthurus* Pierce and Gibron 1962 (California, Miocene) | *Paleosminthurus*: Najt (1987) |
| **Formicidae Incertae Sedis** | |
| †*Baikuris* Dlussky 1987 (USSR, Cretaceous) | |
| †*Camponotites* Dlussky 1981 (USSR, Miocene) | |
| †*Formicites* Dlussky 1981 (USSR, Miocene) | |
| †*Kotshkorkia* Dlussky 1981 (USSR, Miocene) | |
| †*Ponerites* Dlussky 1981 (USSR, Miocene) | |
| †*Promyrmicium* Baroni Urbani 1971, new name for *Myrmicium* Heer (Spitzbergen, Miocene) | *Promyrmicium:* Baroni Urbani (1971b) |

**TABLE 2–3** Regional faunistic studies and checklists.

| | | |
|---|---|---|
| **Worldwide** | Generic revision and species list | Emery (1910–1925a) |
| | Key to the genera (outdated and of historic interest only) | Wheeler (1922) |
| | Numbers of species in various localities worldwide | Kusnezov (1957b) |
| **Palaearctic Region, including Atlantic islands and Middle East** | | |
| Balkan countries | List | Agosti and Collingwood (1987a,b) |
| China | List | Wheeler (1930–31) |
| England | Distributions and keys | Collingwood (1958a,b, 1964), Collingwood and Barrett (1964), Barrett (1977, 1979) |
| Europe exclusive of Spain | Key to species | Agosti and Collingwood (1987b) |
| Fennoscandia and Denmark | Annotated list | Collingwood (1979) |
| Germany | Monographs | Gösswald (1932), Stitz (1939) |
| Ireland | Annotated list | Collingwood (1958c) |
| Italy | Monographs | Baroni Urbani (1968b, 1971a,c) |
| Japan | Generic synopsis, key to Ponerinae | Ogata (1987) |
| Northern Europe | Biogeography | Baroni Urbani and Collingwood (1977) |
| Saudi Arabia | Annotated list | Collingwood (1985) |
| Scottish Highlands | Annotated list | Collingwood (1961) |
| Soviet Union | Annotated list | Marikovsky (1979) |
| Spain | Annotated list | Collingwood and Yarrow (1969) |
| St. Helena | Annotated list | Taylor and Wilson (1961) |
| Switzerland | Monograph | Kutter (1977) |
| Temperate Asia, Afghanistan to Japan and Korea | List | Chapman and Capco (1951) |
| Tunisia (Galita Archipelago) | Annotated list | Baroni Urbani (1976a) |
| Turkey | Annotated lists | Kutter (1975), Aktaç (1976) |
| Western Europe and Mediterranean Basin | Monograph | Bernard (1968) |
| **Ethiopian and Malagasy Regions** | | |
| Africa south of the Sahara, and Malagasy Region | List and partial monograph | Wheeler (1922) |
| Central Sahara | Descriptions and keys; ecology | Bernard (1948, 1951, 1953), Délye (1968) |
| Madagascar | Monograph | Forel (1891) |
| Southern Africa | Monograph | Arnold (1915–1926) |
| West Africa | Generic synopsis and key to genera | Bolton (1973a) |
| **Oriental Region** | | |
| Asia | List | Chapman and Capco (1951) |
| Borneo | Annotated list | Wheeler (1919b) |
| India, Burma, and Sri Lanka | Monograph | Bingham (1903) |
| Korea | Annotated list | Kim and Kim (1983a,b, 1986), Kim (1986) |
| Melanesia | Monographs and annotated lists | Emery (1914), Mann (1919, 1921), Chapman and Capco (1951), Wilson (1958c,d, 1959a,c,e,f, 1962c, 1964), Taylor (1976b, 1987) |
| **Australian Region, including Oceania** | | |
| Australia | Brief discussion and key to subfamilies | Brown and Taylor (1970) |
| | Checklist | Taylor (1987) |
| New Zealand | Monographs, checklist | Brown (1958a), Taylor (1962b, 1971, 1987) |

*continued*

**TABLE 2–3** *(continued)*

| | | |
|---|---|---|
| Polynesia | Monographs | Taylor and Wilson (1961), Wilson and Taylor (1967a,b), Taylor (1967b), Wilson and Hunt (1967) |
| South Australia | Keys to genera | Greenslade (1979) |
| **Nearctic Region** | | |
| Alaska | Annotated catalog | Nielsen (1987) |
| Arizona | List | Hunt and Snelling (1975) |
| California | Monograph, Deep Canyon | Wheeler and Wheeler (1973) |
| Colorado | Monograph | Gregg (1963) |
| Florida | Local faunal list and ecology | Van Pelt (1956), Deyrup et al. (1988) |
| Iowa | Annotated list | Buren (1944) |
| Michigan | List | Talbot (1975) |
| Nevada | Monograph | Wheeler and Wheeler (1986a) |
| North America north of Mexico | Monograph with keys | Creighton (1950) |
| | Annotated catalog | Smith (1979) |
| North Dakota | Monograph | Wheeler and Wheeler (1963) |
| Quebec | Introduction and monograph of Ponerinae | Francoeur (1979) |
| South Dakota | List | Wheeler and Wheeler (1987) |
| Tennessee | Annotated list | Cole (1940) |
| **Neotropical Region** | | |
| All of Neotropical Region, including West Indies | Checklist | Kempf (1972b) |
| | Generic review of Solenopsidini | Kusnezov (1957a) |
| Argentina | Monographs | Bruch (1916), Gallardo (1916–1932), Kusnezov (1953) |
| Bahamas | Annotated list | Smith (1954) |
| Chile | List, monograph | Kempf (1970a), Snelling (1975), Snelling and Hunt (1975) |
| Cuba | Annotated lists | Wheeler (1913, 1937a), Alayo (1974), Baroni Urbani (1978c) |
| Hispaniola | Annotated lists | Wheeler and Mann (1914), Wheeler (1936b), Wilson (1988) |
| Puerto Rico | Monograph | Smith (1936b), Wilson (1988) |
| Suriname | Annotated list | Kempf (1961b) |
| West Indies | Lists and general description | Wilson (1988) |

Creighton's (1950) review of the ants of North America north of Mexico remains one of the most useful as a convenient guide and quick reference, although it has now been replaced by revisions of more than half the Nearctic genera. It has the distinction of being the first major work to dispense with the clumsy and meaningless polynomials that plagued ant taxonomy for a hundred years. Creighton substituted a much simpler and more efficient system of binomials and trinomials based on modern population concepts (for example, "*Camponotus herculeanus pennsylvanicus* var. *whymperi*" became a synonym under *Camponotus herculeanus*). Through his influence, and that of William L. Brown and a few others working on the entire world fauna, taxonomic procedures have been thoroughly modernized since 1945. Also, the number of generic monographs and faunistic studies has increased conspicuously during the 1980s with the entry of more young investigators into the field. Yet, like a mosaic lacking just enough pieces so that the pattern remains obscure, the classification of the world fauna still lacks satisfying coherence and practical utility.

Ant larvae have been systematically described by Wheeler and Wheeler (1951–1986; syntheses in 1976 and 1979), with a supplementary analysis supplied by Picquet (1958). The basic anatomy and a classification of larval body forms are presented in Figures 2–12 and 2–13. Excellent comparative accounts have been given of the proventriculus (Eisner, 1957), adult mouthparts (Ettershank, 1966; Gotwald, 1969; Buren et al., 1970); wing venation (Brown and Nutting, 1949); antennal sensillae (Masson, 1974; Walther, 1981a,b); adult body sculpturing (Harris, 1979); myrmicine sting apparatus (Kugler, 1978a, 1986); grooming behavior (Farish, 1972); adult carrying behavior (Möglich and Hölldobler, 1974; Duelli, 1977); strigil (cleaning comb) of the foreleg pretarsus (Francoeur and Loiselle, 1988); and physiology of digestion and colonial food flow (Abbott, 1978). The generic characteristics of male ants in the North American fauna have been analyzed by Smith (1943). Otherwise, males have been neglected in most taxonomic studies—usually because of

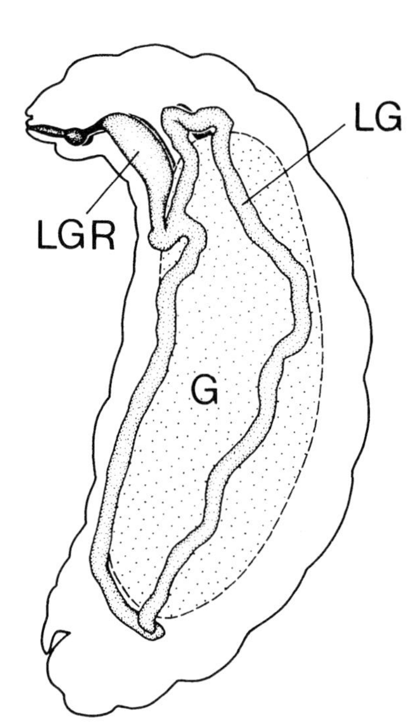


**FIGURE 2–12** A male larva of *Formica* sp., showing the location of the gut (*G*) and large labial (salivary) glands (*LG*); *LGR*, labial gland reservoir. (From Emmert, 1968.)

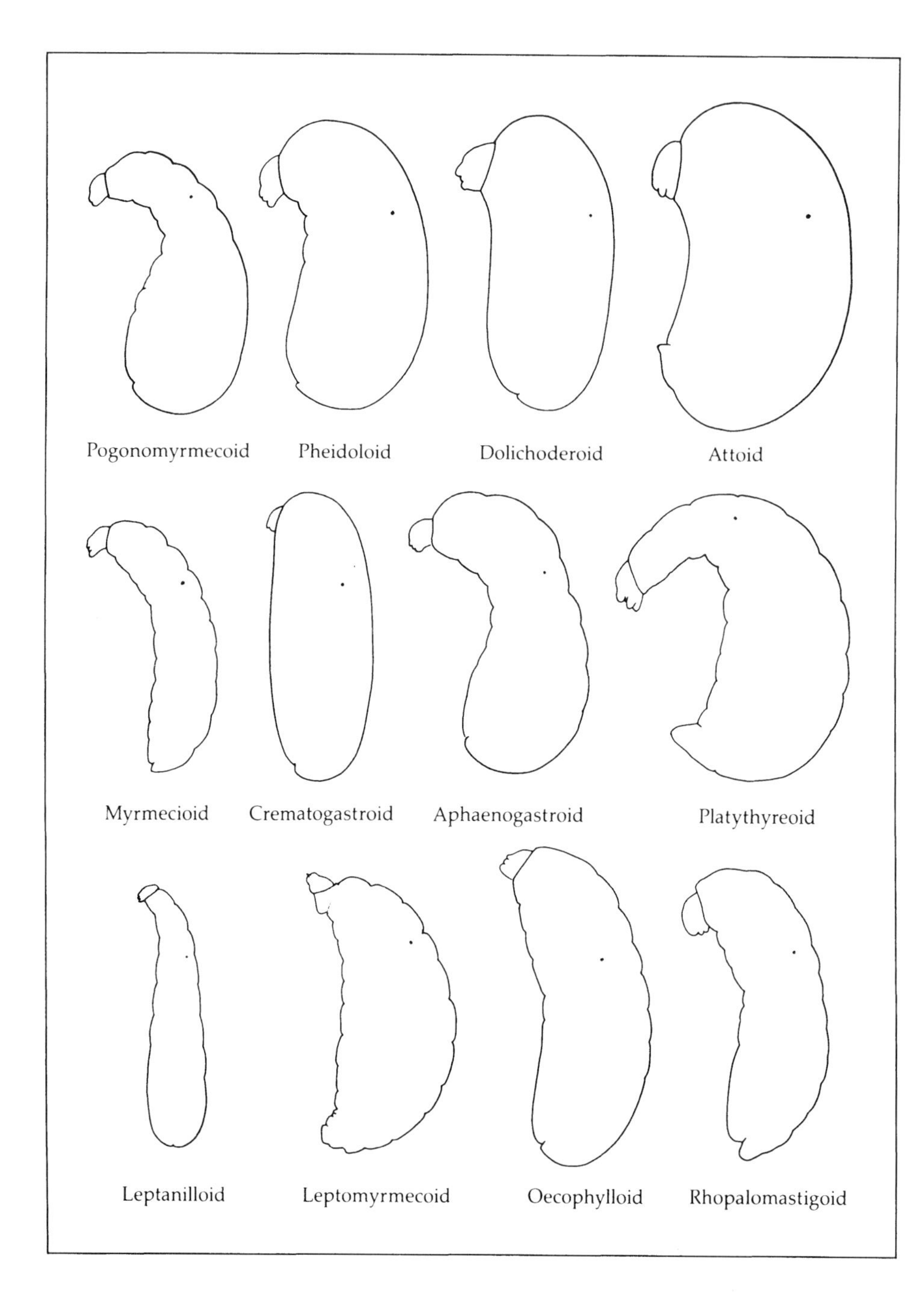


**FIGURE 2–13** The classification of body forms in ant larvae offered by Wheeler and Wheeler (1976).

their greater scarcity in collections in comparison with workers. The cytotaxonomy of ants, which is still a young but potentially very important subject, has been reviewed by Crozier (1975, 1987b), Imai et al. (1977, 1984), Sherman (1979), Hauschteck-Jungen and Jungen (1976, 1983), and Taber (1986). A remarkable case of a karyotype with only one pair of chromosomes has been reported in the primitive bulldog ant *Myrmecia pilosula* by Crosland and Crozier (1986). The possible relation between chromosome numbers and social evolution will be discussed in Chapter 4. Finally, a perceptive and entertaining account of the early history of ant taxonomy has been written by Brown (1955b).

## ORIGIN OF THE ANTS

Until recently the search for the ancestry of the ants always ended in frustration. For more than a hundred years large numbers of fossil ants were recovered from Oligocene and Miocene deposits, but all were members of living subfamilies. Even at the generic level they possessed a distinctly modern aspect. Myrmecologists were forced to consider the Eocene Epoch or even more remotely, the Cretaceous Period, where no certain fossils belonging to the Formicidae were yet known. In 1967 Wilson et al. (1967a,b) obtained the first ant remains of Cretaceous age. The species, *Sphecomyrma freyi* (see Plate 1), and the new subfamily founded on it (Sphecomyrminae) were described from two well-preserved workers in New Jersey amber dating to the late middle (Late Santonian) portion of the Cretaceous Period. The age of the specimen, first estimated to be about 100 million years, has been adjusted to 80 million years (Donald Baird and F. M. Carpenter, personal communication).

*Sphecomyrma freyi* proved to be the nearly perfect link between some of the modern ants and the nonsocial aculeate wasps. In particular, the Cretaceous ants had the following primitive wasp-like traits: mandibles very short and with only two teeth, gaster unconstricted, sting extrusible, and middle and hind legs furnished with double tibial spurs. The *Sphecomyrma* were in fact a mosaic, for they also possessed distinctively ant-like character states: thorax reduced in size and wingless, petiole or "waist" pinched down posteriorly at its juncture with the rest of the abdomen (but still primitive in form in comparison with later ants), and—most important—an apparent metapleural gland, the possession of which is the key diagnostic trait of modern ants. The *Sphecomyrma* were intermediate between most modern aculeate wasps and almost all modern ants in the form of the antennae, which combined a proportionately short first segment with a long, flexible funiculus.

Additional specimens of *Sphecomyrma* of about the same age were next discovered in amber from Alberta, Canada (Wilson, 1985f). In the interim Dlussky (1975, 1983) described an important collection of ant-like forms from several time horizons in the Upper Cretaceous of the Taymyr Peninsula (extreme north-central Siberia), southern Kazakh S.S.R., and the Magadan region of extreme eastern Siberia. He established ten new genera to accommodate this material. In his second report (1983) he also created a new family, the Armaniidae, to receive some of these new forms, while elevating the Sphecomyrminae to family rank (hence, Sphecomyrmidae) to accommodate *Sphecomyrma* and a few related Soviet fossils. In a later analysis, however, Wilson (1987c) marshaled new morphological evidence to assemble all of the Cretaceous formicoids into a single subfamily, the Sphecomyrminae, within the Formicidae and into at most two genera, *Sphecomyrma* and *Cretomyrma*, rather than two families and numerous genera. The females appear to have been differentiated as queen and worker castes belonging to the same colonial species instead of winged and wingless solitary females belonging to different species. This conclusion is supported by the fact that the abdomens of workers of modern ant species and extinct Miocene ant species (chosen to include fossil forms) are smaller relative to the rest of the body than are the abdomens of modern wingless solitary wasps. The wingless Cretaceous formicoids, including the original *Sphecomyrma freyi*, conform to the proportions of ant workers rather than to those of wasps and are therefore reasonably interpreted to have lived in colonies. Representatives inferred to belong to the male and female castes by this interpretation are depicted in Figure 2–14.

In 1986 Jell and Duncan (1986) described *Cretacoformica explicata*, a supposed ant, from Lower Cretaceous beds in Victoria, Australia. If verified, this species would be the earliest known formicid and, because of its geographic origin, extraordinarily important in reconstructing the origin of the ants. However, the single poorly preserved specimen, a male, cannot be assigned with certainty to either the Formicidae or a pre-formicid line. The region of the body that might contain a petiole is covered by the abdomen, which was folded over the rear portion of the alitrunk during preservation. The wings are rounded at their tips and reduced in venation. Overall, the specimen seems more likely to be an aculeate wasp than an ant, but judgment must be reserved until more material becomes available.

The Mesozoic fossils unearthed to date seem to present us with the following picture. During middle and late Cretaceous times representatives of a few species belonging to the very primitive subfamily Sphecomyrminae ranged widely across the northern hemisphere in what was then the supercontinent Laurasia. They were evidently scarce in comparison with later ants in Tertiary and modern times. Only 2 individuals (*Sphecomyrma canadensis*) have been found so far among thousands of insects in amber from Alberta (J. F. McAlpine, personal communication). Formicoids constituted just 13 of the 1,200 insect impressions in the Magadan collection and 5 of the 526 impressions among the Kazakhstan fossils, in other words about 1 percent in both cases (Dlussky, 1983). These figures contrast sharply with the proportionately high representation of ants in Oligocene and Miocene deposits. In the Florissant and other shales of North America (Carpenter, 1930), as well as the Baltic amber of northern Europe (Wheeler, 1914b) and the amber of the Dominican Republic (Wilson, 1985c–e,h), ants are the most abundant insects, making up a large minority of all specimens.

The adaptive radiation destined to propel the ants to dominance took place no later than the beginning of the Tertiary Period, about 65 million years ago. *Eomyrmex guchengziensis*, a species apparently combining traits of *Sphecomyrma* and the living Ponerinae, has been recorded from the early Eocene Fushan deposits of Manchuria (Hong et al., 1974). Amber of mid-Eocene age from Arkansas has yielded representatives of the Myrmicinae, Dolichoderinae, and Formicinae (Wilson, 1985f). In addition, Dlussky (personal communication) has recently found representatives of four living subfamilies (Ponerinae, Aneuretinae, Dolichoderinae, and Formicinae) in

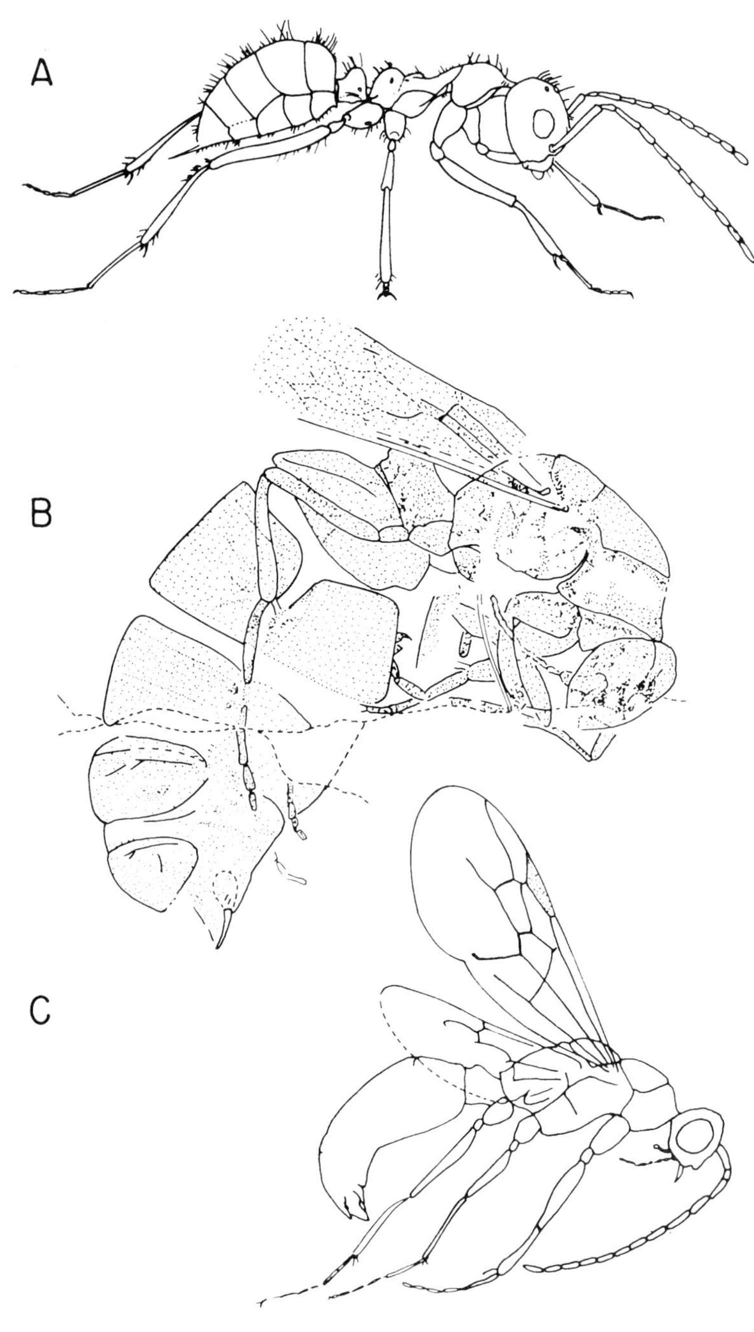


**FIGURE 2–14** The three castes of *Sphecomyrma*, the most primitive known ants, as provisionally associated. *A*, worker: the holotype of *Sphecomyrma freyi*, Cretaceous (Santonian) of New Jersey. *B*, Winged queen: the holotype of *Armania robusta*, Cretaceous (Cenomanian) near Magadan, northeastern Siberia. *C*, Male: the holotype of *Paleomyrmex zherichini*, Cretaceous (Santonian) of the Taymyr Peninsula, north central Siberia. (From Wilson, 1987c.)

Eocene amber from Sakhalin. The exact age cannot be determined, because the amber pieces were not collected in the original deposits, but it is quite likely that the material dates to the early Eocene.

Finally, Lutz (1986) has recognized a new subfamily of gigantic ants, the Formiciinae, from the Lower Eocene of Tennessee and the Middle Eocene of England and Germany. The single genus *Formicium*, not to be confused with *Formica* (the type genus of the Formicinae), is evidently the same as *Eoponera* and *Pseudosirex*, which hymenopterists had previously consigned to the family Pseudosiricidae and classified as a siricoid wasp. The new material described by Lutz makes placement of the Formiciinae plausible. The key traits of the specimens, which are all winged queens and males, are the following:

1. Petiole produced into an erect, thin scale of the kind found in many species of the Formicinae.
2. Reduced sting, also similar to the Formicinae.
3. Wing venation primitive, closely resembling the "idealized" pattern of our Figure 2–11 and unique to the Formiciinae.
4. Huge size, with the forewing length of the queen from 25 to 65 millimeters according to species; the largest species, *Formicium giganteum*, exceeds in size any other known ant, living or extinct.
5. Spiracles of the gaster proportionately very large and slit-shaped.
6. Distinctive forewing venation, which we interpret as follows (differently from Lutz): cuticle of the stigma thin and transparent, making it appear to be an additional cell.

Only 1 of the approximately 10 genera thus far recorded from these Cretaceous and Eocene deposits is extant (*Iridomyrmex*, from the Arkansas amber). Yet no fewer than 24 genera, or 56 percent of the 43 total represented in the early Oligocene Baltic amber fossils, still survive, including such currently abundant and widespread forms as *Ponera, Tetraponera, Aphaenogaster, Monomorium, Iridomyrmex, Formica*, and *Lasius* (Wheeler, 1914b). At least one species, *Lasius schiefferdeckeri*, is so close to living species of the *L. niger* group of North America and Eurasia that it can be distinguished only by minor average differences in antennal and mandibular form (Wilson, 1955a). This modern facies is even more evident in the Dominican amber, which apparently dates from the early Miocene. Here no fewer than 35 genera, or 92 percent of the total 38, still survive. Further, the great majority of species analyzed to date have been placed in modern species groups. In a few instances they are difficult to separate from modern forms even at the species level (Wilson, 1985c–e,h).

From what taxonomic family did the Sphecomyrminae evolve? In a preliminary phenetic analysis of the original *Sphecomyrma freyi*, Wilson et al. (1967b) placed the Mesozoic subfamily closest to the methochine Tiphiidae among living aculeate wasps. In a later study Brothers (1975) applied a cladistic analysis of 92 characters to all of the living aculeate families including the ants (Formicidae) as a whole, and arrived at a different result. As shown in Figure 2–15, he derived the ants from a clade later than that giving rise to the Tiphiidae but earlier than the branching that led to the modern Eumenidae, Masaridae, Scoliidae, and Vespidae. Among the 12 families of the Vespoidea in this scheme, the Formicidae are so distinctive that Brothers felt compelled to place them in an informal section of their own, the Formiciformes, with the remaining 11 families composing the Vespiformes. A case could be made for the retention of the superfamily Formicoidea, but according to the logic of cladistic classification this is not permissible—unless families antecedent to the Formicidae were also split off to create several other superfamilies.

Our current view of phylogeny within the Formicidae is summarized in Figure 2–16. This arrangement is the latest revision in a succession of earlier cladograms by Brown (1954a), Wilson et al. (1967b), and Taylor (1978c). The principal new feature is the more

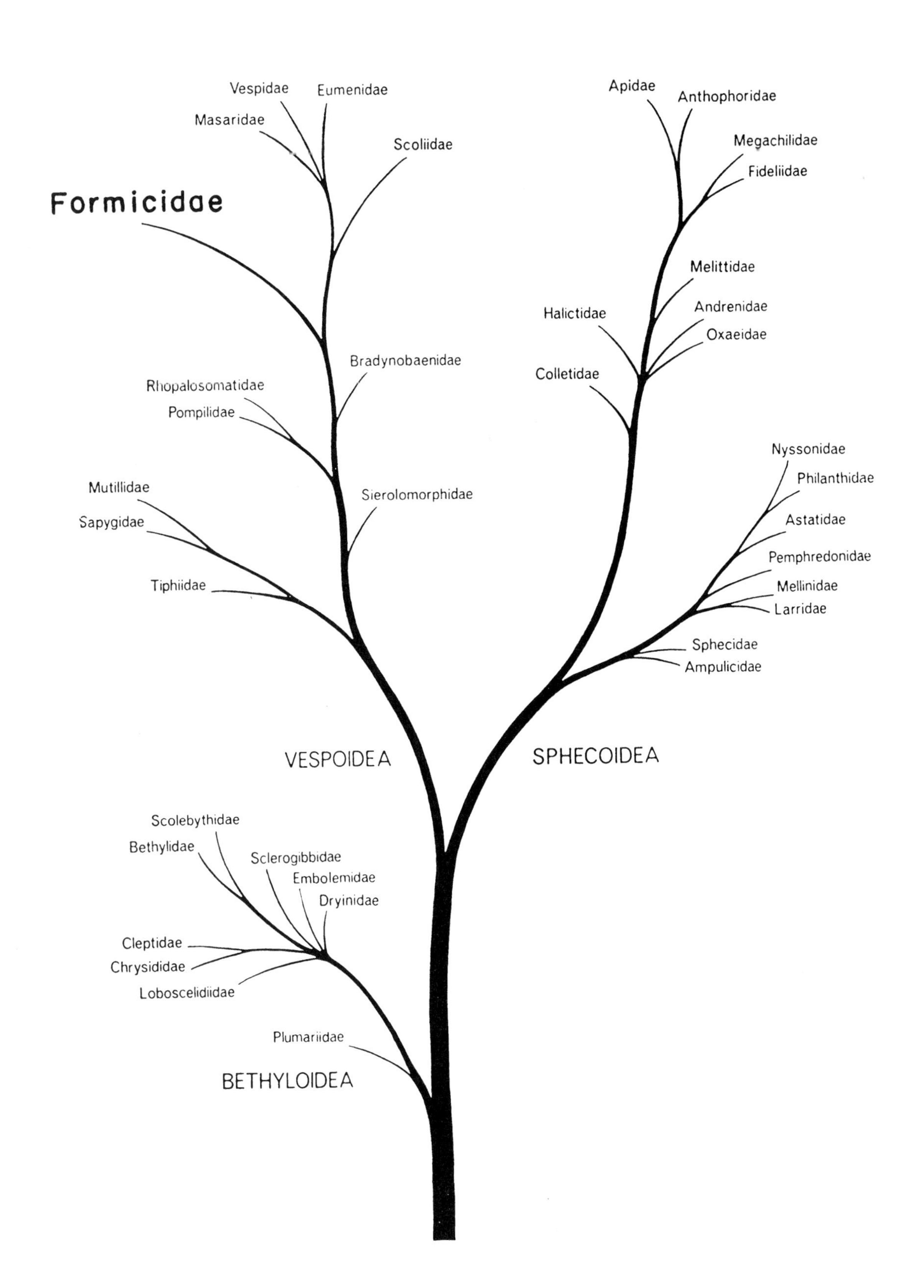


**FIGURE 2–15** Cladogram of the living aculeate families based on the analysis by Brothers (1975).

extensive case of exocrine glands, about which a great deal has been learned during the 1980s. One important change from earlier schemes is the very early divergence of the subfamily Formicinae, probably in late Cretaceous or earliest Tertiary times. A similar conclusion was reached independently by W. L. Brown (personal communication).

In reconstructing phylogenies in this manner, we must keep in mind the distinction between primitive character states and primitive taxa. A primitive character state (the "plesiomorphous state" of many authors) is simply one that precedes a more advanced one ("apomorphous state") in evolution. The two states can be major, as in absence of the metapleural gland changing to presence of the gland; or they can be quite trivial, as in 10 hairs on the pronotum changing to 20 hairs. A primitive taxon, on the other hand, is judged to be so on a much more subjective basis. It is a species, or a genus, or a taxon belonging to some other, higher category that possesses a relatively large number of primitive character states in comparison with other taxa of the same rank. Thus we speak of the Sphecomyr-

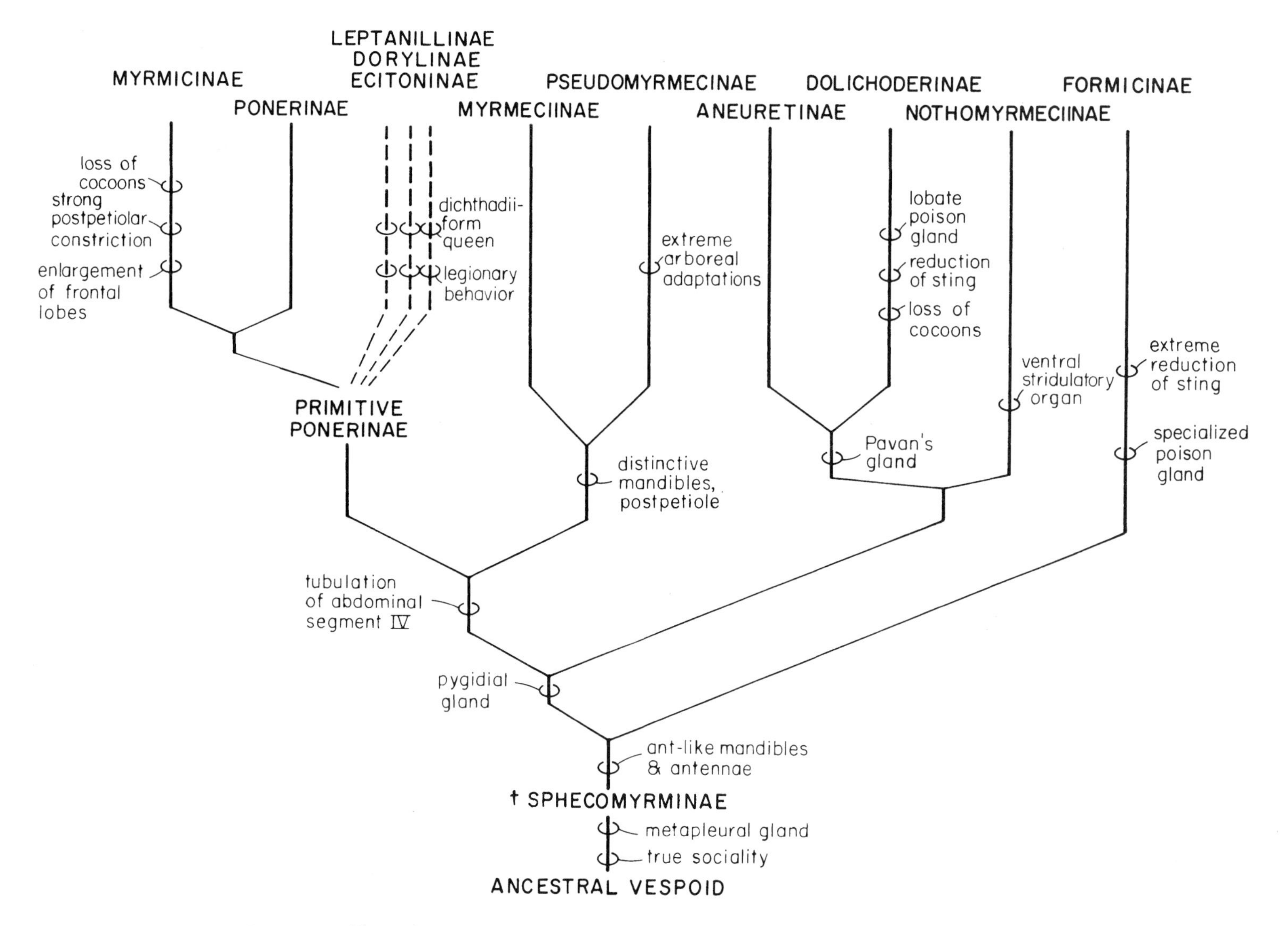


**FIGURE 2–16** The authors' current view of the phylogeny of the subfamilies within the Formicidae.

**FIGURE 2–17** Diagrams of exploded abdominal plates of primitive ants, illustrating anatomical differences in segment IV, which has its sclerites associated to form a tubulate structure in *Amblyopone* (*A*) and *Myrmecia* (*B*), but separate in *Nothomyrmecia* (*C*). (From Taylor, 1978c.)

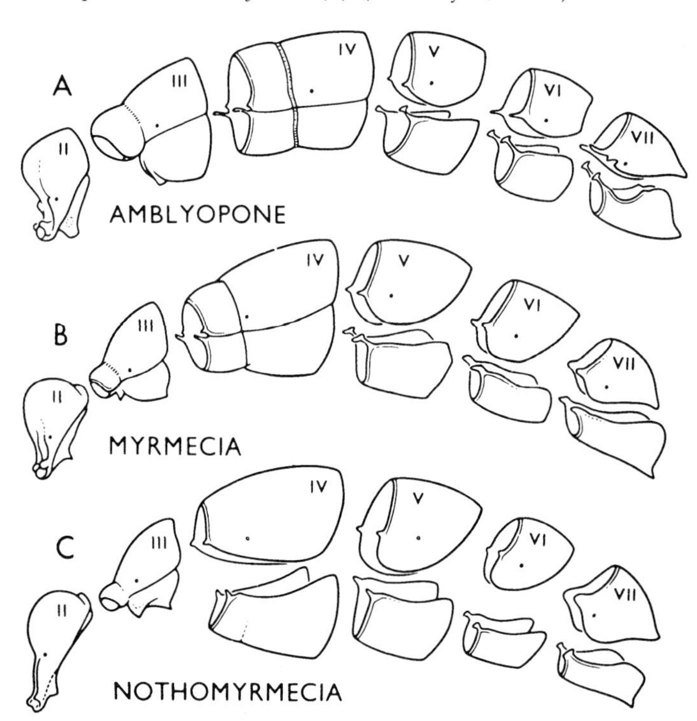


minae as a primitive subfamily relative to the Myrmicinae because sphecomyrmine workers are characterized by short, wasp-like mandibles, a single symmetric petiole, and other important (i.e., complex) primitive character states.

Entomologists will no doubt alter our ant cladogram further as new species are discovered, especially fossil forms from Cretaceous, Paleocene, and Eocene deposits, and as new characters and dates are added to the analysis. They will also disagree on which are the most primitive taxa. A number of difficulties already exist in our own version. One of the most troubling is the relation of *Nothomyrmecia macrops,* a very primitive ant with reference to the living Formicidae, to *Amblyopone,* a worldwide genus of primitive ponerines thought to be derived from the sphecomyrmine-nothomyrmeciine clade. *Amblyopone* has a tubulated abdominal segment, clearly a derived character state (see Figure 2–17). But *Amblyopone* also has a petiole (abdominal segment II) that attaches broadly to the gaster (abdominal segments III–VII), which is presumably a primitive state. In addition, it possesses reduced eyes, proportionately short appendages, and a heavily sclerotized exoskeleton. *Amblyopone* is clearly an ant adapted for hypogaeic foraging, that is, for hunting underground and within enclosed spaces in rotting wood and leaf litter. *Nothomyrmecia,* in contrast, is epigaeic, hunting aboveground and in the open (see Plate 12). *Sphecomyrma,* to judge from its appearance (large eyes, proportionately long appendages, thin exoskeleton) was also epigaeic. Either the broad petiolar attachment of *Amblyopone* and related genera in the tribe Amblyoponini represents an evolutionary reversion from the primary narrow petiole of the Ur-Formicidae, or the Amblyoponini (and possibly the remainder of the Ponerinae, Myrmicinae, and related higher subfamilies) are an independent clade of ants predating the Nothomyrmeciinae and Sphecomyrminae. We favor the first, more conservative hypothesis. The matter will, however, remain open until more evidence is obtained. For example, the discovery of *Amblyopone* like forms with fully constricted petioles from post-Cretaceous deposits would favor the first hypothesis. If, on the other hand, *Amblyopone*-like forms are discovered in deposits at least as old as those containing *Sphecomyrma,* the second, more radical hypothesis will receive strong support.

## BEGINNINGS OF SOCIAL BEHAVIOR

All known living ants are eusocial, with strong physical differences separating the queen and worker castes. Thus a large gap in social behavior remains between even the most primitive ants, including *Amblyopone, Myrmecia,* and *Nothomyrmecia,* and their closest living relatives among the vespoid wasps. Without further evidence it would be very difficult to infer the steps that led to eusocial behavior in these insects. The living eusocial species of Vespoidea *are* connected, however, to solitary species of other vespoids and aculeate wasps by finely graded steps, providing an independent but otherwise solid base for inference. Howard E. Evans (1958), drawing on his own extensive knowledge of the solitary Hymenoptera and on studies by Richards and Richards (1951) and other contemporary students of the social wasps, proposed an ethocline that has stood up well under the test of more recently accumulated field data. His schema collates two independent forms of information: first, the morphological similarity and the inferred direction of morphological change, which are expressed in the branching pattern of cladograms; and second, the sequence of grades that can be logically envisioned to have occurred during behavioral evolution. It is notable that several families of wasps (Pompilidae, Sphecidae, Eumenidae, and Vespidae) have to be included in order to tell the whole story. This in no way vitiates the theory. Indeed, if we believe that behavioral evolution is even loosely correlated with morphological evolution, it follows that single taxa such as families and genera should encompass less behavioral variation than all the wasps taken together. Evans' 13 grades, starting with the simplest and presumably most primitive, are as follows:

1. Female stings prey, lays egg.
2. Female stings prey, places it in a convenient niche, lays egg.
3. Female stings prey, constructs a nest on the spot, lays egg.
4. Female builds a nest, stings prey, transports it to nest, lays egg.
5. Female builds a nest, stings and transports a prey item, lays egg, then mass provisions with several more prey (added quickly, before egg hatches).
6. As in (5) but prey items are progressively provided, as the larva grows.
7. As in (6) but progressive provisioning occurs from the start.
8. In addition to progressive provisioning in a preconstructed nest, female macerates prey items and feeds the pieces directly to the larvae.
9. Founding female is long-lived, so that offspring remain with her in the nest, add cells, and lay eggs of their own.
10. Little colony of cooperating females engages in trophallaxis (liquid food exchange), but there is still no division into reproductive and worker castes.
11. Behavioral division between a dominant queen caste and subordinate worker caste appears; unfertilized workers may still lay male-destined eggs.
12. Larvae are fed differentially; queen and workers that result are physically distinct, but intermediates remain common.
13. Worker caste is physically strongly differentiated, and intermediates are rare or absent.

These grades are said to follow the *subsocial* route, through which a single foundress gains enough longevity to coexist in the same nest as her female offspring. In this case the most primitive colony is an extended family: the founding female is accompanied by her daughters, sons, and grandsons, but not under ordinary circumstances by her granddaughters, because the unfertilized eggs of her daughters produce only males.

A relatively minor variation on the 13 grades is the *parasocial* route (Michener, 1969, 1974), which starts with members of the same generation using the same composite nest and also cooperating in brood care, instead of a single foundress carrying through on her own. It is possible for one of the females to dominate her contemporaries and to become the de facto queen, as occurs, for example, in founding aggregations of the paper wasp *Polistes.* On the basis of more recent field studies, West-Eberhard (1978) has argued that this was the prevalent route followed during the origin of the eusocial wasps. She has proposed the "polygynous family group hypothesis," in which the aggregating foundresses are typically sisters or at least close cousins. Their mutualistic association is followed by serial queendom, in which first one female or small group of females and then another takes over oviposition. Both foundresses and offspring can lay eggs in such family groups. In later evolution progressively fewer females participate in reproduction. Finally, a sharply demarcated, mostly nonreproductive worker caste emerges. West-Eberhard's scheme has been supported by a cladistic study of anatomical and behavioral traits performed by J. M. Carpenter (1989), and it has been widely favored among other wasp specialists.

Against this backdrop we can now place existing knowledge of the anatomically most primitive ants. As summarized in Table 2–4, all of the key traits are sufficiently similar across *Amblyopone, Myrmecia,* and *Nothomyrmecia* to justify placing them on the Evans' vespoid *scala.* All three genera are in the most advanced (thirteenth) stage, yet their behavior is in every way reminiscent of a clade that began as tightly knit, soil-dwelling families of medium-sized vespoid wasps. Haskins and Haskins were essentially correct when they concluded, on the basis of their studies of *Myrmecia:*

> The existent wealth and variety of Formicid social structures, with their tremendous range of variation from group to group, took their

**TABLE 2-4** Primitive behavior and social traits in three anatomically primitive genera. X, present; O, absent (that is, a derived trait is present); XO, present in some species, absent in others; —, no data. (Based on data from Wheeler, 1933b; Haskins and Haskins, 1950a, 1951; Freeland, 1958; Traniello, 1978, 1982; and Hölldobler and Taylor, 1983.)

| Trait | *Amblyopone* spp. | *Myrmecia* spp. | *Nothomyrmecia macrops* |
|---|---|---|---|
| Multiple queens in colony founding | XO | XO | X |
| Queens forage during colony founding | X | X | X |
| Queens revert to foraging outside nest when deprived of workers | — | X | — |
| Eggs lie separately on nest floor | O | X | X |
| Larvae are fed directly with insect fragments | X | X | X |
| Adults highly nectarivorous and collect prey mainly to feed larvae | O | X | X |
| Larvae active and able to crawl short distances | — | X | X |
| Trophic eggs absent | X | O | O |
| Workers do not assist larvae at pupation or new adults at eclosion | O | O | O |
| Colonies typically small at maturity (<100 adults) | X | O | X |
| Nests very simple, in soil or rotting vegetation | X | O | X |
| Workers monomorphic (only one subcaste) | X | O | X |
| Workers and queens closely similar in size and anatomy | X | XO | X |
| Alarm communication absent | O | O | O |
| Recruitment to food sources absent | O | X | X |
| Coordinated foraging absent | X | X | X |
| Regurgitation among colony members absent | X | O | O |
| Adults do not groom brood | O | O | O |
| Adults do not groom one another | O | O | O |
| Adult transport awkward and rare, or absent | — | X | X |
| Nest odors absent | — | O | O |
| Roles remain the same regardless of age (age polyethism) | X | — | — |

evolutionary beginnings in the activities of solitary, winged, ground-dwelling wasplike types in which the female, having dealated herself after fertilization, constructed a shelter in the ground and reared a small family to maturity. The larvae were provided with freshly killed prey captured and supplied through a behavior pattern intermediate in its complexity between the simple provisioning of paralyzed insects characteristic of the modern solitary Sphecoid wasps and the malaxated pellets of insects prepared for the larvae by their nurses in such primitive social Vespids as *Polistes*. (Haskins and Haskins, 1951: 444)

It is further possible that the original foundresses gathered in small associations as in some *Polistes*. In the language of our formal lexicon, the evolution was parasocial rather than strictly subsocial. In this case the polygynous-group hypothesis of West-Eberhard could also apply to the origin of the ants. Colonies of *Nothomyrmecia macrops* are in fact sometimes founded by multiple queens, although well-developed colonies typically have only a single queen. When Hölldobler and Taylor (1983) introduced two queens to a group of queenless workers, the queens behaved amicably toward each other at first, but later one began to dominate the other by standing above her at frequent intervals (see Figure 6-7). Later the workers expelled the subordinate queen by repeatedly dragging her outside the nest. This pattern of the steady eviction or execution of multiple foundresses until only one remains is also widespread in the phylogenetically more advanced subfamilies of ants, as will be shown in Chapter 6.

An opposing hypothesis of the origin of social behavior was advanced by the late Soviet entomologist S. I. Malyshev (1960, 1968). He postulated that specialization on big prey resulted in the mother wasp's staying in the vicinity of her young long enough for them to get to know her and to cooperate with her. Malyshev further postulated that the precursors of the ants must have fed on fungi growing in the nest wall. Otherwise, he contended, the young colonies would have had no way of tiding over the period of scarcity after the initial large prey had been consumed. But this suggestion ignores what we know of colony founding throughout the ant world and is wholly unsupported by any evidence of fungus eating in the lower ants.

The model for the large prey theory is provided by members of the bethylid genus *Scleroderma*, particularly *S. immigrans* and *S. macrogaster*, which were studied in detail by Bridwell (1920) and Wheeler (1928). The female *S. macrogaster*, for example, is only 2.5–3 millimeters long, and she attacks beetle larvae that are hundreds or thousands of times greater in bulk. In a typical sequence, the female first crawls over the surface of her prey, pausing from time to time to grip little folds of the cuticle. She stings the larva at any point where the muscles show signs of contraction. Finally, after one to four days, the larva becomes completely paralyzed. The *Scleroderma* now feeds for several days by making little punctures in the cuticle and drinking the hemolymph. Her abdomen then begins to swell as the ovaries develop, and after a time she lays eggs on the surface of the prey. The remainder of the life cycle has been described by W. M. Wheeler in the following striking passage:

> The eggs laid on a larva or young pupa produce minute larvae which at first lie on the surface but later become spindle-shaped and erect, so that the host bristles with them like a porcupine. The older larvae acquire the colour of the juices of the prey; those feeding on the pink larvae or pupae of *Liopus* becoming red. They are always spotted with white, owing to the large masses of urate crystals in their fat bodies. The mother *Scleroderma* remains with the larvae, often stands over them and may sometimes lick them, holding them meanwhile in her fore feet. She also continues occasionally to drink the host's blood, which exudes about the deeply inserted heads of her larval offspring. Although she will sometimes eat her eggs I have never seen her attack one of her larvae. The devouring of some of the eggs seems to be due to a tendency to regulate their number according to the volume of the prey. When the larvae are mature they fall away from its shrivelled and exhausted remains and spin snow-white cocoons in a cluster. Pupation covers a period of fourteen to thirty days. The males emerge first from their cocoons, at once eat their way into the female cocoons and fecundate the pupae. They also mate readily with the same individual five to eight times after brief intervals. The same females may also mate with several males in succession. So great is the ardour of the latter that they often attempt to mate with one another. The mother being a long-lived insect may mate with one of her sons and will readily paralyze another beetle larva, rear another brood and mate again with one of her grandsons. (Wheeler, 1928: 63)

Although Malyshev, inspired by Wheeler's observations, spoke of the sclerodermoid ancestors of the ants, morphological evidence rules against any of the ant groups having been derived from *Scleroderma*-like progenitors or any of the other known Bethylidae. Furthermore, as suggested by the family-level cladogram of Figure 2–15, the bethyloids diverged from the early vespoids well before the origin of the ants from a vespoid stock. The accumulating behavioral evidence also militates against the Wheeler-Malyshev route to eusociality in favor of the vespoid route, in which prey are brought to previously constructed nests.

Wilson (1971) suggested that amblyoponine ants might have evolved in the Wheeler-Malyshev manner, because *Amblyopone* were known occasionally to transport their larvae to centipedes and other large prey rather than the other way around. It is now clear, however, that one species at least, *A. pallipes*, more commonly transports prey back to the brood chambers of the nest as part of a stereotyped and efficient predatory sequence (Traniello, 1982). Wilson also found that the amblyoponine *Myopopone castanea* carries larvae to large wood-boring larvae that cannot possibly have been transported any significant distance from the spot where they were disabled by the workers. These ants spread their nests widely beneath the bark of rotting logs, however, and they also forage there. It can be argued that the entire subcortical surface constitutes the "nest" of the ants. Finally, the small amblyoponine *Prionopelta amabilis* has now been shown to be closer in some important respects to higher ants than to *Amblyopone* and *Myopopone*. The colonies are large, consisting of hundreds of workers; there is a marked size difference between queen and worker; the nests are more elaborate, with pupal chambers being "wallpapered" with fragments of discarded cocoons; and, perhaps most important, the workers specialize on campodeid diplurans, small flightless insects that they capture and carry back to their nests (Hölldobler and Wilson, 1986a).

## THE CAUSES OF SUCCESS

What unusual or unique biological traits led to the remarkable diversification and unchallenged success of the ants for over 50 million years? The answer appears to be that they were the first group of predatory eusocial insects that both lived and foraged primarily in the soil and in rotting vegetation on the ground. Although many ant species are specialized for arboreal existence, the great majority of these distinctive forms live in tree boles, hollow twigs, and moist subcortical cavities that simulate an earthen environment; arboreal life appears to represent a secondary, minority adaptation. To an exquisite degree ants are creatures of the ground. The wingless workers can easily penetrate small, remote cavities less accessible to flying wasps, which are burdened with wings and bulky thoraces. Armed with stings and toxic chemical weapons, the ant workers are formidable predators. They orient in part by odor cues on the ground, and most species are able to recruit foraging parties with a high degree of efficiency through the use of odor trails laid over the surface. After entering the adaptive zone of social, terrestrial predators, no later than the Upper Cretaceous, the ants apparently preempted its occupation by other candidate groups among the insects.

Eusocial behavior is a rare evolutionary achievement among insects. The evidence of living groups indicates that it has developed about twelve times within the Hymenoptera and once in the protoblattoid line that gave rise to the termites (Wilson, 1971; Michener, 1974). Richly organized colonies of the kind made possible by eusociality enjoy several key advantages over solitary individuals.

Under most circumstances groups of workers are better able to forage for food and defend the nest, because they can switch from individual to group response and back again swiftly and according to need. Individual ant workers for the most part can perform as competently as individual solitary wasps—except, of course, in the case of reproduction. When a food object or nest intruder is too large for one worker to handle, nestmates can be assembled by alarm or recruitment signals. Of equal importance is the fact that the execution of multiple-step tasks is accomplished in a series-parallel sequence instead of a parallel-series sequence (Oster and Wilson, 1978). That is, individual ants can specialize on particular steps, moving from one object (such as a larva to be fed) to another (a second larva to be fed). They do not need to carry each task to completion from start to finish—for example, to check the larva first, then collect the food, then feed the same larva. Hence, if each link in the chain has many workers in attendance, a series directed at any particular object (a given hungry larva) is less likely to fail. Moreover, ants specializing on particular labor categories typically constitute a caste specialized by age or body form or both. There has been some documentation of the superiority in performance and net energetic yield of various castes for their modal tasks, but careful experimental studies are still relatively few (e.g., Wilson, 1980b; Porter and Tschinkel, 1985).

What makes ants unusual even in the select company of eusocial insects is the fact that they are the only eusocial *predators* occupying the soil and ground litter. Termites live in the same places and also have wingless workers, but they feed almost exclusively on dead vegetation.

Ants have a number of adaptations fitting them for their special way of life. One of the most striking is the elongation of the mandibles into working tools. The primitive formicid mandible is a blade whose inner border is lined with a row of sharp teeth used for gripping and cutting. This basic shape is found in most species of both primitive and advanced subfamilies, but it has been altered in a few to resemble a sickle and other shapes, which serve either for the capture of unusual prey or as fighting instruments in the defense of the colony.

A second important innovation is the metapleural gland, a pair of cell clusters that open into chambers located at the extreme rear corners of the mesosoma, the major middle portion of the body (see Figure 7–27). The gland produces phenylacetic acid, which is active against fungi and bacteria, and possibly other antibiotic substances as well. The body of the average worker of the leafcutter ant *Atta sexdens* contains 1.4 micrograms of phenylacetic acid at any given time (Maschwitz et al., 1970). Fungistatic activity has recently been demonstrated in the metapleural gland of the primitive ant *Myrmecia nigriscapa* (Beattie et al., 1986). Where bees and wasps protect their immature forms by constructing antibiotic-impregnated brood cells, ants appear to disseminate antibiotic secretions diffusely through the nest from the metapleural gland. This innovation is likely to have played a role in the successful colonization of the moist, microorganism-ridden environment in which the great majority of ant species live.

The metapleural gland comes closest to being a single diagnostic character separating the Formicidae from all other aculeate Hymenoptera, but it is far from universal. The gland has been secondarily lost in a few phyletic lines, especially genera such as *Camponotus*, *Dendromyrmex*, *Oecophylla*, and *Polyrhachis*, which specialize in the occupation of arboreal (hence drier and cleaner) environments (Hölldobler and Engel-Siegel, 1984). In addition it is reduced or absent in the males of many ant species as well as in the more extreme social parasites (Brown, 1968).

**TABLE 2–5** A glossary of anatomical and other specialized terms commonly used in ant taxonomy (see also diagrams in Figures 2–1 to 2–11).

**Pilosity (body hairs)**

*Appressed*. Referring to a hair that runs parallel, or nearly parallel, to the body surface.
*Decumbent*. Referring to a hair that stands 10 to 40 degrees from the surface.
*Erect*. Referring to a hair that stands straight up, or nearly straight up, from the cuticle.
*Hair*. A seta; see *Pilosity*.
*Lanuginous*. Woolly or down-like.
*Piligerous*. Bearing a hair, as in the bottom of a fovea or on the summit of a tubercle.
*Pilosity*. The longer, stouter hairs, or setae, which are outstanding above the shorter, usually finer hairs that constitute the pubescence.
*Plumose*. Referring to hairs that are multiply branched and hence feather-like in overall appearance.
*Pubescence*. Exceptionally short, fine hairs, typically forming a second layer beneath the pilosity.
*Recumbent*. Referring to a hair lying on the body surface.
*Squamate*. Scale-shaped.

**TABLE 2–5** *(continued)*

*Subdecumbent*. Referring to a hair that stands about 45 degrees from the body surface.
*Suberect*. Referring to a hair that bends about 10 to 20 degrees from the vertical.

**Sculpturing (of body surface)**

*Aciculate*. Finely striate, as if scratched by a needle.
*Alveolate*. Honeycombed; furnished with alveoli, or cup-shaped depressions, each of which often contains a hair.
*Areolate*. Divided into a number of small, irregular cavities or spaces.
*Carinate*. Possessing carinae (elevated ridges), especially in parallel rows.
*Carinulate*. Possessing carinulae (small elevated ridges), especially in parallel rows.
*Corrugated*. Wrinkled, especially with alternative and parallel ridges and channels.
*Costate*. Bearing costae (elevated ridges rounded at the crest), especially in parallel rows.
*Costulate*. With costulae (small elevated ridges), especially in parallel rows.
*Foveate*. With multiple foveae, or deep pits with well-marked sides.
*Foveolate*. With multiple small, deep pits which are deeper and larger than punctures; hence coarser than punctate.
*Glabrous*. Smooth, hairless, and shining.
*Pruinose*. Having the appearance of being frosted or lightly dusted in the manner of a plum; for example, the body surface of *Platythyrea*.
*Punctate*. Bearing fine punctures like pinpricks; compare with *Foveate* and *Foveolate*.
*Reticulate*. Covered with a network of carinae, striae, or rugae.
*Reticulate-punctate*. Covered by a network of carinae, striae, or rugae with punctures in the interspaces.
*Rugoreticulate*. With rugae (wrinkles) forming a network or grid.
*Rugose*. With multiple wrinkles, especially running approximately in parallel. See also *Rugoreticulate* and *Scabrous*.
*Rugulose*. With multiple small wrinkles, especially running in parallel.
*Scabrous*. Roughly and irregularly rugose.
*Shagreened*. Covered with a fine but close-set roughness, like shark leather.
*Striate*. Marked with striae, or multiple impressed lines.
*Striate-punctate*. With rows of punctures.
*Strigate*. Transversely striate.
*Sulcate*. Deeply furrowed or grooved, in other words bearing a sulcus (singular) or sulci (plural); compare with *Striate*.
*Tuberculate*. Covered with tubercles (small thick spines or pimple-like structures).
*Verrucose*. Covered with irregularly shaped lobes or wart-like protuberances.

**Shapes and locations**

*Acuminate*. Tapering to a fine point.
*Acute*. Sharply angulate, less than 90 degrees.
*Angulate*. Forming an angle, as the rear corners of the head or pronotal border.
*Apical*. At or near the tip.
*Arcuate*. Curved, bow-like.
*Attenuated*. Drawn out, tapered, as a portion of a waist segment.
*Basad*. Located near or toward the base.
*Biconvex*. Convex on opposite sides, lens-shaped, as a segment of the waist.
*Bidentate*. Bearing two teeth, as on the anterior clypeal border.
*Clavate, claviform*. Thickened, especially toward the tip.
*Cordate*. Heart-shaped.

*continued*

**TABLE 2–5** *(continued)*

*Dentate.* Toothed, as the dentate inner borders of the mandibles.
*Denticulate.* Furnished with minute teeth or tooth-like structures.
*Depressed.* Flattened down as if pressed.
*Distad.* Located toward the distal or farthest end.
*Distal.* Farthest away from the body, as the distal portion of a spine; compare with *Proximal.*
*Dorsal.* Pertaining to the dorsum or upper surface, versus *Ventral.*
*Dorsoventral.* Along a line drawn from the upper to lower surface.
*Dorsum.* The upper surface.
*Emarginate.* Notched, with a piece of any shape such as rounded or square, seemingly cut from the margin; for example, an emarginate anterior clypeal border. Versus *Entire.*
*Entire.* Referring to a smoothly unbroken margin, versus *Emarginate.*
*Excised.* With a deep cut or notch, as on the margin of a segment; an extreme form of emargination.
*Falcate.* Sickle-shaped.
*Flagellate.* Whip-like, as an antenna.
*Gibbose.* Humpbacked.
*Incrassate.* Conspicuously swollen, especially near the tip, as an incrassate scape.
*Lobiform.* Lobe-shaped.
*Median.* Pertaining to the middle.
*Mesad.* In the direction of the middle or median.
*Pectinate.* Comb-shaped.
*Pedunculate.* Stalk-like, or set on a stalk or peduncle, as the waist of many ant species.
*Phragmotic.* Sharply truncated, as the overall front of the head in some stem-dwelling ants.
*Proximad.* Located toward the proximal or nearest end with reference to the body.
*Proximal.* Closest with reference to the body.
*Replete.* Conspicuously swollen with liquid food; said of the abdomen.
*Serrate.* With teeth along the edge, saw-like.
*Sternal.* Pertaining to the sternum or lower portion of the body or a body part.
*Tergal.* Dorsal (pertaining to the upper surface).
*Ventral.* Pertaining to the lower surface.

**Body parts and special terms**

*Abdomen.* The third, posteriormost major division of the body.
*Aedeagus.* The penis.
*Alate.* Winged.
*Alitrunk.* The second, middle major division of the body, consisting of the true thorax to the rear of which is fused the first segment of the true abdomen (called the propodeum). This division is sometimes called the mesosoma.
*Allotype.* The paratype of opposite sex to the holotype.
*Anal cell.* The space between the anal veins.
*Antennal condyles.* The narrowed, neck-like portions of the first antennal segment that connect to the head surface.
*Antennal fossa.* The cavity or depression of the head into which the antenna is articulated.
*Antennomere.* An antennal segment.
*Apodeme.* An ingrowth or other rigid process of the exoskeleton, typically serving for muscle attachment.
*Arolium* (plural: *arolia*). A pad-like structure between the tarsal claws.
*Atrium* (plural: *atria*). A chamber at the entrance of a body opening, as for example the cavity over the metapleural gland.

**TABLE 2–5** *(continued)*

*Basitarsus.* The first segment of the tarsus ("foot") outward from the body.
*Buccal cavity.* The mouth cavity.
*Bulla.* A blister-like structure, as for example the thin, convex roof of the metapleural gland chamber.
*Bursa.* A sac or pouch, as the bursa copulatrix or modified vaginal area of the female.
*Carina* (plural: *carinae*). An elevated ridge or keel, not necessarily acute.
*Carinula* (plural: *carinulae*). A small carina.
*Clypeus* (plural: *clypei*). The foremost section of the head capsule, just back of the mandibles, demarcated posteriorly by a transverse suture (see Figures 2–1 and 2–2).
*Condyle.* A structure that articulates an appendage to the body surface, e.g., the condyle of the antenna.
*Costa* (plural: *costae*). A ridge or keel rounded at the top.
*Costula* (plural: *costulae*). A small costa.
*Cotype.* Old expression for *syntype* (*q.v.*).
*Coxa.* The basalmost segment of the leg, by which it is attached to the body.
*Cribellum.* A sieve-like plate.
*Dealate.* Having shed the wings, as an inseminated female.
*Declivity.* A downward-sloping surface, as the posterior face of the propodeum.
*Description.* A shorthand expression for *Original description* of a species, genus, or other newly recognized taxon.
*Diagnosis.* A brief description of a species or other taxon that states the most important distinguishing features.
*Diastema.* A relatively large and conspicuous gap between two adjacent teeth on the mandible.
*Dimorphic.* Occurring in two distinct forms without connecting intermediates, as in certain ant caste systems.
*Dorsum.* The upper surface.
*Endemic.* Limited to and native to a particular region.
*Epinotum.* Alternative term for propodeum, the first abdominal segment fused to the rear of the thorax.
*Exudatoria.* Papillae or other rounded appendages of larvae thought to be excretory in function.
*Facet.* The ommatidium, one of the basic units of the compound eye.
*Family.* A higher taxonomic category, comprising a set of similar genera that are descended uniquely from a common ancestor.
*Femur* (plural: *femora*). The "thigh," or third segment of the leg away from the body, after the coxa and trochanter.
*Fossa* (plural: *fossae*). A relatively large and deep pit, as the antennal fossa of the head into which the first antennal segment is inserted.
*Fovea* (plural: *foveae*). In sculpturing, a deep depression with well-marked sides.
*Foveola* (plural: *foveolae*). In sculpturing, a small deep pit.
*Frons.* The area above the clypeus, approximately in the center of the front of the head; it often includes the frontal triangle, which is roughly triangular in form and demarcated by grooves.
*Frontal area.* Same as *Frons.*
*Frontal triangle.* See *Frons.*
*Funiculus* (plural: *funiculi*). All of the antenna other than the first, basal segment (*Scape, q.v.*).
*Gaster.* The globular terminal four or five segments of the abdomen, immediately posterior to the waist. This is a functional unit of anatomy, based on whatever segments constitute the terminal portion of the body, rather than a unit based on homology.
*Gena* (plural: *genae*). The "cheek" of the head, the area just ventral to the eyes, that is, on the opposite side from the antennal insertions.

*continued*

**TABLE 2–5** *(continued)*

*Genera.* Plural of genus.
*Genotype.* In taxonomy, the "typical" species on which the name of the genus is based.
*Genus* (plural: *genera*). A set of similar species uniquely possessing a single common ancestor.
*Gula.* The central part of the lower surface of the head.
*Habitus.* The general appearance. The subjective overall "look" of a particular species.
*Hair.* See "Pilosity" section.
*Holotype.* The single specimen on which the name of a species is based.
*Humerus* (plural: *humeri*). The shoulder, i.e., the anterior corners of the pronotum, or first segment of the thorax.
*Hypopygium.* The last sternite (lower plate) of the abdomen.
*Incertae sedis.* In classification, of uncertain taxonomic placement.
*Inferior.* In a lower anatomical position.
*Infradental lamella.* One of a pair of flanges extending from the lower posterior corners of the propodeum.
*Joint.* Segment, as a joint of the antenna.
*Labial palps.* The pair of jointed appendages originating from the labium.
*Labium.* The second maxilla, forming a lower lip beneath the maxillae.
*Labrum.* A broad lobe suspended from the clypeus above the mouth, forming an upper "lip."
*Lamella.* A thin, plate-like process.
*Laterad.* Toward the side.
*Lectotype.* A specimen selected as the ultimate standard for the name of a species out of a series of syntypes or a set of type specimens previously treated as equivalent in this regard.
*Mandibles.* The first pair of jaws.
*Maxilla.* The second pair of jaws, usually kept folded beneath the principal pair of jaws, or mandibles.
*Maxillary palps.* The pair of jointed appendages originating from the maxillae.
*Mesosoma.* The middle of the three principal body parts; also called the alitrunk.
*Monograph.* A detailed treatise on a single species, genus, or other taxonomic group.
*Node.* A rounded, knob-like structure, such as the petiolar node, the upper rounded portion of the petiole.
*Nomen nudum* (plural: *nomina nuda*). A scientific name proposed without an accompanying description.
*Occipital lobes.* The rear corners of the head.
*Occiput.* The rearmost portion of the head.
*Ocellus* (plural: *ocelli*). One of the three simple, bead-like eyes located in the rear central portion of the head.
*Ommatidium* (plural: *ommatidia*). A facet, one of the units of the compound eye.
*Original description.* The description of a newly recognized species, genus, or other taxon, including the assignment of a formal scientific name to the taxon.
*Parameres.* Two lateral processes sheathing the male aedeagus.
*Paratype.* Any type, in addition to the holotype, on which the description of a species is based.
*Pectinate.* Supplied with a comb-like structure, as the tibial spurs.
*Pedicel.* Either the "waist" (petiole and postpetiole) or the second segment of the antenna from the base outward.
*Petiole.* The first (and many times the only) segment of the ant waist; see also *Postpetiole.*
*Postpetiole.* The second segment of the ant waist (not all kinds of ants have this segment).

**TABLE 2–5** *(continued)*

*Pretarsus.* The terminal segment of the foot, supplied with a pair of claws and (usually) an arolium, or central pad.
*Propodeum.* The epinotum, or first abdominal segment fused to the thorax to form the central of the three main body parts (also known as the alitrunk).
*Proventriculus.* The narrow, sclerotized posterior part of the crop; the gizzard.
*Psammophore.* A basket-like array of long, curved hairs beneath the head of some desert ants, used as an aid in carrying sand.
*Punctures.* Pinpoint impressions in the exoskeleton.
*Pygidium.* The last complete tergite (upper plate) of the abdomen.
*Ruga* (plural: *rugae*). A wrinkle.
*Rugula* (plural: *rugulae*). A small wrinkle.
*Scape.* The first, elongated segment of the antenna next to the head.
*Scrobe.* A large groove for the reception of an appendage; especially, the antennal scrobe of the head.
*Segment.* A joint; a transverse subdivision of the body or of an appendage.
*Seta.* Hair.
*Spur.* A spine-like appendage, often paired and/or pectinate, at the end of the tibia.
*Stria* (plural: *striae*). Any fine, impressed line, especially (but not necessarily) longitudinal in orientation; compare with *Striga* and *Sulcus.*
*Striga* (plural: *strigae*). A narrow, transverse line; a special term for transverse striae.
*Subgenus* (plural: *subgenera*). One or more distinctive species of common phylogenetic origin within a genus.
*Sulcus.* A deep furrow or groove.
*Suture.* A seam or line separating two body plates.
*Synonym* (denoted by =). A name applied to a species or other taxon that is superseded by an older, hence more valid name.
*Syntype.* One of two or more specimens from which the species is described and where no holotype has been selected.
*Taxon* (plural: *taxa*). Any systematic entity, such as a particular species or genus.
*Tibia* (plural: *tibiae*). The fourth division of the leg, between the femur ("thigh") and the tarsus ("foot").
*Tribe.* The taxonomic category between the genus and the subfamily.
*Trochanter.* The short second division of the leg, between the coxa and the femur.
*Venter.* The lower surface.
*Vertex.* The upper surface of the head between eyes, frons, and occiput.
*Volsella* (plural: *volsellae*). The median or inner pair of appendages of the male copulatory organ, between the paramere and aedeagus.

**Standard measurements**

*Cephalic index* (*CI*). Head width × 100/head length.
*Eye length* (*EL*). The maximum measurable length of the eye.
*Eye width* (*EW*). The maximum width of the eye measured at a right angle to the long axis.
*Head length* (*HL*). The length of the head, held in perfect full face, measured from the midpoint of the anterior border of the median clypeal lobe to the midpoint of the occipital border (or, if it is greater, a line drawn between the anteriormost points of the anterior clypeal border and the posteriormost points of the occipital lobe).
*Head width* (*HW*). Worker and queen: the maximum width of the head seen in perfect full face and excluding the eyes. If the eyes extend beyond the lateral borders of the head in this position, the measurement is taken across whatever part of the lateral borders are left exposed. Male: the maximum head width across and including the eyes.

*continued*

**TABLE 2–5** *(continued)*

*Mandibular index (MI).* Mandible length × 100/head length.

*Palp formula.* The number of segments in the maxillary and labial palps, respectively. Thus 6,4 would mean 6 segments in the maxillary palp and 4 in the labial palp.

*Paramere length.* The length measured exactly parallel to the long axis from the level of the distalmost part of the basiparamere to the level of the tip of the paramere.

*Pronotal width (PW).* The maximum width of the pronotum measured from directly above and a right angle to the long axis of the alitrunk.

*Scape index (SI).* Scape length × 100/head width.

*Scape length (SL).* The maximum measurable length of the scape exclusive of the basalmost condyle (or "neck" of the segment attaching it to the head).

**Color**

*Bicolored.* With two contrasting colors, e.g., reddish brown and yellow.

*Concolorous.* Of a uniform color.

*Ferrugineous.* Rusty reddish brown.

*Fuscous.* Brownish gray.

*Piceous.* Pitch black or black with a slight reddish tinge.

*Testaceous.* Brownish yellow.

## TAXONOMIC KEYS

**TABLE 2–6** A key to the **living subfamilies** of ants, based on the worker caste. (Modified slightly from B. Bolton, personal communication; used with permission.)

**1** Body with a single reduced or isolated segment (the petiole) between alitrunk and gaster; first gastral segment either entirely confluent with the second or separated from it by only a narrow girdling impression; if the latter, the first gastral segment not markedly reduced in size . . . . . . . . . . . . . . . . . . . . . . . **2**

Body with 2 reduced or isolated segments (the petiole and postpetiole) between alitrunk and gaster; either both segments much reduced or the second somewhat larger than the first; if the latter, the postpetiole distinctly smaller than the first gastral segment and separated from it by an extensive deep girdling constriction . . . . . . . . . . . . . . . . . . . . . . . . . . . . . . . . . . . . . . **10**

**2(1)** Apex of gaster with a semicircular to circular acidopore formed from the hypopygium (last lower plate of the gaster), this structure often projecting as a nozzle and fringed with setae; sometimes the acidopore concealed by a projection of the pygidium (last upper plate of the gaster), but if so the antennal insertions located well behind the posterior clypeal margin. Sting absent, replaced by an acid-projecting system of which the acidopore is the orifice. (Worldwide) . . . . . . . . . . . . . . . . . **Formicinae**

Apex of gaster with hypopygium lacking an acidopore. Sting present or absent; when present, usually visible; but when reduced or vestigial, the hypopygium forms a smooth posterior margin (not modified as above), and the antennal sockets abut the posterior margin of the clypeus . . . . . . . . . . . . . . . . . . . . . . **3**

**3(2)** Pygidium or hypopygium armed with peg-like teeth or short spines. If the pygidium is armed, it is transversely flattened to impressed and has either a single pair of short posterolaterally situated spines or a marginal row of short spines or peg-like teeth; hypopygium in these forms unarmed. If the pygidium is unarmed and convex, then the hypopygium has a marginal row of teeth or spines that project dorsally outside the pygidium . **4**

Pygidium and hypopygium both unarmed. Pygidium transversely convex and rounded, lacking either a posterolateral pair of short spines or a marginal row of short spines or peg-like teeth. Hypopygium with margins smooth and without spines . . . . . . . **5**

**4(3)** *Driver ants.* Armament of pygidium consisting solely of a single pair of posteriorly directed short spines situated posterolaterally, combined with frontal lobes that are vestigial or vertical and do not at all conceal the antennal sockets; hypopygium always unarmed, and promesonotal suture present. (Old World tropics and subtropics) . . . . . . . . . . . . . . . . . . **Dorylinae** (in part)

Either (1) the pygidium is armed with a single pair of spines (as above), in which case the frontal lobes cover the antennal sockets, or else the hypopygium possesses a marginal row of short spines or teeth, in which case the pygidium is unarmed and the frontal lobes present; or (2) the pygidium possesses a lateral or posterior row (or both) of short spines or peg-like teeth, in which case the promesonotal suture is absent. (Tropics and subtropics) . . . . . . . . . . . . . . . . . . . . . . . . . . . **Ponerinae** (in part)

**5(3)** Sting vestigial or absent, in any case not visible without dissection. (Worldwide) . . . . . . . . . . . . . . . . . . . . . . **Dolichoderinae**

Sting present and functional, often projecting in dead specimens; in many species the sting shaft visible through the cuticle of the gastral apex even when fully retracted . . . . . . . . . . . . . . . **6**

**6(5)** Pretarsal claws with a tooth, sometimes several teeth, on the inner curvature . . . . . . . . . . . . . . . . . . . . . . . . . . . . . . . . . . . **7**

Pretarsal claws simple, without teeth on the inner curvature . . . **9**

**7(6)** *Rare, very primitive ants from southern Australia.* Stridulatory system present ventrally on gaster, between first and second sternites; maxillary palp 6-segmented, labial palp 4-jointed; first gastral segment confluent with second, without an impression between them; mandibles elongate-triangular, blade-like and multidentate, the masticatory margins meeting along their entire length. (Southern Australia only) . . . . . . **Nothomyrmeciinae**

Stridulatory system absent or present dorsally on gaster between first and second tergites; palp formula variable but usually lower than 6,4; if the palp formula is 6,4, then the gaster has an impression between the first and second segments, or the mandibles are not constructed as above, or both . . . . . . . . . . . . . . . . . . **8**

**8(7)** *New World army ants.* Frontal lobes vestigial and vertically aligned, not at all covering the antennal sockets; antennal sockets very close to anterior margin of head, the clypeus extremely narrow in front of them. (New World tropics to warm temperate zone) . . . . . . . . . . . . . . . . . . . . . . . . . . . **Ecitoninae** (in part)

Frontal lobes present and horizontally aligned, partially to conceal completely the antennal sockets. Antennal sockets a notable distance behind the anterior margin of the head, the clypeus well developed in front of them. (Old and New World tropics) . . . . . . . . . . . . . . . . . . . . . . . . . . . . . **Ponerinae** (in part)

**9(6)** *Rare, small, yellowish-brown ants from Sri Lanka.* Major tibial spur of hind leg simple or with a few minute barbules; metathoracic spiracles on dorsal surface when alitrunk viewed from the side; palp formula 3,4; petiole with a long, narrow anterior peduncle; propodeum armed with a pair of spines. (Sri Lanka only) . . . . . . . . . . . . . . . . . . . . . . . . . . . . . . . . . . . . . . . **Aneuretinae**

*continued*

*Worldwide*. Major tibial spur of hind leg broadly and distinctly pectinate, or the metathoracic spiracles not on the dorsal surface of the alitrunk when viewed from the side (either not visible or on the lateral surface), or both; palp formula variable, generally not 3,4 but if so, then either the petiole lacks a long narrow anterior peduncle, or the propodeum lacks spines, or both . . . . . . . . . . . . . . . . . . . . . . . . . . . . . . . . . . . . . . **Ponerinae** (in part)

10(1) Pygidium transversely flattened or impressed and armed laterally, posteriorly, or both, with a row of short spines or peg-like teeth that usually project vertically. (Tropics and subtropics) . . . . . . . . . . . . . . . . . . . . . . . . . . . . . . . . . **Ponerinae** (in part)

Pygidium transversely rounded, not armed laterally or posteriorly with a row of short spines or peg-like teeth . . . . . . . . . . . **11**

11(10) Frontal lobes either absent or very reduced and vertical; in either case the antennal sockets completely open in full-face view and not at all concealed or covered by the frontal lobes . . . . . . . **12**

Frontal lobes present, horizontal to somewhat elevated; the antennal sockets always partially or completely covered by the frontal lobes in full-face view and never completely open . . . . . . . **16**

12(11) Eyes present and conspicuous, with many distinct ommatidia . **13**

Eyes absent or at most represented by a single ommatidium or small featureless blister . . . . . . . . . . . . . . . . . . . . . . . . **14**

13(12) Hind tibia with a conspicuous pectinate apical spur. Posterior margin of median portion of clypeus not projecting back between antennal sockets. Promesonotal suture present. Antenna always with 12 segments . . . . . . . . . . . **Pseudomyrmecinae** (in part)

Hind tibia without a pectinate apical spur; spur either simple or (usually) absent. Posterior margin of median portion of clypeus projecting back between antennal sockets. Promesonotal suture absent. Antenna usually with fewer than 12 segments . . . . . . . . . . . . . . . . . . . . . . . . . . . . . . . . . . **Myrmicinae** (in part)

14(12) *Old World army ants*. Antenna with 10 segments . . . . . . . . . . . . . . . . . . . . . . . . . . . . . . . . . . . . . . . . **Dorylinae** (in part)

Antenna with 12 segments . . . . . . . . . . . . . . . . . . . . . . . **15**

15(14) Minute ants, total length at most 2.5 mm, usually less than 2.0 mm; labial palp with 1 segment; genae ("cheeks") not carinate outside the depressions housing the antennal sockets; propodeal spiracle circular, its orifice directed laterally; eyes completely absent. (Old World, temperate and tropical regions) . . . . . . . . . . . . . . . . . . . . . . . . . . . . . . . . . . . . . . . . **Leptanillinae**

*New World army ants*. Larger ants, only very rarely as small as 2.5 mm; labial palp with 2, or more commonly 3, segments; genae carinate outside the depressions housing the antennal sockets; propodeal spiracle usually oval, D-shaped, elliptical or slit-like, only very rarely circular, its orifice directed posteriorly to some extent; eyes usually represented by a single ommatidium or small featureless blister, but only rarely completely absent. (New World, temperate and tropical regions) . . . **Ecitoninae** (in part)

16(11) Posterior margin of the median portion of the clypeus straight to weakly arcuate and more or less level with the anterior margins of the antennal sockets, the median portion of the clypeus not projecting strongly backward between the frontal lobes and antennal sockets; slender, mostly arboricolous ants. (Tropics and subtropics) . . . . . . . . . . . . . . . **Pseudomyrmecinae** (in part)

Posterior margin of median portion of clypeus arcuate to triangular and extending well behind the level of the anterior margins of the antennal sockets, thereby separating them . . . . . . . . . **17**



17(16) *Bulldog ants of Australia and New Caledonia*. Pretarsal claws with a tooth at or close to the midlength of each, on the inner curvature; ocelli always present; palp formula always 6,4. (Australia and New Caledonia) . . . . . . . . . . . . . . . . . . . . . . **Myrmeciinae**

Pretarsal claws simple, without teeth on inner curvature; ocelli usually absent, only very rarely present; palp formula usually less than 6,4, only very rarely with this count, in which case ocelli are absent. (Worldwide) . . . . . . . . . . . . . **Myrmicinae** (in part)

**TABLE 2–7** A key to the ant genera endemic to the **Palaearctic Region** and those elements of the Afrotropical and Oriental regional faunas that penetrate the southern portions of the Palaearctic. Because other genera of these extralimital faunas may occur in the southern Palaearctic but have not yet been detected there, this key may be run through with reference to the Ethiopian (Afrotropical)-Malagasy and Oriental keys. Based on the worker caste. (By B. Bolton, previously unpublished; used with permission. The description embodied in the key of the genus *Anomalomyrma* derives from R. W. Taylor.)

### SUBFAMILY PONERINAE

1 Petiole broadly attached to first gastral segment, the two separated dorsally and laterally only by a constriction; petiole without a free posterior face. Mandibles elongate, multidentate and linear, articulated at corners of anterior margin of head, not closing against the clypeus . . . . . . . . . . . . . . . . . . . . *Amblyopone*

Petiole narrowly attached to first gastral segment, the two joined via a slender articulatory junction; petiole usually with a free posterior face. Mandibles usually triangular or subtriangular; if linear, they articulate in the middle of the anterior margin of the head and have a vertical apical fork of teeth . . . . . . . . . . . . **2**

2(1) Pygidium transversely flattened or impressed and armed laterally, posteriorly, or both, with a row of short spines, peg-like teeth, or denticles that project dorsally . . . . . . . . . . . . . . *Cerapachys*

Pygidium transversely rounded, not armed laterally or posteriorly with a row of short spines, peg-like teeth, or denticles . . . . . **3**

3(2) Mandibles long and linear, inserted in the middle of the anterior margin of the head, with an apical armament of 3 teeth arranged in a vertical series . . . . . . . . . . . . . . . . . . . . . . . . . . . . **4**

Mandibles linear to triangular, inserted at sides of anterior margin of the head and not armed apically with a vertical series of 3 teeth . . . . . . . . . . . . . . . . . . . . . . . . . . . . . . . . . . . . . . . . **5**

4(3) Nuchal carina (separating dorsal from posterior surfaces of head) converging in a V at the midline, and also receiving a pair of prominent dark posterior apophyseal lines that converge to form the sharp median-dorsal groove of the vertex . . *Odontomachus*

Nuchal carina forming a broad uninterrupted curve across the posterodorsal extremity of the head; posterior surface without paired dark apophyseal lines; on vertex, median groove absent or ill defined and shallow . . . . . . . . . . . . . . . . . . . . . . . *Anochetus*

5(3) Tergite of second gastral segment strongly arched and vaulted so that the remaining segments point anteriorly. Alitrunk devoid of sutures . . . . . . . . . . . . . . . . . . . . . . . . . . . . . . . . . . . **6**

*continued*

Tergite of second gastral segment not arched and vaulted, remaining segments directed posteriorly. Alitrunk usually with at least 1 suture visible dorsally, only very rarely without sutures . . . 7

**6(5)** Mandibles edentate, overhung by the projecting clypeus. Apical funicular segment strongly bulbous . . . . . . . . . . ***Discothyrea***

Mandibles with 3 or more teeth, not overhung by the clypeus. Apical funicular segment moderately enlarged but not strongly bulbous . . . . . . . . . . . . . . . . . . . . . . . . . . . . . . ***Proceratium***

**7(5)** Frontal lobes absent. With head in full-face view the antennal sockets entirely visible and directed vertically. Frons and clypeus fused into a broad shelf, which projects forward over the mandibles; antennae articulated on dorsum of this shelf . . . . . . . . . . . . . . . . . . . . . . . . . . . . . . . . . . . . . . ***Probolomyrmex***

Frontal lobes present. With head in full-face view the antennal sockets mostly or entirely concealed by the frontal lobes. Frons and clypeus not fused into a shelf that projects forward over the mandibles . . . . . . . . . . . . . . . . . . . . . . . . . . . . . . . . 8

**8(7)** Sides of head with broad, deep antennal scrobes present . . . . . . . . . . . . . . . . . . . . . . . . . . . . . . . . . . . . . . . . . ***Aulacopone***

Sides of head without antennal scrobes . . . . . . . . . . . . . . . . 9

**9(8)** Pretarsal claws pectinate on the inner curvature behind the apical point . . . . . . . . . . . . . . . . . . . . . . . . . . . . . . ***Leptogenys***

Pretarsal claws unarmed or at most with a single tooth on the inner curvature behind the apical point . . . . . . . . . . . . . . . . . . 10

**10(9)** Petiole node armed dorsally with a pair of spines. Alitrunk laterally with a conspicuous pocket-like excavation above the mesopleuron . . . . . . . . . . . . . . . . . . . . . . . . . . . . . . . ***Diacamma***

Petiole node unarmed dorsally. Alitrunk laterally without a pocket-like excavation above the mesopleuron . . . . . . . . . . . . . . 11

**11(10)** Mandibles elongate-triangular and armed with 5 long but slender spiniform teeth. Apical tooth particularly long, saber-like, broadly curved, and strongly crossing over its opposite number when mandibles closed . . . . . . . . . . . . . . . . . . ***Belonopelta***

Mandibles triangular and closing tightly against the clypeus, not armed with 5 spiniform teeth. Apical tooth not saber-like, overlapping, but not strongly crossing with its opposite number when mandibles closed . . . . . . . . . . . . . . . . . . . . . . . 12

**12(11)** Basal portion of mandible with a distinct circular or near-circular pit located on top and to the outer side . . . . . . . . ***Cryptopone***

Basal portion of mandible lacking a distinct pit or fovea . . . . . 13

**13(12)** Tibia of hind leg equipped apically with only 1 spur, this spur always pectinate. Without a second spur . . . . . . . . . . . . . . 14

Tibia of hind leg equipped apically with 2 spurs. Median spur always large and pectinate, the other never absent but highly variable, ranging from almost the same size as the first and pectinate, to much smaller and smooth . . . . . . . . . . . . . . . . . . . . . 15

**14(13)** Subpetiolar process in profile with an acute angle (in fact a pair of bilateral teeth) posteroventrally and with a translucent fenestra or thin spot anteriorly . . . . . . . . . . . . . . . . . . . . . . ***Ponera***

Subpetiolar process in profile a simple lobe, without an acute posteroventral angle and lacking an anterior fenestra or thin spot . . . . . . . . . . . . . . . . . . . . . . . . . . . . . . . . . . . ***Hypoponera***

**15(13)** Tibiae of middle and hind legs each with 2 pectinate spurs. Sculpture universally of fine dense shagreening with associated larger punctures. Pretarsal claws each with a tooth on inner curvature, some distance from the apex . . . . . . . . . . . . . . . ***Platythyrea***

**TABLE 2–7** Palaearctic (*continued*)

Tibiae of middle and hind legs each with 1 pectinate and 1 simple spur. Sculpture not of universal fine dense shagreening with associated larger punctures. Pretarsal claws lacking teeth on the inner curvature . . . . . . . . . . . . . . . . . . . . . ***Pachycondyla***

### SUBFAMILY DORYLINAE

**1** Waist of 2 segments; antenna 10-segmented; gena carinate; pygidium not impressed and not armed with spines . . . . . ***Aenictus***

Waist of a single segment; antenna 9- to 11-segmented; gena not carinate; pygidium impressed, armed with a short spine or tooth at each side posteriorly . . . . . . . . . . . . . . . . . . . . ***Dorylus***

### SUBFAMILY LEPTANILLINAE

**1** Mandibles scoop-shaped, inner surfaces lined with conspicuous peg-like structures. (Known only from Japan) . . . . . . . . . . . . . . . . . . . . . . ***Anomalomyrma*** (R. W. Taylor, pers. comm.)

Mandibles linear, more typically formicid in shape, inner surfaces not lined with peg-like structures. (Very sparsely distributed in southern France, Sardinia, Corsica, Algeria, Israel, and Japan) . . . . . . . . . . . . . . . . . . . . . . . . . . . . . . . . . . . . . ***Leptanilla***

### SUBFAMILY MYRMICINAE

[The genus *Phacota*, which may be based on an ergatoid female rather than a worker (Bolton, 1987), is omitted from the key. Known only from a single specimen from Spain (Malaga), *Phacota* would run out at couplet 12 but would fail to satisfy either lug of the couplet.]

**1** Postpetiole articulated on dorsal surface of first gastral segment; the gaster in dorsal view roughly heart-shaped and capable of being bent forward over the alitrunk. Petiole dorsoventrally flattened and without a node . . . . . . . . . . . . . . ***Crematogaster***

Postpetiole articulated on anterior surface of first gastral segment; the gaster in dorsal view not roughly heart-shaped, not capable of being bent forward over the alitrunk. Petiole not dorsoventrally flattened, with a node of some form . . . . . . . . . . . . 2

**2(1)** Apical and preapical antennal segments much larger than preceding funicular segments and forming a conspicuous 2-segmented club . . . . . . . . . . . . . . . . . . . . . . . . . . . . . . . . . . . . . 3

Either apical plus 2 preapical funicular segments of antennae enlarged and forming a conspicuous 3-segmented club, or (less commonly) the club with more than 3 segments. Rarely the funiculus filiform and without a developed apical club . . . . . 13

**3(2)** Antennae with 4 or 6 segments . . . . . . . . . . . . . . . . . . . . . 4

Antennae with 9–12 segments . . . . . . . . . . . . . . . . . . . . . 11

**4(3)** Mandibles elongate and linear, produced into narrow projecting blades. Mandibles never triangular or subtriangular, never serially multidentate or denticulate . . . . . . . . . . . . . . . . . . . . 5

Mandibles triangular or subtriangular, not produced into narrow projecting blades; apical (masticatory) margins usually serially multidentate or denticulate, but teeth sometimes reduced . . . 7

**5(4)** Apex of each mandibular blade either with a single long tooth at the dorsal apex subtended by a series of minute denticles, or with a series of minute denticles only; always lacking an apical fork of 2 spiniform teeth. Labral lobes long and conical, in full-face view visible between the mandibles when the latter are closed . . . . . . . . . . . . . . . . . . . . . . . . . . . . . . ***Epitritus***

*continued*

Apex of each mandibular blade armed with a fork of 2 spiniform teeth set in a vertical series, with or without intercalary denticles between the spiniform fork teeth. Labral lobes not long and conical, not visible between the mandibles when the latter are closed . . . . . . . . . . . . . . . . . . . . . . . . . . . . . . . . . . . . 6

**6(5)** Antennae with 4 segments . . . . . . . . . . . . . . . ***Quadristruma***

Antennae with 6 segments . . . . . . . . . . . . . . . . ***Strumigenys***

**7(4)** Spongiform appendages absent from petiole and postpetiole. Frontal lobes confluent, situated centrally and high on dorsum of head. Mandibles with 4 teeth. Antennal scrobes absent. Anterior coxae much smaller than the massively developed middle and hind coxae . . . . . . . . . . . . . . . . . . . . . ***Melissotarsus***

Spongiform appendages present on petiole, postpetiole, or both. Frontal lobes widely separated, situated laterally on anterior half of head. Mandibles with more than 4 teeth or denticles. Antennal scrobes present. Anterior coxae as large as or larger than the middle and hind coxae . . . . . . . . . . . . . . . . . . . . . . . . 8

**8(7)** Fully closed mandibles with a strongly defined transverse basal border, which is separated from the anterior clypeal margin by a conspicuous impression or gap. Basal lamella of mandible situated ventral to the basalmost tooth, in a plane almost at right angle to the anterior portion of the mandible, not visible in full-face view with the mandibles open . . . . . . . . . . ***Trichoscapa***

Fully closed mandibles without a strongly defined basal border, the basal region of the mandible overlapped by the anterior clypeal margin, the two not separated by an impression or gap. Basal lamella of mandible following basalmost tooth in the same plane (sometimes separated by a long diastema), visible in full-face view with the mandibles open . . . . . . . . . . . . . . . . . . . 9

**9(8)** Propodeum unarmed. Dorsal alitrunk with 2 distinct convexities, promesonotal and propodeal, separated by a transverse groove . . . . . . . . . . . . . . . . . . . . . . . . . . . . . . . . . . . . ***Kyidris***

Propodeum armed with a pair of spines or teeth. Dorsal alitrunk not raised into 2 separate convexities separated by a transverse groove . . . . . . . . . . . . . . . . . . . . . . . . . . . . . . . . . 10

**10(9)** Standing hairs of some form present on head, alitrunk, petiole, or all these areas . . . . . . . . . . . . . . . . . . . . . . ***Smithistruma***

Standing hairs completely absent from head, alitrunk, and petiole . . . . . . . . . . . . . . . . . . . . . . . . . . . . . . . . . ***Pentastruma***

**11(3)** Antennae with 12 segments. Palp formula 5,3. Mandibles with 5 teeth. Lateral portions of clypeus flattened and prominent, fused with raised projecting median portion of the clypeus to form a shelf that projects forward over the mandibles . . . . . . . . . . . . . . . . . . . . . . . . . . . . . . . ***Cardiocondyla*** (in part)

Antennae with 9–11 segments. Palp formula 1,2 or 2,2. Mandibles with 4 or 5 teeth. Lateral portions of clypeus not flattened and prominent, not fused with median portion of clypeus to form a shelf that projects forward over the mandibles . . . . . . . . . 12

**12(11)** Anterior clypeal margin with a single long, anteriorly projecting, unpaired median seta at the midpoint of the margin. Propodeum always unarmed and rounded. Antennae always with 10 segments. Mandible 4-dentate . . . . . . . . . . . . . . ***Solenopsis***

Anterior clypeal margin lacking a single unpaired median seta, instead with a pair of hairs that straddle the midpoint of the margin. Propodeum with spines or teeth, or sharply angulate. Antennae with 9–11 segments. Mandible 5-dentate . ***Oligomyrmex***

**13(2)** Antennae with 11 segments . . . . . . . . . . . . . . . . . . . . . 14

Antennae with 12 segments . . . . . . . . . . . . . . . . . . . . . 24

**14(13)** Propodeum armed with a pair of spines that curve upward and forward. Postpetiole-gaster junction strongly dorsoventrally compressed and very narrow in profile. Basal tooth of mandible broad and with 2 points . . . . . . . . . . . . . . . . ***Trigonogaster***

Propodeum unarmed or with a pair of teeth or spines directed posteriorly. Postpetiole-gaster junction not strongly dorsoventrally compressed. Basal tooth of mandible with a single point, or sometimes mandible edentate . . . . . . . . . . . . . . . . . . . 15

**15(14)** Frontal lobes absent so that the antennal articulations are exposed and the depressed area containing the antennal sockets clearly visible. Anterior clypeal margin irregular, crenulate to denticulate . . . . . . . . . . . . . . . . . . . . . . . . . . . ***Pristomyrmex***

Frontal lobes present, covering most or all of the antennal articulations; the antennal sockets not visible in dorsal view. Anterior clypeal margin unarmed or with a pair of small teeth . . . . . 16

**16(15)** Petiole sessile to subsessile. In profile the petiole lacking an anterior bar-like peduncle between the portion that articulates with the alitrunk and the ascending (anterior) face of the node . . 17

Petiole distinctly pedunculate. In profile the petiole with a longitudinal anterior bar-like peduncle between the portion that articulates with the alitrunk and the ascending (anterior) face of the node . . . . . . . . . . . . . . . . . . . . . . . . . . . . . . . . . . . 20

**17(16)** Antennal scrobes present on sides of head above the eyes. Mandibles edentate . . . . . . . . . . . . . . . . . . . . ***Harpagoxenus***

Antennal scrobes absent. Mandibles usually with teeth, only very rarely edentate . . . . . . . . . . . . . . . . . . . . . . . . . . . . 18

**18(17)** Sternite of petiole expanded ventrally into a massive process, lamella, or both. Postpetiole ventrally lacking a median tooth-like projecting process. Palp formula 3,2 or 4,2 . . . . . . . ***Epimyrma***

Either sternite of petiole not expanded ventrally into a massive process or lamella, or postpetiole ventrally with a median tooth-like projecting process. Palp formula 4,3 or 5,3 . . . . . . . . . 19

**19(18)** Postpetiole ventrally with a median tooth-like projecting process. Palp formula 4,3 . . . . . . . . . . . . . . . . . . . . ***Formicoxenus***

Postpetiole ventrally without a median tooth-like projecting process. Palp formula 5,3 . . . . . . . . . . . . . ***Leptothorax*** (in part)

**20(16)** Propodeum unarmed. Mandibles with only 3–4 teeth. Midpoint of anterior clypeal margin with a single long unpaired seta . . . . . . . . . . . . . . . . . . . . . . . . . . . . . . . ***Monomorium*** (in part)

Propodeum armed with a pair of spines or teeth. Mandibles with 5 or more teeth. Midpoint of anterior clypeal margin without a single long unpaired seta, instead usually a pair of setae that straddle the midpoint . . . . . . . . . . . . . . . . . . . . . . . . 21

**21(20)** Palp formula 2,2. Metapleural lobes minute to absent. Pronotal dorsum a flat plateau that is sharply marginate laterally, the marginations terminating anteriorly in projecting flat acute tooth-like or triangular processes, above and behind the true humeral angles of the pronotum . . . . . . . . . . . . . . . . ***Lophomyrmex***

Palp formula 3,2 or more (up to 5,3). Metapleural lobes conspicuous. Pronotal dorsum usually rounded, only rarely otherwise, in which case marginations and processes as described above are absent . . . . . . . . . . . . . . . . . . . . . . . . . . . . . . . . . . 22

**22(21)** Sting with an apicodorsal lamellate appendage projecting from the shaft. Lateral portions of clypeus raised into a sharp ridge or shield wall on each side, in front of the antennal insertions. Propodeal spiracle low on the side and distinctly behind the mid-length of the segment . . . . . . . . . . . . . ***Tetramorium*** (in part)

*continued*

Sting without an apicodorsal lamellate appendage projecting from the shaft. Lateral portion of clypeus not raised into sharp ridges or shield walls in front of the antennal insertion. Propodeal spiracle high on the side, at or close to the midlength of the segment . . . . . . . . . . . . . . . . . . . . . . . . . . . . . . . . . . . 23

**23(22)** With head in profile the large eyes drawn out anteroventrally in a broad lobe that runs down the side and almost onto the ventral surface of the head close to the mandibular insertions . . . . . . . . . . . . . . . . . . . . . . . . . . . . . . . . . . . . . *Oxyopomyrmex*

With head in profile the eyes roughly oval, not drawn out anteroventrally in a broad lobe that runs almost onto the ventral surface of the head; anterior margin of the eye a considerable distance from the mandibular insertion . . . *Leptothorax* (in part)

**24(13)** Palp formula 6,4. Tibial spurs of middle and hind legs usually pectinate; rarely the spurs simple or absent . . . . . . . . . . . . . 25

Palp formula 1,2 to 5,3. Tibial spurs of middle and hind legs usually simple, sometimes absent; only rarely the spurs pectinate . . . . . . . . . . . . . . . . . . . . . . . . . . . . . . . . . . . . . . 26

**25(24)** Propodeum bidentate to bispinose. Mandibles with 6–10 teeth. Metasternal process a closely approximated pair of raised flanges or plates, the ventral midline not visible between them . . . . . . . . . . . . . . . . . . . . . . . . . . . . . . . . . . *Myrmica*

Propodeum unarmed and rounded. Mandibles with more than 12 teeth. Metasternal process a pair of crudely arched, convex, thickened lobes, the ventral midline visible between them . . . . . . . . . . . . . . . . . . . . . . . . . . . . . . . . . . . . . . . *Manica*

**26(24)** Sting with an apicodorsal triangular to pennant-shaped lamellate appendage projecting from the shaft. Lateral portions of clypeus raised into a sharp-edged ridge or shield wall on each side in front of the antennal insertions . . . . . . . . . . . . . . . . . . . 27

Sting without an apicodorsal triangular to pennant-shaped lamellate appendage projecting from the shaft. Lateral portions of clypeus not raised into sharp-edged ridges or shield walls in front of the antennal insertions . . . . . . . . . . . . . . . . . . . 29

**27(26)** Mandibles narrow and falcate, edentate or at most with a single minute denticle close to the acute apex . . . . *Strongylognathus*

Mandibles triangular or subtriangular, dentate, with 2 or 3 larger teeth apically, followed by a row of 4 or more small teeth or denticles . . . . . . . . . . . . . . . . . . . . . . . . . . . . . . . . . . . 28

**28(27)** Head heart-shaped in full-face view. Ventral margin of petiole convex and keel-like. Anterior clypeal margin strongly arcuate and prominent. Eyes behind midlength of sides of head. Median clypeal and median cephalic carinae vestigial or absent. Palp formula 3,2 . . . . . . . . . . . . . . . . . . . . . . . *Rhoptromyrmex*

Head not heart-shaped in full-face view. Ventral margin of petiole not convex and keel-like. Anterior clypeal margin not strongly arcuate. Eyes usually at or in front of midlength of sides of head, only extremely rarely otherwise. Either median clypeal carina or median cephalic carina usually present, or both present; only infrequently both absent. Palp formula predominantly 4,3; rarely reduced . . . . . . . . . . . . . . . . . *Tetramorium* (in part)

**29(26)** Sides of head with broad deep scrobes, which can accommodate the entire antenna when folded. Masticatory margin of mandible continuously dentate from apex to base, without a long diastema . . . . . . . . . . . . . . . . . . . . . . . . . . . . *Lordomyrma*

Antennal scrobes absent or weakly present; if the latter then masticatory margin of mandible with 2 apical teeth followed by a long diastema and 1 or 2 smaller basal teeth . . . . . . . . . . . 30

**TABLE 2–7** Palaearctic (*continued*)

**30(29)** Ventrolateral margin of head delineated by a sharp longitudinal carina on each side. The carina starts close to the inner ventral mandibular base, runs the length of the head below the eye, and ascends the occipital surface posteriorly . . . . . . . *Myrmecina*

Ventrolateral margin of head without a longitudinal carina on each side . . . . . . . . . . . . . . . . . . . . . . . . . . . . . . . . . . . . 31

**31(30)** With head in profile the large eyes drawn out anteroventrally in a broad lobe that runs down the side and almost onto the ventral surface of the head close to the mandibular insertions . . . . . . . . . . . . . . . . . . . . . . . . . . . . . . . . . . . . . . . *Goniomma*

With head in profile the eyes not drawn out anteroventrally in a broad lobe that runs almost onto the ventral surface of the head; anterior margin of the eye a considerable distance from the mandibular insertions . . . . . . . . . . . . . . . . . . . . . . . . . . 32

**32(31)** Median portion of clypeus (in front of level of frontal lobes) sharply raised and with a relatively narrow dorsal area, the raised portion with a pair of longitudinal carinae, which usually arise between the frontal lobes and usually diverge anteriorly. Median portion of clypeus never with a single unpaired median longitudinal carina . . . . . . . . . . . . . . . . . . . . . . . . . 33

Median portion of clypeus (in front of level of frontal lobes) broad, evenly convex to more or less flat, not sharply raised, without carinae or frequently with an unpaired median carina. Never with a median pair of anteriorly divergent carinae upon a sharply raised narrow dorsal area . . . . . . . . . . . . . . . . . 35

**33(32)** Petiole sessile, lacking an anterior peduncle. Petiole node ventrally with a large and strongly projecting plate-like process . . . . . . . . . . . . . . . . . . . . . . . . . . . . . . . . . . . . . . . *Vollenhovia*

Petiole pedunculate. Petiole node ventrally without a large plate-like process (usually small anteroventral process present on the peduncle) . . . . . . . . . . . . . . . . . . . . . . . . . . . . . . . . 34

**34(33)** Propodeum unarmed and rounded. Mandibles with total of 3–5 (usually 4) teeth or denticles. Petiolar spiracle at the node or on the peduncle very close to the node. Midpoint of anterior clypeal margin with a single unpaired long seta, which projects forward over the mandibles . . . . . . . . . . . . . *Monomorium* (in part)

Propodeum bidentate. Mandibles with total of 6 or more (usually more) teeth or denticles, although sometimes the basalmost teeth poorly defined. Petiolar spiracle on the peduncle very close to the articulation with the alitrunk. Midpoint of anterior clypeal margin straddled by a pair of setae, without an unpaired long median seta . . . . . . . . . . . . . . . . . . . . . . . *Stenamma*

**35(32)** Petiole sessile. In profile the petiole lacking a longitudinal anterior bar-like peduncle between the section which articulates with the alitrunk and the ascending (anterior) face of the node . . . . 36

Petiole pedunculate. In profile the petiole with a longitudinal anterior bar-like peduncle between the section which articulates with the alitrunk and the ascending (anterior) face of the node . . . . . . . . . . . . . . . . . . . . . . . . . . . . . . . . . . . . . . . 37

**36(35)** Sternite of petiole expanded ventrally into a massive process . . . . . . . . . . . . . . . . . . . . . . . . . . . . . . . . . . . . . . *Myrmoxenus*

Sternite of petiole with at most a small anteroventral tooth . . . . . . . . . . . . . . . . . . . . . . . . . . . . . . . *Leptothorax* (in part)

**37(35)** Either the apical (masticatory) margin of the mandible with total of more than 5 teeth or denticles (dental count usually more than 7), or the teeth variously worn down to blunt stubs, or the margin functionally edentate. In the last two cases the original basal outline of each tooth may or may not be discernible . . . . . . 38

*continued*

Apical (masticatory) margin of mandible with total of 3–5 well-defined teeth or denticles; apical margin not showing the remains of numerous teeth worn down to stubs or functionally edentate . . . . . . . . . . . . . . . . . . . . . . . . . . . . . . . . . . **40**

**38(37)** Palp formula 2,2 or 3,2. Antennal funiculi terminating in a strongly defined 3-segmented club . . . . . . . . . . . . ***Pheidole*** (in part)

Palp formula 4,3 or 5,3. Antennal funiculi terminating in a weakly defined 4-segmented club or without a differentiated club, the segments gradually increasing in size toward the apex . . . . **39**

**39(38)** Metasternal process large or very large. Head massive and broad in media and major workers (Cephalic Index greater than 90). Mandibles short and powerful, massively constructed, outer margins strongly curved toward the midline. Mostly polymorphic species, some of which may have the mandibular teeth much worn down . . . . . . . . . . . . . . . . . . . . . . . ***Messor***

Metasternal process minute to absent. Head narrow in all workers (CI 90 maximum, usually much less). Mandibles elongate-triangular and not massively constructed, outer margins not strongly curved toward the midline. Monomorphic species . . . . . . . . . . . . . . . . . . . . . . . . . . . . . . . . . . . . . . . . . ***Aphaenogaster***

**40(37)** Mandibles powerfully constructed, armed with 2 large apical teeth followed by a long diastema and then with 1 or 2 (rarely 3) basal teeth. 2–4 hypostomal teeth usually present on the posterior margin of the buccal cavity. Palp formula 2,2 or 3,2 . . . . . . . . . . . . . . . . . . . . . . . . . . . . . . . . . . . . . ***Pheidole*** (in part)

Mandibles delicately constructed, armed with 5 teeth, serially dentate, and the teeth decreasing in size from apex to base, not arranged as above. Hypostomal teeth absent from the posterior margin of buccal cavity. Palp formula 5,3 . . . . . . . . . . . . . **41**

**41(40)** Midpoint of anterior clypeal margin with an unpaired long seta, which projects forward over the mandibles. Lateral portions of clypeus flattened and projecting over the mandibles; sometimes the lateral portions projecting farther forward than the median portion of the clypeus . . . . . . . . . . . ***Cardiocondyla*** (in part)

Midpoint of anterior clypeal margin without an unpaired long seta; instead the midpoint usually straddled by a pair of setae. Lateral portions of clypeus not flattened and projecting over the mandibles; never projecting farther forward than the median portion of the clypeus . . . . . . . . . . . . . . . . . . . . . . . . . . . . . . **42**

**42(41)** Frontal carinae absent . . . . . . . . . . . . . . ***Leptothorax*** (in part)

Frontal carinae present but weak, running back from the frontal lobes as a pair of finely raised lines . . . . . . . . . ***Chalepoxenus***

### SUBFAMILY DOLICHODERINAE

**1** Petiole in profile usually a simple transversely flattened strip, sometimes slightly swollen anterodorsally but never equipped with a standing scale. Petiole overhung by first gastral segment and not visible in dorsal view when alitrunk and gaster are in the same plane . . . . . . . . . . . . . . . . . . . . . . . . . . . . . . **2**

Petiole in profile surmounted by a scale, which may be high and erect or lower and somewhat inclined forward, but scale always present and conspicuous. Petiole scale not or only weakly overhung by first gastral segment, usually visible in dorsal view when alitrunk and gaster are in the same plane . . . . . . . . . **3**

**2(1)** In dorsal view only 4 gastral tergites visible. Fifth tergite bent forward below the fourth, visible in ventral view where it forms a transverse plate abutting the fifth sternite; the anal and associated orifices thus situated ventrally . . . . . . . . . . ***Tapinoma***

In dorsal view 5 gastral tergites visible, the fifth small but continuing the line of the gaster and not bent forward below the fourth; the anal and associated orifices thus situated apically . . . . . . . . . . . . . . . . . . . . . . . . . . . . . . . . . . . . . . . . ***Technomyrmex***

**3(1)** Palp formula 4,2 or 2,3 . . . . . . . . . . . . . . . . ***Bothriomyrmex***

Palp formula 6,4 . . . . . . . . . . . . . . . . . . . . . . . . . . . . . . **4**

**4(3)** Integument thick, hard, and strongly sculptured; the head with foveolate punctures present. Propodeal declivity strongly concave and overhung by the posterodorsal propodeal angle. Metanotal groove broad, conspicuous, and shallowly U-shaped in profile . . . . . . . . . . . . . . . . . . . . . . . . . . . ***Hypoclinea***

Integument thin and flexible, densely but weakly sculptured; the head without foveolate punctures. Propodeal declivity not concave or overhung by the posterodorsal propodeal angle. Metanotal groove not a broad U-shaped impression in profile . . . **5**

**5(4)** With alitrunk in profile metanotal groove impressed and metathoracic spiracles dorsal. Ocelli absent. Anterior face of first gastral tergite without a deep concavity in the area immediately behind the petiole scale . . . . . . . . . . . . . . . . ***Iridomyrmex***

With alitrunk in profile metanotal groove not impressed and metathoracic spiracles lateral. Ocelli sometimes present. Anterior face of first gastral tergite with a deep concavity in the area immediately behind the petiole scale . . . . . . . . . . ***Liometopum***

### SUBFAMILY FORMICINAE

**1** Antennae with 9 segments . . . . . . . . . . . . . . . ***Brachymyrmex***

Antennae with 11–12 segments . . . . . . . . . . . . . . . . . . . . . **2**

**2(1)** Antennae with 11 segments . . . . . . . . . . . . . . . . . . . . . . . **3**

Antennae with 12 segments . . . . . . . . . . . . . . . . . . . . . . . **6**

**3(2)** Propodeum armed with a pair of spines, teeth, or tubercles. Dorsal edge of petiole usually armed with a pair of teeth or spines but sometimes only emarginate . . . . . . . . . . . ***Acantholepis***

Propodeum and petiole unarmed, without trace of spines, teeth, or tubercles . . . . . . . . . . . . . . . . . . . . . . . . . . . . . . . . . **4**

**4(3)** Palp formula 5,3 or less . . . . . . . . . . . . . . . . . . . . ***Acropyga***

Palp formula 6,4 . . . . . . . . . . . . . . . . . . . . . . . . . . . . . **5**

**5(4)** With alitrunk in dorsal view the mesonotum separated from the metanotum by a conspicuous transverse groove or impression, so that the metanotum forms a distinctly isolated sclerite . . . . . . . . . . . . . . . . . . . . . . . . . . . . . . . . . . . . . . . . . ***Plagiolepis***

With alitrunk in dorsal view the mesonotum fused with the metanotum, the two not separated by a transverse impression or groove, the metanotum not forming an isolated sclerite . . . . . . . . . . . . . . . . . . . . . . . . . . . . . . . . . . . . . . . . ***Anoplolepis***

**6(2)** Metapleuron with a distinct wide orifice for the metapleural gland, situated above the hind coxa and below the level of the propodeal spiracle. Orifice of metapleural gland protected by a line or tuft of guard-hairs, usually very conspicuous. Antennal sockets situated close to posterior margin of clypeus . . . . . . **7**

Metapleural gland orifice absent, the surface of the metanotum uninterrupted by a gland orifice above the hind coxae and below the level of the propodeal spiracle. Antennal sockets situated well behind the posterior margin of the clypeus . . . . . **15**

**7(6)** Orifice of propodeal spiracle elongate-oval, elliptical, or an elongate slit that is near-vertical or inclined from the vertical. With alitrunk in absolute profile, the propodeal spiracle well in front

*continued*

of the point where the propodeal side rounds into the declivity . . . . . . . . . . . . . . . . . . . . . . . . . . . . . . . . . . . . . . . . . . **8**

Orifice of propodeal spiracle circular to subcircular. With alitrunk in absolute profile, the propodeal spiracle bordering or actually on the curvature where the propodeal side rounds into the declivity . . . . . . . . . . . . . . . . . . . . . . . . . . . . . . . . . **13**

**8(7)** Mandibles with narrow sickle-like to saber-like curved blades, which taper to a sharp apical point. Inner border of each blade with a single tooth or with a minutely denticulate or weakly jagged margin, without conspicuously differentiated serial teeth . . . . . . . . . . . . . . . . . . . . . . . . . . . . . . . . . . . . . . . . . **9**

Mandibles triangular to elongate-triangular and armed with 5 or more conspicuously differentiated serial teeth . . . . . . . . . **10**

**9(8)** Palp formula 6,4. Maxillary palpi extremely long and very distinct . . . . . . . . . . . . . . . . . . . . . . . . . . . ***Cataglyphis*** (in part)

Palp formula 4,2. Maxillary palpi short and inconspicuous . . . . . . . . . . . . . . . . . . . . . . . . . . . . . . . . . . . . . . ***Polyergus***

**10(8)** Apical (masticatory) margin of mandible usually with 8 teeth but sometimes with more. Third tooth of mandible, counting from the apex, always distinctly smaller and shorter than the fourth; the fourth tooth larger than all the remaining teeth to the basal angle . . . . . . . . . . . . . . . . . . . . . . . . . . . . . . . ***Formica***

Apical (masticatory) margin of mandible usually with 5–7 teeth, only very rarely with more. If more than 5 teeth present then the third tooth, counting from the apex, larger and longer than the fourth; teeth after the fourth either irregular or evenly decreasing to the basal angle . . . . . . . . . . . . . . . . . . . . **11**

**11(10)** With the head in full-face view the occipital margin concave and the occipital corners projecting as a pair of blunt posteriorly directed processes. Mandibles with 7–8 teeth. Clypeus in profile steeply sloping and with an evenly concave outline . . . . . . . . . . . . . . . . . . . . . . . . . . . . . . . . . . . . . . . . ***Rossomyrmex***

With the head in full-face view the occipital margin not concave and the occipital corners bluntly angular to rounded, not projecting into posteriorly directed processes. Mandibles with 5–7 teeth on apical margin. Clypeus in profile shallowly sloping and without an evenly concave outline . . . . . . . . . . . . . . . . **12**

**12(11)** Posterior (basal) area of maxillae and labium with long and very obvious anteriorly curved or J-shaped hairs. Propodeal spiracle a long narrow ellipse or an elongate slit. Petiole a node or a thick scale . . . . . . . . . . . . . . . . . . . . . . . ***Cataglyphis*** (in part)

Posterior (basal) area of maxillae and labrum hairless or with short inconspicuous hairs. Propodeal spiracle oval or short broad ellipse. Petiole a thin scale . . . . . . . . . . . . . . . . . ***Proformica***

**13(7)** With the head in full-face view the eyes at or in front of the midlength of the sides. Head and alitrunk with stout bristles arranged in distinct pairs . . . . . . . . . . . . . . . . . ***Paratrechina***

With the head in full-face view the eyes distinctly behind the midlength of the sides. Hairs on head and alitrunk not distinctly paired . . . . . . . . . . . . . . . . . . . . . . . . . . . . . . . . . . . **14**

**14(13)** Mandibles with 6 teeth, very rarely with 7. Anterior face of first gastral segment broadly transversely concave throughout its height. Antennal scapes relatively very long; when laid straight back from their insertions at least half their length projects beyond the occipital margin . . . . . . . . . . . . . . . . . ***Prenolepis***

Mandibles with at least 7 teeth, usually with more than 7. Anterior face of first gastral segment with a small concave area immediately above the petiole-gaster articulation, but the face not broadly transversely concave throughout its height. Antennal scapes much shorter; when laid straight back from their insertions much less than half their length projects beyond the occipital margin . . . . . . . . . . . . . . . . . . . . . . . . . . . . ***Lasius***

**15(6)** Polymorphic species. Petiole a node or scale, never armed with teeth or spines. Propodeum and humeral angles of pronotum unarmed . . . . . . . . . . . . . . . . . . . . . . . . . ***Camponotus***

Monomorphic species. Petiole armed with 1 or 2 pairs of teeth or spines. Propodeum and humeral angles of pronotum usually armed with teeth or spines . . . . . . . . . . . . . . . ***Polyrhachis***

**TABLE 2–8** A key to the ant genera of sub-Saharan Africa, Madagascar, and the Seychelles, Mascarene, and other western islands of the Indian Ocean, referred to generally as the **Ethiopian (Afrotropical) and Malagasy regions.** Based on the worker caste. (Except for Dorylinae, the keys to all subfamilies are by B. Bolton; used with permission.)

### SUBFAMILY PONERINAE

**1** Petiole broadly attached to first gastral segment, the two separated dorsally and laterally only by a constriction; petiole without a free posterior face . . . . . . . . . . . . . . . . . . . . . . . . **2**

Petiole narrowly attached to first gastral segment, the two joined via a slender articulatory junction; petiole usually with a free posterior face . . . . . . . . . . . . . . . . . . . . . . . . . . . . . . **5**

**2(1)** Mandibles elongate and usually linear, multidentate and not closing tightly against the clypeus, always with more than 3 teeth . . . . . . . . . . . . . . . . . . . . . . . . . . . . . . . . . . . . . . . . . **3**

Mandibles short, triangular or subtriangular, with relatively few teeth (1–3) and closing tightly against the clypeus . . . . . . . . **4**

**3(2)** Mandibles pointed at apex, not as long as head; tooth row on inner margin of mandible single; spatulate hairs absent from head . . . . . . . . . . . . . . . . . . . . . . . . . . . . . . . . . ***Amblyopone***

Mandibles blunt at apex and very long, longer than head; tooth row on inner margin of mandible double; spatulate hairs present on head . . . . . . . . . . . . . . . . . . . . . . . . . . ***Mystrium***

**4(2)** Apical tooth of mandible followed by a short cleft, remainder of masticatory margin edentate; alitrunk in dorsal view marginate and denticulate anteriorly, and strongly constricted from side to side in front of metanotal groove; antenna 9-segmented with a strongly defined 4-segmented club . . . . . . . . . . . ***Concoctio***

Apical tooth of mandible followed by 2 more teeth, the second tooth small, the third larger than the second; alitrunk in dorsal view not marginate anteriorly nor constricted from side to side in front of metanotal groove; antenna 8–12-segmented with a 3–4-segmented club . . . . . . . . . . . . . . . . . . . . ***Prionopelta***

**5(1)** Pygidium transversely flattened or impressed and armed laterally, posteriorly, or both, with a row of short spines, peg-like teeth, or denticles that project dorsally . . . . . . . . . . . . . . . . . . . **6**

Pygidium transversely rounded, not armed laterally or posteriorly with a row of short spines, peg-like teeth, or denticles . . . . . **8**

*continued*

6(5) Each gastral segment separated from adjoining segments by distinct girdling constrictions; eyes absent . . . . ***Sphinctomyrmex***

Gaster constricted only between first and second segments, sometimes very deeply, so that a postpetiole is delimited; eyes present, varying from large to minute . . . . . . . . . . . . . . . . . . 7

7(6) Tibial spurs absent from middle legs; claws usually with a single preapical tooth . . . . . . . . . . . . . . . . . . . . . . . ***Simopone***

Tibial spurs present on middle legs; claws always simple . . . . . . . . . . . . . . . . . . . . . . . . . . . . . . . . . . . . . . . ***Cerapachys***

8(5) Tergite of second gastral segment strongly arched and vaulted so that the remaining segments point anteriorly; alitrunk devoid of sutures . . . . . . . . . . . . . . . . . . . . . . . . . . . . . . . . . . 9

Tergite of second gastral segment not arched and vaulted, remaining segments directed posteriorly; alitrunk usually with at least one suture visible dorsally, rarely without sutures . . . . . . . 10

9(8) Mandibles edentate, overhung by the projecting clypeus; apical funicular segment strongly bulbous . . . . . . . . . . ***Discothyrea***

Mandible with 3 or more teeth, not overhung by the clypeus; apical funicular segment moderately enlarged but not strongly bulbous . . . . . . . . . . . . . . . . . . . . . . . . . . . ***Proceratium***

10(8) Mandibles long and linear, inserted in the middle of the anterior margin of the head, with an apical armament of 2 or 3 teeth arranged in a vertical series . . . . . . . . . . . . . . . . . . . . . 11

Mandibles linear to triangular, inserted at sides of anterior margin of head, not armed apically with a vertical series of 3 teeth . 12

11(10) Nuchal carina (separating dorsal from posterior surfaces of head) converging in a V at the midline, and also receiving a pair of prominent dark posterior apophyseal lines that converge to form the sharp median-dorsal groove of the vertex . . . . . . . . . . . . . . . . . . . . . . . . . . . . . . . . . . . . . . ***Odontomachus***

Nuchal carina forming a broad uninterrupted curve across the posterodorsal extremity of the head; posterior surface without paired dark apophyseal lines; on vertex, median groove absent or ill defined and shallow . . . . . . . . . . . . . . . . ***Anochetus***

12(10) Frontal lobes absent; with head in full-face view, the antennal sockets entirely visible and directed vertically . . . . . . . . . 13

Frontal lobes present; with head in full-face view, the antennal sockets mostly or entirely concealed by the frontal lobes . . 14

13(12) Promesonotal suture absent; frons and clypeus fused into a broad shelf, which projects forward over the mandibles, with the antennae articulated on the dorsum of this shelf . . . . . . . . . . . . . . . . . . . . . . . . . . . . . . . . . . . . . . . ***Probolomyrmex***

Promesonotal suture present; frons and clypeus not fused into a broad shelf nor projecting over the mandibles, which close against the anterior clypeal margin . . . . . . . . . . ***Apomyrma***

14(12) Tibia of hind leg equipped apically with only 1 spur, this spur always pectinate and unaccompanied by a second spur . . . 15

Tibia of hind leg equipped apically with 2 spurs; median spur always large and pectinate; the other spur highly variable, ranging from almost the same size as the first and as strongly pectinate to much smaller and smooth, but this second spur never absent . . . . . . . . . . . . . . . . . . . . . . . . . . . . . . . . . . 22

15(14) Mandibles elongate, linear and weakly curved, blunt apically, the inner margin with 0–2 blunt teeth; mandibular articulation associated with a marked semicircular excavation of the dorsal anterior margin of the head in front of the eyes . . . ***Plectroctena***

Mandibles triangular to elongate-triangular and with a sharp apical tooth; the masticatory margin sometimes edentate but usually with several to many teeth; mandibular articulation not associated with a semicircular excavation of the dorsal anterior margin of the head in front of the eyes . . . . . . . . . . . . . . 16

16(15) Basal portion of mandible with a distinct circular or near-circular pit or fovea dorsolaterally . . . . . . . . . . . . . . . . ***Cryptopone***

Basal portion of mandible without a dorsolateral pit or fovea . . . . . . . . . . . . . . . . . . . . . . . . . . . . . . . . . . . . . . . . 17

17(16) Dorsal (outer) surfaces of middle tibiae, and middle and hind basitarsi, equipped with numerous stiff conical spines or very conspicuous peg-like teeth . . . . . . . . . . . ***Centromyrmex*** (in part)

Dorsal (outer) surfaces of middle tibiae, and middle and hind basitarsi, with hairs but lacking spines or teeth . . . . . . . . . . . 18

18(17) Gaster in profile and in dorsal view without an impression or constriction between the first and second segments . . . . . . . . . . . . . . . . . . . . . . . . . . . . . . . . . . . ***Pachycondyla*** (in part)

Gaster in profile and in dorsal view with a distinct impression or constriction between the first and second segments . . . . . . 19

19(18) Mandibles elongate-falcate, with an extremely long apical tooth so that the tips cross over at rest; masticatory margin edentate or crenulate; labrum prominent, in dorsal view projecting beyond the anterior clypeal margin as a striated lobe. Palp formula 3,4. Larger ants, total length 9–16 mm . . . . . . . . ***Psalidomyrmex***

Mandibles short and triangular, lacking an extremely long apical tooth; masticatory margin multidentate; labrum not projecting beyond clypeus as a striated lobe in dorsal view. Palp formula less than 3,4 (unknown in *Dolioponera*). Smaller ants, total length less than 6 mm . . . . . . . . . . . . . . . . . . . . . . . . 20

20(19) Frontal lobes massive, produced anteriorly and overlapping the clypeus; median portion of clypeus projecting as a broad truncated lobe; basal angle of mandible evenly rounded . . . . . . . . . . . . . . . . . . . . . . . . . . . . . . . . . . . . . . . . ***Dolioponera***

Frontal lobes small, not produced anteriorly and not overlapping the clypeus; median portion of clypeus not forming a broad truncated lobe; basal angle of mandible angulate to dentate . 21

21(20) Subpetiolar process in profile with an acute posteroventral angle and with a fenestra or thin-spot anteriorly that is translucent . . . . . . . . . . . . . . . . . . . . . . . . . . . . . . . . . . . . . . . ***Ponera***

Subpetiolar process in profile a simple lobe, without an acute posteroventral angle and lacking an anterior fenestra or thin-spot . . . . . . . . . . . . . . . . . . . . . . . . . . . . . . . . . ***Hypoponera***

22(14) Pretarsal claws of middle and hind legs armed on the inner curvature with a tooth, either close to the midlength or near the base, or else the entire inner curvature dentate to pectinate . 23

Pretarsal claws of middle and hind legs simple, the inner curvature without a tooth medially or near the base, never dentate or pectinate . . . . . . . . . . . . . . . . . . . . . . . . . . . . . . . . . 28

23(22) Pretarsal claws of middle and hind legs pectinate or with 1–3 small teeth behind the apex. If only 1 preapical tooth present, then mandibles with only 1–3 teeth and clypeus with a sharp median longitudinal carina . . . . . . . . . . . . . . . ***Leptogenys*** (in part)

Pretarsal claws of middle and hind legs never pectinate, the claws always with only a single preapical tooth. Mandibles usually with more than 3 teeth but sometimes edentate, in which case the clypeus without a median longitudinal carina . . . . . . . 24

24(23) Tibiae of middle and hind legs each with 2 pectinate spurs. Sculpture universally of fine dense shagreening with associated larger punctures. Eyes never positioned well behind the midlength of the sides of the head . . . . . . . . . . . . . ***Platythyrea***

*continued*

Tibiae of middle and hind legs each with 1 large pectinate spur and 1 small simple spur. Sculpture usually not of universal fine dense shagreening with associated larger punctures, but if such is present then the eyes are positioned well behind the midlength of the sides of the head . . . . . . . . . . . . . . . . . . . 25

**25(24)** Median portion of clypeus raised in front of the frontal lobes, the dorsum of the raised section transversely concave, and anteriorly the raised section projecting forward over the mandibles as a short truncated lobe . . . . . . . . . . . . . . . . ***Paltothyreus***

Median portion of clypeus transversely convex in front of the frontal lobes, not projecting forward over the mandibles as a truncated lobe . . . . . . . . . . . . . . . . . . . . . . . . . . . . . . . . 26

**26(25)** Gena with a carina running longitudinally between the eye and the mandibular articulation . . . . . . . . . . . . . . . ***Megaponera***

Gena without a carina running between eye and mandibular articulation . . . . . . . . . . . . . . . . . . . . . . . . . . . . . . . . . . 27

**27(26)** With the head in full-face view, the eye slightly in front of the midlength of the sides; dorsal surface of mandible, near the base, with an elongate transverse impression or trench . . . . . . . . . . . . . . . . . . . . . . . . . . . . . . . . . . . . . . . . . ***Hagensia***

With the head in full-face view, the eye far behind the midlength of the sides; dorsal surface of mandible entire, without an elongate transverse impression or trench near the base . . . . . . . . . . . . . . . . . . . . . . . . . . . . . . . . . ***Ophthalmopone*** (in part)

**28(22)** Eyes absent; dorsal (outer) surface of middle tibiae, and middle and hind basitarsi, with numerous stiff spine-like hairs or peg-like teeth . . . . . . . . . . . . . . . . . . . ***Centromyrmex*** (in part)

Eyes present, varying from large to insignificant; middle tibiae, and middle and hind basitarsi, without cuticular spines or teeth, although stiff bristles may be present . . . . . . . . . . . . . . . 29

**29(28)** Petiole dorsally with a comb of 5 long spines which curve backward over the base of the first gastral segment . . . . . . . . . . . . . . . . . . . . . . . . . . . . . . . . . . . . . . . . . ***Phrynoponera***

Petiole dorsally without a comb of 5 spines . . . . . . . . . . . . 30

**30(29)** Sides of petiole converging dorsally into a sharp longitudinal crest, which runs the length of the segment. Posterolateral margins of petiole also sharply angulate in the dorsal half, these sharp angles meeting the dorsal crest at its posterior end. Anterior clypeal margin broadly concave, the concavity terminating at each side in prominent angle or tooth-like projection . . . . . . . . . . . . . . . . . . . . . . . . . . . . . . . . . . ***Streblognathus***

Petiole scale-like to nodiform but without a sharp longitudinal crest running the length of the dorsum. Clypeus usually prominent, but if shallowly concave in the middle then the concavity not terminating in prominent angles or teeth . . . . . . . . . . 31

**31(30)** Mandible armed with only 1–3 teeth (usually 2) . . . . . . . . . . . . . . . . . . . . . . . . . . . . . . . . . . . . . . . ***Leptogenys*** (in part)

Mandible armed with 5 or more teeth . . . . . . . . . . . . . . . . 32

**32(31)** Maxillary palp with only 2 segments, the second segment very reduced. (Malagasy only) . . . . . . . . . . . . . . . . ***Euponera***

Maxillary palp with 4 segments . . . . . . . . . . . . . . . . . . . . 33

**33(32)** Eyes very large, in full-face view situated well behind the midlength of the sides of the head . . . . . ***Ophthalmopone*** (in part)

Eyes small to moderate, in full-face view situated in front of the midlength of the sides of the head . . . . ***Pachycondyla*** (in part)



### SUBFAMILY DORYLINAE

**1** Waist of 2 segments; antenna 10-segmented; gena carinate; pygidium not impressed and not armed with spines . . . . . ***Aenictus***

Waist of a single segment; antenna 9- to 11-segmented; gena not carinate; pygidium impressed, armed with a short spine or tooth at each side posteriorly . . . . . . . . . . . . . . . . . . . . ***Dorylus***

### SUBFAMILY LEPTANILLINAE

(Represented by the single genus *Leptanilla*.)

### SUBFAMILY PSEUDOMYRMECINAE

(Represented by the single genus *Tetraponera*, which includes the synonymous genera *Pachysima* and *Viticicola*.)

### SUBFAMILY MYRMICINAE

**1** Postpetiole articulated on dorsal surface of first gastral segment; the gaster in dorsal view roughly heart-shaped and capable of being bent forward over the alitrunk. Petiole dorsoventrally flattened and without a node . . . . . . . . . . . . . . ***Crematogaster***

Postpetiole articulated on anterior surface of first gastral segment; the gaster in dorsal view not roughly heart-shaped, not capable of being bent forward over the alitrunk. Petiole not dorsoventrally flattened, with a node of some form . . . . . . . . . . . . . 2

**2(1)** Apical and preapical antennal segments much larger than preceding funicular segments and forming a conspicuous 2-segmented club . . . . . . . . . . . . . . . . . . . . . . . . . . . . . . . . . . . . . . 3

Either apical plus 2 preapical funicular segments of antennae enlarged and forming a conspicuous 3-segmented club, or less commonly the club with more than 3 segments; rarely the funiculus filiform and without a developed apical club . . . . . . . 21

**3(2)** Mandible elongate and linear, produced into narrow projecting blades each one of which is much longer than broad; mandible never triangular or subtriangular, never serially multidentate or denticulate . . . . . . . . . . . . . . . . . . . . . . . . . . . . . . . . . 4

Mandible triangular or subtriangular, not produced into narrow projecting blades; apical (masticatory) margin usually serially multidentate or denticulate, but teeth sometimes reduced . . . 8

**4(3)** Apex of each mandibular blade armed with a fork of 2 or 3 spiniform teeth set in a more or less vertical series, with or without intercalary denticles between the spiniform fork teeth . . . . . 5

Apex of each mandibular blade either with a single long tooth at the dorsal apex subtended by a series of minute denticles, or with a series of minute denticles only; never possessing an apical fork of 2–3 spiniform teeth . . . . . . . . . . . . . . . . . . . . . . 7

**5(4)** Apical fork of mandible with 3 spiniform teeth; blades of mandible without preapical teeth. Maxillary palp 3-segmented. Antennal scrobes absent, the eyes dorsolateral. Petiole node with a pair of teeth or short spines, postpetiole with lateral lamellate appendages . . . . . . . . . . . . . . . . . . . . . . . . . . . ***Microdaceton***

Apical fork of mandible with 2 spiniform teeth; blades of mandible usually with preapical teeth. Maxillary palp 1-segmented. Antennal scrobes present, the eyes ventrolateral. Petiole node unarmed, postpetiole with spongiform appendages . . . . . . . 6

**6(5)** Antenna with 4 segments . . . . . . . . . . . . . . . . ***Quadristruma***

Antenna with 6 segments . . . . . . . . . . . . . . . . . ***Strumigenys***

*continued*

**7(4)** Antennal scape with a broad, anteriorly projecting subbasal lobe. Clypeal margin with spatulate or strap-like projecting hairs. Head with large orbicular hairs present; the head broad, wider than long . . . . . . . . . . . . . . . . . . . . . . . . . . . *Epitritus*

Antennal scape linear, without a projecting lobe. Clypeal margin without spatulate or strap-like projecting hairs. Head with only simple hairs present; the head longer than wide . . . . . . . . . . . . . . . . . . . . . . . . . . . . . . . . . . . . . . . . . . *Cladarogenys*

**8(3)** Antenna with 4–6 segments . . . . . . . . . . . . . . . . . . . . . . . . **9**

Antenna with 8–12 segments . . . . . . . . . . . . . . . . . . . . . **13**

**9(8)** Spongiform or lamellate appendages absent from petiole and postpetiole. Frontal lobes confluent, situated centrally and high on dorsum of head. Mandibles with 4 teeth. Antennal scrobes absent. Propodeum unarmed. Anterior coxae much smaller than the massively developed middle and hind coxae . . . . . . . . . . . . . . . . . . . . . . . . . . . . . . . . . . . . . . . . . . *Melissotarsus*

Spongiform or lamellate appendages present on petiole, postpetiole, or both. Frontal lobes widely separated, situated laterally on anterior half of head. Mandibles with more than 4 teeth. Antennal scrobes present. Propodeum armed. Anterior coxae as large as or larger than the middle and hind coxae . . . . . . . **10**

**10(9)** Differentiated prominent basal lamella of mandible absent. Apical (masticatory) margin of mandible with more than 20 denticles, the basal 4–8 of which may be enlarged. Mandibles relatively long, mandibular index more than 25 . . . . . . . . . *Serrastruma*

Differentiated prominent basal lamella of mandible present. Apical (masticatory) margin of mandible with 17 or fewer teeth or denticles of varying size. Mandibles relatively short, mandibular index less than 25 . . . . . . . . . . . . . . . . . . . . . . . . . . **11**

**11(10)** Fully closed mandibles with a strongly defined transverse basal border, separated from the anterior clypeal margin by a conspicuous impression or gap. Basal lamella of mandible situated ventral to the basalmost tooth, in a plane at almost a right angle to the anterior portion of the mandible, not visible in full-face view with the mandibles open . . . . . . . . . . . . . . . . . *Trichoscapa*

Fully closed mandibles without a strongly defined basal border, the basal region of the mandible contiguous with or overlapped by the anterior clypeal margin, the two not separated by an impression or gap. Basal lamella of mandible following basalmost tooth in the same plane, visible in full-face view with the mandibles open . . . . . . . . . . . . . . . . . . . . . . . . . . . . . **12**

**12(11)** Viewed in profile, the mandible increasing in width from base to apex and the distal portion of the blade passing into a strongly down-curved arc so that part or most of the apical margin is at right angle to the long axis of the head. Masticatory margin of mandible armed with a basal lamella plus 8–11 teeth, the basal 5–8 of which may be very strong . . . . . . . . *Glamyromyrmex*

Viewed in profile, the mandible with upper and lower margins approximately parallel for most of its length or evenly tapering anteriorly. At most, the extreme tip of the mandible down-curved, without the major part of the apical margin at right angle to the long axis of the head. Masticatory margin of mandible armed with a basal lamella plus 12–17 teeth or denticles, the apicalmost group of which are minute . . . . . *Smithistruma*

**13(8)** Mandible with 7 large teeth, which increase in size from apex to base; between each pair of teeth is a minute denticle. Mesopleuron with a depressed circular organ filled with fine, radially arranged hairs. Antennal scrobes present; antenna with 8 segments. (Madagascar only) . . . . . . . . . . . . . . . *Pilotrochus*

**TABLE 2–8** Ethiopian (*continued*)

Mandible with 4–6 teeth, which decrease in size from apex to base; without denticles between the teeth. Mesopleuron lacking hair-filled circular organ. Antennal scrobes usually absent; antennae with 9–12 segments, only extremely rarely with 8 . . . . . . . **14**

**14(13)** Antenna with 12 segments . . . . . . . . . . . . . . . . . . . . . . . . **15**

Antenna with 8–11 segments . . . . . . . . . . . . . . . . . . . . . . **16**

**15(14)** Palp formula 5,3 (maxillary palp 5-segmented, labial palp 3-segmented). Frontal lobes separated and median portion of clypeus broadly inserted between them. Lateral portions of clypeus flattened and prominent, fused to the raised projecting median portion of the clypeus to form a shelf, which projects forward over the mandibles. Metapleural lobes low and rounded, not connected to propodeal spines (when present) by broad projecting lamellae . . . . . . . . . . . . . . . . . . . . *Cardiocondyla* (in part)

Palp formula 2,2. Frontal lobes closely approximated and median portion of clypeus reduced to an extremely narrow strip between them. Lateral portions of clypeus not prominent, not fused to median portion and not forming a shelf; instead, median portion of clypeus sharply raised centrally and in the form of a narrow longitudinal ridge. Metapleural lobes large and prominent, connected to propodeal spines by broad conspicuous lamellae . . . . . . . . . . . . . . . . . . . . . . . . . *Baracidris*

**16(14)** Anterior clypeal margin with a single long, anteriorly projecting unpaired median seta at the midpoint of the margin. Propodeum always unarmed and rounded. Antenna always with 10 segments . . . . . . . . . . . . . . . . . . . . . . . . . . . *Solenopsis*

Anterior clypeal margin lacking a single unpaired median seta, instead usually with a pair of hairs which straddle the midpoint of the margin. Propodeum sometimes unarmed and rounded but usually with spines or teeth, or sharply angulate. Antenna with 8–11 segments . . . . . . . . . . . . . . . . . . . . . . . . . . . **17**

**17(16)** Antenna with 8–9 segments . . . . . . . . . . . . . . . . . . . . . **18**

Antenna with 10–11 segments . . . . . . . . . . . . . . . . . . . . . **20**

**18(17)** Propodeum bidentate, bispinose or sharply angulate in profile. Worker caste dimorphic, without intermediates . . . . . . . . . . . . . . . . . . . . . . . . . . . . . . . . . . . *Oligomyrmex* (in part)

Propodeum unarmed. Worker caste monomorphic . . . . . . . . **19**

**19(18)** Eyes absent; mandible with 5–6 teeth; promesonotum not marginate laterally . . . . . . . . . . . . . . . . . . . . . . . . . . *Carebara*

Eyes present; mandible with 4 teeth; promesonotum marginate laterally . . . . . . . . . . . . . . . . . . . . . . . . . . . . *Paedalgus*

**20(17)** Clypeus longitudinally bicarinate on median portion. Workers dimorphic without intermediates . . . . . . *Oligomyrmex* (in part)

Clypeus not bicarinate on median portion. Workers polymorphic with a graded series of intermediates connecting minor to major workers . . . . . . . . . . . . . . . . . . . . . . . . *Pheidologeton*

**21(2)** Antenna with 7 segments . . . . . . . . . . . . . . . . . . *Myrmicaria*

Antenna with 9–12 segments . . . . . . . . . . . . . . . . . . . . . **22**

**22(21)** Antenna with 9 segments. Petiole sessile, without an anterior peduncle. Pronotum and mesonotum fused into a shield, which overhangs the sides of the alitrunk and sometimes also the propodeum. Antennal scrobes present above the eyes . . . . . . . . . . . . . . . . . . . . . . . . . . . . . . . . . . . . . . . . . *Meranoplus*

Antenna with 10–12 segments. Petiole usually with an elongate anterior peduncle; if not, then either the pronotum and mesonotum do not form a shield or antennal scrobes are present below the eyes . . . . . . . . . . . . . . . . . . . . . . . . . . . . . **23**

*continued*

23(22) Median portion of clypeus vertical, with a clypeal fork (a conspicuous anteriorly projecting bilobed appendage) projecting out over the mandibles . . . . . . . . . . . . . . . . . . . . . . . . . . 24

Median portion of clypeus not vertical, without a bilobed appendage projecting out over the mandibles . . . . . . . . . . . . . . 25

24(23) Antenna with 11 segments. Peduncle of petiole short and very thick in profile. All body hairs simple, without bizarre pilosity . . . . . . . . . . . . . . . . . . . . . . . . . . . . . . . . . . . . **Dicroaspis**

Antenna with 12 segments. Peduncle of petiole elongate and narrow in profile. Body hairs bizarre, either spatulate, squamate, clavate, star-shaped, or very short thick and stubbly with abruptly tapered points . . . . . . . . . . . . . . . . **Calyptomyrmex**

25(23) Propodeal spiracle long and narrow, its orifice slit-like. Mesothoracic spiracles opening on dorsum of alitrunk. Third tooth of mandible, and usually also fourth tooth, paired internally on masticatory margin . . . . . . . . . . . . . . . . . . . . **Ocymyrmex**

Propodeal spiracle circular or subcircular, not long and narrow. Mesothoracic spiracles concealed by a pronotal flap on the sides of the alitrunk. No teeth paired internally on masticatory margin . . . . . . . . . . . . . . . . . . . . . . . . . . . . . . . . . . . . . . 26

26(25) Antenna with 10 segments . . . . . . . . . . . . . . . . . **Decamorium**

Antenna with 11–12 segments . . . . . . . . . . . . . . . . . . . . . . 27

27(26) Antenna with 11 segments . . . . . . . . . . . . . . . . . . . . . . . . 28

Antenna with 12 segments . . . . . . . . . . . . . . . . . . . . . . . . 37

28(27) Antennal scrobes present below the eyes. First gastral tergite enormously enlarged and constituting almost all of the gastral dorsum . . . . . . . . . . . . . . . . . . . . . . . . . . . . . . . . **Cataulacus**

Antennal scrobes either absent or present above the eyes. First gastral tergite not enormously enlarged . . . . . . . . . . . . . 29

29(28) Frontal lobes reduced or absent so that the antennal articulations are exposed and the depressed area containing the antennal sockets clearly visible. Anterior clypeal margin armed with denticles . . . . . . . . . . . . . . . . . . . . . . . . . . . **Pristomyrmex**

Frontal lobes present, covering most or all of the antennal articulations; the antennal sockets not visible in dorsal view. Anterior clypeal margin unarmed or with a pair of small teeth . . . . . 30

30(29) Eyes placed behind midlength of sides of head. Median portion of clypeus raised and produced forward as a large shield-like lobe that projects strongly over the mandibles. Tibiae and basitarsi of middle and hind legs terminating in a number of peg-like stout spines . . . . . . . . . . . . . . . . . . . . . . . . . . . . . **Metapone**

Eyes placed at or in front of the midlength of the sides of the head, or sometimes absent. Median portion of clypeus not produced forward as a large shield-like lobe that projects strongly over the mandibles. Tibiae and basitarsi of middle and hind legs not terminating in peg-like stout spines . . . . . . . . . . . . . . . . . . 31

31(30) Mandible with 4 teeth. Maxillary palp with 1 or 2 segments. Propodeum rounded to angulate, never armed with differentiated teeth or spines. Antennal scrobes always absent . . . . . . . . 32

Mandible with 5 or more teeth. Maxillary palp with 3 or 4 segments. Propodeum usually bidentate or bispinose, rarely otherwise. Antennal scrobes frequently present . . . . . . . . . . . . 34

32(31) Eyes absent. Propodeal spiracle enormously enlarged. Frontal lobes closely approximated . . . . . . . . . . . . . . . **Bondroitia**

Eyes present. Propodeal spiracle small. Frontal lobes widely separated . . . . . . . . . . . . . . . . . . . . . . . . . . . . . . . . . . . 33

33(32) Median portion of clypeus distinctly raised, strongly to weakly longitudinally bicarinate. Postpetiole node less voluminous than petiole node in profile and narrowly attached to the gaster . . . . . . . . . . . . . . . . . . . . . . . . . . . . . **Monomorium** (in part)

Median portion of clypeus evenly transversely convex, not distinctly raised or longitudinally bicarinate. Postpetiole node much more voluminous than petiole node in profile and very broadly attached to gaster . . . . . . . . . . . . . . . **Diplomorium**

34(31) Mandible with only 5 teeth or denticles, the basal tooth generally concealed by the anterior clypeal apron. Sting acute apically, not terminating in a lamellate, spatulate, or dentiform appendage. (Minute yellow ants introduced from Neotropical region) . . . . . . . . . . . . . . . . . . . . . . . . . . . . . **Wasmannia**

Mandible with 6 or more teeth or denticles, usually at least 7. Sting terminating in an apical or apicodorsal lamellate, spatulate, or dentiform appendage . . . . . . . . . . . . . . . . . . . . . . . . . 35

35(34) Lateral portions of clypeus not raised into a narrow ridge or wall in front of the antennal insertions. Median portion of clypeus narrow and bicarinate, narrowly inserted between frontal lobes. Mandible armed with 10–14 teeth, which decrease in size from apex to base. Promesonotum in profile with a swollen, dome-like outline . . . . . . . . . . . . . . . . . . . . . . . . . **Cyphoidris**

Lateral portions of clypeus raised into a narrow ridge or wall in front of the antennal insertions. Median portion of clypeus broad, not bicarinate, broadly inserted between frontal lobes. Mandible armed with 2–3 enlarged teeth apically, followed by a row of at least 4 smaller denticles. Promesonotum in profile without a swollen, dome-like outline . . . . . . . . . . . . . . . 36

36(35) Palp formula 3,2. Head heart-shaped and median portion of clypeus with a prominent arcuate anterior margin, which overlaps the basal angle of the mandible. Antennal scrobes absent. Ventral margin of petiole keel-like. Eyes behind midlength of sides of head . . . . . . . . . . . . . . . . . . . . **Rhoptromyrmex** (in part)

Palp formula usually 4,3, only rarely reduced. If less than 4,3 then head not heart-shaped and median portion of clypeus without a prominent arcuate anterior margin. Antennal scrobes usually present. Ventral margin of petiole not keel-like. Eyes at or somewhat in front of midlength of sides of head . . . . . . . . . . . . . . . . . . . . . . . . . . . . . . . . . . . . . . **Tetramorium** (in part)

37(27) Petiole sessile, lacking an anterior peduncle, and the venter of the petiole node with a large and very strongly projecting plate-like process. Median portion of clypeus longitudinally bicarinate. (Malagasy only) . . . . . . . . . . . . . . . . . . **Vollenhovia**

Petiole distinctly pedunculate to subsessile, in either case the venter of the petiole node lacking a large plate-like process (usually a small anteroventral process present on the peduncle). If petiole subsessile then median portion of clypeus not bicarinate . . . 38

38(37) Dorsum of petiole node armed with a pair of sharp spines . . . 39

Dorsum of petiole node unarmed or indented medially, lacking sharp spines . . . . . . . . . . . . . . . . . . . . . . . . . . . . . . 41

39(38) All of visible portion of gaster consisting of the first tergite, which is massively enlarged and subcircular, ball-like but with an anteroventral orifice within which the remaining gastral segments are telescoped. Eyes at extreme posterior corners of head. Clypeus projecting far forward and almost concealing the mandibles . . . . . . . . . . . . . . . . . . . . . . . . . . . . **Ankylomyrma**

Gaster composed of 4 visible tergites and sternites, which decrease in size posteriorly, the gaster with the first tergite not massively

*continued*

enlarged or ball-like. Eyes not at extreme posterior corners of head. Clypeus not projecting far forward over the mandibles . . . . . . . . . . . . . . . . . . . . . . . . . . . . . . . . . . . . . . . . **40**

**40(39)** Occipital corners of head evenly rounded in full-face view. Ventral surface of alitrunk with a very deep broad pit between the hind coxae. Ventral margin of sides of metapleuron eroded in front of the metapleural gland bulla. Propodeum armed with a pair of long spines. Polymorphic species . . . . . . . . . ***Atopomyrmex***

Occipital corners of head sharply angulate to denticulate in full-face view. Ventral surface of alitrunk without a broad deep pit between the hind coxae. Ventral margin of sides of metapleuron not eroded in front of metapleural gland bulla but with a conspicuous broad groove running forward to the mesopleuron. Propodeum either bidentate or unarmed. Monomorphic species . . . . . . . . . . . . . . . . . . . . . . . . . . . . . ***Terataner*** (in part)

**41(38)** Lateral portions of clypeus raised into a sharp-edged ridge or shield wall in front of the antennal insertions . . . . . . . . . . **42**

Lateral portions of clypeus not raised into a sharp-edged ridge or shield wall in front of the antennal insertions . . . . . . . . . . **44**

**42(41)** Sting lacking a spatulate or dentiform lamellate appendage dorsally at or close to the apex of the shaft. Anterior clypeal margin with a small triangular point medially. (Madagascar only) . . . . . . . . . . . . . . . . . . . . . . . . . . . . . . . . . . . ***Eutetramorium***

Sting with a spatulate or dentiform lamellate appendage dorsally at or close to the apex of the shaft. Anterior clypeal margin without a small triangular point medially . . . . . . . . . . . . . . . **43**

**43(42)** Head heart-shaped in full-face view. Ventral margin of petiole convex and keel-like. Anterior clypeal margin strongly arcuate and prominent. Eyes behind midlength of sides of head. Propodeum unarmed . . . . . . . . . . . . . . . . . . ***Rhoptromyrmex*** (in part)

Head not heart-shaped in full-face view. Ventral margin of petiole never convex and keel-like. Anterior clypeal margin not strongly arcuate or prominent. Eyes only rarely behind midlength of sides of head. Propodeum usually armed . . . . . . . . . . . . . . . . . . . . . . . . . . . . . . . . . . . . . . . . . ***Tetramorium*** (in part)

**44(41)** Occipital corners of head distinctly tuberculate to sharply denticulate in full-face view. Pronotum marginate laterally. Frontal carinae present . . . . . . . . . . . . . . . . . . . . ***Terataner*** (in part)

Occipital corners of head usually rounded, rarely angular, but never tuberculate or denticulate in full-face view. Pronotum usually lacking lateral margination. Frontal carinae absent . . . . **45**

**45(44)** Apical (masticatory) margin of mandible with more than 5 teeth or denticles, usually 7 or more altogether; very rarely, the apical margin may be worn down and entirely edentate . . . . . . . **46**

Apical (masticatory) margin of mandible with 3–5 teeth or denticles altogether, never more; apical margin never entirely edentate . . . . . . . . . . . . . . . . . . . . . . . . . . . . . . . . . . . . . **48**

**46(45)** Palp formula 2,2 or more rarely 3,2. Antennal funiculi terminating in a strongly defined 3-segmented club . . . . ***Pheidole*** (in part)

Palp formula 4,3 or 5,3. Antennal funiculi terminating in a weakly defined 4-segmented club or, lacking a differentiated club, the segments gradually increasing in size toward the apex . . . . **47**

**47(46)** Ventral surface of head with a psammophore. Head massive and broad (Cephalic Index greater than 90); mandibles massive, their outer margins strongly curved toward the midline. Metasternal process large or very large, conspicuous. Polymorphic species . . . . . . . . . . . . . . . . . . . . . . . . . . . . . . . . . . . . . ***Messor***

**TABLE 2-8** Ethiopian (*continued*)

Ventral surface of head without a psammophore. Head narrow (Cephalic Index 90 at maximum, usually less); mandibles elongate triangular, not massive, their outer margins not curved toward the midline. Metasternal process minute to absent. Monomorphic species. (Madagascar only) . . . . . . . ***Aphaenogaster***

**48(45)** Mandibles powerfully constructed, armed with 2 large apical teeth followed by a long diastema and then 1 or 2 (rarely 3) basal teeth. Two to 4 hypostomal teeth usually present on posterior margin of buccal cavity. Palp formula 2,2 or 3,2; clypeus lacking a long unpaired median seta on the anterior margin . . . . . . . . . . . . . . . . . . . . . . . . . . . . . . . . . . . . . . . . . . ***Pheidole*** (in part)

Mandibles delicately constructed, armed with 3–5 teeth, serially dentate and decreasing in size from apex to base, not arranged as above. Hypostomal teeth absent from posterior margin of buccal cavity. If palp formula 2,2 then clypeus with a long unpaired median seta on the anterior margin . . . . . . . . . . . . **49**

**49(48)** Midpoint of anterior clypeal margin with a single, unpaired, elongate seta, which projects forward over the mandibles and is very conspicuous . . . . . . . . . . . . . . . . . . . . . . . . . . . . . . . **50**

Midpoint of anterior clypeal margin lacking a single, unpaired, elongate seta; instead, usually with a pair of short hairs, one on each side of the midpoint . . . . . . . . . . . . . . . . . . . . . . **51**

**50(49)** Maxillary palps usually with 1 or 2 segments, rarely more. Mandible with 3 or 4 teeth. Propodeum always unarmed and evenly rounded. Median portion of clypeus concave to prominent anteriorly, usually overhanging the mandibles; lateral portions of clypeus not expanded forward or fused with the median portion to form a broad projecting shelf . . . . . ***Monomorium*** (in part)

Maxillary palps always 5-segmented. Mandible with 5 teeth. Propodeum usually bidentate to bispinose, only extremely rarely unarmed. Lateral portions of clypeus flattened and prominent, fused to the raised projecting median portion of the clypeus to form a shelf, which projects forward over the mandibles . . . . . . . . . . . . . . . . . . . . . . . . . . . . . . ***Cardiocondyla*** (in part)

**51(49)** With the alitrunk viewed in profile the anterior margin of the mesonotum is suddenly and very steeply raised above the level of the pronotum; mesonotal free anterior face near-vertical and somewhat concave, projecting and prominent. (Known only from Mauritius) . . . . . . . . . . . . . . . . . . . . . . ***Ireneopone***

With the alitrunk viewed in profile the mesonotum follows the line of the pronotum and is not suddenly and steeply raised above the level of the pronotum; mesonotum lacking a near-vertical and concave free anterior face . . . . . . . . . . . . . . ***Leptothorax***

### SUBFAMILY DOLICHODERINAE

(Not including *Ecphorella*, known only from the single type specimen from Angola. It appears closest to *Tapinoma*, distinguished by thick antennae, but palpal formula and gastral form not discernible.)

**1** Antenna with 11 segments . . . . . . . . . . . . . . . . . . . ***Semonius***

Antenna with 12 segments . . . . . . . . . . . . . . . . . . . . . . . . **2**

**2(1)** Propodeum armed with a pair of teeth, tubercles or acute angles posterolaterally, these sometimes linked across by a narrow carina. Between the tubercles, teeth or angles, or somewhat more dorsally, the propodeum at its midwidth with a prominent tubercle, projecting plate, longitudinal ridge, or tooth . ***Axinidris***

Propodeum unarmed, without posterolateral teeth, tubercles or

*continued*

acute angles, without a median prominence of any sort on the dorsum . . . 3

3(2) Scale of petiole well developed, in profile very distinct. Scale somewhat inclined forward but not reduced and not overhung by an anterior projection of the first gastral segment above it . . . ***Iridomyrmex***

Scale of petiole very reduced or vestigial, in profile very small and low or even entirely absent. Petiole overhung and concealed by an anterior projection of the first gastral segment above it . . . 4

4(3) Palp formula 4,3 . . . ***Engramma***

Palp formula 6,4 . . . 5

5(4) In dorsal view only 4 gastral tergites visible. Fifth tergite bent forward below the fourth, visible in ventral view where it forms a transverse plate abutting the fifth sternite; the anal and associated orifices are thus situated ventrally (see also *Ecphorella*, in note above) . . . ***Tapinoma***

In dorsal view 5 gastral tergites visible, the fifth small but continuing the line of the gaster and not bent forward below the fourth; the anal and associated orifices are thus situated apically . . . ***Technomyrmex***

## SUBFAMILY FORMICINAE

1 Antenna with 9 segments . . . 2

Antenna with 10–12 segments . . . 4

2(1) Palp formula 6,4. Basal border of mandible edentate. (Malagasy only) . . . ***Brachymyrmex***

Palp formula less than 6,4 (either 5,3 or 3,3). Basal border of mandible with a small tooth . . . 3

3(2) Palp formula 5,3. Polymorphic species with roughly rectangular head capsule in full-face view. In this view the eyes situated well in from the lateral margins of the head . . . ***Aphomomyrmex***

Palp formula 3,3. Monomorphic species with heart-shaped head capsule in full-face view. In this view the eyes situated very close to the lateral margins of the head . . . ***Petalomyrmex***

4(1) Antenna with 10 segments . . . ***Agraulomyrmex***

Antenna with 11–12 segments . . . 5

5(4) Antenna with 11 segments . . . 6

Antenna with 12 segments . . . 9

6(5) Propodeum armed with a pair of spines, teeth, or tubercles. Dorsal edge of petiole usually armed with a pair of teeth or spines but sometimes only emarginate . . . ***Acantholepis***

Propodeum and petiole unarmed, without trace of spines, teeth, or tubercles . . . 7

7(6) Palp formula 5,3 or less . . . ***Acropyga***

Palp formula 6,4 . . . 8

8(7) Seen in dorsal view the mesonotum separated from the metanotum by a conspicuous transverse groove or impression, so that the metanotum forms a distinctly isolated sclerite . . . ***Plagiolepis***

Seen in dorsal view the mesonotum fused with the metanotum, the two not separated by a transverse groove or impression, the metanotum not forming an isolated sclerite . . . ***Anoplolepis***

9(5) Eyes enormous, occupying almost the entire side of the head. Ventrolateral margin of head with a tooth at each side . ***Santschiella***

Eyes smaller, occupying less than one half the side of the head.

**TABLE 2-8** Ethiopian (*continued*)

Ventrolateral margin of head unarmed, without a tooth at each side . . . 10

10(9) Metapleuron with a distinct wide orifice for the metapleural gland, situated above the hind coxa and below the level of the propodeal spiracle. Orifice of metapleural gland protected by a line or tuft of guard-hairs, which are usually very conspicuous. Antennal sockets situated close to posterior margin of clypeus . . . 11

Metapleural gland orifice absent, the surface of the metapleuron uninterrupted by a gland orifice above the hind coxa and below the level of the propodeal spiracle. Antennal sockets situated well behind the posterior clypeal margin . . . 13

11(10) Propodeal spiracle an elongate vertical or near-vertical ellipse or slit. Ocelli present . . . ***Cataglyphis***

Propodeal spiracle circular to subcircular, usually small. Ocelli absent . . . 12

12(11) Palp formula 6,4. Eyes large and very conspicuous. Dorsal surfaces of head and body with distinctly paired coarse setae . . . ***Paratrechina***

Palp formula 3,4, 3,3, or 3,2. Eyes small to absent. Dorsal surfaces of head and body without distinctly paired coarse setae . . . ***Pseudolasius***

13(10) Palp formula 5,4. Mandible with 10 or more teeth or denticles altogether. Apical tooth disproportionately large and the fourth tooth from the apex much larger than the third and the fifth teeth. Petiole reduced to an elongate low node, which allows the gaster to be bent forward over the alitrunk . . . ***Oecophylla***

Palp formula 6,4. Mandible usually with 7 teeth at most, sometimes fewer and only very rarely with more. Whatever the number, the teeth decrease in size from apex to base; the fourth tooth is not enlarged as above. Petiole an erect node or scale, the gaster not capable of being bent forward over the alitrunk . . . 14

14(13) Polymorphic species. Petiole a node or scale, never armed with teeth or spines. Propodeum unarmed. Humeral angles of pronotum unarmed . . . ***Camponotus***

Monomorphic species. Petiole armed with teeth, spines, or prominent angles. Propodeum usually bidentate or bispinose but may be unarmed. Humeral angles marginate or armed with teeth or spines . . . 15

15(14) Acidopore formed at apex of sclerotized portion of hypopygium, always open and usually fringed with hairs; not concealed by pygidium when inactive. Anterior clypeal margin broadly and shallowly concave . . . ***Phasmomyrmex***

Acidopore formed from the invaginated membranous area of the hypopygium, not fringed with hairs, and concealed by the pygidium when not in use. Anterior clypeal margin not broadly and shallowly concave . . . ***Polyrhachis***

**TABLE 2–9** Key to the ant genera of tropical and warm-temperate Asia, as well as Micronesia and Melanesia east to Fiji, referred to generally as the **Oriental Region.** Australia, New Caledonia, and New Zealand (the Australian Region) are excluded. Based on the worker caste. (Except for Dorylinae and Leptanillinae all keys are by B. Bolton, previously unpublished; used with permission. The descriptions of *Anomalomyrma* and *Protanilla* under the Leptanillinae are the responsibility of R. W. Taylor with reference to priority.)

SUBFAMILY PONERINAE

1 Petiole broadly attached to first gastral segment, the two separated dorsally and laterally only by a constriction; petiole without a free posterior face . . . . . . . . . . . . . . . . . . . . . . . . . . . **2**

Petiole narrowly attached to first gastral segment, the two joined via a slender articulatory junction; petiole usually with a free posterior face . . . . . . . . . . . . . . . . . . . . . . . . . . . . . **5**

2(1) Mandibles elongate and usually linear, multidentate, and not closing tightly against the clypeus, always with more than 3 teeth . . . . . . . . . . . . . . . . . . . . . . . . . . . . . . . . . . . . . **3**

Mandibles short and narrow, closing tightly against the clypeus, and armed with only 3 teeth of which the median tooth is the smallest . . . . . . . . . . . . . . . . . . . . . . . . . . ***Prionopelta***

3(2) With head in full-face view, the frontal lobes approximately even with, or slightly surpassing, the anterior clypeal border beneath them. Antennal funiculi (outer segments) markedly compressed in cross section . . . . . . . . . . . . . . . . . . . . . . . ***Myopopone***

With head in full-face view, the frontal lobes distinctly posterior to the anterior clypeal border. Antennal funiculi not compressed, approximately round in cross section . . . . . . . . . . . . . . . . **4**

4(3) Mandible pointed at apex, in the form of an acute tooth. Spatulate hairs absent from head . . . . . . . . . . . . . . . . . ***Amblyopone***

Mandible blunt at apex, rounded or subtruncate in full-face view. Spatulate hairs present on head . . . . . . . . . . . . . ***Mystrium***

5(1) Pygidium transversely flattened or impressed and armed laterally, posteriorly, or both, with a row of short spines, peg-like teeth, or denticles that project dorsally . . . . . . . . . . . . . . . . . . . **6**

Pygidium transversely rounded, not armed laterally or posteriorly with a row of short spines, peg-like teeth, or denticles . . . . . **8**

6(5) Each gastral segment separated from adjoining segments by distinct girdling constrictions . . . . . . . . . . . . . ***Sphinctomyrmex***

Gaster constricted only between first and second segments, sometimes very deeply, so that a postpetiole is delimited . . . . . . . **7**

7(6) Tibial spurs absent from middle legs. Claws usually with a single preapical tooth . . . . . . . . . . . . . . . . . . . . . . . ***Simopone***

Tibial spurs present on middle legs. Claws always simple . . . . . . . . . . . . . . . . . . . . . . . . . . . . . . . . . . . . . . ***Cerapachys***

8(5) Mandibles long and linear, inserted in the middle of the anterior margin of the head, each with an apical armament of 2–3 teeth arranged in a vertical series . . . . . . . . . . . . . . . . . . . . . . **9**

Mandibles linear to triangular, in all cases inserted at anterolateral corners of head and not armed apically with a vertical series of 2–3 teeth . . . . . . . . . . . . . . . . . . . . . . . . . . . . . . . . . . **10**

9(8) Nuchal carina (separating dorsal from posterior surface of head) converging in a V at the midline, and also receiving a pair of prominent dark apophyseal lines that converge to form the sharp median-dorsal groove of the vertex . . . . ***Odontomachus***

Nuchal carina forming a broad uninterrupted curve across the posterodorsal extremity of the head; posterior surface without paired dark apophyseal lines; on vertex, median groove absent or ill defined and shallow . . . . . . . . . . . . . . . . . ***Anochetus***

10(8) With head in full-face view, horizontal frontal lobes appear to be absent. Either the frontal lobes actually are completely absent, in which case the antennal sockets are entirely visible and set on a shelf-like projection that overhangs the mandibles, or the sockets at extreme anterior margin of head; or else (rarely) strongly elevated strip-like frontal lobes are present, in which case the sockets are at the extreme anterior margin of the head and the head capsule has a median carina running its length . . . . . **11**

With head in full-face view, horizontal frontal lobes are clearly apparent; they usually cover and conceal the antennal sockets, but if the sockets are partially visible then either the sockets are well behind the anterior margin of the head, or else a median longitudinal carina is absent from the head capsule, or both; antennal sockets never on a shelf-like projection overhanging the mandibles . . . . . . . . . . . . . . . . . . . . . . . . . . . . . . . . . . . **13**

11(10) Tergite of second gastral segment not arched and vaulted, remaining segments directed posteriorly. Eyes absent . . . . . . . . . . . . . . . . . . . . . . . . . . . . . . . . . . . . . . . . . ***Probolomyrmex***

Tergite of second gastral segment strongly arched and vaulted so that the remaining segments point anteriorly. Eyes present, even if very small . . . . . . . . . . . . . . . . . . . . . . . . . . . . . . . **12**

12(11) Mandibles edentate, overhung by the projecting clypeus. Apical funicular segment strongly bulbous . . . . . . . . . . ***Discothyrea***

Mandibles with 3 or more teeth, not overhung by the clypeus. Apical funicular segment moderately enlarged but not strongly bulbous . . . . . . . . . . . . . . . . . . . . . . . . . . . . . . ***Proceratium***

13(10) Hind tibiae each with 2 distinctly pectinate spurs, the median spur usually much larger than the lateral . . . . . . . . . . ***Platythyrea***

Hind tibiae either with only one spur, which is pectinate, or with a large pectinate median spur and a much smaller simple lateral spur . . . . . . . . . . . . . . . . . . . . . . . . . . . . . . . . . . . . **14**

14(13) Frontal lobes widely separated throughout their length. Posterior clypeal margin usually broadly rounded or truncated between anterior ends of frontal lobes; the lobes themselves never separated by only a slender triangle anteriorly or a narrow median strip of cuticle throughout. Frontal lobes usually elongate and generally more or less straight-sided, not consisting of simple semicircles or blunt triangles and not having a distinct pinched-in appearance posteriorly . . . . . . . . . . . . . . . . . . . . . . **15**

Frontal lobes closely approximated or even partially to entirely confluent. Frontal lobes separated by only a slender triangle of cuticle anteriorly or by a longitudinal line or narrow median strip of cuticle throughout. Frontal lobes usually consisting of simple short semicircles or blunt triangles, and having a distinct pinched-in appearance posteriorly . . . . . . . . . . . . . . . . . . **16**

15(14) With alitrunk viewed in profile, the anteroventral pronotal angle, just in front of the anterior coxa, with a distinct and usually acute tooth (rarely this tooth may be reduced or missing in individual specimens). Posterior pretarsal claws always with a distinct median tooth; posterior coxae unarmed above . . . . . . . . . . . . . . . . . . . . . . . . . . . . . . . . . . . . . . . ***Rhytidoponera***

With alitrunk viewed in profile, the anteroventral pronotal angle unarmed or forming an obtuse angle. In rare cases where the angle is present or more nearly tooth-like, then the posterior pretarsal claws lack a median tooth, or else the posterior coxae are toothed above . . . . . . . . . . . . . . . . . . ***Gnamptogenys***

*continued*

**16(14)** Basal portion of mandible with a distinct circular or near-circular pit or fovea dorsolaterally . . . . . . . . . . . . . . . . *Cryptopone*

Basal portion of mandible without a dorsolateral pit or fovea . . . . . . . . . . . . . . . . . . . . . . . . . . . . . . . . . . . . . . . . . . **17**

**17(16)** Ventral apex of hind tibia, when viewed from in front with the femur at right angle to the body, with a single large pectinate spur; without a second simple spur in front of the pectinate, main spur in the direction of the observer . . . . . . . . . . . . **18**

Ventral apex of hind tibia, when viewed from in front with the femur at right angle to the body, with two spurs, consisting of a large pectinate spur and a second simple spur, which is in front of the pectinate main spur in the direction of the observer . . **21**

**18(17)** Mandible elongate-triangular and armed with 5 long, slender, spiniform teeth. Apical tooth particularly long, saber-like, broadly curved, and strongly crossing over its opposite number when mandibles closed . . . . . . . . . . . . . . . . . . *Belonopelta*

Mandible triangular and closing tightly against the clypeus, not armed with 5 spiniform teeth. Apical tooth not saber-like, overlapping but not strongly crossing over its opposite number when mandibles closed . . . . . . . . . . . . . . . . . . . . . . . **19**

**19(18)** Dorsal (outer) surfaces of middle tibiae, and middle and hind basitarsi, equipped with numerous strong cuticular spines or peg-like teeth . . . . . . . . . . . . . . . . . . . . . . . . *Centromyrmex*

Dorsal (outer) surfaces of middle tibiae, and middle and hind basitarsi, with hairs but lacking cuticular spines or teeth . . . . . **20**

**20(19)** Subpetiolar process in profile with an acute angle (actually, a pair of bilateral teeth) posteroventrally and with a fenestra (i.e., thin-spot) anteriorly, which is translucent . . . . . . . . . . . *Ponera*

Subpetiolar process in profile a simple lobe, without an acute posteroventral angle and lacking an anterior fenestra (i.e., thin-spot) . . . . . . . . . . . . . . . . . . . . . . . . . . . . . . . . *Hypoponera*

**21(17)** Pretarsal claws of hind leg either pectinate or equipped with one or more teeth on the inner curvature behind the apical point . . . . . . . . . . . . . . . . . . . . . . . . . . . . . . . . . . . . . . . **22**

Pretarsal claws of hind leg neither pectinate nor armed with preapical teeth . . . . . . . . . . . . . . . . . . . . . . . . . . . . . . . . . **23**

**22(21)** Ocelli present. Pretarsal claws each with a large stout preapical tooth. Mandibles forceps-like, each blade with a double longitudinal row of teeth and with more than 25 teeth in each row, so that each mandibular blade has 50 or more teeth. Ventral surface of each mandible close to base, with a large triangular flange whose inner margin forms an extension of the main gripping edge of the mandible . . . . . . . . . . . . . . . . *Harpegnathos*

Ocelli absent. Pretarsal claws usually pectinate, less commonly the pectination reduced to 1–3 small teeth on the basal half of the claws. Mandibles highly variable in shape but never with 2 rows of teeth on each blade and with fewer than 30 teeth on each mandible; often with only 1–3 teeth on each blade. Ventral surface of mandible close to base without a triangular flange . . . . . . . . . . . . . . . . . . . . . . . . . . . . . . . . . . . . . *Leptogenys*

**23(21)** Alitrunk with a conspicuous pocket-like excavation on the lateral surface above the mesopleuron. Petiole a node, armed dorsally with a pair of spines . . . . . . . . . . . . . . . . . . . . *Diacamma*

Alitrunk without a pocket-like excavation above the mesopleuron. Petiole usually unarmed but sometimes forming a scale that is emarginate dorsally, sometimes a node with a tridentate to multidentate posterodorsal margin . . . . . . . . . . . . . . . . . . . **24**

**24(23)** Antennal sockets very close to or at the anterior clypeal margin. With head in full-face view, frontal lobes reaching or overhanging the anterior clypeal margin on each side. Medially, the frontal lobes usually with a narrow truncated clypeal lobe projecting freely in front of them; very rarely this lobe absent . . *Myopias*

Antennal sockets located well behind the anterior clypeal margin. With head in full-face view, frontal lobes far behind anterior clypeal margin on each side. Medially, the frontal lobes usually without a narrow truncated freely projecting clypeal lobe in front of them, very rarely such a lobe is present . . . . . . . . **25**

**25(24)** Pronotum with a pair of laterally directed triangular teeth. Mandible with 5 large, stout teeth and usually also with a minute basal denticle. Anterior clypeal margin with 7–9 acute to blunt projecting teeth . . . . . . . . . . . . . . . . . . . . . . . *Odontoponera*

Pronotum unarmed. Mandible usually with 7 or more teeth, rarely with 6. Anterior clypeal margin unarmed, without projecting teeth . . . . . . . . . . . . . . . . . . . . . . . . . . *Pachycondyla*

### SUBFAMILY DORYLINAE

**1** Pedicel ("waist") of 2 segments; gena (cheek) carinate; pygidium not impressed and not armed with spines . . . . . . . . *Aenictus*

Pedicel of a single segment; gena not carinate; pygidium impressed and armed with a short spine or tooth at each side posteriorly . . . . . . . . . . . . . . . . . . . . . . . . . . . . . . *Dorylus*

### SUBFAMILY LEPTANILLINAE

**1** Mandible scoop-shaped, with a single long spike-like tooth on the lower margin and the inner surfaces lined with conspicuous peg-like structures. (Known only from Japan) . . . . . . . . . . . . . . . . . . . . . . . . . . . . . . . . . . . . . . . *Anomalomyrma*

Mandible linear, more typically formicid in shape, its inner surface not lined with peg-like structures, although such structures may occur along the anterior border. (Widespread in Oriental Region, including Japan) . . . . . . . . . . . . . . . . . . . . . . . . . **2**

**2(1)** Labrum lined with conspicuous forward-directed trigger hairs; mandibles can be opened 180 degrees, their anterior borders lined with slender peg-like structures (see Figure 16–18) . . . . . . . . . . . . . . . . . . . . . . . . . . . . . . . . . . . . . . *Protanilla*

Labrum lacking especially conspicuous hairs on its anterior margin; mandibles not capable of being opened 180 degrees and lacking a row of peg-like structures on their anterior borders . . . . . . . . . . . . . . . . . . . . . . . . . . . . . . . . . . . . . *Leptanilla*

### SUBFAMILY PSEUDOMYRMECINAE

(Genus *Tetraponera* only, comprising abundant arboricolous ants.)

### SUBFAMILY MYRMICINAE

**1** Antennal scrobes present and running below the eyes; eyes usually distinct but rarely may be minute and situated on the underside of the upper scrobe margin, not visible in full-face view . . . . . . . . . . . . . . . . . . . . . . . . . . . . . . . . . . . . . . . . . **2**

Either antennal scrobes absent, or present but running above the eyes; in some genera both eyes and scrobes are absent . . . . **5**

**2(1)** Antenna with 11 segments. Petiole sessile, without an anterior peduncle. Dorsum of gaster consisting entirely of the first tergite, the remaining tergites visible in profile below the posterior margin of the first . . . . . . . . . . . . . . . . . . . . . . . *Cataulacus*

*continued*

Antenna with 6–7 segments. Petiole with an anterior peduncle. Dorsum of gaster not consisting entirely of the first tergite, the remaining tergites continuing the line of the first and visible in dorsal view . . . . . . . . . . . . . . . . . . . . . . . . . . . . . . . . 3

**3(2)** Antenna with 6 segments. Petiole and postpetiole with extensive foliaceous or membranous outgrowths. Palp formula 5,3 . . . . . . . . . . . . . . . . . . . . . . . . . . . . . . . . . . . . ***Colobostruma***

Antenna with 7 segments. Petiole and postpetiole without foliaceous or membranous outgrowths. Palp formula 2,2 or less . . 4

**4(3)** Mandibles triangular, with serially dentate masticatory borders engaging directly along their entire length at full closure . . . . . . . . . . . . . . . . . . . . . . . . . . . . . . . . . . . . . . ***Eurhopalothrix***

Mandibles linear, with insertions remote so that their masticatory borders cross or engage only near their apices . . ***Rhopalothrix***

**5(1)** Postpetiole articulated on dorsal surface of first gastral segment; the gaster in dorsal view roughly heart-shaped and capable of being bent forward over the alitrunk. Petiole dorsoventrally flattened and without a node; eyes present . . . . . ***Crematogaster***

Postpetiole articulated on anterior surface of first gastral segment; the gaster in dorsal view not roughly heart-shaped, and usually not capable of being bent forward over the alitrunk. Petiole not dorsoventrally flattened, possessing a node of some form. If postpetiole articulated high on anterior face of first gastral segment, then petiole with a node, and eyes absent . . . . . . . . . 6

**6(5)** Apical and preapical antennal segments much larger than preceding funicular segments and forming a conspicuous 2-segmented club, or the apical and preapical segments preceded by an elongate bar-like fusion segment . . . . . . . . . . . . . . . . . . . . . . 7

Either apical plus 2 preapical funicular segments of antennae enlarged and forming a conspicuous 3-segmented club, or less commonly the club with more than 3 segments. Rarely, the funiculus filiform and without a developed apical club . . . . . 30

**7(6)** Antennae with 4–6 segments . . . . . . . . . . . . . . . . . . . . . . 8

Antennae with 8–12 segments . . . . . . . . . . . . . . . . . . . . . 18

**8(7)** Mandible elongate and linear, produced into a narrow projecting blade which is much longer than broad, and which always lacks a cluster of 3 stout teeth near the midlength of the blade. Mandible never triangular or subtriangular, never serially multidentate or denticulate . . . . . . . . . . . . . . . . . . . . . . . . . . . 9

Mandible triangular to subtriangular or with a cluster of 3 stout teeth near the midlength of the blade. Mandible usually serially multidentate or denticulate but sometimes with diastemata . 12

**9(8)** Apex of each mandibular blade either with a single long tooth at the dorsal apex subtended by a series of minute denticles, or with only a series of denticles; always lacking an apical fork of 2–3 spiniform teeth. Labral lobes long and conical, visible between the mandibles in full-face view when the latter are closed . . . . . . . . . . . . . . . . . . . . . . . . . . . . . . . . . . . . . ***Epitritus***

Apex of each mandibular blade armed with a fork of 2–3 spiniform teeth set in a more or less vertical series, with or without intercalary denticles between the fork teeth. Labral lobes not long and conical, not visible between the mandibles in full-face view when the latter are closed . . . . . . . . . . . . . . . . . . . . . . 10

**10(9)** Antennal scrobe absent. Antenna with 5 segments; of the 4 funicular segments the second is bar-like and elongate. Eyes lateral. Palp formula 5,3 . . . . . . . . . . . . . . . . . . . . ***Orectognathus***

Antennal scrobe present, although reduced in some species. Antenna with 4 or 6 segments, never 5; if antenna 6-segmented then second funicular segment never bar-like and elongate. Eyes ventrolateral, placed on or near the ventral margin of the scrobe. Palp formula 1,1 . . . . . . . . . . . . . . . . . . . . . . . . . . . 11

**11(10)** Antenna with 4 segments . . . . . . . . . . . . . . . . ***Quadristruma***

Antenna with 6 segments . . . . . . . . . . . . . . . . . . ***Strumigenys***

**12(8)** Closed mandibles with a strongly defined transverse basal border which is separated from the anterior clypeal margin by a conspicuous impression or gap . . . . . . . . . . . . . . . . . . . . 13

Closed mandibles without a strongly defined basal border; the basal region of the mandible contiguous with or overlapped by the anterior clypeal margin, the two not separated by an impression or gap . . . . . . . . . . . . . . . . . . . . . . . . . . . . . . 14

**13(12)** Mandible with 12 teeth or denticles. At full closure the entire length of the masticatory margin engaging directly, from basal to apical tooth . . . . . . . . . . . . . . . . . . . . . . . ***Trichoscapa***

Mandible with more than 20 teeth or denticles. At full closure only the apical halves of the masticatory margins engaging, with a distinct gap between the basal portions of the margins when the apical halves are engaged . . . . . . . . . . . . . ***Dysedrognathus***

**14(12)** Propodeum unarmed. Dorsum of alitrunk with two distinct convexities, promesonotal and propodeal, which are separated by a transverse groove . . . . . . . . . . . . . . . . . . . . . . . . ***Kyidris***

Propodeum armed with a pair of spines or teeth. Dorsum of alitrunk not raised into two convexities separated by a transverse groove . . . . . . . . . . . . . . . . . . . . . . . . . . . . . . . . . . 15

**15(14)** Masticatory margin of mandible with a cluster of 3 stout teeth near the midlength. Between this cluster and the clypeal apron the mandibular margin is unarmed. Diastema and a single small tooth occur distal to the 3 stout teeth before the apical series of small teeth and minute denticles . . . . . . . . . . . ***Asketogenys***

Masticatory margin of mandible serially dentate or denticulate, lacking a cluster of 3 stout teeth near the midlength. Diastema between basal tooth and clypeal apron variably developed but usually absent . . . . . . . . . . . . . . . . . . . . . . . . . . . . 16

**16(15)** Mandibles short, powerful, and shaped like a bear trap, armed with relatively few teeth (total usually 8 or less), most or all of which are large and strongly developed . . . . ***Glamyromyrmex***

Mandibles triangular to elongate-triangular, not shaped like a bear trap, armed with many teeth or denticles (total 10 or more), all of which are short and triangular to peg-like . . . . . . . . . . 17

**17(16)** Standing hairs of one form or another present on the head, alitrunk, petiole, or all of these areas . . . . . . . . . . ***Smithistruma***

Standing hairs completely absent from head, alitrunk, and petiole . . . . . . . . . . . . . . . . . . . . . . . . . . . . . . . . . ***Pentastruma***

**18(7)** Antenna with 12 segments . . . . . . . . . . . . . . . . . . . . . . 19

Antenna with 8–11 segments . . . . . . . . . . . . . . . . . . . . 21

**19(18)** Dorsum of alitrunk with a series of pairs of conical tubercles or prominences. Frontal carinae present . . . . . . ***Proatta*** (in part)

Dorsum of alitrunk without conical tubercles or prominences. Frontal carinae absent . . . . . . . . . . . . . . . . . . . . . . . . . 20

**20(19)** Frontal lobes closely approximated; median portion of clypeus, where it is inserted between the lobes, narrower than either frontal lobe. Basal border of mandible with a tooth close to or behind its midlength. Palp formula 2,2 or less . . ***Adelomyrmex***

Frontal lobes widely separated; median portion of clypeus, where it is inserted between the lobes, much broader than either frontal

*continued*

lobe. Basal border of mandible unarmed. Palp formula 5,3 . . . . . . . . . . . . . . . . . . . . . . . . . . . . . *Cardiocondyla* (in part)

**21(18)** Antennal scrobes present, varying from long, broad, but shallow indentations bounded above by the frontal carinae to extensive excavations in the sides of the head above the eyes . . . . . . **22**

Antennal scrobes and frontal carinae absent . . . . . . . . . . . . **24**

**22(21)** Mandible with 12 teeth, which alternate in size and which become larger basally. Antenna with 9 segments . . . . . . . . *Ishakidris*

Mandible with 4–5 teeth, which decrease in size from apex. Antenna with 10 or 11 segments . . . . . . . . . . . . . . . . . . . . **23**

**23(22)** Median portion of clypeus with a near-vertical anterior face and forming a bilobed or bidentate process which projects forward over the mandibles. Antenna with 10 segments . . . *Mayriella*

Median portion of clypeus convex but lacking a near-vertical anterior face and not produced into a bilobed or bidentate process which projects forward over the mandibles. Antenna with 11 segments . . . . . . . . . . . . . . . . . . . . . *Wasmannia* (in part)

**24(21)** Anterior clypeal margin with a single long, anteriorly projecting, unpaired median seta at the midpoint of the margin . . . . . . . . . . . . . . . . . . . . . . . . . . . . . . . . . . . . . . . . *Solenopsis*

Anterior clypeal margin lacking a single median unpaired seta, instead usually with a pair of setae that straddle the midpoint . . . . . . . . . . . . . . . . . . . . . . . . . . . . . . . . . . . . . . . . **25**

**25(24)** Antenna with 8–10 segments . . . . . . . . . . . . . . . . . . . . . **26**

Antenna with 11 segments . . . . . . . . . . . . . . . . . . . . . . . **29**

**26(25)** Median portion of clypeus not longitudinally bicarinate. Postpetiole very broadly attached to gaster. Frontal lobes very closely approximated, touching or separated by only an extremely narrow impression. Alitrunk box-like, flattened dorsum finely and densely longitudinally striate . . . . . . . . . . . *Rhopalomastix*

Median portion of clypeus longitudinally bicarinate. Postpetiole narrowly attached to gaster. Frontal lobes separated by median portion of clypeus. Alitrunk usually not box-like, but if so then dorsum not finely and densely longitudinally striate . . . . . **27**

**27(26)** Propodeum bidentate, bispinose, or sharply angulate in profile. Worker caste dimorphic, without intermediates . . . . . . . . . . . . . . . . . . . . . . . . . . . . . . . . . . . . . *Oligomyrmex* (in part)

Propodeum unarmed. Worker caste monomorphic . . . . . . . . **28**

**28(27)** Eyes absent. Mandibles with 5–6 teeth. Promesonotum not marginate laterally . . . . . . . . . . . . . . . . . . . . . . . . . *Carebara*

Eyes present. Mandibles with 4 teeth. Promesonotum marginate laterally . . . . . . . . . . . . . . . . . . . . . . . . . . . . *Paedalgus*

**29(25)** Median portion of clypeus longitudinally bicarinate. Workers dimorphic without intermediates . . . . . . *Oligomyrmex* (in part)

Median portion of clypeus not longitudinally bicarinate. Workers polymorphic with a graded series of intermediates between minors and majors . . . . . . . . . . . . . . . . . . . . *Pheidologeton*

**30(6)** Antenna with 7 segments . . . . . . . . . . . . . . . . . . *Myrmicaria*

Antenna with 9–12 segments . . . . . . . . . . . . . . . . . . . . . **31**

**31(30)** Masticatory border of mandible with a long edentate edge apically and 4 small teeth basally, the edentate portion of the margin longer than the dentate section. Entire gastral dorsum formed by the first tergite, which curves strongly downward posteriorly so that tergites 2–4 are on the ventral surface. Eyes vestigial. Very rare. (Borneo) . . . . . . . . . . . . . . . . . . *Secostruma*

**TABLE 2–9** Oriental (*continued*)

Masticatory border of mandible not divided into a long edentate apical portion and shorter, 4-toothed basal section. Gastral dorsum not formed entirely of the first gastral tergite. Eyes present or absent . . . . . . . . . . . . . . . . . . . . . . . . . . . . . . . . **32**

**32(31)** Eyes completely absent . . . . . . . . . . . . . . . . . . . . . . . . **33**

Eyes present, varying from large and conspicuous to very small and bearing a single ommatidium . . . . . . . . . . . . . . . . . **34**

**33(32)** Frontal lobes very closely approximated, the posterior portion of the clypeus passing between them narrower than the width of either lobe. Petiole without an anteroventral process. Postpetiole articulated at top of anterior face of first gastral segment . . . . . . . . . . . . . . . . . . . . . . . . . . . . . . . . . . . . . *Anillomyrma*

Frontal lobes widely separated, the posterior portion of the clypeus passing between them much broader than the width of either lobe. Petiole with a large to very large anteroventral process. Postpetiole articulated close to center of anterior face of first gastral segment . . . . . . . . . . . . . . . . . . . . *Liomyrmex*

**34(32)** Median portion of clypeus vertical, with a conspicuous, anteriorly projecting bilobed appendage above (the clypeal fork), which projects out over the mandibles from about the same level as the frontal lobes . . . . . . . . . . . . . . . . . . . . . . *Calyptomyrmex*

Median portion of clypeus not vertical, lacking a bilobed appendage that projects out over the mandibles from about the same level as the frontal lobes . . . . . . . . . . . . . . . . . . . . . . **35**

**35(34)** Antenna with 9 segments . . . . . . . . . . . . . . . . . . *Meranoplus*

Antenna with 10–12 segments . . . . . . . . . . . . . . . . . . . . . **36**

**36(35)** Antenna with 10 segments . . . . . . . . . . . *Tetramorium* (in part)

Antenna with 11–12 segments . . . . . . . . . . . . . . . . . . . . . **37**

**37(36)** Antenna with 11 segments . . . . . . . . . . . . . . . . . . . . . . **38**

Antenna with 12 segments . . . . . . . . . . . . . . . . . . . . . . **52**

**38(37)** Frontal lobes absent or reduced and elevated so that the antennal articulations are exposed; in either case the anterior clypeal margin denticulate or sharply crenulate. Distal portion of mandible suddenly broadened, broader than proximal portion. Mandible rotated on its long axis so that at full closure the masticatory margin is vertical or near-vertical below the anterior clypeal margin . . . . . . . . . . . . . . . . . . . . . . . . . . . . . *Pristomyrmex*

Frontal lobes present and covering most or all of antennal articulations. Anterior clypeal margin at most with a pair of teeth, usually unarmed. Distal portion of mandible not suddenly broadened. Mandible not rotated on its long axis; at full closure the masticatory margin not vertical below the anterior clypeal margin . . . . . . . . . . . . . . . . . . . . . . . . . . . . . . . . . **39**

**39(38)** Lateral portions of clypeus raised into a narrow ridge or wall in front of the antennal insertions. Sting either with a lamellate appendage, which projects dorsally, close to the sting apex but at an angle to the shaft, or rarely the appendage continuing the line of the shaft and upcurved at its apex. Mandible with 7 teeth, consisting of 3 larger teeth apically, followed by 4 smaller teeth . . . . . . . . . . . . . . . . . . . . . . . . . . . *Tetramorium* (in part)

Lateral portions of clypeus not raised into a narrow ridge or wall in front of the antennal insertions. Sting shaft usually simple and lacking a lamellate appendage; rarely the sting straight-spatulate apically. Mandibles usually with fewer than 7 teeth, if 7 or more present then they are not arranged as above . . . . **40**

**40(39)** Masticatory margin of mandible with 8 or more teeth, denticles, or crenulations . . . . . . . . . . . . . . . . . . . . . . . . . . . . . **41**

*continued*

Masticatory margin of mandible with 4–6 teeth or denticles altogether . . . . . . . . . . . . . . . . . . . . . . . . . . . . . . . . . . . **43**

**41(40)** With alitrunk in profile, the propodeal spiracle well in front of the margin of the declivity. Antennal scrobes completely absent . . . . . . . . . . . . . . . . . . . . . . . . . . . . . ***Lophomyrmex*** (in part)

With alitrunk in profile, the propodeal spiracle at or extremely close to the margin of the declivity. Antennal scrobes present, varying from weakly defined broad shallow longitudinal impressions to deep excavations . . . . . . . . . . . . . . . . . . . . **42**

**42(41)** Spongiform tissue present ventrally on petiole, postpetiole, and base of first gastral sternite. Petiole without a differentiated peduncle between node and articulation with alitrunk . . . . . . . . . . . . . . . . . . . . . . . . . . . . . . . . . . . . . . . . . . . ***Dacetinops***

Spongiform tissue absent from petiole, postpetiole, and first gastral sternite. Petiole with a peduncle between node and articulation with alitrunk . . . . . . . . . . . . . . . . . . . . ***Indomyrma***

**43(40)** Narrow but deep antennal scrobes present, which are bordered above by very broad horizontal laterally directed or downcurved frontal carinae; the frontal carinae extensively overhanging the scrobe and concealing it from dorsal view. Eyes situated below posterior ends of scrobal impressions . . . . . . . . . . ***Metapone***

Antennal scrobes completely absent or shallowly present. If the latter, then the scrobal impressions bounded above by narrow frontal carinae, which do not extensively overhang and conceal the scrobe, and the eyes not situated below the posterior ends of the scrobal impressions . . . . . . . . . . . . . . . . . . . . . . . . **44**

**44(43)** Maxillary palp with 4 or 5 segments . . . . . . . . . . . . . . . . . **45**

Maxillary palp with 1–3 segments . . . . . . . . . . . . . . . . . . . **48**

**45(44)** Petiole subsessile, with an extremely short and very broad anterior peduncle. Petiole node high and thickly scale-like . . . . . . . . . . . . . . . . . . . . . . . . . . . . . . . . . . . . . . . ***Stereomyrmex***

Petiole either with an elongate narrow anterior peduncle or the petiole subcylindrical and armed above with a single tooth; in either case, the node not high and thickly scale-like . . . . . . **46**

**46(45)** Propodeum armed with a pair of spines, which curve upward and forward. Junction of postpetiole and gaster strongly dorsoventrally compressed and very narrow in profile. Basal tooth of mandible broad and with two points, or else basal margin of mandible with a single tooth . . . . . . . . . . . . . ***Trigonogaster***

Propodeum unarmed or with a pair of teeth or spines, which are more or less straight and directed posteriorly or posterodorsally. Junction of postpetiole and gaster not strongly dorsoventrally compressed. Basal tooth of mandible with a single point and basal margin of mandible edentate . . . . . . . . . . . . . . . . . **47**

**47(46)** Femora of middle and hind legs grossly swollen medially. Pronotal humeri at least sharply angulate, usually dentate or spinose. Standing pilosity present on dorsum of the alitrunk . . . . . . . . . . . . . . . . . . . . . . . . . . . . . . . . . . . . . . . . ***Podomyrma***

Femora of middle and hind legs slender, not grossly swollen medially. Pronotal humeri rounded or at most bluntly angled. Standing pilosity absent from dorsum of alitrunk . . . . . . . . . . . . . . . . . . . . . . . . . . . . . . . . . ***Cardiocondyla*** (in part)

**48(44)** Broad but shallow antennal scrobes present on sides of head above the eye, the scrobes running almost the length of the side of the head capsule . . . . . . . . . . . . . . . . . . . ***Wasmannia*** (in part)

Antennal scrobes completely absent . . . . . . . . . . . . . . . . . **49**

**49(48)** Petiole sessile to subsessile and with a large to very large ventral process. Petiole not enormously more voluminous than postpetiole in dorsal view and in profile . . . . ***Vollenhovia*** (in part)

Petiole pedunculate and at most with a small tooth-like ventral process on the peduncle. If peduncle short and stout then petiole enormously more voluminous than postpetiole in dorsal view and in profile . . . . . . . . . . . . . . . . . . . . . . . . . **50**

**50(49)** Pronotal dorsum flat and sharply marginate laterally, the marginations terminating anteriorly in projecting flat acute tooth-like or triangular processes, which are above and behind the true humeral angles of the pronotum . . . . . . . ***Lophomyrmex*** (in part)

Pronotal dorsum convex, without lateral marginations and lacking acute tooth-like or triangular processes above and behind the true humeral angles of the pronotum . . . . . . . . . . . . . . . **51**

**51(50)** Propodeum unarmed, evenly rounded . . ***Monomorium*** (in part)

Propodeum with a pair of stout, posteriorly directed spines . . . . . . . . . . . . . . . . . . . . . . . . . . . . . . . . . . . . . . ***Willowsiella***

**52(37)** Mandibles hook-shaped, curving apically into a single sharp point, with an additional large tooth near the base; a deep groove running the length of the masticatory border; very rare. (Sulawesi and Sabah) . . . . . . . . . . . . . . . . . . . . . . . . ***Chimaeridris***

Mandibles of a different conformation, not hook-shaped and with the masticatory border usually lined with many serial teeth or denticles . . . . . . . . . . . . . . . . . . . . . . . . . . . . . . . . . **53**

**53(52)** Palp formula 6,4. Spurs on posterior tibiae usually pectinate . . . . . . . . . . . . . . . . . . . . . . . . . . . . . . . . . . . . . . . . ***Myrmica***

Palp formula less than 6,4 (up to 5,3). Spurs on posterior tibiae usually simple or absent, only rarely pectinate . . . . . . . . . **54**

**54(53)** Sting with an apicodorsal triangular to pennant-shaped lamellate appendage projecting from the shaft at an angle to its long axis. Lateral portions of clypeus raised into a sharp-edged ridge or shield wall on each side, in front of the antennal insertions . . . . . . . . . . . . . . . . . . . . . . . . . . . . . . . . . . . . . . . . **55**

Sting without an apicodorsal triangular to pennant-shaped lamellate appendage projecting from the shaft at an angle to its long axis, although sometimes the sting apex may be straight-spatulate. Lateral portions of clypeus not raised into a sharp-edged ridge or shield wall on each side in front of the antennal insertions . . . . . . . . . . . . . . . . . . . . . . . . . . . . . . . . . . **56**

**55(54)** Head heart-shaped in full-face view. Ventral margin of petiole convex and keel-like. Anterior clypeal margin strongly arcuate and prominent. Eyes behind midlength of sides of head. Median clypeal and median cephalic carinae vestigial or absent. Palp formula 3,2 . . . . . . . . . . . . . . . . . . . . . . . . ***Rhoptromyrmex***

Head not heart-shaped in full-face view. Ventral margin of petiole not convex and keel-like. Anterior clypeal margin not strongly arcuate. Eyes usually at or in front of midlength of sides of head, only extremely rarely otherwise. Either median clypeal carina or median cephalic carina usually present, or both present; only infrequently both absent. Palp formula usually 4,3, rarely reduced . . . . . . . . . . . . . . . . . . . . . . ***Tetramorium*** (in part)

**56(54)** Ventrolateral margin of head delineated by a sharp longitudinal carina on each side. Carina starts close to the inner-ventral mandibular base, runs the length of the head below the eye, and ascends the occipital surface posteriorly . . . . . . . ***Myrmecina***

Ventrolateral margin of head without a longitudinal carina on each side . . . . . . . . . . . . . . . . . . . . . . . . . . . . . . . . . . . **57**

*continued*

**57(56)** Alitrunk armed with spines at the pronotal humeri and the propodeal angles; in addition, dorsum has 4 pairs of elongate conical to thickly spiniform prominences on the promesonotum, and the propodeum with a single anteromedian spine. Occipital region of head with 3 pairs of similar prominences . . . . . . . . . . . . . . . . . . . . . . . . . . . . . . . . . . . . . . . . ***Proatta*** (in part)

Alitrunk at most with spines at pronotal humeri and propodeal angles; pronotum usually unarmed and propodeum sometimes unarmed; in either case, the dorsal alitrunk never with 4 pairs of elongate conical to thickly spiniform prominences on the promesonotum and never with 3 pairs of similar prominences on the occipital region of the head . . . . . . . . . . . . . . . . . . . **58**

**58(57)** Mandibles edentate. Either structurally so, with a sharp-edged toothless masticatory margin from apex to base, or functionally so, the teeth having been worn down to nothing from a previously multidentate condition. (Major workers of some dimorphic or polymorphic species.) . . . . . . . . . . . . . . . . . **59**

Mandibles dentate. Masticatory margin with a total of 3 or more teeth or denticles present . . . . . . . . . . . . . . . . . . . . . . **61**

**59(58)** Tergum of first gastral segment medially overlapping onto the anteroventral surface of the same segment, the suture between tergite and sternite of the first segment basally in the form of a rounded M-shape and the postpetiole articulated in the base of the M. In profile, the postpetiole attached on the apparent anteroventral surface of the gaster . . . . . ***Acanthomyrmex*** (in part)

Tergum of first gastral segment medially not overlapping onto the anteroventral surface of the same segment, the suture between tergite and sternite of the first segment basally transverse and not a rounded M-shape; postpetiole articulated in the middle of the anterior surface. In profile, the postpetiole not attached anteroventrally on the gaster . . . . . . . . . . . . . . . . . . . . . . **60**

**60(59)** With the head in full-face view, the eyes usually behind the midlength of the sides, or more rarely at the midlength. Head generally approximately transversely rectangular. Maxillary palp with 4–5 segments. Metasternal process large to massive . . . . . . . . . . . . . . . . . . . . . . . . . . . . . . . . . . . . ***Messor*** (in part)

With the head in full-face view, the eyes in front of the midlength of the sides, usually very obviously so. Head not transversely rectangular. Maxillary palp with 2–3 segments. Metasternal process vestigial to absent . . . . . . . . . . . . . . ***Pheidole*** (in part)

**61(58)** Masticatory margin of mandible with only 3–6 teeth or denticles in total, the dentition usually sharply defined and the teeth decreasing in size from the apical to the basalmost. Masticatory margin never with a series of ill-defined crenulations or semi-effaced denticles near the basal angle or with teeth radically alternating in size along the length of the margin . . . . . . . . **62**

Masticatory margin of mandible with 7 or more teeth or denticles in total, the dentition sometimes decreasing in size from the apex to the base but often the masticatory margin with ill-defined crenulations or denticles between the main teeth, or with a series of ill-defined crenulations or denticles near the basal angle. Sometimes teeth alternating in size along the length of the margin . . . . . . . . . . . . . . . . . . . . . . . . . . . . . . . . . **72**

**62(61)** Petiole sessile or subsessile, without a roughly horizontal anterior peduncle between the portion that articulates with the alitrunk and the ascending anterior face of the strongly developed node . . . . . . . . . . . . . . . . . . . . . . . . . . . . . . . . . . . . . . . . . . **63**

Petiole pedunculate, with a roughly horizontal anterior peduncle between the portion that articulates with the alitrunk and the ascending anterior face of the node, or the entire petiole roughly cylindrical to claviform and lacking a sharply defined node . **64**

**63(62)** Elongate frontal carinae and shallow antennal scrobes present. Palp formula 5,3. Ventral process of petiole a minute denticle. Propodeum armed with a pair of stout long spines . . . . . . . . . . . . . . . . . . . . . . . . . . . . . . . . . . . . . . . . . . ***Romblonella***

Frontal carinae and antennal scrobes absent. Palp formula 2,2. Ventral process of petiole large and angular to plate-like. Propodeum unarmed or at most with a pair of small triangular teeth . . . . . . . . . . . . . . . . . . . . . . . . . . . ***Vollenhovia*** (in part)

**64(62)** Basal border of mandible with 2 posteriorly directed broad rounded lobes, the first lobe close to the basalmost tooth of the 5 on the masticatory margin, the second lobe near the trulleum (basin-shaped depression at mandibular base) . . . ***Epelysidris***

Basal border of mandible without 2 posteriorly directed lobes; basal border usually unarmed or (rarely) with a single small tooth . . . . . . . . . . . . . . . . . . . . . . . . . . . . . . . . . . **65**

**65(64)** With head in full-face view, the occipital corners seen to be acutely angulate to dentate. Frontal carinae and antennal scrobes present, propodeum unarmed, and petiole usually subcylindrical to claviform . . . . . . . . . . . . . . . . . . . . . . . . ***Dilobocondyla***

With head in full-face view, the occipital corners broadly to narrowly rounded. If the latter then either frontal carinae and antennal scrobes are absent, or the propodeum is armed with a pair of spines or teeth, or the petiole has a definitive node; or sometimes all of these . . . . . . . . . . . . . . . . . . . . . . . . **66**

**66(65)** Midpoint of anterior clypeal margin with a long unpaired median seta, which projects forward over the mandibles. Median portion of clypeus longitudinally bicarinate, or lateral portions of clypeus flattened and strongly prominent, fused to the raised median portion and forming a shelf, which projects forward over the mandibles . . . . . . . . . . . . . . . . . . . . . . . . . **67**

Midpoint of anterior clypeal margin without a long unpaired median seta; instead either with a pair of setae, which straddle the midpoint, or with an unbroken row of long strong setae, or hairless. Median portion of clypeus not longitudinally bicarinate and lateral portions of clypeus not flattened and prominent . . . **69**

**67(66)** Propodeum unarmed and rounded or at most with minute denticles; if the latter then eyes with only a single ommatidium . . . . . . . . . . . . . . . . . . . . . . . . . . . . ***Monomorium*** (in part)

Propodeum armed with a pair of teeth or spines; eyes always with many ommatidia . . . . . . . . . . . . . . . . . . . . . . . . . . . **68**

**68(67)** Maxillary palp with 5 segments. Lateral portions of clypeus dorsoventrally flattened and thin, strongly prominent over the mandibles and sometimes projecting farther than the median clypeal portion. Median portion of clypeus not longitudinally bicarinate . . . . . . . . . . . . . . . . . . .***Cardiocondyla*** (in part)

Maxillary palp with fewer than 5 segments (usually 2–3). Lateral portions of clypeus not dorsoventrally flattened or projecting over the mandibles or even as far forward as the median clypeal portion. Median portion of clypeus longitudinally bicarinate . . . . . . . . . . . . . . . . . . . . . . . . . . . . . . . ***Rogeria*** (in part)

**69(66)** Dorsal profile of alitrunk simple, more or less flat to evenly shallowly convex from front to back, without breaks in the outline; at most the metanotal groove is shallowly present. Pronotum and mesonotum usually indistinguishable. Palp formula 5,3 . . . . . . . . . . . . . . . . . . . . . . . . . . . . . . . . . . . . . . . . . . **70**

*continued*

Dorsal profile of alitrunk complex, the pronotum or pronotum plus anterior mesonotum forming a high dome-like or markedly convex arc. Behind this convexity the mesonotum may or may not form a second eminence before sloping steeply to the metanotal groove. Propodeum forming a separate convexity or flat plateau behind metanotal groove. Pronotum and mesonotum usually distinguishable. Palp formula 2,2 to 4,3 . . . . . . . . 71

**70(69)** Metapleural lobes elongate, sharply narrowly triangular and directed almost vertically. (Endemic to Fiji) . . . . ***Poecilomyrma***

Metapleural lobes short and rounded, not projecting almost vertically as sharp narrow triangles . . . . . . . . . . . . . ***Leptothorax***

**71(69)** With head in full-face view, the eyes at or behind the midlength of the sides, usually the latter. Masticatory margin of mandible with up to 6 teeth arranged along its length, without a long central diastema. Maxillary palp with 4 segments. Metasternal process large to massive . . . . . . . . . . . . . . . . ***Messor*** (in part)

With head in full-face view, the eyes in front of the midlength of the sides. Masticatory margin of mandible with 2 teeth apically, a long diastema, and 0–3 teeth basally. Maxillary palp with 2–3 segments. Metasternal process vestigial to absent . . . . . . . . . . . . . . . . . . . . . . . . . . . . . . . . . . . . . ***Pheidole*** (in part)

**72(61)** Tergum of first gastral segment medially overlapping onto the anteroventral surface of the same segment, the suture between tergite and sternite of the first segment basally in the form of a rounded M-shape and the postpetiole articulated in the base of the M. In profile, the postpetiole attached on the apparent anteroventral surface of the gaster . . . . . ***Acanthomyrmex*** (in part)

Tergum of first gastral segment medially not overlapping onto the anteroventral surface of the same segment, the suture between tergite and sternite of the first segment basally transverse and not a rounded M-shape; postpetiole articulated in the middle of the anterior surface. In profile, the postpetiole not attached anteroventrally on the gaster . . . . . . . . . . . . . . . . . . . . . . 73

**73(72)** Petiole armed dorsally with a pair of narrow acute spines, which are directed posteriorly (possible synonym of *Lordomyrma*) . . . . . . . . . . . . . . . . . . . . . . . . . . . . . . . . . . . . . ***Ancyridris***

Petiole unarmed dorsally or with a single tooth or spine; sometimes dorsum of node emarginate but never with a pair of posteriorly directed narrow spines . . . . . . . . . . . . . . . . . . . 74

**74(73)** Median portion of clypeus narrow and longitudinally bicarinate, the surface between the 2 carinae usually transversely concave. Frontal lobes relatively close together so that the posteromedian portion of the clypeus, where it projects between the frontal lobes, is at most only slightly broader than one of the lobes. Frontal lobes themselves usually flat and transverse, not sharply elevated . . . . . . . . . . . . . . . . . . . . . . . . . . . . . . . . . 75

Median portion of clypeus broad and not longitudinally bicarinate. Frontal lobes relatively far apart so that the posteromedian portion of the clypeus, where it projects between the frontal lobes, is usually very much broader than one of the lobes. If posteromedian clypeus not distinctly much broader than one of the lobes (very rare), then the frontal lobes are markedly elevated . . . . . . . . . . . . . . . . . . . . . . . . . . . . . . . . . . . . . . 78

**75(74)** Elongate frontal carinae present, running back from the posteriormost points of the frontal lobes. Antennal scrobes variously developed, ranging from broad but shallow impressions to extensive excavations in the side of the head . . . . . . . ***Lordomyrma***

Frontal carinae absent. Antennal scrobes absent . . . . . . . . . 76

**TABLE 2-9** Oriental (*continued*)

**76(75)** Petiole sessile to subsessile, lacking a roughly horizontal anterior peduncle between the portion that articulates with the alitrunk and the anterior ascending face of the node. Petiole with a large to enormous ventral process . . . . . . . . ***Vollenhovia*** (in part)

Petiole pedunculate, with a roughly horizontal anterior peduncle between the portion that articulates with the alitrunk and the anterior ascending face of the node. Petiole with at most a small dentiform anteroventral process . . . . . . . . . . . . . . . . . . 77

**77(76)** Petiole with a very long anterior peduncle, much longer than the height of the node; petiolar node small, low and conical to subconical in profile. Antennal club of 4 segments . . . . ***Stenamma***

Petiole with a shorter anterior peduncle, shorter than or at most about equal to the height of the node; petiolar node not low and conical in profile. Antennal club of 3 segments . . . . . . . . . . . . . . . . . . . . . . . . . . . . . . . . . . . . . . . . . ***Rogeria*** (in part)

**78(74)** With alitrunk in profile, the dorsum of the promesonotum flat or forming a single very shallowly convex curve from front to back. Dorsal surface of propodeum on approximately the same level as the promesonotum, at most only fractionally lower. Petiole node rectangular or rounded-rectangular, block-shaped . . . . . . . . . . . . . . . . . . . . . . . . . . . . . . . . . . . . . . ***Paratopula***

With alitrunk in profile, the pronotum or pronotum plus anterior mesonotum forming a high dome-like or markedly convex arc. Behind this the mesonotum may or may not form a second eminence before sloping steeply, and sometimes sinuously, to the metanotal groove. Dorsal surface of propodeum depressed below level of promesonotum, usually considerably so. Petiole node usually conical to subconical, only rarely otherwise . . 79

**79(78)** Palp formula 2,2 or 3,2. Masticatory margin of mandible with the third tooth (counting from the apex) smaller than the fourth, or the reduced third tooth followed by a minute denticle before the larger fourth tooth (all minor workers); or mandible with 2 large apical and 1–2 enlarged basal teeth, the margin between these teeth irregularly crenulate or bluntly dentate (major workers of a few species) . . . . . . . . . . . . . . . . . . . . ***Pheidole*** (in part)

Palp formula 4,3 or 5,3. Masticatory margin of mandible with the third tooth (counting from the apex) larger than the fourth. Mandible never with dentition as described above for major workers . . . . . . . . . . . . . . . . . . . . . . . . . . . . . . . . . . . . . . 80

**80(79)** Metasternal process large or very large. Head massive and broad in media and major workers. Mandibles short and powerful, massively constructed, their outer margins strongly curved toward the midline. Mostly polymorphic species . . . . . . . . . . . . . . . . . . . . . . . . . . . . . . . . . . . . . . . . . ***Messor*** (in part)

Metasternal process minute to absent. Head elongate and narrow in all workers. Mandibles elongate-triangular and not massively constructed, their outer margins not strongly curved toward the midline. Monomorphic species . . . . . . . . . . . ***Aphaenogaster***

**SUBFAMILY ANEURETINAE**

(Genus *Aneuretus* only: known only from moist forests of central Sri Lanka.)

**SUBFAMILY DOLICHODERINAE**

1 Antennae with 11 segments . . . . . . . . . . . . . . . . . . ***Semonius***

Antennae with 12 segments . . . . . . . . . . . . . . . . . . . . . . . . **2**

*continued*

2(1) Petiole in profile usually a simple transversely flattened strip, sometimes slightly swollen anterodorsally but never equipped with a standing scale. Petiole overhung by first gastral segment and usually not visible in dorsal view when alitrunk and gaster are in the same plane . . . . . . . . . . . . . . . . . . . . . . . . . 3

Petiole in profile surmounted by a node or scale, which may be high and erect or lower and somewhat inclined forward, but scale (or node) always present and conspicuous. Petiole not or only weakly overhung by first gastral segment, usually visible in dorsal view when alitrunk and gaster are in the same plane . . 5

3(2) In dorsal view 5 gastral tergites visible, the fifth small but continuing the line of the gaster and not bent forward below the fourth; the anal and associated orifices are thus situated apically . . . . . . . . . . . . . . . . . . . . . . . . . . . . . . . . . . . *Technomyrmex*

In dorsal view only 4 gastral tergites visible. Fifth tergite bent forward below the fourth, visible in ventral view where it forms a transverse plate abutting the fifth sternite; the anal and associated orifices are thus situated ventrally . . . . . . . . . . . . . . 4

4(3) Monomorphic species. Maximum diameter of eye usually distinctly greater than maximum width of antennal scape . . . . . . . . . . . . . . . . . . . . . . . . . . . . . . . . . . . . . . . *Tapinoma*

Markedly dimorphic species. Maximum diameter of eye about equal to the maximum width of the antennal scape in minor worker; major worker with disproportionately large rectangular head and bilobed, abruptly truncated clypeus . . *Zatapinoma*

5(2) Palp formula 4,2 or 2,3 . . . . . . . . . . . . . . . . . *Bothriomyrmex*

Palp formula 6,4 . . . . . . . . . . . . . . . . . . . . . . . . . . . . . . . 6

6(5) Head and alitrunk extremely elongate and slender; appendages narrow and extremely long. Petiole nodiform. Mandibles elongate-triangular and slender, outer margins shallowly concave through the central section of their length. Masticatory margin of mandible with 17 or more (usually about 20) teeth or denticles in total . . . . . . . . . . . . . . . . . . . . . . . . . . *Leptomyrmex*

Head and alitrunk usually broad and stocky, not extremely elongate; appendages not extremely long. Petiole usually a scale of some form, only rarely nodiform. Mandibles triangular, their outer margins generally convex but sometimes straight or slightly concave through the central section of their length. Masticatory margin of mandible usually with 12 or fewer teeth, rarely with up to 15 . . . . . . . . . . . . . . . . . . . . . . . . . . 7

7(6) Integument thick, hard and armor-like, the surface varying from smooth (very rare) to strongly and coarsely sculptured . . . . . 8

Integument thin and flexible, not armor-like, the surface usually very finely and densely sculptured, only extremely rarely more or less smooth . . . . . . . . . . . . . . . . . . . . . . . . . . . . . 9

8(7) Propodeum drawn out into a single long, horn-like protuberance. (New Guinea) . . . . . . . . . . . . . . . *Monoceratoclinea*

Propodeum not bearing a long, horn-like protuberance, although in some species it is angulate in side view. (Widespread in tropical Asia) . . . . . . . . . . . . . . . . . . . . . . . *Hypoclinea*

9(7) Posterodorsal angles of propodeum drawn out into short tubercles or prominences, the propodeal spiracles situated at the apices of the tubercles or prominences . . . . . . . . . . . . . . . . *Turneria*

Posterodorsal areas of propodeum not drawn out into short tubercles or prominences. Propodeal spiracles lateral . . . . . . 10

10(9) With alitrunk viewed in profile, metanotal groove impressed and metathoracic spiracles dorsal. Anterior face of first gastral tergite without a deep concavity in the area immediately behind the petiolar scale . . . . . . . . . . . . . . . . . . . . . . . *Iridomyrmex*

With alitrunk in profile, metanotal groove not impressed and metathoracic spiracles lateral. Anterior face of first gastral tergite with a deep concavity in the area immediately behind the petiolar scale . . . . . . . . . . . . . . . . . . . . . . . . . *Liometopum*

### SUBFAMILY FORMICINAE

1 Antennae with 8 segments . . . . . . . . . . . . . . . . . . . . . . . 2

Antennae with 9–12 segments . . . . . . . . . . . . . . . . . . . . . 3

2(1) Antennal scape, when laid back in its natural resting position, passing below the eye. Masticatory margin of mandible with more than 4 teeth . . . . . . . . . . . . . . . . . . . *Gesomyrmex*

Antennal scape, when laid back, passing above the eye. Masticatory margin of mandible with 4 teeth . . . . . . . . *Cladomyrma*

3(1) Antennae with 9–11 segments . . . . . . . . . . . . . . . . . . . . . 4

Antennae with 12 segments . . . . . . . . . . . . . . . . . . . . . . 8

4(3) Palp formula 5,3 or less . . . . . . . . . . . . . . . . . . . *Acropyga*

Palp formula 6,4 . . . . . . . . . . . . . . . . . . . . . . . . . . . . . . 5

5(4) Antennae with 9 segments . . . . . . . . . . . . . . . *Brachymyrmex*

Antennae with 11 segments . . . . . . . . . . . . . . . . . . . . . . 6

6(5) Propodeum armed with a pair of spines, teeth, or tubercles. Dorsal edge of petiole usually armed with a pair of teeth or spines but sometimes only emarginate . . . . . . . . . . . . . . *Acantholepis*

Propodeum and petiole unarmed, without spines, teeth, or tubercles . . . . . . . . . . . . . . . . . . . . . . . . . . . . . . . . . . . 7

7(6) With alitrunk in dorsal view, the mesonotum seen to be separated from the metanotum by a conspicuous transverse groove or impression, so that the metanotum forms a distinctly isolated sclerite . . . . . . . . . . . . . . . . . . . . . . . . . . . . . *Plagiolepis*

With alitrunk in dorsal view, the mesonotum seen to be fused with the metanotum, the two not separated by a transverse groove or impression, the metanotum not forming an isolated sclerite . . . . . . . . . . . . . . . . . . . . . . . . . . . . . . . . . . . . . *Anoplolepis*

8(3) Mandibles extended into extremely long, slender, linear blades, which project far in front of the anterior clypeal margin; the mandibles at least 0.85 times the head length and often exceeding the head length . . . . . . . . . . . . . . . . . . . . *Myrmoteras*

Mandibles subtriangular to elongate-triangular, not extended into long, slender blades and usually very obviously less than 0.85 times the head length . . . . . . . . . . . . . . . . . . . . . . . . 9

9(8) Antennal sockets situated close to the posterior clypeal margin, and metapleuron with a distinct metapleural gland orifice, the orifice situated above the hind coxa and below the level of the propodeal spiracle . . . . . . . . . . . . . . . . . . . . . . . . . 10

Either antennal sockets situated far behind the posterior clypeal margin, or the metapleuron lacking a metapleural gland orifice in the location described above, or sometimes both . . . . . . 16

10(9) Maxillary palp with 2–4 segments . . . . . . . . . . . . . . . . . . 11

Maxillary palp with 6 segments . . . . . . . . . . . . . . . . . . . . 12

11(10) With alitrunk in profile, the mesonotum and anepisternum seen to form together a roughly triangular oblique wedge between pronotum and remainder of alitrunk. Posterolateral angle of pronotum very nearly touching the katepisternal anterior margin. An-

*continued*

terior clypeal margin convex or indented medially, not broadly and evenly concave. Outer margin of mandible shallowly curved in apical half; at full closure the apical tooth directed laterally or anterolaterally . . . . . . . . . . . . . . . . . . . . . . . ***Pseudolasius***

With alitrunk in profile, the mesonotum and anepisternum seen not to form a roughly triangular oblique wedge between pronotum and remainder of alitrunk. Instead this region narrow and elongated, with a distinct horizontal border ventrally between the posterolateral angle of the pronotum and the katepisternal anterior margin. Anterior clypeal margin broadly and evenly concave across its entire width. Outer margin of mandible strongly curved in apical half; at full closure the apical tooth directed posterolaterally or posteriorly . . . . . . . . ***Euprenolepis***

**12(10)** Orifice of propodeal spiracle either elongate-oval or elliptical in shape, or an elongate slit, and near-vertical or inclined from the vertical. With alitrunk in absolute profile, the propodeal spiracle well in front of the point where the propodeal side rounds into the declivity . . . . . . . . . . . . . . . . . . . . . . . . . . . . . . . **13**

Orifice of propodeal spiracle circular to subcircular. With alitrunk in absolute profile, the propodeal spiracle bordering or actually on the curvature where the propodeal side rounds into the declivity . . . . . . . . . . . . . . . . . . . . . . . . . . . . . . . . . . . . **14**

**13(12)** Apical (masticatory) margin of mandible usually with 8 teeth but sometimes with more. Third tooth of mandible, counting from the apex, always distinctly smaller and shorter than the fourth; the fourth tooth larger than all the remaining teeth to the basal angle . . . . . . . . . . . . . . . . . . . . . . . . . . . . . . . ***Formica***

Apical (masticatory) margin of mandible with 5–7 teeth. If more than 5 teeth present, then the third tooth, counting from the apex, larger and longer than the fourth; teeth after the fourth decreasing in size to the basal angle . . . . . . . . . . ***Cataglyphis***

**14(12)** With the head in full-face view, the eyes at or in front of the midlength of the sides. Head and alitrunk with stout bristles arranged in distinct pairs . . . . . . . . . . . . . . . . . ***Paratrechina***

With the head in full-face view, the eyes distinctly behind the midlength of the sides. Hairs on head and alitrunk not distinctly paired and usually not stout bristles . . . . . . . . . . . . . . . . . **15**

**15(14)** Mandibles with 6 teeth, very rarely with 7. Anterior face of the first gastral segment broadly and transversely concave throughout its height. Antennal scapes relatively very long; when laid straight back from their insertions at least half their length projects beyond the occipital margin . . . . . . . . . . . . . . . . . ***Prenolepis***

Mandibles with at least 7 teeth, usually with more than 7. Anterior face of first gastral segment with a small concave area immediately above the petiole-gaster articulation, but the face not broadly transversely concave throughout its height. Antennal scapes much shorter; when laid straight back from their insertions much less than half their length projects beyond the occipital margin . . . . . . . . . . . . . . . . . . . . . . . . . . . ***Lasius***

**16(9)** Mandible with 10 or more teeth or denticles in total. Apical tooth disproportionately large and the fourth tooth, counting from the apical, larger than the third and fifth teeth. Petiole reduced to an elongate low node, which allows the gaster to be bent forward over the alitrunk. Palp formula 5,4 . . . . . . . . . . . ***Oecophylla***

Mandible usually with 5–7 teeth at most, only very rarely with more. If 7 or more teeth present, they decrease in size from apex to base; the fourth tooth is not enlarged as above. Petiole an erect node or scale, the gaster not capable of being bent forward over the alitrunk. Palp formula 6,4 . . . . . . . . . . . . . . . . . **17**

**TABLE 2–9** Oriental (*continued*)

**17(16)** Eyes very large and in an extreme posterolateral position on the head. In full-face view, the occipital corner is formed by the curvature of the eye on each side, and posteriorly the eyes form the lateral portions of the occipital margin. The eyes often project slightly farther posteriorly than the true occipital margin that runs between them . . . . . . . . . . . . . . . . . . . . ***Opisthopsis***

Eyes moderate to large and usually situated behind midlength of sides, but not occupying the occipital corners or constituting a part of the occipital margin . . . . . . . . . . . . . . . . . . . . **18**

**18(17)** With alitrunk in profile, the metathoracic spiracles forming tuberculiform prominences that project beyond the outline of the dorsum. Propodeum posteriorly with a raised transverse ridge, which appears as a tooth in profile. Pronotum and petiole node unarmed. (One species in Java, very rare) . . . . ***Forelophilus***

With alitrunk in profile, the metathoracic spiracles usually not forming tuberculiform prominences that project beyond the outline of the dorsum. When such spiracles are present (rare), the propodeum posteriorly does not have a raised transverse ridge resembling a tooth in profile, or the pronotum and petiole are armed with teeth or spines, or both . . . . . . . . . . . . . . . . **19**

**19(18)** Antennal funiculus with apical segments gradually but strongly broadening, forming a club. Proventriculus with sepals of calyx short, scarcely longer than the basal bulb. (One species in Singapore, rare) . . . . . . . . . . . . . . . . . . . . . . ***Overbeckia***

Apical segments of antennal funiculus not forming a club. Proventriculus with sepals of calyx much longer than the basal bulb . . . . . . . . . . . . . . . . . . . . . . . . . . . . . . . . . . . . . . . . **20**

**20(19)** Metapleural gland orifice present on side of metapleuron above the hind coxa and below the level of the propodeal spiracle. Orifice usually preceded by a longitudinal impression, which is overhung by a projecting rim of cuticle, the orifice itself usually with a conspicuous tuft of downward-directed guard hairs . **21**

Metapleural gland orifice absent from side of metapleuron. An oblique impression separating metapleuron from propodeum frequently present, but gland orifice as described above is absent . . . . . . . . . . . . . . . . . . . . . . . . . . . . . . . . . . . . . . . . **23**

**21(20)** Tergite of first gastral segment extremely large, accounting for considerably more than half the length of the gaster in dorsal view or in profile; sometimes the entire gastral dorsum consists of the first tergite alone. Dorsum of petiole usually armed with spines, teeth, or tubercles, the dorsolateral angles frequently dentate or spinose and the sides sometimes spinose; petiole only very rarely nodiform . . . . . . . . . . . . . . . . . . . . . . . . ***Echinopla***

Tergite of first gastral segment much smaller, accounting for much less than half the length of the gaster in dorsal view or in profile. Petiole nodiform to thickly scale-like, never armed with spines or teeth . . . . . . . . . . . . . . . . . . . . . . . . . . . . . . . . . . **22**

**22(21)** Mandible with 5 teeth. Median portion of clypeus usually shorter and less prominent than the lateral portions, not forming a narrow lobe that projects forward over the mandibles. Metanotal groove deeply impressed . . . . . . . . . . . . . . . ***Calomyrmex***

Mandible with more than 5 teeth. Median portion of clypeus much longer than lateral portions, forming a narrow lobe that projects forward over the mandibles. Metanotal groove an unimpressed transverse line . . . . . . . . . . . . . . . . ***Camponotus*** (in part)

**23(20)** Tergite of first gastral segment large, accounting for at least half the length of the gaster in dorsal view or in profile; the first tergite distinctly much longer than the second. Spines or teeth

*continued*

present on pronotum, propodeum, petiole, or on two or all of these . . . . . . . . . . . . . . . . . . . . . . . . . . . . . ***Polyrhachis***

Tergite of first gastral segment shorter, accounting for distinctly less than half the length of the gaster in dorsal view or in profile; the first tergite at most only slightly longer than the second. Spines or teeth usually absent from pronotum, propodeum, and petiole; very rarely one of these locations armed . . . . . . . . . . . . . . . . . . . . . . . . . . . . . . . . . . . . ***Camponotus*** (in part)

**TABLE 2–10** A key to the ant genera of Australia, New Caledonia, and New Zealand, referred to generally as the **Australian Region.** Based on the worker caste. (Modified slightly from R. W. Taylor, previously unpublished; used with permission.)

### SUBFAMILY PONERINAE

1 Petiole broadly attached to first gastral segment, without free posterior face, mandibles often elongate and narrow, their entire inner margins dentate, the teeth frequently double ranked; anterior clypeal border often denticulate . . . . . . . . . . . . . . . . 2

Petiolar-gastral junction narrow, petiole with a distinct free posterior face, mandibles and clypeus not as above . . . . . . . . . . 6

**2(1)** Mandibles short, narrow, closing tightly against clypeus, their apical borders distinct, completely occupied by 3 large teeth; basal borders of mandibles unarmed . . . . . . . . . . . . . ***Prionopelta***

Mandibles otherwise; usually strongly projecting beyond clypeus when closed, and with more than 3 teeth . . . . . . . . . . . . . 3

**3(2)** Antennal funiculi markedly compressed; when head is viewed full-face, lobes of frontal carinae approximately even with or extending beyond the anterior clypeal border below them. (Northern Cape York Peninsula) . . . . . . . . . . . . . . . . ***Myopopone***

Antennal funiculi not compressed, approximately round in cross section; in full-face view, lobes of frontal carinae distinctly behind the median clypeal border . . . . . . . . . . . . . . . . . . . 4

**4(3)** Mandibular apex bluntly rounded or subtruncate as seen from above; many body hairs clavate or spatulate . . . . . . ***Mystrium***

Apex of mandible an acute tooth; body hairs simple, fine and tapered . . . . . . . . . . . . . . . . . . . . . . . . . . . . . . . . . . . . 5

**5(4)** Hind tibia lacking an apical spur, or at most with a nonpectinate vestige; small, shining, slender ants with greatly enlarged middle and posterior tarsal claws; queens wingless, small-eyed ("dichthadiiform"); legionary ants of Queensland rain forest . . . . . . . . . . . . . . . . . . . . . . . . . . . . . . . . ***Onychomyrmex***

Tibia of hind leg bearing a well-developed apical spur with a curved, broadly pectinate inner margin; species of diverse size and form, tarsal claws rarely enlarged; queens normally winged, sometimes ergatoid . . . . . . . . . . . . . . . . . . . ***Amblyopone***

**6(1)** Pygidium impressed, the laterapical borders with raised edges bearing a row or field of minute denticles, sometimes reduced to 4 or 6, and small in size . . . . . . . . . . . . . . . . . . . . . . . 7

Pygidium with rounded, convex contours, lacking such denticles . . . . . . . . . . . . . . . . . . . . . . . . . . . . . . . . . . . . . . . . 8

**7(6)** Principal segments of gaster (true abdominal segments IV, V, VI) separated by marked annular constrictions, the sternite and tergite of each segment fused laterally and together forming an anterior articular boss within the preceding segmental exoskeleton, i.e., "tubulate" . . . . . . . . . . . . . . . . . . . ***Sphinctomyrmex***



Gastral segments not separated by constrictions, following that behind the postpetiole (segment IV). Segments behind IV not tubulated, sternites and tergites each separate . . . . . ***Cerapachys***

**8(6)** Tarsal claws pectinate, each bearing a series of minute comb-like bristles (including "*Prionogenys*") . . . . . . . . . . . . ***Leptogenys***

Tarsal claws simple, or else with a single median tooth . . . . . . **9**

**9(8)** Mandibles linear, long and straight, with apical armament of 3 teeth, lacking other dentition; the jaws inserted at the middle of the anterior margin of head, their bases closely approximate . . . . . . . . . . . . . . . . . . . . . . . . . . . . . . . . . . . . . . . . . **10**

Mandibles inserted at the sides of the anterior margin of head, usually triangular, sometimes elongate, but with the bases well separated, the inner margins usually serially dentate . . . . . **11**

**10(9)** Nuchal carina (separating dorsal from posterior surfaces of head) converging in a V at the midline, and also receiving a pair of prominent dark posterior apophyseal lines, which converge to form a sharp median-dorsal groove of the vertex; petiole surmounted by a slender drawn-out tooth . . . . . . ***Odontomachus***

Nuchal carina forming a broad uninterrupted curve across the posterior extremity of the head; posterior surface lacking dark apophyseal lines; on vertex, median groove absent or ill defined; petiole a transverse scale, sometimes with angulate dorsolateral corners, or with its apex produced as a blunt apical knob . . . . . . . . . . . . . . . . . . . . . . . . . . . . . . . . . . . . . . ***Anochetus***

**11(9)** Mesosomal dorsum uninterrupted by transverse sutures, its surface continuous except for sculpturation . . . . . . . . . . . . . **12**

Mesosomal dorsum clearly divided by a distinctly incised promesonotal suture, often also by a metanotal groove . . . . . . . . **15**

**12(11)** Frontal carinae and clypeus fused and projecting anteriorly over the mandibles; the antennae inserted close together on the anterior part of the clypeo-carinal process, with their bases exposed in frontal view . . . . . . . . . . . . . . . . . . . . . . . . . **13**

Frontal carinae and clypeus not as above, their structure conventional . . . . . . . . . . . . . . . . . . . . . . . . . . . . . . . . . . . . **14**

**13(12)** Antennae with 7 to 10 free segments; the tubulate IV abdominal segment exoskeleton strongly reflexed so that its dorsum is strongly arched in side view, and the planes of its anterior and posterior apertures forming approximately a right angle . . . . . . . . . . . . . . . . . . . . . . . . . . . . . . . . . . . . . . ***Discothyrea***

Antennae 12-segmented; slender, elongate, eyeless ants lacking a reflexed IV abdominal segment exoskeleton . . ***Probolomyrmex***

**14(12)** Eyes large, multifaceted; posterior coxae each surmounted by a slender acute spine. (Known only from rain forests of far northern Cape York Peninsula) . . . . . . . . . . . ***Gnamptogenys***

Eyes minute, single-faceted; posterior coxae without slender acute spines . . . . . . . . . . . . . . . . . . . . . . . . . . . . ***Proceratium***

**15(11)** Tarsal claws each with a single median tooth, which may be very small, but always distinct at appropriate magnifications . . . **16**

Tarsal claws simple . . . . . . . . . . . . . . . . . . . . . . . . . . . **17**

**16(15)** Ventrolateral borders of the pronotum on each side drawn out as an acute tooth-like process. In the rare exceptions the petiolar apex is produced as posteromedian tooth . . . . ***Rhytidoponera***

Ventrolateral pronotal teeth lacking; petiole with an extensive dorsal surface, never produced as a tooth-like process . . . . . . . . . . . . . . . . . . . . . . . . . . . . . . . . . . . . . . . . . . ***Platythyrea***

**17(15)** Extreme base of outer border of mandible with a small but distinct circular or elliptical pit . . . . . . . . . . . . . . . . . . . . . . . **18**

*continued*

Mandibular pit lacking . . . . . . . . . . . . . . . . . . . . . . . . . 19

18(17) Eyes minute or vestigial, their diameter much less than that of the antennal scapes . . . . . . . . . . . . . . . . . . . . . . ***Cryptopone***

Eyes larger, their maximum diameter at least approximating that of the antennal scapes . . . . . . . . . . . . . . . . . ***Brachyponera***

19(17) Large species (length ca. 1 cm) with relatively heavy costate sculpturation; mesosoma on each side with a large pit near its midlength; petiolar node bearing a pair of acute posterodorsal spines on its summit . . . . . . . . . . . . . . . . . . . ***Diacamma***

Usually much smaller species, more lightly sculptured, without mesosomal pits or petiolar spines . . . . . . . . . . . . . . . . . 20

20(19) Frontal carinae more or less parallel, bordering an extensive median portion of the frons . . . . . . . . . . . . . . . . ***Heteroponera***

Frontal carinae lacking, the antennal bases protected by small frontal lobes, which converge sharply behind the antennal insertions . . . . . . . . . . . . . . . . . . . . . . . . . . . . . . . . . . . . . . 21

21(20) Mandibles slender, elongate, enclosing a broad triangular space in front of the clypeus; clypeus with a truncated median process projecting anteriorly between the frontal lobes . . . . . ***Myopias***

Mandibles broadly triangular, closing against the clypeal border . . . . . . . . . . . . . . . . . . . . . . . . . . . . . . . . . . . . . . . 22

22(21) Larger species, seldom less than 3 mm in length, often much larger. Maxillary palpi 3- or 4-segmented . . . . . ***Pachycondyla***

Small species, length seldom exceeding 3 mm. Maxillary palpi 1- or 2-segmented . . . . . . . . . . . . . . . . . . . . . . . . . . . . 23

23(22) Subpetiolar process low, anteriorly with a minute fenestra, posteriorly with a pair of minute parallel denticles; palpal formula 2,2 . . . . . . . . . . . . . . . . . . . . . . . . . . . . . . . . . . . . . ***Ponera***

Subpetiolar process a simple lobe without fenestra or posteroventral teeth; maxillary palpi 1-segmented, labials 1- or 2-segmented . . . . . . . . . . . . . . . . . . . . . . . . . . ***Hypoponera***

SUBFAMILY NOTHOMYRMECIINAE

(Represented by the single species *Nothomyrmecia macrops* of south central and southwestern Australia.)

SUBFAMILY MYRMECIINAE

(Represented by the single genus *Myrmecia*, the bulldog ants of Australia and New Caledonia.)

SUBFAMILY DORYLINAE

(Genus *Aenictus* only: tropical habitats of eastern Australia.)

SUBFAMILY LEPTANILLINAE

(Genus *Leptanilla* only: rare and widely dispersed in Australia.)

SUBFAMILY PSEUDOMYRMECINAE

(Genus *Tetraponera* only: tropical and subtropical Australia.)

SUBFAMILY MYRMICINAE

1 Antennae 4-segmented . . . . . . . . . . . . . . . . . . . . . . . . . 2

Antennae 5-segmented . . . . . . . . . . . . . . . . . . . . . . . . . 3

**TABLE 2–10** Australian (*continued*)

Antennae 6-segmented . . . . . . . . . . . . . . . . . . . . . . . . . 4

Antennae 7-segmented . . . . . . . . . . . . . . . . . . . . . . . . . 8

Antennae 9-segmented . . . . . . . . . . . . . . . . . . . ***Meranoplus***

Antennae 10-segmented . . . . . . . . . . . . . . . . . . . . . . . . 9

Antennae 11-segmented . . . . . . . . . . . . . . . . . . . . . . . . 12

Antennae 12-segmented . . . . . . . . . . . . . . . . . . . . . . . . 24

2(1) Mandibles broad, triangular, engaging continuously along their serially dentate masticatory borders when closed, without an intervening gap . . . . . . . . . . . . . . . . . ***Colobostruma*** (in part)

Each mandible a short curved shaft, edentate except for an apical cluster of 3 spine-like teeth; a large subcircular gap is framed between the shafts when the jaws are closed . . ***Quadristruma***

3(1) Mandibles triangular; when fully closed there is no gap between their minutely serially dentate masticatory borders . . . . . . . . . . . . . . . . . . . . . . . . . . . . . . . . . ***Colobostruma*** (in part)

Mandibles linear, elongated, each armed only with an apical cluster of 2 to 4 spine-like teeth; these alone engage when the jaws are closed and, with the exception of major workers of one species, a broad open gap is framed between the shafts at full closure . . . . . . . . . . . . . . . . . . . . . . . . . . ***Orectognathus***

4(1) Mandibles short or elongate, basically triangular in shape, and without a significant gap between their serially dentate masticatory borders when closed . . . . . . . . . . . . . . . . . . . . . . . 5

Mandibles elongated, linear, armed only with 2 to 5 spine-like teeth at or near each of their apices; these engage when the jaws are closed, leaving a broad open gap between the shafts . . . . 7

5(4) Petiole and postpetiole bearing massive lateral and ventral blocks of pale foam-like cuticular material . . . . . . . ***Glamyromyrmex***

Petiole and postpetiole lacking foam-like material . . . . . . . . . 6

6(5) Both petiole and postpetiole with lateral wing-like extensions, sometimes reduced, but always distinct, and often with transparent sections . . . . . . . . . . . . . . . . ***Colobostruma*** (in part)

Lateral alary extensions lacking on petiole although, with the exception of one species, they are present on the postpetiole . . . . . . . . . . . . . . . . . . . . . . . . . . . . . . . . . . . . ***Mesostruma***

7(4) Petiole and postpetiole bearing massive lateral and ventral blocks of pale foam-like cuticular material . . . . . . . . . . ***Strumigenys***

Petiole often bilaterally spinose; postpetiole usually with lateral wing-like extensions, spines, etc.; both nodes lack foam-like material . . . . . . . . . . . . . . . . . . . . . . . . . . ***Epopostruma***

8(1) Mandibles triangular, their masticatory borders serially dentate and fully engaging at closure . . . . . . . . . . . . . ***Rhopalothrix***

Mandibles linear, their insertions remote, so that their masticatory borders cross or engage only near the apices . . ***Eurhopalothrix***

9(1) Antennal funiculus without a segmentally differentiated club . . . . . . . . . . . . . . . . . . . . . . . . . . . . ***Monomorium*** (in part)

Antennal club 2-segmented . . . . . . . . . . . . . . . . . . . . . 10

10(9) Strong antennal scrobes present above eyes; the latter relatively large, elongate-oval (roughly kidney-shaped), each drawn to a point extending almost to the base of the adjacent mandible; eyes almost as long as anterior tarsi; propodeum bispinose . . . . . . . . . . . . . . . . . . . . . . . . . . . . . . . . . . . ***Mayriella***

Antennal scrobes and propodeal spines lacking; eyes smaller, much shorter than the anterior tarsi, subcircular in outline . . 11

11(10) Mesosomal dorsum planar, straight in profile, lacking all trace of

*continued*

transverse sutures, minutely longitudinally striate; apical antennomere flattened, narrowly oval in end view . . ***Rhopalomastix***

Mesosomal dorsum not planar or flat in profile, broken by a (usually deep) metanotal groove, and differently sculptured; apical antennomere basically cylindrical, circular in end view . . . . . . . . . . . . . . . . . . . . . . . . . . . . . . . . . . . . . . . . . ***Solenopsis***

**12(1)** Postpetiole attached anterodorsally to first segment of gaster; gastral dorsum in side view flat or concave, contrasting with the convex ventral face; gaster more or less heart-shaped in dorsal view, capable of being reflexed forward over the mesosoma . . . . . . . . . . . . . . . . . . . . . . . . . . . . . . . . . . . ***Crematogaster***

Postpetiole attached to anterior end of first gastral segment; dorsal and ventral outlines of gaster approximately equally convex; gaster more or less oval in dorsal view, not capable of flexure over the mesosoma . . . . . . . . . . . . . . . . . . . . . . . . **13**

**13(12)** Well-marked antennal scrobes present above the eyes, normally capable of receiving the full length of the associated scape . . **14**

Antennal scrobes lacking or at most weakly developed and shallow . . . . . . . . . . . . . . . . . . . . . . . . . . . . . . . . . . . **17**

**14(13)** Mesosomal dorsum more or less planar, its profile straight, broken by a finely incised metanotal groove, which is not impressed in side view; propodeal spines lacking; eyes minute, often almost indiscernible, much shorter than maximum thickness of the scape; antennal apices flattened, each narrowly oval in end view . . . . . . . . . . . . . . . . . . . . . . . . . . . . . . . . . . . ***Metapone***

Mesosomal dorsum transversely arched, usually not straight in profile; metanotal groove often vestigial or lacking; eyes larger, usually as long as or longer than the scapes are thick; antennal apices not flattened, circular in end view . . . . . . . . . . . . . **15**

**15(14)** Petiole and postpetiole each extended dorsally as a high conical turret, which is somewhat compressed laterally, with the pointed apex inclined posterodorsally . . . . . . ***Peronomyrmex***

Petiole and postpetiole not so peculiarly elaborated . . . . . . . **16**

**16(15)** Infradental lamellae at the lower, posterior end of the propodeum well developed, each usually extended as a strong spine or denticle, which may be almost as long as the associated propodeal spine above it; sting with a spatulate or lamelliform appendage apicodorsally; antennal scrobes not bordered by lateral genal carinae . . . . . . . . . . . . . . . . . . . . . . . ***Tetramorium*** (in part)

Infradental lamellae small, rounded, not acutely extended; sting without an apicodorsal appendage; antennal scrobes each bordered by a lateral genal carina. (Neotropical genus introduced onto New Caledonia) . . . . . . . . . . . . . . . . . . . . ***Wasmannia***

**17(13)** Mesosoma with well-developed armament, comprising a pair of spines or strong denticles each on the pronotal humeri, propodeum, and infradental lamellae . . . . . . . . . . . ***Pristomyrmex***

Propodeal spines alone sometimes present, usually absent; pronotal spines lacking . . . . . . . . . . . . . . . . . . . . . . . . . . **18**

**18(17)** Antennal club distinctly 2-segmented . . . . . . . . . . . . . . . . **19**

Antennal club constituted differently . . . . . . . . . . . . . . . . **20**

**19(18)** Worker caste markedly dimorphic, without intermediates between the minute minors and the large-headed majors; minors sometimes eyeless; majors with a pair of short, forwardly directed, horn-like projections on the occipital area of the head. (Cryptobiotic and widespread, nesting in soil or rotting wood) . . . . . . . . . . . . . . . . . . . . . . . . . . . . . . . . . . . . . . ***Oligomyrmex***

Workers strongly and continuously polymorphic, the smallest and the largest individuals connected by a series of graded intermediates, which show allometric enlargement of the head and increase in size; eyes distinct in all variants; majors without horn-like projections on the head. (Surface-active army ants in northern Australian wet forests, with bivouac nests in soil or leaf-litter) . . . . . . . . . . . . . . . . . . . . . . . ***Pheidologeton***

**20(18)** Workers highly polymorphic; antennal club distinctly 3-segmented . . . . . . . . . . . . . . . . . . . . . . . . . . . . . . . . . . **21**

Workers monomorphic; antennal club not or only barely segmentally differentiated, not distinctly 3-segmented . . . . . . . . . **22**

**21(20)** Integument of head and mesosoma coarsely striate-rugose (longitudinally on head) and subopaque; clypeus longitudinally bicarinate; eyes relatively large and distinctly darkly pigmented; their maximum length about the same as that of the penultimate antennal segment; palpal formula 4,3 . . . . . . . . . . . . ***Adlerzia***

Head and mesosoma without sculpture, smooth and strongly shining; clypeus not longitudinally bicarinate; eyes lacking in smallest minors, minute and depigmented in large majors; palpal formula 2,2 . . . . . . . . . . . . . . . . . . . . . ***Machomyrma***

**22(20)** Sting with a spatulate or lamelliform appendage apicodorsally; infradental lamellae strongly developed, acutely dentate or spinose, the spines sometimes as long as the propodeals; the latter present except in one New Caledonian species . . . . . . . . . . . . . . . . . . . . . . . . . . . . . . . . . . . . ***Tetramorium*** (in part)

Sting without an apicodorsal appendage; infradental lamellae absent or vestigial; propodeal spines often lacking in Australian species . . . . . . . . . . . . . . . . . . . . . . . . . . . . . . . . . . . **23**

**23(22)** Genus endemic to New Zealand; median area of clypeus bordered behind by a broad, deeply impressed transverse suture between the frontal carinae; propodeum bispinose; petiolar node rounded in all directions, without denticles or other elaboration; maximum length about 5 mm . . . . . . . . . . . . . . . . ***Huberia***

Genus endemic to Australia and Melanesia, absent from New Zealand and New Caledonia; clypeus less strongly defined behind; propodeum usually not bispinose; petiolar node often bispinose, crested, or otherwise elaborated, especially in smaller species within the *Huberia* size range; many species are of larger size, some with total length exceeding 10 mm . . . ***Podomyrma***

**24(1)** Sting with an apicodorsal spatulate or lamelliform appendage . **25**

Sting lacking an apicodorsal appendage . . . . . . . . . . . . . . . **26**

**25(24)** Infradental lamellae well developed, each usually produced as an acute angle or strong spine; shallow antennal scrobes usually present . . . . . . . . . . . . . . . . . . . . . . ***Tetramorium*** (in part)

Infradental lamellae and antennal scrobes lacking . . . . . . . . . . . . . . . . . . . . . . . . . . . . . . . . . . . . . . . ***Rhoptromyrmex***

**26(24)** Strong antennal scrobes present above the eyes . . . . . . . . . . **27**

Antennal scrobes lacking . . . . . . . . . . . . . . . . . . . . . . . . **28**

**27(26)** Antennal scrobes exceptionally large and deep, but lacking lateral genal carinae; eyes minute, with only a few obscure facets, almost enclosed by antennal scrobes; frontal lobes projecting very strongly forward to crowd the median portion of the clypeus, which forms a narrow bifurcate, anteriorly directed projection, and to deeply obscure antennal sockets; hairs spatulate or otherwise elaborated; mesosoma disproportionately short, in dorsal view little longer than the head is wide . . . . . ***Calyptomyrmex***

Antennal scrobes strong, but not exceptionally so; eyes larger (maximum diameter exceeding thickness of scape), separated

*continued*

from antennal scrobes by well-developed genal carinae; clypeus bicarinate, but not unduly crowded by the frontal lobes, which are not exceptionally produced and do not greatly obscure the antennal sockets; hairs unexceptional. Mesosoma of more typical proportions, in dorsal view at least twice as long as the head is wide . . . . . . . . . . . . . . . . . . . . . ***Lordomyrma*** (in part)

**28(26)** Head on each side with a closely parallel pair of minute but distinct carinae separated by a fine groove, which extends posteriorly from the base of the mandible to pass below the eye and terminate obliquely at the similarly fine but usually medially unpaired occipital carinae; the lateral carinae may delimit a marked transition in the sculpturation . . . . . . . . . . . . . . . . . . . . . **29**

Area on each side of head between the eye and the genal suture, which marks the ventral midline of the cranium, totally uninterrupted by carinae, grooves, or other structures; occasional species have a line of sculptural transition in this area, but without accompanying carinae . . . . . . . . . . . . . . . . . . . . . . . **30**

**29(28)** Petiole subcylindrical, barrel-shaped, nearly quadrate in form when viewed from above, without a distinct anterior peduncle; mesosomal dorsum longitudinally sharply costulate . . . . . . . . . . . . . . . . . . . . . . . . . . . . . . . . . . . . . . . . ***Myrmecina***

Petiole with a well-developed anterior peduncle, the node distinct, rather quadrate in side view; sculpturation not as prescribed above . . . . . . . . . . . . . . . . . . . . . . . . . . . . ***Leptothorax***

**30(28)** Antennal club distinctly 2-jointed; known in the Australian Region (as defined for this key) only from New Caledonia . . . . . . . . . . . . . . . . . . . . . . . . . . . . . . . . . . . . . . . . . ***Adelomyrmex***

Antennal club differently constituted . . . . . . . . . . . . . . . . **31**

**31(30)** Polymorphic species complete with a caste of large-headed soldiers . . . . . . . . . . . . . . . . . . . . . . . . . . . . . . . . . . . **32**

Monomorphic species lacking a soldier caste . . . . . . . . . . . . **33**

**32(31)** Workers markedly dimorphic, without intermediates between the very different major and minor castes; pronotal humeri sometimes acutely extended or spinose; mesosoma in side view with the promesonotal section elevated above the propodeal dorsum . . . . . . . . . . . . . . . . . . . . . . . . . . . . . . . . . . . ***Pheidole***

Workers continuously polymorphic, with a graded series of intermediates connecting the allometrically very different smallest and largest individuals; pronotal humeri unarmed; profile of mesosomal dorsum almost straight, the promesonotum barely elevated above the propodeum . . . . . . . . . . . ***Anisopheidole***

**33(31)** Petiole subtended by a large, plate-like, ventrally directed, apically rounded process; propodeum totally lacking spines; small, slender, usually dark-colored, heavily sculptured species. (New Caledonia and far northern Queensland rain forests) . . . . . . . . . . . . . . . . . . . . . . . . . . . . . . . . . . . . . . . ***Vollenhovia***

Subpetiolar process absent or rudimentary; propodeum sometimes with posterodorsal spines; mostly light-colored, less heavily sculptured species. (Relevant species throughout the region) . . . . . . . . . . . . . . . . . . . . . . . . . . . . . . . . . . . . **34**

**34(33)** Dorsum of body devoid of standing hairs; median portion of clypeus convex, not longitudinally bicarinate, its anterior margin entire; postpetiole relatively large, subcircular in dorsal view, and almost twice as wide as the petiole; small species, total length around 2 mm . . . . . . . . . . . . . . . . . . . . . . ***Cardiocondyla***

Dorsum of body usually hirsute, rarely lacking at least a few standing hairs; in all but a few species the median portion of the clypeus is longitudinally concave or flat, and defined laterally by a pair of longitudinal carinae, which sometimes terminate anteriorly as a pair of forwardly directed denticles; postpetiole usually scarcely wider than petiole in dorsal view; similarly small to much larger species . . . . . . . . . . . . . . . . . . . . . . . . . **35**

**TABLE 2-10** Australian (*continued*)

**35(34)** Antennal club distinctly 4-segmented; median portion of clypeus not longitudinally bicarinate, strongly convex, its anterior border entire; relatively large species, total length about 5 mm . . . . . . . . . . . . . . . . . . . . . . . . . . . . . . . . . . . . ***Aphaenogaster***

Antennal club not clearly segmentally defined, or with 3 segments; median portion of clypeus longitudinally bicarinate, its anterior border medially concave or emarginate, often bidentate; usually much smaller species . . . . . . . . . . . . . . . . . . . . . . . . **36**

**36(35)** Australian, New Zealand, or New Caledonian species; anterior margin of clypeus with a median, forwardly-directed seta; propodeum seldom with a pair of posterodorsal teeth; infradental lamellae of propodeum at most weakly developed and not spinose; mesosomal structure unexceptional; maxillary palps 2-jointed, labials 1- to 3-jointed . . . . . . . ***Monomorium*** (in part)

New Caledonian species; anteromedian clypeal seta lacking; propodeum armed with a pair of posterodorsal teeth or spines; the infradental lamellae also spinose; mesosoma often of bizarre form; palpal formula 3,2 . . . . . . . . . . ***Lordomyrma*** (in part)

### SUBFAMILY DOLICHODERINAE

**1** Propodeum armed with a distinct pair of posterodorsal spines or spine-like processes . . . . . . . . . . . . . . . . . . . . . . . . . . **2**

Propodeum unarmed, its contours generally broadly rounded, lacking paired processes, even if produced posterodorsally . . **5**

**2(1)** Pronotal shoulders armed with a pair of forwardly-directed spines ("*Acanthoclinea*": may be a distinct genus) . . . . . . . . . . . . . . . . . . . . . . . . . . . . . . . . . . . . . . . . ***Hypoclinea*** (in part)

Pronotal shoulders rounded and unarmed . . . . . . . . . . . . . . **3**

**3(2)** Larger species, total length generally exceeding 5 mm; integument of head and mesosoma thick and strongly sculptured, generally densely foveolate and opaque ("*Diceratoclinea*": may be a distinct genus) . . . . . . . . . . . . . . . . . . . . . . ***Hypoclinea*** (in part)

Smaller species, total length generally less than 2–3 mm; integument of head and mesosoma thin, often collapsing in dried specimens, lacking strong sculpturation . . . . . . . . . . . . . . . . . **4**

**4(3)** Propodeum armed with a pair of narrow, elongate, but somewhat lobate spine-like processes. Eyes of normal size, their maximum diameter less than the length of the 2 apical antennal segments together. (Widespread on mainland Australia) . . ***Froggattella***

Propodeum armed with a pair of broad, pointed tumosities. Eyes very large, maximum diameter well exceeding the length of 2 apical antennal segments together. (Generally restricted to northern tropical coastal or insular areas) . . . . . . . . ***Turneria***

**5(1)** Large, exceptionally slender species (total length exceeding 10 mm); the head constricted behind the eyes; legs and antennae exceptionally elongated and slender; scapes exceeding occipital border by at least half their length . . . . . . . . . . ***Leptomyrmex***

Smaller species, usually much smaller, and not agreeing with most or all of the above remaining traits . . . . . . . . . . . . . . . . . . **6**

**6(5)** Integument of head and mesosoma thick, strongly sculptured, usually densely foveolate and opaque; propodeum in profile usually with the declivitous face concave, its apex overarching the petiolar articulation . . . . . . . . . . . . . ***Hypoclinea*** (in part)

Integument of head and mesosoma thin, often collapsing in dried specimens, and lacking strong sculpturation . . . . . . . . . . . **7**

*continued*

7(6) Petiole low, overhung dorsally by first gastral segment, and with the scale rudimentary or vestigial . . . . . . . . . . . . . . . . . . 8

Petiole with a well-developed scale, which is sometimes inclined forward but seldom indistinct and not overhung by the gaster . . . . . . . . . . . . . . . . . . . . . . . . . . . . . . . . . . . . . . 9

8(7) Fifth gastral tergite protruding from beneath the fourth, so 5 segments are visible in dorsal view; cloacal orifice terminal . . . . . . . . . . . . . . . . . . . . . . . . . . . . . . . . . . . ***Technomyrmex***

Only 4 gastral segments visible in dorsal view; cloacal orifice ventral . . . . . . . . . . . . . . . . . . . . . . . . . . . . . . . . ***Tapinoma***

9(7) Maxillary palp 6-segmented . . . . . . . . . . . . . . . ***Iridomyrmex***

Maxillary palp 2-segmented . . . . . . . . . . . . . . ***Bothriomyrmex***

## SUBFAMILY FORMICINAE

1 Antenna 9-segmented . . . . . . . . . . . . . . . . . . ***Brachymyrmex***

Antenna 10- to 12-segmented . . . . . . . . . . . . . . . . . . . . . . 2

2(2) Antenna 10-segmented . . . . . . . . . . . . . . . ***Acropyga*** (in part)

Antenna 11- to 12-segmented . . . . . . . . . . . . . . . . . . . . . . 3

3(2) Antenna 11-segmented . . . . . . . . . . . . . . . . . . . . . . . . . . 4

Antenna 12-segmented . . . . . . . . . . . . . . . . . . . . . . . . . . 7

4(3) Slender ants with exceptionally long legs and antennae; scapes exceeding occipital border by more than two-thirds their length; midline length of pronotum in dorsal view exceeding its maximum width . . . . . . . . . . . . . . . . . . . . . . . . . ***Anoplolepis***

Ants of more conservative proportions; scapes often failing to meet occipital border, never surpassing it by more than half their length; pronotum shorter than wide . . . . . . . . . . . . . . . . 5

5(4) Propodeum complex in shape, with one or more pairs of spines, bosses, or tumosities; at least with a pair of posterolateral spines near the spiracles. Petiole often bispinose . . . . . . ***Stigmacros***

Propodeum simple in shape, its basic contours broadly rounded without trace of spines or other armament . . . . . . . . . . . . . 6

6(5) With *both* of the following features: eyes well developed, their outlines and faceting clearly defined, minimum diameter clearly exceeding maximum diameter of scape; first funicular segment longer than the 2 or 3 following segments (because the latter are relatively short) . . . . . . . . . . . . . . . . . . . . . . . ***Plagiolepis***

Eyes usually poorly defined, usually very small, often minute, vestigial, or lacking. Eyes clearly defined with diameter approximating that of scape in one tropical species, but here the second funicular segment is as long as the first . . . . ***Acropyga*** (in part)

7(3) Ocelli present (the facets sometimes small to minute, but distinct at appropriate magnifications) . . . . . . . . . . . . . . . . . . . . 8

Ocelli lacking . . . . . . . . . . . . . . . . . . . . . . . . . . . . . . . 14

8(7) Propodeum with a pair of slender, acute posterodorsal spines . . 9

Propodeum lacking slender acute spines, though sometimes with inflated tumosities . . . . . . . . . . . . . . . . . . . . . . . . . . 10

9(8) Propodeum with a pair of ventrolateral spines additional to the posterodorsals. Petiolar dorsum similarly bispinose. Pronotal humeri unexceptional . . . . . . . . . . . . . . . . ***Pseudonotoncus***

Ventrolateral propodeal and petiolar spines lacking. Pronotal humeri extended laterally as exceptional wing-like projections . . . . . . . . . . . . . . . . . . . . . . . . . . . . . . . . . ***Teratomyrmex***

10(8) Propodeal spiracles slit or comma-shaped. Clypeus, exposed surfaces of labiomaxillary complex, and underside of head with psammophore hairs. (Fast-running diurnal foragers, even at high temperatures, mainly in arid and semi-arid habitats) . . . . . . . . . . . . . . . . . . . . . . . . . . . . . . . . . . . . . . ***Melophorus***

Propodeal spiracles circular or oval; psammophore lacking. (Cryptic and usually nocturnal foragers in humid and subhumid areas) . . . . . . . . . . . . . . . . . . . . . . . . . . . . . . . . . . . 11

11(10) Large dimorphic *Camponotus*-like species, head width exceeding 2 mm, often greatly. Flagellar antennomeres each much longer than wide. (Rain forests of northeast New South Wales and eastern Queensland) . . . . . . . . . . . . . . . . . . . ***Notostigma***

Smaller species, seldom *Camponotus*-like, head width usually considerably less than 1.5 mm. Flagellar antennomeres seldom more than twice as long as wide . . . . . . . . . . . . . . . . . . 12

12(11) Mesosomal structure elaborate, pronotal humeri and corners of propodeum often inflated or extended. Mesonotum inflated, sometimes elevated behind as a chisel-like or bifurcated process . . . . . . . . . . . . . . . . . . . . . . . . . . . . . . . . . . . ***Notoncus***

Mesosoma not so elaborated . . . . . . . . . . . . . . . . . . . . . . 13

13(12) Polymorphic; antennal scapes seldom attaining occipital border in majors, exceeding it by less than twice their width in minors . . . . . . . . . . . . . . . . . . . . . . . . . . . . . ***Myrmecorhynchus***

Monomorphic; scapes (usually much) longer, exceeding occipital border by more than three times their width, usually by at least one-third their length . . . . . . . . . . . . . . . . . . . . ***Prolasius***

14(7) External bulla of metapleural gland lacking (since the gland itself is absent) . . . . . . . . . . . . . . . . . . . . . . . . . . . . . . . . 15

External bulla of metapleural gland present on each side, at the posteroventral corners of the mesosoma, above the hind coxae . . . . . . . . . . . . . . . . . . . . . . . . . . . . . . . . . . . . . . . 17

15(14) Petiole an elongate low node, allowing the gaster to be reflexed over the mesosoma. Mesothorax elongate and cylindrical, the mesosoma strongly constricted in dorsal view . . . ***Oecophylla***

Petiole block- or scale-shaped. Mesothorax of more conventional configuration, mesosoma not constricted in dorsal view . . . 16

16(15) Petiolar node always armed with 4, 3, or 2 spines or denticles, which may be very long or very reduced. Mesosoma usually with one or more pairs of spines or denticles (variously on the pronotum, mesonotum, propodeum, or combinations of these parts); if lacking spines, the laterodorsal mesosomal borders are marginate, or the whole tagma subspherical in shape . . . . . . . . . . . . . . . . . . . . . . . . . . . . . . . . . . . . . . . ***Polyrhachis***

Petiole a simple scale or node, never armed. Mesosoma lacking all trace of spines or other armament, or laterodorsal margination . . . . . . . . . . . . . . . . . . . . . . . . . . . . . . . . ***Camponotus***

17(14) Petiolar node strongly transverse, with several small denticles at each lateral extremity. First gastral tergite covering most of the tagma. Promesonotal suture lacking, metanotal groove present or absent . . . . . . . . . . . . . . . . . . . . . . . . . . . ***Echinopla***

Petiole a simple, unarmed scale or node; other features not as above . . . . . . . . . . . . . . . . . . . . . . . . . . . . . . . . . . . 18

18(17) Eyes very large, in frontal view occupying the extreme posterior corners of the head. Propodeum somewhat longitudinally constricted above; metanotal suture often vestigial. Usually brightly multicolored in shades of orange, brown, and black . . . . . . . . . . . . . . . . . . . . . . . . . . . . . . . . . . . . . . . . ***Opisthopsis***

Eyes more conventionally proportioned and positioned; coloration and propodeal structure other than above . . . . . . . . . . . . 19

19(18) Large (8–10 mm long) species with profuse silver or white pilosity;

*continued*

usually with coarsely granular sculpturation, dark in color and often iridescent deep green to purple. Metanotal groove a deep, transverse incision . . . . . . . . . . . . . . . . . . . . *Calomyrmex*

Much smaller species, usually lightly colored and noniridescent. Metanotal groove a broad, shallow, transverse depression . . 20

20(19) Most dorsal body surfaces and the legs with numerous black or dark-colored stout hairs, which are narrowed only at the extreme apex . . . . . . . . . . . . . . . . . . . . . . . . *Paratrechina*

Body hairs abundant, but slender, light in color, and tapered. (Rain forests of extreme northern Cape York Peninsula) . . . . . . . . . . . . . . . . . . . . . . . . . . . . . . . . . . . . *Pseudolasius*

**TABLE 2–11** A key to the ant genera of the Polynesian islands from Wallis-Futuna and Samoa to Hawaii, referred to generally as the **Polynesian Region.** The fauna is a mixture of native species and "tramp" species introduced by human commerce from both Old and New World tropics. Based on the worker caste. (Modified from Wilson and Taylor, 1967.)

### SUBFAMILY PONERINAE

1 Mandibles linear and very long, inserted in the middle of the anterior margin of the head, with an apical armament of 3 teeth . . . . . . . . . . . . . . . . . . . . . . . . . . . . . . . . . . . . . . 2

Head differently shaped, with mandibles inserted at its anterior corners . . . . . . . . . . . . . . . . . . . . . . . . . . . . . . . . . . 3

2(1) Large (head width about 2 mm), dark reddish-brown with petiolar apex drawn into an acute conical spine; nuchal carina (ridge across the neck) V- or wedge-shaped . . . . . . . *Odontomachus*

Smaller (head width about 1 mm), golden brown with petiolar summit a narrow transverse ridge; nuchal carina evenly, continuously curved across midline . . . . . . . . . . . . . . . *Anochetus*

3(1) Petiole depressed, articulated over its whole posterior surface with postpetiole . . . . . . . . . . . . . . . . . . . . . . . . . . . . . . . . 4

Articulation between petiole and postpetiole narrow, petiole usually with a distinct transverse posterior face . . . . . . . . . . . . 5

4(3) Mandibles short, closing tightly against clypeus, their apical borders distinct and occupied by 3 large teeth, of which the middle is shortest; basal border of mandible edentate . . . . *Prionopelta*

Mandibles linear, strongly projecting beyond clypeus when closed, their inner borders armed with a number of bipartite teeth . . . . . . . . . . . . . . . . . . . . . . . . . . . . . *Amblyopone*

5(3) Mandibles falcate, very slender and strongly curved, lacking distinct teeth; when closed, there is an extensive gap (with an area much greater than that of mandibles themselves) between their inner borders and anterior clypeal border; tarsal claws pectinate . . . . . . . . . . . . . . . . . . . . . . . . . . . . . . . . . *Leptogenys*

Mandibles differently shaped, usually triangular, and with distinct teeth; when closed, there is little if any gap between them and clypeus; tarsal claws simple, or with a single median tooth . . 6

6(5) Head, mesosoma, and node very roughly punctate-rugose, dorsal surfaces of postpetiolar and first gastral tergites densely and finely arched-striate; entire body with strong greenish or purplish metallic reflections; lower margins of pronotum each armed with a strong acute tooth . . . . . . . . . . *Rhytidoponera*

Sculpturation unlike that described above, dorsal aspects of postpetiole and first gastral segments never striate; color ranging



from black to pale yellowish-brown, without metallic reflections; lower pronotal margins rounded . . . . . . . . . . . . . . . . . . . 7

7(6) Petiolar node distinctly longer than broad in dorsal view; body almost entirely lacking erect hairs—none break its dorsal outline except at gastral apex; tarsal claws each with a distinct median tooth . . . . . . . . . . . . . . . . . . . . . . . . . . . . . *Platythyrea*

Petiolar node usually distinctly broader than long in dorsal view, occasionally almost as long as broad; body with abundant erect or suberect hairs breaking its dorsal outline; tarsal claws simple, lacking a median tooth . . . . . . . . . . . . . . . . . . . . . . . . 8

8(7) Posterior faces of propodeum and node heavily striate, striae usually transverse, although sometimes partly longitudinal on upper parts of node; mesepisternum divided by a transverse suture into anepisternal and katepisternal plates ("*Ectomomyrmex*") . . . . . . . . . . . . . . . . . . . . . . . . . . . . . . . . . *Pachycondyla*

Posterior faces of propodeum and node smooth and shiny, at most with a few transverse striae on their lower parts; mesepisternum entire, not divided by a horizontal suture . . . . . . . . . . . . . 9

9(8) Mandible elongate triangular, the angle between its posterior and masticatory borders obtuse, approximating 120 degrees; masticatory border with 5 or 6 distinct strong teeth . . . . . . . . . 10

Mandible broadly triangular, the angle between its posterior and masticatory borders approximating 90 degrees; masticatory border with more numerous small teeth or minute denticles . . 11

10(9) Small (head width less than 0.5 mm), pale brown species, entirely lacking compound eyes; antenna bearing distinctly 4-segmented club . . . . . . . . . . . . . . . . . . . . . . . . . . . . . . *Cryptopone*

Larger (head width greater than 1 mm), dark brown species, with small but distinct compound eyes and lacking a distinctly segmented antennal club ("*Trachymesopus*") . . . . . *Pachycondyla*

11(9) Subpetiolar process well developed as a lobe, with a translucent circular depression on either side . . . . . . . . . . . . . . *Ponera*

Subpetiolar process weakly developed or absent; if present, not impressed laterally . . . . . . . . . . . . . . . . . . . . *Hypoponera*

### SUBFAMILY MYRMICINAE

1 Antenna with 7 segments or less; head cordate (heart-shaped) in front view and often bearing 2 or more conspicuous scale-like hairs; postpetiole commonly (but not always) with bunches of whitish, fungus-like material . . . . . . . . . . . . . . . . . . . . . 2

Antenna with at least 8 segments; except in large-headed soldier caste of *Pheidole* (in which the head is somewhat cordate), the head in front view is always subrectangular to elliptical; scale-like hairs not present on head and spongiform appendage never developed on postpetiole . . . . . . . . . . . . . . . . . . . . . . . 6

2(1) Mandibles short, thick, and serially dentate; when fully closed, they engage along their entire masticatory margins and leave no appreciable interspace . . . . . . . . . . . . . . . . . . . . . . . . 3

Mandibles linear, elongate, with only 3 or 4 spiniform teeth at or near their apices; when closed, only these teeth engage; and a broad, open space is framed between mandibular shafts . . . . 5

3(2) Eyes large, dorsolaterally placed, and conspicuous when head is viewed directly from front; large, head width greater than 1.2 mm; dark brown . . . . . . . . . . . . . . . . . . . . *Eurhopalothrix*

Eyes very small, laterally placed, and not visible when head is viewed directly from front; head width less than 0.8 mm; light to medium reddish-brown . . . . . . . . . . . . . . . . . . . . . . . 4

4(3) Mandibles short-triangular, with distinct, transverse basal borders;

*continued*

hairs of dorsum of head capsule limited to a single short, erect, clavate pair on vertex . . . . . . . . . . . . . . . . . . . *Trichoscapa*

Mandibles long-triangular, without transverse basal borders; vertex and occiput with more than 1 pair of specialized erect spatulate or clavate hairs (8 in undamaged specimens), along with a ground pilosity of subreclinate spatulate hairs . . *Smithistruma*

**5(2)** Antennal funiculus with only 3 distinct segments (small species with strongly bowed mandibles; head covered with large orbicular or squamiform hairs) . . . . . . . . . . . . . . *Quadristruma*

Antennal funiculus with 5 distinct segments, of which second and third are small . . . . . . . . . . . . . . . . . . . . . . . *Strumigenys*

**6(1)** Antenna 9-segmented, including a robust 2-segmented club longer than entire remainder of funiculus; propodeum bluntly dentate . . . . . . . . . . . . . . . . . . . . . . . . . . . . . . . . *Oligomyrmex*

Antenna 10-, 11-, 12-segmented; if terminal club is very large and 2-jointed, then propodeum is unarmed . . . . . . . . . . . . . . . 7

**7(6)** Antenna 10-segmented, with a distinct 2-segmented club . . . . . . . . . . . . . . . . . . . . . . . . . . . . . . . . . . . . . . . . *Solenopsis*

Antenna 11- or 12-segmented, with a distinct 1- or 3-segmented club, or else terminal joints not forming a distinct club . . . . . 8

**8(7)** Anterior clypeal border bearing 4 conspicuous teeth; a large, recurved, accessory tooth present near base of mandible and well behind masticatory border; antennal club 1-segmented; small, robust, heavily sculptured, dark brown. (Samoa) . . . . . . . . . . . . . . . . . . . . . . . . . . . . . . . . . . . . . . . *Adelomyrmex*

Anterior clypeal border with at most 2 teeth; accessory, basal tooth lacking on mandible; antennal club either 3-segmented or not distinct from remainder of funiculus . . . . . . . . . . . . . . . . 9

**9(8)** Petiole subtended by a smooth, very flat, ventrally rounded, translucent flange about as broad as the depth of anterior peduncle of petiole; slender, heavily sculptured species . . . *Vollenhovia*

Petiole at most subtended by a small, thin knob placed at anteriormost part of ventral surface of anterior peduncle . . . . . . . . 10

**10(9)** Frontal lobes fused with median third or fourth of clypeus, which forms a distinct shelf raised sharply from lateral portions of clypeus, or else the portion of the clypeus in front of the antennal lobes reduced to the ridge alone (in which case the pronotum bears a pair of spines) . . . . . . . . . . . . . . . . . . . . . . . . . 11

Frontal lobes clearly demarcated from clypeus in front; center of clypeus well developed and not conspicuously raised as a separate element from remainder of sclerite . . . . . . . . . . . . . . 12

**11(10)** Pronotum armed with a conspicuous pair of spines. (Tonga only; R. W. Taylor, personal communication) . . . . . . *Pristomyrmex*

Pronotum unarmed . . . . . . . . . . . . . . . . . . . . *Tetramorium*

**12(10)** Propodeum either smoothly rounded or, at most, armed with a pair of blunt processes forming angles of not less than 90 degrees; clypeus bicarinate . . . . . . . . . . . . . . . . *Monomorium*

Propodeum armed with a pair of acute teeth or spines; clypeus not bicarinate . . . . . . . . . . . . . . . . . . . . . . . . . . . . . . . . 13

**13(12)** Seen from directly above, postpetiole nearly twice as broad as petiole; dorsum of body devoid of standing hairs; small (head width about 0.4 mm), slender, monomorphic species . *Cardiocondyla*

Seen from directly above, postpetiole at most 1.3 times broader than petiole; dorsum of body bearing numerous standing hairs . . . . . . . . . . . . . . . . . . . . . . . . . . . . . . . . . . . . . . . . 14

**14(13)** Monomorphic; pronotum coarsely rugoreticulate and never armed with spines; scape in repose not reaching occipital border . . . . . . . . . . . . . . . . . . . . . . . . . . . . . . . . . . . . . . . *Rogeria*



Dimorphic; pronotum of small-headed minor worker either smooth and shining or finely "shagreened," or armed with a pair of spines; scape of minor worker in repose exceeding occipital border . . . . . . . . . . . . . . . . . . . . . . . . . . . . . . *Pheidole*

### SUBFAMILIES DOLICHODERINAE AND FORMICINAE

**1** Petiole armed with 2 large, laterally directed horn-like spines; monomorphic, medium-sized, black. (Island of Rotuma only) . . . . . . . . . . . . . . . . . . . . . . . . . . . . . . . . . . . *Polyrhachis*

Petiole either unarmed or else bearing a single median tooth-like protuberance . . . . . . . . . . . . . . . . . . . . . . . . . . . . . . 2

**2(1)** Antenna 9-segmented; small, robust, brown . . . . *Brachymyrmex*

Antenna 11- or 12-segmented . . . . . . . . . . . . . . . . . . . . . 3

**3(2)** Juncture of dorsal and basal faces of propodeum of workers (or minor workers if workers are polymorphic) drawn into an acute tooth-like protuberance; medium-sized species . . . . . . . . . . . . . . . . . . . . . . . . . . . . . . . . . . . **Camponotus** (in part)

Juncture of dorsal and basal faces of propodeum of minor workers rounded or at most obtusely angulate . . . . . . . . . . . . . . . 4

**4(3)** Polymorphic; medium-sized to large, with head width (exclusive of compound eyes) of smallest worker greater than 0.80 mm . . . . . . . . . . . . . . . . . . . . . . . . . . . . **Camponotus** (in part)

Monomorphic; small to medium-sized, with head width of largest worker not greater than 0.75 mm . . . . . . . . . . . . . . . . . 5

**5(4)** Antenna 11-segmented; minute, robust, yellow formicine ants with almost completely hairless mesosoma and large abdomen (gaster) distinctly longer than mesosoma in undistended state . . . . . . . . . . . . . . . . . . . . . . . . . . . . . . . . . . *Plagiolepis*

Antenna 12-segmented; not combining all of other characters cited above . . . . . . . . . . . . . . . . . . . . . . . . . . . . . . . . . . . . 6

**6(5)** Minute, head width approximately 0.40 mm, with mesosoma (alitrunk) completely devoid of standing hairs . . . . . . *Tapinoma*

Either much larger (head width greater than 0.60 mm), or else mesosoma bears numerous standing hairs . . . . . . . . . . . . . . . 7

**7(6)** Body extremely thin and elongate; antennal scape at least 1.5 times as long as head including closed mandibles . . . . . . . . . . . 8

Body of "average" to somewhat robust proportions; antennal scape not more than 1.5 times as long as head including closed mandibles . . . . . . . . . . . . . . . . . . . . . . . . . . . . . . . . . 9

**8(7)** Dorsum of mesosoma almost completely devoid of standing pilosity; color yellow; mesonotum viewed from side weakly concave . . . . . . . . . . . . . . . . . . . . . . . . . . . . . . . . . . *Anoplolepis*

Dorsum of mesosoma bearing numerous long, erect hairs; color grayish-brown with occasional weak purplish reflections; mesonotum viewed from side weakly convex . . . . . . . . . . . . . . . . . . . . . . . . . . . . . . . . . . . . . . . . **Paratrechina** (in part)

**9(7)** Mesosoma devoid of standing pilosity. (*Iridomyrmex humilis*: Hawaii only) . . . . . . . . . . . . . . . . . . . **Iridomyrmex** (in part)

Mesosoma bearing at least several prominent standing hairs . . 10

**10(9)** Anterior clypeal border whole and smoothly convex . . . . . . . . . . . . . . . . . . . . . . . . . . . . . . . . . . . **Paratrechina** (in part)

Anterior clypeal border emarginate (notched in middle) . . . . 11

**11(10)** Petiolar node well developed; mesosoma feebly shining . . . . . . . . . . . . . . . . . . . . . . . . . . . . . . . . . **Iridomyrmex** (in part)

Petiolar node rudimentary; mesosoma densely shagreened and opaque . . . . . . . . . . . . . . . . . . . . . . . . . . *Technomyrmex*

**TABLE 2–12** A key to the ant genera of North America and northern and central Mexico, known as the **Nearctic Region.** Based on the worker caste. (Subfamily Ponerinae by S. P. Cover and authors; subfamilies Myrmicinae and Formicinae by S. P. Cover, published with permission; subfamily Dolichoderinae by S. P. Cover and S. O. Shattuck, published with permission; additional data provided by B. Bolton, A. Francoeur, and W. P. MacKay; Ecitoninae based on multiple sources.)

**SUBFAMILY PONERINAE**

1 Mandibles slender, elongate, abruptly bent inward near apex, and articulated near middle of anterior margin of head; postpetiole not differentiated from remainder of gaster . . . . . . . . . . . . **2**

Mandibles variously shaped, but not abruptly bent inward near apex, and articulated near anterolateral corners of head; postpetiole often slightly to moderately constricted posteriorly, and thus more or less differentiated from remainder of gaster . . . **3**

**2(1)** Nuchal carina (ridge delimiting the occiput, or extreme rear of head) continuously curved across midline; apophyseal lines absent. (Currently known from the Florida Keys and Mexico) . . . . . . . . . . . . . . . . . . . . . . . . . . . . . . . . . . ***Anochetus***

Nuchal carina V- or wedge-shaped, narrowed toward median into a middorsal groove; apophyseal lines present as a pair that converge from head insertion up to nuchal carina. (Southern United States and Mexico) . . . . . . . . . . . . . . ***Odontomachus***

**3(1)** Median lobe of clypeus broad, its anterior margin lined with multiple denticles. [*Note*: In small species less than 3 mm in total length, the denticles may not be clearly visible unless high magnification (> 50X) is used. In addition, it is helpful if the mandibles of the specimen are open.] Petiole narrowed only slightly behind, so that its posterior face attaches broadly to the gaster in dorsal view . . . . . . . . . . . . . . . . . . . . . . . . . . . . . . **4**

Median lobe of clypeus variously shaped, but anterior margin never multidenticulate; petiole usually strongly constricted behind in dorsal view . . . . . . . . . . . . . . . . . . . . . . . . . . **5**

**4(3)** Mandibles short and narrow, with 3 apical teeth that close tightly against the clypeus. (Central Florida and Mexico) . . . . . . . . . . . . . . . . . . . . . . . . . . . . . . . . . . . . . . . . ***Prionopelta***

Mandibles longer, strongly projecting beyond clypeus when closed, and having many teeth along the inner borders. (Widespread in the United States, probably also in Mexico) . . . . . . . . . . . . . . . . . . . . . . . . . . . . . ***Amblyopone***

**5(3)** Pygidium bearing rows of erect spines or teeth along both sides that converge toward each other at rear; antennal sockets nearly touching; body elongate and cylindrical . . . . . . . . . . . . . . **6**

Pygidium lacking rows of erect spines or teeth; separation between antennal sockets variable, often greater than maximum diameter of a single antennal socket; body form variable, rarely elongate and cylindrical . . . . . . . . . . . . . . . . . . . . . . . . . . . . **7**

**6(5)** Each gena margined ventrally by a carina or ridge that runs forward from posteroventral corner of head for a distance of one-quarter to one-half the head length in side view; antennal sockets bordered laterally by a second carina or ridge that extends to lateral lobe of clypeus or projects past anterior clypeal border (sometimes slightly over dorsal surface of closed mandibles) as a blunt tooth. (South central United States and Mexico) . . . . . . . . . . . . . . . . . . . . . . . . . . . . . . . . . . . . . ***Cerapachys***

Lacking both carinae described above . . . . . . . . . . . . . . . . . . . . . . . . . . . . . . . . . . . . . . . . . ***Ctenopyga/Acanthostichus***

[*Note:* Currently these two genera are separated by differences in the queens (born winged in *Ctenopyga* and wingless in *Acanthostichus*). Workers of *Ctenopyga* are very poorly known and, at present, cannot be distinguished from those of *Acanthostichus*. *Ctenopyga*-type queens are known from southwestern United States and Mexico. *Acanthostichus* records are from Central and South America, but the genus might occur much farther north; see Brown (1975) for a further discussion.]



**7(5)** Tarsal claws finely pectinate; antennal sockets close together, separated by less than the maximum diameter of a single socket; anterior margin of clypeus sharply triangular in form, its center projecting far forward; mandibles elongate and slender, usually without conspicuous subapical teeth. (Florida, southern Texas, and Mexico) . . . . . . . . . . . . . . . . . . . . . . . . . ***Leptogenys***

Tarsal claws simple, or with 1 or 2 subapical teeth, but not finely pectinate; separation of antennal sockets variable; anterior margin of clypeus broad, and rounded or subtriangular in form . **8**

**8(7)** Lobes of frontal carinae more or less sharply raised above plane of front of head, or vertical and fused, so that condylar bulbs of antennae are mostly or completely exposed in frontal view; eyes greatly reduced (less than 15 facets) or apparently absent. [*Note*: Dorsal surface of first gastral segment often strongly convex so that apex of gaster points anteriorly.] . . . . . . . . . . . . . . . . **9**

Lobes of frontal carinae horizontal or raised, but wholly or mostly covering the condylar bulbs of antennae in front view; eyes variable in development, but if condylar bulbs seem less than half covered by the frontal lobes, then eyes well developed (more than 30 facets) . . . . . . . . . . . . . . . . . . . . . . . . . . . . **10**

**9(8)** Lobes of frontal carinae vertical and fused; antennal sockets placed low on shelf-like extension of the clypeus that projects forward, partly or completely overhanging mandibles; apical segment of antenna greatly enlarged and approximately equal to or longer than second to penultimate segments combined; mandibles inconspicuous and strap-like. (Southeastern United States and Mexico) . . . . . . . . . . . . . . . . . . . . . . . . . . ***Discothyrea***

Lobes of frontal carinae raised, not vertical and fused; antennal sockets normally placed; apical segment of antenna not greatly enlarged; mandibles easily visible in front view, with several teeth, at most only slightly overhung by the clypeus. (Widespread in forests of eastern and southern United States and Mexico) . . . . . . . . . . . . . . . . . . . . . . . . . . . ***Proceratium***

**10(8)** Erect hairs absent from head and body; tarsal claws on middle and hind legs each bearing a submedian tooth; body dull or feebly shining, bearing sparse, shallow punctures, and entirely covered with very fine, appressed, pruinose pubescence that gives surface a pronounced grayish or silvery cast. (Southern Florida and Mexico) . . . . . . . . . . . . . . . . . . . . . . . . ***Platythyrea***

Erect hairs present on head and/or body, although sometimes short and sparse; tarsal claws on middle and hind legs simple, or if bearing a submedian tooth, body not appearing as described above . . . . . . . . . . . . . . . . . . . . . . . . . . . . **11**

**11(10)** Tarsal claws on front legs (and other legs as well) simple; frontal carinae close together posteriorly, just prior to flaring outward anteriorly to produce distinct lateral lobes . . . . . . . . . . . . **12**

Tarsal claws on front legs (and often middle and hind legs as well) with a subapical tooth that can be nearly basal to median in position. [*Note:* In a few small species the subapical teeth are inconspicuous unless examined at high power (> 50X) with good illumination.] Frontal carinae well separated and subparallel posteriorly, usually subparallel or feebly diverging anteriorly, sometimes diverging more strongly to produce lateral lobes . **16**

*continued*

**12(11)** Middle and hind tibiae with 2 apical spurs, the outer spur often no more than half as long as the inner . . . . . . . . . . . . . . . . 13

Middle and hind tibiae each with a single apical spur . . . . . . 15

**13(12)** Outer face of mandible at base with oval, pit-like depression; outer face of middle tibia with several spine-like bristles; eyes present, although small (less than 15 facets); small (2–5 mm), cryptobiotic ants. (Southeastern United States, perhaps Mexico also) . . . . . . . . . . . . . . . . . . . . . . . . . . . . . . . . . . . . . ***Cryptopone***

Outer face of mandible at base lacking an oval, pit-like depression; spine-like bristles on outer face of middle tibia generally absent . . . . . . . . . . . . . . . . . . . . . . . . . . . . . . . . . . . . . . . . 14

**14(13)** In profile, mesonotum convexly swollen and surrounded by deeply impressed sutures; neither genae (cheeks) nor pronotum laterally marginate. (*Brachyponera chinensis*, = *solitaria*, from Japan and China has been introduced into the southeastern United States) . . . . . . . . . . . . . . . . . . . . . . . ***Brachyponera***

In profile, mesonotum feebly convex or flat, not set off by deeply impressed sutures; genae and/or pronotum may be laterally marginate. (Florida and Texas south) . . . . . . ***Pachycondyla***

**15(12)** Subpetiolar process prominent, impressed anteriorly on each side to form a roughly circular, central, translucent window, and projecting posteriorly as a pair of right angles or teeth. [*Note*: Maxillary palps 2-jointed. Common in the United States and Mexican highlands.] . . . . . . . . . . . . . . . . . . . . . . . . . . . . ***Ponera***

Subpetiolar process a small, relatively simple lobe lacking a central window. [*Note:* Maxillary palps 1-jointed. Abundant in southern United States and Mexico.] . . . . . . . . . . . . . . ***Hypoponera***

**16(11)** Viewed in profile, the mesonotum prominent, conspicuously set off from the propodeum by a deep transverse fissure; the mesonotum and propodeum forming distinct, separate convexities; coxa of hind leg without dorsal tooth or spine. [*Note*: Known from Mexico and may also occur in southern Texas.] . . . . . . . . . . . . . . . . . . . . . . . . . . . . . . . . . . . . . . . . ***Ectatomma***

Viewed in profile, the mesonotum not unusually prominent; it and the propodeum form one continuous or near-continuous profile that is interrupted at most by a suture-like groove at their juncture (rarely deep), or by an ill-defined impression in this region; coxa of hind leg often armed with a dorsal tooth or spine. (Texas and Mexico south) . . . . . . . . . ***Gnamptogenys***

### SUBFAMILY ECITONINAE

**1** Waist 1-segmented. (Rare: southern Mexico) . . . ***Cheliomyrmex***

Waist 2-segmented . . . . . . . . . . . . . . . . . . . . . . . . . . . . . 2

**2(1)** Tarsal claws simple, lacking median tooth . . . . . ***Neivamyrmex***

Tarsal claws each with a single, median tooth in addition to the terminal point . . . . . . . . . . . . . . . . . . . . . . . . . . . . . . . . 3

**3(2)** Flagellum robust; apical width of scape more than one-third its entire length. (Southern Texas, Mexico) . . . . ***Nomamyrmex***

Flagellum slender; apical width of scape less than one-third its entire length . . . . . . . . . . . . . . . . . . . . . . . . . . . . . . . . . 4

**4(3)** Propodeum with teeth or lamellae; highly polymorphic; major subcaste usually with falcate (sickle-shaped) mandibles. (Southern Mexico) . . . . . . . . . . . . . . . . . . . . . . . . ***Eciton***

Propodeum lacking teeth or lamellae; moderately polymorphic; major subcaste with shorter mandibles not shaped like sickles. (Arkansas, Louisiana, Oklahoma, Texas, Mexico) . . . ***Labidus***



### SUBFAMILY PSEUDOMYRMECINAE

(Represented only by the genus *Pseudomyrmex*, found widely through the southernmost United States as well as all of Mexico.)

### SUBFAMILY MYRMICINAE

**1** Antenna with 4, 6, or 7 segments . . . . . . . . . . . . . . . . . . . . 2

Antenna with 10 or 11 segments . . . . . . . . . . . . . . . . . . . . 7

Antenna with 12 segments . . . . . . . . . . . . . . . . . . . . . . . 25

**2(1)** Antenna with 4 segments. [*Note:* Mandible curvilinear, bearing a long, apical fork, its masticatory border otherwise mostly or completely bare. Head cordate, strongly narrowed anteriorly, and bearing scale-like hairs. One species, *Q. emmae*, introduced (probably from Africa) into southern Florida and possibly Mexico.] . . . . . . . . . . . . . . . . . . . . . . . . . . . . ***Quadristruma***

Antenna with 6 or 7 segments . . . . . . . . . . . . . . . . . . . . . . 3

**3(2)** Antenna with 7 segments. Antennal scrobe present and with carinate margins. Eye located on upper scrobal margin at or close to midlength of head capsule. No sponge-like growth on petiole or postpetiole. Mandibles triangular, their masticatory borders serially dentate and fully engaging at closure. Some hairs scale- or strap-like. Antennal scape abruptly bent near insertion and flattened. (One species, *E. floridana*, present in Florida, other species in Mexico) . . . . . . . . . . . . . . . . . . . . ***Eurhopalothrix***

Antenna with 6 segments. Antennal scrobe absent or, if present, the eye is located on or below the lower scrobal margin at or below the midlength of the head capsule. Sponge-like growth often present on petiole or postpetiole. Mandibles sometimes linear with principal teeth at apex, often engaging only at apex (tribe Dacetini in part) . . . . . . . . . . . . . . . . . . . . . . . . 4

**4(3)** Mandibles elongate and linear, frequently with distinctive apical fork of inwardly directed spiniform teeth . . . . . . . . . . . . . 5

Mandibles triangular or subtriangular, their masticatory margin serially dentate and lacking an apical fork of inwardly directed spiniform teeth . . . . . . . . . . . . . . . . . . . . . . . . . . . . . . . . . 6

**5(4)** Mandibles with distinctive apical fork of two (rarely 3) spiniform teeth, lowest tooth equal to or larger in size than the upper (or uppermost) tooth. Labral lobes short and not visible between the mandibles as seen from above. (Southern United States and Mexico) . . . . . . . . . . . . . . . . . . . . . . . . . . ***Strumigenys***

Mandibles with the lower tooth of apical fork much reduced or rudimentary. Labral lobes long, conical and clearly visible between mandibles in dorsal view. (*Epitritus hexamerus*, native to Japan, is known from 1 collection in Marion Co., Florida: M. Deyrup, personal communication) . . . . . . . . . . . . . ***Epitritus***

**6(4)** Pronotum strongly margined along the front and sides. Pilosity of head limited to 2 short, erect, clavate hairs on the vertex. (*T. membranifera*, an Old World species, is established in the Gulf Coast states, California, and probably in Mexico) . ***Trichoscapa***

Pronotum not marginate. Cephalic hairs usually abundant and conspicuous, often clavate, scale-like, long and sinuous, or otherwise unusual in shape. (Diverse and common, although inconspicuous, in temperate and subtropical North America and Mexico; a few species present in mesic habitats in the Southwest and California) . . . . . . . . . . . . . . . . . . . . . ***Smithistruma***

**7(1)** Antenna with 10 segments, including a distinct 2-jointed club.

*continued*

Clypeus usually bicarinate, the carinae diverging toward the anterior clypeal margin and frequently terminating as teeth projecting beyond the margin. Workers polymorphic or monomorphic. Propodeum always lacking spines or teeth. (Common south of boreal Canada, abundant throughout southern United States and Mexico) . . . . . . . . . . . . . . . . . ***Solenopsis***

Antenna with 11 segments . . . . . . . . . . . . . . . . . . . . . . . . **8**

**8(7)** Postpetiole attached to anterior dorsal surface of first gastric segment. Gastric dorsum in profile concave, flat, or at most feebly convex, in contrast with strongly convex ventral surface. In dorsal view gaster more or less heart-shaped and is capable of being flexed so that it points forward over the alitrunk. Petiole dorsoventrally flattened; node inconspicuous or absent. (Occurs south of boreal Canada, abundant and diverse in southern United States and Mexico) . . . . . . . . . . . . . . ***Crematogaster***

Postpetiole attached to anterior end (not to the anterior dorsal surface) of the first gastric segment. Dorsal and ventral surfaces approximately equally convex when viewed in profile, or ventral surface less convex than dorsal. Gaster more or less oval or teardrop-shaped in dorsal view and not capable of being flexed forward over the alitrunk. Petiole usually with obvious node . . . **9**

**9(8)** Frontal carinae greatly expanded forward and laterally, forming a more or less flat surface that completely covers the cheeks in dorsal view, and forming a deep, distinctive antennal scrobe anterior to the eye, which is located at or near the posterior corner of the head. Workers more or less dimorphic; majors often with saucer-shaped heads. Body broad and notably flattened; petiole and postpetiole each with pair of lateral spines or teeth; body hairs often scale- or strap-shaped; sculpture of head and alitrunk sometimes foveate. (Arboreal species in southern Florida, Texas, and Arizona, and in Mexico) . . . . . . . . ***Zacryptocerus***

Frontal carinae generally not expanded forward and laterally so as to completely cover the cheeks. In cases where they are so expanded, the eye is located more or less at head mid-length, and the antennal scrobes (when present) continue past it to the occipital corners . . . . . . . . . . . . . . . . . . . . . . . . . . . . . . **10**

**10(9)** A single, more or less longitudinal carina present above and sometimes slightly anterior to the eye; if an antennal scrobe is present, the carina may form part or all of its lateral border; longitudinal rugae absent on head and/or pronotal dorsum; antenna lacking distinct 2- or 3-jointed apical club (tribe Attini) . . . . **11**

Carina as described above usually absent. In cases where a carina is present, at least some longitudinal rugae are present on the sides and rear of head and/or the pronotal dorsum, and the antenna has a distinct 2- or 3-jointed apical club . . . . . . . . . . **15**

**11(10)** Pilosity on upper body surface absent or short and strongly appressed; hairs flattened, scale- or strap-like. Simple, erect, or recurved hairs extremely scarce (usually absent) on dorsum of head and alitrunk. Frontal carinae moderately to strongly expanded laterally (and often forward) forming distinct lobes that, in some cases, cover most or all of the cheeks in dorsal view. Gaster never tuberculate. (Two species found in Gulf coast states, one in Texas and southwestern states; more in Mexico) . . . . . . . . . . . . . . . . . . . . . . . . . . . . . . ***Cyphomyrmex***

Pilosity on upper body surfaces variable: several to many simple, erect, or recurved hairs present on dorsum of head and alitrunk. Frontal carinae expanded laterally (and sometimes forward) to form lobes, but these do not ever cover most or all of the cheeks in dorsal view. Gaster often tuberculate . . . . . . . . . . . . . **12**

**TABLE 2–12** Nearctic (*continued*)

**12(11)** Workers monomorphic or at most very weakly polymorphic . . **13**

Workers moderately to extremely polymorphic . . . . . . . . . . **14**

**13(12)** Frontal carinae projecting forward (as well as laterally) beyond the antennal insertions, forming conspicuous lobes that overhang the sides of the clypeus. (At present *M. hartmanni*, found in Texas and Louisiana, is the only species known from the region) . . . . . . . . . . . . . . . . . . . . . . . . . . . . . . . . ***Mycetosoritis***

Frontal carinae expanded laterally to form more or less conspicuous lobes, but not projecting forward beyond antennal insertions. Upper body surfaces conspicuously tuberculate; alitrunk frequently with teeth or spines. (Long Island, New York, and Illinois south, and Mexico. See note under *Acromyrmex* below) . . . . . . . . . . . . . . . . . . . . . . . . . . . . . . . . ***Trachymyrmex***

**14(12)** Workers moderately polymorphic; dorsum of alitrunk with 4 pairs of teeth, spines, or large tubercles. Occiput and/or first gastric tergite tuberculate. (One species, *A. versicolor*, is present in western Texas, and the desert southwest into Mexico; other species also present in Mexico) . . . . . . . . . . . . . . ***Acromyrmex***

[*Note:* There are currently no reliable characters to separate *Acromyrmex* and *Trachymyrmex* workers other than the presence or absence of worker polymorphism.]

Workers extremely polymorphic: minims with head widths less than 1 mm and majors with head widths often greater than 4.5 mm. Dorsum of alitrunk armed with 3 pairs of spines or teeth. Occiput and first gastric tergite smooth, not tuberculate. (*A. texana* present in Texas and Louisiana south, *A. mexicana* in deserts of southern Arizona and Mexico) . . . . . . . . . . . . . ***Atta***

**15(10)** Propodeum in profile evenly rounded or angulate; no teeth or spines present at juncture of basal and declivitous faces. Petiole in side view subcylindrical; node absent or rudimentary. (Arboreal; Florida and Mexico) . . . . . . . . . . . . . . . . . ***Xenomyrmex***

Propodeum armed with teeth (sometimes small) or spines, rarely evenly rounded. Petiole with node . . . . . . . . . . . . . . . **16**

**16(15)** Frontal carinae extending posteriorly past the eye and almost reaching the vertex . . . . . . . . . . . . . . . . . . . . . . . . . **17**

Frontal carinae short, not reaching close to the vertex . . . . . . **20**

**17(16)** Humeri distinctly angulate. Mandibles each with 5 teeth including the basal tooth or angle, which is usually hidden from view when the mandibles are closed. [*Note:* Antennal scrobe more or less distinct, a carina sometimes present above the eye and forming part or all of the scrobe's lateral border. Propodeal spines long and divergent in dorsal view. Pilosity on head and alitrunk long and sparse.] (Small yellow or orange ants, only 1.5–2.0 mm long, found in Florida and Mexico, also introduced into southern California) . . . . . . . . . . . . . . . . . . . . ***Wasmannia***

Humeri rounded, not angulate. Dentition variable, but number of mandibular teeth other than 5 . . . . . . . . . . . . . . . . . . . **18**

**18(17)** In dorsal view lateral portions of posterior clypeal border with a distinctive, roughly semicircular emargination adjacent to each antennal fossa. The emargination is actually a ridge or lamella that drops off sharply on its posterior side, creating the impression that the antenna is inserted at the bottom or side of a deep pit. Mandibles with at least 6 (often 7) teeth. Dorsum of head and alitrunk completely covered with coarse rugoreticulate sculpture. (Four species known from Texas, the desert southwest, and Mexico, formerly *Xiphomyrmex*) . . . . . . . . . . . . . . . . . . . . . . . . . . . . . . . . . . . . . . . . ***Tetramorium*** (in part)

Lateral portions of posterior clypeal border lacking semicircular

*continued*

emarginations at each antennal fossa as described above. Antennae not appearing to be inserted in deep pits. Mandibles with 4 teeth or lacking teeth. At least vertex and occiput of head largely smooth and shining . . . . . . . . . . . . . . . . . . . . . . . . . . **19**

**19(18)** Mandibles with flattened dorsal surfaces and lacking teeth. Clypeus with pronounced median notch. (Dulotic ants enslaving *Leptothorax* species: one species, *H. canadensis*, present in Canada and extreme northern United States) . . . . . . . ***Harpagoxenus***

Mandibles with strongly convex dorsal surfaces and 4 teeth. Entire anterior clypeal border, not just median area, moderately concave. (One species, *P.*—formerly *Harpagoxenus—americanus*, enslaves 3 *Leptothorax* species in eastern North America) . . . . . . . . . . . . . . . . . . . . . . . . . . . . . . . . . . ***Protomognathus***

**20(16)** Workers dimorphic; minors small (total length less than 3 mm) with a 2-jointed apical club and minute eyes (fewer than 5 facets, often rudimentary); majors much larger, with enormous quadrate heads and a much less distinct antennal club. (One rare species, *E. longi*, known from Texas and Oklahoma; other species probably in Mexico) [*Note:* Majors often not collected.] . . . . . . . . . . . . . . . . . . . . . . . . . . . . . . . . . . . ***Erebomyrma***

Antennal club 3-jointed or indistinct. Eyes usually larger, with more than 10 facets, rarely rudimentary. Workers monomorphic or weakly polymorphic . . . . . . . . . . . . . . . . . . . . . . . . **21**

**21(20)** Combining the following traits: Eyes rudimentary or apparently absent. Propodeum evenly rounded in side view. Mandibles with 4 teeth (the basal tooth somewhat offset) and a notably oblique cutting margin. Antenna with 3-segmented apical club. Workers slightly polymorphic and minute compared to the queen. (At present known only from two recent collections in the Chiricahua Mountains of Arizona. Apparently close to the Old World genus *Anillomyrma*, possibly a new genus: S. Cover, personal communication) . . . . . . . . . . . . . . New genus (?)

Eyes well developed, though sometimes small, with 10 or more facets. Propodeum with spines or teeth (sometimes small), rarely angulate or rounded. Mandibles with 5 or more teeth, the basal tooth not offset, and having a more or less transverse cutting margin . . . . . . . . . . . . . . . . . . . . . . . . . . . . . . . **22**

**22(21)** Humeri distinctly angulate. Frequently some of the hairs on the petiole and postpetiole are swollen at the base or borne on tubercles, giving the surface a tuberculate appearance. (Arboreal; southern Texas and Arizona south through Mexico) . . . . . . . . . . . . . . . . . . . . . . . . . . . . . . . . . . . . . . . ***Nesomyrmex***

Humeri rounded or at most feebly angulate. Petiole and postpetiole not tuberculate . . . . . . . . . . . . . . . . . . . . . . . . **23**

**23(22)** Mandibles with 5 teeth. Alitrunk weakly convex in profile. Strigil without a basal tooth. Petiole with a more or less distinct anterior peduncle. ("*Myrafant*," a possibly distinct genus, common south of boreal Canada) . . . . . . . . . . . ***Leptothorax*** (in part)

Mandibles with 6 teeth. Alitrunk flattened in profile. Strigil with a basal tooth. Petiole lacking a distinct anterior peduncle . . . **24**

**24(23)** Eyes with erect hairs. [*Note:* Commensals of *Myrmica, Manica,* and *Formica,* often with ergatoid males and queens.] (Found principally in boreal and mountainous North America) . . . . . . . . . . . . . . . . . . . . . . . . . . . . . . . . . . . . . . . . . ***Formicoxenus***

Eyes lacking erect hairs. ("Typical" *Leptothorax:* free-living species, abundant in boreal and mountainous North America, possibly south into Mexican highlands) . . . . ***Leptothorax*** (in part)

**25(1)** In profile petiole short and subcylindrical, lacking an anterior peduncle and with a rudimentary node (node often absent); humeri moderately to sharply angulate; propodeum armed with spines or short teeth. (Eastern and southern United States, California, and Mexican highlands; at least 3, probably 4 species; one is a social parasite) . . . . . . . . . . . . . . . . . . ***Myrmecina***

Petiole decidedly nodiform; humeri usually rounded, rarely angulate . . . . . . . . . . . . . . . . . . . . . . . . . . . . . . . . . . . . **26**

**26(25)** Combining the following traits: Petiolar node set off sharply from the long, distinctive anterior peduncle. The node in side view roughly triangular, usually with a short, steep anterior face and a longer, gradually sloping posterior face, and reaching its highest point at or toward its anterior surface. [*Note:* Where the node apex is rounded care may be needed to determine its highest point.] Psammophore usually well developed; if absent, erect hairs are often present on the gula and the alitrunk is extensively and coarsely rugoreticulate (see the *Pogonomyrmex* and *Ephebomyrmex* generic figures) . . . . . . . . . . . . . . . . . . . . . . . . **27**

Anterior peduncle sometimes long, often short or absent. Node in side view variable in shape, often roughly rectangular or quadrate. If triangular, it generally reaches its highest point at its longitudinal midpoint or toward its posterior surface. In rare cases where the peduncle is long and the node is triangular and reaches its highest point toward its anterior surface, then the alitrunk is not coarsely and extensively rugoreticulate (although longitudinal or transverse rugae or weak reticulate sculpture may be present). A true psammophore is rarely present, although scattered erect hairs on the gula occur fairly often . . **28**

**27(26)** Mandible with 6 teeth, the ultimate basal tooth reduced in size (rarely with 5 teeth and no reduced ultimate basal tooth). Psammophore usually absent, although long, erect hairs are often present on the gula. Dorsum of alitrunk extensively and coarsely rugoreticulate. (Texas, the desert southwest, and Mexico; in xeric, open habitats) . . . . . . . . . . . . . . . . . . ***Ephebomyrmex***

Mandible usually with 7 teeth, the ultimate basal tooth not reduced in size (rarely with 6 teeth and the ultimate basal tooth not reduced). Psammophore strongly developed. Alitrunk not extensively and coarsely rugoreticulate. (Coast to coast in southern United States, north to Washington state and the Dakotas, also in Mexico; very abundant in prairies, deserts, and other xeric habitats) . . . . . . . . . . . . . . . . . ***Pogonomyrmex***

**28(26)** Head and alitrunk without visible erect hairs on dorsal surfaces. [*Note:* Head and alitrunk dull and densely punctulate. In side view the anterior clypeal margin elevated above the mandibles and extending slightly over them, forming a small "shelf." Postpetiole often enlarged in dorsal view. Metanotal suture slightly to moderately impressed. Small, monomorphic ants often found in early successional and disturbed habitats. Most species dwell in soil but one in the region is arboreal.] (Old World genus introduced in California, the Gulf Coast states, and Mexico) . . . . . . . . . . . . . . . . . . . . . . . . . . . . . . . . . . ***Cardiocondyla***

Head and alitrunk bearing at least a few (often many) erect or suberect hairs . . . . . . . . . . . . . . . . . . . . . . . . . . . . . . **29**

**29(28)** Dorsum of alitrunk flattened or convex, but without impressed sutures. [*Note:* In some cases a small transverse ridge may be present on the propodeal dorsum.] . . . . . . . . . . . . . . . . . . . **30**

Dorsum of alitrunk variously shaped in profile, but never forming a continuous surface; its outline always interrupted by one or more sutural impressions . . . . . . . . . . . . . . . . . . . . . **36**

*continued*

30(29) Clypeus with 2 longitudinal carinae that sometimes fail to reach the anterior margin and never form clypeal teeth . . . . . . . **31**

Clypeus not bicarinate, although sometimes longitudinally rugulose . . . . . . . . . . . . . . . . . . . . . . . . . . . . . . . . . . . **32**

31(30) Petiole lacking an anterior peduncle and bearing a large, translucent plate-like process projecting downward from its anterior ventral surface. Alitrunk in profile flattened. (A Japanese species, *V. emeryi*, is well established in moist stream valleys in the western half of the District of Columbia: S. Cover, personal communication) . . . . . . . . . . . . . . . . . . . . . . . . . ***Vollenhovia***

Petiole with distinct anterior peduncle, and lacking a plate-like subpetiolar process. Alitrunk in profile notably convex. (Southern Texas and Arizona, and in Mexico) . . . . . . . . . . ***Rogeria***

32(30) Frontal carinae long, extending rearward past the eye and reaching or almost reaching the vertex, and/or the clypeus nearly completely covered by conspicuous longitudinal rugulae . . . . . **33**

Frontal carinae short, not extending past the eye, and never almost reaching the vertex. Clypeus variously sculptured or smooth and shining, but not covered by conspicuous longitudinal rugulae . . . . . . . . . . . . . . . . . . . . . . . . . . . . . . . . . . . **34**

33(32) Frontal carinae long, extending rearward past the eye and reaching or almost reaching the vertex, and/or the clypeus longitudinally rugulose, the rugulae clearly continuous with those between the frontal carinae and thus extending all the way to the rear of the head. Posterior lateral portions of clypeus that border the antennal sockets forming a thin, vertical ridge to create the impression of a deep pit surrounding the socket. Antennal club 3-segmented. (Several Old World species with long frontal carinae occur in the Gulf Coast states and Mexico; *T. caespitum*, with short frontal carinae, has been naturalized in the eastern United States and introduced into California) . . . ***Tetramorium*** (in part)

Frontal carinae always short, never closely approaching the vertex. Clypeus longitudinally rugulose, its rugulae often not continuous to rear of head. Posterior lateral portions of clypeus bordering the antennal sockets do not form a vertical ridge; thus the socket does not appear to be surrounded by a deep pit, especially anteriorly. [*Note:* Two aberrant *Myrmica (M. colax* and *M. rugiventris)* from the southwestern states have a clypeal structure as described for *Tetramorium*, but may be distinguished by conspicuous, fine longitudinal rugae on the basal third of the first gastric tergite that are entirely lacking in *Tetramorium caespitum*.] Antennal club indistinct or absent. (Canada, temperate and mountainous North America, and the Mexican highlands) . . . . . . . . . . . . . . . . . . . . . . . . . . . . . . . . . ***Myrmica*** (in part)

34(32) Humeri distinctly angulate. Frequently some hairs on the petiole and the postpetiole are swollen at the base or borne on tubercles, giving the surface a tuberculate appearance. (Southern Texas, Arizona, and Mexico; arboreal or nesting in twigs or hollow stems in scrubby, secondary vegetation) . . . . . . ***Nesomyrmex***

Humeri rounded or at most weakly angulate. Petiole and postpetiole never tuberculate . . . . . . . . . . . . . . . . . . . . . . . . **35**

35(34) Alitrunk smoothly and strongly convex in side view. Anterior peduncle of petiole frequently long and thin and set off sharply from the node. In dorsal view, the postpetiole often bell-shaped and/or broad in relation to the gaster (in some cases two-thirds to three-fourths or more of maximum gastric diameter). Propodeal spines, when present, often long. (Arizona, Texas, southern Florida and Mexico. See note under *Leptothorax* just below) . . . . . . . . . . . . . . . . . . . . . . . . . . . . . . . . ***Macromischa***

**TABLE 2–12** Nearctic (*continued*)

Alitrunk weakly convex in profile. Anterior peduncle short, thick, and not sharply set off from the node. Postpetiole usually two-thirds or less maximum diameter of the gaster. Propodeal spines or teeth short, or at most only moderately long. ("*Myrafant*" group of species, possibly a distinct genus. More or less abundant south of boreal Canada, a few species in the Mexican highlands.) [*Note:* A few species possess unusually broad postpetioles and may be confused with some small *Macromischa* that have very broad postpetioles and relatively short anterior petiolar peduncles. Major possibility of confusion exists in southern Florida, where 2 yellow *Macromischa* and 1 black *Leptothorax* are present.] . . . . . . . . . . . . . . . . . . . . . ***Leptothorax*** (in part)

36(29) Mandibles with 3 or 4 teeth. Clypeus almost always with 2 longitudinal carinae (sometimes weak, rarely absent) that often end as teeth on the anterior margin. Antenna with 3-segmented apical club. Propodeum lacking teeth or spines. (Eastern and central North America south, and Mexico; several introduced species more or less naturalized in portions of the southern United States) . . . . . . . . . . . . . . . . . . . . . . . . . . ***Monomorium***

Mandibles with 5 or more teeth. Propodeum frequently bearing teeth or spines . . . . . . . . . . . . . . . . . . . . . . . . . . . . **37**

37(36) Antenna with 3- or 4-segmented apical club . . . . . . . . . . . . **38**

Antenna lacking an apical club, the terminal segments gradually enlarging toward the apex . . . . . . . . . . . . . . . . . . . . . **40**

38(37) Antennal club 4-segmented. Clypeus usually with 2 longitudinal carinae that do not form teeth on the anterior margin (carinae rarely indistinct or absent). Workers monomorphic. Propodeum usually armed with small teeth, rarely sharply angulate. (Throughout region in forest soils and litter) . . . . . ***Stenamma***

Antennal club almost always clearly 3-segmented and often thicker than the remainder of the funiculus. [*Note:* If club 4-segmented, then the worker caste is dimorphic.] Clypeus always lacking 2 longitudinal carinae . . . . . . . . . . . . . . . . . . . . . . . . . **39**

39(38) Workers monomorphic; metanotal impression prominent, forming a conspicuous valley between the promesonotum and the propodeum, which are at about the same elevation in profile (or else propodeum very slightly lower); basal face of propodeum in profile convex; mandibles with only 5 teeth or denticles; promesonotal profile always continuous. ("Subgenus *Dichothorax*": one variable species, *L. pergandei*, present in southern United States west to Arizona; may be a distinct genus) . . . . . . . . . . . . . . . . . . . . . . . . . . . . . . . . . . . . ***Leptothorax*** (in part)

Workers dimorphic (rarely polymorphic). Majors with large heads and (usually) hypostomal teeth. Minors with the propodeum usually distinctly lower in elevation than the pronotum, the mesonotum often as high as the pronotum and separated from the propodeum by a distinct "step" at the metanotal region, or else the mesonotum sloping gradually (interrupted by a notch or depression at the metanotal region) to the level of the propodeum. Basal face of propodeum in profile flat or weakly concave, rarely convex. Mandibles with 6 or more teeth or denticles. Promesonotal profile sometimes clearly discontinuous. (Present south of boreal Canada, abundant in central and southern United States and Mexico) . . . . . . . . . . . . . . . . . ***Pheidole***

40(37) Propodeum not armed with teeth or spines, almost always evenly rounded but rarely with small, blunt protuberances. Metanotal region strongly impressed; promesonotum and propodeum forming decidedly separate convexities in profile. (Mountains of western North America) . . . . . . . . . . . . . . . . . ***Manica***

Propodeum armed with teeth (sometimes small) or spines. Metanotal impression variable, sometimes absent . . . . . . . . . . **41**

*continued*

41(40) In profile, the metanotal region weakly to moderately impressed. Propodeum barely differentiated from remainder of alitrunk and at most slightly depressed below the level of the promesonotum in profile. Antennal scape often bent abruptly near the base (sometimes 90 degreees) and bearing a more or less obvious lamina at the bend. (Canada, temperate and mountainous North America, and the Mexican highlands) . . . . . ***Myrmica*** (in part)

Metanotal impression variable. Propodeum usually strongly differentiated from remainder of the alitrunk and, in profile, always substantially depressed below the elevation of the pronotum, between which the mesonotum forms a more or less gradually sloping link. Antennal scape not abruptly bent at base, rarely with lamina near base . . . . . . . . . . . . . . . . . . . . . . . . . **42**

42(41) Head quadrate, not noticeably narrower behind the eyes than in front of them. Psammophore often present. Workers sometimes polymorphic. (Western and southwestern United States and Mexico, desert harvesting ants formerly placed in "*Veromessor*") . . . . . . . . . . . . . . . . . . . . . . . . . . . . . . . . . . . . . ***Messor***

Head clearly longer than broad and often noticeably narrower behind the eyes than in front of them. Psammophore absent. Workers monomorphic or at most very feebly polymorphic. (Widespread in North America and Mexico. Mostly in forests and other mesic habitats; 2 species common in deserts—see note below) . . . . . . . . . . . . . . . . . . . . . ***Aphaenogaster***

[*Note:* Two large species (8–12 mm) common in deserts in southwestern United States and Mexico. Usually referred to under the generic name *Novomessor* and may be distinguished from other *Aphaenogaster* by the rudimentary metanotal impression (often absent) that does not interrupt the smooth, sinuous profile of the alitrunk. In other *Aphaenogaster* the metanotal impression is significant and the profile of the alitrunk is not smooth and sinuous.]

### SUBFAMILY DOLICHODERINAE

1 Hypostomal tooth-like projection present adjacent to the ventral surface of mandibular insertions; declivitous face of propodeum strongly concave in profile, and meeting the basal face to form a conspicuous, narrow convexity or sharp angle projecting rearward; scale of petiole erect. (Canada and United States east of Rockies, also Mexico) . . . . . . . . . . . . . . . . . . . ***Hypoclinea***

Hypostomal tooth-like projection not present; declivitous face of propodeum convex, straight, or weakly concave; propodeal profiie evenly convex, sometimes angulate, or bearing a tooth-like protuberance projecting more or less upward; scale of petiole erect, flattened and inconspicuous, or absent . . . . . . . . . . . **2**

2(1) In side view, propodeum bearing a tooth-like protuberance projecting vertically (and sometimes slightly rearward) from the juncture of the basal and declivitous faces; third joint of maxillary palps unusually elongate, commonly as long as or longer than the 3 succeeding distal joints combined. (Arid habitats in the eastern and central United States, common throughout southern states and Mexico) . . . . . . . . . . . . . . . ***Conomyrma***

Propodeum unarmed, evenly convex, or angulate at juncture of basal and declivitous faces; third joint of maxillary palp not unusually long, notably shorter than 3 succeeding distal joints combined . . . . . . . . . . . . . . . . . . . . . . . . . . . . . . . . . . . **3**

3(2) Metanotal region (narrow transverse band between mesonotum and propodeum) not impressed dorsally; mesonotum and propodeum forming a smooth, continuous, flat or convex profile; workers moderately polymorphic. (Western United States, Mexico) . . . . . . . . . . . . . . . . . . . . . . . . . . ***Liometopum***

Metanotum region slightly to profoundly impressed dorsally, forming a shallow, concave depression, an angle, or a notch between the mesonotal and propodeal profiles; workers monomorphic . . . . . . . . . . . . . . . . . . . . . . . . . . . . . . . . . . **4**

4(3) Petiolar scale absent (in rare instances present but minuscule). Basal face of propodeum shorter than the declivitous face; the juncture of the two faces in profile is more or less angular, rarely evenly convex. Apical 3 or 4 mandibular teeth relatively large, succeeding teeth progressively much smaller toward the base (often becoming mere denticles). Basal tooth or angle lacking . . . . . . . . . . . . . . . . . . . . . . . . . . . . . . . . . . . . . . **5**

Petiolar scale present, though sometimes small. Basal face of propodeum as long as or longer than the declivitous face; juncture of the two faces in profile usually evenly convex, rarely weakly angular. Mandibles possessing basal tooth or angle (although it is sometimes small and inconspicuous); masticatory margins possessing either 5 obvious teeth (several denticles may also be present) or with the apical and first subapical teeth large and succeeding teeth notably smaller, and with the third, fourth, and basal tooth or angle separated by several to numerous denticles, these teeth often little larger than the denticles themselves . **6**

5(4) Four gastric tergites visible in dorsal view. The fifth (terminal) tergite is reflexed forward ventrally so that the anus is not located at the terminus of the gaster viewed in profile, but under the "shelf" of the reflexed fifth tergite. [*Note:* This character works well with fresh and alcohol material, and also with pointed specimens in which the gaster is not distorted or shriveled. When gaster is shriveled or distorted, rehydration and dissection may be necessary.] . . . . . . . . . . . . . . . . . . . . . . . . . ***Tapinoma***

Five gastric tergites visible in dorsal view (terminal one is often small). The terminal tergite is not reflexed ventrally and thus the anus is located at the terminus of the gaster in side view. (*Technomyrmex albipes*, a common Indo-Pacific tramp species, has been established in California and it may also occur in Florida) . . . . . . . . . . . . . . . . . . . . . . . . . ***Technomyrmex***

6(4) Mandibles with 5 obvious teeth (including basal tooth or angle), with 1 to 4 denticles separating fourth tooth from third and basal tooth. [*Note:* Denticles often inconspicuous or absent.] In profile clypeus with 2 or more long, straight, erect hairs projecting forward and/or upward. Also present are 2 or more hairs (as a rule longer than the straight hairs) originating at or near the anterior margin that curve downward over anterior surface of the closed mandibles. Metanotal region slightly to moderately impressed; propodeal profile slightly or moderately distinct from outline of remainder of alitrunk. (Xeric habitats from Great Lakes south, common throughout southern states and Mexico) . . . ***Forelius***

Mandibles usually bearing 5 teeth, with the apical and first subapical teeth prominent and notably larger than the succeeding subapical teeth. Third, fourth, and basal tooth or angle separated by several to numerous denticles (basal tooth or angle sometimes small and inconspicuous), and are often little larger than the denticles themselves. Clypeus in profile with 2, occasionally several, long, straight, erect hairs projecting forward and/or upward, usually on either side of the clypeal center. Long, downward-curving hairs at or near anterior margin commonly absent, but if present, shorter than straight hairs. Metanotal area notably impressed; the propodeal profile forms a prominent, rounded convexity that is relatively distinct from the

*continued*

profile of the remainder of the alitrunk. (*Iridomyrmex humilis*, native to temperate South America, is well established in California and the Gulf Coast states; "*Iridomyrmex*" *pruinosus* and its variants of North America are *Forelius*.) . . . . . . . ***Iridomyrmex***

### SUBFAMILY FORMICINAE

1 Antenna with 11 or fewer segments . . . . . . . . . . . . . . . . . . 2

Antenna with 12 segments . . . . . . . . . . . . . . . . . . . . . . . . 6

2(1) Antennal scape greatly elongate, over 1.5 times as long as head length excluding mandibles. Dorsum of alitrunk usually lacking erect hairs. Petiolar node rounded dorsally, not scale-like. Body elongate. (Old World tramp not yet recorded from the region but to be looked for in southern California, Mexico, and the Gulf Coast states) . . . . . . . . . . . . . . . . . . . . . . . . ***Anoplolepis***

Antennal scape little (if at all) longer than head length. Dorsum of alitrunk almost always with some conspicuous, erect hairs. Petiole sometimes scale-like. Body not notably elongate . . . . . . . 3

3(2) Mandible slender, nearly straight, with a strongly oblique masticatory margin bearing 3 teeth. Eye minute, sometimes rudimentary, and positioned anterior to the middle of the head. (One species, *A. epedana*, known from southern Arizona, more in Mexico; subterranean, yellowish ants) . . . . . . . . . ***Acropyga***

Mandible triangular with a more or less transverse masticatory margin, bearing 4 or more teeth or denticles. Eye large and usually located near the middle of the side of the head . . . . . . . 4

4(3) Antenna with 11 segments. Alitrunk lacking erect hairs, though some decumbent or appressed hairs may be present. Dorsum of alitrunk without sutural impressions (a faint metanotal impression rarely present), its profile flattened and continuous. (An Old World tramp, *P. alluaudi*, is established in California, perhaps in Mexico, and to be looked for in the Gulf Coast states) . . . . . . . . . . . . . . . . . . . . . . . . . . . . . . . . . . ***Plagiolepis***

Antenna with 9 or 10 segments. Dorsum of alitrunk often bearing 1 or more sutural impressions . . . . . . . . . . . . . . . . . . . . 5

5(4) Antenna 9- or 10-segmented with a 2- or 3-segmented apical club. Scale of petiole erect, prominent, and fully exposed. Alitrunk in profile with a distinct metanotal impression. (Central Florida—probably introduced—and Mexico; arboreal) . . ***Myrmelachista***

Antenna 9-segmented and lacking a distinct apical club. Scale of petiole strongly inclined forward and hidden beneath the base of the first gastric tergite. Metanotal impression often absent or faint. (Southern Canada south through Mexico; Nearctic species are terrestrial) . . . . . . . . . . . . . . . . . . . ***Brachymyrmex***

6(1) Mandibles falcate, with numerous denticles. Petiole with prominent, rounded node (not scale-like). (Temperate and mountainous North America south into Mexican highlands; a dulotic genus found in mixed colonies with *Formica* spp.) . . . . . . . . . . . . . . . . . . . . . . . . . . . . . . . . . . . . . . . . . . ***Polyergus***

Mandibles more or less triangular, masticatory margin with 5–12 teeth. Petiole usually scale-like, sometimes with a rounded node. Mostly free-living species . . . . . . . . . . . . . . . . . . . 7

7(6) Maxillary palp 3-segmented and very short. Yellow or orange subterranean ants, whose eye width is generally less than that of the last antennal segment. (Throughout most of North America and extending into the Mexican highlands; most or all species are probably temporary social parasites of *Lasius* species) . . . . . . . . . . . . . . . . . . . . . . . . . . . . . . . . . . . ***Acanthomyops***

Maxillary palp 6-segmented and moderately to exceptionally long . . . . . . . . . . . . . . . . . . . . . . . . . . . . . . . . . . . . . . . . . 8

8(7) Maxillary palp longer than the head length (excluding the mandibles), the third and fourth segments *each* as long or longer than 2 terminal segments combined. Psammophore usually present, though sometimes weakly developed. Workers monomorphic or moderately polymorphic. (Western North America and Mexico, in semi-arid or arid habitats) . . . . . . . . . ***Myrmecocystus***

Maxillary palp not longer than head length and usually distinctly shorter, its third and fourth segments not disproportionately long. Psammophore absent . . . . . . . . . . . . . . . . . . . . . 9

9(8) Profile of alitrunk continuous and evenly convex, with the propodeum not depressed below the level of the promesonotum and the metanotal region at most slightly impressed (usually not at all impressed). Alitrunk in dorsal view usually wedge-shaped and tapering posteriorly. Workers often polymorphic, sometimes dimorphic . . . . . . . . . . . . . . . . . . . . . . . . . . . 10

Profile of alitrunk clearly discontinuous and not evenly convex, the metanotal area moderately to strongly impressed, and the propodeum often distinctly depressed below the level of the promesonotum. Alitrunk in dorsal view *not* wedge-shaped, usually constricted to some degree in the middle. Workers usually monomorphic, sometimes weakly polymorphic . . . . . . . . . . . 11

10(9) Workers completely dimorphic (medias absent). Head of major obliquely truncate in front with the borders of the truncated zone sharply marginate. Mandible of major with 3 distinct apical teeth, followed by basal teeth that are fused into a single, broad cutting edge. Mesepisternal carina ending shortly below level of lower corner of pronotum, not turned mesad in front of middle coxae. (Arboreal: southern United States and in Mexico) . . . . . . . . . . . . . . . . . . . . . . . . . . . . . . . . . . . . . ***Colobopsis***

Workers continuously polymorphic, rarely dimorphic. Head of major usually not obliquely truncate in front (if somewhat truncate, the truncated zone *not* sharply marginate). Mandible of major dentate along entire masticatory margin. Mesepisternal carina reaching to the inner base of the middle coxae. (Abundant throughout region) . . . . . . . . . . . . . . . . . . . ***Camponotus***

11(9) Frontal carinae short but distinct, each a small ridge with a moderately to sharply angulate summit that is sometimes slightly reflected upward. [*Note:* The frontal carinae are best examined by looking perpendicularly down on the summit of the ridge.] Lower rim of antennal socket usually nearly touching the posterior border of the clypeus, the distance between them less than one-fourth the maximum diameter of the antennal socket. The basal face of the propodeum longer than (sometimes as long as, but only rarely somewhat shorter than) the declivitous face. [*Note:* In a few cases propodeal profile is so evenly rounded that it is hard to distinguish the 2 faces.] Mandibles with 7 or more teeth or denticles. (Abundant and diverse throughout North America, also occurring in Mexican highlands) . . . . . ***Formica***

Frontal carinae are indistinct or absent. If present, each carina is a small ridge with a distinctly (and often broadly) *rounded* summit. Lower rim of the antennal socket usually not nearly touching the posterior clypeal border, the distance between them commonly one-third the maximum diameter of the antennal socket or more. Either declivitous face of propodeum markedly longer than basal face and mandibles with 7 or more teeth, or propodeal faces nearly equal in length (or declivitous face slightly longer than the basal face) and mandibles with 5 or 6 teeth . 12

*continued*

**12(11)** Mandibles with 7 or more teeth. Antennal scapes passing occipital border by no more than 2–3 times maximum diameter of the scape, usually less. Declivitous face of propodeum decidedly longer than the basal face, both faces meeting so that the propodeal profile resembles a distinct upward-facing "peak" with a more or less rounded apex. (Abundant throughout North America, present in Mexican highlands) . . . . . . . . . . ***Lasius***

Mandibles with 5 or 6 teeth. Antennal scapes frequently passing the occipital border by 4 or 5 times maximum diameter of the scape or more . . . . . . . . . . . . . . . . . . . . . . . . . . . . . . **13**

**13(12)** Mesonotum, viewed from directly above, severely constricted and giving the alitrunk a distinctive hourglass-like shape. Pilosity not conspicuously coarse or bristle-like; erect hairs mostly slender and golden or brownish. (Widespread in temperate and mountainous United States and in the Mexican highlands) . . . . . . . . . . . . . . . . . . . . . . . . . . . . . . . . . . . . . . ***Prenolepis***

Mesonotum, viewed from directly above, weakly constricted; alitrunk lacking obvious hourglass-like shape. Erect hairs usually coarse and bristle-like, and often dark brown or black. (Temperate North America and south, common throughout southern United States and in Mexico) . . . . . . . . . . . . . ***Paratrechina***

**TABLE 2-13** A key to the ant genera of Central and South America, the West Indies, and lowland tropical Mexico, generally referred to as the **Neotropical Region.** Based on the worker caste. (Written by S. P. Cover, assisted by B. Bolton, J. E. Lattke, S. O. Shattuck, J. C. Trager, and G. J. Umphrey, and incorporating some elements of an unpublished key to the Central American genera by R. R. Snelling; used with permission.)

### SUBFAMILY PONERINAE

**1** Mandibles slender, elongate, abruptly bent inward near apex, articulated near middle of anterior margin of head; postpetiole not differentiated from remainder of gaster . . . . . . . . . . . . . . . **2**

Mandibles variously shaped, but not abruptly bent inward near apex, and articulated near anterolateral corners of head; postpetiole often slightly to moderately constricted posteriorly, and thus more or less differentiated from the remainder of the gaster . . . . . . . . . . . . . . . . . . . . . . . . . . . . . . . . . . . . . . . . **3**

**2(1)** Nuchal carina (the ridge delimiting the occiput, or extreme rear of head) continuously curved across midline; apophyseal lines absent . . . . . . . . . . . . . . . . . . . . . . . . . . . . . . . . ***Anochetus***

Nuchal carina V- or wedge-shaped, narrowed toward the median into a middorsal groove; apophyseal lines present as a pair that converge from the head insertion up to nuchal carina . . . . . . . . . . . . . . . . . . . . . . . . . . . . . . . . . . . . . ***Odontomachus***

**3(1)** Median lobe of clypeus broad, its anterior margin lined with multiple denticles. [*Note:* In small species (workers less than 3 mm total length) the denticles may not be clearly visible unless high magnification (> 50X) is used. In addition, it is helpful if the mandibles of the specimen are open.] Petiole narrowed only slightly behind, so that its posterior face attaches broadly to the gaster in dorsal view . . . . . . . . . . . . . . . . . . . . . . . . . **4**

Median lobe of clypeus variously shaped, but anterior margin never multidenticulate. Petiole usually strongly constricted behind in dorsal view . . . . . . . . . . . . . . . . . . . . . . . . . . **5**



**4(3)** Mandibles short and narrow, with 3 apical teeth that close tightly against clypeus . . . . . . . . . . . . . . . . . . . . . . . ***Prionopelta***

Mandibles longer, strongly projecting beyond clypeus when closed, and having many teeth along the inner borders . . . . . . . . . . . . . . . . . . . . . . . . . . . . . . . . . . . . . . ***Amblyopone***

**5(3)** Each gena (cheek) projecting forward as a simple, blunt tooth over the dorsal surface of the closed mandibles on either side of the antennal insertions. Frontal carinae vertical and fused; condylar bulbs of antennae completely exposed in dorsal view. Head lacking conspicuous carinae or ridges that laterally border the antennal sockets or ventrally border the genae running from the posteroventral corners of the head for a distance of one quarter to one half the head length in side view. Ants eyeless, with stout antennal scapes less than one half the length of the funiculus. Very small (less than 5 mm total length), slender ants. (Rare: at present known only from Bolivia, Colombia, and Ecuador) . . . . . . . . . . . . . . . . . . . . . . . . . . . . . . . . . . . ***Leptanilloides***

Each gena lacking a blunt tooth projecting forward over the dorsal surface of the mandibles. (In rare cases where such teeth are present, they are extensions of prominent carinae or ridges laterally bordering the antennal sockets.) Antennal scapes longer than one half the length of the funiculus. Ants frequently possessing eyes, even if comprising only 1 or 2 facets . . . . . . . . **6**

**6(5)** Giant ants: head width usually greater than 4.8 mm. Antennal scrobes present, extending above the eye, and thence downward between the eye and ventral margin of the gena. Pronotal dorsum with large lateral tubercles . . . . . . . . . . ***Paraponera***

Head width usually less than 2.5 mm. Antennal scrobes often absent, but when present not curving downward between the eye and ventral margin of the gena. Pronotal dorsum almost never bearing large lateral tubercles . . . . . . . . . . . . . . . . . . . . . **7**

**7(6)** Pygidium (last dorsal plate of abdomen) bearing rows of erect spines or teeth along both sides that converge toward one another at the rear. Antennal sockets nearly touching. Body elongate and cylindrical . . . . . . . . . . . . . . . . . . . . . . . . . . **8**

Pygidium lacking rows of erect spines or teeth (rarely bearing a single pair of prominent teeth, or numerous spine-like setae). Separation between antennal sockets variable, often greater than the maximum diameter of a single antennal socket. Body form variable, rarely elongate and cylindrical . . . . . . . . . . . . . **10**

**8(7)** Each gena margined ventrally by a carina or ridge that runs forward from the posteroventral corner of the head for a distance of one-quarter to one-half the head length in side view. Antennal sockets bordered laterally by a second carina or ridge that extends to the lateral lobe of the clypeus or projects past the anterior clypeal border (sometimes slightly over the dorsal surface of the closed mandibles as well) as a blunt tooth . . . . ***Cerapachys***

Genae lacking both carinae as described above . . . . . . . . . . **9**

**9(8)** Eye conspicuous, consisting of more than 15 facets. Middle and hind tibiae each with 2 apical spurs, the outer spur half (or less) as long as the inner . . . . . . . . . . . . . . . . ***Cylindromyrmex***

Eye consisting of a single, usually unpigmented facet. Middle and hind tibiae each with a single apical spur . . . . . . . . . . . . . . . . . . . . . . . . . . . . . . . . . . . . . . ***Acanthostichus/Ctenopyga***
[*Note:* Currently these genera are separated by differences in the queens (born winged in *Ctenopyga*, wingless in *Acanthostichus*). Workers of *Ctenopyga* are very poorly known and indistinguishable from those of *Acanthostichus*. *Ctenopyga*-type queens have been found only in the southern United States and in Mexico.

*continued*

*Acanthostichus* records are from Central and South America, but some species might occur much farther north. See Brown (1975) for a more detailed discussion.]

**10(7)** Anterior margin of clypeus bearing a tooth projecting forward, one each side of a median notch (emargination). Giant black ants of South American forests and wetter savanna woodlands . . . . . . . . . . . . . . . . . . . . . . . . . . . . . . . . . . . . . . . ***Dinoponera***

Anterior margin of clypeus not bidentate as described . . . . . . **11**

**11(10)** Tarsal claws finely pectinate. Antennal sockets close together, separated by less than the maximum diameter of a single socket. Anterior margin of clypeus sharply triangular in form and projecting far forward in the middle. Mandibles elongate and slender, usually without conspicuous subapical teeth . ***Leptogenys***

Tarsal claws simple, or with 1 or 2 subapical teeth, but not finely pectinate. Separation of antennal sockets variable. Anterior margin of clypeus usually broad, and rounded or subtriangular in form, rarely sharply triangular and projecting far forward (if so, the mandibles bear 3–5 large teeth and are broadly triangular in shape) . . . . . . . . . . . . . . . . . . . . . . . . . . . . . . . . . . **12**

**12(11)** Mandibles long and slender, and curved, attaining the opposite anterolateral corners of the head when closed, and armed with very long, narrow teeth (each mandible reminiscent of a pitchfork). Antennal sockets widely separated, the distance between them greater than one-third the head width exclusive of the eyes . . . . . . . . . . . . . . . . . . . . . . . . . . . . . ***Thaumatomyrmex***

Mandibles usually broad, but if slender not curved to opposite anterolateral corners of head, and without bizarre, long, slender teeth. Separation between antennal sockets variable; often the sockets are close together . . . . . . . . . . . . . . . . . . . . . . **13**

**13(12)** Front of head bearing a distinct median costa that runs from the anterior clypeal border across the frontal triangle and reaches or almost reaches the vertex. Antennal scapes relatively short, rarely reaching or barely surpassing the occipital angles . . . **14**

Front of head usually without a median costa or with an incomplete costa. If bearing a complete median costa as described, the antennal scapes are long, surpassing the occipital angles by at least the maximum diameter of the scape and often by twice that amount . . . . . . . . . . . . . . . . . . . . . . . . . . . . . . . . . . **15**

**14(13)** Tarsal claws each with a prominent basal lobe and a large submedian tooth. Petiole with single, large posterapical spine projecting rearward; propodeum bispinose . . . . . . . . ***Acanthoponera***

Tarsal claws lacking prominent basal lobe, often lacking submedian tooth as well. Petiole with or without posterapical tooth; propodeum angulate, or having short, blunt teeth, rarely bispinose . . . . . . . . . . . . . . . . . . . . . . . . . . . . . ***Heteroponera***

**15(13)** Lobes of frontal carinae more or less sharply raised above the line of the front of the head, or vertical and fused, so that the condylar bulbs of the antennae are mostly or completely exposed in frontal view. Eyes greatly reduced (fewer than 15 facets) or apparently absent . . . . . . . . . . . . . . . . . . . . . . . . . . . . **16**

Lobes of frontal carinae horizontal or raised but wholly or mostly covering the condylar bulbs of the antennae in front view. Eyes variable in development, but if condylar bulbs seem less than one-half covered by the frontal lobes, then eyes well developed (more than 30 facets) . . . . . . . . . . . . . . . . . . . . . . . . **18**

**16(15)** Lobes of frontal carinae raised, not vertical and fused. Antennal sockets normally placed. Mandibles easily visible in frontal view, with several teeth, at most only slightly overhung by the clypeus. [*Note*: Dorsal surface of first gastric segment often strongly convex so that apex of gaster points anteriorly, rarely ventrally.] . . . . . . . . . . . . . . . . . . . . . . . . . . . . . . . . . ***Proceratium***

Lobes of frontal carinae vertical and fused. Antennal sockets placed low on shelf-like extension of the clypeus that projects forward, partly or completely overhanging the mandibles. Mandibles inconspicuous and strap-like . . . . . . . . . . . . . . . . **17**

**17(16)** Apical segment of the antenna greatly enlarged and approximately equal to or longer than the second to penultimate segments combined. Eyes very small, consisting of only one to several facets. Dorsal surface of first gastral segment very strongly convex so that the gastral apex points anteriorly . . . . . . ***Discothyrea***

Apical segment of antenna not greatly enlarged, much shorter than the second to penultimate segments combined. Eyes indistinct or apparently absent. Dorsal surface of first gastral segment slightly convex; gastral apex normally pointing downward or rearward, not anteriorly . . . . . . . . . . . . . . ***Probolomyrmex***

**18(15)** Erect hairs absent from head and body. Tarsal claws on middle and hind legs each bearing a submedian tooth. Body dull or feebly shining, bearing sparse, shallow punctures, and entirely covered with very fine, appressed, pruinose pubescence that gives the surface a pronounced grayish or silvery cast . . ***Platythyrea***

Erect hairs present on head and/or body, though sometimes short and sparse. Tarsal claws on middle and hind legs simple, or if bearing a submedian tooth, body not appearing as described above . . . . . . . . . . . . . . . . . . . . . . . . . . . . . . . . . . . . **19**

**19(18)** Antenna with apical club consisting of 3 (rarely 4) segments. Petiole with distinct anterior peduncle and bearing a large subpetiolar process that reaches maximum size under the peduncle or under the juncture of the peduncle and the node. The subpetiolar process is subtriangular or tooth-like (rarely rounded) and projects forward or downward (sometimes pointing downward and slightly deflected posteriorly). Eyes small (less than 15 facets) or apparently absent. Upper surface of head dull or feebly shining. Small, light-colored, cryptobiotic ants . ***Typhlomyrmex***

Antenna only rarely with 3- or 4-jointed apical club. In many cases the terminal antennal segments are enlarged, but they do nc form a distinct club due to a gradual increase in segment size from base to apex of funiculus. Petiole usually lacking a distinct peduncle; often lacking a subpetiolar process as well, but if present it extends over most of the ventral surface of the node, is subrectangular in shape, or points posteriorly. (In cases where the petiole appears to have a peduncle and the subpetiolar process is prominent, subtriangular or tooth-like, and points downward, either the head and alitrunk are smooth and strongly shining or the eyes are well developed, with more than 30 facets) . . . . . . . . . . . . . . . . . . . . . . . . . . . . . . . . . . . . . . . **20**

**20(19)** Tarsal claws on front legs (and other legs as well) simple. Frontal carinae close together posteriorly, just prior to flaring outward anteriorly to produce distinct lateral lobes . . . . . . . . . . . . **21**

Tarsal claws on front legs (and often on middle and hind legs as well) with a subapical tooth that can be nearly basal to median in position. [*Note*: In a few small species the subapical teeth are inconspicuous unless examined at high power (greater than 50–60X) with good illumination.] Frontal carinae well separated and subparallel posteriorly, usually subparallel or feebly diverging anteriorly, sometimes diverging more strongly to produce lateral lobes . . . . . . . . . . . . . . . . . . . . . . . . . . . . . . . . . . . **27**

**21(20)** Mandible elongate, its cutting margin strongly oblique and bearing

*continued*

3-5 large teeth. Eye consisting of a single enlarged facet. Median lobe of clypeus roughly triangular in shape and projecting far forward of the rest of the clypeus . . . . . . . . . . . . . . . . . **22**

Mandible more or less triangular, with a distinct cutting margin armed with numerous teeth or denticles. Eye often multifaceted, or consisting of a single, minute facet, or (rarely) absent altogether. Median lobe of the clypeus rarely triangular in shape or projecting far forward . . . . . . . . . . . . . . . . . . . . . . . . . **23**

**22(21)** Mandible bearing 5 teeth. Frontal lobes in profile not elevated above plane of front of head (includes *Leiopelta*) . . ***Belonopelta***

Mandible bearing 3 or 4 teeth; the 2 apical teeth separated from the basal tooth (or pair) by a large gap. Frontal lobes in profile abruptly elevated above the plane of the front of head . . . . . . . . . . . . . . . . . . . . . . . . . . . . . . . . . . . . . . . . ***Simopelta***

**23(21)** Middle and hind tibiae with 2 apical spurs, the outer spur often only half as long (or less) as the inner . . . . . . . . . . . . . . . **24**

Middle and hind tibiae each with a single apical spur . . . . . . **25**

**24(23)** Outer face of mandible at base with an oval, pit-like depression. Outer face of middle tibia with several spine-like bristles. Eyes present, although small, with fewer than 15 facets. Small (2–5 mm), cryptobiotic ants . . . . . . . . . . . . . . . . . . ***Cryptopone***

Outer face of mandible at base lacking an oval, pit-like depression. Spine-like bristles on outer face of middle tibia generally absent. [*Note*: 2 small cryptobiotic species have numerous spine-like bristles on outer faces of their middle tibiae but lack mandibular pits, and their eyes are indistinct or absent.] Often medium-sized to large ants, generally abundant in tropical forests . . . . . . . . . . . . . . . . . . . . . . . . . . . . . . . . . . . . ***Pachycondyla***

**25(23)** Outer face of middle tibia with numerous spine-like bristles. Head capsule slightly to notably broader than long. Eyes absent. Pubescence on upper head and body surfaces extremely dilute or lacking; integument smooth, highly polished, and strongly shining . . . . . . . . . . . . . . . . . . . . . . . . . . . ***Centromyrmex***

Outer face of middle tibiae lacking spine-like bristles. Head capsule slightly to notably longer than broad. Eyes usually present, consisting of single, small facet (rarely multifaceted). Pubescence on upper body surfaces often moderately dense, sometimes dilute; integument dull, or weakly to moderately shining, very rarely highly polished and strongly shining . . . . . . . . **26**

**26(25)** Subpetiolar process prominent, impressed anteriorly on each side to form a roughly circular, central, translucent window, and projecting posteriorly as a pair of right angles or teeth. [*Note*: Maxillary palps 2-jointed.] (Common in Mexican highlands, not yet known from Central or South America) . . . . . . . . . . ***Ponera***

Subpetiolar process a small, relatively simple lobe lacking a central window. [*Note*: Maxillary palps one-jointed.] (Abundant throughout Neotropical Region) . . . . . . . . . . . ***Hypoponera***

**27(20)** Viewed in profile, the mesonotum prominent, conspicuously set off from the propodeum by a deep transverse fissure; mesonotum and propodeum forming distinct, separate convexities. Coxa of last leg without dorsal tooth or spine . . . . ***Ectatomma***

Viewed in profile, the mesonotum not prominent. It and the propodeum form one continuous or near-continuous profile interrupted at most by a suture-like groove at their juncture (rarely deep or trench-like), or by an ill-defined impression in this region. Coxa of hind leg often armed with a dorsal tooth or spine . . . . . . . . . . . . . . . . . . . . . . . . . . . . . . ***Gnamptogenys***



### SUBFAMILY ECITONINAE

#### Worker

**1** Waist 1-segmented . . . . . . . . . . . . . . . . . . . . ***Cheliomyrmex***

Waist 2-segmented . . . . . . . . . . . . . . . . . . . . . . . . . . . . **2**

**2(1)** Tarsal claws simple, lacking a median tooth in addition to the terminal point . . . . . . . . . . . . . . . . . . . . . . . ***Neivamyrmex***

Tarsal claws with a single median tooth in addition to the terminal point . . . . . . . . . . . . . . . . . . . . . . . . . . . . . . . . . . **3**

**3(2)** Flagellum robust; width of scape near its apex more than one-third its entire length . . . . . . . . . . . . . . . . . . . . . ***Nomamyrmex***

Flagellum slender; apical width of scape less than one-third its entire length . . . . . . . . . . . . . . . . . . . . . . . . . . . . . . . . **4**

**4(3)** Propodeum with teeth or lamellae; major subcaste usually with falcate mandibles . . . . . . . . . . . . . . . . . . . . . . . . . ***Eciton***

Propodeum without teeth or lamellae; major subcaste without falcate mandibles . . . . . . . . . . . . . . . . . . . . . . . . ***Labidus***

#### Male

**1** Flagellum only slightly longer than width of head . . . . . . . . . . . . . . . . . . . . . . . . . . . . . . . . . . . . . . . . . . ***Cheliomyrmex***

Flagellum much longer than width of head . . . . . . . . . . . . . . **2**

**2(1)** Legs short, apex of hind femur not reaching posterior margin of second gastral segment . . . . . . . . . . . . . . . . ***Neivamyrmex***

Legs long, apex of hind femur reaching or surpassing posterior margin of second gastral segment . . . . . . . . . . . . . . . . . . **3**

**3(2)** Apex of penis valve without setae . . . . . . . . . . . . . . . . ***Eciton***

Apex of penis valve with setae . . . . . . . . . . . . . . . . . . . . . . **4**

**4(3)** Gastral terga with lateral clusters of long setae . . . ***Nomamyrmex***

Gastral terga without lateral clusters of setae . . . . . . . . ***Labidus***

### SUBFAMILY LEPTANILLINAE

(No true leptanillines have been found in the New World. A remarkably convergent genus *Leptanilloides* occurs in the Ponerinae, and can be keyed out in that subfamily.)

### SUBFAMILY PSEUDOMYRMECINAE

**1** All castes with 12 antennal segments; hind basitarsus lacking a longitudinal groove or sulcus. (Widespread throughout warmer parts of the Neotropical Region) . . . . . . . . . . ***Pseudomyrmex***

Worker and queen with 10–11 antennal segments, male with 13 antennal segments; hind basitarsus of worker and queen with a distinct longitudinal sulcus on the anterior face. (Brazil) . . . . undescribed genus (P. S. Ward, ms., personal communication)

### SUBFAMILY MYRMICINAE

**1** Antenna with 4 segments . . . . . . . . . . . . . . . . . . . . . . . . . **2**

Antenna with 6 or 7 segments . . . . . . . . . . . . . . . . . . . . . . **3**

Antenna with 8 segments . . . . . . . . . . . . . . . . . . . . . . . . . **17**

Antenna with 9 or 10 segments . . . . . . . . . . . . . . . . . . . . **19**

Antenna with 11 segments . . . . . . . . . . . . . . . . . . . . . . . **28**

Antenna with 12 segments . . . . . . . . . . . . . . . . . . . . . . . **62**

*continued*

2(1) Mandible subtriangular, its masticatory border lined with small, serial teeth. Cephalic hairs simple. Head capsule subrectangular, about 1.5 times longer than broad, narrowed only slightly anteriorly. Head surface shining, with scattered piligerous foveae. (Known only from Cuba; very rare) . . . . ***Codioxenus***

Mandible curvilinear, bearing a long, apical fork; its masticatory border otherwise mostly or completely bare. Head cordate, strongly narrowed anteriorly, and bearing scale-like hairs. (One species, *Q. emmae*, has been widely introduced throughout the Caribbean, and more sporadically throughout the rest of the Neotropics) . . . . . . . . . . . . . . ***Quadristruma***

3(1) Antenna with 6 segments (tribe Dacetini in part) . . . . . . . . . . **4**

Antenna with 7 segments . . . . . . . . . . . . . . . . . . . . . . . . **13**

4(3) Broad, thin, semitransparent lamellae surround the clypeus, antennal scrobes, posterior border of the head, and anterior border of alitrunk. (Known only from Bolivia; rare) . . . ***Tingimyrmex***

Head and anterior portion of alitrunk lacking semitransparent lamellae as described above . . . . . . . . . . . . . . . . . . . . . . **5**

5(4) Mandible elongate, linear or sublinear, with a distinct apical fork of 2 (rarely 3) inwardly directed spiniform teeth, sometimes with 1 or more minute denticles between the teeth . . . . . . . . . . . **6**

Mandible triangular or subtriangular, the masticatory margin serially dentate or denticulate and lacking an apical fork . . . . . . **8**

6(5) Both principal teeth of apical fork reduced and about equal in size. Subapically, a short spiniform tooth is present at or anterior to the midpoint, flanked front and rear by a short series of minute, separated denticulae . . . . . . . . . . . . . . . . . . . ***Neostruma***

At least 1 principal tooth of apical fork prominent and elongate. Subapical teeth not as described above . . . . . . . . . . . . . . . **7**

7(6) Head in dorsal view oblong, with nearly parallel sides and slightly convex occipital border. Mandible broad at base, possessing 3 subapical teeth. Integument smooth. (Known only from Cuba; rare) . . . . . . . . . . . . . . . . . . . . . . . . . . . . . . . . ***Dorisidris***

Head in dorsal view strongly narrowed anteriorly. Occipital border concave to distinctly excised. At least the dorsal surface of the head sculptured and opaque. (Abundant and widespread throughout Neotropics) . . . . . . . . . . . . . . . . . ***Strumigenys***

8(5) Cephalic hairs abundant and conspicuous, often strap-shaped, scale-like, or otherwise bizarre. Anterior margin of clypeus bearing fringe of prominent hairs . . . . . . . . . . . . . . . . . . . . . **9**

Cephalic hairs reduced and inconspicuous; a few on the vertex may be larger. Clypeus lacking a fringe of prominent hairs . **10**

9(8) Mandible viewed from the side not notably swollen, its dorsal profile nearly flat. Head elongate, more than 1.2 times longer than broad, and as a rule finely, densely, and evenly sculptured over its dorsal surface. Hairs often clavate, scale-like, long and sinuous, or otherwise unusual in shape. (Abundant and widespread in Neotropics) . . . . . . . . . . . . . . . . . . ***Smithistruma***

Mandible notably swollen; its dorsal surface in profile strongly bowed. Head not elongate, only about 1.2 times longer than broad; its dorsum strongly rugose at least in posterior region. Pilosity fleece-like, composed of abundant, long, soft filiform hairs. (Rare, local) . . . . . . . . . . . . . . . . . . . ***Codiomyrmex***

10(8) Head strikingly elongate, its capsule more than 2 times longer than broad. Head and alitrunk completely lacking erect hairs, smooth and shining. (Rare, Brazil) . . . . . . . . . . . . . ***Gymnomyrmex***

Head not strikingly elongate, its capsule less than 2 times longer than broad. Head and trunk with at least some standing strap-like or filiform hairs . . . . . . . . . . . . . . . . . . . . . . . . . **11**

11(10) Dorsal surfaces bearing conspicuous longitudinal costulae. Postpetiole in dorsal view broad and kidney-shaped, with anterolateral portions embracing posterior part of petiole . . . . . . . . . . . . . . . . . . . . . . . . . . . . . . . . . . . ***Glamyromyrmex*** (in part)

Dorsal surfaces variously sculptured but never conspicuously costulate in longitudinal direction. Postpetiole ordinary in dorsal view, not embracing petiole . . . . . . . . . . . . . . . . . . . . **12**

12(11) Head densely and finely sculptured, opaque. Vertex with 2 short, erect clavate hairs. Pronotum strongly margined anteriorly and laterally . . . . . . . . . . . . . . . . . . . . . . . . . . . ***Trichoscapa***

Head weakly or not at all sculptured. Vertex lacking a single pair of erect, clavate hairs. Pronotum not strongly marginate . . . . . . . . . . . . . . . . . . . . . . . . . . . . . . ***Glamyromyrmex*** (in part)

13(3) Antennal scrobes absent. (Obligate symbionts of several tropical trees, including some *Cordia*, *Tococa* and *Duroia* species) . . . . . . . . . . . . . . . . . . . . . . . . . . . . . . . . . ***Allomerus*** (in part)

Antennal scrobes present, often deep and conspicuous. Hairs often scale- or strap-like . . . . . . . . . . . . . . . . . . . . . . . **14**

14(13) Mandibles triangular, their masticatory borders serially dentate or denticulate, and fully engaging when closed . . . . . . . . . . **15**

Mandibles linear (rarely subtriangular), with their insertions remote so that their masticatory borders cross or engage only near their apices . . . . . . . . . . . . . . . . . . . . . . . . . . . . . . **16**

15(14) Eye located at or just above dorsal margin of antennal scrobe more or less at scrobe midlength. Antennal scape abruptly bent near insertion, reaching its maximum diameter at or near the bend, and strongly flattened from the bend to its apex. Dorsal surface of head completely sculptured and very feebly shining or dull. Pilosity on upper body surfaces erect to subappressed (rarely absent); at least some hairs prominently scale- or strap-like. Propodeum usually armed with spines or teeth, rarely sharply angulate or unarmed. (Widespread in Neotropics) . . . . . . . . . . . . . . . . . . . . . . . . . . . . . . . . . . . . . . . . . ***Eurhopalothrix***

Eye located at extreme posterior border of antennal scrobe. Antennal scape not notably bent at base, increasingly flattened toward the apex, and reaching its maximum diameter at one-half to two-thirds its total length from the insertion. Dorsal surface of head without sculpture except for scattered punctures, and strongly shining. Pilosity on upper body surfaces, when present, consisting only of simple, delicate erect to subappressed hairs. Propodeum unarmed. [*Note:* Mandibles, head, alitrunk unusually robust. Ventral surfaces of mandibles, if visible, with dense brush of heavy setae near masticatory margins.] (Rare: known only from tropical Mexico, El Salvador, Honduras, and Costa Rica) . . . . . . . . . . . . . . . . . . . . . . . . . . . . . . . . . . . . ***Tatuidris***

16(14) Mandibles each with a long, conspicuous, spiniform tooth near the apex. (Cuba and continental Central and South America) . . . . . . . . . . . . . . . . . . . . . . . . . . . . . . . . . . . ***Rhopalothrix***

Mandible apices with interlocking teeth, none notably long or spiniform. [*Note:* In side view, the masticatory margin of each mandible is oriented at about 120-degree angle to the longitudinal plane of the head, meaning that the ventral end of the masticatory margin is closer than the dorsal end to the anterior border of the head. In almost all other ants, the reverse is true.] (Rare: known from Trinidad and northern South America) . . ***Talaridris***

*continued*

17(1) Antenna with 8 segments . . . . . . . . . . . . . . . . . . . . . . . . 18

Antenna with 9 or 10 segments . . . . . . . . . . . . . . . . . . . . . 19

18(17) Antennal scrobes present (rarely indistinct or absent); eye located at or just above dorsal margin of scrobe more or less at scrobe midlength; antennal scape bent abruptly near insertion and usually strongly flattened from bend to apex; pilosity on dorsal body surfaces and antennal scapes, when present, with many hairs scale- or strap-like; propodeum armed with teeth or spines, rarely sharply angulate or unarmed; upper body surface heavily sculptured and dull. (Mostly ground dwelling, widespread in Neotropics) . . . . . . . . . . . . . . . . . . . . . . . . . ***Octostruma***

Antennal scrobes absent; antennal scapes gradually and evenly bent at base and not strongly flattened; pilosity on upper body surfaces and scapes consisting of simple hairs; propodeum unarmed; dorsal surfaces of head and alitrunk largely free of sculpture and shining. (Obligate symbionts of several tropical trees, including species of *Cordia, Tococa,* and *Duroia,* in South American tropical forests) . . . . . . . . . . . . . . . ***Allomerus*** (in part)

19(1) Antenna with 9 segments . . . . . . . . . . . . . . . . . . . . . . . . 20

Antenna with 10 segments . . . . . . . . . . . . . . . . . . . . . . . 24

20(19) Eyes small, with fewer than 10 facets, or indistinct, or apparently absent . . . . . . . . . . . . . . . . . . . . . . . . . . . . . . . . . . 21

Eyes well developed, with more than 15 facets . . . . . . . . . . 23

21(20) Mandibles linear, elongate, and meeting at closure only at tips of 2 pairs of stout, spiniform teeth present on inner margin; mandibles cross near apices as sinuous spines that curve upward in side view. Antennal scapes bent strongly (about 90 degrees) near their insertions. Antennomeres enlarging gradually in size (terminal antennomere greatly enlarged), but not forming a 2-jointed club distinct from rest of the funiculus. Propodeum armed with distinct teeth. (Known only from Colombia and Ecuador) . . . . . . . . . . . . . . . . . . . . . . . . . . ***Protalaridris***

Mandible triangular or subtriangular, without spiniform teeth. Antennal scapes not strongly bent at base. Antenna with prominent 2-jointed apical club distinct from remainder of the funiculus. Propodeum unarmed or bearing inconspicuous teeth . . 22

22(21) Worker caste dimorphic: minors with dorsum and sides of petiolar node (and usually at least half the sides of the propodeum as well) dull and covered with coarsely granulose or irregularly rugose sculpture, and with the propodeum sometimes bearing small teeth. Majors far larger than minors, with enormous quadrate or rectangular heads, alitrunk often somewhat queen-like in shape and suturation. [*Note:* Major workers frequently not collected.] (Widespread in Neotropics) . . . ***Erebomyrma*** (in part)

Workers varying slightly in size but essentially monomorphic. Dorsum and sides of petiolar node without sculpture except for coarse piligerous punctures that do not obscure the shining surface. Propodeum unarmed, and sides lacking coarsely granulose or irregularly rugose sculpture. Little known: at least some species associated with soil-dwelling termites. Queens are spectacularly larger than conspecific workers. (So far known only from South America within Neotropics; also occurs in African and Oriental tropics) . . . . . . . . . . . . . . . . . . . . . ***Carebara***

23(20) Anterior margin of clypeus without teeth, median lobe strongly and evenly convex. Dorsal surfaces of head and alitrunk without conspicuous sculpture and shining. Propodeum unarmed. (Obligate symbionts of several tropical trees, including species of *Cordia, Tococa,* and *Duroia;* South American tropical forests) . . . . . . . . . . . . . . . . . . . . . . ***Allomerus*** (in part)



Anterior margin of clypeus with several conspicuous, blunt, irregular teeth. Mandibles with 4 teeth, the basal tooth separated from the others by a wide gap and clearly pointing toward the clypeus at closure. Dorsum of head and mandibles covered with coarse longitudinal rugae; interrugal sculpture largely absent so that surface is strongly shining. Head and antennal scapes with moderately abundant, long, somewhat coarse erect hairs . . . . . . . . . . . . . . . . . . . . . . . . . . . . . . . . . . . ***Perissomyrmex***
[*Note:* A major myrmecological mystery: known only from one series taken in quarantined begonia tubers in Guatemala; *Perissomyrmex* may not be native to Neotropics.]

24(19) Clypeus lacking any trace of longitudinal carinae, the median lobe of the anterior margin strongly and evenly convex; propodeum unarmed; antennal club doubtfully distinct and may seem vaguely 2- or 3-jointed; workers monomorphic. (Obligate symbionts of several tropical trees, including species of *Cordia, Tococa,* and *Duroia* in South American forests) . . . . ***Allomerus*** (in part)

Clypeus bicarinate (with 2 longitudinal carinae), or without such carinae; the median lobe of the clypeal anterior margin emarginate, flat, truncate, or feebly rounded, not strongly and evenly convex; propodeum armed or unarmed; antennal club very distinct, either 2- or 3-jointed; workers polymorphic, dimorphic, or monomorphic . . . . . . . . . . . . . . . . . . . . . . . . . . . . . 25

25(24) Antennal club 3-jointed. Worker caste dimorphic. Minors with dorsum of head and pronotum sculptured and dull or feebly shiny, the sculpture usually coarsely granulose. Majors with hypostomal teeth. ("Subgenus *Decapheidole*" of Central and South America) . . . . . . . . . . . . . . . . . . . . . . . ***Pheidole*** (in part)

Antennal club 2-jointed. Much or all of cephalic and pronotal dorsum smooth and shiny. If completely sculptured and dull, the sculpturing is rugulose, not coarsely granulose . . . . . . . . . 26

26(25) Workers dimorphic. Minors with entire sides of petiolar node (and commonly all or most of the sides of the propodeum) covered with coarse, granulose sculpture, and dull or feebly shiny. Majors much larger than minors, with enormous quadrate heads. [*Note:* Majors often not collected.] (Widespread in Neotropics) . . . . . . . . . . . . . . . . . . . . . . . . . . . ***Erebomyrma*** (in part)

At least some portion of the sides of the petiolar node and/or the propodeal sides smooth or very weakly sculptured with the surface moderately to strongly shiny. Workers polymorphic or monomorphic, rarely dimorphic . . . . . . . . . . . . . . . . . . . . . 27

27(26) Clypeus with 2 subparallel longitudinal carinae (i.e., bicarinate) that curve sharply to meet at the longitudinal clypeal midline and form an abruptly raised, flat, oblong area. Anterior clypeal border resembles 3 sides of a hexagon in dorsal view and lacks teeth or any trace of a central emargination. Workers monomorphic. (Scattered records from tropical Central and South America, including Trinidad) . . . . . . . . . . . . . . . . . . . . . ***Carebarella***

Clypeus variable, but not as described above. Clypeus strongly or feebly bicarinate, in some cases lacking carinae. If bicarinate, carinae usually moderately to strongly divergent toward the anterior clypeal margin and often terminating as teeth projecting beyond the margin. One to several additional teeth may also be present, and the section of the margin between the carinae is frequently emarginate (sometimes weakly so). Area between the longitudinal carinae is commonly somewhat concave, not flat. Workers polymorphic, monomorphic, or rarely dimorphic. (Abundant and diverse throughout region) . . . . . . . ***Solenopsis***

*continued*

28(1) Head strongly narrowed anteriorly; mandibles elongate (not triangular), sometimes extremely so, and abruptly bent mesad near the apex; in dorsal view the occipital border of the head concave to deeply excised . . . . . . . . . . . . . . . . . . . . . . . . . . . **29**

Head often not strongly narrowed anteriorly; mandibles more or less triangular, not abruptly bent mesad near the apex; occipital border variable, frequently flat or convex . . . . . . . . . . . . . **30**

29(28) Mandibles very slender, usually almost as long as head capsule and sometimes longer, each bearing a prominent spine-like process on the ventral surface near the mandibular insertions, and with an apical fork consisting of 3 spiniform teeth. Workers monomorphic. (Small, beautifully delicate ants of moist tropical forests in Central and South America) . . . . . . . . ***Acanthognathus***

Mandibles stout, thick, notably flattened dorsoventrally, less than two-thirds head length, lacking basoventral spine-like processes, and with an apical fork of 2 stout teeth. Workers polymorphic. [*Note:* Large yellowish ants with prominent eyes found in canopies of South American forests.] . . . . . . . . . . . . . . ***Daceton***

30(28) Postpetiole attached to the anterior *dorsal* surface of the first gastral segment. Gastral dorsum in profile feebly convex, flat, or concave in contrast to the strongly convex ventral surface. In dorsal view the gaster is more or less heart-shaped and is capable of being flexed forward over the alitrunk. Petiole dorsoventrally flattened; node inconspicuous or absent. (Abundant and diverse throughout region) . . . . . . . . . . . . . . . ***Crematogaster***

Postpetiole attached to the anterior end (not the *dorsal* surface) of the first gastral segment. Dorsal and ventral surfaces approximately equally convex when viewed in profile, or ventral surface less convex than the dorsal. Gaster more or less teardrop-shaped in dorsal view, and not capable of being flexed so that it points forward over the alitrunk . . . . . . . . . . . . . . . . . . . . . . **31**

31(30) Frontal carinae greatly expanded, covering the cheeks in dorsal view, and subparallel or weakly diverging anteriorly. [*Note:* Sometimes the carinae form a saucer-shaped plate that constitutes dorsal surface of the head.] Propodeum armed with spines or prominent teeth . . . . . . . . . . . . . . . . . . . . . . . . . . **32**

Frontal carinae commonly not greatly expanded and covering cheeks in dorsal view. In cases where they are so expanded the propodeum lacks spines or prominent teeth. [*Note:* The propodeum may be angulate, bear rounded bumps, or have small, inconspicuous teeth at juncture of basal and declivitous faces.] . . . . . . . . . . . . . . . . . . . . . . . . . . . . . . . . . . . . . . . . . . . **36**

32(31) Eye situated beneath the antennal scrobe, which continues past the eye toward the occipital corner . . . . . . . . . . . . . . . . . . . **33**

Eye situated, at least in part, behind or slightly above the scrobe, which terminates in front of the eye and does not continue past it toward the occipital corner . . . . . . . . . . . . . . . . . . . . . **35**

33(32) Head and alitrunk lacking spines, with the exception of conspicuous propodeal teeth. Eyes small, fewer than 25 facets. Petiole nodiform. Mandible with alternating long and much shorter teeth along most of masticatory margin. (Rare, smooth, shiny ants known only from southern Brazil) . . . . . . . ***Phalacromyrmex***

Pronotum bearing prominent spines or at least a pair of teeth at the humeral angles. Petiole lacking a node, or else node rudimentary. Mandibles without alternating long and much shorter teeth along most of masticatory margin . . . . . . . . . . . . . . . . . **34**

34(33) Pair of prominent spines present at each posterior corner of the head, and at least 2 prominent spines present on pronotal dorsum. Eyes large, with more than 100 facets. (Large, black, arboreal, polymorphic ants, Panama and south) . . . . ***Cephalotes***

Lacking prominent spines near corners of head. Prominent spines also absent on pronotum, but with a tooth present at each humeral angle. Propodeal spines long. (Small to medium-sized, brownish, heavily sculptured, soil-dwelling ants in South America) . . . . . . . . . . . . . . . . . . . . . . . ***Blepharidatta*** (in part)

35(32) Occipital angles of the head each bispinose. Sides of petiole and postpetiole lacking spines or teeth. Eyes globose. (Arboreal; Amazon Basin, the Guianas) . . . . . . . . . . . . . ***Eucryptocerus***

Occipital angles each bearing a single spine, tooth, or lamina. Sides of petiole and postpetiole armed with spines or teeth. Eyes not globose. (Arboreal ants; abundant and diverse throughout region. Workers almost always dimorphic. Includes some of the most bizarre forms in the Neotropics. Many species are strongly flattened, are conspicuously spinose and/or have prominent laminae on the head, or have majors with saucer-shaped heads; includes *Hypocryptocerus* of Hispaniola, a likely synonym) . . . . . . . . . . . . . . . . . . . . . . . . . . . . . . . . . . . . . . ***Zacryptocerus***

36(31) Head in dorsal view roughly circular in outline (including closed mandibles), with deep antennal scrobes that diverge strongly toward the posterior margin of the head; eyes large; pronotum laterally marginate; humeri angulate; propodeal spines conspicuous; upper surfaces of head and alitrunk usually covered with coarse rugose and/or foveate sculpture, never smooth and strongly shining. (Medium-sized, robust, blackish, arboreal ants occurring throughout region) . . . . . . . . . ***Procryptocerus***

Not matching the above description . . . . . . . . . . . . . . . . . . **37**

37(36) A single carina present above (and/or sometimes slightly anterior to) the eye. [*Note:* Look carefully at proper magnification; the carina is weak but nevertheless present in some species.] If an antennal scrobe is present this carina may form part or all of its lateral (i.e., outer) border. Longitudinal rugae absent on sides and rear of head or dorsum of alitrunk. Antenna lacking a distinct 2- or 3-jointed club. (Tribe Attini in part) . . . . . . . . . **38**

Carina as described above usually absent; in cases where a carina is present, at least some longitudinal rugae are present on the sides and rear of head and/or the pronotal dorsum, and/or the antenna has a distinct 2- or 3-jointed club . . . . . . . . . . . . . . . . . . **45**

38(37) Pilosity on upper body surface absent or short and strongly appressed; the hairs usually flattened, strap- or scale-like . . . . **39**

Pilosity on upper body surface variable: several to many erect and/or recurved simple hairs present on dorsum of head and alitrunk, or else upper body surface covered with long, flexuous, decumbent hairs. Gaster often tuberculate . . . . . . . . . . . . . . . . **41**

39(38) Frontal carinae moderately to strongly expanded laterally (and often forward) forming lobes that, in some cases, may cover all or most of the cheeks in dorsal view. Antennal scrobes present, often prolonged to the occipital corners, which in a few cases form conspicuous blunt or triangular protuberances. Dorsum of petiole lacking spines or teeth. (Abundant and diverse throughout region) . . . . . . . . . . . . . . . . . . . . . . . ***Cyphomyrmex***

Frontal carinae modestly expanded laterally (and sometimes slightly forward) to form small lobes. Antennal scrobes absent or, in rare instances, very weakly developed. [*Note:* Carina above eye sometimes very weak.] Dorsum of petiole sometimes with spines or teeth . . . . . . . . . . . . . . . . . . . . . . . . . . . . . . . . . **40**

*continued*

**40(39)** Occipital corners armed with 1–3 short teeth. Petiolar dorsum bearing a pair of finger-like teeth projecting upward and rearward. Dorsum of alitrunk with conspicuous teeth or spines. (2 species, known only from Brazil and Argentina) . . . ***Mycetarotes***

Occipital corners and petiolar dorsum unarmed. Alitrunk unarmed or at most bearing 2 small pronotal teeth and low, rounded bumps on the mesonotum. (*M. conformis* occurs in sandy beach habitats throughout the region; several other species occur in South America) . . . . . . . . . . . . . . . . . . . . . ***Mycetophylax***

**41(38)** Pilosity and pubescence on head and alitrunk abundant, long, flexuous, and lying more or less close to the body, which usually appears "silky." Head and gaster lacking spines, teeth, or tubercles. Alitrunk lacking spines, teeth or tubercles except for a pair of blunt, cone-like teeth on the mesonotum (a second, smaller pair may also be present just behind the first pair). Small propodeal teeth also sometimes present; rarely the humeri take the form of small, blunt teeth. (Widespread in Central and South America) . . . . . . . . . . . . . . . . . . . . . ***Sericomyrmex***

[*Note:* Several *Trachymyrmex* species exhibit long, flexuous pilosity but may be distinguished by head shape, as follows: their head is square or slightly longer than broad and narrows slightly toward the mandibular insertions. In *Sericomyrmex* the head is broader than long, more or less cordate, and narrows significantly toward the mandibles.]

Pilosity on head and alitrunk erect and/or consisting of coarse recurved hairs, which sometimes occur primarily on tubercles and teeth. Head often bearing spines, teeth, or tubercles; gaster frequently tuberculate. Alitrunk bearing spines, teeth, or tubercles on the pronotum and often on the propodeum as well . . . . **42**

**42(41)** Workers monomorphic or at most very feebly polymorphic . . . **43**

Workers moderately to extremely polymorphic . . . . . . . . . . . **44**

**43(42)** Frontal carinae projecting forward beyond the antennal insertions (as well as laterally), forming conspicuous lobes that overhang the sides of the clypeus in dorsal view. (Uncommon, small species known only from south central United States, Brazil, and Argentina) . . . . . . . . . . . . . . . . . . . . . . . . ***Mycetosoritis***

Frontal carinae expanded laterally to form more or less conspicuous lobes, but not projecting forward beyond the antennal insertions. Alitrunk and gaster often conspicuously tuberculate; alitrunk frequently with teeth or spines. (Abundant and diverse throughout region) . . . . . . . . . . . . . . . . . . . . . . . . ***Trachymyrmex***

**44(42)** Workers moderately polymorphic. Dorsum of alitrunk bearing 4 pairs of teeth, spines, or large tubercles. Occiput and/or first gastral tergite tuberculate. (Southwestern United States south throughout region, including some Caribbean islands, such as Guadeloupe and Tobago) . . . . . . . . . . . . . . . . ***Acromyrmex***

Workers extremely polymorphic: head width of minimas often under 1 mm and those of majors often over 4.5 mm. Dorsum of alitrunk bearing 3 pairs of teeth or spines. Occiput and first gastral tergite smooth, not tuberculate. (Louisiana and southwestern United States south through continental Central and South America, as well as Cuba, Trinidad, and Tobago) . . . . . . ***Atta***

**45(37)** Pronotum and anterior of mesonotum bearing a conspicuous circlet of 8 teeth or spines (sometimes with 2 teeth in middle of the circlet), or else head and alitrunk tuberculate and with pilosity consisting of distinctive erect or recurved clavate or scale-like hairs. (Tribe Attini in part) . . . . . . . . . . . . . . . . . . . . . **46**

Lacking both the above-described traits . . . . . . . . . . . . . . . **47**

**46(45)** Pronotum and anterior of mesonotum bearing a conspicuous circlet of 8 (very rarely 6) upward-directed teeth or spines (sometimes 2 teeth present in middle of the circlet). (Several species: inconspicuous but apparently common throughout most of region) . . . . . . . . . . . . . . . . . . . . . . . . . . . . . . . . . ***Mycocepurus***

Head and alitrunk moderately to weakly tuberculate. Pilosity on head and alitrunk consisting of distinctive erect or recurved clavate or scale-like hairs that often are concentrated on the tubercles and occurring also (without tubercles) on the gaster. (Continental Central and South America, as well as Trinidad) . . . . . . . . . . . . . . . . . . . . . . . . . ***Myrmicocrypta***

**47(45)** Frontal carinae long, extending posteriorly past the eye, surpassing, reaching, or almost reaching the vertex . . . . . . . . . . . **48**

Frontal carinae short, not reaching or almost reaching the vertex . . . . . . . . . . . . . . . . . . . . . . . . . . . . . . . . . . . . . . **51**

**48(47)** Antennal scrobe very distinct, with a pronounced carinate lateral border. Humeri rounded or toothed, not angulate. Antenna with distinct 2-jointed apical club . . . . . . . . . . . . . . . . . . . . . **49**

Antennal scrobe lacking or weakly developed with a feebly carinate or indistinct lateral border. In rare cases where scrobe looks distinct, the humeri are angulate. Antennae variable, often with 2- or 3-jointed club . . . . . . . . . . . . . . . . . . . . . . . . . . . **50**

**49(48)** Antennal scrobe deep (basal two-thirds to three-fourths of scape partly sheltered by overhanging edge of frontal carina) and extending to the occipital corner, which is prolonged to form a blunt protuberance. Area within scrobe dull or weakly shiny. Pilosity on upper body surface erect, long, coarse, and sparse. (Propodeal spines long. Small to medium-sized, brownish, heavily sculptured South American ants) . . . . . . . . . . . . . . . . . . . . . . . . . . . . . . . . . . . . . . ***Blepharidatta*** (in part)

Antennal scrobe shallow (basal third or less of scape slightly sheltered by overhanging edge of frontal carina) and not reaching the occipital corners, which are rounded and do not form protuberances. Area within scrobe smooth and strongly shining in contrast to remainder of head, which is heavily sculptured and dull or weakly shiny. Pilosity on head and alitrunk (and sometimes gaster) long, fine, erect, and abundant. (Central and South American tropical forests) . . . . . . . . . . . . . . . ***Lachnomyrmex***

**50(48)** Mandibles with 5 teeth, including the basal tooth or angle, which is usually hidden from view when the mandibles are closed. Humeri of pronotum angulate. Sculpture on dorsum of alitrunk variable, usually weakly rugose (longitudinal or reticulate). (Small, yellow or orange—rarely brown—ants abundant throughout region in both disturbed and undisturbed habitats) . . ***Wasmannia***

Mandibles with at least 6 (usually 7) teeth. Humeri well rounded. Sculpture on dorsum of alitrunk strongly and coarsely rugoreticulate. (4 species known from the southwestern United States and Mexico; formerly *Xiphomyrmex*) . . . . . . . . . ***Tetramorium***

**51(47)** Clypeus with 2 longitudinal carinae terminating as teeth that project beyond the anterior border; each clypeal tooth accompanied by a small lateral denticle. Petiole nodiform. Antenna with 3-jointed club. Propodeum angulate or dentate. Mandibles with 4 teeth. (Little-known, terrestrial, South American ants; a few species appear to be symbiotic with *Pheidole* colonies) . . . . . . . . . . . . . . . . . . . . . . . . . . . . . . . . . . . . . . . ***Oxyepoecus***

Clypeus generally not bicarinate, and lacking clypeal teeth. In in-

*continued*

stances where the clypeus appears bicarinate, clypeal teeth are either absent or else the anterior margin lacks the accompanying denticles and the petiolar node is absent or rudimentary . . . **52**

**52(51)** Propodeum lacking teeth or spines at juncture of basal and declivitous faces, in profile rounded or angulate . . . . . . . . . . . . **53**

Propodeum armed with teeth (sometimes small) or spines at juncture of basal and declivitous faces . . . . . . . . . . . . . . . . . . **56**

**53(52)** Upper body surfaces covered with pilosity that is moderate to dense, long, flexuous, and mostly decumbent to appressed. Mandibles with at least 8 teeth (usually more). In dorsal view frontal carinae forming prominent, horizontal, rounded lobes strongly elevated over the plane of the cheeks. (Abundant throughout region in moist, tropical forests; colonies small; ants slender, frequently slow moving, and prone to feigning death) . . . . . . . . . . . . . . . . . . . . . . . . . . . . . . . . ***Apterostigma***

Pilosity not as described above, usually short and sparse or long and mostly erect. Mandibles with 6 or fewer teeth. Frontal lobes not particularly prominent, only modestly elevated above the plane of the cheeks . . . . . . . . . . . . . . . . . . . . . . . . . . **54**

**54(53)** Petiole in side view subcylindrical, the node absent or rudimentary. Clypeus with median emargination that is sometimes between 2 broad clypeal teeth or blunt lobes. Metanotal region moderately to profoundly impressed, forming trough between the promesonotum and the propodeum, which appear to be separate convexities in profile. Workers monomorphic. (Arboreal; most records are from Mexico, Central America, and the Caribbean, with only a few from South America) . . . . . . . . . . . . ***Xenomyrmex***

Petiole in side view not subcylindrical and possessing a distinct node. Anterior margin of clypeus usually more or less evenly convex and always without teeth or blunt lobes; rarely a very weak median emargination. Workers usually weakly polymorphic . . . . . . . . . . . . . . . . . . . . . . . . . . . . . . . . . . . **55**

**55(54)** Eyes very small but distinct, consisting of 6–10 or more discrete facets. Mandible with 4–5 teeth, the basal tooth not offset. In profile the promesonotum and the propodeum form a flattened, continuous profile interrupted only by a weak notch or impression at the metanotal region. [*Note:* In one species the metanotal region is profoundly impressed, resulting in a discontinuous alitrunk profile.] Petiolar node in profile set off from a distinct anterior peduncle, the node occupying one-half to two-thirds of the total petiolar length. (Largely subterranean yellow or orange ants with 3-jointed antennal club common throughout Central and South America, including Trinidad. Workers very small compared to queens) . . . . . . . . . . . . . . . . . . . . . . ***Tranopelta***

Eyes rudimentary, not consisting of discrete facets, or apparently absent. Mandibles with 4 teeth, the basal tooth somewhat offset. Profile of alitrunk not continuous, the promesonotum and propodeum forming separate convexities divided by a slight to moderate impression at the metanotal region. Petiolar node not notably set off from the short anterior peduncle, occupying three-fourths or more of the total petiolar length. (Subterranean, yellow ants with a 3-jointed antennal club known only from 2 recent collections in the Chiricahua Mountains of Arizona. Genus almost certainly present in Mexico. Workers minute compared to the queen. Apparently close to the Old World genus *Anillomyrma*, possibly a new genus: S. Cover, personal communication) . . . . . . . . . . . . . . . . . . . . . . . . . . New genus ( ? )

**56(52)** Workers dimorphic; minors small, less than 3 mm in total length; antennae with a distinctive 2-jointed apical club; eyes minute (fewer than 5 facets, often indistinct); mandibles with 4 teeth. In addition, majors very much larger, with enormous quadrate heads and a less distinct antennal club. (Widespread in continental Mexico, Central and South America, and Trinidad) [*Note:* Majors often not collected.] . . . . . . . . . . ***Erebomyrma*** (in part)

Antennal club clearly 3-jointed or indistinctly clubbed; eyes usually with more than 20 facets . . . . . . . . . . . . . . . . . . . . . . . **57**

**57(56)** In profile the propodeum notably depressed below the common level of the pronotum and the mesonotum and separated by an abrupt "step" at the metanotal region. Antennal club clearly 3-jointed. Workers dimorphic: majors with large heads and hypostomal teeth. ("Subgenus *Hendecapheidole*" of Central and South America) . . . . . . . . . . . . . . . . . . . . . . . ***Pheidole*** (in part)

Propodeum not abruptly depressed below level of pronotum and mesonotum. Metanotal region without an impression or impressed, but not as described above. Workers monomorphic. Antennal club 3-jointed or indistinct . . . . . . . . . . . . . . . . . . **58**

**58(57)** In profile the metanotal region (located between the mesonotum and the propodeum) slightly (but distinctly) or moderately impressed; pronotal dorsum largely smooth and moderately to strongly shining . . . . . . . . . . . . . . . . . . . . . . . . . . . . **59**

In profile metanotal region generally not impressed; if a faint impression is present the pronotal dorsum is sculptured and dull or feebly shining . . . . . . . . . . . . . . . . . . . . . . . . . . . . . **60**

**59(58)** Mandibles with 4 teeth; the second subapical tooth distinctly smaller than the other 3 teeth, rarely reduced to a denticle. Humeri of pronotum moderately to strongly angulate. (Little-known South American genus with 2 species) . . . . . . . . . . . . . . . . . . . . . . . . . . . . . . . . . . . . . . . . . . . ***Ochetomyrmex***

Mandibles with 4 or 5 teeth; the subapical teeth subequal in size or successively decreasing in size. Humeri well rounded. (Obligate symbionts of several South American trees, including species of *Cordia*, *Tococa*, and *Duroia*) . . . . . . . . . . . ***Allomerus*** (in part)

**60(58)** Humeri distinctly angulate. Frequently some hairs on the petiole and/or the postpetiole are swollen at the base or borne on tubercles, giving the surface a tuberculate appearance. Arboreal or nesting in twigs or hollow stems in scrubby secondary vegetation. (Common throughout much of region) . . . . . . . . . . . . . . . . . . . . . . . . . . . . . . . . . . . . . ***Nesomyrmex*** (in part)

Humeri rounded or at most weakly angulate. Petiole and postpetiole not tuberculate . . . . . . . . . . . . . . . . . . . . . . . . **61**

**61(60)** Mandibles with 5 teeth. Alitrunk weakly convex in profile. Petiole with a more or less distinct anterior peduncle. Strigil lacking a basal tooth. ("*Myrafant*," possibly a distinct genus; a few species in Mexican highlands) . . . . . . . . . . . . . . . . . . . . . . . . . . . . . . . . . . . . . . . . . ***Leptothorax (= Myrafant)*** (in part)

Mandibles with 6 teeth. Alitrunk flattened in profile. Petiole lacking a distinct peduncle. Strigil with a basal tooth. (*Leptothorax* strict sense; possibly occurs at high elevations in the Mexican highlands) . . . . . . . . . . . . . . . . . . . . . . . ***Leptothorax*** (in part)

**62(1)** Frontal carinae greatly expanded and forming lobes that project forward over the mandibles, or else antennal scape abruptly bent (almost 90 degrees) near base to form an "elbow," strongly flattened (at least near the bend), and with prominent clavate or strap-like hairs . . . . . . . . . . . . . . . . . . . . . . . . . . . . . **63**

Frontal carinae not forming lobes that overhang mandibles. Antennal scape straight or gradually and evenly bent near base, or if abruptly bent and somewhat flattened, the scape lacks conspicuous clavate or strap-like hairs . . . . . . . . . . . . . . . . . . . . **65**

*continued*

**63(62)** Frontal carinae greatly expanded to form prominent lobes that project forward over the mandibles. Antennal scrobes extremely deep; eye located on or near lower scrobal margin. Scape not conspicuously flattened and is widest toward distal end. Body covered with abundant, long, erect, clavate hairs. (Rare: records from Panama, Ecuador, and Brazil) . . . . . . . . . . ***Stegomyrmex***

Frontal carinae variable, but not forming lobes as described above. Antennal scrobes distinct, sometimes deep; eye located on or near upper scrobal margin. Scape abruptly bent near base (almost 90 degrees), conspicuously flattened at least near the bend, and with abundant clavate or strap-like hairs . . . . . . . . . . **64**

**64(63)** Dorsal surface of the basal half of each mandible bearing a deep, transverse, oblique groove running outward from the masticatory border. Apical portion of mandible curved sharply downward. (Very rare, known only from Brazil) . . . ***Creightonidris***

Mandibles lacking a conspicuous dorsal, transverse groove; dorsal surfaces only moderately convex to their apices. (Central and South America, including Trinidad) . . . . . . . . . . . ***Basiceros***

**65(62)** In profile petiole short and subcylindrical, lacking an anterior peduncle, and with a rudimentary node (node often absent); humeri moderately to sharply angulate. Propodeum armed with spines or short teeth. (Mexican highlands) . . . . . ***Myrmecina***

Petiole distinctly nodiform; if the node is very low and rounded, an anterior peduncle is also present; humeri usually rounded, rarely angulate; propodeal armature variable . . . . . . . . . . . . . . . **66**

**66(65)** Combining the following traits: Mandibles with strongly oblique cutting edge; the node in profile very low, much rounded, and much longer than high (see the *Hylomyrma* generic figure). [*Note:* Rarely the node is vaguely triangular and reaches its highest point at or toward its anterior border.] Alitrunk in profile usually evenly convex. (Mostly found in reasonably moist tropical forests in Central and South America, including Trinidad) . . . . . . . . . . . . . . . . . . . . . . . . . . . . . . . . . . . . . . . ***Hylomyrma***

Mandibles with transverse cutting margin. Petiolar node variable, but not as described above. Profile of alitrunk often discontinuous or flattened, sometimes evenly convex . . . . . . . . . . . **67**

**67(66)** Combining the following traits: Petiolar node sharply set off from the long, distinctive anterior peduncle. Node in side view roughly triangular, usually with a short, steep anterior face and a longer, gradually sloping posterior face, and reaching its highest point at or toward its anterior surface. [*Note:* Where the node apex is rounded, care may be needed to determine highest point.] Psammophore usually well developed; if absent, erect hairs are often present on the gula and alitrunk is extensively and coarsely rugoreticulate (see the *Pogonomyrmex* and *Ephebomyrmex* generic figures). (Mostly deserts and other dry habitats from southern United States to Argentina) . . . . . . . . . . . . . . . **68**

Anterior peduncle sometimes long, often short or absent. Node in side view variable in shape, often roughly rectangular or quadrate. If triangular, it generally reaches its highest point at its longitudinal midpoint or toward its posterior surface. In rare cases where the peduncle is long and the node is triangular and reaches its highest point toward its anterior surface, then the alitrunk is not coarsely and extensively rugoreticulate (although longitudinal or transverse rugae or weak reticulate sculpture may be present). True psammophore is rarely present, although scattered erect hairs on the gula may occur . . . . . . . . . . . . . . **69**

**68(67)** Mandibles with 6 teeth, the ultimate basal tooth reduced in size (rarely with 5 teeth and no reduced ultimate basal tooth). Psammophore usually weakly developed or absent. Head and alitrunk extensively and coarsely rugoreticulate. . . . . . . ***Ephebomyrmex***

Mandibles usually with 7 teeth, the ultimate basal tooth not reduced in size (rarely with 6 teeth with the ultimate basal tooth not reduced). Psammophore strongly developed. Head and alitrunk not extensively and coarsely rugoreticulate, usually finely rugose . . . . . . . . . . . . . . . . . . . . . . . . . ***Pogonomyrmex***

**69(67)** Head and alitrunk lacking visible erect hairs on their dorsal surfaces. Head and alitrunk dull and densely punctate. In side view the clypeal anterior margin elevated above the mandibles and extending slightly over them, forming a small "shelf." Postpetiole often enlarged in dorsal view. Metanotal suture present and slightly to moderately impressed. Small monomorphic ants often found in early successional and disturbed habitats. Most species are soil-dwelling, a few are arboreal. (An Old World genus widely introduced throughout the region, especially in the Caribbean) . . . . . . . . . . . . . . . . . . . . . . . . . . . ***Cardiocondyla***

Head and alitrunk bearing at least a few (often many) erect or suberect hairs . . . . . . . . . . . . . . . . . . . . . . . . . . . . . . . . **70**

**70(69)** Median lobe of clypeus bicarinate and strongly elevated, forming a narrow plate fused with the frontal carinae. This plate terminates as 2 blunt teeth on anterior clypeal margin. The sides of the clypeus descend sharply to the level of the mandibular insertions where, in some cases, 2 more clypeal teeth are present. Dorsum of head and usually of alitrunk conspicuously rugoreticulate or longitudinally rugose. (Widespread throughout continental Mexico, Central America, and South America. Soil and litter ants of tropical forests) . . . . . . . . . . . . . . . . . . . ***Adelomyrmex***

Clypeus not bicarinate, or if bicarinate either the head and alitrunk are not conspicuously rugoreticulate or longitudinally rugose, or else the carinae do not terminate as teeth on the anterior clypeal margin . . . . . . . . . . . . . . . . . . . . . . . . . . . . . . . . . **71**

**71(70)** Dorsum of alitrunk flattened or convex in profile, but without impressed sutures. [*Note:* Small transverse ridge may be present on the propodeal dorsum.] . . . . . . . . . . . . . . . . . . . . . . . . **72**

Dorsum of alitrunk variously shaped in profile, but never forming a continuous surface; its outline always interrupted by 1 or more sutural impressions . . . . . . . . . . . . . . . . . . . . . . . . . **77**

**72(71)** Clypeus with 2 longitudinal carinae that diverge toward the anterior margin. Carinae often fail to reach anterior margin and never form clypeal teeth. Alitrunk in profile often strongly convex. (Widespread throughout region. Inconspicuous soil- and litter-dwelling ants of tropical forests) . . . . . . . . . . . ***Rogeria***

Clypeus not bicarinate, though sometimes longitudinally rugulose . . . . . . . . . . . . . . . . . . . . . . . . . . . . . . . . . . . . . . . **73**

**73(72)** Frontal carinae long, extending posteriorly past the eye and reaching or almost reaching the vertex, and/or the clypeus conspicuously longitudinally rugulose . . . . . . . . . . . . . . . . . . . . **74**

Frontal carinae short, not extending past the eye and never approaching the vertex closely. Clypeus variously sculptured or smooth and shining, but never with conspicuous longitudinal rugulae . . . . . . . . . . . . . . . . . . . . . . . . . . . . . . . . . . . **75**

**74(73)** Frontal carinae long, extending posteriorly past the eye and reaching or almost reaching the vertex, and/or the clypeus longitudinally rugulose, its rugulae clearly continuous with those between the frontal carinae and thus extending all the way to the rear of the head. Posterolateral portions of clypeus bordering the antennal sockets forming a thin, vertical ridge to create the impression

*continued*

of a deep pit surrounding the socket. Antennal club 3-segmented. (Several Old World species with long frontal carinae introduced widely throughout the region. *T. caespitum*, with short frontal carinae, has been introduced, but may not be established, in Chile, Belize, and Mexico) . . . . . . . . . ***Tetramorium***

Frontal carinae always short and never almost reaching the vertex. Clypeus longitudinally rugulose, its rugulae often not continuous to rear of head. Posterior lateral portions of clypeus that border antennal sockets do not form a vertical ridge; thus socket does not appear to be surrounded by a deep pit, especially anteriorly. Antennal club indistinct or absent. (Mexican highlands) . . . . . . . . . . . . . . . . . . . . . . . . . . . . ***Myrmica*** (in part)

**75(73)** Humeri distinctly angulate. Frequently some hairs on the petiole and/or postpetiole are swollen at the base or borne on tubercles, giving the surface a tuberculate appearance. (Arboreal or nesting in twigs or hollow stems in scrubby, secondary vegetation. Common throughout most of region) . . . ***Nesomyrmex*** (in part)

Humeri rounded or at most weakly angulate. Clypeus in side view usually somewhat rounded. Petiole and postpetiole never tuberculate . . . . . . . . . . . . . . . . . . . . . . . . . . . . . . . . . . **76**

**76(75)** Alitrunk smoothly and strongly convex in side view. Anterior peduncle of petiole frequently long and thin (sometimes extremely so), and often set off sharply from the node. In dorsal view, postpetiole often bell-shaped and/or broad in relation to the gaster (in some cases two-thirds to three-fourths or more of the maximum gastral diameter). Propodeal spines, when present, often long. (Most diverse and abundant in Cuba, common on the larger Caribbean islands, a few species in Mexico and Central America. Some species are brightly colored and strikingly beautiful) . . . . . . . . . . . . . . . . . . . . . . . . . . . . ***Macromischa***

Alitrunk weakly convex in profile. Anterior peduncle of petiole short, thick, and not sharply set off from the node. Postpetiole usually less than two-thirds the maximum gastral diameter in width. Propodeal spines or teeth short or, at most, only moderately long. ("Subgenus *Myrafant*," possibly a distinct genus; a few species in Mexican and Guatemalan highlands) . . . . . . . . . . . . . . . . . . . . . . . . . . . . . . . . . ***Leptothorax*** (in part)

**77(71)** Mandibles with 3 or 4 teeth. Clypeus almost always with 2 longitudinal carinae (sometimes weak, rarely absent) that often end as teeth on the anterior clypeal border. Antennae with 3-segmented club. Propodeum lacking teeth or spines, evenly rounded or weakly angulate in profile. (Throughout region; several Old World tramp species widespread and abundant) . . . . . . . . . . . . . . . . . . . . . . . . . . . . . . . . . . . . . . . . . . . ***Monomorium***

Mandibles usually with 5 or more teeth (if only 4 teeth are present, propodeum with short but distinct teeth). Propodeum frequently bearing teeth or spines . . . . . . . . . . . . . . . . . . . . . . . . **78**

**78(77)** Antenna with 3- or 4-segmented apical club . . . . . . . . . . . . . **79**

Antenna lacking a distinct antennal club, the terminal segments gradually enlarging toward the apex . . . . . . . . . . . . . . . . **83**

**79(78)** Antennal club 4-segmented. Clypeus usually with 2 longitudinal carinae that do not form teeth on the anterior margin (carinae rarely indistinct or absent). Workers monomorphic. Propodeum usually armed with small teeth, rarely sharply angulate or unarmed. (Mexico and Central America in forest soils and litter) . . . . . . . . . . . . . . . . . . . . . . . . . . . . . . . . . . ***Stenamma***

Antennal club almost always clearly 3-segmented, and often thicker than the remainder of the funiculus. [*Note:* If 4-segmented, then the worker caste is dimorphic.] Clypeus sometimes bicarinate . . . . . . . . . . . . . . . . . . . . . . . . . . . . . . . . . . . . . . . . **80**

**TABLE 2–13** Neotropical (*continued*)

**80(79)** Clypeus longitudinally and conspicuously bicarinate (the carinae sometimes forming ridges marked by transverse rugules), its upper surface moderately to strongly truncated. Monomorphic. (Genera limited to Argentina and Chile) . . . . . . . . . . . . . **81**

Clypeus usually not bicarinate, its upper surface more or less evenly convex, not truncated. If bicarinate, then inconspicuously so, and in addition the workers are strongly dimorphic, with a large-headed major caste. (Widespread throughout Neotropics) . . . . . . . . . . . . . . . . . . . . . . . . . . . . . . . . . . . . **82**

**81(80)** Mandibles with 5 teeth; propodeum weakly angulate or with small broad teeth that point vertically; workers weakly polymorphic. (Very poorly known: Chile) . . . . . . . . . . . . ***Nothidris***

Mandibles with 4, 5, or 6 teeth; propodeum with small teeth pointing rearward and slightly upward; workers monomorphic. (Known only from Argentina and Chile) . . . . ***Antichthonidris***

**82(80)** Workers dimorphic (rarely polymorphic). Majors with large heads and hypostomal teeth. Minors commonly with propodeal teeth or spines (rarely propodeum angulate or evenly convex), and with at least the sides of the propodeum sculptured and feebly shiny to opaque (rarely smooth and shiny). Dorsum of head and alitrunk often sculptured and dull or feebly shiny, sometimes smooth and shiny. Mandibles with 6 or more teeth or denticles. (Ubiquitous and spectacularly diverse throughout region) . . . . . . . . . . . . . . . . . . . . . . . . . . . . . ***Pheidole*** (in part)

Workers monomorphic. Propodeum in profile usually evenly convex or angulate, very rarely with small, broad teeth. Propodeal sides smooth and shiny. Dorsum of head and alitrunk smooth and moderately to strongly shiny. Mandibles often with 5 teeth, sometimes with more teeth and/or numerous denticles. (Central and South America) . . . . . . . . . . . . ***Megalomyrmex*** (in part)

**83(78)** Propodeum not armed with teeth or spines, usually rounded or angulate, very rarely with small, broad teeth; metanotal impression distinct, notch-like, and often deep, clearly separating the propodeal profile from that of the rest of the alitrunk; body predominantly smooth and shiny, the propodeal sides always smooth and shiny; postpetiole in dorsal view somewhat constricted at gastral juncture. (Central and South America) . . . . . . . . . . . . . . . . . . . . . . . . . . . . ***Megalomyrmex*** (in part)

Propodeum usually armed with teeth (sometimes small) or spines; propodeal sides usually sculptured and not strongly shiny; metanotal impression variable; postpetiole in dorsal view either strongly constricted or else not at all constricted at gastral juncture . . . . . . . . . . . . . . . . . . . . . . . . . . . . . . . . . . . . **84**

**84(83)** In profile, the metanotal region weakly to moderately impressed. Propodeum barely differentiated from remainder of alitrunk and scarcely (sometimes slightly) depressed below the level of the promesonotum in profile. Antennal scape often bent abruptly near the base (sometimes 90 degrees) and bearing a more or less obvious lamina at the bend. Postpetiole in dorsal view strongly constricted at gastral juncture. (Mexican highlands) . . . . . . . . . . . . . . . . . . . . . . . . . . . . . . . . . . . . ***Myrmica*** (in part)

Metanotal impression variable. Propodeum usually strongly differentiated from the remainder of the alitrunk and, in profile, always substantially depressed below elevation of promesonotum. Antennal scape not abruptly bent at base, rarely possessing a lamina near base . . . . . . . . . . . . . . . . . . . . . . . . . . . . **85**

**85(84)** Head quadrate, not notably narrower behind the eyes than in front. Psammophore often present. Workers sometimes polymorphic. (Desert harvesting ants, "*Veromessor*"; Mexico) . . . . . . . . . . . . . . . . . . . . . . . . . . . . . . . . . . . . . . . . . . . ***Messor***

*continued*

Head longer than broad and often notably narrower behind the eyes than in front. Psammophore absent. Workers monomorphic or at most very feebly polymorphic. (Mexico and Central America. Mostly in forests and other mesic habitats) . . . . . . . . . . . . . . . . . . . . . . . . . . . . . . . . . . . . . . . . . . . . . . . . . . . . ***Aphaenogaster***

[Note: Two large (10–12 mm) species of *Aphaenogaster* common in Mexican deserts. Frequently referred to under the generic name *Novomessor* and distinguishable from other *Aphaenogaster* by the rudimentary metanotal impression (often absent) that does not interrupt the smooth, sinuous profile of the alitrunk. In other *Aphaenogaster* the metanotal impression is significant, and the profile of the alitrunk is not smooth and sinuous.]

### SUBFAMILY DOLICHODERINAE

1 Workers eyeless, pale yellow in color, and about 2 mm long. Maxillary and labial palps reduced: 3- and 2-jointed respectively. Body lacking teeth or spines. Petiolar scale distinct and strongly inclined forward. (Rare: known only from Argentina) . . . . . . . . . . . . . . . . . . . . . . . . . . . . . . . . . . . . . . . . . . ***Anillidris***

Workers possessing eyes; color and size variable. Maxillary palps frequently 6-jointed, very rarely 3-jointed. Body sometimes bearing teeth or spines . . . . . . . . . . . . . . . . . . . . . . . . . **2**

2(1) Hypostomal tooth-like projection adjacent to ventral surface of mandibular insertions lacking. Integument thin and flexible, rarely conspicuously sculptured. Thoracic spines generally absent, although a tooth-like protuberance may be present at the juncture of the propodeal faces. (Tribe Tapinomini) . . . . . . **3**

Hypostomal tooth-like projection present adjacent to ventral surface of mandibular insertions. Integument thickened, sometimes conspicuously sculptured. Thoracic spines often present. (Tribe Dolichoderini) . . . . . . . . . . . . . . . . . . . . . . . . . **10**

3(2) Viewed in profile, the propodeum bearing a tooth-like protuberance that projects more or less vertically and/or rearward from the juncture of the basal and declivitous faces. Third joint of the maxillary palps unusually elongate, commonly as long as or longer than the 3 succeeding distal joints combined . . . . . . . **4**

Propodeum unarmed; the juncture of the basal and declivitous faces evenly rounded or angulate. Third joint of maxillary palps not unusually long, notably shorter than the 3 succeeding distal joints combined . . . . . . . . . . . . . . . . . . . . . . . . . . . . . **5**

4(3) In profile the propodeal tooth projects rearward (and in some instances somewhat upward). Petiole nodiform, not scale-like, its crest rounded and blunt. (Most records are from arid areas in southern South America) . . . . . . . . . . . . . . . ***Dorymyrmex***

In profile the propodeal tooth projects upward (and sometimes slightly forward or rearward). Petiole scale-like, not nodiform, its crest moderately to very sharp. (Throughout region in dry, open habitats; includes "*Araucomyrmex*") . . . . . . . ***Conomyrma***

5(3) Metanotal region (narrow, transverse band between mesonotum and propodeum) not impressed dorsally; mesonotum and propodeum forming a smooth, continuous, flat or convex profile. Workers moderately polymorphic. (Mexico) . . . ***Liometopum***

Metanotal region slightly to profoundly impressed dorsally, forming a shallow, concave impression, an angle, or a notch between the mesonotal and propodeal profiles. Workers monomorphic or polymorphic . . . . . . . . . . . . . . . . . . . . . . . . . . . . . . . . **6**

6(5) Petiolar scale or node in side view often moderately conspicuous and longer than tall. Mesonotum almost always raised slightly above level of the pronotum. Workers moderately polymorphic. Pronotum often with numerous long, erect hairs. Occipital margin moderately to strongly concave. (Arboreal; throughout region) . . . . . . . . . . . . . . . . . . . . . . . . . . . . . . . . . ***Azteca***

Petiolar scale usually small and thin and taller than long, or absent altogether. Mesonotum seldom raised above level of pronotum. Workers monomorphic. Pronotum commonly with 6 or fewer erect hairs. Occipital margin flat or weakly to moderately concave . . . . . . . . . . . . . . . . . . . . . . . . . . . . . . . . . . . . . . **7**

7(6) Petiolar scale absent (in rare cases present but minuscule); basal face of propodeum shorter than the declivitous face; the juncture of 2 propodeal faces in profile more or less angular, rarely evenly convex; apical 3 or 4 mandibular teeth large, succeeding teeth progressively much smaller toward the base (often becoming mere denticles), with basal tooth or angle lacking . . . . . . . . **8**

Petiolar scale present, though sometimes small. Basal face of propodeum as long as or longer than the declivitous face; the juncture of the two faces in profile usually evenly convex, rarely weakly angular; mandibles possessing basal tooth or angle, though sometimes small and inconspicuous; the masticatory margins with either 5 prominent teeth including the basal tooth or angle (1 to several denticles may also be present) or else with the apical and first subapical teeth large and the succeeding teeth notably smaller; the third, fourth, and basal tooth or angle separated by several to numerous denticles, and often little larger than the denticles themselves . . . . . . . . . . . . . . . . . **9**

8(7) Four gastral tergites visible in dorsal view. Fifth (terminal) tergite reflexed ventrally, so that the anus is not located at the terminus of the gaster viewed in profile, but under the "shelf" of the reflexed fifth tergite. [*Note:* This character works well with fresh and alcoholic material, and also with pointed specimens in which the gaster is not shriveled or distorted. Where the gaster is shriveled, rehydration and dissection may be necessary.] (Occurs throughout region) . . . . . . . . . . . . . . . . ***Tapinoma***

Five gastral tergites visible in dorsal view (the terminal one often small). The terminal tergite is not reflexed ventrally, thus the anus is located at the true terminus of the gaster in side view. (*Technomyrmex fulvum,* formerly *Tapinoma fulvum,* is known from Barro Colorado Island in Panama. *T. albipes,* a common Indo-Pacific tramp species, is likely to have been introduced somewhere in the region) . . . . . . . . . . ***Technomyrmex***

9(7) Mandibles with 5 prominent teeth (including the basal tooth or angle), with 1–4 denticles separating the fourth from the third and the basal tooth or angle. [*Note:* Denticles often inconspicuous or absent.] In profile clypeus bearing 2 or more long, straight, erect hairs projecting forward and/or upward, and 2 or more hairs (longer as a rule than the straight hairs) originating at or near the anterior margin that curves downward over the upper surfaces of the closed mandibles. Metanotal region slightly to moderately impressed; the propodeal profile slightly or moderately distinct from remainder of the alitrunk. (Xeric habitats throughout the region) . . . . . . . . . . . . . . ***Forelius***

Mandibles usually bearing 5 teeth, with the apical and first subapical teeth prominent and notably larger than the succeeding subapical teeth. Third, fourth, and basal tooth or angle separated by several to numerous denticles and often little larger than the denticles themselves. [*Note:* Basal tooth or angle sometimes small and inconspicuous.] Clypeus in profile with 2, occasionally several, long, straight, erect hairs projecting forward and/or upward, usually on either side of the clypeal center.

*continued*

Long, downward-curving hairs at or near the anterior margin commonly absent, but if present, shorter than straight hairs. Metanotal area notably impressed, and the propodeal profile forms a prominent, rounded convexity relatively distinct from the remainder of the alitrunk. (Mostly mesic habitats, patchily distributed through most of region) . . . . . . . . . ***Iridomyrmex***

10(2) Mesonotum in dorsal view markedly longer than broad. (Arboreal; South American forests) . . . . . . . . . . . . ***Dolichoderus***

Mesonotum in dorsal view at most slightly longer than broad . **11**

[*Note:* Genera near *Dolichoderus* are currently being revised; and substantial changes are likely (S. O. Shattuck, personal communication). Couplets 10 and 11, adapted from a preliminary key by Shattuck, will allow specimens to be sorted into the traditional groups.]

11(10) Scale of petiole ending above in a sharp angle or a single spine. Pronotum usually with 2 spines or teeth, and marginate on the sides. (Arboreal, sometimes in old *Nasutitermes* nests; Central and South America) . . . . . . . . . . . . . . . . . . . . . . ***Monacis***

Scale of petiole rounded. Pronotum variable. Dorsum of alitrunk lacking spines, teeth, or denticles (minute teeth). (Arboreal, sometimes litter dwelling. Diverse in Mexico, Central and South America) . . . . . . . . . . . . . . . . . . . . . . . . . . ***Hypoclinea***

### SUBFAMILY FORMICINAE

1 Antenna with 11 or fewer segments . . . . . . . . . . . . . . . . . . **2**

Antenna with 12 segments . . . . . . . . . . . . . . . . . . . . . . . . **6**

2(1) Antennal scape elongate, over 1.5 times as long as head length excluding mandibles. Dorsum of alitrunk usually lacking erect hairs. Petiolar node rounded dorsally, not scale-like. Body elongate. (An Old World tramp species, *A. longipes*, introduced sporadically throughout region) . . . . . . . . . . . . . . ***Anoplolepis***

Antennal scape little (if at all) longer than head length. Dorsum of alitrunk almost always with some conspicuous, erect hairs. Body not notably elongate . . . . . . . . . . . . . . . . . . . . . . . . . **3**

3(2) Mandible slender, nearly straight, with a strongly oblique masticatory margin bearing 3 teeth. Eye minute, sometimes rudimentary, and positioned anterior to the middle of the head. (Subterranean, yellowish ants; occur throughout region) . . . . . . . . . . . . . . . . . . . . . . . . . . . . . . . . . . . . . . . . . . ***Acropyga***

Mandible triangular with a more or less transverse masticatory margin, bearing 4 or more teeth or denticles. Eye large and usually located near the middle of the side of the head . . . . . . . **4**

4(3) Antenna with 11 segments. Alitrunk lacking erect hairs, although some decumbent or appressed hairs may be present. Dorsum of alitrunk without sutural impressions (a faint metanotal impression rarely present), its profile flattened and continuous. (An Old World tramp, *P. alluaudi,* has been introduced widely throughout the region.) . . . . . . . . . . . . . . . . . . ***Plagiolepis***

Antenna with 9 or 10 segments. Dorsum of alitrunk often bearing 1 or more sutural impressions . . . . . . . . . . . . . . . . . . . . **5**

5(4) Antenna 9- or 10-segmented with a 2- or 3-segmented apical club. Scale of petiole erect, prominent, and fully exposed. Alitrunk in profile with a distinct metanotal impression. (Arboreal: fairly common throughout the region) . . . . . . . . . . ***Myrmelachista***

Antenna 9-segmented and lacking a distinct apical club. Scale of petiole strongly inclined forward and hidden beneath base of the first gastral tergite. Metanotal impression often absent or faint. (Mostly terrestrial, occasionally arboreal. Common throughout region; its classification is badly in need of revision) . . . . . . . . . . . . . . . . . . . . . . . . . . . . . . . . . ***Brachymyrmex***

6(1) Eye spectacularly large, occupying nearly the entire side of the head. Rear legs unusually long with basally swollen femorae. Petiolar node in profile triangular. (South American forests, 1 species only: *G. destructor*) . . . . . . . . . . . . . . . . ***Gigantiops***

Eye medium to small in size, not occupying more than half of side of the head. Rear legs normal in length; their femorae not abnormally swollen . . . . . . . . . . . . . . . . . . . . . . . . . . . . . . **7**

7(6) Mandibles falcate, with numerous denticles. Petiole with prominent, rounded node (not scale-like). (Mexican highlands: a dulotic genus found in mixed colonies with *Formica* spp.) . . . . . . . . . . . . . . . . . . . . . . . . . . . . . . . . . . . . . . . . . ***Polyergus***

Mandibles more or less triangular, the masticatory margin with 5–12 teeth. Petiole usually scale-like, sometimes with more or less rounded node. Mostly free-living species . . . . . . . . . . . . . **8**

8(7) Maxillary palp 3-segmented and very short. (Yellow or orange subterranean ants with eye width generally less than that of the last antennal segment. Extends from North America into Mexican highlands. Most or all species are probably temporary social parasites of *Lasius* species) . . . . . . . . . . . . . . ***Acanthomyops***

Maxillary palp 6-segmented and moderately to exceptionally long . . . . . . . . . . . . . . . . . . . . . . . . . . . . . . . . . . . . . . . . **9**

9(8) Maxillary palp longer than head length (excluding the mandibles), the third and fourth segments each as long as or longer than 2 terminal segments combined. Psammophore usually present, though sometimes weakly developed. (Workers monomorphic or moderately polymorphic. Mexico, in semi-arid or arid habitats) . . . . . . . . . . . . . . . . . . . . . . . . . . . ***Myrmecocystus***

Maxillary palp not longer than head length and usually distinctly shorter, its third and fourth segments not disproportionately long. Psammophore absent . . . . . . . . . . . . . . . . . . . . . **10**

10(9) Profile of the alitrunk continuously and evenly convex, with the propodeum not depressed below level of the promesonotum and the metanotal region at most slightly impressed (usually not at all impressed). Alitrunk in dorsal view usually wedge-shaped and tapering posteriorly. Workers often polymorphic, sometimes dimorphic or monomorphic . . . . . . . . . . . . . . . . . **11**

Profile of alitrunk clearly discontinuous and not evenly convex, the metanotal area moderately to strongly impressed, and the propodeum often distinctly depressed below the level of the promesonotum. Alitrunk in dorsal view not wedge-shaped, usually constricted to some degree in the middle. Workers usually monomorphic, sometimes weakly polymorphic . . . . . . . . . . . **13**

11(10) Workers monomorphic. Metanotal suture almost always absent across dorsum of alitrunk. Mesepisternal carina abruptly elevated and slightly reflexed immediately behind base of coxa of foreleg. Eyes large and hemispherical. (Arboreal, living in silk-impregnated carton nests, Central and South American tropical forests) . . . . . . . . . . . . . . . . . . . . . . . . . . ***Dendromyrmex***

Workers polymorphic or dimorphic. Metanotal suture usually present across dorsum of alitrunk. Mesepisternal carina rarely abruptly elevated and reflexed near base of coxa of foreleg . **12**

12(11) Workers completely dimorphic (medias absent). Head of the major obliquely truncate in front with the borders of the truncated zone sharply marginate. Mandible of major with 3 distinct apical teeth, and with the basal teeth fused into a single, broad cutting edge. Mesepisternal carina ending shortly below level of lower

*continued*

corner of pronotum, not turned mesad in front of middle coxae. (Arboreal: throughout region) . . . . . . . . ***Colobopsis***

Workers continuously polymorphic, sometimes dimorphic. Head of major usually not obliquely truncate in front; if somewhat truncate, the truncated zone is *not* sharply marginate. Mandible of major dentate along entire masticatory margin. Mesepisternal carina reaching to inner base of middle coxae. (Diverse and abundant throughout region) . . . . . . . . . . . . . ***Camponotus***

**13(10)** Frontal carinae short but distinct, each a small ridge having a moderately to sharply angulate summit, sometimes slightly reflected upward. [*Note:* Frontal carinae best examined by looking perpendicularly down on the summit of the ridge.] Lower rim of the antennal socket usually nearly touching the posterior border of the clypeus, the distance between them less than one-fourth the maximum diameter of the antennal socket. The basal face of the propodeum usually longer (sometimes as long as, but only rarely slightly shorter) than the declivitous face. [*Note:* In a few cases propodeal profile is so evenly rounded that the 2 faces are hardly distinguishable.] Mandibles with 7 or more teeth or denticles. (Mexican highlands) . . . . . . . . . . . . . . . . ***Formica***

Frontal carinae are indistinct or absent. If present, each is a small ridge with a distinctly (and often broadly) rounded summit. Lower rim of the antennal socket often not almost touching the posterior clypeal border, the distance between them commonly one-third the maximum diameter of antennal socket or more. Declivitous face of the propodeum often decidedly longer than the basal face, or 2 faces about equal in length. Mandibular dentition variable, but often consisting of 5 or 6 teeth. [*Note:* If ants from southern South America, this lug of couplet is certainly correct.] . . . . . . . . . . . . . . . . . . . . . . . . . . . . . . . . **14**

**14(13)** Mandible with 7 or more teeth. Antennal scape surpassing occipital border by no more than 2–3 times maximum diameter of the scape, usually much less . . . . . . . . . . . . . . . . . . . . . . **15**



Mandible with 5 or 6 teeth. Antennal scape frequently surpassing the occipital border by 4 or 5 times maximum diameter of the scape . . . . . . . . . . . . . . . . . . . . . . . . . . . . . . . . . . . **16**

**15(14)** Declivitous face of the propodeum decidedly longer than the basal face, both faces meeting so that the propodeal profile resembles a distinct upward-facing "peak" with a more or less rounded apex. Lower margin of antennal socket usually separated from the posterior clypeal margin by about one-third the maximum diameter of the antennal socket. (Mexican highlands and north) . . . . . . . . . . . . . . . . . . . . . . . . . . . . ***Lasius***

Declivitous face of the propodeum as long as or somewhat longer than basal face. Propodeal profile usually evenly rounded and forming a convexity, rather than a distinct upward-facing peak. Lower margin of antennal socket almost always nearly touching the posterior margin of the clypeus (less than one-fourth the maximum diameter of the antennal socket separating these structures). (Poorly understood genus closely convergent to *Lasius* and best separated from it by substantial differences in the structure of the proventriculus. Known at present only from temperate Argentina and Chile) . . . . . . . . . . . ***Lasiophanes***

**16(14)** Mesonotum, viewed from directly above, severely constricted, giving the alitrunk a distinctive hourglass-like shape. Pilosity not conspicuously coarse or bristle-like; erect hairs mostly slender and golden or brownish. Eyes never strongly reduced (fewer than 20 facets). (Mexican highlands, and one species endemic to Cuba) . . . . . . . . . . . . . . . . . . . . . . . . . . . ***Prenolepis***

Mesonotum, viewed from directly above, weakly constricted; alitrunk lacking obvious hourglass-like shape. Erect hairs usually coarse, and often dark brown or black. [*Note:* Erect hairs sometimes fine and yellowish, particularly in a few subterranean species with greatly reduced eyes having fewer than 20 facets.] (Abundant throughout region) . . . . . . . . . . . . ***Paratrechina***

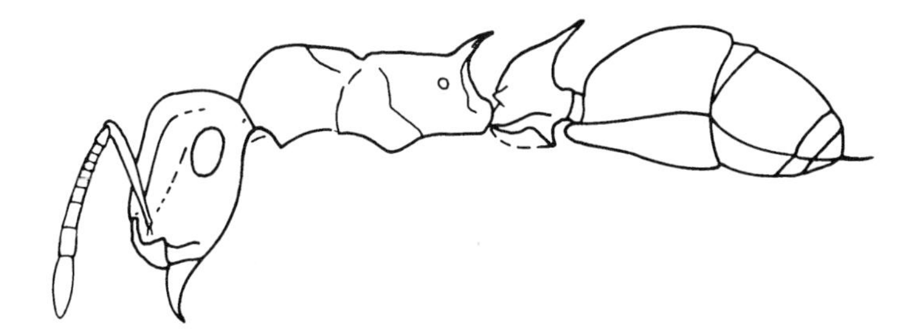

*Acanthoponera*

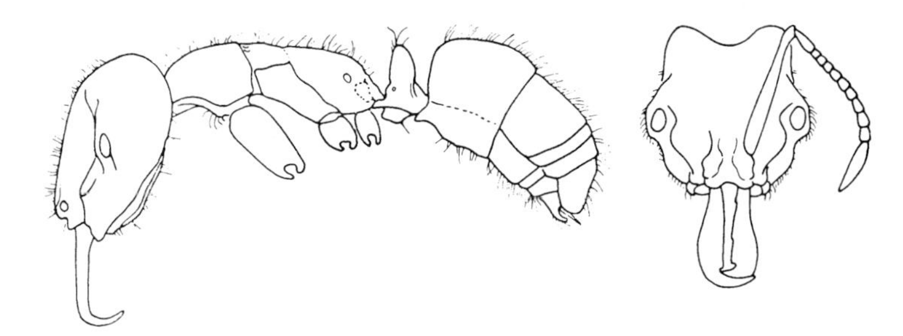

*Anochetus*

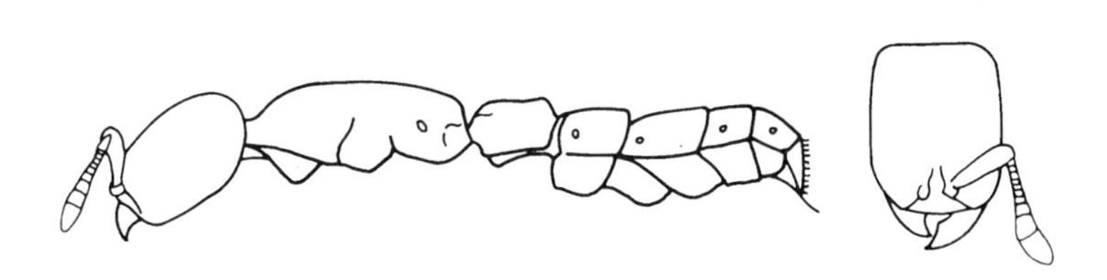

*Acanthostichus*

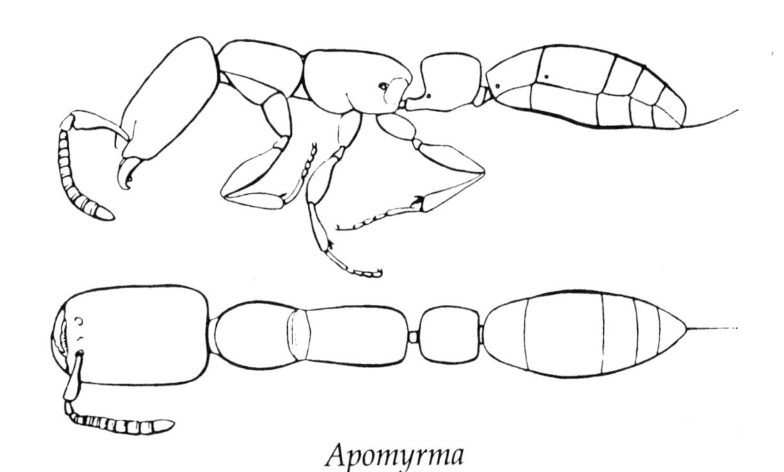

*Apomyrma*

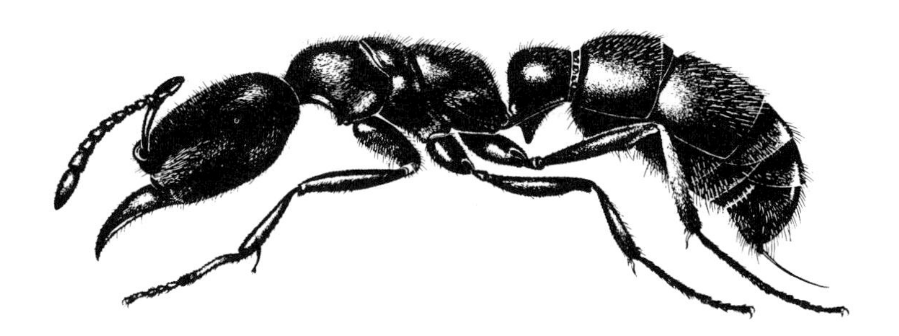

*Amblyopone*

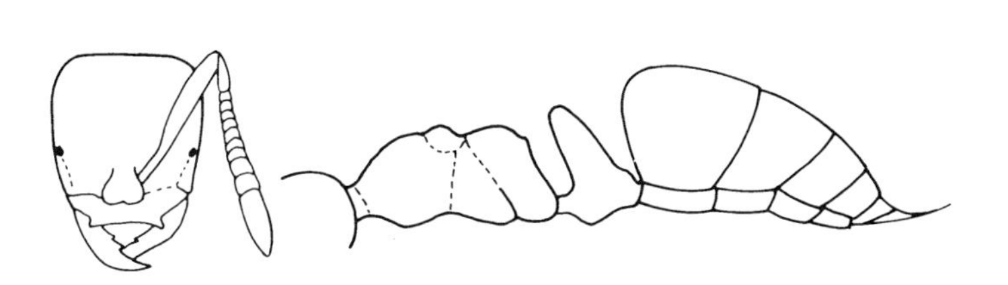

*Asphinctopone*

*Acanthoponera minor*, Costa Rica

*Acanthostichus femoralis*, Argentina

*Amblyopone* sp., Australia (R. W. Taylor, unpublished; F. Nanninga, artist)

*Anochetus graeffei*, Samoa (Wilson and Taylor, 1967b)

*Apomyrma stygia*, Ivory Coast (Brown et al., 1970a)

*Asphinctopone lucidus*, Zaïre

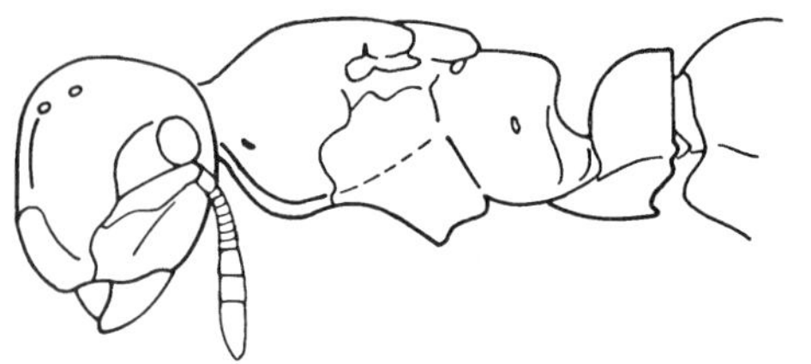

*Aulacopone*

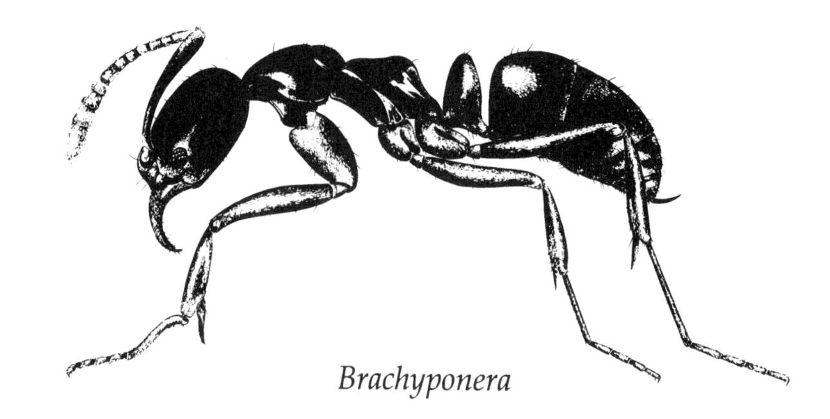

*Brachyponera*

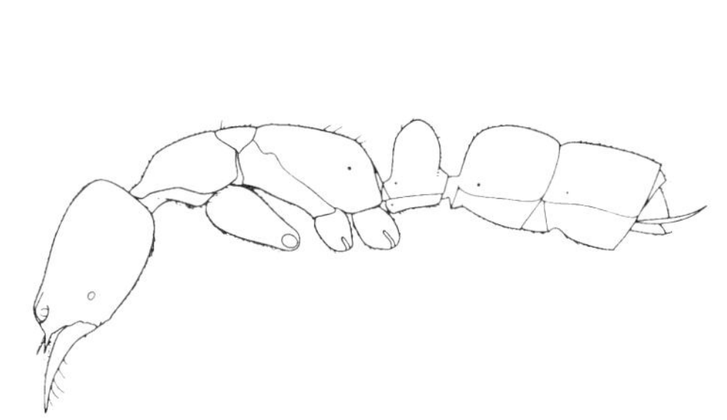

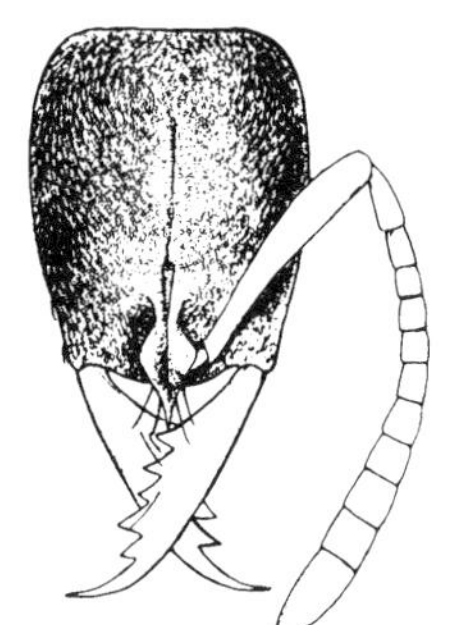

*Belonopelta*

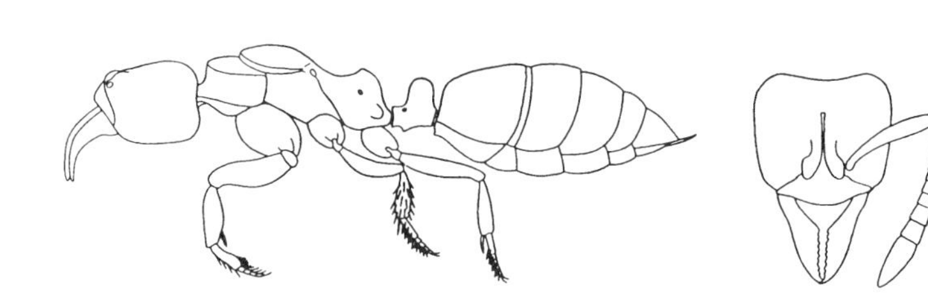

*Centromyrmex*

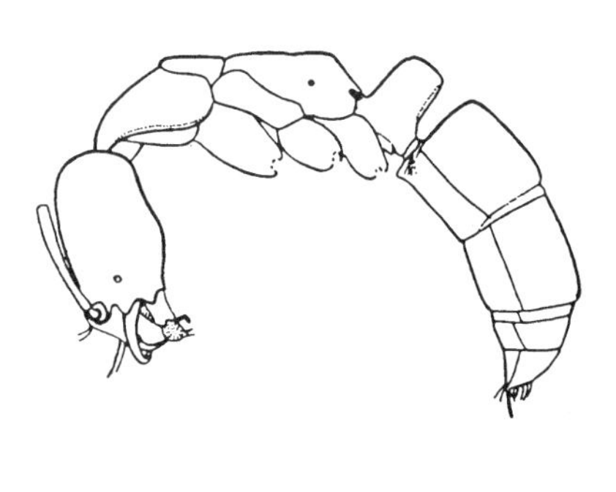

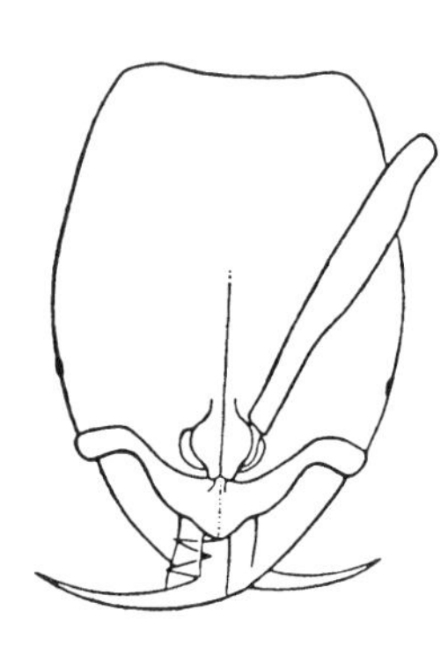

*Belonopelta (= Leiopelta)*

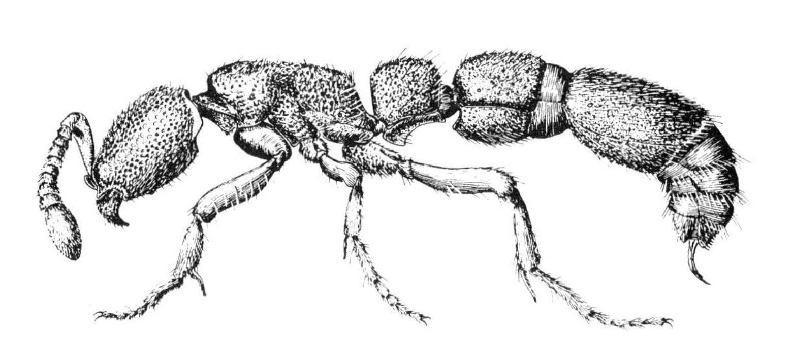

*Cerapachys*

*Aulacopone relicta,* USSR

*Belonopelta attenuata,* Colombia (Baroni Urbani, 1975a)

*Belonopelta (= Leiopelta) deletrix,* Mexico (Wilson, 1955a)

*Brachyponera croceicornis,* Australia (R. W. Taylor, unpublished; R. J. Kohout, artist)

*Centromyrmex feae,* Java (Wheeler, 1936c)

*Cerapachys augustae,* United States (Smith, 1947a)

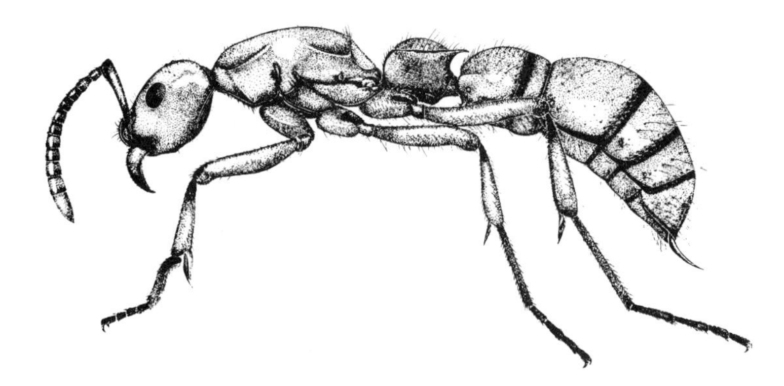

*Cerapachys*

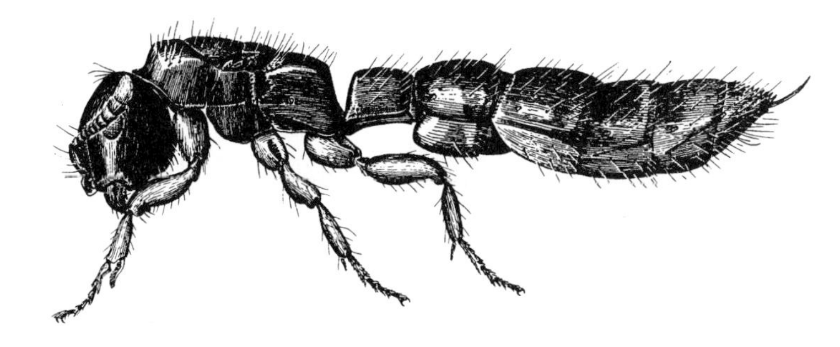

*Ctenopyga*

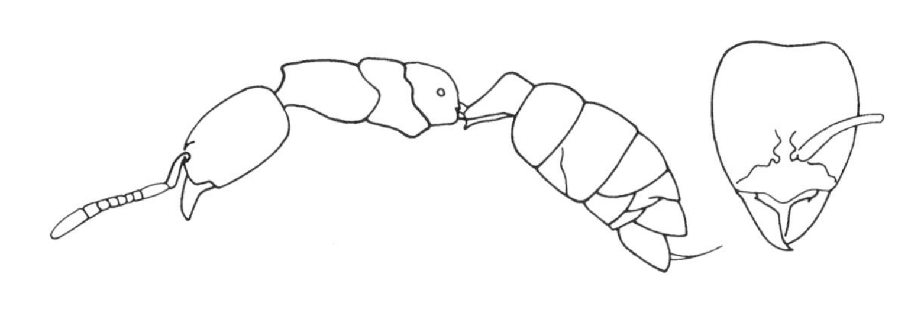

*Concoctio*

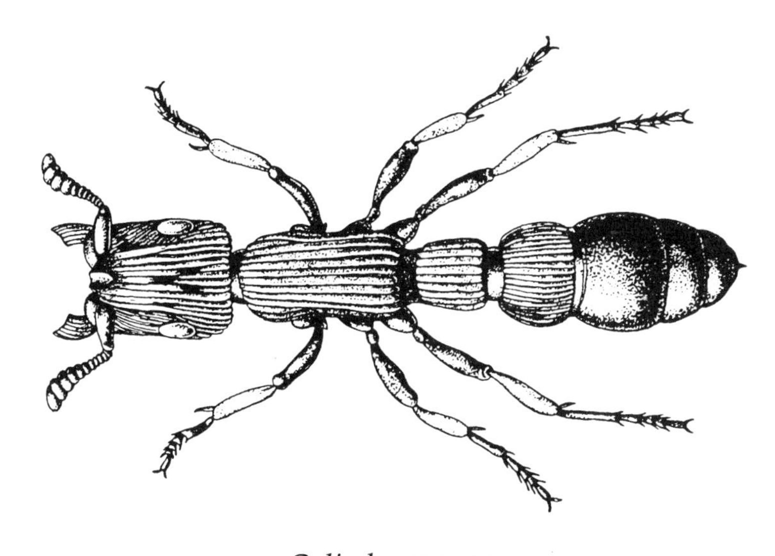

*Cylindromyrmex*

*Cryptopone*

*Diacamma*

*Cerapachys* sp., Australia (R. W. Taylor, unpublished; R. J. Kohout, artist)

*Concoctio concenta*, Gabon (Brown, 1974c)

*Cryptopone mjobergi*, Australia (R. W. Taylor, unpublished; F. Nanninga, artist)

*Ctenopyga texanus*, United States (Smith, 1947a)

*Cylindromyrmex whymperi*, Ecuador (Wheeler, 1910a)

*Diacamma australe*, Australia (R. W. Taylor, unpublished; F. Nanninga, artist)

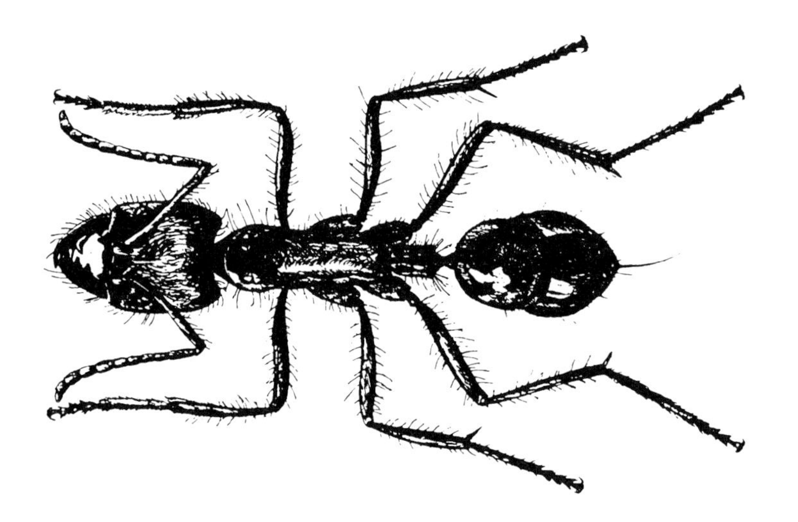

*Dinoponera*

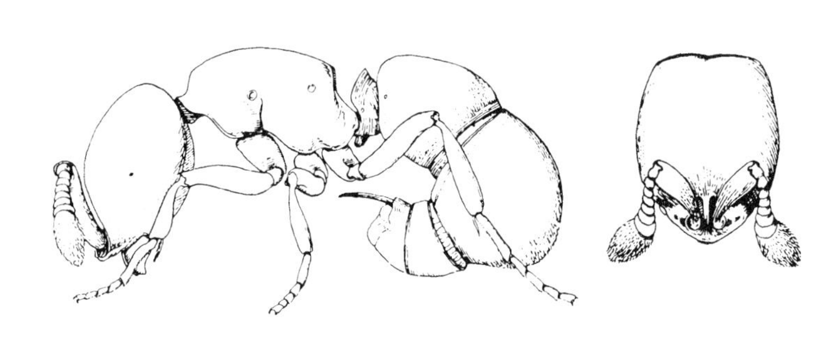

*Discothyrea*

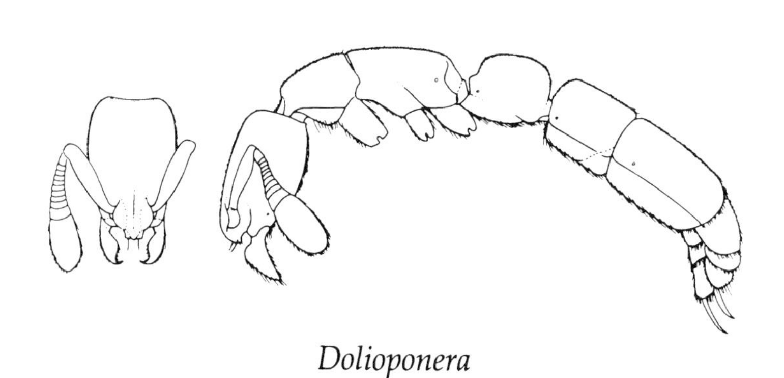

*Dolioponera*

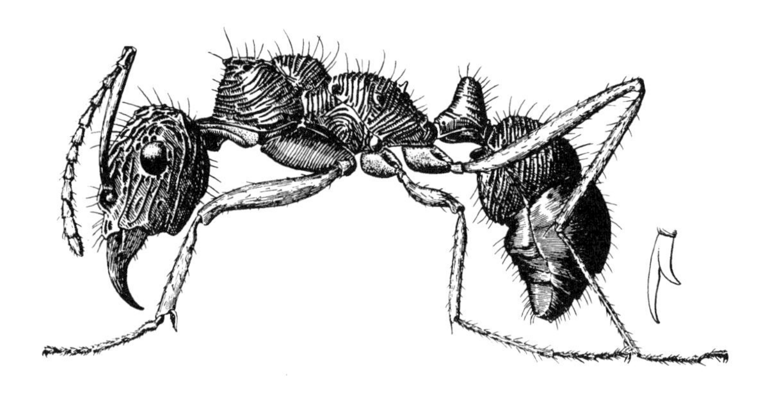

*Ectatomma*

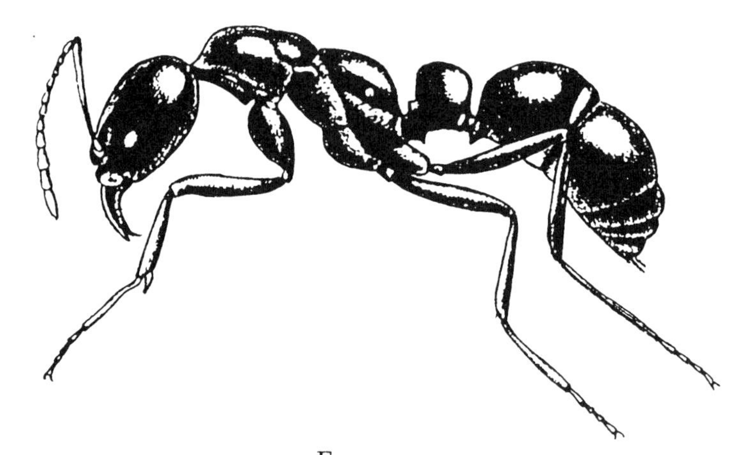

*Euponera*

*Gnamptogenys*

*Dinoponera lucida,* Brazil (Kempf, 1971)

*Discothyrea testacea,* United States (Smith and Wing, 1954)

*Dolioponera fustigera,* Gabon (Brown, 1974d)

*Ectatomma tuberculatum,* United States (Smith, 1947a)

*Euponera sikorae,* Madagascar (Forel, 1891)

*Gnamptogenys biroi,* Australia (R. W. Taylor, unpublished; R. J. Kohout, artist)

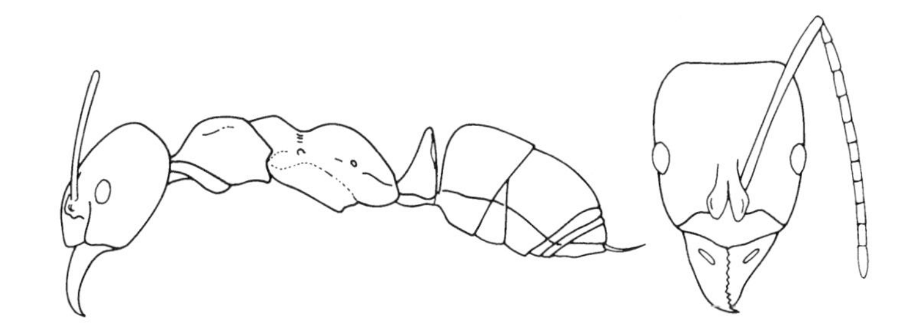

*Hagensia*

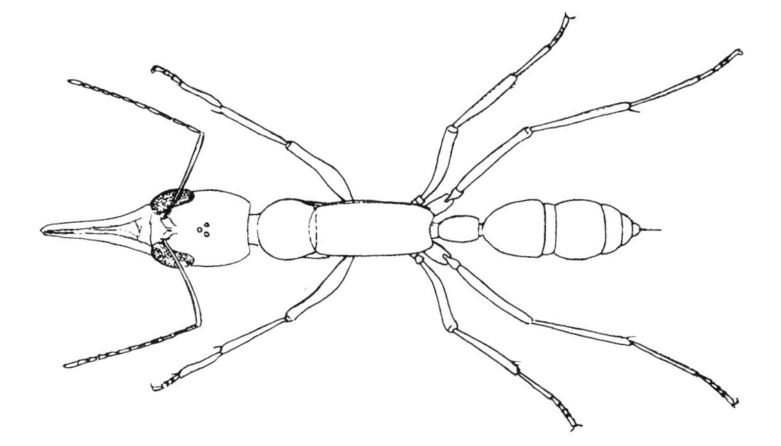

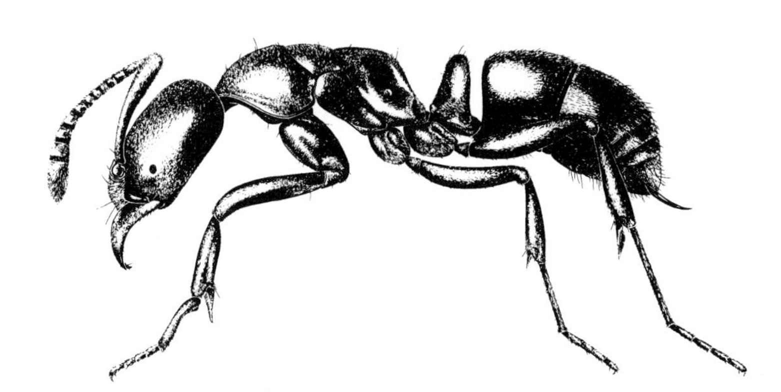

*Hypoponera*

*Harpegnathos*

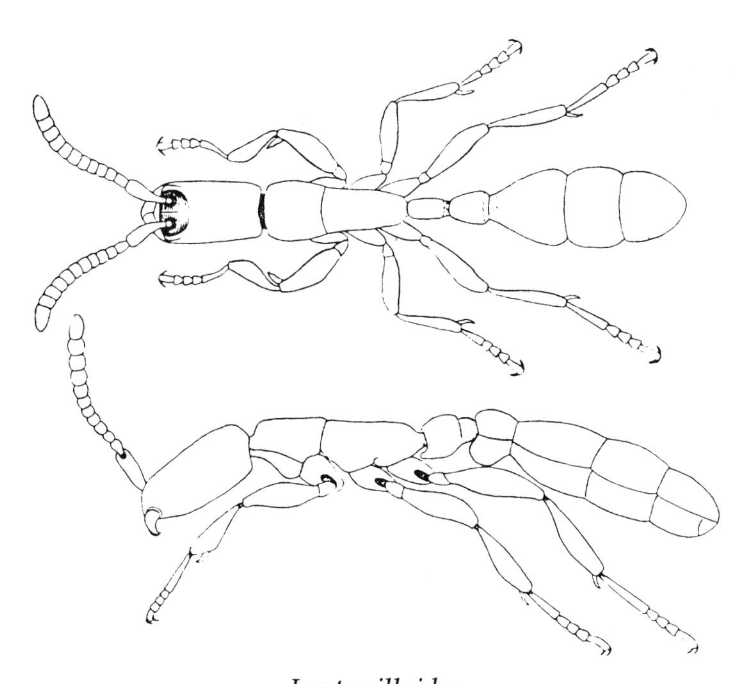

*Heteroponera*

*Leptanilloides*

*Hagensia peringueyi*, South Africa

*Harpegnathos saltator*, tropical Asia (Emery, 1911b)

*Heteroponera leae*, Australia (R. W. Taylor, unpublished; F. Nanninga, artist)

*Heteroponera monticola*, Colombia (Kempf and Brown, 1970)

*Hypoponera* sp., Australia (R. W. Taylor, unpublished; R. J. Kohout, artist)

*Leptanilloides biconstricta*, Bolivia (Mann, 1923a)

*Leptogenys*

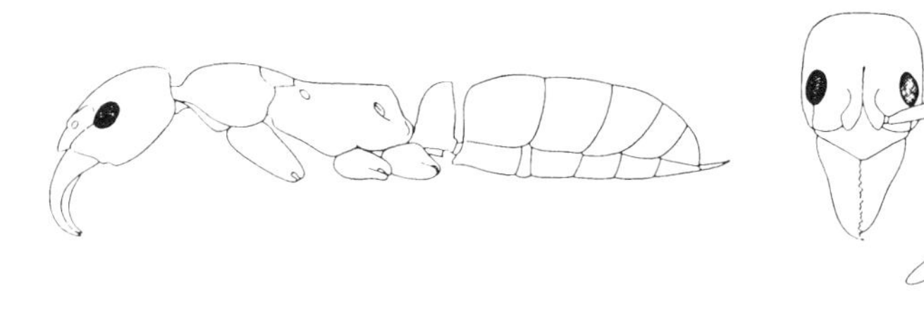

*Megaponera*

*Leptogenys*

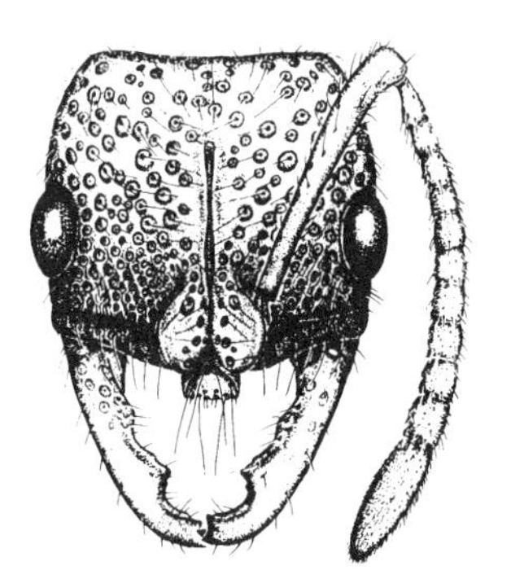

*Myopias*

*Leptogenys (= Prionogenys)*

*Myopopone*

*Leptogenys diminuta,* Australia (R. W. Taylor, unpublished; F. Nanninga, artist)

*Leptogenys longensis,* Australia (R. W. Taylor, unpublished; F. Nanninga, artist)

*Leptogenys (= Prionogenys) podenzanai,* Australia (R. W. Taylor, unpublished; F. Nanninga, artist)

*Megaponera foetens,* Zimbabwe (Wheeler, 1936c)

*Myopias concava,* Papua New Guinea (Willey and Brown, 1983)

*Myopopone castanea,* Australia (R. W. Taylor, unpublished; F. Nanninga, artist)

*Mystrium*

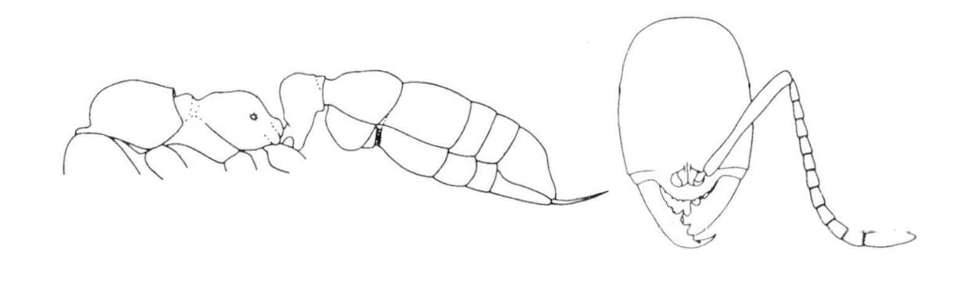

*Onychomyrmex*

*Odontomachus*

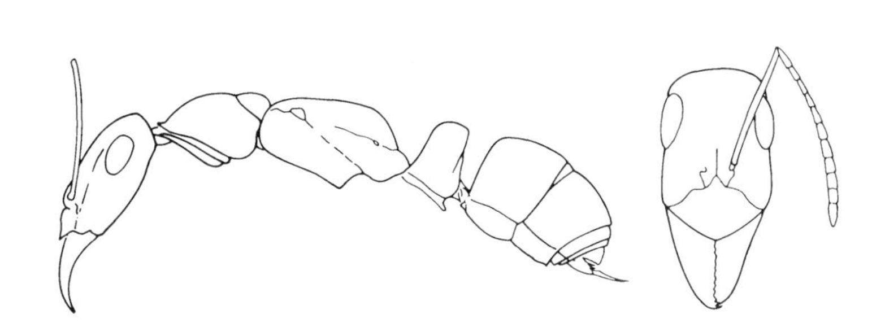

*Ophthalmopone*

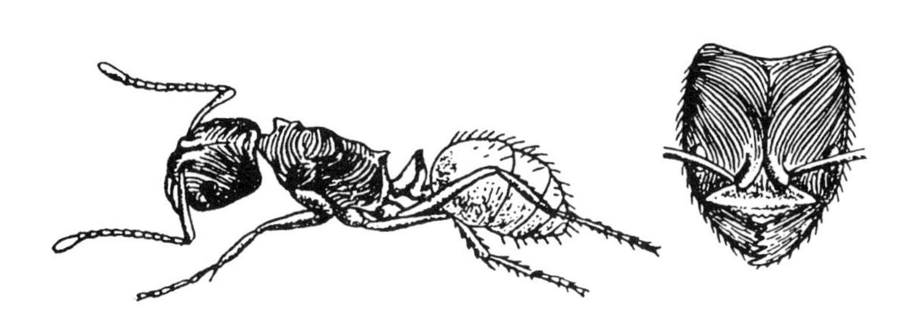

*Odontoponera*

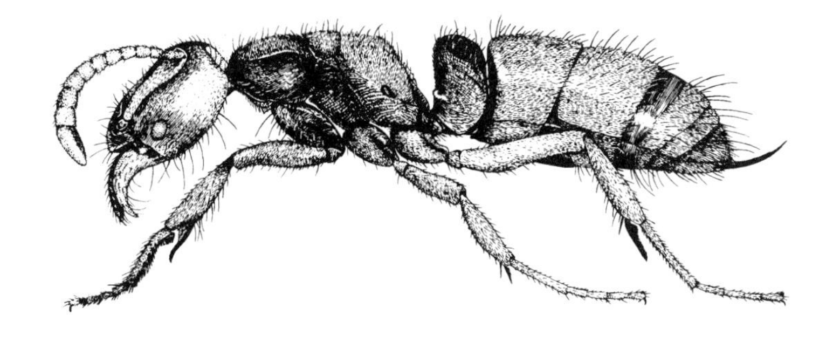

*Pachycondyla*

*Mystrium camillae,* Australia (R. W. Taylor, unpublished; F. Nanninga, artist)

*Odontomachus* sp., Australia (R. W. Taylor, unpublished; F. Nanninga, artist)

*Odontoponera transversa,* India (Wheeler, 1910a)

*Onychomyrmex hedleyi,* Australia (Brown, 1960b)

*Ophthalmopone berthoudi,* Zimbabwe

*Pachycondyla harpax,* United States (Smith, 1947a)

*Pachycondyla (= Bothroponera)*

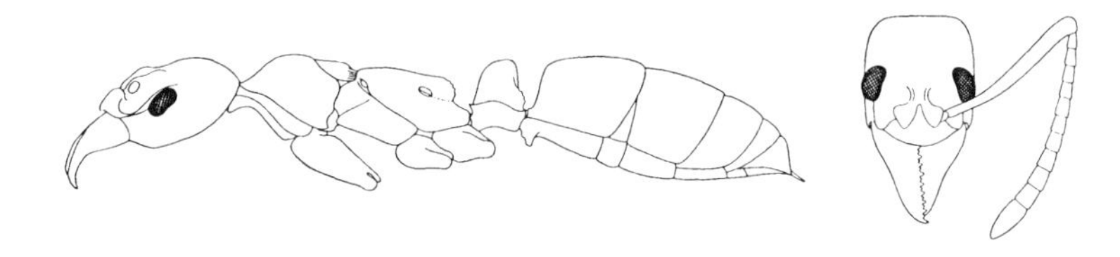

*Pachycondyla (= Termitopone)*

*Pachycondyla (= Mesoponera)*

*Pachycondyla (= Trachymesopus)*

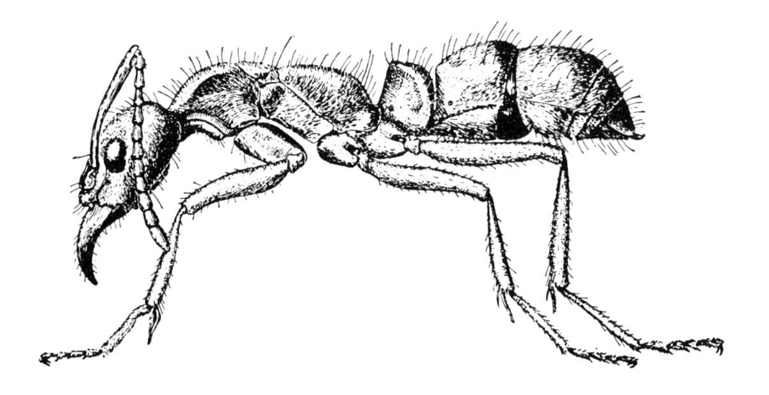

*Pachycondyla (= Neoponera)*

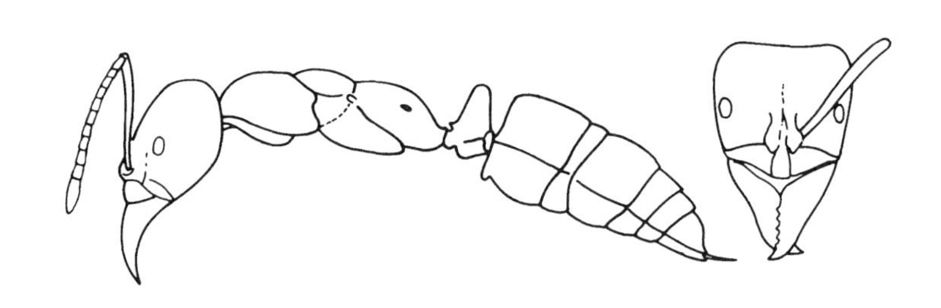

*Paltothyreus*

*Pachycondyla (= Bothroponera)* sp., Australia (R. W. Taylor, unpublished; F. Nanninga, artist)

*Pachycondyla (= Mesoponera) australis,* Australia (R. W. Taylor, unpublished; R. J. Kohout, artist)

*Pachycondyla (= Neoponera) villosa,* United States (Smith, 1947a)

*Pachycondyla (= Termitopone) commutata,* South America (Wheeler, 1936c)

*Pachycondyla (= Trachymesopus)* sp., Australia (R. W. Taylor, unpublished; R. J. Kohout, artist)

*Paltothyreus tarsatus,* Kenya

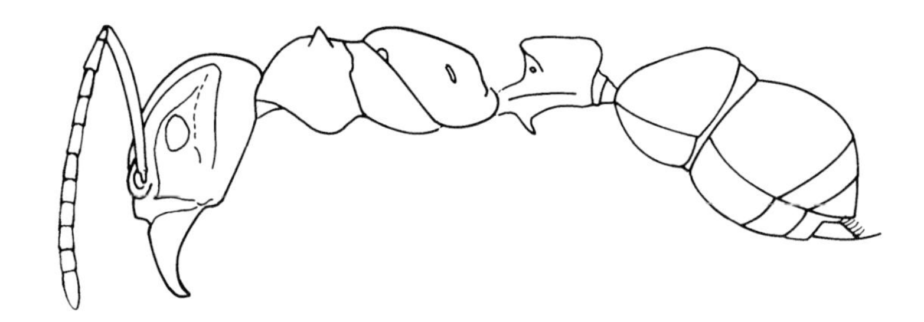

*Paraponera*

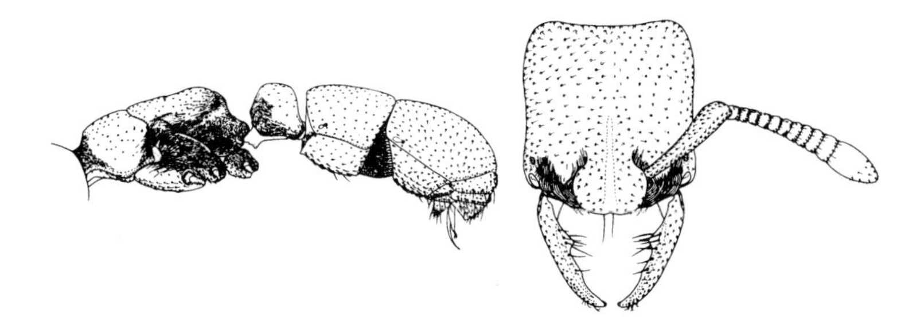

*Plectroctena*

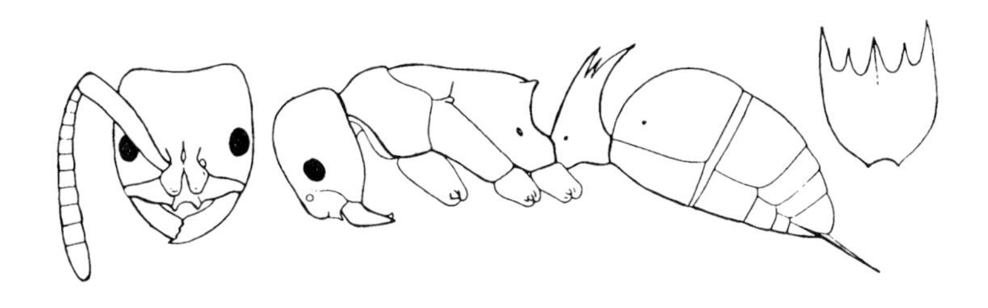

*Phrynoponera*

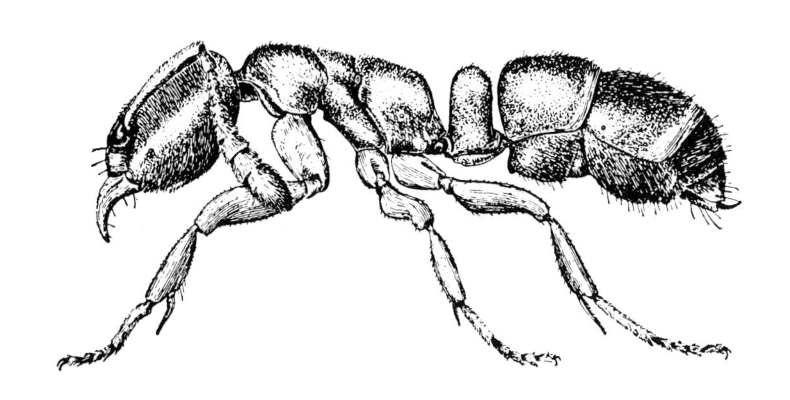

*Ponera*

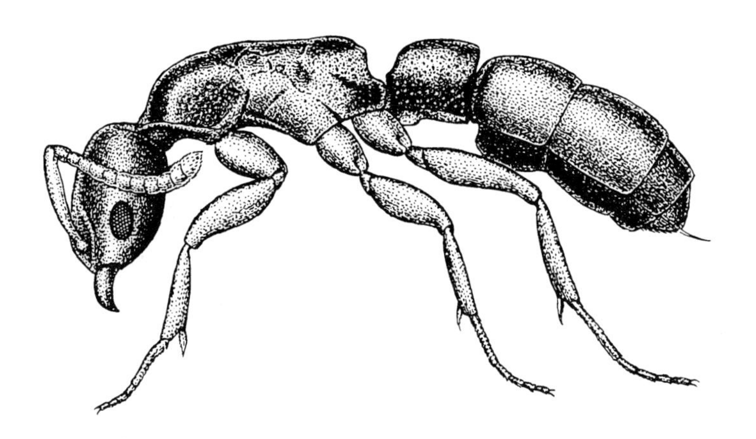

*Platythyrea*

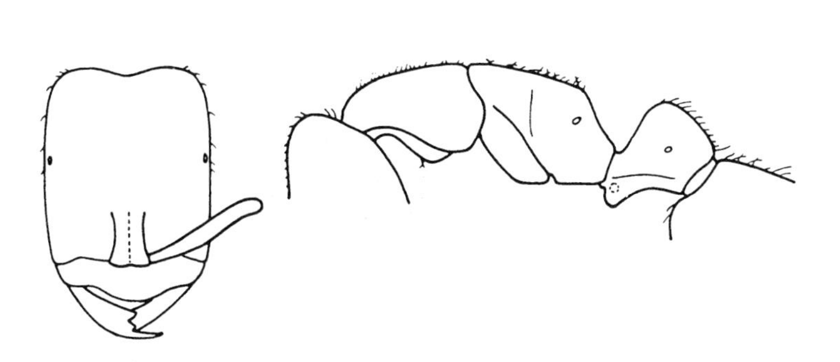

*Prionopelta*

*Paraponera clavata*, Costa Rica

*Phrynoponera gabonensis*, Zaïre (Wheeler, 1922)

*Platythyrea parallela*, Papua New Guinea (Wilson and Taylor, 1967b)

*Plectroctena lygaria*, Ivory Coast (Bolton et al., 1976)

*Ponera pennsylvanica*, United States (Smith, 1947a)

*Prionopelta kraepelini*, Samoa (Wilson and Taylor, 1967b)

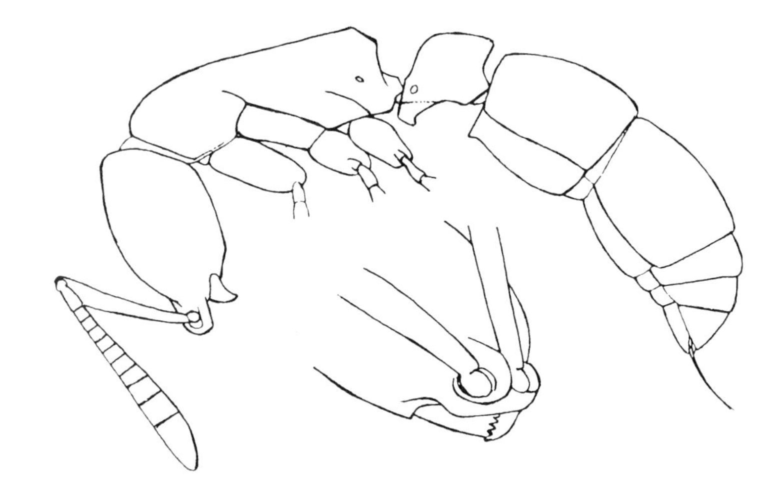

*Probolomyrmex*

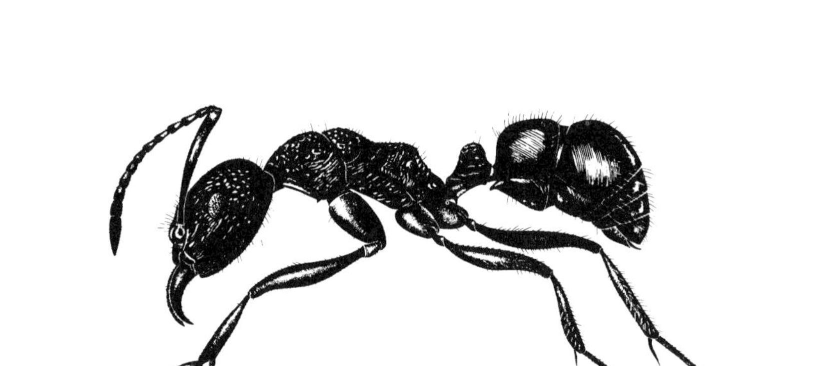

*Psalidomyrmex*

*Rhytidoponera*

*Proceratium*

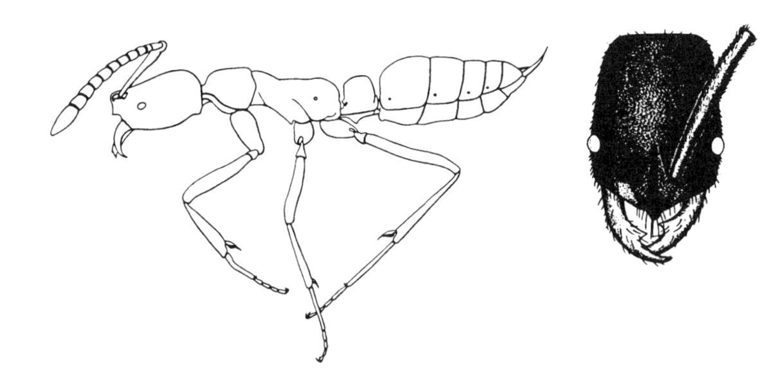

*Simopelta*

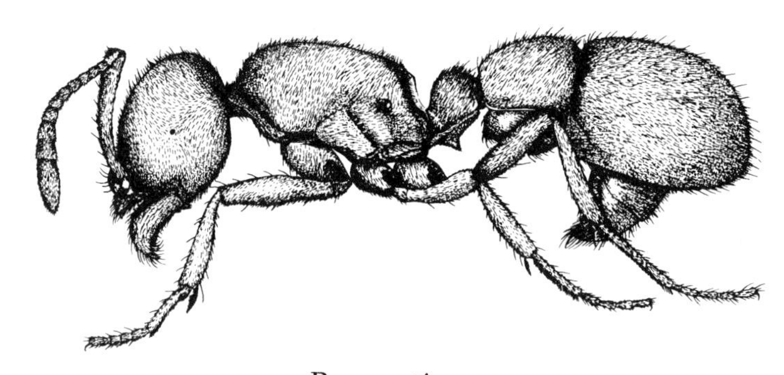

*Proceratium*

*Probolomyrmex filiformis*, South Africa (Emery, 1911b)

*Proceratium papuanum*, Australia (R. W. Taylor, unpublished; F. Nanninga, artist)

*Proceratium pergandei*, United States (Smith, 1947a)

*Psalidomyrmex procerus*, Zaïre (Wheeler, 1922)

*Rhytidoponera purpurea*, Australia (R. W. Taylor, unpublished; R. J. Kohout, artist)

*Simopelta oculata*, Costa Rica (Gotwald and Brown, 1966)

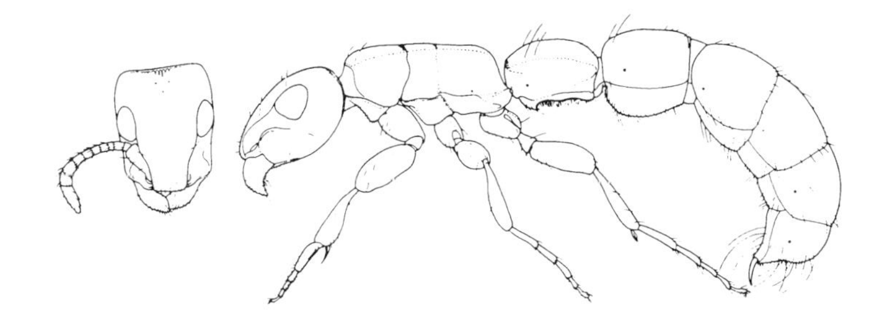

*Simopone*

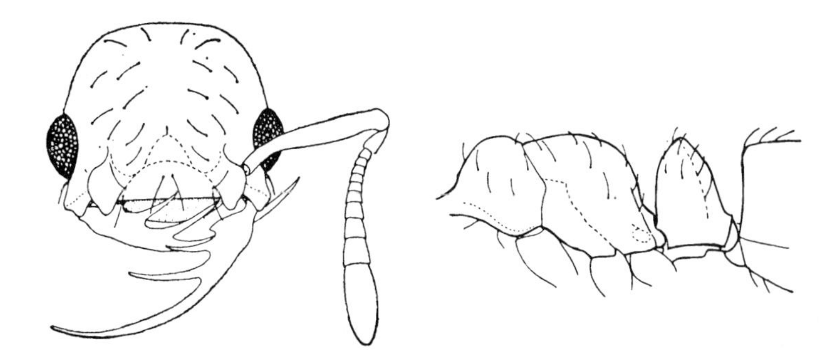

*Thaumatomyrmex*

*Sphinctomyrmex*

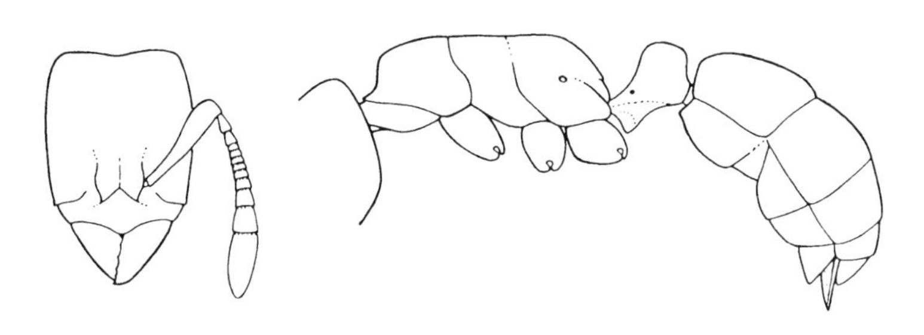

*Typhlomyrmex*

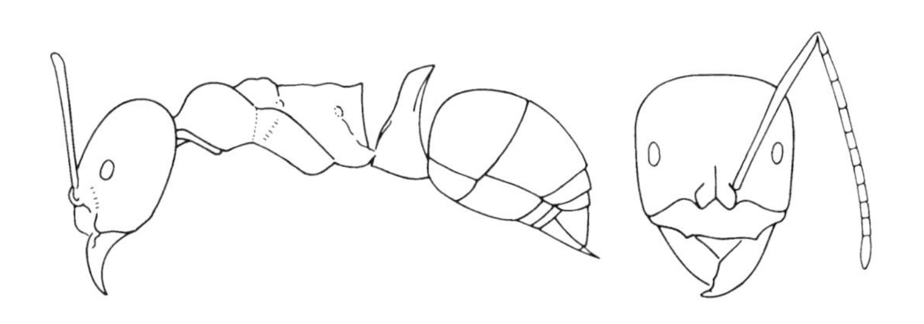

*Streblognathus*

*Simopone gressitti,* Papua New Guinea (Taylor, 1965d)

*Sphinctomyrmex* sp., Australia (R. W. Taylor, unpublished; R. J. Kohout, artist)

*Streblognathus aethiopicus,* South Africa

*Thaumatomyrmex contumax,* Brazil (Kempf, 1975)

*Typhlomyrmex pusillus,* Argentina (Brown, 1965)

## SUBFAMILY NOTHOMYRMECIINAE

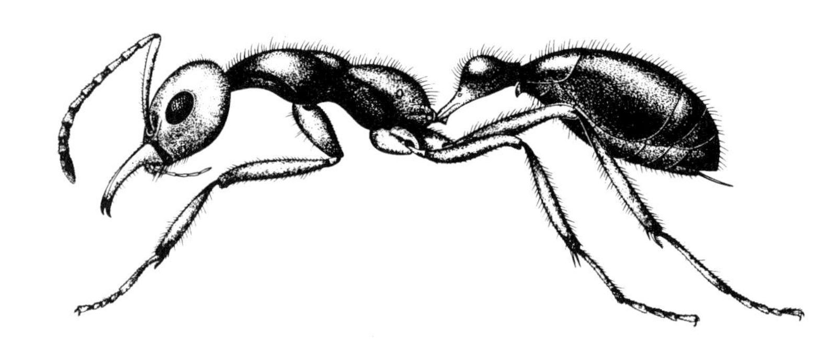

*Nothomyrmecia*

*Nothomyrmecia macrops,* Australia (R. W. Taylor, unpublished; F. Nanninga, artist)

## SUBFAMILY MYRMECIINAE

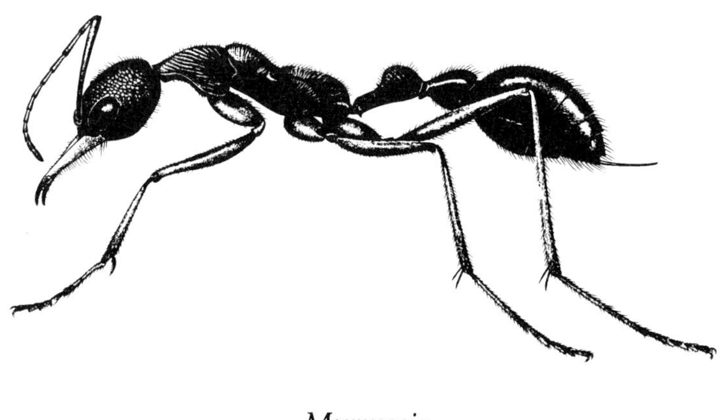

*Myrmecia*

*Myrmecia nigriceps,* Australia (R. W. Taylor, unpublished; F. Nanninga, artist)

## SUBFAMILY DORYLINAE

*Aenictus*

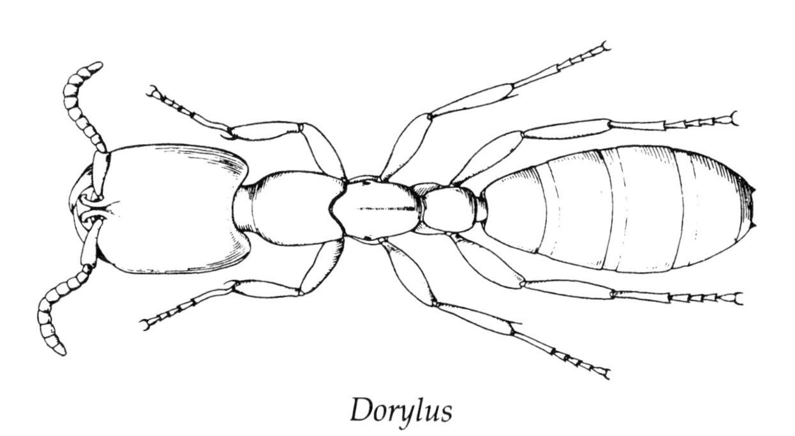

*Dorylus*

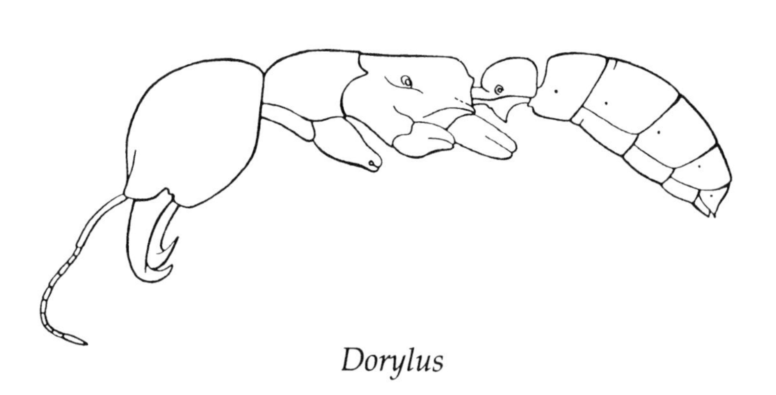

*Dorylus*

*Aenictus* sp., tropical Asia (Gotwald, 1978)

*Dorylus fulvus,* Zaïre (Wheeler, 1922)

*Dorylus* sp., tropical Africa (Gotwald, 1978)

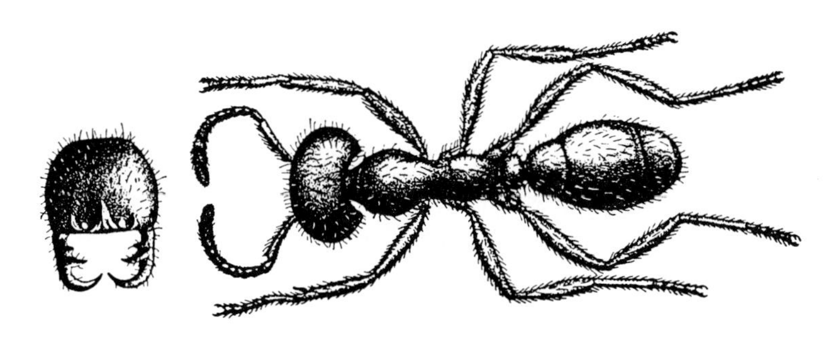

*Cheliomyrmex*

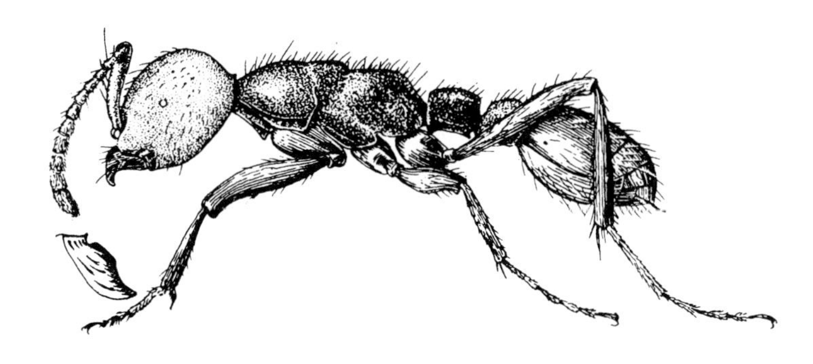

*Neivamyrmex*

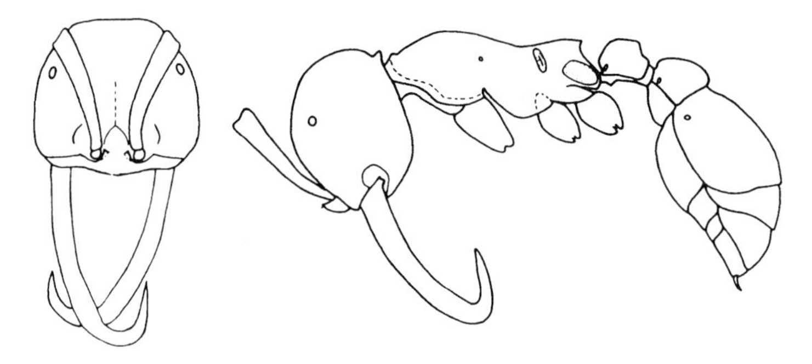

*Eciton*

*Nomamyrmex*

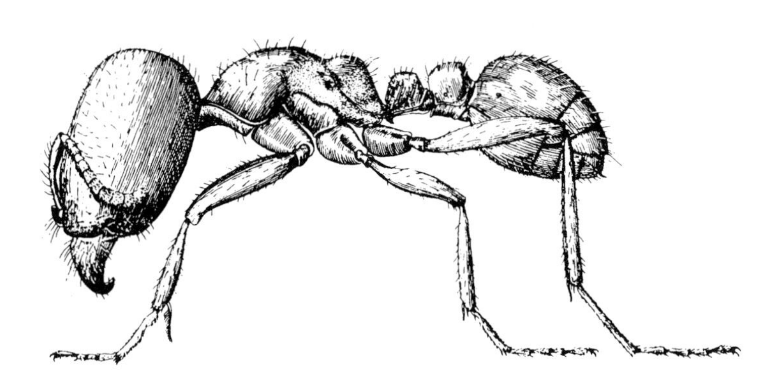

*Labidus*

*Cheliomyrmex nortoni*, Central America (Wheeler, 1910a)

*Eciton burchelli*, Mexico (Watkins, 1982)

*Labidus coecus*, United States (Smith, 1947a)

*Neivamyrmex opacithorax*, United States (Smith, 1947a)

*Nomamyrmex esenbecki*, Brazil

## SUBFAMILY LEPTANILLINAE

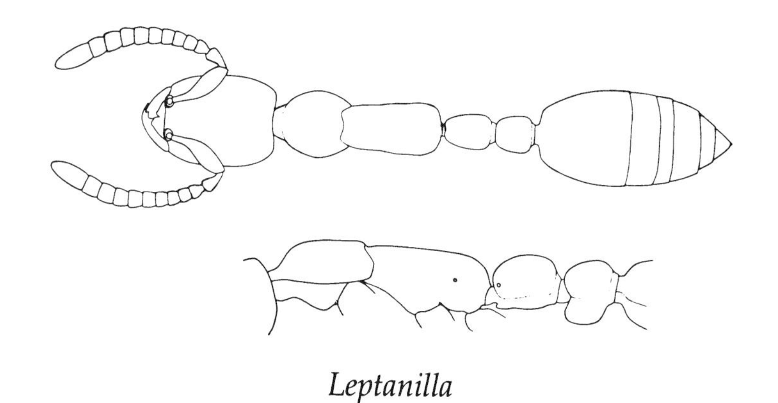

*Leptanilla*

*Leptanilla kubotai*, Japan (Baroni Urbani, 1977)

## SUBFAMILY PSEUDOMYRMECINAE

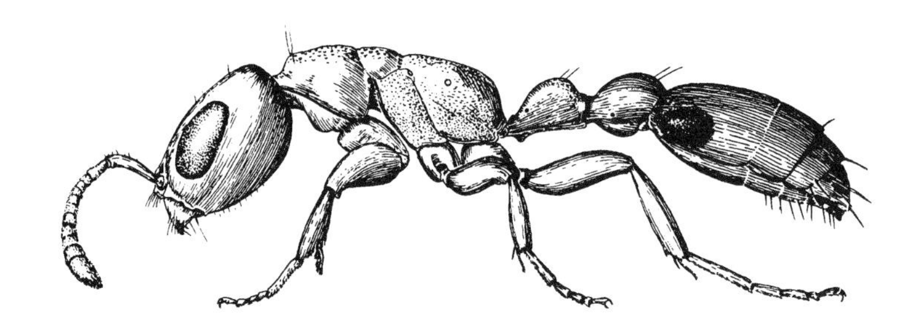

*Pseudomyrmex*

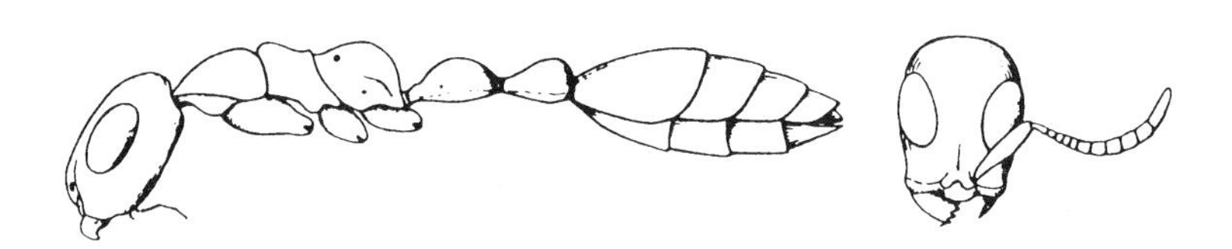

*Tetraponera*

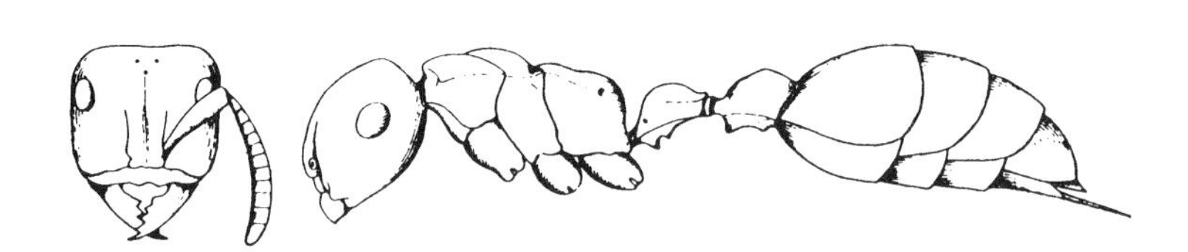

*Tetraponera (= Pachysima)*

*Pseudomyrmex pallidus*, United States (Smith, 1947a)

*Tetraponera ophthalmica*, Zaïre (Wheeler, 1922)

*Tetraponera (= Pachysima) aethiops*, Zaïre (Wheeler, 1922)

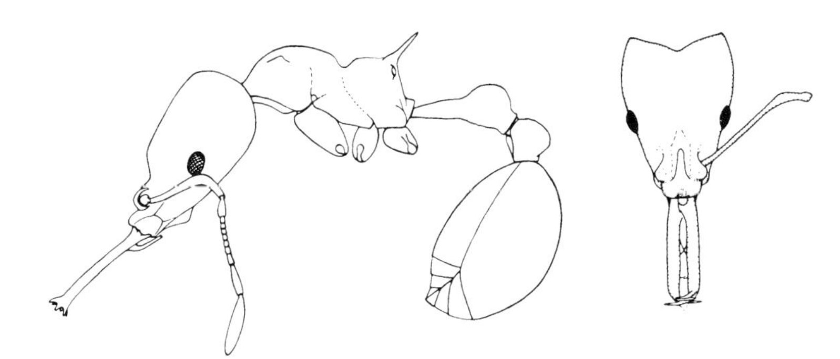

Acanthognathus

Acromyrmex (= Pseudoatta)

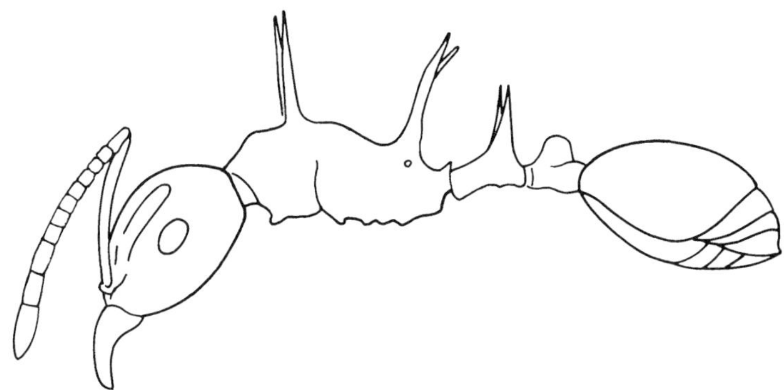

Acanthomyrmex

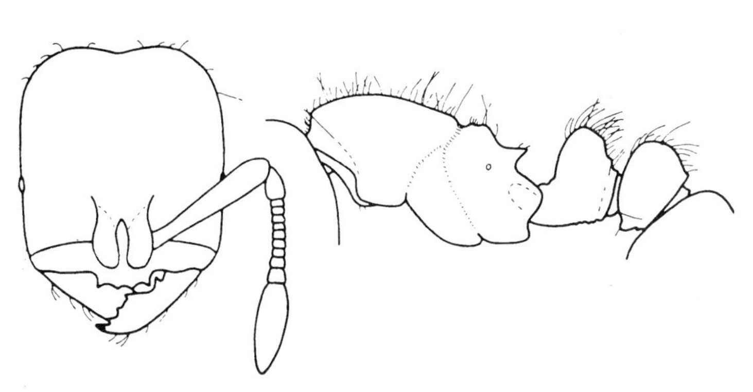

Adelomyrmex

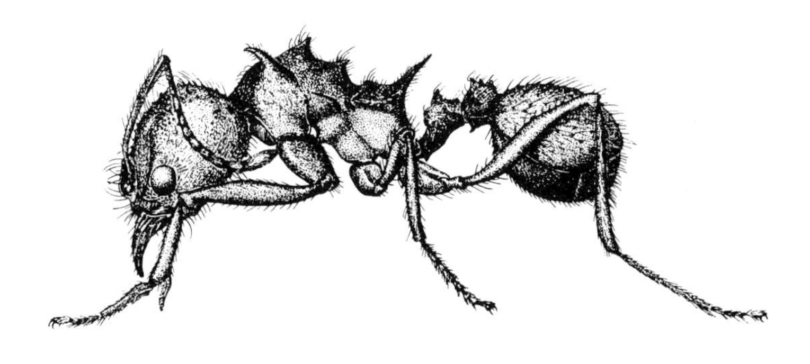

Acromyrmex

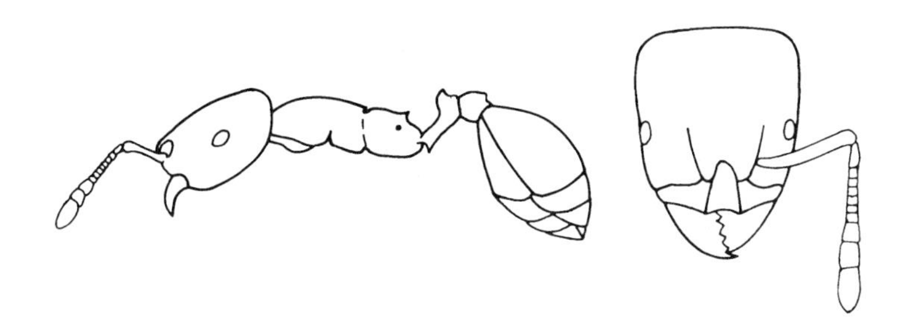

Adlerzia

*Acanthognathus rudis*, Brazil (Brown and Kempf, 1969)

*Acanthomyrmex ferox*, Indonesia (based on Moffett, 1986b)

*Acromyrmex versicolor*, United States (Smith, 1947a)

*Acromyrmex (= Pseudoatta) argentina*, Argentina

*Adelomyrmex samoanus*, Samoa (Wilson and Taylor, 1967b)

*Adlerzia froggatti*, Australia

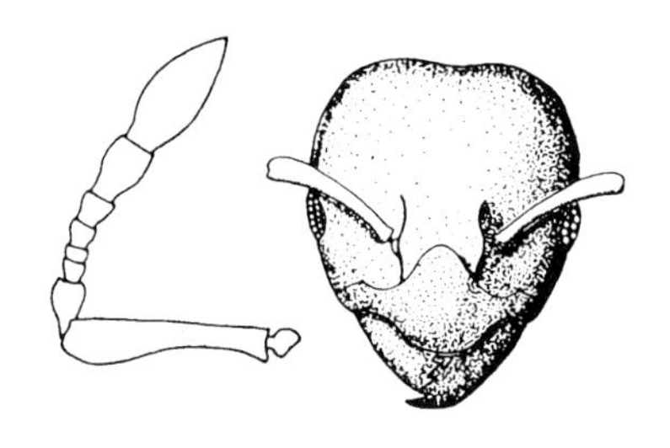
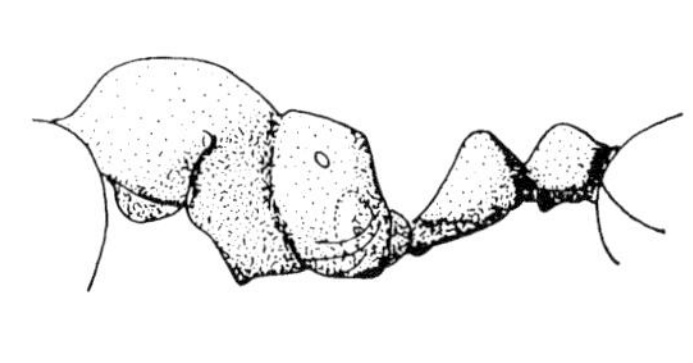

*Allomerus*

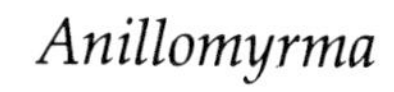

*Anillomyrma*

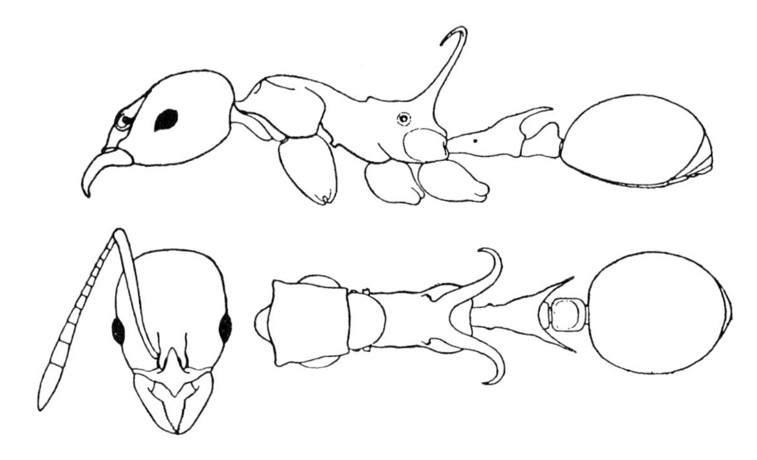

*Ancyridris*

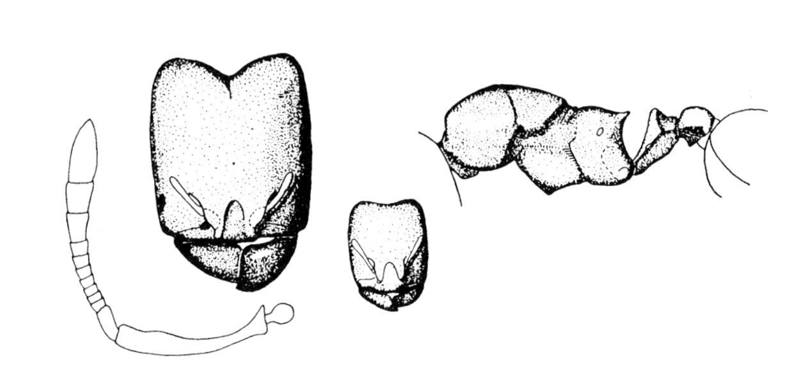

*Anisopheidole*

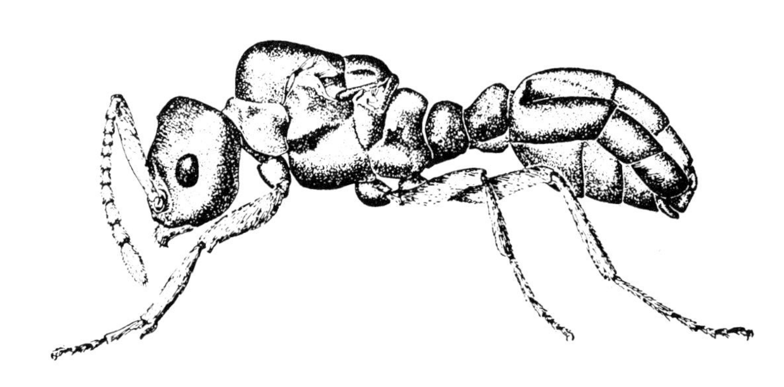

*Anergates*

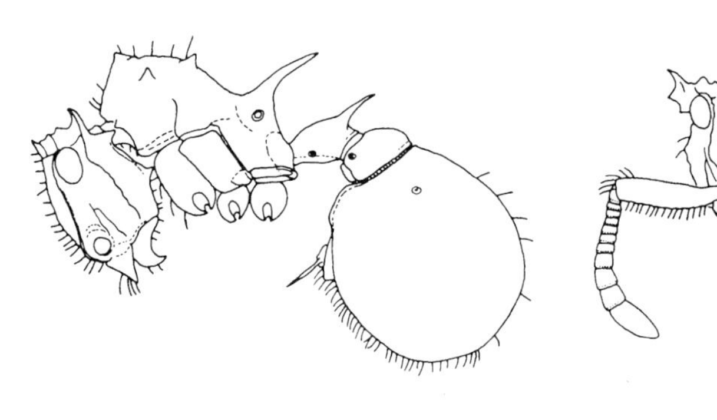

*Ankylomyrma*

*Allomerus angulatus,* Bolivia (Ettershank, 1966)

*Ancyridris polyrhachioides,* Papua New Guinea (Wheeler, 1935)

*Anergates atratulus,* United States (Smith, 1947a)

*Anillomyrma tridens,* Malaysia (Bolton, 1987)

*Anisopheidole froggatti,* Australia (Ettershank, 1966)

*Ankylomyrma coronacantha,* Gabon (Bolton, 1981b)

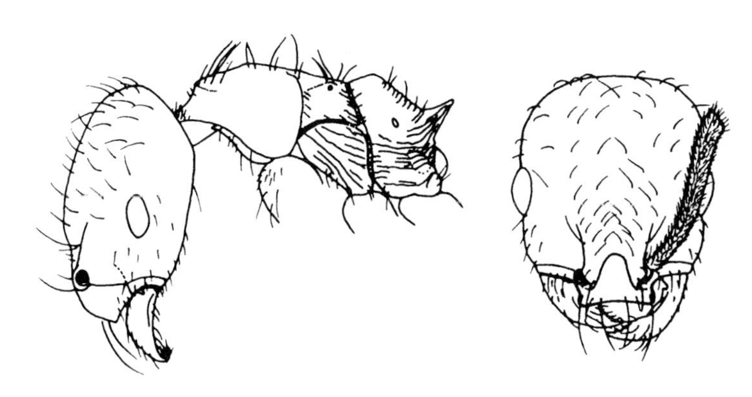

*Antichthonidris*

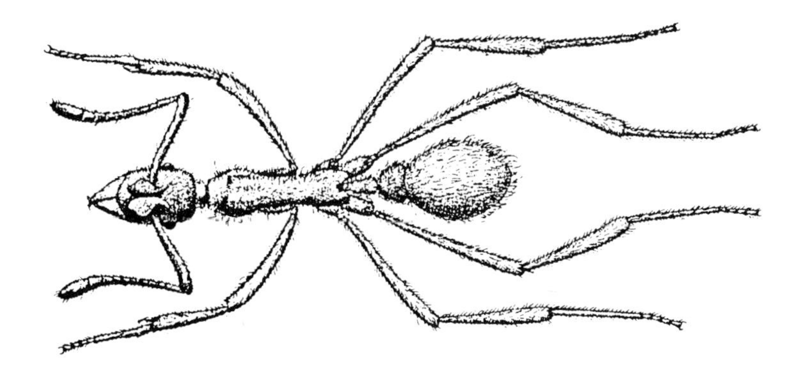

*Apterostigma*

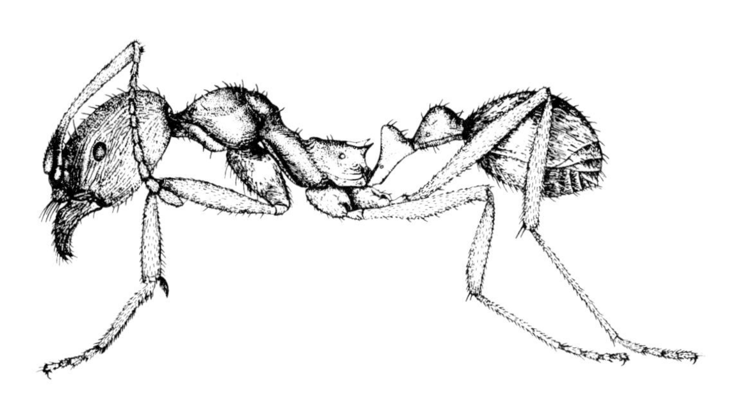

*Aphaenogaster*

*Asketogenys*

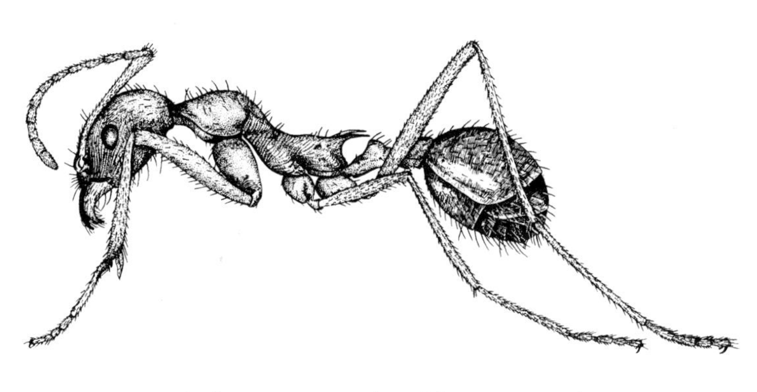

*Aphaenogaster (= Novomessor)*

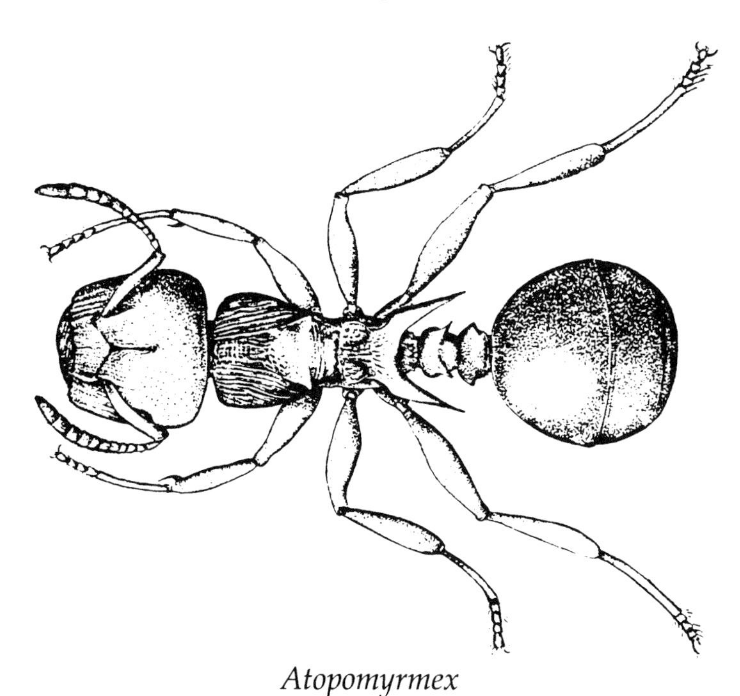

*Atopomyrmex*

*Antichthonidris bidentatus,* Chile (Snelling, 1975)

*Aphaenogaster treatae,* United States (Smith, 1947a)

*Aphaenogaster (= Novomessor) cockerelli,* United States (Smith, 1947a)

*Apterostigma pilosa,* Brazil (Wheeler, 1910a)

*Asketogenys acubecca,* Malaysia (Brown, 1972)

*Atopomyrmex mocquerysi,* Zaïre (Wheeler, 1922)

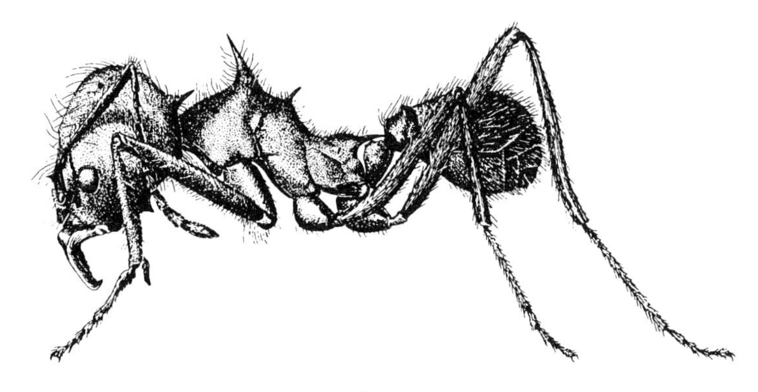

*Atta*

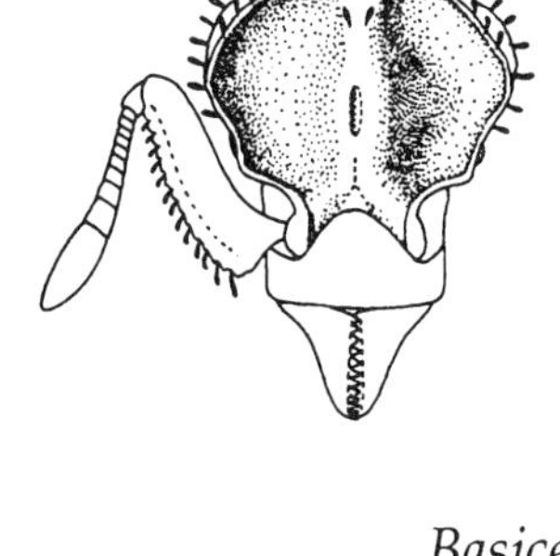

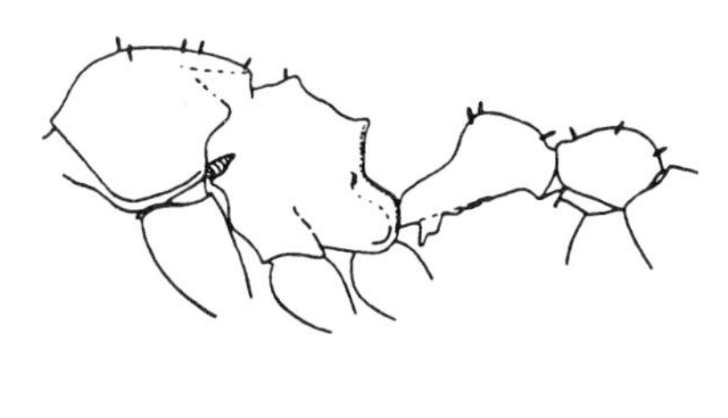

*Basiceros (= Aspididris)*

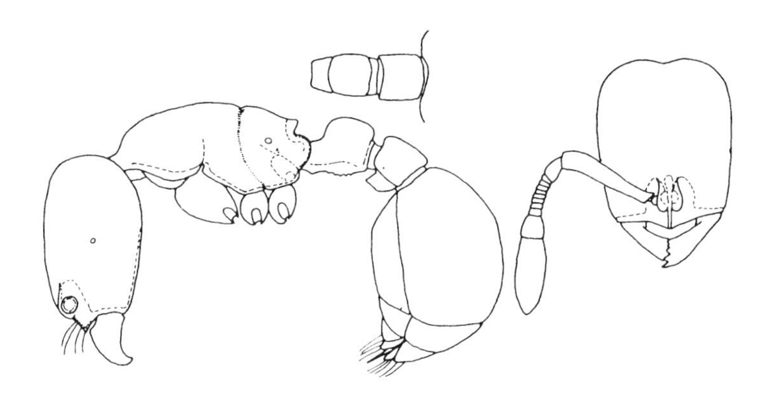

*Baracidris*

*Blepharidatta*

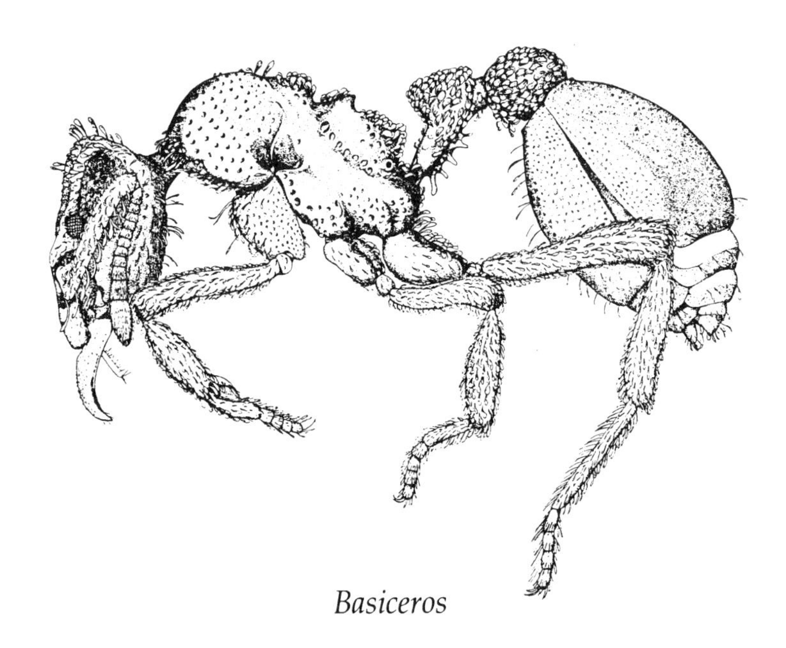

*Basiceros*

*Bondroitia*

*Atta texana,* United States (Smith, 1947a)

*Baracidris meketra,* West Africa (Bolton, 1981b)

*Basiceros conjugans,* Ecuador (Brown, 1974b)

*Basiceros (= Aspididris) discigera,* Brazil (Brown and Kempf, 1960)

*Blepharidatta brasiliensis,* Brazil (Wheeler, 1915)

*Bondroitia lujae,* Zaïre (Bolton, 1987)

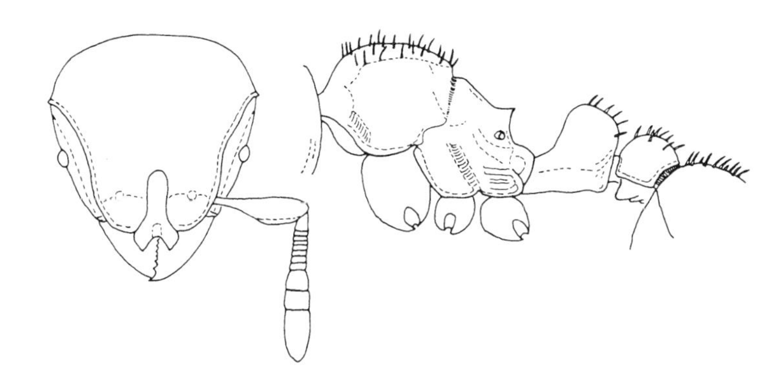

*Calyptomyrmex*

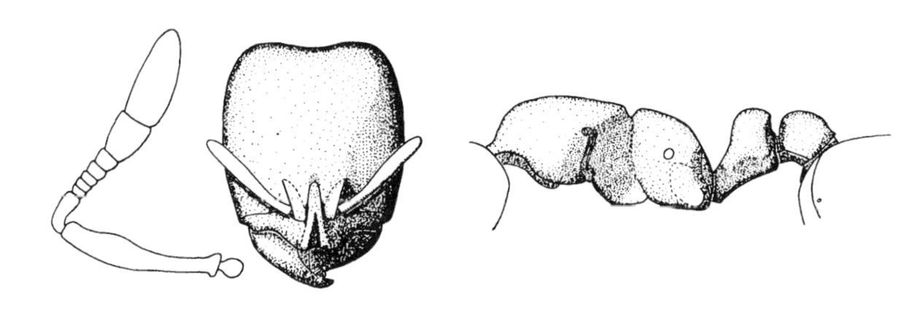

*Carebara*

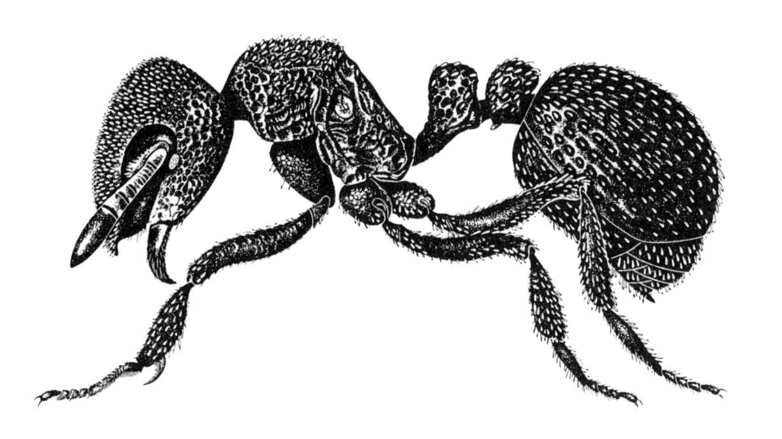

*Calyptomyrmex*

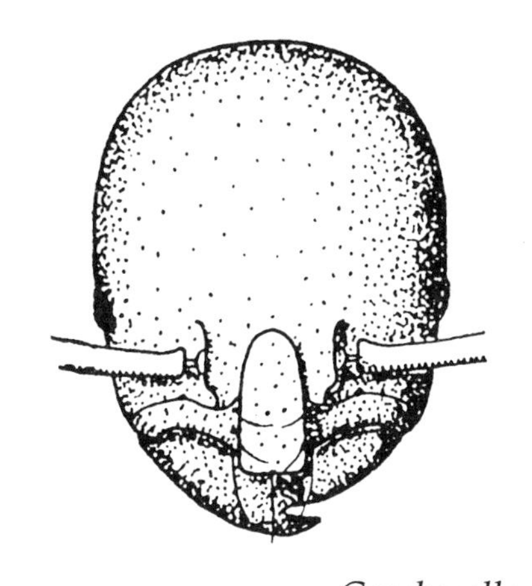

*Carebarella*

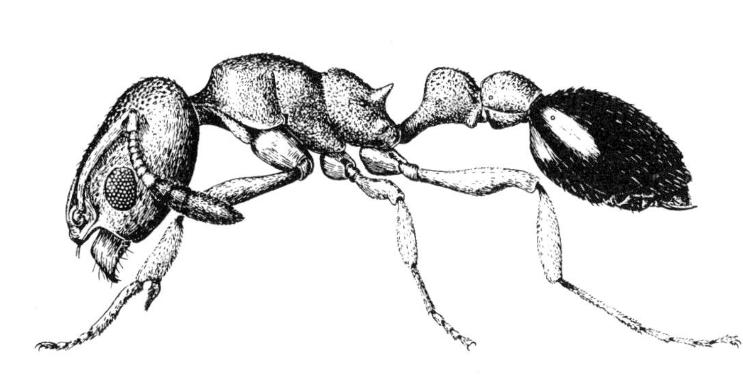

*Cardiocondyla*

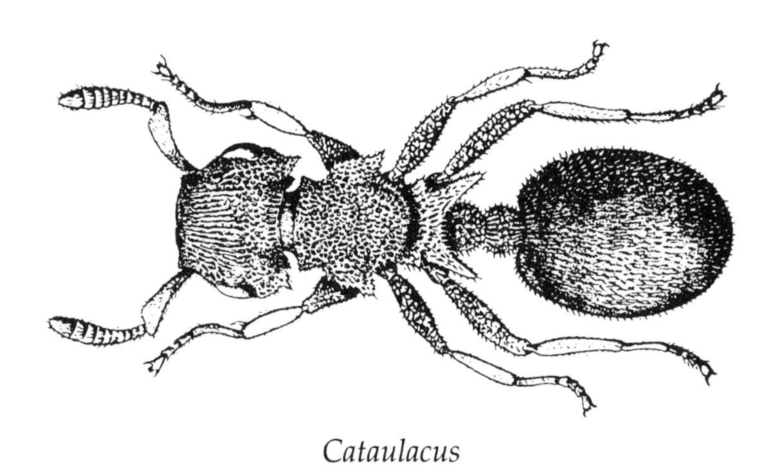

*Cataulacus*

*Calyptomyrmex barak,* Ghana (Bolton, 1981a)

*Calyptomyrmex* sp., Australia (R. W. Taylor, unpublished; F. Nanninga, artist)

*Cardiocondyla emeryi,* United States (Smith, 1947a)

*Carebara winifredae,* Guyana (Ettershank, 1966)

*Carebarella bicolor,* Argentina (Ettershank, 1966)

*Cataulacus erinaceus,* Zaïre (Wheeler, 1922)

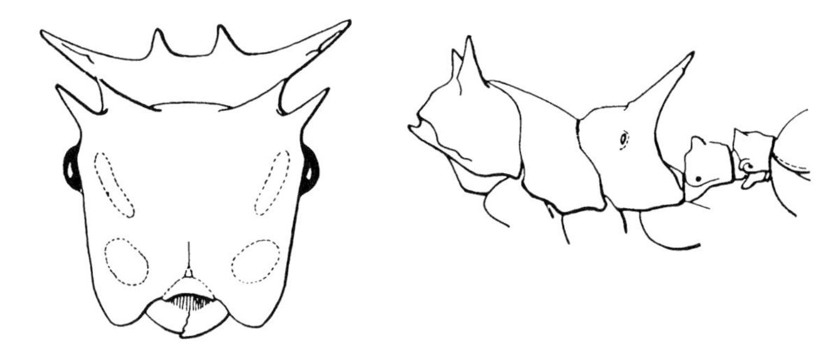

*Cephalotes*

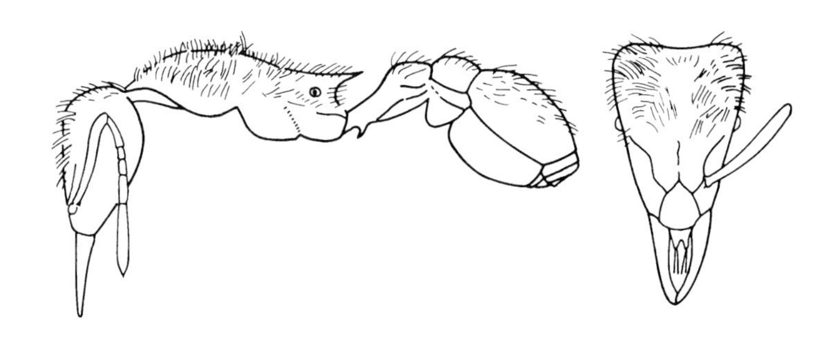

*Cladarogenys*

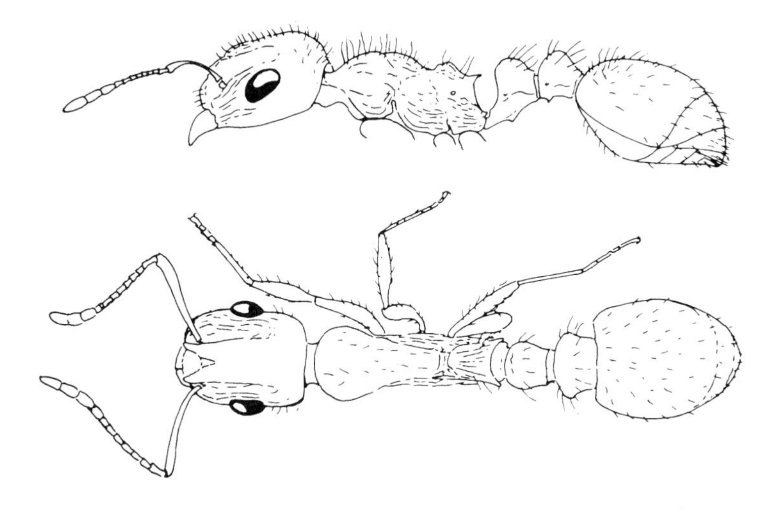

*Chalepoxenus*

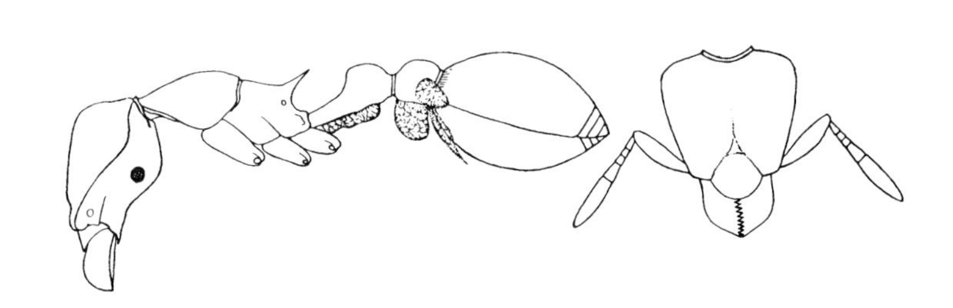

*Codiomyrmex*

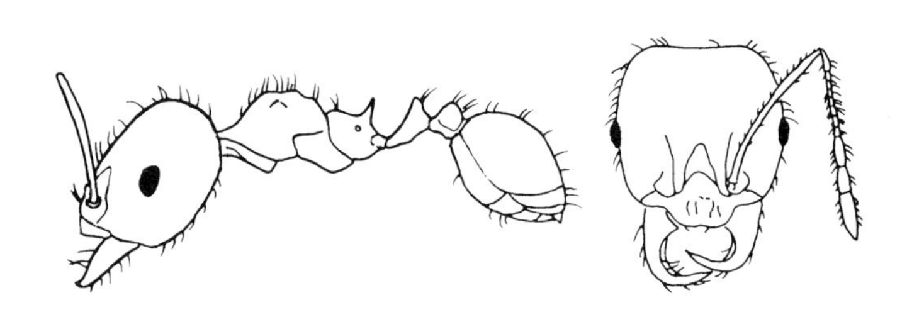

*Chimaeridris*

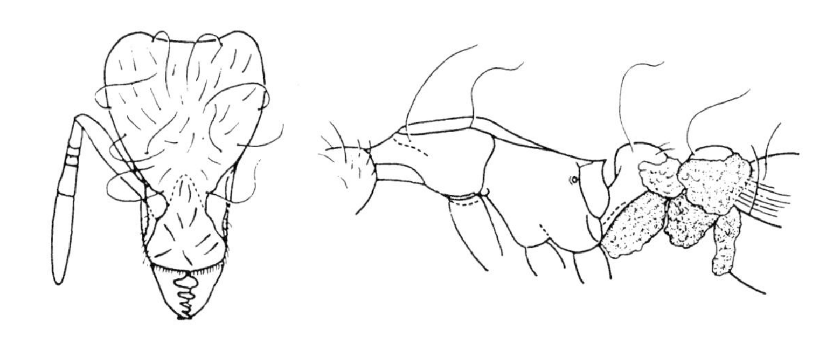

*Codiomyrmex*

*Cephalotes decemspinosus,* French Guiana (Kempf, 1951)

*Chalepoxenus muellerianus,* Switzerland (Kutter, 1978)

*Chimaeridris boltoni,* Indonesia (Wilson, 1989)

*Cladarogenys lasia,* Gabon (based on Brown, 1976a)

*Codiomyrmex flagellatus,* Australia (Taylor, 1962a)

*Codiomyrmex thaxteri,* Trinidad (Wheeler, 1916b)

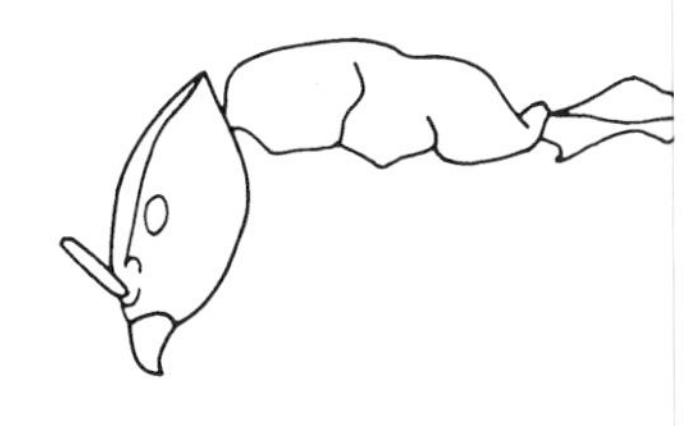

*Dil*

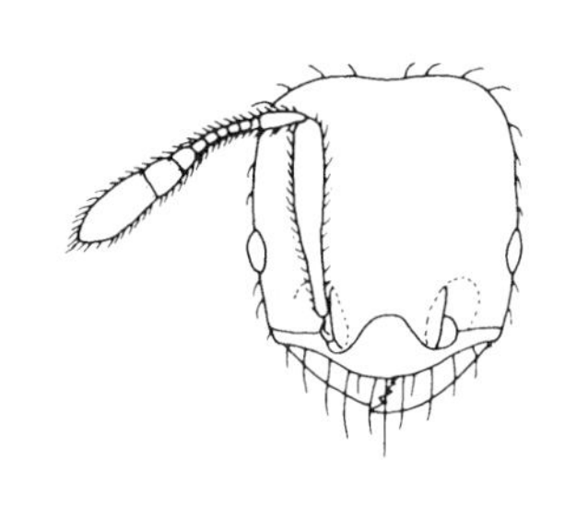

*Dip*

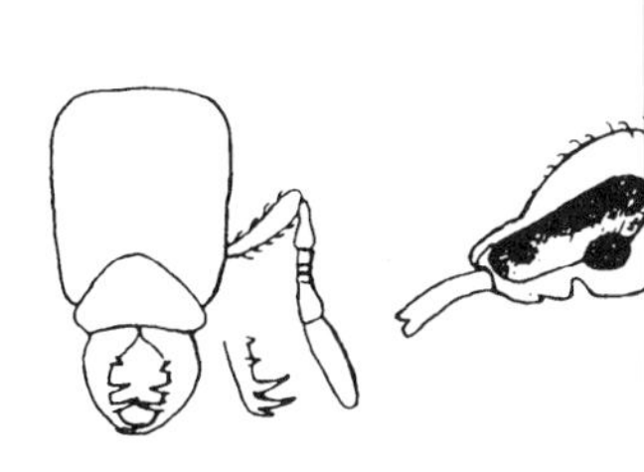

*D*

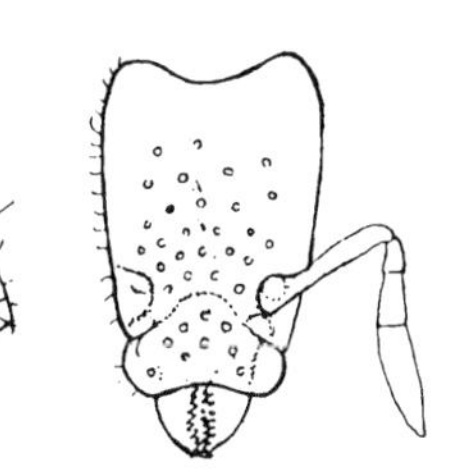

*Codioxenus*

*Crematogaster*

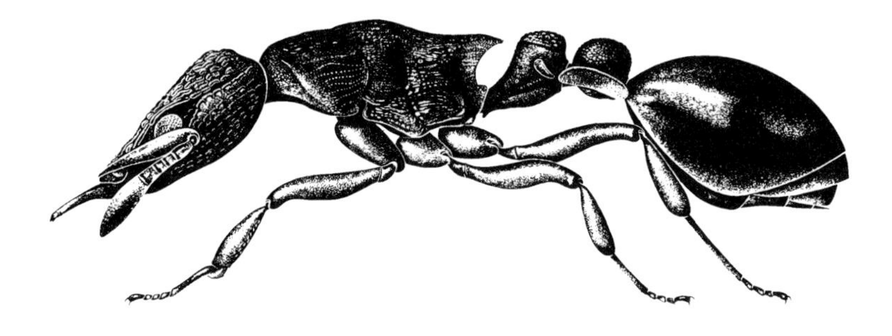

*Colobostruma*

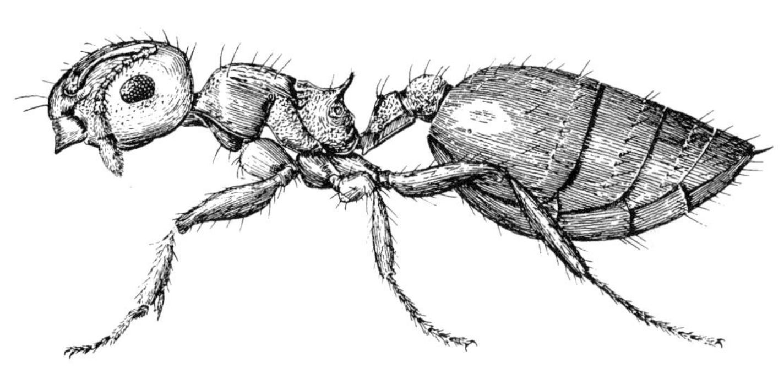

*Crematogaster (= Orthocrema)*

*Creightonidris*

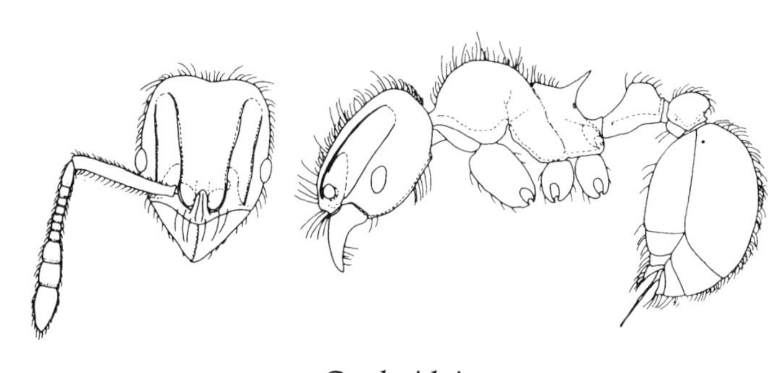

*Cyphoidris*

*Dilobocondyla cataulacoides,* Papua N

*Diplomorium longipenne,* South Afri

*Dorisidris nitens,* Cuba (Santschi, 1

*Codioxenus simulans,* Cuba (Santschi, 1931)

*Colobostruma* sp., Australia (R. W. Taylor, unpublished; F. Nanninga, artist)

*Creightonidris scambognatha,* Brazil (Brown and Kempf, 1960)

*Crematogaster clara,* United States (Smith, 1947a)

*Crematogaster (= Orthocrema) minutissima,* United States (Smith, 1947a)

*Cyphoidris spinosa,* Zaïre (Bolton, 1981b)

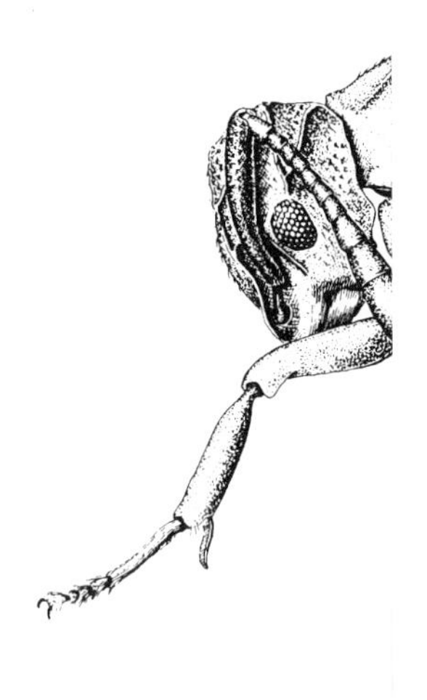

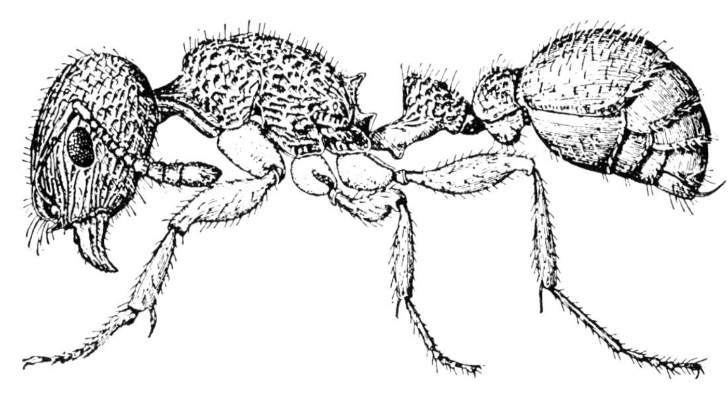

*Ephebomyrmex*

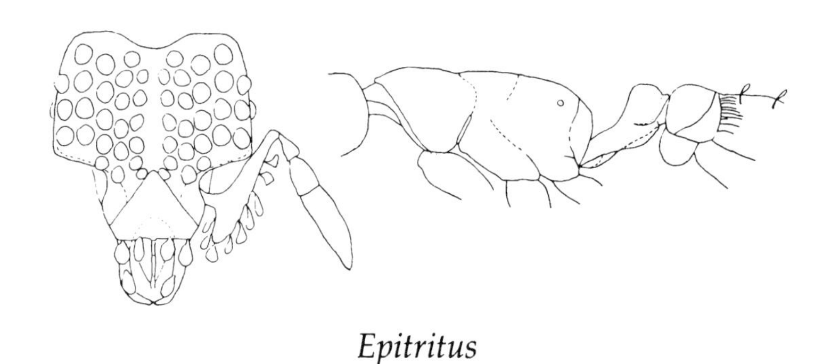

*Epitritus*

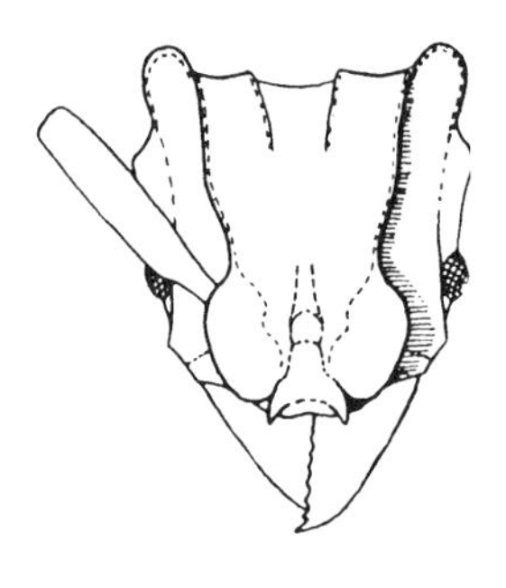

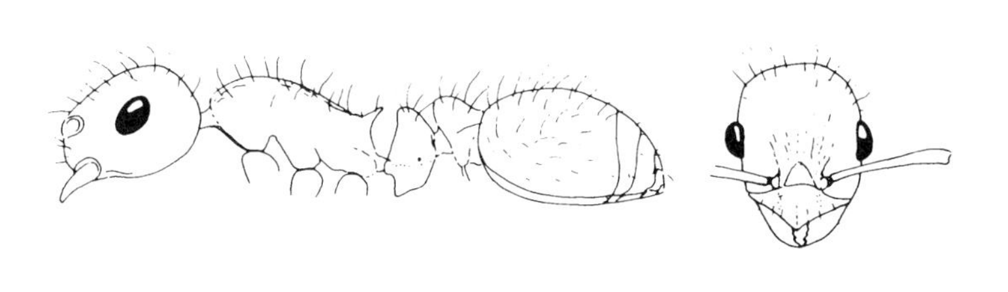

*Epimyrma*

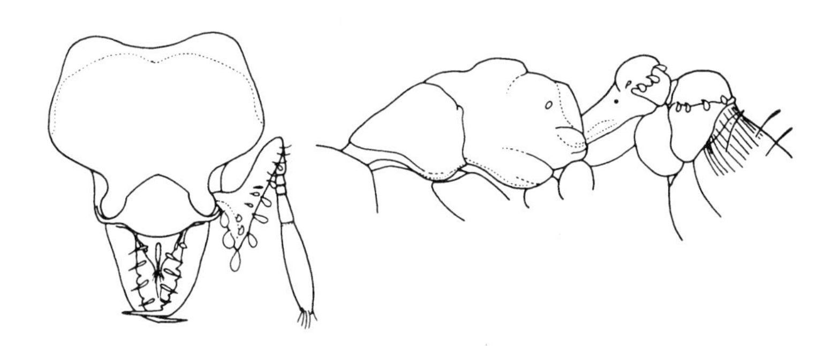

*Epitritus*

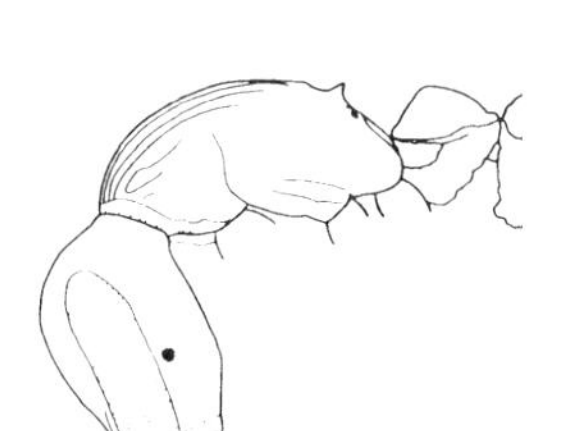

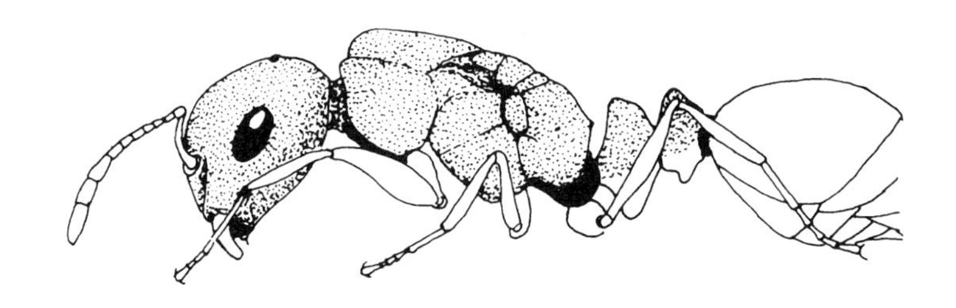

*Epimyrma*

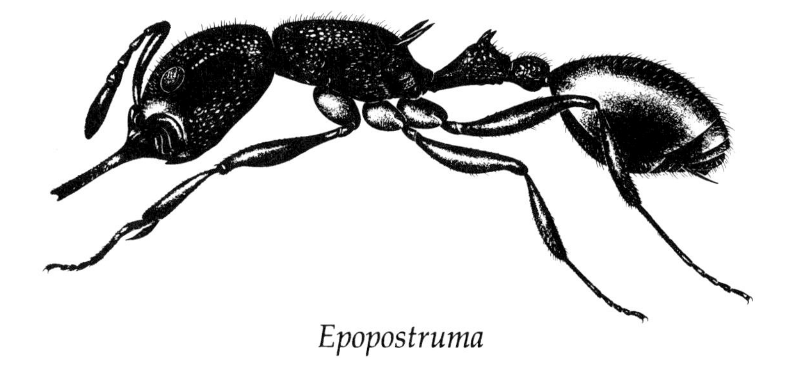

*Epopostruma*

*Cyphomyrmex rimosus*, Unit

*Cyphomyrmex strigatus*, Bra

*Dacetinops cibdela*, Papua N

*Ephebomyrmex imberbiculus*, United States (Smith, 1947a)

*Epimyrma kraussei*, Europe (Kutter, 1973b)

*Epimyrma stumperi*, Switzerland (Kutter, 1978)

*Epitritus minimus*, Ghana (Bolton, 1971)

*Epitritus murphyi*, tropical Asia (Taylor, 1968)

*Epopostruma frosti*, Australia (R. W. Taylor, unpublished; F. Nanninga, artist)

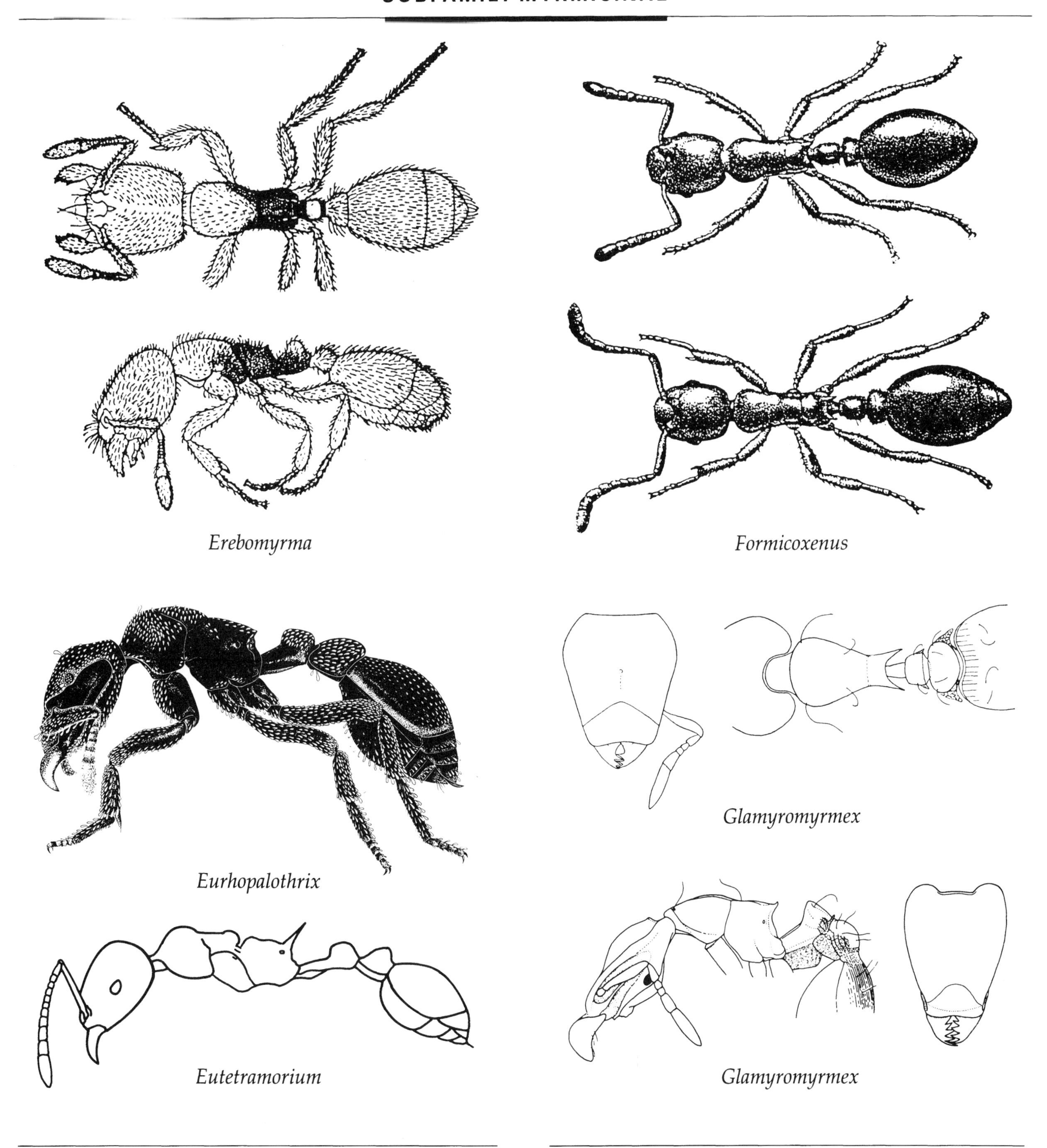


*Erebomyrma longi*, United States (Wheeler, 1910a)

*Eurhopalothrix procera*, Australia (R. W. Taylor, unpublished; R. J. Kohout, artist)

*Eutetramorium mocquerysi*, Madagascar

*Formicoxenus nitidulus*, Europe (Wheeler, 1910a); see also Figure 12–17

*Glamyromyrmex appretiatus*, Brazil (Borgmeier, 1954)

*Glamyromyrmex tetragnathus*, Angola (Taylor, 1965c)

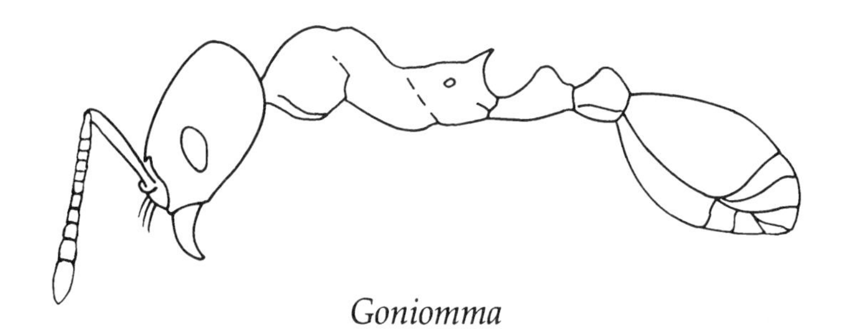

*Goniomma*

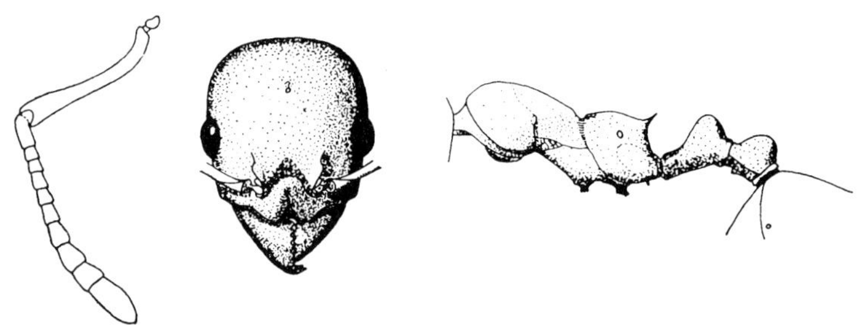

*Huberia*

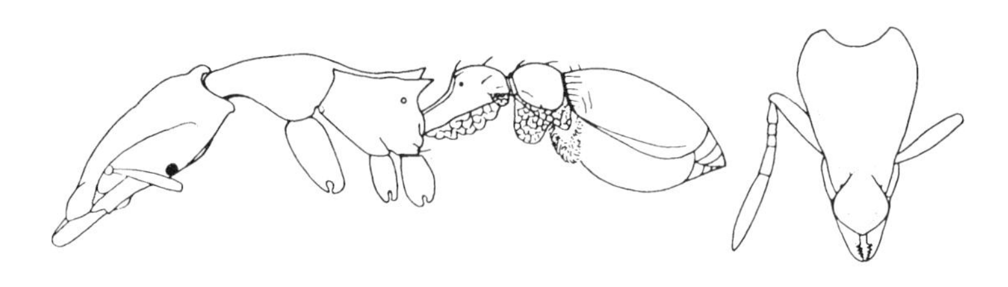

*Gymnomyrmex*

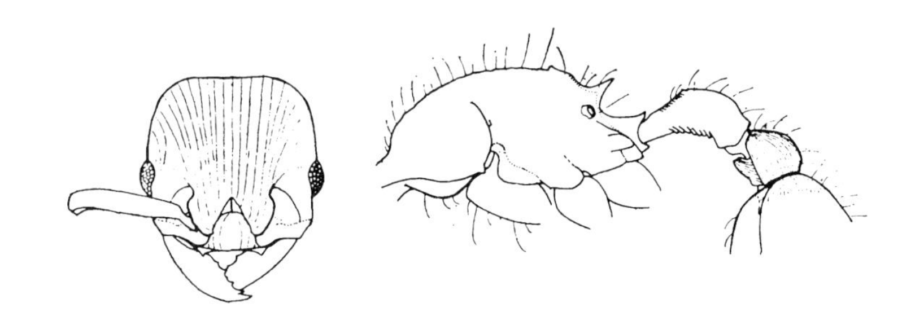

*Hylomyrma*

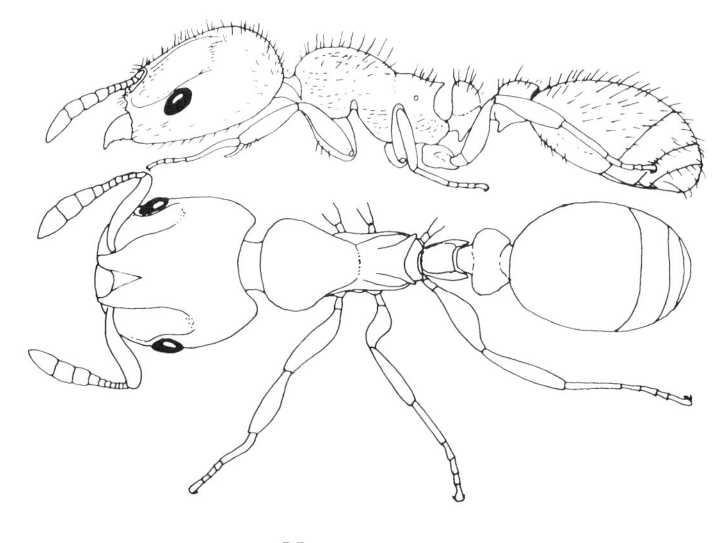

*Harpagoxenus*

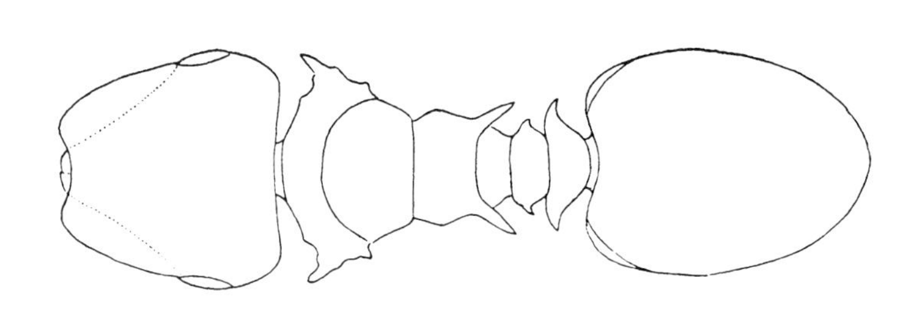

*Hypocryptocerus*

*Goniomma blanci*, Tunisia

*Gymnomyrmex splendens*, Brazil (Borgmeier, 1954)

*Harpagoxenus sublaevis*, Europe (Kutter, 1977); see also Figure 12–11

*Huberia striata*, New Zealand (Ettershank, 1966)

*Hylomyrma dentiloba*, Panama (Kempf, 1973)

*Hypocryptocerus haemorrhoidalis*, Haiti (Wheeler and Mann, 1914)

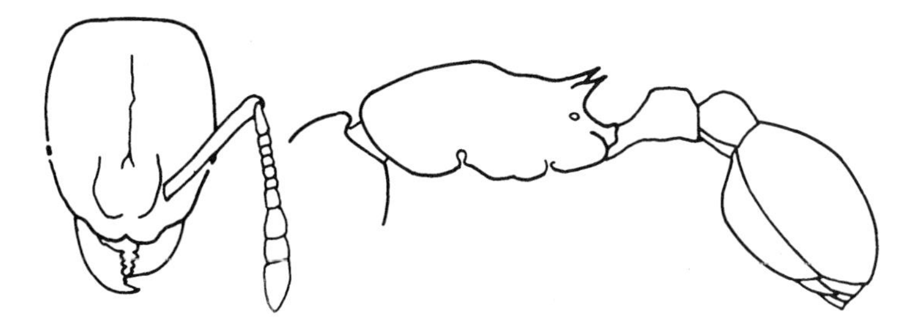

*Indomyrma*

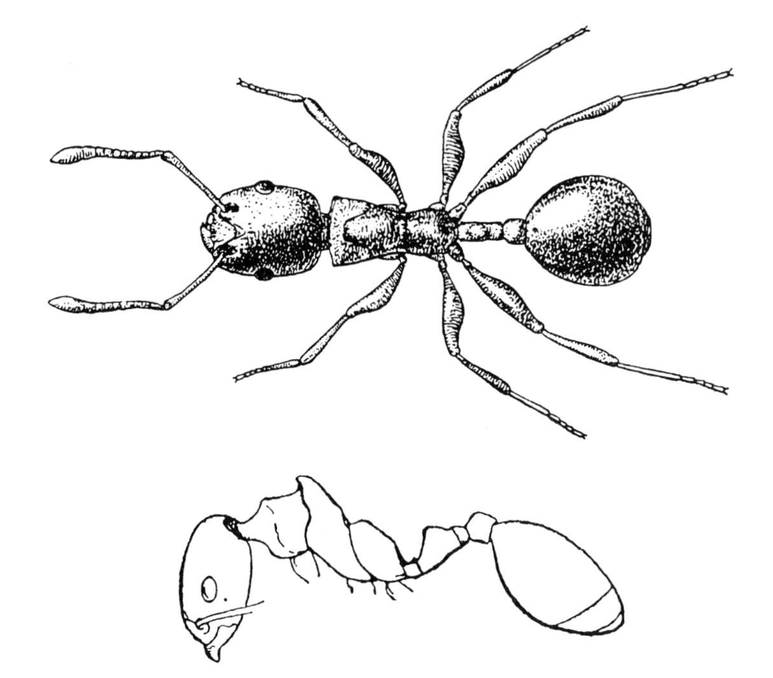

*Ireneopone*

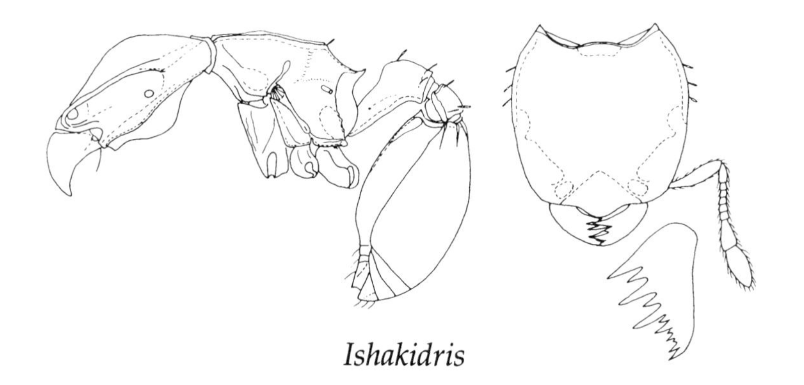

*Ishakidris*

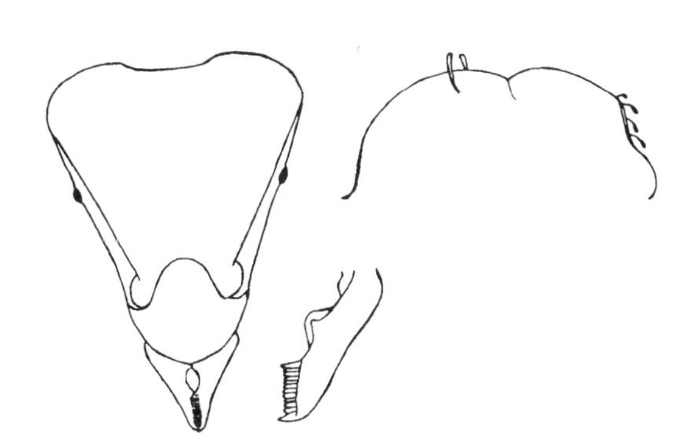

*Kyidris*

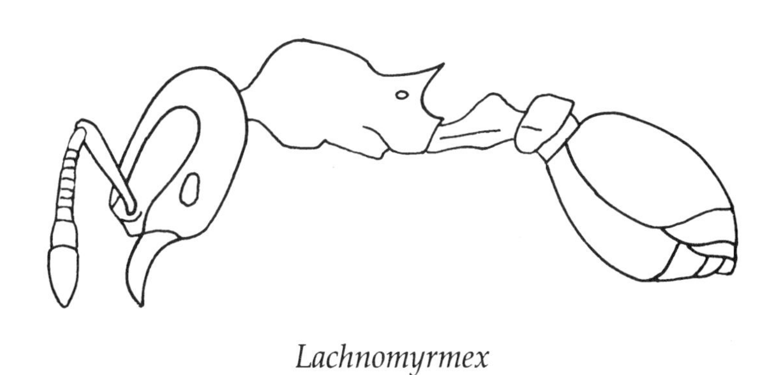

*Lachnomyrmex*

*Leptothorax*

*Indomyrma daspyx*, India

*Ireneopone gibber*, Mauritius (Donisthorpe, 1946)

*Ishakidris ascitaspis*, Sarawak (Bolton, 1984)

*Kyidris mutica*, Japan (Brown, 1949)

*Lachnomyrmex plaumanni*, Brazil

*Leptothorax muscorum*, Canada (Smith, 1947a)

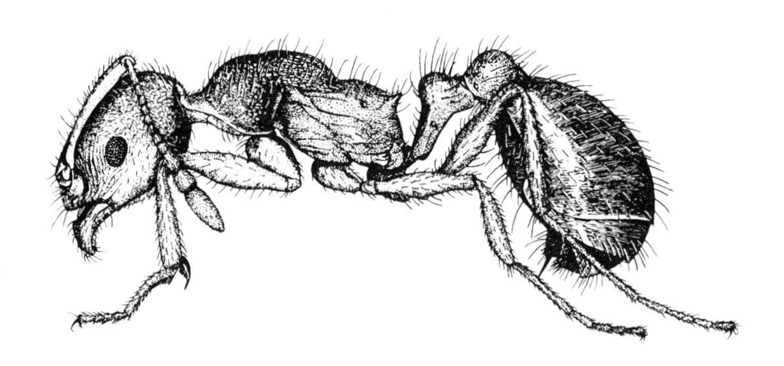

*Leptothorax (= Dichothorax)*

*Lordomyrma*

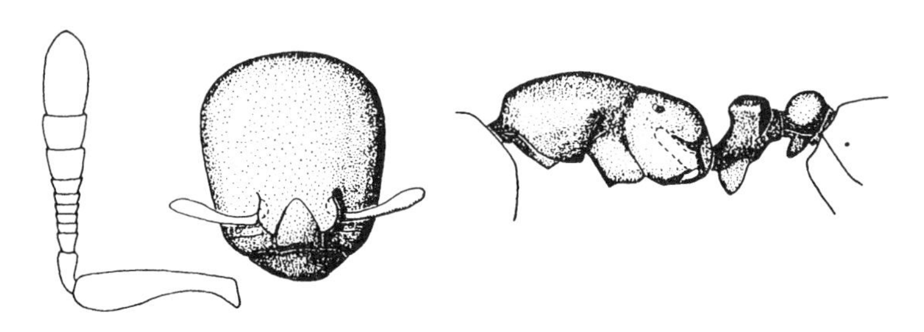

*Liomyrmex*

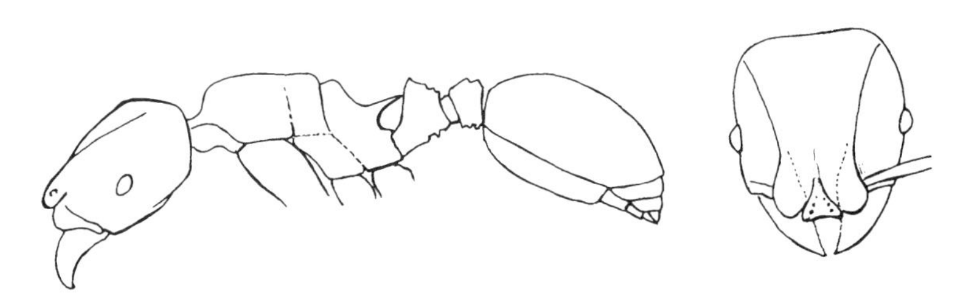

*Lordomyrma (= Prodicroaspis)*

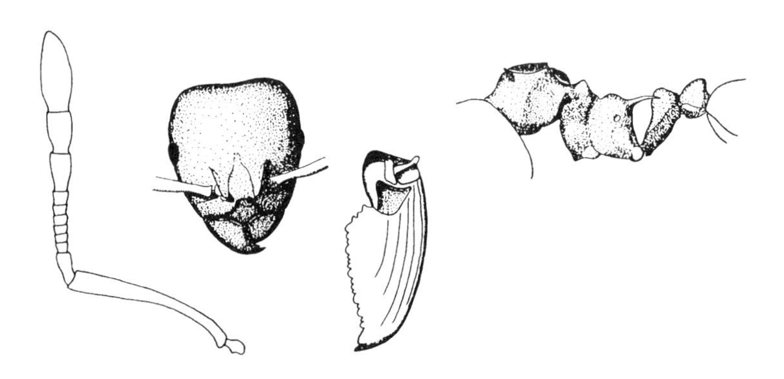

*Lophomyrmex*

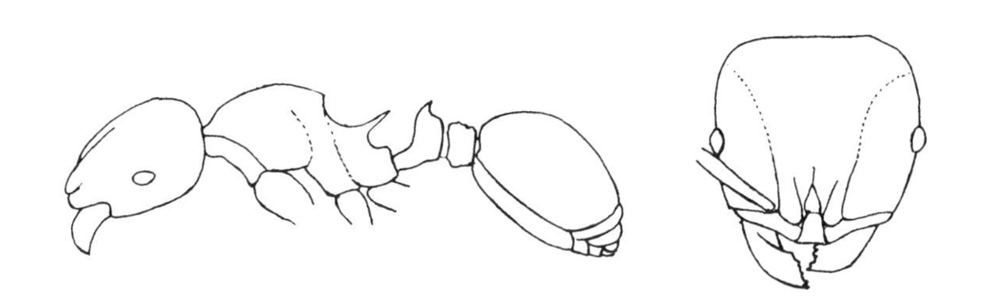

*Lordomyrma (= Promeranoplus)*

*Leptothorax (= Dichothorax)* sp., United States (Smith, 1947a)

*Liomyrmex aurianus*, Philippines (Ettershank, 1966)

*Lophomyrmex quadrispinosus*, India (Ettershank, 1966)

*Lordomyrma* sp., Australia (R. W. Taylor, unpublished; R. J. Kohout, artist)

*Lordomyrma (= Prodicroaspis) sarasini*, New Caledonia (Emery, 1914)

*Lordomyrma (= Promeranoplus) rouxi*, New Caledonia (Emery, 1914)

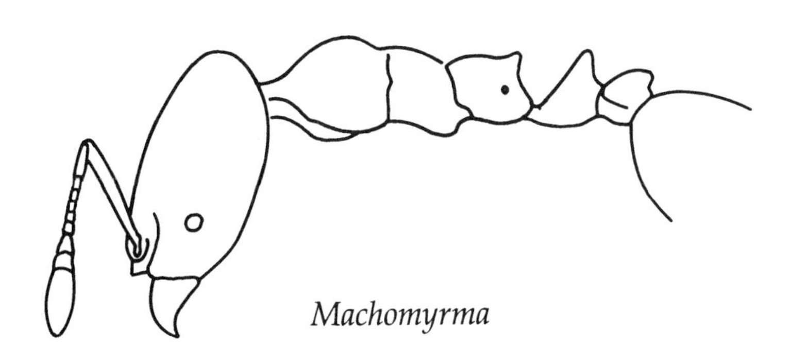

*Machomyrma*

*Mayriella*

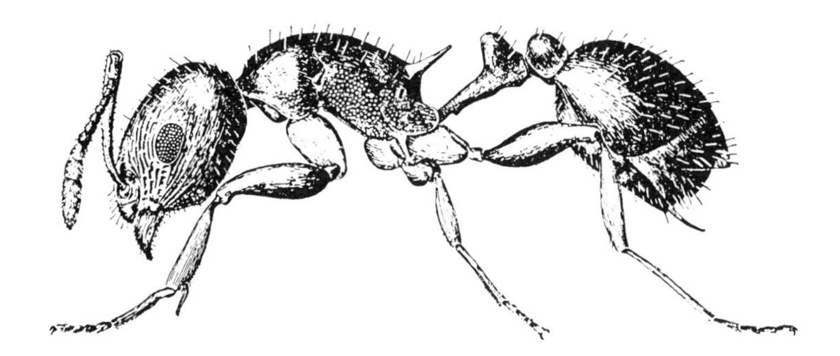

*Macromischa*

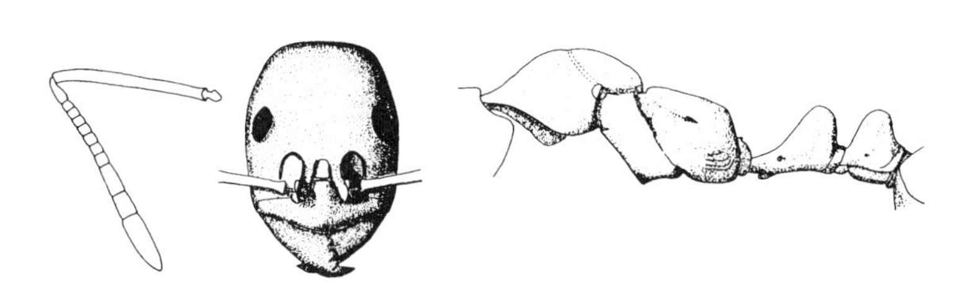

*Megalomyrmex*

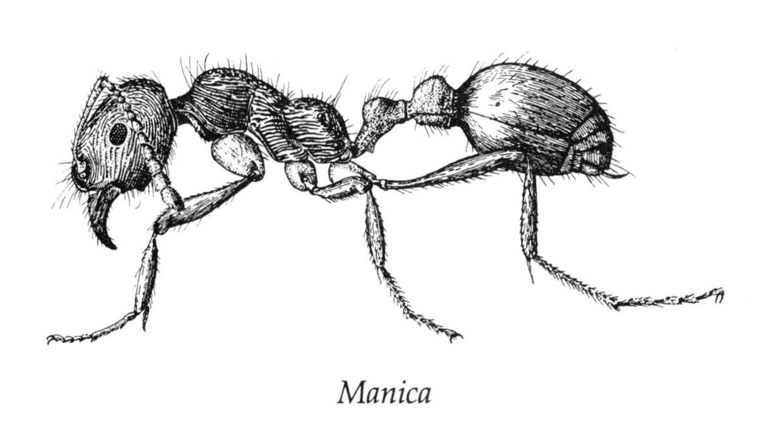

*Manica*

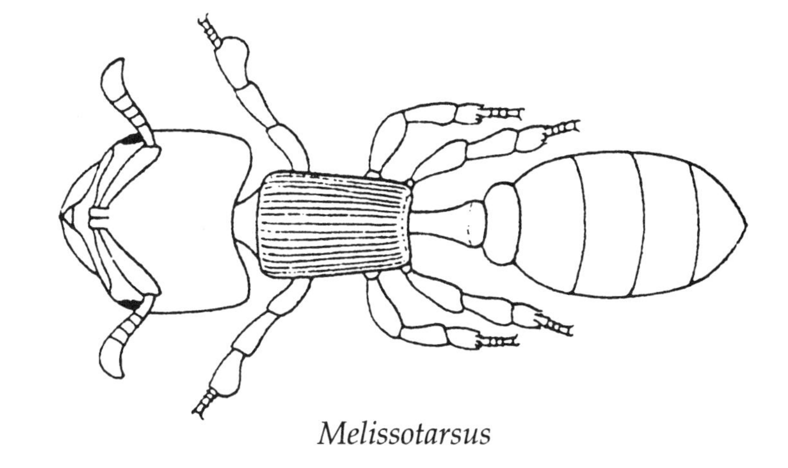

*Melissotarsus*

*Machomyrma* sp., Australia

*Macromischa subditiva*, United States (Smith, 1947a)

*Manica mutica*, United States (Smith, 1947a)

*Mayriella abstinens*, Australia (R. W. Taylor, unpublished; R. J. Kohout, artist)

*Megalomyrmex duckei*, Brazil (Ettershank, 1966)

*Melissotarsus beccarii*, Zimbabwe (Arnold, 1915–26)

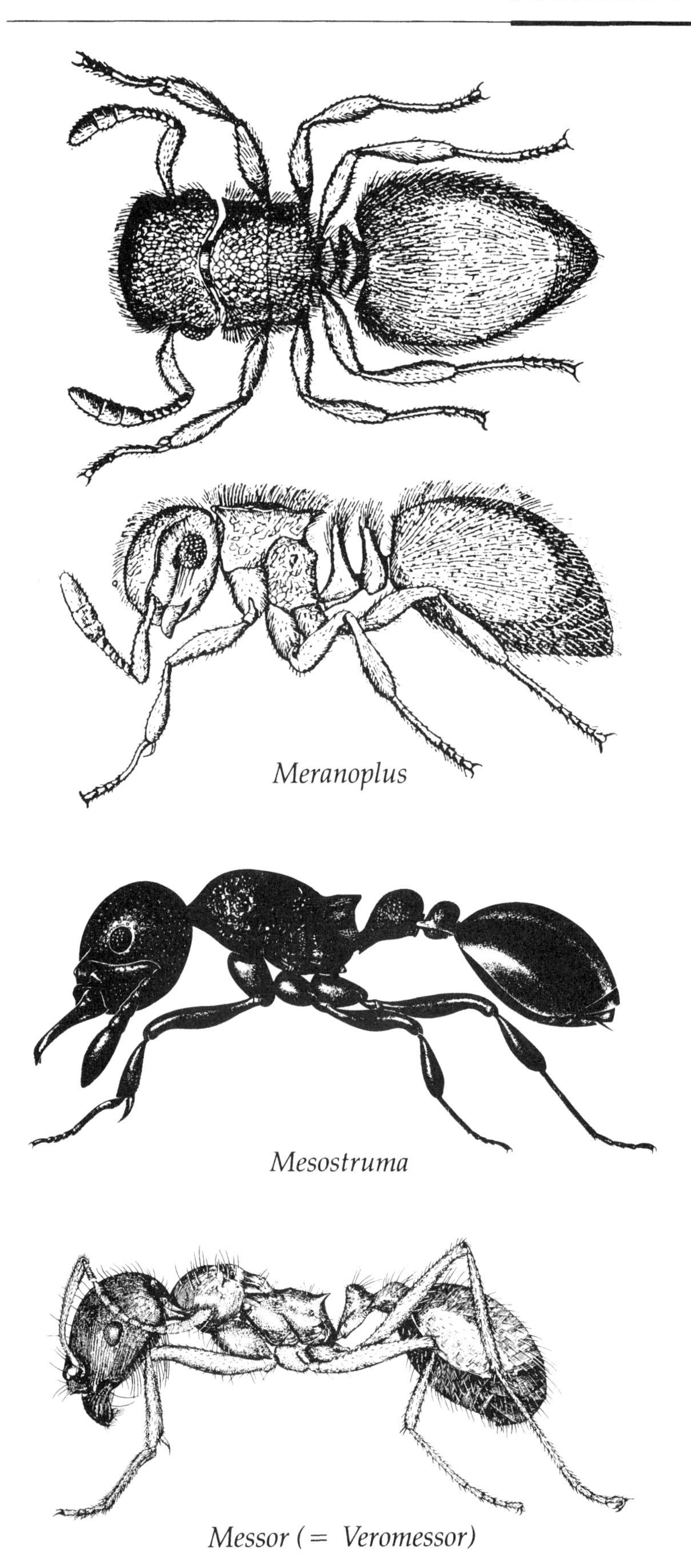

*Meranoplus*

*Mesostruma*

*Messor (= Veromessor)*

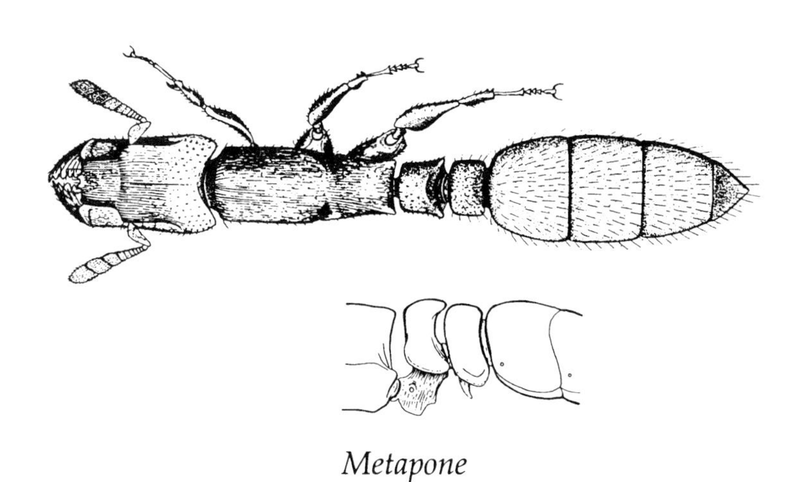

*Metapone*

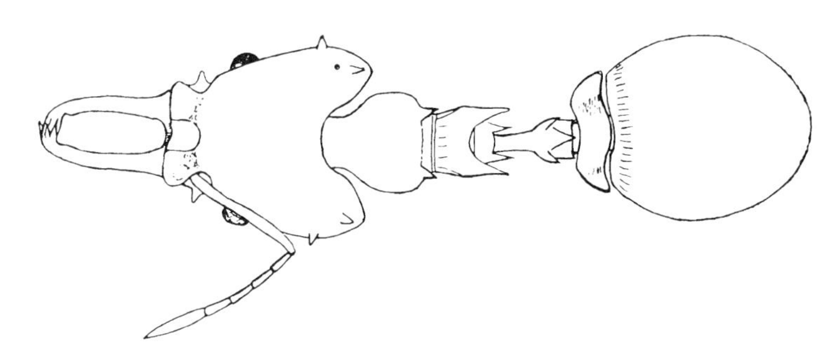

*Microdaceton*

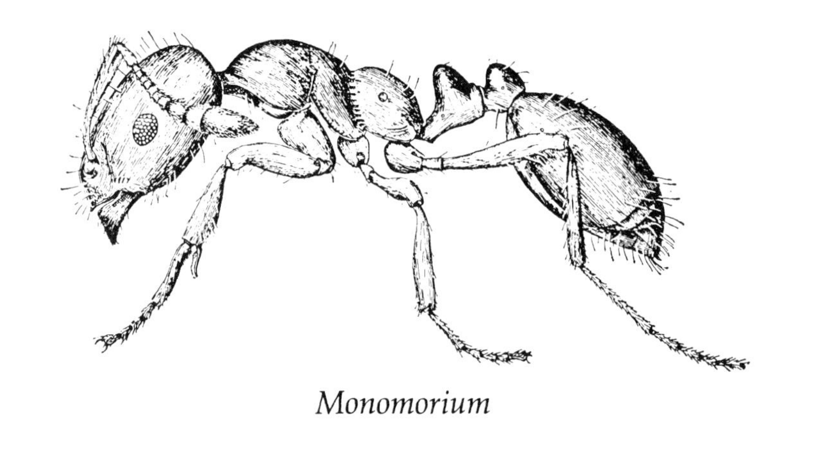

*Monomorium*

*Meranoplus nanus*, Zaïre (Wheeler, 1922)

*Mesostruma browni*, Australia (R. W. Taylor, unpublished; F. Nanninga, artist)

*Messor (= Veromessor) pergandei*, United States (Smith, 1947a)

*Metapone madagascarica*, Madagascar (Gregg, 1958b)

*Microdaceton exornatum*, South Africa (Santschi, 1914)

*Monomorium minimum*, United States (Smith, 1947a)

Monomorium (= Chelaner)

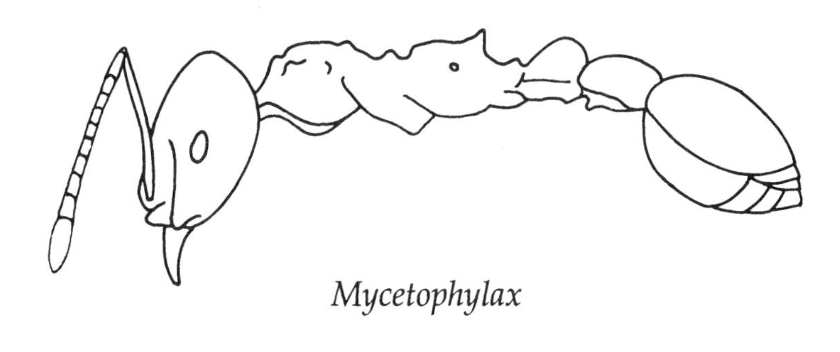

Mycetophylax

Monomorium (= Xenhyboma)

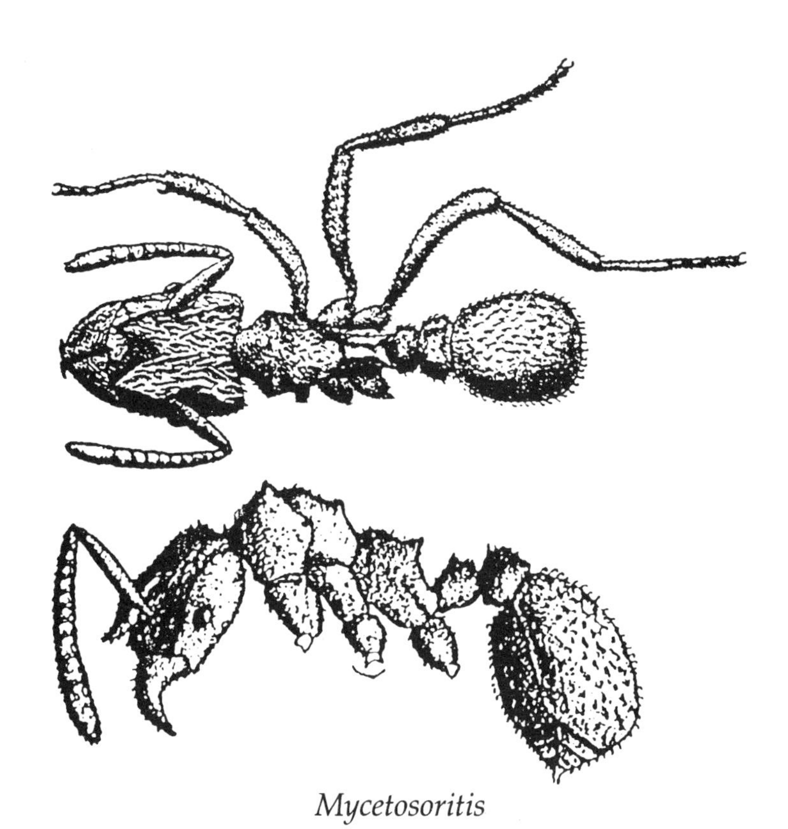

Mycetosoritis

Mycetarotes

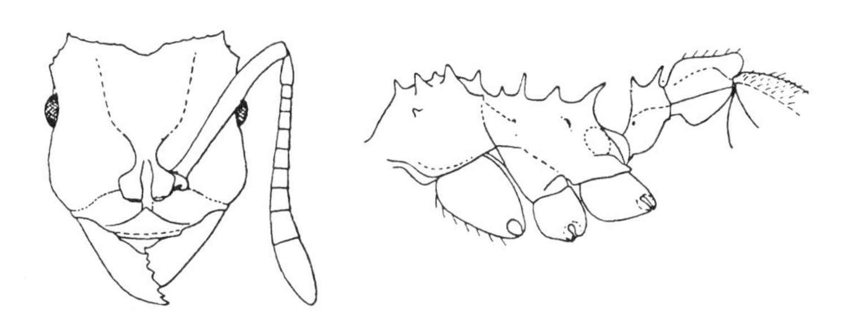

Mycocepurus

*Monomorium (= Chelaner)* sp., Australia (R. W. Taylor, unpublished; R. J. Kohout, artist)

*Monomorium (= Xenhyboma) mystes*, Canary Islands (Kutter, 1972)

*Mycetarotes parallelus*, Brazil

*Mycetophylax emeryi*, Argentina

*Mycetosoritis hartmani*, United States (Wheeler, 1910a)

*Mycocepurus goeldii*, Brazil (Kempf, 1963a)

*Myrmecina*

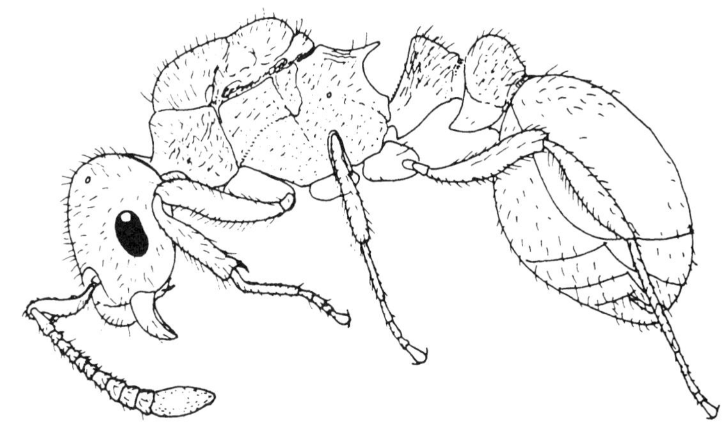

*Myrmica (= Sifolinia)*

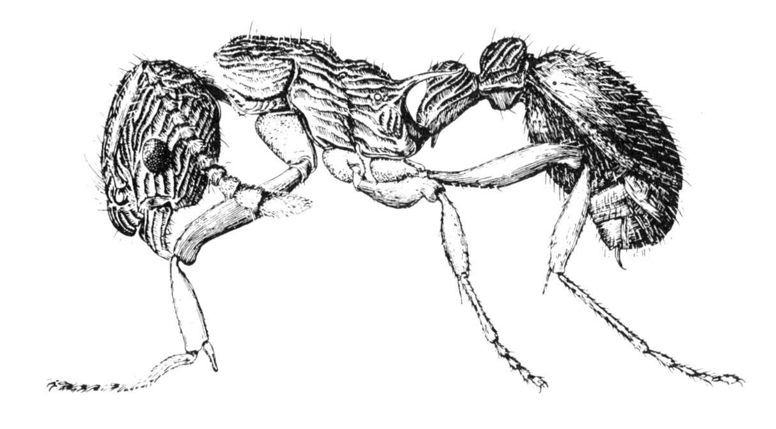

*Myrmica*

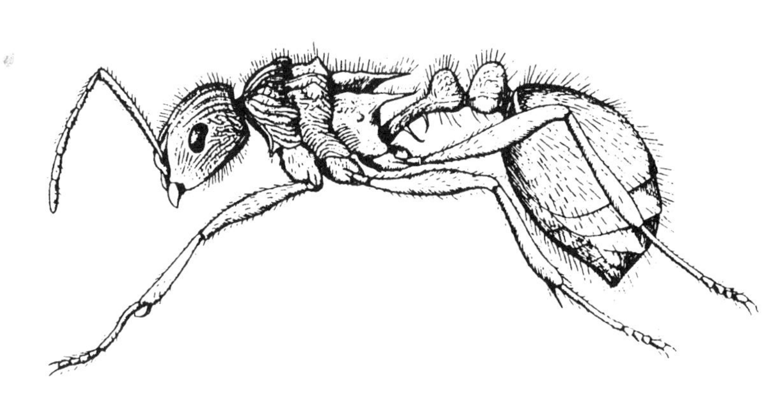

*Myrmicaria*

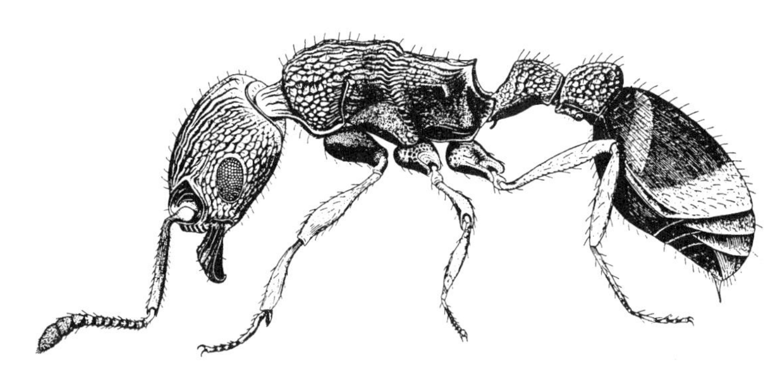

*Myrmica (= Paramyrmica)*

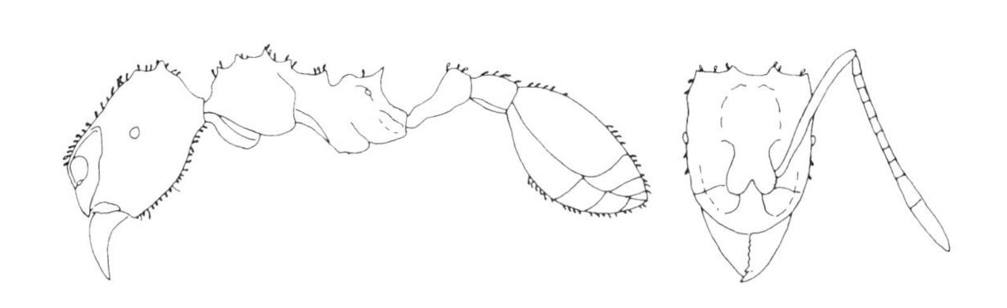

*Myrmicocrypta*

*Myrmecina americana,* United States (Smith, 1947a)

*Myrmica punctiventris,* United States (Smith, 1947a)

*Myrmica (= Paramyrmica) rugiventris,* United States (Smith, 1947a)

*Myrmica (= Sifolinia) lemasnei,* France (Kutter, 1973a)

*Myrmicaria salambo,* Zaïre (Wheeler, 1922)

*Myrmicocrypta spinosa,* Guyana

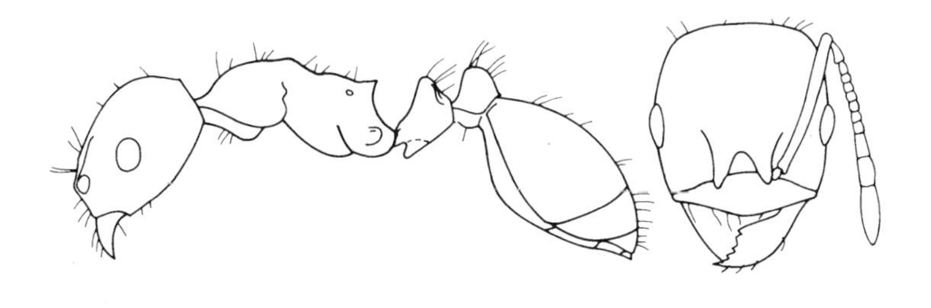

*Myrmoxenus*

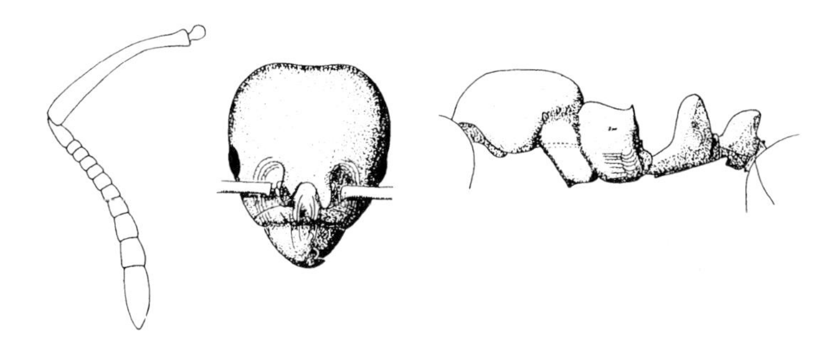

*Nothidris*

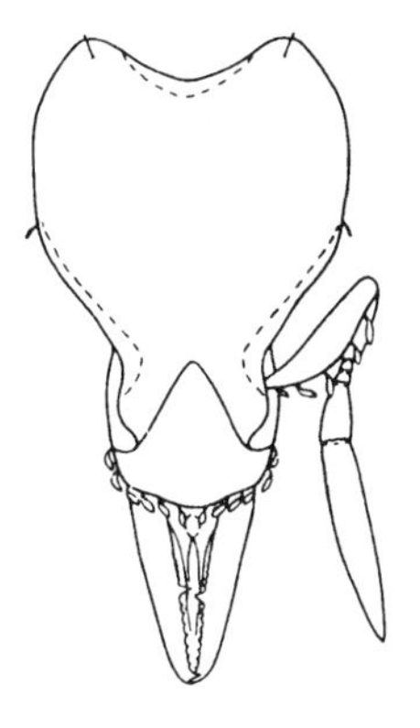

*Neostruma*

*Ochetomyrmex*

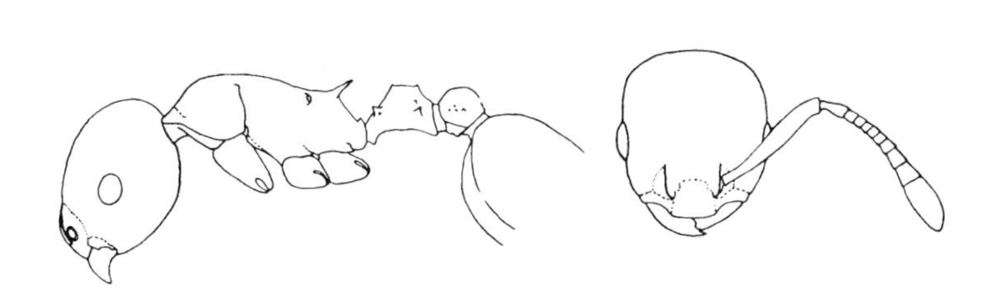

*Nesomyrmex*

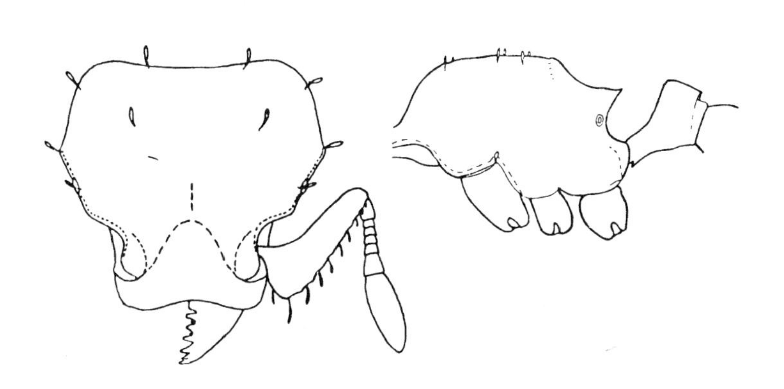

*Octostruma*

*Myrmoxenus gordiagini*, Yugoslavia

*Neostruma zeteki*, Panama (Brown, 1959)

*Nesomyrmex rutilans*, Guyana (Kempf, 1958d)

*Nothidris latastei*, Chile (Ettershank, 1966)

*Ochetomyrmex subpolita*, Colombia

*Octostruma balzani*, Central America (Brown and Kempf, 1960)

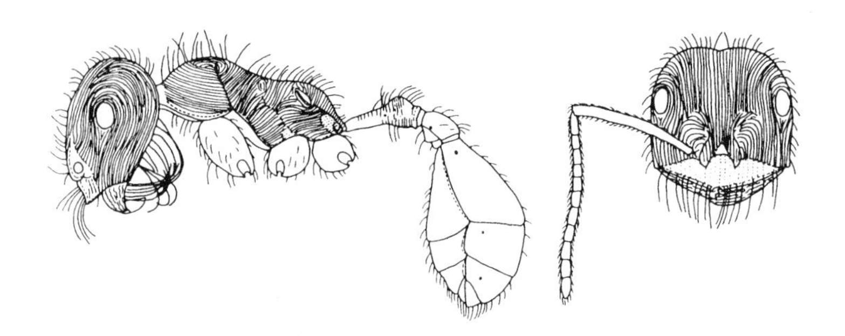

*Ocymyrmex*

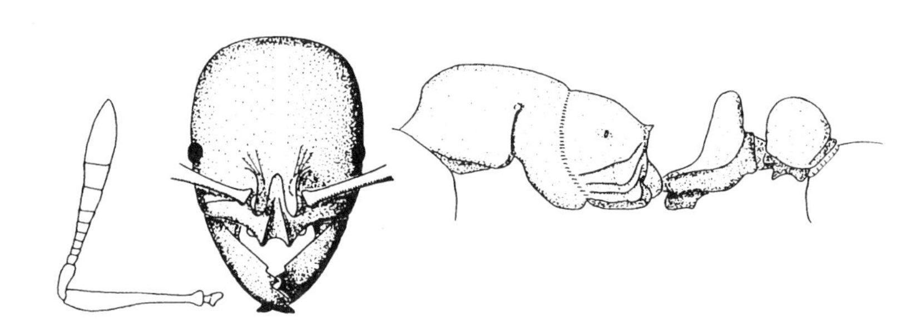

*Oxyepoecus*

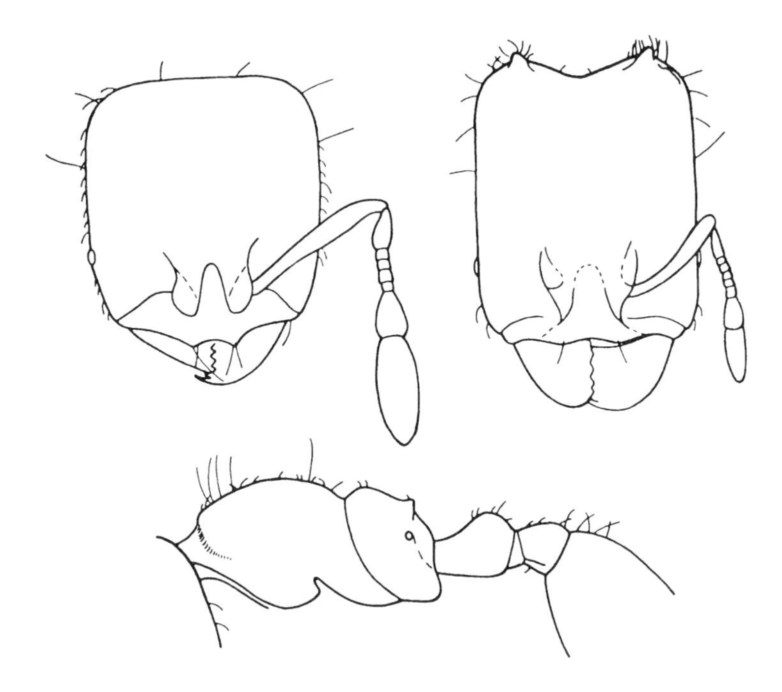

*Oligomyrmex*

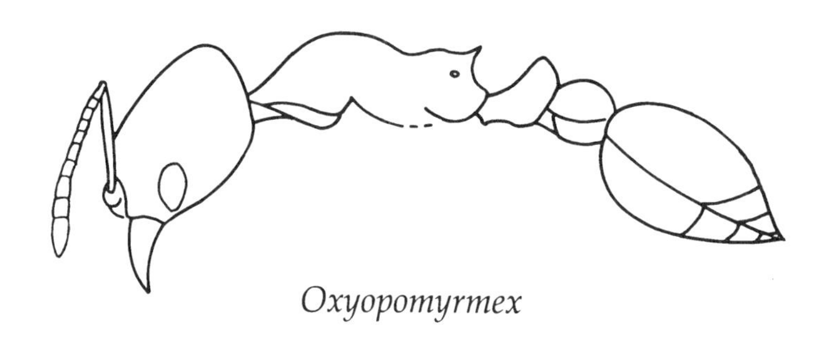

*Oxyopomyrmex*

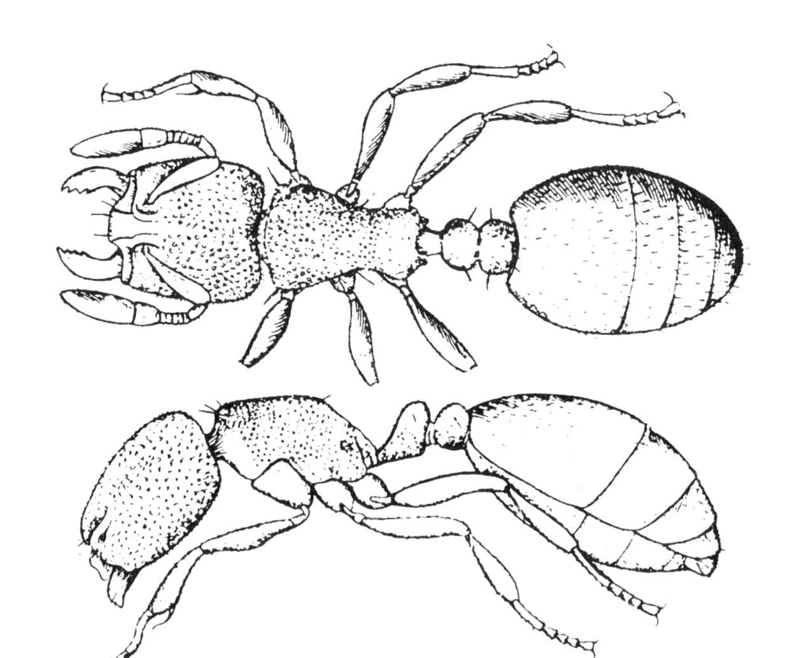

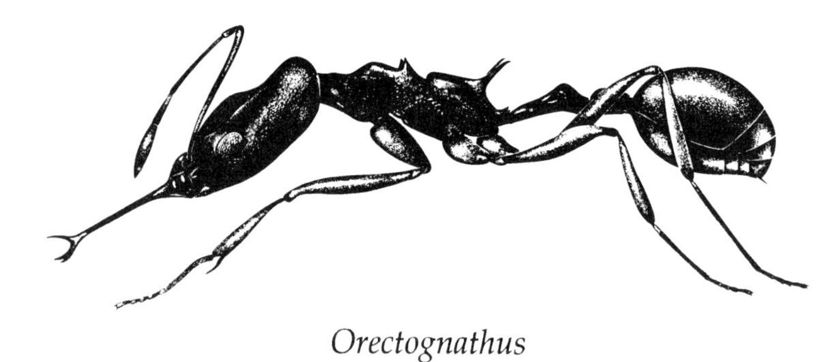

*Orectognathus*

*Paedalgus*

*Ocymyrmex nitidulus*, Somalia (Bolton, 1981b)

*Oligomyrmex atomus*, Samoa (Wilson and Taylor, 1967b)

*Orectognathus antennatus*, Australia (R. W. Taylor, unpublished; F. Nanninga, artist)

*Oxyepoecus vezenyii*, Paraguay (Ettershank, 1966)

*Oxyopomyrmex santschii*, Sicily

*Paedalgus termitolestes*, Zaïre (Wheeler, 1922)

*Paratopula*

*Peronomyrmex*

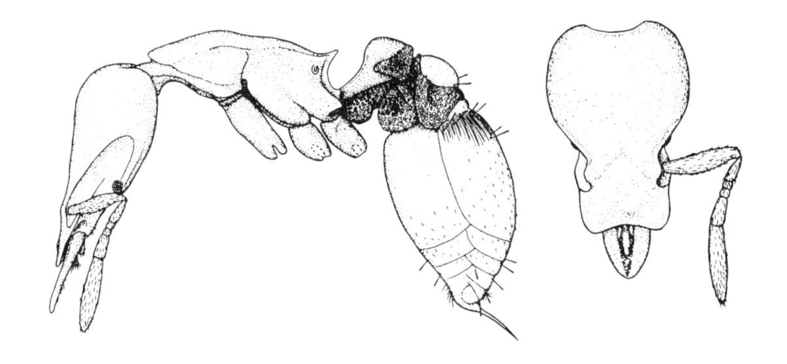

*Pentastruma*

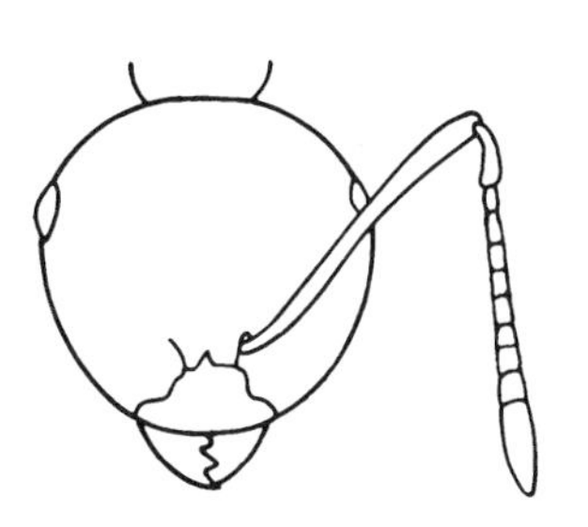

*Phacota*

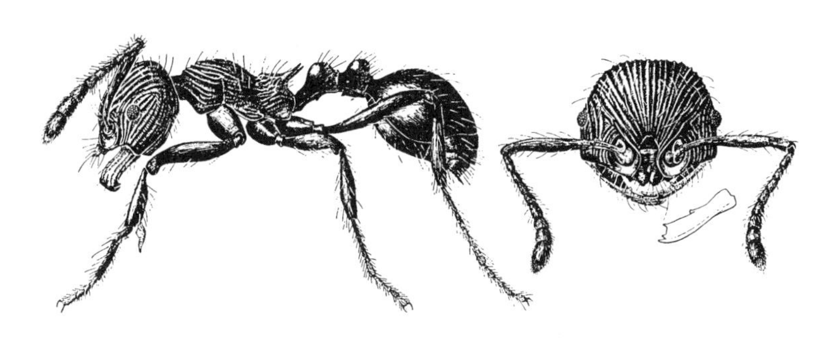

*Perissomyrmex*

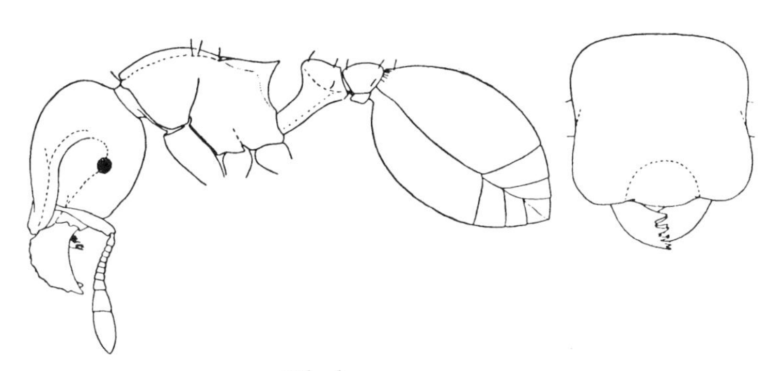

*Phalacromyrmex*

*Paratopula catocha*, Indonesia

*Pentastruma sauteri*, Taiwan (Brown and Boisvert, 1978)

*Perissomyrmex snyderi*, Guatemala (Smith, 1947b)

*Peronomyrmex overbecki*, Australia (Taylor, 1970b)

*Phacota sicheli*, Spain (modified from Roger, 1862)

*Phalacromyrmex fugax*, Brazil (Kempf, 1960b)

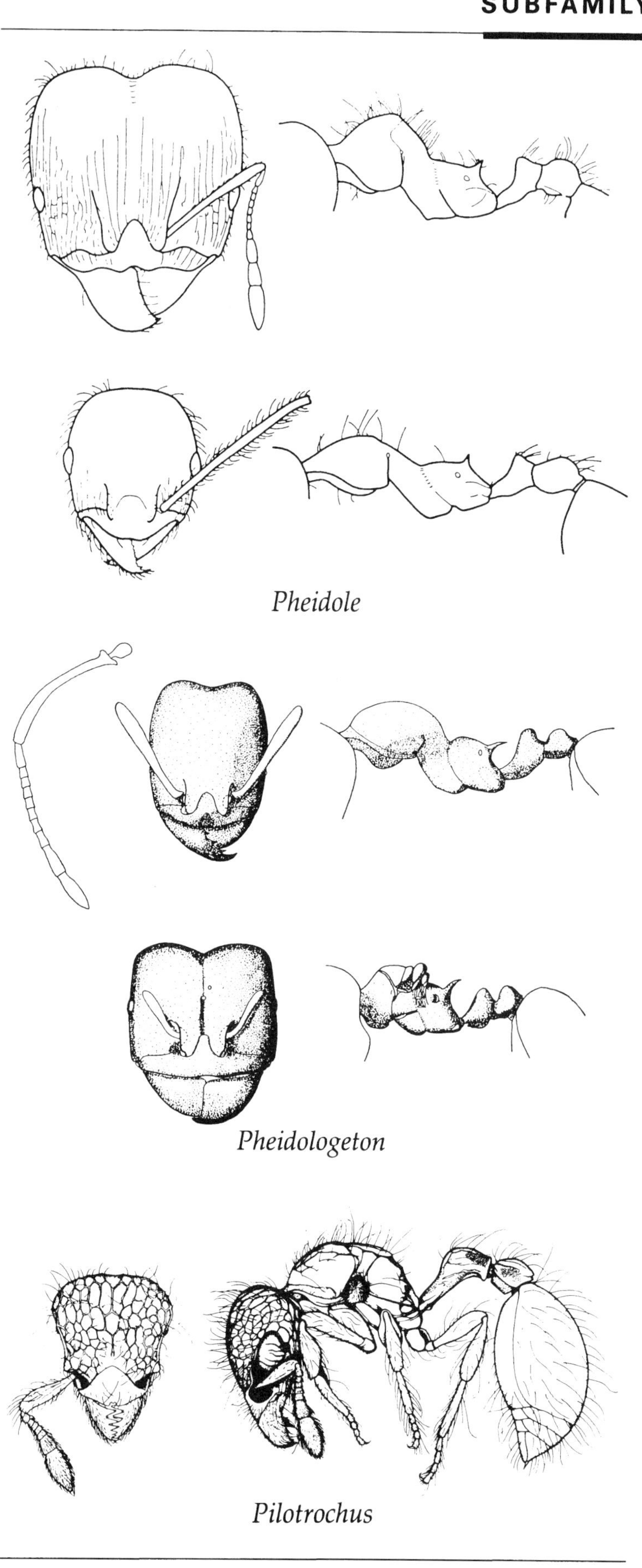

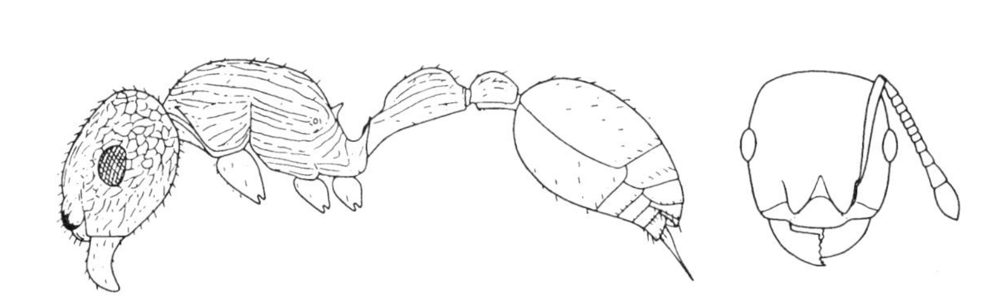

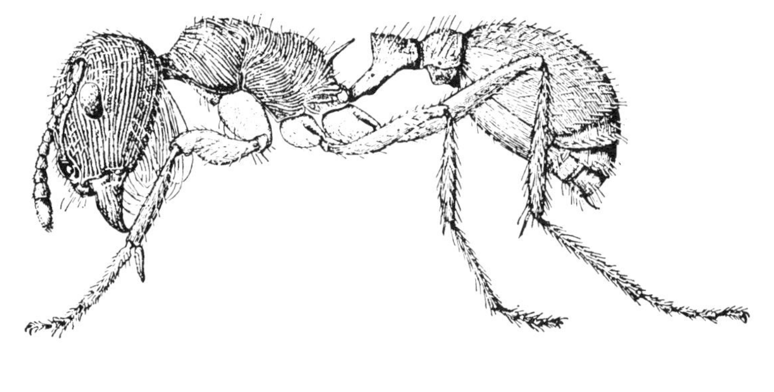

Pogonomyrmex

*Pheidole megacephala*, Samoa (Wilson and Taylor, 1967b)

*Pheidologeton* sp. minor worker (*above*), *P. diversus* major worker (*below*), Southeast Asia (Ettershank, 1966)

*Pilotrochus besmerus*, Madagascar (Brown, 1977a)

*Podomyrma* sp., Australia (R. W. Taylor, unpublished; F. Nanninga, artist)

*Poecilomyrma senirewae*, Fiji (Mann, 1921)

*Pogonomyrmex occidentalis*, United States (Smith, 1947a)

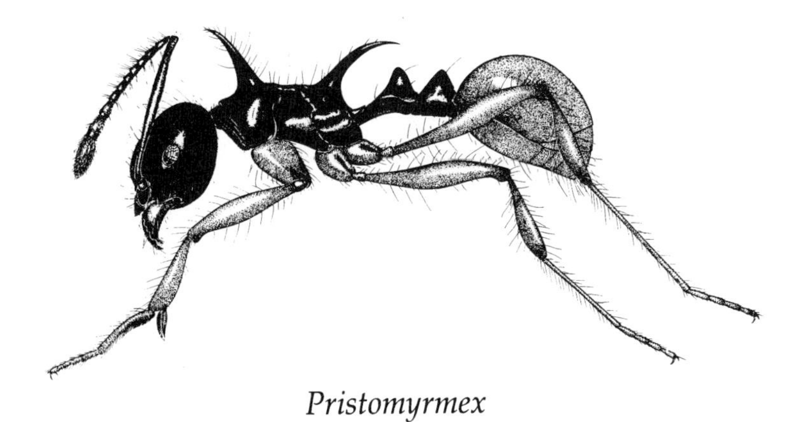

*Pristomyrmex*

*Protalaridris*

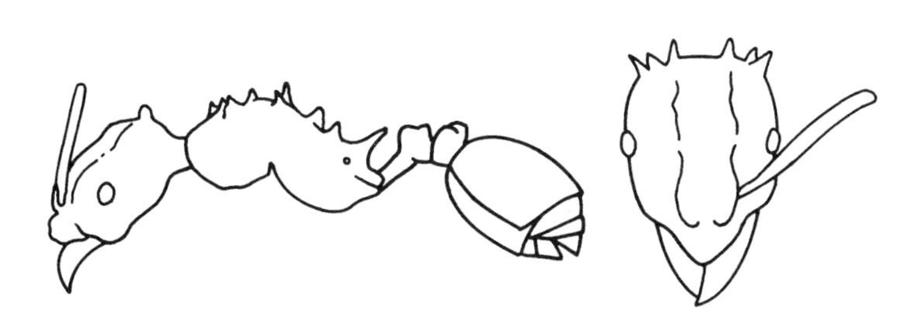

*Proatta*

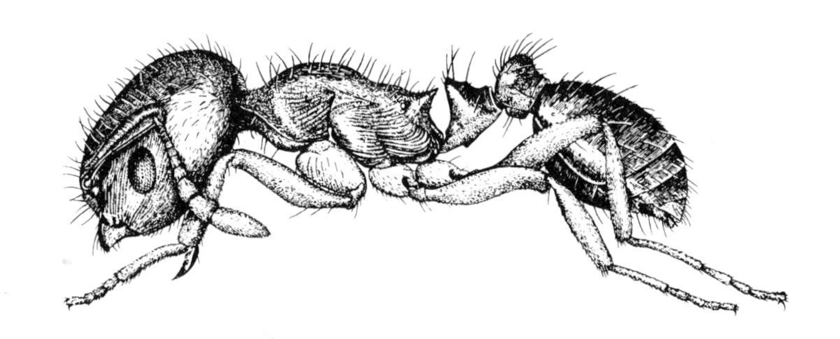

*Protomognathus*

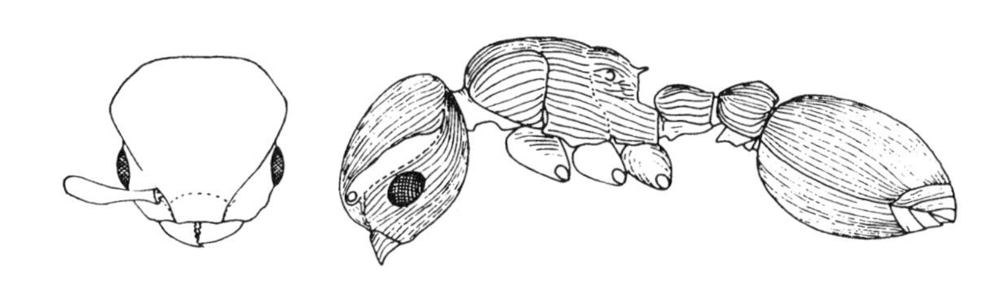

*Procryptocerus*

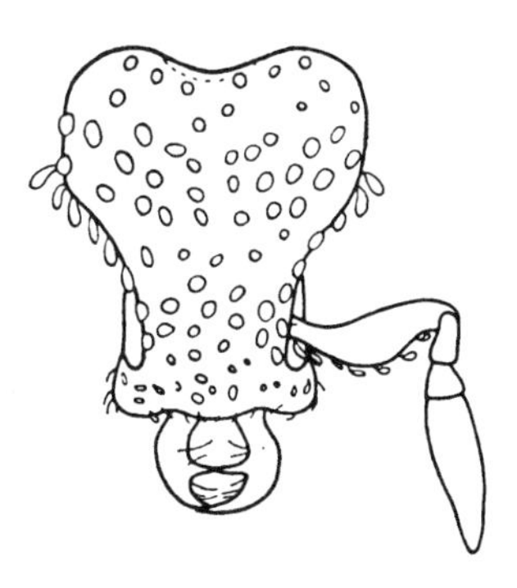

*Quadristruma*

*Pristomyrmex wilsoni*, Australia (R. W. Taylor, unpublished; R. J. Kohout, artist)

*Proatta butteli*, Singapore

*Procryptocerus lenkoi*, Brazil (Kempf, 1969)

*Protalaridris armata*, Ecuador (based on Brown, 1980)

*Protomognathus americanus*, United States (Smith, 1947a)

*Quadristruma emmae*, Samoa (Wilson and Taylor, 1967b)

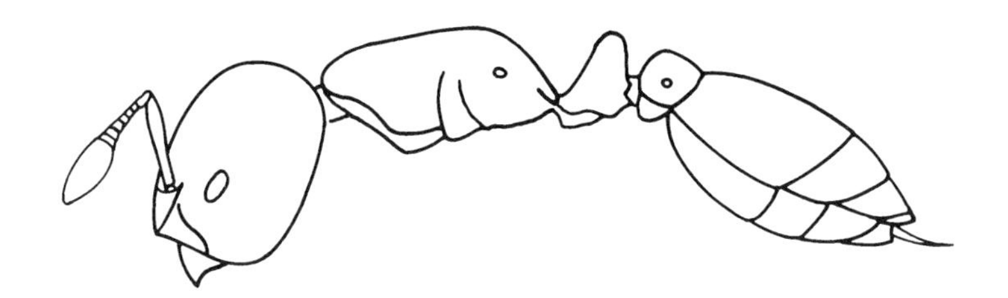

*Rhopalomastix*

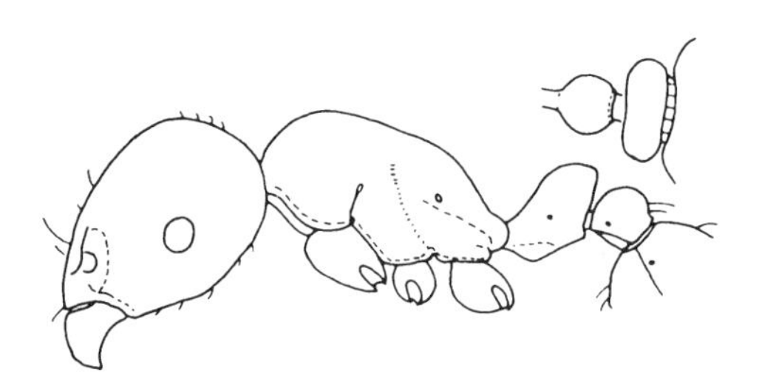

*Rhoptromyrmex*

*Rhopalothrix*

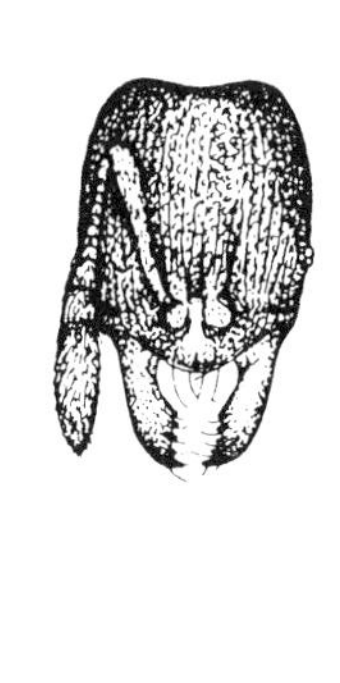

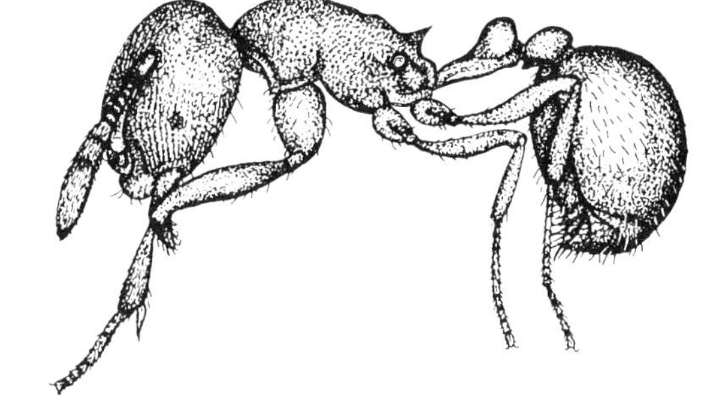

*Rogeria*

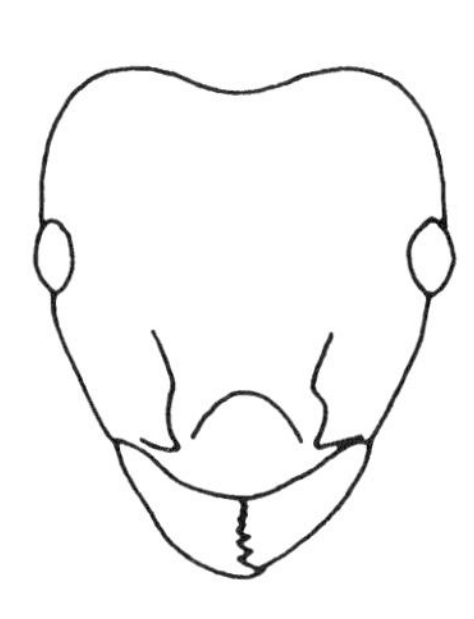

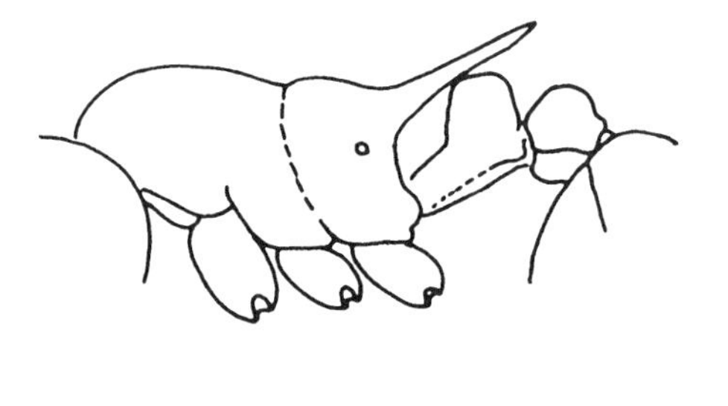

*Rhoptromyrmex*

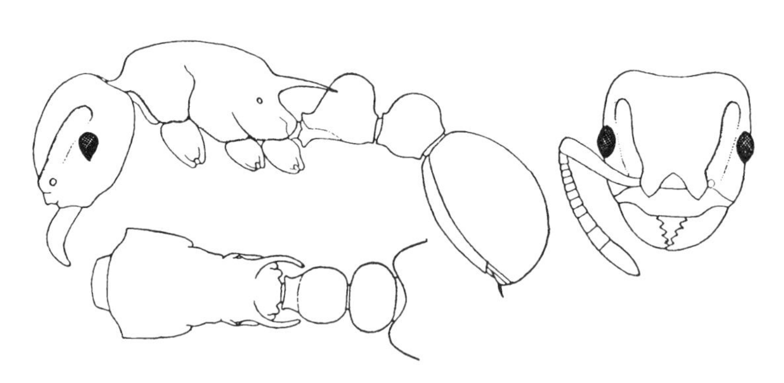

*Romblonella*

*Rhopalomastix rothneyi,* Singapore

*Rhopalothrix orbis,* Australia (R. W. Taylor, unpublished; R. J. Kohout, artist)

*Rhoptromyrmex melleus,* Papua New Guinea (Bolton, 1976)

*Rhoptromyrmex transversinodis,* South Africa (Bolton, 1986a)

*Rogeria huachucana,* United States (Snelling, 1973a)

*Romblonella grandinodis,* Philippines (Wheeler, 1935)

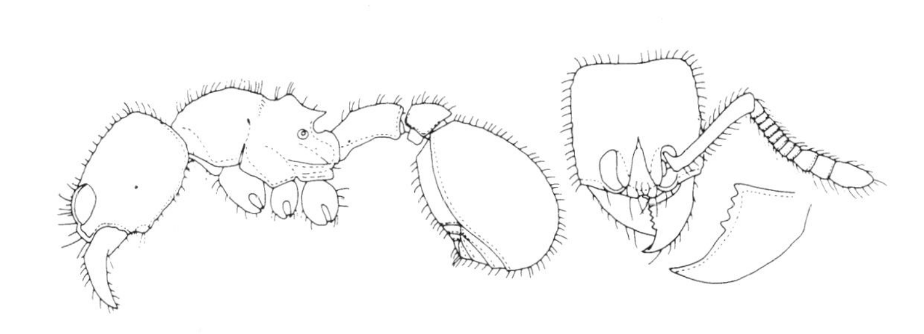

Secostruma

Smithistruma

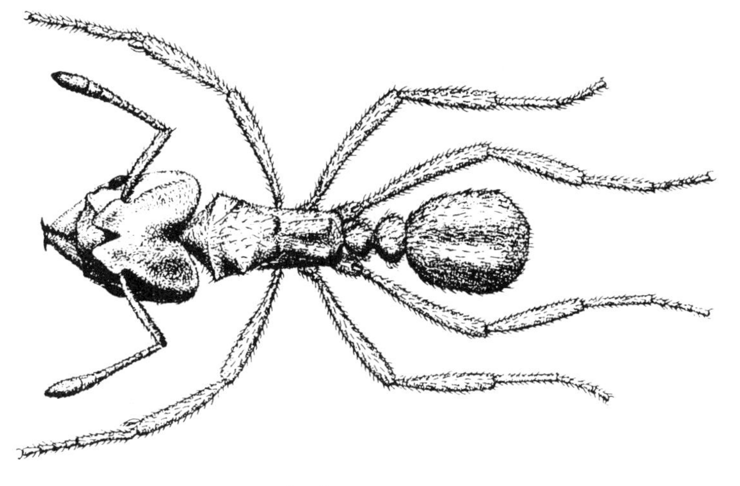

Sericomyrmex

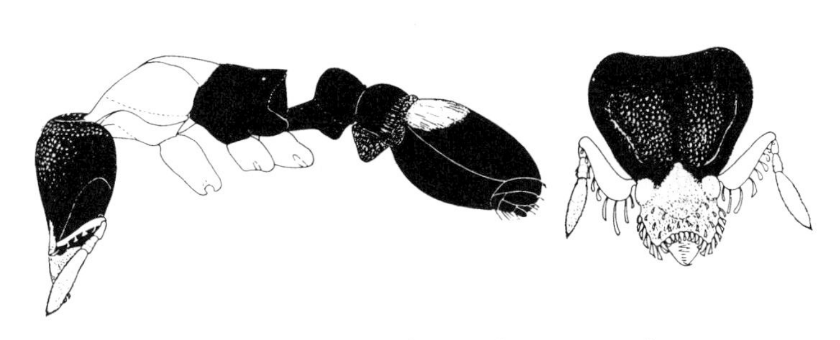

Smithistruma (= Miccostruma)

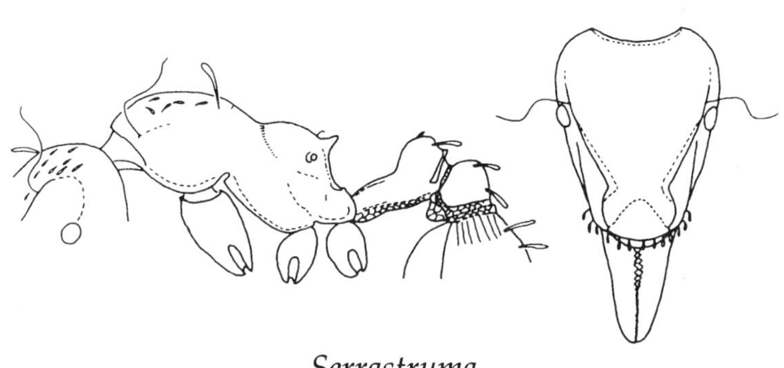

Serrastruma

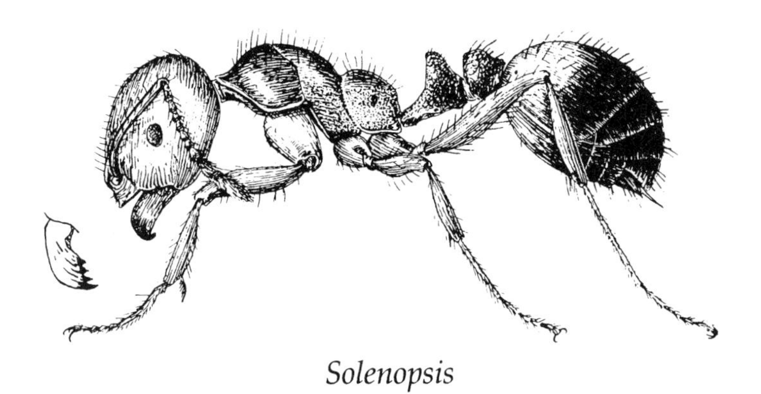

Solenopsis

*Secostruma lethifera*, Indonesia (Bolton, 1988c)

*Sericomyrmex opacus*, South America (Wheeler, 1910a)

*Serrastruma ludovici*, tropical Africa (Bolton, 1983)

*Smithistruma rostrata*, United States (Smith, 1947a)

*Smithistruma (= Miccostruma) tigrilla*, Ivory Coast (Brown, 1973b)

*Solenopsis invicta*, United States (Smith, 1965)

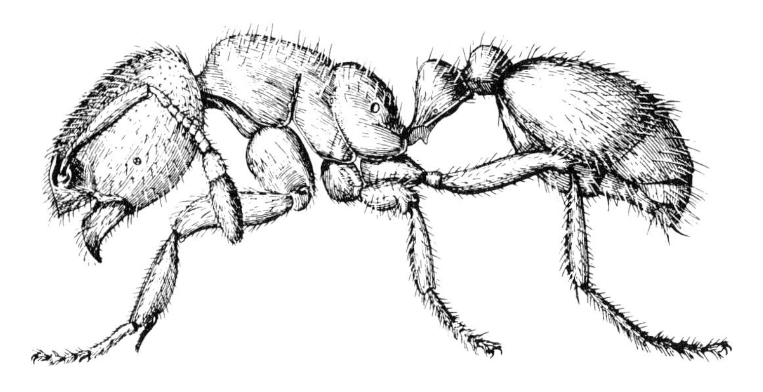

*Solenopsis (= Diplorhoptrum)*

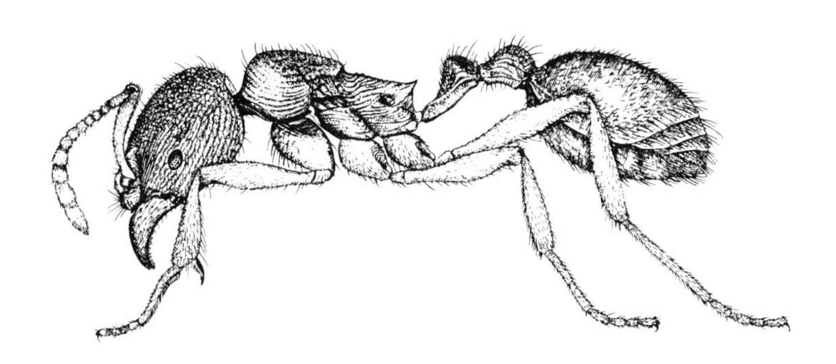

*Stenamma*

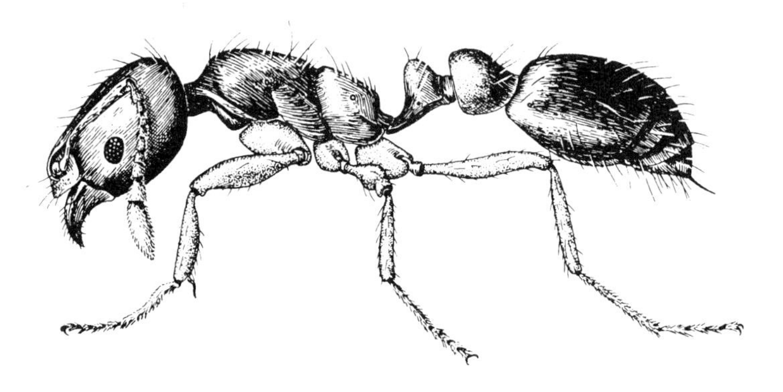

*Solenopsis (= Euophthalma)*

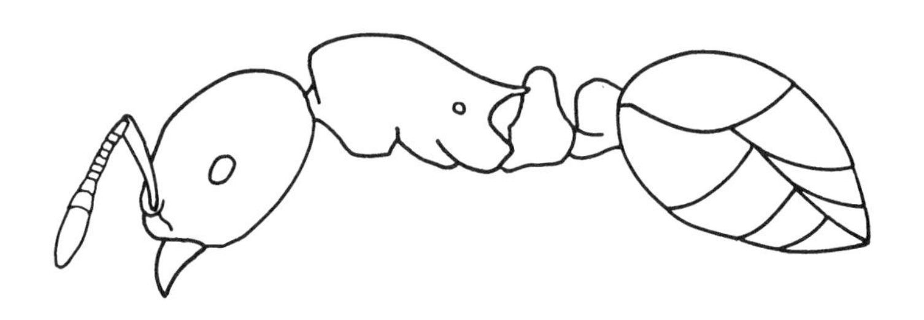

*Stereomyrmex*

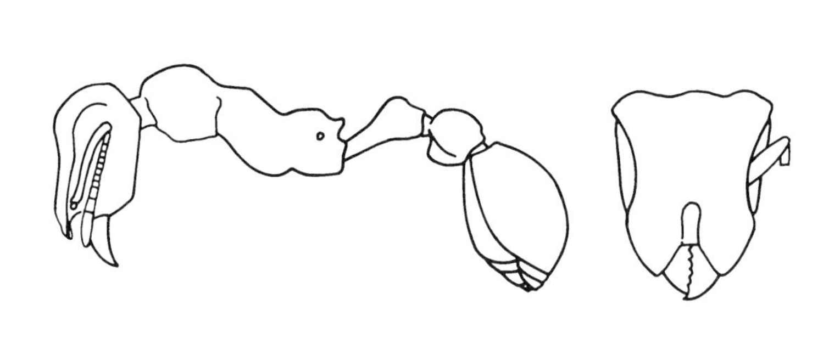

*Stegomyrmex*

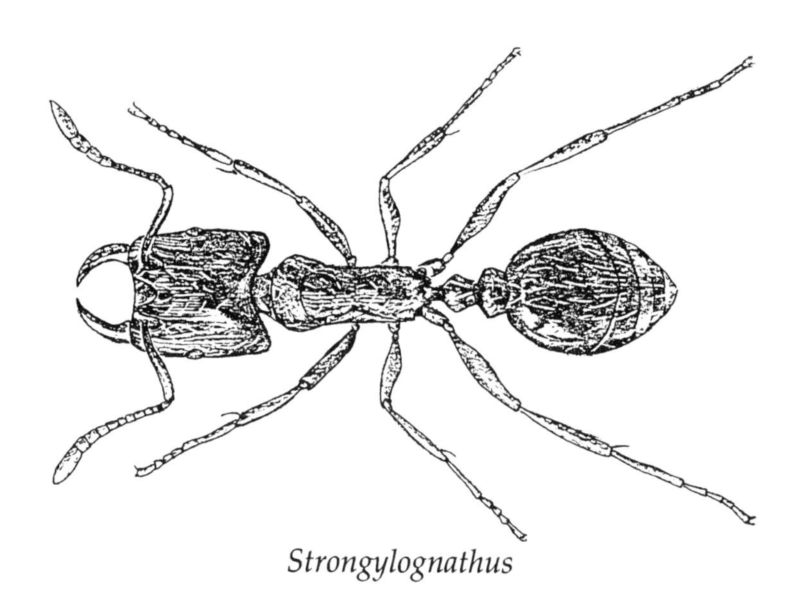

*Strongylognathus*

*Solenopsis (= Diplorhoptrum) pergandei,* United States (Smith, 1947a)

*Solenopsis (= Euophthalma) globularia,* United States (Smith, 1947a)

*Stegomyrmex connectens,* Ecuador

*Stenamma foveolocephalum,* United States (Smith, 1947a)

*Stereomyrmex horni,* Sri Lanka

*Strongylognathus testaceus,* Germany (Gösswald, 1985)

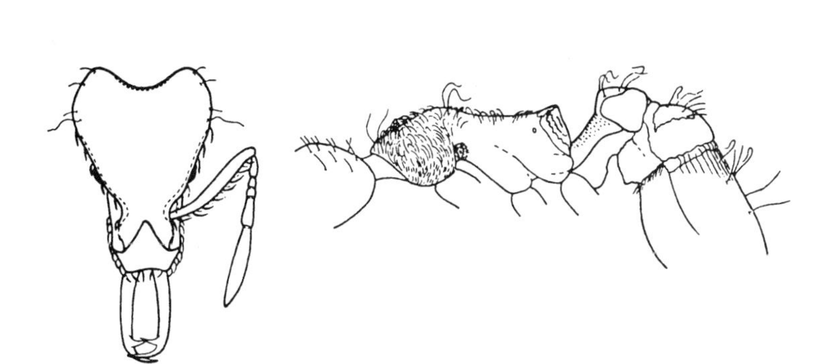

*Strumigenys*

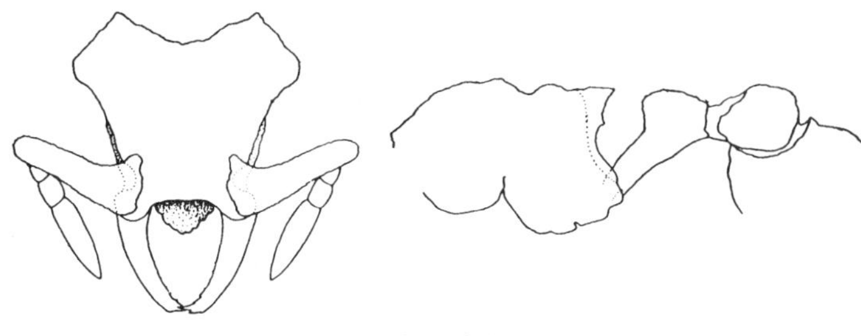

*Talaridris*

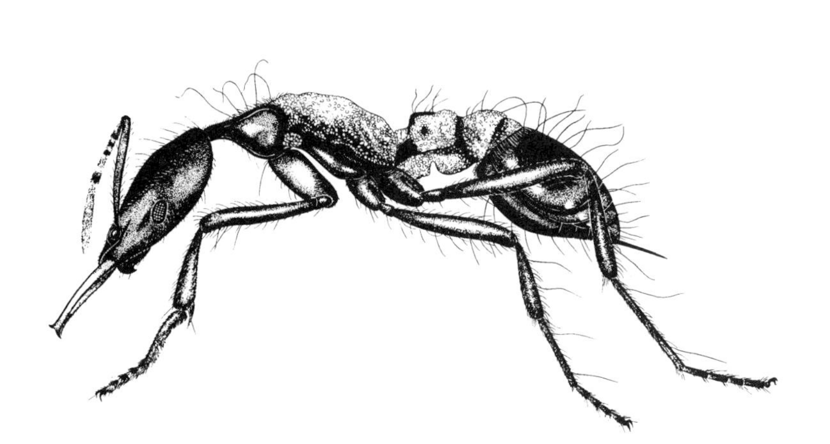

*Strumigenys*

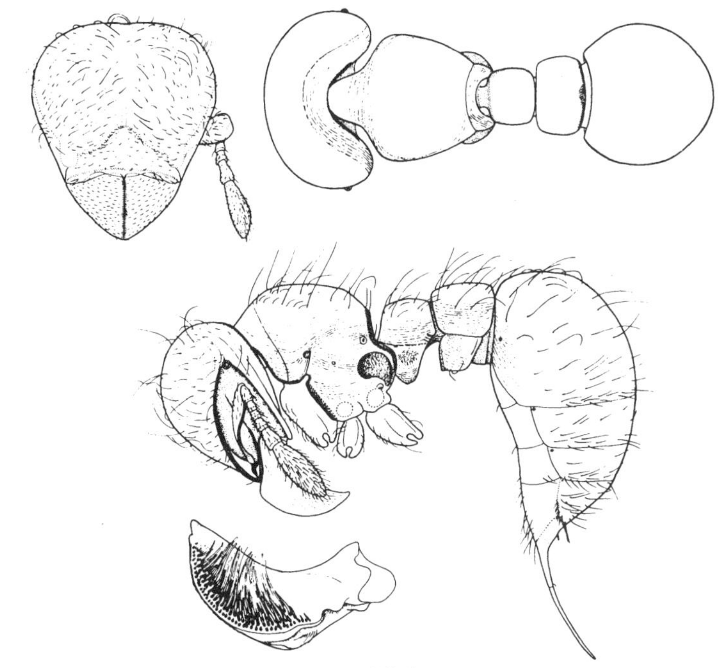

*Tatuidris*

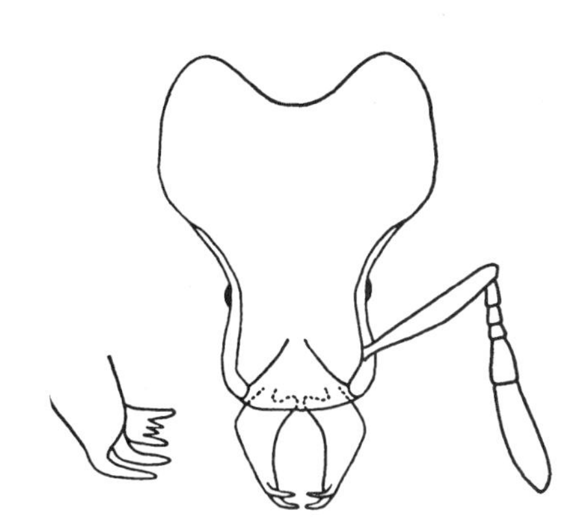

*Strumigenys (= Labidogenys)*

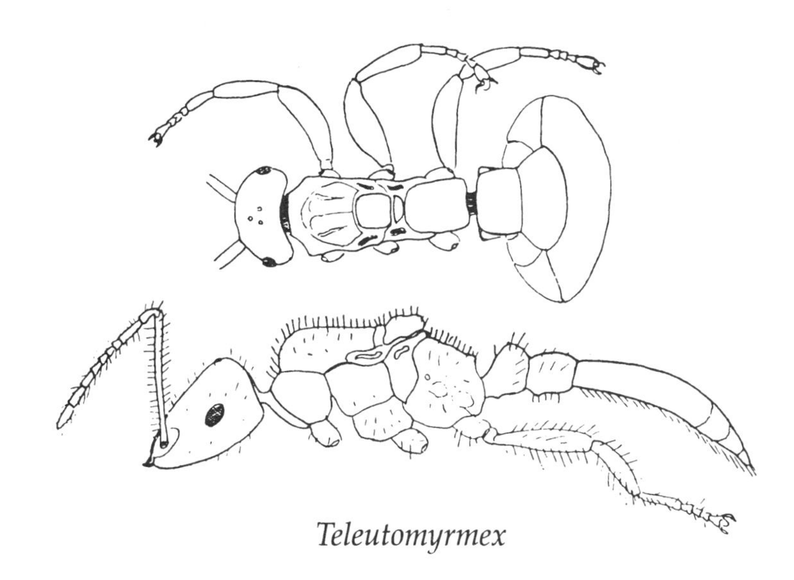

*Teleutomyrmex*

*Strumigenys godeffroyi*, Samoa (Wilson and Taylor, 1967b)

*Strumigenys* sp., Australia (R. W. Taylor, unpublished; R. J. Kohout, artist)

*Strumigenys (= Labidogenys)* sp., tropical Asia (Brown, 1948)

*Talaridris mandibularis*, Trinidad (Weber, 1941a)

*Tatuidris tatusia*, El Salvador (Brown and Kempf, 1967)

*Teleutomyrmex schneideri*, Switzerland (Kutter, 1950)

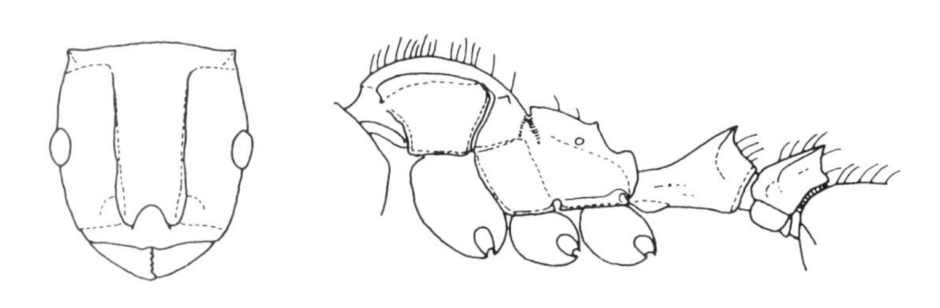

*Teratane*r

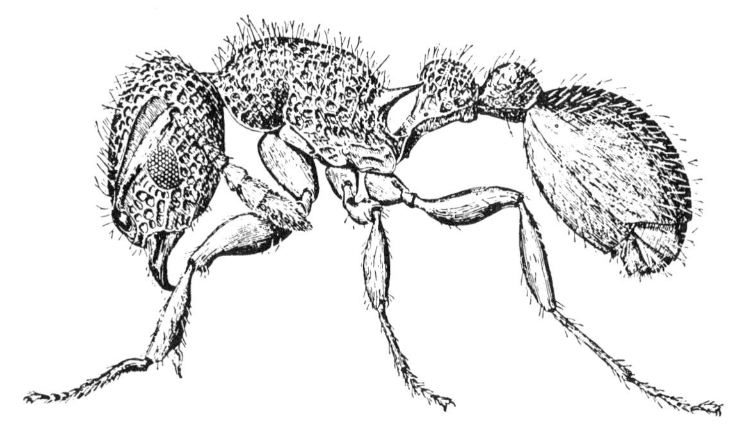

*Tetramorium (= Triglyphothrix)*

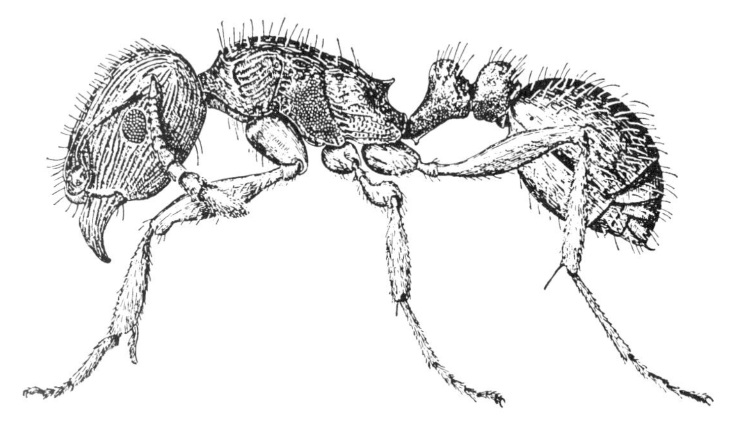

*Tetramorium*

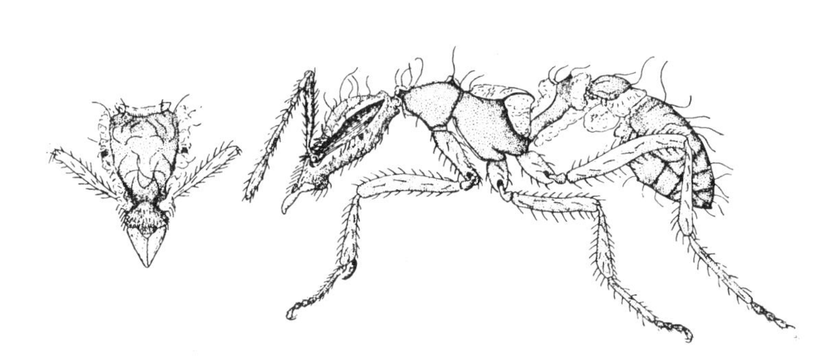

*Tingimyrmex*

*Tetramorium*

*Trachymyrmex*

*Erataner transvaalensis*, South Africa (Bolton, 1981b)

*Tetramorium caespitum*, United States (Smith, 1947a)

*Tetramorium pacificum*, Australia (R. W. Taylor, unpublished; F. Nanninga, artist)

*Tetramorium (= Triglyphothrix) striatidens*, United States (Smith, 1947a)

*Tingimyrmex mirabilis*, Bolivia (Mann, 1926)

*Trachymyrmex septentrionalis*, United States (Smith, 1947a)

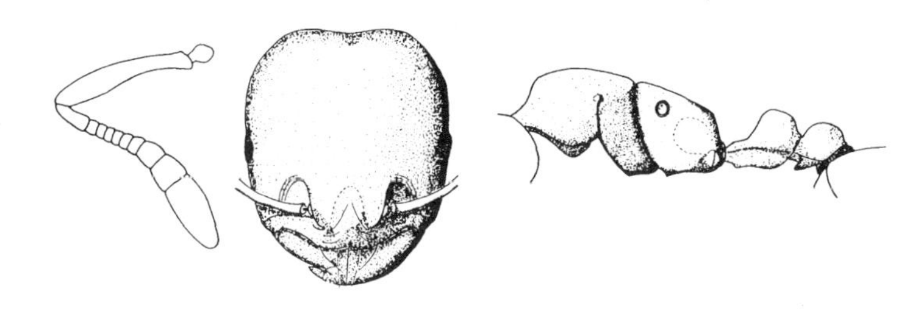

*Tranopelta*

*Vollenhovia*

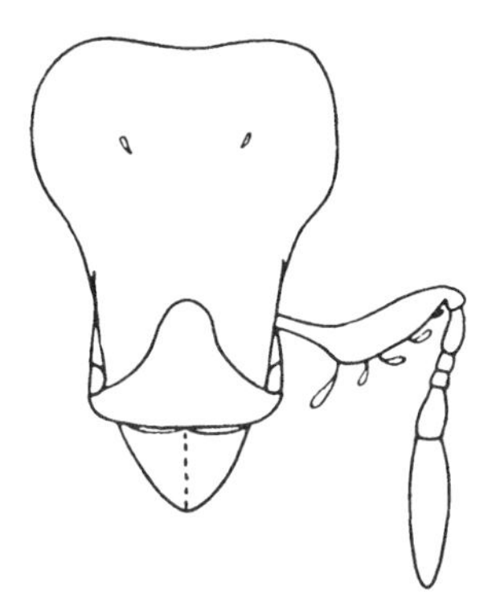

*Trichoscapa*

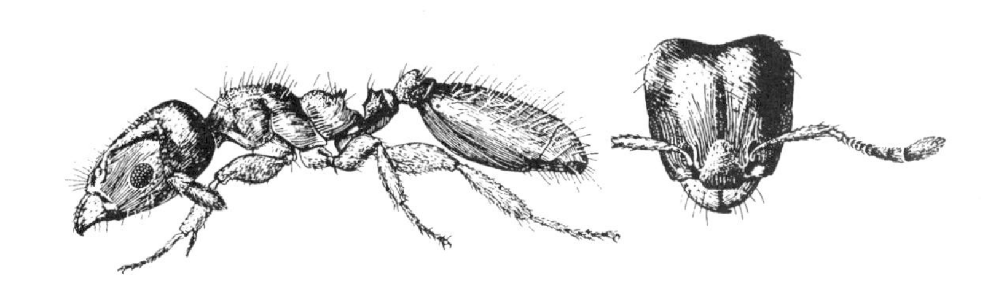

*Vollenhovia (= Acalama)*

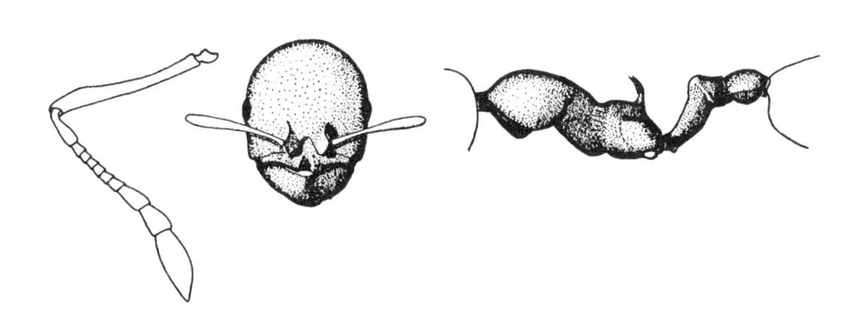

*Trigonogaster*

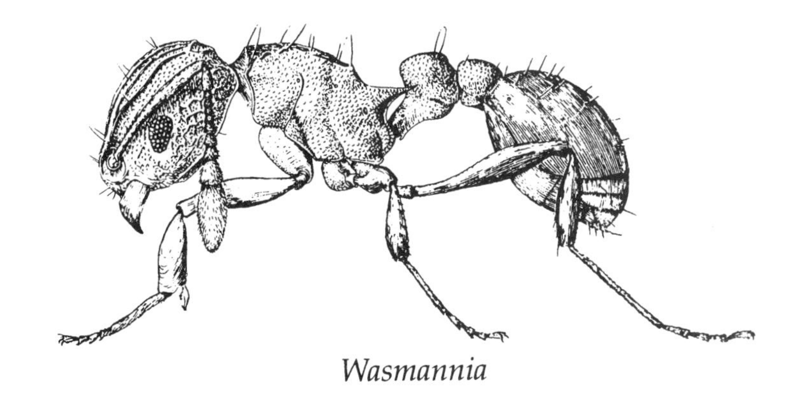

*Wasmannia*

*Tranopelta gilva*, South America (Ettershank, 1966)

*Trichoscapa membranifera*, Samoa (Wilson and Taylor, 1967b)

*Trigonogaster recurvispinosa*, tropical Asia (Ettershank, 1966)

*Vollenhovia* sp., Australia (R. W. Taylor, unpublished; R. J. Kohout, artist)

*Vollenhovia (= Acalama) donisthorpei*, India (Smith, 1948)

*Wasmannia auropunctata*, United States (Smith, 1947a)

## SUBFAMILY MYRMICINAE

*Willowsiella*

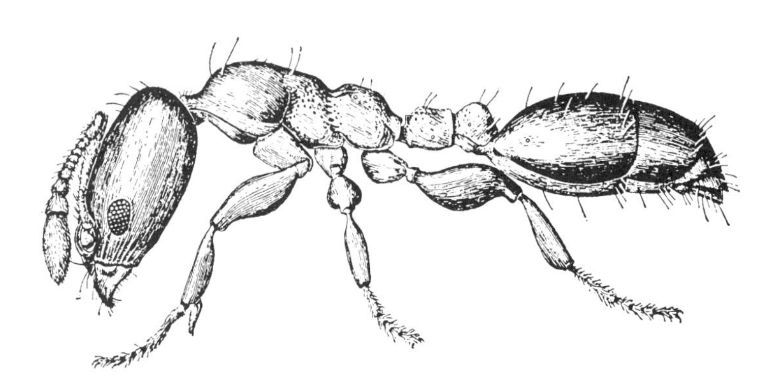

*Xenomyrmex*

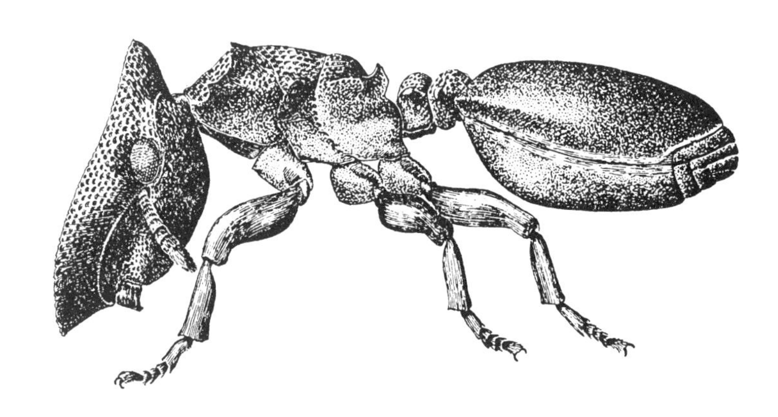

*Zacryptocerus (= Cyathomyrmex)*

*Willowsiella dispar,* Solomon Islands (redrawn from Wheeler, 1934b)

*Xenomyrmex floridanus,* United States (Smith, 1947a)

*Zacryptocerus (= Cyathomyrmex) varians,* United States (Smith, 1947a)

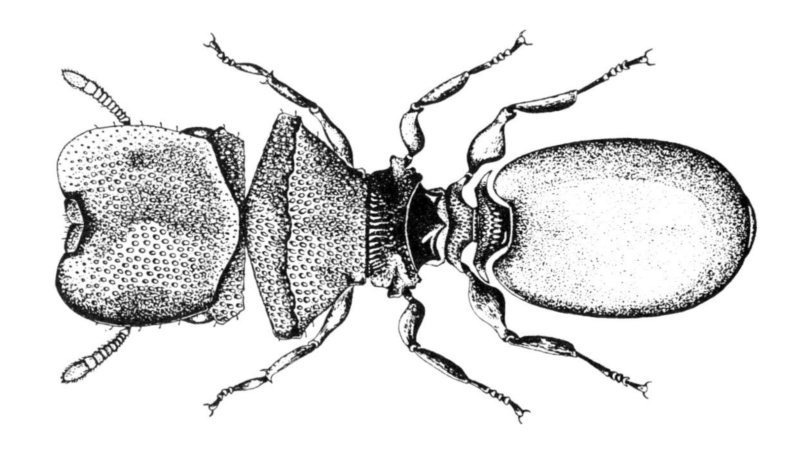

*Zacryptocerus (= Paracryptocerus)*

*Zacryptocerus (= Paracryptocerus) texanus,* United States (Creighton and Gregg, 1954)

## SUBFAMILY ANEURETINAE

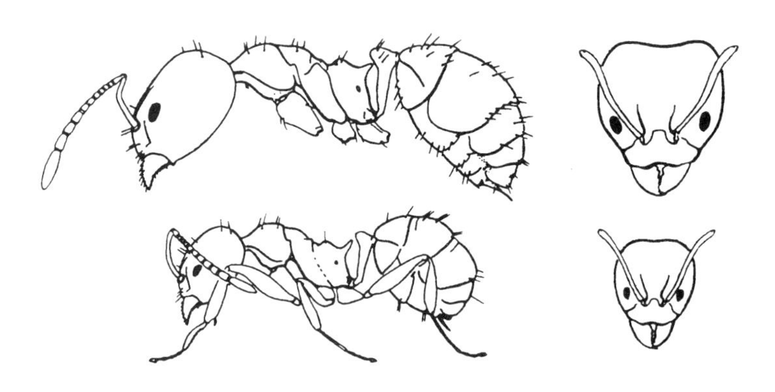

*Aneuretus*

*Aneuretus simoni,* Sri Lanka (Wilson et al., 1956)

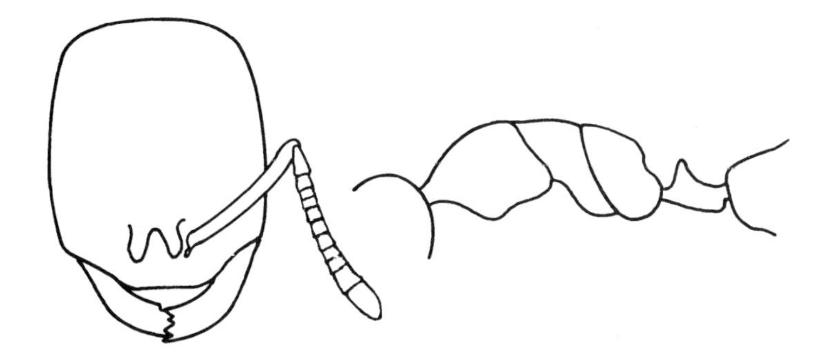

*Anillidris*

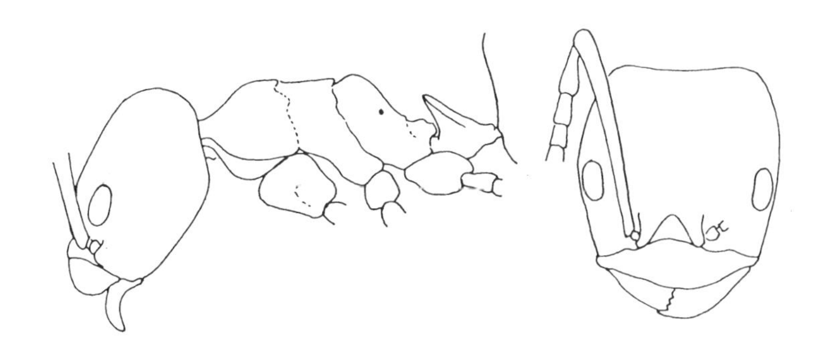

*Bothriomyrmex*

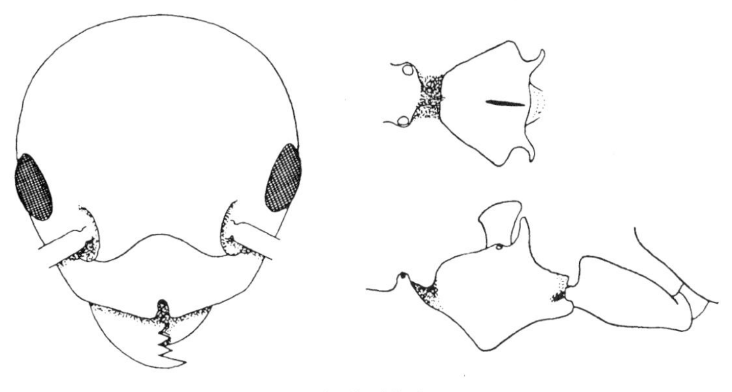

*Axinidris*

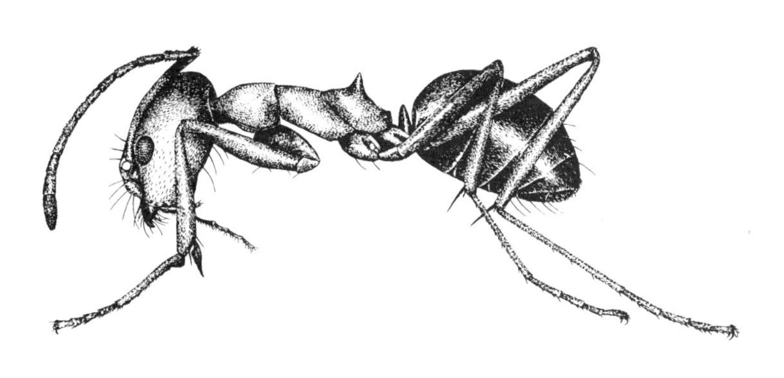

*Conomyrma*

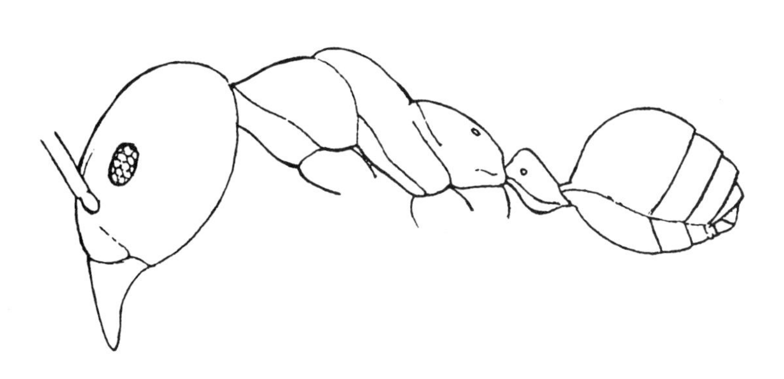

*Azteca*

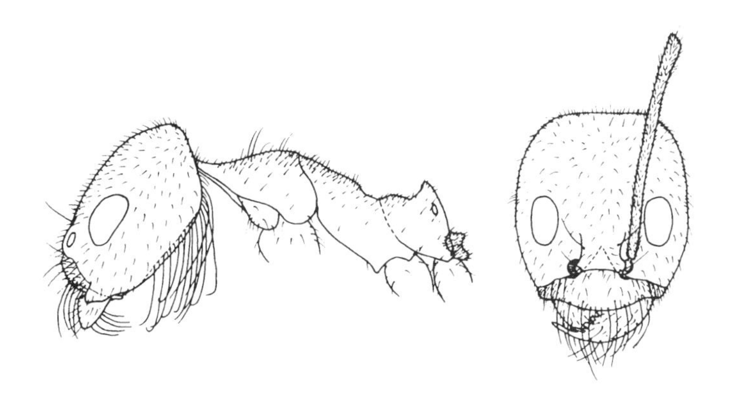

*Conomyrma (= Araucomyrmex)*

*Anillidris bruchi*, Argentina (Santschi, 1936)

*Axinidris acholli*, Sudan (Weber, 1941a)

*Azteca coeruleipennis*, Central America (Emery, 1912)

*Bothriomyrmex menozzii*, southern Europe (Kutter, 1971)

*Conomyrma flavopecta*, United States (Smith, 1947a)

*Conomyrma (= Araucomyrmex) pappodes*, Chile (Snelling, 1975)

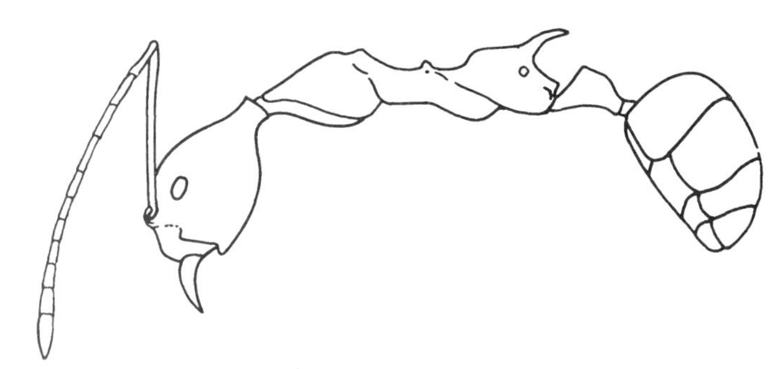

*Dolichoderus*

*Engramma*

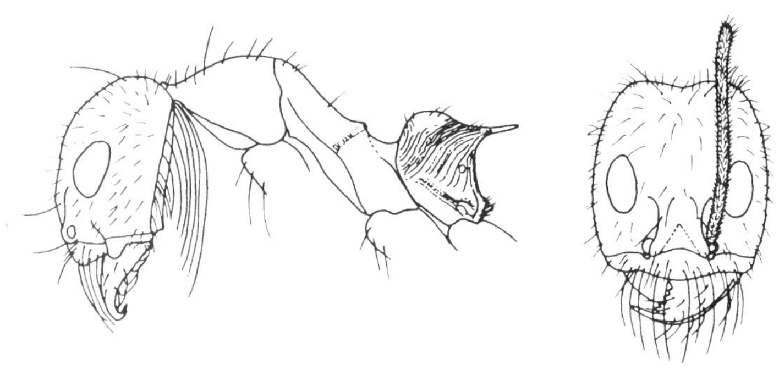

*Dorymyrmex*

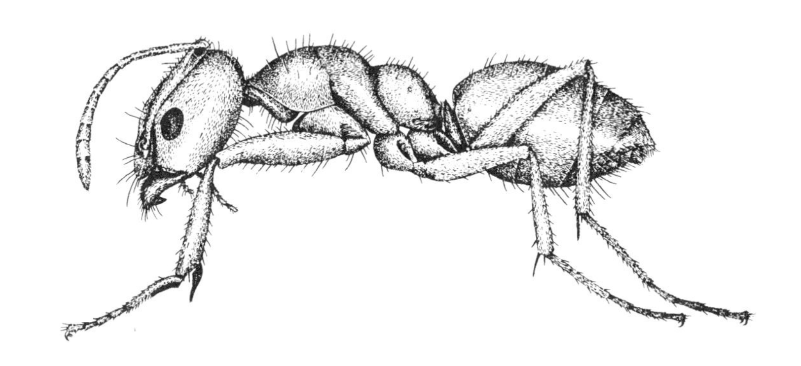

*Forelius*

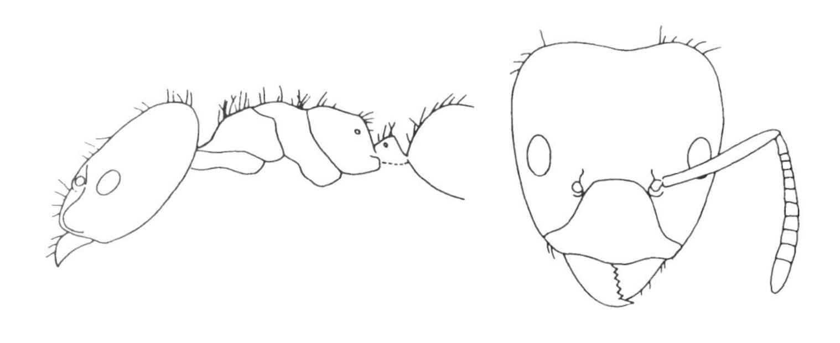

*Ecphorella*

*Froggattella*

*Dolichoderus attelaboides*, Guyana

*Dorymyrmex agallardoi*, Chile (Snelling, 1975)

*Ecphorella wellmani*, Angola

*Engramma lujae*, Zaïre (Wheeler, 1922)

*Forelius foetidus*, United States (Smith, 1947a)

*Froggattella kirbyi*, Australia (R. W. Taylor, unpublished; R. J. Kohout, artist)

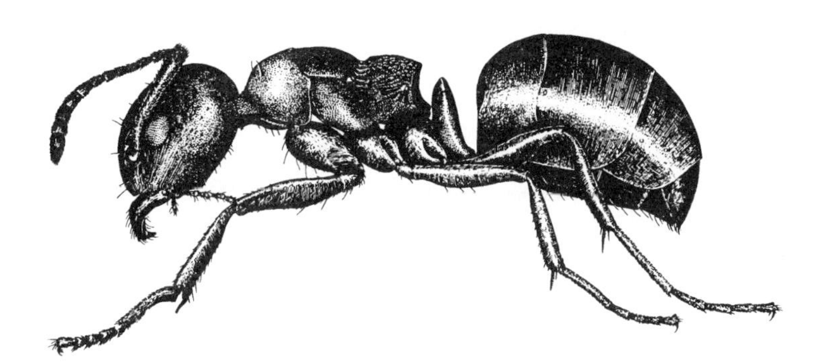

*Hypoclinea*

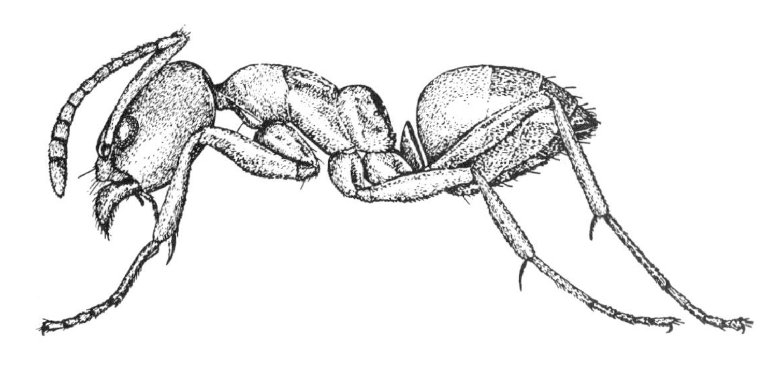

*Iridomyrmex*

*Hypoclinea* (= *Acanthoclinea*)

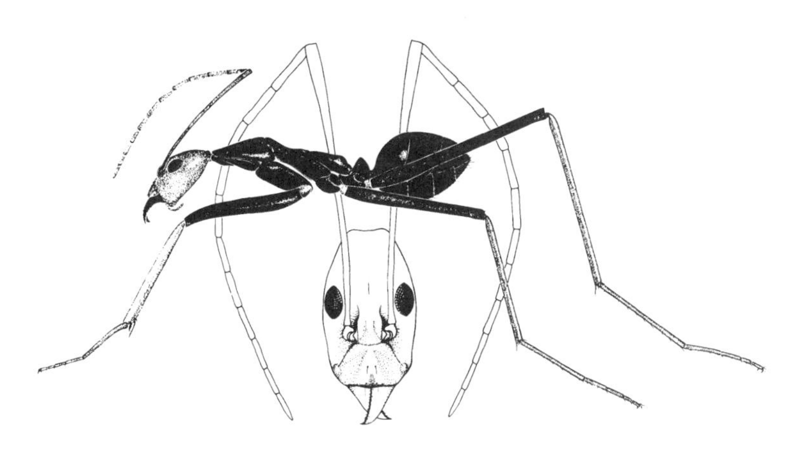

*Leptomyrmex*

*Hypoclinea* (= *Diceratoclinea*)

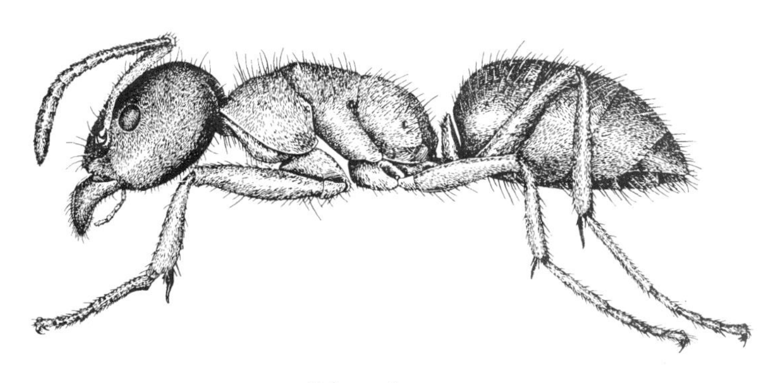

*Liometopum*

*Hypoclinea taschenbergi*, United States (Smith, 1947a)

*Hypoclinea* (= *Acanthoclinea*) sp., Australia (R. W. Taylor, unpublished; R. J. Kohout, artist)

*Hypoclinea* (= *Diceratoclinea*) sp., Australia (R. W. Taylor, unpublished; R. J. Kohout, artist)

*Iridomyrmex humilis*, United States (Smith, 1947a)

*Leptomyrmex erythrocephalus*, Australia (R. W. Taylor, unpublished; R. J. Kohout, artist)

*Liometopum occidentale*, United States (Smith, 1947a)

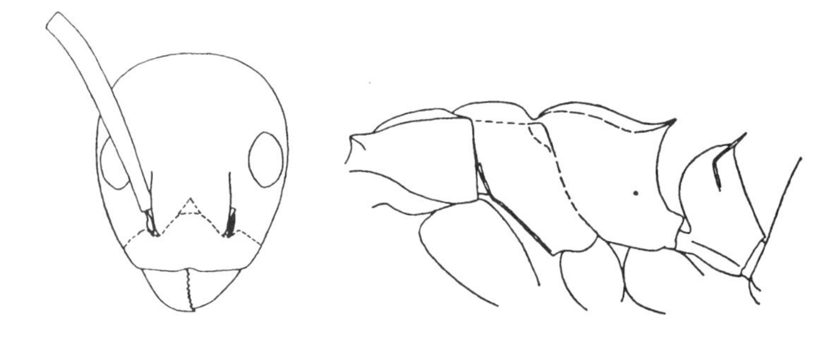

*Monacis*

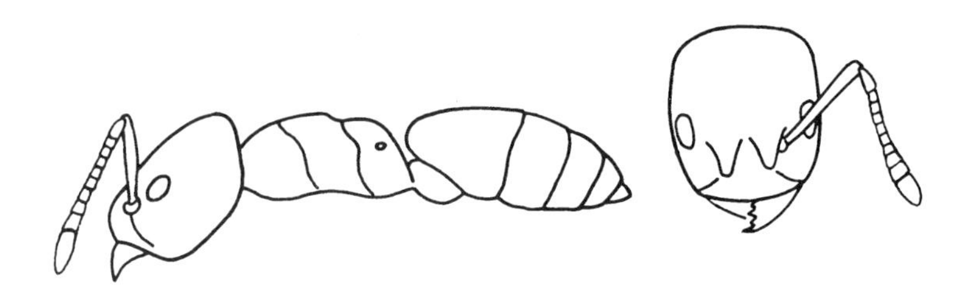

*Semonius*

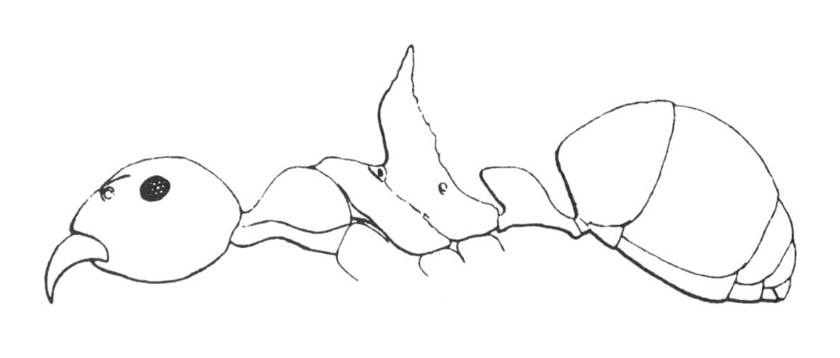

*Monoceratoclinea*

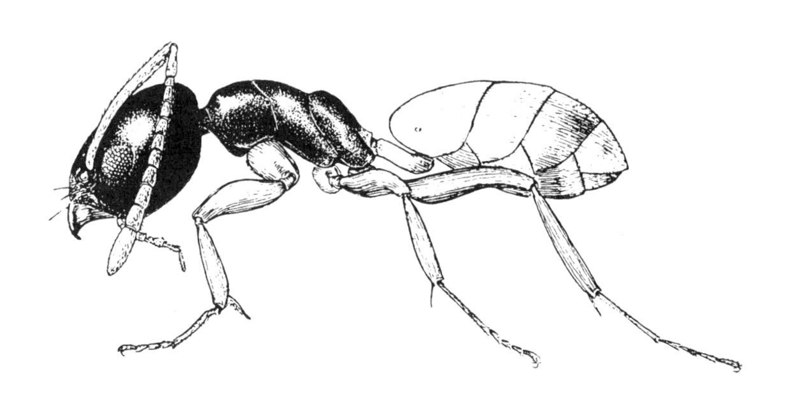

*Tapinoma*

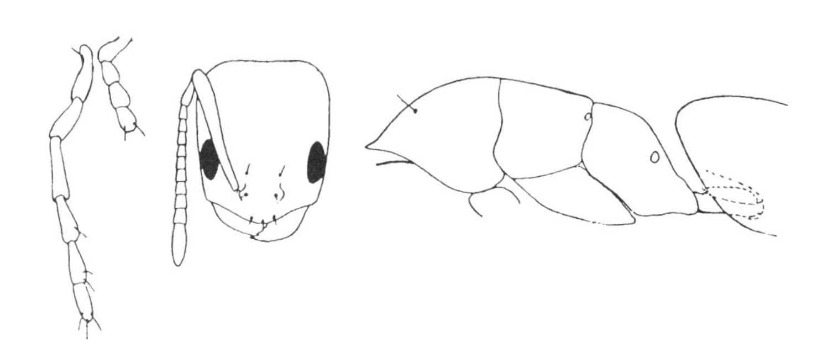

*Neoforelius*

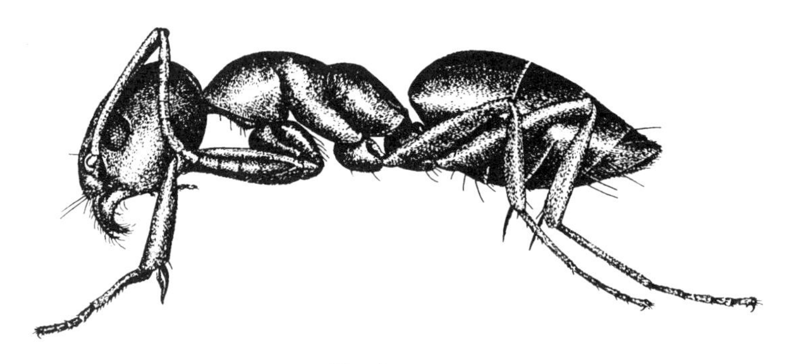

*Tapinoma*

*Monacis laminata*, South America (Kempf, 1959a)

*Monoceratoclinea monoceros*, Papua New Guinea (Emery, 1897)

*Neoforelius tucumanus*, Argentina (Kusnezov, 1953)

*Semonius schultzei*, Zaïre (Wheeler, 1922)

*Tapinoma melanocephalum*, United States (Smith, 1965)

*Tapinoma sessile*, United States (Smith, 1947a)

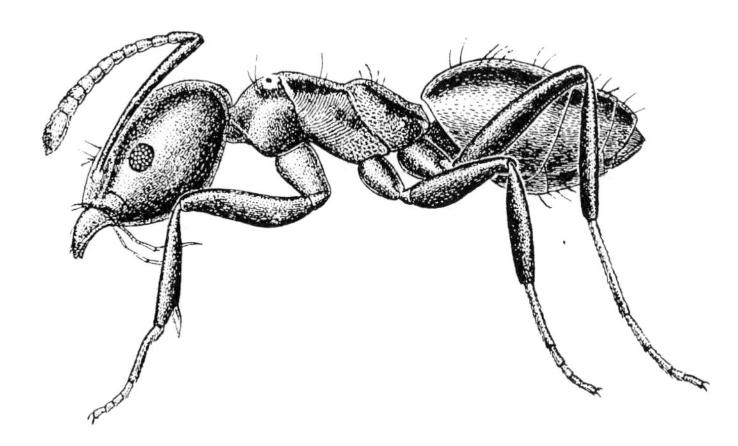

*Technomyrmex*

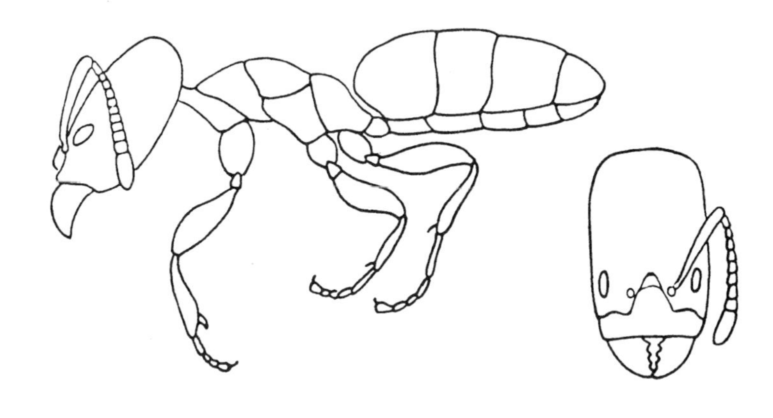

*Zatapinoma*

*Turneria*

*Technomyrmex albipes*, Palau (Wilson and Taylor, 1967b)

*Turneria pacifica*, Vanuatu

*Zatapinoma wheeleri*, Samoa(?) (Wilson and Taylor, 1967b)

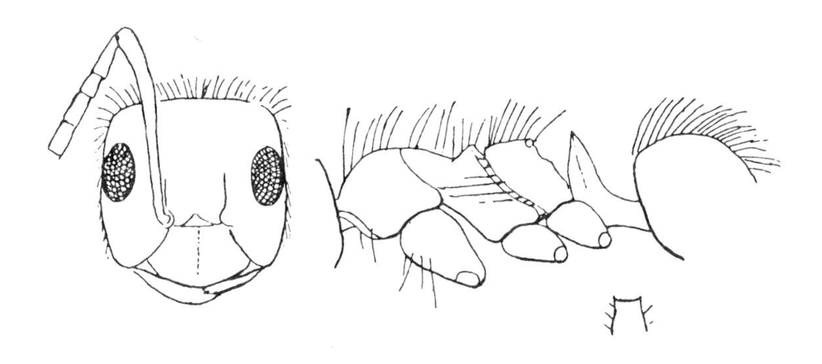

*Acantholepis*

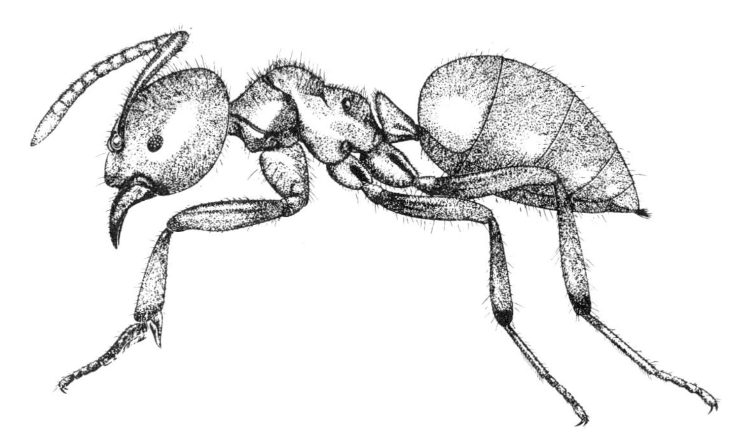

*Acropyga*

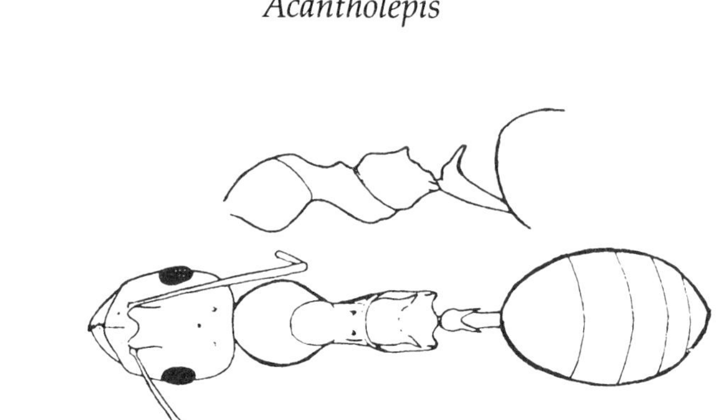

*Acantholepis*

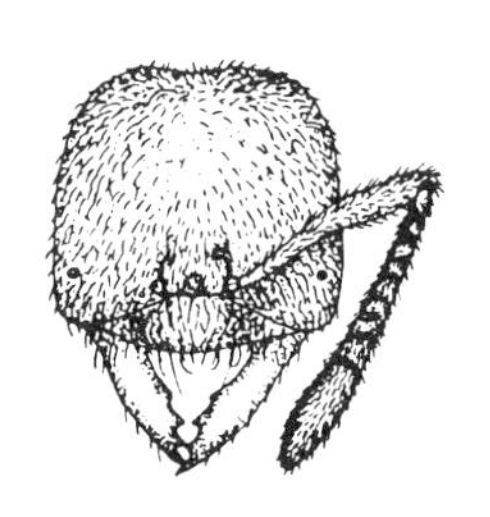

*Acropyga*

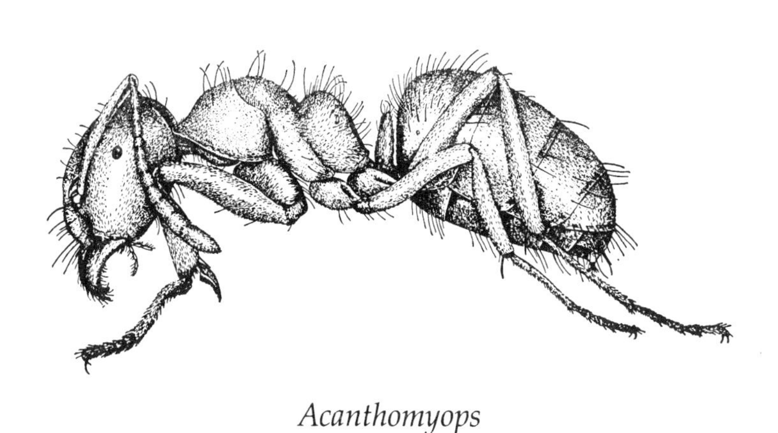

*Acanthomyops*

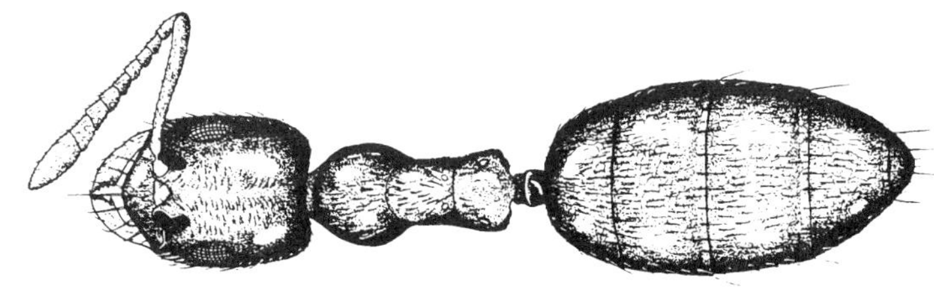

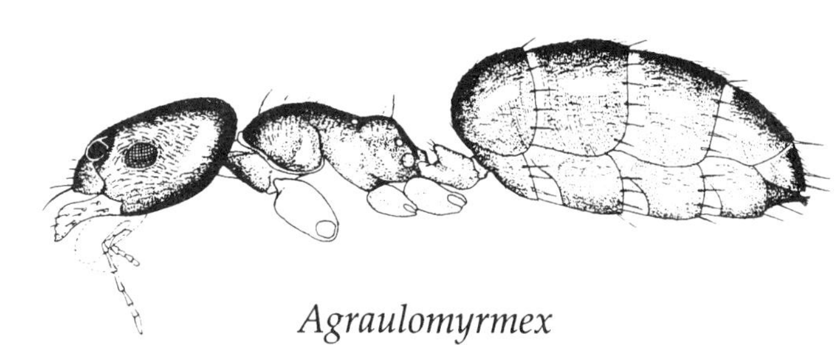

*Agraulomyrmex*

*Acantholepis albata,* Zaïre (Santschi, 1935)

*Acantholepis arenaria,* Zimbabwe (Arnold, 1915–26)

*Acanthomyops interjectus,* United States (Smith, 1947a)

*Acropyga acutiventris,* Australia (R. W. Taylor, unpublished; R. J. Kohout, artist)

*Acropyga epedana,* United States (Snelling, 1973a)

*Agraulomyrmex meridionalis,* South Africa (Prins, 1983)

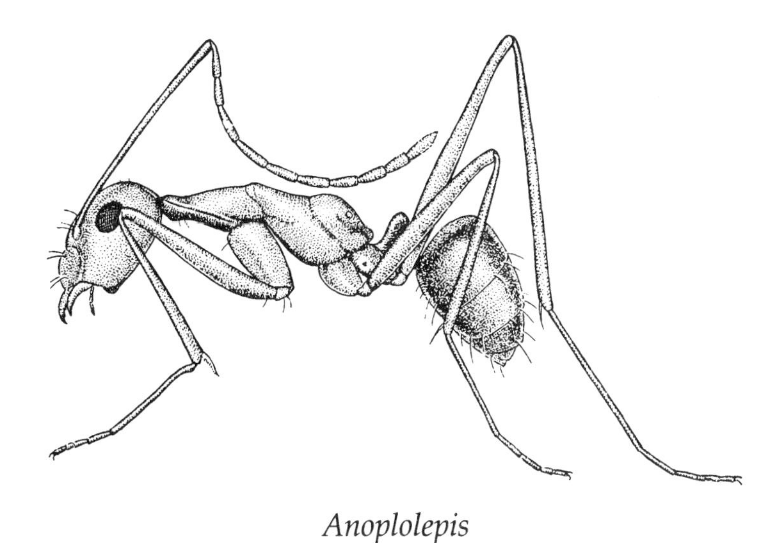

*Anoplolepis*

*Bregmatomyrma*

*Aphomomyrmex*

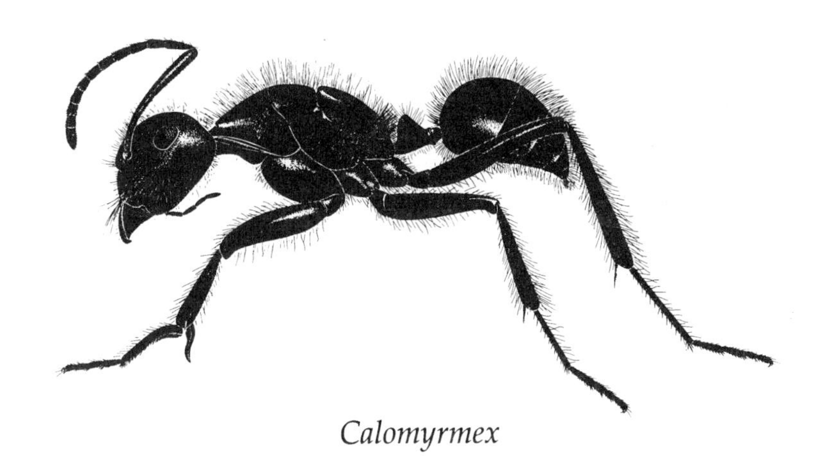

*Calomyrmex*

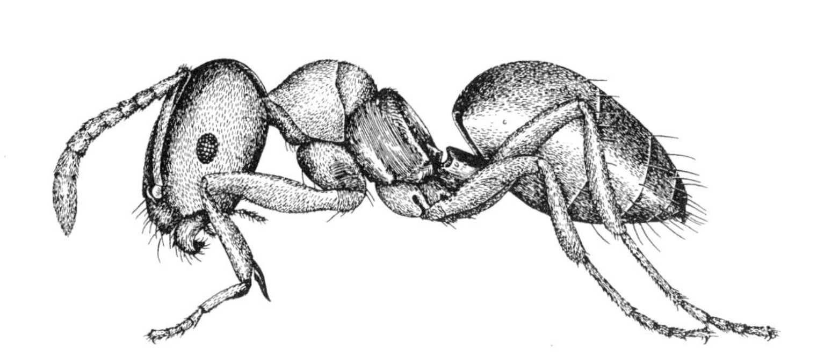

*Brachymyrmex*

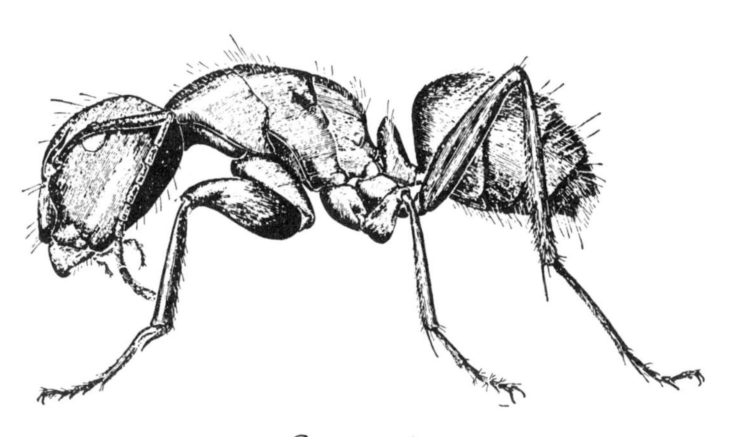

*Camponotus*

*Anoplolepis longipes*, Micronesia (Wilson and Taylor, 1967b)

*Aphomomyrmex afer*, Cameroon

*Brachymyrmex* sp., United States (Smith, 1947a)

*Bregmatomyrma carnosa*, Indonesia (Wheeler, 1929b)

*Calomyrmex* sp., Australia (R. W. Taylor, unpublished; R. J. Kohout, artist)

*Camponotus pennsylvanicus*, United States (Smith, 1947a)

*Camponotus*

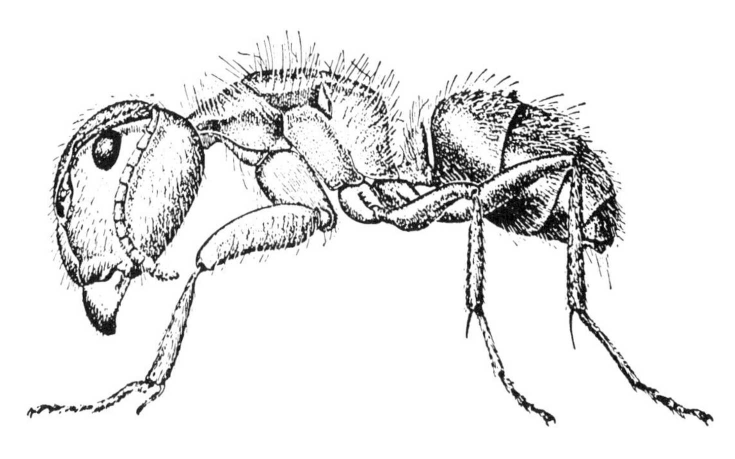

*Camponotus (= Myrmothrix)*

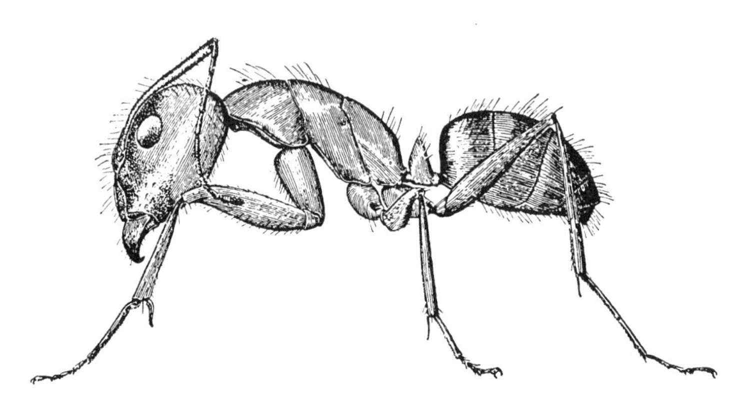

*Camponotus (= Myrmentoma)*

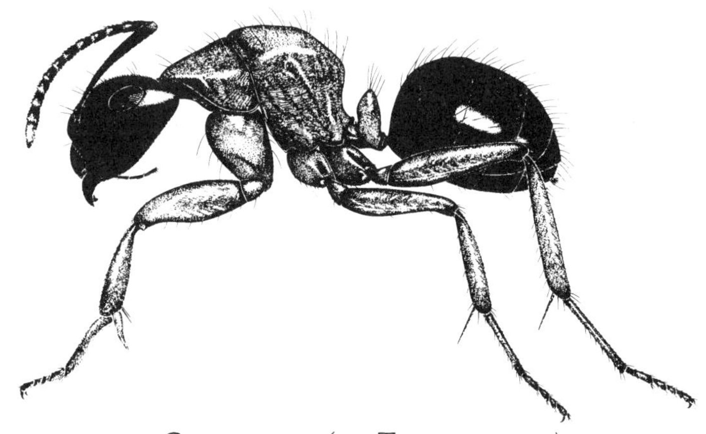

*Camponotus (= Tanaemyrmex)*

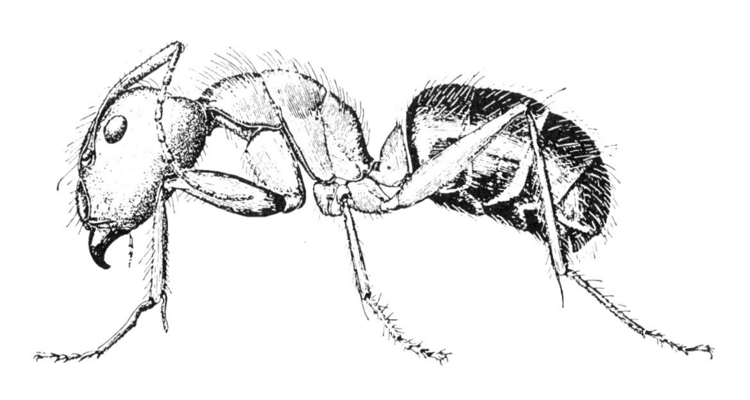

*Camponotus (= Myrmobrachys)*

*Cataglyphis*

*Camponotus* sp., Australia (R. W. Taylor, unpublished; R. J. Kohout, artist)

*Camponotus (= Myrmentoma) nearcticus*, United States (Smith, 1947a)

*Camponotus (= Myrmobrachys) planatus*, United States (Smith, 1947a)

*Camponotus (= Myrmothrix) floridanus*, United States (Smith, 1947a)

*Camponotus (= Tanaemyrmex) fumidus*, United States (Smith, 1947a)

*Cataglyphis fortis*, North Africa (Wehner, 1983b)

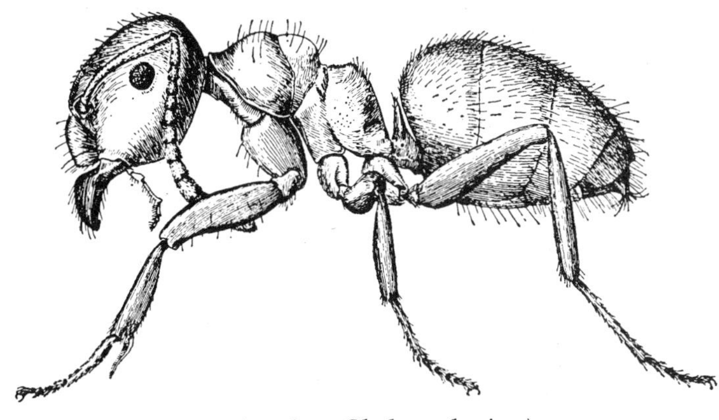

*Lasius ( = Chthonolasius)*

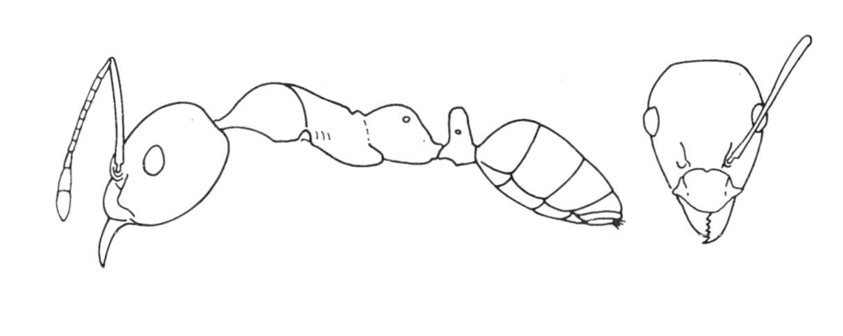

*Myrmecorhynchus*

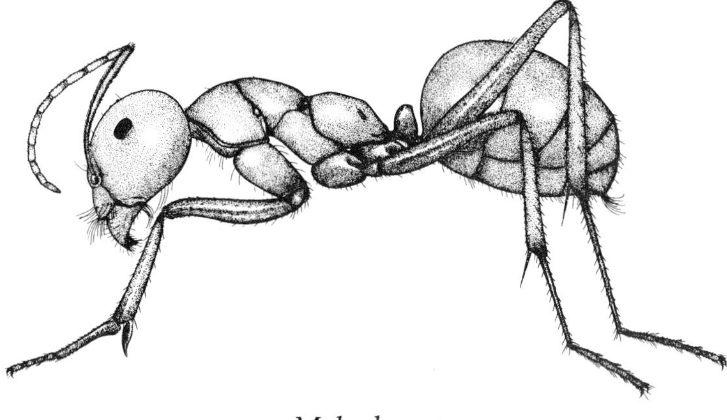

*Melophorus*

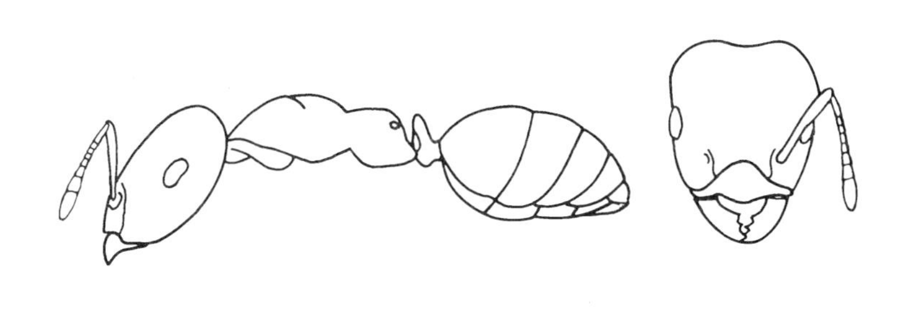

*Myrmelachista*

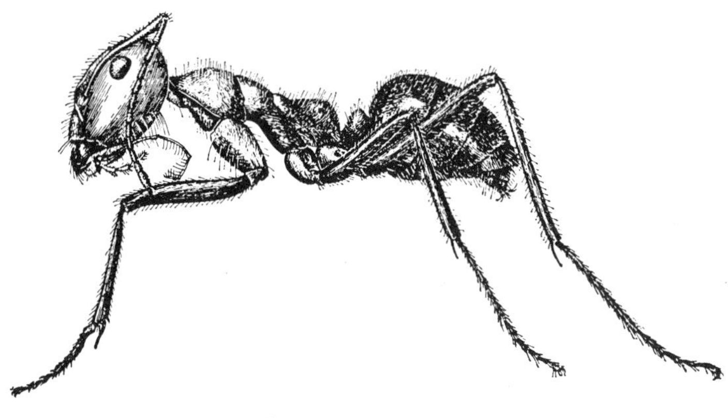

*Myrmecocystus*

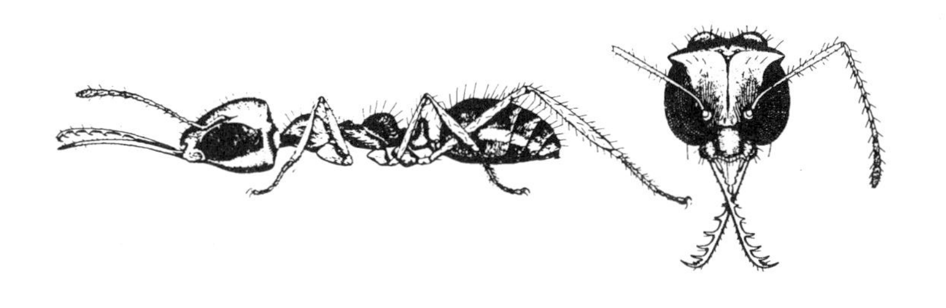

*Myrmoteras*

*Lasius ( = Chthonolasius) umbratus*, United States (Smith, 1947a)

*Melophorus* sp., Australia (R. W. Taylor, unpublished; R. J. Kohout, artist)

*Myrmecocystus* sp., United States (Smith, 1947a)

*Myrmecorhynchus emeryi*, Australia

*Myrmelachista ramulorum*, Mexico

*Myrmoteras karnyi*, Indonesia (Gregg, 1954)

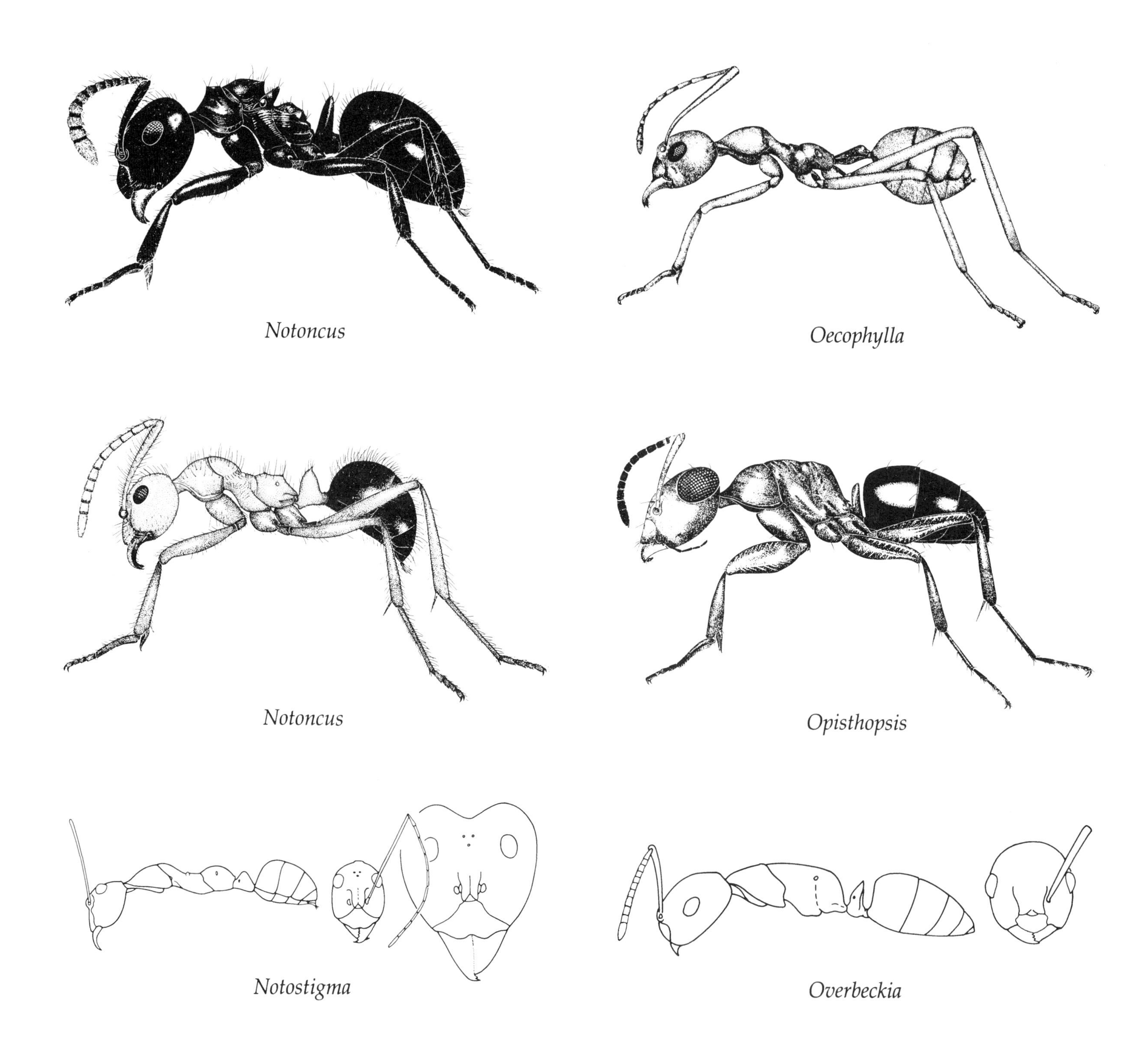


*Notoncus ectatommoides,* Australia (R. W. Taylor, unpublished; F. Nanninga, artist)

*Notoncus spinisquamis,* Australia (R. W. Taylor, unpublished; R. J. Kohout, artist)

*Notostigma carazzii,* Australia

*Oecophylla smaragdina,* Australia (R. W. Taylor, unpublished; R. J. Kohout, artist)

*Opisthopsis* sp., Australia (R. W. Taylor, unpublished; R. J. Kohout, artist)

*Overbeckia subclarata,* Singapore

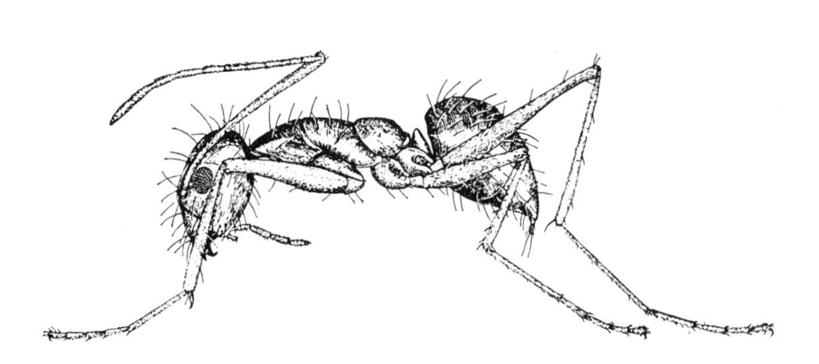

Paratrechina

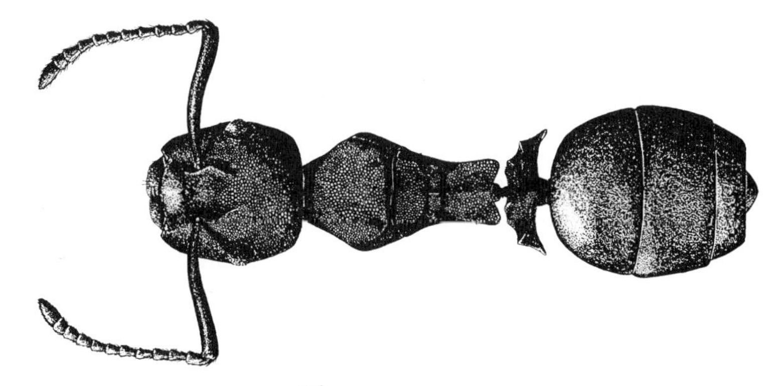

Phasmomyrmex

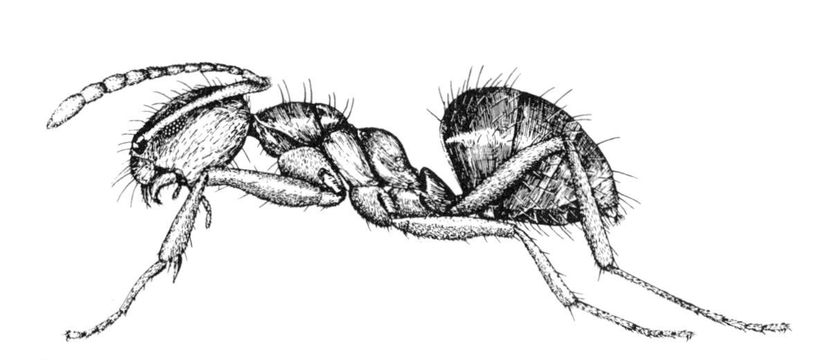

Paratrechina (= Nylanderia)

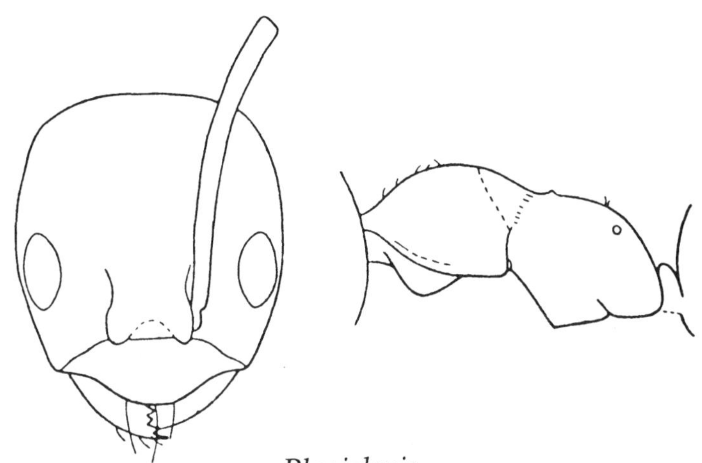

Plagiolepis

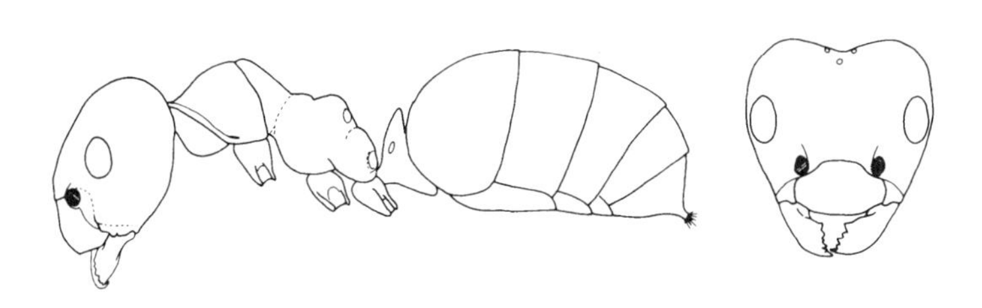

Petalomyrmex

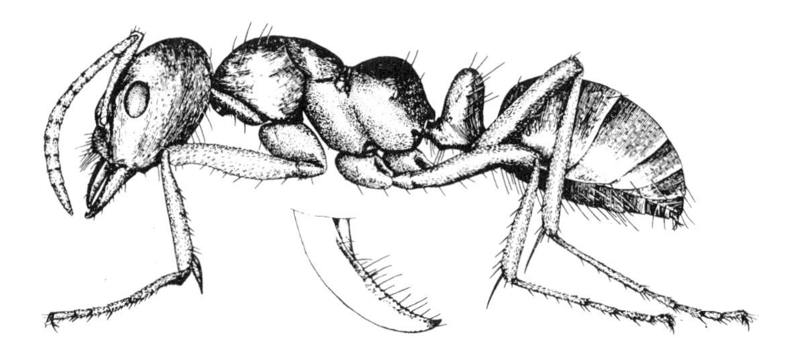

Polyergus

*Paratrechina longicornis*, United States (Smith, 1947a)

*Paratrechina* (= *Nylanderia*) *parvula*, United States (Smith, 1947a)

*Petalomyrmex phylax*, Cameroon (Snelling, 1979)

*Phasmomyrmex aberrans*, Ghana (B. Bolton, unpublished)

*Plagiolepis alluaudi*, Society Islands (Wilson and Taylor, 1967b)

*Polyergus lucidus*, United States (Smith, 1947a); see also Figure 12–7

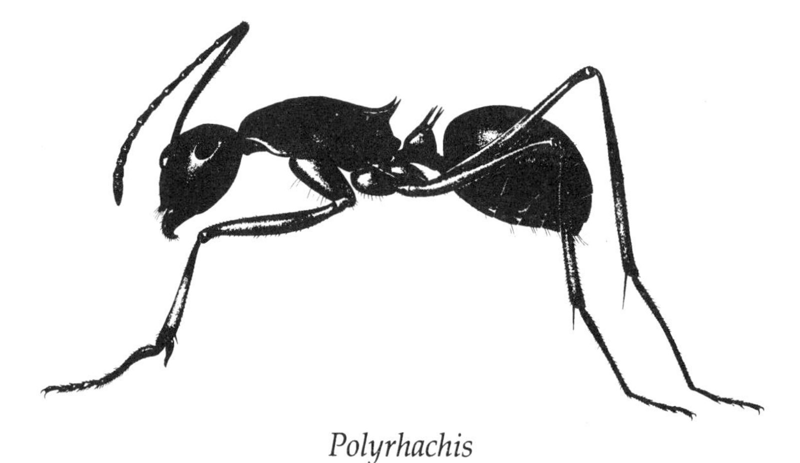

Polyrhachis

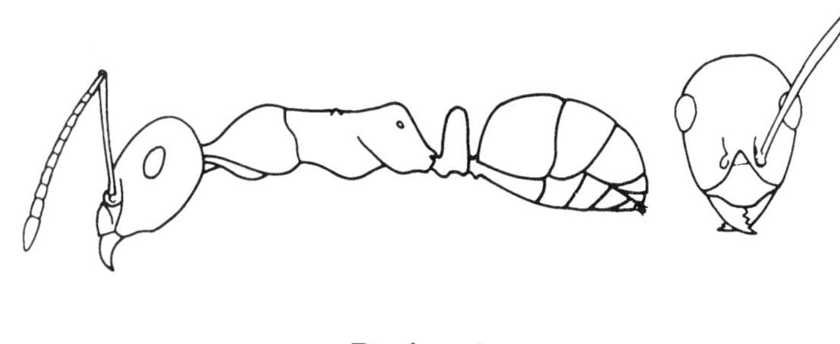

Proformica

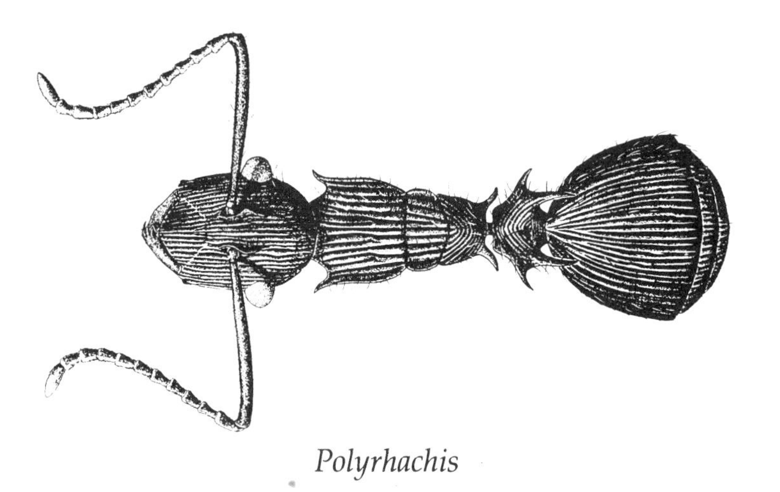

Polyrhachis

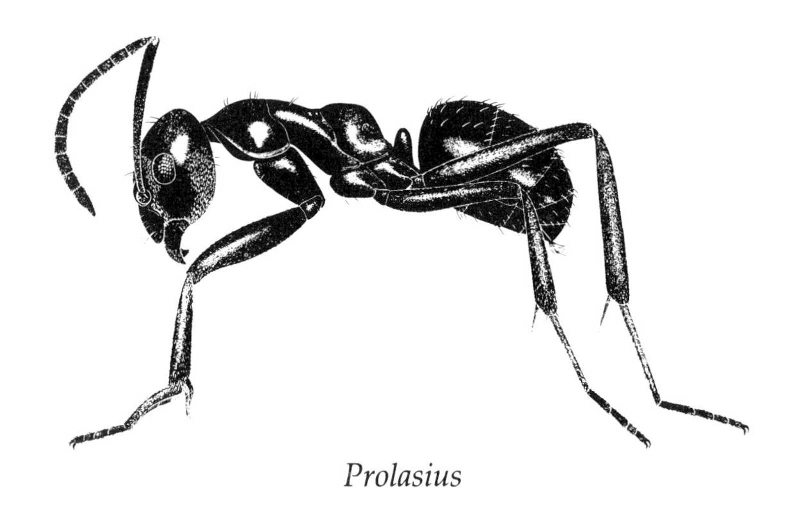

Prolasius

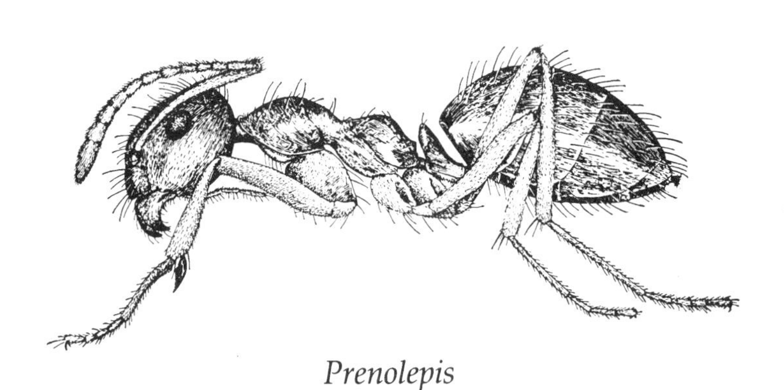

Prenolepis

Pseudaphomomyrmex

*Polyrhachis sulcata*, Ghana (Bolton, 1973)

*Polyrhachis* sp. nr. *macropus*, Australia (R. W. Taylor, unpublished; F. Nanninga, artist)

*Prenolepis imparis*, United States (Smith, 1947a)

*Proformica coriacea*, USSR

*Prolasius* sp., Australia (R. W. Taylor, unpublished; R. J. Kohout, artist)

*Pseudaphomomyrmex* sp., Vietnam (determination tentative)

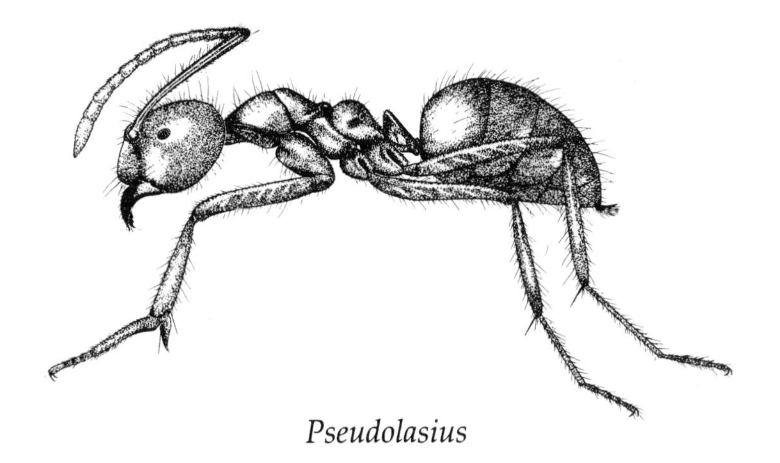

Pseudolasius

Santschiella

Pseudonotoncus

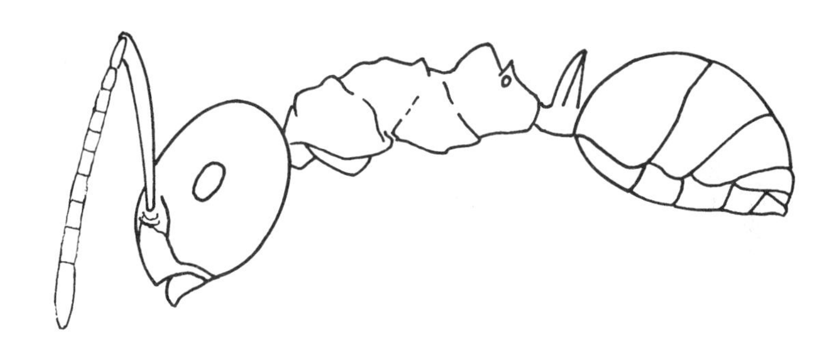

Stigmacros

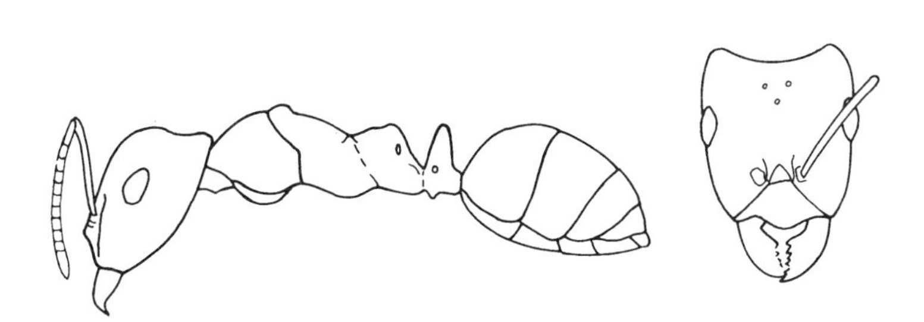

Rossomyrmex

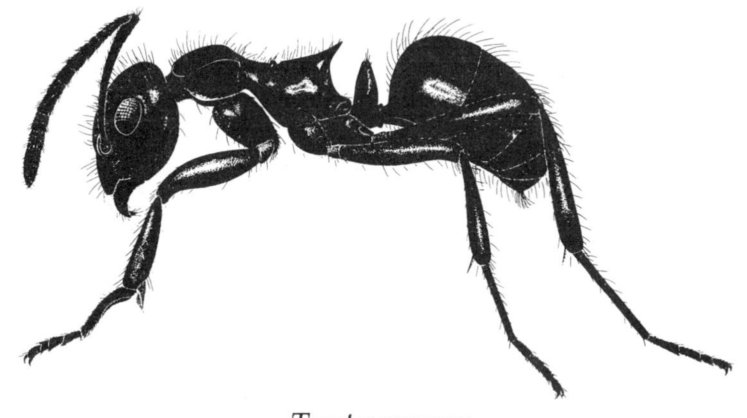

Teratomyrmex

*Pseudolasius australis*, Australia (R. W. Taylor, unpublished; R. J. Kohout, artist)

*Pseudonotoncus hirsutus*, Australia (R. W. Taylor, unpublished; R. J. Kohout, artist)

*Rossomyrmex proformicarum*, USSR

*Santschiella kohli*, Zaïre (redrawn from Emery, 1925a)

*Stigmacros occidentalis*, Australia

*Teratomyrmex greavesi*, Australia (R. W. Taylor, unpublished; R. J. Kohout, artist)

# CHAPTER 3

# The Colony Life Cycle

The ant colony is an almost exclusively female society with the males remaining in the nest only until the time of their invariably fatal nuptial flight. Also, the entire activity of the colony can be said to pivot on the welfare of the queen. It is, to paraphrase Samuel Butler's remark about the hen and the egg, the procedure by which a queen makes more queens. Seen in yet another way, the colony life cycle can be fruitfully analyzed as an orchestration of energy investments, in which workers are multiplied until such time as it is profitable to convert part of the net yield into new queens and males. In some extreme cases, this maturation point comes with the accumulation of only a few tens of workers, which are organized by means of the simplest caste and communication systems. For example, the average size of a colony of the fungus-grower *Apterostigma dentigerum* producing queens and males is 35 (Forsythe, 1981). The rare Central American myrmicine *Basiceros manni* reaches maturity at 50 workers (Wilson and Hölldobler, 1986). In other species maturity is not attained until the worker population reaches tens of thousands and develops complex caste and communication systems. The extreme examples are the army ants, whose colonies do not divide until the worker populations exceed hundreds of thousands or even a million (Raignier and van Boven, 1955; Rettenmeyer, 1963a).

The life cycle of a particular species can be viewed as the story of how the maturation point is attained with maximum combined speed and freedom from risk. Only by studying it as one strategy out of a great many possible strategies can we expect to understand more deeply the way a given species has adapted by social means to the particular environment in which it lives.

## STAGES OF COLONY GROWTH

Like the life cycle of the individual ant, the life cycle of an ant colony can be conveniently divided into three parts (Oster and Wilson, 1978). The *founding stage* begins with the nuptial flight. The virgin queen departs from the nest in which she was reared, leaving behind her mother, who is the queen of the colony, and her sisters, who are either sterile workers or virgin reproductives like herself. She meets one or more males and is inseminated. The males soon die without returning home, while the queen finds a suitable nest site in the soil or plant material and constructs a first nest cell. Here she rears the first brood of workers, drawing on her own tissue reserves to produce eggs and feed the growing larvae. Soon after reaching the adult stage, the workers take over the tasks of foraging, nest enlargement, and brood care, so that the queen may confine herself to egg laying. Over the coming weeks and months the population of workers grows, the average size of the workers increases, and new physical castes are sometimes added. The colony is now in the *ergonomic stage:* its activities are exclusively concerned with work devoted to colony growth, rather than with colony-level reproduction or dispersal (Figure 3–1). After a period that ranges according to species from a single warm season to five or more years, the colony begins to produce new queens and males (*reproductive stage*). The sexual forms go forth to start new colonies, and the new colony life cycle has begun. As depicted in Figure 3–2, colonies of all known ant species are perennial. Like flowering plants, they issue a crop of seeds, then return to an interval of purely vegetative (i.e., worker) growth.

Substantial variation has been elaborated out of this elementary theme, especially with reference to details in the mode of colony founding and the number of egg-laying queens that coexist during the several stages of the life cycle. Figure 3–3 presents a classification of the variations and the relevant terminology. *Monogyny* refers simply to the possession by a colony of a single queen, as opposed to *polygyny,* which is the possession of multiple queens. The founding of a colony by a single queen is referred to as *haplometrosis;* when multiple queens start a colony the condition is called *pleometrosis.* The term *metrosis* refers generally to this biological variable. Monogyny can be *primary,* meaning that the single queen is also the foundress; or it can be *secondary,* meaning that multiple queens start a colony pleometrotically but only one survives. In a symmetric fashion, polygyny can be primary, in which multiple queens persist from a pleometrotic association, or secondary, in which the colony is started by a single queen and supernumerary queens are added later by adoption or fusion with other colonies. (The patterns of queen numbers will be discussed in greater detail in Chapter 6.)

Next, the mode of colony founding is subject to complicated variation among species. It can be accomplished by *swarming,* a process also called budding, hesmosis, or sociotomy, in which two or more forces of workers separate in the company of queens. We prefer to divide swarming into two types: the more common *budding,* in which a group of workers departs from the main nest with one or more queens and starts a new nesting unit; and *fission* of the kind used by army ants, in which portions of the colony containing fertile queens separate from each other and go their own ways. Colony founding in ants is frequently *claustral,* meaning that the queen seals

**FIGURE 3–1** A young colony of the North American carpenter ant *Camponotus pennsylvanicus,* shortly after claustral founding by the queen and in the early ergonomic stage of growth. Alongside the large queen can be seen the first generation of workers, cocoons enclosing pupae, and grub-like larvae in various stages of growth.

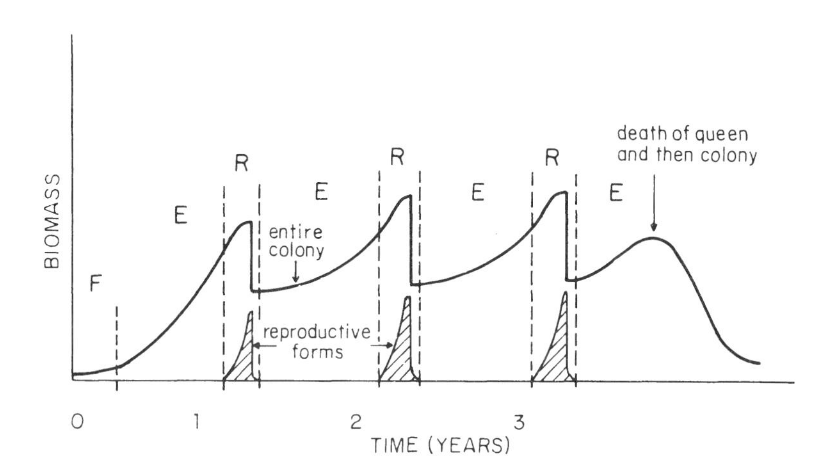


**FIGURE 3–2** The colony life cycle of most ant species can be conveniently divided into the three stages depicted here and labeled as follows: *F,* founding stage; *E,* ergonomic stage; *R,* reproductive stage. With the release of the reproductive forms (virgin queens and males) during the breeding season, the colony is diminished in size and reenters the ergonomic stage. (From Oster and Wilson, 1978.)

**FIGURE 3–3** The basic life cycle is modified in various species of ants by variations in the mode of colony foundation and the number of egg-laying queens that coexist in various stages of the life cycle. This diagram presents the several possibilities and the special terms employed to describe them. (From Hölldobler and Wilson, 1977b.)

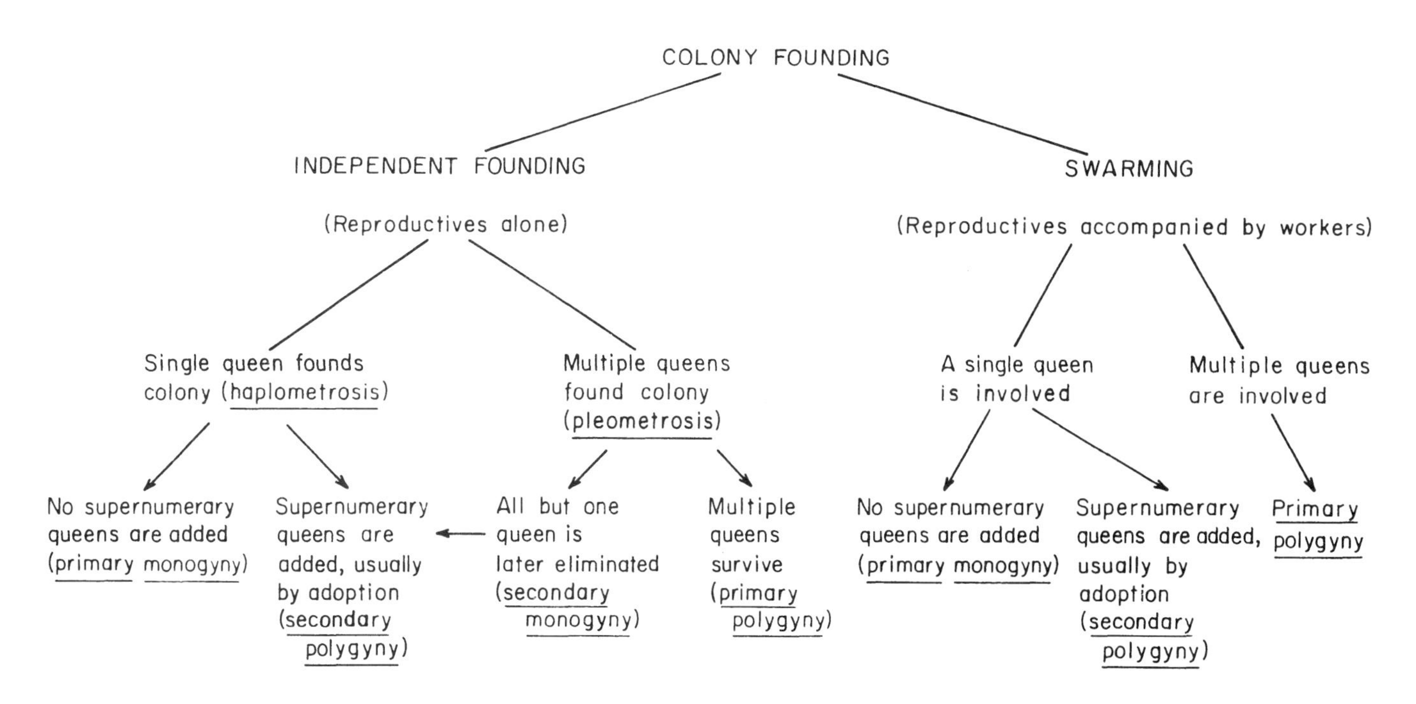

herself off in a chamber and rears the first brood in isolation. This is in fact the prevailing mode of independent colony formation in ants. However, the queens of such very primitive forms as *Amblyopone* and *Myrmecia*, as well as of some more advanced ponerine genera, still forage outside their cells for food, a condition known as *partially claustral colony founding* (Wheeler, 1933b; Haskins and Haskins, 1950a,b, 1951). The same behavior has been observed in the myrmicines *Acromyrmex, Manica,* and some species of *Pogonomyrmex,* and in the formicine *Cataglyphis* (Cordero, 1963; Le Masne and Bonavita, 1967; Fridman and Avital, 1983; B. Hölldobler, unpublished observations).

## NUPTIAL FLIGHTS AND MATING

The vast majority of virgin queens die within hours of leaving the mother nest. Most are destroyed by predators (Figure 3–4) and hostile workers of alien nests, while the others are variously drowned, overheated, and desiccated. In species with large nest populations, such as the leafcutter ants (*Atta*) and fire ants (*Solenopsis*), it is not uncommon for one colony to release hundreds or thousands of the young winged queens in less than an hour. If the surrounding area is dominated by stable, mature colonies, only one or two of the queens may become the progenetrices of new colonies. Most of the rest will die before they can construct a first shelter—or even before they can find a mate. In an unusual study of its kind, Whitcomb et al. (1973) have produced a catalog of the many kinds of predators that decimate young queens of the red imported fire ant *Solenopsis invicta*. The few individuals that navigate all the dangers must also avoid breeding with males of other species, thereby producing nonviable or sterile offspring.

It follows that the brief interval between leaving the home nest and settling into a newly constructed nest is a period of intense natural selection among queens, a dangerous odyssey that must be precisely timed and executed to succeed. We should expect to find an array of physiological and behavioral mechanisms that enable the young queens simultaneously to avoid enemies, to get to the right habitat on time in order to build a secure nest, and to mate with males of the same species. Field studies have shown that such specialized traits exist in abundance.

As also expected from the evolutionary argument, mating patterns vary greatly from one species to the next. Most of the patterns thus far studied, however, fall into one or the other of two broad classes, or "syndromes" (Hölldobler and Bartz, 1985). In the first, the *female-calling syndrome*, the females, which are often wingless and sometimes just fertile workers, do not travel far from the nest. Standing on the ground or low vegetation, they release sex pheromones to "call" the winged males to them (Figure 3–5). This pattern

**FIGURE 3–4** The mortality of colony-founding queens is extremely high. In this case a recently dealated queen of *Pogonomyrmex maricopa* has been captured by a crab spider. (From Hölldobler, 1976b.)

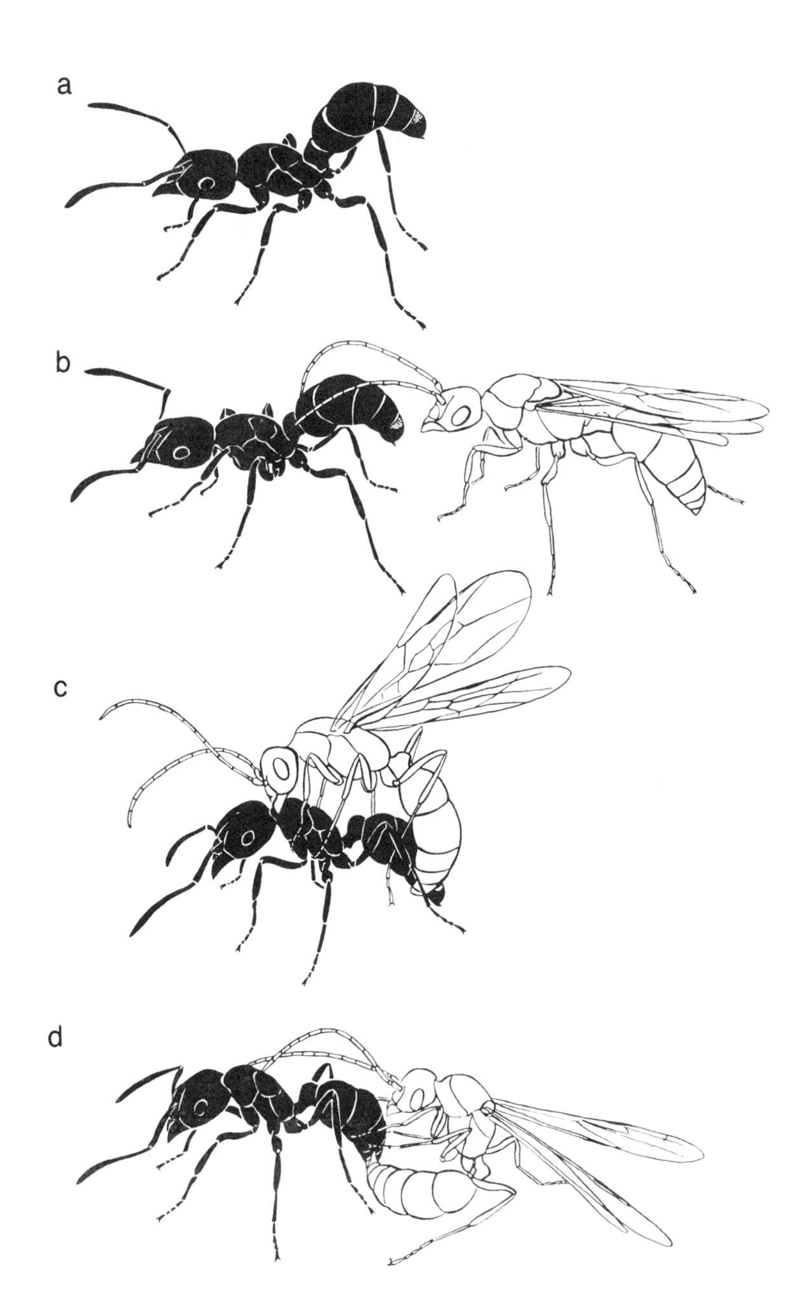


**FIGURE 3–5** An example of one of the two major categories of mating behavior in ants, the female-calling syndrome, in the Australian ponerine *Rhytidoponera metallica*. (*a*) An ergatoid or worker-like female (*black*) assumes the calling posture, during which she releases a sex pheromone from the pygidial gland located between the VIth and VIIth abdominal tergites. (*b*) A male approaches her and touches her with his antennae. (*c*) The male mounts the female, grasps her by the prothorax, and extrudes his copulatory organ in search of the female's genitals. (*d*) Copulation occurs. (From Hölldobler and Haskins, 1977; drawing by T. Hölldobler-Forsyth.)

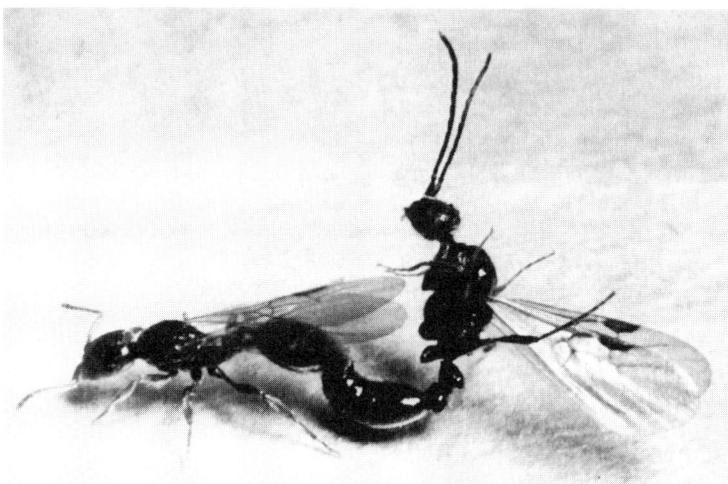

**FIGURE 3–6** Sexual behavior of the socially parasitic ant *Doronomyrmex pacis*. *Above:* a virgin female assumes the calling position, during which she extrudes her sting and releases a pheromone attractive to males. *Below:* a pair in copulation. (From Buschinger, 1971b.)

is displayed by *Amblyopone* and *Rhytidoponera*, which are members of the phylogenetically primitive subfamily Ponerinae (Haskins, 1978); presumably also by the very primitive *Nothomyrmecia macrops* (Hölldobler and Taylor, 1983); at least one pseudomyrmecine, the Neotropical acacia ant *Pseudomyrmex ferruginea* (Janzen, 1967); and the socially parasitic species of the myrmicine genera *Doronomyrmex, Formicoxenus, Harpagoxenus,* and *Leptothorax* (Buschinger, 1968a,b, 1971a,b, 1975b; see Figure 3–6). The colonies of female-calling species are typically small at maturity, with 20 to 1,000 workers, and produce relatively few reproductives. As far as we know, the females mate only once. An unusual variation on this pattern is followed by the Florida harvester ant *Pogonomyrmex badius*. Females gather on the surface of their home nest and are inseminated by males; afterward they fly off to start new colonies. Van Pelt (1953) thought that the males come from the same nest as the females with whom they copulate, but S. D. Porter (personal communication) observed that males usually fly for about a quarter-hour first before settling on a nest different from their own. Porter observed one case in which a male mated with two females after alighting.

The second combination of traits during mating is the *male-aggregation syndrome*. Males from many colonies gather at specific mating sites, usually prominent features of the landscape such as sunflecked clearings, forest borders, hilltops, crowns of trees, and even tops of tall buildings. Sometimes, as in some species of *Lasius* and *Solenopsis,* the males cruise in large numbers at characteristic heights above the ground. The females fly into the swarms, often from great distances, in order to mate (see Figures 3–7 through 3–9 and Plate 2), and afterward they typically disperse widely before shedding their wings and excavating a nest. The winged queens and males of the fire ant *Solenopsis invicta,* for example, fly to heights of 250 meters or more; 99 percent then descend to the ground within a 2-kilometer radius of their origin, while a very few travel as far as 10 kilometers. The ability of a single mature colony to disseminate fertile queens in many directions over long distances is one of the reasons the fire ant is so difficult to eradicate (Markin et al., 1971). Male-aggregation species typically differ from those utilizing female calling in two other key respects: the mature colonies are large, containing from several thousand to over a million workers and producing hundreds to thousands of reproductive adults yearly; and multiple insemination is common. An unusual reversal of the swarming procedure was recently discovered in some *Pheidole* species of the southwestern United States: the winged queens gather in aerial swarms, where they maintain a more or less uniform distance from each other while attracting males with pheromones. The males fly into the female swarms and mate with individual females (B. Hölldobler, unpublished). Swarms of variable composition, some predominantly male and others predominantly female (occasionally exclusively female), have been reported by Eberhard (1978) in the coccid-tending formicine *Acropyga paramaribensis* of northern South America.

Ant species can be classified another way into two broad types. When the males alight on the surface of the mating site, either in response to female calling or in swarms to compete directly with one another, they are often large and robust in form and possess well-developed mandibles. In contrast, males that gather in aerial swarms are usually (but not invariably) smaller relative to the queen than are males of the first type. Also, their mandibles are reduced in size and dentition, sometimes consisting of nothing more than vestigial lobate or strap-shaped organs. An example of this type is the small myrmicine *Pheidole sitarches* of the southwestern United States. Up to 50 males form circular swarms that hover from a few centimeters to 2 meters above the surface of woodland clearings. Each virgin queen flies in slow, even circles through the aggregations until mounted in midair by a male, whereupon the pair cease flying and spiral to the ground together to complete the copulation (Wilson, 1957b).

The swarms of some ant species are among the more dramatic spectacles of the insect world. W. W. Froggatt describes the flight of the giant Australian bulldog ant *Myrmecia sanguinea* as follows:

> On January 30th, after some very hot, stormy weather, while I was at Chevy Chase, near Armidale, N.S.W., I crossed the paddock and climbed to the top of Mt. Roul, an isolated, flat-topped, basaltic hill, which rises about 300 feet above the surrounding open, cleared country. The summit, about half an acre in extent, is covered with low "black-thorn" bushes (*Bursaria spinifera*). I saw no signs of bull-dog ant nests till I reached the summit. Then I was enveloped in a regular

**FIGURE 3–7** The second of the two categories of mating behavior in ants, the male-aggregation syndrome, exhibited by species of American harvester ants (*Pogonomyrmex*). (*a*) Winged *Pogonomyrmex maricopa* queens emerge from the nest before taking flight. (*b*) A mating aggregation of *P. desertorum* at an acacia tree, with males and females approaching upwind. (*c*) A mating cluster of *P. desertorum*, in which a young queen is surrounded by two males. (From Hölldobler, 1976b.)

**FIGURE 3–8** Behavioral sequence during mating in *Pogonomyrmex*, from top to bottom: (*a*) A male (*black*) approaches a female (*white*) and touches her with his antennae. (*b*) The male grasps the female's thorax and attempts to insert his copulatory organs into the female's cloaca. (*c*) After successfully inserting his copulatory organs, the male releases his mandibular grip on the female's thorax; a second male grasps the female's thorax. (*d*) The copulating male massages the female's gaster with his mandibles and forelegs; the female begins to gnaw at the male's gaster. A third male has arrived and grasps the second male's gaster. The entire sequence is based on film recordings. (From Hölldobler, 1976b; drawing by T. Hölldobler-Forsyth.)

> cloud of the great winged ants. They were out in thousands and thousands, resting on the rocks and grass. The air was full of them, but they were chiefly flying in great numbers about the bushes where the males were copulating with the females. As soon as a male (and there were hundreds of males to every female) captured a female on a bush, other males surrounded the couple till there was a struggling mass of ants forming a ball as large as one's fist. Then something seemed to give way, the ball would fall to the ground and the ants would scatter. As many as half a dozen of these balls would keep forming on every little bush and this went on throughout the morning. I was a bit frightened at first but the ants took no notice of me, as the males were all so eager in their endeavors to seize the females.
> (in Wheeler, 1916c: 72)

Donisthorpe (1915), from the distinctively British viewpoint of an earlier observer, tells of the mass flights of the abundant *Myrmica rubra:*

> Farren-White in 1876 observed a swarm of ants near Stonehouse rising and falling over a small beech tree. The effect of those in the air—gyrating and meeting each other in their course, as seen against the deep blue sky—reminded him of the little dodder, with its tiny clustered blossoms and its network of ramifying scarlet threads, over the gorse or heather at Bournemouth. He noticed the swarm about thirty paces off, and it began to assume the appearance of curling smoke; at forty paces he could quite imagine the tree to be on fire. At fifty paces the smoke had nearly vanished into thin air. (p. 108)

A still different mating pattern was described in the Australian formicine species *Notoncus ectatommoides* by Brown (1955a):

> In a cropped lawn at Montville, numerous small holes appeared, each opened by workers and accompanied by a minute pile of dark earthen particles. From these holes, males began to issue almost immediately in numbers, until within a few minutes there had accumulated on the surface a surprisingly large number of this sex and also a few workers. The males travelled aimlessly over the sward in low, flitting flight from one blade of grass to another, never rising more than a foot or so from the ground. Movement seemed to take place at random in all directions. Suddenly, however, the males of one area all rushed simultaneously to a single focal point, which proved to be a winged female emerging from a small hole. In a few seconds, the female was surrounded by a dense swarm of males in the form of a ball, which at times must have exceeded 2 cm in diameter. This ball moved in a half-tumbling, half-dragging motion over and among the densely packed grass blades, and held together for perhaps 20 seconds, after which the female escaped, flying straight upward. She appeared not to be encumbered by a male, and no males were seen to follow her for more than a foot above the ground; she flew steadily, and soon passed out of sight.
>
> Meanwhile, the lawn had become dotted with similar balls of frenzied males, each surrounding a female in a fashion similar to the first. Obviously, many more males than females were involved in this particular flight. On each occasion, the female left the ball after 20–30 seconds and flew straight upward. (pp. 487–488)

In a similar fashion males and females of *Formica obscuripes* conduct nuptial swarms on the ground. Talbot (1972) observed them flying to "swarming grounds" near their nests, which were maintained throughout the nuptial flight season and perhaps even from year to year. The males fly back and forth above the ground searching for females, which "stand on grasses, forbs or bushes," and apparently signal their presence to the males by pheromones.

No encompassing theory exists to explain the extreme variation in the patterns of mating behavior so far observed. A close exami-

**FIGURE 3–9** A mating cluster of *Pogonomyrmex rugosus,* in which more than ten males compete for access to a young female.

nation of individual species, however, reveals details that clearly contribute to the greater success of the sexual castes. For example, flying queens of the formicine *Lasius neoniger* stay strictly within open fields, the exclusive habitat of the earthbound colonies. Fewer than 1 percent make the mistake of venturing into adjacent woodland, a habitat dominated by the closely similar *Lasius alienus.* In one experimental study (Wilson and Hunt, 1966), newly inseminated and flightless queens were labeled with radioactive material for easy tracking and displaced to woodland sites. They attempted to crawl out but were unable to do so; thus the *Lasius* queens depend on controlled flight patterns to survive.

Like the orientation, the timing of the flights is important for successful mating and colony foundation. Flights conducted as part of the female-calling syndrome do not appear to be well synchronized at the level of either the colony or the population of colonies. The search by airborne males for solitary calling females in fact resembles that of many solitary wasps (Buschinger, 1975b; Haskins, 1978). In contrast, flights leading to male aggregation are tightly synchronized within the colony as well as among colonies of the same species.

The manner in which this coordination is achieved is typified by *Pogonomyrmex* harvester ants of the southwestern United States (Hölldobler, 1976b). Just prior to take-off, males and females move restlessly in and out of the sandy crater nests or cluster around the entrance, as shown in Figure 3–7. This preflight activity is especially pronounced in *P. maricopa,* a morning flyer whose queens and males evidently need more time to warm up before taking wing. As the time of departure approaches, the reproductives run back and forth in mounting intensity. In a frenzy, they climb up and down on grass leaves or small bushes around the nest. At this point many more workers pour out of the nest, running excitedly around it and attacking any moving object encountered (including the careless myrmecologist). When the first reproductives try to take flight, the workers delay many of them by pulling or carrying them back to the nest. Once the flight is in full progress, however, workers cease to interfere. Although the timing of the take-off overlaps considerably between the two sexes, the males generally fly from the nest first. Once aloft, both sexes appear initially to drift with the wind, but after a few seconds they take a course upwind or across the wind. Soon afterward they arrive at the swarm sites, centered on conspicuous landmarks such as tree crowns and hilltops or (in the case of *P. rugosus*) merely flat local areas in the desert.

A similar marching order is observed by the carpenter ant *Camponotus herculeanus,* which nests in the trunks of both living and dead trees in the boreal forests of Eurasia and North America. Males leave before queens, although the periods broadly overlap. The early departure of the winged forms is inhibited by the workers, who drag or carry many back to the nest entrance (Figure 3–10). When the males do succeed in taking flight, they discharge a pheromone from their mandibular gland. The concentration of this substance is highest at the peak of male activity—the gland emission can be smelled readily by humans—enough to trigger the mass take-off of the females (Figure 3–11). Blum (1981b) reports methyl 6-methylsalicylate and mellein as two of the three components of the secretion. This pleasantly aromatic combination is shared by most other species of *Camponotus,* but considerable differentiation is nevertheless achieved by the addition of other substances, such as octanoic acid and methyl anthranilate, according to species (see also Lloyd et al., 1984). A similar function may be accomplished in *Pogonomyrmex* harvester ants by vibrational signals rather than by pheromones. Both males and virgin queens stridulate just before and during take-off, running the sharp posterior rim of their postpetiole over the actively moving, striated file on the first gastric tergite (Markl et al., 1977).

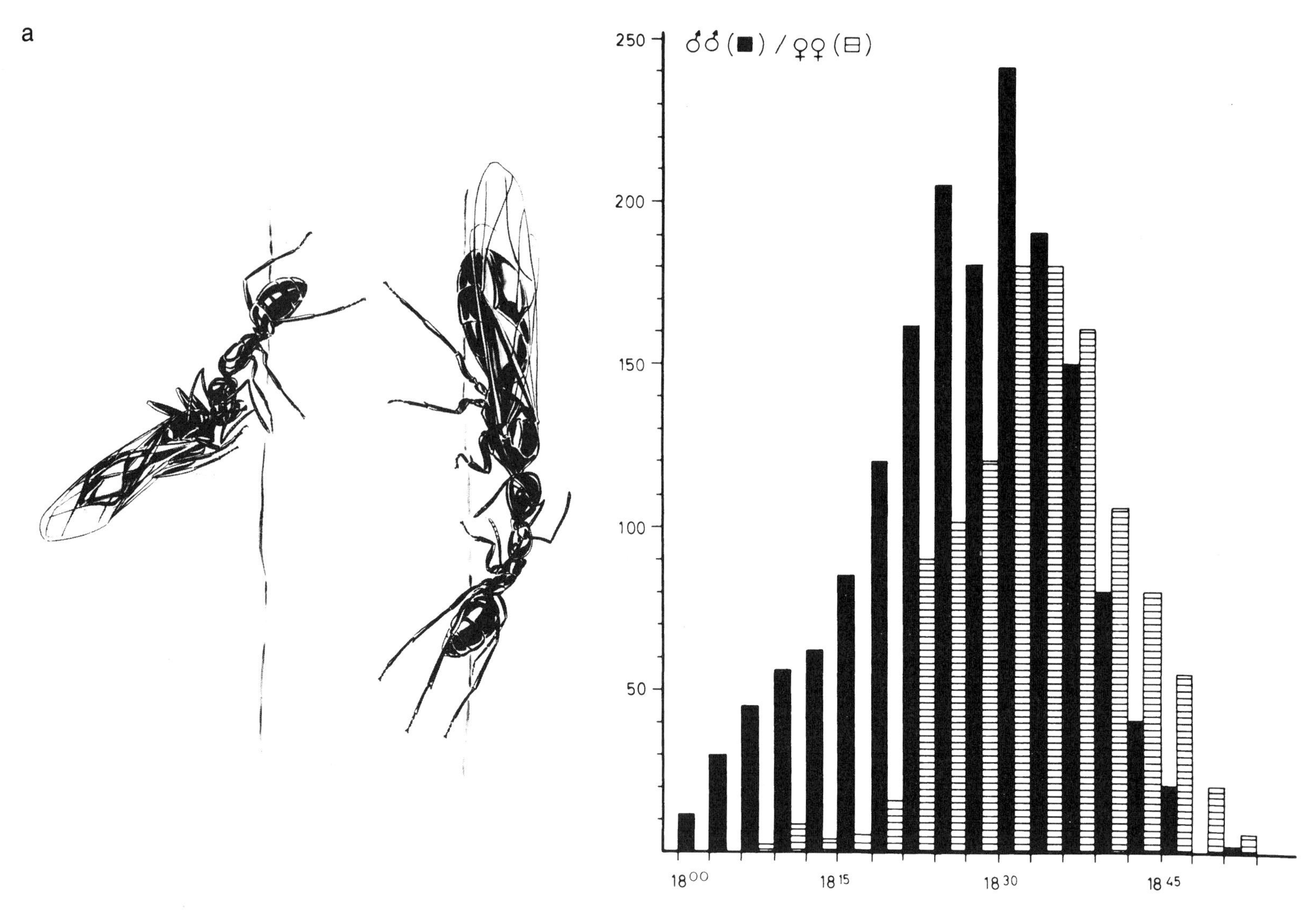


**FIGURE 3–10** Synchronization of the nuptial flight in the carpenter ant *Camponotus herculeanus.* (*a*) The early take-off of both males (on the left) and virgin queens (on the right) is hindered by the workers, who carry or drag them back to the nest entrance. (*b*) On the average, males depart earlier than do females. The ordinate represents the number of ants leaving the nest; the abscissa gives the time of day. (From Hölldobler and Maschwitz, 1965.)

Many entomologists, including especially Kannowski (1959a, 1963) and Weber (1972), have observed that each ant species, at least each displaying the male-aggregation syndrome, swarms at a precise time in the 24–hour diel cycle; and the time differs among species. Under controlled laboratory conditions, McCluskey (1958, 1965, 1967, 1974) and McCluskey and Soong (1979) demonstrated in fact that the rhythms of males are generally if not universally circadian and endogenous. Once set in a laboratory regime of 12 hours light alternating with 12 hours dark, the rhythms persist for up to a week in total darkness. They are also quite precise. McCluskey found that males of the harvester ant *Messor (= Veromessor) andrei* increase in movement during the last hour of darkness, then peak during the first hour of light. Throughout the remainder of the 24-hour cycle they are quiescent, usually stirring themselves only to groom, solicit food from the workers, or walk sluggishly about the nest. Males of the Argentine ant *Iridomyrmex humilis,* in contrast, are most active at the very end of the light period. Similarly distinctive rhythms, each spanning only 1 or 2 hours, have been documented by McCluskey and his co-workers across a wide diversity of species from four subfamilies (Ponerinae, Myrmicinae, Dolichoderinae, and Formicinae), including some that are wholly nocturnal.

Queens of at least two species, *Pogonomyrmex californicus* and *Messor (= Veromessor) pergandei,* also display circadian rhythms, and these are more or less synchronous with those of the males (McCluskey, 1967; McCluskey and Carter, 1969). In the case of *P. californicus,* at least, the rhythm persists even after the female has flown and lost her wings. But it ceases when she is mated.

In summary, the time of day at which flights occur is programmed by a species-specific diel rhythm. But what determines the particular *day* on which the flights occur? Several studies, including the one by Boomsma and Leusink (1981), have shown that weather conditions play a major role in the timing of nuptial flights. One of the commonest triggering stimuli is rain, especially in species that occupy dry habitats such as deserts, grasslands, and forest clearings. A typical species in this respect is *Lasius neoniger,* one of the most abundant ants in abandoned fields and other open environments in eastern North America. This small formicine emerges in immense swarms in the late afternoon in the second half of August or the beginning of September. The flights almost always occur within 24 hours of moderate or heavy rainfall on warm, humid days with little wind. For an hour or so the air seems filled with winged ants, rising from the ground like snowfall in reverse. After mating, the queens find themselves on moistened soil that is easier to excavate. They are also protected from desiccation due to overheating (Wilson, 1955a). A very similar pattern is followed by the North American leafcutter ant *Atta texana,* except that the flights occur well before dawn, between 0300 and 0415 hours (Moser, 1967a).

Because there are relatively few "best days" during which the young queens can be successfully launched, species belonging to the same genus are likely to swarm at the same time and location. In one respect this is a favorable result, since an apparent function of mass emergence and swarming in cicadas, termites, and other insects is the reduction of mortality by overloading predators (Wilson, 1975b). But in another respect it can be detrimental. In the tumult of the swarms, with males struggling to copulate with each female encountered, there is a strong likelihood of interspecific hybridization resulting in either sterility or the production of less viable hybrids. The standard argument from natural selection theory suggests that this circumstance favors the evolution of premating isolating mechanisms. This conventional explanation does seem compatible with a great deal of evidence. Species belonging to genera as phylogenetically diverse as *Myrmecia, Pheidole, Solenopsis,* and *Lasius* have been observed to conduct their nuptial flights within the major habitats occupied by the colonies, thus automatically avoid-

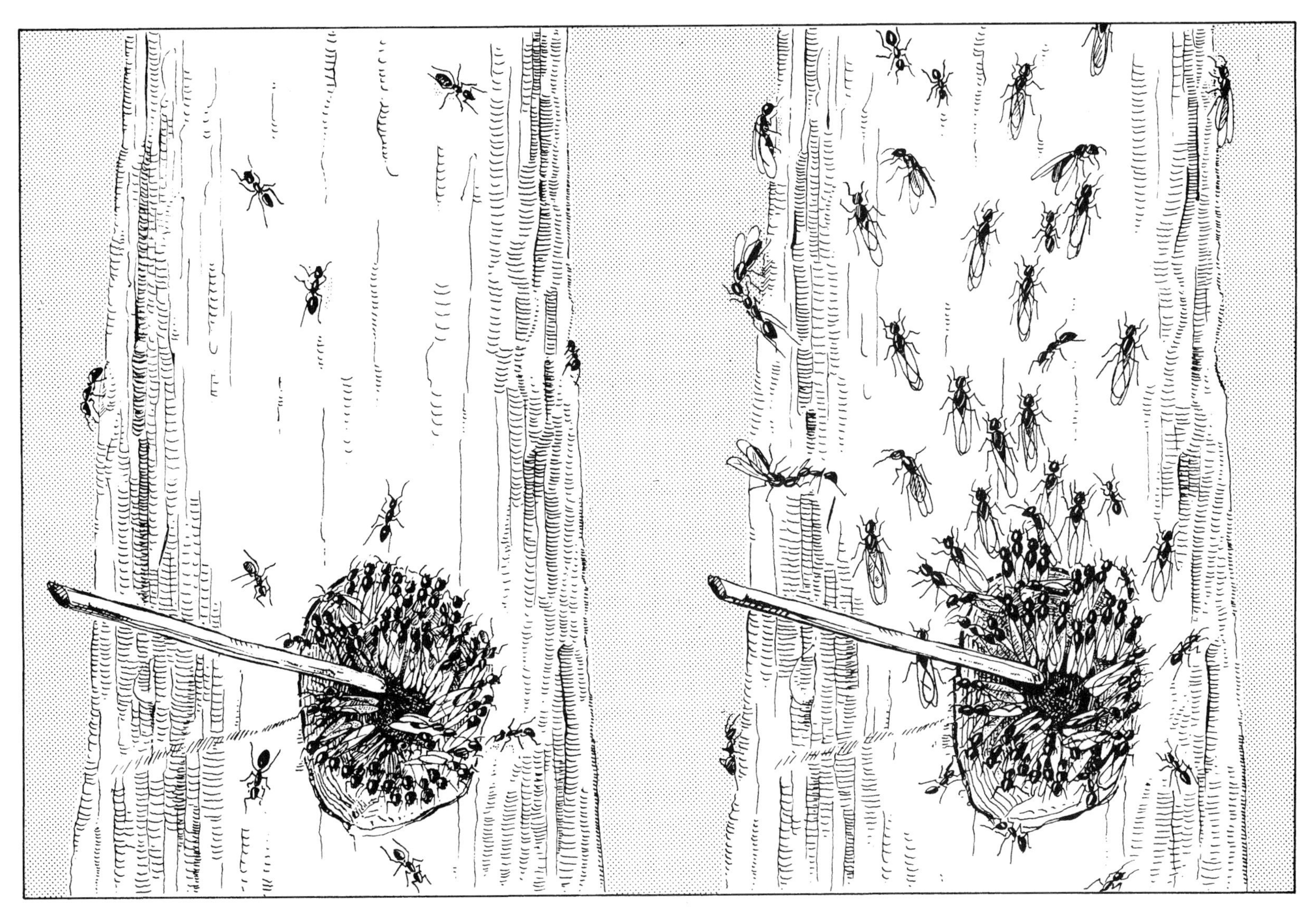

**FIGURE 3–11** An experiment showing the role of a male mandibular gland pheromone in synchronizing the nuptial flight of the carpenter ant *Camponotus herculeanus.* (*Left*) An applicator stick bearing a crushed mandibular gland on the tip is presented to winged reproductives clustered at the nest entrance. (*Right*) Moments after being stimulated by the pheromones, the reproductives rush from the entrance and prepare to take flight. (Modified from Hölldobler and Maschwitz, 1965.)

ing sexual contact with closely related species limited to other major habitats. How widespread and efficient this isolating mechanism is among ants in general has not been determined, but it cannot be the sole device in deserts, savannas, and tropical moist forests, where large numbers of congeneric species nest closely together. To take an extreme case, in many forest localities in the Amazon Basin, 30 or more species of *Pheidole* can be found within a single plot of a few square kilometers.

Another intrinsic isolating mechanism is differentiation in the preferred mating site within the major habitat. Among the sympatric species of *Pogonomyrmex* of Arizona, *P. desertorum* and *P. maricopa* congregate on bushes and trees, whereas *P. barbatus* and *P. rugosus* gather at different sites on the ground. In addition, males mark the sites with secretions from their mandibular glands, and apparently the females and other males are attracted by volatile pheromones contained in the material (Hölldobler, 1976b). It is possible (but not yet experimentally verified) that the pheromones are species specific and serve as an additional isolating device.

Many congeneric species are further separated by the timing of their mating flight, either the season of the year or the hour of the day. In Figures 3–12 and 3–13 we have presented two sets of data from army ants that suggest just such a mutually repulsing spread of flight times across the seasons and the daily cycle respectively. The males of army ants, on which the data were based, fly for an unknown distance before entering the columns or bivouacs of alien colonies belonging to the same species. If the receptiveness of the workers is synchronized by the same circadian rhythm, even the hours of flight can serve as an effective barrier to "mistakes" and interspecific hybridization. Such staggering in the diel flight schedule appears to be common among ants. In Michigan, for example, *Myrmica emeryana* flies between 0600 and 0800 hours, *M. americana* between 1230 and 1630 hours, and *M. fracticornis* between 1800 and 1930 hours (Kannowski, 1959a). Similarly, in Arizona *Pogonomyrmex maricopa* flies between 1000 and 1130 hours, *P. barbatus* between 1530 and 1700 hours, and *P. rugosus* between 1630 and 1800 hours. As morning flyers, the *P. maricopa* queens appear to be at some disadvantage. The heat of midday prevents them from beginning nest excavation for three or four hours, during which time they are subject to higher predation than the other species (Hölldobler, 1976b). Some of the most closely related European species of *Leptothorax* swarm at different times of the day; others come into contact and occasionally hybridize (Plateaux, 1978, 1987).

Another potential advantage of synchronous nuptial swarming is the increase in the numbers of colonies participating and hence the degree of outbreeding. The sparse data on allozyme variation in ants collected so far indicate that outbreeding is indeed nearly total (Craig and Crozier, 1979; Pamilo and Varvio-Aho, 1979; Pearson, 1983; Ward, 1983a). Hence mating is either effectively at random, as demonstrated in experimental choice tests with *Pogonomyrmex californicus* by Mintzer (1982a), or disassortative, that is, directed away from nestmates.

The glandular sources of sex pheromones produced by female ants have been identified only for a few species. The reproductive females of *Rhytidoponera metallica* call males with a sex attractant from the pygidial gland, an intersegmental structure between the VIth and VIIth abdominal tergites (Hölldobler and Haskins, 1977). Although some of the contents of this gland have been chemically identified (Meinwald et al., 1983), the specific behavior-releasing

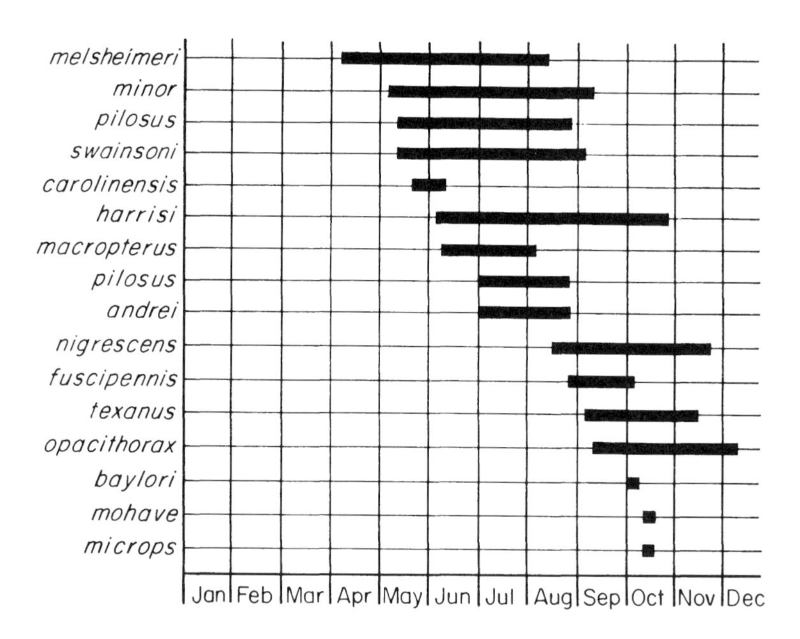


**FIGURE 3–12** The seasons of flight by the males of 16 North American species of the army ant genus *Neivamyrmex*. The differences appear to reduce or even to prevent the possibility of interspecific hybridization. (Modified from Baldridge et al., 1980.)

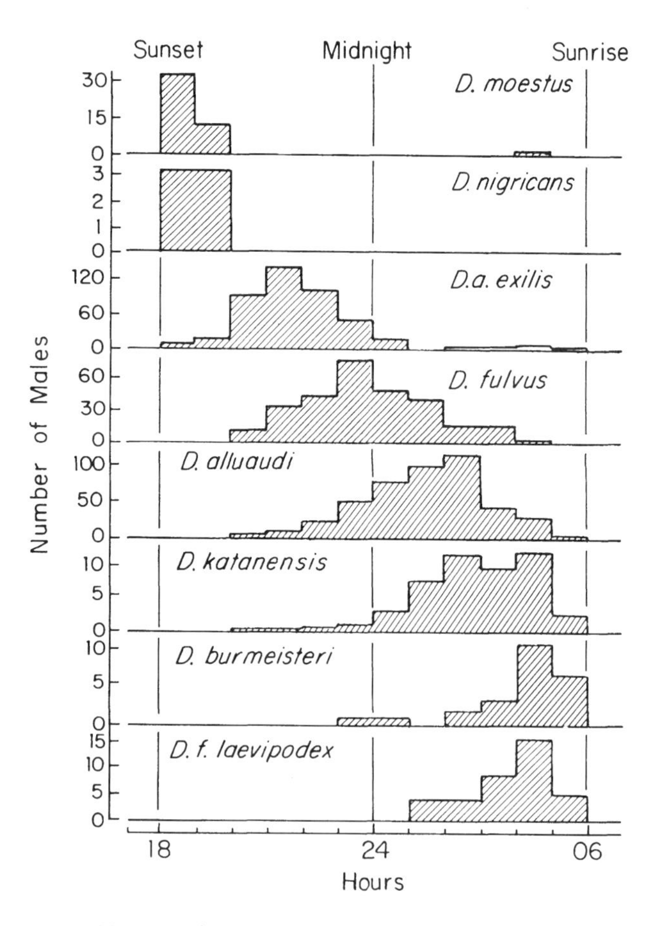


**FIGURE 3–13** Hours in the evening during which Ugandan driver ant males (*Dorylus*) were attracted to light traps. The differences in flight time could contribute to the reproductive isolation of species and hence the avoidance of mating errors between species. (Modified from Haddow et al., 1966.)

components have not yet been established experimentally. In several myrmicine species glands associated with the sting apparatus have been pinpointed as the sources of female sex pheromones. Virgin queens release a male-attracting pheromone from the poison gland in the myrmicines *Xenomyrmex floridanus* (Hölldobler, 1971), *Harpagoxenus sublaevis* (Buschinger, 1972a), *Doronomyrmex kutteri* and *D. pacis* (Buschinger, 1975b; see Figure 3–6), and *Formicoxenus nitidulus* (Buschinger, 1976a,b).

Buschinger (1972b) was also able to demonstrate that males of *Doronomyrmex kutteri* and *D. pacis* react to the other species' female sex pheromones, and that hybridization is possible in laboratory experiments. In the field, however, both species, which occur sympatrically, appear to be sexually isolated by different diel rhythms in mating activity. In general, specificity in sexual communication is consistent with phylogenetic relationships among the leptothoracines. The Canadian slavemaker *Harpagoxenus canadensis* shows the same mating behavior as the European *H. sublaevis*, and males of both species respond to the other species' female sex pheromones. Very similar sexual behavior and responses to sex pheromones have been described in several other social parasites of the "subgenus *Mychothorax*" of *Leptothorax*, whose hosts, like those of *H. canadensis* and *H. sublaevis*, also belong to the "subgenus *Mychothorax*." The same is true of at least some non-parasitic members of the subgenus. In fact, there appears to be no pheromone specificity among the *Leptothorax* species. In contrast, *Protomognathus americanus* males do not respond to *H. canadensis* or *H. sublaevis* pheromones. This anomaly suggests that *P. americanus* may be more closely related to its host of the "subgenus *Leptothorax*" than to *Harpagoxenus* or the "subgenus *Mychothorax*" (Buschinger, 1975b, 1981; Buschinger and Alloway, 1979).

Poison gland secretions of *Pogonomyrmex* females also attract males (Hölldobler, 1976b). In *Monomorium pharaonis*, on the other hand, the female sex pheromone is derived from the Dufour's gland and the bursa pouches (Hölldobler and Wüst, 1973).

Male ants are richly endowed with exocrine glands (Hölldobler and Engel-Siegel, 1982), but little is known about their function. One important fact, noted earlier, is that *Camponotus herculeanus* males discharge mandibular gland contents when departing from the nest that stimulate the virgin reproductive females to launch their nuptial flight as well. A variety of compounds of the mandibular gland secretions of several *Camponotus* species have been identified (for review see Blum, 1981b), but it is not yet clear which substance or combination of compounds elicits the behavior. Similarly, the males of *Lasius neoniger* discharge their mandibular gland contents sometime during the nuptial flight (Law et al., 1965), but the precise timing and function remain unknown.

Males of *Pogonomyrmex* discharge mandibular gland secretions when arriving at the mating sites. The collectively discharged pheromone appears to attract the virgin females to the lek (Hölldobler, 1976b). It is possible that in other species where males have well-developed mandibular glands and distinct blends of compounds, the secretions also promote aggregation and competition. Examples include *Lasius* and *Acanthomyops* (Law et al., 1965), *Camponotus* (Brand et al., 1973b,c; review by Blum, 1981b), *Calomyrmex* (Brown and Moore, 1979), *Myrmecocystus* (review by Blum, 1981b), *Tetramorium caespitum* (Pasteels et al., 1980), and *Polyrhachis ?doddi* (Bellas and Hölldobler, 1985).

A hypothesis concerning a possible novel role of male pheromones in sexual selection in army ants has been proposed by Franks and Hölldobler (1987). A detailed morphological examination of the reproductives has shown a close resemblance of conspecific males and females. Males are remarkably queen-like, large and robust, with long, cylindrical abdomens partially filled with an impressive battery of exocrine glands similar in form and location to those of females. Because queens are flightless and never leave their colony, males must fly between colonies and run the gauntlet of the workers before they approach the queen. For this reason, the workers can choose which males will be admitted and which virgin queens will be inseminated by the males. Army ant workers may therefore be involved in a unique form of sexual selection in which they choose both the matriarch and patriarch of new colonies. If this interpretation is correct, males resemble queens not because they are deceitful mimics but because, under the influence of sexual selection, they have come to use the same channels of communication as those used by queens to demonstrate their potential fitness to the workers.

Worker involvement in sexual selection may not be restricted to the army ants. Longhurst and Howse (1979a) observed that males of *Megaponera foetens* enter the nests of alien colonies, after utilizing recruitment pheromone trails laid by workers to guide them to the nests. Wheeler (1910a) noted that males of *Leptogenys elongata* are also accepted into alien colonies to mate with the wingless ergatoid females, and Maschwitz and Mühlenberg (1975) observed that males run along permanent foraging trails of *Leptogenys ocellifera*, apparently in an attempt to find access to ergatoid females. It may therefore be significant that Hölldobler and Engel-Siegel (1982) discovered very large exocrine sternal glands in *Leptogenys* males. Some other ponerines have ergatoid queens and therefore are not likely to engage in ordinary nuptial swarms, including species of *Diacamma*, *Dinoponera*, and *Ophthalmopone*. Males of *Ophthalmopone berthoudi* also enter strange nests after dispersal flights in the same manner as those of *Megaponera*, but as far as we know do not follow odor trails—*O. berthoudi* workers in fact forage in an exclusively solitary manner and hence are less likely to lay recruitment trails of any kind (Peeters and Crewe, 1986a, 1987).

Male ants compete for females rigorously, whether they are orienting to calling females in the primitive manner, flying in aerial swarms, or massing on the surface of the ground and vegetation. The competitive nature of mating is vividly illustrated by *Pogonomyrmex rugosus* (see Plate 2). The males gather in what can properly be called leks of the vertebrate kind. That is, the males occupy the same site year after year, use pheromones to attract other reproductives of the same species, and then compete with one another for access to the females. In the desert near Portal, Arizona, Hölldobler (1976b) was able to locate only one such site in an area of approximately 120,000 square meters. The mating arena covered 4,800 square meters of completely flat land unmarked by any distinctive physical features. The winged reproductives approached the arena upwind, which may suggest the presence of an olfactory cue. The first individuals to arrive (at around 1630 hours) were males, who alighted and began to race about in a frenzied manner. Soon afterward the first females alighted. They were immediately surrounded by 3 to 10 males, as shown in Figures 3–8 and 3–9. At the height of the activity thousands of such mating clusters carpeted the ground, in densities as high as 50 per square meter. The queens actively terminated mating after several copulations and stridulated when pre-

**FIGURE 3–14** A young mated *Pogonomyrmex* queen, shortly after shedding her wings, begins to excavate the founding nest chamber. (From Hölldobler, 1984d; drawing by J. D. Dawson reprinted with permission of the National Geographic Society.)

vented from leaving by other suitors. This stridulatory vibration evidently served as a "female liberation signal" that communicated the female's non-receptivity to approaching males and induced them to cease pursuit (Markl et al., 1977). The females then climbed onto grass leaves to launch their flights or else flew directly from the ground. Some landed a short distance away, but others flew at least 100 meters and possibly much farther. Each then shed her wings and began to excavate a nest chamber in the soil (Figure 3–14).

The general activity at the *Pogonomyrmex rugosus* mating site lasted about two hours, ending completely by 1900 hours as darkness approached. The males then withdrew into shelters around the mating site, such as crevices beneath grass clumps or little cavities in the soil. There they remained clustered overnight and through the following day until 1500–1600 hours, when they resumed activity. As on the previous day, new males flew in to the site to swell the population, and shortly afterward females began to arrive. This cycle was repeated on three more consecutive days.

The ant leks differ from those of the sage grouse, hammerheaded bats, and Hawaiian *Drosophila* (see for example Bradbury, 1985) in one important respect. Ant males are constrained in a way that vertebrates and fruit flies are not: each male ecloses from the pupa into full maturity with all of the sperm that he will ever possess. Dissections of males from phylogenetically divergent genera such as *Nothomyrmecia, Camponotus, Lasius, Myrmica,* and *Pogonomyrmex* reveal that the males' testes have degenerated and all of the sperm have migrated to the expanded vas deferens (Hölldobler, 1966 and unpublished data). When the male mates, he discharges most or all of the sperm together with the secretions of the mucus gland; he is thus incapable of additional inseminations (see Figure 3–15). As a result, reproductive success in male ants does not increase with repeated copulations, as it does with other kinds of insects whose males continuously replenish their sperm supply. Furthermore, it does not appear, from the few cases known, that males have enough sperm to inseminate more than one female. In cases where the queen is destined to produce very large numbers of offspring, one male is not even able to supply all of her needs. In *Atta sexdens*, for example, each newly eclosed male has between 40 million and 80 million spermatozoans, while each newly mated female contains between 200 million and 310 million spermatozoans in her spermatheca (Kerr, 1962). Fire ant queens *(Solenopsis invicta)* receive a supply of about 7 million sperm initially, which they gradually parcel out over a period of almost seven years until the supply is exhausted (Tschinkel and Porter, 1988). Male ants are thus under strong pressure in natural selection to husband their sperm carefully.

One obvious question concerning the ultimate reproductive success of males is whether it is better for a male to invest all of his sperm in a single female or else to copulate with several females. As Hölldobler and Bartz (1985) pointed out, it is important to note that in ants, unlike other nonsocial species, a male's sperm does not all go toward effective reproduction. This is because for an ant colony to begin to produce any reproductive forms, it first must produce many workers. In most advanced ant societies workers are rarely reproductive, and because workers are females derived from fertilized eggs, a substantial portion of a male's sperm is used in colony growth and maintenance rather than in direct production of new queens. The consequent trade-off for a male is obvious. If he delivers all his sperm into a single female, and if she mates with no other male, then the male is certain to father any reproductives that she eventually produces. Mortality of colony-founding females is extremely high, however. Hence a male that inseminates only a single female puts all of his sperm in one fragile basket. If he inseminates several females, on the other hand, surrendering each instant only a fraction of his sperm supply, he increases the chance that his sperm will end up in a successful foundress of a colony. In this case, however, he may decrease the chance that his sperm will be used by the queen to make alates. On the other hand, if males do inseminate several females, there may be selection favoring males whose sperm mixes with other males' sperm in the females' spermathecae. Mixing sperm increases the chance that each male will have at least some offspring among the new crop of alate queens. Allozyme variation studies in multiple-mating ant species do in fact indicate that workers in colonies are fathered by several males (Pamilo, 1982b,c; Pearson, 1983; Ward, 1983a).

In species with large mature colonies, whose females must mate with several males to acquire sufficient sperm, males seldom attempt to monopolize females (Cole, 1983b). In many other kinds of insects, and other organisms as well, sperm competition is an important selective force, and a male is often favored to ensure that no other male copulates with his mate (Parker, 1970). In multiple-mating ant species, however, a male who prevents his mate from mating again may well prevent her from acquiring enough sperm to generate a mature colony, that is, one large enough to produce reproductives. Males in these species therefore should be selected to mate with females that are already mated. The active vying for position in waiting lines behind copulating pairs in *Pogonomyrmex* species indicates at least that males do not discriminate against previously mated females, but it does not prove the optimum multiple-mating hypothesis.

To summarize, male ants are faced with two limiting resources: a restricted number of females available for mating, and a finite supply of sperm, which suffices for only one or at most several matings. An expected consequence in evolution is the fierce competition of the kind observed in the *Pogonomyrmex* leks. Davidson (1982) observed that *P. barbatus* and *P. desertorum* males indiscriminately seize

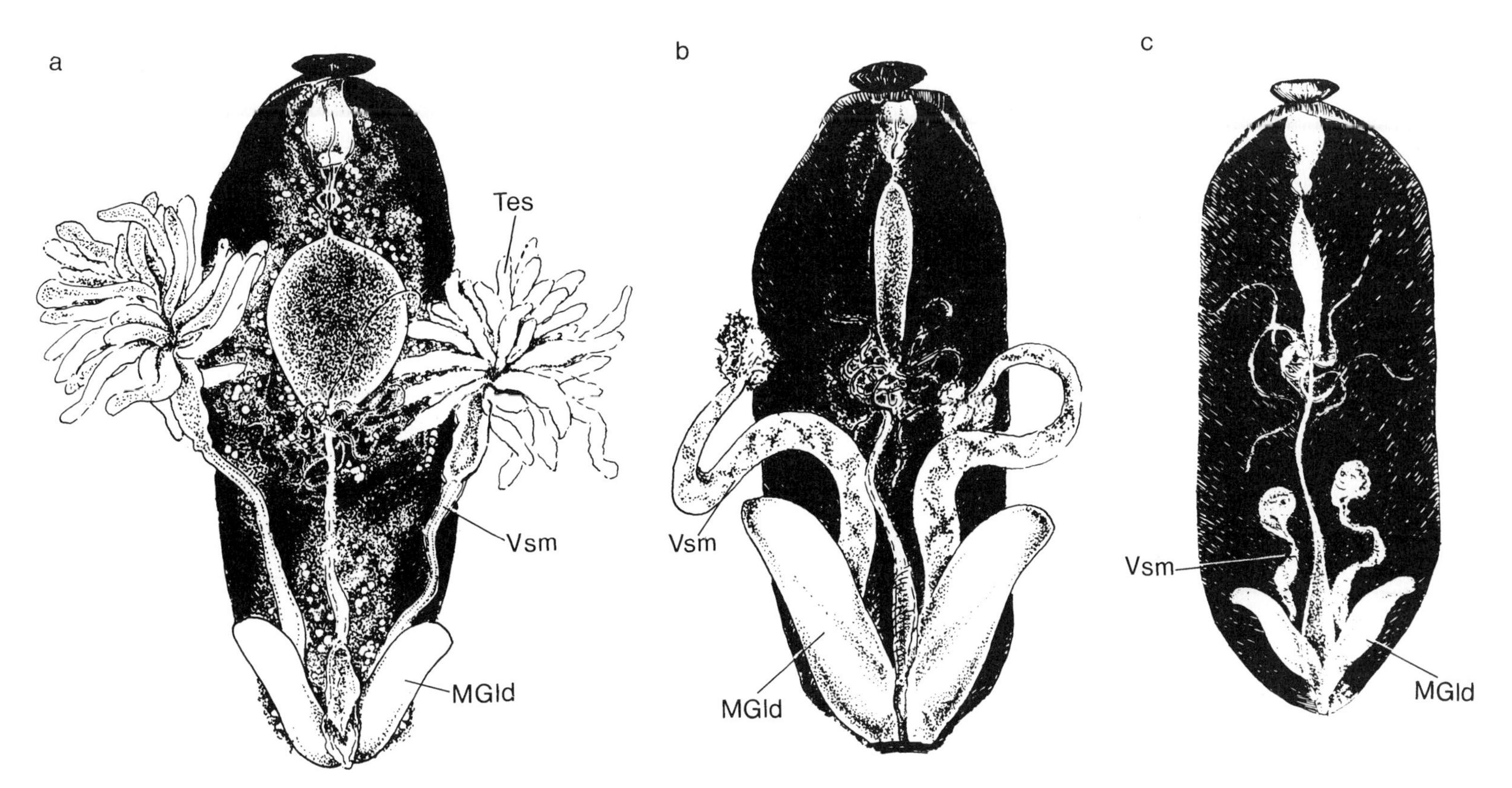


**FIGURE 3–15** The males of most ant species are physiologically programmed to mate during a single nuptial flight, and they die soon afterward. These dissections of the abdomens of *Formica polyctena* males show a typical sequence: (*a*) production of sperm in the testes; (*b*) transfer of all sperm to the vas deferens or "staging organs" for copulation; (*c*) depletion of the sperm and mucus gland supplies after copulation. *Tes,* testis; *Vsm,* seminal vesicle; *MGld,* mucus gland. (From Hölldobler and Bartz, 1985.)

females and attempt to mate, while the females actively resist copulation. As a result, large males are disproportionately successful at gaining access to mates. In addition, large females mate even more disproportionately with large males. And still further, the average size of males produced by an individual colony depends on the total number of reproductives reared in a given season, which in turn is a function of the size and vigor of the colony. In short, the bigger the colony, the more likely its individual males are to succeed in the mating arenas. Why has this selection pressure not created ever larger males in evolution? Davidson offers two reasons: larger males mean fewer males per colony, an obvious trade-off in colony fitness, and *very* large males have been observed to lose some of their advantage to those slightly smaller. The result is the existence of an optimum male size in *Pogonomyrmex*.

A confounding bit of data reported by Davidson (1982) is that not only do larger females tend to mate with larger males, but smaller females tend to mate with smaller males. If sexual selection is operating in such a way that females choose larger males, why do small females not also choose to mate with larger males? The answer to this question may be that males are selected to be choosy as well. As we have pointed out, male ants have only so much sperm at their disposal, and they cannot afford to be profligate. Selection may favor males who compete for larger females because there is a better chance that large females will survive to produce a mature colony. A result of the competition would be that the smaller, less competitive males must settle for the smaller, less desirable females.

The evolution of male biology has been subjected to few rigorous studies, and most questions concerning trends and optimality in its evolution remain unanswered. We are in a somewhat better position with reference to both data and theory on the number of female matings. As documented in Table 3–1, which includes most or all of the information available, some fraction of the queens of fully three-quarters of all species copulate with more than one male. It is also true, as revealed by allozyme marker studies (Pamilo, 1982b,c; Pearson, 1983; Ward, 1983a), that the sperm from different fathers contribute randomly to fertilization. Cole (1983b) established that multiple mating (polyandry) occurs more frequently in species with large colony size. He concluded, as West-Eberhard (1975) and a few other writers had suggested earlier on more intuitive grounds, that polyandry was therefore likely to be a response to the need on the part of queens in large colonies for more spermatozoans than one male could provide. In a study of 25 species in 5 subfamilies, Tschinkel (1987a) added stronger evidence from the number of sperm acquired by queens. In comparisons across species, the number of sperm increases very rapidly with the number of ovarioles. It ranged in Tschinkel's sample from a few tens of thousands in *Ponera* and *Hypoponera,* which form small, slow-growing colonies, to 400 million in the leafcutter *Atta texana,* which attains populations of over a million workers at a time. At another level, the number of sperm stored per ovariole (as opposed to per queen) increased from 2,000 for queens with only 6 ovarioles to about 30,000 for queens with about 200 ovarioles.

**TABLE 3–1** The number of matings by females of various ant species. Mode of study: A, allozyme markers; D, dissection; O, observation of behavior. (From Page, 1986, based on numerous sources, and Ross and Fletcher, 1985a, for *Solenopsis invicta*.)

| Species | Number of matings | Mode of study |
|---|---|---|
| **Single mating** | | |
| *Aphaenogaster rudis* | 1 | A |
| *Formica dakotensis* | 1 | O |
| *F. obscuripes* | 1 | O |
| *F. transkaucasica* | 1 | A |
| *Harpagoxenus canadensis* | 1 | O |
| *H. sublaevis* | 1 | O |
| *Iridomyrmex purpureus* | 1 | A |
| *Myrmica americana* | 1 | O |
| *Pheidole sitarches* | 1 | O |
| *Rhytidoponera chalybaea* | 1 | A |
| *R. confusa* | 1 | A |
| *Solenopsis invicta* | 1 | A |
| **Multiple mating** | | |
| *Acromyrmex landolti* | >1 | O |
| *Atta laevigata* | >3 | D |
| *A. sexdens* | 3–8 | D |
| *A. texana* | >1 | D |
| *Brachymyrmex depilis* | 2–3 | O |
| *Eciton burchelli* | 1–5 | O |
| *Formica aquilonia* | >1 | O |
| *F. bradleyi* | 2–3 | O |
| *F. exsecta* | 1–2 | A |
| *F. montana* | >1 | O |
| *F. opaciventris* | >2 | O |
| *F. pergandei* | >1 | O |
| *F. pressilabris* | >1 | A |
| *F. rufa* | >1 | O |
| *F. sanguinea* | >1 | A |
| *F. subintegra* | 2–4 | O |
| *F. yessensis* | >1 | O |
| *Lasius flavus* | >1 | O |
| *L. niger* | >1 | O |
| *Monomorium salomonis* | >1 | O |
| *Mycocepurus goeldii* | 4 | O |
| *Myrmica rubra* | 5–6 | O |
| *Pogonomyrmex badius* | 2–4 | O |
| *P. barbatus* | 4–5 | O |
| *P. californicus* | 1–6 | O |
| *P. desertorum* | 2–3 | O |
| *P. maricopa* | 2–3 | O |
| *P. occidentalis* | >1 | O |
| *P. rugosus* | 4–5 | O |
| *Polyergus lucidus* | 6 | O |
| *Prenolepis imparis* | >1 | O |
| *Solenopsis lou* | >1 | O |

Not satisfied with the intuitively simplest explanation, however, Crozier and Page (1985) went on to employ the method of multiple competing hypotheses to test the adaptiveness of polyandry. They constructed no fewer than eight such explanations (some admittedly very improbable) to account for the trend documented by Cole. The explanation of limited male contribution favored by Cole and earlier authors was downgraded, because "males of species with big females are generally larger than those with small females, so there is no *absolute* bar to male size (within reason!)." This does not seem to be a very strong counterargument. Aerial swarmers can potentially benefit from smaller size, which confers greater agility during the approach to incoming queens. Also, as we have noted with reference to mating in *Pogonomyrmex*, there is a trade-off between male size and male numbers, still poorly analyzed, that might contribute to the preferred production by colonies of smaller males. Hence it is prudent to keep alive the limited-sperm hypothesis of polyandry.

Crozier and Page, after discarding most of the other competing explanations, hold on to three (not counting the limited-sperm hypothesis, which we favor) as both inherently plausible and compatible with the correlation between colony size and polyandry. The first is that caste determination may be genetic, and if so polyandry would allow fuller expression of the caste system in each colony. It follows that species with more complex caste differentiation (a trait associated with large colonies) should be more polyandrous than species with simpler caste systems. As Crozier and Page note, there is no evidence for genetic caste determination in ants to the present time, although recent evidence suggests some kind of genetic predisposition toward various forms of labor specialization in honey bees (Calderone and Page, 1988; Frumhoff and Baker, 1988; Robinson and Page, 1988). All of the many substantial studies to date have implicated a single genotype with multiple developmental pathways controlled by nutritive and other environmental factors (see Chapter 8).

The second surviving explanation in the Crozier-Page analysis is that polyandry maximizes the production of divergent worker genotypes, quite apart from caste phenotypes, and hence the range of environmental conditions that a colony can tolerate. Broad-niche species, most often those possessing large colonies, should be more polyandrous than species with narrow niches. This relationship has not yet been tested empirically.

The third favored hypothesis is that multiple matings reduce the chances of disaster due to the production of diploid males. Males of Hymenoptera, it will be recalled, ordinarily come from unfertilized eggs and are determined as males simply by being haploid, that is, having only one set of sex-determining genes. When one locus or a very few loci are involved in the process, and recessive male-determining alleles exist, it is also possible to get males from fertilized eggs, the so-called diploid males. The queen of an ant colony can ordinarily control male egg production precisely by opening or closing her spermathecal valve "at will," thus determining whether an egg in the vaginal passage is fertilized. But she has no control whatever over the production of diploid male eggs, because the effort to produce females will still result in a fixed percentage of males by Mendelian chance alone. This circumstance does not matter much if the strategy of the colony is to produce males during early stages of colony growth (beyond the very earliest, fragile stage of colony founding), a not uncommon event in species with a small mature colony size. But it can lay a substantial energetic burden on species whose strategy is to delay production of drones until the colony is large. By mixing sperm from multiple males, the variance of such a load is reduced. In other words, more colonies are likely to have *some* diploid males, but on the average they are less likely to produce large numbers of diploid males.

Reasoning in another mode, Woyciechowski and Łomnicki (1987) proposed that multiple matings prevent workers from producing male offspring. According to their model of kin selection, workers are at an advantage if they produce sons and care for nephews in the presence of a mother queen who has mated only once, but they should avoid personal reproduction and care for brothers in the presence of a mother queen who has mated several times. The existing data on queen mating patterns and worker reproduction are not adequate to test the hypothesis.

More recently, Sherman et al. (1988) have argued that the role of polyandry is to increase genetic variation within colonies, thereby reducing the likelihood that parasites or pathogens can decimate the worker force by overcoming all of its physiological and behavioral defenses at once. A balanced portfolio of investments in genetic variation, in other words, is more likely to produce the highest long-term probability of survival and successful growth. This argument is logical, but in our opinion does not accord with the remarkable correlations that exist between polyandry, sperm count, and colony size, which favor the limited-sperm hypothesis but no other.

In any case, the reproductive behavior of ants is still a poorly explored domain with rich possibilities for general evolutionary biology. More studies are needed on all fronts, including the comparative natural history of nuptial flights, the detailed analysis of individual males and females during mating, genetic studies of sex determination, and more sophisticated models of reproductive competition at individual and colony levels.

## COLONY FOUNDING AND GROWTH

### The Founding Stage

As soon as the queens are inseminated, they shed their membranous wings by raking their middle and hind legs forward and snapping the wings free at the basal dehiscent sutures. Over the coming weeks the alary muscles and fat bodies are metabolized and converted into eggs, as well as food to rear the first batch of larvae. The nutrient materials are packaged as either trophic eggs (eggs that cannot develop but are used exclusively as food), specialized salivary secretions, or both. The basic process was first described in the classic study of the formicine ant *Lasius niger* by Janet (1907). More recently, it has been found that the esophagus of the queen expands into a "thoracic crop" in which the converted tissues are temporarily held in liquid form. In Pharaoh's ant (*Monomorium pharaonis*), the esophagus diameter widens from 7–10 micrometers to 265 micrometers. The thoracic crop has been demonstrated in five genera of Myrmicinae and Formicinae so far (Petersen-Braun and Buschinger, 1975).

In the case of *Solenopsis invicta* at least, the conversion process is mediated by the corpora allata. Allatectomized queens fail to cast their wings or undergo wing muscle histolysis, whereas treatment of these operated individuals with juvenile hormone causes both processes to proceed (Barker, 1979).

The external trigger for wing shedding and histolysis is not exclusively insemination, as one might guess. When virgin queens of some species are taken from the presence of other queens, they drop their wings after about a day and begin to behave more in the manner of inseminated nest queens. In the case of *Solenopsis invicta* in particular, the crucial signal is a relatively nonvolatile pheromone produced by queens and conveyed to the virgin alates (Fletcher and Blum, 1981; Fletcher et al., 1983).

The conversion of body tissues into food for the larvae was a vital evolutionary advance in the Formicidae. Wheeler (1933b) suggested that partially claustral colony founding (where the queen still leaves the nest to obtain some of the food) is the primitive state and fully claustral colony founding was derived from it. This inference is based on sound logic: if the ants did originate from predatory vespoid wasps, as the anatomical evidence suggests, they or their immediate ancestors were likely to have passed through a stage in which foundresses still captured insect prey and transported them to preexisting nests. In other words, the earliest ants are likely to have been partially claustral.

It is further true that species of the relatively primitive subfamily Ponerinae display finely graded steps leading from the partial to the fully claustral mode, from dependence on outside foraging to more or less complete freedom from it. The queens of *Pachycondyla* (= *Bothroponera*) *soror*, for example, are in an exactly intermediate stage. They forage outside the nest, but their wing muscles are still reduced and metabolized in the manner of the higher ants (Haskins, 1941). The queens of *Odontomachus haematodus* are capable of rearing their first brood at least partially with their own oral secretions, but in one experiment performed by Haskins and Haskins (1950b), the larvae still failed to reach maturity. Finally, these authors found that the unusually bulky queens of *Pachycondyla* (= *Brachyponera*) *lutea* are able to rear the first brood all the way through solely with their own secretions, even though they continue to forage outside the nest when given the opportunity. From *P. lutea* it is but a short step to the condition typifying most myrmicines, formicines, and other "higher" ants, in which complete claustral colony founding is the mode.

At least one substantial advantage of claustral colony founding is obvious and may in fact have played a role in its general adoption among the phylogenetically more advanced ants. Social insect workers suffer their highest mortality during foraging trips (Porter and Jorgensen, 1981; Schmid-Hempel, 1984), and it is probable that the same is true of founding queens forced to leave their nests to search for food.

### The Ergonomic (Exponential) Stage

The first workers produced by the queen are typically "nanitics" or "minims," that is, miniature forms somewhat smaller than the smallest workers encountered in older colonies of the same species (Figure 3–16). They are characteristically timid in behavior but otherwise perform the same repertory of tasks as do workers in older colonies. In the fire ant *Solenopsis invicta* at least, they differ from other worker castes in venom composition, specifically in the relative proportions of piperidine alkaloids (Vander Meer, 1986b). In the case of dimorphic species, the first-brood nanitics possess the basic anatomical structure of the minor caste. Major workers usually do not appear until later, and even then are initially smaller in average size. Minims are a general, perhaps a universal phenomenon in

**FIGURE 3–16** A physogastric queen of *Myrmecocystus mexicanus* with her first worker offspring.

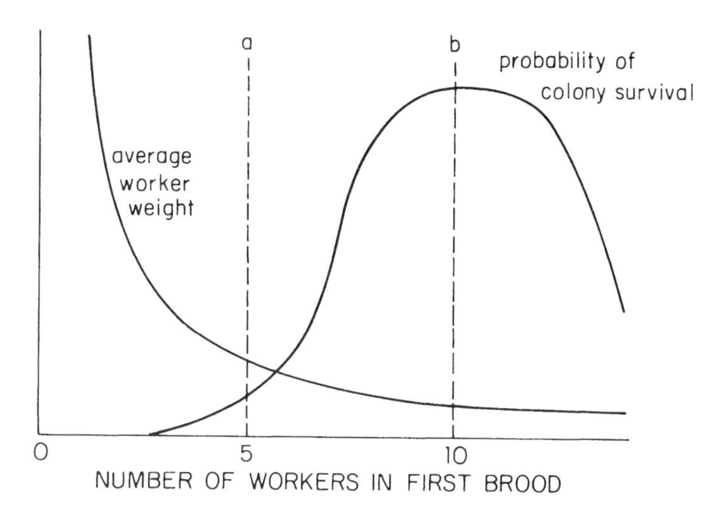


**FIGURE 3–17** The inferred evolutionary determinants of size and number of workers in the first brood of an ant colony. The first workers are characteristically very small (nanitics). The hypothesis represented in the diagram is that there exists a threshold number required to perform all of the vital tasks, and in addition a higher optimum number that compensates for average adult mortality prior to the maturing of the second brood. By producing very small workers at first, the incipient colony divides the available biomass of the first adult brood into the optimum number of nanitic workers. (*a*) The number of workers in the first brood if the workers are as large as those in later broods. In this case, the number would bring fewer individuals in the second brood to maturity. (*b*) The optimum number. (Modified from Oster and Wilson, 1978.)

ants, occurring not only in the "higher" subfamilies and genera, but also in the primitive Australian genus *Myrmecia* (see review in Wood and Tschinkel, 1981).

Ergonomic models designed to calculate the net energetic yield and hence growth rates of colonies support the intuition that the small size and timidity of the first workers represent prudent features built into the investment strategy of colonies as a whole (Oster and Wilson, 1978). A newly founded colony should strive to maximize the number of workers and their initial survival rate at the expense of everything else. The intuited reasons are as follows. With the queen's internal resources exhausted, a *minimum number* of workers is needed to accomplish an adequate performance in each of the vital tasks—a certain number to enlarge the nest, a certain number to nurse the second brood, a certain number to forage, and so forth. There should also exist an *optimum number,* above this minimum, since adult mortality is probable before the second brood reaches adulthood. The optimum number of nanitics can be defined as that above which the survival probability of the queen can no longer be significantly increased and in fact is likely to be decreased. Because the biomass of adult workers that the founding queen can produce is very limited, it is efficient for the queen to divide it into many small workers, as documented by Porter and Tschinkel (1986) in the fire ant *Solenopsis invicta*. But this advantage is easily reversed, because to raise a great many such individuals would necessitate the production of excessively small nanitics unable to exploit the food items and nest sites for which the species is anatomically and behaviorally adapted. As a result, there should be an optimum number of nanitic workers, determined by the balance between the advantages of a larger initial worker force and the disadvantages of a smaller body size (see Figure 3–17). This result has been confirmed experimentally in *Solenopsis invicta* by Porter and Tschinkel (1986). They found that nanitics are less efficient on an individual basis than ordinary minor workers in rearing brood, but more efficient as a group than a group of minor workers of equal combined weight—because nanitics are more numerous. On the other hand, they are energetically more expensive to maintain and thus are superior only for the brief period of colony founding.

Moreover, the small size of the incipient colony seems to dictate that its members be relatively timid in behavior. Suppose that an encounter with a single enemy such as a group of foragers from an alien colony results in the loss of five workers. For a mature colony containing thousands of members, this sacrifice is not only tolerable but desirable, if it clears enemy scouts from the territory on which the population depends for food. But for an incipient colony of only ten workers, the loss could be fatal. Furthermore, the potential gain from expelling territorial intruders is expected to be less, because the incipient colony is still living on a fraction of the available food supply yielded by the surrounding terrain.

The nanitics appear to owe their miniature size at least partially to the meager nutrients supplied them by the founding queen in their larval stage. Pheromonal or other programmed stimuli from the queen may also be important. When Wood and Tschinkel (1981) introduced newly inseminated *Solenopsis invicta* queens into groups of workers with differing numbers, the workers in the first brood increased in average size according to the number of attending workers in the adoptive group. Yet none of these sets of offspring were as small in average size as the nanitics of normal incipient colonies.

When the first brood of workers reaches the adult stage, the new colony undergoes a radical transformation. If the queen has been performing the ordinary chores of the colony, she now stops in order to devote herself exclusively to egg-laying. The workers take over all of the remaining tasks, including the feeding of the queen herself. For a few worker generations, the average number of which varies among species, no new reproductive forms are reared. Also, with the exception of raiding species such as *Myrmecocystus mimicus* and *Solenopsis invicta,* few if any reproductive individuals or alien workers are adopted from the outside. Thus the colony is a semiclosed system devoted to its own exponential growth (Brian, 1957b, 1965b, 1983; Wilson, 1971; Oster and Wilson, 1978). The colony can in fact be viewed in this middle, ergonomic stage as a growth machine: its hypothesized "purpose" is to proliferate workers as quickly and safely as possible. The growth function is implemented chiefly by division of labor—the right number of foragers to harvest the surrounding terrain, the right number of nurses to stoke larval growth, and a sufficient but not excessive number of defenders and auxiliaries standing by for emergencies. The focus of the colony is not yet reproduction or dispersal. New nest sites are sought only when the old ones become environmentally untenable or too small to hold the expanding colony.

During the ergonomic stage competition within the colony is at a minimum (see our analysis of competition in Chapter 6). The beginning of the stage, however, or more precisely the transition to this stage from the preceding, founding stage, is sometimes accompanied by hostile interactions. In *Lasius flavus, Messor pergandei, Myrmecocystus mimicus,* and *Solenopsis invicta,* pleometrotic laboratory groups revert to monogyny when the first brood or mature workers appear. The *Lasius* queens fight one another and then break apart into single-queen units (Waloff, 1957). Those of *Myrmecocystus* form dominance hierarchies in which the supernumerary individuals are eventually driven out by the workers (Bartz and Hölldobler, 1982). When multiple *Solenopsis invicta* queens are introduced to queenless workers, the latter usually execute all but one, both in the laboratory and under natural conditions (Wilson, 1966; Fletcher and Blum, 1983; Tschinkel and Howard, 1983). In the carpenter ants *Camponotus herculeanus* and *C. ligniperda,* large colonies often contain several queens, but these individuals are intolerant of one another and maintain territories within the diffuse nests, a condition referred to by Hölldobler (1962) as oligogyny. The same phenomenon occurs in the Australian meat ant *Iridomyrmex purpureus:* queens that cooperated amicably during nest founding become antagonistic after the first workers appear, and in the end permanently separate within the nest (Hölldobler and Carlin, 1985).

If the colony survives the precarious period during which the first and second worker broods are being reared, it is likely to enjoy an interval of sustained exponential growth. This growth, however, like that in all populations, can be expected to slow with time and eventually to come to a halt. Most data on the course of colony growth in social insects generally suggest curves that are roughly sigmoidal (hence "logistic") in form (Brian, 1965b, 1983). In a recent, thorough study of the fire ant *Solenopsis invicta,* Tschinkel (1988a) was able to show that colonies under natural conditions grow logistically, attaining the maximum worker population of about 220,000 in four to six years. This is the expected result, but the underlying density-dependent controls are more complex than those determining typical logistic growth in nonsocial insects. The theory of colony growth, based on the concepts of economies of scale and evolutionary optimization, has been developed in some detail by Oster and Wilson (1978). We will return to the subject repeatedly in later chapters as part of our analysis of caste, division of labor, foraging strategies, and defense.

## The Reproductive Stage

If a monogynous colony were to maintain its worker population at zero population growth, it would be unable to reproduce, since total investment means by definition that no production of virgin queens and males is possible. Consequently, at some point short of its maximum possible size, the colony should devote part of its production to the creation of virgin queens and males. The timing of the conversion varies among species according to the special adaptations the species have otherwise made to their environments (Wilson, 1971; Oster and Wilson, 1978). Within species, the production of reproductives increases as a function of colony size. Among species of the genus *Myrmica* at least, the average worker stature (related to colony size) also has a particularly strong influence on queen production but not male production (Elmes and Wardlaw, 1982). It is a common event, as documented in *Myrmica rubra* by Brian (1957a,b), for males to appear in the nest before females—and sometimes at erratic intervals, possibly as a consequence of workers laying eggs in competition with the queen. Brian has termed the interval of early male production the "adolescent" period of colony growth, coming between the "juvenile" (= ergonomic) period during which the worker population expands and the "mature" period during which new, virgin queens are produced. In truth very little information has been published on the timing of these key events in the colony life cycle, so it is impossible to generalize about the sequence in which males and queens appear. In many, perhaps even most, ant species,

**TABLE 3–2** Representative numbers of adults in ant colonies. *a* = average number; other data pertain to individual colonies or range of multiple colonies.

| Species | Locality | Number of adults | Number of colonies censused | Authority |
|---|---|---|---|---|
| **PONERINAE** | | | | |
| *Amblyopone pallipes* | Massachusetts, USA; Quebec, Canada | 12[a] (1–35) | 37 | Francoeur (1965), Traniello (1982) |
| *A. pluto* | Ivory Coast | 33 (30–40) | 3 | Gotwald and Lévieux (1972) |
| *Apomyrma stygia* | West Africa | 13–92 | — | Lévieux (1976b) |
| *Centromyrmex sellaris* | Ivory Coast | 408 | 1 | Lévieux (1976b) |
| *Cerapachys opaca* | New Guinea | 100 | 1 | Wilson (1959c) |
| *C. polynikes* | New Guinea | 20 | 1 | Wilson (1959c) |
| *Cryptopone motschulskyi* | New Guinea | 20 | 1 | Wilson (1959c) |
| *Diacamma rugosum* | New Guinea | 30; 50 | 2 | Wilson (1959c) |
| *Ectomomyrmex striatulus* | New Guinea | 10; 20 | 2 | Wilson (1959c) |
| *Gnamptogenys macretes* | New Guinea | 40 | 1 | Wilson (1959c) |
| *Leptogenys bituberculata* | New Guinea | 300 | 1 | Wilson (1959c) |
| *L. diminuta* | New Guinea | 90–400 | 4 | Wilson (1959c) |
| *L. ocellifera* | Sri Lanka | >16,000 | 1 | Maschwitz and Mühlenberg (1975) |
| *L. purpurea* | New Guinea | 500; 2,000 | 2 | Wilson (1959c) |
| *Megaponera foetens* | Ivory Coast | 442; 593 | 2 | Lévieux (1976b) |
| *Myopias* sp. 1 | New Guinea | 60 | 1 | Wilson (1959c) |
| *M.* sp. 2 | New Guinea | 40; 70 | 2 | Wilson (1959c) |
| *M.* sp. 3 | New Guinea | 30 | 1 | Wilson (1959c) |
| *Mystrium* sp. | Sarawak | 29 | 1 | Moffett (1986a) |

*continued*

some colonies rear only males in a given season, others rear only queens, while still others rear a mixture (Nonacs, 1986a,b). Whether individual colonies change their strategy from one year to the next is not known.

Table 3–2 gives most of the available data on colony size in various species of ants. A few additional data have been compiled by Baroni Urbani (1978a). These raw numbers tell us nothing directly about the growth rates or factors limiting mature colony size. Also, the numbers often represent underestimates of the typical mature colony size, because colonies of all ages were censused and colonies in many wild populations tend to be young. Nevertheless, the data do permit some inferences when comparisons are made between major groups:

1. No apparent correlation is yet evident between the size of the mature colony and the longevity of the colony, at least as measured in monogynous species by the life span of the queen (see also Table 3–3). However, the data are so few, especially of colony life spans, that it would be premature to draw any firm conclusion. There is a need for more longevity studies of ants of all castes because of the relevance of such data to population dynamics.

2. There is no clear relation between climate and colony size. If anything, temperate species tend to have somewhat larger colonies on the average. This is because of a special ecological effect connected with constraints in the nest site of many tropical species, as follows.

3. There is a strong relation between preferred nest site and mature colony size. Among the ants of New Guinea rain forests, for example, species that nest in rotting logs and other pieces of decaying wood on the ground (almost all Ponerinae and the majority of Myrmicinae) form smaller colonies than those living in less restricted nest sites, such as the open soil of the forest floor (*Acidomyrmex, Pheidologeton, Leptomyrmex, Pseudolasius, Acropyga,* most *Paratrechina*), open air at the ground surface (*Aenictus*), and various parts of the tree canopy (most *Crematogaster, Iridomyrmex, Camponotus, Polyrhachis,* and *Oecophylla*).

4. The most elaborate caste and communication systems occur in species with large, perennial colonies, for example the legionary ants (*Dorylus, Eciton*), the leafcutter ants (*Acromyrmex, Atta*), and the marauding ants of the genus *Pheidologeton.*

5. The great variation in colony size among species belonging to the same taxonomic group (for example, the Myrmicinae) attests to the capacity of this population trait to evolve with relative speed. Small alterations in the physiological parameters of individual ants such as mean worker life span and thresholds in queen determination of larvae can bring about major differences in mature colony size. Given this ability to adapt colony size to local environmental conditions at the species level, we should feel encouraged about the possibility of inferring which conditions have been critical in evolution.

**TABLE 3–2** *(continued)*

| Species | Locality | Number of adults | Number of colonies censused | Authority |
|---|---|---|---|---|
| *Paltothyreus tarsatus* | Ivory Coast | 200–1,000 | ? | Lévieux (1976b) |
| *Paraponera clavata* | Costa Rica | 708; 1,355 | 2 | Janzen and Carroll (1983) |
| *Platythyrea conradti* | Ivory Coast | 300–500 | ? | Lévieux (1976b) |
| *P. parallela* | New Guinea | 50 | 1 | Wilson (1959c) |
| *Prionopelta amabilis* | Costa Rica | 282; 709 | 2 | Hölldobler and Wilson (1986a) |
| *P. opaca* | New Guinea | 20 | 1 | Wilson (1959c) |
| *Proceratium silaceum* | United States | 28[a] (9–60) | — | Kennedy and Talbot (1939) |
| *Rhytidoponera araneoides* | New Guinea | ~50[a] | 3 | Wilson (1959c) |
| *R. chalybaea* | Australia | 271[a] | 68 | Ward (1981b) |
| *R. confusa* | Australia | 203[a] | 132 | Ward (1981b) |
| *R. laciniosa* | New Guinea | 100; 150 | 2 | Wilson (1959c) |
| **NOTHOMYRMECIINAE** | | | | |
| *Nothomyrmecia macrops* | Australia | 50–70 | 5 | Taylor (1978c) |
| **MYRMECIINAE** | | | | |
| *Myrmecia dispar* | Australia | 125 (15–329) | 20 | Gray (1971a) |
| *M. gulosa* | Australia | 188; 1,586 | 2 | Haskins and Haskins (1950a) |
| *M. nigrocincta* | Australia | 455; 1,187 | 2 | Gray (1971a) |
| *M. pilosula* | Australia | 553; 862 | 2 | Haskins and Haskins (1950a) |
| *M. vindex* | Australia | 109; 272 | 2 | Haskins and Haskins (1950a) |
| **DORYLINAE** | | | | |
| *Aenictus currax* | New Guinea | >100,000 | 1 | Wilson (1959c) |
| *A. laeviceps* | Philippine Islands | 60,000–110,000 | — | Schneirla (1971) |
| *Dorylus wilverthi* | Africa | $2 \times 10^6$–$22 \times 10^6$ | — | Raignier and van Boven (1955) |
| **ECITONINAE** | | | | |
| *Eciton burchelli* | Central America | 150,000–700,000 | — | Rettenmeyer (1963a), Schneirla (1971) |
| *E. hamatum* | Central America | 100,000–500,000 | — | Rettenmeyer (1963a), Schneirla (1971) |
| *Neivamyrmex nigrescens* | United States | 80,000–140,000 | — | Rettenmeyer (1963a), Schneirla (1971) |
| **LEPTANILLINAE** | | | | |
| *Leptanilla japonicus* | Japan | 100–200 | 11 | Masuko (1987) |
| **MYRMICINAE** | | | | |
| *Acanthomyrmex ferox* | Kalimantan | 57 | 1 | Moffett (1985b) |
| *A. notabilis* | Sulawesi | 40 | 1 | Moffett (1985b) |
| *Adelomyrmex biroi* | New Guinea | 10 | 1 | Wilson (1959c) |
| *Aphaenogaster dromedarius* | New Guinea | 100 | 1 | Wilson (1959c) |
| *A. rudis* | Michigan, USA | 326[a] (26–2,079) | 72 | Talbot (1951) |
| *A. rudis* | Ohio, USA | 280[a] (11–950) | 46 | Headley (1949) |

*continued*

**TABLE 3–2** *(continued)*

| Species | Locality | Number of adults | Number of colonies censused | Authority |
|---|---|---|---|---|
| *A. treatae* | United States | 682[a] (65–1,662) | 30 | Talbot (1954) |
| *Apterostigma angulatum* | Panama | 155 | 1 | Weber (1941b) |
| *Atopomyrmex mocquerysi* | Ivory Coast | 65,000 | ? | Lévieux (1976b) |
| *Atta colombica* | Central America | Between $10^6$ and $2.5 \times 10^6$ | 1 | M. Martin (personal communication) |
| *Basiceros manni* | Costa Rica | 24[a] (8–49) | 3 | Wilson and Hölldobler (1986) |
| *Cardiocondyla paradoxa* | New Guinea | 50 | 1 | Wilson (1959c) |
| *C. thoracica* | New Guinea | 70 | 1 | Wilson (1959c) |
| *Crematogaster dohrni artifex* | India | 56,947 | 1 | Ayyar (1937) |
| *C. dohrni rogenhoferi* | India | 5,690 | 1 | Roonwal (1954) |
| *C. elegans* | New Guinea | 300 | 1 | Wilson (1959c) |
| *C. larreae* | Mexico | 789[a] (163–1,727) | 13 | MacKay et al. (1984) |
| *C. subtilis* | New Guinea | >5,000 | 1 | Wilson (1959c) |
| *Cyphomyrmex rimosus* | Venezuela | 300 | 1 | Weber (1947b) |
| *Dacetinops cibdela* | New Guinea | 10 | 1 | Wilson (1959c) |
| *Daceton armigerum* | Suriname; Trinidad | 300; 5,000 | 2 | Wilson (1962a) |
| *Erebomyrma nevermanni* | Costa Rica | 180 | 1 | Wilson (1986a) |
| *E. urichi* | Suriname; Trinidad | 568 (500≥1000) | 2 | Wilson (1962d) |
| *Eurhopalothrix biroi* | New Guinea | 50 | — | Wilson (1959c) |
| *Leptothorax curvispinosus* | United States | 84[a] (8–368) | 38 | Headley (1943) |
| *L. longispinosus* | United States | 47[a] (2–142) | 97 | Headley (1943) |
| *Lordomyrma* sp. | New Guinea | 10; 15 | — | Wilson (1959c) |
| *Meranoplus spinosus* | New Guinea | 150 | 1 | Wilson (1959c) |
| *Monomorium rothsteini* | Australia | 15,000–58,000 | — | Davison (1982) |
| *M. whitei* | Australia | 450–40,000 | — | Davison (1982) |
| *Myrmecina transversa* | New Guinea | 100 | 1 | Wilson (1959c) |
| *Myrmica ruginodis* | Scotland | 1,216[a] (305–2,855) | 12 | Brian (1950) |
| *M. schencki emeryana* | United States | 255[a] (35–561) | 36 | Talbot (1945) |
| *M. sulcinodis* | England | 120[a] | Many | Elmes (1987a) |
| *Myrmicaria eumenoides* | Ivory Coast | 18,320; 21,477 | 2 | Lévieux (1983a) |
| *Oligomyrmex pygmaeus* | Tropical Asia | 2,000–20,000 | 6 | Moffett (1986g) |
| *Pheidole* sp. | New Guinea | 150 | 1 | Wilson (1959c) |
| *Pheidologeton* sp. | New Guinea | >3,000 | 1 | Wilson (1959c) |
| *Pogonomyrmex badius* | United States | 4,000–6,000 | 6 | Golley and Gentry (1964) |
| *P. badius* | United States | 4,736[a] | 25 | Gentry and Stiritz (1972) |
| *P. barbatus* | United States | 12,358 | 1 | Wildermuth and Davis (1931) |
| *P. californicus* | United States | 4,536[a] | 11 | Erickson (1972) |
| *P. carbonarius* | Argentina | 400–500 | ? | Kusnezov (1951) |
| *P. desertorum* | United States | 400–600 | ? | Whitford and Bryant (1979) |
| *P. laticeps* | Argentina | 40–50 | ? | Kusnezov (1951b) |
| *P. longibarbis* | Argentina | 200–300 | ? | Kusnezov (1951b) |
| *P. magnacanthus* | United States | 100–225 | ? | Cole (1968) |
| *P. mayri* | Colombia | 603[a] | 8 | C. Kugler (in MacKay, 1981) |
| *P. montanus* | United States | 1,665[a] | 70 | MacKay (1981) |
| *P. occidentalis* | United States | 3,024[a] | 33 | Lavigne (1969) |
| *P. occidentalis* | United States | 2,676[a] | 11 | Rogers et al. (1972) |
| *P. rugosus* | United States | 7,740[a] (2,586–14,742) | 20 | MacKay (1981) |

*continued*

**TABLE 3–2** *(continued)*

| Species | Locality | Number of adults | Number of colonies censused | Authority |
|---|---|---|---|---|
| *P. subnitidus* | United States | 5,934[a] (1,850–13,056) | 26 | MacKay (1981) |
| *Pristomyrmex* sp. | New Guinea | 100 | 1 | Wilson (1959c) |
| *Pristomyrmex pungens* | Japan | 22,000[a] (maximum > 300,000) | Many | A. Mizutani (personal communication) |
| *Proatta butteli* | Singapore | 1,000–10,000 | Several | Moffett (1986d) |
| *Rhoptromyrmex melleus* | New Guinea | >5,000 | 2 | Wilson (1959c) |
| *Sericomyrmex amabilis* | Central America | to 300 | Many | Wheeler (1925) |
| *S. urichi* | Central and South America | 200–1,691 | Many | Weber (1967) |
| *Solenopsis invicta* | United States | 220,000[a] (100,000–250,000, varying seasonally) | Many | Wilson and Eads (1949), Morrill (1974b), Tschinkel (1988a) |
| *Stenamma diecki* | Quebec, Canada | 41[a] (5–103) | 7 | Francoeur (1965) |
| *Strumigenys bajarii* | New Guinea | 400 | 1 | Wilson (1959c) |
| *S. frivaldszkyi* | New Guinea | 15 | 1 | Wilson (1959c) |
| *S. loriai* | New Guinea | 300; 500 | 2 | Wilson (1959c) |
| *S. mayri* | New Guinea | 100 | 1 | Wilson (1959c) |
| *Tetramorium caespitum* | England (1963) | 14,068[a] (2,603–29,571) | 23 | Brian et al. (1967) |
| *T. caespitum* | England (1964) | 7,881[a] (1,395–30,943) | 26 | Brian et al. (1967) |
| *Trachymyrmex ruthae* | Panama | 331 | 1 | Weber (1941b) |
| *T. septentrionalis* | United States | 200–1,400 | 99 | Lenczewski (1985) |
| *Vollenhovia brachycera* | New Guinea | 150 | 1 | Wilson (1959c) |
| **DOLICHODERINAE** | | | | |
| *Iridomyrmex purpureus* | Australia | 11,000–64,000 | Many | Greaves and Hughes (1974) |
| *I. purpureus* | Australia | 300,000 | 1 | Ettershank (1971) |
| *I. scrutator* | New Guinea | 500–>3,000 | 1 | Wilson (1959c) |
| *Leptomyrmex fragilis* | New Guinea | 350 | 1 | Wilson (1959c) |
| **FORMICINAE** | | | | |
| *Acropyga* sp. | New Guinea | >1,000 | 1 | Wilson (1959c) |
| *Calomyrmex laevissimus* | New Guinea | 250 | 1 | Wilson (1959c) |
| *Camponotus confusus* | New Guinea | 200 | 1 | Wilson (1959c) |
| *C. herculeanus* | Ontario, Canada | 13,376 (max) | — | Sanders (1970) |
| *C. noveboracensis* | Michigan, USA | 10,800 (max) | — | Sanders (1970) |
| *C. papua* | New Guinea | 300 | 1 | Wilson (1959c) |
| *C. pennsylvanicus* | United States | 1,943–2,500 | Many | Pricer (1908) |
| *C. vitreus* | New Guinea | >4,000 | 1 | Wilson (1959c) |
| *Formica exsectoides* | United States | 41,366; 238,510 | 2 | Cory and Haviland (1938) |
| *F. "incerta"* | United States | 714[a] (107–1,668) | 24 | Talbot (1948) |
| *F. japonica* | Japan | 2,495[a] (657–4,639) | 4 | Kondoh (1968) |
| *F. pallidefulva nitidiventris* | United States | 2,352[a] (541–7,050) | 24 | Talbot (1948) |
| *F. rufa* | Europe | to >100,000 | Many | Brian (1965b) |
| *F. yessensis* | Japan | 307 million | 1 | Higashi and Yamauchi (1979) |

*continued*

**TABLE 3–2** *(continued)*

| Species | Locality | Number of adults | Number of colonies censused | Authority |
|---|---|---|---|---|
| *Myrmoteras barbouri* | Singapore | 11 | 1 | Moffett (1986c) |
| *M. toro* | Sulawesi, Indonesia | 23 | 1 | Moffett (1986c) |
| *Oecophylla longinoda* | Africa | ca. 500,000 | Many | Way (1954a) |
| *Paratrechina pallida* | New Guinea | 500 | 1 | Wilson (1959c) |
| *P.* sp. 1 | New Guinea | 200 | 1 | Wilson (1959c) |
| *P.* sp. 2 | New Guinea | 150 | 1 | Wilson (1959c) |
| *Polyrhachis debilis* | New Guinea | ~325[a] (300–350) | 3 | Wilson (1959c) |
| *P. dives* | Japan | ca. 1 million | Many | Yamauchi et al. (1987) |
| *P. hirsutula* | New Guinea | 150 | 1 | Wilson (1959c) |
| *P. limbata* | New Guinea | 100 | 1 | Wilson (1959c) |
| *P. omymyrmex* | New Guinea | 60 | 1 | Wilson (1959c) |
| *P. rufiventris* | New Guinea | 200 | 1 | Wilson (1959c) |
| *Prenolepis imparis* | United States | 1,582[a] (48–2,208) | 11 | Talbot (1943a) |
| *P. imparis* | United States | 14–10,300 | 10 | Tschinkel (1987c) |
| *Pseudolasius breviceps* | New Guinea | 200; 500 | 2 | Wilson (1959c) |

## BROOD CARE AND LARVAL RECIPROCATION

The workers of all ant species thus far investigated lavish care on all of the immature stages, from egg to larva to pupa. This is true even of *Nothomyrmecia* and *Amblyopone*, both of which are very primitive anatomically and have the simplest social organizations known in the ants (Figure 3–18). In the case of the North American species *A. pallipes*, Traniello (1982) recorded the following behaviors by the workers on behalf of the brood: lick and carry all stages, place larva on prey, assist larval molt by licking ecdysial skin free, bank mature larva with soil to facilitate cocoon spinning, assist the removal of the meconium at the commencement of pupation, and remove empty cocoon after eclosion. All of these behaviors have also been observed in *Nothomyrmecia* (B. Hölldobler and R. W. Taylor, unpublished; Figure 3–19).

The workers of most species belonging to the phylogenetically more advanced subfamilies Myrmicinae, Aneuretinae, Dolichoderinae, and Formicinae engage in "trophallaxis" (Wheeler, 1918a). That is, they regurgitate liquid food to other members of the colony, including the larvae. In some species, such as *Formica sanguinea* or *Solenopsis invicta*, the larvae elicit the response by rocking their heads back and forth, flexing their mandibles, and "swallowing" rapidly (Hölldobler, 1968b; O'Neal and Markin, 1973). In the phylogenetically more primitive subfamilies Ponerinae, Nothomyrmeciinae, and Myrmeciinae, the existence of oral trophallaxis is less certain. Reports have been published of occasional trophallaxis among workers of *Nothomyrmecia macrops* (Taylor, 1978c) and those of some species of the Ponerinae and Myrmeciinae (Le Masne, 1953; Haskins and Whelden, 1954; Freeland, 1958). However, these observations need confirmation by experimental evidence (Hölldobler, 1985).

Myrmecologists have long believed that special nutrients needed for the rearing of reproductive females are produced in one or the other of the exocrine glands in the head or thorax of the worker. The two most frequently mentioned by past authors have been the labial and postpharyngeal glands (see Figure 7–2). Early experimental investigations of the two glands in *Formica polyctena* were contradictory. Gösswald and Kloft (1960a,b), using radioactive phosphorus as a tracer, concluded that the labial gland is the principal source of nutrition. In contrast Naarmann (1963), using a similar technique, assigned the nutritive role to the postpharyngeal gland.

The discrepancy was resolved by more detailed studies by Paulsen (1969) of the labial, propharyngeal, and postpharyngeal glands in *Formica polyctena*. It is worth mentioning that the propharyngeal

**FIGURE 3–18** A worker of the primitive Australian ant *Amblyopone australis* attends larvae in the brood nest.

**FIGURE 3–19** Brood-tending behavior of the primitive Australian ant *Nothomyrmecia macrops*. (*a*) A worker licks larvae. (*b*) When carrying a larva the worker frequently holds it by the neck, as shown in this figure. (*c*) A worker banks a mature larva with soil to facilitate cocoon spinning. (*d*) Workers remove the soil particles from the cocoon.

gland of ants is homologous to the pharyngeal gland in honey bees (Emmert, 1969). In bees the pharyngeal gland is the source of important nutritional substances used in the nursing of larvae (Rembold, 1964; Weaver, 1966). Paulsen found no indication, however, that the propharyngeal gland serves the same function in *Formica polyctena*. Instead, as Ayre (1963a) suggested earlier for *Camponotus herculeanus*, it is evidently a source of digestive enzymes. The postpharyngeal gland, as then noted by Paulsen, is found only in ants, and it occurs in all adult castes, including the queens, workers, and males. The organ is glove-shaped, with two symmetrical halves terminating in a varying number of finger-shaped projections. Its lumen is filled with a complex mixture of lipids. When Paulsen injected triolein labeled with radioactive carbon into the hemolymph of "storage workers," who have a large abdominal fat body, radioactivity accumulated in the postpharyngeal gland but not in the propharyngeal gland or labial gland. Some of the labeled material was subsequently transported to other workers as well as to queens. In addition, Paulsen found that larvae of *Formica polyctena* collected in the field had droplets in their gut microscopically identical to those in the postpharyngeal gland lumen of the nursing workers.

These and subsequent investigations suggest that the postpharyngeal gland is the source not only of basic nutrients but also of "profertile substances" (Gösswald and Bier, 1953) used by workers in the determination of reproductive females. The fundamentally nutrient role of the gland is supported further by the following findings:

1. There exists a strong positive correlation between the secretion of lipophilic substances in the postpharyngeal gland and the storage of lipids in the fat body of *Formica* workers.

2. *Formica* workers store lipids in their fat body during the autumn and metabolize them following the winter dormancy (Kirchner, 1964). There is a parallel increase in the volume of the cell nuclei in the glandular epithelium of the postpharyngeal gland in the autumn and a decrease in size in the late spring and summer. The nuclear volume is believed to be an index of secretory activity (Bausenwein, 1960).

3. A similar association has been found in freshly eclosed workers, who have larger fat bodies and postpharyngeal gland nuclei than older workers have (Bausenwein, 1960).

The caste-determining connection involving profertile substances was made as follows. Bier (1958a) found that only recently eclosed workers or young workers that served as storage workers during winter dormancy have the ability to rear reproductive females. Data provided later by Paulsen (1969) strongly suggest that this capacity is due to the high secretory activity of the postpharyngeal gland in these workers.

How general is the nutritive role of the postpharyngeal gland in ants? After demonstrating that dyes fed to *Messor* workers are transferred directly from the pharynx into the lumen of the postpharyngeal gland, Barbier and Delage (1967) postulated that the postpharyngeal gland is an organ for digestion and nutrient resorption. In later studies, however, one of the authors (Delage-Darchen, 1976) demonstrated that materials from the postpharyngeal lumen serve as nutrients and are fed to the larvae. Markin (1970) provided circumstantial evidence that the postpharyngeal gland of the Argentine ant *Iridomyrmex humilis,* a dolichoderine, produces important nutrients fed primarily to queens and small larvae.

Thus a substantial amount of evidence points to a nutritive function of the postpharyngeal gland across at least three of the major ant subfamilies, the Myrmicinae, Dolichoderinae, and Formicinae. In the case of *Formica* in particular, both Naarmann (1963) and Paulsen (1969) concluded that the contents of the gland are first transferred to the crop, where they are mixed with other foods, and then regurgitated to larvae and adult nestmates.

This is not the whole story, however. Work on the biochemistry of the postpharyngeal gland in fire ants (*Solenopsis*) indicates that the organ serves additional, still unknown functions in at least some species. Attygalle et al. (1985) found that the postpharyngeal gland of *Solenopsis geminata* is hypertrophied, forming four huge lobes that completely fill the upper part of the head cavity. The glandular secretion is an oil containing a complex mixture of hydrocarbons, of which the major constituents are heneicosane, tricosane, and tricosene (linear $C_{21}$ and $C_{23}$ hydrocarbons). Interestingly, these substances differ in proportions from one worker to the next, for reasons that have not been explained. Thompson et al. (1981) showed that the postpharyngeal gland of *Solenopsis invicta* queens also contains hydrocarbons as major components (63 percent), accompanied by triglycerides (32 percent) and some free fatty acids. The hydrocarbons are all methyl-branched and include 13-methylheptacosane, 3-methylheptacosane, 13,15-dimethylheptacosane, and 3,9-dimethylheptacosane. When these queen-derived substances were presented to a *Solenopsis invicta* colony, workers clustered around them. On the other hand, Attygalle and his colleagues failed to obtain a similar response in *Solenopsis invicta* with the contents of the worker postpharyngeal gland.

The labial gland, unlike the postpharyngeal gland, can discharge its contents directly through an opening on the labium (see Figure 7–2). Data from Ayre (1963a) on *Camponotus* and from Delage (1968) on *Messor* indicate that the labial gland serves primarily as a source of digestive enzymes, and amylase in particular. This sugar-generating role of the organ was confirmed in *Formica polyctena* by Paulsen (1969), who found that the labial gland secretions are rich in glucose (0.46 M). This is not all. When Paulsen injected radioactivity-labeled glucose or fructose into the hemolymph of workers, he detected radioactivity 24 hours later in the labial gland but not in the postpharyngeal gland. Paulsen also found that the labial gland secretions, which are mostly carbohydrates obtained from resorbed food or mobilized reserves from the fat body, are fed to queens, workers, and larvae. The labial glands are active in older workers and their secretory activity does not display the same strong annual cycle characterizing the postpharyngeal glands.

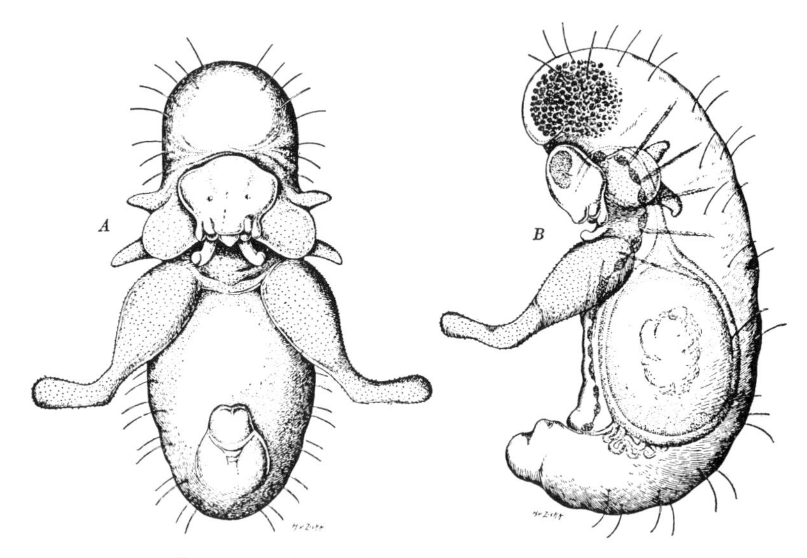

**FIGURE 3–20** First-instar larvae of the African arboreal ant *Tetraponera* (= *Pachysima*) *latifrons,* showing the peculiar thoracic and abdominal organs considered by Wheeler to be the source of substances attractive to workers: (*A*), front view; (*B*), side view. (From Wheeler, 1918a.)

Stomodeal trophallaxis by larvae has been observed in nearly all of the major ant subfamilies (Wheeler, 1928; Le Masne, 1953; Haskins and Whelden, 1954; Maschwitz, 1966). The single exception is the Dolichoderinae, a negative generalization seemingly reinforced by the fact that Athias-Henriot (1947) found no evidence of secretory tissue in the labial glands of the dolichoderine species *Tapinoma nigerrimum.* The workers also actively seek the stomodeal fluid, since they lick the head region more frequently than the remainder of the body and occasionally employ the same kind of antennal stroking used to offer food to larvae (Stäger, 1923; Le Masne, 1953). The larval secretions of at least one species, *Leptothorax curvispinosus,* are avidly consumed by queens who "graze" from one larva to another (Wilson, 1974a), while final-instar larvae of the Australian harvesting ant *Monomorium* (= *Chelaner*) *rothsteini* convert seeds given to them into secretions that are fed back to the workers (Davison, 1982). The origin of these liquids is not known, but the most likely source is the paired salivary glands, the only well-developed exocrine glands that open into the mouth. Wheeler (1918a) pointed out that these glands are hypertrophied in the myrmicine *Paedalgus termitolestes,* and he suggested that they serve to produce secretions attractive to the workers. Wheeler also speculated that the unusual thoracic abdominal appendages of larvae of the pseudomyrmecine *Tetraponera* (= *Pachysima*), which he called exudatoria, produce liquids that attract and bind the affection of workers (see Figure 3–20). Similar structures are found in certain species of *Crematogaster,* including *C. rivai* (Menozzi, 1930).

The larvae of the migratory ant *Leptanilla japonica* have a specialized duct organ on each side of the third abdominal segment. Masuko (1987 and personal communication) demonstrated that adult ants imbibe larval hemolymph directly through this organ. The queen in particular seems to feed exclusively on larval hemolymph. In addition, Masuko (1986) found that larval hemolymph provides an important source of nutrients for the queens of *Amblyopone silves-*

*trii.* In this case, however, no special organ is involved. The queen punctures the larval skin and imbibes hemolymph from the bleeding wound.

Wheeler's mutual attraction hypothesis was put forth as evidence favoring the Roubaud-Wheeler theory of the origin of insect sociality through trophallaxis. As beguiling as this idea may be, it has not yet been subjected to critical experimental investigation. Maschwitz (1966) found that the stomodeal contents of *Tetramorium* larvae have much higher concentrations of amino acids than their hemolymph. He concluded that the stomodeal fraction "very probably originates from the salivary secretion," and thus the secretion must have nutritive value. Sorenson et al. (1983) obtained the same result in larvae of the fire ant *Solenopsis invicta;* the concentration of amino acids, to be precise, is two to three times greater in the oral secretions of the larvae than in their hemolymph. On the other hand, the total protein concentration in the oral secretions is only 60–70 percent that in the hemolymph. The material consists substantially of amylases and proteinases, which are used (in the final instar) partly outside the body to digest solid food placed below the head by the nurse workers (Petralia et al., 1980). At the present time almost no information exists on the nature and function of the abdominal exudatoria or other possible glandular tissue located elsewhere on the bodies of larvae, with the single exception of the abdominal glands of *Leptanilla* just mentioned.

Ant larvae periodically produce yet another kind of liquid from the anal region. The workers often solicit this material, which appears likely to contain waste substances and to originate from the Malpighian tubules, by stroking the tip of the larva's abdomen with their antennae (Le Masne, 1953; Ohly-Wüst, 1977). In the case of *Solenopsis invicta,* O'Neal and Markin (1973) have distinguished two forms of the anal liquid. One is milky and sought by the workers, who gently squeeze the posterior sternites up to five times in a row in an attempt to obtain it. Produced by the late second instar as well as third and fourth (final) instars, the milky liquid excites and attracts nearby workers, and it appears to serve as a supplemental food for them. The second kind of liquid is clear and less viscous and never solicited. Whenever a droplet is extruded, a passing worker picks it up in half-opened mandibles, carries it to the edge of the nest, and deposits it. This material, unlike the milky liquid, appears to consist wholly of true excreta.

The possibility that larvae serve as specialized digestive castes has received substantial support from a series of studies on ants belonging to the subfamily Myrmicinae. In a pioneering analysis of the European harvester *Messor capitatus,* Delage (1968) showed that the stomodeal secretions of larvae (presumably originating in the salivary gland) contain lipases and proteases, but no carbohydrases. In contrast, workers produce carbohydrases in their cephalic glands (including the labial gland, which extends into the thorax), but no proteases. Delage therefore proposed that the larvae provided proteases to the workers while receiving carbohydrases in return. *Messor* feeds heavily on seeds, and such a mutualistic exchange between adults and larvae might contribute to a more efficient digestion of the starch and other carbohydrates.

A similar symbiotic relationship was proposed by Went et al. (1972) for the American harvesting ant *Messor (= Veromessor) pergandei,* following the discovery of gluconeogenesis in the wasp *Vespa orientalis* by Ishay and Ikan (1968). Went and his co-workers speculated that *Messor* larvae digest solid particles of seeds given them by the workers, who (it was suggested) are unable to digest solid food and hence depend on the larvae for this service. The hypothesis is plausible, especially in view of Delage's results on the European *Messor,* but it has not yet been tested experimentally.

In the next principal development, a decisive analysis was performed by Margarete Ohly-Wüst (1977) on *Myrmica rubra (= M. laevinodis)* and *Monomorium pharaonis.* The larvae of both of these myrmicine species respond to tactile stimulation from the workers by discharging saliva from their labial glands as well as proctodeal liquid from the rectal bladder. Ohly-Wüst found that the larval saliva of both species contains amino acids, proteases, carbohydrases, and, at least in the case of *Myrmica,* lipases as well (no tests for lipases were made on *Monomorium*). A remarkable quantity of fluid is discharged by the larva during a single "milking." The amounts in volume and percentage of body weight obtained for *Myrmica rubra* were as follows:

Second instar: 0.026 microliter (14.4 percent)
Third instar: 0.055 microliter (14.5 percent)
Fourth instar: 0.067 microliter (5 percent)

The rectal fluid of both *Myrmica rubra* and *Monomorium pharaonis* contains uric acid, amino acids, and traces of proteins—but no carbohydrates or lipids. The quantity of discharges measured by Ohly-Wüst was as impressive as in the case of the salivary discharges:

Second instar: 0.050 microliter (27.8 percent)
Third instar: 0.044 microliter (11.6 percent)
Fourth instar: 0.054 microliter (4.3 percent)

The total nitrogen content of the two kinds of fluid is also substantial; for *Myrmica rubra* it was measured at 2.2 micrograms per microliter for the saliva and 2.1 micrograms per microliter for the rectal fluid. Nitrogenous material is potentially valuable to the workers both as a nutrient and as a source of enzymes. In fact, Ohly-Wüst found that the larvae use salivary enzymes to predigest their own food, then donate some to the workers, who pass it around among themselves by regurgitation. In *Monomorium pharaonis,* the larval secretions also serve as an emergency food supply during periods of starvation.

The workers of many ant species have functional ovaries that play a key role in colony integration (Eidmann, 1928; Goetsch, 1938; Ledoux, 1949; H.-J. Ehrhardt, 1962). In some species worker-laid eggs are usually consumed by the queen, by the larvae, and, less commonly, by other workers shortly after they are deposited; thus they deserve to be called "trophic" eggs (*oeufs alimentaires*). Trophic eggs have been observed in a wide diversity of both primitive and advanced ant genera: *Prionopelta* (Hölldobler and Wilson, 1986a), *Nothomyrmecia* (B. Hölldobler and R. W. Taylor, unpublished), *Myrmecia* (Freeland, 1958), *Aphaenogaster (= Novomessor)* (B. Hölldobler, N. Carlin, and E. P. Scovell, unpublished), *Atta* (Bazire-Benazet, 1957), *Basiceros* (Wilson and Hölldobler, 1986), *Leptothorax* (Gösswald, 1933; Le Masne, 1953; Wilson, 1975a), *Myrmica* (Brian, 1953, 1969), *Pogonomyrmex* (Wilson, 1971), *Zacryptocerus* (Wilson, 1976a), *Hypoclinea* (Torossian, 1959, 1965), *Iridomyrmex* (Torossian, 1961, 1965), *Oecophylla* (where the queen is fed chiefly with trophic eggs; Hölldobler and Wilson, 1983a), *Plagiolepis* (Passera, 1966), and *Formica* (Weyer, 1929). Trophic eggs are characteristically flaccid or differ in shape from reproductive eggs, and, in some cases at least, they lack detectable DNA (for the latter point see Voss, 1981). The frequency with which such structures are laid varies enormously among ant species. They do not occur at all in the myrmicine genera *Pheidole*, *Pheidologeton*, and *Solenopsis*, the workers of which completely lack ovaries. In *Iridomyrmex humilis* trophic eggs are a rare event, while in *Pogonomyrmex badius*, *Aphaenogaster* (= *Novomessor*) *cockerelli*, and *Hypoclinea quadripunctata* they form the usual diet of the larger larvae and nest queens. As a rule among species, the more frequent the exchange of trophic eggs, the less frequent the exchange of liquid food by regurgitation. The two systems often occur in the same species, but their overall pattern of distribution among species suggests that they tend to replace one another in evolution. In *Pogonomyrmex badius* and *Aphaenogaster cockerelli* the trophic eggs are misshapen and more flaccid than reproductive eggs. Those of *Plagiolepis pygmaea* are smaller in size as well, and they are formed when the trophocytes and follicular epithelium degenerate prematurely so that the oocyte remains small and fails to acquire a chorion (Passera et al., 1968). In *Atta rubropilosa* the oocytes even fuse together in masses, creating an extraordinary trophic "omelette" (Bazire-Benazet, 1957). It is tempting to speculate that these drastic alterations, which appear to be regular programmed processes of worker physiology, are induced by primer pheromones passed from the queen or larvae, with the effects being mediated by changes in the endocrine system.

Trophic eggs are also laid by colony-founding queens belonging to at least the following genera: *Atta*, *Solenopsis*, and *Tetramorium* among the Myrmicinae; and *Lasius* and *Paratrechina* among the Formicinae (Taki, 1987). When deliberate searches are made, this phenomenon will probably be found to occur in other subfamilies and genera. Finally, unmated queens of *Pheidole pallidula* and *Solenopsis invicta* lay trophic eggs, at least under certain conditions (Passera, 1978; Voss, 1981). It may be significant that *Pheidole* and *Solenopsis* are among the very few ant genera in which the workers lack ovaries and hence cannot lay trophic eggs.

In general, first-instar larvae are fed by regurgitation, trophic eggs, or both. They are kept by the workers with the eggs, so that the two characteristically form an "egg-microlarva pile." At this point the larvae sometimes consume reproductive eggs next to them, a form of cannibalism. Liquid food exchange by regurgitation continues through the life of the larva. The larvae of predatory, scavenger, and granivorous ant species (hence the great majority of all ant species) are also fed fragments of insects or seeds, especially during the later instars (Plate 3). The workers cut up this material and often chew it as well before placing it directly on the head or in the feed basket (just below the "chin") of the larva. Workers of the primitive genera *Amblyopone* and *Myopopone* also carry the larvae to freshly caught centipedes and insects when these arthropods are large. A similar behavior has been observed in the myrmicine ant *Aphaenogaster subterranea* (Buschinger, 1973a; Figure 3–21).

Another novel and surprising category of brood care has been suggested in the fire ant *Solenopsis invicta* by Obin and Vander Meer (1985). Nurse workers raise and then vibrate their abdomens while extruding the sting and dispensing about 1 nanogram of venom as an aerosol. Because the fire ant venom possesses antibiotic activity, and the metapleural glands (the source of antibiotics in other ants) do not, Obin and Vander Meer have proposed that the secretion serves the secondary function of protecting the brood from microorganisms. A different form of this "gaster flagging," dispensing up to 500 nanograms of venom, is used to repel other species of ants from the nest area.

**FIGURE 3–21** The alternative mode of serving prey to larvae: a few species, such as *Aphaenogaster subterranea* shown here, sometimes carry larvae to the prey rather than the reverse. (From Buschinger, 1973a.)

Although a great deal of information has accumulated on brood care, especially during the past fifteen years, the subject remains largely unexplored. There is a need for close comparative studies across the subfamilies and tribes, including histological and biochemical analyses. We can reasonably anticipate many surprising new discoveries, some of which may force changes in our thinking about colony organization.

## DEMOGRAPHY OF COLONY MEMBERS

The demography of individual colony members has received remarkably little attention, even though it is basic to the understanding of the ergonomics and population dynamics of colonies. Relatively few data exist on the life span, the survivorship schedule, and the reproduction of the members of any of the various castes.

Data on the longevity of the adult stage of queens and workers are summarized in Table 3–3. An unusually thorough study of longevity in *Solenopsis invicta* workers was conducted by Porter and Tschinkel (1985 and personal communication), who followed marked workers from near eclosion to death. Longevity increased with size while decreasing with rising ambient temperatures: at 17°C the averages in various laboratory colonies ranged from 60 weeks in minors (head width 0.6 mm) to 70 weeks in majors (head width 0.9–1.3 mm); at 24°C the averages were 18 weeks for minors and 36 weeks for majors; and at 30°C the averages ranged from 10 weeks for minors to 16 weeks for majors. Thus the higher costs of manufacturing larger workers were partially amortized by the greater longevity of these insects.

The following additional generalizations can be drawn from the meager data on other ant species:

1. As expected, mother queens live much longer than workers in all groups of ants. The astounding figures for longevity of *Campo-*

**TABLE 3–3** Longevity of individual ant species. Only the adult life spans are given; the immature stages usually add one to several months.

| Species | Caste | Locality | Average longevity or range | Maximum recorded longevity | Authority | Comments |
|---|---|---|---|---|---|---|
| *Aphaenogaster rudis* | Queen | United States | 8.7 yrs | 4.6–13 yrs | Haskins (1960) | Based on 11 queens in laboratory nests |
| *A. rudis* | Worker | United States | ? | >3 yrs | Fielde (1904b) | |
| *Atta sexdens* | Queen | Guyana | ? | 14 yrs | K. M. Horton and E. O. Wilson (unpublished) | Based on 1 queen in laboratory nest |
| *A. sexdens rubropilosa* | Queen | Brazil | ? | 15.3 yrs | Autuori (1950b) | Based on 1 queen in laboratory nest |
| *Camponotus consobrinus* | Queen | Australia | >7 yrs | | B. Hölldobler (unpublished) | Based on 1 queen in laboratory nest |
| *C. herculeanus* | Queen | Germany | ? | >10 yrs | B. Hölldobler (unpublished) | Based on 1 queen in laboratory nest |
| *C. lateralis* | Queen | Italy | 8 yrs | | K. Hölldobler (unpublished) | Based on 1 queen in laboratory nest |
| *C. lateralis* | Queen | France | >5 years | | Palma-Valli and Délye (1981) | Based on 1 queen in laboratory nest |
| *C. perthiana* | Queen | Australia | ? | 21 yrs | Haskins and Haskins (personal communication) | Based on 1 queen in laboratory nest |
| *Ectatomma ruidum* | Queen | Australia | ? | 8.8 yrs | Haskins and Haskins (1980) | Based on 1 queen in laboratory nest |
| *E. ruidum* | Queen | ? | ? | 9 yrs | Haskins and Haskins (1980) | Based on 1 queen in laboratory nest |
| *Formica rufibarbis* | Queen | Germany | ? | 14 yrs | H. Appel (in Kutter and Stumper, 1969) | Based on 1 queen in laboratory nest |
| *F. sanguinea* | Queen | Germany | ? | 20 yrs | H. Appel (in Kutter and Stumper, 1969) | Based on 1 queen in laboratory nest |
| *Lasius alienus* | Queen | France | ? | 9.25 yrs | Janet (1904) | Based on 1 queen in laboratory nest |
| *L. flavus* | Queen | Germany | 18 yrs | 18 yrs | H. Appel (in Kutter and Stumper, 1969) | Based on 3–4 queens in laboratory nests |
| *L. flavus* | Queen | England | ? | 22.5 yrs | Prescott (1973) | |
| *L. niger* | Queen | Germany | ? | 29 yrs | H. Appel (in Kutter and Stumper, 1969) | Based on 1 queen in laboratory nest |
| *Leptothorax lichtensteini* | Worker | France | 2.5 yrs | ? | Plateaux (1986) | Based on laboratory observations |
| *L. lichtensteini* | Queen | France | ? | 12–15 yrs | Plateaux (1986) | Based on laboratory observations |
| *L. nylanderi* | Worker | France | 3 yrs | ? | Plateaux (1986) | Based on laboratory observations |
| *L. nylanderi* | Queen | France | ? | 15 yrs | Plateaux (1986) | Based on laboratory observations |
| *Messor semirufus* | Queen | Lebanon | ? | 9 yrs | Tohmé and Tohmé (1978) | |
| *Monomorium pharaonis* | Queen | England | ? | 39 wks | Peacock and Baxter (1950) | |
| *M. pharaonis* | Worker | England | ? | 9–10 wks | Peacock and Baxter (1950) | |
| *Myrmecia gulosa* | Worker | Australia | 1.7 yrs | 1.3–2.2 yrs | Haskins and Haskins (1980) | Based on 3 workers in laboratory nest |
| *M. nigriceps* | Worker | Australia | 2.2 yrs | 2.1–2.4 yrs | Haskins and Haskins (1980) | Based on 2 workers in laboratory nest |
| *M. nigrocincta* | Worker | Australia | 1.2 yrs | 1.1–1.3 yrs | Haskins and Haskins (1980) | Based on 5 workers in laboratory nest |
| *M. pilosula* | Worker | Australia | 1.3 yrs | 1.12–1.6 yrs | Haskins and Haskins (1980) | Based on 6 workers in laboratory nest |
| *M. vindex* | Worker | Australia | 1.9 yrs | 1.4–2.6 yrs | Haskins and Haskins (1980) | Based on 5 workers in laboratory nest |
| *Myrmecocystus mimicus* | Queen | United States | ? | >11 yrs | B. Hölldobler (unpublished) | Based on 1 queen in laboratory nest |
| *Myrmica rubra* (= *laevinodis*) | Worker | England | ? | 2 yrs | Brian (1951b) | |
| *Odontomachus* sp. | Queen | ? | ? | 4 yrs | Haskins and Haskins (1980) | Based on 2 queens in laboratory nests |
| *Pogonomyrmex badius* | Queen | Florida, USA | ? | 17 yrs | K. M. Horton and E. O. Wilson (unpublished) | Inferred from colony longevity in laboratory |
| *P. owyheei* | Queen, | Idaho, USA | 17 yrs | 30 yrs | Porter and Jorgensen (1988) | Age of colonies in the field; evidence presented that colonies last only as long as founding queen |

*continued*

**TABLE 3–3** *(continued)*

| Species | Caste | Locality | Average longevity or range | Maximum recorded longevity | Authority | Comments |
|---|---|---|---|---|---|---|
| *Rhytidoponera purpurea* | Worker | Australia | 2.6 yrs | 2.1–2.8 yrs | Haskins and Haskins (1980) | Based on 4 workers in laboratory nest |
| *R. purpurea* | Queen | Australia | 12.5 yrs | 12.5 yrs | Haskins and Haskins (1980) | Based on 1 queen in laboratory nest |
| *R. purpurea* | Queen | ? | ? | 12 yrs | Haskins and Haskins (1980) | Based on 1 queen in laboratory nest |
| *Solenopsis invicta* | Worker | United States | 10–70 weeks | 97 weeks (17°C) | Porter and Tschinkel (1985 and personal communication) | Based on many workers in laboratory nests; depends on temperature and worker size |
| *S. invicta* | Queen | United States | 5.8–6.8 yrs (by locality) | — | Tschinkel (1987b) | Based on sperm depletion curves from 97 queens |
| *Stenamma westwoodi* | Queen | England | ? | 17 or 18 yrs | Donisthorpe (1936) | Based on 1 queen in laboratory nest |

*notus, Formica,* and *Lasius* queens, ranging from 18 to 29 years, make these ants the most long-lived insects ever recorded.

2. Males have a shorter adult life span than either queens or workers.

3. No correlation is yet apparent between the degree of sociality of a species and the longevity of its queens. On the other hand, the workers of *Myrmecia* and the ponerines monitored to date do have greater natural longevities than those in many higher myrmicine genera, including *Monomorium, Pheidole,* and *Solenopsis.* This difference, if it holds, may indicate a larger phylogenetic trend, at least within certain clades of the Myrmicinae. It may also merely reflect the larger size of workers in *Myrmecia* and the Ponerinae. The distinction can be tested by obtaining data on small ponerines such as *Ponera* and *Cryptopone.*

The duration of the immature period, from the birth of the egg to the eclosion of the adult ant from the pupa, is ordinarily only a small fraction of the adult life span and highly dependent on temperature. It is about 20 to 45 days for workers of the fire ant *Solenopsis invicta* (Wilson and Eads, 1949; S. D. Porter, personal communication), 35 to 45 days for males and queens of the European wood ant *Formica polyctena* (Schmidt, 1974a,b), 40 to 60 days for workers of the South American leafcutter *Atta sexdens* (Autuori, 1956), 44 to 61 days for the Australian meat ant *Iridomyrmex purpureus* (Hölldobler and Carlin, 1985), 48 to 74 days for seven species of *Camponotus* in several subgenera (Mintzer, 1979a), 60 days for workers of the American harvester *Messor (= Veromessor) pergandei* (Wheeler and Rissing, 1975a), 70 to 90 days in *Prenolepis imparis* (Tschinkel, 1987c), and 100 days for workers of the Australian bulldog ant *Myrmecia forficata* (Haskins and Haskins, 1950a). Army ants go through brood cycles that are tightly synchronized and therefore easily timed during field studies. The full span from egg through pupa is 50 days in the Neotropical army ant *Eciton hamatum,* 65 days in the small Asian army ant *Aenictus laeviceps,* and 30 days in the African driver ant *Dorylus wilverthi* (Schneirla, 1971).

These figures are in accord with impressions from our own laboratory cultures of many genera representing all of the subfamilies except the Dorylinae and Leptanillinae. The duration of the immature period ranges according to species from about one month to no more than three or four months, except in those cases where growth is temporarily halted during the winter—as in boreal species of the carpenter ant genus *Camponotus* (Hölldobler, 1961). It also depends crucially on temperature, in ways that have not yet been precisely measured.

The number of larval instars has only recently been established firmly in ant species, and can be summarized as follows: three in *Crematogaster scutellaris* (Casevitz-Weulersse, 1984), *C. stadelmanni* (Delage-Darchen, 1972b), and *Pheidole pallidula* (Passera, 1973); three or possibly four in *Oecophylla longinoda* (Wilson and Hölldobler, 1980); four in *Cephalotes atratus* (D. E. Wheeler, personal communication), *Formica polyctena* (A. Maidhoff, cited in Schmidt, 1974), *Myrmica rubra* (Ohly-Wüst, 1977), *Pheidole bicarinata* (Wheeler and Nijhout, 1984), *Solenopsis invicta* (O'Neal and Markin, 1973; Petralia and Vinson, 1979b), and *Zacryptocerus minutus* (D. E. Wheeler, personal communication); five in *Eciton burchelli* and *E. hamatum* (Wheeler and Wheeler, 1986b), *Plagiolepis pygmaea* (Passera, 1968a), and the workers of *Camponotus aethiops* (Dartigues and Passera, 1979b); and six in the queen of *Camponotus aethiops* (Dartigues and Passera, 1979b). The basic number of larval instars in ants generally appears to be four (D. E. Wheeler, personal communication).

The recorded egg production of queens varies enormously according to species, from as few as 400 per queen each year in *Myrmica rubra* (Brian and Hibble, 1964) to 2 million in *Eciton burchelli* (Schneirla, 1971) and (probably the absolute world record for all insects!) 50 million in *Dorylus nigricans* (Raignier and van Boven, 1955). The rate increases within and across species as colony size increases, and across species as the average worker longevity declines.

## COLONY MOVEMENTS

Since the mid 1970s a startling new picture has begun to emerge concerning the stability of ant nests. Colonies move from one site to another more frequently than previously imagined, and the emigrations are organized by sophisticated systems of communication among the workers that entail motor and other tactile signals, release of pheromones, division of labor, and bodily transport. These topics have become the objects of what is virtually a small field of investigation in itself (see Chapter 7).

Herbers (1985) found that all of 14 species she monitored in New York State woodland regularly shifted their nest sites as part of an annual cycle. During her study colonies of these species, including the primitive ponerine *Amblyopone pallipes* and representatives of the Myrmicinae and Formicinae, dispersed from hibernation sites in the spring and summer and began to contract again as fall approached. The emigrations of some of the notoriously polydomous forest species, such as *Leptothorax longispinosus* and *Myrmica punctiventris*, were part of a cycle of summer expansion of outposts and winter contraction to hibernation sites. In the cases of *Myrmica punctiventris*, *Aphaenogaster rudis*, and *Lasius alienus*, between 77 and 100 percent of the nest sites were evacuated at least once.

Additional studies have shown that other kinds of ants have very different waiting times, in concert with their principal lifeways (Table 3–4). The Japanese queenless ant *Pristomyrmex pungens* does not construct elaborate subterranean nests and frequently relocates its nests. The mean duration of nests in two different study sites was 15.7 and 17.4 days (K. Tsuji, personal communication). Some army ants (*Aenictus*, *Eciton*) emigrate as often as once a day during the nomadic part of their brood cycle (or even more than once a day in the case of *Aenictus*), when larvae are present and raiding for insect prey is most intense.

At the opposite extreme, mound-builders and a few other forms with deep, secure nests often remain at the same site for many years. Species such as leafcutters of the genus *Atta*, *Myrmecocystus* honey ants, *Camponotus* carpenter ants, and members of the *Formica rufa* group invest huge amounts of energy and time in the construction of nests. Nests of *Camponotus herculeanus* and *Myrmecocystus mimicus* have been observed at the same location for more than ten years (B. Hölldobler, unpublished). They enjoy relative security and excellent microclimate control, and hence rarely encounter enough stress to trigger emigration. Yet even these sedentary forms move when sufficiently disturbed, either mechanically or following insecticide treatment, and their emigration procedures do not differ fundamentally from those of the more agile species (Wilson, 1980a; Fowler, 1981).

The causes of colony emigration are numerous and complex and can be understood only by attention to the natural history of particular species. The following factors have been identified in various studies:

*Nest disturbance.* A mechanical disturbance of the nest, admitting light and air currents to the brood chambers, causes an immediate retreat by the queen and nurse workers, with the latter hastily transporting brood pieces and fragments of food into the intact nest interior. Even soldiers and extremely young, callow workers participate as brood carriers. In all but the most timid species, some of the workers—typically older individuals and members of the soldier (major) caste—run excitedly over the nest surface and the surrounding terrain. The entire response is frequently enhanced by the release of alarm substances, especially in larger, more aggressive colonies. Colonies of some genera, for example *Leptothorax* (Möglich, 1978), then organize emigration to new nest sites.

*Flooding.* Some ant species respond in a dramatic manner to even partial flooding of the nest. Minor workers of the Neotropical forest ant *Pheidole cephalica* react to as little as a single drop placed in the nest entrance by making alarm runs through the nest, which often end at alternate entrances (Wilson, 1986c). They use odor trails to lead nestmates into the unobstructed entrance galleries and sometimes out of the nest altogether. With this procedure one or two workers can mobilize a large fraction of the colony in 30 seconds or less and even initiate colony emigration. Other *Pheidole* species react to flooding with both alarm runs and alarm waves, in which short loops generate broader and more slowly advancing fronts of excitement. Still others respond with alarm waves alone. Both social patterns together constitute what Wilson has termed the flood evacuation response. Alarm runs occur more frequently in *Pheidole* species that nest in pieces of rotting wood, and hence typically excavate a linear array of nest galleries and chambers. They tend to be absent in species that nest in soil and hence excavate a broader array of nest galleries and chambers.

An equally remarkable phenomenon is the living rafts of fire ants. Those of *Solenopsis invicta* have been observed in the floodplains of the southeastern United States, and those of *S. saevissima* have been observed in the savannas of northern South America. As flood waters rise, the ants move upward through their nests to ground level and then form large masses that float on the water surface. Both brood and queens have been found alive in the centers. The masses eventually anchor to grass stems or bushes sticking out of the water, and apparently return to the soil when the flood recedes (Morrill, 1974a; E. O. Wilson, unpublished).

*Nest microclimate change.* A more gradual alteration in nest microclimate induces emigration even in the absence of mechanical disturbance or flooding. In a long-term study of the Australian meat ant *Iridomyrmex purpureus*, Greenslade (1975a,b) found that shading by the encroachment of overhanging vegetation was a principal cause of colony movements. Several field experiments with artificial shading, including those by Carlson and Gentry (1973) on the harvester ant *Pogonomyrmex badius* and Smallwood (1982) on the American woodland ant *Aphaenogaster rudis*, have confirmed more directly the response of colonies to this ubiquitous microclimatic change.

*Predation.* One response to invasion of the nest by hostile ants or other enemies is "panic alarm," during which the workers incite nestmates to flee into the nest interior or out of the nest altogether (Wilson and Regnier, 1971). *Pheidole dentata* colonies have a three-stage strategy of defense against enemies, especially the formidable fire ants (*Solenopsis*), that culminates in evacuation and emigration (Wilson, 1976b). At the lowest level of stimulation, in which a few *Solenopsis* scouts are contacted away from the nest, the minor workers recruit nestmates over considerable distances. The major work-

**TABLE 3–4** Residence times of colonies in particular nest sites.

| Species | Mean residence times (days) | Minimum (days) | Maximum (days or years) | Reference |
|---|---|---|---|---|
| **DORYLINAE** | | | | |
| *Aenictus* spp. | — | ≤1 | — | Schneirla (1971) |
| **ECITONINAE** | | | | |
| *Eciton* spp. | — | 1 | — | Schneirla (1971) |
| **MYRMICINAE** | | | | |
| *Aphaenogaster cockerelli* | — | — | 8 yrs | Chew (1987) |
| *A. rudis* | 20 | 19 | 37 days | Smallwood (1982), Herbers (1985) |
| *Myrmica americana* | 48 | — | — | Talbot (1946) |
| *M. punctiventris* | — | 5 | 26 days | Herbers (1985) |
| *Pheidole desertorum* | 6 | 1 | 28 days | Droual and Topoff (1981) |
| *Pogonomyrmex badius* | 234 | — | — | Gentry (1974) |
| *P. badius* | — | — | >2 yrs | B. Hölldobler (unpublished) |
| *P. barbatus* | 1,054 | — | — | Van Pelt (1976) |
| *P. barbatus* | — | 10 | — | Hölldobler (1976a) |
| *P. barbatus* | — | — | >5 yrs | B. Hölldobler (unpublished) |
| *P. occidentalis* | — | — | >7 yrs | B. Hölldobler (unpublished) |
| *P. owyheei* | — | — | >17 yrs | S. D. Porter (personal communication) |
| *Pristomyrmex pungens* | 8.4 | — | — | K. Tsuji (personal communication) |
| **DOLICHODERINAE** | | | | |
| *Tapinoma sessile* | 13 | — | — | Smallwood (1982) |
| **FORMICINAE** | | | | |
| *Camponotus herculeanus* | — | — | >10 yrs | B. Hölldobler (unpublished) |
| *Formica difficilis* | 27 | — | — | Smallwood (1982) |
| *F. exsecta* | — | 1 | — | Dobrzańska (1973) |
| *F. exsectoides* | — | — | 30 yrs | Andrews (1926) |
| *F. obscuripes* | 1,132–2,847 | — | — | King and Sallee (1953, 1956) |
| *F. pallidefulva* | 72 | — | — | Smallwood (1982) |
| *F. polyctena* | — | — | >5 yrs | B. Hölldobler (unpublished) |
| *F. subsericea* | 90 | — | — | Smallwood (1982) |
| *Lasius alienus* | — | 6 | 30 days | Herbers (1985) |
| *L. fuliginosus* | — | — | >16 yrs | B. Hölldobler (unpublished) |
| *Myrmecocystus depilis* | — | — | 17 yrs | Chew (1987) |
| *M. mexicanus* | — | — | 40 yrs | Chew (1987) |
| *M. mimicus* | — | — | >11 yrs | B. Hölldobler (unpublished) |
| *Polyergus rufescens* | — | — | >2 yrs | B. Hölldobler (unpublished) |

ers ("soldiers") attracted in this manner then take over the main role of destroying the intruders. If the fire ants invade in larger numbers, fewer trails are laid, and the *Pheidole* meet the enemy close to the nest along a shorter perimeter. Finally, if the invasion becomes more intense, the *Pheidole* abscond with their brood and scatter away from the nest in all directions.

A similarly complex but otherwise different avoidance response is utilized by the desert species *Pheidole desertorum* and *P. hyatti* when approached by their chief predators, army ants of the genus *Neivamyrmex* (Droual and Topoff, 1981; Droual, 1983, 1984). The colonies normally have multiple nests, only one of which is used at a time. When *Neivamyrmex* raiders are detected near the occupied nest, the colony enters an "alert" phase, in which *Pheidole* workers carry brood pieces outside the nest but remain in the close vicinity. If the nest is not discovered by the *Neivamyrmex* column, the alert ends and the *Pheidole* go back inside. If on the other hand the raiders come close, the alert escalates into a full evacuation. *P. desertorum* workers then scatter in all directions, in the manner of *P. dentata*, while *P. hyatti* workers follow recent recruitment trails out of the vicinity. In both cases the colonies rendezvous in the surplus nests.

This odd method appears to work very well. Of 46 *Neivamyrmex* raids observed during a three-month period, only five *Pheidole* colonies were seen to be destroyed (Droual and Topoff, 1981). When a few *P. desertorum* colonies were denied access to their extra nests, significantly fewer brood and alates survived (Droual, 1984).

It is evident that the sign stimuli and organization of response to enemies have been crafted separately through natural selection in each major phyletic line, and in many cases the response does not include emigration. In the same area as the *Pheidole* observations, LaMon and Topoff (1981) found that species of *Camponotus* differ from these species and from each other in their response to the approach of *Neivamyrmex*. Workers of *C. festinatus* evacuate the nest with brood and run up nearby vegetation. *C. ocreatus* and *C. vicinus*, in contrast, stand their ground and fight back. In none of the three species has a full-scale emigration been observed to accompany a *Neivamyrmex* raid.

*Competition.* Seen from low-flying airplanes, the craters of *Pogonomyrmex* harvesting ants in the western United States appear remarkably uniform in distribution. They are, in the parlance of ecology, statistically overdispersed. There is substantial evidence that this pattern emerges at least in part from emigrations induced by too frequent contact among colonies of the same or congeneric species. Hölldobler (1976a) observed such forced movement of colonies on ten occasions, one by a *P. barbatus* colony that came into successive hostile contacts with two colonies of *P. rugosus* (Figure 3–22). A similar tendency to move away from nearest neighbors has been observed in *P. badius* by Harrison and Gentry (1981) and in *P. californicus* by De Vita (1979).

The impact of competition and interference is not always so definite. In laboratory experiments Bradley (1972, 1973) found that *Hypoclinea taschenbergi* regularly caused *Formica obscuripes* to emigrate. But in the field the interaction between the Canadian forest species was less decisive. *Formica* colonies transplanted into *Hypoclinea* territories were forced to emigrate, just as in the laboratory. In the reverse transplantation, however, the *Formica* shifted their foraging trails to avoid the *Hypoclinea* nests but otherwise held their position.

At the opposite extreme among ant species, Longhurst and Howse (1979b) could find no evidence of the role of competition in the frequent emigrations of *Megaponera foetans*, a giant termite-raiding ponerine in tropical Africa. They reasoned that because nine colonies occurred per acre at the study site while the raiding columns extended as far as 95 meters, exploitation of termites was "likely to be so intense as to render emigration fruitless." They suggested that some other factor might be responsible for colony movements, such as flooding or attacks from their own predators, subterranean driver ants of the genus *Dorylus*.

*Colony satellite formation and budding.* Many ant species expand their foraging domain by dividing into subcolonies that disperse to extra nest sites, maintaining contact through an exchange of foraging workers as well as the transport of immature forms back and forth. This is the case, for example, in the huge supercolony of *Formica yessensis* found along the Ishikari Coast of Hokkaido (Higashi and Yamauchi, 1979). Often the satellite colonies are inhabited by supernumerary queens, but in many instances only workers and brood are present. A similar polycalic colony organization with multiple queens has recently been described in the weaver ant *Polyrhachis dives* (Yamauchi et al., 1987). There has been a tendency for some investigators to treat each such unit as a separate colony, but this is clearly an error in those cases where the fragmentation is only part of a fluid and often cyclical process entailing the formation of temporary satellite nests.

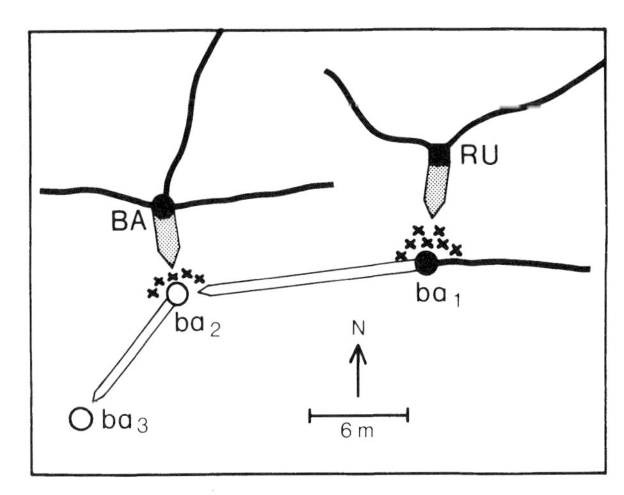


**FIGURE 3–22** Schematic illustration of forced emigrations by a colony of the American harvester ant *Pogonomyrmex barbatus*. A *P. rugosus* colony (*RU*) attacks a *P. barbatus* colony ($ba_1$); $ba_1$ emigrates to a new nest site ($ba_2$). There it is attacked by a *P. barbatus* colony (*BA*) and emigrates a second time to $ba_3$. (From Hölldobler, 1976a.)

An entirely different circumstance obtains, however, in the process of colony multiplication by budding. In this case a cohort of workers departs the main nest in the company of one or more inseminated queens and establishes a peripheral nest that eventually grows into an independent colony, with few or no contacts with the mother colony. This is often the case in the European red wood ant *Formica polyctena* (Mabelis, 1979a). It is invariably the case when the queens are flightless or, as in the large ponerines *Dinoponera gigantea* and *Ophthalmopone berthoudi* and species of the myrmicine *Megalomyrmex leoninus* group, have been wholly replaced by fertile workers (Overal, 1980; Peeters and Crewe, 1985; C. R. F. Brandão, personal communication). But budding also occurs in some species with normal alate queens. In the North American slavemaker *Polyergus lucidus* it has evidently replaced invasion of host colonies of other species by the queen, the usual procedure of slave-making species (Marlin, 1968). It is apparently the exclusive mode of colony founding in the Mediterranean formicine *Cataglyphis cursor* (Lenoir et al., 1987a).

*Improvement of predation.* It has long been assumed that the dramatic emigrations of army ants during the nomadic phase improve predatory efficiency. That is, they serve ultimately to find the new crops of arthropod prey needed to fuel the huge colonies (Wilson, 1958e; Schneirla, 1971). This is the case even though, as Schneirla discovered in his pioneering study of *Eciton*, the proximate triggering signal to the workers is the emergence of callow workers and larvae.

Two recent lines of evidence support the hypothesis of programmed emigration directed toward improved harvests. Franks and Fletcher (1983) have documented that the effects of the raids of *E. burchelli* on arthropod faunas are so devastating that raids in new directions produce higher yields. In the laboratory, Topoff and Mirenda (1980a,b) found that underfed colonies of *Neivamyrmex nigres-*

*cens* emigrated twice as frequently as overfed colonies. This cause-and-effect relation can be extended only with caution to other kinds of ants, however. Smallwood (1982), for example, observed higher emigration rates in underfed colonies of the woodland myrmicine *Aphaenogaster rudis* one year, but lower rates the following year.

*Social integration and resource distribution.* When colonies disperse to multiple sites, the exchange of colony members typically ensues, with a few specialists carrying fellow workers and immature forms back and forth. As Økland (1934) showed in the course of his early studies on the wood ant *Formica rufa,* this phenomenon can be an important means of colony integration. In the closely related *F. polyctena,* adult transport among multiple nest sites is seasonal, reaching a maximum in spring and autumn. In one German colony of approximately a million workers studied by Kneitz (1964a), between 200,000 and 300,000 transportations occurred during the course of a year. Most of the transporting workers were older foragers, and most of the workers being carried were younger individuals of the kind that engage principally in nursing and ingluvial food storage. The adaptive value of such strenuous activity remains to be demonstrated in any direct or otherwise convincing manner. It is conceivable that the transporters are engaged in allocating labor resources accordingto overall colony needs which, by virtue of these insects' more extensive wanderings, they are the most qualified to sense.

## ALTERNATIVE STRATEGIES IN COLONY LIFE CYCLES

A well-established principle of evolutionary biology is that when different species are faced with similar challenges in the environment, they often develop very different mechanisms, even when they are closely related genetically at the outset. The reasons can only be guessed in most cases, but it appears that even subtle differences in preadaptations (traits already existing that are incorporated in the new evolution) and environmental pressures can cause a divergence of the evolving lines. The ants amply support this generalization. Some of the examples are so extreme as to verge on the bizarre. Within the large genus *Pheidole* there exists a unique species, *P. embolopyx* of the Brazilian Amazon, in which the queen's abdomen is posteriorly truncated, the pronotum dilated laterally into flanges, and the scapes, anterior clypeal border, and frontal carinae covered from time to time with gelatinous sheaths secreted by hypertrophied glands in the head (Brown, 1967; Wilson and Hölldobler, 1985). The combination of traits appears to protect the queen from enemies by letting her pull into a tight, turtle-like defensive posture. The genus *Chimaeridris* from Indonesia, evidently derived from *Pheidole,* is distinguished by the possession in the minor worker of saber-shaped mandibles, which differ radically from the broad, multiply toothed mandibles found in all of the hundreds of other species. Why these two particular clades departed in such extreme, unexpected directions within *Pheidole* is still unknown.

The principle of alternative strategic solutions applies in the case of colony life cycles, in which striking differences often occur among ant species occupying the same habitats. An especially well documented example is provided by two European species, the mound-building wood ant *Formica polyctena* and the carpenter ant *Camponotus herculeanus.* The distinction between them is essentially as follows. Because of metabolic heat production combined with physical properties of the *Formica* mound (see Plate 4) that enhance the collection of heat from sunlight, the inhabitants are able to raise the internal nest temperatures rapidly during the early spring (Kneitz, 1964b). This head start permits them to rear queens and males from egg to adult in only five to six weeks and to avoid keeping immature individuals during the winter (Gösswald, 1951a; Schmidt, 1964, 1974a,b; Cherix, 1986). In contrast, a *C. herculeanus* colony has no special adaptation for thermoregulation, and its springtime development is much slower. Yet it attains a similar schedule of production by keeping larvae and virgin sexual forms in the nest through the winter (Hölldobler, 1962, 1964, 1966). This difference in strategy between the two species has profound repercussions in other aspects of their sociobiology.

It is appropriate to begin the *Formica polyctena* annual cycle in August, when most of the larvae of the summer crop have grown to maturity (Figure 3–23). No further worker generation is in sight, because the queens stopped laying eggs in July. However, foraging is still heavy and will remain so through the last warm days of October and November. Much of the food gathered is fed to the youngest workers, whose abdomens begin to swell with fat bodies. These individuals become temporary "repletes" (as they are called in the myrmecological literature), a caste specialized for food storage. With the onset of cooler weather during the fall, the repletes and the queens move into the deep chambers below the level of the ground surface. They cluster together tightly, with as many as 20 queens collected inside a single cell. The older workers, who have functioned as *Aussendiensttiere,* or outside foragers, and nest-builders during the summer, also move to the lower chambers, but they can still venture upward and even forage outside the nest if the temperature rises sufficiently.

During mild winters the *Formica polyctena* colony can be periodically activated in this manner as early as late January, but usually significant awakening does not begin until March. On the first warm, sunny days large numbers of workers and queens migrate out onto the mound surface, where they stand almost motionless in the sunlight. This exposure evidently activates their repletes, whose metabolism adds to the heat of the inner nest chambers. Thus the mound temperature is effectively regulated, so that a *Wärmezentrum* (warm core) near the ground surface stays at approximately 27°C even when the temperature on the nest surface falls to freezing, during the early weeks of spring.

The young workers gather in the *Wärmezentrum,* where their nutrient-producing glands and especially the large postpharyngeal gland grow more active, converting lipids and proteins derived from the fat bodies into a complex mixture of lipoproteins and holding them in readiness to be fed to queens and larvae (Figure 3–24). The queens remain in the *Wärmezentrum* for about eight days, during which each individual lays approximately 500 "winter eggs" (Figure 3–25). They then retire to the lower nest chambers where, after another three to four weeks, they begin to lay "summer eggs." This second reproductive activity lasts until the middle of July.

As Bier (1954a) first showed, the winter eggs are richer in RNA and capable of developing into reproductive forms. However, this development can be realized only if a sufficient amount of high-grade food is received by the larvae during the first 72 hours after they hatch. Female larvae not exposed to the specialized nurse workers during this critical period complete their development as

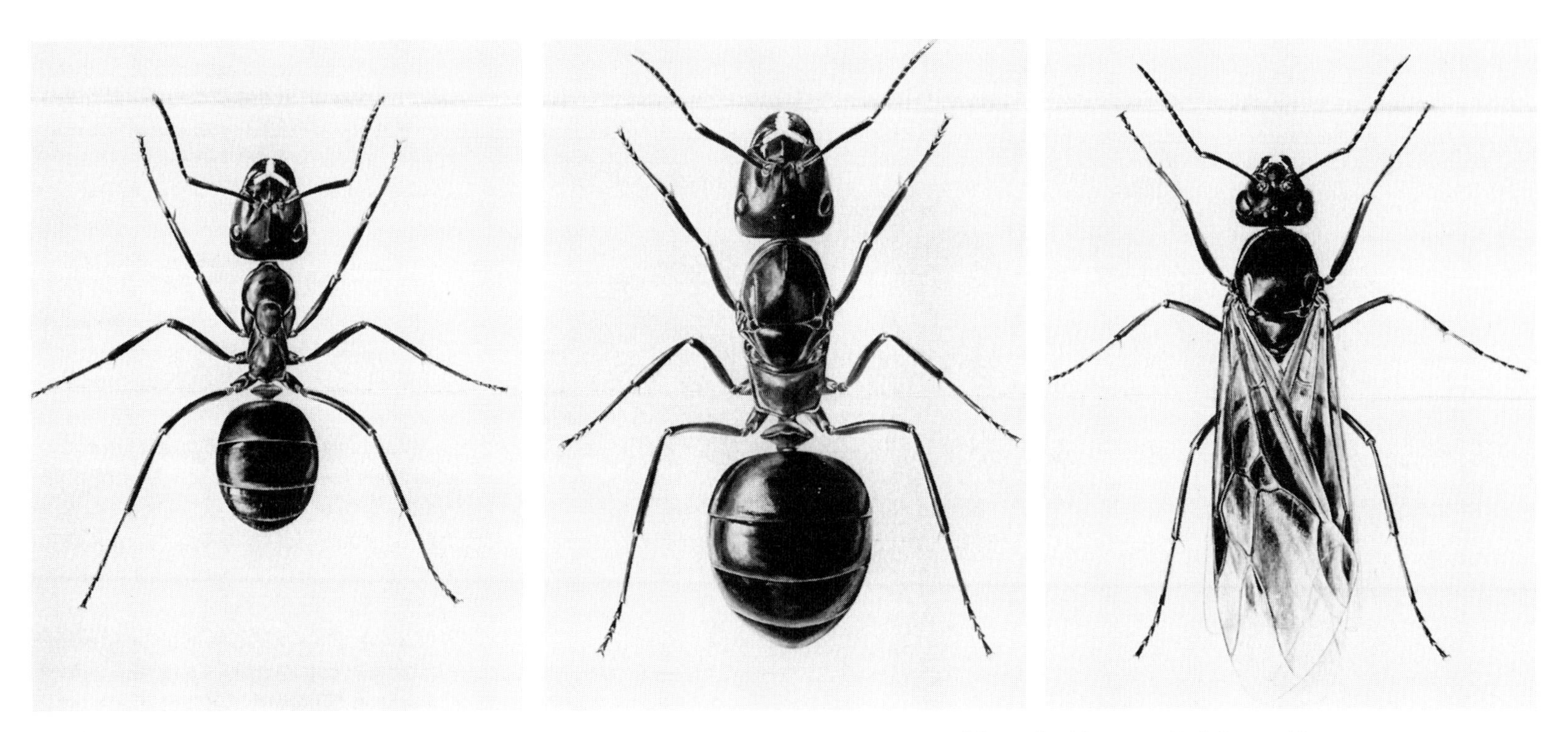

**FIGURE 3–23** The *Formica polyctena* castes: (*left*) worker; (*center*) queen; (*right*) male. (Courtesy Karl Gösswald; drawing by T. Hölldobler-Forsyth.)

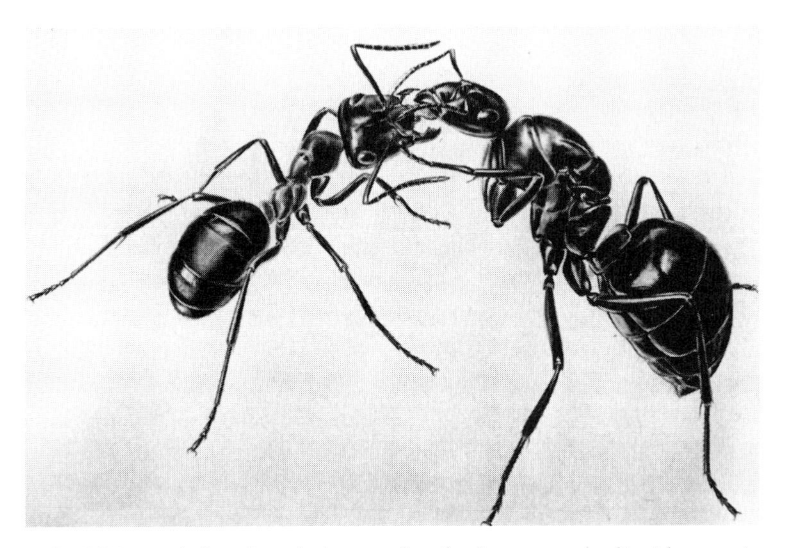

**FIGURE 3–24** A *Formica polyctena* worker feeds a queen by liquid regurgitation. (Original painting by T. Hölldobler-Forsyth.)

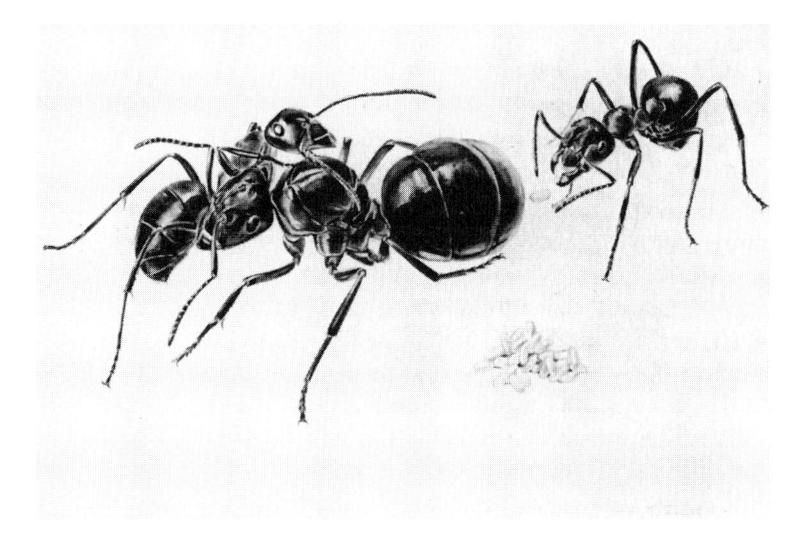

**FIGURE 3–25** A *Formica polyctena* queen lays an egg, which will be immediately carried by an attending worker to a nearby egg pile. (Courtesy K. Gösswald; original drawing by T. Hölldobler-Forsyth.)

workers, while male larvae so deprived are killed and eaten. Hence caste is determined by the combination of a "blastogenic" factor (properties of the egg cytoplasm, including at least RNA density) and a "trophogenic" factor (quality of food received during the early part of the first larval instar) (Schmidt, 1974a; Schmidt and Winkler, 1984).

The key feature of the *Formica polyctena* annual cycle is the relative environmental independence of the reproductive crop, which is reared from egg to adult at a time of year when little or no food is being gathered from outside the nest. The colony depends instead on food reserves harvested during the previous summer and fall and stored in the late summer crop of young workers, and on the early rousing of activity the following spring made possible by the mound nests.

At the conclusion of the development of the reproductive forms (virgin queens and males) from egg to eclosion of adult (Figure 3–26), the nurse workers have used up almost all of their body food reserves. The spring weather has now turned consistently warm, and new batches of worker-destined summer eggs have been laid and are hatching. The mode of production has changed: the new

**FIGURE 3–26** *Formica polyctena* nurse workers aid young alate females as they eclose from their pupal cocoons. (From Gösswald, 1964; painting by T. Hölldobler-Forsyth.)

**FIGURE 3–27** Copulation in *Formica polyctena*. (Courtesy K. Gösswald; painting by T. Hölldobler-Forsyth.)

crop of larvae are being nourished with food collected by foragers outside the nest.

Within a few days of eclosion from the pupa, the males are ready to mate. In this brief interval they have undergone a profound physiological transformation. Spermatogenesis is completed, the sperm are transferred from the follicles into the sperm vesicles, the small fat bodies are mostly used up, and the corpora allata, oenocytes, and midgut have begun to degenerate. In essence, the males have been quickly converted into single-purpose sexual missiles. The queens are also primed to mate within several days, but they retain a large fat body and other forms of physiological robustness that foreshadow a natural longevity of many years.

The reproduction itself is a simple affair. In the case where both sexes have been reared in the same nest, some of the males and females copulate on the mound surface without benefit of a nuptial flight (Figure 3–27). Others fly away to find partners outside the nest vicinity. Soon afterward the males die, while the newly inseminated queens shed their wings and attempt to commence a new colony life cycle (Figure 3–28). Some of them are adopted into the home nest or other *Formica polyctena* nests in the area. In later stages they may serve as pioneers when the colony subdivides through budding and emigrates to new auxiliary nests. During the emigrations the queens are usually carried by the workers (Figure 3–29).

Returning to August, we can now trace the very different annual

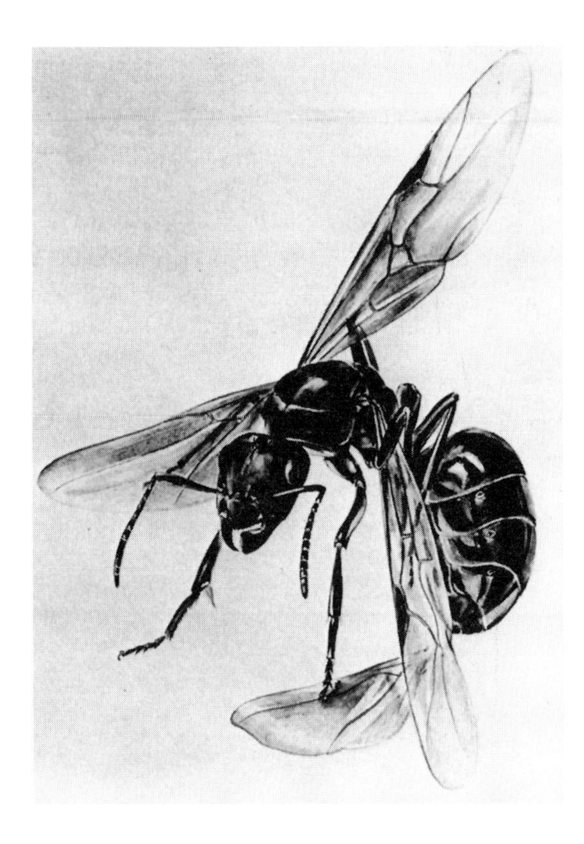

**FIGURE 3–28** After mating, the young queen of *Formica polyctena* sheds her wings. (From Gösswald, 1964; painting by T. Hölldobler-Forsyth.)

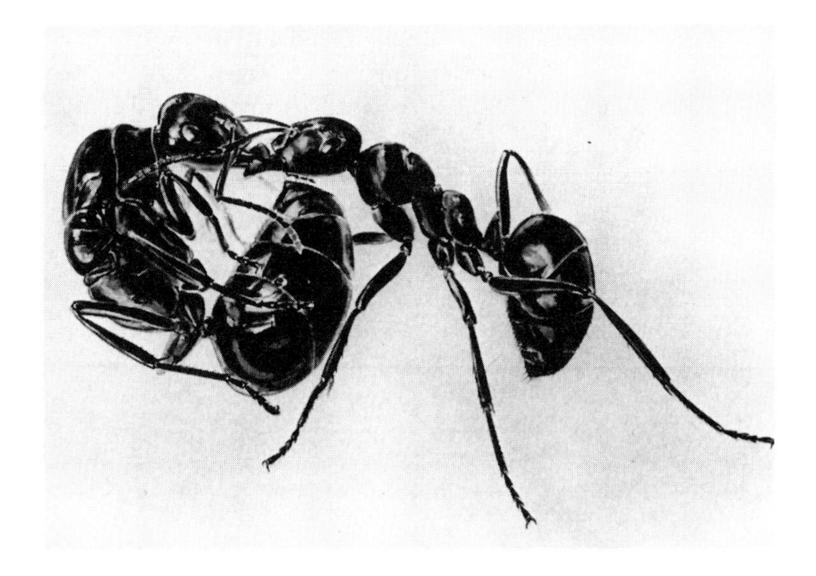

**FIGURE 3–29** During colony division through budding *Formica polyctena* workers carry their queen to a new nest site. (Courtesy K. Gösswald; painting by T. Hölldobler-Forsyth.)

cycle of *Camponotus herculeanus,* a species known in Germany as the *Rossameise* or "horse ant." Unlike *Formica polyctena, C. herculeanus* produces both workers and reproductive forms in late summer (it may be noted in passing that a large majority of colonies produce both males and queens, in contrast to *F. polyctena,* where all-queen and all-male colonies are common). This "eclosion guild" remains together until the nuptial flight the following year. The young workers stay close to the queens and males, feeding them food that they themselves have received from returning foragers. All of the members of the guild accumulate fat reserves as the summer draws to a close. It is useful to refer to the males to mark the annual cycle because, in this social phase, the males both receive and donate liquid food to their nestmates, a rare instance within the ants generally of males (i.e., drones) displaying altruistic, worker-like behavior. At the same time the mother queen lays a clutch of late-summer eggs. They hatch and develop as far as the second instar before the entire colony shuts down for the winter. Unlike the *F. polyctena* response, this inactivity is based on a true physiological diapause, because it continues even if the colony is transferred to a laboratory interior kept at 22–25°C.

The males, to trace the cycle again from the point of view of this sex, are now in the hibernation phase. Sufficient chilling is attained by late January or early February to break the diapause. If the nest temperature is maintained below 18°C the colony will remain relatively inactive, and the virgin queens and males then stay within the nest for an entire additional year. Thus the *Camponotus herculeanus* males are the longest-lived members of this sex known in all the ants. But if the temperature is sustained above 22°C, a common event in the field in late March and early April, the hibernation phase ends. The males become "socially emancipated" in the sense that they now participate much less in grooming and food exchange, start to use up their fat bodies, and transfer the sperm from the testes to the seminal vesicles. With their bodies light in weight and their sperm poised for ejaculation, they are primed for the nuptial flight. The workers continue to fatten the virgin queens as well as the overwintered larvae, which are destined to mature during the late spring and summer.

By May the stage is at last set for the *Camponotus herculeanus* nuptial flights. In the afternoon the queens and males start to move to the nest entrance to sun themselves. The excursions become more and more extended, until finally, aided by a synchronization pheromone from the mandibular glands of the males, a mass flight occurs (see Figure 3–10). The newly inseminated queens then start new colonies by sealing themselves into old beetle burrows or other preformed cavities in wood. They live on their body reserves until the first brood of workers is reared (see Figure 3–1).

To conclude this exposition of alternative life histories, we will turn to two remarkable species in which the ordinary cycle is short-circuited by the abrogation of the queen caste. In the large African ponerine *Ophthalmopone berthoudi,* a specialized predator on termites, queens have been completely replaced by inseminated, laying workers, or "gamergates" (Peeters and Crewe, 1985). These reproductives appear not to differ anatomically in any way from other workers engaged in foraging, nursing, and other quotidian tasks. Their ovaries are small and contain only a few oocytes. They do not receive more food during termite meals or enjoy any other apparent preferred treatment. Up to a hundred workers are found in each colony, but not all of them have a chance to be inseminated and hence serve as gamergates. Only those males who are young and hence sexually attractive during their brief life span, mostly in the late South African summer (February–March), play this role. Each year a new crop of gamergates is created, gradually replacing the dwindling supply that survived from earlier mating seasons. New colonies are formed by budding.

An even more radical reduction has occurred in the myrmicine *Pristomyrmex pungens,* one of the most abundant ants of the Japanese countryside (Itow et al., 1984). The normal queen caste has been completely eliminated. Males are rare (2–3 percent of the entire adult crop during June and July) and nonfunctional. An ergatogyne caste, distinguished by an enlarged abdomen and possession of ocelli, is extremely unusual and plays at most an insignificant role. Reproduction is almost exclusively parthenogenetic by unfertilized workers, who do not differ in any apparent manner from other workers in the colony. In short, the *P. pungens* colony is asexual. It has come to resemble a vegetatively reproducing plant in its organization. The concept of "queen" cannot be applied in this case, and it is difficult even to classify the species as truly eusocial.

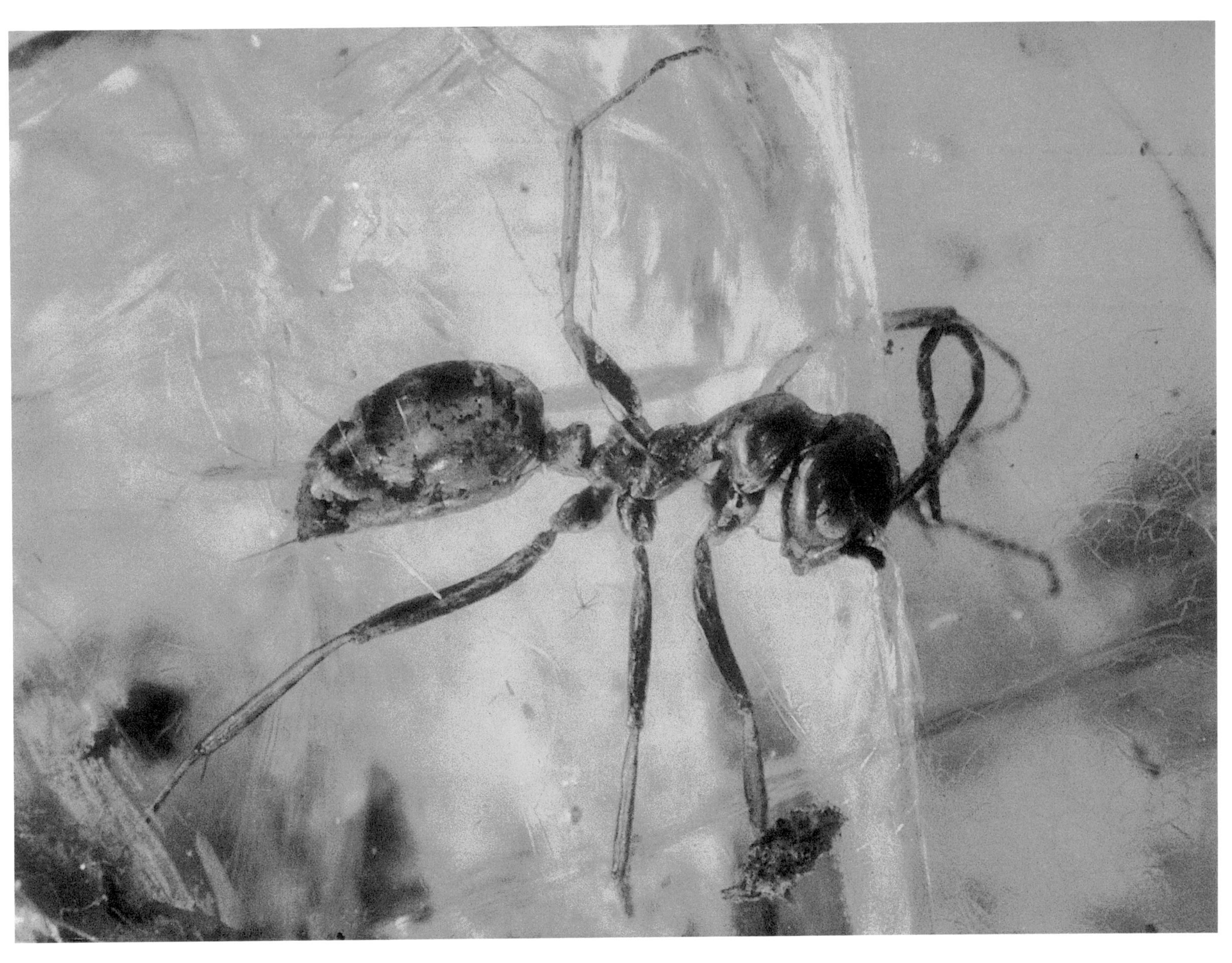

**PLATE 1.** A worker of the subfamily Sphecomyrminae, the oldest and most primitive known group of ants. This worker, the type specimen of *Sphecomyrma freyi* and the Sphecomyrminae, is in sequoia amber from New Jersey. Its age is the lower part of the Upper Cretaceous, or approximately 80 million years. (Photograph by F.M. Carpenter.)

**PLATE 2.** A mating aggregation of the harvester ant *Pogonomyrmex rugosus*. Each year winged males fly out from many different nests and gather by the thousands in traditional assembly areas. Once on the ground they discharge secretions from their mandibular glands that appear to attract virgin queens and other males. (From Hölldobler, 1984d; painting by J.D. Dawson reprinted with permission of the National Geographic Society.)

**PLATE 3.** A *Nothomyrmecia macrops* worker feeds a group of half-grown larvae with a recently captured fly.

**PLATE 4.** A mound of the wood ant *Formica polyctena* in a German forest. In the foreground workers kill a sawfly larva *(Diprion)*. This is only one of some 100,000 such prey items a single colony, which contains a million or more workers and hundreds of queens, can consume in a single day. (From Hölldobler, 1984d; painting by J.D. Dawson reprinted with permission of the National Geographic Society.)

**PLATE 5.** The queen of the Neotropical ant *Daceton armigerum* is surrounded by workers who protect, groom, and feed her. There is always a large number of callow workers (the light-colored individuals in the photograph) in the vicinity of the queen.

**PLATE 6.** The African weaver ant, *Oecophylla longinoda,* establishes large territories in tree canopies. The maintenance and defense of the territories are organized by a complex communication system. Confronting a stranger *(left foreground),* a worker displays hostility with gaping mandibles and the gaster cocked over the forward part of the body. Another pair in the background are clinched in combat. Rushing toward the leaf nest *(upper right),* another ant lays an odor trail with secretions from the rectal gland at the abdominal tip. The chemical substances in this trail will lead reinforcements to the fray. When capturing a prey object, such as a giant black African stink ant *(Paltothyreus tarsatus),* ants organize cooperation by means of chemical short-range recruitment signals from the sternal gland and alarm pheromones from the mandibular gland. (From Hölldobler, 1984d; painting by J.D. Dawson reprinted with permission of the National Geographic Society.)

**PLATE 7.** Food exchange among workers of the Australian sugar ant *Camponotus ephippium.* On the right two minor workers exchange liquid by regurgitation. On the left a minor feeds a major. (From Haskins, 1984; painting by J.D. Dawson reprinted with permission of the National Geographic Society.)

**PLATE 8.** A view inside the nest of *Myrmecocystus mimicus,* a honeypot ant found in the desert of the southwestern United States. In more favorable times the foraging workers gather termites, nectar from desert plants, and honeydew from homopterous insects, and store them in the crops of the repletes, nestmates specialized to serve as food storage receptacles. In the foreground a forager regurgitates liquid from her crop into an expanding replete. Other repletes, about the size of peas, hang from the ceiling of the nest chamber. The queen can be seen beyond the pile of cocoons and larvae. (From Hölldobler, 1984d; painting by J.D. Dawson reprinted with permission of the National Geographic Society.)

# CHAPTER 4

# Altruism and the Origin of the Worker Caste

By almost any conceivable standard, the single most important feature of insect social behavior is the existence of the nonreproductive worker caste. The altruistic actions of this caste integrate the colony tightly and make possible advanced forms of labor specialization. The baseline for the role of the worker is provided by the queen, who in most species still behaves in a primitive, totipotent manner resembling that of a solitary aculeate wasp. She alone traverses the whole life cycle of the species. Acting like a solitary insect, she leaves the mother colony, mates, and builds a nest. During this time her anatomy and physiology are essentially those of a solitary wasp, and her behavioral patterns are equally complicated. Only when the first brood of workers arrives does she become specialized, narrowing her repertory to an almost exclusively egg-laying role. In contrast, workers are specialized throughout their lives, with a large part of their repertory devoted, start to finish, to the welfare of the queen and their siblings.

## ALTRUISM

Is it correct to call the nonreproductive workers "altruistic"? Some authors have begun to drop this admittedly value-laden word. They point out that the prescribing genes are "selfish" rather than altruistic, because if our conception of evolution by natural selection is true, it must follow by definition that genes persisting at the expense of others are selfish—even if they prescribe altruism. Alternative expressions have been suggested for outwardly selfless behavior, such as social donorism (Williams and Williams, 1957), nepotism (Alexander, 1974), and reciprocation (Trivers, 1971). But if altruism is defined in the original lexical manner as self-denying behavior performed for the benefit of others, its application to ants and other social insects is justified, at least at the levels of the organism and the colony. The key question, as we shall see, is how natural selection can produce selfish genes that prescribe unselfishness.

Altruistic behavior has been documented in many ways. To start, the great majority of ant workers make no effort to reproduce at all. Although the ovaries of young individuals are often active, the eggs they produce are more often than not trophic, used to feed the larvae and queen, and unable to develop even if left unharmed. Older workers typically leave the nest to search for food outside it, where they find life very dangerous. Such foragers in the Idaho harvester *Pogonomyrmex owyheei*, which make up less than 10 percent of the worker population at any given time, undergo a weight loss of 40 percent and increased mandible wear. They are subject to intense predation and live an average of only 14 days after starting their forays (Porter and Jorgensen, 1981). In a study conducted by De Vita (1979), individual colonies of the California harvester *Pogonomyrmex californicus* were observed to suffer an average of 0.06 death per worker foraging hour due to fighting with neighboring colonies. The level of sacrifice while foraging approaches suicide in the formicine *Cataglyphis bicolor*, a scavenger of dead arthropods in the North African desert. At any given time about 15 percent of the workers are engaged in long, dangerous searches away from the colony, where they are downed mainly by spiders and robber flies. They have a life expectancy of only 6 days, but during that time each one retrieves food weighing 15 to 20 times her own body weight (Schmid-Hempel in Wehner et al., 1983; Schmid-Hempel and Schmid-Hempel, 1984). Porter and Jorgensen have referred to foragers in such extreme cases as constituting a "disposable" caste, since they exchange their lives for a high productivity on behalf of the colony.

The trade-off between individual sacrifice and colony welfare is even more evident in the case of defense. Aging workers of the green tree ant of Australia (*Oecophylla smaragdina*), who are distinguished by reduced fat bodies and ovaries, emigrate to special "barrack nests" located at the territorial boundary of the colony. When *Oecophylla* workers from neighboring nests or other invaders cross the line, these guards are the first to rush to the attack (Hölldobler, 1983). (It can be said that a principal difference between human beings and ants is that whereas we send our young men to war, they send their old ladies.)

Altruistic behavior is sometimes accompanied by anatomical specialization. Workers of some *Pogonomyrmex* species possess reverse barbs on the sting that cause the venom apparatus and other portions of the viscera to come free when the ants move away from the sting site. This device appears to be used to defend the colony against vertebrates (Hermann, 1971). Sting autotomy or autothysis, as it is sometimes called, also occurs in honey bees and in some genera of the social polistine and polybiine wasps, and thus constitutes a remarkable example of convergence in social behavior (Hermann and Blum, 1981). An even more bizarre suicidal defense is mounted by workers of a tropical Asian species belonging to the *Camponotus saundersi* group. The worker mandibular gland is hypertrophied, occupying not only a large part of the head capsule but extending all the way back to the tip of the abdomen. When the ants are sufficiently provoked, they contract their abdomens violently until their body wall bursts along the intersegmental membranes, discharging large quantities of the sticky mandibular gland secretion, which traps the attackers (Maschwitz and Maschwitz, 1974; see Figures 7–25 and 7–26).

West-Eberhard (1979, 1981) has tempered the interpretation of worker altruism as the dominant mode of ant life by pointing out that competition among nestmates occurs more commonly than was recognized in the earlier literature. This is particularly the case in those species in which workers are able to lay eggs and thus to compete reproductively. In other words, workers are less than perfectly altruistic. Furthermore, natural selection at the level of the "selfish" individual might have played a role even in the evolution of division of labor, a social arrangement considered the exemplar of cooperation and harmony. West-Eberhard argues the case especially for the centrifugal pattern of temporal castes, in which older workers move away from the queen and brood and devote themselves more to outside work. The "selfish" worker, by staying close to the brood chambers while still young and while her personal reproductive value is highest, maximizes her potential to contribute personal offspring. As she ages and her fertility declines because of programmed senescence, her optimum strategy for contributing genes to the next generation is to enhance colony welfare through more dangerous occupations such as foraging and defense. This is an appealing hypothesis, and it should be kept in mind, especially when considering ant species with more primitively organized, smaller colonies. It loses most or all of its force, however, in phylogenetically advanced species with extreme worker specialization for foraging and defense. In some species workers completely lack ovaries and hence are constitutionally exempt from individual selection. Examples include the members of *Monomorium, Pheidole, Pheidologeton,* and *Solenopsis.*

## KIN SELECTION

To explain the more clear-cut cases of worker altruism, we must turn to kin selection, which can be defined as the alteration of the frequencies of genes shared by relatives through actions that favor or disfavor their relatives' survival and reproduction. In other words, it is selection mediated by interactions among kin. Kin selection is inferentially a powerful force in evolution. To take an extreme imaginary case, if an allele (that is, an alternate form of a particular gene) appears in a population that causes its bearer to act so as to triple the reproduction of one or more of the bearer's brothers and sisters, the allele will spread rapidly through the population. This will occur even if the allele-bearer sacrifices herself in the process, because many of her siblings also carry the altruistic gene. Thus undeniable altruistic behavior can become the norm in the population.

Although the roots of the theory of kin selection go back to Darwin, the concept was developed most originally and most forcefully by W. D. Hamilton in the 1960s (his now-famous seminal article was published in 1964) and then taken to new levels of sophistication over the past twenty years by Bartz, Charlesworth, Charnov, Craig, Crozier, Feldman, Oster, Trivers, and others. An especially important extension was made by Trivers and Hare (1976), who predicted a 3:1 ratio of energy investment in queens as opposed to males during the production of reproductives, at least in cases where workers control the sex allocation, a rare instance of a quantitative prediction in evolutionary biology. In short, the theory's progress has been such that kin selection in ants can now be considered fundamental to general sociobiology.

Kin selection is actually one of three hypotheses that have been constructed at various times to explain the origin and evolution of eusociality:

*Kin selection.* By reducing personal survival and reproduction, workers nevertheless increase the survival and reproduction of genes they share with other members of the colony by common descent. Individuals suffer, but the colony flourishes and so do the genes (including the altruistic genes).

*Mutualism.* In some fashion, individuals do better in personal survival and reproduction when they live in groups than when they live alone, even though they defer to other colony members and sacrifice on their behalf to some extent.

*Parental manipulation.* One or both parents (actually, the mother in ants and other social hymenopterans) are able to neuter and control some of their offspring so as to produce a larger total number of offspring. The parents' personal fitness is raised even though that of some of the offspring is lowered.

The case for kin selection can be most clearly made by comparing it to the hypothesis of social mutualism, first developed by Michener (1958) and subsequently elaborated by Lin and Michener (1972). The idea of mutualism is simple to the point of seductiveness: cooperation will evolve if a group of individuals can reproduce better than a single individual under the same circumstances. Exactly this association has been documented in bees and wasps, as follows: the larger the colony, the more nest cells, eggs, larvae, and pupae it produces. The relationship is not linear, however. As first noted by Michener (1964, 1969), the production of nest cells and immature forms per individual member falls off with an increase in group size. This reproductivity effect, as it is known, appears to occur generally in the social insects (Wilson, 1971).

Why has social existence evolved at all if belonging to a group diminishes personal reproduction? The answer is clearly the enhancement of group survival that promotes individual survival. If an insect has a longer life as a member of a group than as a solitaire, the advantage can more than compensate for its decrease in reproductive potency. And again, the evidence demonstrates that such enhancement exists. In the paper wasp *Polistes canadensis* (Pickering, 1980) and the honeypot ant *Myrmecocystus mimicus* (Bartz and Hölldobler, 1982), average lifetime reproduction per individual, taken as the summed products of survival probability and reproduction in each interval of time, increases with group size during colony founding.

There are at least two reasons why increased survival should compensate for lowered reproduction among insects. First, groups can fend off competitors and other enemies more effectively, an advantage that has been well documented in the literature (Wilson, 1975b, and Chapter 6 of this book). Second, as we noted earlier, it is far safer to stay at home than to forage. A large percentage of colony members are able to remain in the nest, whereas all solitary hymenopterans must forage.

Thus the mutualism hypothesis by itself does appear adequate to account for the origin of colonial existence. However, one of the three diagnostic traits of eusociality, the existence of a sterile worker caste, cannot be explained in such a manner. This is why kin selection has risen to importance in evolutionary theory. The concept was actually originated, in a very general form, by Charles Darwin in *The Origin of Species*. Darwin, whose interest in social insects was strong in his later life (he consulted the early myrmecologist Frederick Smith at the British Museum about slave-making ants), had found in them the "one special difficulty, which at first appeared to me insuperable, and actually fatal to my whole theory." How, he asked, could the worker castes of insect societies have evolved if they were sterile and left no offspring? This paradox proved truly fatal to Lamarck's theory of evolution by the inheritance of acquired characters, for Darwin was quick to point out that the Lamarckian hypothesis required characters to be developed by use or disuse of the organs of individual organisms and then to be passed directly to the next generation, an impossibility when the organisms were sterile.

To save his own theory, Darwin introduced the idea of natural selection operating at the level of the family rather than that of the single organism. In retrospect his logic seems impeccable. If some of the individuals of the family are sterile and yet important to the welfare of fertile relatives, as in the case of insect colonies, selection at the family level is inevitable. With the entire family serving as the unit of selection, it is the capacity to generate sterile but altruistic relatives that becomes subject to genetic evolution. To quote Darwin, "Thus, a well-flavoured vegetable is cooked, and the individual is destroyed; but the horticulturist sows seeds of the same stock, and confidently expects to get nearly the same variety; breeders of cattle wish the flesh and fat to be well marbled together; the animal has been slaughtered, but the breeder goes with confidence to the same family" (Darwin, 1859: 237). Employing his familiar style of argumentation, Darwin noted that intermediate stages found in some living species of social insects connect at least some of the extreme sterile castes, making it possible to trace the route along which they evolved. Speaking of the soldiers and minor workers of ants, he wrote, "With these facts before me, I believe that natural selection, by acting on the fertile parents could form a species which regularly produce neuters, either all of a large size with one form of jaw, or all of small size with jaws having a widely different structure; or lastly, and this is the climax of our difficulty, one set of workers of one size and structure, and simultaneously another set of workers of a different size and structure" (Darwin, 1859: 24).

Although J. B. S. Haldane, the noted British population geneticist, pointed out the implications of Darwin's insight, the modern genetic theory of kin selection and sterile castes was inaugurated by Hamilton. He recognized that there are two ways for alleles (alternative forms of a gene found on the same locus) to be passed to future generations. The first is by personal reproduction, in other words the production of sons and daughters. The measure of personal reproductive success, which preoccupied the earlier theoreticians of population genetics, has come to be known as *classical fitness*. The second mode of gene descent is collateral, promoting the welfare of brothers, sisters, and other relatives besides offspring who possess the same alleles by reason of common descent. Hamilton recognized the importance of a measure he called *inclusive fitness*, which incorporates both the individual's personal reproduction (classical fitness) and its influence on the reproduction of collateral relatives. To avoid confusion, we need to use the expression "kin selection" to refer to circumstances involving the reproduction of collateral relatives. Yet the complete effects of kin selection on evolution cannot be evaluated without including the effects on personal reproduction. This notion and the terminology expressing it have been put in the most nearly standard form by Pamilo and Crozier (1982) and Pamilo (1984a,b).

The ordinary (nonsocial) measure of classical fitness is

$$W = \frac{E(RS)}{\text{Average } RS \text{ for Population}}$$

where $E(RS)$ is the average direct reproductive success of individuals possessing the genotype of interest. It measures the number of offspring the individual injects into the population, in comparison with the contribution from the remainder of the population. This is the most common measure of fitness encountered in the literature of population genetics. Inclusive fitness, on the other hand, incorporates two components:

$$IF = \frac{E(RS) + \Sigma[b_j E(RS)]}{\text{Average } IF \text{ for Population}}$$

where the second term, $\Sigma[b_j E(RS)]$, is the effect on the reproduction of all of the collateral relatives. The quantity $b_j$ is the *coefficient of relatedness*, the probability that the relative $j$ of the focal individual also possesses the allele of interest. In ordinary diploid systems, for example, $b_j$ is ½ for brothers and sisters; ¼ for uncles, aunts, grandparents, and grandchildren; and ⅛ for first cousins. Outside this tight circle of close relatives, $b_j$ continues to fall swiftly, so that kin selection becomes a proportionately negligible force.

After Hamilton's original formulation, a great deal of confusion arose over the best definition and the usefulness of measures of relatedness, but this appears now to be largely resolved. Michod and Hamilton (1980) demonstrated that the five principal different coefficients of relatedness invented to account variously for different degrees of penetrance, gene frequency, and inbreeding, are equivalent. Seger (1981) generalized the result to cases in which the average gene frequencies of the altruists differ from those of the recipients. The best intuitive way to think about the degree of relatedness and to approximate the coefficient of relatedness is to ask the following question: if the focal individual has an allele $a$, what is the chance that a relative also possesses it?

The pivotal idea can now be put as follows. If the allele $a$ affects altruism in some manner, then self-sacrificing propensities have the potential to evolve. The allele is always "selfish"; it spreads through the population by promoting itself via the increased success of collateral relatives. The necessary minimal condition for this to happen is stated by "Hamilton's rule": $C/B < b$. This says that the cost $C$ (which is the loss in expected personal reproductive success through the self-sacrificing behavior) divided by the benefit $B$ (the increase in the relatives' expected reproductive success) must be less than $b$, the probability that the relatives have the same allele.

Another way of expressing Hamilton's rule is to say that the benefit to relatives is discounted by their degree of relationship, so that the less the relatedness, the greater the benefit must be to counterbalance the cost. Consider, for example, a highly simplified network consisting solely of an individual ant and her sister. If the focal ant is altruistic she will perform some sacrifice for the benefit of her sister. She may share food, labor more in nest construction, or place herself between the sister and some enemy. The important consequence for the focal ant, from an evolutionary point of view, is her loss of personal genetic fitness, due to a reduced life span, fewer offspring, or both, leading in turn to less representation of her personal genes in the next generation. But three-fourths of the sister's genes are identical to those of the altruist by virtue of common descent. Suppose, to take the extreme case, that the altruist leaves no offspring. If her altruistic act enlarges the sister's personal representation in the next generation to a sufficient degree (in this case, by more than 50 percent), it will increase the three-quarters of the genes identical to those in the altruist, and the altruist will actually gain representation in the next generation. Some of the genes shared by such sisters will be the very ones that encode the tendency toward altruistic behavior. The inclusive fitness, in this case determined solely by the sister's contribution, will be sufficient to cause the spread of the altruistic genes through the population.

This result can be changed somewhat by certain restrictive conditions, but in general it is surprisingly robust. One way to bend it a little is to regard the costs and benefits as multiplicative or "interactive" instead of additive (Uyenoyama and Feldman, 1981). In the original additive model,

$$IF = 1 - C + bB$$

This quantity must be greater than one if the allele of interest is to spread, or

$$1 - C + bB > 1$$

which is rearranged to produce Hamilton's rule: $C/B < b$.

In the interactive model proposed by Uyenoyama and Feldman the components are multiplied:

$$IF = (1 - C)(1 + bB)$$

In this case the fitness is greater than one and the altruism allele will spread if

$$\frac{C}{B(1-C)} < b.$$

The interactive model is more restrictive than the additive. In other words, some conditions exist under which the altruism allele would spread under the additive relation but not under the multiplicative model. The difference is not great, however, and it is minimal when the costs and benefits are both low. Furthermore, the additive model appears intuitively preferable to the interactive model. The cost to the altruist and the benefit to the relative are distinct events, unlike many interactive physiological processes within the same organism, and it is difficult to imagine how they could operate on inclusive fitness in a multiplicative way (Bartz, 1983). Again, Hamilton's rule is robust as a theoretical proposition.

## PARENTAL MANIPULATION

Although kin selection successfully reaches beyond social mutualism to explain the origin of sterile workers, it is not the only conceivable explanation. Several writers have noted that a worker caste can arise if the mother is forceful enough to dominate and manipulate her own offspring. In other words, if the mother can rear enough additional daughters ($b = \frac{1}{2}$) by enslaving some of them, it is to her advantage to sterilize them rather than to let them depart to create a larger crop of granddaughters ($b = \frac{1}{4}$). Such an arrangement, with sisters or other cogenerational females enslaving each other, was proposed to explain the origin of eusocial bees by Michener (1974) and Michener and Brothers (1974). These authors noticed that queens of the primitively eusocial bee *Lasioglossum zephyrum* control other adult females by two simple behaviors. Other females are systematically nudged, an act that appears to be aggressive and may have the effect of inhibiting ovarian development. The individuals most frequently nudged are the ones with the largest ovaries, who are best able to compete with the queen. Nudging is followed by backing, in which the nudger retreats down the nest galleries, apparently attempting to draw the other bee after her. The effect is to maneuver the follower closer to the brood cells, where she can assist in the construction and provisioning of the cells used by the queen. It is not difficult to imagine, with Michener and Brothers, that sterile castes can evolve if certain allelic combinations arise that are very powerful in controlling nestmates. Alexander (1974) independently suggested that the exploitation of offspring by their parents has been a general force in the social evolution of insects. Any conflict between parents and their children, he argued, is likely to be resolved in favor of the parents, who are bigger, stronger, and more forceful in any episode of conflict.

To summarize, we have two strong competing hypotheses for the origin of the sterile castes and, following that, the more complex forms of social organization. Which one is correct? Were the parents (meaning, in the case of the social Hymenoptera, the mothers) molded by natural selection to enslave their offspring? Or were the offspring shaped by natural selection to be willing helpers—or even further, to manipulate their mothers to produce more brothers and sisters?

Some potent arguments have been raised in favor of the parental manipulation hypothesis. One stresses the fact that some of the offspring become workers while others become queens. Such plasticity might suggest control on the part of the queen, because if there is no intrinsic advantage to a female to be a worker, all should choose to be queens. However, this argument is blunted by the evidence, examined in Chapters 3 and 8, that the ergonomic phase of colony growth postpones queen production only to allow a very large crop at a later date. The outcome serves both the classical fitness of the mother queen and the inclusive fitness of the workers.

A second argument, advanced by Charlesworth (1978), Charnov (1978), and Craig (1979) in connection with models for the spread of rare alleles for eusociality, identifies a circumstance that appears to make daughter enslavement easier: from the mother queen's point of view, the offspring need only be half as good at raising their brothers and sisters as would be required if allowed to raise their own sons and daughters. The reason is that the queen is related to her own offspring (the brothers and sisters of the enslaved workers) by ½, but she is related to her grandsons and granddaughters (the would-be offspring of the workers) by only ¼. As a consequence she can afford to degrade her offspring in the process of making them into workers, for example by starvation or hormone-mediated sterilization, even if it reduces their competence as helpers. Hence the matrix of relatedness favors the origin of a worker caste by the evolution of manipulative behavior on the part of the queen—all other things being equal.

The difficulty with the argument, and the parental manipulation theory generally, is that all things are not equal. If offspring have lower inclusive fitness as a consequence of being workers, any allele that prescribes resistance to the queen's machinations would be favored. Hence parental manipulation can be invaded by genes that prescribe its reversal as a social trend. On the other hand, as Craig has pointed out, mothers can also manipulate their offspring so that it is in the offspring's best interest to stay around and help. Parental manipulation of this kind devolves to kin selection and brings us to the possibility that offspring "consent" to become workers.

## OFFSPRING CONSENT

The opposing view to crude parental manipulation is of course offspring consent. Worker castes arise and are maintained because under certain circumstances inclusive fitness is enhanced by surrendering reproduction. Hamilton's original formulation of this idea permitted a surprisingly detailed prediction of some hitherto unexplained features of behavior in ants and other social Hymenoptera. Its explanatory power gave the model a great deal of initial appeal and, more generally, launched kin selection as an important idea in general sociobiology. The key step was the connection between the haplodiploid method of sex determination, which is universal in the Hymenoptera, and the remarkable prevalence of eusociality in this order.

Haplodiploidy is the mode of sex determination in which males are derived from unfertilized (haploid) eggs and females from fertilized (diploid) eggs. The ultimate basis, however, is not the mere presence or absence of chromosomes but rather the one or more sex-determining genes they bear. Haplodiploidy has a number of odd effects (Andersson, 1984). The parthenogenetic origin of hymenopteran males means that all alleles will be expressed in a homozygous condition (or, more precisely, hemizygous condition). As a result, lethal and subvital alleles will be exposed each generation, and total genetic variability in the population will tend to be reduced. Hymenopteran species have between one-tenth and one-half the heterozygosity per individual that nonsocial insects have. Also, the more advanced social hymenopteran species have less heterozygosity than do the solitary ones, possibly because of a higher degree of inbreeding caused by reduced gene flow among the population of colonies (Graur, 1985).

The negative effects of haplodiploidy are true only for genes expressed in the male, however. Those limited in expression to female characters are theoretically expected to behave as though they existed in wholly diploid populations, enjoying the same potential variability and obeying the same equilibrium laws (Kerr, 1967). Another curious effect of haplodiploidy is that characters that are both under polygenic control and not sex limited should be more variable

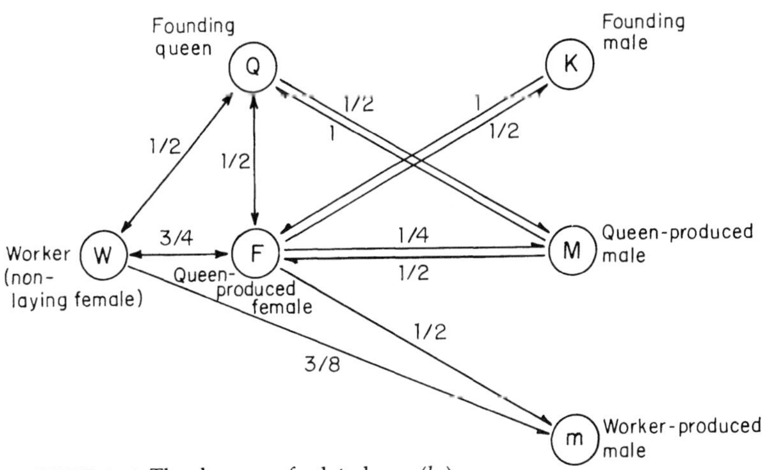


**FIGURE 4–1** The degrees of relatedness ($b_{ij}$) among the members of an ant colony. (From Oster and Wilson, 1978.)

among males in sibling groups than among females. In fact, under the simplest possible conditions (panmixia and an absence of dominance and epistasis), the theory of polygenic inheritance predicts a genetic variance in males four times that of their biparental sisters. Since most characteristics are under polygenic control, it should be a rule that males are more variable than virgin queens collected from the same colony. Eickwort (1969) found this proposition to be true for ten external morphological characters that she measured in the paper wasp *Polistes exclamans*.

But the strangest of all consequences of haplodiploid sex determination is the asymmetries it creates in the relatedness among close relatives (Figure 4–1). The coefficient of relationship between sisters is ¾, whereas between mother and daughter it is ½, the same as in diploid organisms. Sisters are exceptionally close because they share all of the genes they receive from their father (since their father is homozygous and produces genetically uniform spermatozoans), and they share on the average one-half of the genes they receive from their mother. Each sister receives one-half of her genes from her father and one-half from her mother, so that the probability that a gene possessed by the focal females will be shared by a sister is

| From father | | From mother | | Shared overall |
|---|---|---|---|---|
| (1 x 1/2) | + | (1/2 x 1/2) | = | 3/4 |

Hamilton reasoned that when the mother lives as long as the eclosion of her female offspring, her daughters can increase their inclusive fitness more by caring for their younger sisters than by caring to an equal degree for their own offspring.

During the twenty years that followed Hamilton's publication, the basic model received two major adjustments that were to strengthen its precision and improve its falsifiability. The first, by Trivers and Hare (1976), noted that whereas the advantage accruing to the care of sisters is inevitable, it is counterbalanced by an equal disadvantage that results from the rearing of brothers. This is because brothers and sisters are related by only ¼. The male comes from an unfertilized egg, whereas his sister comes from a fertilized egg. The coefficient of relatedness between the male and the haploid egg that produces his sister is ½. This is diluted when a sperm is added to the female-destined egg by an outside male, so that the

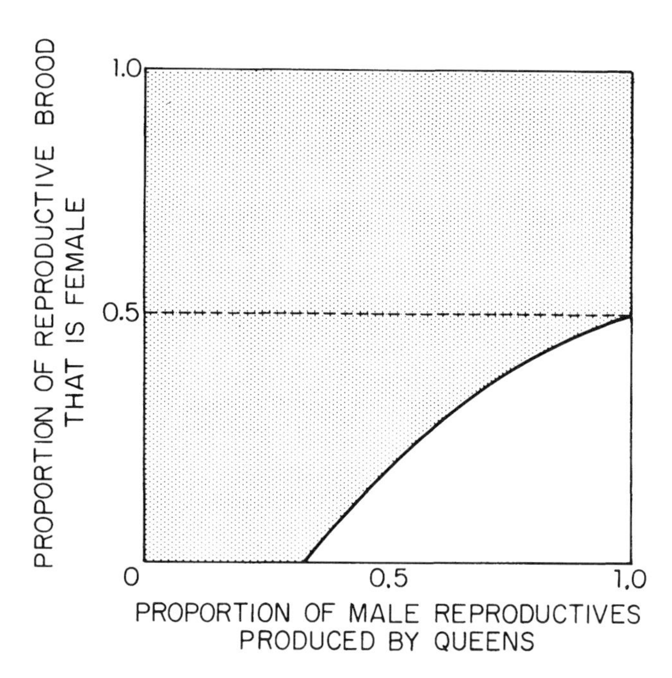


**FIGURE 4–2** Bartz's rule in the evolution of the hymenopteran worker caste: societies can be expected to have male workers or female workers but not both; in addition, conditions are generally favorable to the evolution of female workers. (Modified from Bartz, 1982.)

relation between the female and her brother drops to ¼. More precisely, the probability that an allele chosen in a focal female will also be present in the brother by immediate common descent is ¼. In ordinary diploid organisms with a 1:1 sex ratio, the degree of relatedness between a focal female and all her sisters and brothers is

| Related to sister | | Related to brother | | Related overall |
|---|---|---|---|---|
| (1/2 x 1/2) | + | (1/2 x 1/2) | = | 1/2 |

This turns out to be identical to the haplodiploid system when that system includes a 1:1 sex ratio, in spite of the asymmetries in relationship:

| Related to sister | | Related to brother | | Related overall |
|---|---|---|---|---|
| (3/4 x 1/2) | + | (1/4 x 1/2) | = | 1/2 |

Trivers and Hare pointed out that the advantage to rearing sisters as opposed to daughters could nevertheless be restored if the ratio of investment (approximated by dry weight) is altered to 3:1 in favor of sister production over brother production. The 3:1 ratio should be at equilibrium because the expected reproductive success of the males will then be three times that of the queen on a per-gram basis, balancing the one-third initial investment. This important result can of course be tested; it is one of the few cases in which evolutionary theory actually predicts a specific quantity rather than merely a trend or inequality. The substantial amount of research it has stimulated appears to favor the kin selection hypothesis, a conclusion we will discuss in more detail shortly.

The second adjustment deserving special mention is due to Bartz (1982), who deduced that in a haplodiploid system it is theoretically possible to evolve either female workers or male workers but not both. As shown in Figure 4–2, male workers will be favored only if half or less of the reproductive investment made by the colony results in females, while at the same time the queen produces a large fraction of the males. Furthermore, the less the queen contributes to male production, the higher the proportion of females in the reproductive brood must be in order to compensate. "Bartz's rule" is in close enough accord with the facts to support the kin selection hypothesis. Pamilo (1984a) has confirmed the main result in a separate analysis, but he notes that it depends upon a population-wide sex ratio of 0.5. When the population of colonies deviates from this value, the result is less certain.

## TESTING THE KIN SELECTION THEORY

Few ideas in biology have been probed as aggressively and from so many directions as the theory of kin selection in the eusocial insects. The testing has been made still more rigorous by the presence of a strong alternative explanation of the evolution of eusocial behavior, the hypothesis of parental manipulation. The key critiques and reviews include those of Alexander and Sherman (1977), Andersson (1984), Bartz (1982, 1983), Craig (1979, 1980), Crozier (1977, 1979, 1982), Michod (1982), Page (1986), Pamilo (1982a, 1984a,b), Starr (1979), West-Eberhard (1982), and Wilson (1975b). How well has kin selection held up? In the sections to follow we will examine the most important features of ant biology that bear on the origin of the nonreproductive caste and attempt to weigh the empirical evidence accumulated to the time of writing. We will come out on the side of kin selection, but cautiously—because some new twist in theory or important empirical finding might yet overturn it.

*Which insects have become eusocial?* The kin selection theory predicts that haplodiploid insects should show a higher incidence of eusociality than completely diploid insects, and this appears to be the case. Haplodiploidy is a characteristic of all of the Hymenoptera but is shared by only a few other arthropod groups, including certain mites, thrips, and whiteflies; iceryines and possibly other scale insects; and the beetle genera *Micromalthus, Xylosandrus,* and, perhaps, *Xyleborus.* Outside the Arthropoda, some nematodes and most rotifers are also haplodiploid. At the same time true eusociality is very nearly confined to the Hymenoptera, where it has arisen at least eleven times independently: twice in the wasps (more precisely at least once each in the stenogastrine and vespine-polybiine wasps and probably a third time in the sphecid *Microstigmus*), eight or more times in the bees, and at least once or perhaps twice in the ants. Quite probably this lower estimate will increase with the growth in knowledge of hymenopteran biology, especially that of tropical bees.

Throughout the entire remainder of the Arthropoda, true eusociality is known to have originated in only one other living group, the termites or order Isoptera. Among the rest of the higher animal phyla, it is known only in the remarkable African mole rat *Heterocephalus glaber* (Jarvis, 1981). Aphids have evolved nonreproductive soldier castes no fewer than four times independently. They have also achieved an overlap of generations—but not cooperative brood care (Aoki, 1977, 1982, 1987). This dominance of the social condition by the Hymenoptera cannot be a coincidence. Of the 751,000 living insect species described to 1985 (Arnett, 1985), only about 103,000 or 14 percent belong to the Hymenoptera. Something close to this partition has persisted throughout at least the Cenozoic, further diminishing the possibility that the bias can be explained as a mere historical accident.

Several leading hymenopterists, namely Michener and Brothers (1974), West-Eberhard (1975), and Evans (1977a), have questioned the primacy of kin selection even while acknowledging its importance. They have urged the equivalent importance of at least two other preadaptations that exist in the Hymenoptera: the universal existence of mandibulate mouthparts and the frequent building of nests, especially in the aculeate wasps (as opposed to parasitoid wasps) and bees. Their case with reference to nest building is quite strong. Starr (1985) has added a third preadaptation: the aculeate hymenopteran sting, which is an effective defense against vertebrate predators. Only between 50,000 and 60,000 aculeate hymenopteran species have been described, yet they include all of the eusocial insects with the exception of the termites. The sting was needed in the earlier stages of eusocial evolution, Starr argues, because groups of individuals are generally more vulnerable to large predators that preferentially seek aggregations of prey and have the size and strength to overcome them. This seems to us to be a much less persuasive argument, but it cannot easily be rejected or confirmed.

What is needed is a kind of accounting system in the origin of eusociality, perhaps in the form of a multiple regression equation that incorporates each of the factors, including kin selection, in order to provide a more precise measure of the inclusive fitness of the would-be queens and workers. When the averaged relative fitnesses rise to a certain "eusociality threshold" (Wilson, 1975c), the species is likely to evolve all three of the basic traits of a higher social (eusocial) insect, that is, cooperative brood care, overlap of at least two generations, and division of the group into reproductive and sterile castes. It would appear that a precondition that carried the 11 hymenopteran phyletic lines across the eusociality threshold is the enhancement of kin selection by haplodiploidy. Mandibulate mouthparts and nest building are not enough by themselves. Other kinds of arthropods, including many beetles, spiders, and orthopterans, build nests, manipulate objects skillfully with their mandibles and legs, and care for their young, occasionally in elaborate fashion, but none has attained the eusociality threshold. It is equally true that haplodiploid enhancement is insufficient on its own, because it occurs in other, nonsocial arthropod species, including all of the symphytan and most of the parasitoid hymenopterans as well as a majority of the bees and aculeate wasps. (Diploid males have been reported in the parasitoid genera *Bracon*, *Nasonia*, and *Neodiprion*, as well as *Apis*, *Bombus*, *Melipona*, and *Trigona*; see Page and Metcalf, 1982.)

*Do males ever serve as castes?* In a manner consistent with Bartz's argument (see Figure 4–2), the worker caste of ants is universally female. Alexander (1974) and a few other writers have argued in contrary fashion that hymenopterous males, being stingless and otherwise highly specialized for reproduction throughout the order, simply are not able to evolve a worker-like anatomy and behavioral repertory. Hence phylogenetic inertia, rather than kin selection, can account for their lack of involvement.

This reasonable-sounding explanation is considerably vitiated by the evidence of substantial evolutionary lability among male ants. In at least two species males have assumed a partial worker role, under special environmental circumstances that indicate that when selection pressures are strong enough to countervail the conventional components of inclusive fitness, evolution away from strict male idleness does occur. Males of the carpenter ant *Camponotus herculeanus* are exceptionally long-lived, because they are produced in the fall, overwinter in the nest, and if kept sufficiently cool through the following spring and summer, live on through a second annual cycle. Unlike most other ant males studied to date, they store food in their crops and regurgitate some of it back to workers and other males, thus participating in a key homeostatic role for the entire colony (Hölldobler, 1964, 1966).

Male larvae of the weaver ant *Oecophylla longinoda* have well-developed silk glands. Like the worker-destined female larvae, they contribute the silk to nest construction rather than to construction of their own cocoons. Because the behavior is very specialized and clearly derived in evolution with reference to the ants as a whole, participation in nest building represents an important shift on the part of the males toward a new social role (Wilson and Hölldobler, 1980). Male anatomy is at least as malleable as behavior.

The males of *Formicoxenus* and some species of *Hypoponera*, *Cardiocondyla*, and *Technomyrmex* are wingless and "ergatomorphic," convergent to the worker of the species in overall body form (see Figures 12–17, 12–18). The modification appears to be associated with the loss of between-colony dispersal by the males and with a tendency toward pairing with female nestmates (Le Masne, 1956a; Terron, 1972b; Hamilton, 1979). In some species of *Cardiocondyla*, such as *C. papuana* and *C. wroughtoni*, males have evolved saber-shaped mandibles (Kugler, 1983; see Figure 4–3). The ergatomorphic males of *C. wroughtoni* fight among themselves until only one remains in the colony (Kinomura and Yamauchi, 1987; Stuart et al., 1987a,b). Fighting ergatoid males have also been recorded in the small ponerine *Hypoponera punctatissima* (Hamilton, 1979). No evidence has ever been adduced, however, of participation by ergatomorphic males belonging to the three genera in any worker-like social function. Finally, it is probably relevant that termites, which are not haplodiploid, have both male and female workers.

The social status of males is complicated somewhat by the coexistence in a few species of two morphological forms, the small "micraner" and the large "macraner." The males of *Solenopsis invicta* have a bimodal size-frequency distribution; the small individuals are haploid and the large males are diploid (Ross and Fletcher, 1985b). Size dimorphisms also exist in *Formica naefi* (Kutter, 1957), *F. exsecta* (Pamilo and Rosengren, 1984), *F. sanguinea* (Agosti and Hauschteck-Jungen, 1987), and species of the *Rhytidoponera impressa* complex (Ward, 1983a), but the genetic basis has not been thoroughly studied. In the case of *Formica exsecta* at least, micraners and macraners are not always haploid and diploid, respectively. However, the micraners have a somewhat lower percentage of haploid brain cells, whereas chromosome numbers higher than $2N$ occur only in micraners (Agosti and Hauschteck-Jungen, 1987). The behavioral and ecological significance of the size dimorphism in all of the species remains unknown.

*Do males suffer higher mortality?* Smith and Shaw (1980) have pointed out that because of their homozygous state, hymenopterous males always express lethal and subvital alleles in their genome and hence inevitably suffer higher mortality than females. Even if selection has completely "cleaned out" such alleles by passing them through the male haploid filter, they will still appear each generation in very low levels through the occurrence of new mutations. Hence the biased sex ratio attributed to kin selection might be due in part to new lethal mutations, provided the differential mortality occurs during or after the stage at which parental care is invested. The effect can be expressed as the probability that a male dies either because of a new mutation in his own soma or because of a new mutation picked up from his mother. This probability is

$$P = 1 - \exp(-3m)$$

where $m$ is the mutation rate per genome (not colony) per generation. The effect on the female:male ratio is displayed in Figure 4–4. The value of $m$ for parasitoid wasps in the genus *Apanteles* is 0.035, giving a differential mortality of 10 percent. If a similar value holds for ants, the mutation-induced distortion of the ratio of investment would only be a small fraction of that expected from kin selection. The effect is further attenuated by the fact that the equation applies to the ratio of individuals, rather than to the ratio of biomass. Since individual queens are usually much heavier than males at eclosion, and receive proportionately larger amounts of food prior to the nuptial flight, the mortality ratio translates to a narrower biomass ratio. The importance of the equation is that it forms a baseline or null hypothesis against which kin selection can be more reliably tested. It is rendered much less tractable, however, by the fact that $m$ must be considered separately for each phylogenetic group, and no values are yet available for ants.

*How closely related are the colony members?* Here we encounter a potential difficulty of considerable significance for the kin selection hypothesis. Four key factors can be distinguished that determine relatedness within a colony: the number of laying queens, the degrees of relatedness among the laying queens, the number of males with which the laying queens mate, and the intensity of egg laying by the workers. In some species a fifth, at least episodic, factor is queen succession, which occurs in species that reproduce by colony fission and queen supersedure. Prominent examples include ecitonine army ants. John Tobin (personal communication) argues:

> When colony fission occurs one of the daughter colonies will be headed by the mother queen. However, the other daughter colony (or colonies, if there are more than two daughter colonies after fission) will get a new queen that is a sister of the workers, rather than of the mother; the effect will be that the workers in this colony will be raising not their sisters (or half-sisters) but their nieces (or half-nieces). As the younger generation of workers (the daughters of the new queen) replaces the older generation (the daughters of the original queen), average intracolony relatedness will decrease, and will reach a minimum when the proportion of the two generations, or subfamilies, of workers is 1:1. As the older workers continue to die away, average relatedness will increase back towards the pre-fission levels.

In Table 4–1 we have summarized the information collected to date on the degrees of relatedness. These data are mostly estimates based on the electrophoretic separation and identification of allozymes (different forms of the same enzyme) encoded by multiple alleles and treated as representative samples from the larger genotype.

One important finding is the distant relatedness among queens in polygynous colonies. In *Myrmecia pilosula* and *Formica sanguinea*, the species for which data are available, the queens are less closely related than full sisters; in other words their $b$ values are significantly

**FIGURE 4–3** The ergatomorphic males of *Cardiocondyla wroughtoni*, a small myrmicine of Asiatic origin, fight among themselves with the aid of saber-shaped mandibles. The existence of this bizarre variant illustrates the potential for anatomical evolution in male ants, an important consideration for the theory of the origin of social life. (*Above*) A more darkly pigmented ergatomorph male with a lethal fighting grip on a recently eclosed ergatoid male nestmate. (*Below*) A *C. wroughtoni* ergatomorph male rests on the brood pile within its nest. The unusual mandibles are clearly visible. (From Stuart et al., 1987b; photographs by Mark Moffett.)

**FIGURE 4–4** Because male ants are derived from unfertilized eggs and hence homozygous, they are subject to higher mortality due to new lethal mutations alone. As shown in this figure, the female:male ratio (of individuals, not biomass) will rise with the lethal mutation rate ($m$) and mimic at least in part the biased investment expected from kin selection. (Modified from Smith and Shaw, 1980.)

**TABLE 4–1** The degree of genetic relatedness (estimated coefficient of relatedness, or *b*) in various ant species. When one species was sampled from multiple localities, the localities are indicated in parentheses after the scientific name. The measurement given is mean *b* ± s.e.

| Species | Social organization | Relationship measured | Method of measurement | Degree of relatedness | Authority |
|---|---|---|---|---|---|
| **PONERINAE** | | | | | |
| *Rhytidoponera chalybaea*, type A colonies | Monogynous, singly mated | Worker-worker | Allozymes | 0.756 ± 0.042 | Ward (1983a) |
| *R. chalybaea*, type B colonies | Queenless; multiple egg-laying workers | Worker-worker | Allozymes | 0.337 ± 0.068 | Ward (1983a) |
| *R. chalybaea*, type A colonies | Monogynous, singly mated | Worker-male | Allozymes | 0.296 ± 0.139 | Ward (1983a) |
| *R. chalybaea*, type B colonies | Queenless; multiple egg-laying workers | Worker-male | Allozymes | 0.189 ± 0.013 | Ward (1983a) |
| *R. confusa*, type A colonies | Monogynous, singly mated | Worker-worker | Allozymes | 0.679 ± 0.100 | Ward (1983a) |
| *R. confusa*, type B colonies | Queenless; multiple egg-laying workers | Worker-worker | Allozymes | 0.278 ± 0.159 | Ward (1983a) |
| *R. confusa*, type A colonies | Monogynous, singly mated | Worker-male | Allozymes | 0.157 ± 0.115 | Ward (1983a) |
| *R. confusa*, type B colonies | Queenless; multiple egg-laying workers | Worker-male | Allozymes | 0.137 ± 0.072 | Ward (1983a) |
| *R.* "sp. 12" (= *mayri* in part) | Queenless; multiple egg-laying workers | Worker-worker | Allozymes | 0.158 ± 0.018 | Crozier et al. (1984) |
| **NOTHOMYRMECIINAE** | | | | | |
| *Nothomyrmecia macrops* | Mostly monogynous, occasionally 2–several queens | Worker-worker | Allozymes | 0.172 ± 0.068 | Ward and Taylor (1981) |
| **MYRMECIINAE** | | | | | |
| *Myrmecia pilosula* | Primary polygyny | Queen-queen<br>Worker-worker | Allozymes<br>Allozymes | 0.243 ± 0.056<br>0.172 ± 0.026 | Craig and Crozier (1979) |
| **MYRMICINAE** | | | | | |
| *Aphaenogaster rudis* | Monogynous, singly mated | Worker-worker | Inferred from social organization | 0.75 | Crozier (1973) |
| *Harpagoxenus sublaevis* | Monogynous, singly mated | Worker-worker | Allozymes | 0.73 ± 0.07 | A. Bourke (personal communication) |
| *Myrmica rubra* (site A-1975) | Polygynous | Worker-worker | Allozymes | 0.106 | Pearson (1983) |
| *M. rubra* (site A-1977) | Polygynous | Worker-worker | Allozymes | 0.022 | Pearson (1983) |
| *M. rubra* (site A-1978) | Polygynous | Worker-worker | Allozymes | 0.083 | Pearson (1983) |
| *M. rubra* (site B-1977) | Polygynous | Worker-worker | Allozymes | 0.543 | Pearson (1983) |
| *Solenopsis invicta* | Monogynous strain, singly inseminated | Females, same generation | Allozymes | 0.75 ± 0.05<br>0.68 ± 0.06 | Ross and Fletcher (1985a) |
| *S. invicta* | Polygynous strain, singly inseminated | Foundress queens, same nest | Allozymes | 0.02 ± 0.02<br>0.01 ± 0.02 | Ross and Fletcher (1985a) |
| *S. invicta* | Polygynous strain, singly inseminated | Worker-worker | Allozymes | 0.08 ± 0.06<br>0.04 ± 0.04 | Ross and Fletcher (1985a) |
| **FORMICINAE** | | | | | |
| *Formica aquilonia* (Espoo) | Polygynous | Worker-worker | Allozymes | 0.09 ± 0.09 | Pamilo (1982b) |
| *F. aquilonia* (Vantaa) | Polygynous | Worker-worker | Allozymes | -0.02 ± 0.14 | Pamilo (1982b) |
| *F. polyctena* | Polygynous | Worker-worker | Allozymes | 0.19 ± 0.34 | Pamilo (1982b) |
| *F. polyctena* | Polygynous | Worker-worker | Allozymes | 0.30 ± 0.23 | Pamilo (1982b) |

*continued*

**TABLE 4–1** *(continued)*

| Species | Social organization | Relationship measured | Method of measurement | Degree of relatedness | Authority |
|---|---|---|---|---|---|
| *F. sanguinea* (2 successive years) | Polygynous, multiply mated | Worker-worker<br>Worker-worker | Allozymes<br>Allozymes | 0.420 ± 0.098<br>0.311 ± 0.124 | Pamilo and Varvio-Aho (1979) |
| *F. sanguinea* | Polygynous, multiply mated | Worker-worker | Allozymes | 0.19 | Pamilo (1981) |
| *F. sanguinea* | Polygynous, multiply mated | Queen-queen | Allozymes | 0.21 | Pamilo (1981) |
| *F. sanguinea* | Polygynous, multiply mated | Worker-queen | Allozymes | 0.38 ± 0.17 | Pamilo and Varvio-Aho (1979) |
| *F. sanguinea* | Polygynous, multiply mated | Worker-male | Allozymes | 0.62 ± 0.40 | Pamilo and Varvio-Aho (1979) |
| *F. transkaucasica* | Polygynous, occasionally monogynous; singly mated | Worker-worker | Allozymes | 0.33 ± 0.07 | Pamilo (1981, 1982b) |

below the theoretically expected 0.75. They are nevertheless as close as half-sisters or first cousins. The other measurements taken so far match expectations. In strains of *Rhytidoponera chalybaea* and *R. confusa* that possess only one singly mated queen per colony, $b$ among workers does not depart significantly from 0.75, as anticipated. In queenless colonies of the same species, with fertile workers serving as reproductives, the worker-worker $b$ values are highly variable from one colony to another but always well below 0.75, as expected from the diverse provenance of the worker population in each nest.

The difficulty raised for the kin selection theory by the estimates in Table 4–1 is the following. In polygynous colonies the degree of relatedness between selected individual workers and the cogenerational females and males in the same nest is less than would be the case if the individual workers had produced their own sons and daughters. If kin selection is a powerful force, what prevents evolution from leading to a more competitive state in which the workers (who have ovaries) try to take over reproduction? One obvious explanation is that polygynous colonies are so successful as smoothly operating units as to make the inclusive fitness of focal workers higher than it would be if workers were egg-layers. On the other hand, evolution of egg-laying workers has occurred in some cases and is even associated with the disappearance of the queen caste in a few phyletic lines. All species of *Diacamma*, *Dinoponera*, and *Ophthalmopone*, as well as some species of *Rhytidoponera* and *Pristomyrmex*, have discarded the queen caste and reverted to reproduction by workers (Peeters and Crewe, 1986b). In some cases a large percentage of the workers lay viable eggs during early stages of their lives. From one point of view, expressed with reference to *Pristomyrmex pungens* by Itow et al. (1984), the worker caste has been lost. We may even ask whether the colonies are classifiable as eusocial.

Another escape from the difficulty posed by polygyny would be provided if workers in polygynous colonies could be shown to distinguish close kin from more distant kin. This ability has been demonstrated in social bees and ants, a subject to be reviewed in Chapter 5. Is the discriminatory activity strong enough to divide a polygynous colony into cliques of closely related individuals? All that would be required is for individual workers to favor their siblings over half-sisters and more distantly related nestmates among the reproductive larvae. Whether this occurs in ants remains to be learned.

The occasional occurrence of distant relatedness raises questions about the origin of eusociality in ants, because some species of the very primitive ant genera *Amblyopone* and *Myrmecia* are polygynous. If true polygyny were the primitive condition for the Formicidae, it is difficult to see how the simple model of kin selection can explain the origin of eusociality. We need much more information on the reproductive structure and relatedness among the colony members in primitive ant species to help resolve these important questions.

*Do males from the same colony compete?* Hamilton (1967) pointed out that another selection force that might bias sex allocation ratios in favor of new queens is local mate competition. Alexander and Sherman (1977) agreed and argued that this might be a more harmonious explanation than kin selection, when combined with control on the part of the queen. If a queen's sons habitually inseminated their own sisters, the queen could maximize her own fitness by producing only enough sons to ensure the insemination of all her daughters. In fact, Hamilton showed that such extraordinary sex ratios occur elsewhere in the animal kingdom. The extreme cases include certain mites and parasitoid wasps, where total incest is accompanied by very high female:male ratios.

The local mate competition hypothesis is not favored, however, by the natural history of most ant species, whose reproductive forms disperse widely and often gather in large, mixed swarms assembled from hundreds or thousands of colonies (Hölldobler and Bartz, 1985; Woyciechowski and Łomnicki, 1987). Among the several species studied so far that possess large colonies and limited dispersal, mating has been shown by allozyme analysis either to be random (in *Formica sanguinea* and *F. transkaucasica*) or to depart from randomness (in *F. pressilabris*). No information is yet available concerning the degree of mating within single nest units (Pamilo, 1982c). In a study of equal importance, Nonacs (1986a) has systematically tested kin selection in opposition to local mate competition by examining the various phenomena in which they predict different outcomes. Kin selection is throughout the more reliable. In particular, the numerical sex ratios (as opposed to the biomass sex ratios) are close enough to 1:1 to be inconsistent with local mate competition, while the biomass sex ratios (as opposed to numerical ratios) are close enough in monogynous species to 3:1 to be harmonious with kin selection. Also, the biomass ratios of parasitic species, in which inbreeding does occur, is close to the 1:1 value expected from kin selection and not consistent with local mate competition. Finally, the evidence is overwhelming that ant males are capable of mating only once or at most two or three times. Furthermore, in some species of ants the queen requires the output of more than one male to fill her spermatheca (see Chapter 3). In short, the conditions do not exist for male ants to inseminate numerous closely related monandrous females.

*How reliable are biomass measurements?* Reproductive forms have been collected, dried, and weighed with little reference to timing in the adult phase of the life cycle. Ignoring the life cycle is risky, because queens (gynes) and males undergo major body alterations from the time they eclose to the moment of the nuptial flight. Even worse, the time courses of the two sexes are exactly opposite: queens are fattened by the workers and grow steadily heavier, while males fast and grow lighter. Hence timing makes a considerable difference in measuring the two sexes. In the European mound-building formicine *Lasius flavus,* for example, individual queens triple their weight during the first six days after emerging from the pupa. At the same time their fat content rises from 20 percent to 60 percent, while their water content falls from 75 percent to 45 percent (Nielsen et al., 1985a,b). In addition, their respiratory rate doubles. Except for falling weight, males of the same species show little change during the same period. In addition, the weight-specific respiratory rate (μl $O_2$/mg dry weight/hour) of queen-destined larvae is higher than in worker-destined larvae (Peakin et al., 1985). Similar results have been obtained independently for *Lasius niger* by Boomsma and Isaaks (1985; see also van der Have et al., 1988).

Ideally, then, measures of energy investments should include both construction (that is, growth in biomass) and maintenance (respiration during the tenure in the nest) over the life spans of the two sexes. This has been achieved in only one instance of which we are aware. In their audit of *Lasius niger,* Boomsma and Isaaks estimated that it takes 689 joules to make a queen and 90 joules to make a male. They found that the population means of energy investment in queens are lowered by the adjustment but remain in good agreement with the theoretical kin selection optimum of 3:1. The closest fit was in colonies taken from environments considered most favorable for *Lasius niger.* We suggest that without exacting metabolic studies of this kind, the best time to collect data is in the middle period of adult life, when the queens have been fattened and the males still possess most of their reserves and original body weight. Failing even that, the aim should be to collect large enough samples to wash out the differences in the time courses of energy investment in females and males. Most investigators of the subject have in fact attempted to follow this second procedure.

*Is there sperm competition?* Multiple matings are widespread in the ants, occurring most commonly in species with large mature colonies (Cole, 1983b). If the sperm of two or more males are used extensively to fertilize the eggs, the degree of relatedness among the workers and between the workers and newly produced sexual forms will be diminished accordingly. If on the other hand there is sperm precedence, so that the sperm of one male or another dominates in fertilization, at least for intervals as long as the average longevity of females, the effect will be negligible. The relatively sparse evidence accumulated to date for solitary Hymenoptera and honey bees and *Polistes* wasps (reviewed by Crozier and Brückner, 1981; Page, 1986) indicates that sperm mixing is a general phenomenon, although there is some bias in utilization. Bias occurs in ants of the *Formica rufa* group; whether through true competition or precedence is not yet known (Pamilo, 1982c).

*Who lays the eggs?* The provenance of the eggs, whether from one queen, multiple queens, workers, or some numerical combination of these alternative sources, makes a great difference in the organization of the ant colony and the degree of relatedness of its members. The available data on worker-produced offspring in species with queens or ergatogynes (intermediates between queens and workers) are presented in Table 4–2. This information needs to be treated cautiously. Queens are sometimes difficult to find even when present, because they often crouch in hidden recesses of the nest. Also, the immature stages are sometimes prolonged by diapause, so it is easy to misidentify queen-produced workers and males as worker-produced. Some confusing controversies have resulted in the literature (for recent reviews see Choe, 1988, and Bourke, 1988b).

For example, when proposing a *cycle évolutif* in the African weaver ant *Oecophylla longinoda,* Ledoux (1949, 1950) reported that many new colonies are begun by workers that leave the territory of the mother colony and start colonies on their own. Some of the workers then lay small eggs 0.6 millimeter in length, most of which develop into workers; a few develop into queens. Ledoux suggested that the parthenogenesis in this case is apomictic, resulting in diploid eggs, and that the eggs are small because they are ejected before normal meiosis can begin. On the other hand, Way (1954a), working in the field with free colonies, and Crozier (1971) and Wilson and Hölldobler (1980), experimenting on laboratory colonies, could find no evidence of thelytoky in workers. Groups of workers separated from the queen produced only males.

The general picture so far is that reproduction by workers is common but far from universal, and in most instances it is limited to the production of males in compliance with the ordinary working of haplodiploid sex determination. Thelytoky is relatively rare and it is often an infrequent facultative process in the species that display it. When Haskins and Enzmann (1945) made a careful attempt to obtain offspring from virgin queens of *Aphaenogaster rudis,* only 18 of 100 such individuals reared brood (mostly male) to the pupal stage, and only 2 females were brought to maturity among them. Bier (1952) found that a similar difficulty in rearing workers from worker-laid eggs of *Lasius niger* is due to the fact that the great majority of larvae coming from such eggs are actually male determined and die at an early age. Female-determined eggs are viable but constitute only a tiny fraction of the total number laid.

In spite of the considerable uncertainties in most cases of reported thelytokous parthenogenesis in ant workers, there exist at least two species where worker thelytoky appears to have been clearly demonstrated. One is *Cataglyphis cursor,* where Cagniant (1979, 1982) and Lenoir and Cagniant (1986) provided solid experimental evidence for worker thelytoky, and the other is *Pristomyrmex pungens,* where worker thelytoky seems to be the predominant mode of re-

**TABLE 4–2** Worker reproduction in ants. 1 = workers become inseminated and produce females; 2 = workers are reported to produce females thelytokously; 3 = involved in social parasitism; 4 = polygynous species; 5 = male eggs are laid, but resulting larvae die before pupation. (Modified slightly from Choe, 1988.)

| Taxon | Workers reproduce— In queenless nests | In queenright nests | Authority |
|---|---|---|---|
| **PONERINAE** | | | |
| *Diacamma rugosum* | X[1] | | Wheeler and Chapman (1922) |
| *Hypoponera eduardi* | X[1] | | Le Masne (1953) |
| *Odontomachus haematodes* | X | X | Colombel (1972, 1974) |
| *Ophthalmopone berthoudi* | X[1] | | Peeters and Crewe (1984) |
| *Rhytidoponera chalybaea* | X[1] | | Ward (1981b) |
| *R. confusa* | X[1] | | Ward (1981b) |
| *R. impressa* | X[1] | | Ward (1981b) |
| *R. inornata* | X[1] | | Haskins and Whelden (1965) |
| *R. mayri* | X[1] | | Crozier et al. (1984) |
| *R. metallica* | X[1] | | Haskins and Whelden (1965), Ward (1986) |
| *R. purpurea* | | X | Haskins and Whelden (1965) |
| *R. tasmaniensis* | X[1] | | Haskins and Whelden (1965) |
| *R. victoriae* | X[1] | | Haskins and Whelden (1965) |
| *R. violacea* | X[1] | | Haskins and Whelden (1965) |
| *R.* "sp. 12" | X[1] | | Pamilo et al. (1985) |
| **NOTHOMYRMECIINAE** | | | |
| *Nothomyrmecia macrops* | | X | Taylor (1978c) |
| **MYRMECIINAE** | | | |
| *Myrmecia forficata* | X | X | Haskins and Haskins (1950a) |
| *M. gulosa* | X | X | Haskins and Haskins (1950a) |
| *M. mandibularis* | X | X | Haskins and Haskins (1950a) |
| *M. nigrocincta* | X | X | Haskins and Haskins (1950a) |
| *M. piliventris* | X | X | Haskins and Haskins (1950a) |
| *M. pilosula* | X | X | Haskins and Haskins (1950a) |
| *M. regularis* | X | X | Haskins and Haskins (1950a) |
| *M. swalei* | X | X | Haskins and Haskins (1950a) |
| *M. tarsata* | X | X | Haskins and Haskins (1950a) |
| *M. vindex*[3] | X | X | Haskins and Haskins (1950a) |
| **DORYLINAE** | | | |
| *Dorylus nigricans* | X[5] | | Raignier (1972) |
| *D. wilverthi* | X[5] | | Raignier (1972) |
| **PSEUDOMYRMECINAE** | | | |
| *Tetraponera anthracina*[3] | X | | Terron (1970) |
| **MYRMICINAE** | | | |
| *Acromyrmex* spp. | X? | | Haskins and Enzmann (1945) |
| *Aphaenogaster cockerelli* | X | | Hölldobler and Bartz (1985) |
| *A. lamellidens* | X[2]? | | Haskins and Enzmann (1945) |
| *A. rudis*[4] | X[2]/X? | | Haskins and Enzmann (1945), Crozier (1974) |
| *A. senilis* | X[2]? | | Ledoux and Dargagnon (1973) |
| *A. subterranea* | X | | Bruniquel (1972) |
| *Apterostigma dentigerum* | X | | Forsyth (1981) |
| *Atta cephalotes* | X[2]? | | Tanner (1892) |
| *Atta* spp. | X? | | Haskins and Enzmann (1945) |
| *Crematogaster auberti* | X[2]? | | Soulié (1960) |
| *C. impressa* | X | | Delage-Darchen (1974a) |

*continued*

**TABLE 4–2** *(continued)*

| Taxon | Workers reproduce— In queenless nests | In queenright nests | Authority |
|---|---|---|---|
| *C. scutellaris* | X[2]? | | Soulié (1960) |
| *C. skounensis* | X[2]? | | Soulié (1960) |
| *C. vandeli* | X[2]? | | Soulié (1960) |
| *Epimyrma ravouxi*[3] | X | | Winter and Buschinger (1983) |
| *Harpagoxenus canadensis*[3] | X | X | Buschinger and Alloway (1978) |
| *H. sublaevis*[3] | X[1] | X | Buschinger (1968b), Buschinger and Winter (1978) |
| *Leptothorax allardycei* | X | X | Cole (1981) |
| *L. ambiguus*[3,4] | X | | Alloway et al. (1982) |
| *L. curvispinosus*[3,4] | X | | Alloway et al. (1982) |
| | | X | T. Graham (personal communication) |
| *L. longispinosus*[3,4] | X | | Alloway et al. (1982) |
| | | X | T. Graham (personal communication) |
| *L. nylanderi* | | X | Plateaux (1970, 1981) |
| *L. recedens*[3] | | X | Dejean and Passera (1974) |
| *L. tuberum* | X | X[1]? | Gösswald (1933), Bier (1954b) |
| *Messor capitatus* | X | | Delage (1968) |
| *M. ebeninus* | X | | Tohmé (1972) |
| *M. semirufus* | X | | Tohmé (1972) |
| *Myrmica rubra*[4] | X | X | Brian (1953), Smeeton (1981) |
| *M. ruginodis*[4] | X | | Brian (1953) |
| *M. sabuleti*[3,4] | X | | Brian (1972) |
| *Pristomyrmex pungens* | X[2] | | Itow et al. (1984) |
| *Protomognathus americanus*[3] | X | X | Wesson (1939), Buschinger and Alloway (1977) |
| *Stenamma fulvum* | X | | Fielde (1905a) |
| *Zacryptocerus varians* | X | | Wilson (1976a) |
| **DOLICHODERINAE** | | | |
| *Dolichoderus quadripunctatus* | | X | Torossian (1968, 1974) |
| *Iridomyrmex humilis*[4] | X | | Torossian (1974) |
| *I. purpureus*[4] | X | | N. Carlin (personal communication) |
| *Technomyrmex albipes* | X[2]? | | Terron (1972) |
| **FORMICINAE** | | | |
| *Camponotus acvapimensis* | X | | Lévieux (1973) |
| *C. aethiops*[3] | X | | Dartigues and Passera (1979a) |
| *C. ferrugineus* | X | | Goetsch and Käthner (1937) |
| *C. herculeanus*[4] | X | | Fielde (1905) |
| *C. lateralis* | X | | Goetsch and Käthner (1937) |
| *C. pennsylvanicus* | X | | Fielde (1905a) |
| *C. pictus* | X | | Fielde (1905a) |
| *C. vagus* | X | | Benois (1969, in Brian, 1979a) |
| *Cataglyphis cursor* | X[2] | | Cagniant (1979, 1982) |
| *Formica argentata* | X | | Fielde (1905a) |
| *F. canadensis*[3] | X | | Hung (1973) |
| *F. cinerea* | X | | Lubbock (1894) |
| *F. exsecta*[3,4] | X | | Pamilo and Rosengren (1983) |
| *F. fusca*[3] | X | | Lubbock (1894) |
| *F. pallidefulva* | X | | Fielde (1905a) |

*continued*

**TABLE 4–2** *(continued)*

| Taxon | Workers reproduce— In queenless nests | In queenright nests | Authority |
|---|---|---|---|
| *F. pergandei*[3] | X | | Hung (1973) |
| *F. polyctena*[4] | X[2]/X? | | Otto (1960), Schmidt (1982) |
| *F. pratensis*[3] | X | | Goetsch and Käthner (1937) |
| *F. rufa*[3] | X[2]/X? | | Goetsch and Käthner (1937), Bier (1956, 1958b), Schmidt (1974a) |
| *F. rufibarbis* | X | | Goetsch and Käthner (1937) |
| *F. sanguinea*[3] | X | | Forel (1874) |
| *Lasius alienus*[3] | X[2]? | | Wheeler (1903b) |
| *L. brunneus*[3] | X[2]? | | Goetsch and Käthner (1937) |
| *L. niger*[3] | X[2]? | | Reichenbach (1902) |
| *Oecophylla longinoda* | X | | Hölldobler and Wilson (1983a) |
| *O. smaragdina* | X | | Hölldobler and Wilson (1983a) |
| *Plagiolepis pygmaea*[3,4] | X | | Passera (1966) |
| *Polyergus breviceps*[3] | X | | Hung (1973) |
| *P. rufescens*[3] | X | | Lubbock (1894) |

production (Itow et al., 1984; Tsuji and Itō, 1986). In both cases the experiments were carefully controlled: the workers were reared from pupae and kept isolated from males. In addition, histological and macroscopic dissections were employed to investigate possible inseminations, but no signs of inseminations were revealed (Suzzoni and Cagniant, 1975; K. Tsuji, personal communication).

Male production by workers exists mostly in queenless colonies, or, more precisely, in orphaned colonies. Even when it takes place in the presence of the queen, the rate of production is usually small and limited to special circumstances. For example, workers of *Odontomachus haematodes* and *Hypoclinea quadripunctata* produce males only when separated to some degree within the nest from the mother queen (Colombel, 1974; Torossian, 1978). *Leptothorax recedens* lay male-destined eggs only during the first several weeks after hibernation, when the queen's inhibitory power is weak or absent (Dejean and Passera, 1974). Usually the sole *Myrmica rubra* workers who produce males are the callows, and this activity is most likely to occur in the late summer in England, when there is a flush of such young individuals in the nest (Smeeton, 1981). "True" workers of the European slavemaker *Harpagoxenus sublaevis* exist, and Buschinger and Winter (1978) found 11 fertile individuals among 230 dissected; of these 7 had been laying eggs in the presence of the queen. But the workers in this unusual species differ from the much more fertile ergatogynes only in their lack of the seminal receptacle necessary to store semen for the normal generation of female offspring. A stronger exception to the general rule appears to be *Leptothorax allardycei*, whose workers produce 20 percent of the eggs in the presence of the queen (Cole, 1986).

The overall picture of worker reproduction is one of marginal productivity and nearly universal inhibition by the laying queens. This circumstance is difficult to interpret with reference to the two major competing hypotheses, but it appears to favor parental manipulation. There would seem, prima facie, to be strong selection for individual worker reproduction in the presence of the queen. By producing males, the egg-layer trades brothers for sons and hence a degree of relatedness (*b*) of ¼ for one of ½. The worker's sisters should be compliant, at least in the presence of a single queen who mated once, because by exchanging brothers for nephews they trade a relatedness of ¼ for one of ⅜.

On the other hand, both the kin selection and parental manipulation hypotheses predict some conflict among the workers for whatever reproductive rights they assume. In fact, dominance hierarchies have been reported among the workers of the little stem-dwelling myrmicine *Leptothorax allardycei* (Cole, 1981) and the slavemaker *Harpagoxenus americanus* (Franks and Scovell, 1983). Otherwise, however, observations of conflict among workers have been limited to queenless colonies, in which not only has the queen's contribution ended but the entire reproductive future of the colony is at risk.

*Who controls the investments?* Because the predicted ratios of investment are quantitative, they provide the most rigorous comparison of kin selection and parental manipulation. Trivers and Hare (1976) distinguished three kinds of societies that should possess characteristic investment ratios if kin selection is operating.

In *monogynous colonies,* the ratio of investment should be 3:1 in favor of the female reproductives, provided the workers have evolved so as to control the investment in a way that maximizes their inclusive fitness. Put another way, the workers should see to it that the male biomass constitutes 25 percent of the total invested in reproductive adults as a whole. The reason: workers are weighing degrees of relatedness of ¾ for their sisters against ¼ for their brothers.

In *polygynous colonies,* the investment ratio can be expected to subside toward 1:1 (50 percent male investment), because the workers are no longer all sisters. Indeed, as revealed by recent allozyme studies (Table 4–1), the average degree of relatedness among the workers from individual colonies is usually well below ½ and often below ¼.

FIGURE 4–5 Support for the kin selection theory is provided by investment in biomass and numbers of males as opposed to investment in biomass and numbers of females. The hypothesis generated by the theory predicts a biomass investment of 25 percent in males in monogynous (single-queen) species and 50 percent in polygynous (multiple-queened) species, while both kinds of species should produce numerical investment of 50 percent. As summarized here, the predictions are upheld by data from 34 principally monogynous species and 15 polygynous species. (Modified from Nonacs, 1986a.)

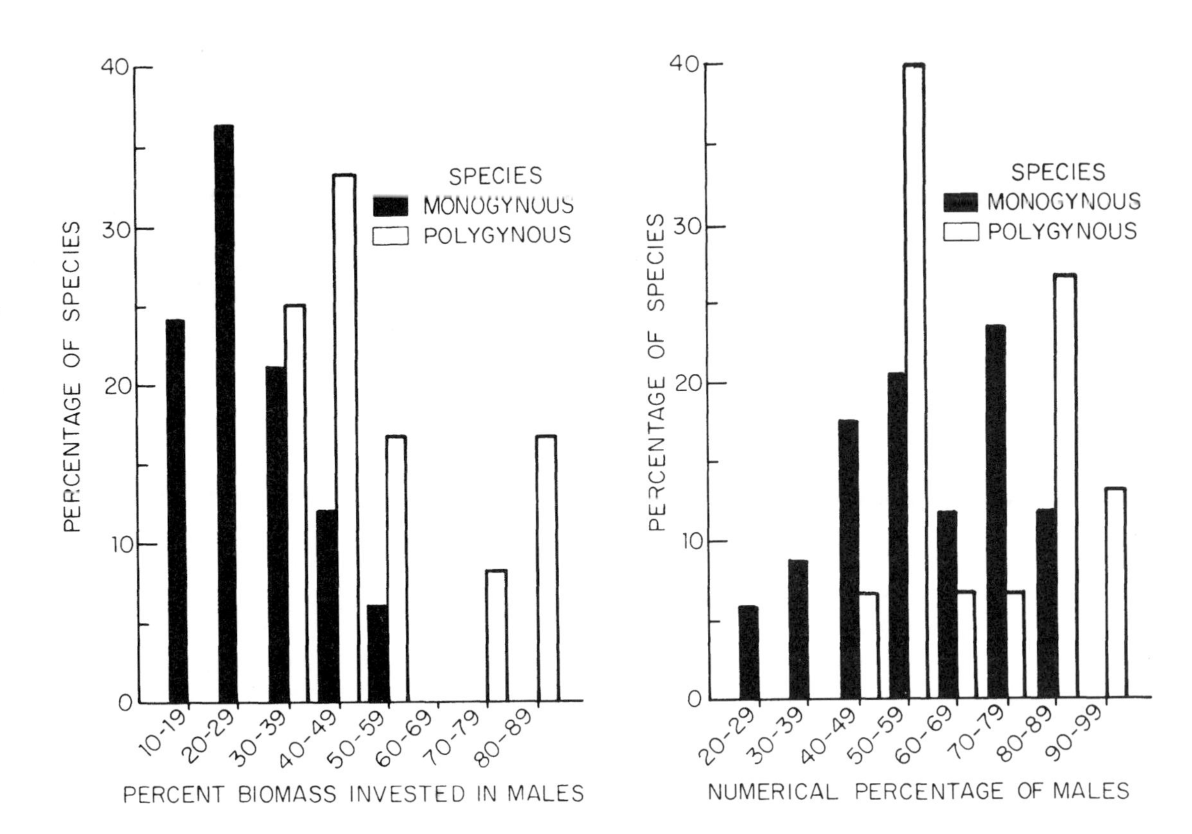


In *mixed colonies* of parasitic and host species, the ratio should also be 1:1, because the parasite queens are selected to control the ratio, whereas the captive and usually queenless workers of the host species cannot be selected to resist manipulation by the parasite. They are unable to evolve resistance because they leave behind no offspring and cannot otherwise aid in the continuance of their own species.

In a recent analysis of the rapidly accumulating studies on investment ratios, Nonacs (1986a) found the Trivers-Hare predictions based on kin selection to be "remarkably robust," whereas the parental manipulation hypothesis fell short in several key areas. The principal results are depicted in Figure 4–5. The percentages of biomass investment in males, with the standard error, are 0.282 ± 0.06 for the monogynous species, 0.522 ± 0.09 for the polygynous species, and 0.515 ± 0.02 for the parasitic species, all quite close to the equilibrium figures predicted by the kin selection model. The numerical percentages are 0.579 ± 0.09, 0.687 ± 0.09, and 0.628 ± 0.03 for the same three groups of species. These are again respectably close to the predicted value of 50 percent. It will be recalled that the numerical sex ratio is expected to be close to unity in a fully outbreeding system, because males can usually inseminate only one queen.

It is a curious feature of the investment ratio strategies that they apply in a given species to the population of colonies as a whole rather than to individual colonies. As illustrated in Figure 4–6, colonies tend to specialize on one sex or the other in any given reproductive season. Nonacs (1986a,b) found that the proportion of males decreases with an increase in the biomass of new reproductives in the colony as a whole. The correlation is not primarily with the size of the colony, however. When Nonacs parceled out the biomass of reproductives, there was almost no relationship between the number of workers in the colony and the percentage of investment in males. These trends led him to conclude that colonies invest mostly in males when resources are low (and only a small reproductive crop

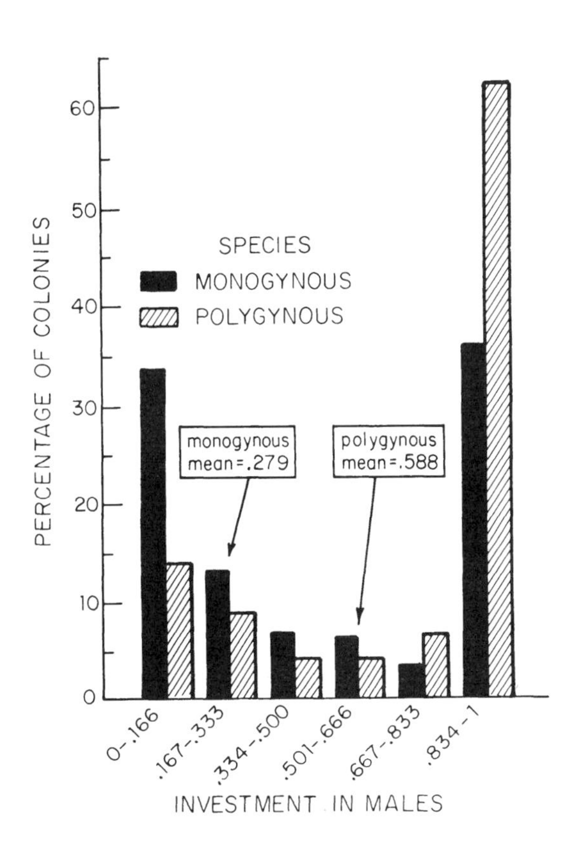


FIGURE 4–6 The trends in investment ratios by ants are population-wide and not particular to individual colonies, because colonies tend to specialize at various times in the production of males or queens. The result is a bimodal curve when individual colonies rather than species are assessed. The curves illustrated here are based on 22 monogynous species and 6 polygynous species. (Redrawn from Nonacs, 1986a.)

can be raised) and mostly in females when resources become more plentiful. This conclusion seems supported by the independent finding of Rosengren and Pamilo (1986) that colonies of *Formica aquilonia* and *F. rufa* bias the sex ratio investment toward females when the environment is most favorable. The input of the resources thus serves as a proximate cue to the colony on where to place its surplus energy income. The effects summed over the population of colonies as a whole yield the 3:1 or 1:1 ratio. If a large population of colonies is at equilibrium, then all sex ratios will be equally fit, as shown in a general proof by Taylor and Sauer (1980). Hence even colonies that produce all males or all females cannot be assumed to be acting against the inclusive fitness of their workers.

The intuitive basis for this argument is as follows. In an equilibrial population, with either a 3:1 or a 1:1 ratio overall, the colony that produces mostly males will find itself compensated during mating because some other colonies are producing excess females. What evolves over a period of time, in theory, is the sensitivity of the individual colony to resource levels. These thresholds are selected so that the colony acts "appropriately" with respect to the population as a whole. In short, the sex ratios chosen at different times by a single colony represent an evolutionary stable strategy of mixed responses of the kind suggested by Maynard Smith (1982).

Two detailed studies of variation among colonies of the same species have shed further light on the evolution of investment ratios. Herbers (1984) analyzed colony units of *Leptothorax longispinosus* with variable numbers of queens. She found that the fewer the queens, the closer the investment ratio approached 3:1 in favor of females. Queenless groups had the highest ratio of all. The result can be interpreted as kin selection operating amid a conflict between the queens and the workers. When no queens are present, the workers boost the ratio without interference. When many queens are present, they push the ratio back to a level more in their favor. The 1:1 ratio tends to be more in the interest of the workers in this case, however, so conflict might well be attenuated at the extreme multiple-queen end of the scale. Nonacs (1986b) interprets the result as a response of the polygynous colonies to fewer resources, supplied by a smaller proportion of workers. Whichever of the two interpretations is correct, Herbers and Nonacs agree that kin selection is an important force in *Leptothorax longispinosus* and that some amount of conflict between the queens and the workers does occur.

Ward (1983b) has conducted an equally instructive study on two species of Australian ponerine ants belonging to the *Rhytidoponera impressa* group (*chalybaea* and *confusa*). Each of these species has two types of colonies. Type A colonies have a single queen, and their overall mean proportional investment is 0.82 in favor of the females. Type B colonies are queenless and instead are serviced entirely by reproductive workers. They are therefore usually polygynous. Their mean proportionate investment in queens ranges from 0.35 to 0.72. The two types of colonies are intermingled. The higher the relative frequency and density of the type B (queenless) colonies in particular local populations, the higher the proportionate investment by the neighboring type A (queenright) colonies in new queens. In fact, their average investment has risen above the 0.75 level predicted by kin selection theory in some localities, presumably as a local adaptive response to the larger number of males being generated in nearby type B colonies. This interpretation is supported by the fact that one population of *Rhytidoponera purpurea*, which also belongs to the *impressa* group and consists entirely of type A colonies, has a proportionate investment of 0.74, in close accordance with the kin selection model.

As Ward points out, there are two ways by which type A colonies could compensate their investment ratios in the manner observed. First, there could be some form of communicating by which they gauge the density of type B colonies. Such an assessment among colonies is not impossible. Hölldobler (1976c, 1981a) has shown that it occurs during the ritualized tournaments among colonies of the honeypot ant *Myrmecocystus mimicus*. Alternatively, the investment ratios of type A colonies might have evolved in local populations in response to stable environments and equilibrial densities of type B colonies. This explanation is made plausible by the fact that the rain forest populations of *chalybaea* and *confusa* are in habitats that have persisted for thousands of years.

Finally, Bourke et al. (1988) provided solid evidence supporting the prediction made by Trivers and Hare that mixed colonies of parasitic and host species invest at a ratio of 1:1. They found that the mean proportion of dry weight investment in queens belonging to populations of the monogynous slavemaker *Harpagoxenus sublaevis* is 0.54. Using allozyme analysis, they also confirmed that each queen mates with only one male, that female nestmates are full sisters (coefficient of relatedness 0.73 ± 0.07), that inbreeding does not occur, and that queen and worker siblings are not genetically differentiated.

Frank (1987) presented an alternative model, which attempts to explain the phenomenon that small colonies of some ant species tend to produce predominantly males, whereas large colonies produce mostly female alates. This model, which Frank calls the constant male hypothesis, suggests that "when there is any local mate competition, however slight, small colonies are favored to make mostly males, and large colonies are expected to increase their investment in females as their total brood increases." As yet, however, no convincing behavioral or genetical evidence exists indicating even a weak trend toward local mate competition in mating swarms of ants. In an important study of *Lasius niger,* van der Have et al. (1988) excluded mate competition in favor of kin selection. They found that in marginal populations, investment ratios tend to drop from 0.75 toward 0.50, in part because of higher production of males by workers. Similarly, Elmes (1987b) found that the largest and most reproductively successful colonies of *Myrmica sulcinodis* have investment ratios close to 0.75, which drop toward male bias in marginal, polygynous populations.

Finally, it is surely significant that the larger and more complexly organized the monogynous ant colony, and hence the less preponderant the physical presence of the queen, the more care the workers devote to her. Put another way, as the colony grows larger the workers appear to treat the queen more as a valuable resource and less as a rival. Among the more arresting spectacles of the ant world are the dense retinues of certain species that guard the mother queen as she moves from one nest site to another. In extreme cases, such as species of the legionary ponerine *Simopelta*, the doryline and ecitonine army ants, the leptanilline army ant *Leptanilla*, *Solenopsis* fire ants, *Pheidologeton* marauder ants, and *Oecophylla* weaver ants, the entire body of the queen is covered by a seething shell of these guards (see Figure 4–7). In ecitonine army ants, the retinue extends as much as a meter in front of the queen and 2 meters behind her

**FIGURE 4–7** Retinues of workers form around the mother queens of large monogynous colonies, in a manner suggesting that the workers treat this caste as a valuable resource. (*a*) A mass of major workers of the African weaver ant *Oecophylla longinoda* covers the queen during an emigration from one nest site to another. Many members of the retinue have raised their abdomens in a defensive posture. (*b*) Inside a nest chamber, the *Oecophylla* queen is surrounded, licked, and fed continuously by her retinue. The queen's head and intersegmental protrusions (the latter bearing pores through which special glands discharge secretions) are especially attractive to the attendants. (Modified from Hölldobler and Wilson, 1983a.)

and contains up to five times the number of workers found in an ordinary 2-meter segment of the column of emigrating ants. Epigaeic species, in other words those whose colonies move mostly above ground and in sight of predators, have the largest retinues (Rettenmeyer et al., 1978). In retinue-forming ant species investigated so far, the queen possesses unusually extensive exocrine glands, which evidently produce pheromones attractive to the workers.

A fruitful direction to take in ant biology will be to examine investment priorities among different classes of colony members. In principle, we should expect workers to maximize their personal inclusive fitness by favoring particular castes and life stages over others. They might provide special care to their own sisters as opposed to half-sisters and more distantly related relatives. A few tantalizing pieces of evidence concerning such discrimination have surfaced already. Petersen-Braun (1982) found that workers of the house ant *Monomorium pharaonis* respond most aggressively to the queens who are less productive, with attacks turning fatal when the colony breeding cycle is threatened. In this study, however, no specific discrimination based on relatedness could be detected. A similar phenomenon has been reported in the fire ant *Solenopsis invicta* by Fletcher and Blum (1983a,b). Lenoir (1981) reports that when workers of *Lasius niger* discover lost brood, they retrieve the different stages in the following order: large larvae and pupae, then small larvae, and finally eggs. This is the sequence predictable from the differing amounts of payoff (in healthy adults produced) expected from these stages. In particular, pupae have already received the most energy and will enter the adult generation with the least additional care, whereas eggs still require large amounts of energy investment and future time. Unfortunately, the tests were not calibrated in a manner that allowed choices by the ants to be made on a per-gram basis.

## EUSOCIALITY AND CHROMOSOME NUMBERS

Sherman (1979) pointed out that colonies of social insects tend to be more harmonious when the members are more closely related; therefore mechanisms to reduce within-colony genetic variance should be favored by natural selection operating at the level of the colony—or more precisely, at the level of the queen. One way of reducing variance is to increase the number of chromosomes, so that large blocks of genes held together by linkage are broken up and their constituents distributed more evenly among colony members during meiosis and fertilization. Increased cross-over should produce the same result. The data assembled by Sherman appear to confirm that species of eusocial insects, including ants and termites, do have higher chromosome numbers than the solitary species phylogenetically closest to them. Crozier's summary (1987b) supports this result for ants and other social insects, even though one species, a bulldog ant in the *Myrmecia pilosula* complex, has a haploid number of only 1 (*Myrmecia* is exceedingly variable in this respect, with another species, *M. brevinoda*, possessing a haploid number of 42). As Crozier points out, there is no way at present to judge which came first, that is, whether high chromosome numbers predisposed certain phylogenetic lines toward eusociality or the reverse. He is certainly correct in suggesting that "sociogenetics," the blending of genetic and sociobiological analysis at the chromosome and genic levels, is a still inchoate field of great promise. Ultimately it will change our view of the way in which eusociality originated and is sustained.

## OVERVIEW

Over the past twenty years, the analysis of kinship and investment strategies in ants and other social insects has matured into one of the more sophisticated enterprises of evolutionary biology. During this time opinion concerning the dominant selective force of colonial evolution has shifted twice. Throughout the 1960s and into the 1970s, enthusiasm prevailed for the pure kin selection hypothesis, which appeared to explain the origin of the sterile worker caste in a uniquely robust manner. Sentiment in this direction was reinforced when Trivers and Hare (1976) added the concepts of conflict among colony members and the optimum 3:1 queen-to-male ratio of investment. Then the fortunes of kin selection declined as Alexander and Sherman (1977) pressed the opposing parental manipulation model. It seemed that theoreticians might dispense with a complex calculus of investments if the mother could be shown to force some of her daughters into a slave-like worker status with a reduced inclusive fitness.

Doubts about kin selection were reinforced by flaws discovered in the statistical analysis of the early investment data, along with a battery of new theoretical arguments that made parental dominance seem easier to attain in the course of evolution. Alexander and Sherman argued further that biasing toward female production could be explained as the result of local mate competition, wherein the queen produces fewer males because a smaller force is needed to inseminate the cogenerational crop of sisters and other close kin. Crozier (1977) pointed out that for kin selection to create a sterile worker caste, as opposed to merely maintaining it, the ancestral preformicids would have to have been able to tell the sex of larvae *before* such species could create the favorable 3:1 ratio. In other words, the sensory cart must be put in front of the behavioral horse. Charlesworth, Charnov, and Craig independently noted that the mother need only make offspring half as efficient at rearing other brood in order to serve her own interests, so that worker formation was easier by means of parental manipulation than had been earlier imagined. Altogether, kin selection took a heavy beating during the late 1970s.

Then, in the 1980s, the tide turned back. As Crozier himself cautioned (in 1982), "Don't discard kin selection in favor of parental manipulation." More and more pieces began to fall into place to create a pattern favorable to kin selection within a setting of parent-offspring conflict. From Bartz's rule to the cumulative investment measurements analyzed by Nonacs, as well as other contributions we have reviewed in this chapter, details of colony life histories and caste structure were elucidated that seemed difficult to interpret by any other existing model, including parental manipulation. Furthermore, it began to be appreciated that parental manipulation could take a form concordant with positive kin selection rather than opposed to it, if the mother changed the environment and rearing conditions so as to improve the inclusive fitness of her "enslaved" daughters.

This interesting saga, reminiscent of some of the best historical controversies in physics and molecular biology, is far from over. Too many surprising theoretical and empirical discoveries have been made during the past ten years for us to suppose that all competing explanations, including parental manipulation, can be discarded for good. One of the more desirable targets for future research is the close comparison of primitive and advanced ant taxa. Anatomically primitive genera with relatively simple social organizations, especially *Amblyopone* and other members of the Amblyoponini, as well as *Myrmecia* and *Nothomyrmecia,* should be compared in detail with evolutionarily advanced genera such as *Formica* and *Pheidole.* It is likely that differences in the degree of conflict and investment patterns will be found and will be used to discriminate more confidently among the competing hypotheses.

The basis of this last prediction is as follows. Charnov (1978) first pointed out that the conditions for the beginning of eusociality must have been very different from the conditions for its maintenance in advanced eusocial species. Indeed, many of the difficulties envisioned by Alexander, Crozier, Pamilo, and others centered on the origin of the worker caste. It is entirely possible that parental manipulation played a role in spreading the first rare genes that made worker development possible. Alternatively, the first workers could have been females parasitizing their mothers by choosing to settle in the home environment rather than to risk creating new nests. As Bartz (1982) pointed out, the stay-at-homes were not required to bias their reproductive siblings into a 3:1 investment ratio in order to evolve into workers themselves. It was only when the queen produced all of the males herself that a female-biased ratio was necessary to create degrees of relatedness favorable to the evolution of female workers. If daughters produced any male reproductives at all, then the requirement for a biased investment ratio was sidestepped. If a female raised a nephew rather than a brother, she would have traded a degree of relatedness of $1/4$ for one of $3/8$. Consequently her average relatedness to a reproductive brood composed of sisters, brothers, and nephews would have been greater than the relatedness to a brood composed of only sisters and brothers, and under many combinations, greater than her relatedness to a brood she could have expected to raise had she started her own nest.

Once eusocial species reached a more advanced stage, with large colonies and specialized worker subcastes, the rules changed. It is likely that worker reproduction and conflict lost most of their profit in comparison with the maintenance and fine-tuning of colony organization. As far back as the late Cretaceous Period, the ants as we know them may have reached a point of no return. The earliest stages of eusociality have not been found in any living species of ant and no solitary aculeates are known that might have originated from ants through a secondary loss of eusociality.

The bees and wasps remain by far the best insect groups for studying the origin of the worker caste. They offer a graded series of living species that range from completely solitary to completely eusocial. But the extraordinary phylogenetic spread of the ants and the immense diversity of their social systems make them the most favorable group for reconstruction of the middle and advanced grades of eusocial evolution.

# CHAPTER 5

# Colony Odor and Kin Recognition

When a worker inspects a nestmate she seems to do nothing more than casually sweep her antennae over the other's body. The true intensity of the inspection is revealed, however, when an alien ant enters the nest. An intruder who belongs to a different species is almost always violently attacked. On the other hand, if the new ant is a member of the same species but from a different colony, the hostility falls somewhere along a broad gradient of responses. At one extreme, intruders are accepted but offered less food until they have time to acquire the colony odor. At the other extreme, the residents attack strangers with extreme violence, locking their mandibles on body and appendages while stinging or spraying with formic acid, citronellal, or some other toxic substance. Intermediate degrees of rejection include avoidance, mutual threatening with open mandibles, and nipping and leg pulling. This array of responses from aversion to violence has been used as the basis of sensitive bioassays in studies of colony odor by a number of authors from Lange (1960, 1967) to Carlin and Hölldobler (1983, 1986, 1987), as summarized in Table 5–1 and Figure 5–1.

## KIN RECOGNITION: GENERAL PRINCIPLES

The ability to distinguish nestmates from strangers is vital to social life among the ants. Waldman (1987) recommends that the term *kin discrimination* be used to denote differential treatment of conspecifics correlated with kinship, and that *recognition* be defined more narrowly as "the processes by which individuals assess the genetic relatedness of conspecifics to themselves or others, based upon their perception of traits expressed by or associated with these individuals." Waldman et al. (1988) reemphasize the distinction between the two terms, because "recognition—a series of internal and essentially unobservable physiological events—may occur without any behavioral response."

Throughout the evolutionary history of the ants and other social insects there has been an intense selection pressure to sharpen recognition ability, because favors bestowed on an unrelated individual are wasted in the remorseless crucible of natural selection. Consequently, behavioral discrimination of relatives from non-relatives must be intimately tied to kin selection. If kin selection works as effectively as the evidence (reviewed in Chapter 4) suggests, in other words if it constitutes a strong "ultimate" factor in evolution, then mechanisms of kin recognition would be advantageous as a means of directing nepotism correctly. In studies across a broad diversity of animals, kin recognition of one kind or another has been implicated in most kinds of social behavior, from simple aggregations of tadpoles to the most complex colonial organizations of ants and termites (reviews by Hölldobler and Michener, 1980; Gadagkar, 1985; Fletcher and Michener, 1987). Where a discriminating capacity was expected from kin selection theory, it has almost always been found to exist in fact. Moreover, study of the linkage has proved heuristic for biology. The mediating behavior is often complex and highly effective, and new physiological processes are more easily discovered when research is animated by kin selection theory.

When considering recognition labels, keep in mind that two forms of kin discrimination can be distinguished in insect societies: exclusion of non-kin from the colony, and preferential aid for kin of higher relatedness within the colony. As Hölldobler and Carlin (1987) point out, labels involved in recognition at the colony level are simultaneously specific and anonymous. That is, workers are able to discriminate between nestmates and intruders, but they also tend to treat all nestmates as fellow colony members, regardless of the degree of their relatedness. This view of anonymity among genetically varying nestmates (Jaisson, 1985, called it the fellowship concept) does not preclude specificity at the within-colony level. Generally, though, it appears that workers encountering one another in the context of territorial defense or nest guarding respond to cues that indicate colony membership rather than kinship.

Useful classifications of the labeling phenomena have been prepared by Holmes and Sherman (1983) and Sherman and Holmes (1985) for animals generally and by Hölldobler and Michener (1980) for social insects in particular. Four principal strategies exist, each employing a different kind of cue or stimulus by which one individual classifies other members of the same species.

The first strategy is based on *purely spatial distribution*. It occurs in species with a high degree of site fidelity. Adult bank swallows (*Riparia riparia*), to cite one well-studied example, learn the location of the nest holes they excavate and feed any chicks found in these retreats (including alien swallow chicks introduced by the investigator) up until the time their own offspring fledge at about two weeks of age. Conversely, they ignore their own nestlings when the experimenter transfers them to nearby burrows. No example of reliance on purely spatial distribution has yet been documented in the ants or any other social insects, although Klahn (1979) demonstrated the use of spatial as well as phenotypic cues in *Polistes* wasps.

In *purely allelic recognition*, as conceived in theory at least, animals would depend on the innate capacity to discriminate other individ-

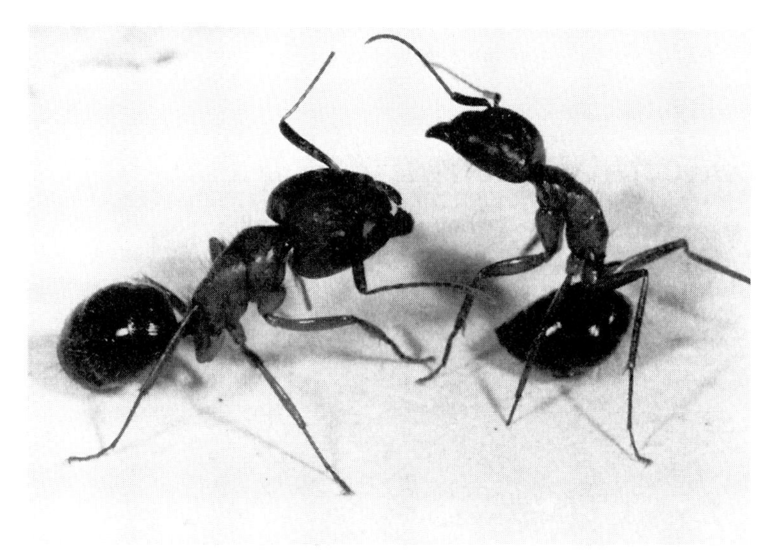

**FIGURE 5–1** Three degrees of aggressive behavior among workers of *Camponotus floridanus*. (*Top*) Threat display. (*Center*) Grappling and pulling appendages, in this case the antennae. (*Bottom*) Full attack, which usually ends with the death of one or both of the adversaries.

**TABLE 5–1** Seven-level aggression scale used to score the responses of unfamiliar workers of the carpenter ant genus *Camponotus*. (From Carlin and Hölldobler, 1986.)

| | |
|---|---|
| 0 | Casual tolerance; huddling together; allogrooming; exchanging food |
| 1 | Initial jerking back, then tolerance; initial or weak avoidance; weak open-mandible threat |
| 2 | Intense antennation ("investigation"); rapid mutual antennation; jerking back at each encounter; strong open-mandible threat |
| 3 | Strong avoidance or flight; light mandible-mandible nipping ("nibbling"); aggressive regurgitation ("spitting fight"); standing atop the opponent |
| 4 | Repeated, rapid forward-and-back jerking with open mandibles; stilt-legged posture; "advance-retreat"; carrying the intruder |
| 5 | Strong mandible-mandible nipping ("sparring"); seizing and dragging; lunging (weak charge); nipping antennae, body limbs; chasing; gaster twisted forward to spray formic acid |
| 6 | Charge and attack; brief locking together; prolonged biting/spraying fight |

uals with which they share certain alleles. The alleles could be expressed in the phenotype in one of several ways, as for example by odor or by physical appearance. The important qualifier in this extreme category is that the phenotypic difference is innately recognized, not learned. Dawkins (1976) has called the hypothetical phenomenon the green beard effect: if the focal animal were to have a green beard it would classify all green-bearded strangers as kin. The possibility is intriguing and may be relevant to within-colony discrimination, but most investigators agree that allelic discrimination is unlikely to evolve as the primary system in discrimination at the colony level (Crozier, 1987a). It requires the possession of individually distinctive sets of genes that prescribe *both* the recognition cue and the neurosensory apparatus to recognize it.

The principal difficulty in proving such an ultrasimple arrangement is that learning has been implicated in most cases of colony discrimination, even though it is often of a very restricted and predictable nature. No one has invented a way to control for, and thereby to eliminate, all conceivable learning possibilities, including the familiarization of an animal with its own cues. Even so, we may confidently look for cases in the simplest of organisms, where interactions take the form of growth and tissue rejection rather than conventional learning in neuronal systems. Examples include corals, sponges, and other colonial invertebrates that fuse bodily with genetically identical organisms but reject those that are even slightly different.

The third strategy of kin recognition is discrimination "of specific individuals with which one has previously interacted." This form can be called *recognition by association*. As Waldman et al. (1988) describe it, "individuals presumably learn one another's traits in a setting where only relatives are likely to be present, later distinguishing them from non-kin in other potentially more ambiguous settings . . . If the labels expressed by every member of a circumscribed kin group (e.g., within a nest, burrow, or insect colony) are distinct, individual recognition may result."

Waldman (1987) points out that this kind of kin recognition mechanism is nevertheless very similar and perhaps even identical to *phe-*

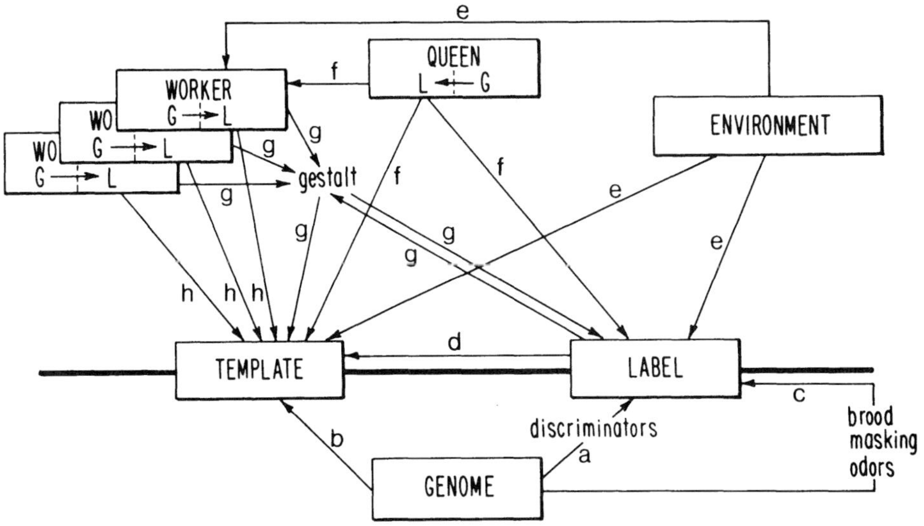

**FIGURE 5–2** A general model of a kin recognition system in the social insects, utilizing the terminology applied to phenotype matching by Holmes and Sherman (1983). The specific mechanisms are labeled *a–h*, and their hierarchical ranking in the ant genus *Camponotus* is described in the text. The thick horizontal line represents the external surface of the worker; *G* denotes the genome and *L* the olfactory label by which comparisons are made against the template, or standard of acceptable odors learned by the workers. (From Carlin and Hölldobler, 1986.)

*notype matching,* the fourth strategy of kin recognition proposed by Sherman and Holmes (1985). In this case kinship identity may be inferred from the perceived overlap in cues between conspecifics, whereby animals are able to discriminate between unfamiliar kin and unfamiliar non-kin, or among familiar kin with different degrees of relatedness.

The evidence to date from social insects exclusively implicates phenotype matching. Furthermore, only chemical signals are known to be employed. This being the case, we have to consider that nestmate recognition consists of both perceptive and expressive components, and the ontogeny of both must be included in any complete explanation of the recognition system (Gamboa et al., 1986; Hölldobler and Carlin, 1987). For variation in labels to be functional, an individual must have some criteria for determining whether to respond to a given variant. These decision-making rules can be innate or learned or, to use less problematic terminology, they can be determined by closed or open ontogenetic programs (Mayr, 1974). Hölldobler and Carlin (1987) noted that the rules of perception for both anonymous and specific communication signals can be either genetically encoded or acquired by experience, depending on the predictability of signal expression. When the expression of a semiochemical is highly predictable, the genome of the receiver can "know" in advance what characteristics to expect and can program an efficient, hard-wired neural mechanism for recognizing them. This is clearly true for certain chemical communication signals, such as anonymous sex pheromones detected by specialist receptors. Conversely, when the expression of a semiochemical is unpredictable, the receiver's genome cannot dictate a perception mechanism in advance, and the criteria for responding must be derived from experience.

In all of the species of social insects studied thus far, nestmate recognition cues appear to be learned shortly after eclosion into the adult stage. New workers eclosing into a colony whose queen has mated more than once cannot know which heritable recognition signals to expect among their half-siblings. In addition, they must learn cues that were acquired from other colony members and the external environment. Masson and Arnold (1984) and Gascuel et al. (1987) suggest that young adult honey bees learn odors in an imprinting-like manner, through restrictive timing of olfactory center development in the brain, which would admirably fit Mayr's (1974) definition of an open ontogenetic program.

The following general scheme of phenotype matching in social insects can be reasonably suggested. Each individual possesses both phenotypic recognition cues—a recognition label—and a sensory template specifying a learned set of cues likely to be borne by kin. An observing individual determines whether or not an unfamiliar conspecific is a relative by matching the latter's label to its own template. Carlin and Hölldobler (1986) constructed a flow-diagram model of possible extrinsic and intrinsic inputs to the label and template of a social insect worker, which is summarized in Figure 5–2. The thick horizontal line represents the body surface and sensory receptors of the observing individual. Intrinsic inputs (below the line) are genetic in origin, while extrinsic inputs to the label (above the line) presumably occur on the body surface. Inputs to the template are learned.

Genetically determined discriminators are known, or at least strongly suspected, in a wide range of species of bees (Greenberg, 1979; Breed, 1981, 1983) and ants (Jutsum et al., 1979; Haskins and Haskins, 1983; Mintzer and Vinson, 1985b; Stuart, 1987a). Direct genetic specification of both labels and templates has not been documented in any species. Such a mechanism risks false-negative discrimination against kin when colonies contain even a moderate genetic diversity, a circumstance that arises when queens mate with more than one male. The need to accept nestmates who are unpredictably varied, thus preserving colony cohesiveness, favors the evolution of mechanisms based on learning. Callows whose discriminators differ from the ambient cues of the colony in which they emerge may be protected by special pheromones, such as brood-masking substances, while they acquire labels and learn templates (Jaisson, 1972a,b; Hölldobler, 1977). Environmental input to recognition labels is known in a number of social Hymenoptera (Kalmus and Ribbands, 1952; Jutsum et al., 1979; Mabelis, 1979b; Boch and

Morse, 1981; Haskins and Haskins, 1983; Gamboa et al., 1986; Obin, 1986; Breed and Bennett, 1987; Stuart, 1987b,c; Obin and Vander Meer, 1988, 1989). Purely extrinsic recognition risks false-positive acceptance between neighboring colonies that share environmental cues, however, and it is precisely these neighbors that nestmates will encounter and need to oppose. In most cases, therefore, recognition is not likely to rely solely on differences in soil and nest-material odors, although a prominent environmental component has been demonstrated in the recognition system of the fire ant *Solenopsis invicta* (Obin, 1986; Obin and Vander Meer, 1988) and *Polistes* wasps (Gamboa et al., 1986).

Queen discriminators have been demonstrated in honey bees (Breed, 1981; Boch and Morse, 1982) and inferred in some ant species (Watkins and Cole, 1966; Jouvenaz et al., 1974). Their transferability has been demonstrated in *Camponotus* species (Carlin and Hölldobler, 1986, 1987, 1988; Carlin and Vander Meer, unpublished data). Effects of queens on worker nestmate discrimination have been documented in several *Camponotus* species (Carlin and Hölldobler, 1983, 1986, 1987) as well as in *Leptothorax lichtensteini* (Provost, 1985, 1987) and interspecific mixed colonies of *Myrmica* (Brian, 1986a). In other *Leptothorax* species (Stuart, 1987a–c), *Pseudomyrmex ferruginea* (Mintzer, 1982b), and *Solenopsis invicta* (Obin and Vander Meer, 1988, 1989), however, queens have been shown to contribute little or nothing to nestmate discrimination.

A worker gestalt mechanism, according to the criteria in the flow-diagram model (Figure 5–2), has been clearly demonstrated in some *Leptothorax* species, which have small and frequently polygynous colonies (Stuart, 1987a–c). It is noteworthy, however, that the queen input in label and template formation in *Camponotus* may also be considered part of the "gestalt model" as originally formulated by Crozier and Dix (1979). In fact the queen can dominate a collective gestalt. Carlin and Hölldobler (1986, 1987) demonstrated that, in *Camponotus pennsylvanicus* and *C. floridanus*, queens, worker genotype, and environmental cues all contribute to the nestmate discrimination labels of adult workers, but not all are of equal importance. Breed (1987) reports that honey bee workers acquire extrinsic recognition cues in the presence of queens and environmental odors, although there is some uncertainty over the possible formation of collective gestalt labels among queenless workers (Breed et al., 1985; Getz and Smith, 1986; Getz et al., 1986).

## DISCRIMINATORY ABILITY

The existence of a finely tuned recognition system based on degrees of genetic similarity has been demonstrated in the social bees, who are relatively amenable to breeding experiments. Greenberg (1979) utilized *Lasioglossum zephyrum*, a primitively eusocial sweat bee that constructs simple burrows in the soil. One of the colony members serves as a guard by sitting at the entrance to the outer burrow and blocking alien *L. zephyrum* and other intruders. Greenberg bred bees in the laboratory to produce colonies having 12 different levels of genealogical relationship to one another, from completely unrelated groups to colonies whose members were sisters in inbred brother-sister lines. Bees were tested in pairs by introducing an individual from one nest to a guard bee of a different nest, under regimes in which the two individuals could never previously have seen or smelled each other. In the course of 1,586 such introductions there was a strong positive correlation between the degree of relatedness between the two bees and the frequency with which guards permitted introduced bees to pass (Figure 5–3). The experiments were reinforced during subsequent trials by Buckle and Greenberg (1981), who provided several lines of new evidence indicating that odor cues are not transferred among nestmates. Their additional conclusion, that the guard bee knows only the odors of her nestmates, and not her own, is questionable, however (Getz, 1982).

Similar experiments by Breed (1981, 1983) revealed the existence of a genetic component in the colony odor of honey bees. He found that both queens and workers of *Apis mellifera* were accepted into strange nests with a probability that rose with the degree of relatedness, where three degrees of relatedness were employed (inbred sisters, outbred sisters, and unrelated). Getz and Smith (1983) performed similar experiments but used genetic markers to measure genetic relatedness. By this means they discovered that worker bees can distinguish full sisters from half-sisters. Further evidence has been adduced by Getz and Smith (1986), as well as by Breed et al. (1985), that in some contexts honey bee workers perceive their own labels and use this "self-awareness" as a template for recognition of nestmates and full sisters within the colony. Page and Breed (1987) suggest that self-awareness is probably learned, a process sometimes called self-matching.

The common occurrence of multiple insemination and polygyny makes necessary a mechanism for anonymously identifying all nestmates as colony members as opposed to intruders. In the context of interactions among colony members, especially the rearing of reproductive brood, intracolony genetic heterogeneity also reduces the level of inclusive fitness considered important for the maintenance of hymenopteran eusociality. A worker who indiscriminately rears half-sisters (patrilines) or the offspring of other queens (matrilines) does not have the same proxy reproductive success that haplodiploid workers obtain by rearing three-quarters-related full sisters. Kin discrimination has been proposed as a solution to this difficulty, which would preserve eusociality by kin selection despite low relatedness within colonies (Gadagkar, 1985). Even if eusociality is maintained by some other means, workers that find themselves among nestmates of varied relatedness can improve their inclusive fitness by discriminating on the basis of kinship. In either case sufficient cue specificity must be retained within colonies to permit discrimination of full sisters from other patrilines and matrilines if cohorts defined by three-quarters relationship are to cooperate preferentially (Hölldobler and Carlin, 1987).

The studies with bees just cited indicate that within-colony discrimination among workers is possible. In addition Getz et al. (1982) reported segregation of worker patrilines during swarming, Evers and Seeley (1986) observed aggressive discrimination between patrilines in queenless colonies containing ovipositing workers, and Frumhoff and Schneider (1987) found that workers prefer to exchange food with full sisters rather than with half-sisters. They also prefer to groom full sisters. All of these investigators used genetic color markers to identify the patrilineal cohorts, which were produced by artificial insemination. Later results indicated that the color markers exaggerate the specificity of half-sister discriminators, however (Frumhoff, 1987).

Although discrimination among patrilines has not been tested in

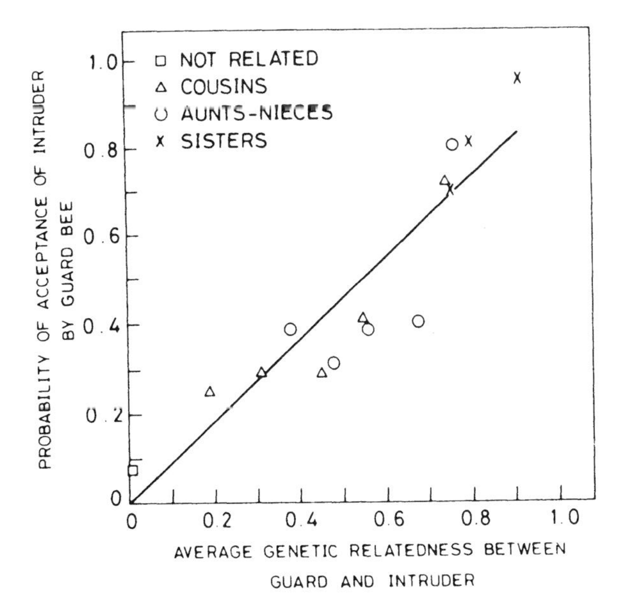


**FIGURE 5-3** In the sweat bee *Lasioglossum zephyrum*, the probability that a guard bee will admit a strange bee into the nest burrow is correlated with the degree of the guard's relatedness to the intruder. (Based on Greenberg, 1979.)

**FIGURE 5-4** Two workers of *Camponotus floridanus* exchanging food. The workers are adoptive nestmates; in other words they originated from different colonies. The worker on the left is marked with a tiny wire ring around the waist.

ants, Carlin et al. (1987b) found that *Camponotus floridanus* workers originating from unrelated colonies and introduced into mixed nests antennate familiar non-kin more frequently than familiar sisters but fail to discriminate consistently in food exchange and grooming (Figure 5-4). The higher frequency of antennation toward non-kin is apparently an inquiring rather than a soliciting behavior. Yet the absence of such discrimination in experimental colonies does not preclude the possibility that the *Camponotus* favor siblings when rearing reproductive brood. Indeed, kin-based rearing of queens by honey bees has been tested experimentally by several investigators. Breed et al. (1984) found no evidence that workers preferred nestmate queen larvae over unrelated, non-nestmate larvae. On the other hand, Page and Erickson (1984) observed a preference for nestmate larvae that were full sisters (three-quarters related) over approximately one-quarter-related larvae from another colony. The latter may have been additionally distinguishable because of their origin in a different nest, however. In still other experiments, Visscher (1986) transferred brood pieces among colonies in controlled experiments and obtained the same result. After separating out nest-specific cues by presenting workers with non-nestmate unrelated and non-nestmate related eggs, he obtained preferences for siblings (mixed full sisters and half-sisters) over non-kin. Noonan (1986) performed the only reported test of worker interactions with queen larvae of different patrilines, all originating in the same nest. She found that workers significantly preferred to visit, inspect, and feed full-sister larvae. Unfortunately, these results may also have been influenced by genetic color markers used to identify the patrilines. Thus the evidence accumulated to date indicates that the signal variation correlated with kinship required for within-colony kin recognition does exist alongside the cues used for between-colony discrimination. Whether this specificity is utilized in adaptive nepotistic behavior under natural conditions remains to be conclusively demonstrated.

Individual odors have been implicated in a few other cases of ant behavior. Nest-founding queens of the honeypot ant *Myrmecocystus mimicus* and the Australian meat ant *Iridomyrmex purpureus*, as well as workers of *Leptothorax* and *Harpagoxenus*, set up aggressive relationships very similar in outward appearance to vertebrate dominance hierarchies (Cole, 1981; Bartz and Hölldobler, 1982; Franks and Scovell, 1983; Hölldobler and Carlin, 1985). Queens of *Leptothorax curvispinosus* do not dominate one another by overt aggression, but they do consume some of the eggs of other queens. At least part of the basis of this differential oophagy is rough treatment. One queen was observed to handle the eggs of all of her rivals in an aggressive manner, licking them vigorously and chewing on them until the shells crumpled slightly before springing back into shape. Eggs from one of her rivals were more fragile and as a consequence were ruptured and eaten (Wilson, 1974a).

In a remarkable development, Jessen and Maschwitz (1986) discovered that workers of the tropical Asian ponerine *Pachycondyla tesserinoda* recognize and follow their own personal odor trails from a still unknown glandular source while returning from food discoveries or leading nestmates to new nest sites. They guide other workers along these trails by means of tandem running to the target locations. The follower ants fix on the general body odor of the leader and run close behind her. By using such personal trails the leaders can return to food or nest sites they themselves have personally selected. Individual specific trails have also been discovered in *Leptothorax affinis* (Maschwitz et al., 1986b).

## BROOD DISCRIMINATION

Other, less direct pieces of evidence suggest that widespread discriminatory ability exists in the ants (Carlin, 1988). In the relatively primitive myrmicine genus *Myrmica*, workers seem not to be capable of distinguishing the tiny first-instar larvae from eggs, so that when eggs hatch the larvae are left for a time in the midst of the egg pile. The larvae feed during the first instar by breaking into a single adjacent egg. As soon as they molt and enter the second instar, however, they are removed by the workers and placed in a separate pile (Weir, 1959a). Third-instar larvae vary greatly in size; the smaller ones are destined to metamorphose into workers, while the larger ones retain the ability to develop into queens. If a nest queen is present, the large larvae receive proportionately less food, and they are licked and bitten more by the workers, an action that may reduce their growth still further. The ultimate effect of the presence of a nest queen is the production of fewer new queens (Brian and Hibble, 1963).

The workers of the fire ant *Solenopsis invicta* are able to distinguish sexual pharate pupae ("prepupae") from worker brood, and a key component used by the ants in recognition is triolein, the triglyceride of oleic acid, which is present in many vegetable oils and in substantial amounts in the pupae. During experiments by Bigley and Vinson (1975), the ants placed filter-paper disks treated with triolein with groups of sexual pharate pupae rather than with worker-destined brood. These results have been criticized by Morel and Vander Meer (1987), however, on the grounds that the experimental design was likely to confound brood retrieval and food retrieval.

The tendency to segregate eggs, larvae, and pupae into separate piles is a nearly universal trait in ants, and probably has a basis in chemical differentiation like that demonstrated for the fire ant sexual pupae. Furthermore, workers of many species can distinguish larvae of two or more size classes (Le Masne, 1953). Workers and larvae are also able to distinguish trophic eggs (special eggs produced for consumption) from the normal eggs destined to develop into larvae. *Monomorium pharaonis* workers can even tell male eggs from female eggs (Peacock et al., 1954).

As Carlin (1988) has pointed out, kin recognition in brood discrimination by workers need not be limited to the differential feeding, grooming, and retrieval of close relatives. Brood cannibalism is well known in ants and the eating of viable eggs has been reported among queens in *Leptothorax curvispinosus* (Wilson, 1974a). Preliminary results on oophagy in this species and *L. longispinosus* suggest that queens may recognize their own eggs (T. P. Graham and N. F. Carlin, personal communication). Discrimination of eggs based on kin has also been suggested in oligogynous colonies of *Iridomyrmex purpureus* (Hölldobler and Carlin, 1985).

Although brood can easily be transferred among conspecific, and, within limits, even among interspecific colonies, the evidence clearly demonstrates that discrimination between nestmate and non-nestmate brood does occur in ants. Carlin (1988) observes that "this behavior differs strikingly from the antagonism and rejection of alien adults by the same species. In choice situations (such as the retrieval bioassay), non-nestmate conspecific brood are not rejected; rather, both are accepted, but nestmate brood are favored (retrieved earlier or in greater numbers). That is, preference for nestmate adults is usually absolute and exclusive, while preference for nestmate brood is probabilistic and non-exclusive."

Nestmate brood discrimination has been demonstrated in a phylogenetically diverse group of ant species (see Table 5–2). The most detailed investigation of the phenomenon was conducted on the desert ant *Cataglyphis cursor.* Lenoir (1984) discovered that when newly eclosed workers are transferred into an alien conspecific colony, they engage in less brood care than do control workers of equal age in their own colony. Furthermore, Isingrini et al. (1985) found evidence suggesting that *Cataglyphis* workers respond differentially

**TABLE 5–2** Species in which discrimination between nestmate and non-nestmate conspecific brood has been demonstrated.

| Species | References |
|---|---|
| **MYRMICINAE** | |
| *Acromyrmex octospinosus* | Febvay et al. (1984) |
| *Atta cephalotes* | Robinson and Cherrett (1974) |
| **DOLICHODERINAE** | |
| *Tapinoma erraticum* | Meudec (1973, 1977) |
| **FORMICINAE** | |
| *Camponotus floridanus* | Carlin and Schwartz (1989) |
| *C. pennsylvanicus* | Carlin (1988) |
| *C. tortuganus* | N. Carlin and R. Eitel (unpublished observations) |
| *C. vagus* | Bonavita-Cougourdan et al. (1988) |
| *Cataglyphis cursor* | Lenoir (1984), Isingrini et al. (1985), Isingrini (1987) |
| *Lasius niger* | Lenoir (1981) |

to nestmate and non-nestmate brood because they learn to recognize nestmate brood during larval life. Since workers of this species are known to produce female eggs parthenogenetically and also to migrate frequently among neighboring nests, the adaptive significance of nestmate and kin recognition in *C. cursor* is of special interest (Lenoir et al., 1987b). Finally, Rosengren and Cherix (1981) made an interesting observation concerning differential brood treatment in *Formica lugubris.* Colonies from populations in Switzerland preferred to retrieve nestmate pupae over conspecific pupae from Finland, but did not discriminate against pupae from geographically closer Swiss and Italian populations. In symmetric fashion, workers from Finland discriminated strongly against pupae from Switzerland and Italy.

Since the early work of Forel (1874) and Fielde (1903), we have known that brood can be transferred not only between nests of the same species but also between those of different species. Hölldobler (1977) hypothesized that brood pheromones are not colony specific and often not even species specific, but are so highly attractive to worker ants in general as to override or mask the response to colony-specific odors. Indeed, Brian (1975) demonstrated that the recognition signals indicating reproductive potential in *Myrmica rubra* and *M. scabrinodis* remain effective when queen larvae are cross-fostered between species. Many other examples of cross-fostering within a genus and even among different genera exist. For example, Plateaux (1960a–c) was able to exchange larvae among several myrmicine genera. The responsiveness of ant workers to generalized brood signals is the basis for social parasitism and interspecific slavery in ants (see Chapter 12), and it may also be the essential prerequisite for parasitic symbiosis in many myrmecophilous arthropods (Chapter 13).

As N. F. Carlin (1988) has emphasized, however, interspecific brood transfers in general tend to be more successful when phylogenetically close taxa are used.

> Emery's (1909) rule, that social parasites are closely related to their hosts, may be attributed to the greater similarity of brood (and adult)

cues among species with common ancestors; however this pattern applies equally to species uninvolved in parasitism. Jaisson (1985) reported that workers of *Formica* spp. may recognize and care for pupae of other formicine genera (*Camponotus* and *Lasius*), but reject those genera in other subfamilies (*Atta* and *Eciton*). Founding queens of *Camponotus pennsylvanicus* have narrower tolerance (Carlin, unpublished results). Queens will adopt larvae and pupae from 14 congeneric species (*C. abdominalis, americanus, castaneus, consobrinus, ferrugineus, festinatus, floridanus, herculeanus, nearcticus, noveboracensis, ocreatus, socius, tortuganus,* and *vicinus*), but not those of *C. planatus* and *C. ulcerosus*—which belong to the morphologically aberrant subgenera *Myrmobrachys* and *Maniella*, respectively. *C. pennsylvanicus* foundresses destroy or reject brood of the formicine genera *Acanthomyops, Formica,* and *Lasius* (though a few *Lasius neoniger* larvae may be retained for nearly a month before finally being eaten), and always reject brood from other subfamilies (Ponerinae: *Amblyopone;* Myrmicinae: *Aphaenogaster, Crematogaster, Novomessor* [= *Aphaenogaster*]; Dolichoderinae: *Iridomyrmex*). (p.277)

Foreign brood pieces that have been adopted are occasionally allowed to mature into adults, which leads to the formation of mixed worker forces (Fielde, 1903; Carlin and Hölldobler, 1983; Errard and Jaisson, 1984). Nonetheless, the adoptees are then often attacked and killed by their hosts (Fielde, 1904a; Haskins and Haskins, 1950a; Plateaux, 1960c; Wilson, 1975a). A further complication was observed by Carlin (1988): "When heterospecific brood was adopted and reared to adulthood, the adoptee workers may tend host brood but kill host workers as they eclose (e.g., *Camponotus (Myrmothrix) floridanus* workers adopted into *C. (Camponotus) pennsylvanicus* colonies; *C. (Tanaemyrmex) tortuganus* adoptees of a *C. (Hypercolobopsis) paradoxus* queen). These results together suggest that the interspecific similarity of brood pheromones is greater than that of adult (species-level) recognition cues."

In discriminating against alien brood, worker ants do not have an "all or nothing" response, but rather exhibit a preferential choice behavior. For example, workers of *Camponotus floridanus* and *C. tortuganus* reared with conspecific brood picked up and retrieved nonnestmate conspecific pupae significantly earlier than pupae of the other species. The heterospecific pupae were also retrieved, but only 35 percent of them were retained and tended after five days, whereas 69 percent of the conspecific pupae were still alive in the nest after that period (Carlin et al., 1987a). Similar results were obtained by Hare and Alloway (1987) for *Leptothorax ambiguus* and *L. longispinosus.* Workers of each species preferred to accept conspecific larvae, but they also adopted a smaller proportion of the alien brood. Finally, Elmes and Wardlaw (1983) reported that they were often, but not always, able to introduce larvae of several *Myrmica* species into heterospecific colonies; larval survival was usually better in conspecific colonies.

Carlin (1988), in his extensive review of brood discrimination in ants, describes experiments by different investigators that indicate an absolute preference on the part of some ant species:

Exclusive conspecific preferences have been reported in three species of *Formica*. When *Formica polyctena* workers that had been reared with brood of their own species were offered the choice of conspecific and heterospecific pupae (*F. sanguinea* or *Camponotus vagus*), the former were retrieved first, but heterospecific pupae were also collected. Exclusivity was revealed in their subsequent treatment, as conspecific pupae were retained while heterospecifics were rapidly eaten (Jaisson, 1975; Jaisson and Fresneau, 1978). *Formica rufa* workers similarly offered conspecific pupae and those of *F. lugubris* retrieved and tended only the former in one experiment (Le Moli and Passetti, 1977), but in another test they brought both species into the nest and neglected and destroyed the latter (Le Moli and Passetti, 1978). Conversely, Le Moli and Mori (1982) found that *F. lugubris* workers retrieved only conspecifics, and ate *F. rufa* pupae outside the nest. Thus exclusive preferences were manifested consistently by destruction of the non-preferred species, though variable retrieval responses also occurred.

These findings bear re-examination, however. In a review, Jaisson (1985) reports that the same basic results were obtained with a species of *Camponotus*, suggesting methodological dissimilarities between these studies and that of Carlin et al. (1987a). One potentially relevant difference is the use of cocoons killed by freezing by Jaisson, Le Moli and their respective colleagues in all experiments that yielded exclusive preferences. The studies of Hare and Alloway (1987) and Carlin et al. (1987a), demonstrating non-exclusive biases in both retrieval and retention, employed live larvae and pupae respectively. Ants are well known to eat dead or injured brood (Wilson, 1971), and brood killed by freezing have been shown to decline in attractiveness (Robinson and Cherrett, 1974; Walsh and Tschinkel, 1974). Having retrieved two species of dead pupae, the *Formica* spp. workers might cannibalize first the less familiar . . . ; the thorough destruction of all heterospecifics needs to be confirmed by offering live pupae. In addition, *F. polyctena* is sometimes the host of a temporary social parasite, *F. truncorum* (Gösswald, 1957; Kutter, 1969), indicating that heterospecific parasite brood must be sufficiently attractive to adult workers reared under normal field conditions. This same difficulty applies to the report that workers of *Lasius niger* respond only to conspecific brood and destroy all others (Le Moli and Angeli, cited in Le Moli, 1980), as this species is the natural host of temporary social parasite *L. umbratus* (K. Hölldobler, 1953). (pp. 278–279)

Additional discussions of the brood discrimination problem, with special reference to social parasitism, are presented by Le Moli and Mori (1987) and in Chapter 12 of this book.

## SOURCE OF THE COLONY ODOR

In most if not all cases studied in ants so far, identification of nestmates and life stages is by antennal contact. This fact in itself suggests chemoreception, even though Brian (1968) has speculated that several age classes of *Myrmica* larvae may also be distinguished by certain differences in hairiness that are quite apparent to the human observer. At least one example has been reported in which the chemoreception appears to operate over a distance. When workers of the harvester *Pogonomyrmex badius* lay trophic eggs, they search for hungry larvae while sweeping their antennae through the air. When they come within about a centimeter of the head of the larva they move directly to it and unerringly place the egg on its mouthparts (Wilson, 1971). This response appears to be exceptional, however. The employment of close-range olfaction in the discrimination of life stages has been proved by the identification of several of the surface pheromones involved: triolein in the sexual pharate pupae of *Solenopsis invicta* as described above, two pyranones and dihydroactinidiolide as components of the queen attractant of *S. invicta* (Vander Meer et al., 1980; Glancey, 1986), and neocembrene as a queen attractant in *Monomorium pharaonis* (Edwards and Chambers, 1984).

No colony odor by which kin are recognized has yet been identified with certainty, but some tantalizing clues exist. One is the assimilation of the cuticular hydrocarbons of fire ants by the scarabaeid beetle living with them, *Myrmecaphodius excavaticollis.* The relations of the two insects are harmonious, despite the fact that the beetles prey on the ants. Vander Meer and Wojcik (1982) discovered that the *Myrmecaphodius* are coated with a series of hydrocarbons identical to those of one of their fire ant hosts, *Solenopsis richteri.* The beetles also possess a second set of high-molecular-weight hydrocarbons not shared with the ants and evidently peculiar to themselves. When *Myrmecaphodius* adults are isolated from the ant hosts, they lose the *S. richteri* hydrocarbons but retain their own, heavier hydrocarbons. When subsequently introduced into colonies of *Solenopsis invicta,* a second fire ant host species, the beetles acquire the *S. invicta* hydrocarbons. On this basis Vander Meer and Wojcik postulated that the ants use the hydrocarbons for odor identification among themselves, at least to the level of species, and that the beetles exploit this behavior by acquiring the molecular signature. When *Myrmecaphodius* first enter *Solenopsis* nests, they rely on their heavily armored exoskeleton and a death-feigning behavior for protection. After a few days, according to the hypothesis, enough hydrocarbons are adsorbed by the beetles to gain fuller acceptance into the ant society.

Nelson et al. (1980), who first identified the surface material of the fire ant as saturated normal, mono-, and dimethyl branched hydrocarbons, found that *Solenopsis invicta* and *S. richteri* have the same substances (as evidenced by their chromatographic peaks), but that the relative amounts are so different as to make the species easily distinguishable. Several years later, Vander Meer (1986b) showed that *S. geminata* and *S. xyloni* also possess distinctive patterns based on proportionality. Thompson et al. (1981) discovered that the five major hydrocarbons occur in the postpharyngeal glands of queens of *S. invicta,* while Vander Meer et al. (1982) found that the quantities stored there are quite large (between 15 and 50 μg), increasing substantially and shifting in proportion shortly after the ants mate. The postpharyngeal glands are believed to function primarily as a source of food for larvae. In fire ant queens these organs are fully engorged with oils just prior to the nuptial flight (Phillips and Vinson, 1980). Whether they also provide odor signatures for the colony remains an untested possibility.

The first evidence of differences in hydrocarbon proportions existing between colonies of ants, and not just between species, was adduced by Clément et al. (1987), Bonavita-Cougourdan et al. (1987a,b), and Morel and Vander Meer (1987) in the carpenter ants *Camponotus floridanus* and *C. vagus.* By direct chemical analysis they found that at least some colonies of the same species differed among themselves in the hydrocarbon blends. When live workers of the focal colony, dead workers, and dummies were washed with solvents containing extracts of workers from alien colonies of the same species, other workers from the focal colony reacted aggressively to them, whereas the responding workers reacted in a neutral or at most a mildly aggressive manner to workers and dummies washed with extracts of their nestmates. This experiment was suggestive but did not constitute sufficient proof that the critical components in the washes were hydrocarbons. A complicating factor is the additional finding by Bonavita-Cougourdan et al. (1987b) that the hydrocarbon proportions change with age. Such shifts could serve as the basis of subcaste recognition, such as discrimination of foragers from nurses, but they undoubtedly make recognition of nestmates in general more difficult. Similar results have been obtained by Morel and Vander Meer (1987) and Morel et al. (1988) for *Camponotus floridanus.* Blum (1987) has reviewed the findings on the possible role of cuticular hydrocarbons for species and nestmate recognition in social insects.

The first to claim the discovery of a glandular source of colony odors were Jaffe and Sanchez (1984a,b) in the course of studies of the Neotropical formicine ant *Camponotus rufipes.* They concluded that the signature comes from the mandibular glands, which are also well known to produce alarm substances, and offered two proofs. First, they reported that isolated heads of workers from alien colonies were bitten significantly more and elicited more alarm responses than did isolated heads from the same colonies, whereas isolated thoraces and gasters did not cause this differential response. The result cannot be taken as conclusive, however, since some additional substance in the head, including the mandibular gland substances themselves, may simply have excited the target ants and lowered their attack threshold in response to colony odor found uniformly over the entire body. The results of a second bioassay employed by Jaffe and Sanchez seem stronger but are still less than definitive. Plastic dummies were contaminated either with mandibular gland secretions of alien workers or with the secretions of nestmates of the target workers. The alien-treated models elicited more bites ($3.5 \pm 1.8$, mean $\pm$ standard deviation) than did the nestmate-tested dummies ($1.2 \pm 0.9$). The result is compromised by the possibility that the secretions could have been contaminated with body odors originating elsewhere in the body. Furthermore, as noted in Chapter 7, colony-specific blends in the mandibular gland secretions may also function as modulators of alarm signals, and do not need to be the source of the colony-specific nestmate recognition label.

Leaving for the moment the chemical identity and anatomical source of the colony odor, we can take up the separate question of its intrinsic as opposed to its extrinsic control. The existence of an important hereditary component in at least some ant species has been established by several lines of evidence. Haskins and Haskins (1979) found that different colonies of the Australian ponerine *Rhytidoponera metallica* retained the same level of incompatibility from the time they were collected in the field through 13 years of identical laboratory conditions. When worker pupae were placed with workers of other laboratory colonies, allowed to eclose into active adults, and then placed with their home colony, they were fully accepted. An equally convincing experiment was performed by Mintzer and Vinson (1985b) with the arboreal ant *Pseudomyrmex ferruginea,* an obligate symbiont of acacia in Mexico. The *Pseudomyrmex* are wholly bound to the acacia trees, nesting on them and feeding in large part on secretions from the foliar nectaries and on Beltian bodies, special food organs that grow at the tips of the leaves. Mintzer and Vinson raised a series of *Pseudomyrmex* colonies on clones of *Acacia hindsii* under uniform conditions. Although the environmental variance had thus been reduced to a minimum, the *Pseudomyrmex* colonies retained their mutual incompatibility. Of equal significance, colonies belonging to inbred lines of the ants underwent a slight increase in compatibility when compared with outbred controls.

Another experiment, this time utilizing the leafcutter ant *Acromyrmex octospinosus,* demonstrated the existence of endogenous factors, probably genetic in origin, as well as exogenous factors origi-

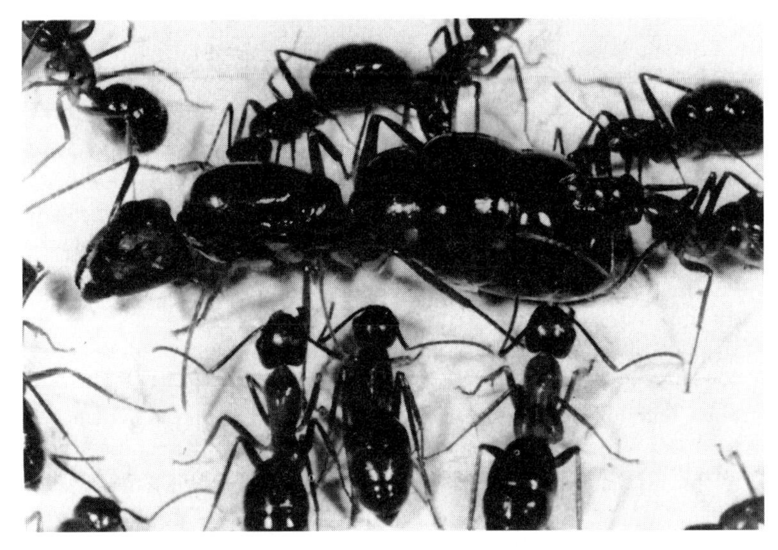

FIGURE 5–5 The queen of *Camponotus floridanus* is surrounded by workers, who lick and groom her almost continuously. During this process they are believed to pick up queen labels that serve as an important component of the colony odor.

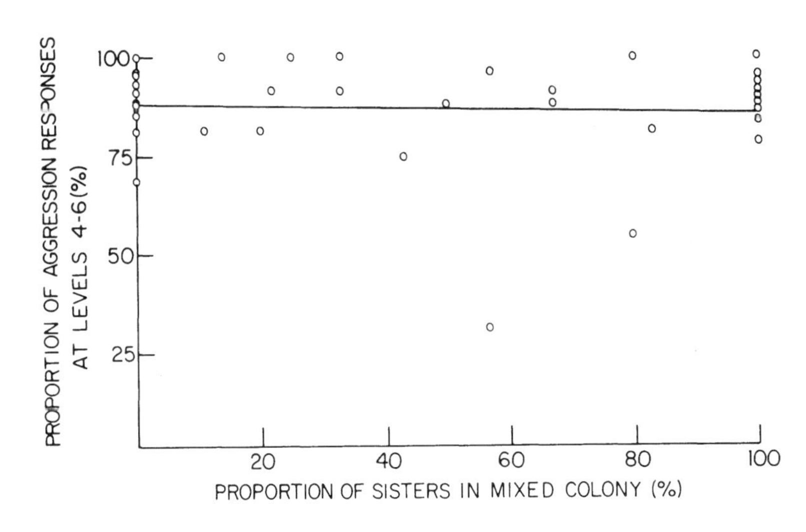


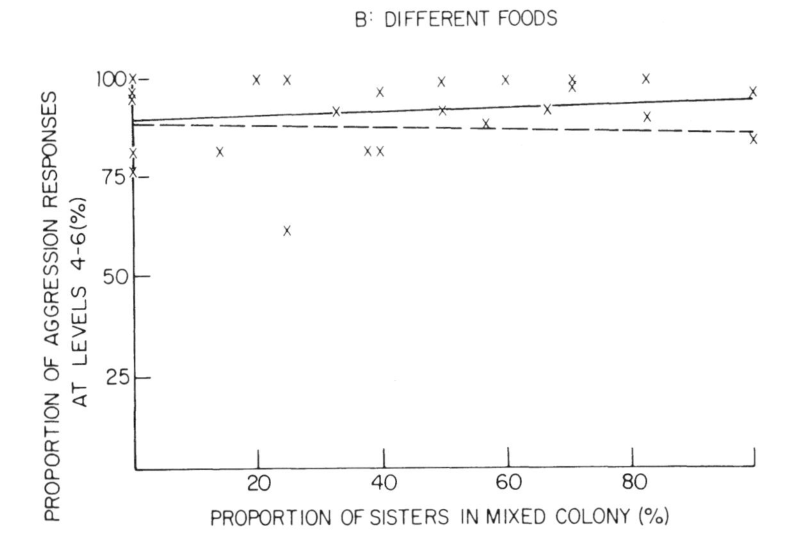


FIGURE 5–6 Proportion of aggressive responses at levels 4 to 6 (see Table 5–1) between adoptees. The adoptees were reared in queenright intraspecific mixed colonies of *Camponotus pennsylvanicus* that contained varying proportions of adoptee sisters and true siblings. Each point represents aggregated aggression data for one colony of a given percentage of adoptees. (*A*) Mixed and stock colonies were fed the same diet ($n = 777$ encounters). The regression slope is not significantly different from 0 ($p > 0.05$), indicating no effect of proportion of kin in the colony on the nest-mate recognition. (*B*) Mixed and stock colonies were fed different diets ($n = 452$ encounters). The slope is also not significantly different from 0 ($p > 0.05$). The dashed line represents sisters fed the same diet, repeated from *A*; the slopes of the two lines are not significantly different from each other, indicating that different environmental cues do not enhance discrimination. (From Carlin and Hölldobler, 1986.)

nating in the food (Jutsum et al., 1979). *A. octospinosus* colonies subsist entirely on a symbiotic fungus grown on freshly cut leaves and other vegetation, possibly along with sap released from fresh incisions. Prior to the tests various colonies were maintained on vegetation from different species of plants. Jutsum et al. used as a measure of hostility the time spent by workers investigating strange workers. Reintroductions of ants into their own nests provided the control tests. The results demonstrated that both endogenous and exogenous factors contribute to the differentiation of colony odor (see Table 5–2).

Given that at least part of the colony odor is endogenous, the question now arises as to which individuals produce it. Is it generated by one dominant member, such as the queen, or is it a gestalt drawn from many or all of the colony members?

Carlin and Hölldobler (1983, 1986, 1987) addressed this question by creating a series of experimental colonies of *Camponotus* species, variously containing (1) a queen and workers, (2) queenless worker groups, and (3) pairs of worker groups between which a queen was repeatedly switched. Some colonies of the first and second type also received different diets. The goal of this regime was to assess the relative magnitude of the contributions of the queen, the workers, and the environment under strictly controlled laboratory conditions. Queenless workers, removed from the same colony serving as the source of pupae but reared separately, were relatively tolerant of one another but more aggressive toward non-relatives. Dietary differences slightly enhanced aggression among separately reared kin. However, workers reared in queenright colonies (Figure 5–5) attacked all non-nestmates, whether kin or non-kin, and with equal violence. This response was unaffected by diet odors or by the proportion of different kin groups in the colony (Figure 5–6). Non-kin sharing a switched queen were as tolerant of one another as were kin, and cues derived from healthy queens were sufficient to label colonies of approximately 190 workers each. The workers' own discriminators became more important in large groups with queens that had less active ovaries and were relatively infertile. From these results Carlin and Hölldobler proposed a hierarchy of colony-specific label sources in *Camponotus* colonies, with cues derived from fertile queens most important, followed by worker discriminators, and then environmental odors.

A different system, or at least a different order in the hierarchy of cues, appears to be employed by the Mexican acacia ant *Pseudomyrmex ferruginea*. Mintzer (1982b) separated worker brood from a stock colony into groups of larvae and pupae, then allowed each group to be reared by a different foster queen. The workers from the separate groups were not antagonistic toward each other, which suggests that their own odors (rather than those of the foster queens) were the dominant discriminators. In a follow-up study Mintzer and Vinson (1985b) found that nestmates were often treated differently by colonies into which they were introduced. This variation in response favors an individualistic rather than a gestalt mode of colony odor production. Experiments with inbred colonies further indicated that the odor labels are under multilocus rather than single-locus genetic control. The experiments by Mintzer and Vinson were not designed to learn whether the odor identifications in *Pseudomyrmex* are entirely genetic or, as in the case of *Camponotus*, require some form of learning.

A still different system has been discovered in *Leptothorax curvispinosus*, a myrmicine species that forms facultatively polygynous and polydomous colonies. R. J. Stuart (1987c) measured the acceptance of workers among nests of three categories: (1) the nests were relatively near one another (0.09–1.87 m); (2) the nests were spaced farther apart (1.52–4.65 m); and (3) the nests were very far apart (7 km). In the first group the workers sometimes accepted each other without any aggression, and they were generally significantly less hostile to one another than were workers from sites 7 kilometers apart. The aggression between foreign workers within the second group was initially also very strong. After being cultured for several weeks under uniform environmental conditions, however, the workers grew significantly more friendly. The results indicate that polydomous colonies of *L. curvispinosus* occur within multicolonial populations, and that colony segregation within local populations is largely maintained by transient environmentally based nestmate recognition cues. Stuart proposes that in fact "the maintenance of colony autonomy within genetically highly interrelated populations may be the prime function of environmentally-based nestmate recognition cues. Colony autonomy under such circumstances may be important to maintain a relatively small but optimal colony size, or because the mechanisms which regulate colony growth development, etc., require a limited colony size."

Stuart (1987a) also demonstrated more stable cues of possible genetic origin that contribute to discrimination even after extended periods of uniform laboratory environment. In one of the studies worker pupae were removed from colonies and allowed to eclose and age for 38–157 days in isolation. When these workers were introduced into their parental colonies, they were readily accepted. But they were frequently attacked and even killed when introduced to non-parental conspecific colonies. These results demonstrate that individual workers produce persistent, colony-specific recognition cues, even when completely isolated from the parental nest. Stuart suggests that the cues involved may be either genetically based or acquired prior to pupation.

Finally, using interspecific mixed *Leptothorax* colonies, Stuart (1985a, 1987b) demonstrated that workers are labeled not only by transient environmental odors and individually produced cues, but also by a colony gestalt odor. His was the first demonstration of the existence of a gestalt label in ants. Supporting evidence that labels are transferred among nestmates was provided by the work of Errard and Jallon (1987), who used interspecific mixed colonies of *Manica rubida* and *Formica selysi*. A gestalt label had previously been demonstrated only in the desert woodlouse *Hemilepistus reaumuri* (Linsenmair, 1985, 1987). In this collective system, as Stuart (1988a,b) also called it, nestmates share recognition cues, and each colony member bears some kind of mixture of cues representative of the variation among the member's nestmates. Although the gestalt label does not appear to be completely uniform in *Leptothorax* colonies, Stuart (1988b) concludes that the

> collective system may have evolved as one means of reducing the degree of individual variation in individually produced nestmate recognition cues among colony members. Nonetheless, as evidenced by the *Leptothorax* data, this solution to the problem of individual variation may be far from perfect. Although this system might lessen individual variation to some extent and permit larger colony sizes, the dynamics of odor sharing and the relative degree of heterogeneity in individually borne cues might still impose a fairly small limit on colony size and compromise nestmate discrimination and colony defense when larger colony sizes are achieved. This might explain the relatively small size typical of *Leptothorax* colonies, and reports of relatively low levels of aggression between colonies under some circumstances (Provost, 1979).

The work by Stuart (1988a,b), Jutsum et al. (1979), Obin (1986), and others on ants, together with that by Gamboa and his collaborators on social wasps (Gamboa et al., 1986), suggests that social insects are programmed to respond to odor differences without regard to ultimate origin. Thus whereas the acquisition of unifying extrinsic labels seems common, the particular sources involved (or their hierarchy of importance in recognition) varies widely as a function of the relative strengths of ambient odors available to each species. These strengths are fixed in turn by the general biology, ecology, and social organization of the species. Among ant genera, for example, it may prove significant that *Camponotus* exhibits greater queen-worker dimorphism and stronger queen suppression of worker oviposition than does *Leptothorax*, and that *Solenopsis* workers do not require queen suppression for the simple reason that they lack ovaries. On the other hand it is possible that colonies of the same species employ different strategies suited to their particular ontogenetic stage. In *Leptothorax*, for example, Stuart (1987c) suggests that "special integrating mechanisms associated with newly eclosed workers may be required only under certain circumstances, perhaps during early colony growth, after young reproductives have been adopted into established colonies, or when newly eclosed adults must acquire additional cues to achieve acceptability." It seems also possible that in smaller colonies of *Camponotus* (with several hundred workers) the queen has a stronger influence on the colony label than in colonies with several thousand workers. The different components could also be weighted differently according to age and location of the workers. For example the queen's influence could be much stronger in the "core area" around her, where it would affect primarily younger workers, whereas in the peripheral zone of the colony, occupied by older workers, discrimination might be based more on gestalt labels derived from older nestmates.

A *Camponotus*-like hierarchical system, with queen odor overriding that of the workers, may be found to prevail in species that are either monogynous, oligogynous (two to several queens with approximate reproductive parity), or polygynous with one queen predominant in egg laying. On the other hand truly polygynous species, in which many queens are active egg-layers, can be expected to rely on a gestalt odor contributed in greater measure by the workers. In addition, extremely polygynous species in large colonies may lose their capacity to create colony odors of any kind, since the large variety of genotype represented in each unit is likely to be duplicated in adjacent units by the law of large numbers and the inevitable minimization of between-unit variance. The ultimate result would be a supercolony, or unicolonial population, in which the local population constituted just one vast colony. Supercolonies with large populations of queens or reproductive workers do occur in some species of *Pheidole, Monomorium, Pristomyrmex, Wasmannia, Iridomyrmex,* and *Formica.* In such cases the generation of intrinsic colony odors has virtually ceased, but some odor differences arising from different extrinsic influences may still exist.

## THE ONTOGENY OF RECOGNITION

It is now generally accepted that learning plays a major role in the acquisition of a template for labels used in colony-level recognition. Carlin (1988) has reviewed the work on the significance of learning in brood recognition in ants; we can do no better than to quote his concise analysis of the diverse studies published:

> Jaisson (1975) demonstrated that newly eclosed *Formica polyctena* workers learn to recognize any brood species with which they are familiarized. A process comparable to imprinting takes place in a sensitive period of approximately one week following emergence (Jaisson and Fresneau, 1978). Young workers exposed during this period only to pupae from another species (*F. pratensis, F. sanguinea, Camponotus vagus* or *Lasius niger*) retrieved the familiar heterospecific pupae first and destroyed the unfamiliar conspecific after retrieval. Provided with conspecific pupae, they preferred familiar conspecifics and destroyed unfamiliar heterospecifics; deprived of any experience with brood, they ignored or destroyed all species impartially. These results were replicated with young *Formica lugubris* and *F. rufa* workers, exposed to their own or one another's pupae or deprived of brood experience, by Le Moli and Passetti (1977, 1978) and Le Moli and Mori (1982). Le Moli (1978) also reported apparent social facilitation of the brood-tending response. *F. rufa* workers deprived of brood through the sensitive period, then placed together with experienced workers and pupae from their colony of origin, subsequently retrieve and tend both conspecific and *F. lugubris* pupae, rather than destroying them.
>
> In the genus *Leptothorax,* by contrast, preference for conspecific brood can be eradicated by early experience with heterospecifics, but not reversed. Alloway (1982), examining the effect of enslavement on pupa acceptance in three *Leptothorax* species, found that free-living, previously and currently enslaved workers always preferred conspecific pupae to heterospecifics. The experience of enslavement did increase the proportion of heterospecific pupae accepted by the workers; behavioral interactions were also implicated, as slaves accepted more heterospecific pupae in the presence of the slave-makers (*Harpagoxenus americanus*) than if the latter were removed. Hare and Alloway (1987) exposed newly eclosed *L. ambiguus* and *L. longispinosus* workers either to conspecific or to each others' larvae, or deprived them of brood (isolation). Workers familiar with conspecific larvae retrieved and retained significantly more conspecifics, but those exposed to heterospecific larvae or isolated accepted larvae of both species without preference. Carlin *et al.* (1987a) independently obtained the same results using *Camponotus floridanus* and *C. tortuganus* workers, which preferred conspecific pupae if familiar with them, but retrieved both species indiscriminately if naive or exposed only to heterospecifics. However, naive young workers were not entirely incapable of brood species recognition, as they retained unfamiliar conspecifics significantly longer than unfamiliar heterospecific pupae. In addition, the behavior of older workers (20 days post-eclosion), already exposed to conspecific pupae in their natal nest, proved to be malleable to a limited extent. Older adults retrieved conspecific pupae preferentially irrespective of recent experience, but their retention of heterospecific pupae was significantly increased by recent familiarization; since the effect of early learning can be subsequently modified, this behavior does not qualify as imprinting strictly defined.
>
> Thus while newly-eclosed workers of *Formica* spp. respond to experience with any species, those of *Camponotus* and *Leptothorax* spp. are apparently equipped with a somewhat modifiable bias that favors the learning of their own species, and cannot be made to prefer the brood of others. These genera may possess some innate "template" mechanism that restricts the range of effective learning stimuli, predisposing them to learn conspecific brood signals, or the workers may simply respond more strongly to odors that resemble their own. Alternatively, pre-imaginal learning may be involved . . .
>
> The ontogeny of inter-colony recognition [of brood] has been addressed in *Cataglyphis cursor,* and early learning is clearly involved (Isingrini *et al.*, 1985). The data are in fact strikingly similar to those on species recognition in *Camponotus* and *Leptothorax:* newly eclosed workers that were exposed to nestmates licked and carried nestmate larvae significantly more than conspecific alien larvae. Those provided post-eclosion experience with non-nestmate larvae or kept in isolation exhibited significantly weakened preferences for nestmate larvae, but non-nestmates were not preferred. Pre-imaginal learning, that is learning of ambient larval recognition cues by larvae, was implicated by transferring eggs from their natal colony into alien colonies, allowing them to develop there, returning them to the natal nest as pupae and then testing them after eclosion. Larvae from the colony in which the workers spent their own larval life were preferred to larvae from their natal nest. Transferring large larvae rather than eggs did not alter the preference for natal nest brood, perhaps because the larvae spent only 2–5 days in the adoptive colony before pupating. Pre-imaginally induced biases declined as the workers aged, a factor which must be investigated further before the significance of larval learning can be evaluated. However, this is an intriguing possibility, which may be important in other forms of recognition as well. (pp. 280–283)

Similar results were obtained with *Camponotus floridanus* by Carlin and Schwartz (1989).

Because colony odor is learned in part, yet another property of colony and species discrimination is the time at which the workers acquire the odor and the time at which they learn the odor of their group. Fielde (1903, 1904a,b, 1905b) discovered that the colony odor changes with age. *Aphaenogaster rudis* workers, segregated at the pupal stage and isolated as a group until they were 16 to 20 days into adult life, accepted additional alien *rudis* workers up to 40 days after they emerged from the pupa, but attacked workers older than that as well as queens. Indeed, Bonavita-Cougourdan et al. (1987a,b) discovered that different age classes of *Camponotus vagus* workers bear distinct blends of cuticular hydrocarbons. Fielde had proposed a progressive odor hypothesis, which suggested that certain hereditary nestmate recognition odors in adults change progressively with age, so that workers often do not accept older siblings unless they have eclosed among nestmates of that age class and learned their recognition cues. In a reexamination of this hypothesis, however, Stuart (1987b,c) was unable to confirm it for three species of *Leptothorax.* Workers automatically acquire a distinctive odor no later than 38 days after eclosion, even when isolated from the mother colony as far back as the pupal stage. They are then readily accepted by their own colony but not by alien colonies.

In fact, very few studies have been made of the precise timing of

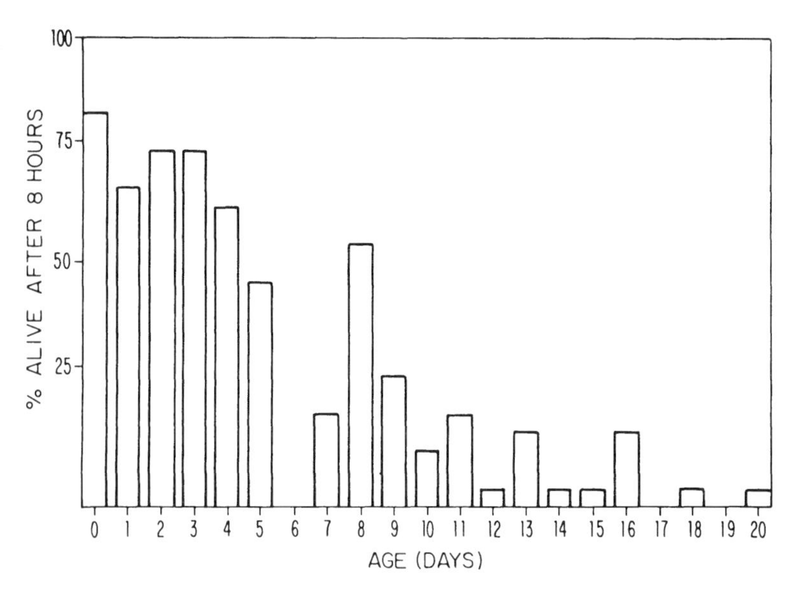


**FIGURE 5–7** The time course of adoptability of young adult *Camponotus pennsylvanicus* workers. Age: number of days between eclosion and introduction into the alien colony. Percentage alive: proportion of 25 callow workers of each age class accepted inside an alien nest eight hours after introduction ($n$ = 525 introductions). (From Carlin and Hölldobler, 1986.)

the odor acquisition or the sensitive period at which odor is learned. The mixed-colony experiments of Carlin, Fielde, Hölldobler, Jaisson, Lenoir, Morel, Vander Meer, and others have demonstrated that brood are far more readily adopted by alien colonies than adults, while very young (callow) workers are more acceptable than their older nestmates. More detailed information on the ontogeny of acceptability of young adults has been provided by the work of Lenoir et al. (1987b) on *Cataglyphis cursor,* Morel and Vander Meer (1987) and Morel et al. (1988) on *Camponotus floridanus,* Clément et al. (1987) and Bonavita-Cougourdan et al. (1987) on *Camponotus vagus,* and Carlin and Hölldobler (1986) on *Camponotus pennsylvanicus* (Figure 5–7). In this last study *C. pennsylvanicus* workers a few days old, like brood, were transferred among colonies without suffering much mortality. Adoptability then declined, from about the fifth day following eclosion onward. Workers accepted initially were not attacked at any time afterward; they presumably acquired cues of the recipient colony before their own discriminators developed or their immunity to attack wore off. Most callows over one week old were killed within eight hours of introduction into a queenright colony. More than 98 percent of all introductions, including those involving the youngest callows, provoked at least initial threats at level 1 (see Table 5–1). Only nine callows were immediately accepted at level 0, of whom eight were less than two days old and one was nine days old. These results represent discrimination against callows by the recipient colonies exclusively, and not discrimination on the part of the callows, who had not yet become aggressive.

## PROSPECT

Entomologists have scarcely begun to explore the intricacies of learning in kin recognition among social insects. Identification and discrimination are major features in biological systems, from embryogenesis and immune responses to communities and ecosystems, and it is no surprise to find them developed to a high degree in ants (Hölldobler and Carlin, 1987). All forms of recognition require distinguishable signals, whether antigens or colony odors, varying in ways that are correlated with the evolutionary advantages they confer. The pattern of diversity in recognition systems of ants is still mostly unknown and is one of the principal challenges in future research.

# CHAPTER 6

# Queen Numbers and Domination

The number of queens profoundly alters several of the key features of colonial organization among the ants, including the kinship of nestmates, the rate of colony growth, and the number and distribution of nests. The 1980s witnessed an explosive growth of knowledge about this complex subject, with investigators addressing the following questions:

- Why does the number of queens vary among species? The colonies of some species always have a single egg-laying queen. Those of others have up to thousands or, as in the case of the "supercolonial" *Formica yessensis* of Japan, millions of queens. In addition, the number often changes through different stages of the colony life cycle.
- Which colony members control the number of queens? Is the number a consequence of dominance and elimination among queens, or do the workers regulate it?
- How is the number controlled? Regulation of the queen population can be achieved by physical elimination, reproductive castration (perhaps mediated by pheromones), or emigration of supernumeraries.
- What are the consequences of varying queen numbers on the growth and genetic structure of colony populations?

It will be useful to begin by reviewing the generally accepted terminology of queen numbers in relation to the life cycle. Monogyny refers simply to the possession by a colony of a single egg-laying queen, or reproductive female, or "gyne," as she is often called in the myrmecological literature. Polygyny is the possession of multiple queens. Oligogyny is a special case of polygyny, in which two to several queens coexist in the same nest but remain well apart from one another (Hölldobler, 1962; Buschinger, 1974a). As a rule, oligogyny in ants is characterized by the workers' tolerance of supernumerary queens combined with intolerance among the queens, so that the queens space out in the same nest (Hölldobler and Carlin, 1985). The founding of a colony by a single queen is called haplometrosis; multiple-queen founding is referred to as pleometrosis (Wasmann, 1910b; Wheeler, 1933b). The term "metrosis" can be used to refer generally to this biological variable. Monogyny can be primary, meaning that a single queen is also the foundress; or it can be secondary, as when multiple queens start a colony pleometrotically but only one of them survives. In a symmetric fashion, polygyny can be primary (multiple queens persist from a pleometrotic association) or secondary (the colony is started by a single queen and others are added later by adoption or fusion with other colonies).

True polygyny, in which two or more queens contribute to egg-laying, has been conclusively demonstrated in many genera in the Ponerinae, Myrmeciinae, Myrmicinae, Dolichoderinae, and Formicinae, both directly, by observation of egg laying, and indirectly, through the electrophoretic identification of allozymes in the workers and second generation of reproductives (Pamilio, 1982b; Berkelheimer, 1984). Without such proof the mere presence of multiple queens in the same nest does not necessarily mean that the colony is polygynous. Winged queens are invariably virgin and not active egg-layers. These alate individuals are young and native to the nests in which they are found. They await the mating flight and, in most cases, initiate new colonies on their own. Even if supernumerary queens are wingless (called dealate if they have shed wings they earlier possessed), they still might not be egg-layers. In some species such individuals are inseminated but their oviposition is suppressed by the presence of a major egg-laying queen in the same nest. This form of reproductive latency, also called functional monogyny, has been demonstrated in *Formicoxenus* and *Leptothorax* (Buschinger, 1967, 1968c, 1979a; Buschinger and Winter, 1976; Buschinger et al., 1980b) and in *Solenopsis invicta* (Tschinkel and Howard, 1978). Functional monogyny has also been reported in *Myrmecina graminicola* by Baroni Urbani (1968a, 1970). However, pointing out that functional monogyny had not been observed in colonies collected fresh from the field, Buschinger (1970b) cautioned that these observations might have resulted from laboratory manipulations. In other species, as for example *Leptothorax acervorum*, the dealate supernumeraries are unfertilized. Buschinger (1967) calls this condition pseudopolygyny to distinguish it from true, functional polygyny and latent polygyny. He notes that in *Leptothorax* the infertile queens often act like workers by sharing ordinary labor in the nest. On the other hand, Ehrhardt (1970) found that dealated virgin queens of *Formica polyctena* often contribute to the male brood. Similar observations have been made in polygynous colonies of *Solenopsis invicta* (D. J. C. Fletcher, personal communication).

## THE ORIGINS OF POLYGYNY AND MONOGYNY

Polygyny can arise by one of three means: pleometrosis, with the founding queens remaining together after the first workers appear; adoption of extra inseminated queens after their nuptial flights; and fusion of colonies. Different ant species have used these devices in various ways to produce a remarkable diversity of statistical pat-

**FIGURE 6–1** Workers from mature colonies of the Australian meat ant *Iridomyrmex purpureus* often join newly inseminated queens and help them excavate their first nest. As a result, satellite colonies are founded around the periphery of the older established nest. (From Hölldobler and Carlin, 1985.)

terns of queen numbers. Many species employ them to create mixed strategies of colony founding and structure.

The most versatile species studied to date in this regard is the Australian meat ant *Iridomyrmex purpureus.* Most new colonies are founded by single queens after the nuptial flight in the spring month of October. Sixty-five of 72 newly founded nests excavated near Canberra contained a single queen, 6 contained 2 queens, and 1 contained 3 queens (Hölldobler and Carlin, 1985). As the nearby inseminated queens scurry over the ground and then start to dig burrows in the soil, they are attacked by many enemies. Greaves and Hughes (1974) reported losses of 80 percent or higher from ground-feeding birds alone. Many queens are killed by hostile workers of their own species when they stray too near the established colonies. Surprisingly, however, other queens succeed in digging their nest chambers in the immediate vicinity of mature *I. purpureus.* Not only are most of these females tolerated by the resident workers, but they are often attended and protected, and the workers even help them dig the chambers (see Figure 6–1). Such actions are likely to protect the foundresses from hostile ants, birds, and other predators. Meat ant workers are extremely aggressive, biting enemies and spraying them with poisonous secretions from their pygidial glands. Hölldobler and Carlin (1985) believe that the queens protected in this manner are the ones fortunate enough to settle near their natal nests, so that the foundresses are also likely to be absorbed into the mother colonies to become supernumerary queens. Thus variation in queen numbers appears to be due at least in part to the vicissitudes of foundress association and tolerance by neighboring mature colonies.

Variation in the number of queens, usually less complexly organized than in *Iridomyrmex purpureus,* is widespread but not universal in the ants. Among 46 nests of *Aphaenogaster rudis* excavated by Headley (1949), for example, no queens were found in 6 of the nests, a single queen was found in each of 38 nests, and 2 queens each were found in 2 nests. Talbot (1951) obtained similar proportions in 71 nests of *A. rudis.* In 20 nests of *Prenolepis imparis* excavated in Missouri, Talbot (1943a) found no queens in 3 of the nests, a single queen in each of 15 nests, and 2 queens in each of 2 nests. In Florida, Tschinkel (1987c) found most colonies of *P. imparis* to be polygynous, with a range of 1–6 and a mean of 4 queens per nest. In *Lasius flavus* colonies with more than 1 queen occur but are rare (Waloff, 1957). The same is true of harvester ants in the genus *Pogonomyrmex;* in one sample of 70 *P. rugosus* incipient nests studied by Pollock and Rissing (1985), only 4 contained more than 1 queen. Mature colonies, however, invariably contained only a single queen (B. Hölldobler, unpublished observations). The level of polygyny can likewise vary geographically within the same species. Some local populations of the fire ant *Solenopsis invicta* are primarily monogynous, others primarily polygynous (Tschinkel and Howard, 1978; Fletcher et al., 1980). Unfortunately, most species reported to be polygynous have not been studied carefully enough to establish that the queens are all fully fertile and inseminated.

Although precise data are not available for enough species to assess the Formicidae as a whole, it is our impression from a good deal of field experience that the mature colonies of the majority of species are strictly monogynous. It is reasonable to suppose that properties in colony organization tend to bias species toward monogyny in the course of evolution, and that the tendency is reversed only when special ecological constraints are imposed on the species (Hölldobler and Wilson, 1977b; Oster and Wilson, 1978).

Two such properties can be expected as a consequence of evolution by natural selection. First, queens of all kinds of social insects should prefer to retain personal reproductive rights and surrender none to their sisters or daughters, because they are more closely related to their daughters and sons than to their nieces, nephews, and grand-offspring. Furthermore, queens living in multiples are likely to contribute fewer offspring than they would if they were the sole egg-layers, an effect in fact documented experimentally in *Leptothorax curvispinosus* by Wilson (1974b) and in *Plagiolepis pygmaea* by Mercier et al. (1985a,b). Second, workers should prefer to have only one queen serving as the colony progenitrix. In species with colonies of small to moderate size, the rate of colony growth and hence the degree of colony genetic fitness are limited primarily by the number of workers and not by the number of queens, one queen usually being able to supply as many eggs as a worker force can rear. This effect has been demonstrated in the myrmicine genera *Tetramorium* (Brian et al., 1967), *Myrmica* (Elmes, 1973), and *Leptothorax* (Wilson, 1974b). Since extra queens would be an unnecessary energetic burden on the colony, an especially significant factor during the colony's early growth, it should be of advantage to the workers to eliminate them. In short, one can reasonably expect the dominant queen and the workers to conspire to eliminate supernumerary queens during the early phase of colony growth and thus to attain a state of monogyny.

Under a wide range of conceivable conditions, independent colony foundation should be a second trait favored by natural selection. If entire colonies can be started by one or a few queens, mother colonies producing such females can deploy far more of them over greater distances than can otherwise comparable colonies that reproduce by swarming. Each swarm drains off a part of the original worker force, and its dispersal range is limited by the difficulties inherent in mass orientation and mobility. Swarming is likely to be of advantage only if the survival rate of queens is overwhelmingly greater when they are accompanied by workers than when they proceed alone.

In addition, one can expect haplometrosis to be the preferred mode of independent colony foundation. Unless circumstances give a large advantage to founding in groups, each queen should attempt to start a colony well away from all possible rivals. The tendency just cited toward the restitution of monogyny by both the dominant queen and the first worker force means that a queen choosing to become a member of a group of $n$ founding queens has a $1/n$ chance of surviving to be the nest queen.

Finally, colony foundation should be claustral whenever possible. The highest mortality of social insect workers occurs during foraging trips, and it is probable that the same is true of founding queens forced to leave their nests to search for food.

A first logical exploration of the more abstract general properties of insect societies thus indicates that natural selection will lead species to monogyny, haplometrosis, and claustral nest founding.

Yet deviations from this expected pattern are many and exceedingly diverse. They probably can originate rather quickly in evolution as well. In the case of the fire ant *Solenopsis invicta*, for example, local strains of polygyne colonies have originated and flourished repeatedly in various parts of the United States since 1940 (Greenberg et al., 1985). An examination of the trends toward polygyny can, if made in a broader theoretical context, shed light on the evolution of other aspects of social behavior. Let us begin with a comparison of two of the major groups of social Hymenoptera. Most ant species are monogynous and obligatorily haplometrotic. Within most major phyletic lines, polygyny and swarming appear to be evolutionarily derived conditions. Among the wasps, in contrast, only the temperate zone species of *Vespa* and *Vespula* are known to be obligatorily haplometrotic. *Belonogaster, Mischocyttarus,* and *Polistes* are sometimes haplo- and sometimes pleometrotic, while most or all of the Polybiini, containing 20 of the 26 known social wasp genera, are polygynous and reproduce by swarming (Evans and West-Eberhard, 1970; Spradbery, 1973). We suggest the following simple explanation for the difference. When an ant colony moves to a new nest site, as for example after a disturbance by a predator, it must walk. The workers are wingless, and the queen must travel on the ground with them. Thus queens can afford to shed their wings following the nuptial flight and initiate claustral colony foundation, in which, in all but a few species, they histolyze their alary muscles to nurture the first worker brood within a completely closed cell. Both wide dispersal and independent, claustral colony founding are within their reach. Since these are the optimum techniques under almost all conceivable conditions, most ant species have evolved to acquire them.

When wasp colonies are disrupted, they fly to a new nest site. Lengthy ground travel is not only unnecessary but would be disadvantageous for insects so fully adapted to life in the air. Because the queens must fly with them, they must "stay in shape" by not losing their flight muscles. (The nest queens of *Vespa* lose the power of flight as they become older, presumably because of the weight of their ovaries, but this is the exception rather than the rule in social wasps.) Thus wasp queens are less well equipped to be solitary foundresses, since independently founding individuals would find it disadvantageous to convert the alary muscles into energy for the brood and must engage in the risky process of foraging for food. It follows that wasp species should be more likely to rely on pleometrosis or even swarming, which is in fact the case.

Why, then, does polygyny occur at all in ants? It is evidently derived in evolution and has arisen repeatedly during the hundred-million-year history of these insects, despite the fact that it has certain intrinsic disadvantages. When a phenomenon displays this kind of pattern, the biologist is justified in searching for unusual circumstances that have promoted its deviant development.

### Endangered Populations

Population extinction rates are likely to be highest in rare or locally distributed species. In ants and other social hymenopterans, in which the males are derived from unfertilized eggs, the effective size of the population of colonies ($\tilde{N}$ in the parlance of population genetics; see Li, 1955) is

$$[(4.5N_t)/N_c](\overline{MQ}/1+2\overline{M})$$

where $N_t$ is the total number of adults of all castes in the population of colonies, $N_c$ is the average number of adults per colony, $\overline{Q}$ is the average number of queens contributing offspring to individual colonies, and $\overline{M}$ is the average number of males that fertilized each queen (these males no longer need to be living).

$\tilde{N}$ is the equivalent of the number of breeding individuals in an idealized nonsocial population with equal numbers of males and females, and it provides an exact measure for the estimate of inbreeding, gene loss through random drift, and probability of extinction.

In ants as in nonsocial organisms the effective breeding size will be exactly equal to the term given if breeding is panmictic, that is, completely random among the reproductive individuals, and less than the term if breeding is not panmictic (Wilson, 1963, 1971).

Examination of this formula shows that the most efficient means of enlarging effective population size is by increasing the number of queens. Merely adding a single queen to a monogynous colony system, for example, has the effect of doubling the effective population size. And to double the population size will enormously increase the mean survival time of the population under a wide range of environmental conditions and demographic constraints (MacArthur and Wilson, 1967). Consequently polygyny alone might "rescue" rare populations from quick extinction. In other words, group selection could establish polygyny in systems of populations small enough to be subject to frequent extinction. If there are multiple rare species or rare populations belonging to the same species, those containing polygynous colonies would be more likely to survive than those composed exclusively of monogynous colonies.

The effective population size of some ant species is indeed very low. The *Erebomyrma urichi* population occupying Trinidad's Oropouche Cave in 1961 was estimated not to exceed 400 (Wilson, 1962d). The population of *Lasius minutus* at Hidden Lake, Michigan, nested in only 700 mounds during the 1950s. Since a single colony occupied an average of about 4.4 mounds (Kannowski, 1959a,b), the population could be estimated to contain approximately 160 colonies and an effective population size as low as 320 outside the reproductive season.

The extreme rarity of many social parasites, coupled with patchy distributions, is also well known to myrmecologists. A typical example is *Manica parasitica*, recorded only from nests of *M. bradleyi* on top of Polly Dome, Yosemite National Park, California, and apparently absent from immediately surrounding areas (Creighton, 1934; Wilson, 1963). A second population was discovered in the Stanislaus National Forest of California by Wheeler and Wheeler (1968). Chamberlin (in Wheeler, 1910a) found only three colonies of *Formicoxenus* (= *Symmyrmica*) *chamberlini* "in several parts of a ten-acre field" near Salt Lake City, Utah, in 1902. After an intensive search eighty years later Buschinger and Francoeur (1983) rediscovered a sparse population in the same area. The extreme parasite *Teleutomyrmex schneideri* has one population apparently limited to the east side of the isolated Saas-Fee Valley of Switzerland, between 1,800 and 2,300 meters elevation (Kutter, 1950a); a second local population has been discovered near Briançon in the French Alps (Collingwood, 1956).

A second means of enlarging the effective population size, which might be favored by group selection (or more precisely, interdemic selection), is reduction in the average colony size, which increases the number of colonies for a fixed total number of individual ants. Still another is the promotion of outbreeding by such mechanisms as the appearance of the males in advance of the reproductive females in individual colonies (protandry), fully developed nuptial flights that carry the reproductive forms away from the natal nests, and the synchronization of nuptial flights to bring together males and females from different colonies at the time of mating.

Have rare species in fact taken these steps? The evidence presented in Table 6–1 shows that a significantly higher number of rare species have multiple queens in comparison with related species. The inference is that selection is strong enough at the level of entire populations of colonies (as opposed to individual colonies), because of the danger of extinction through small population size, to force polygyny on the colonies. Since polygyny also occurs in many ant species that are abundant and widespread, this inference is weak, however.

Thus one way in which polygyny might have arisen in ants is by group selection acting on rare species, where interdemic selection is generally most likely to be potent. An alternative explanation for polygyny in rare species is available, however: because of the small number of individuals, associated queens might be more closely related than is the case in larger, more typical populations. Hence, as originally suggested for *Formica rufa* by Williams and Williams (1957), altruistic cooperation among queens would be favored through sister-group selection. At present there is no clear way to decide between these two competing hypotheses.

An examination of Table 6–1 shows that decrease in colony size, another means of raising effective population size, does not occur among the rare species. Furthermore, the data show no evidence of an increase in exogamy, the tendency of the sexual forms to mate with members of other colonies. Quite the contrary, in fact: the rare species show a reduced level of exogamy. Mating among members of the same colony, which has the effect of turning the colony into something resembling a self-fertilizing hermaphrodite, has long been recognized by myrmecologists as a trait of the rarest, parasitic species. The males are often apterous or subapterous, mating takes place in or near the nest, and the fecundated, winged queens then either disperse in search of new host colonies or else return to the old.

Whether intracolonial mating characterizes other categories of rare species is an open question. Since the trait must cause a decrease in the effective breeding size of the population, just the reverse of what purely logical considerations concerning population size alone dictate, it is necessary to consider other possible advantages of such a design feature. At least one can be deduced: intracolonial mating certainly eliminates the loss of virgin reproductives that would normally occur during dispersal and ensures that the queens will be inseminated, however scarce the species. This advantage can easily outweigh disadvantages from inbreeding. Passing from random mating to perfect inbreeding reduces the effective breeding size by only half, a deficit that can be balanced merely by doubling the average number of nest queens. The exact extent of true brother-sister mating is unknown. Because of the additional trait of polygyny in these same species, the offspring of several matings probably breed with each other as a matter of course. In fact, Wesson (1939) did find that the queens and males of *Protomognathus* (= *Harpagoxenus*) *americanus* prefer to mate with unrelated individuals. It is even conceivable that parasitic species are less "adelphogamous" than they were previously assumed to be, for it has not been established with certainty that true brothers and sisters within polygynous colonies mate with each other at all. In polygynous colonies the opposite may be true.

### Specialized Nest Sites

A much stronger correlation exists between polygyny and the manner in which colonies occupy nest sites. Truly polygynous species, most or all of whose colonies contain multiple inseminated queens, fall into one of three sets characterized by very different adaptation

**TABLE 6–1** Characterization of rare species with respect to certain social features that affect reproductive population size. +, the species shows a significant difference from commoner members of the genus or from the nearest related taxa in the indicated trait. –, the reverse is true, that is, the species departs from related taxa. 0, no difference. —, information available but comparison meaningless because worker caste is lacking. ?, insufficient information.

| | Predicted devices to increase population size | | | Predicted devices to increase exogamy | | | |
|---|---|---|---|---|---|---|---|
| Species | Decreased colony size | Polygyny | Polyandry | Special behaviors | Dioeciousness, or protandry | Increased synchronization of mating activity among colonies | Reference |
| **Bog "endemics"** | | | | | | | |
| *Formica neorufibarbis* | 0 | + | ? | ? | ? | 0? | Kannowski (1959a) |
| *Lasius minutus* | 0 | 0? | ? | ? | ? | 0 | Kannowski (1959a) |
| *L. speculiventris* | 0 | 0? | ? | – | ? | 0 | Kannowski (1959a) |
| *Leptothorax muscorum* | 0 | + | ? | ? | ? | ? | Kannowski (1959a) |
| **Cave species (troglophilic)** | | | | | | | |
| *Erebomyrma urichi* | 0 | + | ? | ? | ? | ? | Wilson (1962d) |
| **Parasitic species** | | | | | | | |
| *Anergates atratulus* | — | + | ? | – | 0 | 0 | Wheeler (1910a), Gösswald (1932) |
| *Epimyrma stumperi* | — | + | ? | ? | ? | ? | Kutter (1951), Stumper and Kutter (1951) |
| *Formicoxenus chamberlini* | +? | + | ? | – | ? | 0 | Buschinger and Francoeur (1983) |
| *Kyidris yaleogyna* | 0 | + | ? | ? | 0 | ? | Wilson and Brown (1956) |
| *Monomorium santschii* | — | 0 | ? | – | 0 | ? | Forel (1906) |
| *Pheidole (= Bruchomyrma) acutidens* | — | + | ? | – | 0 | 0 | Bruch (1932a) |
| *Plagiolepis grassei* | — | + | ? | – | 0 | 0 | Le Masne (1956b) |
| *P. xene* | — | + | ? | – | 0 | 0 | Kutter (1952) |
| *Teleutomyrmex schneideri* | — | + | + | – | 0 | 0 | Kutter (1950a), Stumper (1950) |

syndromes. The first type is specialized on exceptionally short-lived nest sites. Such species are opportunistic in the sense employed by ecologists—they occupy local sites that are too small or unstable to support entire large colonies with life cycles and behavioral patterns dependent on monogyny. The second type is specialized on scarce nest sites, which evidently confer advantages on queens willing to group together in lasting associations. The third type is specialized on entire habitats, as opposed to nest sites, that are long-lived, patchily distributed, and extensive enough to support large populations. The three forms of specialization are not mutually exclusive; some ant species, for example *Iridomyrmex humilis* and *Pheidole megacephala*, possess features of both the first and the third types.

Consider first the class of opportunistic nesters. Colonies of *Tapinoma melanocephalum, T. sessile, Paratrechina bourbonica*, and *P. longicornis* often occupy tufts of dead grass, plant stems, temporary cavities beneath detritus in urban environments, and other local sites that sometimes remain habitable for only a few days or weeks. Colony fragments of *T. sessile* observed by Smallwood (1982) moved an average of every 12.9 days. Some species of *Cardiocondyla* excavate shallow tunnels and chambers in patches of soil in such places as the edges of palm trunks, sidewalks, and street gutters. *Leptothorax curvispinosus* typically nests in confined and relatively unstable preformed cavities in acorns, hickory nuts, galls, stems, and twigs. *Lasius sakagamii* favors unstable river banks, sandy and sparsely vegetated areas that are frequently flooded (Yamauchi et al., 1981). On the rocky islets of the Gulf of Finland, *Formica truncorum* occupies flimsy nest sites in which temperature varies in an unreliable manner (Rosengren et al., 1985). Colonies of *Monomorium pharaonis* and *Tapinoma melanocephalum* go so far as to invade houses to occupy cracks in walls, linings of instrument cases, spaces in piles of discarded clothing and between leaves of books, and similarly unlikely microhabitats.

Colonies of these species are characterized by extreme vagility—a readiness to move when only slightly disturbed and an ability swiftly to discover new sites and to organize emigrations. Their colonies are also typically broken into subunits that occupy different

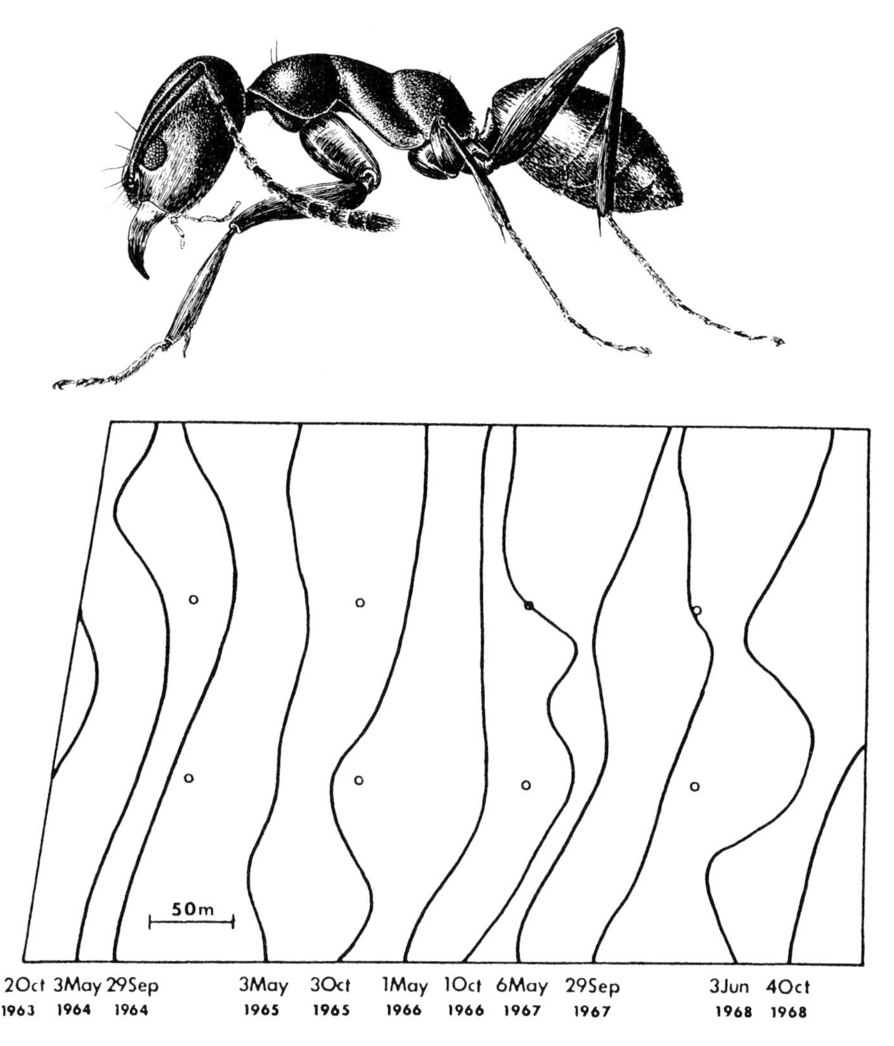


**FIGURE 6-2** The forward advance of a unicolonial species, the Argentine ant *Iridomyrmex humilis,* across an old field in San Luis Rey, California. The small circles are 100-meter reference points. As they spread, the Argentine ants displaced a native harvesting ant, *Pogonomyrmex californicus.* (Modified from Erickson, 1971.)

**FIGURE 6-3** *(facing page)* Part of the range of a single supercolony (unicolonial population) of the mound-building ant *Formica lugubris* in the Swiss Jura. The extent of the area mapped is 12 hectares. (Modified from Cherix, 1980.)

nest sites and exchange individuals back and forth along odor trails. Colonies of opportunistic nesters bud, disperse, and fuse again, as documented in *Tapinoma erraticum* by Dubuc and Meudec (1984) and in various species of *Leptothorax* by Buschinger (1974a), Möglich (1978), Alloway et al. (1982), and Stuart (1985b). We suggest that it is the tendency to emigrate swiftly and efficiently that gives polygyny a premium in opportunistic nesting and ties it closely to polydomy, the occupation of multiple nest sites. Because of the inevitable frequent fragmentation of the colonies, subunits probably lose contact with one another for long periods of time and occasionally forever. Having enough reproductive females to service most or all of the subunits means that the colony as a whole can exploit the rapidly fluctuating environment in which it lives. Other kinds of ants are not fragmented in this manner, and consequently a single queen suffices as the colony progenitrix.

The second class of polygynous nesters has been defined by Herbers (1986a,b) and has only a single known example, the small North American myrmicine *Leptothorax longispinosus.* The species prefers hollow acorns and other preformed cavities on the ground, sites that are usually in short supply. Herbers found that the percentage of colonies that are secondarily polygynous increases with the local density of colonies, and she was able to exclude nest site fragility as a potent factor in polygyny.

The third major class of fully polygynous ants contains only a few species, but these include the great majority of examples that have been well studied and do not qualify as opportunistic nesters. Thus most fully polygynous species whose natural history is known to us are accounted for by the simple three-way classification proposed here. The habitats favored by species in the third category are first of all patchily distributed; they have distinctive qualities and are more or less isolated from each other. They are also extensive enough to support substantial populations of ants. Hence propagules of species that are specially adapted for such places encounter a potential bonanza when they succeed in colonizing one. The habitats are also relatively long lived, giving a premium to the type of slow but thorough occupation made possible by polygyny and budding.

The Allegheny mound-builder *Formica exsectoides,* for example, typically occupies persistent grassy or heath-like clearings. Such habitats are relatively scarce and patchily distributed, and many are fully occupied by dense unicolonial populations of *F. exsectoides.* The *"microgyna"* form of *Myrmica ruginodis* (which itself is almost certainly a distinct species; see Pearson, 1981) shows a similar preference for scattered, very stable, open habitats in England, some of which are known to have persisted at the same localities for as long as two hundred years. The typical, or *"macrogyna"* form, which is haplometrotic and monogynous, favors less stable but more widespread habitats (Brian and Brian, 1955). *Pseudomyrmex venefica* is spe-

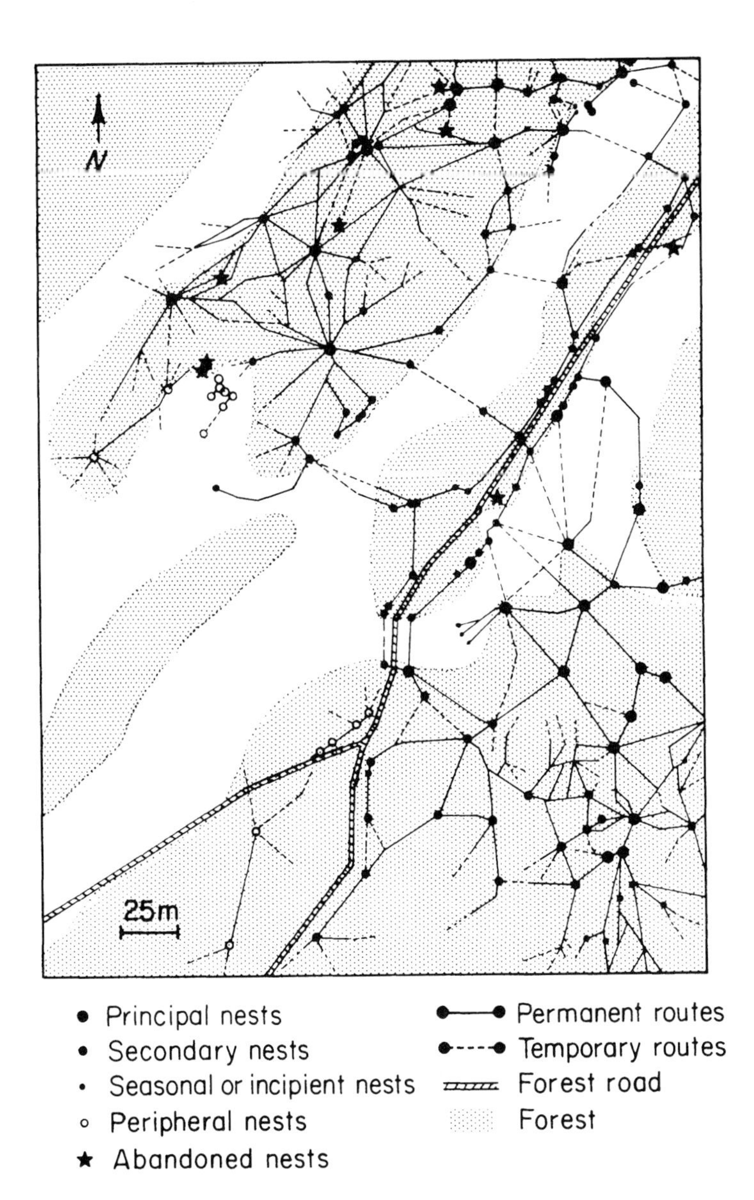


cialized to occupy species of swollen-thorn acacia that grow for long periods of time in areas of slow floristic succession. Because the acacias are able to expand into extensive thorn forests, the *P. venefica* colonies have the opportunity to build large unicolonial populations over a period of many years. Single populations may contain 20 million or more workers, rivaling *Dorylus* ants for the possession of the largest ant "colonies" in the world (Janzen, 1973b). Both are exceeded easily, however, by *Formica yessensis*, one supercolony of which in Hokkaido was estimated to contain 306 million workers and over one million queens (Higashi and Yamauchi, 1979; Higashi, 1983). "Tramp" species, those ants distributed widely by human commerce and living in close association with man, are typically polygynous. In a sense they can be said to have been preadapted for patchy but persistent and species-poor habitats created within man-made environments. Some of the best-known species comprise unicolonial populations and spread largely or entirely by the budding off of groups of workers accompanied on foot by inseminated queens. Examples include *Monomorium pharaonis*, *Pheidole megacephala*, *Iridomyrmex humilis*, and *Wasmannia auropunctata*.

The isolated-habitat specialists, as distinguished from the nest-site opportunists, are species that have difficulty locating their preferred habitats; having once found a suitable place, however, they are opposed by relatively less competition from colonies of their own or other ant species. Thus it is to the advantage of the founding colony to spread out as a continuous unicolonial population, occupying all of the habitable nest sites and foraging areas. An example involving the Argentine ant *Iridomyrmex humilis* is shown in Figure 6–2. In contrast, most other ant species have no trouble finding a suitable habitat, but such places are typically already saturated with ant colonies. The best strategy of these ants is to send out large numbers of flying queens capable of founding new colonies independently. A tiny fraction of the propagules will locate some of the rare nest sites and foraging areas left unfilled; the colonies they produce will not find it profitable to spread outward through the surrounding occupied territories by budding in the unicolonial manner.

The full life cycle of unicolonial populations remains to be thoroughly worked out, although details of the structure and secular changes over a period of up to a few years have been studied in the huge colonies of *Formica lugubris* in Switzerland by Cherix (1980), *F. polyctena* in the Netherlands by Mabelis (1986), and *F. aquilonia* in the Soviet Union by Zakharov et al. (1983). A map of part of the *F. lugubris* colony is presented in Figure 6–3. Among the interesting variables to be studied as functions of the age and condition of the populations are the queen:worker ratios and the behavior of queens during nuptial flights. Scherba (1961) reported that most of the queens of *F. opaciventris*, a North American, mound-building, unicolonial species, mate within a few meters of their nest of origin and return at once, while a small minority, originating entirely from mounds near the edge of the population, fly out of sight away from the population. A very similar pattern occurs in at least one European equivalent species, *F. polyctena*, whose life cycle was presented in Chapter 3 (see also Gösswald and Schmidt, 1960). We predict that in the early stages of population growth, the queen:worker ratio will start high (immediately following independent nest founding by one or more newly inseminated queens, usually by adoption of an alien host species), drop to a steady state as the population comes to saturate the habitat, then rise again as the habitat quality declines and a higher premium is placed on dispersal away from the habitat. It also seems likely that the fraction of the queens engaging in dispersal flights will increase when the habitat is saturated, and increase still more as the habitat declines.

Thus in a curious fashion the extreme polygynous species have evolved unicolonial populations as a means for trading dispersibility for potential colony immortality. Where competitive pressure is lower, the species can afford to gamble on great longevity. The colonies can be fused into single units, and newly mated queens can more safely return to the parental nests. And since colonies are on the average longer lived, fewer empty nest sites will be available at any given time, which makes the dissemination of queens away from the parental nests less profitable. The results will be a dual pressure for the unicolonial species to suspend claustral nest founding altogether. The adoption of unicolonialism, with its increased degree of inbreeding and reliance on budding as the principal means of reproduction, is an evolutionary step that parallels the adoption of apomixis and vegetative reproduction in nonsocial organisms. Both permit the rapid growth of population in sparsely distributed habitats that are relatively free of competition.

An interesting phenomenon for which no explanation yet exists is the frequent coexistence of pairs of closely related species, one of which is monogynous and the other polygynous. Examples include the *"macrogyna"* and *"microgyna"* forms of *Myrmica ruginodis*, probably at least a few other European species of *Myrmica* (Pearson,

1981), the acacia ant *Pseudomyrmex venefica* and an undescribed polygynous sibling form in Mexico (Janzen, 1973b), two apparent species placed under *Crematogaster minutissima* (E. O. Wilson, unpublished), *Conomyrma insana* and *C. flavopectus* (Nickerson et al., 1975), *Formica incerta* and *F. nitidiventris* (Talbot, 1948), and two apparently distinct species of *F. neorufibarbis* in the White Mountains of New Hampshire (Wilson, 1971). A similar duality appears to exist in *F. rufa,* which apparently consists of two sibling species, one strictly monogynous and the other polygynous (Kutter, 1977). It is also possible for at least the rough equivalents of monogynous and polygynous strains to occur as genetic polymorphs within the same species, as in the A and B colony types of *Rhytidoponera confusa* and *R. chalybaea* in Australia (Ward, 1983a,b).

### Variable Food Supply

A third principal clue to the origin of polygyny, as well as to mixed strategies among and within species with respect to the number of queens, has been provided by Briese (1983) in his study of an Australian species of *Monomorium* (= *Chelaner*) belonging to the *rothsteini* group. He followed two colonies over a two-year period in a semi-arid saltbush steppe in New South Wales. One, containing winged queens capable of flight, started new colonies by the conventional means of nuptial flights followed by claustral nest founding. The other, containing brachypterous (short-winged) queens unable to fly, started new colonies by fission. The brachypterous queens were either adopted back into the natal nest or escorted to new nest sites by their worker sisters. Both procedures led to polygyny with brachypterous queens. Flights by the normal-winged queens occurred after a favorable period of substantial rainfall that increased the food supply (principally seeds). Fission by brachypterous females followed a drought and a reduced food supply (Figure 6–4). Briese considered the two kinds of colonies to belong to the same species, and their founding methods to constitute alternative responses of a mixed strategy:

> When conditions are favourable, the production of fully alate queens, with sufficient body reserves to raise a first brood by themselves, would allow a colony to extend its genetic material over a much wider area with maximised chances of successful establishment. When stress is very severe queen production may cease altogether. However, under conditions of less severe food stress, the mode of colony foundation might alter, and brachypterous queens with less body reserves be produced. During colony foundation, these would be accompanied by groups of worker ants which act as food gatherers to support the initial brood. Such daughter colonies would remain in the same, relatively favourable area, still capable of supporting a colony, rather than facing the risk of failure through the alternative mode of random dispersal to areas of unknown favourability.

The attractiveness of Briese's hypothesis is that it can be tested, whether the two queen morphs represent different species or variants of the same species. Further studies of this and other species can help determine whether ants do have a propensity to evolve fission in areas of wide and erratic fluctuations in climate such as occur in the arid Australian interior, also whether species that are truly polyethic favor fission during hard times.

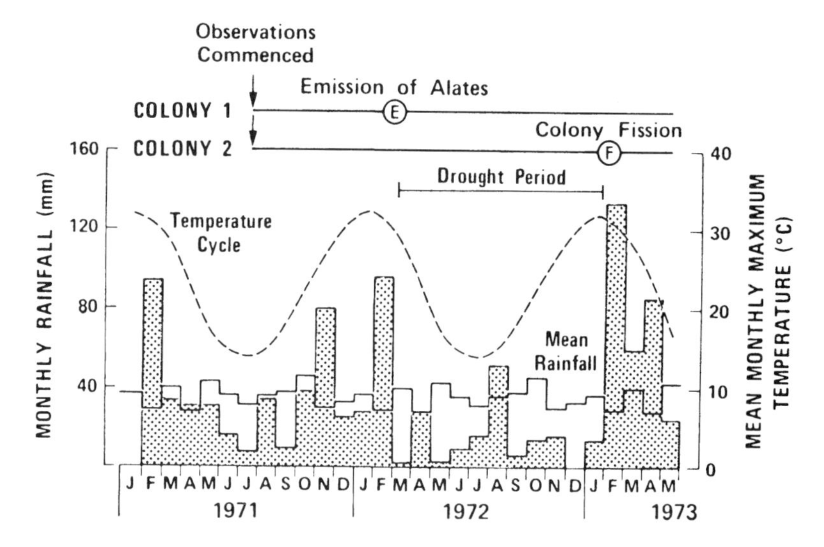


**FIGURE 6–4** Monogyny and polygyny appear to be favored under different rainfall regimes in the Australian steppe ant *Monomorium* (= *Chelaner*) sp. Emission of fully winged queens (alates), which produces new monogynous colonies, followed a period of rainfall, while fission of the natal colony, involving flightless queens and promoting polygyny, followed a period of drought. (From Briese, 1983.)

## GYNY AND SPECIES DIVERSITY

Having reviewed the possible ecological prime movers that lead to monogyny or polygyny, we may now briefly consider some of the broader implications of variation in the number of queens. Monogyny is closely associated with colony distinctness. Each colony occupies a separate nest site, and its workers either avoid or attack the members of other colonies they encounter while foraging. The loss of territorial boundaries in the case of highly polygynous, unicolonial species changes the rules drastically. Unicolonial species are notable for their high local abundance and the degree to which they appear to dominate the environment. Introduced populations of *Pheidole megacephala, Wasmannia auropunctata,* and *Iridomyrmex humilis* extirpate many other kinds of ants, although it is not known whether species occurring within the native ranges of the unicolonial ants have evolved competitive resistance to the point of being able to coexist. Habitats containing populations of *Formica exsectoides* have notably sparse ant faunas, and the related *F. exsecta* and *F. opaciventris* are known to exclude some other territorial ants, including other species of *Formica,* by aggression (Scherba, 1964; Pisarski, 1972). The same is true of *F. aquilonia* and *F. polyctena,* the "large-scale conquerors" of the *F. rufa* group in Europe (Rosengren and Pamilo, 1983). It is not known to what extent the general occurrence of unicolonial species in species-poor habitats is a cause and to what extent it is an effect. Have the unicolonial ants simply adapted to habitats that were species poor in the first place, or has the formation of supercolonies provided them with a decisive competitive edge in habitats that would otherwise be species rich? This problem seems eminently tractable to field analysis (see Chapter 11).

It is possible that gyny and the nature of territoriality also affect the patterns of species diversity and quite possibly the mechanism of speciation itself. Local faunas consisting of multicolonial ant populations, which comprise chiefly monogynous and oligogynous colonies, seem to be more susceptible to increases in within-habitat species diversity, for three reasons. First, as previously noted, they appear to suppress other ant species less severely than do unicolonial populations, permitting the buildup of larger numbers of coexisting species. Second, because the colonies are much smaller in size, multicolonial species are able to specialize more on nest sites and food. For example, a colony of one species may do well with a single hollow stem, whereas a colony of another species can occupy the space beneath a nearby stone. Some species prey exclusively on small arthropods, others tend scale insects, and so forth. The relatively huge populations of unicolonial species cannot afford such a narrowing of their niche. To survive they must remain broad generalists, which brings them into competition with a large number of specialists as well as with other generalists.

The third reason why multicolonial species can be expected to exhibit higher within-habitat diversity is more subtle. Many studies have shown that colonies of ants are hostile to a degree inversely proportional to the degree of similarity to their competitors. That is, they are most aggressive to other colonies of the same species, somewhat less to other species in the genus, and least of all to forms that not only belong to other genera but differ strongly in size and behavior. This being the case, within-habitat diversity can be expected to be enhanced by the divergence of species odors between closely related species. Suppose that two newly formed, cognate species have arisen by genetic divergence during geographical isolation, and have recently come into contact along the boundaries of their respective geographical ranges. Suppose further that the species are sufficiently divergent in ecological requirements that neither would exclude the other by means of nonaggressive preemption of resources. Yet the two forms are likely to remain in separate geographical ranges as long as their species odors are too similar, so that interspecific territorial exclusion is as strong as intraspecific exclusion. The two species can penetrate one another's ranges only if one or both undergo a divergence in the species-specific components of the colony odor. This is a form of character displacement comparable to the divergence of identifying songs by which territorial bird species penetrate one another's ranges (Murray, 1971). The result is an increase in the within-habitat species diversity.

The penetration of territories belonging to alien monogynous species is made easier by the very fact that territorial colonies repel one another so effectively—to the extent that colonies are overdispersed in their statistical pattern of distribution. In many kinds of ants, such as members of the genus *Pogonomyrmex*, this pattern is geometrically very regular and is maintained by high levels of intercolonial aggression (Hölldobler, 1974, 1976a). If colonies of other species are different enough in species-specific components of the colony odor to escape such aggressive response, and if they are also distinct enough in their foraging habitats not to be replaced through competitive exclusion, they might easily slip into nest sites located between those of the resident species.

In an earlier analysis (Hölldobler and Wilson, 1977b) we hypothesized that the degree of odor specificity associated with monogyny and intercolony territoriality lends itself to interspecific recognition, the reduction of interspecific territorial aggression by means of that recognition, and an increase in numbers of species that can coexist in the same habitat. Many polygynous ant species, and particularly those that are unicolonial, have surrendered some of this discriminatory power. As a result they remain aggressive toward a broader range of species, and it is not surprising to find very few unicolonial species occupying the same habitat, as well as an overall decrease in species diversity. Although it is theoretically possible for unicolonial ant species to build up between-habitat diversity by means of specialization on habitats as opposed to niches within habitats, very little of such specialization appears to have occurred in nature—quite possibly because of the preemption of most kinds of habitats by monogynous, multicolonial ant species.

## PLEOMETROSIS

The alliance of two or more queens during colony founding is a widespread but still far from universal habit in ants. Within particular species it is more or less an optional procedure. For example, 10 percent of founding queens have been recorded as being in pleometrotic groups in *Iridomyrmex purpureus* (Hölldobler and Carlin, 1985), 80 percent in *Lasius flavus* (Waloff, 1957), 82 percent in *Acromyrmex versicolor* (Rissing et al., 1986), 89 percent in *Messor (= Veromessor) pergandei* (Rissing and Pollock, 1986), and 97 percent in *Myrmecocystus mimicus* (Bartz and Hölldobler, 1982).

These data suggest only coarse differences between species. In fact, the incidence of pleometrosis is usually flexible within particular species, rising under environmental conditions that promote crowding of the newly dealated queens. In the honeypot ant *Myrmecocystus mimicus*, queens are strongly attracted to each other after they have mated, and the pleometrosis is enhanced further by the queens' tendency to start nests in close proximity (Bartz and Hölldobler, 1982). Even after the nest digging has begun, the foundresses aggregate still further, until the great majority of nests contain more than one queen, as illustrated in Figure 6–5. A closely similar pattern has been described in the fire ant *Solenopsis invicta* by Tschinkel and Howard (1983). After the fire ant queens have mated and cast their wings, they tend to settle on higher ground, away from rain-washed areas and puddles. Within the favored spots they also aggregate in a strongly non-random manner. A 64-fold increase in queen density across various sample areas near Tallahassee, Florida, was found to result in a 2.2-fold increase in the average number of queens per nest, with most burrows still containing one to three queens but a few containing ten or more. Variation in queen density is quite potent in this aspect of social behavior. In the Tschinkel-Howard study it accounted for 70 percent of the variation in the mean number of queens per nest and for 86 percent of the overall aggregation of queens in local areas. An even stronger initial clumping as an effect of microhabitat selection has been recorded in the leafcutter ant *Acromyrmex versicolor* by Rissing et al. (1986). Newly mated queens strongly prefer to settle under the shade of the scattered trees of the Arizona desert, where close proximity leads to a preponderance of joint occupancy in the newly excavated nests. The shaded, cooler environment permits a much higher survival of the incipient colonies.

Although pleometrosis is very common among ants, it seldom leads smoothly to polygyny. In most cases studied to date, multiple queens are reduced to a single egg-laying queen, at least within local areas of the nest, shortly after the first brood of workers ecloses. The workers eliminate the supernumerary queens (*Nothomyrmecia ma-*

FIGURE 6–5 Founding queens of the honeypot ant *Myrmecocystus mimicus* dig burrows in the Arizona soil in clumped rather than randomly dispersed patterns, as illustrated on the left. The queens also tend to aggregate still more after the burrows have been started, as shown on the right. (Modified from Bartz and Hölldobler, 1982.)

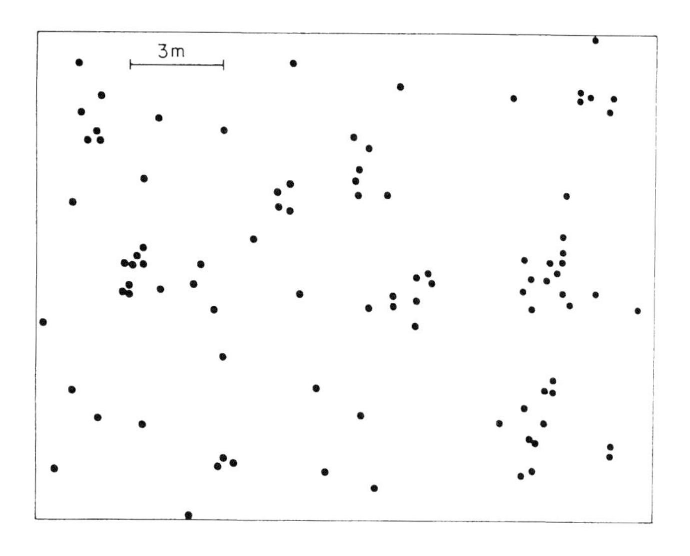


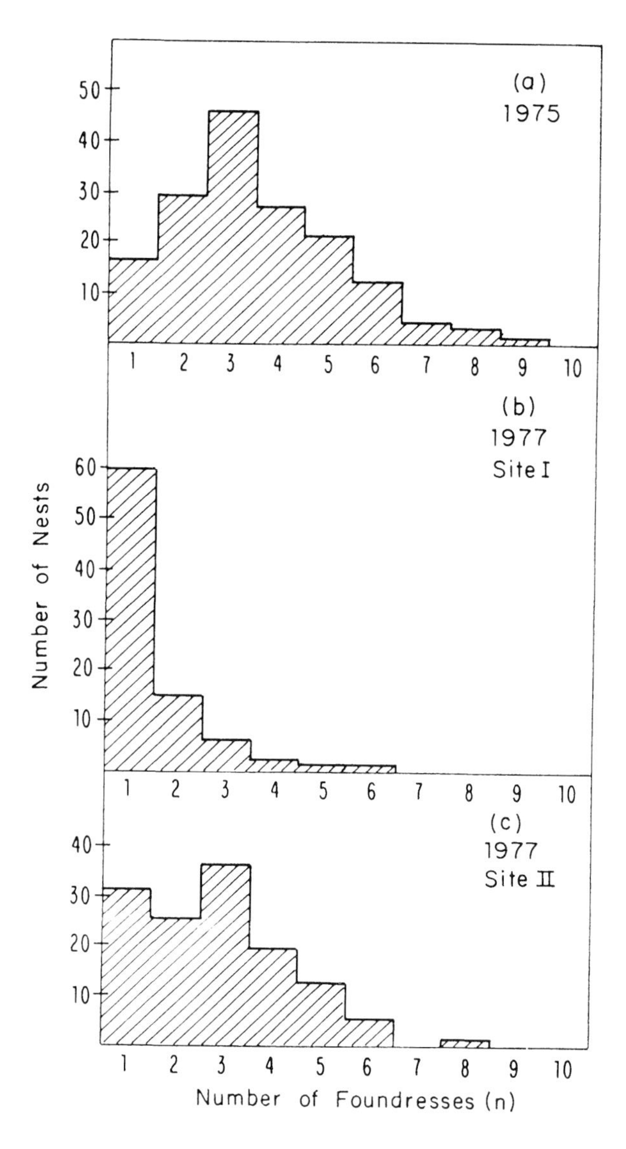


*crops, Myrmecocystus mimicus,* some strains of *Solenopsis invicta*), or the queens fight among themselves and are reduced further by worker aggression (*Messor pergandei*), or the queens begin to fight and disperse to different parts of the nest, creating a condition of oligogyny (*Iridomyrmex purpureus, Camponotus herculeanus, C. ligniperda, Lasius flavus*). Even when extra inseminated queens are tolerated in close association with the primary egg-layer, their ovarian development is usually suppressed (species of *Leptothorax* and *Formicoxenus,* at least one strain of *Solenopsis invicta*). Primary polygyny, in which associations of founding queens survive to become multiple egg-layers at close quarters in mature colonies, appears to be rare. In fact we know of only two cases: the leafcutter *Atta texana* (Mintzer and Vinson, 1985a; Mintzer, 1987) and *Pheidole morrisi* (S. Cover, personal communication). In all other species whose colony ontogeny has been thoroughly studied, polygyny is secondary, that is, it originates when additional inseminated queens are adopted by an already existing worker force.

The association of queens during founding does not appear to be due to kin selection of the kind observed in foundress females of *Polistes* wasps. These insects prefer to associate with their former nestmates, who are likely to be full sisters (Ross and Gamboa, 1981; Post and Jeanne, 1982; Strassmann, 1983). The evidence so far points to no such discrimination in ants. In particular, foundress queens of *Myrmecocystus mimicus* and *Messor* (= *Veromessor*) *pergandei* associate freely with females who could not have come from the same nest. Rissing and Pollock (1986) combined newly inseminated queens of *M. pergandei,* collected 1–6 kilometers apart. In all of nine replicates, the queens combined with no outward aggression to start a single cooperative nest. Another clue has been provided by the discovery that multiple queens of older *Solenopsis invicta* colonies are no more closely related to one another than to queens collected outside the nest (Ross and Fletcher, 1985a).

The key advantage that does accrue to multiple founding queens is the fact that in comparison to solitary founding queens they produce larger initial broods and worker force in less time and with less individual weight loss. This effect has been documented across a wide variety of ant genera, including *Lasius* (Waloff, 1957), *Solenopsis* (Wilson, 1966; Markin et al., 1972; Tschinkel and Howard, 1983),

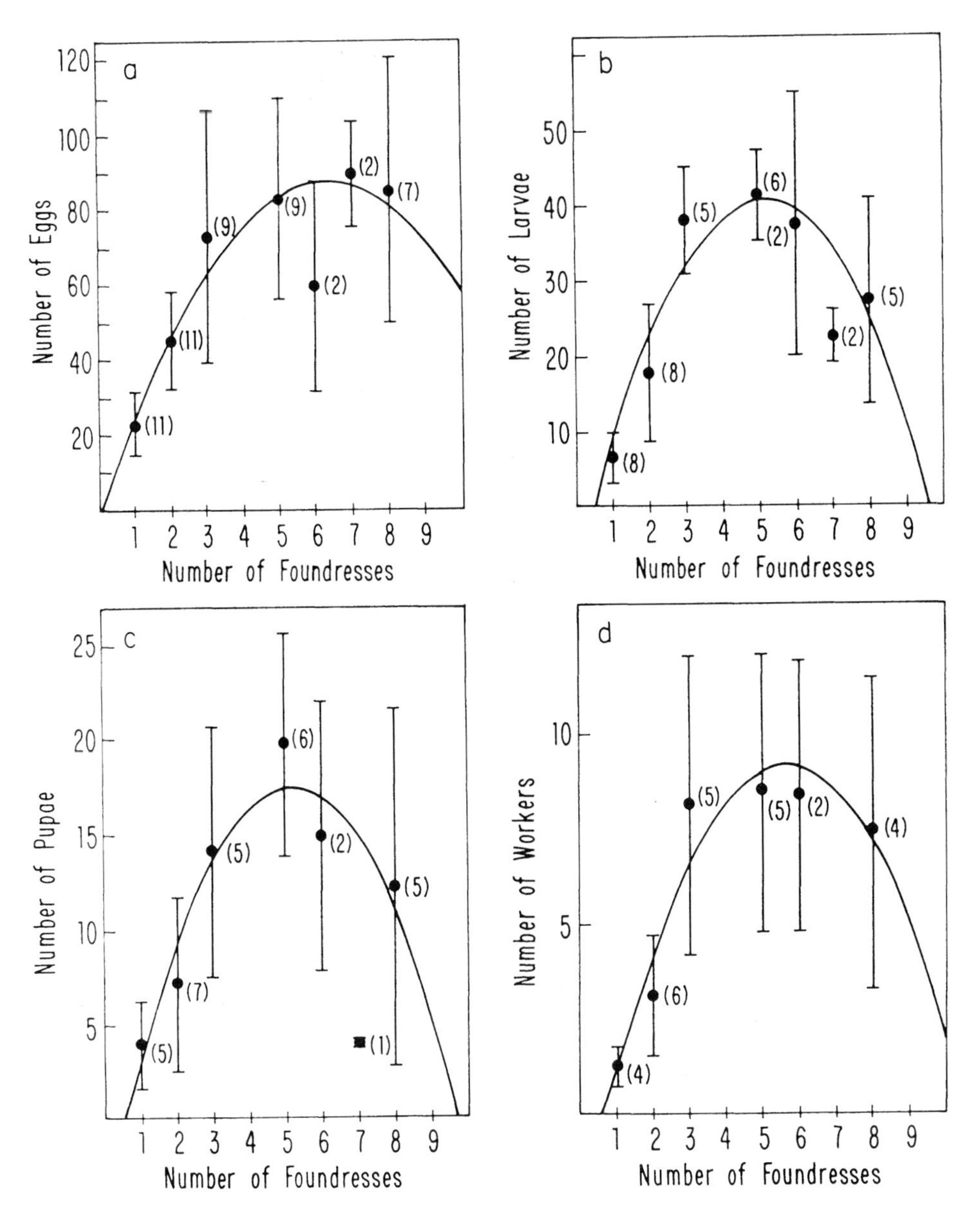


**FIGURE 6–6** Optimization during colony founding is attained by the honeypot ant *Myrmecocystus mimicus*. Under laboratory conditions an intermediate number of queens is most successful in rearing the first worker brood. The group size that maximizes the number of workers per female (as opposed to the number produced by the whole group) is between 2 and 3, closely corresponding to the modal founding group in nature—as shown in Figure 6–5. (From Bartz and Hölldobler, 1982.)

*Camponotus* (Stumper, 1962; Mintzer, 1979a), *Tapinoma* (Hanna, 1975), *Messor* (Taki, 1976), and *Myrmecocystus* (Bartz and Hölldobler, 1982). In addition, colonies starting with larger initial worker forces are more successful in brood raids or territorial fights directed against other incipient colonies. This advantage has been documented in *Myrmecocystus mimicus* by Bartz and Hölldobler (1982) and *Messor* (= *Veromessor*) *pergandei* by Rissing and Pollock (1986). No contrary cases are known.

Experiments by Bartz and Hölldobler (1982) on *Myrmecocystus mimicus* have allowed a more precise assessment of the advantage of pleometrosis as well as a test of optimization. As shown in Figure 6–5, the most frequent group size of founding queens in newly dug burrows is 2 to 4. Under laboratory conditions, the maximum production of workers per founding queen—as opposed to workers produced by the entire founding group—is attained when the number of associated queens is about 3 (Figure 6–6), corresponding very closely to the modal group size in nature. It also turns out that the mortality rate among the founding queens is lowest when the group size is 3 to 4, a second selection factor seemingly favoring this intermediate level. The explicit reasons for this optimal group size have not been worked out. They may include convex curves of efficiency in brood care, along with steadily rising curves of mutual egg cannibalism (some of which was observed) and maximal resistance to disease.

Apart from optimal group size, Bartz and Hölldobler discovered an additional behavior favoring group formation over solitary founding in general. After the first *Myrmecocystus* workers appear, they start to raid nearby incipient colonies by transporting brood to their own nests. In laboratory enclosures that simulated the natural dispersal of incipient nests, all of the brood eventually ended up in a single nest. Workers frequently abandoned their own mothers in favor of the winning colonies. In 16 of 23 such experimental arenas, colonies starting with 5 queens prevailed in the end, while in 7 cases the winning group consisted of 3 queens. In no case did colonies founded by 1 or 2 queens prevail. Furthermore, in 19 cases the winning colonies were the ones with the largest initial worker force. This martial behavior foreshadows an even more remarkable competition that develops among mature *Myrmecocystus mimicus* colonies. These colonies conduct ritualized tournaments as part of the defense of their foraging territories. Opposing colonies summon their worker forces to the tournament area, where hundreds of ants perform highly stereotyped display fights. When one colony is considerably stronger than the other, in other words able to summon a larger worker force, the tournaments end quickly and the weaker colony is sacked. During these final incursions, the queen is killed or driven off and the larvae, pupae, callow and honeypot workers are transported to the raiders' nest (Hölldobler, 1976c, 1981a).

A closely similar pattern of prevalence of groups over solitaires was independently discovered in *Solenopsis invicta* by Tschinkel and Howard (1983). The mean number of founding queens per nest varied according to site from 1.1 to 3.4. In laboratory tests reproductive performance was compared across groups of founding queens containing 1, 5, 10, and 15 individuals. Groups with 5 queens produced the most pupae and workers, while productivity per queen was greatest at a lower number, somewhere between 1 and 5. The advantages of group life do not stop at the founding stage, however. When the colonies were maintained over a period of 100 days, with the worker populations reaching 200 to 1,000 individuals, the most rapid growth occurred in the polygynous colonies. This differential persisted even after the workers had executed all but 1 of the founding queens. And as in *Myrmecocystus mimicus*, workers of *Solenopsis invicta* tend to move into whatever neighboring nest has the largest number of workers, leaving their mothers to die of starvation. Tschinkel and Howard suggest four advantages of pleometrosis in the fire ants, with accompanying documentation from laboratory and field studies: (1) earlier maturation and reproduction lead to a shorter generation time and higher population growth rate (Tschinkel and Howard, 1983); (2) fire ants are territorial and war with conspecific colonies along their mutual territorial boundaries, with the advantage presumably going to the larger colonies (Wilson et al., 1971); (3) the coalition of queens in an incipient colony benefits those with the larger worker force, as just described (Tschinkel and Howard, 1983); (4) a colony's ability to survive adverse physical conditions is enhanced by larger size. Markin et al. (1973) found that *S. invicta* colonies not attaining a certain size by the onset of cold weather fail to survive the winter, and the same could be true of survival in the prolonged dry season of the Brazilian homeland of this species.

An effect similar to this last survival advantage has been discovered in *Myrmecocystus mimicus*. Young colonies require a large number of workers in order to produce repletes. In the arid conditions of the American Southwest, the advantage provided by repletes in surviving the harshest seasons may be large enough to favor cooperation among founding queens and their workers (Bartz and Hölldobler, 1982).

Tschinkel and Howard have further addressed the puzzling circumstance that founding queens readily join oversized groups of 15 or more in which there is no hope of reproduction. They point out that *Solenopsis invicta* is far less abundant in southern Brazil, where it evolved, and where populations are mostly restricted to disturbed or seasonally flooded habitats. Consequently, mating flights and post-flight queen density are low, pleometrosis much less frequent, and average group size low, so (as is not the case for *Myrmecocystus mimicus*) there may be little opportunity for selection to operate against excessively large groups. Yet selection for joining an available group should be strong.

Raiding and coalition of incipient colonies may be a commoner phenomenon in ants than previously realized. A third case has been reported in the desert granivore *Messor (= Veromessor) pergandei* by Rissing and Pollock (1986). Once again, as in *Myrmecocystus* and *Solenopsis*, pleometrotic colonies produce earlier brood and are more successful at raiding than are haplometrotic colonies.

## DEMOGRAPHIC CONSEQUENCES OF GYNY AND OF DOMINANCE ORDERS

The effects of variation in the number of queens will be fully understood only when the consequences of the variation on colony growth and survival are subject to quantitative measurements. Only a few studies thus far have been designed to acquire such data. Mercier et al. (1985a,b) found that the little European formicine ant *Plagiolepis pygmaea* is truly polygynous, averaging 17 laying queens per nest. The productivity of individual queens is very uneven, varying directly with their weight, their attractiveness to the workers, and the number of workers in the colony. When experimentally combined in pairs, the queens' oviposition rate is approximately half what it is when they are kept alone with a fixed number of workers (200). The implication from the combined studies is that the number of workers, rather than the number of queens, is the prime determinant of the total oviposition rate and hence of the potential for colony growth. The same result was obtained by Wilson (1974b) for colonies of *Leptothorax curvispinosus*. In *Solenopsis invicta*, on the other hand, polygynous colonies have higher brood:worker ratios than monogynous colonies, which implies that under some circumstances the number of queens is important (W. R. Tschinkel, personal communication).

By employing enzyme genetic markers in the polygynous *Solenopsis invicta*, Ross (1988) was able to assess directly the contributions of individual queens to the relative numbers of workers and female reproductives in the nest. He discovered that some queens who contributed substantial numbers to the worker pool often added few if any daughters to the pool of female sexuals. Offspring of other queens developed primarily into reproductive females. The mechanism of this remarkable differential reproduction among associating queens is not yet known. Ross's results strongly suggest, however, that "significant variability in short- as well as long-term reproductive success may occur among the distantly related queens associating in polygyne *S. invicta* nests." Thus the common earlier view that queens of polygynous colonies perform about equally will probably have to be revised.

Interference among competing queens and reproductive workers is yet another factor that must be entered into the equations. In queenless *Leptothorax allardycei* groups, at least, the most competitive workers spend more time on dominance exchanges than on brood care. Cole (1986) has estimated that the cost to worker reproduction is 15 percent of the total in the case of time spent on brood care and 13 percent of brood care when measured as brood pieces tended per unit time.

Aggression and dominance, which have proved far more common within ant colonies than was believed as recently as 1970, are very important in determining the numbers of reproductive females. Methods of control range from periodic all-out attacks to mammal-like dominance posturing, differential eating of the eggs of

**FIGURE 6–7** Fighting and dominance behavior in the primitive ant *Nothomyrmecia macrops*. (*a*) Two workers fight after one was experimentally introduced into a foreign laboratory nest. (*b*) Dominance behavior between two queens; the dominant individual stands over her crouching subordinate. (From Hölldobler and Taylor, 1983.)

rivals, and the use of primer pheromones to suppress ovarian development.

An extreme case of fighting was recorded by Wilson and Brown (1984) in a laboratory colony of the Asian termite predator *Eurhopalothrix heliscata*. The mother queen had evidently been lost during the transport of the colony from the field, and unfertilized dealated queens were contending for control. Battles ensued almost continuously for a week; they then became increasingly intermittent and finally ceased altogether. On several occasions two pairs were locked in combat at the same time, and at various times either fully dealate or partially (but never fully) winged individuals were engaged. At least five individuals were evidently involved on different days, and the total number during the 17–day period may have been far higher. Not all interactions were aggressive, however. Most of the time queens simply moved on past when they encountered each other, and on two occasions a dealate queen was observed grooming another.

In a typical aggressive episode the two individuals were locked together like wrestlers, with one trying to maintain a hold on the body of the other with all her legs while her opponent struggled to escape. Sometimes the aggressor rolled her own body around the head and anterior portion of the opponent's alitrunk; on other occasions she centered her attack on the alitrunk and occasionally farther back, on the gaster. Usually the aggressor pressed her mandibles downward and against the body of her opponent, evidently to gain added purchase. Several times Wilson and Brown saw one queen biting an antenna or a portion of the vertex of her opponent. In most episodes the aggressor also pressed the tip of her gaster against the body of her adversary in an additional attempt to gain purchase. No evidence of stinging was observed at this time, although the sting could easily have been extruded periodically and jabbed into the body of the adversary without being visible.

In most instances one queen was clearly dominant and the other subordinate. The dominant attempted to hold the subordinate in place, while the subordinate either struggled to work free or else lay quietly in a ball-like "pupal" posture. Several times the two appeared to be more evenly matched, so that neither could get a commanding position on top of the other. In both cases the combatants jockeyed and rolled around sluggishly for up to 15 minutes or longer. In one case three queens rather than two fought briefly; one then walked away and left the struggle to the remaining pair.

Subordinates who managed to break free usually walked briskly a short distance from the scene. One was seen to depart from the brood chambers to the outer nest chambers, only to return after a few minutes—and suffer another attack. There was one piece of indirect evidence that damage was being inflicted during the battles. The mortality of the queens was much higher than that of the workers, the reverse of the usual case in ant colonies. By the end of the 17–day period during which fighting occurred, the number of queens had dropped from 25 to 5. During this entire time no episodes of aggression were witnessed among workers or between workers and queens.

A less extreme dominance hierarchy was demonstrated among founding queens of the American carpenter ant *Camponotus ferrugineus* by Fowler and Roberts (1983) and in the resident queens of the polygynous *Myrmica rubra* by Evesham (1984a). The *Camponotus* queens showed consistent (but at most marginally significant) differences among themselves in lunging behavior, displacement from original position, and consumption of eggs laid by rivals. When secretions of the resting queens were collected on absorbent paper and presented to separate groups of queenless workers, the order of attractiveness corresponded to the dominance order of the three queens. A strikingly mammal-like ritual was observed by Hölldobler and Taylor (1983) in the very primitive Australian ant *Nothomyrmecia macrops:* one queen stood above a rival, occasionally stepping on top of her while pointing downward with her head. The subordinate remained still, in a crouching posture. The contact did not lead to fighting of the kind observed among *Nothomyrmecia* workers from different colonies (see Figure 6–7), but it was followed by the expulsion of the subordinate queen from the nest by the

**FIGURE 6–8** Dominance behavior between two founding queens of *Myrmecocystus navajo*. The dominant queen steps onto the subordinate individual, who responds by crouching and opening her mandibles.

workers. Similar ritual postures have been observed among queens of *Myrmecocystus* (Figure 6–8) and workers of *Leptothorax allardycei* by Cole (1981) and among wasps of the genera *Polistes* and *Mischocyttarus* by Pardi (1948), West (1967), and Jeanne (1972).

More subtle indicators of rivalry have been discovered in other ant species. Queens of *Leptothorax curvispinosus* do not attack or threaten each other, but do consume one another's eggs to varying degrees, so that a hierarchy of sorts emerges (Wilson, 1974a). Dominant queens of the Neotropical arboreal *Procryptocerus scabriusculus* are distinguished solely by the higher rates at which they solicit food from workers and larvae and groom themselves (Wheeler, 1984).

A still subtler but pervasive phenomenon is the inhibition by the egg-laying queen of ovarian development in nestmates through the release of inhibitory pheromones. Substances producing this effect have been demonstrated in *Plagiolepis pygmaea* (Passera, 1980b), *Oecophylla longinoda* (Hölldobler and Wilson, 1983a), and *Aphaenogaster* (= *Novomessor*) *cockerelli* (B. Hölldobler, N. F. Carlin, and E. P. Scovell, unpublished). The inhibition of reproductive egg laying by workers through the presence of the laying queen has been widely demonstrated in other ants, and the experimental evidence suggests that it is likely to be mediated at least to some extent by inhibitory pheromones. Examples include *Odontomachus haematodus* (Colombel, 1972), *Leptothorax tuberum* (Bier, 1954b), *Leptothorax* (= *Myrafant*) *recedens* (Dejean and Passera, 1974), various species of *Formicoxenus* (Buschinger, 1979a), and *Myrmica rubra* (Brian and Rigby, 1978). A second kind of pheromone produced by mated queens of *Solenopsis invicta* acts directly on virgin queens to prevent them from shedding their wings, histolyzing the wing muscles, and undergoing rapid oogenesis, all of which would convert them into reproductive rivals (Fletcher and Blum, 1981, 1983b; Fletcher and Ross, 1985).

Aggression can lead not only to dominance orders and despotism but to mutual separation of the queens into different parts of the nest, in other words spatial oligogyny. Hölldobler and Carlin (1985), who followed the transition from pleometrosis to large oligogynous colonies of *Iridomyrmex purpureus*, found that the process could be divided into three phases correlated with the size of the worker population. Soon after the first worker in laboratory nests emerged, the previously amicable co-foundress queens commenced aggressive displays. In this phase I, they faced each other head-on while engaging in stereotyped bouts of rapid mutual antennation (see Figure 6–9), and occasionally shifted to threats with open mandibles and even biting. The encounters resulted in clear-cut dominance, with the winning queen holding her ground while the loser backed or turned away. The dominant queen also spent more of her time on the egg pile and contributed more of her own eggs to it. Phase I, during which the queens remained together in an uneasy hierarchy, lasted for about a year, until the worker population reached 200 to 400 workers. Toward the end of the period the antennation bouts became more frequent, and the queens began to leave the founding nest tubes for intervals of up to 3 minutes. In phase II, which lasted approximately 9 months, the separations became gradually longer, lasting from several hours at a time to over 30 days. Meanwhile, the size of the single colony kept under observation increased to 2,000 workers and dispersed out over many more nest tubes. During phase III, when the colony had reached a size of 2,500 to 3,000 workers, one of the queens moved permanently to a separate part of the nest area. The structure of colonies in the wild is consistent with this pattern of social change. Eleven mature nests were excavated in Australia, containing 10,000 or more workers; two queens were found in each of two of the nests, one in each of five nests, and none in the remaining four nests. The queens were clearly separated in different galleries within both of the colonies that contained two queens. When these two pairs were placed in small artificial nests with 500 of their workers, they soon initiated ritualized antennation bouts and subsequently moved apart in the same manner as observed in the laboratory colony.

Aggression and dominance hierarchies, sometimes involving outright battles, have been described among workers of a few species when the queens were removed. Examples include the ponerines *Platythyrea cribrinoda* (B. Hölldobler and E. O. Wilson, unpublished observations) and *Rhytidoponera metallica* (Ward, 1986) and the myrmicines *Leptothorax duloticus* (Wilson, 1975a) and *Aphaenogaster cockerelli* (B. Hölldobler, N. F. Carlin, and E. P. Scovell, unpublished). The *Novomessor* workers most frequently attacked were the ones with the greatest degree of ovarian development.

Dominance hierarchies among workers have also been recorded in queenright colonies of the myrmicines *Leptothorax allardycei* (Cole, 1981, 1986), *Protomognathus americanus* (Franks and Scovell, 1983), and *Harpagoxenus sublaevis* (A. F. G. Bourke, personal communication). In each species the high-ranking workers also have the greatest ovarian development and receive more food. Even in the presence of the queen the *L. allardycei* dominants produce 20 percent of the eggs, which are unfertilized and hence destined to produce males if they develop. In the aggressive reactions the workers antennate each other, and the dominant ant often climbs above her opponent while (in *Leptothorax* at least) pummeling her with her mandibles. The subordinate responds by freezing, crouching, and drawing her mandibles back to the side of her head. Bourke (1988a) discovered that in queenless worker groups of *H. sublaevis* worker dominance behavior inhibits egg laying by subordinates. In queenright colonies, on the other hand, the queen appears to restrict both dominance behavior and oviposition by workers, probably by chemical means.

**FIGURE 6–9** The beginning of aggression among colony-founding queens of the Australian meat ant *Iridomyrmex purpureus* is marked by stereotyped bouts of mutual antennation. Occasionally the exchanges escalate into threats with open mandibles and biting. In time the aggression leads to a separation of the rivals into different parts of the nest. (From Hölldobler and Carlin, 1985.)

## WHO IS IN CHARGE, QUEENS OR WORKERS?

The picture that has emerged in contemporary studies is of a moderate amount of struggle within ant colonies. Queens and workers appear to be in general conflict over the management of the ratio of investment in new queens and males. In some species, under appropriate conditions, queens battle queens for principal reproductive rights. Workers compete with their nestmates for the same privileges in the absence of the queen and, in a few cases, even when the queen is present. As we have seen, all of this conflict falls into patterns. What do the patterns tell us about social organization, in particular about whose reproductive interest and genetic fitness are being served? The question can be focused to some extent by returning to the question partially explored in Chapter 4: which caste controls the reproductive activities of the colony, the queen or the workers?

A remarkable general phenomenon of ant biology that might be interpreted as dominance by the queen over the workers is the queen-tending pheromone. Queens are generally attractive to workers, who often form entourages around them, licking their bodies, feeding them with trophic eggs and regurgitated liquids, and guarding them against intruders. The queen of a monogynous colony, or at least the queen who prevails as the egg-layer, is more attractive than are the queens in a functionally polygynous colony. In comparisons both of species and of growing colonies belonging to the same species, the queens are most attractive in the largest colonies. In *Solenopsis invicta,* the best-studied ant in this regard, no fewer than five behaviors are evoked by the secretions of the physogastric queens when the queens are displaced by the experimenter outside the nest (Glancey et al., 1982; Glancey, 1986):

1. Intense initial attraction toward the queen.
2. Formation of a dense cluster of workers around the queen.
3. Transport of brood to the queen and deposit of the brood next to her.
4. Formation of odor trails to the nest, often with several branches coalescing into a very wide trail terminating at the nest.
5. Guidance of the queen along one of the trails; if the queen does not walk along the trail under her own power, the workers drag her into the nest.

At least some of the components of the pheromone inducing this remarkable series of responses originate in the poison gland sac of the queen. Those characterized so far are two pyranones and a dihydroactinidiolide, the structures of which are illustrated in Figure 6–10. The queen pheromone of a second myrmicine species, *Monomorium pharaonis,* has been identified as neocembrene. Found only in mated queens, it exercises a strong attraction on the workers, but no other responses have been reported thus far (Edwards and Chambers, 1984). In the European wood ant *Formica polyctena,* methyl-3-isopropylpentanoate, which is produced in the cephalic glands of the queen, inhibits aggression in workers (Francke et al., 1980).

These immediate and overt effects are only part of the total regime of influence inseminated queens have on workers, at least among some of the phylogenetically more advanced subfamilies of ants. As Brian and Hibble (1963) first showed, the presence of laying queens in colonies of *Myrmica rubra* also reduces the likelihood that larvae will become queens and hastens their development into workers. Some of these results are achieved by an increased tendency of the workers to bite and scar the growing larvae when a queen is present. Further studies by Brian and his colleagues have revealed a complex repertory of variable worker responses to different kinds of

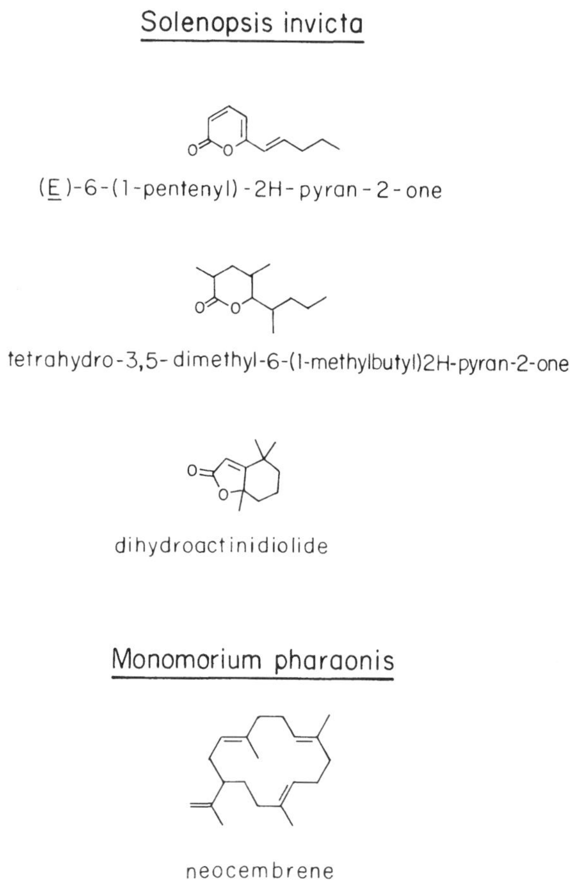

Solenopsis invicta

(E)-6-(1-pentenyl)-2H-pyran-2-one

tetrahydro-3,5-dimethyl-6-(1-methylbutyl)2H-pyran-2-one

dihydroactinidiolide

Monomorium pharaonis

neocembrene

**FIGURE 6–10** Queen substances in the red imported fire ant *Solenopsis invicta* and in Pharaoh's ant (*Monomorium pharaonis*). These materials by themselves are very attractive to workers. In the case of *S. invicta* they also trigger a complex of behaviors that result in the care and protection of egg-laying queens. (Modified from Glancey, 1986; Edwards and Chambers, 1984.)

larvae, which are sensitive to the presence or absence of queens. The workers' propensity to lay eggs of their own is also strongly affected (see Table 6–2). The overall result is a system of checks and balances that inhibits worker reproduction and production of new queens when laying queens are present and promotes these two activities when laying queens are absent. The system sharpens the division of labor between the queens and the workers.

In view of the considerable evidence of queen control by pheromones, it is surprising that little attention has been given to the queens' exocrine glands. Whelden (1963) and B. Hölldobler and C. W. Rettenmeyer (unpublished observations) found that queens of New World army ants belonging to the genus *Eciton* are much more richly endowed with such structures than are workers of the same species. A similar difference has been noted in weaver ants of the genus *Oecophylla* (Hölldobler and Wilson, 1983a), the amblyoponine army ants of the genus *Onychomyrmex* (B. Hölldobler and R. W. Taylor, unpublished), and the leptanilline army ants of the genus *Leptanilla*. It is to be expected that the exocrine development of the queens will be greatest in species with the largest colonies, because the amount of pheromonal material required to reach the entire worker force is correspondingly greater.

Yet another indication that queens can exert control comes from the extraordinary discovery by Masuko (1986) that queens of the primitive ant *Amblyopone silvestrii* are vampires: they depend exclusively on hemolymph (blood), which they obtain by biting the larvae. *Amblyopone* workers do not lay trophic eggs or exchange liquid food by regurgitation, the two ordinary means by which ant queens obtain nourishment. Nor do the queens feed directly on the geophilomorph centipedes and other arthropod prey captured by the workers. Instead they use their sharp mandibles to pierce the dorsal integument of the forward part of the abdomens of older larvae, then drink the hemolymph leaking from the puncture. The bodies of the donors are scarred, but they appear otherwise unharmed. Masuko has observed similar behavior in queens of three Japanese species of the aberrant ponerine genus *Proceratium*, which ordinarily prey on the eggs of arthropods. In another study, Masuko (1987) found that queens of *Leptanilla japonica*, a minute army-ant-like species, feed from special exudatory organs of the larvae. These structures, unique to *Leptanilla* and other Leptanillinae, resemble spiracles and are located on either side of the third abdominal segment.

Yet another apparent mode of queen control is the aversion to nest queens displayed by workers of *Leptothorax curvispinosus* (Wilson, 1974a). The workers regularly feed the queens by regurgitation, but they respond more commonly to them by withdrawing for a distance of 2 millimeters or more after making antennal contact. Sometimes they back off or turn aside and walk away at a normal gait. But with equal frequency they run away rapidly. When a queen moves into and through a crowd of workers, they "explode," scattering away from the queen and clearing a path in front of her. The workers seem to avoid her head in particular and to pay little attention to the rest of her body. The result is that the laying queens are able to move more freely around the nest. In particular, they gain easy access to the larvae, whose salivary secretions form part of their diet.

To summarize, we have reviewed a wide diversity of anatomical structures and behaviors in ants that may be interpreted as mechanisms by which queens control workers. This functional explanation is by no means certain, however. In each case it could be argued that the workers are not really "controlled," in the sense that a conflict exists between them and the queens that is resolved in favor of the queens. Quite the contrary: the workers may in fact simply be monitoring the presence of the queens and responding in ways that increase their own inclusive fitness. It may even be said that *Amblyopone* and *Proceratium* larvae surrendering their blood to the queen are not acting in any fundamentally different way from *Leptothorax* larvae surrendering their personal salivary secretions. Overall, the question of queen control is moot with reference to conflict and the queen's effect on the rest of the colony.

We now turn to evidences of worker control. In Chapter 4 we saw how a conflict does arise in monogynous colonies between the queen and her daughter workers concerning the optimal ratio of investment in new male and female reproductives, and how the workers settle this dispute to their own advantage. There are many other circumstances in which the workers are far removed from any possible control by the queen other than pheromonal signals, and are hence able to make decisions by themselves on a moment-by-moment basis. One of the most evident is the harvesting and distribution of food. Fire ant workers (*Solenopsis invicta*) are probably typ-

**TABLE 6–2** The effects of inseminated queens on worker behavior toward larvae and on worker oviposition in the ant genus *Myrmica*. (Modified from Brian, 1983.)

| Larval type or behavior | Queens present | Queens absent |
|---|---|---|
| Worker-biased larvae | Nurse workers feed such larvae in preference to other larvae, and these immature stages then metamorphose into workers. | The larvae are neglected and they stop growing halfway through the third instar. |
| Labile female larvae | Workers (mostly foragers) attack these larvae during the spring, at a time when the larvae show signs of developing into queens. The larvae betray their queen-prone tendency by means of secretions on their lower surfaces. Attacks include biting and scarring of cuticle. | Both nurse and foraging workers feed the labile larvae with solid, protein-rich foods, focusing on a few at a time. |
| Male larvae | These larvae are neglected but not attacked. | These larvae are fed more often than worker-biased larvae but less often than queen-biased larvae. |
| Egg formation and laying | As long as the queens are laying or a cluster of reproductive eggs of either sex exists, young nurse workers lay trophic eggs used as food but incapable of development. Also, very young workers (less than 3 weeks old) lay some reproductive eggs. | Young workers lay reproductive eggs that can develop into males. |

ical of this caste in ants in the way that they partition the different kinds of food. During experiments conducted by Howard and Tschinkel (1981), fire ant foragers and nurses passed sugar to other workers, amino acids to larvae and the queen, and soybean oil to larvae and other workers in equal amounts. In other experiments (Petersen-Braun, 1982; Fletcher and Blum, 1983a), during the elimination of supernumerary queens of *S. invicta* and *Monomorium pharaonis* the workers selected the losing contenders and executed them.

A particularly interesting case of queen selection by workers seems to occur in army ants. Franks and Hölldobler (1987) argued that during colony fission, workers should, in theory at least, select the queen that will enable them to maximize their own inclusive fitness (Macevicz, 1979; Franks, 1985). This calculation is complicated by three factors: (1) workers that accompany sister queens rather than their mother queen suffer the fate of raising nieces and nephews rather than brothers and sisters; (2) at some stage in the life of a colony, workers have to reject their maternal queen on the grounds of her senility; and (3) virgin queens may not be full sisters of the workers, because army ant queens probably mate more than once in their lifetime. If all the virgin queens and all the workers are full sisters, then the workers should unanimously select the potentially most fertile queens, preferring their mother to a sister. If, on the other hand, the maternal queen mates more than once in her lifetime, as has been suggested for both *Eciton burchelli* (Rettenmeyer, 1963) and species of *Dorylus* (Raignier and van Boven, 1955), then each patrilineal group of workers should prefer that their own full sister become one of the new queens. This alternative will be possible only if workers can discriminate between full and half-sisters, however. That degree of kin recognition has been suggested in honey bees (Getz and Smith, 1983; Page and Erickson, 1984; Noonan, 1986; Visscher, 1986). It has also been claimed that colony division in honey bees is associated with the segregation of workers into sororities (Getz et al., 1982). Furthermore, worker bees are known to distinguish between individual queens on the basis of their odors (Boch and Morse, 1974, 1979). Breed (1981) has shown that the rate of acceptance of foreign queens is correlated with the degree of genetic relationship among the queens involved in the transfers. Still, few studies have been conducted on whether ants can discriminate between individual nestmates on the basis of relatedness, so that very little can be concluded (see Carlin et al., 1987b). Furthermore, although army ant queens may mate with two or more males, they may do so only once each year (Rettenmeyer, 1963a). It is therefore possible that new queens and the majority of the worker population are full sisters.

Thus we do not know if kinship influences how workers choose new queens. What is evident, however, is that whether workers are able to recognize their full sisters or not, there will be very strong selection for workers to discriminate between queens on the basis of potential fertility and survivorship.

Army ant queens have to be exceptionally vigorous and productive. *Eciton burchelli* queens, for example, may live six years, during which time they produce some 3 million workers and walk between successive bivouacs a total distance of 60 kilometers (Franks, 1985). Because of the huge worker mortalities during foraging, army ant colonies grow relatively slowly, and three years on the average elapse between bouts of sexual reproduction. The workers must select highly fecund and long-lived queens in order to realize any inclusive fitness at all.

Studies on colony foundation by multiple queens in other species of ants suggest that workers are skilled at choosing the most fertile and attractive queens, and that kinship is of minor importance in the rejection of supernumerary queens (Bartz and Hölldobler, 1982). Although almost nothing is known about the genetic constraints on worker choice of queens in army ants, a considerable amount is known about the proximate mechanisms by which workers discriminate between queens.

Schneirla (1956a) observed that conflicts arise between workers who are associated with different virgin queens even when the latter are still maturing larvae. He noted that individual queen larvae are often separated by considerable distances within the bivouacs and that each larva is surrounded by a cluster of "satellite workers." Adjacent worker groups can come into conflict, which occasionally results in the death of some potential queens. Schneirla (1971) further observed that the first queens to emerge are more attractive and more likely to be successful. The first young females will have had longer to produce their pheromones and to win the allegiance of the workers. That workers may come into conflict over the allegiance to queens does not necessarily mean that they are forming sororities. Such conflict may be a mechanism by which workers can compare the strength of their advocacy for certain queens and hence the queens' attractiveness. Therefore such competition may result purely from colony-level selection.

The phenomenon of workers changing their queen allegiances has been demonstrated by experiments in which the old queen is removed from the bivouac when the sexual larvae are mature. Under these circumstances the parental queen will be readmitted to her colony in an unequivocal manner only if she is met by the group of workers previously affiliated with her. Otherwise she may be segregated in a tight cluster of workers and eventually abandoned (Schneirla and Brown, 1952; Schneirla, 1956a,b). On the other hand, *Eciton* colonies will accept workers of alien colonies only if the transplanted workers are isolated from their queen for a number of days (Schneirla, 1971). After a few days a colony deprived of its queen will fuse with a colony of the same species possessing a viable queen. Thus it appears that queen pheromones unite and coordinate the huge army ant society. The chemical basis of the queen's attractiveness and signature has been demonstrated by a simple experiment in which workers were more attracted to paper discs upon which the queen had been resting than to control discs (Watkins and Cole, 1966).

These observations provide strong circumstantial evidence that army ant workers can recognize and choose particular queens during colony fission, making their selection on the basis of relative queen attractiveness. Still, it is difficult to conclude that the workers are really exerting their will over that of the queens. Even when they execute supernumeraries they serve the interest of the winning queens as well as their own.

Conflict within ant colonies may be summarized as follows. Under many but not all circumstances there is direct aggression among queens, resulting in either dominance hierarchies, oligarchies of a few queens spaced through the nest, or the eviction or execution of supernumeraries. Dominance is not always based on physical domination. It is sometimes mediated by pheromones or some other still undiscovered, circuitous signals. Often it takes the form of the inhibition of ovarian development in subordinates who are otherwise left unmolested. When the laying queen is removed in a few species, virgin queens present at the time then contend for dominance.

Similarly, there is often definitive dominance behavior among workers. In rare cases hierarchies leading to differential reproduction develop among members of this caste even in the presence of laying queens, but more often the dominance hierarchies emerge when laying queens are absent.

Conflict between queens and workers is less certain. In monogynous species, an inherent conflict arises between the two castes over the ratio of investment in new queens and males, and the difference has been resolved evolutionarily in favor of the workers. Interactions between queens and workers are often too ambiguous, however, to be clearly labeled as queen control or worker control. Even such apparently explicit behaviors as ovarian suppression in workers and execution of supernumerary queens may well benefit the inclusive fitness of both castes simultaneously. It is a mistake to interpret individual forms of interaction in terms of vertebrate behavior, where dominant individuals are able to raise their genetic fitness. Control and dominance in ant colonies, at least between the queen and worker castes, must be viewed within a far tighter, more complex social context where (as experience has taught us) appearances are often very deceptive.

CHAPTER 7

# Communication

The intensive study of communication among the ants over the past thirty years has yielded a profusion of results that deeply affect our understanding of social organization. The demonstrated modes of communication are extremely diverse. There exist the expected tappings, stridulations, strokings, graspings, nudgings, antennations, tastings, and puffings and streakings of chemicals that evoke various responses from simple recognition to recruitment and alarm. To this list can be added other, often bizarre activities, such as the exchange of pheromones that inhibit caste development, the soliciting and exchange of trophic eggs and special secretions from the anal region, the acceleration or inhibition of work performance by the presence of other colony members, and programmed execution.

Researchers on communication in ants and other social insects have come to recognize the following twelve broad functional categories of responses:

1. Alarm.
2. Simple attraction.
3. Recruitment, as to a new food source or nest site.
4. Grooming, including assistance at molting.
5. Trophallaxis (the exchange of oral and anal liquid).
6. Exchange of solid food particles.
7. Group effect: either facilitating or inhibiting a given activity.
8. Recognition, of both nestmates and members of particular castes, including (broadly) discrimination of injured and dead individuals.
9. Caste determination, either by inhibition or by stimulation.
10. Control of competing reproductives.
11. Territorial and home range signals and nest markers.
12. Sexual communication, including species recognition, sex recognition, synchronization of sexual activity, and assessment during sexual competition (see Chapter 3).

## CHEMICAL COMMUNICATION

If any single generalization applies to all of these categories, it is that chemical signals pervade them. In 1958, on the basis of early results then just emerging, Wilson predicted the dominance of chemoreception in ant behavior as follows:

> The complex social behavior of ants appears to be mediated in large part by chemoreceptors. If it can be assumed that "instinctive" behavior of these insects is organized in a fashion similar to that demonstrated for the better known invertebrates, a useful hypothesis would seem to be that there exists a series of behavioral "releasers," in this case chemical substances voided by individual ants that evoke specific responses in other members of the same species. It is further useful for purposes of investigation to suppose that the releasers are produced at least in part as glandular secretions and tend to be accumulated and stored in glandular reservoirs. (1958a)

As each improvement in bioassay design and organic microanalysis has permitted the separation and bioassay of secretory substances, new evidence reinforces this early impression.

A generally accepted terminology has evolved to classify the functions of the chemical substances (Nordlund, 1981). A *semiochemical* is any substance used in communication, whether between species (as in symbioses) or between members of the same species (Law and Regnier, 1971). A *pheromone* is a semiochemical, usually a glandular secretion, used within a species; one individual releases the material as a signal and the other responds after tasting or smelling it (Karlson and Lüscher, 1959). An *allomone* is a comparable substance employed in communication across species, as for example a lure used by a predator in attracting prey. It evokes a response that is adaptively favorable to the emitter but not to the receiver (Brown, 1968; Brown et al., 1970a). In contrast, the term *kairomone* was proposed by Brown et al. in 1970 to cover chemicals emitted by an organism that elicit a response adaptively favorable to the receiver but not to the emitter. Semiochemicals can be classified as olfactory or oral according to the site of their reception. Also, their various actions can be distinguished as either *releaser effects* (then we speak of releaser pheromones), comprising the classical stimulus-responses mediated wholly by the nervous system, or *primer effects* (induced by primer pheromones), in which endocrine and reproductive systems are altered physiologically. In the latter case the body is truly primed for new biological activity, responding afterward with an altered behavioral repertory when presented with appropriate stimuli (Wilson and Bossert, 1963).

The sum of current evidence, which will be described in the remainder of this chapter, indicates that pheromones play the central role in the organization of ant societies. In general, it appears that the typical ant colony operates with somewhere between 10 and 20 kinds of signals, and most of these are chemical in nature. This rule is illustrated very well by the fire ant *Solenopsis invicta*, perhaps the most thoroughly studied ant species in this respect. As summarized in Table 7–1, 13 signals of a communicative or quasi-communicative nature are employed. Of these, all but one or two are mediated through chemoreception.

**TABLE 7–1** Known categories of communication and similar interactions within colonies of the fire ant *Solenopsis invicta*. Categories unequivocally concerned with communication are numbered sequentially. (Based on Wilson, 1962b; O'Neal and Markin, 1973; Howard and Tschinkel, 1976; Fletcher and Blum, 1981; Vander Meer, 1983; LaMon and Topoff, 1984; Obin and Vander Meer, 1985; Obin, 1986.)

| Stimulus | Transmission | Response |
|---|---|---|
| **Worker to worker** | | |
| 1. Colony odor | Chemical | None, if odor is similar to that of focal worker |
| 2. Caste odor | Chemical | Recognition of caste |
| 3. Casual antennal or bodily contact | Tactile | Turning-toward movement or increased undirected movement |
| 4. Body surface attractants | Chemical | Grooming |
| 5. Carbon dioxide | Chemical | Clustering and digging |
| 6. Ingluvial food solicitation | Probably tactile (tapping or stroking of labium) | Regurgitation |
| 7. Regurgitation | Chemical, at least in part | Feeding |
| 8. Emission of Dufour's gland substances as trails | Chemical | Attraction followed by movement along trail (used in mass foraging and emigration) |
| 9. Emission of Dufour's gland substances during attack | Chemical | Attraction to disturbed worker |
| 10. Emission of cephalic substance | Chemical | Alarm behavior |
| Oleic acid and oleates among body decomposition products | Chemical | Necrophoric behavior: recognition and disposal of corpse |
| **Worker to queen** | | |
| Colony odor, casual contact, body surface attractants, carbon dioxide, regurgitation, trail substances, and necrophoric substances, the same as in worker-to-worker communication above. | | |
| **Queen to worker** | | |
| Colony odor, caste odor, casual contact, body surface attractants, carbon dioxide, ingluvial food solicitation, and necrophoric substances, the same as in worker-to-worker communication above. | | |
| 11. Queen recognition pheromones | Chemical | Recognition of and attraction to the queen |
| **Queen to queen** | | |
| 12. Dealation inhibiting pheromone | Chemical | Inhibition of wing shedding by winged virgin queens |
| **Worker to brood** | | |
| Regurgitation | Chemical or tactile | Feeding |
| Dispersal of venom droplets by vertical abdominal "flagging" | Chemical | Inhibition of microorganism growth (?) |
| **Brood to worker** | | |
| Colony odor, body surface attractants, carbon dioxide, and necrophoric substances, the same as in worker-to-worker communication above. | | |
| 13. Life stage odors | Chemical | Recognition of eggs, larvae of two or more stages, pupae, and eclosing workers as distinct life stages |

## Glandular Sources

The typical ant worker is a walking battery of exocrine glands, developed to a degree well beyond that typifying nonsocial hymenopterans. More than ten of the organs have been implicated thus far in the production of semiochemicals. They vary greatly in form and distribution among the major groups of ants, as illustrated in Figures 7–1 through 7–20. A few, such as the sternal and rectal glands of *Oecophylla,* the independently evolved sternal gland of *Onychomyrmex,* the pygidial gland of *Polyergus,* and the cloacal gland of *Camponotus,* appear to be unique to particular genera and to have arisen *de novo* during the course of social evolution (Hölldobler and Wilson, 1978; Hölldobler, 1982b,d, 1984a). Others, such as the poison gland of the Formicinae and Pavan's gland (sternal gland) of the Aneuretinae and Dolichoderinae, are peculiar to higher groups and thus provide valuable clues for the reconstruction of ant phylogeny. At least one structure, the metapleural glands, characterizes the ants as a whole (Maschwitz, 1974; Hölldobler and Engel-Siegel, 1984). Still other glands are shared with aculeate bees and wasps, including those that are nonsocial. Their versatile employment as sources of pheromones illustrates the economy of evolution, or Romer's rule, as it sometimes is called, whereby organs and new functions tend to arise as modifications of preexisting organs and functions rather than as true novelties. This evolutionary process has been repeated many times to create a confusing pattern of glandular form and function across the Formicidae.

The repeating pattern of communicative evolution can be partially deciphered by focusing on six of the key exocrine glands that occur widely through the ants and serve a variety of functions in different phylogenetic groups. These structures are Dufour's gland, the poison gland, the pygidial gland, the sternal glands, the mandibular glands, and the metapleural glands. Our knowledge concerning them has been thoroughly summarized at successive intervals by Maschwitz (1964), Bergström and Löfqvist (1970, 1973), Blum and Hermann (1978a,b), Hölldobler and Engel (1978), Parry and Morgan (1979), Hölldobler (1982a, 1984c), Vander Meer (1983), Bradshaw and Howse (1984), Buschinger and Maschwitz (1984), Morgan (1984), Attygalle and Morgan (1985), and Schmidt (1986). The treatment of anatomy and biochemistry by Blum and Hermann is close to being exhaustive for the earlier period of this research.

*Dufour's gland* is typically a small gland, usually finger shaped but sometimes bulbous or bifurcate in form, that opens at the base of the sting very near the egress of the poison gland (Figures 7–17 and 7–18). On morphological grounds it has been assumed that doryline, ponerine, and myrmicine ants can discharge the contents of Dufour's gland and the poison gland independently (Whelden, 1960; Hermann and Blum, 1967a,b). In contrast, formicines were believed to release the contents of the two glands simultaneously, because no mechanisms were known that could close either one separately (Percy and Weatherston, 1974). Billen (1982b), however, discovered a closing apparatus of the Dufour's gland in several formicine ants (*Formica sanguinea, F. fusca,* and *Lasius fuliginosus*). Four sets of muscles play a part, two of which are directly attached to the slit-like glandular duct (Figure 7–19). Billen suggested that the opening of the Dufour's gland duct is achieved by active muscular contractions, while its closure is achieved by a passive return to the rest position of the thickened cuticular intima. A similar structure had been previously found by Beck (1972) in *F. sanguinea* and *Polyergus rufescens,* but was not recognized as a closing mechanism.

In a comparative ultrastructural study of the glandular epithelium of the Dufour's gland in ants, Billen (1986a) discovered remarkable differences in the cellular organization in eight ant subfamilies (Figure 7–20). Most of the Myrmicinae and Ponerinae possess a rather simple epithelium without special modifications. "In the African Dorylinae, the epithelium has a crenellated appearance and numerous basal invaginations, while the New World Ecitoninae have a very uniform epithelium with a basal layer of membrane foldings. Myrmeciinae, Pseudomyrmecinae and Dolichoderinae, each shows a different kind of apical microvilli, whereas Formicinae exhibit a characteristic subcuticular layer of mitochondria and a very thick basement membrane." In the Myrmicinae the Dufour's gland produces only aliphatic hydrocarbons, but in dazzling variety among the various species, including such compounds as methylundecane, tridecane, hexadecane, hexadecene, and an array of farnesenes. In formicine ants the Dufour's gland is even more versatile. Aliphatic hydrocarbons are produced in abundance in most species, with *n*-undecane and *n*-tridecane typically present as major components and longer chain hydrocarbons as minor components. The alkanes are often accompanied by their corresponding alkenes, and in several species their dienes are also present. In addition, a great many oxygenated compounds occur in various combinations with the alkanes, especially in *Lasius.* They include alcohols, ketones, esters, acids, and lactones. The evolutionarily primitive function of the Dufour's gland and its basic set of alkanes is still uncertain. At least some of the compounds mediate alarm, recruitment, and sexual attraction among various species of ants. These communicative roles are clearly a derived condition within the Hymenoptera in general and the Formicidae in particular.

The *poison gland apparatus* typically consists of paired filamentous glands that converge into a single convoluted gland, which in turn empties into a thin-walled, sac-like reservoir or "poison sac." The most evolved version is that found in the Formicinae. The convoluted gland proper is located on the dorsum of the poison sac, a condition unique within the Hymenoptera. The sac as a whole is also exceptionally large and it produces sizable quantities of formic acid by biosynthetic pathways that are now known (Hefetz and Blum, 1978). This simplest of all organic acids has for historical reasons been popularly regarded as characteristic of all ants, perhaps because it was one of the first natural products isolated in pure form, from the distillate of *Formica* workers, in 1670. Nevertheless, it is evidently limited to the subfamily Formicinae.

The primary function of the poison gland in ants is the production of formic acid (in the Formicinae) or venom used in predation and defense. The primitive components, shared as a class with other aculeate hymenopterans, are proteinaceous. They are also neurotoxic, histolytic, or both in their effect—hence crippling to small invertebrate enemies and painful to human beings. This type of venom is the most common form in the anatomically more primitive ant subfamilies, namely the Ponerinae, Myrmeciinae, Pseudomyrmecinae, Dorylinae, and Ecitoninae, and is widespread among the tribes and genera of the Myrmicinae as well. Its effects are enhanced in bulldog ants of the genus *Myrmecia* by the addition of histamine

**FIGURE 7–1** Schematic illustration of a sagittal section through a *Formica* worker, showing the major internal organs and exocrine glands. (Based on Otto, 1962, and Gösswald, 1985.)

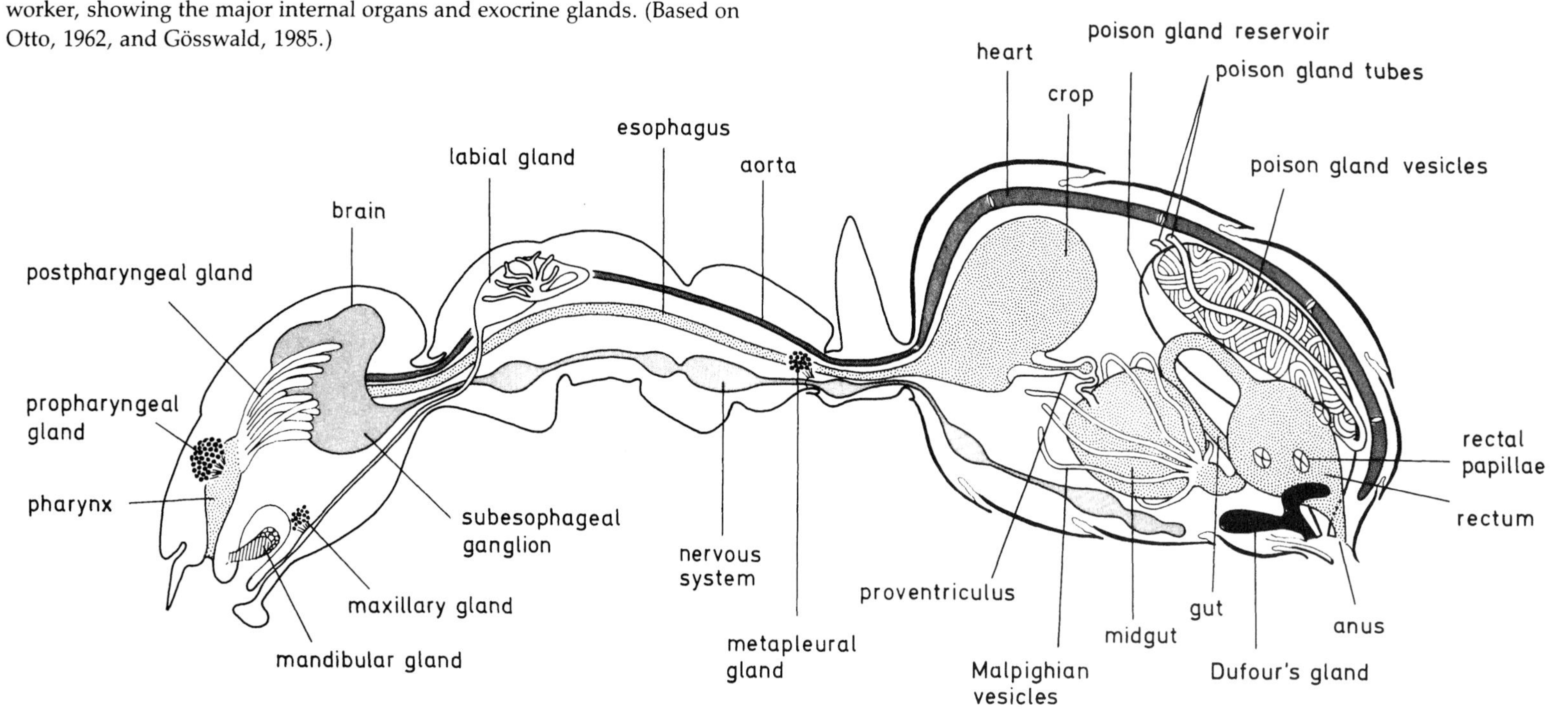


**FIGURE 7–2** Exocrine glands in the head and thorax of a *Formica* worker. (Modified from Emmert, 1968.)

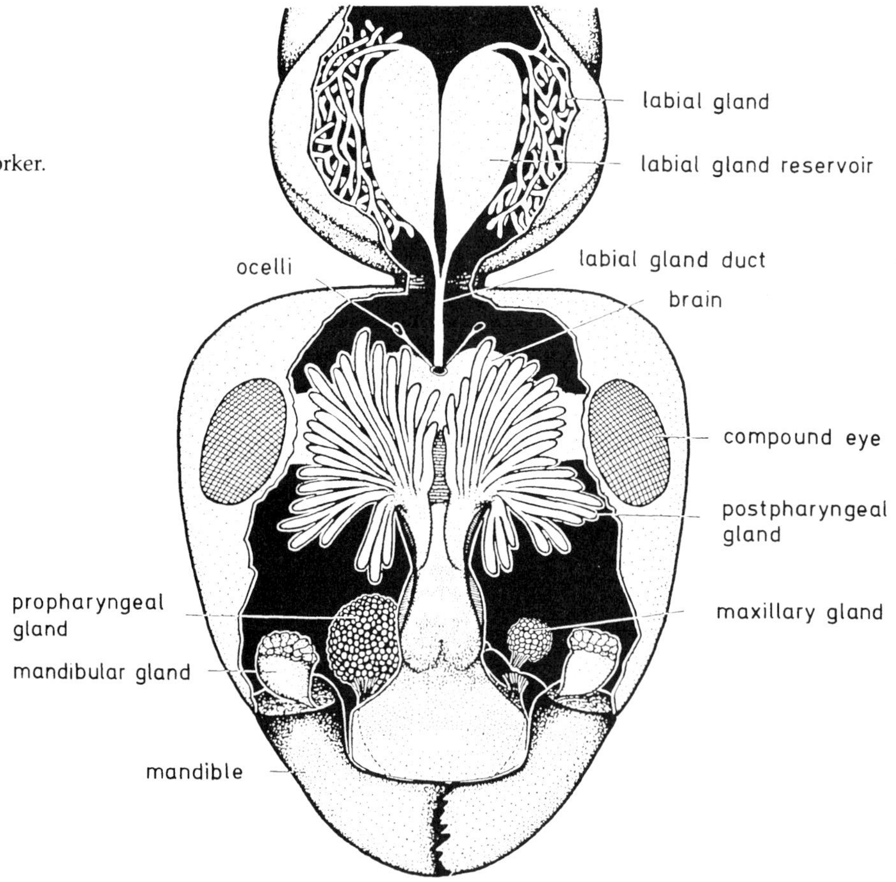

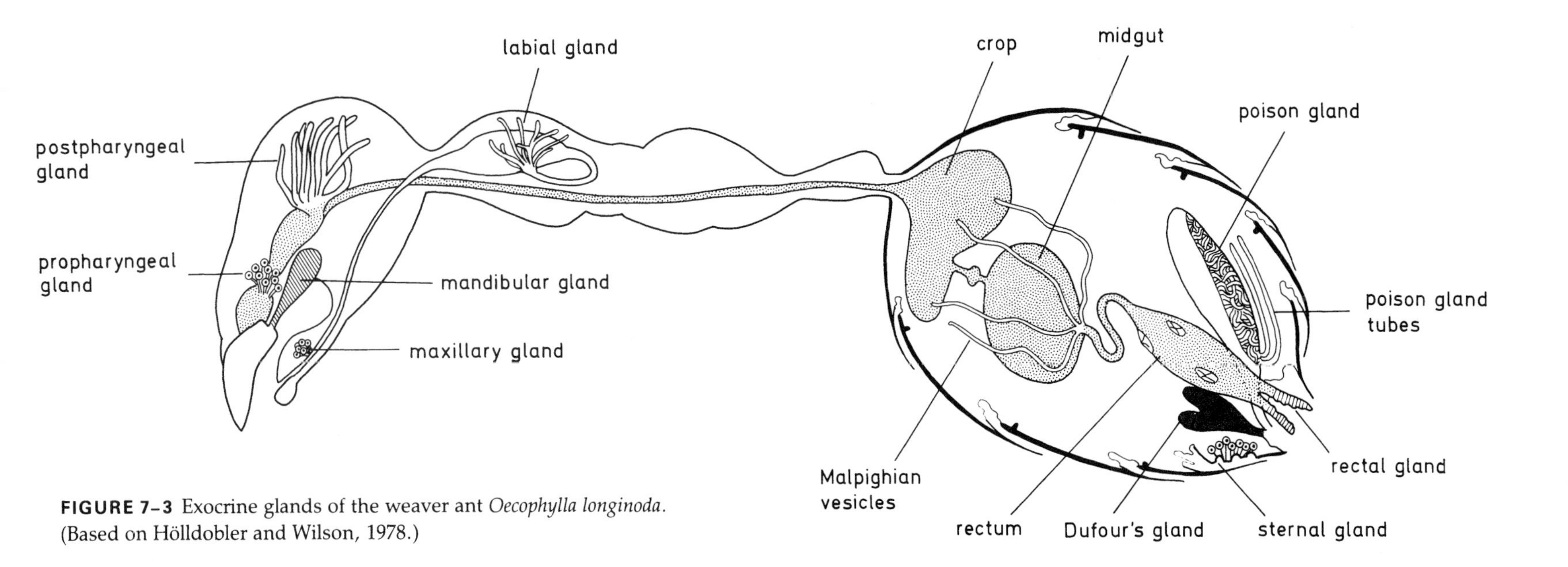


**FIGURE 7–3** Exocrine glands of the weaver ant *Oecophylla longinoda*. (Based on Hölldobler and Wilson, 1978.)

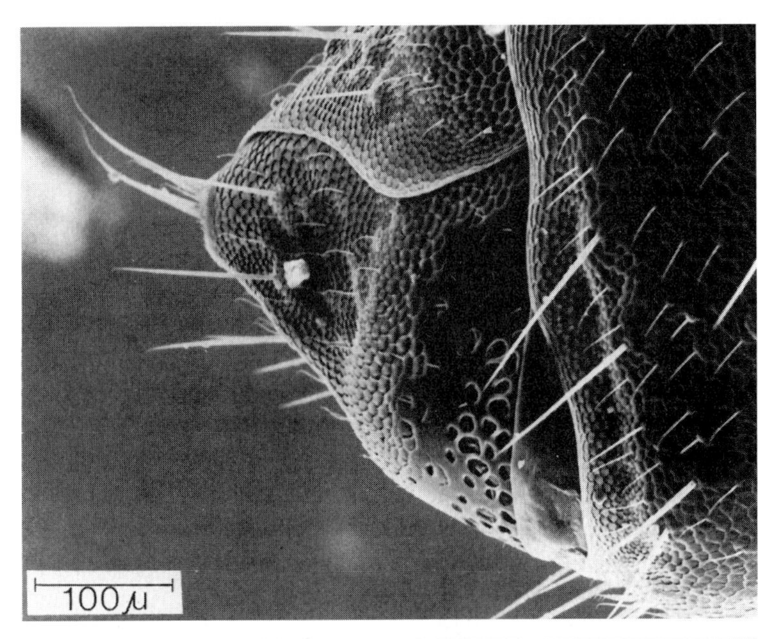


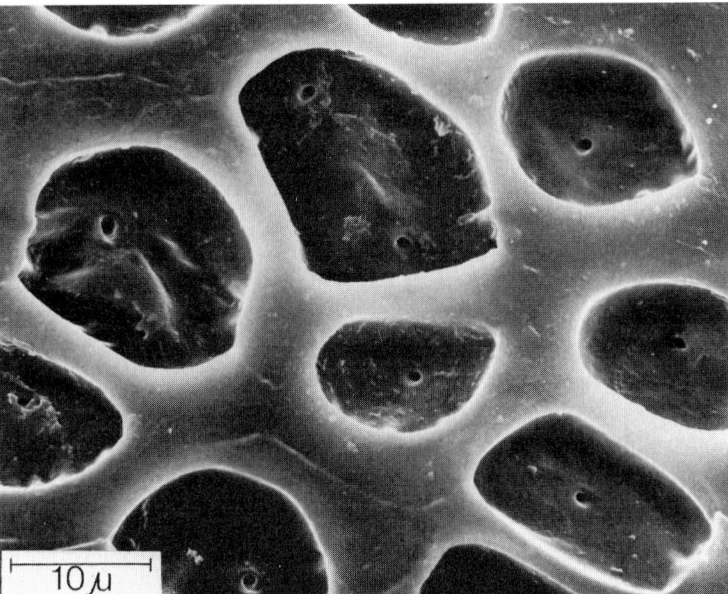


**FIGURE 7–4** External structure of the sternal gland of an *Oecophylla longinoda* worker, revealed by scanning electron micrographs. (*a*) The terminal abdominal segment of a major worker has been rotated upward to expose its anterior lower surface and the cup-like receptacles in which the sternal gland pheromone is accumulated. (*b*) Closer view of the cups, showing the minute openings to the underlying glandular cells, which are connected by intracellular channels to the lower cup surfaces. (From Hölldobler and Wilson, 1978.)

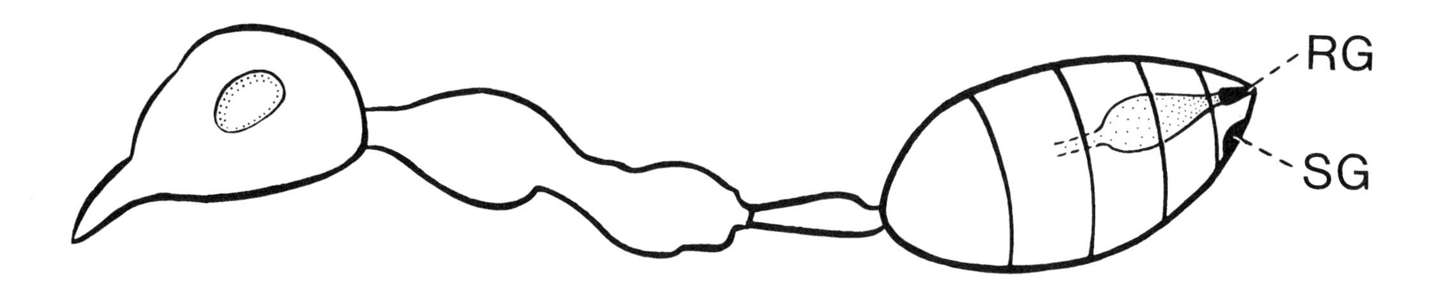


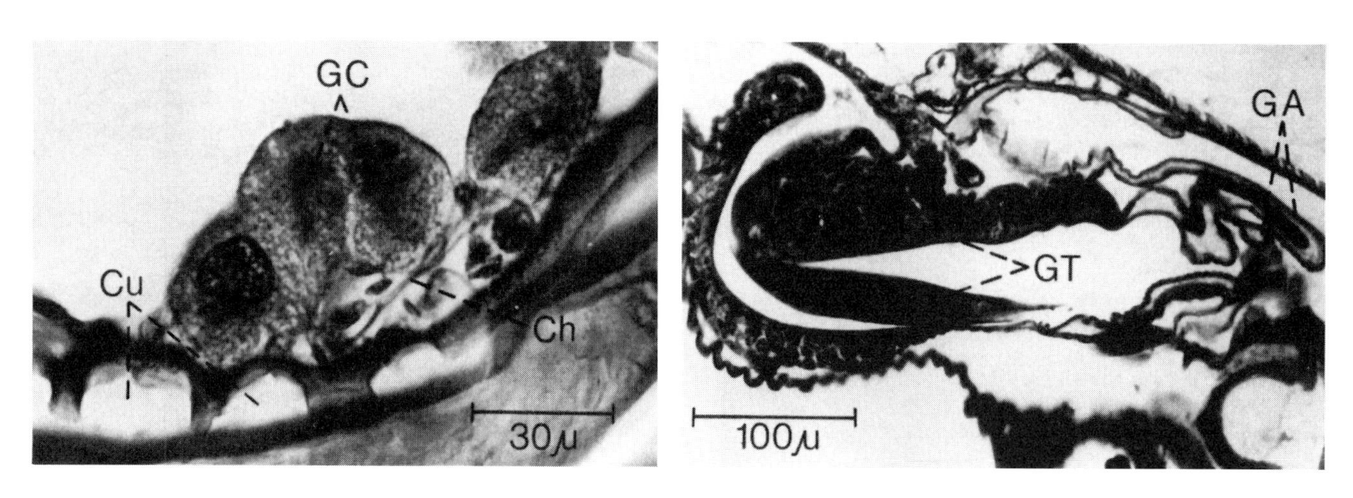


**FIGURE 7–5** Sagittal section through the sternal gland (*SG*) and the rectal gland (*RG*) of an *Oecophylla longinoda* major worker. *Cu*, receptacle cup of sternal gland; *GC* and *Ch*, cells and channels of sternal gland. The rectal gland is created from an infolding of the rectal wall comprising glandular tissue (*GT*) and an eversible gland applicator (*GA*). (From Hölldobler and Wilson, 1978.)

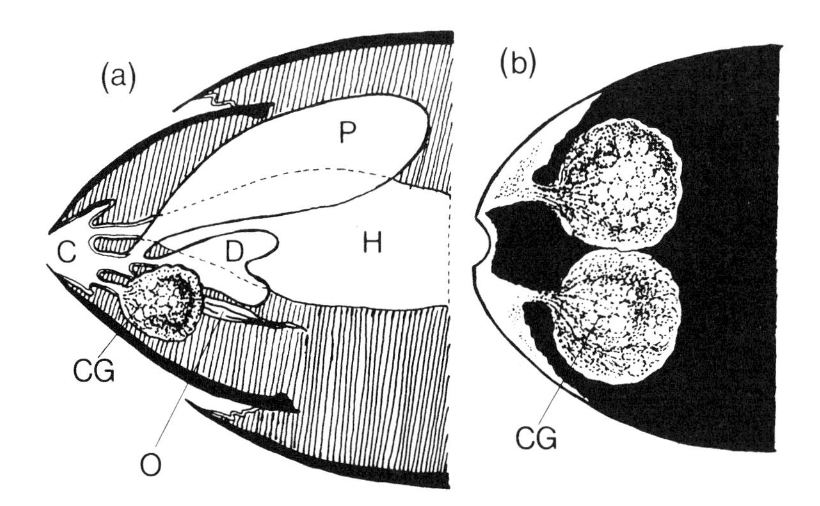


**FIGURE 7–6** (*a*) Schematic drawing of a sagittal section through the gaster tip of a *Camponotus ephippium* worker, including *C*, cloacal chamber; *CG*, cloacal gland; *D*, Dufour's gland; *H*, hindgut; *O*, ovaries; *P*, poison gland. (*b*) View of the VIIth sternite from above. The ventral membranous wall of the cloacal chamber has been partly removed to reveal the paired cell clusters of the cloacal gland. (From Hölldobler, 1982d.)

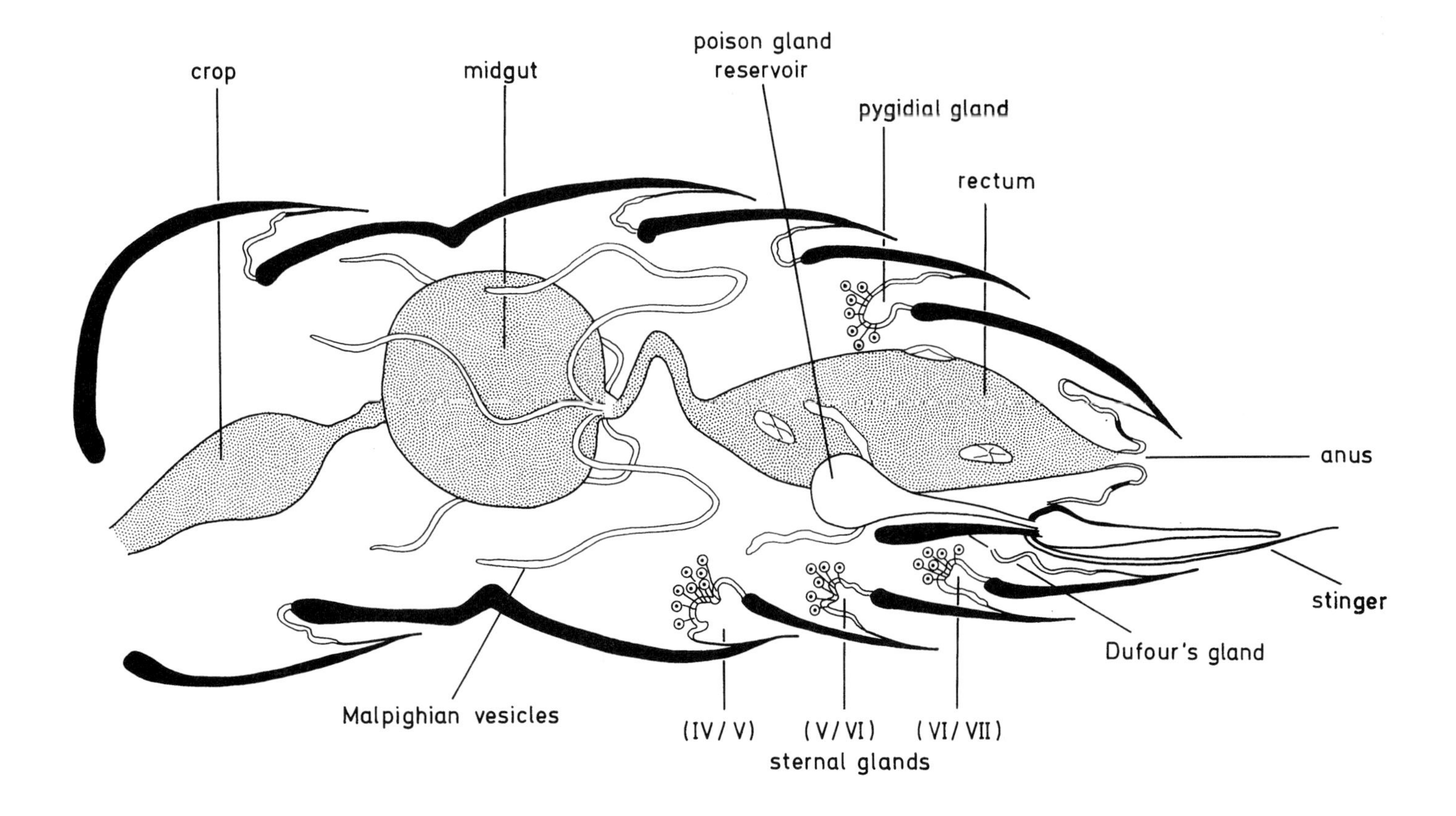


**FIGURE 7–7** Gaster of a worker of the African stink ant, *Paltothyreus tarsatus,* showing the major exocrine glands. (Based on Hölldobler, 1984b.)

**FIGURE 7–8** Gaster of a worker of the ponerine ant genus *Onychomyrmex,* showing the major exocrine glands. (Based on Hölldobler et al., 1982.)

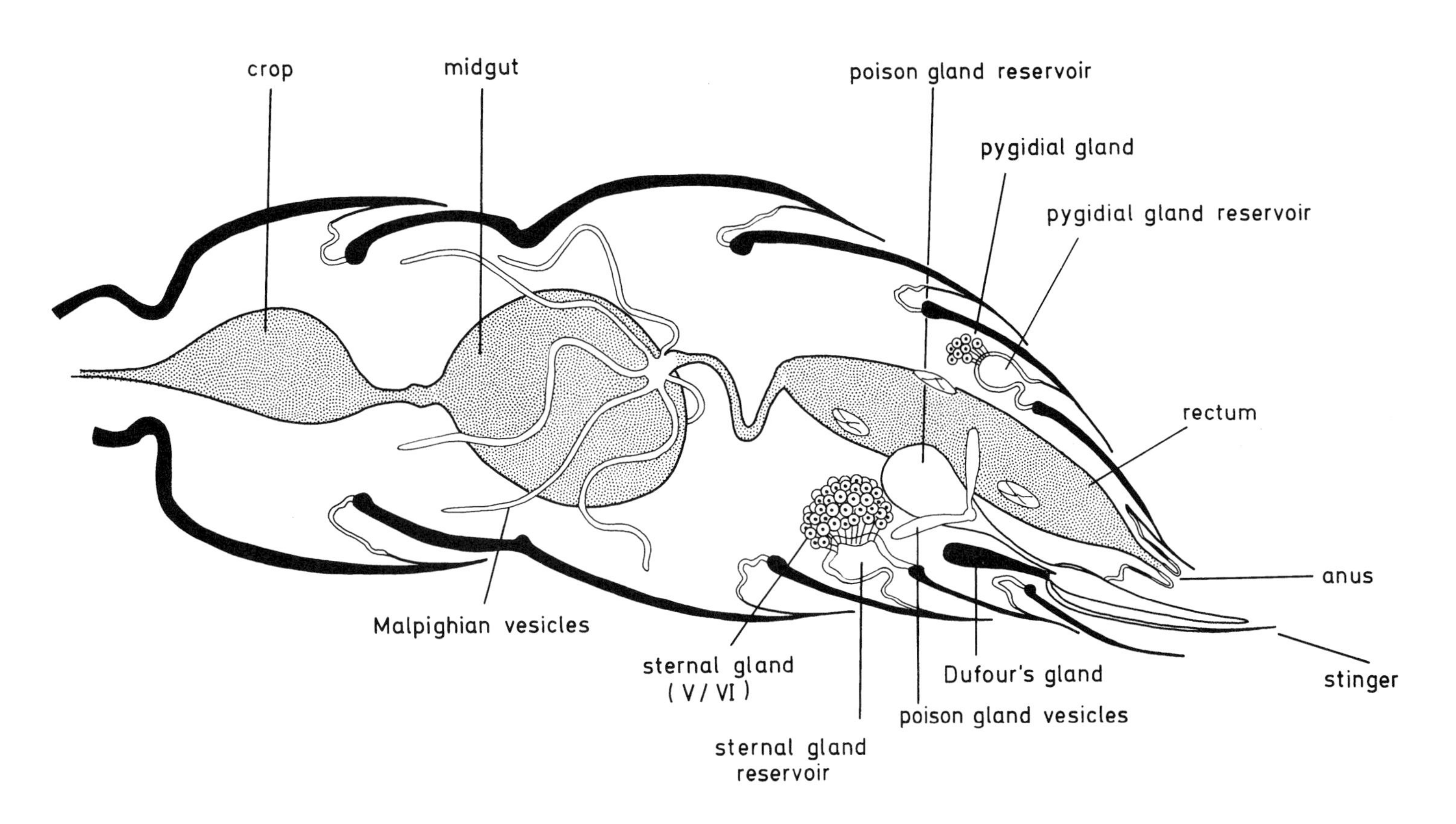

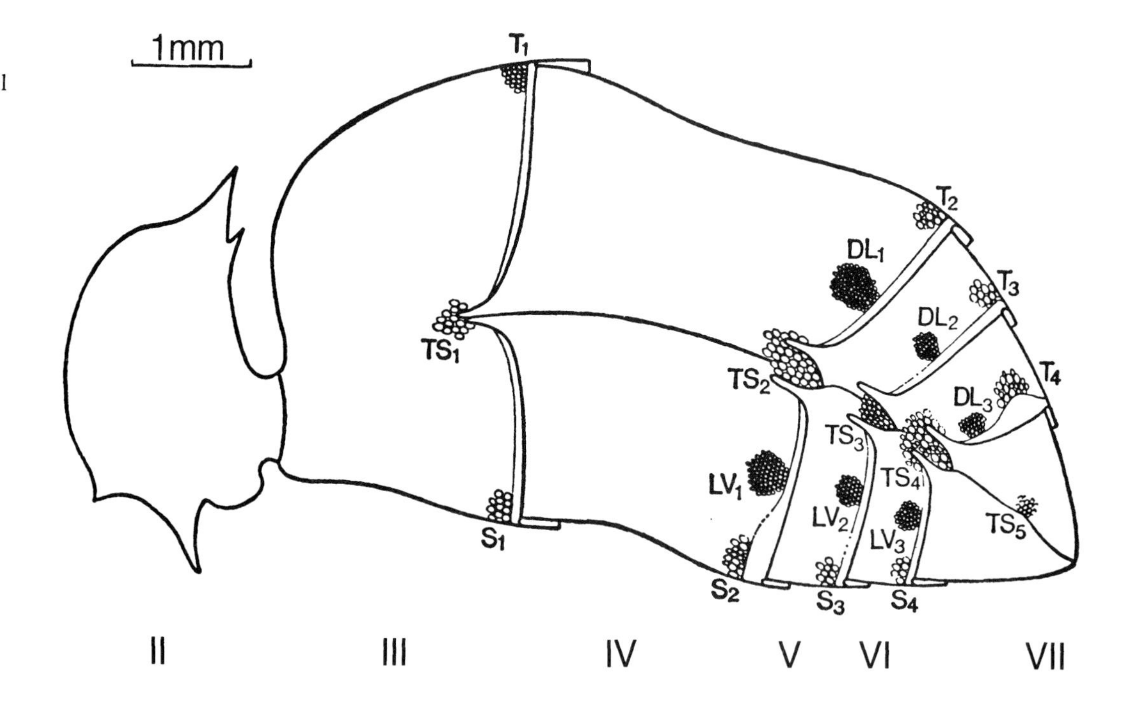


**FIGURE 7–9** Exocrine tergal and dorsal glands in the gaster of *Pachycondyla tridentata*. $T_{1-4}$, tergal glands; $DL_{1-3}$, dorsolateral glands; $TS_{1-5}$, tergasternal glands; $LV_{1-3}$, lateroventral glands; $S_{1-4}$, sternal glands. (From Jessen and Maschwitz, 1983.)

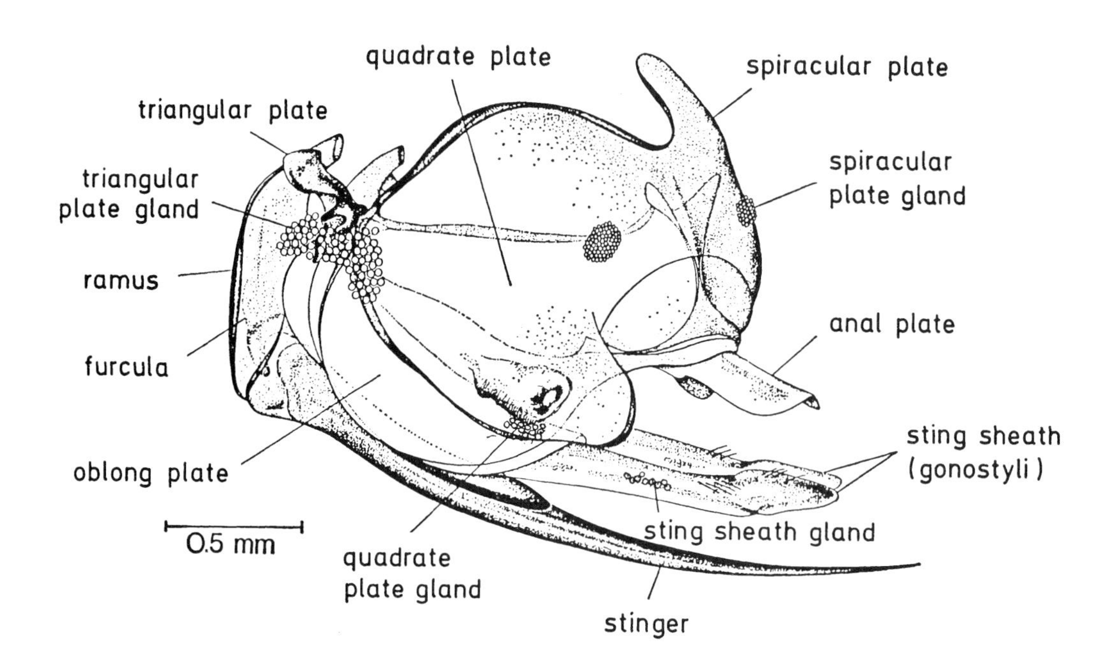


**FIGURE 7–10** The sting apparatus of *Pachycondyla tridentata* and the glands most closely associated with it. Omitted are the poison gland, the Dufour's gland, and the layer of epithelial glandular cells in the gonostyli and on the base of the gonostyli. (Modified from Jessen and Maschwitz, 1983.)

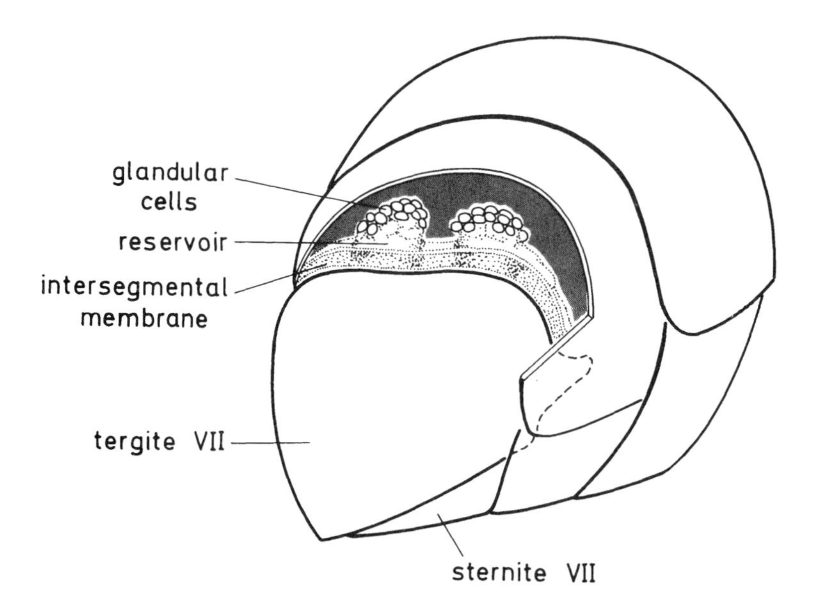


**FIGURE 7–11** The paired pygidial glands of *Leptogenys ocellifera*. The reservoirs are formed by an invagination of the intersegmental membrane between the VIth and VIIth tergites. (Based on Jessen et al., 1979.)

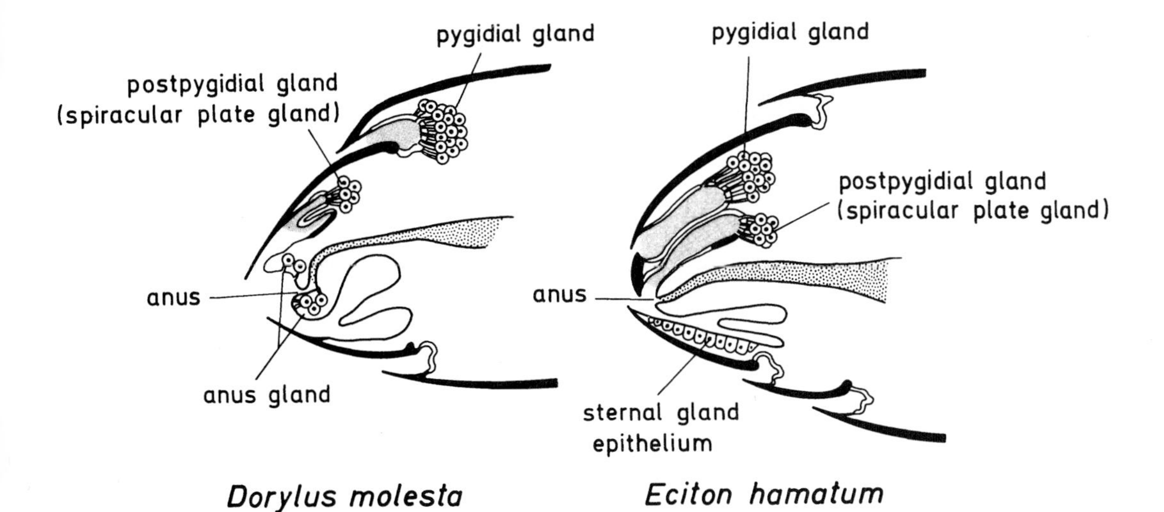


**FIGURE 7-12** Sagittal sections through the gaster tips of Old World driver ant *Dorylus molesta* and New World army ant *Eciton hamatum*. Both genera have well-developed pygidial and postpygidial glands, but the anatomical arrangement is strikingly different. Note that *E. hamatum* lacks the anus gland, but possesses a well-developed glandular epithelium on the VIIth sternite. (Modified from Hölldobler and Engel, 1978.)

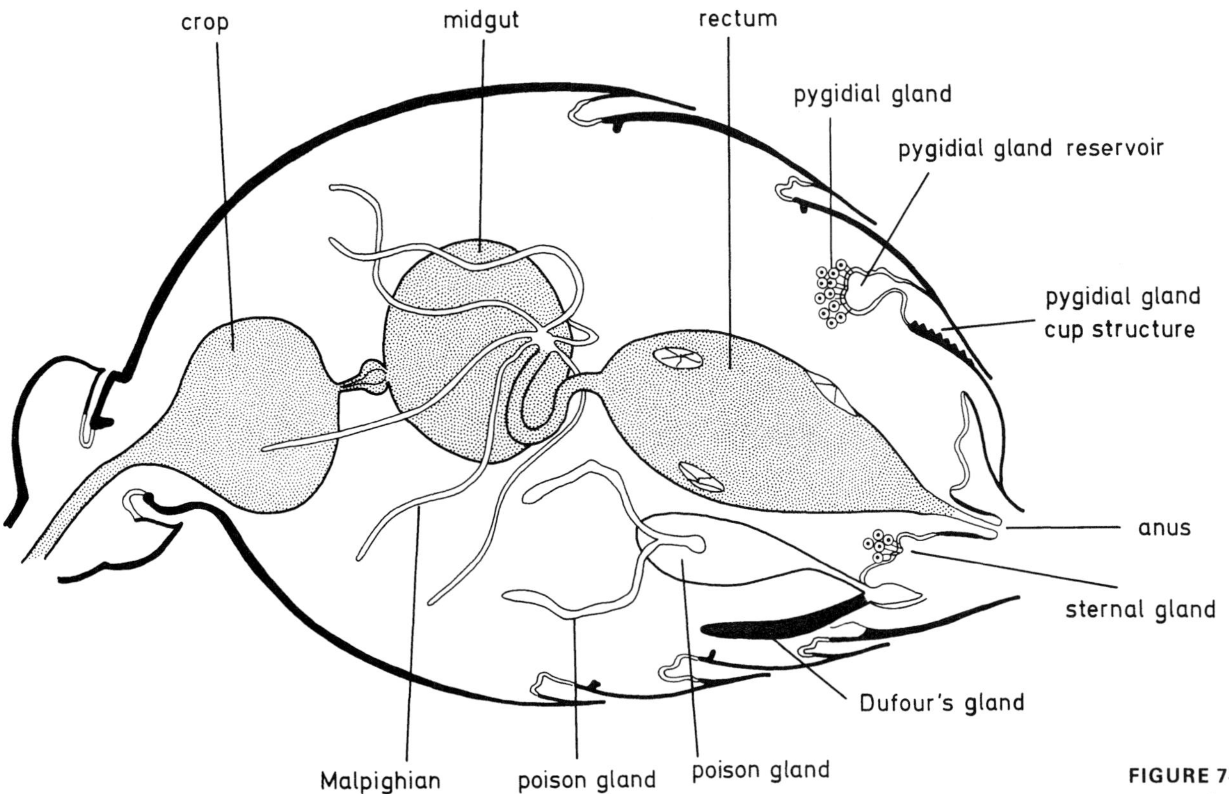


**FIGURE 7-13** Sagittal section through the gaster of an *Aphaenogaster* (= *Novomessor*) worker showing the major exocrine glands. (Based on Hölldobler et al., 1976.)

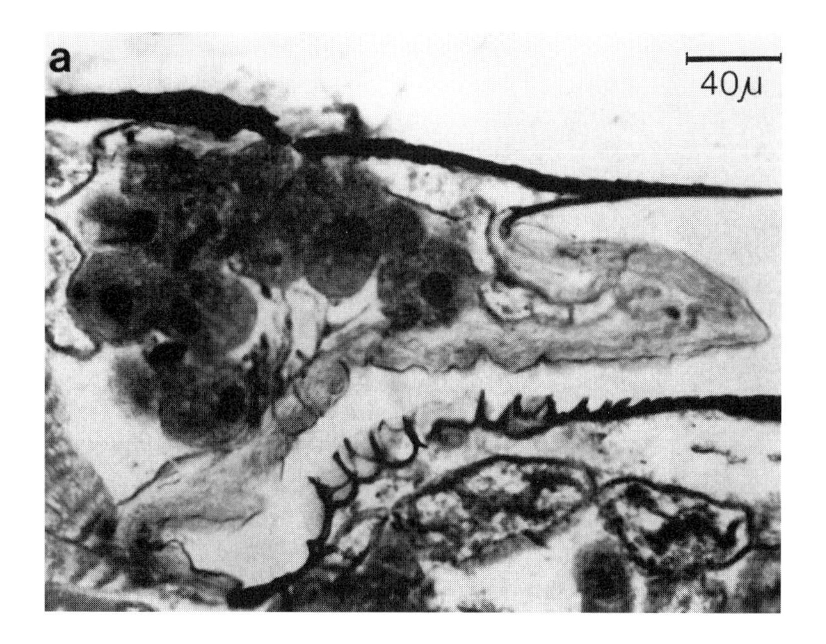


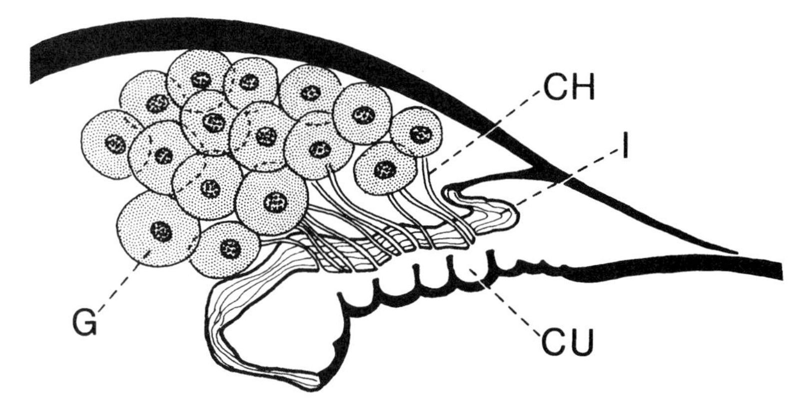


**FIGURE 7–14** (*a*) Sagittal section through the abdominal tergites VI and VII of a worker of *Aphaenogaster* (= *Novomessor*) *albisetosus.* (*b*) Schematic drawing of the histological section, showing the glandular cells (*G*) and the glandular channels (*CH*) through which the secretion is drained into the cuticular cup structure (*CU*) on tergite VII. The glandular ducts penetrate the intersegmental membrane (*I*). (From Hölldobler et al., 1976.)

**FIGURE 7–15** The exocrine gland system of a worker of the Argentine ant *Iridomyrmex humilis: 1,* mandibles; *2,* labium; *3,* labrum; *4,* opening of labial gland; *5,* infrabuccal pocket; *6,* base of maxillary gland; *7,* mandibular gland; *8,* pharynx; *9,* postpharyngeal gland; *10,* brain; *11,* duct of labial gland; *12,* labial gland; *13,* metapleural gland; *14,* reservoir of metapleural gland; *15,* intestine; *16,* rectal bladder; *17,* ovaries; *18,* pygidial gland cells (anal gland); *19,* pygidial gland reservoir; *20,* sternal gland (Pavan's gland); *21,* poison gland; *22,* Dufour's gland. (From Pavan and Ronchetti, 1955.)

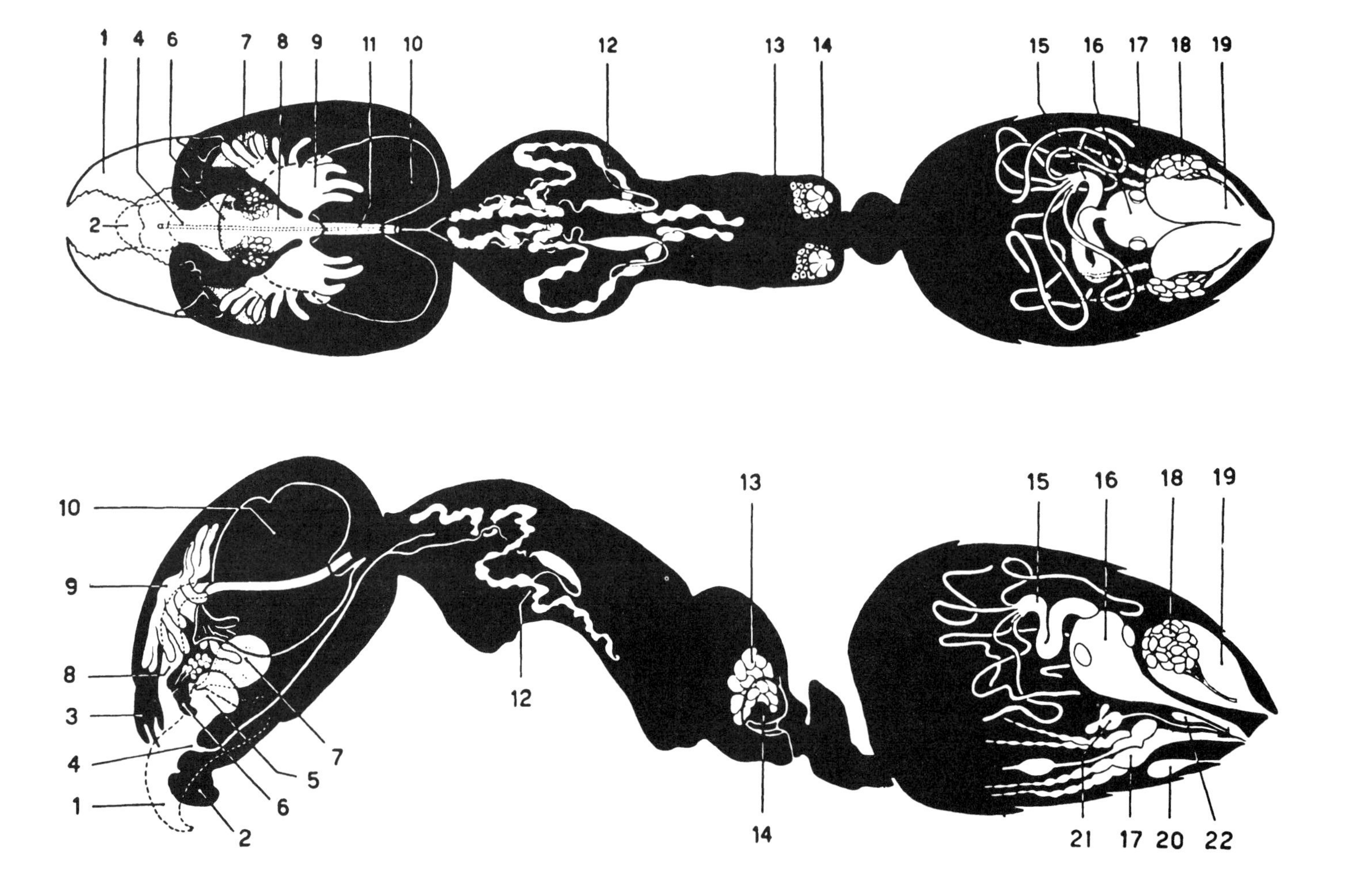

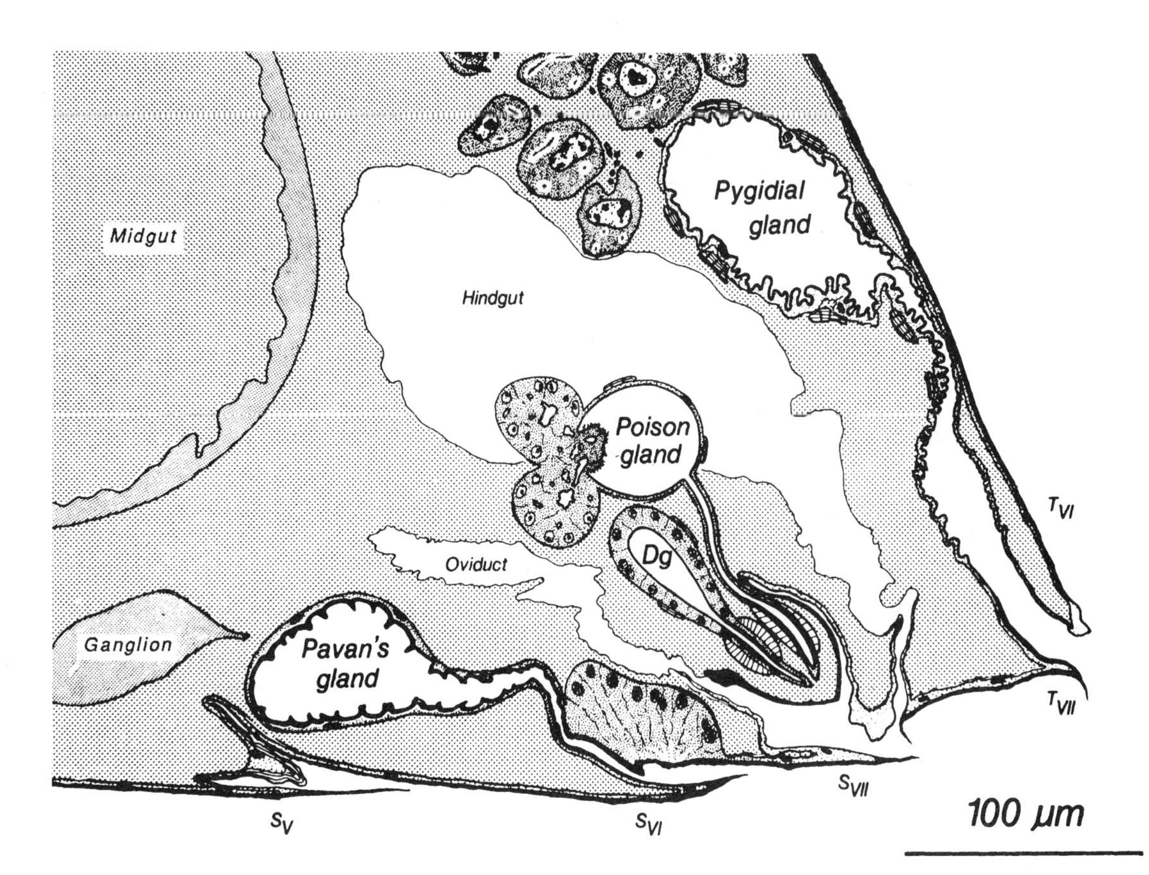


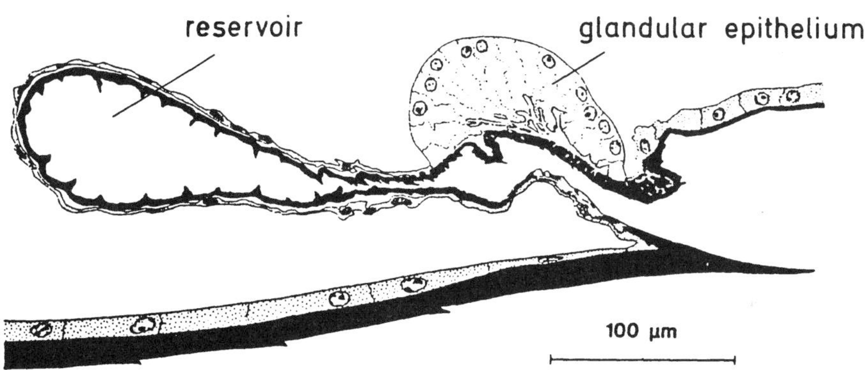


**FIGURE 7–16** Sagittal section through the abdominal tip of dolichoderine ant workers. *Above, Iridomyrmex humilis,* showing the four major exocrine glands. *Dg,* Dufour's gland. (From Billen, 1986c.) *Below,* section through the VIth and VIIth sternites of *Hypoclinea quadripunctata,* showing the reservoir and glandular epithelium of the sternal gland (Pavan's gland) in greater detail. (From Billen, 1985b.)

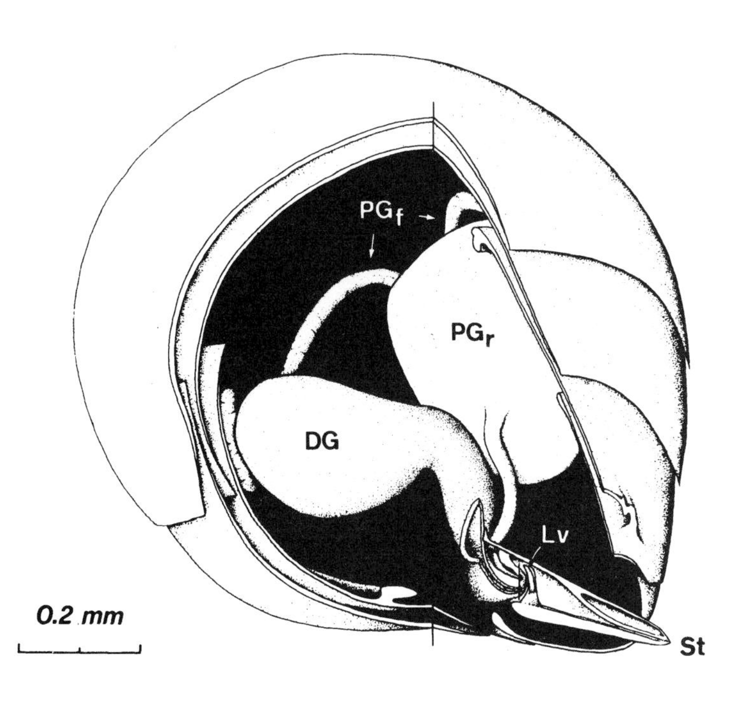


**FIGURE 7–17** Three-dimensional reconstruction of the abdomen of a *Myrmica rubra* worker. The left hind quarter of the abdominal wall has been removed to show the poison and Dufour's glands. *DG,* Dufour's gland; *Lv,* lancet valves; $PG_f$, poison gland free filaments; $PG_r$, poison gland reservoir; *St,* sting. (From Billen, 1986b.)

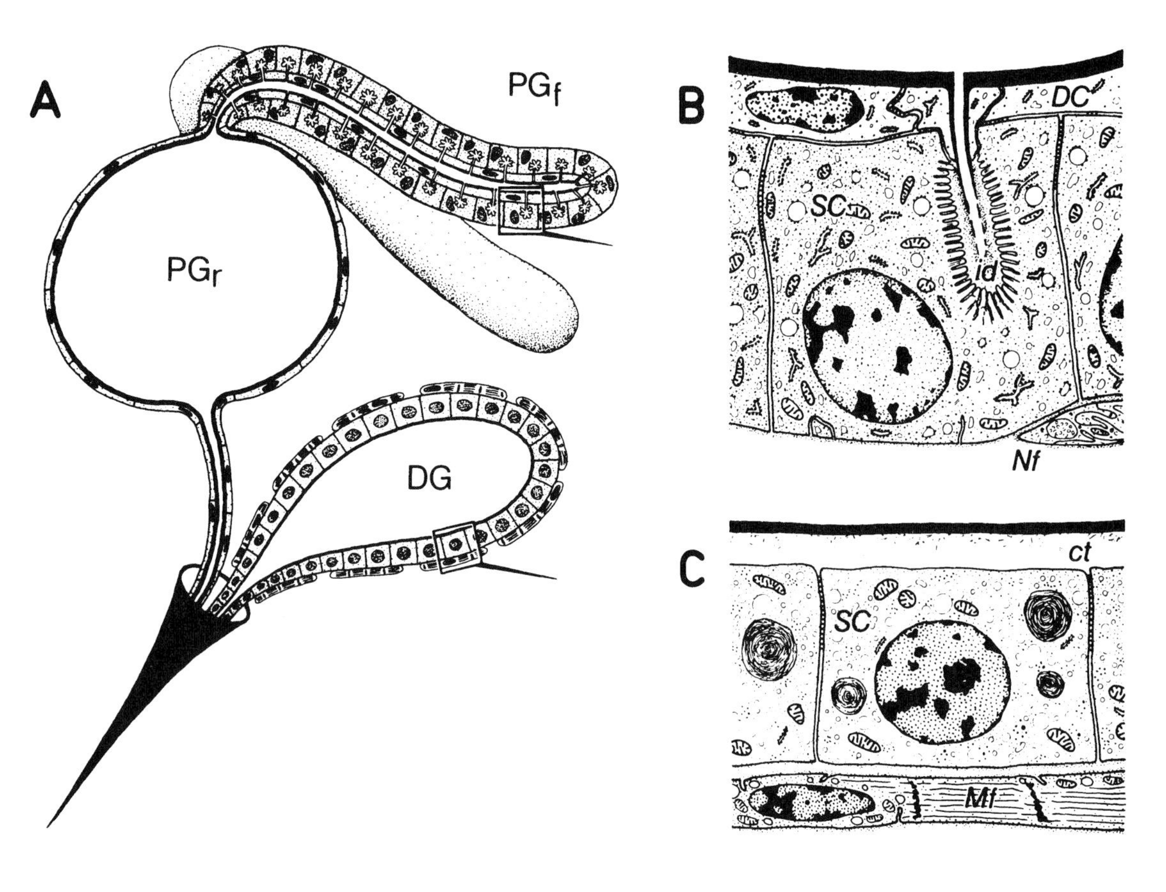


**FIGURE 7–18** The two major sting glands of *Myrmica rubra*. (*A*) General morphology showing the poison gland with its reservoir ($PG_r$) and glandular filaments ($PG_f$), and the Dufour's gland (*DG*). Small portions of the glandular epithelia of the poison gland (*B*) and the Dufour's gland (*C*) are enlarged to show the cytological organization of the secretory cells. *SC*, secretory cells; *id*, intracellular ductule; *DC*, duct cell; *Nf*, nerve fiber; *ct*, cuticle; *Mf*, muscle fiber. (From Billen, 1986b.)

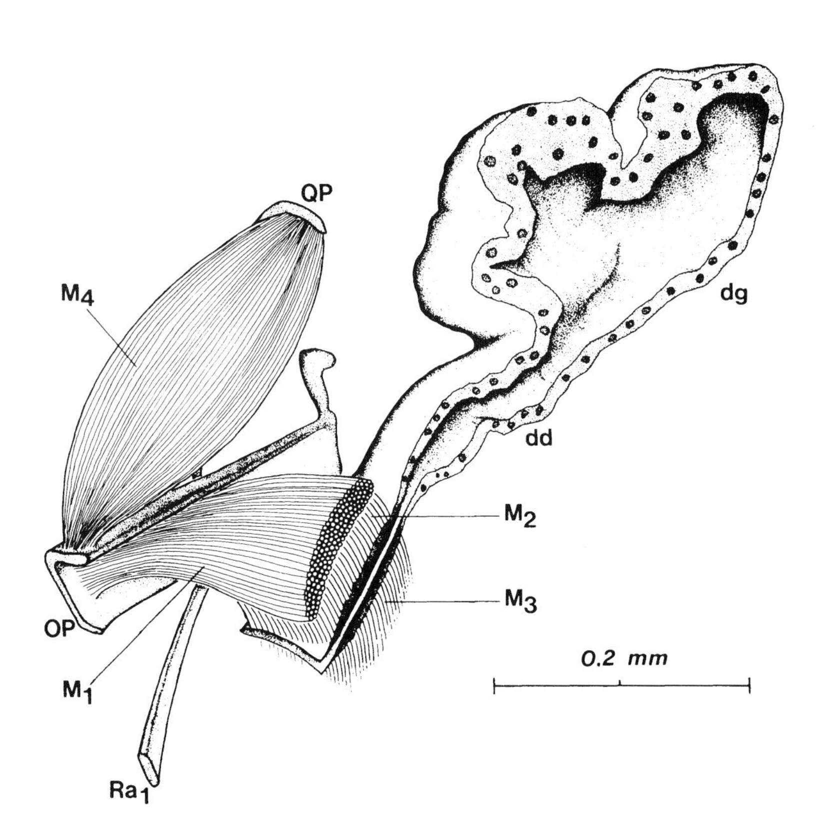


**FIGURE 7–19** Microreconstruction of the Dufour's gland (*dg*) and the adjacent elements showing the four muscle sets ($M_{1-4}$) affecting the gland duct (*dd*). Only the parts to the left of the medial body axis are represented; muscles surrounding the gland itself are omitted. *OP*, oblong plate; *QP*, quadrate plate; $Ra_1$, ramus of the first valvifer. (From Billen, 1982b.)

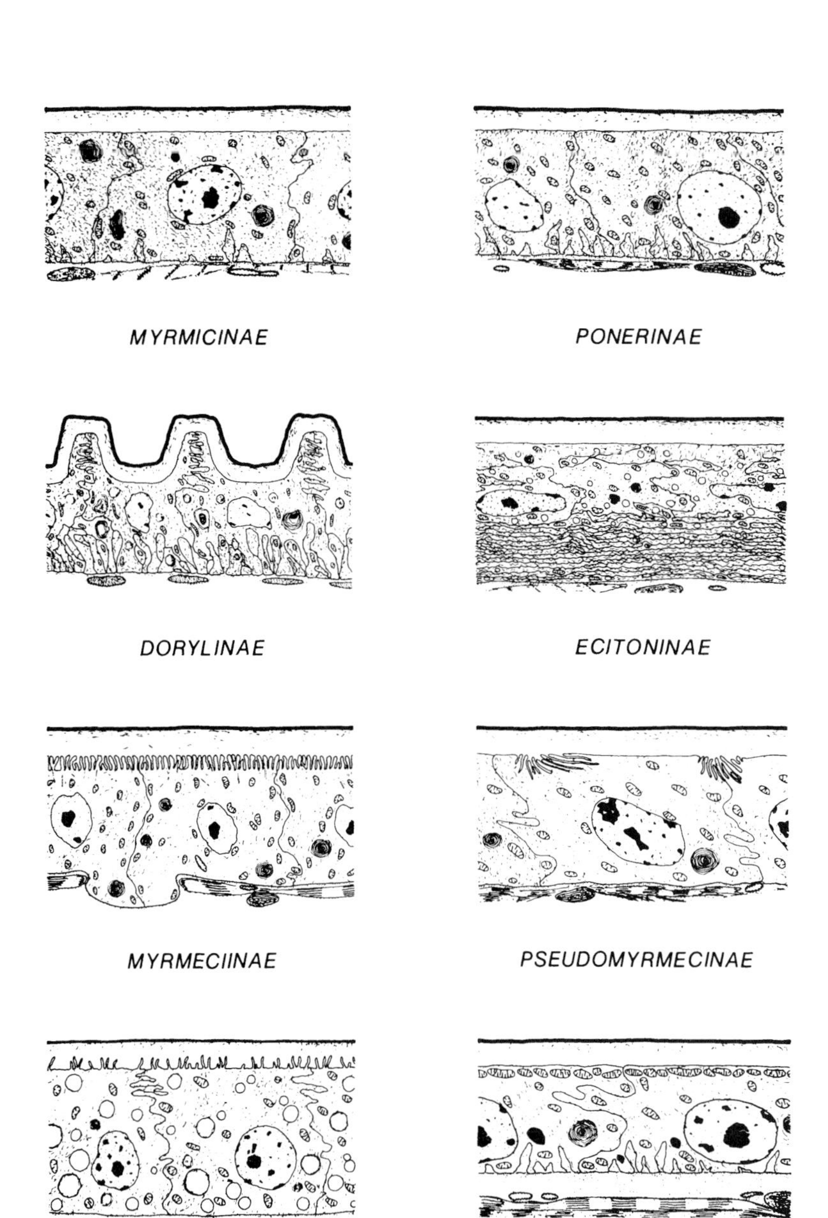


**FIGURE 7–20** Schematic representation of the Dufour's gland epithelium in eight ant subfamilies. (From Billen, 1986a.)

**FIGURE 7–21** Fragment of a colony of *Aphaenogaster* (= *Novomessor*) *cockerelli*. In the middle, the queen sits close to a cluster of eggs and larvae.

and histamine-releasing factors. The "fire" in the venom of fire ants (*Solenopsis*), which indeed feels like a pinpoint burn, is caused by an unusual class of alkaloids, the piperidines, with 2,6-dialkylpiperidines composing the major components. In some species of myrmicines and formicines, a few constituents of the poison gland serve as recruitment or alarm substances. In *Monomorium* and *Solenopsis* at least, they are effective repellents against enemy ants and other arthropods.

In the Dolichoderinae the poison gland is typically reduced, and its function is replaced at least in part by the abundant toxic secretions of the *pygidial gland*. The homology of the pygidial gland has only recently been determined by anatomical studies. Its importance for ant biology has been enhanced by its newly discovered ubiquity and diversity in subfamilies other than the Dolichoderinae. The history of research on the pygidial gland provides a cameo of the often haphazard way in which knowledge of anatomy and behavior is acquired. In his magisterial study of the *Myrmica rubra* worker, Janet (1898) discovered the gland as a cluster of a few cells under the VIth abdominal tergite, with ducts leading to the intersegmental membrane between the VIth and VIIth tergites. After eighty years it was rediscovered as a well-developed organ in *Aphaenogaster* (= *Novomessor*) by Hölldobler et al. (1976), and subsequently shown to be widespread among other genera of the Myrmicinae by Hölldobler and Engel (1978) and Kugler (1978b). Because the gland opens at the VIIth tergite (the pygidium), the name given it by Kugler, the pygidial gland, is now generally accepted (see Figure 7–11).

Substances from the gland have been shown to function as alarm pheromones in three myrmicine genera. The large desert harvesters *Aphaenogaster* (= *Novomessor*) *albisetosus* and *A.* (= *N.*) *cockerelli* (Figure 7–21) release strong-smelling components to evoke a form of panic alarm, which evidently serves to organize swift evacuations during the approach of army ants. Workers of *Orectognathus versicolor*, a highly predaceous Australian dacetine species, lay alarm recruitment trails to prey (Hölldobler, 1981b). Yet another evolutionary direction has been taken by the South American *Pheidole biconstricta:* minor workers produce large quantities of a secretion from their hypertrophied pygidial gland that are used in both chemical defense and aggressive alarm (Kugler, 1979). In *Pheidole embolopyx*, a Brazilian species, the major workers discharge alarm pheromones from the pygidial gland (Wilson and Hölldobler, 1985).

Once defined anatomically in the Myrmicinae, the pygidial gland was quickly located as well in the subfamilies Ponerinae, Myrmeciinae, Dorylinae, Pseudomyrmecinae, Aneuretinae, and Dolichoderinae (Hölldobler and Engel, 1978). Only the Formicinae lack the gland altogether, except in the slave-raiding genus *Polyergus*, where it appears as an independent evolutionary development (Hölldobler, 1984a). A surprising find, however, was the reidentification of the structure in the Dolichoderinae. Generations of researchers had diagnosed this subfamily in part by the possession of the supposedly unique "anal gland," which produced strongly odorous secretions often referred to informally as the Tapinoma odor after the dolichoderine genus *Tapinoma*. Now it is recognized that the anal gland is homologous to the pygidial gland of other ant groups. The finding has bearing on the phylogeny of several subfamilies. It has long been thought that the Aneuretinae are ancestral to the Dolichoderinae, on the basis of common features in external anatomy (Wheeler, 1914b; Wilson et al., 1956). Recent studies by Traniello and Jayasuriya (1981a,b) of the pygidial gland, as well as of the ster-

**FIGURE 7–22** Trail-laying behavior of the ponerine *Pachycondyla laevigata,* a predator of termites. (*Above*) Normal locomotory behavior. (*Below*) A worker carries a termite prey and simultaneously lays a trail by dragging the surface of the pheromone-laden tergite VII over the ground. (From Hölldobler and Traniello, 1980b.)

nal gland, in the sole living species *Aneuretus simoni,* lend further support to this hypothesis. On the other hand W. L. Brown (quoted by Kugler, 1978b) suggests that "the Aneuretinae might just be closer to the Myrmicinae than has been thought." The anatomy and functions of the pygidial gland are at least consistent with this additional linkage. Furthermore, Blum and Hermann (1978b), noting similarities in the chemistry of the secretions of the mandibular glands (quite apart from the pygidial gland) in several myrmicine and dolichoderine species, concluded that "from an exocrinological standpoint, the Dolichoderinae have far more in common with the Myrmicinae than any other formicid subfamily." Because Taylor (1978c) considers the Nothomyrmeciinae ancestral to the Aneuretinae, it is noteworthy that the pygidial gland secretions of the very primitive *Nothomyrmecia macrops* elicit an aggressive alarm response in nestmates as well as a repellent effect on some other ant species occurring sympatrically with it (Hölldobler and Taylor, 1983). Thus in the Nothomyrmeciinae, Aneuretinae, Dolichoderinae, and Myrmicinae, the pygidial gland appears to produce alarm pheromones, defensive substances, or both.

The pygidial gland is both widespread and functionally diverse in the Ponerinae (Hölldobler and Engel, 1978; Jessen et al., 1979; Fanfani and Dazzini Valcurone, 1986). The secretions also play different roles from those of the nothomyrmeciine-myrmicine complex: in several species thus far studied, they elicit either recruitment or sexual attraction. In some species of *Pachycondyla* they are used in either tandem running or trail laying (Hölldobler and Traniello, 1980a,b; Traniello and Hölldobler, 1984; see Figures 7–22 and 7–23). In species of *Leptogenys, Cerapachys,* and *Sphinctomyrmex,* the pygidial gland substances are mixed with poison gland pheromones to produce odor trails (Maschwitz and Schönegge, 1977, 1983; Hölldobler, 1982b and unpublished data). Finally, the results of preliminary experiments suggest that the pygidial gland is at least one of the sources of the trail pheromones in ecitonine army ants (Hölldobler and Engel, 1978).

A plethora of *sternal glands,* representing several independent evolutionary origins, have been discovered in ants. Pavan's gland, a well-developed, often paddle-shaped structure located beneath the VIIth sternite, is the source of the trail pheromone in the Aneuretinae and Dolichoderinae (see Figure 7–16). It consists of a medioventral sac between the VIth and VIIth abdominal sternites, which serves as the gland's reservoir, and a thick glandular epithelium on the anterior margin of the VIIth sternite (Traniello and Jayasuriya, 1981a; Fanfani and Dazzini Valcurone, 1984; Billen, 1985b). It may well have originated in the primitive aneuretines, which in turn gave rise to the dolichoderines in late Cretaceous or early Eocene times (Wilson et al., 1956; Traniello and Jayasuriya, 1981b). Many myrmicine species possess paired clusters of cells beneath the VIIth sternite, but their anatomy is so different as to suggest that they are not homologous with Pavan's gland (Hölldobler and Engel, 1978). Nothing is known at the present time concerning their function, although some circumstantial evidence reported by Cammaerts (1982) suggests that a secretion from the VIIth abdominal sternite serves as an auxiliary trail pheromone in *Myrmica.*

The greatest variety of sternal glands has been encountered in the Ponerinae (Hölldobler and Engel, 1978; Jessen et al., 1979; Fanfani and Dazzini Valcurone, 1986). In the termite-hunting *Paltothyreus tarsatus* of Africa these structures occur beneath the intersegmental membranes that connect the terminal three abdominal sternites, and they produce pheromones for both the recruitment and orientation trails (see Figure 7–7). Workers of *Onychomyrmex,* an Australian genus unique among the amblyoponine Ponerinae for its legionary (army ant) behavior, have a single large gland that opens between the Vth and VIth abdominal sternites (see Figure 7–8). Their secretions serve as a powerful trail and recruitment pheromone during predatory raids and colony emigrations (Figure 7–24). Other, nonlegionary amblyoponines investigated thus far (in the genera *Amblyopone, Myopopone, Mystrium,* and *Prionopelta*) lack the gland. Thus both the gland and the communication it serves appear to have evolved *de novo* in *Onychomyrmex* as part of the army ant syndrome.

Sternal glands found in some species of the subfamily Formicinae are also unique. One such structure, apparently limited to *Oecophylla* weaver ants, occurs beneath the VIIth sternite (Hölldobler and Wilson, 1977d, 1978). This gland consists of an array of single cells, which send short channels into cuticular cups on the outer surface of the sternite (see Figure 7–4). Its original function might have been to secrete lubrication for the seventh abdominal segment, which is frequently rotated when the ant raises the gaster to spray venom through the acidopore. The secretions also function as a

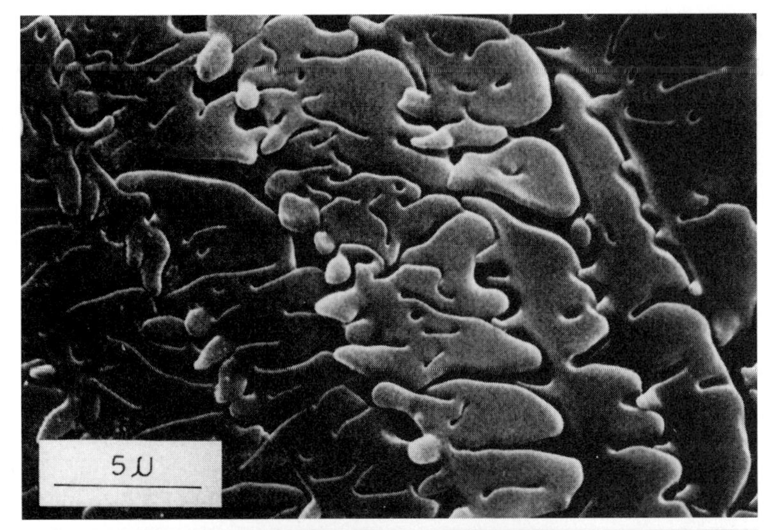


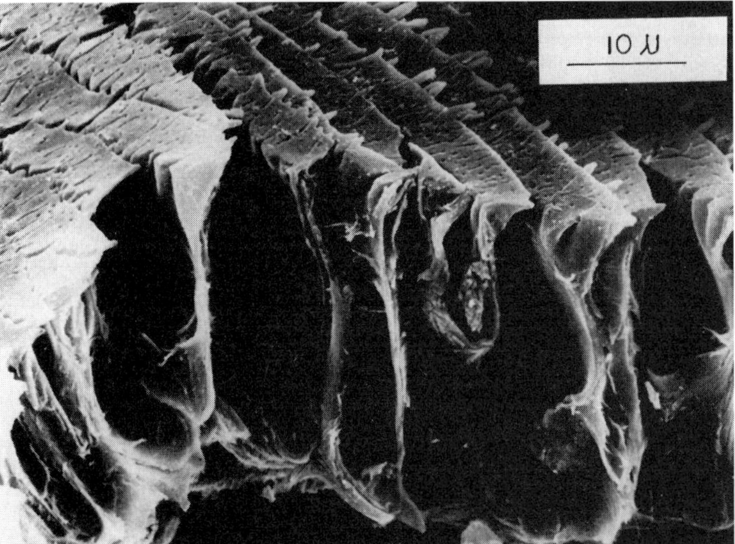


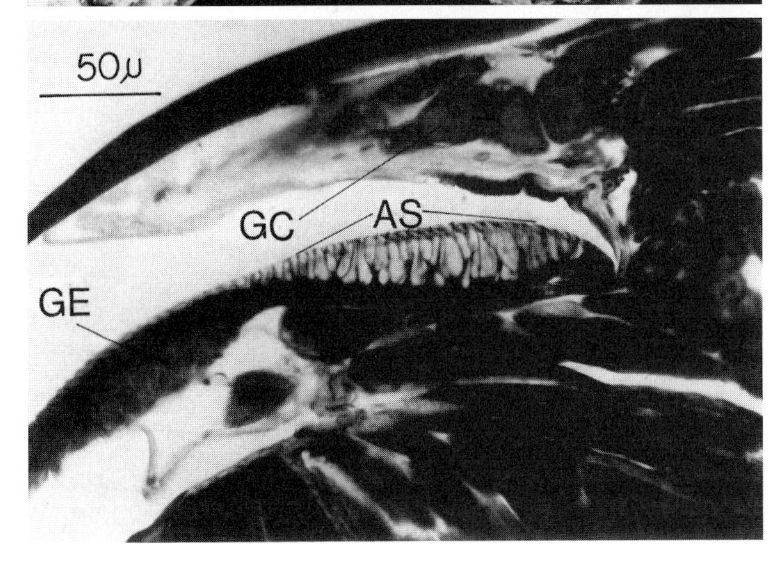


**FIGURE 7–23** The pygidial gland of *Pachycondyla laevigata*, which is the source of the trail pheromone in this termite-raiding ponerine species. *Top:* scanning electron micrograph of the gland-applicator surface on the VIIth tergite of a *P. laevigata* worker. *Center:* scanning electron micrograph of the applicator surface, with part sheared away to reveal the large cavities associated with the structure. *Bottom:* sagittal section through the pygidial gland, showing the glandular epithelium (*GE*), glandular cells (*GC*), and cuticular structure of the applicator surface (*AS*) on tergite VII. (From Hölldobler and Traniello, 1980b.)

short-range recruitment signal. A very different organ is the "cloacal gland" found in several *Camponotus* species, consisting of a paired cluster of glandular cells located at the base of the VIIth abdominal sternite. Each cluster is associated with a major duct elaborated from an invagination of the cloacal chamber. The channels of the glandular cells of each cluster open in dense bundles into these two major ducts (see Figure 7–6). Experiments on *C. ephippium* suggest that the secretions of the cloacal gland serve as recruitment pheromones (Hölldobler, 1982d).

The *mandibular glands* are a pair of thin-walled sacs filled typically with mixtures of alcohols, aldehydes, and ketones. Each of the two structures consists of a flattened glandular mass on the surface of a reservoir. The exit ducts are always connected to the mesal side of the mandibles and open near the anterior edge of the preoral cavity (Blum et al., 1968b). The glands vary relatively little through the Formicidae, although they are generally small in the Ponerinae and large in the Formicinae. In a few species they are hypertrophied in connection with special functions. For example, in a Malaysian species of the *Camponotus saundersi* group, they extend posteriorly all the way into the abdomen and are burst by muscle contractions during combat (Maschwitz and Maschwitz, 1974; Figures 7–25 and 7–26).

When the mandibles are carefully torn away from the head capsule of ant workers, the gland often (but not always!) pulls free intact, which makes its study much easier. Buren et al. (1970) pointed out that longitudinal mandibular grooves are widespread in ants and other aculeate Hymenoptera. This observation led to the oft-cited suggestion that these structures serve as channels for the outward flow of mandibular gland secretions. The grooves do not extend to the gland orifice, however, and in any case are on the opposite side of the mandible from the glandular orifices, making a guiding function unlikely (Hermann et al., 1971).

The mandibular gland secretions of the ants as a whole are so chemically diverse as to preclude any generalization at this time. The substances manufactured by ponerines are especially diverse, including (according to species) organic sulfides, ketones, pyrazines, and a salicylate ester. The glands of myrmicines are a "veritable storehouse of ethyl ketones" and are further often accompanied by their corresponding carbinols, according to Blum and Hermann (1978a). Those of the Formicinae are dominated by terpenoid constituents. The mandibular gland secretions appear to function primarily if not exclusively in defense and in alarm communication. In most species of ant the two roles are combined, but their relative importance varies greatly from one species to the next. Thus in a few species the glands are large, produce copious quantities of toxic secretions, and appear to have little behavioral impact on the ants. In other species the glands are small, yet contain behaviorally very active components.

The *metapleural glands* (also called metasternal or metathoracic glands) are complex structures located at the posterolateral corners of the alitrunk (Figure 7–27). Each consists of a cluster of glandular cells, with each cell draining through a duct into a common membranous collecting sac. The collecting sac leads directly into the storage chamber or reservoir, which is a simple sclerotized cavity. Externally the metapleural glands are often marked by a pronounced vault or "bulla," and a slit-shaped opening to the outside (Figure 7–27a). Brown (1968) suggested that the glands produce

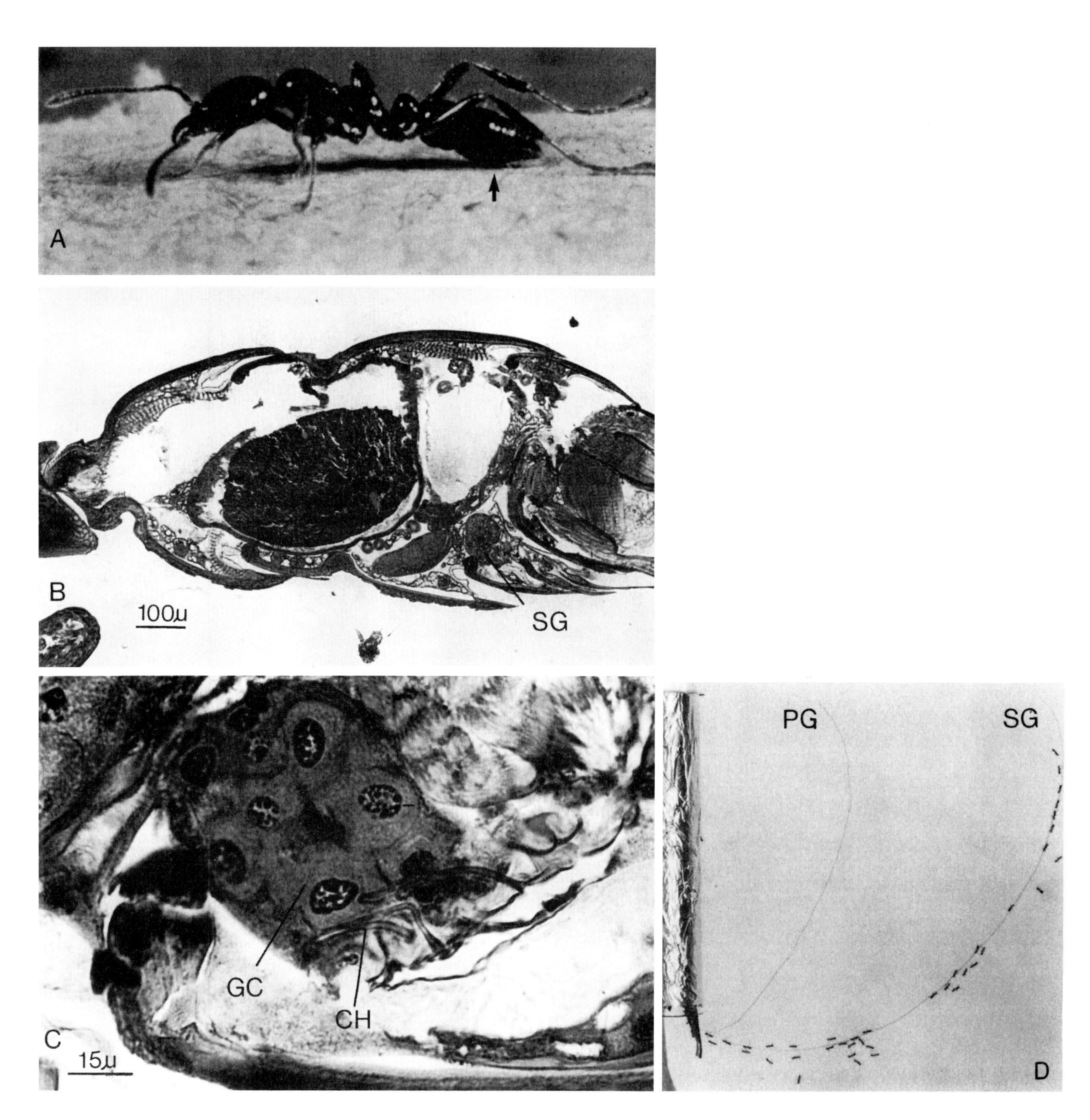


**FIGURE 7–24** Ants of the Australian ponerine genus *Onychomyrmex*, who exhibit army ant behavior, have a unique sternal gland used in the production of trail pheromones. (*A*) Worker lays an odor trail by dragging the gland over the surface. Arrow indicates the position of the gland. (*B*) Sagittal section through the gaster of an *Onychomyrmex* worker (*SG*, sternal gland). (*C*) Sagittal section through sternal gland (*GC*, glandular cells; *CH*, glandular channels leading to surface). (*D*) When two artificial trails are drawn with a crushed poison gland (*PG*) and sternal gland (*SG*), the *Onychomyrmex* workers follow only the sternal gland trail. (From Hölldobler et al., 1982.)

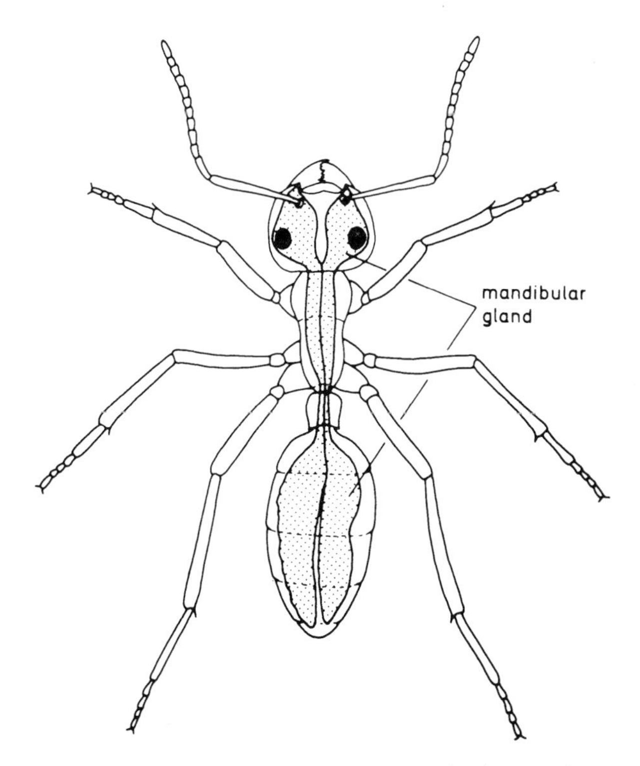


**FIGURE 7–25** In workers of the *Camponotus saundersi* group, hypertrophied mandibular glands extend all the way into the abdomen. (Modified from Maschwitz and Maschwitz, 1974.)

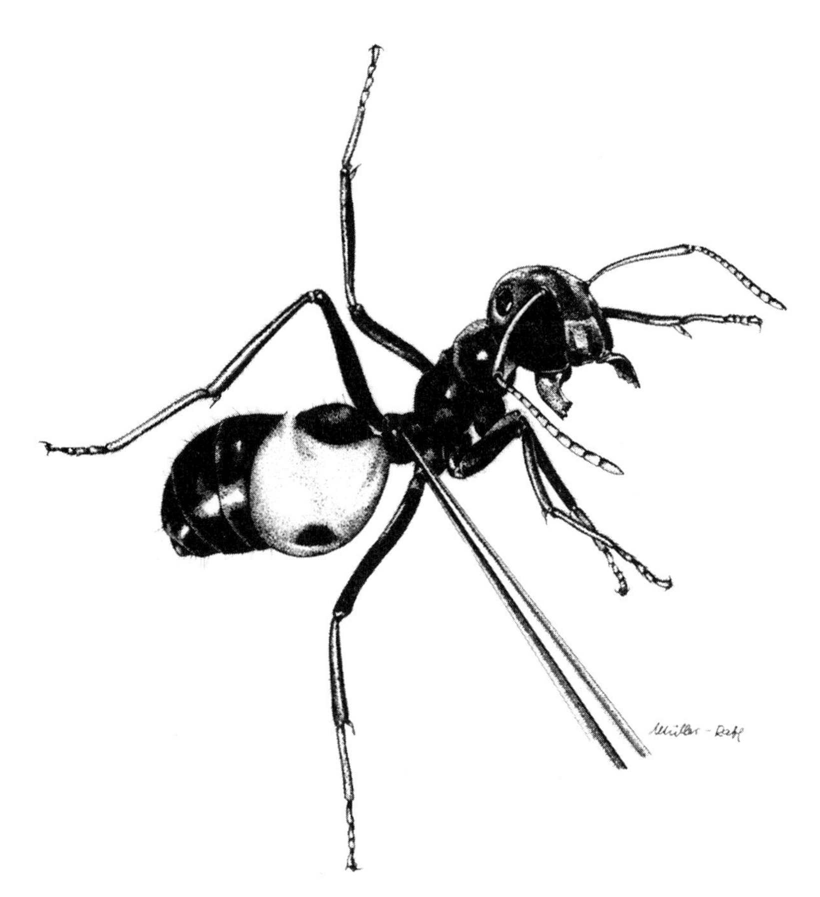

**FIGURE 7–26** An ant "bomb." When a worker of the *Camponotus saundersi* group is seized with a pair of forceps, it contracts its abdominal wall violently, finally bursting open to release secretions from its hypertrophied mandibular glands. (From Buschinger and Maschwitz, 1984.)

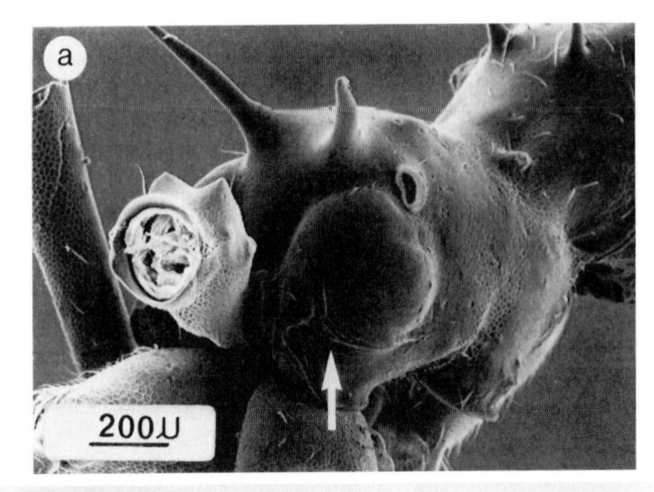


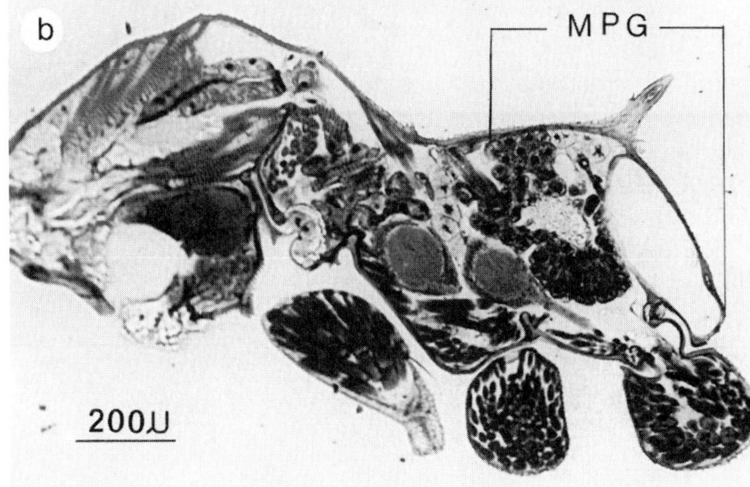


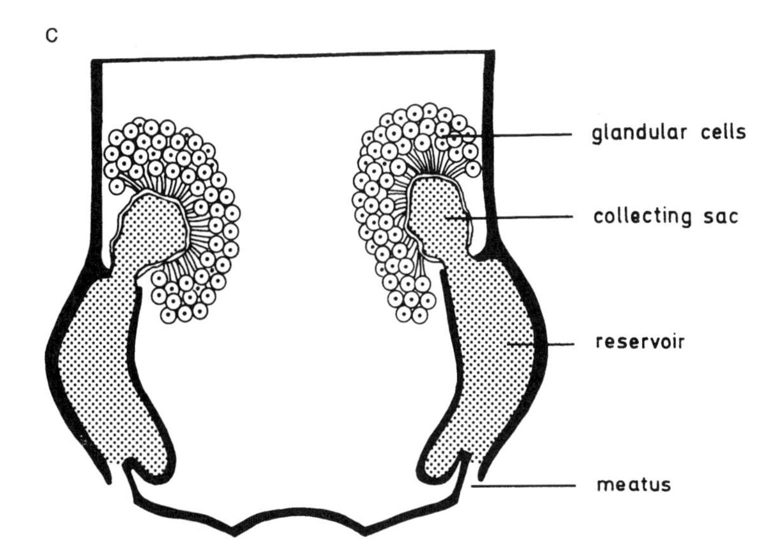


**FIGURE 7–27** The metapleural glands of leafcutter ants of the genus *Atta*, which illustrate the major anatomical features of this organ in ants generally. (*a*) Scanning electron micrograph of the mesosoma of *A. cephalotes*, showing the large pronounced vault (bulla) which covers the storage chamber. Arrow points to the slit-shaped opening (meatus). (*b*) Sagittal section through the mesosoma of *A. sexdens*, showing the large internal region of the metapleural gland (*MPG*). (*c*) Schematic illustration of a dorsal view of the paired metapleural glands. (Modified from Hölldobler and Engel-Siegel, 1984.)

**FIGURE 7–28** When a worker of *Crematogaster inflata* is seized with a pair of forceps, she discharges droplets of sticky fluid from her hypertrophied metapleural glands. (From Buschinger and Maschwitz, 1984.)

pheromones for recognition and identification of nestmates and alien species, and Jaffe and Puche (1984) claimed that in *Solenopsis geminata* metapleural gland secretions serve as territorial markers. This general explanation seems unlikely, because other investigations have found no evidence that secretions from the metapleural glands are involved in communication at all (Maschwitz et al., 1970; Maschwitz, 1974). Maschwitz and his collaborators did, however, demonstrate that in a number of ant species the metapleural gland secretions serve as powerful antiseptic substances that protect the body surface and nest against microorganisms. One active antibiotic component of *Atta sexdens*, for example, is phenylacetic acid, of which one ant carries an average of 1.4 micrograms at any given moment. In *Crematogaster deformis* the hypertrophied metapleural glands contain a mixture of phenols, including mellein (Attygalle et al., 1989). The worker regularly releases small amounts of this mixture that serve as an antiseptic. But when she is attacked by enemy ants, particularly at the highly vulnerable petiolar-postpetiolar region of the abdomen, she suddenly discharges large quantities of the metapleural gland secretions, which now function as a powerful repellent. Finally, in *C. inflata*, which also possess hypertrophied metapleural glands, the sticky secretions serve primarily as an alarm-repellent substance (Maschwitz, 1974; Figure 7–28).

It is generally assumed that the metapleural glands are a universal and phylogenetically old character of the Formicidae. Even the extinct species *Sphecomyrma freyi* of Cretaceous age appears to have possessed one (Wilson et al. 1967a,b). The organ is well developed in the Ponerinae, Myrmeciinae, and in *Nothomyrmecia macrops*, the only living species of the primitive subfamily Nothomyrmeciinae. The species of only a few genera, such as *Oecophylla*, *Polyrhachis*, and *Dendromyrmex*, as well as most *Camponotus* and certain socially parasitic ants, have secondarily atrophied or completely lost the metapleural glands (Brown, 1968; Hölldobler and Engel-Siegel, 1984).

### Design Features of Ant Pheromones

It is not always the "purpose" of animal communication systems to maximize the information transmitted. In many cases, a simple yes-or-no signal is sufficient, for example, when nestmates are distinguished from aliens or workers broadcast a state of alarm. In others, such as the pinpointing of food discoveries by means of odor trails and waggle dances, the precision and hence the quantity of spatial information are at a premium. The optimal *gain* in transmission, in other words the number of group members contacted (Markl, 1985), also varies according to circumstance. Alarm signals are typically local, whereas caste-inhibitory signals are colony-wide.

Research on ant pheromones has revealed these and other design features of signals to be adaptations to the moment-by-moment needs of the colony. The theory of design is based on the concept of the *active space*, which is the zone within which the concentration of a pheromone (or any other behaviorally active chemical substance) is at or above threshold concentration (Bossert and Wilson, 1963; Wilson and Bossert, 1963). The active space is, in fact, the chemical signal itself. According to need, the space can be made large or small; it can reach its maximum radius quickly or slowly; and it can endure briefly or for a long period of time. These adjustments have been made in the course of evolution by altering the $Q{:}K$ ratio, the ratio of the amount of pheromone emitted ($Q$) to the threshold concentration at which the receiving animal responds ($K$). $Q$ is measured in number of molecules released in a burst, or in number of molecules emitted per unit of time, whereas $K$ is measured in molecules per unit of volume. Where location of the signaling animal is relevant, the rate of information transfer can be increased either by lowering the emission rate $Q$ or by raising the threshold concentration $K$, or both. This adjustment achieves a shorter fade-out time and permits signals to be more sharply pinpointed in time and space by the receiver. A lower $Q{:}K$ ratio characterizes both alarm and trail systems. The mathematical models based on diffusion and plume formation can be used to predict the form and duration of the active space or, conversely, either $Q$ or $K$ when the other parameter is known along with the elementary dimensions of the active space (Bossert and Wilson, 1963).

If part of the message is the location of the signaler, as it typically is in alarm, recruitment, and sexual communication, the information in each signal increases as the logarithm of the square of the distance over which the signal travels. In chemical systems it is the active space that must be expanded. An increase in active space can be achieved either by increasing $Q$ or decreasing $K$. The latter option is far more efficient, since $K$ can be altered over many orders of magnitude by changes in the sensitivity of the chemoreceptors, whereas a comparable change in $Q$ requires enormous increases or decreases in pheromone production as well as large changes in the capacity of the glandular reservoirs. The reduction of $K$ has been especially prevalent in the evolution of trail systems and airborne sex pheromones, where threshold concentrations are sometimes on the order of only hundreds of molecules per cubic centimeter.

The duration of the signal can be shortened by an enzymatic deactivation of the molecules. When Johnston et al. (1965) traced the metabolism of radioactive queen substance, (*E*)-9-keto-2-decenoic acid, fed to worker honey bees, they found that within 72 hours more than 95 percent of the pheromone had been converted into inactive substances consisting principally of 9-ketodecanoic acid, 9-hydroxydecanoic acid, and (*E*)-9-hydroxy-2-decenoic acid. No comparable investigations have been conducted on the pheromones of ants, but such deactivations are likely to occur in systems requiring both a long reach and a rapid fade-out.

Communication can be enriched by variation in the response according to the concentration of the pheromone. In workers of the Florida harvester *Pogonomyrmex badius,* the principal alarm pheromone is 4-methyl-3-heptanone, which is stored in quantities of 0.2–34.0 micrograms (average: about 16 μg) in the mandibular gland reservoir (Vick et al., 1969; N. Lind, personal communication). Workers near the nest respond to threshold concentrations averaging $10^{10}$ molecules per cubic centimeter by moving toward the odor source; when a zone of concentration one or more orders of magnitude greater than this amount is reached, the ants switch into an aggressive alarm frenzy (Wilson, 1958a). The active space of the alarm can therefore be envisioned as a concentric pair of hemispheres. As the ant enters the outer zone she is attracted inward toward the point source; when she next crosses into the central hemisphere, she is excited into a frenzy. A very similar pattern of response to the same pheromone occurs in the leafcutter ant *Atta texana* and is illustrated in Figure 7–29.

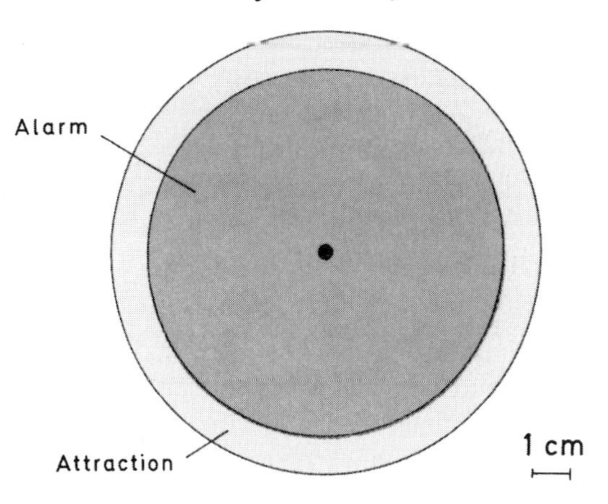


**FIGURE 7–29** The active space of the alarm pheromone 4-methyl-3-heptanone in the leafcutter ant *Atta texana.* At low concentrations the ants are attracted to the substance; at higher concentrations they are launched into aggressive alarm behavior. The two zones are more accurately envisioned as hemispheres above the ground surface than as circles as shown here. (Modified from Bradshaw and Howse, 1984, projected from the data of Moser et al., 1968.)

The size of the pheromone molecules transmitted through air can be expected to conform to certain broad physical rules (Wilson and Bossert, 1963). In general, they should possess a carbon number between 5 and 20 and a molecular weight between 80 and 300. The a priori arguments that led to this inference are essentially as follows. Below the lower limit, only a relatively small number of molecules can be readily manufactured and stored by glandular tissue. Above it, molecular diversity increases very rapidly. In at least some insects, and for some homologous series of compounds, olfactory efficiency also increases steeply. As the upper limit is approached, molecular diversity becomes astronomical, so that further increase in molecular size confers no further advantage in this regard. The same consideration holds for intrinsic increases in stimulative efficiency, insofar as they are known to exist. On the debit side, large molecules are energetically more expensive to make and to transport across membranes, and they tend to be far less volatile. However, differences in the diffusion coefficient caused by reasonable variation in molecular weight do not cause much change in the properties of the active space, contrary to what one might intuitively expect. The large number of ant pheromones identified to date conform to this rule of molecular size variation. Wilson and Bossert (1963) further predicted that alarm substances, which have no requirements for specificity and can be "read" by other species without harm to the sender, should have lower molecular weights than trail substances and other kinds of pheromones in which privacy is at a premium. The reason is that the smaller the molecule, the less likely it is to be unique. For example, there are vastly fewer variations possible on a 6–carbon alcohol than on a 12-carbon alcohol, or on a 6-carbon alkane than on a 6-carbon nitrogen heterocycle. So far, this prediction has been vindicated in the Myrmicinae but not in the Formicinae. In the latter group the alarm and trail substances overlap very broadly in their molecular weight, and they show no additional design features that conspicuously enhance or diminish their molecular specificity. The molecular design in these substances remains a puzzle.

Because of the large numbers of species of ants and other social insects, and the natural constraints on biosynthesis limiting molecular diversity far below the theoretical maximum, a considerable amount of convergence has occurred in pheromone chemistry. Examples of identical pheromones across species are given in Table 7–2. Since the insects listed are phylogenetically so remote from one another, every one of the pairings can be regarded as a result of convergent evolution rather than of homology.

Some biochemical matches are nevertheless probably due to homology, with particular compounds having persisted over long periods of time through conservative biosynthesis and function. Possibly the most stable of all glands in this respect is the Dufour's gland, which often contains mixtures of terpenoid and straight-chain hydrocarbons that vary little from one genus to the next. *Z,E*-α-farnesene, for example, is the principal recruitment pheromone laid down in trails of the fire ant *Solenopsis invicta,* while two of its homofarnesene homologs serve as synergists (Vander Meer, 1986a,b). *Myrmica lobicornis* and *M. scabrinodis* also produce *Z,E*-α-farnesene and homofarnesenes in their Dufour's glands, but these substances do not function as trail pheromones and their role remains unknown (Attygalle et al., 1983). The trail pheromone of *Myrmica,* produced in the poison gland, is in fact 3-ethyl-2,5-dimethylpyrazine; it is also a poison gland product and a trail pheromone of *Tetramorium caespitum* and two species of *Atta* (see Table 7–5). Finally, the Dufour's gland contents of the large, primitive dacetine *Daceton armigerum* are followed by *Solenopsis invicta* (Wilson, 1962a), while the poison gland contents are followed by species of *Acromyrmex* and *Atta* (Blum and Portocarrero, 1966), which suggests that the Dufour's gland of *Daceton armigerum* contains the farnesene and its poison gland the pyrrole, one of the trail pheromones identified in attine ants (Tumlinson et al., 1971, 1972). The comparative biochemistry of ant exocrine glands and their primitive and derived functions are fascinating but still relatively unexplored subjects.

**TABLE 7–2** Biochemical convergence of pheromones among ants (Formicidae), bees (Apidae), and termites (Termitidae and Rhinotermitidae). (From Blum, 1982.)

| Compound | Function | Occurrence | |
|---|---|---|---|
| | | Family | Genus |
| Benzaldehyde | Trail pheromone | Apidae | *Trigona* |
| | Defense | Formicidae | *Veromessor* |
| Citral | Trail pheromone | Apidae | *Trigona* |
| | Alarm pheromone | Formicidae | *Acanthomyops* |
| Citronellol | Aggregation pheromone | Apidae | *Bombus* |
| | Defense | Formicidae | *Atta* |
| 2,5-Dimethyl-3-isopentylpyrazine | Sex pheromone | Formicidae | *Camponotus* |
| | Alarm pheromone | Formicidae | *Odontomachus* |
| Mellein | Sex pheromone | Formicidae | *Camponotus* |
| | Defense | Termitidae | *Cornitermes* |
| Methyl anthranilate | Sex pheromone | Formicidae | *Camponotus* |
| | Alarm pheromone | Formicidae | *Xenomyrmex* |
| Methyl 6-methylsalicylate | Sex pheromone | Formicidae | *Camponotus* |
| | Alarm pheromone | Formicidae | *Gnamptogenys* |
| 2-Tridecanone | Alarm pheromone | Formicidae | *Acanthomyops* |
| | Defense | Rhinotermitidae | *Schedorhinotermes* |

## Efficiency of Semiochemicals

Possibly the chief advantage of semiochemicals over signals in other sensory modalities is the extreme economy of their manufacture and transmission. The sensory apparatus has evolved in some cases to respond to particular substances at a virtually quantal level, with only a few molecules striking the receptive membranes in each antennal sensillum every few seconds. The process is abetted by the existence of isomerism, in which relatively minor differences in the configuration of the same molecule generate new physical or chemical properties that are discernible by the ants. The most extreme form is optical isomerism, the existence of pairs of chemical compounds (enantiomorphs) whose molecules are nonsuperimposable mirror images. One configuration is capable of rotating plane-polarized light to the right and constitutes the dextro or (+) form, and the other, rotating to the left, constitutes the levo or (−) form. In leafcutter ants of the genus *Atta,* workers are 100 to 200 times more sensitive to the natural, (+) enantiomer of 4-methyl-3-heptanone, an alarm pheromone, than they are to its (−) enantiomer (Riley et al., 1974a). *Pogonomyrmex* harvester ants are similarly more sensitive to the (+) enantiomer (Benthuysen and Blum, 1974).

A principal consequence of such acute sensitivity is that only minute amounts of the pheromones are needed at any given time. The extreme cases recorded thus far occur in the trail substances. The amounts of methyl 4-methylpyrrole-2-carboxylate found in each worker of the leafcutters *Acromyrmex* and *Atta* range according to species from 0.3 nanogram to 3.3 nanograms (Evershed and Morgan, 1983). Workers of *Myrmica rubra* contain 5.8 ± 1.7 nanograms of the trail substance 3-ethyl-2,5-dimethylpyrazine (Evershed et al., 1982). Even such trace amounts, while wholly undetectable to human beings without the aid of elaborate instrumentation, are sufficient to convey complete messages among ants. Tumlinson et al. (1971), the discoverers of methyl 4-methylpyrrole-2-carboxylate as the trail substance of *Atta texana,* estimated that one milligram of this substance (roughly the quantity in a single colony), if laid out with maximum efficiency, would be enough to lead a column of ants three times around the world.

The chief disadvantage of such chemical systems is the slowness of fade-out. When using pheromones alone, ants cannot transmit a rapid sequence of signals in the manner of vocalizations or quickly changing visual signals. In order to replace signals, they must wait until the active space of the pheromones expands to maximum diameter, then shrinks back to the point of emission or is blown away by air currents. In many cases this property has been turned to the advantage of the insects. A long-standing active space is needed, for example, in the employment of colony odors and caste-identification substances, alarm pheromones, and trail substances. It is also possible to create sequential and compound messages either by a graded reaction to different concentrations of the same substance, as illustrated in the case of the *Atta texana* alarm system (see Figure 7–29), or by blends of signals. Let us now consider the latter very interesting elaboration in some detail.

## Pheromone Blends

The following rule has emerged from surveys of the natural product chemistry of ants: individual exocrine glands usually produce mixtures of substances, which are moreover often complex in both constitution and function. A typical example is provided by the subterranean "citronella ant" *Acanthomyops claviger* of the eastern United States, depicted in Figure 7–30. The highly modified poison gland, typical of the Formicinae, appears to produce only formic acid, used in defense. But the multiple terpenoid aldehydes and alcohols of the hypertrophied mandibular glands serve in both defense and alarm. Among the homologous alkanes and ketones of the Dufour's gland, undecane is an alarm pheromone, whereas the remaining components serve mostly or entirely in defense.

The identification of components has out-paced an understanding of their function, and differences among closely related species of ants have compounded the mystery. For example, the sibling species *Tetramorium caespitum* and *T. impurum* of Europe can be identified morphologically only by differences in the male genitalia, but their Dufour's gland secretions are quite distinct. The gland in *T. caespitum* is of moderate size and contains about 70 nanograms of $C_{13}$ to $C_{17}$ linear hydrocarbons together with a mixture of pentadecenes as major components; that of *T. impurum* is smaller, containing 40 nanograms of the same mixture but with a prevalence of *n*-pentadecane and a sesquiterpenoid compound. *T. semilaeve,*

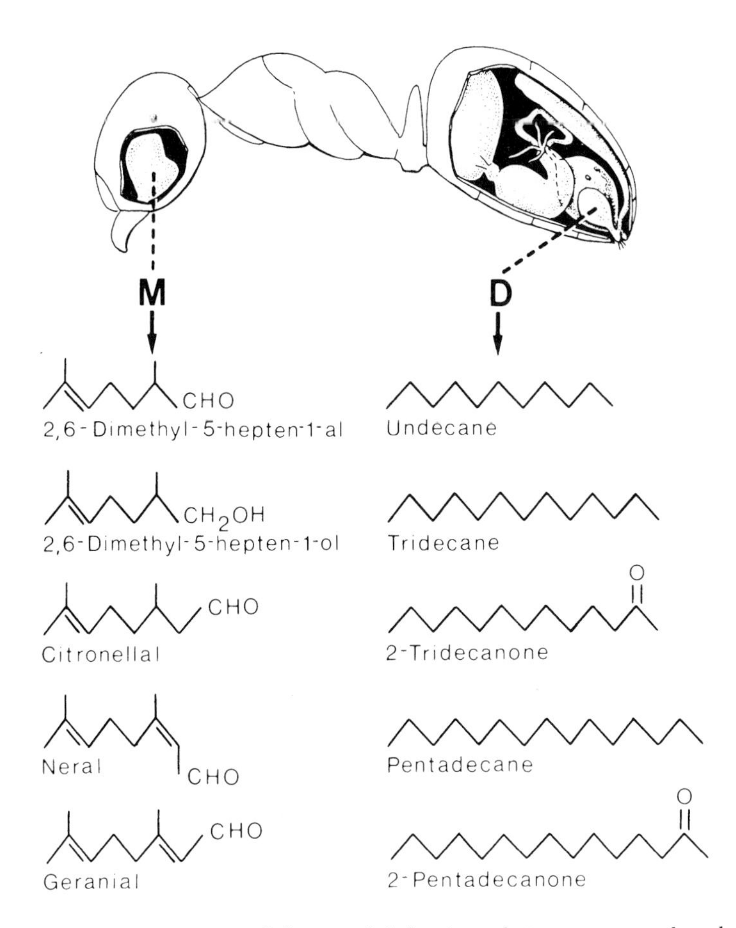


**FIGURE 7–30** Mixtures of alarm and defensive substances are produced by two glands in workers of the subterranean formicine ant *Acanthomyops claviger. M,* mandibular gland; *D,* Dufour's gland. (Modified after Regnier and Wilson, 1968, from Hölldobler, 1978.)

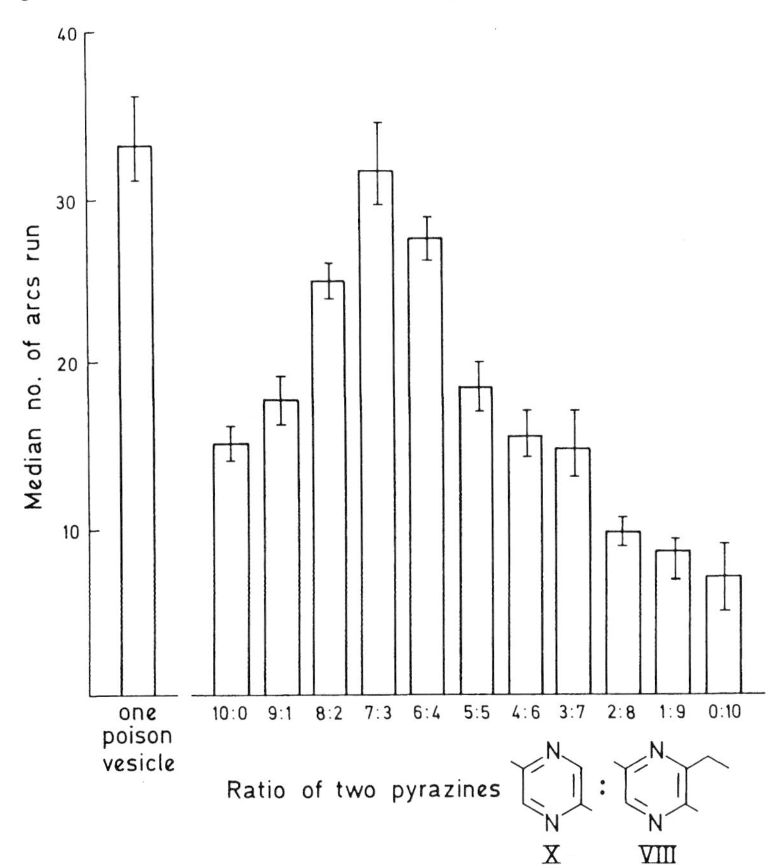


**FIGURE 7–31** Optimum blending of pheromones: the response of *Tetramorium caespitum* workers to artificial trails drawn with different proportions of the two pyrazines that are the principal products of the poison gland. (Modified from Attygalle and Morgan, 1983.)

a species anatomically more distinct from *caespitum* and *impurum,* has a still smaller gland (30 ng capacity) with a simpler mixture of hydrocarbons and the pentadecane present in much higher proportion (Billen et al., 1986). The functional significance of the mixtures and the differences discovered among these *Tetramorium* species remains unknown.

In only a few ant species has the significance of chemical blending been clarified. The functions fall into one of two categories. Either an increase in specificity allows one species to distinguish its own pheromones from those of others, or the production of multiple simultaneous signals allows the transmission of more complex messages.

An example of the first role, the promotion of privacy during communication, is provided by the trail substances of leafcutter ants. All of the species of *Atta* and *Acromyrmex* analyzed so far either produce methyl 4-methylpyrrole-2-carboxylate in their poison glands, or react to this substance, or both. Nevertheless, *Acromyrmex octospinosus* actively avoids trails of *Atta cephalotes,* an effect that appears to be due to components that occur in blends with the pyrrole (Blum et al., 1964; Blum, 1982). This hypothesis has been verified in the case of *Atta sexdens,* which possesses the pyrrole (in addition to some minor components) but utilizes yet another substance, 3-ethyl-2,5-dimethylpyrazine, as its major trail pheromone (Cross et al., 1979).

In a closely parallel manner, the fire ant *Solenopsis invicta* relies primarily on *Z,E*-α-farnesene as its recruitment trail pheromone, with two homofarnesenes acting as synergists. All three substances are emitted from the Dufour's gland. It also produces two isomeric tricyclic homosesquiterpenes in small quantities. The closely related *S. richteri,* in contrast, relies entirely on the tricyclic homosesquiterpenes, producing neither alpha-farnesenes nor homofarnesenes. Yet each species responds weakly to the trails of the other. The reason is that *S. invicta* is sensitive to some extent to its own tricyclic homosesquiterpenes, and it produces enough (over 50 femtograms per gland) to activate *S. richteri* (Vander Meer, 1986a,b).

A more precise delineation of the effects of pheromone blends has been made in *Tetramorium caespitum* by Attygalle and Morgan (1983). This myrmicine lays trails comprising two pyrazines, designated as VIII and X. As shown in Figure 7–31, workers respond maximally to a blend with a ratio of 3:7 of VIII to X. Pheromone blending of this kind has not always arisen in evolution, however, even in the case of pyrazines. In eight species of *Myrmica,* the trail consists only of component VIII, which has been identified as 3-ethyl-2,5-dimethylpyrazine.

Thus research on specificity indicates that related species, for example those similar enough to be placed within the same genus, have diversified repeatedly during evolution by creating variable mixtures of pheromones in the same exocrine gland. Some manufacture and rely wholly on single components, whereas others generate combinations that can be easily shifted during evolution to create optimal mixes peculiar to individual species. One important result is the enhanced privacy of communication within species, a feature that is of clear adaptive significance at least in the case of trail systems.

In a wholly different dimension, precision of communication has been improved by evolving mixtures of pheromones with different effects. When laying odor trails, workers of the imported fire ant discharge a medley of substances from the Dufour's gland, some of which are illustrated in Figure 7–32. The principal component for recruitment is *Z,E*-α-farnesene (I), which nevertheless requires the two homofarnesenes (III, IV) and a still unidentified component in order to attain the full activity observed from complete Dufour's gland extracts. Oddly, these substances remain inactive unless the ants have been primed by yet another, still unidentified component of the gland. Once the ants have encountered this unknown pheromone, they respond fully to artificial trails made entirely from the alpha-farnesene and homofarnesenes (Vander Meer, 1983, 1986a). A similar but less fully investigated Dufour's gland system has been discovered in the household ant *Monomorium pharaonis*. The principal component is faranal, or (6*E*,10*Z*)-3,4,7,11-tetramethyl-6,10-tetradecadienal. Several nitrogen heterocycles serve as supplementary attractants. Moreover, indolizines and pyrrolidines from the poison gland are attractive to the workers and may play a special role of their own, although their presence in odor trails remains to be proven (Ritter et al., 1977a,b).

The partitioning of foraging areas among sympatric species of the harvester ant genus *Pogonomyrmex* illustrates the involvement of both anonymous and specific semiochemicals in inter- and intraspecific territoriality. The relatively short-lived recruitment signal from the poison gland is, so far as we know, invariant among *Pogonomyrmex* species. In addition to these anonymous recruitment trails, persistent trunk routes are established by clearing vegetation and marking with Dufour's gland secretions, which contain species-specific mixtures of hydrocarbons (Regnier et al., 1973; Hölldobler, 1976a, 1986b). The trunk routes also contain colony-specific chemical markers which, together with species-specific cues from the Dufour's gland, serve to channel the foragers of neighboring nests in diverging directions, effectively partitioning limited food resources. These examples illustrate that not all constituents of chemical communication signals need have the same functional significance. In many cases, one or several components act as key stimuli, triggering a basic anonymous response, while additional components add specificity (Hölldobler and Carlin, 1987). Undecane, for example, is apparently the active alarm signal in most ant species of the subfamily Formicinae, and it is also usually the most abundant product in formicine Dufour's glands. Other hydrocarbons are also present, however, and the total mixture is often species specific (Morgan, 1984). In *Oecophylla longinoda*, further specificity is added by droplets originating from the rectal bladder, which are used in colony-specific territorial marking (Hölldobler and Wilson, 1978). Because rectal marking alters the probability of winning territorial conflicts (ants are more aggressive on ground that they have previously marked, and less so on ground marked by another colony), it fulfills the criteria for a *modulator* of the alarm response (Markl, 1985).

To date the rigorous investigation of modulatory communication signals, defined as those that do not themselves release behavioral responses but that influence reactions to other signals, has been limited to cases in which one signal modulates another of a different modality (Markl, 1983, 1985; Hölldobler, 1984c). Different elements of cues in a single modality can also interact in this fashion, however; thus the paradigm also applies to multicomponent semiochemicals in which the additional information of specificity may be seen as modulating the response to anonymous chemical releasers (Hölldobler and Carlin, 1987).

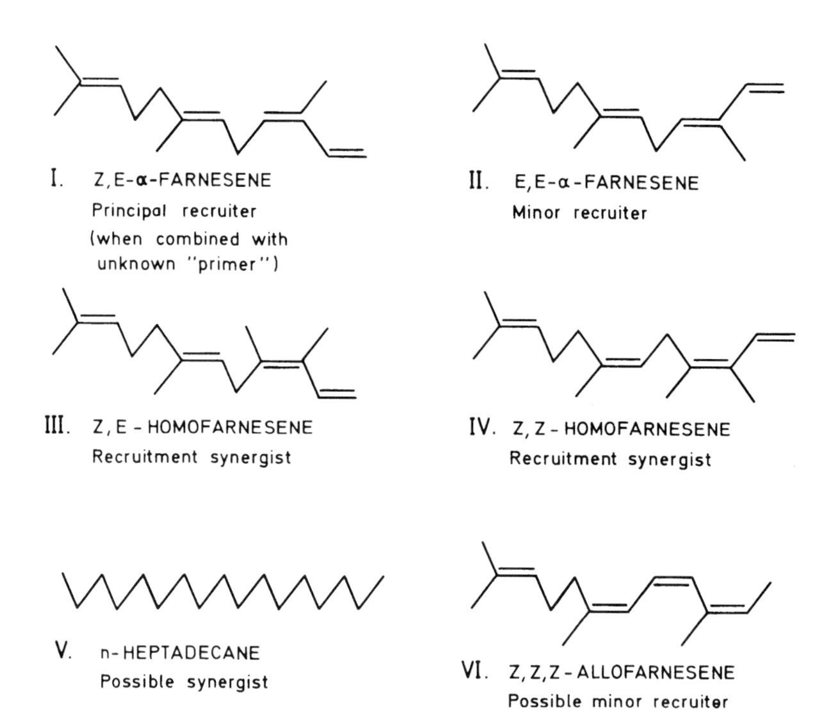


**FIGURE 7–32** Components of the Dufour's gland of the fire ant *Solenopsis invicta*, together with their known functions. (Based on Vander Meer, 1983, 1986.)

If specificity is considered as a form of modulation, and modulatory functions presuppose the existence of the behavior being modulated, a possible evolutionary route to signal specificity can be envisaged. Hölldobler and Carlin (1987) argued that the production of simple semiochemicals releasing elementary anonymous reactions is subject to the inevitable impression of all biosynthetic processes. The resulting degree of variation may well be perceptible to the receivers' sensory system, but will have no effect on the response to the signal. Should an adaptive advantage happen to correlate with any of the available variants, however, selection will favor individuals that respond differentially on the basis of these specific characteristics, in other words, that modulate the original response. For example, other Dufour's gland hydrocarbons will be released along with undecane. If, say, genetically similar colony members tend to produce similar hydrocarbon patterns, the signal may come to be modulated by this added specificity, informing workers whether nestmates or aliens are sending the alarm signal. Once the presence or proportions of additional components significantly affect the response to the basic releaser in an adaptive manner, selection can be expected to improve their distinctiveness and stereotypy. In fact, the exploration of variation in communication among colonies of the same species should prove fruitful in the future. Cherix (1983) found that two adjacent colonies of *Formica lugubris* in the Swiss Jura mountains possessed both qualitative and quantitative differences in their pheromones, including the presence or absence and proportionality of undecane, tridecane, and nonadecanol. Such within-species variation may come about as a result of genetic differences, or the succession of stages in colony growth, or previously unsuspected factors in nest environment.

Of equal importance, multiple pheromones permit the spread of different messages across space, especially when the mixtures are

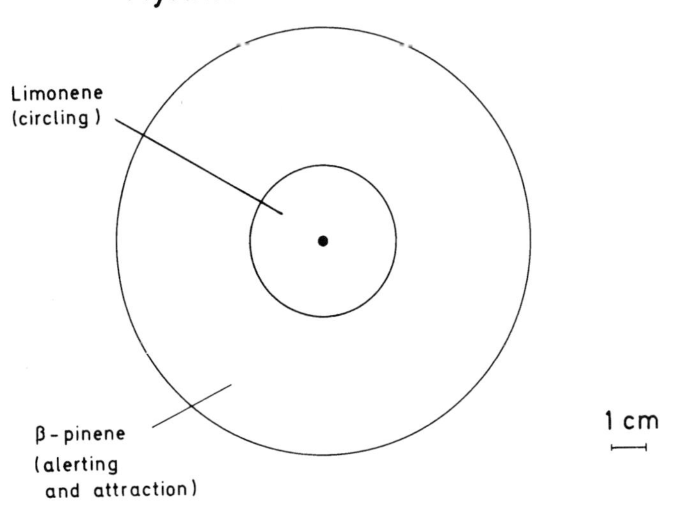


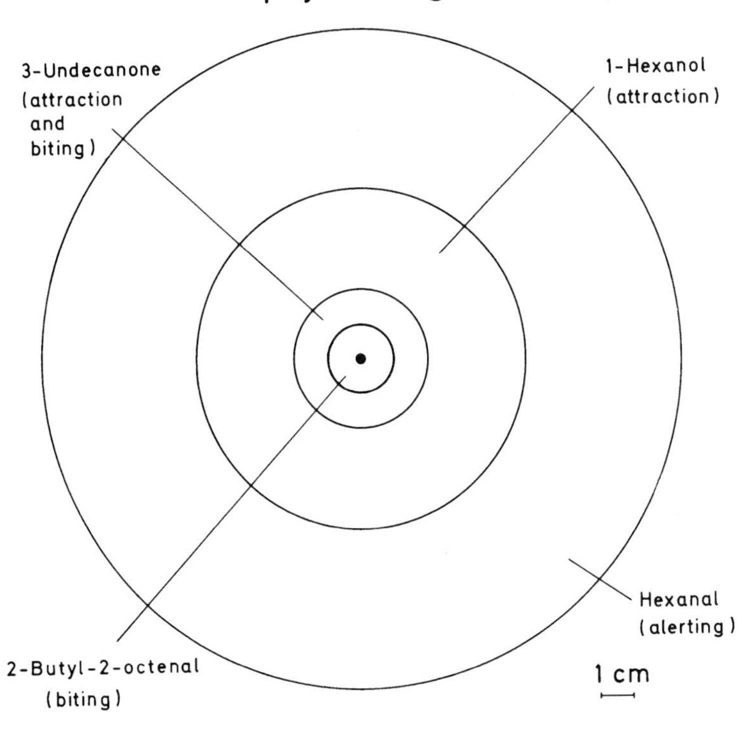


**FIGURE 7–33** Concentric active spaces of multiple pheromones released from the same exocrine gland in two species of ants. The spaces are depicted as circles but are actually overlapping hemispheres that spread above the ground from the point source. (*Above*) Poison gland substances used in recruitment by *Myrmicaria eumenoides*, 40 seconds after deposition on a flat surface at the point in the center. (*Below*) Mandibular gland substances used in alarm recruitment by *Oecophylla longinoda*, 20 seconds after deposition on a flat surface at another point in the center. (Modified from Bradshaw and Howse, 1984. Based in part on Brand et al., 1974; Bradshaw, 1981; Bradshaw et al., 1975, 1979; as well as C. Longhurst as cited by Bradshaw and Howse.)

released from a single point. This paradoxical effect is made possible by the fact that chemical substances produce different active spaces. For example, if pheromone A is produced in larger quantities than pheromone B or is behaviorally more active, it will generate a larger hemispherical space that encompasses the similarly shaped active space of pheromone B. As the receiver ant approaches the point source, she first receives signal A and then signal B, and responds in a sequence of actions.

Two cases of this interesting phenomenon are illustrated in Figure 7–33. Workers of the African myrmicine *Myrmicaria eumenoides* each deposit a single droplet of poison gland secretion near potential prey items. Unlike the venoms of most other myrmicine ants, which are proteinaceous toxins, the gland contents of *M. eumenoides* are monoterpene hydrocarbons, including β-myrcene, β-pinene, and limonene. The β-pinene generates the larger active space, which causes nestmates to move toward the droplet. At closer range the limonene induces circling behavior, which deploys the workers around the prey so that they approach from many directions during the attack itself (Bradshaw and Howse, 1984). In the African weaver ant *Oecophylla longinoda* the multiple components of the mandibular gland secretion trigger a stepwise escalation of responses as the ants approach an enemy. In the outermost space, hexanal alerts the workers. Then 1-hexanol attracts them, and finally 3-undecanone and 2-butyl-2-octenal induces them to attack and bite any alien object in the vicinity (Bradshaw et al., 1975, 1979).

In addition to such multicomponent pheromones, *multisource systems* are commonplace in the ants. In such systems various compounds are released from multiple glandular sources. The substances may serve the same essential functions, as in the case of the alarm pheromones of *Acanthomyops claviger* (see Figure 7–30), but often the roles are different. In the Florida harvester ant *Pogonomyrmex badius*, for example, the recruitment pheromones are voided from the poison glands, whereas the homing pheromones originate at least in part in the Dufour's gland (Hölldobler and Wilson, 1970). Workers of the primitive Australian ant *Myrmecia gulosa* induce territorial alarm behavior *in toto* by pheromones from three sources: an alerting substance from the rectal sac, an activating pheromone from the Dufour's gland, and an attack pheromone from the mandibular glands (Robertson, 1971).

In what may be the ultimate evolutionary development, communication can be part of *multimodal systems*, which transmit signals through more than one sensory modality. The species with the most elaborate organization discovered thus far is *Oecophylla longinoda*, in which four of five recruitment systems incorporate pheromones (from the anal and sternal glands) together with specialized tactile signals. This example will be examined in some detail in the following section on ritualization.

## RITUALIZATION

In the vast majority of cases the origin of communicative systems in animals is based on *ritualization*, the evolutionary process by which a phenotypic trait is altered to serve more efficiently as a signal. Commonly, the process begins when some movement, anatomical feature, or physiological process that is functional in another context acquires a secondary value as a signal. For example, members of a

**TABLE 7–3** Properties of the five recruitment systems of the African weaver ant *Oecophylla longinoda*. (From Hölldobler and Wilson, 1978.)

| System | Chemical signals | Tactile signals | Pattern of movement | Apparent function |
|---|---|---|---|---|
| Recruitment to food | Odor trail from rectal gland; regurgitation of liquid crop contents | Antennation; head waving; mandible opening associated with food offering | Occasional signpost marking with looping trails laid around food source; main trail directly to nest | Recruitment of major workers to immobile food source, especially sugary materials |
| Recruitment to new terrain | Odor trail from rectal gland | Antennation; occasional body jerking | Main trail directly to nest; broad looping movements resembling signpost marking, but only after foragers physically contact terrain; increase in frequency of anal spotting | Recruitment of major workers to new terrain |
| Emigration | Odor trail from rectal gland | Antennation; physical transport of nestmates and tactile invitation of signals leading to transport | Main trail directly to nest site; no signpost marking; predictable sequence of categories of nestmates carried | Emigration of colony to new nest site |
| Short-range recruitment to enemies | Short looping trails from sternal gland and exposure of gland surface with abdomen lifted in air | None | Trails short, looping, and limited to vicinity of contact with enemy | Attraction and arrest of movement of nestmates; inducement of clumping and quicker capture of invaders and prey |
| Long-range recruitment to enemies | Odor trail from rectal gland | Antennation; at higher intensities, body jerking | Main trail directly to nest; no signposts laid | Attraction of major workers to vicinity of invaders and prey; operation in conjunction with short-range recruitment; especially intense during territorial wars with conspecifics |

species may recognize the opening of mandibles or the release of an odor as a threat. Alternatively they can interpret the turning away of an opponent's body in the midst of conflict as an intention to flee. During ritualization such movements (or odors, or visual features) are altered in a way that makes their communicative function still more effective. They acquire support in the form of additional anatomic structures or biochemical changes that enhance the distinctiveness of the signal. The movements also tend to become stereotyped and exaggerated in form. Finally, the receiving apparatus is modified to detect such ritualized signals with less ambiguity. In the case of trail systems among ants, the chemoreceptors have been modified to detect minute traces of the appropriate pheromone, which often occur in nanogram or even femtogram amounts.

The classic example of ritualization in the behavior of social insects is the waggle dance of the honey bee. The dance, first "decoded" by von Frisch in 1945, is easy to understand if one thinks of it as a ritual flight, a scaled-down version of the journey from the nest to the food source. The essential element in the maneuver is the straight run, the middle piece in the figure-eight pattern. (The remainder of the figure eight consists of a doubling back to repeat the straight run.) The dancing bee has just returned from several back-and-forth journeys to the target. The straight run she performs on the vertical surface of the comb is a miniaturized version of the outward flight that she now invites her nestmates to undertake. The angle between the straight run and the vertical line of the comb surface (in other words the line pointing straight up) indicates the direction of the target relative to the position of the sun. The duration of the straight run indicates the distance to the source: the longer the straight run takes to complete, the farther away the target. The straight run is rendered more conspicuous by a rapid waggling motion of the body, a typical case of an enhancing embellishment during ritualization. The dancing bee also produces a distinctive sound.

Relatively few communicative systems in ants have been analyzed explicitly with reference to their evolutionary origin, but suggestive evidence of ritualization has been adduced. One clear use is the invitational movements of *Camponotus* workers recruiting nestmates to new nest sites. Workers of *C. sericeus* jerk their bodies back and forth vigorously in front of other workers, then seize them by the mandibles and pull them forward a short distance (see Figures 7–56 and 7–57). Those of *C. socius* have taken the next step. They employ body jerking alone, evidently having entirely deleted the rudiments of physical transport during the early stages of recruitment. Food offering is also highly ritualized in *C. socius*. After filling her crop with liquid food and returning to the nest, the scout shakes her body from side to side with her mandibles wide open, allowing nestmates to scan the lower mouthparts and odor of the recently ingested food (Hölldobler, 1971c; Hölldobler et al., 1974; see Figure 7–53). The ponerine *Pachycondyla* (= *Bothroponera*) *tesserinoda* of Sri Lanka, representing a separate evolutionary development, uses mandible pulling as a signal to initiate tandem running both to new

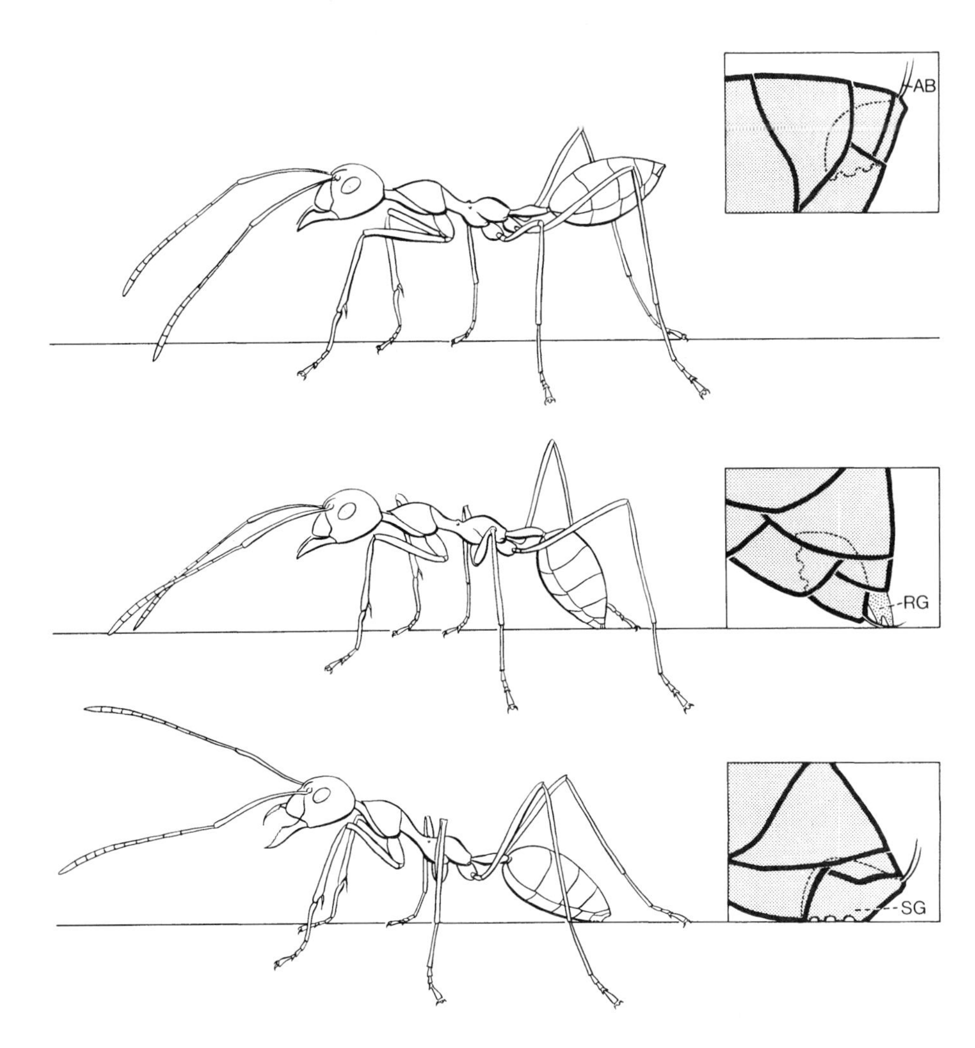


**FIGURE 7–34** These diagrams show how pheromones are voided during the five forms of recruitment in the African weaver ant *Oecophylla longinoda* (see Table 7–3 for descriptions of the systems). (*Top*) Ordinary running posture of a worker. As shown in the inset to the right, the terminal abdominal segment is held so that the sternal gland surface is covered by the penultimate abdominal sternite, and the rectal gland remains retracted within the wall of the rectal vesicle. (*Center*) As a worker lays an odor trail from the extruded rectal gland to food, new terrain, or enemies, the rectal gland (*RG*) rides on the paired bristles (*AB*) of the acidopore, located just beneath the anus (see top figure). While the rectal gland is exposed, the sternal gland surface remains covered. (*Bottom*) A worker deposits sternal gland (*SG*) substance onto the substratum during short-range recruitment to enemies; the terminal abdominal segment has been rotated upward to expose the gland openings, and the rectal gland remains retracted. (From Hölldobler and Wilson, 1978.)

nest sites and to food finds (Maschwitz et al., 1974).

Additional evidence of ritualization can be found in the multiple recruitment systems of the African weaver ant *Oecophylla longinoda*, which are the most complex form of communication thus far discovered in the ants. Workers of this species, who construct arboreal nests in part from larval silk, utilize no fewer than five recruitment systems to draw nestmates from the nests to the remainder of the nest tree and to the foraging areas beyond (Hölldobler and Wilson, 1978; see Plate 6). These include (1) recruitment to new food sources, under the stimulus of odor trails produced by the scout from her rectal gland, together with tactile stimuli presented while the scout engages in mandible gaping, antennation, and head waving; (2) recruitment to new terrain, employing pheromones from the rectal gland and tactile stimulation by antennal play; (3) emigration to new sites under the guidance of rectal gland trails; (4) short-range recruitment to territorial intruders, during which the terminal abdominal sternite is maximally exposed and dragged for short distances over the ground to release an attractant from the sternal gland; and (5) long-range recruitment to intruders, mediated by odor trails from the rectal gland and by antennation and intense body jerking. These systems exist in addition to the elaborate pheromone-mediated alarm communication described by Bradshaw et al. (1975).

The organization of the five recruitment systems is summarized in Table 7–3, and some of the behavior is illustrated in Figures 7–34 and 7–35. The forward jerking movement used during recruitment to enemies closely resembles maneuvering during the actual attack maneuvers themselves, and we have therefore interpreted the signals to be a ritualized version, "liberated" during evolution to serve as a signal when a nestmate is encountered rather than an enemy. When workers recruit nestmates to food, they use a wholly different set of movements. They wave their heads laterally while opening their mandibles. The effect is evidently to waft food odors from the lower mouthparts to the antennae of the potential recipient.

Ritualization is not limited to tactile signaling. Chemical alarm communication evidently evolved from chemical defense behavior. Like many solitary insects, ants and other social insects use chemical secretions to repel predators and other enemies. In social insects, however, defensive reactions are closely linked with alarm communication, and quite often a single substance serves both functions. A well-documented example is citronellal, a mandibular gland product of *Acanthomyops claviger* (see Figure 7–30).

*Acanthomyops* and other formicine ants use hindgut contents as trail pheromones (Hangartner, 1969c), a procedure that might have evolved as a gradual ritualization of the defecation process. The final development, exemplified by the extraordinary rectal gland of the *Oecophylla* weaver ants, is the origin of a wholly new structure to generate the trail substances. In fact, *Oecophylla* workers employ the hindgut in two ways that could have evolved from defecation. The second application is the use of fecal material directly in territorial

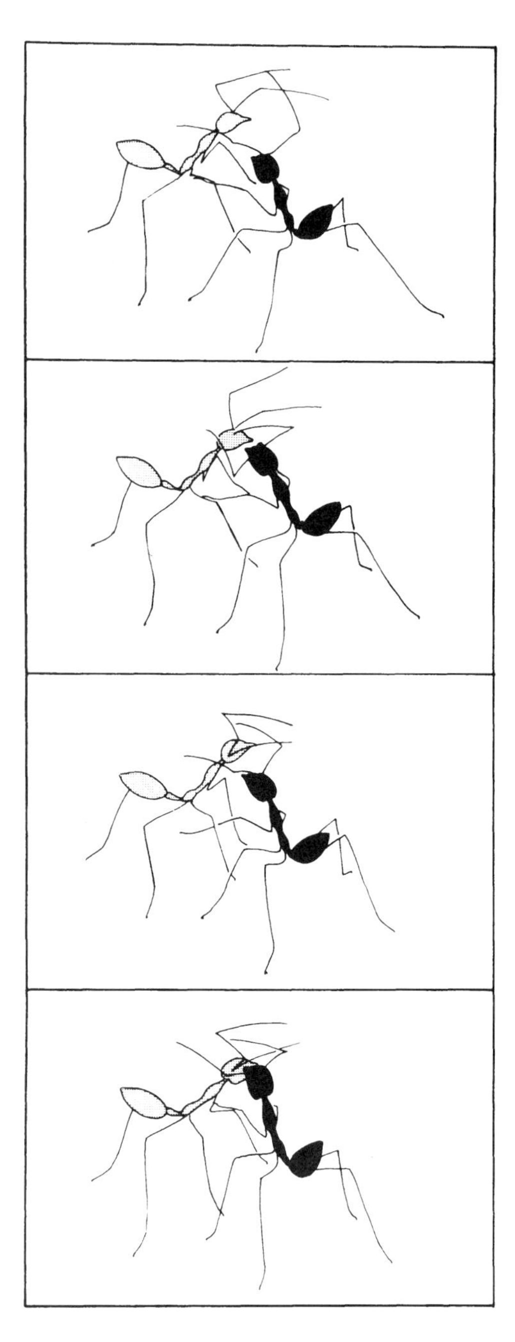

**FIGURE 7–35** Alarm-recruitment in weaver ants alerts nestmates to the presence of enemies. Following an encounter with alien ants, an excited worker (*black*) greets a nestmate with a rapid back-and-forth jerking movement. The movement is thought to have evolved as a ritualized version of attack behavior. This figure was drawn from 4 successive frames in a motion picture taken at 25 frames per second. (From Hölldobler and Wilson, 1978.)

marking. The ants deposit fecal droplets more or less uniformly over the surface of the vegetation around their nests, rather than in refuse piles or other special zones. The droplets contain substances that are specific to their colony, and they permit the ants to determine from moment to moment whether they are in the vicinity of their own nests or on foreign terrain (Hölldobler and Wilson, 1978). The general topic of territorial marking will be taken up again later in this chapter.

## SIGNAL ECONOMY AND "SYNTAX"

For two reasons ants can be intuitively expected to practice economy in the evolution of their communication systems, that is, to use a small number of relatively simple signals derived from a limited number of ancestral structures and movements. First, the small brain and short life span of ant workers limit the amount of information these insects can process and store. Second, the tendency toward signal evolution through ritualization restricts the range of potential evolutionary pathways.

The five recruitment systems of *Oecophylla longinoda* just described illustrate signal economy in a striking manner. Although the messages differ from one another strongly (see Table 7–3), they are built out of pheromones from just two organs, the rectal and sternal glands, together with a modest array of stereotyped movements and tactile stimuli. The specificity of each of the recruitment systems comes principally from the combinations of chemical and tactile elements. There is a primitive sort of syntax in this differentiation. For example, both recruitment to food and recruitment to new terrain are stimulated and guided by pheromones from the rectal gland. But food is further specified by head waving and mandible opening, while fresh terrain is specified (occasionally) by body jerking. There is no proof that the ants themselves perceive the difference, but it seems unlikely that such stereotyped movements would have been incorporated into the behavior if it was wholly lacking in function. The parallel nature of the recruitment systems is accompanied by a lack of any clear distinction on the part of the *Oecophylla* workers between colony defense and predation. When defenders vanquish invading ants, they remove them to the nest interior and convert them into food. The full sequence of responses does not appear to differ significantly from that following the encounter of ordinary prey insects within the territorial boundaries, although the communal response to conspecific invaders is considerably more massive.

A comparable degree of parsimony is widespread among behavioral categories of the most diverse kinds. Workers of *Formica subintegra*, a North American slave-making ant, spray acetates from the Dufour's gland when they invade the nests of other *Formica* species. These substances serve to attract the *F. subintegra* workers, but act to alarm the assaulted workers, who succumb to panic and disperse (Regnier and Wilson, 1971). Workers of the fire ant *Solenopsis invicta* disperse venom as an aerosol by forming a droplet on the tips of their extruded stings and vibrating ("flagging") the abdomen vertically. They perform this unusual act in two radically different circumstances, using different quantities of venom. When repelling other species of ants from the foraging arena, each worker dispenses up to 500 nanograms of the substance. Inside the nest, brood attendants dispense about 1 nanogram over the surface of the brood, possibly as an antibiotic. Defensive flagging of the first kind is distinguished either by a vertical orientation of the entire body with the head pointing downward—the "head stand"—or by a horizontal orientation with the sting pointed at the intruder. Brood flagging is accomplished with an essentially horizontal orientation (Obin and Vander Meer, 1985). In yet another category of behavior, some species of *Leptothorax* recruit nestmates initially by "tandem calling," in which the worker slants her abdomen upward and discharges poison gland secretions from the extruded sting. Nestmates

**FIGURE 7–36** Recruitment in the European myrmicine ant *Leptothorax acervorum* illustrates the principle of economy in the evolution of recruitment systems. (*Above*) After discovering a new food source, a worker assumes the calling posture, extrudes her sting, and voids pheromone-laden material from the poison gland. (*Below*) A nestmate arrives and touches the abdomen of the caller. Soon the caller will move forward, causing the nestmate to follow close behind in the typical tandem running behavior. An apparently identical behavior is employed during sexual calling by reproductive workers of *Harpagoxenus*, a genus phylogenetically derived from *Leptothorax*. (From Möglich et al., 1974.)

are attracted, and as soon as one of them touches the calling ant, tandem running begins and the recruiter leads the nestmate to the target area (Möglich et al., 1974; Figure 7–36). An apparently identical calling behavior is employed by the reproductive workers of the slavemaker *Harpagoxenus*, a phylogenetic derivative of *Leptothorax*, in order to attract males (Buschinger, 1968a,b; Buschinger and Alloway, 1979; phylogeny discussed by Buschinger, 1981). Finally, as we will show in describing trophallaxis later in this chapter, ponerine ants employ closely similar antennal signals in social greeting, recruitment invitation, and food solicitation.

## MODULATORY COMMUNICATION

Communication in complex social systems is seldom characterized by a direct, all-or-nothing response. Signals do not always merely "release" behavioral responses of a particular kind, but instead often appear to adjust the behavior of nestmates toward one another in a manner appropriate to the surrounding environment. In such instances the communication may have relatively low informational content when measured simply by the number of bits transferred. According to this interpretation, advanced by Markl and Hölldobler (1978) for social insects and subsequently elaborated in later studies by Markl (1983, 1985) and Hölldobler (1984c), outwardly inefficient communication systems serve different but no less important purposes than more direct, deterministic systems. They influence the behavior of receivers, not by forcing them into narrowly defined behavioral channels but by slightly shifting the probability of the performance of other behavioral acts. Put another way, the signals of such "modulatory communication" do not release specific fixed-action patterns in the familiar manner envisaged by ethologists, but instead alter the probability of reactions to other stimuli by influencing the motivational state of the receiver. Modulatory communication can be expected to be more frequent in the most complex animal societies, where many members perform many different tasks at the same time. In this arrangement, a flexible program is required if the work force is to distribute its energy investment among the different tasks in an effective manner.

The best-analyzed form of modulatory communication in ants is stridulation in the enhancement of short-range recruitment by workers of *Aphaenogaster* (= *Novomessor*) *albisetosus* and *cockerelli* (Hölldobler et al., 1978; Markl and Hölldobler, 1978). These gracefully slender ants of the American desert are adept at retrieving large prey objects, such as dead insects, within short periods of time. After discovering an object too large to be carried or dragged by a single ant, a scout worker releases poison gland secretion into the air. Nestmates as far away as 2 meters are attracted and move toward the source. When a sufficient number of foragers has assembled around the prey, they gang-carry it swiftly to the nest (Figure 7–37). Time is of the essence, because the *Aphaenogaster* must remove food from the scene before formidable but less agile ant competitors, including fire ants and *Iridomyrmex pruinosum*, arrive in large numbers on the scene. *Aphaenogaster* workers, in addition to releasing the poison gland pheromone, also regularly stridulate by rubbing the sharp posterior edge of the postpetiolar tergite (scraper) against a file of horizontally arrayed ridges on the anterior end of the first gastric tergite. The chirping sounds generated last up to 200 milliseconds each with frequencies between 0.1 and 10 kilohertz. Once the foragers encounter the vibration, they remain in the vicinity for up to twice as long as when no stridulation occurs. Ants perceiving the signals also start to encircle the prey sooner, and they are likely to release the attractive poison gland pheromone earlier. Overall, both the recruitment of workers and the retrieval of the food object are advanced by 1–2 minutes as a consequence of stridulation. The vibration thus serves as a *communication amplifier* in this particular circumstance, conferring a considerable advantage on the *Aphaenogaster*, who must race to acquire food in the highly competitive desert environment.

A similar enhancement of recruitment by stridulation has been noted in the European harvesting ant *Messor rufitarsis*, which uses the sound in conjunction with odor trails from her Dufour's gland (Hahn and Maschwitz, 1985). Other studies suggest the phenomenon may be widespread in *Messor* (Schilliger and Baroni Urbani, 1985; Buser et al., 1987). In the tropical Asiatic predatory ant *Leptogenys chinensis* stridulation is combined with pygidial gland secretions during colony emigration (Maschwitz and Schönegge, 1983).

A second category of modulatory communication is drumming in

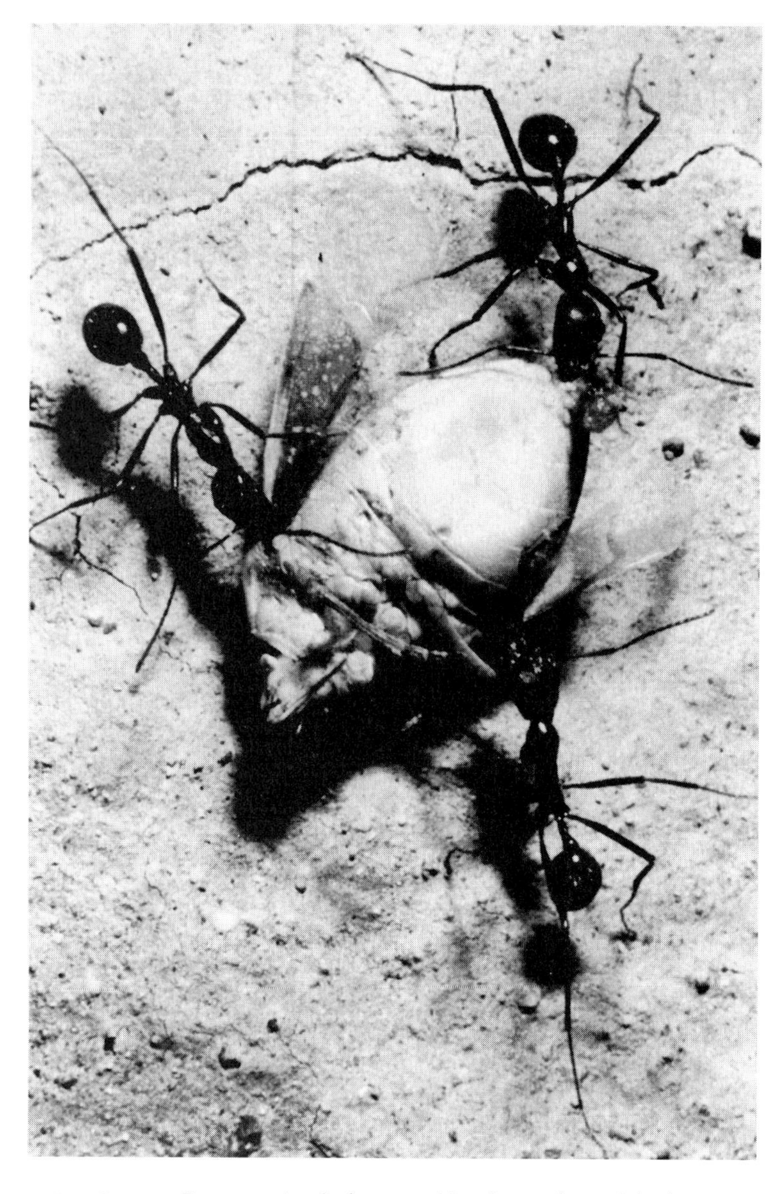

**FIGURE 7–37** Group retrieval of a prey object by workers of the large desert ant *Aphaenogaster* (= *Novomessor*) *cockerelli.* Three workers are carrying a coreid bug with swift, cooperative movements. (From Hölldobler et al., 1978.)

the European carpenter ants *Camponotus herculeanus* and *C. ligniperda* (Fuchs, 1976a,b). Workers strike the surface of the wooden chambers and galleries in which they live with their mandibles and gasters, producing vibrations that can be perceived by nestmates for 20 centimeters or more. Much of the behavior is classifiable as direct alarm communication, which we will describe shortly. But Fuchs showed that the drumming alters other behavioral responses as well, in a manner that becomes clear only with the aid of statistical analysis of transition probabilities between the different responses. In short, the behavior of some categories is "tightened up." Transition probabilities are raised, and hence uncertainty reduced, when the ant is in particular initial states when the signal is received. The particular initial states include antennal waving (probably a condition of monitoring the environment), grooming, and running. No effect was observed by Fuchs when the *Camponotus* workers were either inactive or feeding. The meaning of the modulatory changes remains to be elucidated.

Modulatory communication appears to be a primitive phenomenon in ants and other social insects. Other communicative motor patterns in ants, including the abbreviated fast runs and body jerking employed during recruitment, may have evolved from early forms that merely modulated the behavior of nestmates in other categories. Some of them may then have been ritualized into specialized signals with direct effects of their own, usually working in combination with other signals—such as trail or alarm pheromones.

In the case of chemical signaling, modulatory communication resembles the primer effects of pheromones. One of the best-documented examples of primer effects in ants and other social insects is caste inhibition, in which the detection of substances secreted by one caste (such as a queen or soldier) induces a larva or nymph to mature into a different caste. Primer pheromones also inhibit wing shedding in queens and ovarian development in both queens and workers. In both of these categories, the pheromones have a profound influence on behavior and communication in the receptor individuals. The time scale is far greater than that of ordinary modulatory communication, however. It spans a large part of the life cycle rather than just seconds or minutes.

## SYNERGISM

We have shown how some signals can modulate the response of ants to others in unexpected ways. Many cases have also been discovered of multiple signals, sometimes transmitted through two or more sensory modalities, that operate in concert to evoke complex responses. A third, related phenomenon is true synergism, in which two signals have the same or a closely similar effect but a combination of the two causes a stronger reaction than comparable magnitudes of either one presented singly. An apparent example of this phenomenon has been discovered by Cammaerts et al. (1982) in the European species of the ant genus *Myrmica.* The mandibular glands of the workers contain mixtures of low-molecular-weight alcohols and ketones that serve as alarm pheromones, inducing nestmates to increase the linear speed and decrease the sinuosity (angular velocity) of their running. In *M. schencki* a mixture of 3–octanol and 3-octanone evoked a stronger response when presented in natural concentrations than did either component alone. The difference is relatively slight, however, and its meaning not yet clear. Species of the arboreal dolichoderine genus *Azteca* produce mixtures of three cyclopentyl ketones in their pygidial glands. Each component causes some alarm response, but a mix of all three is far more effective than any one pheromone alone. Some species also add 2-heptanone, another powerful pheromone. Other *Azteca* use the heptanone exclusively, and a few rely entirely on the cyclopentyl ketones (Wheeler et al., 1975). In general, few efforts have been made to test for synergism in the communicative systems of ants. The effort might prove profitable, once the complete compositions of more pheromone complexes are known.

## CASTE AND COLONY DIFFERENCES IN COMMUNICATION

A large part of the ant social organization is based on differences among the various castes and subcastes in patterns of communication. The Brazilian myrmicine *Pheidole embolopyx* is typical in this regard. The minor and major workers, illustrated in Figure 7–38, display different responses during recruitment and defense of the nest and food sources. The caste-specific behaviors are nevertheless coordinated to create efficient reactions at the level of the colony. Only the minor workers lay odor trails, which originate from the poison gland and are used to direct the remainder of the colony to new nest sites and food finds. Both castes cooperate in defending food objects too large to transport to the colony, by forming clusters that encircle these bonanzas for periods of hours or even days. The minor workers protect the food from intruder ants by seizing their legs and pinning them down, while the majors attack the intruders' bodies directly. When the nest is disturbed, both castes communicate alarm by means of abdominal pheromones, which in the case of the major worker have been traced to the pygidial gland. The queen behaves in still another, radically different, manner. She can be recruited with odor trails only when the colony is emigrating to new nest sites. While the colony is in a state of alarm, she relies for personal defense on her uniquely turtle-like body form, which includes a truncated abdomen and paired flange-like protrusions of the pronotum and first gastric segments. She crouches tightly into small cavities in a way that exposes a minimum of vulnerable body surface. The defense is enhanced by gelatinous sheaths secreted from enlarged cephalic glands onto the scapes and anterior portions of the head (Wilson and Hölldobler, 1985).

The differentiation in communicative behavior among ant castes is often underlain by strong biochemical differences. The soldier of *Pheidole fallax* produces skatole in a hypertrophied poison gland that fills about one-third of the entire abdominal cavity. This substance is used in defense and possibly also in alarm communication. As in *P. embolopyx,* the majors follow odor trails but do not lay them. The trails are voided instead by the minors from their poison glands. Conversely, the minors do not produce skatole (Law et al., 1965). In a third *Pheidole* species, *P. biconstricta,* the minor workers produce large quantities of odorous secretion from greatly enlarged pygidial glands. This material, which appears peculiar to the caste, is used in both alarm and defense (Kugler, 1979). Queen and male ants are well known to synthesize an array of distinctive pheromones used during mating and, in the case of the queens, for the attraction of workers (see Chapters 3 and 6).

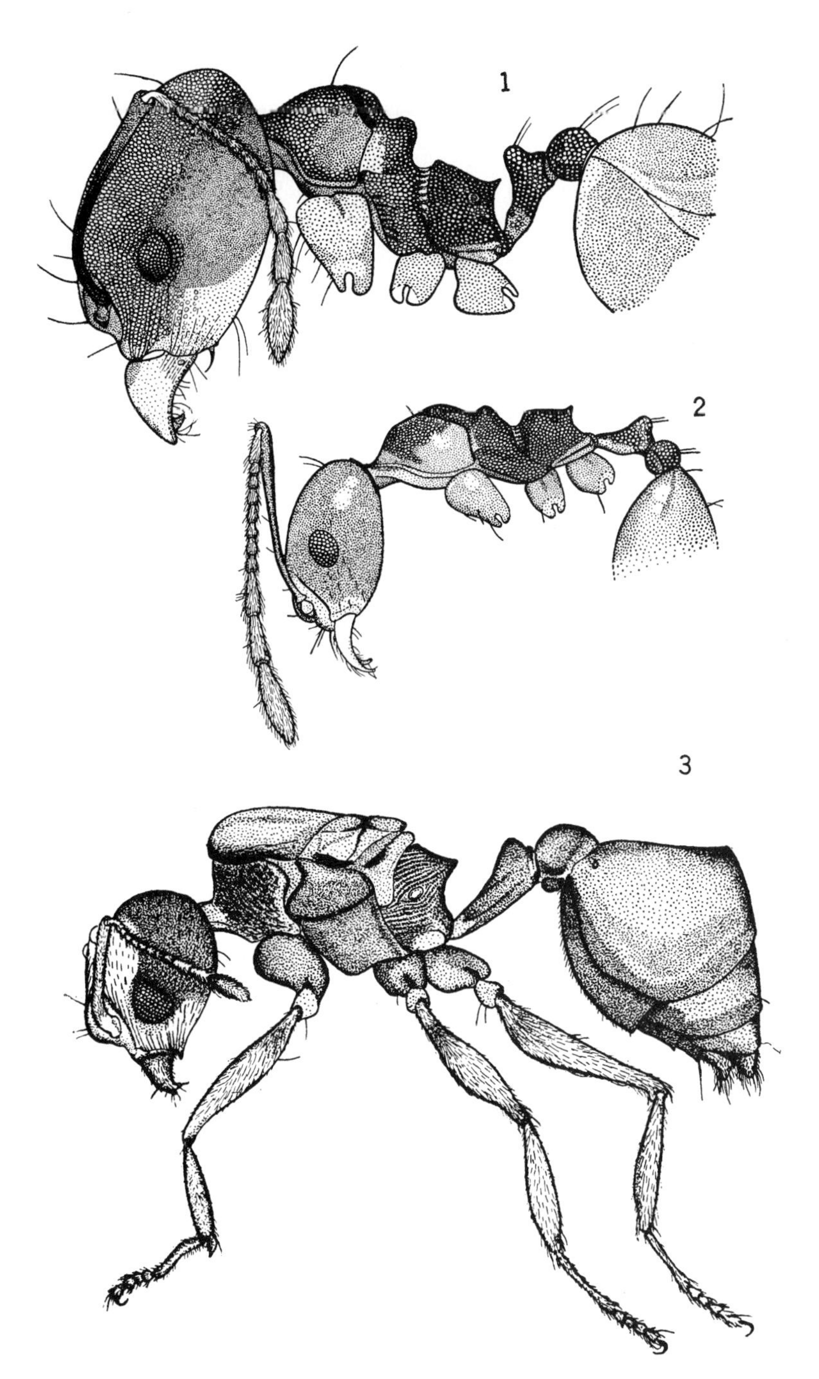


**FIGURE 7–38** The three female castes of the myrmicine ant *Pheidole embolopyx,* which display radically different responses in recruitment and defensive communication: (*1*) major worker ("soldier"); (*2*) minor worker; (3) queen. (From Brown, 1967.)

## ACOUSTICAL COMMUNICATION

The use of vibrational signals is weakly developed in ants in comparison with communication by pheromones. It often occurs in conjunction with chemical signals. Most but probably not all acoustical signals are transmitted primarily through the soil, nest wall, or some other solid substratum rather than through the air. Two forms of sound production have been identified, body rapping against the substratum and stridulation, the latter employing files and scrapers clearly evolved for a communicative purpose. According to species, one or the other of four functions is served by sound signals in ants: alarm, recruitment, termination of mating by females, and, as noted previously, modulation of other communication and forms of behavior.

The production of sound signals by body rapping or drumming in social insects occurs most commonly in colonies that occupy wooden or carton nests, where substrate vibrations are transmitted with high efficiency in comparison with otherwise similar nest structures in soil. It is widespread among the arboreal species of the dolichoderine genera *Dolichoderus* and *Hypoclinea* and the formicine genera *Camponotus* and *Polyrhachis* (Markl, 1973). Workers of the carpenter ants *Camponotus herculeanus* and *C. ligniperda,* the best-studied species to date (Markl and Fuchs, 1972; Fuchs, 1976a,b), can be launched into drumming by any moderate disturbance to their nests, including air currents (a sign that the nest has been breached), sound, touch, or chemical contaminants. The drumming ant strikes the substrate with mandibles and gaster while rocking her entire body violently back and forth. Up to seven such strikes are delivered at 50-millisecond intervals. The signaling pattern is independent of the triggering stimulus. That is, the ants do not modify the drumming to identify the category of danger to the nest. The signals have a maximum energy content at 4–5 kilohertz and attenuate at the rate of 2 decibels per centimeter. They carry through the thin wooden shells of the nest for several decimeters or more.

As we noted earlier, drumming is a modulatory signal, altering the transition probabilities of other kinds of behavior. The vibration is also an effective alarm signal. The response of workers in or close to the nest depends on their initial level of activity, as measured by their running speed. Those less active "freeze" into immobility, while those more active do the exact opposite—they increase their running speed, move toward the source of the vibrations, and attack any moving object in the vicinity. The acoustical signals are also modulatory in nature, in that they alter the transition probability from antennal waving, grooming, and running to other behavioral states. It is not yet known whether this latter alternation also serves alarm and defense or is merely "noise" in the social system.

Stridulation, a more sophisticated sound-producing behavior than drumming, is found in the ant subfamilies Ponerinae, Nothomyrmeciinae, Pseudomyrmecinae, and Myrmicinae. Entomologists define it generally as the rubbing of specialized body parts together to produce a "chirp." The sound has been characterized by Broughton (1963) as "the shortest unitary rhythm-element that can be readily distinguished as such by the unaided human ear." In almost all ant species studied to date the chirps are specifically produced by raising and lowering the gaster (the most posterior discrete section of the abdomen) in such a way as to cause a dense row of fine ridges (the "file") located on the middle portion of the first segment of the gaster, that is, the fourth abdominal segment, to rub against a scraper ("plectrum") situated near the border of the preceding third abdominal segment (Haskins and Enzmann, 1938; Forrest, 1963; Markl, 1968; Schilliger and Baroni Urbani, 1985). The basic details are illustrated in Figures 7–39 and 7–40. Two exceptions to the position of the stridulatory organ are known. The first is provided by the Australian ant *Nothomyrmecia macrops* (Taylor, 1978c). The workers of this very primitive species bear a scraper on the third abdominal segment and a file on the fourth abdominal segment. This location is the same as in other ants, except that in *Nothomyrmecia* the scraper and file are on the ventral segment of the abdomen instead of the dorsal. The other exception occurs in the ponerine genus *Rhytidoponera,* which possesses a dorsal and ventral file on the fourth abdominal segment (H. Markl, personal communication). The arrangement is unique within the Hymenoptera as a whole, where the file is otherwise invariably on the fourth or fifth tergite (Brothers, 1975).

Spangler (1967) gave a characterization of stridulation in the harvester ant *Pogonomyrmex occidentalis.* As the gaster is pulled down, a relatively weak sound is produced that lasts about 100 milliseconds and generates its principal energy at 1–4 kilohertz. The gaster is then jerked back up, producing a second and similar sound, also

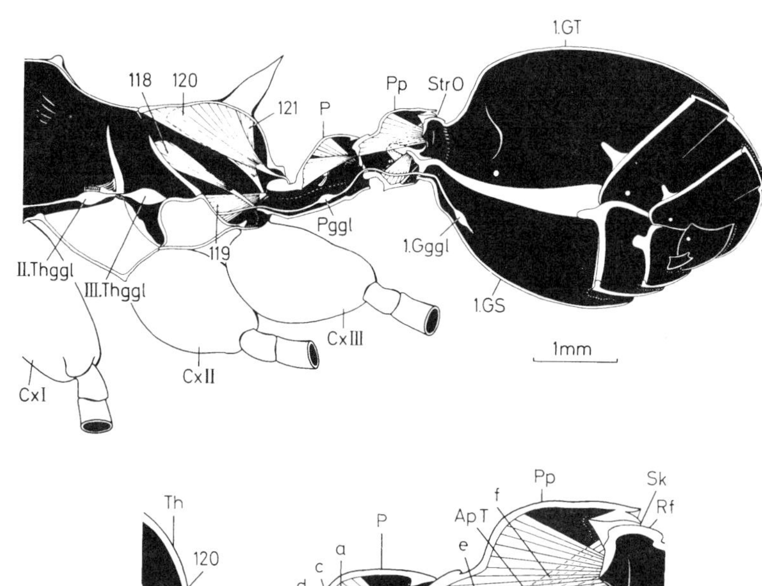


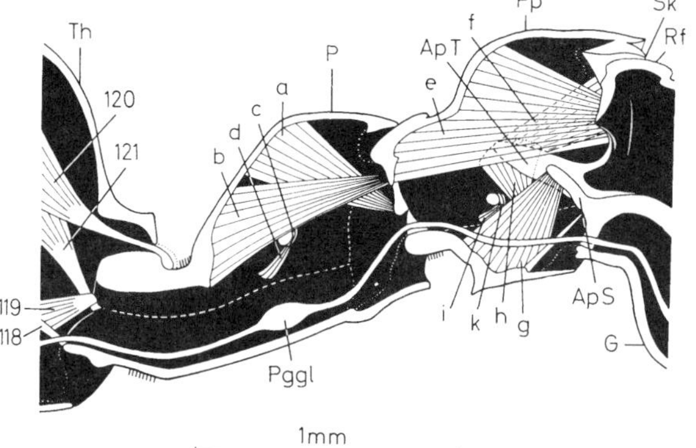


**FIGURE 7–39** Sagittal section of the rear portion of the body of a worker of *Atta cephalotes,* showing the location of the stridulatory organ. (*Above*) The organ (*StrO*) consists of a sharp scraper on the posterior border of the postpetiole (*Pp*), which rubs against a file of transverse ridges located on the upper surface of the anteriormost part of the first gastric segment. The rubbing occurs when the gaster is jerked up and down. (*Below*) The petiole (*P*) , postpetiole (*Pp*), and anterior face of the gaster (G) are shown in magnified view. The scraper is labeled *Sk* and the file surface *Rf* (for the German *Rippenfeld*). (From Markl, 1968.)

**FIGURE 7–40** The file surface, a part of the stridulatory organ, of a worker of the leafcutter *Acromyrmex octospinosa.* (From Markl, 1968.)

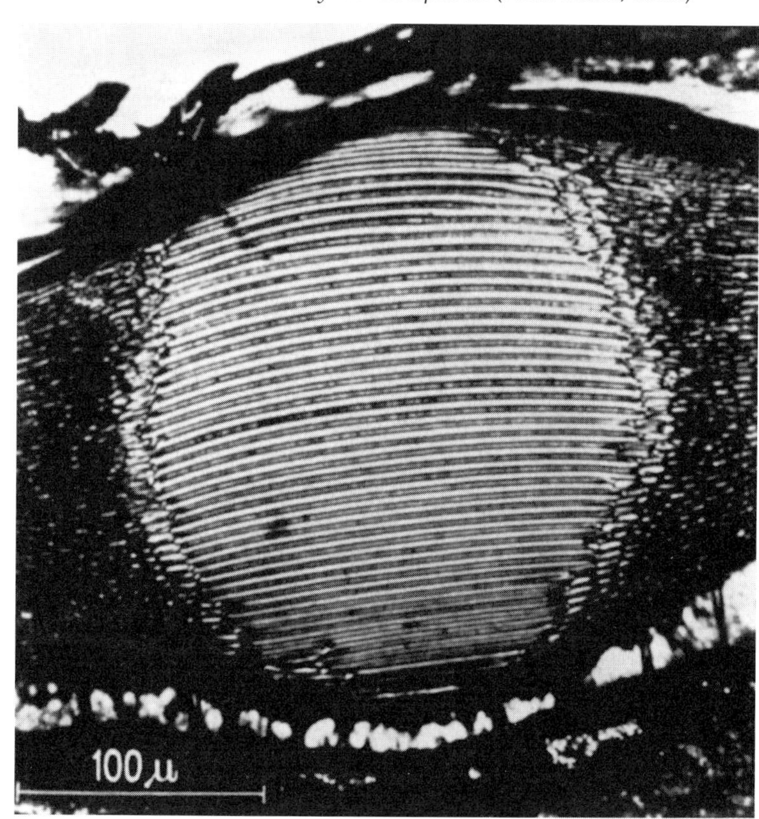

lasting about 100 milliseconds but overall louder and containing in addition some higher frequencies (7–9 kHz). All of the instrumentally detected frequencies are within the range of human hearing, with the result that the stridulating *Pogonomyrmex* worker, when held close to the ear, seems to squeak faintly and almost continuously. In his independent studies of *Atta cephalotes,* Markl (1965–1973) found that the chirp produced differs in quality from that of *Pogonomyrmex.* It is much louder, as much as 75 decibels at 0.5 centimeter from a major worker of *A. cephalotes,* whereas the intensity is less than 2 decibels in *P. occidentalis.* It is also higher pitched. In a later study Masters et al. (1983), employing laser-Doppler vibrometry, determined that the gaster, on which the file is located, appears to be the principal sound-radiating part of the ant. In their analysis of *A. sexdens,* these investigators divided the energy of the radiated airborne sound into three frequency regions, the first two corresponding to low frequency (LF) and high frequency (HF) peaks of the vibration spectrum, and the third extending above about 20 or 30 kilohertz. They note that

> in the lower two regions energy in the sound spectrum can be interpreted rather directly in terms of the measured vibration. Energy in the range 0.5 to 1.5 Hz can be attributed to the LF oscillation of the tooth impact rate. This low-frequency component has been found in other stridulatory insects and may be an important carrier of information. . . . In the range from about 5 to 20 or 30 kHz the energy comes mainly from HF oscillation. Generally the vibration signal peak at about 10 kHz appears at a slightly higher frequency in the airborne sound pressure spectrum. The shift to higher frequencies is attributable to the increasing radiation efficiency of the gaster as frequency rises. Radiation efficiency reaches a plateau value at a wavelength of $2\pi a$, where $a$ is the gaster radius, representing a frequency of about 40 kHz for a typical ant, and above this frequency radiation is about equally efficient at all frequencies, although of course the amplitude of vibration will be small at such high frequencies.

It has long been known, thanks to the experiments of Fielde and Parker (1904) and Haskins and Enzmann (1938), that ants are nearly deaf to airborne vibrations but extremely sensitive to vibrations carried through the substratum—sensitive enough, in fact, to detect stridulation sounds even through well-packed soil. Markl found that workers of leafcutter ants are attracted to nestmates pinned to the ground as far away as 8 centimeters, when sound was the only signal that could be transmitted. Workers buried under the soil to a depth of 3 centimeters induced digging with their stridulation, while those buried more deeply, to a depth of 5 centimeters, only attracted workers to the vicinity.

Markl and his co-workers have discovered three functions of stridulation in various species and castes:

1. In *Atta* at least, the stridulation serves as an underground alarm system, employed most frequently when a part of the colony is buried by a cave-in of the nest. The activity varies greatly among the castes. The large majors contribute very little, while nearly half of the medias most active in nest excavation also participate in rescue digging in response to stridulation (Markl, 1985). A similar function is implied for *Pogonomyrmex badius,* the workers of which readily stridulate when they are pinned by forceps, pressed beneath some object, or restricted to a small space in a container (Wilson, 1971). In *Pogonomyrmex* the signal works in conjunction with alarm pheromones of the mandibular gland. Higher concentrations of the latter release digging behavior (Wilson, 1958a).

2. Stridulation is used during mating by young queens of *Pogonomyrmex* (Markl et al., 1977). These ants gather in nuptial swarms on the ground or vegetation, where substratal sound transmission is possible. The young queens stridulate vigorously when the spermatheca has been filled, and they struggle to escape from the swarms of males chasing them. This "female liberation signal" benefits both sexes by saving the males lost mating chances, while allowing the females to escape and commence the highly competitive business of colony founding—where time is of the essence. The signal is not used during courtship itself; in other words, it does not aid the males in either finding or selecting their mates.

3. In some species of *Aphaenogaster* (= *Novomessor*), *Leptogenys,* and *Messor,* stridulation enhances the effectiveness of pheromones during recruitment of nestmates to food finds, new nest sites, or both (Markl and Hölldobler, 1978; Maschwitz and Schönegge, 1983; Hahn and Maschwitz, 1985).

No evidence exists to rank the chirps of stridulation as anything more than simple unitary signals. In other words, ants do not "talk" by modulating sound through time. Although the sounds differ greatly from one species to the next, they do not appear to vary much within species or within the repertory of one worker ant through time (Forrest, 1963). Spangler (1967) showed that the two phases of the *Pogonomyrmex occidentalis* chirp, caused by the up and down swings of the gaster, are interrupted at regular intervals, but that the pulses thus created do not form any discernible temporal pattern. It is not even certain that such patterns could be preserved during transmission through the ground. A similar result was obtained in Markl's analysis of sound production in *Atta cephalotes.* So far as we know, then, stridulation in ants produces a monotonous series of chirps with limited meaning.

In a survey of 1,354 ant species belonging to 205 ant genera, Markl (1973) demonstrated that the taxonomic distribution of the stridulatory organ is the result of the interplay of phylogenetic inertia at the level of the subfamily and environmental adaptation at the level of the genus and species. On the one hand, the organ appears to be limited to 4 subfamilies, the Ponerinae, Nothomyrmeciinae, Pseudomyrmecinae, and Myrmicinae. The extinct, ancestral subfamily Sphecomyrminae bears no trace of the organ, so that stridulation may well be a derived trait within the ants as a whole. Within the Ponerinae, Nothomyrmeciinae, and Myrmicinae, the stridulatory organ tends to be present in genera and species that nest in the soil, and hence are both subject to cave-ins and able to perceive substratum vibrations. It tends to be absent in ants nesting in plants and rotten logs, leaf litter, and other soft ground materials that are poor sound transmitters. The Pseudomyrmecinae, which are almost all completely arboreal, are a puzzling exception to this rule. All 36 species examined by Markl possess a stridulatory organ. A similar paradox is presented by members of the myrmicine genus *Crematogaster,* all of which appear to possess the organ but many of which are primarily or exclusively arboreal. *Procryptocerus,* the most primitive genus of the exclusively arboricolous tribe Cephalotini, has a stridulatory organ, but the structure is lacking in the remainder of the cephalotines. It would thus appear, in accordance with Markl's principle, that the organ was lost secondarily as a consequence of arboreal life at an early period in the history of the Cephalotini.

## TACTILE COMMUNICATION

Several independent investigators, including especially Bonavita-Cougourdan, Hölldobler, Jaisson, Lenoir, and Torossian, have concluded that tactile signals serve several communicative functions but also, despite their outward complexity, convey only a limited amount of information. This relatively modest assessment of the role of touch is a considerable departure from that of Wasmann (1899a), who interpreted the play of antennae on the bodies of sister ants as a complex "antennal language" playing a basic role in colonial organization. We now understand that a great deal of such antennation serves to receive information rather than to send it. Ants antennate the bodies of nestmates in order to smell them, not to inform them.

Nevertheless, a role for touch by the antennae and forelegs has been well established in a few categories of communication (Lenoir and Jaisson, 1982). They have been firmly implicated in invitation behaviors that entrain most forms of recruitment. Typically, one ant runs up to a nestmate and beats the nestmate's body very lightly and rapidly with her antennae, often raising one or both forelegs to touch the nestmate with these appendages as well. The recruiter then turns and follows a recently laid odor trail, or lays a new one, or commences tandem running. In the case of tandem running, experiments have shown that the leader ant requires the touch of the follower ant before proceeding back toward the food find or new nest site (Hölldobler, 1971c, 1984b; Hölldobler et al., 1974; Maschwitz et al. 1974).

Because of the superficial complexity of tactile behavior, the observer is always at risk of overinterpreting it. When one ant rushes up to another and partially mounts her, the function could be to bring chemical signals as close to the other ant as possible. And when an ant antennates her nestmate, the function could easily be chemoreception instead of signal transmission. Even so, the ritualized nature of many of the movements, including body jerking and intense antennal beating, suggests a true signal function even when direct experimental confirmation is lacking.

The best-documented mode of tactile communication is in stomodeal (i.e., oral) trophallaxis, the exchange of liquid food from the crop of one ant to the alimentary tract of another. As Hölldobler found (1967, 1968b, 1970a,c), the ability possessed by some myrmecophilous staphylinid beetles and other social parasites to induce ant workers to regurgitate to them, despite their outlandishly different anatomies, suggests that there must be some simple trick involved in the soliciting procedure. He was able experimentally to induce *Myrmica* and *Formica* workers to regurgitate by touching parts of their bodies with the tip of a human hair and to augment the result with close observations of natural regurgitation. The whole procedure can be summarized as follows. The most susceptible worker ant is one that has just finished a meal and is searching for nestmates with whom to share her crop content. In order to gain her attention, a nestmate (or myrmecophile) has only to tap the do-

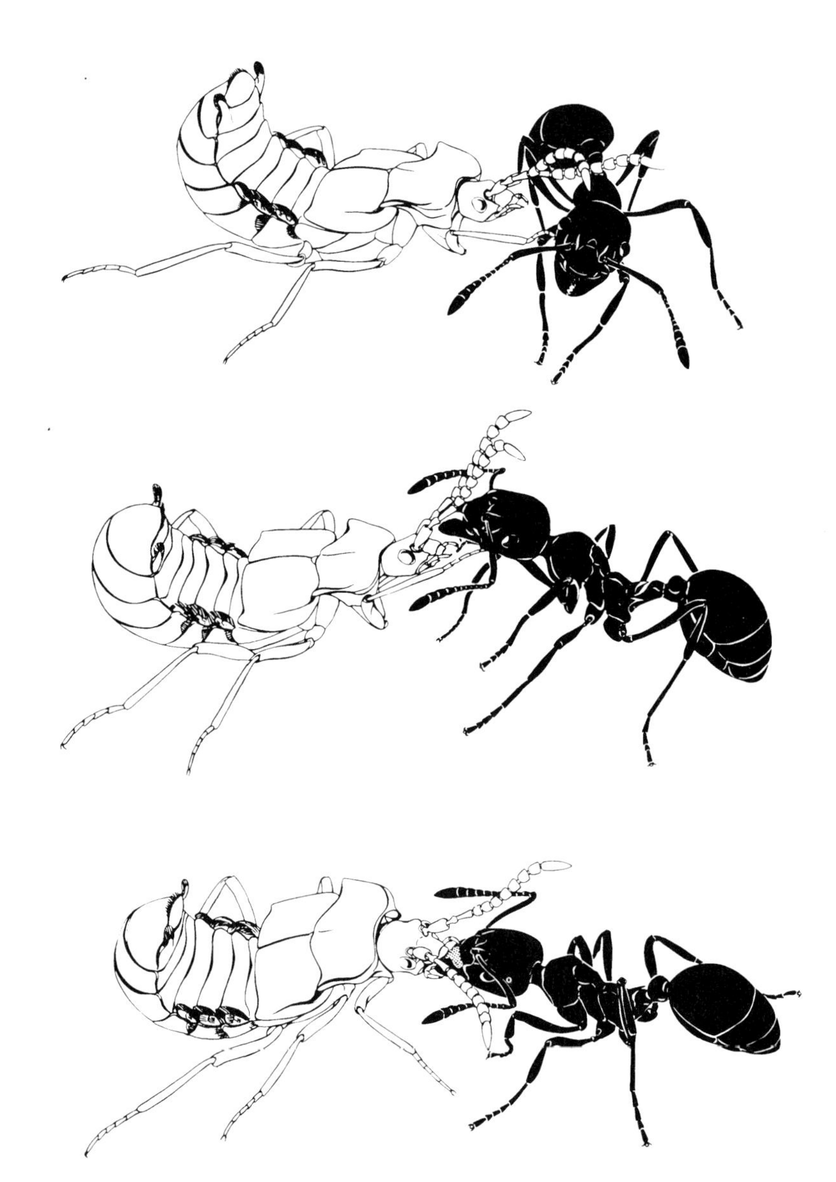

**FIGURE 7–41** Tactile communication in an ant colony: the soliciting of regurgitation by the socially parasitic staphylinid beetle *Atemeles pubicollis*. (*Top*) The beetle gains the attention of the ant worker (*Myrmica* sp.) by tapping her on the side with its antennae. The ant turns (*center*) and is induced to regurgitate when the beetle taps its fore tarsi on her labium (*bottom*). (From Hölldobler, 1970c; drawing by T. Hölldobler-Forsyth.)

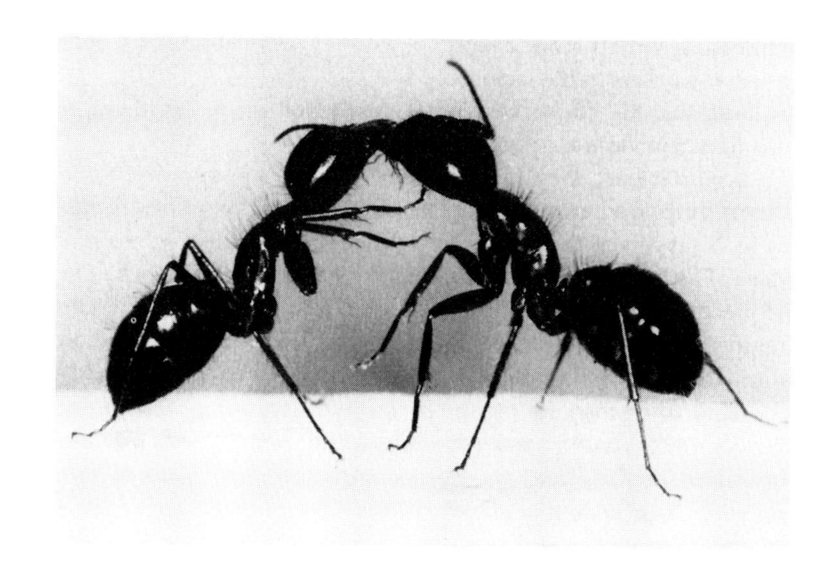

**FIGURE 7–42** Liquid food exchange (oral trophallaxis) between two workers of *Camponotus floridanus*. The worker on the left is inducing regurgitation from the worker on the right by touching her forelegs and antennae to her nestmate's head.

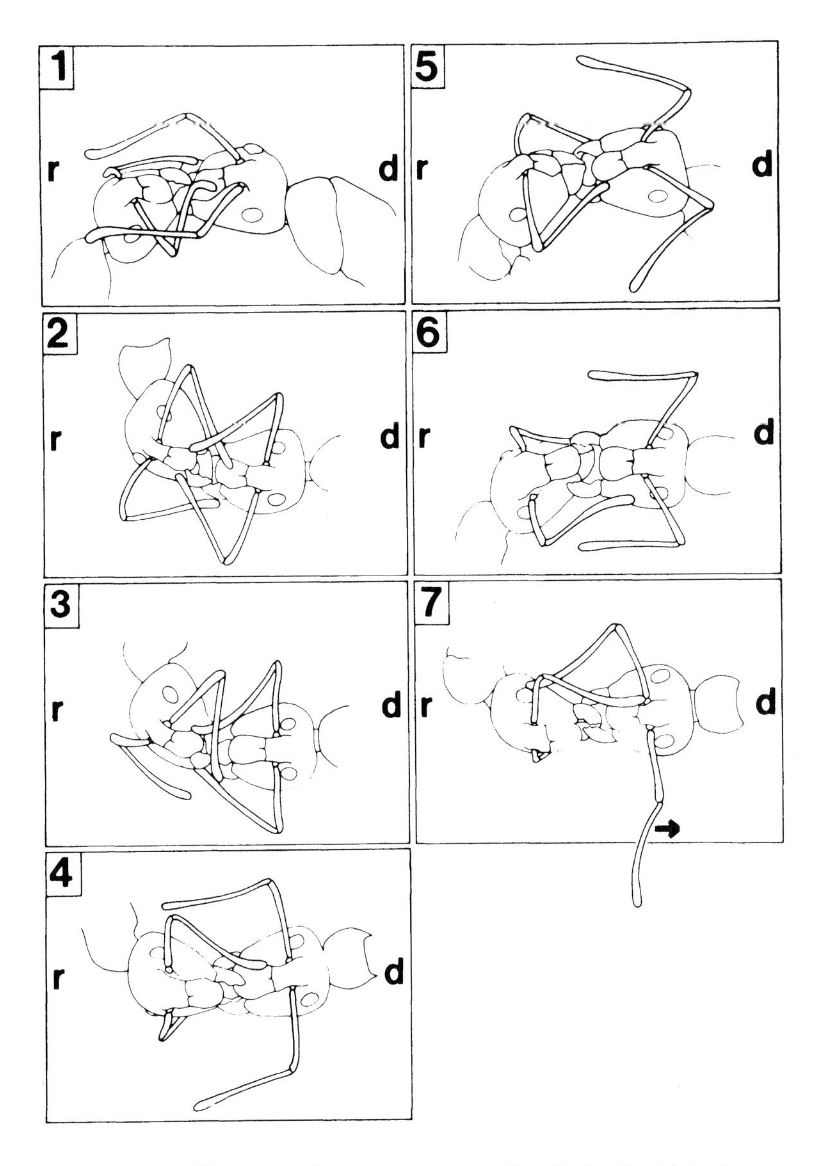


**FIGURE 7–43** The antennal movements occurring during liquid food exchange by regurgitation between workers of *Camponotus vagus: r,* recipient; *d,* donor. Seven positions and associated movements were recognized during this study, but all appear to convey about the same information. (From Bonavita-Cougourdan, 1983.)

nor's body lightly with antennae or forelegs. This causes the donor to turn and face the individual that gave the signal. If tapped lightly and repeatedly on the labium (lower mouth plate), the ant will regurgitate (see Figure 7–41). Some ants use their fore tarsi for this purpose, although the antennae may suffice (Figure 7–42). The myrmecophiles also employ their tarsi or antennae.

Once the food starts to flow, there is a heavy play of antennae between the donor and recipient, with that of the donor being more complex and variable. In both *Myrmica rubra* (Lenoir, 1982) and *Camponotus vagus* (Bonavita-Cougourdan, 1983; Bonavita-Cougourdan and Morel, 1984), the donor's paired funiculi are moved backward and forward, sometimes together and sometimes singly, to touch various parts of the dorsum of the recipient's head and, on occasion, the front ventral half of the recipient's head. The recipient directs her own funicular stroking to the anterior portion of the head, both the dorsal and ventral surfaces (Figure 7–43). The French investigators performed information analyses by measuring transition probabilities in the shifts from one antennal posture to another in both the donor and recipient ants. By this criterion very little transmission of information was recorded. One antennal posture did not lead to another within or across the repertories of the two ants in a regular sequence. In other words signal A did not reliably cause response B, and so forth. In an important additional experiment, Bonavita-Cougourdan used radioactive gold to monitor the flow of liquid food between *Camponotus* pairs while recording the antennal postures of the donor. Again, varying positions brought no change in the flux. Overall it seems unlikely that any of the postures convey particular messages to the receiver ants.

What then is the meaning of the elaborateness of antennal play? The paradox of complexity's exceeding function is not unique to ants. Some other kinds of animal communication, for example the alternative courtship and territorial displays of many bird species, also appear to invest little individual meaning in the particular signals that are switched back and forth. We suggest that the chief role of variety in such signaling is dishabituation. Animals tend to habituate when they receive the same stimulus repeatedly; that is, their response threshold rises and the contact is thus more likely to be broken off. By constantly changing the parts of the head stroked, both the donor and the recipient are more likely to sustain the trophallactic exchange. It follows that when the donor has a large load she should use a more intense and variable repertory than when her load is small. The same is true of the recipient if she is very hungry. Both correlations appear to be the case in nature. This fact alone does not prove dishabituation, however, because the correlations could also be simply a result of the general intensification of behavior with rising motivation.

## VISUAL COMMUNICATION

The use of visual signals in ants is at best very minor and in fact not a single example has yet been solidly documented. This negative generalization is not simply the outcome of a physiological constraint. Many large-eyed ants have excellent form vision and are very good at detecting moving objects (Jander, 1957; Ayre, 1963b; Voss, 1967; Via, 1977; Wehner, 1981; Wehner et al., 1983). The South American formicine *Gigantiops destructor* (Figure 7–44) and the African formicine *Santschiella kohli* have huge eyes that cover much of the sides of the head. *Gigantiops* workers are difficult to catch because they see approaching human observers from several meters away and flee by swiftly running and jumping away (the behavior of *Santschiella,* a very rare ant, has never been recorded). Workers of large-eyed ants generally do not respond to prey insects that are standing still, but run toward them as soon as they begin to move. Stäger (1931) noticed that when lone workers of *Formica lugubris* foraging in a field encounter an insect, they dash in erratic circles around it and attract other workers that happen to be in the vicinity. Stäger believed that the communication is mediated by vision alone and labeled the phenomenon "kinopsis." Later Sturdza (1942) showed that the sight of a running *F. nigricans* worker, apparently without reinforcement of other kinds of stimuli, is enough to set another worker running. Workers of *Daceton armigerum,* a large-eyed predatory ant living in the canopy of South American rain for-

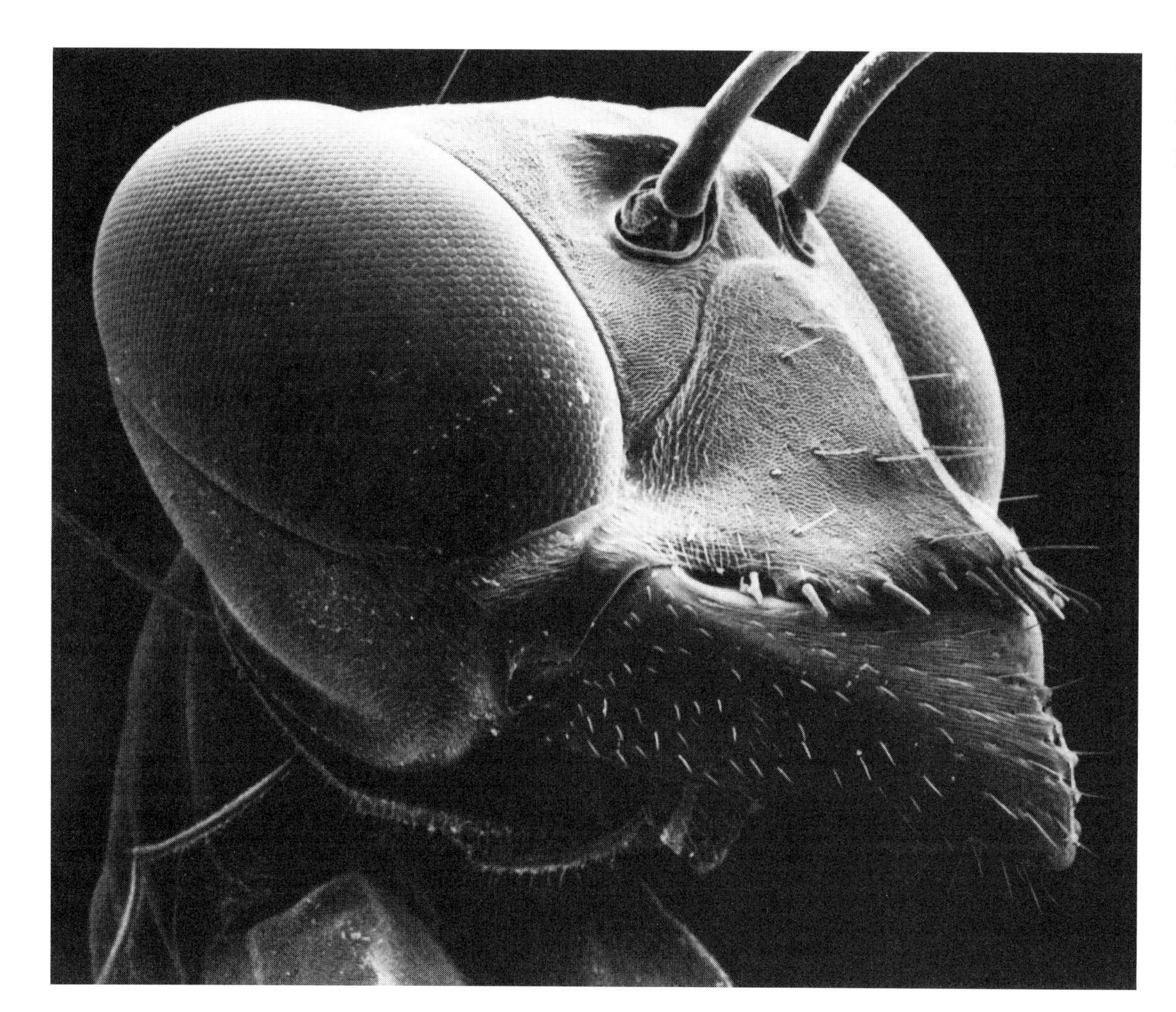

**FIGURE 7–44** Head of the South American formicine ant *Gigantiops destructor,* which is characterized by huge compound eyes.

ests, rush to join other workers running around prey insects (Wilson, 1962a; see Plate 5). Unfortunately, neither Stäger nor Wilson eliminated the possibility of short-range chemical recruitment signals. It is now well known that workers of the *cockerelli* group of *Aphaenogaster* (Hölldobler et al., 1978), *Oecophylla longinoda* (Hölldobler and Wilson, 1978), *Lasius neoniger* (Traniello, 1983), and species of *Formica* (Horstmann et al., 1982; Horstmann and Bitter, 1984) are able to spread short-range recruitment pheromones over distances of at least several centimeters within a few seconds. This invisible chemical communication is accompanied by conspicuous running that could easily be misinterpreted as visual signaling.

### ALARM COMMUNICATION

Alarm is the most difficult of behavioral responses to define, because investigators have used it as a portmanteau category for all responses to danger. In the broad sense worker ants are said to be in a state of alarm when they move away from a potentially dangerous stimulus, either calmly or in panic, or charge toward it aggressively, or simply mill about in a state of heightened alertness. Nevertheless, for each particular species in turn it is usually easy to devise a bioassay based on the precise reactions of the colonies, without having to take into account all the variations of behavior displayed by other species.

Most alarm signals are multicomponent, typically consisting of two or more pheromones, which often serve simultaneously to alert, attract, and evoke aggression. Acoustical signals are sometimes added to the pheromones, including especially stridulatory chirps that enhance attraction. The ants also occasionally touch nestmates with their antennae and forelegs during ritualized motor displays.

Alarm behavior is made additionally difficult to characterize because it so often blends into other major behavioral responses. Most ant species engage in some form of *alarm-defense,* in which the same chemicals are used to repel enemies and to alert nestmates. Examples of alarm-defense substances that have been identified in various ant species are citronellal, dendrolasin, dimethyl sulfate, and undecane. Some secretions are probably purely defensive, such as the formic acid employed by at least some formicine species. It is also possible that other compounds are purely communicative, especially those produced in minute quantities, but this specialized function remains to be documented in many cases. Probably the great majority of compounds used in alarm communication also serve in defense. Placed in the sequence more likely to have occurred in evolution, many chemicals employed in defense have also been ritualized to some extent into alarm signals.

A second intergradient category that has been documented is *alarm-recruitment.* Alarm signals both alarm and attract in some species, while in others the alarm pheromones are combined with odor trails that lead nestmates to or from the source of the danger. Many species employ a single alarm-recruitment procedure to alert nestmates to both enemies and prey, and in fact the distinction between the two may be wholly blurred with reference to communication.

Whether joined with recruitment or not, alarm behavior can be conveniently classified into one of two broad categories (Wilson and Regnier, 1971). In *aggressive alarm* some of the colony members (often the majors or "soldiers") are drawn toward the threatening stimulus and seek to attack it. In *panic alarm* the colony as a whole flees from the stimulus or dashes around in erratic patterns. If disturbed strongly enough by the waves of alarm, individuals may even evacuate the nest.

A wide range of stimuli evoke alarm communication. Oddly, substratal vibrations and air currents disturb the entire colony and may even induce evacuation if severe enough, but they seldom trigger the release of alarm pheromones. Alarm communication is initiated far more predictably when an enemy penetrates the close environs of the nest. Certain stimuli work better than others in this context. In the phenomenon called enemy specification, dangerous species are more effective than less threatening ones in evoking a response. For example, minor workers of *Pheidole dentata* recruit majors to only a single fire ant forager approaching the nest (Wilson, 1976b). Those of *P. desertorum* and *P. hyatti* initiate panic alarm leading to nest evacuation when they become aware of the approach of army ants of the genus *Neivamyrmex* (Droual and Topoff, 1981). Such hair-trigger responses are not limited to living enemies. A single drop of water in the nest entrance of *P. cephalica* is often enough to cause alarm-recruitment and the full retreat of the colony away from that part of the nest (Wilson, 1986c).

Alarm is technically one of the most difficult forms of communication to study with precision. At lower stimulus intensities the responses may be subtle, entailing nothing more than increased alertness or an increase in running velocity combined with a reduced sinuosity in the direction taken (Cammaerts et al., 1982). It is further possible to get an alarm response from any volatile compound extracted from ants, if the concentrations used are high enough. A positive reaction cannot be used as evidence that the substance is an alarm pheromone if the experimental concentrations are higher than those that could be generated by the ants themselves. An essential part of the analysis is to create active spaces of the size and geometry resembling those expected under natural conditions. Investigators usually accomplish this end crudely by crushing single glands containing the pheromone.

Let us now examine a relatively uncomplicated aggressive alarm system. When a worker of the subterranean formicine ant *Acanthomyops claviger* is severely threatened, say attacked by a member of a rival colony or an insect predator, she reacts strongly by simultaneously discharging the contents of the reservoirs of her mandibular and Dufour's glands. After a brief delay, other workers resting a short distance away display the following responses: the antennae are raised, extended, and swept in an exploratory fashion through the air; the mandibles are opened; and the ants begin to walk, then run, in the general direction of the disturbance (Regnier and Wilson, 1968). Workers sitting a few millimeters away begin to react within seconds, whereas those a few centimeters distant take a minute or longer. Thus the signal appears to obey the laws of gas diffusion. Experiments have implicated some of the terpenes, hydrocarbons, and ketones as the alarm pheromones; these are shown in Figure 7–30. Undecane and the mandibular gland substances (all terpenes) evoke the alarm response at concentrations of $10^9$–$10^{12}$ molecules per cubic centimeter, reflecting a moderate amount of sensitivity as far as pheromones go. These same substances are individually present in amounts ranging from as low as 44 nanograms to as high as 4.3 micrograms per ant, and altogether they total about 8 micrograms. Released as a vapor during experiments, similar quantities of the synthetic pheromones produce the same responses. Apparently the *A. claviger* workers rely entirely on these pheromones for alarm communication. Their system seems designed to bring workers to the aid of a distressed nestmate over distances of up to 10 centimeters. Unless the signal is reinforced by additional emissions, it dies out within a few minutes. The $Q{:}K$ ratios are on the order of $10^3$–$10^4$. If the entire contents of the Dufour's gland, containing about 2.5 micrograms of undecane, are discharged as a puff from the poison gland, the diffusion model of Bossert and Wilson (1963) predicts that the pheromone signal will reach a maximum of about 20 centimeters in still air. If on the other hand only 0.1 percent is discharged, this active space can still reach a maximum of 2 centimeters. Hence the match with the observations of natural behavior is reasonably close.

The alerted *Acanthomyops* workers approach their target in a truculent manner. This aggressive defensive strategy is in keeping with the structure of their colonies, which are large and often densely concentrated in narrow subterranean galleries. It would not pay the colonies to try to disperse when their nests are invaded, and the workers have apparently evolved so as to meet danger head-on.

A different strategy, based on panic and escape, is employed in the chemical alarm-defense system of the related ant *Lasius alienus* (Regnier and Wilson, 1969). Colonies of this species are smaller and normally nest under rocks or in pieces of rotting wood on the ground; such nest sites give the ants ready egress when the colonies are seriously disturbed. *L. alienus* produce the same volatile substances as *Acanthomyops claviger*, with the exception of citronellal and citral. The alarm substances' principal volatile component is undecane. When the *Lasius* workers smell this pheromone, they scatter and run frantically in an erratic pattern. They are more sensitive to undecane than the *Acanthomyops* workers, being activated by only $10^7$–$10^8$ molecules per cubic centimeter. Thus, in contrast to *A. claviger*, *L. alienus* utilizes an "early warning" system and subsequent evacuation to cope with serious intrusion.

Alarm systems are nearly universal in ants, having been discovered in every species in which a test for the phenomenon was performed. This is true even of the species considered to be among the most primitive, *Amblyopone pallipes* (Traniello, 1982), *Myrmecia gulosa* (Robertson, 1971), and *Nothomyrmecia macrops* (Hölldobler and Taylor, 1983). By far the most common mode of alarm communication is chemical. Many species, such as the members of *Acanthomyops* and *Lasius*, employ alarm pheromones without the accompaniment of acoustical or tactile displays. In general it appears that where acoustical and tactile signals exist they have been added to chemical alarm and recruitment as modulators. This does not mean, however, that such signals are specialized or derived with reference to the evolution of the ants as a whole. Stridulatory organs are present in *Nothomyrmecia* and many members of the primitive subfamily Ponerinae. Among members of the latter group, *Amblyopone australis* and *A. pallipes* utilize a tactile signal during a vibratory display in which the ants vigorously jerk the head and thorax up and down while coming into contact with nestmates (Hölldobler, 1977; Traniello, 1982). In *A. pallipes* at least, mandibular gland pheromones also serve as weak attractants. They might serve in alarm, recruitment, or both—not enough experiments have been made to justify a distinction.

A remarkable alarm behavior in the Australian bulldog ant *Myrmecia* has been discovered by Hubert Markl (unpublished observations). On the basis of models of the two-dimensional random alarm process, Markl and his collaborators (in Frehland et al., 1985) give the following account:

> Around an undisturbed nest one usually finds a small number of workers randomly distributed over a few square meters, sitting still or moving about slowly. If an intruder disturbs one of these "sentinels," it may trigger it into a frantic, erratic run for a number of seconds. If it thereby comes within sight of a second nestmate, this will lunge forward as if in attack. Upon direct contact, however, combat is avoided and now the second ant starts the same alarm-run while the first one may continue its own. Depending on density and distribution of ants over the guarded area and on number of guards initially stimulated, the alarm can spread two-dimensionally over an extended area. Soon, one or the other of the alarmed workers will return to the nest and there recruit additional forces to join the excited crowd. Intruders into the nest territory will thus be reliably detected and attacked by ants (who can sting ferociously) notwithstanding the stochasticity of the environment. After intruders have been driven out, the alarm subsides by not receiving more stimulating input and/or by having spread out over too large an area. Thus, with a small number of widely distributed guards, the colony can efficiently control a fairly large territory which cannot completely be overrun by any single individual, without engaging more work force than necessary at any given time. (p. 198)

Most of the alarm pheromones identified to the present time are listed in Table 7–4. Their great structural diversity is a reflection of the antiquity and phylogenetic complexity of the ants themselves. Additional variety comes from the fact that most alarm pheromones also serve as defensive substances and have probably obtained their communicative function through ritualization many times independently. In some instances certain exocrine glands initially involved in alarm communication are secondarily hypertrophied and function as the major defensive devices of the ants (Buschinger and Maschwitz, 1984; Figure 7–45). Defensive chemicals, basic to the biology of social insects, are in turn tied into varying strategies that involve entire syndromes of anatomical structure and behavior. These syndromes are known to evolve across species in correlation with colony size, nest site, and other aspects of natural history. In short, the diversity of alarm pheromones is an expected consequence of phylogeny and the close linkage that exists in ants between defense and alarm. The one common feature the substances seem to share is molecular weight. The great majority of the compounds are in the $C_6$ to $C_{10}$ range. This is in accord with the prediction (Wilson and Bossert, 1963) that alarm pheromones should evolve in the lower molecular size range because of the need for substantial volatility and a lower $Q:K$ ratio, which would permit the rapid expansion and fade-out of the active space. Further, there is little need for privacy in communication, so that molecular complexity and size have not been at a premium during evolution.

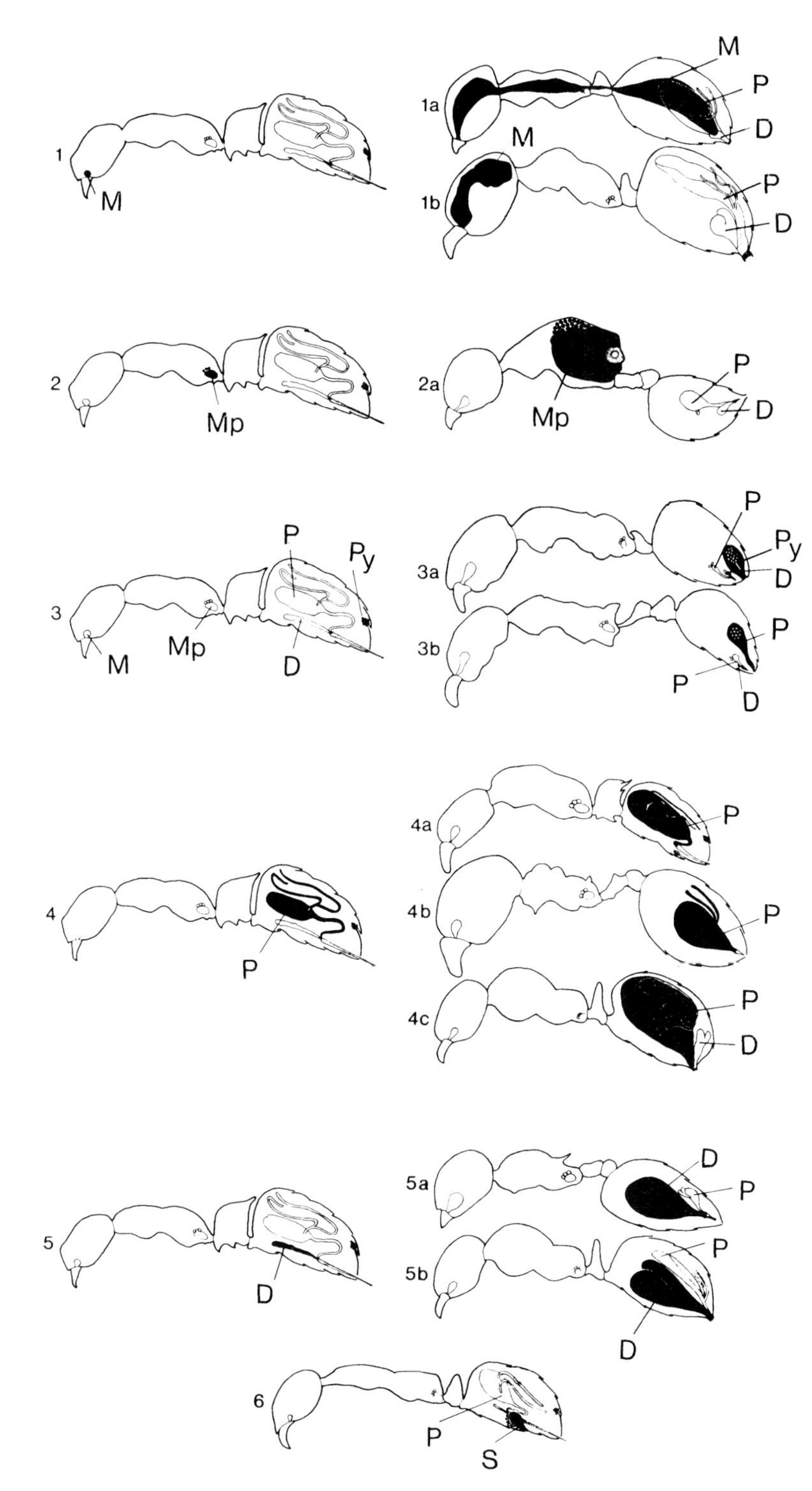


**FIGURE 7–45** Exocrine glands in ants which are employed in producing defensive secretions. In developing as defensive devices many of the glands have become hypertrophied. *1*. Mandibular gland (*M*): *1a, Camponotus* sp. near *saundersi; 1b, Lasius fuliginosus*. *2*. Metapleural gland (*Mp*): *2a, Crematogaster inflata*. *3*. Pygidial gland (*Py*): *3a, Iridomyrmex humilis; 3b, Pheidole biconstricta*. *4*. Poison gland (*P*): *4a, Pachycondyla tridentata; 4b, Pheidole fallax; 4c, Formica polyctena*. *5*. Dufour' s gland (*D*): *5a, Crematogaster scutellaris; 5b, Formica subintegra*. *6*. Sternal gland (*S*): *Leptogenys ocellifera*. (Modified from Buschinger and Maschwitz, 1984.)

**TABLE 7–4** Chemicals recognized as alarm pheromones in ants. In some cases a substance has been found to be a pheromone in one species and present in the same gland in other related species. Only those species are listed for which activity has been demonstrated. (Based principally on the review by Parry and Morgan, 1979; with additions by Wilson, 1971; Duffield et al., 1980; Olubajo et al., 1980; Cammaerts et al., 1981, 1982, 1983; and Tomalski et al., 1987.)

| Compound | Molecular formula | Glandular source | Species and reference |
|---|---|---|---|
| **Alcohols** | | | |
| 3-Decanol | $C_{10}H_{22}O$ | Mandibular | *Myrmica lobicornis* (Cammaerts et al., 1983) |
| 4-Heptanol | $C_7H_{16}O$ | Mandibular | *Zacryptocerus varians* (Olubajo et al., 1980) |
| 1-Hexanol | $C_6H_{14}O$ | Mandibular | *Oecophylla longinoda* (Bradshaw et al., 1975) |
| 4-Methyl-3-heptanol | $C_8H_{18}O$ | Mandibular | *Pogonomyrmex badius* (McGurk et al., 1966) |
| 6-Methyl-5-hepten-2-ol | $C_8H_{16}O$ | Pygidial | *Iridomyrmex* sp. (Blum et al., 1966; Crewe and Blum, 1971); *Tapinoma melanocephalum* (Tomalski et al., 1987) |
| 3-Octanol | $C_8H_{18}O$ | Mandibular | *Myrmica lobicornis, M. rubra, M. scabrinodis* (Cammaerts-Tricot, 1973, 1974b; Morgan et al., 1978; Cammaerts et al., 1981, 1983) |
| **Aldehydes** | | | |
| 2-Butyl-2-octenal | $C_{12}H_{22}O$ | Mandibular | *Oecophylla longinoda* (Bradshaw et al., 1975) |
| Hexanal | $C_6H_{12}O$ | Mandibular | *Oecophylla longinoda* (Bradshaw et al., 1975) |
| 2-Hexenal | $C_6H_{10}O$ | Cephalic | *Crematogaster* (*Atopogyne*) sp. (Blum et al., 1969) |
| **Aliphatic ketones** | | | |
| 3-Decanone | $C_{10}H_{20}O$ | Mandibular | *Manica bradleyi, M. mutica* (Fales et al., 1972) |
| 4,6-Dimethyl-4-octen-3-one | $C_{10}H_{18}O$ | Mandibular | *Manica bradleyi, M. mutica* (Fales et al., 1972) |
| 2-Heptanone | $C_7H_{14}O$ | Pygidial | *Conomyrma pyramica* (Blum and Warter, 1966); *Forelius* (=*Iridomyrmex*) *pruinosus* (Wilson and Pavan, 1959; Blum et al., 1963) |
| 4-Heptanone | $C_7H_{14}O$ | Mandibular | *Zacryptocerus varians* (Olubajo et al., 1980) |
| 2-Methyl-4-heptanone | $C_8H_{16}O$ | Pygidial | *Tapinoma nigerrimum* (Trave and Pavan, 1956) |
| 4-Methyl-3-heptanone | $C_8H_{16}O$ | Mandibular | *Atta texana* (Moser et al., 1968); *Manica bradleyi, M. mutica* (Fales et al., 1972); *Pachycondyla villosa* (Duffield and Blum, 1973); *Pogonomyrmex badius, P. californicus, P. desertorum, P. occidentalis, P. rugosus* (McGurk et al., 1966); *Trachymyrmex seminole* (Crewe and Blum, 1972) |
| 5-Methyl-3-heptanone | $C_8H_{16}O$ | Mandibular | *Atta rubropilosa* (Butenandt et al., 1959) |
| 6-Methyl-5-hepten-2-one | $C_8H_{14}O$ | Pygidial | *Conomyrma pyramica* (McGurk et al., 1968); *Dolichoderus* sp. (Cavill and Hinterberger, 1960b); *Iridomyrmex* sp. (Blum et al., 1966; Crewe and Blum, 1971), *I. conifer* (Cavill et al., 1956); *I. nitidiceps, I. rufoniger* (Cavill and Hinterberger, 1960b); *Tapinoma melanocephalum* (Tomalski et al., 1987); *T. nigerrimum* (Pavan and Trave, 1958) |
| 4-Methyl-2-hexanone | $C_7H_{14}O$ | Pygidial | *Dolichoderus clarki* (Cavill and Hinterberger, 1960a) |
| 4-Methyl-3-hexanone | $C_7H_{14}O$ | Mandibular | *Manica bradleyi, M. mutica* (Fales et al., 1972) |
| 3-Nonanone | $C_9H_{18}O$ | Mandibular | *Myrmica rubra, M. scabrinodis* (Cammaerts-Tricot, 1974b) |
| 3-Octanone | $C_8H_{16}O$ | Mandibular | *Acromyrmex octospinosus, A. versicolor* (Crewe and Blum, 1972); *Crematogaster peringueyi* (Crewe et al., 1969, 1970); *Manica bradleyi, M. mutica* (Fales et al., 1972); *Myrmica lobicornis, M. rubra, M. ruginodis, M. sabuleti, M. scabrinodis* (Tricot et al., 1972; Cammaerts-Tricot, 1973; Morgan et al., 1978; Cammaerts et al., 1981, 1982, 1983); *Trachymyrmex seminole, T. septentrionalis, T. urchii* (Crewe and Blum, 1972) |
| 2-Tridecanone | $C_{13}H_{26}O$ | Dufour's | *Acanthomyops claviger* (Regnier and Wilson, 1968) |
| **Carboxylic acids** | | | |
| Formic acid | $CH_2O_2$ | Poison | *Camponotus* spp. (Hefetz and Orion, 1982); *C. pennsylvanicus* (Ayre and Blum, 1971); *Cataglyphis* spp. (Hefetz and Orion, 1982); *Formica aquilonia* (Dlussky and Chernyshova, 1975); *F. rufa* (Löfqvist, 1976); *Oecophylla longinoda* (Bradshaw et al., 1975); *Polyrhachis* spp. (Hefetz and Orion, 1982) |

*continued*

**TABLE 7–4** *(continued)*

| Compound | Molecular formula | Glandular source | Species and reference |
|---|---|---|---|
| **Cyclic ketones** | | | |
| *cis*-1-Acetyl-2-methylcyclopentane | $C_8H_{14}O$ | Pygidial | *Azteca* sp. (Wheeler et al., 1975) |
| 2-Acetyl-3-methylcyclopentene | $C_8H_{12}O$ | Pygidial | *Azteca* sp. (Wheeler et al., 1975) |
| 2-Methylcyclopentanone | $C_6H_{10}O$ | Pygidial | *Azteca* sp. (Wheeler et al., 1975) |
| **Esters** | | | |
| Decyl acetate | $C_{12}H_{14}O_2$ | Dufour's | *Formica pergandei, F. sanguinea, F. subintegra* (Regnier and Wilson, 1971) |
| Dodecyl acetate | $C_{14}H_{18}O_2$ | Dufour's | *Formica pergandei, F. sanguinea, F. subintegra* (Regnier and Wilson, 1971) |
| Methyl anthranilate | $C_8H_5NO_2$ | Cephalic | *Aphaenogaster fulva, Xenomyrmex floridanus* (Duffield et al., 1980) |
| Methyl 6-methylsalicylate | $C_9H_{10}O_3$ | Cephalic | *Gnamptogenys pleurodon* (Duffield and Blum, 1975); *Pachycondyla* (= *Bothroponera*) *soror* (Longhurst et al., 1980) |
| Tetradecyl acetate | $C_{16}H_{22}O_2$ | Dufour's | *Formica pergandei, F. subintegra* (Regnier and Wilson, 1971) |
| Undecyl acetate | $C_{13}H_{16}O_2$ | Dufour's | *Formica sanguinea* (Regnier and Wilson, 1971) |
| **Hydrocarbons** | | | |
| *n*-Decane | $C_{10}H_{22}$ | Dufour's | *Formica rufa* (Löfqvist, 1976) |
| *n*-Undecane | $C_{11}H_{24}$ | Dufour's | *Acanthomyops claviger* (Regnier and Wilson, 1968); *Camponotus ligniperda* (Bergström and Löfqvist, 1972); *Cataglyphis* spp. (Hefetz and Orion, 1982); *Formica aquilonia* (Dlussky and Chernyshova, 1975); *F. rufa* (Löfqvist, 1976); *Lasius alienus* (Regnier and Wilson, 1969); *L. fuliginosus* (Dumpert, 1972); *Oecophylla longinoda* (Bradshaw et al., 1975) |
| **Nitrogen heterocycles** | | | |
| 2,6-Dimethyl-3-ethylpyrazine | $C_8H_{12}N_2$ | Mandibular | *Odontomachus brunneus* (Wheeler and Blum, 1973) |
| 2,5-Dimethyl-3-isopentylpyrazine | $C_{11}H_{18}N_2$ | Mandibular | *Odontomachus clarus, O. hastatus* (Wheeler and Blum, 1973) |
| 2,6-Dimethyl-3-pentylpyrazine | $C_{11}H_{18}N_2$ | Mandibular | *Odontomachus brunneus* (Wheeler and Blum, 1973) |
| 2,6-Dimethyl-3-propylpyrazine | $C_9H_{14}N_2$ | Mandibular | *Odontomachus brunneus* (Wheeler and Blum, 1973) |
| **Sulphur compounds** | | | |
| Dimethyl disulphide | $C_2H_6S_2$ | Mandibular | *Paltothyreus tarsatus* (Casnati et al., 1967) |
| Dimethyl trisulphide | $C_2H_6S_3$ | Mandibular | *Paltothyreus tarsatus* (Casnati et al., 1967) |
| **Terpenoid compounds** | | | |
| Citral | $C_{10}H_{16}O$ | Mandibular | *Acanthomyops claviger* (Chadha et al., 1962); *Atta rubropilosa* (Butenandt et al., 1959) |
| Citronellal | $C_{10}H_{18}O$ | Mandibular | *Acanthomyops claviger* (Ghent, 1961; Regnier and Wilson, 1968); *Lasius spathepus, L. umbratus* (Blum, 1969) |
| Dendrolasin | $C_{15}H_{22}O$ | Mandibular | *Lasius fuliginosus* (Pavan, 1955, 1956; Quilico et al., 1956, 1957) |
| 2,6-Dimethyl-5-heptenal | $C_9H_{16}O$ | Mandibular | *Acanthomyops claviger* (Regnier and Wilson, 1968) |
| 2,6-Dimethyl-5-hepten-1-ol | $C_9H_{18}O$ | Mandibular | *Acanthomyops claviger* (Regnier and Wilson, 1968); *Lasius alienus* (Law et al., 1965; Regnier and Wilson, 1969) |
| Geraniol | $C_{10}H_{18}O$ | Mandibular | *Cataglyphis* spp. (Hefetz, 1985); *Oecophylla longinoda* (Bradshaw et al., 1975) |
| Limonene | $C_{10}H_{16}$ | Poison | *Myrmicaria natalensis* (Quilico et al., 1960) |
| Nerol | $C_{10}H_{18}O$ | Mandibular | *Oecophylla longinoda* (Bradshaw et al., 1975) |

## PROPAGANDA

Workers of the slave-making ant *Formica subintegra* possess an enormously enlarged Dufour's gland loaded with approximately 700 micrograms of a mixture of decyl, dodecyl, and tetradecyl acetates (Regnier and Wilson, 1971). The substances are sprayed at defending colonies of other species of *Formica* during slave raids. They attract the *F. subintegra* workers, but alarm and disperse the defenders. Hence they act at least in part as "propaganda substances." The acetates are well designed for this purpose in accordance with the engineering rules for the evolution of pheromones. Having a higher molecular weight than most ordinary alarm substances, the *F. subintegra* pheromones evaporate at a slower rate and exert their influence for longer periods of time. The larger size of the molecules also confers the potential for lower response thresholds, although this effect, demonstrated in *Acanthomyops claviger* (Regnier and Wilson, 1968), has not been tested explicitly in *Formica*.

In a remarkably parallel manner, workers of the myrmicine slavemaker *Harpagoxenus sublaevis* employ propaganda substances from their Dufour's glands to subdue workers of target *Leptothorax acervorum* colonies. One or more of the components causes the *Leptothorax* to fight their own nestmates, throwing the colony into greater confusion and rendering it more vulnerable to raids by the *Harpagoxenus* workers. An identical effect is produced by Dufour's gland secretions voided by the queens of *L. kutteri*, a workerless social parasite of *L. acervorum* (Allies et al., 1986), and queens of the slave-raiding *Polyergus breviceps* (Topoff et al., 1988).

## RECRUITMENT

Recruitment is defined as communication that brings nestmates to some point in space where work is required (Wilson, 1971). The ants have evolved an astonishing array of devices to assemble workers for joint efforts in food retrieval, nest construction, colony defense, and emigration to new nest sites. Various species employ idiosyncratic combinations of touch, stridulation, and chemical cues to achieve these ends. The category of recruitment is thus a very loose one, and in particular cases it often cannot be clearly distinguished from alarm and simple assembly. The Australian dacetine ant *Orectognathus versicolor* is typical in the way it mixes the functions. Workers employ trails laid from their poison gland to organize colony emigrations, during which their pheromones serve as a stimulative recruitment signal as well as a longer-lasting orientation cue. On the other hand, trails laid from the *Orectognathus* pygidial gland appear to function as a short-lasting alarm-recruitment signal, channeling workers to areas of disturbance near the nest. There is also some evidence that the pygidial gland pheromone is discharged by successful foragers as they enter the nest, increasing the rate at which nestmates exit and follow the poison gland trail (Hölldobler, 1981b).

The study of recruitment has grown into a substantial field of inquiry. During the 1950s and 1960s investigators emphasized the straightforward identification of the glandular source of the pheromones, while paying some attention to the details of trail-laying behavior (see reviews in Wilson, 1971; Blum, 1974b; Hölldobler, 1977; Parry and Morgan, 1979). Later they began to focus on an additional topic: the analysis of the organizational levels and ecological significance of recruitment. The possession of one kind of recruitment as opposed to another clearly constitutes an adaptation by individual species to particular long-term conditions in the environment. Researchers generally agree that the recruitment strategy makes little sense except with reference to the ecology of the species. Conversely, the ecology of most species cannot be fully understood without a detailed knowledge of their recruitment procedures.

The most prevalent form of recruitment behavior is chemical trail communication. Carthy (1950, 1951a,b) was one of the first to conduct an experimental study of trail laying in ants. He obtained strong circumstantial evidence that *Lasius fuliginosus* produces the trail pheromone in the hindgut, a result later confirmed by Hangartner and Bernstein (1964). Wilson (1959d, 1962b), working with the red imported fire ant *Solenopsis invicta* (= *S. saevissima* in part), provided the first bioassay methods to test trail-laying behavior even in the absence of a recruiting ant. He laid artificial trails made from different glandular extracts away from the nest entrance and from worker aggregations around food finds. By comparing the trail-following response of worker ants, he was able to identify the Dufour's gland as the source of the trail pheromone in fire ants. This technique has subsequently been used by many investigators and has led to the discovery of a number of trail pheromone glands in different taxonomic groups (Table 7–5 and Figure 7–46). There has also been rapid progress in the chemical identification of trail pheromones (Attygalle and Morgan, 1985; Table 7–6).

Wilson's (1962b) analysis also revealed for the first time the existence of chemical mass communication in ants, one of the most complex forms of social behavior occurring in the social insects. The details of the communication can be summarized as follows. When fire ant workers leave the nest in search of food, they may follow odor trails for a short while, but they eventually separate from each other and begin to explore singly. When alone they orient visually, at least in part, a fact that was established by the following simple experiment. Single workers were first permitted to explore a foraging table in the laboratory for distances of up to a meter from the nest. The only source of illumination was a lamp beamed from one side of the table. The workers were next allowed to discover a drop of sugar water. When they had fed and started home, laying an odor trail behind them, the light was switched off and a second light located on the opposite side of the table was simultaneously switched on. After the direction of the light source had thus been changed by 180 degrees, the ants almost invariably performed a complete about-face. By alternately switching the light source from one side to the other, individual workers could be marched back and forth like puppets and finally brought home at will.

When a fire ant discovers a food source, she heads home at a slower, more deliberate pace. Her entire body is held closer to the ground. At frequent intervals the sting is extruded, and its tip is drawn lightly over the ground surface, much as a pen is used to ink a thin line (Figure 7–47). The key recruitment pheromone in *Solenopsis invicta* is an alpha-farnesene, with synergistic effects being added by two homofarnesenes, possibly *n*-heptadecane, and a still unidentified component from the Dufour's gland. A second alpha-farnesene and an allofarnesene may also play minor roles as recruiters (Vander Meer, 1983, 1986a; see Figure 7–32). Each worker possesses less than a nanogram at any given moment. Artificial trails made from a single Dufour's gland induce following by dozens of

**TABLE 7–5** Glandular sources of trail and tandem-running substances.

| Species | Behavior | Glandular source | Authority |
|---|---|---|---|
| **PONERINAE** | | | |
| *Cerapachys* sp. (*turneri* group) | Trail | Poison (orientation, some stimulatory effects); pygidial (stimulatory) | Hölldobler (1982b) |
| *Diacamma rugosum* | Trail | Hindgut (orientation, in trail) | Maschwitz et al. (1986c) |
| *Leptogenys ocellifera* and other *Leptogenys* spp. | Trail | Poison (orientation, some stimulatory effects); pygidial (stimulatory and orienting effects) | Fletcher (1971), Maschwitz and Mühlenberg (1975), Maschwitz and Schönegge (1977), Hölldobler (1982a) |
| *Megaponera foetens* | Trail | Poison | Longhurst et al. (1979b) |
| *Onychomyrmex* spp. | Trail | Sternal (intersegmental, abdominal sternites V–VI) | Hölldobler et al. (1982) |
| *Pachycondyla* (= *Neoponera*) *obscuricornis* | Tandem running | Pygidial | Traniello and Hölldobler (1984) |
| *P.* (= *Termitopone*) *laevigata* | Trail | Pygidial | Hölldobler and Traniello (1980b) |
| *Paltothyreus tarsatus* | Trail | Sternal (intersegmental, abdominal sternites IV–V, V–VI, VI–VII) | Hölldobler (1984b) |
| *Sphinctomyrmex steinheili* | Trail | Poison (orientation, some stimulatory effects); pygidial (stimulatory) | Hölldobler (1982b) |
| **DORYLINAE** | | | |
| *Eciton hamatum* | Trail | Hindgut; pygidial (stimulatory effect: provisional finding) | Blum and Portocarrero (1964), Hölldobler and Engel (1978) |
| *Neivamyrmex* spp. | Trail | Hindgut | Watkins (1964) |
| **MYRMICINAE** | | | |
| *Acromyrmex octospinosus* | Trail | Poison | Blum et al. (1964) |
| *Aphaenogaster* (= *Novomessor*) *albisetosus* and *A.* (= *N.*) *cockerelli* | Trail | Poison | Hölldobler et al. (1978) |
| *Atta* spp. | Trail | Poison | Moser and Blum (1963), Blum et al. (1964), Tumlinson et al. (1971), Cross et al. (1979), Jaffe and Howse (1979) |
| *Crematogaster ashmeadi, C. peringueyi* | Trail | Tibial | Fletcher and Brand (1968), Leuthold (1968a,b), Billen (1984a) |
| *Cyphomyrmex* sp. | Trail | Poison | Blum et al. (1964) |
| *Daceton armigerum* | Trail | Poison | B. Hölldobler and M. Moffett (unpublished) |
| *Harpagoxenus sublaevis* | Tandem running | Poison | Buschinger and Winter (1977) |
| *Huberia* sp. | Trail | Poison | Blum (1966) |
| *Leptothorax* spp. | Tandem running | Poison | Möglich et al. (1974), Möglich (1979) |
| *Manica* sp. | Trail | Poison | Blum (1974a) |
| *Meranoplus* spp. | Trail | Poison (stimulatory effect); Dufour's (long-lasting orientation effect) | B. Hölldobler (unpublished) |
| *Messor rufitarsis* | Trail | Dufour's (recruitment and orientation) | Hahn and Maschwitz (1985) |
| *M.* (= *Veromessor*) *pergandei* | Trail | Poison | Blum (1974a) |
| *Monomorium minimum* | Trail | Dufour's | Adams and Traniello (1981) |
| *M. pharaonis* | Trail | Dufour's | Hölldobler (1973a) |
| *Myrmica rubra* and other *Myrmica* spp. | Trail | Poison (principal trail); Dufour's (short-range attraction) | Cammaerts-Tricot (1974a,b), Cammaerts and Cammaerts (1980, 1981), Evershed et al. (1982) |

*continued*

**TABLE 7–5** *(continued)*

| Species | Behavior | Glandular source | Authority |
|---|---|---|---|
| *Orectognathus versicolor* | Trail | Poison and pygidial | Hölldobler (1981b) |
| *Pheidole dentata* and other *Pheidole* spp. | Trail | Poison | Wilson (1976b), Hölldobler and Möglich (1980), Wilson and Hölldobler (1985) |
| *Pheidologeton diversus* | Trail | Poison (recruitment, orientation); Dufour's (orientation) | Moffett (1987b) |
| *Podomyrma* sp. | Trail | Poison | Hölldobler (1983) |
| *Pogonomyrmex badius* and other *Pogonomyrmex* spp. | Trail | Poison (recruitment); Dufour's (long-lasting orientation and homing cue) | Hölldobler and Wilson (1970), Hölldobler (1971b, 1976a) |
| *Proatta butteli* | Trail | Poison | Moffett (1986d) |
| *Sericomyrmex* sp. | Trail | Poison | Blum and Portocarrero (1966) |
| *Solenopsis invicta* and other *Solenopsis* (*S.*) spp. | Trail | Dufour's | Wilson (1959d, 1962b), Vander Meer (1983, 1986a) |
| *S.* (*Diplorhoptrum*) *fugax* | Trail | Dufour's | Hölldobler (1973a) |
| *Tetramorium* spp. | Trail | Poison | Blum and Ross (1965), Attygalle and Morgan (1983) |
| *Trachymyrmex septentrionalis, T. urichi* | Trail | Poison | Blum and Portocarrero (1966), Jaffe and Villegas (1985) |
| **ANEURETINAE** | | | |
| *Aneuretus simoni* | Trail | Sternal (Pavan's) | Traniello and Jayasuriya (1981a,b) |
| **DOLICHODERINAE** | | | |
| *Conomyrma insana, C. bicolor* | Trail | Sternal (Pavan's) | B. Hölldobler (unpublished) |
| *Forelius foetidus* | Trail | Sternal (Pavan's) | B. Hölldobler (unpublished) |
| *F. pruinosus* | Trial | Sternal (Pavan's) | Wilson and Pavan (1959) |
| *Iridomyrmex humilis* | Trail | Sternal (Pavan's) | Wilson and Pavan (1959), Robertson et al. (1980), Hölldobler (1982c) |
| *Liometopum apiculatum* | Trail | Sternal (Pavan's) | B. Hölldobler (unpublished) |
| *Monacis bispinosa* | Trail | Sternal (Pavan's) | Wilson and Pavan (1959) |
| *Tapinoma* sp. | Trail | Sternal (Pavan's) | Wilson (1965a) |
| **FORMICINAE** | | | |
| *Acanthomyops claviger* | Trail | Hindgut | Hangartner (1969c) |
| *Camponotus ephippium* | Trail | Cloacal (stimulatory effect); hindgut (orienting cue) | Hölldobler (1982d) |
| *C. consobrinus, C. pennsylvanicus, C. perthiana, C. rufipes, C. sericeus, C. socius* | Trail | Hindgut (orientation, in association with motor display; in some cases poison as additional stimulus) | Hölldobler (1971c), Hölldobler et al. (1974), Barlin et al. (1976), Traniello (1977), Jaffe and Sanchez (1984a), B. Hölldobler (unpublished) |
| *Formica fusca* | Trail | Hindgut | Möglich and Hölldobler (1975) |
| *F. polyctena* | Trail | Hindgut (trail); Dufour's (possible source of food-recruitment "alarm" in trails) | Horstmann (1982b), Horstmann et al. (1982) |
| *Lasius fuliginosus, L. neoniger* | Trail | Hindgut | Carthy (1950), Hangartner and Bernstein (1964), Huwyler et al. (1975), Traniello (1980, 1983) |
| *Myrmecocystus mimicus, M. depilis* | Trail | Hindgut (in association with motor display and perhaps poison gland secretions) | Hölldobler (1981a) |
| *Myrmelachista ramulorum* | Trail | Hindgut | Blum and Wilson (1964) |
| *Oecophylla longinoda* | Trail | Rectal gland (orientation in association with motor displays) | Hölldobler and Wilson (1978) |
| *Paratrechina longicornis* | Trail | Hindgut | Blum and Wilson (1964) |

**FIGURE 7–46** Diversity of trail and recruitment pheromone glands in ants. Representative species in each subfamily are listed from top to bottom. **Ponerinae:** *Onychomyrmex hedleyi, Pachycondyla laevigata, Leptogenys chinensis* or *Cerapachys* sp., *Paltothyreus tarsatus;* **Ecitoninae** (formerly in subfamily Dorylinae): *Eciton hamatum;* **Myrmicinae:** *Orectognathus versicolor, Pogonomyrmex badius* or *Myrmica rubra, Solenopsis invicta* or *Monomorium pharaonis, Crematogaster ashmeadi;* **Aneuretinae:** *Aneuretus simoni;* **Dolichoderinae:** *Monacis bispinosa* or *Iridomyrmex* spp.; **Formicinae:** *Lasius* or *Formica* spp., *Oecophylla* spp., *Camponotus ephippium*. *C*, cloacal gland; *D*, Dufour's gland; *H*, hindgut; *PO*, poison gland; *PY*, pygidial gland; *R*, rectal gland; *S*, sternal glands; *T*, tibial gland. All sternal glands are labeled *S;* the morphology and anatomy of these organs are very diverse, however. Circumstantial evidence reported by Cammaerts-Tricot (1982) suggests that secretions obtained from the seventh abdominal sternite serve as an auxiliary trail pheromone in *Myrmica*. Most myrmicine ants possess a sternal gland in this region; therefore, we have tentatively marked this area with an *S*. In several species it has been demonstrated that more than one gland is simultaneously involved in trail and recruitment communication. For example, in *Pogonomyrmex* the Dufour's gland secretions appear to serve as long-lasting species-specific orientation cues, while the poison gland secretions function as stimulating, short-lasting (not species-specific) recruitment signals. (From Hölldobler, 1984c.)

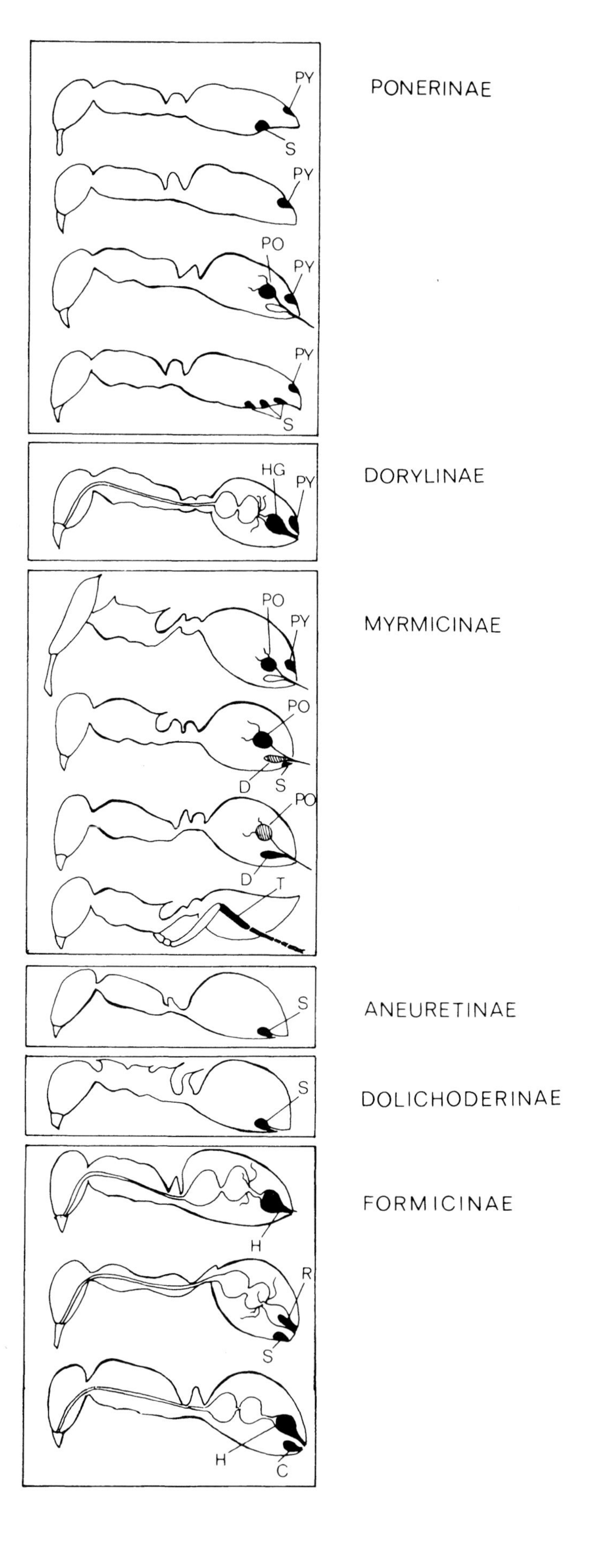


individuals over a meter or more. When the concentrated pheromone is allowed to diffuse from a glass rod held in the air near the nest, workers mass beneath it, and they can be led along by the vapor alone if the rod is moved slowly enough. When Wilson presented large quantities of the substance at the entrances of laboratory nests, he was able to draw out most of the inhabitants, including workers carrying larvae and pupae and, on a single occasion, the mother queen.

While laying a trail, the fire ant worker sometimes loops back in the direction of the food find, but only for short distances, before turning toward the nest again. If another worker is contacted, the homeward-bound worker turns to face her. She may do no more than rush against the encountered worker before moving on again, but sometimes the reaction is stronger: the recruiter climbs on top of the worker and, in some instances, shakes her body lightly but vigorously in a vertical plane. The movement may represent a modulatory signal that enhances the effect of the pheromone, but it does not appear to impart any essential information about the food find, because contacted workers do not exhibit trail-following behavior obviously different from those not contacted. Moreover, the pheromone is by itself sufficient to induce immediate and full trail following when laid down in artificial trails.

Most workers encountering a freshly laid trail respond at once by following it outward from the nest. They can detect the farnesene pheromones by smelling the vapor over distances as great as a centimeter. The workers do not follow a liquid odor trace on the ground. Instead, they move through the vapor created by diffusion of the substances through the air. There is an active space, which theoretical calculations show to be semi-ellipsoidal in shape, within which the pheromone is detected by the ants (Figure 7–48). As follower workers travel through this "vapor tunnel," they sweep their antennae from side to side, evidently testing the air for odorant molecules. In fact, they are able not only to detect these molecules in the gaseous state but also to move up gradients of molecular concentration, a process of orientation referred to as osmotropotaxis (Martin, 1964). When fire ants are presented with trail substance evaporating

**TABLE 7–6** Trail pheromones identified in ants.

| Compound | Molecular formula | Glandular source | Species and references |
|---|---|---|---|
| **Alcohols** | | | |
| 4-Methyl-3-heptanol | $C_8H_{18}O$ | Pygidial | *Leptogenys diminuta* (Attygalle et al., 1988) |
| **Aldehydes** | | | |
| (Z)-9-Hexadecenal | $C_{16}H_{30}O$ | Pavan's | *Iridomyrmex humilis* (Cavill et al., 1979, 1980; Van Vorhis Key et al., 1981; Van Vorhis Key and Baker, 1982a,b) |
| **Esters** | | | |
| Methyl 6-methylsalicylate | $C_9H_{10}O_3$ | Poison | *Tetramorium impurum* (Morgan and Ollett, 1987) |
| **Fatty acids** | | | |
| Decanoic acid | $C_{10}H_{20}O_2$ | Hindgut | *Lasius fuliginosus* (Huwyler et al., 1975) |
| Dodecanoic acid | $C_{12}H_{24}O_2$ | Hindgut | *Lasius fuliginosus* (Huwyler et al., 1975) |
| Heptanoic acid | $C_7H_{14}O_2$ | Hindgut | *Lasius fuliginosus* (Huwyler et al., 1975) |
| Hexanoic acid | $C_6H_{12}O_2$ | Hindgut | *Lasius fuliginosus* (Huwyler et al., 1975) |
| Nonanoic acid | $C_9H_{18}O_2$ | Hindgut | *Lasius fuliginosus* (Huwyler et al., 1975) |
| Octanoic acid | $C_8H_{16}O_2$ | Hindgut | *Lasius fuliginosus* (Huwyler et al., 1975) |
| Higher-molecular-weight fatty acids | — | Unknown | *Pristomyrmex pungens* (Hayashi and Komae, 1977) |
| **Nitrogen heterocycles** | | | |
| 2,5-Dimethylpyrazine | $C_6H_8N_2$ | Poison | *Tetramorium caespitum* (Attygalle and Morgan, 1983) |
| 3-Ethyl-2,5-dimethylpyrazine | $C_8H_{12}N_2$ | Poison | *Atta rubropilosa* (Cross et al., 1979); *A. sexdens* (Evershed and Morgan, 1983); *Manica rubida* (Attygalle et al., 1986); *Myrmica* spp., including *M. rubra* (Evershed et al., 1982); *Tetramorium caespitum* (Attygalle and Morgan, 1983) |
| Methyl 4-methylpyrrole-2-carboxylate | $C_7H_8NO_2$ | Poison | *Acromyrmex octospinosus* (Cross et al., 1982); *Atta cephalotes* (Riley et al., 1974b); *A. octospinosus* (Cross et al., 1982); *A. texana* (Tumlinson et al., 1971) |
| Monomorine I | $C_{13}H_{25}N$ | Poison | *Monomorium pharaonis* (Ritter et al., 1977b) |
| Monomorine III | $C_{14}H_{27}N$ | Poison | *Monomorium pharaonis* (Ritter et al., 1977b). True trail pheromone of *M. pharaonis* appears to be faranal from the Dufour's gland; monomorine I and III may act as additional attractants. |
| **Terpenoids** | | | |
| Z,Z,Z-Allofarnesene | $C_{15}H_{24}$ | Dufour's | *Solenopsis invicta* (Vander Meer, 1983, 1986a) |
| Faranal | $C_{17}H_{30}$ | Dufour's | *Monomorium pharaonis* (Ritter et al., 1977a) |
| Z,E-α-Farnesene, E,E-α-farnesene | $C_{15}H_{24}$ | Dufour's | *Solenopsis invicta* (Vander Meer et al., 1981; Vander Meer, 1983, 1986a) |
| Z,E-Homofarnesene, E,E-homofarnesene | $C_{16}H_{26}$ | Dufour's | *Solenopsis invicta, S. richteri* (Vander Meer, 1983, 1986a) |

in still air from the tip of a glass rod, they run directly to the glass rod. The sensory mechanism enabling fire ants to orient in this fashion is still unknown. However, in a set of ingenious experiments, summarized in Figure 7–49, Hangartner (1967) demonstrated the basis of osmotropotaxis in another trail-following ant species, *Lasius fuliginosus.* The method of following odor trails disclosed by these experiments makes it very unlikely that directional signals can be built into the trails. In other words, the odor streaks may or may not be tapered or shaped in some other way so as to point the way home—as discovered for example in *Myrmica ruginodis* trails by Macgregor (1948)—but it would be difficult for the follower ant to "read" this information. Additional experiments performed on *Las-*

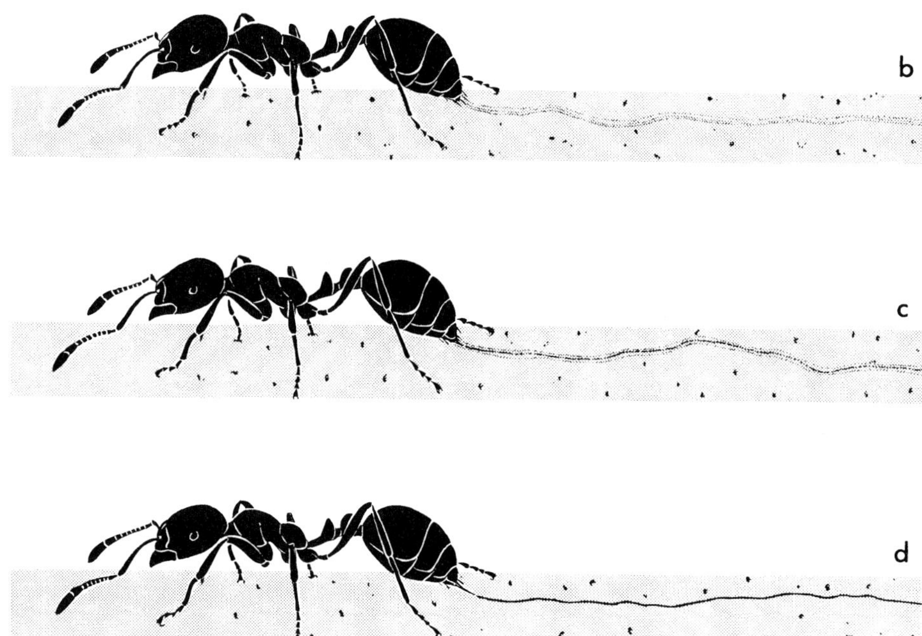


**FIGURE 7–47** A fire ant worker (*Solenopsis geminata*) lays an odor trail from her extruded sting. By allowing the ant to walk over pieces of smoked glass the tracks can be recorded of the feet (*a*), hairs at the tip of the abdomen (*b* and *c*), and the tip of the sting itself (*d*) from which the pheromone is emitted. (From Hölldobler, 1970a; drawing based on photographs by Hangartner, 1969b.)

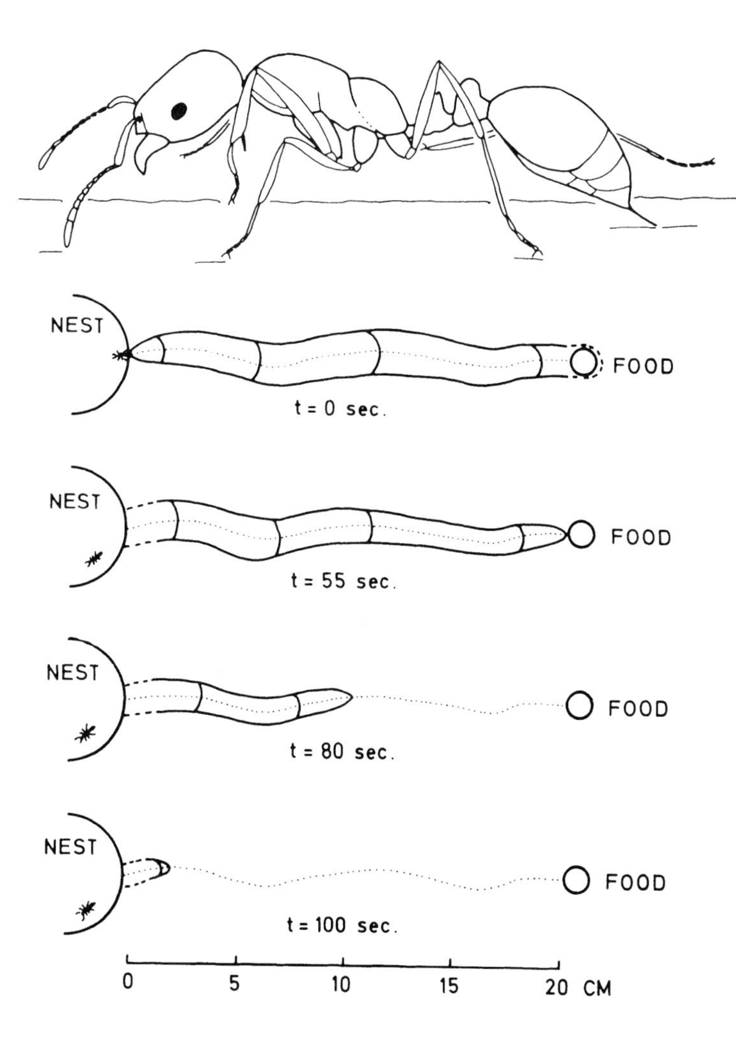


**FIGURE 7–48** The form of the odor trail of *Solenopsis invicta* laid on glass. As the pheromone (a mix of farnesenes) diffuses from its line of application on the surface, it forms a semiellipsoidal active space within which it is at or above threshold concentration. This space, and therefore the entire recruitment signal, fades after about 100 seconds. The times shown here are given from the moment a worker reaches the nest after laying a trail in a straight or slightly wavering line from a food source 20 centimeters away. The trail is shown in this model as continuous. In nature it is irregularly segmental, but the dimensions and fade-out times remain nearly the same. At the top a worker is shown laying a trail from right to left. (Redrawn from Wilson, 1962b, and Wilson and Bossert, 1963.)

*ius, Acanthomyops,* and *Solenopsis* have indicated in fact that no such information is transmitted (Carthy, 1951b; Wilson, 1962b). These findings refute the early hypothesis of Bethe (1898) that trails are effectively polarized, as well as the conjecture of Piéron (1904) that ants find their way back and forth by a special kinesthetic sense. Much more recently, Moffett (1987a) discovered that food-laden workers of *Pheidologeton diversus* orient along chemical trails but determine the home direction according to the directions taken by other laden nestmates.

Our present knowledge makes it desirable to reassess the value of Auguste Forel's mysterious theory of the "topochemical sense." Although this notion is frequently mentioned in the older literature, especially in connection with odor trails, there is general confusion over what it really means. Forel seems to have been talking about the perception of form and the spatial relation of discrete objects by means of smell, but his pronouncements on the subject were discouragingly obscure. Here is the clearest we have encountered: "By topochemical I mean a sense of smell which informs the ant as to the topography of the places surrounding it by means of chemical emanations, which give an odor to objects" (Forel, 1928: I, 116). The sense was said to be located in the terminal segments of the antennae. The perception of the spatial relation of objects implies an integration of sensory input in the central nervous system considerably more complex even than the following of an odor trail by osmotropotaxis. This has been well documented in the case of visual input (Jander, 1957; Wehner, 1981) but not yet in the case of olfactory input.

A more important finding from the work on *Solenopsis invicta* involves the way in which mass communication is achieved. By mass communication Wilson (1962b) meant information that can be transmitted only from one group of individuals to another group of individuals. In the case of *S. invicta,* the number of workers leaving the nest is controlled by the amount of trail substance being emitted by foragers already in the field. Tests involving the use of enriched trail pheromone showed that the number of individuals drawn outside the nest is a linear function of the amount of the substance presented to the colony as a whole. Under natural conditions this quantitative relation results in the adjustment of the outflow of workers to the level needed at the food source. An equilibration is then achieved in the following manner. The initial buildup of workers at a newly discovered food source is exponential, and it decelerates toward a limit as workers become crowded on the food mass because workers unable to reach the mass turn back without laying trails and because trail deposits made by single workers evaporate within a few minutes. As a result, the number of workers at food masses tends to stabilize at a level that is a linear function of the area of the food mass. Sometimes, for example when the food find is of poor quality or far away or when the colony is already well fed, the workers do not cover it entirely, but equilibrate at a lower density. This mass communication of quality is achieved by means of an "electorate" response, in which individuals choose whether to lay trails after inspecting the food find. If they do lay trails, they adjust the quantity of pheromone according to circumstances (Hangartner, 1969b; see Figure 7–47). The more desirable the food find, the higher the percentage of positive responses, the greater the trail-laying effort by individuals, the more trail pheromone presented to the colony and hence the more newcomer ants emerging from the nest.

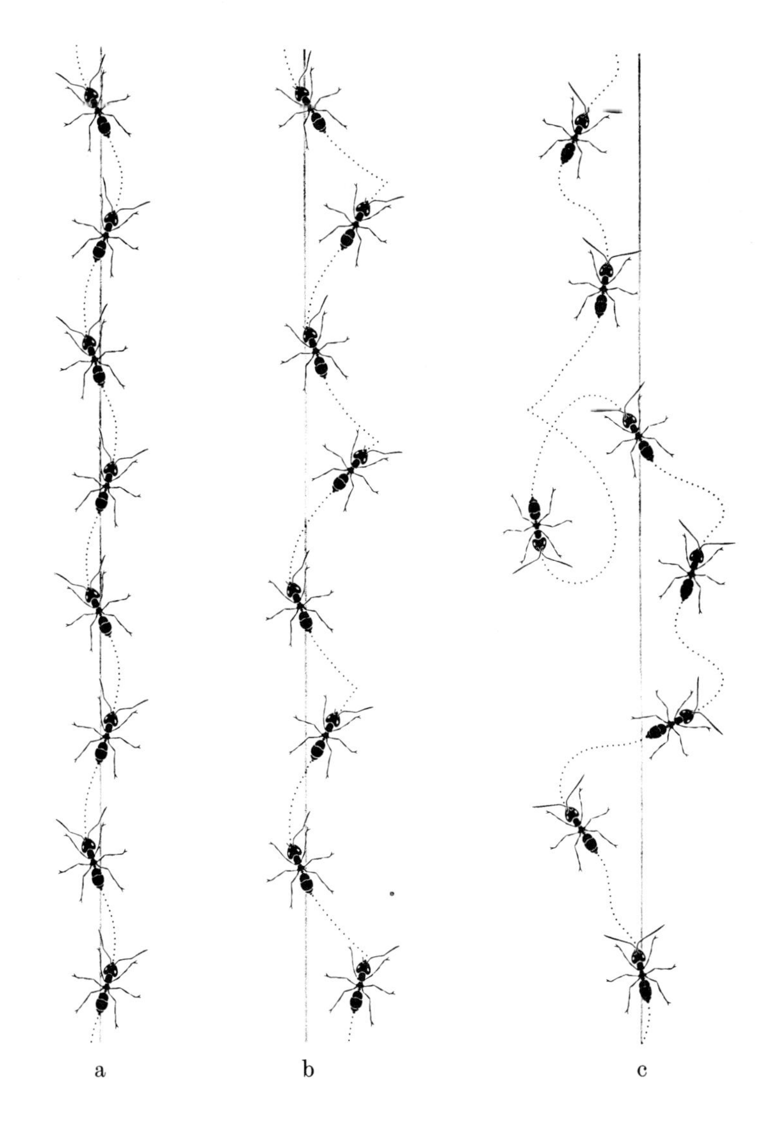


**FIGURE 7–49** Hangartner's demonstration that worker ants (*Lasius fuliginosus*) orient along odor trails by detecting varying concentrations of the trail pheromone in vapor form. The straight line represents (in idealized form) the point of application of the pheromone to the ground. (*a*) As a worker follows the trail, she tends to move out of the active vapor space, first to one side and then to another. As one antenna, containing the odor receptors, leaves the active space the ant swings back in the opposite direction. Consequently, she moves in a weaving pattern. (*b*) When the left antenna is amputated, the ant repeatedly overcorrects on the right side. (*c*) When the antennae are crossed and glued in this position, the ant is disoriented and recovers the active space only with difficulty. Nevertheless she is still able to move along in one direction, presumably with the aid of light-compass orientation. (From Hangartner, 1967.)

Thus the trail pheromone, through the mass effect, provides a control that is more complex than might have been assumed from knowledge of the relatively elementary individual response alone. This complexity is increased still further by the fact that the pheromone assumes different meanings in at least two different contexts. When colonies move from one nest site to another, a common event in the life of fire ants as well as many other kinds of ants, the new site is chosen by scout workers, who then lay odor trails back to the old nest. Other workers are drawn out by the pheromone. They investigate the new site and, if satisfied, add their own pheromone to the trail. In this fashion the number of workers traveling back and

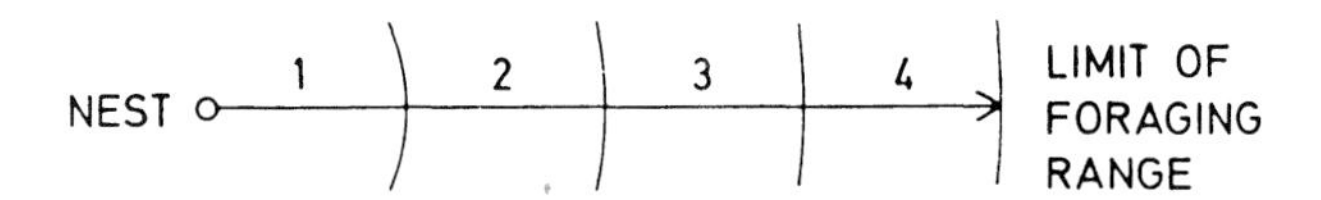


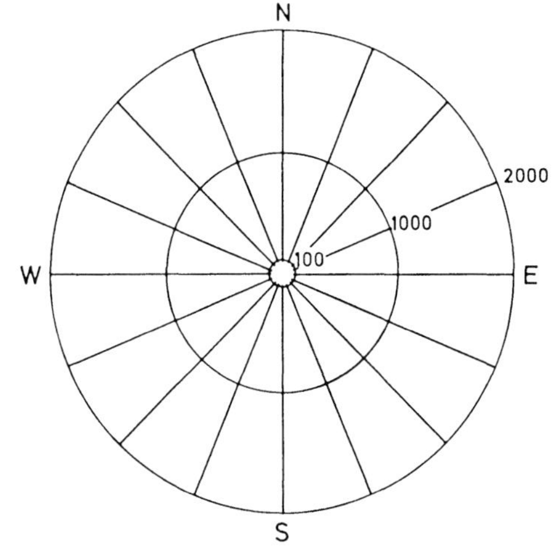


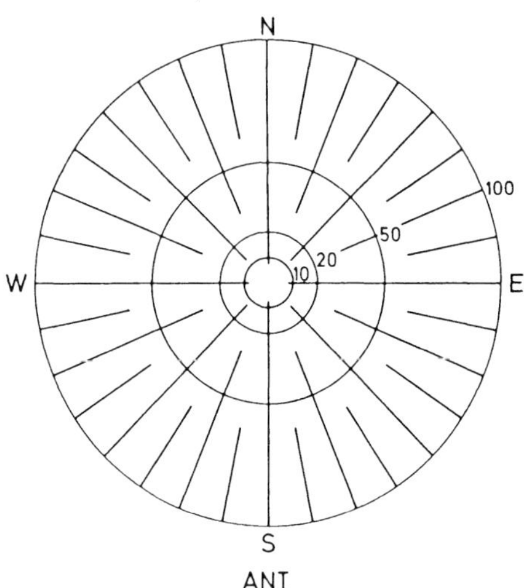


**FIGURE 7–50** These abstract figures represent the amounts of information transmitted by the honey bee (race *carnica*) around the time she performs the waggle dance, and by the fire ant when she lays an odor trail. (*Center*) The "bee compass" indicates that the worker honey bee receives up to 4 bits of information with respect to distance, or the equivalent of information necessary to allow her to pinpoint a target within 1 of 16 equiprobable angular sectors. The compass lines are represented arbitrarily as bisecting the sectors. The amount of direction information remains independent of distance, given here in meters. This last estimate is probably subject to revision, as explained in the text. (*Top*) The "distance scale" of both bee and fire ant communication shows that approximately 2 bits are transmitted, providing sufficient information for the worker to pinpoint a target within 1 of 4 equal concentric divisions between the nest and the maximum distance over which a single message can apply. (*Bottom*) The "fire ant compass" shows approximately how direction information increases with distance, given here in millimeters. (From Wilson, 1962b.)

forth builds up exponentially. In time the brood is transferred, the queen walks over, and the emigration is completed. The pheromone also functions as an auxiliary signal in alarm communication. When a worker is seriously disturbed, she releases some of the trail substance simultaneously with alarm substance from her head, so that nearby workers are not only alarmed but also attracted to the threatened nestmate.

Wilson (1962b) was able to measure the information transmitted by the odor trail of *Solenopsis invicta* and compare it with that transmitted by the waggle dance of the honey bee, which had been analyzed previously by Haldane and Spurway (1954). The method is straightforward. Workers were allowed to lay odor trails away from a food find (a drop of sugar water), and the starting point of the trail was marked in order to measure its distance from the food find—and hence, the initial error committed by the trail-laying ant. Similarly, the paths taken by the outward-bound follower ants and the distances of their nearest approaches on the trail to the food find were each recorded. The spatial precision of the two actions was then correlated and condensed into transmitted information, which was measured in bits. The bit is the amount of information needed to make a choice between two equiprobable alternatives. If $n$ alternatives are present, a choice provides the following quantity of information: $H = \log_2 n$. Thus in a most elementary communication system, the sending of one of $n$ equiprobable messages reduces $\log_2 n$ amount of uncertainty. By definition, it provides that much information.

Suppose we ourselves were communicating information about direction, and suppose our system allowed us to transmit any one of 16 directions with perfect accuracy. Then the message "go north by northwest," one of 16 equiprobable messages, conveys $\log_2 16 = 4$ bits of information. The essential results for honey bees and fire ants are shown in Figure 7–50. The key result is that both systems transmit about the same amount of information concerning both direction and distance. Also, perhaps coincidentally, they convey a degree of precision comparable to that of human beings when we think about spatial locations and express them verbally. Thus we can just about place direction to "north by northwest" mentally, and to no compass direction of finer degree; this also happens to be the performance of the honey bee and fire ant.

Why don't the social insects do better? As Haldane and Spurway noted in the case of the honey bee, "Natural selection is always acting so as to reduce the error of the mean direction, while acting less intensely, if at all, to reduce individual errors which lead to a scatter round this direction." They cite the analogous problem in naval gunnery, "where a superior force pursuing ships with less fire power should fire salvoes with a considerable scatter, in the hope that at least one shell will hit a hostile ship and slow it down." The same analogy at first glance appears even more relevant in the case of the fire ant. One of the chief problems faced by this species in the course of foraging is the recruitment of sufficient numbers of workers in time to immobilize small prey animals detected while passing through the colony territory. In laboratory arenas workers often succeed in capturing insects only because they deviate from odor trails that had been rendered inaccurate by the continued movement of the prey. Deneubourg et al. (1983) have also pointed out the importance of "noise" in chemical trail recruitment in ants for simultaneously directing and spreading the foraging efforts.

There is an alternative and simpler explanation of the quantity of information transmitted by ant recruitment trails. It is possible that the level of accuracy of the pheromone system has been arrived at as a compromise between the utmost effort of the ants' chemosensory apparatus to follow trails accurately and the simultaneous need

**TABLE 7–7** Species that use tandem running in recruitment.

| Species | Target | Authority |
|---|---|---|
| **PONERINAE** | | |
| *Diacamma rugosum* | Nest site | Fukumoto and Abe (1983), Maschwitz et al. (1986c) |
| *Hypoponera* spp. | Food, nest site | Agbogba (1984), Hölldobler (1985) |
| *Pachycondyla caffraria* | Food | Agbogba (1984) |
| *P. constricta* | Probably food | S. Levings (personal communication) |
| *P. crassa* | Probably food, nest site | W. L. Brown (personal communication), B. Hölldobler (unpublished) |
| *P. harpax* | Probably food | S. Levings (personal communication) |
| *P. impressa* | Probably food | S. Levings (personal communication) |
| *P. obscuricornis* | Nest site | Hölldobler and Traniello (1980a), Traniello and Hölldobler (1984) |
| *P.* (= *Bothroponera*) *tesserinoda* | Food, nest site | Hölldobler et al. (1973), Maschwitz et al. (1974) |
| *Paltothyreus tarsatus* | Nest site | Hölldobler (1984b) |
| *Ponera coarctata* | Nest site | N. Carlin (personal communication) |
| *P. eduardi* | Food?, nest site | Le Masne (1952), U. Maschwitz, B. Hölldobler and K. Böhringer (unpublished) |
| **MYRMICINAE** | | |
| *Cardiocondyla emeryi* | Food | Wilson (1959b) |
| *C. venustula* | Food | Wilson (1959b) |
| *Chalepoxenus muellerianus* | Slave raids | Buschinger et al. (1980a) |
| *Harpagoxenus canadensis* | Slave raids | Buschinger et al. (1980a), Stuart and Alloway (1983) |
| *H. sublaevis* | Slave raids | Buschinger and Winter (1977) |
| *Leptothorax acervorum* | Food, nest site | Dobrzański (1966), Möglich et al. (1974), Möglich (1978) |
| *L. ambiguus* | Food, nest site, nest defense | T. M. Alloway (personal communication) |
| *L. crassipilis* | Food, nest site | Möglich (1978) |
| *L. curvispinosus* | Food, nest site, nest defense | Möglich (1978), T. M. Alloway (personal communication) |
| *L. longispinosus* | Food, nest site, nest defense | Möglich (1978), T. M. Alloway (personal communication) |
| *L. muscorum* | Food, nest site | Möglich (1978) |
| *L. nylanderi* | Food, nest site | Möglich (1978) |
| *L. rugatulus* | Food, nest site | Möglich (1978) |
| *L. unifasciatus* | Food, nest site | Lane (1977) |
| **FORMICINAE** | | |
| *Camponotus consobrinus* | ? | B. Hölldobler (unpublished) |
| *C. ocreatus* | ? | G. Alpert (personal communication), B. Hölldobler (unpublished) |
| *C. perthiana* | ? | B. Hölldobler (unpublished) |
| *C. sericeus* | Food, nest site | Hingston (1929), Hölldobler et al. (1974), Möglich et al. (1974) |

to reduce the quantity and to increase the volatility of the trail pheromones in order to minimize overcompensation in the mass response. The amount of the trail substances is indeed microscopic, which seems logical if the mass response is to be finely governed. Which of the two evolutionary hypotheses is closer to the truth cannot be determined until deeper physiological analyses have been made.

The mass communication system exemplified by *Solenopsis invicta* clearly represents an advanced evolutionary grade. In order to identify the more primitive forms of recruitment from which this system may have evolved, we must discover and characterize less sophisticated modes of recruitment communication. Tandem running, used variously during emigration and recruitment to food, is generally considered to be one such mode. Only a single nestmate is recruited at a time, and the follower has to stay in direct antennal contact with the leader. The pair then proceed to the target site while remaining tightly linked together. The phenomenon was described for the first time, albeit sketchily, by the Swedish entomologist Gottfrid Adlerz in 1896, who saw a *Leptothorax* slave worker leading a *Harpagoxenus* (= *Tomognathus*) *sublaevis* slavemaker to a new nest site (cited by Stuart, 1986; confirmed by Buschinger and Winter, 1977). It was encountered a second time in the tropical Asian ant *Camponotus sericeus* by Hingston (1929), and then described in detail for the first time in *Cardiocondyla* by Wilson (1959b), who also first applied the expression "tandem running." The ensuing thirty years witnessed the discovery of this form of communication across a wide range of additional genera in the Ponerinae, Myrmicinae, and Formicinae (Table 7–7). Also, the first experimental analyses have been conducted on the signals employed by the ants.

One of the best-understood species is the little European myrmicine *Leptothorax acervorum* (Möglich et al., 1974; Möglich, 1978, 1979). When a successful scout returns to the colony, she first regurgitates food to several nestmates. Then she turns around and tilts her gaster upward into a slanting posture (see Figure 7–36). Simultaneously she exposes her sting and extrudes a droplet of liquid. Nestmates are attracted by this "tandem calling" behavior, as Möglich and his co-workers have labeled it. When the first ant arrives at the calling ant, she touches the caller on the hind legs or gaster with her antennae, and tandem running commences. The recruiting ant leads the nestmate to the newly discovered food source. During tandem running the leader ant lowers the gaster while the sting remains extruded. But the sting is not dragged over the surface, as in the case of *Solenopsis invicta* and other species that lay chemical trails from their stings. Instead, the *L. acervorum* follower keeps close antennal contact with the leader, repeatedly touching her hind legs and gaster. Whenever this contact is interrupted, as for example when the follower accidentally loses her leader or is removed experimentally, the leader immediately stops and resumes the calling posture. Once again she stands still and points her abdomen obliquely upward. She may remain in this posture for several minutes, continuously discharging the calling pheromone. Under normal circumstances the lost follower quickly orients back to the leader ant and tandem running is resumed. Closely similar behavior has been observed in *L. muscorum* and *L. nylanderi*.

Experiments on this interesting recruitment behavior have revealed tandem running to be mediated by signals in two sensory modalities:

1. If a *Leptothorax acervorum* tandem pair has been separated, the leader immediately stops and resumes the calling posture. However, when the ant is carefully touched with a hair at the hind legs or gaster with a frequency of at least two contacts per second, the leader continues running to the target area. This experiment shows that the presentation of the tactile signals normally provided by the follower ant is sufficient to release "tandem running" by a leader ant.

2. The calling pheromone originates from the poison gland. In the studies by Möglich and his co-workers, ants were strongly attracted to dummies that had been contaminated with poison gland secretions but not to those contaminated with Dufour's gland secretions. Further experiments revealed that the poison gland substance not only functions as a calling pheromone but also plays an important role during tandem running itself by binding the follower ant to the leader. Möglich et al. (1974) found that the leader could easily be replaced by a dummy contaminated with poison gland secretions. Gasters of freshly killed ants from which the sting and venom glands had been removed could not replace a leader ant. When these body parts were contaminated with secretions of the poison gland, however, they functioned effectively as leader dummies.

The discovery of tandem calling with pheromones in *Leptothorax* throws considerable light on the evolution of chemical recruitment techniques in myrmicine ants generally. It now seems very plausible that the highly sophisticated chemical mass recruitment performed by *Solenopsis, Monomorium, Pheidole, Pheidologeton,* and certain other myrmicine ants was derived from chemical tandem-calling behavior of the *Leptothorax* mode. With the exception of *Crematogaster,* which produces a trail pheromone in the tibial glands of the hind legs, all other myrmicine species generate the pheromone from one of the sting glands. As Morgan (1984) has pointed out, the trail pheromone is found mostly in the poison gland of species with a proteinaceous venom, and mostly in the Dufour's gland in species with strongly alkaloidal venom (as in *Solenopsis* and *Monomorium*). Some ant species have lost the ability to sting but still retain a proteinaceous poison secretion. It is likely that pheromonal calling, during which an alerting and attracting pheromone is discharged through the sting into the air, was one of the first steps that led to chemical trail laying and mass communication in the later evolution of myrmicine ants.

Chemical calling, which as we noted in the discussion of signal economy closely resembles tandem calling, is also employed by reproductive castes of the social parasites *Doronomyrmex kutteri* and *Harpagoxenus sublaevis* (Buschinger, 1968a, 1971a,b). *Harpagoxenus* uses tandem running during slave raids (Buschinger and Winter, 1977; Buschinger et al., 1980a; see Figure 12–12). The manufacture of sex pheromones in the sting glands is widespread among myrmicine ants, having been documented in *Harpagoxenus, Leptothorax, Monomorium, Xenomyrmex,* and *Pogonomyrmex* (for review see Hölldobler and Bartz, 1985). It is therefore reasonable to suppose that in at least some myrmicine phyletic lines, sex attractants and recruitment pheromones had a common evolutionary origin (Hölldobler, 1978). In fact, the same substances may have originally served as sex pheromones in one context and as recruitment signals in another. Nearly identical modes of recruitment communication have been discovered in other ant subfamilies. But since the anatomical origins of the pheromones are different, we have to assume that the patterns themselves, however outwardly similar, have evolved independently several times. The parallel is especially striking between the Ponerinae and Myrmicinae.

Even some ponerine species that are exclusively solitary foragers employ tandem running recruitment during nest relocations. In fact, it is possible that tandem running recruitment, like social carrying, first evolved as a device to lead stray nestmates back to the nest. Many ponerine ants construct relatively simple nests in soil that are friable and prone to cave-ins. It would be advantageous for workers to be able to lead nestmates to intact portions of the nests with little delay. Several species of the ponerine genus *Pachycondyla* employ tandem running when recruiting nestmates to new nest sites or food sources (Maschwitz et al., 1974; Traniello and Hölldobler, 1984). A recruiting worker of *P. obscuricornis* "invites" a nestmate with a ritualized display that leads to the formation of the tandem pair. The nestmate then follows the recruiter to the new nest by keeping in close contact with her hind legs. If the contact is broken, as for example when the follower accidentally loses the leader or is removed experimentally, the leader immediately stops. Only after she receives the tactile signal on her hind legs or gaster does she continue to travel toward the target area. During tandem running, secretions from the pygidial gland released by the leader ant provide a chemical bond between her and the follower (Hölldobler and Traniello, 1980a). In the workers of some other ponerine species, including at least a few *Pachycondyla* (= *Bothroponera*), *Hypoponera,* and *Diacamma,* which also recruit by tandem running, the pygidial gland secretion is evidently not involved (Jessen and Maschwitz, 1986; Hölldobler, unpublished data).

*Pachycondyla obscuricornis* is a solitary forager that uses tandem running only during nest emigration. In contrast, *P. laevigata* is an obligate predator on termites, and conducts massive group raids when foraging. The raids are organized by a powerful trail pheromone discharged from the pygidial gland by scout ants (Hölldobler and Traniello, 1980b; see Figures 7–22 and 7–23). The shift in diet from evenly dispersed to strongly clumped food sources (termites) apparently led to the use of the pygidial gland secretions in mass communication and the transformation of the pheromones into stimulatory and orienting signals. As we showed earlier, queens of some myrmicine species display sexual calling behavior identical to the tandem calling behavior of the workers. Again, the parallel to the Ponerinae in this respect is striking: here the pygidial gland appears to have a functional repertory identical to that of sting glands in the Myrmicinae. Depending on the species and the behavioral context, the secretions of the ponerine pygidial glands can function

**FIGURE 7–51** Recruitment to food in the tropical Asiatic ant *Camponotus sericeus* by a form of tandem running evolved independently from that of *Leptothorax* and other myrmicines. (*1–5*) A recruiting ant (*black*) approaches nestmates head-on and performs food-offering rituals, each of which lasts only a few seconds. Many of the ants encountered try to keep close antennal contact with the recruiter. But when this worker finally leaves the nest to return to the food source, only the single nestmate maintaining closest contact is successful in tandem following. (*Bottom panel*) Diagram of the time course of a typical recruitment episode. •, Worker returning from the food source to the nest; Δ, single worker leaving the nest; the paired and shaded triangles indicate tandem pairs leaving the nest. (From Hölldobler et al., 1974.)

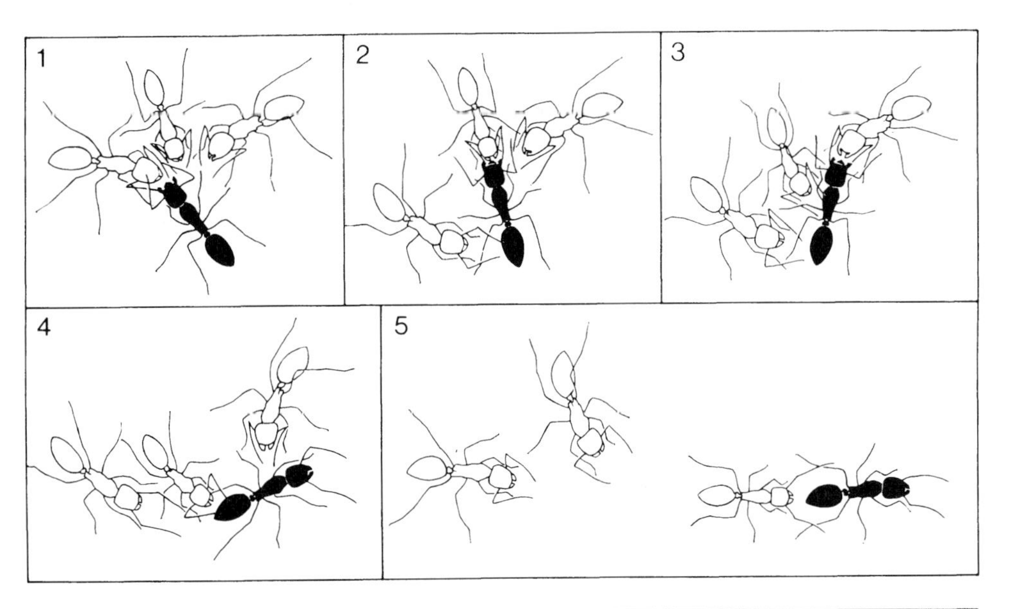


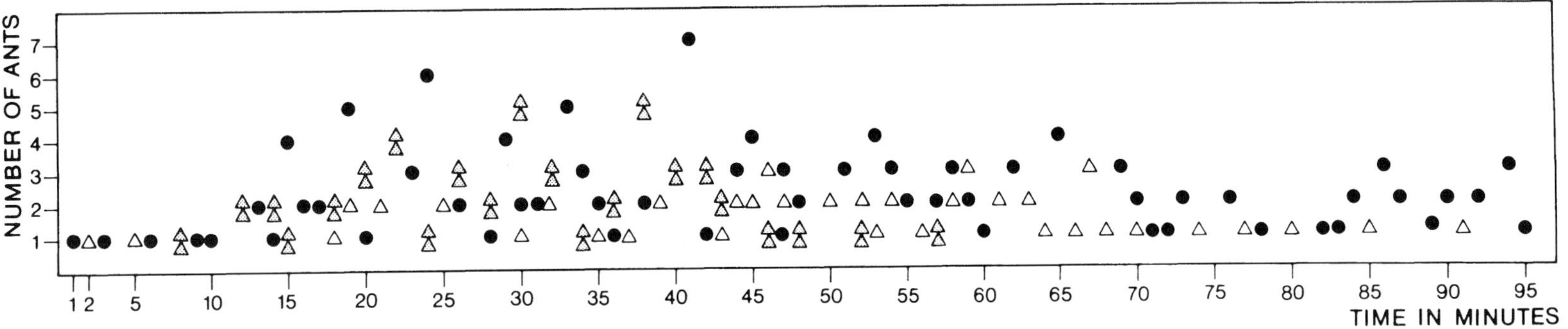


as tandem running pheromones, recruitment trail pheromones, or sex pheromones (Hölldobler and Haskins, 1977).

In short, comparative studies have revealed several strikingly convergent pathways in the evolution of mass communication in ants (see also Jaffe, 1984). On the other hand, morphological and behavioral findings indicate that communication by chemical trails in ants is considerably more diverse overall than researchers had assumed. As illustrated in Figure 7–46, no fewer than ten different anatomical structures have been identified in various species of ants as sources of trail pheromones. In the Ponerinae alone four different trail pheromone glands have been identified. It is obvious that trail communication has evolved many times independently. Even in the same subfamily the mechanisms and anatomical structures for trail communication have diverged considerably. In the termitophagous species *Megaponera foetens* and in several group-raiding *Leptogenys* species, the trail pheromone originates from the poison gland (Fletcher, 1971; Maschwitz and Mühlenberg, 1975; Longhurst et al. 1979b). In some *Leptogenys* species at least, a second recruitment pheromone originates in the pygidial gland (Maschwitz and Schönegge, 1977; B. Hölldobler, unpublished data). In the myrmecophagous group-raiding species *Cerapachys turneri*, the trail pheromone originates from the poison gland, but appears to be reinforced by a chemical recruitment signal from the pygidial gland (Hölldobler, 1982b). In several species of the legionary and group-raiding genus *Onychomyrmex*, nest emigrations and raids are organized by trails laid with secretions from a large sternal gland located between the Vth and VIth abdominal sternites (Hölldobler et al., 1982; see Figures 7–8 and 7–24). Finally, the African stink ant *Paltothyreus tarsatus*, a scavenger and termite predator, employs tandem running during nest relocation and trail communication during foraging. The trail pheromone, which functions mostly as an orientation cue, originates from intersegmental sternal glands located between abdominal sternites IV and V, V and VI, and VI and VII (Hölldobler, 1984b; see Figure 7–7).

Trail communication appears to be much more common in the Ponerinae than researchers had previously believed. Overal (1986) has provided circumstantial evidence that the Neotropical species *Ectatomma quadridens* employs chemical trails in nest emigrations and recruitment of foragers. S. Pratt (personal communication) discovered trail communication in *E. ruidum* and identified the Dufour's gland as the source of the trail pheromone. Similarly, Breed et al. (1987) demonstrated that the giant Neotropical ponerine *Paraponera clavata* exhibits graded recruitment responses, depending on the type, quantity, and quality of the food source. Their observations suggested that a trail pheromone is used in orientation by individual ants, independent of recruitment, very similar to that documented in *Pachycondyla tesserinoda* by Jessen and Maschwitz (1985, 1986). On the other hand *P. tesserinoda* recruits by tandem running, while *Paraponera clavata* appears to have a recruitment system similar to that of *Paltothyreus*. In other words, the recruitment signal by which nestmates are stimulated to leave the nest and follow the orientation trail appears to be separate from the trail pheromone (Hölldobler, 1984b).

The trail pheromones of most formicine ants originate from the hindgut (see Table 7–5). Because of this distinctive trait, the peculiar chemical nature of the pheromones (see Table 7–6), and the strongly divergent position of the Formicinae suggested by their anatomy, it is likely that the recruitment systems of the Formicinae evolved independently from those of the Ponerinae and the Myrmicinae. Yet studies of the genus *Camponotus* have led to the conclusion that mass recruitment methods originated stepwise from a convergently evolved tandem running (Hölldobler et al., 1974; Hölldobler, 1978, 1981c).

The recruitment system of *Camponotus sericeus* exemplifies the more elementary evolutionary grade of tandem running in the Formicinae. The first scouting ant to discover the food source typically fills her crop and returns to the nest. As the worker heads home, she touches the tip of her abdomen to the ground for short intervals. Tracer experiments have shown that the ant is depositing chemical signposts with material from her hindgut. Inside the nest she performs short-lasting fast runs, which are interrupted by food exchange and grooming. During individual recruitment episodes these rituals are repeated 3 to 16 times. The behavior evidently serves to keep nestmates in close contact with the successful scout. When the scout finally leaves the nest to return to the food source, most of the nestmates that she encountered try to follow her. However, only the ant maintaining the closest antennal contact succeeds in accompanying her out of the nest (see Figure 7–51). Most of the followers who reach the food source in this manner soon return to the nest and commence to recruit nestmates on their own. Experiments have shown that the hindgut trail laid by homing scouts has no recruiting effect by itself. Only experienced ants follow the trail, and they appear to use it exclusively for orientation. Similarly, during tandem running the trail pheromone appears to serve no significant communicative function. The leader and follower are bound together by a continuous exchange of tactile signals and the perception of persistent pheromones on the surface of the body. Tandem running is also used in *Camponotus sericeus* to recruit nestmates to new nest sites (see Figure 7–55).

We now come to the next higher organizational level of recruitment communication in formicine ants, in a system that Hölldobler has called group recruitment. In this case one ant summons from 5 to 30 nestmates at a time, and the recruited ants follow closely behind the leader ant to the target area (Figure 7–52). This behavior has been observed in *Camponotus compressus* (Hingston, 1929), *C. beebei* (Wilson, 1965b), *C. socius* (Hölldobler, 1971c), *C. ephippium* (Hölldobler, 1982d), *C. gigas* (U. Maschwitz, personal communication), and in several *Polyrhachis* species (B. Hölldobler, unpublished).

In the case of *C. socius*, a strikingly colored terrestrial ant from the southern United States, Hölldobler found that scouts use chemical signposts around newly discovered food sources and lay a trail with hindgut contents from the food source to the nest. The trail pheromone alone, however, does not induce recruitment to any significant extent. Inside the nest the recruiting ant performs a waggle display when facing nestmates head-on (see Figure 7–53). The vibrations last 0.5–1.5 seconds and comprise 6–12 lateral strokes per second. Nestmates are alerted by this behavior and subsequently follow the recruiting ant to the food source. Hölldobler demonstrated the significance of the motor display inside the nest by closing the gland openings of recruiting ants with wax plugs. With the

**FIGURE 7–52** Group recruitment in *Camponotus socius*. (*Above*) The leader ant (*L*) lays a hindgut odor trail to the target site, and the pheromone trace is followed closely by a group of workers alerted in the nest by a waggle display. (*Below*) Close-up view of a leader ant laying an odor trail while it is followed closely by a nestmate. (From Hölldobler, 1971c.)

waggle display thus separated from the chemical signals, it was proved that only workers stimulated first by the waggle display follow an artificial trail drawn with hindgut contents. The presence of a leader ant is still essential for a complete performance, however. In the experiments freshly recruited ants without a leader followed a hindgut trail through a distance of only about 100 centimeters. Similar behavioral patterns are employed during recruitment to new nest sites. The main difference is that the motor display leading to emigration is more frequently a back-and-forth jerking movement (see Figure 7–53). Also, in contrast to recruitment to food sources, males respond to the emigration signals and hence are recruited (see Figure 7–59).

The next higher organizational level within the formicine ants is represented by species in which workers stimulated by a motor display follow the trail to the food source even in the absence of the recruiting ant. In *Formica fusca*, for example, successful scouts lay a hindgut trail from the food source to the nest. The pheromone has no primary stimulating effect. After the scout has performed a vigorous waggle display inside the nest, however, frequently interrupted by food exchanges, nestmates rush out and follow the trail to the food source without receiving additional cues from the recruiter (Möglich and Hölldobler, 1975).

*Camponotus pennsylvanicus* represents the next most advanced level. Scouts returning from newly discovered food sources also lay odor trails. These individuals further stimulate nestmates by a waggle motor display. When nestmates are alerted by the display, they follow the previously laid trail. The scout does not usually guide the recruited group to the target area (Traniello, 1977). However, *C. pennsylvanicus* workers encountering the odor trail follow it even without being mechanically stimulated by the scout ant. Barlin et al. (1976) have identified a chromatographic peak, evidently corresponding to a single substance, whose harvested fraction releases trail-following behavior. Nevertheless, motor displays remain an integral part of the recruitment process in *C. pennsylvanicus*. The number of ants responding to an artificial trail consisting of hindgut contents plus poison gland secretion is higher if a scout is allowed first to stimulate her nestmates with motor displays.

Finally, from the *Camponotus pennsylvanicus* level it is only a short step to chemical mass communication of the *Solenopsis* kind, where the trail pheromone is the overwhelmingly prevalent recruitment signal and the outflow of foragers is controlled by the amount of pheromone discharged. It appears that this evolutionary grade has been attained within the Formicinae as within the Myrmicinae, but studies to date have not established that fact beyond doubt. Good candidates are *Paratrechina longicornis* and *Lasius fuliginosus*, both of which readily follow artificial trails made from the rectal sac (Blum and Wilson, 1964; Hangartner, 1967).

The cumulative studies have made it clear that motor displays and tactile signaling play an important role during recruitment communication in many ant species (see also Sudd, 1957; Szlep and Jacobi, 1967; Leuthold, 1968b; and Szlep-Fessel, 1970). It appears, however, that during the course of evolution these signals became less important with the increasing sophistication of the chemical recruitment system. The correlates of this advance remain to be worked out, but it is evident that they include the population size of colonies. The larger the mature colony size among ant species, the more reliance is placed on trail pheromones as opposed to motor displays in the initiation of recruitment. Also, chemical trails become more important than tandem running in the orientation of the follower ants.

It is further evident that formicine ants evolved chemical trails by means of a ritualization of defecation. Hindgut contents are discharged by ants at necessarily frequent intervals. A comparative study has revealed that in many species workers do not defecate randomly but visit specific locations preferentially. Besides certain sites within the nest, they favor nest borders, garbage dumps ("kitchen middens" as they are called by entomologists), and the borders of trunk trails that lead to permanent food sources or connect multiple nest entrances. Such disposal areas also seem to be ideally suited to receive chemical cues used in home range orientation. In fact, this phenomenon has been documented in a number of species (Hölldobler, 1971b; Hölldobler et al., 1974). The *Camponotus sericeus* pattern, in which the scout first deposits the pheromone and then uses it to lead nestmates to food sites, is evidently primitive. It occurs mostly in monomorphic species with smaller colonies and is widespread in the primitive subfamily Ponerinae. We may reasonably suppose that in the Formicinae hindgut material first became an important cue in home range orientation and then was transformed into a more specific orienting and stimulating signal used during recruitment. In fact, Traniello (1980) discovered colony specificity in the hindgut-derived trail pheromone of *Lasius neoniger*. This may be a more common phenomenon in ants generally, because colony specificity in the orientation trails has also been found

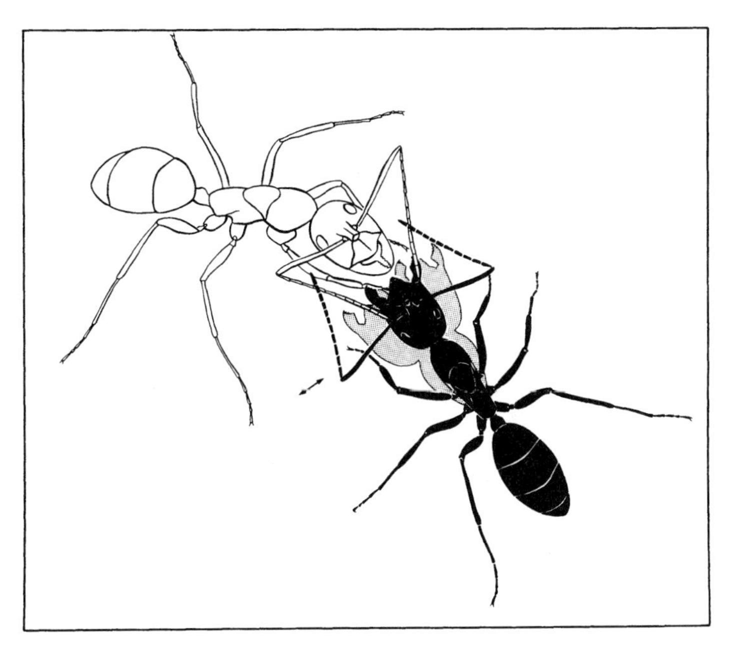

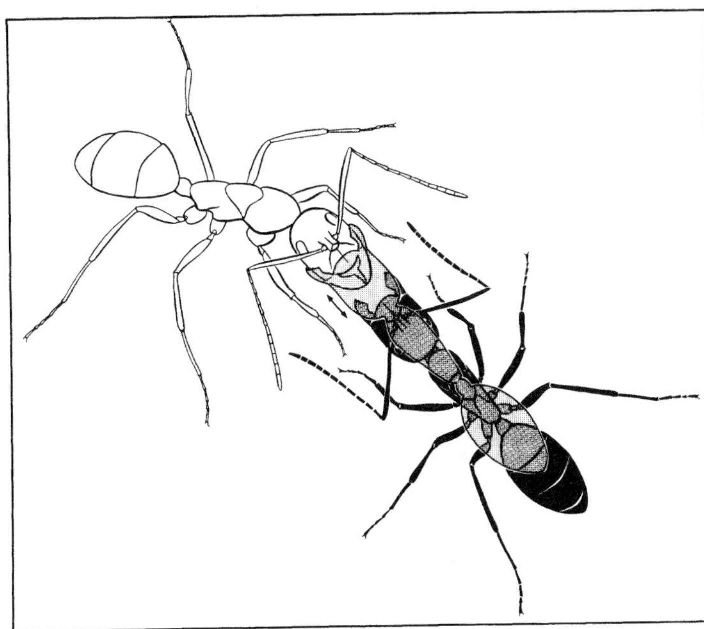

**FIGURE 7–53** In *Camponotus socius* the invitation signal employed during recruitment to food (*above*) is a lateral wagging of the body that appears to be a ritualization of food offering. It differs from the invitation signal employed during emigration (*below*), which is a back-and-forth jerking of the body, apparently a ritualization of the more primitive mandible pulling displayed by *C. sericeus* (see Figure 7–57). (From Hölldobler, 1971c.)

in *Pogonomyrmex* (Regnier et al., 1973; Hölldobler, 1976a) and *Oecophylla* (Hölldobler, 1979). Even more surprising is the recent discovery that workers of certain ant species lay down individual-specific trail markers. This level of specificity occurs in *Pachycondyla tesserinoda* (Jessen and Maschwitz, 1985, 1987), *Leptothorax affinis* (Maschwitz et al., 1986b), and possibly also in *Paraponera clavata* (M. D. Breed, personal communication).

The ritualization hypothesis of the origin of recruitment trails can be tested by broad comparative analyses of the many formicine tribes and genera that remain unstudied. This is especially true of the ones that are relatively primitive by anatomical criteria, includ-

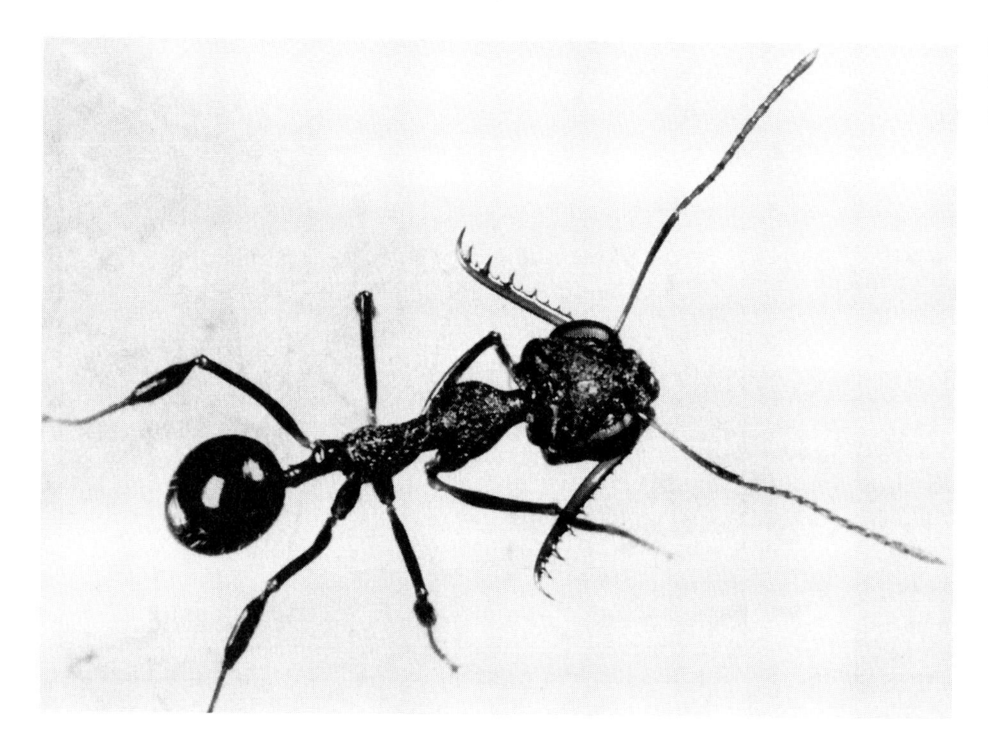

**FIGURE 7–54** Worker of the formicine *Myrmoteras toro,* with her specialized trap mandibles folded back. (Photograph by M. W. Moffett.)

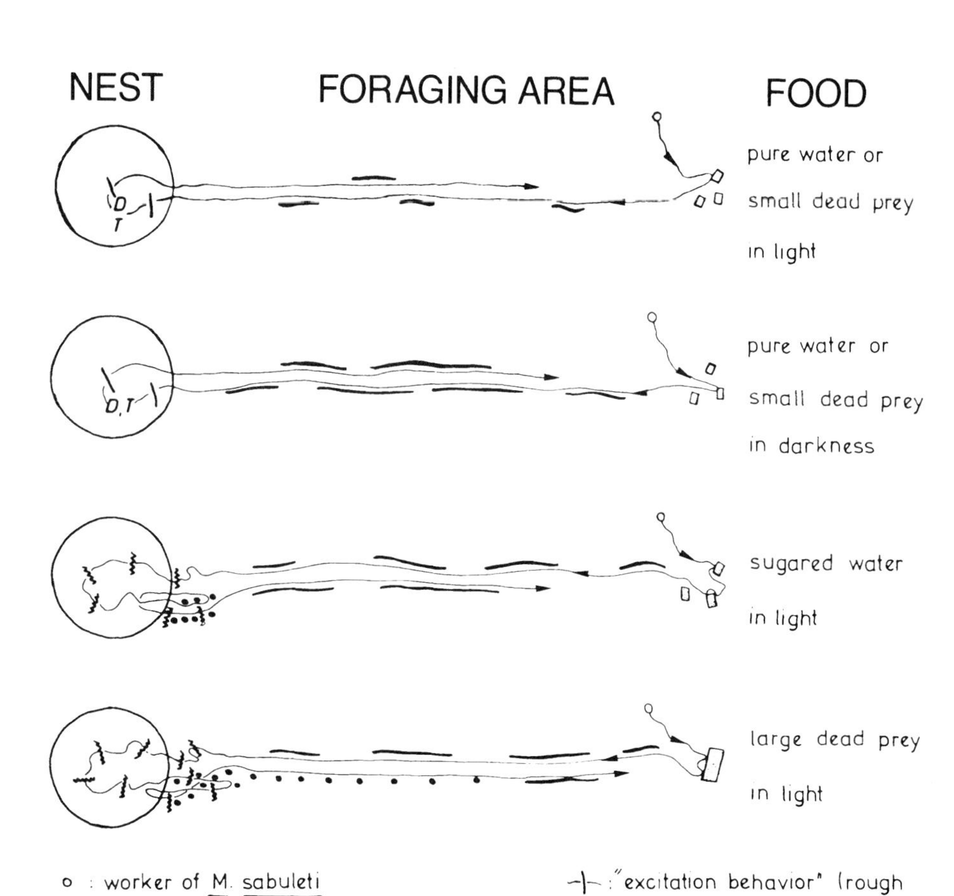


**FIGURE 7–55** The graded recruitment system of *Myrmica sabuleti,* a species inhabiting woodlands and meadows in Europe. (Modified from Cammaerts and Cammaerts, 1980.)

ing *Myrmecorhynchus* and *Notoncus* of Australia and *Gesomyrmex* of tropical Asia. It is possibly significant in this respect that the tropical Asiatic *Myrmoteras*, which has highly specialized predatory trap jaws (Figure 7–54) but is primitive in many other physical and sociobiological features, lacks a recruitment system altogether (Moffett, 1986c). At the opposite extreme is the African weaver ant *Oecophylla longinoda*, which has evolved a novel organ, the rectal gland, to replace the hindgut. This arboreal species uses hindgut material as a territorial pheromone (Hölldobler and Wilson, 1978).

Another trend of increasing sophistication among the ants generally is the use of a mix of chemical components, each of which serves a different role in the recruitment process. We noted earlier that the principal trail pheromone of the fire ant *Solenopsis invicta* is *Z,E*-α-farnesene, which is responsible for most of the following response of nestmates. Foraging workers do not follow trails made from this substance alone, however. The combination of the pheromone with other Dufour's gland farnesenes and *n*-heptadecane improves the response, but the blend still does not work as well as the complete Dufour's gland extract. It appears that still another, unidentified "priming" compound exists in the natural secretion (Vander Meer, 1983, 1986). Why a multiplicity of elements is required to obtain the full response is not clear, but it may serve to enhance the specificity of the trails. The North American ant species *S. invicta*, *S. geminata*, *S. richteri*, and *S. xyloni* respond to each other's trails weakly or not at all (Blum, 1982), despite the fact that at least two of the species, *S. invicta* and *S. richteri*, share some of the components. The crucial difference lies in the blend.

Yet another trend toward complexity is to vary the signals given and their intensity in order to produce graded messages. We have seen how two *Aphaenogaster* species, *A. albisetosus* and *A. cockerelli*, enhance pheromonal recruitment by adding stridulation. It is a common observation that the motor displays of *Camponotus*, *Pheidole*, and other ant genera increase in intensity as the colony grows hungrier or in greater need of new nest sites. Whether information is conveyed by the gradation in these displays remains to be established, but such an effect seems likely. Individual *Solenopsis invicta* workers adjust the amount of pheromone in their odor trails according to the quality of the food find (Hangartner, 1969b). A similar modulation occurs in the odor trails of *Formica oreas* (Crawford and Rissing, 1983) and in the recruitment behavior of *Paraponera clavata* (Breed et al., 1987).

Another system is employed by the species of *Myrmica*, exemplified by *M. sabuleti* of Europe (see Figure 7–55). When workers find water or smell insect prey, they collect the material individually. If the ants are working in a lighted environment, they employ no pheromones during the back-and-forth trips. If on the other hand the ants are in darkness, they lay odor trails from their poison gland. The substance, 3–ethyl–2,5–dimethylpyrazine, serves only as an orientation marker. Emitted alone, it does not arouse nestmates. If the scout *M. sabuleti* discovers sugar water or prey, she not only lays the orientation trail, but also recruits nestmates by strong motor displays, and in addition deposits Dufour's gland substances just outside the nest entrance. If the food discovery is a large insect, the scout distributes Dufour's gland material along the length of the orientation trail as well (Cammaerts and Cammaerts, 1980).

## ADULT TRANSPORT

The modes of transport from one nest site to another are based upon some of the most highly evolved communicative techniques in the ants, and they vary greatly among species in ways that have begun to be mapped across the phylogenetic groups (see Table 7–8). A pattern useful for comparison is that employed by the tropical Asiatic ant *Camponotus sericeus* (see Figures 7–56 and 7–57). Colony emigrations can be induced easily by keeping the old nest under unfavorable conditions such as excessive light or inadequate moisture and providing a superior nest nearby. When a scout discovers the second site she inspects it briefly and returns to the old nest. Approaching a nestmate she employs an "invitation behavior": she first jerks her entire body back and forth, seizes the nestmate by her mandibles, and pulls her forward a short distance. She then turns back in the direction of the new nest site, releasing a pheromone from her hindgut. Now the two ants engage in tandem running: the recruiter leads the nestmate forward along the pheromone trail, while the other ant stimulates the recruiter by playing her antennae over the recruiter's abdomen and hind legs. On arriving at the new site, the nestmate inspects the premises and may return to become a recruiter herself. Soon workers are bodily carrying other colony members by the sequence shown in Figure 7–58. The activity builds up exponentially and in short order a large portion of the colony is on the move. This sequence is very different from the previously described recruitment to food by workers of the same colony. Alate females are mostly led by tandem running to the new nest site, but males are usually carried by the workers in a fashion quite different from worker transportation.

A variation on this theme, illustrating the evolutionary lability of transport communication, is shown by the North American *Camponotus socius* (Hölldobler, 1971c). The invitation behavior is closely similar to that of *C. sericeus*, except that the recruiter usually does not pull on the mandibles of nestmates prior to pheromone release. Also, when she returns to the new nest site, she lays an odor trail that causes her to be followed at a short distance by one to several colony members (Figure 7–59). As already shown in Figure 7–53, *C. socius* also uses sharply different invitation signals when recruiting to food and when recruiting to nest sites. The first is a lateral wagging of the body with the mandibles wide open, permitting a scan of the lower mouthparts and the recently ingested food. The second is a back-and-forth jerking, an apparently ritualized version of the primitive dragging behavior. A remnant of the latter, directly functional behavior still exists in the mandible pulling of *C. sericeus*.

One of the striking features of ant biology is that the complex communication mediating colony emigration is more widespread among the phylogenetic groups than is recruitment to food. It also appears to be more primitive, in other words precedent to the latter behavior, since it occurs in *Myrmecia* unaccompanied by any form of food recruitment (Haskins and Haskins, 1950a). As already mentioned, the same asymmetry occurs in the predaceous dacetine ant *Orectognathus versicolor* and in some species of *Pachycondyla* (= *Neoponera*), which employ recruitment communication during colony emigration but not for recruitment to food.

Since the writings of Escherich (1917) and Arnoldi (1932), it has been recognized that the postures assumed during adult transport

**TABLE 7–8** Social techniques used by worker ants during colony movements from one nest site to another.

| Species | Carry adult | Tandem run | Group recruit-ment | Lay odor trail for orienta-tion only | Lay recruitment trail | References |
|---|---|---|---|---|---|---|
| **PONERINAE** | | | | | | |
| *Amblyopone australis* | | | X | X | | B. Hölldobler and H. Markl (unpublished) |
| *A. longidens* | | | X | X | | B. Hölldobler and H. Markl (unpublished) |
| *Cerapachys* sp. | | | X | | X | Hölldobler (1982b) |
| *Diacamma rugosum* | | X | | X | | Maschwitz et al. (1974), Fukumoto and Abe (1983), Maschwitz et al. (1986c) |
| *Dinoponera gigantea* | X | X | | | | Overal (1980) |
| *Ectatomma ruidum* | X | | | ? | ? | S. Pratt (personal communication) |
| *Gnamptogenys* sp. | X | | | ? | ? | B. Hölldobler (unpublished) |
| *Hypoponera* sp. | | X | | ? | | Hölldobler (1985), U. Maschwitz (personal communication) |
| *Leptogenys attenuata* | | | X | | X | Fletcher (1971) |
| *L. chinensis* | | | X | | X | Maschwitz and Schönegge (1983) |
| *L. nitida* | | | X | | X | Fletcher (1971) |
| *Odontomachus* sp. | X | | | ? | ? | Möglich and Hölldobler (1974) |
| *Onychomyrmex* sp. | | | X | | X | Hölldobler et al. (1982), B. Hölldobler and R. W. Taylor (unpublished) |
| *Pachycondyla obscuricornis* | X | X | | ? | | Traniello and Hölldobler (1984) |
| *P. tesserinoda* | | X | | X | | Maschwitz et al. (1974), Jessen and Maschwitz (1986) |
| *Paltothyreus tarsatus* | | X | | ? | | Hölldobler (1984b) |
| *Ponera coarctata* | | X | | | | N. Carlin (personal communication) |
| *Rhytidoponera impressa* group | X | | | ? | | Ward (1981a) |
| *R. metallica* | X | | | ? | | Möglich and Hölldobler (1974) |
| **MYRMECIINAE** | | | | | | |
| *Myrmecia* sp. | X | | | | | Haskins and Haskins (1950a) |
| **DORYLINAE** | | | | | | |
| *Eciton* spp. | | | X | | X | Topoff (1984) |
| *Neivamyrmex* spp. | | | X | | X | Topoff (1984) |
| **PSEUDOMYRMECINAE** | | | | | | |
| *Pseudomyrmex* spp. | X | | | ? | ? | Duelli (1977) |
| *Tetraponera* sp. | X | | | | | P. Duelli (personal communication) |
| **MYRMICINAE** | | | | | | |
| *Aphaenogaster albisetosus* | X | | X | | X | B. Hölldobler (unpublished) |
| *A. cockerelli* | X | | X | | X | Möglich and Hölldobler (1974), B. Hölldobler (unpublished) |
| *A. floridanus* | X | | X | | X | Möglich and Hölldobler (1974), B. Hölldobler (unpublished) |
| *A. rudis* | X | | X | | X | B. Hölldobler (unpublished) |
| *Leptothorax* spp. | X | X | | X | | Möglich (1978), Maschwitz et al. (1986b) |

*continued*

**TABLE 7–8** *(continued)*

| Species | Carry adult | Tandem run | Group recruit-ment | Lay odor trail for orienta-tion only | Lay recruitment trail | References |
|---|---|---|---|---|---|---|
| *Meranoplus* spp. | | | X | | X | B. Hölldobler (unpublished) |
| *Myrmica rubra* | X | | X | | X | Abraham and Pasteels (1980) |
| *Orectognathus versicolor* | X(males) | | X | | X | Hölldobler (1981b) |
| *Pheidole dentata* | | | X | | X | Wilson (1976b) |
| *P. desertorum* | | | X | | X | Droual (1983) |
| *P. hyatti* | | | X | | X | Droual (1983) |
| **DOLICHODERINAE** | | | | | | |
| *Forelius pruinosus* | | | X | | X | B. Hölldobler (unpublished) |
| *Iridomyrmex purpureus* | X | | X | | X | B. Hölldobler (unpublished) |
| *Tapinoma sessile* | | | X | | X | S. Beshers (personal communication) |
| **FORMICINAE** | | | | | | |
| *Camponotus consobrinus* | X | X | | X | ? | B. Hölldobler and D. Perlman (unpublished) |
| *C. pennsylvanicus* | X | | X | | X | B. Hölldobler (unpublished) |
| *C. perthiana* | X | X | | X | ? | B. Hölldobler (unpublished) |
| *C. sericeus* | X | X | | X | | Hölldobler et al. (1974) |
| *C. socius* | X | | X | | X | Hölldobler (1971c) |
| *Cataglyphis bicolor* | X | | | | | Duelli (1973), Harkness (1977b) |
| *C. iberica* | X | | | | | De Haro (1983) |
| *Formica fusca* | X | | X | | X | Möglich and Hölldobler (1975) |
| *F. polyctena* | X | | ? | | ? | Möglich and Hölldobler (1974) |
| *F. sanguinea* | X | | X | | X | Möglich and Hölldobler (1974) |
| *Oecophylla longinoda* | X | | X | | X | Hölldobler and Wilson (1978) |
| *O. smaragdina* | X | | X | | X | B. Hölldobler (unpublished) |
| *Polyergus lucidus* | X | | X | | X | Kwait and Topoff (1983) |

vary among the subfamilies and in some cases the genera of ants to such a consistent degree that they serve as taxonomically useful characters. Reviews of this subject, which is still in an early stage of investigation, have been provided by Wilson (1971), Möglich and Hölldobler (1974), and Duelli (1977). The most striking difference is between the Formicinae and Myrmicinae, as illustrated in Figure 7–60. Formicine species are the most stereotyped of any major ant group, with the transporter providing invitation signals and the nestmate rolling into a backward-facing "pupal" posture—as described earlier for *Camponotus*. An exception is *Oecophylla*, where the transportee is either grasped on the petiole or held on the head and curled over the body of the transporting ant. This latter position is very unusual for formicine ants, but it is the common mode of adult transportation in the Myrmicinae, as illustrated in Figure 3–29, and has also been observed in the Neotropical cryptobiotic myrmicine ant *Basiceros manni* during nest emigrations induced in the laboratory (Wilson and Hölldobler, 1986). *Crematogaster* workers hold nestmates by the petiole so that their head and legs point forward. In some species of *Pogonomyrmex* (*badius, barbatus, rugosus*), workers grasp nestmates at virtually any accessible part of the body, but in at least two species (*maricopa* and *californicus*) the posture is typical of other myrmicines.

Among the primitive ants, *Myrmecia* workers usually transport only aged and ailing individuals, callow workers, nest queens, and males. The carrying worker faces the nestmate, seizes her by the mandibles or (in the case of males) the antennae, and drags her over the ground while walking backward. The transported individual does not fold her appendages in the pupal position or cooperate actively in any other way. A similar crude behavior has been reported in the ponerines *Odontomachus* and *Pachycondyla (= Bothroponera) tesserinoda:* the transporter seizes the nestmate by any convenient part of the body and lifts her off the ground, and the transportee assumes the pupal response. On the other hand the Neotropical ponerine *Pachycondyla obscuricornis* exhibits a stereotyped mode of adult transport resembling the formicine pattern. The ponerine genera *Ectatomma, Gnamptogenys,* and *Rhytidoponera,* and the pseudomyrmecine genera *Pseudomyrmex* and *Tetraponera* show the myrmicine adult transport mode. In the New World army ants (Ecitoninae), adults are carried in the same strange fashion as the larvae and pupae, that is, slung back beneath the body and between the long legs of the workers. In general, the modes of adult transport have only been partially explored. It would be valuable for future investigators to include descriptions of the phenomenon routinely in sociobiological accounts of individual species.

Emigration is usually organized by a minority of the workers, who are either "elites," that is, individuals who work hard at many tasks (Wilson, 1971; Meudec, 1973; Abraham, 1980) or else moving specialists—no studies have been conducted to establish this two-

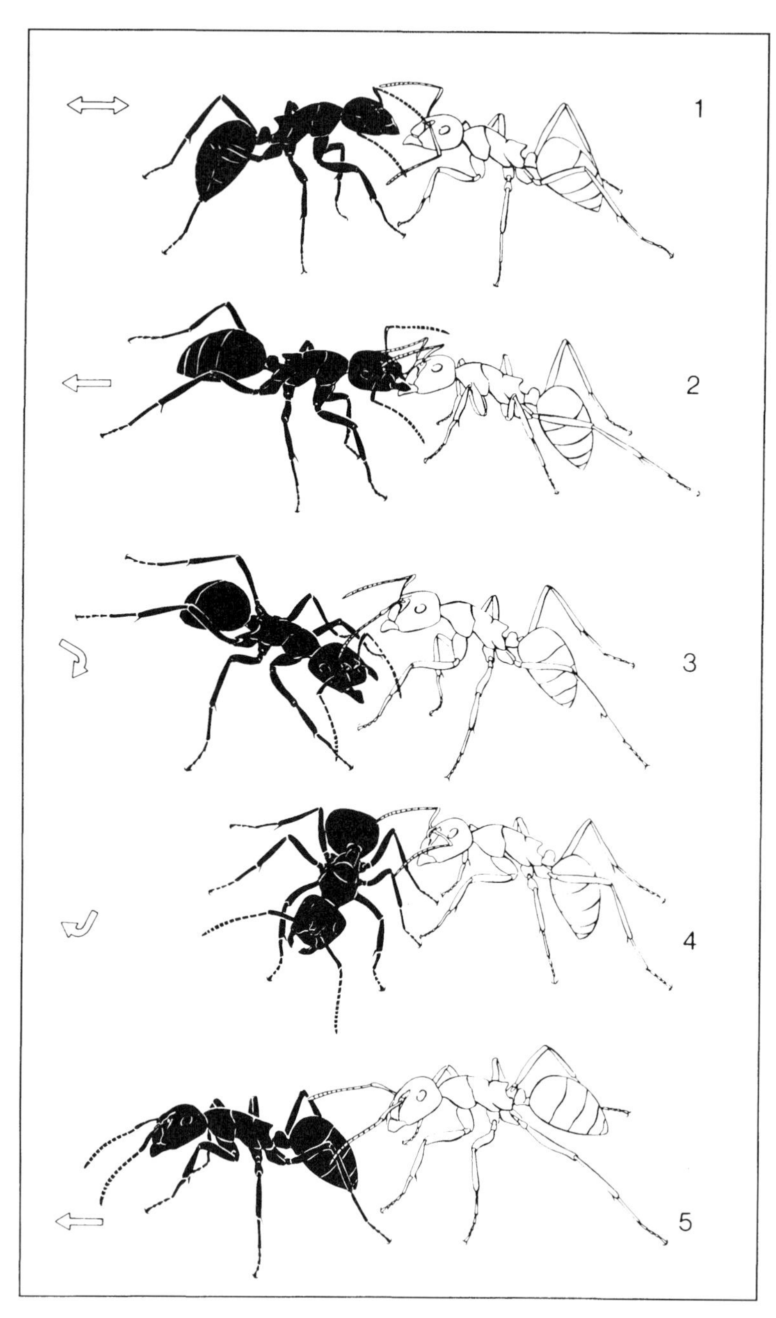


**FIGURE 7–56** Colony emigration in the Asiatic species *Camponotus sericeus* begins when scouts employ a ritualized invitation behavior, causing nestmates to follow them to the new nest site. (*1*) The recruiter (*black*) approaches a nestmate (*white*) and jerks her body back and forth for 2–3 seconds. (*2*) She then grasps the nestmate by the mandibles and pulls her through a distance of about 2–20 centimeters. (*3*) Next the recruiter loosens her grip and (*4*) turns around. (*5*) Finally, releasing a pheromone from the hindgut, she leads the other worker toward the new nest site, while the nestmate maintains direct tactile contact by playing her antennae over the leader's abdomen and hind legs. (From Hölldobler et al., 1974; drawing by T. Hölldobler-Forsyth.)

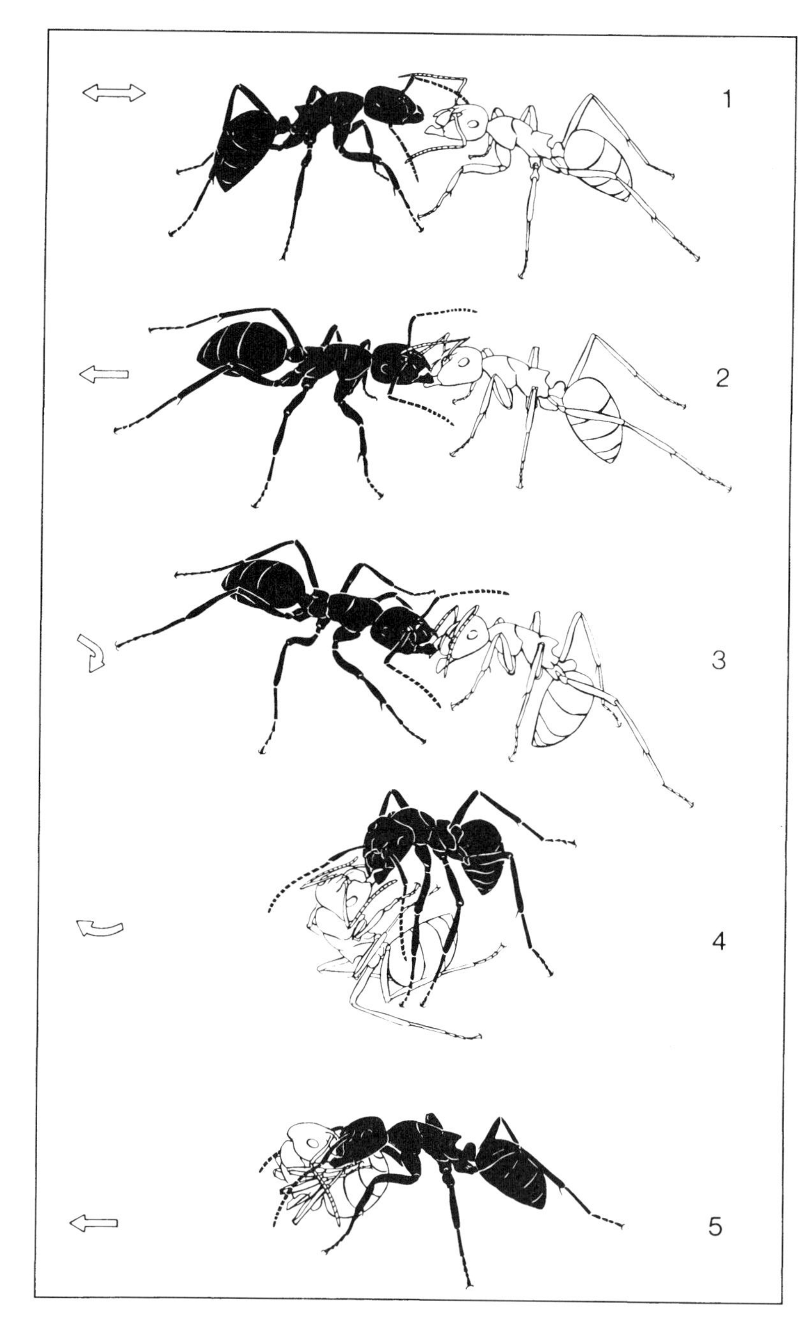


**FIGURE 7–57** Adult transport during colony emigration of *Camponotus sericeus* is depicted in this sequence. (*1*) The recruiting worker (*black*) approaches a nestmate (*white*) and jerks her body back and forth for 2–3 seconds. (*2*) She then grasps the nestmate on the mandibles and pulls her a distance of 2–20 centimeters. (*3*) Holding on to mandibles of the nestmate and lifting her slightly, the recruiter begins to turn. (*4*) The nestmate folds her antennae and legs tightly against the body and rolls her gaster inward. (*5*) The recruiter now carries the nestmate to the new nest site. (From Hölldobler et al., 1974; drawing by T. Hölldobler-Forsyth.)

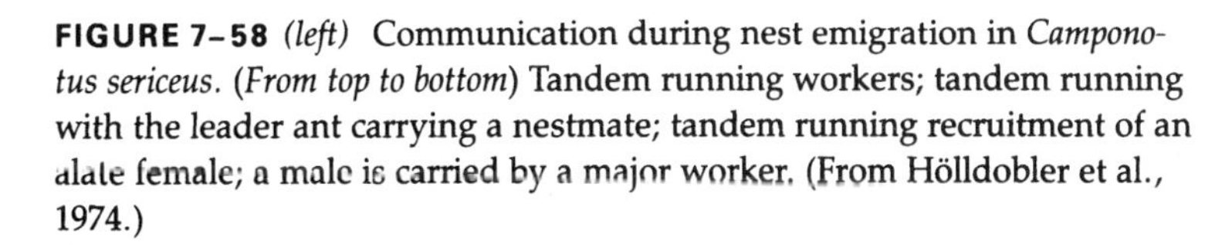

**FIGURE 7–58** *(left)* Communication during nest emigration in *Camponotus sericeus.* *(From top to bottom)* Tandem running workers; tandem running with the leader ant carrying a nestmate; tandem running recruitment of an alate female; a male is carried by a major worker. (From Hölldobler et al., 1974.)

**FIGURE 7–60** *(below)* The posture assumed during adult transport varies markedly among the major phylogenetic groups of ants. *(Above) Formica sanguinea,* a member of the subfamily Formicinae. *(Below) Aphaenogaster* (= *Novomessor*) *cockerelli,* a member of the subfamily Myrmicinae. (From Möglich and Hölldobler, 1974.)

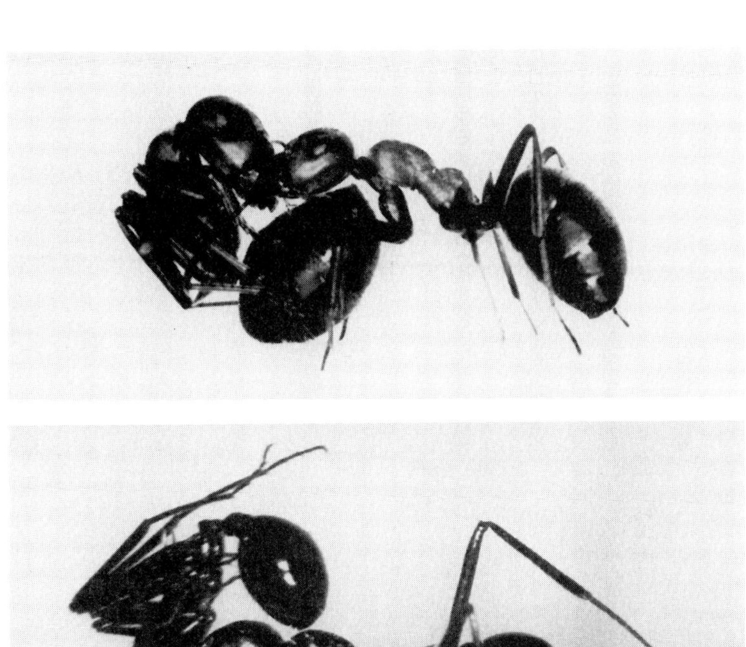

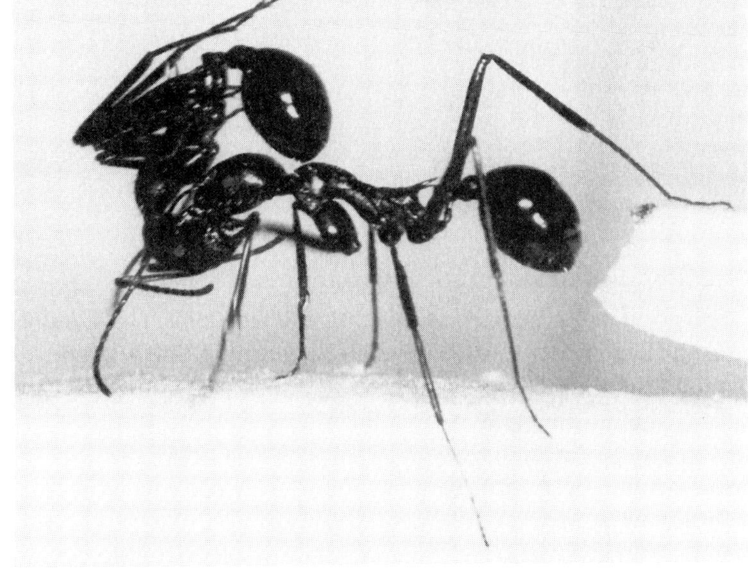

**FIGURE 7–59** Colony emigration in *Camponotus socius,* a ground-dwelling formicine of the southeastern United States, is mediated by the laying of a trail pheromone. The recruiting worker *(black)* carries a nestmate, simultaneously dragging the tip of her abdomen over the ground to deposit secretions from the hindgut. She is followed by a male and another worker. (From Hölldobler, 1971c; drawing by T. Hölldobler-Forsyth.)

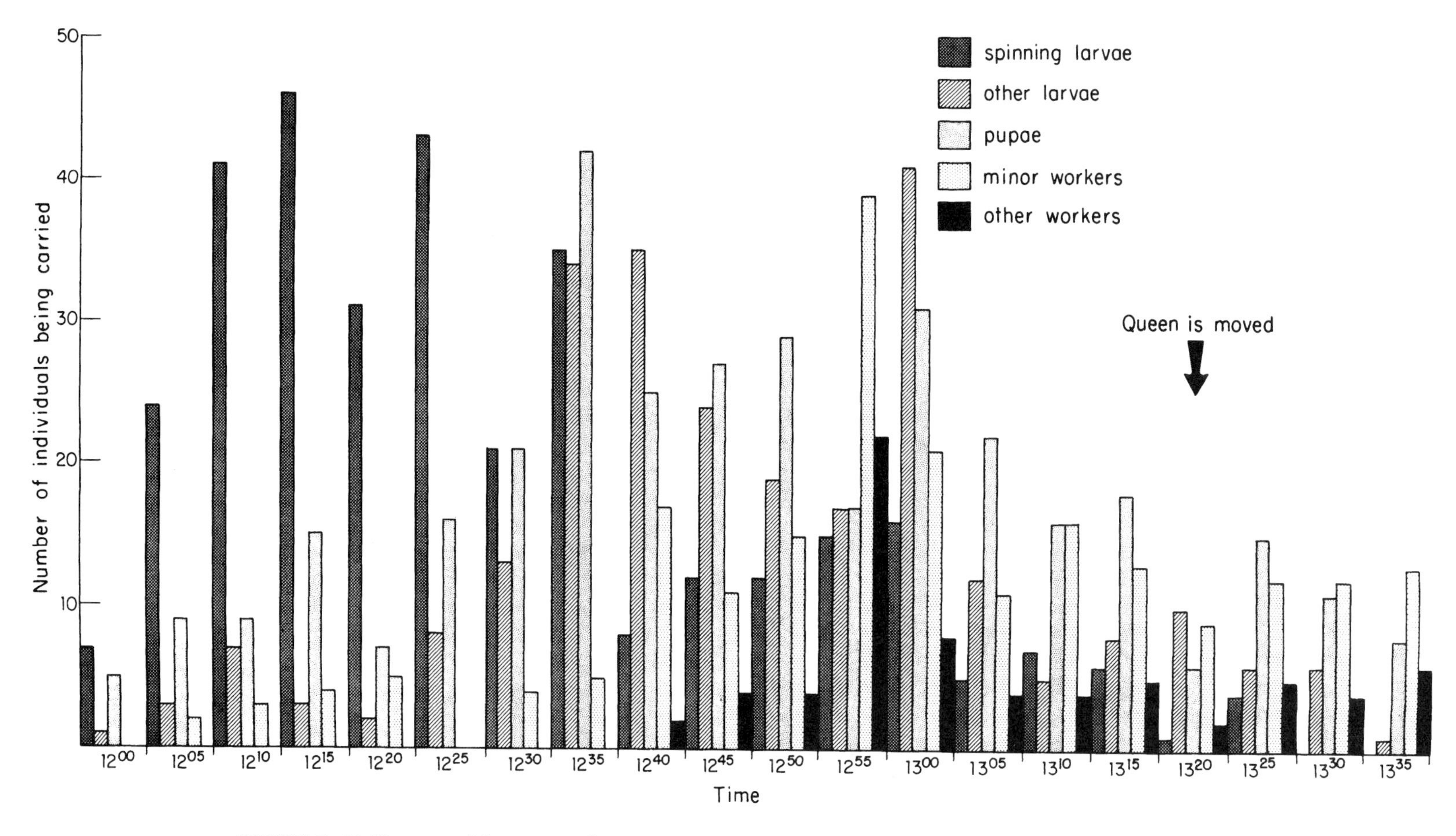


**FIGURE 7–61** The most elaborate marching order during emigration thus far discovered occurs in the African weaver ant *Oecophylla longinoda*. The sequence illustrated here is typical of several recorded in a laboratory colony. The data shown are the numbers of individuals of various classes carried by major workers from an artificial nest onto a nearby tree. (From Hölldobler and Wilson, 1978.)

way distinction for any particular species. In a *Camponotus sericeus* colony containing 81 workers, Möglich and Hölldobler (1974) found that the 8 most active workers led in 91 percent of the instances of tandem running, the top 3 workers led in 52 percent of the runs, and the single "champion" ant led in 24 percent of the runs. Meudec (1977) and Meudec and Lenoir (1982) discerned lesser degrees of specialization in the dolichoderine *Tapinoma erraticum*, with the percentage of brood carriers varying according to the number of brood pieces to be transported and the severity of the stress inducing emigration. In ants generally, the older workers and especially the foragers usually take the lead during emigration (Möglich and Hölldobler, 1974; Möglich, 1978; Abraham and Pasteels, 1980). But in at least one species, the slavemaker *Polyergus lucidus*, this does not appear to be the case (Kwait and Topoff, 1983).

A close examination of the emigration process in the most socially complex species reveals idiosyncratic traits that clearly contribute to the survival of the colonies. For example, workers of the harvester ant *Pogonomyrmex badius* surround their crater nests with irregular rings of charcoal fragments. When some of these middens were removed during experiments conducted by Gordon (1984a), the frequency of invasions of the nests by other species of ants increased, interfering with the daily round of activities of the harvesters. The value placed on such material may be general in *Pogonomyrmex*. When colonies of *P. barbatus* emigrate, they carry midden debris from the old nest site to the new (Van Pelt, 1976). Similarly, when workers of the leafcutter species *Atta sexdens* evacuate a chamber following a major disturbance, they first remove the brood, then tear off and transport pieces of the fungus garden (Wilson, 1980a). In some cases ants also transport their homopteran symbionts and even their parasitic guests when emigrating (see Chapter 13).

Some remarkable variations in the strategy of emigration have been discovered. When colonies of the slavemaker *Polyergus lucidus* move to their winter nests, most of the transport is ordinarily conducted by their *Formica* slaves. Very swift "rapid transit" emigrations also occur, however, which closely resemble slave raids both in the timing and in the form of communication. On such occasions the *Polyergus* usually carry their *Formica* slaves—a further simulation of raiding, although in emigrations only adults are carried, not larvae and pupae. This special type of emigration appears to be well adapted as a quick response to a season of worsening and unpredictable weather (Kwait and Topoff, 1983).

The most elaborate and specialized strategy of emigration thus far recorded occurs in the African weaver ant *Oecophylla longinoda* (Hölldobler and Wilson, 1978). When newly captured colonies are given access in the laboratory to potted trees suitable as nest sites, they respond decisively. A well-organized emigration is initiated within minutes and is all but completed several hours later. The process begins when exploring workers investigate the trees and return to recruit nestmates with rectal gland odor trails and antennation signals. Masses of major workers accumulate on certain leaves and branch tips and proceed to fold the leaves and pull them together, often making living chains of their bodies to span the large gaps in the vegetation. Very soon major workers start to bring in final-instar larvae as a source of silk threads to bind the leaves together. Now the whole colony moves in a fixed marching order (Figure 7–61). After the major workers make the new leaves secure, they carry an increasing number of older larvae and pupae, then a rising proportion of minor workers, and finally other major workers. At its height the parade closely follows a trunk trail. Additional recruitment through antennation and trail laying continues throughout the remainder of the emigration but at a lower overall intensity. Thus emigration is achieved through a combination of recruitment and physical transport. Finally, the queen always leaves the old nest after a large part of the remainder of the colony has emigrated. She travels under her own power but is covered almost to the point of invisibility by a dense retinue of major workers (see Figure 4–7).

The research of the past twenty years has established colony emigration as basic in ant biology: surprisingly frequent in occurrence, vital to survival, and mediated by stereotyped techniques of communication and transportation that vary from one species to the next. In the future it will be exciting to explore the phenomenon more extensively, in order to correlate the many forms emigration takes with exigencies faced by species in their individual environments. A growing knowledge of the stereotyped behaviors occurring during emigration will also provide valuable new data for the reconstruction of phylogenetic relationships for the larger taxonomic groups of ants.

## TRUNK TRAILS AND "HIGHWAYS"

A great many species in the Myrmicinae, Dolichoderinae, and Formicinae lay trunk trails, which are traces of orientation pheromones enduring for periods of days or longer. In the case of leafcutter ants of the genus *Atta,* harvesting ants of the genus *Pogonomyrmex,* the European shining black ant *Lasius fuliginosus,* and the polydomous wood and mound-building ants in the *Formica exsecta* and *rufa* groups, the trails can last for months at a time. The workers clear them of vegetation and debris to form veritable highways along which large numbers of ants travel easily (Figure 7–62).

Trunk trails used for foraging are typically dendritic in form. Each one starts from the nest vicinity as a single thick pathway that splits first into branches and then into twigs (Figure 7–63). This pattern deploys large numbers of workers rapidly into foraging areas. In the species of *Atta* and the remarkable desert seed-eating *Pheidole militicida,* tens or even hundreds of thousands of workers move back and forth on a daily basis. A few workers drift away from the main trunk

**FIGURE 7–62** The "highway" system of *Pogonomyrmex rugosus,* a harvesting ant found in the deserts of the southwestern United States; *N,* nest; *T,* cleared trunk trails. (From Hölldobler, 1976a.)

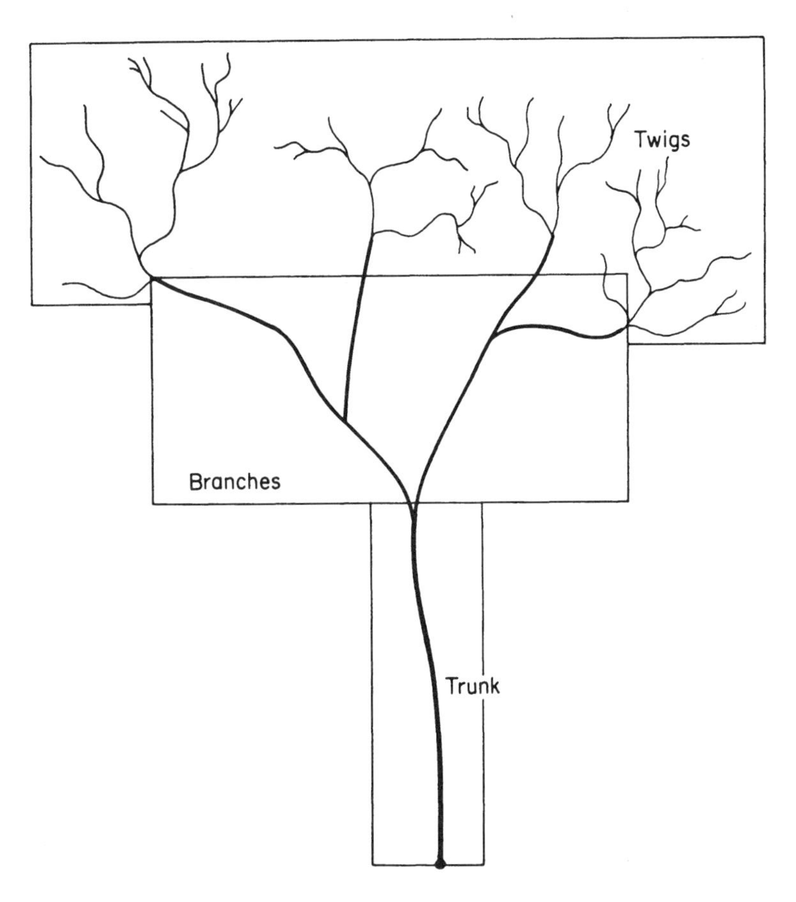


**FIGURE 7–63** A schematic representation of the complete foraging route of *Pheidole militicida,* a harvesting ant of the southwestern U.S. deserts. Each day tens of thousands of workers move out to the dendritic trail system, disperse singly, and forage for food. (From Hölldobler and Möglich, 1980.)

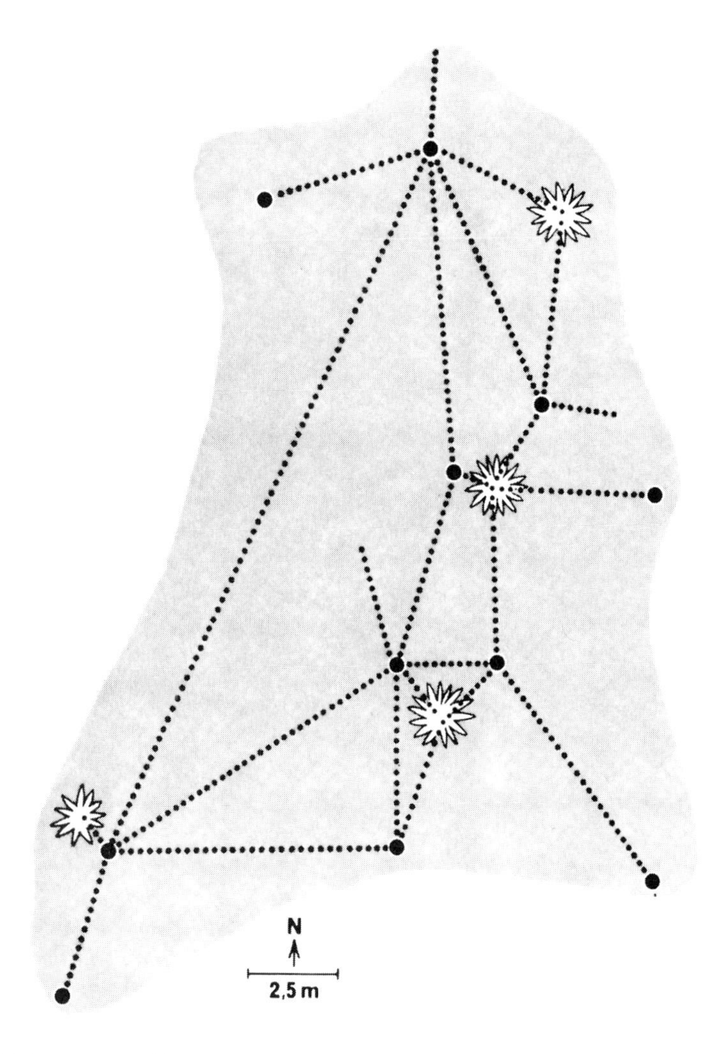


**FIGURE 7–64** Nest area of a colony of *Camponotus socius* in Tampa, Florida. The dotted lines between the nest entrances (●) and the palmetto bushes (white symbols) indicate the paths along which the ants were observed moving on a regular basis. (From Hölldobler, 1971c.)

route, but most do not disperse on a solitary basis until they reach the terminal twigs. When small food items are encountered, the workers carry them back into the outer branches of the system and then homeward. The twigs and branches can now be envisioned as tributaries of ant masses flowing back to the nest. When rich deposits of food are found, on the other hand, the foragers lay recruitment trails to them. In time new deposits of orientation pheromone accumulate along which foragers move with the further inducement of recruitment pheromones. By this process the outer reaches of the trunk-trail system shift subtly from day to day. The orientation pheromones comprising the trunk trails are often secreted by glands different from those that produce the recruitment pheromones. In *Pogonomyrmex* harvester ants, for example, orientation substances are produced in the Dufour's gland and recruitment substances in the poison gland (Hölldobler and Wilson, 1970).

Trunk trails also provide rapid transit to persistent food sources such as seedfalls and herds of aphids and other honeydew-producing homopterans. Some ants that occupy multiple nests use trunk trails as connecting routes. Conspicuous examples include arboreal species of *Crematogaster;* many species in the dolichoderine genera *Azteca, Hypoclinea,* and *Iridomyrmex;* mound-building ants in the *Formica exsecta* and *F. rufa* groups; and the polydomous colonies of *Camponotus socius* (Figure 7–64).

## INTERSPECIFIC TRAIL FOLLOWING

Trail systems are not wholly private. A few cases have been reported in which ants utilize the odor traces of other species. The parabiotic ant species of the tropical forests in Central and South America follow one another's trails, with certain forms dominating and exploiting others (see Chapter 12). In southern Europe, workers of *Camponotus lateralis* sometimes follow the trails of *Crematogaster scutellaris* in large numbers to the *Crematogaster* feeding grounds and share their food resources (Kaudewitz, 1955). In Trinidad, workers of *Camponotus beebei* regularly if not invariably utilize the trunk trails of a locally dominant dolichoderine species, *Azteca chartifex*. The *Camponotus* "borrow" the *Azteca* trails during the day, when *Azteca* foraging is at a low ebb. The *Camponotus* are treated as enemies by the *Azteca,* but they are too swift and agile to be caught (Wilson, 1965b). The European guest ant *Formicoxenus nitidulus* follows odor trails laid by its *Formica* host (Elgert and Rosengren, 1977).

It is likely that many other such social parasites have developed this capacity and hence are able to emigrate with their hosts from one nesting site to another. The reverse is true in the case of slavery: the host species are sometimes able to follow the recruitment trails of their captors. For example, *Formica neorufibarbis* workers were observed to accompany workers of the slavemaker *F. wheeleri* on a raid that was being waged simultaneously against colonies of *Formica subsericea* and *F. lasioides* (Wilson, 1955c). In a more ambiguous case, Starr (1981) has reported a case of non-parasitic species of the formicine genus *Polyrhachis* using the same trail in the Philippines. *Polyrhachis bihamata,* which evidently laid the trail, shared it with *P. armata* in apparently complete amity. The adaptive significance of this symbiosis, if any, has not been determined. Finally, two species may accidentally follow one another's trails if they use identical or closely similar pheromones. An example is the common trail used by leafcutter ants *Acromyrmex versicolor* and *Atta mexicana* observed by Mintzer (1980) in the desert of Sonora, Mexico. The workers of the two species displayed no hostility except near one of the *Acromyrmex* nest entrances. The phenomenon appears to be rare and perhaps lacks adaptive significance.

## MARKING OF HOME RANGES, TERRITORIES, AND NEST ENTRANCES

Workers of a few ant species lay odorous materials outside the nest in spots or short streaks, in patterns that serve neither to recruit nestmates nor to direct them away from the nest, and which in fact are difficult to interpret. Although too few cases have been analyzed to draw general conclusions about the function of the behavior, three categories of roles seem likely. These are (1) *home-range marking,* in which newly opened terrain is flecked with pheromones that appear to label the surface as hospitable and available for foraging; (2) *territorial marking,* in which colony-specific pheromones are used to mark the terrain as belonging to the colony and subject to defense; and (3) *nest-entrance marking,* in which colony-specific substances label the entrances as belonging to the colony and assist foragers in the final stages of homing.

The first of these categories, home-range marking, is still a vaguely defined category of communication. Cammaerts et al.

**FIGURE 7–65** Chemical territorial marking by weaver ants (*Oecophylla longinoda*). *Above:* anal spots on a paper surface in a laboratory foraging arena. Experiments have shown that the spots contain a true territorial pheromone. *Below:* two workers in combat on the marked paper. The ant on the left is fighting on an area marked by her own colony, and enjoys an advantage as is typical in such cases. (From Hölldobler and Wilson, 1978.)

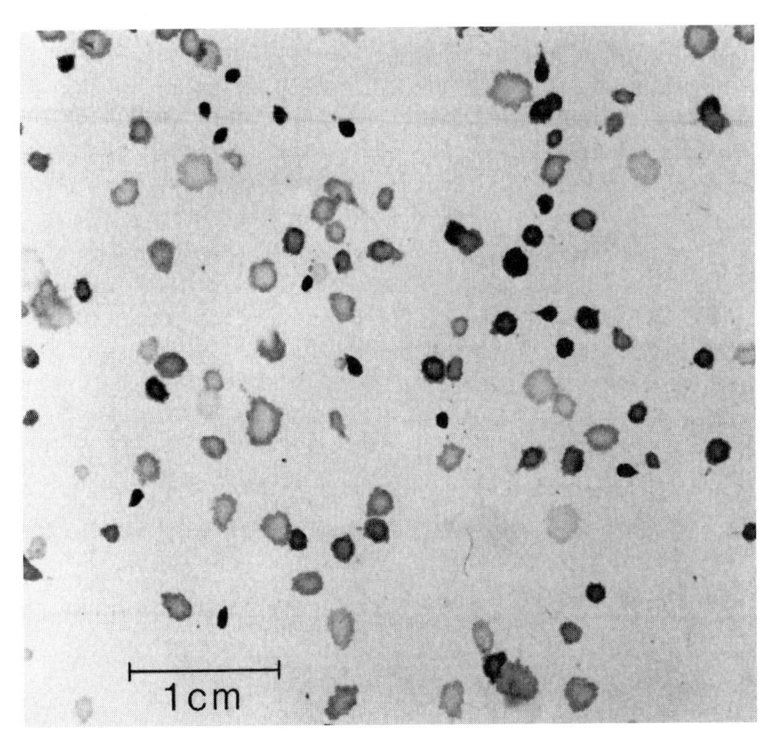

(1977) found that when *Myrmica rubra* foragers are allowed onto new terrain near the nest, they move very slowly while laying down spots of material in short rows. The deposits, which apparently include secretions from the Dufour's gland, are not linked together in the form of a trail back to the nest. Instead, the odor induces nestmates to approach the area and to explore the vicinity thoroughly. Most of the attraction disappears after about three minutes, but a residual effect, causing increased rates of locomotion, persists for longer periods. We have observed a closely similar phenomenon in *Pogonomyrmex badius* (Hölldobler and Wilson, 1970; Hölldobler, 1971b) and in the fire ant *Solenopsis invicta*, in which workers admitted to new glass platforms in the laboratory lay short, haphazardly directed odor trails over the surface.

Gordon (1988a) found that the exploration of new terrain by fire ant workers consists in a good deal more than random wandering. Four classes of individuals can be distinguished as follows. (1) Some ants, perhaps those serving as recruiters into new terrain, engage in more frequent antennal contact with nestmates they encounter. (2) Others come into the new region and then remain stationary, seemingly serving as sentinels. (3) Still others spend more time moving slowly through the region, inspecting other slow or stationary ants, perhaps to gather information. (4) Members of a fourth class move rapidly and directly through the region, as though exploring farther out.

The function of this marking behavior is not yet entirely clear. Cammaerts and her co-workers have referred to the phenomenon in *Myrmica rubra* as territorial marking, but they have put too much weight on the data. There is no evidence yet that the deposits contain components that allow individual colonies to distinguish them from those of other colonies. Nor has it been demonstrated that

**FIGURE 7–66** *Oecophylla longinoda* worker in typical aggressive display.

**FIGURE 7–67** Workers of *Meranoplus hirsutus* mark their nests with anal spots, which are densest near the entrance (*N*). When the nest is removed and the foraging arena placed in a new position to avoid visual orientation, the homing workers still assemble at the previous nest entrance location, as shown here. When the nest markers are covered by a sheet of paper, the ants are unable to find the nest entrance area.

workers are repelled or moved to aggression when they encounter the substances of alien colonies.

All of these territorial criteria are met, however, in the African weaver ant *Oecophylla longinoda* (Hölldobler and Wilson, 1977a, 1978). The behavior is part of a complex system of land tenure. The workers lay true trails from the rectal gland to newly opened terrain, attracting nestmates in large numbers and permitting the colony to explore and occupy the surface within a short period of time. This first step is recruitment and not territoriality. When *O. longinoda* workers enter such newly opened space, for example a potted tree placed adjacent to their nest tree in the laboratory, they periodically touch the tip of the abdomen to the substrate and extrude large drops of brown fluid from the anus. This material quickly soaks into the surface or else hardens into shiny, shellac-like, shallowly convex solids (Figure 7–65). At first the rate of deposition is very high. One colony containing several thousand workers deposited approximately 500 drops onto the surface of a fresh 71-by-142-centimeter arena during just the first hour. Thereafter the marking rate declined to a much lower, constant rate. It is further notable that the anal spots were not concentrated in a "kitchen midden" or in some remote corner of the arena, the pattern used by workers of many other ant species when defecating.

The ability of the *Oecophylla* workers to recognize deposits from their own colony was tested by the following method. A colony was allowed to mark the papered floor of an arena for a period of several days. Then the ants were removed overnight, and the arena was shifted slightly to one side to make room for a second, identical arena that had been marked by an alien colony of *Oecophylla.* The first colony was next given access to both arenas simultaneously by the emplacement of their wooden bridges. The first workers to enter the alien odor field displayed greater caution and a significantly higher rate of aggressive posturing, which consisted of opening the mandibles and raising the abdomen above the remainder of the body (Figure 7–66). This response was obtained even though no alien workers had been in the arena for more than 12 hours. The exploring ants showed a special interest in the anal deposits, often stopping to inspect individual spots with their antennae. After a few minutes, some of the foragers then returned to their nest tree while laying odor trails, and a full-scale recruitment to the alien arena began. Some recruitment to the familiar arena occurred simultaneously but at a significantly lower level. In other experiments, squares of paper marked by alien colonies were placed near those marked by the home colony. The results showed unequivocally that *Oecophylla* workers distinguish their own deposits from those of aliens. Moreover, artificial spots made from the rectal sac contents of *Oecophylla* workers yielded the same result, although the difference in the response to alien material was less strong.

Jaffe et al. (1979) reported that a territorial pheromone is deposited by workers of the leafcutter *Atta cephalotes.* Jaffe and his colleagues provided evidence that the substance comes from the valves gland (at the base of the sting apparatus), is colony specific, and reduces the level of aggressive posturing when workers encounter material from their own nestmates. Although we could confirm that *Atta* workers mark their nest entrance area with long-lasting colony-specific secretions, we were unable to verify some of the Jaffe group's other results, even though we used several methods that included a close duplication of their own bioassays. In fact, we found no evidence that the *Atta* foragers use valves gland secretions as a territorial pheromone. Rather, the valves gland produces a typical alarm pheromone and, with the Dufour's gland, appears to be the principal source of this kind of signal in the abdomen (Hölldobler and Wilson, 1986b).

In yet another category, workers of some ant species mark the substrate in the vicinity of their nest entrances and use the odor to orient homeward, or to distinguish their own nest from the nests of other colonies, or both. When workers of the Florida harvester ant *Pogonomyrmex badius* are placed in a circular olfactometer and given a choice between purified sand, sand from the nest entrances of alien *P. badius* colonies, and sand from their own nest entrances,

**FIGURE 7–68** Surface attraction in the Australian myrmicine ant *Meranoplus hirsutus*. Workers cluster around the mother queen (*left*), and the large larva of a reproductive female (*right*).

they orient chiefly to their own material (Hangartner et al., 1970). A similar power to find the nest entrance has been demonstrated in *Eurhopalothrix heliscata*, a cryptobiotic predaceous ant of Malaysia. Whether the foragers also distinguish their own deposits from those of other colonies has not been determined (Wilson and Brown, 1984).

We have noted that workers of the leafcutter *Atta cephalotes* also deposit such "nest exit pheromones" around their nest entrances. Lasting for 24 hours or longer, these substances orient the workers to the nest openings and increase the rate of trail laying, leaf cutting, and leaf retrieval. Their perception by the workers forms part of a cognitive map by which the ants adjust the intensity and pattern of their activity while foraging. The material is voided at least in part from the poison gland. It is considerably more persistent as an orienting stimulus than the poison gland pyrrole that serves as the primary recruitment substance. The nest exit pheromones may also come in part from hindgut fluid, which contains arrestants of the ant's locomotory activity. In any case, they are colony specific. When workers encounter deposits by alien colonies, they increase the rate of abdominal dipping, whereby they evidently add colony-specific chemicals of their own (Hölldobler and Wilson, 1986b).

Recent studies have revealed that colony-specific nest marking is a widespread phenomenon in ants. In addition to those species already mentioned, it has been demonstrated in *Nothomyrmecia macrops* (Hölldobler and Taylor, 1983), *Paltothyreus tarsatus* (Hölldobler, 1984b), *Odontomachus troglodytes* (Dejean et al., 1984), *Pseudomyrmex termitarius* and *P. triplarinus* (Jaffe et al., 1986), *Ectatomma ruidum* (Jaffe and Marquez, 1987), and in species of *Pachycondyla, Leptogenys, Hypoponera, Ponera, Diacamma, Leptothorax, Meranoplus, Aphaenogaster,* and *Myrmecocystus* (Jessen and Maschwitz, 1986; B. Hölldobler and U. Maschwitz, unpublished data). In arena experiments, ants of these species when searching for their nests were attracted by colony-specific nest markers. When the nest was experimentally removed, the ants settled where the density of the pheromone deposits was greatest (Figure 7–67).

## ATTRACTION AND SURFACE PHEROMONES

Ants, like other social insects, have a universal tendency to aggregate. If a group of workers are taken from their nest and placed in a separate container, most will soon coalesce into tight clusters. The brood and queen are especially attractive as nuclei around which entire colonies readily gather (Figure 7–68). Several of the key queen attractants have been identified in the myrmicine genera *Monomorium* and *Solenopsis* (see Chapter 6).

An exceptionally simple system of attraction exists in fire ants of the genus *Solenopsis*. When away from the nest and in close quarters, workers attempt to move up carbon dioxide gradients, hence in the direction of the largest nearby clusters of ants, and they attempt to dig through soil and other barriers placed in their way (Wilson, 1962b; Hangartner, 1969a). Carbon dioxide has thus been implicated as the simplest chemical signal used in any known animal communication system. It is likely that the carbon dioxide gradients are used by the ants for orientation in the immediate vicinity of their nests. High concentrations of the order used in the experiments have been reported in ant nests (Portier and Duval, 1929; Raffy, 1929). The ability to detect carbon dioxide and differences in its con-

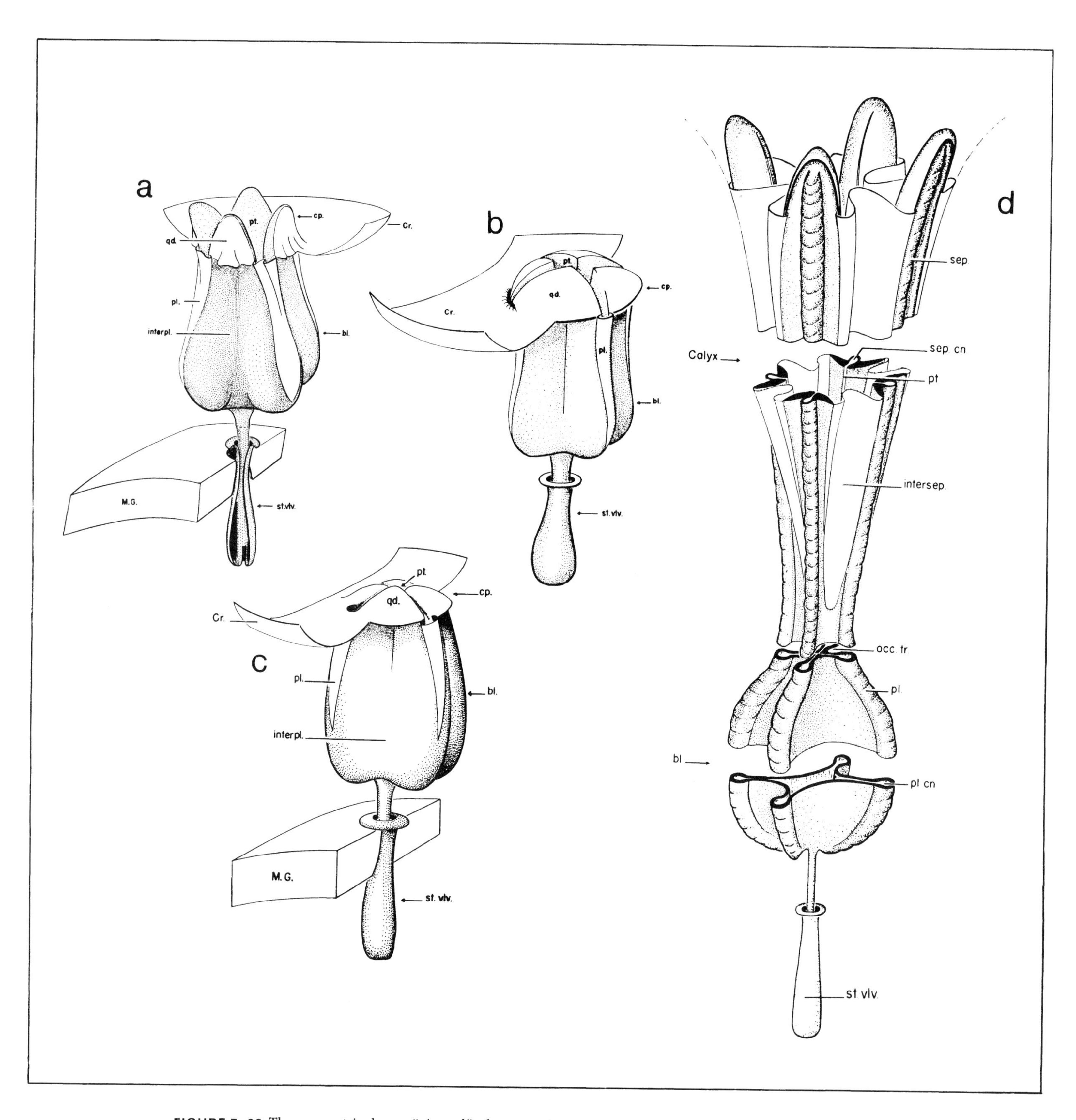


**FIGURE 7–69** The proventriculus or "gizzard" of ants, an important organ in trophallaxis. Cuticular framework of the proventriculus of (*a*) *Myrmecia regularis,* (*b*) *Pseudomyrmex pallidus,* and (*c*) *Aneuretus simoni.* (*d*) Exploded diagram of the generalized formicine proventriculus, based on *Camponotus. pt,* portal; *cp,* cupola; *Cr,* crop; *qd,* quadrant; *pl,* plica; *bl,* bulb; *interpl,* interplicary plate of bulb; *intersep,* intersepalary cuticle; *MG,* midgut; *st vlv,* stomodeal valve; *sep,* sepal; *sep cn,* sepal canal; *pl cn,* plicary canal; *occ tr,* occlusory tract. (From Eisner, 1957.)

centration has also been reported in the honey bee (Lacher, 1967), but we do not know whether this strange capacity is used in orientation, monitoring air quality, or some other still unsuspected function. Too high a concentration of carbon dioxide immobilizes ants, but they can be revived even after hours of immersion in the gas. This is probably the explanation of why many ants can survive long periods under water. When dried out, they appear to have drowned, but revive after an hour or so. The same is true of ants kept overlong in small containers for shipment. Colonies thought to be "lost" in transit often revive within an hour or two after being exposed to fresh air. This curious phenomenon also allows continuous anesthesia of ants during laboratory experiments.

It is of course probable that other pheromones are involved in the clustering phenomenon in ants. In studies of several *Camponotus* species, Ayre and Blum (1971) demonstrated that small amounts of the Dufour's gland secretions have a strong attracting and settling effect on workers. A similar response is elicited by the sternal gland secretions of *Oecophylla longinoda* (Hölldobler and Wilson, 1978). Here the pheromone acts as a short-range recruitment signal and arrestant, causing a greater clustering of patrolling foragers. We know of the existence of "surface pheromones" (Wilson, 1971) of moderate or high molecular weight and low volatility that are located in the epicuticle. These substances generate a shallow active space, so that they come into play only when two nestmates are in virtual physical contact. In addition to colony odors, which may be mixtures of hydrocarbons, it is likely that there exist less specific attractants as well as releasers of other forms of behavior.

## TROPHALLAXIS

One of the elementary bonds of insect colonies is the sharing of food among nestmates. Prey objects and seeds brought into the nest by a few individuals are usually freely consumed by many other individuals. Liquid food, stored in the forager's crop (the "social stomach") is regurgitated to nestmates and thus distributed over large portions of the colony. This latter form of food transmission is called stomodeal (or oral) trophallaxis. When food is passed out from the rectum, it is called abdominal (or anal) trophallaxis.

Stomodeal trophallaxis is by far the most common type of liquid food exchange. The crops of most ant species that feed on nectar and homopteran-secreted honeydew are capable of considerable distention. Individual foragers are consequently able to carry home large loads of carbohydrates. Some groups of workers serve as living reservoirs during lean periods. The storage of liquid food in the crop has been carried to great heights by the repletes of certain ant species, individuals whose abdomens are so distended they have difficulty moving and are forced to remain permanently in the nest as "living honey casks" (see Plate 8). The liquid, digested only to a limited extent while held in the crop, is freely passed from one ant to another. Thus the crops of all the workers taken together serve as a social stomach from which the colony as a whole draws nourishment.

Eisner (1957), adding extensively to the original discoveries of Forel (1878), showed how the proventriculus has evolved in ants to facilitate this communal function. The proventriculus forms a tight constriction at the posterior end of the crop (see Figure 7–1). It regulates the flow of liquid back to the midgut where the food is digested, and thus serves to segregate the communal supply in the crop from the personal supply in the midgut. The proventriculus of *Myrmecia* (Figure 7–69) is typical of primitive ants, while that of *Camponotus* represents an advanced form found in formicine ants. So distinctive are the structures that they provide useful characters for the study of phylogeny at the generic and tribal levels (Eisner and Brown, 1958). The peculiar infrabuccal pocket, a sizable cavity located just beneath the tongue of worker ants, filters out and compacts most of the solid material that would otherwise clog the narrow, rigid proventriculus channels (Eisner and Happ, 1962; Figure 7–70). From time to time workers of most ant species disgorge the infrabuccal waste material in the form of a pellet (see Febvay and Kermarrec, 1981). This waste material is then carried out of the nest and discarded. In arboreal ants of the genus *Pseudomyrmex*, it is routinely fed to the larvae (Wheeler and Bailey, 1920). Typically, a soliciting ant elicits regurgitation from a nestmate by stimulating her with stereotyped tactile signals with the antennae and forelegs. During this episode the mechanical stimulation of the donor's mouthparts, especially her labium, serves as the important releaser of the regurgitator reflex (see discussion of tactile communication earlier in this chapter; also Figures 7–42, 7–43, 7–71, and 7–72).

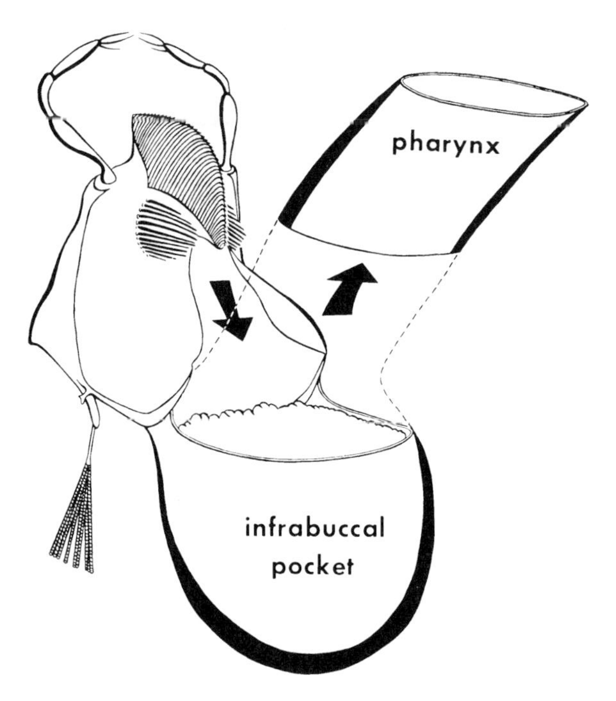


**FIGURE 7–70** Diagram of the labium (tongue), infrabuccal pocket, and pharynx. Arrows indicate path of materials along labial groove, past or into infrabuccal pocket, and through mouth into the pharynx. (From Gotwald, 1969.)

Liquid food exchange by regurgitation, which distributes food through colonies with remarkable rapidity, is a highly evolved form of behavior. It is much more common among species belonging to phylogenetically advanced subfamilies (Wilson, 1971). A likely precursor exists in the distinctive mode of liquid food transport practiced by some ponerine species. Most ponerine ants are primarily predators and scavengers, but some species collect liquid material as well. Evans and Leston (1971) discovered that workers of a West African species of *Odontomachus* gather honeydew from aphids and

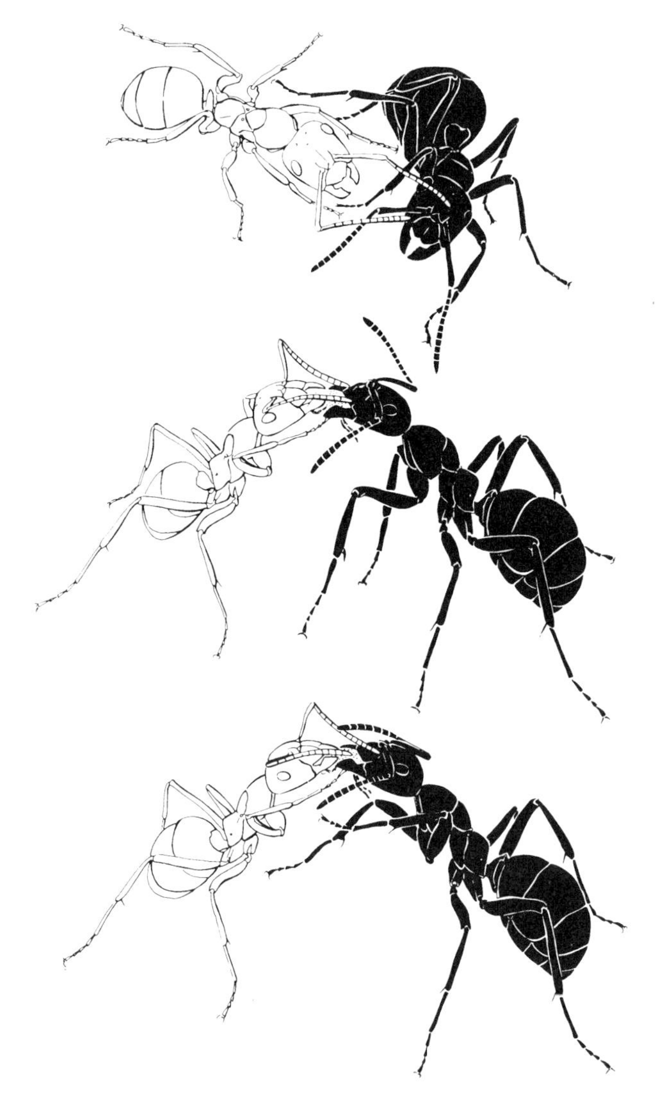

**FIGURE 7–71** The initiation of stomodeal trophallaxis, or exchange of liquid food by regurgitation, between two workers of *Formica sanguinea.* (*From top to bottom*) The soliciting worker, on the left, approaches a potential donor and antennates her, causing her to turn and face the solicitor. When the solicitor stimulates the donor's head and lower mouthparts with her antennae and forelegs, the donor regurgitates liquid as a droplet, on which the other ant feeds. (Hölldobler,1973b; drawing by T. Hölldobler-Forsyth.)

**FIGURE 7–72** Stomodeal trophallaxis between workers of *Formica sanguinea,* with the alimentary tract shown schematically. The donor, on the right, passes liquid food from her crop or "social stomach" (*K*) through her esophagus and into the mouth and crop of the recipient. Small amounts of food are also passed from the crop into the midgut (*M*) to serve as nourishment for the donor. *R,* rectal bladder. (From Hölldobler, 1973b; drawing by T. Hölldobler-Forsyth.)

coccids and carry the liquid homeward as droplets between their mandibles. Other large ponerines transport liquid in a similar manner. They include *Ectatomma tuberculatum* (Weber, 1946b) and *Paraponera clavata* (McCluskey and Brown, 1972; Hermann, 1975). The giant *Paraponera* workers appear to gather most of the liquid from extrafloral nectaries, standing water, and fruit, and this material composes a substantial portion of the harvested food (Hermann, 1975; Young, 1977).

Recent studies have revealed not just the transport but also the transmission of the liquid droplets by workers of the ponerine species *Pachycondyla obscuricornis* and *P. villosa,* large ants found in the New World tropical forests (Hölldobler, 1985). When a forager enters the nest laden with liquid food, she stands still for a period of time, swinging her head from side to side while waiting for a nestmate to approach; or else she moves directly toward nestmates and presents them with the food droplet held between her widely opened mandibles. If the colony is well fed, a forager may have to wait as long as 30 minutes before a nestmate responds. Sometimes she is wholly ignored and is not able to share her booty. In this case she imbibes a portion of the droplet herself and wipes off the residue on the floor and walls of the nest. Most of the time, however, nestmates readily accept the liquid food and even actively solicit it from the forager. While jerking her head rapidly up and down, the solicitor approaches the food carrier head-on and intensively antennates the front of her head and mandibles (see Figure 7–73). She makes a "spooning" or licking motion with the labium, and slowly transfers part of the standing drop to the space between her own mandibles. All the while she continues to antennate the head and mandibles of the donor. When about half of the liquid has been transferred, the ants pull apart. After the separation, the solicitor appears to imbibe a small fraction of the liquid. The remainder she shares with other nestmates, until as many as ten or more have received a portion.

In short, the *Pachycondyla* workers do not share food by regurgitation in the characteristic manner of most other ants. Rather they employ a "social bucket" system in which they first collect liquid food, then spoon portions into the gaping mandibles of nestmates. The bucket itself is formed by the mandibles on the side and by inwardly curving setae and the extruded labium underneath. The liquid is held in place by surface tension.

The whole social bucket procedure, while crude, nevertheless bears a striking similarity to liquid food exchange by regurgitation as it is employed by the Formicinae and other phylogenetically advanced ants (see Figure 7–71). In the latter case the food is collected in the crop. In response to very similar antennal signals and the mechanical stimulation of her labium, the food carrier regurgitates a droplet of liquid from her crop. Simultaneously she opens her mandibles widely, extrudes the labium, and folds the antennae backward. Occasionally, when a large droplet is regurgitated all at once, it is held between the mandibles in the ponerine manner (Figure 7–74). In contrast to the typical ponerine exchange, however, the soliciting ant imbibes all the food she receives and stores it in her crop. Small amounts of this food pass through the proventriculus into the midgut, where they are digested. The major portions, however, are distributed by regurgitation to nestmates.

With this evidence at hand, it is quite reasonable to suppose that the social bucket method of liquid food exchange is a precursor to

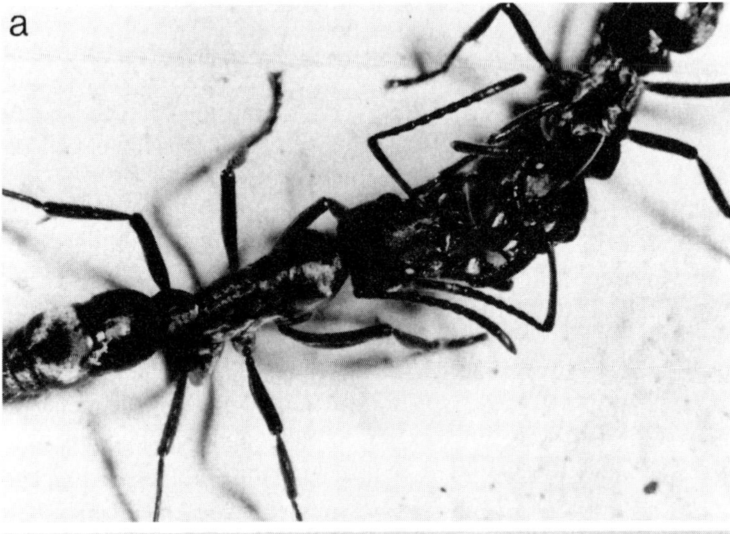

**FIGURE 7–73** The elementary form of "social bucket" technique used in liquid food transmission by the large ponerine *Pachycondyla villosa*. (*a*) The soliciting ant, on the right, approaches the food carrier and antennates her head and mandibles. (*b*) Food exchange proceeds by a direct unloading of part of the standing droplet from the donor to the solicitor. (*c*) After about half of the liquid has been transferred, the two ants pull apart. (From Hölldobler, 1985.)

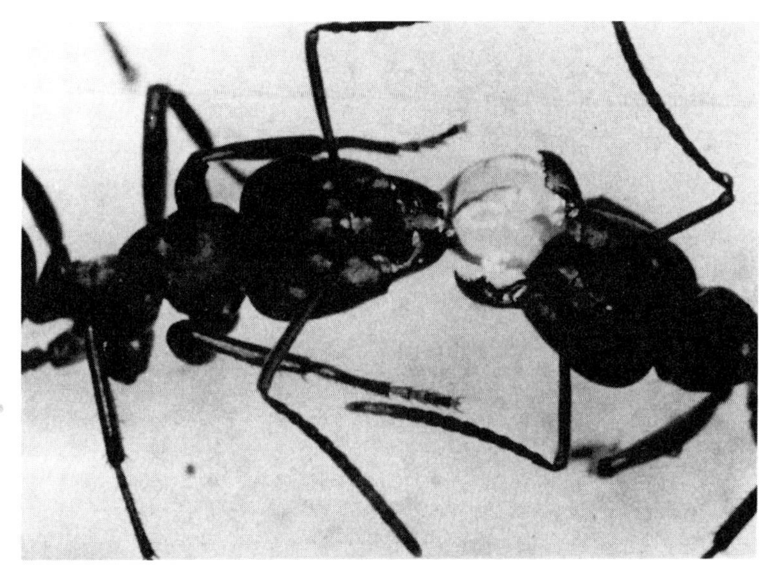

**FIGURE 7–74** Food exchange between two *Formica* workers. The ant on the right has regurgitated a large droplet, which is now held between the mandibles while it is slowly imbibed by the ant on the left. (From Hölldobler, 1973b.)

stomodeal regurgitation. It is not the only evolutionary entrée conceivable but for the moment it seems the most plausible one. The hypothesis gains further support from the fact that *Ectatomma* and *Paraponera*, which employ the social bucket, are members of the tribe Ectatommini. This taxonomic group is generally considered to be close to the stock that gave rise to the Myrmicinae, among the master users of regurgitation.

It is interesting to take one more step back in time and inquire about the evolutionary origin of the antennal signals used in both the social bucket and regurgitation. A clue is provided by the similarity of the antennal signals used in widely different behavioral categories within the Ponerinae. They include food begging, recruitment initiation, and social greeting, in which nestmates are recognized and alerted into examining the greeter. For example, when a patrolling worker of the African ponerine *Paltothyreus tarsatus* meets a stray nestmate, both ants first engage in mutual antennation. This behavior closely resembles the antennation pattern preceding food exchange in many other ant species, but no trophallaxis or liquid food exchange of any kind occurs. In fact, *Paltothyreus* apparently never practices liquid food exchange. In this species the stereotyped antennation is part of a greeting and invitation behavior, by which the nestmate is solicited to follow in tandem back to the nest (Hölldobler, 1984b). The invitation behavior is even more striking in an Australian species of the ponerine genus *Hypoponera* (Hölldobler, 1985). After a pair of workers meet face to face, the recruiter tilts her head sideways almost 90 degrees and strikes the upper and lower surfaces of the nestmate's head with her antennae. Often the solicited ant responds with similar antennation. The recruiting ant then turns around and tandem running starts (Figure 7–75). Similar behavior also occurs inside the nest, but it has never been observed to elicit food exchange—only the enticement of nestmates to travel from one nest site to another.

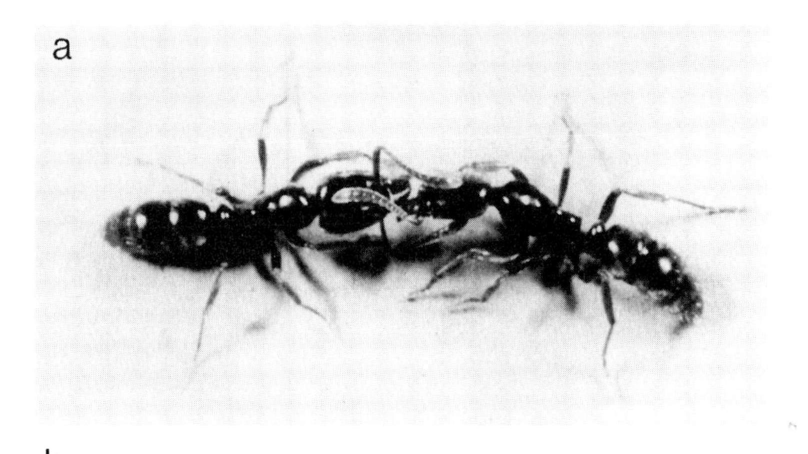

**FIGURE 7–75** (*a*) A worker of *Hypoponera* (*right*) invites a nestmate to follow in tandem. (*b*) Schematic illustration of the antennation behavior employed during invitation for tandem running recruitment. (*c*) Tandem running in *Hypoponera*. (From Hölldobler, 1985.)

In addition to antennation, the head-jerking movements often associated with food solicitation in *Pachycondyla villosa* have been found to be part of the invitation behavior in other species of *Pachycondyla* (Maschwitz et al., 1974; Traniello and Hölldobler, 1984).

Thus solicitation signals employed by ponerine ants in recruitment are similar if not identical to those employed in food exchange. Because the signals are employed by many ponerine species exclusively for invitation, yet no species uses them exclusively for food soliciting, invitation is reasonably interpreted to be the more primitive of the two functions. It would appear that the repertory of some ponerine phylogenetic lines was expanded by ritualization of invitation signals to encompass food-soliciting signals.

A very different form of exchange that has been reported in adult workers of several myrmicine genera is *abdominal trophallaxis*, the extrusion of a droplet of rectal liquid that is consumed by nestmates. In general form it resembles the donation of anal droplets from larvae to workers. The phenomenon was discovered in *Zacryptocerus varians*, an arboreal cephalotine species from the West Indies and Florida (Wilson, 1976a; Cole, 1980). Corn (1980) was not able to observe abdominal trophallaxis in the giant species *Cephalotes atratus*, but Wheeler (1984) found it in the cephalotine *Procryptocerus scabriusculus*. In *P. scabriusculus* the behavior is usually initiated by newly eclosed workers, who seek out older workers and solicit the droplets by licking their abdominal tips. Abdominal trophallaxis was also observed between older workers and between a worker and a queen, but the bouts were much shorter than those between callows and older workers. The function of the behavior remains unknown. However, Wheeler has pointed out its similarity to proctodeal feeding in termites, by which symbiotic protozoans and bacteria are transferred from older to younger colony members. There may be a connection with a second striking peculiarity associated with digestion in cephalotine ants: a mushroom-shaped, sclerotized cap on the proventriculus, an organ that intervenes between the crop and midgut and is thought to filter food as part of the social function of crop storage (Eisner, 1957).

Outside the Cephalotini, abdominal trophallaxis has been observed between the myrmicine slavemaker *Harpagoxenus americanus* and its hosts species, *Leptothorax ambiguus* and *L. longispinosus* (Stuart, 1981). Workers and queens of the *Harpagoxenus* occasionally assume a stereotyped posture, standing quietly with the abdomen raised, and extrude a droplet of liquid, which is eaten by the slaves. This behavior is doubly remarkable because it is a rare instance of a social parasite donating something to her host. Its function remains unknown, and may prove to be a form of dominance or other exploitative behavior.

## FACILITATION AND GROUP EFFECTS

In 1946 Grassé proposed to classify all the effects of "social physiology" into two categories: mass effects, in which the surrounding medium is modified by the population; and group effects, in which the members of the population affect one another directly by sensory stimulation. The phrase *effet de groupe* was then used repeatedly in the French literature on social insects. But Grassé's terminology did not catch on elsewhere because, like Allee's earlier exposition of "group behavior" (1931, 1938), it was too amorphously formulated. From the beginning there has been no clear boundary between mass and group effects. It is also very difficult to make a sharp distinction between group in the Allee-Grassé sense and the rest of communicative behavior. Like the word "trophallaxis," the expression "group effect" can, with little effort, be stretched to become synonymous with communication in the broadest sense.

Good use can nevertheless be made of the group effect label to cover a particular set of communicative phenomena that are of considerable importance in insect societies. A group effect can be usefully defined as an alteration in behavior or physiology within a species brought about by signals that are directed in neither space nor time (Wilson, 1971). Alarm signals, odor trails, and sex attractants obviously do not qualify. On the other hand, most primer pheromones, which act on animals over long periods of time without necessarily evoking a directed response, do qualify. In addition, there exist a wide range of communicative phenomena in social insects that are undirected and long lasting but not necessarily pheromonal in nature. They are the co-actions that the French investigators have intuitively called group effects. Most are examples of what has been termed social facilitation in the psychological literature, meaning communication that promotes rather than inhibits activity. The conception of social facilitation as a discrete phenomenon began in studies of human social psychology. Allport (1924) defined it as "an increase of response merely from the sight or sound of others making the same movement." To complete the terminology, the opposite effect from social facilitation should be labeled social inhibition.

A relatively clear example of facilitation from the ants is the behavior of workers of *Lasius emarginatus,* who excavate the soil and attend larvae at a higher rate when in large groups. When Francfort (1945) separated small groups of this European soil-dwelling species consisting of four to six workers from larger groups by only a gauze barrier, their activity rate remained high, but it dropped when he inserted a glass barrier. Francfort concluded that the facilitating stimulus is an odor. Hangartner (1969a), following up this result, discovered that fire ant workers (*Solenopsis geminata*) attempted to dig through porous barriers put up to separate them from other members of their colony. He was able to induce the same effect by substituting tubes containing slightly higher concentrations of carbon dioxide. Hence it is likely that Francfort's result was due to carbon dioxide or some other general metabolic product rather than a specialized facilitation pheromone.

The relation between activity and group size is not simple, however. Chen (1937a,b) reported that workers of the carpenter ant *Camponotus japonicus aterrimus* placed in groups of two or three in earth-filled jars began digging sooner, moved more earth per ant, and displayed less variation in individual effort than workers placed alone in the same kind of containers. "Leader" ants, that is, ants that worked best when alone, had a stimulating effect on nestmates, whereas the slower "follower" ants had a retarding effect. Leader ants also had a higher metabolic rate, as evidenced by their greater vulnerability to starvation, drying, and poisoning by chloroform and ether fumes. A qualitatively similar result was obtained by Klotz (1986) in *Formica subsericea* and Imamura (1982) in *F. yessensis.* Imamura suggested that the effect could be due to the release of small quantities of alarm pheromone from the mandibular glands when the mandibles are opened. In a puzzling development, contrary results were obtained by Sudd (1972) for *F. lemani* and Sakagami and Hayashida (1962) for *F. fusca.* When the number of *F. fusca* workers was increased to eight, for example, the average digging performance either remained the same or began to fall off. The difference in these results may be due to experimental design or to sampling variations in the proportion of leader (or elite) workers used in the various replicates. In general, however, it appears well established that the facilitation of digging behavior does occur in at least some ant species under appropriate conditions.

Facilitation and group effects remain relatively unexplored subjects in the sociobiology of ants. There may well be important communicative signals and social phenomena awaiting discovery. At the very least, a diversity of behavior patterns other than nest building is modulated in some manner by the size of the group. The aggressiveness of an individual ant increases as the size of the crowd of nestmates around her grows. When workers of *Acanthomyops claviger,* a subterranean formicine ant of the eastern United States, are kept in solitude, they are nearly insensitive to the natural alarm substances of the species, including undecane and citronellal. In contrast, those placed in the same nests with a few hundred nestmates respond normally to the pheromones (Wilson, 1971).

The opposite effect also occurs. When worker ants are placed in groups and not otherwise stimulated, they cluster and "calm down," expending less energy. In *Camponotus* and *Formica,* for example, the change is reflected in a decline of oxygen consumption as measured in microliters per milligram dry weight per hour. The curve approaches an asymptote, such that the oxygen consumption is reduced by half or more when the group size reaches ten (Gallé, 1978).

Perhaps the most fundamental form of facilitation is the high-frequency pulsing of patrolling, brood care, and other activities demonstrated in *Leptothorax* by Franks and Bryant (1987) and *Macromischa* by B. J. Cole (personal communication). Individual ants are quiescent most of the time, and when they become active it is usually in pulses that occur 2 to 4 times an hour. The pulses tend to be coupled among nestmates, a phenomenon that appears to be based on some form of facilitation.

## AUTOSTIMULATION

Is it possible for ants to communicate with themselves? In effect, foraging workers do just that when they dispense orientation pheromones in their odor trails and then follow the traces during the return journey to the target area. A striking example of this kind of autostimulation is seen in the recruitment of *Camponotus sericeus* to food finds and new nest sites. The successful scout lays an orientation trail back to the nest, recruits a single nestmate by a motor display, and leads her back along the trail by means of tandem running. The recruitment does not involve the odor trail, which serves only for orientation of the worker that laid it (Hölldobler et al., 1974). An even more extreme development exists in *Leptothorax affinis,* which uses odor trails during emigration. Each worker recognizes and prefers her own trail and ignores those of her nestmates (Maschwitz et al., 1986b).

After workers of *Atta cephalotes* cut leaves and before they transport them back to the nest for use as a fungal substrate, they mark them with an abdominal secretion. The marked fragments are more readily picked up by nestmates and by the cutting ants themselves than unmarked pieces. The effect has been duplicated by *n*-tridecane and (Z)-9-nonadecene, which are components of the Dufour's gland (Bradshaw et al., 1986).

## MEDIATION OF LARVAL DIAPAUSE

Colonies of most ant species in the north temperate zone undergo some form of diapause during the late fall and winter. The metabolic rate and locomotor activity slow down drastically and reproduction ceases. In the genus *Camponotus,* the species of which usually nest in fresh and decaying wood and hence are called carpenter ants, the adults and brood enter diapause in the cold season (Hölldobler,

1961). Similarly, most myrmicine species overwinter with larvae in the nest, and these immature stages also pass through a true diapause of their own.

In most organisms diapause is entrained by a shortening of the daily photoperiod, which is by far the most reliable "calendar" available. Workers of *Myrmica rubra* have been proved to use photoperiod (Brian, 1986b), and there is no reason to expect otherwise for adults of ants generally. On the other hand a mystery remains. Ant larvae are hidden by the adults in soil or rotting vegetation, and thus live in permanent darkness. How do they measure the change in season and choose to enter diapause? Weir (1959b) proved that fall ("serotinal") workers of *Myrmica* tend to induce diapause in terminal-instar larvae, whereas spring ("vernal") workers cannot. This result was achieved by keeping workers in the laboratory at the warm temperature of 25°C for 11 weeks after lengthy chilling to simulate in them the physiological state of wild fall workers, and by keeping other workers at the same temperature for only 5 weeks after chilling to simulate the wild spring condition. The "fall" workers could induce diapause; the "spring" workers could not. Closely similar experiments were performed by Kipyatkov (1979, 1988) on *M. rubra* in the Soviet Union, with the same result. Weir further guessed that diet might be the key, since *Myrmica* workers are known to increase the proportion of protein in their diet as the season progresses, and dormant larvae have a higher nitrogen-to-carbon content in their meconia and fat bodies than do nondormant larvae. When Weir fed spring workers a sufficiently increased amount of protein (by means of a pure diet of *Drosophila*), it turned out that they too were able to induce dormancy. Weir, after considering the matter at length, remained uncertain whether the larval dormancy is true diapause in the purest sense, in other words a shut-down mediated by the endocrine system. Still, it qualifies as diapause in the broad sense that, once initiated, it is persistent in its effect, even at raised temperatures. Kipyatkov is in substantial agreement. In addition, he has provided evidence of a remarkable new volatile pheromone from spring workers that reactivates diapausing larvae and queens.

## NECROPHORESIS OR CORPSE REMOVAL

Identification of the dead is not communication in the strict sense, but it has some features in common with communication, particularly in its dependence on stereotyped responses triggered by narrowly specific chemical stimuli. The removal of dead nestmates (a behavior called necrophoresis) and other decomposing material from the nest also serves the hygiene of the colony as a whole. The interiors of the nests of ants, and particularly the brood chambers, are kept meticulously clean. Workers drag alien objects, including particles of waste material and defeated enemies, out of the nest and dump them onto the ground nearby. They carry waste liquid and meconia (the pellets of accumulated solid waste voided by larvae at pupation) to the nest perimeter or beyond. They respond to disagreeable but immovable objects by covering them with pieces of soil and nest material.

The same behavior has been modified to serve a new function in some of the species that keep aphids and other honeydew-producing "cattle" outside the nest. The ants enclose their charges in chambers irregularly built from soil or vegetable matter. In a few species, such as some of the members of *Crematogaster*, the behavior has advanced to the point where an elaborate carton shelter is constructed from chewed vegetable fibers (Wheeler, 1919a). Ant workers also occasionally try to cover small pools of water or other liquid in the nest vicinity. A casual observation of this phenomenon has misled some authors to report erroneously that ants construct "bridges" to cross obstacles. If honey is placed in a small pool outside the nest, workers sometimes treat it in the same manner, covering the honey with particles of soil or bits of vegetable matter. They may then carry some of the honey-dipped particles back to the nest, giving the erroneous impression that they are using "tools" (Fellers and Fellers, 1976).

Ants, like other social insects, are especially fastidious when dealing with corpses. The dead of some species of ants are eaten by their own nestmates. This occurs in varying frequency in some species of the myrmicine genera *Pheidole* and *Solenopsis*, and in the weaver ants *Oecophylla* that we have studied in the laboratory. In the spring, colonies of the red wood ant *Formica polyctena* regularly war on one another and eat their dead enemies. Mabelis (1979b) proposed that the cannibalism is adaptive, serving to tide colonies over during periods of prey shortage. This idea has been further supported by the field studies of Driessen et al. (1984), who found that the raids are concentrated during periods when insect prey is in short supply. In most cases, however, corpses of nestmates and other arthropods are carried out and discarded.

When nesting on a level surface, workers of the red imported fire ant *Solenopsis invicta* carrying their dead proceed outward from the nest entrance in randomly distributed directions. On a slope, they tend to walk downward, and the headings reach a constant level in this bias at 15 degrees inclination or greater. Whatever the slope, the corpses are dropped at unpredictable distances and hence do not accumulate in piles (Howard and Tschinkel, 1976). Other kinds of ants, for example army ants of the genus *Eciton* (Rettenmeyer, 1963a), pile the dead among the general refuse in kitchen middens located a short distance from the nest or bivouac. Still others, including leafcutters of the genus *Atta*, place them in special refuse chambers (Stahel and Geijskes, 1939; Moser, 1963). One species, the small predatory myrmicine *Strumigenys lopotyle* of New Guinea, pile fragments of corpses of various kinds of insects in a tight ring around the entrance of their nest in the soil of the rain forest floor (E. O. Wilson, reported in Brown, 1969). On the other hand, despite the claim of some authors in both ancient and early modern times (Wilde, 1615), there is no creditable evidence of the existence of "ant cemeteries," to which only the bodies of fallen nestmates are consigned. Nor is there any documented case of ants burying their dead in anything approaching a ritualistic or organized fashion.

The transport of dead nestmates from the nest is nevertheless one of the most conspicuous and stereotyped patterns of behavior exhibited by ants. A full description of the behavior is given by McCook (1879) in his classic monograph on the harvesting ant *Pogonomyrmex barbatus*. Wilson et al. (1958) analyzed the stimuli that trigger this "necrophoric" pattern in *Pogonomyrmex* and *Solenopsis*. The results have been confirmed and extended in key respects in *Solenopsis invicta* by Blum (1970) and Howard and Tschinkel (1976) and in the primitive bulldog ants *Myrmecia* by Haskins (1970).

When a corpse of a *Pogonomyrmex badius* worker is allowed to decompose in the open air for a day or more and is then placed in the nest or outside near the nest entrance, the first sister worker to encounter it ordinarily investigates it briefly by repeated antennal contact, then picks it up and carries it directly away toward the refuse piles. In the laboratory nests employed by Wilson in his *Pogonomyrmex* study, the most distant walls of the foraging arenas were less than a meter from the nest entrances, and the ants had built the refuse piles against them. The distance was evidently inadequate to allow the rapid consummation of the corpse removal response because workers bearing corpses frequently wandered for many minutes back and forth along the distant wall before dropping their burdens on the refuse piles. Others were seen to approach the distant wall unburdened, pick up the corpses already on the piles, and transport them in similarly restless fashion before redepositing them. This behavior constituted a distinctive and easily repeated bioassay. It was soon established that bits of paper treated with acetone extracts of *Pogonomyrmex* corpses were treated just like intact corpses by both *P. badius* and *Solenopsis invicta* workers. Separation and behavioral assays of principal components of the extract implicated long-chain fatty acids and their esters. Furthermore, it turned out that oleic acid, a common decomposition product in insect corpses, is fully effective. Blum (1970) has subsequently identified this substance in *Solenopsis* corpses. On the other hand, many other principal products of insect decomposition, including short-chain fatty acids, amines, indoles, and mercaptans, were ineffective. When *Pogonomyrmex* corpses were thoroughly leached in solvents, dried, and presented to colonies, they were seldom transported as corpses, but were more commonly eaten instead.

Thus the worker ants appear to recognize corpses on the basis of a limited array of chemical breakdown products. They are, moreover, very "narrow-minded" on the subject. Almost any object possessing an otherwise inoffensive odor is treated as a corpse when daubed with oleic acid. This classification even extends to living nestmates. When a small amount of the substance is daubed on live workers, they are picked up and carried, unprotesting, to the refuse pile. After being deposited, they clean themselves and return to the nest. If the cleaning was not thorough enough, they are sometimes mistaken a second or third time for corpses and taken back to the refuse piles.

Perhaps even more remarkable than the simplicity of this control of necrophoric behavior is the tendency of the workers of some ant species to remove themselves from the nest when they are about to die. We have repeatedly observed that injured and dying ants loiter more in the vicinity of the nest entrance or outside the nest than do normal workers. Injured *Solenopsis invicta,* particularly those that have lost their abdomens or one or more appendages, tend to leave the nest more readily when the latter is disturbed. Marikovsky (1962a) reports that workers of *Formica rufa* fatally infected with the fungus *Alternaria tenuis* leave the nest and cling fast to blades of grass with their mandibles and legs just before dying.

# CHAPTER 8

# Caste and Division of Labor

It is possible to distinguish two very general problems in the study of social insects, to which all other topics and theoretical reflection play a tributary role. The first is the origin of social behavior itself. Biologists are especially interested in the advanced state of eusociality, which characterizes all of the ants and a small percentage of aculeate wasps and bees. This evolutionary grade is defined as the combination of three features: care of the young by adults, overlap in generations, and division of labor into reproductive and nonreproductive castes.

The last of the eusocial traits, the existence of a subordinate or even completely sterile worker caste, is the rarest of the three. It is also by far the most significant with reference to the further evolutionary potential of social life, for when individuals can be turned into specialized working machines, an intricate division of labor can be achieved and a complicated social organization becomes attainable even with a relatively simple repertory of individual behavior. To use a rough metaphor, we may say that it does not matter if the separate pieces of a clock are simple in construction, so long as they can be shaped to serve particular roles and then fitted together into a working whole. You cannot construct a proper clock if every little piece must also keep some sort of time on its own. By the same token natural selection cannot readily assemble a complex colony if every member is designed to complete a reproductive life cycle of its own in the vertebrate manner.

This brings us to the second major problem in the study of social insects, which can be called *strategic design,* or, more fully, the strategic design of colonies. Once a sterile worker is in place, that is, once the equivalent of the less-than-independent levers and wheels of the clock can be manufactured, the important consideration becomes the best arrangement of castes and division of labor for the functioning and reproduction of the colony as a whole. The colony can be most effectively analyzed if it is treated as a factory within a fortress. Natural selection operates so as to favor colonies that contribute the largest number of mature colonies in the next generation, so that the number of workers per colony is only of incidental importance. In other words the key measure is not how big, or strong, or aggressive a colony of a particular genotype can become but how many successful new colonies it generates. Hence the functioning of the workers in gathering energy and converting it into virgin queens and males is vital. This part of the colony's activities constitutes the factory. But the colony is simultaneously a tempting target for predators. The brood and food stores are veritable treasure houses of protein, fat, and carbohydrates. As a consequence colonies must have an adequate defense system, which often takes the form of stings and poisonous secretions and even specialized soldier castes. This set of adaptations constitutes the fortress.

The ultimate currency in the equations of colony fitness is energy. The workers appear programmed to sweep the nest environs in such a way as to gain the maximum net energetic yield. Their size, diel rhythms, foraging geometry, recruitment techniques, and methods of food retrieval are the qualities most likely to be shaped by natural selection. But even as energy is being collected and distributed to the queen and brood, the colony is subject to predation and competition. A certain number of individuals must be sacrificed in periodic defenses of the colony and during the riskier steps of the daily foraging expeditions. The loss of energy required to replace them is entered on the debit side of the fitness ledger.

Colony design then becomes effectively a problem in economics. The more exact expression is *ergonomics,* to acknowledge that work and energy are the sole elements of calculation, and also that nothing resembling human transactions with credit and money is involved (Wilson, 1968; Oster and Wilson, 1978).

## CASTE, TASK, ROLE

The core of any ergonomic (as well as any human economic) analysis is the prescription of a complete balance sheet, and to this end it is necessary to define as precisely as possible the agents entailed. The fundamental distinctions among individual workers in a colony are based on the actual tasks the workers perform and the various roles they play in accomplishing the functions of the colony as a whole. An example of the behavioral repertories of two physical castes is given in Table 8–1.

An ergonomic analysis is started with a listing of distinct behavioral acts from protracted observations of individuals, selected sets of individuals, or the entire colony. If physical and temporal groups can be distinguished, separate repertories for each are compiled. In the unlikely event that some of these arbitrarily defined categories prove to have identical repertories, they cannot be listed as separate castes—and the data concerning their repertories can be subsequently combined. The next step is to compute the relative frequencies of the acts and to fit them to standard distributions such as the lognormal Poisson or negative binomial. An observer thus can obtain a measure of the adequacy of his or her sample size. This method was perfected and used with success by Fagen and his collaborators (especially Fagen and Goldman, 1977) to make estimates of total repertory size even before a complete catalog was actually compiled. For example, after 1,222 separate behavioral observations had been made on the minor worker caste of a colony of *Pheidole dentata*, 26 kinds of behavioral acts were recognized; these acts constituted the known behavioral repertory of this caste (Wilson, 1976c). By fitting the frequency data (see Table 8–1) to a lognormal Poisson distribution, the researcher could estimate the actual number of kinds of behavioral acts to be 27, with a 95 percent confidence interval of (26, 28). The major workers were observed performing 8 kinds of behavioral acts; the true number was estimated to be 9, with a confidence interval of (8, 10).

As analyzed by Wilson and Fagen (1974) this method of tabulating behavioral repertories, which has gained wide usage in the study of behavior, has notable strengths and weaknesses. One obvious advantage of the technique is that it improves unaided intuition without forcing any new, unsupportable assumptions on the analysis. Another is that the tabulation allows a more precise judgment of the point of diminishing return during ethological studies. That point comes surprisingly early in the case of the ants. A typical example is the study of worker behavior in *Leptothorax curvispinosus* conducted by Wilson and Fagen. After only 51 hours of observation, during which 1,962 separate acts were recorded, the mode of the frequency curve emerged and the estimated sample coverage attained was 99.95 percent. Thus the effort required to secure a nearly complete repertory is at least a full order of magnitude less than is the effort required in the case of vertebrates, which typically consumes months or years of arduous field research. The result is that comparative behavioral studies have proceeded much more rapidly in ants and other insects.

The tractability of ant studies has a feature of considerable biological interest. This is the small number of rare behavioral acts that exist relative to common acts (see Figure 8–1). Whatever ants do they do rather frequently in comparison with vertebrates; few if any rare behaviors exist to surprise the investigator in the late stages of the study. It is possible that the small size of the ant brain precludes the storage of acts that are not used commonly or at least are not of crucial importance to the colony. This interpretation is consistent with the principle of economy in the evolution of ant communication systems, which we documented in Chapter 7. In briefest form the concept of economy holds that a characteristic of ant behavior is the repeated use of the same communicative signals and responses in different contexts to achieve multiple purposes.

There are two disadvantages of the method, which we do not regard as particularly serious. The first is the probability that the repertories and especially the frequency distributions change in different contexts. It remains for the biologist to define those contexts and to repeat the analysis within them. In the case of ants distinguishable contexts are not only finite but also probably quite limited in number. By far the greatest part of an ant's life is conducted in the homeostatic environment of the nest interior. Thus the lifetime sample coverage in the *Pheidole dentata* study, defined as the cumulative probability of all behavior for all contexts, was probably very high in spite of the fact that it was limited to one environment. We suggest that the following list might exhaust the remaining contexts for the worker caste: extended foraging periods and territorial patrolling and defense; major disturbances of the nest, including invasion by alien colonies, flooding, and overheating; emigration to a new nest site; and assisting during the initiation of nuptial flights on the part of the reproductive forms.

**TABLE 8–1** A behavioral repertory: behavioral acts by the two physical castes of the ant *Pheidole dentata* in an undisturbed colony, with their relative frequencies. *N*, total number of behavioral acts recorded in each column. (From Wilson, 1976c.)

| Behavioral act | Frequency of behavioral acts: Minor workers ($N$ = 1,222) | Major workers ($N$ = 208) |
|---|---|---|
| Self-grooming | 0.18003 | 0.56373 |
| Allogroom adult | | |
| Minor worker | 0.04992 | 0 |
| Major worker | 0.00573 | 0 |
| Alate or mother queen | 0.01146 | 0 |
| Brood care | | |
| Carry or roll egg | 0.01391 | 0 |
| Lick egg | 0.00245 | 0 |
| Carry or roll larva | 0.12357 | 0 |
| Lick larva | 0.09984 | 0.02941 |
| Assist larval ecdysis | 0.00409 | 0 |
| Feed larva solid food | 0.00573 | 0 |
| Carry or roll pupa | 0.03601 | 0 |
| Lick pupa | 0.01882 | 0 |
| Assist eclosion of adult | 0.00818 | 0 |
| Regurgitate | | |
| With larva | 0.02128 | 0 |
| With minor worker | 0.03764 | 0.22059 |
| With major worker | 0.00573 | 0 |
| With alate or mother queen | 0.00327 | 0 |
| Forage | 0.12111 | 0.02941 |
| Feed outside nest | 0.04337 | 0.01471 |
| Carry food particles inside nest | 0.05237 | 0 |
| Feed inside nest | 0.05810 | 0.01471 |
| Lick meconium | 0.00573 | 0 |
| Carry dead nestmate | 0.01882 | 0.04902 |
| Carry or drag live nestmate | 0.00246 | 0 |
| Eat dead nestmate | 0.06383 | 0.07843 |
| Handle nest material | 0.00655 | 0 |
| Total | 1.0 | 1.0 |

The second difficulty in repertory estimation is the arbitrariness of the definition of the behavioral act. One observer might see three distinct neuromuscular patterns where another sees only one. Thus "foraging" as defined in Table 8–1 could easily be broken down into several acts. This is essentially a problem of language, and different observers can solve it by a straightforward mapping procedure. One observer's acts *a*, *b*, and *c* will be recognized as constituting the second observer's act *a*, the first observer's act *h* will be seen as representing the second observer's acts *m* and *n*, and so forth. No great difficulty should occur when the same species is considered or closely related species are compared. Serious conceptual problems

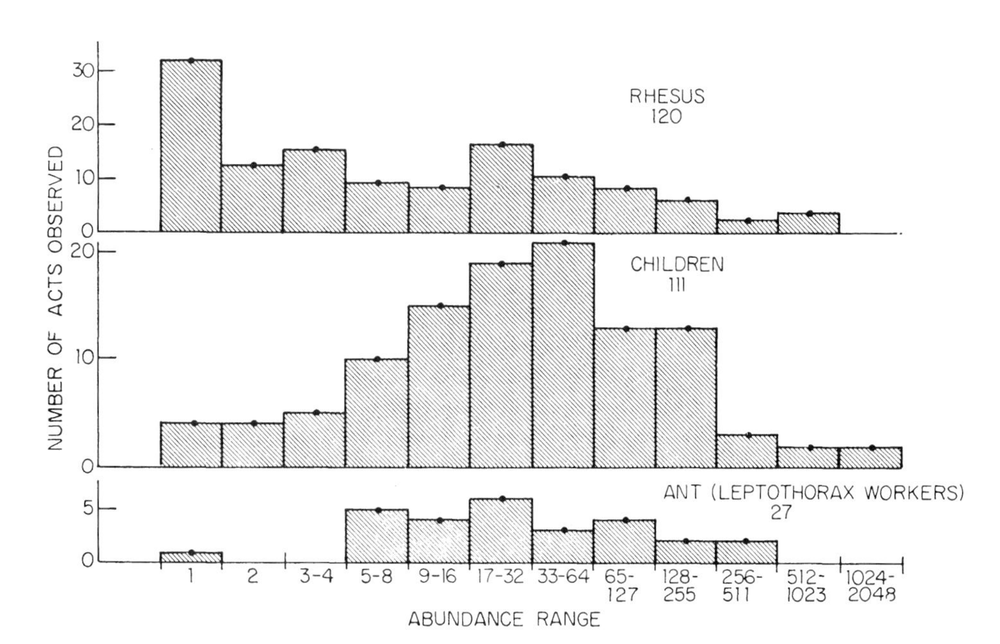


**FIGURE 8–1** The number of types of acts having a given abundance in 27 observed behaviors of *Leptothorax curvispinosus* workers, 111 behaviors of playing children, and 120 behaviors of rhesus monkeys. The mode has clearly emerged in the ants and children, indicating that most kinds of behaviors have been cataloged. This is particularly true of the ants, in which relatively few rare categories have so far been discovered. (Human and rhesus data from Fagen and Goldman, 1977; ant data from Wilson and Fagen, 1974.)

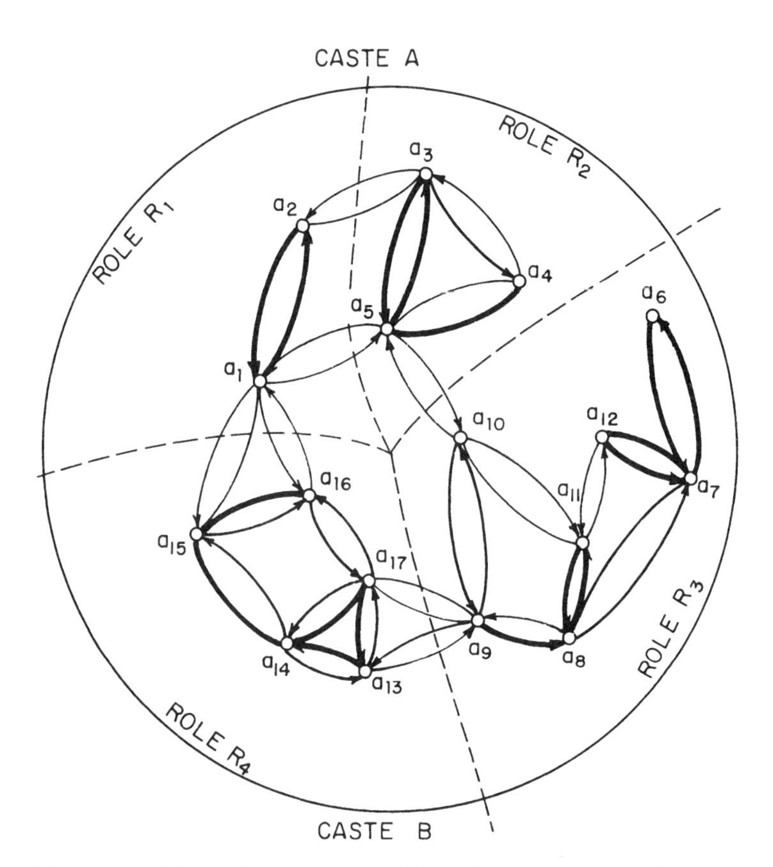


**FIGURE 8–2** A form of ethogram by which roles and castes can be more precisely defined. Each set of behavioral acts $a_i$ linked by relatively high transition probabilities constitutes a role. The thicknesses of the connecting lines indicate the magnitudes of the transition probabilities. When a distinct group (for example, a size or age cohort) attends preferentially to one or more roles, it is defined as a caste. Roles can also consist in overlapping sets of behavioral acts; for convenience of graphical representation they are shown here as disjoint sets. (From Oster and Wilson, 1978.)

might develop, however, when an attempt is made to compare the size and frequency characteristics across radically different species.

The next step beyond the behavioral repertory is the construction of an *ethogram*, which incorporates not only the repertory of a caste but also the transition probabilities connecting individual acts and the time distributions spent in each act. When the ethogram also takes into account the interactions of parents and offspring, dominant and subordinate individuals, and other members of the society, the description is referred to as a *sociogram*. Ethograms can cover all of the repertory or certain well-defined portions of it. Such quantitative studies have been conducted on ants, honey bees, hermit crabs, mantis shrimps, and rhesus monkeys, with promising results (see reviews by Dingle, 1972; Wilson, 1975b: 194–200).

Figure 8–2 illustrates an imaginary example of an ethogram with role and castes delimited. For any individual certain of the behavioral acts ($a_i$) will be linked together by relatively high transition probabilities: in *Pheidole dentata*, for example, pupal licking is associated with high frequency to pupa carrying, egg licking, and egg transport; nest construction gives way with high probability to foraging outside the nest; and so on. A set of closely linked behavioral acts can be defined as a *role*, even if the acts are otherwise quite different. It is generally true, for example, that the act of grooming the queen is closely linked with the very different acts of regurgitating to the queen and removing freshly oviposited eggs. All of these responses can be considered part of the single role of queen care. We can define a *caste* as a group that specializes to some extent on one or more roles. Broadly characterized, a caste is any set of a particular morphological type, age group, or physiological state (such as inseminated versus barren) that performs specialized labor in the colony. A natural classification follows from this definition. A *physical caste* is distinguished not only by behavior but also by distinctive anatomical traits. A *temporal caste*, in contrast, is distinguished by age. A *physiological caste* is distinguished by a principal physiological state that is frequently but not necessarily linked to anatomy or age, such as insemination. Sometimes all of the workers together are referred to as a caste, as distinct from the queen, and subsets of workers are called subcastes.

The term *task* is used to denote a particular sequence of acts that accomplishes a specific purpose, such as foraging or nest repair. Ordinarily a task is identical to a role or is composed of the subset of acts within a role, but it might conceivably consist of acts distributed across two or more roles. Finally, the division of labor by the allocation of tasks among various castes is often referred to as *polyethism,* a term apparently first employed by Weir (1958a,b). We further speak of physical caste polyethism as opposed to temporal caste polyethism.

Few studies have probed to the depth necessary to yield quantitative measures of castes and tasks. Wilson (1976c) recognized four worker subcastes of *Pheidole dentata,* comprising a unified age category in the major worker and three age categories in the minor worker. Together the castes perform a known total of 26 acts in undisturbed colonies, of which 23 can be classified as true social tasks. Similarly, a total of 29 tasks have been identified in the leafcutter ant *Atta sexdens* (Wilson, 1980a). These are performed by an estimated four physical castes, of which three are further subdivided into temporal castes to make a total of at least seven castes overall. The most thorough studies of labor division have been conducted on *Lasius niger* (Lenoir, 1979a,c) and on species of *Leptothorax* (Herbers, 1983; Herbers and Cunningham, 1983). These studies combined matrix analysis with methods of multivariate analysis to cluster the tasks into roles. Two or three roles, or "components," as they were called by Lenoir, could be recognized in *Lasius.* In the *Leptothorax,* the workers were divided into three behavioral castes filling a total of four roles.

It will be useful at this point to introduce a short glossary of terms applied to castes that have gained at least moderate employment by myrmecologists. This is a difficult procedure, for two reasons. First, an extraordinary number of technical terms have been proposed. Many of them were suggested by William Morton Wheeler, who was not only a great entomologist but a classical scholar fluent in Greek with an inordinate fondness for neologisms. In 1907 he proposed a caste system based on anatomy with no fewer than 30 categories. He recognized pathological forms such as the phthisaner, a pupal male with appendage growth caused by an *Orasema* wasp larva; the pseudogyne, a worker-like form with swollen mesothorax in the genus *Formica,* whose abnormality is caused by the presence of lomechusine staphylinid beetles in the nest; and the pterergate, a worker ant with vestiges of wings. During his lifetime and especially in his last work on the subject in 1937, Wheeler believed he had found good reasons for multiplying and naming every qualitatively distinguishable form. On the one hand he considered the parasitogenic forms to be different enough and rare enough to fit comfortably into the system. More important, however, he believed (erroneously) that most nonparasitogenic castes arise by genetic mutations. He saw no fundamental difference between normal functional castes and true anomalies. All except the queen and typical males were basically anomalous forms to Wheeler, and he referred to his categories alternatively as castes, phases, and anomalies.

In a reevaluation of Wheeler's system, Wilson (1953b) concluded that some of the names are superfluous, some are virtually synonymous with others, and some are no more than stages in an allometric progression. Over the years a substantial literature of teratology has accumulated on gynandromorphs and other developmental anomalies, some of which possess traits intermediate between normal castes, but it has not been considered useful to recognize any of these forms as castes; descriptions and reviews have been provided variously by Wheeler (1937b), Novak (1948), Kusnezov (1951a), Buschinger and Stoewesand (1971), Wilson and Fagen (1974), Sokolowski and Wiśniewski (1975), Peru (1984), and Taber and Francke (1986).

A second cause of ambiguity in classification is the dispute among leading students of the subject over whether castes should be defined primarily by role or by anatomy. In particular, Buschinger (1987b) has proposed that "queen" and "worker" be used to designate females with reproductive and nonreproductive roles, regardless of whether they differ anatomically in a particular species, whereas various other terms such as gynomorph and intermorph should be used to denote the anatomical variants within these two principal categories. Peeters (1987a,b), in opposition, prefers a morphological definition of the castes "queen" and "worker," and reserves the term "gamergate" for the relatively uncommon condition in which mated, completely worker-like forms have replaced the queen caste in "queenless" societies. The two positions are conceptually not far apart and can be semantically resolved. Our opinion is that the classification should be kept somewhat loose, incorporating either anatomy or roles in a manner that maximizes convenience, precision, and clarity of expression. It must be recognized that anatomy and reproductive role are related across the various ant species not by a one-to-one linkage but by a many-to-many linkage. That is, anatomical workers often join the anatomical queens in reproduction although usually to a limited extent. In such species the reproductive role can be said to be assumed by both anatomical castes. And queens in their turn assume many roles other than reproduction. Such many-to-many relationships are nearly universal among the anatomical worker subcastes and labor roles. It is also true that anatomically distinct (i.e., physical) castes are almost always biased toward certain roles, whereas individual roles are almost always filled preferentially by one anatomical caste or another in the case of polymorphic species. With these qualifications, the following annotated glossary is efficient and for the most part reflects current usage (see also Figure 8–3).

1. *Male.* In the great majority of ant species, males (designated by the symbol ♂) fill no role in the colony that generated them, being content to receive food from their sisters while awaiting the nuptial flight that will end their lives. In such cases it is misleading to refer to them as a caste. In some species of *Camponotus,* the males are long-lived and serve as donors in food exchange within the colony; hence they qualify marginally as castes. To take another, very different phenomenon, worker-like or "ergatomorphic" males occur in several genera (*Hypoponera, Cardiocondyla, Formicoxenus, Technomyrmex*). So far as we know they do not contribute to colony labor, however.

2. *Queen.* This caste is often designated by the symbol ♀ (contrast with symbols for worker, to follow). In the broad sense, the principal female reproductive type, that is, any form anatomically distinguishable from the worker caste and responsible to a disproportionate extent for reproduction. In the narrow sense, employed in more technical literature on caste biology, the fully developed reproduc-

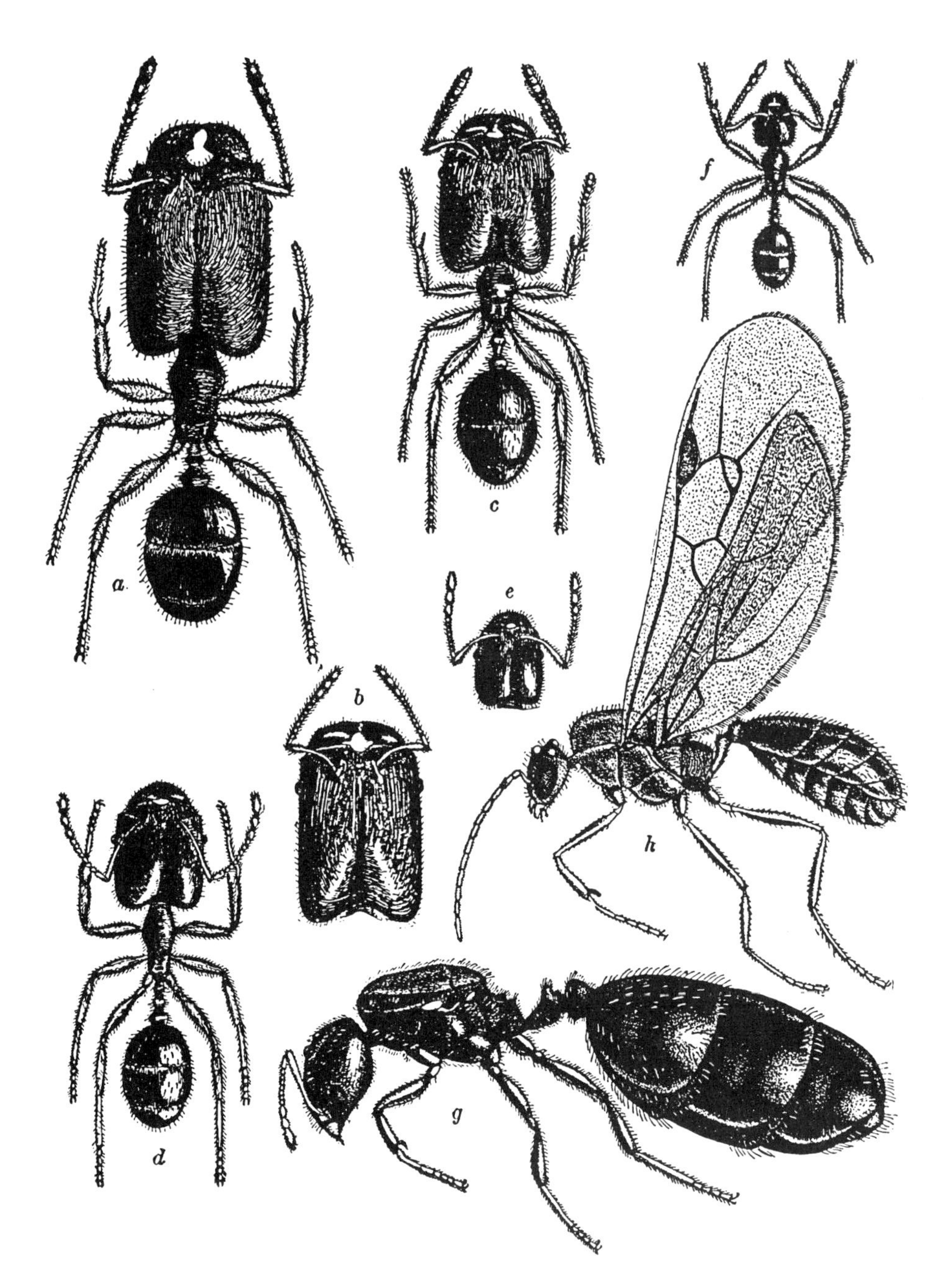


**FIGURE 8–3** The female castes and the male of the myrmicine ant *Pheidole tepicana*. In this classic drawing by Wheeler, the worker caste is shown to be composed of subcastes of successively diminishing size from the major workers (*a*) through media workers (*b–e*) to the minor worker (*f*). The queen (*g*) and male (*h*) are also shown. (From Wheeler, 1910a.)

tive female, possessing a generalized hymenopterous thorax and functional but deciduous wings. The queen is sometimes referred to loosely as the "female" of the colony. The term *gyne* was used synonymously with queen by Wheeler (1907a). Brian (1957b) employed it to denote more specifically "a sexual female that is not socially a functional reproductive." During the past thirty years the word has gained wider usage in conformity with Wheeler's original meaning. It is employed most consistently as a neutral expression, to cover any reproductive female whether virgin or inseminated, laying or infertile. In current writings, "queen" then denotes a functioning gyne.

In a few species such as the primitive Australian *Nothomyrmecia macrops*, the wings are reduced in size and probably nonfunctional (Figure 8–4). Such individuals are designated *micropterous queens* (or micropterous females) in the literature. Occasionally, queens are exceptionally small, even smaller than the workers in some dimensions. Such microgynes, as they are called, are usually social parasites. Examples are found in the South American myrmicine *Pheidole microgyna* (Wilson, 1984c), the parasitic and microgynous form of the European *Myrmica rubra* complex (Pearson, 1981), and members of the North American *Formica microgyna* group (Wheeler, 1910a; Creighton, 1950). Microgynes coexist in the same colonies with normal, "macrogynous" queens in the arboreal *Pseudomyrmex venefica* (Janzen, 1973b) and an arboreal Australian species of *Polyrhachis* subgenus *Cyrtomyrma* (Bellas and Hölldobler, 1985; see Figure 8–5).

3. *Worker*. This caste is often represented by the symbols ♀ or ☿. The ordinarily sterile female, possessing reduced ovarioles (or none at all) and a greatly simplified thorax, the nota of which typically consist of no more than a single sclerite each (Tulloch, 1935). Even when ovarioles are present, the workers usually lack spermathecae in which sperm can be stored. Hence when they become reproductive they cannot produce daughters except by thelytokous parthenogenesis—an uncommon event in ants generally (Crozier, 1975; Buschinger and Winter, 1978; Bourke, 1988b; Choe, 1988). The disparity between the reproductive apparatus of workers and queens varies considerably among ant species. In *Nothomyrmecia macrops*, a

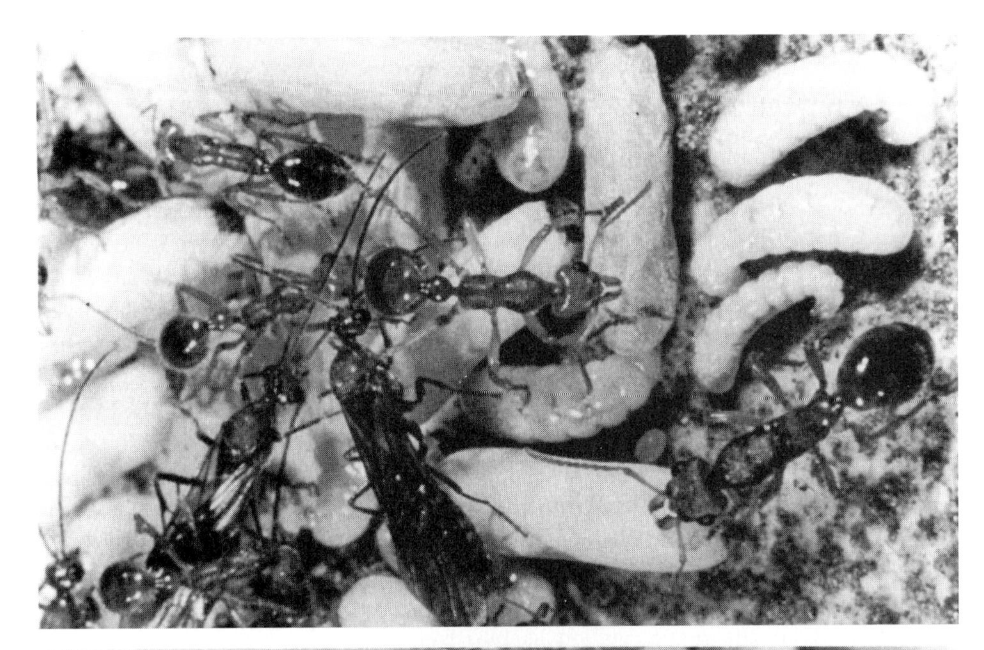

**FIGURE 8–4** Part of a colony of the primitive Australian ant *Nothomyrmecia macrops* including workers, nest queen, males, larvae, and pupae (*above*) and virgin brachypterous queens, workers, and pupae (*below*). (Photograph courtesy of R. W. Taylor.)

**FIGURE 8–5** Dimorphism in the queen caste of *Polyrhachis* (*Cyrtomyrma*) *?doddi*. A large queen is indicated by an arrow in *a* and small queen by an arrow in *b*. (From Bellas and Hölldobler, 1985.)

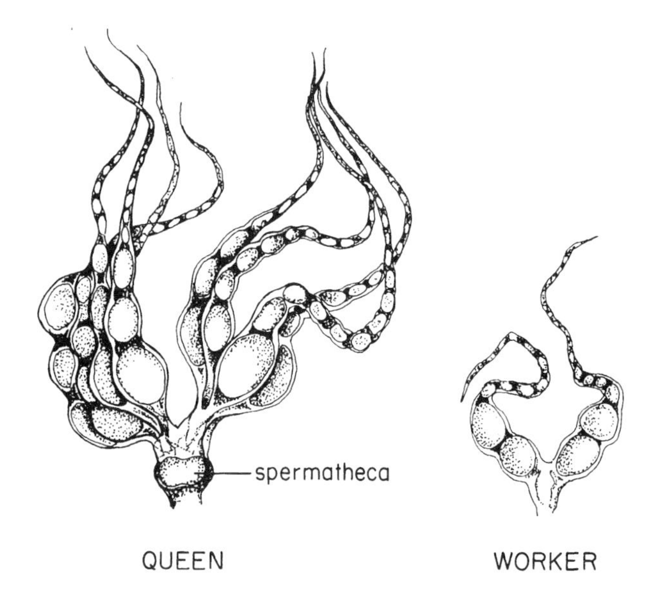


**FIGURE 8–6** Ovarian development in a queen and a worker of *Leptothorax nylanderi.* The queen has eight ovarioles, most of which are manufacturing eggs in series; she also has a spermatheca in which sperm from the original nuptial flight can be stored for years. The fecund worker has only two ovarioles. She lacks a spermatheca, which renders her incapable of fertilizing the eggs. (Based on Plateaux, 1970.)

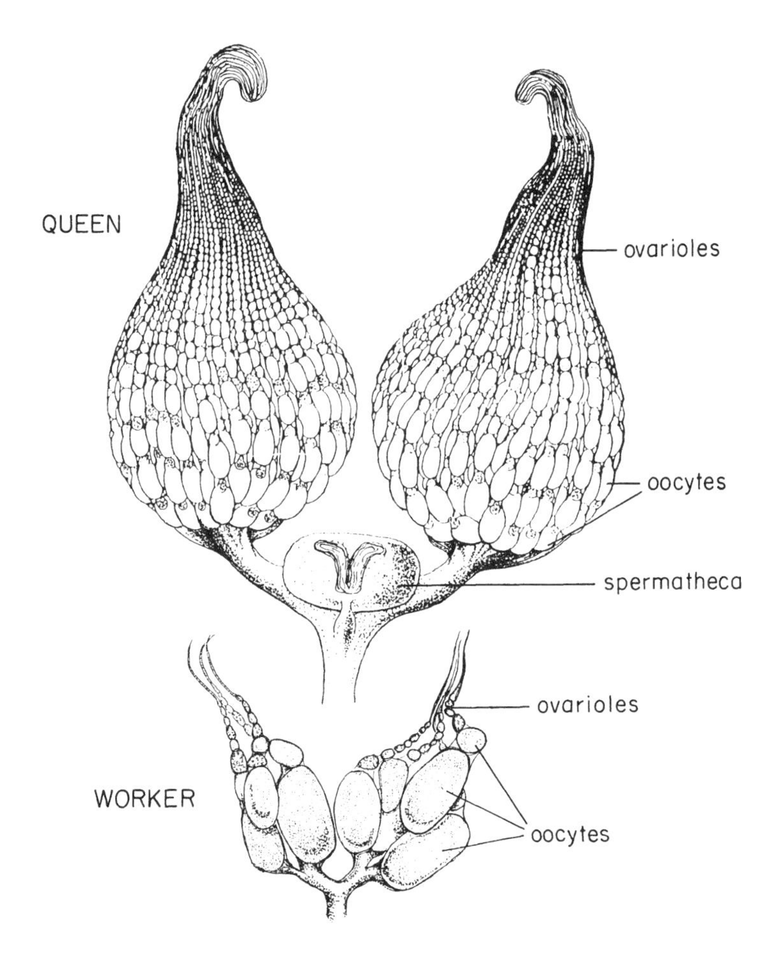


**FIGURE 8–7** Reproductive system of females in the European wood ant *Formica polyctena.* The ovaries of the worker are those of a young individual. (Modified from Otto, 1962.)

very primitive ant, workers possess 4 to 7 ovarioles and queens 8 to 10 (Hölldobler and Taylor, 1983). In *Leptothorax nylanderi* only 2 ovarioles are found in workers and 8 in queens (Plateaux, 1970; Figure 8–6), whereas in *Formica polyctena* workers have 4 to 6 ovarioles and queens have 90 to 270 (Otto, 1958, 1962; Figure 8–7). On the other hand in some species of *Rhytidoponera* whose mated workers have replaced the queen as the functional egg-layers, there is no significant difference between the ovaries of reproductives and those of nonreproductives. Only inseminated workers appear to lay eggs, however (Peeters, 1987b). The queen of *Eciton hamatum* has approximately 4,600 ovarioles (Hagan, 1954), and the queen of *E. burchelli* has at least twice as many (Whelden, 1963; Schneirla, 1971). In some other species, such as the members of *Pheidole* and *Solenopsis*, workers possess no ovarioles at all (Wilson, 1971). The worker caste is often subdivided into additional castes, or subcastes, as follows: *minor, media, major.* Castes in a size-variable worker series are designated according to ascending size. In some species the media subcaste mostly or entirely drops out, and the minors and majors are easily distinguished as two frequency distributions with distinct modes. In this case, the symbol ♃ designates the major, and the symbol ♀ usually designates the minor. When their function is partly or wholly fighting on behalf of the colony, the majors are often referred to as *soldiers.*

4. *Ergatogyne.* This caste is a reproductive form anatomically intermediate between the worker and the queen. A variety of relatively informal terms have been used to denote the intermediates. These terms range from "apterous females" (Bolton, 1986b) to "gynomorph" or "gynomorphic workers," "ergatomorph" or "ergatomorphic queens," and "intermorphs" (Buschinger and Winter, 1976). We are inclined to favor a bipartite classification suggested to us by C. P. Peeters (personal communication). It makes a distinction between *intercastes,* forms that are intermediate in anatomy between workers and queens, commonly lack a spermatheca (and hence are incapable of mating), and occur in conjunction with typical workers and queens; and *ergatoid queens,* also intermediate in anatomy but possessing spermathecae, who typically replace the ordinary queen caste as the female reproductive.

Intercastes as just defined occur most commonly in social parasites of the leptothoracine genera *Formicoxenus, Harpagoxenus,* and *Protomognathus* (see Figure 8–49), but they also occur in some free-living species of *Leptothorax* and *Monomorium.* They appear to arise from a failure of endocrine regulation in the differentiation of queens and workers. In the most detailed study to date, Plateaux (1970) showed that there is a continuous anatomical progression in intercastes of the free-living *Leptothorax nylanderi,* manifested in body size, ocelli, segmentation of the thorax, number of ovarioles, the spermatheca, and elsewhere. Although the character states are strongly concordant in the workers and queens, they vary independently of one another in the intercastes. For example, the thorax and gaster do not increase in size in a well-coordinated manner, and some individuals combine large ovaries with worker-like thoraces. Intercastes of most of the species, such as *Harpagoxenus canadensis, Leptothorax nylanderi,* and *Protomognathus americanus,* lack spermathecae. On the other hand the intercastes of *Formicoxenus* species and *H. sublaevis* do have spermathecae and hence often mate and lay eggs.

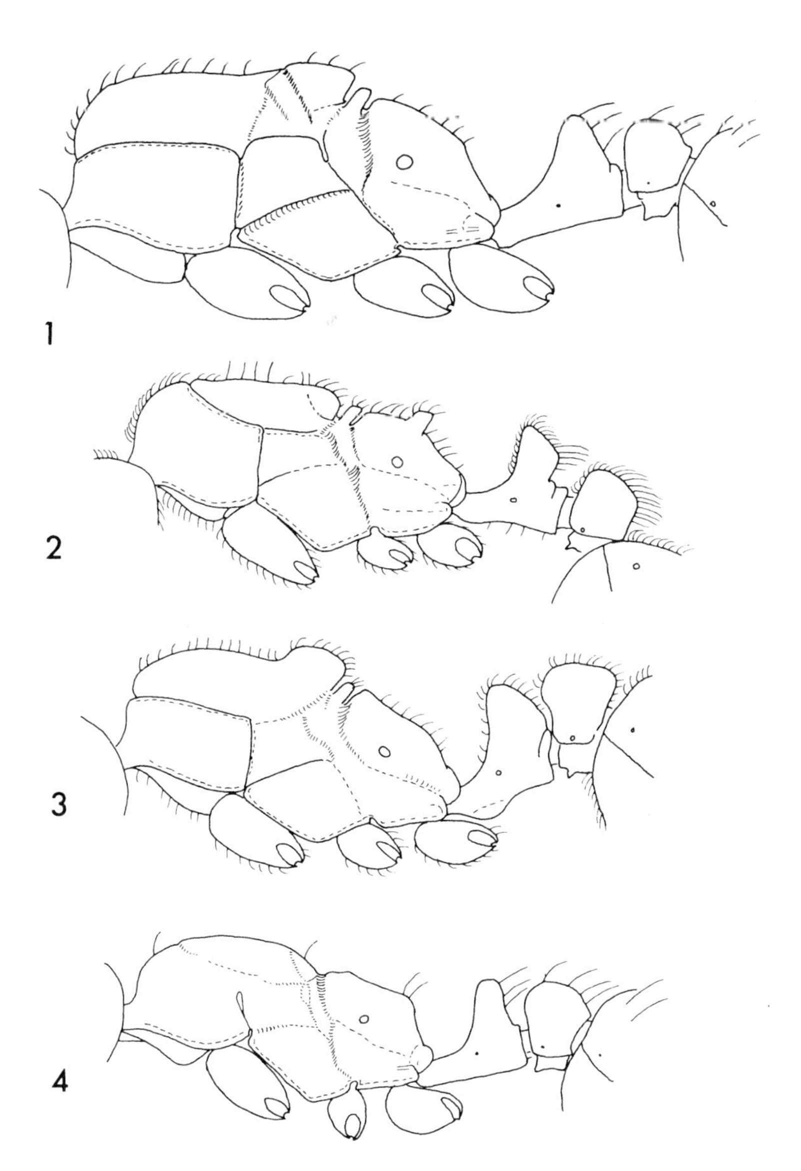


**FIGURE 8–8** A graded series of ergatoid queens occurs across species of the *Monomorium salomonis* group. In each species one or the other of intermediate forms depicted here replaces the full, winged queen as the reproductive. (1) *Monomorium rufulum;* (2) *M. hesperium;* (3) *M. medinae;* (4) *M. advena.* (From Bolton, 1986b.)

Ergatoid queens, on the other hand, replace the typical queen entirely and are not connected to the worker caste by a graded series. This second, truly functional type of ergatogyne is especially common in the primitive genus *Myrmecia* and in several genera of the subfamily Ponerinae. Haskins and Haskins (1955) have pointed out that the tendency of a species to develop ergatoid queens is reflective of the habit of queens of primitive ants to forage for food outside the nest during colony founding. In higher ants claustral founding is the rule, and queens usually must rear their first brood entirely on the reserves contained in their own fat bodies and degenerating flight muscles. Hence fully differentiated queens are a necessity when colonies are founded claustrally.

Ergatoid queens nevertheless do occur in some free-living higher ants. They are the rule, for example, in the aberrant dolichoderine genus *Leptomyrmex* (Wheeler, 1934a) and in the legionary cerapachyines (a tribe of the Ponerinae) and dorylines (Wilson, 1958e). They occur in *Aphaenogaster phalangium* of Central America and in *Blepharidatta brasiliensis* (E. O. Wilson, unpublished). They are common in species of the Old World *Monomorium salomonis* group, where a graded series on the queen-worker gradient can be found among various species but not within species (Bolton, 1986b; see Figure 8–8). Ergatoid queens are also the sole form of reproductive in a high percentage of the endemic species of New Caledonia, belonging to such phylogenetically advanced genera as *Monomorium* (= *Chelaner*), *Lordomyrma, Prodicroaspis,* and *Promeranoplus* (Wilson, 1971). New Caledonia is an old, very isolated island, and ergatogyny in its ants corresponds to the flightless state found so commonly among the endemic species of birds and insects on oceanic islands. Where ergatoid and micropterous queens have replaced true queens in higher species, it seems likely either that workers accompany the reproductives during colony founding, or that the reproductives revert to foraging on their own during colony founding. Supporting evidence of the former alteration has been adduced in the case of *Monomorium* (= *Chelaner*) (Briese, 1983). Many ergatoid queens show a divergent trend away from the queen-worker gradient in that the development of the gaster and postpetiole outpaces the development of the thorax and head. Such variants possess a gaster that approaches (or surpasses) in size that of the typical queen, whereas the thorax and head are more typically worker-like.

5. *Gamergates.* A growing number of ponerine ants has been discovered in which the queen caste is replaced by reproductives anatomically indistinguishable from the worker caste. The species in this category include *Diacamma rugosum* (Wheeler and Chapman, 1922; Fukumoto, 1983; Moffett, 1986h), *Dinoponera grandis* (Haskins and Zahl, 1971), *Leptogenys schwabi* (Peeters, 1987a), *Ophthalmopone berthoudi* (Peeters and Crewe, 1984, 1985), *Pachycondyla krugeri* (Peeters and Crewe, 1986b), *Platythyrea schultzei* (Peeters, 1987a), and several species of *Rhytidoponera* (Haskins and Whelden, 1965; Ward, 1981a,b, 1983a,b; Peeters, 1987b). The only examples known from the Myrmicinae are apparently a few species belonging to the South American *Megalomyrmex leoninus* group (Brandão, 1987). Peeters has suggested the term *gamergate* for such inseminated workers, which seems a useful and clarifying step. It is the one case known to us of a true physiological caste, in which a particular physiological state (insemination and enhanced oogenesis), rather than anatomy or age, distinguishes a labor group. On the other hand it may be pushing the term too far to include the reproductive workers of *Pristomyrmex pungens,* which are not fertilized and produce females by parthenogenesis. In this extreme, aberrant case the colony can be said to lack a physical caste system (Itow et al., 1984; Tsuji and Itō, 1986). On the other hand *P. pungens* have a typical age caste system: young workers work inside the nest and lay eggs, whereas older workers have reduced ovaries and tend to work outside the nest (K. Tsuji, personal communication).

6. *Dichthadiiform ergatogyne.* The dichthadiiform female is the extreme stage of the phylogenetic trend toward enlargement of the gaster in the ergatogyne and is usually recognized as a distinct category. In fact, the dichthadiiform female is just an aberrant queen and is never accompanied by ordinary queens. The total size of the female is greatly increased, the gaster is huge, and the postpetiole is expanded to the extent that it has come secondarily to resemble the first gastric segment (see Figure 8–9). In addition, the head is broad-

**FIGURE 8–9** The queen of the army ant *Eciton burchelli*, a dichthadiiform ergatogyne, is shown in the egg-laying phase. Note the extremely large gaster, broad postpetiole appressed to the anterior face of the gaster, and the worker-like thorax. (Photograph courtesy of C. W. Rettenmeyer.)

**FIGURE 8–10** A queen of *Onychomyrmex hedleyi*, an Australian legionary ant in the ponerine tribe Amblyoponini, is surrounded by workers during nest emigration. The queen, like the queen of the more familiar ecitonine army ants (see Figure 8–9), is a dichthadiiform female.

ened and rounded, the mandibles are often falcate, and the petiole is commonly bilaterally cornulate. Dichthadiiform females are limited to ants with a legionary mode of life. The extreme development is seen in the subfamilies Dorylinae, Ecitoninae, and Leptanillinae and the ponerine genus *Onychomyrmex* (Wilson, 1958e; see Figure 8–10). An intermediate stage in the evolution of this form occurs in the ergatogynes of the ponerine *Acanthostichus quadratus* and species of the ponerine genus *Simopelta* (Gotwald and Brown, 1966). All of these phyletic lines, both intermediate and advanced, evidently evolved their dichthadiiform form independently.

We now take up a different dimension in caste formation. A *temporal caste* is a set of individuals distinguished both by behavior and by age. It is thus wholly different from physical castes, which are marked by some anatomical feature, and physiological castes, which may in a few cases be distinguished by insemination, or some other principal physiological feature alone, in a manner unrelated to anatomy and age. In *Pheidole dentata*, for example, one temporal caste consists of the youngest workers, from the moment of emergence from the pupa to about two days into adult life. These individuals care preferentially for the queen, eggs, and young larvae. The second temporal caste, which tends to care for medium-sized larvae, extends to about the sixteenth day of adult life. The third temporal caste, which prefers nest work and foraging to a greater extent, consists of still older workers (Wilson, 1976c).

## ADAPTIVE DEMOGRAPHY

A key principle of caste evolution is the adaptive nature of colony demography (Wilson, 1968; Oster and Wilson, 1978). Ordinary demography, of the kind found in nonsocial organisms, in social vertebrates, and in more primitively social insects such as subsocial wasps and bees, is a function of the individual parameters of growth, reproduction, and death. Substantial documentation from free-living and laboratory populations supports the general belief that growth and natality schedules are direct adaptations shaped by natural selection at the level of the individual (Krebs and Davies, 1984). On the other hand the size and age structure of the population as a whole are epiphenomena in the sense that they reflect the individual-level adaptations but do not constitute adaptations in their own right. Thus a sharply tapered age distribution in any species of fishes and birds results from a high birth rate and a high mortality schedule throughout the life span, but in itself does not contribute to the survival of the population or the individual members. The exact reverse is the case of the eusocial insect colony defined as a population. The demography, not its causal parameters, is directly adaptive. The workers are for the most part sterile; their birth and death schedules have meaning only with reference to the survival and reproduction of the queen. Hence the unit of selection is the colony as a whole. The traits of the colony are the larger features of demography, the age- and size-frequency distributions. What matters are such higher-level traits as the number of very large adults available to serve as soldiers, the number of small young adults functioning as nurses, and so on through the entire caste roster and behavioral repertory. Each species has a characteristic age-size frequency distribution, and the evidence is strong that this colony-wide trait is not an epiphenomenon, but has been shaped by natural selection and constitutes a direct adaptation (see Figure 8–11).

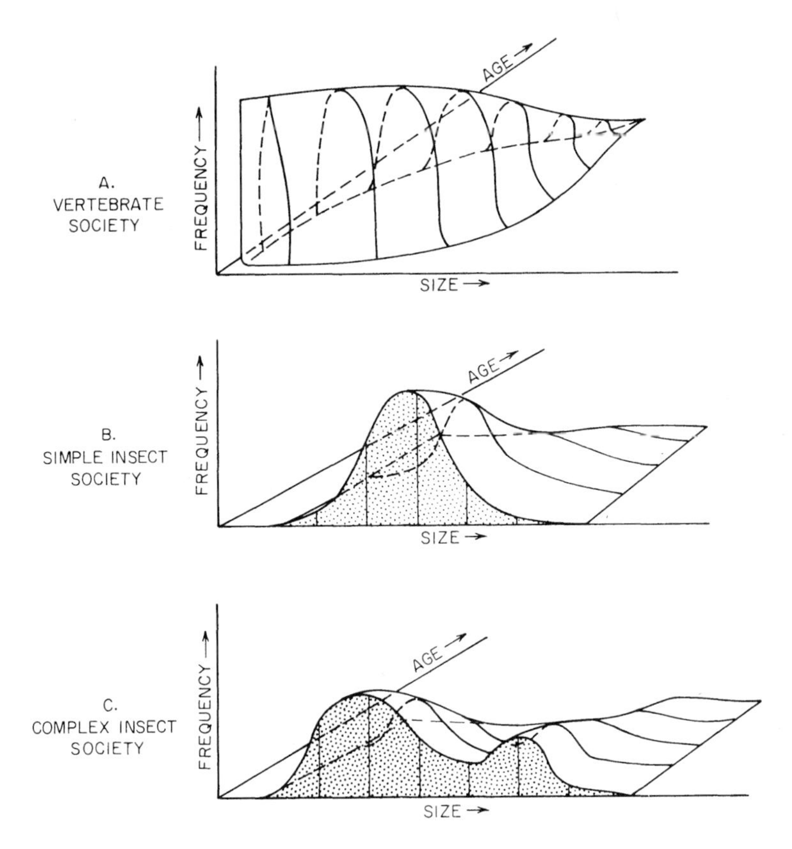


**FIGURE 8–11** The concept of adaptive demography in ant colonies and other complex insect societies. In vertebrate societies the overall size and frequency distributions are nonadaptive at the level of the population. In the simplest insect societies, such as those of the primitively eusocial bees, this remains the case. In complex insect societies, however, the proportion of individuals of various sizes and ages determines the efficiency of the division of labor and hence is adaptive at the level of the entire colony. In these imaginary but realistic examples, the ages shown for the two insect societies apply only to the final, adult stage, during which most or all of the labor is performed. Hence no further increase in size occurs with aging. (From Wilson, 1975b.)

A striking example of adaptive demography is provided by developmental changes in the caste system of the leafcutter ant *Atta cephalotes* (Wilson, 1983b). Beginning colonies, started by a single queen from her own body reserves, have a nearly uniform size-frequency distribution across a relatively narrow head-width range of 0.8–1.6 millimeters. The key to the arrangement is that workers in the span 0.8–1.0 millimeter are required as gardeners of the symbiotic fungus on which the colony depends, whereas workers with head widths of 1.6 millimeters are the smallest that can cut vegetation of average toughness. This range also embraces the worker size groups most involved in brood care. Thus, remarkably, the queen produces about the maximum number of individuals who together can perform all of the essential colony tasks. As the colony continues growing, the worker size variation broadens in both directions, to head width 0.7 millimeter or slightly less at the lower end to more than 5.0 millimeters at the upper end, and the frequency distribution becomes more sharply peaked and strongly skewed to the larger size classes (see Figures 8–12 and 8–13).

A more physiological question immediately arises from the observed sociogenesis (colony ontogeny) of *Atta:* which is more important in determining the size-frequency distribution, the size of the colony or its age? To learn the answer, Wilson selected four colonies three to four years old with about 10,000 workers and reduced the population of each to 236 workers, giving them a size-frequency distribution characteristic of natural young colonies of the same size collected in Costa Rica. The worker pupae produced at the end of the first brood cycle possessed a size-frequency distribution like that of small, young colonies rather than larger, older ones. Thus colony size and, indirectly, the amount of food produced are more important than age.

The *Atta* example is just an extreme case of what appears at least on the surface to be programmed colony demography among ant species. The physiological control mechanisms remain almost entirely unexplored, however. The "rejuvenation" effect in *A. cephalotes* indicates that a feedback loop of some kind is involved, as opposed to an irreversible maturation of the size-frequency distribution; but its nature has not been investigated.

The ontogeny of the physical caste system has also been traced in the red imported fire ant *Solenopsis invicta,* as illustrated in Figure 8–14. In this species the first adult generation of workers consists entirely of "nanitics" or "minims," with head widths about 0.50 millimeter. These individuals, as Porter and Tschinkel (1986) have demonstrated experimentally, are more effective as a group in rearing the second generation of adults, but less efficient energetically on an individual basis than the slightly larger "minor" workers.

**FIGURE 8–12** A young colony of the leafcutter *Atta cephalotes* from Costa Rica. The huge queen rears both a fungus garden and the first brood of workers from energy reserves in her own body. The workers at first have head widths that span 0.8–1.6 millimeters, the minimum range required to gather and process vegetation to sustain the fungus garden. As the colony population grows, the head width range increases, as shown in Figure 8–13. (Photograph courtesy of C. W. Rettenmeyer.)

When the population grows by the addition of the second and later broods, it is to be expected on the basis of ergonomic selection at the colony level that the minims will be quickly replaced by minors and still larger workers. Just such a transition actually occurs, as shown in Figure 8–14. Subsequent field studies by Tschinkel (1988a) revealed that as the fire ant colonies grow, they change from a monomorphic to a polymorphic worker force. The production of larger, major workers causes the overall size-frequency curve to become skewed in the manner just illustrated (Figure 8–13) for *Atta.* The skewed distributions are actually bimodal. They consist of two slightly overlapping normal distributions, a narrow one defined as the minor workers and a much broader one defined as the major workers. The "media" workers have no clear developmental definition and are simply individuals in the zone of overlap between the minor and major frequency curves. In full-sized colonies, the majors make up about 35 percent of the total worker force and 68 percent of its biomass.

It is evident that each ant species possesses a particular programmed ontogeny in the ratios of its subcastes (size-frequency distribution). In other words the ratios are species specific and hence genetically prescribed. Even differences encountered among colonies of the same species may have a partial genetic basis, as suggested by laboratory studies conducted on the North American myrmicine ant *Pheidole dentata* by Johnston and Wilson (1985). The

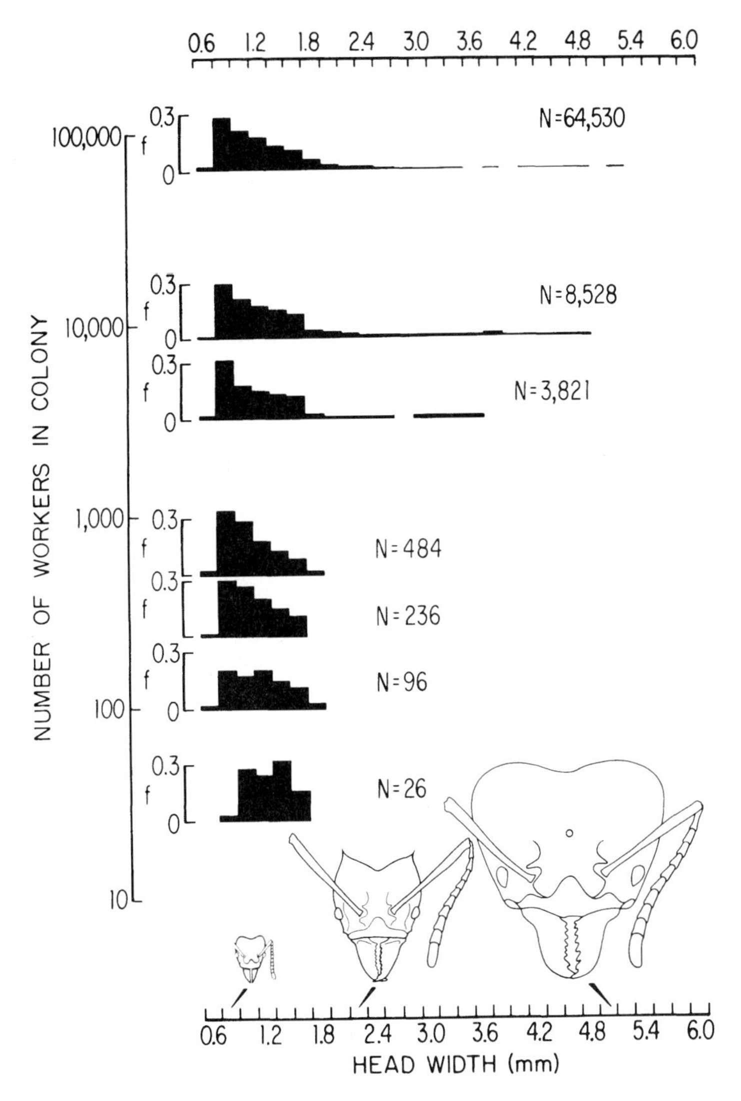


**FIGURE 8–13** Sociogenesis of the leafcutter ant *Atta cephalotes:* the ontogeny of the caste system is illustrated here by seven representative colonies collected in the field or reared in the laboratory. The worker caste is differentiated into subcastes by continuous size variation associated with disproportionate growth in various body parts. The number of workers in each colony (*N*) is based on complete censuses; *f* is the frequency of individuals according to size class. The heads of three sizes of workers are shown to illustrate the disproportionate growth. (From Wilson, 1985g.)

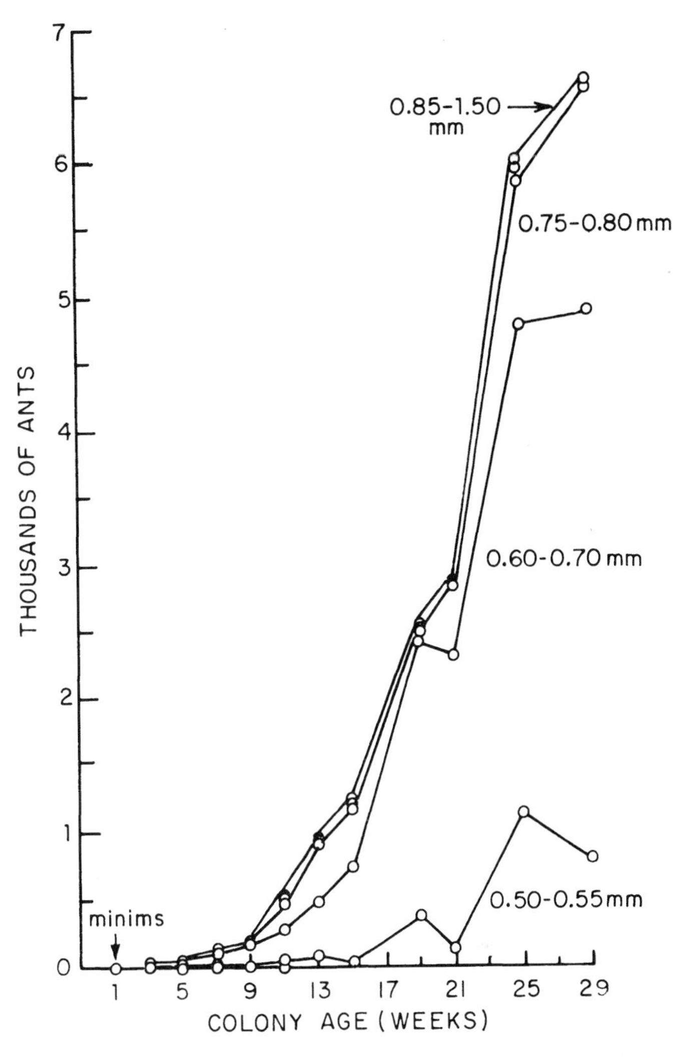


**FIGURE 8–14** The ontogeny of physical castes in laboratory colonies of the red imported fire ant *Solenopsis invicta,* expressed cumulatively along the vertical axis as the sum of the mean numbers of ants in each size class. As predicted from ergonomic theory, the tiny minim caste drops out after the first worker brood. (Redrawn from Wood and Tschinkel, 1981.)

worker force of colonies of *P. dentata,* like those of other species of *Pheidole,* is divided into two discrete castes: small-headed minor workers, who conduct most of the quotidian tasks of the colony, and large-headed major workers, who are specialized for defense (see Table 8–1 and Figure 8–20). In natural populations of *P. dentata* in northern Florida, majors make up between 5 and more than 50 percent of the worker force. The higher figures are rare and almost certainly pathological, and the great majority of colonies in the field and laboratory maintain a representation of majors between 5 and 15 percent. When Johnston and Wilson altered laboratory colonies displaying this latter range of variation to a uniform percentage of 5 percent and then allowed them to grow freely, the colonies changed to the original rank ordering, in most cases not far from their original subcaste percentages. Wheeler and Nijhout (1984) have shown that excess numbers of majors in *P. bicarinata* (= *vinelandica*) colonies inhibit further production of majors in a manner that restores the species-characteristic ratio. The inhibition is evidently due to a pheromone that affects the endocrine system of presumptive soldiers. It seems likely that a similar mechanism operates in *P. dentata,* and that the variation in ratios among colonies arises from genetic differences in major pheromone production, sensitivity to the pheromone, or both. The result is potentially important for an understanding of the evolution of caste ratios, since genetic variation in any trait is a prerequisite for its modification through time.

In summary, there appears to be a species average or "norm" in caste ratios around which individual colonies vary as a result of both genetic and environmental differences. The environmental factors most important to the variance have only begun to be explored. For example, it is a common observation that majors of ant species live longer than minors, so that when the queen dies or ceases oviposition for long periods of time the major:minor ratio rises. It also increases above ordinary levels when the colony is starved or desiccated, because larger ants withstand such stress longer than their nestmates. The differential is commonly observed in the laboratory and has been documented explicitly in *Camponotus* and *Formica* by Kondoh (1977). In *Messor* (= *Veromessor*) *pergandei*, a seed-harvesting myrmicine living in the deserts of the southwestern United States, there is an annual cycle in average worker size. Smaller workers appear in the foraging force as a result of the wintertime "triple crunch" caused by a reduced seed crop, shorter suitable foraging time, and the added expense of production of reproductive forms (Rissing, 1987). In a reverse trend, major workers of the fire ant *Solenopsis invicta* become relatively most common in the late winter and early spring, probably because of their lower mortality under adverse conditions (Markin and Dillier, 1971). Such unusual fluctuations may well be just epiphenomena, that is, "noise" in the system of no adaptive significance to the colonies.

## THE EVOLUTION OF PHYSICAL CASTES

In ants the diversification of the female castes is based mostly on allometry. During larval development the imaginal discs (patches of undifferentiated tissue destined to be transformed into adult organs at the pupal state) grow at different rates, a process that swiftly accelerates during pupal development (Brian, 1957a, 1979a; Schneirla et al., 1968). The principal result of differential growth rate in the imaginal discs is that various organs end up with different sizes relative to one another according to how large the individual is at the termination of the larval period. That is, the final adult size determines how much overall growth the various organs have attained. Thus, if the disc destined to transform into part of the head is growing faster than the disc destined to transform into part of the thorax, it will finish proportionately larger in an ant that attains a larger overall size. In short, big ants will have proportionally even bigger heads. If each disc grows exponentially, and if the disc growth rates do not change much in the course of development, the sizes of two parts will be related by a simple power law: $\log y = \log b + a \log x$, or, equivalently, $y = bx^a$ where $y$ and $x$ are linear measures of the two body parts and $a$ and $b$ are fitted constants the values of which depend on the nature of the measurement taken.

This simple relation is referred to as allometry or heterogonic growth. Its study in ants and other organisms was pioneered by Huxley (1932) and has been reviewed extensively by Gould (1966). On a double logarithmic plot the curve is a straight line. Its slope $a$ is determined by the rate of divergence of the two body parts with increasing total size and can be referred to as the allometric constant. If $a$ is equal to unity, no divergence takes place with an overall increase in size; the growth is then referred to as isometric. The greater the departure of $a$ from unity, the more striking the differential growth. Skellam, Brian, and Proctor (1959) found that in *Myrmica rubra* the imaginal discs of the wings and legs actually grow in this elementary fashion during development of adult queens and males. The organs predictably conform to the basic allometry equation in their final adult form.

It is possible to produce a wide array of deviations from elementary allometric growth simply by speeding or slowing the growth rates of different discs according to different time schedules. Further complexity can be introduced by making the rates dependent on the total size of the larva reached by certain ages. This last effect is crucial in the determination of the queen and worker castes in *Myrmica*, as documented by Brian (1979a, 1983). It also occurs during the differentiation of the worker caste of the *Eciton* army ants, as discovered by Tafuri (1955). These modifications are crucial for the discretization of physical castes within the adult instar of ants, a process that will now be examined in some detail.

Wilson (1953b) demonstrated that the allometry equation, or relatively simple modifications of it, can be applied almost universally to continuous variation in the hard parts of ants. The comparative study of allometry has proved fruitful in tracing the evolution of castes. Polymorphism, as this research has led us to understand it, embraces three variable characters in the adult females within colonies of any species: size variation, shape variation through allometric growth at the time of adult formation, and the frequency distribution of different-sized workers. A physical caste system (or *polymorphism* as it is also frequently called) is defined as growth occurring over a sufficient range to produce individuals of distinctly different sizes, body proportions, or both. Where polymorphism exists, it has always been found to be closely linked to division of labor.

By comparing a large fraction of the 8,800 known living ant species, Wilson (1953b) inferred five major steps in the evolution of physical castes:

1. *Monomorphism.* The workers of the normal mature colony display isometry (with a log-log curve slope of approximately 1.0) and very limited size variability. A plot of their size-frequency distribution is symmetrical and has only a single mode. In other words the properties of variation are not basically different from those in a typical random collection for nonsocial insects. The worker castes of most ant genera and species are monomorphic. Also, within the majority of genera and higher taxonomic groups monomorphism is evidently the primitive state.

2. *Monophasic allometry.* The relative growth is nonisometric, meaning that the allometric constant $a$ is greater or less than unity to a subjectively noticeable degree. In the most elementary form of monophasic allometry, and hence of worker polymorphism generally, feeble nonisometric variation is displayed over a short span of size variation; this variation in turn is grouped around a single mode with possible skewing in the direction of the major caste. A more advanced stage, involving an increased variation in size together with an apparent bimodal size-frequency distribution, is exemplified by the fire ant *Solenopsis invicta* (Figure 8–15). The two modal groups are still overlapping, and hence the majors and minors are connected by intermediate-sized workers, or medias.

3. *Diphasic allometry.* The allometric regression line, when plotted on a double logarithmic scale, "breaks" into two segments of different slopes that meet at an intermediate point. In the several species known to possess this condition, including the leafcutter ant *Atta*

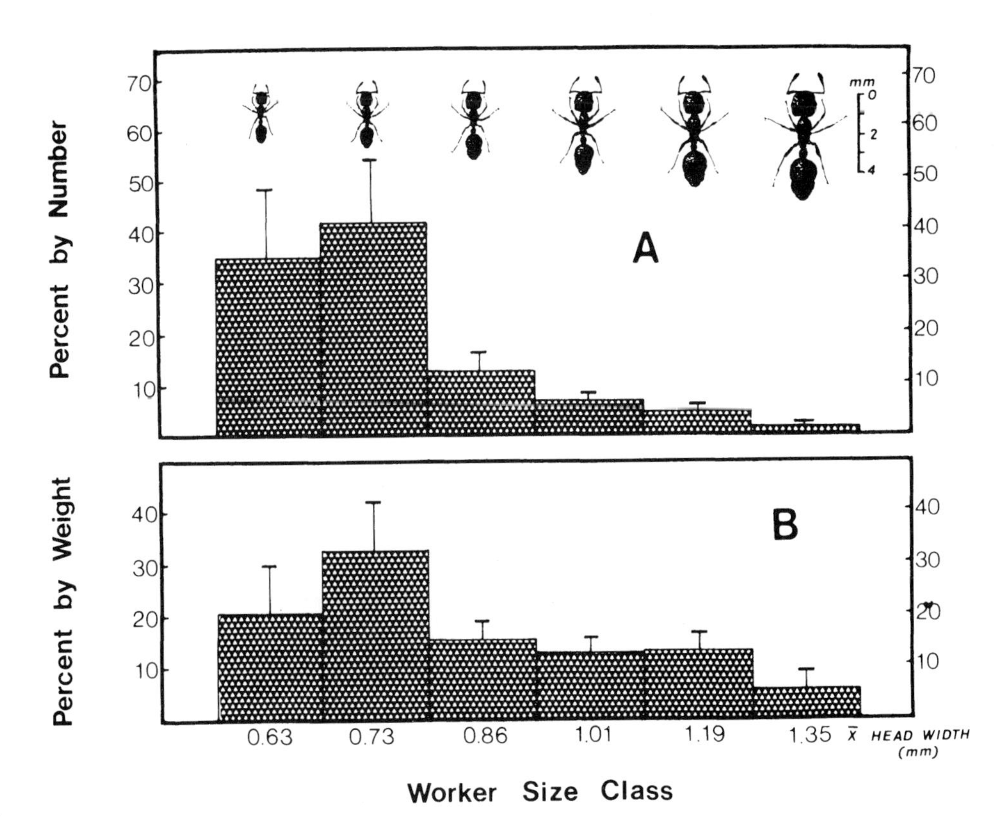


**FIGURE 8–15** Physical caste systems in ants are based on three qualities: an increased size variation among the adult females of each colony, allometric growth, and a tendency toward multimodality in the size-frequency distribution. The workers of the fire ant *Solenopsis invicta* depicted here possess an elementary form of polymorphism. Most of the body parts are isometric but a few are weakly allometric—in other words, they increase or decrease in relative size as total body size is enlarged. Head width increases faster toward the posterior border, and the antennae grow relatively shorter, whereas pronotal width is isometric with reference to most of the rest of the body. The allometry of *S. invicta* is "monophasic," meaning that the slope of the relative growth curve remains constant or nearly so. The size-frequency curve is bimodal, or at least skewed toward the larger workers. The small-headed individuals clustered around the smaller mode are called minor workers, those with large heads around the larger mode are the major workers, and those around the midpoint between the modes are the media workers. The data are averages from 34 mature colonies; standard deviations are indicated on each bar. (From Porter and Tschinkel, 1985.)

*texana* (Wilson, 1953b), the African driver ant *Dorylus nigricans* (Hollingsworth, 1960), and the carpenter ant *Camponotus sericeiventris* (Busher et al., 1985), the size-frequency curve is bimodal. Also, the saddle between the two frequency modes falls just above the bend in the allometry curve. Diphasic allometry permits the stabilization of the body form in the small caste while providing for the production of a markedly divergent major caste by means of a relatively small increase in size. The lower segment of the relative growth curve is nearly isometric, so that individuals falling within a large segment of the size range are nearly uniform in structure; but the upper segment leading to the major caste is strongly nonisometric, with the result that a modest increase in size yields a new morphological type.

4. *Triphasic and tetraphasic allometry.* The allometric line breaks at two points into three straight segments. The two terminal segments, representing the minor and major castes, deviate only slightly from isometry while the middle segment, encompassing the media caste, has a very steep slope. The effect of triphasic allometry is the stabilization of body proportions in the minor and major castes. An example from the weaver ant genus *Oecophylla* is presented in Figures 8–16 and 8–17. Baroni Urbani (1976b) has reported a case of tetraphasic allometry in the antennal length of the West African ant *Camponotus maculatus*. The curve resembles that of triphasic species except that in the largest size classes a high degree of allometry is resumed. The biological significance of this extreme pattern is wholly unknown.

5. *Complete dimorphism.* Two morphologically very distinct size groups exist, separated by a gap in which no intermediates occur. Each class is either isometric or nonisometric, but the allometric regression curves are not aligned, a condition suggesting that complete dimorphism can arise directly from triphasic allometry. Examples include most queen-worker differences in ants. They also include minor-major divisions in no fewer than eight phylogenetically scattered genera: the myrmicines *Acanthomyrmex, Oligomyrmex, Pheidole,* and *Zacryptocerus;* the aneuretine *Aneuretus;* the dolichoderine *Zatapinoma;* and the formicines *Camponotus* (subgenera *Colobopsis* and *Tanaemyrmex*) and *Pseudolasius* (Plate 7).

By arranging species of ants along a gradient from what appear to be the simplest to the most advanced systems, we can find worker castes that display virtually every conceivable step in a transition from perfect monomorphism to a complete dimorphism (see Figure 8–18). Certain large taxonomic groups, such as the subfamilies Myrmicinae and Formicinae, embrace the entire evolutionary sequence within themselves. The evolution has thus occurred repeatedly within multiple phyletic lines in the ants and has produced a remarkable degree of convergence between the lines.

Of the three principal qualities of polymorphism (size variation, allometry, and size frequency), the size-frequency distribution has been subject to the stricter and more notable convergence. When individual colonies of a given species show a slight increase in size variance, the frequency curve is almost always skewed toward the larger size classes. When the intracolonial size variance is still greater, the frequency curve is bimodal. In at least one species of army ant with extreme size variation, data published by Topoff (1971) and da Silva (1972) show the existence of three size modes. Three modes also occur in the extremely polymorphic myrmicine *Pheidologeton diversus* (Moffett, 1987b). Otherwise enlarged intracolonial size variation is typically associated with bimodality, with the large workers being less common. In weaver ants of the genus *Oecophylla,* the reverse is true: the majors are more common than the minors, and they fill most of the labor roles of the colony. This unusual form of polymorphism has existed since at least as far back as Miocene times, as shown by Wilson and Taylor (1964). These authors described the extinct species *Oecophylla leakeyi* from a colony

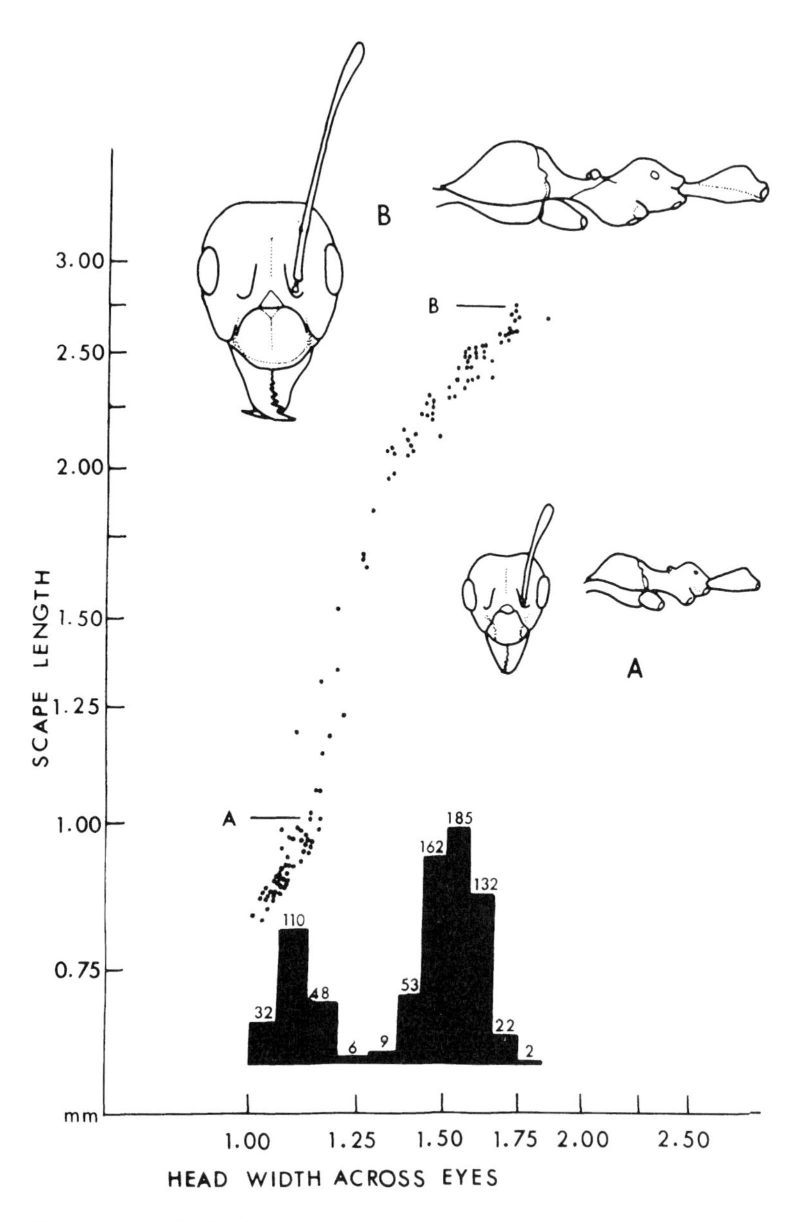


**FIGURE 8–16** In the Asiatic weaver ant *Oecophylla smaragdina* the allometry is triphasic. Three different slopes are possessed by such allometric characters as the length of the scape (first antennal segment) taken as a function of body size, in this case represented by the maximum width of the head. The size-frequency curve (based on head width) is strongly bimodal, with majors predominating. The heads and middle body parts of selected minor (*A*) and major (*B*) workers are also depicted, and their positions on the allometric curve indicated. (From Wilson, 1953b.)

**FIGURE 8–17** A major of *Oecophylla longinoda* carries a minor. Minor workers are seldom seen outside the leaf-nests, and they are usually transported by majors to places where their work is required.

fragment that had been preserved intact, including even pupae and packets of larvae. With this extraordinary material, the only fossil insect colony found to date, it was possible to measure allometry and size frequency and to compare these traits with those of the modern species.

The ways in which ant species create castes out of the adult instar are few. It is reasonable to ask how a colony limited to simple skewing and a maximum of two or three modes is able to produce castes in proportions that match the frequencies of numerous environmental contingencies. The answer appears to be that the evolutionary sequence unfolds within narrow physiological constraints. Species can only improve their situation to the extent of increasing the intracolonial size range and arranging something close to the optimum numbers of majors, minors, and medias. Physical caste systems, in a word, are coarse-tuned rather than fine-tuned (Oster and Wilson, 1978). Moreover, as Franks and Norris (1987) have shown using cartesian transformation analysis, the allometric relationships of various pairs of body measurements (as employed by Wilson, 1953b) are linked over the entire body by the imposition of simple developmental rules. Hence the evolution of bizarre major morphology "is severely constrained by minor morphology and vice versa."

## TEMPORAL CASTES

The adult workers of almost all kinds of social insects change roles as they grow older, ordinarily progressing from nurse to forager. Each species has its own distinctive pattern of temporal polyethism, and in many the behavioral changes are accompanied by patterned shifts in the activity of exocrine glands. For example, as honey bee workers shift during a two-week period from a preoccupation with brood care and nest work to an emphasis on foraging, the activity of the hypopharyngeal and wax glands declines somewhat while that of the labial glands increases. For the purposes of ergonomic analysis it is useful to consider different age groups as constituting distinct temporal castes (Wilson, 1968). Just as a species may manipulate its own developmental biology in the course of evolution to produce optimum ratios of physical castes, it can adjust the program of role change during adult life to approach optimum ratios of temporal castes.

Two extreme alternatives are open to a species in the process of evolving temporal castes. These are represented in Figure 8–19. The aging period depicted begins at the moment of the worker's eclosion from the pupal skin and terminates at the moment of her death by senescence. The division of the life span into six periods in this imaginary example is arbitrary. The worker undergoes physiological change with age in such a way that her responsiveness to various environmental stimuli changes. For example, suppose that $T_1$ is the responsiveness to a misplaced egg: the curve indicates that when the worker is very young (age I) she is likely both to be in the vicinity of the egg and to react by picking the egg up and putting it on an egg pile. Her location and/or behavioral responsiveness change as she ages in such a way that her probability of response to the contingency drops off rapidly after ages I or II. Such age-dependent responsiveness has been abundantly documented in ants (Topoff et al., 1972; Cammaerts-Tricot, 1974b; Jaisson, 1975; Topoff and Mirenda, 1978; MacKay, 1983b).

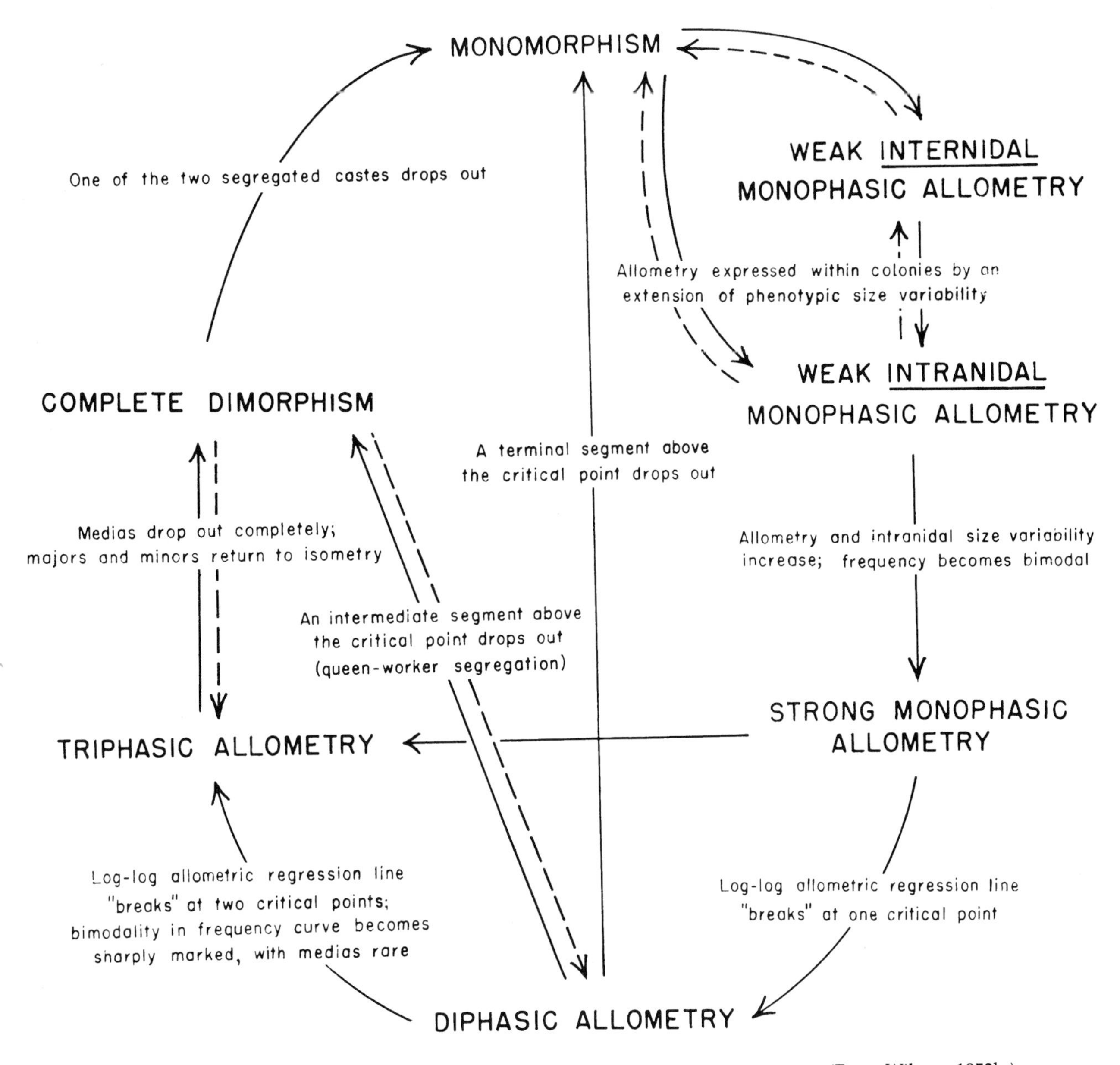


**FIGURE 8–18** Inferred pathways in the evolution of physical caste systems in ants. (From Wilson, 1953b.)

Let us now consider the possibilities. In the upper diagram of Figure 8–19, labeled Model 1, the response curves to four contingencies ($T_1$–$T_4$) are all out of phase. The curve of response to $T_1$ (misplaced egg) is different from the curve of response to $T_2$ (say, a hungry larva), and so on. As a result the ensemble of age groups represented on the right-hand side by the frequency distribution of workers in different age groups that attend to task $T_1$ is different from that attending to $T_2$, and so on. As the number of contingencies is increased, and their response curves are all made discordant, there will be one age-group ensemble for each task. Let us now define an age-group ensemble as a *temporal caste.* In the extreme case represented in Model 1 there is a caste for each task. The distinction between age-group ensembles will be blurred, however, as more tasks are added. The overlap in the age-group frequency curves is so extensive that after ten or so contingencies are added, the system becomes effectively continuous. For this reason we suggest that such an arrangement be called a *continuous caste system.*

The approach to continuity in Model 1 is marked by complexity and subtlety. The evolving ant species can easily adjust the programming to individual worker responsiveness to attain discordance that in turn yields one caste specialized for each task. Thus only a relatively elementary alteration in physiology is needed to produce a complex caste system.

Next consider the alternative option, depicted in the lower half of Figure 8–19. Here various of the response curves are concordant, or at least approximately so. As a result the same statistical ensemble of workers attends to more than one task. As the number of tasks increases the number of castes does not keep pace; conceivably it could remain low, say corresponding to as few as two or three distinct ensembles. Thus the species has chosen to operate with a *discrete caste system*—comprising relatively few, easily recognized age-group ensembles. The evolutionary process leading to such a system can be called *behavioral discretization* (Wilson, 1976c). It can operate through physiological alterations as potentially simple as those that yield continuous caste systems.

Few studies of temporal polyethism have been designed in a way that permits a consideration of the hypothesis of discretization. An analysis of the ant *Pheidole dentata* by Wilson (1976c) showed that the system has been discretized. Virtually all of the 26 behavioral acts recorded in the minor worker caste can be divided according to three age periods in which they are performed approximately in concert (see Figures 8–20 and 8–21). Data presented by Higashi (1974) on the Japanese wood ant *Formica yessensis* indicate that temporal castes have also been discretized in this species, but the behavioral categories followed through time were too few to be certain.

The observed discretization in *Pheidole dentata* appears to represent an adaptation that increases spatial efficiency. It is obviously more efficient for a particular ant grooming a larva to regurgitate to it as well. Similarly, a worker standing "guard" at the nest entrance can be expected to be especially prone to excavate soil when the entrance is buried. The other juxtapositions in the *P. dentata* repertory make equal sense when the spatial arrangement of the colony as a whole is considered. The queen, eggs, first-instar larvae (microlarvae), and pupae are typically clustered together and apart from the older larvae, although the positions are being constantly shifted, and pupae are often segregated for varying periods of time well away from other immature stages. Thus, the *A*-ensemble of workers can nicely care for all of these groups, moving from egg to pupa to queen with a minimum of travel. The mean free path of a patrolling worker is minimized by such versatility so that the least amounts of time and energy are consumed. It makes equal sense for *A* workers to assist the eclosion of adults from pupae, since pupae are already under their care.

It will be of interest to learn to what extent and by which patterns the temporal polyethism of other species of social insects has been discretized. Calabi et al. (1983) found the phenomenon lacking in the Asian *Pheidole hortensis;* in other words, the caste system is continuous. Although it is too early to put the matter on a sound empirical basis, we suspect that the recognizable age ensembles—that is, temporal castes—will prove to be related functionally to nest architecture. Species with complex nest structure, which provide a wider array of housekeeping tasks and the opportunity for a more precise distribution of brood stages into chambers with differing microclimates, can be expected to have more temporal castes than those with relatively simple structures. For example, leafcutter ants of the genus *Atta,* which construct elaborate chambers for the gardening of symbiotic fungi, are expected to possess a correspondingly complex system of temporal polyethism. The patterns of temporal polyethism may be related to the dietary specialization of the species and the external environment of its nests. An ant species that forages widely outside the nest for small particles of food, while simultaneously defending its nest entrances from frequent attacks by predators, is likely to have an unusually discrete polyethic division at the end of worker life. The *C* age ensemble of *P. dentata* represents just such a case. Finally, the more dispersed the tasks for which temporal castes are specialized, the more likely is the labor to be discretized.

To the extent that temporal castes are discretized, all of the castes of a species can be counted. *Pheidole dentata,* for example, has five adult castes: three temporal stages of the minor worker, a major worker (which has only one temporal stage), and the queen. The male does not occupy any known labor role. The larva may provide gluconeogenesis or some other metabolic service and hence constitute a sixth caste, but this possibility remains uninvestigated in *Pheidole.* There is every reason to suspect that the modest temporal caste system of *P. dentata* is typical for the majority of ant species.

More generally, the total characterization of an insect society appears more feasible than it did. It is likely that the number of physical and temporal castes will not exceed 10 in ants and 20 in termites. The categories of behavior recorded in individual physical castes of ants have so far ranged between 20 and 42, with a broad overlap of

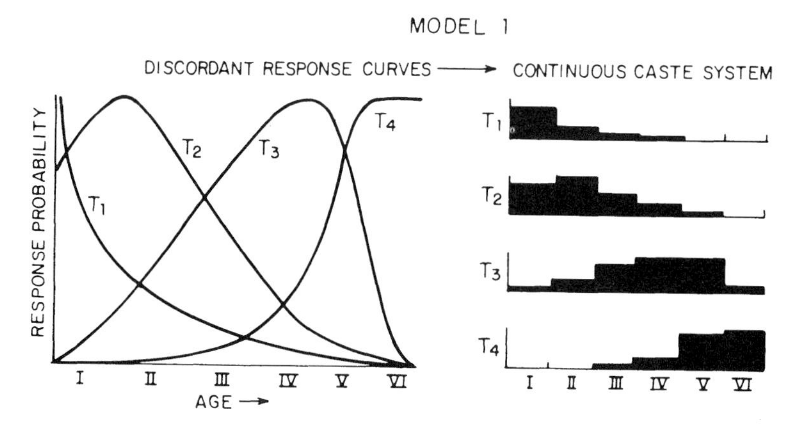


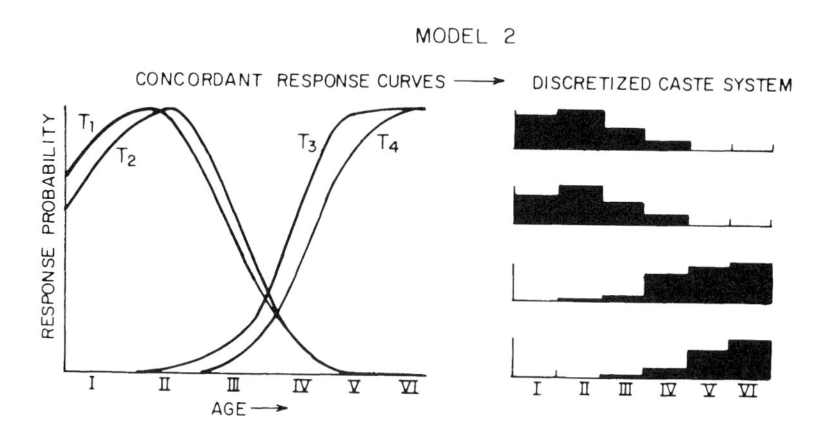


**FIGURE 8–19** These two schemas represent the extreme alternatives open to an ant species in the evolution of temporal castes. The age of the adult worker (or of any instar that can function as a worker) is arbitrarily divided into six periods. In the first model (*above*) the responsiveness of the worker to each of four tasks ($T_1$–$T_4$) changes markedly out of phase with reference to the worker's responsiveness to the other contingencies. As a consequence each of the four tasks is addressed by a distinct ensemble of age groups (temporal castes) which are represented on the right by the frequency distributions of workers in different age groups attending to the contingency. If the number of contingencies (tasks) were increased substantially, the overlap of the age-group ensembles would increase to a corresponding degree and the resulting system would approach a continuous transition. In the second model (*below*) the response curves are clustered into groups that are approximately in phase, resulting in more than one contingency being addressed by the same age-group ensemble (caste). Two ensembles perform two tasks each. Even if the number of contingencies were increased substantially, the number of ensembles would remain small. (From Wilson, 1976c.)

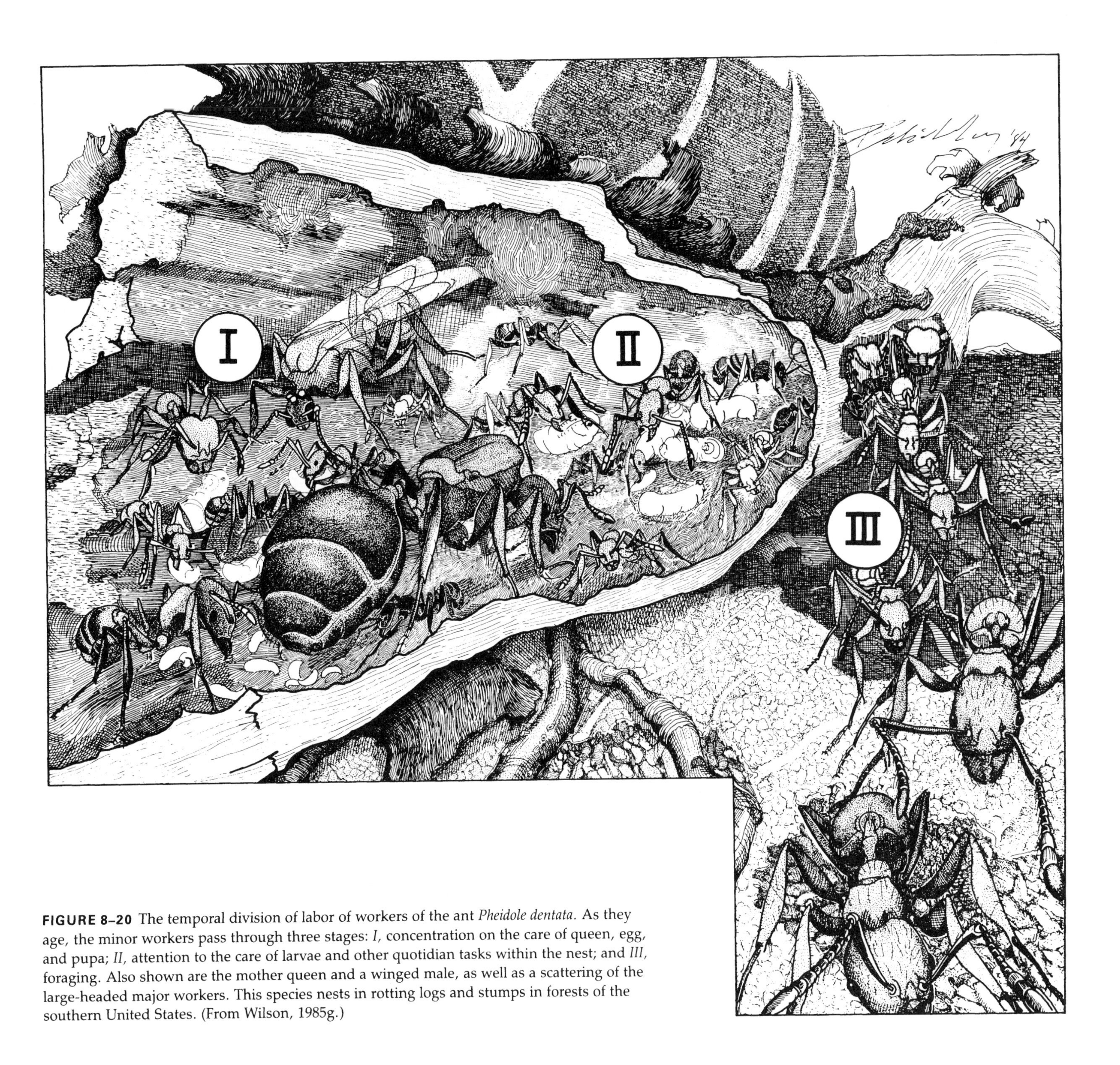


**FIGURE 8–20** The temporal division of labor of workers of the ant *Pheidole dentata.* As they age, the minor workers pass through three stages: *I,* concentration on the care of queen, egg, and pupa; *II,* attention to the care of larvae and other quotidian tasks within the nest; and *III,* foraging. Also shown are the mother queen and a winged male, as well as a scattering of the large-headed major workers. This species nests in rotting logs and stumps in forests of the southern United States. (From Wilson, 1985g.)

categories among castes and a total species repertory probably not much exceeding 50 (Oster and Wilson, 1978). Moreover, the smaller the size of the worker characterizing a species, the smaller the behavioral repertory of the species. The largest ants have a repertory size somewhat more than 1.5 times that of the smallest ants (Cole, 1985). The number of categories of signals used in communication, mostly chemical, is likely to fall between 10 and 20 (Wilson, 1971).

The study of the relation of age and behavior in ants has proceeded slowly because it is a technically difficult task. A few authors, including Otto (1958) and Lenoir (1979c), have simply marked and followed individual ants through portions of their lives. This brute-force procedure, although unassailably accurate, is also very tedious. Other investigators have sought shortcuts that allow the rapid estimation of the ages of masses of workers engaged in particular tasks. In other words instead of piecing together many individual trajectories and from those data inferring the age bias among workers generally with reference to particular tasks, researchers do the reverse, with approximately the same results: they ascertain the age bias among the bulk of the workers and from this information infer "typical" individual repertories. The age marker used most

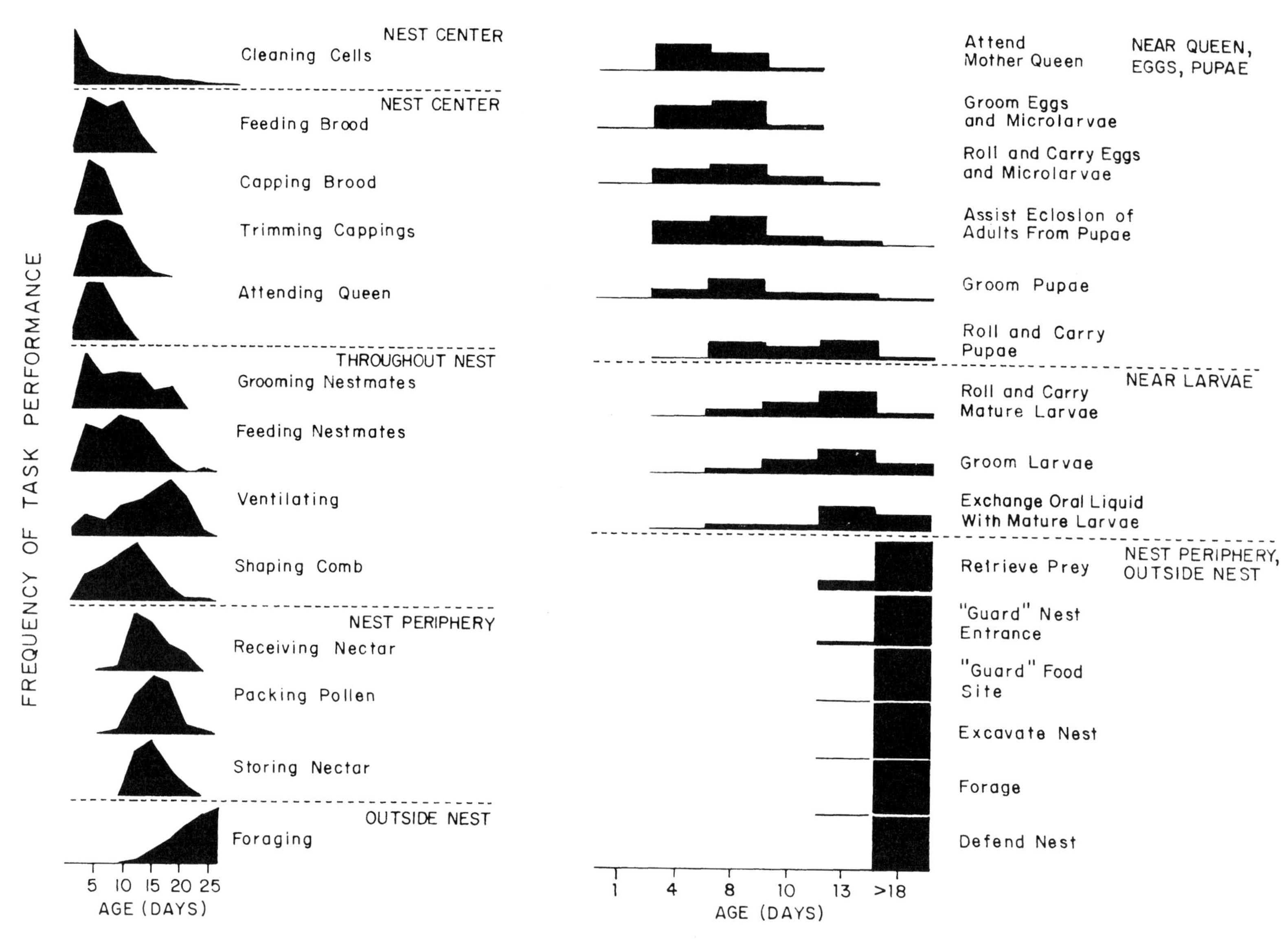


**FIGURE 8–21** The temporal division of labor, based on changes of behavior in the adult workers with aging, is shown in the ant *Pheidole dentata* and honey bee *Apis mellifera:* the insects shift from one linked set of tasks to another as they move their activities outward from the nest center (see Figure 8–19). The similarities between the two species are due to evolutionary convergence. The sum of the frequencies in each histogram is 1.0. (From Wilson, 1985g, based on Wilson, 1976c, and Seeley, 1982.)

commonly is the darkening color of workers as they grow older. In *Pheidole dentata,* for example, newly eclosed adult minor workers are uniformly clear yellow; after 2 days the gaster darkens to a contrasting yellowish-brown; when the ant reaches about a week in age the head also darkens to contrast with the middle of the body; and so on until after about 16 days the body is nearly uniformly dark brown (Wilson, 1976c). A closely similar series has been effectively employed for *Myrmica* by Weir (1958a,b) and by Cammaerts-Tricot (1974b). In the Neotropical cryptobiotic ant *Basiceros manni* we recognized four stages in coloration and cumulative body deposits that were correlated with age and hence useful in the study of temporal division of labor (Wilson and Hölldobler, 1986).

Another kind of marker, employed by Higashi (1974) in his studies of *Formica yessensis,* is the degree of mandible wear. The three apical teeth of the masticatory border are needle-sharp in newly eclosed workers and gradually wear down through the course of adult life.

A third widely used marker is the condition of the ovaries, which typically reach maximal development early in the adult stage and gradually atrophy as the ants age and undertake work away from the brood chambers (Figure 8–22). In *Formica sanguinea,* for example, yolk deposition commences about the tenth day of adult life and reaches a maximum between the twenty-sixth and thirty-fifth days (Billen, 1982a). Oogenesis by the young workers produces either trophic eggs that are fed to the queen and larvae or viable eggs that almost always develop into males. The progression of ovarian development has been used in other studies of the division of labor in *Formica* by Otto (1958), Kneitz (1964b, 1970a), Hohorst (1972), and

**FIGURE 8–22** Different stages in the ovarian development of *Formica* workers. (*a–d*) Young workers (*Innendiensttiere*). (*e–g*) Older workers (*Aussendiensttiere*). (Based on Otto, 1962.)

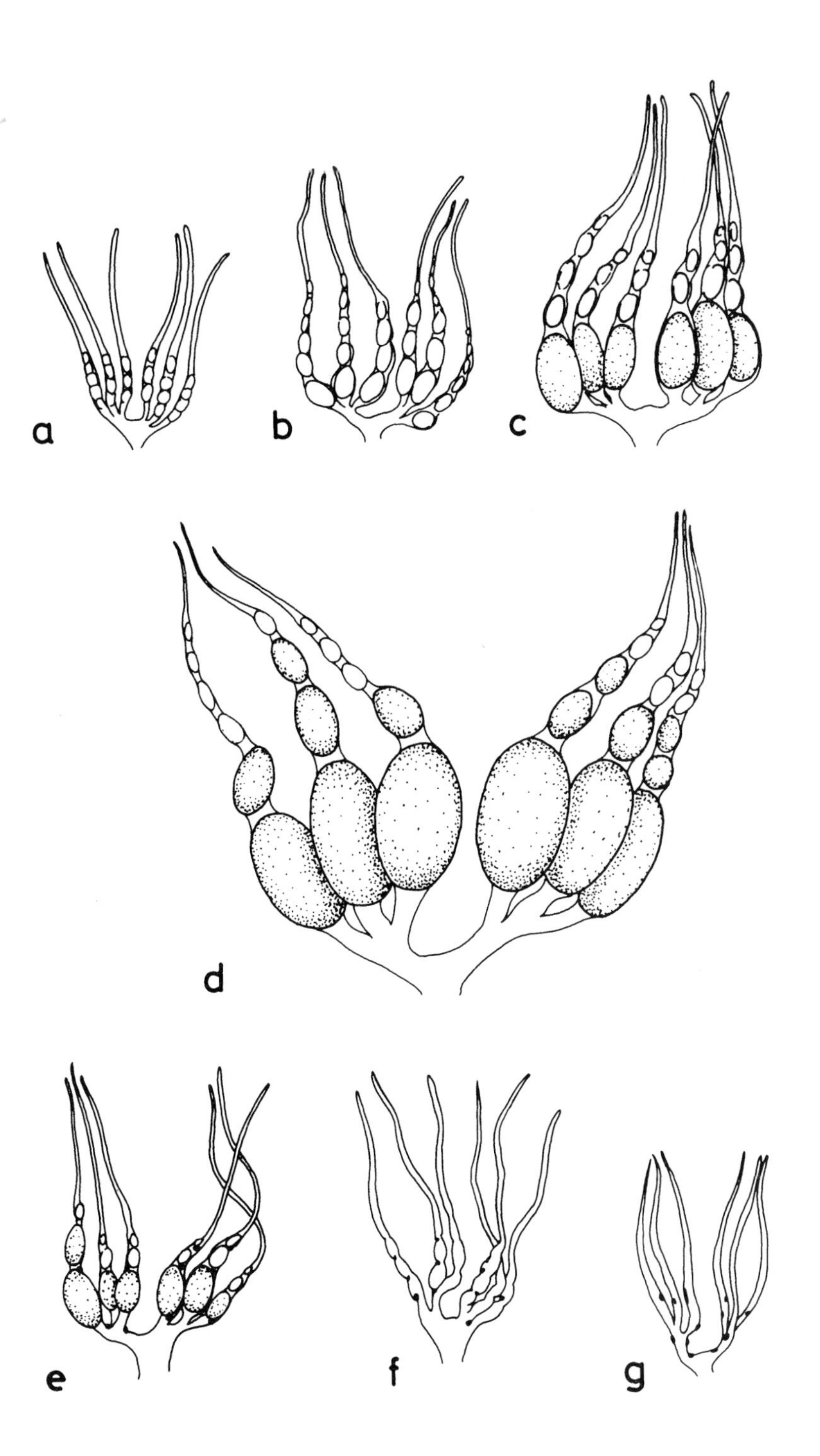


Möglich and Hölldobler (1974), and in *Oecophylla* by Hölldobler (1983). These authors have all noted the close correspondence between the position of the ants in the nest and the state of their ovaries.

Each of these methods of age estimation is subject to considerable error due to individual differences among the workers and fluctuations in the colony environment. Still, if large samples of workers are taken at least the main course of age polyethism can be charted. Its reliability increases when the colonies are maintained under controlled laboratory conditions and the marker is calibrated by following marked workers through the early part of their lives.

The technical advance most needed in studies of temporal division of labor is a physiological or biochemical "clock," one that changes steadily as the ant grows older and is subject to minimal error through environmental covariance. One candidate is the density of particles in the cytoplasm of oenocytes (Buschinger, 1967; Ehrhardt, 1970). In *Pheidole dentata* it increases at a seemingly steady rate (E. O. Wilson, unpublished). Another is the accretion of daily rings in the endocuticle, a process that works well in the honey bee for up to 11 days and less reliably thereafter (Menzel et al., 1969). Endocuticular rings occur in the harvester ant *Pogonomyrmex badius* but have not been calibrated (E. O. Wilson, unpublished).

## CASE HISTORIES

Temporal polyethism, in other words division of labor by age, is far more prevalent in ants than caste polyethism (the more easily recognizable division of labor correlated with anatomical differences). Only a single species, *Amblyopone pallipes*, is known that lacks temporal polyethism altogether. All of the remaining studies to date display some variation on "typical" age polyethism, in which young workers devote themselves to brood care and other forms of labor inside the nest, whereas older workers tend to travel outside the nest for nest construction, defense, and foraging. As the German investigators have succinctly put it, ants pass from *Innendienst* (inside work) to *Aussendienst* (outside work). In contrast, only a minority of ant genera contain one or more polymorphic species, that is, species in which the workers are divided into two or more physical subcastes. Of the 297 living genera recognized during the preparation of this book (Table 2–2), only 46, or 15 percent, contain at least some polymorphic species. These taxa are listed in Table 8–2, along with the small number of genera whose caste systems are unknown. One of the remarkable features of the phylogenetic distribution revealed by this tabulation is the scarcity of physical worker subcastes within the large and relatively primitive subfamily Ponerinae. Only a single genus (*Megaponera*), out of 58 known living ponerine genera, or 2 percent of the total, is known to be polymorphic. According to Breed and Harrison (1988) another case may be the giant Neotropical ponerine *Paraponera clavata*, where worker variation appears to be mildly allometric and the size-frequency distribution of mature colonies is unimodal and skewed to the right. Although the sizes of workers performing different tasks overlap widely, there is a statistical association between size and task performance. Large workers function more often as guards and foragers. One other ponerine genus, *Aenictogiton*, is of unknown status with respect to caste.

Most of the studies of caste and division of labor conducted to the present time are listed in Table 8–3. It is informative to compare this compilation with Table 2–2, which contains all of the living genera, and Table 8–2, which lists the genera known to display polymorphism; and then to note that almost all ant species probably have some form of division of labor, even if it is only based on age. It will then be obvious that although a large amount of information is now at our disposal, it applies to only a small fraction of genera and species. The exploration of caste and division of labor in ants is in an early stage, and many surprises surely lie ahead.

The best way to encompass existing knowledge of division of la-

**TABLE 8–2** Genera of ants with at least one species in which the worker caste is divided into physical subcastes, along with genera of still unknown status. All other ant genera (not listed here; see Table 2–2) either have a single worker subcaste or, in the case of extreme social parasites, lack workers altogether.

**At least one species polymorphic, possessing two worker subcastes**

Ponerinae: *Megaponera*
Myrmeciinae: *Myrmecia*
Dorylinae: *Dorylus*
Ecitoninae: *Cheliomyrmex, Labidus, Nomamyrmex*
Pseudomyrmecinae: *Tetraponera*
Myrmicinae: *Acanthomyrmex, Acromyrmex, Adlerzia, Anisopheidole, Cephalotes, Crematogaster, Machomyrma, Messor, Monomorium, Oligomyrmex, Pogonomyrmex, Solenopsis, Strumigenys, Zacryptocerus*
Aneuretinae: *Aneuretus*
Dolichoderinae: *Azteca, Iridomyrmex, Liometopum, Zatapinoma*
Formicinae: *Camponotus, Cataglyphis, Euprenolepis, Formica, Gesomyrmex, Melophorus, Myrmecocystus, Myrmecorhynchus, Notostigma, Oecophylla, Proformica, Pseudaphomomyrmex, Pseudolasius*

**At least one species polymorphic, possessing three subcastes**

Ecitoninae: *Eciton*
Myrmicinae: *Atta, Daceton, Orectognathus, Pheidole* (most *Pheidole* spp. with two subcastes only), *Pheidologeton*

**Degree of polymorphism, if any, unknown**

Ponerinae: *Aenictogiton*
Leptanillinae: *Noonilla, Phaulomyrma, Scyphodon*
Dolichoderinae: *Linepithema*
Formicinae: *Forelophilus, Phasmomyrmex*

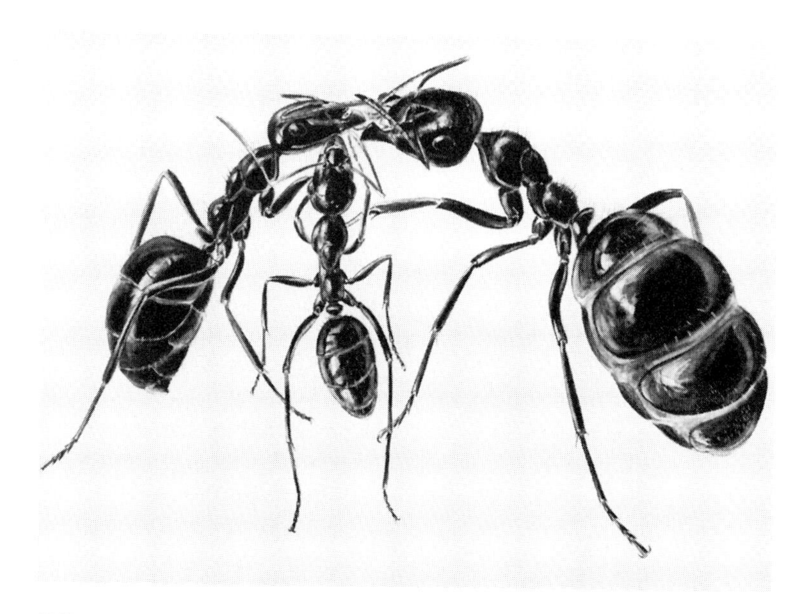

**FIGURE 8–23** A media of *Proformica nasuta* feeds a major from the same colony, as a minor attempts to intrude. (Painting by T. Hölldobler-Forsyth.)

bor is to compare several grades of complexity across the entire range of the living ants. A well-studied example of weak polymorphism correlated with polyethism has been provided by Brandão (1978) for *Formica perpilosa.* This American montane species displays weak polymorphism of a kind widespread in ants. There is a modest size variation within the worker caste of individual colonies, accompanied by a slight but distinct allometry in which the bodies grow more robust and the heads broader relative to length with an overall increase in size. Medium-sized workers have the broadest repertories, covering all of the quotidian tasks of brood and queen care, nest work, and foraging. The smallest workers are very similar in overall behavior while performing slightly fewer acts. The largest workers, in contrast, constitute a true "major" caste specialized for liquid food storage. They engage in few other social acts. Overall, *F. perpilosa* closely parallels caste and division of labor in the species of *Proformica,* a formicine genus found in dry habitats in Europe and Asia that also possesses weak polymorphism and a major subcaste specialized for food storage (Figure 8–23). The similarity is almost certainly due to convergent evolution.

In almost all species for which differences in size have been found to be correlated with division of labor, the size variation is also distinctly allometric. In addition to this first rule of caste evolution we encounter a second rule: monomorphic ant species rarely divide labor according to size. More precisely, the greater the combined amounts of size variation and allometry, the more pronounced the division of labor. Considerable size variation often occurs, but unless it is accompanied by allometry strong enough to be evident by casual inspection, it is not correlated with marked role differences (Wilson, 1971; Herbers et al., 1985). No deviation is known from the first rule relating polymorphism to division of labor. One exception has been discovered to the second rule, however. *Leptothorax longispinosus* is a typically monomorphic species, yet the large workers forage more than the smaller ones, a foreshadowing of the division of labor characterizing the most simply allometric ant species (Herbers and Cunningham, 1983). In contrast, the monomorphic *L. ambiguus,* when subjected to the same form of analysis, evinced no sign of size bias in behavior. Herbers (1983) suggested that the difference between the two species might be due to the greater size variation that occurs within colonies of *L. longispinosus.* Another exception is the formicine honey ant *Myrmecocystus mimicus,* whose larger workers, Hölldobler (1981a) found, are significantly more engaged in tournament behavior than smaller workers.

Successive grades in the early evolution of polymorphism and division of labor are nicely demonstrated among species of fire ants. The more primitive pattern of *Solenopsis invicta* is typical of almost all of the fire ant species, which compose the subgenus *Solenopsis* of *Solenopsis* (as opposed to the thief ants, which compose the subgenus *Diplorhoptrum*). The allometry of *S. invicta* is very slight (see Figures 8–15 and 8–24). The majors have heads that are only somewhat shorter, broader, and more quadrate with reference to the remainder of the body. The behavioral differences are also modest, since the majors perform the same tasks as the minors with the exception of caring for the eggs and young larvae. Workers in these larger size classes simply handle pieces of brood, prey, and soil particles that are correspondingly larger in size. Hence transport and nest excavation appear to be enhanced by the polymorphism, although that subjective impression has not yet been put to a quanti-

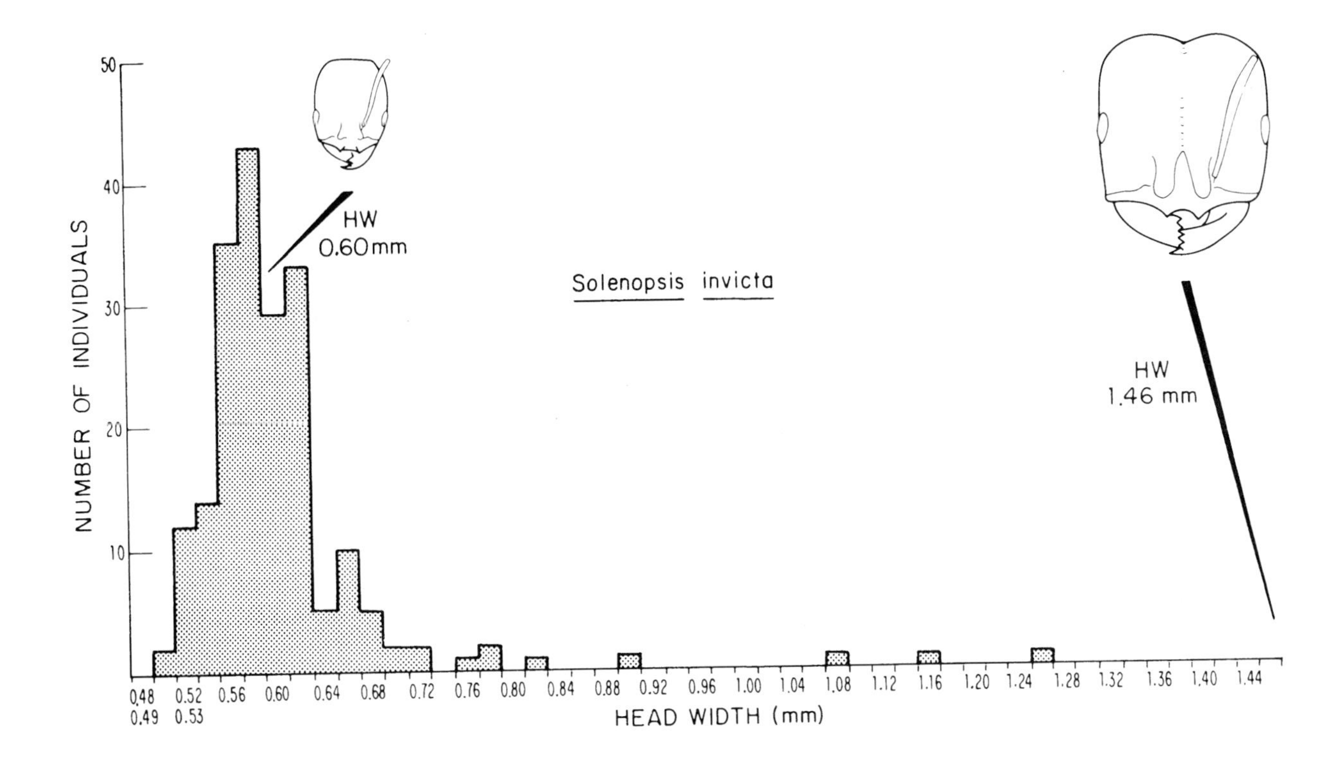


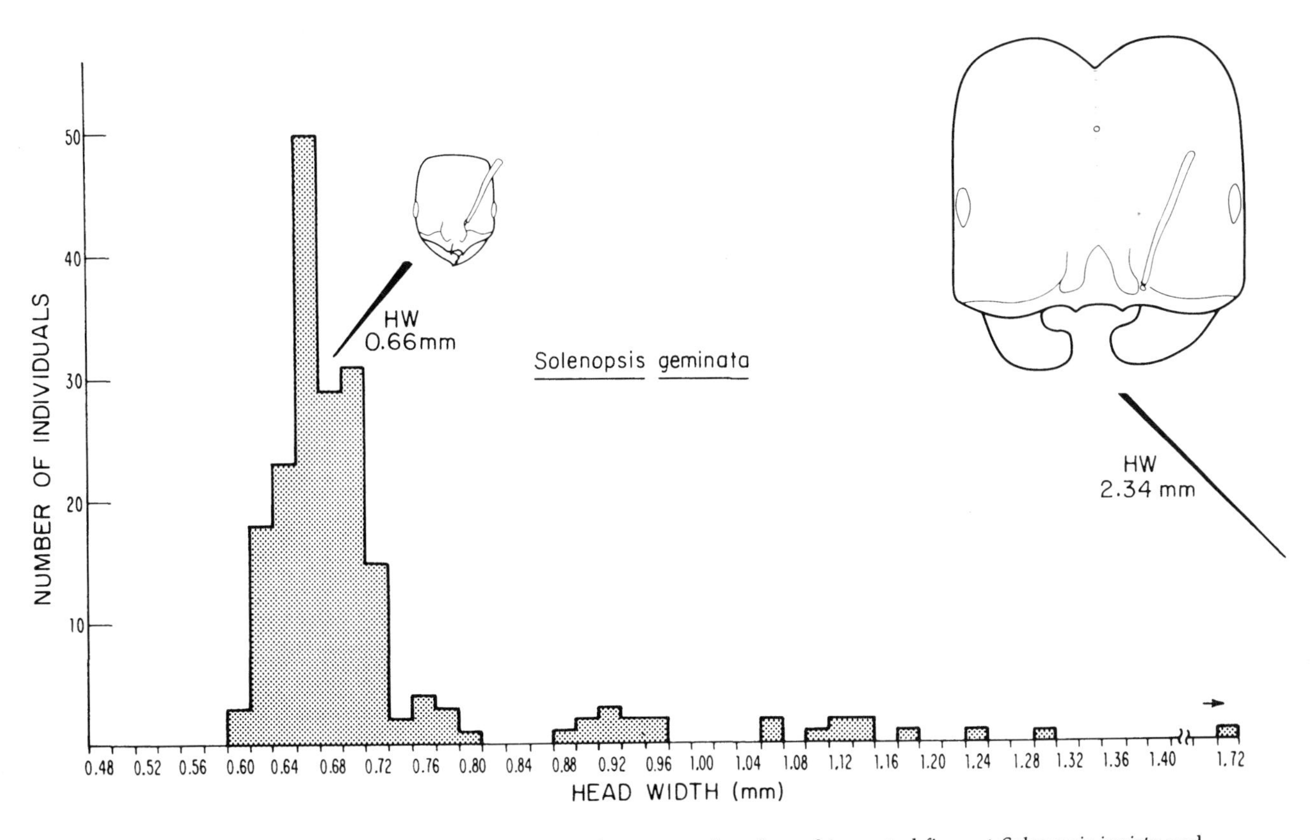


**FIGURE 8–24** The worker caste systems of two fire ant species, the red imported fire ant *Solenopsis invicta* and the native granivorous form *S. geminata*. The outlines shown are the heads of workers at the two extremes of size variation; that is, a small minor worker contrasted with a large major worker. Also presented are the size-frequency distributions of random samples of workers from large laboratory colonies of the two species. (From Wilson, 1978.)

**TABLE 8–3** Studies of caste and division of labor. "Typical age polyethism" means that the younger workers attend the brood and queen, with the older workers changing to nest work (usually) and then foraging.

| Species | Principal results | Reference |
|---|---|---|
| **PONERINAE** | | |
| *Amblyopone pallipes* | No caste or division of labor | Traniello (1978, 1982) |
| *Diacamma rugosum* | Foragers are emigration leaders; queenless, with gamergates as reproductives | Wheeler and Chapman (1922), Fukumoto and Abe (1983), Moffett (1986h) |
| *Ectatomma ruidum* | Automated record reveals strong individual labor differences in this polymorphic species; typical age polyethism also demonstrated | Corbara et al. (1986), Lachaud and Fresneau (1987) |
| *Odontomachus affinis* | Increase in worker behavioral repertory with colony growth | Brandão (1983) |
| *Ophthalmopone berthoudi* | Queens absent; reproduction by gamergates | Peeters and Crewe (1984, 1985) |
| *Pachycondyla* (= *Neoponera*) spp. | Typical age polyethism | Fresneau (1984), Fresneau and Lachaud (1985), Pérez-Bautista et al. (1985) |
| *Paraponera clavata* | Larger workers are guards and foragers; smaller workers tend to remain inside the nest. Ovaries of foragers and guards are smaller than those of brood workers | Breed and Harrison (1988) |
| *Prionopelta amabilis* | Typical age polyethism; young workers lay trophic eggs | Hölldobler and Wilson (1986a) |
| **MYRMECIINAE** | | |
| *Myrmecia* spp. | Typical age polyethism | Freeland (1958) |
| **DORYLINAE** | | |
| *Dorylus* spp. | Moderate diphasic allometry and bimodal size-frequency distribution linked to complex division of labor | Raignier and van Boven (1955), Hollingsworth (1960), Raignier et al. (1974), Gotwald (1978, 1982) |
| **ECITONINAE** | | |
| *Eciton* spp. | Strong polymorphism associated with complex division of labor, even to the formation of "teams"; size-frequency curve trimodal | Rettenmeyer (1963a), Schneirla (1971), Silva (1972), Franks (1986) |
| *Neivamyrmex nigrescens* | Early participation by young workers in predator raids, even earlier in emigrations | Topoff and Mirenda (1978) |
| **PSEUDOMYRMECINAE** | | |
| *Tetraponera nasuta* | A major caste described, unique within the subfamily; role unknown | Terron (1971) |
| **MYRMICINAE** | | |
| *Acanthomyrmex* spp. | Majors specialized for milling of seeds and defense | Moffett (1986b) |
| *Aphaenogaster* (= *Novomessor*) *albisetosus* | Typical age polyethism | McDonald and Topoff (1985) |
| *A. rudis* | Very weak size polyethism and role persistence; seasonal change in behavior patterns | Herbers et al. (1985) |
| *Atta* spp. | Very complex division of labor, with "assembly-line" processing of substrate according to worker size; majors exclusively defensive | Eibl-Eibesfeldt and Eibl-Eibesfeldt (1967), Weber (1972), Autuori (1974), Wilson (1980a,b, 1983a,b) |
| *Basiceros manni* | Typical age polyethism; young workers probably lay trophic eggs | Wilson and Hölldobler (1986) |
| *Cephalotes atratus* | Wide size variation; moderate allometry with division of labor | Corn (1976) |
| *Daceton armigerum* | Moderate continuous polymorphism associated with complex division of labor; smallest workers are nurses | Wilson (1962a) |

*continued*

**TABLE 8–3** *(continued)*

| Species | Principal results | Reference |
|---|---|---|
| *Erebomyrma nevermanni* | Majors defend, subdue large prey, are semirepletes; young minors tend brood, old ones form queen retinue | Wilson (1986a) |
| *Eurhopalothrix heliscata* | Typical age polyethism | Wilson and Brown (1984) |
| *Leptothorax acervorum* | Non-inseminated queens function as auxiliary workers, both as slaves of *Harpagoxenus* and in independent colonies | Buschinger (1968c, 1983) |
| *L. longispinosus* | A "monomorphic" species, yet strong size polyethism exists; transition matrix provided | Herbers and Cunningham (1983) |
| *Messor* spp. | Allometry with overlapping bimodal frequency distribution; typical age polyethism | Goetsch and Eisner (1930), Ehrhardt (1931), Delage-Darchen (1974b) |
| *Monomorium* spp. (*salomonis* group) | Evolution of queen-worker intercastes | Bolton (1986b) |
| *Myrmica* spp. | Typical age polyethism; physiology, pheromone production, and aggressiveness change | Heyde (1924), Ehrhardt (1931), Weir (1958a,b, 1959b), Cammaerts-Tricot (1974b, 1975), Cammaerts-Tricot and Verhaeghe (1974) |
| *Oligomyrmex* spp. | Dimorphism with rare intergrades; majors with very limited repertory; both castes form semirepletes; typical age polyethism in minors | Kusnezov (1951a), Moffett (1986g) |
| *Orectognathus versicolor* | Strong polymorphism; minors specialize in brood care, majors defend nest by "bouncing" with mandibles | Carlin (1981) |
| *Pheidole dentata* | Dimorphism with soldiers recruited by minors to defend especially against *Solenopsis;* typical age polyethism with roles clumped (discretized) | Wilson (1976b,c) |
| *P. embolopyx* | Majors defend nest and food, are recruited by workers | Wilson and Hölldobler (1985) |
| *P. hortensis* | Typical age polyethism, but not discretized | Calabi et al. (1983) |
| *P. lamia* | Majors with phragmotic heads defend against *Solenopsis* | Buren et al. (1977) |
| *P. pubiventris* | Basis for division of labor at brood pile | Wilson (1985a) |
| *Pheidole* spp. | Majors defend nest and food, serve as repletes, according to species; minors with typical age polyethism; majors can substitute for minors in emergencies | Creighton and Creighton (1959), Wilson (1971, 1976b,c, 1984b, 1985a), Hölldobler and Möglich (1980), Calabi et al. (1983), Wilson and Hölldobler (1985), Calabi (1986) |
| *Pheidologeton diversus* | Extreme physical caste variation accompanied by strong division of labor; 4 worker subcastes and 2 worker temporal castes; only minors lay trails | Moffett (1987a–c, 1988a,b) |
| *Pogonomyrmex badius* | Typical age polyethism, at least from midden work to patrolling to foraging; larger workers with lower per-gram respiration | Hölldobler and Wilson (1970), Gordon (1984b), Porter (1986) |
| *P. barbatus* | Perturbations in one role-group cause responses in other role-groups; workers of small colonies switch more readily from one task to another | Gordon (1987, 1988b) |
| *P. owyheei* | Disposable caste: old workers forage, suffer high mortality | Porter and Jorgensen (1981) |
| *Pogonomyrmex* spp. | Stratification in the nests: youngest (nurse) ants nearest bottom | MacKay (1983b) |
| *Pristomyrmex pungens* | Queen absent; unfertilized workers are sole reproductives, and all workers appear to lay eggs when young; typical age polyethism | Itow et al. (1984), Tsuji and Itō (1986), K. Tsuji (personal communication) |
| *Procryptocerus scabriusculus* | Typical age polyethism; callows engage in extensive abdominal trophallaxis; comparisons with other species of cephalotines | Wheeler (1984) |
| *Solenopsis geminata* | Large medias and majors highly specialized for milling seeds | Wilson (1978) |

*continued*

**TABLE 8–3** *(continued)*

| Species | Principal results | Reference |
|---|---|---|
| *S. invicta* | Small workers care for brood more, large ones forage; particle size matching during foraging; typical age polyethism; large number of behaviorally flexible reserve workers; polymorphism aids brood production; size-frequency distribution changes as colony grows; majors can serve as repletes | Markin et al. (1972, 1973), Glancey et al. (1973), Wilson (1978), Mirenda and Vinson (1981), Wood and Tschinkel (1981), Porter and Tschinkel (1985), Sorenson et al. (1985) |
| *Zacryptocerus* spp. | Caste dimorphism associated with strong division of labor | Creighton and Gregg (1954), Creighton (1963), Wilson (1976a), Wheeler and Hölldobler (1985) |
| **ANEURETINAE** | | |
| *Aneuretus simoni* | Dimorphism with marked division of labor, although with no evidence of defense by majors; typical age polyethism | Wilson et al. (1956), Traniello and Jayasuriya (1985) |
| **DOLICHODERINAE** | | |
| *Azteca laticeps* | Larger workers defend nest and forage on ground; smaller workers forage arboreally | Wheeler (1986c) |
| *Hypoclinea quadripunctata* | Queen-worker differentiation | Torossian (1974) |
| *Tapinoma erraticum* | Typical age polyethism; but transport not age dependent; elitism in brood transport | Meudec (1973), Torossian (1974), Lenoir (1979b), Meudec and Lenoir (1982) |
| **FORMICINAE** | | |
| *Camponotus ephippium* | Major workers guard nest, do not forage | B. Hölldobler (unpublished) |
| *C. herculeanus* | Typical age polyethism | Hölldobler (1965) |
| *C. impressus* | Test of ergonomic theory in major:minor ratios (see text) | Walker and Stamps (1986) |
| *C. ligniperda* | Typical age polyethism | Hölldobler (1965) |
| *C. maculatus* | Worker allometric curve breaks into 4 segments | Baroni Urbani (1976b) |
| *C. pennsylvanicus* | (See under *Camponotus* spp.) | |
| *C. sericeiventris* | Strong polymorphism correlated with division of labor | Busher et al. (1985) |
| *C. socius* | Majors do not scout, are recruited by minors and medias | Hölldobler (1971c) |
| *C. vagus* | Allometry and size-frequency distribution; aging associated with increase in aggression and changes in food-exchange behavior | A. Benois in Passera (1984), Morel (1984), Bonavita-Cougourdan and Morel (1985) |
| *Camponotus* spp. | Evolution of allometry and size-frequency distribution. As a rule, small workers attend brood, work in nest; larger workers guard and forage | Pricer (1908), Buckingham (1911), Heyde (1924), Kiil (1934), Lee (1938), Sanders (1964), Baroni Urbani (1974), Jaffe and Sanchez (1984a) |
| *Cataglyphis bicolor* | Typical age polyethism | Schmid-Hempel (1982) |
| *C. bombycina* | Majors have saber-shaped mandibles and serve as nest guards | Délye (1957) |
| *Colobopsis* spp. | Majors guard nest entrances with plug-shaped heads; serve as repletes | Forel (1874), Wheeler (1910a), Szabó-Patay (1928), Wilson (1974c) |
| *Formica neorufibarbis gelida* | Small workers darker in color, able to forage earlier at high elevations | Bernstein (1976) |
| *F. obscuripes* | Large workers hunt prey, medias gather honeydew, etc. Predator stress reduces size variance, does not shift mean | Herbers (1979, 1980) |
| *F. perpilosa* | Weak allometry, strong division of labor; larger workers semirepletes | Brandão (1978) |
| *F. polyctena* | Allometric variation in worker ovarioles; ovariole degeneration with age (see also *rufa* group below) | Otto (1958), Schmidt (1964), Horstmann (1973), Billen (1982a) |
| *F. sanguinea* | Typical age polyethism; ovarian regression with age; also age-dependent hibernation, physiology, and aggressiveness among workers | Dobrzańska (1959), Billen (1984b) |

*continued*

**TABLE 8–3** *(continued)*

| Species | Principal results | Reference |
|---|---|---|
| *F. yessensis* | Typical age polyethism; small workers remain more in nest | Higashi (1974) |
| *Formica* spp. (*rufa* group) | Typical age polyethism, with great variability among individuals; older workers establish trunk trails; larger workers venture farther | Økland (1930, 1934), Kiil (1934), Jander (1957), Otto (1958), Kneitz (1964), Lange (1967), Rosengren (1977a, 1987) |
| *Lasius niger* | Typical age polyethism, but highly variable in expression. Principal component analysis of roles; log-normal distribution of activity defining elites; difference in brood retrieval | Lenoir and Mardon (1978), Lenoir (1979c), Lenoir and Ataya (1983) |
| *Lasius* spp. | Typical age polyethism | Heyde (1924) |
| *Myrmecocystus mimicus* | Primarily larger workers involved in territorial tournaments | Hölldobler (1981a) |
| *Myrmecocystus* spp. | "Honey pot" ants: some larger workers become extreme repletes and store liquid food | Wheeler (1910a), Rissing (1984) |
| *Oecophylla longinoda* | Dimorphic: minors attend brood and homopterans; majors perform other functions; typical age polyethism | Cole and Jones (1948), Ledoux (1950), Hölldobler and Wilson (1978), E. O. Wilson (unpublished) |
| *O. smaragdina* | As majors age, they move toward periphery of nest trees, as guards; majors are principal nest weavers | Hemmingsen (1973), Hölldobler (1983) |
| *Prenolepis imparis* | Basis of "replete" formation in workers: adaptation to low-temperature foraging | Tschinkel (1987c) |
| *Proformica* spp. | Larger workers become repletes and store liquid food for the colony; in *P. epinotalis* large workers forage at night, small ones by day | Stumper (1961) |

tative test. The more important point is that *S. geminata* has added something entirely new. Allometry is steep, so that the larger medias and majors are grotesque creatures possessing massive heads and blunt, toothless mandibles. These two features together contribute to a substantial increase in the crushing power of the ants. Wilson (1978) was able to show that the largest workers were in fact specialized for milling seeds, the majors so much so that they have become restricted to the smallest behavioral repertory thus far recorded in the social insects—milling and self-grooming. *S. geminata* has undergone this striking revision in caste system and division of labor as part of its trend toward granivory, which is the most extensive known among the fire ant species.

Fire ants illustrate a further principle in the division of labor, which can be summarized by the expression *alloethism:* the regular and disproportionate change in behavior in a particular category of behavior as a function of worker size (Wilson, 1978). Alloethism parallels allometry, or the regular and disproportionate change in one anatomical dimension relative to other dimensions as a function of worker size. In both species in which sufficiently detailed measurements have been taken, the fire ant *Solenopsis geminata* and the leafcutter *Atta sexdens,* the curves were found to possess steeper slopes than the corresponding allometric curves of the anatomical structures employed in the behavior (Wilson, 1978, 1980a). Put succinctly, alloethism amplifies allometry. Part of the fire ant pattern is illustrated in Figure 8–25.

Passing to the opposite extreme, we may next consider one of the most advanced grades of caste and division of labor discovered so far in ants, that of *Pheidologeton diversus* of Southeast Asia. Colonies of this "marauder ant" contain hundreds of thousands of workers and conduct wide-ranging raids to collect insects and a variety of vegetable materials. The workers display only a moderate amount of allometry, which is largely confined to the proportions of the head. But their size variation is the greatest ever recorded within single ant colonies: a full 10-fold range in head width and a 500-fold range in dry weight (see Figure 8–26). The size-frequency distribution of the workers is trimodal. Superimposed on this pattern are the following four size-related roles:

*Minors* (head widths less than 0.8 mm) conduct most of the tasks of the colony, including brood care, much of the foraging, construction of the nest and trail arcades, and defense.

*Medias* (head widths 1.0–2.6 mm) assist the minors in foraging and construction.

*Majors* (head widths greater than 2.6 mm) hoist obstructions and smooth the surface of the trails.

*Repletes* are medias and majors that remain in the nest and acquire food until their abdomens are swollen; they share this stored food with other members of the colony.

The medias and majors act as the colony bulldozers. They move obstructions on the trails too large for their small nestmates to handle, pack the soil with their massive heads, and chew through the tougher fibers of fruit and the more massive appendages of captured prey. In addition to the physical castes, Moffett (1987b) has distinguished two temporal castes in the small workers that conform to the usual pattern in ants. That is, the youngest workers associate more with the brood, whereas the older workers presumably

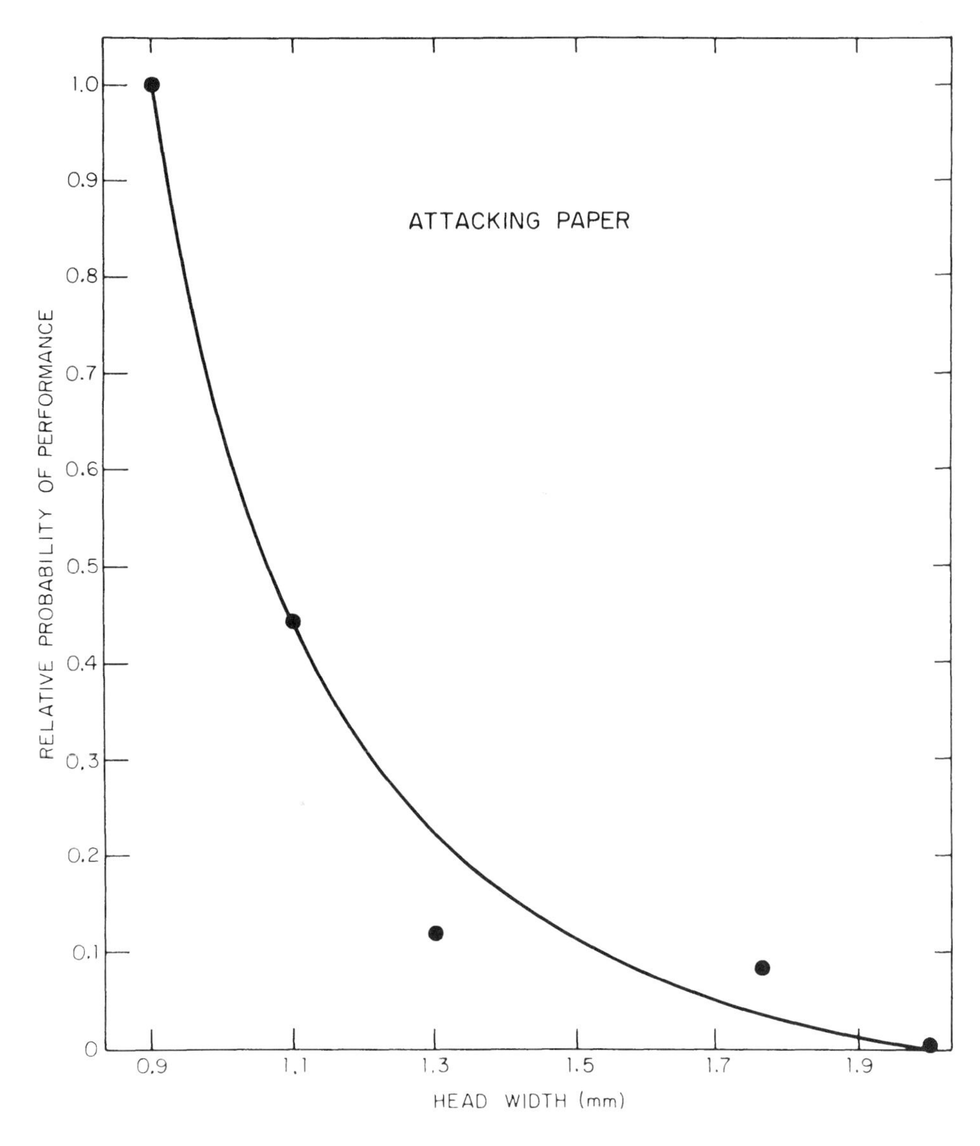


**FIGURE 8–25** Part of the alloethic curve for aggressive response by workers of the fire ant *Solenopsis geminata* to a disturbance of the nest. "Relative probability" is defined as the fraction of workers responding in a size group relative to the highest fraction responding in any size group (in this case the group possessing head width 0.9 mm). (From Wilson, 1978.)

**FIGURE 8–26** The greatest size variation of nestmates ever recorded in ants occurs in the Asian marauder ant *Pheidologeton diversus.* The minor worker depicted in this scanning electron micrograph has a head width exactly 1/10 that of the major on which it sits, and a dry weight only about 1/500 that of the larger ant. (From Moffett, 1987b.)

**FIGURE 8–27** The caste systems of leafcutter ants in the genus *Atta* are among the most complex in the social insects, involving extreme size variation accompanied by strong diphasic allometry. The workers illustrated here are from a single colony of *A. laevigata.* (From Oster and Wilson, 1978; drawing by T. Hölldobler-Forsyth.)

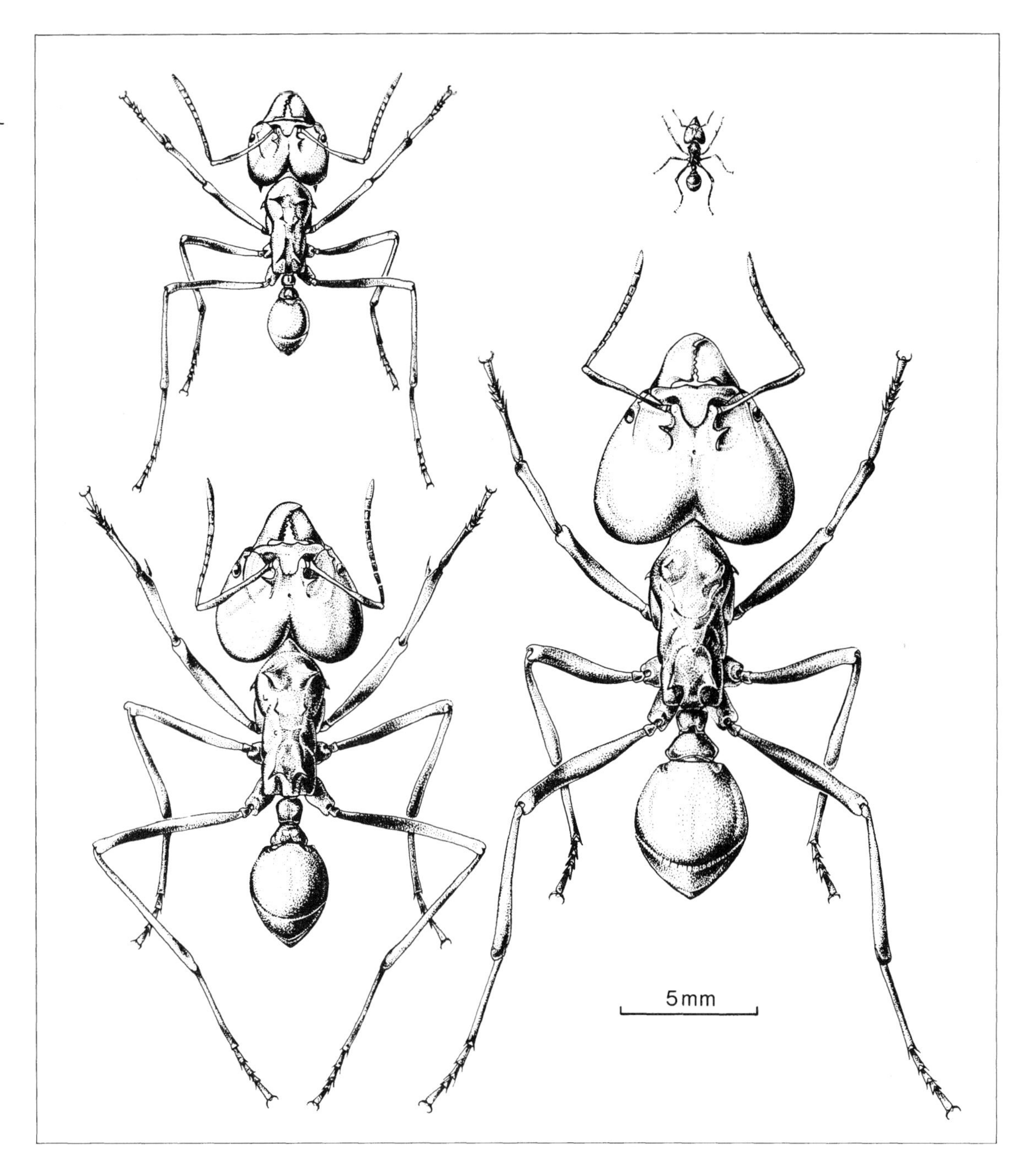


expand their repertory to fill the entire range of tasks. In addition, exceptionally dark-colored, hence old or wounded, workers are disproportionately represented among the workers guarding the trails. The caste system of *Pheidologeton diversus,* when combined with its large colony size and ability to field a large army of workers, has enabled it to utilize an extraordinarily wide variety of foods.

Precisely the opposite dietary adaptation has been achieved by the equally complex system of the leafcutter ants in the genus *Atta,* illustrated in Figure 8–27. Like *Pheidologeton, Atta* generates a broad array of physical types by combining extreme size variation with moderate allometry. In *Atta sexdens,* the head width varies 8-fold and the dry weight 200–fold from the smallest minors to the huge majors; the polymorphism thus runs a close second to *Pheidologeton diversus.* However, the *Atta* caste diversity is not directed at broadening the diet of the ants, as in *Pheidologeton,* but at narrowing it severely. Colonies of *Atta* subsist entirely on the sap of plants and a symbiotic fungus they grow on fragments of vegetation. *Atta* and its sister genus *Acromyrmex* are unique in the animal kingdom in their ability to utilize freshly cut leaves and other vegetation for the rearing of fungus. Their unusual caste system has made the adaptation possible.

The fungus-growing ants of the ant tribe Attini, to which *Atta* belongs, are of unusual interest in biology because (if we may cite a familiar metaphor) they alone among the ants have achieved the transition from a hunter-gatherer to an agricultural existence. But this major shift did not require an elaborate caste system. The great majority of attine genera and species are monomorphic, including the presumably most primitive forms belonging to *Cyphomyrmex.* The complex caste system and division of labor of *Atta* represent a much narrower, more idiosyncratic adaptation to the collecting of

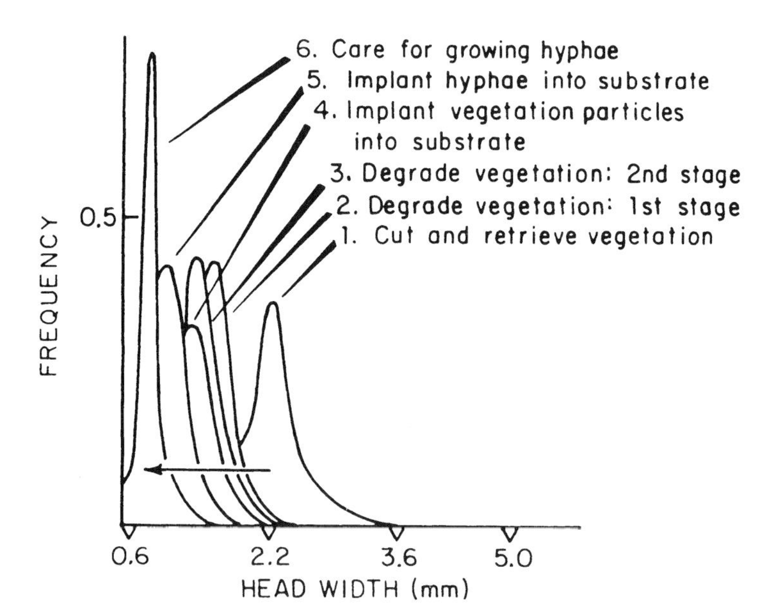


**FIGURE 8–28** The assembly-line processing of vegetation by leafcutter ants (*Atta sexdens*) depends on an intricate division of labor among the minor and media workers. Medias gather the vegetation, after which a succession of ever smaller media and minor workers process it and use it to cultivate fungal hyphae. The polyethism curves are smoothed versions of histograms of head width distributed over 0.1-mm intervals. (From Wilson, 1980a.)

fresh vegetation as a novel form of fungal substrate. Most of the monomorphic attines such as *Cyphomyrmex* utilize decaying vegetation, insect remains, or insect excrement, all of which are materials ready-made for fungal growth. Fresh leaves and petals, in contrast, require a whole series of special operations before they can be converted into substrate. They must first be cut down, then chopped into fine pieces, next chewed and treated with enzymes, and finally incorporated into the garden comb. Beyond the harvesting process, the fungus must also be provided with constant care after it sprouts on the substratum.

The *Atta* workers organize the gardening operation in the form of an assembly line, which is illustrated in Figures 8–28 and 8–29 (see also Plates 9 through 11). Tough vegetation can be cut only by workers with head widths of 1.6 millimeters or greater. The most frequent size group among foragers consists of workers with head widths 2.0–2.2 millimeters. At the opposite end of the line, the care of the delicate fungal hyphae requires very small workers, and this task is filled within the nests by workers with head widths predominantly 0.8 millimeter. The intervening steps in gardening are conducted by workers of graded intermediate size. After the foraging medias drop the pieces of vegetation onto the floor of a nest chamber, they are picked up by workers of a slightly smaller size, who clip them into fragments about 1–2 millimeters across. Within minutes, still smaller ants take over, crush and mold the fragments into moist pellets, and carefully insert them into a mass of similar material. The resulting "comb" ranges in size from that of a clenched fist to that of a human head and is riddled with channels. Resembling a gray cleaning sponge, it is light and fluffy and crumbles under slight pressure. On its surface the fungus spreads out like a frost, sinking its hyphae into the leaf paste to digest the abundant cellulose and proteins held there in partial solution.

The remainder of the gardening cycle proceeds. Worker ants even smaller than those just described pluck loose strands of the fungus from places of dense growth and plant them on the newly constructed surfaces. Finally, the very smallest—and most abundant—workers patrol the beds of fungal strands, delicately probing them with their antennae, licking their surfaces clean, and plucking out the spores and hyphae of alien species of mold. These colony dwarfs can travel through the narrowest channels deep within the garden masses. From time to time they pull loose tufts of fungal strands resembling miniature stalked cabbage heads and carry them out to feed their larger nestmates.

Although the assembly line of fungal cultivation is the core of caste and division of labor in the leafcutter colony, it is far from the entire story. The defense of the colony is also organized to some extent according to size. All of the size groups attack intruders, but in addition there is a true soldier caste. Among the scurrying ants can be seen a few extremely large majors. Their sharp mandibles are powered by massive adductor muscles that fill the swollen, 5-millimeter-wide head capsules. Working like miniature wire clippers, these specialists chop enemy insects into pieces and easily slice through the skin of vertebrate intruders. The giants are especially adept at repelling large enemies. When entomologists digging into a nest grow careless, their hands become nicked all over as if they had been pulled through a thorn bush.

When brood care, nest construction, and other tasks are added, the ants perform a total of 22 such social functions. The workers of *Atta sexdens* fall into four size groups defined by the clusters of these tasks on which they specialize, together with size. These four physical castes can be characterized respectively as gardeners-nurses, within-nest generalists, foragers-excavators, and defenders. As shown in Figure 8–30, this analysis starts with the drawing of polyethism, or size-frequency, curves of the workers engaged in each of the 22 tasks in turn. The polyethism curves are then clustered to define the roles. Three of the four physical castes recognized in this fashion have also been shown to pass through changes of behavior with aging, with two temporal castes in each. The total number of castes recognizable in *Atta sexdens*, physical plus temporal, is therefore seven (Wilson, 1980a).

The absolute commitment of the *Atta* worker caste to the social life of the species is reflected by its intricate polyethism. It is also underscored by the distinctive patterns of size-related variation in the exocrine glands and body spines. In every instance where the function of an organ is known, the organ proves to be maximally developed in the size classes that specialize on the tasks it serves (see Figure 8–31). Hence the pronotal spines, taken as representative of armament over the entire body, are proportionately longest in the size classes (with head widths 1.6–2.6 mm) that spend the most time foraging outside the nest and are exposed most frequently to danger from predators and competitor ants. The same size classes do most of the trail laying, and sure enough, they have the proportionately largest poison glands, the source of the trail pheromone. The postpharyngeal gland is the source of larval food in species of ants that have been studied in this respect, and it turns out that in *A. sexdens* the postpharyngeal gland is largest in the smallest size class, with head widths 0.6–1.2 millimeters; these workers are the ones that feed larvae by regurgitation. The mandibular gland is known to produce the alarm substances citral and 5-methyl-3-heptanone in species of *Atta*, including, at least in the case of the methylheptanone, *A. sexdens* (Blum et al., 1968a). As expected, this paired organ is proportionately most massive in the

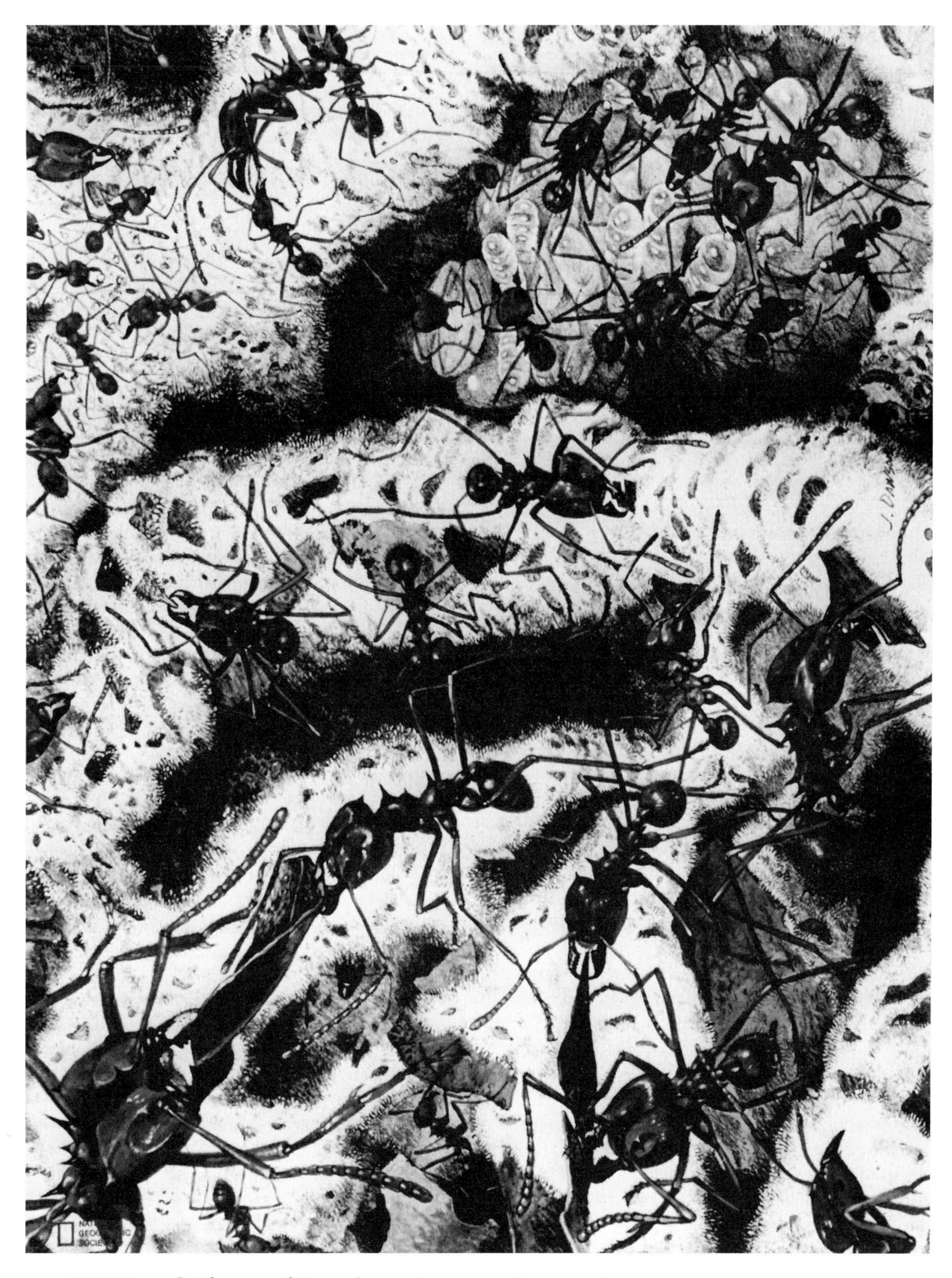

**FIGURE 8–29** Inside a nest of *Atta sexdens,* workers convert fragments of vegetation into moist clumps of substrate. They also cultivate the symbiotic fungus on which they subsist, a basidiomycete in the genus *Leucocoprinus.* The fungal growth superficially resembles that of bread mold. The processing of the substrate and fungal gardening follows an assembly-line sequence involving successively smaller workers, a procedure documented by the frequency curves of Figure 8–28. (From Hölldobler, 1984d; painting by J. D. Dawson, reprinted with permission of the National Geographic Society.)

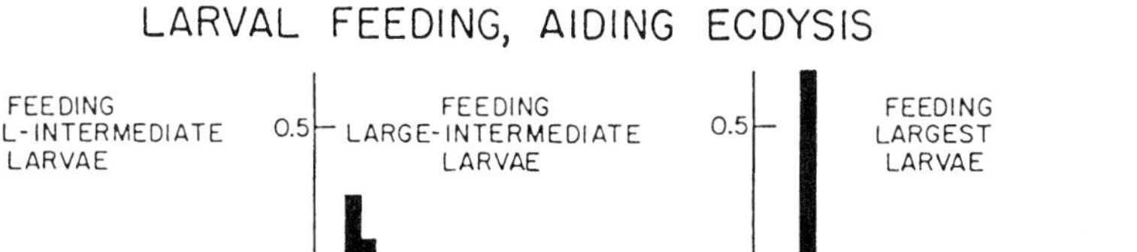


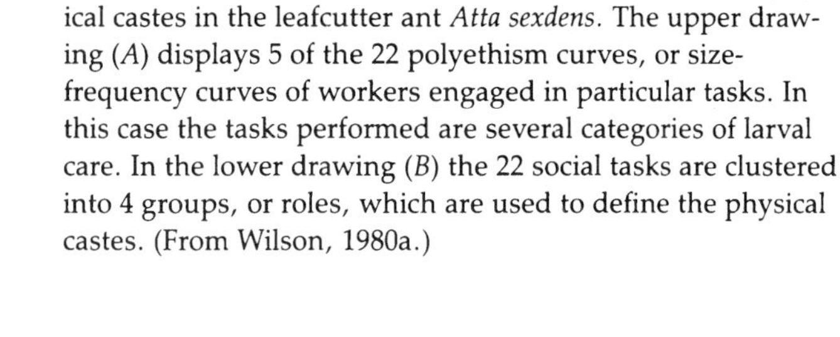

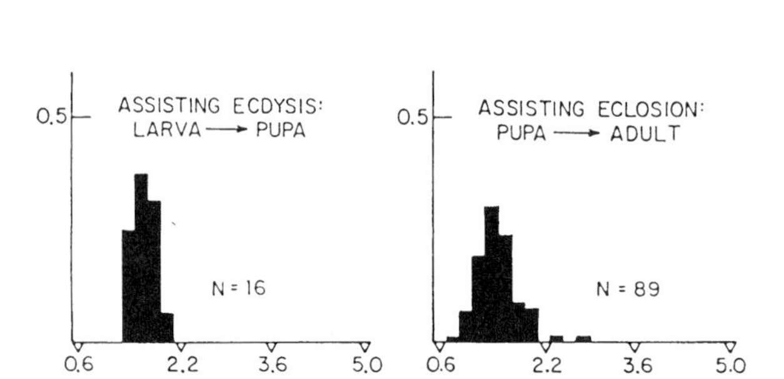


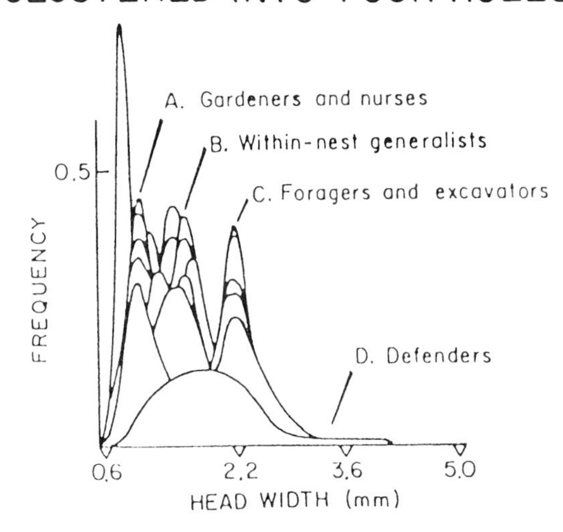


**FIGURE 8–30** The definition of roles and the attending physical castes in the leafcutter ant *Atta sexdens*. The upper drawing (*A*) displays 5 of the 22 polyethism curves, or size-frequency curves of workers engaged in particular tasks. In this case the tasks performed are several categories of larval care. In the lower drawing (*B*) the 22 social tasks are clustered into 4 groups, or roles, which are used to define the physical castes. (From Wilson, 1980a.)

largest size classes of *A. sexdens*, which are specialized for colony defense. The consistency of correlation suggests that proportionate size alone may be used in the future as a first clue concerning the roles of previously unstudied or experimentally less tractable organs. For example, it can be predicted that secretions of the labial and hypopharyngeal glands in *A. sexdens* function in brood care or fungus gardening, whereas those of the Dufour's gland are used either in defense or in alarm communication. The technique can also be used to infer the roles of sensory organs. Jaisson (1972c), for example, reports that the sensilla ampulacea and sensilla coeloconica of the antennae vary according to worker size in *A. laevigata*. This important aspect of sensory physiology awaits closer study.

Although caste and division of labor in *Atta* are very complex in comparison with other ant systems, they are still derived from surprisingly elementary processes of increased size variation, allometry, and alloethism. This important point is made graphic by Figure 8–32, in which the properties of the true *A. sexdens* system are compared with those of an imaginary, more complex arrangement that could be generated by only slightly more elaborate rules. Ant species in general and *A. sexdens* in particular have thus been remarkably restrained in the elaboration of their castes. They have relied on a single rule of deformation to create physical castes, which translates into a single allometric curve for any pair of specified dimensions such as head width versus pronotal width or, as illustrated in Figure 8–31, exocrine gland size versus pronotal width. The *Atta* could have created a more complex and precise array of castes by programming an early divergence of developing larval lines along with differing allometry among those lines at pupation, producing the effect illustrated in the upper diagram of Figure 8–32. But neither the *Atta* nor to our knowledge any of the many kinds of polymorphic ants have ever done so.

Behavior follows similarly elementary rules. The polyethism curves presented in Figure 8–30 are all of relatively simple form. They are unimodal and show only limited amounts of skewing. These properties suggest the existence of underlying alloethism functions of fundamental simplicity: for each behavior, a monotonic rise of responsiveness to a peak along the size range is followed by a monotonic decline.

Thus, even though *Atta sexdens* possesses one of the most complicated caste systems found in the ants, it has not evolved anywhere close to the conceivable (and we believe evolutionarily attainable) limit. There are far more tasks than castes: by the first crude estimate seven castes cover a total of between 20 and 30 tasks. Furthermore, one can discern another important phenomenon in *A. sexdens* that constrains the elaboration of castes: polyethism has evolved further than polymorphism. As we showed in the case of the fire ant *Solenopsis geminata*, the alloethic curves rise and descend more steeply than the size-frequency distributions and they are generally steeper than the allometric curves drawn from any selected pair of physical measurements. Consequently ensembles specialized on particular tasks are more differentiated by behavior than by age or anatomy. In the course of evolution *Atta* created its division of labor primarily by greatly expanding the size variation of the workers while adding a moderate amount of allometry and a relatively much greater amount of alloethism.

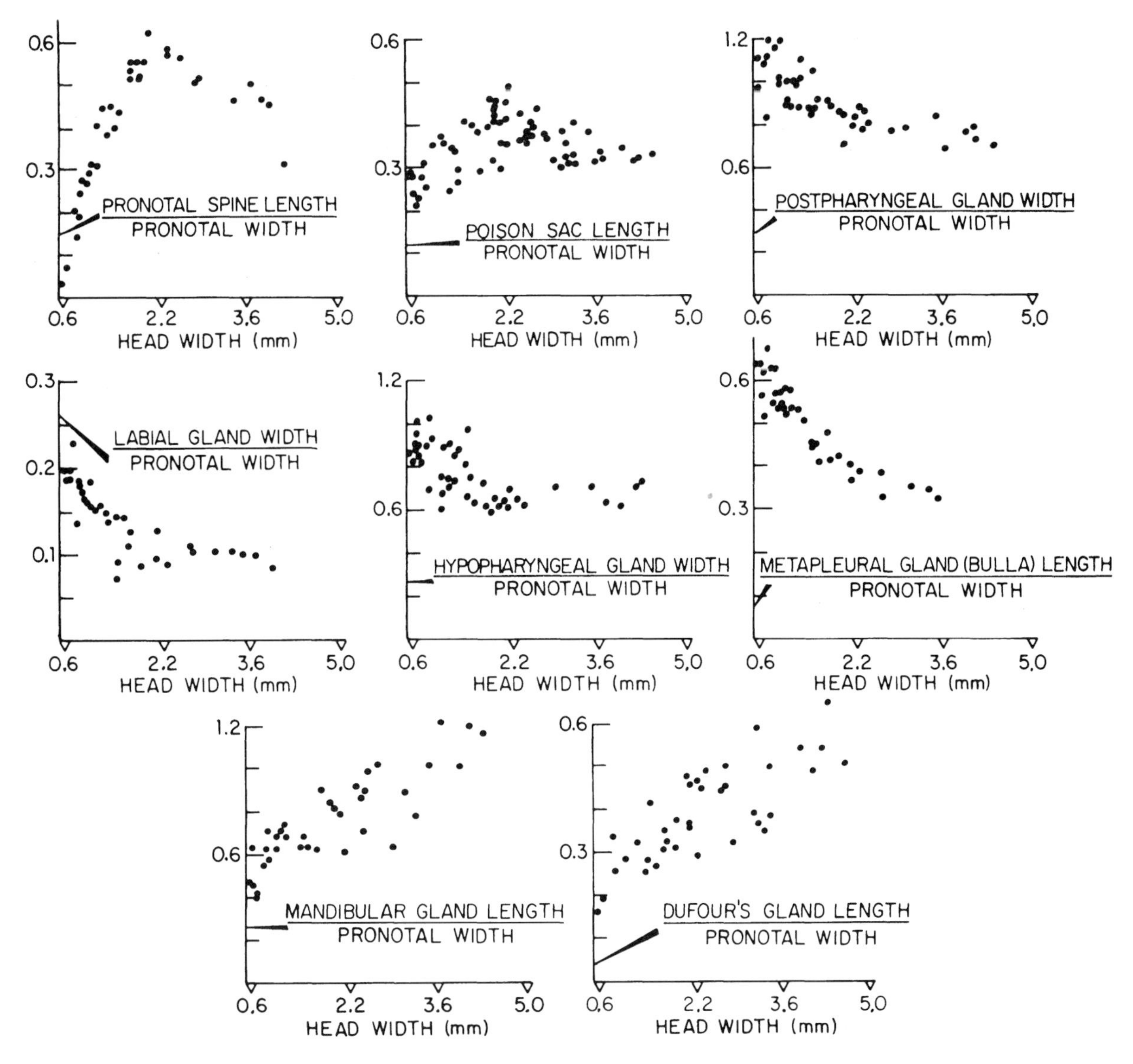


**FIGURE 8-31** Division of labor in the leafcutter ant *Atta sexdens* is served by complex variation in exocrine glands and body spination, where each organ is maximally developed in the worker size classes that specialize on the task served by the organ. (From Wilson, 1980a.)

**FIGURE 8-32** A conceivable *Atta* caste system compared with the actual system in *A. sexdens*. As illustrated in the upper, imaginary figure, ant species may evolve multiple allometric curves that generate physical castes precisely fitted to various tasks. Even in the unusually complex *A. sexdens* system, however, all physical castes are in fact generated by the single allometric curve illustrated in the lower figure. (From Wilson, 1980a.)

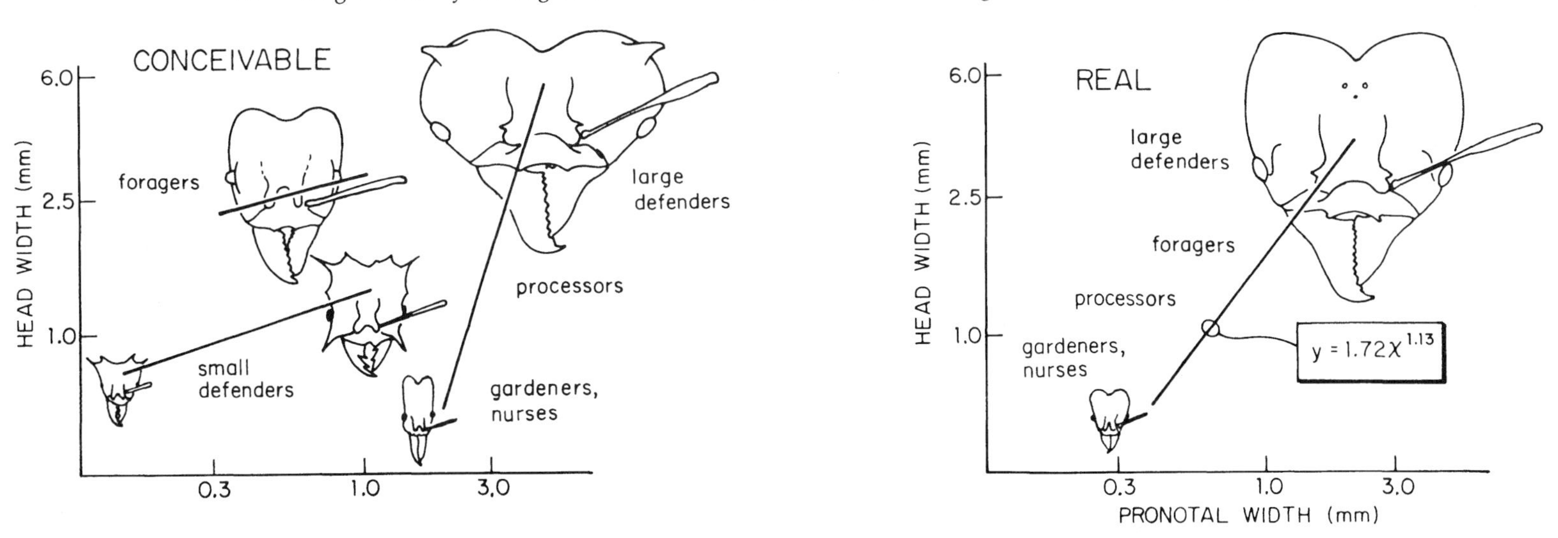

## SPECIALIZATION BY MAJORS: DEFENSE

It is remarkable that the anatomically more deviant major workers of various ant species have evolved solely as specialists for one or the other of three primary tasks: defense, milling (chewing and comminution) of seeds, and food storage (by the extreme distension of the crop to create repletes). In cases of advanced polymorphism, especially complete dimorphism where intermediates have dropped out and the major and minor workers have begun to evolve allometric patterns of their own, the greatest modifications in the major caste are found in the head and mandibles. The majors often look like members of an entirely different species.

The dominant role of majors, first noted by Westwood (1838) and since repeatedly confirmed, is defense of nest and food. The majors, or "soldiers" as they are properly designated, are adapted to one of four basic fighting techniques: shearing, piercing, blocking, and bouncing.

1. *Shearing.* The mandibles are large but otherwise typical, the head is massive and cordate, and the soldiers are adept at cutting the integument and clipping off the appendages of enemy arthropods. Examples are found in *Pheidole* (Figure 8–3), *Pheidologeton* (Figure 8–26), *Atta* (Figure 8–27), *Oligomyrmex, Aneuretus, Zatapinoma, Camponotus* (see Plate 7), and a few other genera of diverse relationships. Wheeler (1927) pointed out that the peculiar head shape of this form of soldier is simply due to an enlargement of the adductor muscles, which imparts greater cutting or crushing power to the mandibles.

2. *Piercing.* The mandibles are pointed and sickle-shaped or hook-shaped. Soldiers of the most highly evolved army ants (*Dorylus* and *Eciton*) often line up along the flanks of the moving columns, heads facing outward and mandibles gaping. An identical guard posture is assumed around nest entrances by the saber-jawed soldiers of the formicine *Cataglyphis bombycina* in the Sahara Desert (Délye, 1957). These formidable-looking individuals rush at any moving object when the nest is disturbed. They seldom perform other tasks. Contrary to an earlier suggestion by Felix Santschi, Délye found that the *Cataglyphis* soldiers are not well suited to carting particles during the excavation of nests.

3. *Blocking.* The behavior of soldiers that block nest entrances is the most specialized of all. The members of *Colobopsis* can be taken as typical of this category. *C. truncatus* of Europe has been studied in detail by Forel (1874) and later authors. Wheeler (1910a) described the American *C. etiolatus* under natural conditions, and Wilson (1974c) made observations on colonies of the American *C. fraxinicola* housed in glass tubes. The soldiers seldom leave the nests, which consist of narrow cavities in the dead wood of standing trees and shrubs. One or more of them stand guard at the nest entrances, where they serve literally as living doors. When minor workers approach them from either end and give the right signal (presumably a combination of simple touch and colony-recognition scent, although the matter has never been experimentally investigated) the soldiers pull back into the nest to allow their nestmates free passage. The nest entrances are cut into wood or plastered with carton so as just to accommodate the head of a soldier. It is a remarkable fact that it is the minors, not the soldiers, who engineer this fit. In those instances when the entrance is larger, several soldiers join to plug it with the combined mass of their heads. Both arrangements are shown in the illustration of *C. truncatus* by Szabó-Patay (1928; Figure 8–33).

The specialized role of the *Colobopsis* soldiers was demonstrated still more convincingly in experiments conducted by Wilson (1974c). Undecane, stored in the Dufour's gland of the abdomen, is a general formicine alarm pheromone. When small quantities of this substance were allowed to evaporate near the nest entrance, all members of the *C. fraxinicola* colony were thrown into the typical excited running movements of the *fraxinicola* alarm response. But some of the soldiers moved to the nest entrances, filling even holes that had been unattended prior to the alarm reaction. On the other hand the soldiers were not especially adept in combat. When twigs containing *fraxinicola* colonies were first broken open, both minor workers and soldiers rushed out. Many attacked any accessible alien objects, such as the observer's hand or a bit of cloth offered to them, biting it and spraying it with formic acid. The same response is obtained in the laboratory by permitting fire ant workers (*Solenopsis invicta*) to invade the nests. Individuals of both castes are about equally aggressive and effective in repelling these invaders. The total population of minor workers, by virtue of its greater size, was more effective than that of the soldiers. In the Australian species *Camponotus ephippium* the soldiers appear to function solely in nest defense. They have never been observed foraging. They are fed by minors (Plate 7) but rarely if ever regurgitate food to nestmates. The nest entrance in the soil, which is slightly larger than the head of a soldier, is always blocked by the head of one member of this caste. When a stick is touched to the nest entrance the guarding soldier attacks it with her powerful mandibles. When the stick is removed the soldier usually does not release her grip and is pulled out. Immediately another soldier takes her place at the entrance (B. Hölldobler, unpublished; Figure 8–34). *Camponotus ephippium* often nests within the territories of the dominant species *Iridomyrmex purpureus*. One of the major functions of the soldiers of *C. ephippium* appears to be to protect the colony against raids by these aggressive "meat ants."

A different form of blocking behavior is exhibited by the North American cephalotine *Zacryptocerus texanus*. The entrance hole to the arboreal nest is somewhat larger than the head of the soldier and is blocked by the combined mass of the head and expanded prothorax, the latter structure being heavily armored and pitted like the head. The head is held obliquely, rather like the animated blade of a miniature bulldozer. This posture, combined with the thrust and pull of the short, powerful legs, allows the soldier to press the intruders right out of the nest—a kind of nonviolent defense. When a minor worker returns to the entrance hole, the following sequence unfolds.

> The returning minor may or may not touch the antennae of the guard, although it usually does so. Thereafter the guard crouches down. This brings the anterior rim of the head below the level of the floor of the passage or, if the guard stands completely inside the passage, the front of the head is raised as the guard crouches. The dorsum of the guard's thorax is now no longer close to the roof of the passage and the minor can, if it is sufficiently active, wriggle between the dorsum of the thorax and the roof of the passage . . . If the passageways are made large enough to accommodate two majors simul-

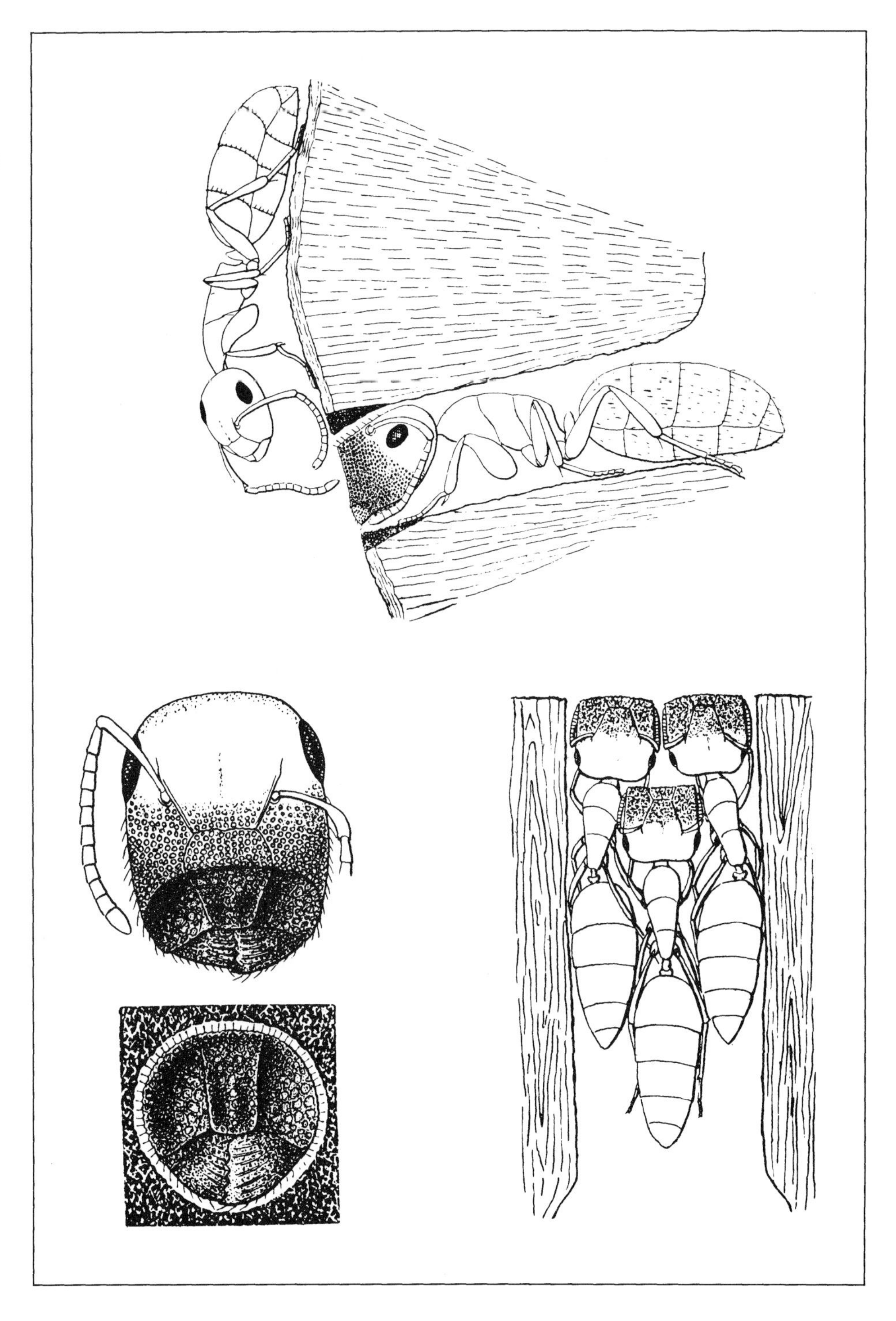

**FIGURE 8–33** Nest guarding by soldiers of the European ant *Colobopsis truncatus:* (*above*) a minor worker approaches a soldier that is blocking a nest entrance with her plug-shaped head; (*lower left*) two views of the head of a soldier, the lower (frontal) view framed by a small circular entrance to the nest, demonstrating how the head serves as a living "gate" to the nest; (*lower right*) a group of workers assume the position they use in plugging a large entrance to their nest. (From Szabó-Patay, 1928.)

taneously, they ordinarily assume a position where they are back to back. Under such circumstances the two opposed cephalic discs form a V-shaped area. The bottom of this V is open but the space behind it is closed by the closely approximated thoracic dorsi of the two guards. When minors are admitted to the nest both majors crouch and the entering worker struggles through the narrow space between the thoraces of the guards. It seems scarcely necessary to state that there is no part of this behavior which at all resembles that of the *Colobopsis* major, which must back away from the nest entrance to admit the returning minor. (Creighton and Gregg, 1954)

Closely similar blocking maneuvers occur in the Neotropical *Zacryptocerus varians* (Wilson, 1976a; see Figure 8–35). But this is only part of the total strategy of defense. In fact, both minor and major workers are very active. The minor workers respond at a lower threshold. Thus they form the "early warning system" of the colony and can dispose of less formidable intruders without help. Soldiers respond less readily, but once activated they are individually more effective.

The specialization of the *Zacryptocerus* soldiers is augmented by a unique encrusting layer of filamentous material that covers the surface of their heads. This odd material, which resembles a mass of fungal mycelia, gives a grimy appearance to the part exposed to the outside during entrance guarding. Wheeler (1942) suggested that it serves as camouflage for the nest entrance. He commented that the cephalic disc of older soldiers and the queen were often coated with "dirt and extraneous particles so that it closely resembled the bark of the plant." Clean, shiny head surfaces might be more easily spotted by visual predators, such as birds and lizards, which could break

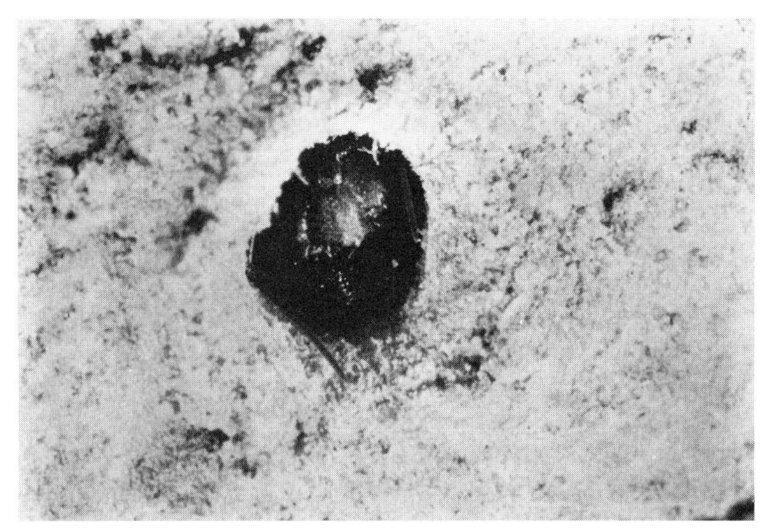

FIGURE 8–34 The use of the head of major workers to block nest entrances in the Australian *Camponotus ephippium*. (*Above*) A major blocks the entrance with her head. Often the tips of the antennae reach out. When nestmates approach the guard backs off and lets them enter. (*Below*) A major enters the nest. The entrance is only slightly larger than the head of the ant.

open twigs for the rich reward of ant brood. The resemblance of head to bark would effectively conceal the location of the nest entrance. Research by D. E. Wheeler and Hölldobler (1985) has disclosed that the filaments are secreted through a large number of secretory pores only 1–3 micrometers in diameter and scattered over the disc surface.

A case of abdominal phragmosis has been reported in the ponerine *Proceratium melinum* by Poldi (1963). The second gastral segment is strongly convex, so much so that the succeeding gastral segments point obliquely forward. The posterior surface of the second segment is therefore the posteriormost part of the body, and it is used to block the nest galleries against intruders. The trait is possessed by all of the workers, rather than by a specialized major caste.

4. *Bouncing*. A wholly different mode of defense occurs in the Australian dacetine ant *Orectognathus versicolor* (Carlin, 1981). As depicted in Figure 8–36, the mandibles of the largest individuals (the soldiers) are broadened and flattened from bottom to top. The apical teeth, which the minor and media workers use to capture prey, are shortened and blunted. When an alien ant enters the *Orectognathus* nest, the soldiers spread their peculiar mandibles about 120 degrees apart. At the moment the intruder's body comes within range of a soldier's gape, the soldier snaps her mandibles shut, pinching the intruder with such force as to shoot her away—like a slippery seed pressed hard between the fingers. Only the largest workers, with their broad, nearly toothless mandibles and powerful adductor muscles, are equipped to perform the bouncing maneuver. They are very effective in this specialized role, successfully propelling enemy ants up to 10 centimeters through the air. The workers of *Odontomachus* species use a similar bouncing technique to repel invaders from their nest entrances.

FIGURE 8–35 The three female castes of the ant *Zacryptocerus varians* are included in this colony fragment, which occupies a typical nest cavity in a dead stem of the red mangrove *Rhizophora mangle*. The nest queen rests on the floor of the cavity to the left, while on the right a large major worker blocks the nest entrance with her saucer-shaped head (an eye and an antenna can be seen just beneath the left margin of the expanded, circular frontal lobes). To the rear of the queen another major worker receives regurgitated liquid from a minor worker. (From Wilson, 1976a; drawing by T. Hölldobler-Forsyth.)

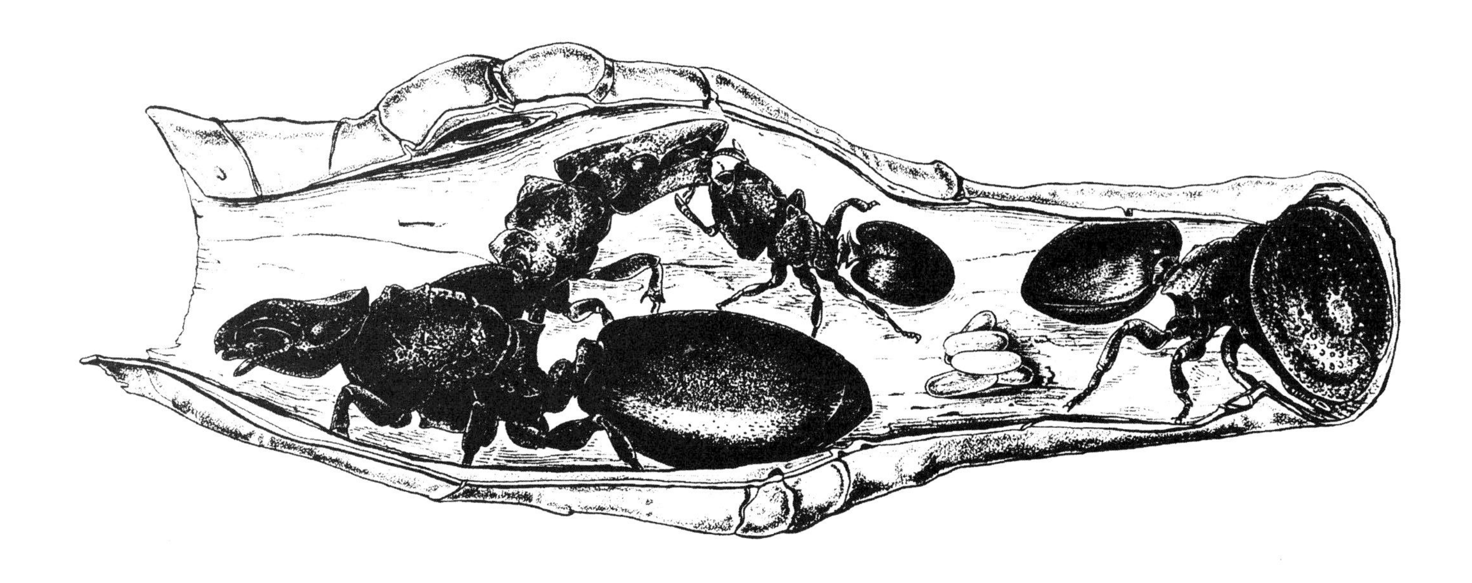

**FIGURE 8–36** A colony of the dacetine ant *Orectognathus versicolor*, a species with an unusual allometry and mode of nest defense. The queen is at the far left. On her immediate right are two soldiers (major workers), whose heads are larger than the queen's and whose mandibles are broadened and flattened in an unusual manner; another soldier is at the extreme upper right. The soldiers use the mandibles to "bounce" intruders away from the nest. The smaller individuals are minor workers. (From Carlin, 1981.)

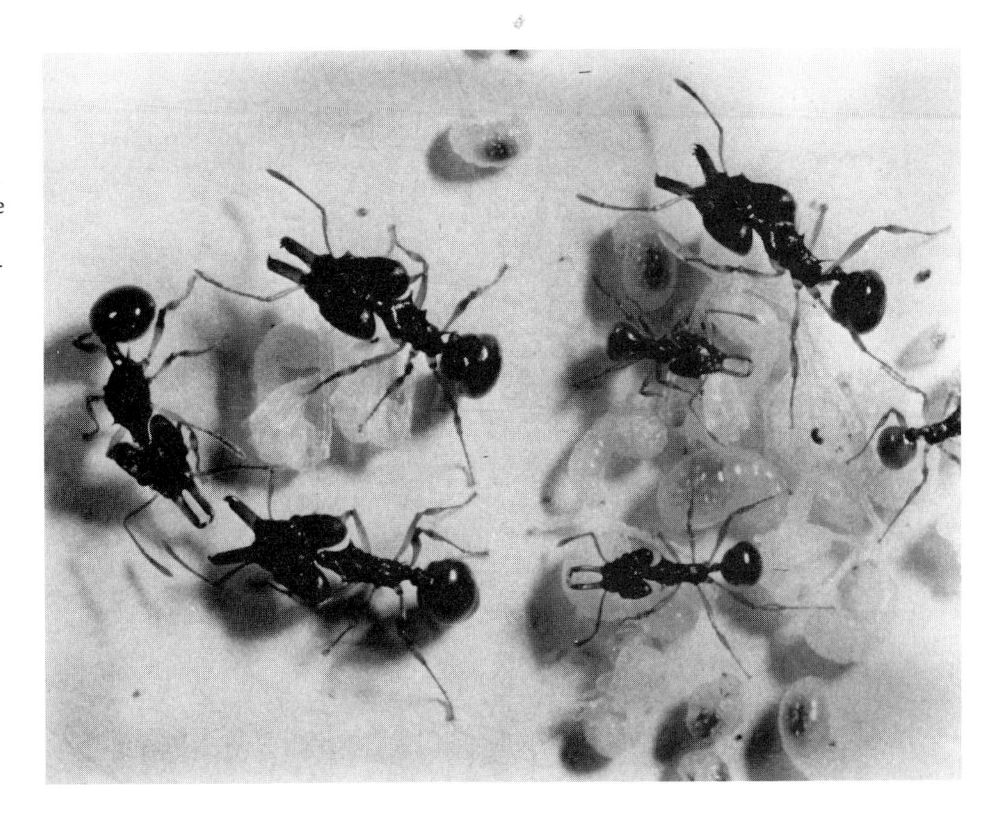

## SPECIALIZATION BY MAJORS: MILLING AND FOOD STORAGE

The next principal specialization, the addition of a major caste for milling, has arisen in evolution only when seeds constitute a substantial (but less than exclusive) supplement to the diet. Milling occurs in *Solenopsis geminata* (Figure 8–24), *Pogonomyrmex badius*, the species of *Acanthomyrmex* (Figure 8–37) and granivorous members of the large, cosmopolitan genus *Pheidole* (Wilson, 1978, 1984b; Moffett, 1985b). The majors employ their massive, blunt-edged mandibles to strip away the coat and break apart the endosperm of the seed. They are demonstrably more efficient at this task than the much smaller and weaker minor workers. Paradoxically, where seeds are the principal or sole dietary items, the species tend to be monomorphic or weakly polymorphic. In other words all of the workers are adapted to milling, yet remain generalists in other tasks and hence are more "normal" in appearance. Examples of the second category are believed to include species of *Monomorium* ("subgenus *Holcomyrmex*") and the North American members of the genus *Messor*, that is, the species formerly placed in the separate genus *Veromessor*.

The storage of liquid food in the crop has been carried to great heights by the repletes of certain ant species, individuals whose abdomens are so distended they have difficulty moving and are forced to remain permanently in the nests as "living honey casks" (Plate 8). The extreme examples are ground-dwelling species that live in arid habitats: species of *Myrmecocystus*, a genus confined to the western United States and Mexico; *Camponotus inflatus*, *Melophorus bagoti*, and *M. cowlei* of the deserts of Australia; some species of *Leptomyrmex* in Australia, New Guinea, and New Caledonia; and *Plagiolepis trimeni* of Natal (McCook, 1882; Wheeler, 1910a; Creighton, 1950). *Leptomyrmex* is a dolichoderine, and the remainder belong to the Formicinae. Australian aborigines dig up and eat repletes of *Camponotus* and *Melophorus* as a kind of candy. Intermediate stages of repletion are seen in such diverse genera as *Erebomyrma*, *Pheidologeton*, *Prenolepis*, *Proformica*, and *Oligomyrmex*.

Repletes are usually drawn from the ranks of the largest workers, who apparently begin playing their servile role as callows, while their abdomens are still soft and elastic (McCook, 1882; Wheeler, 1908a, 1910a; Rissing, 1984). No fewer than 1,500 such individuals were recovered from the nest of a colony of *Myrmecocystus melliger* by Creighton and Crandall (1954). The complete population of a mature *Myrmecocystus* colony typically comprises a single queen and about 15,000 workers (Hölldobler, 1984d). Crandall, with the aid of professional gravediggers, followed the nest galleries through 5 meters of Arizona desert soil until he recovered the nest queen from a small chamber at the very bottom. Thus the earlier conjecture by Wheeler (1908a) that the *Myrmecocystus* nests are shallow and the population of replete workers is small has been thoroughly refuted. Rissing (1984) found that the repletes are drawn from the largest workers in colonies of *M. mexicanus*. This appears to be generally true for *Myrmecocystus*, because we have confirmed it in numerous laboratory colonies of several other *Myrmecocystus* species, including *depilis*, *mimicus*, *navajo*, and *placodops*. The repletes of *Myrmecocystus* often vary in color from clear yellowish-brown to dark amber. Conway (1977) analyzed the crop contents of the repletes of *M. mexicanus* and found that the dark fluid "contained more dissolved solids and that they were mainly glucose and fructose. In the more dilute clear sample, sucrose made up the bulk of the solids and there

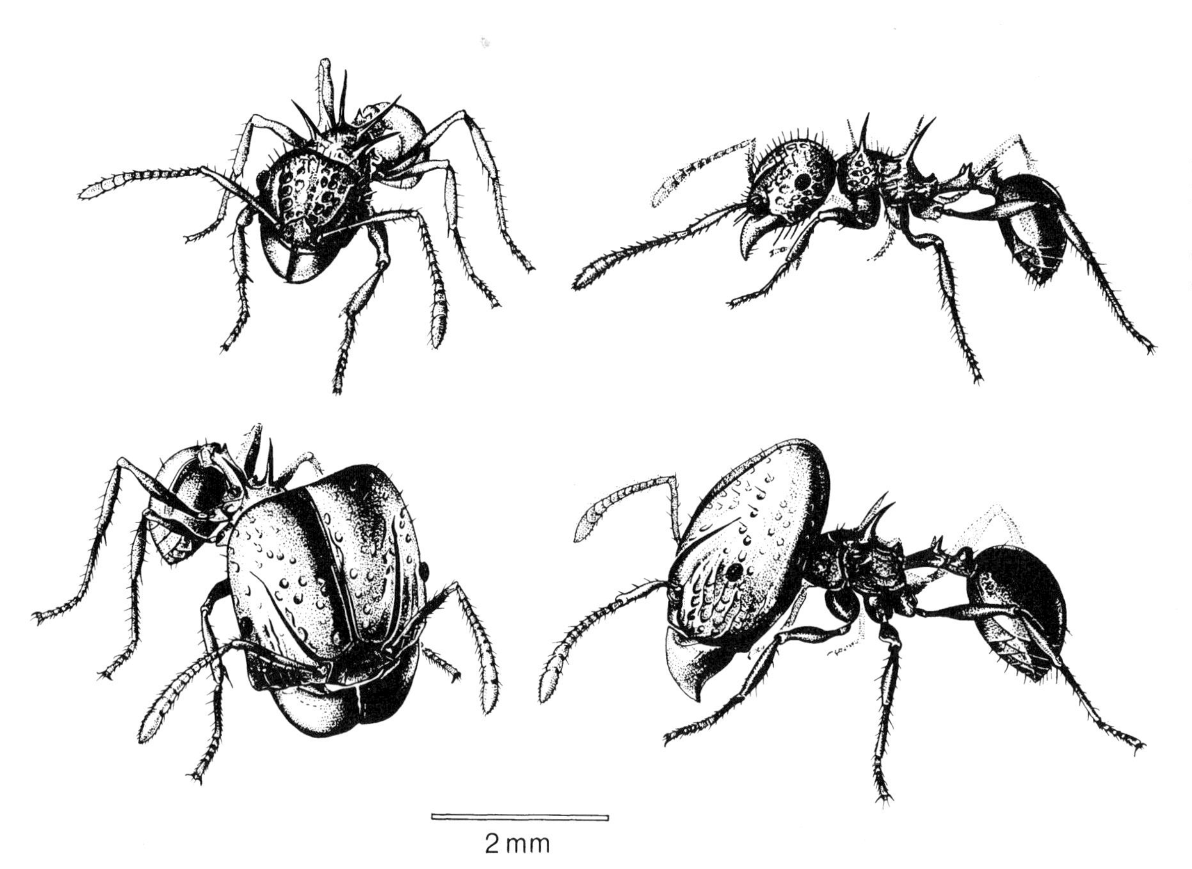


**FIGURE 8–37** The minor and major workers of a Celebes species of *Acanthomyrmex* are shown here in scale. The majors, who are virtually "walking heads," are specialized for the milling of seeds, but they also participate in colony defense. (From Oster and Wilson, 1978; drawing by T. Hölldobler-Forsyth.)

were only traces of glucose and fructose. Clear repletes may function primarily as water-storage vesicles, an adaptation well-suited to their semi-arid habitats."

Stumper (1961) used radioactive tracers in honey to study the behavior of the repletes of the European ant *Proformica nasuta.* His results went far to resolve the long-standing controversy about the adaptive significance of repletism (Creighton, 1950). At moderate temperatures food passes chiefly from foraging workers to the repletes, but at 30–31° C, when the metabolism of the colony sharply increases, the direction of flow is reversed. Stumper inferred that in nature the communal supply in the crops of the *Proformica* repletes is built up in relatively cool, moist weather and tapped in hot, dry weather. These results perhaps explain why the species that produce the extreme replete forms are also for the most part desert dwellers. A parallel phenomenon, "adipogastry," involves the exceptional development of the abdominal fat bodies of certain nocturnal, deserticolous species of *Camponotus* (Emery, 1898). The same phenomenon has been described in *Prenolepis imparis* of North America, which forages during cool weather and estivates during the hottest months (Tschinkel, 1987c).

Additional experimental evidence of the adaptiveness of the replete condition was obtained by Wilson (1974c) in the case of *Colobopsis fraxinicola.* The majors of this species serve two of the four specialized roles of the major caste of ants. Specifically, they defend the nest by using their heads to block the entrances and they store substantial quantities of liquid in their disproportionately large gasters. When Wilson first starved a colony of *C. fraxinicola* and then fed it to satiety with sugar water, the minors registered an average weight gain of 50.6 percent whereas the majors gained 63.4 percent. The disparity in storage capacity can be seen even more clearly by examining the colony as a whole. Although the majors made up a little less than 16 percent of the worker population and contained 28.4 percent of the wet weight in the starved condition, they stored 38.2 percent of the liquid at repletion.

## CASTE OPTIMIZATION

The simple fact of the occurrence of evolution is easy to document at the level of both single genes and complex traits controlled by genes at multiple loci. It is also relatively easy to demonstrate the adaptive nature of the evolving traits in most cases by careful analysis of the function of the phenotypic traits and their effect on survival and reproduction. It is an entirely different matter, however, to judge such cases of evolution with reference to optimization. In practical terms, we are required to ask how well a species has performed in the course of its evolution, and whether there exists an attainable adaptive peak toward which the species is still moving. Such questions, despite their abstract flavor, are not merely philosophical exercises but truly scientific in nature, that is, solvable by means that can be objectively repeated and verified. The answers to them, if well formulated, can help to predict such important phenomena as the rate and direction of evolution, the impact of a newly imported species on a host environment, and the outcome of competitive interactions between pairs of species.

Under most circumstances, and in the biology of most species, the measure of optimization is obviously going to be technically forbidding. In biology, as in engineering, optimization implies an optimum, a goal with reference to which the system may or may not have been ideally designed. Only when the nature of the goal is precisely identified may we presume to use such familiar terms as "suboptimum," "transitional," and "maladaptive" with any degree of confidence. The complete analysis of optimization, or strategic analysis as it might equally well be called, requires the following four components: (1) a state space, designating the conceivable parameters such as size and behavioral response and their relevance to the goal; (2) a set of conceivable strategies, such as foraging procedures, caste structure, division of labor, and pattern of colony development; (3) the goal itself, consisting in one or more optimization criteria, or fitness functions, including net energetic yield during foraging; (4) a set of constraints, defined by those states and strategies beyond the reach of the species and hence outside the scope of analysis (we know, for example, not to expect ants that weigh 10 kilograms or have metal jaws) (Oster and Wilson, 1978).

In most studies of behavioral ecology and sociobiology, optimization models are very difficult to design in a way that satisfies the formal and rigorous demands of the four criteria. But this is not an unusual circumstance for biological disciplines in general. It is equally impracticable to test Mendelian genetics with redwoods or to advance the cellular biology of learning with butterflies. The study of animal behavior seeks paradigmatic species that are easily managed in the field and laboratory, as well as phenomena that are relatively unambiguous with reference to the question asked. The biologist should always keep in mind August Krogh's rule of biological research: for every problem there exists an organism ideally suited to its solution. Moreover, an often unappreciated bonus awaits the successful completion of an optimization study. By appropriate experimental design it is possible to learn which fitness criteria actually hold in nature, in other words, which selection pressures have been most active in shaping the trait under consideration. Indeed, the data may indicate that *no* feasible criterion is approached closely, so that the trait can be identified with some confidence as maladaptive or at least substantially suboptimal. So much for the baseless charge sometimes heard that natural selection theory is "panglossian," circular in logic, and not susceptible to rigorous testing.

Ant castes are exceptionally well suited for optimization studies in sociobiology. The reason is that individual worker ants are full organisms with ordinary, whole patterns of social behavior, yet they are also clearly specialized for particular well-defined tasks. This restriction is extreme in certain highly specialized castes and in particular in the major workers of the dimorphic species, who may perform only one or several tasks during their entire lifetime and who possess bodies and behavioral repertories clearly modified to that end. Thus well-defined anatomical structures and behavioral acts can be more readily assayed with reference to the four elements of optimization models; we really are able to specify a restricted and relatively easily defined state space, a set of strategies, testable fitness functions, and measurable constraints.

Consider the case of the prime foraging caste of the leafcutter ants of the genus *Atta*. These specialists, who possess head widths of 1.8–2.4 millimeters (and a modal head width between 2.0 and 2.2 millimeters), represent less than 10 percent of the work force and do little except forage and cut vegetation. To learn whether the *Atta* colony as a whole, and the leafcutting size class in particular, was optimized with reference to foraging, Wilson (1980b) conceived three alternative a priori criteria of evolutionary optimization consistent with his understanding of *Atta* biology. These are the reduction of predation by means of evasion and defense during foraging, the minimization of foraging time through skill and running velocity during foraging, and energetic efficiency, which must be evaluated on the basis of both the energetic construction costs of new workers and the energetic cost of maintenance of the existing worker force.

To measure the performance of various size groups within the *Atta sexdens* worker caste in isolation, Wilson devised a "pseudomutant" technique. At the start of each experiment, groups of foraging workers were thinned out until only individuals of one size class were left outside the nest. Measurements were then made of the rate of attraction, initiative in cutting, and performance of each size group at head-width intervals of 0.4 millimeter. Other measurements were made of body weight, oxygen consumption, and running velocity for each of the size classes.

The pseudomutant technique has the great advantage of permitting the same colony to be used over and over again, with the experimenter's modifying it in a precise manner as though it were a mutant in a certain trait, but with little or no alteration of the remainder of its traits. There is no genetic noise or pleiotropy to confuse the analysis. The procedure used in the *Atta* experiments is roughly comparable to an imaginary study of the efficiency of various conceivable forms of the human hand in the employment of a tool. In the morning we painlessly pull off a finger and measure performance of the four-fingered hand, then restore the missing finger at the end of the day; the next morning we cut off the terminal digits (again painlessly) and measure the performance of a stubby-fingered hand, restoring all the fingers after the experiment; and so on through a wide range of variations and combinations of hand form. Finally, we are able to decide whether the natural hand form is near the optimum.

The data from the pseudomutant scanning revealed that the size-frequency distribution of the leafcutter caste in *Atta sexdens* conforms closely to the optimum predicted by the energetic efficiency criterion for harder varieties of vegetation, such as thick, coriaceous leaves. The distribution is optimum with reference to both construction and maintenance costs (see Figure 8–38). It does not conform to other criteria conceived prior to the start of the experiments.

Wilson next constructed a model in which the attraction of the ants to vegetation and their initiative in cutting were allowed to "evolve" genetically to uniform maximum levels. The theoretical maximum efficiency levels obtained by this means were found to reside in the head-width 2.6–2.8-millimeter size class, or 8 percent from the actual maximally efficient class. In the activity of leaf cutting, *Atta sexdens* can therefore be said not only to be at an adaptive optimum but also, within a relatively narrow margin of error, to have been optimized in the course of evolution. In other words there appears to be no nearby adaptive peak that is both higher and attainable by gradual microevolution.

Not all such studies reveal optimal responses or even adaptive mechanisms of the sort intuitively expected. For example, what would be the physiological response of an *Atta* colony, as opposed to the purely behavioral reaction described earlier, if a large part of its foraging specialists were suddenly removed by some catastrophe outside the nest? One prudent response on the part of a colony would be to manufacture a higher proportion of workers in the 1.8–

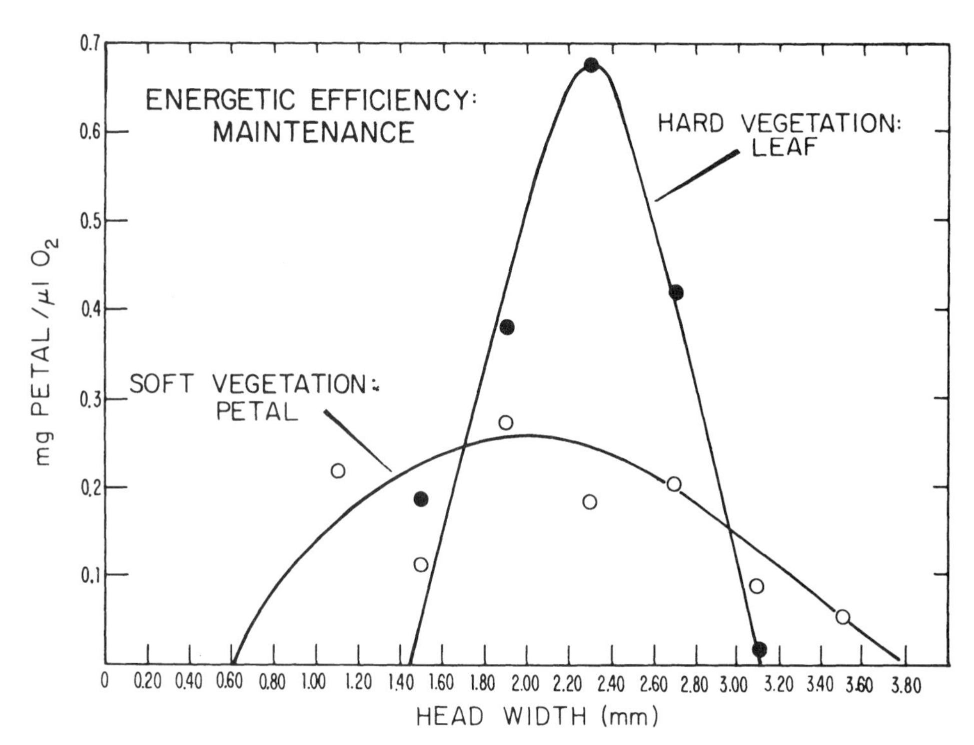


**FIGURE 8–38** The principal foraging caste of leafcutter ants (*Atta sexdens*), comprising workers in the size range 1.8–2.4 millimeters, is also the most efficient in net energetic yield during foraging. The cost criterion used in the measurements depicted here is oxygen consumption, reflecting the energetic cost to maintain workers of various sizes within the colony. A similar result was obtained when the criterion employed was construction cost, indexed by the dry weight of workers of different sizes. (From Wilson, 1980b.)

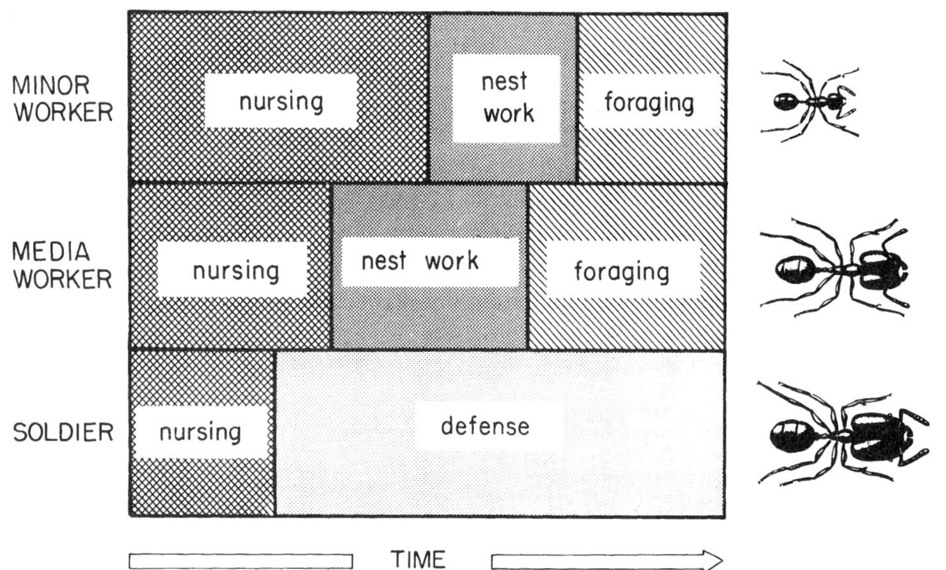


**FIGURE 8–39** This diagram illustrates the principal work periods traversed in the life spans of three worker subcastes of a generalized polymorphic ant species. The work periods are those in which the indicated task is the one most frequently performed; other tasks may be performed, but less often. The forms of the castes and the sequences of work periods within each caste are based on real species, but the precise durations of the periods are imaginary. In this case each of the eight periods, the total (arbitrary) of periods encountered in all three castes together, is treated as a separate "caste." The optimal mix can be evolved by varying both the relative numbers in each subcaste and the relative time spent in each work period (the ants represented in this figure belong to the myrmicine genus *Pheidole* and are shown only as an intuitive aid). (After Wilson, 1968.)

2.2-millimeter size class during the next brood cycle to make up the deficit. When over 90 percent of this group in experimental colonies were removed, however, no differential increase in the production of 1.8–2.2-millimeter workers could be detected in comparison with sham-treated control colonies. As a consequence, this group remained underrepresented in the foraging arenas by about 50 percent at the end of the first brood cycle, a period of eight weeks (Wilson, 1983a).

Another striking example of a feedback loop that did not evolve is stress-induced change in the production of major workers in the ant *Pheidole dentata*. What happens when a colony is attacked repeatedly over a long period by fire ants, the enemy to which *P. dentata*'s alarm-defense system is specially tuned? An obvious adaptation would be to raise the production of majors (increase defense expenditures, to put it in more familiar human terms) until the stress is relieved. But this response could not be induced by Johnston and Wilson (1985) in laboratory colonies. They stressed *P. dentata* colonies heavily through four brood cycles by regularly forcing them to fight and destroy *Solenopsis invicta* workers. Contrary to their expectation, the proportion of majors did not change significantly from that in colonies stressed with another species of ant (*Tetramorium caespitum*) or from the proportion that prevailed prior to the experiments, when the colonies were free of any pressures through many brood cycles.

The optimization theory pursued in the case of caste ratios assumes selection at the colony level. In fact, colony selection in the advanced social insects does appear to be the one example of group selection that can be accepted unequivocally so long as we are careful to bear in mind that the group in this case is the colony and not the population of colonies. To be sure, it is the queen, the mother of all other workers and second generation reproductives in the colony, who transmits the gametes and is the ultimate focus of selection. In this special sense colony selection differs from group selection in the hypothetical Wynne-Edwards sense, where most or all of the mature individuals of the populations are involved in reproduction (Wynne-Edwards, 1962, 1986; Williams, 1966). Yet it remains true that the colony is selected as a whole, and its members contribute to colony fitness rather than to individual fitness. It therefore seems a sound procedure to accept colony selection as a mechanism and to press on in search of optimization theory based on the assumption that the mechanism operates generally. For, if selection is mostly at the colony level, workers can be altruistic with respect to the remainder of the colony, and their numbers and behavior can be regulated in evolution to achieve maximum colony fitness. What is required is a theory of group behavior, a way of abstracting our empirical knowledge of caste and colony ergonomics into a form that can be used to analyze optimality.

The term ergonomics was borrowed from human sociology (e.g., Murrell, 1965) to identify the quantitative study of the distribution of work, performance, and efficiency in insect societies (Wilson, 1968). Wilson attempted a first formulation by means of the techniques of linear programming and obtained some surprising but still largely theoretical results. The essential arguments and results are presented here in a simplified form.

First, consider the concept of cost in colony reproduction. As colonies grow, their caste ratios change. Very young colonies founded by single queens typically consist only of the queens and minor workers. As they approach maturity, these same colonies may add medias and soldiers. Finally, they produce males and new, virgin queens. Here we will consider ergonomics and cost in the mature colony only. A mature colony is defined as a colony large enough to produce new, virgin queens. Also, for convenience, the category "caste" will include both physical castes, such as minor workers and soldiers, and temporal castes, the various periods of labor specialization through which most ants pass in the course of their lives. What determines the efficiency of the mature colony is the number of workers in each temporal caste at any given moment. This conception is spelled out in the example given in Figure 8–39.

In the mature colony, depending on the species, the adult force may contain anywhere from a few tens of workers to several millions. The number is a species characteristic. It has been evolved as an adaptation to ultimate limiting factors in the environment. An ultimate limit may be imposed by the constraints of a peculiar kind of nest site to which the species is adapted, by the restricted productivity of some prey species on which the species specializes, or, conversely, by a prey species or competitor so physically formidable as to require a larger worker force as a minimum for survival. The mature colony, on reaching its predetermined size, can be expected to contain caste ratios that approximate the *optimal mix*. This mix is simply the ratio of castes that can achieve the maximum rate of production of virgin queens and males while the colony is at or near its maximum size.

Entrenched in the nest site and harassed by enemies and capricious changes in the physical environment, the colony must send foragers out to gather food while converting the secured food inside the nest into virgin queens and males as rapidly and as efficiently as possible. The rate of production of the sexual forms is an important, but not an exclusive, component of colony fitness. Suppose we are comparing two genotypes belonging to the same species. The relative fitness of the genotypes could be calculated if we had the following complete information: the survival rates of queens and males belonging to the two genotypes from the moment they leave the nest on the nuptial flights, their mating success, the survival rate of the fecundated queens, and the growth rates and survivorship of the colonies founded by the queens. Such complete data would, of course, be extremely difficult to obtain. For the sake of developing an initial theory of ergonomics, however, we can get away with restricting the comparisons to the mature colonies. To do this and still retain precision, we must take the difference in survivorship between the two genotypes outside the period of colony maturity and reduce it to a single weighting factor. But we can sacrifice precision without losing the potential for general qualitative results by taking the difference as zero. Now we are concerned only with the mature colony, and the production of sexual forms becomes (keep in mind the artificiality of our convention) the exact measure of colony fitness. The role of colony-level selection in shaping population characteristics within the colony can now be clearly visualized. If, for example, colonies belonging to one genotype contain on the average 1,000 sterile workers and produce 10 new, virgin queens in their entire life span, and colonies belonging to the second genotype contain, on the average, only 100 workers but produce 20 new, virgin queens in their life span, the second genotype has twice the fitness of the first, despite its smaller colony size. As a result selection would reduce colony size. The lower fitness of the first could be due to a lower survival rate of mature colonies, or to a smaller average production of several forms for each surviving mature colony, or to both. The important point is that the rate of production in this case is the measure of fitness, and evolution can be expected to shape mature colony size and organization to maximize this rate.

The production of sexual forms is determined in large part by the number of "mistakes" made by the mature colony as a whole in the course of its fortress-factory operations. A mistake is made when some potentially harmful contingency is not met—a predator successfully invades the nest interior, a breach in the nest wall is tolerated long enough to desiccate a brood chamber, a hungry larva is left unattended, and so forth. The cost of the mistakes for a given category of contingencies is the product of the number of times a mistake is made times the reduction in queen production per mistake. With this formal definition, it is possible to derive in a straightforward way a set of basic theorems on caste. In the special model, the average output of queens is viewed as the difference between the ideal number made possible by the productivity of the foraging area of the colony and the number lost by failure to meet some of the contingencies. (The model can be modified to incorporate other components of fitness without altering the results.) The evolutionary problem postulated to have been faced by social insects can be solved as follows: the colony produces the mixture of castes that maximizes the output of queens. In order to describe the solution in terms of simple linear programming, we must restate the solution in terms of the dual of the first statement: the colony evolves the mixture of castes that allows it to produce a given number of queens with a minimum quantity of workers. In other words the "objective" is to minimize the energy cost. Meeting the objective confers higher genetic fitness to the colony members, both collectively and individually.

This form of ergonomic theory leads to at least two results that appear at first to be counterintuitive but in fact are straightforward consequences of selection at the colony level. One, illustrated in Figure 8–40, involves the relation between the efficiency and the numerical representation of a given caste. If in the course of evolution one caste increases in efficiency while others do not, the proportionate total weight of the improving caste will decrease. That is, the expected result of colony-level selection is precisely the opposite of that of individual selection, which would be an increase in the more efficient form.

The first prediction can be tested in the following way. Increased efficiency at a given task implies increased specialization on that task, a relation that has been documented abundantly and virtually without exception throughout the ants. The theoretical prediction can therefore be translated to the simple statement that the less the members of a caste do, the fewer there are. *Pheidole* is an excellent group with which to test such relationships. With the exception of several parasitic species, the members of this myrmicine genus are consistently dimorphic, with distinctive, large-headed major workers that are specialized for defense, milling, food storage, or some combination of these three roles. Yet there is variation: the majors differ from species to species in size and degree of anatomical and behavioral specialization, as well as in numerical representation within the colony. *Pheidole* is also the largest of all ant genera, with hundreds of species available for sampling and comparison. Finally, most species of *Pheidole* have proved relatively easy to culture in the laboratory. Wilson (1984b) obtained the behavioral repertories and major/minor ratios of ten species, representing an equal number of phylogenetically distinct species groups from various localities around the world. The proportionate representation of the majors was found to decrease significantly as their behavioral repertory decreased among the species, in accordance with ergonomic theory. This effect is shown in Figure 8–41.

A second counterintuitive result of ergonomic theory is the predicted generalization that species with initially unspecialized castes will have on the average fewer castes and more variable caste ratios, and this effect will be enhanced in fluctuating environments. In other words generalized castes are prone to extinction. The more specialized the castes become in evolution, the more entrenched they become, in the sense that they are more likely to be represented in the optimal mix regardless of long-term fluctuations in the environment (Wilson, 1968, 1971). In classical evolutionary theory, which entails individual selection, it is the generalized genotypes and species, and not the specialized ones, that are more likely to survive in the face of long-term fluctuation in environment. This result will be hard to test in any definitive manner. Well-developed major castes have persisted with little change in *Camponotus*, *Oecophylla*, and *Pheidole* since at least Miocene times, in other words about 15 million years, but data are lacking on the longevity of less differentiated caste systems.

The ergonomic models lead to the result that in a perfectly constant environment and in the absence of developmental constraints of any kind, it is of advantage to colonies to evolve so that there is one caste specialized to respond to each kind of contingency. That is, one caste should come into being that perfects the appropriate response, even at the expense of losing proficiency in other tasks. This is very far from what prevails in nature, however. Behavioral repertories prepared for a variety of species in *Atta*, *Camponotus*, *Cephalotes*, *Leptothorax*, *Pheidole*, *Solenopsis*, and other genera (references in Table 8–3) indicate that as a rule workers of each colony perform between 20 and 50 distinct tasks, the precise number varying according to species. Yet most ant species have only a single worker physical subcaste plus three or four discernible temporal subcastes. Even species possessing the most complex polymorphism and division of labor, such as the members of *Atta* and *Pheidologeton*, appear to have no more than seven combined physical and temporal castes.

Oster and Wilson (1978) considered at some length the possible existence of constraints on the proliferation of physical and temporal castes. Through a combination of deduction and inference based on empirical evidence they suggested a restrictive role for the following seven properties:

1. *Holometabolism*. In ants, bees, and wasps adult size is fixed permanently at the end of larval growth, and most of the labor is performed later, following the attainment of the adult instar. As a consequence the size-frequency curve of the existing adult population cannot be altered to meet new contingencies imposed by the environment. There exists a time lag between the onset of the new conditions and the attainment of a more efficient size-frequency distribution that minimally equals the period between the moment of cessation of larval growth and the eclosion of new adults, in other words roughly the duration of the pupal stage. The coarse nature of this adjustment to the environment reduces the number of castes to be expected in the optimal mix below that predicted by the elementary linear programming models.

2. *Allometry*. In the ants, which are the only social hymenopterans that display a significant amount of worker polymorphism, the physical differentiation is based on allometry—the regular disproportionate increase of body parts relative to one another. Thus there is usually only a single rule of deformation, or (in the case of diphasic and triphasic allometry) two or three, which substantially diminishes the number of castes that can be generated over a given amount of total size variation.

3. *Fidelity costs during development*. In order to regulate not only allometry but also the characteristic size-frequency distributions that fix the proportions of physical castes, feedback mechanisms must be employed that correct repeatedly for the dispersion in growth of each age cohort. As precision is increased, the energetic costs of the regulatory mechanisms are likely to grow in a greater than linear relation. Above some size the costs must become prohibitive, and species that reach this limit are expected to compromise accordingly in the degree of complexity of their caste systems. Wheeler (1986a) has argued further that species relying on larval nutrition as the principal trigger for queen determination will find it more difficult to add worker physical subcastes to the prior and requisite differentiation of queens and workers. The physiological systems needed to maintain both levels at the same time would be difficult to operate with any precision. On the other hand systems of queen determination based on direct pheromonal or hormonal in-

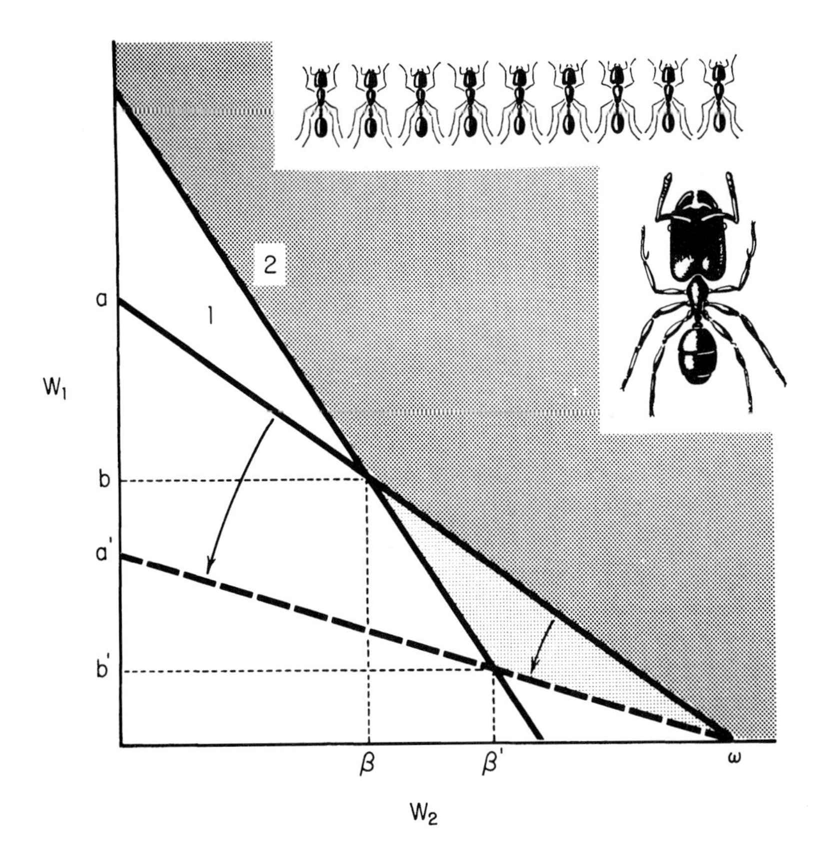


**FIGURE 8-40** A counterintuitive principle from ergonomic theory is derived in this diagram. If one caste increases in efficiency during the course of evolution and others do not, the proportionate total weight of the improving caste will decrease. This theoretical result of colony-level selection is the opposite of that expected from individual-level selection, which tends to increase improving phenotypes. (From Wilson, 1968.)

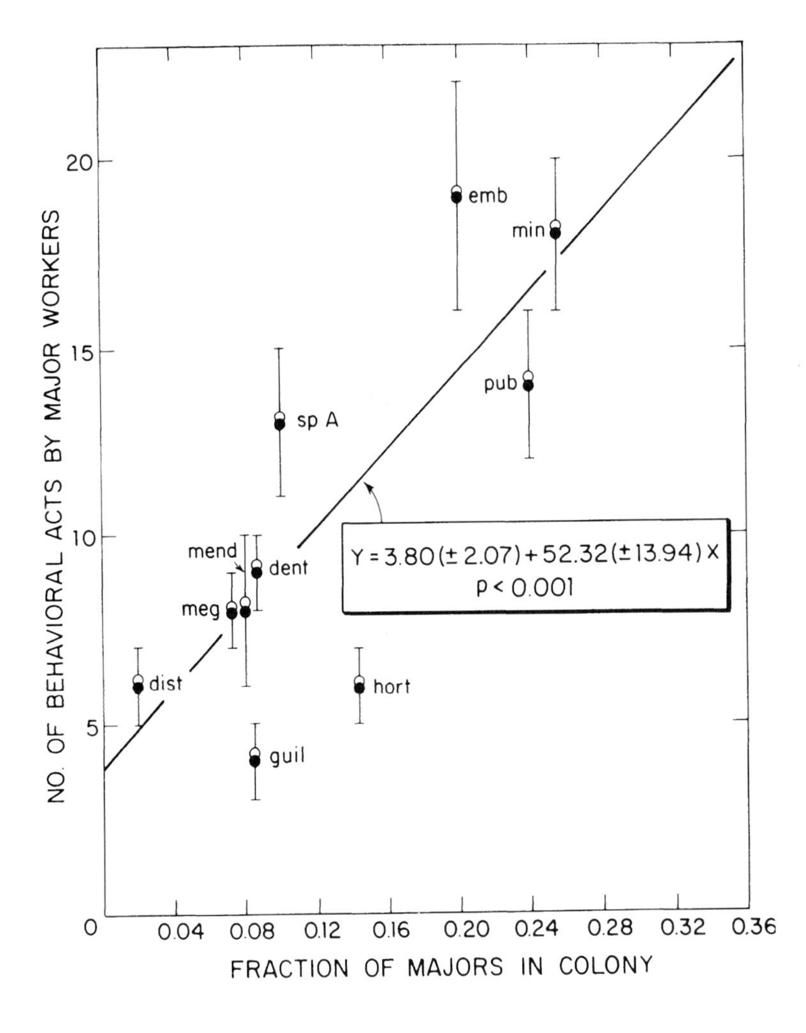


tervention will pose less of a barrier to the evolution of worker polymorphism. Pheromonal and hormonal intervention is in fact the rule in ants.

4. *Environmental variance.* Prey captured by the colony foragers and soil particles moved by the excavators have a substantial size variance. It should be of advantage for the worker force of colonies to have a built-in size variance to compensate for this environmental uncertainty. If true, the result will be a decrease in the number of discrete castes that can be maintained.

5. *Task overlap.* Tasks that are very different from one another may, nevertheless, require anatomical features that are closely similar. For example, prey items and the pupal stages of the ants' own nestmates often resemble each other in size and shape, so that the same worker caste could be employed to handle both categories of objects.

6. *Behavioral plasticity.* Workers specialized for one task are usually required to perform other rare but essential tasks. When ant nests are broken open, to take one familiar example, most workers stop what they are doing and either attack the intruder or carry larvae and other immature forms into deeper chambers. This minimal flexibility limits the degree of specialization of individual castes and perhaps also the number that can be differentiated within a single colony population.

7. *Ergonomic costs.* Soldiers, millers, and other exceptionally large castes are energetically expensive to manufacture and maintain. Although one of these forms might be "ideal" in its capacity to function, it would never evolve because the energy it would expend is greater than the energy its presence would gain for the colony. Put another way, it is necessary for majors to perform some special service of exceptional importance to make their creation profitable to the colony. It is not surprising, then, to find that majors of strongly polymorphic species are characteristically very specialized in behavior.

In addition, the existence of large numbers of reserve workers, a common feature of ant colonies, is likely to inhibit the proliferation of castes (Porter and Jorgensen, 1981; Porter, personal communication). It is probably better, for example, to utilize 10 or 20 unemployed generalist workers to accomplish a task that occurs only occasionally than it is to create 1 or 2 workers specialized for the task.

There are two ways to alleviate the burden of majors—by discounting the cost of either their manufacture or their maintenance. Manufacture costs can be reduced in two ways. The first would be to lower the per-gram metabolism of major-destined larvae over the same developmental period as minor-destined larvae. The existing data do not yet indicate the existence of this phenomenon. Nielsen (1986) measured the per-gram oxygen consumption of larvae in two myrmicine and three formicine species, including polymorphic aspects of *Camponotus,* and found that they do not vary with changes in total body weight. Still, the evidence is incomplete with reference to the ants as a whole. More important, it cannot be applied to er-

**FIGURE 8-41** Among species of the ant genus *Pheidole,* the numerical representation of the major workers decreases when the number of social acts they perform decreases and hence their degree of specialization increases. The result is consistent with selection at the colony level but not at the individual level. (From Wilson, 1984b.)

gonomic calculations until comparable within-species measurements are made of major-destined and minor-destined larvae in the most extreme polymorphic ants such as *Atta* and *Pheidologeton*.

The second way to reduce manufacture costs would be to reduce turnover by extending longevity of the majors, in other words by cutting the rate of manufacture itself. This difference does occur in many polymorphic species, at least in colonies of *Camponotus, Pheidole,* and *Solenopsis,* whose majors are longer-lived than the minors.

Maintenance costs can be reduced by lowering the per-gram metabolic rate of larger adult members of the colony—as opposed to growing larvae. Substantial evidence exists that this is indeed the case. In his general treatment of ant metabolic rates, Nielsen (1986) provided evidence that at first glance might seem to weigh against metabolic discounting. He showed that per-gram oxygen consumption does not vary significantly across the myrmicine and formicine species thus far sampled, even those from different climates. Moreover, Jensen and Holm-Jensen (1980) found that the energy cost of running is $7.2 \pm 1.8$ times the energy cost at rest in both the monomorphic *Formica* and polymorphic *Camponotus*. Nielsen et al. (1982) showed that loads carried by *C. herculeanus* increased energy expenditure at a linear rate. However, these evidences of linearity do not eliminate the possibility of differences among castes within the same species, as Nielsen (1986) has been careful to point out. Such differences do exist and are consistent with ergonomic theory. In *Atta sexdens*, which Wilson (1980b) studied with respect to caste differences, the larger workers consume less oxygen on a per-gram basis, with the following relation holding at 30°C:

$$\log [O_2 \text{ consumption } (\mu l/mg/hr)] = 0.63(\pm 0.06) - 0.53(\pm 0.13)[\text{head width}]$$

Beraldo and Mendes (1982) independently found differences among the size groups of *Atta laevigata* and *A. rubropilosa,* but the correlation coefficients were low and the results difficult to evaluate. Lighton et al. (1987) confirmed that oxygen consumption per gram weight falls off with total worker size in the worker caste of *A. colombica,* and Bartholomew et al. (1988) established the same result for the army ant *Eciton hamatum*. Porter and Tschinkel (1985) found a decline in per-gram oxygen consumption with size among workers of the polymorphic fire ant *Solenopsis invicta,* and Porter (1986) obtained a similar result with *Pogonomyrmex badius,* the only polymorphic member of its genus. Altogether, the evidence of metabolic discounting of the major caste is persuasive, but there is room for a great deal more study. Furthermore, variation within and between the reproductive castes is an intriguing possibility. MacKay (1982a) found that males of *Pogonomyrmex* double or triple their respiratory rates prior to the nuptial flights, but no such increases occur in the winged queens. He also discovered a size dimorphism in the males, wherein each size has a different respiratory rate.

## THE ALLOMETRIC SPACE

The ergonomic analysis of insect societies is in its earliest stages. It will be difficult to tease apart and to estimate the relative importance of the various evolutionary constraints on caste proliferation. Oster and Wilson have devised a number of models that attempt to define the agents involved and in some cases crudely to measure them. An

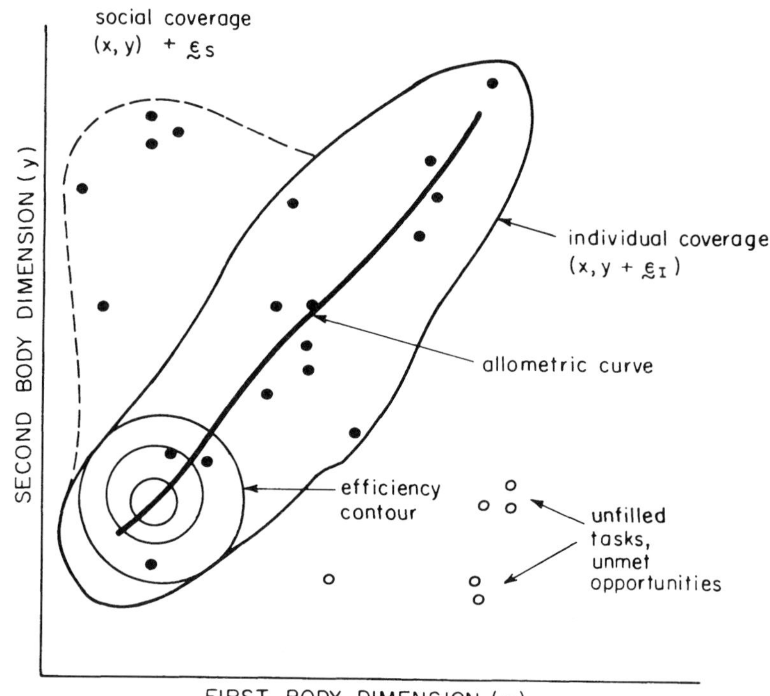


**FIGURE 8–42** The concept of the allometric space is one of the models that can be used to test the hypothesis of evolutionary optimization. The worker caste of each ant species has a characteristic allometric curve, drawn here for two anatomical dimensions only (for example, thorax width and head width). The allometric curve exists within the broader allometric space of all possible physical forms. In the local environment where the colonies live there exists a set of contingencies, consisting of opportunities such as food items and nest sites and perils such as predators and cave-ins. It is postulated that for each task by which the colony meets the contingency there is a point in the allometric space that corresponds to the physical caste ideally suited to perform the task. There is also a zone in the allometric space, with radius ε, within which the task is performed with at least adequate proficiency. The species can cover a greater number of task points by altering the allometric curve, by increasing individual coverage, or by increasing social coverage through cooperative effort. (From Oster and Wilson, 1978.)

example is the conceptualization of the *allometric space,* illustrated in Figure 8–42. Each ant species has a characteristic allometric curve displayed by the worker force of its colonies; this curve is a minute subset of all of the possible anatomical forms that constitute the allometric space. The colonies live in a particular microhabitat, such as rotting logs, hollow twigs, desertic soil, or any of a number of other narrowly defined parts of local environments. The habitat challenges the colony with a set of contingencies: certain kinds of food items available, certain types of nest sites open, certain enemies present, a given probability of the nest's being flooded, and so on. To meet the contingencies successfully requires the performance of a task (defined as a set of appropriate behaviors) and for each task there is a particular physical caste (a point in the allometric space) ideally suited for its performance.

It seems reasonable to postulate that because of the ergonomic costs and the accidents of task overlap the species evolves in such a way that some of the colony members can perform more than one task. There are three ways in which this multiple coverage can be improved: by lengthening and twisting the allometric curve, thus changing the array of physical castes; by expanding the capacity of individual workers to perform various tasks, that is, by increasing behavioral flexibility; and by adding communication and cooperative behavior so that the task can be performed by groups.

Let us suppose that the task points are distributed randomly over an allometric space (Oster and Wilson, 1978). For convenience and without much loss of generality the task points can be regarded as Poisson distributed over two-dimensional space. Because task points cluster in groups, and the allometric curve only has to come within a certain distance (ε) of each task point, the number of castes required to do all necessary tasks is not expected to increase as a linear function of the number of tasks as our attention shifts from simply organized to complexly organized species. In fact, it should increase no faster than the logarithm of the number of tasks:

$$N_C \precsim \frac{1}{\varepsilon^2} \ln N_T$$

where $N_C$ is the number of castes required to provide adequate coverage of a given number ($N_T$) of tasks and ε is the maximum distance on the allometric plane between the task point and the nearest actual caste found in the colony.

By estimating the number of tasks and castes in a group of closely related species, as Wilson (1976c) has done for *Pheidole dentata*, we should be able to characterize the task-caste model and more sophisticated derivatives of it. One of the most promising groups is the tribe Attini, a tightly knit phylogenetic group of New World tropical ants within which occur species that are variously monomorphic, weakly polymorphic, and extremely polymorphic. As we demonstrated earlier, this variation is linked to the form of substrate on which the workers grow fungi and hence to the number of tasks required to obtain and process the substrate.

## ELITES

Workers of some ant species, especially those belonging to the phylogenetically advanced subfamilies Dolichoderinae and Formicinae, vary greatly in their readiness to work. In the original study of this phenomenon, Chen (1937a,b) found that "leader" workers of the carpenter ant *Camponotus japonicus aterrimus* begin to dig earth sooner when placed in earth-filled jars, move more earth per individual, and show less variation in effort than others. They furthermore have a stimulating effect on their more sluggish nestmates. Similar "elitism" has been observed in *Tapinoma erraticum* during brood transport (Meudec, 1973); in *Formica fusca, F. sanguinea*, and *C. sericeus* during adult transport (Möglich and Hölldobler, 1974, 1975); in *Diacamma rugosum* during tandem running (Fukumoto and Abe, 1983); and in *C. vagus* during aggression and food exchange (Bonavita-Cougourdan and Morel, 1988). When Möglich and Hölldobler removed the small fraction of *C. sericeus* workers transporting their nestmates during a change in nest site, the emigration virtually ceased.

Without further studies across several behavioral categories, it is impossible to say whether the most active ants in one category are generally elites or mere specialists in the category under observation. A striking degree of specialization has been recorded in some formicine ants, especially in wood ants of the *Formica rufa* species complex. For example, Horstmann (1973) observed that foraging workers of *F. polyctena* fall into one of three groups: arboreal foragers that collect honeydew primarily and search for prey secondarily, ground foragers that hunt prey almost exclusively, and collectors of nest materials. Individual ants remain in one category or another for periods of at least two weeks. A remarkable specialization also occurs in *Myrmecocystus mimicus:* certain individuals specialize in attending guard posts at the territorial boundaries, whereas others forage exclusively by robbing prey from *Pogonomyrmex* foragers (Hölldobler, 1981a, 1986a). Within-caste specialization may even be idiosyncratic in degree. Workers of *Lasius fuliginosus* patrol certain portions of the foraging ground over periods of weeks or longer, during which time they become familiar with specific portions of the terrain, the phenomenon of *Ortstreue* (Dobrzańska, 1966). Similar learning and particularization of behavior have been noted in the leafcutter ants of the genus *Atta*, harvester ants of the genus *Pogonomyrmex*, and wood ants of the *F. rufa* group, and the desert ant *Cataglyphis bicolor* (Jander, 1957; Rosengren, 1971; Lewis et al., 1974a; Hölldobler, 1976a; Herbers, 1977; Rissing, 1981a; Wehner et al., 1983).

Not all apparent elitism can be explained away as specialization, idiosyncracy, and *Ortstreue.* In the detailed protocols published by Otto (1958) on *Formica polyctena*, it is apparent that a few workers are much more active than others in the pursuit of a multiplicity of tasks. Furthermore, some are more catholic in their choice of roles over most or all of the adult life span. Hence substantial variation can exist among individual colony members within the broad role sectors of particular age-size classes.

Some of the mystery of elitism and leadership has been solved by several careful statistical studies of the individual contributions of worker ants. Abraham (1980) measured the transporting activity of *Myrmica rubra* workers during repeated emigrations, then plotted the individual contributions in descending rank order. The resulting curve was smooth. In other words it reflected not just elites and sluggards but a graded series of intermediates. It can be described approximately as follows:

$$\text{number of workers} = 30\, e^{-0.04x}$$

where $x$ is the measure of transporting activity. Qualitatively similar results were obtained across a wide range of roles by Fukumoto and Abe (1983) in the ponerine *Diacamma rugosum* and by Lenoir and Ataya (1983) in the formicine *Lasius niger.* For *Lasius* at least the distribution of total activity proved to be log-normal. The latter result in particular suggests that labor in ants, like wealth in human societies, is derived from early developmental causes that might at first be normally distributed but are then skewed toward the upper classes by an enhancement effect in which prior possession raises the likelihood of obtaining more with the passage of time. The rich get richer, so to speak, successful athletes train still harder, and labor-prone ants are more likely to assume still more duties.

Of course such a quasi-mathematical explanation still tells us nothing about the physiological causation underlying the initial differences separating workers. It merely helps to elucidate why there are outstanding performers even if the initial differences are relatively modest.

Very little is known concerning this ultimate basis of elitism but it appears that both innate and learned components can be important.

Bernstein and Bernstein (1969) reported that the ability to run a maze by *Formica rufa* workers is positively correlated with the size of the head, the diameters of the compound eyes and median ocellus, and the dimensions of the calyxes of the corpora pedunculata, the latter structures being the part of the brain most conclusively implicated in the control of complex behavior. Whether this apparent variation is genetic or merely the outcome of random developmental variation is not known. The important point is that in either case it is fixed at the beginning of the adult instar. Changes induced by experience can also be major and long-lasting: when Jaisson (1975) prevented adult *F. polyctena* workers from contacting cocoons during the first fifteen days following eclosion, they proved incapable of tending cocoons in later life.

## PATROLLERS AND RESERVES

Students of social insects generally agree with Elton (1927) that "all cold-blooded animals spend an inordinately large proportion of their time doing nothing at all, or at any rate, nothing in particular." At any given moment only a very small fraction of workers in the typical ant colony engage in foraging. Inside the nest a majority of the workers are standing still, grooming themselves, or just walking around, apparently aimlessly. Individual workers of *Leptothorax acervorum*, a species typical in this respect, are inactive 78 percent of the time (Allies, 1984). Some ant species may even exhibit sleep-like behavior. Workers of the nocturnal Australian carpenter ant *Camponotus perthiana* were observed lying on their sides in the pupal position during the day (B. Hölldobler, unpublished; see Figure 8–43). The ants were "wakened" by gentle probing with a pair of forceps. For a few seconds the awakened ant moved sluggishly before commencing the typical swift movements of a worker ant. Thus ant societies do not differ much from human societies in the apportionment of time devoted to labor and idleness.

As Herbers (1981b) has pointed out in an elegant analysis, such apparent laziness is not a character flaw, as it may be considered in human societies, but is likely to have survival value. Under a wide range of conceivable environmental conditions and foraging, a perfectly adequate strategy is a thermostat feeding process whereby the animal initiates foraging and simply continues until it reaches a certain level of acquired energy. For many animal species, including ants, this rule of thumb will result in the idling of a large percentage of the population at any particular time.

Still, idleness is a far more complicated and subtle matter in ant colonies and other insect societies, because work is set in a social context. What each worker does affects not just her own physiological state but that of many others around her, in a way that strongly affects her genetic fitness. Being largely sterile, workers are programmed to respond to the needs of the colony as a whole, even at the risk to their personal lives. This broader involvement helps to explain the curiously undirected behavior of patrolling ants, whose aim can be most generously described as an inspection tour. Gordon (1984b, 1986, 1987) has characterized patrolling in *Pogonomyrmex* harvesting ants as a distinctive set of activities that include (1) walking around the nest exterior while inspecting the terrain with the antennae and making more frequent stops and changes of direction than during foraging; (2) pawing at the ground with the forelegs and

**FIGURE 8–43** A major worker of *Camponotus perthiana* in the sleeping position.

inspecting the resulting small depressions with the antennae; (3) standing with mandibles open at the site of a disturbance, for example around a new object brought in by nestmates. Patrollers are more likely than other labor groups to encounter threats to the nest, and they are immediately available for defense. Furthermore, patrolling precedes foraging in the daily round of undisturbed colonies, and it appears that patrollers regularly recruit foragers to new food sources (Gordon, 1983a).

Masses of ants that rest or idly patrol also function as reserves. They constitute a backup for the colony in such emergencies as flooding of the nest or invasion by a predator that require the simultaneous engagement of many individuals. Moreover, superabundance of individuals performing any given task, as for example nest construction or larval feeding, forces some into other, less crowded functions, and the division of labor tends to approach proportions of laborers that match the needs of the colony as a whole. Many of these displaced workers are searching for new roles or at least are prepared to assume them. They impart a flexibility to the labor schedules that allows the colony to adapt quickly to capricious changes in the environment. They also permit fine-tuning to the inevitable wide shifts that occur in the requirements of the brood. The worker force of large fire ant colonies (*Solenopsis invicta*) is probably typical in this regard. In addition to two main castes, nurses and foragers, there is at any given time a large reserve group of generalists, heterogeneous in age and size, who variously nurse, forage, store liquid food, and relay food from the nurses to the foragers (Mirenda and Vinson, 1981). The idle workers at any given time may also serve as a "metabolic caste," converting raw food into glandular secretions for the eventual feeding of the queen and brood (A. Buschinger, personal communication).

Finally, idleness can be exacerbated by "selfishness" among workers (Schmid-Hempel, 1990). If colonies are polygynous (with multiple queens) or polyandrous (queens inseminated by more than one male), so that workers are not always full sibs, it will be advantageous for workers to "hold back" and to protect their lives and health more carefully. The result will be that as the colony approaches the reproductive phase, during which new reproductive forms and colonies are created, a higher proportion of workers with selfish genes will exist than at the outset of colony growth. These workers will then be in a position to lay eggs to contribute a higher proportion of males. Alternatively, they may be able to distinguish full sibs as opposed to half-sibs and unrelated individuals among the growing reproductive larvae and to give preferential treatment to these relatives. As a result, Schmid-Hempel has argued, laziness will tend to increase in species with higher degrees of polygyny, polyandry, or ability to recognize kin. At present there is no way to evaluate the importance of this selfish factor in evolution relative to the ergonomic factors previously discussed.

## CLIQUES AND TEAMS

The relation of the members of an ant society to one another can be characterized as one of impersonal intimacy. With the possible exception of limited dominance orders among reproductive females in some species, ants do not appear to recognize one another as individuals. Their classificatory ability is limited to the recognition of nestmates, different castes such as majors and minors, the various growth stages among immature nestmates, and possibly also kin groups within the colony.

A consequence of this coarse grade of discrimination is that members of colonies do not form what can be conveniently termed *cliques:* groups of workers whose members recognize one another as individuals within larger castes and regularly come together as individuals to accomplish some task. For example, a clique would exist if worker "alpha" regularly met with ants "delta" and "gamma" to transport larvae. Of course, ants assemble to capture prey, excavate soil, and fulfill other functions requiring mass action. Also, odor trails and other sophisticated techniques exist that allow the rapid recruitment of nestmates to the work sites. But within castes the participants are entirely interchangeable. There is no evidence that they come and go in personalized cliques.

Until recently there has also been no evidence for the existence of *teams,* which can be defined as members of different castes that come together for highly coordinated activity in the performance of a particular task. A team would not consist of particular ants, but rather of interchangeable members of particular castes. Thus it would be less specific in arrangement than a clique. The general lack of team organization is not necessarily the outcome of the limited brain power of social insects. It can be shown at a very general level that processes are less efficient when conducted by redundant teams than when conducted by redundant parts not organized into teams (Oster and Wilson, 1978). This disparity can be overcome or reversed, as in fact it is in human beings, only if the degree of coordination among the members of the teams or between the teams is sufficiently great to compensate for the shortcomings inherent in the system redundancy.

An exceptional case of team organization has been reported by

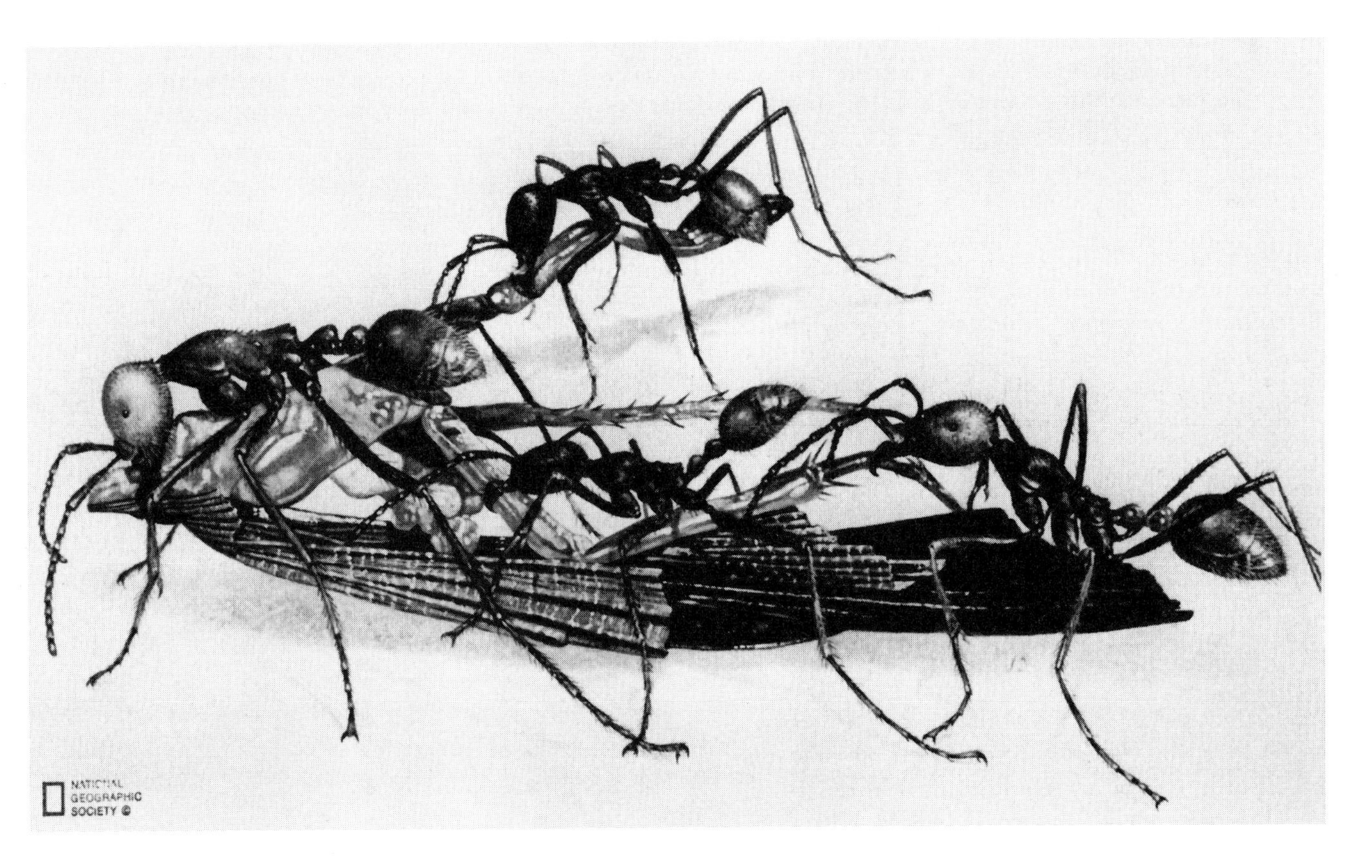


**FIGURE 8-44** True teams are formed by the army ant *Eciton burchelli* during the group transport of prey. A single submajor (*left*), a member of the caste specialized for this function, is assisted by smaller workers in carrying part of the carcass of a large cockroach. (From Hölldobler, 1984d; illustration by J. D. Dawson reprinted by permission of the National Geographic Society.)

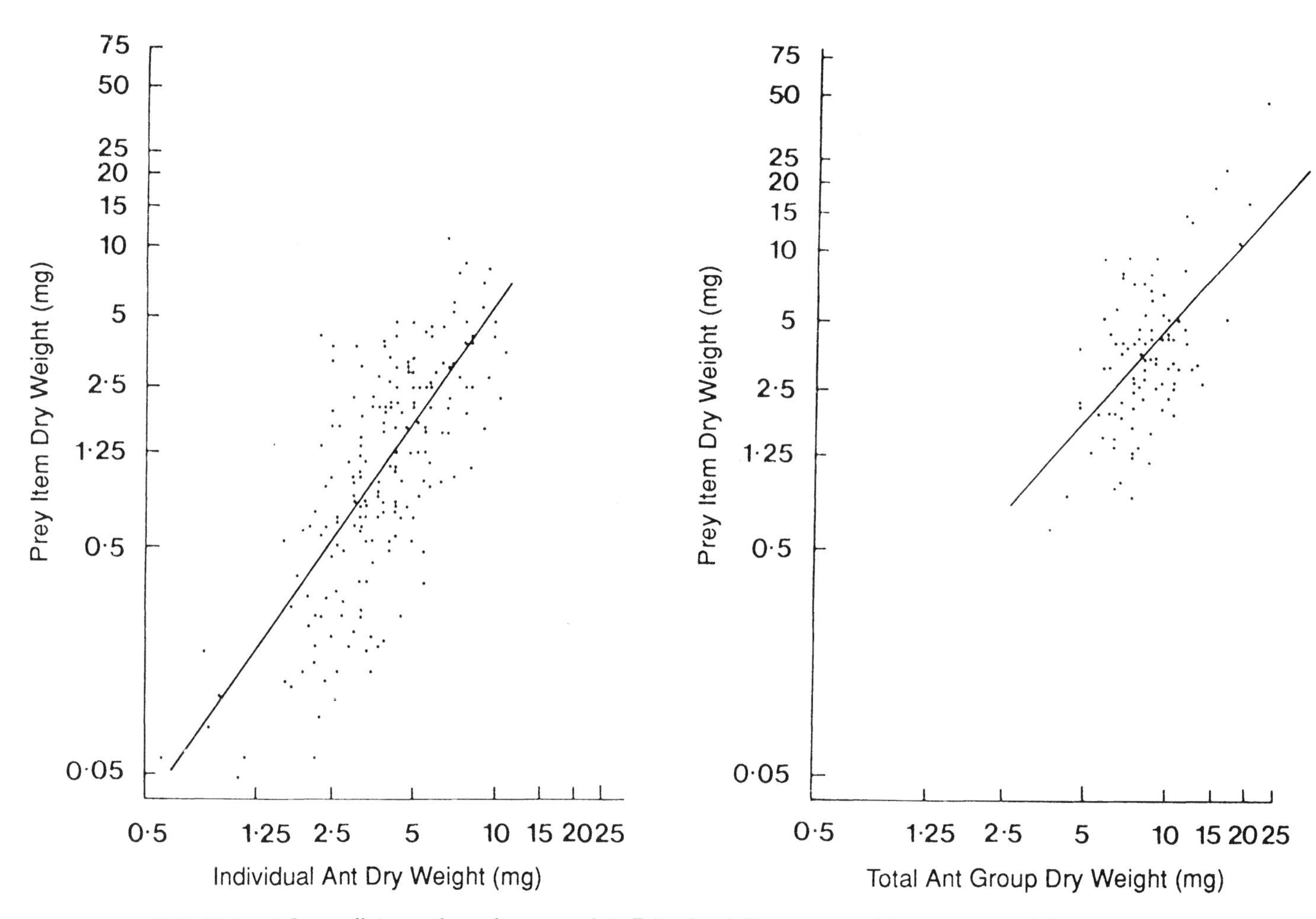


**FIGURE 8–45** Superefficiency through teamwork in *Eciton burchelli*, a swarm-raiding army ant of Central and South America. These diagrams show the relation between the weight of ants and the weight of prey items they are carrying. Ants working in teams (*right*) manage a higher ratio than ants of the same weight carrying items singly (*left*). (From Franks, 1986.)

Franks (1986) in the group retrieval of prey by *Eciton burchelli*, the swarm-raiding army ants of Central and South America. Large prey items are regularly carried by groups of foragers, among which is a single submajor, the specialized "porter caste" of *E. burchelli*. Submajors are the size group just below the very large majors; their mandibles are disproportionately long compared with those of smaller nestmates but not nearly so long as the bizarre, sickle-shaped mandibles of the majors (see Figure 8–44). The submajors do not form cliques with particular groups of smaller ants in performance of their task; they fall in with whatever individuals are nearby and of appropriate size. When Franks removed and then replaced large prey items in the raiding column, the *Eciton* followed a predictable sequence. First, workers formed a tight group over the item. They often included a large major that stood guard. The item was never immediately picked up by workers acting in concert. Rather, a transport gang was gradually formed after the prey item had already been put in motion by one individual. Because such large pieces would be moved by a large ant, the first member of the group was typically a submajor. After this prime mover started carrying or dragging the item, the smaller ants joined in until the group reached the standard retrieval speed in the raiding column, after which no more ants were added.

Franks showed that the structured teams of *Eciton burchelli* are "superefficient." They can carry items that are so large that if they were fragmented, the original members of the group would be unable to carry all the fragments. This surprising effect is documented in Figure 8–45. It is explained at least in part by the ability of the teams to overcome rotational forces. Cooperating individuals can support an object so that these forces are balanced and disappear.

Teamwork needs closer study in *Eciton* and other ant species that employ group transport. Two weaknesses are evident in Franks's analysis that need to be addressed. First, his data on large groups are few: he sampled only two prey items being carried by more than three ants. Second, the energetic consumption has not yet been measured. Workers in teams carry more weight relative to their own weight than do solitary foragers, and they are probably expending more energy. Does the total energy expense of the team exceed the energy gained by using fewer workers? This possibility seems intuitively unlikely, but it needs to be tested.

## PULSATILE ACTIVITY

For many years myrmecologists noticed that activity within ant nests tends to spread in waves, but little thought was given to the frequency or meaning of the phenomenon. In 1987, however, Franks and Bryant discovered that in *Leptothorax acervorum* at least, the activity is not only synchronized to an unexpected degree but also is periodic, regularly attaining 3 to 4 peaks an hour. An example from their protocols is given in Figure 8–46. A closely similar result was independently obtained for *Macromischa allardycei* by B. J. Cole (personal communication), who recorded an average period of 26 minutes and a rate of spread in activity through the colony of 0.08 centimeter per second. *Macromischa* is a close relative of *Leptothorax* and often placed as a synonym within it.

What is the significance of this strange periodicity? N. Franks and S. Bryant (personal communication) offer the following explanation, tying it to the idleness effect in ant labor we discussed earlier:

> The activity rhythms we have observed may represent integrative phenomena which might influence the subsequent behaviour of particular individuals allowing them to maximize their inactivity and so be the basis of greater ergonomic efficiency. With few exceptions (Cole, 1981; Franks and Scovell, 1983) what an individual ant does is not dependent on what it alone requires, but is instead based on the needs of many other members of the colony. For example, it has long been established that certain individuals forage, effectively for the rest of the colony (Wilson and Eisner, 1957). The decisions of such individuals to forage or to remain in the nest need to be based not only on their own nutritional status but on those of many other individuals in the nest. This in turn requires many individuals to be active, as defined in this paper, in order to communicate and determine task priorities. It is clear that many individuals may need regularly to exchange information about their own energy or other requirements in order to build up a picture of the distribution of resources in the colony as a whole. If a long time is taken acquiring such information, resource levels will have changed and it will be less reliable. If an individual must acquire enough information to attain a "critical reliability" with regard to colony status, it would need to be involved in more interactions in a short time. Therefore, many individuals should be active at the same time and colonies should exhibit short bursts of activity interspersed with longer periods of rest. Randomly organised workers would need to be active for longer periods and more energy would be consumed within the nest, causing a probable reduction in overall productivity and inclusive fitness. Interpreted in this way the extensive inactivity of ants within their nests can be seen as being possible only because periods of activity are synchronized. Rhythms of activity may facilitate the flow of accurate, regularly updated information through the nest whilst maximizing inactivity and energy conservation.

B. J. Cole (personal communication) is in independent agreement with the essentials of this argument. He regards the periodicity itself as an epiphenomenon of the more basic synchronization that serves the needs of the colony. Cole has also begun a comparison of synchronization of behavior across species, beginning with a study of the very primitive *Amblyopone pallipes*. The colonies, which are small and loosely organized in most respects, show less synchronization than those of *Leptothorax* and *Macromischa*. The workers are more spontaneous in their activity and less dependent on the tactile stimulation of working nestmates than the myrmicines.

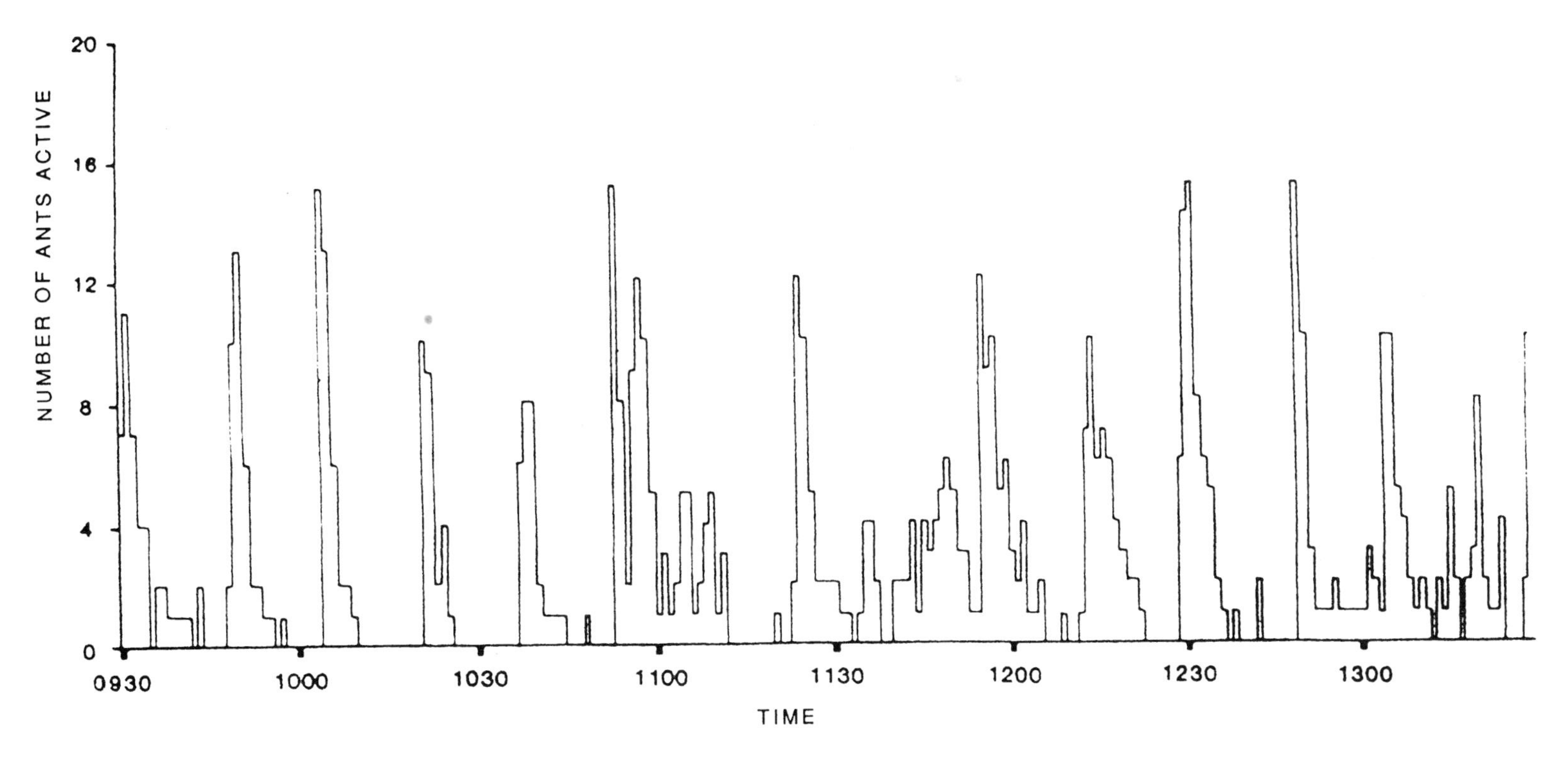


**FIGURE 8–46** Pulsatile activity in a colony of the myrmicine ant *Leptothorax acervorum*. Each column represents the number of individuals active in a sample of 20 workers at 1-minute intervals. The activity is synchronized and occurs at periods of 3 to 4 times an hour. (N. Franks and S. Bryant, personal communication.)

## MALES AND LARVAE AS CASTES

Despite an intensive search for examples over many years, investigators have merely confirmed that adult males fail to contribute services to the remainder of the colony. Within the nest their repertory is almost always limited to grooming themselves and receiving food from the workers. In the great majority of species the males leave the nest to mate, after which they die. The near absoluteness of male deferment from social life, or put more positively, the relatively powerful commitment of females to social life, is in accord with Bartz's principle (1982; see our Figure 4–2) that species of Hymenoptera can evolve female workers or male workers, but not both. The only known exception to the egocentricity of adult males is a feeble one associated with extraordinary environmental circumstances. Adult males of *Camponotus herculeanus* and *C. ligniperda,* species that range north in Europe and North America to the arctic circle, are the longest lived known in the ants as a whole. They emerge from pupae in the fall, overwinter in the nest, and swarm the following late

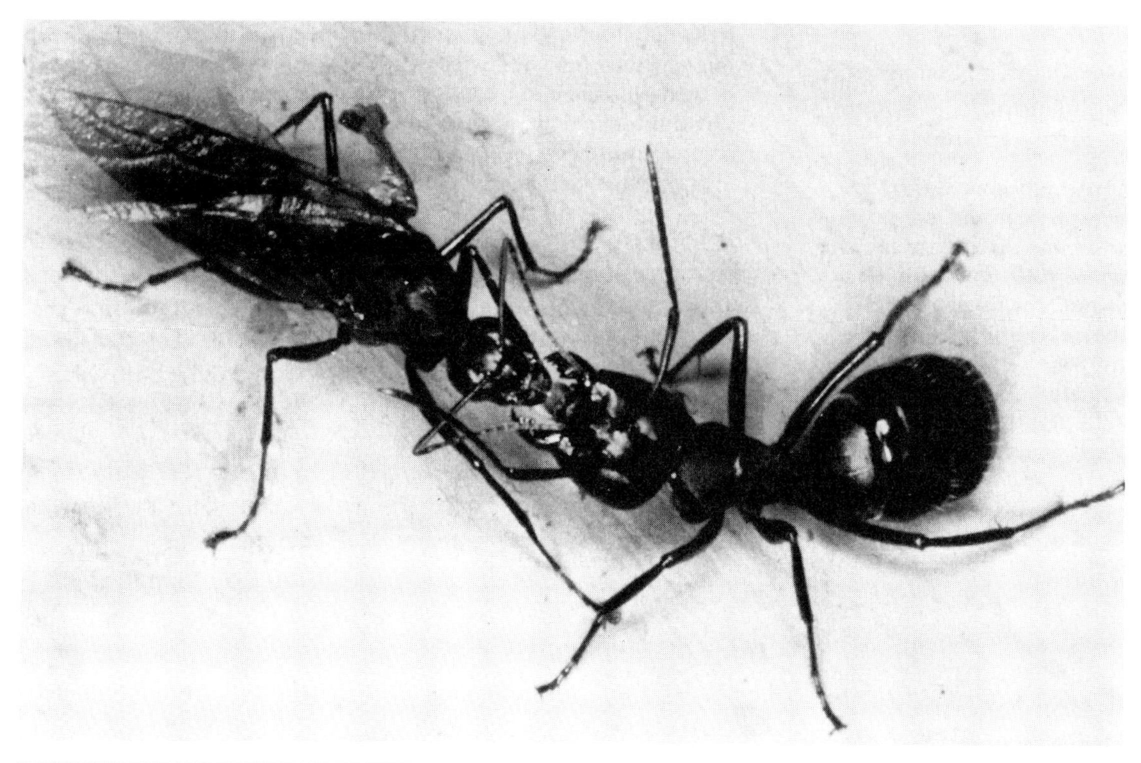

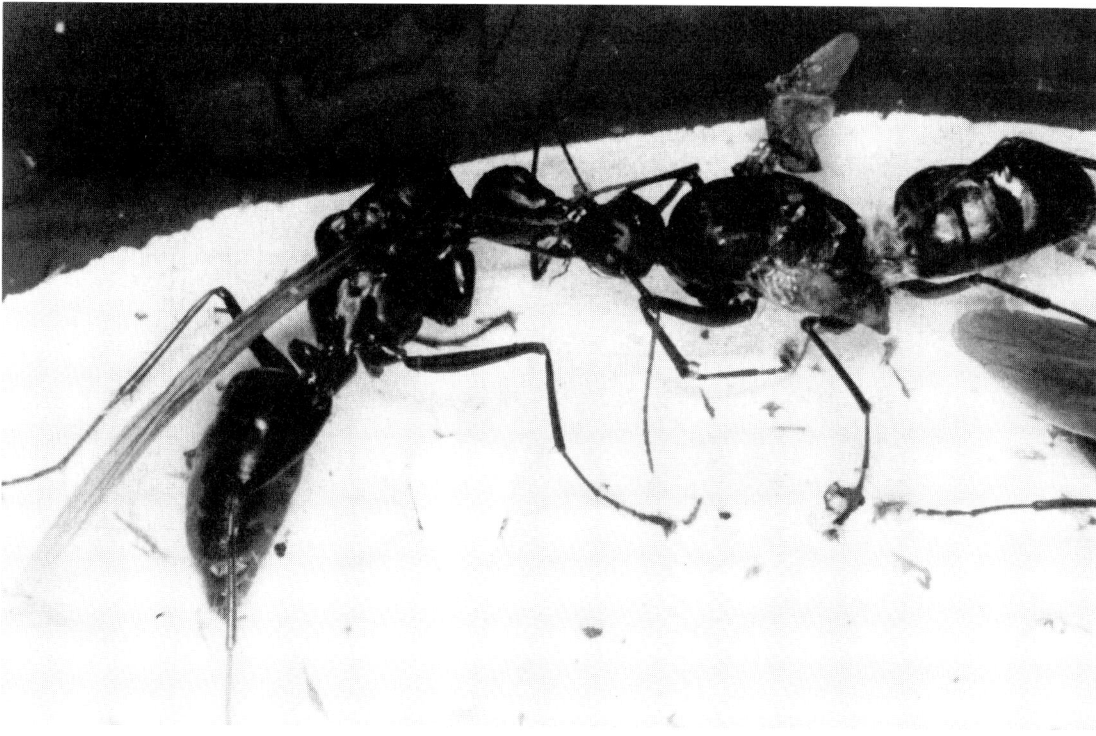

**FIGURE 8–47** Social behavior of males of *Camponotus ligniperda.* (*Above*) A worker feeds a male. (*Below*) Food is exchanged between two males, one of whom has not yet unfolded his wings. (From Hölldobler, 1965.)

**FIGURE 8–48** Larvae of the bulldog ant *Myrmecia* eating a *Camponotus* worker.

spring. If kept cool enough, they remain in the nest through a second annual cycle. They are also unique in their habit of storing liquid food in their crops and sharing it with workers and other males (Hölldobler, 1964, 1966; Figure 8–47). A. Buschinger (personal communication) has evidence that long life also characterizes several Mediterranean species of *Camponotus* as well as *Colobopsis truncatus*. Similar observations were made with *Camponotus socius*, a common species in the southeastern United States (B. Holldobler, unpublished).

Two forms of males, distinguished by size, occur in *Formica exsecta* and *F. sanguinea* in Finland (Fortelius et al., 1987). The small males, or "micraners," appear to mature later, display sharper circadian activity peaks, and disperse farther than the larger males, or "macraners." The small forms evidently occur more prevalently in crowded, polydomous colonies characterized by strongly biased male sex ratios. Fortelius and his co-workers suggest that the larger forms are more common during the early stages of the growth of local populations, when there is an advantage to the reduction of dispersal to enhance the buildup of the local population. Later, the shift to small forms favors dispersal, which is the better strategy when local resources have become strained by large colony size. As impressive as it is, however, this size dimorphism is still basically different from a true caste system, since it adds nothing to labor organization within the colony. Similarly, two forms of males occur together in colonies of *Cardiocondyla wroughtoni:* a normal winged morph that evidently disperses and a worker-like morph with saber-shaped mandibles that remains behind and fights for dominant status and access to the virgin queens (Kinomura and Yamauchi, 1987; Stuart et al., 1987a). As in the *Formica,* the diversification is part of a breeding strategy rather than a true caste system (see our discussion of this phenomenon in Chapter 4).

A very different—indeed startling—picture has emerged from studies of ant larvae. The anatomy of larvae bears many modifications that enable closer care by their adult nurses (Wheeler and Wheeler, 1976; Petralia and Vinson, 1979a). This general class of social adaptation is especially conspicuous in the conformation of the ventral surface of the body. In the case of larvae fed only with regurgitated food from the workers, as in the genera *Crematogaster* and *Camponotus,* the head is too closely appressed to the thorax for the larva to reach its own venter with its mouthparts. The venter is correspondingly simplified in form, with a smooth surface, unspecialized hair pattern, and few if any small spines. The forward part of the body is thick and nearly immobile. In contrast, larvae that are fed solid food possess anatomical features that are adaptive to this end. Those in the most primitive ant subfamilies, including the Nothomyrmeciinae, Myrmeciinae, and most Ponerinae, as well as the "true" army ants (Dorylinae and Ecitoninae) are primitive in form. They have tapered, flexible necks and strong mandibles, and are able to bend and stretch to reach prey items placed near them (Figure 8–48). The simplest anatomical modification for holding solid food, found in some species of *Monomorium* and *Solenopsis,* is a set of anteriorly directed spinules set on the venter just beneath the head. The spinules and hairs of *S. invicta* form a "food basket" surrounding a bare central area on which food particles are placed (Petralia and Vinson, 1979b). The basket reaches its most elaborate form in larvae of *Pogonomyrmex* harvester ants, which can thrust their heads deep into the enclosure to feed on the fragments of seeds and insects placed there by the nurse workers.

A second modification for feeding is the "food platter" of the ponerines *Odontomachus* and *Pachycondyla,* a wide, flattened region toward the rear of the body venter. Wheeler (1918a) described its use in *Odontomachus:* "These larvae are placed by the ants on their broad backs, and their heads and necks are folded over onto the concave ventral surface, which serves as a table or trough on which the food is placed by the workers."

The most elaborate structure possessed by larvae for feeding on solid food is the "food pocket," hellenistically designated a trophothylax by W. M. Wheeler and Bailey (1920) in *Pseudomyrmex* and a praesaepium by G. C. and J. Wheeler (1953a) in *Camponotus.* It is an invagination of the ventral thoracic body wall into which the food particles are inserted by the workers. The food pocket was almost certainly evolved independently in these two groups of ants. It also appears to be closely associated with arboreal life. The pocket is best developed in *Pseudomyrmex* and *Colobopsis,* both of which mostly nest in hollow twigs, cavities in dead standing branches, and other dry arboreal habitats.

The hairs on body parts other than the food baskets and platters are often long and dense. In many species they also take peculiar shapes—sinuous, spiral, hook-like, and anchor-tipped (Y-shaped with the tips curved downward). The hairs help to hold the smaller larvae together in conveniently managed packets, allowing them to be more efficiently transported by the workers. The anchor-tipped hairs of *Crematogaster* and probably the other myrmicine genera are used by the workers to suspend the larvae, papoose-like, from the nest walls (Wheeler and Wheeler, 1976).

One of the most bizarre larval structures is found in the rare and obscure ponerine genus *Probolomyrmex.* It is a large tubercle shaped like a doorknob (or a toadstool on a stalk) protruding conspicuously from the dorsal surface of the last abdominal segment. Workers of *P. angusticeps* of Panama, the only species of the genus studied in life to date, carry the larvae by the tubercles. They also attach the larvae to the roof of the nest with the apparently adhesive outer surface of the knob-like endings. The larvae then hang down like bats over the huddled masses of workers (Taylor, 1965a). In similar manner the larvae of *Leptanilla japonica,* the only member of the equally rare and obscure subfamily Leptanillinae thus far studied, have peculiar appendages projecting from the prothorax, which the workers grip with their lower mouthparts. In this unique manner the larvae are transported during colony emigrations (Masuko, 1987).

Although most of the social modifications of ant larvae are for their benefit, a few serve the adult members of the colony. Final-instar larvae of *Oecophylla* weaver ants, both female and male, contribute all of their silk to the construction of the leaf nests of the colony (Wilson and Hölldobler, 1980). Larvae of at least some ponerines, myrmicines, and formicines donate small quantities of apparently nutritious liquid from their mouthparts to the adults (Maschwitz, 1966; Wilson, 1974a). Those of *Solenopsis invicta* void an apparently nutritive liquid from their anus (O'Neal and Markin, 1973). Salivary liquid provided by larvae of *Monomorium pharaonis* during periods of extreme dryness allow the workers to survive longer (Wüst, 1973; Ohly-Wüst, 1977). Larvae of the ponerines *Amblyopone* and *Proceratium* involuntarily provide hemolymph to the queens, who bite small incisions in their cuticles and drink the oozing liquid, vampire-like. The activity does not seriously impair the larvae, and in large colonies of *Amblyopone silvestrii* at least, it is the exclusive source of the queen's food (Masuko, 1986). *Leptanilla japonica* has taken larval hemolymph donation a step further. The larvae bear a single pair of organs resembling spiracles on the third abdominal segment. These structures exude hemolymph on which the queen subsists exclusively (Masuko, 1987). Final-instar larvae of *Monomorium (= Chelaner) rothsteini,* a harvester ant of Australia, convert fragments of seeds into cephalic secretions that are fed back to the workers (Davison, 1982).

Thus a wealth of new information has tended to confirm the view of Wheeler (1918a) that ant larvae contribute significantly to the life of the colony. The picture has emerged slowly because of the generally passive nature of the larvae and their heavy reliance on the exchange of liquids, making their behavior difficult to analyze. It is now sufficiently clear, however, to show that larvae often serve as a distinct caste and must be included in future studies of division of labor. In this context it is interesting to note that *Oecophylla* male larvae also contribute silk for the communal nest construction, but they do so significantly less than worker larvae. We do not know whether such sex-biased asymmetries in larval trophallaxis exist in other ant species.

## CASTE DETERMINATION

A wealth of studies on caste determination has followed the pioneering experiments by Brian (1951a), who first suggested the key steps on the divergence of queens and workers in the genus *Myrmica.* All of the research has come together in an overriding principle: the female castes of ants are differentiated by physiological rather than genetic factors. In other words allelic differences almost never separate the female castes of a colony, even to the extent of slightly biasing individuals to develop into one caste as opposed to another.

The single exception known applies to alternative forms of reproductive females among colonies and not to the ordinary complement of castes within single colonies. In the European slavemaking ant *Harpagoxenus sublaevis,* ergatomorphic reproductives (wingless and resembling workers) differ from fully winged (gynomorphic) queens by a single recessive allele. The ergatomorph intercastes possess the genotypes *EE, Ee,* or *ee,* whereas winged queens are always *ee* (Buschinger, 1975a; Buschinger and Winter, 1975; see Figure 8–49). The ratios of the workers to queens produced are also affected by the alleles (Winter and Buschinger, 1986). Control in this case is not absolute. Larvae of all three genotypes develop into workers under stringent feeding regimes. When the larva makes the physiological commitment to become a reproductive, it adheres strictly to the type of reproductive prescribed by its genotype, in other words an intercaste or a gynomorph, but the likelihood that it will make this decision in the first place is affected by the *E* and *e* alleles. When a larva has an *E* allele, its development is slowed. It is also more sensitive to inhibition by adult gynomorphs and hence is more likely to become a worker instead of an ergatomorph. Conversely, *ee* larvae grow faster and are less sensitive to the presence of gynomorphs. They are consequently more likely to turn into gynomorphs themselves (Winter and Buschinger, 1986).

Because caste determination is almost entirely environmental in ants as a whole, the observed caste systems can be usefully interpreted as the consequence of variations among species on *growth transformation* during larval development (Oster and Wilson, 1978). With the aid of this concept, evolutionary scenarios can be constructed that are subject to experimental testing. Consider first the production of a small class of majors by skewing of the adult worker size-frequency distribution, the lowest grade of worker polymorphism known in the ants. As suggested in the schema of Figure 8–50, newly hatched larvae vary in body weight, probably according to a normal distribution. It is likely that they also vary in other qualities,

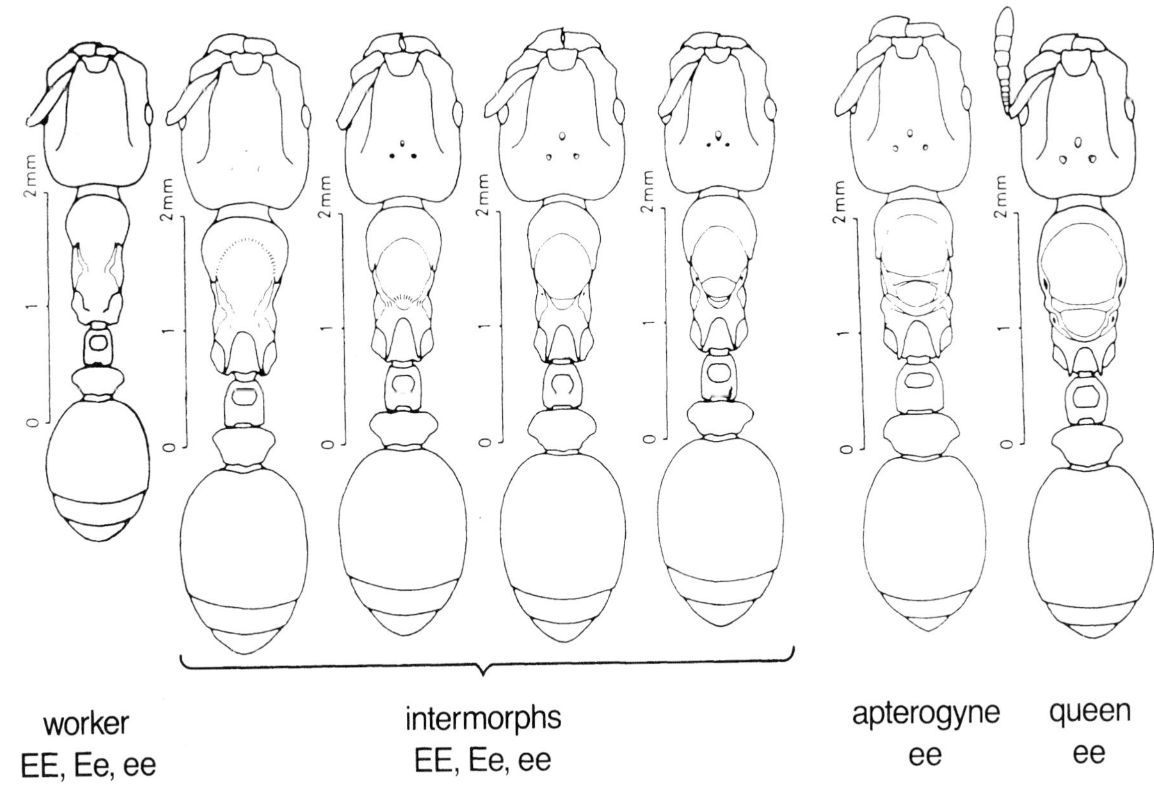


**FIGURE 8–49** A rare case of genetic control of caste, in the European slave-making ant *Harpagoxenus sublaevis*. This illustration depicts a morphological worker on the far left, a fully developed queen on the far right, and a series of intermorphs in between. The fully developed queen is originally winged; the specimen shown here has shed her wings. The presence of the allele *E* prevents the development of a fully developed queen. An *ee* larva can become a queen if nutrition is good and there is minimal inhibition from her mother queen; it can also become a worker or an intermorph if nutrition is poor or queen inhibition strong. The "apterogyne," second from right, is only known to originate from *ee* larvae, and occurs very rarely. *EE* and *Ee* larvae become workers or fall somewhere on the intermorph gradient according to nutrition, intensity of queen inhibition, and whether or not hibernation was experienced. Larger workers than the one shown on the left also occur. In contrast to true workers, intermorphs function as reproductives. They possess spermathecae and usually leave the nest for mating and founding of new colonies. (Modified from Passera, 1984; adapted from Buschinger and Winter, 1975, with qualifications by Buschinger, personal communication.)

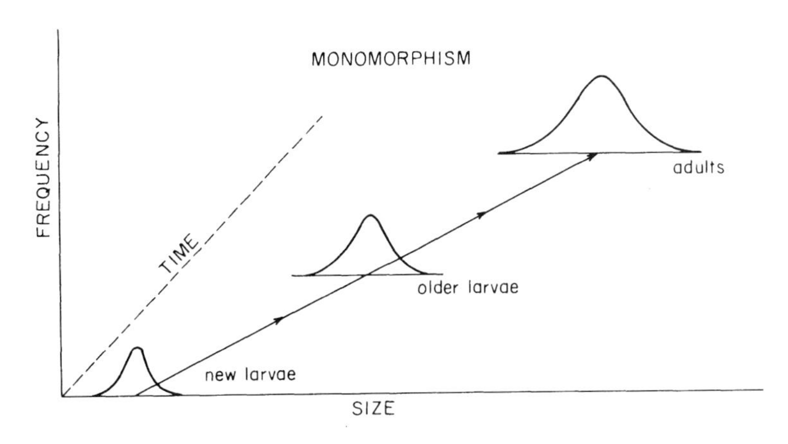


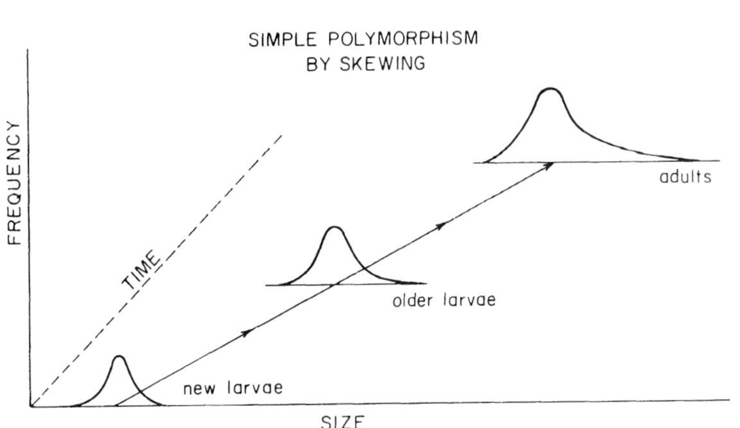


**FIGURE 8–50** Simple caste systems are created during the evolution of ant species by the skewing of the size-frequency curve toward the larger size classes, accompanied by allometric variation. This can be achieved by a growth transformation function that converts small differences in initial larval size, or any other factor affecting caste, into disproportionately greater differences in the final larval size. (From Oster and Wilson, 1978.)

such as the amount of yolk available to them during embryonic development, the temperature at which they developed, and so forth. Suppose that some of these factors influence the final body size attained by the larvae and hence by the adult ants. If this growth transformation remains constant with increases in initial size, the result will be a set of workers whose size-frequency distribution is approximately normal. Such distributions are typically accompanied by isometry (non-allometric growth), and consequently the entire worker caste will be monomorphic.

Suppose, however, that the transformation is such that the larger the initial size of the larva (or the greater the quantity of other caste-biasing factors present), the more rapid its growth, so that individuals starting large finish proportionately larger. Such a relationship might be described as follows: $y_1 = F[y_0]$ where $y_0$ and $y_1$ are the body size (or quantity of other caste-biasing factors) at the start and

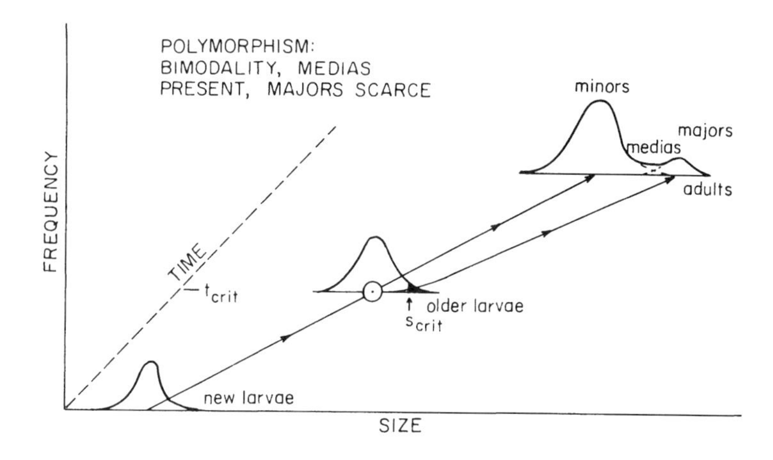


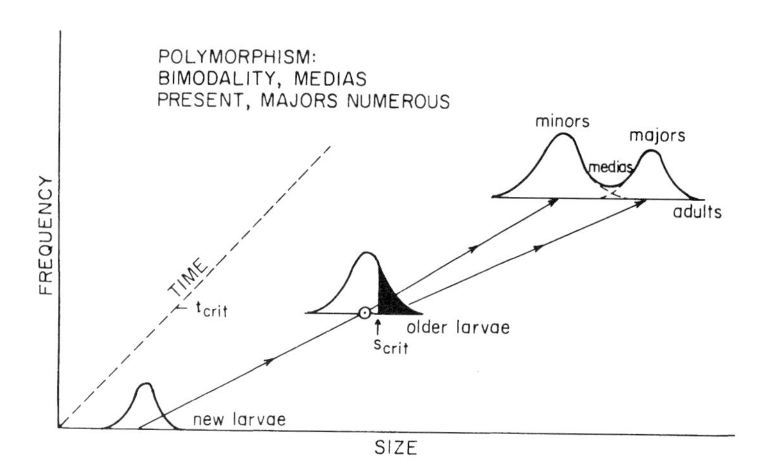


**FIGURE 8–51** When a switching point is introduced at a critical developmental time ($t_{crit}$), larvae that have attained a threshold size ($s_{crit}$) increase their growth rate and move toward the major worker mode of the final adult size-frequency distribution. (The fraction of larvae destined to travel this divergent pathway are indicated by the shaded portion of the middle frequency curves.) By adjusting $s_{crit}$, species can set the percentage of major workers, as illustrated by a comparison of the upper and lower diagrams. In both of these imaginary examples, the divergence of the two developing pathways is slight enough so that the final minor and major frequency distributions overlap, producing a media class. (From Oster and Wilson, 1978.)

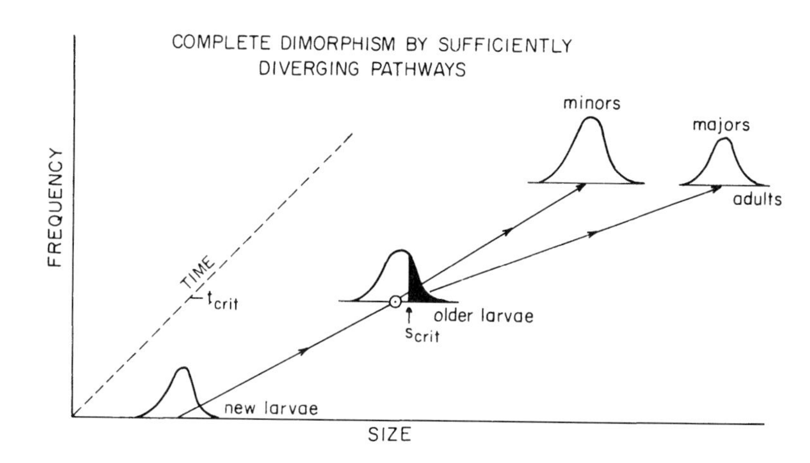


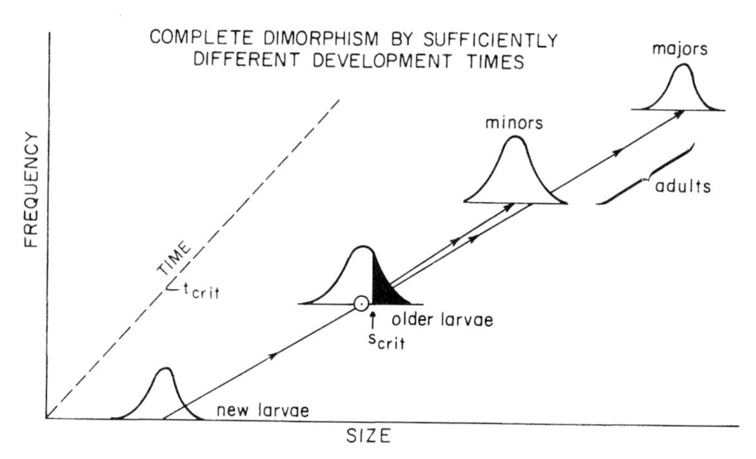


**FIGURE 8–52** This schema is an extension of that presented in Figure 8–51. When the two developmental pathways diverge rapidly enough (*above*) or the larger larvae continue their development for a sufficiently longer period of time in comparison with the smaller larvae (*below*), the final size-frequency distributions are disjunct and the worker caste is completely dimorphic. (From Oster and Wilson, 1978.)

finish of larval growth. If the transformation is linear, the result will be monomorphism, in conformity with the conditions just described. Where the transformation is exponential, so that the subsequent rate of growth increases as a function of the starting point 0, the result will be a size-frequency distribution skewed toward the upper size classes. The magnitude of the exponential constant represents the sensitivity of larval growth to the initial conditions encountered by the young larva. Alternatively, the constant can depend on conditions encountered later in larval life, with similar final results. By relatively small adjustments in the transformation function, the caste system of a species can be conspicuously altered.

The next simplest conceivable step in the elaboration of size-frequency distributions is the introduction of decision points during development, as illustrated in Figures 8–51 and 8–52. The decision point, or control point as it is sometimes called, is a time in development at which one of two sets of growth constants is acquired by the immature form. Thereafter the individual will proceed in its development toward one subcaste or another, with little chance of deviation. Decision points are efficient devices for sorting colony members into two or more independent populations. The rules for shunting individuals into one direction as opposed to another can be adjusted to regulate the relative sizes of the two populations. The examples given in Figure 8–51 utilize threshold size at a critical developmental time as the shunting rule; this is the situation in queen-worker determination in the ant *Myrmica ruginodis* (Brian, 1955). Larvae that attain a certain size by a critical time continue rapid development toward a still larger ultimate size; those that fall short proceed at a distinctly slower rate and are even more behind at the completion of adult development. By setting the threshold size low or high relative to the usual size-frequency distribution of larvae at the critical development time, species can arrange a lower or higher proportion of the larger caste.

Shunting rules are based on a variety of token stimuli in addition to size. No fewer than six classes of such stimuli have been identified in various ant species as influencing the determination of individual females to the worker caste as opposed to the queen caste.

Two of these are also known to be potent in minor-major worker determination; the latter process is still relatively unexplored and will probably be shown to be subject to other influences as well (see reviews in Wilson, 1971; Schmidt, 1974a,b; Brian, 1979a, 1983, 1985; Nijhout and Wheeler, 1982; de Wilde and Beetsma, 1982; Passera, 1984; Vargo and Fletcher, 1986a,b; Wheeler, 1986a).

1. *Larval nutrition*. Competition among larvae alone may produce the bimodal size-frequency curves that underlie most queen-worker and minor-major distinctions in ant species. Larvae that attain a threshold size by a critical developmental time are shunted toward the larger caste.

2. *Winter chilling*. Intraovarian eggs of *Formica* and larvae of *Myrmica* that have been chilled have a greater tendency to develop into queens, an apparent device for timing the emergence of queens in spring or early summer. Other responses to temperature, humidity, or photoperiod could produce crops of queens (or even worker subcastes) at other times. These would depend on local climatic conditions and idiosyncratic features of the colony life cycle in each species.

3. *Temperature*. The larvae of *Formica* and *Myrmica* tend to develop into queens more readily if reared at temperatures close to the optimal for larval growth.

4. *Caste self-inhibition*. The presence of a mother queen inhibits production of new queens in *Myrmica*, *Monomorium*, and *Solenopsis*; likewise the presence of soldiers inhibits the production of soldiers in *Pheidole*. This negative feedback loop could obviously serve to stabilize caste ratios when there is a need to fix the ratio instead of making it flexible in response to short-term needs.

5. *Egg size*. In *Formica*, *Myrmica*, and *Pheidole* the larger the egg, the more yolk, and the more likely the larva is to develop into a queen as opposed to a worker. No information is available on the relation of egg size to minor-major determination.

6. *Age of queen*. Young queens of *Myrmica* tend to produce more workers; the queen's age could, of course, be reflected in the size of the eggs she lays. Smaller egg size in the batch produced by a nest-founding queen might further explain the occurrence of nanitic workers in the first brood; in other words, egg size could evolve to ensure the production of nanitics at this stage in the colony life cycle.

Some of these factors are merely biasing in their effects, rendering an individual more likely to take one direction as opposed to another upon reaching the point of bifurcation. Others exert their influence directly at the decision points themselves. Often one factor can override another in the following manner: if condition *a* prevails earlier instead of *a*′, then the larva can respond to either *b* or *b*′, but *a*′ forces the same response regardless of whether *b* or *b*′ is present. For example, mature larvae of *Myrmica ruginodis* subjected to winter chilling have the capacity to develop into either queens or workers. But only those that subsequently reach a weight of 3.5 milligrams within approximately eight days after the start of posthibernation development actually become queens. Larvae not exposed to chilling always develop into workers, regardless of their size.

The scheduling of receptiveness to caste-biasing stimuli almost certainly represents an idiosyncratic genetic adaptation on the part of each individual species. The later the decision point, the more flexible is the system, in the sense that it permits the colony to make rapid adjustments in the caste ratios. This would seem a priori to be of special advantage to species that possess a soldier caste subject to

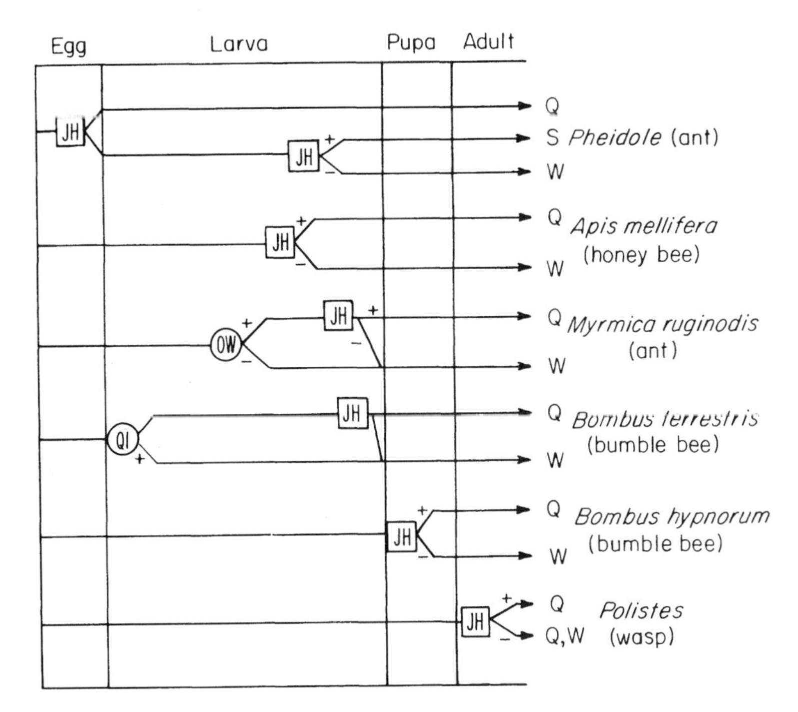


**FIGURE 8–53** Variation in the timing of the decision points in caste determination among species of social insects. Points in the larval and pupal stadia are probably associated with nutritional switches. *OW*, overwinter causing chilling; *QI*, inhibition by the queen; *Q*, queen; *S*, soldier (major); *W*, worker, including minor worker when soldiers are also present; *JH*, juvenile hormone titer, which influences the decision. (Modified from Wheeler, 1986a.)

occasional heavy mortality. The older the larvae are when shunted to the soldier developmental pathway, the shorter the time required to fill gaps created by casualties. It is consequently of interest that in *Pheidole* the point of soldier-vs.-minor worker determination is in the final instar (Passera, 1974a; Wheeler and Nijhout, 1984). It is also true, as stressed by Wheeler (1986a), that a wide separation of the queen-worker and major-minor decision points minimizes confusion of the two and allows a more precise regulation of caste ratios.

The timing of queen-worker bifurcation in fact varies greatly among species. In *Myrmica ruginodis* it is very late—about a week prior to the cessation of larval growth. But in *Formica polyctena* the opposite is true: a larva is determined to queen or worker within 72 hours of hatching from the egg (Bier, 1958a,b; Schmidt, 1974a,b; Weaver, 1957). A similarly wide variation in timing has been observed among species of bumblebees (Röseler, 1974; Röseler et al., 1984). The full range of timing in decision points is illustrated in Figure 8–53. The relation of these differences to the ecology of the species awaits investigation.

*Queen determination in* Myrmica. The most thoroughly studied system of caste determination in ants is that of the genus *Myrmica*, pursued by M. V. Brian and his associates in England through various ramifications from 1950 to the present time. The species employed has been primarily *M. ruginodis*, with supplementary studies made on *M. rubra*. Both species are polygynous. In the earlier work Brian sometimes referred to *M. rubra* as *M. laevinodis* and to *M. ruginodis* variously as *M. rubra*, *M. rubra macrogyna*, or just *Myrmica*. Fortunately, the two species appear to be similar in key aspects of caste determination.

*Myrmica ruginodis* and *M. rubra*, like most north temperate myr-

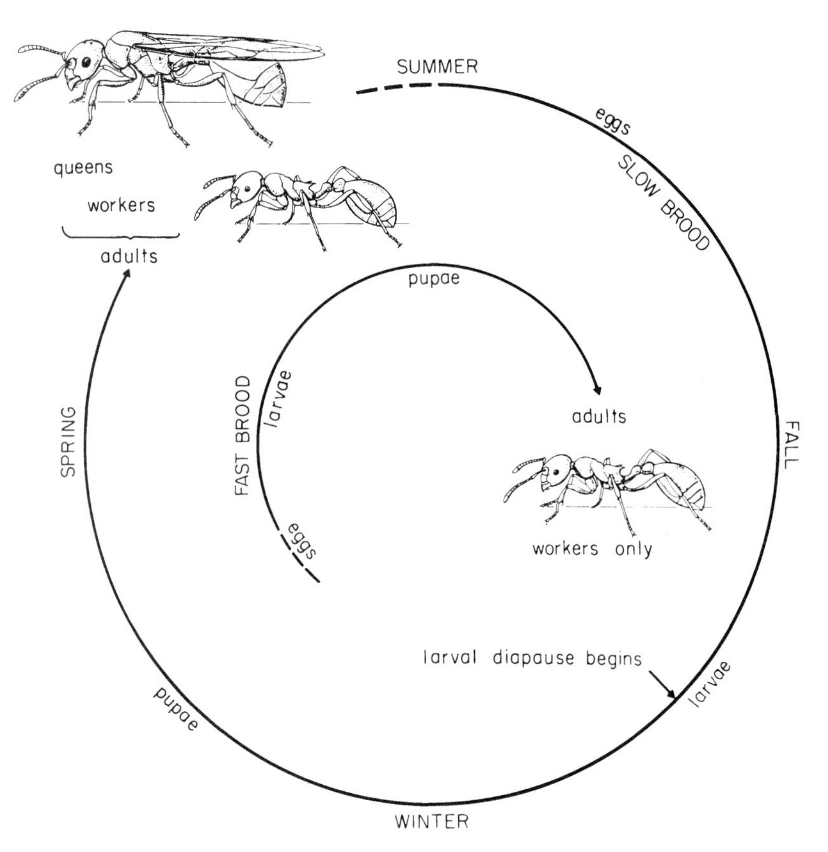


**FIGURE 8–54** The annual cycle of brood development in a mature colony of *Myrmica ruginodis*. The mother queen continues to lay eggs intermittently through the spring and summer. Many of the larvae that hatch early in the season are able to complete development by the end of the summer and become workers (fast brood). Others persist as larvae through the winter and can become workers or queens the following spring (slow brood). The full development of fast brood requires about three months, that of slow brood almost a year. (From Wilson, 1971, based on data from M. V. Brian.)

micines, overwinter with their brood in the larval stage. In most mature colonies the following spring, some of the female larvae develop into queens and others develop into workers. The annual cycle is illustrated in simplified form in Figure 8–54. As the "slow" larvae, hatched in early summer, reach a certain point in their development in the late summer or fall, they become dormant. This crucial juncture is in the terminal (third) instar. More precisely, the larvae are halted at a stage of early pupal development when the brain has migrated halfway back into the larval prothorax. Many larvae do not proceed beyond this 0.5 point until the following spring, a condition referred to as primary diapause. Others can continue developing until the brain is 0.8 into the prothorax, but they always halt there in a condition of secondary diapause. Still other, younger larvae do not reach the 0.5 stage at all before cold weather; these pass the winter in a nondiapause state. In any case all the brood overwinters in the larval stage, and it is from some of these vernalized larvae that the yearly crop of queens is matured in the spring.

Diapause, as pointed out by Brian (1956b), has the effect of holding over until winter all of the late brood in the larval stage. The dormant state can be broken by sustained high temperatures and handling by spring workers. Both of course are normally encountered by larvae in wild colonies in the spring. The important point is that chilling in winter temperatures confers on the larvae the capacity to sustain a high growth rate in critical periods of the final instar and, as a result, to metamorphose into queens. Dormancy itself is not a prerequisite. Some small larvae, as we have noted, are immobilized by winter cold before they enter stage 0.5 dormancy; yet the following spring they too have the capacity to transform into queens.

Final caste determination occurs in the spring larvae late in their terminal instar. The critical periods were revealed by experiments in which sets of larvae were starved at different periods in pupal development. Of course, larvae are not normally steered into one path of development or another by any such regimen of sudden starvation or overfeeding. In order to learn about the natural course of determination, it was necessary for Brian to follow the development of many larvae being reared individually under relatively undisturbed conditions. When this was done, a clearer picture of the role of the growth rates was obtained. Brian learned first that the time required for spring larvae to develop into worker pupae did not differ from the time required for development into queen pupae. In both cases duration of development ranged from 9 to 21 days and averaged about 13 days. But the final weight attained by larvae transforming into the two castes differed greatly, averaging about 4.5 milligrams in the worker and 8 milligrams in the queen. Clearly the larvae destined to be queens either must grow at a faster rate or else start at a higher weight. And either can be the case. In general, queens come from spring larvae that either start relatively large and maintain a moderate to high growth rate or else start at a medium weight and maintain a consistently high growth rate. For the most part larvae that have failed to reach a weight of about 3.5 milligrams by stage 0.8 are destined to become workers.

An important finding from Brian's studies is that caste determination in *Myrmica ruginodis* is essentially worker determination, or the failure of this event to occur. This means that the worker is an individual diverted from a normal female (that is, queen) course of development by having part of her adult system shut down. For convenience, the imaginal discs of a larva can be divided into a dorsal set, containing wing buds, gonad rudiments, and ocellar buds, and a ventral set, containing leg buds, mouthpart buds, and central nervous system (Brian, 1957a, 1965a). In the case of queen development leading to a more or less typical hymenopteran female, both sets maintain full growth and development at the onset of pupal development. In the case of larvae destined to become workers, however, the dorsal set stops growth and development for the most part, and only the ventral set continues on. The abruptness of the shutdown is quite striking. It is interesting that the dorsal organs do not always shut down together. One of the anomalous "castes" described by Wheeler (1905) was the pterergates, or otherwise normal workers bearing external vestiges of wings. One worker of *M. scabrinodis* figured by Wheeler had wings as long as the thorax itself.

Growth studies, including the starvation experiments, disclosed the important role of chilling in developing queen potential in larvae, and of nutrition in leading *Myrmica* larvae to growth beyond the thresholds required to sustain development as queens. Subsequent experimentation has revealed the existence of at least four additional factors influencing caste determination in the closely related species *Myrmica rubra* (Brian, 1963, 1979a, 1983). First, an increase in temperature from 22 to 24°C, that is, from the optimal temperature for larval survival to slightly above it, results in an eightfold increase in the proportion of larvae metamorphosing into workers. Second, the presence of nest queens results in a fourfold increase in the proportion of workers. The latter effect is caused at least in part by a change in the behavior of the workers evoked by the perception of queen pheromones. When two sizes of larvae are presented to adult workers in the presence of a queen, the small larvae are fed more and the larger larvae less than in the case of control groups lacking a queen. Also, the larger larvae are bitten and licked more in the presence of the queen, which presumably lowers their chances of survival (Brian and Hibble, 1963; Brian, 1983; Elmes and Wardlaw, 1983). Similar changes are obtained when dead queens are presented daily to queenless colonies (Carr, 1962). Workers make sharper discriminations if they are exposed to queens during the first several weeks after they emerge as adults than they make if kept away from this caste (Brian and Evesham, 1982; Evesham, 1984b). The presence of a primer pheromone is suggested by the fact that when the sterol fraction of extracts of the heads of queens is fed to larvae or applied topically, larval growth is inhibited (Brian and Blum, 1969).

A third influence is blastogenic. Small eggs laid during periods of most rapid oviposition yield higher numbers of workers. When queens are allowed to emerge from hibernation at 20°C, the rate of oviposition rises during the first three weeks to a maximum that persists through the following three to four weeks. Then the rate gradually declines toward zero, reaching a very low level after about the sixteenth week. Simultaneously the size of the eggs changes, declining rapidly in the first three weeks and then remaining approximately constant. Eggs laid during the first three weeks show a greater capacity for transforming to queens than do eggs laid after three weeks when both kinds are cultured under identical conditions.

Finally, a most interesting additional blastogenic effect discovered by Brian and Hibble (1964) is that eggs laid by different queens differ markedly in their tendency to produce queens or workers. Most of the variation appears to be the result of age; younger queens have a higher tendency to lay worker-biased eggs. This effect was foreshadowed by the results of early experiments on *Pheidole pallidula* by Goetsch (1938), who introduced eggs from colony-founding queens into larger colonies in which normal-sized workers were being reared. Despite the richer environment, dwarf-sized minor workers typical of first brood were obtained. If confirmed, the Goetsch effect appears to be the first demonstration of some form of ovarian, or truly blastogenic, influence on the determination of worker subcastes.

To summarize, we can list at least six factors that determine whether a *Myrmica* female will become a worker or a queen: larval nutrition, winter chilling, posthibernation temperature, queen influence, egg size, and queen age. The next question should logically be, what is their relative importance in nature? The clearest way to view the entire caste-determining system is to regard it metaphorically as a series of checkpoints arranged more or less in sequence. An egg "aspires" to develop into an adult queen. This ambition is "approved" by the colony, providing the following two checkpoints are passed. First, has the larva been through diapause and chilled to resume full development? Second, has the larva reached the requisite size by the onset of adult development in the final larval instar? In addition, are the mother queens nearby and potent, and is the colony young? If so, borderline cases are more likely to fail the queen test and be consigned to workerhood. Taken together, the caste-biasing factors make it more likely that the *Myrmica* colony will produce new queens in the spring and also when it is large and robust—the conditions under which it can most profitably invest in reproduction.

*Queen determination in* Formica. Work on the *Formica rufa* group, paralleling in many respects that on *Myrmica*, was published by K. Gösswald and K. Bier (1953–1957) and Schmidt (1974a). As in *Myrmica*, there was originally much confusion in the taxonomy of this difficult species group, but it has been largely elucidated by Yarrow (1955), Lange (1958), and Betrem (1960). Two species were used in the Gösswald-Bier study: *F. nigricans* (= *rufa pratensis*) and *F. polyctena* (= *rufa rufopratensis minor*). The following account applies primarily to *F. polyctena*.

These formicine ants hibernate without brood. In the spring, when the nest temperature rises to 13°C or above, the queens migrate to the warmest part of the nest near the surface, lay batches of eggs (referred to as the "winter" eggs), and afterward retreat to cooler parts of the nest. Eggs laid at temperatures under 19.5°C remain unfertilized, and as a consequence those first produced in the spring, when the nest temperatures are between 13 and 19.5°C, are male. Also, the smaller the colony, the poorer its thermoregulation, and, hence, the higher the proportion of males produced (Gösswald and Bier, 1955). Later eggs in this first "winter" batch are fertilized and capable of producing either queens or workers. Eggs laid still later—the "summer" eggs—are capable of producing only workers.

The winter and summer eggs differ strikingly from each other. Viewed in ovarian preparations, the winter eggs contain more RNA around their nuclear membranes and have a much more voluminous polar plasm than is the case in summer eggs; in addition, the nurse cells have larger nuclei (Bier, 1953, 1954a). It has been postulated, quite reasonably, that these cytological differences are in some way intimately connected with the later biasing of larval growth, but the relation has not been proved experimentally.

In any case it is evident that chilling of the eggs in the ovaries makes them bipotent with respect to caste. Final determination, however, occurs during about the first 72 hours of larval life. This was proven in experiments in which Gösswald and Bier (1957 and contained references) introduced eggs of *F. polyctena* into colonies of *F. nigricans* in order to permit the tracing of individual development. When 30 or more host *nigricans* workers, deprived of their own queen, were given small numbers of *polyctena* eggs, they reared queens. Groups of less than 30 workers reared workers. At 27°C the young larvae remained plastic for 72 hours.

Two other factors were discovered that match those in *Myrmica*. The presence of *F. nigricans* queens inhibits development of the *polyctena* eggs into queens. If a queen is present with a large group of workers, the winter eggs transform into either queens or workers, but if no queen is present the eggs always transform into queens. Part of the difference is due to the propensity of workers to direct the flow of food to queens differentially. Workers just emerged from hibernation, as well as young workers, increase the tendency of larvae hatched from winter eggs to develop into queens.

In summary, although caste determination in *Formica* differs from that in *Myrmica* in several important details, there is a close resemblance in general pattern. Multiple controls exist; most of the six factors of *Myrmica* also occur in *Formica*. A close interplay of responses to hibernation and nutrition characterizes both. In *Formica*, as in *Myrmica*, the relative weight and precise degree of interaction of the factors under natural conditions are still unknown.

*Queen determination in other ants.* The queen-determining systems of other ant species have so far proved to be variations on the themes exemplified by *Myrmica* and *Formica*. The differences discovered are in the timing of the determination during development of the egg and larva and in the relative importance of the six environmental factors.

The formicine *Plagiolepis pygmaea* resembles *Myrmica ruginodis* in the key events of queen determination. The spring larvae that have overwintered in the final instar are bipotent. If they are exposed to inhibitory queen pheromones during the first nine days after diapause is broken, they develop into workers. When the queen is absent or less potent, the better-nourished larvae become queens, whereas those less well fed turn into workers. The queen secretes the pheromone primarily or entirely from her head (Passera, 1974b, 1980b, 1984; Suzzoni et al., 1980). The *Myrmica* pattern is also repeated in *Leptothorax nylanderi*. Larvae that overwinter and emerge from diapause are bipotent; in the absence of the queen and with sufficient nourishment, they develop into queens. The critical period, like that of *Myrmica* and *Plagiolepis*, is in the third and final instar (Plateaux, 1971).

A more *Formica*-like determination has been recorded in the Mediterranean myrmicine *Pheidole pallidula*. Bias toward queen development originates during the egg stage. Eggs laid by fat, heavy queens after hibernation are strongly prone to yield queens (Passera, 1980b, 1984).

*Monomorium pharaonis*, a cosmopolitan house ant of tropical Old World origin, has the most aberrant queen-determining system yet discovered in ants. The colonies are unusual in being potentially immortal; as old queens die, they are replaced by young queens that mate in the nest. The typical longevity of the queens is no more than 200 days, the shortest known in the Formicidae as a whole. As Petersen-Braun (1975, 1977) demonstrated, inhibition by the queen pheromone is decisive, and the brevity of queen life regulates bursts of new queens at intervals of 3 to 4 months owing to what Petersen-Braun called a state of "physiological queenlessness." Queens pass through three periods in their 28-week life span: a juvenile phase which lasts about a month, a fertile phase of 2 to 3 months, and a final senile phase. During both the juvenile and senile periods, the queen lays eggs destined to become workers regardless of larval nutrition. During the lengthy intervening fertile phase, as the queens approach senescence, their eggs are bipotent. If the larvae hatched from these eggs are given sufficient food they become queens, and if not, they turn into workers. As a consequence a crop of new queens arrives about the time the old queens die off. In the intermediate period, spanning four to five developmental periods from egg to adult, an "avalanche" of new workers is eclosed. These ants had been initiated during the juxtaposed senile and juvenile periods of the two queen generations. The inhibitory pheromones are secretions known to be picked up from the cuticle of the queens by the workers, but their ultimate glandular source remains unknown (Berndt and Nitschmann, 1979).

Still incomplete information indicates that queen inhibition is a very widespread if not universal phenomenon, at least in ant species possessing a queen caste in the first place (Vargo and Fletcher, 1986a,b, 1987). In the ponerine *Odontomachus "haematodes" troglodytes* (= *O. "haematodes"*) larvae are sensitive to queens during their final one or two instars. Once determination has occurred by this means it does not appear to be reversible by changes in nutrition (Colombel, 1978). In colonies of the pseudomyrmecine *Tetraponera anthracina*, the sensitive period is very late in larval life. Worker-destined larvae commence metamorphosis soon after this point, but those determined as queens continue growing for a while (Terron, 1977).

*Soldier determination in* Pheidole. The mode of regulation of ratios of minor and major workers is now relatively well understood in the genus *Pheidole*. When Passera (1974a) altered the percentages of *P. pallidula* majors (or "soldiers") from the usual 3–5 percent, most returned to this level within 75 days through the differential production of minors and majors. Even colonies containing 50 or 100 percent majors at the outset dropped to an average level of 20–25 percent and were still declining at the end of the experiment. A similar regulation has been demonstrated in *P. morrisi* by Gregg (1942), *P. bicarinata* (= *vinelandica*) by Wheeler and Nijhout (1981, 1984), and *P. dentata* by Johnston and Wilson (1985). These results show that major:minor ratios are not set entirely by an automatic shunting of larvae at the critical developmental time consistent with the model of simple determination depicted in Figures 8–51 and 8–52. Feedback controls have been added that can temporarily increase or decrease the proportion directed to development as majors.

The nature of the controls was elucidated in *Pheidole bicarinata* by Wheeler and Nijhout (1981, 1983, 1984). Worker larvae have four instars. They molt to the final instar when they are about 0.6 millimeter in length, and soldier determination takes place when the larvae grow to 0.9–1.2 millimeters in length. Minor worker larvae usually begin metamorphosis at 1.3 millimeters, whereas major worker larvae continue to grow beyond that point. Methoprene, a juvenile hormone analogue, delays metamorphosis (as expected) and yields an abnormally large crop of soldiers. Conversely, the presence of an excessive number of soldiers reduces the production of new soldiers, an effect evidently the result of a contact inhibitory pheromone. The model suggested by Wheeler and Nijhout to explain these results incorporates a feedback loop from the colony to the individual. When the colony is growing and foraging workers bring in sufficient food, individual larvae are well fed, and as a consequence their juvenile hormone titer is elevated in some fashion, predisposing them toward soldier determination in the final instar. As the proportion of soldiers in the colony rises, however, so does the amount of inhibitory pheromone to which the still-growing larvae are exposed. The inhibitory pheromone has the effect of making the larvae less sensitive to juvenile hormone, so even though many have an elevated titer of the hormone, it is no longer sufficient to trigger soldier development—and the larva transforms into a minor worker. In essence, a balance is struck between the proportion of soldiers in the colony and the sensitivity of the developing larvae so that the proportion of soldiers remains more or less constant.

We noted earlier that each species of *Pheidole* has a characteristic norm in its major:minor ratio, and that, at least in the case of *P. dentata*, differences among colonies of the same species appear to have a partial genetic basis. The Wheeler-Nijhout model explains how such variation can arise during evolution. All that would be required are mutations affecting production of the juvenile hormone, production of the soldier inhibitor pheromone, or sensitivity to either of these substances.

CHAPTER 9

# Social Homeostasis and Flexibility

What makes social systems most appealing intellectually is the existence of hierarchies. When organization of this kind occurs, there is often more to a whole society than the sum of its parts. Yet even the most emergent properties of the society's behavior remain strictly ordered by devices of communication and regulation operating at the level of organism. This perception is not a mere truism. There is no way to predict which devices have evolved in particular social species, even though general principles can be adduced that affect all of the species together.

A hierarchy is a system "composed of interrelated subsystems, each of the latter being in turn hierarchic in structure until we reach some lowest level of elementary subsystem" (Simon, 1981). Hierarchies are said to be decomposable, which means that the linkages within the components at each level are much stronger than the linkages between the components at the same level. Thus an ant colony can be decomposed into the occupants of different nests when the colony is polydomous. That is, the occupants of each nest are more tightly connected to one another within each short interval of time than are the occupants of different nests. Moving on down, the investigator can decompose the occupants of a nest into various castes, which can then be decomposed into individual ants. An individual ant is in turn a hierarchy of organs, tissues, cells, and so on. Often hierarchies are run by a central command structure, in other words a "boss" at the top, but this is not necessary and in fact does not occur in ant colonies. All that is needed to create a hierarchy is the property of decomposability.

In two respects the ant colony is a special kind of hierarchy, which can be usefully called a "dense heterarchy" (Wilson and Hölldobler, 1988). The colony is dense in the sense that each individual insect is likely to communicate with any other. Groups of workers specialize as castes of particular tasks, and their activities are subordinated to the needs of the whole colony. They do not act by a chain of command independent of the other groups of workers, however. They are open at all times to influence by most or all of the membership of the colony. An ant colony thus differs in basic organization from the "partitioned" hierarchies of human armies and factories, in which instructions flow down parallel, independent groups of members through two or more levels of command. The colony is also a heterarchy, a hierarchy-like system of two or more levels of units with activity in the lower units feeding back to influence the higher levels. Finally, the highest level of the ant colony is the totality of its membership rather than a particular set of superordinate individuals who direct the activity of members at lower levels.

An excellent example of organization by a dense heterarchy is the pattern of food flow into a colony of the fire ant *Solenopsis invicta* as demonstrated in the experiments performed by Sorenson et al. (1985). When foragers were first starved as a group, they collected disproportionately more honey. When they were well fed but the nurse workers or larvae were starved, the foragers collected more vegetable oils and egg yolk. Previous research had shown that sugars are used mainly by adult ant workers, lipids by workers and some larvae, and proteins by larvae and egg-laying queens (Wilson and Eisner, 1957; Echols, 1966; Lange, 1967; Abbott, 1978; Howard and Tschinkel, 1981). Hence the foragers, which are older workers, respond to the nutritional needs of the colony as a whole and not just to their own hunger.

How do they monitor this generalized demand? The answer is that they rely on a combined system and individual decisions joined on a massive scale (Sorenson et al., 1985). A large additional group of workers, the reserves, receive most of the food as it comes into the fire ant nest and then pass it on to other colony members, including the nurses. When the demand they encounter declines in any sector of the colony, the reserve workers accept the corresponding food less readily from the foraging workers. The foragers are unable to dispose of their loads as quickly as before, and they reduce their efforts to collect more food of the same kind. As a result they shift as a group in their emphasis from carbohydrates to oils or proteins, or in the reverse direction, according to the needs of the colony. A similar mediating effect based on the rate of acceptance of newly harvested food has been reported in honey bees and other ant species (Wilson, 1971; Brian and Abbott, 1977; Seeley, 1985).

A second example from the same species is found in the phenomenon of mass communication during recruitment (Wilson, 1962b, 1971). Individual workers arriving at food finds such as an animal carcass lay odor trails composed of secretions from the Dufour's gland (as reviewed in Chapter 7). They decide whether to recruit in this manner according to the nutritional needs of the colony and the richness and appropriateness of the food. Individual workers also vary the amount of pheromone emitted according to these variables. The quickness with which a forager inspects the food and returns to the nest, laying a trail, depends on these same variables, as well as on the amount of crowding by ants already at the food site. Food retrieval proceeds as the number of recruiting workers sets the amount of recruitment pheromone in the trail, which in turn sets the number of workers answering the call. As the colony approaches satiety, or the food site becomes saturated, the likelihood of trail laying by individual workers diminishes. The flow of food correspondingly stabilizes and finally declines.

A third and very different form of dense heterarchy occurs in the regulation of egg laying by fire ants. The queen is the sole progenitrix of new colony members, and much of the colony's activity is devoted to her protection and nurture. It is reasonable to suppose that her production of eggs is related in some manner to the capacity of the colony to rear additional members. When too few eggs are laid, the growth of the colony is slowed below capacity. When too many are laid, energy is wasted, again reducing colony growth—at least in small colonies. How then does the queen monitor the condition of the colony in order to manufacture approximately the correct number of eggs? More accurately, how do nurse workers monitor the condition of the colony and adjust the feeding of the queen to this desirable end? In a series of ingenious experiments, Tschinkel (1988c) showed that the control is based on mass communication. Workers and pupae do not by themselves stimulate queen oviposition. However, the rate of egg laying by the queen increases in a log-log relationship as the number of fourth- (and final-) instar larvae and earliest pupal stages increases. Pharate pupae ("white larvae") are the most stimulating of all, becoming maximally so during the 24-hour period after passing the meconium. Tschinkel found that vital dyes placed in food were transferred rapidly from the fourth-instar larvae to the nurse workers and thence to the queen's eggs, indicating a substantial passage of food along this route. General observations on the behavior of *Solenopsis invicta* and other kinds of ants indicate that such larvae probably serve as a metabolic caste, and are often active in processing protein for egg production by the queen. The queen and fourth-instar larvae are therefore linked in a positive feedback loop.

It is becoming evident that much of the structure of the ant heterarchies is based on order parameters, defined as the proportion and physical distribution of individuals existing in one physiological state or another. Thus the frequency of fourth-instar fire ant larvae relative to young larvae determines the rate of queen egg-laying and consequently the number of additional young larvae due to appear during the following month. The number of hungry larvae in comparison with hungry workers back in the nest determines the emphasis placed by foragers on the kind of food they collect, independent of their own hunger. In a second example, the proportion of major workers in the adult population of *Pheidole bicarinata* affects the ability of final-instar larvae to transform into majors themselves. As the proportion rises, the level of inhibiting pheromone produced by the majors also rises. The pheromone evidently has the effect of making the larvae less sensitive to juvenile hormone, so that they enter metamorphosis more quickly and are more likely to become minor workers (Wheeler, 1986a).

In conclusion, the organization of ant colonies into dense heterarchies permits relatively efficient social regulation with the loosest of command structures. The most striking social effects turn out to be the holistic outcomes of mass communication combined with the rise and decline of pheromones and foodstuffs.

## SERIES–PARALLEL OPERATIONS

The dense heterarchy also provides the colony with the ability to act locally and to blanket a large area in quick response. The colony does not depend on the transmission of information up and down chains of command before decisive local action can be taken. It thus meets important contingencies as they arise, such as the appearance of an enemy or the hunger signal of a larva. The full system is also more secure, because when one colony member fails to complete a task another is likely to succeed. In the design of control devices, engineers commonly utilize such series-parallel operations because the breakdown of one unit will not cause the failure of the whole device. This procedure is more reliable than a parallel-series operation, a simpler system in which sequences of tasks are performed in isolation. When performance in one line of a parallel-series operation breaks down, it is not fixed but merely replaced by another, parallel sequence (see Figure 9–1).

This conception of behavioral acts and sequences, introduced by Oster and Wilson (1978) and refined by Herbers (1981a), sheds new light on the origins of insect social behavior. It can be shown that the reliability of the colony as a whole ($P$) is related to the reliability of its individual components ($p$) by a sigmoidal curve of the kind illustrated in Figure 9–2. When the competence of individuals is low, the performance of the group is lower than if the individuals acted independently. There is, however, a threshold ($p^*$) of individual reliability above which group reliability exceeds that of individual reliability. In this region, colony-level selection favors cooperative behavior.

The concept of a threshold of individual competence may seem paradoxical, but an example will make the relation intuitively clear. Suppose that several ants are attempting to move larvae from one part of the nest to another part with a more favorable temperature. If each ant tries the entire operation alone, the group result will equal the average individual result. Now suppose that two workers start working together, as in fact they often do when the larvae are large (Figure 9–3). If their competence is low, they will tend to cancel out one another's actions. They may pull in opposite directions, or start and stop independently so as to stall the operation indefinitely. The probability that incompetent ants will make the right combination of correct movements ($k/n$), such as gripping the larva in the right position, lifting it to the proper height, and selecting the right direction to walk, will be even lower than the chance of doing one thing correctly. In the end fewer larvae will have been transported. As individual competence increases, however, the group will reach a point at which they are able to put together the right combination of actions and speedily move a larger number of larvae. The improvement will be greater still if they divide the labor. For example, some might concentrate on laying odor trails or leading tandem runs while others concentrate on transporting the larvae.

The potency of cooperation in a dense heterarchy is enhanced in another way predicted from reliability theory. This second theorem states that replication at the subunit level is more efficient than redundancy at the system level. In other words, if a designer is given two sets of parts, it is better for him or her to build a single system with redundant components than to build two separate systems. The reliability of the redundant system, hence the chance that it will work at all, is greater than the reliability of the separate systems. This relationship is illustrated in Figure 9–4.

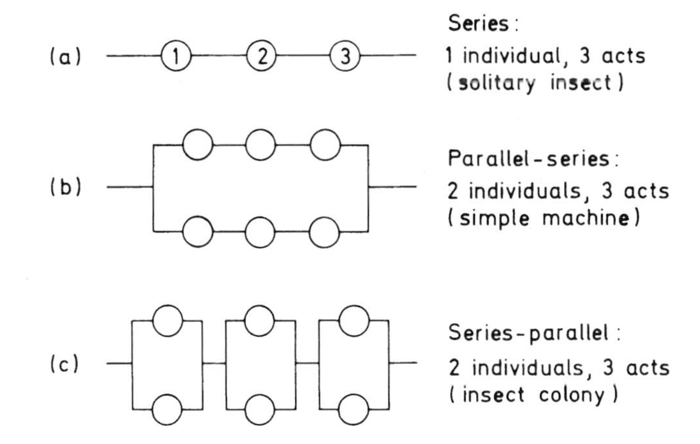


**FIGURE 9–1** A comparison of the reliability of behavioral sequences. The series-parallel operation is the most reliable. For example, if the probabilities of both the correct performance at each step and the correct transition from one step to the next are all set at 0.2, then the probability of completing the sequence is 0.01 in the elementary series (*a*), 0.02 in the parallel series (*b*), and 0.05 in the series-parallel operation (*c*). (Modified from Oster and Wilson, 1978.)

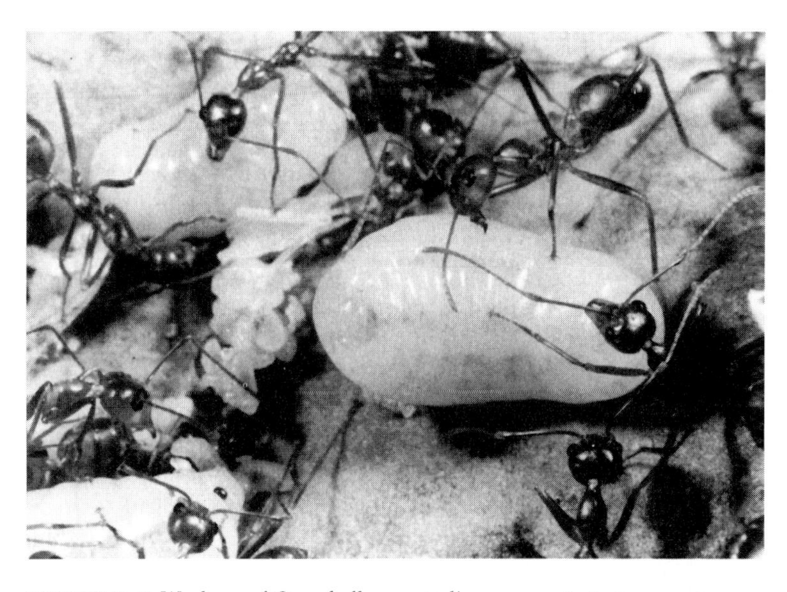

**FIGURE 9–3** Workers of *Oecophylla smaragdina* cooperate to transport a large queen larva.

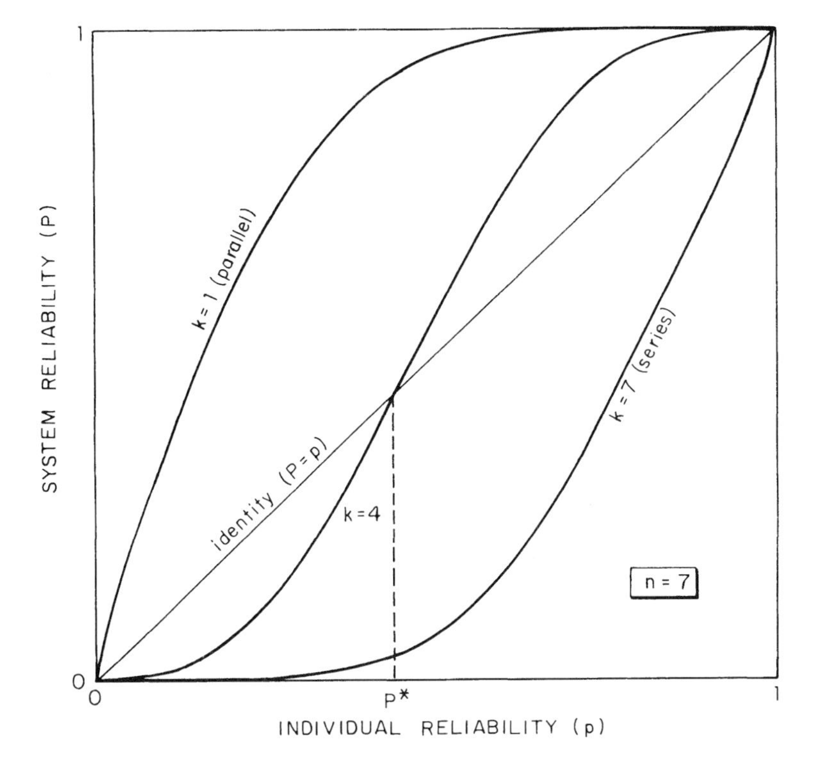


**FIGURE 9–2** The reliability of a system (such as an ant colony), defined as the probability that the system can act correctly or in some designated number of times, depends on whether its units operate in series or in parallel. $P$ is the probability that the entire system operates successfully (i.e., the system reliability) and $p$ is the probability that an individual carries out her task successfully (i.e., the individual reliability). If the requirement is to perform the act correctly during at least $k$ attempts out of $n$, the overall success ($P$) increases with individual competence ($p$) sigmoidally. For $n = 7$, the value $k = 7$ corresponds to a completely serial operation, whereas $k = 1$ corresponds to a completely parallel operation. The typical ant colony, organized as a dense heterarchy, conducts parallel operations. (From Oster and Wilson, 1978.)

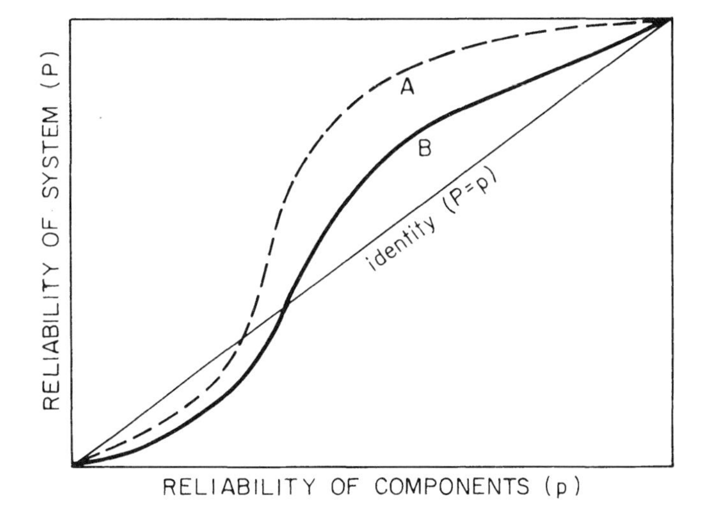


**FIGURE 9–4** The reliability of an ant colony or similar whole system depends on the reliability and redundancy of its components. As illustrated here, a system with redundancy at the level of the individual workers (*A*) does better than a system with redundancy at higher levels, such as entire groups of workers (*B*). The ant colony, which acts as a dense heterarchy, approaches redundancy at the level of the individual worker. (From Oster and Wilson, 1978.)

## MULTIPLIER EFFECTS

The reliability theorems illustrated in Figures 9–2 and 9–4 lead us to believe that the responses of the colony as a whole are generally amplified. That is, not only do the parallel operations of redundantly acting workers increase the chance that a given task will be performed, but the speed and effectiveness of the colony will be improved.

An example is provided by mass communication during trail laying in the myrmicine *Tetramorium impurum,* as described by Verhaeghe (1982). Like scouts of fire ants, the foraging workers of *T. impurum* vary the amount of pheromone paid into the trail according to the quality of the food find and the degree to which food is needed by the colony. The probability that a follower ant succeeds in following the trail to the end increases as the logarithm of the amount of pheromone deposited. As a result, variations in the quantity of pheromone voided by individual workers translate into large differences in the number of ants traveling back and forth to the target. As more scouts lay trails during their parallel efforts, the segments they individually contribute are more likely to unite into a complete trail. The result is a rapid exploitation of food and new food sites. This particular form of amplification is likely to be widespread in trail-laying ants. In most cases scouts act more or less independently in laying trails. Also, in at least the species of *Acanthomyops, Myrmica, Solenopsis,* and *Tetramorium,* workers individually vary the quantity of pheromone (Hangartner, 1969b; M.-C. Cammaerts, 1977; Verhaeghe, 1982).

Other multiplier effects can arise as a consequence of the statistical properties of caste systems. The tournaments of the honeypot ants of the genus *Myrmecocystus,* for example, are conducted by major workers. This caste is not a discrete group but comprises the largest individuals in a population of workers whose size-frequency curve is unimodal and skewed toward the larger end. The size-frequency curve expands as the colony grows in size, so that larger colonies have a higher percentage of majors and a much larger absolute number of individuals belonging to the extreme size class. As a result they possess an overwhelming advantage over smaller colonies during tournaments, and they are more likely to win territorial disputes (B. Hölldobler, unpublished).

## RULES OF THUMB

The total behavioral repertory of an individual ant worker is relatively simple, consisting according to species of no more than 20 to 45 acts. Yet the behavior of the colony as a whole is vastly more complex in all but the most primitive species. As we showed in Chapter 8, some of this superstructure is built on the physical caste system. Small and large workers have different behavioral repertories, which fit together to form larger wholes. Of even greater consequence are the social feedback loops operating through mass communication. The individual ant need operate only with "rules of thumb," elementary decisions based on local stimuli that contain relatively small amounts of information. The examples of rules of thumb we have reviewed thus far in this chapter could be expressed as follows: continue hunting for a certain foodstuff if the present foraging load is accepted by nestmates; follow a trail if sufficient pheromone is present; feed the queen more if final-instar larvae are present; and attend the larvae and other immature stages if regular nurse workers are absent. Each of these rules is easily handled by the individual worker, even when we allow for brains as small as a tenth of a cubic millimeter. Each action is also performed in a probabilistic manner with limited precision. Yet when the actions are put together in the form of dense heterarchies involving large numbers of workers, the whole pattern that emerges is strikingly different and more complicated in form, as well as more precise in execution.

## THE SUPERORGANISM

The new perception of heterarchical organization has revived the venerable idea of the ant colony as superorganism. William Morton Wheeler, like many of his contemporaries, was guided by this concept. In his celebrated essay "The Ant Colony as an Organism" (1911) he stated that the animal colony is really an organism and not merely the analogue of one. Of course, one has to pay close attention to his definition of organism:

> An organism is a complex, definitely coordinated and therefore individualized system of activities, which are primarily directed to obtaining and assimilating substances from an environment, to producing other similar systems, known as offspring, and to protecting the system itself and usually also its offspring from disturbances emanating from the environment. The three fundamental activities enumerated in this definition, namely nutrition, reproduction and protection, seem to have their inception in what we know, from exclusively subjective experience, as feelings of hunger, affection and fear respectively.

These malleable criteria are easily applied to the ant colony and other insect societies. Wheeler saw several important qualities of the ant colony that qualified it as an organism:

1. It behaves as a unit.
2. It shows some idiosyncrasies in behavior, size, and structure that are peculiar to the species.
3. It undergoes a cycle of growth and reproduction that is clearly adaptive.
4. It is differentiated into "germ plasm" (queens and males) and "soma" (workers).

In *The Social Insects* (1928), Wheeler began to call the insect colony a superorganism rather than the more obviously metaphorical organism. His ideas on the subject of colonial organization had actually changed little by that time, although he had begun to conceive in a vague fashion of the phenomenon that was later to be called homeostasis: "We have seen that the insect colony or society may be regarded as a super-organism and hence as a living whole bent on preserving its moving equilibrium and integrity."

Wheeler's imagery grew from the *Zeitgeist* prevailing during his most productive years. From about 1900 to 1950, many biologists and philosophers besides Wheeler developed a keen interest in holism and emergent evolution. Although a considerable amount of mysticism was generated, the most famous example being Maurice Maeterlinck's "spirit of the hive," the core of scientifically oriented writing concentrated on analogies between the organism and the superorganism. Trophallaxis (the exchange of liquid food among colony members) was cited as the equivalent of circulation, whereas soldiers were thought of as paralleling the immune system, and so on. Although initially stimulating, even inspirational, the entire elaborate exercise eventually exhausted its possibilities. The limitations of a primarily analogical approach became increasingly apparent when investigators turned more to the fine details of communication and castes. By the early 1960s the expression "superorganism" had all but disappeared from the technical vocabulary (see the historical review by Wilson, 1971: 317).

Old ideas in science never really die, however. They only sink to mother Earth, like the mythical giant Antaeus, to rise again with renewed vigor. The time may have arrived for a revival of the superorganism concept. We see two reasons, both stemming from the increase of information and technical competence in its analysis since 1960. The first is the beginning of a sound developmental biology of the insect colony, and the second is the rapid improvement of optimization analysis in behavioral ecology and sociobiology.

The new developmental biology has provided an understanding of the ways in which castes are determined, their ratios regulated, and their actions coordinated through heterarchical forms of communication. There is a need now for drawing analogies at a deeper level, in which the organizational processes of societies are more precisely measured and compared with their equivalents in the growth and differentiation of tissues of organisms. It is not too much to suggest that insect sociobiology will contribute to a future general theory of biological organization based on quantitatively defined principles of feedback, multiplier and cascading affects, and optimal spatial arrangements. The new effort will prove additionally useful in calling attention to poorly understood organizational processes and the techniques by which they are more precisely analyzed. For example, Meyer (1966), Hofstadter (1979), Markl (1983, 1985), and Minsky (1986) have cited similarities between the organization of neurons in the brain and that of workers in an insect colony.

In *The Insect Societies* (1971), Wilson observed that faith in reductionism rides on the ability of researchers to use many piecemeal analyses to reconstruct the full colonial system in vitro. Such a reconstruction will mean the full explanation of social behavior by means of integrative mechanisms experimentally demonstrated and the proof of that explanation by the artificial induction of the complete repertory of social responses on the part of isolated members of insect colonies. Three achievements appear to be central to the realization of this dream. None seemed within reach in 1971, but now, nearly twenty years later, all have been partly attained:

1. The activation of social behavioral responses, including the more intricate and delicate aspects of brood care, by exposure of isolated colony members to synthetic pheromones, sounds, and other stimuli emanating from lifeless dummies or else presented wholly in vacuo. These procedures are now routine in research on ants (see Chapter 7).

2. The rearing of isolated ant and honey bee larvae or termite nymphs and the determination of their caste at the will of the investigator by appropriate manipulation of food, pheromones, hormones, and other caste-biasing factors. Although immature forms have not yet been raised in isolation, the experimental conditions have been adequately controlled and the key factors of caste determination identified in some species of ants (see Chapter 8).

3. The cybernetic simulation of nest building, incorporating into the model only those behavioral elements and sequences of elements actually observed in individual workers, leading in turn to the successful prediction of the responses of isolated workers presented with synthetic nests in various stages of construction. Preliminary steps have been taken in this direction during studies of *Nasutitermes* termites by Jones (1979, 1980), but we are still a long way from in vitro duplication for any species of social insect.

The historical cycle of research in sociobiology has run from an imagery of emergent properties in complicated societies to the closer scrutiny of these properties by experimental studies and finally to the reconstitution of the facts into a new whole. The reconstituted imagery will inevitably differ in important details from the original, and will provide still more distant perceptions that invite experimental investigation.

A partial explanation of ant colonies may be built in such a manner, but it requires in addition an evolutionary perspective in which the whole system is examined as an adaptation to the environment. The adaptationist explanation is not a dogma, in which the investigator sees each newly discovered process as increasing the fitness of the society. Rather it is a mode of framing hypotheses, a heuristic device used to design field studies and laboratory experimentation.

Let us take a concrete example. The relative frequencies of the castes are expected to be adjusted by natural selection, and so the caste system is conjectured to be adaptive. That is, the numbers of minors, majors, callow workers, and so forth are postulated to compose a mix enabling the colonies to survive and reproduce better than would be the case if they possessed a different mix. The optimal mix is attained by adaptive demography, a genetically programmed schedule of birth, growth, and mortality that yields standing crops of caste members in the appropriate numbers. Moving down to the next level of explanation, we are challenged to ask why certain mixes are superior. Lumsden (1982) has pointed to three determining properties of the colony members: (1) the ergonomic gaps between the castes, in other words those tasks not exactly covered by particular castes which have evolved to perform them; (2) the short, Markovian memory of colony members, requiring virtually moment-by-moment perception and adjustment of the colony as a whole to environmental exigencies; and (3) interactions among the colony members. In theory at least, the organization of the colony can be described as the matrix of interaction of the members of the colony both within and across castes, subject to the constraints of ergonomic gaps and limited memory. Some of the workers will interact at very frequent intervals, others seldom. Most of the interactions will be cooperative and productive for the colony whereas others will consist mostly of interference and reduce production. The ergonomic matrix presumably evolves toward higher fitness states by the genetic alteration of the relative frequencies and behavioral patterns of the castes and the details of their interactions. The matrix is therefore the scaffolding of the superorganism. Our growing knowledge of caste determination, communication, and feedback loops will make complete sense only when these individual-level processes are correctly placed on the scaffolding.

Thus, the new holism subsumes explanations of three kinds that can be aligned with one another: the relative adaptiveness of the colonies as superorganismic operating units within their natural environments, the ergonomic matrix that determines an optimal or at least evolutionarily stable mix of castes and communication systems, and the details of the castes and communication systems themselves.

## REGULATION OF HOMEOSTASIS

The concept of the colony as an organized whole also implies a *norm of reaction* in each of the categories of social organization. That is, the genotype prescribes a particular caste system and a particular communication system that emerge in response to the environment in which the colony grows up. In some instances the norm of reaction is narrow, so that only one social response is to be expected regardless of the environment. We then speak of the response as rigid or stereotyped. In other instances the response differs from one environment to another but in a consistent and evidently adaptive manner. In the latter case the norm of reaction is broad, based upon a flexibility in worker behavior or colony development. A useful corollary distinction can be made between "noise," the irregular fluctuation around the norm of reaction caused by changes in the environment and accidents in development, and the regularized shift in social organization that adapts the colony to new circumstances. In the case of noisy perturbations, the colony tends to return to the original state by means of negative feedback loops in physiology and behavior. In the case of adaptive shifts to new positions in the norm of reaction, the colony uses physiological and behavioral responses to move to alternative modes of social organization.

Both procedures constitute homeostasis in the broad sense. Whether a social process constantly readjusts itself back to the status quo ante or shifts to a new adaptive state, it permits the basic life functions such as regulation of temperature and care of the larvae to proceed with little interruption. In the traditional expression, the life functions are homeostatic, and the social responses have evolved as either stereotyped or flexible enabling devices to achieve their end.

An example of social homeostasis is provided by the caste systems of the ant genus *Pheidole*. Each species of this large cosmopolitan genus has a characteristic ratio of small-headed minor workers to large-headed major workers (see Figures 7–38 and 8–21). When the ratio is altered in a particular colony by an excess of birth or mortality in one of the castes, the colony converges back toward the original ratio within one or two worker generations, extending across one to three months. The feedback loop is an inhibitory pheromone produced by majors that lowers sensitivity to juvenile hormone, so that larvae surrounded by an excess of majors curtail growth and tend to become minors. Those present during a shortage of majors become more sensitive to juvenile hormone, extend growth, and turn into majors.

A second form of social homeostasis is displayed by the behavior (as opposed to the physiology) of the *Pheidole* major workers. When the relative frequencies of adult minors and majors are anywhere near the species norm, the majors are inactive most of the time. When they do work they keep to the specialized roles of nest defense, ingluvial storage of liquid food, and milling of seeds. In contrast, the minors take responsibility for virtually all nursing of immature forms, nest construction, and other quotidian tasks. However, when the minor:major ratio is lowered to below 1:1 (from 3:1 to 20:1, the norm according to species), so that minors are much scarcer than usual, the repertory of the majors suddenly and dramatically expands to cover most of the tasks of the minors. Within less than an hour, majors can be seen attending brood, handling nest material, and in general acting like rather clumsy, overweight minors. Under these transformed circumstances, the number of kinds of acts has been observed to increase by 1.4 to 4.5 times according to species. The rate of activity, measured in number of acts of all categories performed in each unit of time, increases by 15 to 30 times (Wilson, 1984b; see Figure 9–5). By this means the majors restore about 75 percent of the missing minors' activity. Even when all of the minors are taken away, the majors can rear brood to maturity and otherwise keep the colony going until a substantial minor force is restored.

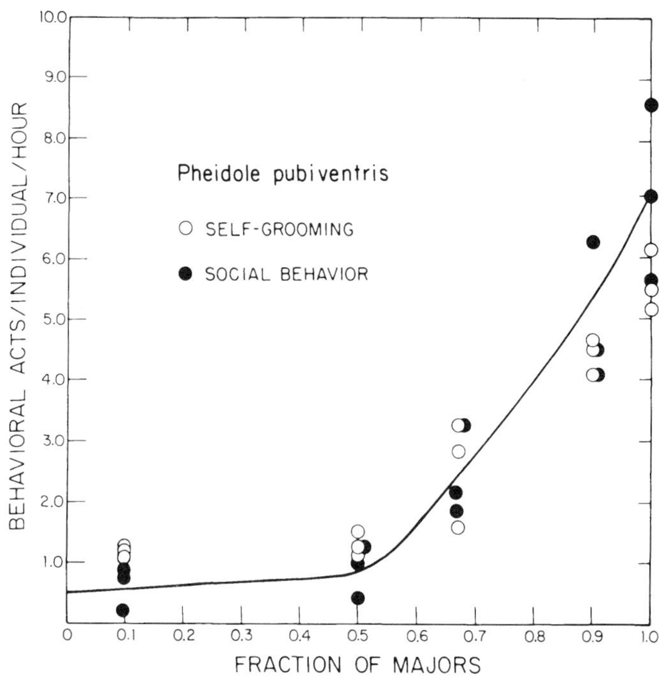


**FIGURE 9–5** Elasticity in the behavior of the major workers in a colony of *Pheidole pubiventris*, a species found in South American rain forests. When the colony is modified so that majors outnumber minors, the rate of their social behavior increases by approximately tenfold. The photograph shows minor and major workers on a pile of larvae and pupae. The diagram below depicts the rapid increase of grooming and social activity by the majors as the minors become scarce. (From Wilson, 1984b.)

In such cases as the standby role of the *Pheidole* majors, a distinction can be made between the programmed *elasticity* in the repertory of individual workers and the *resiliency* of the colony as a whole, which depends on the pattern of caste-specific elasticity. The homeostasis achieved is in the vital functions of the colony, sustaining a near-normal load of brood care and other tasks until the normal caste ratios are restored. The signal that informs the *Pheidole* majors when to remain quiescent and when to spring into action is a ritualized form of aversion on the part of the majors (Wilson, 1985a). They actively avoid minors while in the vicinity of the larvae and other immature states. Majors do not turn away from each other near the immature stages, however, and they avoid neither minors nor majors while in other parts of the nest. For their part, minors do not avoid either minors or majors anywhere in the nest. When minors are scarce enough, majors move onto the piles of eggs, larvae, and pupae and commence attending to them. At the same time they commence other tasks and greatly increase their rate of self-grooming. The result of this seemingly simple rule of procedure is a striking division of labor when colonies have a normal caste composition, followed by a homeostatic shift of behavior by the majors when minors become unusually scarce.

A similar aversive phenomenon occurs in the leafcutter ants of the genus *Atta*. Most of the foragers engaging in active cutting of leaves and other vegetation are in the media size group, with head widths between 1.8 and 2.2 millimeters (see Figures 8–29 and 8–30). When 90 percent of these medias were removed experimentally from laboratory colonies by Wilson (1983a), the rate of harvesting remained unaffected, owing to the fact that excess workers in the adjacent size classes were already present on a standby basis in the foraging area. In addition, the survivors in the 1.8–2.2-millimeter group increased their individual activity rate by approximately five times. In a manner reminiscent of dominance by the *Pheidole* minor workers, the prime foragers in the *Atta* 1.8–2.2-millimeter group tended to displace the others from the edges of the leaves, where most of the cutting takes place. When these individuals were removed, the auxiliaries participated more freely.

Cases of social homeostasis that return the colony repeatedly to its previous state have been discovered with increasing frequency. When a colony is disturbed, to cite another category, only a small percentage of the workers respond with all-out attacks on the intruder. Le Roux and Le Roux (1979) found that only about 20 percent of *Myrmica laevinoda* workers attacked an approaching worker of *Myrmica ruginodis*. Another 30 percent approached it without violence, and the remainder did not visibly respond. When the ants were then rearranged into groups more nearly homogeneous with respect to aggressivity, the behavior of enough ants changed so that the percentage of aggressive workers approached that in the original ensembles. Finally, when the ants were restored to their original groups, there was a tendency for individuals to retain the level of aggression they had assumed in the altered groups. In the end the proportions of violent and nonviolent ants were not greatly changed.

The speed with which workers alter their individual "idiosyncratic" behavior in response to stress varies according to both the behavioral category and the context. Queens and very young workers of *Myrmica rubra*, like those of most other ant species, do not participate in brood transport during colony emigration. They readily do so, however, when isolated from the older workers that normally take responsibility for this task, and they accomplish the task in only a little less time. When the older workers are separated into elite and non-elite groups with reference to this task, the two ensembles converge in their levels of activity, but only after a period of three months (Abraham et al., 1980, 1984). Moderate flexibility of a similar kind has been documented in *Camponotus vagus* by Bonavita-Cougourdan and Morel (1988).

The degree of flexibility can change with the size or age of the colony. Young workers of *Lasius niger* in founding colonies were found to be more flexible in behavior and slower to specialize than those in more mature colonies (Lenoir, 1979a). Workers of young colonies of *Pogonomyrmex barbatus* are variable in their response to perturbations, but this greater flexibility—or more precisely variability—results in a less rapid return to the status quo ante (Gordon, 1987). In other words individual workers of young colonies appear to be more versatile, but the colonies as a whole are less homeostatic. Thus the phenomenon of adaptive demography, entailing programmed changes in the size and age of workers as the colony grows older, is augmented by programmed changes in the pattern of behavior of the individual workers.

## ADAPTIVE SHIFTS IN BEHAVIOR

The most conspicuous form of adaptive shift in social organization, as opposed to changes in behavior that merely restore preexisting social organization, is in the degrees of polydomy (number of nests per colony) and polygyny (number of queens per colony or local nest). Repeated studies of *Iridomyrmex* and *Leptothorax* have revealed how variation in these traits within single species arises from flexible procedures in the modes of colony founding and subsequent dispersal of colonies to multiple nest sites by means of budding (Buschinger, 1967, 1968c; Alloway et al., 1982; Greenslade and Halliday, 1983; Hölldobler and Carlin, 1985). Most of the analysis has depended on direct observation and the correlation of polygyny and polydomy with environmental variables in the field. However, Herbers (1986b) also used field experimentation in her studies of the North American *L. longispinosus*. She seeded a forest tract with artificial nest sites consisting of wooden rods with holes drilled in them. During the ensuing two years the number of both queens and workers per nest declined in comparison with unseeded plots, a reflection of an expansion of existing colonies into the newly available nests and hence a thinning effect of the overall population. At the same time the total population appeared to increase in the experimental plot, which suggested that it was undergoing growth as the colony fragments multiplied.

Some ant species adjust the time of principal foraging activity dramatically in response to weather, the presence of food, and the availability of foragers at any given time (see also Chapter 10). In tropical habitats daily and seasonal changes in foraging activities of ants are related both to temperature (Torres, 1984a,b) and moisture (Levings, 1983; Levings and Windsor, 1984). Documented examples include *Paraponera clavata* (McCluskey and Brown, 1972; Harrison and Breed, 1987), *Acromyrmex octospinosus* (Therrien et al., 1986), and *Atta cephalotes* (Cherrett, 1968). Other species, such as *Monacis bispinosus*, have been observed to shift their foraging schedules in

**FIGURE 9–6** The spider *Steatoda fulva* builds its webs in front of the nest exits of the Florida harvester *Pogonomyrmex badius*. *Pogonomyrmex* foragers leaving the nest are caught in the web and eaten by the spider. Surviving workers respond by closing the nearest exits with sand. (From Hölldobler, 1971e.)

response to invasion by workers of a competing ant species (Swain, 1977). *Pheidole titanis,* an ant that hunts termites in the deserts and deciduous thorn forests in the southwestern United States and Mexico, changes the diel timing of its raids to avoid a parasitoid phorid fly (Feener, 1988).

Several cases have been reported in which ant species modify their nests in response to parasites or predators. Workers of *Formica subsericea* plug their nest entrances with soil, pebbles, or grass after being raided by the slave-making ant *F. subintegra*. They also remove the traces of excavated soil and discarded cocoons that otherwise typically litter the nest surface (Talbot and Kennedy, 1940). Similarly, colonies of *Myrmecocystus mimicus,* endangered by raids from neighboring colonies of the same species, close their nest entrances and cover the surroundings with sand. In so doing they cover up colony-specific nest markers (Hölldobler, 1981a and unpublished observations). Workers of the harvester ant *Pogonomyrmex rugosus* under persistent attack by the western widow spider (*Latrodectus hesperus*) respond by closing their nest entrances and decreasing or halting their foraging activity, even when less than 0.2 percent of the population is being lost per day (MacKay, 1982b). An identical behavior has been recorded in the Florida harvester ant, *P. badius,* which is beleaguered by the theridiid spider *Steatoda fulva* (Hölldobler, 1971e; Figure 9–6). Workers of the desert genus *Cataglyphis* in North Africa also close their nest with soil when attacked by hunting spiders (*Oxyopis*), which are among their most important enemies. The spiders often overcome the obstacle by removing the earth, particle by particle (Harkness and Wehner, 1977).

## VARIATION AMONG COLONIES

Given even a small amount of such flexible behavior, one is not surprised to find considerable differences in the overall social behavior of ant colonies belonging to the same species. The variance is enhanced by differences in queen fertility, amount of brood present, and other qualities that vary from one week to the next within the same colony. Nor can genetic differences be discounted, as demonstrated by the results of experiments on the caste ratios of *Pheidole dentata* (Johnston and Wilson, 1985). Gordon (1983a) discovered strong intercolony variation in the "daily rounds" of *Pogonomyrmex badius,* including the numbers of workers engaged in patrolling, nest maintenance, and foraging as well as the time of peak activity. Glunn et al. (1981) found striking differences in the dietary preferences of colonies of fire ants in both field and laboratory. Herbers (1983) found so much variation in the ethograms of colonies of *Leptothorax ambiguus* and *L. longispinosus* respectively that the two species could be only tenuously separated on the basis of the behavioral data alone. Research on such variation and its multiple causes is obviously in its earliest stages and will no doubt provide new surprises.

## POSITIVE FEEDBACK AND RUNAWAY REACTIONS

Positive feedback loops, in which the product of a process increases the rate of reaction leading to more of the same product, are relatively rare in biology. Where negative feedback loops decelerate processes in such a way as to regulate the product at a more or less constant level (hence homeostasis), positive feedback tends to produce a runaway reaction that ends only when supervenient negative loops are activated or else the materials needed to create the product are exhausted.

An example of positive feedback is the stimulation cited earlier of queen egg-laying by the presence of mature larvae in the fire ant *Solenopsis invicta* (Tschinkel, 1988c). The presence of mature larvae causes the queen to lay more eggs that lead to more mature larvae, which stimulate the queen to lay even more eggs, and so on. But the system does not run away, because when large larvae become numerous enough, it is expected (although not yet proved) that further production will be slowed by the workers' inability to feed them. Also, the stimulative effect per larva decelerates as more larvae are added.

A possible positive feedback leading to a runaway reaction does occur in the nuptial flights of the carpenter ant *Camponotus herculeanus.* When males become sufficiently aroused, they release a pheromone from their mandibular glands that stimulates both other males and the virgin queens, causing these reproductive forms to fly away from the nest for a short interval of time (Hölldobler and Maschwitz, 1965). Whether detection of the pheromone itself causes males to release still more of the pheromone remains to be established.

Still another form of runaway reaction appears to occur during absconding by colonies of the myrmicine ant *Pheidole dentata* under attack by fire ants, which are among their principal enemies (Wilson, 1976b). When *P. dentata* minor workers encounter no more than one or a few fire ants near their nest, they recruit major workers to

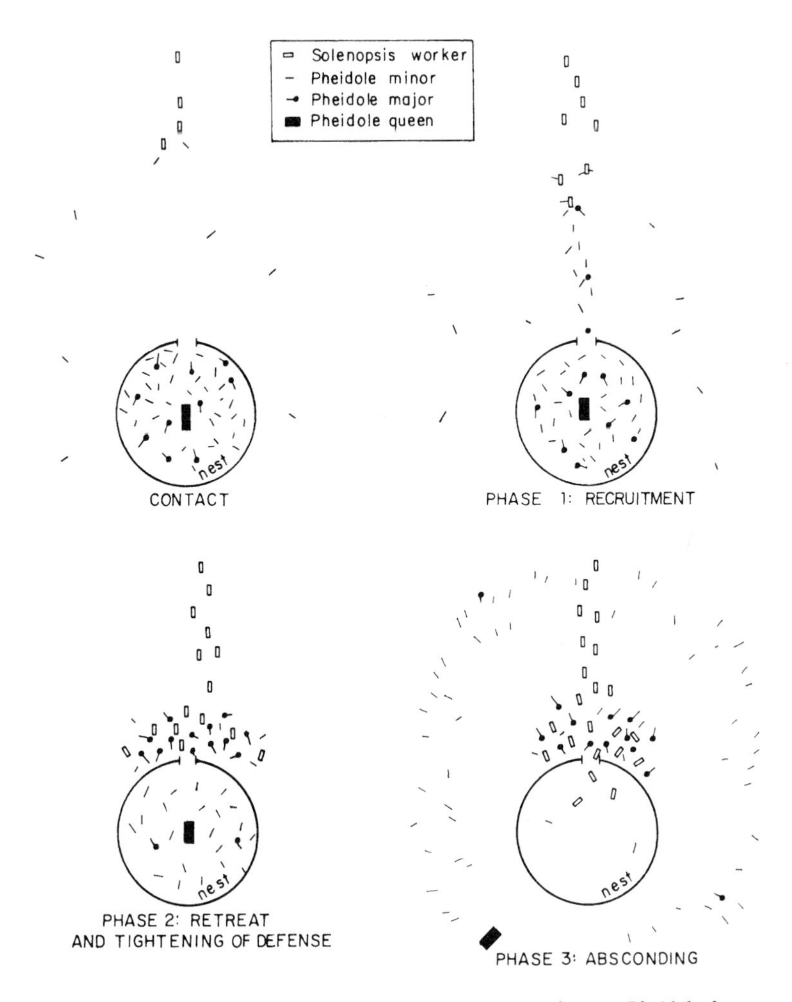


**FIGURE 9–7** The three phases of colony defense in the ant *Pheidole dentata*. The third phase, culminating in desertion of the nest by the colony, appears to be organized by positive feedback in the release of alarm pheromones. (From Wilson, 1976b.)

the site. This reaction usually results in an early destruction of the invaders. But when the minor workers run into large numbers of fire ants, a second phase of defense is initiated. A few grapple with the intruders but most scatter outward in all directions. Only a minority of those fleeing return directly to the nest, and proportionately few odor trails are laid even by these individuals. Consequently the fire ants are able to move quickly to the vicinity of the *Pheidole* nest. There they meet stiffening resistance as more minor workers are encountered, a larger number of trails are laid, and a growing force of major workers is assembled. Thus the second phase of defense is only an extension of the first conducted in another location. It may seem paradoxical that when faced with more enemies the colony should recruit majors less efficiently. But this weaker response applies only to the territory farthest from the nest. The result of importance is that the majors are committed to battle close to the nest, where in fact they are now most needed.

If the second phase of defense begins to break down, the colony absconds, which constitutes the final, desperate, yet very effective maneuver (see Figure 9–7). The following events lead to absconding. As the fire ants crowd closer to the nest entrances, an increasing number of *Pheidole* minor workers lay odor trails all the way back to the brood areas. As a result the excitement of the nest workers increases, and many begin to pick up pieces of brood and carry them back and forth. Meanwhile, the defense perimeter continues to shrink as *Pheidole* majors are immobilized in combat and more *Pheidole* minor workers are either immobilized or flee from the nest area after contacting fire ants. Activity within the brood area builds up, sometimes at an apparently exponential rate, and minor workers then begin to run out of the nest, many laden with pieces of brood. At first there is a strong tendency for the refugees to loop back and reenter the nest, but if they continue to encounter fire ants they break away entirely and flee outward. No particular direction is taken by individuals during the exodus, so that the colony ends up scattered over a wide area. The queen also departs under her own power. Later, after settling down, she attracts minor workers who cluster around her. No direct physical transport of one adult by another has been observed. When the fire ants are removed, the *Pheidole* adults slowly return to their nest and reoccupy it.

In summary, there appear to be two processes contributing to the relative suddenness and *en masse* quality of absconding: As fire ants press in, more and more *Pheidole* of both castes are killed, while other minor workers simply desert the area. This leads to a steadily decreasing ratio of defenders to invaders and a shrinking of the perimeter of defense. A point is reached in which the nest workers are contacting fire ants at a sufficiently high rate to cause them to seize brood pieces and leave. This final critical level is approached steeply. Simultaneously there is an exponential increase in the rate at which minor workers run back and forth between the vicinity of the nest entrance and the brood chambers. Since many lay odor trails, there appears to be a buildup in the concentration of the pheromone.

In short, absconding is an explosive social response created by the reciprocal acceleration of alarm and contact with the invading ants. Each process enhances the other until they attain the threshold level at which absconding occurs. This final action dissolves the buildup that created it.

## CASTE AND BEHAVIORAL FLEXIBILITY

In Chapter 8 we argued from the theory of the allometric space that the number of castes in the optimal mix can be expected to increase as the logarithm of the number of tasks performed by the colony. It is further true that a very sensitive relation exists between the width of the repertory of individual workers, in other words their moment-to-moment behavioral flexibility, and the number of castes in the optimal mix. The greater the flexibility, the fewer the castes. Moreover, the transition is readily made from a polymorphic state, in which multiple physical castes are specialized on sets of tasks, to a monomorphic state, in which a single physical caste addresses itself to all tasks. In some instances all that is required is a slight increase in either behavioral flexibility or cooperation among the caste members (Oster and Wilson 1978; Traniello, 1987b; West-Eberhard, 1987). The theory can be briefly summarized as follows, using the version of the allometric space displayed in Figure 9–8. The shaded area is the "reach" of an ant of a particular caste, where the caste is defined by some pair of physical traits $x$ and $y$ on the allometric plane. As behavioral flexibility increases, and worker anatomy becomes less specialized, or else cooperation increases, the shaded area expands. This means that workers of that caste can perform more and more tasks with at least the minimal required efficiency.

**FIGURE 9–8** The conditions affecting the relation between behavioral flexibility and numbers of physical caste are conjectured in this model. The efficiency function η of a single caste *x,y* is projected over the allometric space. The points shown, both those covered by the minimal competence of the caste (filled in) and those uncovered (empty), are the tasks. The location of each of the tasks is given as the physical properties of ants on the *x,y* plane (e.g., on the plane of head width and head length, two of many properties) that are best suited for the performance of the task. Beyond a certain distance ε, individuals cannot perform the task with the minimum level of efficiency required to make the effort energetically profitable. (From Oster and Wilson, 1978.)

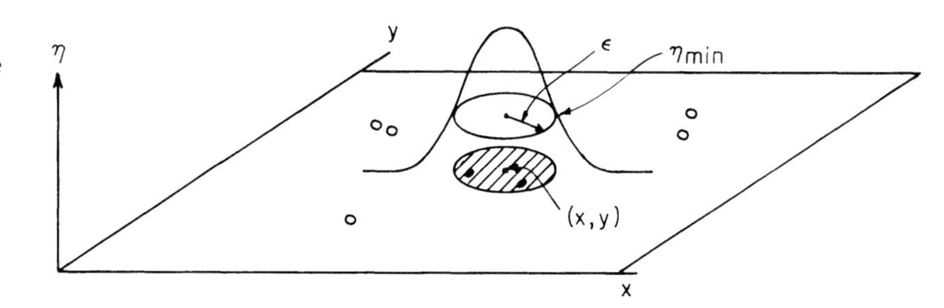


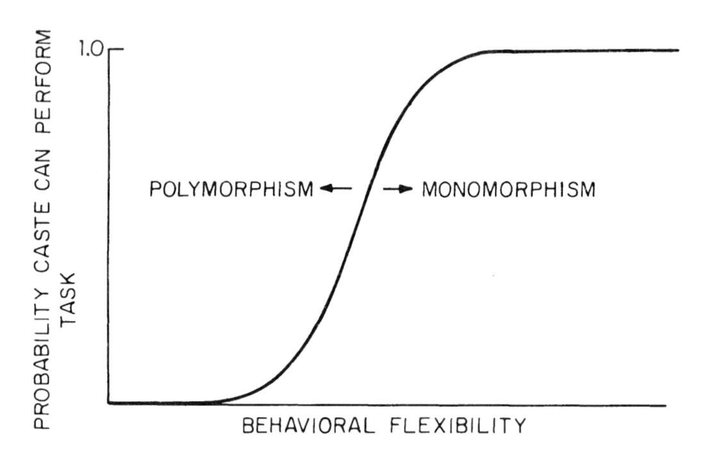


**FIGURE 9–9** The probability that a caste can perform a task with adequate competence increases abruptly when the behavioral flexibility of the caste reaches a certain level. The shift from polymorphism to monomorphism can occur with a relatively small increment in either behavioral flexibility or the capacity to work cooperatively. (Modified from Oster and Wilson, 1978.)

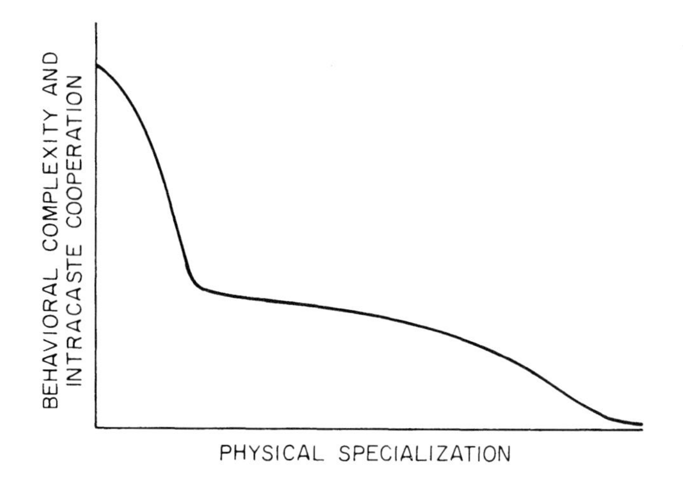


**FIGURE 9–10** The trade-off between the number of physical castes and behavioral flexibility of individual ants is unlikely to be proportional. As suggested in this curve, selected from many possible curves of the same general form, a phase transition exists in which the optimum strategy of a species can shift abruptly from multiple specialized castes to a single flexible worker caste. (From Oster and Wilson, 1978.)

Now suppose that the tasks are distributed at random over the allometric space. As behavioral flexibility or cooperation increases, there is a surprisingly abrupt transition from an optimal state of polymorphism, comprising multiple physical castes, to monomorphism, in which only a single physical caste is optimal (Figure 9–9). The principal inference from this argument is that for a given degree of physical specialization, a small change in behavioral flexibility or cooperativeness can result in a shift in the colony optimum from polymorphism to monomorphism and back again. The phenomenon is reminiscent of "phase transition" curves in physics, which characterize sudden condensations, shifts from order to disorder, and other abrupt changes of important magnitude. As Figure 9–10 suggests, the trade-off between physical and behavioral castes is not proportionate: the strong selective pressures for morphological differentiation should disappear rather quickly as behavioral complexity increases. The same is true as the ability of members of the same caste to cooperate increases, enlarging the social coverage of the allometric space.

If such an evolutionary phase transition truly exists, we should expect genera and other sets of phylogenetically closely related species to display striking variation in the degree of worker polymorphism. This proves to be notably true in the case of ants. Of the 45 genera known to have polymorphic species, at least 11 also contain some monomorphic species (*Azteca, Crematogaster, Formica, Iridomyrmex, Monomorium, Myrmecia, Neivamyrmex, Pogonomyrmex, Solenopsis, Strumigenys,* and *Tetraponera*). Some other genera that are fully polymorphic also possess strong interspecific variation in the degree of polymorphism. The species of *Camponotus,* for example, cover the entire range of possibilities from feeble monophasic allometry and modest size variation to complete dimorphism.

It is possible to envisage three very different means by which the transition occurs in the direction of monomorphism. First, the behavior of workers in particular age-size cohorts can simply become more complex and flexible. The second means is by the addition of temporal castes; the members of a given size cohort now specialize according to age. This second mode, which is followed to some degree by virtually all of the ants and other higher social insects, has the virtue of allowing reversible specialization. Individuals are not frozen within any particular repertory. When the needs of the colony demand it, workers can switch to a new behavioral regime and even, given a few days time, reactivate dormant exocrine glands. Finally, members of the same age-size cohort can add to their repertory by cooperative action, such as moving large prey items in a coordinated manner or defending the nest against enemies too formidable to be subdued by a single member.

## TEMPO

An important property first analyzed by Oster and Wilson (1978) is the tempo of activity, which varies enormously among different species of social insects. The workers of some ant species walk slowly and with seeming deliberation. As they make their rounds among the brood or forage outside the nest, they appear to waste few movements. Examples of such "cool" species include many members of the Ponerinae, most species of the myrmicine tribe Attini, Basicerotini, and Dacetini, and the large Neotropical members of the dolichoderine genus *Dolichoderus*. In contrast, the colonies of army ants, fire ants, and species of the dolichoderine genera *Conomyrma, Forelius,* and *Iridomyrmex* literally seethe with rapid motion. The workers appear to waste substantial amounts of time canceling one another's actions. During colony emigration, one ant in such a "hot" colony may run in one direction with a pupa, another in the opposite direction. Additional time seems to be wasted by colony members who run back and forth to nest sites empty-mandibled.

Although the matter has not been subjected to appropriate measurements, a positive correlation appears to exist between tempo and the mature colony size of species, even when evolutionary "inertia" due to phylogenetic similarities is taken into account. In Chapter 8 we established that a loose correlation exists among species between the mature colony size and the complexity of the physical caste system. Does a causal relation then exist among the three variables—tempo, colony size, and degree of polymorphism? Perhaps species with small colonies must consist of workers that are more "careful." In order not to deplete their numbers to a level fatal to the colony as a whole, they must act with greater deliberateness and precision, even at the expense of less productivity. Such species live in circumstances in which colony size is necessarily small for other reasons, such as the low density of food items or the small size of the preferred nest site. Under these conditions colony growth is expected to be relatively slow and worker longevity greater. Such colonies are the analogues of *K*-selected species. The *r*-selected colonies of other species are better fitted to a niche in which the premium is on large mature colony size, rapid colony growth, and a high turnover of workers. Such colonies can afford some degree of inefficiency as the price of their large size and high rate of exploitation of the environment. They are "labor saturated" in the sense that each task is attended by many individuals. With such increased redundancy, the reliability theorems described earlier ensure that the system reliability will be high even if each individual is performing erratically.

Pasteels et al. (1982) and Deneubourg et al. (1983) have pointed out that "errors" committed by ants during recruitment actually have other adaptive advantages when colonies are labor saturated. Because the responses of individual ants are probabilistic, with a large random component during each moment of time, the mass response of foragers can be adjusted by altering probabilities of individual response through changes in recruitment signaling. In a tightly deterministic system the same mass adjustment would require a much more sophisticated system beyond the brain capacity of insects. Also, errors during recruitment could allow the ants to discover nearby food sources. Finally, a scattering of ants in this manner allows the colony to follow and capture moving prey more efficiently than if the recruitment were highly precise (Wilson, 1962b).

One coadaptation favored by high tempo in colonies is a more differentiated caste system: specialized castes replace the "careful" generalized workers of the low-tempo colonies. Also, since workers of such high-tempo societies live for shorter periods of time, and often will not even survive to the reproductive season, they will be more likely to lack ovaries. Their differentiation into anatomical castes further increases the likelihood of their becoming sterile. These circumstances might account more precisely for the association between large colony size, worker polymorphism, and the lack of worker ovaries that has come to light in separate studies.

Two environmental conditions can lead to the maximization of energy yield at high tempo. The most likely condition is the existence of a relatively rich food source that occurs unpredictably in space. The rate of performance in items retrieved per worker will be the multiple of the frequency of occurrence of the items, the probability of encounter, and the probability of a successful retrieval following each encounter. An increase in tempo will increase the probability of encounter but is likely to decrease the retrieval rate. Such a strategy can be expected in species that capture a wide array of arthropod prey and hence form larger, more polymorphic colonies. In contrast, a species specialized on a few rare, difficult species of prey is likely to adopt a low tempo: the search for the items is careful and aided by special sensory devices, and the assault is deliberate, achieved either by stealth or recruitment of nestmates. Foraging, in other words, is more specially tailored to the prey. A species adapted to predictable and rich food sources, such as aggregations of honeydew-producing aphids, is also likely to operate at a low tempo. The workers need to invest very little in reconnaissance but a great deal in the protection and exploitation of the resource.

The second circumstance leading to a high tempo, in theory at least, is a relatively low loss of energy because of the activities of competitors and predators. Some energy expenditure is of course unavoidable owing to metabolism, and the expenditure will accelerate as a function of tempo. If the additional cost through injury and mortality inflicted by enemies is high, so that the energy curve rises even more steeply as a function of tempo, the optimum tempo of the colony will be correspondingly lower. In other words there will be a tendency to decrease tempo in order to compensate for energy lost to the colony through competition and predation.

## LEARNING

A great deal of the flexibility observed in the behavior of individual ants can be attributed to learning. There is no hard and fast line between innate patterns and learning. It is true that some complicated responses, such as self-grooming and regurgitation, appear to be wholly programmed so that the insect performs them more or less expertly with no prior experience. Other responses are altered according to experience. But even when learning is required, it is genetically "prepared"—that is, certain responses are learned more readily than others. The degree of preparedness in learning varies greatly according to category of behavior. Some forms of behavior, such as the differential aggression toward certain kinds of enemy ants, represent little more than shifts in the intensity of otherwise wholly programmed responses (Carlin and Johnston, 1984). Others, including visual orientation by landmarks, incorporate a sophisticated memorization of details (Jander, 1957; Wehner et al., 1983). Still others entail a complex integration of information on the colony's nutritional status and available food in the environment (Deneubourg et al., 1987; Traniello, 1987b, 1988).

The simplest form of learning in ants, like that in other animals, is *habituation*, the diminution of the intensity of the response as a result of experience. When an ant colony is disturbed by the intrusion of a glass rod into the nest, the workers launch an attack. If the disturbance is repeated at close intervals, say every hour afterward, the ants grow steadily less responsive. In this fashion ant workers can be "tamed," especially if they are kept away from the nest. Even large, venomous species can be picked up, allowed to walk over the hand, and fed (carefully!) with sugar water. Habituation appears to be an adaptive process in which the individual comes to recognize less dangerous stimuli as such, enabling it to adjust at a calmer and more efficient tempo. In Schneirla's experiments on maze learning in *Formica* (1941), the workers first went through a stage of habituation ("generalized stage"). During this time progress occurred through a simultaneous decrease in excited and erratic behavior and an increase in the tendency to continue running when obstacles were encountered.

Ants are also notably capable of *associative learning*. In this second major category of behavioral modification, the ants acquire conditioned responses by the association of rewards (such as sugar water or attractive pheromones) with previously meaningless stimuli. This is the phenomenon used by Karl von Frisch to measure the sensory capabilities of honey bees. He and subsequent investigators recognized that if a worker bee can be trained to associate a given wavelength of light or odor with food, the linkage is ipso facto evidence that the bee can detect that particular stimulus. The great majority of cases of learning that have been demonstrated in ants can be loosely classified as associative learning. Some of the effects also constitute imprinting, which is learning that occurs mostly or exclusively during brief "sensitive periods" in the life cycle and can be reversed later only with difficulty, if at all.

The most striking example of imprinting by simple association is the learning of colony odors (Chapter 5). The period of greatest sensitivity is usually in the first days after eclosion of the young adult from the pupa, although Isingrini et al. (1985) have provided evidence in the case of *Cataglyphis cursor* that the learning can begin during the larval stage. Carlin and Schwartz (1989) obtained a similar result in *Camponotus floridanus*. There is additional evidence that when no contact with nestmates is permitted during the sensitive period, later social behavior can be seriously disturbed. Lenoir (1979a,c) found that young workers of *Lasius niger* not allowed to contact larvae after eclosion tended to become foragers at an early age. A few hours of later exposure to larvae were enough to cause them to transport larvae in a normal manner. Workers kept in complete social isolation were aggressive, however, and could be integrated only with difficulty into their colony of origin. In other experiments, cocoon care was abolished altogether by keeping workers of *Formica polyctena* or *F. rufa* from cocoons for three weeks after eclosion (Jaisson, 1975; Jaisson and Fresneau, 1978; Le Moli, 1978; Le Moli and Passetti, 1978). An imprinting-like mode of learning may also play a role in a different socioecological context. Experimental results obtained by Jaisson (1980) indicate that in certain ant species early experience can induce an environmental preference in colony-founding queens. Jaisson suggests that this may explain specific associations that can be found between ants and plants. Similarly, Goodloe et al. (1987) found that workers of the North American slave-making ant *Polyergus lucidus* prefer to raid nests of the same species of *Formica* as that already represented by slaves in their nest.

Some associative learning appears to be little more than the enhancement of innately programmed responses. Colonies of *Pheidole dentata* react more strongly to invasions by fire ants (*Solenopsis invicta*) than to invasions by most other kinds of ants, even in the absence of earlier experience. After repeated exposure to *S. invicta* and *Tetramorium caespitum*, the *Pheidole* increase their response to both species, but much more strongly in the case of the *Solenopsis* (Carlin and Johnston, 1984). A similar enhancement occurs in species of the "*Novomessor*" group of *Aphaenogaster* after experience with their own archenemies, army ants of the genus *Neivamyrmex* (McDonald and Topoff, 1986). In a parallel category, workers of *Myrmica rubra* kept in the presence of queens from the time of eclosion exert more control over the growth of larvae than do those kept apart from queens. A critical period exists in which the young workers become sensitized, after which they are more likely to use aggression and neglect to suppress development of larvae into the queen caste (Evesham, 1984b).

Ants are capable of feats of learning considerably more sophisticated than simple association when finding their way through the environment. Workers of *Formica pallidefulva* studied by Schneirla (1953a) learned a six-point maze with relative ease at a rate only two or three times slower than that achieved by laboratory rats. Workers of *F. polyctena* can remember their way through mazes for periods of up to four days (Chauvin, 1964). Those of *F. rufa*, operating under more natural circumstances, can simultaneously memorize the position of four separate landmarks and remember them well enough for use in orientation for as long as a week (Jander, 1957). The latter result was greatly extended by Rosengren (1971), who discovered that the ants can remember specific locations outside the nest over winter.

A complex process also takes place in the brains of ants during exploratory foraging trips. An outward-bound worker winds and loops in tortuous searching patterns until she encounters food. But then she takes a direct route, the "beeline," in her return trip to the nest. This feat is made possible by sun-compass orientation, that is, using the sun as the reference point for guided movement. Santschi, who discovered the phenomenon in 1911, was attracted by the problem of how workers of desert ants in Tunisia manage to leave their nests, forage at a distance, and then find their way back home over the featureless desert terrain, even when strong winds make odor trails impracticable. He found the answer by means of his now famous "mirror experiment." When workers of *Aphaenogaster*, *Messor*, and *Monomorium* returning home with booty were shaded from the sun on one side and presented with the image of the sun by means of a mirror held on the opposite side, they reversed their direction 180 degrees and headed confidently away from home. When the shade and mirror were removed, they again turned about 180 degrees and ran homeward. It was apparent that the ants were reckoning the angle subtended by the sun and the nest and holding it constant as they returned home. Later von Buddenbrock (1917) established that this form of orientation occurs widely among insects, and it has since been discovered in many other invertebrate groups as well. For nocturnally foraging ants a moon-compass response is equally feasible, and has, in fact, been demonstrated in *Monomorium* by Santschi (1923) and in *Formica* by Jander (1957).

Santschi also tested the desert ant *Cataglyphis bicolor*, a common species in Tunisia. The workers forage individually and often travel more than 100 meters away from the nest over featureless terrain. Surprisingly, however, the mirror experiment did not work very well with this species and Santschi (cited in Wehner, 1982) concluded that cues in the sky *other* than the sun must serve *Cataglyphis* for celestial navigation. Almost forty years later von Frisch (1950) discovered these cues: he found that insects can perceive the polarized light in the sky and use its pattern for orientation. Subsequently Wehner and his collaborators, using exacting experiments, proved that the *Cataglyphis* use the specific pattern of polarized light from the sky to navigate across a featureless plain (Wehner, 1982, 1983a; Fent and Wehner, 1985; Wehner and Müller, 1985). Previously Vowles (1950, 1954) had also demonstrated orientation to polarized light in the ant *Myrmica ruginodis*. Other investigators established its existence in *Tetramorium, Tapinoma, Lasius*, and *Camponotus* (G. Schifferer in von Frisch, 1950; Carthy, 1951a,b; Jander, 1957; Jacobs-Jessen, 1959).

Since the sun moves through an arc of approximately 15 degrees every hour, what will happen if an ant is trapped on her way home and not permitted to see the sun or portion of the sky for a substantial period of time? If she always keeps a constant angle to the sun regardless of the passage of time—what German investigators call *winkeltreue* orientation—then the error the trapped ant makes on being released should equal the arc through which the sun has passed in the interim. Brun (1914) performed just such an experiment with workers of *Lasius* and found that they did behave as though complying with the *winkeltreue* rule. For example, a worker was confined in darkness from 16:00 hours to 17:30 hours one afternoon, during which time the sun moved through an arc of 22.5 degrees. When released, the ant set off in the direction of her original angle to the sun and consequently deviated from her original, true path by 23.5 degrees, approximately the amount the sun had traveled. This would seem to be a relatively poor way for an ant to get around, especially if she spends hours at a time away from her nest, and Brun's result never seemed to provide a full explanation of visual orientation. In 1957 Jander showed that experienced *Formica rufa* workers are really able to do much better. They duplicate the feat of the honey bee, keeping track of the sun's movement and constantly adjusting the angle of their return journeys. Newly eclosed *rufa* workers and those that have just emerged from overwintering must learn the sun's movement; until they accomplish this, they orient to the sun in the *winkeltreue* manner. It is not known whether Brun's *Lasius* were similarly naive, but at least his results are not inconsistent with the time-compensated orientation demonstrated in *Formica* by Jander.

It is a matter worth further reflection that ants do not need to see the sun in order to perform time-compensated celestial orientation, in other words true compass orientation. Wehner and his collaborators recently worked out the mechanisms by which insects analyze the conspicuous pattern of polarized light displayed on the celestial canopy and use it for compass orientation (Rossel and Wehner, 1982, 1984; Wehner and Rossel, 1985; see Figure 9–11a, b). Wehner (1983a) describes it in the following way:

> In analyzing skylight patterns insects do not behave like human astrophysicists. Neither do they perform spherical geometry in the sky, nor do they come programmed with a set of skylight tables which would inform them correctly about all possible e-vector patterns occurring in the daytime sky. Thus, they do not seem to rely on some abstract knowledge about the laws of atmospheric optics, but get along quite well without such mathematics. What they use instead is a simple celestial map based on the most remarkable feature of skylight polarization: the intrinsic symmetry line of the e-vector patterns which is identical with the line of maximum polarization [Figure 9–11c]. In the real sky, this line is confined to that half of the sky that lies opposite to the sun, and so is the insect's celestial map [Figure 9–11d]. Furthermore, while the actual skylight patterns vary during the course of the day, the insect's celestial map does not change, but is used in the same way at every time of the day. It only rotates about the zenith as the sun moves across the sky (Wehner and Lanfranconi, 1981). (p. 370)

Using the honey bee as an example, but aware that identical results have been obtained with the ant *Cataglyphis*, Wehner continues:

> In operational terms, the insect's celestial map is analogous to what can be called a neural template. Imagine that while steering a particular compass course the bee aligns the symmetry line of this template with the anti-solar meridian in the sky. All it must do in later trying to reestablish the former compass course is to match the template as closely as possible with whatever e-vector pattern it experiences in the sky. Let us recall, however, that the insect's template is a generalized rather than detailed copy of the actual skylight patterns. Using such a generalized map based on a simple rule [Figure 9–11c] implies that under certain skylight conditions navigational errors must occur. On the other hand, such a rule provides the insect with a rather simple strategy of reading compass information from complicated and ever-changing skylight patterns. An intriguing hypothesis is that decisive features of the map are already determined by the geometry of the receptor array within the insect's retina (for the geometry of these arrays in bees and ants see Wehner, 1982). (pp. 370–371)

On the basis of his experiments with *Formica*, Jander suggested that the foraging worker performs a continuous series of calculations analogous to the simplest mathematical operation. As she runs outward the ant continuously perceives the location of the sun and remembers the angle she takes relative to the sun during each of her twists and turns. For each new direction taken, the product of the angle to the sun times the duration of the outward leg of the run is calculated, and the sum of all these products is divided by the total running time to produce the average (weighted) movement angle to the light. When the insect is ready to come home, she needs only to reverse this mean angle by 180 degrees.

In studies of the desert ant *Cataglyphis bicolor*, Wehner and his coworkers found an even more subtle interplay of cognitive processes (Wehner et al., 1983; Fent and Wehner, 1985; Wehner and Müller, 1985). When colonies are located on flat desert terrain, with no distinguishing landmarks in the vicinity, the workers employ dead reckoning in a manner similar to that found in Jander's laboratory studies of *Formica*. That is, they depend on vector navigation by

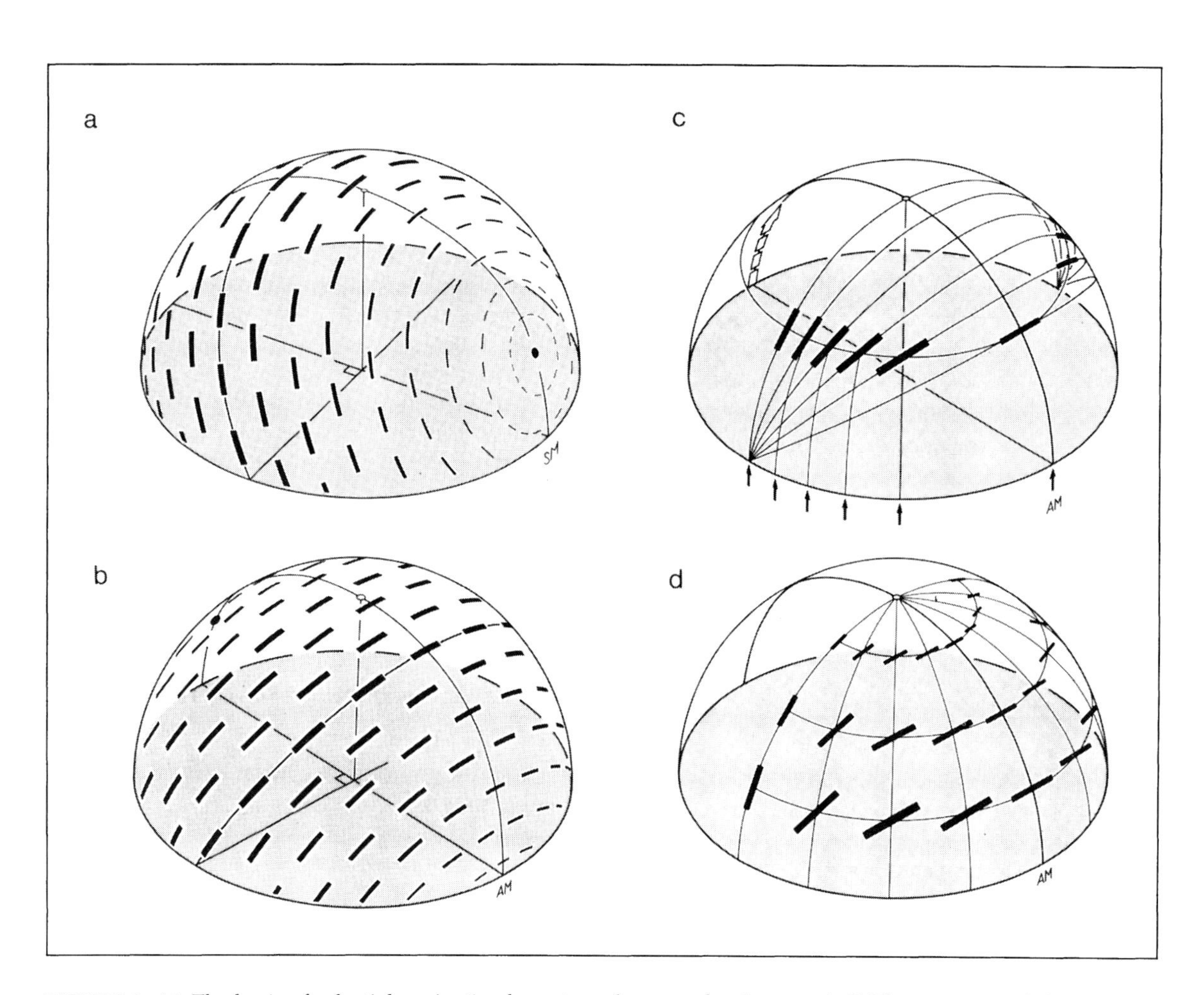


**FIGURE 9–11** The basis of celestial navigation by ants and some other insects. (*a,b*) The geometrical design of the pattern of polarized light in the sky. The **e**-vectors (black bars) are arranged along concentric circles around the sun (black disk: elevation of the sun 24 degrees). The direction and width of each black bar mark the direction and degree of polarization, respectively. Figure *b* is rotated relative to Figure *a*, so that the reader faces either the solar meridian (*SM*) in *a* or else the antisolar meridian (*AM*) in *b*. The other great circle marks the line of maximum polarization. (*c,d*) The celestial map used by the insect is a simplified version of the actual **e**-vector patterns. It is based on just one feature of skylight polarization—the line of maximum polarization. As the sun moves up (white arrows in *c*) the line of maximum polarization tilts down. This results in a distribution of maximally polarized **e**-vectors as shown for an elevation of 45 degrees. The distribution is very similar for all elevations except those close to the horizon. By taking the mean of all these distributions and assigning the resulting mean distribution to all elevations above horizon, one arrives exactly at the insect's celestial map, illustrated in *d*. It shows the position in the sky at which the bee expects any particular **e**-vector to occur. Note that it does not coincide exactly with any **e**-vector pattern present in the sky at any time of the day (*b*), but is a generalized version of these patterns. (From Wehner, 1983a.)

celestial compass. Fent and Wehner also discovered that the *Cataglyphis* employ both the compound eyes and ocelli in this form of orientation. When the nests are located in terrain with rocks, bushes, or other emergent features, the ants rely substantially on these objects as landmarks. Experiments with artificial landmarks by Wehner and his co-workers showed that the workers use two-dimensional memory images of their surroundings as seen from the nest. When moving back and forth between the nest and foraging areas, they employ a sequence of such memory images. There is no evidence that the ants can integrate a true topographic map in the human manner. Nevertheless, each *Cataglyphis* worker is able to remember several such sites simultaneously and to travel to each one of them on separate occasions.

Memories of visual images also play a role in the "canopy orientation" of *Paltothyreus tarsatus*, a large ponerine widely distributed in African forests south of the Sahara (Hölldobler, 1980). The foraging workers memorize particular configurations of branches and leaves in the foliage overhead and use them to find their way back to the nest. To reveal this remarkable phenomenon, Hölldobler took photographs of a Kenyan forest canopy with a wide-angle lens and suspended large prints of it over the foraging arenas of laboratory colonies. When the prints were rotated, homeward-bound workers shifted their compass direction in corresponding degree.

Ants learn more than visual orientation cues during orientation. Many species are arboreal and forage, often exclusively, in a three-dimensional maze of branches and twigs in the tree canopy. In this situation orientation based on gravity perception is at least as important as visual orientation. In 1962 Markl discovered a gravity receptor system in ants. It consists of rather simple-looking clusters of external sensory bristles, one pair on the neck (or, more accurately, on lobes of the episternum that protrude into the neck region), two pairs on the petiole, and similar organs on the legs and antennae (Figure 9-12). As the ant shifts her position, the head and gaster swing on their respective articulations with the thorax and petiole, the segments of the antennae flex, and the body bends slightly on the upper leg articulations. These minor movements press the sensilla of the hair plates to one side or another with the degree varying according to position. The shift causes the underlying neuron clusters to fire differentially. Markl (1962) was able to train workers of *Formica polyctena* to walk up slopes of as little as 2 degrees from the horizontal. He also succeeded in training them to maintain a constant angle with reference to gravity while running up and down a vertical surface.

The associative learning documented in ants thus far has been passive in nature, requiring only that the individual experience a new, unconditioned stimulus in conjunction with an unconditioned, innate stimulus that carries either a positive or a negative reinforcement. Another form of associative learning that contains greater possibilities for behavioral evolution is *operant conditioning*. The process begins when the animal performs a new act by accident or exploratory movement that "pays off" in providing reinforcement. Thereafter the animal either repeats the act or avoids repeating it, depending on whether it was rewarded or punished the first time. Operant conditioning is important because it allows individuals to realize the potential of their basic innate programs of behavior more fully. It also enlarges the possibility of new behavioral evolution through *genetic assimilation*. In the latter phenomenon, the new act provides an advantage in natural selection, so that genotypes with the capacity to perform the act—or still better, the tendency to perform it—spread through the population and alter the future pattern of responses of the member organisms. Operant conditioning appears to occur in foraging honey bees and bumblebees as they perfect their harvesting techniques on different species of flowers (Heinrich, 1984). To the best of our knowledge no examples have been reported in ants, but this lack could be due merely to the failure of past investigators to search for them. Behavioral skills likely to be improved by operant learning include the search and pursuit of prey, the handling and transportation of booty, and the manipulation of nest material and brood.

By the same token it is not known whether ants partition learning into short-term memory and long-term memory. Experiments with honey bees have shown that workers visiting flowers first store information in short-term memory and then transfer it to long-term memory. If a short-term memory is not reinforced by additional experience within minutes, however, it is lost. Erber (1976) and Menzel (1985) speculated that this rapid decay prevents the bee from registering the signals from uneconomical types of flowers into its long-term memory. Menzel (1979) further concluded that short-term memory permits a refinement of incoming information, allowing the bee to continue testing flowers until each type has been encountered enough times to judge its relative productivity. In other words, a single experience is not sufficient to commit the forager permanently to a less desirable flower type. It seems likely that a similar strategy is employed by ants during tasks such as nest excavation, larval feeding, and liquid food transfer, but the possibility has not been experimentally tested.

At least a few ants are capable of learning to perform tasks at particular times of the day. Harrison and Breed (1987) succeeded in training workers of *Paraponera clavata*, a giant ponerine of Central and South America, to come to sugar baits at selected hours during the day or night. Most of the arrivals were accurate to within 30 minutes. When the *Paraponera* were no longer rewarded at particular places or times in the forest where the experiments were conducted, they ceased responding by the third day. The upper bounds of their ability were not measured. Hence *Paraponera* cannot yet be compared with honey bees, which have been shown to learn up to nine different times and places within a 24-hour cycle with a high degree of accuracy (Koltermann, 1974).

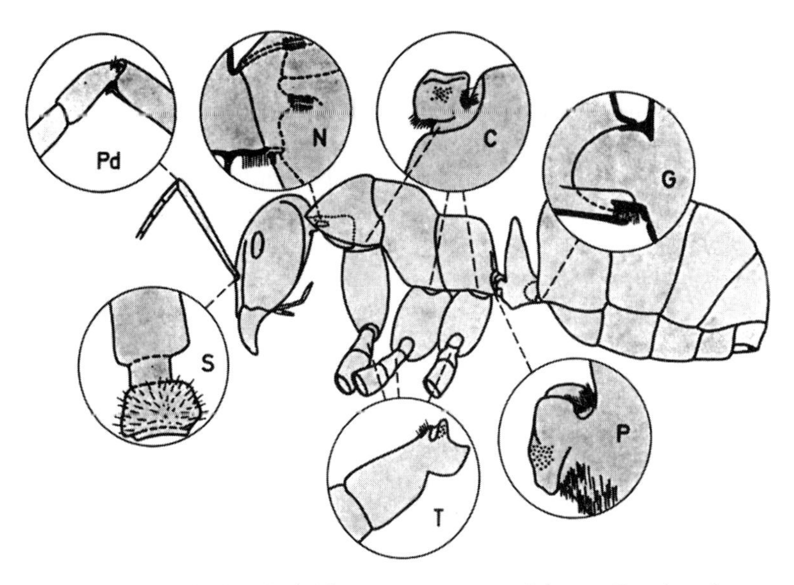


**FIGURE 9-12** The bristle-field gravity receptors of the ant *Formica polyctena*. Within the circles are enlarged views of the different joints bearing hair plates; dashed lines show the location of the sensitive joints. *C*, coxal joint; *G*, gastral joint; *N*, nuchal (neck) joint; *P*, petiole; *S*, joint between antenna and head capsule; *T*, joint between coxa and trochanter; *Pd*, joint between first and second antennal segments. (Based on Markl, 1962.)

In fact temporal learning is likely to prove to be a phylogenetically restricted phenomenon in ants. It appears to be involved in the rapid shifts of foraging times by colonies of *Atta*, *Monacis*, and a few other ants in response to competition and food availability (Cherrett, 1968; Lewis et al., 1974a; Swain, 1977). The shifts occur within one or a few days and persist for days to weeks thereafter. They cannot be easily ascribed to the daily occurrence of cues extraneous to the colony, as distinguished from temporal learning. On the other hand, with the exception of the Harrison-Breed data, evidence of temporal learning in ants has not been forthcoming. Grabensberger (1933) claimed to have trained workers of *Myrmica rubra* to search for food at intervals of 3, 5, 21, 22, 26, and 27 hours respectively, as well as 24 hours. This is an extraordinary result, especially in view that no such non-circadian rhythms have ever been taught to honey bees, and in fact it appears that Grabensberger was in error. Reichle (1943), working mostly with *Myrmica rubra*, and Dobrzański (1956), examining 14 European ant species, were unable to detect temporal learning of any kind. Both Reichle and Dobrzański suggested that none of the species examined is capable of such learning because none is adapted to feeding substantially on nectar. Indeed, *Paraponera clavata* is different from most other ants in that it does regularly collect nectar (Janzen and Carroll, 1983).

Although the documented cases of associative learning imply an impressive cognitive ability, a closer examination of the actual behavior of ants reveals some peculiar shortcomings. Above all, their responses are severely tailored. Learning of any magnitude is limited to the kinds of tasks in which flexibility and fine adjustment are at a premium, such as orientation to and from the nest and the recognition of colony odors. Even in these cases learning is strongly prepared by innate tendencies to lead the ant to quick, accurate responses. Other constraints have been discovered. *Formica pallidefulva* workers "follow their nose" during maze learning. If the passage in a maze turns to the left, the ant tends to follow it around on the outer, or right-hand, side. If the next turn brings the ant to a T-choice, she will usually proceed around the right-hand corner and into the right-hand arm of the choice. Consequently ants can learn a maze much more quickly if the arms they tend to follow by momentum have been set by the experimenter to be the correct ones. Also, when the *Formica* workers do make a wrong turn in the early stages of learning, they usually follow the blind alley to its very end before turning back. Only in later stages do they develop the ability to turn back shortly after a mistake has been made. In short, it takes ants a long time to "realize" that they have been in error (Schneirla, 1943).

It also appears that ants are not capable of insight learning, one of the more advanced categories in the classification by Thorpe (1963). This means that they cannot duplicate the mammalian feat of reorganizing their memories to construct a new response in the face of a novel problem. A dog can, if given time, work purposefully around a transparent barrier instead of trying to push its way through or climb over. With no coaching, a chimpanzee can deduce how to pile boxes in order to reach a banana previously out of reach. No behavior approaching this level of sophistication has been observed in ants or in any other kind of social insect.

## ANTS DO NOT PLAY

Play is generally defined as an activity that makes no immediate contribution to survival or reproduction. Still, there is abundant evidence that the experience gained during play often confers adaptive advantages in later life (Fagen, 1981). Higher mammals, who have evolved play to its highest levels, use it primarily to rehearse the motions of searching, fighting, courtship, hunting, and copulation in a nonfunctional context. Among the young especially, play appears to have two functions: to gain acquaintance with the environment and social partners, and to perfect adaptive responses to both.

Several authors have claimed that ants engage in play or something closely resembling it. Huber (1810) described the following peculiar behavior in a mound-building species of *Formica:*

> One day I approached some of their mounds, which were exposed to the sun and sheltered from the north. The ants had gathered in large numbers and seemed to be enjoying the heat which prevailed on the surface of their nests . . . But when I undertook to follow each ant separately, I saw that they approached one another waving their antennae with astonishing rapidity; with the front feet they lightly stroked the sides of the heads of other ants; and after these preliminary gestures, which resembled caresses, they stood up on their hind legs, two by two, and wrestled one another seizing a mandible, a leg, or antenna, and letting it go immediately, to return to the attack; clinging to one another's waist or abdomen, embracing one another, overturning one another, falling and scrambling up again, and taking revenge for their defeat without appearing to inflict an injury; they did not eject their venom, as they do in their battles, and they did not grip their opponents with the tenacity seen in serious fights; they soon released the ants which they had seized, and tried to catch others; I saw some who were so ardent in their efforts that they pursued several workers in succession, and wrestled with them for a few moments, and the contest was concluded only when the less lively ant, having overthrown her opponent, succeeded in escaping and hiding herself in some gallery. (pp. 170–171)

Huber observed the same behavior repeatedly on the one nest, and he remained especially impressed by the fact that it never seemed to result in death or injury. In 1921 Stumper described a similar form of *combat amical* in the little myrmicine ant *Formicoxenus nitidulus,* which he attributed to play functioning to get rid of excess "muscular energy." But later the same author (1949) observed that the fighting was in deadly earnest and was, in fact, being carried on between members of different colonies! The territorial wars of the pavement ant *Tetramorium caespitum* have much the same appearance as the activity described by Huber: prolonged "wrestling" by pairs of ants in the midst of a struggling mass of contending workers, seldom resulting in injury and death. In short, these activities have a simple explanation having nothing to do with play. We know of no behavior in ants or any other social insects that can be construed as play or social practice behavior approaching the mammalian type.

## THERMOREGULATION

For some reason still unknown, ants are strongly thermophilic. With the exception of a very few cold-temperate species such as *Nothomyrmecia macrops* and *Prenolepis imparis,* they function poorly below 20°C and not at all below 10°C. Even *Myrmica rubra,* a typical northern European species, is unable to produce reproductive adults much below 20°C. Those living in wooded habitats in western Scotland fall below the temperature minimum. Although they can raise workers, they are unable to reproduce (Brian and Brian, 1951; Brian, 1973). This peculiarity is reflected in biogeography. The diversity of ants drops off steadily along a transect drawn from the tropics to the north temperate zones (Kusnezov, 1957b). Colonies are especially sparse in the northern coniferous forests, and there appear to be no native ants at all in Iceland and the Falkland Islands. Ants also largely disappear from the slopes of heavily forested mountains in the tropics above 2,500 meters. In contrast, they are both diverse and relatively abundant in the hottest and driest habitats on earth, including all of the major deserts of the tropics and warm temperate zones. As a group, they seek heat for the rearing of larvae and are able to lower it to tolerable levels in the face of dangerously high ambient temperatures.

Lacking wings, ants are not able to ventilate their nests by fanning in the manner of bees and wasps. And since they are unable to commute rapidly to sources of open water, they cannot readily use droplet evaporation to regulate temperature. Instead, various species rely on combinations of five procedures: (1) correct location of the nests, (2) efficient construction of the nests, (3) migration within the nest, (4) migration among multiple nests, and (5) regulation of metabolic heat, which can be used either to raise the temperature locally by clustering or to lower it by dispersion.

**PLATE 9.** In the first step of vegetation processing, a media worker of the agricultural ant *Atta sexdens* shears a leaf outside the nest.

**PLATE 10.** Two media workers of *Atta sexdens* cooperate to cut the twig of a food plant.

**PLATE 11.** A media worker of *Atta cephalotes* carries a freshly cut piece of leaf to the nest.

**PLATE 12.** A forager of the primitive ant *Nothomyrmecia macrops* retrieving a freshly captured wasp. *Nothomyrmecia* are nocturnal foragers; they can be found outside their nest only after dusk. The painting by J.D. Dawson is based on a photograph taken in the field in South Australia. (From Hölldobler, 1984d; reprinted with permission of the National Geographic Society.)

**PLATE 13.** Foragers of *Oecophylla smaragdina* retrieve an injured hermit crab to one of their leaf nests. (From Hölldobler, 1983.)

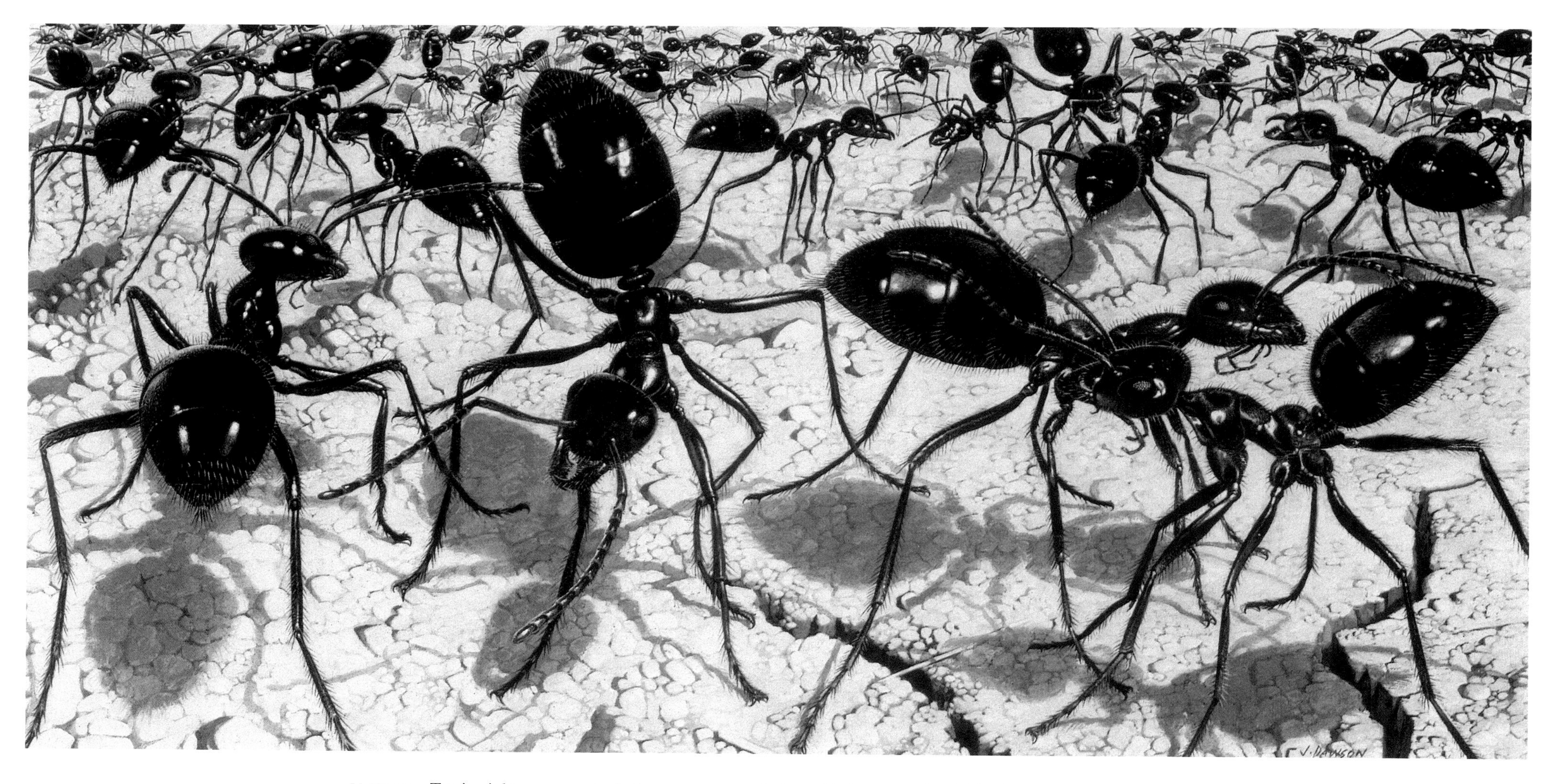

**PLATE 14.** Territorial tournament of *Myrmecocystus mimicus.* Worker ants of neighboring colonies confront each other in ritualized displaying fights during which they walk high with the legs in a stilt-like position while raising the gaster and head. (From Hölldobler, 1984d; painting by J.D. Dawson reprinted with permission of the National Geographic Society.)

**PLATE 15.** The red Amazon ants *(Polyergus rufescens)* invade the nest of *Formica fusca* to capture the pupae. At this moment, the scouts that discovered the site are leading a raiding party into the nest interior. Some defenders grasp the brood and attempt to flee. The mandibles of *Polyergus* are specialized fighting weapons with which they can easily penetrate the Formica worker's cuticle. (From Hölldobler, 1984d; painting by J.D. Dawson reprinted with permission of the National Geographic Society.)

**PLATE 16.** *Amphotis marginata* on the trails of its host ant *Lasius fuliginosus.* In the foreground an *Amphotis* solicits regurgitation from a food-laden forager. In the upper right side an ant is seen attacking *Amphotis,* but the beetle is well protected by its heavily sclerotized carapace. The picture also shows the staphylinid beetle *Pella* hunting L. *fuliginosus* foragers. (From Hölldobler, 1984d; painting by J.D. Dawson reprinted with permission of the National Geographic Society.)

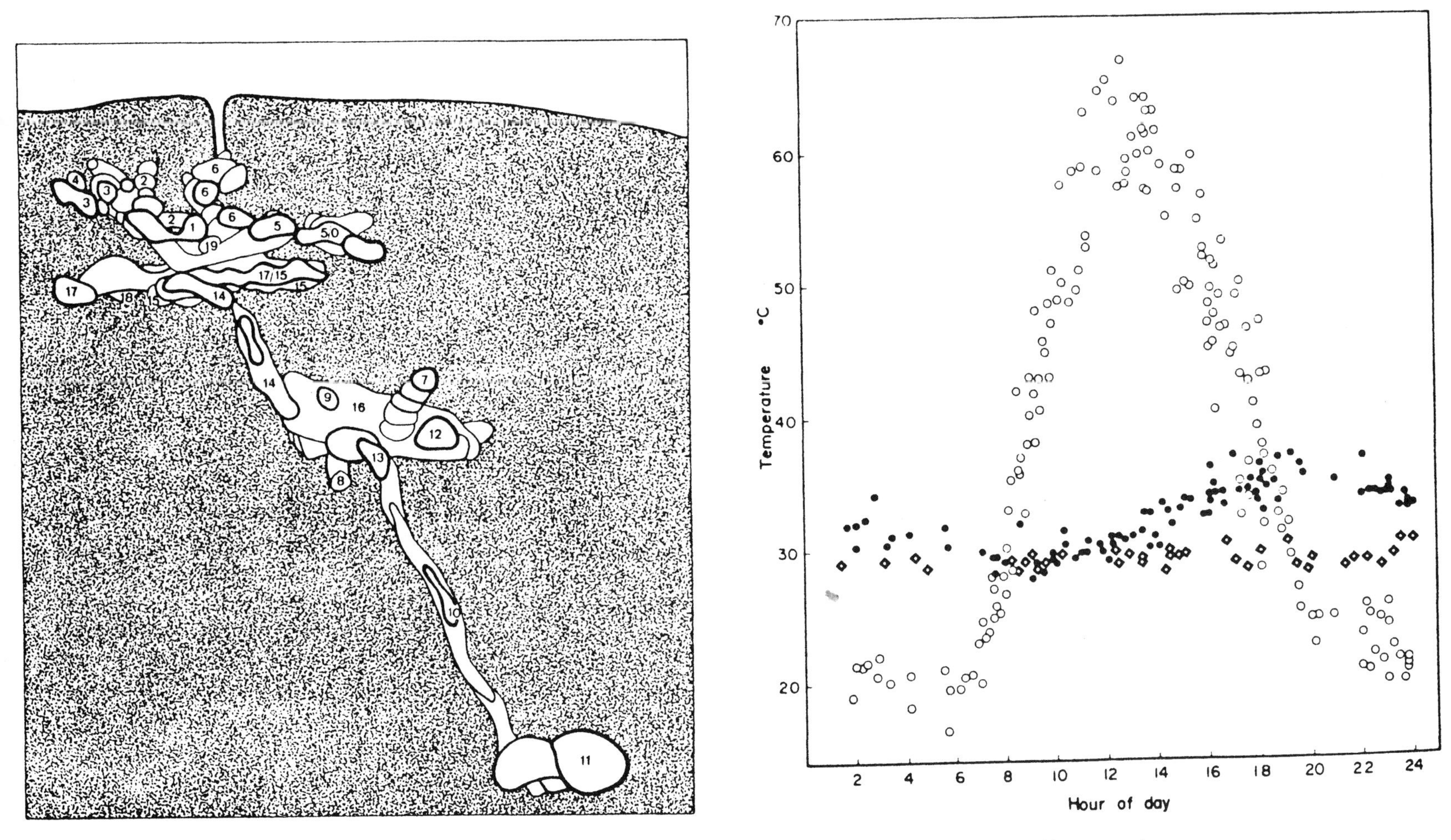


**FIGURE 9–13** *Cataglyphis bicolor,* a desert and dry-forest ant of the Mediterranean region, avoids extremely high temperatures by building soil nests. (*Left*) Diagram of a seven-day-old nest constructed in soil in Tunisia. The numbers indicate the quantities of ants found in each chamber, the deepest of which is approximately 40 centimeters beneath the surface. (*Right*) Temperatures at a nest site in Greece in August. ○, surface temperature; ●, 22-centimeter depth; ◇, 40-centimeter depth. (Modified from Harkness and Wehner, 1977.)

Ant colonies nesting in the soil exploit the universal circumstance that, at depths below a few centimeters, the temperature and humidity vary very little throughout the year. In sandy soil of northern Florida, colonies of *Prenolepis imparis* each excavate a single vertical tunnel to a depth of 2.5–3.6 meters, along which they construct horizontally floored, shallowly domed chambers. No chambers are built higher than 60 centimeters from the ground surface, and most chambers are in the bottom half of these deep nests. As a result most of the colony enjoys temperatures between 16 and 24°C year round. The automatic control is especially important for *Prenolepis imparis,* which prefer cooler temperatures and cannot tolerate the extreme heat of the Florida summers (Tschinkel, 1987c).

Nest structure is at least equally crucial in the case of ants that live in deserts and other dry, open environments. Even extreme desert specialists such as scavenger ants of the genus *Cataglyphis* die if forced to stay aboveground for more than two or three hours. Temperatures above 50°C cause death within minutes or even seconds (S. D. Porter, personal communication). The ants are nevertheless able to nest in locations that are inhospitable at the surface by constructing nests deep within the soil. A striking example is shown in Figure 9–13.

In warm temperate zones most species construct nests in the soil or at the soil surface just beneath rocks or the covering layer of leaf litter and humus, and many occupy pieces of rotting wood. Species in tropical forests have a very different pattern. Most utilize small pieces of rotting wood on the ground, a smaller number nest arboreally or in rotting logs, and still fewer nest entirely in the soil. When rocks occur on the ground in lowland tropical habitats, they seldom serve as primary nesting sites. In contrast, the largest number of ant species in north temperate areas nest in the soil beneath rocks, a tendency that is especially marked in arboreal coniferous forests and other northernmost habitats. A considerable number of north temperate species also occupy either dead logs and stumps or the open soil.

Rocks have thermoregulatory properties, especially those that are flat and set shallowly into the soil. The reason is that when dry they have a low specific heat, meaning that only a relatively small amount of solar energy is required to raise their temperature. Hence in temperate zones during the spring the sun warms the rocks and the underlying soil much more rapidly than it does the surrounding soil. This allows the colonies to initiate egg-laying and brood development earlier than would be otherwise possible. The same phenomenon occurs in the bark of decaying stumps and logs and in the frass-filled spaces underneath. In spring workers, queens, and brood crowd together in such places, retreating to the inner chambers of the nest at the onset of hot, dry weather. Species utilizing rocks, stumps, or logs are also physiologically less vulnerable to low humidity and high temperature than those nesting exclusively in

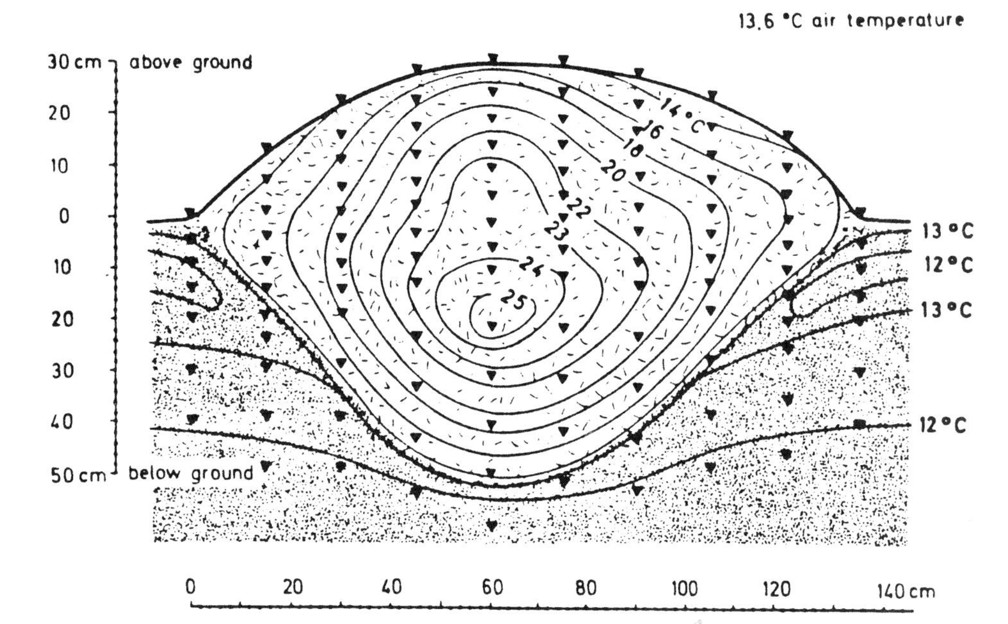


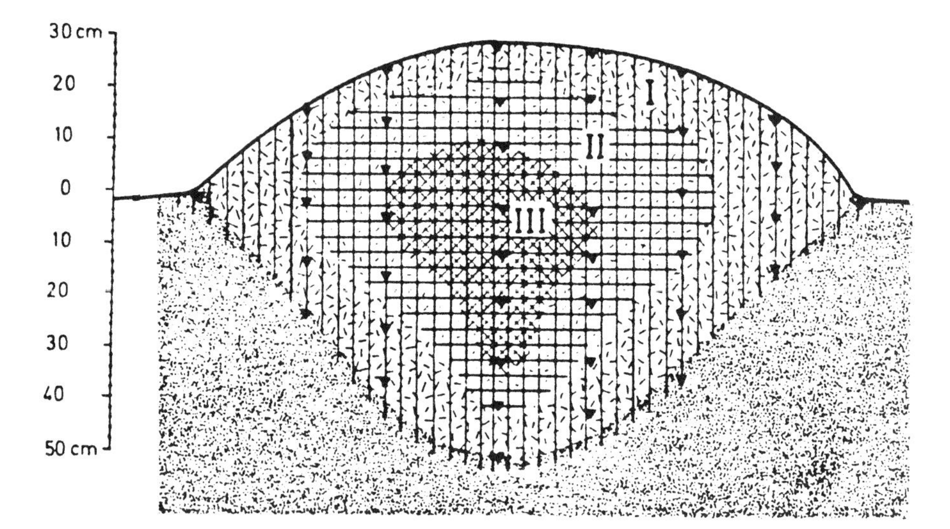


**FIGURE 9–14** Large colonies of the European wood ant *Formica polyctena* warm their mounds with heat from decaying nest materials and their own body metabolism. The upper figure gives the temperature contours and the lower figure the zones of density of ants (increasing from I to III). (Modified from Coenen-Stass et al., 1980.)

the soil. Gösswald (1938c) found that in Germany only the soil-dwelling species require 100 percent relative air humidity for indefinite survival.

Ant nests are typically constructed from beneath rocks or the free ground surface vertically into the soil, or alternatively from spaces beneath the bark of rotting wood fragments inward through the heartwood or around the heartwood surface to encompass portions of the wood facing the soil. This spatial arrangement allows workers to move the brood around at will within the nest to reach the chamber best suited for development. Workers of most ant species keep all stages of brood in the warmest chambers, in the upper range of 25–35°C when these temperatures are available, and pupae are segregated further into the drier parts. When choosing microenvironments inside the nest, the workers can be said to seek one or the other of at least three preferenda: one for themselves when alone, another for eggs and larvae, and still another for pupae. A typical daily cycle in cool climates proceeds as follows: the brood is transported to the rock or bark surface during the morning as the sun raises local temperatures, and back into the lower portions of the nest in the evening as the outer temperatures decline. Steiner (1926, 1929) confirmed that temperature is the critical orientation factor in *Lasius* by alternately heating stones with a flatiron during cool parts of the day and shading them from the sun during warm periods. As a result the workers reversed their usual pattern of migration. In further support, Kondoh (1968) has provided a detailed account of the vertical within-nest migrations of *Formica japonica* that correspond closely to seasonal changes in temperature.

When all else fails, colonies of ants move to nest sites that provide more favorable microenvironments. During the emigrations queens walk under their own power and the brood is carried by the workers. There are even a few "fugitive" species of ants, including *Monomorium pharaonis, Tapinoma melanocephalum, Iridomyrmex humilis,* and *Paratrechina longicornis,* that occupy flimsy, unstable nest sites and rely heavily on frequent colony emigrations to maintain a favorable brood environment. A few species engage in a true, back-and-forth, seasonal migration. On Zanzibar colonies of weaver ants (*Oecophylla longinoda*) shift their silken nests alternately to the northern and southern sides of the trees just after each equinox, in a way that keeps the nests more fully exposed to the sun (Vanderplank, 1960). The most striking seasonal migration of which we are aware is conducted by colonies of the weaver ant *Polyrhachis simplex* at En-Gedi, Israel. During the winter the colonies nest on streamside cliffs, where the temperatures average 17.5–24°C during the daytime. By May the average daily temperatures at these sites rise to an intolerable 30°C, and the ants shift their domicile to small trees at the edge of the stream (Ofer, 1970; E. O. Wilson, personal observation).

A more advanced form of microclimatic regulation has been attained by the small minority of ant species that build mounds. True mounds are not to be confused with simple craters, as they are often called by myrmecologists, which are no more than rings of excavated soil around nest entrances. True mounds are symmetrically shaped piles of excavated soil, rich in organic materials, perforated with dense systems of interconnected galleries and chambers that serve as living quarters for the ants, and often thatched with bits of leaves and stems or sprinkled with pebbles or pieces of charcoal (see Figure 9–15 and Plate 4). The soil beneath the mounds also contains extensive galleries and chambers. During the active season only a fraction of the colony's population occupies the mound itself at any given moment, although in the fire ant *Solenopsis invicta* 60–65 percent of the worker force and 90 percent of the brood is located there during late morning on sunny spring days (S. D. Porter, personal communication). Mound-building species occur in the myrmicine genera *Acromyrmex, Aphaenogaster ("Novomessor" group), Atta, Myrmicaria, Pogonomyrmex,* and *Solenopsis* in the tropics and warm temperate zones in various parts of the world; the dolichoderine genus *Iridomyrmex* in Australia; the formicine genera *Formica* and *Lasius* in Europe, Asia, and North America; and the desert formicine genus *Cataglyphis* in parts of temperate Asia. True mounds are encountered in a wide range of environments, but they are most commonly found in habitats subject to extremes of temperature and humidity, particularly bogs, stream banks, coniferous woodland, and deserts.

Pierre Huber (1810) was the first to suggest that the primary function of mounds is microclimatic regulation. The hypothesis was elaborated by Forel (1874) in his "Théorie des Domes" and has become a subject for investigation by a long line of European, American, and Japanese researchers. The most intensive analyses have been conducted by Andrews (1927) on *Formica exsectoides;* Wellenstein (1928) on *F. rufa;* Steiner (1926, 1929) on species of *Formica* and *Lasius;* Katō (1939) on *F. truncorum;* Scherba (1962) on *F. ulkei;* Orr (1985) on *F. glacialis;* Raignier (1948), Heimann (1963), Kneitz (1970), Coenen-Stass et al. (1980), Horstmann and Schmid (1986), and Rosengren et al. (1986b) on *F. polyctena;* Ettershank (1968, 1971) on *Iridomyrmex purpureus;* and MacKay and MacKay (1985) on *Pogonomyrmex montanus.*

Colonies of *Formica* and *Lasius* maintain much higher temperatures within the core of the mound during cool weather. The temperatures at 20–30 centimeters beneath the surface of the mound apex vary less than those in the surrounding air and soil and stay consistently close to the preferenda of the ants themselves. A striking example from *F. polyctena* is displayed in Figure 9–14. The basis of the thermoregulation is now fairly well understood. The outer, crust-like layer of the mound seems to reduce loss of heat and moisture. The shape of the mound itself exposes it to more sunlight and enhances warming on cool days, especially in the spring and fall. The mounds of some species of *Formica* and *Lasius* also have longer, more gently sloping faces to the south, which increases the amount of exposure still more. For centuries such nests have been used as crude compasses by natives of the Alps. Thatching of the mound surface, a common feature in the nests of many but not all species, appears to reduce erosion of the crust by rain. The trapped air spaces it provides may also improve insulation.

In addition to collecting solar energy with their nests, the mound-building ants rely heavily on metabolic heat production. According to Coenen-Stass et al. (1980), the heat produced by decaying nest material in nests of *Formica polyctena* is seven times that evolved by the ants themselves. Rosengren et al. (1986b) doubt a primary role for nest material, however, noting the lack of correspondence between the heat fluctuation outside the nest and that inside the nest, as well as a sudden jump in inner temperature occurring in July that cannot be linked to decomposition. They attribute most of the internal heating to worker metabolism and patterns of distribution of the ants within the nests.

Small *Formica polyctena* colonies depend more on solar heat than large ones, whereas large colonies that later produce sexual forms maintain a higher temperature during the winter and early spring. Rosengren and his co-workers also found that colonies in Finland exceeding one million workers are "autocatalytic": they are able to commence the final warming of the nest after winter without sunning behavior. Horstmann and Schmid (1986) heated *polyctena* mounds with wires to observe the response of the workers to excessive temperatures. They found that a lowering of temperature is achieved by a reduction of the mound height, an enlargement of the exits on the mound surface, and a dispersion of both workers and brood away from the heat center. It would seem that the minute-by-minute behavior and metabolism of the ants are indeed key factors in thermoregulation by *F. polyctena.*

Some ants, especially *Pogonomyrmex, Iridomyrmex,* and *Cataglyphis,* decorate their nest surfaces with small pebbles, dead fragments of vegetation, or pieces of charcoal (Figure 9–15). These dry materials heat rapidly in the sun and evidently serve as solar energy traps. Harkness and Wehner (1977) note that colonies of *Cataglyphis* on the high plains of Afghanistan build mound nests and decorate them with small stones. They propose a realistic explanation of the legendary gold mining by ants reported by Pliny and Herodotus. The ants were said by these ancient writers to resemble those in Greece that run with great speed, build their homes in the soil, and carry gold to the surface. Herodotus placed the ants near the town of Caspatyros in the country of Pactyike, which has been identified as modern-day Kabul or nearby Peshawar. It is well known that gold is found in the rock and alluvial soil in this part of Afghanistan. Hence it may well be that at least one animal fable involving ants has a basis in fact. In a somewhat similar fashion, fossil hunters in the western United States regularly collect the bones of small mammals from the decoration zones of *Pogonomyrmex* nests. They inspect the mounds early in their expeditions to see if there are any fossils in the region (F. A. Jenkins, Jr., personal communication).

Regardless of which specific features are important in functional design, the mound is certainly no accidental accumulation of excavated earth. It is in a constant state of flux, as workers move materials around to reinforce and to repair the crust and the interior (Cole, 1932; Chauvin, 1959; Kloft, 1959b). When mounds are broken apart

**FIGURE 9–15** A small nest mound of *Pogonomyrmex badius* decorated with dry plant material, stone pebbles, and charcoal.

or their shape is altered experimentally, the ants set to work immediately to restore the original form.

In a few cases ants arrange their own bodies to achieve environmental control in a manner that suggests the winter clustering of honey bees. The most impressive example is temperature control within the nomadic-phase bivouacs of the surface-dwelling army ants (Schneirla et al., 1954; Jackson, 1957). The hundreds of thousands of *Eciton* workers belonging to a single colony form shelters out of nets and chains created from their own bodies, which are hooked together by their tarsal claws and piled layer upon layer to create a single massive cylinder suspended from some log or tree trunk on the rain forest floor (Figure 9–16). In the center of the mass, where the queen and brood are placed, the temperature is consistently 2–5°C higher than the surrounding air (see Figure 9–17). So far as is known, the effect is attained entirely by the trapping of metabolic heat within the air spaces created by the intertwined bodies of the ants. A different kind of direct behavioral thermoregulation has been suggested for *Formica rufa* by Zahn (1958). In cool weather workers leave the nests in large numbers to sun themselves in open air close by. Zahn believes that, when the workers return, they significantly raise the nest temperature by radiating heat from their bodies. This *Wärmeträgertätigkeit* was put forth as a new kind of stereotyped social behavior. The amount of heat flux generated by the movement of such workers has not been measured, however; nor can it be certainly identified as anything more than a felicitous by-product of normal foraging activity.

## HUMIDITY REGULATION

In general, ants in all habitats are subject to stress from desiccation. Even colonies in tropical forests are likely to be displaced from part of their nest sites during the dry season. Arboreal species, much more susceptible to drying than terrestrial ants, have evidently acquired more resistance during evolution. They take much longer to die when placed in dry containers, and they also possess a greater rectal pad area relative to total body size, which suggests that they are able to reclaim fecal water more efficiently (W. G. Hood and W. R. Tschinkel, personal communication).

Ants employ a diversity of techniques, some of them approaching the bizarre, to regulate humidity within their nests. Mounds appear to be constructed to keep not just the temperature but also the moisture of the air and soil within tolerable limits. This form of control was documented convincingly by Scherba (1959) in his study of *Formica ulkei*. Compared with the adjacent soil, the mound nests monitored by Scherba had higher minimum values at 30 centimeters and lower maximum values at 5 centimeters and, consequently, maintained a lesser moisture gradient between these two depths. The ground weekly moisture content of the nests was 29 percent at 30 centimeters and 27 percent at 5 centimeters—in effect, spatially uniform. There can be no doubt that such humidity control is adaptive. The workers regularly shift brood up and down through the dense system of chambers and vertical galleries, with the pupae normally concentrated in the drier upper layers. By means of laboratory experiments, in which workers were permitted to move their brood along artificially produced humidity gradients, Scherba determined the humidity preference for larvae and pupae. The results provided a remarkably close match with the moisture contents at different levels of the natural mound nests: 30 percent for larvae and 28 percent for pupae.

A wholly different form of humidity control is employed by *Pachycondyla villosa*, a giant ponerine that ranges from Mexico to Argentina (R. Mendez, cited in Wilson, 1985b). During the dry season the colonies are in constant danger of desiccation. Many of the workers make repeated trips to beads of dew or any other source of standing water wherever it can be found. They gather drops between their widely opened mandibles and return to the nest. There they often stand still and allow some of the nestmates to drink from the drop. The remainder of the water is fed to larvae, daubed onto cocoons, or placed directly onto the ground. In this fashion the interior of the nest is kept much moister than the surrounding soil. The procedure is at least outwardly identical to the transport and delivery of nectar

**FIGURE 9–16** *(right)* A bivouac nest of the New World army ant *Eciton burchelli.* (Photograph courtesy of C. W. Rettenmeyer.)

**FIGURE 9–17** *(below)* Thermoregulation within a body mass of army ants (*Eciton hamatum*) in a Central American rain forest. The ants form bivouacs consisting entirely of their own intertwined bodies suspended from a tree trunk or log, as indicated in the right-hand portion of the figure (see also Figure 9–16). Within this cluster temperatures are maintained at a level several degrees higher than in the surrounding air. (From Schneirla et al., 1954.)

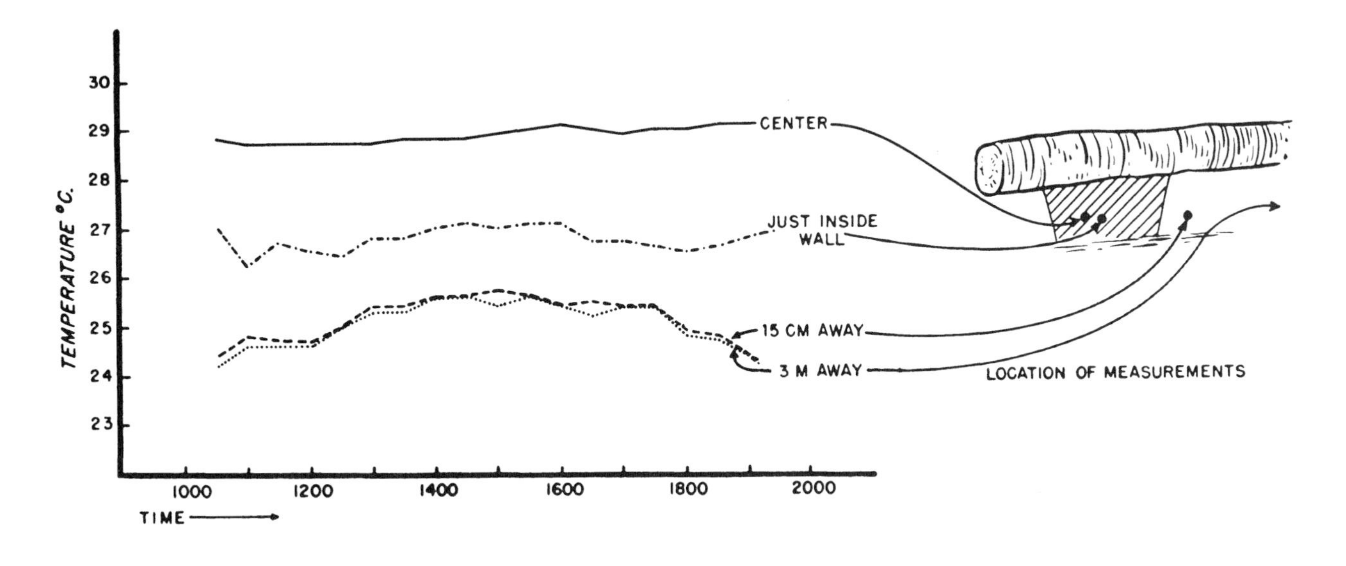

observed in *Pachycondyla* and other ponerine ants, a phenomenon studied independently by Hölldobler (1985). Another peculiar method of obtaining water has been discovered in the ponerine ant *Diacamma rugosum.* Moffett (1985a) observed that workers of *D. rugosum* in southern India decorate the entrances of their nests with relatively hygroscopic objects such as dead ants and bird feathers. In the early morning hours a light dew forms on the material and is subsequently collected by the *Diacamma*. Thus the objects serve as water traps (Figure 9–18). According to Moffett, during the dry season the droplets of dew appear to be the only external source of humidity available to the ants.

Still another and equally bizarre form of humidity control is "wallpapering" by the ponerine *Prionopelta amabilis* (Hölldobler and Wilson, 1986a; Figure 9–19). This primitive rain forest species typically constructs nests in logs and other fragments of rotting wood on the forest floor. The walls of the galleries occupied by the queen, eggs, larvae, and resting groups of workers are bare and moist. Those housing pupae, on the other hand, are usually papered by fragments of pupal cocoons from which adults have previously emerged. Sometimes the fragments are added on top of one another to form several layers (see Figure 9–20). As they decay they become darker in color, staining the substrate of rotting wood beneath. The surfaces of the special wallpapered galleries are drier than those of other nest spaces. The adaptiveness of the modification is suggested by the observation that *Prionopelta* workers move pupae to the drier portions of laboratory nests.

**FIGURE 9–18** Water harvesting by the Indian ponerine ant *Diacamma rugosum.* (*Above*) A nest entrance has been "decorated" by bird feathers. The roll of film indicates the scale. (*Below*) A worker of *D. rugosum* collects droplets of dew that have accumulated on the feathers. (Photographs courtesy of M. W. Moffett.)

**FIGURE 9–19** Portion of a *Prionopelta amabilis* nest with the queen surrounded by workers and cocoons that contain both worker and queen pupae. (From Hölldobler and Wilson, 1986a.)

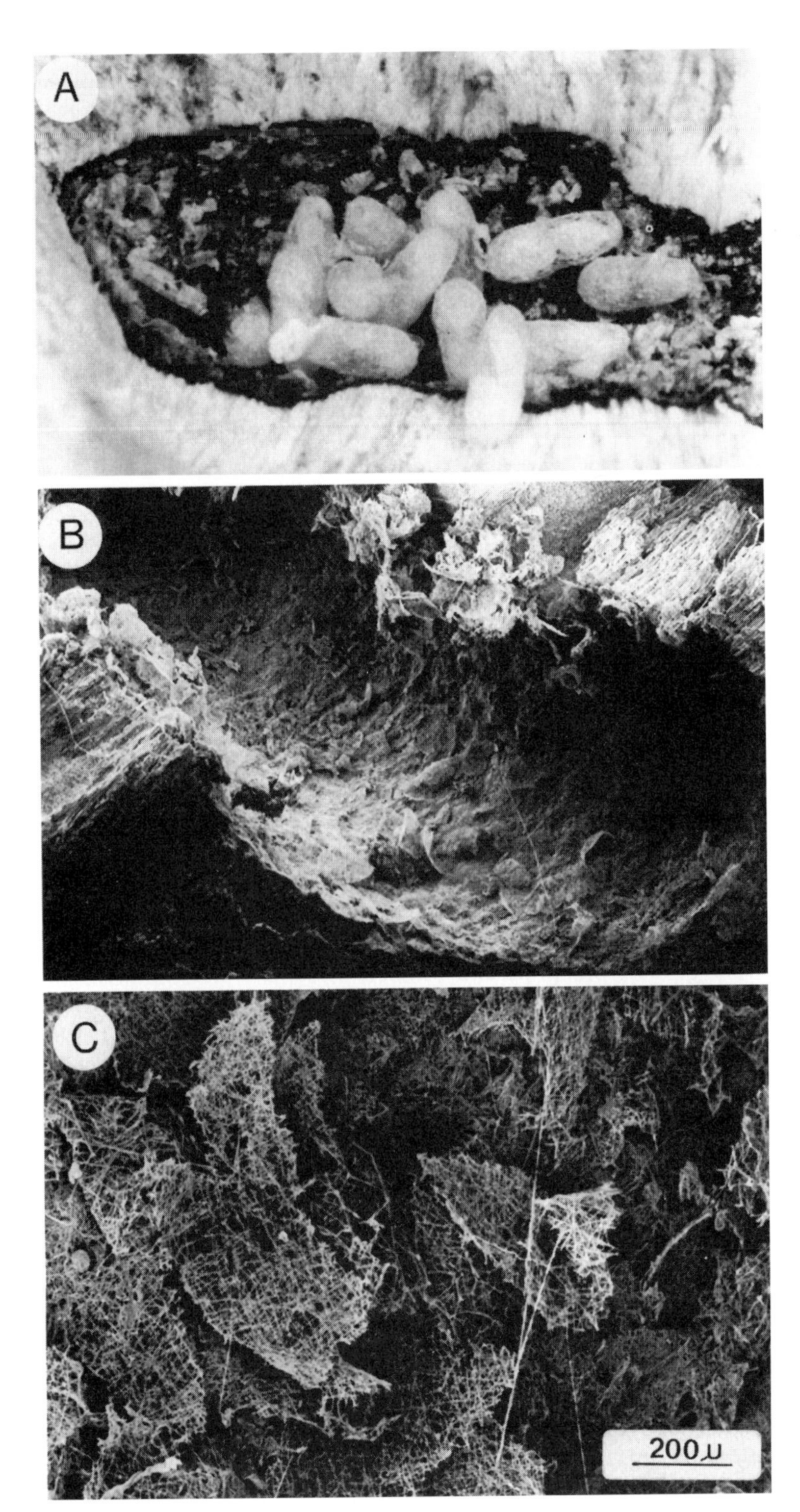


**FIGURE 9–20** The "wallpapered" galleries of the pupal chambers of the ponerine ant *Prionopelta amabilis* evidently serve an unusual form of humidity control in the rain forests where these ants nest. (*A*) The walls of a pupal chamber are lined with fragments of discarded cocoons. (*B* and *C*) Scanning electron micrographs show the old cocoon fragment forming a drier surface on which the pupa-containing cocoons are placed. (From Hölldobler and Wilson, 1986a.)

# CHAPTER 10

# Foraging Strategies, Territory, and Population Regulation

Ant colonies exploit the environment wholly by social means. They deploy aging workers into the field as expendable probes to blanket the terrain around the nest in a nearly continuous and instantaneous manner. In many species scouts that encounter a large food supply recruit nestmates in numbers appropriate to the size and richness of the discovery. When they meet enemies they can recruit defenders or else withdraw, closing the nest entrance and lasting out the danger with the aid of food that has been conveniently stored in the crops and fat bodies of the younger workers.

In the formal framework of a theoretical ecologist an ant colony resembles a plant as much as it does a solitary animal. It is sessile, in other words rooted to a nest site. Its growth is indeterminate, so that under stress it can revert to the size and caste composition of younger colonies. As a consequence size and age of colonies are only weakly correlated. As Rayner and Franks (1987) have pointed out, close parallels also exist between ant colonies and mycelia of fungi: "both are collectives of genetically related or identical semi-autonomous units, consisting respectively of discrete multicellular individuals and hyphae." The patterns created by the growing hyphae are remarkably similar in geometric design to the foraging columns and swarms of army ants and trunk trail systems of wood ants. In both cases the probes of genetically different cultures avoid one another, creating an interdigitating pattern that thoroughly exploits the available resources.

To continue the comparison, we may note that many ant colonies adjust to environmental change as plants do, more by flexibility of response than by movement from place to place. They attain this adaptability through a broad range of social behaviors, including adjustable alarm and recruitment, differential clustering, alterations of nest architecture, and shifts in foraging pattern.

## FORAGING THEORY

Foraging theory can be said to have begun, at least in mathematical metaphor, with Emlen (1966) and MacArthur and Pianka (1966). It has been extended greatly by Schoener (1971), Pyke et al. (1977), Orians and Pearson (1979), Stephens and Krebs (1986), and others. The special case of social insects has been theoretically treated by Oster and Wilson (1978) and advanced by many investigators through excellent field and laboratory studies.

Foraging theory is composed mostly of optimality models. The marginal value utilized, or "currency," is usually energy. The modelers visualize the organism as seeking to maximize its net energetic yield. Evolution by natural selection, they reason, has modified behavior in one of four basic ways so as to increase this currency: (1) choice of food items (optimum diet), (2) choice of food patch (optimum patch choice), (3) allocation of time invested in different patches (optimum time budget), and (4) regulation of the pattern and speed of movement. In the case of ants and other animals with a nest or other permanent retreat, the foraging patterns are analyzed in terms of sallies from a central location. This special case is the subject of "central-place foraging theory," about which we will say more later.

Optimal foraging patterns are generally thought to be constrained by two forces, one external to the organism and the other internal. The external constraint is mortality due to accidents, disorientation, and attacks by predators and competitors. Many field studies have shown that worker populations of ant colonies are subject to terrible losses through predation and battles with competing colonies. It is also clear that the foraging patterns of individual species are profoundly influenced by these pressures. The internal constraints, on the other hand, are particular physiological constraints and the limited sensory and psychological capacities of organisms. Animals, especially small-brained creatures like ants, can rarely if ever maximize their net energy yield by calculating exact costs, benefits, and mortality risks for each new situation in turn (of course, even human beings are hard put to accomplish that much). Instead, they are more likely to employ "rules of thumb," which are quick decisions triggered by relatively simple stimuli or stimulus configurations. The rules of thumb work adequately for individual worker ants most of the time, and they may result in considerable precision when added up across multiple workers during the process of mass communication. In Chapter 9, to illustrate colony homeostasis, we introduced rules of thumb used by fire ants, which include the following: "continue hunting for a certain foodstuff if the present foraging load is accepted by nestmates" (it will probably satisfy the needs of the colony); "follow a trail if enough pheromone is present" (it will probably lead to food); and "feed the queen more if final-instar larvae are present" (the colony developmental pipeline is in good working order).

In adapting foraging theory to the social insects, Oster and Wilson (1978) treated risk of worker mortality not as a constraint but as an energy cost that must be borne to varying degrees by the colony. When a worker is killed in action, the game is far from over; the colony has merely lost a packet of energy that had been invested earlier through the processes of egg laying, larva rearing, and pupa care. The ultimate payoff in colony-level selection is the summed production of reproductives over a colony generation. In general ant colonies appear to act so as to build as large a population of workers as possible during the ergonomic phase, then to "cash in" with a partial or full conversion to reproductives during the breeding season. Consequently a reasonable measure of fitness for a colony in its ergonomic phase of growth is energy, with the growth rate of the adult biomass expressed as calories per unit of time and the full colony biomass achieved at the end of the ergonomic phase expressed in net caloric accumulation. The analytic advantage of employing energy as the basic currency is that it can be folded one way or the other into all colony processes. The key particulars are as follows.

- The biomass and biomass growth of the colony or any portion of it can be translated to energy units.
- Foraging success, brood care, and other nurturant activities can be translated into energy gained.
- Protective activities, including nest defense and construction, can be converted into calorie equivalents saved because of the reduction in mortality and impaired colony functions that such activity avoids.
- Metabolism and mortality can be converted directly into caloric loss.

It is of course an oversimplification to measure ergonomic success entirely by the energy equivalents of biomass. Reproductive success depends not just on colony size at the end of the ergonomic phase but also on the proportion of immature and mature forms as measured by the relative frequency of castes at this critical time. It also depends on the forms and kinds of "capital" at the disposal of the colony, particularly the amount of stored resources and the structure and location of the nest. Nevertheless energy equivalence is probably the overriding determinant of reproductive success. The optimal caste-frequency distribution and nest structure at the onset of the reproductive phase are unlikely to differ from those built up during the ergonomic phase, whereas few species store large amounts of food materials near the end of the ergonomic phase. In addition to such an audit of the total life span, analyses of particular components of foraging appear to be energy efficient in the case of ants. As we shall see shortly, the time spent selecting food items and the degree of selectivity are consistent with the models of central-place foraging theory, which are based on the postulate of net energy maximization. Also, in the case of leafcutting ants, net energy yield appears to be the key function among all those that might have influenced the evolutionary choice of leafcutting ant medias over minors and majors as foraging specialists (Wilson, 1980b).

Foraging behavior has two principal interlocking components: the search for and retrieval of food items, often accompanied by the recruitment of nestmates, and the avoidance or defeat of enemies. It is feasible in theory at least to assess the relative importance of the two components by equating them in terms of energetic cost and gain.

## THE TEMPERATURE–HUMIDITY ENVELOPE

Every ant species operates within ranges of temperature and humidity that can be depicted as a two-dimensional space; in other words every species has a temperature-humidity envelope. Also, the tolerance of an isolated foraging worker is very different from that of an entire colony. The forager is more swiftly affected by the preexisting ambient temperature and humidity of the microenvironment, whereas the colony can control the microenvironment by shifting into deeper reaches of the nest or by clustering to retain metabolic heat and moisture. Thus no nuptial flights of the fire ant *Solenopsis invicta* occur when morning temperatures of the soil (from the surface to 10 cm deep) are below 18°C, and colony founding by newly inseminated queens occurs only if the soil temperatures at 5–7 centimeters below the surface are equal to or greater than 24°C. In addition, young colonies survive winter poorly even in southern Mississippi (Rhoades and Davis, 1967; Markin et al., 1973, 1974). As a consequence the imported fire ant appears already to have reached its geographic range limit in the northern part of the coastal states. To the west the colonies are likely to be able to spread across desert areas only by colonizing stream beds, irrigated agricultural fields, and urban areas. Foraging fire ant workers, on the other hand, have even less tolerance for extreme cold, heat, and low relative humidity, and they are unable to survive in ambient conditions in which colonies as a whole do well (Francke and Cokendolpher, 1986). The main point to consider is that if fire ants were solitary insects, the geographic distribution of the species would be far more restricted than it is in fact.

A large portion of the available data on temperature tolerances of foraging workers is summarized in Table 10–1. Not surprisingly, a rough correlation exists between the tolerances and the environment, with desert species at the upper end and some (but not all) species from cold temperate forests at the lower end. The extreme thermophiles of the world fauna belong to the myrmicine genus *Ocymyrmex* of southern Africa and the formicine genus *Cataglyphis* of North Africa and Eurasia. The two are biogeographic vicariants of each other. In particular, both are specialized in similar manner for diurnal foraging on extremely hot desert terrain. The workers hunt singly for the corpses of insects and other arthropods that have succumbed from the heat (Harkness and Wehner, 1977; Marsh, 1985a,b; Wehner, 1987). Their behavior and even outward physical appearance are remarkably similar. *O. barbiger* forages in the sun at surface temperatures up to 67°C, which must be close to the record for insects generally. *C. fortis* is a specialist of the extremely hot, dry, and food-impoverished terrain of the Saharan salt pans. At the opposite end the cryophilic (cold-loving) species *Camponotus vicinus* and *Prenolepis imparis* start foraging at just above freezing and cease when the temperature reaches about 20°C. Similarly, the Australian species *Nothomyrmecia macrops* forages exclusively after dusk, and the workers seem to be more active at times when temperatures are low (5–10°C). Hölldobler and Taylor (1983) suggested that low temperatures hamper the escape of potential prey encountered in the tree tops, increasing the hunting success of the *Nothomyrmecia* foragers (Plate 12).

Most diurnal desert ants pass through two periods in the day when their preferred temperatures occur, in the morning and late afternoon. As a consequence they have a bimodal distribution of

**TABLE 10–1** Limits of temperature within which ant workers forage.

| Species | Study location | Temperature (°C) at which foraging— Begins | Peaks | Ends | Authority |
|---|---|---|---|---|---|
| **PONERINAE** | | | | | |
| *Odontomachus* sp. | Semi-arid Australia | 10 | — | 30 | Briese and Macauley (1980) |
| *Rhytidoponera convexa* | Semi-arid Australia | 13 | — | 31 | Briese and Macauley (1980) |
| *R. metallica* | Semi-arid Australia | 14 | — | 32 | Briese and Macauley (1980) |
| **MYRMICINAE** | | | | | |
| *Aphaenogaster rudis* | Maryland | 15 | 25–30 | — | Lynch et al. (1980) |
| *A. senilis* | France | 10 | — | 35 | Ledoux (1967) |
| *A.*(= *Novomessor*) *cockerelli* | New Mexico | 20 | — | 40 | Whitford et al. (1980) |
| *Atta mexicana* | Mexico | 12 | — | — | Mintzer (1979) |
| *A. texana* | Louisiana | 12 | — | 30 | Moser (1967a) |
| *Crematogaster emeryana* | Desert, western United States | 14 | — | 62 | Bernstein (1979b) |
| *C. scutellaris* | France | 11 | — | 40 | Soulié (1955) |
| *Messor aegypticus* | North Africa | 24 | — | 39 | Sheata and Kaschef (1971) |
| *M.* (= *Veromessor*) *pergandei* | California | 17 | — | 39 | Bernstein (1974) |
| *Monomorium minimum* | Massachusetts | 32 | 32 | 40 | Adams and Traniello (1981) |
| *M.* (= *Chelaner*) sp. | Semi-arid Australia | 12 | 22 | 33 | Briese and Maccauley (1980) |
| *Ocymyrmex barbiger* | Namibia | 25 | 52 | 67 | Marsh (1985a,b) |
| *Pheidole gilvescens* | Desert, western United States | 23 | — | 29 | Bernstein (1979b) |
| *P. militicida* | New Mexico, Arizona | 16 | 31 | 24–33 | Hölldobler and Möglich (1980) |
| *P. xerophila* | Desert, western United States | 14 | — | 33 | Bernstein (1979b) |
| *Pheidole* sp. | Semi-arid Australia | 10 | 20 | 30 | Briese and Macauley (1980) |
| *Pogonomyrmex californicus* | California | 32 | — | 53 | Bernstein (1974) |
| *P. californicus* | New Mexico | 20 | 45 | 60 | Whitford et al. (1976) |
| *P. desertorum* | New Mexico | 25 | 45 | — | Whitford and Ettershank (1975) |
| *P. occidentalis* | Colorado | 28 | — | 50 | Rogers (1974) |
| *P. rugosus* | California | 22 | — | 46 | Bernstein (1974) |
| *P. rugosus* | New Mexico | — | 45 | — | Whitford and Ettershank (1975), Whitford et al. (1976) |
| *Solenopsis invicta* | Florida | 15 | 22–36 | 43 | Porter and Tschinkel (1987) |
| *S. xyloni* | Desert, western United States | 21 | — | 41 | Bernstein (1979b) |
| *Tetramorium caespitum* | England | 10 | — | 40 | Brian et al. (1965) |
| *Trachymyrmex smithi* | New Mexico | 14 | — | 48 | Schumacher and Whitford (1974) |
| **DOLICHODERINAE** | | | | | |
| *Araucomyrmex antarcticus* | Chile | 10 | — | 48 | Hunt (1974) |
| *Conomyrma bicolor* | Desert, western United States | 26 | — | 39 | Bernstein (1979b) |
| *C. insana* | Desert, western United States | 20 | — | 41 | Bernstein (1979b) |

*continued*

**TABLE 10–1** *(continued)*

| Species | Study location | Temperature (°C) at which foraging— Begins | Peaks | Ends | Authority |
|---|---|---|---|---|---|
| *Forelius pruinosus* | Desert, western United States | 19 | — | 52 | Bernstein (1979b) |
| *Iridomyrmex purpureus* form *viridiaeneus* | Arid Australia | 14 | — | 43.5 | Greenaway (1981) |
| *Tapinoma antarcticum* | Chile | 16 | — | 50 | Hunt (1974) |
| *T. sessile* | Desert, western United States | 6 | — | 35 | Bernstein (1979b) |
| **FORMICINAE** | | | | | |
| *Camponotus vicinus* | California | 2 | — | 23 | Bernstein (1979b) |
| *Formica perpilosa* | New Mexico | 17 | — | 55 | Schumacher and Whitford (1974) |
| *F. schaufussi* | Massachusetts | 15 | — | 40 | Traniello et al. (1984) |
| *Lasius neoniger* | Massachusetts | 29 | 29 | 36 | Adams and Traniello (1981) |
| *Melophorus* sp. | Semi-arid Australia | 37 | — | 55 | Briese and Macauley (1980) |
| *Paratrechina melanderi* | Maryland | 15 | 25–30 | — | Lynch et al. (1980) |
| *Prenolepis imparis* | Maryland | 5 | 15–19 | 26 | Lynch et al. (1980) |
| *P. imparis* | Missouri | 0 | 7–18 | 18.5 | Talbot (1943b) |

foraging activity, with a period of decline around midday. In mesic temperate habitats such as northern coniferous woodland, the reverse pattern occurs among diurnal species, with foraging peaking in the middle of the day. These patterns are probably determined by external environmental cues. Hunt (1974) was able to shift the desert bimodal pattern of the Chilean dolichoderine *Dorymyrmex antarcticus* to a mesic unimodal pattern merely by shading the nest at midday.

Although the temperature tolerances of ant species are correlated with climate and major habitat, the relation is only loose. Added variance comes from two sources, microhabitat specialization and competition avoidance. Forest species, such as *Aphaenogaster rudis* and *Paratrechina melanderi*, have lower tolerance than other species, such as *Monomorium minimum* and *Tetramorium caespitum*, which are adapted to clearings in the same immediate area. It is also of advantage for ants to use unusual activity regimes to escape competitors. The thermophilic species of *Cataglyphis* and *Ocymyrmex* have the desert terrain virtually to themselves at midday. In the grassy meadows of Massachusetts the little black ant *Monomorium minimum* recruits at higher temperatures than does its three closest competitors, *Lasius neoniger*, *Myrmica americana*, and *Tetramorium caespitum*. It combines this specialization with an effective form of venom dispersal to appropriate a substantial fraction of dead insects and other food finds (Adams and Traniello, 1981).

The subtle differences in microhabitat that characterize species are very well illustrated in *Formica perpilosa* and *Trachymyrmex smithi*, two ant species common in the Chihuahuan desert of southern New Mexico. As illustrated in Figure 10–1, *F. perpilosa* is active in a much broader area around its nest than is *T. smithi*. The reason is that the *Formica* are more resistant to water loss, and they also tend to forage on mesquite and other forms of vegetation, where the relative humidity is higher (Schumacher and Whitford, 1974).

Quantitative studies of humidity preference by foraging ants of the kind conducted by Schumacher and Whitford are rare in the literature, but abundant anecdotal evidence exists to show that the higher the relative humidity, the greater the temperature tolerance. An increase in foraging activity when humidity rises at high temperatures has been observed in *Pheidole militicida* (Hölldobler and Möglich, 1980), *Formica polyctena* (Rosengren, 1977b), *Prenolepis imparis* (Talbot, 1943b, 1946), and a wide range of semi-desert ant genera in Australia (Briese and Macauley, 1980); and it is probably a general phenomenon in ants as a whole. Another source of differential foraging in the face of varying humidity is simple physiological resistance to desiccation, a phenomenon that has been observed in *Tetramorium caespitum* by Brian (1965b), *Pogonomyrmex* by Hansen (1978), and diurnal ant species in Australia by Briese and McCauley (1980).

On the other hand, rain halts most foraging in places where the drops pelt the ground and form small puddles and rivulets (Hodgson, 1955; Lewis et al., 1974b; Skinner, 1980a). Every collector is familiar with this phenomenon, to his frequent frustration and especially during tropical wet seasons. On the other hand there are a few long-legged ants, including the species of the dolichoderine genus *Leptomyrmex* of Australia and New Guinea and the myrmicine "giraffe ants" *Aphaenogaster* (*Deromyrma*) *phalangium* of Central America, that use their unusual stature to navigate water films. As a result they are among the first insects to forage after the rain has ceased.

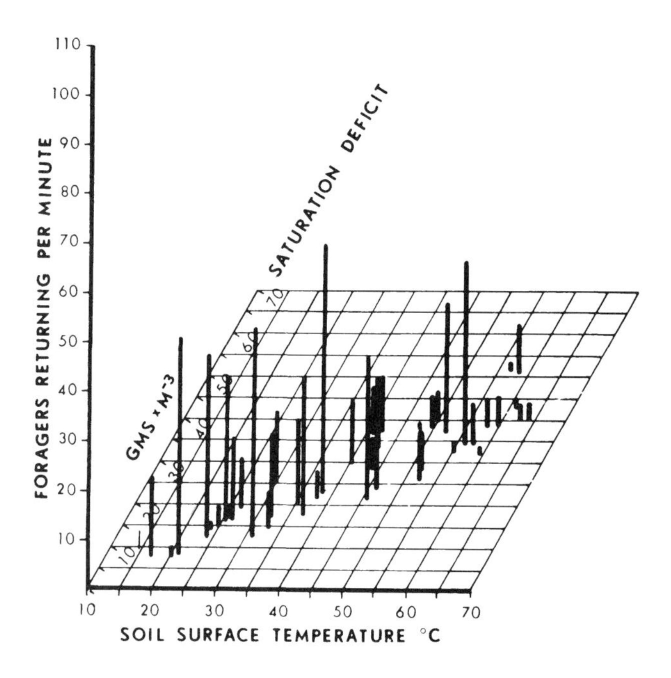


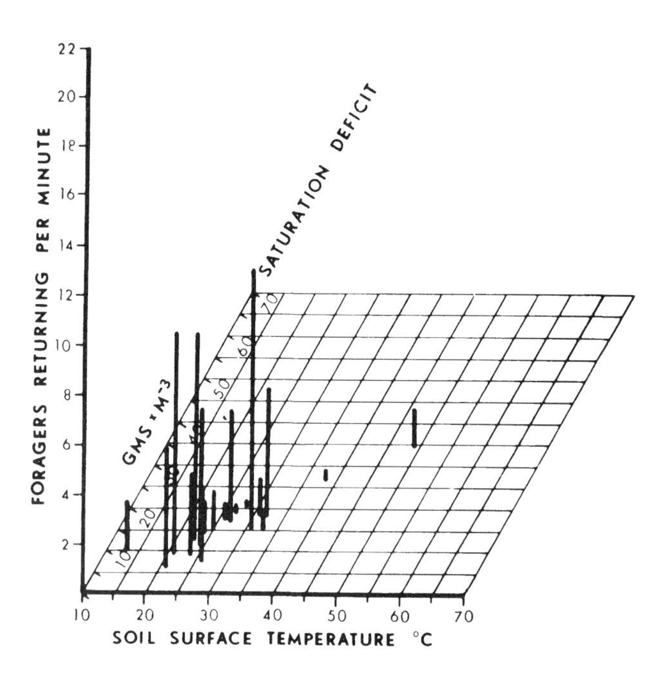


**FIGURE 10–1** The temperature-humidity envelopes of foraging activity by workers of two American desert ants, *Formica perpilosa* (*above*) and *Trachymyrmex smithi* (*below*). The measurements were taken around the nest surfaces of the colonies. The envelope of the *Formica* is broader in part because the workers forage on plants, where the temperature is lower and humidity higher than on the ground. The humidity saturation deficit is expressed in grams of water per square meter. (From Schumacher and Whitford, 1974.)

## DAILY CYCLES OF ACTIVITY

Each ant species has a distinctive daily foraging schedule. An extreme example is the remarkable degree of precision in the changeover of ant species at dusk in the heath of southwestern Australia (Wilson, 1971). In midafternoon the ground and the branches and leaves of the low bushes that dominate the vegetation contain hordes of workers, mostly brown, red, or black, with medium-sized compound eyes, belonging to ten or so species of *Myrmecia, Rhytidoponera, Dacryon, Iridomyrmex,* and other typically Australian genera. As dusk falls, first one species, then another, begins to pull back into its nest, while the nocturnal species—pale-colored, mostly large-eyed species in *Colobostruma, Iridomyrmex,* and *Camponotus*—make their appearance in a regularly staggered succession. So orderly is the changeover that approximately the same number of foraging workers remain on the bushes throughout.

Are such daily cycles based on circadian rhythms or are they guided hour by hour by changes in temperature and other external stimuli? The answer appears to be that circadian rhythms affecting foraging behavior are widespread, but in many ant species they can be overridden—or at least frame-shifted—by colony hunger or certain environmental changes. McCluskey and Soong (1979), for example, established diel rhythms in four species that occur together in southern California. *Pogonomyrmex californicus* and *P. rugosus,* which are harvesters, commence foraging at early midday and remain very active through the afternoon. Two other harvesters, *Messor* (= *Veromessor*) *andrei* and *M. pergandei,* as well as *Formica pilicornis* and *Myrmecocystus mimicus,* which are general insectivores, have bimodal activity schedules, declining toward midday and picking up again from late afternoon through dusk. When colony fragments were placed in the laboratory at a constant temperature with alternating light and dark, they maintained their natural schedules to a significant degree. A similar result was obtained by Rosengren (1977b) with Finnish colonies of the wood ant *F. polyctena.* Workers responded to a 12:12 light:dark cycle in the laboratory with a rise in activity toward the end of the dark period, as though anticipating daybreak. The ants could be entrained to new 12:12 cycles out of phase with that occurring in nature. When kept in complete darkness on a 12:12 warm-cool cycle, the workers grew most active in the middle of the warm period. Thus light alone is not crucial for the entrainment of diel cycles.

In later work Rosengren and Fortelius (1986a) extended this result to a variety of species in the *exsecta* and *rufa* groups of *Formica.* They observed a tendency of workers to increase activity before the artificial dawn, but there was also a great deal of variation in the overall activity pattern among colonies. Ambiguity is also the rule in the giant ponerine ant *Paraponera clavata* (Figure 10–2). McCluskey and Brown (1972) observed that workers on Barro Colorado Island emerged regularly at dusk and apparently foraged until dawn. When transferred to an artificial nest in the laboratory they displayed striking peaks of activity at dawn and dusk. But when placed in a dark room with alternating light and dark and constant temperature, they became diurnal rather than crepuscular and nocturnal. In Costa Rica, over a period of two years, we observed considerable variation in the diel pattern of activity by *Paraponera* workers, although these giant ants appear to be primarily nocturnal. The same phenomenon was observed by Harrison and Breed (1987), who succeeded in training *Paraponera* workers to come to the sites of sugar baits at fixed times during both the day and night.

In fact many ant species seem quite capable of shifting the time of peak foraging back and forth as an adjustment to vagaries of the environment. We have seen *Atta cephalotes*, a dominant leafcutter of the moist lowland forests in Central and South America, change from predominantly diurnal to nocturnal foraging over a period of a few days in the Brazilian Amazonian forest. Lewis et al. (1974a) found that in Trinidad the foraging columns emerge decisively and reach peak traffic within 2 hours. This event usually occurs at night and lasts for an average of 12 hours; when it unfolds during the day it persists for 7.5 hours on the average. Colonies in the same locality are often unsynchronized, and it seems likely that they are entrained in an independent manner by food discoveries rather than by some more pervasive cue from the physical environment.

This form of adaptive plasticity has in fact been demonstrated in the case of *Messor galla* and *M. regalis*, two granivores inhabiting the savanna of northeastern Ivory Coast (Lévieux, 1979). During the dry season, when seeds are abundant, the colonies forage mostly at night and in tight columns. During the rainy season, or in periods of the dry season with poor seed crops, the ants shift to diurnal foraging and the columns are less organized. It is very common for desert ants to switch from nocturnal foraging in the summer to diurnal foraging in the winter. The phenomenon has been observed in species of *Aphaenogaster (= Novomessor)* (Whitford and Ettershank, 1975), *Atta* (Mintzer, 1979b), *Messor (= Veromessor)* (Tevis, 1958), *Pheidole* (Hölldobler and Möglich, 1980), and *Pogonomyrmex* (Hölldobler, 1976a). Nearly the reverse pattern is displayed by *Aphaenogaster rudis* and *Paratrechina melanderi* in the hardwood forests of Maryland: minimal activity occurs around midnight during the summer months, shifting to more nearly uniform round-the-clock activity during the spring and fall and quiescence during the winter (Lynch et al., 1980). Authors who have reported these patterns generally consider them to be adaptations to promote thermoregulation. Where rodents can remain strictly nocturnal year round, at least in deserts, ants are more sensitive to ambient temperature changes and must alter their daily activity accordingly (Davidson, 1977b; Brown et al., 1979a,b).

Differences in foraging rhythms among sympatric species of ants can serve in temporal partitioning of significant resources. Such activity differences may be based proximately on different humidity and temperature ranges tolerated by the species, yet be an ultimate, evolutionary result of interspecific competition. Although only a few studies directly address this question, some of the observations reported by Talbot (1946), Greenslade (1972), Bernstein (1979b), Swain (1977), Klotz (1984), and others strongly suggest temporal partitioning of resources. Klotz reports that in his study area in Kansas both *Camponotus pennsylvanicus* and *Formica subsericea* utilize the same aphid honeydew sources, but in a temporally displaced manner. *Camponotus* is more active at cool temperatures and predominantly nocturnal, whereas *Formica* is active at higher temperatures and is primarily diurnal. We made similar observations on *C. socius* and *C. floridanus* in central Florida. Where both species coexisted and used the same honeydew sources, *socius* was primarily diurnal and *floridanus* primarily nocturnal.

**FIGURE 10–2** A forager of the giant ponerine ant *Paraponera clavata* carrying a droplet of nectar between her mandibles. (Photograph by Dan Perlman.)

## FORAGING STRATEGIES

The approximately 8,800 known ant species use a dazzling variety of procedures for the discovery and retrieval of food. Several authors have attempted classifications of this important category of behavior, including Oster and Wilson (1978), Passera (1984), and Moffett (1987b). We will provide a synthesis here of what we regard as the most useful elements of these schemes. The resulting classification breaks all of the phenomena into three categories: hunting (3 kinds), retrieving (4 kinds), and defense (4 kinds). The elements in each can be combined to form $3 \times 4 \times 4 = 48$ possible three-state foraging techniques. This arrangement provides an informal and convenient framework for the description of the behavior of individual species.

*Hunting:* (1) by solitary workers; (2) by solitary workers directed to specific trophophoric fields by trunk trails; or (3) by groups of workers searching in concert, in the manner of army ants.

*Retrieving:* (1) by solitary workers who return home on their own; (2) by individual workers who return home along persistent trunk trails; (3) by individual workers recruited to the food sites by scouts; or (4) by groups of workers who gang-carry the food items.

*Defense:* (1) by guard workers during hunting or (2) the absence of such defense; (3) by guard workers during harvesting and retrieval of food; or (4) the absence of such defense.

Although no complete accounting has been attempted across all of the ants, it is our impression that virtually all of the 48 possibilities in this broad space of foraging techniques are employed by one species or another. At one extreme are the completely solitary foragers, including scavengers and specialized predators in all subfamilies other than the army ants of the subfamilies Dorylinae and Ecitoninae. The workers hunt singly and retrieve food items entirely on their own. At the other extreme are the army and driver ants of the genera *Eciton* and *Dorylus,* who hunt in groups, often gang-retrieve large prey items, and deploy specialized guard workers along the perimeters of the foraging columns.

The *total* foraging strategy of an ant species often comprises two or more three-state techniques, each chosen according to the quality and nature of the food of the moment. For example, when the solitary foragers of *Pogonomyrmex* encounter a seed or dead insect small enough to be carried singly, they carry it home alone. When the object is too large to be transported by one ant, or there is a cluster of small food items, the worker lays a recruitment trail with secretions from her poison gland. When the food is persistent at a particular site, as is the case for a continuing seedfall, the ants deposit trunk trails with secretions that come at least in part from the Dufour's gland (Hölldobler and Wilson, 1970; Hölldobler, 1971b; Hölldobler, 1976a). The minor workers of the Amazonian myrmicine *Pheidole embolopyx* also search in solitary fashion and single-load the small insects they encounter. When the food item is too large to move, the scouts recruit other minor and major workers from the nest. The majors spend most of their time carving up the insect and defending it from intruders, while the minors concentrate on drinking the hemolymph and carrying fragments back to the nest (Wilson and Hölldobler, 1985). Minor workers of very small *Pheidole dentata* colonies spend more time feeding away from the nest and transferring the food to nestmates within the nest, as well as less time recruiting, than do minor workers in larger colonies. Evidently the effort of a single worker can more nearly satisfy the requirements of the small colony (Burkhardt, 1983).

Recruitment thus permits a multiple strategy of foraging and permits the exploitation of a wider range of food items. When workers of *Monomorium, Myrmica, Formica,* and *Lasius* forage singly, prey size is positively correlated with worker size. Yet when cooperative foraging based on recruitment is used, the correspondence breaks down. The ants are able to handle items far larger than their individual size (Traniello, 1987a). Although the workers in Traniello's study varied in fresh weight from 0.1 to 6.8 milligrams, their arthropod prey varied from 1 milligram to over 2 grams, a 2,000-fold range from the lightest to the heaviest.

## SOLITARY FORAGERS

The extreme category of solitary hunting combined with solitary retrieval is illustrated by the desert scavengers of the genera *Cataglyphis* and *Ocymyrmex* (Harkness and Wehner, 1977; Wehner et al., 1983; Schmid-Hempel, 1984, 1987; Schmid-Hempel and Schmid-Hempel, 1984; Wehner, 1987). The complete foraging pattern of *C. bicolor* is described in a striking manner by Wehner in his 1987 review:

> When the daily foraging activity of a desert ant's colony is compressed in both space and time by assuming a bird's eye point of view and using a quick-motion camera, an impressive picture emerges: a constant stream of particles radiates out from a centre and spreads evenly over a roughly circular area. Within this field of flow each (individually labelled) particle follows its own radial path, lingers at some distance from the start, and returns to the centre. Some time later, the same particle reappears, follows a similar path, disappears again in the centre, reappears, and continues to do so until at the end of the day the whole stream of particles ceases to flow. Now assume that the time window is expanded from one to several days and the time machine speeded up. One would then become aware of particles disappearing from the scene forever after having performed a number of moves, and new particles taking their place. It is as though one were observing a gigantic slime mold in which cells continually moved in and out from an aggregation centre, but in which the inward and outward movements of the cells were largely desynchronized. Scaling up the scene by about five orders of magnitude and returning to real time, the central-place foraging behaviour of desert ants unfolds in front of one's eyes, and a number of questions about the spatial and temporal organization of this behaviour immediately spring to mind. (p. 16)

The workers of *Cataglyphis bicolor* (Figure 10–3) make about 5 to 10 forays each day, and an average of 290 foragers of a colony conduct 1,500 to 3,000 such expeditions. The entire effort is made during daytime. The ants all retreat into the nest at night and often close the entrance with soil. As morning approaches they dig out the entrance again. At sunrise, a worker comes out slowly and stands about for a minute or so within a few centimeters of the entrance. As Harkness and Wehner (1977) describe it, the first foragers next begin their runs:

> Then one or two more come out and will set off away from the nest in more or less a straight line in some direction, which varies all round the compass with different individuals. Each individual, however, keeps to a constant foraging direction, that is maintained at least over a period of weeks. The frequency of ants leaving is low at first but eventually rises to 100–200 per hour (according to the size of the nest) later in the morning. In half-an-hour or more the first ants begin to come back, usually carrying the corpse or part of the corpse of a dead insect. The arrival of an ant with a burden seems often to be followed by the exit of a group of half-a-dozen or more ants . . . After an ant returns from a foray, commonly it goes out again in a matter of minutes. By midday there is a constant traffic in and out. Although the frequency of exits and entries stays high for several hours, in general the activity of an individual ant is restricted to a limited period of time that is the same every day and lasts 2–3 hours. Towards sunset the traffic in and out of the nest falls away. (p. 115)

The solitary foragers are guided by sun-compass orientation and the location of bushes and other landmarks, the latter being remembered as successive frames of configuration (Wehner, 1982, 1987).

**FIGURE 10–3** A solitary forager of the desert ant *Cataglyphis bicolor.* (Photograph by R. Wehner.)

The foragers tend to persist in only one or a very few directions for their lifetime, if for no other reason than that travel outside the nest is very dangerous and life is short. Most of the workers are soon picked off by spiders and robber flies, in spite of their ability to run at great speed, up to a meter per second. In southern Tunisia they enjoy a half-life time of only 4.2 days and a life expectancy of 6.1 days (Schmid-Hempel and Schmid-Hempel, 1984). Yet the system is so efficient that the average forager retrieves a food weight during her lifetime 15 to 20 times greater than her own body weight (Wehner et al., 1983).

When occasionally changing direction, both *Cataglyphis* and *Ocymyrmex* appear to follow a simple rule of thumb: continue to forage in the direction of the preceding foraging trip whenever the trip proved successful. Otherwise abandon that direction and select a new one at random, but decrease the probability of doing so as the number of previously successful runs increases. Following these rules, Wehner (1987) designed a model that closely described the directional choices actually made by individual foragers. The result was in accord with the observation that colonies exhibiting low foraging efficiencies simultaneously develop low sector fidelities. The overall effect of this method of foraging is that each colony blankets the area around the nest entrance so evenly that maps of foraging activity do not provide a clue to the location of the nest entrance (see Figures 10–4 and 10–5).

As Schmid-Hempel (1984) has pointed out, the rule of thumb used by *Cataglyphis* and other desert ants creates a kind of polymorphism in the population of foragers. When some of the ants encounter aggregated items, so that they are repeatedly rewarded, they concentrate their efforts. Other workers enter terrain with more scattered resources and disperse their efforts to a corresponding degree. In the ensuing total pattern the entire foraging domain of the colony is covered efficiently, with a temporary focus on the most productive sites.

The efficiency of deployment is enhanced by "noise" in the repetitive journeys. That is, unpredictable deviations regularly occur away from paths taken on previous sallies. Such errors increase the chance that the workers will strike food items missed in earlier efforts. The same explanation applies to the scatter of fire ants around odor trails, a variance that sometimes allows the workers to locate moving prey (Wilson, 1962b, 1971).

Pasteels, Deneubourg, and their co-workers have taken the next step of reasoning and developed a sophisticated theory of error adjustment to optimize net energetic yield during recruitment (Pasteels et al., 1982; Deneubourg et al., 1983; Verhaeghe and Deneubourg, 1983; Champagne et al., 1984; Pasteels et al., 1987). Each species concentrates on food items with a certain pattern of dispersion and size. When the items are extremely clumped and large, as in the case of a termite colony, the recruitment needs to be very precise, as it is in the case of termitophages *Pachycondyla* (= *Termitopone*) *laevigata* and *Megaponera foetans.* When food is less aggregated and more easily handled by single workers, the error in recruitment is typically larger. Such seems to be the case in the *Messor* and *Pogonomyrmex* harvester ants, for example. Furthermore, the error level may be adjusted within the repertories of individual species in ways that accord with the Pasteels-Deneubourg models. Food items that are small or of poor quality elicit less recruitment pheromone and tactile signaling and more error in the response, which leads to a wider exploration of the environment, as in the single-forager strategy of the desert ants (Wilson, 1962b; Hangartner, 1969b; Hölldobler, 1976a; Verhaeghe, 1982; Crawford and Rissing, 1983). The same is true of food sources that are distant from the nest (Wilson, 1962a; Ayre, 1969; Hölldobler, 1976a; Taylor, 1977).

## *ORTSTREUE* AND MAJORING

Workers belonging to the same colony display a striking degree of specialization in their foraging activity. A growing body of research has revealed the widespread occurrence of individual persistence in foraging site, the phenomenon called *Ortstreue.* We have seen how *Cataglyphis* workers return repeatedly to the same approximate area for as long as they are rewarded with food. Similar learning and particularization of behavior have been documented in the ponerine ants *Diacamma rugosum* and *Pachycondyla* (= *Neoponera*) *apicalis,* both of which are solitary hunting ants (Uezu, 1977; Fresneau, 1985); leafcutter ants of the genus *Atta* (Lewis et al., 1974a); harvester ants of the genera *Messor, Monomorium* (=*Chelaner*), *Pheidole,* and *Pogonomyrmex* (Hölldobler, 1976a; Hölldobler and Möglich, 1980; Onoyama, 1982; Onoyama and Abe, 1982; Davison, 1982); the carton-building "shining black ant" *Lasius fuliginosus* of Europe (Dobrzańska, 1966; Hennaut-Riche et al., 1980); the North American species *Lasius neoniger* (Traniello and Levings, 1986); and ants of the *Formica rufa* group (Økland, 1930; Jander, 1957; Rosengren, 1971, 1977a; Herbers, 1977; Rosengren and Fortelius, 1986b).

Experiments on *Cataglyphis, Formica,* and *Pogonomyrmex* suggest that the memories of location in site fidelity are primarily visual. In

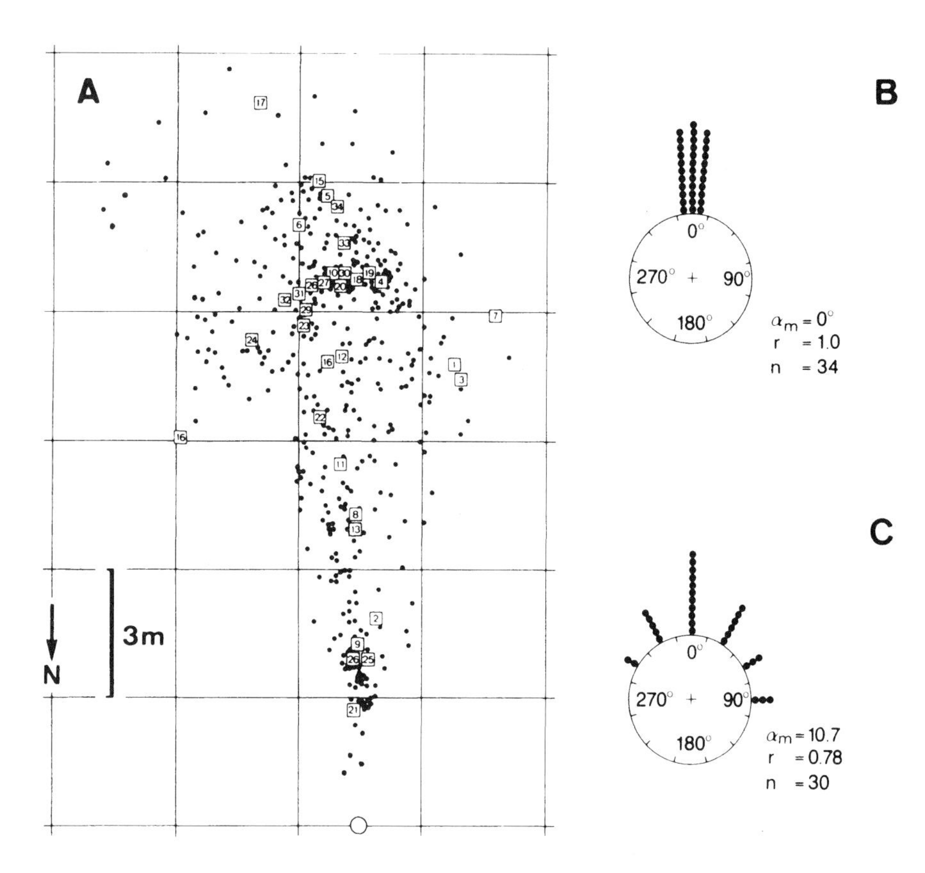


**FIGURE 10–4** The foraging pattern of a desert ant species (*Cataglyphis bicolor*) that searches for and retrieves food items in a solitary fashion. (*A*) This plot shows 34 foraging paths of one typical individual over a period of 5 days. The ant's position was recorded every 30 seconds. The sites where the ant found a piece of food are indicated by squares containing the code numbers of the foraging trips. (*B,C*) Foraging directions of one worker (*B, Cataglyphis bicolor; C, Ocymyrmex velox*) recorded over a period of 5 days. (From Wehner, 1987.)

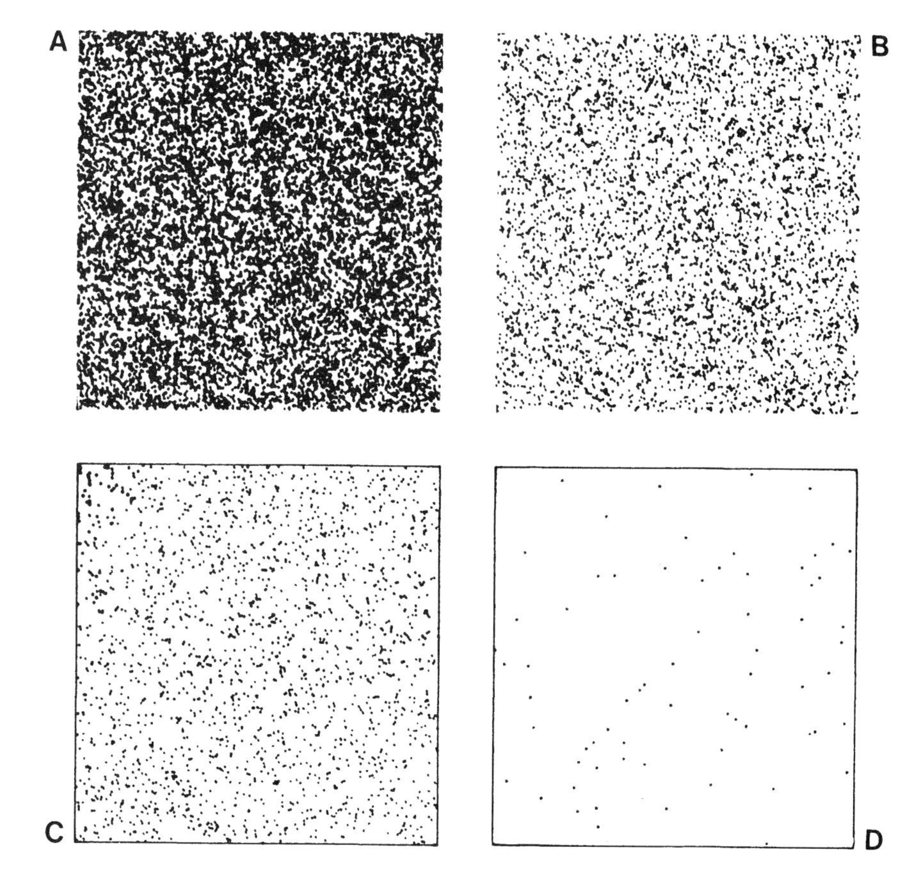


**FIGURE 10–5** When the foraging paths of all of the foragers of desert ant colonies are plotted together, they uniformly blanket a wide area around the nest entrance. The four species whose patterns are depicted here are: *A, Cataglyphis bicolor* (possibly sibling species no. 1); *B, C. albicans; C, C. bicolor* (possibly sibling species no. 2); *D, Ocymyrmex velox*. Each plot represents an area of 25 × 25 m. The position of each ant is given every 10 minutes over a 1-day period. (From Wehner, 1987.)

extreme cases the learning persists for weeks or months. Rosengren and Fortelius were able to show that in *F. aquilonia* at least, olfactory orientation is used to the same purpose in the dark. Specifically, the workers appear to follow scent marks used in home range marking. Thus the *Formica* evidently depend on a hierarchy of cues of the kind demonstrated in homing *Pogonomyrmex* workers by Hölldobler (1971b, 1976a). A cue in one sensory modality (such as vision) is used predominantly until it fails, then a cue in a second modality (such as olfaction) is employed, and so on. *Pheidole militicida*, which forages extensively during the night, also employs primarily olfactory cues in home range orientation and *Ortstreue* (Hölldobler and Möglich, 1980).

It is reasonable to expect that memory and fidelity to particular sites improve, through natural selection, as food sources become richer and more persistent. Such appears to be the case in herds of honeydew-producing aphids and other homopterans (Pasteels et al., 1987; Goss et al., 1989). When mistakes are made, so that the ants arrive at the wrong tree, the result can be a substantial loss of time and energy. Just such an unproductive leakage has been documented in *Formica lugubris* by Sudd (1983), who found that a small percentage of workers wander from a species of pine supporting large populations of aphids to another species with few or no aphids.

Another way to divide the environment among foragers belonging to the same colony is by majoring, or specialization on different kinds of food. It is well known that individual honey bees and bumblebees persist in visiting one species of flower for days at a time, even when the favored blossoms are intermingled with those of other flower species (Heinrich, 1979; Seeley, 1985). A similar phenomenon occurs in ants, although it has been documented less extensively. In field experiments by Rissing (1981a), for example, workers of *Messor* (= *Veromessor*) *pergandei* and *Pogonomyrmex rugosus* continued to harvest one species of grass seed mixed with two other species, even when larger seeds in the medley were available and being collected by nestmates. The specialization sometimes lasted for several days but occasionally shifted rapidly. The choices made were not just a function of body size, with larger ants selecting larger seeds. This variable accounted for only 4 percent of the variance in seed size. It is not known how the initial selections are made and why preferences change, but the process is far from a random first choice: *Messor* and *Pogonomyrmex* workers examine as many as 60 seeds before carrying a single one home (Hölldobler, 1976a; Rissing, 1981a).

## CENTRAL-PLACE FORAGING

When a forager is bound to a nest or sleeping site, she faces a different set of problems in harvesting energy than she would if she merely rested at intervals while conducting a search. How these problems can be solved so as to maximize the net energy yield is the subject of the special set of models that make up central-place foraging theory (Orians and Pearson, 1979; Schoener, 1979; Stephens and Krebs, 1986). The basic reasoning must be modified to accommodate the social insects, which commit expendable "energy packets" in the form of nonreproductive workers (Oster and Wilson, 1978; Seeley, 1985). The number of workers and hence the energy costs can be fitted through mass communication to the spatial distribution of food items and thus to the yield of energy moment by moment. Also, the workers can individually specialize on particular sectors of the terrain and on different kinds of food items, which makes possible a nearly simultaneous coverage of a wide area.

Most of the predictions made from the foraging models are intuitively clear and have a double heuristic value in the study of ant ecology. First, they allow a test of the basic proposition that natural selection shapes behavior and, in the case of social insects, does so by acting at the level of colonies. Second, by framing research in terms of these models, the ecologist asks questions and searches for predicted phenomena that may otherwise easily be missed.

Let us now consider some of the predictions and see how they have fared so far in empirical studies. The key overall assumption is that the greater the energy expenditure to get to a patch of food items, the more choosy the animal should be in selecting the item out of the patch to carry home. In other words the more energy you spend to get there, the larger the energy package you try to bring home. This general conception leads to a series of explicit predictions, as follows.

*The greater the distance the ant travels from the nest to a food patch, the longer she should take to select the food item.* This prediction was confirmed by Schmid-Hempel (1984) for workers of *Cataglyphis bicolor*, who spend more time searching around a productive site (where dead insects were found) when the site is farther from the nest. It was also confirmed by the finding of Rissing and Pollock (1984) that workers of *Messor* (= *Veromessor*) *pergandei* take longer to search through more distant patches of seeds once they have arrived on the site. It is not supported by data of Shepherd (1982) on the leafcutter *Atta colombica*, the workers of which fail to increase their search time on leaf baits when they have traveled farther to get there. The *Atta* data are ambiguous, however, because the large variance in individual times obscured possible real differences in a one-way analysis of variance conducted by Shepherd. For ants generally, then, this test of the basic proposition of energy conservation appears to be supported by the available data, but more experimental studies with different species are needed to make the test convincing.

*The greater the distance the ant travels to a food patch, the more selective the ant should be in choosing a food item.* The expected relation was obtained by Davidson (1979), who found that the farther from the nest *Pogonomyrmex rugosus* foragers journeyed, the narrower the range of barley seeds they accepted in a patch, the smaller seeds being ignored more consistently. Rissing and Pollock (1984) discovered that whereas *Messor pergandei* workers generally take larger seeds from patches, the size of the seed is not correlated with the time spent selecting it and hence the time spent in traveling to the patch. Possible nutritive cues other than size could not be discounted. Overall the available data appear to favor the energy model, but more studies are needed.

*The higher the temperature, the more selective the forager should be in food patches.* To our knowledge no experiment has been performed to test this expected correlation. However, Traniello et al. (1984) have examined the behavior of *Formica schaufussi*, at varying temperatures, in the pursuit of scattered prey as opposed to variable prey aggregated in patches. As the temperature rose, the ants became less selective, accepting smaller and less profitable prey. Ants, like other insects, increase their metabolism and hence pay higher energy costs at higher temperatures (Nielsen, 1986). We would therefore expect that if the ants encounter food items in groups, and can pick and choose, they will be more selective at higher temperatures. If they encounter prey items singly, however, they will be inclined to be less selective, because "a bird in the hand is worth two in the bush." This is the result obtained by Traniello and his co-workers.

## ENERGY MAXIMIZERS VERSUS TIME MINIMIZERS

Schoener (1971) made an important distinction between animals that maximize their net energy yield and those that have a fixed quota of energy required for body maintenance and reproduction. Probably only bacteria and other very simple organisms are pure energy maximizers, but even within the ranks of the more complex animals some species appear to follow an energy maximization program more closely than others. A model of conversion of energy to genetic fitness in social insects has been offered by Oster and Wilson (1978). Their key result can be stated as follows: so long as the production of new queens remains a linear function of the amount of energy harvested, ant colonies can be expected to maximize their net energy yield. This inference, along with the thrust of information from natural history, indicates that ant foraging procedures are based predominantly on energy maximization. Indeed, the celebrated industriousness of ant colonies stems largely from their seemingly incessant foraging activity and rearing of young. And there is a direct connection to genetic fitness. Numerous studies have shown that the larger the colony, the more likely it is to extirpate neighboring colonies and to extend its foraging range (see Chapter 3), and consequently, the larger the number of queens it produces.

Nevertheless the ant colony is far from a mere growth machine. Prudence would seem to dictate that the colony commit only a small fraction of the worker force at any one time to foraging. Workers expend up to seven times more energy while running than while resting (Nielsen et al., 1982; Lighton et al., 1987), so that a point of diminishing return in energy harvesting is quickly reached. To this can be added the high construction costs of replacing lost foragers, along with the need to have a substantial force available to defend the colony against attack at any moment. The trade-offs among these countervailing selection forces remain to be quantified in both theory and experiment.

## SPECIAL STRATAGEMS THAT IMPROVE HARVESTING

One of the most distinctive devices evolved by ants in foraging is *group retrieval*, in which two or more workers cooperate to bring home a food object too large to be managed by a single individual. The behavior occurs in elaborate form in the Asiatic marauder ant *Pheidologeton diversus* (Moffett, 1987b). When individual workers carry prey or other food items, they lift them from the ground in their mandibles and hold them forward while walking back toward the nest. Group-retrieving ants carry burdens differently. They lay one or both forelegs on the burden, an action that appears to assist in lifting it. The ants also open their mandibles and press them against the object, although they seldom attempt to grip it in the manner of solitary *Pheidologeton* foragers carrying smaller burdens. Workers utilize different movement patterns corresponding to their positions around the perimeter of the object and the direction of transport. Those at the forward margin walk backward, pulling the burden, while those along the trailing margin walk forward and push it. The ants along the sides of the burden shuffle their legs more or less laterally in the direction of transport (Figure 10–6).

This group-retrieval technique permits the *Pheidologeton* to carry food at many times more the weight per worker than is possible through retrieval by solitary workers. One of the heaviest burdens recorded by Moffett was an earthworm 10 centimeters long with a dry weight of 0.55 gram, or over 5,000 times the weight of a single minor worker. The worm was borne by about 100 ants, who transported it at about 0.4 centimeter per second on level ground. Comparisons with transport of smaller objects showed that the workers carrying the worm handled at least 10 times more weight per ant than did solitary ants carrying a summed equivalent in small loads, with a loss of only a little more than half the velocity. In another extreme case a large piece of cereal was carried by 14 ants. If the ants had first gnawed the chunk into pieces just small enough to be handled by single ants, at least 60 workers would have been needed to retrieve the same amount of food. In other words group transport reduced the required worker force by 75 percent.

Group transport has evolved independently among most of the subfamilies of ants (see Table 10–2). It is limited to species that practice recruitment to food sources. Among various phyletic groups it is developed to variable degrees of skill, with the greatest sophistication and efficiency occurring in *Pheidologeton, Aphaenogaster, Eciton,* and *Oecophylla* (Plate 13). At least three genera, *Camponotus, Leptogenys,* and *Pheidole,* contain some species that utilize it conspicuously and others that do not practice it at all. Franks (1986) has shown that workers of the swarm-raiding army ant *Eciton burchelli* actually form teams to carry large burdens, in which medias specialized for transport get the process started. Medias also frequently join the transport gangs of *Pheidologeton diversus* minor workers, but media participation is much less important here than in *Eciton*. Finally, the species of *Pheidologeton, Eciton,* and *Neivamyrmex* use groups to transport food, immature stages, and (in the case of *Pheidologeton*) large pieces of nest material and refuse.

A second highly evolved technique that enhances productivity, namely the frequent rotation of foraging direction, has been discovered in two species of ants that possess large colonies and send out columns or swarms of foragers. During the nomadic phase of the *Eciton burchelli* 35-day cycle, when colonies emigrate to a new site following the daily raid, the ants run little risk of depleting the arthropod populations on which they depend. During the statary phase, however, when the colonies remain in the same site for an extended period of time, the arthropod populations are decimated and require as much as a week to recover. Other ant and wasp species, which are favored targets, take much longer. Franks and Fletcher (1983) showed that the *E. burchelli* solve the problem by rotating their central swarm direction around the central bivouac site. Each day they shift an average of 126 degrees, which is significantly greater than the 90 degrees that would be expected if raids were oriented at random. Thus the ants increase their harvesting efficiency by a design similar to the spiral leaf arrangement used by many plant species, to minimize self-shading. The raiding fronts are

**FIGURE 10-6** Foragers of the Asiatic marauder ant *Pheidologeton diversus* join forces to carry large seeds. (Photograph by M. W. Moffett.)

**TABLE 10-2** Genera in which examples of group transport are known for at least some of the constituent species. An asterisk (*) indicates that the transport is poorly executed. (Provided by M. W. Moffett.)

| Taxon | Items carried by groups of workers | Selected references |
|---|---|---|
| **PONERINAE** | | |
| *Leptogenys* | Food | Fletcher (1973), Maschwitz and Mühlenberg (1975) |
| *Mesoponera* | Food | Agbogba (1984) |
| *Onychomyrmex* | Food | B. Hölldobler and R. Taylor (unpublished) |
| **Paltothyreus* | Food | Hölldobler (1984b) |
| *Plectroctena* | Food | Fletcher (1973) |
| *Rhytidoponera* | Food | R. H. Crozier and P. J. M. Greenslade (personal communication) |
| **DORYLINAE** | | |
| *Aenictus* | Food | Chapman (1964) |
| *Dorylus* | Food | Gotwald (1974) |
| **ECITONINAE** | | |
| *Eciton* | Food, brood | Wilson (1971), Franks (1986) |
| *Labidus* | Food | Rettenmeyer (1963a) |
| *Neivamyrmex* | Food, brood | Rettenmeyer (1963a) |
| *Nomamyrmex* | Food | C. W. Rettenmeyer (personal communication) |
| **LEPTANILLINAE** | | |
| **Leptanilla* | Food | K. Masuko (personal communication) |
| **PSEUDOMYRMECINAE** | | |
| **Pseudomyrmex* | Enemies, adult workers | D. H. Janzen (personal communication) |

*continued*

**TABLE 10–2** *(continued)*

| Taxon | Items carried by groups of workers | Selected references |
|---|---|---|
| **MYRMICINAE** | | |
| *Aphaenogaster* (including "*Novomessor*") | Food | Hölldobler et al. (1978) |
| **Daceton* | Food | Wilson (1962a) |
| **Leptothorax* | Food | S. Cover (personal communication) |
| *Myrmica* | Food | Sudd (1965) |
| **Oligomyrmex* | Food | Moffett (1986g) |
| *Pheidole* | Food | Sudd (1960), E. O. Wilson (unpublished) |
| *Pheidologeton* | Food, brood, adult workers, building material, refuse, enemies | M. W. Moffett (personal communication) |
| *Tetramorium* | Food | J. F. A. Traniello (personal communication) |
| *Wasmannia* | Food | Clark et al. (1982) |
| **ANEURETINAE** | | |
| *Aneuretus* | Food | Traniello and Jayasuriya (1981b) |
| **DOLICHODERINAE** | | |
| *Azteca* | Food | C. W. Rettenmeyer (personal communication) |
| **FORMICINAE** | | |
| *Anoplolepis* | Food | M. W. Moffett (personal communication) |
| *Camponotus* | Food | B. Hölldobler (unpublished) |
| *Formica* | Food | Sudd (1965) |
| *Lasius* | Food | Traniello (1983) |
| *Myrmecocystus* | Food, honeypot slaves | Hölldobler (1981a) |
| *Oecophylla* | Food, enemies | Hölldobler (1983) |
| **Paratrechina* | Food | M. W. Moffett (personal communication) |
| *Prenolepis* | Food | B. Hölldobler (unpublished) |

in fact analogous to very long, thin leaves, in which a spiral phyllotaxy can be less than exact and yet still avoid overlap.

A similar foraging rotation has been claimed in the desert ant *Messor (= Veromessor) pergandei*, which sends out tens of thousands of foragers in lengthy columns to collect seeds. Went et al. (1972) reported the average change between consecutive foraging periods to be 15 degrees, whereas Ruth Bernstein (in Carroll and Janzen, 1973) gave it as about 20 degrees. Rissing and Wheeler (1976), however, found the shifts to be very irregular in magnitude and direction, sometimes proceeding consistently clockwise, sometimes counterclockwise, and sometimes reversing to head in the opposite direction. Where the *Eciton* "phyllotaxy" is relatively precise and endogenous, the *Messor* pattern is evidently more dependent on the day-to-day vagaries of seedfall.

Yet another method that improves foraging efficiency is to deploy workers in outposts, shortening the length of their sallies and allowing them to store food temporarily at way stations or even to feed larvae carried there. The behavior is displayed by the termite-hunting workers of the Malaysian *Eurhopalothrix heliscata*, who tend to settle in groups and rest for prolonged periods of time away from the brood chambers. The pattern enhances the effectiveness of their predation. *Eurhopalothrix*, like other members of the tribe Basicerotini, do not rely on mass raiding to overcome termites as do some species of *Leptogenys*, *Megaponera*, and *Pachycondyla*. They are able to recruit nestmates, but the communication is relatively short-range and imprecise. It therefore seems to be of advantage for colonies with these traits to have staging areas from which forays against termite colonies can be conducted over relatively short distances whenever workers discover foraging columns of the insects or holes in the termite nests. The prey items can then be relayed at a more leisurely pace back to the brood chambers (Wilson and Brown, 1984). A similar system of outposts is used by the strange little myrmicine *Proatta butteli* of tropical Asia. The workers prey on termites, isopods, and other arthropods, some of which are quite large and can be subdued only by rapidly recruited workers (Moffett, 1986d).

The outpost phenomenon probably accounts for many cases of polydomy, defined as the dispersal of the colony into multiple nest sites. Polydomy may occur in colonies with either single or multiple queens. Traniello and Levings (1986) found that in *Lasius neoniger*, a monogynous formicine that nests in the soil in fields and other open habitats, polydomy follows an annual cycle and is close to optimal for foraging efficiency. During early and middle summer workers construct increasing numbers of little crater nests, expanding the foraging territory of the colony as a whole. The nests at the Massachusetts study site were more uniformly distributed than they would have been had chance alone dictated their distribution. Workers emerging from a given nest entrance were most successful at retrieving prey items over distances of less than 20 centimeters. The average distance between nest entrances was 38 centimeters, that is, about twice the most effective foraging range, as the investigators had expected. In late summer, when the season drew to a close, the number of nests declined and the colony contracted toward a central core.

## POPULATION REGULATION

The population of workers in an ant colony grows in a roughly sigmoidal (S-shaped) pattern. It accelerates at first, then decelerates as some of its resources are diverted to the production of virgin queens and males. During the dry season or winter, when local conditions are harshest, the worker population declines still further, because mortality is not adequately offset by larval growth and new worker production. The reproductive forms leave the nest after the growth season, so that the workers are free to increase their own numbers during the first part of the next growth season. As a consequence the population of workers in a mature colony fluctuates annually around a mean, roughly approximating a stable limit cycle (Brian 1965b, 1983; Wilson, 1971).

Superimposed on the growth of individual colonies is the increase in the population of colonies. The standing density of colonies is determined by the rate at which queens start new colonies, balanced by their mortality rate and that of the colonies they found. At the next lower level, the density of workers belonging to a given species in a particular area is the summed product of the number of colonies in various growth stages and the workers present in each growth stage.

It is convenient at this stage of our knowledge of ant ecology to make a rough distinction between density-independent effects and density-dependent effects in the determination of the local numbers of colonies and workers. Independent effects are those that alter birth and death rates in a manner not related to the density of the colonies or workers. Some features of the physical environment, for example, limit numbers by restricting the replacement rate of the colonies and workers regardless of the density attained in the growth curves. Such is the case for flooding and severe cold waves. In contrast, density-dependent effects either lower the birth rate or raise the death rate as population density increases. In time the braking effect slows the growth to zero. Depending on the initial conditions and the lag time in which the density-dependent effects take place, the population stabilizes at a more or less constant level (the "carrying capacity" of that particular environment), or enters a stable limit cycle around a constant level, or enters a chaotic regime in which the density fluctuates in an unpredictable manner. The totality of these effects constitutes the population regulation.

Put briefly, the existing evidence implicates territorial aggression as an important and possibly premier mode of population regulation in ants. Other factors, including climate, predation, food, and availability of nest sites, also clearly affect the density of ant colonies, but so far they have been only sparsely documented. No experiments have been performed that permit an evaluation of all of the controls together. It is possible, however, to describe the effects of separate factors separately, at least in qualitative terms.

Starting then with density-independent factors, the role of temperature is most evident in the biology of the ants of cold temperate zones. In northern Europe the distribution of *Myrmica* species, for example, is highly correlated with the number of hours of bright sunshine per day (Baroni Urbani and Collingwood, 1977). Barrett (1979) found that portions of England lacking *M. sabuleti* average only 4.6 hours of bright sunshine per day, those with sparse records of the species have 5.2 hours per day, and those with abundant records average 5.8 hours per day. In general *Myrmica* colonies of various species receiving the most solar heat acquire the largest worker populations, and those with the most workers produce a disproportionate share of virgin queens and males (Brian and Brian, 1951; Brian and Elmes, 1974; Elmes and Wardlaw, 1982). Elmes and Wardlaw fitted the known correlates to a model in which a chain of annual events determines the production of these reproductive forms; the key factors they intuited include the number and size of larvae held over winter and the amount of solar heat acquired in the following spring.

In milder portions of the temperate zones, population densities are more likely to be reduced not only by cold stress in the winter but also by heat and drought stress in the summer. In fact, studies of the meat ant *Iridomyrmex purpureus* in Australia across terrain of variable drainage showed that summer stress affects the growth of worker populations within individual colonies, whereas winter stress affects the survival of colonies as a whole (Greenslade, 1975a,b). A similar bracketing control through the annual cycle has limited the spread of the imported fire ant *Solenopsis invicta* in the southern United States (Francke and Cokendolpher, 1986).

For some kinds of ants a limitation in the number of suitable nest sites is important in population control. This is especially true in regions with marginal climatic conditions, including northern Europe, where exposure to the sun is a critical limiting factor (Brian, 1952b, 1956c; Sudd et al., 1977). It also appears to be the case in the lowland rain forests of New Guinea, where most of the ant species are specialized for living in rotting tree branches and other small fragments of wood lying on the ground, and where most such nest sites are fully occupied by colonies of either ants or termites (Wilson, 1959a,c). Hence the rotting-wood specialists appear largely to have filled the space available to them. A majority of species studied by Room (1971) in a cocoa farm in Ghana occupy the same kind of nest sites. In Sri Lankan forests a similar near saturation occurs, but with termites preempting a larger percentage of the nest sites. *Daceton armigerum*, a large predatory ant of South America, is specialized for life in preformed cavities in the canopy of moist tropical forests. The relative scarcity of such retreats appears to be an important factor in determining the density of *Daceton* colonies, but this hypothesis has not been subject to verification due to the relative inaccessibility of the habitat (Wilson, 1962a). A similar specialization appears to restrict the European "shining black ant" *Lasius fuliginosus*, which nests preferentially in large preformed cavities in tree trunks (Maschwitz and Hölldobler, 1970).

On the other hand persuasive experimental confirmation has been provided by Herbers (1986b) in the case of *Leptothorax longispi-*

*nosus*, an ant that favors preformed cavities in small pieces of rotting wood on the floor of North American deciduous forests. When Herbers added additional nest sites in the form of hollow dowels, the density of colony fragments increased, and the total worker population appeared to grow as well. In an independent study Brian (1956c) has provided evidence that in the west of Scotland the density of colonies is controlled by the density of available nest sites. In England as a whole, which has only a marginal environment for ants, "nest sites need to have high insolation, to be moist but not too wet, to be soft enough for excavation but mechanically stable, and they need to be within reach of plants to supply sugar and water through Homoptera and protein through the many small insects that feed on the plants and their litter." The constraining condition of having to nest in a place with few plants and yet still be near places with many plants is offset by the extreme abundance of insect food during the spring and summer. In Brian's study area in western Scotland, much of the land is covered by higher vegetation of one form or another. Consequently nest sites are in short supply and the ant colonies compete heavily for them. When Brian laid out slabs of rocks, which make ideal nest sites, in an open, food-rich area populated by *Myrmica ruginodis*, the number of colonies increased. In a parallel study he found that the gradual growth of trees in a glade, which increased shading and reduced the number of warm, dry nest sites on the ground, resulted in a reduction in the number of ant colonies.

The same general conclusion was independently drawn by Gösswald (1951b) on the basis of his long-term studies of the same ant genera in Germany. He pointed to the curious fact that dense ant populations characteristically occur around quarries, where fragments of stones lying on the ground provide an unusual number of nest sites. Yet in spite of such evidence nest site limitation is probably far from universal. It appears not to be a significant factor in Puerto Rican forests (Torres, 1984a,b), for example, and it is obviously of limited importance for the soil-dwelling ants of deserts and grasslands.

The food supply of colonies has been implicated in colony regulation in at least two studies. According to Davison (1982), colonies of *Monomorium* (= *Chelaner*) in arid New South Wales are sensitive to fluctuations in the supply of seeds on which they depend. In times of scarcity the number of larvae declines. Because the adult workers depend on larval secretions for food (the larvae consume and metabolize the raw seeds for the colony), the worker population also declines. In another study, by excluding rodents that compete with seed-eating ants and counting the seeds available in the soil, Brown et al. (1979a) obtained evidence that the colony density of the granivorous ants in Arizona is limited by food.

As noted earlier, swarm-raiding army ants (*Eciton burchelli*) devastate the arthropod fauna on the ground and low vegetation over which they conduct their daily forays. On Barro Colorado Island in Panama, approximately 50 colonies harvest the island in a relatively efficient manner, seldom if ever colliding with one another (Franks and Bossert, 1983). Nearby Orchid Island has only one-eightieth the area of Barro Colorado, and for many years no *E. burchelli* lived there. Since both islands were created by the rise of Gatun Lake, Franks and Fletcher (1983) concluded that the army ants had previously existed on Orchid Island but had disappeared because of an inadequate food supply; Orchid Island has less area than the average portion of land shared by each *Eciton* colony on Barro Colorado Island.

In a reverse direction, predation can also depress colony and worker density of the ant species taken as prey. Franks and Fletcher found that the ant colonies on Orchid Island were denser and more mature than those on Barro Colorado. When these investigators introduced an *Eciton burchelli* colony onto the small island, it preyed much more heavily on ants during the first few weeks than did the *Eciton* colonies remaining on Barro Colorado and as a result reduced the densities of the prey ants substantially.

## OFFENSE AND DEFENSE BY FORAGERS

The ants possess an enormous variety of offensive and defensive techniques and employ them in almost every imaginable circumstance in the search for food and the protection of nests. The existing information on this very eclectic category of behavior is briefly summarized in Table 10–3.

Hunt (1983) correctly noted that predators, especially visually searching vertebrates, play a key role in the shaping of foraging behavior, perhaps rivaling the maximization of net energetic yield as a natural selection factor. Separate species have "chosen" whether to commit only a few foragers or a large force, whether to hunt stealthily or in the open, whether to recruit soldiers to the food site, and so forth. All of these options are important in the evolution of foraging strategies. In spite of their potential significance, studies of the phenomenon have been curiously incomplete. A great deal of information has been acquired on the vertebrates, spiders, assassin bugs, and other animals specialized to prey on ants, as reviewed for example by Wilson (1971), Hunt (1983), Oliveira and Sazima (1984, 1985), and Redford (1987). Yet there are virtually no systematic audits of the entire set of predators of individual ant species, nor have field experiments been performed to determine the long-term effects of predation on population growth.

### Competition

Our presentation on foraging strategies so far tells us that each ant species lives within a particular temperature-humidity envelope. Its workers employ foraging procedures that evidently represent tradeoffs between the optimal food-gathering strategy and the need to avoid being eaten by predators. Population density is constrained by the kinds of food and nest sites utilized by the species, as well as by predation. Each of these factors is known to limit geographic ranges and local densities in particular species. Some of the published accounts documenting their effect have been detailed and definitive, others anecdotal and less persuasive. In either case, studies of population control in ants have been for the most part unifactorial. Assessments have not been made of the relative importance of multiple factors and their interactive, second-order effects. Long-term studies remain very scarce.

Yet in spite of the inchoate, not to say incoherent state of population regulation studies in ants, one process has asserted itself with undeniable force. Competition, by its ubiquity and the ease with which it is detected, has become the dominant theme in studies of ant ecology. The trend cannot be written off as fashion or an artifact

**TABLE 10–3** Offensive and defensive techniques.

| Technique | Representative genera and species | Employment | Source |
|---|---|---|---|
| Armor: heavily sclerotized exoskeleton | *Cerapachys, Sphinctomyrmex,* and other members of the tribe Cerapachyini, subfamily Ponerinae; *Cylindromyrmex; Cataulacus; Cephalotes; Dolichoderus* s.str. | Preying on other ants and termites; defense against predators, especially when combined with spines | Wheeler (1936c), Wilson (1957a), Brown (1975), Hölldobler (1982b), Hunt (1983), Buschinger and Maschwitz (1984) |
| Large size | *Dinoponera, Megaponera, Paraponera,* and other giant ponerine ants | Combined with powerful sting, effective defense against most other animals; some also termite raiders | Wheeler (1936c), Lévieux (1972, 1982), Weber (1972) |
| Trap mandibles | *Anochetus* and *Odontomachus* in the Ponerinae; *Daceton, Strumigenys,* and other members of the tribe Dacetini in the Myrmicinae; *Myrmoteras* in the Formicinae | Elongated mandibles snap shut convulsively, impaling enemy on sharp teeth at end | Brown and Wilson (1959), Brown (1976b, 1977b, 1978), Moffett (1985c) |
| Saber- and hatchet-shaped mandibles | Slavemaking species: *Harpagoxenus, Polyergus, Strongylognathus* | Mandibles efficient at piercing exoskeleton or chopping off appendages of opponents | Buschinger (1986) |
| Bouncing | *Orectognathus* | Elongated, blunt mandibles of major caste snap shut, "squirting" opponent some distance away | Carlin (1981) |
| Cryptic coloration, "playing dead" | *Basiceros, Eurhopalothrix,* and other members of the tribe Basicerotini; *Stegomyrmex; Myrmicocrypta, Trachymyrmex,* and other small members of the tribe Attini; many other small myrmicines | Earth-colored integument; workers freeze motion when disturbed; specialized body hairs for collecting soil on basicerotines and *Stegomyrmex* | Brown and Kempf (1960), Weber (1972), Hölldobler and Wilson (1986c), Wilson and Hölldobler (1986) |
| Spines | A wide range of Myrmicinae; some Ponerinae, Dolichoderinae, and Formicinae | Spines most commonly on thorax, next on head and petiole, rarely on gaster (which may bear flanges or tubercles); apparently most effective against predators, especially in conjunction with large size and thick exoskeleton | Dumpert (1978), Hunt (1983), Buschinger and Maschwitz (1984), Oliveira and Sazima (1984) |
| Antennal scrobes (longitudinal grooves in side of head to receive the antennae) | Developed to variable degree in many myrmicine genera and a few ponerine genera, including *Aulacopone* and *Paraponera* | Antennae fold back into scrobes for protection: in most species only scape is covered, but in others (e.g., members of *Aulacopone, Cephalotes, Glamyromyrmex, Phalacromyrmex,* and most Cephalotini) entire antenna is received | Wheeler (1910a), Kempf (1951, 1958a), Corn (1976, 1980), Wilson (1976a), Taylor (1979), Hunt (1983) |
| Venomous stings, poisonous droplets and secretions | Universal in ants. Stings prevail in anatomically primitive groups, including Sphecomyrminae, Nothomyrmeciinae, Myrmeciinae, and almost all Ponerinae; in a majority of Myrmicinae (especially primitive genera); in all Pseudomyrmecinae, Ecitoninae, and Dorylinae; and in Aneuretinae. Many of these species also employ poisonous droplets and sprays; stings are vestigial or absent in the Dolichoderinae and Formicinae, which use often elaborate systems of droplets and sprays | Employed to varying degrees in offense and defense, according to species. Most crippling sting against human beings may be from *Paraponera clavata* of South and Central America | Weber (1937, 1939, 1959), Hölldobler (1973a), Blum and Hermann, (1978a,b), Blum et al. (1980), Adams and Traniello (1981), Jones et al. (1982a,b), Schmidt (1982, 1986), Hermann (1984a,b), Buschinger and Maschwitz (1984), Blum (1985), Merlin et al. (1988); see also Chapter 7 |
| Other pungent secretions from exocrine glands (see Figure 7–45) | Many species and genera | Secretions sprayed or wiped on opponent, or else allowed to evaporate from body surface | Blum and Hermann (1978a,b), Blum (1981a), Buschinger and Maschwitz (1984) |

*continued*

**TABLE 10–3** *(continued)*

| Technique | Representative genera and species | Employment | Source |
|---|---|---|---|
| Propaganda pheromones | Slavemaking species, *Formica subintegra* and *Harpagoxenus sublaevis*; also the inquiline *Leptothorax kutteri* | | Wilson and Regnier (1971), Allies et al. (1986) |
| Warning coloration | Some species of *Macromischa, Myrmecia*, and *Pseudomyrmex* are brightly colored (e.g., red and black or even metallic blue and green), forage in the open, and possess strong stings; warning coloration of defensive secretions of mandibular glands in *Calomyrmex* | Coloration presumably a warning to predators, especially vertebrates | Brough (1978), Brown and Moore (1979), B. Hölldobler and E. O. Wilson (personal observations) |
| Phragmosis (heads with specialized shapes used to block nest entrances) | Plug-shaped heads in some *Camponotus* (subgenera *Colobopsis* and *Hypercolobopsis*), *Colobostruma, Crematogaster* (= *Colobocrema*), *Pheidole* (*P. lamia*), and *Oligomyrmex* (*"Crateropsis" elmenteitae*; see Ettershank, 1966). Shield-shaped heads in *Zacryptocerus* | Head fills nest entrance; used to block intruders or push them out of entrance galleries | See Chapter 8 |
| Reverse phragmosis (gaster truncated or strongly rounded and capable of plugging nest entrance) | Rounded first gastral segment in *Proceratium*; severely flattened and plug-shaped terminal gastral segments in *Pheidole embolopyx* | Role documented in *Proceratium*; not yet proved in *Pheidole* | Poldi (1963), Wilson and Hölldobler (1985) |
| Soldiers | Specialized physical castes, with enlarged or otherwise modified heads, cutting or crushing mandibles, enlarged glands, etc.; used to defend nests and (in some species) foraging columns and food finds | Defend nest and occasionally also foraging columns and food finds. Often recruited by other workers for one of these purposes | See Chapter 8 |
| Evasive movements | Jumping and bouncing in some *Myrmecia* and *Odontomachus*; deliberate free fall from trees and bushes in many formicine species; hiding under leaves and debris in many ponerine ants, such as *Pachycondyla* and *Paltothyreus* | | Wheeler (1910a), K. Hölldobler (1965), B. Hölldobler and E. O. Wilson (unpublished) |
| Closing nest entrance | See column at right | At sunset species of *Cataglyphis, Harpegnathos, Messor*, and *Pogonomyrmex* close nest entrance with soil. *Proformica epinotalis* do so when attacked by slavemaker *Rossomyrmex proformicarum*. When attacked by fire ants, *Crematogaster* close nest entrance with their bodies, abdomens pointed outward with poison drops extruded. Small *Myrmecocystus* and *Pogonomyrmex* colonies close nest when attacked by foraging ants. *Pogonomyrmex* close nest when preyed on | Hölldobler (1981a), Marikovsky (1974), Harkness and Wehner (1977), Buschinger and Maschwitz (1984), B. Hölldobler and E. O. Wilson (unpublished), H. Markl and B. Hölldobler (unpublished) |
| Covering trails and food sites with soil and carton, partly as walls or wholly as tunnels | *Crematogaster, Dorylus, Pheidologeton, Solenopsis* | Protection of foraging workers and persistent food sources | Wheeler (1910a), Wilson (1971), Gotwald (1982), Moffett (1986f, 1987b) |
| Absconding | *Aphaenogaster, Formica, Meranoplus, Oecophylla, Pheidole, Pogonomyrmex, Solenopsis* | Rapid and well-organized evacuation of nest in presence of formidable enemies or flooding | Forel (1874), Maidl (1934), Hölldobler (1976a), Buschinger and Maschwitz (1984), Wilson (1986c); see Chapter 7 |

of the procedures of field study. It is evidently a very important phenomenon in the biology of ants.

Before we lay out the results of research on the subject to date, it will be worthwhile to provide a brief theoretical background. The role of competition has been challenged on largely methodological and statistical grounds in recent years (for example Connell, 1975; and Simberloff, 1983, 1984), but most of the evidence has stood up well under critical examination, while additional field studies have supported it further (Schoener, 1983, 1986; articles in Diamond and Case, 1986). It has been convenient to divide competition into two modes: *interference competition*, in which individuals exclude one another by threats, fighting, or poisons; and *exploitative competition*, in which individuals use resources and hence deprive others of their use, but without direct aggression. The distinction is made more vivid by the occasional use of the expressions "contest" for interference competition and "scramble" for exploitative competition, and is analogous to the difference between small boys running a race to see who wins a pile of coins (contest competition) and the same small boys racing to pick up as many coins as possible thrown in front of them (scramble competition).

The intensity of competition both within and between species differs greatly according to habitat and the position of the species in the food web. In 1960 Hairston, Smith, and Slobodkin presented a model that postulated major differences among species as a function of their trophic level. They proposed that carnivores, at or near the top of the food web, should actively compete, since their principal density-dependent control would be the exhaustible food supply of herbivores on which they prey. The same should be true of primary producers (plants), which have a limited supply of incident solar energy, and scavengers and other decomposers, which are limited by the finite amount of dead tissue and waste material made available by the other trophic levels. In contrast, herbivores should not compete as much because, occupying the intermediate position in the food web, they are more likely to be held down by predators. This trophic-level rule of competition has held up well in numerous field studies of competition during the past twenty-five years. It is well marked in terrestrial and fresh-water systems, where producers, as well as granivores, nectarivores, carnivores, and scavengers taken together, display more competition than herbivores and filter feeders. In marine systems, however, no trend is yet detectable in one direction or the other.

In two notable reviews of the subject, Schoener (1983, 1986) stressed not only the trophic levels, but also other properties of particular species within each trophic level. Perhaps most important, we should expect competition to be most intense in large animals, since they are closest to the top of the food chain as well as long lived and resistant to other predators. Such organisms are likely to saturate the environment in a way that reduces food or fills nest sites until these requisites are in short supply and competition becomes a principal density-limiting force. Large size and long life are of course most familiarly displayed by carnivorous vertebrates. But they are also hallmarks of ant colonies. An ant colony, the replicating unit of these social insects, is a large "organism." It is long lived, in some species the record holder of all insects and longer lived than most vertebrates. Beyond its founding stage, the colony is also typically sheltered and heavily defended in specialized nests, so that the queen is usually immune to predators. Finally, the ant colony weighs heavily on the surrounding terrain, regardless of whether it preys on insects and other organisms, scavenges, or gathers seeds. In a nutshell, ants have all of the traits expected to generate competition within and between species, and competition is what we find.

The evidence for the pervasive role of competition in ants is complex and multilayered. It can be arranged into four categories:

1. Local communities of ant species are often overdispersed with reference to body size, food type, or weight of food item.
2. Nest sites and foraging columns of the same or closely related species are overdispersed.
3. Experimental removal of some colonies (of the same or different species) causes increased growth in other colonies.
4. Displacement of individual foragers or colonies by other foragers or colonies is commonly observed.

## Overdispersion of Biological Traits

The single most prominent pattern on which animal ecologists fasten their attention is the diversification of closely related species occupying the same locality. Bernstein (1974), for example, noted that three dominant seed-eating ants in the Mojave Desert of California, *Messor (= Veromessor) pergandei* and two species of *Pogonomyrmex*, forage at different surface temperatures, which in turn are correlated with the time of day at which foraging occurs. In arid environments of southwestern France, the granivore *M. capitatus* is limited to dry, calcareous grassland slopes, whereas *M. structor* occupies a wide range of surrounding habitats (Delage, 1968). In the savanna of the Ivory Coast, *Crematogaster heliophila* builds carton nests in the canopy and defends arboreal territories that extend through several trees. *C. impressa* occupies a complementary position, nesting mostly in hollow twigs in bushes near the ground (Delage-Darchen, 1971). In a wholly different dimension studied by Chew and Chew (1980), ant species occupying an evergreen woodland in Arizona and belonging to the same feeding guild (granivore or fluid-feeding) were most different in body size. The differences between a given pair of species were less marked when one of the species was uncommon, and they were smallest of all when the species occupied different guilds. Comparable results were obtained by Whitford (1978b) in a study of the granivore and omnivore guilds of the New Mexican desert, and by Chew and De Vita (1980) in granivore and insectivore guilds of the Arizona desert.

The divergence of species along many axes of the potential niche open to ants is of course to be expected if species are competing with each other. In order to coexist—at least in the classical view—species must differ from one another to a sufficient degree in their utilization of limiting resources such as food and nest sites. This minimal critical divergence should be reflected in the phenotypic traits that make it possible, such as body size and hence size of food items, or the time of foraging. The mere existence of the pattern is not proof in itself of competitive displacement, however. Correlative studies in general prove to be suggestive but not definitive. There is always the possibility of some third, still undisclosed class of causative factors. Another complication is the fact that ants of very different sizes often fight and exclude one another locally, or steal food from each other in a way that tends to neutralize the phenotype differences that might otherwise circumvent competition. It is therefore desirable to seek other, more rigorous forms of verification, and, happily, these are easily obtained.

### Overdispersion of Colonies

As a rule mature colonies of social insects belonging to the same or closely related species are overdispersed, that is, spaced so that the distances between them are too nearly uniform to have been randomly set. Most entomologists who have examined local distributions of ant colonies, among them Elton (1932), Talbot (1943a, 1954), Brian (1956a, 1965b), Wilson (1959a,c), and Yasuno (1964, 1965b), have arrived at this conclusion. In an important review Levings and Traniello (1981) statistically analyzed 80 data sets gathered by themselves and previous authors. In many cases the distributional data had been independently assembled by different authors for the same species, especially in the genera *Formica, Lasius,* and *Myrmica.* The studies covered both terrestrial and arboreal nest sites in a wide range of habitats in both temperate regions and the tropics. For the 80 cases that Levings and Traniello were able to test statistically, 67 showed overdispersed nest distributions or tended toward such a pattern. In another 80 cases that could not be tested statistically, all appeared nevertheless to be overdispersed. Thus it is safe to say that a majority of species so far studied tend to have regular nest arrays. This pattern holds across the principal ant subfamilies with persistent nest sites (Ponerinae, Pseudomyrmecinae, Myrmicinae, Dolichoderinae, and Formicinae), as well as across many habitats and food types, not to mention investigators. One group that could not be assessed in this manner, because of inadequate data, are the ants that defend only their nests. Many of these, including the collembolan-feeding members of *Strumigenys* and other Dacetini, are cryptic and difficult to census in the field.

In a second study Levings and Franks (1982) examined the nest distribution of 16 ecologically similar species that live on the floor of the moist tropical forest on Barro Colorado Island, Panama. When all the species were grouped into a single sample, the nests were found to be overdispersed from each other. Each of the more abundant species treated separately was also overdispersed. Levings and Franks concluded that the aversive interaction, whatever its nature, is stronger among the colonies belonging to the same species than among colonies belonging to different species, but both types are potent enough to create a relatively easily detected effect in the fauna as a whole.

What of army ants, which have no fixed nest site but wander from one place to another? It might seem that their paths would be randomly distributed, but this appears not to be entirely the case, at least not in the swarm raider *Eciton burchelli.* Franks and Bossert (1983) developed a computer simulation model incorporating all of the known aspects of *E. burchelli* foraging behavior. With a density of model colonies similar to that actually occurring on Barro Colorado Island, collisions occurred in the computer at a frequency of approximately once per colony per 250 days. Yet during the thousands of raids observed since 1929 by Schneirla (1933b), Willis (1967), and Franks (in Franks and Fletcher, 1983), not a single collision has been recorded in nature. A likely explanation for this nonevent is that colonies reject areas still contaminated with the trail pheromones of other *E. burchelli* colonies. The hypothesis is supported by Willis' observation that one colony withdrew its raid from an area recently visited by another colony. Accordingly Franks and Bossert programmed their computer colonies to stop raiding when they encountered trail systems that had been laid during the past 20 days—in accordance with field observations of trail durability. With this and several other realistic alterations in raiding frequency and direction, the model colonies collided only once per colony per 600 days.

Overdispersion is most simply explained as the outcome of some form of territorial behavior. The mutual exclusion can come about by an active aversion in which the colonies emigrate until they are spaced on all sides from their nearest competing neighbors. It can also result from colonies mutually annihilating one another until only one remains in the minimal defensible space. Finally, it can arise through preemption, in which the first colony to become established destroys incoming foundress queens before they can establish strong colonies. All of these processes occur widely in the ants, as we shall see, which makes the simplest unitary explanation of overdispersion also the most plausible.

The competition hypothesis of colony overdispersion gains added credence from the observation by Waloff and Blackith (1962) that nests of *Lasius flavus,* a species known from direct behavioral observations to be territorial, are randomly distributed in areas of low population density but overdispersed in areas of high population density. A comparable result has been reported by Byron et al. (1980), who found that during periods of food shortage the desert ants *Aphaenogaster (= Novomessor) cockerelli* and *Messor (= Veromessor) pergandei* broaden their diets, increase their foraging space, and overdisperse their nests with reference to each other. Ryti and Case (1984, 1986) found that not only are colonies of *Messor pergandei* and *Pogonomyrmex californicus* overdispersed, but large colonies are separated by greater distances than small colonies. By careful quantitative studies Ryti and Case were able to eliminate virtually every conceivable cause of overdispersion except intraspecific competition. A study of *Myrmica lemani* in Poland led Pętal (1980) independently to a similar conclusion. During a reduction of food over two years, the ants enlarged the spectrum of the food items they collected while foraging farther from their nests. The dispersal of the nests simultaneously shifted from a clumped to a uniform pattern.

### Experiments on Competition

A series of field and laboratory experiments have left no doubt of the powerful role of competition in determining the structure of ant populations and community structure. One of the simplest but most persuasive was Brian's 1952 analysis of habitat selection by Scottish ants. He noticed that a rank order exists among species in the appropriation of favored nest sites by competing queens. In the cool, moist woodland of western Scotland, ant colonies are limited to sunny places where higher temperatures persist long enough to permit the rearing of brood. As newly mated ant queens enter rotting pine stumps, they move to the south side of the vertical surface just beneath the bark. When individuals belonging to the same species encounter one another, they group together or space out at very short intervals. When queens of different species meet, however, they space out at much greater distances. Under such conditions *Formica fusca* occupies the southern face of the stump, which is the warmest area, whereas *Myrmica ruginodis* (= *M. "rubra"*) moves, for the most part, to the east face, which is the second warmest. *M.*

*scabrinodis* takes what is left. The tiny *Leptothorax acervorum* avoids conflict altogether by occupying galleries in the core of the stump too narrow to admit the other species.

Brian was then able to demonstrate through a laboratory experiment that the segregation of the species is due, at least in large part, to the repulsion of the *Myrmica* by the *Formica* from the favored southerly position. He placed newly inseminated queens in glass vessels heated to 30°C on one side. When joined with members of their own species alone, both *F. fusca* queens and *M. ruginodis* queens invariably clumped together in the warmest part of the chambers. When the two species were mixed together, however, the *M. ruginodis* were displaced to the cooler part. The segregation matches that observed in dead stumps under natural conditions. In later stages of their population growth, the colonies of the different species occupy different positions, and in some cases they remain there into maturity.

Every ant researcher has observed the displacement of one ant species by another at baits, and the simplicity of the phenomenon invites its use in experimental tests of competition. In the forests of the eastern United States, for example, the formicine *Prenolepis imparis* is dominant over the myrmicine *Aphaenogaster rudis* in this respect. When Lynch et al. (1980) experimentally removed *Prenolepis* foragers from baits made of small pieces of tuna, the number of *Aphaenogaster* foragers at the baits dramatically increased. They were also able to harvest more food, as Lynch and his co-workers expected.

An alternative experimental procedure is to remove foragers of a colony entirely from the field and to observe the response of its presumptive competitors. All such procedures known to us have resulted in an expansion of the released colonies, as we would expect if competition is occurring. In the remainder of this section we will present some of the most persuasive examples of such experiments.

When Kugler (1984) removed a colony of the Colombian harvester ant *Pogonomyrmex mayri*, the nearest *P. mayri* colony nearly doubled its foraging area within four days, sending scouts deep within the territory of its former neighbor. In another locality two *P. mayri* nests were surrounded by the multiple nest entrances of an apparent competitor species, the abundant ponerine *Ectatomma ruidum*. When Kugler removed the *Ectatomma* from the field by plugging up their entrances, both *Pogonomyrmex* colonies expanded the area of their foraging activity. When their fields met, however, one colony retreated as the other advanced, with the result that the areas occupied by the two colonies remained mutually exclusive.

In Ghana, as elsewhere in tropical Africa, the weaver ant *Oecophylla longinoda* forms huge colonies that suppress many other arboricolous ant species. When Majer (1976a–c) removed the ant from local areas in a cocoa farm, *Crematogaster striatula* moved in. When *C. striatula* was also deleted, *C. depressa* moved in and *Tetramorium aculeatum* spread its nests into the thinner vegetation of the kind earlier preempted by the dominant species.

On the Siberian steppes, *Formica pratensis* dominates *F. cunicularia*. When Reznikova (1982) isolated nests of *pratensis* by means of oil-smeared obstacles, so that the workers could no longer forage, the weight of prey collected by the newly liberated *cunicularia* workers increased threefold and the average weight of the prey increased from 1.4 milligrams to 3.1 milligrams.

Only four species of ants occupy small mangrove islets in the Florida Keys with areas of 12 square meters or less. When Cole (1983a) introduced colonies of *Pseudomyrmex elongatus* and *Zacryptocerus varians* onto empty islets, they persisted. When placed on similar islets containing *Crematogaster ashmeadi* or *Xenomyrmex floridanus*, they soon disappeared. (This experiment, which is important for the understanding of the assembly of island faunas, will be discussed in fuller detail in Chapter 11.)

Three sympatric species in the desert of the southwestern United States, *Pogonomyrmex rugosus*, *Pogonomyrmex desertorum*, and *Pheidole xerophila*, feed on seeds of different but partially overlapping size ranges. *Pogonomyrmex rugosus* and *Pogonomyrmex desertorum* resemble each other substantially in this regard, as do *Pogonomyrmex desertorum* and the much smaller *Pheidole xerophila*, but there is little overlap between *Pogonomyrmex rugosus* and the *Pheidole*. If the three species are competing for seeds in the zones of overlap to an extent affecting their population size, general theory (see Levins, 1976) predicts that *Pogonomyrmex rugosus* actually aids *Pheidole xerophila* by impacting *Pogonomyrmex desertorum*, their common competitor. If *Pogonomyrmex rugosus* were absent, *Pogonomyrmex desertorum* should increase and *Pheidole xerophila* should decline. When Davidson (1985) destroyed *Pogonomyrmex rugosus* colonies with insecticides, she got precisely the predicted effect in the two surviving species.

In the deserts of the American Southwest, ants and rodents are the principal consumers of seeds (Davidson et al., 1980). When Brown et al. (1979a) excluded rodents from small experimental plots near Tucson, Arizona, the number of ant colonies increased by 71 percent. In the reverse experiment, where ants were excluded, the numbers of rodents increased 20 percent and their biomass rose by 29 percent. A second series of experiments near Portal, Arizona, yielded a similar result for one species of ant, *Pheidole xerophila*, but not for another, *Pogonomyrmex desertorum*, or for rodents (Davidson et al., 1985).

The converse of the exclusion experiments just cited are packing experiments, in which species are added to localities to observe their effects on population structure and growth. Those performed to date have strongly implicated competition.

The overdispersion of colonies due to mutual repulsion of different species was demonstrated in experiments by Bradley (1972). In the jack pine stands of Manitoba, Canada, mature colonies of *Formica obscuripes* and *Hypoclinea taschenbergi* reach very large size, with an estimated minimum of 40,000 adult workers; and they appear to be spaced at a minimum distance from each other. Bradley transplanted colonies into previously unoccupied pine plantations. When nest sites of the two species were alternated in position and spaced at intervals of 20 meters, few of the colonies emigrated. But when the sites were spaced at only 5 meters, a large percentage of the colonies emigrated, often repeatedly, and in sufficiently divergent directions to create greater spacing.

*Formica truncorum* is an aggressive ant species that forms huge polydomous colonies in the drier habitats of northern Europe. Detailed biogeographic data collected in Finland suggest that it competes successfully with other ants and is most effective in suppressing populations of the smaller formicine *Lasius niger* (Rosengren et al., 1985, 1986a; Rosengren, 1986). *L. niger* colonies in turn compete with those of the more subterranean *L. flavus* (see Pontin, 1960, 1961, 1963), although less intensely than with one another. *F. truncorum* and *L. flavus*, it appears, interact to a lesser degree. Rosengren and his associates introduced colonies of *F. truncorum* to tiny islands in the Gulf of Finland already occupied by the *Lasius* species but lacking the *Formica*. The introduced colonies prospered as the *L. niger* colonies declined. On the other hand, *L. flavus* colonies increased their populations. Nearby "control" islets, on which no introduction of *Formica truncorum* was attempted, experienced little change in the *Lasius* populations.

### Direct Observation of Competition

One of the behaviors of ants most commonly observed in nature is overt aggression, often resulting in injury, death, and displacement of one colony by another. Auguste Forel was correct in observing that "the greatest enemies of ants are other ants, just as the greatest enemies of men are other men."

Interference competition among ants can be demonstrated within hours or even minutes almost anywhere in the world merely by placing food baits on the ground. This ultrasimple method has become a standard tool for analysis (for example, De Vroey, 1979; Lynch et al., 1980). The baits most commonly used by entomologists are sugar water and pieces of tuna. In the faunas of the West Indies, which have relatively few species, consistent interactions follow easily perceived rules (Wilson, 1971; Levins et al., 1973). *Paratrechina longicornis,* the swift-running *hormigas locas* (crazy ants) are "opportunists." The workers are very adept at locating food and often are the first to arrive at newly placed baits. They fill their crops rapidly and hurry to recruit nestmates with odor trails laid from the rectal sac of the hindgut. But they are also very timid in the presence of competitors. As soon as more aggressive species begin to arrive in force, the *Paratrechina* draw back and run excitedly in search of new, unoccupied baits. The species appears to survive by arriving and recruiting early enough to appropriate a share of the food, without, however, exposing themselves to any great risk in battle. A second class of species, the "extirpators," recruit by odor trails and fight it out with competitor species. Examples include some species of *Pheidole, Crematogaster steinheili,* the fire ant *Solenopsis geminata,* the "little fire ant" *Wasmannia auropunctata,* and *Camponotus sexguttatus.* Some of the species have a well-developed soldier caste which, in the case of *Pheidole* and *Solenopsis* at least, plays a key role in the fighting. Injury and death are commonplace, and one species (or one colony in competition with other colonies of the same species) usually comes to dominate the bait. Preemption is the rule: the colony whose foragers arrive first typically wins, because the foragers recruit nestmates, who surround the bait and prevent scouts of competing colonies from sampling the food and recruiting on their own. When a few scouts encounter large numbers of a competitor colony they are more timid and easily repulsed. Large colonies also enjoy a local advantage over small colonies.

Finally, a third class of species, the "insinuators," are exemplified by *Tetramorium simillimum* and species of *Cardiocondyla.* The colonies are small, lack a soldier caste, and evidently rely on small body size and stealthy behavior to reach the baits. When a *Cardiocondyla* scout worker discovers food, she recruits only one nestmate at a time by means of tandem running. The small numbers of workers reaching the baits by this means are usually able to ease their way to the edge of the bait through crowds of "extirpator" workers without provoking an aggressive response.

Territorial fighting is very general in the ants, sometimes widespread over the foraging areas and sometimes confined to the immediate vicinity of the nest. Among the most dramatic battles witnessed among colonies of the same species are those conducted by the common pavement ant *Tetramorium caespitum.* First described by the Reverend Henry C. McCook (1880), these "wars" occur in abundance on sidewalks and lawns in towns and cities of the eastern United States throughout the warm season. Masses of hundreds or thousands of the small, dark brown workers lock in combat for hours at a time, tumbling, biting, and pulling each other, while new recruits are guided to the melee along freshly laid odor trails.

When major workers from two laboratory colonies of the African weaver ant *Oecophylla longinoda* are permitted to enter a new arena simultaneously, they quickly join in a territorial battle (Hölldobler and Wilson, 1978). As individual pairs meet, they rear up by straightening their legs and dodge around each other like prancing, stiff-legged dancers, then thrust back and forth in an attempt to seize each other. When one succeeds in grasping a mandible, antenna, leg, or other part of the adversary's body to secure a good grip, she bites down and pulls back on her legs (Figure 10–7). Sometimes the mandibles cut directly through, lopping off one of the appendages or slashing open the abdominal wall. Often nestmates recruited to the site join in the attack, spread-eagling the enemy by pulling her legs in all directions and finally dispatching her with further mandibular cuts. Encounters with alien members of the same species are exceptionally exciting to the *Oecophylla* workers. Some of those not immediately locked in combat return to the nest, laying odor trails by dragging their rectal glands over the ground. As they encounter nestmates they antennate them vigorously while jerking their bodies back and forth in swift, exaggerated movements. As a result substantial forces of new workers are recruited and a veritable war between the two colonies ensues. The colony that wins is the one able to assemble the heaviest battalions of major workers, and this is typically the one that meets the enemy on ground previously marked with its own territorial pheromone. To use the celebrated phrase of General Nathan Bedford Forrest during the American Civil War, the winners arrive at the battle scene "first with the most." The colony with the highest density of workers present is also victorious in a majority of local duels, since small groups are able to gang up more frequently on single opponents. As the combat by attrition proceeds and fatalities mount, the more weakly represented colony is gradually eliminated from the arena.

The same pattern of combat and exclusion has been observed under natural conditions in *Oecophylla longinoda* in East Africa and in its sister species *O. smaragdina* in Australia (Hölldobler, 1979, 1983; Hölldobler and Lumsden, 1980). It also occurs regularly when *Oecophylla* meets other species of ants, including members of *Anoplolepis, Iridomyrmex,* and *Pheidole* (Way, 1953; Steyn, 1954; E. S. Brown, 1959). This leads to a mosaic distribution of territories of codominant ant species (Leston, 1973a–c; Majer, 1976a; Jackson, 1984). Similar interspecific territorial patterns have been reported from the Neotropical ant fauna (Leston, 1978) and from Europe (Pisarski, 1982), and will be treated in more detail in Chapter 11.

Ant wars are in fact a commonplace among species with large colonies. They have been especially well documented in the mound-building ants *Formica lugubris* (Cherix and Gris, 1978) and *F. polyctena* (Mabelis, 1979b; Driessen et al., 1984). In *F. polyctena* at least, they are cannibal wars. The bodies of the enemy dead are carried back to the nest and eaten; battles are especially intense during

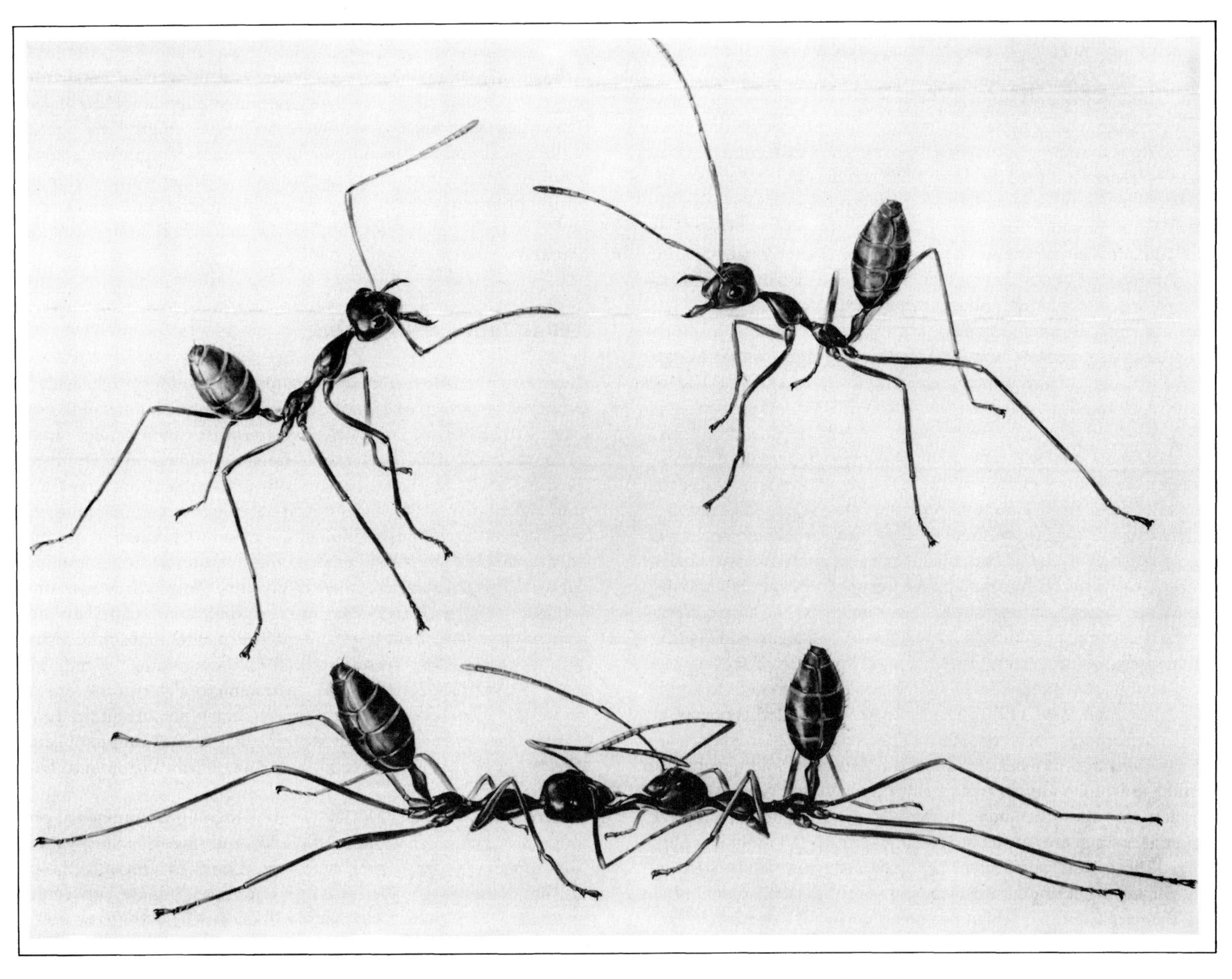

**FIGURE 10–7** Combat in weaver ants. Workers from different colonies of *Oecophylla longinoda* fight by rearing up on their legs and mandibles in a threat display (*above*). They also dodge around each other and seize one another with the mandibles (*below*). Such encounters often end in injury or death. (From Hölldobler and Wilson, 1978; painting by T. Hölldobler-Forsyth.)

times of general food shortage. *F. polyctena* has also been directly observed extirpating colonies of *Lasius, Myrmica,* and *F. fusca* for use as food (Vepsäläinen and Pisarski, 1982). In the case of the supercolonial *F. yessensis* in Japan, the combination of warfare and local population saturation results in the virtual elimination of many other species within the local range of the dominant species (Higashi and Yamauchi, 1979). Another case is the spread of the unicolonial *Wasmannia auropunctata* on Santa Cruz Island in the Galápagos, where it was introduced by human commerce. By 1976 the ant had expanded across approximately the southern half of the island, drastically reducing or even extinguishing the populations of every other ant species it encountered, including two endemics of the Galápagos (Clark et al., 1982).

The most extreme episodes of species exclusion, in which colonies can be observed extirpating colonies, occur when species possessing large colonies are introduced by human commerce into previously undisturbed native ant faunas. Among the best-known cases, in addition to *Wasmannia,* are the notorious *Pheidole megacephala,* a principal insect pest and destroyer of the Hawaiian native insect fauna, the fire ant *Solenopsis invicta,* which has invaded almost the entire southern United States, and the so-called Argentine ant *Iridomyrmex humilis* (Illingworth, 1917; Smith, 1936a; Haskins, 1939; Wilson and Brown, 1958a; Haskins and Haskins, 1965; Wilson and Taylor, 1967b; Ward, 1987). The process of extirpation has been observed by many authors. Fluker and Beardsley (1970) found that in Hawaii an *I. humilis* colony takes about 10 to 14 days to eliminate all of the workers from an *S. geminata* nest. When they invade *P. megacephala* nests, the *Pheidole* abscond and move to a new, presumably less desirable site.

It is entirely possible that given enough time, evolutionary or ecological change will result in the "taming" of such species and their increasingly harmonious assimilation into the indigenous faunas. Wheeler (annotations to Réaumur, 1926) reminds us that in the early 1500s a stinging ant appeared in such huge numbers as to cause the near abandonment of the early Spanish settlements on Hispaniola and Jamaica. The colonists of Hispaniola called on their patron saint, St. Saturnin, to protect them from the ant and conducted religious processions through the village streets to exorcise the pest. What was evidently the same species, which came to be known by the Linnaean name *Formica omnivora,* appeared in plague proportions in Barbados, Grenada, and Martinique in the 1760s and 1770s. The legislature of Grenada offered a reward of £20,000 for anyone who could devise a way of exterminating the pest, to no avail. It now appears that *F. omnivora* was none other than the familiar "native" fire ant, *Solenopsis geminata,* which at the present time is a peaceable and only moderately abundant member of the West Indies fauna.

*Iridomyrmex humilis* and *Pheidole megacephala* are wholly incompatible species, with *humilis* the usual if not invariable winner in subtropical to warm temperate localities between 30° and 36° latitude north and south, and *megacephala* the winner in the tropics. Thus *humilis* is dominant in disturbed habitats in California, the Mediterranean area, southwestern Australia, and the island of Madeira, but only above 1,000 meters in Hawaii, the zone cool enough to favor it (Fluker and Beardsley, 1970). The *Iridomyrmex* populations penetrate new environments on foot, with raiding columns of workers clearing the way and pioneer communities of workers and queens, like Zulu bands, following into the freshly opened nest sites. New populations, on the other hand, spring from little groups of workers and queens that are carried inadvertently in human freight and baggage. The gradual replacement of *Pheidole megacephala* by *Iridomyrmex humilis* on Bermuda, which began with the establishment of the *Iridomyrmex* around 1953, has been recorded across a decade by Haskins and Haskins (1965), Crowell (1968), and Lieberburg et al. (1975). The most recent survey, by Lieberburg and his co-workers, suggests that the two contestants may be approaching a dynamic equilibrium. Between 1966 and 1973 the *Iridomyrmex* established two new local populations while the *Pheidole* finished reclaiming three other sites.

As the red imported fire ant *Solenopsis invicta* progressed across the southern United States, it virtually eliminated the native fire ant *S. xyloni* and reduced the population densities of *S. geminata* and the harvesting ant *Pogonomyrmex badius* (Wilson and Eads, 1949; Wilson and Brown, 1958a). Throughout this newly conquered range the *S. invicta* population spread principally by nuptial flights and the establishment of additional colonies from newly inseminated queens. As a result the populations are patchy in local distribution and often form mosaics with those of the chief competing species (Tschinkel, 1988b). At the Brackenridge Field Laboratory of the University of Texas in Austin, the local population has become unicolonial. In other words it has abandoned its colony territorial boundaries and is now composed of huge numbers of workers and queens that emigrate over short distances. The continuity and high density of this peculiar population allows it to penetrate microhabitats not ordinarily available to the species, and it has been able to devastate most of the resident native species (S. D. Porter, personal communication).

Finally, De Vita (1979) has gone so far as to measure the amount of aggression among colonies in a natural population of the Mojave Desert harvester ant *Pogonomyrmex californicus.* Eighty-one percent of the encounters between foraging workers from different colonies resulted in some form of aggression and 7 percent were fatal to one or the other participant. The cost of this interference competition was 0.06 death per ant foraging hour, a deficit that has a substantial impact on colony size and growth.

## TERRITORIAL STRATEGIES

A territory is defined as an area occupied more or less exclusively by an animal or group of animals by means of repulsion through overt defense or advertisement. Any given territory can be *absolute,* in the sense that an established terrain is defended all the time, whether it contains food, nest sites, or any other resource at any particular time. Alternatively, the territory can be *spatiotemporal,* meaning that only limited sectors of the foraging area are defended at any moment, and there are usually places where resources are temporarily located. Thus the large colonies of fire ants, *Oecophylla* weaver ants, wood ants of the *Formica rufa* species group, and Australian meat ants of the species *Iridomyrmex purpureus* maintain absolute territories (Yasuno, 1965b; Wilson et al., 1971; Greenslade, 1975a,b; Mabelis, 1979b; Hölldobler, 1983). Spatiotemporal territories are defended by colonies of *Prenolepis imparis,* honeypot ants of the genus *Myrmecocystus,* and most species of the diverse and abundant genus *Pheidole* (Talbot, 1943a; Hölldobler, 1981a, 1986b; Wilson and Hölldobler, 1985).

Territorial behavior is in fact a very general if not universal phenomenon, ranging according to species and stage of colony growth from defense of large absolute foraging domains down to the immediate nest vicinity. Because of the division of labor between reproductive individuals and sterile worker castes, fatalities caused by territorial defense have a qualitatively different significance for ants and other social insects as compared to solitary animals. The death of worker ants represents only an energy and labor debit, not the destruction of a reproductive agent. The acts performed at risk of death can more than offset the cost of death by providing or maintaining resources and colony security. In this sense frequent death can even become a positive element in the colony's adaptive repertory, as suggested by Oster and Wilson (1978) and documented by Porter and Jorgensen (1981) for *Pogonomyrmex* and Schmid-Hempel and Schmid-Hempel (1984) for *Cataglyphis.* Hence strong territorial defense is more likely to have evolved in ants than in most solitary animals.

In more general terms, natural selection theory suggests that an animal should establish a territory only when the territory's size and design make it defensible in economic terms. That is, the territorial defense should gain more energy than it expends (J. L. Brown, 1964; Brown and Orians, 1970; Oster and Wilson, 1978; Davies and Houston, 1984).

It follows that the analysis of territorial strategies in ants must include the study of the design and spatiotemporal structure of the territory, as well as the social mechanisms through which the strategies are achieved. Important differences in the use of space exist among ant species. Species foraging on relatively stable resources dispersed over a wide area defend territories that are different in geometry from those of species exploiting stable but patchily distributed resources. Distinct strategies are also associated with food resources that are patchy, relatively stable, and predictable, or else frequently changing in location in an unpredictable manner. In the following sections we will briefly examine ant species that display much of this variation among themselves. Each of the species comprises ecologically dominant animals that have been studied extensively both in the laboratory and in their natural habitats. Adaptation to food sources that are relatively uniform in space and stable in time is exemplified by the African weaver ant *Oecophylla longinoda*. In contrast, certain species of harvester ants in the genus *Pogonomyrmex* utilize food sources that tend to be stable but are often distributed patchily in space. Finally, the honeypot ant *Myrmecocystus mimicus* harvests resources that for the most part occur randomly in space and time.

### African Weaver Ants: Absolute Territories

For blackbirds (*Euphagus cyanocephalus*) and many other solitary animal species, resources are uniformly distributed and continuously renewing. It is advantageous for them to maintain a complete defense of whatever portions of the foraging area can be patrolled in a reasonably short period of time (Horn, 1968). A striking analogy exists in the absolute foraging territories of the African weaver ants (*Oecophylla longinoda*), although the mechanisms of establishing and maintaining the territories are very different.

*Oecophylla longinoda* is a dominant species in the forest canopy in a large part of tropical Africa. Workers bind leaves into tight nest compartments with silk spun by the final-instar larvae. One colony usually builds hundreds of such leaf nests, which are distributed over several nest trees and concentrated in the peripheral canopy of the trees. Weaver ant colonies are monogynous; the single reproductive queen resides in one of the leaf nests, completely surrounded and protected by workers. Most of the other leaf nest compartments are filled with brood of all stages and hundreds of workers. The worker force of a mature *O. longinoda* colony sometimes consists of more than 500,000 individuals.

Weaver ants are strongly predaceous, using their cooperative ability to capture a wide range of large insect prey that ventures onto their territory (Figure 10–8). Although the foragers tend to remain on the trees and surrounding low vegetation, they also hunt extensively on the ground; their territories are three dimensional. *Oecophylla* workers patrol every part of their territory, tolerating only very few ant species on the trees they occupy. But they exclude foreign colony members of their own species in aggressive interactions (Figures 10–9 and 10–10) so severely that they create narrow, unoccupied corridors, which are in effect a "no ant's land."

In an area of the Shimba Hills Reserve in Kenya, Hölldobler (1979) studied individual territories, sometimes covering an area of up to approximately 1,600 square meters across the canopies of 17 large trees. He was able to determine the territorial borders by repeatedly transplanting sets of 20 *Oecophylla* workers from one nest tree to another. As long as the trees belonged to the same territory, no conspicuous response was observed. But when the transplanted ants were released within the territory of a neighboring colony, a massive defensive recruitment response was elicited in the resident ants. Individual ants that encountered the foreign intruders ran back to some of the nearest leaf nest compartments in the tree canopy, laying a trail with secretion from the rectal gland. When they encountered nestmates, they performed a vigorous jerking display similar to the initial fighting behavior between two individual opponents in the manner described earlier in laboratory colonies (see Chapter 7). Nestmates approached by a jerking recruiter ant became very excited and ran along the trail toward the battle site. After 30 minutes, often many more than 100 ants had assembled in this area and remained there long after all the "intruder" ants had been killed.

The weaver ant colony finds considerable advantage in excluding competitor colonies of all but the smallest or most dietarily different species. Indeed, *Oecophylla longinoda* is highly aggressive not only to conspecific aliens but also to many other ant species. In the Shimba Hills, only a few ant species were found to coexist with *Oecophylla* on the same tree. Some other arboreal ant species (for example, *Camponotus* sp.) that were never found to coexist with *Oecophylla* on the same tree elicited as vehement a defensive recruitment response in *Oecophylla* as conspecific aliens when they were introduced into a weaver ant territory. It is interesting to note that the *Camponotus* also responded with a defensive recruitment of hundreds of nestmates when *Oecophylla* workers were released on a *Camponotus* tree. These results suggest that both *Oecophylla* and *Camponotus* react with a massive defensive response only to certain intruding ants, particularly the most serious competitors for essential resources (nesting sites and food), and predators including the army ant *Dorylus* and the formicine *Anoplolepis longipes*. Thus *Oecophylla longinoda* employs "enemy specification" in a selective form of absolute territory. *Oecophylla* diverges from the classic central-place model in the mechanisms by which it maintains territorial control. Its nests are not aggregated in one central location but are effectively decentralized throughout much of the territory that it controls. This large territory can be patrolled and exploited over much of its volume simultaneously without the colony's incurring the costs of transporting prey from distant capture points to a single central nest deep within the territory. These factors induced Hölldobler and Lumsden (1980) to formulate a different general model concerning the geometry of the "nest decentralization" and the economic defensibility of the *Oecophylla* territory.

A solitary animal, or even a social group with relatively few members, faces a challenge if a territory of considerable size is to be established and maintained: the defender cannot be everywhere at once. Invaders can often penetrate deep into the territory and even harvest resources before they are detected and expelled. This qualitative observation suggests that the solitary holder of a territory faces serious constraints in maintaining a large territory, especially when the density of the invader is high. In contrast, *Oecophylla* is virtually everywhere at once throughout its territory. Invaders are detected at or very close to a colony's boundary. Consequently, the *Oecophylla* territorial system allows harvesting benefits from throughout the interior of the territory while restricting the defensive costs almost purely to the boundary or margin.

It is interesting to note that the Nearctic species *Lasius neoniger*, although much smaller, appears to have similar decentralized territories. Like those of *Oecophylla*, societies of *L. neoniger* are monogyn-

**FIGURE 10–8** Cooperative combat by the African weaver ant *Oecophylla longinoda*. The workers subdue a large worker of the ponerine *Paltothyreus tarsatus* by a "rope pulling" technique, in which they form chains of their own bodies to provide added power. (From Hölldobler, 1986b.)

**FIGURE 10–9** Three *Oecophylla longinoda* workers attack an intruder of the same species. (From Hölldobler and Lumsden, 1980.)

ous and occupy multiple nests that are distributed throughout the colony's foraging area. Traniello and Levings (1986) hypothesize that this nest structure of *L. neoniger* partitions "the territory within a colony by spatial subdivision of its foragers, and thus may reduce loss of prey to competitors."

The strategic implications of such a system are far-reaching. To explore some of the consequences, consider a circular territory of radius $r$ with uniformly distributed, temporally stable food sources (see Figure 10–11; basically the same argument as the one about to be made holds for territories that approximate convex regions, as in the case of *Oecophylla* depicted in Figure 10–10). A cohort of foragers, dispersed throughout the interior of the territory at density $\rho_w$, returns net foraging benefit $B_2 > 0$ per unit area per unit time. Because resources and workers flow locally into many leaf nests dispersed throughout the territory and not back to one central place located deep inside the territory, $B_2$, which includes the cost of food transport, does not decrease as $r$ increases. Thus $B_2$ can be approximated by a constant independent of $r$.

Since costs arising from intercolony aggression occur primarily at the edge of the territory, let $C_2$ be the cost per unit time per unit length of boundary. Under these circumstances a territory of radius $r$ yields benefit $B_2\pi r^2$ per unit time to a colony and costs it $2C_2\pi r$. A net ergonomic profit rate $\pi r(B_2 r - 2C_2)$ is produced by this strategy. Thus we find economic defensibility threshold $r^* = 2C_2/B_2$ beyond which the territory yields net positive return. Moreover, the net return is an increasing function of $r$. Once beyond $r^*$, the net profit rate is positive and continues to increase (Figure 10–10a). The ratio of defense cost to net foraging benefit is just $r^*/r$ and thus decreases monotonically with $r$. Colonies that use such a strategy should prefer the largest possible territory. Eventually, sufficiently large $r$ is reached such that the egg-laying capacity of the queen is reached and the colony, in addition, begins to control an unusable surplus of

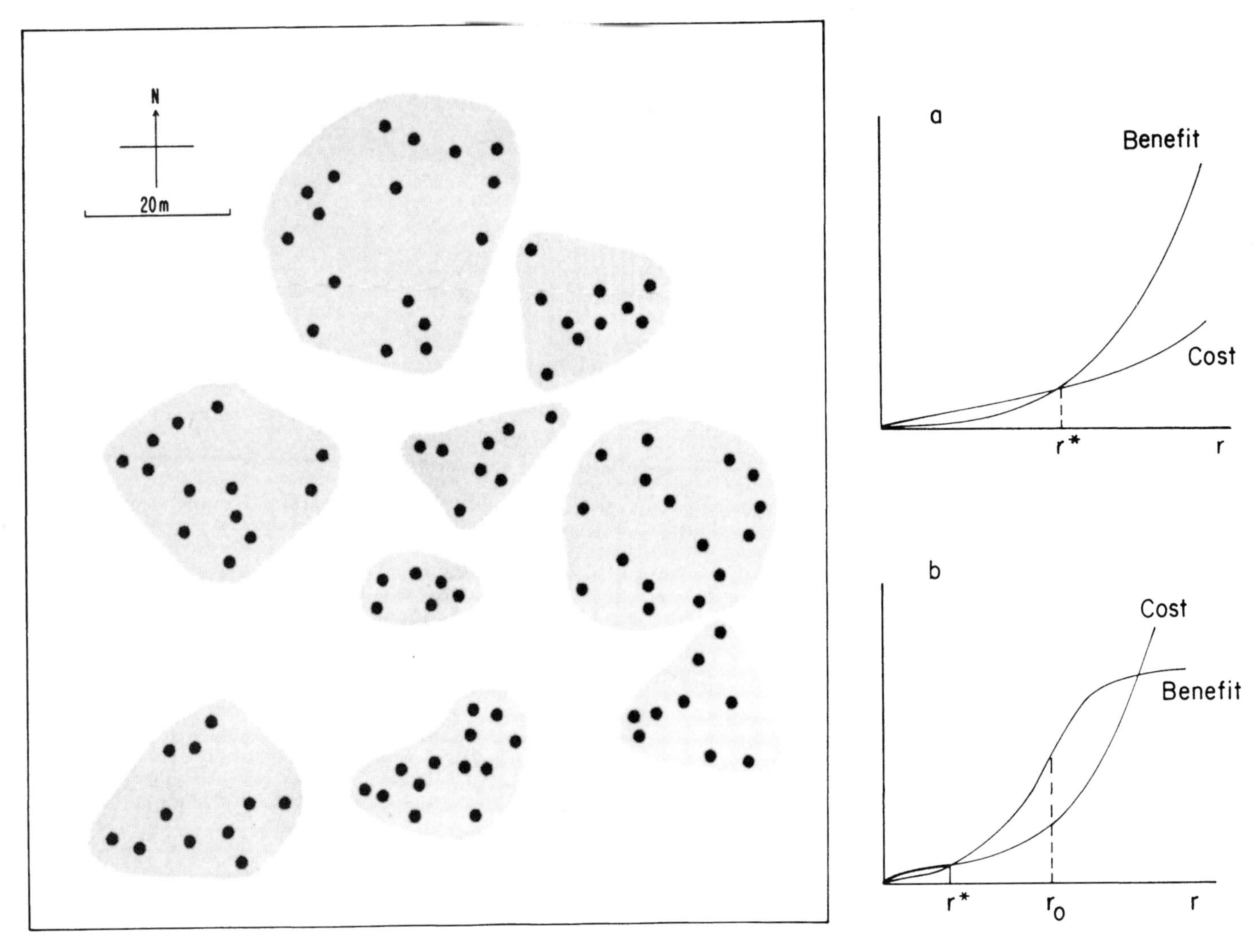


**FIGURE 10–10** (*Left*) Territories of *Oecophylla longinoda* in the study area at the Shimba Hills Reserve, Kenya. The solid circles (●) represent trees occupied by *Oecophylla*, and the shaded areas delineate the individual territories. (*Right*) (*a*) Theoretical cost and benefit curves for boundary defense strategy on spherical territories of the *Oecophylla* type. The symbol $r^*$ represents the economic defensibility threshold. Maximum of benefit-cost is obtained for $r \to \infty$. (*b*) Toward large $r$, the benefit curve tips over, as a result of decreasing returns to scale in the economics of territorial expansion. Maximum rate of return occurs for territories of radius $r_0$. (From Hölldobler, 1979; Hölldobler and Lumsden, 1980.)

resource. Thus the increment to benefit eventually approaches zero, and optimal economic defensibility occurs at an intermediate $r$, say $r_0$ (Figure 10–10b). The very large number of workers that mature *Oecophylla* colonies can maintain suggests a large laying capacity in the queen and the potential for very large territories.

The territories of *Oecophylla* are actually three dimensional rather than two dimensional. But so long as the territorial strategy is one of decentralized nests, with workers monopolizing the interior volume of the territory while confining costs of defense to the boundary, there are economic defensibility thresholds; and the ratio of cost to benefit continues to decline along the curve $r^*/r$ as the territory expands in size below saturation of the egg-laying capacity of the queen and the surplus limits of resources. Differences between the geometry of two and three dimensions and deviations from this idealized strategy impose quantitative differences in the details of territorial design.

There may also be territorial size limits beyond which such factors as transport of brood from the queen nest to other leaf nests in the colony becomes ergonomically impractical or the system of inter-nest coordination becomes too complex to maintain. The observed sizes of *Oecophylla* territories suggest, however, that the way stations offered by established leaf nests and the rather local nature of the nest-nest interactions within the territory make such limits relatively weak. It appears that there is a high selection pressure for expanding territorial space and thus increasing access to essential resources while simultaneously improving the relative economic defensibility of the territory. Indeed, the three-dimensional territories of *O. longinoda* are probably the largest territories known for all invertebrates. They are approached in structure and size by a few other arboreal ants, including some species of *Crematogaster* and *Pseudomyrmex*, which have evolved parallel social systems.

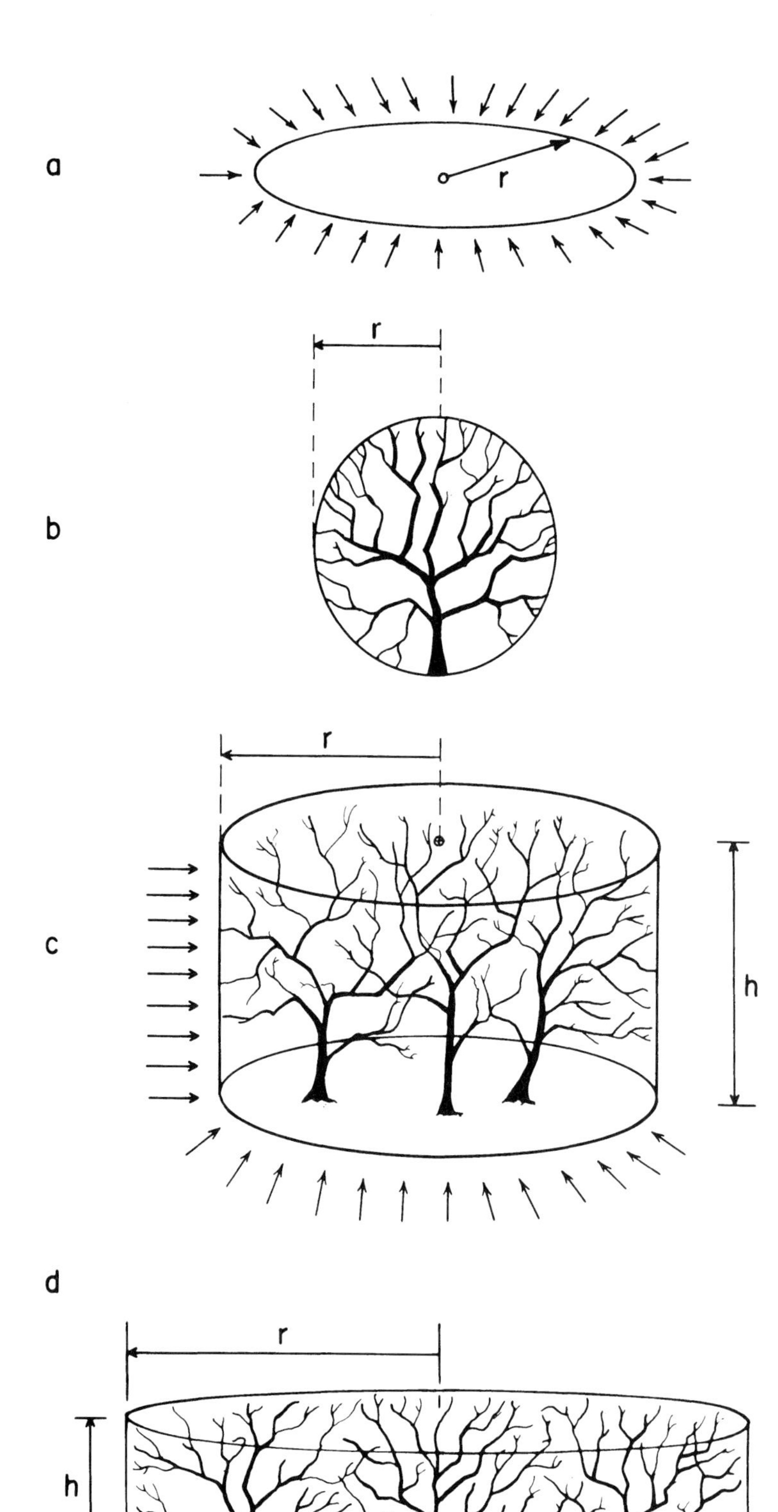


**FIGURE 10–11** A comparison of four models of boundary defense for *Oecophylla*. (*a*) The two-dimensional model is that analyzed in the text. Its territory is circular with radius *r*. In three-dimensional models we have to consider several possibilities. (*b*) "Spheroid" is a spherical territory with radius *r*. This is appropriate to thick canopy systems in which the colony can expand outward from the queen nest in all directions for significant distances. (*c*) "Cylindrical 1" is a cylindrical territory of radius *r* and height *h*. All boundaries are defended and challenges come over the cylinder surface (arrows) but not the end faces, which are contained within the territory. In (*d*) "cylindrical 2," only the trunk surfaces leading up into the occupied canopy are defended. Trunk radius is $r_1$ and the defense occurs along a length $h_1$ of the trunks. (From Hölldobler and Lumsden, 1980.)

## Harvester Ants: Trunk Trail Territories

Generally, the partitioning of space between ant colonies is effective in reducing aggression between individuals belonging to the same species but to different colonies. The pattern of space partitioning can be very different, however, and depends largely on the foraging strategies of the species. In contrast to the African weaver ants, the harvester ants *Pogonomyrmex badius, P. barbatus,* and *P. rugosus,* which are among the most abundant species of ants in the southern United States, tend to exploit patchy food supplies and accordingly show a more complex partitioning structure (Hölldobler, 1974, 1976a; Davidson, 1977b; Brown et al., 1979a,b; Harrison and Gentry, 1981).

Foragers of the harvester species travel on well-established trunk trails before diverging onto individual excursions. After foraging, the workers return to these routes for homing. Such trunk trails sometimes extend for more than 40 meters; they are remarkably persistent over long periods of time and even survive heavy rainfalls. The trunk routes originate from recruitment trails laid to newly discovered seedfalls. The recruitment pheromone, which is relatively short lived, is discharged from the poison gland and deposited on the ground through the extruded sting of foraging ants. Because the seed patches are frequently quite stable, the ants continue to travel along the former recruitment trail to these foraging sites. Persistent chemical signposts are also deposited along the trail and, together with the visual markers, serve as orientation cues long after the recruitment signal has vanished.

Trunk trails of intraspecific neighboring colonies are overdispersed and almost never cross (Figure 10–12). On the contrary, they usually diverge and channel the mass of foragers of hostile neighboring nests into very different directions. The trunk trails, together with the immediate vicinity of the nest entrance of mature colonies, can be considered the core area of the colony's territory. Although foraging areas of nearby colonies can overlap, aggression in the overlapping zone is usually limited to individual confrontations between two foragers. When two trunk trails of neighboring colonies are brought into contact, however, aggressive mass confrontations occur; and they continue until the trunk trails have diverged again (Figure 10–13). Although the major function of the trunk route foraging system seems to be to facilitate long-distance orientation and the exploitation of patchily distributed and relatively stable food sources, the topographic design of the route system of each colony in turn depends greatly on the route maps of its neighboring colonies.

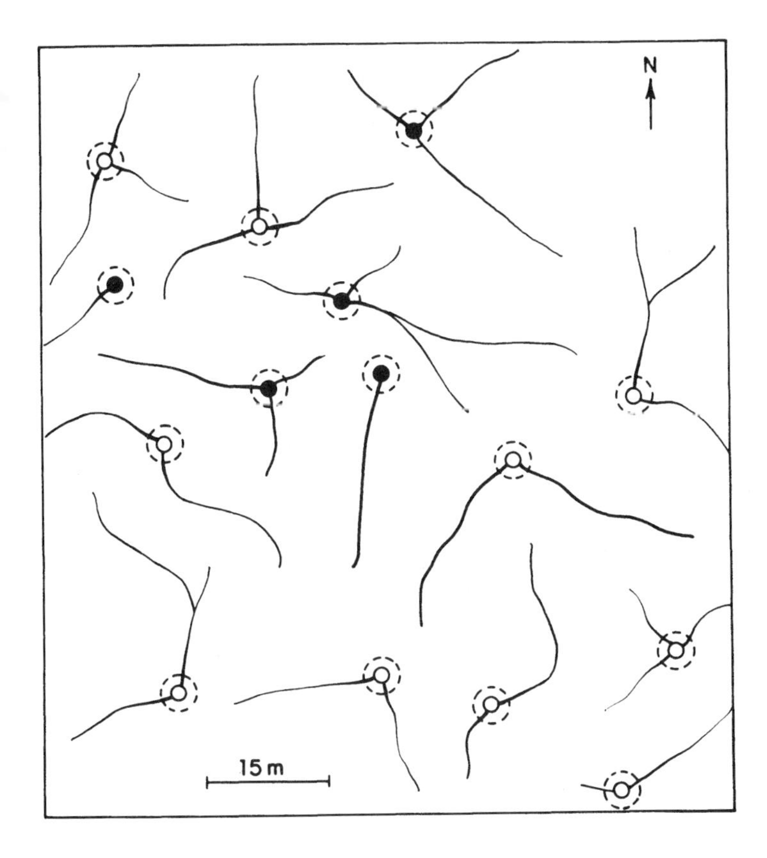


**FIGURE 10–12** Territorial displacement between the ecologically very similar harvester ants *Pogonomyrmex barbatus* and *P. rugosus* is revealed in the map of nests and trunk routes in one section of the study site in New Mexico. Solid black circles (●): *P. barbatus;* open circles (○): *P. rugosus.* (From Hölldobler and Lumsden, 1980.)

**FIGURE 10–13** Territorial combat in the American harvesting ant *Pogonomyrmex barbatus.* (*a*) Fighting among foragers of two neighboring colonies. (*b*) The head of a former opponent is still attached to the petiole of a forager. Such signs of heavy intraspecific aggression are frequently encountered in the field. (From Hölldobler, 1986b.)

Up to this point we have considered only the partitioning of foraging grounds by the trunk route system between colonies belonging to the same species. *P. barbatus* and *P. rugosus,* however, seem to be identical with regard to food type, nesting site, and foraging activity period, as though they were ecologically the same species. Although populations of both species are largely separate geographically, zones of overlap exist (Hölldobler, 1976a; Davidson, 1977b). It is not surprising therefore that territoriality is strongly developed both within and between these two species, and that the foraging area is subdivided interspecifically by the trunk route system. These findings lead to the question of the cues used by the ants to discriminate between colonies of their own and other species (see Chapter 5). The foragers not only have to be able to recognize members of their own colony and those belonging to another conspecific colony, they also need to identify other species that are potential competitors for essential resources.

Although not much is known about the mechanisms of species discrimination in ants, there is some analytical evidence of species specificity in the mixtures of hydrocarbons of the Dufour's gland in *Pogonomyrmex*. It is interesting that *P. rugosus* and *P. barbatus,* which exhibit strong interspecific territoriality, have almost identical patterns in the mixtures of the compounds from their Dufour's glands, whereas other *Pogonomyrmex* species, whose territories overlap with those of *P. rugosus* and *P. barbatus,* have very different patterns (Regnier et al., 1973; Regnier and Hölldobler, unpublished; see Hölldobler, 1986b, 1987).

When designing a mathematical model of trunk trails and territories, Hölldobler and Lumsden (1980) focused on species such as the *Pogonomyrmex* harvester ants, which utilize spatially patchy but long-lived resources. The investigators visualized the foraging area as a system of sectors that converge in a polar coordinate system on the nest (see Figure 10–14). The foragers are deployed in such a way as to maximize net energetic yield, measured as the benefit in calories of seeds harvested minus the costs incurred because of metabolism and mortality during seed retrieval. The resulting territorial design approximates a trunk trail system. The model incorporates such details as the efficiencies of food retrieval over narrow trails, the effects of nearby alien colonies, the presence of multiple patches in each sector, and time-varying costs. In simplest terms, the width of each sector is likely to be determined by these several biological factors. The colony should avoid empty sectors altogether except for occasional patrols by scouts. And whenever the cost:benefit ratio of a formerly productive sector falls below unity, the colony should close it down. Such would be the case when the patch is not suffi-

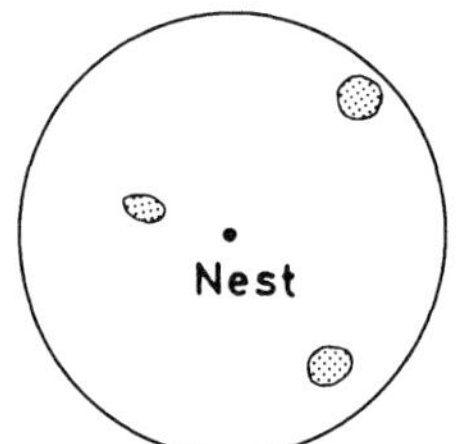


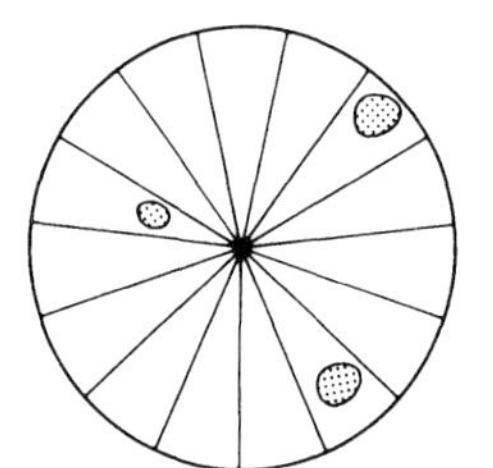


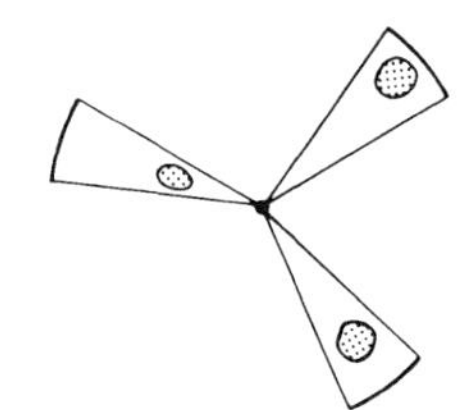


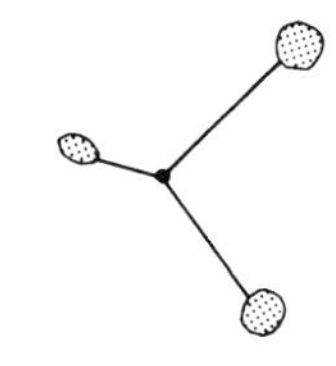


**FIGURE 10–14** The basis of the model of trunk trail defense. (Modified from Hölldobler and Lumsden, 1980.)

ciently rich or close to the nest, or when a rival nest already occupies the patch and usurpation costs would be greater than the acquired benefit.

Experimental field data agree well with these economic considerations. New foraging sites, which have not yet been discovered by competitors, are rapidly explored and occupied with the aid of an effective chemical recruitment system. In fact the significance of the recruitment communication in harvester ants becomes especially apparent in view of the strong intra- and interspecific competition for the same foraging areas. Hölldobler (1976a) found that the forager recruitment activity not only depends on a number of parameters of the food source—such as distance to the nest, density of the seed fall, and size of the grains—but also on the presence or absence of foreign foragers at the resource patch. Seed sites previously occupied by competing foragers were considerably less attractive than unoccupied seed sites.

Although most *Pogonomyrmex barbatus* and *P. rugosus* colonies studied by Hölldobler (1976a) and Davidson (1977b) had extensive trunk trail systems, a number of nests, especially in areas with sparser vegetation, were without trunk trails. This is a general trait of *P. maricopa,* which has never been observed to produce long-lasting trunk trails. In these cases ants usually leave the nest on their individual foraging excursions and disperse in all directions. Yet as Hölldobler's experiments demonstrated, the foragers also show a high directional fidelity, even though all sectors appear to be more or less equally frequented by the worker force as a whole. This suggests that the seeds do not accumulate in patches but instead are randomly dispersed. In other words the individual sectors are more or less equally productive. This foraging system obviously does not allow as subtle a partitioning of foraging grounds between neighboring colonies as does the trunk trail system. Such colonies invariably show a much wider spacing pattern (Hölldobler, 1976a).

In short, the trunk routes of *Pogonomyrmex* are partitioning devices which curtail aggressive confrontations between neighboring colonies, while at the same time enlarging the foraging area for patchy food supplies. A similar foraging and partitioning system has been reported for several species of *Formica* (Rosengren, 1971; Bruyn and Mabelis, 1972; Bruyn, 1978), *Lasius* (Traniello, 1980), *Messor* (Lévieux and Diomande, 1978a), and the leafcutter ants of the genus *Atta* (Rockwood, 1973; S. Hubbell, personal communication).

### Other Trunk Trail Territories

The trunk routes of the seed harvesting ant *Pheidole militicida* share many traits with those of *Pogonomyrmex* (Hölldobler and Möglich, 1980). They, too, originate from chemical recruitment trails and are stabilized by enduring chemical orientation cues and visual markers. Nevertheless they are not as persistent as the foraging pathways of *Pogonomyrmex.* In fact the route foraging system of *Pheidole militicida* seems to be intermediate between that of *Pogonomyrmex* and that of *Messor pergandei. Messor* employs a foraging strategy in which various sectors of the foraging area around the nest are successively exploited by the ants through rotational changes in direction (Rissing and Wheeler, 1976). Like *Messor, Pheidole militicida* workers shift the direction of the foraging pathway or establish a new route when the seed supplies in a foraging area diminish. The shifts do not occur as regularly as in *Messor,* however, and no geometric pattern in the change of the foraging columns has been detected. There is some recent evidence that the trunk trails of *Messor* also serve in territorial partitioning (Ryti and Case, 1986; Rissing, 1987), but this appears not to be the case in *Pheidole militicida.* Although little is known about the territorial behavior of these species, it seems that they have less rigid territorial boundaries and defend only those areas being heavily used by foragers.

### Honeypot Ants: Spatiotemporal Territories

When resident ants defend only those portions of their territory in which they happen to forage, as opposed to the trunk trails or entire foraging area, and encounter intruders at close range, the territorial defense can be said to be truly spatiotemporal. As shown by the work of Hölldobler (1981a, 1986b), a spatiotemporal territorial strategy of honeypot ants (*Myrmecocystus mimicus*) appears to have evolved as part of a foraging system designed to utilize patchily distributed but unpredictable or unstable food sources. *M. mimicus* is abundant in the mesquite-acacia community of the southwestern United States (Snelling, 1976). Like other members of its genus, it has a special honeypot caste, the members of which function as living storage containers.

One of the major food sources of *Myrmecocystus mimicus* is termites. Hölldobler found that when a scouting ant discovers a rich supply of termites, for example under a piece of dried cattle dung, she directs a group of nestmates to this food supply by means of special recruitment signals, including motor displays and odor trails laid with hindgut liquid. If another colony of *M. mimicus* is located near the food source and is detected by the foragers of the first colony, some of these individuals rush home and recruit an army of 200 or more workers to the foreign colony. They swarm over the nest and engage all of the workers emerging from the alien nest entrance in an elaborate display tournament, thus blocking this colony's access to the food supply (Figure 10–15). Frequently scouts leave the tournament to return to their colony in order to recruit reinforcements, while the other group of nestmates continues to retrieve the termite prey. Once the food source has been exhausted, and the foraging activity in the area declines, the tournament activity at the neighboring nest site also declines and the intruding army finally retreats to its own nest.

Although hundreds of ants are often involved in the territorial tournaments, almost no physical fights occur. Instead individual ants engage each other in highly stereotyped aggressive displays (Plate 14). Each group of displaying ants breaks up after 10–30 seconds, but the ants continue to run about on stilt-like legs. When they meet a nestmate they respond with a brief jerking display, but when they meet another opponent the whole aggressive display ceremony is repeated.

Field observations have shown how it is possible for scouting ants to venture very deeply into the foraging area of a neighboring colony. Although individual foragers of *Myrmecocystus mimicus* frequently disperse in all directions when leaving the nest, they tend to swarm out of the nest at intervals. Such departures are usually spaced irregularly in time, and an interval between two departures can last as long as several hours (Figure 10–16). Sometimes only very few foragers leave a nest during a whole day, while in contrast the foragers of a neighboring nest may be very active. Then within a few days the situation is reversed. If a scout of a relatively inactive colony discovers a rich food source, however, such as an access to a gallery of a termite nest, she can quickly galvanize the colony into foraging activity by recruiting nestmates to the food source (Figure 10–17). But the termites may have already been discovered by foreign scouts of the very active neighboring colony, which then sends an army of ants to the competing nest in order to interfere with the first colony's foraging activity.

Territorial tournaments also occur in the zone between two adjacent *Myrmecocystus mimicus* nests, especially when both colonies are active at the same time. Alien foragers are then blocked from the respective foraging areas of each colony (Figure 10–15b). These tournaments sometimes last for several days, being interrupted only at night when workers of this species are normally inactive.

When one colony is considerably stronger than the other, that is, when it can summon a much larger worker force, the tournaments end quickly and the weaker colony is raided. During these raids the queen is killed or driven off. The larvae, pupae, callow workers, and honeypot workers are carried or dragged to the nest of the raiders (Figure 10–18). Field observations and laboratory experiments have led to the discovery that the surviving workers as well as the honeypots and brood of the raided colony are incorporated to a large extent into the raiders' colony as full members.

Hölldobler (1981a) analyzed a total of 34 raids conducted by *Myrmecocystus mimicus* on conspecific neighboring nests in the field. These episodes constituted only about 8 percent of all tournament interactions actually observed. A total of 9 raiding events was observed from beginning to end, enabling Hölldobler to make a fairly accurate count of the number of larvae, pupae, honeypots, and workers abducted into the raiders' nest. From these data he estimated that the raiding colony is at least ten times larger than the raided colony. Thus raiding seems to be primarily directed against

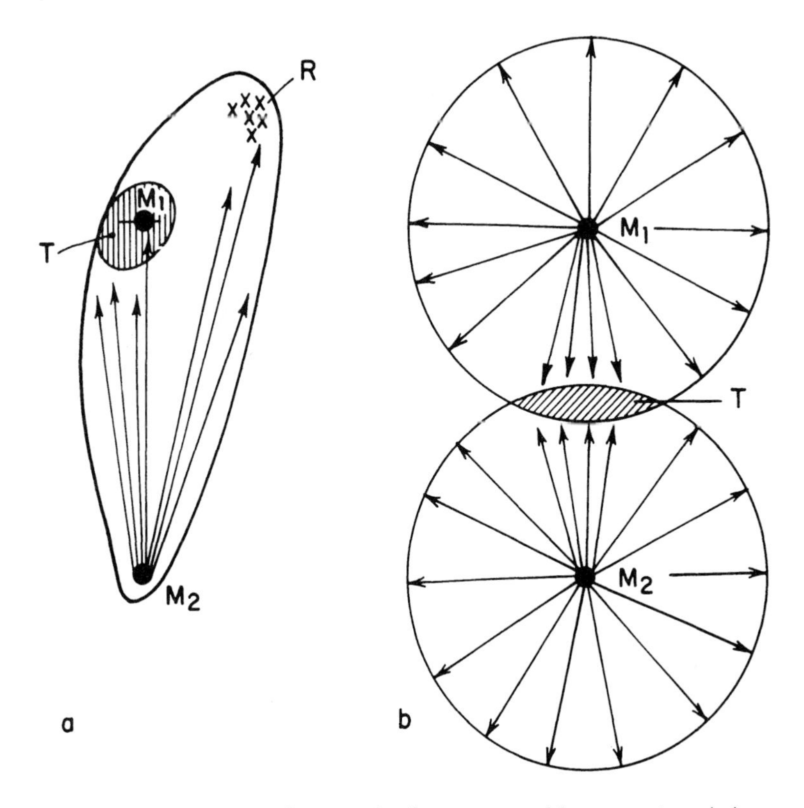


**FIGURE 10–15** Territorial interaction between two *Myrmecocystus mimicus* nests ($M_1$ and $M_2$). (*a*) Nest $M_2$ forages at food source *R*; simultaneously it engages $M_1$ in a display tournament (*T*) directly at the nest of $M_1$, thus interfering with the foraging activity of $M_1$. (*b*) $M_1$ and $M_2$ foragers disperse in all directions around their nests. This can lead to territorial tournaments in an area located between both neighboring nests. (From Hölldobler and Lumsden, 1980.)

younger, still developing colonies in the neighborhood. On the other hand, Hölldobler later observed a successful raid against a mature colony that had diluted its defensive forces by being simultaneously aggressively engaged with three other neighboring colonies. In this case the victor colony captured approximately 1,000 immatures.

### The Territorial Logic of *Myrmecocystus* Colonies

The adaptive significance of this peculiar territorial strategy is of interest. As already mentioned, one of the major food sources of *Myrmecocystus* is termites. The temporal and spatial distributions of the termite colonies are highly unpredictable. Since there is little point in defending an area that is unlikely to provide adequate food in a given time, *Myrmecocystus* does not establish fixed territorial borders around its entire foraging range. Instead it defends only areas into which it is currently conducting intensive forays. This procedure obviously enables it to extend its foraging range considerably and leads to frequent incursions into potential foraging ranges of neighboring *Myrmecocystus* colonies. Since there are no well-established territorial borders, aggressive mass confrontations with conspecific competitors are much more common in *M. mimicus* than in *Oecophylla* or *Pogonomyrmex*. If these confrontations were as violent as physical combat in the latter species, they would result in a constant and heavy drain on the worker force. Thus the tournaments of fighting display seem to be the most economical strategy in defense of

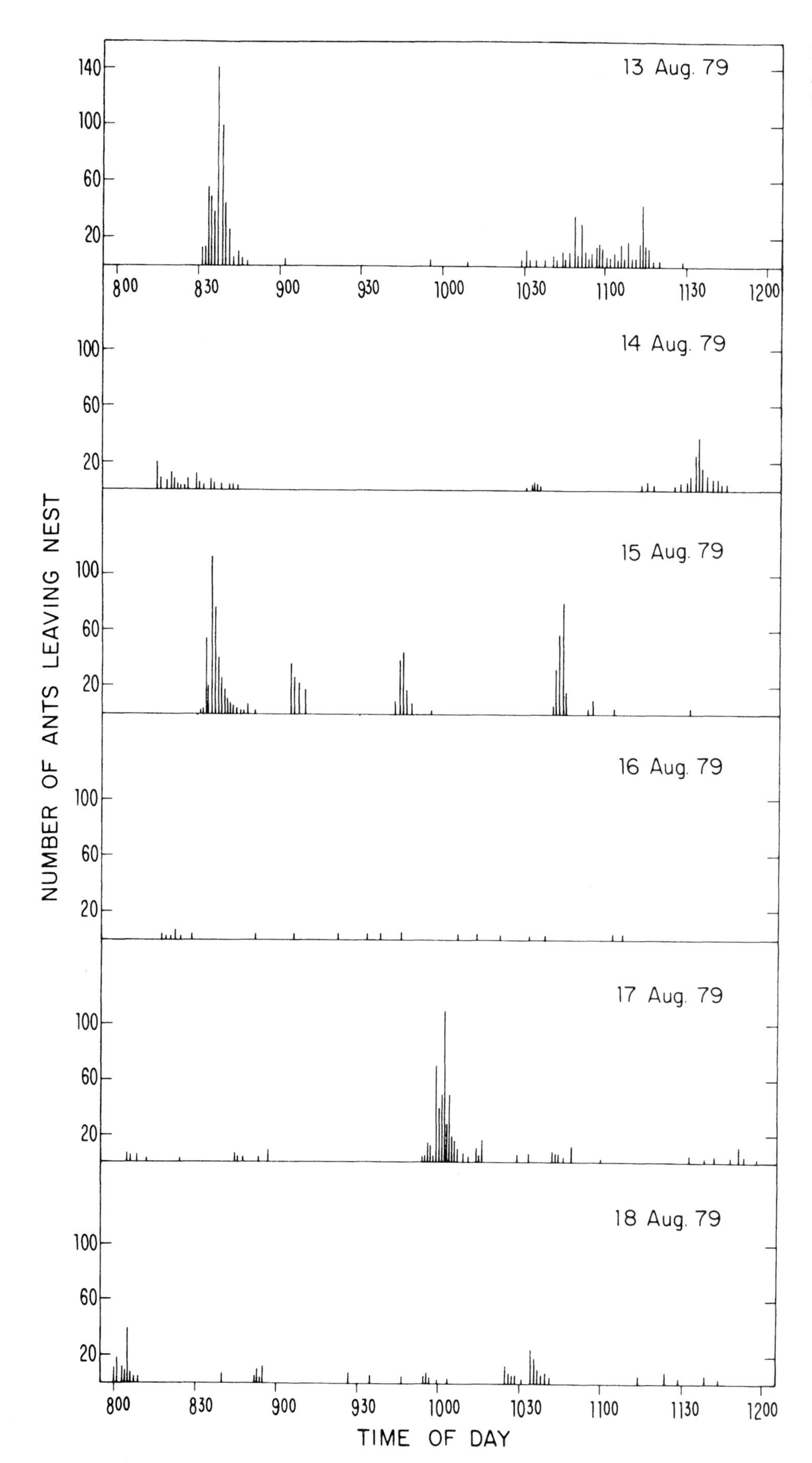


**FIGURE 10–16** Activity of workers departing from a *Myrmecocystus mimicus* nest during six consecutive mornings. (From Hölldobler, 1981a.)

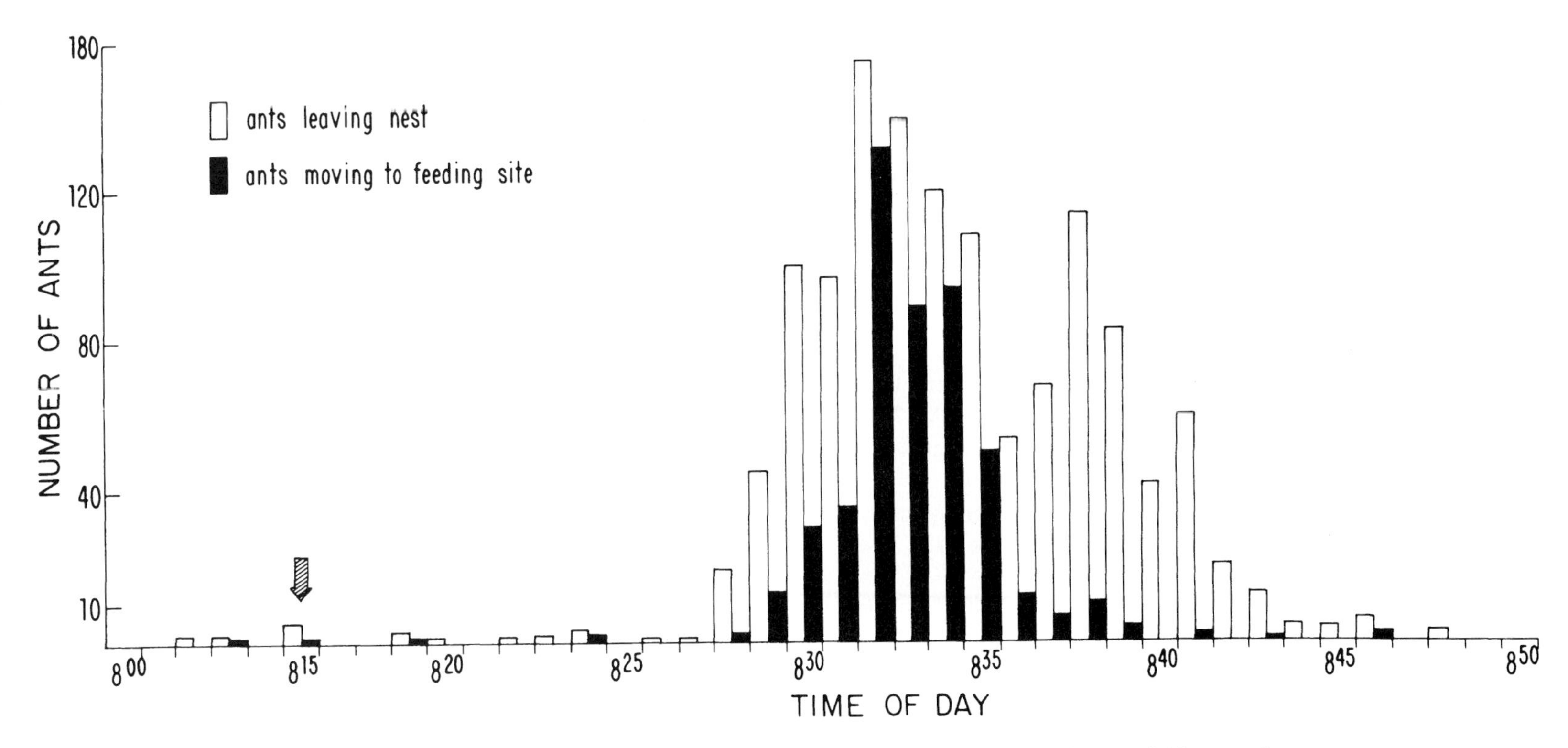


**FIGURE 10–17** Recruitment response of *Myrmecocystus mimicus* to a rich food source (termites) 10 meters from the nest entrance. The arrow indicates the time at which the termites were presented. (From Hölldobler, 1981a.)

the temporal territorial borders. Only when one colony is considerably weaker does it risk being overrun by the stronger colony, losing its queen, and being enslaved (Hölldobler, 1976c, 1981a).

The territorial logic of the *Myrmecocystus* colonies can be further analyzed with the aid of a geometric model (Hölldobler and Lumsden, 1980). We may consider a group of neighboring colonies located in an area $A$ in which resources occur patchily in space and time in an essentially stochastic manner. A honeypot ant colony pursuing a rigid monopolization strategy would maintain exclusive use of an area $\alpha$ around the nest. If monopolizing a territory of area $\alpha$ costs the colony $C(\alpha)$ per unit time, then the total cost of defense for a given interval $T$ is $C(\alpha)T$.

During $T$ a number of resource patches (fractions of termite nests) will appear in the area $A$, patterned in a spatial distribution. In the simplest case this distribution is random and each point in $A$ is just as likely to receive a resource patch as is any other point. Let $R$ be the total number of resource patches which, on the average, appear in $A$ during the period $T$. Then the average density of resource patches is $\rho = R/A$ patches per unit area, and the probability that a honeypot ant colony will receive $n$ of these on its area is

$$P(n) = e^{-\alpha\rho}\frac{(\alpha\rho)^n}{n!} \qquad (1)$$

For clarity let each patch return benefit $B_0$ to the colony. During $T$ the colony acquires on the average a total of $\alpha\rho$ resource patches on its territory and thus harvests an average benefit $B = B_0\alpha\rho$. As re-

**FIGURE 10–18** True slavery in honeypot ants. A pupa (*above*) and honeypot worker (*below*) of *Myrmecocystus mimicus* are pulled out of the nest entrance by victorious *M. mimicus* workers during a raid. (From Hölldobler, 1986b.)

sources become scarcer, ρ decreases and the size of territory required to harvest a given benefit $B$ increases.

If the colony requires a minimum return $B^*$ on its territorial investment during the period $T$, then the fixed territory of area α is economically defensible on the average if

$$\rho B_0 - \frac{B^*}{\alpha} \geq C(\alpha)\frac{T}{\alpha} \tag{2}$$

The surplus yield per unit area must exceed the costs of defense per unit area during $T$. For environments with scarce resources, ρ will be small, and these conditions are unlikely to be met.

The expected return $B$ characterizes the average benefit accrued over a very large number of repeats of the period $T$. For any single period $T$, the appearance of resource patches in a fixed territory is a sampling from $P(n)$ and will lie close to $B$ each period only if $P(n)$ peaks sharply around its mean αρ. For $P(n)$ with the Poisson structure (Eq. 1), however, the variance $\sigma^2$ is as large as the mean itself

$$\sigma^2 = \alpha\rho \tag{3}$$

and the harvestable return to the colony can vary significantly from period to period. Similar conclusions follow for related distributions. In such an environment holding fixed territories is a high-risk strategy in the sense that variance on return is of the same magnitude as the return itself. For many periods $T$ there will be either a resource surplus on the territory, which can exceed the colony's handling capacity, or a substantial resource deficit, which will make the defense effort a heavy drain. If the colony can handle surplus and ride out hard times (which is evidently possible because of the honeypot caste), the effects of this short-term stochasticity can be smoothed out somewhat and net benefit $B$ predicted by Eq. 1 will be more closely realized. Economic problems of defensibility will then again be addressed by Eq. 2. But a colony locked into a fixed territory is at the mercy of the fluctuations in the resource. A colony struck by chance with a long chain of "hard time" periods would be eliminated if it is unable to search further afield for resources.

A general rule of thumb is that animals faced with chains of deficits and surpluses on fixed territories should dissolve such boundaries and let foraging ranges overlap. We have seen that this condition applies in part for honeypot ants. Furthermore, for territorial risk-prone colonies there is a basic economic formula that will determine whether fixed territories pay off in the long run. For scarce, quickly exhausted resources this appears unlikely. The alternative to fixed territories is a foraging system with floating, temporary territorial boundaries, as has been observed in honeypot ants. Although the colony will lose a fraction of termite clusters in its region to competitors, it will gain others in the region of the competitor colony. Since costs of defense are now much reduced, and even further reduced by the ritualized fighting display strategy, floating boundaries and overlapping ranges become a preferred option.

## Intercolony Communication and Assessment

The dynamics of tournament interactions indicate that a kind of dominance order can exist among neighboring colonies. For a period of three weeks in July, when the rainy season had already started and the foraging activity of *Myrmecocystus mimicus* was very high, Holldobler (1981a) observed the interactions among three neighboring nests. In the sequence depicted in Figure 10–19 (I), nest 3 was very active and interfered with nest 1 and nest 5, foraging in the immediate vicinity of these two nests. Several days later (II) nest 3 was still engaging nest 5 in a tournament directly at nest 5, but the tournament between nest 1 and nest 3 had shifted toward nest 3, and nest 1 was now actively foraging. Seven days later (III) nest 1 and nest 3 did not show any outside activity, but nest 5 was highly active for the first time since the three nests were first monitored. The pattern suggests that nest 3 was the most dominant colony, and frequently interfered aggressively with the foraging activity of its conspecific neighboring colonies. Nest 5 seemed to be the weakest of the three colonies, because its foraging activity was most frequently suppressed by neighboring colonies. This is a representative example of what Hölldobler observed many times in his study area during the peak foraging period. Although on occasion one colony previously dominant over another was later suppressed by the same colony (still later the situation could be reversed yet again), in most cases the direction of domination remained constant throughout the active season.

During tournaments scouting ants repeatedly return to their nest, where they recruit reinforcement of 100 or more workers to the tournament area. Usually the recruitment of a new army of major workers by one colony is countered by a similar action by the opposing party. The relative sizes of the opposing forces appear to reflect the relative sizes of the two colonies. If the colonies are markedly different in size, so that one has a significant majority of the ants engaged in tournament interactions, the tournament is shifted more and more toward the nest of the weaker colony. When one of the opposing colonies is still incipient and therefore unable to summon a large defending worker force, the tournament ends quickly and the weaker colony is raided by the mature colony.

These field data suggest that the tournaments of *Myrmecocystus* provide the colonies with a means of assessing one another's strength. Lumsden and Hölldobler (1983) postulated that numerous threat displays between individual workers are integrated into a massive group display between opposing colonies. In parallel to the procedure followed by solitary animals, the groups' "strategic decision" whether to retreat, to recruit, to continue to fight by display or convention, or to launch an escalated attack depends on information about the strength of the opposing colony, and this information is somehow obtained during the ritualized combats at the tournament site. During the tournament contest the ants walk on stilt-like legs while raising the gaster and head. When two hostile workers meet, they initially turn to confront each other head-on. Subsequently they engage in a more prolonged lateral display, during which they raise the gaster even higher and bend it toward the opponent (Figure 10–20a,b). Simultaneously they drum intensively with their antennae on and around each other's abdomen, and frequently kick their legs against the opponent. This is almost the only physical contact, although each ant seems to push sideways as if she were trying to dislodge the other. After several seconds one of the ants usually yields and the encounter ends. The ants continue to move on stilt-like legs. They soon meet other opponents, and the whole procedure is repeated. In a tournament situation encounters with nestmates last only 1–2 seconds and are terminated by a brief jerking movement of the body. The exchanges usually do not develop into a lateral display. One other feature that appears to be im-

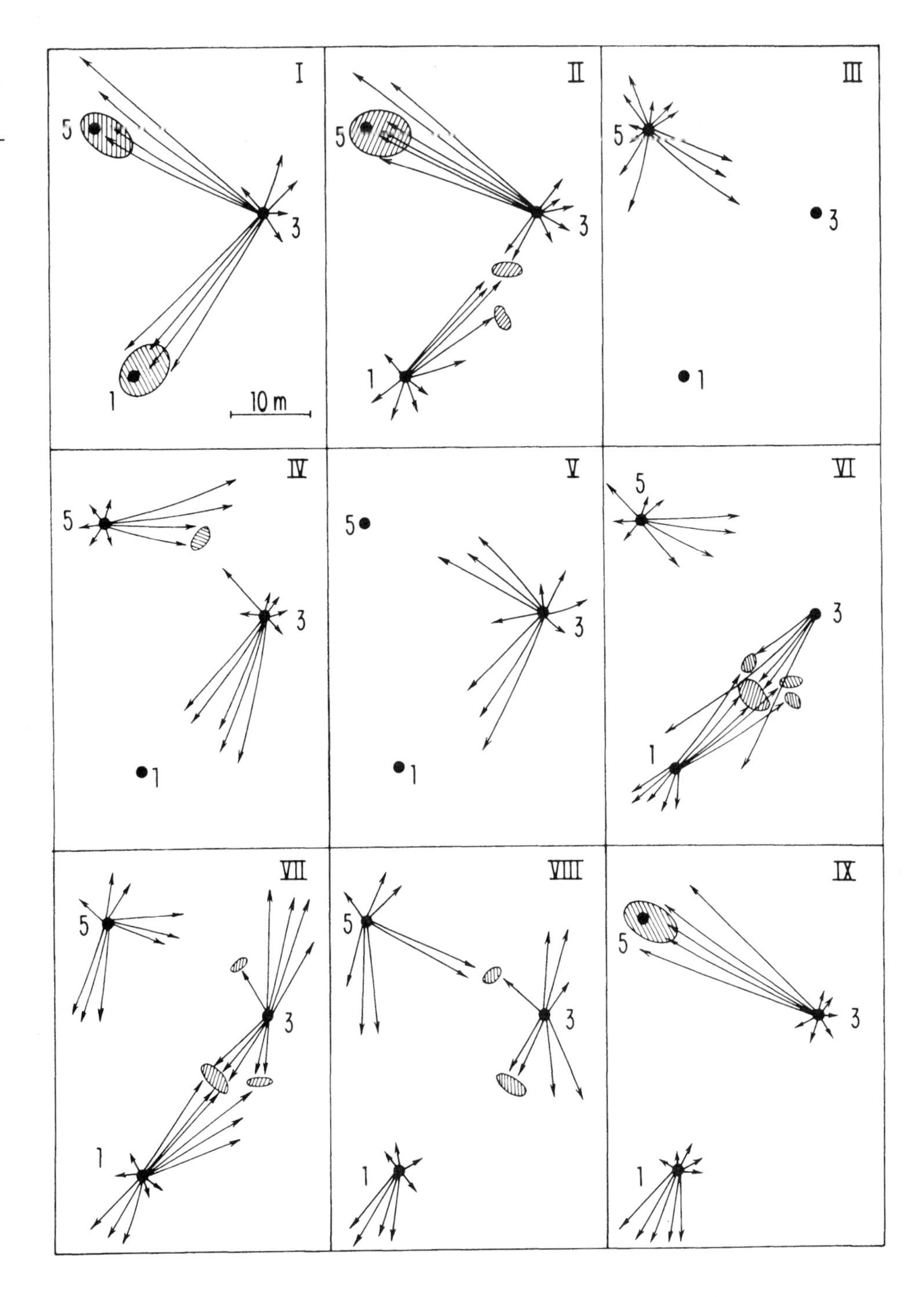


**FIGURE 10–19** Territorial expansion by tournaments in honeypot ants (*Myrmecocystus mimicus*). These diagrams follow interactions among three neighboring colonies during a period of three weeks in July 1977. The shaded portions represent the tournament areas. The long arrows indicate the principal routes followed by the ants; small arrows indicate minor traffic. (From Hölldobler, 1981a.)

portant during the displaying tournament is the size of the individual ants. If a large and a small ant are matched in a displaying encounter, usually the smaller ant yields. Displaying ants not only walk with their legs in a stilt-like manner while raising the gaster and head, but in addition they sometimes appear to inflate the gaster, so that the tergites are raised and the whole gaster appears considerably larger. There is also a tendency for the tournament ants to mount little stones and pebbles and display down to their opponents (Figure 10–20c). In fact the behavioral analysis of the displaying patterns suggests that during encounters the contestants gauge each other's size, and that there is a tendency among the ants to bluff, that is, pretend to be larger than they really are.

From these observations Lumsden and Hölldobler (1983) developed two models of ways in which *Myrmecocystus* may assess one another's strength during the tournaments. Individual workers may "count heads" during the encounters to gain a rough measure of the enemy's strength. Alternatively, individuals may determine whether a low or high percentage of the opponents are major workers and use it to estimate the opposing colonies' strength, since a high percentage is a reliable index of large colony size (Figure 10–20d). Hölldobler showed that among tournamenting ants majors are more frequently represented than among groups of foragers. Among colonies reared in the laboratory from founding queens, those younger than four years have a disproportionately small group of majors in the worker population.

Recent experiments indicate that both assessment mechanisms are involved in intercolony communication, and the data suggest that in particular small, immature colonies rely on the "caste poll-

**FIGURE 10–20** The tournament of the honeypot ants. *Myrmecocystus mimicus* workers are shown here in display fights. (*a*) A worker attempts to dislodge her opponent by bending sideways. (*b*) Lateral display between two opponents. (*c*) Large workers of *M. mimicus* in a display fight with small workers. (*d*) *M. mimicus* workers occasionally mount little stones in an apparent enhancement of the display ritual. (From Hölldobler, 1986b.)

ing" technique, which enables them quickly to assess whether or not the opponent is a mature colony. When confronted with large workers small colonies immediately retreat into the nest and close the nest entrance. This tactic enables small colonies to prevent larger ones from mounting a raid (Hölldobler, unpublished data).

Finally, still another method of assessment may be queuing: a long wait before meeting an unengaged opponent means a small colony, whereas a short wait means a large colony. However it is accomplished, whether by head counting, assessment of major percentages, or queuing, such colony-to-colony communication amounts to a kind of "negotiation" short of violent warfare. It entails behavior no more complex than that already demonstrated for the assessment and regulation of mass communication of many ant species during recruitment to food finds and new nest sites.

In addition to intercolony communication by means of tournament interactions, there is another, more subtle aspect of intercolony communication. Often no major antagonistic interactions occur for weeks between neighboring *Myrmecocystus* colonies. What Hölldobler (1981a) has called guard contingents are positioned daily at the same spots, however. The contingents consist of a few workers (usually not more than a dozen) that stand on stilt-like legs, often posing on top of little stones. Neighboring colonies also dispatch guard contingents to these posts, where they conduct "mini-tournaments" with the foreign guards. These mini-tournaments represent demonstrations of temporarily stabilized territorial borders. If the number of workers of one party at the guard-post suddenly increases, however, the other party responds by recruiting a large counterforce, and a full tournament often ensues.

## Ritualized Territorial Behavior in Other Ant Species

Ritualized aggression of the *Myrmecocystus* type or at least approaching it may occur more widely among ants than observers had previously suspected. It occurs for example in *Prenolepis imparis* (B. Hölldobler and J. F. A. Traniello, unpublished). The display may have the same ecological function as in *Myrmecocystus*, because as Talbot (1943a) first noted, *P. imparis* defend spatiotemporal territories. Le Moli and his collaborators reported that fighting among workers of different colonies of *Formica rufa* never leads to overt aggression (Le Moli and Parmigiani, 1982; Le Moli et al., 1982). These authors stated that alien workers instead confront one another by lifting their bodies up on stilt-like legs and bringing the abdomen forward between the posterior legs in what appears to be a ritualized version of formic acid spraying. A similar behavior has been observed as a threat display in *Camponotus americanus* (Carlin and Hölldobler, 1983; Figure 10–21). Actual spraying of the acid during such exchanges has rarely been noticed. According to Le Moli and Mori (1986), the absence of overt fighting with injury and death among colonies of the same species appears to be a common trait of the members of the *F. rufa* group, including *F. rufa*, *F. cunicularia*, and *F. lugubris*. In contrast, workers belonging to different species typically respond violently to one another and often fight to the death. Such generalizations must be made very cautiously, however, because seasonal intraspecific ant wars with many deaths have been reported in *F. polyctena* (Mabelis, 1979b) and *F. lugubris* (Cherix and Gris, 1978). Furthermore, Pisarski (1982) demonstrated that the degree of aggressivity in *F. (Coptoformica) exsecta* depends on the social organization of the colony.

According to Ettershank and Ettershank (1982), a form of ritualized territorial behavior leading to tournaments of the *Myrmecocystus* type occurs in the Australian meat ant *Iridomyrmex purpureus*. Ants from different colonies perform a series of acts more or less in sequence: they gape at one another with their mandibles, raise their bodies in stilt-like posture, "box" with their front legs, grab at one another with their mandibles, stand side by side, circle around one another while presenting the tips of their abdomens, and kick with their hind legs. The workers do not actually fight. Rather, they either part at some point of the display or else appease their opponents by grooming them. The winner in the exchanges remains in the stilt-like posture and the loser lowers her body, often while leaning away (Figure 10–22). A similar ritualized territorial display has been observed in *Forelius* (= *Iridomyrmex*) *pruinosus*, a common species in southwestern North America and Mexico. Members of separated nest populations usually do not engage in open physical aggression but instead perform a stereotyped display. This behavior consists of a rapid shaking motion as the ants confront each other, followed by a brief lateral display during which each ant holds her gaster up with the tip pointed toward the head of the other (Hölldobler, 1982c).

Workers of both *Formica* and *Iridomyrmex* use their distinctive behaviors to defend absolute territories, rather than spatiotemporal territories as in the case of the *Myrmecocystus* honeypot ants. In fact, Greenslade (1975a,b) and Greenslade and Halliday (1983) have demonstrated that the territories of meat ants are quite stable. Nevertheless real tournaments occur in *I. purpureus*. Several hundred ants occasionally display at the same time for up to several days, during which period some workers return to the same spot repeatedly. In one case observed by the Ettershanks the larger colony pressed steadily toward the weaker colony and eventually destroyed it. The myrmecocide—to coin a term—was similar to that common in *Myrmecocystus*.

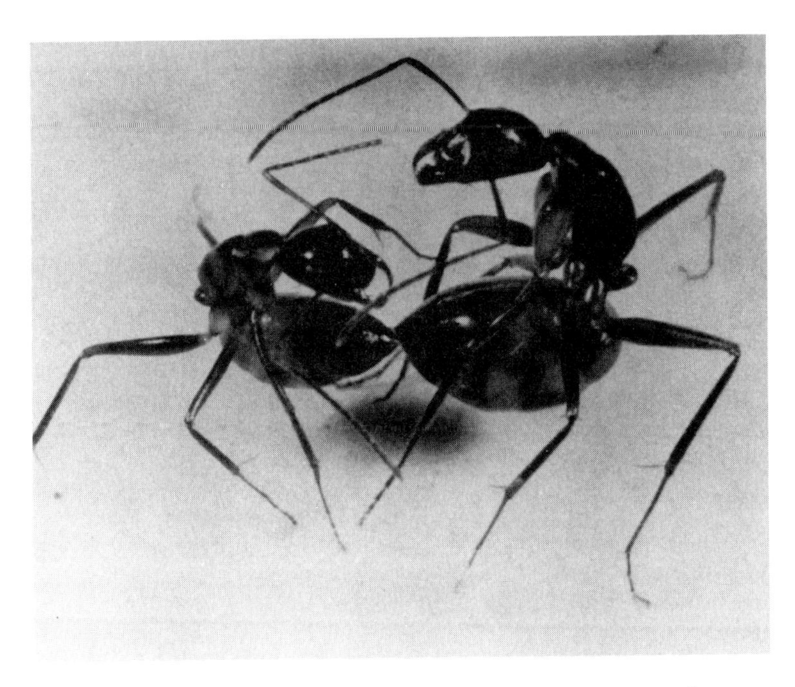

**FIGURE 10–21** The threat display of the American carpenter ant *Camponotus americanus*. (From Carlin and Hölldobler, 1983.)

Another form of ritualized aggressive interaction with appeasement behavior may be intercolony trophallaxis. In fact Ettershank and Ettershank (1982) point out that the origin of the agonistic display of the meat ant "is seemingly the more benign interaction of solicitation of food. The initial contact, in which each ant tries to get the other to fold its mandibles and submit is very similar to the begging of a solicitor from a donor. The ultimate appeasement gesture of the loser lowering the body resembles the attitude adopted by the donor in feeding; similarly stilting is an exaggeration of the superior posture adopted by a successful solicitor so that it can feed from the droplet." Intercolonial and interspecific food exchange during aggressive confrontations staged in a laboratory arena have been observed among conspecific alien workers, and even among workers belonging to different genera and subfamilies (Gösswald and Kloft, 1963; Bhatkar, 1979a,b, 1983; Kloft, 1983, 1987). This behavior has also been interpreted as a form of appeasement behavior regulating territorial interactions among neighboring colonies. These observations are based primarily on laboratory experiments, however, and the full understanding of the functional significance of intercolonial trophallaxis has to await more extensive field studies.

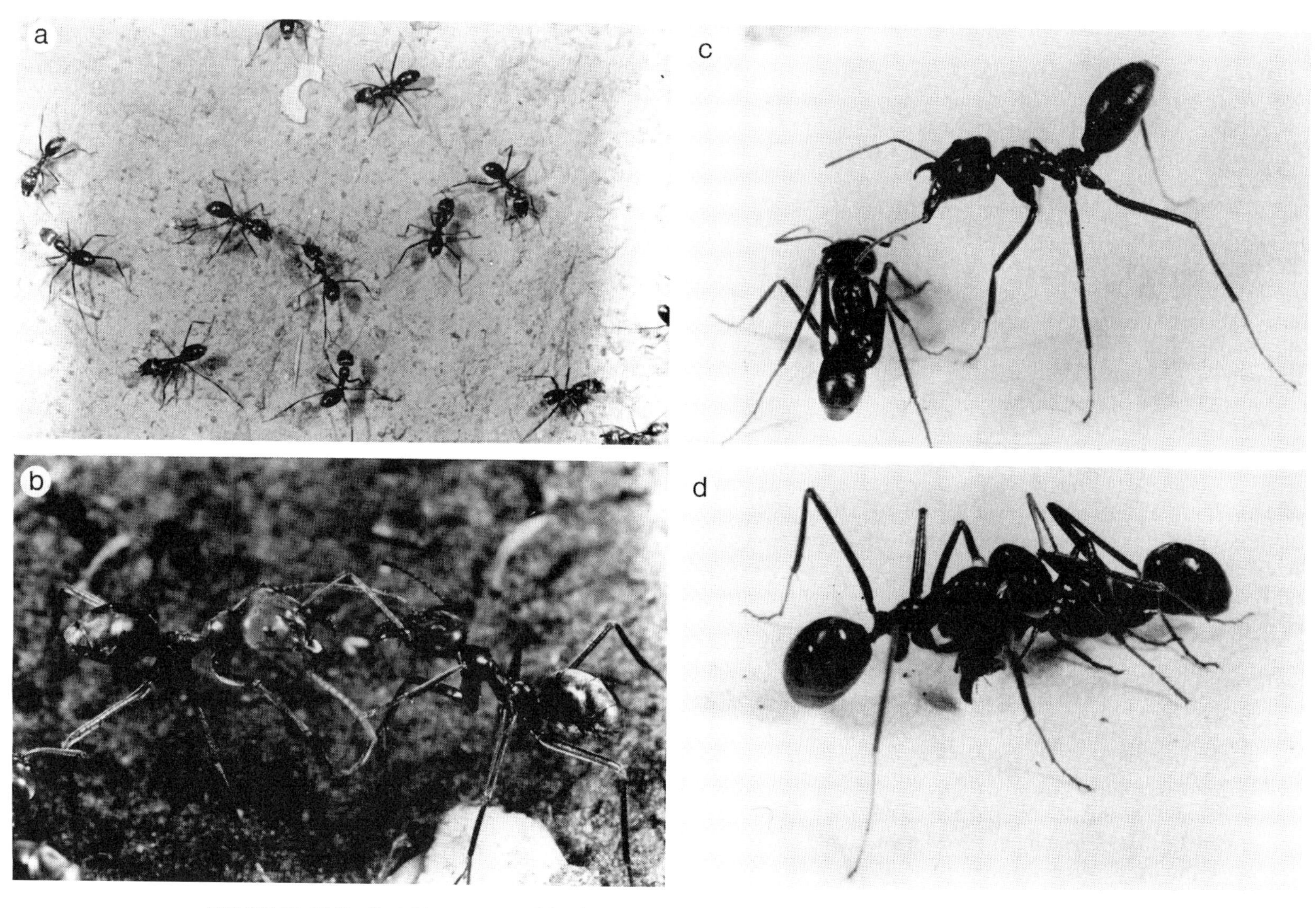


**FIGURE 10–22** Territorial tournament of the Australian meat ant *Iridomyrmex purpureus:* (*a*) group of opposing workers in display; (*b*) confrontational display between two workers, who threaten with open mandibles and "box" with their forelegs; (*c*) displays by submissive ant (*left*) and her dominant opponent (*right*); (*d*) the dominance display sometimes leads to carrying behavior, in which the dominant ant picks up the submissive opponent and carries her away from the tournament site.

## TERRITORY, PREDATION, AND TRUE SLAVERY

In Chapter 7 we introduced the idea of economy in the evolution of ant communication. Species have often combined the same or very similar elements of behavior in different ways to create new messages. They have also enlarged their repertories by ritualization, in which behavioral acts that are functional in one context are modified for use as signals in another context. A similar evolutionary parsimony exists in the relation between territorial aggression and other major categories of behavior, particularly predation and parasitism.

Many ant species are known to carry enemies killed in territorial battles back to their nests for use as food. This is true for at least a few within-species conflicts, as in the cannibal wars of *Formica polyctena* (Mabelis, 1979b, 1984; Driessen et al., 1984). It is more commonly the case for between-species aggression. For example, the fire ant *Solenopsis invicta* consumes *Pheidole* colonies that it displaces (Wilson, 1976b). The weaver ant treats alien workers of conspecific colonies and of other species like prey to be killed and carried back to the nest, but the effect as territorial defense is just the same (Hölldobler, 1983; Figure 10–23). *F. polyctena* invades the nests of *F. fusca* and species of *Lasius* and *Myrmica* and carries the inhabitants back as prey. In the process *F. polyctena* largely eliminates these species for distances of up to tens of meters from their nest entrances, an effect that could equally well be called overutilization of prey or territorial elimination of competitors. It is easy to see how specialized predators of ants may originate by the territorial route. Such specialists are the rule in the army ant genera *Aenictus, Eciton,* and *Neivamyrmex,* as well as in the ponerine tribe Cerapachyini. They also occur in derived species of *Gnamptogenys* and *Myopias,* which are ponerine genera whose members prey primarily on other kinds of arthropods.

Territorial raids also appear to have led to slavemaking in a few species. The colony conquests of *Myrmecocystus mimicus,* in which worker and brood of alien colonies are captured and adopted as nestmates, result in true slavery—the colonies use the labor of captives belonging to the same species (Hölldobler, 1981a). In addition

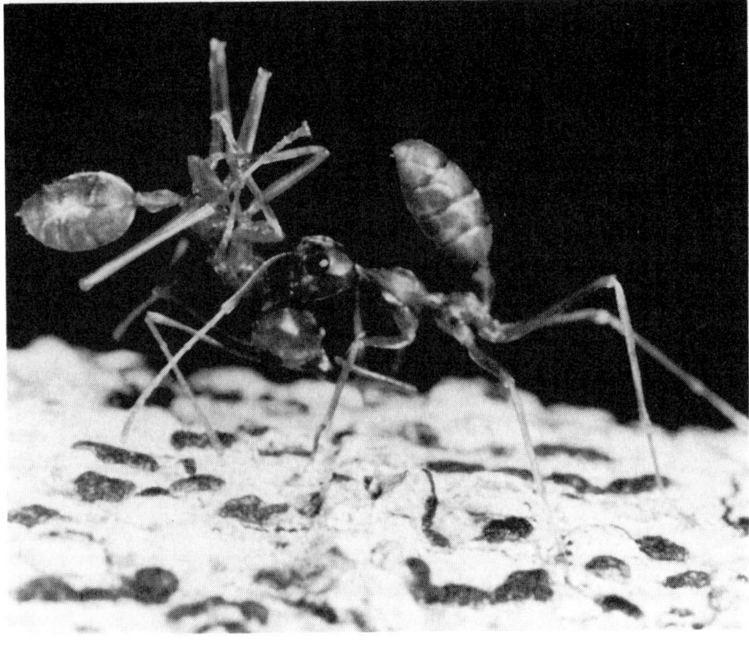

FIGURE 10–23 Territoriality and predation in weaver ants of the genus *Oecophylla*. (*Above*) A worker of *O. smaragdina* in Australia has captured a minor worker of *Pheidole megacephala* and is carrying her as prey into the nest. (*Below*) A worker of *O. longinoda* in Kenya carries a newly killed enemy worker from a territorial battlefield to the nest, where the defeated ant will be used as food.

*Myrmecocystus* utilizes all injured enemy ants as prey, especially the honeypots. The largest proportion of the pillaged honeypots do not survive the raid and are eaten by the victors. Thus a successful raid is a true bonanza for the victorious colony. It not only gains a large labor force, without having to spend energy and time to raise the workers, but it also obtains highly valuable food by robbing and preying on the living storage containers of the victim colony. A closely similar phenomenon occurs in the *curvispinosus* group of *Leptothorax*, colonies of which attack related species during territorial expansion. The workers consume most of the larvae and pupae they capture but allow some to mature and remain as nestmates (Wilson, 1975a; Alloway, 1980; Stuart and Alloway, 1982). In Finland, Rosengren and Pamilo (1983) observed a case in which workers from a *Formica polyctena* colony raided a smaller colony of the same species and carried away its young, semi-replete workers used in food storage (*Speichertiere*). They killed the queens but did not harm the workers, who evidently were incorporated as slaves.

## OTHER TECHNIQUES OF COMPETITION AND EXPLOITATION

In competition for food sources and nest sites, ants employ a rich repertory of interference techniques, of which territoriality is only the most commonly observed form. Many ant species monopolize food discoveries by the mass recruitment of foragers that use powerful chemical repellents. Their competitors often counter these actions by employing chemical assaults of their own or by relying on the rapid cooperative retrieval of prey. Still other ant species invade the immediate nest area of their competitors and use threats and chemical weapons to prevent them from foraging. Let us examine some of the better-studied examples in these several categories.

The European thief ant *Solenopsis* (*Diplorhoptrum*) *fugax* usually lives close to colonies of other species and preys on their brood. Scouting workers of the *Solenopsis* build an elaborate tunnel system leading into the neighbor's brood chambers. As soon as the construction of the tunnels is completed, the scouts lay chemical trails back to their own nest and recruit masses of nestmates, who then mount an invasion. While invading the brood chambers and preying on the brood, the *Solenopsis* workers discharge a highly effective and long-lasting repellent from the poison gland. The principal component of this secretion is trans-2-butyl-5-heptyl-pyrrolidine. The substance prevents the brood-tending ants from defending their own larvae and enables the *Solenopsis* to rob brood virtually unopposed (Hölldobler, 1973a; Blum et al., 1980).

Similar chemical techniques are used in interference competition in *Solenopsis* (Baroni Urbani and Kannowski, 1974; Hölldobler et al., 1978), *Monomorium* (Hölldobler, 1973a; Adams and Traniello, 1981; Jones et al., 1988), *Meranoplus* (Hölldobler, unpublished), and *Forelius* (= *Iridomyrmex*) (Hölldobler, 1982c) (see Figure 10–24). The last case shows how effective the methods can be. In the Arizona desert the dolichoderine ant *F. pruinosus* and the formicines *Myrmecocystus depilis* and *M. mimicus* overlap widely in their nesting and food habits. Although the individual *Forelius* workers are only one-fourth the length of *Myrmecocystus* workers, they usually succeed in displacing the larger ants from food bait. They also block the nest entrance of their competitors, preventing them from leaving the nest altogether. They accomplish this feat by swiftly channeling large numbers of workers to food sources and nest entrances of the *Myrmecocystus*, using a pheromonal mass communication system, while chemically repelling their competitors with secretions from the pygidial gland (Hölldobler, 1982c; Figure 10–25).

A bizarre variation of the nest-blocking technique is used by workers of the dolichoderine *Conomyrma bicolor*, which interferes with sympatric species of *Myrmecocystus*. The *Conomyrma* workers surround the nests of their competitors, pick up pebbles and other small objects in their mandibles, and drop them down the nest entrances (Möglich and Alpert, 1979; see Figure 10–26). Although it is not known exactly how the stone-dropping alters the behavior of

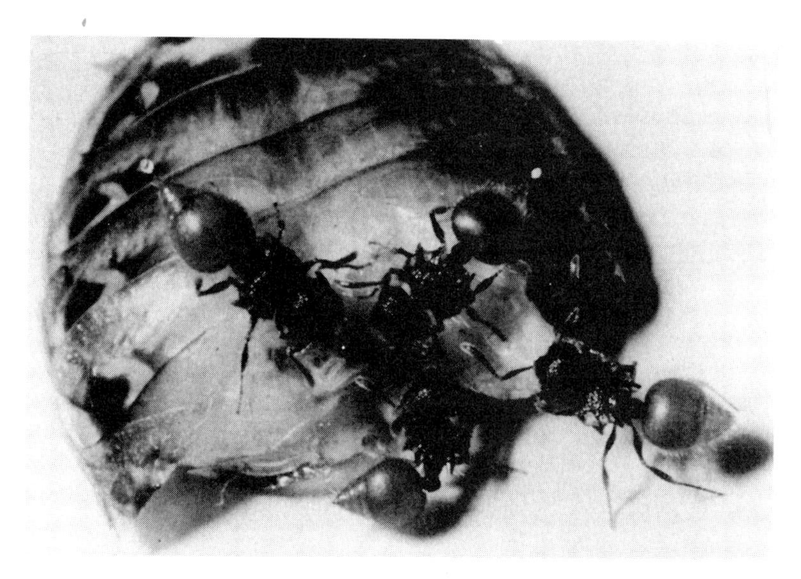

**FIGURE 10-24** The use of chemical defense by ants to ward off competitors at food sites. (*Left*) Workers of the fire ant *Solenopsis xyloni* defend a severed abdomen of a honeypot ant *Myrmecocystus mimicus* by raising their abdomens and extruding their stings to release a repellent venom into the air. (From Hölldobler, 1986b.) (*Right*) Foragers of *Meranoplus* sp. guard a cockroach abdomen by leaving the sting extruded and releasing a sticky white substance from the tip. The defensive material originates in the hypertrophied Dufour's gland.

**FIGURE 10-25** Competition by aggressive displacement between species. (*a*) In Arizona, a worker of *Forelius pruinosus* displays in front of the nest entrance of *Myrmecocystus mimicus*. (*b*) Histogram illustrating the activity of *M. mimicus* foragers at their nest. The arrow indicates the first appearance of *F. pruinosus* workers at the nest entrance. From this point on the activity of *Myrmecocystus* foragers declined rapidly. (*c*) As soon as the first *F. pruinosus* workers appeared at the *Myrmecocystus* nest (arrow), they were continuously removed by the experimenter. In the absence of their small competitors, the foraging activity of *Myrmecocystus* continued at a high level. (From Hölldobler, 1982c, 1986b.)

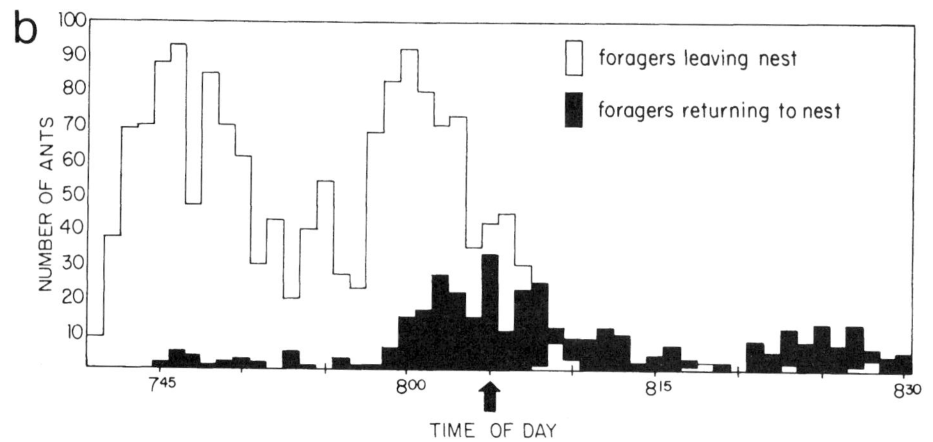


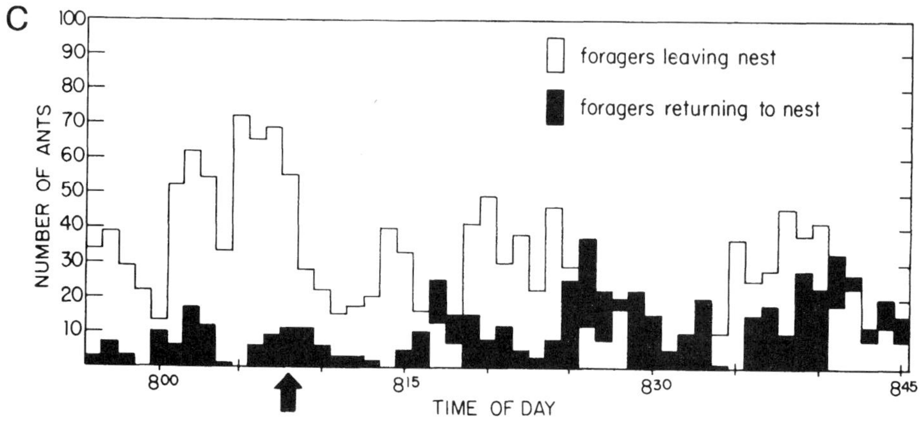

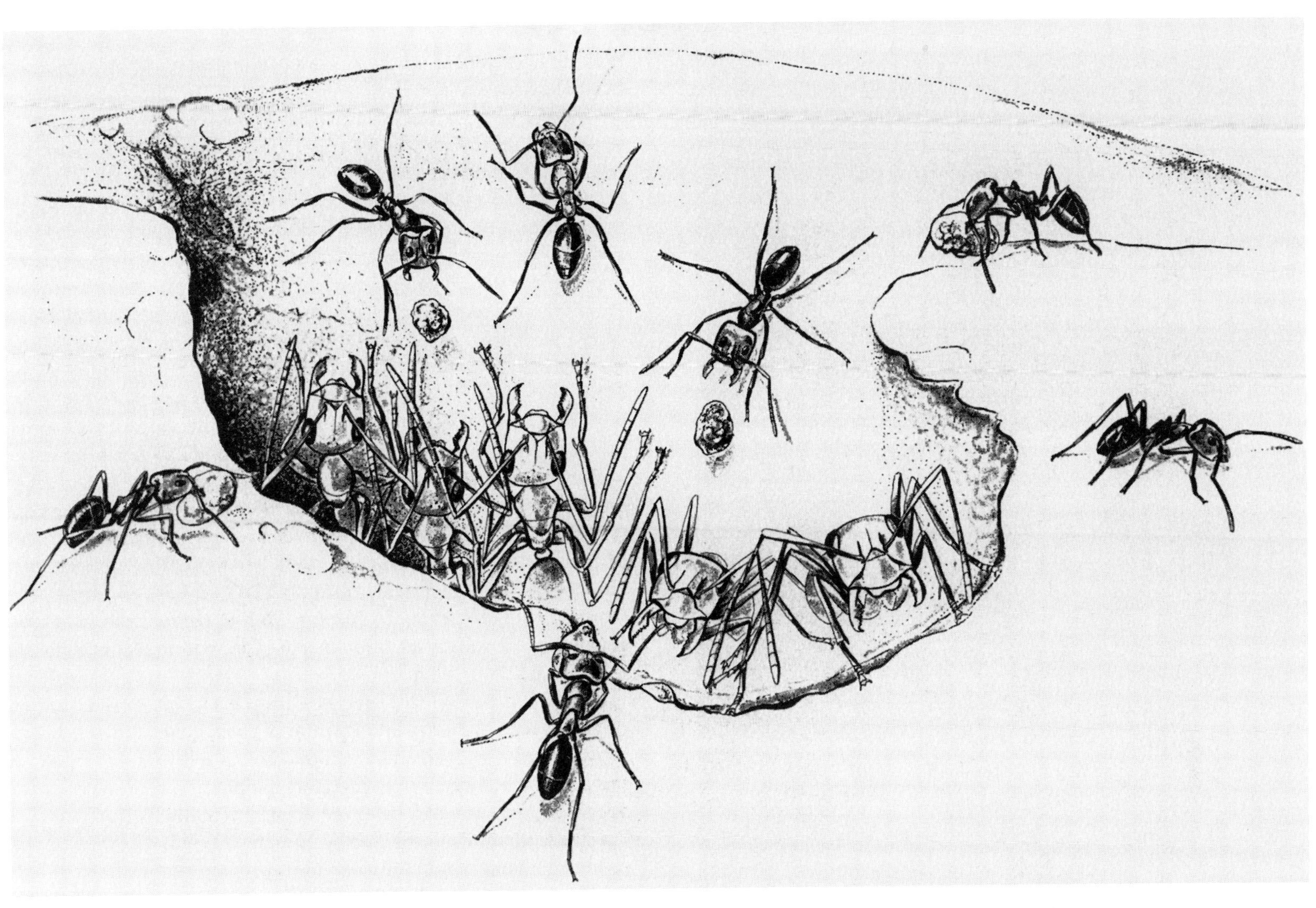

**FIGURE 10–26** In the deserts of Arizona, workers of *Conomyrma bicolor* inhibit foraging of a species of *Myrmecocystus* by dropping pebbles down the nest entrances of the *Myrmecocystus.* (From Möglich and Alpert, 1979.)

the workers in the target nests, the effect is to reduce their foraging by a significant amount. A closely similar behavior was observed by Gordon (1988b) in *Aphaenogaster (= Novomessor) cockerelli,* which blocks the nest entrances of *Pogonomyrmex barbatus* and thereby shortens the daily foraging period of this harvesting species by one to three hours.

Oster and Wilson (1978) inferred from elementary mathematical models that as prey size increases, so does the probability of interference competition. The prediction was confirmed in field studies on *Monomorium* by Adams and Traniello (1981) and on *Lasius* by Traniello (1983). In the deserts of the southwestern United States, the species of *Aphaenogaster (= Novomessor)* partially solve this problem by their large size, which enables single foragers to retrieve large prey items without delay (Hölldobler et al., 1978). When the items are very large, however, the *Aphaenogaster* are still in danger of losing them to colonies of *Forelius, Monomorium,* and *Solenopsis,* which can assemble large forces of workers by mass communication. The *Aphaenogaster* foragers do not fight back very effectively, at least not with chemical weapons of their own. Instead they use a highly efficient system of recruitment and retrieval. When a lone scout finds a prey object too large for her to carry, she discharges a poison-gland pheromone into the air as a short-range recruitment signal. Other scouts are soon attracted from as far away as 2 meters. The effect is enhanced by vibrational signals that lead to a more efficient group retrieval of the prey item (Markl and Hölldobler, 1978). Whenever possible the prey is not dissected on the spot but lifted up by the gang of workers and carried immediately back to the nest. In this fashion *Aphaenogaster* greatly reduces the number of confrontations with its more formidable, mass-communicating competitors.

A very widespread form of competitive exploitation in ants is food robbing or "cleptobiosis." Although the phenomenon has been described by many authors over the past century (Wroughton, 1892; Forel, 1901; Wheeler, 1910a, 1936c; Abe, 1971; Maschwitz and Mühlenberg, 1973), relatively little attention has been given to its ecological significance. A particularly interesting case involving desert ants in the southwestern United States was analyzed in this regard by Hölldobler (1986a). Harvester ants of the genus *Pogonomyrmex,* which harvest mostly seeds and other vegetable material, at first seem unlikely competitors for honeypot ants of the genus *Myrmecocystus,* which are predators and scavengers that also collect floral nectar and honeydew from homopterans. *Pogonomyrmex* workers also occasionally retrieve dead insects, however, and they are especially prone to join other ant species in preying on termites when

**FIGURE 10–27** Competition through robbing. (*a*) In an Arizona desert, *Myrmecocystus mimicus* workers stop and inspect a *Pogonomyrmex maricopa* worker. (*b*) An *M. mimicus* worker robs a termite prey from a *Pogonomyrmex* forager. (From Hölldobler, 1986b.)

these insects appear on the surface of the desert soil in large numbers following rain.

Workers of *Myrmecocystus mimicus* waylay returning foragers of several *Pogonomyrmex* species and rob them of insect prey, especially termites. As the *Pogonomyrmex* workers return to the nest, they are often stopped by the *Myrmecocystus* workers and thoroughly inspected. In a typical sequence, a *Myrmecocystus* climbs on the harvester ant's back and nibbles her head, mandibles, and mouthparts. Sometimes two or even three *Myrmecocystus* gang up to harass one *Pogonomyrmex* worker in this manner. If the harvester ant does not carry insect prey, she normally stands still and tolerates the inspections, which last only a few seconds. But if she has a termite or any other insect prey in her mandibles, the *Myrmecocystus* workers attack ferociously, vigorously yanking the prey away from the *Pogonomyrmex*. At most, the *Pogonomyrmex* responds to her tormentors with open mandibles and an occasional forward lunge in an apparent attempt to bite the *Myrmecocystus* (Figure 10–27). The *Myrmecocystus* are easily able to avoid the attack by swift, evasive movements of their own. In a survey of 76 nests belonging to *Pogonomyrmex*, Hölldobler found that between 5 and 39 percent are affected by the *Myrmecocystus* according to species. Thus it seems likely that food robbing has a substantial influence on the food intake of both the robbers and their victims.

Finally, it is possible for species to invade a nearly closed community of species by heavy immigration of propagules. Either a beachhead is finally attained and the population is able to sustain itself, or the permanent supply of immigrants maintains the presence of the species in spite of rapid turnover. A general mathematical model of the process has been provided by Loeschcke (1985). Documentation of the phenomenon in ants has been provided by Rosengren and Pamilo (1983), who call it the Austerlitz effect, after the 1805 battle in which Napoleon's forces vigorously attacked superior forces on both sides of the Pratzen Plateau. *Formica truncorum*, which is a "beachland specialist," assails habitats surrounding its established colonies with numerous inseminated queens evidently capable of long-distance dispersal. It also spreads rapidly by encroachment into vacant nesting sites by portions of the colony that emigrate over the ground. The result is a patchy distribution of the species in the midst of the more nearly continuous distribution of *F. aquilonia* and *F. polyctena* populations.

# CHAPTER 11

# The Organization of Species Communities

Species belonging to the same community are said to be organized when they interact in such a way as to make certain subsets more frequent than would be expected as a result of chance alone. In Elton's words (1933) the community has "limited membership." At the outset only some of the species in the surrounding pool have the dispersal ability to reach the habitat, so that admission is restricted. And of those that colonize, some fit in less well than others and pass into extinction sooner, making expulsion from membership discriminatory. Organization is based on the latter process of differential extinction. But the addition and loss of species is not the only way that organization can occur. It is also possible for species to affect one another's abundance, spatial distribution, and behavior.

By definition the more numerous the connections among the species, and the more complex the hierarchies they create, the more organized the community (Diamond et al., 1986). The most obvious type of effective species interaction is competition. But other such processes occur, including predation, parasitism, commensalism, and mutualism. In addition it is possible for species to alter the physical environment in a way that allows other species to colonize it more readily. All of these phenomena have been demonstrated at one time or another in the ants.

## INTERSPECIFIC COMPETITION

In Chapter 10 we saw that competition is the hallmark of ant ecology. Many, perhaps most, ant species employ aggressive techniques up to and including organized warfare. What are the consequences of this fierce competition for community organization? Since the early 1950s myrmecologists have recognized the existence of dominance hierarchies among species, in which some species regularly displace others at territorial sites. These hierarchies are often linear through three or more links. Documentation has been provided in a series of detailed studies by Kaczmarek (1953), Marikovsky (1962b), Yasuno (1963, 1965b), Brian (1965b, 1983), Dlussky (1965), Wilson (1971), Zakharov (1972, 1977), Pisarski (1973, 1982), Reznikova (1975), Czechowski (1979), Jutsum (1979), Levings and Traniello (1981), Pisarski and Vepsäläinen (1981), Vepsäläinen and Pisarski (1982), Rosengren and Pamilo (1983), Fellers (1987), Ward (1987), Savolainen and Vepsäläinen (1988), and others.

Vepsäläinen and Pisarski have devised a useful three-tiered classification from general winners at the top to general losers at the bottom. On the Tvärminne Archipelago, a cluster of small islands off the Finnish Baltic coast, this hierarchy is as follows:

I. The lowest level comprises species that defend only their nests; examples include *Formica fusca* and the three Finnish species of *Leptothorax*.

II. At the intermediate level, species defend their nests and food finds; in other words they maintain spatiotemporal territories. Examples include *Tetramorium caespitum*, *Camponotus herculeanus*, and *C. ligniperda*, as well as smaller colonies of *Lasius niger*.

III. At the top level species successfully defend their nests and all of their foraging areas as absolute territories. Examples include *Formica exsecta*, *F. sanguinea*, *F. truncorum*, and members of the *F. rufa* group of species, as well as large colonies of *Lasius niger*. They are the "large-scale conquerors" recognized by Rosengren and Pamilo (1983) for Fennoscandia generally. The most prevalent of all species on the mainland is *F. aquilonia*.

Some of the population and behavioral characteristics of representative species are given in Table 11–1. Intermediate (level II) species are often extirpated entirely from the extensive domains of colonies belonging to level III. On the other hand colonies of level I species, which fight with alien workers only in the immediate vicinity of their own nests, are able to survive better in the presence of the dominants. The compatibility they enjoy is nevertheless far from perfect, because *Formica polyctena* workers periodically invade nests of the low-ranking *F. fusca* and *Lasius flavus* and carry out adults and brood as prey. As a result of this pressure, few colonies of these two species survive within tens of meters of the *F. polyctena* nests.

Since the dominant species patrol such large foraging areas, local faunas can be profoundly influenced by relatively few of their colonies. On the island of Joskär, for example, a single colony of *Formica polyctena* containing over a million workers dominated 20,000 square meters of terrain (the size of four football fields). Entire communities can be closed against some species. *F. rufibarbis* is absent from the Tvärminne Archipelago, even though it is abundant on the adjacent mainland. The reason appears to be that its favored habitat, open dry pine forests, is preempted by colonies of the more aggressive *F. sanguinea*.

The species hierarchies offer an unusual opportunity to analyze the role of aggressive behavior and interference competition in the assembly of communities. As Vepsäläinen and his co-workers have demonstrated, it is possible to translate upward from experimental studies of the encounters of individual ants on baits to local faunal lists. It is also possible to make qualitative predictions of relative abundance. The lowest-ranking, level I species, for example, appear to be the same as the "opportunistic" species recognized by Wilson (1971) at food baits. Their workers are adept at finding new food but

**TABLE 11–1** Formicine ant species of the Tvärminne Archipelago, Finland, ranked into the levels of the competition hierarchy; the numbers given are the magnitudes of the main biological parameters that affect rank order. (From Savolainen and Vepsäläinen, 1988; based in part on supplementary data from Collingwood, 1979, and Zakharov, 1975.)

| Level of hierarchy | Species | Size of society (number of workers) | Size of workers (mm) | Dynamic density (individuals/sq dm/min) | Radius of foraging area (m) | Recruitment to food | Defense of— | | |
|---|---|---|---|---|---|---|---|---|---|
| | | | | | | | Nest | Food | Foraging area |
| III | *Formica polyctena* | $10^5$–$10^6$ | 4.0–8.5 | ca. 3 or more | <100–200 | Effective | + | + | + |
| | *F. truncorum* | $10^3$–$10^4$ | 3.5–9.0 | — | <15 | Effective | + | + | + |
| | *F. exsecta* | $10^3$ | 4.5–7.5 | — | <10 | Effective | + | + | + |
| II | *Camponotus* sp. | $10^2$–$10^3$ | 6.0–14.0 | — | <15–30 | Moderate | + | + | – |
| | *Lasius niger* | $10^2$–$10^3$ | 3.5–5.0 | 0.5–0.6 | <3–5 | Effective | + | + | – |
| I | *Formica fusca* | <500 | 4.7–7.0 | <0.5 | <15–20 | Poor | + | – | – |

abandon it when confronted by more aggressive species. The exemplar in the Tvärminne Archipelago is *Formica fusca*. The workers of this species utilize a wide range of food items, which they locate quickly, and they shift to less desirable food when put under pressure from higher-ranking species.

As one proceeds from small to large islands in various archipelagos around the world, the number of ant species increases roughly as the fifth to third root of the island area (Wilson, 1961; MacArthur and Wilson, 1967). It is also true that the growing assemblages often form nested sets. That is, on extremely small islands only species A–C might be present, on somewhat larger islands A–F, on still larger islands A–H, and so on. The sequence is seldom if ever of the form A–C, D–I, K–X, and so forth (Wilson and Hunt, 1967). This regularity, which is far from perfect and evidently subject to considerable stochasticity, reflects the fact that larger islands provide a greater array of microhabitats in which specialized ant species can settle. It is also likely to be shaped in part by the kind of rank-ordering and preemption illustrated in the Finnish studies, however. The nested-set phenomenon is unusually evident in very small mangrove islands in the Florida Keys (Cole, 1983a). As depicted in Figure 11–1, the species of arboreal ants build up in a regular manner with an increase in volume of the foliage. Each species has a minimum island volume it requires for indefinite survival, which for some species at least is that amount of plant surface needed to protect the colony from excessive stress due to wind and wave action. The nests are especially vulnerable, because the mangrove trees are rooted in tidal mud. Two species, *Crematogaster ashmeadi* and *Xenomyrmex floridanus*, are dominant. When a colony of either one is established first, it precludes the invasion by a colony of the other species. When one of these two species is present on islands with volumes less than 5 cubic meters, it precludes an invasion by the remaining, subordinate species. Workers of the two dominant species are consistently aggressive toward workers of all other species, whereas those of the subordinate species (*Camponotus* sp., *Pseudomyrmex elongatus*, and *Zacryptocerus varians*) almost invariably run from enemies. When Cole removed the dominants from islands less than 5 cubic meters in volume, the subordinates were able to invade the trees and persist indefinitely.

Dominance hierarchies among ant species are a worldwide phenomenon. In the arid and semi-arid regions of Australia, the overwhelming alpha species are members of the dolichoderine genus *Iridomyrmex* (Greenslade, 1979; Greenslade and Halliday, 1983; Andersen, 1986a,b). As described by Greenslade (1979) in his study of the fauna of South Australia,

> *Iridomyrmex* are successful competitors and generally they dominate associations of ant species. The more abundant, diurnal *Iridomyrmex* species compete with each other and since they tend to be mutually exclusive their colonies often form a patchwork on the ground. The distribution of dominant *Iridomyrmex* then sets up patterns in space and time with which other ants must conform. At high temperatures, above-ground activity by the *Iridomyrmex* ceases and they are replaced by *Melophorus* . . . Some *Camponotus* species are nocturnal, foraging at night when many dominant *Iridomyrmex* are inactive. Others are much larger than the *Iridomyrmex* with which they coexist and reduce interaction by the size difference. Others again nest in no-man's land between colonies of *Iridomyrmex*. The behaviour of *Camponotus* is normally evasive when they encounter other ants and they seem to have succeeded as subordinate competitors. Other common genera are *Rhytidoponera*, *Monomorium* and *Pheidole*. As a rule the species are not closely adapted to the rest of the ant community. Instead they tend to be rather unspecialized, catholic feeders, flexible in their time of foraging and able to occupy a wide variety of habitats. Together with some species of *Iridomyrmex* and *Paratrechina* they often seem to be opportunists, exploiting areas or resources that are not intensively used by other ants. (p. 4)

An example of the way in which the subordinate species fit into the *Iridomyrmex* community is provided by a formicine of the genus *Melophorus* that lives in unoccupied chambers of nests of *I. purpureus*, emerging onto the surface and foraging in the middle of the day in hot weather when the meat ants are inactive. This general adaptation to extreme heat, paralleling the thermophily of *Cataglyphis* and *Ocymyrmex* in Africa, has evidently been a key factor in the evolution of *Melophorus* into a large number of species that fill many nesting and feeding niches. The genus includes general predators, predators on termites and other ants, and seed harvesters. One species has a minor worker with a conspicuously flattened head and alitrunk that allows it to forage under the peeling bark of *Eucalyptus* trees.

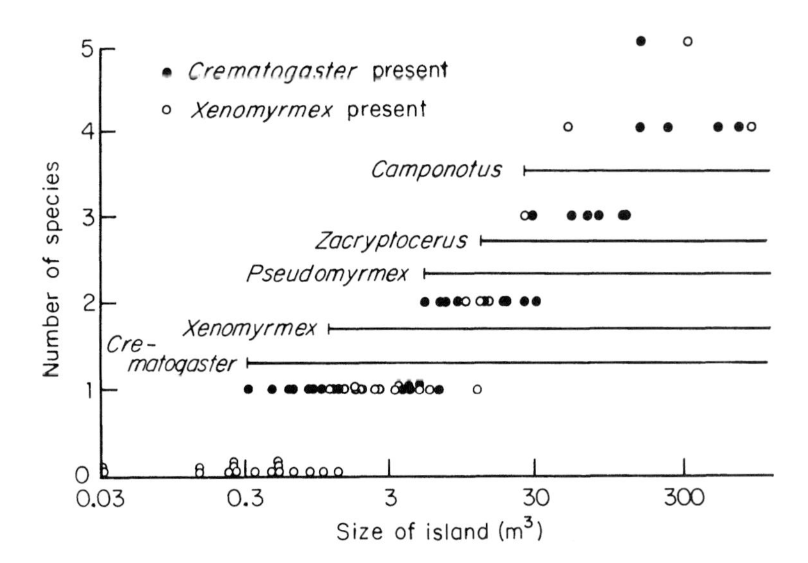


**FIGURE 11–1** Assembly rules in the ant faunulae of very small mangrove islands in the Florida Keys. The lines represent the range of island sizes on which the common species can occur in the absence of dominant competitors. When either *Crematogaster ashmeadi* or *Xenomyrmex floridanus* are present on the very smallest islands, they exclude the other, more timid species. As a consequence, species communities are built up in an orderly manner. (Modified from Cole, 1983a; note that space is measured in volume rather than area.)

**FIGURE 11–2** Community organization as a series of ant mosaics in West Africa. The species shown are arboreal ants found on a cocoa farm in Ghana. The dominant species, in *Crematogaster* and *Oecophylla*, are centrally located within the enclosure lines, and the subordinate species associated at the 95-percent confidence level with each dominant is connected to it by a line. The genera are abbreviated as follows: *A.*, *Acantholepis*; *At.*, *Atopomyrmex*; *Camp.*, *Camponotus*; *Cat.*, *Cataulacus*; *Crem.*, *Crematogaster*; *O.*, *Oecophylla*; *Odont.*, *Odontomachus*; *Phas.*, *Phasmomyrmex*; *Platy.*, *Platythyrea*; *Poly.*, *Polyrhachis*; *T.*, *Tetramorium*. (Modified from Room, 1971.)

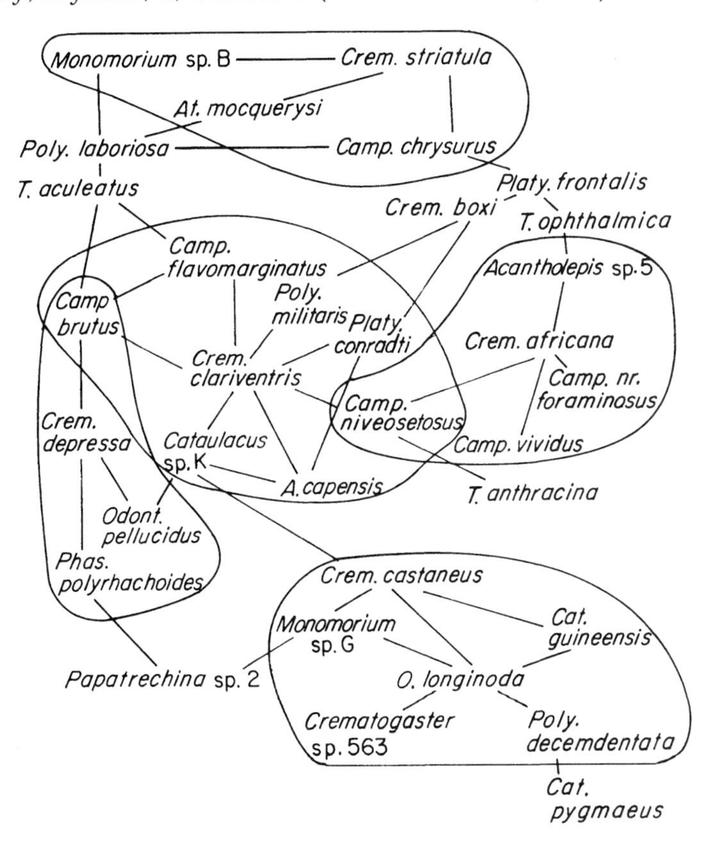


The dominant species form the core of the local community. They affect the composition and abundance not only of other ant species but even of plants and other arthropods. Their pervasive influence is clearly marked in the canopies of tropical forests of West Africa, where they spread out into mutually exclusive territories that cover from one to several trees each. The resulting "ant mosaic" was originally described by Leston (1973a) as follows.

> The humid tropics fauna include certain dominant ants. Where they occur these are more numerous than other ants and to the exclusion of other dominants, *i.e.*, those ants which elsewhere are more abundant. Dominants are usually non-nomadic, arboreal, multi-nested, saccharophilic and predatory, practicing mutualism with Homoptera: they have a potential for rapid population growth. Optimally dominants are spaced out in a three-dimensional mosaic with few lacunae but forest degradation leads to a two-dimensional structure. Three mosaic patterns are recognized: (1) the forest mosaic, occurring too but simplified in the crowns of cocoa, coffee, coconut and oil palm, etc.; (2) the forest understorey pattern; (3) the *Crematogaster striatula* pattern, peculiar to West Africa . . . Each dominant is the centre of a positive association of other insects, spiders and non-dominant ants, and of a negative association. The positive associates are not merely those insects with which the dominant is in mutualistic symbiosis. In the Old World tropics many tree crop pests have patchy distributions through their negative association with some dominants, their predators, and positive association with others. "Flitting" insects are more affected than "flying" or endophytic species: insects can be arranged in an "ant impact" hierarchy. Manipulation of the mosaic leads to changes in the overall population levels of many pests. Mosaics occur throughout the Old World tropics and probably in the New World too: it is likely that *Azteca* and its allies are dominant, to some extent limiting leaf-cutting Attini.

The details of the African mosaics were further elucidated by Room (1971), Leston (1973b,c), Majer (1976a–c), and Jackson (1984). Room studied more than a hundred species of ants in cocoa farms in Ghana and plotted their pairwise occurrences. The five dominant species and their associates are shown in Figure 11–2, while the negative associations are given in Figure 11–3. The reason why some species in the African canopy repulse one another is clear enough, but the basis of the positive association between many pairs is unknown. A strikingly similar mosaic pattern occurs in tropical Australia, with *Oecophylla smaragdina* and a species of *Crematogaster* playing key roles (Figure 11–4). Majer established the dominance of *Oecophylla longinoda*, *Crematogaster striatula*, *C. depressa*, and *Tetramorium* (= *Macromischoides*) *aculeatus* in that order. He proved competitive displacement by removing colonies of some of the dominants and watching the competitor species move in, along with their favored ant and other arthropod associates.

## CHANGING AND REMOVING DOMINANTS

If dominant ant species are really organizing agents, then changing from one dominant species to another should have marked effects on the remainder of the ant community. We have seen that this is indeed the case in passing from one site to another in the African canopy. The ants and other insects associated with *Oecophylla longinoda* are very different from those associated with the species of *Crematogaster*, creating a spatial mosaic across the tree tops. Fox and Fox (1982) described a changeover through time of the dominant ants in

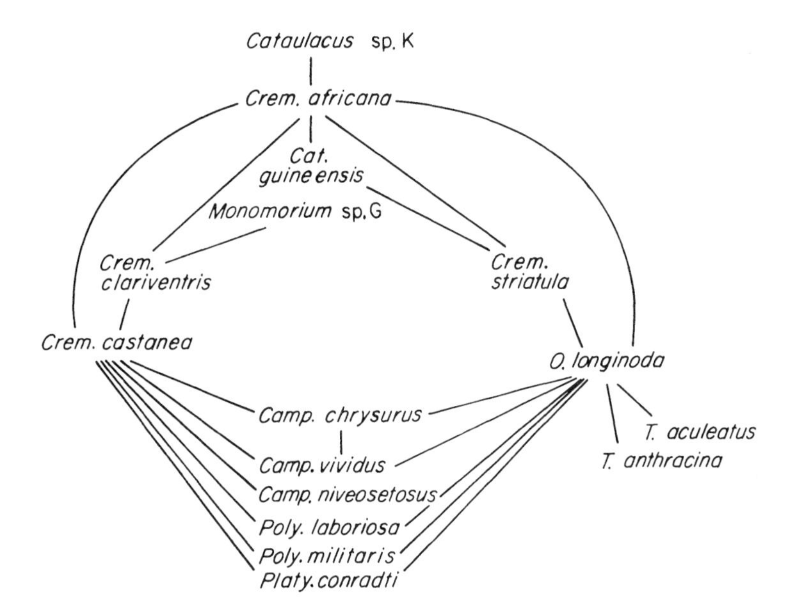


**FIGURE 11–3** Additional structure in the ant community of an African forest canopy is revealed in the negative associations at the 95-percent confidence level among species. Abbreviations of the scientific names are the same as in Figure 11–2. (Modified from Room, 1971.)

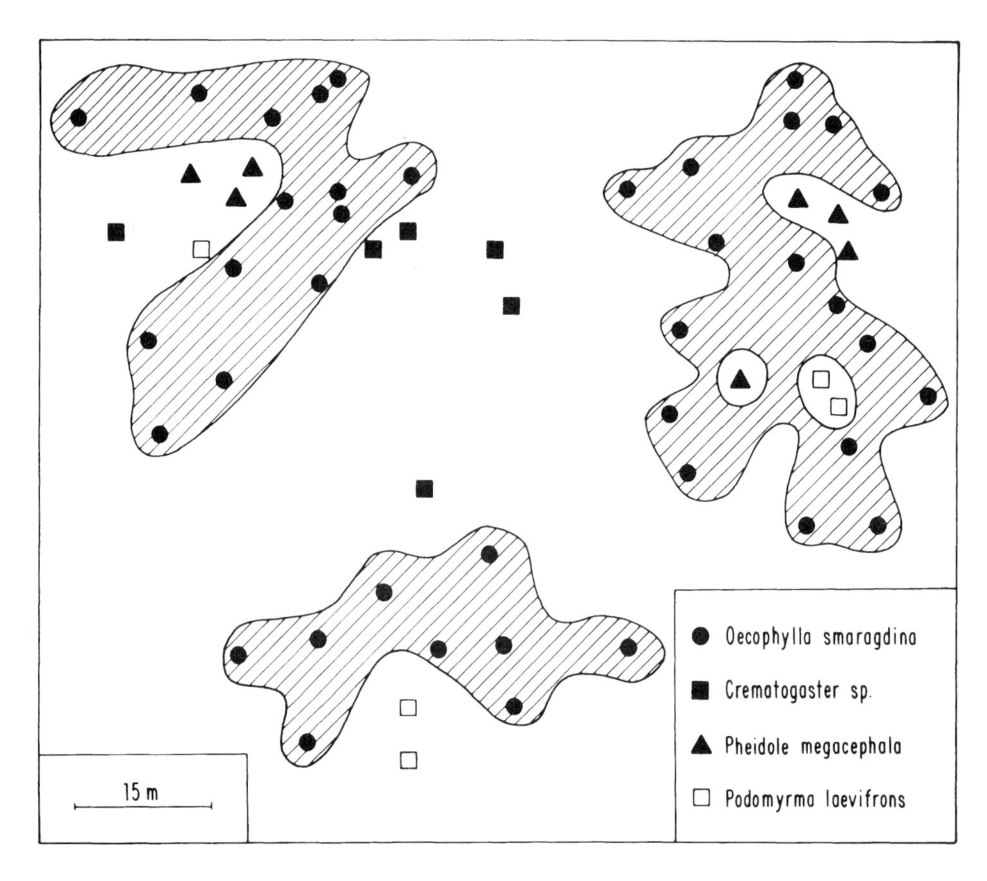


**FIGURE 11–4** An ant mosaic in Queensland, Australia. The four species represented in this map occupy mutually exclusive foraging and nesting areas in forest canopies. The territories of *Oecophylla smaragdina* are cross-hatched. (From Hölldobler, 1983.)

regenerating heath at Hawks Nest, Australia. There was a shift in only two years from a community dominated by three species of *Iridomyrmex* to one dominated by two other species of *Iridomyrmex* and *Tapinoma minutum*. Not only were the dominant species replaced (all dolichoderines), but also all the communities of ants associated with them. The Foxes observed a change in species diversity and density of individual ants across the changeover point. Similarly, major shifts in the local ant fauna were observed by Greenslade (1971) in changeovers of the dominant species *Anoplolepis longipes* and *Oecophylla smaragdina* in coconut plantations of the Solomon Islands. When *A. longipes* flourished, species diversity fell sharply; and when it declined, diversity increased.

In communities lacking such dominant species, or "large-scale conquerors" of level III importance, the hierarchical arrays of remaining species are much less predictable (Dobrzański and Dobrzańska, 1975). Some of this change is due in various localities to the harshness of the local environment and an increase in diversity of microhabitats (Gallé, 1975; Boomsma and van Loon, 1982), but the relaxation of competition and predatory pressure almost certainly also plays a role (Savolainen and Vepsäläinen, 1988). In the grasslands of southern England, the three principal species are *Tetramorium caespitum* (the overall dominant), *Lasius alienus*, and *L. niger*. *Formica fusca* is a marginal species, able to live within the territories of the *Tetramorium* but avoiding contact by building a single small entrance to its nests and foraging more widely in places less frequented by the *Tetramorium*. Farther north, in the vicinity of Strathclyde, *T. caespitum* is absent and *Formica fusca* is replaced by the related *F. lemani*. Under these circumstances *F. lemani* is a dominant of sorts, taking the favored nesting sites and preying on two species of *Myrmica* (Brian, 1983).

Community structure is clearly affected by climate, the diversity of nest sites, the diversity of food, and interspecific competition. But the relative importance of these factors, despite excellent and strongly suggestive studies conducted on several continents, remains to be definitively measured. To this end comparative quantitative analyses of different communities are a challenging task for students of ant ecology.

## THE DOMINANCE–IMPOVERISHMENT RULE

We have noticed a worldwide tendency in the relation between behavior and species diversity, as follows: *the fewer the ant species in a local community, the more likely the community is to be dominated behaviorally by one or a few species with large, aggressive colonies that maintain absolute territories.* The relation holds in the relatively species-poor canopies of Africa (Room, 1971; Leston, 1973a), Australia (Hölldobler, 1983), and the Solomon Islands (Greenslade, 1971); the boreal faunas of northern Europe (Vepsäläinen and Pisarski, 1982; Rosengren and Pamilo, 1983); the mangrove islets of the Florida Keys (Cole, 1983a); the small islands of the West Indies (Levins et al., 1973); valley riparian woodland in California (Ward, 1987); and the arid and semi-arid as opposed to mesic habitats of Australia (Greenslade, 1976; Greenslade and Halliday, 1983; Andersen, 1986a,b). It does not hold in at least some terrestrial faunas of Australia, such as those in the tropical part of the Northern Territory (A. N. Andersen, personal communication).

What is cause and what is effect in this rule? Are some ant faunas impoverished because of the suppressing effect of the large-scale conquerors, or have the large-scale conquerors originated in environments with impoverished ant faunas? At first it might seem that the first alternative is more likely to be true, because the dominant species have been repeatedly shown to reduce species diversity and abundance within their territories. When the populations of colonies of such species are dense, the effect can be widespread. Indeed, Carroll (1979) has postulated that the low biomass of stem-dwelling ants he found in Liberia in comparison with that of stem-dwellers in Costa Rica is due to the greater prevalence of dominant ant species (such as *Oecophylla*) in Liberia.

Initial intuition can be wrong in this case, however. We have concluded that the opposite is true, that impoverished faunas promote dominant species rather than the other way around. Our reasoning is as follows. If the appearance of dominant species promotes impoverished faunas (the first alternative), the loss should occur only in alpha diversity, that is, the number of species found in the particular sites where dominant species are present, but not in beta diversity, the number of species occurring across many localities with and without dominants. In other words, the faunas of whole regions and habitats in which dominants prevail should be rich even though the local, individual sites where they occur are poor. But this is not the case. In the regions where the dominants occur, such as boreal Europe and small tropical islands, the faunas as a whole are small. Furthermore, arboreal ant faunas in tropical forests are generally much less diverse than the terrestrial ant faunas that lie just below them, and they are also the ones dominated by species with large aggressive colonies and absolute territories. This appears particularly to be true in West African forests, which were hard hit during the Pleistocene dry periods (Moreau, 1966; Carroll, 1979). This difference exists even though the leaves and branches of the arboreal zone are of the same geologic age as the litter and humus of the terrestrial zone.

We suggest, therefore, that the primary causal chain is from impoverishment to dominance rather than from dominance to impoverishment. Some habitats have relatively few species because they are physically harsh, are restricted to a limited area, or are geologically young, or because of some combination of these features. As a consequence there is a relatively limited number of specialists available to preempt the narrower niches of nest sites and food items. The opportunity exists for a few other, generalist species to expand ecologically and to occupy a wide range of nest sites and food items. They will tend to evolve a large colony size and behavioral mechanisms, such as absolute territories, that vouchsafe control of the larger niches into which they have moved.

## THE CONDITIONS FOR COEXISTENCE OF SPECIES

Interference between colonies belonging to the same species has the important effect of increasing the numbers of competing species that can coexist. This result is predicted in the graphical models devised by Gause and Witt (1935) on the basis of the Lotka-Volterra competition equations. In words, the Gause-Witt theory states that, if two species interfere with one another to any extent, one will always replace the other unless the following condition is met: the population densities of the two species must be self-limiting in such a way that each will stop increasing before the other species becomes extinct. The most familiar way in which such an equilibrial coexistence can come into being is if the two species occupy sufficiently different niches. Then one will tend to reach a limit in the part of the habitat to which it is specialized and to reach maximal density before it is able to crowd out the second species in the part of the habitat optimal for the second species. This special case has become so familiar in ecological writing as to be frequently referred to as *the* Gause hypothesis, Gause's law, and so forth, but the Gause-Witt model embraces other potential mechanisms as well. Consider, for example, the possibility that the population density of each species is under the control of a parasite specialized for feeding on it. This, too, could easily lead to stable coexistence of the two prey species—as well as of the two parasitic species. It follows that any density-dependent control peculiar to a species will contribute to the stable coexistence of competing species.

The greater the difference between species in their respective niches, comprising nest sites, time of foraging, food, and so forth, the more likely it is that each species will be independent in its population controls. In this light the work of Pontin (1960, 1961, 1963) takes on a particular significance. Pontin made careful studies of the ecology of two related species of formicine ants, *Lasius flavus* and *L. niger*, with special reference to the ways in which each affects the survival and reproduction of the other. In calcareous grassland near Wytham, England, the two species are dominant, and their colonies are intermingled at saturated densities. In order to measure the consequences of interaction, Pontin first placed newly mated *Lasius* queens in tubes with openings large enough to admit workers but too small to permit the escape of the queens, and seeded them within the territories of mature colonies. He found that the queens were attacked and destroyed preferentially by workers of their own species. Pontin then made studies of the relation between the productivity of new queens and the distance between the nest of the colony and the nearest nests belonging to both species. In a related experiment colonies of *L. flavus* were transplanted to new positions in a circle around nests of *L. niger* in order to increase the competitive pressure on them. The results showed conclusively that queen productivity is reduced more by intraspecific than by interspecific interference. Therefore, through both the depression of the production rate of new queens and their destruction following the nuptial flights, each of the two species controls its own population densities to a greater extent than it does those of its competitor. This effect fulfills, at least in principle, the essential condition of the Gause-Witt equilibrium. The behavioral basis of the effect can only be guessed. Perhaps the reason why workers attack alien queens of their own species preferentially is that, as Brian (1956a) has claimed to be the case in *Myrmica*, they tend to be repulsed at a distance by the odors of both queens and workers of alien species. Such a response, which serves primarily as an adaptation to avoid injurious conflict, could not be extended to members of the same species without interfering with normal communication within the colonies.

Interference between mature colonies seems to be reduced by innate ecological differences between the two species. *Lasius niger* is a versatile ant that nests in rotting stumps, beneath stones, or in the open soil (often in mounds), and forages both below and above ground and up onto low vegetation. *L. flavus* is a primarily subterranean species that builds mounds in the open soil. Where the two species live together in the Wytham grassland, *niger* inserts itself in suitable nest sites between the *flavus* mounds. By competition for space and food (and limited predation on *flavus*, which is not reciprocated) *niger* depresses the queen production of *flavus*. Symmetrically, *flavus* takes away space and food from *niger*; it also interferes with *niger* by using stones as props for the mounds of excavated earth, thus covering them and denying them to the *niger* for use as nest covers. The degree of interference between the two species seems to be superimposed on a larger degree of interference among colonies belonging to the same species, however. The latter, intraspecific interference is not only enough to stabilize the populations of colonies, but is also sufficiently greater than interspecific interference to permit the permanent coexistence of the two species. A similar relationship appears to exist between two species of honeypot ants, *Myrmecocystus depilis* and *M. mimicus*, which have much the same ecological requirements and frequently coexist in the same habitats (Hölldobler, 1981a).

Once the nest site of an individual ant colony has been selected, and if the area of its foraging ground is fixed by the colonies in residence around it, the productivity of the colony will probably depend on the food yield of the territory. It follows that in truly territorial species the production rate of new queens in mature colonies will increase as a function of the size of the territory. The relation has been verified in the two analyses in which the proposition has been tested: that of Pontin (1961) on *Lasius* and that of Brian et al. (1966) on *Myrmica*.

Actually, the factors controlling the density of populations of social insects are probably what ecologists call intercompensatory. This means that, in a given environment at a given time, one factor is usually limiting, and, if it were removed, the population would increase until a second factor became limiting, and so on. If nest sites became unlimited in a food-rich area, the colony density would increase until food became scarce. If food were then presented in unlimited amounts, the populations would presumably increase until territorial behavior (to be sure, centered around smaller territories) stabilized the population at a new, still higher level. This simple sequence is based on only a few empirical observations, mostly those of Brian, and it is speculative when applied to social insects as a whole. It is altogether probable that other sequences and other factors are at work. It is even likely that different populations belonging to the same species equilibrate under different schedules of controls.

Against this theoretical background it is possible to make more sense of the diverse interaction phenomena of ant species, and to gain some understanding of the mechanisms that organize species communities. In briefest form, species use an amazingly wide range of procedures to push back competitors, fit among them unobtrusively, or escape into marginal environments. Those documented to date will now be examined.

## Niche Differentiation

The reduction of interference by foraging at different times of the day is well known (Baroni Urbani and Aktaç, 1981; Hölldobler, 1981a; Klotz, 1984). Yet even when this occurs a dynamic tension between the species can "fine-tune" the adjustment. In Australia *Iridomyrmex purpureus* and *Camponotus consobrinus* utilize the same food sources, and they often nest in close association (Greaves and Hughes, 1974). As Hölldobler (1986a) found, the *Iridomyrmex* forage mostly during the day and the *Camponotus* mostly at night, with the two species replacing each other at particular homopteran aggregations and other persistent food sources (Figure 11–5). Where either of the species occurs alone, its foraging period is usually longer by one to two hours. Where the *Iridomyrmex* and *Camponotus* occur together, they shorten each other's foraging period by direct interference. In the morning *I. purpureus* workers gather around the nest exits of *C. consobrinus* and close them with pebbles and clumps of soil. At dusk the situation is reversed: the *Camponotus* gather to prevent the *Iridomyrmex* from leaving the nest (Figure 11–6).

An equally striking displacement of diel schedules has been discovered in several other ant species. In Costa Rica Swain (1977) observed that workers of the dolichoderine *Monacis bispinosa* ceased tending scale insects at night when a large yellow formicine of the

**FIGURE 11–5** Diel replacement among ant species at the same food source: a nymph of *Eurymela* is tended on its eucalyptus host tree by *Iridomyrmex purpureus* during the day (*above*) and by *Camponotus consobrinus* at night (*below*). Both ant species collect honeydew excreted by homopteran insects through the anus.

genus *Camponotus* appeared. In order to recreate the displacement experimentally, he lured *Monacis* workers to a sugar bait in a new nearby site. At first the *Monacis* workers continued feeding after dusk, the normal activity pattern of the species in other localities. As soon as *Camponotus* workers found the bait, however, the *Monacis* retreated. Those who hesitated were attacked and killed, and the diel displacement was quickly established. In the Siberian steppes colonies of *Formica subpilosa* are usually most active during the mid-afternoon. In the presence of *F. pratensis,* however, they shift the peak of their activities to the evening. Stebaev and Reznikova (1972) were able to induce the change by moving nests of *F. pratensis* to the vicinity of those of *F. subpilosa*. In forests of Ghana the arboreal myrmicine *Tetramorium* (= *Macromischoides*) *aculeatus* spaces out its foraging time and becomes more nocturnal in the immediate presence of *Crematogaster clariventris* and *Oecophylla longinoda* (Leston, 1973c; Majer, 1976a).

Comparable cases of displacement leading to dynamic equilibria exist in the differentiation of space. On Hicacos Island near Puerto

**FIGURE 11–6** Aggressive display in the 24-hour cycle of Australian ants: workers of the meat ant *Iridomyrmex purpureus* close the nest entrance of the sugar ant *Camponotus consobrinus* in the morning (*above*). At dusk workers of *C. consobrinus* guard the nest exits of *I. purpureus*, preventing foragers of the competitors from leaving (*below*).

Rico, Levins et al. (1973) observed that when food baits were shaded, they were dominated by *Pheidole megacephala* workers. When the same baits were flooded by sunlight, the *Pheidole* retreated and the baits were taken over by *Brachymyrmex heeri*. When the baits were again shaded, the *Brachymyrmex* left and the *Pheidole* returned. The turnover occurred within a half hour and was accompanied by sporadic fighting.

### Density Specialization

Davidson (1977b) suggested that ant species may divide the environment by specializing on different densities of seeds and other food items. Those that forage in large numbers along trunk trails are likely to enjoy an advantage when food is denser. They are especially effective in rich patches where cooperating groups operate most efficiently. At lower food densities the energetic cost of foraging in this manner may exceed the energetic yield, allowing solitary foragers to take over. Davidson provided some empirical evidence instantiating this model of niche division. Among granivorous ants of the southwestern American desert, trunk trail foragers such as *Solenopsis xyloni* and *Messor (= Veromessor) pergandei* concentrate on high-density patches of seeds. In contrast, individual foragers like *Aphaenogaster (= Novomessor) cockerelli* and *Pheidole desertorum* harvest seeds that are mostly dispersed and hence require independent discovery. They also spend more time searching than trunk trail foragers. A mixed strategy was used by *Pogonomyrmex rugosus*, which use trunk trails during peak seed abundance and revert more to individual foraging when seeds are scarcer. In experiments trunk trail foragers were more selective in their choice of food when food was abundant than were individual foragers. In addition Hölldobler et al. (1978) demonstrated that *Aphaenogaster cockerelli* is less efficient in recruiting nestmates to seed patches than is *Pogonomyrmex rugosus*.

The interesting consequence of density specialization, to the extent it occurs in nature, is that it permits the coexistence of species that are ecologically identical in all respects except in the distribution of their food items. It is also a form of niche division that can profoundly affect the modes of communication and foraging techniques of ant species.

The concept of density specialization can be fitted at least loosely to the phenomenon of dominance hierarchies among species. In some environments the trunk trail, high-density specialists would be the same as the highest ranking competitors in the dominance orders. Such species not only utilize rich food sources but also monopolize them by means of territorial aggression (Hölldobler, 1976a). If we return briefly to the Tvärminne Archipelago, we note that *Formica polyctena* and *F. lugubris* are approximately equal dominants. Their colonies attain very large size, and the workers monopolize aphid clusters and other rich food sites as part of their absolute territories. Their large mound nests represent considerable energetic investments and are seldom abandoned. The two species are separated by a slight difference in habitat preference: *F. polyctena* favors older pine forest with undergrowth and *F. lugubris* newly grown pine forest. At the next competitive level is *F. truncorum*, whose colonies have smaller populations and occupy less food-rich and stable habitats, thus partially escaping competition from *F. polyctena* and *F. lugubris*. *F. truncorum* also builds smaller, less "expensive" nests. Close behind in the hierarchy is *F. exsecta*, which is similar in most respects to *F. truncorum* but has a smaller worker size. Below both of these species is *Camponotus ligniperda*, which defends only the food sources it happens to find. It nests in sites outside the territories of the dominant *Formica*, such as open, rocky pine forest. At the bottom of the hierarchy are *F. fusca* and the species of *Leptothorax* and *Myrmica*, characterized by much smaller colonies and marginal nest sites. The workers search individually, and most frequently in areas with low food density where none of the dominant *Formica* occur (Pisarski and Vepsäläinen, 1981; Vepsäläinen and Pisarski, 1982). In southern California a similar relation between rank, nest site, and foraging pattern has been noted in the dominance of *F. haemorrhoidalis* over *C. laevigatus* (MacKay and MacKay, 1982).

### Size Differences and Worker Polymorphism

Size differences have often been implicated in the reduction of competition among closely related animal species. Ants offer some persuasive examples. In a study of three *Pogonomyrmex* species at Portal, Arizona, Hansen (1978) found a close correlation between the size of the workers and the size of the seeds they collected. The mean wet weight of the foraging workers was 5.8 milligrams in *P. desertorum*, 8.9 milligrams in *P. maricopa*, and 15.1 milligrams in *P. rugosus*, whereas the mean weights of seeds collected by them were 1.0, 2.2, and 2.8 milligrams, respectively, with pairwise differences all significant at the 99-percent confidence level or higher. Size differences are a striking feature of the western North American species of *Pogonomyrmex* generally, as we have illustrated in Figure 11–7. It is a notable fact that the only species occurring in eastern North America, the Florida harvester *Pogonomyrmex badius*, is strongly polymorphic, with the continuous size variation of workers from a single mature colony spanning most of that covered by many species in western North America (Figure 11–8). It seems possible that *P. badius* workers also collect a wider range of seeds than those of individual species in the west. In other words, in the absence of competition *P. badius* might have evolved into a generalist. The full diets of *P. badius* and other *Pogonomyrmex* species remain to be described, however, and J. F. A. Traniello (personal communication) has found no correlation of forage load weight and worker size in *P. badius*.

A strong case for the evolutionary relation of polymorphism to competition has been made for the harvester *Messor (= Veromessor) pergandei* by Davidson (1977a,b, 1978, and as a co-author in Brown et al., 1979b). This "supreme specialist" is the most abundant ant in the least productive deserts of the American West. Its huge colonies use a mixed strategy of trunk trail and individual foraging to exploit seeds that are either clumped in patches or independently scattered. At different localities the degree of size variation within single *M. pergandei* colonies is inversely correlated with the presence or absence of competing granivorous ant species. At the eastern edge of the species range, the moderately productive Sonoran Desert of south central Arizona, *M. pergandei* coexists with up to four potential competitors that are either larger or smaller but approximately monomorphic. Here the *Messor* workers display less within-colony polymorphism, and those with intermediate size predominate. In desert habitats with still greater resource productivity and a correspondingly higher diversity of ant species, *M. pergandei* is replaced by species whose colonies possess workers of nearly uniform body sizes and forage on narrower ranges of seed densities.

Whether the correlation of body size and seed size is always real-

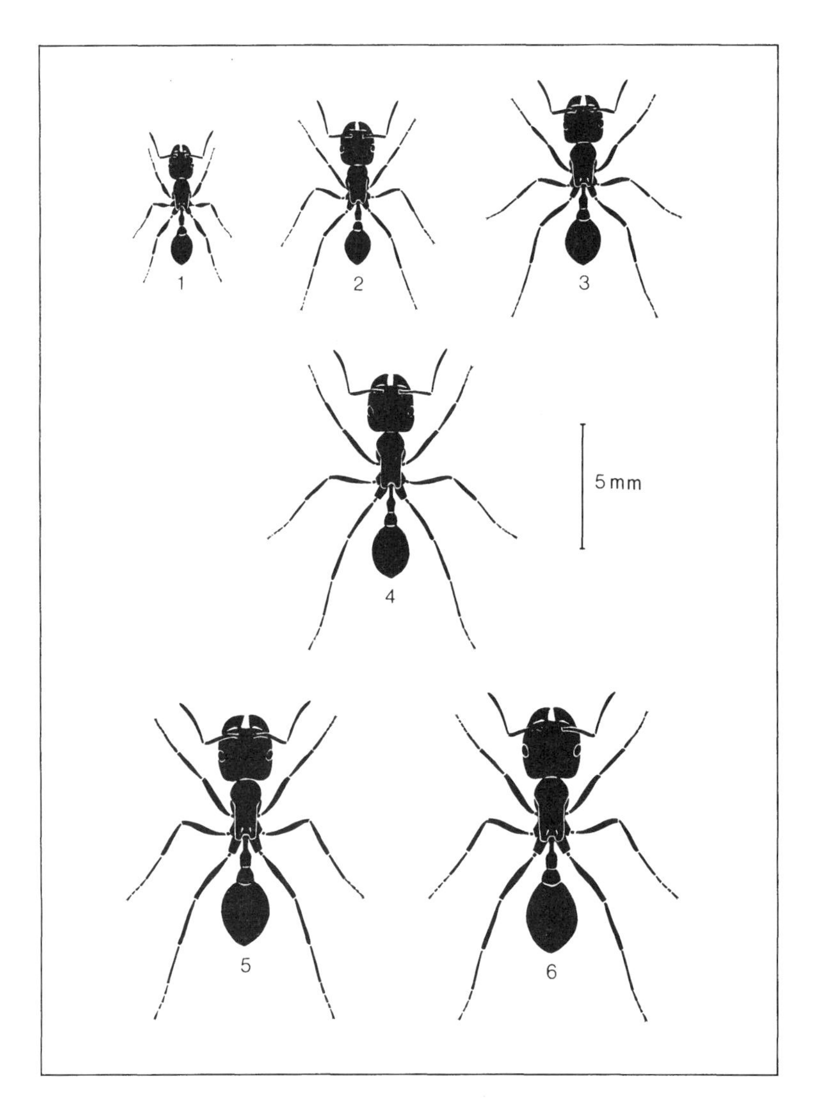


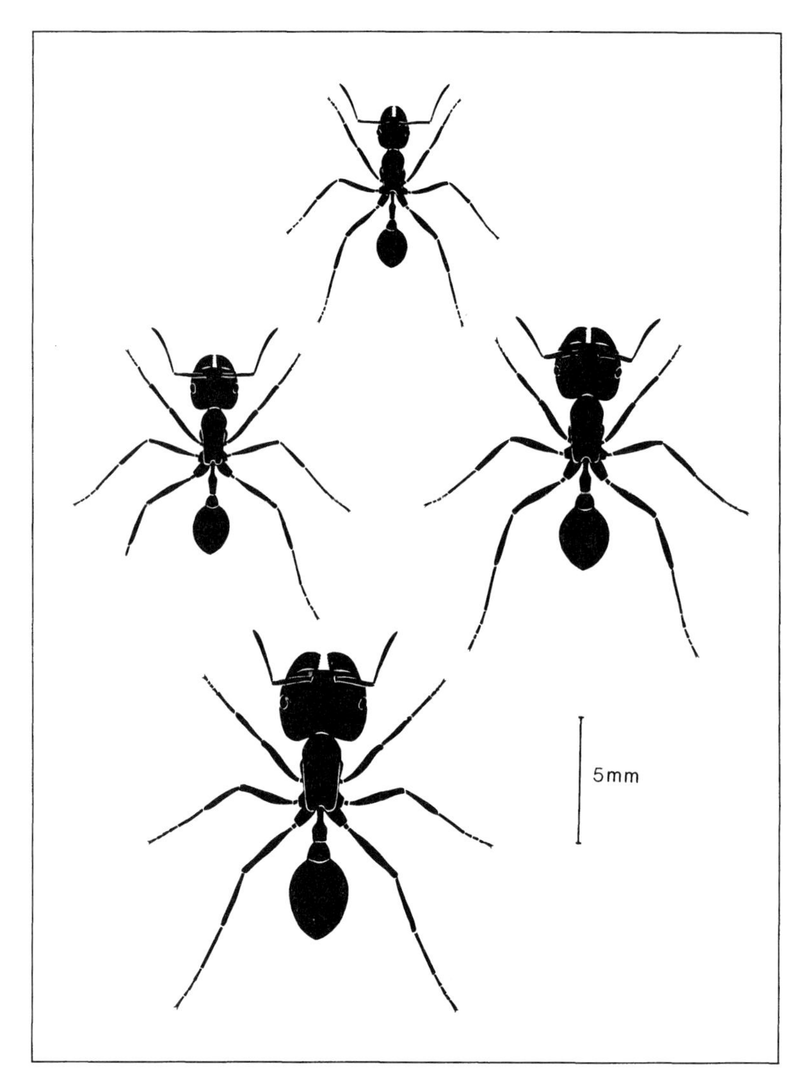


**FIGURE 11–7** *(left)* The worker caste of six species of *Pogonomyrmex,* illustrating the often strong size differences among members of this genus. In at least some cases the variation is correlated with differences in the size of seeds collected by the workers. (1) *Pogonomyrmex imberbiculus;* (2) *P. californicus;* (3) *P. desertorum;* (4) *P. maricopa;* (5) *P. barbatus;* (6) *P. rugosus.*

**FIGURE 11–8** *(right)* Size variation in the Florida harvester ant *Pogonomyrmex badius.*

ized in polymorphic harvester ants is still an open question. Davidson (1978) reported such correlations in *Messor (= Veromessor) pergandei* at four separate locations, but not at a fifth site. On the other hand Rissing and Pollock (1984) working in different habitats were unable to detect size matching in this species. The phenomenon was demonstrated in the fire ant *Solenopsis invicta* (Wilson, 1978). Rissing and Pollock (1984), however, point out that in this case the prey size taken by small workers (head width less than 0.8 mm) is much smaller than that taken by large workers. This is most likely due to physiological limitation on the carrying ability of small individuals. *Messor pergandei* workers are

> substantially larger, with most workers measuring 1.0–2.0 mm in head width. Size-matching is absent in *V. pergandei* and in *S. invicta* workers of the same size range as *V. pergandei* (Wilson, 1978). An analysis of the effect of burden on velocity of *V. pergandei* foragers (Rissing, 1982) indicates these individuals are substantially larger than necessary to carry seeds commonly found in their habitat. This suggests that physiological limitations on harvestable item size does not occur in *V. pergandei* . . . To the extent that size-matching does not occur in *V. pergandei,* worker size variability must be a weak to nonexistent factor in determining diet breadth or foraging efficiency in this species.

There is one other aspect which might in part account for the sometimes conflicting results. Rissing (1987) recently found in *M. pergandei* a "distinct annual cycle in mean worker body size that replicates across colonies and habitats; this cycle occurs through alteration of the worker size distribution." Rissing argues that the worker size variance over the year is a mechanism to maintain a constant, large worker force, in which smaller workers appear in the foraging force following periods of reduced seed availability, reduced favorable times to forage, and alate production during winter months. Since *M. pergandei* exhibits interspecific territoriality, a large and constant worker force appears of selective advantage. If it is too costly to make large workers, the colony chooses to make smaller workers rather than to reduce the size of the worker force.

## TROPHALLACTIC APPEASEMENT

One mechanism by which ant species may mutually adjust to one another is through some form of appeasement whereby dominance is recognized and halted short of fatal aggression. Interactions between competing colonies are often thought to consist entirely of threat, fighting, and avoidance. These are indeed the prevailing responses observed under natural conditions. Yet it also appears that some ant species regurgitate liquid food to adversaries during hostile encounters, and that they benefit to some extent by stopping or delaying physical attacks. Kutter (1963b, 1964) found that when he placed colonies of *Lasius fuliginosus* and various species of *Formica (exsecta, pratensis, rufa, truncorum)* in containers close together and connected them by wooden bridges, intense fighting broke out as expected. But eventually the surviving workers grew more friendly, engaging in mutual grooming and feeding. The *Lasius* workers nevertheless remained hostile to the *Formica* queens and hunted them down, so that in time the former "alliance colony" turned into a pure *Lasius* colony. Similarly, workers of *Pheidole dentata* and *P. morrisi* placed with fire ants (*Solenopsis geminata* and *S. invicta*) close together in plastic cells or larger laboratory nests appeased the more aggressive fire ants by regurgitating food to them. The same proved true of the two fire ant species when they were placed together, as well as various combinations of *Formica*. These encounters consistently lowered the frequency of overtly aggressive acts (Bhatkar and Kloft, 1977; Bhatkar, 1979a,b, 1983; Kloft, 1987).

Is trophallactic appeasement, to use Bhatkar's phrase (1979b), just an artifact of laboratory confrontations? Evidence that it occurs under more natural conditions was obtained by Bhatkar (1983), who observed the behavior of ant species attracted to sugar-water baits in a lightly wooded area of northern Florida. He found that whenever *Pheidole dentata*, *P. morrisi*, or *Solenopsis geminata* workers arrived at the baits before *S. invicta* workers, the *invicta* scouts challenged them by raising the gaster, secreting a droplet of venom, vibrating the abdomen (probably to dispense the venom as an aerosal, as described by Obin and Vander Meer, 1985), and advancing slowly with mandibles open. The food-laden workers of *Pheidole* and *S. geminata* often responded by opening their mandibles, regurgitating a droplet of the sugar water, and offering it to the attacking *S. invicta*. The *invicta* workers became less aggressive, holding their own mandibles open to receive the food, or stopping the interaction entirely to groom themselves. Similar exchanges were observed among different species visiting the same aphid associations to collect honeydew.

Thus trophallactic appeasement occurs in nature, but we have too little information to assess its extent among ants generally or its role in community organization. The exchange appears to be specialized as an adjustment to aggressive, dominant species and is most likely to occur around long-lasting food sources that attract large numbers of foragers.

## ENEMY SPECIFICATION

The major predators of ant species are frequently other ant species, including many specialized to prey on ants (see Chapters 15 and 16). It is equally true that the most serious interspecific competitors in ant communities are dominant territorial species. It is therefore not surprising to find that at least some ant species have evolved defensive maneuvers directed in a precise way to identify and confound their most dangerous adversaries.

This phenomenon of "enemy specification" was first discovered in *Pheidole dentata*, a small myrmicine abundant in woodland over most of the southern United States (Wilson, 1976b). The native fire ant *Solenopsis geminata* occurs in many of the same habitats and to some extent utilizes the same nest sites as *P. dentata*. It forms large and aggressive colonies that are strongly territorial. This is also true of the red imported fire ant *S. invicta*, which has spread throughout much of the southern United States during the past forty years. Fire ant scouts recruit masses of workers to food sites. They also treat *Pheidole dentata* as food and can destroy a colony within an hour. The *Pheidole* can avoid this fate by intercepting the scouts before the *Solenopsis* are able to mount an invasion. The *Pheidole* minor workers respond to the presence of only one or two *Solenopsis*. Within seconds, some of them start to run swiftly back and forth to the nest, dragging the tips of their abdomens over the ground. The trail pheromone thus deposited attracts both minor and major workers from the nest in the direction of the invaders. The majors are especially excited by the combination of the fire ant odor adhering to the bodies of the returning minor worker scouts and to movement while they are being contacted. They do not lay odor trails of their own. Upon arriving at the battle scene they become even more excited, rushing about and snapping at the fire ants with their powerful mandibles and chopping them to pieces (Figure 11–9). The recruited minor workers also join in the fighting, but they are less persistent and remain in the area for much shorter periods of time. As a result the majors increase in proportion, and for all but the most transient invasions they eventually outnumber the minors, despite the fact that they constitute only 8–20 percent of the worker population in most nests. The majors remain in the battle area for an hour or more after the last *Solenopsis* has been dispatched, restlessly patrolling back and forth.

In this manner the *Pheidole* are able to "blind" colonies of fire ants by destroying the scouts of these dangerous enemies. But they do not react to other potential adversaries in the same way. Ants of a wide variety of species in other ant genera tested by Wilson were neutral or required a large number of workers to induce the response. In a more recent study Carlin and Johnston (1984) demonstrated that *P. dentata* colonies can be sensitized to react more swiftly with a defensive recruitment of majors when encountering other ant genera, such as *Tetramorium*, *Crematogaster*, and *Atta*, provided they have previously been repeatedly exposed to these genera. Even then, however, they never exhibit the same immediate and trigger-like response as when exposed to *Solenopsis*. Similar reactions to *Solenopsis* have been reported in *P. militicida*, a seed-harvesting species of the southwestern United States, and *P. morrisi*, an omnivorous species of eastern U.S. woodlands (Feener, 1986, 1987a).

Enemy specification has also been discovered in African and Asian weaver ants of the genus *Oecophylla* (Hölldobler, 1979, 1983). At the Shimba Hills in Kenya only a few ant species coexist with *O. longinoda* on the same tree. Some of these, including a large species of *Polyrhachis*, are occasionally hunted and attacked by the weaver

**FIGURE 11–9** Enemy specification by the ant *Pheidole dentata* (black), in which the workers respond much more aggressively to fire and thief ants of the genus *Solenopsis* (gray) than to other kinds of ants. After contacting fire ant workers near the nest, minor workers of *Pheidole* run back and forth to the nest, dragging the tips of their abdomens over the ground to lay odor trails (depicted in upper left). The trail pheromone attracts both minor and major workers to the battle ground. The majors are especially effective in destroying the invaders, which they chop to pieces with their powerful, clipper-like mandibles. Some of the *Pheidole* are themselves crippled or killed by the venom of the fire ants. (From Wilson, 1976b, drawing by S. Landry.)

ants. Most of the time, however, the *Oecophylla* are relatively indifferent, and the ants are able to avoid capture by quick, skillful movements. Other species, such as a common *Camponotus* of the region, are never found on the same trees as the *Oecophylla*. When workers are placed on an *Oecophylla* territory, the weaver ants react by recruiting masses of defending workers. The response is as strong as that to other colonies of *Oecophylla* and is of the kind that sometimes leads to major warfare. *O. smaragdina* reacts with equal violence to *Pheidole megacephala*, *Podomyrma* sp., and *Iridomyrmex* sp., some of its principal competitors in Melanesia and Australia. On the other hand Hölldobler (1983) observed in Queensland, Australia, that where *O. smaragdina* workers ventured onto a tree occupied by *Podomyrma laevifrons*, they were immediately attacked by the *Podomyrma* workers. In fact, where only 20 *Oecophylla* workers were released on a *Podomyrma* tree, *Podomyrma* reacted with an effective defensive recruitment. Alarming scouts summoned nestmates by laying trails with poison gland secretions. Using their massive mandibles, the *Podomyrma* quickly dismembered the *Oecophylla* workers and carried the body parts into their nests. These results suggest that *Oecophylla*, and possibly *Camponotus*, *Podomyrma*, and other dominants in the arboreal ant mosaic, recruit large defensive forces only when they are confronted with the most dangerous enemies. According to Vanderplank (1960) a limited number of ant species are effective predators of *Oecophylla*, including *Pheidole megacephala*, *P. punctulata*, *Anoplolepis longinoda*, and perhaps some species of *Crematogaster*.

Enemy specification of one form or another appears to be a very wide occurrence in the ants. Species of *Camponotus* and *Aphaenogaster* (= *Novomessor*), for example, rapidly evacuate their nests when army ants of the genus *Neivamyrmex* approach (LaMon and Topoff, 1981; McDonald and Topoff, 1986). The response, so far as is known, is triggered exclusively by army ants, and it is evidently based on the recognition of chemical cues specific to *Neivamyrmex*. The approach of *Neivamyrmex* also causes *Pheidole desertorum* and *P. hyatti* to evacuate their nests (Droual and Topoff, 1981; Droual, 1983, 1984).

Enemy specification may also prove to be a key process in the organization of communities of ant species. On the one hand it provides the means whereby a vulnerable species can live alongside a dominant one, in the way *Pheidole dentata* manages to coexist with the *Solenopsis* fire ants. On the other hand it can lead to the opposite effect by excluding one species from within the territories of other species, thereby sharpening the patterns of the ant mosaics. A case of this second effect is provided by the species of *Oecophylla* and their co-dominants.

## CHARACTER DISPLACEMENT

There is yet another consequence of competitive interaction on which ecological analysis can profitably focus. We have seen evidence that for species to coexist, it is necessary that they be sufficiently different to reach their equilibrial densities before eliminating their competitors, and the usual way this occurs is through differences in critical dimensions of the "niche," namely, those parameters of habitat, nest site, diet, foraging periodicity, and other factors capable of limiting populations. Now when the ranges of two species first meet, it may be that the species are already so different that competition is negligible, and the ranges come to overlap with no difficulty. If interference is considerable, however, it will be of adaptive value for the species to diverge ecologically in the zone of overlap. In the case of social insects such "character displacement" (Brown and Wilson, 1956; Futuyma, 1986; Grant, 1986) will have a dividend measurable in increased colony survival or queen production or both.

In view of the kind of competition revealed by Pontin's analysis it is appropriate that one of the first and best-documented examples of ecologically based character displacement occurs in the genus *Lasius* (Wilson, 1955a). *L. flavus,* one of the protagonists in Pontin's study described earlier, occurs throughout the temperate portions of Europe, Asia, and North America. Over most of this range it occupies a wide array of principal habitats, including grassland and both deciduous and coniferous woodland with varying degrees of shade and drainage. In the forested portion of eastern North America, however, it encounters a very closely related species, *L. nearcticus.* There it is more restricted in habitat choice, being limited chiefly to open woodland and fields, whereas *nearcticus* occupies the darker, moister woodland habitats. Where they occur together, the two species can be distinguished by no fewer than five morphological characters, including differences in eye size, color, maxillary palp development, antennal length, and head shape. At least the first three of these reflect a greater adaptation on the part of *flavus* to a less subterranean existence. From the Great Plains of North America, where *nearcticus* is left behind, westward to the Pacific coast and beyond across Asia and Europe, *flavus* displays both a wider ecological range and a greater variation in each of the morphological characters sufficient to incorporate the *nearcticus* as well as the eastern *flavus* traits. In short, where *flavus* overlaps the range of *nearcticus,* it is displaced to more open habitats, and its populations display morphological characteristics correlated with specialization to these habitats.

A case of character displacement in actual progress has been recorded in fire ants by Wilson and Brown (1958a). The South American species *Solenopsis invicta* was introduced into the port of Mobile, Alabama, around 1918, and in the 1940s began a rapid expansion that was to extend its range over most of the southern United States by 1970. The species builds up very dense populations in open habitats, but is largely absent from woodland. *S. xyloni,* a closely related native species which also favors open environments, was mostly eliminated from its old range in the southern United States in only twenty years. *S. geminata,* on the other hand, has been only partially displaced by *invicta;* whereas previously *geminata* occurred in both open and woodland environments, now it is limited mostly to woodland where the *invicta* do not penetrate. There has been a concurrent morphological change in the *geminata* population inside the *invicta* range. Previously the open environment was occupied chiefly by a reddish form and the woodland by a dark brown form; with the advent of *invicta,* the reddish form has been mostly eliminated.

The evolutionary phenomenon of character displacement may or may not be of general occurrence in social insects. It can usually be detected only by extensive studies of geographic variation and so far has been recorded only in several ant genera, including *Odontomachus, Rhytidoponera, Pristomyrmex, Solenopsis,* and two species groups of *Lasius* (see also Taylor, 1965b). It deserves closer study because of the likelihood that interspecific competition sets constraints not only on the distribution, ecology, and morphology of particular social insect species, but on their social characteristics as well. A species restricted to pieces of dead wood or some other cramping nest site will, by interspecific competition, tend to evolve a smaller mature colony size and lower its production rate of sexual forms. If the reduction is great enough, it may abandon odor trails as a form of communication.

There is another dimension to the displacement phenomenon, revealed in biogeographic analyses of entire faunistic regions. Wilson (1959a, 1961) showed that ant species invade and subsequently evolve to endemicity within New Guinea and other parts of Melanesia through a *taxon cycle* of range expansion and contraction. Most of the Melanesian fauna has been derived ultimately from Asiatic stocks entering by way of New Guinea; some invading species are able to spread beyond this great island to Queensland, to the Solomon Islands, and to other parts of outer Melanesia. A smaller part of the fauna has been derived from old Australian stocks that have entered by way of New Guinea or New Caledonia. Faunal flow from New Guinea through outer Melanesia has been largely unidirectional, with an ever-diminishing number of species groups found outward from the Bismarcks to the Fiji Islands (see Figure 11–10).

From the totality of these distribution patterns, Wilson inferred the cyclical pattern of expansion depicted in Figure 11–11. Following the invasion of Melanesia (Stage I, primary), the pioneer populations may then diverge to species level (Stage II) and further diversify. Eventually the source populations outside Melanesia contract, leaving the species group as a whole peripheral and Melanesian-centered (Stage III). Endemic Melanesian species occasionally enter upon a secondary phase of expansion (Stage I, secondary) but are rarely if ever able to push beyond Australia or the Philippines.

The expanding, Stage I species are characterized on New Guinea by their greater concentration in "marginal" habitats, which have the lowest species diversity. These habitats include open lowland forest, savanna, and seashore. They are evidently the most favorable beachheads for invasion, as well as launching areas for further range expansion. Stage I species are also characterized by their individual occurrence in a greater range of major habitats. These species also make up a significantly higher proportion of the faunas of the archipelagos of central Melanesia, including the Bismarck Archipelago, Solomon Islands, and the New Hebrides. Stage II and Stage III species, in contrast, are concentrated in the "central," high-diversity habitats of the lowland and montane rain forests.

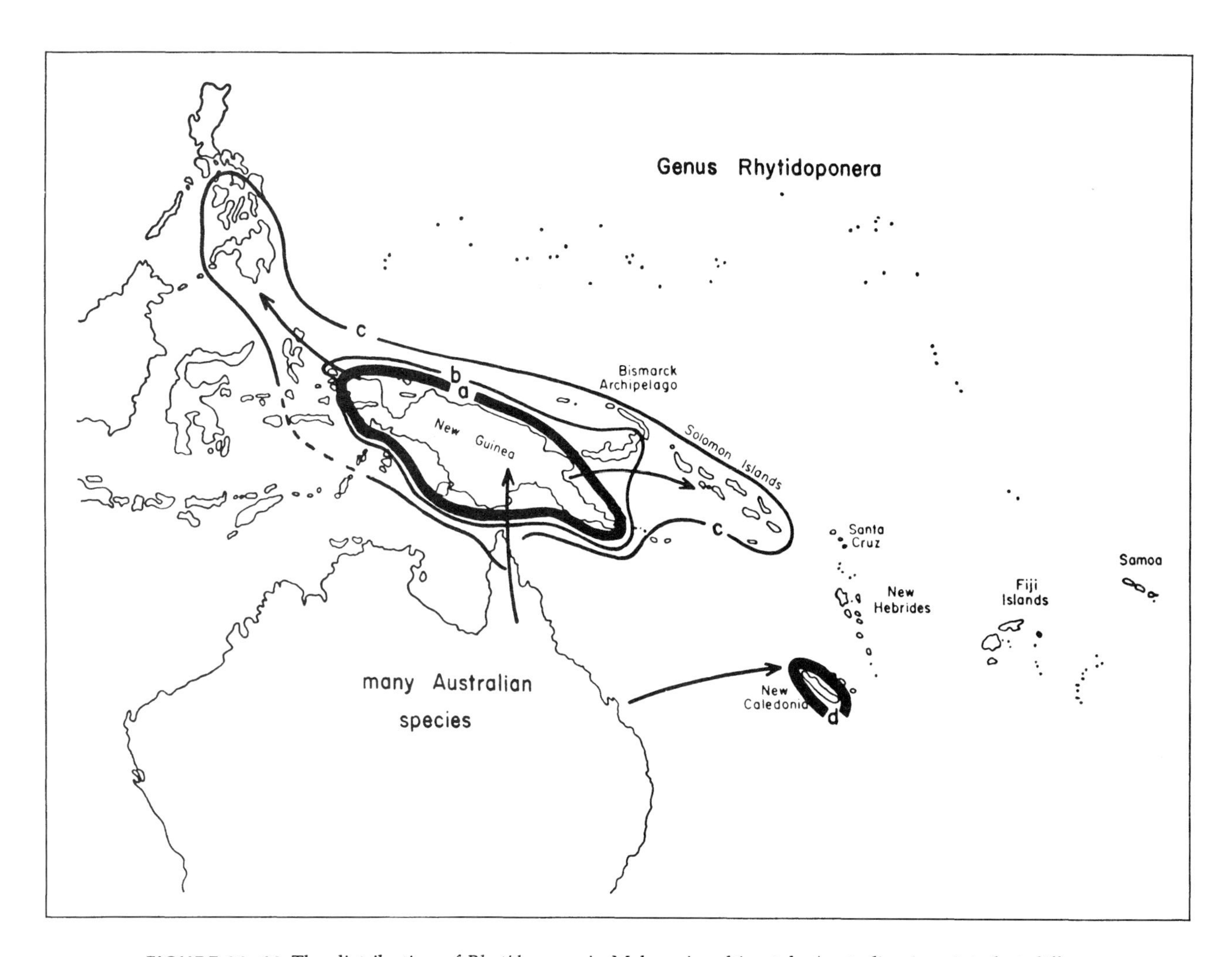


**FIGURE 11–10** The distribution of *Rhytidoponera* in Melanesia, ultimately Australian in origin but differentiating and expanding secondarily out of New Guinea and New Caledonia. The thick lines around the latter two islands represent large concentrations of endemic species. (From Wilson, 1959a.)

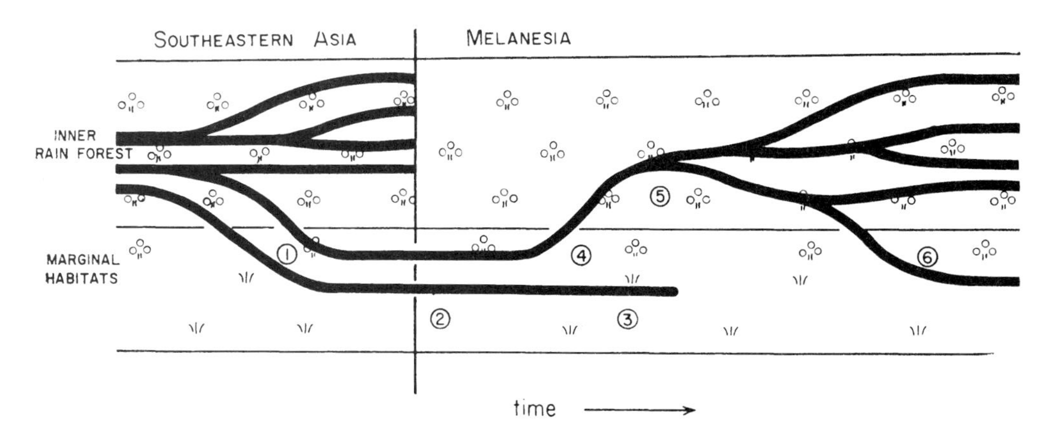


**FIGURE 11–11** The inferred taxon cycle of ant species groups in New Guinea and other islands of Melanesia that are ultimately Asian in origin. *(1)* Species or infraspecies populations adapt to marginal habitats in southeastern Asia, then cross the water gaps to New Guinea and colonize marginal habitats there *(2)*. In time these colonizing populations either become extinct *(3)* or invade the inner rain forests of New Guinea and surrounding islands *(4)*. If they succeed in adapting to the inner rain forests, they eventually diverge to species level *(5)*. As diversification progresses in Melanesia, the species in the group remaining in Asia may contract, so that in time the group as a whole becomes Melanesia-centered. A few of the Melanesian species, especially those in New Guinea, may readapt to the marginal habitats *(6)* and expand secondarily. (From Wilson, 1959a.)

It appears that ant species invade New Guinea by way of the marginal habitats. Evolutionary opportunity is nevertheless limited in the marginal habitats, and there is strong selection pressure favoring re-entry into the inner forest habitats. In general Stages II and III, leading to the origin of the great bulk of the Melanesian fauna and its most distinctive elements, are played out primarily in the inner rain forest. Stage I species are characterized on the average by larger colony size and the use of trunk trails (Wilson, 1961). These, it will be recalled, are also traits of the dominant species of the boreal forests and other species-poor habitats in other parts of the world. Passage through the taxon cycle appears to entail changes in biological traits consistent with the dominance-impoverishment principle described earlier. In other words species with large colonies dominating parts of their foraging range do best in species-poor environments.

## ECOLOGICAL EXPANSION AND CONTRACTION

The entry of a species into a smaller fauna is often accompanied by ecological release. On New Guinea some of the most widespread of the Indo-Australian ant species, notably *Rhytidoponera araneoides, Odontomachus simillimus, Pheidole oceanica, P. sexspinosa, P. umbonata, Iridomyrmex cordatus,* and *Oecophylla smaragdina,* are mostly or entirely limited to species-poor marginal habitats such as grassland and gallery forest. But in the Solomon Islands, which have a smaller native fauna, these same species also penetrate the rain forests, where they are among the most abundant species. In Vanuatu (New Hebrides), which has a truly impoverished ant fauna, the species of *Odontomachus* and *Pheidole* just listed almost wholly dominate the rain forests as well as the marginal habitats. Ecological release in the opposite direction, from central to marginal habitats, has also occurred. In Queensland and New Guinea, *Turneria* is a genus of rare species mostly confined to rain forests. It is also the only genus of the subfamily Dolichoderinae to have reached the northern islands of Vanuatu. On Espiritu Santo, for example, two species of *Turneria* are among the most abundant arboreal insects in both marginal habitats and virgin rain forest.

The degree of compression or release in new environments varies among species and is difficult to predict in advance. A case in point is the marked difference in behavior between 2 of the 13 ant species that have succeeded in colonizing the Dry Tortugas, the outermost of the Florida Keys. In the presence of such a sparse fauna, *Paratrechina longicornis* has undergone extreme expansion. In most other parts of its range it nests primarily under and in sheltering objects on the ground in open environments. On the Dry Tortugas it is an overwhelmingly abundant ant and has taken over nest sites that are normally occupied by other species in the rest of southern Florida: tree boles, usually occupied by species of *Camponotus* and *Crematogaster,* which are absent from the Dry Tortugas; and open soil, normally occupied by the crater nests of *Conomyrma* and *Iridomyrmex,* which genera are also absent from the Dry Tortugas. In striking contrast is the behavior of *Pseudomyrmex elongatus.* This ant is one of 10 species that commonly nest in hollow twigs of red mangrove in southern Florida. It tends to occupy the thinnest twigs near the top of the canopy and is only moderately abundant. *Pseudomyrmex elongatus* is also the only member of the arboreal assemblage that has colonized the Dry Tortugas, where it has a red mangrove swamp on Bush Key virtually all to itself. Yet it is still limited primarily to thinner twigs in the canopy and, unlike *Paratrechina longicornis,* has not increased perceptibly in abundance (MacArthur and Wilson, 1967).

## SPECIES PACKING AND EQUILIBRIUM

It is generally thought that the numbers of species inhabiting islands and other more or less closed habitats are at equilibrium, with the rate of species extinction approaching the rate of immigration of new species. The result is the more or less regular area-species curve, in which according to taxonomic group and location the number of species on individual islands increases as the fifth to third root of the land area. As a rule of thumb, the number of species doubles with every tenfold increase in area. This relation has been documented in ant faunas of different parts of the world by Wilson (1961), Baroni Urbani (1971c), Goldstein (1975), Pisarski et al. (1982), Vepsäläinen and Pisarski (1982), Ranta et al. (1983), and Boomsma et al. (1987). There exist universal qualities of biology at the species level, however, which dictate that equilibria must always be quasi-equilibria. If we observe a well-established fauna for short periods of time, we will note that the species extinction rate at least approximately equals the species immigration rate; but if we continue to watch it for a long period of time we will probably note a steady and very slow drift in the average species number, most likely in an upward direction. Wilson (1969) postulated the four stages of quasi-equilibria described below.

1. *Non-interactive equilibrium.* This approximate balance is reached in certain cases prior to sufficiently high population densities, which make competitive exclusion and other forms of species interaction major factors in species survivorship. The condition and the following stage were documented by Simberloff and Wilson (1969) in ants and other arthropods colonizing small islands in the Florida Keys.

2. *Interactive species equilibrium.* When the populations of individual species become dense enough to make competitive exclusion and other forms of species interaction major factors in species survivorship, or at least significantly more important factors than when densities are very low, the equilibrium can be said to be interactive. There is no way to predict from theory the direction or velocity of change in attaining this stage. In the Florida Keys it occurred within several months and entailed changes of less than 20 percent in species numbers.

3. *Assortative species equilibrium.* After interactive equilibria were reached on the islands observed by Simberloff and Wilson (1969) in the Florida Keys, the composition of arthropod species continued to change rapidly. Thus new combinations of species were being generated. Inevitably, combinations of longer-lived species must accumulate by this process on island systems in general. Such species persist longer either because they are better adapted to the peculiar conditions of the local environment or else because they are able to coexist longer with the particular set of species among which they find themselves. As a rule, we would expect such assortative equilibria to consist of a greater number of species than their antecedent interactive equilibria.

4. *Evolutionary species equilibrium.* If the community persists for a sufficiently long period of time, its member species can be expected

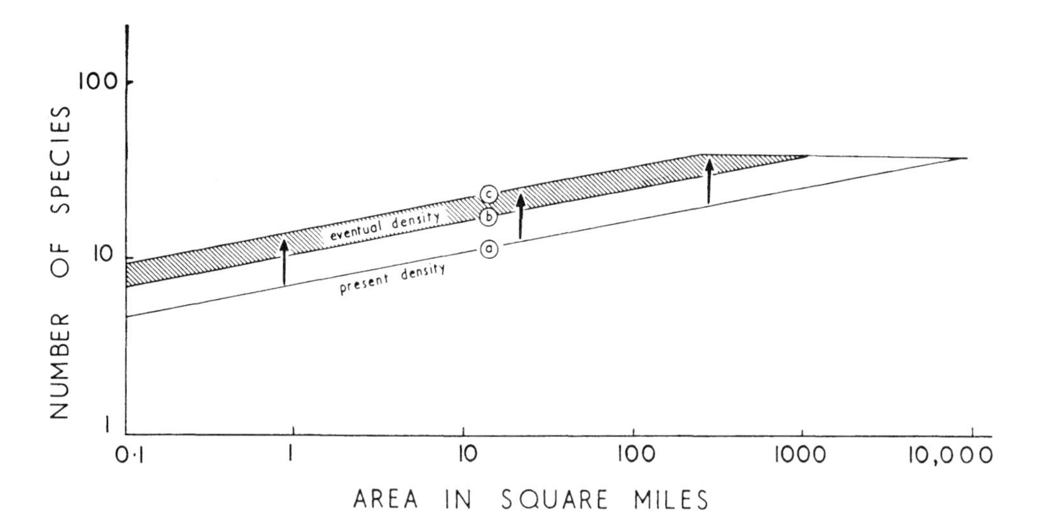


**FIGURE 11–12** The predicted increase in species numbers in Polynesia during passage from the present interactive or assortative equilibria to the future evolutionary equilibrium. It is assumed that the pool of species being introduced by human commerce remains about 38. (*a*) The present area-species curve based on the faunas of Upolu, Samoa, and Fakaofo of the Tokelau group. (*b*) and (*c*) The curves derived from older, native faunas suggest that an evolutionary increase in species numbers of 1.5× to 2× will eventually be attained. (From Wilson and Taylor, 1967a.)

to adapt genetically to local environmental conditions and to each other. The result should be the lowering of the extinction rate and consequently an increase in species diversity.

The existence of evolutionary species equilibria implies a tighter packing of species and a greater compatibility based on such mechanisms as enemy specification, species dominance hierarchies, and niche division leading to separate density-dependent controls. Although the process cannot yet be treated in the context of quantitative theory, some idea of the amount of change in species numbers can be gained by comparing newly assembled biotas with older ones containing at least some endemic species.

Wilson and Taylor (1967a) took this approach in their analysis of the Polynesian ant faunas. There are no native ant species in the far-flung archipelagos east of Tonga. The islands are inhabited instead by "tramp" species carried inadvertently by man from many different parts of the world. Most were introduced by shipping during the past two hundred years. The total number of such species known to occur in the Pacific area is 38. Remarkably, this synthetic fauna behaves as though it is equilibrial on the separate islands, because it follows a fairly regular area-species curve. In particular, the number of species found on individual islands increases approximately as the cube root of the area. Islands the size of Upolu, in the Samoan group, contain 15 to 25 tramp species. The islands of extreme western Polynesia also possess old, partly endemic ant faunas. Altogether, 43 undoubted native species occur in this area, of which 35 are found on Upolu. Thus the pools of tramp and native species are similar in size (38 versus 43 species), but the equilibrial number of tramp species on Upolu (22) and islands of similar size is much lower than the equilibrial number of native species occurring on Upolu (35). Wilson and Taylor concluded that the transition from the early interactive or assortative equilibria, roughly duplicated at the present time by the synthetic tramp faunas, to the evolutionary equilibria of the native faunas resulted in an increase in the number of species of 1.5 to 2 times. And since the area-species relation is a simple power function, the same result should apply to other Polynesian islands, as depicted in Figure 11–12.

## THE INFLUENCE OF PARASITES AND PREDATORS

Students of ant ecology have recently begun to document a role of parasitism and predation in the structure of species communities. In essence, differential mortality caused by parasites and predators can put some ant species at a disadvantage with reference to others, altering not only their abundance but also their foraging behavior as the workers attempt to meet the threat.

Feener (1981) found phorid flies in Texas of the genus *Apocephalus* that attack soldiers of *Pheidole dentata* when they come out of the nests to defend food baits against *Solenopsis texana*. In an attempt to lay their eggs, the flies stampede the *Pheidole* majors, which retreat back to the nests, leaving the minor workers to confront the *Solenopsis*. Neither the *Pheidole* minor workers nor the *Solenopsis* are parasitized by the flies. Under these circumstances the balance is tipped in favor of the *Solenopsis*, which dominate the baits. When the flies are absent, in the early spring and late fall, the *Pheidole* soldiers are able to operate in full strength, and the *Pheidole* largely replace the *Solenopsis* at the food baits. It is not difficult to see how the presence of just one effective parasite or predator can contribute to the coexistence of two otherwise incompatible ant species.

Such complex interactions among hosts, parasites, and competitors may be more common than we had previously realized. Feener (1981) has observed parasitic phorid flies in all warm temperate woodlands in which he has searched. Workers of *Camponotus pennsylvanicus*, one of the most abundant ants of the southern and eastern United States, were seen to abandon food baits on the approach of the phorid *Apocephalus pergandei*. These baits were then quickly occupied by other ants, including *Crematogaster punctulata*, *Pheidole dentata*, and *Solenopsis geminata*. Ants may go so far as to change their foraging times in response to phorids. *Pheidole titanis* is a specialized predator on termites in the southwestern United States and western Mexico. During the dry season it conducts its raids during the daytime. In the wet season, under pressure from *Apocephalus* flies that attack both soldiers and minor workers, *P. titanis* raids only at night, when the flies are inactive. As a result fewer termites are harvested during the wet season, a differential that presumably increases opportunities for other ant species feeding on termites (D. H. Feener, personal communication).

Vertebrate predators can similarly exercise mediating effects in community organization. In the Mojave Desert the horned lizard *Phrynosoma platyrhinos* preys on the individually foraging harvester ant *Pogonomyrmex californicus* 10 to 100 times more frequently than on the trunk trail harvesters *Pogonomyrmex rugosus* and *Messor pergandei*. The trunk trail foragers are able to mob the lizards in large numbers, causing the predators to retreat. *Pogonomyrmex californicus* appears to use other devices to avoid the lizards, including the construction of small, inconspicuous nests, the storage of refuse in underground chambers, and the cessation of aboveground foraging when lizards appear (Rissing, 1981b).

## THE FUTURE OF COMMUNITY STUDIES

Although still in an early stage, the study of the behavioral ecology of ants has revealed a startling variety of social mechanisms that appear to adjust species to one another and hence to organize local communities. We are persuaded that the mechanisms we know of today represent only a fraction of the processes that actually exist. We also believe that the bottom-to-top approach is the best way to understand communities. In other words it is best to start with the identification of the processes in individual species and proceed to a simulated synthesis of the community of species within the context of the entire local fauna and flora. Ants, with their great abundance and easily observed social behavior, are superb organisms for the study of community ecology.

# CHAPTER 12

# Symbioses among Ant Species

The ant society is a decidedly more open system than is a lower unit of biological organization such as the organism or cell. In the course of evolution the tenuous lines of communication among members of ant colonies have been repeatedly opened and extended to incorporate alien species. Many kinds of ants, for example, adopt aphids, mealybugs, and other homopterans as cattle to provide a steady source of honeydew; a few raid colonies of other species to acquire workers as domestic slaves, or utilize the odor trails of other species, or defend common nest sites. Just as frequently the lines of communication have been tapped by other alien species that have insinuated themselves into the colony as inconspicuous social parasites. Taken together, the hundreds of cases of interspecific symbioses among ant species that have come to light encompass almost every conceivable mode of commensalism and parasitism. True cooperation, however, is rare or nonexistent. No verified examples are yet known of mutualism, in which two species cooperate to the benefit of both. All of the relationships carefully analyzed to date are unilateral, with one species profiting and the other species either remaining unaffected or, in the great majority of cases, suffering from the attentions of its partner.

## THE ULTIMATE SOCIAL PARASITE

There is no better way to begin a survey of the social symbioses than by considering the most extreme example known, that of the "ultimate" parasitic ant *Teleutomyrmex schneideri*. This remarkable species was discovered by Heinrich Kutter (1950a) at Saas-Fee, in an isolated valley of the Swiss Alps near Zermatt. Its behavior has been studied by Stumper (1950) and Kutter (1969), its neuroanatomy by Brun (1952), and its general anatomy and histology by Gösswald (1953). A second population has been reported from near Briançon in the French Alps by Collingwood (1956), a third in the French Pyrenees by Buschinger (1987c), and still others in the Spanish Sierra Nevada by Tinaut Ranera (1981). Appropriately, the name *Teleutomyrmex* means "final ant."

The populations of *Teleutomyrmex schneideri*, like those of most workerless parasitic ant species (Wilson, 1963), are small and isolated. The Swiss population appears to be limited to the eastern slope of the Saas Valley, in juniper-*Arctostaphylos* woodland ranging 1,800–2,300 meters in elevation. The ground is covered by thick leaf litter and sprinkled with rocks of various sizes, providing an ideal environment for ants. The ant fauna is of a typically boreal European complexion, comprising the following free-living species listed in the order of their abundance (Stumper, 1950): *Formica fusca, F. lugubris, Tetramorium caespitum, Leptothorax acervorum, L. tuberum, Camponotus ligniperda, Myrmica lobicornis, Myrmica sulcinodis, C. herculeanus, F. sanguinea, F. rufibarbis, F. pressilabris,* and *Manica rubida*. For some unexplained reason this little assemblage is extremely prone to social parasitism. *F. sanguinea* is a facultative slave-making species, preying on the other species of *Formica*. *Doronomyrmex pacis*, a workerless parasite living with *L. acervorum*, was discovered by Kutter as a genus new to science in the Saas-Fee forest in 1945. In addition Kutter and Stumper found *Epimyrma stumperi* in nests of *L. tuberum*, as well as two parasitic *Leptothorax*, *goesswaldi* and *kutteri*, in nests of *L. acervorum* (Kutter, 1969).

*Teleutomyrmex schneideri* is a parasite of *Tetramorium caespitum* and *Tetramorium impurum*. Like so many other social parasites, it is phylogenetically closer to its host than to any of the other members of the ant fauna to which it belongs. In fact, it may have been derived directly from a temporarily free-living offshoot of this species, since *Tetramorium caespitum* and *Tetramorium impurum* (the host species at Briançon and in the Pyrenees) are the only non-parasitic tetramoriines known to exist at the present time through most of central Europe. It is difficult to conceive of a stage of social parasitism more advanced than that actually reached by *Teleutomyrmex schneideri*. The species occurs only in the nests of its hosts. It lacks a worker caste, and the queens contribute in no visibly productive way to the economy of the host colonies. The queens are tiny compared with most ants, especially other tetramoriines; they average only about 2.5 millimeters in total length. They are unique among all known social insects in being ectoparasitic. In other words they spend much of their time riding on the backs of their hosts (Figure 12–1). The *Teleutomyrmex* queens display several striking morphological features that are correlated with this peculiar habit. The ventral surface of the gaster (the large terminal part of the body) is strongly concave, permitting the parasites to press their body close to that of the host. The tarsal claws and arolia are unusually large, permitting the parasites to secure a strong grip on the smooth chitinous body surface of the hosts. The queens have a marked tendency to grasp objects. Given a choice, they will position themselves on the top of the body of the host queen, either on the thorax or the abdomen. Deprived of the nest queen, they will then seize a virgin *Tetramorium* queen, or a worker, or a pupa, or even a dead queen or worker. Stumper observed a case in which six to eight *Teleutomyrmex* queens simultaneously grasped one *Tetramorium* queen, immobilizing her.

The mode of feeding of the *Teleutomyrmex* is not known with certainty. The adults are evidently either fed by the host workers

**FIGURE 12–1** The extreme social parasite *Teleutomyrmex schneideri* with its host *Tetramorium caespitum*. The two *Teleutomyrmex* queens sitting on the thorax of the host queen have not yet undergone ovarian development, and their abdomens are consequently flat and unexpanded. One still bears her wings and is, therefore, almost certainly a virgin. The third *Teleutomyrmex* queen, who rides on the abdomen of the host queen, has an abdomen swollen with hyperdeveloped ovarioles. A host worker stands in the foreground. (Drawing based on a painting by W. Linsenmaier; published in Wilson, 1971, by courtesy of R. Stumper.)

through direct regurgitation or else permitted to share in the liquid regurgitated to the host queen. In any case they are almost completely inactive most of the time. The *Teleutomyrmex* adults, especially the older queens, are highly attractive to the host workers, who lick them frequently. According to Gösswald (1953), large numbers of unicellular glands are located just under the cuticle of the thorax, pedicel, and abdomen of the queens; these are associated with glandular hairs and are believed to be the source of a special attractant for the host workers. The abdomens of older *Teleutomyrmex* queens become swollen with fat body and ovarioles, as Figure 12–1 shows. This physogastry is made possible by the fact that the intersegmental membranes are thicker and more sclerotized than is usually the case in ant queens and can therefore be stretched more. Also, the abdominal sclerites themselves are widely overlapping in the virgin queen, so that the abdomen can be distended to an unusual degree before the sclerites are pulled apart. The ovarioles increase enormously in length, discard their initial orientation, and infiltrate the entire abdomen and even the postpetiolar cavity.

From one to several physogastric queens are found in each parasitized nest, usually riding on the back of the host queen. Each lays an average of one egg every thirty seconds. The infested *Tetramorium* colonies are typically smaller than uninfested ones, but they still contain up to several thousand workers. The *Tetramorium* queens also lay eggs, and these are capable of developing into either workers or sexual forms (A. Buschinger, personal communication). Consequently the brood of a parasitized colony consists typically of eggs, larvae, and pupae of *Teleutomyrmex* queens and males mixed with those of *Tetramorium* workers.

The bodies of the *Teleutomyrmex* queens bear the mark of extensive morphological degeneration correlated with their loss of social functions. The labial and postpharyngeal glands are reduced, and the maxillary and metapleural glands are completely absent. The mandibular glands, on the other hand, are apparently normal. In addition the queens possess a tibial gland, the function of which is unknown. The integument is thin and less pigmented and sculptured in comparison with that of *Tetramorium;* as a result of these reductions the queens are shiny brown, a contrast to the opaque blackish brown of their hosts. The sting and poison apparatus are reduced; the mandibles are so degenerate that the parasites are probably unable to secure food on their own; the tibial-tarsal cleaning apparatus is underdeveloped; and, of even greater interest, the brain is reduced in size with visible degeneration in the associative centers. In the central nerve cord, ganglia 9–13 are fused into a single piece. The males are also degenerate. Their bodies, like those of the males of a few other extreme social parasites, are "pupoid," meaning that the cuticle is thin and depigmented, actually grayish in color; the petiole and postpetiole are thick and provided with broad articulating surfaces; and the abdomen is soft and deflected downward at the tip.

In its essentials the life cycle of *Teleutomyrmex schneideri* resembles that of other known extreme ant parasites. Mating takes place within the host nest. The fecundated queens then either shed their wings and join the small force of egg-layers within the home nest or else fly out in search of new *Tetramorium* nests to infest. Stumper found that the queens could be transferred readily from one *Tetramorium* colony to another, provided the recipient colony originated from the Saas-Fee. *Tetramorium* colonies from Luxembourg were hostile to the little parasites, however. Less surprisingly, ant species from the Saas-Fee other than *Tetramorium caespitum* always rejected the *Teleutomyrmex*. A. Buschinger (personal communication) has pointed out, however, that the Saas-Fee population could be *caespitum* or *impurum*, or a mixture of both. In other words the transfer may have been attempted across species.

## THE KINDS OF SOCIAL PARASITISM IN ANTS

Social parasitism in ants is complicated, and its study has become virtually a little discipline of entomology in itself. The source of the complexity is first the large number of ant species that have entered into some form of parasitic relationship with one another. Second, at least two and possibly three major evolutionary routes lead to the ultimate stage of permanent, workerless parasitism. Finally, no two species are exactly alike in the details of their parasitic adaptation. Table 12–1 contains a list of the known parasitic ants, together with certain essential data concerning each of them. With this informa-

**TABLE 12–1** The known parasitic ants, their hosts, their distribution, and their form of parasitism.

| Parasite | Host | Nearest related free-living genus | Form of parasitism | Range | Authority |
|---|---|---|---|---|---|
| **SUBFAMILY MYRMECIINAE** | | | | | |
| *Myrmecia inquilina* | *Myrmecia nigriceps, M. vindex*; queen survives | *Myrmecia* | Inquilinism; workerless | Australia | Douglas and Brown (1959), Haskins and Haskins (1964) |
| **SUBFAMILY PSEUDOMYRMECINAE** | | | | | |
| *Tetraponera ledouxi* | *Tetraponera anthracina* | *Tetraponera* | Temporary | West Africa | Terron (1969) |
| *Pseudomyrmex leptosus* | *Pseudomyrmex ejectus* | *Pseudomyrmex (P. pallidus)* | Inquilinism; workerless | Florida | Ward (1985) |
| **SUBFAMILY MYRMICINAE** | | | | | |
| *Anergates atratulus* | *Tetramorium caespitum*; queen eliminated | *Tetramorium* | Extreme inquilinism; workerless | Europe, introduced into United States | Wheeler (1910a), Creighton (1950), Buschinger (1974b), Bolton (1976) |
| *Antichthonidris bidentatus* | *Antichthonidris bidenticulatus* | *Antichthonidris* | Possible dulosis; mixed colonies, with *bidentatus* predominating; relationship uncertain | Argentina | Ettershank (1966), Snelling (1975), Bolton (1987) |
| *Aphaenogaster tennesseensis* | *Aphaenogaster fulva* | *Aphaenogaster* | Temporary | Eastern United States | Creighton (1950) |
| *Cardiocondyla zoserka* | *Cardiocondyla ? schuckardi* | *Cardiocondyla* | Suspected inquilinism with workerless queen | Nigeria | Bolton (1982) |
| [*Cardiocondyla* (= *Xenometra*) *gallica* and *C.* (= *X.*) *monilicornis*, considered by Wheeler (1910a) and Bernard (1968) to be workerless parasites of *Cardiocondyla*, are actually ergatoid males of *Cardiocondyla elegans* and *C. emeryi* respectively; see Baroni Urbani (1973) and Kugler (1983).] | | | | | |
| *Chalepoxenus brunneus* | *Leptothorax* (= *Myrafant*) sp. | *Leptothorax* (= *Myrafant*) | Inquilinism; workerless | North Africa | A. Buschinger (personal communication) |
| *C. insubricus, kutteri, mullerianus* (= *gribodoi*), *siciliensis, tramieri* | *Leptothorax* (= *Myrafant*) spp.; queen stung to death | *Leptothorax* (= *Myrafant*) | Dulosis; workers present | Southern Europe, Yugoslavia, Sicily | Kutter (1950b), Bernard (1968), Le Masne (1970b,c), Ehrhardt (1982), Cagniant (1983), Buschinger (1987a), Buschinger et al. (1988) |
| *Crematogaster atitlanica* | *Crematogaster sumichrasti* | *Crematogaster* | Inquilinism; workerless | Guatemala | Wheeler (1936a) |
| *C. creightoni, kennedyi* | *Crematogaster lineolata*; queen survives | *Crematogaster* | Inquilinism; workerless | Eastern United States | Wheeler (1933a), Creighton (1950) |
| *Doronomyrmex goesswaldi, kutteri, pacis* | *Leptothorax acervorum*; queen survives with *kutteri*, but *goesswaldi* kills her by cutting off appendages | *Leptothorax* | Inquilinism; workerless | Switzerland, Austria, northern Italy, southern France, Sweden, and Estonia (USSR) | Kutter (1945, 1969), A. Buschinger (1981 and personal communication), Buschinger et al. (1981), Allies (1984), Allies et al. (1986) |

*continued*

**TABLE 12–1** *(continued)*

| Parasite | Host | Nearest related free-living genus | Form of parasitism | Range | Authority |
|---|---|---|---|---|---|
| *D. pocahontas* | *Leptothorax muscorum*; queen may not survive | *Leptothorax* | Inquilinism?; workers produced in laboratory colonies | Western Canada | A. Buschinger (1979b, personal communication) |
| *Epimyrma africana, algeriana, bernardi, corsica, kraussei* (= *foreli* = *vandeli*), *ravouxi* (= *goesswaldi*), *stumperi, tamarae, zaleskyi* | *Leptothorax* (= *Myrafant*) spp.; host queen killed by parasite queen | *Leptothorax* (= *Myrafant*) | Dulosis; worker caste present, or degenerate dulosis with rare workers | Europe, Turkey, North Africa | Kutter (1951, 1969, 1973b), Bernard (1968), Cagniant (1968), Arnoldi (1968), Buschinger et al. (1981, 1986), Buschinger (1985, 1986, and personal communication), Ehrhardt (1982), Espadaler (1982a), Buschinger and Winter (1983b) |
| *E. corsica* | *Leptothorax* (= *Myrafant*) *exilis*; host queen killed by parasite queen | *Leptothorax* (= *Myrafant*) | Inquilinism; workerless | Corsica, Italy, Yugoslavia | Buschinger and Winter (1985), Buschinger (1986 and personal communication) |
| *Formicoxenus diversipilosus, hirticornis* | *Formica haemorrhoidalis, integroides, obscuripes* | *Leptothorax* | Xenobiosis | Western United States | Snelling (1965), Alpert and Akre (1973), Buschinger (1981), Francoeur et al. (1985) |
| *F. nitidulus, sibiricus* (= *orientalis*) | *Formica aquilonia, lugubris, pisarskii, polyctena, pratensis, rufa* | *Leptothorax* | Xenobiosis | Europe, USSR | Stäger (1925), Stumper (1949, 1950), Buschinger (1976a,b), Buschinger and Winter (1976), Francoeur et al. (1985) |
| *F. provancheri* (= *Leptothorax emersoni*) | *Myrmica incompleta* | *Leptothorax* | Xenobiosis | North America | Wheeler (1910a), Buschinger et al. (1980b) |
| *F. quebecensis* | *Myrmica alaskensis* | *Leptothorax* | Xenobiosis | Canada | Francoeur et al. (1985) |
| *Formicoxenus* (= *Symmyrmica*) *chamberlini* | *Manica mutica* | *Leptothorax* | Xenobiosis | Utah | Wheeler (1910a), Buschinger and Francoeur (1983), Francoeur et al. (1985) |
| *Harpagoxenus canadensis* | *Leptothorax muscorum* | *Leptothorax* | Dulosis | Eastern North America | Creighton (1950), Buschinger and Alloway (1978), Stuart and Alloway (1983), Stuart (1984) |
| *H. sublaevis* | *Leptothorax acervorum, muscorum* | *Leptothorax* | Dulosis | Europe | Buschinger (1966a,b, 1968a,b, 1978, 1987a, and personal communication), Kutter (1969), Buschinger and Winter (1975, 1977, 1978), Winter (1979a) |

*continued*

**TABLE 12–1** *(continued)*

| Parasite | Host | Nearest related free-living genus | Form of parasitism | Range | Authority |
|---|---|---|---|---|---|
| *H. zaisanicus* | *Leptothorax* sp. | *Leptothorax* | Dulosis (?) | Mongolia | Pisarski (1963) |
| *Kyidris media, yaleogyna* | *Strumigenys loriae;* queen survives | *Smithistruma* (?) | Inquilinism; workers numerous | New Guinea | Wilson and Brown (1956) |
| *Leptothorax buschingeri* | *Leptothorax acervorum* | *Leptothorax* (*L. acervorum*) | Possible inquilinism; known from males only; status uncertain, may be teratological *acervorum* (A. Buschinger, personal communication) | Switzerland | Kutter (1967, 1969) |
| *L. ergatogyna* | *Leptothorax recedens* | *Leptothorax* | Status uncertain; possible intermorph of *L. recedens* | France | Bernard (1968), A. Buschinger (personal communication) |
| *L. faberi* | *Leptothorax muscorum;* queen survives | *Leptothorax* | Inquilinism; workers present | Western Canada | Buschinger (1982b) |
| *Leptothorax* (= *Myrafant*) *duloticus* | *Leptothorax* (= *Myrafant*) *curvispinosus, L.* (= *M.*) *ambiguus, L.* (= *M.*) *longispinosus* | *Leptothorax* (= *Myrafant*) | Dulosis | Eastern United States | Wesson (1940), Talbot (1957), Wilson (1975a), Alloway (1979, 1980) |
| *Leptothorax* (= *Myrafant*) *minutissimus* | *Leptothorax* (= *Myrafant*) *curvispinosus;* queen survives | *Leptothorax* (= *Myrafant*) | Inquilinism; workerless | Eastern United States | Smith (1942a) |
| *Manica parasitica* | *Manica bradleyi* | *Manica* | Inquilinism; workers present | California | Creighton (1950), Wheeler and Wheeler (1968) |
| *Megalomyrmex symmetochus* | *Sericomyrmex amabilis* | *Megalomyrmex* | Xenobiosis | Panama | Wheeler (1925) |
| *M. wheeleri* | *Cyphomyrmex costatus* | *Megalomyrmex* | Xenobiosis | South America | Weber (1940) |
| *Monomorium hospitum* | *Monomorium floricola* | *Monomorium* | Inquilinism; probably workerless | Singapore | Bolton (1987) |
| *M. inquilinum* | *Monomorium cyaneum* | *Monomorium* | Inquilinism; workerless | Mexico | Dubois (1981, 1986) |
| [*Monomorium metoecus*, described by Wilson and Brown (1958b) as a parasite of *Monomorium minimum*, is probably only an aberrant form of the host species; see Ettershank (1966) and DuBois (1986).] | | | | | |
| *Monomorium noualhieri* | *Monomorium subnitidum* | *Monomorium* | Xenobiosis (?); worker caste present | Algeria | Wheeler (1910a), Ettershank (1966), Bolton (1987) |
| *M. talbotae* | *Monomorium minimum* | *Monomorium* | Inquilinism; workerless | Michigan | Dubois (1981, 1986) |
| [*Monomorium* (= *Epixenus*) *andrei, biroi,* and *creticus*, thought by Tohmé and Tohmé (1979) and other authors to be possible inquilines, are almost certainly only apterous, ergatoid queens of the "host" species (Bolton, 1987).] | | | | | |
| *Monomorium* (= *Epoecus*) *pergandei* | *Monomorium minimum;* queen survives | *Monomorium* | Inquilinism; workerless | Washington, D.C. | Creighton (1950), Ettershank (1966), Dubois (1986) |
| *Monomorium* (= *Wheeleriella* ) *santschii* (= *adulatrix* ) | *Monomorium salomonis;* queen eliminated | *Monomorium* | Inquilinism; workerless | Tunisia | Forel (1906), Wheeler (1910a), Ettershank (1966), Bolton (1987) |
| *Monomorium* (= *Wheeleriella* ) *effractor* (= *wroughtoni*) | *Monomorium indicum* | *Monomorium* | Inquilinism; workerless | India | Forel (1910), Ettershank (1966), Bolton (1987) |

*continued*

**TABLE 12–1** *(continued)*

| Parasite | Host | Nearest related free-living genus | Form of parasitism | Range | Authority |
|---|---|---|---|---|---|
| *Myrmica bibikoffi* | *Myrmica sabuleti* | *Myrmica* | Facultative inquiline; worker caste present | Switzerland | Kutter (1963a) |
| *M. ereptrix* | *Myrmica rugosa* | *Myrmica* | Inquilinism; workerless | India | Bolton (1988a) |
| *M. faniensis* | *Myrmica scabrinodis* | *Myrmica* | Inquilinism; workerless? | Belgium | van Boven (1970) |
| *M. hirsuta* | *Myrmica sabuleti*; queen survives | *Myrmica* | Inquilinism; workerless | England | Elmes (1983) |
| *M. lampra* | *Myrmica alaskensis* | *Myrmica* | Inquilinism; workerless | Canada | Francoeur (1968, 1981) |
| *M. lemasnei* | *Myrmica sabuleti* | *Myrmica* | Inquilinism; status of worker caste uncertain | Pyrenees-Orientales, France; Spain | Bernard (1968), Espadaler (1981) |
| [*Myrmica microgyna* of England was considered by Pearson (1981) to be a workerless inquiline of *M. rubra* but it is considered by Barry Bolton (personal communication) and others to be a polygynous morph of the "host" *M. rubra*.] | | | | | |
| [*Myrmica myrmecophila* was thought to be a workerless inquiline of *M. sulcinodis* by Bernard (1968), but it is evidently only a gynecoid worker of the "host" *M. sulcinodis* (van Boven, 1970).] | | | | | |
| *Myrmica myrmecoxena* | *Myrmica lobicornis* | *Myrmica* | Inquilinism; workerless | Switzerland | Kutter (1969), van Boven (1970) |
| *M. quebecensis* | *Myrmica alaskensis* | *Myrmica* | Inquilinism; workerless | Canada | Francoeur (1981) |
| *M. samnitica* | *Myrmica sabuleti* | *Myrmica* | Inquilinism; workerless | Italy | Mei (1987) |
| *Myrmica* (= *Paramyrmica*) *colax* | *Myrmica striolagaster* | *Myrmica* (*M. striolagaster*) | Probably temporary, with worker caste | Texas | Cole (1957) |
| *Myrmica* (= *Sifolinia*) *kabylica* | *Myrmica aloba* | *Myrmica* | Inquilinism; workerless | Algeria | Cagniant (1970) |
| *Myrmica* (= *Sifolinia* = *Symbiomyrma*) *karavejevi* (= *M. pechi*) | *Myrmica rugulosa*, *sabuleti*, *scabrinodis* | *Myrmica* | Inquilinism; workerless | England, Sweden, Czechoslovakia, Poland, USSR | Bernard (1968), Kutter (1969), Douwes (1977) |
| *Myrmica* (= *Sifolinia*) *laurae* | Unknown; holotype swept from flowers | *Myrmica* | Inquilinism; workerless (?) | Italy | Emery (1921–22), Wilson (1984c) |
| *Myrmica* (= *Sifolinia*) *winterae* | *Myrmica rugulosa* | *Myrmica* | Inquilinism; workerless | Switzerland | Kutter (1973a), B. Bolton (personal communication) |
| *Myrmica* (= *Sommimyrma*) *symbiotica* | *Myrmica rubra* | *Myrmica* | Inquilinism; workerless | Italy | Kutter (1969) |
| *Myrmoxenus gordiagini* (= *Myrmetaerus microcellatus*, provisional synonymy) | *Leptothorax* (= *Myrafant*) spp., queen killed | *Leptothorax* (= *Myrafant*) | Dulosis; worker caste present | USSR (eastern Siberia), Yugoslavia | Ruzsky (1902), Buschinger et al. (1983) |
| *Oxyepoecus bruchi*, *daguerri*, *inquilina*, *minuta* | *Pheidole* spp. and *Solenopsis* spp. | *Solenopsis* | Apparent inquilinism; worker caste present | Argentina | Kusnezov (1952), Ettershank (1966), Kempf (1974) |
| *Parapheidole oculata* | Unknown | *Pheidole* | Inquilinism (?) | Madagascar | Wilson (1984c) |
| *Pheidole lanuginosa* | *Pheidole indica* | *Pheidole* | Inquilinism; workerless | India | Wilson (1984c) |

*continued*

**TABLE 12–1** *(continued)*

| Parasite | Host | Nearest related free-living genus | Form of parasitism | Range | Authority |
|---|---|---|---|---|---|
| *P. microgyna* | *Pheidole* species near *minutula* | *Pheidole* | Apparently workerless inquilinism | Guyana | Wilson (1984c) |
| *P. parasitica* | *Pheidole indica* | *Pheidole* | Inquilinism; workerless | India | Wilson (1984c) |
| *Pheidole* (= *Anergatides*) *neokohli* | *Pheidole megacephala melancholica* | ? | Extreme inquilinism; workerless | Zaire | Wasmann (1915a), Emery (1921–22), Wilson (1984c) |
| *Pheidole* (= *Bruchomyrma*) *acutidens* | *Pheidole nitidula* | *Pheidole* | Extreme inquilinism; workerless | Argentina | Bruch (1932a), Wilson (1984c) |
| *Pheidole* (= *Epipheidole*) *inquilina* | *Pheidole pilifera* | *Pheidole* (*P. pilifera*) | Inquilinism; workers present but scarce | Colorado to Nebraska | Creighton (1950), Wilson (1984c) |
| *Pheidole* (= *Eriopheidole*) *symbiotica* | *Pheidole obscurior* | *Pheidole* | Inquilinism; workerless | Argentina | Kusnezov (1951c), Wilson (1984c) |
| *Pheidole* (= *Gallardomyrma*) *argentina* | *Pheidole nitidula* | *Pheidole* | Extreme inquilinism; workerless | Argentina | Bruch (1932b), Wilson (1984c) |
| *Pheidole* (= *Sympheidole*) *elecebra* | *Pheidole ceres* | *Pheidole* | Inquilinism; workerless | Colorado | Creighton (1950); Wilson (1984c) |
| *Pogonomyrmex anergismus* | *Pogonomyrmex rugosus* | *Pogonomyrmex* | Inquilinism; workerless | New Mexico | Cole (1968), MacKay and Van Vactor (1985) |
| *P. colei* | *Pogonomyrmex rugosus*; queen survives | *Pogonomyrmex* | Inquilinism; workerless | Nevada | Rissing (1983) |
| *Protomognathus americanus* | *Leptothorax* (= *Myrafant*) *longispinosus* and *L.* (= *M.*) *curvispinosus* | *Leptothorax* (= *Myrafant*) in part | Dulosis | Eastern North America | Wesson (1939), Creighton (1950), Alloway (1979, 1980), Alloway and Del Rio Pesado (1983), Del Rio Pesado and Alloway (1983), Buschinger (1987a) |
| *Pseudoatta argentina* | *Acromyrmex lundi* | *Acromyrmex* (*Pseudoatta* a likely synonym of *Acromyrmex*) | Inquilinism; workerless | Argentina | Gallardo (1916b), Kusnezov (1954) |
| *Rhoptromyrmex* spp. | ? | ? | Possible temporary parasites | Old World tropics | Bolton (1976, 1986a) |
| *R. mayri* | *Pheidole latinoda* | *Rhoptromyrmex* | Inquilinism | India | Brown (1964a), Wilson (1984c), Bolton (1986a) |
| *Rhoptromyrmex* (= *Hagioxenus*) *schmitzi* | *Tapinoma erraticum* (?) | *Rhoptromyrmex* | Inquilinism (?) | Jerusalem | Wilson (1984c), Bolton (1986a) |
| *Serrastruma inquilina* | *Serrastruma lujae* | *Serrastruma* | Inquilinism; workerless (?) | Zaire | Bolton (1983) |
| *Solenopsis* (= *Labauchena*) *acuminata* | *Solenopsis invicta* and *S. clytemnestra* | *Solenopsis* | Inquilinism; workerless | Argentina | Borgmeier (1949), Kusnezov (1957a), Ettershank (1966) |

*continued*

**TABLE 12–1** *(continued)*

| Parasite | Host | Nearest related free-living genus | Form of parasitism | Range | Authority |
|---|---|---|---|---|---|
| *Solenopsis* (= *Labauchena*) *daguerrei* | *Solenopsis invicta* | *Solenopsis* | Inquilinism; workerless | Argentina, Uruguay | Santschi (1930), Ettershank (1966) |
| *Solenopsis* (= *Paranamyrma*) *solenopsidis* | *Solenopsis* ? *clytemnestra* | *Solenopsis* | Inquilinism; workerless | Argentina | Kusnezov (1954, 1957a), Ettershank (1966) |
| *Strongylognathus afer, alboini, alpinus, arnoldi, bulgaricus* (= *kratochvili*), *caeciliae, cecconii, cheliferus, christophi, dalmaticus, destefanii, emeryi, foreli, huberi, insularis, italicus, karawajewi, kervillei, koreanus, palaestinensis, rehbinderi, ruzskyi, silvestrii, testaceus* | *Tetramorium caespitum* and *T. jacoti*; queen tolerated by *Strongylognathus testaceus* | *Tetramorium* | Dulosis | Temperate parts of Europe and Asia | Wheeler (1910a), Stumper (1950), Pisarski (1966), Kutter (1969), Baroni Urbani (1969), Bolton (1976), Radchenko (1985) |
| *Strumigenys xenos* | *Strumigenys perplexa*; queen survives | *Strumigenys* (*S. perplexa*) | Inquilinism; workerless | Victoria, Australia; New Zealand | Brown (1955c), Taylor (1967c) |
| *Teleutomyrmex schneideri* | *Tetramorium caespitum*; queen survives | *Tetramorium* | Extreme inquilinism; workerless | France, Spain, Switzerland | Stumper (1950), Gösswald (1953), Kutter (1969), Buschinger (1987c) |
| *Tetramorium microgyna* | *Tetramorium sepositum, T. sericeiventre* | *Tetramorium* (*sericeiventre*) | Extreme inquilinism; workerless | South Africa, Zimbabwe | Bolton (1980) |
| *T. parasiticum* | *Tetramorium avium* | *Tetramorium* | Extreme inquilinism; workerless | South Africa | Bolton (1980) |
| **SUBFAMILY DOLICHODERINAE** | | | | | |
| *Bothriomyrmex* (34 spp.) | Some species at least: *Tapinoma* spp.; queen killed; biology of most species unknown | *Tapinoma* | Temporary, during colony formation | Old World | Santschi (1906), Wheeler (1910a), Emery (1925b), Bernard (1968), B. Bolton (personal communication) |
| *Conomyrma medeis* | *Conomyrma bureni* | *Conomyrma* | Temporary; host queen apparently eliminated; also transient slavemaking following territorial invasion | Southeastern United States | Buren et al. (1975), Trager (1988) |
| *C. reginicula* | *Conomyrma bossuta, bureni* | *Conomyrma* | Temporary | Florida | Trager (1988) |
| **SUBFAMILY FORMICINAE** | | | | | |
| *Acanthomyops latipes, murphyi* | *Lasius* (*L.*) spp. | *Lasius* | Temporary, during colony foundation | North America | Wing (1968) |
| *Anoplolepis nuptialis* | *Anoplolepis custodiens* | *Anoplolepis* | Inquilinism; workerless | South Africa | Santschi (1917) |

*continued*

**TABLE 12–1** *(continued)*

| Parasite | Host | Nearest related free-living genus | Form of parasitism | Range | Authority |
|---|---|---|---|---|---|
| *Camponotus universitatis* | *Camponotus aethiops* | *Camponotus* | Inquilinism (?); worker caste present | France, Switzerland | Bernard (1968), Kutter (1969) |
| *Formica dirksi* | *Formica* species near *fusca* | *Formica* (*microgyna* group) | Possible inquilinism; workerless | Maine | Wing (1949), Wilson (1976d) |
| *F. talbotae* | *Formica obscuripes* | *Formica* (*microgyna* group) | Inquilinism; workerless | Iowa, Michigan, North Dakota | Wilson (1976d), Talbot (1976) |
| *Formica* (*Coptoformica*), also called the "*exsecta* group" of subgenus *Formica: bruni, exsecta, foreli, forsslundi, goesswaldi, naefi, pressilabris,* and *suecica* in Europe; *exsectoides, opaciventris,* and *ulkei* in North America | *Formica fusca* group; queen does not survive | *Formica* | Temporary, during colony foundation | Europe, North America | Wheeler (1910a), Creighton (1950), Bernard (1968), Kutter (1969) |
| *Formica* (*F.*) *microgyna* group (14 species) | *Formica fusca* group spp. | *Formica* | Temporary; host queen does not survive | North America | Wheeler (1910a), Creighton (1950) |
| *Formica* (*F.*) "*rufa* group," some but not all of the species, namely: *aquilonia, lugubris, pratensis, rufa, truncorum,* and *uralensis* in Europe; *dakotensis* and *reflexa* in North America | *Formica fusca* group ("subgenus *Serviformica*") and *Formica* (*Neoformica*); queen does not survive | *Formica* | Temporary, during colony foundation; one species (*reflexa*) apparently permanent, with workers | Europe, North America | Wheeler (1910a), Creighton (1950), Gösswald (1951a,b, 1957), Bernard (1968), Kutter (1969) |
| *Formica* (*Raptiformica*), also called the "*sanguinea* group" of subgenus *Formica* (2 spp. in Europe and Asia and about 10 spp. in North America) | *Formica* spp. | *Formica* | Dulosis (facultative in *sanguinea*) | Europe, Asia, North America | Creighton (1950), Bernard (1968), Buren (1968a), Kutter (1969) |
| *Lasius fuliginosus* group ("subgenus *Dendrolasius*") (5 species) | *Lasius* spp.; queen eliminated | *Lasius* | Temporary, during colony foundation | Europe, Asia | Wilson (1955a), Andrasfalvy (1961), Yamauchi and Hayashida (1968) |
| *Lasius umbratus* group ("subgenus *Chthonolasius*") (11 species) | *Lasius* (*L.*) spp.; queen eliminated | *Lasius* | Temporary, during colony foundation | Europe, Asia, North America | Crawley (1909), Gösswald (1938a), Hölldobler (1953), Wilson (1955a), Cole (1956), Andrasfalvy (1961), Kutter (1969) |
| *Lasius* ("subgenus *Austrolasius*") *reginae* | *Lasius alienus;* queen killed | *Lasius* | Temporary, during colony foundation | Austria | Faber (1967) |
| *Paratrechina* sp. | *Paratrechina parvula* | *Paratrechina* | Inquilinism; workerless | Massachusetts | S. Cover (personal communication) |
| *Plagiolepis grassei* | *Plagiolepis pygmaea;* queen survives | *Plagiolepis* (*P. pygmaea*) | Inquilinism; worker caste rare | Pyrenees-Orientales, France | Le Masne (1956b) |

*continued*

**TABLE 12-1** *(continued)*

| Parasite | Host | Nearest related free-living genus | Form of parasitism | Range | Authority |
|---|---|---|---|---|---|
| *P. xene* | *Plagiolepis pygmaea;* queen survives | *Plagiolepis* (*P. pygmaea*) | Inquilinism; workerless | Hungary to Pyrenees-Orientales | Le Masne (1956b), Passera (1964, 1966, 1968b) |
| *Plagiolepis* spp. | *Plagiolepis* | *Plagiolepis* | Inquilinism; workerless | Algeria, Morocco | A. Buschinger (personal communication) |
| *Plagiolepis* (= *Aporomyrmex*) *ampeloni* | *Plagiolepis vindobonensis;* queen survives | *Plagiolepis* | Inquilinism; workerless | Austria | Faber (1969) |
| *Plagiolepis* (= *Aporomyrmex*) *regis* | *Plagiolepis* sp. | *Plagiolepis* | Inquilinism; workerless | Daghestan, USSR | Faber (1969) |
| *Polyergus breviceps, lucidus, nigerrimus, rufescens, samurai* | *Formica* spp. | *Formica* | Dulosis | Europe, Asia, North America | Wheeler (1910a), Creighton (1950), Marikovsky (1963), Bernard (1968), Kwait and Topoff (1983), Topoff et al. (1984, 1985a–c, 1987, 1988), Goodloe and Sanwald (1985), Topoff (1985), Goodloe and Topoff (1987), Goodloe et al. (1987) |
| *Rossomyrmex minuchae* | *Proformica longiseta* | *Formica* | Dulosis | Spain | Tinaut Ranera (1981) |
| *R. proformicarum* | *Proformica nasuta* | *Formica* | Dulosis | Transcaucasia, USSR | Arnoldi (1932), Marikovsky (1974) |

tion readily at hand for constant reference, we will now present what is deliberately a rather didactic review of the entire subject, attempting to make it as orderly and clear as possible from the outset.

Wasmann (1891) distinguished two classes of consociations, or myrmecobioses as Stumper (1950) later dubbed them, that occur between different species of ants. These are the *compound nests,* in which two or more species live very close to each other, in some cases even running their nest galleries together, but keep their brood separated; and the *mixed nests,* in which the brood are mingled and cared for communally. Compound nests are very common. They reflect relationships that range, depending on the species involved, all the way from the accidental and trivial to total parasitism. Mixed colonies, on the other hand, almost always come about as a result of social parasitism. Forel (1898, 1901), who was the first to use the expression "social parasitism," and Wheeler (1901a,b, 1910a) devoted a great deal of attention to compound and mixed nests and provided a useful classification of the underlying relationships, complete with a somewhat less useful set of Hellenistic terms to label the various categories. Let us examine this classification briefly. Then we will make more interesting use of it in tracing the evolution of parasitism and other forms of symbioses.

## COMPOUND NESTS

*Plesiobiosis.* In this most rudimentary association, different ant species nest very close to one another, but engage in little or no direct communication—unless their nest chambers are accidentally broken open, in which case fighting and brood theft may ensue. The less similar the species are, morphologically and behaviorally, the more likely they are to cluster together in an accidental, truly "plesiobiotic" relationship. Conversely, closely related species of ants are the least likely to tolerate one another's presence.

*Cleptobiosis.* Some species of small ants build nests near those of larger species and either feed on refuse in the host kitchen middens or rob the host workers when they return home carrying food. R. C. Wroughton (quoted by Wheeler, 1910a) has described a species of *Crematogaster* in India whose workers "lie in wait for *Holcomyrmex,* returning home, laden with grain, and by threats, rob her of her load, on her own private road and this manoeuvre was executed, not by stray individuals, but by a considerable portion of the whole community." Workers of *Conomyrma pyramica* in the southern United States collect dead insects discarded by colonies of *Pogonomyrmex,* including corpses of the *Pogonomyrmex* themselves. Our impression in the field has been that some *Conomyrma* colonies obtain a large part of their food in this way, to the point of preventing kitchen middens from building up near the *Pogonomyrmex* nests.

*Lestobiosis.* Certain small species, most belonging to *Solenopsis* and related genera, stay in the walls of large nests built by other ants or termites and enter the nest chambers of their hosts to steal food and prey on the inhabitants. For example, colonies of the "thief ants" of the subgenus *Solenopsis* (*Diplorhoptrum*), including especially *S. fugax* of Europe and *S. molesta* of the United States, often nest next to larger ant species, stealthily enter their chambers, and prey on their brood. Species of *Carebara* in Africa and tropical Asia frequently construct their nests in the walls of termite mounds and are believed to prey on the inhabitants (see Chapter 15). The relationship is parasitic with respect to nest sharing and predatory with respect to brood theft.

*Parabiosis.* In this peculiar form of symbiosis two or more species use the same nest and sometimes even the same odor trails, but they keep their brood separate. The situation is similar to the mixed foraging flocks of birds so prevalent in tropical forests, except that in some instances, at least, one species dominates and exploits another.

*Xenobiosis.* This symbiotic state falls just short of a truly mixed colony. One species lives in the walls or chambers of the nests of the other and moves freely among its hosts, obtaining food from them by one means or another, usually by soliciting regurgitation. The brood is still kept separate. This relationship is truly parasitic.

## MIXED COLONIES

The following phenomena are vital in the later stages of parasitic evolution. In a sense they form categories comparable to those just cited for compound nests, although they are less than ideal because they are not mutually exclusive. Nevertheless, we favor continuing to distinguish them on the grounds that the associated terminology is the familiar one in literature dating back over nearly a century and, more important, the classification can still be relied on to serve as a guide through the complex relationships as we understand them.

*Temporary social parasitism.* This symbiosis was first clearly recognized by Wheeler (1904b) as a result of his studies of the life cycle of members of the *Formica microgyna* group, especially *F. difficilis.* It has since been discovered in a diversity of genera belonging to several subfamilies. The newly fecundated queen finds a host colony and secures adoption, either by forcibly subduing the workers or by conciliating them in some fashion. The original host queen is then assassinated by the intruder or by her own workers, who somehow come to favor the parasite. With the development of the first parasite brood, the worker force soon becomes a mixture of host and parasite species. Finally, since the host queen is no longer present to replenish them, the host workers die out, and the colony comes to consist entirely of the parasite queen and her offspring. Temporary social parasitism is generally considered to be preceded in evolution by the re-adoption of queens by colonies of their own species following the nuptial flights. Bolton (1986b) has referred to this condition as autoparasitism.

*Dulosis* (*slavery*). Certain ant species have become dependent on workers of other species which they keep as slaves. The slave raids of the evolutionarily advanced species are dramatic affairs in which the slave-making workers go out in columns, penetrate the nests of colonies belonging to other, related species, and bring back pupae to their own nests. The pupae are allowed to eclose, and the workers become fully functional members of the colony. The workers of most slave-making species seldom if ever join in the ordinary chores of foraging, nest building, and rearing of the brood, all of which are left to the slaves. Facultative inter- and intraspecific slavemakers also occur. These less specialized forms provide an illuminating glimpse into the likely early stages of dulosis.

*Inquilinism* (*permanent parasitism*). In this final, degenerate stage, the parasitic species spends its entire life cycle in the nests of the host species. Workers may be present, but they are usually scarce and display atrophied behaviors. In many of these species, as for example in *Teleutomyrmex schneideri,* the worker caste has been lost altogether. Wilson (1971) suggested the use of the term "inquilinism" in preference to the somewhat more familiar expression "permanent parasitism," since obligatorily dulotic species are also permanent parasites. Inquilinism and dulosis, on the other hand, form exclusive categories; they are meant to be the streamlined equivalents of Kutter's (1969) "permanent parasitism without dulosis" and "permanent parasitism with dulosis." The queens of some inquiline species permit the host queen to live, whereas others either assassinate her or else somehow, in a procedure yet to be firmly established by experiments, induce her own workers to accomplish the task.

## THE OCCURRENCE OF SOCIAL PARASITISM THROUGHOUT THE ANTS

A rich variety of new parasitic species, representing almost every conceivable evolutionary stage, have been added since the time of Wheeler's classic synthesis in 1910. They continue to be discovered at such a consistently high rate as to suggest that at this moment only a small fraction of the total world fauna of social parasites is known. The reason for the slow uncovering of the world fauna seems clear: parasitic species tend to be both rare and locally distributed. As a rule, moreover, the more advanced the stage of parasitism, the rarer the species. Thus we find (Table 12–1) that temporary social parasites, such as members of the *Formica exsecta* and *Lasius umbratus* groups, are often nearly as widely distributed as their free-living congeners, and a few of the species are also very abundant. Species in which dulosis is weakly developed or even facultative, as, for example, the representatives of the *F. sanguinea* group, are also relatively abundant and widespread. On the other hand extreme dulotic species, such as the members of *Strongylognathus, Polyergus,* and *Rossomyrmex,* exist in more restricted, sparser populations. Finally, the extreme workerless parasites are, as a rule, both very rare and very locally distributed. *Anergates atratulus* comes closest to being an exception. It has been collected over a wide area from southern France to Germany, and it has even been accidentally introduced into the United States with its host *Tetramorium caespitum.* Yet everywhere within this range it is still a comparatively rare ant. The great majority of other workerless parasites have been found at only one or two localities and are extremely difficult to locate, even when a deliberate search is made for them in the exact spots where they were first discovered. Usually they give the impression, quite possibly false, that they have no more than a toehold on their host populations and that they exist close to the edge of extinction.

Most of the known parasitic species have been recorded exclusively from the temperate areas of North America, Europe, and South America. Almost certainly this reflects at least in part the strong bias of ant collectors, most of whom reside in these areas and devote a large part of their lives to a meticulous examination of local faunas. Switzerland, for example, is the present "capital" of parasitic ants for the simple reason that both Auguste Forel and Heinrich Kutter lived there. About one-third of the 110 Swiss species are parasitic (Kutter, 1969). Europe has received the attention of the expert collector Alfred Buschinger and his students for more than twenty years. The United States has benefited similarly from the efforts of W. M. Wheeler, the Wesson brothers, and other more recent gifted collectors, while the rich trove of species in Argentina was uncovered by three men who spent a large part or all of their lives in the country—Carlos Bruch, Angel Gallardo, and Nicolás Kusnezov.

We believe that as the huge and still little-known tropical ant faunas are more carefully worked (there are no resident myrmecologists on the Amazon!), many more parasitic species will come to light. Some evolutionarily advanced forms are already known from tropical regions. Wilson (1984c) recognized four tropical parasites in the genus *Pheidole,* including two new species and the Zaïrian *Pheidole* (= *Anergatides*) *neokohli,* which rivals *Teleutomyrmex* in the extremeness of its degeneration. Equally impressive are the strange postxenobiotic *Kyidris* parasites of New Guinea (Wilson and Brown, 1956). Wheeler (1925) pointed out that females of the numerous species of *Crematogaster* belonging to the "subgenera" *Atopogyne* and *Oxygyne,* groups widely distributed in Africa, Madagascar, and tropical Asia, have all of the morphological characteristics of northern ants known to be temporary parasites in that they tend to be small and shining and to possess falcate or very oblique mandibles and large postpetioles which are attached broadly to the gaster. The last of these characteristics is usually associated with physogastry, also a common but not diagnostic feature of social parasitism. Emery (1899) recorded a highly physogastric nest queen of *Crematogaster* ("*Oxygyne*") *ranavalonae* from Madagascar. At least two species of the Neotropical dolichoderine genus *Azteca* (*aurita* and *fiebrigi*) possess some of these traits. The species of *Rhoptromyrmex,* found in South Africa, Asia, New Guinea, and Australia also possess them (Brown, 1964a; Bolton, 1986a). A special study of such species, and any others that can be found to possess various of the "temporary parasite syndrome" of characters, may prove very rewarding to future students of tropical myrmecology.

Even so, the vast differences in quality of sampling from around the world render the matter inconclusive, and there remains the possibility that life in certain climates and environments actually does predispose ant species toward parasitism. It is true, for example, that a disproportionate number of parasitic species, especially the complete inquilines, occur in mountainous and arid regions. We have already mentioned the extraordinary diversity of parasites found in the little forest of the Saas-Fee. Among numerous other examples that can be cited are the montane species *Pheidole* (= *Epipheidole*) *inquilina, Pheidole* (= *Sympheidole*) *elecebra, Manica parasitica, Pogonomyrmex anergismus, Pogonomyrmex colei, Doronomyrmex pocahontas,* and *Leptothorax faberi,* which together make up about half of the known inquiline fauna of North America. Temporary social parasites, along with species that can be tentatively placed in this category by virtue of their morphology, are more abundant in the colder portions of Europe and North America than in the warm temperate and subtropical portions, even though the faunas of the two climatic zones are otherwise not radically different. Even more impressive, dulosis is a common phenomenon in the colder parts of Europe and Asia but rare in the warmer parts; and not a single example has ever been reported from the tropical or south temperate zones.

It is conceivable that cooler temperatures facilitate the introduction of parasitic queens in the early evolution of the phenomenon by dulling the responses of the host colonies. We have found, in general, that if ant colonies are first chilled in the laboratory they are more likely to adopt queens of their own species, whom they would otherwise attack and destroy. In nature parasite queens need not wait for winter to utilize this effect. Some degree of chilling, say to 10 or 15°C, occurs commonly during the cool summer nights in mountainous regions, right in the middle of the season of nuptial flights. It should prove instructive to study the effects of various degrees of cooling of potential host colonies on the success of introduction of queens belonging to species at any early stage of inquilinism, such as *Leptothorax faberi* and *Manica parasitica.* Useful information may also be obtained from an analysis of the behavioral effects of cooling on ant groups that most commonly serve as hosts, such as the genera *Leptothorax* and *Formica,* as opposed to those that are relatively immune to social parasitism, such as the genus *Camponotus.*

Other clues to the origin of social parasitism may be found in the phylogenetic distribution of the phenomenon, which is remarkably patchy. The more advanced forms of parasitism, namely dulosis and inquilinism, are almost wholly limited to the subfamilies Myrmicinae and Formicinae and are furthermore heavily concentrated in certain genera, including *Pheidole, Myrmica, Leptothorax, Tetramorium, Plagiolepis, Lasius,* and *Formica,* and in the satellite parasitic genera derived from them. Two inquilines (*Myrmecia hirsuta* and *M. inquilina*) have been described from the primitive subfamily Myrmeciinae. In view of the relatively small number of species known in the Myrmeciinae (about 120) and the limited amount of field study devoted to it to date, parasitism in this group may eventually be found to occur at about the same level of frequency as in the Myrmicinae and Formicinae. The only parasites known with certainty among the Dolichoderinae, on the other hand, are the temporarily parasitic species of *Bothriomyrmex.* This relative immunity is puzzling since the dolichoderines are a relatively large, numerically abundant group of advanced phylogenetic rank. Perhaps the explanation lies in the fact that very few dolichoderine species range into the cooler portions of the north temperate zone where parasitic species are most likely to evolve. Yet it is also true that a rich dolichoderine fauna exists in subtropical and temperate Argentina, where many myrmicine parasites have been discovered. No parasitic species of any kind are yet known in the Ponerinae, Cerapachyinae, and Dorylinae. One can speculate almost endlessly on why this is the case. One reason may be, for example, that the Ponerinae are primitive—but so are the Myrmeciinae, and in any case many ponerine species form large colonies with advanced social traits. Similarly, the Dorylinae engage in frequent nest changes—but many parasitic beetles, millipedes, wasps, and other arthropods emigrate with them along their odor trails.

An important lead from the phylogenetic distribution has emerged from studies by A. Buschinger, T. M. Alloway, and their co-workers on the myrmicine tribe Leptothoracini. Although leptothoracines represent fewer than 3 percent of the 8,800 described ant species, they contain 30 (15 percent) of the 200 known parasitic species. From the research of Buschinger, Francoeur, and their co-workers, which utilizes a combination of cytological, morphological, and behavioral traits, we can be reasonably sure that slave making alone has arisen a minimum of six times within the Leptothoracini: once each in the lines leading to *Leptothorax* (= *Myrafant*) *duloticus*, *Harpagoxenus*, *Protomognathus* ("*Harpagoxenus*" *americanus*), *Chalepoxenus*, *Epimyrma*, and *Myrmoxenus*. The Leptothoracini, as Buschinger has said, include "an astoundingly rich variety of socially parasitic genera and species," in which "new species can be found nearly everywhere when populations of independent species are closely examined."

What is the cause of the vulnerability of the leptothoracines to social parasites? What inclines so many to turn into parasites? Buschinger (1970a, 1986) and Alloway et al. (1982) believe that the key predisposing traits are polygyny, the regular occurrence of multiple laying queens in colonies, and polydomy, the spread of colonies to multiple nest sites. All these traits are developed in *Leptothorax* (= *Myrafant*) and *Leptothorax* (*s.str.*), the preeminent northern hemisphere representatives of the Leptothoracini and the stock group from which most of the parasitic genera and species have arisen. To be structured in this manner means that the colonies are relatively "open," in other words they are more easily invaded by alien queens of the same or different species. Polygyny usually arises by the readoption of queens after they have mated outside the nest. This habit thus being fixed in the workers' repertory, colonies are more susceptible to invasion by "cuckoo" queens able to provide the right chemical cues. Polydomy often results in the creation of outlier nests containing only workers and immature forms. It is possible that these queenless fragments add to the general vulnerability of the colonies.

If we accept the view that colony structure can predispose species for or against social parasitism, how does this explanation apply to the apparent scarcity of parasitism in the tropics? It is possible that for some reason ant species with leptothoracine-type biology are rare in warmer climates. However, our knowledge of the social organization and life cycle of the vast majority of tropical ant species is too meager to search for this correlation. Buschinger (personal communication) has offered one promising hypothesis:

> In my opinion a very important factor might be that parasitism occurs most frequently when the host species form dense, large, and homogeneous populations. This is the case in many temperate-zone species, whereas in warmer areas a high species diversity is often combined with rarity and wide dispersal of nests of a given species. I had presumed this for a long time, and recently in Australia I found a perfect confirmation of this idea (I also did not find any parasites there, but did not check *Myrmecia* nests, which often form dense populations—they were too aggressive!).

Another factor to examine is the means by which workers recognize colony odor at the species level. If they have a fully innate, "hard-wired" recognition separating individuals of the same species from those of different species, they will be very resistant in evolution to the intrusion of social parasites. If, on the other hand, they learn the species odor early in life, they can be more easily duped. A worker captured by a slavemaker while still in the pupal or callow (newly eclosed) adult stage can be imprinted on the odor of the captor. She will serve automatically as a slave thereafter.

The relative flexibility of early learning of conspecific brood labels may also play a role. Extensive research has been conducted on the recognition of brood (immature stages) by ants of the genus *Formica* (Jaisson, 1975, 1985; Le Moli and Passetti, 1977, 1978; Jaisson and Fresneau, 1978; Le Moli and Mori, 1982). The results demonstrate that young adult workers learn to recognize whatever species of brood they encounter within a period of approximately one week after their emergence from the cocoon. Unfamiliar brood pieces, whether of another species or their own, are rejected or destroyed. It has been suggested that an early learning of brood labels has favored the repeated evolution of slave making and social parasitism among ants. Le Moli (1980) and later Brian (1983) argued that the only species suitable as hosts or slaves are those in which brood recognition is based on learning without any bias favoring individuals of the same species. This preadaptive flexibility would ensure that immature slave or host ants, when eclosing, would learn the brood labels of their own species as well as those of their parasites.

The Brian-Le Moli hypothesis is only partially supported by the evidence, however. *Formica rufa* and *F. lugubris*, which do not exhibit a bias for learning conspecific brood labels, are not themselves victims of parasitism, although they are temporary social parasites of other *Formica* species (Gösswald, 1951a; Kutter, 1969). *F. polyctena*, which is usually free-living, is occasionally parasitized by *F. truncorum* (Kutter, 1969). Preference for early learning of conspecific brood labels has been found in *Camponotus* (Carlin et al., 1987a,b). It is tempting to attribute the "immunity" to social parasitism of the genus *Camponotus* to this bias of learning conspecific brood labels. These facts appear to support the Brian-Le Moli hypothesis. Since the bias discovered in *Camponotus* is not exclusive, however, colonies should still be vulnerable to potential parasites. Le Moli cites as supporting evidence his discovery that *Lasius niger* does not learn brood labels at an early age, and he uses the trait to explain why *L. niger* is evidently immune against social parasitism. He overlooks the fact, however, that *L. niger* serves as the host of the temporary social parasites *L. umbratus* (Crawley, 1909; Gösswald, 1938a; Hölldobler, 1953) and *L. fuliginosus* (Andrasfalvy, 1961).

To summarize, we can recognize several predisposing features toward social parasitism that might explain why it occurs in some ant genera and not in others. Those most likely to take the step (1) live in cool or arid climates; (2) have multiple queens as a result of readoption of newly mated queens; (3) occupy multiple nests, some of which are at least temporarily without a resident queen; (4) live in dense populations; and (5) learn the species odor early in life.

## THE EVOLUTION OF SOCIAL PARASITISM IN ANTS

In 1909 Emery formulated what is perhaps the single most important generalization concerning social parasitism: "the dulotic ants and the parasitic ants, both temporary and permanent, generally originate from the closely related forms that serve them as hosts." What he meant, of course, was that the parasitic species tend to resemble their host species more closely than they do any other free-

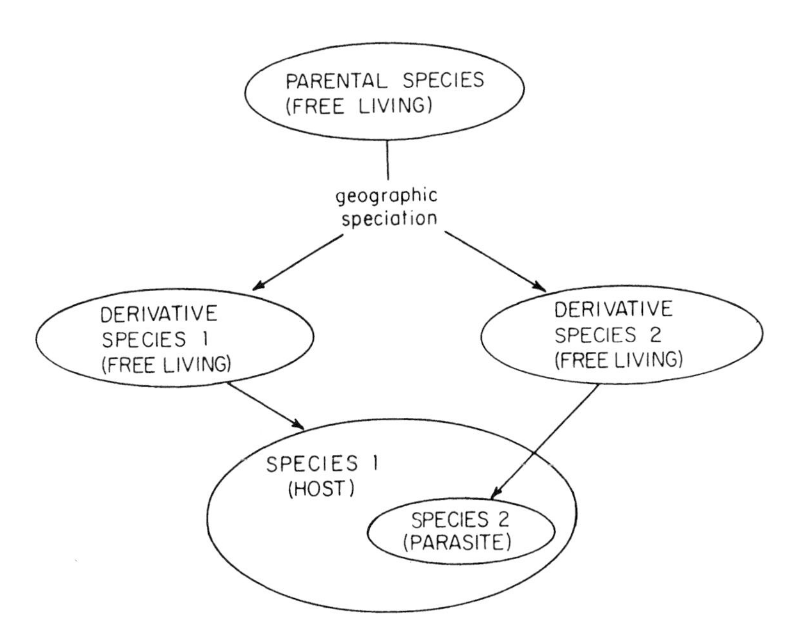


**FIGURE 12–2** The means by which a species can originate by geographic speciation and come to live as a social parasite with its closest living relative, in accord with Emery's rule. (From Wilson, 1971.)

living form. Emery's rule, as it has been called (Le Masne, 1956b), has continued to hold well for the true inquilines. Taxonomists have stressed that certain of the parasites—for example, *Myrmica* (= *Paramyrmica*) *colax*, *Pheidole* (= *Epipheidole*) *inquilina*, *Leptothorax buschingeri*, and *Strumigenys xenos*—really are morphologically more similar to their hosts than to any other known species. At first glance this relation seems to create a paradox: how can a species generate its own parasite? As long ago as 1919 Wheeler experienced difficulty in even conceiving of a mechanism by which it could occur. In 1971 Wilson proposed a scheme based on the known process of geographic speciation in other organisms (see Figure 12–2). In what is generally regarded as the prevalent sequence of animal speciation, a single "parental" species can be divided into two "daughter" species by, first, fragmentation through geographic barriers and, second, genetic divergence of the populations thus isolated geographically until they acquire intrinsic isolating mechanisms. If and when the newly formed species reinvade one another's ranges, the isolating mechanisms prevent them from interbreeding. And if, in addition, one of the species then becomes specialized as a parasite on the other, the condition of Emery's rule is fulfilled.

This model suggests that, all other circumstances being equal, the frequency of parasitism within a given genus should increase as a function of the rate of speciation. A corollary is that the more taxonomically "difficult" the genus, in other words, the larger the percentage of newly formed, indistinctly defined species in it, the higher should be the percentage of parasitism. This prediction does indeed seem to be met by such genera as *Pheidole*, *Leptothorax*, *Plagiolepis*, *Lasius*, and *Formica*, although the correlation through all ant groups taken together is far from perfect. But at least it is clear that the speciation rate is one additional factor that must be considered in future evolutionary analyses of the subject.

An alternative mode of evolution that must always be kept in mind is sympatric speciation, in which certain homogamous mutants or recombinants (forms that breed only with their own kind) arise in sufficient numbers at the same place and time to segregate a distinct breeding population. In other words the new parasitic species may arise *in situ* directly from the host species, without the intervening step of geographic isolation. Buschinger (1986) expressed a simple model as follows:

> Ant species or populations may change from monogyny to polygyny (and vice versa) depending on various environmental conditions, and polygyny is a sufficiently frequent phenomenon in ants as to represent a condition from which parasitism may have evolved convergently several times. A conceivable mutation, *e.g.*, causing certain sexuals to mate at a different time of day, might quickly produce a subpopulation of individuals genetically isolated from the original form but still living in its polygynous colonies. Once this isolation has been achieved, development in the inquiline or the dulotic direction may depend upon the ability of such a form to produce further workers. And when the host species evolves toward monogyny, or the parasite spreads into monogynous populations of its host species, selection will favor those parasite queens which are able to replace the host colony queens by force, as in dulotic or temporary parasitic species.

Pearson (1981) has added the view that inquilines, whether created by allopatric or sympatric speciation, are more likely to arise when competition occurs prominently. The species that is subordinated must survive on smaller amounts of food and is more likely to be miniaturized to fit into its marginal ecological role. Once miniaturized, it can more easily enter nests of the dominant species.

Buschinger's model, with or without miniaturization, is basically the same that has been elaborated many times in studies of other, free-living organisms. It is difficult to disprove, but it is also rendered less probable by the fact that disruptive selection, the basic mode of selection involved, must be severe to create incipient species. Furthermore, species-isolating mechanisms are usually genetically complex, a circumstance lessening the likelihood of sympatric speciation (Futuyma, 1986).

Many exceptions to Emery's rule exist. The most notable ones fall into the special categories of xenobiosis and parabiosis, in which the parasitic species typically belongs to a different genus and sometimes even to a different subfamily. The explanation for these two classes of exceptions is simple enough. When members of different genera associate at all, they are not likely to combine their brood, for they tend to be very different in biology and mutually incompatible. Any association achieved will be of a more tenuous sort, involving grooming, food exchange, trail sharing, or combinations of these relations—xenobiosis and parabiosis. The more closely related the two species, the more likely they are to enter into the more intimate forms of parasitism, producing the effect that is generalized in Emery's rule.

How do different ant species come together in symbiosis in the first place? The evolutionary schema presented in Figure 12–3 is an extension of one evolved in a long sequence of contributions by Wheeler (1904b, 1910a), Emery (1909), Escherich (1917), Stumper (1950), Dobrzański (1965), Wilson (1971), and Buschinger (1986), with additions of our own. The single most important idea embodied in this diagram is that inquilinism is a convergent phenomenon, reached independently by many different species following one of at least two available pathways in evolution. Also, complete inquilinism is viewed as an evolutionary sink; a return to free life or even to a partially parasitic existence by reversed evolution seems impossible. For convenience we have arranged known cases of social parasitism according to these hypothesized sequences.

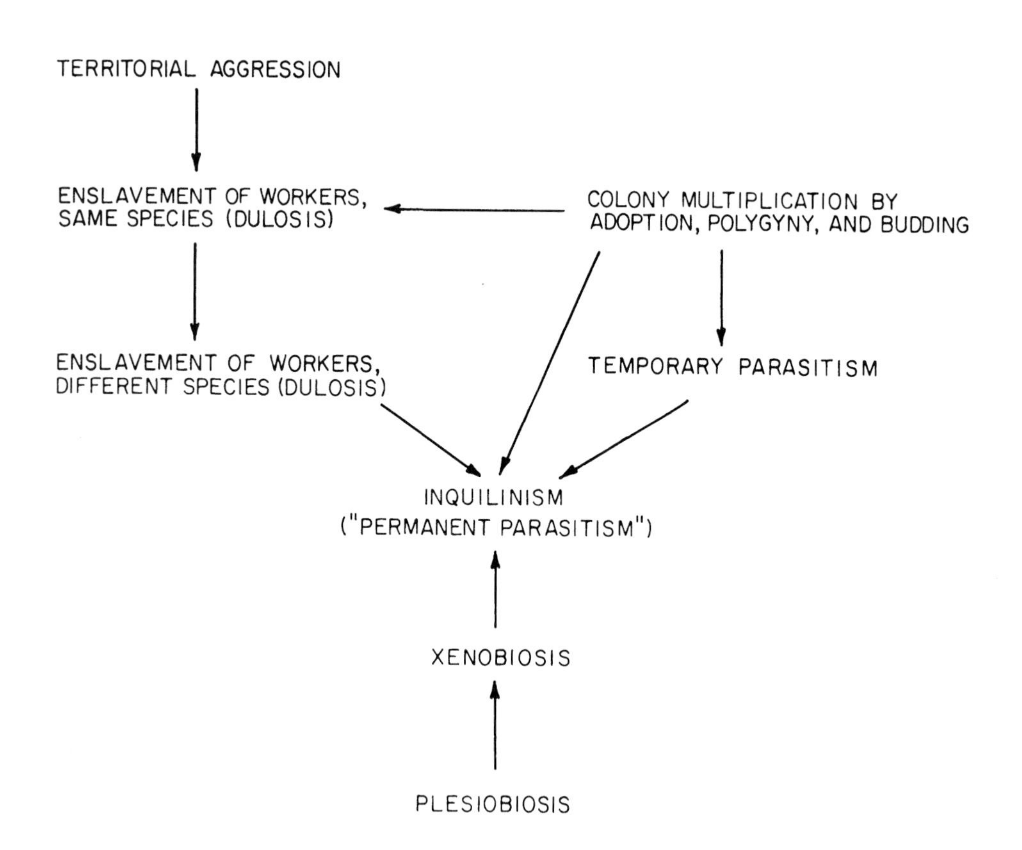


**FIGURE 12–3** Hypothesized evolutionary pathways of social parasitism in ants.

## THE TEMPORARY PARASITISM ROUTE

The earliest stages of temporary parasitism are displayed by members of the *Formica rufa* group (Gösswald, 1951a,b; Kutter, 1913, 1969). Several of the members, *F. lugubris, F. polyctena,* and *F. pratensis,* form colonies with multiple queens. New colonies are usually created by budding, or hesmosis, as it has occasionally been called. After the nuptial flights the newly fecundated queens normally return to the home nest, and at some later date some of them may move to a new site nearby with a group of workers. The new unit thus created is a colony only in the purely spatial sense because it may exchange workers with the mother nest for an indefinite time afterward. Multiplication by budding creates the pattern, so characteristic of these species of *Formica,* of dense aggregations of interconnecting nests that dominate local areas. Occasionally young queens do not find their way back to a nest of their own species. They may then seek adoption in a colony of the *F. fusca* group. Whenever one of them succeeds in penetrating such an alien nest, the host queen is somehow eliminated; the intruder takes over the role of egg laying exclusively, and eventually the host workers die off. The final result is that the colony consists entirely of the intruder and her offspring. Such temporary parasitism is regarded as a secondary mode of colony founding for these ants since mixed host-parasite colonies are rarely encountered in nature.

However, a closely related species, *Formica rufa,* has taken the step of founding its colonies predominantly by temporary parasitism, then forming monogynous colonies or relying on budding in a minority of instances, where polygynous supercolonies build up. Its host species in Europe include *F. fusca* and *F. lemani.* The *rufa* queen is still a rather inept parasite. On approaching the host colony she does not hide, play dead, conciliate, or display any of the other dissembling tricks ordinarily used by parasitic queens; instead, she plunges right into the nest. Such intrusions frequently result in the death of the queen at the hands of hostile host workers, but enough attempts succeed to maintain *F. rufa* as one of the more abundant and widespread ant species of Europe.

Most European students of *Formica,* starting with Emery (1909), have argued that loss of the ability to found nests in the usual claustral manner, with the resulting dependence on adoption and budding, preadapts members of the *rufa* group to temporary parasitism on other species. Also, the fact that *F. rufa* itself is usually monogynous (its colonies each tolerate only one egg-laying queen) predisposes this species even further to incursions on other species.

The species of the *exsecta* group of *Formica* (collectively referred to by earlier European writers as the subgenus *Coptoformica*) have a life cycle very similar to that of the *rufa* group species, except that the queens have become more skillful at penetrating host colonies (Kutter, 1956, 1957). The European species of *F. exsecta,* for example, depend chiefly on homospecific adoption and budding, but a few queens seek colonies of the *fusca* group of species (formerly called the subgenus *Serviformica*). The *exsecta* queens stalk the host colonies and either enter the nests by stealth or else permit themselves to be carried in by host workers. The *exsecta* queens are smaller and shinier than those of most members of the *rufa* group, and they seem to be treated with less hostility by the host workers. This is also the case for *F. pressilabris,* a second member of the *exsecta* group found

**FIGURE 12–4** (*Above*) A newly mated female of the social parasite *Epimyrma stumperi* has entered the nest of the host species *Leptothorax tuberum* and is strangling the host queen to death. (Modified from Kutter, 1969.) (*Below*) The queen of the temporary social parasite *Lasius reginae* exhibits almost identical behavior when invading the host nest of *Lasius alienus* and killing the queen. (Modified from Faber, 1967.)

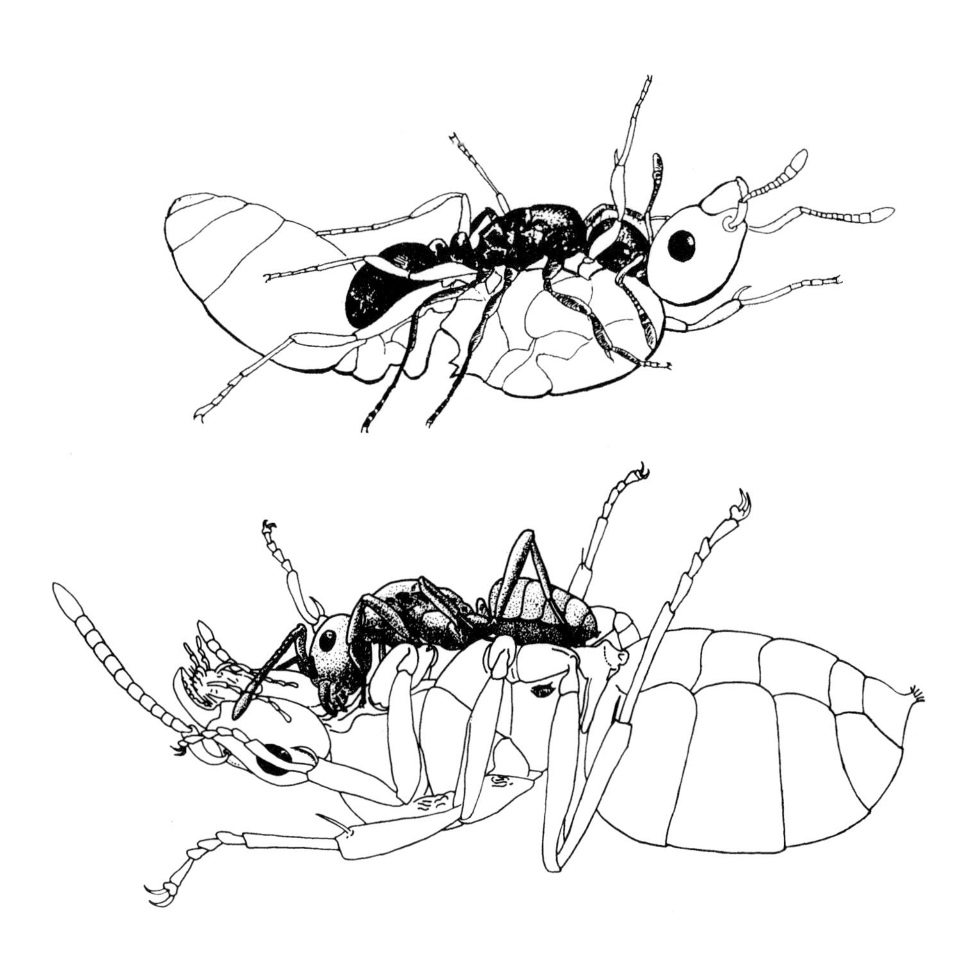

in Europe. Queens approached by host workers lie down and "play dead" by pulling their appendages into the body in the pupal posture. In this position they are picked up by the host workers and carried down into the nests without any outward show of hostility. Later they somehow manage to eliminate the host queen and take over the reproductive role. Similar life histories have been described for the North American species of the *exsecta* group (Wheeler, 1906; Creighton, 1950; Scherba, 1958, 1961) and have been postulated on the basis of limited laboratory experiments for members of the North American *microgyna* group (Wheeler, 1910a).

The transition from temporary social parasitism to full inquilinism, depicted in Figure 12–3, has been achieved by *Formica talbotae*, a workerless species of the *F. microgyna* group that lives with *Formica obscuripes* in the north central United States (Talbot, 1976; Wilson, 1976d). A second species evidently in the same category is the closely related *F. dirksi* of Maine (Wing, 1949; Wilson, 1976d).

Further subtleties have been developed by the related genus *Lasius*. Apparently all of the species of the *fuliginosus, reginae,* and *umbratus* groups are temporary parasites on members of the *L. niger* group (see Wilson, 1955a). This relationship is obligatory, not optional as in the *rufa* and *exsecta* groups of *Formica*. The colonies are monogynous for the most part, and homospecific adoption is not practiced. When newly mated queens of *L. umbratus* are searching for a host colony, they first seize a worker in their mandibles, kill her, and run around with her for a while before attempting to penetrate the nest (Hölldobler, 1953). Apparently all of the parasitic *Lasius* get rid of the host queens, but the exact means employed are still unknown in most cases. The queens of *L. reginae,* a species discovered in Austria by Faber (1967), eliminate their rivals by rolling them over and throttling them (Figure 12–4).

Assassination is also the technique employed by the queens of the dolichoderine species *Bothriomyrmex decapitans* and *B. regicidus* in gaining control of colonies of *Tapinoma* (Santschi, 1906, 1920). These temporary parasites occur in the deserts of North Africa. After the nuptial flight the *Bothriomyrmex* queen sheds her wings and searches over the ground until she finds a *Tapinoma* nest. She allows herself to be accosted by the aroused *Tapinoma* workers and dragged by them into the interior of their nest. There she takes refuge among the brood or on the back of the *Tapinoma* queen. In time she settles down for good on the back of the host queen and begins the one act for which she is uniquely specialized: slowly cutting off the head of her victim. When this is accomplished, sometimes only after many hours, the *Bothriomyrmex* takes over as the sole reproductive, and the colony eventually comes to consist entirely of her offspring and herself. Lloyd et al. (1986) discovered that the pygidial glands of *B. syrius* queens and of the *T. simrothi* host workers contain the same ketone. They speculate that the odor identity assists the *Bothriomyrmex* queens when they attempt to penetrate the host colonies.

A similar mode of entry into host nests is employed by the myrmicine *Monomorium* (= *Wheeleriella*) *santschii,* also a native of North Africa and a permanent workerless parasite of *M. salomonis*. In this case, however, it is the *salomonis* workers who destroy their own queen. They then adopt the *santschii* queen as the sole reproductive in her place (F. Santschi, in Forel, 1906).

## THE DULOSIS ROUTE

Slavery in ants, particularly as practiced by *Polyergus rufescens* and the species of the *Formica sanguinea* group, has been a favorite subject of myrmecologists in Europe and the United States ever since it was originally described by Huber in 1810. Latreille (1805) had been first to note the large number of workers emerging from *Polyergus* nests in "une espèce d'ordre de bataille" but apparently did not recognize the nature of the slavery that resulted. Darwin was fascinated by the phenomenon, and in *On the Origin of Species* he offered the first hypothesis of how it originated in evolution. The ancestral *Formica,* he proposed, began by raiding other species of ants in order to obtain their pupae for food. Some of the pupae survived in the storage chambers long enough to eclose as workers, whereupon they were accepted by their captors as nestmates. This fortuitous addition to the work force helped the colony as a whole, and consequently there was an increasing tendency, propelled by natural selection, to raid other colonies solely for the purpose of obtaining slaves. If Darwin's explanation seems at first a bit farfetched, it is only commensurate with the phenomenon itself. Several authors, most notably Erich Wasmann (1905), rejected Darwin's hypothesis on various grounds, chiefly a priori. The years have brought an increasing amount of confirmation of an evolutionary sequence approximately consistent with Darwin's scheme, however, through the discovery of species whose behavior collectively bridges the gap between free-living and slavemaker species in ever shorter, more plausible steps.

Before examining the details of behavior across the dulotic species, let us review the existence today of three hypotheses concerning the origin of slavery:

1. *Predation.* As suggested by Darwin, the predulotic ancestral ants raided the nests of other species to obtain prey. The first slaves were prey items allowed to live.

2. *Territory.* Part of territorial exclusion practiced by predulotic ants was the invasion of the nests of rivals and the robbing of the brood. The brood pieces, including eggs, larvae, and pupae, were routinely eaten for the most part, but some survived long enough to join their captors as slaves. The first step in this sequence was territorial aggression among colonies belonging to the same species. The next step, leading to the earliest stage in social parasitism as traditionally conceived, was territorial aggression directed at colonies of other species. The more closely related the rival species, the more often the captives were tolerated, leading to Emery's rule. The territorial hypothesis was developed by Wilson (1975a), Alloway (1979, 1980), and Stuart and Alloway (1982, 1983).

3. *Transport.* In the suggestion originally made by Buschinger (1970a), slave raiding evolved as the outcome of regular brood transport among the nests of single polydomous colonies. If the habit pattern is extended to less familiar populations of the same species or to other species, it will create an early version of dulosis.

These hypotheses are not mutually exclusive. They merely draw attention to the existence of three propensities that predispose the species to the practice of slavery. They are logically linked as follows: territorial raids combined with a strong propensity to transport brood lead to the regular retrieval of alien brood back to the raiders' nest; the raiders destroy and eat most of the captives, but a few survive to join the colony as slave workers.

This synthetic model, combining the three hypotheses, identifies territorial behavior rather than predatory raids as the prime mover leading to slavery, with food being a secondary benefit to the raiders. A substantial amount of evidence appears to support this view. First, none of the ant groups that specialize in predation on other ants, including one branch each in *Myrmecia* and *Gnamptogenys,* the entire tribe Cerapachyini of the subfamily Ponerinae, and *Aenictus, Eciton,* and *Neivamyrmex* among the army ants, has produced a single slave-making species. On the other hand, slave making is rampant in the tribe Leptothoracini, the species of which are generalized insectivores and honeydew collectors. Leptothoracines limit their myrmecophagy (if indeed it occurs in nature) mostly to other colonies of their own species, in what appears to be an incidental outcome of territorial aggression. Wilson (1975a) showed that when colonies of *Leptothorax* (= *Myrafant*) *curvispinosus* were placed close to one another in the laboratory, workers from the larger colony attacked the smaller colony, expelling the queen and workers. They carried away the brood and allowed at least some of the pupae to eclose into adults and join them as nestmates. *L.* (= *M.*) *ambiguus* colonies raided smaller *L.* (= *M.*) *curvispinosus* colonies in the same manner. They allowed *curvispinosus* pupae to eclose, and at first licked them and treated them like sister workers. But within one or two days they dragged them out of the nest and killed them. No *curvispinosus* worker was permanently adopted. Even so, the behavior of *L. ambiguus,* a typical free-living leptothoracine, was revealed to be but one short step away from an elementary form of dulosis.

Later Alloway (1980) discovered mixed colonies of *Leptothorax* (= *Myrafant*) in Canada and the northern United States. Most contained a queen and majority of workers of *L. longispinosus* with a scattering of workers belonging to either *L. ambiguus* or *L. curvispinosus.* One was a colony of *L. ambiguus* with a single *longispinosus* worker, another a *curvispinosus* colony with both *ambiguus* and *longispinosus* workers. When cultured, almost all the mixed colonies produced additional workers of the majority species. Alloway was able to create similar mixed colonies by placing pure colonies of the *Leptothorax* species close enough in the laboratory to trigger territorial raids of the kind described earlier by Wilson. In other words a natural condition of low-intensity, facultative slavery already exists in free-living species of *Leptothorax,* and it is evidently the outcome of territorial behavior. Alloway's result must be treated with caution, however, because his colonies were fed only Bhatkar diet. Buschinger and Pfeifer (1988) have shown that the diet is deficient in protein, causing the ants to consume more brood and perhaps increasing their propensity to raid neighboring colonies.

The connection between territoriality and predation is also well established. Even specialized slave-making species consume substantial portions of the prey, as documented in *Leptothorax* (= *Myrafant*) *duloticus* by Alloway (1979) and *Polyergus breviceps* by Topoff et al. (1984).

A second line of evidence supporting the primacy of territoriality was found by Stuart and Alloway (1983) in a laboratory study of *Harpagoxenus canadensis,* a specialized slave raider, and its slave *Leptothorax (s.str.) "muscorum."* The two species display similar raiding behavior, and both carry brood back to the nests after the raids. *H. canadensis* rear the captured brood to produce slave workers, how-

**FIGURE 12–5** *Formica sanguinea*, the slave-making "sanguinary ant" of Europe: (*a*) dealated queen; (*b*) pseudogyne, an abnormal female form found in colonies of this (and other) species of *Formica* infested by lomechusine staphylinid beetles; (*c*) worker; (*d*) head of worker, showing the notched clypeus that characterizes members of the *sanguinea* group of *Formica*. Colonies of this species are not dependent on slaves for survival, and they engage only occasionally in raids. The morphology is not especially well adapted for slave making. (From Wheeler, 1910a.)

ever, whereas *L. "muscorum"* colonies mutilate the immature forms and feed them to their own larvae. The *Leptothorax* slaves then join their *Harpagoxenus* mistresses in raids against other *Leptothorax* colonies. The only important difference between free and enslaved *Leptothorax* workers was the willingness of the enslaved individuals to care for captured brood instead of destroying it. Not only are the *Harpagoxenus* and *Leptothorax* very similar in this respect, but their method of recruitment is identical. Both recruit nestmates to the scene of fighting among workers, rather than to newly discovered nest sites, the usual stimulus triggering raids by other species of leptothoracine slavemakers. It is possible that *H. canadensis* arose directly from *L. muscorum* or from an immediately common stock (Buschinger, 1981).

The same facultative, low-level slavery appears to occur in *Formica*, another genus already known to be prone to advanced dulosis. Scherba (1964) found that in Wyoming the dense, polydomous colonies of *F. opaciventris*, a member of the *exsecta* group, commonly oust colonies of *F. fusca* from their nest sites by laying siege to them and robbing larvae and pupae when they get the chance. When Kutter (1957) placed colonies of *F. naefi*, also a member of the *exsecta* group, near colonies of species belonging to the *fusca* group, the *naefi* attacked their neighbors, penetrated their nest, and carried away both the brood and the adult workers. Kutter was unsure whether such behavior occurs in nature, but he noted that all larger *naefi* colonies observed in the field contained a few *fusca*-group workers. It seems reasonable to suggest that *naefi* represents the first interspecific dulotic stage envisioned in the territorial hypothesis.

It is also possible that the phenomenon described by Kutter occurs in other *Formica*. King and Sallee (1957, 1962) reported the puzzling existence of natural mixed colonies of *F. clivia* and *F. fossaceps* that persisted over a period of up to 16 years in Iowa. Workers and sexuals of both forms were produced in the nests. King and Sallee believed that the two forms are either genetic morphs or distinct species linked in some aberrant and unexplained symbiosis. The field data strongly suggest the second alternative. In laboratory experiments small homospecific groups of workers readily accepted queens of the opposite species combined with alien worker groups. The significance of this permissiveness and the nature of the interaction of colonies of the same species in nature are promising subjects for future study.

It is traditional to use the expression "slavery" for the exploitation of one species by another. In the human sense this is not slavery but more akin to the forcible domestication of dogs and cattle by humans. Does *true* slavery—the use of captives of the same species—exist in the ants? Evidence from the laboratory experiments just cited indicates that it occurs as an accidental outcome of territorial raiding in *Leptothorax* ( = *Myrafant*). On the other hand, true slavery is also practiced as a highly organized, evidently adaptive behavior in the honeypot ant *Myrmecocystus mimicus* (Hölldobler, 1976c, 1981a). The foraging grounds of neighboring *mimicus* colonies often overlap, setting off massive territorial confrontations. The conflicts do not consist of deadly physical fights but rather of elaborate tournaments in which very few ants are injured. The rival workers stilt-walk on extended legs, lift their abdomens and point the tips at each other, and drum their antennae on one another's abdomens (see Figure 10–20). The *Myrmecocystus* tournaments sometimes last for days. If one colony is considerably weaker than the other and therefore unable to recruit a large enough worker force to the tournament area, it is eventually overrun and raided by the larger colony. Workers of the winning colony kill or drive off the queens and carry or drag the larvae, pupae, and callow workers to their own nests (see Figure 10–18). Raids do not originate exclusively from territorial interactions. New evidence shows that scout ants recruit nestmates to newly discovered small conspecific colonies, which are subsequently raided by the larger colony. Similar behavior has been observed among colonies of *M. depilis*, which often occur sympatrically with *M. mimicus*. No cases have been recorded, however, in which *M. mimicus* raided *M. depilis* or the reverse.

The raiding of smaller colonies by larger conspecific or congeneric colonies is probably much more common in territorial interactions in ants than investigators assumed. In a recent survey of colony interactions in populations of *Pogonomyrmex*, H. Markl and B. Hölldobler (unpublished) observed several incidents of intraspecific and interspecific raids, which in some cases clearly led to the enslavement of the captured immature stages. This explains the occurrence of *Pogonomyrmex* colonies with mixed-species worker populations that has occasionally been noted.

The next step in the dulotic progression has been well documented in *Formica sanguinea* (Figure 12–5), a European species that has been thoroughly studied by Huber (1810), Forel (1874), Wasmann (1891), Dobrzański (1961, 1965), and others. The "sanguinary ants" are very aggressive and territorial, dominating the local spots near their nests that are richest in food. They are "facultative slaveholders," in Wasmann's terminology, since colonies are sometimes found with no slaves present. Also, workers isolated in laboratory nests are able to conduct all of the affairs of colony life, including nest construction, in a competent manner. According to Wheeler (1910a) the percentage of slaveless colonies in different populations of *sanguinea* varies enormously, from about 2 percent to 98 percent. Thus the *sanguinea* are far more committed to dulosis than the species of *Leptothorax* ( = *Myrafant*), which take captives only rarely and apparently by accident. The commonest slaves taken by the *sanguinea* belong to the *fusca* group ("*Serviformica*") and include *fusca*, *lemani*, and *rufibarbis;* less commonly exploited are *gagates*, *cunicularia*, *transkaukasica*, and *cinerea*, all of which are also members of the *fusca* group as conceived in the broadest sense. On rare occasions workers of the *rufa* group, in particular *nigricans* and *rufa*, have been found in *sanguinea* nests, but always in the company of *fusca*-group slaves (Bernard, 1968). As a rule *sanguinea* colonies enslave the *fusca*-group species nearest their nest, and the seeming preferences are merely a reflection of local relative abundance of the slave species. Two or even three slave species are sometimes present in a given *sanguinea* nest simultaneously, and the composition of slaves may change from year to year.

The raids of *sanguinea* have been lucidly described by Wheeler:

> The sorties occur in July and August after the marriage flight of the slave species has been celebrated and when only workers and mother queens are left in their formicaries. According to Forel the expeditions are infrequent—"scarcely more than two or three a year to a colony." The army of workers usually starts out in the morning and returns in the afternoon, but this depends on the distance of the *sanguinea* nest from the nest to be plundered. Sometimes the slavemakers postpone their sorties till three or four o'clock in the afternoon. On rare occasions they may pillage two different colonies in succession before going home. The *sanguinea* army leaves its nest in a straggling, open phalanx sometimes a few meters broad and often in several companies or detachments. These move to the nest to be pillaged over the directest route permitted by the often numerous obstacles in their path. As the forefront of the army is not headed by one or a few workers that might serve as guides, but is continually changing, some dropping back while others move forward to take their places, it is not easy to understand how the whole body is able to go so directly to the nest of the slave species, especially when this nest is situated, as is often the case, at a distance of 50 or 100 m . . .
>
> When the first workers arrive at the nest to be pillaged, they do not enter it at once, but surround it and wait till the other detachments arrive. In the meantime the *fusca* or *rufibarbis* scent their approaching foes and either prepare to defend their nest or seize their young and try to break through the cordon of *sanguinea* and escape. They scramble up the grass-blades with their larvae and pupae in their jaws or make off over the ground. The sanguinary ants, however, intercept them, snatch away their charges and begin to pour into the entrances of the nest. Soon they issue forth one by one with the remaining larvae and pupae and start for home. They turn and kill the workers of the slave-species only when these offer hostile resistance. The troop of cocoon-laden *sanguinea* straggle back to their nest, while the bereft ants slowly enter their pillaged formicary and take up the nurture of the few remaining young or await the appearance of future broods. (Wheeler, 1910a: 456–457)

The communicative signals that trigger and orient the raids of colonies belonging to the *sanguinea* group of slave-making ants were identified, at least in part, by F. E. Regnier and Wilson (in Wilson, 1971). They found that workers of the American species *Formica rubicunda* readily followed artificial odor trails made from whole body extracts of *rubicunda* workers and applied with a camel's hair brush over the ground in the vicinity of the nest. When the trails were drawn away from the nest opening in the afternoon, at about the time raids are usually conducted, the *rubicunda* workers showed behavior that was indistinguishable from ordinary raiding sorties. They ran out of the nest and along the trail in an excited fashion, and, when presented with colony fragments of a slave species (*F. subsericea*), they proceeded to fight with the workers and to carry the pupae back to their nest. It seems likely that under normal circumstances *rubicunda* scouts lay odor trails from the slave colonies they discover to the home nest, and the raids result when nestmates follow the trails out of the home nest back to the source. In addition the scouts sometimes travel at the head of the raiding column. Chemical trails are probably the general mode of communication among slave-making ants. As we shall see shortly, they are employed by the evolutionarily more advanced amazon ants of the genus *Polyergus*, as well as some myrmicine slavemakers (Buschinger et al., 1980a). The tendency of *F. sanguinea* to fan out into "phalanxes" in their outward march does not conflict with this interpretation; there could be several odor trails involved, around which orientation is less than perfect, or else the recruits may swarm loosely around the leader ant—as in *P. breviceps* (Topoff et al., 1984, 1985a–c).

The general biology and raiding behavior of *Formica subintegra*, an American member of the *sanguinea* group, have been studied by Wheeler (1910a) and by Talbot and Kennedy (1940). The latter investigators, by keeping a chronicle over many summers of a population on Gibraltar Island, Lake Erie, were able to show that raiding is much more frequent in *subintegra* than in *sanguinea*. Some colonies raided almost daily for weeks at a time, striking out in any one of several directions on a given day. Occasionally the forays continued into the night, in which case the *subintegra* workers remained in the looted nest overnight and returned home the next morning. In other details the raiding behavior resembled that of *sanguinea*. Subsequently Regnier and Wilson (1971) discovered that each *subintegra* worker possesses a grotesquely hypertrophied Dufour's gland, which contains approximately 700 micrograms of a mixture of decyl, dodecyl, and tetradecyl acetates. These substances are sprayed at the defending colonies during the slave raids. They act at least in part as "propaganda substances" because they evaporate slowly and help to alarm and to disperse the defending workers (see Figure 12–6).

Little is known about the other nine or so American species of the *sanguinea* group (Creighton, 1950; Buren, 1968a), and their study is likely to reveal new behavioral phenomena related to dulosis. For example, a colony of *Formica wheeleri* that Wilson (1955c) observed in Yellowstone Park, Wyoming, divided its labor in a remarkable fashion between two species of slaves. *F. neorufibarbis* accompanied the *wheeleri* on a raid (against colonies of *F. fusca* and *F. lasioides* simultaneously) and assisted them in excavating and breaking into the plundered nests. Later, when the mixed nest was excavated for closer examination, the *neorufibarbis* were found to be concentrated in the middle and upper layers. They were very aggressive and joined the *wheeleri* in defending the nest. The workers of the second slave species, *F. fusca*, did not accompany the slavemakers on the

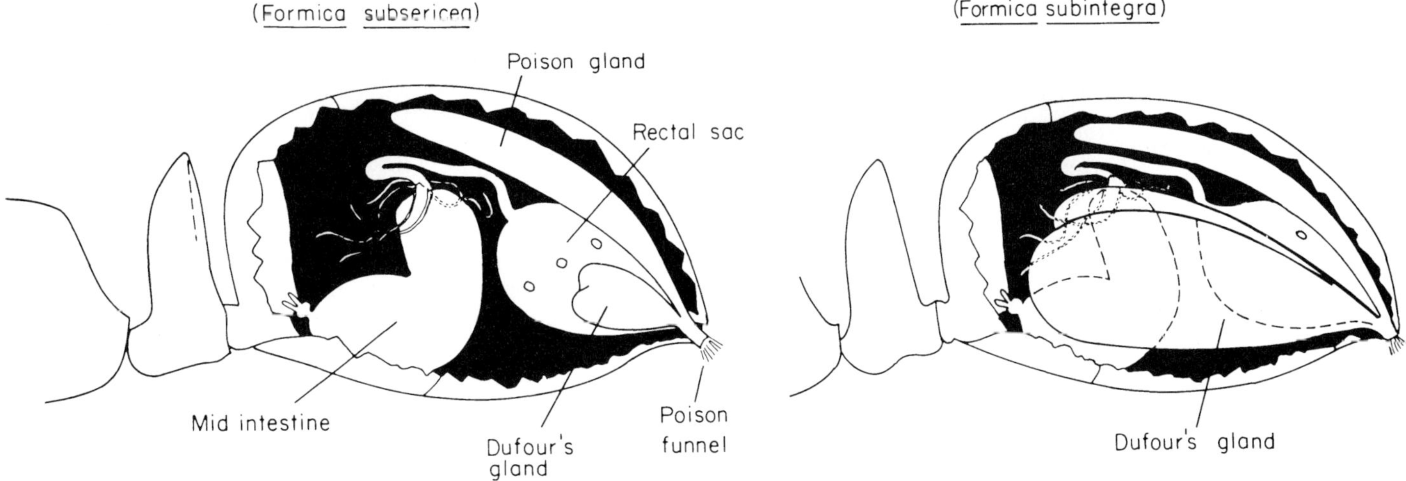


**FIGURE 12–6** The Dufour's gland of the slavemaker ant *Formica subintegra* is the source of large quantities of "propaganda substances" that scatter other *Formica* colonies during raids. It is contrasted here with the Dufour's gland of *F. subsericea*, which is of ordinary size. (From Regnier and Wilson, 1971.)

raid, and later they made only feeble attempts to defend the nest. Instead they were found concentrated in the lower layers of the nest close to the brood, and most had their crops distended with liquid food. These circumstances suggest that the *fusca* workers were specializing on food storage and brood care. A deeper significance of the dulotic habit is indicated by this example. It is apparent that the slavemaker colony not only adds to its labor force quantitatively by taking slaves, but can also incorporate specialists that increase the efficiency of the colony in a fashion analogous to that seen in normal worker polymorphism.

The mode of colony founding by queens of the *Formica sanguinea* group has not been observed in nature, and this surprising gap in our information continues to prevent a secure understanding of the evolutionary origins of dulosis. Wheeler (1906) conducted a series of laboratory experiments on the American species *F. rubicunda* which strongly indicate that the queens can function at least facultatively as temporary parasites. When he placed newly dealated (but still virgin) *rubicunda* queens in nests containing workers and brood of *F. fusca,* they responded in an aggressive and effective manner. They advanced on the *fusca* colonies, fighting and killing *fusca* workers that attacked them, then seizing and sequestering the *fusca* pupae, until finally all of the *fusca* workers were dead and the *rubicunda* queens stood guard over the confiscated brood. When new *fusca* workers emerged from the brood pile at a later date, they accepted the *rubicunda* queens and began to lick and to feed them. Viehmeyer (1908) and Wasmann (1908) subsequently repeated Wheeler's experiment with young mated queens of *F. sanguinea* and obtained the same result. The behavior of the intruding queens differs markedly from those belonging to the *fusca* and *microgyna* groups used in parallel experiments. There is no reason to doubt that, at least under certain conditions, the *sanguinea*-group queens do start new colonies by this form of unaided assault on colonies of slave species.

Wheeler, in his early writings, and later Santschi (1906) and Wasmann (1908), believed that such temporary parasitism not only characterized the ancestors of the slave-making *Formica* but was a prerequisite for the evolution of the dulotic habit itself. Together they postulated this explanation as an alternative to the Darwinian predation hypothesis, believing that, once predatory habits evolved in the queen during nest founding, it was far easier for the species to extend such behavior to the worker caste in the form of raiding for slaves. Later, Wheeler (1910a) saw the incongruity in his position, namely, that dulosis represents a wholly new behavior pattern that cannot be viewed simply as a variant of the temporarily parasitic mode of colony founding. He concluded, "In my opinion both temporary parasitism and dulosis have arisen independently from the practice of *F. rufa* and *F. sanguinea* of adopting fertilized queens of their own species." This opinion seems about right at the present time. Neither the predation nor territorial hypotheses are excluded by the demonstration of temporary social parasitism in the *sanguinea* group; they are moreover considerably strengthened by the growing evidence of raiding and accidental dulosis in *F. naefi* and *F. sanguinea,* as we pointed out earlier. It is still not known to what extent the various species of the *sanguinea* group rely on temporary parasitism to start new colonies, as opposed to homospecific adoption followed by budding.

The pinnacle (or nadir if you prefer) of the slave-holding way of life is reached in the formicine genus *Polyergus,* a totally dulotic group of species that have evidently been phylogenetically derived from *Formica* (see Plate 15). Five species are known: *rufescens* of Europe and North America, *breviceps* and *lucidus* of North America, *nigerrimus* of the Soviet Union, and *samurai* of Japan and eastern Siberia (Figures 12–7 and 12–8). These "amazon ants" are nowhere very common, but their striking appearance (large size, bright red or black coloration, and shiny body surface), the extraordinary degree of their behavioral specialization, and the spectacular qualities of their slave raids have placed them among the most frequently studied of all the ants from the time of Latreille (1805) and Huber (1810) onward. As usual, no one has ever approached Wheeler's ability to distill the important information in the form of a gripping narrative,

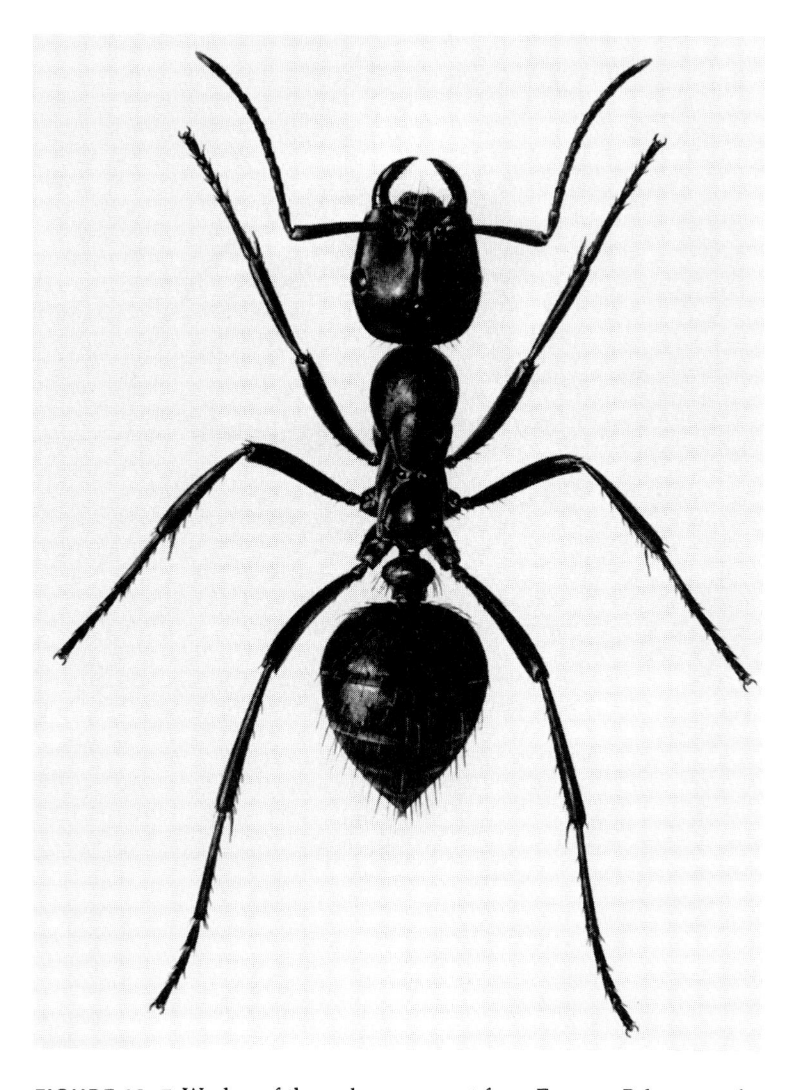

**FIGURE 12–7** Worker of the red amazon ant from Europe, *Polyergus rufescens.* (From Gösswald, 1985; painting by T. Hölldobler-Forsyth.)

and we will again defer to him. In the following passage he describes *P. rufescens:*

> The worker is extremely pugnacious, and, like the female, may be readily distinguished from the other Camponotine [formicine—authors] ants by its sickle-shaped, toothless, but very minutely denticulate mandibles. Such mandibles are not adapted for digging in the earth or for handling thin-skinned larvae or pupae and moving them about in the narrow chambers of the nest, but are admirably fitted for piercing the armor of adult ants. We find therefore that the amazons never excavate nests nor care for their own young. They are even incapable of obtaining their own food, although they may lap up water or liquid food when this happens to come in contact with their short tongues. For the essentials of food, lodging and education they are wholly dependent on the slaves hatched from the worker cocoons that they have pillaged from alien colonies. Apart from these slaves they are quite unable to live, and hence are always found in mixed colonies inhabiting nests whose architecture throughout is that of the slave species. Thus the amazons display two contrasting sets of instincts. While in the home nest they sit about in stolid idleness or pass the long hours begging the slaves for food or cleaning themselves and burnishing their ruddy armor, but when outside the nest on one of their predatory expeditions they display a dazzling courage and capacity for concerted action compared with which the raids of *sanguinea* resemble the clumsy efforts of a lot of untrained militia. The amazons may, therefore, be said to represent a more specialized and perfected stage of dulosis than that of the sanguinary ants. In attaining to this stage, however, they have become irrevocably dependent and parasitic. Wasmann believes that *Polyergus* is actually descended from *F. sanguinea*, but it is more probable that both of these ants arose in pretertiary times from some common but now extinct ancestor. The normal slaves of the European amazons are the same as those reared by *sanguinea*, viz: *F. fusca, glebaria, rubescens, cinerea,* and *rufibarbis;* and of these *fusca* is the most frequent. But the ratio of the different components in the mixed nests is the reverse of that in *sanguinea* colonies, there being usually five to seven times as many slaves as amazon workers. The simultaneous occurrence of two kinds of slaves in a single nest is extremely rare, even when the same amazon colony pillages the nests of different forms of *fusca* during the same season. This is very probably the result of the slaves' having a decided preference for rearing only the pupae of their own species or variety and eating any others that are brought in. Two slave forms may, however, appear in succession in the same nest. Near Morges, Switzerland, Professor Forel showed me an amazon colony which during the summer of 1904 contained only *rufibarbis* slaves, but during 1907 contained only *glebaria.*
>
> Unlike *sanguinea, rufescens* made many expeditions during July and August, but these expeditions are made only during the afternoon hours. One colony observed by Forel (1874) made 44 sorties on thirty afternoons between June 29 and August 18. It undoubtedly made many more which were not observed, as Forel was unable to visit the colony daily . . . Forel estimated the number of amazons in the colony at more than 1,000 and the total number of pupae captured at 29,300 (14,000 *fusca*, 13,000 *rufibarbis*, and 2,300 of unknown provenience, but probably *fusca*). The total number for the summer (1873) was estimated at 40,000. This number is certainly above the average, as the amazon colony was an unusually large one. Colonies with only 300 to 500 amazons are more frequent, but a third or half of the above number of pillaged cocoons shows what an influence the presence of a few colonies of these ants must have on the *Formica* colonies of their neighborhood. Of course, only a small proportion of the cocoons are reared. Many of them are undoubtedly injured by the sharp mandibles of the amazons and many are destroyed and eaten after they have been brought home.
>
> The tactics of *Polyergus*, as I have said, are very different from those of *sanguinea*. The ants leave the nest very suddenly and assemble about the entrance if they are not, as sometimes happens, pulled back and restrained by their slaves. Then they move out in a compact column with feverish haste, sometimes, according to Forel, at the rate of a meter in 33⅓ seconds or 3 cm. per second. On reaching the nest to be pillaged, they do not hesitate like *sanguinea* but pour into it at once in a body, seize the brood, rush out again and make for home. When attacked by the slave species they pierce the heads or thoraces of their opponents and often kill them in considerable numbers. The return to the nest with the booty is usually made more leisurely and in less serried ranks. (Wheeler, 1910a: 472–473)

The means by which the *Polyergus* workers are able to mobilize themselves within minutes and run in a compact column straight for the target colony was for a long time one of the classic problems of entomology. While watching *P. lucidus* colonies in Michigan, Talbot (1967) noticed that, prior to the onset of each raid, several scout workers explored the surrounding terrain, including the vicinity of the specific nest later raided. By monitoring the *Polyergus* nests carefully, she saw that the beginning of the raid was often signaled by the appearance of a scout returning from the direction of the target nest. As she describes it:

> On other days the departure of scouts was less conspicuous, and seldom was one lucky enough to spot a scout coming in. But whenever an ant came in hurriedly from the grass and went directly into the

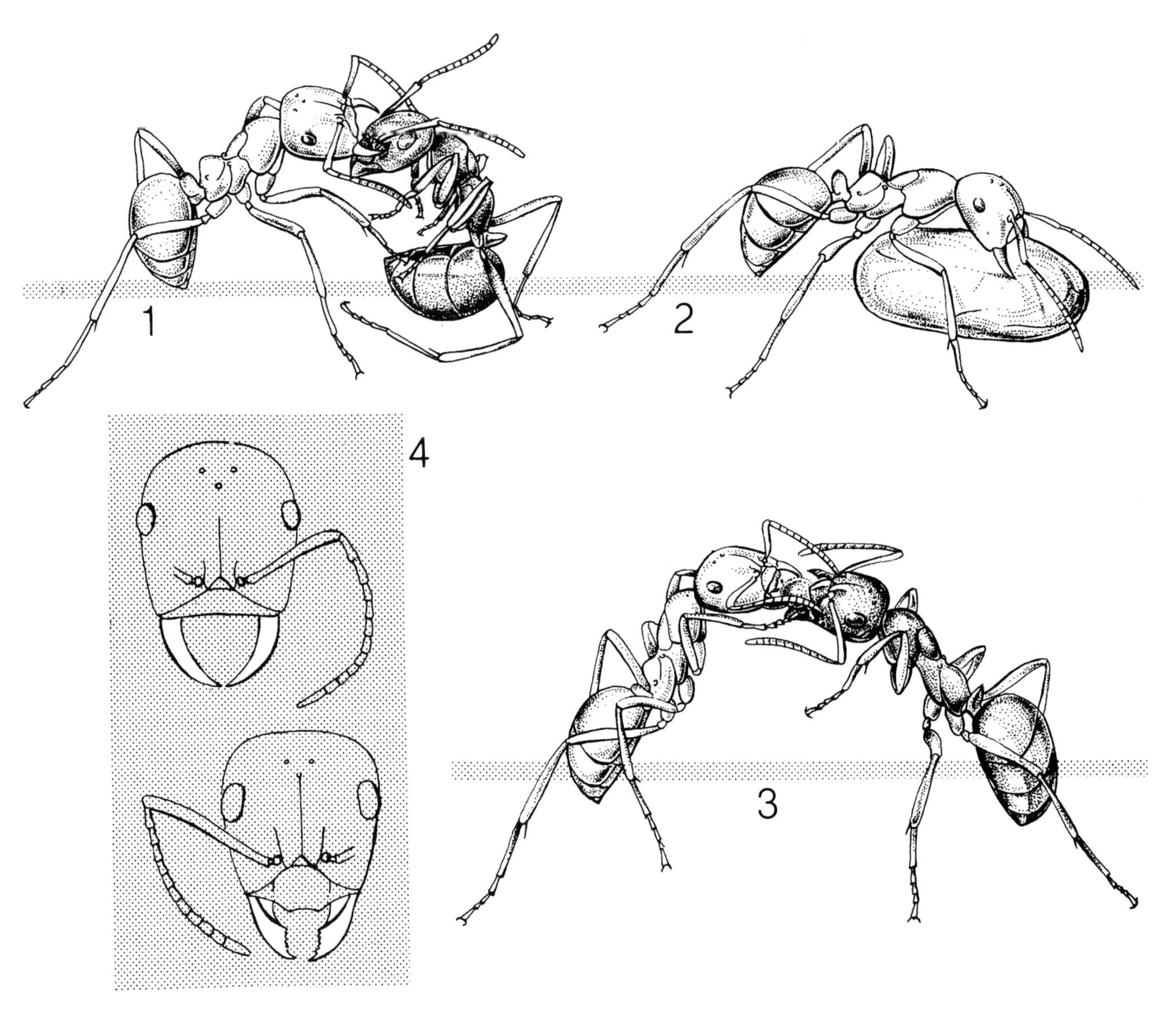


**FIGURE 12–8** Adaptations of the European amazon ant *Polyergus rufescens* for slave making. *(1)* A *P. rufescens* worker attacks a worker of *Formica fusca* during a slave raid, then *(2)* returns to her own nest with a *fusca* pupa (encased in a cocoon). *(3)* A *P. rufescens* worker is fed by an *F. fusca* slave that has eclosed from a captured pupa. *(4)* The sickle-shaped mandibles of *P. rufescens* are contrasted with the ordinary mandibles of the slavemaker *F. subintegra,* which has wider masticatory blades lined with teeth. (From Hölldobler, 1973c; based on Wilson, 1971; drawing by T. Hölldobler-Forsyth.)

nest, there was an outpouring of ants. It was thus assumed that whenever a sudden emergence occurred it was in response to a messenger arriving with news of a located colony. If this was correct and if the scouting ant, which found a colony, laid down an odor trail on its way home, then the odor must have been quite long lasting, for it sometimes took an ant 30 to 45 minutes to return from a raided nest. It seemed unlikely that a raiding group could be following anything but an odor trail, for it moved rapidly, did not maintain leaders, and usually stopped at exactly the right place.

The next logical step was to try to induce false raids by means of the artificial trail test. Talbot accomplished this in a manner that decisively favored her hypothesis. When she laid down dichloromethane extracts of whole *Polyergus* bodies over the ground along an arbitrary path away from the nest and at the time of day raids normally occur, *Polyergus* workers poured from the nest and followed the trails to the end. Thus Talbot was able to activate the raid swarms at will and lead them to targets of her choosing. Finally, she induced a complete raid on a colony of *Formica nitidiventris* by placing it in a box 2 meters from a *Polyergus* colony and drawing an artificial *Polyergus* trail to it. Talbot concluded that the *Polyergus* raid phalanxes do not contain leaders, but this may not be correct. Although it is true that excited *Polyergus* are capable of moving out along trails without further guidance, Topoff et al. (1984, 1985a,b) recently demonstrated that in the case of *P. breviceps* at least, naturally occurring raids are always led by scout ants. Experiments further revealed that visual cues are more important to the leaders than are the chemical trails. After the attack, however, *P. breviceps* workers orient homeward by a combination of visual guideposts and the chemical trails laid by the outward-bound leaders.

Emery (1911a), by employing introduction experiments of the kind invented by Wheeler for studies of temporary parasitism, discovered that the queens of *Polyergus rufescens* act essentially like those of other parasitic groups during colony founding. When presented with a colony of *Formica fusca* in the laboratory, the *rufescens* queen works her way into the nest by submissive posturing and secures adoption by the *fusca* workers and queen. Then, after a week or so has elapsed, she kills the *fusca* queen by piercing her head with her sharp mandibles. The frequency with which this mode of colony formation is used in nature is not known. It must occur at least occasionally since single dealated *P. rufescens* queens have been found alone in small *F. fusca* nests on at least three occasions, the most recent at Aosta, Italy, by A. Buschinger (personal communication).

The American amazon ant *Polyergus lucidus* is evidently similar in most respects to the European *P. rufescens* in colony foundation. Goodloe and Sanwald (1985) observed newly mated queens penetrate and secure adoption in queenless colony fragments of *Formica nitidiventris* and *F. schaufussi*. The queens were given a choice of the two host species, and in all of 13 penetrations they entered a colony fragment of the same species used as slaves by their colony of origin. In addition, colonies of *Polyergus lucidus* can reproduce by budding (Marlin, 1968). Some of the queens return to their home nests following the nuptial flights. Later they accompany workers on a raid and remain behind with a few of them in a plundered *Formica* nest or in some neighboring nest site.

H. R. Topoff and his co-workers (personal communication, 1987) conducted a series of laboratory experiments in which newly mated queens of *Polyergus breviceps* were allowed to invade colonies of their host species, *Formica gnava*. The *Polyergus* queens quickly detected the *Formica* queen and killed her. Surprisingly, the aggressive behavior of the *Formica* workers toward the invading *Polyergus* queen subsided shortly after their own queen was killed. Topoff and his co-workers suggest that the *Polyergus* queen releases an appeasement substance from her enlarged Dufour's gland. When *Pogonomyrmex* workers, which served as test objects, were contaminated with the Dufour's gland secretions of *Polyergus* queens, they were ignored by the *Formica* workers, whereas untreated *Pogonomyrmex* were attacked.

In 1932 Arnoldi reported the discovery of a new and equally spectacular kind of formicine slave-making ant. The species, *Rossomyrmex proformicarum*, superficially resembles *Polyergus*, but has evidently been derived from *Formica*-like ancestors in a line separate from that leading to the amazon genus. His observations were later supplemented in considerable detail by Marikovsky (1974). *R. proformicarum* is locally common in the semi-deserts and dry *Artemisia-Festuca* steppes over a 1,000-kilometer-wide area of Soviet Central Asia. It enslaves *Proformica epinotalis* and *P. nasuta*, which are closely related to the genus *Formica* and abundant in xeric habitats. The method of raiding is unique. After a long reconnaissance one or a few scouts begin to transport some of their nestmates from the home nest to the *Proformica* nest. The scout seizes a fellow worker by her mandibles, whereupon the latter folds up her legs, tucks under her abdomen, and allows herself to be carried in the typical formicine fashion. Fifty or more such pairs set off in a loose file for distances of as much as 50 meters. They then halt, uncouple, and search for the entrances to the *Proformica* nests. As soon as the *Rossomyrmex* approach, the *Proformica* close the exits with particles of earth. The slavemakers may require several hours to break through the barrier. They then make short work of the defenders and begin to carry away brood. Whereas other slave-making species, including *Formica* and *Polyergus*, usually steal only pupae, the *R. proformicarum* raiders steal all stages of brood—eggs, larvae, and pupae. The pairwise carrying behavior used in the raids is basically the same as that used by *Formica* during colony movements from one nest site to another. The *Rossomyrmex* have simply adapted it to a new function. It is reminiscent of the "emigration raids" described by Kwait and Topoff (1983) in *Polyergus lucidus*, which occur at the end of the raiding season. The *Polyergus* workers gather into a swarm, "raid" their old nest, and carry the adult *Formica* slaves to a new site. An intermediate stage in the evolution of the *Rossomyrmex* habit is displayed by the slavemaker *F. wheeleri*, whose workers have been observed carrying one another back to the home nest following a raid (Wilson, 1955c).

Even more remarkable in another sense is the existence of a phylogenetically independent form of dulosis in the myrmicine genera *Chalepoxenus*, *Epimyrma*, *Harpagoxenus*, *Leptothorax* (= *Myrafant*), *Myrmoxenus*, *Protomognathus* (recently separated as a genus from *Harpagoxenus*), and *Strongylognathus*. The *Protomognathus* case is the most specialized and also by far the best understood. The single known species, *P. americanus* of North America, has been closely studied by Sturtevant (1927), Creighton (1929), and especially Wesson (1939), whose analysis of the life cycle and behavior is a model of its kind. Alloway (1979) and Alloway and Del Rio Pesado (1983) have added important details of the raiding behavior. The *Protomognathus* workers are small, blackish-brown ants superficially resembling some of the *Leptothorax* (= *Myrafant*) species they enslave. Their most distinguishing feature is the presence of "antennal scrobes"—long, deep pits along the sides of the head into which the antennae are folded for protection during the raids (Figure 12–9). *P. americanus* is a relatively widespread species, existing in very local but dense populations from Ontario south to Virginia and west to Ohio. It enslaves three of the commonest *Leptothorax* (= *Myrafant*) species of eastern North America, *L. ambiguus*, *L. curvispinosus*, and *L. longispinosus*. In the populations studied by Wesson, the ratio of *P. americanus* mixed colonies to local pure *Leptothorax* colonies was about 1:15. The mixed colonies contained up to 50 *Protomognathus* workers and 300 *Leptothorax* worker slaves. Most colonies were much smaller than this, the medians being about 6 *Protomognathus* and 30 *Leptothorax*, respectively.

Following the nuptial flight in early or middle July, the newly fecundated queen sheds her wings and crawls about on the ground or low vegetation in search of a *Leptothorax* nest. On encountering a nest, she begins quite literally to throw the *Leptothorax* adults out. As each worker approaches her in turn, she seizes her by an antenna, drags her out of the nest entrance, and flings her to one side. She avoids attacks on her own body by very rapid, shifting movements. After she has savaged the colony in this manner for a while, the *Leptothorax* queen and workers finally panic and desert the nest. The *Protomognathus* queen then appropriates the larvae and pupae left behind.

When *Leptothorax* workers later eclose from the pirated brood, they adopt the *Protomognathus* queen without hesitation, and soon afterward she begins to lay eggs. The *Protomognathus* workers that develop from these eggs are degenerate in behavior. They spend almost all of their time in the nest grooming each other and "loafing." They are fed with liquid food regurgitated to them by the *Leptothorax* slaves, who also assume the thankless tasks of foraging, nest construction, and brood care. When the *Protomognathus* depart on the slave raids, however, they reveal themselves to be efficient little fighting machines. The raids are initiated by scouts, who hunt singly or in small groups for *Leptothorax* colonies in the vicinity of the *Protomognathus* nest. When a *Protomognathus* scout encounters a

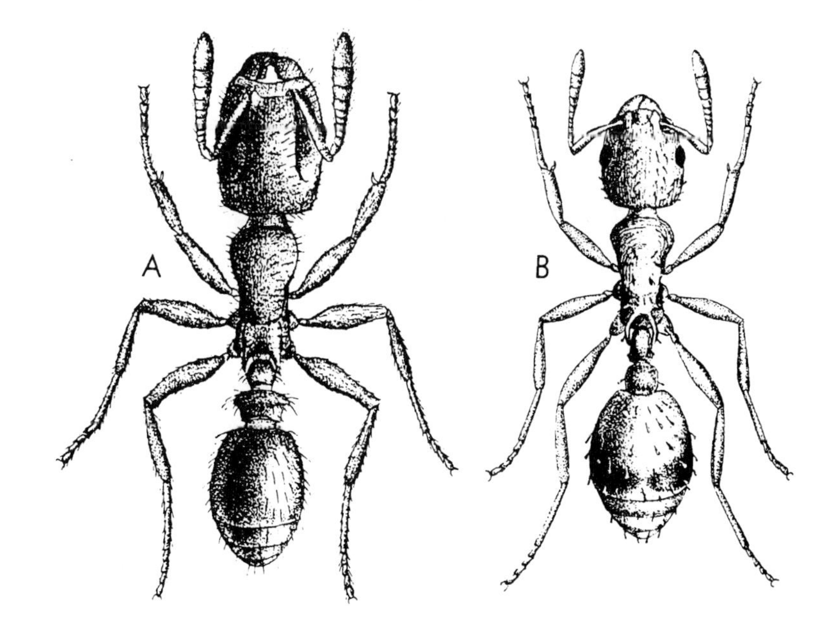


**FIGURE 12–9** Workers of *Protomognathus americanus* (*A*) and one of its slave species, *Leptothorax* (= *Myrafant*) *curvispinosus* (*B*). The fighting ability of the *Protomognathus* is enhanced by clipper-like mandibles, which are powered by large adductor muscles filling most of the elongate head. During combat the slavemakers are also able to fold their antennae back into the grooves running the length of each side of the head. The genus *Protomognathus* has recently been split off taxonomically from *Harpagoxenus*. (From Wheeler, 1910a.)

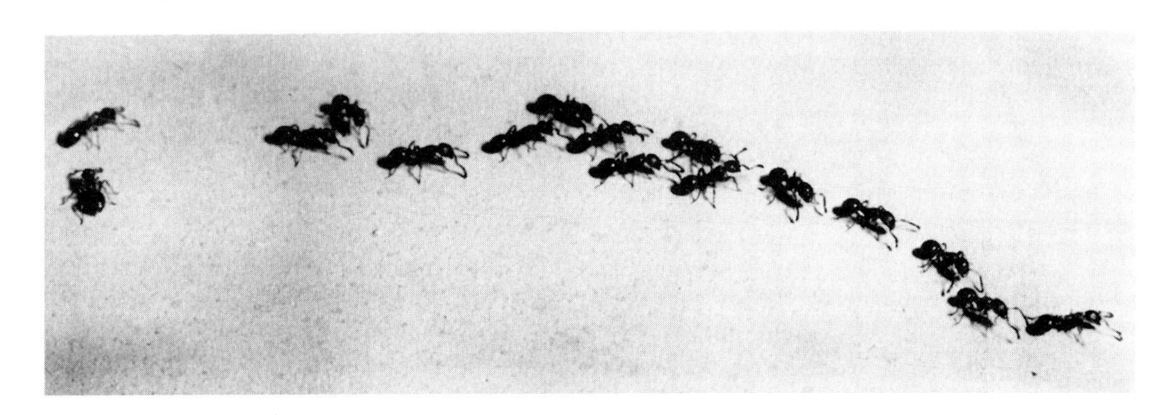

**FIGURE 12–10** Slave raid of *Epimyrma ravouxi* (= *goesswaldi*). A raiding column is led by a scout ant. (From Winter, 1979b.)

*Leptothorax* nest, she normally attempts to penetrate the entrance without hesitation. If the colony is small and weak, she may succeed in capturing it single-handedly, after which she begins to transport the *Leptothorax* brood back to her own nest. If, on the other hand, the scout is repulsed, she returns to her nest, excites her nestmates (evidently by release of a pheromone), and soon sets out again for the newly discovered *Leptothorax* nest. This time she lays down a short-lived odor trail which draws out a tight little column of *Protomognathus* workers and *Leptothorax* slaves. If this group is still not sufficient to breach the *Leptothorax* nest, some of the *Protomognathus* workers return to the home nest and bring out auxiliary columns. A similar recruitment behavior during slave raids has been observed in *L. duloticus* (Alloway, 1979), *Epimyrma ravouxi* (= *E. goesswaldi*) (Buschinger et al., 1980a; see Figure 12–10), and *Myrmoxenus gordiagini* (Buschinger et al., 1983).

Toward the end of the summer the raids are transmuted into an unusual form of colony multiplication. An increasing tendency develops for some of the *Protomognathus* workers to remain behind in the conquered nests, where they stand guard over the *Leptothorax* brood. When this happens the expatriates soon lose contact with the home nest, and they are treated as queens by the *Leptothorax* workers who subsequently eclose from the pupae. As Buschinger and Alloway (1977) showed, the eggs laid by the *Protomognathus* workers are unfertilized (the workers in fact lack spermathecae) and give rise to males, possibly along with a few workers. Such secondary colonies are very common and even rival in number the primary colonies started by single *Protomognathus* queens. Of 32 colonies censused by Wesson in Maryland, Ohio, and Pennsylvania, no fewer than 16 were populated exclusively by *Protomognathus* workers together with their slaves.

The genus *Harpagoxenus* is outwardly very similar to *Protomognathus* in physical appearance but evidently independently derived in evolution. It is closest to species constituting *Leptothorax* in the strict sense and enslaves some of them, whereas *Protomognathus* resembles and enslaves species of *Leptothorax* sometimes distinguished as the genus or subgenus *Myrafant*. The European *H. sublaevis* (Figure 12–11) has been examined in successively greater detail by Adlerz (1896), Viehmeyer (1921), Buschinger (1966a,b, 1968a,b), Buschinger and Winter (1977), and Winter (1979a), while the North American *H. canadensis* has been studied by Stuart and Alloway (1983). Both *Harpagoxenus* species, as well as *Chalepoxenus muellerianus*, recruit nestmates by tandem running. As soon as the scout ant returns from a newly discovered host nest, she displays an invitation behavior. She then leads a single nestmate to the host nest (Figure 12–12). "The follower ant may stay near the host nest, or return, like the scout, to the mother nest in order to recruit further nestmates. This system often involves a rather long 'siege' of the host nest, until the number of warriors is high enough to risk a direct

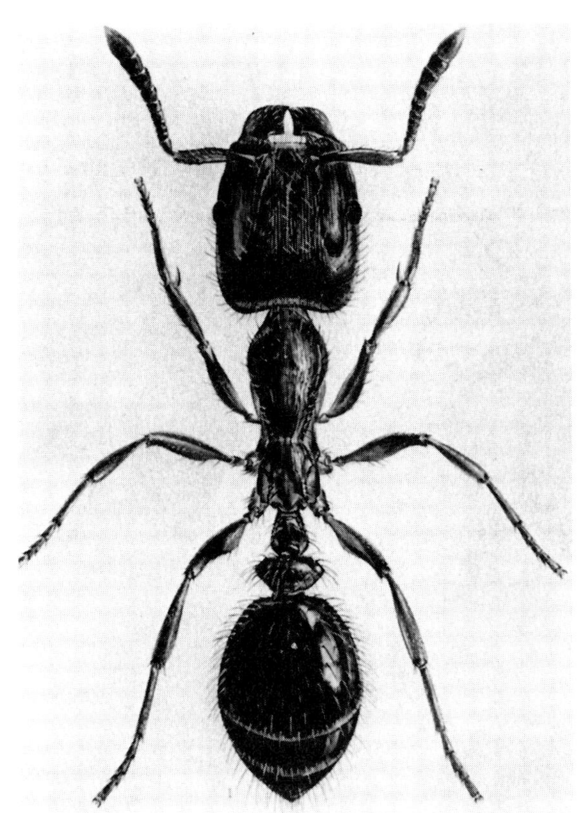

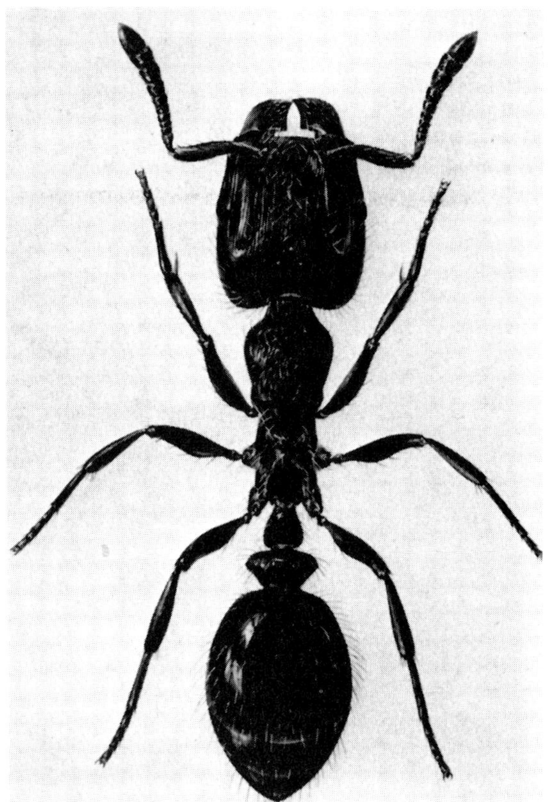

**FIGURE 12–11** A worker (*left*) and intermorph of *Harpagoxenus sublaevis*. The specimen on the right has well-developed ocelli. (Paintings by T. Hölldobler-Forsyth, courtesy of K. Gösswald.)

**FIGURE 12–12** *Harpagoxenus sublaevis* workers recruit by tandem running during slave raids. (Photograph courtesy of A. Buschinger.)

attack" (Buschinger et al., 1980a). The slave-making *Leptothorax* (= *Myrafant*) *duloticus* was discovered by Wesson (1937) in Ohio and later studied by Talbot (1957), Wilson (1975a), and Alloway (1979). Its basic parasitic behavior is similar to that of *Protomognathus* and *Harpagoxenus*. The same is true of the European leptothoracines *Chalepoxenus* (Ehrhardt, 1982), *Epimyrma* (Buschinger and Winter, 1985; Buschinger, 1986), and *Myrmoxenus* (Buschinger et al., 1983).

Raiding workers and colony-founding queens of *Harpagoxenus sublaevis* appear to employ chemical weapons when invading the nest of the host species *Leptothorax acervorum*. Buschinger (1968b, 1974d) found that the *Harpagoxenus* have hypertrophied Dufour's glands (Figure 12–13). He observed *L. acervorum* workers attacking each other following contact with *H. sublaevis* that were either conducting slave raids or founding colonies (Figure 12–14). He suggested that *Harpagoxenus* discharges a propaganda pheromone from the Dufour's gland. Recently Allies et al. (1986) confirmed Buschinger's hypothesis and also demonstrated in laboratory experiments that the workerless inquiline ant *Doronomyrmex kutteri* (which also possesses hypertrophied Dufour's glands; see Buschinger, 1974b) uses Dufour's gland secretions as a chemical weapon in defense against hostile workers of *L. acervorum*. Host workers, contaminated with Dufour's gland secretions of *D. kutteri* queens, attack each other. They evidently no longer recognize each other as nestmates. It is interesting to note that another inquiline parasite of *L. acervorum, D. pacis,* also possesses a hypertrophied Dufour's gland. A third inquiline species, *D. goesswaldi,* has only a slightly enlarged Dufour's gland (Buschinger, 1974b).

Thanks largely to the outstanding efforts of Buschinger and his collaborators, we now know that the species of *Epimyrma* form a series of steps from full dulotic behavior to workerless inquilinism—or perhaps more precisely, degenerate dulosis, since unlike true inquilines the parasite queens continue to assassinate the host queens (Buschinger, 1982a, 1986; Buschinger and Winter, 1983b). The earlier view (Wilson, 1971) that the genus followed the temporary parasitism route to inquilinism is incorrect. Eight species in this European and North African genus are known at the present time. *Myrmoxenus gordiagini,* a dulotic ant of the Soviet Union and Yugoslavia, is very close morphologically and behaviorally and very likely constitutes a ninth species. Three of the *Epimyrma* species, including *algeriana, ravouxi* (= *goesswaldi*), and *stumperi,* conduct well-organized

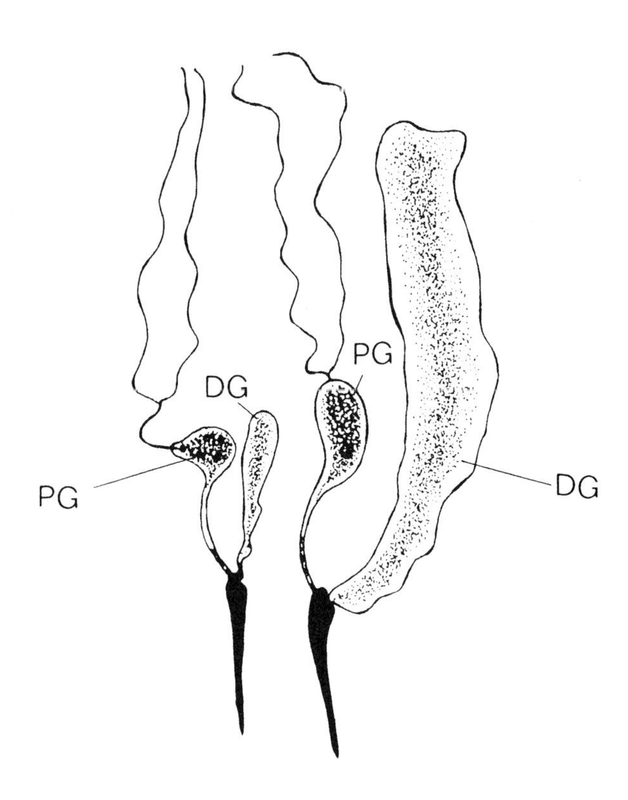


**FIGURE 12–13** Poison gland (*PG*) and Dufour's gland (*DG*) of *Leptothorax acervorum* (*left*) and *Harpagoxenus sublaevis* (*right*), showing the hypertrophied condition of the gland in the latter, slave-making species. (Modified from Buschinger, 1968b.)

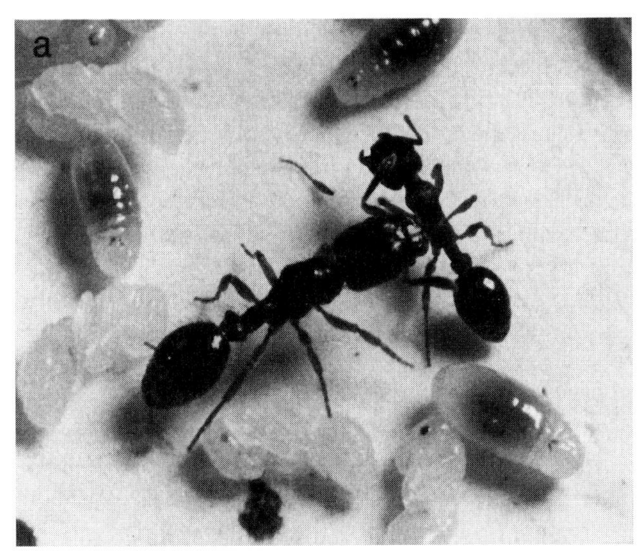


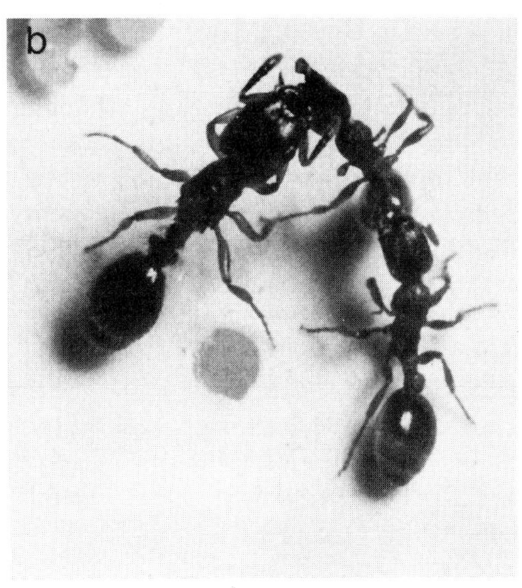


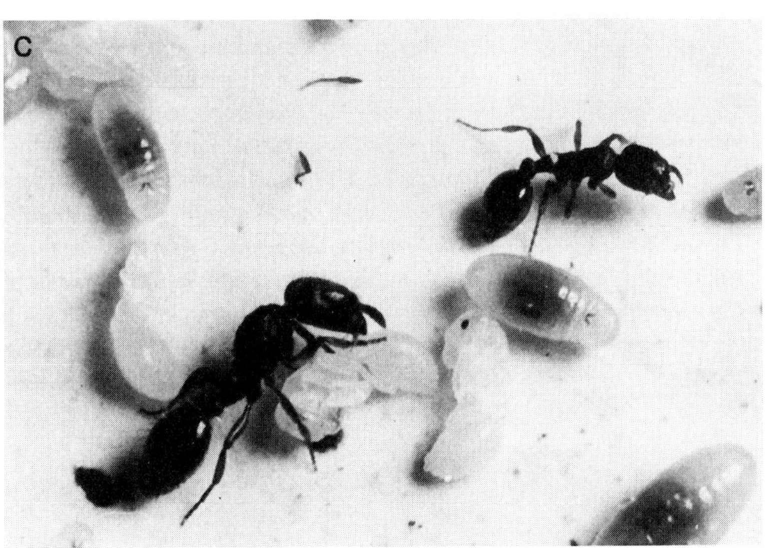


**FIGURE 12–14** A newly mated female of the European slavemaker *Harpagoxenus sublaevis* invades a host colony of *Leptothorax acervorum*. (*a*) The *H. sublaevis* queen dismembers a worker of *L. acervorum*. (*b*) After contact with the *Harpagoxenus* the *L. acervorum* workers attack each other. A *Leptothorax acervorum* worker bites a nestmate, who is simultaneously attacked by the *Harpagoxenus* queen. (*c*) An injured *L. acervorum* worker stands aside as the *Harpagoxenus* queen takes over the *L. acervorum* brood. (From Buschinger, 1974d.)

slave raids with group recruitment and fight with a powerful sting. In a second group, including *kraussei,* the number of workers is reduced drastically to five or fewer. The workers can still raid, but their efforts are ineffectual. Finally, at least two species, *corsica* and a still undescribed species from Greece, are completely workerless (A. Buschinger, personal communication). A simultaneous evolution in sexual behavior has occurred, from normal nuptial flights to mating within the nest and hence inbreeding. Three of the fully dulotic species conduct the flights. A fourth, *algeriana,* and all of the degenerate slavemakers and workerless species mate inside the nest.

It is a remarkable fact that so far as we now know, the queens of all of the *Epimyrma* species, including the workerless *E. corsica,* found new colonies by forcibly entering the host nest and throttling the resident queen. The details have been worked out over many years in studies by Gösswald (1933), Kutter (1951, 1969), and Buschinger and his associates. After shedding her wings, the queen searches for a new host colony. The mode of entry and subsequent behavior vary greatly among the various species. The queen of *E. kraussei,* approaching a *Leptothorax recedens* colony, makes repeated hostile approaches to the host workers and "intimidates" them, to use Kutter's expression. If she succeeds in entering the nest, she kills the host queen and secures complete adoption by the rest of the colony. The queen of *E. ravouxi* (= *goesswaldi*), on the other hand, calms the host workers (*L. unifasciatus*) by stroking them with her antennae and lower mouthparts. Once inside the nest, she mounts the host queen from the rear, seizes her around the neck with her saber-shaped mandibles, and kills her.

*Epimyrma stumperi,* studied in Switzerland by Kutter, uses still another variation to enter the nests of its host, *Leptothorax* (= *Myrafant*) *tuberum.* The queen first stalks the host colony with slow, deliberate movements. When approached by the *Leptothorax* workers, she crouches down, "freezes," and seems to feign death. After a time she begins to mount the workers from the rear, strokes their bodies with her foreleg combs, and grooms herself, perhaps thereby passing nest odors back and forth. With this display of sophistication in evidence, it is not surprising to find that queens of *E. stumperi* (like those of other *Epimyrma*) are able to penetrate host colonies relatively quickly. Once inside the nest, the *E. stumperi* queen begins an implacable round of assassinations directed at the host queens, of which there are usually between two and five in the *L. tuberum* colonies. She mounts each queen in turn, forces her to roll over, then seizes her by the throat with her mandibles. The sharp tips of the mandibles squeeze the soft intersegmental membrane of the neck of the victim. The *Epimyrma* maintains her grip for hours or even days, until the paralyzed *Leptothorax* queen finally dies. Then the invader moves on to the next queen and repeats her behavior until no host queen is left. It is a matter of more than ordinary interest that the *E. stumperi* workers also occasionally mount *Leptothorax* workers and go through an ineffectual rehearsal of the assassination behavior, but without harming their "victims" and with no visible benefit to themselves. This seems best interpreted as a partial transfer of the queen's behavioral pattern to the worker caste where it has neither positive nor harmful effects.

The Palaearctic myrmicine genus *Strongylognathus* provides a second possible example of the transition from dulosis to inquilinism. The natural history of the genus has been gradually explored over a period of many years by Forel (1874), Wasmann (1891), Wheeler (1910a), Kutter (1923, 1969), Pisarski (1966), and others. *Strongylognathus* is closely related to *Tetramorium,* and its species enslave members of the latter genus. The most favored slave species is *T. caespitum,* one of the most abundant and widespread ant species of Europe. *S. alpinus* has a life cycle more or less typical of the majority of *Strongylognathus* species. It is at an evolutionary level somewhat less advanced than that of *Harpagoxenus* and *Protomognathus* in the one special sense that the behavior of its workers is less degenerate. The workers, like those of most parasitic ant species, do not forage for food or care for the immature stages; nevertheless, they still feed themselves and assist in nest construction. The raids of *alpinus* are notoriously difficult to observe. They are thought to occur in the middle of the night and to take place, for the most part, along underground galleries. The *alpinus* workers are accompanied by *T. caespitum* slaves, who, true to the aggressive nature of their species, join in every phase of the raid. Warfare against the target colony is total: the nest queen and winged reproductives are killed, and all of the brood and surviving workers are carried back and incorporated into the mixed colony. This union of adults should not be too surprising when it is recalled that *T. caespitum* colonies, even in the absence of *Strongylognathus,* frequently conduct pitched battles that sometimes terminate in colony fusion. The *S. alpinus* workers are well equipped for lethal fighting. Like some other dulotic and parasitic ant species, they possess saber-shaped mandibles adapted for piercing the heads of their resisting victims (see Figure 12–15). The mode of colony multiplication is not known, but it is evident that the host queen is somehow eliminated in the process.

One member of the genus, *Strongylognathus testaceus* (Figure 12–16), has evolved at least partway to inquilinism. The *Tetramorium* queen is tolerated and lives side by side with the *S. testaceus* queen. There are fewer *testaceus* than host workers, the usual situation found in advanced dulotic species. The *testaceus* workers do not engage in ordinary household tasks and are wholly dependent on the host workers for their upkeep. They have never been observed to conduct raids, although R. Johann (cited by Buschinger, 1986) found that laboratory colonies can somehow gain slaves from *Tetramorium* colonies—perhaps by some form of chemical warfare. The key fact is that the *S. testaceus* queen can depend upon the host queen to supply her with slave workers, without the necessity of raids. Furthermore, the parasites soon control the reproductive activity of the host queens. The *Tetramorium* queen generates only workers and no reproductives, whereas the *S. testaceus* queen is privileged to produce both castes. Nevertheless, the presence of the *Tetramorium* queens permits the mixed colonies to attain great size. Wasmann found one comprising 15,000 to 20,000 *Tetramorium* workers and several thousand *Strongylognathus* workers. The brood consisted primarily of queen and male pupae of the inquiline species. It is evident that *S. testaceus* is in a stage of parasitic evolution just a step beyond that occupied by *S. alpinus.* The worker caste of *testaceus* has been retained, and it still has the murderous-looking mandibles dating from the species' dulotic past, but it has evidently lost all of its former functions and is in the process of being reduced in numbers. Probably *S. testaceus* is on the way to dropping the worker caste altogether, a final step that would take the species into the ranks of the extreme inquilines.

An interesting and still puzzling phenomenon is "revolt" among the slaves. On three occasions Wilson (1975a) saw *Leptothorax curvi-*

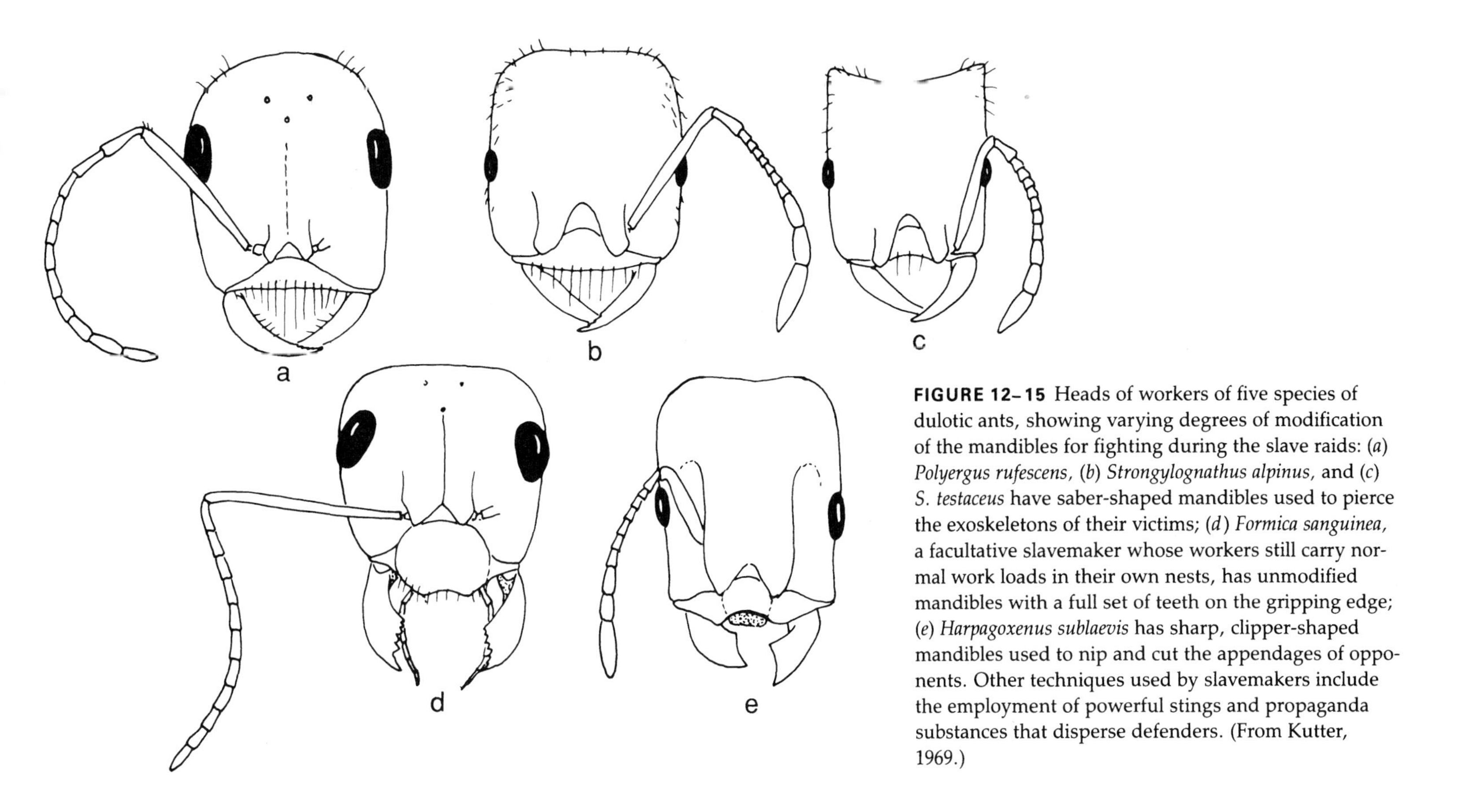


**FIGURE 12–15** Heads of workers of five species of dulotic ants, showing varying degrees of modification of the mandibles for fighting during the slave raids: (*a*) *Polyergus rufescens,* (*b*) *Strongylognathus alpinus,* and (*c*) *S. testaceus* have saber-shaped mandibles used to pierce the exoskeletons of their victims; (*d*) *Formica sanguinea,* a facultative slavemaker whose workers still carry normal work loads in their own nests, has unmodified mandibles with a full set of teeth on the gripping edge; (*e*) *Harpagoxenus sublaevis* has sharp, clipper-shaped mandibles used to nip and cut the appendages of opponents. Other techniques used by slavemakers include the employment of powerful stings and propaganda substances that disperse defenders. (From Kutter, 1969.)

*spinosus* slave workers approach the queen of *L. duloticus,* the mother slavemaker, and bite at her head and thorax. Simultaneously or immediately afterward the worker laid an egg. In two of the incidents the worker safely placed the egg in one of the egg piles; but in the third case the queen seized the egg, pulled it back and forth with the worker holding on to the other end, and finally ruptured and ate it. Slave hostility was not limited to the moment of oviposition. Once, as the *duloticus* queen wandered away from the egg pile, she was seized on the right hind tarsus by a young *curvispinosus* slave, who then alternately tried to drag her backward and to sting her. From time to time during this incident, which lasted 20 minutes, the worker stridulated. All of these actions are typical of *curvispinosus* workers engaged in fighting alien ants. Alloway and Del Rio Pesado (1983) witnessed comparable aggression of *L. ambiguus* and *L. longispinosus* slaves against *Protomognathus americanus.* On many occasions they saw the *Leptothorax* biting their mistresses and dragging them out of the slavemaker nests. A few *Protomognathus* workers lost parts of appendages as a result of these attacks. On the other hand no slavemaker was ever seen to attack a slave. Nor was the revolt generalized. The same slave that attacked one slavemaker would typically feed and groom another, and any slavemaker attacked by one slave was cared for by others.

Another trait of slave-making ants of interest to theoreticians is the degree of degeneracy of the worker caste. We have seen that at

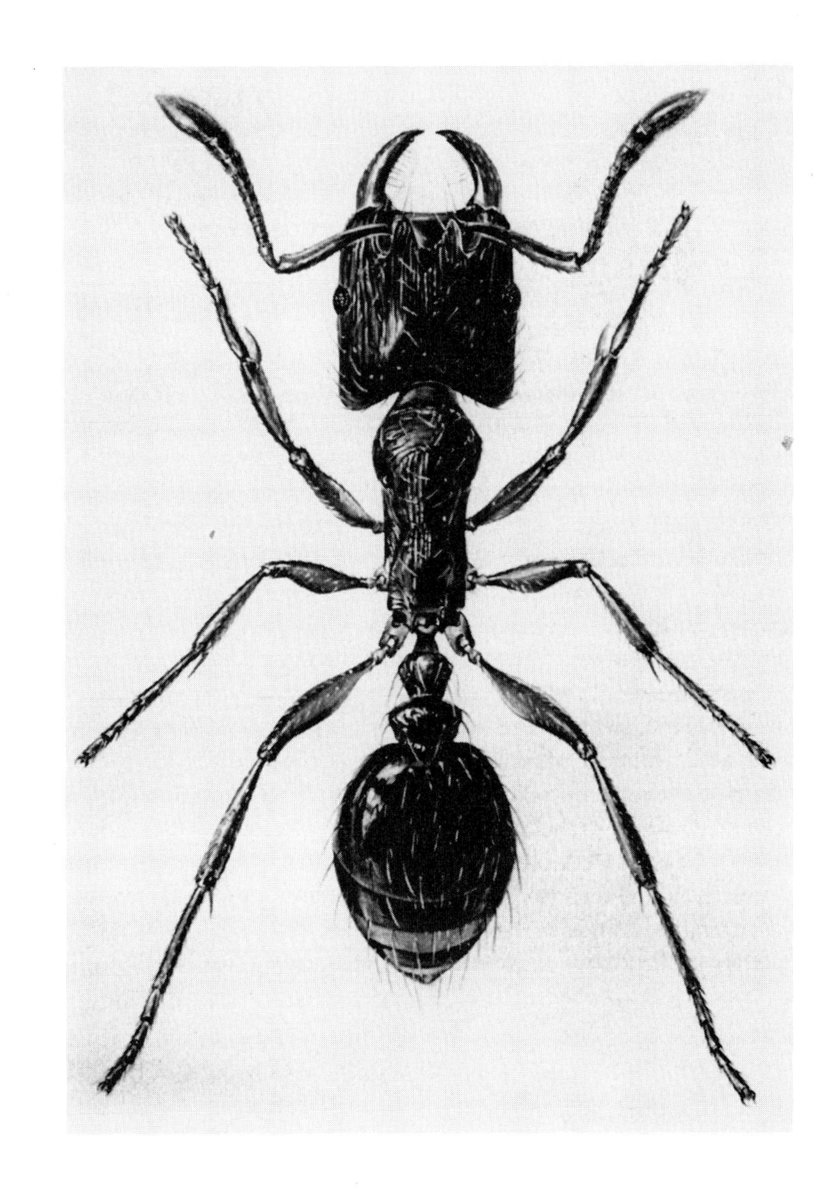

**FIGURE 12–16** A worker of the social parasitic ant *Strongylognathus testaceus.* (From Gösswald, 1985; painting by T. Hölldobler-Forsyth.)

one end of the spectrum, represented by *Formica sanguinea*, the workers are self-sufficient. They conduct all of the quotidian tasks of the colony on their own, and they can easily survive without the support of slaves. At the other extreme, represented by the degenerate slavemakers of the genus *Epimyrma*, the workers have a very limited behavioral repertory and are apparently completely helpless without their slaves.

Wilson (1975a), in his study of *Leptothorax duloticus*, asked the question, If such an obligate slavemaker were artificially deprived of its slaves, would it expand its repertory to assume the essential tasks? When *L. curvispinosus* slaves were present, the *duloticus* did not gather or feed on food items directly, depending instead on regurgitation from the slaves for their nourishment. They also failed to function as nurses; that is, they licked the larvae and collected secretions in what might well be a "selfish" harvesting of secretions from these immature forms. When the *curvispinosus* slaves were taken away, however, the repertory of the *duloticus* expanded dramatically. Entire behaviors appeared for the first time. The workers moved away from the nest entrance and elsewhere and gravitated toward the brood, which they now attended much more intensively than before. The workers also began to feed the larvae solid materials such as collapsed eggs or pupal skins. But they were generally inept as nurses. Larvae and pupae were allowed to sit in their partly shed skins for hours at a time, something previously never permitted by the *curvispinosus* slaves. A lack of competence also characterized nest-building behavior. Workers carried pieces of nesting material around in their mandibles but did not succeed in placing them together to form a plug at the nest entrance. The *duloticus* workers began to feed on honey for the first time but took as much as ten times longer to drink the same quantity as the *curvispinosus* slaves. They never retrieved solid food. The result of all this ineptness was a rapid deterioration of the slaveless colony. When the original *curvispinosus* slaves were returned to the nest following the experiment, they quickly displaced the *duloticus* in the brood area, and the earlier patterns of activity and division of labor were restored.

A similar experimental approach to behavioral degeneracy was taken by Stuart and Alloway (1985) in a comparative study of *Harpagoxenus* and *Protomognathus*. *H. canadensis* proved more self-sufficient than either *H. sublaevis* or *P. americanus* when its slaves were taken away. It was able both to forage and to organize emigrations by use of the tandem-running method characteristic of leptothoracine ants.

The concept of the early evolution of dulosis proposed by Wilson (1975a) seems increasingly borne out by recent research. With aggressive territorial behavior already part of their common repertory, few physiological and behavioral changes separate free-living ant species from their slavemaker relatives. Only two relatively slight quantitative changes in the behavior of the *Leptothorax ambiguus* or *L. muscorum* would be required to turn them into facultative slavemakers. First, the tolerance toward adult captives would have to be increased. Instead of accepting newly eclosed adults for a few hours, slavemakers would have to extend their tolerance for days or even the lifetime of the captives. Second, the raiding distance would have to be increased to encompass not just adjacent nests but those as much as a meter away. Both of these modifications would involve quantitative changes in the response thresholds of existing behavior patterns. The theory of population genetics allows that the changes could occur in a few tens of generations, given moderate selection pressures.

To pass from facultative to obligatory slave making is a more drastic step. Now the geographic range of the species would be altered to fit within those of the combined slave species, and its population densities would be reduced or held to lower levels merely by the necessity to "harvest" continuously from surrounding host colonies. The obligatory state of dulosis implies some degree of behavior decay in the slavemaker. It is also reasonable to suppose that as the dulotic workers become more specialized raiding machines, with martial anatomy and behavior patterns, the decay will increase. We have seen that the dulotic species of *Leptothorax*, *Harpagoxenus*, and *Protomognathus* do show varying degrees of incompetence in nest building, foraging, and brood care. The idea of a progressive loss of behavior causes no theoretical difficulties. A single gene can block a behavioral pattern, and the loss or severe reduction of behavioral elements has in fact occurred in laboratory populations of *Drosophila* and *Peromyscus* mice during as few as ten generations.

Finally, dulosis always involves relatively closely related species. No solid case is known of ants enslaving other ants belonging to another tribe or subfamily. Bernstein (1978) claimed to have found the dolichoderines *Conomyrma bicolor* and *C. insana* enslaving the myrmicine *Crematogaster emeryana* and formicine *Myrmecocystus kennedyi* in the western United States. Her documentation is sparse, however, and the records are to be doubted until corroboration can be obtained.

## XENOBIOSIS AND TROPHIC PARASITISM

The classic example of xenobiosis is the relation of the shampoo ant *Formicoxenus provancheri* to its host *Myrmica incompleta* as studied by Wheeler (1903a, 1910a), who described the *Formicoxenus* under the synonymic name of *Leptothorax emersoni* (and called the host species *M. brevinoda*). Ants in the tribe Leptothoracini, to which *Formicoxenus* belongs, generally form small colonies that nest in tight little places, for example the interior of hollow twigs lying on the ground, cavities in rotted acorns, or abandoned beetle galleries in the bark of trees. The workers forage singly, and when they encounter other ants they usually avoid them by moving in a stealthy, unobtrusive manner. Because of these traits colonies of leptothoracines are often found close to the nests of larger ants, and their workers are able to forage freely among their large neighbors.

The trend has been extrapolated into parasitism by *Formicoxenus provancheri*. This species has been found living only in close association with colonies of *Myrmica incompleta*. Both species occur widely throughout the northern United States and southern Canada. Colonies of *M. incompleta* construct their nests in the soil, in clumps of moss, and under logs or stones, especially in wet meadows and bogs. Smaller *F. provancheri* colonies excavate their nests near the surface of the soil and join them to the host nests by means of short galleries. They keep their broods strictly apart. The *Myrmica* are too large to enter the narrow *Formicoxenus* galleries, but the *Formicoxenus* move freely through the nests of their hosts. The *F. provancheri* workers do not forage for their own food. They depend almost entirely on crop liquid obtained from the host workers, using begging movements to induce the incoming *Myrmica* foragers to regurgitate

**FIGURE 12–17** The guest ant *Formicoxenus nitidulus:* (*left*) the wingless male; (*right*) queen. (Paintings by T. Hölldobler-Forsyth, courtesy of K. Gösswald.)

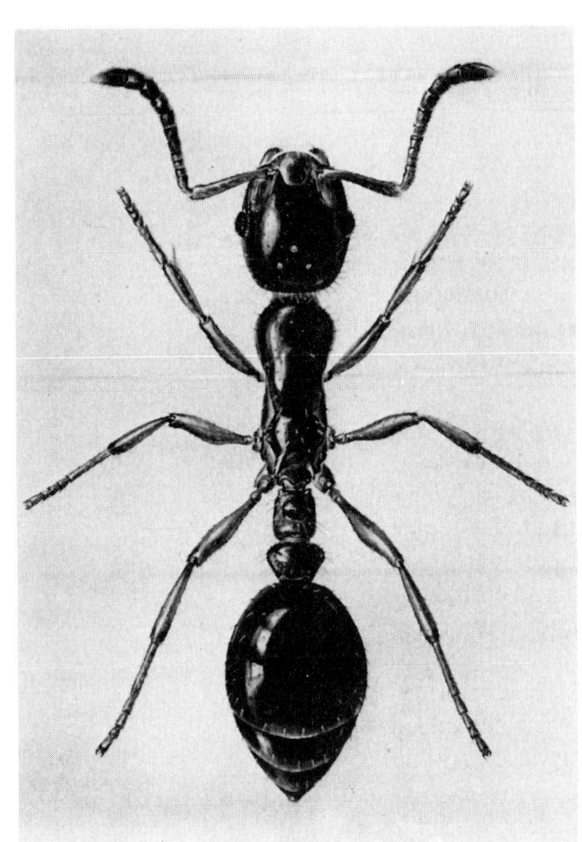

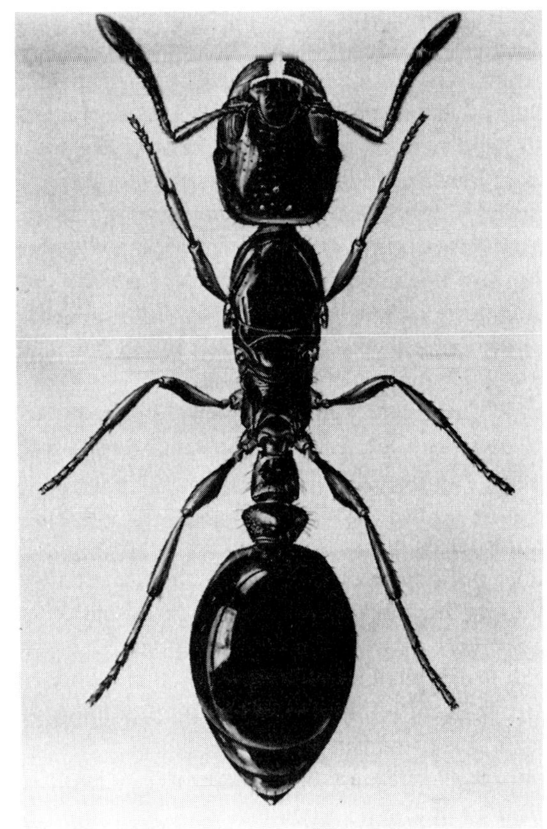

to them. They also mount the *Myrmica* adults and lick them in what Wheeler (1903a) described as "a kind of feverish excitement," to which the hosts respond with "the greatest consideration and affection." Wheeler was under the impression that the *Myrmica* sought the *Formicoxenus* in order to obtain a "shampoo," and he believed at first that the relationship might be mutually beneficial. Later (1910a) he conceded that the *Formicoxenus* are probably no more than parasites. They are nevertheless far from being totally dependent on *M. incompleta.* Not only do they construct their own nests and rear their own brood, but they are also able to feed themselves, albeit awkwardly, when isolated in artificial nests in the laboratory.

*Formicoxenus nitidulus* is a northern and central European species with habits closely similar to *F. provancheri.* This small reddish ant closely resembles *Leptothorax* and may have been derived from it in evolution (Figure 12–17). It is specialized for life inside the large mound nests of members of the *Formica rufa* group, particularly *F. lugubris, F. polyctena, F. pratensis,* and *F. rufa.* It has also been found occasionally in nests of other *Formica* species, including *F. exsecta* and *F. fusca,* and even *Polyergus rufescens.* The relation of *Formicoxenus* to its hosts has been studied over a period of many years by Forel (1874, 1886), Adlerz (1884), Wasmann (1891), Janet (1897a), Wheeler (1910a), Stäger (1925), Stumper (1950), and Buschinger and his co-workers (see Buschinger, 1976a,b). The colonies, which contain 100 to 500 workers and multiple queens, appear to be functionally monogynous (Buschinger and Winter, 1976) and to nest exclusively within the host nests. They excavate their own chambers and keep their brood strictly segregated. Like the *Formicoxenus provancheri* shampoo ants, *Formicoxenus nitidulus* build narrow galleries that open directly into the interior of the *Formica* nests, and from these they periodically emerge to forage among the host workers. Unlike the *provancheri,* however, they do not lick their hosts. In fact it has been difficult to observe interactions of any kind between the *Formicoxenus nitidulus* and the *Formica.* Although Stäger reported regurgitation from *Formica* workers to *Formicoxenus nitidulus,* and this exchange was confirmed by Buschinger (personal communication), Stumper concluded that this must be uncommon since most of the time the *nitidulus* workers appear to keep strictly to themselves. *Formicoxenus nitidulus* nevertheless displays at least two remarkable adaptations to its commensal existence. First, the males are wingless and highly worker-like in appearance (Figures 12–17 and 12–18). They can be distinguished externally only by their longer antennae, which contain one more segment than those of the workers, by an additional abdominal segment, and of course by the extrusible portions of their genitalia. The matings take place on top of the host nests. The second adaptation to life with *Formica* is the ability to emigrate in the columns of the host workers when the latter change nest sites. Forel (1928) and later Elgert and Rosengren (1977) demonstrated that the *Formicoxenus* follow the scent trails of their *Formica* hosts.

In 1925 Wheeler reported the discovery of a new and thoroughly surprising case of xenobiosis between a myrmicine guest ant, *Megalomyrmex symmetochus,* and a fungus-growing host species, *Sericomyrmex amabilis,* also a myrmicine. The *Sericomyrmex* found modest-sized colonies, comprising 100 to 300 workers and a queen, that nest in the wet soil of the laboratory clearing on Barro Colorado Island, Panama. They subsist entirely on a special fungus raised on beds of dead vegetable material. The *Megalomyrmex* form smaller colonies, consisting of 75 or fewer adults, that live directly among the fungus gardens of the host. Since the *Sericomyrmex* also place their brood in the gardens, the young of both species become mixed to a limited extent. The *Megalomyrmex* tend to segregate their brood in little clumps, however, each of which is closely attended by a few workers, and neither species feeds or licks the brood of the other. The most remarkable fact is that the *Megalomyrmex* appear to subsist exclusively on the fungus. This represents a major dietary shift which must have occurred relatively recently in the evolution of the genus. Because liquid food exchange is either uncommon or completely lacking in fungus-growing ants, the *Megalomyrmex* do not secure nutriment from the *Sericomyrmex* in this way. They do, however, lick the body surfaces of their hosts.

An important common feature of the three examples of xenobiosis just cited is what German writers call *Futterparasitismus*, which can perhaps best be translated into the rather formal expression "trophic parasitism." This is intrusion into the social system of one species by another in order to steal food. Trophic parasitism does not by itself require a close association of nests or even entry into the host nest by foraging workers. In other words it can occur apart from xenobiosis. A weak, nonxenobiotic form of such parasitism is exhibited by *Camponotus lateralis* toward *Crematogaster scutellaris* in Europe. Goetsch (1953) and Kaudewitz (1955) have described instances in which *Camponotus* workers followed the *Crematogaster* odor trails to their feeding grounds and exploited the same food resources during the same time of day. The *Crematogaster* were hostile to the *Camponotus*, which assumed a crouching, conciliatory posture when they met the legitimate users of the trails. Unlike the xenobionts, the *Camponotus lateralis* nest separately. Moreover, the relationship is not obligatory on *lateralis*, since colonies and foraging workers of that species are often found far from *Crematogaster* nests.

On the island of Trinidad Wilson (1965b) discovered an instance of trail sharing that approaches a neutral, or commensalistic, relationship. Each of the several colonies of the formicine *Camponotus beebei* encountered were in close association with a large colony of the dolichoderine *Azteca chartifex*, one of the dominant ant species of the island forests. The *Camponotus* nested in cavities in tree branches near the arboreal carton nests of the *Azteca*, and their workers followed the *Azteca* odor trails down the branches and tree trunks to foraging areas on the nearby ground and weedy vegetation. The diets of the two species were not determined, but regardless of the degree of similarity, potential interference between the two species was reduced by the existence of opposite daily schedules. The *Camponotus* therefore "borrowed" the *Azteca* trails when the owners were putting them to minimal use. The *Azteca* workers were hostile to the *Camponotus* workers and attacked them on the rare occasions when the latter slowed in their running, but the *Camponotus* were larger and faster and usually easily avoided their hosts without causing any visible disturbance. On a single occasion a *Camponotus* worker was seen to lead out a tight column of six other *Camponotus* workers, guiding them along by means of a short-lived odor trail. The pheromone was laid directly on top of the *Azteca* odor trail, yet the *Azteca* workers did not fall in line or show any other response to the passage of the *Camponotus* group. Thus it appears that the *Camponotus*, while "eavesdropping" on the *Azteca* odor trails, have reserved a special recruitment trail system of their own which they do not share with their hosts.

Two phenomena have been discovered in very different parts of

**FIGURE 12–18** A wingless male of the xenobiotic European ant *Formicoxenus nitidulus* mates with a winged female. (From Buschinger, 1976b.)

the world that constitute alternative strategies of trophic parasitism. In the Siberian steppes *Formica pratensis* is a territorial dominant over *F. cunicularia*, driving its workers away from favored nest sites and food finds. Occasionally *F. cunicularia* foragers can steal food items that have been temporarily laid on the ground by *pratensis* foragers, but in general they must depend on quickness and luck to gather food before the *pratensis* arrive on the scene. For their part the *pratensis* workers use the *cunicularia* as scouts. They are attracted to the movements of these particular ants (and not to other species), so that when the *cunicularia* discover food the *pratensis* are often able to appropriate it for themselves. Reznikova (1982), who discovered this relationship, found that when *pratensis* colonies were prevented from foraging under otherwise natural conditions, nearby *cunicularia* colonies increased their food intake. Yet when *cunicularia* colonies were constrained, the *pratensis*, now deprived of their scouts, harvested significantly less food.

In Panama hundreds of workers of the little myrmicine *Crematogaster limata* were observed to file into the nests of the large ponerine *Ectatomma tuberculatum*. They climbed up the legs of the *Ectatomma* onto their bodies, which they proceeded to lick. The big ponerines did not respond aggressively to these little intruders; they even occasionally opened their mandibles and let the *Crematogaster* lick their extended mouthparts. Then the *Crematogaster* climbed down and quickly left (Wheeler, 1986b). This odd relationship resembles the xenobiosis of *Formicoxenus*, but its significance can only be guessed on the basis of present evidence.

An interesting evolutionary question can now be raised: do xenobiosis or the more tenuous forms of trail parasitism ever lead to full inquilinism? The evidence to look for is the coexistence in the same genus of xenobiotic species and fully inquiline species, both of which parasitize other species belonging to the same genus. This is the criterion, it will be recalled, by which inquilinism in *Epimyrma corsica* was inferred to be of dulotic origin. So far, no such examples have been found. Even so, some of the traits of *Kyidris*, an inquilinous genus which will be described in a moment, at least suggest the possibility of a xenobiotic origin. The same is true of the species of *Formicoxenus*.

## PARABIOSIS

Forel (1898) designated as parabiosis the following complex behavior which he discovered in Colombia. Colonies of the arboreal, rain forest ant species *Crematogaster limata parabiotica* and *Monacis debilis* (called *Dolichoderus debilis* var. *parabiotica* by Forel) commonly nest in close association, with the nest chambers kept separate but connected by passable openings. They are the principal occupants of ant gardens in the rain forests of South America and as such rank among the most abundant arboreal ants (see Chapter 14). The workers of the two species also run together along common odor trails. Wheeler (1921a) confirmed the phenomenon in his Guyana studies and showed that the two species collect honeydew together from membracids. Wheeler also discovered a similar association between *Crematogaster limata parabiotica* and *Camponotus femoratus*. Both species were observed utilizing common trails and gathering honeydew from jassids and membracids on the same plants as well as nectar from the same extrafloral nectaries of *Inga*. Wheeler believed that the *Crematogaster* and *Camponotus* workers were tolerant of each other in this potentially confrontational situation. He saw them "greet" each other with calm antennation on the trails. On three occasions he observed *Camponotus* workers regurgitating to individuals of *Crematogaster*.

In a subsequent and more detailed study of the same three species in Brazil, Swain (1980) obtained a very different picture. He learned first of all that parabiotic Crematogasters actually belong to two species, one found with the *Monacis* and one with the *Camponotus*. The parabionts do not mix their odor trails. Rather, both the *Monacis* and *Camponotus* follow trails laid by their respective *Crematogaster* associates. Moreover, the parabionts are not always amicable at food sites. At sugar and insect baits set out by Swain, the *Camponotus femoratus* workers drove the *Crematogaster limata* away and fed exclusively by themselves. So during foraging at least, the relationship is not mutualistic, as formerly believed, but parasitic.

It has also been commonly thought (Wheeler, 1921a; Weber, 1943) that the larger ants of each pair provide protection for the *Crematogaster* at the nest site. It is at least true that *Camponotus femoratus* plays this role against vertebrates. Its workers are among the most formidable ants in the forest canopy, swarming out to bite and spray formic acid on any intruder and at the slightest disturbance. *Monacis debilis*, on the other hand, are timid ants that appear to offer even less defense than the *Crematogaster*. But of course they may be very effective against ants and other invertebrate enemies. It is entirely possible that the losses the *Crematogaster* suffer from domination by *Camponotus femoratus* at food sites (their relation to *Monacis* is not known) is counterbalanced by the added protection they receive at the nest. The *Crematogaster* seem well served by the parabiotic association, because they usually if not always exist within it.

Other forms of parabiosis probably exist. One marginal case has been found in South Australia by Greenslade and Halliday (1983). Three species of *Camponotus* belonging to the *ephippium* and *innexus* groups outwardly resemble species in the dominant, more abundant species of meat ants in the *Iridomyrmex detectus* group. These apparent mimics nest within the meat ant territories and forage among the *Iridomyrmex* workers, using speed and agility to avoid their formidable neighbors. The majors of at least one species of the *C. ephippium* group have huge heads with which they block the nest entrances (Figure 8–34). An even more intimate parabiotic relationship exists between the Australian "sugar ant" *C. consobrinus* and *C. perthiana* on the one side and the meat ant *Iridomyrmex purpureus* on the other. The *Camponotus* often nest directly inside the large mounds of the *Iridomyrmex*, yet without connecting their nest galleries and chambers to those of their hosts (Greaves and Hughes, 1974; B. Hölldobler, unpublished observations).

## THE DEGREES OF INQUILINISM

Once an ant species enters complete inquilinism, whether by temporary parasitism, dulosis, or xenobiosis, it seems to evolve quickly on down into a state of abject dependence on its host. It acquires some of what can be termed the "inquiline syndrome," a set of characteristics found in varying combinations in all of the relatively specialized inquiline species (Wilson, 1971):

1. The worker caste is lost.
2. The queen is either replaced by an ergatogyne, or ergatogynes appear together with a continuous series of intergrades connecting them morphologically to the queens.
3. There is a tendency for multiple egg-laying queens to coexist in the same host nest.
4. The queen and male are reduced in size, often dramatically; in some cases (for example, *Teleutomyrmex schneideri, Plagiolepis* (= *Aporomyrmex*) *ampeloni, P. xene*) the queen is actually smaller than the host worker.
5. The male becomes "pupoid": his body is thickened, the petiole and postpetiole become much more broadly attached, the genitalia are more externally exposed when not in use, the cuticle becomes thin and depigmented, and the wings are reduced or lost. The extreme examples of this trend are seen in *Anergates atratulus, Pheidole* (= *Anergatides*) *neokohli*, and *P.* (= *Bruchomyrma*) *acutidens* (see Figures 12–19 and 12–20).
6. There is a tendency for the nuptial flights to be curtailed, and to be replaced by mating activity among nestmates ("adelphogamy") within or near the host nest. Dispersal of the queen afterward is very limited.
7. Probably as a consequence of the curtailment of the nuptial flight just cited, the populations of inquiline species are usually very fragmented and limited in their geographic distribution.
8. The wing venation is reduced.
9. Mouthparts are reduced, with the mandibles becoming smaller and toothless and the palps losing segments. Concomitantly, the inquilines lose the ability to feed themselves and must be sustained by liquid food regurgitated to them by the host workers.
10. Antennal segments are fused and reduced in number.
11. The occiput, or rear portion of the head, of the queen is narrowed.

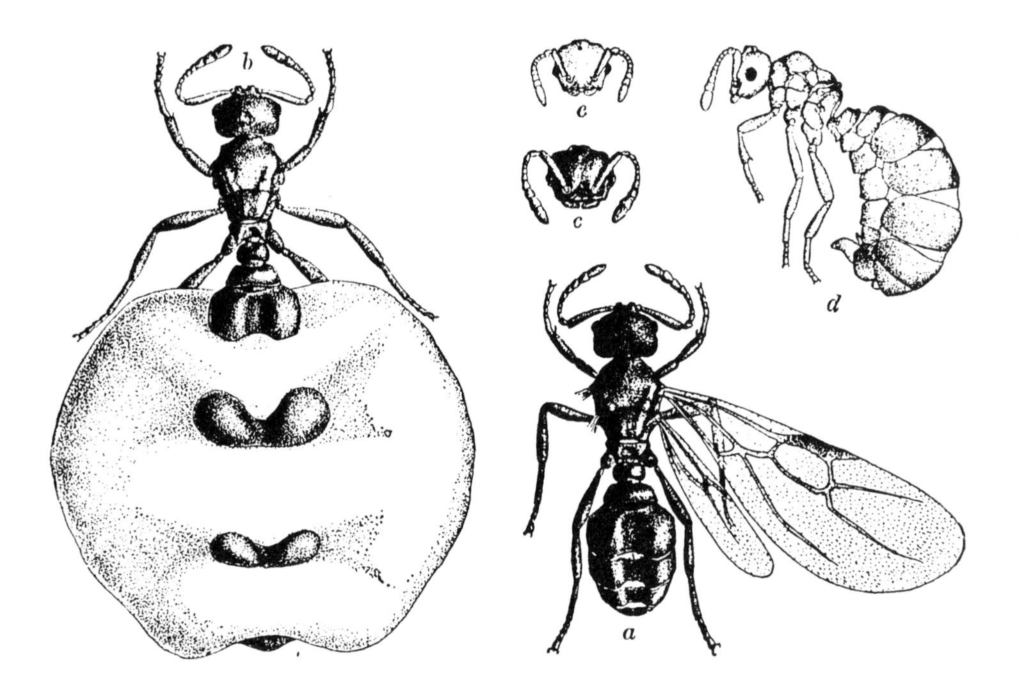


**FIGURE 12–19** *Anergates atratulus,* an extreme workerless parasite that lives with *Tetramorium caespitum* in Europe: (*a*) virgin queen; (*b*) old queen with typically physogastric abdomen; (*c*) head of queen, showing the reduced mandibles and narrowed occipital region characteristic of extreme inquilines; (*d,e*) male, a pupoid form of the kind found in a few other extreme inquilinous species. (From Wheeler, 1910a.)

**FIGURE 12–20** *Anergates atratulus:* (*a*) a wingless male; (*b*) a physogastric queen; (*c*) a male attempting to mate with a pupa; (*d*) a male mating with a winged queen. (Modified from Buschinger, 1974b.)

12. The central nervous system is reduced in size and complexity, usually through reduction of associative centers.
13. The petiole and especially postpetiole are thickened, and the postpetiole acquires a broader attachment to the gaster.
14. A spine is formed on the lower surface of the postpetiole (the *Parasitendorn* of Kutter).
15. The propodeal spines (if present in the ancestral species) "melt," that is, they thicken and often grow shorter, and their tips are blunted.
16. The cuticular sculpturing is reduced or lost altogether over most of the body; in extreme cases the body surface becomes very shiny.
17. The exoskeleton becomes thinner and less pigmented.
18. Many of the exocrine glands are reduced or lost, a trait already described in some detail in the earlier account of *Teleutomyrmex schneideri*.
19. The queens become highly attractive to the host workers, who lick them frequently. This is especially true of the older, physogastric individuals, and it appears to be due to the secretion of special attractant substances which are as yet chemically unidentified.

Let us examine more closely one of the most interesting of these trends, namely, reduction of the worker caste. The inquiline species *Kyidris yaleogyna* of New Guinea represents the most primitive known level in this evolutionary regression since it retains an abundant and partly functional worker caste (Wilson and Brown, 1956). Colonies of *K. yaleogyna* are parasitic on *Strumigenys loriae*, one of the more abundant and ecologically widespread of the Papuan ants. Both *Kyidris* and *Strumigenys* are members of the tribe Dacetini of the subfamily Myrmicinae, but they are otherwise very different from one another. *Kyidris* is a short-mandibulate form, closer to *Smithistruma* and *Serrastruma* than to the highly distinctive, long-mandibulate *Strumigenys*. Four mixed colonies were discovered nesting in pieces of decaying wood on the floor of rain forests. In each the *Strumigenys* slightly outnumbered the parasites. One large colony collected *in toto* contained 1,622 workers and 16 dealated queens of *Strumigenys*, in combination with 1,170 workers, 4 dealated queens, 84 alate queens, and 51 males of *Kyidris*. A second colony contained 243 workers and 4 dealated queens of *Strumigenys* and 64 workers, 2 dealated queens, and 31 males of *Kyidris*.

These large groups lived in completely harmonious mixtures in which the parasitic nature of the *Kyidris* was only subtly evident. The *Kyidris* workers foraged for food. One group was found attending coccids near the nest. Others engaged in hunting for small insects, but compared with the *Strumigenys* they were quite ineffectual. They wandered through the food chambers of the artificial nests like typical restless dacetines, but rarely tried to catch prey. Even when they tried they usually failed, in sharp contrast to the highly efficient performances of their *Strumigenys* nestmates. One *Kyidris* worker was seen to seize a symphylan, pull it backward, hold it for about thirty seconds without trying to sting it, and finally release it when it began to struggle. Another seized an entomobryid collembolan, pulled it back vigorously, then lost it when the entomobryan kicked with its furcula. Still another was seen actually to carry an entomobryid at a brisk clip across the food chamber floor; it reached the entrance to the brood chamber only to have a *Strumigenys* meet it and take the insect away. The general impression created is that the predatory behavior has regressed, but not completely disappeared, in *Kyidris* workers. In the artificial nests, at least, the *Strumigenys* workers did most of the productive hunting. The *Kyidris* studied by Wilson and Brown (1956) also aided in brood care much less frequently than the *Strumigenys*, and their efforts seemed ineffectual. *Kyidris* workers were never observed in nest construction. They received regurgitated liquid food from the *Strumigenys* workers, and sometimes obtained food by inserting their mouthparts between those of two *Strumigenys* exchanging regurgitated material, but they were never seen to offer anything in return. In sum the New Guinea *Kyidris* appear to represent inquilinism at a very early stage when the worker caste has only begun to reduce its behavioral repertory. Probably the degeneration has proceeded past the point of no return since it is doubtful that *Kyidris* colonies could survive without their hosts.

A somewhat more advanced stage of behavioral decay is shown by the workers of *Strongylognathus alpinus*, a European species already mentioned in the previous section on dulosis. The workers are able to conduct raids, they participate in nest building, and they can feed themselves. Unlike the *Kyidris* workers, however, they have lost the capacity either to hunt or to care for brood. In the great majority of all other dulotic and inquilinous ant species that still possess a worker caste, the workers appear to have lost entirely the ability to carry on the ordinary functions of nest construction, food gathering, and queen and brood care.

Small wonder, then, that most truly inquilinous species have taken still one more step in evolution and discarded the worker caste altogether. The stages leading to this final abrogation have been beautifully documented in the genus *Plagiolepis* by Le Masne (1956b) and Passera (1966, 1968b). The two species *P. grassei* and *P. xene* are parasitic on the closely related free-living form *P. pygmaea*. In certain key characteristics, namely loss of worker caste, size reduction, and alteration of the male form, *xene* qualifies as an extreme inquiline, whereas *grassei* occupies an almost exactly intermediate position between it and the free-living *Plagiolepis* (see Table 12–2). The most interesting annectant feature of *grassei* is in the status of the workers. This caste is almost extinct, and it appears in a given host nest only after the winged parasitic sexuals are produced—the reverse of the order that is universal in free-living ant species.

In a separate analysis Wilson (1984c) examined traits of the inquiline syndrome in the nine known parasitic species of *Pheidole*. He assumed that when most or all the species possessed a given trait, the change producing the trait had appeared in evolution prior to other, less widespread traits. Conversely, he viewed a rare trait, such as the existence of pupoid males, as a late event in evolution. If this assumption is correct, then the earliest changes to occur in the *Pheidole* parasites were loss of the worker caste, reduction of size, rounding of the occiput (associated with reduction of the mandibular adductor muscles and loss of mandibular strength), loss of body sculpture (associated with a thinning of the exoskeleton), and broadening of the postpetiole. These shifts were followed by reduction of the antennal segments (including the three-segmented club), reduction of the mandibles, and development of a pupoid body form in the male.

The majority of inquiline species for which adequate information is available permit the host queens to live. This is the situation one

**TABLE 12–2** Characteristics of a host species (*Plagiolepis pygmaea*) and two closely related species that live as social parasites with it. (Based on data from Le Masne, 1956b.)

| Characteristic | *P. pygmaea* (host) | *P. grassei* (intermediate parasite) | *P. xene* (extreme parasite) |
|---|---|---|---|
| Length of queen (mm) | 3.4–4.5 | 2.0–2.4 | 1.2–1.3 |
| Development of queen wings | Normal | Normal | Variable, often rudimentary |
| Condition of worker caste | Normal, abundant; appears before sexual forms | Rare; only 1 to every 10 fecundated queens; appears after sexual forms | Absent |
| Condition of male | Normal, with functional wings | Slightly female-like; wings occasionally rudimentary | Very female-like; wings always rudimentary |

would intuitively expect. It seems to make good sense for the parasites to insure themselves a long-lasting supply of host workers. This reasoning is also consistent with the fact that species of *Strongylognathus* which obtain host workers by slave raids also destroy the host queens, whereas *S. testaceus*, the one species of the genus that does not obtain host workers in this fashion, tolerates the host queens.

But what of the minority of inquilines whose presence causes the death of the host queens, in particular the queen-killing *Epimyrma* species that do not conduct slave raids? For some species, in which either the host colonies or the parasites themselves are relatively short lived, it may be advantageous to get rid of the host queen and invest all the efforts of the host workers in the production of as many parasite queens and males as soon as possible. Buschinger and Winter (1983b) have provided strong evidence in *Epimyrma* supporting this hypothesis. Colonies of the slave-raiding *E. ravouxi* have a maximum life span of ten years. In contrast, the queens of *E. kraussei*, whose daughter workers conduct slave raids only rarely, produce mainly sexual forms and a scattering of workers until they run out of *Leptothorax* host workers to support them. This is the "big bang" strategy of reproduction, essentially the same as that employed by such fishes as the migratory eels and salmon and such plants as the bamboos (Gadgil and Bossert, 1970; Oster and Wilson, 1978). Other parasitic ant species, with longer lives and more stable host colonies, would find it advantageous to let the host queens live and to employ the host colony in the production of parasite queens and males at a lower rate—but for a longer period of time. We should bear in mind that even the continuous reproducers inhibit host reproduction to some degree. In general the production of host sexual forms in the presence of inquilines is a rare event even in those cases where the host queen is permitted to live. Furthermore, Passera (1966) has discovered that in *Plagiolepis pygmaea* even the ability to produce workers is partially inhibited by the presence of parasitic *P. xene* queens.

The theoretical implications raised by these various observations were formalized by Wilson (1971) in the following conjecture: *The degree of reproductive repression inflicted by a given inquiline species is such as to maximize the total production of parasite queens and males per host colony under the particular ecological conditions in which the mixed colonies occur.* To test this hypothesis and more generally to advance a population theory of social parasitism, investigators need data on the population dynamics of both parasitized and unparasitized host colonies.

CHAPTER 13

# Symbioses with Other Arthropods

For more than a century we have known that many species of insects and other arthropods live with ants and have developed thriving symbiotic relationships with them. Most do so only occasionally, functioning as casual predators or temporary nest commensals. A great many others, however, are dependent on the ant society during part or all of their life cycles. These ant guests, commonly known as myrmecophiles, include a great variety of beetles, mites, collembolans, flies, and wasps, as well as less abundant representatives of a wide range of other insect groups (see Table 13–1).

Myrmecophily is almost exclusively an invertebrate phenomenon. A respectable list of vertebrate species, including synbranch eels, frogs, lizards, snakes, birds, small and medium-sized mammals and even primates occasionally live with social insects or prey upon them. Yet very few are specialized for such an association (Myers, 1929, 1935; Hindwood, 1959; Scherba, 1965; Chew, 1979; Vogel and von Brockhusen-Holzer, 1984; Redford, 1987). The vast majority of obligate invertebrate symbionts are moreover arthropods, and it is among these organisms that the most striking adaptations for life with social insects have taken place. A number of these myrmecophiles make their homes in nests of the ants and enjoy all the social benefits of their hosts. Although the interlopers in some cases eat the brood, the ants treat the guests with astonishing tolerance: they not only admit the invading species to the nest, but often feed, groom, and rear the guest larvae as if they were the ants' own young.

An ant colony possesses a complex system of communication (see Chapters 5 and 7) that enables it not only to carry out its collaborative activities in food gathering, brood care, and other social activities, but makes possible instant recognition of nestmates and discrimination of foreigners. This identification and discrimination system functions like a social immune barrier: only colony members are allowed to enter the ant society, and alien individuals are harshly rejected. Nevertheless, by using various techniques, a considerable number of solitary arthropods have managed to penetrate ant nests. The fact that the ants treat many of these alien guests amicably suggests that the guests have broken the ants' communication and recognition code. In other words, they have attained the ability to speak the ants' language of mechanical and chemical cues.

## HISTORY OF MYRMECOPHILE STUDIES

Before we address this phenomenon in greater detail, it will be useful to review briefly the historical background of the study of myrmecophilous symbiosis. Erich Wasmann initiated the modern study of the subject. In 1894 and in subsequent publications he developed a classification that divides species into five behavioral categories, which reflect increasing levels of integration into the social system of their hosts:

1. *Synechthrans.* These arthropods are treated in a hostile manner by their hosts. They are predators for the most part, managing to stay alive by means of greater speed and agility or by using defensive mechanisms such as repellent secretions and retraction beneath shell-like cuticular shields.
2. *Synoeketes.* These arthropods, which are also primarily scavengers and predators, are ignored by their hosts because they are either too swift or else very sluggish and apparently neutral in odor.
3. *Symphiles.* Also referred to occasionally as "true" guests, these symbionts are accepted to some extent by their hosts as though they were members of the colony.
4. *Ectoparasites and endoparasites.* These arthropods are conventional parasites. They live on the body surfaces of their hosts, or lick up their oily secretions, or bite through the exoskeleton and feed on their hemolymph, or penetrate the body itself.
5. *Trophobionts.* This category includes phytophagous homopterans, heteropterans, and lycaenid caterpillars that are not dependent on the social insects for food but instead supply their hosts with honeydew and nutritive glandular secretions. In exchange they receive protection from parasites and predators.

With the accumulation of more detailed information on the behavior of the symbionts in recent years, the Wasmannian classification has turned out to be considerably less than perfect. Many symbionts fit into more than one category. Symphiles, for instance, not only exist on the charity of their hosts, grooming and soliciting food from them, but also prey simultaneously on the hosts and their brood.

Several alternative schemes have been proposed to categorize the many life styles of myrmecophiles, by Delamare Deboutteville (1948), Paulian (1948), Akre and Rettenmeyer (1966), and Kistner

(1979). In a more modern, ecological twist, Kistner distinguishes two major categories: integrated species, "which by their behavior and their hosts' behavior can be seen as incorporated into their hosts' social life"; and non-integrated species, "which are not integrated into the social life of their hosts but which are adapted to the nest as an ecological niche." When we consider the complex diversity of myrmecophilous adaptations it is often difficult, however, for us to draw a clear distinction between "integrated" and "non-integrated" symbionts. Thus, despite its considerable shortcomings, the original Wasmann nomenclature continues to be useful in labeling the majority of cases, and it is frequently employed as a kind of shorthand in the literature on social symbioses.

## DIVERSITY OF MYRMECOPHILES

The literature on myrmecophiles is enormous and growing each year, much of it consisting of incidental notes buried in taxonomic and ecological studies of selected genera and higher taxa. A very extensive review of solitary symbionts in insect societies was published by Kistner in 1982. Much of the information on myrmecophiles is summarized in Table 13–1, which is based on the original table compiled by Wilson (1971), together with new information provided by Kistner's review and other publications.

Not surprisingly, the cumulative data reveal that certain taxa are much more preadapted for life as symbionts than others. The mites (Acarina), for example, are the foremost representatives among ectosymbiont species in terms of sheer numbers of individuals. Although there exist only a few quantitative faunistic investigations of symbionts in ant colonies, Rettenmeyer's study (1962a) of 150 colonies of army ants in Kansas and Panama is characteristic: he counted almost twice as many ectosymbiotic mites as all the other myrmecophiles taken together. Also, Kistner (1979) reports that he collected 3,288 mites from a single colony of the army ant *Eciton burchelli*. Mites find easy entry into nests either as scavengers too small and quiescent to be evicted by their hosts or as ectoparasites adapted for life on the body surface of the ants.

Myrmecophilous mites are rivaled in diversity by staphylinid beetles, a family of approximately 28,000 species. Staphylinids, like acarines, are predisposed to life in ant and termite nests by virtue of their preference for moist, hidden environments and the role they commonly assume as generalized scavengers and predators. The greatest variety of staphylinid symbionts is found in colonies of the "true" army ants, that is, the dorylines and ecitonines (Seevers, 1965; Akre and Rettenmeyer, 1966; Kistner and Jacobson, 1975a,b; Kistner, 1979). But they also commonly occur in nests of nonlegionary ant species. Kistner (1982) lists 19 staphylinid genera recorded from nests of attine ants, 17 genera found with *Solenopsis* fire ants, and approximately 15 genera known to occur in nests of formicine ants. Diversity aside, probably the most abundant insect guests of social insects in general and army ants in particular are phorid flies. When all species are combined, as many as 4,000 adults occur in a single ecitonine army ant bivouac (Rettenmeyer and Akre, 1968).

Another safe generalization is that by far the greatest diversity of species of myrmecophiles, measured either per host species or per host colony, is found with species that form exceptionally large mature colonies. The ultimate in this trend is found in the great faunas of symbionts that live with ecitonine and doryline army ants, meat ants of the genus *Iridomyrmex*, and leafcutters of the genus *Atta*, whose nests normally contain from hundreds of thousands to millions of inhabitants. An exceptional variety of guests has also been recorded from large colonies of *Hypoclinea* in tropical Asia and the north temperate species of the *Formica rufa* group. By contrast, very few symbionts are known from nests of species with the smallest mature colony sizes, including the great majority of ponerine, dacetine, and leptothoracine ants.

This last rule of population size lends itself readily to theoretical explanation. The insect colony and its immediate environment can be thought of as an ecological island, partitioned into many microhabitats that symbiotic organisms are continuously attempting to colonize (Wilson, 1971; Hölldobler, 1972). In general, species with the largest mature colony size possess the greatest diversity of ecological niches and also enjoy the longest average span of mature colony life. The more diverse the microhabitats presented by the host colony, the greater the potential diversity of symbiont species. Furthermore, a long colony life increases the probability that symbiotic propagules will penetrate a given colony at some time or other. It is also true that if colony size is large, the equilibrial population size of its symbionts will be proportionately large and their species extinction rate (measured as the number of symbiont populations going extinct per colony per unit of time) will be correspondingly low. In general, large colony size enhances three factors—long colony life, high microhabitat diversity, and low symbiont extinction rates—that reinforce one another to produce a higher diversity and abundance of symbiotic species.

The evidence reviewed by Wilson (1971) and Kistner (1979, 1982) shows that many species can maintain a viable population size only under the protection of large colonies. The great majority of very specialized species are uncommon, and many are rare and local in distribution.

## THE GUESTS OF ARMY ANTS: AN EXTREMELY RICH COMMUNITY

The species of mites associated with army ants, according to Rettenmeyer (1962a), can be conveniently divided into two main groups: those that live in the refuse piles or "kitchen middens" of the ants, and those that live on the bodies of the ants or within their bivouacs. The latter group, the only true myrmecophiles in this case, have finely apportioned the environment through amazing feats of specialization. Several of the extreme adaptations are illustrated in Figure 13–1. Of special interest are the Circocyllibanidae, which are phoretic on adult workers but occasionally ride on larvae, adult males, and queens (Elzinga and Rettenmeyer, 1975). They appear to live primarily on certain parts of the host bodies such as the mandibles, head, thorax, and gaster of adult workers. Of equal interest are the Coxequesomidae, which are specialized to live on the antennae or coxae of their *Eciton* hosts.

The most extraordinary adaptation of all, however, is exhibited by the macrochelid, *Macrocheles rettenmeyeri*, which is a true ectoparasite. This mite feeds on blood taken from the terminal membranous lobe (arolium) of the hind tarsus of large media workers of its exclusive host species, *Eciton dulcius*. In the process it allows its entire

**TABLE 13–1** Arthropods that are symbiotic with ants.

| Taxon | Host | Biology | Selected references |
|---|---|---|---|
| **ISOPODA** | | | |
| **Armidillidiidae** | | | |
| *Typhloschizidium* | *Tetramorium* | | Argano and Pesce (1974) |
| **Oniscidae** | | | |
| *Exalloniscus* | *Leptogenys* | Accompany ants on emigrations; often ride on ant brood | Ferrara et al. (1987) |
| **Porcellionidae** | | | |
| *Metaponorthrus* | *Messor* | Feed on stored grain | Bernard (1968) |
| **Squamiferidae** | | | |
| *Plathyarthrus* | Many ant genera | Scavenge, occasionally attend aphids, eat expelled pellets of infrabuccal chambers of ants | Donisthorpe (1927), Hölldobler (1947), Bernard (1968), Williams and Franks (1988) |
| **Trachelipidae** | | | |
| *Nagurus* | *Myrmecia* | Scavenge | Gray (1971a, 1974b) |
| **PSEUDOSCORPIONIDA** | | | |
| *Pychnochermes* | *Atta* | Prey on small synoeketes in debris chambers | Eidmann (1937) |
| *Syndeipnochernes* | *Atta, Camponotus* | | Beier (1970), Weber (1972) |
| **ARANEAE** | | | |
| **Agelenidae** | | | |
| *Tetrilus* | *Formica* | | Donisthorpe (1927) |
| **Aphantochilidae** | | | |
| *Aphantochilus* | *Zacryptocerus* | Specialized predator | Oliveira and Sazima (1984) |
| **Clubionidae** | | | |
| *Phrurolithus* | *Crematogaster* | | Emerton (1911) |
| **Gnaphosidae** | | | |
| *Eilica* | *Camponotus* | Host ants carry spider eggs to safety when nest is disturbed | Noonan (1982) |
| **Linyphiidae** | | | |
| *Acartauchenius* | *Tetramorium* | | Donisthorpe (1927) |
| *Cochlembolus* | *Formica* | | Dondale and Redner (1972) |
| *Evansia* | *Formica, Lasius* | | Donisthorpe (1927) |
| *Masoncus* | *Pogonomyrmex* | Prey on collembolans inside host ant nest; emigrate with host colony | Porter (1985) |
| *Thyreosthenius* | *Formica* | | Donisthorpe (1927) |
| **Oonopidae** | | | |
| *Brucharachne, Myrmecoscaphiella* | *Eciton* | | Fage (1938) |
| *Gamasomorpha* | *Myrmecia* | | Gray (1971) |

*continued*

**TABLE 13–1** *(continued)*

| Taxon | Host | Biology | Selected references |
|---|---|---|---|
| **Salticidae** | | | |
| *Cotinusa* | *Tapinoma* | | Shepard and Gibson (1972) |
| *Habrocestum* | | Prey on ants | Cutler (1980) |
| **Theridiidae** | | | |
| *Euryopis, Latrodectus* | *Pogonomyrmex* | Specialized predator | MacKay (1982b), Porter and Eastmond (1982) |
| *Steatoda* | *Pogonomyrmex* | Specialized predator | Hölldobler (1971e) |
| **Thomisidae** | | | |
| *Strophius* | *Camponotus* | Specialized ant predator | Oliveira and Sazima (1985) |
| **Zodariidae** | | | |
| *Zodarium* | *Cataglyphis* | Specialized ant predator | Harkness (1977a) |
| **ACARINA** | | | |
| **Acaridae** | *Eciton*; nonlegionary ants | Possible commensals; phoretic | Samsinak (1960), Rettenmeyer (1960, 1961) |
| **Anoetidae** | *Eciton* | Possible commensals | Rettenmeyer (1960) |
| **Antennophoridae** | | | |
| *Antennophorus* | Nonlegionary ants | Phoretic on adults, live on regurgitated food | Janet (1897a,b), Wheeler (1910b), Park (1932), Bernard (1968) |
| **Circocyllibanidae** | *Eciton* | Phoretic on adults | Rettenmeyer (1960) |
| **Coxequesomidae** | *Eciton* | Phoretic on adults | Rettenmeyer (1960) |
| **Ereynetidae** | *Eciton* | Possible commensals | Rettenmeyer (1960) |
| **Gamasidae** | Several ant genera | | Bernard (1968) |
| **Hypochthoniidae** | *Eciton* | Possible commensals | Rettenmeyer (1960) |
| **Laelaptidae** | *Eciton* | Phoretic on adults | Rettenmeyer (1960) |
| *Cosmolaelaps* | Several ant genera | Scavengers on dead ants | Kistner (1982) |
| *Hypoaspis* | *Formica, Lasius, Myrmecia, Tetramorium* | Feed on immature stages of other mites | Donisthorpe (1927), Gray (1974b) |
| *Laelaps* | *Formica* | | Hölldobler (1947) |
| *Laelaspis* | *Aphaenogaster, Crematogaster, Tetramorium* | Phoretic on adult ants | Hunter (1964), Kistner (1982) |
| *Oolaelaps* | *Formica* | Live among the eggs of ants; eat secretions of ants' salivary glands, which are deposited over each egg | Kistner (1982) |
| *Sphaerolaelaps* | *Solenopsis (Diplorhoptrum)* | | Hölldobler (1928) |
| **Macrochelidae** | *Eciton* | Phoretic and ectoparasitic on adults and larvae | Rettenmeyer (1960, 1962b) |
| **Neoparasitidae** | *Eciton* | Phoretic and ectoparasitic on adults and larvae | Rettenmeyer (1960) |

*continued*

TABLE 13–1 (*continued*)

| Taxon | Host | Biology | Selected references |
|---|---|---|---|
| **Planodiscidae** | *Eciton* | Phoretic on legs of adults | Rettenmeyer (1960, 1970), Elzinga (1978) |
| **Pyemotidae** | *Eciton* | Phoretic and ectoparasitic on adults and larvae | Rettenmeyer (1960) |
| **Scutacaridae** | *Eciton* | Phoretic on legs of adults | Rettenmeyer (1960) |
| DIPLOPODA | | | |
| **Pyrgodesmidae** | | | |
| *Calymmodesmus, Cynedesmus, Rettenmeyeria, Yucodesmus* | Ecitonine army ants | Scavengers, feed on dirt, follow ant trails, tolerated by ants, sometimes carried by ants | Rettenmeyer (1962c), Akre and Rettenmeyer (1968) |
| **Other families** | | | |
| Possibly other species of *Blaniulus, Nopoiulus, Polydesmus, Polyxenus* | Nonlegionary ants | Scavengers | Donisthorpe (1927), Manfredi (1949), Bernard (1968) |
| COLLEMBOLA | | | |
| **Entomobryidae, Isotomidae, Onychiuridae** | | | |
| Many genera | Many ant species | Commensals, probably mostly scavengers | Donisthorpe (1927), Delamare Deboutteville (1948), Ellis (1967), Bernard (1968), Gray (1971a,b, 1974b) |
| THYSANURA | | | |
| **Lepismatidae, Nicoletiidae** | | | |
| *Atelura* | Many ant species | Groom hosts, steal regurgitated food | Janet (1896), Wheeler (1910a), Pohl (1957), Gray (1971a,b, 1974b) |
| *Attatelura* | *Atta* | | Weber (1972) |
| *Grassiella, Trichatelura* | Ecitonine army ants; *Atta, Solenopsis* | Groom hosts, feed on their prey | Eidmann (1937), Rettenmeyer (1963b), Torgerson and Akre (1969) |
| Probably other genera: *Allatelura, Assmuthia, Atopatelura, Braunsina, Crypturella, Ecnomatelura, Lepisma, Lepismina, Metriotelura, Neatelura, Platystylea, Proatelura* | Many ant species and termites | | Escherich (1905), Folsom (1923), Wygodzinsky (1961), Joseph and Mathad (1963), Hocking (1970), Kistner (1982) |
| ORTHOPTERA | | | |
| **Gryllidae** | | | |
| *Myrmecophila* | Many ant species | Lick host secretions, prey on host ants' eggs, solicit regurgitated food; adult and immature crickets follow trails of ants; feed on prey brought in by ants | Wheeler (1900b, 1910a), Schimmer (1909), Hölldobler (1948), Bernard (1968), Graves et al. (1976), Henderson and Akre (1986a,b,c) |

*continued*

**TABLE 13–1** *(continued)*

| Taxon | Host | Biology | Selected references |
|---|---|---|---|
| **BLATTARIA** | | | |
| **Attaphilidae** | | | |
| *Attaphila* | Fungus-growing ants, *Acromyrmex, Atta* | Lick host secretions | Wheeler (1900a, 1910a), Princis (1960), Roth and Willis (1960), Moser (1964), Weber (1972), Brossut (1976) |
| **Atticolidae** | | | |
| *Atticola, Myrmeblattina, Myrmecoblatta, Phorticolea* | Fungus-growing ants | | Princis (1960), Roth and Willis (1960) |
| **Blattellidae** | | | |
| *Blattella* | *Crematogaster* | | Hocking (1970) |
| *Escala* | *Myrmecia* | | Gray (1974b) |
| **Nothoblattidae** | | | |
| *Nothoblatta* | Fungus-growing ants | | Princis (1960), Roth and Willis (1960) |
| **HOMOPTERA** | | | |
| **Aphididae, Coccidae, Pseudococcidae, Chermidae, Cercopidae, Membracidae, Fulgoridae** | Many ant species | Trophobiosis | See "The Trophobionts" section in Chapter 13; reviews by Way (1963), Kloft et al. (1985) |
| **Cixiidae** | | | |
| *Mnemosyne* | *Odontomachus* | Presumably trophobiosis | Myers (1929) |
| **Diaspididae** | | | |
| *Andaspis* | *Melissotarsus* | Presumably maintained as prey | Ben-Dov (1978) |
| **Eurymelidae** | | | |
| *Eurymela* | *Camponotus, Iridomyrmex* | Trophobiosis | B. Hölldobler (unpublished) |
| **HETEROPTERA** | | | |
| **Alydidae** | | | |
| *Hyalymenus* | *Camponotus, Ectatomma* | Probably Batesian mimicry | Oliveira (1985) |
| **Coreidae** | | | |
| *Hygia* | *Meranoplus* | Trophobiosis | Maschwitz et al. (1987) |
| Many species | Several ant species | Nymphal stages are ant-like, probably predaceous on ants | Donisthorpe (1927), Bernard (1968) |
| **Miridae** | | | |
| Many species | Several ant species | Probably predaceous on ants; ant mimics (probably Batesian mimicry) | Donisthorpe (1927), Schuh (1973), McIver (1987) |
| **Plataspididae** | | | |
| *Coptosoma* | *Crematogaster* | Trophobiosis | Green (1900) |
| *Tropidotylus* | *Crematogaster, Meranoplus* | Trophobiosis | Maschwitz et al. (1987) |

*continued*

**TABLE 13–1** *(continued)*

| Taxon | Host | Biology | Selected references |
|---|---|---|---|
| **Reduviidae** | | | |
| *Acanthaspis* | *Solenopsis* | Prey on ants | Mühlenberg and Maschwitz (1976) |
| *Eidmannia* | Fungus-growing ants, *Atta* | Prey on ants | Eidmann (1937) |
| *Eremocoris* | *Formica* | Prey on ants | Jordan (1937) |
| *Ptilocerus* | *Hypoclinea* | Attract workers from odor trails and prey on them | Jacobson (1911) |
| **Tingidae** | | | |
| *Allocader* | *Amblyopone* | | Kistner (1982) |
| *Anommatocoris, Thaumamannia, Vianaida* | Several ant species | Found in ant nests but not yet proven to be obligate guests | Drake and Davis (1959), R. C. Froeschner (personal communication) |
| *Lasiacanta* | *Iridomyrmex* | | Kistner (1982) |
| **PSOCOPTERA** | *Atta, Crematogaster, Lasius, Oecophylla* | | Eidmann (1937) |
| **NEUROPTERA** | | | |
| **Chrysopidae** | | | |
| *Italochrysa* | *Crematogaster* | Larvae stay close to ant nest and feed on ant larvae | Principi (1946) |
| *Nadiva* | *Camponotus* | Larvae carried by and presumably tolerated by host ants | Weber (1942) |
| **COLEOPTERA** | | | |
| **Alleculidae** | | | |
| *Jophon* | *Rhytidoponera* | | Lea (1910) |
| *Lobopoda* | Fungus-growing ants, *Acromyrmex, Atta* | | Eidmann (1937, 1938), Moser (1963), Campbell (1966) |
| *Pseudocistela* | *Formica* | | Wiśniewski (1963) |
| **Anthicidae** | | | |
| *Anthicus, Formicoma, Temnopsophus* | | | Schwarz (1890), Lea (1910), Clark (1921) |
| **Brenthidae** | | | |
| *Amorphocephalus* | *Camponotus*, probably other genera | Solicit regurgitated food from hosts; also feed hosts by regurgitation | Lea (1910, 1912), Kleine (1924), Le Masne and Torossian (1965) |
| *Cordus, Eremoxemus, Myrmecobrenthus* | Several ant species | | Kleine (1924), Schedl (1975) |
| **Byrrhidae** | | | |
| *Microchaetes* | *Iridomyrmex, Rhytidoponera* | | Lea (1912) |
| **Carabidae** | | | |
| *Adelotopus, Anillus, Elaphropus, Harpalus, Illaphanus, Nototarus, Philophlaeus, Pseudotrechus, Tachys (sens. lat.)* | *Formica, Iridomyrmex, Lasius, Messor, Myrmecia, Orectognathus, Pheidole, Solenopsis* | | Brues (1902), Lea (1910, 1912), Clark (1921), Park (1929), Ayre (1958), Kolbe (1969) |

*continued*

**TABLE 13–1** *(continued)*

| Taxon | Host | Biology | Selected references |
|---|---|---|---|
| *Amara* | *Pogonomyrmex* | | MacKay (1983a) |
| *Helluomorphoides* | Army ants (*Neivamyrmex*) | Follow odor trails of host ants, prey on host brood | Plsek et al. (1969), Topoff (1969) |
| *Notiophilus* | *Pogonomyrmex* | | MacKay (1983a) |
| *Pseudomorpha* | *Camponotus*, *Myrmecocystus* | Larvae prey on ant brood | Erwin (1981) |
| *Scarites*, *Ophryognathus*, *Tachyura* | *Atta* | Live in debris chambers of *Atta* nests | Eidmann (1937) |
| *Sphallomorpha* | *Iridomyrmex* | Larvae burrow near ant nest, prey on adult ants | Moore (1974) |
| Tribe Paussinae: | | | |
| Probably all species | Many species of ants | Prey on adult ants and larvae; many species tolerated and licked by ants | Escherich (1907), Wheeler (1910a), Reichensperger (1922, 1939, 1954, 1958), Mou (1938), Janssens (1949), Darlington (1950), Le Masne (1961a,b) |
| *Physea* | *Atta* | Live in debris chambers of *Atta* nests | van Emden (1936), Eidmann (1937), Darlington (1950) |
| **Cerylonidae** | | | |
| *Aculagnathus* | *Amblyopone* | | Oke (1932), Besuchet (1972) |
| *Hypodacne* | *Camponotus* | | Lawrence and Stephan (1975) |
| *Lapethus* | *Atta* | Live in refuse deposits | Lawrence and Stephan (1975) |
| **Chrysomelidae** | | | |
| Tribe Clytrinae | | | |
| *Clytra*, *Hockingia*, *Saxinis* | *Atta*, *Camponotus*, *Formica*, other nonlegionary ants | Larvae feed on vegetable material in host ants' nests | Jolivet (1952), Selman (1962), Hocking (1970) |
| Tribe Cryptocephalinae | | | |
| *Isnus* | *Crematogaster*, other arboreal species in *Acacia* | | Selman (1962) |
| **Cleridae** | | | |
| *Clerus* | *Formica* | Predaceous on hosts | Wiśniewski (1963) |
| **Coccinellidae** | | | |
| *Coccinella*, *Hyperaspis* | *Formica*, *Tapinoma* | Feed on aphids and fulgorids tended by ants | Silvestri (1903), Donisthorpe (1927) |
| *Scymnus* | *Pogonomyrmex* | | MacKay (1983a) |
| **Colydiidae** | | | |
| *Bothrideres*, *Ditoma*, *Euclarkia*, *Euxestus*, *Kershawia*, *Myrmecoxenus* | *Camponotus*, *Crematogaster*, *Formica*, *Iridomyrmex* | | Lea (1910, 1919), Clark (1921), Dajoz (1977), Stephens (1968) |
| **Cryptophagidae** | | | |
| *Catopochrotus*, *Emphylus* | *Crematogaster*, *Formica* | | Reitter (1889), Bernard (1968) |
| **Cucujidae** | | | |
| *Cryptamorpha* | *Rhytidoponera* | | Lea (1912) |
| *Nepharinus*, *Nepharis* | *Crematogaster*, *Formica* | | Reitter (1889), Bernard (1968) |

*continued*

**TABLE 13–1** *(continued)*

| Taxon | Host | Biology | Selected references |
|---|---|---|---|
| **Curculionidae** | | | |
| *Crematogasterobius, Myrmecorhinus* | *Crematogaster* | | Wasmann (1894b), Hustache and Bruch (1936) |
| *Dryophthorus* | *Lasius* | | Donisthorpe (1927) |
| *Letopius* | *Myrmecia* | | Gray (1974b) |
| *Liometophilus* | *Liometopum* | | Fall (1912) |
| **Dacoderidae** | | | |
| *Tetrothorax* | *Leptogenys, Odontomachus* | | Lea (1910), Watt (1967) |
| **Elateridae** | | | |
| *Cardiophorus* | *Pogonomyrmex* | | MacKay (1983a) |
| **Endomychidae** | | | |
| *Rhymbillus, Symbiotes, Trochoideus* | *Formica, Lasius,* other species | Probably scavengers | Reichensperger (1915), Lawrence and Reichardt (1969) |
| **Erotylidae** | | | |
| *Tritomidea* | *Amblyopone* | | Lea (1910) |
| **Histeridae** | | | |
| *Abraeus, Acritus, Attalister, Chlamydopsis, Epiglyptus, Geosaprimus, Hetaerius, Mesynodites, Myrmetes, Orectoscelis, Paramyrmetes, Poneralister, Psiloscelis, Sternocoelis, Wasmannister,* many other genera | Many nonlegionary ant genera | Some feed on detritus, others feed on hosts and on prey captured by hosts; some solicit regurgitated food | Wasmann (1886, 1905), Lea (1910, 1919), Wheeler (1908c), Bruch (1929, 1930, 1932c), Eidmann (1937), Kistner (1982), MacKay (1983a), Helava et al. (1985) |
| *Acritus, Coelocraera, Hypocacculus, Paratropus, Tribalus,* many other genera | Doryline army ants | Some carried by ants; food habits not known | Paulian (1948), Kistner (1982) |
| *Chrysetaerius, Daitrosister, Euxenister, Neolister, Oaristes, Pselaphister, Pulvinister, Symphilister, Synodites, Xylostega,* many other genera | Ecitonine army ants | Groom hosts and prey on host brood | Reichensperger (1924), Bruch (1926), Borgmeier (1948), Rettenmeyer (1961), Akre (1968), Torgerson and Akre (1970), Kistner (1982), Helava et al. (1985) |
| **Hydrophilidae** | | | |
| *Oosternum* | *Atta* | | Weber (1972) |
| **Lagriidae** | | | |
| *Lagria* | *Brachyponera, Myrmecia* | | Lea (1910, 1912) |
| **Lathridiidae** | | | |
| *Cartodere* | *Formica, Lasius, Tetramorium* | | Bernard (1968) |

*continued*

**TABLE 13–1** *(continued)*

| Taxon | Host | Biology | Selected references |
|---|---|---|---|
| **Leiodidae** | | | |
| *Attaephilus, Attumbra, Catopomorphus, Choleva, Dissochaeta, Echinocoleus, Eocatops, Myrmicholeva, Nargus, Nemadus, Philomessor, Preposcia, Ptomaphagus, Synaulus* | *Aphaenogaster, Camponotus, Formica, Lasius, Leptogenys, Messor, Pogonomyrmex* | | Lea (1910), Hatch (1933), Jeannel (1936), Fall (1937), Peck (1973, 1976), MacKay (1983a) |
| **Leptinidae** | | | |
| *Leptinus* | *Formica* | Tolerated by ants; food habits unknown | Park (1929) |
| **Limulodidae** | | | |
| Most or all species | Many ant species, especially army ants | Groom hosts, possibly other food habits | Lea (1910), Park (1933), Seevers and Dybas (1943), Wilson et al. (1954), Dybas (1962) |
| **Merophysidae** | | | |
| *Coluocera* | *Messor, Monomorium, Paratrechina, Pheidole* | | Wasmann (1899b, 1905), Bernard (1968), Kistner (1982) |
| **Nitidulidae** | | | |
| *Amphotis* | *Crematogaster, Formica, Lasius* | Solicit regurgitated food | Schwarz (1890), Wasmann (1892), Donisthorpe (1927), Hölldobler (1968a) |
| **Pselaphidae** | | | |
| Many genera and species | Many ant species | Predators on hosts | Brauns and Bickhardt (1914), Lea (1919), Donisthorpe (1927), Hölldobler (1948), Park (1964), Akre and Hill (1973), R. Cammaerts (1974, 1977), Kistner (1982) |
| Tribe Clavigerinae: All species | Many ant species | Most or all are symphiles that prey on larvae, solicit regurgitated food, and lick hosts; they are carried by host ants | Brauns and Bickhardt (1914), Lea (1919), Donisthorpe (1927), Hölldobler (1948), Park (1964), Akre and Hill (1973), R. Cammaerts (1974, 1977), Kistner (1982) |
| **Ptiliidae** | | | |
| Probably many species | *Atta, Eciton, Formica* | | Donisthorpe (1927), Eidmann (1937) |
| **Ptinidae** | | | |
| About 12 genera | Several ant genera | Probably scavengers; some species have reduced mouthparts and may be fed by ants | Lawrence and Reichardt (1969) |
| **Rhizophagidae** | | | |
| *Monotoma* | *Formica* | | Bernard (1968) |

*continued*

**TABLE 13–1** *(continued)*

| Taxon | Host | Biology | Selected references |
|---|---|---|---|
| **Scarabaeidae** | | | |
| *Canthon, Cartwrightia, Chaeridium, Coelosis, Euparixia, Euphoria, Lamanoxia* | Fungus-growing ants, *Acromyrmex, Atta* | Some are predators on hosts, others feed as larvae on fungus | Eidmann (1937, 1938), Ritcher (1958), Woodruff and Cartwright (1967), Krikken (1972), Kistner (1982) |
| *Cetonia* | *Formica* | Larvae live in ant nests; probably eat detritus and vegetable material there | Gösswald (1985) |
| *Cremastocheilus* | Several ant genera | Predators on host larvae; also feed occasionally on crop contents of dead workers | Wheeler (1908b), Cazier and Statham (1962), Cazier and Mortenson (1965), Alpert and Ritcher (1975), Krikken (1976), Alpert (1981) |
| *Cryptodus, Euphoria* | *Camponotus, Formica, Leptomyrmex* | | Lea (1910), Wheeler (1910a) |
| *Euparia, Rhyssemus* | *Solenopsis* | | Summerlin (1978), Wojcik et al. (1978) |
| *Euphoriaspis* | *Formica* | Probably predators on host larvae | Ratcliffe (1976) |
| *Haroldius* | *Diacamma, Pheidole* | | Ritcher (1958) |
| *Myrmecaphodius* | *Iridomyrmex, Solenopsis* | | Collins and Markin (1971), Wojcik et al. (1977, 1978), Vander Meer and Wojcik (1982) |
| *Paracotalpa* | *Pogonomyrmex* | | MacKay (1983a) |
| **Scydmaenidae** | | | |
| Many species | Many ant species | | Lea (1910, 1912), Clark (1921), Costa Lima (1962), Bernard (1968) |
| **Staphylinidae** | | | |
| Many tribes and genera | Doryline, ecitonine and aenictine army ants | Depending on the species, predatory on hosts or scavengers; at least some also groom the hosts | Wasmann (1917), Patrizi (1948), Paulian (1948), Kistner (1958, 1966a,b, 1968, 1979, 1982), Seevers (1965), Akre and Rettenmeyer (1966), Akre and Torgerson (1968), Kistner and Jacobson (1975a, 1979), Jacobson et al. (1987) |
| Many tribes and genera | Nonlegionary ants | Some are symphiles; depending on the species, either predators on hosts or insects living with them, as scavengers, or inducers of regurgitation, or host groomers, or a combination of these roles | Lea (1910, 1912), Wheeler (1910a), Wasmann (1915b, 1925, 1930), Donisthorpe (1927), Hölldobler (1967, 1968b, 1969, 1970b,c), Bernard (1968), Kistner (1972, 1982), Hölldobler et al. (1981) |
| **Tenebrionidae** | | | |
| *Alaudes, Anepsius, Blapstinus, Conibius, Metoponium, Notibius* | *Pogonomyrmex* | Probably seed predators in storage chambers of host ants' nests | MacKay (1983a) |
| *Chariophenus, Scleropatrum* | *Pheidologeton* | | Blair (1929) |
| *Cossyphodes, Cossyphodites, Dichillus, Lenkous,* | Many ant genera | Some are seed predators in storage chambers of harvester ant hosts | Lea (1905, 1910), Clark (1921), Basilewsky (1952), Bernard (1968), Le Masne (1970a), |

*continued*

**TABLE 13–1** *(continued)*

| Taxon | Host | Biology | Selected references |
|---|---|---|---|
| *Myrmecopeltoides, Oochrotus, Thorictosoma, Tribolium* | | | Kaszab (1973) |
| **Thorictidae** | | | |
| *Thorictus* | Many ant genera | Attack legs or antennae of hosts to be carried into nests, then scavenge on refuse and dead ants | Wasmann (1895), Wasmann and Brauns (1925), Banck (1927) |
| **LEPIDOPTERA** | | | |
| **Agrotidae** | | | |
| *Epizeuxis* | *Formica* | Larvae are scavengers | Smith (1941) |
| **Arctiidae** | | | |
| *Crambidia* | *Formica* | Larvae are protected in ant nest from parasites | Ayre (1958) |
| **Batrachedridae** | | | |
| *Batrachedra* | *Polyrhachis* | Larvae feed on host brood | Hinton (1951) |
| **Cyclotornidae** | | | |
| *Cyclotorna* | *Iridomyrmex* | Larvae prey on cicadellids and subsequently on ant larvae inside ant nest | Dodd (1912) |
| **Lycaenidae** | | | |
| Many species | Many ant genera and species | Trophobiotic: ants attend larvae on food plant; larvae of some species enter nest to pupate; some also feed on host brood or solicit food from host workers | See special section in Chapter 13; Hinton (1951), Downey (1966), Ross (1966), Malicky (1969, 1970a,b), Maschwitz et al. (1975, 1984, 1985b,c), Callaghan (1977), Schremmer (1978), Atstatt (1981a,b), Henning (1983a,b), Kitching (1983, 1987), Pierce (1983, 1984, 1985), Cottrell (1984), Kitching and Luke (1985), Pierce and Elgar (1985), DeVries et al. (1986), Pierce and Easteal (1986), Pierce (1987) |
| **Nymphalidae** | | | |
| *Mechanitis* | Ecitonine army ants | Males and females attracted to raiding swarms by ants' odor; search out bird droppings, which are abundant at army ant swarms because of accompanying ant birds | Drummond (1976), Ray and Andrews (1980) |
| **Papilionidae** | | | |
| *Gramphium* | Ecitonine army ants | Males and females attracted to raiding swarms by ants' odor; search out bird droppings, which are abundant at army ant swarms because of accompanying ant birds | Drummond (1976) |

*continued*

**TABLE 13–1** *(continued)*

| Taxon | Host | Biology | Selected references |
|---|---|---|---|
| **Pieridae** | | | |
| Several genera; obligatory status not certain | Several ant genera | Possess attractive glands and are attended by ants on food plants | Hinton (1951) |
| **Pyralidae** | | | |
| *Pachypodistes* | *Hypoclinea* | Larvae feed on nest carton | Hagmann (1907) |
| *Stenachroia* | *Crematogaster* | | Turner (1913) |
| *Wurthia* | *Oecophylla, Polyrhachis* | Larvae prey on host brood | Kemner (1923), Hinton (1951) |
| **Tineidae** | | | |
| *Ardiosteres, Iphierga* | *Iridomyrmex* | Larvae are scavengers in host nests | Hinton (1951) |
| *Atticonviva* | *Acromyrmex, Atta* | Larvae feed on plant material brought into nest by host ants | Eidmann (1937, 1938) |
| *Hypophrictoides* | *Hypoclinea, Plagiolepis* | Predaceous on ant pupae | Roepke (1925) |
| *Myrmecozela* | *Formica* | Larvae feed on nest material of host nest | Hinton (1951) |
| **Tortricidae** | | | |
| *Semutophila* | Species of Formicinae, Dolichoderinae, Myrmicinae | Larvae live in trophobiosis with ants | Maschwitz et al. (1986a) |
| **DIPTERA** | | | |
| **Bombyliidae** | | | |
| *Glabellula* | *Formica* | Last larval instar lives in *Formica* nests; probably predaceous on ant larvae | Andersson (1974) |
| **Calliphoridae** | | | |
| *Bengalia* | African driver ants, *Dorylus* | Rob prey and larvae from columns of ants | Bequaert (1922a), Thorpe (1942) |
| *Bengalia* | *Crematogaster, Leptogenys, Monomorium, Pheidole, Pheidologeton* | Rob prey and possibly larvae from ants | Bequaert (1922a) |
| *Bengalia* | *Camponotus, Meranoplus, Myrmicaria, Oecophylla, Pachycondyla (= Bothroponera), Plagiolepsis, Polyrhachis, Solenopsis, Technomyrmex* | Rob prey from foraging ants outside nest | Maschwitz and Schönegge (1980) |
| **Ceratopogonidae** | | | |
| *Ceratopogon* | *Formica* | Larvae live in host nests; obligatory relationship uncertain | Long (1902), E. Séguy in Bernard (1968) |
| **Chironomidae** | | | |
| *Forcipomyia* | *Aphaenogaster, Formica, Labidus* | Presumably larvae prey on host larvae | Long (1902), Wheeler (1928), E. Séguy in Bernard (1968) |
| **Conopidae** | | | |
| *Stylogaster* | Ecitonine army ants, doryline driver ants | Hovering females follow raids of ecitonine army ants and parasitize cockroaches | Rettenmeyer (1961), Smith (1969) |

*continued*

**TABLE 13–1** *(continued)*

| Taxon | Host | Biology | Selected references |
|---|---|---|---|
| | | flushed out by ants; probably also parasitize adult tachinids that follow ants | |
| **Culicidae** | | | |
| *Malaya (Harpagomyia)* | *Crematogaster* | Adults solicit regurgitated food from workers on odor trails | Jacobson (1909), Farquharson (1918), Wheeler (1928) |
| **Milichiidae** | | | |
| *Milichia* | *Crematogaster, Lasius* | Adult flies solicit regurgitated food from ants on trunk trails; larvae inside nests | Donisthorpe (1927) |
| *Pholeomyia* | Fungus-growing ants, *Atta* | Eat detritus in refuse chambers of host ant nests | Sabrosky (1959), Moser (1963), Moser and Neff (1971) |
| *Phyllomyza* | *Formica, Lasius* | Larvae live in refuse pile of host ant nests | Donisthorpe (1927) |
| *Prosoetomilichia* | *Hypoclinea* | Adults feed on anal droplets of workers | Jacobson (1909) |
| **Muscidae** | | | |
| *Fannia* | Fungus-growing ants, *Atta* | Larvae probably feed on detritus in debris chambers of host ant nests | Chillcott (1965) |
| *Stomoxys* | African driver ants, *Dorylus* | | Thorpe (1942), Kistner (1982) |
| **Mydidae** | | | |
| *Mydas* | Fungus-growing ants, *Atta* | Live in debris chambers of host ant nests | Zikán (1942) |
| **Phoridae** | | | |
| Six subfamilies, all of which have been recorded with social insects | Especially abundant in and around ecitonine army ants | Usually scavengers; some species also share prey of their hosts or feed on hosts themselves | Schmitz (1950), Borgmeier (1968, 1971), Rettenmeyer and Akre (1968) |
| *Aenigmatias* | *Formica* | Apterous female flies (males are winged) lay eggs on host larvae | Donisthorpe (1927) |
| *Aenigmatopoeus* | African driver ants, *Aenictus, Dorylus* | Apterous female flies live in debris of host ants, accompany ants at end of raiding and emigration trails | Borgmeier (1963) |
| *Apocephalus* | *Pheidole* | Adult flies lay eggs on workers | Burges (1979), Feener (1981) |
| *Cataclinusa* | *Pachycondyla* | Adult flies lay eggs on larvae of host ants | Wheeler (1901b) |
| *Metopina* | *Solenopsis (Diplorhoptrum)* | Adult flies receive regurgitated food from hosts | Hölldobler (1928) |
| *Neodohrniphora* and other genera | Fungus-growing ants, *Atta* | Adult flies lay eggs on workers | Borgmeier (1925), Eidmann (1937), Eibl-Eibesfeldt and Eibl-Eibesfeldt (1967), Weber (1972) |
| *Plastophora* | *Solenopsis* | Adult flies lay eggs on workers; one undescribed species parasitizes alate females | Wojcik et al. (1987) |
| *Pseudacteon* | *Solenopsis* | Oviposit on worker ants | Feener (1987b) |
| *Tubicera* | *Plagiolepis* | Larvae tended by host ants; ants regurgitate liquid to larvae; adult flies not tolerated by ants | Le Masne (1941) |

*continued*

**TABLE 13–1** *(continued)*

| Taxon | Host | Biology | Selected references |
|---|---|---|---|
| **Sciaridae** | | | |
| *Sciara* | *Formica, Lasius* | Larvae live in host nests and are groomed by workers | E. Séguy in Bernard (1968) |
| **Sphaeroceridae** | | | |
| *Aptilotella, Hemalomitra* | Ecitonine army ants | | Kistner (1982) |
| *Limosina* | *Formica, Lasius* | | Donisthorpe (1927) |
| *Safaria, Leptocera* | Doryline driver ants | | Richards (1968), Kistner (1982) |
| **Syrphidae** | | | |
| *Microdon* | Many ant genera | Larvae live as scavengers in host ant nests; adults attacked by ants; predation on ant brood recorded in some species | Hocking (1970), Van Pelt and Van Pelt (1972), Akre et al. (1973), Lopez and Bonaric (1977), Duffield (1981), Thompson (1981), Garnett et al. (1985) |
| *Trichopsomyia* | *Polyrhachis* | Larvae live as scavengers and brood predators in ant nests | B. Hölldobler (unpublished) |
| *Xanthogramma* | *Lasius* | Larvae tended by ants | Hölldobler (1929a) |
| **Tachinidae** | | | |
| *Androeuryops, Calodexia* | Ecitonine army ants | Females follow ant raids and parasitize cockroaches and crickets flushed out by ants | Rettenmeyer (1961) |
| *Strongygaster* (= *Tamiclea*) | *Lasius* | Larvae live endoparasitically in young queens; pupae tended by ant queen; adult fly attacked | Gösswald (1950) |
| **HYMENOPTERA** | | | |
| **Bethylidae** | | | |
| *Dissomphalus* | Fungus-growing ants, *Acromyrmex, Atta* | | Bruch (1917), Eidmann (1937) |
| *Dissomphalus* | Ecitonine army ants | | Evans (1964) |
| *Epyris, Rhabdepyris* | *Formica, Lasius, Tetramorium* | | Evans (1964), Bernard (1968) |
| *Pseudisobrachium* | *Aphaenogaster, Solenopsis* | Larvae probably endoparasites of host brood | Mann (1915) |
| **Braconidae, Paxylommatidae** | | | |
| *Aclitus* | *Pheidole* | Adult females antennate host ants frequently; oviposition behavior on aphids observed; adults live on honeydew extruded by aphids | Takada and Hashimoto (1985) |
| *Elasmosoma, Hybrizon, Neoneurus* | *Formica* and other genera | Internal parasites: females insert egg into intersegmental membranes of host ants; larvae develop inside gasters of ants | Wheeler (1910a), Donisthorpe (1927), Bernard (1968) |
| *Paralipsis* | *Lasius* | Adult females live inside host nests and solicit regurgitated food from ants; are parasites of aphids tended by host ants. When entering ant colony females attempt to ride on host ant workers, rubbing their legs on ant. Soon afterward, parasite is | Maneval (1940), Starý (1970), Takada and Hashimoto (1985) |

*continued*

**TABLE 13-1** *(continued)*

| Taxon | Host | Biology | Selected references |
|---|---|---|---|
| | | accepted by colony and presumably obtains colony odor thereby | |
| **Diapriidae** | | | |
| *Ashmeadopria, Auxopaedeutes, Basalys, Bruchopria, Bruesopria, Geodiapria, Lepidopria, Loxotropa, Plagiopria, Solenopsia, Tetramopria, Trichopria,* others | Nonlegionary ants: *Plagiolepis, Solenopsis (Diplorhoptrum), Tetramorium* | *Lepidopria* and *Solenopsia* solicit regurgitation; larvae probably endoparasites of host brood | Donisthorpe (1927), Hölldobler (1928), Borgmeier (1939), Wing (1951) |
| *Mimopria, Myrmecopria, Phaenopria* | Ecitonine army ants | Some species mimetic, run in host columns; some species treated by hosts like ants, probably because of chemical mimicry; many species cast their wings when entering host colony. Larvae probably endoparasites of host brood | Holmgren (1908), Borgmeier (1939), Mann (1912b, 1923b), Masner (1959, 1976, 1977) |
| **Eucharitidae** | | | |
| *Eucharis, Orasema, Pheidoloxenus* | *Camponotus, Ectatomma, Formica, Messor, Myrmecia, Pheidole* | Larvae endoparasites of host brood | Wheeler (1910a), Mann (1914), Gösswald (1934), Bernard (1968) |
| **Ichneumonidae** | | | |
| *Pesomachus* | *Formica, Myrmica* | Adults wingless, ant-like in appearance; have been found inside ant nests | Donisthorpe (1927) |
| **Pompilidae** | | | |
| *Dimorphagenia* | *Azteca* | | Evans (1973) |
| *Iridomimus* | *Iridomyrmex* | Adults wingless and resemble host ants | Evans (1970) |
| **Pteromalidae** | | | |
| *Spalangia* | *Lasius* | Parasitizes fly (*Phyllomyza*) that lives in *Lasius fuliginosus* nests | Donisthorpe (1927) |
| **Sphecidae** | | | |
| *Aphilanthops* | *Formica* | Specialized predator on specific ant species | Evans (1962) |
| *Clypeadon* | *Pogonomyrmex* | Specialized predator on specific ant species | Evans (1962, 1977b) |

body to be used by the host ant as a substitute for the terminal segment of the foot. Rettenmeyer (1962a) described its behavior:

> The mite inserts its chelicerae into the membrane, but it is not known if the palpi assist in holding to the ant. Presumably the mite feeds in this position and may be considered a true ectoparasite rather than a strictly phoretic form . . . Army ants characteristically form clusters by hooking their tarsal claws over the legs or other parts of the bodies of other workers. Temporary nests or bivouacs of *Eciton dulcius crassinode* may have clusters several inches in diameter hanging in cavities in the ground. In small laboratory nests, when a worker hooked the leg with the mite onto the nest or another worker, the hind legs of the mite usually served in place of the ant's tarsal claws. The hind legs of the mite were never seen in a straight position but were always curved. In all cases the entire hind legs of the mite served as claws, not just the mite's claws. No difference was noted in the behavior of the ant whether its own claws or the hind legs of the mite were hooked onto some support. When no other part of the ant's leg was touching any object, adequate support was provided by the mite's legs for at least five minutes.

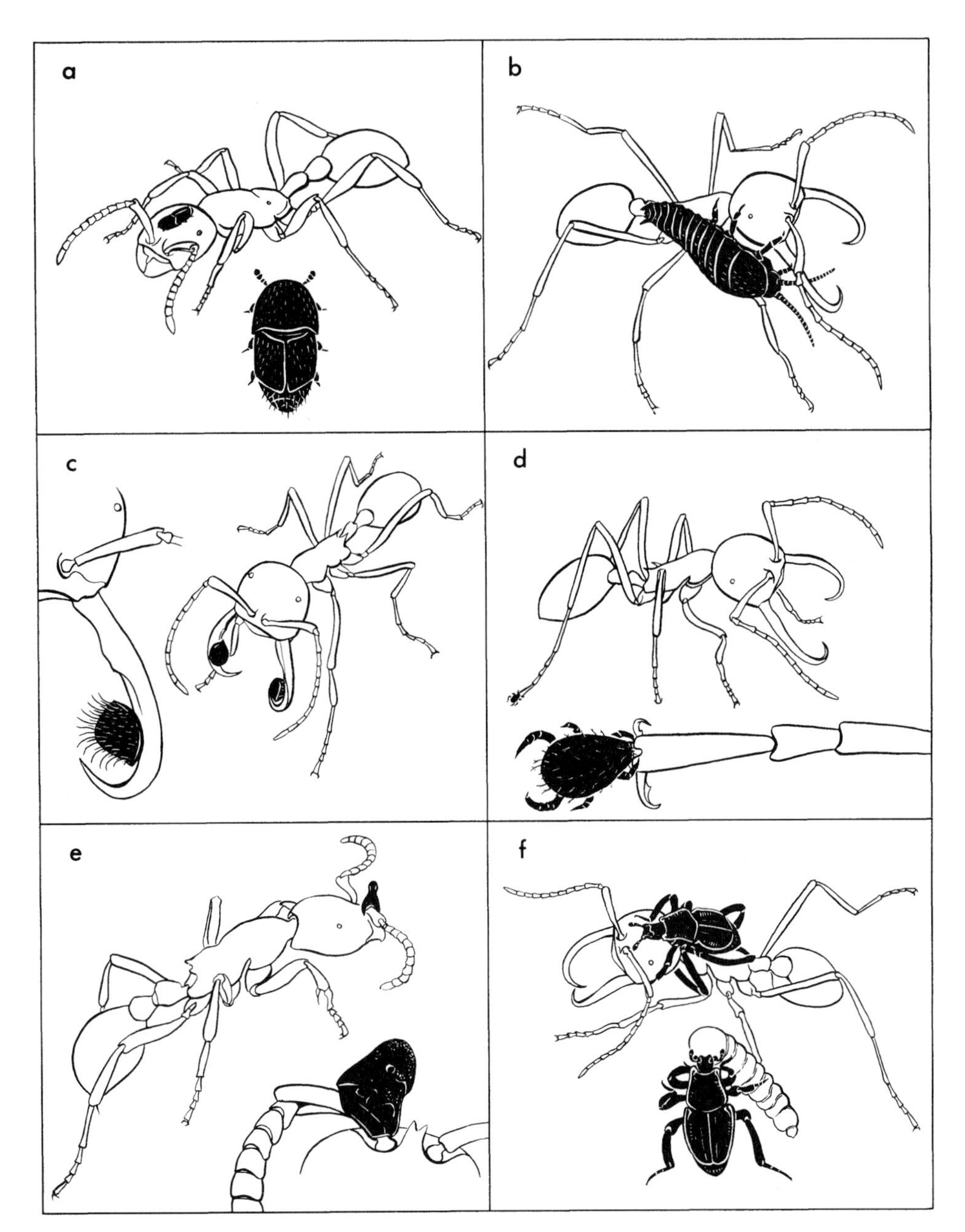


**FIGURE 13–1** Six arthropod guests of army ants, which display a few of the many kinds of symbiotic adaptations to these hosts. (*a*) *Paralimulodes wasmanni* is a limulodid beetle that spends most of its time riding on the bodies of host workers (*Neivamyrmex nigrescens*). (*b*) *Trichatelura manni* is a nicoletiid silverfish that scrapes and licks the body secretions of its hosts (*Eciton* spp.) and shares their prey. (*c*) The *Circocylliba* mite shown here belongs to a species specialized for riding on the inner surface of the mandibles of major workers of *Eciton*. (*d*) *Macrocheles rettenmeyeri* is another mite species which is normally found attached in the position shown, serving as an extra "foot" for workers of *E. dulcius*. (*e*) *Antennequesoma* is a mite genus highly specialized for attachment to the first antennal segment of army ants. (*f*) The histerid beetle *Euxenister caroli* grooms the adults of *E. burchelli* and feeds on their larvae. (From Wilson, 1971; original drawings by T. Hölldobler-Forsyth.)

A second species, *M. dibanos*, was observed to parasitize *E. vagans* in the same manner.

Many of the mites found with Neotropical army ants are endowed with particular holdfast mechanisms, such as modified claws, empodia, specialized ridges and teeth on their enlarged dorsa, and highly modified posterior idiosomas that adapt the mites to specific regions of the hosts' bodies (Elzinga, 1978).

Mites, being much smaller than their hosts, function primarily as ectoparasites or scavengers within the army ant nests. Many of them, on the other hand, also either scavenge or obtain part of their nutriment by licking the oily secretions of the bodies of the hosts, a specialized behavior referred to by Wheeler and a few other authors as strigilation. The nutritive value of strigilation is open to some question, at least in the case of the ecitophiles. Both Rettenmeyer and Akre have suggested that it also serves to transfer the host odor to the parasites. This hypothesis received support from the recent finding by Vander Meer and Wojcik (1982) that the scarabaeid beetle *Myrmecaphodius excavaticollis* acquires colony-specific cuticular hy-

**FIGURE 13–2** An unidentified myrmecophilous staphylinid attacks a worker of the army ant *Nomamyrmex esenbecki*. (Photograph courtesy of C. W. Rettenmeyer.)

**FIGURE 13–3** A worker of *Neivamyrmex sumichrasti* is holding a male of the staphylinid species *Ecitosius robustus* by his leg. Note the remarkable resemblance in body shape and surface sculpture of the myrmecophile to its host. (Photograph courtesy of C. W. Rettenmeyer.)

drocarbons from its *Solenopsis* hosts. Other insects feed on the booty brought in by the army ant host workers, or else prey on the hosts themselves, especially the defenseless larvae and pupae. As we saw in Figure 13–1, the insect guests vary greatly in size and in the positions they take on the bodies of the host workers. They also vary in the locations they prefer within the bivouac as a whole and in the ant columns during colony emigrations and raids.

A typically instructive case of positional specialization is provided by the aleocharine staphylinid *Tetradonia marginalis*. Akre and Rettenmeyer (1966) note that *T. marginalis*

> is the most common staphylinid found in refuse deposits of *Eciton burchelli* and in the emigrations. Sometimes more than 100 individuals can be found in a single large deposit of this host. Many of these myrmecophiles do not run along the emigration columns but apparently fly from one colony to another. *Tetradonia marginalis* was the species most frequently attacking uninjured, active workers as well as injured workers of both *E. burchelli* and *E. hamatum* along edges of columns and near bivouacs [Figure 13–2]. The staphylinids were not seen to attack workers running along in the middle of raid or emigration columns or those running quickly along the edges.

During field work in La Selva, Costa Rica, in 1985, we were able to confirm some of the observations of Akre and Rettenmeyer. We were especially intrigued by the beetles' spatial positions at the end of the emigration columns and by the hunting techniques they employed while there. For an hour late one March evening, we watched as a large part of the emigration column of *Eciton hamatum* crossed a forest path. At the very end of the column, with no more than a hundred stragglers on that portion of the trail, we spotted ten or more *Tetradonia marginalis* running swiftly along the trail, abdomen curled up over the thorax. Most were attacking *Eciton* media workers, and one was attacking a major. All of the victims were slow and partly disabled, unable to walk swiftly. We speculated that they had fallen back because of injuries or sickness, and had been singled out by the beetles for attack. The staphylinids circled around, seizing the legs or tip of the abdomen, in the latter case seeming to bite the anal region of the ant. Some dragged their victims swiftly along the trail by a leg and even those appearing to bite the anal region started off in this manner within a few minutes. In one instance two beetles attacked an ant together, running along with it in a coordinated fashion (unpublished observations).

Akre and Rettenmeyer (1966) made a distinction between "generalized" species of symbiotic staphylinids, which have the typical appearance of most Staphylinidae and no obvious modifications for life with ants, and "specialized" species, which include mimetic forms (to be described later in this chapter). The specialized staphylinids are distinguished by rigid abdomens that are held in a single position (Figure 13–3). Paradoxically, these insects are also less common than the nonmimetic forms in the host colonies. Akre and Rettenmeyer compiled an impressive list of consistent behavioral differences between the two kinds of species, which we have reproduced in Table 13–2. These data reveal a gradually increasing degree of behavioral integration with the host society, which may constitute the most obvious of the evolutionary pathways along which progress in symbiosis can be measured. The two kinds of species can be reasonably interpreted as progressive stages in an adaptation to the host ants and their way of life.

## THE ANT COLONY AS AN ECOSYSTEM

In more general terms the concept of an insect colony as an ecological island, or, to use a slightly more precise language, a partially isolated ecosystem, can be extended to gain a better understanding of certain aspects of the biology of the symbionts. The ant colony and its surroundings are richly structured into many diverse microhabitats, such as the foraging trunk routes, refuse areas, peripheral nest chambers and guard nests, storage chambers, brood chambers with separate areas for pupae, larvae, and eggs, queen chambers, and the bodies of adult and immature inhabitants of the nest. These microhabitats are occupied by a diversity of symbionts, which show special adaptations to each of the niches in turn.

**TABLE 13–2** Behavioral differences between morphologically "generalized" and "specialized" Staphylinidae associated with army ants. (From Akre and Rettenmeyer, 1966.)

| Generalized species | Specialized species |
|---|---|
| 1. Live around periphery of bivouacs or in refuse deposits | 1. Live within bivouacs; rarely seen in refuse deposits or outside bivouacs (except on emigrations) |
| 2. Found at start of emigration before brood is carried; may be found in smaller numbers during emigration; also found at end of emigration, including after ants have disappeared from column | 2. Most frequently found when first part of brood is carried; may be present throughout time when brood is seen; rare or absent at other times |
| 3. Run along edges of emigration columns and short distances from columns; sometimes in centers of columns | 3. Run in centers of emigration columns |
| 4. Sometimes found at bivouac sites shortly after ants have emigrated | 4. Absent or rare at bivouac sites shortly after ants have emigrated |
| 5. Do not ride on brood or booty carried by workers in raid or emigration columns | 5. May ride on booty and brood in emigration columns |
| 6. Not carried by ants (excluding attacking workers) | 6. Carried by workers (few species only) |
| 7. Frequently attacked by ants in nest | 7. Usually not attacked by ants in nest |
| 8. Attack adult ants but usually kill only injured or weak workers | 8. Do not attack adult ants |
| 9. Can tolerate a wider range of ecological conditions; live with a few or no ants for a day to weeks | 9. Can tolerate only a narrow range of ecological conditions; usually die within a few hours of removal from colony |
| 10. Never groom living workers; sometimes appear to feed on surfaces of dead workers; workers do not groom staphylinids | 10. Often groom surfaces of living workers; workers tolerate grooming and may groom staphylinids |
| 11. Never rub their legs on workers | 11. May rub their legs on workers |
| Examples: *Microdonia, Tetradonia, Ecitana, Ecitodonia* | Examples: *Ecitomorpha, Ecitophya, Probeyeria, Ecitosius* |

## The Bodies of the Host Ants

The bodies of ants are occupied internally by a diversity of endoparasites, including nematodes (Gösswald, 1938b), trematodes (Hohorst and Lämmler, 1962; Schneider and Hohorst, 1971), and cestodes (Muir, 1954; Plateaux, 1972; Buschinger, 1973b). This form of parasitism can affect the individual host's morphology (Gösswald, 1938b; Kutter, 1958), metabolism (Kloft, 1949), and behavior (Hohorst and Graefe, 1961), but it usually has little impact on the social behavior of the host ants. We will therefore not discuss this subject in greater detail.

One endoparasite, however, deserves special mention, namely the tachinid fly *Strongygaster globula* (formerly called *Tamiclea globula*), whose larvae develop as endoparasites inside the gaster of colony-founding queens of *Lasius niger* and *L. alienus* (Gösswald, 1950). The behavior of the infected queens is not noticeably affected, except that they cease to lay eggs. When the last-instar larva of the parasite leaves the host's abdomen through the cloaca, it quickly pupates and is groomed and tended by the host ants as if it were a member of the ants' own brood (Figure 13–4). In contrast, the fly imago is not treated amicably by the ants and has to leave the incipient ant nest quickly. The ant queens infected with *Strongygaster* die after the parasites leave the nest.

The bodies of immature and adult army ants, as noted earlier, provide special microhabitats for a diversity of ectosymbionts. The same is true for nonlegionary ant species. Probably the best-known examples are species of the antennophorid mite genus *Antennophorus*, which live with formicine ants of the genera *Lasius* and *Acanthomyops* (Janet, 1897b; Wasmann, 1902; Karawajew, 1906; Wheeler, 1910b). The mites ride on the bodies of the ants, shifting their positions when two or more are present on the same ant so as to produce a balanced load. When three mites are present, two ride on either side of the gaster, and the third on the lower surface of the head (Figure 13–5). When four are present two position themselves on either side of the abdomen and two on either side of the head. The workers first try to remove the mites from their bodies, but seem to give up after several unsuccessful attempts.

The *Antennophorus* live on food regurgitated by the ant hosts, either imbibing it as it passes between workers or soliciting it directly by stroking the mouthparts of workers with long, antenna-like forelegs. In the latter case the mites evidently imitate the tactile signals used by the ants themselves. It has even been observed that mites attached to the gaster of one worker are able to solicit food from another. The mites prefer to mount newly eclosed workers. This habit confers a distinct advantage: young workers usually stay inside the nest, and they are attractive to older workers who frequently feed them by regurgitation, affording the mites plenty of opportunity to participate in the trophallaxis.

A remarkable analogy to *Antennophorus* in its ectosymbiotic adaptations was discovered by K. Hölldobler (1928) in the phorid fly *Metopina formicomendicula*, which lives in nests of the thief ant *Solenopsis* (= *Diplorhoptrum*) *fugax*. The *Metopina* adult mounts a thief ant

**FIGURE 13–4** A myrmecophilous fly of Europe. (*Above*) A young queen of *Lasius niger* tends a pupa of the parasitic tachinid fly *Strongygaster globula*. (*Below*) The adult fly after eclosion. (Photograph by T. Hölldobler-Forsyth.)

worker and rapidly strokes the head and mouthparts of the ant with its forelegs. The worker usually responds by slightly raising her head, opening her mandibles, and regurgitating a droplet of food, which is then rapidly imbibed by the fly (Figure 13–6). The ants occasionally attack the phorids when they move through the nest but they rarely harm them because the flies are too swift and elusive. When resting the *Metopina* frequently ride on the queen, where they are usually ignored by the workers. Hölldobler (1951) categorized *Metopina* as a "synoekete of the evasive type."

## Trails and Kitchen Middens

Many guests of army ants, especially the staphylinid beetles, regularly accompany raiding columns, where they prey on the captured booty of their hosts or on the host ants themselves. These myrmecophilous scavengers and predators, as exemplified by *Tetradonia marginalis,* often bring up the "rear guard" of the raiding columns, sprinting deftly along each twist and turn of the newly vacated trails. It has been known for a long time that guests of army ants can orient by the odor trails without further cues provided by the hosts. Since army ants frequently change the location of their bivouacs, it is important for the myrmecophiles to be able to follow the emigration trails. Akre and Rettenmeyer (1968) investigated this phenomenon systematically by exposing a large variety of guests of ecitonine army ants to natural odor trails laid over the floor of the laboratory arena. They found that nearly all of the species tested, including a haphazardly chosen sample of Staphylinidae, Histeridae, Limulodidae, Phoridae, Thysanura, and Diplopoda, were able to orient accurately by means of the trails alone. Some myrmecophiles did better than others, however. For example, the staphylinids *Ecitomorpha, Ecitophya,* and *Vatesus,* which regularly accompany ant columns in close formation, followed experimental trails readily and accurately, whereas phorid flies, which under natural conditions exhibit a kind of meandering movement along emigration trails, deviated more frequently from the experimental trails. Other species, such as histerid beetles, which normally ride on host ants or booty in ant columns, did relatively poorly in the trail-following tests.

In some instances the guests were more sensitive to trail pheromones than the ants themselves. In choice tests the symbionts preferred trails laid by their own host species to those of other army ants, and sometimes they were repelled by the trails of the wrong species. However, a few instances were recorded in which the ecitophiles were less specific in their response. Interestingly, the behavior was partly correlated with the myrmecophiles' acceptance of several host species in nature. None of the symbionts showed an ability to distinguish the trails of its own host colony in competition with trails laid by other colonies belonging to the same species.

Trail following may also be a means by which parasites locate the host colony. Kistner (1979) observed a limuloid beetle *Mimocetes* as it flew into a raiding column of *Dorylus,* then zigzagged from one side to another for about 10 centimeters, and finally entered the thick swarm of driver ants. The cockroach *Attaphila fungicola* lives with the fungus-growing ant *Atta texana,* whose workers depend strongly on odor trails for orientation during foraging expeditions and colony emigrations. Moser (1964) demonstrated that *Attaphila fungicola* follows the trail pheromone of its host ants, which he extracted directly from the poison gland of the workers and laid along arbitrarily selected routes in the laboratory. This experiment substantiates an earlier observation by Bolivar (1905) of *Attaphila schuppi* following trails of *Acromyrmex* in the field in Brazil. Moser also noted that the cockroaches do not depend solely on the ability to disperse from nest to nest, because individuals have been observed riding on the backs of *Atta texana* queens during the nuptial flights of these ants. The myrmecophilous cricket *Myrmecophila manni* probably also disperses along the foraging trails of its host, the mound-building ant *Formica obscuripes.* Henderson and Akre (1986a,b) observed a total of 63

FIGURE 13–5 Ectoparasitic mites of ants. (*a*) The mite *Antennophorus pubescens*. (Photograph by K. Hölldobler.) (*b*) Schematic illustration of *Antennophorus* riding on its host ant *Lasius*. (Based on Janet, 1897b; drawing by E. Kaiser.)

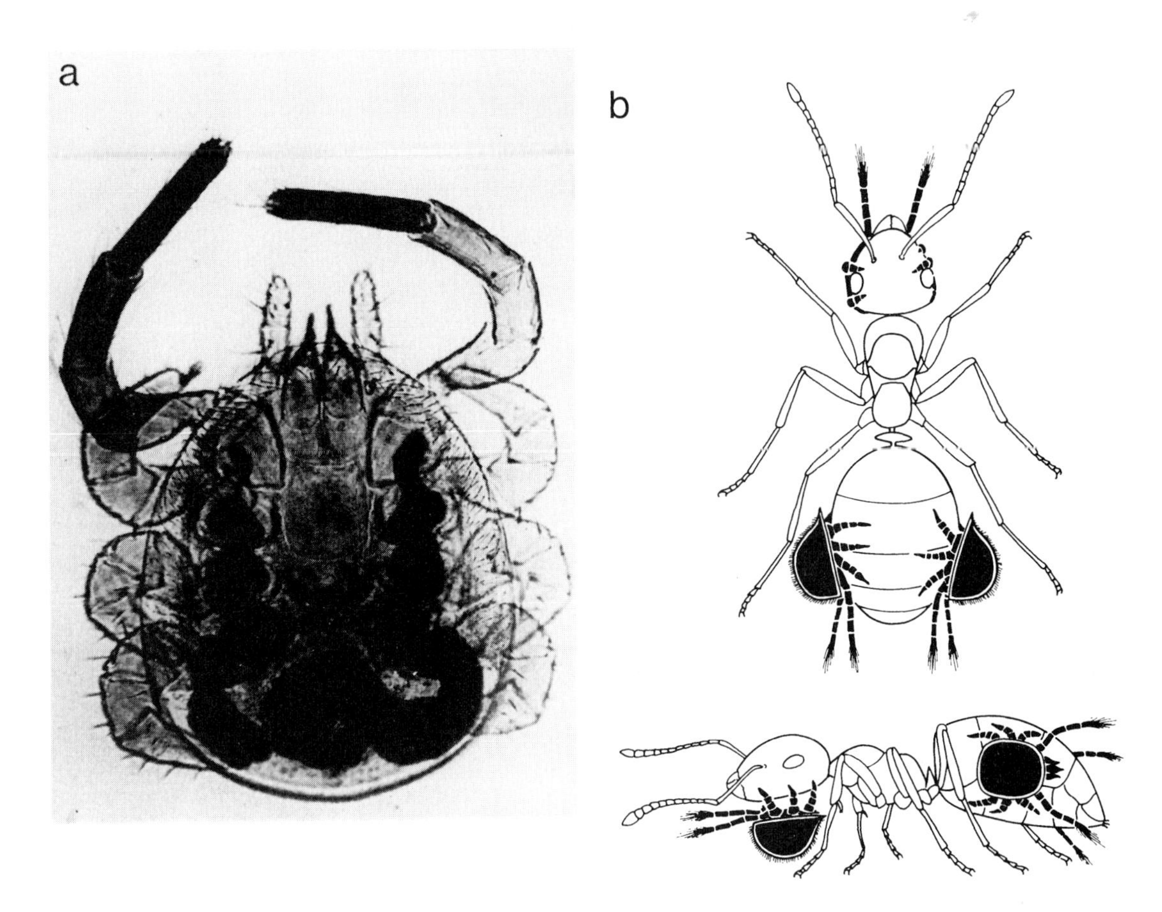


crickets, including both sexes and all nymphal instars, on foraging trails of *F. obscuripes* more than 20 meters distant from the nests. During June and July these insects were seen running on the trails nearly every evening from dusk to almost midnight.

It is well known that caterpillars of the lycaenid *Maculinea teleius* change from phytophagous to myrmecophilous habits after casting the skin of the third larval instar. It was previously assumed that the myrmecophilous stage is carried by its host ants, *Myrmica rubra*, into the host nest, where the *Maculinea* larvae prey on the ant brood (Malicky, 1969). Schroth and Maschwitz (1984) found, however, that fourth-instar larvae of *Maculinea teleius* actively move into the host ants' nest, thus confirming an earlier observation by Chapman (1920). In laboratory experiments Schroth and Maschwitz demonstrated that the lycaenid larvae follow trail pheromones of their host ants when searching for the nests.

Trails and ant paths not only serve as orientation cues, they also inadvertently create a diversity of ecological niches within which the symbionts specialize. The *Eciton* army ants and their attending ecitophiles provide some of the most striking examples of modes of adaptive radiation. Another, wholly unexpected case involves the adult females of three nymphalid butterfly species, *Mechanitis isthmia, Mechanitis doryssus*, and *Melinaea imitata*, which regularly follow raiding swarms of the army ant *E. burchelli*. Ray and Andrews (1980)

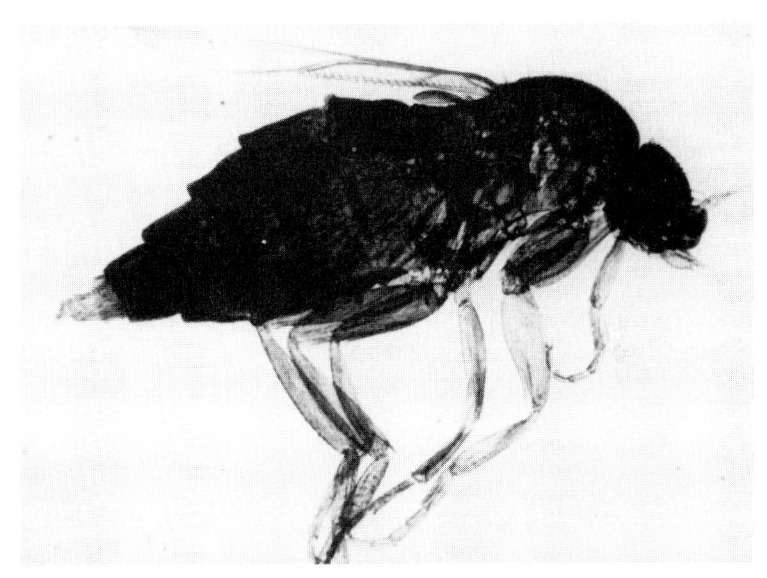

FIGURE 13–6 The phorid fly *Metopina formicomendicula* is illustrated above and shown riding on its host ant *Solenopsis* (= *Diplorhoptrum*) *fugax* below. The parasite rapidly strokes the ant's mouthparts with its forelegs to elicit regurgitation of food. (Photograph by K. Hölldobler, 1948; drawing by E. Kaiser based on Hölldobler.)

**FIGURE 13–7** The nitidulid beetle *Amphotis marginata,* a myrmecophile of the ant *Lasius fuliginosus.* (Painting by T. Hölldobler-Forsyth.)

**FIGURE 13–8** *Ptilocerus ochraceus* (*center*) and the ant *Hypoclinea bituberculata* (*right*), on which it feeds. At left is the undersurface of the bug's abdomen with the trichomes that dispense a tranquilizing attractant to the ants. (From Wilson, 1971; based on China, 1928.)

found that these butterflies feed on bird droppings in the swarm vicinity. Many species of birds, mostly in the antbird family Formicariidae, deposit droppings as they follow raiding swarms of *E. burchelli* to feed on insects flushed from the leaf litter by the ants (Willis and Oniki, 1978). Thus army ant swarms provide an indirect yet reliable and relatively rich source of nutrients in the form of bird droppings. Although no experimental proof has yet been attempted, it is most likely that odors associated with the swarms allow the butterflies to orient to the general vicinity of the droppings.

The trails and paths of nonlegionary ant species also harbor a diversity of myrmecophiles. One of the most fascinating examples is the nitidulid beetle *Amphotis marginata*, the "highwayman" of the local ant world (Figure 13–7). These beetles occupy shelters along the foraging trails of the formicine ant *Lasius fuliginosus* during the day. At night they patrol the trails and successfully stop and obtain food from ants returning to the nest. Ants that are heavily laden with food are easily deceived by the beetles' simple solicitation behavior. The *Amphotis* adult induces an ant to regurgitate food droplets by using its short antennae to tap the ant's labia and rapidly drum on her head (Plate 16). Soon after the beetle begins to feed, however, the ant seems to realize she has been tricked and attacks the beetle. The beetle then is able to defend itself simply by retracting its appendages and flattening itself on the ground. With the aid of special setae on its legs it firmly attaches its lower body surface to the ground. The ant is then unable to lift the beetle or turn it over (B. Hölldobler, 1968a, unpublished observations). This mechanism appears to give the beetle nearly complete protection. A somewhat similar parasitic pattern is exhibited by the calliphorid fly genus *Bengalia*. This paleotropical form has been reported to steal prey and larvae from columns of doryline driver ants and trails of a large number of nonlegionary ant genera (Bequaert, 1922; Maschwitz and Schönegge, 1980).

A diversity of insect predators settle near ant trails and prey on the ants. They employ an astonishing repertory of hunting techniques to capture their prey without being attacked or killed by the usually very aggressive ants. Some of the most specialized predators occur among the Hemiptera (Jordan, 1937). One remarkable hunting technique is employed by the reduviid bug *Acanthaspis concinnula* (Mühlenberg and Maschwitz, 1976). This species lives near the nests of the fire ant *Solenopsis geminata*. The bug captures *Solenopsis* workers, immobilizes them by injecting poison, and then sucks out the hemolymph. Then it places the shriveled corpses on its back. The accumulated shield of dead victims provides an excellent disguise and even attracts other ant workers, which approach to inspect their freshly killed nestmates. A similar shielding behavior has been described in some hunting spiders (Oliveira and Sazima, 1984). A wholly different method is employed by the reduviid bug *Ptilocerus ochraceus* (Figure 13–8) of Java, which feeds to a large extent on the dolichoderine ant *Hypoclinea biturberculata*, an extremely abundant ant in southeastern Asia (Jacobson, 1911). The predators take up a position in or near an ant path and offer an intoxicating attractant secreted from trichome glands on the underside of the bug's abdomen. According to Jacobson, when it encounters a *Hypoclinea* worker the bug presents the trichomes by raising the front of its body, simultaneously lifting up its front legs,

> folding them in such a manner that the tarsi nearly meet below the head. The ant at once proceeds to lick the trichomes, pulling all the while at the tuft of hairs, as if milking the creature, and by this manipulation the body of the bug is continually moved up and down. At this stage of the proceedings the bug does not yet attack the ant; it only takes the head and thorax of its victim between its front legs, as if to make sure of it; very often the point of the bug's beak is put behind the ant's head, where it is joined to the body, without, however, doing any injury to the ant. It is surprising to see how the bug can restrain its murderous intention as if it was knowing that the right moment had not yet arrived. After the ant has indulged in licking the tuft of hair for some minutes the exudation commences to exercise its paralyzing effect. That this is only brought about by the substance which the ants extract from the trichomes, and not by some thrust from the bug, is proved by the fact, that a great number of ants, after having licked for some time the secretion from the trichomes, leave the bug to retire to some distance. But very soon they are overtaken by the paralysis, even if they have not been touched at all by the bug's proboscis. In this way a much larger number of ants is destroyed than actually serves as food to the bugs, and one must wonder at the great profligacy of the ants, which enables them to stand such a heavy draft on the population of one community. As soon as the ant shows signs of paralysis by curling itself up and drawing in its legs, the bug at once seizes it with its front legs, and very soon it is pierced and sucked dry. (Jacobson, 1911)

A similar waylaying technique is employed by larvae of the carabid beetles *Sphallomorpha colymbetoides* and *S. nitiduloides,* which prey on the Australian meat ants *Iridomyrmex purpureus*, also a dolichoderine species. Moore (1974) discovered that the beetle larvae dig burrows near the nests and paths of the meat ants and wait for passing *Iridomyrmex* workers. When an ant approaches, the larva lunges at her, grasping the ant's legs and holding her for one or two minutes. The struggling ant soon relaxes, apparently in a state of paralysis. Moore speculates that the larva may apply toxic secretions, possibly from its maxillary glands. This, however, has not yet been tested experimentally. Once the prey is motionless, the larva sucks out the hemolymph of the ant and then discards the shriveled corpse. Often several *Sphallomorpha* burrows are found grouped together. Moore believes that the aggregation improves the predators' hunting success, because captured meat ants discharge an alarm pheromone that attracts other ants from nearby.

A similar ecological niche near ant nests or ant paths is occupied by ant lions, which are members of the order Neuroptera. The larvae of these insects build funnel-shaped pit traps in loose soil, where they wait on the bottom for ants to slip and fall down to them (Wheeler, 1930). The ant lion larvae are not exclusively predators on ants, however. Of 222 prey of the species *Myrmeleon immaculatus* identified by Heinrich and Heinrich (1984), only 36 percent were ants.

Yet another, very different technique exploiting trails and kitchen middens is employed by the European staphylinid beetle *Pella funesta* (Hölldobler et al., 1981). In early spring the adult beetles deposit eggs near the kitchen middens of the formicine ant *Lasius fuliginosus*. The larvae develop in the refuse pile and pupate sometime during the period from May to July. In July or August the adult beetles eclose. The young adult beetles then apparently emigrate, as indicated by a short phase of high diurnal locomotory and flight activity. After this period the beetles hunt the ants near the *L. fuliginosus* nest during the night, while remaining hidden in convenient shelters during the day. The *Pella* adults overwinter in dormancy inside the *Lasius* nest. At the end of the winter they enter a second diurnal activity phase during which mating takes place. After reproduction the beetles die, normally a few weeks before the new adult beetle generation ecloses in June.

The larvae of *Pella funesta* are as closely associated with the ants as the adults. They are found almost exclusively near or in the kitchen middens of the *Lasius fuliginosus* nest, where they subsist primarily on dead ants. While feeding, the larvae always attempt to stay out of sight, either by remaining entirely beneath the booty and devouring it from below, or by crawling inside the corpses. Occasionally two to four larvae feast on the same corpse. When the beetles become overly crowded, however, they attack and cannibalize each other (Figure 13–9).

When ants encounter the *Pella* larvae they usually attack them. Almost invariably the larvae raise their abdominal tips toward the heads of the ants in a defensive maneuver. *Pella* larvae possess a complex tergal gland near the abdominal tip, which may secrete appeasement or diversion substances. At least the ants respond to the movement by stopping their attack and palpating the abdominal tip of the larvae. In most cases this brief interruption is sufficient to allow the larvae to escape. In general, *Pella* larvae are able to come into close contact with workers of *Lasius fuliginosus* without being attacked, and this is especially the case when the temperature is relatively low (14–17°C). Under these circumstances larvae have been observed licking the cuticle of live ants, including even the mandibles and mouthparts. Tracer experiments with radioactive-labeled food did not, however, indicate that the larvae are able to solicit regurgitation from the ants. Their main food source seems to be dead ants and nest debris. They can easily be raised in the laboratory entirely apart from living ants, provided they are given a regular diet of freshly killed ants.

The genus *Pella* was originally combined with *Zyras* and *Myrmedonia,* but raised to full generic rank by Kistner (1972). Wasmann (1886, 1920, 1925, 1930) reported that all species he had studied, including *cognata, funesta, humeralis, laticollis, lugens,* and *similis,* live with *Lasius fuliginosus.* Of these, only *P. humeralis* uses alternate hosts, in this case species of the *Formica rufa* group. According to Wasmann all *Pella* species prey on ants, concentrating especially on disabled adults. In addition Wasmann observed that the beetles are active primarily during the night. In a more recent study Kolbe (1971) failed to find predatory behavior in *P. humeralis* and concluded that this species primarily feeds on dead ants. Similar observations were made with *P. japonicus,* which lives with *Lasius spathepus* (Yasumatsu, 1937; Kistner, 1972). Kistner also observed that these beetles "ate small insects that are being transported by the ants." He could not see the *Pella* eating live ants, however, or fighting any of the ants on the trail.

Hölldobler et al. (1981) confirmed that adult *Pella funesta, P. laticollis,* and *P. humeralis* live as scavengers, feeding on dead or disabled ants and debris discarded by the ants. They also observed, however, that these beetles, in concert with Wasmann's claim, act as very effective predators on the ants. In the field and laboratory *Pella* beetles

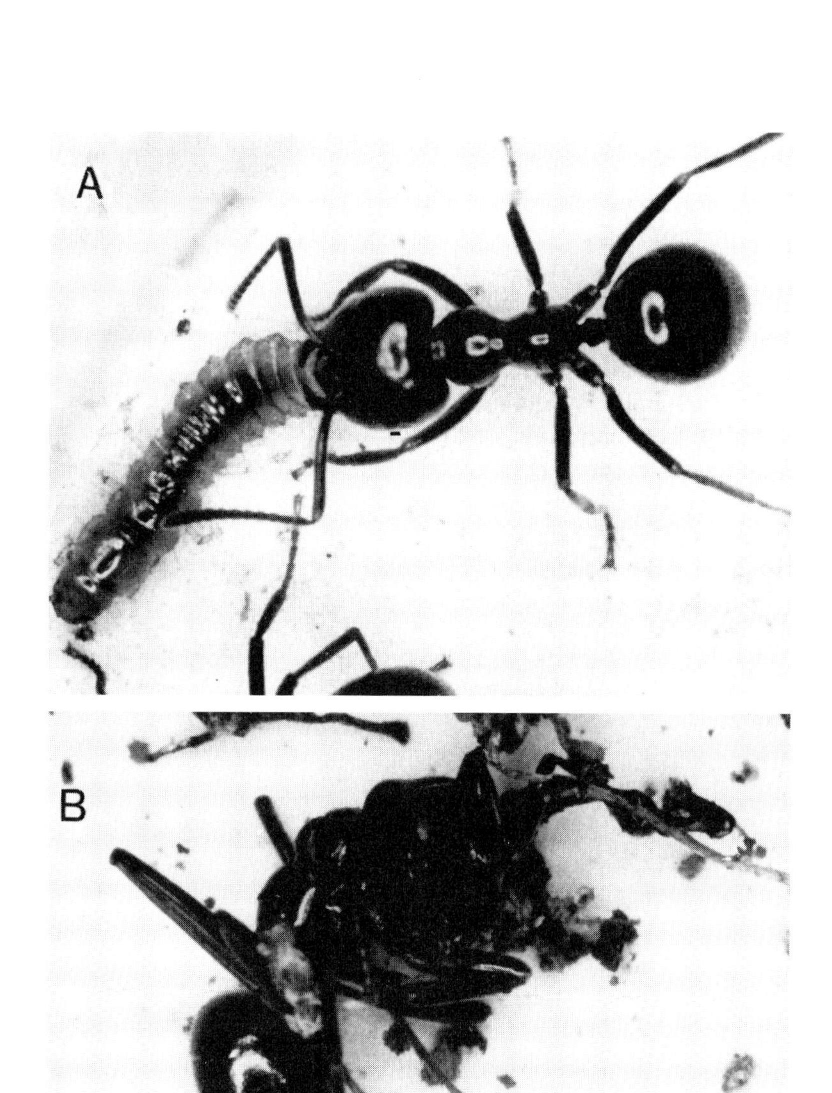

**FIGURE 13–9** Behavior of the larvae of the staphylinid beetle *Pella funesta.* (*A*) The larva presents its abdominal tip to an attacking worker of *Lasius fuliginosus* in an appeasement gesture, causing the ant to interrupt her attack and lick the larva. (*B*) The larva feeds on a dead ant. (*C*) Cannibalistic behavior of *Pella* larvae. (From Hölldobler et al., 1981.)

were seen hunting *L. fuliginosus*. The beetles chased individual ants and pursued them for distances of up to 6 centimeters. When a beetle made a rear approach, it attempted to mount the ant and insert its head between the victim's head and thorax (Figure 13–7). When attacked the ant usually reacted by stopping abruptly and pressing her femurs rapidly and tightly against her body. Often this reaction threw the beetle off the back of the ant, allowing the ant to escape. Of 178 beetle onslaughts on *L. fuliginosus* workers observed during a period of three hours, only 9 were successful. The beetles partially decapitated their victims, at least to the extent of severing their esophagus and nerve cord. Occasionally two or three beetles were seen chasing an ant. Once the ant was caught by a beetle the other beetles joined in subduing and killing it. Although individual *Pella* often tried to drag the prey away from the others, all the members of the pack usually ended up feeding on it simultaneously. No aggression among the beetles was observed on such occasions. When the beetles were starved for several days, they occasionally chased each other, even jumping on one another's backs, but they never engaged in cannibalism.

When foraging on the ants' garbage dumps or running along the ants' trails, *Pella* adults frequently encounter ants and are attacked by them. How do the beetles manage to escape their aggressive hosts? Usually they run around with their abdomens curved slightly upward. When encountering an ant, the beetles flex the abdomen even more strongly. This is a typical and frequently described behavior of many staphylinid myrmecophiles and is commonly considered a defensive response. It has been suggested that during abdominal flexing the beetles discharge secretions from their tergal gland (Jordan, 1913; Kistner and Blum, 1971). The gland is located between the sixth and seventh abdominal tergites and is unique to the subfamily Aleocharinae, to which most myrmecophilous and termitophilous staphylinids belong (Jordan, 1913; Pasteels, 1968). The chemical composition of the tergal gland secretions of several species has been investigated and found to be extraordinarily diverse (Blum et al., 1971; Brand et al., 1973a; Kolbe and Proske, 1973; Hölldobler et al., 1981). Kistner and Blum (1971) suggested that *P. japonicus* and possibly also *P. comes*, both of which live with *L. spathepus*, produce citronellal in their tergal glands. This substance is also a major compound of the mandibular gland secretions produced by the host ants, for which it functions as an alarm pheromone.

*Pella japonicus* tergal gland contents have not yet been chemically identified. Because irritated beetles seemed to smell like the ants' mandibular gland secretions, Kistner and Blum speculated that *P. japonicus* produce citronellal in the tergal glands and thereby mimic the alarm pheromone of their host ants. They suggested that in this way the beetles can "cause the ants to reverse their direction, a reaction which allows the myrmecophiles to escape." However, the investigations by Hölldobler et al. (1981) of the defensive strategy employed by the European *Pella* toward their host ants *Lasius fuliginosus* lead to a different interpretation. When irritated mechanically,

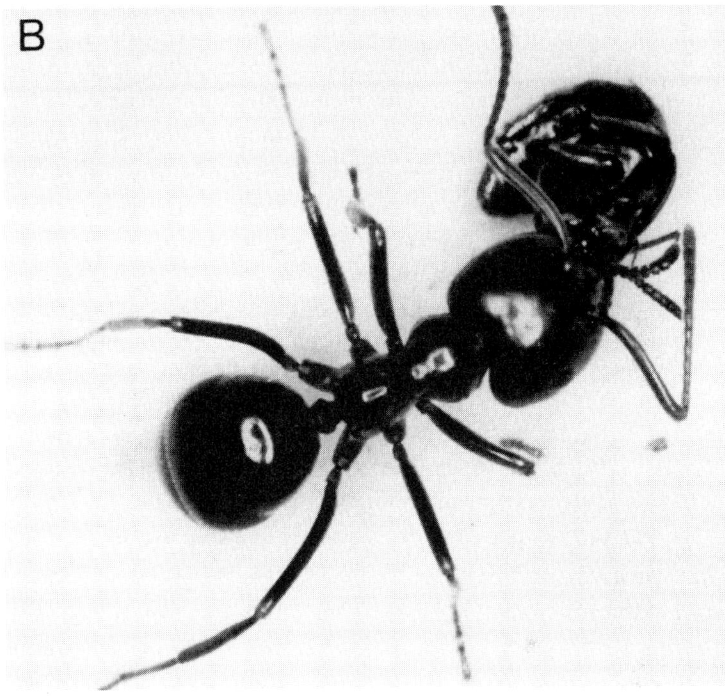


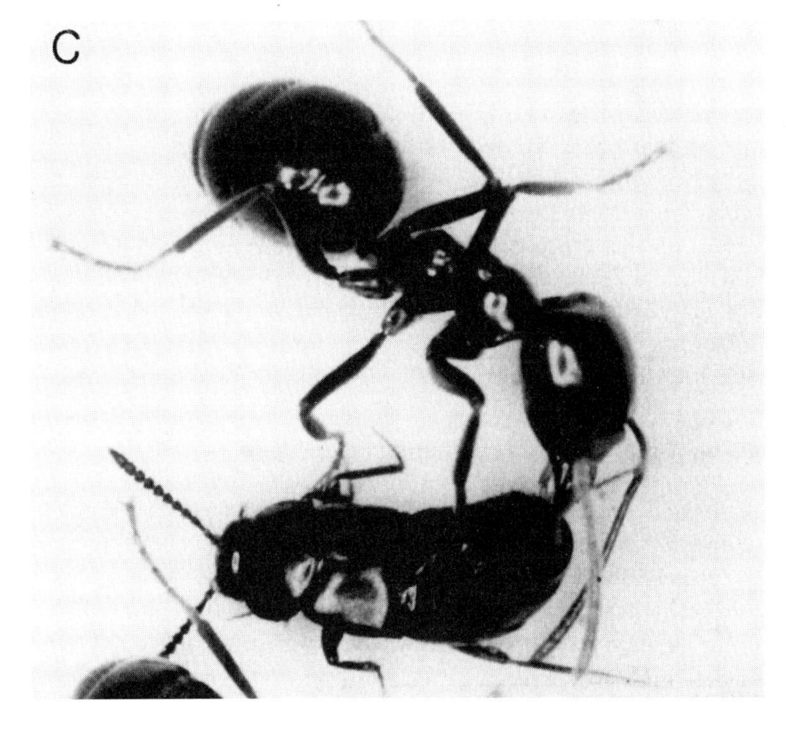


**FIGURE 13–10** Defensive behavior of the myrmecophilous staphylinid *Pella* when attacked by host ants *Lasius fuliginosus*. (*A*) *Pella* in death-feigning position. (*B*) A death-feigning beetle is carried around by an *L. fuliginosus* worker. (*C*) *Pella* presents abdominal tip to the attacking worker, and the ant licks the abdominal tip. (From Hölldobler et al., 1981.)

*P. laticollis* discharged a pungent brownish secretion from its tergal gland. A chemical analysis of the secretions did not reveal any resemblances to the mandibular gland secretions of the host ants. Furthermore, the beetles never employed the tergal gland secretions when they were attacking ants. They did so only when severely attacked themselves, and in particular when the ants seized their appendages. Ants contaminated with tergal gland secretions in these episodes usually displayed an aversive reaction, releasing the grip on the beetles and grooming and wiping their mouthparts and antennae on the substrate. The beetles then had to escape quickly, however, because other ants nearby were alerted and rapidly converged on the scene, apparently in response to the alarm pheromones of their nestmates.

Although Hölldobler and his collaborators could not find any resemblance of the bulk of *Pella* tergal gland secretions to the mandibular gland secretions of *Lasius fuliginosus,* it is noteworthy that the *Pella* secretions did contain undecane, a hydrocarbon commonly found in the Dufour's glands of formicine ants and considered to be an alarm pheromone in *L. fuliginosus* (Dumpert, 1972). Even with this element, however, isolated tergal gland contents of *P. laticollis* elicited a repellent reaction rather than an alarm response in *L. fuliginosus.* Apparently the repellent effect of the quinones, the major components of the tergal gland secretions, was stronger than the attractant effect of undecane. When the ants' antennae were directly contaminated with the beetles' tergal gland secretions, they hung almost motionless, and the ants were disoriented for several minutes. Hence it is unlikely that the beetle uses its powerful repellent defensive system each time it makes a routine encounter with its host ants. Hölldobler and his co-workers concluded that in such situations the beetles rely on an appeasing or diverting defensive strategy, and employ the repellent defense only as a last resort.

In early spring, when most of the *Pella* adults are close to the entrance of the ants' nest, they feign death when attacked by ants. The ants then either ignore the motionless beetles or carry them around and finally discard them on the refuse piles. Later in the season, when the activity of ants and beetles is much higher, the beetles employ a different appeasement technique. Each time they encounter ants they flex their abdomen forward and point the abdominal tip toward the head of their adversaries. Usually the ants respond by antennating the tip and licking it briefly (Figure 13–10). This ordinarily damps the ants' aggressions, allowing the beetle to escape. When on occasion the ants become persistent, the beetle extrudes a white, viscous droplet from the abdominal tip, which the ants eagerly lick up. Histological investigations revealed the abdominal tips of *Pella* to contain exocrine glandular structures, which Hölldobler has called the appeasement gland complex. It is not yet possible to assign one specific gland to the appeasement function.

### Kitchen Middens and Peripheral Nest Chambers

We have described the genus *Pella* in some detail as representative of an early evolutionary grade in the myrmecophilous evolution of the aleocharine staphylinids. Species of *Pella* are specialized predators and scavengers, but they are less advanced in their myrmecophilic relationships than aleocharine species specialized on the kitchen middens, peripheral nest chambers, and brood chambers of the host ants' nests. A representative of the latter, more advanced

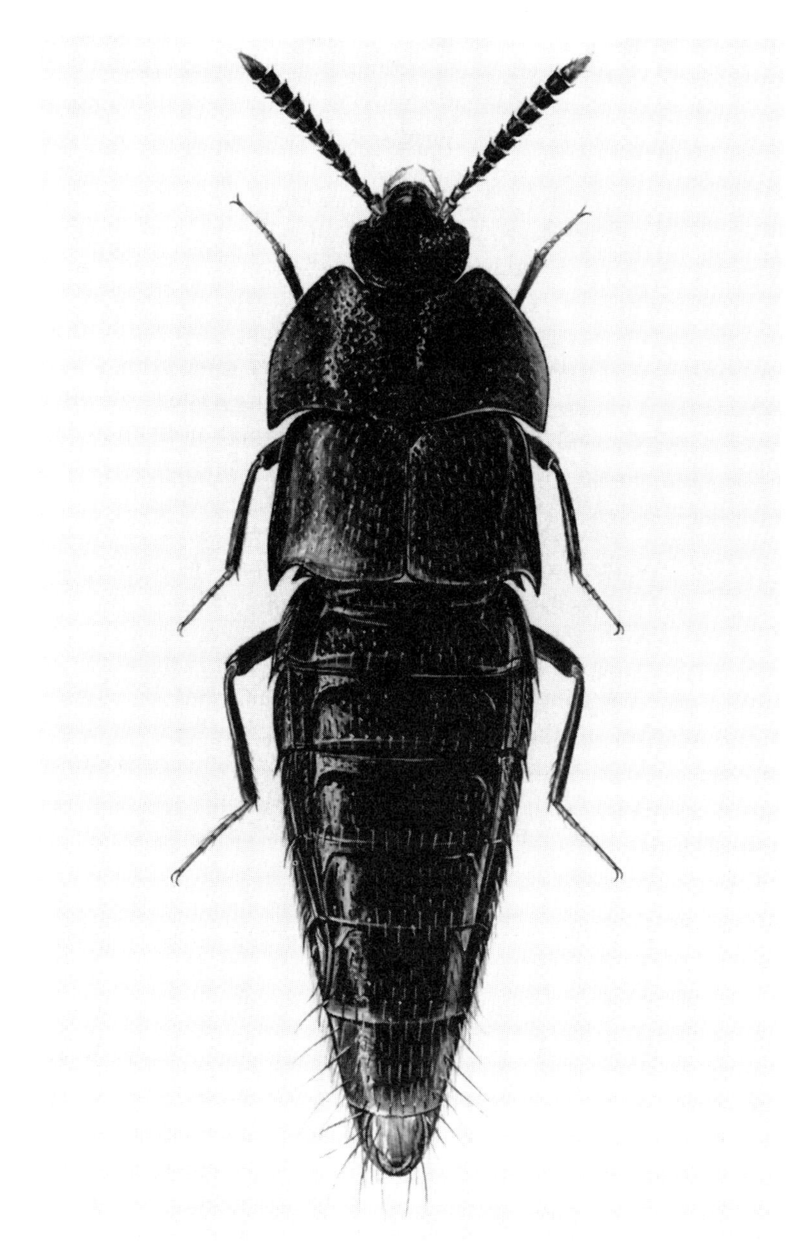

**FIGURE 13–11** *Dinarda dentata* is a regular guest in nests of the ant *Formica sanguinea.* (From Hölldobler, 1973c; painting by T. Hölldobler-Forsyth.)

evolutionary grade is *Dinarda.* The myrmecophilous habits of this genus have been known since Wasmann's first account published in 1894, and the genus has since been recorded with many species of *Formica* (Bernard, 1968). Kistner (1982) points out that the species-level taxonomy is still in poor condition, so that the best-studied form, *D. dentata,* may in fact comprise several species. The following account is based mostly on the form (or species) of *D. dentata* that lives in nests of the red slave-making ant *F. sanguinea* of Europe (Figure 13–11).

The larvae of *Dinarda* are concentrated in the kitchen middens of the *Formica* hosts, where they feed on dead ants and debris. When they encounter worker ants they exhibit the same appeasingly defensive behavior described for *Pella* larvae (Figure 13–12). Not surprisingly, they also possess a similar complex glandular structure in the tip of their abdomen. Some adult *Dinarda* beetles are found in

FIGURE 13–12 A staphylinid beetle (*Dinarda dentata*) appeases a host ant (*Formica sanguinea*) by offering appeasement substances from the tip of its abdomen.

the kitchen middens, where they exist as scavengers. Most, however, patrol through the peripheral nest chambers, where they feed on arthropod prey brought in by the host workers. Wasmann (1894a) reported that *Dinarda* also eat mites and occasionally even ant eggs and larvae.

Finally, studies of laboratory colonies have revealed that *Dinarda* taps into the liquid food flow of its hosts (Hölldobler, 1971a, 1973c). This parasitism occurs mostly in the peripheral chambers of the nest, where the bulk of regurgitation occurs between newly returning foragers and nest workers. The *Dinarda* beetles sometimes insinuate themselves between two workers exchanging food and literally snatch the droplet of food from the donor's mouth. They also use a simple begging behavior to obtain food directly from the returning foragers. The beetle approaches an ant and furtively touches her labium, causing the ant to regurgitate a small droplet of food (Figure 13–13). The ant, however, soon recognizes the beetle as an alien and commences to attack it. At the first sign of hostility the beetle raises its abdomen and offers the ant the appeasement secretions at the

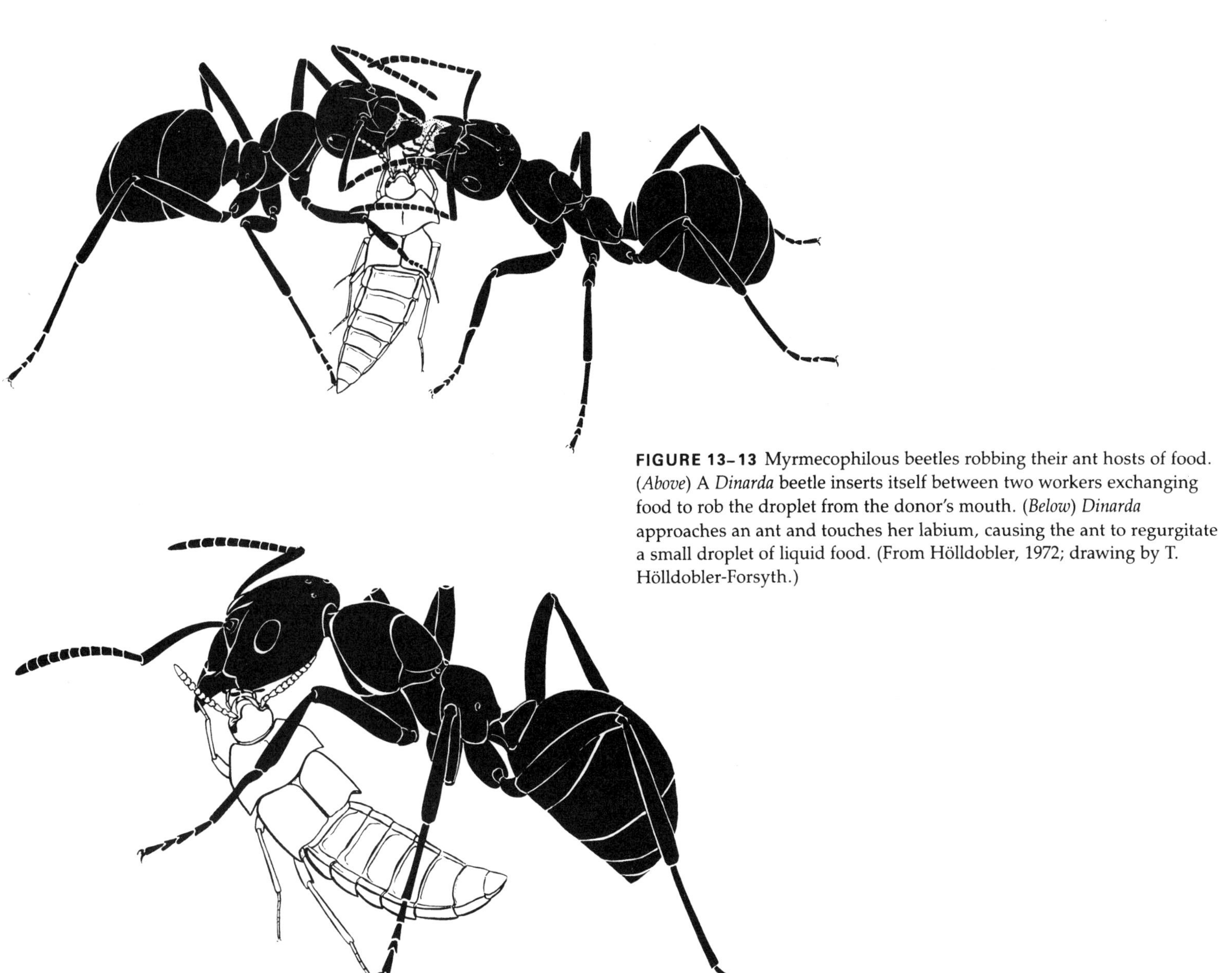

FIGURE 13–13 Myrmecophilous beetles robbing their ant hosts of food. (*Above*) A *Dinarda* beetle inserts itself between two workers exchanging food to rob the droplet from the donor's mouth. (*Below*) *Dinarda* approaches an ant and touches her labium, causing the ant to regurgitate a small droplet of liquid food. (From Hölldobler, 1972; drawing by T. Hölldobler-Forsyth.)

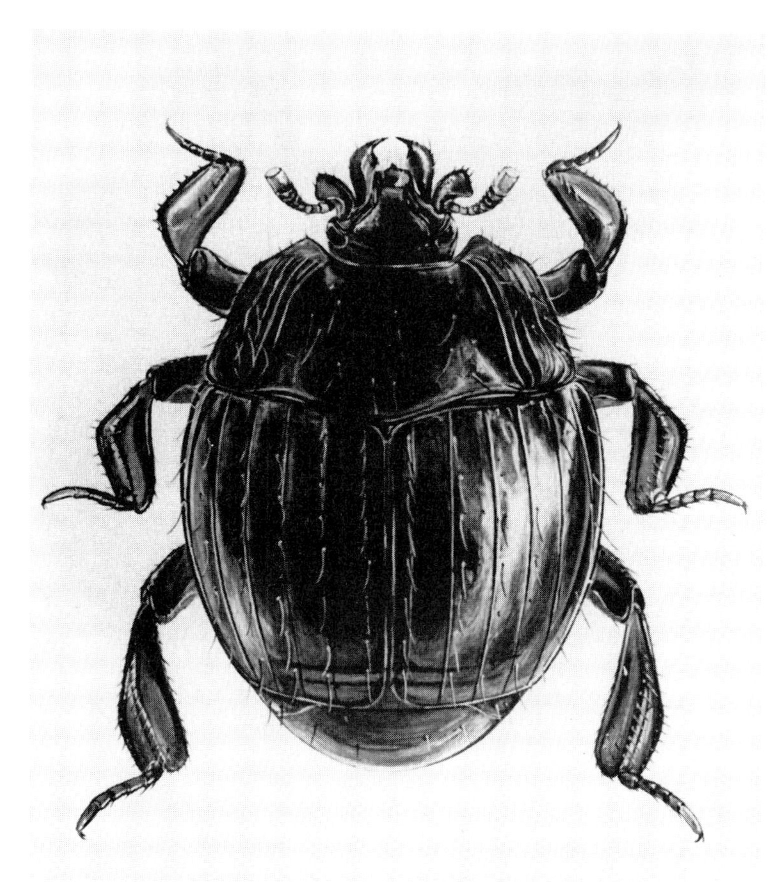

**FIGURE 13–14** *Hetaerius ferrugineus,* a histerid myrmecophile most frequently found in nests of the ant *Formica fusca.* (From Hölldobler, 1973c; painting by T. Hölldobler-Forsyth.)

abdominal tip. The abdominal tip is quickly licked by the ant, and almost instantly the attack ceases. During this brief interlude the beetle makes its escape. *Dinarda* also possesses a well-developed tergal gland, but its repellent secretions are applied only as a last resort against ant attacks.

It is noteworthy that a myrmecophile belonging to a radically different taxonomic group obtains food in a similar manner. Janet (1896) discovered that the thysanuran *Atelura formicaria,* which is a very common symbiont of many European ant genera (Bernard, 1968), occasionally snatches food during food exchanges between worker ants. Like the *Dinarda* beetles, *Atelura* favors the peripheral chambers of the host nests.

Another example of a myrmecophile adapted especially to kitchen middens and peripheral nest chambers is the histerid beetle genus *Hetaerius* (Figure 13–14). As Wasmann (1886, 1905) first reported, the European species *H. ferrugineus* is most frequently found in nests of *Formica fusca,* where it feeds on dead and wounded ants. Occasionally it also consumes ant larvae and pupae. Most of the time the ants pay no attention to the *Hetaerius.* When they do attack an adult, the beetle exhibits death-feigning behavior by holding perfectly still with its legs closely appressed to its body. The ants frequently react to this nonviolent resistance by carrying the beetle around, licking it, and finally releasing it. It has been suggested that *Hetaerius* has special trichome glands opening on the margins of the thorax (Escherich and Emery, 1897; Wasmann, 1903; Wheeler, 1908c). No histological investigations have been performed to confirm this assumption, however.

Wheeler (1908c) observed that adults of the North American *Hetaerius brunneipennis* solicit regurgitated food from the host ants. The beetle sometimes waves its forelegs toward passing ants, and by this action appears to attract their attention. A very similar behavior has been more recently recorded in *H. ferrugineus.* The soliciting beetle takes an upright position, stretching its forelegs widely apart and waving them slightly toward the approaching ant. The ant antennates and licks the beetle with her mandibles usually held nearly closed (Figure 13–15). Tracer experiments have revealed that the beetle also obtains small amounts of regurgitated food (B. Hölldobler, unpublished results). Thus, contrary to a suggestion originally made by Wheeler, *H. brunneipennis* does not represent a more advanced evolutionary state over *H. ferrugineus* (where food solicitation was previously unknown).

## Brood Chambers

The brood chambers constitute the optimal niche within an ant nest for a social food-flow parasite, because there the food of the highest quality is concentrated to be fed to the developing larvae, callow workers, and queen(s). Moreover, the immature ant stages housed in the brood chambers provide the most valuable prey for specialized ant predators. The brood chambers are nevertheless difficult for parasites to penetrate, because they are fiercely defended by the ants. Very special adaptations are thus needed for myrmecophiles to exploit these sites as an ecological niche. This has been accomplished by some of the evolutionarily most advanced myrmecophilous staphylinids, of which the aleocharine beetles *Lomechusa strumosa* and several species of the genus *Atemeles* are the premier examples.

*Lomechusa strumosa* (Figure 13–16) lives with *Formica sanguinea* in Europe. *Atemeles pubicollis* (Figure 13–16), also a European species, is normally found with the mound-building wood ant *F. polyctena* during the summer, but in the winter it inhabits the nests of ants of the myrmicine genus *Myrmica.* We know from the observations of Wasmann of more than seventy years ago that these beetles are both fed and reared by their host ants. The behavioral patterns of the larvae of the beetles are similar for *Lomechusa* and the various *Atemeles* species; in particular the larvae prey to a certain extent on their host ants' larvae. It is therefore astonishing that nurse ants not only tolerate these predators but also feed and protect them as readily as they do their own brood (Figure 13–17).

B. Hölldobler (1967, 1968b) demonstrated that the interspecific communication is both chemical and mechanical in nature. The beetle larvae show a characteristic begging behavior toward their host ants. As soon as they are touched by an ant, they rear upward and try to make contact with the ant's head. If they succeed, they tap the ant's labium with their own mouthparts. This apparently releases regurgitation of food by the ants. The ant larvae beg for food in much the same way but usually less intensively than the beetle parasites. With the aid of experiments using food mixed with radioactive sodium phosphate it has been possible to measure the social exchange of food in a colony. The results show that myrmecophilous beetle larvae present in the brood chamber obtain a proportionately greater share of the food than do the host ant larvae. The pres-

**FIGURE 13–15** Food solicitation behavior of *Hetaerius ferrugineus*. The beetle assumes an upright position, stretches its forelegs apart and waves them in the direction of an approaching ant (*above*). The ant antennates and licks the beetle with her mandibles held almost closed (*below*).

ence of ant larvae does not affect the food flow to the beetle larvae, whereas ant larvae always receive less food when they compete with beetle larvae. This disparity suggests that the releasing signals presented by the beetle larvae to the nurse ants may be more effective than those presented by the ant larvae themselves.

The *Lomechusa* and *Atemeles* larvae are also frequently and intensively groomed by the brood-keeping ants. Thus it appears probable that chemical signals are involved in the interspecific relationship as well. The transfer of substances from the larvae to the ants was detected with the aid of radioactive tracers. These substances are probably secreted by glandular cells beneath the integument of the dorsolateral area of each segment. The biological significance of the secretions was elucidated by the following experiments (Hölldobler, 1967). Beetle larvae were first completely covered with shellac to prevent the release of the secretion. They were then placed outside the nest entrance, together with freshly killed but otherwise untreated control larvae. The ants quickly carried the control animals into the brood chamber. The shellac-covered larvae, on the other hand, were either ignored or carried to the garbage dump. It was found that for adoption to be successful, at least one segment of the

**FIGURE 13–16** Two species of staphylinid myrmecophiles that have become fully integrated in their host ants' society. (*Left*) *Lomechusa strumosa,* which lives with *Formica sanguinea*. (*Right*) *Atemeles pubicollis,* which is normally found with *F. polyctena* during the summer, but inhabits nests of the genus *Myrmica* during the winter. (From Hölldobler, 1973c; painting by T. Hölldobler-Forsyth.)

larva had to be free of shellac. Furthermore, after all the secretions had been extracted with solvents, the larvae were no longer attractive. When the extracted larvae were then contaminated with secretions from normal larvae, they regained their attractiveness. Even filter paper dummies soaked in the extract were carried into the brood chambers.

These experiments show that the adoption of the *Lomechusa* and *Atemeles* larvae and their care within the ant colony depend on chemical signals. It may be that the beetle larvae imitate a pheromone which the ant larvae themselves use to release brood-tending behavior in the adult ants. This inference is supported by the fact that the beetle larvae and host ant larvae can be transplanted from the host species to closely related *Formica* species. On the other hand, the ant species that do not accept the larvae of the ant hosts also reject the beetle larvae.

The question next arises as to how the ant colony manages to survive the intensive predation and food parasitism by the beetle larvae. The answer appears to be very simple. The beetle larvae are cannibalistic, and this factor alone appears to be effective in limiting the number of beetle larvae in the brood chambers at any time.

The larvae of both genera pupate in the summer, and they eclose as adult beetles at the beginning of autumn. The young *Lomechusa* adults leave the ant nest and after a short period of migration seek adoption in another nest of the same host ant species. *Atemeles* adults, on the other hand, emigrate from the *Formica* nest, where they have been raised, to nests of the ant genus *Myrmica*. They winter inside the *Myrmica* brood chambers and in the springtime return to a *Formica* nest to breed (Wasmann, 1910a; Hölldobler, 1970c). The fact that the adult beetle is tolerated and is fed in the nest of ants belonging to two different subfamilies suggests that it is able to communicate efficiently in two different "languages."

The *Atemeles* face a major problem in finding their way from one host species to another. *Formica polyctena* nests normally occur in woodland, whereas those of *Myrmica* are found in grassland around the woods. Experiments have revealed that when *Atemeles* leave the *Formica* nest they show high locomotor and flight activity and orient toward light. This may well explain how they manage to reach the relatively open *Myrmica* habitat. Once they reach the grassland, the beetles must distinguish the *Myrmica* ants from the other species present and locate their nests. Choice experiments in the laboratory have revealed that they identify the *Myrmica* nests innately by specific odors. Windborne species-specific odors are equally important in the springtime movement back to *Formica* nests.

Having found the hosts, the beetles must then secure their own

**FIGURE 13–17** (*Above*) A *Formica* worker regurgitates to an *Atemeles* larva. Glands believed to be the source of a false brood identification odor are located in pairs on the dorsal surface of each of the body segments of the *Atemeles* larva. (*Below*) A larva of *Atemeles pubicollis* feeds on a larva of one of its host ants (*Formica*). (From Hölldobler, 1971a.)

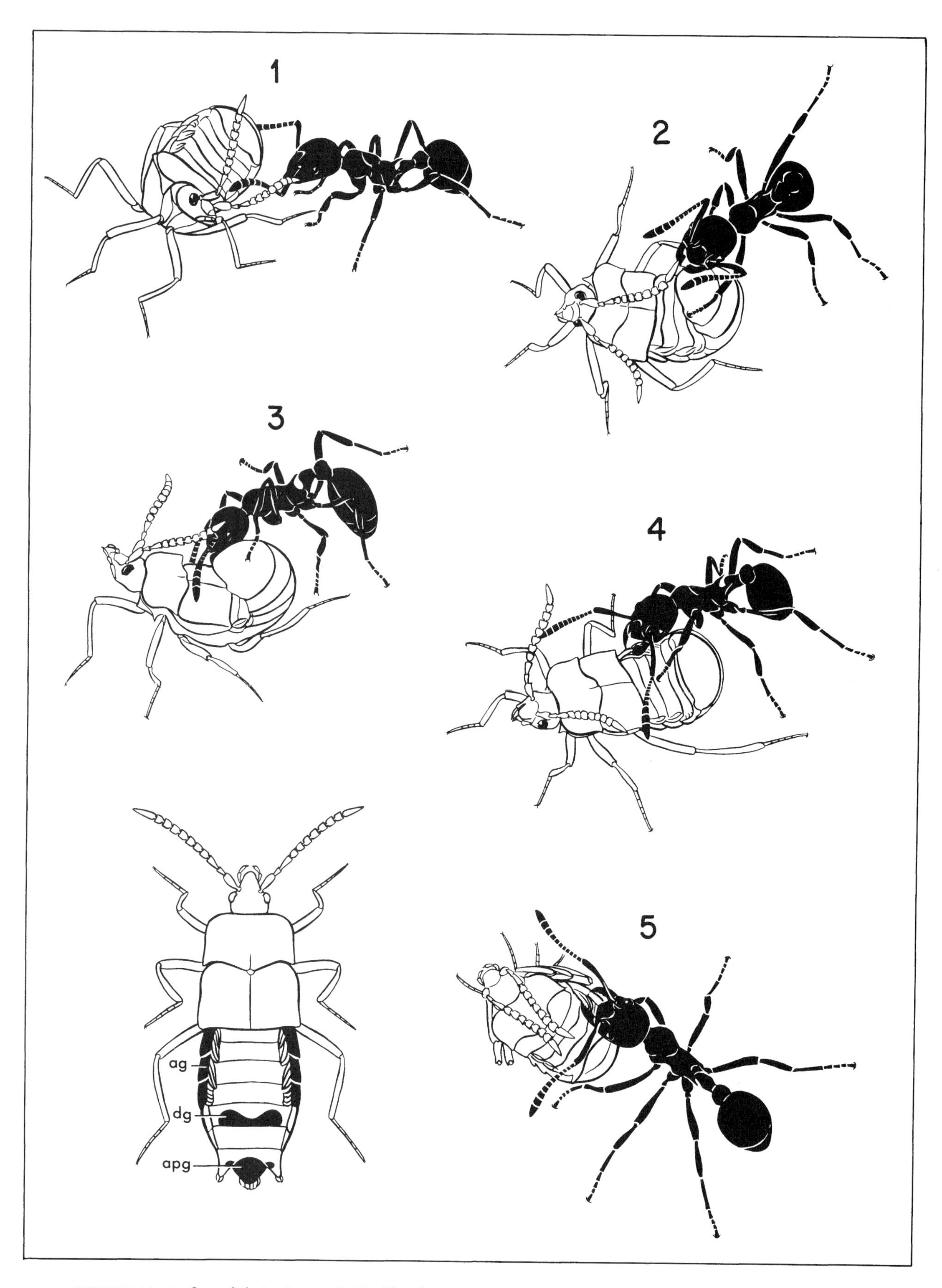


**FIGURE 13–18** Symphily in the staphylinid beetle *Atemeles pubicollis*. The figure at the lower left indicates the location of the three principal abdominal glands of the parasite: (*ag*) adoption glands; (*dg*) defensive glands; (*apg*) appeasement gland. The beetle presents its appeasement gland to a worker of *Myrmica* that has just approached it (*1*). After licking the gland opening (*2*), the worker moves around to lick the adoption glands (*3, 4*), after which she carries the beetle into the nest (*5*). (From Hölldobler, 1970b; drawing by T. Hölldobler-Forsyth.)

**FIGURE 13–19** Food solicitation behavior of the myrmecophilous beetle *Atemeles pubicollis*. The beetle taps the ant's mouthparts (*Formica polyctena, above; Myrmica rubra, below*) with its forelegs, triggering the regurgitation of a droplet of food from the ant. (From Hölldobler, 1971a.)

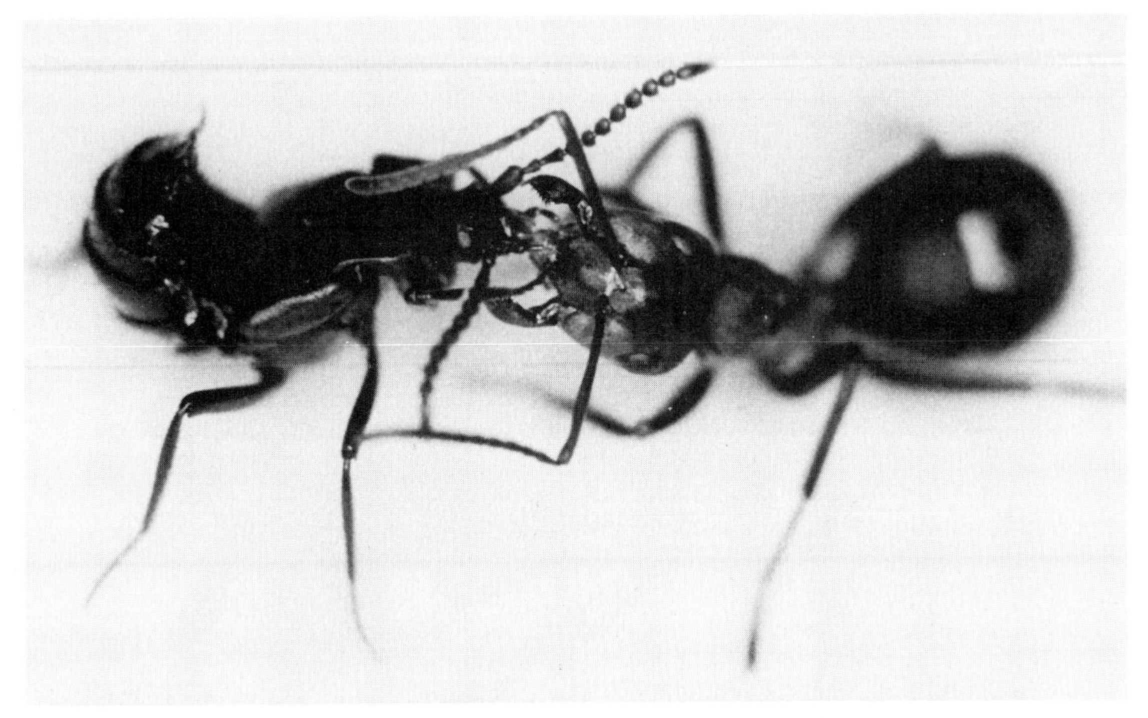

adoption. The process comprises the four sequential steps depicted in Figure 13–18. First, the beetle taps the ant lightly with its antennae and raises the tip of the abdomen toward it. The latter structure contains the appeasement glands, the secretions of which are immediately licked up by the ant and appear to suppress aggressive behavior. The ant is then attracted by a second series of glands along the lateral margins of the abdomen. The beetle lowers its abdomen in order to permit the ant to approach this part of its body. The glandular openings are surrounded by bristles, which are grasped by the ant and used to carry the beetle into the brood chambers. By experimentally occluding the openings of the glands, Hölldobler found that their secretion is essential for successful adoption, and for this reason he called them adoption glands. Thus the adoption of the adult beetle, like that of the larva, depends on chemical communication, and it most probably entails an imitation of specific brood pheromones (Hölldobler, 1970c).

Before leaving the *Formica* nest the *Atemeles* beetle must obtain enough food to enable it to survive the trek to the *Myrmica* nest. This it obtains by vigorous solicitation from its hosts. The begging behavior is essentially the same toward both *Formica* and *Myrmica* (Figure 13–19; see also Figure 7–41). The beetle attracts the ant's attention by rapidly drumming on her with its antennae. Using its maxillae and forelegs it taps the mouthparts of the ant, thus inducing regurgitation. As we noted previously, the ants themselves employ a similar mechanical stimulation of the mouthparts to obtain food from one another. It is thus clear that *Atemeles* is able to obtain food by imitating these very basic tactile food-begging signals.

What is the significance of the seasonal changes in hosts in *Ate-*

**FIGURE 13–20** Mating in the myrmecophilous staphylinid beetle *Lomechusa strumosa*.

**FIGURE 13–21** (*facing page*) A *Lomechusa strumosa* adult is fed by a worker of its host ant *Formica sanguinea*. The beetle simultaneously appeases another worker by offering a special calming secretion from its abdominal tip. (From Hölldobler, 1973c.)

*meles?* There are good reasons for believing that *Atemeles* first evolved myrmecophilic relationships with *Formica* rather than with *Myrmica*. It seems likely that the ancestral *Atemeles* beetles hatched in *Formica* nests in the autumn and then dispersed, returning to other *Formica* nests only to overwinter. This simpler life cycle is followed by *Lomechusa* today (Wasmann, 1915b; Hölldobler, 1972). In the *Formica* nest, however, the immature stages disappear during the winter, and consequently social food flow is reduced. In contrast, the *Myrmica* colony maintains brood throughout the winter. Thus in *Myrmica* nests larvae and nutrients from the social food flow are both available as high-grade food sources to the myrmecophiles. These circumstances, coupled with the fact that the beetles are sexually immature when they hatch, suggest why it is advantageous for them to overwinter in *Myrmica* nests. While there the *Atemeles* complete gametogenesis, so that when spring comes the beetles are sexually mature. They then return to the *Formica* nest to mate and lay their eggs. At this time the *Formica* are just beginning to raise their own larvae and the social food flow is again maximized. The life cycle and behavior of *Atemeles* is thus synchronized with that of its host ants in such a manner as to take greatest advantage of the social life of each of the two species in turn.

The North American staphylinid myrmecophiles of *Xenodusa* seem to have a life history similar to that of *Atemeles*. The larvae are found in *Formica* nests and the adults overwinter in the nests of the carpenter ants of the genus *Camponotus* (Wheeler, 1911). It is undoubtedly significant that *Camponotus*, like *Myrmica*, maintains larvae throughout the winter. It may well be that the host-changing behavior of *Xenodusa* has the same significance as that inferred for *Atemeles*.

Wasmann considered *Lomechusa strumosa* to be the most evolutionarily advanced of the myrmecophilous staphylinids. We agree with this assessment, which is supported by a more recent analysis of the myrmecophilous behavior of *Lomechusa* (Hölldobler, 1971a and unpublished results). *Lomechusa* also appears to have at least one migratory phase in spring, when, shortly after overwintering as an adult inside the *Formica sanguinea* nest, it travels to another nest of the same host species, where mating takes place (Figure 13–20). The adoption procedure into the new host ants' nest is even more complicated than that of *Atemeles*.

Like other aleocharine myrmecophiles, *Lomechusa* is equipped with well-developed tergal glands that produce defensive substances. The secretion consists of benzoquinone, methyl-benzoquinone, ethyl-benzoquinone, and *n*-tridecane, which last compound accounts for more than 80 percent of the volatiles (Blum et al., 1971). This powerful repellent mixture is normally employed only when the beetle is attacked by non-host ants, however. When approached by host ants *Lomechusa* behaves much more gently. It first presents the trichome structures on its legs, particularly the femora. The beetle then slowly circles around on the spot, with its legs widely extended so that the femoral trichomes are easily accessible. Simultaneously the beetle antennates the ants, bending its body sideways or backward in order to reach the ants with its antennae. Next it slowly points its abdominal tip at the ants. *Lomechusa* possesses a battery of exocrine glands at the abdominal tip, and dur-

ing the early adoption phase it frequently extrudes a whitish, viscous droplet which is eagerly licked up by the ants. The secretion is proteinaceous and contains no appreciable amounts of carbohydrates. The ants that feed on the material seem to grow calmer in the process. The *Lomechusa* now permits the ants access to trichomes along the lateral margins of the abdomen that project above the adoption glands. The ants lick and grasp the trichome bristles and finally carry or drag the beetle into the brood chambers of their nests.

Inside the nest the *Lomechusa* continues to be tended by its host ants, despite the fact that it preys on the ants' brood (Figure 13–21). In addition the adult *Lomechusa,* like the larva, receives food directly from the nurse ants. Its soliciting behavior is markedly different from that of adult *Atemeles. Lomechusa* possesses trichomes at the labrum, which are frequently licked by the ants. During this procedure the beetle uses its maxillae to stimulate the mouthparts of the ants, an action that seems to elicit the regurgitation of liquid food. In contrast to *Atemeles,* however, *Lomechusa* does not use its forelegs when begging for food. One gets the impression, as Wasmann first suggested, that the ants' feeding behavior toward the adult *Lomechusa* resembles more the nursing behavior toward ant larvae than the trophallactic food exchange between adult ants. It is possible, but not yet proved, that *Lomechusa* is fed primarily with high-grade secretions from the postpharyngeal and labial glands of nurse workers, in other words with the same food given the larvae.

In short, *Lomechusa strumosa* presents its host ants with a complex variety of chemical signals, including imitations of brood-tending pheromones and other kinds of pseudopheromones. The species is thereby superbly well adapted to live with *Formica sanguinea,* especially in the brood chambers where it finds both food and a protected and regulated environment.

Most of the more casual observations of regurgitation between host and symbionts suggest that the exchange is exploitative, meaning that the liquid flows exclusively from the host to the symbiont. Experiments with radioactive tracers have proved this to be the case for the larvae and adults of *Atemeles* and *Lomechusa* (Hölldobler, 1967, 1968b, 1970c, and unpublished data). An exception has been reported, however, in the case of *Amorphocephalus coronatus.* Adults of this brenthid beetle live with *Camponotus* in southern Europe (Figure 13–22). According to Le Masne and Torossian (1965) they receive food from some of the host workers and regurgitate part of it back to other host workers. This is the first reported example that could be construed as altruistic behavior on the part of the symbiont beetles (Wilson, 1971). Quantitative investigations are needed, however, to see how much of the received food the beetles share with other host ants. It could very well be that *Amorphocephalus* return only very small portions of the food to the colony and that this "pseudoaltruistic" behavior is one other mechanism used by symbionts to become fully integrated into the host colony.

FIGURE 13–22 Myrmecophilous brenthid beetles (*Amorphocephalus coronatus*) with host ant *Camponotus herculeanus*. (Painting by T. Hölldobler-Forsyth.)

## Other Adaptations of Predators in the Ants' Brood Chambers

Entomologists generally agree that a select few of the aleocharine staphylinids, including *Atemeles* and *Lomechusa*, have reached the evolutionary pinnacle of myrmecophilous adaptations. In mimicking the ants' important communication signals they are able to integrate themselves into the ant societies and exploit the hosts' brood chambers. There nevertheless exist a diversity of other arthropods that also prey on ant brood but use a very different set of sophisticated techniques to invade the ants' nests. An especially instructive example is provided by the scarabaeid beetle genus *Cremastocheilus* (Figure 13–23).

*Cremastocheilus* is a North American genus comprising about fifty species (Krikken, 1976), various of which have been studied by Wheeler (1908b), Cazier and Mortenson (1965), Alpert and Ritcher (1975), and Kloft et al. (1979). Most recently Alpert (1981) has provided an insightful review of the ecology, behavior, and evolution of the genus.

Various species of *Cremastocheilus*, the greatest number of which are found in the deserts of the southwestern United States and Mexico, use the formicine ants *Formica*, *Myrmecocystus*, *Lasius*, and *Camponotus*, and the myrmicines *Pogonomyrmex*, *Messor* (= *Veromessor*), and *Aphaenogaster* as hosts. Ant nests in this arid area provide a high concentration of food resources and a refuge from predators and severe climatic change. *Cremastocheilus* larvae, like other scarabaeid larvae, feed primarily on decaying vegetable material. They develop within piles of vegetable matter inside ant nests, despite the apparent lack of any morphological adaptations for defense from their hosts. Alpert's field observations revealed few interactions between beetle larvae and ants. Experiments showed that the larvae are generally ignored by ant workers, whereas unrelated scarabaeid larvae of similar size used as controls were attacked immediately. Only highly excited ants were observed to attack *Cremastocheilus* larvae, which then defended themselves with powerful "mandibular strikes" and escape maneuvers. Alpert concluded that although *Cremastocheilus* larvae are not nutritionally dependent on ants (they can be raised in complete isolation), they gain protection from predators and desiccation simply by virtue of their residency in ant nests.

Cazier and Statham (1962) suggested that the adult beetles are first brought into the ant nests as booty by the foraging ants. The

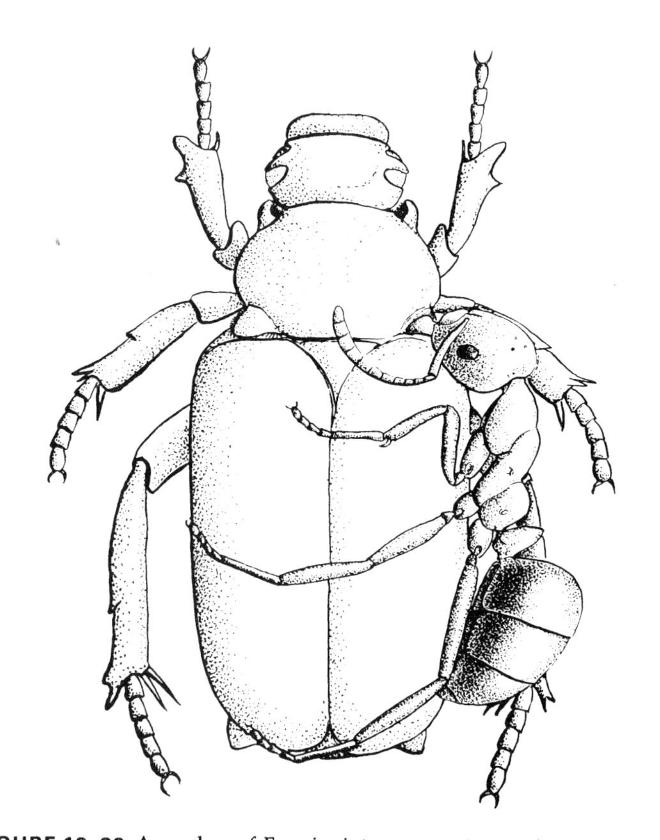

FIGURE 13–23 A worker of *Formica integra* gnawing at the thoracic trichomes of the scarab beetle *Cremastocheilus castaneae*. (From Wheeler, 1910a.)

beetles, however, are so well protected by their heavily sclerotized cuticle that the ants are unable to kill them. Once inside the nest the beetles themselves prey on the ant larvae (Cazier and Mortenson, 1965). This scenario has been basically confirmed by Alpert's (1981) field and laboratory observations. Although *Cremastocheilus* adults are capable of feeding on any species of ant larvae, only one or a few ant species are selected as hosts under natural conditions. This specificity appears to be a product of the host-searching process.

It was originally believed that secretions from glandular trichomes, located on prominent pronotal angles in all species of *Cremastocheilus,* attracted ants to beetles outside the nest and induced adoption (Cazier and Mortenson, 1965); but experiments conducted by Alpert did not confirm this assumption. Alpert released dead *Cremastocheilus* and other scarabaeid beetles with live *C. stathamae* near *Myrmecocystus* colonies. There was no significant preference by the ants for live *C. stathamae,* suggesting that special exocrine gland secretions are not involved in the adoption of the beetles by their ant hosts. Nor did glandular extracts elicit special attraction on the part of the ants. All the evidence supports the hypothesis that certain *Cremastocheilus* species are mistaken for prey items and thereby gain access to the nest. In fact, *Cremastocheilus* beetles commonly feign death when approached by ants, retracting their antennae into protective grooves and sticking out their heavily sclerotized legs.

Whereas some species of *Cremastocheilus* gain entry into the host ants' nest by mimicking prey items, other species are able to march directly through the nest entrance or burrow through thatch piles into the interior. Often the beetles are attacked by ants inside the nest. Occasionally they are forcibly ejected, only to be carried in again by other ants. Attacks usually persist for several minutes, until the ants give up and the *Cremastocheilus* move to concealed corners in chambers or passageways. In the end the beetles move slowly toward the brood chambers, where they are completely ignored by the ants. Alpert reports, "When the adult beetles *(C. stathamae)* approached ant brood, or while beetles were feeding, the host ants clearly licked the surface of the beetle. Although pronotal angles were specifically licked, most of the dorsal surface of the beetles was also involved." Similar observations were made of other *Cremastocheilus* species. Alpert concludes that while adults of *Cremastocheilus* feed on ant brood, trichome secretions may distract workers, reduce aggression, and prevent workers from evacuating brood, and other glands serve to absorb colony odors.

Kloft et al. (1979) proposed that secretions from trichomes at the prosternal apophysis are sought by *Camponotus castaneus* and evoke food-sharing behavior in the ants. They claimed that by effecting the transfer of *Cremastocheilus* trichome substances from worker to worker, the beetles could appease the entire colony. In addition they suggested that the enlarged and projecting propygidial spiracles are glandular and "a possible source of secretions" attractive to the ants. By carefully sectioning *C. castaneus* and members representing most of the other species groups, Alpert (1981) has shown that this conclusion is incorrect. The prosternal apophysis and propygidial spiracles are not glandular in the genus *Cremastocheilus.* Alpert nevertheless did find many new areas that bear exocrine glands. In a series of experiments with radioactive tracers, he further discounted the claim by Kloft et al. (1979) that trophallaxis occurs between host ants and *Cremastocheilus.* Furthermore, he was unable to find evidence of a "peace-making" allomone transferred from trichomes to the colony.

Entomologists continue to be intrigued by the possible evolutionary pathways that led to the specialized predaceous behavior of *Cremastocheilus.* One hypothesis is that *Cremastocheilus* has changed from what was almost certainly a herbivorous diet to become a fully carnivorous scarabaeid (Wilson, 1971; Kistner, 1982). This route may have started with the adaptation of larvae for development in ant nests. In typical scarabaeid fashion *Cremastocheilus* larvae eat vegetable matter in the soil, and because of the concentration of vegetation in ant nests and protection from predators, ant colonies became the new oviposition site. Such behavior is known in cetonine scarabaeids, such as *Euphoria inda,* which develops in *Formica obscuripes* nests in North America (Windsor, 1964), and the European species *Cetonia curprea,* the larvae of which also develop in *Formica* nests (Donisthorpe, 1927). Alpert (1981), on the other hand, argues that the adults of *Euphoria* and *Cetonia* are not preadapted for survival in ant colonies and did not evolve in this direction. *Cremastocheilus* adults, in contrast, have many morphological adaptations for integration into ant nests and an examination of other genera of Cremastocheilini gives the best understanding of how the genus *Cremastocheilus* might have evolved. In Alpert's view, "the tribe Cremastocheilini reflects an evolutionary route from adult predation on soft bodied insects to specialized feeding on ant brood and subsequent development of larvae in ant colonies." The Indian cremastocheiline *Spilophorus maculatus,* in fact, is a predator on membracids and lives independently of ant colonies (Ghorpade, 1975).

The natural history of related beetles provides an added perspective. *Genuchinus ineptus,* for example, develops in plants of the liliaceous genus *Dasylirion* of Mexico and the southwestern United States, and any association with ants is entirely accidental. The adults are general predators, feeding on many different soft-bodied insects in the laboratory. *Dasylirion* plants are inundated with fly larvae, the most probable source of food for *G. ineptus* in the field. *G. ineptus,* in this interpretation, is preadapted for myrmecophilous existence as a predator on ant brood. The predaceous mouthparts of *G. ineptus* adults are almost identical to those of *Cremastocheilus.* There are other morphological features common to *Genuchinus* and *Cremastocheilus* which are apparently adaptations for predation. These include the flattened body shape, hard, pockmarked integument, retractable antennae, and protective mentum. The larvae of *G. ineptus* develop in leaf compost of *Dasylirion* plants, and the pupae are sheltered by protective cases. The *Genuchinus* differ from *Cremastocheilus* by remaining inside their cases until spring.

According to Alpert's reconstruction, the major evolutionary step taken by the genus *Cremastocheilus* was to specialize on ant brood. The development of pronotal angles and trichome hairs is a consequence of this specialization. Other genera of New World Cremastocheilini, such as *Centrocheilus* (Krikken, 1976), appear to occupy an intermediate level leading to this stage. Other species of *Genuchinus* from South America occur commonly in banana leaves and bromeliads. Ant nests are also frequently found at these sites, and beetles may have moved from the leaf litter into the ant colonies to promote their larval development. A study of the life history of *Paracyclidius bennetti,* which is found with ants on bromeliads in Trinidad, might provide insight into this evolutionary stage. The genus *Cremastocheilus* adaptively radiated into ant colonies inhabiting the southwestern deserts of North America. Today members from all five subgenera are found in Arizona, New Mexico, and Colorado. All species groups have at least one member from the Southwest, except for the

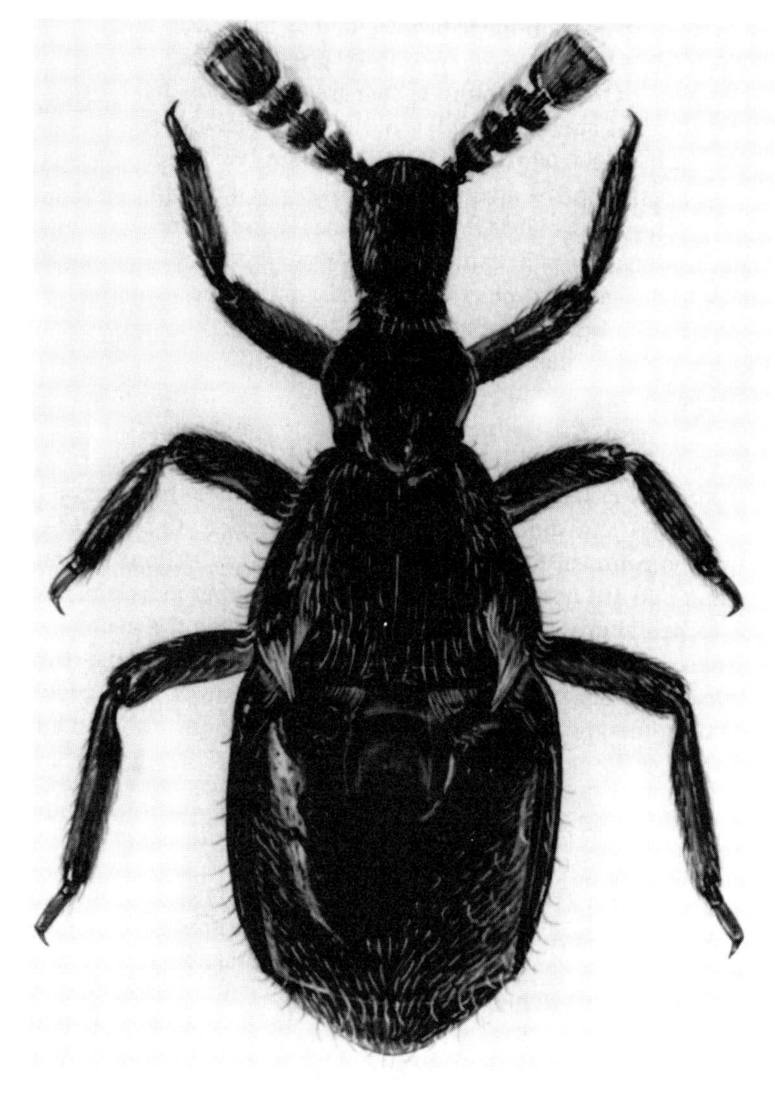

**FIGURE 13–24** *Claviger testaceus*, a frequent guest in *Lasius* nests in Europe. (From Hölldobler, 1973c; painting by T. Hölldobler-Forsyth.)

*castaneus* group, which is strictly eastern in distribution.

A major question remaining is whether after speciation and adaptive radiation species of *Cremastocheilus* are now at different evolutionary stages along a pathway toward full integration. Alternatively, differences in behavior and morphology may simply reflect adaptive radiation to the behavioral ecology of different species of ants. Alpert argues and we agree that the latter hypothesis is more likely, since all species of *Cremastocheilus* have the same basic relationship with ants.

Most species of the beetle family Pselaphidae are carnivorous, with members of the subfamilies Pselaphinae and Clavigerinae being additionally myrmecophilous. According to Park (1964) all of the approximately 250 species of the clavigerines are especially adapted to live in ant nests. The best-studied genera are *Adranes* (Park, 1932; Akre and Hill, 1973; Hill et al., 1976; Kistner, 1982) and *Claviger* (Donisthorpe, 1927; Hölldobler, 1948; R. Cammaerts, 1974, 1977). All observations to date indicate that these clavigerine beetles live among the brood of their host ants, either preying on it or exploiting it in some other manner. Kistner (1982) has provided a table listing all the behavioral acts thus far observed in *Adranes* and *Claviger*. From these data we learn that *Claviger testaceus* (Figure 13–24), a frequent myrmecophile of the formicine genus *Lasius*, preys on ant eggs, larvae, and pupae while at the same time receiving food from larvae and adult ants by means of trophallaxis (Figure 13–25). The beetles occasionally eat dead ants or booty brought into the brood nest by their hosts. Finally, *Claviger* adults are also frequently groomed by the ants. The host workers lick secretions from the numerous trichomes and other exocrine glands, the anatomy of which has been carefully studied by Cammaerts (1974). The ants rarely treat *Claviger* aggressively. When transplanted experimentally to new nests the beetles are accepted by a diversity of ant species—even those genera with which they have never been found to occur naturally (Donisthorpe, 1927; Hölldobler, 1948). When attacks do occur, almost invariably during the initial phase of adoption, the *Claviger* usually show death-feigning behavior. The ants often respond by grasping the beetles on their slender thoraces and carrying them around for a short while before releasing them again, usually in the midst of the ant brood. *Claviger* rarely suffer any harm from this treatment.

R. Cammaerts (1977) observed that *Lasius* workers regurgitate liquid food to *Claviger testaceus* after they have licked them intensively. He noted a similarity of the movements to those occasionally observed when ants lick prey objects before they place them as food among the ant larvae. From this observation he proposed the interesting hypothesis that *Claviger* produce substances from the trichome glands that smell like insect corpses, which cause the ants to place the beetles as "pseudoprey" among the larvae. It is noteworthy in this connection that Akre and Hill (1973) observed that ant larvae lick the trichomes of the clavigerine *Adranes taylori*, although in contrast to the species of *Claviger* the *Adranes* do not eat ant larvae. Instead they depend on food regurgitated to them by the larvae. In both cases, however, the mimicking of prey signals would be an efficient way to gain access to the brood chambers.

One other category of myrmecophilous adaptation by brood predators deserves special mention. The larvae of the syrphid fly genus *Microdon*, which develop in ant nests, have unusual shapes (Figure 13–26). They resemble slugs or limpets and were originally described as mollusks, later as coccids, before they were correctly identified as syrphid larvae (for recent reviews of the literature see Duffield, 1981; Kistner, 1982). Also, the larvae of at least two other nest parasites are convergent in this slug or limpet-like form: the larvae of the predaceous Australian lycaenid butterfly *Liphyra brassolis* and the larvae of several species of the Australian parasitic moths in the family Cyclotornidae (Dodd, 1902; Hinton, 1951). The genus *Microdon* is cosmopolitan and contains about a hundred species, most of which have been recorded from nests of such diverse ant genera as *Camponotus, Formica, Lasius, Polyergus, Aphaenogaster, Crematogaster, Monomorium, Pheidole, Iridomyrmex*, and *Tapinoma*. *Microdon* larvae either feed on detritus (Akre et al., 1973) or prey on the host ants' larvae (Hocking, 1970; Van Pelt and Van Pelt, 1972; Duffield, 1981; Garnett et al., 1985). Duffield (1981) reports that the first-instar larvae of *M. fuscipennis*, which develop in nests of *Forelius* (= *Iridomyrmex*) *pruinosus*, have never been observed eating ant larvae, although they may obtain food from the ants by larval trophallaxis. Second- and third-instar larvae, on the other hand, consume small ant larvae but never pupae. Frequently the adult ants pull their larvae away from the *Microdon*. Successful *Microdon* larvae move up

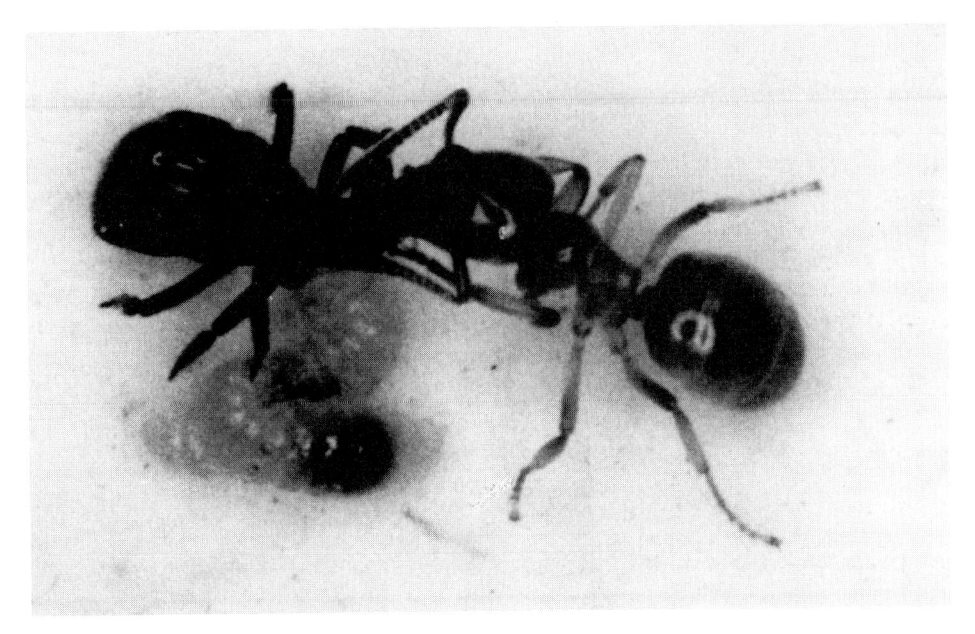

**FIGURE 13–25** The feeding and grooming of *Claviger testaceus* beetles by its *Lasius* hosts. (Photograph by T. Hölldobler-Forsyth.)

and over the ant larvae, piercing the larval skin, emptying the body contents, and discarding the empty shell. Duffield observed third-instar larvae consuming eight to ten ant larvae each during various 30-minute periods of observation.

How are the *Microdon* larvae able to survive inside the ants' brood chambers? The fly larvae are usually bulky and have a heavily reticulated skin, which appears to confer mechanical protection from ant attacks. Garnett et al. (1985) made an interesting discovery that suggests a more subtle form of defensive adaptation on the part of certain *Microdon* species. These investigators studied three Nearctic species, *Microdon albicomatus* and *M. cothurnatus* in nests of *Formica* species, and *M. piperi* in *Camponotus* nests. The larvae are obligate predators on the brood of their hosts. However, they appear to feed almost exclusively on cocoon occupants, that is, on larvae already encased by a cocoon, or prepupae, or pupae. Attacks by ant workers on the *Microdon* larvae were extremely rare, but obvious acts of aggression did occur when second and third instars of *M. albicomatus* and *M. cothurnatus* were introduced into the nest of an inappropriate host, *Camponotus modoc*. This result indicates the presence of a pseudopheromone that mimics the colony odor of the hosts. Garnett et al. propose that at least the first- and second-instar larvae mimic ant cocoons. They observed that when an ant nest is exposed to sunlight *Microdon* larvae compress their bodies laterally into the rough shape of an ant cocoon. They are then quickly picked up by workers and transported into undisturbed parts of the nest along with real cocoons. Third-instar larvae are evidently too large and bulky to exhibit curling behavior, and they are never transported.

> Considering that first and second instars of the three species of *Microdon* studied are physically similar and yet workers appear species-specific in their aggression, we assume that the aggressive mimicry is chemical as well as physical. If the first instar of *Microdon* gains access to a defenseless pupa in cocoon, it will not only find nourishment and protection from attack, but an opportunity to acquire recognition chemicals characteristic of the host. Upon its exit from the

**FIGURE 13–26** A larva and adult of the myrmecophilous syrphid fly *Microdon* with their host ants, *Formica fusca*. (Photograph by T. Hölldobler-Forsyth.)

cocoon the first instar has gained the necessary chemicals to be accepted and even transported as brood. The possible dilution of such chemicals by growth of the larvae is partially compensated by the second instar continuing to attack and enter cocoons.

Another syrphid parasite has been discovered in nests of the Australian weaver ant *Polyrhachis (Cyrtomyrma) ?doddi* (B. Hölldobler, unpublished). It is an undescribed species of the genus *Trichopsomyia* (F. C. Thompson, personal communication). Its larvae, which resemble slugs even more strikingly than those of *Microdon* (Figure 13–27), live inside the ant nest where they prey on the ant brood. The parasitic larvae are ignored by the ants, but the adult fly, when eclosing from the pupa, is attacked by the ant workers and quickly leaves the nest.

**FIGURE 13–27** A myrmecophilous syrphid fly of the genus *Trichopsomyia* that lives with a *Polyrhachis* weaver ant in Australia: larva (*top*), pupa (*center*), adult (*bottom*).

## WASMANNIAN MIMICRY

Among the battalions of parasitic staphylinids that march with army ants are many species that strikingly resemble their hosts (Figure 13–28). This formicoid habitus is found almost nowhere else among the Staphylinidae. It has originated many times over through modifications of the abdomen and thorax that create an ant-like "petiole." In addition, there is a strong tendency to resemble the hosts in the overall slender body form, in color, and even in the sculpturation of the body surface (see Figure 13–3). Seevers (1965) showed that "petioles" have been created in no fewer than seven ways in various groups of the Staphylinidae, with several of the modifications having been chosen by two or more groups independently. Although the ant-like species are accompanied in nests by staphylinids of a more generalized body shape, their large numbers and the remarkable degree of evolutionary convergence they encompass leave little doubt that some kind of mimicry is involved. The question is, exactly who is being fooled? Wasmann (1889, 1903, 1925), in first documenting the phenomenon from museum specimens, was persuaded that the form of the body, including the false petiole, is tactile mimicry that deceives the host ants. He further believed that the coloration is visual mimicry that deceives birds and other predators, which might otherwise be able to pick the beetles out of the columns of running ants. The most prominent vertebrate predators along the columns of army ants are formicariid antbirds that follow the raiding swarms of *Eciton* or *Labidus*. They feed on arthropods flushed from leaf litter by the ants but avoid the army ants.

Kistner (1979) describes *Ecitomorpha nevermanni* as one of the outstanding examples of such Batesian mimicry among ecitophilous staphylinids, whose color varies to match the color variation of its host, *Eciton burchelli*: "In Costa Rica and Panama, where *E. burchelli* is usually reddish-brown in color, so is *Ecitomorpha nevermanni*. At Tikal, Guatemala, where *E. burchelli* is nearly black, so is *Ecitomorpha nevermanni*. In Ecuador, both were bicolored." Since the army ants have extremely poor vision, it is most likely that this remarkable color adjustment is an evolutionary adaptation to vertebrate predation, including that by the formicariid antbirds. Nevertheless, Kistner (1966a,b; 1979), reflecting on his own field observations of staphylinid guests of the African driver ants, agrees with Wasmann's interpretation that visual mimicry evolved as a deceiving mechanism to fool the host ants in order to be accepted into their colonies. Species of *Dorylomimus*, *Dorylonannus*, *Dorylogaster*, and *Mimanomma* run with the ants in their columns. When a host worker encounters one of the beetles, she touches it lightly with her antennae. Kistner remarks:

> This action is identical with that of an ant when it meets another ant. Typically, the antenna rubs along the ant and lingers at the petiole, then both ants move on. I have interpreted this palpation as a signal which tells the blind ants that the passing insect is another ant. I have further interpreted this with regard to the Dorylomimini, that the constriction of the abdomen is such that the same signal is evolved when palpated by the ants and both ants and myrmecophiles go their separate ways. Thus, the morphological constriction permits the myrmecophile to function within the colony as though it were an ant.

Kistner and Zimmerman (1986) made a similar observation on the staphylinid genus *Pheigetoxenus*, a guest of the myrmicine swarm raiders of the genus *Pheidologeton*. This special form of tactile decep-

**FIGURE 13–28** Myrmecophiles that resemble their army ant hosts. In each pair the host is depicted on the lower right. *(Top) Mimanomma spectrum* with its host *Dorylus nigricans. (Center) Crematoxenus aenigma* with *Neivamyrmex melanocephalum. (Bottom) Mimeciton antennatum* with *Labidus praedator.* (From Hölldobler, 1971a.)

tion was designated Wasmannian mimicry by Rettenmeyer (1970).

Does Wasmannian mimicry really exist? We know that chemical identification is paramount in ants generally. When the surface odor of a worker ant is disturbed only slightly, the ant is swiftly attacked by her own sisters even though her morphology remains unchanged. Furthermore, there exist many more myrmecophiles which are fully integrated into the hosts' social system but do not resemble the host workers' shape in the least. Hölldobler (1970c, 1971a) found that when he artificially altered the shape and color of *Atemeles* or *Lomechusa,* the relationship of the beetles to their hosts was not affected. Thus it appears that communicative behavior and chemical mimicry are the essential ingredients for acceptance of social parasites.

Although the hypothesis of tactile mimicry must not be dismissed out of hand, an alternative hypothesis is equally promising (Wilson, 1971). It is noteworthy that all cases of morphological resemblances of staphylinids occur in myrmecophiles that live with legionary ants. These ants and their guests spend most of their time on the surface, either in raiding or emigration columns. It is possible that predators watching the ant columns for edible morsels are fooled by both the color of the beetles and their shape. Thus all of the mimicry apparent to taxonomists may be visual and directed at predators outside the host colony. In fact there are numerous cases of arthropods that are not socially associated with ants yet strikingly resemble them in morphology and locomotion. Both circumstantial and experimental evidence strongly suggest that this similarity protects the mimics from predators (Reiskind, 1977; Oliveira, 1985; McIver, 1987). Only by employing carefully designed experiments will we be able to resolve this question.

Some evidence exists meanwhile to suggest that tactile mimicry can serve as an ancillary mechanism for social integration in the host colony. We have already mentioned the case of the *Microdon* larvae that resemble the shape of ant cocoons and are transported by the ants in the same manner as these brood objects. Another remarkable case is that of the phoretic mite *Planodiscus* which attaches itself to the tibia of its host, *Eciton hamatum.* Kistner (1979) used scanning electron micrographs to demonstrate that the cuticular sculpturing of the mite's body is nearly identical to that of the ant's leg. Furthermore, the arrangement and number of setae on the mite approximates the arrangement and number of setae on the leg. "Thus when the ant grooms its leg," Kistner observed, "the tactile stimulation will be similar to that of the leg itself. Since the mite clings to the leg by grasping setae, small movements of the mite during grooming would translate into small movements of the ant's setae. These would approximate movements caused by grooming activities themselves."

Another form of tactile mimicry has been described in the myrmecophilous cricket *Myrmecophila acervorum* (Figure 13–29) by Karl Hölldobler (1947). Incidentally, this is the European symbiont with the longest history of investigation; the first description of its behavior was by Paolo Savis (1819), and it was examined in increasing detail by Wasmann (1901), Schimmer (1909), and Hölldobler (1947). In North America, *Myrmecophila nebrascensis* and *M. manni* were studied by Wheeler (1900) and by Henderson and Akre (1986a–c). In many ways the findings on *M. manni* parallel those made on *M. acervorum.* Henderson and Akre (1986b), however, found that *M. manni* reproduces sexually, with males exhibiting dominance hierarchies during mating. In contrast, no males at all are known from *M. acervorum,* and it has therefore been suggested that this species reproduces by parthenogenesis (Schimmer, 1909; Hölldobler, 1947).

*Myrmecophila acervorum,* the species with possible tactile mimicry, lives with many different host species. Karl Hölldobler discovered two morphs of different sizes. The larger "major" morph is found primarily in nests of ant species with larger workers, such as *Formica, Camponotus,* and *Myrmica,* whereas the smaller "minor" morph occurs with species that have smaller workers, such as *Tetramorium* and *Lasius.* All developmental stages live in ant nests. The adults and nymphs are extraordinarily versatile in their feeding habits. They prey variously on ant eggs, lick adult ants and other myrmecophiles such as *Hetaerius* and *Claviger,* snatch food from food-exchanging ants, and solicit the regurgitation of liquid food by ant workers.

When the crickets are newly introduced into an ant nest, or when the colony to which they belong is disturbed, they are usually treated in a hostile manner by the worker ants. They are then able to escape death only through swift and nimble running. But the ant aggression usually subsides as soon as the crickets adjust their locomotory pattern to the movement patterns of the undisturbed host ants. Hölldobler (1947) noted that when *Myrmecophila* were transferred from an ant species with a relatively slow movement pattern, such as *Myrmica rubra,* to a species that generally exhibits a more vivacious "temperament," such as *Formica fusca,* the crickets alter their own locomotory behavior to resemble that of their new hosts. The transformation usually takes a few days, during which the crickets stay mostly in hiding and apparently acquire the host-specific colony odor. Hölldobler proposed that in addition to this locomotory convergence, the *Myrmecophila* employ a tactile mimicry. Although the cricket does not look like an ant overall, portions of its body resemble parts of the ants' bodies. Hölldobler elaborated his tactile mimicry concept with a metaphor. Suppose, he said, that we live in a completely dark room and orient primarily by means of the tactile sense in our hands. Among hundreds of us dwells one creature that is very differently constructed but has appendages resembling human hands, and it also manages to mimic our body movements and to touch us with a humanoid caress. This creature is perceived by us as a fellow human being until some crucial behavioral mistake unmasks it as an alien. Thus the ants do not tolerate the crickets because they perceive the presence of the intruders as pleasant or comfortable, as Wasmann suggested, or because the crickets are too swift and elusive, as Wheeler and Schimmer proposed. Instead, the host workers are fooled into classifying the crickets as fellow ants. As soon as they notice the deception they hunt the crickets, even after having tolerated them for weeks or months.

Kistner and Jacobson (1975a) redefined Wasmannian mimicry to cover all mimicry of social releasers, including behavioral imitation of solicitation signals and the imitation of ant pheromones by myrmecophiles. Wasmann himself would probably not have accepted this usage, because it contradicts his theory—now abandoned—about amical selection and symphilic instincts (see Hölldobler, 1948). Nevertheless, we agree that the expanded concept of Wasmannian mimicry is reasonable and honors the name of a great myrmecologist, who was the most important pioneer in discovering the world of symbiosis in insect societies.

An especially well-studied species displaying Wasmannian mimicry, by the modernized definition, is *Myrmecaphodius excavaticollis,* a scarabaeid beetle that lives principally with *Solenopsis* fire ants and secondarily with dolichoderine ants of the genus *Iridomyrmex.* Vander Meer and Wojcik (1982) noted that when these little beetles first move into a host nest, they exhibit a passive defensive behavior that

**FIGURE 13–29** The myrmecophilous cricket *Myrmecophila acervorum.* (Photograph by T. Hölldobler-Forsyth.)

allows them to acquire cuticular hydrocarbons specific to their host species. These adsorbed substances then apparently enable the beetles to become integrated into the host colony. This conclusion is supported by an increasingly convincing body of evidence that cuticular hydrocarbons are part of the species and colony recognition system in social insects (see Chapter 5).

All developmental stages of *Myrmecaphodius excavaticollis* have been found within the nests of the host ants. The adults variously prey on ant larvae, feed on dead ants and booty, and induce workers to regurgitate liquid food to them. They conduct dispersal flights throughout the year. After alighting they seek access to a suitable host colony, which may not belong to the same species they originally came from (Wojcik et al., 1978).

Vander Meer and Wojcik (1982) report that the hydrocarbons, which constitute 65–75 percent of the cuticular lipids of *Solenopsis invicta* and *S. richteri,* differ conspicuously among the four host species of *Myrmecaphodius excavaticollis.* But the beetles are able to shed the hydrocarbons of one *Solenopsis* species and acquire the pattern of another. In addition to species-specific characteristics, the hydrocarbon mixtures probably also contain colony-specific patterns. This, in part, explains how *Myrmecophodius* is able to infest a variety of species and colonies. Vander Meer and Wojcik demonstrated this effect with the following experiment. Beetles from *Solenopsis richteri* colonies were isolated for two weeks, and then introduced into colonies of *S. invicta.* After five days, the beetles were removed and analyzed for cuticular hydrocarbons. It was found that *M. excavaticollis* had acquired the cuticular hydrocarbons of its new hosts (Figure 13–30). Vander Meer and Wojcik observed:

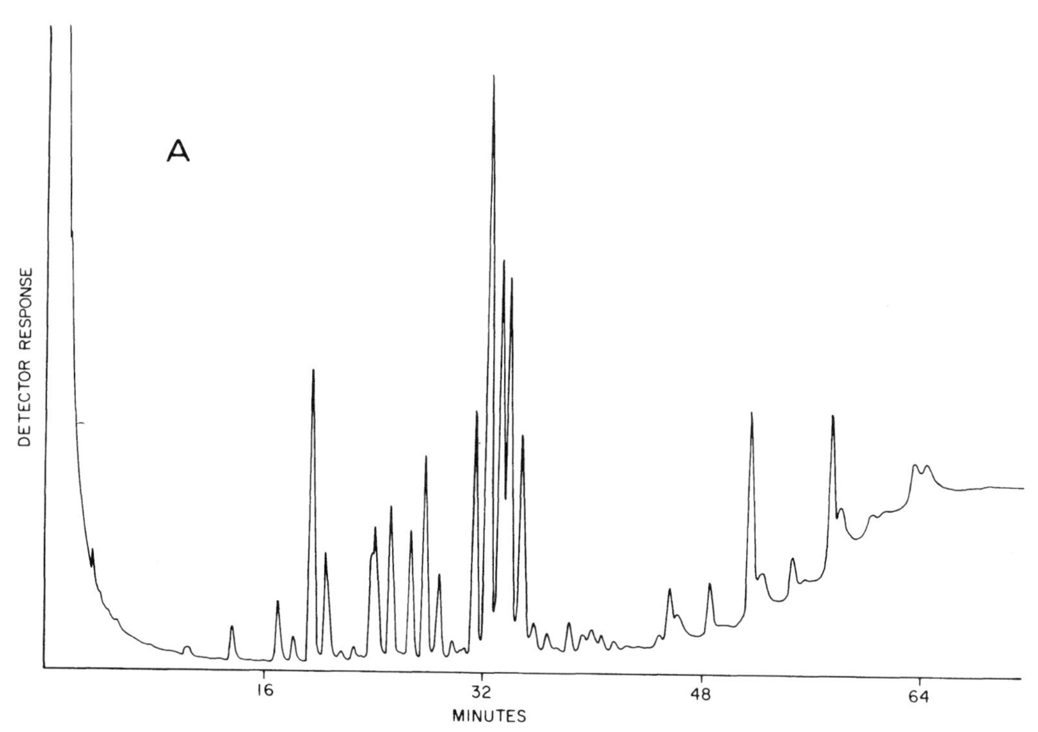


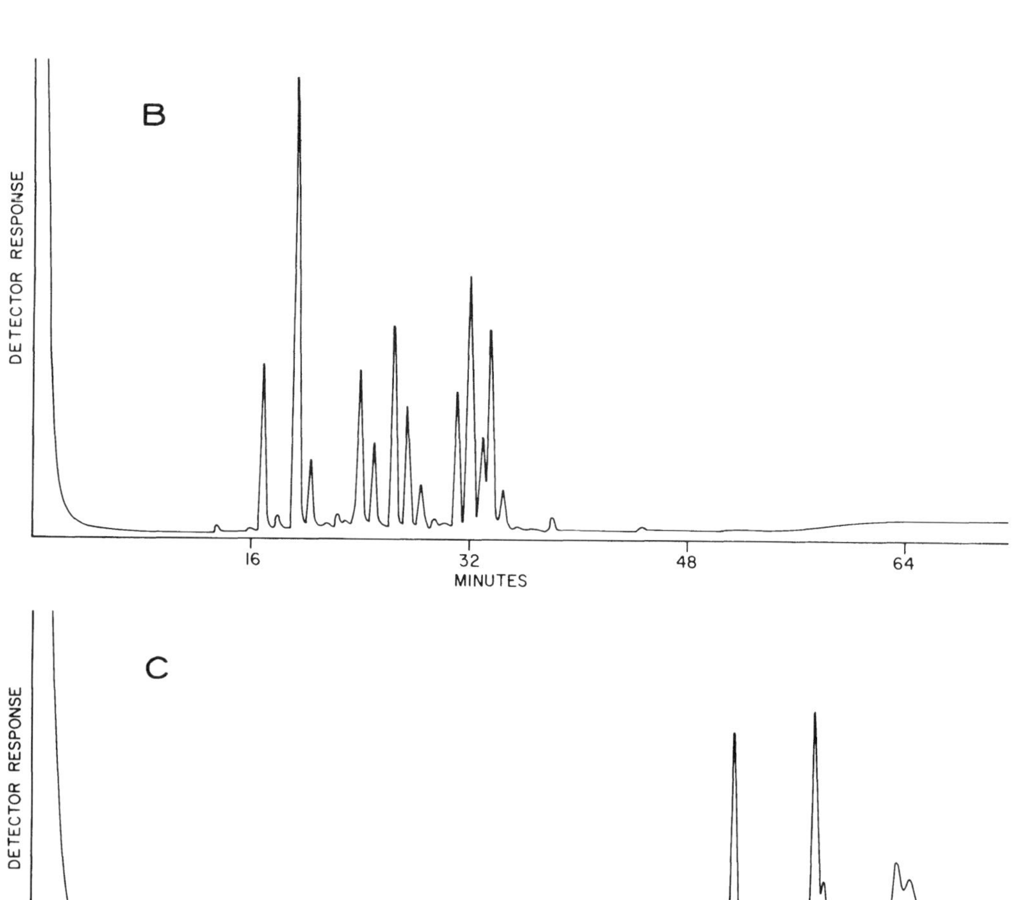


**FIGURE 13–30** Gas chromatographic traces of (*A*) cuticular hydrocarbons from *Myrmecaphodius excavaticollis* taken directly from a *Solenopsis richteri* colony, (*B*) cuticular hydrocarbons from *S. richteri*, and (*C*) cuticular hydrocarbons from *M. excavaticollis* two weeks after the beetles were removed from an *S. richteri* colony. (From Vander Meer and Wojcik, 1982.)

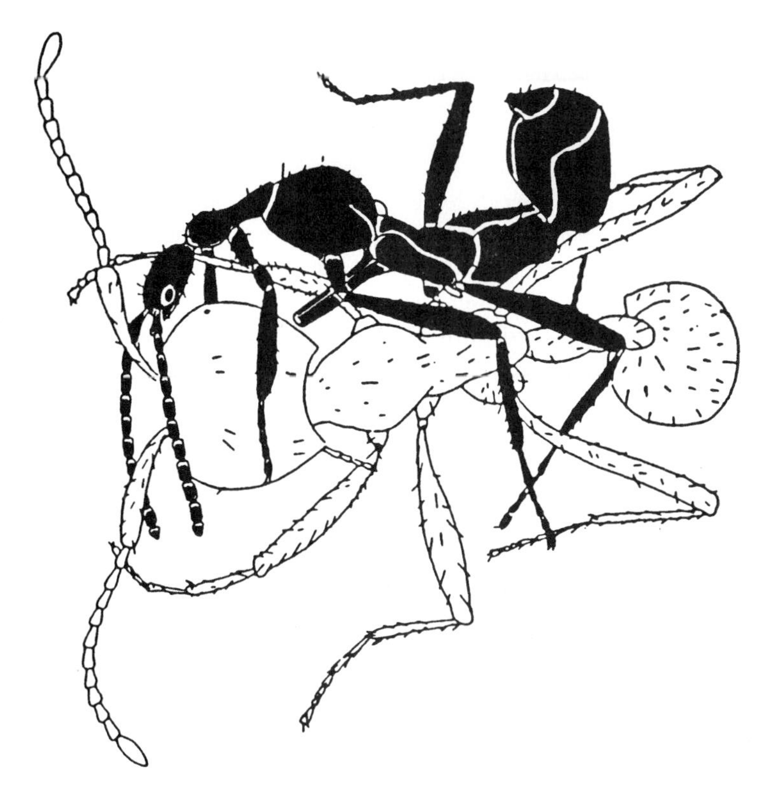

**FIGURE 13–31** *Diploeciton nevermanni* grooming a worker of the host ant *Neivamyrmex pilosus.* (Modified from Akre and Torgerson, 1968.)

> The same phenomenon occurred when previously isolated beetles were introduced into *S. geminata* and *S. xyloni* colonies. Although the switching of hydrocarbon patterns from one host to another weakens the likelihood that they are synthesized by the beetle, we also found that freshly killed isolated beetles had acquired *S. invicta* hydrocarbons within two days after exposure to the ant colony. These data eliminate biosynthesis as a possibility and support a passive mechanism of hydrocarbon acquisition. When initially introduced into a host colony, the *M. excavaticollis* were immediately attacked. The response of the beetles was to play dead and wait for the attacks to cease, or they moved to an area less accessible to the ants. Within two hours after introduction into a host colony, the beetles' cuticle contained 15 percent of host hydrocarbons. The accumulation of hydrocarbon continued up to four days until the beetles' cuticle contained about 50 percent host hydrocarbons. Beetles surviving this long were generally no longer attacked.

In a suggestively parallel manner, K. Hölldobler (1947) observed that about four days are required for the cricket *Myrmecophila acervorum* to become fully integrated into the host colony, at which time it is able to groom the host ant workers. Furthermore, Rettenmeyer (1961) raised the interesting possibility that the guests of army ants pick up the colony odor in an active fashion while grooming their hosts. It is certainly true that grooming by symphilic staphylinids and histerids cannot serve entirely for the ingestion of edible cuticular material. Akre and Torgerson (1968) have described the elaborate grooming rituals of a staphylinid guest of *Neivamyrmex* as follows.

> While *Probeyeria* and *Ecitophya* straddled their host across the longitudinal axis of the body, *Diploeciton* assume a position parallel to, but slightly to one side of and on top of the ant. To position itself, a beetle grasps with its mandibles the scape of an antenna of an ant close to its base. It then positions its body parallel to the body of the ant and uses the first and third legs on the lower side of the body to brace against the substrate. The three legs on the other side of the body then straddle the ant. The mesothoracic leg on the bottom curls under and around the thorax of the ant. This places the sternum of the beetle's thorax against the side of the thorax of the ant as though riding "sidesaddle." In this position the beetle rubs the ant with its legs. The mesothoracic lower leg rubs the bottom of the thorax and the upper legs rub on the dorsal area of the ant; the prothoracic leg usually rubs the head of the ant, the mesothoracic leg rubs on the thorax and gaster, while the metathoracic leg is used sparingly to rub the gaster of the ant. The rubbing strokes are rather slow and alternate between stroking the body of the ant and the staphylinid's own body. The front leg is rubbed on the head and thorax, both middle legs are rubbed on the elytra and the globular portion of the myrmecoid abdomen, while the metathoracic leg was rubbed only rarely on the abdomen [Figure 13–31].

The stroking movements of the staphylinid and histerids calm the ants and in some instances, according to Akre and Torgerson, seem to paralyze them partially. This is not another case of a fatal seduction of the kind worked by the bug *Ptilocerus,* however, for the ecitophiles do not attack their adult hosts. They feed only on the immature stages, which interestingly enough are not recipients of the grooming ritual.

We infer that this kind of integration into the ant societies represents an evolutionarily less advanced grade than that of *Atemeles* or *Lomechusa*. Although these symbionts may also acquire host colony-specific cuticular hydrocarbons, they can be easily transferred from one colony to the other so long as both belong to at least the same species group. This trait probably is based on the ability of both the larvae and the adults of *Atemeles* and *Lomechusa* to produce pseudopheromones that mimic the brood pheromones of their hosts. It has long been known that brood pheromones dominate or mask colony-specific recognition cues in ants (Hölldobler, 1977; Carlin and Hölldobler, 1986), making it easy to transfer ant brood from one colony to another belonging to the same species. The mimicry of brood pheromones thus appears to be the most advanced mechanism of social-parasitic integration, one that opens up the brood chambers, the richest ecological niche within an ant colony.

## SYMBIOSES BETWEEN ANTS AND LYCAENID BUTTERFLIES

The larvae of many species of lycaenid butterflies are closely associated with ants (Hinton, 1951; Atstatt, 1981a; Henning, 1983a; Cottrell, 1984; Pierce, 1987). The majority appear to be mutualistic, although a number are parasitic. An example of a parasitic species is the large blue (*Maculinea arion*) of Europe. The larva feeds on wild thyme and is attended by ants until it reaches the fourth instar. Then it crawls down onto the ground and wanders about until it meets a worker of the ant *Myrmica sabuleti* (Thomas, 1980). When touched by the ant the larva deforms its body into a striking new shape: it retracts its head and swells its thoracic segments up while constricting its abdominal segments, which gives its body a hunched, tapered look (Figure 13–32). Apparently the altered body form serves as a signal to the ant, which may or may not work in concert with substances that resemble larval pheromones of the ants.

Whatever the nature of the key lycaenid stimuli, the ant now picks the caterpillar up and transports it into the nest. Once en-

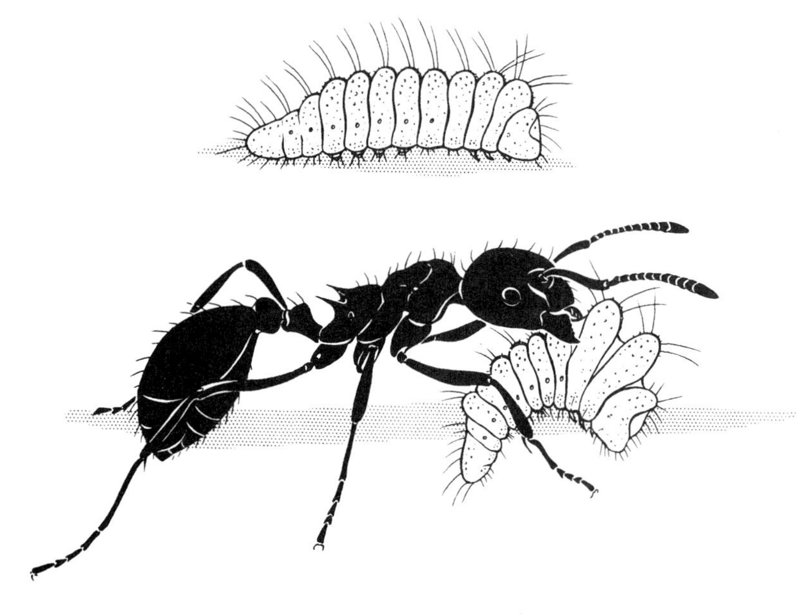

**FIGURE 13–32** The adoption procedure of the third-instar larva of the large blue (*Maculinea arion*). The individual above is searching for a host ant and still has the typical shape of a lycaenid larva. The larva at the bottom has been milked by the *Myrmica* worker, and, after hunching its body, it is being transported back to the nest by the ant. (From Wilson, 1971; redrawn from Frohawk, 1916.)

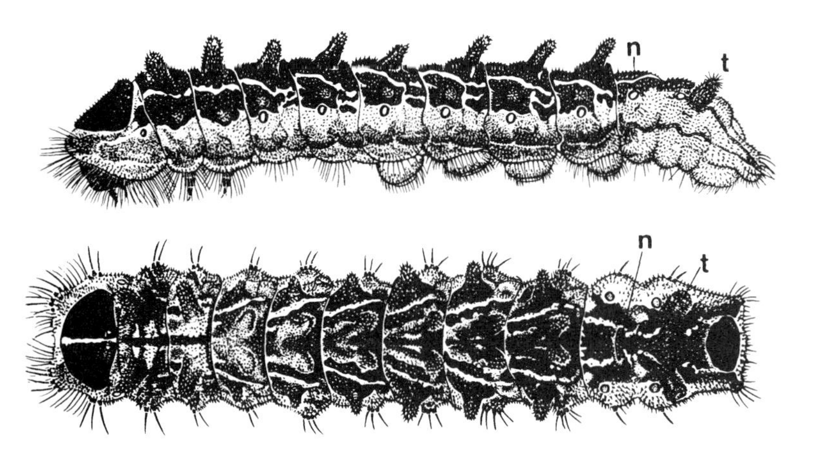


**FIGURE 13–33** Fifth-instar larva of the Australian lycaenid *Jalmenus evagoras*. *n*, Newcomer's gland; *t*, tentacular organ. (From Kitching, 1983.)

sconced in the brood chambers, the caterpillar turns carnivore, feeding heavily on the host ant larvae. When it reaches full maturity, it pupates and overwinters in the nest, finally emerging as an imago the following June (Frohawk, 1916; Cottrell, 1984). A very similar myrmecophilous behavior occurs in species of the lycaenid genera *Lepidochrysops* and *Aloeides*, but only *Lepidochrysops* larvae appear to consume ant brood (Henning, 1983a,b).

A remarkable variation on the *Maculinea* pattern was discovered by Schroth and Maschwitz (1984). The freshly eclosed fourth instars of *Maculinea teleius* leave their host plants, *Sanguisorba officinalis*, and actively search on the ground for the trails of their host ants, *Myrmica rubra*. Once they encounter the chemical trails they follow them until the host ants' nests are reached. The *M. teleius* larvae seem to enter the *Myrmica* nests on their own. Schroth and Maschwitz never observed the lycaenid caterpillars being carried by the ants. Only the fourth-instar larvae respond to the ants' trail pheromone and then only during a brief "sensitive phase" after eclosion. Caterpillars that have already lived two days inside the *Myrmica* nest no longer follow the ants' chemical trails.

Other parasitic species occur among the Asiatic and Australian lycaenids. For example, the caterpillars of *Liphyra brassolis* prey on larvae of the green tree ant *Oecophylla smaragdina* (Dodd, 1902). *Liphyra* larvae, unlike other lycaenid parasites, do not appear to be welcome guests in the host ants' nests, but rather defend themselves by their remarkable morphological adaptation against ant aggression. As we mentioned earlier, the larvae look like mollusks, and they are protected by an extremely hard and thick cuticle (Johnson and Valentine, 1986). When the butterfly hatches, it is covered by a dehiscent cloak of white, gray, and brown scales. As the *Oecophylla* attempt to seize the butterfly during its egress from the nest, they get only mandibles and antennae full of scales. According to Hinton (1951), a similar adaptation occurs in several other lycaenids, including species of *Maculinea*, as well as the South American pyralid *Pachypodistes goeldii*.

The majority of myrmecophilous lycaenid species apparently live in exclusively mutualistic interactions with ants, and it seems likely that the parasitic life pattern has evolved from ancestral mutualistic relationships. Many species of New World riodinid butterflies have similar ecological relationships with ants (Ross, 1966; Callaghan, 1977, 1979, 1981; Schremmer, 1978; Robbins and Aiello, 1982; Horvitz and Schemske, 1984), and for the purposes of our discussion, we have included the Riodininae as a subfamily of the Lycaenidae (Ehrlich, 1958; Kristensen, 1976; Vane-Wright, 1978; but see Eliot, 1973). As more is learned about the morphology and ecology of the riodinids, however, it may be found that they are a distinct taxonomic group whose ant-associated characteristics are convergently derived with those of the Lycaenidae.

Several morphological structures in lycaenid larvae are important in the ant-lycaenid associations (Figures 13–33 and 13–34). Scattered over the surface of the larvae and pupae are many epidermal glands, called lenticles or pore cupolas, the significance of which was first noted by Hinton (1951) and subsequently described in detail by Malicky (1969), Downey and Allyn (1979), Kitching (1983), Pierce (1983), Wright (1983), Franzl et al. (1984), and Kitching and Luke (1985). These perforated structures have been found in all lycaenid subfamilies investigated, including the Riodininae. They are derived from innervated glandular setae that are believed to exude either ant attractants or appeasement substances. Epidermal extracts of several species have been bioassayed and shown to contain substances that are highly attractive to ants (Henning, 1983b; Pierce, 1983).

The larvae of many species also possess the so-called Newcomer's gland (called the dorsal nectar organ or honey gland by some authors), which is located on the dorsum of the seventh abdominal segments. This organ occurs in most species of the Polyommatinae and Theclinae but is absent in other subfamilies, including the Lycaeninae, Miletinae, Liphyrinae, and Poritiinae. The Newcomer's gland presumably developed from dorsal epidermal pores of a kind common in the lycaenids (Malicky, 1969; Kitching and Luke, 1985; Fiedler, 1987). In some groups possibly analogous honey-gland structures have evolved from epidermal pores, such as the "perfo-

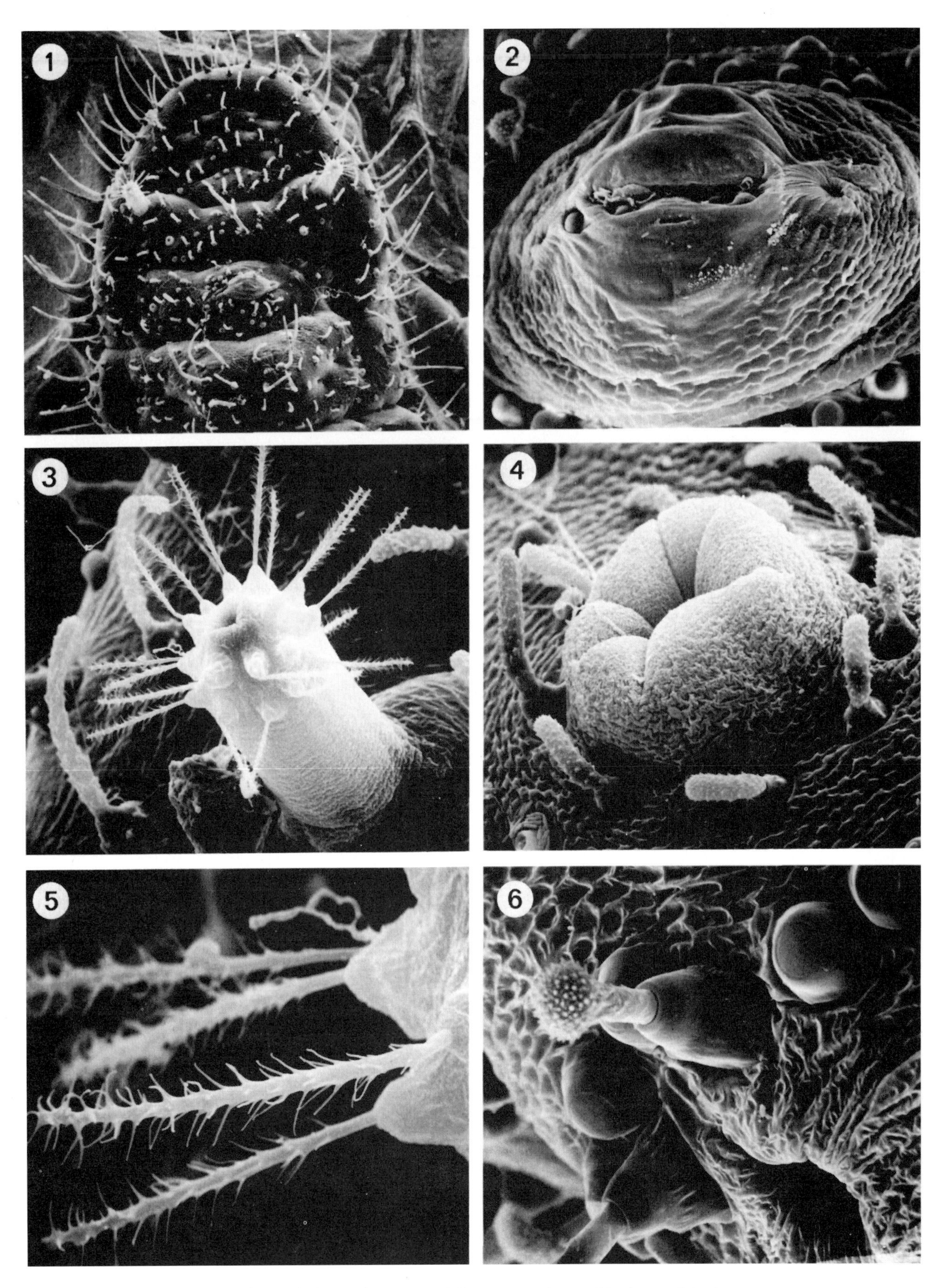


**FIGURE 13–34** External morphology of myrmecophilous organs in an older larva of the lycaenid *Lysandra coridon*. (*1*) posterior and dorsal view; (*2*) Newcomer's organ; (*3*) everted tentacular organ; (*4*) inverted tentacular organ; (*5*) spinose hair on tentacular organ; (*6*) pore cupolas. (From Kitching and Luke, 1985.)

rated chambers" of the Curetinae (DeVries et al., 1986), the "pseudo-Newcomer's gland" of the Miletinae (Kitching, 1987), and the nectar organs of the Riodininae (Ross, 1966; Callaghan, 1977; Schremmer, 1978; Cottrell, 1984).

From its first description by Newcomer (1912), the lycaenid gland was thought to secrete sweetish liquid which was imbibed by the ants. Maschwitz et al. (1975) provided the first analytical proof for this hypothesis. They found that the secretions of the Newcomer's gland of the European species *Lisandra hispana* consist of a 13.1–18.7 percent solution of fructose, sucrose, trehalose, and glucose, as well as minor quantities of protein and a single amino acid, methionine. The hemolymph of the lycaenid caterpillar has a total carbohydrate concentration of only about 2 percent. Thus the Newcomer's gland provides the ants with a highly enriched sugar solution.

The Newcomer's gland of most species of Curetinae, Polyommatinae, and Theclinae is flanked on either side by a pair of eversible tentacles, which are also called lateral organs or tentacular organs. Similar structures exist in species of the subfamily Riodininae, although they are absent in the Lycaeninae. Their function is unclear. It has been variously supposed that they produce substances attractive to the ants, signal the presence of the Newcomer's gland, or serve in defense when the dorsal organ is depleted (Downey, 1962; Claassens and Dickson, 1977). In the 1980s evidence has been adduced to indicate that tentacular gland secretions serve at least in part to evoke alarm behavior in the attending ants (Henning, 1983b; DeVries, 1984; DeVries et al., 1986; Fiedler and Maschwitz, 1987).

Other intriguing anatomical structures await investigation. Setae have been discovered on the surface of lycaenid larvae and pupae that are variously shaped like clubs, mushrooms, trumpets, clover blossoms, and multibranched hydroids (Downey and Allyn, 1973, 1979; Wright, 1983; Kitching and Luke, 1985; Fiedler, 1987). No connection has yet been experimentally established, however, between these structures and the lycaenid-ant associations. In addition the pupae of a number of species in the Lycaenidae possess stridulatory organs that can cause vibrations of the substrate on which the pupae are attached (Downey, 1966). In some instances an audible sound is produced. Late-instar larvae of certain species also possess stridulatory organs (Pierce et al., 1987; DeVries, personal communication). It seems likely that these stridulations play some role in interspecific communication with ants, and, for those species of Lycaenidae whose larvae aggregate, in intraspecific communication between aggregating larvae.

Aside from the case of *Maculinea* and a few other well-studied genera, the exact nature of the symbiosis between lycaenids and ants remains problematic. Malicky (1969, 1970a,b), for example, believed that there is no mutualistic symbiosis at all between ants and lycaenids, and that the so-called myrmecophilous organs such as the pore cupolas and Newcomer's glands serve solely as defensive devices against ant predation. Malicky argued that one of the most important features determining whether a lycaenid associates with ants or not is the "biotope" in which the larva lives. Both Malicky and Atstatt (1981b) pointed out that the structural form of plants may influence the distribution of ants and consequently the distribution of lycaenids. Hence Atstatt suggested that woody perennials, or "apparent" plants, are more likely to be frequented by ants, and are therefore more likely to serve as host plants for myrmecophilous lycaenids. Atstatt (1981a) suggested that the capability of lycaenids to appease ants and thereby avoid predation opened up habitats which are dominated by ants. He considered this "selection for enemy free space" one of the major evolutionary forces that has shaped the lycaenid-ant associations.

This pathway may well have been followed by the Lycaeninae, Miletinae, and some species of the Riodininae, whose larvae are equipped with the pore cupola organs but lack the Newcomer's gland or other nectar organs. This is a relatively minor group within the Lycaenidae, however, comprising few species in comparison with the highly diverse Polyommatinae and Theclinae (Eliot, 1973). Members of the latter two subfamilies possess both Newcomer's gland and cupola organs. Most enjoy a very close symbiotic relationship with ants, which appears to have contributed in turn to their remarkable species multiplication (Pierce, 1984, 1985, 1987). Increasingly strong evidence now exists that many of the associations are mutualistic. Evidence to that end was first provided by Ross (1966), who pointed out that larvae of the riodinine *Anatole*, which are normally associated with *Camponotus* colonies, are preyed upon by other ant species if left unattended. A similar relation was reported in the larvae of the Australian lycaenid *Ogyris genoveva*, which are attended by *Camponotus consobrinus* and preyed upon by *Iridomyrmex purpureus* (Samson and O'Brien, 1981).

It remained for Pierce and her collaborators, however, to demonstrate conclusively the ants' protective effect on the survival of lycaenid larvae (Figure 13–35; Pierce and Mead, 1981; Pierce and Easteal, 1986). When larvae of *Glaucopsyche lygdamus* were experimentally isolated from their attending ants in the field, they were far more likely to disappear from the host plants. More exactly, they were more prone to drop off the plants or succumb to parasitoids. Ant-tended larvae during the study were 4 to 12 times more likely to survive to pupation than an otherwise similar group of untended larvae.

The females of some myrmecophilous lycaenids actively seek, for oviposition, plants already frequented by host ants. This curious behavior was first indicated in the field studies by Atstatt (1981b) and laboratory experiments by Henning (1983b) and has been demonstrated conclusively in a series of field experiments by Pierce and Elgar (1985) on the Australian species *Jalmenus evagoras*. This lycaenid is known to feed on at least 16 species of *Acacia* while being tended by several species of the dolichoderine ant *Iridomyrmex*. *Jalmenus* females are far more likely to lay egg clusters on plants that contain their attendant ants than on plants kept free of ants. Ovipositing females respond to the presence or absence of ants before they alight on the potential food plant. Once they have landed, however, they are equally likely to lay eggs whether or not they encounter ants. Pierce and Elgar describe the symbiosis as follows:

> If newly colonized plants are near nests of the attendant ant *Iridomyrmex* sp. 25, as the plants in our experiments were, then newly hatched clusters of pioneer larvae may be able to attract a sufficient number of tending ants quickly enough to survive and thus establish a new food plant. This process may be facilitated in several ways. First, *A. irrorata*, like many other acacias, has extrafloral nectaries (three near the tip and one at the base of each leaf), and plants are patrolled regularly by nectar seeking ants, including workers of *Iridomyrmex* sp. 25. Larvae of *J. evagoras* that hatch out on these plants are thus likely to be discovered quickly by ants. Second, eggs of *J. evagoras* are laid in clusters that hatch synchronously, thereby creating an aggregation that is a potentially more attractive food source

**FIGURE 13–35** (*Above*) *Formica fusca* tending a final-instar larva of *Glaucopsyche lygdamus*. The ant is imbibing from the honey gland. (*Below*) *F. fusca* defending a final-instar larva. The ant has seized an attacking braconid wasp in her mandibles. (From Pierce and Mead, 1981.)

than only a single larva. Third, like many other social insects, workers of *Iridomyrmex* sp. 25 recruit nestmates to food resources in numbers commensurative with the quality of those resources (see Wilson, 1971). A single worker discovering a cache of *J. evagoras* larvae may be able to recruit enough nestmates sufficiently quickly to tend the larvae and successfully protect them against predators.

Nevertheless commencing life on a new food plant without the benefit of pre-existing conspecific juveniles that have attracted attendant ants involves considerable risk for young larvae of *J. evagoras*. Ant exclusion experiments performed in 1981 (Pierce, 1983) revealed that the first and second instar larvae were preyed upon far more rapidly when they occurred on plants by themselves than when they occurred on plants that contained final instar larvae and pupae that had attracted large numbers of ants. One way *J. evagoras* may circumvent this problem is by ovipositing adjacent to ant tended homopterans. Atstatt (1981b) suggested a similar function for ant tended homopterans on food plants utilized by *Ogyris amaryllis*, although McCubbin (1971) and Das (1959) described situations in which myrmecophilous homopterans appeared to exclude lycaenids from potential host plants.

In addition to acquiring an immediate attentive ant guard, larvae of *J. evagoras* that hatch out beside myrmecophilous membracids gain a further advantage in the form of food. On numerous occasions we observed larvae of *J. evagoras* feeding on the honeydew secretions of homopterans. The first and second instars sometimes even ride on the backs of adult membracids. This phenomenon of honeydew feeding has been described for several lycaenids (e.g., Hinton, 1951). Although other lycaenids actually prey on homopterans (e.g., Cottrell, 1984) we found no evidence of this in *J. evagoras*.

Investigators over the years have described or suggested the existence of ant-dependent oviposition behavior in no fewer than 47 species of Lycaenidae. As a rule the more dependent a lycaenid species is on its attendant ants, the more likely it is to possess ant-dependent oviposition behavior. It would seem to follow that "the propensity of female lycaenids to oviposit in response to myrmecophilous homopterans, when they occur on novel plants that are not the same species as the butterfly's original food plant, could have important implications for the host plant range of species that use ants as well as plants as cues in oviposition" (Pierce and Elgar, 1985). Indeed, a comparison of 282 species of Lycaenidae reported in the literature confirms that species tended by ants feed on a wider range of plants than those not tended by ants.

An observation by U. Maschwitz (personal communication) is consistent with this generalization. During studies in Malaysia, he saw females of the polyommatine lycaenid *Anthene emolus* depositing large clusters of eggs (50 to 100) directly upon or close to the silk pavilions constructed by the weaver ants *Oecophylla smaragdina*. Within a few hours the larvae hatched and moved into the pavilions, where they fed on the surface tissue of the leaves. The second- and third-instar larvae left the pavilions and were subsequently carried by the ants to "grazing sites," usually young shoots or blossoms of the host plants that are rich in carbohydrates and amino acids. The caterpillars were continuously attended by *Oecophylla* workers, which chased off approaching parasitic flies and wasps and frequently milked the secretions from the Newcomer's gland of the caterpillars. *A. emolus* is a common lycaenid in Malaysia which feeds on a wide variety of plants, including *Nephelium litchi* (Sapindaceae) and *Cassia fistula* (Caesalpiniaceae). They are, however, always associated with *O. smaragdina*.

Pierce discovered another important correlation between ant attendance and diet. A comparative survey from the literature showed that ant-tended lycaenids usually feed on plants that fix nitrogen and are rich in proteins, whereas untended lycaenids feed on other kinds of plants (Pierce, 1984, 1985). Pierce and her collaborators demonstrated that the larvae of *Jalmenus evagoras* and two of its congeners *J. ictinus* and *J. pseudictinus* secrete concentrated amino acids in addition to carbohydrates as rewards for attendant ants, and that these amino acids are produced in species-specific profiles (Pierce et al., 1989). The larvae and pupae of *J. evagoras* in particular produce at least 14 different free amino acids among which serine is dominant. Secretions from the dorsal organ contain serine in concentrations of 19–50 millimoles, which is at least an order of magnitude greater than that of amino acids found in most extrafloral nectaries and is comparable to concentrations found in the salivary glands of many social insects (Maschwitz, 1966; Hunt et al., 1982). Even when larvae are raised on four different host plants, the amino acid profile from the dorsal organ remains the same, with serine the primary component. Radioactive labeling experiments and bioassays with synthetic mixtures confirmed that the amino acids secreted by both the larvae and the pupae are an extremely attractive food source for workers of the attendant ant species (Pierce et al., 1989). Nitrogen is a limiting component in the diet of most insects (Mattson, 1980). The need to secrete surplus protein to attract attendant ants could explain why lycaenids associated with ants feed predominantly on protein-rich plants such as legumes, and even on protein-rich parts of plants, including flowers, terminal foliage, and seed pods. Thus a major constraint to reliance upon ants as a mobile defense force may well be the nutritional quality of the host plant.

Fiedler and Maschwitz (1989) witnessed host ants (*Tetramorium caespitum, Plagiolepis pygmaea*) recruiting nestmates to third- and fourth-instar larvae of the lycaenid *Polyommatus coridon*. This recruitment effect could be significantly weakened by closing off the Newcomer's gland with glue. Fiedler and Maschwitz measured the quantities of secretions produced by the Newcomer's gland and provided energetic calculations which strongly indicate that the symbiotic relationship between *Polyommatus* larvae and their attending ants is mutualistic. In the case of *Tetramorium caespitum*, their reasoning was as follows. They noted that an average-sized colony of *T. caespitum* contains 11,000 workers (Brian et al., 1965; Brian, 1979b), whose monthly basal metabolic rate at 15°C is approximately 47.8 kilojoules (Peakin and Josens, 1978). The monthly metabolic rate of one worker is therefore 4 joules. The foraging area of a *T. caespitum* colony covers approximately 40 square meters. In a dense growth of the host plants, *Hippocrepis comosa*, Fiedler and Maschwitz found about 20 *Polyommatus* larvae per square meter. This corresponds to a monthly production of approximately 70–140 milligrams sugar per square meter, or a chemical energy of 1.1–2.2 kilojoules. Assuming that about 25 percent of the colony's foraging area is populated with *Polyommatus* larvae, in accordance with field observations, the energy amount obtained from lycaenid larvae is estimated to be 11–22 kilojoules, which is approximately a quarter to a half of the total monthly basal metabolic rate of the colony. Thus 100 *Polyommatus coridon* larvae could deliver enough sugars to cover the monthly energy needs (at 15°C) of 1,400 to 2,800 workers of *Tetramorium*. Similar calculations were obtained for the ants *Lasius alienus* and *Plagiolepis pygmaea*.

Pierce et al. (1987) examined other costs and benefits of ant associations for the lycaenid *Jalmenus evagoras*. They demonstrated that the degree of mortality caused by predation and parasitism is so severe as to force an obligate dependency on the part of the lycaenids: larvae and pupae deprived of attendant ants cannot survive. In addition to providing protection, attendant ants shorten larval development substantially while extending the duration of the pupal phase only slightly, thereby reducing the time that larvae are exposed to the threat of predators and parasitoids. On the other hand, energetic costs of associating with ants result in a reduction of adult weight and size, traits important to mating success in males and fecundity in females of *J. evagoras* (Elgar and Pierce, 1988).

Pierce et al. (1987) have provided two lines of evidence suggesting that the host ants receive substantial rewards for their efforts.

> First, pupae that are tended by ants for only 5 days lose 25% more weight than their untended counterparts, and develop significantly more slowly. This indicates that pupae are supplying rewards for ants by diverting metabolic resources from metamorphosis. Second, the mean dry weight of an individual worker of *Iridomyrmex anceps* is about 0.4 mg, and our estimate of biomass removal from a tree containing 62 juveniles was about 405 mg. If there is a 10% rate of biomass conversion from one trophic level to the next, then foraging on a single tree containing about 60 juveniles of *J. evagoras* can result in the equivalent production of almost 100 new workers of *I. anceps* in one day.

Finally, Pierce and her collaborators point out that aggregation supplies several benefits to larvae and pupae of *J. evagoras*.

> If a threshold number of ants is necessary to protect the larvae and pupae, then aggregating is one mechanism by which *J. evagoras* could simultaneously increase its collective defense and decrease the amount of food that each individual would need to produce to attract that defense. For example, first instars can gain more ants by joining a group of any size than by remaining alone, and solitary second and third instars can have a higher number of attendant ants by joining the mean size group of about 4 larvae. Moreover, aggregation is not automatic, but occurs in response to ants: young instars that are not tended by ants are less likely to form groups than their tended counterparts. It is interesting and probably significant that most lycaenid species that lay eggs in clusters and whose larvae aggregate have complex and apparently obligate associations with ants (Kitching, 1981; Pierce and Elgar, 1985).

Larval aggregations in lycaenids are mainly known from Australian and South African species (Clark and Dickson, 1971; Common and Waterhouse, 1972). One North American species was discovered by Webster and Nielsen (1984), who found that the third- and fourth-instar larvae of *Satyrium edwardsii* aggregate during the day at the base of the host plant within conical shelters of detritus constructed by the ant *Formica integra*. The lycaenid larvae feed nocturnally on the leaves and are usually attended by the ants. Other *Satyrium* species are also tended by ants but do not form larval aggregations.

The main cost of aggregation for the lycaenids may be competition. Indeed, Pierce and her co-workers found many larvae of *Jalmenus evagoras* that had starved on their host plants after consuming all the available foliage. Furthermore, larvae that occur in groups are likely to be more conspicuous to their predators and parasitoids. The costs and benefits of larval aggregation from the ants' perspective remain to be assessed experimentally, although the Pierce group note: "Ants benefit energetically from aggregation in *J. evagoras:* larvae and pupae that occur in clusters are easier to collect secretions from and to defend. However, if larvae and pupae can regulate the amount of secretion they produce then individuals in groups may be able to provide less food for ants than they would on their own while still receiving the same degree of defense. Furthermore, aggregations may become so attractive to predators (such as other ant species) that the ants themselves are endangered."

Thus it has now been convincingly documented that in the case of at least several myrmecophilous lycaenids, ants protect the larvae

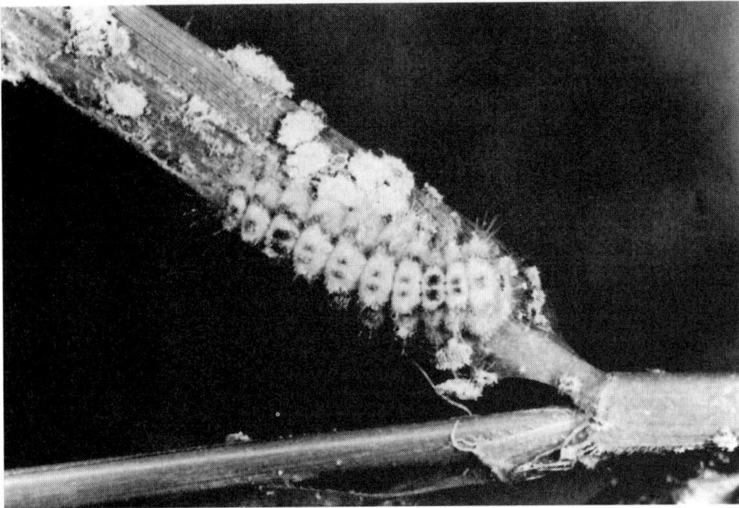

**FIGURE 13–36** A myrmecophilous butterfly (*Taraka hamada*) that feeds on aphids. (*Above*) A female oviposits beside a cluster of her aphid prey, *Ceratovacuna japonica*. The aphids are feeding on bamboo grass (*Pleioblastus china*). (*Below*) A late-instar larva feeds on aphids. *T. hamada* is entirely limited in diet to the aphids, never feeding on the aphid host plant. (Photographs courtesy of N. Pierce.)

and pupae against parasitoids and predators and in return they are rewarded with nutritious secretions provided by these guests.

The evolution of the myrmecophilous anatomical structures and behavior in the lycaenids has been discussed by Hinton (1951) and Malicky (1970a,b). Myrmecophily is judged a primitive trait among the living lycaenids, whereas its absence in some species is considered to be a secondary loss in evolution. The unusually thick cuticle of lycaenid larvae may be a first adaptation to myrmecophily. The Newcomer's gland is also primitive, having been lost in a considerable number of species as the result of decreased interaction with ants. Malicky proposed that the pore cupolas evolved after the Newcomer's gland. When pore cupolas became increasingly efficient as an appeasement device, the Newcomer's gland regressed. The overall evolutionary scheme by Hinton and Malicky has been endorsed at least in part by more recent authors, including Fukuda et al. (1978), Pierce (1983, 1987), and Henning (1983a,b), and challenged at least in part by Maschwitz and Fiedler (1988).

In addition to associating with ants, a number of lycaenids have changed from phytophagous to carnivorous habits, or have come to parasitize ants in other ways (Figure 13–36). Phytophagy in the Lycaenidae has been reviewed by Cottrell (1984), who argues that the switch from phytophagy to carnivory has arisen at least eight separate times within the Lycaenidae. Many members of the Miletinae, for instance, prey on homopterans. Several authors, including Fukuda et al. (1978), Cottrell (1984), and Maschwitz and Fiedler (1988), have suggested that the proclivity of lycaenids for areas inhabited by ants and their homopteran cultures may have led to the evolution of carnivorous life patterns.

Maschwitz et al. (1985b,c) studied three species of the miletine genus *Allotinus* in Malaysia, and Kitching (1987) observed a fourth species in Sulawesi. Maschwitz and his collaborators found adult males and females of the butterfly imbibing honeydew from ant-attended aphids, coccids, and membracids, and they observed females depositing eggs into the midst of loose aphid associations. The ants did not behave aggressively toward the honeydew-stealing butterflies, which suggests that the adult *Allotinus* enjoy some form of chemical appeasement protection. The larvae prey on the homopterans, and the older larvae of *A. unicolor* also imbibe honeydew. Miletine caterpillars in general are also endowed with chemical appeasement substances secreted from the pore cupolas, but they lack the nectar organs (Malicky, 1969; Kitching, 1987). In some species the larvae are visually camouflaged, resembling the substrate in which they live. This feature is more likely to be a protection against vertebrate predators than against ants, however.

Another mode of parasitic behavior in lycaenids has been described by Maschwitz et al. (1984). Many species of the paleotropic plant genus *Macaranga* (Euphorbiaceae) live in symbiosis with the myrmicine genus *Crematogaster.* The ants protect the plants from many herbivorous enemies, and in return the plants provide food bodies and nesting space, and they make possible the culturing of honeydew-producing homopterans (see Chapter 14). Maschwitz et al. discovered that several species of the lycaenid genus *Arhopala* intrude parasitically into this symbiosis system. With the aid of their myrmecophilic organs (pore cupolas, nectar organ, and tentacle organ) the caterpillars overcome the aggressiveness of the ants and feed on the *Macaranga* leaves without disturbance. Moreover, the caterpillars and their pupae are protected against predators and parasitoids by the ants. Maschwitz et al. describe the intricate balance of costs and benefits as follows.

> The secretion of the Newcomer's gland fluid seems to play only a minor role when compared to the honeydew excretion of the scales. This means that these larvae are pests of the plant and by that they harm the scales as well as the protecting ants. There is no severe damage of the plants because the eggs are only oviposited in low numbers. This is a typical pattern of well adapted parasites, which do not destroy their hosts. Against vertebrate enemies which are searching for their prey the larvae are protected by camouflage. There is a strong species specific host plant selection by the butterflies.

The evolution and maintenance of truly mutualistic symbioses between lycaenids and ants depend on the costs and benefits of this relationship for both partners (Pierce and Young, 1986; Pierce, 1987). Mutualism has been a conspicuously successful strategy for the butterflies. By far the majority of all lycaenid species have evolved in those subfamilies where the larvae are endowed with pore cupolas, nectar organs, tentacular organs, and other myrmecophilous adaptations.

## THE TROPHOBIONTS

A great majority of the members of the three phylogenetically most advanced ant subfamilies, the Myrmicinae, Dolichoderinae, and Formicinae, attend homopterans to some extent. To employ one last term from Wasmann, the ants can be said to have entered into *trophobiosis* with the homopterans. As trophobionts, the homopterans resemble many of the lycaenid symbionts in a basic way: they obtain their own food and pass some of it on to their hosts. Unlike the lycaenids that secrete substances from specialized exocrine glands, however, the honeydew provided for ants by homopterans is an excretion derived from a digestive process (see Plate 17).

When aphids feed on the phloem sap of plants, they pass a complex mixture of nutrients, including sugars, free amino acids, amides, proteins, minerals, and vitamins, through the gut and back out through the anus. During this passage the phloem sap changes as some of its components are absorbed and others are converted or added by the aphid (for reviews see Ziegler and Penth, 1977; Kunkel and Kloft, 1977; Dixon, 1985; Maurizio, 1985). According to Maurizio, 0.2–1.8 percent of the honeydew dry weight is nitrogen and 70–95 percent of the nitrogen is amino acids and amides. The mixture of nitrogen compounds in the honeydew is largely identical to that in the phloem sap. Measurements made on *Tuberolachnus salignus* by Mittler (1958) show that as much as half of the free amino acids are absorbed by the aphid's gut. In a few cases the honeydew contains amino acids which are not present in the phloem sap. Presumably these are metabolic products added by the aphids (Gray, 1952; Ehrhardt, 1962).

By far the largest percentage (90–95 percent) of the honeydew dry weight consists of carbohydrates. Sugars from the phloem sap are partly absorbed or converted, and the diverse mixtures of sugars contained in the honeydew are often species specific. They usually comprise fructose, glucose, saccharose, trehalose, and higher oligosaccharides. Trehalose, which is the blood sugar of insects, composes up to 35 percent of the total sugar amount in the honeydew. Typical honeydew sugars also include the trisaccharides fructomaltose and melezitose, with the latter making up 40–50 percent of the total sugar. Other sugars detected in honeydew are maltose, raffinose, melibiose, turanose, galactose, mannose, rhamnose, and stachyose. In addition the honeydew contains other classes of substances, including organic acids, B-vitamins, and minerals.

When unattended by ants, many aphids dispose of the honeydew droplets by flicking them away with their hind legs or caudae, or by expelling them through contractions of the rectum or entire abdomen. The honeydew then falls upon the vegetation and ground below. Similar substances are excreted by several other groups of sap-feeding Homoptera, including scale insects (Coccidae), mealybugs (Pseudococcidae), jumping plant lice (Chermidae = Psyllidae), treehoppers (Membracidae), leafhoppers (Cicadellidae), froghoppers or spittle insects (Cercopidae), and members of the "lanternfly" family (Fulgoridae). Sometimes honeydew accumulates in large enough quantities to be usable by man. The manna "given" to the Israelites in the Old Testament account was almost certainly the excretion of the coccid *Trabutina mannipara*, which feeds on tamarisk. The Arabs still gather the material, which they call "man." In Australia chermid honeydew, or "sugar-lerp," is collected as food by the aborigines. Up to three pounds can be harvested by one person in a single day. It is no surprise, therefore, to find that ants also gather honeydew of all kinds. Many, perhaps most, species collect it from the ground and vegetation where it falls. But many others have developed the capacity to solicit the honeydew directly from the homopterans themselves.

Most aphid species associated with ants insert their stylets into the phloem of the host plant (Kloft, 1953, 1959a, 1960a–c; Kunkel, 1967). Although they can suck up limited amounts of liquid, the aphids appear to depend chiefly on the turgor pressure of the phloem, which forces sap up their stylets (Mittler, 1957; Kunkel and Kloft, 1985). To process a large volume of phloem sap and discard the excess as honeydew evidently costs the aphid fewer calories than would a more nearly total extraction from smaller quantities of sap. The amounts of honeydew produced by individuals are often prodigious. First-instar nymphs of Mittler's *Tuberolachnus* extracted honeydew at the rate of 7 droplets per hour, each droplet containing 0.065 microliter, and the total output per aphid was 133 percent of the aphid's weight every hour. The extractions of other species that have been analyzed are more modest, with an hourly output ranging from 1.9 to 13.3 percent of body weight per hour (Auclair, 1963).

Most myrmecophilous homopterans, especially aphids, have special structural and behavioral adaptations for life with ants (Way, 1963; Kunkel, 1973; Kunkel and Kloft, 1985). Aphids frequently associated with ants tend to have poorly developed cornicles, a reduced cauda, and at most a thin coating of wax filaments. A few ant-attended species have large cornicles, however, and others are densely covered with wax. In the case of one such species, *Prociphilus fraxini*, the ants simply remove wax from the bodies of the aphids (Zwölfer, 1958; Kunkel, 1973). And where most pseudococcids are covered with wax, the ants often clean the symbionts bare. Kunkel (1973) notes that myrmecophilous aphids generally have more setae on the dorsal body surface and tibiae. Anal setae in particular are very numerous in myrmecophilous aphids. They form a basket ("trophobiotic organ") that holds the honeydew droplet until it is imbibed by the ants (Zwölfer, 1958; Kunkel, 1973).

Experiments by El-Ziady and Kennedy (1956) and Johnson and Birks (1960) revealed that the presence of ants belonging to the genera *Lasius* and *Paratrechina* delays the production of alates in populations of species of *Aphis* and hence postpones their dispersal and, from that, increases their population density. In addition alate aphids have long wings that render it more difficult for the ants to collect the honeydew droplets from the anus. Kleinjan and Mittler (1975) demonstrated that if extracts of the mandibular glands of *Formica fusca* are applied to aphid nymphs, the proportion of the population exhibiting winglessness increases. Since winglessness is caused by juvenile hormone, the ants' mandibular gland secretions may well contain material simulating juvenile hormone. Finally, Kunkel (1973) reported that the wings of alate aphids are sometimes bitten and crumpled by their ant guardians.

Numerous authors, beginning with Pierre Huber (1810), have reported that honeydew ejection of certain homopterans has been

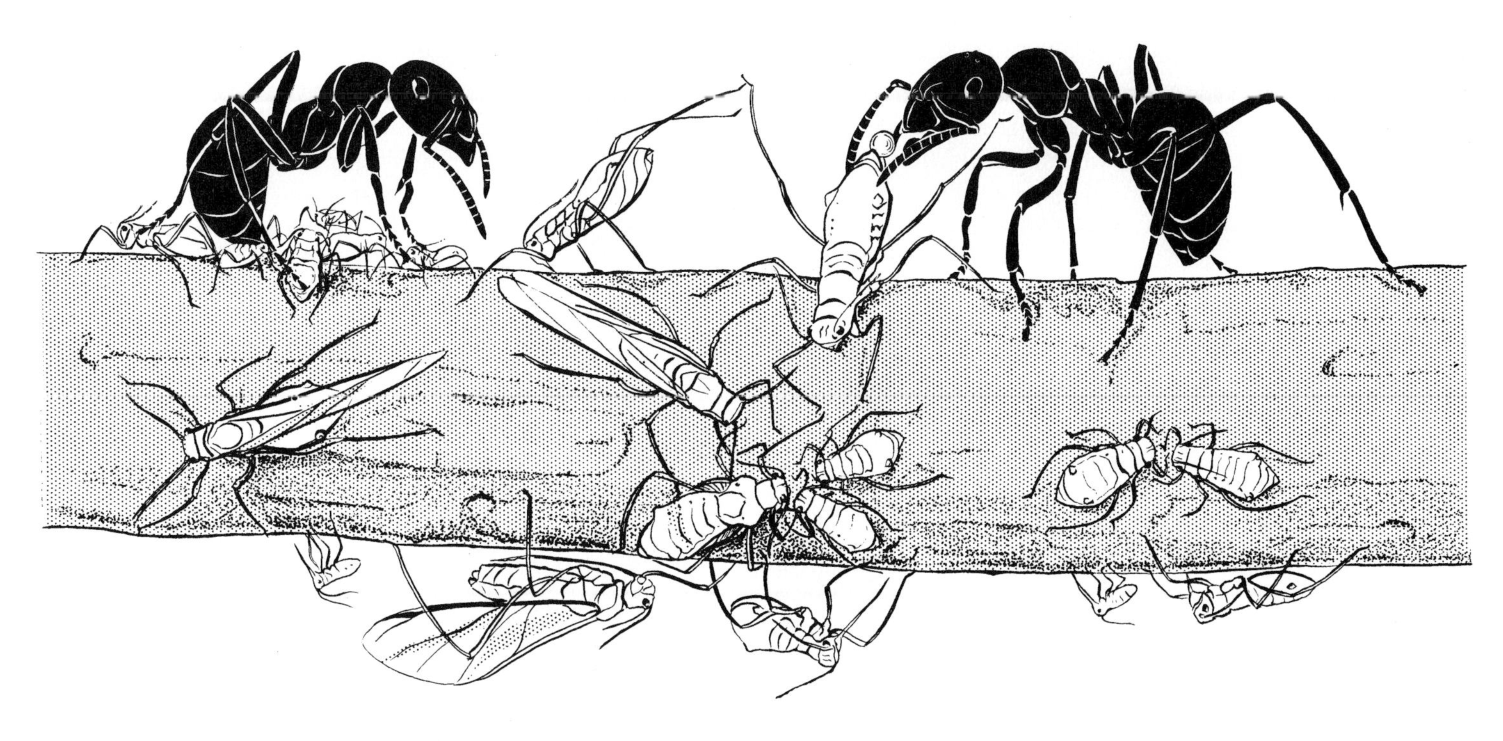

**FIGURE 13–37** Workers of *Formica polyctena* attending aphids (*Lachnus roboris*). (From Wilson, 1971; drawing by T. Hölldobler-Forsyth.)

modified under the attendance of ants (see reviews by Way, 1963; Kunkel, 1973). As a rule symbiotic homopterans ease out the droplets of honeydew when solicited by ants rather than ejecting them away from their bodies. Individuals of the black bean aphid *Aphis fabae* show the following typical specialized responses in the presence of ants: the abdomen is raised slightly, the hind legs are kept down instead of being lifted and waved as in unattended aphids, and the honeydew droplet is emitted slowly and held on the tip of the abdomen while it is being consumed by the ants. If a droplet is not accepted, the aphid will often withdraw it back into the abdomen. Kunkel (1973) interprets the repeated alternate extrusion and withdrawal of honeydew droplets as a signal to the ants that the aphid is ready to defecate. In many cases the required stimulus for honeydew emission is a simple, mechanical one. According to Mordwilko (1907) and Kunkel (1973), many (but not all) of the symbiotic aphid species can be induced to emit a droplet merely by brushing the abdomen with some delicate object in imitation of the caressing movements of the ant's antennae and forelegs.

In other species the aphids appear to exchange mechanical signals with the ants prior to releasing a honeydew droplet. For example *Lachnus roboris* lifts the hind legs when an ant approaches (Figure 13–37). Kloft (1959a, 1960a) has made the intriguing suggestion that the rear of the aphid's abdomen resembles the head of an ant worker offering food, with the cornicles representing the opened mandibles, the cauda symbolizing the ant's extruded labium, and the waving of the aphid's hind legs providing an imitation of the antennal movements of the ant. These stimuli induce the visiting worker to mistake the aphid for a donor ant in the special way that any animal makes a mistake when confronted with a small but vital set of releasers out of context. The solicitation that follows, according to Kloft, is just a slight perversion of the ordinary trophallaxis occurring between sister workers. Kloft hypothesizes that this interspecific communication behavior in aphids evolved from a nonspecific defensive behavior. Approximately 75 percent of all species of the *Aphidina*, when disturbed, kick with their hind legs (Kunkel, 1973). Trophobiotic associations evolved independently many times within the *Aphidina* and in some cases the leg kicking may have been ritualized to become an interspecific trophobiotic communication signal. This ingenious hypothesis still needs to be tested experimentally. In any case interspecific communication is not a general prerequisite of trophobiotic interactions. Coccids and mealybugs are attended with equal fervor and precision, yet their appearance and behavior make them seem wholly different from their ant hosts (Eckloff, 1978; Figures 13–38 and 13–39). It is entirely possible that the subtle resemblances between certain aphids and the heads of ants are just coincidental.

The symbiotic relationship between aphids and ants is based on interspecific communication that is not just tactile but chemical. Certain aphids closely associated with ants appear to possess materials apart from honeydew that are attractive to the host ants. Moreover, there exists some circumstantial evidence that ants mark their aphid populations with colony-specific substances, in effect incorporating them into the nest territory (Eckloff, 1976, cited in Kunkel and Kloft, 1985).

Another form of interspecific chemical communication has been reported by Nault et al. (1976). When aphids are disturbed they discharge defensive and alarm substances from their cornicles. The materials include triglycerides, which are sticky and can impede predators, and trans-β-farnesene, which induces nearby aphids to fall off the plant and escape. The secretions also alert the host ants, which then are quick to attack foreign invaders. In this case aphids do not disperse so readily in response to the alarm substances, but remain close by and benefit from the ants' defensive behavior.

A remarkable example of a predator that goes to great lengths to circumvent ant defense of aphid symbionts is the larva of the green lacewing *Chrysopa glossonae* (Eisner et al., 1978). This insect preys

**FIGURE 13–38** "Stable nest" of the African weaver ant *Oecophylla longinoda*, where the minor workers attend scale insects and collect their honeydew excrement as food. The minors regurgitate the honeydew to major workers, who then transport it into the brood nest for final consumption. (From Hölldobler, 1984d; painting by J. D. Dawson reprinted with permission of the National Geographic Society.)

upon the woolly alder aphid *Prociphilus tesselatus*, a ward of several species of formicine ants. The aphids derive their common name from the filaments of waxy "wool" that cover their bodies. The larvae disguise themselves by "plucking" some of this material from the bodies of the aphids and applying it to their own backs. In other words they employ the "wolf-in-sheep's-clothing" strategy to fool the ant shepherds that guard the aphids.

The life cycle of homopterans, which Zwölfer and others have documented thoroughly, crucially affects the success of the trophobiosis. The ideal trophobiont is one whose life cycle is not tightly synchronized, so that stages capable of producing honeydew are available throughout the year. The evidence shows that not only do taxa possessing these properties as original traits become trophobiotic more frequently, but some species acquire or enhance them secondarily in evolution after the association with ants has begun.

The ant-homopteran symbiosis is probably of ancient origin. Wheeler (1914b) found aphids associated with *Iridomyrmex goepperti* in a block of Baltic amber, which is considered to be of early Oligocene age. *Iridomyrmex*, like many of the other ant genera common in the Baltic amber and other early Tertiary fossil deposits, has survived relatively unchanged in external morphology since as far back as mid-Eocene times, 50 million years ago. Its living species still attend honeydew-producing homopterans.

Whether the homopteran-ant association bestows mutual benefits on both symbiotic partners has long been a subject of controversy. No doubt the extent of honeydew utilization varies greatly among ant species. In the highly predaceous myrmicine *Daceton armigerum*, it is only an occasional event (Wilson, 1962a). In contrast, honeydew collected from aphids is a major part of the diet of the common wood ants of Europe (*Formica rufa* group). Zoebelein (1956) reported that one large colony of *F. rufa* can collect as much as 500 kilograms of honeydew in a year. Gösswald (1954) and Kloft (1953) state that the amino acids in the honeydew are adequate for maintaining colonies of *F. polyctena* during periods when insect prey is scarce. According to Horstmann (1970, 1972, 1974, 1982a), who investigated food intake and energy budgets in *F. polyctena*, honeydew is an essential part of the wood ants' nutrition. The same conclusion has been reached by Skinner (1980b) and Degen et al. (1986). Moreover, aphid colonies attended by wood ants often produce large surpluses of honeydew, which are then harvested by honey bees and other insect species (Wellenstein, 1977).

Because of the prodigious appetites of symbiotic aphids and their potential impact on forests, economic entomologists have conducted intensive research on all conceivable aspects of trophobiosis, and a large and specialized literature on the subject has accumulated through the years. Reviews that mark successive stages in the development of the subject and yet are only partly overlapping in empirical content have been provided by Büsgen (1891), Wheeler (1910a), Jones (1929), Herzig (1937), Nixon (1951), Wellenstein (1952, 1977), Kloft (1960a–c), Auclair (1963), Way (1963), O'Neill and Robinson (1977), Dixon (1985) and Kloft et al. (1985). From this research we can conclude that the ants derive great benefits of some sort from the association with homopterans.

But what do the homopterans gain in turn? As Strickland (1947) and Kunkel (1973) have pointed out, the ants render a hygienic service to their symbionts by constantly removing the large quantities of sugary material. The ants also provide protection from weather by building shelters around homopteran colonies (Way, 1963; Maschwitz et al., 1985a). Their presence stimulates the aphids to grow and mature more rapidly (El-Ziady and Kennedy, 1956; Banks and Nixon, 1958; El-Ziady, 1960). Altogether, the homopterans develop larger, more stable populations and increase their rate of dispersal. Of equal significance, the ants protect aphids against predators and parasitoids. Still, our knowledge concerning this particular benefit is based mostly on anecdotal accounts, and only a few quantitative ecological studies have been conducted to the present time (Way, 1954b, 1963; Brown, 1976; Wood, 1977, 1982; Addicott, 1979; Fritz, 1982; Bristow, 1984). Kunkel (1973) suggested that the mutu-

**FIGURE 13–39** Workers of *Formica polyctena* attend a large scale insect (*Eulecanium coryli*), which extrudes a droplet of honeydew for the ants to consume. (From Wilson, 1975b; photograph by T. Hölldobler-Forsyth.)

alistic relationship between ants and homopterans is specific, that is, the ants are not always interchangeable as guardians. Bristow (1984) experimentally proved this to be the case. She studied colonies of aphids *(Aphis vernoniae)* and membracids *(Publilia reticulata)* that feed on the ironweed *Vernonia noveboracensis* and are tended by the ants *Tapinoma sessile, Myrmica lobicornis,* and *M. americana.* By controlled exclusion experiments she demonstrated that the ants generally had a positive effect on the survivorship of aphid colonies. Aphid colonies tended by *Tapinoma* survived significantly longer than those tended by *Myrmica,* however. Conversely, the membracid colonies benefited more from an association with *Myrmica* than with *Tapinoma.*

Bristow (1983) made another notable discovery concerning ant-membracid symbiosis. It is known that females of some species of treehoppers that tend their own eggs and nymphs abandon their offspring either earlier or later than usual when in the presence of ants (Wood, 1977; McEvoy, 1979). In the case of *Publilia reticulata,* Bristow demonstrated that females tended by ants abandoned their brood significantly sooner than did females from membracid colonies from which ants were excluded. Ten of 25 marked females from ant-tended treehopper colonies were found with new egg masses on other parts of the same plant, but only 1 of 15 females from the colonies in which ants were excluded deserted the first brood to produce a second brood on the same stem. In addition, "ten of 24 marked females from ant-tended colonies were relocated with additional broods on new plants; no females from colonies without ants were found on new plants." Bristow concludes from her experiments that when ants are protecting the *Publilia* broods, females incur no penalty for deserting their own young and may substantially increase their fitness by producing additional clutches. When no ants are present, however, abandonment results in lowered survival of nymphs. In the absence of ants, females find it advantageous to remain with their original brood and provide close protection. As McEvoy (1979) pointed out, treehopper females may also serve to attract ants to the small nymphs, and Bristow proposes that this may be a major role of parental care in *P. reticulata.*

No doubt remains that symbiotic interactions between ants and homopterans are truly mutualistic, in other words that they benefit both partners. There remains one aspect of the symbiosis that has not been given much attention, however: the homopterans' host plants may benefit from the presence of the ants (Gösswald, 1951a, 1985; Nickerson et al., 1977; Laine and Niemelä, 1980; Skinner and Whittaker, 1981). In fact the ants may easily provide one additional service to the homopterans by protecting the host plant from more damaging non-homopteran herbivores.

This effect has been examined by Messina (1981). The membracid *Publilia concava* forms aggregations of nymphs and adults at the base of leaves of the goldenrod plant, *Solidago altissima.* The membracids are tended by several ant species, but the females appear to oviposit preferentially on stems surrounding *Formica* mounds. Two chrysomelid beetles, *Trirhabda virgata* and *T. borealis,* are the dominant herbivores of goldenrod. Both larvae and adults sometimes occur at densities exceeding 35 individuals per stem and can cause complete defoliation. Messina discovered that plants occupied by *P. concava* that were tended by *Formica* escaped defoliation by *Trirhabda.* They attained greater mean height and produced more seeds than their nearest neighbors lacking ants. The plants are probably not truly coevolved partners in the mutualism, however. Indeed, Buckley (1983) found that the interaction between ants in the genus *Iridomyrmex* and the membracid treehopper *Sextius virescens* feeding on *Acacia decurrens* imposed an overall negative effect on plant growth and seed set. Messina himself concluded:

> While plant protection seems to be an extension of the ant-Homoptera mutualism, and is not strictly fortuitous, this phenomenon should not be interpreted as an evolutionarily beneficial relationship between membracids and goldenrod (in the sense of an "indirect mutualism" [Vandermeer, 1980]). In goldenrod stands with few or no beetles, plants with the phloem-feeding membracid colonies may be at a relative disadvantage. Membracid feeding and oviposition cause goldenrod leaves to senesce prematurely and become appressed to the stem, possibly reducing photosynthetic capacity. Unlike plants in typical ant-plant mutualisms, goldenrods possess no adaptations to attract membracids and accompanying ants. Even though ants exclude herbivores from homopteran host plants, they significantly increase the negative impact of the Homoptera themselves, by increasing Homoptera density and feeding rates.

The possibility exists nonetheless that in some of the most advanced tropical ant plants a coadaptation has arisen between the plants on one side and the ants and homopterans that exist in obligatory mutualism with them on the other (Janzen, 1979; see Chapter 14).

Ants are not the only organisms that attend homopterans. According to Salt (1929), stingless bees of the genus *Trigona* collect honeydew directly from membracids in Brazil, and at least one species "tickles" the treehoppers to induce flow. Belt, in *The Naturalist in Nicaragua* (1874), observed social polybiine wasps of the genus *Brachygastra* attending membracids. "The wasp stroked the young hoppers, and sipped up the honey when it was extruded, just like the ants. When an ant came up to a cluster of leaf-hoppers attended by a wasp, the latter would not attempt to grapple with its rivals on the leaf, but would fly off and hover over the ant; then when its little foe was well exposed, it would dart at it and strike it to the ground." Similarly, Letourneau and Choe (1987) observed workers of the polybiine wasp *Parachartergus fraternus* attending the planthopper *Aetalion reticulatum* and the membracid *Aconophora ferruginea* in Costa Rica. The presence of the wasps inhibited the approach of ants, and the planthoppers clearly preferred wasps to ants as attendants.

Silvanid beetles of the Neotropical genera *Coccidotrophus* and *Eunausibius* attend pseudococcids in the hollow leaf petioles of *Tachigalia* (Wheeler, 1928). A few lycaenid butterflies milk homopterans while in the adult or larval stage. Bingham (1907) reports that adults of *Allotinus horsfieldi* solicit honeydew from aphids by stroking them with their forelegs. Maschwitz et al. (1985b) were not able to repeat this observation but found that the imago of the related species *A. unicolor* imitates the mechanical stimulation normally given by ants. It uses the tip of its proboscis, and as a result is able to obtain honeydew. On the other hand, larvae (not the adults) of *Lachnocnema bibulus* depend entirely on honeydew solicited from membracids and jassids and possess modifications in the forelegs that are apparently adapted to this single purpose (Hinton, 1951). Certain flies (*Revellia quadrifuscata*) solicit honeydew from membracids and other homopterans by drumming on the backs of the homopterans with their fore tarsi (Andrews, 1930). Most of these cases, however, are trophobiotic parasites that "steal" the honeydew from the homopterans without rendering any service in return (Kunkel and Kloft, 1985).

Trophobionts occur in a few groups other than the Homoptera and lycaenid butterflies. Green (1900) recorded a case of a hemipteran bug, *Coptosoma* sp. (Plataspididae), being attended in Sri Lanka by *Crematogaster* workers. In Malaysia, Maschwitz and his collaborators discovered two new species of plataspid bugs (*Tropidotylus servus, T. minister*) and a new coreid bug (*Hygia aliens*) associated with the myrmicine ant *Meranoplus mucronatus*. Another coreid species of the genus *Cloresmus* was found on bamboo being attended by a species of *Crematogaster* (Maschwitz and Klinger, 1974; Maschwitz et al., 1987). These several species of "true" bugs imbibe phloem sap and emit honeydew through the anus when tactilely stimulated by the ants. Maschwitz et al. also observed that the ants protect the bugs and their nymphs by biting or repelling potential predators. Finally, a new type of trophobiosis between tortricid larvae of a previously undescribed genus (*Semutophila saccharopa*) and various species of the subfamilies Formicinae, Dolichoderinae, and Myrmicinae has been found in Malaysia (Maschwitz et al., 1986a). The larvae live in silken shelters fixed to the leaves of bamboo. In response to mechanical stimuli from the ants, they discharge an anal liquid that contains sugar and amino acids.

## HERDERS AND NOMADS

The extreme myrmecophilous homopterans have evolved to the status of domestic "cattle." Like artificially bred animals, aphids in particular have reduced or lost the usual defensive structures found in free species, including the modification of the legs for jumping. On the other hand, the fact that ants do not normally attack and eat their homopteran associates is evidence by itself that behavioral evolution has occurred that accommodates them to the mutualism. Their mode of soliciting honeydew is essentially the same as the procedure used to initiate regurgitation within the colony. But the fact that it is directed at totally alien arthropods suggests the existence of a second major behavioral adaptation on the part of the ants. Some species of ants go so far as to care for the homopterans inside their nests.

In an early, classic study, Forbes (1906), discovered that the eggs of the American corn-root aphid *Aphis maidiradicis* are kept by colonies of the ant *Lasius neoniger* in their nests throughout the winter. The following spring the newly hatched nymphs are transported to the roots of nearby food plants. If the host plants are uprooted, the ants move the aphids to undisturbed root systems in the vicinity. During the late spring and summer, some of the aphids transform into alates and disperse on their own. After they have settled and begun to feed, they may be adopted by other ant colonies in whose territories they happen to have fallen. When eggs are tended in this way, they are often mixed with the host brood. Also, when the nest is disturbed, the ants pick up their homopterans and transport them to safety in a manner indistinguishable from the rescue of their own brood. Ant workers chase potential predators and parasites away from the homopterans but, again, in the same way that they protect their nests and inert masses of food.

The similarity of homopteran care to ordinary social behavior led Herzig (1937) and Nixon (1951) to question whether the ant-homopteran symbiosis is really an advanced mutualism involving extensive coadaptation. Herzig went so far as to suggest that most of the behavior of the ants toward aphids can be explained as a compromise between two opposing primitive motivations: avoidance because of unpalatability of their flesh, and attraction because of their honeydew. There is now, however, overwhelming evidence to show that the ants do respond to the aphids in specialized ways that can only represent an adaptation on their part. It has been established that workers of at least some ant species carry their homopteran guests to the appropriate part of the food plant, and at the correct stage of the trophobionts' development. Such behavior has been documented, for example, in the case of *Acropyga* and its root coccids by Bünzli (1935), in *Oecophylla* and *Saissetia* by Way (1954b), and in *Lasius* and *Stomaphis* by Goidanich (1959). Even more impressive is the fact that the queens of certain ant species carry coccids in their mandibles during the nuptial flight. This habit, which has no parallel in the behavior of non-coccidophilous ants, has been observed in species of *Cladomyrma* in Sumatra by Roepke (1930), *Acropyga paramaribensis* in Suriname by Bünzli (1935), and an unidentified formicine (possibly *Acropyga*) in China by Brown (1945). It was recorded most recently from Europe by Buschinger et al. (1987), in what was originally thought to be a species of *Plagiolepis* but now has been identified as *Acropyga nearctica* (Buschinger, personal communication).

Special note must be taken of the extraordinary transport behavior shown by the genus *Hippeococcus*. These pseudococcids, described from Java as a new genus by Reyne (1954), are kept by ants of the dolichoderine genus *Hypoclinea* in underground nests and on trees and shrubs nearby. When disturbed, the small *Hippeococcus* climb onto the bodies of the ants or else are gathered by the ants in their mandibles. Ant riding is made possible by the pseudococcids' long, grasping legs and flat, sucker-like tarsi (Figure 13–40).

Extreme trophobiosis is practiced by certain subterranean ants believed to be totally dependent on root aphids and coccids. The best-known examples are two formicine genera: *Acanthomyops*, which is restricted to North America (Wheeler, 1910a; Wing, 1968), and *Acropyga*, a pantropical genus studied in Central and South America by Bünzli (1935) and Weber (1944a). In neither case has it been established beyond doubt that the colonies live entirely off their "cattle," but the circumstantial evidence is strong that they can do so. As Way (1963) pointed out, the data are not always adequate to determine the matter. Honeydew may or may not suffice by itself. It is possible that *Acanthomyops* and *Acropyga* obtain extra protein by eating some of their homopterans, and we know that cropping of homopterans does occur under at least some circumstances. For example, Way (1954b) found that laboratory colonies of the weaver ant *Oecophylla longinoda*, when presented with excessive members of its trophobiotic coccid, killed and removed individuals until the population had reached the level required for a sufficient but not excessive outflow of honeydew. We have made similar laboratory observations in *O. longinoda*. Finally, Maschwitz et al. (1985a) report that workers of the Southeast Asian weaver ant *Camponotus (Karavaievia) texens* also remove excessive coccids and often transport the symbionts to newly built shelters. These observations and a number of anecdotal reports in the literature suggest that the transport of homopteran symbionts by ants may be quite common worldwide and play a major role in the colonization of new food plants (Janzen, 1969, 1972, 1973a,b; Duviard and Segeren, 1974; Stout, 1979; Schremmer, 1984).

The most remarkable trophobiotic association of all was recently discovered in Malaysia by Maschwitz and Hänel (1985). It entails a completely new mode of life in which the ants can be said to function as migrating herders or nomads. Neither partner can survive for long without the other. Maschwitz and Hänel point out: "Nomads are stock farmers who subsist from their livestock and who closely coordinate their life style with that of their livestock, for instance, by following them to the pastures these need. Several ants with a predatory mode of life, especially the various army ants of the New and Old World tropics, who frequently change their nesting site and thus enter new hunting areas, have been called nomadic. However, these migrating hunters are not true nomads in the genuine sense of the word."

In the following discussion we describe the findings by Maschwitz and Hänel (1985), which document for the first time the existence of true nomads in the animal kingdom. In the rain forest of the Malaysian peninsula the ant *Hypoclinea cuspidatus* lives in obligatory symbiosis with the mealybug *Malaicoccus formicarii* (Pseudococcidae). The ant has also been found with other *Malaicoccus* species, including *M. takahashii, M. moundi,* and *M. khooi,* as well as occasional coccids and membracids. *Malaicoccus* mealybugs, on the other hand, have only been found with *H. cuspidatus,* with one minor exception: *M. khooi* has been partly associated with *H. tubifer,* which has a life pattern similar to that of *H. cuspidatus.* The mealybugs live on many different species of monocotylous and dicotylous angiosperms. They feed exclusively on the phloem sap of young parts of the plant, which are rich in amino acids and sugars. The *H. cuspidatus* workers carry the mealybugs to the feeding sites, some of which are more than 20 meters from the ants' nest. The nests are located in the dense vegetation between leaves or in preformed cavities in wood or soil and consist solely of the bodies of the ants themselves: the workers cling to each other, forming a solid mass that shields the brood and mealybugs. Adult female mealybugs, which are viviparous and give birth to their offspring in these bivouacs, occur commonly among the ant brood.

A mature colony of *H. cuspidatus* contains more than 10,000 workers, about 4,000 larvae and pupae, more than 5,000 mealybugs, and 1 ergatoid queen. The bivouacs and the feeding sites are connected by a trail system, and about 11 percent of the ants running on the trail carry a mealybug between their mandibles. Since young plant shoots are quickly exhausted, new feeding sites have to be located frequently by the ants and the "grazing herds" shifted to them:

> In one instance a main feeding tree broke in two and the ants had to look for new sites. In one day they removed all the mealybugs from the tree and brought them to the nest. Half a day later a completely new trail system branched off from the original route to end up in new trees and twigs. All the sites on leaves and branches which were too old were abandoned by the ants some hours later. On the third day they had all found new feeding sites for their mealybugs. Two days later the whole area was deserted by the colony. Feeding site colonization could be triggered experimentally by cutting off all food plants of one colony and putting them down close to the nest. The ants carried their partners to an intermediate depot within 3 hours and brought them back to the colony later. After a further 2.5 hours there was a new intermediate mealybug assembly in the vicinity of the nest. Within 24 hours they colonized new feeding sites from this point. These were partly deserted after 24 hours and more new sites were colonized.

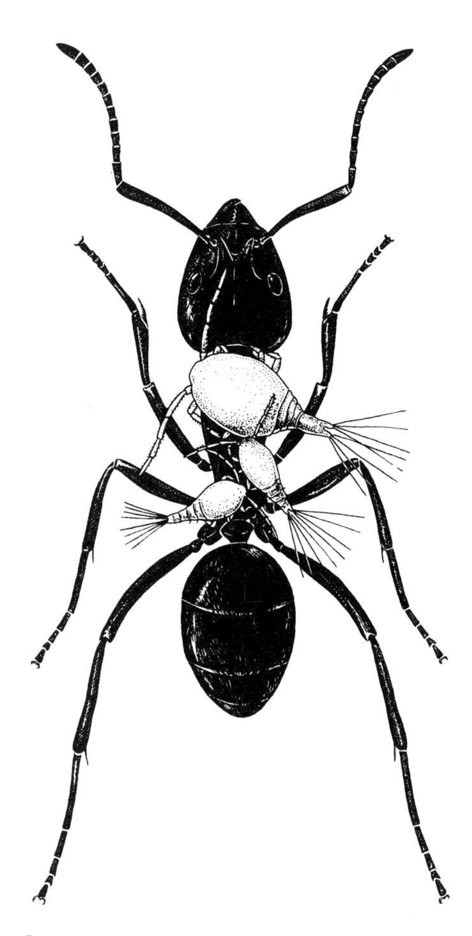

**FIGURE 13–40** Javan pseudococcids of the genus *Hippeococcus* escape from danger by climbing onto the backs of their host ants and allowing themselves to be carried to safety. Their legs and tarsi are apparently specially modified for this purpose. Here three individuals are shown being carried by a worker of *Hypoclinea gibbifer.* (From Wilson, 1971; modified and redrawn from Reyne, 1954.)

When the distance between the nest and the feeding site becomes too large, the colony simply moves to the feeding site, and during the process the brood and the mealybugs are carried along in a well-organized manner. During these emigrations no mealybugs are transported to the feeding sites, but the establishment of intermediate depots is quite common. A shift of nest sites can also be induced by disturbances or by a change in the microclimate in the vicinity of the nest. On the other hand there is no periodicity in nest moving. The frequency varied during the Maschwitz-Hänel study from one or two emigrations per week to none at all during 15 weeks.

At the feeding sites the mealybugs are always attended by the *Hypoclinea* workers, which continuously harvest the honeydew droplets emitted from the anus of the homopterans.

> When we observed the mealybugs at the feeding sites, which was rather difficult because of the permanent layer of ants covering the pseudococcids, we noticed small drops of liquid oozing out of the anus of the mealybugs. They adhered to the numerous long bristles (trophobiotic organ) on the body and were immediately consumed by one of the ants. When tested with a sugar reagent (Dextrostix, Merck) large amounts of sugar obviously originating from the plant phloem were detected. The honeydew was always excreted spontaneously. Its release was not induced by antennal drumming as in many homopterans visited by ants.

When the feeding aggregations are disturbed, both the ants and the mealybugs begin to move about excitedly. Single mealybugs crawl on top of ants, only to be removed quickly with the mandibles. Whereas small mealybugs are merely picked up, often several at the same time, the larger ones raise their bodies, thus inviting the ants to pick them up. During transport the mealybugs remain motionless, except to caress the heads of the ants with light movements of their antennae.

Maschwitz and Hänel observed neither hunting behavior nor active search for any other protein source in the nomadic *Hypoclinea cuspidatus*. Hence these ants appear to depend almost entirely on the nutrition obtained from the honeydew of their trophobiotic partners. When kept without mealybugs, the colonies observed by Maschwitz and Hänel declined rapidly. And conversely, the *Malaicoccus* colonies perished when isolated experimentally from their tending ants. Without ants

> the mealybugs walked around for hours without any feeding or gathering behavior. Later, shortlived assemblies were formed in which some animals produced very large drops of honeydew which accumulated in their anal bristles. As these drops were not removed by *H. cuspidatus* as usual, they fell into the mealybug assemblies contaminating and killing some animals. In the course of the experiment an increasing number of mealybugs left the food plant or fell from it and died on the ground, possibly due to starvation. After 3 days most animals had abandoned the plant. A small number were still actively sucking, which proved that the plants were still in a satisfactory state.

Surprisingly, the *M. formicarii* were not accepted as trophobiosis partners by other ant species. Instead they were attacked and carried into the nest as prey. Workers of a *Crematogaster* species were even observed recruiting fellow huntresses to the abandoned mealybugs. Obviously, then, *M. formicarii* is unable to survive without its trophobiotic partner.

# CHAPTER 14

# *Symbioses between Ants and Plants*

Odoardo Beccari, in his pioneering monograph on myrmecophilous plants (1884), reported that the East Indian pitcher plant *Nepenthes bicalcarata* harbors ant colonies in the hollow stem of the same pitcher-shaped leaf by which it captures and digests other kinds of insects (Figure 14–1). The ants are free to roam over the carnivorous plant and adjacent terrain, gathering insects and other food items of their own. If this relationship is verified, the ants and the plant appear to be engaged in a trade-off of mutual benefit (Jolivet, 1986). The ants risk being eaten by the plant but they get a home; the plants surrender some tissue space and insect prey to the ants but they gain some protection from herbivores.

The *Nepenthes* story is only one, admittedly very peculiar case among hundreds of ant-plant symbioses documented in the past 150 years of research. This topic has been the subject of rich and informative reviews in the 1980s, including systematic accounts of the plants (Huxley, 1980, 1982; Jolivet, 1986), an ecological analysis of proven cases of ant-plant mutualism (Beattie, 1985; Benson, 1985; Huxley, 1986), and a brief summary of all aspects of the symbioses with a bibliography complete through 1981 (Buckley, 1982a–c).

## THE VARIETIES OF ANT–PLANT SYMBIOSES

The angiosperms (flowering plants) and the ants have been closely associated throughout most of their respective histories. By the middle of the Cretaceous Period primitive sphecomyrmine ants were on the scene, while the angiosperms were diversifying and spreading around the world as the newly dominant form of terrestrial vegetation. An intricate coevolution of the two groups probably began during this time. Many of the plant species had come to depend on insects for pollination, and an even greater number of insect species subsisted on nectar and pollen obtained during the pollination process. A legion of other insects fed on the foliage and wood of the angiosperms. Plants responded by evolving various combinations of thick cuticles, dense spines and hairs, and secondary defense substances such as alkaloids and terpenes.

Into this lively theater of coevolution the ants entered. As the Cretaceous drew to a close, the ants increased in diversity and abundance, seized new roles as pollinators and seed dispersers, and appropriated the plants as domiciles. An entomologist returning to

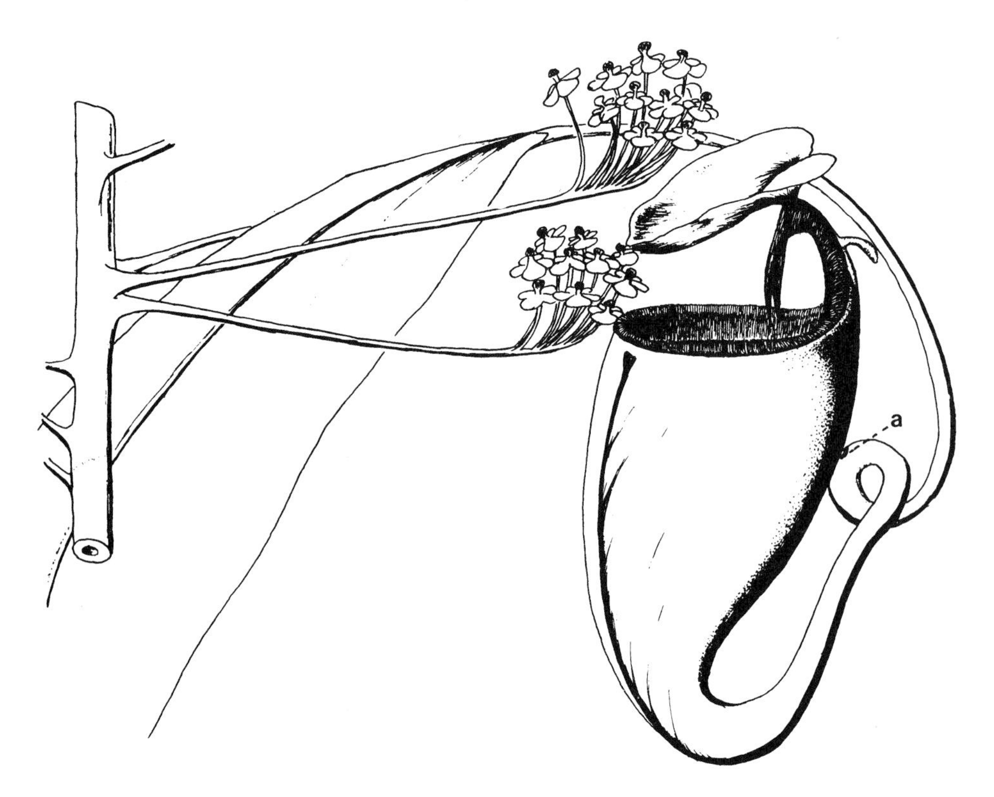


**FIGURE 14–1** The East Indian pitcher plant *Nepenthes bicalcarata,* a presumptive symbiont of ants. On the right is a portion of a pitcher leaf, which is expanded at the end into an urn-like structure that attracts and captures insects. The stem of the leaf coils through a partial spiral which is hollow and harbors an ant colony. The entrance hole (denoted by *a*) is on the dorsal surface of the spiral. A portion of the main stalk and two flower clusters are also illustrated. (From Jolivet, 1986, after Beccari, 1884.)

**PLATE 17.** Trophobiosis between the homopteran *Eurymela* sp. and the Australian meat ant *Iridomyrmex purpureus. (Above)* An adult *Eurymela* excreting a drop of honeydew, which is imbibed by a meat ant worker. *(Below)* A Eurymela nymph surrounded by I*ridomyrmex* workers.

**PLATE 18.** A worker of the African stink ant *Paltothyreus tarsatus* has invaded a termite nest gallery. She first immobilizes the termites by stinging them, then stacks as many termites in her mandibles as she can carry.

**PLATE 19.** The masters of camouflage in the ant world are the members of the Neotropical genus *Basiceros*. Shown here is a portion of a colony of *B. manni* from Costa Rica, including workers and larvae. The older workers are partially encrusted with soil caught in specialized hairs over most of the body.

**PLATE 20.** The early stage of a raid by a colony of the army ant *Eciton burchelli.* In the background a bivouac containing hundreds of thousands of workers and the queen is lodged beneath a fallen tree trunk. The morning swarm is in progress, and columns of the ants have started to spread out across a broad front. In the center workers overwhelm a whip scorpion. A long-mandibled major worker *(left foreground)* stands guard. A bicolored antbird *(upper left)* and a barred woodcreeper *(right background)* watch for insects flushed by the ants. (From Hölldobler, 1984d; painting by J.D. Dawson reprinted with permission of the National Geographic Society.)

**PLATE 21.** During swarm raids *Pheidologeton diversus* workers and soldiers subdue and transport large prey items through well-coordinated group activity. (From Moffett, 1986f; photograph by M.W. Moffett reproduced by permission of the National Geographic Society.)

**PLATE 22.** Nest construction by the Asian weaver ant *Oecophylla smaragdina.* If a leaf is too large or stiff to be turned by a single ant during nest building, groups of workers arrange themselves into multiple chains and pull together to close the gap. The photograph at the right indicated how many individuals can be involved in this stage of the work; at the left, several parallel chains of ants are shown in more detail. (From Hölldobler and Wilson, 1983b.)

**PLATE 23.** Rows of *Oecophylla* workers line up along the edge of a leaf and hold it in place until it can be bound with larval silk. (From Hölldobler and Wilson, 1983b.)

**PLATE 24.** The final stage of nest construction by the weaver ant *Oecophylla.* A worker of the Asian weaver ant *O. smaragdina* holds a last-instar larva between her mandibles and moves it back and forth between the edges of two leaves. The larva, responding to tactile signals from the ant's antennae, releases silk from its sild glands, In this way it serves as a living shuttle responsible for the unique properties of the nest. (Hölldobler and Wilson, 1983b.)

early Eocene times, about 60 million years ago, would find familiar-looking ants swarming over familiar-looking vegetation.

Complex symbioses have been fashioned among the thousands of species of ants and plants. Often these relationships are parasitic, with one exploiting the other and giving nothing in return. In other cases they are commensalistic, with one partner making use of the other but, as in the case of ants occupying hollow stems, neither harming it nor helping it. Of maximum scientific interest, however, is the fact that some symbioses appear to be mutualistic; in other words, both partners benefit from the association. Ants use cavities supplied by the plants for nest sites, as well as nectar and nutritive corpuscles given them as food. They also protect their plant hosts from herbivores, distribute their seeds, and literally pot their roots with soil and nutrients. There is abundant evidence, which we will review shortly, that some pairwise combinations of ants and plants have coevolved so that each is specialized to use the other's services. This mutualistic linkage has produced some of the most elaborate adaptations known in nature.

Before taking up these several categories and the evidence they provide for coevolution, let us define the specialized terms that have grown out of the study of ant-plant mutualism. The definitions below represent the consensus that we believe can be drawn from current usages.

*Ant garden.* A cluster of epiphytic plants inhabited by ant colonies. To qualify as a true ant garden, the plants must benefit from the association. True ant gardens are known from both the Asian and New World tropics.

*Ant plant.* Also known as a myrmecophyte; a species of plant with domatia, or specialized structures for housing ant colonies.

*Arils.* Same as elaiosomes (*q.v.*).

*Bead glands.* Same as pearl bodies (*q.v.*).

*Beccarian bodies.* The pearl bodies (*q.v.*) produced by the stipules or young leaves of the Old World tropical genus *Macaranga*, and consumed by resident ants.

*Beltian bodies.* The food bodies found on the tips of the pinnules and rachises of some New World species of *Acacia*, and consumed by the resident *Pseudomyrmex*.

*Domatia.* Also called *myrmecodomatia;* specialized structures, such as inflated stems, evolved by ant plants for the housing of ants.

*Elaiosomes.* Also called *arils;* specialized nutritive attachments on ant-dispersed seeds. Stimulated by the attractants, the ants transport the seeds to new locations, discard them after feeding on the elaiosomes, and hence aid in the dispersal of the seeds.

*Extrafloral nectaries.* Secretory organs, often no more than small patches of tissue, that produce sugary secretions, which possibly contain amino acids, attractive to ants and other insects. By definition, extrafloral nectaries are not involved in pollination, although they may occur on the flower outside the perianth.

*Food bodies.* Special nutritive corpuscles evolved by ant plants to feed ants; particular kinds of food bodies include Beltian bodies, Müllerian bodies, and pearl bodies.

*Müllerian bodies.* Food bodies produced by *Cecropia* trees on the trichilium (a pad at the base of the petiole) and consumed by the resident ants, which are usually *Azteca* but occasionally *Camponotus balzani* or *Pachycondyla luteola*.

*Myrmecochory.* The dispersal of seeds by ants stimulated by nutritive bodies (elaiosomes) or special seed attractants. Dispersal of seeds by granivorous ants without the aid of such specialized attractants is not included.

*Myrmecodomatia.* See under *domatia*.

*Myrmecophily.* The general condition of encouraging ants. In the system proposed by van der Pijl (1955) and Beattie (1985), myrmecophily is used to denote ant pollination. Jolivet (1986) and other authors, however, have used the expression to refer to the condition of being an ant plant, or myrmecophytism. Since the presence of domatia is so often accompanied by extrafloral nectaries, food bodies, and other coadaptive traits, it seems more appropriate to use the expression "myrmecophilous" in the broadest sense, comprising both pollination and myrmecophytism and all the accouterments plants use to attract and reward ants.

*Myrmecophytism.* The condition of being an ant plant, in other words possessing ant shelters (domatia).

*Myrmecotrophy.* The transport of soil, litter, and other nutrient-bearing materials by ants that results in the feeding of the plant hosts.

*Pearl bodies.* Also called *bead glands;* a heterogeneous group of food bodies with a pearl-like luster and a high concentration of lipids.

## PROTECTIONISM AND THE *ACACIA* CASE

In spite of the obvious intimacy of the associations between many of the tropical plant species and ants, and the extraordinary anatomical features of the plants that seem to have no other function than to serve their guests, biologists for many years disagreed about the significance of the association. On one side stood the "protectionist school," to use Brown's (1960a) expression. It was founded by Belt (1874), whose observations on the ant acacias were the beginning of serious studies of myrmecophytism. Belt, and a majority of subsequent writers, including particularly A. F. W. Schimper, Erich Wasmann, and the leading authority on the bull's-horn acacias, W. E. Safford, agreed that the ants provide the plants protection against their natural enemies. They also postulated that in the course of their evolution the acacias developed hollow thorns, Beltian bodies, and foliar nectaries as devices to promote the welfare of the ants (Figures 14–2 and 14–3). In short, the protectionist authors believed the symbiosis to be mutualistic. The opposing "exploitationist school," represented chiefly by Skwarra (1934) and Wheeler (1942), argued that only the ants benefited and that the various myrmecophilous structures of the acacias serve some other, still unknown function. These opposing viewpoints, rather oversimplified as we have expressed them here, extended to discussions of other genera of ant plants as well.

Brown, crystallizing the issue in 1960, developed new evidence favoring the protectionist hypothesis. He pointed out that *Acacia* is a very old and widespread genus containing no fewer than 700 species. Australia contains a majority of the species as well as the greatest phyletic diversity of any continent. It also contains one of the

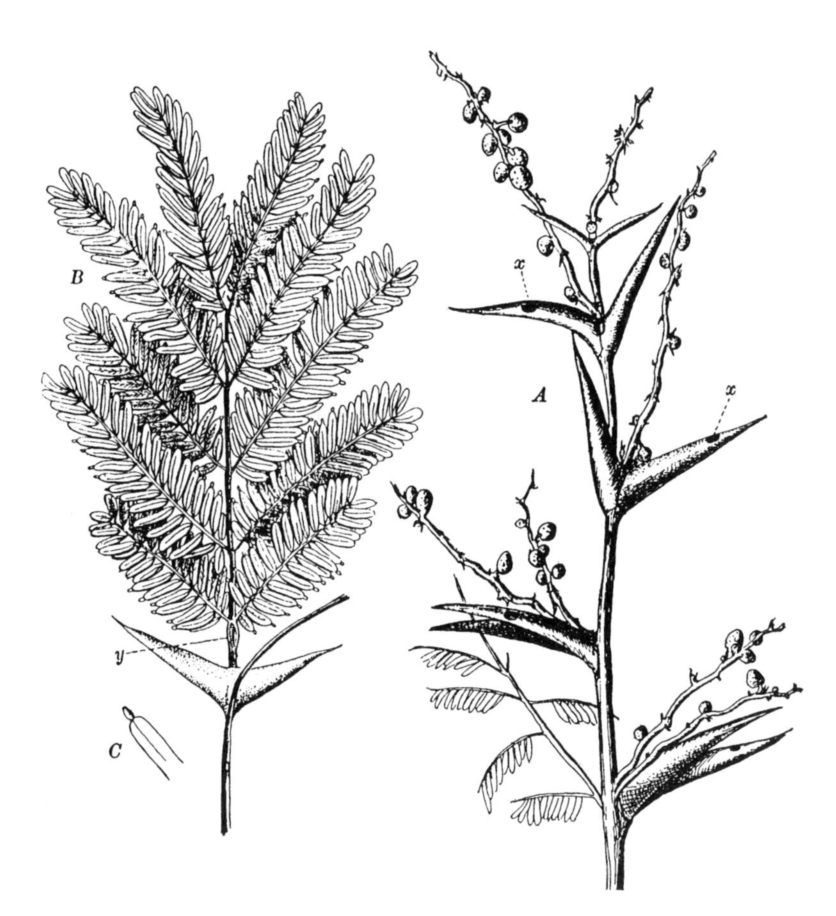


**FIGURE 14–2** *Acacia sphaerocephala*, one of the bull's-horn acacias of the American tropics: (*A*) end of a branch showing pairs of the hollow thorns that are normally occupied by ants of the genus *Pseudomyrmex;* (*x*) holes chewed by the ants to form entrances in the thorns; (*B*) a leaf of the same plant; (*y*) extrafloral nectary; (*C*) tip of a leaflet enlarged to show a Beltian body, the organ which is picked and eaten by the ants. (From Wheeler, 1910a; after A. F. W. Schimper.)

richest ant faunas of any comparable area in the world. Yet not a single potential myrmecophyte has been found among the acacias of Australia. Moreover, the Australian species have mostly lost their spiniform stipules, in striking contrast to their congeners in other parts of the world. This geographic distribution of characteristics agrees with the known occurrence, in the recent geological past and at present, of large and effective faunas of browsing mammals. Brown inferred, in accordance with Belt's hypothesis, that the development of myrmecophytism and spininess in the African and New World *Acacia* species represents an adaptive response to the presence of such mammals, which provide an effective deterrent to browsing. In Australia, where advanced browsing faunas have been unknown (at least in the recent geologic past), the species have either failed to develop myrmecodomatia and spines or lost them secondarily.

It remained for Janzen (1966, 1967, 1969), in a brilliant field experiment in Mexico, to prove directly that the ants do indeed provide vital protection to the bull's-horn acacias. While conducting a pilot survey Janzen noted, as Bequaert (1922) had found earlier for *Barteria fistulosa* in Zaïre, that *Acacia cornigera* shrubs and trees devoid of ants suffer greater damage from attacks by phytophagous insects than do their neighbors harboring ant colonies. They also tend to be overgrown by competing plant species. When Janzen removed the ants (*Pseudomyrmex ferruginea*) with any one of three treatments (spraying with parathion, clipping the thorns, or extirpating entire occupied branches), he found that the acacias became decidedly more vulnerable to attack by their insect herbivores. Coreid bugs and membracids sucked on the shoot tips and new leaves; scarabs, chrysomelid beetles, and assorted caterpillars browsed on the leaves; and buprestid beetle larvae girdled the shoots. Moreover, other plants grew in more closely and shaded the stunted shoots. In nearby control trees, still occupied by *Pseudomyrmex* colonies, Janzen observed that the ants attacked the invading insects, in the great majority of cases successfully driving them off or killing them. Alien plants that sprouted within a radius of 40 centimeters of the occupied acacia trunks were chewed and mauled by the ants until they died. Other plants whose leaves or branches touched the canopy of the acacia were also attacked. Up to one-fourth of the entire ant population were active on the surfaces of the control plants, day and night, constantly patrolling and cleaning them. For the full year during which the experiment was continued, the biomass and growth of the unoccupied acacias steadily fell below that of the occupied ones. In the end it seemed unlikely that they could survive much longer, let alone bear seeds. Thus Belt's view, that the *Pseudomyrmex* "are really kept by the acacia as a standing army," was substantially confirmed.

Although experimental evidence is still lacking, it seems probable that the ants are also effective against browsing mammals. The *Pseudomyrmex* workers are extremely aggressive toward intruders of all sizes. They become alert at the mere smell of a cow or a man, and when their tree is brushed or shaken, they swarm out and attack at once. Their stings are very painful, causing a lasting burning and throbbing effect. To brush against an occupied acacia, and thus to acquire a group of vicious, stinging ants on an arm or a leg, is a sensation very much like walking into a large nettle plant.

According to Janzen, *Pseudomyrmex ferruginea* is an obligate plant ant that occupies at least five species of *Acacia* (*chiapensis, collinsii, cornigera, hindsii,* and *sphaerocephala*). Its life cycle conforms to the basic claustral pattern of ants generally. After the nuptial flight, which can occur in warm weather in any month of the year, the queen alights, sheds her wings, and searches for a nest site. For a *P. ferruginea* there can be only one such place: an unoccupied acacia thorn. If the thorn has not already been opened by a previous occupant, the queen gnaws a circular hole near the tip of the spine and enters. Then she lays 15 to 20 eggs and rears her first brood while remaining secluded in the thorn cavity. Although the exact duration of brood development is not known, it is evidently relatively short for an ant species, and the worker population increases at a rapid rate. Within seven months there are about 150 workers, and, three months later, twice this number. The worker population increases to about 1,100 in two years and to over 4,000 in three years. The largest colony collected by Janzen contained 12,269 workers and a single queen. In old colonies the queen is physogastric, heavily attended by workers, and accompanied by masses of hundreds of eggs and young larvae. The production of males and virgin queens begins during the second year and proceeds continuously thereafter. Workers belonging to the youngest colonies leave the protection of the thorn home only long enough to gather nectar and Beltian bodies and, at rare intervals, to take possession of nearby thorns. When their numbers reach 50 to 100, they begin patrolling

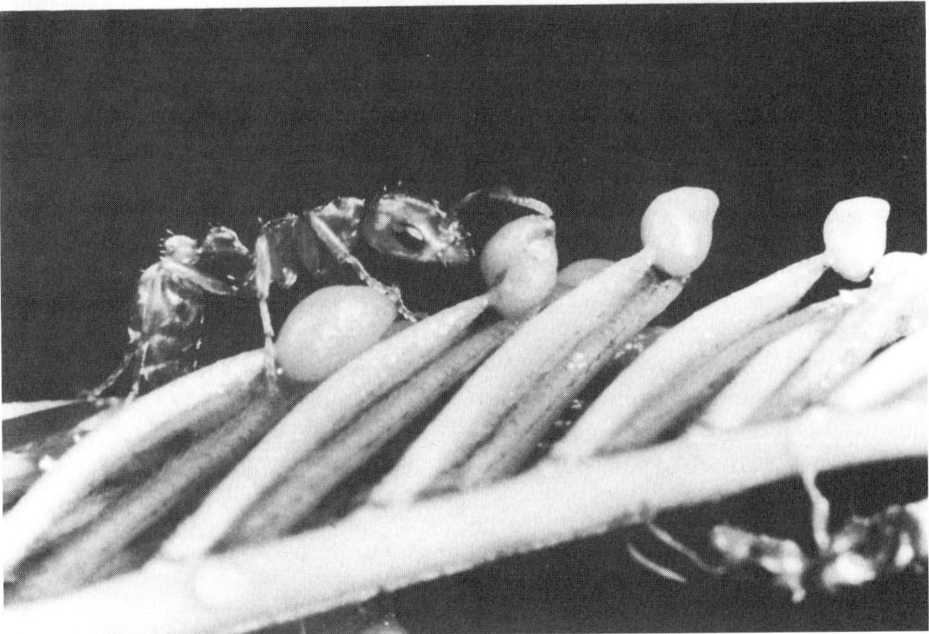

**FIGURE 14–3** (*Above*) A bull's-horn acacia from Costa Rica, showing the hollow thorns with the entrance hole of a *Pseudomyrmex* nest. In the foreground can be seen extrafloral nectaries. (*Below*) Acacia leaf with Beltian bodies being visited by a *Pseudomyrmex* worker. (Photographs courtesy of D. Perlman.)

the open plant surface in the vicinity of the nest thorns. When the population size reaches 200 to 400, the workers become more aggressive and start attacking and destroying other, smaller colonies in nearby thorns. They also become increasingly effective in warding off phytophagous insects that attempt to land in the vicinity. Finally the dominant colony takes possession of the entire tree, wiping out all competitors in the process. A few colonies are also able to extend their territories to other acacias nearby.

The *Pseudomyrmex ferruginea* colonies appear to subsist primarily on the Beltian bodies and foliar nectar obtained from the host trees. The larvae are fed in part on unaltered fragments of Beltian bodies in the following peculiar manner. The nurse worker first pushes the fragment deep within the larva's trophothylax, the special food pouch located on the lower surface of the thorax just behind the head (and found only in pseudomyrmecine larvae; see Petralia and Vinson, 1979a). The larva then starts to rotate its head in and out of the trophothylax, chewing and swallowing the contents. Simultaneously it ejects a droplet of clear fluid, possibly containing a digestive enzyme, into the trophothylax. If the Beltian fragment protrudes from the opening of the pouch, a worker may remove it, cut it up, and redistribute it. From time to time workers also force open the trophothylax and regurgitate droplets of fluid into it. Whether this material consists of elementary crop fluid or some more specialized form of nutritive secretion is unknown. Occasionally the *Pseudomyrmex* workers succeed in capturing insect prey on the nest tree. It is possible that these too are fed to larvae, but they can form a source of protein only secondary in importance to the Beltian bodies.

## THE BALANCE OF MUTUALISM

It is reasonable to conclude from existing evidence that the anatomy and physiology of the ant guests and plant hosts have been modified at least occasionally to enhance the mutualism in a manner consistent with the theory of evolution by natural selection. And as Beattie (1985) has said, the majority of ant-plant mutualisms evolve in response to selection, especially stress selection, on the plants rather than the ants.

> The participating ant species, which vary in time and space, respond to plant rewards on a facultative basis, and traits evolved in specific response to plant rewards are infrequent. Thus, when the definition of coevolution requires reciprocal selection involving heritable traits, coevolution is remarkably difficult to demonstrate in ant-plant mutualisms and appears to be the exception rather than the rule. In fact, the evidence suggests that it is directional selection that drives the vast majority. This does not diminish the importance of their effects. The speed and energy with which ants harvest rewards such as elaiosomes suggest that they are crucial to the economy of ant colonies. This may be especially true when the colony is stressed, either by the external environment or by major internal demographic events such as reproduction. For the plants, we have repeatedly seen that ant services are potent forces that increase plant fitness. In many cases the failure of ant services leads to a variety of demographic failures, including poor seed set and poor seedling recruitment. Populations may contract, crash, or even become extinct as a result. Ant services, either on a continuous basis or as a density response, are crucial to a wide variety of plant species worldwide.

Essentially the same conclusion was reached by Longino (1986) for the ant gardens of Central and South America and Huxley (1980) for epiphytic ant plants generally. Similarly, Davidson et al. (1989) present evidence that ants nesting exclusively on ant plants have restricted diets and refuse baits accepted by many other kinds of ants. In a symmetrical manner, Asian species that have invaded introduced *Cecropia* species (hence are not adapted to them) ignore the Müllerian bodies on which coevolved ant species feed in the American tropics.

## ANTS PROTECT PLANTS

Throughout their ranges ants forage on vegetation in large numbers, searching for arthropod prey. The insects they collect include hemipterans, beetles, sawfly larvae, caterpillars, and other herbivores. Wood ants, in particular *Formica aquilonia, F. polyctena,* and *F. rufa,* are so effective as predators that they have been utilized in Europe for centuries to control forest pests. Entomologists have developed an entire technology since the late 1940s to culture and transplant the ants, and it has often proved cost effective (Gösswald, 1951a, 1985; Grimalski, 1960; Khalifman, 1961; Adlung, 1966; Gösswald and Horstmann, 1966; Smirnov, 1966; Bradley, 1972; Finnegan, 1975, 1977; Skinner and Whittaker, 1981). An ordinary colony of *F. polyctena* was observed to collect about 6 million prey items from one-third of a hectare per year (Horstmann, 1972, 1974); one colony of *F. rufa* monitored by Strokov (1956) gathered 21,700 sawfly larvae and moth caterpillars in a single day. Similarly dramatic data have been published for *polyctena, pratensis,* and *rufa* by Sörensen and Schmidt (1987). *F. polyctena* was so proficient at protecting mountain birches against an outbreak of the geometrid moth *Oporinia autumnata* that green islands of intact trees 40 meters in diameter were left around each nest in the midst of a gray, mostly defoliated forest (Laine and Niemelä, 1980).

There is a potential debit side in the ledger of ant protection. When seed dispersers, such as mammals and birds, are driven away from trees by the resident ants, the fruit and seeds may fall to the ground undispersed. This exact effect was documented by Thomas (1988) in the case of West African fig trees occupied by colonies of weaver ants (*Oecophylla longinoda*).

## PLANTS SHELTER ANTS: MYRMECOPHILY

The strongest evidence for ant-plant mutualism comes from the existence of domatia, or plant structures that serve no evident purpose other than to shelter ant colonies. Domatia increase the density of ants on the plant itself. It is easy to see how the phenomenon could arise in evolution. Ants are always quick to take advantage of whatever hollows and crevices plants have to offer. In most cases the shelters are not true domatia. They are adventitious, and the ants therefore live as parasites or commensals on the plants. These incidental nest sites can be divided into convenient categories as follows:

1. Preformed cavities in live branches and stems, excavated by wood-boring beetles and other insects and later occupied by ant colonies belonging to such diverse genera as *Daceton, Podomyrma, Crematogaster, Azteca, Lasius,* and *Camponotus.*
2. Cavities in stems and branches that are naturally hollow or contain a pith soft enough to be easily excavated by ants. A legion of grasses, sedges, composites, and other herbaceous and shrubby plant forms provide this kind of refuge, and a great variety of ants occupy them. A typical field of sedges and weeds in Florida, for example, contains dense populations of one or more species of *Pseudomyrmex, Colobopsis, Crematogaster,* and *Monomorium.* The low trees and bushes of a Brazilian *cerrado* contain the same genera along with several species of *Zacryptocerus.*
3. Natural or preformed cavities in bark. The bases of pine trees in the southern United States shelter an entire fauna of *Hypoponera, Pheidole, Solenopsis, Crematogaster, Brachymyrmex,* and other ant genera that nest adventitiously in the bark. Higher up can be found colonies of the small myrmicines *Leptothorax bradleyi* and *L. wheeleri,* which are specialized to this environment to some extent (Wilson, 1952). An ant restricted entirely to this microsite is *Melissotarsus titubans,* which occurs in the bark cavities of living trees in tropical Africa and Madagascar. Its body form and locomotory behavior are modified for existence in tight spaces. The workers walk with their middle legs held upright and touching the roofs of the galleries. If placed outside, in the open, they are unable to move around in a normal manner (Delage-Darchen, 1972a).
4. Roots of epiphytes. The tangled root systems of orchids, gesneriads, and other tropical epiphytes are ideal nest sites for ants. One of the most profitable ways to collect ants anywhere in the world is to hold an epiphyte over a pan or ground cloth and strike the root system several times with a trowel. In Central and South America, this technique yields large numbers of colonies of *Hypoponera, Gnamptogenys, Strumigenys, Nesomyrmex,* and other genera,

many belonging to rare or previously undescribed species.

5. Galls formed by cynipid wasp larvae. The phenomenon has been observed in Europe (Torossian, 1972; Espadaler and Nieves, 1983) and North America (Wheeler, 1910a). *Leptothorax obturator* of Texas appears to be a specialist on this nest site (Longino and Wheeler, 1987).

6. Earthen or carton nests constructed vertically against the sunken portions of tree trunks by a few ant species such as the large ponerines *Ectatomma tuberculatum* and *Paraponera clavata* of the New World tropics. The trees provide a partial wall of solid wood that is virtually invulnerable. *Paraponera clavata* prefers trees of the abundant legume *Pentaclethra macroloba*, but the relationship is not obligatory. It is possible that the ants benefit from both the well-formed buttresses of the *Pentaclethra* and the extrafloral nectaries in the foliage of this species (Bennett and Breed, 1985).

None of these diverse structures appears to be "designed" to accommodate ant colonies. All are ordinary anatomical features of the plants that the ants exploit, apparently in a unilateral manner. In contrast, the domatia listed comprehensively in Table 14–1 do appear uniquely to serve as ant nests. They are characterized by cavities that form independently of the ants (even in greenhouses, where no ants are present), adventitious roots and tubercles that absorb nutrients from waste material carried onto the cavities, and even holes or thin windows of tissue through which ants can more conveniently enter and leave. Some of the most distinctive forms are illustrated in Figures 14–4 through 14–8. Domatia are almost always occupied by ant colonies in nature. Furthermore, plant species with the most complex domatia are typically occupied by only one or a small number of species specialized to live with them. Finally, species with domatia usually also manufacture food bodies, which are unique structures with no known function other than the feeding of ants (see also Table 14–2). In short, strong circumstantial evidence indicates that domatia are structures specialized in evolution to promote symbioses with ants. Benson (1985) has suggested that the ant domatia of many ant species evolved from the sheltered feeding sites used by mealybugs and other homopterous insects. Because of the honeydew produced by the homopterans, ants were attracted to these sites, which were enlarged and otherwise structurally modified into ant domatia.

Further evidence of coevolution is provided by the legendary ferocity of many of the guest ants. The vast majority of *Pseudomyrmex* species not occupying domatia are timid and flee even when their nest is broken apart. In sharp contrast, *P. triplarinus*, an obligate resident of *Triplaris americana*, falls upon any intruder touching the nest tree without hesitation or mercy. To be stung by several of these ants within a few seconds is a shocking experience—you pull back at once. Or conversely, if you want to locate *Triplaris* quickly in an Amazonian forest, shake one sapling after another until one produces a swarm of the stinging ants. The *Pseudomyrmex* also attack and remove intruding insects, and they are significantly more efficient than the *Crematogaster* that also occupy *Triplaris* (Oliveira et al., 1987a).

The species of *Camponotus* represent a similar dichotomy. Most retreat or offer limited resistance when the nest is disturbed by a human being. Possibly the shyest ant species in the world is the Amazonian *C. (Hypercolobopsis) paradoxa*, whose workers disperse and hide so quickly that it is difficult to catch any specimens at all. At the opposite extreme is *C. (Myrmothrix) femoratus*, an obligatory resident of epiphytic ant gardens in South America. D. W. Davidson (personal communication) describes its behavior as follows:

> When I approached to within 1–2 m of their nests, workers of this species typically began to run back and forth and frequently jumped or fell onto me. Workers of all size classes of this polymorphic species attempted to bite, but usually only the major castes were capable of breaking the skin with their mandibles and causing a stinging sensation by simultaneously biting and spraying formic acid into the wound. In addition, these workers often exhibited a second type of apparently aggressive behavior, which I will term "coughing" behavior. With mandibles held wide open and the prothoracic legs upraised, they brought their legs down abruptly in a jerking movement that resembled a cough.

*Tetraponera (= Viticicola) tessmanni*, the obligate pseudomyrmecine tenant of the verbenaceous creeper *Vitex staudtii* in West Africa, is "exceedingly vicious and alert," according to Bequaert (1922). "When its host plant is ever so slightly disturbed, the workers rush out of the hollow stalks in large numbers and actively explore the plant. Their sting is extremely painful and sometimes produces vesicles on the skin." Even more redoubtable, Bequaert noted, is the African pseudomyrmecine *Tetraponera (= Pachysima) aethiops*, the obligate tenant of the small flacourtiaceous tree *Barteria fistulosa*: "As soon as any portion of their host plant is disturbed, they rush out in numbers and hastily explore the trunk, branches, and leaves. Some of the workers usually also run over the ground about the base of the tree and attack any nearby intruder, be it animal or man. All observers agree that the sting of the *Tetraponera* is exceedingly painful and is felt for several hours. Its effects can best be compared with those produced by female velvet ants." The ant is feared by the natives of Zaïre, who try to avoid the unpleasant task of cutting the small *Barteria* trees scattered through the forest. As a consequence individuals of *Barteria fistulosa* are often found standing by themselves in the center of clearings or near the sides of forest paths. The species is also abundant in secondary forest growth.

Not all myrmecophyte ants are this formidable. A few, such as the tiny *Pheidole* species that occupy *Maieta* and *Piper*, seem incapable of defeating most herbivores. Many authors have commented on the evident ineptness of these ants in the protection of their adopted plants. Letourneau (1983) points out, however, that such ants can perform their service by removing the eggs and early developmental stages of the herbivores instead of facing down the adults. She showed that *Piper* plants in Costa Rica occupied by *Pheidole bicornis*, their principal ant guest, suffered less damage than those deprived of the ants. The workers preferred to patrol new leaves, which are the most susceptible to insect damage. When Letourneau placed termite eggs on *Piper* bushes, the ants discovered more than 75 percent and dropped them off the plants within an hour. Risch et al. (1977) also found some evidence that the *Pheidole* chew through or push aside alien vines from their host plants, and they postulate that the ants also bring nutrients to the plant cavities as part of their nest material. No one has assayed the relative importance of these benefits to trees.

Competition for the ant plants is intense. Young plants are soon fully occupied by colony-founding queens of plant-ant species. It is common to find different internodes of very young *Cecropia* saplings occupied by one or more queens of *Azteca*. Sometimes the inhabi-

**TABLE 14–1** Plants with domatia, or structures evidently specialized to house ant colonies. Based on Bequaert (1922b), Wheeler (1942), Kostermans (1957), Keay (1958), Steenis (1967), Duviard and Segeren (1974), Stevens (1975), Corner (1976), Risch et al. (1977), Snelling (1979), Huxley (1980, 1982, 1986), Thompson (1981), Buckley (1982a–c), Beattie (1985), Jolivet (1986), Monteith (1986), Davidson and Epstein (1989), McKey (1989a,b), and P. F. Stevens (personal communication).

| Plants | Domatia and food bodies | Resident ants |
|---|---|---|
| **AMERICAN TROPICS** | | |
| **Araceae** (arums) | | |
| *Anthurium gracile* | Cavities among roots with root ball | *Anochetus* sp., *Odontomachus* sp., *Pachycondyla goeldii* |
| **Bromeliaceae** (bromeliads) | | |
| *Tillandsia* spp. | Leaf bases inflated outward to form cavities next to main stem | *Crematogaster* spp. |
| **Chrysobalanaceae** (coco-plums and relatives) | | |
| *Hirtella physophora* | Paired pouches at base of leaf blade | *Allomerus* sp. |
| **Ehretiaceae** | | |
| *Cordia nodosa* | Swollen nodes containing cavity | *Allomerus* spp., *Azteca* spp. |
| **Euphorbiaceae** (euphorbs) | | |
| *Mabea* (50 spp.) | Hollow branches, possibly not domatia | Undetermined ants |
| **Gentianaceae** (gentians) | | |
| *Tachia* spp. | Hollow stems, possibly not domatia | Undetermined ants |
| **Gesneriaceae** | | |
| *Besleria* | ? | *Pheidole* |
| **Lauraceae** (laurels) | | |
| *Pleurothyrium cuneifolium*, other species | Swollen hollow branches | *Azteca* sp., *Myrmelachista* sp. |
| **Leguminosae** (legumes) | | |
| *Acacia* (12 spp.; bull's-horn acacia) | Hollow thorns; nectaries on petioles; edible Beltian bodies on tips of leaflets | *Pseudomyrmex* spp. |
| *Platymiscium* (2 of the 30 spp.) | Hollow stems, possibly not true domatia | Undetermined ants |
| *Sclerolobium odoratissimum* (and up to 34 other known species) | Large sac on upper surface of leaf from petiole to second pair of leaflets | Undetermined ants |
| *Tachigalia* (25 spp.) | Petiole swollen into spindle-shaped sacs | *Pseudomyrmex* spp., *Azteca* spp. |
| **Melastomataceae** (melastomes) | | |
| *Clidemia tococoidea* | Bifid pouch at leaf base | *Azteca* spp. |
| *Maieta* (10 spp.) | Bifid pouch at leaf base | *Pheidole* spp. including *P. minutula; Crematogaster* spp. |
| *Tococa* (50 spp.) | Bifid pouch at leaf base | *Azteca* spp., occasional *Crematogaster* and *Solenopsis* |

*continued*

**TABLE 14–1** *(continued)*

| Plants | Domatia and food bodies | Resident ants |
|---|---|---|
| **Monimiaceae** | | |
| *Siparuna* sp. | Pithy internodes hollowed out by ants | *Pseudomyrmex* sp. |
| **Moraceae** (mulberries and relatives) | | |
| *Cecropia* (100 spp.) | Hollow main-stem internodes; edible Müllerian corpuscles on base of petiole | *Azteca* spp. (usually), *Pachycondyla luteola* in one species (D. W. Davidson, personal communication), occasionally *Camponotus, Crematogaster,* and *Solenopsis* spp. (D. Perlman, personal communication) |
| *Coussapoa* sp. | Similar to *Cecropia* | *Azteca* sp. |
| *Pouruma* spp. | Similar to *Cecropia* | *Allomerus* sp., *Azteca duroiae* |
| **Orchidaceae** (orchids) | | |
| Species of *Caularthron, Epidendrum, Schomburgkia* | Pseudobulbs, hollow roots | *Azteca* spp., *Camponotus abdominalis, C. rectangularis, Crematogaster armandi, C. limata, Ectatomma tuberculatum, Monacis bispinosa, Monomorium floricola* (introduced Asian species), *Odontomachus* spp., *Paratrechina pubens* |
| **Piperaceae** (peppers) | | |
| *Piper cenocladium, P. fimbriulatum, P. sagittifolium* | Flanges of leaf petiole curl over to form cavities (which contain food bodies) | Primarily *Pheidole bicornis* and several other small *Pheidole;* secondarily *Crematogaster* spp. |
| **Polygonaceae** (buckwheats and relatives) | | |
| *Ruprechtia jamesoni* | Hollow internodes and stems | *Pseudomyrmex* sp.; adventitious association? |
| *Triplaris* (25 spp., palo santo) | Hollow main-stem internodes; ants enter by eating axillary bud enclosed by stipules (R. A. Howard, personal communication) | Primarily *Pseudomyrmex dendroicus, P. triplarinus* and other very aggressive species; secondarily *Azteca, Crematogaster, Hypoclinea, Pheidole, Iridomyrmex,* and *Tapinoma* |
| **Polypodiaceae** (ferns) | | |
| *Solanopteris* spp. | Rhizome pseudobulbs | *Azteca filicis, Camponotus* sp., *Pheidole* sp., *Solenopsis* sp. |
| **Rubiaceae** (madders, coffees) | | |
| *Duroia* (20 spp.) | Inflated hollow stems below inflorescence, curved leaf margins, or petiolar pouches | *Allomerus, Azteca, Myrmelachista* |
| *Patima formicaria* | Hollow stem internodes, some of which are inflated | *Azteca traili, Camponotus* spp. |
| *Remijia physophora* | Two pouches at base of leaf blade | Undetermined ants |

*continued*

**TABLE 14–1** *(continued)*

| Plants | Domatia and food bodies | Resident ants |
|---|---|---|
| **AFRICA** | | |
| **Bignoniaceae** (trumpetcreepers) | | |
| *Stereospermum* (15 spp.) | Hollow internodes | *Tetraponera benzigi* |
| **Euphorbiaceae** (euphorbs, spurges) | | |
| *Macaranga saccifera*, other *Macaranga* spp. | Pouch-shaped leaf stipules; extrafloral nectaries at leaf bases | *Crematogaster tibialis* |
| **Flacourtiaceae** (Indian plums and relatives) | | |
| *Buchnerodendron speciosum* | Hollow branches, possibly not domatia | *Crematogaster excisa* |
| *Caloncoba laurentii* | Swollen, hollow branches, possibly not domatia | Unknown |
| **Leguminosae** (legumes) | | |
| *Acacia* | Swollen, hollow spines (in some species nearly globular in shape) | *Cataulacus* spp., *Crematogaster* (many spp.), *Tetraponera* spp. |
| *Leonardoxa africana* | Hollow stem internodes | *Cataulacus mckeyi*, *Petalomyrmex phylax* |
| **Moraceae** (mulberries, figs, and relatives) | | |
| *Musanga cecropioides* | Spaces in internodes filled with pith, which is easily excavated by ants | *Crematogaster* spp. |
| **Passifloraceae** (passion flowers, granadillas) | | |
| *Barteria fistulosa* | Swollen, hollow branches | *Tetraponera* (= *Pachysima*) *aethiops*, *T. latifrons* |
| *Barteria nigritana* | Swollen, hollow branches | *Crematogaster* sp. |
| **Rubiaceae** (madders, coffees) | | |
| *Bertiera simplicicaulis* | Hollow internodes | Undetermined ants |
| *Canthium* (= *Plectronia*) spp. | Swollen, hollow stems | *Cataulacus traegaordhi*, *Crematogaster* spp., *Engramma kohli* |
| *Cuviera* spp. | Swollen, hollow stems | *Cataulacus pilosus*, *Crematogaster* spp., *Engramma denticulatum*, *Technomyrmex hypoclinoides* |
| *Grumilea* (= *Psychotria* = ? *Uragoga* partim) | Swollen, hollow stipules at leaf base | *Crematogaster obstinata* |
| *Nauclea* (= *Sarcocephalus*) (6 spp.) | Hollow branches and stems | *Crematogaster africana winkleri* |
| *Rothmannia* spp. | Swollen, hollow stems | *Camponotus foraminosus*, *Cataulacus weissi*, *Crematogaster rugosa* |
| *Uncaria africana* | Swollen, hollow branches and stems | *Crematogaster excisa andrei* |
| *Vangueriopsis* (18 spp.) | Hollow internodes | Unknown |
| **Sapotaceae** (sapodillas) | | |
| *Delpydora macrophylla* | Pouch at leaf base | Unknown |
| **Sterculiaceae** (cacao trees, sterculias) | | |
| *Cola* spp. | Pairs of pouches at base of leaves | *Engramma* spp., *Plagiolepis* spp. |
| *Scaphopetalum* (2 spp.) | Single club-shaped pouch at base of leaf blade | *Engramma* spp. |

*continued*

**TABLE 14–1** *(continued)*

| Plants | Domatia and food bodies | Resident ants |
|---|---|---|
| **Verbenaceae** (verbenas) | | |
| *Clerodendron* spp. | Hollow internodes | *Crematogaster* spp. |
| *Vitex* spp. | Pith canal in internodes | *Tetraponera* (= *Viticicola*) *tessmanni* |
| **TROPICAL ASIA, OCEANIA, AND AUSTRALIA** | | |
| **Actinidiaceae** (actinidias) | | |
| *Saurauia* spp. | Some species have leaves modified to create apparent domatia | Undetermined ants |
| **Anacardiaceae** (mangos, sumacs, and relatives) | | |
| *Lannea coromandelica* | Hollow stems; status as domatia uncertain | Undetermined ants |
| *Senecarpus australiensis* | Hollow stems | Undetermined ants; adventitious association? (P. F. Stevens, personal communication) |
| **Annonaceae** (custard-apples) | | |
| *Goniothalamus ridleyi* and other plants | Flower masses at base of trunk form spaces invariably occupied by ant colonies; status as domatia uncertain | Undetermined ants |
| **Asclepiadaceae** (milkweeds and relatives) | | |
| *Dischidia* spp. | In certain species of these epiphytes some leaves cup-like, with rim applied to host tree trunk, and with adventive rootlets inside; in other species, leaves flask-like, also with adventitious roots | Species of *Crematogaster, Hypoclinea, Iridomyrmex* |
| *Hoya* spp. | Epiphytes and semi-epiphytes with adaptations similar to those of *Dischidia* | Undetermined ants |
| **Euphorbiaceae** (euphorbs, spurges) | | |
| *Endospermum formicarum, E. moluccanum* | Branches inflated and hollow | *Camponotus angulatus, C. quadriceps* |
| *Macaranga* spp. | Hollow, inflated branches and cavities beneath persistent stipules of buds; edible food bodies (Beccarian bodies) on leaf tips | *Crematogaster* spp. |
| **Flacourtiaceae** (Indian plums and relatives) | | |
| *Gertrudia amplifolia* | Branches inflated beneath leaf buds | Undetermined ants |
| *Ryparosa javanica* | Hollow stems | *Camponotus* ("*Colobopsis*") sp. |
| **Lauraceae** (laurels) | | |
| *Cryptocarya caloneura* | Hollow stems | Undetermined ants |
| **Leguminosae** (legumes) | | |
| *Humboldtia laurifolia* | Flower-bearing stems inflated and hollow | Species of *Crematogaster, Monomorium, Tapinoma, Technomyrmex* |

*continued*

**TABLE 14–1** *(continued)*

| Plants | Domatia and food bodies | Resident ants |
|---|---|---|
| **Melastomataceae** (melastomes) | | |
| *Medinilla loheri* | Pouch on upper leaf surface; role as domatia uncertain | Presence of ants uncertain |
| *Pachycentria macrorhiza, P. microstyla* | Epiphytes: tuberous roots with cavities possibly made by ants; possibly not true myrmecophytes (Janzen, 1974) | Undetermined ants |
| **Meliaceae** (mahoganies and relatives) | | |
| *Aglaia sapindara* | Hollow branches | Undetermined ants |
| *Aphanamixis myrmecophila* | Branches inflated, hollowed by ants | Species of *Crematogaster, Iridomyrmex* (D. W. Davidson, personal communication) |
| *Chisocheton pachyrhachis, C. lasiocarpus,* other species | Branches and bases of petioles inflated and hollow | Species of *Camponotus, Crematogaster, Iridomyrmex; Tapinoma melanocephalum* |
| **Monimiaceae** | | |
| *Kibara ferox,* other species | Hollow, inflated branches | Undetermined ants |
| *Steganthera hospitans,* other species | Hollow, inflated branches | *Iridomyrmex* sp. (P. F. Stevens, personal communication) |
| **Moraceae** (mulberries, figs, and relatives) | | |
| *Ficus inaequalis, F. subinflata* | Some stems hollow, possibly also inflated; status as domatia uncertain | Undetermined ants |
| *Ficus paracamptophylla* | Persistent stipules | Undetermined ants |
| **Myristicaceae** (nutmegs) | | |
| *Myristica heterophylla, Myristica subalulata* (= *M. myrmecophila*) | Somewhat inflated, hollow internodes containing scale insects | *Iridomyrmex* sp. (P. F. Stevens, personal communication) |
| **Myrsinaceae** | | |
| *Tapeinosperma 4-pachycaulum* | Hollow stems | *Pheidole* sp. |
| **Myrtaceae** | | |
| *Syzigium cormifolium* | Hollow stems | *Iridomyrmex* sp. |
| **Nepenthaceae** (East Indian pitcher plants) | | |
| *Nepenthes bicalcarata* | Hollow stem of pitcher leaf | Undetermined ants |
| **Nyctaginaceae** | | |
| *Pisonia longirostria* | Hollow stems; ants accumulate humus in leaf axils | Undetermined ants (P. F. Stevens, personal communication) |
| **Orchidaceae** (orchids) | | |
| *Grammatophyllum speciosum* | Cavities in enlarged basal portion of pseudobulb | Undetermined ants |

*continued*

**TABLE 14–1** *(continued)*

| Plants | Domatia and food bodies | Resident ants |
|---|---|---|
| **Palmae (Arecaceae)** (palms) | | |
| *Calamus amplectens* | Base of leaf inflated, surrounds stem | Undetermined ants |
| *Daemonorops* spp. | Spiny plates form short circular tunnel | Undetermined ants |
| *Korthalsia echinometra*, other species | Inflated, hollow ocrea (stem sheath) | *Camponotus contractus*, *C. korthalsiae*, *Crematogaster deformis* |
| **Piperaceae** (peppers) | | |
| *Piper myrmecophilus* | Leaves modified to provide cavities | Undetermined ants |
| **Polypodiaceae** (ferns) | | |
| *Drynaria* spp. | Cavities in rhizome; extrafloral nectaries | Probably *Pheidole* sp. (Jolivet, 1986) |
| *Lecanopteris* (= *Phymatodes* = *Polypodium* partim) (15 spp.) | Hollow rhizome of epiphyte; *L. mirabilis* with absorptive rootlets in cavities | *Crematogaster deformis*, *C. yappii*, *Iridomyrmex cordatus* (= *I. myrmecodiae*), *Technomyrmex albipes* |
| *Platycerium* (15 spp.) | ? | Undetermined ants |
| **Potaliaceae** | | |
| *Fagraea* (6 spp.) | Auricles on base of petiole form cavity; extrafloral nectaries near leaf base, on blades, and calyces | Undetermined ants |
| **Rubiaceae** (madders, coffees) | | |
| *Anthorhiza* (8 spp.) | ? | Undetermined ants |
| *Hydnophytum* (46 spp.) | Epiphytes: domatia similar to those of *Myrmecodia* | *Iridomyrmex cordatus*, also species of *Camponotus*, *Crematogaster*, *Monomorium*, *Tapinoma*, *Paratrechina* |
| *Myrmecodia* (25 spp.) | Epiphytes: pseudobulb with intricate galleries and chambers | Primarily *Iridomyrmex cordatus* (= *I. myrmecodiae*), also *I. scrutator*, species of *Camponotus*, *Crematogaster*, *Pheidole*, *Poecilomyrma* |
| *Myrmedoma*, *Myrmephytum*, *Squamellaria* (10 spp.) | Epiphytes: domatia similar to *Myrmecodia* | Undetermined ants |
| *Nauclea formicaria*, *N. lanceolata*, *N. strigosa* | Swollen, hollow branches | *Crematogaster* spp. |
| *Psychotria myrmecophila* | Bush with pouch-like stipules | Undetermined ants |
| **Scrophulariaceae** (figworts, foxgloves, and relatives) | | |
| *Wightia borneensis* | Upper branches hollow, possibly excavated by ants; status as true domatia not established | Undetermined ants |
| **Thymelaeaceae** (mezereons, leatherwoods, and relatives) | | |
| *Wikstroemia alata* | Possible domatia | Undetermined ants |
| **Verbenaceae** (verbenas) | | |
| *Callicarpa saccata* | Auriculiform sacs at narrowed lamina bases | Undetermined ants |
| *Clerodendron breviflos*, *C. fistulosum*, *C. myrmecophilum* | Swollen, hollow stems | *Camponotus clerodendri* |

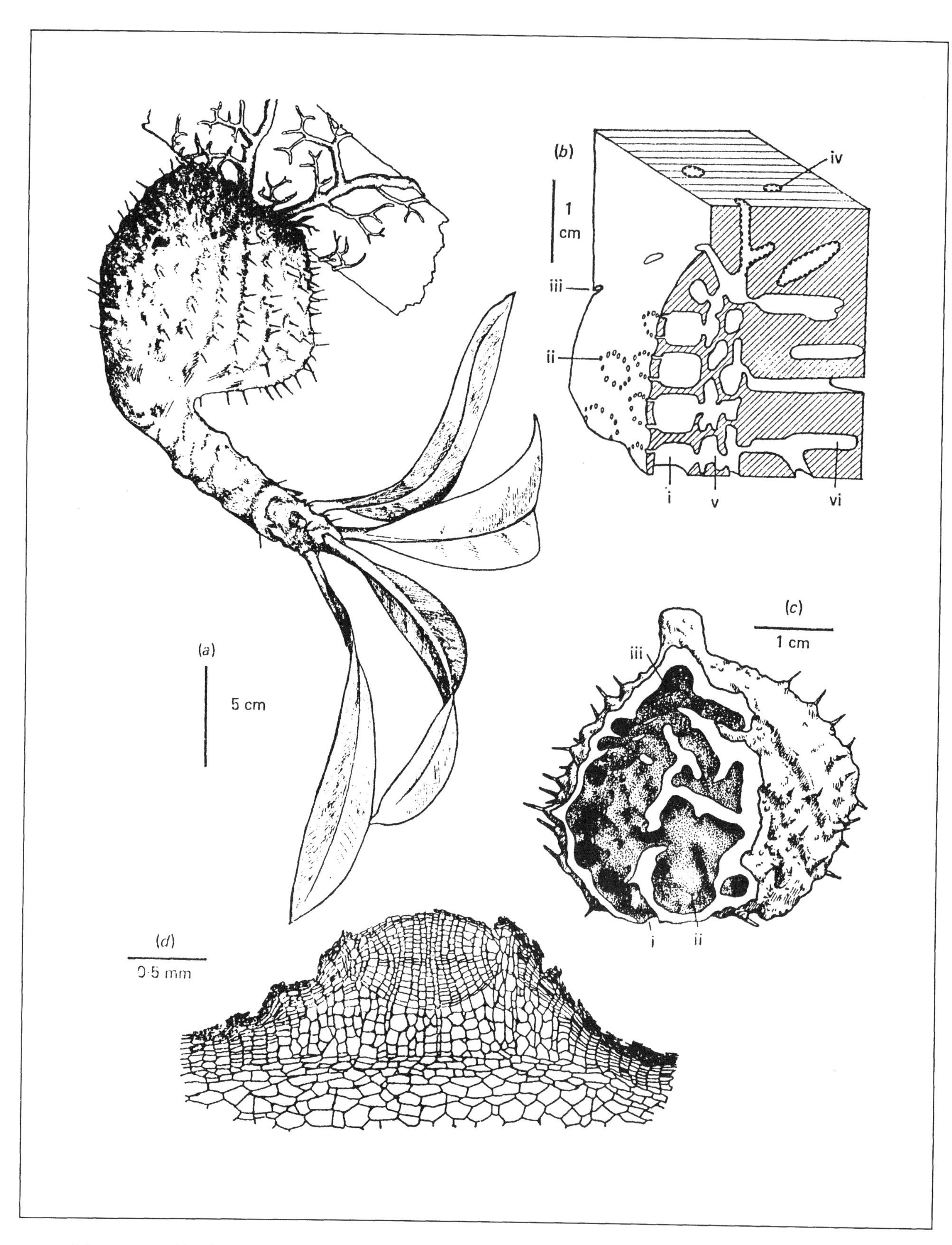


**FIGURE 14–4** Epiphytes of the rubiaceous genus *Myrmecodia* possess some of the most complex domatia (structures housing ants) known in the plant kingdom. (*a*) A whole young plant of *M. tuberosa* of New Guinea is anchored by its root system to a tree branch. (*b*) A diagrammatic block section of the tuber shows the intricate living quarters provided by the plant for its guest ants (*Iridomyrmex cordatus*): (*i*) one cell of the honeycomb; (*ii*) pores leading to the top of a cell, which may aerate and control the cell temperatures; (*iii*) entrance hole used by the ants; (*iv*) tunnel with absorptive warts; (*v*) smooth, dark tunnel; (*vi*) smooth, light-colored tunnel. (*c*) Tuber of young plant to show (*i*) entrance hole, (*ii*) inner smooth chamber favored by the ants, and (*iii*) the dark tunnels with warts that absorb nutrients for the plant. (*d*) Longitudinal section of an absorptive wart, showing cell outlines. (From Huxley, 1978 and 1980.)

**FIGURE 14–5** A species of *Cecropia*, a genus of highly specialized Neotropical myrmecophytes usually occupied by ants of the genus *Azteca*. (*a*) View of entire plants in Costa Rica; (*b*) close-up view of a petiolar base pad bearing Müllerian bodies; (*c*) an *Azteca* worker enters a hole leading to a nest cavity in one of the hollow internodes.

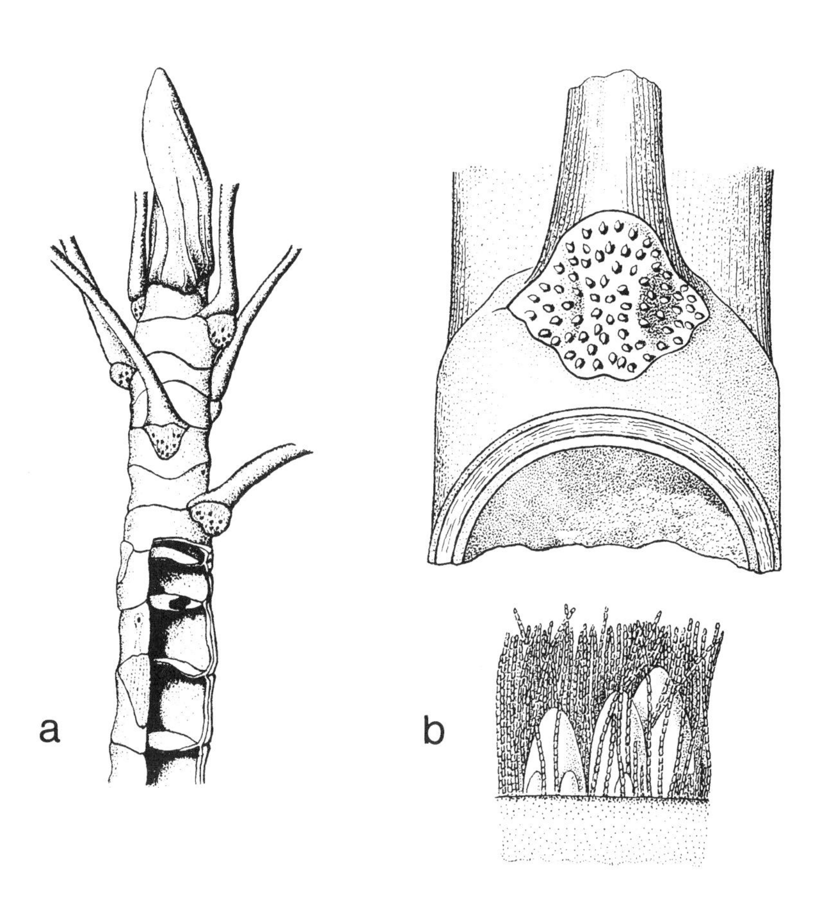


**FIGURE 14–6** *Cecropia adenopus:* (*a*) The growing tip of a young tree, cut away below to show the hollow internodes where the ants live, and the petiole above with the pads bearing Müllerian bodies on which the ants feed. (*b*) A close-up of the petiole base pad with Müllerian bodies, and a magnified side view showing egg-shaped Müllerian bodies in various stages of development. (Modified from Jolivet, 1986; based on Gadeceau, 1907, and Dumpert, 1978.)

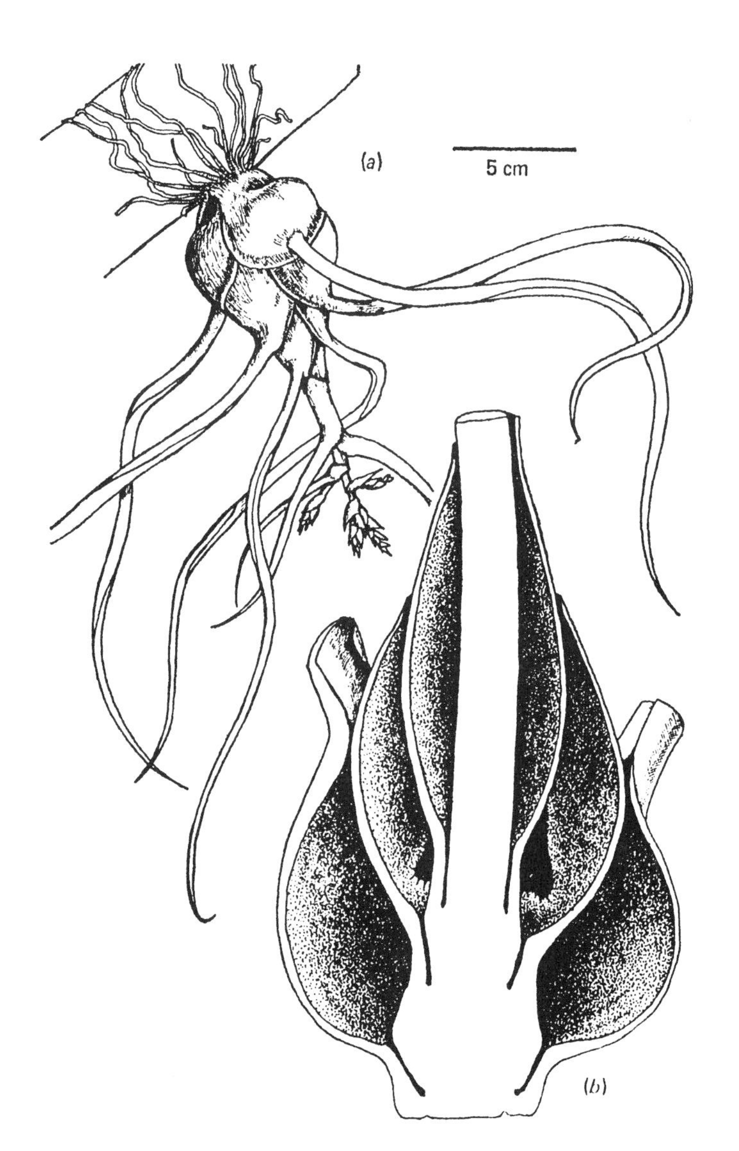


**FIGURE 14–7** *Tillandsia bulbosa,* a New World epiphyte in which domatia for *Crematogaster* ants are created by an inflation of the leaf bases. (*a*) A whole flowering plant; (*b*) a sagittal section of the base of the plant. (From Huxley, 1980.)

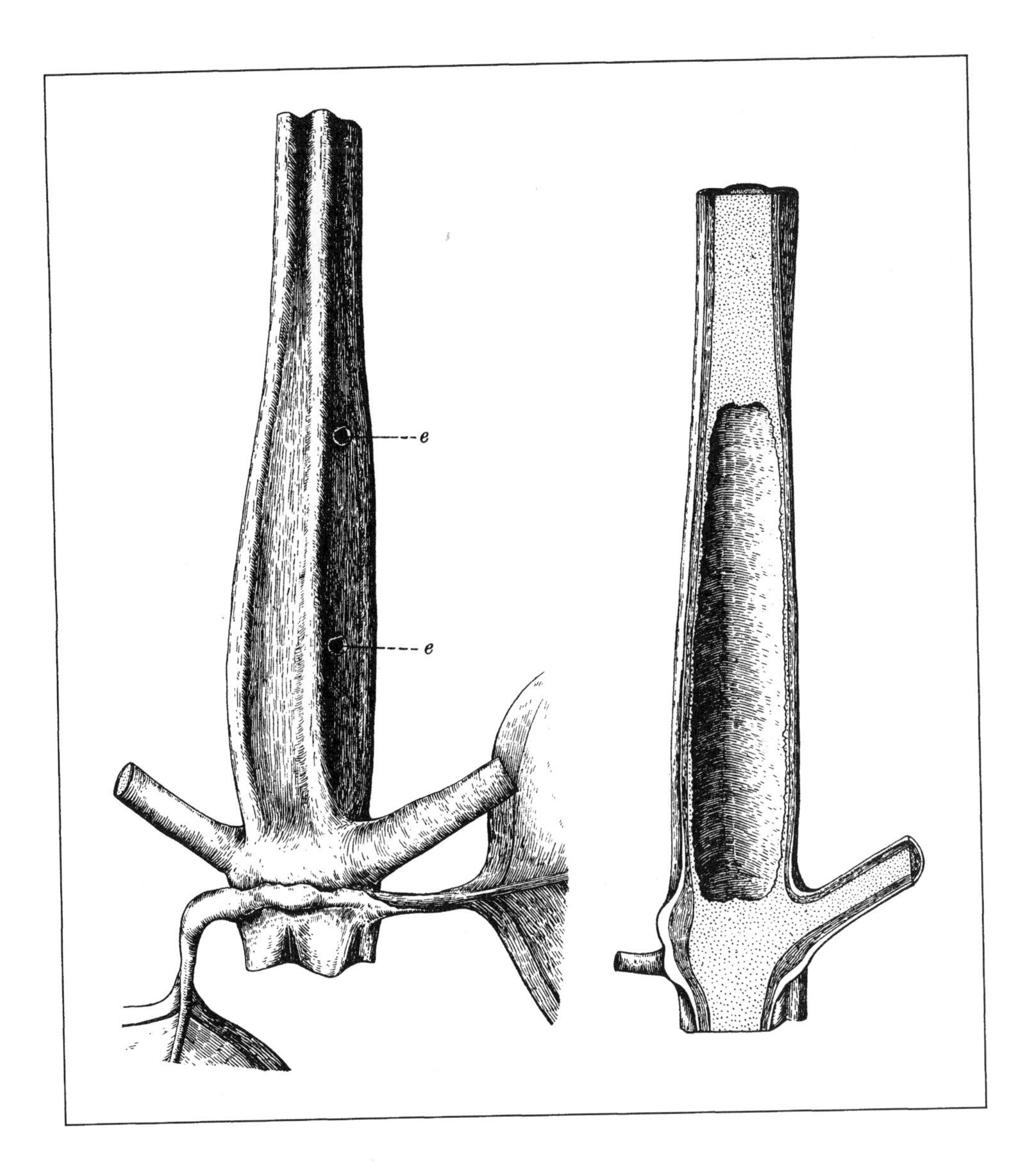


**FIGURE 14–8** *Canthium laurentii*, a tropical African myrmecophyte that harbors ant colonies in its inflated internodes. (*Left*) Intact internode showing entrance holes (*e*) made by the ants. (*Right*) Internode wall cut away to show the natural cavity inside. (Modified from Bequaert, 1922b.)

**TABLE 14–2** Food body–ant associations in five plant genera. (Modified slightly from Beattie, 1985.)

| Characteristics | *Acacia* sp. (Leguminosae) | *Cecropia* sp. (Moraceae) | *Macaranga* sp. (Euphorbiaceae) | *Ochroma* sp. (Bombacaceae) | *Piper* spp. (Piperaceae) |
|---|---|---|---|---|---|
| Major ant associates | *Pseudomyrmex* | *Azteca* | *Crematogaster* | *Solenopsis, Azteca* | *Pheidole* |
| Type of food body | Beltian, tip of pinnule and rachis | Müllerian, produced in large numbers on trichilium (pad of tissue at base of petiole) | Beccarian, on stipules or young leaves | Pearl body, on leaves and stems of sapling | On petiole margins |
| Principal nutrient offered | Protein, lipid | Glycogen, lipid | Lipid, starch, some protein | Lipid, perhaps some starch and protein | Lipid, protein |
| Anatomy | Multicellular, tissues differentiated | Multicellular | Multicellular | Multicellular | Single celled |
| Extrafloral nectaries present? | Yes | No | No | Yes | Yes? |
| Domatia present? | Yes | Yes | Yes | No | Yes? |
| Plant provides complete diet for ants? | Yes | Together with coccid honeydew, probably yes | Together with coccid honeydew, probably yes | No | Yes? |
| Do ants protect plants? | Yes | Yes | Probably | Probably | Probably |

tants belong to two species. As many as a dozen colonies are started in this manner in each tree. When the first workers emerge, they cut holes through the septa separating the internodes. Fighting and other forms of competition ensue, and all of the young colonies except one are either destroyed or perhaps even assimilated, so that in the end only one large colony and a single nest queen survive (D. Perlman, personal communication). The same reductive sequence occurs in *Pseudomyrmex ferruginea*, a resident of swollen-thorn acacias (Janzen, 1967), as well as *P. triplarinus*, an obligate resident of *Triplaris* (Schremmer, 1984).

The pervasiveness of competition is indicated by the patchiness of distribution of the ants among plants of the same species. At the Manu National Park, Peru, Davidson et al. (1989) found eight myrmecophyte species. Among 130 plants dissected, 127 contained ant colonies, and of these 126 belonged to one species only. In many instances the structure of the plant was found to bias which ant species could colonize it. *Maieta guianensis*, for example, has dense epidermal hairs (trichomes) covering the leaves and stems. Its usual guest ants, the tiny *Pheidole minutula*, are able to walk through the hairs without difficulty. A second and less common ant, a *Crematogaster* belonging to the *victima* group, apparently has to cut trails through the trichomes (Davidson, personal communication). In addition, the narrow tunnel entrances leading into the pouch-like domatia are readily entered by the *Pheidole* queens during colony founding but not by the *Crematogaster* queens, who are forced to chew holes into the plant. *Allomerus demararae* enjoy a similar advantage in the occupancy of the myrmecophyte *Cordia nodosa* over species of *Crematogaster* and *Azteca*, which must cut trails through the trichomes just to move around (Davidson et al., 1989). In short, certain ant species occupy individual myrmecophytes predominantly and other species intrude occasionally, but only one species is found in each plant. The final result of this competitive pressure is that the myrmecophytes are saturated with ants.

## ANT GARDENS

Perhaps the most complex mutualism between plants and ants is the ant garden, which is an aggregate of epiphytes assembled by ants. The ants bring the seeds of the epiphytes into their carton nests. As the plants grow, nourished by the carton and detritus brought by the ants, their roots become part of the framework of the nests. The ants also feed on the fruit pulp, the elaiosomes (food bodies) of the seeds, and the secretions of the extrafloral nectaries.

It is a curious fact that although epiphytic myrmecophytes are generally diverse and abundant throughout the tropics, they form ant gardens principally in Central and South America, with some examples having been recently found in tropical Asia by D. W. Davidson (personal communication). First reported by Ule (1902), the gardens are typically round or ellipsoidal, and they range from 6 to more than 60 centimeters in greatest diameter; a typical example is shown in Figure 14–9. The ants construct irregular nest chambers divided by carton walls among their roots (Weber, 1943; Kleinfeldt, 1978, 1986). Specialized garden plants representing no fewer than 16 genera have been identified to the present writing (Buckley, 1982b; Davidson, 1988). At one locality in the Manu National Park alone, Davidson identified 10 such specialists belonging to the following 7 families: Araceae (*Anthurium*, 2 species; *Philodendron*), Bromeliaceae (*Neoregelia*, *Streptocalyx*), Cactaceae (*Epiphyllum*), Gesneriaceae (*Codonanthe*), Moraceae (*Ficus*), Piperaceae (*Peperomia*), and Solanaceae (*Markea*). *Peperomia macrostachya* was by far the dominant species, occurring in 76 percent of the gardens. It was followed by *Anthurium gracile* and *Ficus paraensis*, with 29 and 23 percent occurrence rates respectively. Seven other genera in 6 families were represented among the unspecialized, adventitious species in the gardens. In general the dominant plants of gardens in South and Central America are 5 species of *Codonanthe*, followed in abundance by *Aechmea mertensii* and *Anthurium gracile* (Kleinfeldt, 1986).

The dominant ants in the ant gardens across Central and South America generally are members of the genera *Crematogaster*, *Solenopsis*, *Azteca*, *Monacis*, and *Camponotus*, with *Anochetus* and *Odontomachus* occurring much less commonly (Kleinfeldt, 1986). The overwhelmingly dominant species in the Peruvian Amazon are *Crematogaster parabiotica* (broadly defined) and *Camponotus femoratus*. These forms appear to be the most abundant ant species in the forest canopy generally (Wilson, 1987a; Davidson, 1988). Other, possibly adventitious, ant-garden species belong to the genera *Odontoma-*

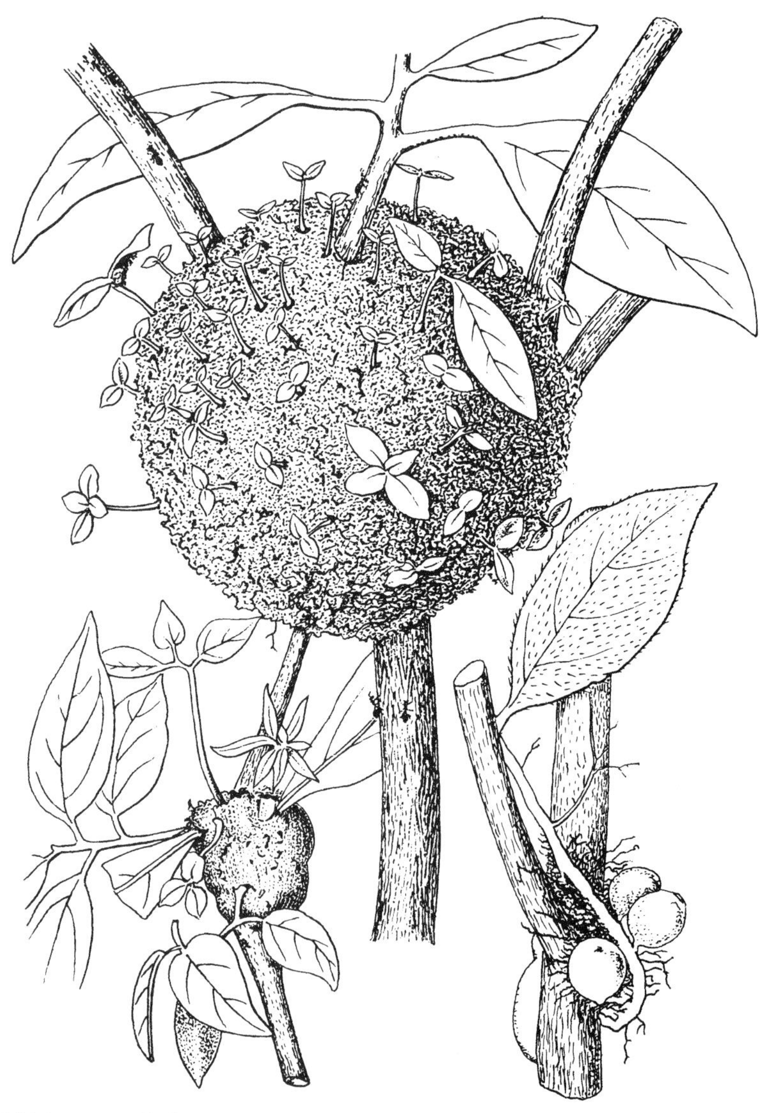

**FIGURE 14–9** Different stages in the development of Amazonian ant gardens. (From Ule, 1902.)

*chus, Anochetus, Monomorium,* and *Solenopsis* (Mann, 1912a; Wheeler, 1921a; Macedo and Prance, 1978). *Monacis debilis* is commonly found in the same garden with one sibling species in the *Crematogaster parabiotica* complex and *Camponotus femoratus* with another. The species pairs nest in separate but contiguous chambers within the same garden; this is the condition originally called parabiosis by Forel (1898). The *Monacis* at least follow the *Crematogaster* out along their odor trails and aggressively displace them at food sites (Swain, 1980).

The tightness of the associations suggests that the plants and ants in the gardens have coevolved, but surprisingly little experimental evidence has been adduced to test this supposition. Kleinfeldt (1978) showed that *Codonanthe crassifolia,* one of the specialized garden plants, grows more quickly when in association with *Crematogaster longispina* than when alone. It is probable that the carton of ant nests provides a physical substrate laced with nutrients preadapting them for epiphytic growth. Longino (1986) observed that *Crematogaster longispina* occurs throughout the Atlantic rain forest of Costa Rica, often as a dominant element of the understory fauna. Its large, diffuse colonies manufacture loose, coarse-fibered carton to build large numbers of nests, shelters for scale insects, and galleries that connect all of these components. The bulk of the carton is placed under and around the roots and stems of aroids and gesneriads running up and down the tree trunks. Whenever new carton is laid down, there is usually a flush of newly sprouted epiphytes in the walls. As Longino points out, it is but a short further step in evolution to nurture an obligate garden species such as *Codonanthe crassifolia.*

How do individual symbiotic gardens get planted in the first place? Ule (1905, 1906) observed that ants retrieve the seeds of the epiphytes, and he proposed that they establish the gardens within the confines of their own nests to start the symbiosis. Wheeler (1921a) argued that the gardens may instead be autonomous, with the ants colonizing the plants later. The evidence has consistently favored Ule. Later authors, including Madison (1979), Kleinfeldt (1978), and Davidson (1988), have repeatedly observed that the symbiont ants are strongly attracted to the fruits and seeds of the garden epiphytes. After the workers consume the adhering fruit pulp and elaiosomes, they place the seeds near their own brood piles. Workers of *Camponotus femoratus* have been observed to build epiphytic gardens over the carton chambers of *Crematogaster parabiotica* in this manner. Davidson noted that the myrmecophyte seeds remain attractive even after they pass through the digestive tracts of frugivorous bats. She suggested that they contain pheromone-mimicking substances rather than just food material. Seidel (1988) has identified several of the compounds as 6-methyl-methylsalicylate, benzothiazole, and a few phenyl derivatives and monoterpenes. Many precedents exist in the behavior of myrmecophilous arthropods that gain entrance to their hosts' nests by pheromonal mimicry (see Chapter 13). It may also be significant that at least some ant species not associated with gardens are repelled by the seeds. Finally, Davidson (personal communication) tested Wheeler's competing hypothesis by setting out garden epiphytes in sites where they could be found by the symbiotic ants. None was colonized.

We may summarize our existing knowledge of the ant gardens by noting that the epiphytes restricted to the gardens appear to be truly adapted to this symbiosis. Their seeds are transported to favorable sites by the ants in response to what appear to be specialized attractive substances, and the subsequent growth of at least some of the species is enhanced by the presence of the ants. For their part the ants are not so clearly adapted to benefit from the ant gardens. They feed on the extrafloral nectar, fruit pulp, and elaiosomes supplied by the plants. This is not a restrictive diet, however, and in fact all of the garden ant species forage away from the gardens. The best evidence that some ants have coevolved with the plants is that the dominant garden species of South America, *Monacis debilis, Camponotus femoratus,* and the two *Crematogaster* in the *parabiotica* group, usually if not invariably nest in the gardens. They are behaviorally specialized to some extent for bringing the seeds of the epiphytes to their carton nests, in effect planting the gardens.

## PLANTS FEED ANTS: FOOD BODIES

The term "food body" or "food corpuscle" applies to any small epidermal structure that is collected and eaten by ants. The principal food bodies discovered so far are named and characterized in Table 14–2. They are extremely diverse in origin and form. For the most part they are best developed in myrmecophytes, that is, plants with domatia, a circumstance reinforcing the judgment that they are beneficial to the ants. Relatively little is known, however, about the actual use of the corpuscles. Ants have been seen collecting them from only a few of the plant species, and little effort has been made to document their consumption by either adults or larvae. Also, the biochemistry of attraction and nutrition of the food bodies remains largely unexplored. Nevertheless, *Pachycondyla luteola, Camponotus balzani,* and species of *Azteca* are wholly dependent on *Cecropia,* and *Pseudomyrmex* on *Acacia,* for their food, making it clear that the mutualism is obligate in at least one direction (Davidson et al., 1988). It is further true that the *Cecropia* and *Acacia* decline when deprived of their ants, so that the relation is truly mutualistic (Janzen, 1966, 1967, 1969; Schupp, 1986). The same is true of the myrmecophyte *Triplaris americana* (Davidson et al., 1989).

Another line of evidence is the evolutionary reduction of the myrmecophytic structures in localities lacking guest ants altogether. At least two species of *Azteca* were abundant on Hispaniola in early Miocene times, but the genus is entirely absent from the Greater Antilles today (Wilson, 1985e). *Cecropia* occurs through the Lesser Antilles to Puerto Rico, where *C. peltata* grows entirely in the absence of resident ants (Janzen, 1973a). And in different populations from Trinidad northward through the Lesser Antilles to Guadeloupe, *C. peltata* shows a progressive reduction of ant-related traits (Rickson, 1977).

In almost all of the principal myrmecophyte genera listed in Table 14–2, the food bodies grow spontaneously. They develop fully on plants reared in greenhouses in the absence of the guest ants. The exception is the genus *Piper.* In *P. cenocladum* at least, the leaf-margin corpuscles are produced only when the plant is occupied by *Pheidole bicornis.* Food-body production declines precipitously when the ants are removed, and it commences again when the ants are restored (Risch and Rickson, 1981).

## PLANTS FEED ANTS: EXTRAFLORAL NECTARIES

Extrafloral nectaries are sugar-producing organs that attract animals but do not promote pollination. They can occur according to species almost anywhere on the plant, including even the flower outside the perianth. When active they attract worker ants, who tend to defend them from other insects. Unlike domatia and food bodies, they are produced by an enormous diversity of plants, occurring in no fewer than 68 families (Elias, 1983). Their anatomical variety is correspondingly great, ranging from small groups of cells that can be located with the naked eye only by detecting the nectar droplets they secrete or the groups of ants that gather around them, to complex cavities filled with trichomes and opening to the outside by a slot or pore. A very large literature exists on the subject, and fortunately this has been well reviewed in recent years by Bentley (1977), Buckley (1982a,b), Elias (1983), Koptur (1984), Beattie (1985), Benson (1985), and Oliveira and Leitão-Filho (1987).

Worker ants treat extrafloral nectaries in the same way that they respond to aggregations of honeydew-producing insects and sugar baits (Figure 14–10). The more aggressive species defend the active nectaries, in some cases extending their territorial zone over the entire plant. In 1889 Wettstein performed a surprisingly modernistic experiment that showed the protective role of the attending ants in the European composites *Jurinea mollis* and *Serratula lycopifolia*. He excluded ants from the plants and recorded an increase in damage from beetles and hemipteran bugs. Many similar studies have yielded the same result, as for example that of Oliveira et al. (1987b) in the cerrado woodland of Brazil. The plant species tested have been variously temperate or tropical in origin; vines, shrubs, or trees; and occupants of either forests or grasslands. The ants tested have belonged to several genera in the Myrmicinae and Formicinae. The kinds of damage averted or reduced by the ants have included destruction of flower parts by grasshoppers, seed predation by bruchid weevils, and withering of shoot tips by psyllid nymphs. In some cases the authors directly observed the ants killing or chasing away the herbivores.

There is also some evidence that plants time their secretions in a way that enhances the protective role of the nectaries. In Michigan the nectaries of the North American black cherry (*Prunus serotina*) are most active during the first three weeks after budbreak, at which time they attract large numbers of *Formica obscuripes* workers. It is probably not a coincidence that the same three-week period is the only time that eastern tent caterpillars (*Malacosoma americana*), which are the major defoliators of the black cherry, are small enough to be captured and killed by the ants. It is a reasonable hypothesis that *P. serotina* has evolved to bring its extrafloral nectaries into play when they can indirectly inflict the greatest amount of damage on the caterpillars (Tilman, 1978).

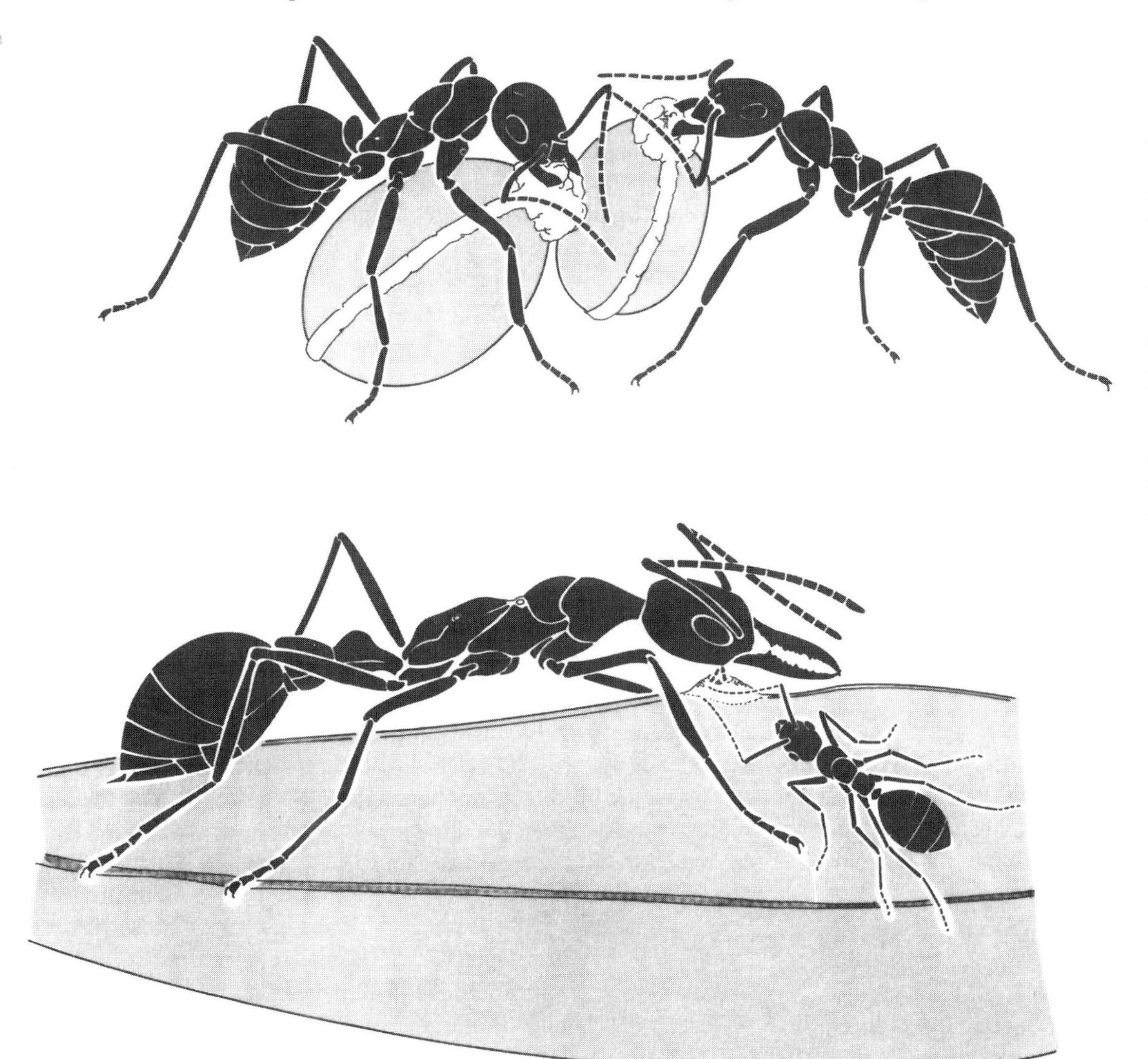

**FIGURE 14–10** The two principal ways in which plants attract and feed ants for their own advantage. (*Above*) Two workers of *Formica podzolica* from the northern United States gather seeds from a violet (*Viola nuttallii*). The ants manipulate the seeds by their elaiosomes, specialized food bodies that will later be eaten by the colony. (*Below*) Two workers in Australia attend an extrafloral nectary on a phyllode of *Acacia*. The large individual is a bulldog ant (*Myrmecia pilosula*) and the small one is a meat ant of the genus *Iridomyrmex*. (Modified from Beattie, 1985; drawing by E. Kaiser.)

## ANTS FEED PLANTS: MYRMECOTROPHY

Ant nests are generally among the most favorable sites for plant growth. The ants turn and aerate the soil, add nutrients in the form of excrement and refuse, and hold the ambient temperature and humidity at moderate levels. Larger nests are often surrounded by more luxuriant and species-rich vegetation than similar but unoccupied sites nearby. The difference is especially conspicuous in deserts, grasslands, and the early successional stages of forests (Gentry and Stiritz, 1972; Beattie and Culver, 1977). And at least in the chalk grasslands of England (King, 1977) and the deserts of Arizona (Rissing, 1986), some plants are much more abundant around the nests and others much scarcer, creating a striking floristic heterogeneity. On the foreshores of salt lakes in the southwestern Baraba Steppe of the Soviet Union, ants construct large hummocks that play a key role in plant succession (Pavlova, 1977).

This general potency of ant nests for the stimulation of plant growth preadapts the myrmecophytes to draw nourishment from their guest ants. Just such a relationship has been documented in the epiphytic myrmecophytes of the genera *Hydnophytum* and *Myrmecodia* (see Figure 14–4). Janzen (1974) found that the workers of *Iridomyrmex cordatus* (= *I. myrmecodiae*) discard the remains of prey in the cavities lined with absorptive tissues, while sequestering their own brood in separate chambers lined with tough, nonabsorptive cells. The absorptive surfaces are dotted with small lenticular warts. Janzen suggested that each of the two zones serves a separate function, namely the feeding of the plant and the housing of the ant brood. Using radioactive tracers, Huxley (1978) and Rickson (1979) demonstrated that this differentiation is indeed the case. The pseudobulbs absorbed ($^{32}$P) phosphate, ($^{35}$S) sulfate, and ($^{35}$S) methionine from waste material deposited by the *Iridomyrmex*, as well as various breakdown products of decomposing *Drosophila* larvae. Most of the activity was concentrated in the warted areas. In short, the ants feed the plants.

Many of the most specialized plants of this kind are tropical epiphytes in open forests and savannas located on nutrient-poor soils. In such areas there are typically few other epiphytes, and when they occur they usually grow on top of the myrmecophyte. Janzen (1974), Huxley (1980), and Thompson (1981) have all suggested that myrmecotrophy allows the plants to penetrate harsh environments otherwise closed to epiphytes.

Further research will no doubt discover new botanical structures and physiological processes that serve the capture and absorption of nutrients supplied by ants. The flask- and bladder-like domatia of *Dischidia*, for example, are penetrated by networks of adventitious roots that almost certainly absorb nutrients. *Crematogaster* and other ants that live and accumulate organic detritus on these roots are in fact literally "potting" the *Dischidia*.

Myrmecotrophy is not limited to plants with domatia. It inevitably occurs whenever ants nest on and around epiphytes, carrying in soil, building carton, and discarding waste materials. An entire fauna of arboreal ant species favor epiphytes for nesting. Only a small fraction of the epiphyte species are myrmecophilous, in other words domatia bearing, and fewer than 1 percent of the ant species are closely associated with domatia. Huxley (1980) has pointed out that the domatia-bearing epiphytes are mostly limited to tropical Asia. Their relative scarcity in the New World tropics may be a result of competition from the immensely successful tank bromeliads and the ant-garden epiphytes. At least in the case of the ant-garden epiphyte *Codonanthe crassifolia*, the study by Kleinfeldt (1978) revealed that the presence of ants promotes growth, very probably as a result of increased nutrient supply. It is a remarkable fact that orchids, the other dominant group of vascular epiphytes, are also dependent on symbiosis for their early nutrition, but in this case the symbiosis is a mycorrhizal association with fungi rather than the harboring of ants.

## ANTS DISPERSE PLANTS: MYRMECOCHORY

Harvesting ants do not manage to carry all the seeds they collect back to their nests, and they do not eat all of the seeds stored in their granaries. The result is that ants are a major, albeit fortuitous, dispersal agent of plants. They are especially effective in deserts and grasslands, but many species, not necessarily specialized harvesters, play some role even in tropical forests.

A wholly different, outwardly "purposeful" category of dispersal is accomplished by plants through myrmecochory, the employment of attractive seed appendages and chemicals that induce the ants to transport the seeds without harming the embryo or endosperm. The phenomenon was first carefully analyzed, and the relevant term proposed, by Sernander (1906). The appendages, which are called arils or elaiosomes, are biochemically distinctive and often large in size (see Figure 14–10). They take various forms according to species, including girdles, sheaths, caps, and finger-shaped terminal extrusions. They have been derived from several kinds of tissues, including the raphe, pericarp, and receptacle (Sernander, 1906; Ridley, 1930; Berg, 1979). Fleshy in consistency, white in color, and containing lipids, protein, starch, sugars, and vitamins, they are carried by foragers back to the nests, where they are eaten by the adult workers and larvae (Beattie, 1985).

Myrmecochory is an almost worldwide phenomenon. Three plant groups have been identified to date in which it is especially common: early-flowering herbs in the understory of north temperate mesic forest (Culver and Beattie, 1978; Pudlo et al., 1980; Beattie and Culver, 1981; Handel et al., 1981), perennials in Australian and southern African dry heath and sclerophyll forest (Berg, 1975; Westoby et al., 1982), and an eclectic assemblage of tropical plants (Horvitz and Beattie, 1980). Of course myrmecochores occur elsewhere; they have been well documented, for example, in the North American desert (Solbrig and Cantino, 1975; O'Dowd and Hay, 1980). Also, many parts of the world, for example tropical Africa, have not been carefully explored for myrmecochory. In spite of such bias in sampling, however, there seems to be little doubt that myrmecochory is disproportionately rich in the three habitats cited. In a New York beech-maple woodland, Handel and his co-workers (1981) found that 13 of 45 herbaceous plant species present were myrmecochores. These plants possessed about half of the stems and 40 percent of the aboveground herbaceous biomass. The proportion of myrmecochorous species in some Swedish habitats is 40 percent (Sernander, 1906). Even more impressive is the occurrence of the phenomenon in sclerophyll vegetation growing on sterile soils.

About 1,500 Australian species are known or thought to be myrmecochores (Berg, 1975) and 1,300 South African (Bond and Slingsby, 1983; Milewski and Bond, 1982), as opposed to only 300 throughout the rest of the world. Botanists who have analyzed myrmecochory have concluded that the shortage of nutrients, particularly phosphorus and potassium, is the key to its irregular geographic distribution (Westoby et al., 1982; Milewski and Bond, 1982). Quite possibly plants rely on the relatively small, inexpensive myrmecochores in preference to the larger, fleshy fruits favored by birds and mammals whenever nutrients are in chronic short supply.

The myrmecochorous plant species are phylogenetically diverse. In Australia alone they are scattered through no fewer than 87 genera in 23 families. That fact, plus the extraordinary variety in the embryological provenance of the myrmecochores themselves, indicates that myrmecochory has originated many times and undergone a great deal of convergence in evolution. A few of the independent phylogenetic pathways have been worked out by Berg (1972, 1975, 1979).

The ants attracted to the myrmecochores are a similarly diverse lot, representing some of the dominant genera peculiar to each local fauna in turn: *Aphaenogaster, Leptothorax, Myrmica, Tapinoma, Formica,* and *Lasius* in north temperate forests (Culver and Beattie, 1978, 1980); *Odontomachus, Pachycondyla, Pheidole, Azteca,* and *Paratrechina* in at least one Neotropical forest locality (Lu and Mesler, 1981); and *Rhytidoponera, Pheidole, Monomorium,* and *Iridomyrmex* in Australia (Buckley, 1982b). At least two ant species, *Pogonomyrmex californicus* and *Messor (= Veromessor) pergandei,* are bivalent; they harvest some seeds for total consumption and disperse others after collecting them for their elaiosomes (O'Dowd and Hay, 1980).

The existence of an elaborate structure with no apparent function other than to attract ants implies that a substantial selective advantage accrues to the myrmecochores. Botanists have given a good deal of thought to what this advantage may be, and they have come up with five possibilities.

1. *Avoidance of interspecific competition.* If the ants carry the seeds of a particular plant species to sites where other, competing species grow less well, myrmecochory will be favored by natural selection. Handel (1978) has adduced some evidence for this effect in forest-dwelling *Carex.*

2. *Avoidance of fire.* As Berg (1975) has noted, myrmecochorous plant species are common in fire climax communities. It is possible that seeds carried into ant nests are protected during the frequent burnovers of their habitat. On the other hand, some species require the high temperatures generated by fires in order to germinate. Majer (1982) has shown that in southwestern Australia ants transport many of the seeds to an intermediate depth where the temperature is high enough to germinate the seeds but not high enough to kill them. A similar finding has been made by Bond and Slingsby (1983) in the heathland of the Cape Province.

3. *Avoidance of parental competition.* Young plants do less well near members of their own species, so that there is an advantage to dispersing well away from the parent. In the case of *Ajuga* at least, ants carry elaiosome-bearing seeds beyond the boundaries of parent clones, eliminating parental competition (Lüönd and Lüönd, 1981).

4. *Avoidance of seed predators.* Bond and Breytenbach (1985) found that seeds not dispersed by ants in the South African heathland are subject to much higher predation by small mammals. When seeds are concentrated they are evidently more attractive and easily discovered. Smith et al. (1986) demonstrated that ants rescue a large percentage of the seeds of the North American myrmecochore *Jeffersonia diphylla* from rodents.

5. *Microsites with superior nutrients.* Nests of ants, especially those belonging to large colonies, usually have higher levels of the nutrients important for plant growth (Gentry and Stiritz, 1972; Haines, 1978; King, 1977; Pętal, 1978). It would seem to be advantageous for ant-dispersed seeds to find their way into the ant nests, which in fact is usually the case.

Beattie (1985) has evaluated these five selection pressures on the basis of the still sparse evidence. Although the factors are likely to vary in relative importance according to species and geographic location, some seem to be more generally potent than others. The interspecific competition model, for example, is compromised by its limitation to related species, some of which are myrmecochores and some not. Fire avoidance has so far been documented only by indirect evidence and is obviously limited to fire-dependent vegetation. Avoidance of parental competition is also a reasonable explanation, but it is easily confounded by any competing hypothesis that postulates an advantage to dispersal, such as the increased probability of predator avoidance. In the absence of optimality models that predict the best distances for particular circumstances, the parental-competition hypothesis is *too* good; it can be applied indiscriminately to almost all situations.

Nutrient enrichment and predator avoidance appear to be both the most plausible and the most readily testable of the suggested selection pressures. Unfortunately, few quantitative studies have been conducted at this writing to assay the effect of nutrient enrichment, and the results so far are mixed. Culver and Beattie (1980) obtained a positive result with two ant-dispersed species of violets in the chalk downland of southern England. Seeds planted in ant nests produced plants that were more numerous and larger in size than those planted in alternative, randomly selected microsites. Three years later almost all of the survivors were in the ant nests. In a similar manner, Davidson and Morton (1981) found that ant-dispersed plants of the family Chenopodeaceae in arid scrubland of Australia do better on ant mounds than elsewhere. The effect varies according to the properties of the soil. In habitats underlain by red, crusty, alluvial loam soils, the chenopods are limited almost entirely to ant mounds, whereas in better-aerated and better-drained sandy soils they form almost continuous stands. It thus appears that the Australian ants alter not only the nutrient levels but also the physical properties of the soils they inhabit. The relative contributions of the two factors will be hard to weigh in most studies. This difficulty is underscored by the finding of Rice and Westoby (1986) that myrmecochorous species in nutrient-poor Australian sclerophyll vegetation grow in soils with no more nitrogen and phosphorus than soils around non-myrmecochorous species. Similarly, Horvitz and Schemske (1986) found that seedlings of the myrmecochore *Calathea ovandensis* of Mexico grow no better in pots of soil from ants' nests than in pots of soil from nearby random sites, despite the fact that the former are richer in nutrients.

In general it appears that nutrient enrichment is an important selection force in some habitats, especially where the soils are relatively sterile, but not in others—for reasons yet to be clarified. Rice and Westoby (1986) say generally of the adaptiveness of myrmecochory:

> Many explanations could account for the particular importance of myrmecochory in Australia and South Africa, or for the importance of myrmecochory in sclerophyll as compared with mesophyll vegetation. But few explanations can account for myrmecochory being more important in sclerophyll than in mesophyll vegetation in Australia and South Africa, but vice versa in North America and Europe. Because sclerophyll shrublands in Australia and South Africa are delimited by low-nutrient soils, while in North America and Europe they are climatically controlled, the nutrient-enriched microsite explanation could have accounted for this distribution. However, this explanation evidently does not hold for a representative cross section of Australian sclerophyll species. This throws open again the problem of finding an adaptive explanation that could account for the geographical distribution of myrmecochory.

## ANTS POLLINATE FLOWERS (SPARINGLY)

Given the abundance and antiquity of ants, it is puzzling to find that they play a relatively minor role as pollinators. Ants are significant for a few flowering plants that display the "ant-pollination syndrome," which comprises the following traits: the plants grow in hot and dry habitats where ants are most abundant and active; the nectaries are morphologically accessible to the flightless ant workers; the plants are short and prostrate; pollen volume per flower is small in order not to stimulate self-grooming in the ant through excessive loading of pollen; and seeds are few per flower, thus requiring less abundant pollen transfers (Hickman, 1974). Examples include plant species occurring on the hot dry slope of the western Cascades of Oregon (Hickman, 1974), granite outcrops of the southeastern United States (Wyatt, 1981), and the Colorado alpine tundra (Petersen, 1977). Ant pollination of other kinds of species occurs, as for example in ground orchids with raised stems (Armstrong, 1979; Brantjes, 1981), but it is evidently much less common. The Australian orchid *Leporella fimbriata* is pollinated by pseudocopulation: winged males of the bulldog ant genus *Myrmecia* mistake the flowers for virgin queens and attempt to mate with them, picking up pollenia in the process (Peakall et al., 1987).

In a striking counterpoint, ants tend to be the "scoundrel in the pollination drama" because they dominate the nectaries while contributing little or nothing to pollination (Faegri and van der Pijl, 1979). Their antibiotic secretions from the metapleural and poison glands, used to suppress bacterial and fungal growth in the nests, are likely to interfere with pollen germination and pollen-tube growth (Iwanami and Iwadare, 1978; Nakamura et al., 1982; Beattie et al., 1984, 1985, 1986). These substances, originally characterized by Schildknecht and Koob (1970, 1971) and Maschwitz et al. (1970), include β-indoleacetic acid, phenylacetic acid, and β-hydroxydecanoic acid ("myrmicacin"). Some plant species are said to have evolved defenses against the ants in the form of sticky belts and repellent substances in the petals and nectar (Faegri and van der Pijl, 1979; Janzen, 1977). The extent and effectiveness of these devices remains to be evaluated, however. Janzen suggested that as a rule ants do not visit flowers in lowland tropical habitats, and that the reason is that floral nectar contains unpalatable substances. This generalization was refuted by Haber et al. (1981), who observed ants feeding on the floral nectar of 27 species of plants in Costa Rica. The ants also fed readily on nectar separated from the flowers and presented to them directly in the field, even though in some cases the liquid contained alkaloids and phenolic compounds.

Beattie (1985) argued that the pollen-suppressing activity of myrmicacin and other ant substances is the fundamental constraint that has limited the evolution of ant pollination. Taking the ants' point of view, Beattie suggested that these insects have bought protection against fungi and bacteria in their underground nests at the cost of the nectar that would otherwise have been given them by the flowering plants. From the plants' point of view, Beattie noted that the only species accepting ants as pollinators are those whose habitats leave them little other choice. Beattie discounted the possibility that the peculiarities of ant foraging prevent ants from serving as efficient pollinators, countering with the statement that "ants have sophisticated sensory and orientation systems and systematically visit plants of all sizes, from low herbs to tall trees, to harvest resources as diverse as insect prey, honeydew, extrafloral nectar, and seeds." This is true as far as it goes, but Beattie overlooked another property of the foraging strategy of many species of ants that may serve as a fundamental constraint. Foragers of many if not most species display *Ortstreue*, returning daily to the same plant and even to the same branch or flower cluster (see Chapter 10). Some, including the omnivores most likely to serve as pollinators, even set up shelters over extrafloral nectaries and honeydew-producing homopteran colonies. Thus plants depending on ants for pollination are likely to find themselves being self-fertilized.

## ANTS PRUNE AND WEED

Some specialized plant-dwelling ants protect their myrmecophyte hosts not only from herbivores but also from other plants that crowd in too closely. *Pseudomyrmex ferruginea* workers attack and destroy any foreign plant that sprouts within 40 centimeters of the trunk of the acacia in which they live, and they cut back vines and foliage of neighboring trees that touch the acacia crown (Janzen, 1967). This pruning action has the effect of promoting the growth and survival of the host plant, but it also removes bridges over which alien ants can attack the resident colony. Davidson et al. (1988) found that in Amazonian Peru, *Crematogaster* workers regularly encroach on *Triplaris americana* trees and attack the resident *Pseudomyrmex dendroicus* colonies by interfering with their foraging and even stealing their brood. The *dendroicus* go so far as to sacrifice valuable leaves of their own host plant when they serve as bridges for major invasions by enemy ants. In a suggestive parallel response, workers of *Allomerus demararae* prune vines on their *Cordia nodosa* host plant when they are pressured by the encroachment of alien ants.

Davidson and her co-workers tested the bridge-demolition hypothesis further by constructing many artificial (and indestructible) bridges of wire between *Triplaris* and *Cordia* trees and neighboring vegetation. The result on *Triplaris* was a significant increase in invasion by alien ants, almost all of which were members of the genus *Crematogaster*. The *Cordia* plants were protected by their trichomes.

**TABLE 14–3** The pruning behavior of plant-associated ants as a function of their defensive mechanisms.

| Ant | Plant | Pruning | Circular plot cleared at base of tree | Authority |
|---|---|---|---|---|
| **ANTS WITH STRONG OR WEAK STINGS** | | | | |
| **Ponerinae** | | | | |
| *Pachycondyla luteola* | *Cecropia* sp. | Facultative | No | D. W. Davidson (personal communication) |
| **Pseudomyrmecinae** | | | | |
| *Pseudomyrmex* cf. *triplarinus* | *Triplaris americana* | Obligate | Yes | Davidson et al. (1988) |
| *Pseudomyrmex* spp. | *Acacia* spp. | Obligate | Yes | Janzen (1966, 1967) |
| *Tetraponera* spp. | *Barteria fistulosa* | Obligate | Yes | Bequaert (1922b), Janzen (1972) |
| **Myrmicinae** | | | | |
| *Allomerus demararae* | *Cordia nodosa* | Facultative | No | Davidson et al. (1988) |
| *Allomerus* sp. | *Duroia* sp. | Either obligate or facultative | No | Davidson et al. (1988) |
| *Pheidole bicornis* | *Piper* spp. | Facultative | No | Risch et al. (1977) |
| *P. minutula* | *Maieta guianensis*, *Clidemia heterophylla* | No | No | Davidson et al. (1989) |
| **CHEMICALLY DEFENDED ANTS LACKING STINGS** | | | | |
| **Myrmicinae** | | | | |
| *Cataulacus mckeyi* | *Leonardoxa africana* | No | No | McKey (1984) |
| *Crematogaster parabiotica* | Ant-garden epiphytes and host plants | No | No | Davidson et al. (1988) |
| *Crematogaster* cf. *victima* | *Tococa* cf. *stephanotricha*, *Maieta guianensis* | No | No | Davidson et al. (1989) |
| *Crematogaster* spp. | *Macaranga triloba* | Yes | No | D. W. Davidson (personal communication) |
| *Crematogaster* spp. | *Acacia drepanolobium* | No | No | Bequaert (1922b), Hocking (1970) |
| **Dolichoderinae** | | | | |
| *Azteca* cf. *alfari* | *Cecropia* spp. | Obligate | No | Janzen (1969) |
| *Azteca* sp. | *Randia ruiziana* | No | No | Davidson et al. (1989) |
| *Azteca* sp. | *Cordia nodosa* | No | No | Davidson et al. (1989) |
| *Azteca* sp. | *Tococa* sp. | No | No | Davidson et al. (1989) |
| **Formicinae** | | | | |
| *Camponotus femoratus* | Host trees of ant-garden | No | No | Davidson et al. (1988) |
| *Petalomyrmex phylax* | *Leonardoxa africana* | No | No | McKey (1984) |

Pruning is not practiced universally by myrmecophyte-dwelling ants. As Davidson and her co-workers discovered (see Table 14–3), the practice is almost entirely limited to species that sting opponents, as opposed to those that have non-penetrating stings and rely on poisonous chemical sprays and droplets. It is possible that stings are generally less effective against other ants than are chemical defenses, so that ants using them are forced to rely more on pruning and bridge demolition to fend off invasions.

## TRADE-OFF AND COMPROMISE

Our account to this point has depicted myrmecophytism as a benefit accruing primarily to plants. The case seems especially strong in the extreme myrmecophytes, which are literally planted, fed, and protected by the ants. Yet there has been an undeniable reluctance on the part of plants generally to enter into such obligatory alliances, because only a small fraction of the genera and species in the world have done so. The explanation of why myrmecophytism is a minority phenomenon may lie in the tendency of arboreal ants to maintain herds of scale insects, mealybugs, and other honeydew-producing homopterous insects (see Chapter 13). So whereas the ants bring gifts to their hosts, they exact a price in energy that the plant must donate to the homopteran sap-feeders.

The depredations of the homopterans are a price the plants can readily pay, however, if the benefits conferred by the ants are high enough. Janzen (1979) has proposed that the maintenance of the herds is best regarded as a more or less fixed cost for the plant, the equivalent of maintaining alkaloids and other secondary substances as a standing defense system against herbivores. Some plants pay energy into the manufacture of the secondary substances, while others, the myrmecophytes, pay the ants to act like secondary substances. In both cases energy is invested to obtain a positive net yield of energy in the end. Janzen expressed this concept of a symbiotic balance sheet as follows:

> In at least two complex and well-developed ant-plant mutualisms, African *Barteria* trees and Neotropical *Cecropia* trees, the ants maintain a standing crop of scale insects or other homopterans inside the hollow stems. These animals feed on the plant and provide a major food source for the ants with their bodies or honeydew exudates. The ants are obligate occupiers of the trees and protect the trees from herbivores and vines. The homopterans are zoological devices used by the plants to maintain an ant colony, the ants being directly analogous to the chemical defenses maintained (and paid for) by more ordinary plants.

Janzen concluded with an intriguing conjecture: "I would not expect there to be selection for traits that reduce the 'damage' done by the Homoptera to the level that would debilitate the ant colony and its protection of the tree." Here we see an application of the concept of coevolution in purest form. The myrmecophyte and the guest ants have evolved by common pressures in natural selection, such that they do not let the homopteran populations explode and kill the tree, and the plants do not defend against the homopterans so vigorously that they starve the ant colony. How might such controls be enacted during evolution? It is well known that ants of various kinds often eat some of the homopterans they attend (Carroll and Janzen, 1973; Hinton, 1977). Among the myrmecophyte-dwelling ants this practice is known to occur in at least one species: *Pseudomyrmex triplarinus* regularly kills some of its symbiotic coccids and feeds them to its larvae (Schremmer, 1984). At the same time homopterans often do not build dense populations under natural conditions, especially those prevailing in tropical forests, so that resistance to them may be relatively easy and inexpensive (Beattie, 1985). Under such conditions, if indeed they occur with consistency, the plant species can "relax" and leave all forms of homopteran control, including that of the myrmecophiles, to the ants themselves.

Ecologists have only begun to investigate the subtle relations that can evolve between plants, their protector ants, and their enemies. In the lowland tropical forest at Los Tuxtlas, Mexico, for example, the perennial marantaceous plant *Calathea ovandensis* attracts ants with extrafloral nectaries, and the ants attack herbivores invading the plant. But the ants also tolerate one herbivore, the larvae of the riodinid butterfly *Eurybia elvina*, which provide them with attractive secretions from eversible glands on the dorsal surface of the eighth abdominal segment. Hence the *Calathea* plants are caught somewhere in the balance between the beneficial control of ordinary herbivores by the ants and the harmful tolerance of the butterfly larvae by the same ants. How do these combinations work out? Horvitz and Schemske (1984) showed that when ants were absent and *Eurybia* butterfly larvae present, seed production was lowered by 66 percent, the greatest loss recorded. When ants were present and *Eurybia* absent, the loss was only 33 percent, the least recorded. The remaining combinations of ants and *Eurybia* both present and ants and *Eurybia* both absent yielded intermediate degrees of production loss. Of equal significance, there was considerable variation in effectiveness among the eight ant species recorded, with *Wasmannia auropunctata* conferring the most protection and *Pheidole gouldi* the least.

It has thus been borne out that some ant species are better symbionts from the plants' point of view than others. This circumstance opens the possibility for some species to intrude as parasites into well-organized mutualistic systems. One such parasite of a mutualism is *Pseudomyrmex nigropilosa* of southern Mexico and Central America (Janzen, 1975). It occupies swollen-thorn acacias but, unlike the other 10 known members of the genus that are obligate residents of these myrmecophytes, *P. nigropilosa* provides no protection for the plants. As a result either the acacias are soon killed by herbivores or the *nigropilosa* are replaced by one of the competent species of *Pseudomyrmex*. As expected of such a parasite, *P. nigropilosa* produces reproductives earlier in the life of the colony than is the case with the mutualistic species. In other words it breeds and disperses before the host dies as a result of its incompetence.

The invasion of a sufficiently dominant but inefficient intruder may endanger the very existence of plants dependent on ant services. According to Bond and Slingsby (1984), something approaching this ecological catastrophe is under way in South Africa. The Argentine ant *Iridomyrmex humilis* has invaded a portion of the fynbos, a local form of scrubland. This unusual habitat possesses a very large number of endemic plants, hundreds of which are myrmecochores and thus dependent on ants for their dispersal and interment in the soil. *Iridomyrmex humilis*, a dominant species of South American origin, has replaced the native ants where it has invaded the fynbos of the Kogelberg State Forest. In field tests Bond and Slingsby found that the *Iridomyrmex* are much slower than the native ants in removing seeds of the proteaceous *Mimetes cucullatus*, a representative myrmecochore of the region. The *Iridomyrmex* also move the seeds shorter distances and then leave them on the soil surface, where they are quickly eaten by invertebrate and small vertebrate granivores. In one trial 35 percent of the *Mimetes* seeds disseminated from depots germinated in *Iridomyrmex*-free sites, but fewer than 1 percent germinated in a nearby infested site.

## PARASITES OF ANTS: MICROORGANISMS AND FUNGI

Relatively little is known about the pathobiology of ants. The species studied most intensely to date is the red imported fire ant *Solenopsis invicta*, for which entomologists have sought—in vain—a biological control agent. *S. invicta* is evidently typical of ants in that microorganisms and fungal parasites are relatively scarce and few in species. They include a virus, one possible species of bacterium, a unicellular fungus (occurring in more than 90 percent of the colonies and not a serious pathogen), two microsporidians, and two neogregarines (Jouvenaz, 1986).

A peculiar haplosporidian parasite was discovered in the European thief ant *Solenopsis (Diplorhoptrum) fugax* by Karl Hölldobler (1929b, 1933), which he described as *Myrmicinosporidium durum*. Because the microorganism takes on the shape of a little bowl when placed in fixative, the disease is called *Näpfchenkrankheit* (small-bowl sickness) in the German literature (Figure 14–11). The parasite was subsequently found in the genus *Leptothorax* (Gösswald, 1932; Buschinger and Winter, 1983a) and in *Pheidole pallidula* (Espadaler, 1982b). Crosland (1988) discovered lemon-shaped objects in the bulldog ant *Myrmecia pilosula*, which appear to be the spores of a protozoan gregarine parasite.

Several kinds of fungi parasitic on ants have been studied by entomologists and mycologists, in some cases for generations. The first is the genus *Cordyceps* (Ascomycetes: Clavicipitaceae), the largest of the entomophagous fungi and often brightly colored as well. The multicellular mycelium pervades the host thoroughly, killing it. It often produces asexual spores or conidia, in which case the specimen has been placed in "*Isaria*," a genus of convenience classified as one of the imperfect fungi. Eventually the mycelium sprouts a boll- or club-shaped organ on a stalk that protrudes as much as 10 centimeters outside the body of the host. The swollen terminus of this structure contains numerous ascocarps, each generating spores within elongate cells or asci (Thaxter, 1888; Rogerson, 1970; Evans and Samson, 1982, 1984; Samson et al., 1982). It is a startling experience to encounter a large ant, dead yet standing rigidly at attention with a *Cordyceps* sporophore raised above it like a flag (see Figure 14–12). At least some species of the genus occur on more than one species of insect host.

The fungus *Desmidiospora myrmecophila* was discovered by Thaxter (1891; cited in Bequaert, 1922) "growing luxuriantly on a large black ant fastened to the underside of a rotting log in Connecticut." Thaxter noted that "the hyphae, much branched and septate, covered the host in a white flocculent mass; they emerged especially from between the abdominal segments, enveloping the insect more or less completely and extending a short distance over the substratum." W. M. Wheeler later identified the ant as *Camponotus pennsylvanicus*.

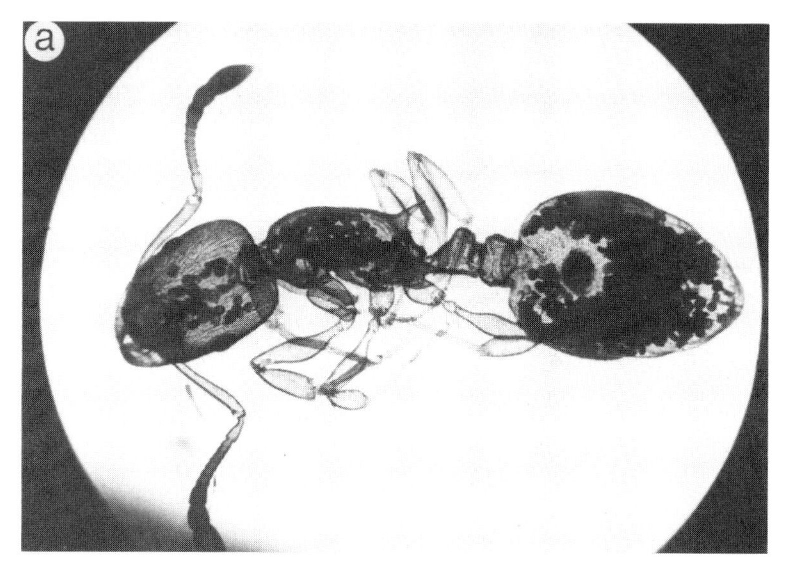


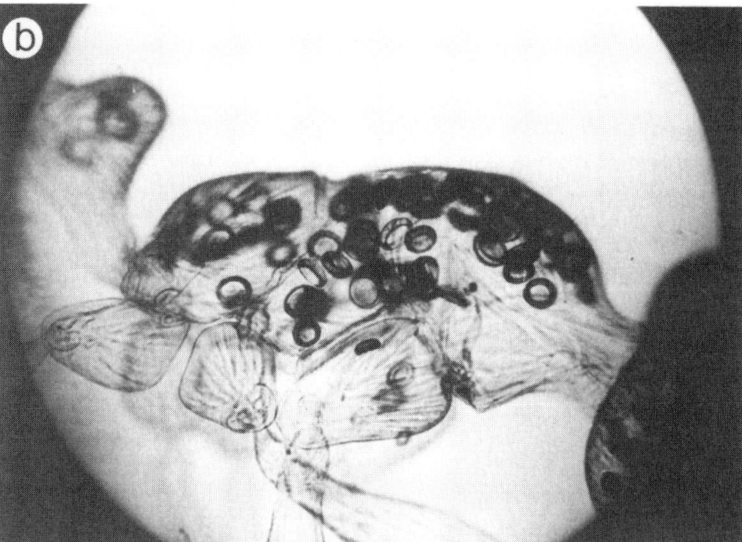


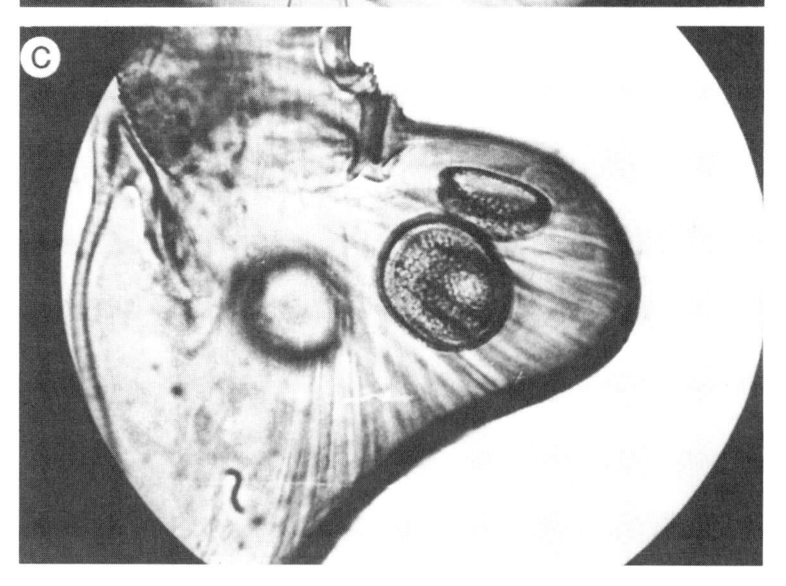


**FIGURE 14–11** *Myrmicinosporidium durum*, a haplosporidian parasite of *Solenopsis fugax* and members of the genus *Leptothorax*. The hosts in these photographs are *Leptothorax* workers: (*a*) a worker completely filled with *Myrmicinosporidium* (from Hölldobler, 1933); (*b*) thorax and (*c*) petiole of another worker, showing the bowl-shaped haplosporidian at higher magnification. (Photographs in *b* and *c* by W. Kloft.)

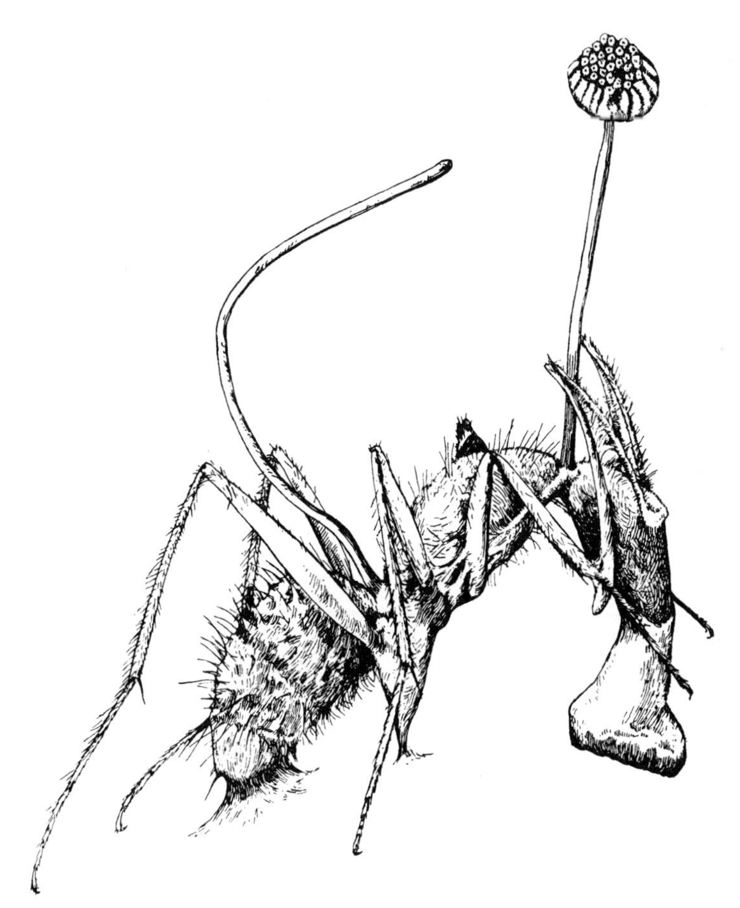

**FIGURE 14–12** A worker of the ant *Camponotus abdominalis* in Guyana has been pervaded and killed by a mycelium of *Cordyceps subdiscoidea*, a parasitic ascomycete fungus. The sporophore of the fungus is borne on a stalk that emerged from behind the ant's head. (From Bequaert, 1922b.)

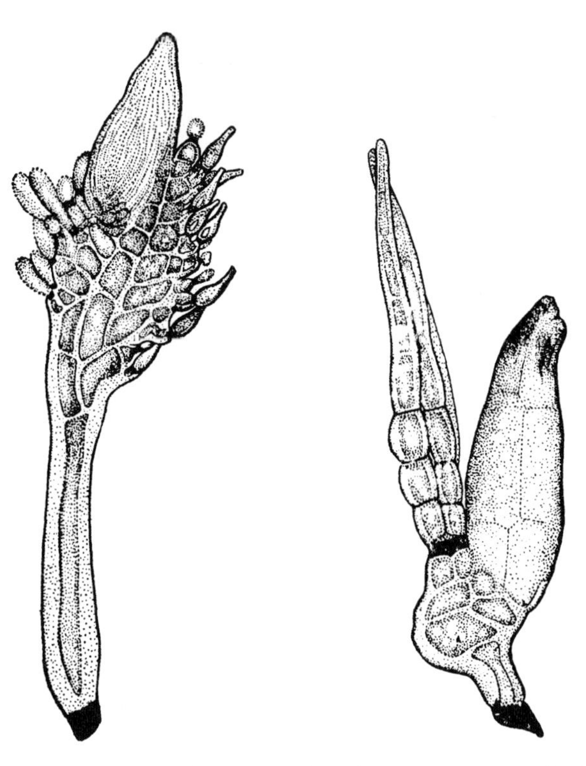

**FIGURE 14–13** Two ant-infesting species of the ascomycete order Laboulbeniales, which neither kill nor significantly harm their insect hosts. (*Left*) *Rickia wasmanni*, a parasite of *Myrmica laevinodis* in Europe. (*Right*) *Laboulbenia formicarum*, a parasite of various North American ant species. The fungi are minute and attach to the outside of the ant's integument. (From Bequaert, 1922b, after Thaxter, 1908.)

In 1986 this fungus was rediscovered in Arizona by Clark and Prusso on a worker of *Camponotus semitestaceus*.

A third group of fungi infecting ants are the members of the ascomycete order Laboulbeniales. These organisms are by far the most specialized of the fungi living on insects. They are transmitted by direct contact from one generation of hosts to the next. They are remarkable in that they neither kill nor noticeably harm the insect hosts. Unlike the *Cordyceps*, they are inconspicuous and, according to Bequaert (1922), resemble minute, usually dark-colored or yellowish bristles or bushy hairs, projecting from the integument either singly or in pairs. Usually scattered, sometimes they are so densely crowded in certain areas that they form a furry coating. The main body of the fungus is external to the insect. It consists of a small number of cells arranged in vertical rows and is attached by a blackened "foot" to the host's cuticle (Figure 14–13). In some species haustoria penetrate the integument and enter the body cavity (Benjamin, 1971).

*Aegeritella roussillonensis* is a new species belonging to the Hyphomycetales. Bałazy et al. (1986), who described it, found the fungus on live *Cataglyphis cursor.* A large number of the workers of a colony and even the queen can be infested. Sometimes the mycelium covers the whole body of the ant, but it is most common on the ant's mouthparts. The behavioral activity of the ants is seriously affected by the fungal infection. Other species of *Aegeritella* have been discovered in *Formica* in Europe and in *Camponotus* in Brazil (Espadaler and Wiśniewski, 1987).

Wheeler (1910c) posed the main theoretical question concerning parasitic fungi in ants in the following way: "At first sight ants would seem to be particularly favorable hosts for such parasites since these insects are in the habit of huddling together in masses in warm subterranean galleries, where the fungi might be supposed to develop luxuriantly and transmit their spores from ant to ant with great facility." Yet fungal parasites are scarce and do not appear to play any significant role in the demography of ant colonies. Wheeler supposed that the fanatical grooming behavior of ants is their principal defense against fungi. Workers are forever cleaning themselves and their nestmates with their tongues and forelegs (Wilson, 1971; Farish, 1972; Figure 14–14). The detritus collected in this way, including fungal spores, is compacted into small masses in the infrabuccal chamber, which is located on the floor of the mouth cavity. The masses are coughed up periodically in the form of "infrabuccal pellets" and carried out of the nest as waste material (Wheeler and Bailey, 1920; Eisner, 1957; Eisner and Happ, 1962). This behavior is undoubtedly an important part of the explanation, but there is more. As we noted earlier (Chapter 7) ants secrete a medley of potent antibiotic secretions from their metapleural gland and other exocrine organs in the body, which they then spread by grooming movements.

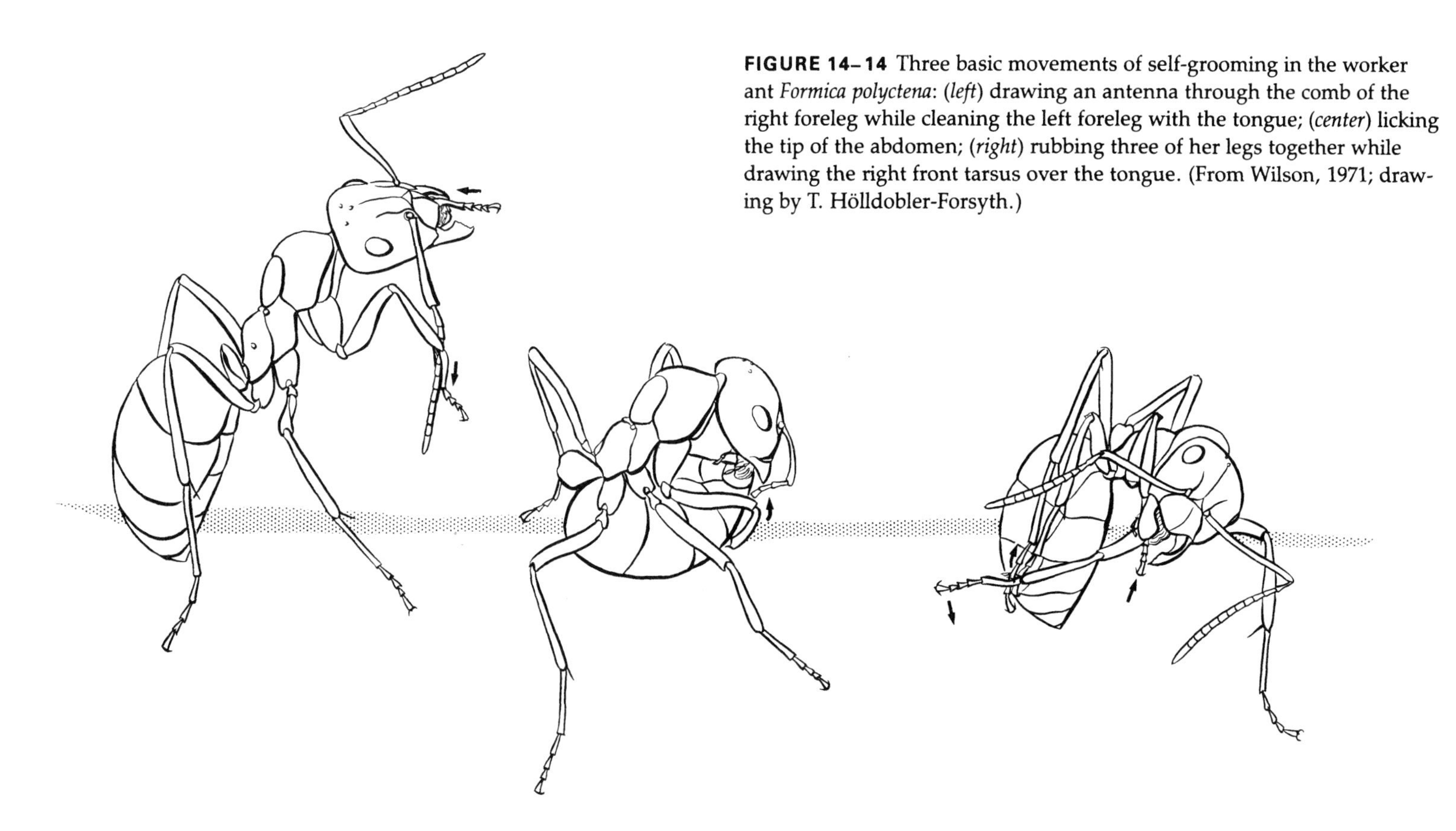

**FIGURE 14–14** Three basic movements of self-grooming in the worker ant *Formica polyctena*: (*left*) drawing an antenna through the comb of the right foreleg while cleaning the left foreleg with the tongue; (*center*) licking the tip of the abdomen; (*right*) rubbing three of her legs together while drawing the right front tarsus over the tongue. (From Wilson, 1971; drawing by T. Hölldobler-Forsyth.)

## POSSIBLE MICROORGANISMIC SYMBIONTS IN ANTS

The tribe Cephalotini, comprising exclusively arboreal ants of the New World tropics, possess a peculiar mushroom-shaped proventriculus or gizzard that has long been thought to strain liquid food from the esophagus and crop posteriorly to the midgut (Eisner, 1957). Myrmecologists have often speculated (at least in conversation) that this unique anatomy is somehow associated with the unusual feeding habits of the cephalotines. Workers of *Zacryptocerus rohweri* and *Z. texanus* do indeed collect and feed on pollen, at least in part (Creighton and Nutting, 1965; Creighton, 1967). On the other hand, *Z. varians* is a general scavenger that can flourish in the laboratory on freshly killed insects and honey (Wilson, 1976a). Caetano and Cruz-Landim (1985) have discovered that in the Brazilian species *Zacryptocerus clypeatus* and *Cephalotes atratus* the anterior portion of the small intestine is nearly filled with fibrillar material. Close examination revealed the mass to consist of bacteria and filamentous, non-septate fungi. Many of the same kind of organisms occur in the posterior region of the midgut, just anterior to the principal aggregate of microorganisms located in the small intestine. The role of the bacteria and fungi is unknown, but Caetano and Cruz-Landim make the entirely reasonable suggestion that they exist in some form of alimentary mutualism with their ant hosts. Perhaps they assist in digesting pollen extracts or some other nutritive materials beyond the ordinary metabolic capabilities of ants.

Intracellular symbiotic bacteria have also been discovered in the ovarioles and midgut epithelium of *Formica fusca* and the carpenter ants *Camponotus herculeanus* and *C. ligniperda* (Hecht, 1924; Lilienstern, 1932; Kolb, 1959; Buchner, 1965). Smith (1944) suggested that the symbiotic microorganisms either contribute vitamins to the hosts or in some way enable *Camponotus* to digest wood. Feeding experiments and biochemical analysis of the gut enzymes of both *C. herculeanus* and *C. ligniperda* have clearly demonstrated, however, that the carpenter ants do not digest wood (Graf and Hölldobler, 1964). If any symbiotic role for the bacteria in the ant's body exists, it remains unknown.

# CHAPTER 15

# The Specialized Predators

Physical adventure and instant scientific rewards await the entomologist who searches for new kinds of ants in unusual food niches or investigates the niches of previously little-known species, the discovery of which in one stroke places these species within the larger context of ant biology. This experience has been especially common in the case of species that hunt unusual prey. Here are several of our favorite examples of past successes:

- Lévieux (1983b), digging below 20 centimeters in the soil of the Ivory Coast savanna, discovered an entire guild of blind, centipede-eating amblyoponine ants. They comprise five or six species of *Amblyopone*, including *A. mutica* and *A. pluto*, as well as *Apomyrma stygia*, the first species of a new genus.
- Working in Guyana (known as British Guiana at the time) in 1920, Alfred E. Emerson found that the large ponerine ant *Pachycondyla commutata* is a specialized hunter of *Syntermes*, the largest termites of the New World. In his notes, published by Wheeler (1936c), he wrote:

> Kartabo, B. G. Oct. 10, 1920. Between 5:30 and 6:00 P.M. I observed a large number of *commutata* (more than 100) raid a trail of *Syntermes territus*. The termites, of which soldiers and two kinds of workers were plentiful, formed an open trail on the ground. The ants, moving either in groups or singly, came down a small hill from their nest in the ground under low bushy vegetation about 150 ft. from the termites and stridulated violently while moving back and forth from their nest. They attacked both the soldier and worker termites by stinging them and then carried them off to the nest.

- Deep in the Oropouche Cave of Trinidad, Wilson (1962d) rediscovered the one supposedly true cave-dwelling ant, "*Spelaeomyrmex*" *urichi*. He found workers gathering arthropod eggs. Soon afterward he encountered the same species nesting in open rain forest. He demonstrated experimentally that *S. urichi* is really a member of the genus *Erebomyrma* and that it is in fact a skilled predator of arthropod eggs.
- When Brown (1974a) explored the native ant fauna of Mauritius, he found a remarkable new species, a discovery that entailed a major extension of the range of the genus. The prime locality to search for the last remnants of this endangered island fauna is Le Pouce, an elongate massif with a topside plateau covered with low, gnarled native forest. After one only partially successful trip to the plateau, Brown decided on a second effort:

> On the first of April, though I was scheduled to leave on an evening flight to Bombay, I tried Le Pouce again. A telephone call to Mr. J. Vinson, who had collected some of Donisthorpe's material, convinced me that the main path on the Le Pouce plateau should not be avoided. I arrived there in the afternoon; the day was heavily overcast, threatening rain on the peaks, and it took me about an hour to walk up to the plateau. Whereas the sunny Sunday in the scrub shade had yielded almost no ants foraging, I now found foragers on foliage and on the hard-packed earth of the trail every few meters of the way. These were mostly *Camponotus aurosus* and *Pristomyrmex* spp. (= *Dodous*), native Mauritian species. Before long, on the trunk of a small tree by the path, I found a sparse trail of bright red ants climbing the bark. Closer examination revealed these to be predominantly the ectatommine I had collected on the previous Sunday, but interspersed with these were workers of *Pristomyrmex bispinosus* which, with their gasters partly curled under, looked remarkably like the ectatommines. It is hard to avoid the impression that some kind of mimicry involves these two species in this habitat. The ectatommines ascending the trunk nearly all carried in their mandibles whitish spherical objects that proved eventually to be arthropod eggs—probably spider eggs. I climbed the tree which was only about 5 meters high, and soon found the nest about 3 meters up. Where two of the gnarled branches crossed, a thick pad of lichens surrounded the place where they touched. Forcing the branches apart, I found a rotted-out pocket, evidently caused by their rubbing together in high winds. The cavity extended downward several centimeters into one of the branches, and it was full of the ectatommine ants with brood and many of the round white arthropod eggs; I estimate that I removed or saw at least 200 workers, and there may have been more. [Afterward, in an incident about which he did not write, Brown attempted to climb the small peak at the end of the plateau. Lightning struck 20 feet upslope; then heavy rain turned the ground into slippery mud. Brown fell and slid down the slope toward the brink of a high cliff. After hanging on to one small bush for a few minutes, he finally regained his purchase and worked his way back down to Port Louis. Another day in the life of the myrmecologist thus came to an end.]

The bright red ant turned out to be an undescribed *Proceratium (P. avium)*, notable for its aboveground foraging for eggs, in marked contrast to the other, wholly subterranean members of the genus. Also, the enlargement of its one-faceted eyes suggests that *avium* emerged as an epigeic forager late in evolution after the arrival of its ancestral stock on the remote island of Mauritius (see Figure 15–1).

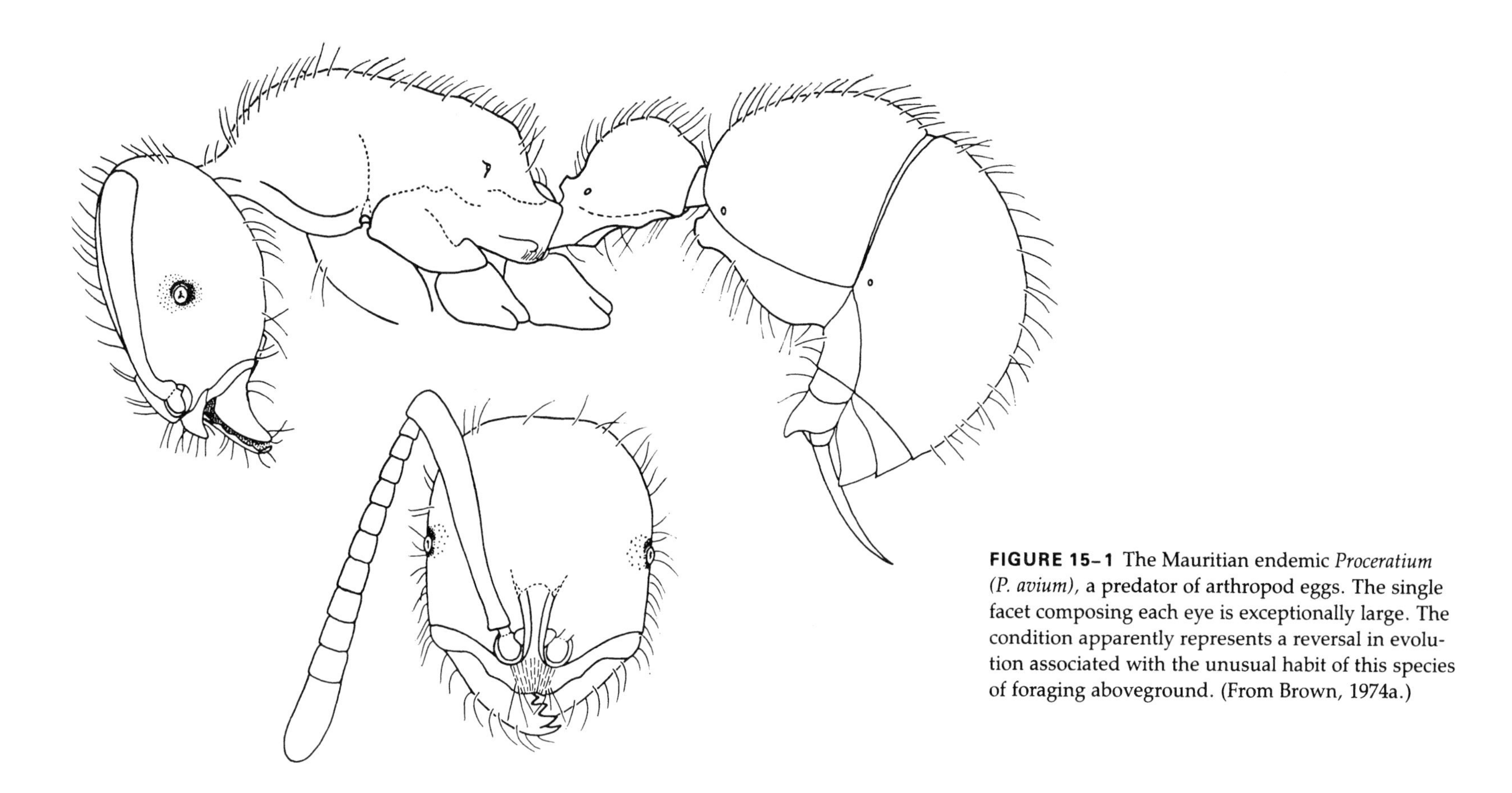

**FIGURE 15–1** The Mauritian endemic *Proceratium* (*P. avium*), a predator of arthropod eggs. The single facet composing each eye is exceptionally large. The condition apparently represents a reversal in evolution associated with the unusual habit of this species of foraging aboveground. (From Brown, 1974a.)

## THE EVOLUTION OF PREY SPECIALIZATION

A large amount of such fieldwork has provided many generalizations of importance in ant biology. The ants display among themselves almost every conceivable degree of prey specialization, from those that accept many kinds of insects and other arthropods to others limited to geophilomorph centipedes or some other, comparably restricted group. The roster of the most specialized species known to us is presented in Table 15–1. The phylogenetic distribution of this extreme class is curious. It is disproportionately concentrated in the Ponerinae, the Leptanillinae, and the army ant subfamilies Dorylinae and Ecitoninae. The Myrmicinae rank a somewhat distant fourth. Remarkably, almost no specialized predators have been recorded in the large subfamilies Dolichoderinae and Formicinae. The single exception is *Myrmoteras barbouri,* which appears to be a specialist on collembolans. It uses its elongate trap jaws and sensitive trigger hairs to snare these elusive prey. A second species, *M. toro,* accepts a wide variety of soft-bodied arthropods. It possesses trap jaws but lacks trigger hairs (Moffett, 1986c; see Figure 7–54).

With reference to the variation in prey utilized from one species to the next (as opposed to variation in prey utilized by single species), the Ponerinae and Myrmicinae are the clear leaders. By and large, the prey utilized by the specialists are ground-dwelling arthropods that are relatively abundant but either swift-moving or armed with formidable defenses. They include collembolans (springtails), centipedes, millipedes, oniscoid isopods, other ants, and termites. As we shall see, various of the specialized predators have developed morphological and behavioral devices to overcome these prey. Eggs of arthropods are an odd class that requires separate explanation.

It must be emphasized that most of the listings in Table 15–1 are based on inadequate data. In a few cases they are inferred from only one to several field records or laboratory experiments. A species is proven oligophagous by one of three methods. First, the species may be seen consistently to capture the preferred prey to the exclusion of almost all other potential prey with which it comes in contact. Such is the case for most of the ant and termite predators. The raids are usually conducted aboveground in such a conspicuous fashion that they have been witnessed repeatedly by entomologists, and specificity is not in doubt. Second, species in a few ant genera, including the ponerines *Gnamptogenys, Leptogenys,* and *Myopias,* store the inedible remains of their prey in kitchen middens contiguous with the nest chambers. The consistent presence of only one or a few kinds of organisms in the middens is prima facie evidence of oligophagy, especially if it can be combined with the existence of at least a few fresh prey among the brood. The third method is the "cafeteria experiment," which has been used to produce persuasive evidence for predatory specificity in *Smithistruma* and other dacetine ants (Wilson, 1953a), *Belonopelta* (Wilson, 1955b), and *Prionopelta* (Hölldobler and Wilson, 1986a). The technique is most useful for species that neither forage aboveground nor accumulate kitchen middens. Colonies of the species are first established in an artificial nest adjacent to a foraging arena. An aspirator is then used to collect arthropods, nematodes, and other small organisms in the terrain in the vicinity of the nest; these are placed live in the foraging arena, along with a scattering of litter and soil. A record is kept of the prey selected by the foraging ants. In the case of dacetines at least, the cafeteria experiment has produced results consistent with observa-

**TABLE 15–1** Specialized predators among the ants.

| Species | Distribution | Prey | Source |
|---|---|---|---|
| **SUBFAMILY PONERINAE** | | | |
| **Tribe Amblyoponini** | | | |
| *Amblyopone pallipes* | North America | Geophilid centipedes primarily, also beetle larvae | Haskins (1928), Wheeler (1936c), Traniello (1982) |
| *A. (mutica, pluto, stygia)* | West Africa | Geophilid centipedes | Gotwald and Lévieux (1972), Lévieux (1983b) |
| *Apomyrma stygia* | West Africa | Geophilomorph centipede (one record only) | Brown et al. (1970b) |
| *Onychomyrmex* spp. | Australia | Centipedes | Hölldobler and Taylor (unpublished) |
| *Prionopelta amabilis* | Mexico, Central America | Campodeidae (diplurans) primarily; geophilomorph centipedes and possibly a limited number of other arthropods secondarily | Hölldobler and Wilson (1986a) |
| **Tribe Ectatommini** | | | |
| *Discothyrea bidens* | Australia | Spider eggs | Brown (1979) |
| *D. oculata* | Tropical Africa | Arthropod eggs | Lévieux (1983b) |
| *D. poweri* | South Africa | Arthropod eggs | Brown (1979) |
| *Discothyrea* sp. nr. *testacea* | Central America | Arthropod (spider?) eggs | Brown (1979) |
| *Gnamptogenys horni* | Central and South America | Ants (specificity uncertain) | E. O. Wilson (unpublished observations) |
| *G. schmitti* | Haiti | Millipedes | Wheeler and Mann (1914) |
| *Gnamptogenys* sp. | Central America | Beetles | W. L. Brown and E. O. Wilson (unpublished notes) |
| *Proceratium avium* | Mauritius | Arthropod (probably spider) eggs | Brown (1974a) |
| **Tribe Platythyreini** | | | |
| *Platythyrea arnoldi* | Southern Africa | Beetles | Arnold (1915–26) |
| *P. conradti* | Tropical Africa | Mostly moth larvae (Noctuidae) | Lévieux (1976d) |
| **Tribe Cerapachyini** | | | |
| *Cerapachys* spp. | New World tropics, tropical Asia, Australia | Ants, especially *Pheidole* | Wheeler (1918c), Wilson (1958b, 1959c), Hölldobler (1982b) |
| *Sphinctomyrmex steinheili* | Australia | Ants | Brown (1975) |
| **Tribe Cylindromyrmecini** | | | |
| *Cylindromyrmex striatus* | South America, Galapagos Islands | Termites | Overal and Bandeira (1985) |
| **Tribe Acanthostichini** | | | |
| *Acanthostichus* sp. | Brazil | Termites | Brown (1975) |
| **Tribe Thaumatomyrmecini** | | | |
| *Thaumatomyrmex contumax* | Central and South America, West Indies | Polyxenid millipedes | Dinitz and Brandão (1989) |
| **Tribe Ponerini** | | | |
| *Belonopelta deletrix* | Mexico, Central America | Campodeidae, Japygidae, young or very small centipedes | Wilson (1955b) |
| *Centromyrmex feae* | Tropical Asia | Termites (probably); live in termite mounds | Wheeler (1936c) |
| *C. sellaris* | Tropical Africa | Termites | Lévieux (1983b) |

*continued*

**TABLE 15–1** *(continued)*

| Species | Distribution | Prey | Source |
|---|---|---|---|
| *Hypoponera* sp. nr. *coeca* | Tropical Africa | Collembolans | Lévieux (1983b) |
| *Leptogenys (aspera, binghami, birmana, chinensis, diminuta, kitteli, ocellifera, processionalis)* | Tropical Asia | Termites | Wroughton (1892), Wheeler (1936c), Maschwitz and Mühlenberg (1975) |
| *L. elongata* | Southern United States | Oniscidae (sowbugs) | Wheeler (1904a) |
| *L. conradti* | Tropical Africa | Oniscidae (sowbugs) | Lévieux (1983b) |
| *L. neutralis* | Western Australia | Ant queens, especially *Crematogaster* | Wheeler (1933b) |
| *L. schwabi* | Southern Africa | Termites | Arnold (1915–26) |
| *L. triloba* | New Guinea | Oniscidae (sowbugs) | Wilson (1959c) |
| *Megaponera foetens* | Tropical Africa | Termites | Wheeler (1936c), Longhurst et al. (1978) |
| *Myopias delta* | New Guinea | Myrmicine ants | Willey and Brown (1983) |
| *M. julivora* | New Guinea | Millipedes | Willey and Brown (1983) |
| *Ophthalmopone ilgi* | Tropical Africa | Termites | Wheeler (1936c) |
| *O. berthoudi* | South and Central Africa | Termites | Peeters and Crewe (1987) |
| *Pachycondyla* (= *Termitopone*) (*commutata, laevigata, marginata*) | Central and South America | Termites | Wheeler (1936c) |
| *Paltothyreus tarsatus* | Tropical Africa | Termites; workers also scavenge for other food in solitary fashion | Hölldobler (1984b) |
| *Plectroctena lygaria* | Tropical Africa | Millipede eggs | Lévieux (1982, 1983b) |
| *P. subterranea* | Tropical Africa | Millipedes | Lévieux (1972, 1983b) |
| *Psalidomyrmex procerus* | Tropical Africa | Earthworms (oligochaetes) | Lévieux (1982 and personal communication) |
| **SUBFAMILY DORYLINAE** | | | |
| *Aenictus* spp. | Old World tropics and warm temperate zone | Other ants primarily | See Chapter 16 |
| **SUBFAMILY ECITONINAE** | | | |
| *Neivamyrmex* spp. | New World tropics and warm temperate zone | Other ants primarily | See Chapter 16 |
| **SUBFAMILY LEPTANILLINAE** | | | |
| *Leptanilla japonica* | Japan | Centipedes | K. Masuko (1987) |
| **SUBFAMILY MYRMICINAE** | | | |
| *Anillomyrma decamera* | Tropical Asia | Possible thief ant in fungus gardens of *Odontotermes* | Wheeler (1936c) |
| *Decamorium uelense* | Africa | Termites, especially *Microtermes*; other insect larvae during dry season | Bolton (1976); Longhurst et al. (1979a) |
| *Epitritus hexamerus* | Japan | Diplurans and centipedes mostly, also entomobryomorph and sminthurid collembolans | Masuko (1984) |
| *Eurhopalothrix biroi* | New Guinea | Soft-bodied arthropods | Wilson (1956) |
| *E. heliscata* | Singapore, Malaysia | Termites primarily; also some soft-bodied arthropods | Wilson and Brown (1984) |
| *Liomyrmex aurianus* | Tropical Asia | Possible thief ant in nests of *Macrotermes* | Wheeler (1936c) |
| *Pentastruma canina* | Japan | Entomobryomorph collembolans; sminthurids not tested | Masuko (1984) |

*continued*

**TABLE 15–1** *(continued)*

| Species | Distribution | Prey | Source |
|---|---|---|---|
| *Pheidole titanis* | Southwestern United States, Mexico | Termites | Creighton and Gregg (1955) |
| *Serrastruma serrula* | Tropical Africa | Entomobryomorph collembolans readily accepted; other possible prey unknown | Dejean (1980a,b) |
| *Smithistruma (brevisetosa, missouriensis, rostrata)* | Eastern United States | Entomobryomorph and sminthurid collembolans | Wilson (1953a) |
| *Smithistruma (clypeata, dietrichi, talpa)* | Eastern United States | Entomobryomorph and sminthurid collembolans primarily, also diplurans and small centipedes | Wilson (1953a) |
| *S. pergandei* | Eastern United States | Collembolans (further specificity not tested) | Wesson (1936) |
| *S. truncatidens* | Tropical Africa | Entomobryomorph collembolans; other choices unknown | Dejean (1985a,b) |
| *Solenopsis fugax* | Europe | Brood of other ant species | Forel (1869, 1874), Wheeler (1936c), Hölldobler (1973a) |
| *Strumigenys louisianae* | Southeastern United States to Argentina | Entomobryomorph and sminthurid collembolans primarily, also other soft-bodied arthropods | Wilson (1950, 1953a) |
| *Strumigenys (lewisi, solifontis)* | Japan | Collembolans, mites, other small soft-bodied arthropods | Masuko (1984) |
| *S. rufobrunea* | Tropical Africa | Entomobryomorph collembolans; other choices unknown | Dejean (1982, 1985a) |
| *Trichoscapa membranifera* | Widespread tropical and warm-temperate tramp species | Entomobryomorph and sminthurid collembolans primarily, other soft-bodied arthropods to limited extent | Wilson (1953a) |
| **SUBFAMILY FORMICINAE** | | | |
| *Myrmoteras barbouri* | Singapore | Entomobryomorph collembolans primarily; possibly also campodeid diplurans | Moffett (1986c) |

tions of undisturbed colonies and foragers in nature. Other more sophisticated techniques of prey identification are possible, including classification of remains in the infrabuccal pockets of the predators and biochemical analysis of stomach contents. With the exception of an early analysis of infrabuccal pellets of *Pseudomyrmex* by Wheeler and Bailey (1920), these approaches have not been attempted, however.

Oligophagy appears to be an evolutionarily derived condition, despite its concentration in the primitive subfamily Ponerinae. In fact the majority of ponerine genera are polyphagous or, as in the case of *Pachycondyla*, comprise both polyphagous and oligophagous species. Some ponerines visit extrafloral nectaries, including *Ectatomma tuberculatum* (Weber, 1946b) and *Paraponera clavata* (Young and Hermann, 1980). At least one species, an African *Odontomachus*, attends aphids and coccids and even builds shelters over these "cattle" in the manner of myrmicine, dolichoderine, and formicine ants (Evans and Leston, 1971). And while it is true that most species of the primitive genus *Amblyopone* thus far studied are specialists on centipedes, at least one, the common Australian species *Amblyopone australis*, is a generalized insectivore (Haskins and Haskins, 1951). *Amblyopone* workers do not accept sugary fluids; possibly their largely subterranean habits keep them from contact with extrafloral nectaries. The very primitive *Nothomyrmecia macrops*, sole living member of the subfamily Nothomyrmeciinae, is a generalized insectivore, readily climbs bushes and trees, and accepts honey in the laboratory (Hölldobler and Taylor, 1983). *Myrmecia*, the sole living genus of the subfamily Myrmeciinae and nearly as primitive in overall anatomy as *Nothomyrmecia*, preys on a wide range of insects and spiders (Gray, 1971a,b, 1974a). Its workers, the fearsome bulldog ants, visit extrafloral nectaries to imbibe sugary secretions (see Chapter 14). They also fit an intuitive conception of primitive huntresses, employing stealth, the excellent vision afforded by their large eyes, and the ability to spring through the air to capture their victims. Gray reports that foragers of *Myrmecia varians* "were observed to remain stationary on branches of *Eucalyptus largiflorens* for as long as 30 minutes with their body crouched close to the branch and mandibles wide apart, ready to spring at unsuspecting prey. One such was seen to capture the fast-moving bush-fly *Musca vetustissima* Walker, which had landed approximately three centimeters away. The worker sprang quickly, clasping the fly's head between its mandibles, and stung the prey to death. The worker then jumped off the branch onto the ground, a distance of nearly one meter, and returned directly to the nest 15 meters from the tree." Limited specificity was observed in *Myrmecia dispar,* which accepts a wide range of insect prey in the winter but turns predominantly to other ants in the summer (Gray, 1971b). Finally, the relatively primitive myrmicine genera *Hylomyrma* and *Myrmica* are also generalized insect predators (E. O. Wilson, unpublished notes).

How can the peculiar phylogenetic distribution of oligophagy be explained? The answer may lie in the interesting fact that all of the specialized predators have well-developed stings, whereas a large percentage of the species that are generalized predators lack functional stings and rely exclusively on biting and the release of poisonous secretions. None of the many members of the subfamilies Dolichoderinae and Formicinae possesses a functional sting, and (with the exception of the very peculiar *Myrmoteras barbouri*) none is a specialized predator. It seems likely that stings are the best weapon to subdue formidable or elusive arthropods in a quick and clean manner. In other words, it seems that the possession of a sting allows the species to be either generalists or specialists, but the absence of a sting precludes the species from being specialists.

The foraging strategies of the great majority of specialized predators fall into one of two broad classes. The members of species belonging to one class raid en masse. Groups of workers are directed along odor trails by scouts to the prey, which they then overwhelm in violent combat. All of the predators of ants use this method, including the army ants (subfamilies Dorylinae and Ecitoninae) and members of the genus *Cerapachys.* Other ants in the same category are termite predators in the genera *Pachycondyla, Leptogenys,* and *Megaponera;* species of *Leptogenys* that capture isopods; and *Leptanilla japonica* and species of *Onychomyrmex,* which attack centipedes. The second principal group of specialists forage singly, most often relying on camouflage and stealth. They include most of the specialist myrmicines, such as *Strumigenys, Smithistruma,* other small members of the tribe Dacetini, and *Eurhopalothrix* in the tribe Basicerotini.

Another useful generalization is that all of the genera of specialized predators are either limited to the tropics or are at least disproportionately represented there. Except for an occasional *Amblyopone, Proceratium, Smithistruma,* or army ant in the genus *Neivamyrmex,* specialists are entirely absent from the northern half of the United States and Europe. *Cerapachys* and smaller dacetines are moderately common in portions of southern, warm-temperate Australia, but they and other specialists are far less well represented there in species and individuals than farther north in Queensland and Melanesia. This ecogeographic rule may in some way be related to another climatic correlation established experimentally by Jeanne (1979): rates of predation by ants, reflected by the harvesting of wasp larvae from bushes and trees, were found to increase gradually southward from Michigan to the Brazilian Amazon. The numbers of ant species appearing at Jeanne's baits also increased, from 22 at the northern limit to 74 at the southern limit. It is also a general rule that the percentage of canopy-dwelling species, as opposed to those that forage onto vegetation from ground nests, also increases. It is well known that in the absence of such arboricolous forms, terrestrial ants forage upward into the vegetation in greater force. On Mont-Ventoux, in the south of France, Du Merle (1982) found that at least 23 of the 40 terrestrial ant species foraged on shrubs and trees, with about 10 doing so intensively.

## THE EVOLUTION OF THE DACETINE ANTS

The analysis by Brown and Wilson (1959) of the Dacetini, supplemented by Wilson (1962a), Hölldobler (1981b), Dejean (1982, 1985a,b), and Masuko (1984), was one of the first attempts directed at any group of animals to correlate the evolution of social behavior with species-level adaptations in feeding. The dacetines are exceptionally well suited for such an analysis. The tribe is worldwide in distribution, and its 24 genera and 250 species vary enormously in size, morphology, and behavior.

Brown and Wilson showed that the main trend within the tribe Dacetini has been the shift from open foraging on the ground and low vegetation to an increasingly cryptic, subterranean existence. The change was associated in the early stages of dacetine evolution with a reduction in the variety of arthropods taken as prey. The most extreme modern forms specialize on entomobryomorph and sminthurid collembolans. There has been a concomitant reduction in body size, a possible shift from polymorphism to monomorphism, and an abandonment of recruitment trails to assemble masses of nestmates. Other anatomical changes include the development of fungus-like growths on the waist and other parts of the body, possibly containing substances that lure the collembolan prey, and a shortening of the mandibles.

The tendency to move from open, even partly arboreal foraging to cryptic, mostly terrestrial and subterranean foraging is not unusual in the ants. It is notable, for example, in the anatomically more primitive ponerine tribes Ponerini and Ectatommini. Haskins (1939) proposed the interesting theory that this trend in the ponerines resulted indirectly from competition with more recently ascendant, dominant aboveground foragers in the Dolichoderinae and Formicinae. A similar explanation is plausible in the case of the Dacetini. The early forms, most closely approached by *Daceton* and *Orectognathus* today, may have retreated when they came up against other myrmicines and the formicines during the great Tertiary radiation of the ants.

Whatever was the prime mover in natural selection, the shift to a more cryptic existence incorporated important changes in hunting behavior of the dacetine ants. The relatively primitive long-mandibulate forms rely on the violent, trap-like action of their mandibles to secure prey (Figures 15–2 through 15–4). Their approach time to the prey is relatively short, and they often do not follow the mandibular snap with a stinging thrust. The short-mandibulate forms, on the other hand, have less shocking power in their mandibles. They rely more on stealth in approaching prey, holding on tenaciously after the mandibular strike, and an immediate, consistent, and efficient use of the sting (Figures 15–5 and 15–6). The essential features of this evolutionary change are illustrated in the contrast between *Strumigenys louisianae,* a relatively primitive long-mandibulate species, and the phylogenetically more advanced, short-mandibulate *Trichoscapa membranifera* (Wilson, 1953a).

**FIGURE 15–2** The Australian *Orectognathus versicolor* represents the primitive body form within the predatory tribe Dacetini as postulated by Brown and Wilson. The long mandibles open and shut like a trap, impaling prey on the sharp apical teeth. The convulsive snap is powered by large adductor muscles that extend into the enlarged corners of the head. (Scanning electron microscope picture by E. Seling.)

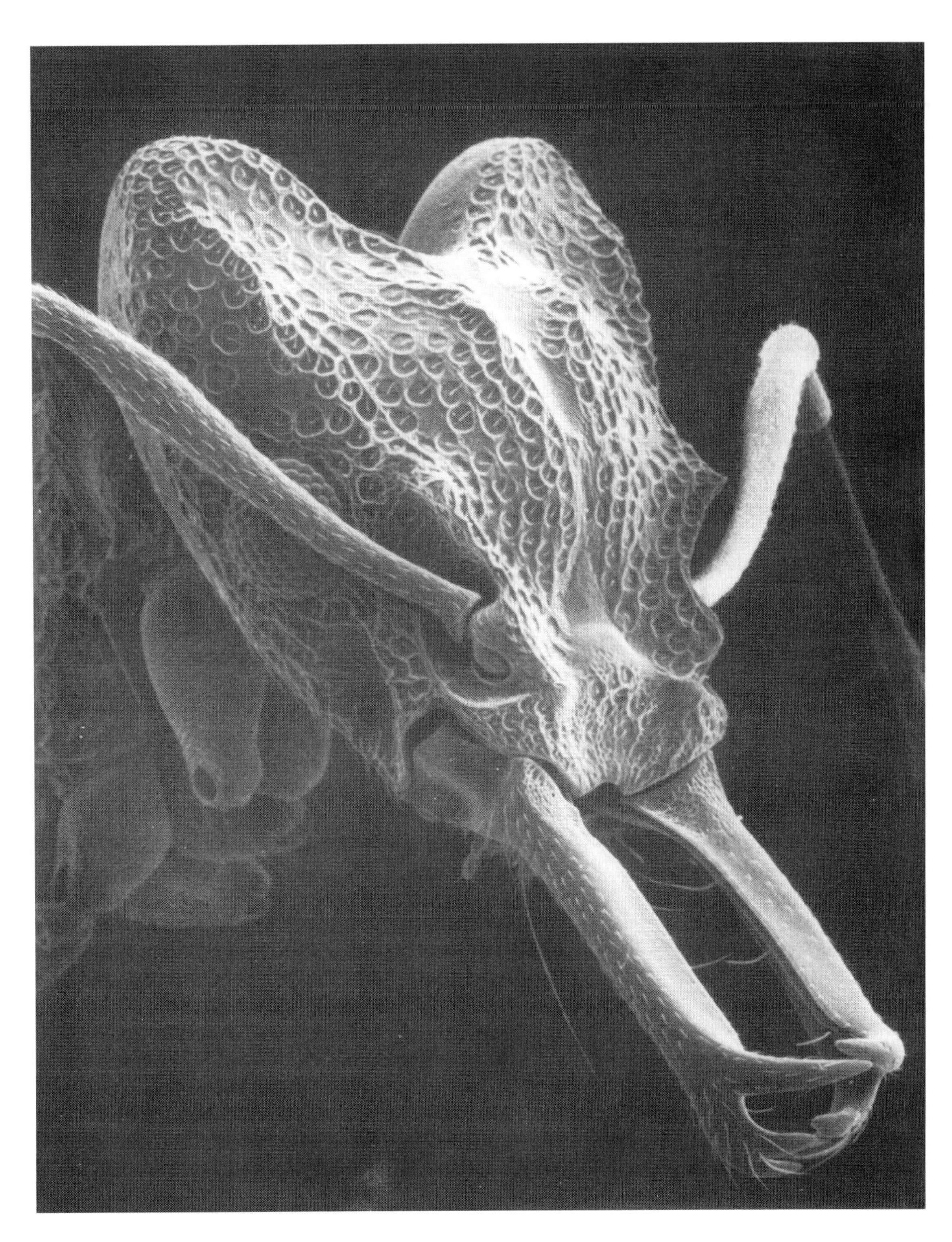

**FIGURE 15–3** A worker of the dacetine *Strumigenys ludia* is shown with mandibles in the open position (*left*) and closed position (*right*). In this more primitive mode of action, the mandibles are first opened to nearly 180 degrees, then snapped shut to impale the prey on the long apical teeth. In the open position a pair of sensitive trigger hairs project forward and assist the ant in locating the prey just prior to the mandibular snap. (Modified from Brown and Wilson, 1959.)

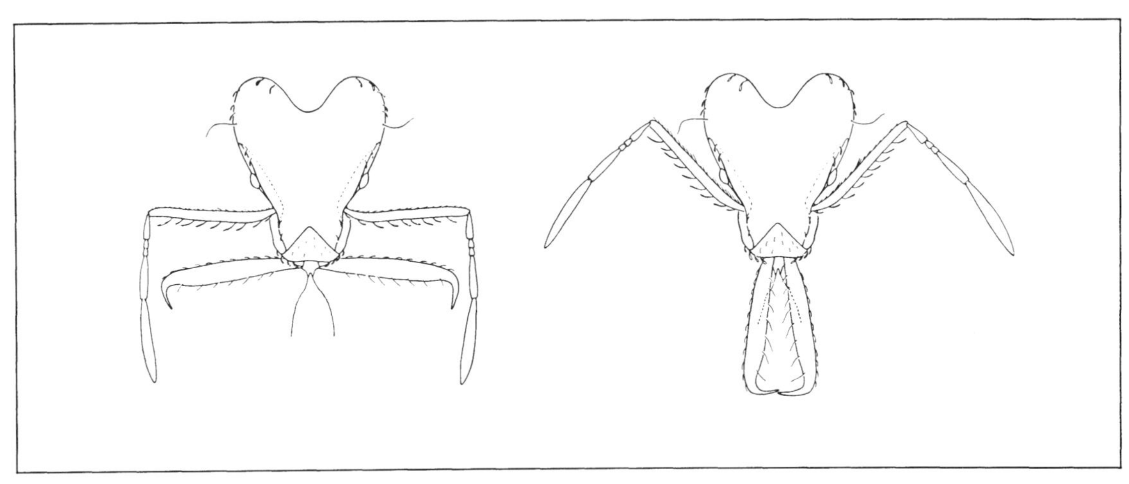

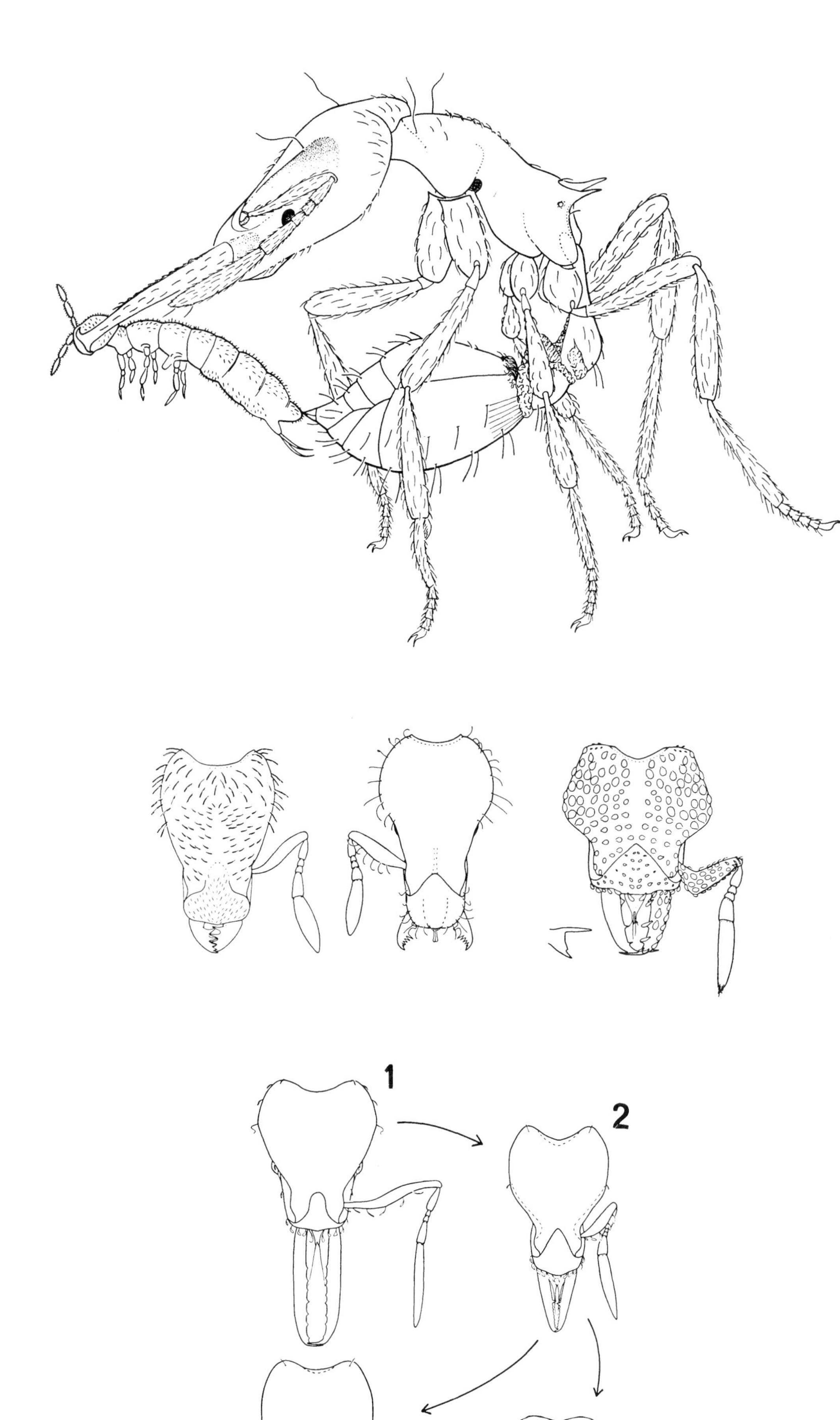


**FIGURE 15–4** A *Strumigenys ludia* worker has caught a small isotomid collembolan with a convulsive snap of her mandibles (see also Figure 15–3). She lifts the prey off the ground and stings it to render it immobile. (From Brown and Wilson, 1959.)

**FIGURE 15–5** Short mandibles are inferred to be an evolutionarily advanced state within the tribe Dacetini. *Left, Codiomyrmex semicomptus; center, Smithistruma weberi; right, Epitritus hexamerus,* with an end-on view of the *Epitritus* mandibular apex shown in the inset. (From Brown and Wilson, 1959.)

**FIGURE 15–6** The inferred evolution of the mandibles and accessory structures in the dacetine genus *Neostruma,* showing the trend toward shortening of the mandibles and a reversal back to the long-mandibulate state. (*1*) *Strumigenys jamaicensis,* a member of the *S. gundlachi* group, which is considered to be ancestral to *Neostruma;* (*2*) *N. zeteki;* (*3*) *N. metopia;* (*4*) *N. myllorhapha.* In *Strumigenys* the paired labral lobes, which are located between the bases of the mandibles, are short, and the trigger hairs, which act as range-finders to set off the spring-snap action of the mandibles, are long. With the shortening of the mandibles in *Neostruma,* the labral lobes have been elongated and the trigger hairs have been reduced, a change believed to lessen damage from the violent struggling of the collembolan prey. In *N. myllorhapha* (*4*), the mandibles have been secondarily elongated, with the labral lobes further lengthened to make up for the shortness of the trigger hairs. In addition the apical segments of the antennae of the four species illustrated are either lengthened or shortened in correspondence with the length of the mandibles, an important additional alteration that improves the timing of the mandibular strike. (From Brown and Wilson, 1959.)

The *Strumigenys,* as exemplified by *S. louisianae,* are bolder and more direct in their manner of stalking prey. This trait is permitted by their more efficient mandibles, which are extremely long and operate very much like a miniature spring trap. Prior to the strike the mandibles are locked at nearly 180 degrees. This position is accomplished by special teeth at their bases that catch on the lateral lobes of the labrum. When the ants tense the retractor muscles alone the mandibles cannot move, but when the labral lobes are also dropped the mandibles snap shut. The worker *Strumigenys* approaching a collembolan moves slowly and cautiously. She spreads her mandibles to the maximum angle and exposes two long hairs that arise from the paired labral lobes. These hairs extend far forward of the ant's head and serve as tactile range finders for the mandibles. When they first touch the prey, its body is well within reach of the apical teeth. A sudden and impulsive snap of the mandibles literally impales it on the teeth, so that drops of hemolymph often well out of the punctures. If the collembolan is small relative to the *Strumigenys,* the ant lifts it into the air and then may sting it. All but the largest collembolans are quickly immobilized by this sequence of actions, and struggling is feeble and brief.

The *Trichoscapa membranifera* worker is much more circumspect than the *Strumigenys* when stalking prey. As soon as the ant becomes aware of the presence of a collembolan, she freezes in a lowered, crouching posture and holds this stance briefly. If the collembolan is to her back or side, the worker now turns very slowly to face it. Once she is aligned, the ant begins a forward movement so extraordinarily slow that it can be detected only by persistent and careful observation. Several minutes may pass before the ant finally maneuvers over less than a millimeter's distance to come within striking distance, and she may remain in this position for as much as a minute or longer. Unlike the *Strumigenys,* the *Trichoscapa* worker opens her mandibles only to about a 60-degree angle. Tactile hairs are present and eventually come to touch the prey. The mandibular strike is as sudden as that of the *Strumigenys,* but since it is usually directed at an appendage, it does not have the same stunning effect on the collembolan. The insect often struggles violently to escape, but the worker is very tenacious and retains a fast grip until she can sting her prey into immobility.

Thus *Strumigenys louisianae* relies on a comparatively swift approach to prey followed by a fixed-action pattern that can be characterized as strike-lift-sting, with the last element occasionally being omitted if the prey is small. In contrast, *Trichoscapa membranifera* employs a more cautious approach followed by strike-hold-sting, with the last element inevitable. The *Strumigenys* pattern is apparently typical for long-mandibulate dacetines generally, whereas that of *Trichoscapa* is typical for the short-mandibulate groups.

The ecological significance of the difference between the two groups of dacetines is evidently the following: the *Trichoscapa* pattern, requiring less space for the operation of the mandibles, is generally associated with cryptic foraging. Masuko (1984) has discovered an extreme version of the short-mandibulate technique in *Epitritus hexamerus* (see Figure 15–5). This bizarre little ant is essentially an ambush hunter. The mandibles are directed slightly upward from the plane of the head, and the dorsal apical tooth is especially long and sharp, enabling the ant to strike with particular effectiveness at objects above her head. The *Epitritus* forager hunts a great deal in small crevices within the soil. Because of the tightness of the passages, she usually encounters the prey in front. The ant immediately crouches and freezes, pulling the antennae completely back into the scrobes that line the sides of her head. The mandibles remain closed. Even though the prey (a collembolan or small centipede) may be very close by, the ant never moves toward it. Instead, she remains perfectly still for periods of 20 minutes or longer, waiting for the prey to step on her head. Then, with a sudden upward snap of her mandibles, she impales the victim on the long apical teeth.

Two recent studies suggest that there is a good deal more to the story of predation by short-mandibulate dacetines than we have just recounted. Masuko (1984) found that the workers of some species smear soil and other detritus on themselves in one of two methods that appear to be unique in the ants generally. In one version, performed by species of *Epitritus, Pentastruma, Smithistruma,* the *"Labidogenys"* group of *Strumigenys,* and *Trichoscapa,* the worker pauses on moist soil and scratches the surface repeatedly with her forelegs. She then rubs the forelegs against her head, alitrunk, and middle and hind legs. The gaster is next stroked with either the hind or fore legs, in the latter case with the abdomen flexed forward beneath the rest of the body. The greatest effort is directed toward the dorsal head surface. In the second version, witnessed by Masuko in a species of *Smithistruma,* the worker picks up minute fragments of organic material (in one case at least, probably arthropod feces). Masuko suggests that body smearing camouflages the odor of the foragers and allows them to close on the prey more efficiently.

Dejean (1985b) found that workers of *Smithistruma emarginata* and *S. truncatidens* are more attractive to entomobryomorph collembolans than are workers of *Pheidole* and blank controls. The source of the attractant is unknown, but a good guess is the spongiform appendage of the waist (illustrated in Figure 15–4), which Dejean showed to be underlain by glandular cells. Another possibility is the head, especially the labrum.

A striking convergence to the Dacetini has occurred in at least two other phylogenetically independent groups of myrmicine ants, the widespread tribe Basicerotini (Brown and Kempf, 1960; Wilson and Brown, 1984) and the Oriental genus *Dacetinops* (Taylor, 1985). These ants are similar to dacetines in body form, pilosity, and predatory behavior. And like the dacetines, they hunt small, soft-bodied arthropods.

## EGG PREDATORS

At least four phyletic lines of ants specialize to some degree on arthropod eggs: the cosmopolitan ectatommine genera *Proceratium* and *Discothyrea,* the African ponerine *Plectroctena lygaria,* and the Neotropical myrmicine genus *Erebomyrma.* The least specialized of the oophagous genera is *Erebomyrma.* The most specialized are *Proceratium* and *Discothyrea,* which, as Brown (1957, 1979) first noted, appear to collect spider eggs exclusively. Workers of at least the species *Proceratium silaceum* use their peculiarly downward-pointing gastral tips to tuck the slippery eggs forward toward the mandibles when transporting them back to the nest.

## THE ANT–TERMITE ARMS RACE

Ants and termites are the superpowers of the insect world. In rain forest near Manaus, Brazil, they make up three-quarters of the entire insect biomass (Fittkau and Klinge, 1973), a preponderance that probably holds roughly in many of the other major habitats of both tropical and temperate regions. Ants, being largely predators, are correspondingly the greatest enemies of termites. The two groups have undoubtedly been locked in struggle for the greater part of the 100 million years of their coexistence, with ants acting as the active aggressors for the most part and termites as the prey and resisters. In addition, both occasionally compete for scarce nest sites in rotting wood and leaf litter. They have engaged in a coevolutionary arms race that has produced the most elaborate weapons and battle strategies known in the animal world.

A large percentage of ant species, perhaps a majority—including *Pheidole* and *Camponotus,* the most speciose of all ant genera—prey on termites if given the opportunity. When a nest of termites is broken open by a large animal, windfall, or some other force, foraging ants nearby rush in to seize workers and nymphs before the termites can pull back and close the breaches in the nest. One of the most effective ways to locate the nests of some wide-ranging ant species, such as *Gigantiops destructor* and the species of *Pachycondyla* in tropical forests, is to scatter portions of a termite nest on the ground and track foraging workers as they carry termites home.

A few ant genera are specialized termite predators, however, and need no assistance from investigators (see Table 15–1). The tropical zone of every continent has a guild of large ponerine ants that organize raids on termite nests. In particular there are three species of *Pachycondyla (= Termitopone)* in Central and South America; *Megaponera, Ophthalmopone,* and *Paltothyreus* in tropical Africa; and *Leptogenys* of the *processionalis* group in tropical Asia. The raids of *Megaponera foetens* are one of the most impressive sights of tropical Africa, if one can scale one's perception down a bit from rhinoceroses and bongos, and they have attracted the attention of explorers and naturalists since the time of Livingstone (1857). The expeditionary force of large black workers is led by a single scout. The raiders proceed along an odor trail in columns two or more workers wide to the termite nest, whereupon, according to Arnold (1915–1926), "the columns break up and pour into every hole and crack which leads into the invaded galleries. The method then adopted is as follows: Each ant brings to the surface one or more termites, and then re-enters the galleries to bring up more victims. This is continued until each has retrieved about half a dozen termites, which, in a maimed condition, are left struggling feebly at the surface. The whole army reassembles again outside, and each marauder picks up as many as it can conveniently carry, usually 3 or 4. The columns are then reformed and march home." While on the move the *Megaponera* frequently stridulate loudly enough to be audible to human observers as much as several meters distant. According to Arnold (cited by Wheeler, 1936c), the *Megaponera* colonies change their nest sites at frequent intervals, apparently in response to the need for fresh supplies of termites.

*Paltothyreus tarsatus,* as Arnold first noted, also raids termites. As in *Megaponera,* each *Paltothyreus* forager stacks as many termites in her mandibles as she can carry before starting home (Figure 15–7; Plate 18). Hölldobler (1984b) found that successful foragers, often carrying termites, recruit nestmates by laying chemical trails with secretions from intersegmental sternal glands located between the Vth and VIth, and VIth and VIIth abdominal sternites. According to Barry Bolton (cited in Longhurst et al., 1979a), *Decamorium uelense* is a small African myrmicine that appears to have arrived at the same specialization. The ants prey mostly on termites during the wet season, but during the dry season take a variety of soft-bodied insect larvae which are encountered underground. *Decamorium uelense* prefers *Microtermes,* a member of the fungus-growing subfamily Macrotermitinae. *Microtermes* species are litter-feeding termites that forage within their food sources of roots and plant debris. When scouts detect foraging *Microtermes,* they return to their nest and recruit a group of 10 to 30 nestmates. These ants attack and immobilize all termites they can capture by stinging them. Later a still larger group is recruited to retrieve the immobilized prey.

A wholly different strategy of termite hunting is employed by the Malaysian basicerotine ant *Eurhopalothrix heliscata.* The workers hunt solitarily through rotting wood housing the termites. They use their peculiar wedge-shaped heads, hard bodies, and short legs to press into tight spaces. They seize the appendages of the termites with their short, sharp-toothed mandibles, clasping these body parts of the prey even more tightly with the aid of heavily sclerotized labra that project like forks from between the bases of the mandibles (Wilson and Brown, 1984).

Beyond observations and a few experiments on the spectacular ponerine raiders, the myrmicine *Decamorium,* and the cryptic *Eurhopalothrix* huntresses, surprisingly little research has been directed at the other specialized termite predators. Many conjectures but few observations have been published on these ants, partly because their behavior is conducted in a secretive manner deep within the intact termite nests. One of the most common and least understood of the adaptive types is the thief ant, whose small workers slip into the termite brood chambers to steal eggs and nymphs. All of the suspected species are myrmicines. They nest close to termite nests, sometimes in the walls of the termitaries themselves. They include *Solenopsis (Diplorhoptrum) laeviceps,* which is known to collect eggs of *Nasutitermes* under natural conditions (A. E. Emerson in Wheeler, 1936c); *Erebomyrma urichi,* which has been observed to steal *Armitermes* eggs in laboratory nests (Wilson, 1962d); the pantropical genus *Carebara;* the Neotropical genus *Carebarella;* the Asian genus *Liomyrmex;* and the African and Asian genus *Paedalgus.*

Except for sparse data on *Solenopsis* and *Erebomyrma,* the case for termite thievery, or "termitolesty," is circumstantial (Forel, 1901; Wheeler, 1936c; Ettershank, 1966). To wit, the ants live close to the termites, the workers are blind and are rarely if ever seen away from the termitaria, they are small enough to slip unobtrusively in and out of the termite nest chambers, and they have no other known source of food. In the case of *Carebara* the colonies produce large masses of queens each year, and it is difficult to see where the food and energy come from if not from the termite hosts. An illustration of the queen and worker castes is presented in Figure 15–8. The queens are vastly larger than the workers. The weight differential is over 4,000 (Wheeler, 1922), the largest known in the ants and matched within the social insects as a whole only by the very same termites that are the hosts of the *Carebara.* Several observers have

**FIGURE 15–7** A forager of the large ponerine species *Paltothyreus tarsatus* emerges from a termite nest gallery with several termites stacked in her mandibles.

found the workers clinging to the tarsal hairs of queens during nuptial flights. Arnold (1915–1926) said of *Carebara vidua:*

> It is probable that the dense tuft of hairs on the tarsi of the female serve an important purpose—that of enabling some of the minute workers to attach themselves to the body of the female when the latter is about to leave the parental nest. Several specimens of the female have been taken by me with one or more workers biting on to the tarsal fimbriae. I am inclined to suspect that the young queen cannot start a new nest without the help of one or more of the workers from the whole nest, on account of the size of her mouthparts, which would probably be too large and clumsy to tend the tiny larvae of her first brood, and that it is therefore essential that she should have with her some workers which are able to feed the larvae by conveying to them the nourishment taken from the mouth of the queen.

Wheeler (1936c) concurred, observing that the problems of a founding queen without the little helpers would be comparable to those of a hippopotamus struggling to feed infants the size of mice. Still, the matter cannot be settled by such logic, however neatly expressed. Anyone who has seen a giant leafcutter queen such as *Atta texana* delicately separating and cleaning her minute eggs during colony founding knows that size alone is not a fatal impediment. Still more impressive is the fact that the founding queen of an Asian *Carebara* kept by Lowe (1948) succeeded in rearing a brood of 16 workers without the aid of helpers.

The unsolved mysteries of *Carebara* are symptomatic of our general ignorance of the myrmicine ants associated with termites. We need detailed studies of the actual behavior of the so-called termitolestic forms as they make contact with their termite hosts. Some of the species, such as *Liomyrmex aurianus* (C. F. Baker in Wheeler, 1914a), have been reported to live in the same chambers as the ter-

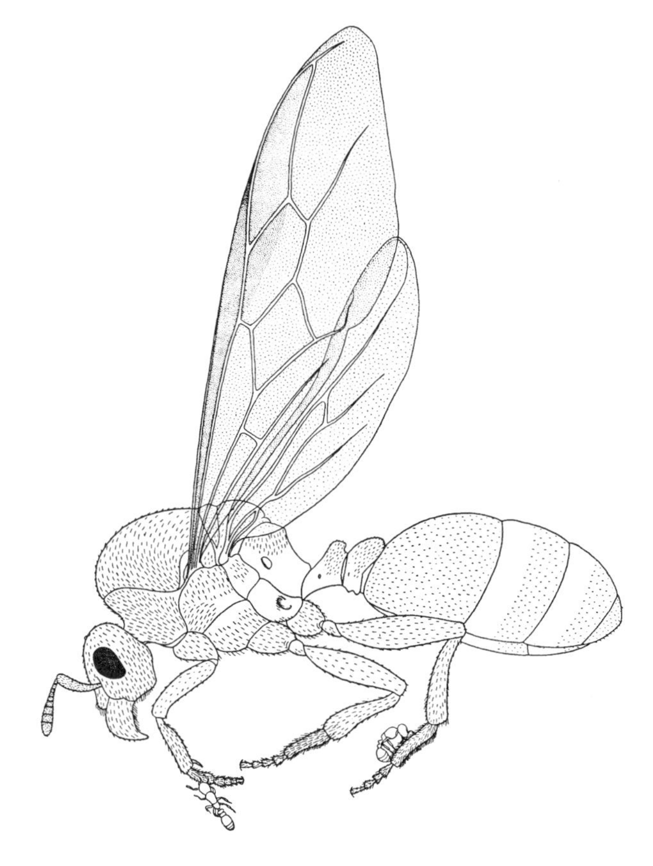

**FIGURE 15–8** A winged, recently inseminated queen of the African ant *Carebara vidua*, carrying two minute, blind workers attached to her dense tarsal hairs by their mandibles. *Carebara* queens settle in the walls of termite nests, and the hitchhiker workers are believed to assist them in rearing the first brood. (After Wheeler, 1936c.)

mites with complete amity. It is even remotely possible that ant species exist that are true termitophiles, accepted by the termites as pseudomembers of their own colony. One example suggested by Wheeler (1936c) is *Stigmacros termitoxenus*, a peculiar Australian formicine ant discovered in a nest of the termite *Tumulitermes peracutus*. Here is Wheeler's account of the single collection made:

> On September 18, 1931, near Mullawa, West Australia, I came upon a colony of diminutive termites nesting under a flat stone in earthen galleries which they had built in a bunch of dry grass. On breaking open one of the galleries I saw several ants of the same size and color as the somewhat more numerous termites and moving about among them. After carefully collecting the occupants of the gallery and making allowance for escaping individuals, the ant-colony was found to comprise only 25 to 30 workers and a single ergatomorphic female (queen). I failed to find any additional ants in the termitary and saw no traces of their brood. The female and more than half the workers attracted my attention because their gasters were enormously distended . . . This distension (physogastry) was not due to liquid stored in their crops but to an unusual accumulation of fat . . . The ants belong to an undescribed species of *Stigmacros*, an exclusively Australian genus of which eleven species are described and of which Mr. John Clark and I have taken quite a number of unpublished forms. Although I have examined hundreds of *Stigmacros* from numerous localities in Eastern, Southern and Western Australia, I have seen no traces of physogastry except in the Mullawa specimens. Since the ants were living in what appeared to be friendly relations with their hosts, I suspect that they are fed by the termite workers and that the physogastry of the female and so many of the workers, like the physogastry of the termite workers and queens, is a result of this feeding.

Wheeler's argument is far from convincing, but the phenomenon he suggests is so extraordinary as to warrant a special effort to rediscover and study in detail the biology of *S. termitoxenus*.

The termites have not stood still in the face of the furious onslaught of the ants. They have responded by evolving an array of weapons and complex tactics of their own (Deligne et al., 1981; Quennedy, 1975, 1984; Grassé, 1986; see Figure 15–9). It is a general characteristic of termite caste systems that the worker caste is morphologically uniform but behaviorally very diverse when different species are compared, whereas the soldier caste is morphologically diverse and behaviorally uniform. The workers construct the nests that vary so drastically among the termite species, and they are also responsible for the greatest part of the phyletic diversity in food habits, foraging styles, and other behavioral properties of the species. The soldiers, on the other hand, are wholly specialized for the single function of defense.

In most species the soldiers are "mandibulate" forms, possessing large, heavily sclerotized heads, powerful adductor muscles, and sharp, elongate mandibles that seem clearly designed for biting and cutting rather than for such nondefensive functions as digging or handling of the brood. In *Termes* and several other genera of the "higher termites" (family Termitidae), the mandibles are shaped like scimitars, with points as sharp as needles. The soldiers of the kalotermitid genus *Cryptotermes* have cylindrical heads that serve as living plugs for the galleries of the nests. The mandibles are short and not particularly suited for defense. The head capsule, on the other hand, is thick-walled, heavily pitted, and truncate in front, so that it forms a barrier across the narrow galleries of the nest that any invader finds very difficult to get around. This "phragmosis" of the

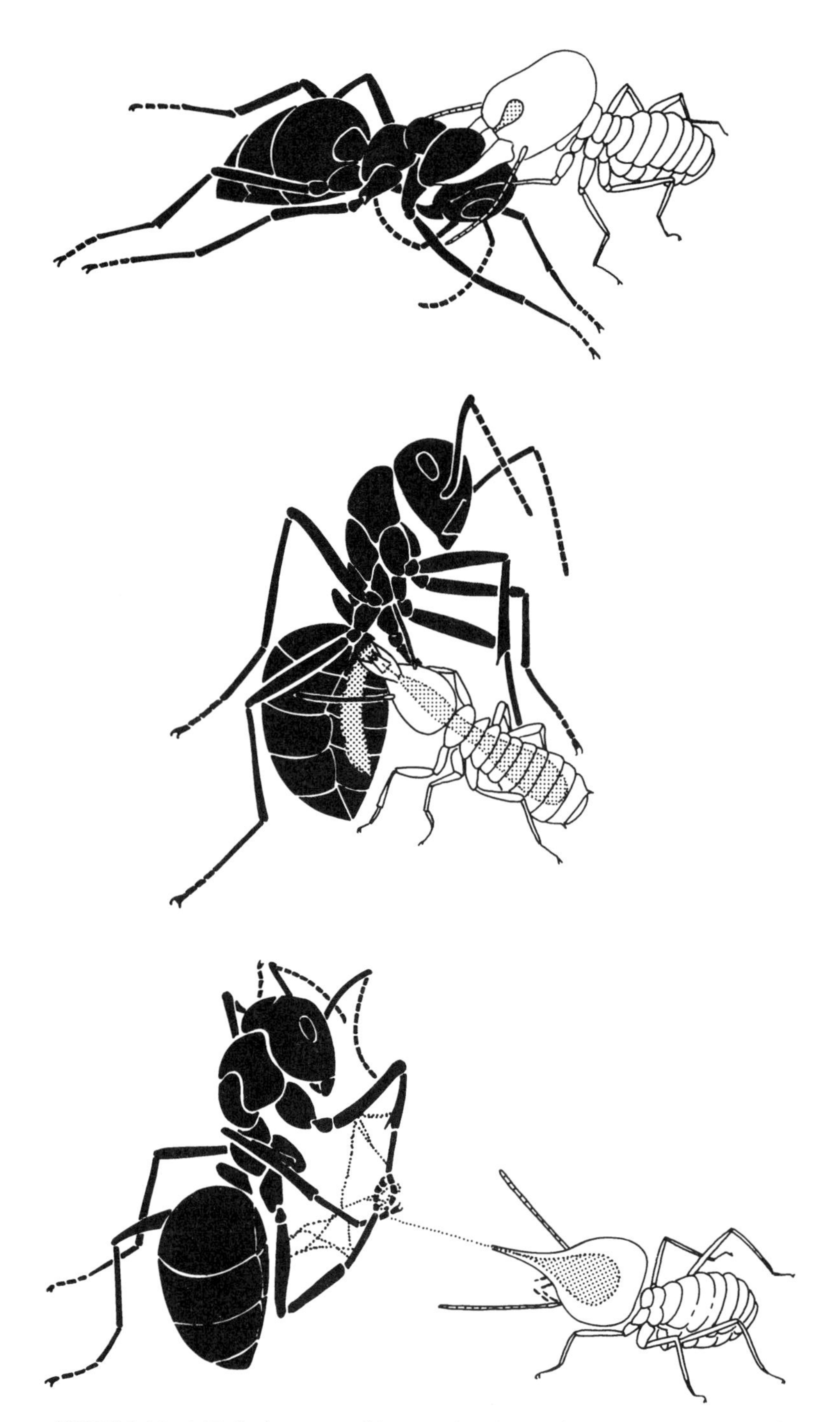

**FIGURE 15–9** Techniques used by termites in combat against ants. (*Top*) A *Cubitermes* soldier seizes an ant behind her head while emitting noxious secretions from the frontal gland, indicated by stippling. (*Center*) A *Schedorhinotermes* soldier rubs its labral brush, which is impregnated with frontal gland secretion, against an ant; it does not employ its mandibles as part of this defensive maneuver. (*Bottom*) A nasute soldier strikes an ant with a jet of sticky fluid from its frontal gland; its mandibles are vestigial. (Modified from Grassé, 1986, after Quennedey, 1975; drawing by E. Kaiser.)

*Cryptotermes* soldiers closely parallels in structure and function that developed by certain ant species in the genus *Camponotus* and tribe Cephalotini (see Chapter 8). Even more bizarre are the "snapping" soldiers of *Capritermes*, *Neocapritermes*, and *Pericapritermes*. The mandibles are twisted, asymmetrical, and arranged so that their flat inner surfaces press against each other as the adductor muscles contract. When the muscles pull strongly enough, the mandibles slip past each other with a convulsive snap, in the same way that we snap our fingers by pulling the middle finger past the thumb with just enough pressure to make it slide off with sudden force. If the mandibles strike an ant or other insect, which seems to be the primary function of the action, a stunning blow is delivered. Even human beings receive a painful flick. The adaptation is similar to that evolved independently by ants of the genus *Mystrium* (Moffett, 1986a).

Finally, the premier combat specialists are the termite soldiers that employ chemical defense. When *Protermes* soldiers attack, for example, they emit a drop of pure white saliva which spreads between the opened mandibles. When they bite, the liquid spreads over the opponent. In general the salivary glands of the soldiers are better developed than those of their worker nestmates, and they sometimes become grotesquely large in proportion to the rest of the body. The salivary reservoirs of *Pseudacanthotermes spiniger* swell out posteriorly to fill nine-tenths of the abdomen. The soldiers of *Globitermes sulphureus,* like those of at least one ant species of the *Camponotus saundersi* group (Buschinger and Maschwitz, 1984), are quite literally walking bombs. Their reservoirs fill the anterior half of the abdomen. When attacking, they eject a large amount of yellow fluid through their mouths, which congeals in the air and often fatally entangles both the termites and their victims. The spray is evidently powered by contractions of the abdominal wall. Occasionally these contractions become so violent that the wall bursts, shedding defensive fluid in all directions.

A separate strategy is the development of a "gun" consisting of a conical organ or "nasus" resembling a large nose on the front of the soldier's head. At its tip is an opening from which secretions of the frontal gland are discharged. Soldiers of some nasutitermitine genera are able to eject the material over distances of many centimeters. Their aim is quite accurate in spite of the fact that they are wholly blind.

Research during the 1980s has disclosed that termites are veritable factories of defensive chemicals. The frontal gland secretions of termitid soldiers, for example, constitute the richest source of monoterpene hydrocarbons known in the insects. Many idiosyncratic compounds have been identified in the secretions of rhinotermitid soldiers. One of them, 1-nitro–1–pentadecene, is the only nitro compound known among insect deterrent substances. This peculiarity of termite biology opens opportunities for new kinds of research. As Deligne and his co-authors (1981) say: "The dazzling variety of novel natural products that characterizes termite defensive secretions provides ideal compounds for studying the toxicology of compounds that may be highly selective insecticides. Hopefully, detailed investigations on the modes of action of these termite defensive products will be forthcoming in the near future."

Untapped opportunities also exist in detailed studies of the relations of ants to termites. For example, Traniello and Beshers (1985) discovered that soldiers of a Neotropical termite, *Nasutitermes costalis,* are strongly recruited in response to intruders of the same species, while termites belonging to other species of *Nasutitermes* are less effective, and other termite genera are ignored. It will be useful to learn whether interspecific enemy specification occurs in termites as it does in ants, that is, whether especially dangerous enemies (in this case particular genera or species of ants) also trigger correspondingly stronger responses. In one suggestive study Traniello (1981) found that the soldier caste of *N. costalis* plays a major role in the foraging-defense system of the species in deterring attacks by ants on *N. costalis* foragers. Conversely, we would also like to know if particular defensive techniques and secretions are broad-spectrum, in other words effective against a wide range of arthropod enemies, or whether they are tailored to combat certain kinds of ants. Might the specificity of predator-prey interactions and of competition between ants and termites play a role in community organization, as it does in ants?

## PREY PARALYSIS AND PREY STORAGE

Workers of the South Asian ponerine *Harpegnathos saltator* are large ants with unusually long mandibles and large, widely separated eyes. Functioning as solitary huntresses, they capture cockroaches and spiders for the most part, but also flies, butterflies, bugs, grasshoppers, cicadas, and beetles (Maschwitz, 1981). *Harpegnathos* workers regularly sting the prey, even when offered a piece of a freshly killed cockroach (Figure 15–10). Maschwitz et al. (1979) provided experimental evidence that *H. saltator* and *Leptogenys chinensis* paralyze prey by stinging and thereby are able to store it for a limited time. In one case the effect was observed to last for two weeks, and in no instance did the stung prey ever recover from the paralysis.

Even more remarkable is the storage of living prey by the Australian ponerines belonging to the *Cerapachys turneri* group. During a raid on a *Pheidole* nest the *Cerapachys* workers briefly sting each *Pheidole* larva and pupa before they pick it up and carry it to the home nest (Hölldobler, 1982b; Figure 15–11). Some prey larvae are stored inside the nest of *Cerapachys* for two months or longer, but they do not pupate or visibly increase in size. Under the microscope the prey larvae can be seen to move their mouthparts slightly, which indicates that they are alive. In a series of additional experiments Hölldobler further confirmed that larvae captured by *Cerapachys* are preserved alive but in a state of metabolic stasis. Approximately 30 *Pheidole* larvae collected from a *Pheidole* colony were put without workers in a small test tube, which was kept moist by a wet cotton plug. A second similar test tube contained 30 *Pheidole* larvae taken from a *Cerapachys* nest. The larvae from the *Pheidole* colony were all dead after two weeks. In contrast, all of the larvae from the *Cerapachys* colony were alive after two weeks, many of them still moving their mouthparts.

There is one other interesting aspect concerning the paralysis of prey larvae by *Cerapachys*. The *Pheidole* larvae are small and tender and the powerful *Cerapachys* sting can easily pierce the larva and thereby kill it. Thus the injection of the paralyzing secretions through the sting has to be very subtle in order to preserve without killing. A clue is presented by the differentiated pygidium with its denticulate margins, which is present in all workers and queens of cerapachyine ants (Brown, 1975; see Figure 15–12). Hölldobler's (1982b) morphological and histological investigations indicate that these denticuliform and spinuliform setae on the pygidium of *Cerapachys turneri* and *Sphinctomyrmex steinheili* are probably sensory setae, in particular mechanoreceptors. It is most likely that during the stinging process the mechanoreceptors signal the gaster tip's touch

**FIGURE 15–10** A worker of *Harpegnathos venetor* stings the head of a freshly killed cockroach offered to her in a laboratory foraging arena. The stinging response automatically follows prey capture.

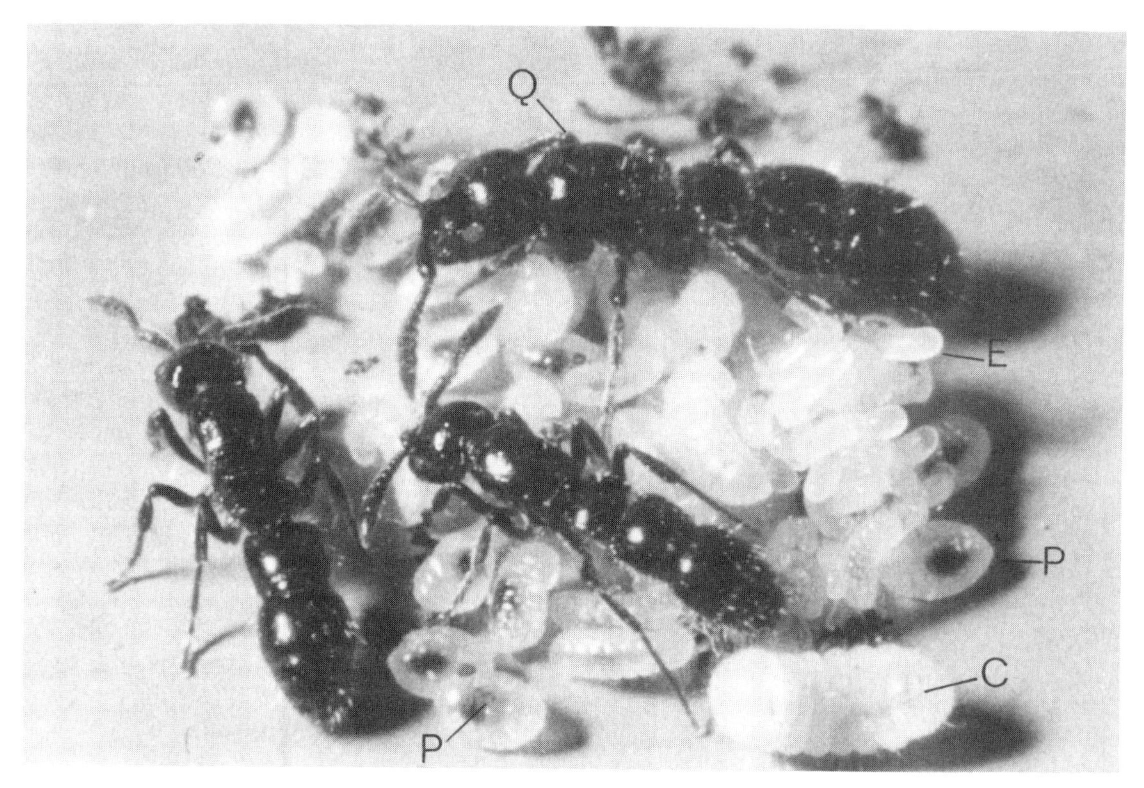


**FIGURE 15–11** A fraction of a colony in the *Cerapachys turneri* group of Australia, including the queen (*Q*), two workers, eggs (*E*), *Cerapachys* larvae (C), and paralyzed *Pheidole* larvae (*P*). (From Hölldobler, 1982b.)

to the prey larva, allowing close control of the sting's penetration. Many of the nonsocial aculeate Hymenoptera that paralyze prey by stinging are equipped with mechanoreceptors on the tip of the sting sheath (Oeser, 1961; Rathmayer, 1978). Hölldobler did not detect similar structures on the tip of the sting sheath of the cerapachyines *Cerapachys* and *Sphinctomyrmex*. The stinging of prey larvae appears to be a very stereotyped behavior. When Hölldobler shook a *Cerapachys* colony containing *Pheidole* larvae out of the nest chamber, *Cerapachys* workers retrieving *Pheidole* larvae almost invariably went through the typical stinging motion. They did not do this, however,

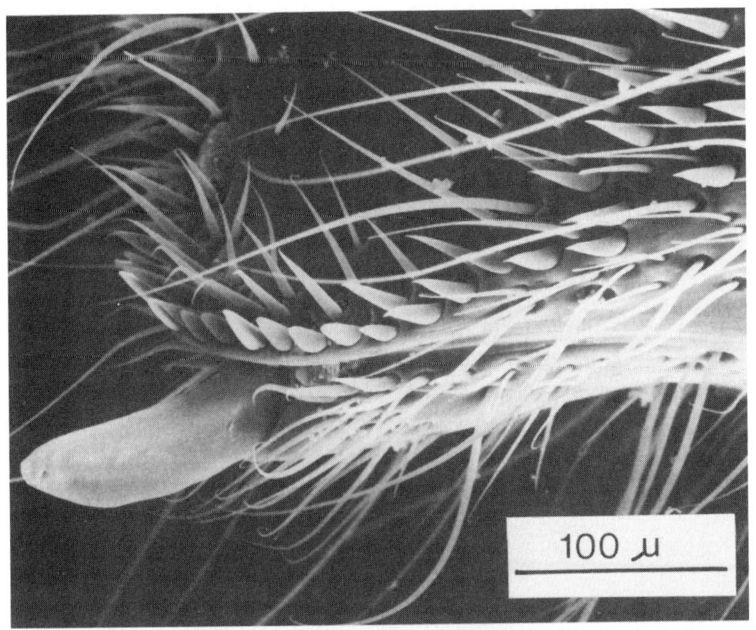


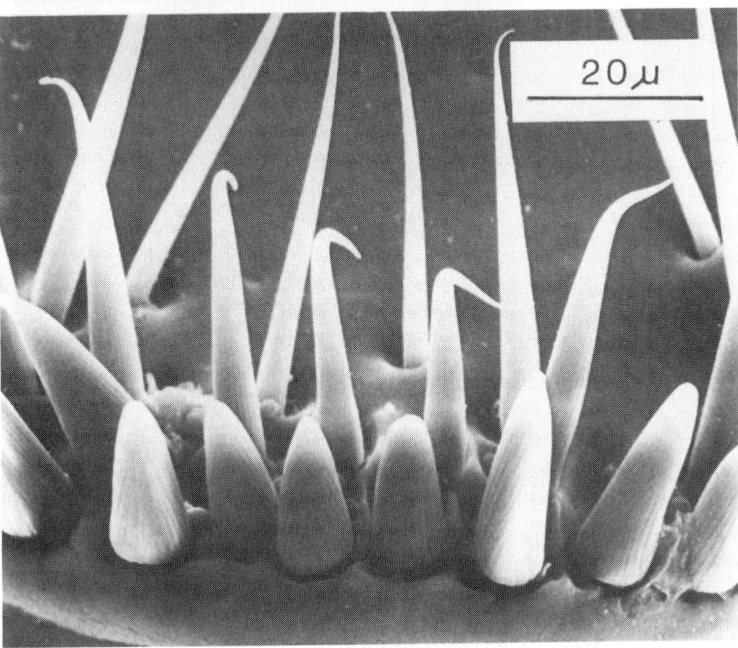


**FIGURE 15–12** Scanning electron microscope picture of the abdominal tip of a *Cerapachys* worker. (*Above*) The partly extruded sting surrounded by the sensory setae at the pygidium and last exposed sternite. (*Below*) An enlarged view of the two kinds of setae on the pygidium. (From Hölldobler, 1982b.)

when they picked up their own larvae. Workers, queens, and larvae of *Cerapachys* all feed on the *Pheidole* larvae. The food storage system appears to enable these specialized predators to stay inside their nests for long intervals. Experiments have shown that they do not conduct raids as long as they have a good supply of prey larvae available.

The paralytic storage of prey appears to be quite common in ponerine ants. *Amblyopone pallipes* and species of *Onychomyrmex* paralyze captured centipedes by stinging (Traniello, 1982; B. Hölldobler and R. W. Taylor, unpublished observations). The immobilized prey can more easily be retrieved into the nest, where it also can be stored for some time. In fact Hölldobler and Taylor found immobilized centipedes in *Onychomyrmex* bivouac nests that did not exhibit any signs of feeding by the ants. Observations made by Masuko (1987 and personal communication) suggest that *Leptanilla japonica* also paralyze large centipedes. If the prey is too large to be transported, the entire colony moves to the immobilized prey to feed on it (see Figure 16–17a). *Paltothyreus tarsatus* workers frequently sting the captured termites before they stack them between their mandibles (Hölldobler, 1984b). It is possible that in this case surplus termite preys are also preserved by paralysis. Finally, *Nothomyrmecia* huntresses regularly sting their prey objects, even when the prey is small and can be easily subdued with the ant's powerful mandibles (Hölldobler and Taylor, 1983). Here too, it is suggestive that the surplus prey can be preserved and stored inside the nest, which makes the colony less vulnerable to climatic irregularities affecting hunting success.

## MASTERS OF CAMOUFLAGE

Until the 1980s, the ants of the Neotropical genus *Basiceros* were thought to be relatively rare. Despite the large size of the worker, no more than ten colonies had been collected and almost nothing was known of the biology of the genus. However, after we had developed a "search image" for *B. manni* in Costa Rica, based in part on the white larvae and pupae that stood out against dark nest material, we began to locate colonies with ease. The reason for the previous obscurity of *Basiceros* is the superb camouflage ability of the workers. To the human eye, and presumably to the eye of many visually orienting predators such as birds and lizards, the ants are difficult to see as they walk over the ground, and virtually invisible when standing still (see Plate 19). The effect is achieved in part by the fact that *B. manni* workers are among the slowest-moving ants we have ever encountered during field experience with more than 180 of the approximately 300 ant genera found worldwide. When observed in an undisturbed state, the entire worker force often stands perfectly still for minutes at a time, even holding their antennae in place. And when moving workers are disturbed by being uncovered or touched by a pair of forceps, they freeze into immobility for up to several minutes (Wilson and Hölldobler, 1986).

The effect is enhanced by the gradual accumulation of fine particles of soil on the bodies of the workers. By the time the workers are old enough to forage, they blend in remarkably well with the soil and rotting litter over which they walk. The collection of the particles is accomplished by two layers of hairs on the dorsal surfaces of the body and outer surfaces of the legs: longer "brush" hairs with splintered distal ends, and shorter, feather-like "holding" hairs. The brush hairs evidently scrape or otherwise capture the soil particles, while the holding hairs help to keep them in place next to the surface exoskeleton (see Figure 15–13). The same phenomenon occurs widely through the other genera of the tribe Basicerotini, including *Eurhopalothrix*, *Octostruma*, and *Protalaridis*, as well as *Stegomyrmex*, the sole member of the tribe Stegomyrmecini (Hölldobler and Wilson, 1986c).

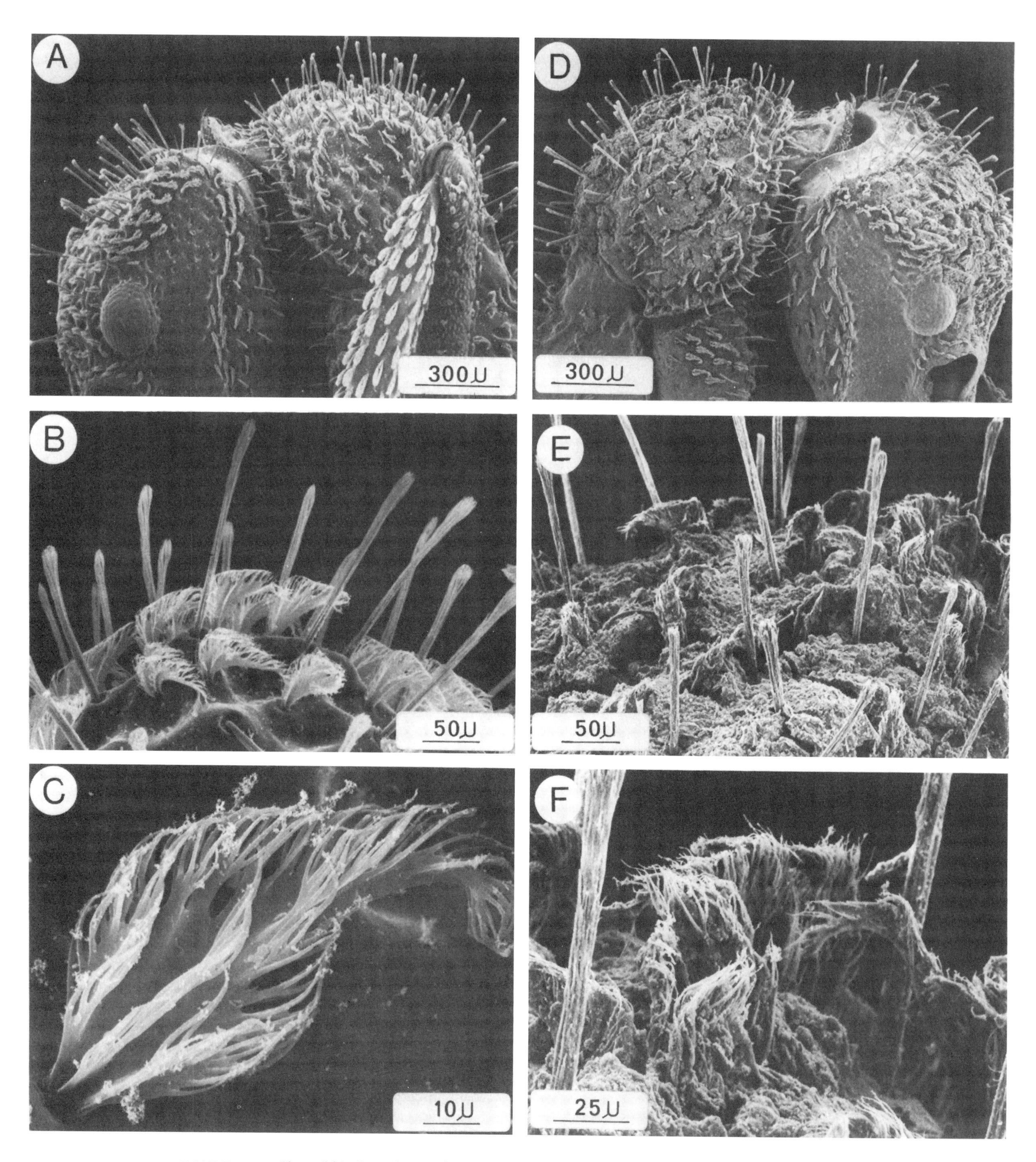


**FIGURE 15–13** The soil-binding pilosity of *Basiceros manni*, part of its camouflage technique. The left row of panels (*A–C*) contains various views of a young worker before she accumulates soil on her body surface; in the row on the right (*D–F*) are the same views of an older, soil-encrusted nestmate. *A* and *D*, portions of the head and thorax; *B* and *E*, close-ups of the prothorax, showing the two layers of soil-collecting hairs; *C* and *F*, plumose hairs from the lower layer. (From Hölldobler and Wilson, 1986c.)

CHAPTER 16

# The Army Ants

No spectacle of the tropical world is more exciting and mystifying than that of a colony of army ants on the march. In *Ants, Their Structure, Development and Behavior,* Wheeler expressed the poetry of the scene:

> The driver and legionary ants are the Huns and Tartars of the insect world. Their vast armies of blind but exquisitely coöperating and highly polymorphic workers filled with an insatiable carnivorous appetite and a longing for perennial migrations, accompanied by a motley host of weird myrmecophilous camp-followers and concealing the nuptials of their strange, fertile castes, and the rearing of their young, in the inaccessible penetralia of the soil—all suggest to the observer who first comes upon these insects in some tropical thicket, the existence of a subtle, relentless and uncanny agency, directing and permeating all their activities. (p. 246)

## THE SWARM RAIDERS

*Eciton burchelli* is one of the best understood of the army ants. This big, conspicuous species is abundant in humid lowland forests from Brazil and Peru north to southern Mexico (Borgmeier, 1955). Its marauding workers, together with those of other species of *Eciton,* are well known to native peoples by such local names as *padicours, tuocas, tepeguas,* and *soldados.* In English they are called army ants, as well as foraging ants, legionary ants, soldier ants, and visiting ants. These insects have, understandably, long been a prime target for study by naturalists, from Lund (1831) through H. W. Bates, T. Belt, H. von Ihering, W. Müller, and Fr. Sumichrast in the last century to W. Beebe, W. M. Wheeler, and many others in more recent times. But it was T. C. Schneirla (1933–1971) who, by conducting patient studies over virtually his entire career, first unraveled the complex behavior and life cycle of this and other species of *Eciton.* His results were confirmed and greatly extended in rich studies conducted by C. W. Rettenmeyer.

A day in the life of an *Eciton burchelli* colony seen through the eyes of Schneirla and Rettenmeyer begins at dawn, as the first light suffuses the heavily shaded forest floor. At this moment the colony is in "bivouac," meaning that it is temporarily camped in a more or less exposed position. The sites most favored for bivouacs are the spaces between the buttresses of forest trees and beneath fallen tree trunks (see Figure 16–1 and Plate 20) or any sheltered spot along the trunks and main branches of standing trees to a height of 20 meters or more above the ground. Most of the shelter for the queen and immature forms is provided by the bodies of the workers themselves. As they gather to form the bivouac, they link their legs and bodies together with their strong tarsal claws (Figure 16–2), forming chains and nets of their own bodies that accumulate layer upon interlocking layer until finally the entire worker force constitutes a solid cylindrical or ellipsoidal mass up to a meter across. For this reason Schneirla and others have spoken of the ant swarm itself as the bivouac. Between 150,000 and 700,000 workers are present. Toward the center of the mass are found thousands of immature forms, a single mother queen, and, for a brief interval in the dry season, a thousand or so males and several virgin queens. The entire dark brown conglomerate exudes a musky, somewhat fetid odor.

When the light level around the ants exceeds about 0.5 lux, the bivouac begins to dissolve. The chains and clusters break up and tumble into a churning mass on the ground. As the pressure builds, the mass flows outward in all directions. Then a raiding column emerges along the path of least resistance and grows away from the bivouac at a rate of up to 20 meters an hour. No leaders take command of the raiding column. Instead, workers finding themselves in the van press forward a few centimeters and then wheel back into the throng behind them, to be supplanted immediately by others who extend the march a little farther. As the workers run on to new ground, they lay down from the tips of their abdomens small quantities of chemical trail substances originating in the hindgut and probably also in the pygidial gland (Hölldobler and Engel, 1978), which guide others forward. Workers encountering prey lay extra recruitment trails that draw nestmates differentially in that direction (Chadab and Rettenmeyer, 1975).

A loose organization emerges in the columns, based on behavioral differences among the castes. The smaller and medium-sized workers race along the chemical trail and extend it at the point, while the larger, clumsier soldiers, unable to keep a secure footing among their nestmates, travel for the most part on either side. The location of the *Eciton* soldiers misled early observers into concluding that they are the leaders. As Thomas Belt put it, "Here and there one of the light-colored officers moves backwards and forwards directing the columns." Actually the soldiers, with their large heads and exceptionally long, sickle-shaped mandibles, have relatively little control over their nestmates and serve instead almost exclusively as a defense force. The minimas and medias, bearing shorter, clamp-shaped mandibles, are the generalists. They capture and transport the prey, choose the bivouac sites, and care for the brood

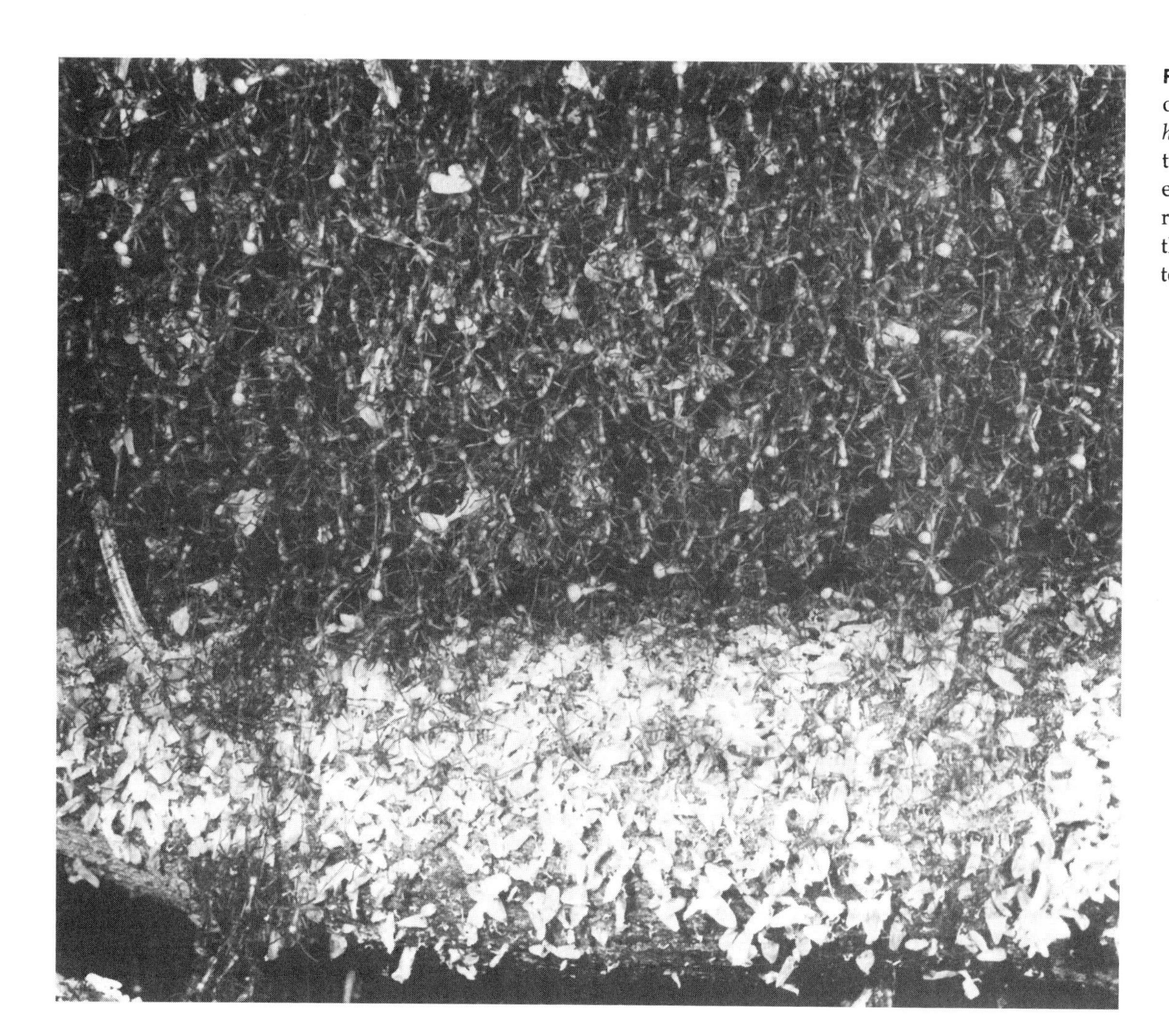

**FIGURE 16–1** The bivouac of this colony of the Central American army ant *Eciton hamatum* is located inside a log. The photograph was made on the first day of the emigration phase; as a consequence recently discarded cocoons are piled up at the bottom of the log. (Photograph courtesy of C. W. Rettenmeyer.)

and queen. Workers often work in teams, with large medias serving as porters. These specialists initiate the transport of large prey items and are joined by workers of equal or smaller size. The teams accomplish their task with greater energetic efficiency than if they cut the prey into small pieces and carried them individually (Franks, 1986; see Figure 8-45).

*Eciton burchelli* has an unusual mode of hunting even for an army ant. It is a "swarm raider," which means that the foraging workers spread out into a fan-shaped swarm with a broad front. Most other army ant species are "column raiders," pressing outward along narrow dendritic odor trails in the pattern exemplified in Figure 16–3. Schneirla (1956b) has described a typical raid as follows.

> For an *Eciton burchelli* raid nearing the height of its development in swarming, picture a rectangular body of 15 meters or more in width and 1 to 2 meters in depth, made up of many tens of thousands of scurrying reddish-black individuals, which as a mass manages to move broadside ahead in a fairly direct path. When it starts to develop at dawn, the foray at first has no particular direction, but in the course of time one section acquires a direction through a more rapid advance of its members and soon drains in the other radial expansions. Thereafter this growing mass holds its initial direction in an approximate manner through the pressure of ants arriving in rear columns from the direction of the bivouac. The steady advance in a principal direction, usually with not more than 15° deviation to either side, indicates a considerable degree of internal organization, notwithstanding the chaos and confusion that seem to prevail within the advancing mass. But organization does exist, indicated not only by the maintenance of a general direction but also by the occurrence of flanking movements of limited scope, alternately to right and left, at intervals of 5 to 20 minutes depending on the size of the swarm.
>
> The huge sorties of *burchelli* in particular bring disaster to practically all animal life that lies in their path and fails to escape. Their normal bag includes tarantulas, scorpions, beetles, roaches, grasshoppers, and the adults and broods of other ants and many forest insects; few evade the dragnet. I have seen snakes, lizards, and nestling birds killed on various occasions; undoubtedly a larger vertebrate which, because of injury or for some other reason, could not run off, would be killed by stinging or asphyxiation. But lacking a cutting or shearing edge on their mandibles, unlike their African relatives the "driver ants," these tropical American swarmers cannot tear down their occasional vertebrate victims. Arthropods, such as ticks, escape through their excitatory secretions, stick insects through repellent chemicals, as tests show, as well as through tonic immobility. The swarmers react to movement in particular as well as to the scent of their booty, and a motionless insect has some chance of escaping them. Common exceptions, which may enjoy almost a community invulnerability in many cases, include termites and *Azteca* ants in their bulb nests in trees, army ants of their own and other species both on raiding parties and in their bivouacs, and leaf-cutter ants in the larger mound communities; in various ways these often manage to fight off or somehow repel the swarmers.
>
> The approach of the massive *burchelli* attack is heralded by three

> types of sound effect from very different sources. There is a kind of foundation noise from the rattling and rustling of leaves and vegetation as the ants seethe along and a screen of agitated small life is flushed out. This fuses with related sounds such as an irregular staccato produced in the random movements of jumping insects knocking against leaves and wood. This noise, more or less continuous, beats on the ears of an observer until it acquires a distinctive meaning almost as the collective death rattle of the countless victims. When this composite sound is muffled after a rain, as the swarm moves through soaked and heavily dripping vegetation, there is an uncanny effect of inappropriate silence.
>
> Another characteristic accompaniment of the swarm raid is the loud and variable buzzing of the scattered crowd of flies of various species, some types hovering, circling, or darting just ahead of the advancing fringe of the swarm, others over the swarm itself or over the fan of columns behind. To the general hum are added irregular short notes of higher pitch as individuals or small groups of flies swoop down suddenly here or there upon some probable victim of the ants which has suddenly burst into view . . . No part of the prosaic clatter, but impressive solo effects, are the occasional calls of antbirds. One first catches from a distance the beautiful crescendo of the bicolored antbird, then closer to the scene of action the characteristic low twittering notes of the antwren and other common frequenters of the raid.

If you wish to find a colony of swarm raiders in Central or South America, the quickest way is to walk quietly through a tropical forest in the middle of the morning, listening. For long intervals the only birds you may hear are in the distance, mostly in the canopy. Then, as Johnson (1954) has expressed it, comes a "chirring, twittering, and piping" of antbirds close to the ground. Mingled in is the murmur or hissing caused by the frantic movements of countless insects trying to escape the raiders, and the buzzing of parasitic flies. Very soon you will see the ants themselves marching in a broad front, hundreds of thousands streaming forward as though drawn toward some goal just out of sight in the forest shadows. Also present may be ithomiine butterflies, which fly over the leading edge of the swarm. First noticed by Drummond (1976), the butterflies are now known to feed on the droppings of the antbirds (Ray and Andrews, 1980; Andrews, 1983).

On Barro Colorado Island, Panama, where Schneirla conducted most of his studies, the antbirds normally follow only the raids of *Eciton burchelli* and those of another common swarm raider, *Labidus praedator.* They pay no attention to the less conspicuous forays of *E. hamatum, E. dulcius, E. vagans,* and other column-raiding army ants. There are at least ten species of antbirds on the island, all members of the family Formicariidae. They feed principally on the insects and other arthropods flushed by the approaching *burchelli* swarms (Johnson, 1954; Willis, 1967). Although a specimen of *Neomorphus geoffroyi* has been recorded with its stomach stuffed with *burchelli* workers, most species appear to avoid the ants completely or at most consume them by accident while swallowing other food.

As one might anticipate from these accounts, the *Eciton burchelli* colonies and their efficient camp followers have a profound effect on the faunas of those particular parts of the forest over which the swarms pass. Williams (1941), for example, noted a sharp depletion of the arthropods at spots on the forest floor where a swarm had struck the previous day. On Barro Colorado Island, which has an area of approximately 17 square kilometers, there exist only about 50 *burchelli* colonies at any one time. Since each colony travels at

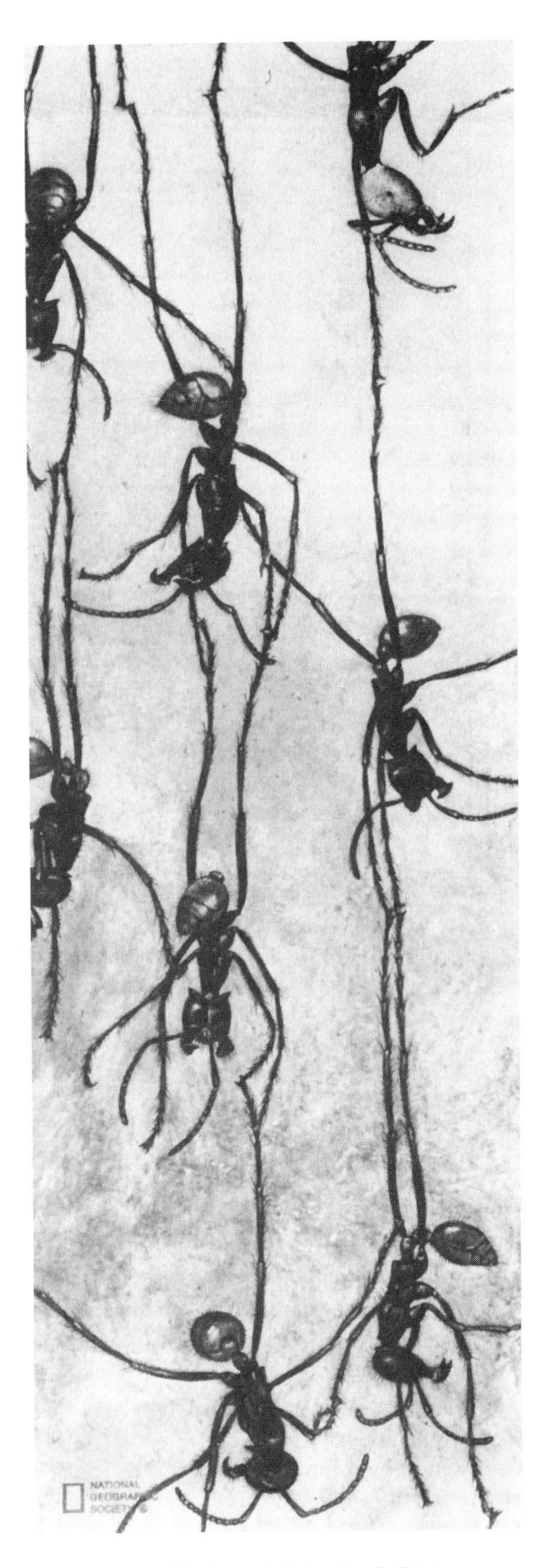

**FIGURE 16–2** Workers of *Eciton burchelli* join together to create shelters out of their own bodies. First several ants choose a log or some other object near the ground and hang from it with their tarsal claws interlocked. Other ants run down the strands and fasten on until strands become ropes that fuse into a meter-wide mass, which constitutes the daily shelter or "bivouac." (From Hölldobler, 1984d; painting by J. D. Dawson reproduced by permission of the National Geographic Society.)

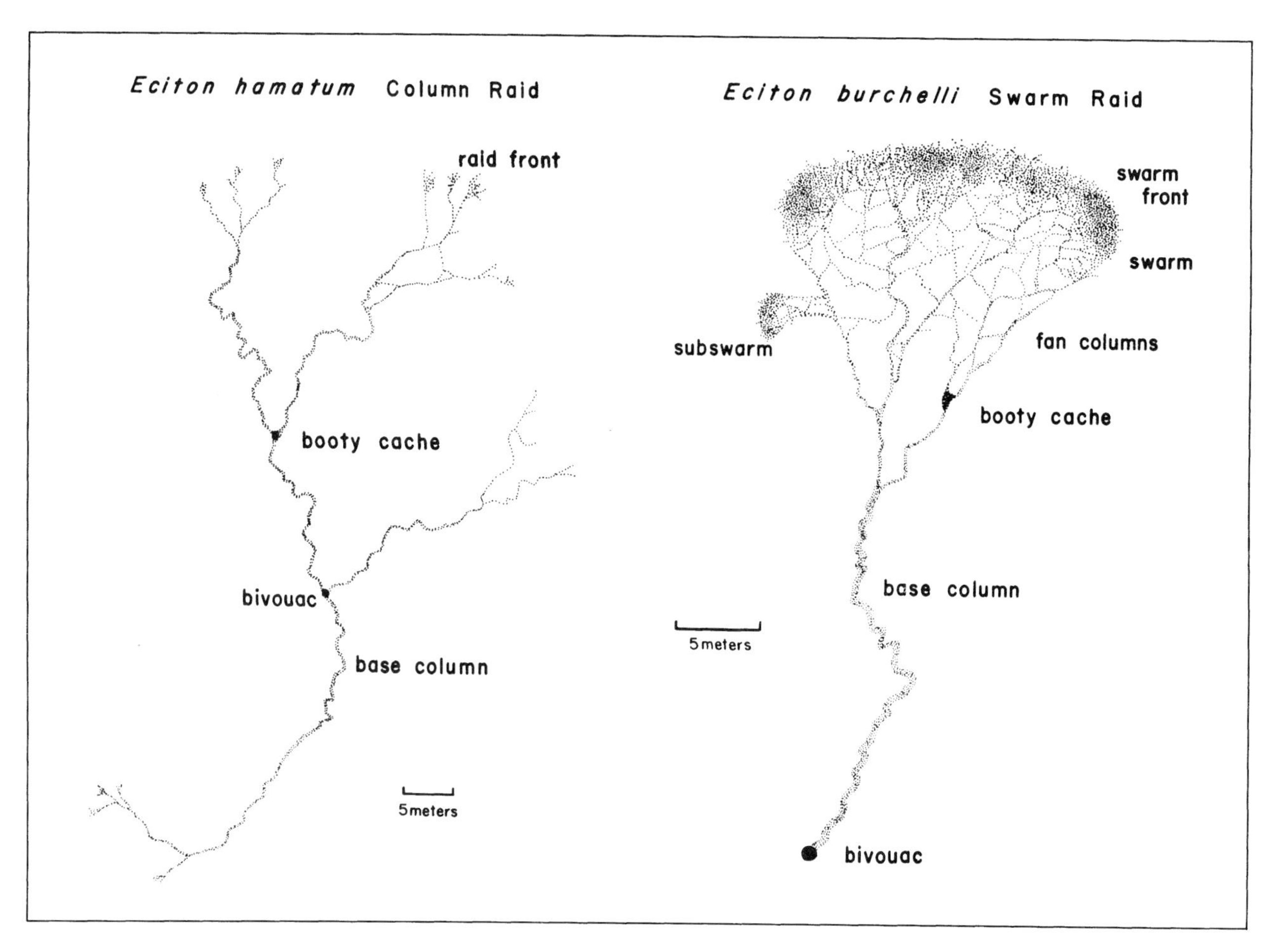


**FIGURE 16–3** The two basic patterns of raiding employed by army ants: (*left*) column raid of *Eciton hamatum* with the advancing front made up of narrow bands of workers (this pattern is displayed by the majority of army ant species); (*right*) swarm raid of *E. burchelli* with the advancing front made up of a large mass or swarm of workers, followed by the anastomosing columns in the fan area which converge toward the base column and bivouac site. (From Rettenmeyer, 1963a.)

most 100–200 meters every day (and not at all on about half the days), the collective population of *burchelli* colonies raids only a minute fraction of the island's surface in the course of one day, or even in the course of one week. But the strikes are probably frequent enough during the course of months to have a significant effect on the composition and age structure of the colonies of ants and social wasps.

The food supply is quickly and drastically reduced in the immediate vicinity of each colony. Early writers, especially Müller (1886) and Vosseler (1905), jumped to the reasonable conclusion that army ant colonies change their bivouac sites whenever the surrounding food supply is exhausted. At an early stage of his work, however, Schneirla (1933b, 1938) discovered that the emigrations are subject to an endogenous, precisely rhythmic control unconnected to the immediate food supply. He proceeded to demonstrate that each *Eciton* colony alternates between a *statary phase,* in which it remains at the same bivouac site for as long as two to three weeks, and a *nomadic phase,* in which it moves to a new bivouac site at the close of each day, also for a period of two to three weeks. (The nomadic phase is better called the migratory phase, since army ants are migratory hunters rather than nomads in the strict sense. That is, they move periodically to areas of fresh prey, rather than guide herds to fresh pastures in the manner of true nomads. True nomadism is known among ants only in certain Malaysian species of *Hypoclinea;* see Chapter 13.) The basic *Eciton* cycle is summarized in Figure 16–4. Its key feature is the correlation between the *reproductive cycle,* in which broods of workers are reared in periodic batches, and the *behavior cycle,* or the alternation of the statary and migratory phases. The single most important feature of *Eciton* biology to bear in mind when trying to grasp this rather complex relation is the remarkable degree to which development is synchronized within each successive brood. The ovaries of the queen begin developing rapidly when the colony enters the statary phase, and within a week her abdomen is greatly swollen by 55,000 to 66,000 eggs (Figure 16–5). Then, in a burst of prodigious labor lasting for several days in the middle of the statary period, the queen lays from 100,000 to 300,000 eggs. By the end of the third and final week of the statary period, larvae hatch, again all within a few days. Several days later the callow workers (so called because they are at first weak and lightly pigmented) emerge from the cocoons. The sudden appearance of tens of thousands of new adult workers has a galvanic effect on their older sisters. The general level of activity increases, the size and intensity of the swarm raids grow, and the colony starts emigrating at the end of each day's marauding. In short, the colony enters the migratory phase. The migratory phase itself continues as long as the brood initiated during the previous statary period remains in the larval stage. As soon as the larvae pupate, however, the intensity of the raids diminishes, the emigrations cease, and the colony (by definition) passes into the next statary phase.

The emigration is a dramatic event requiring sudden complex behavioral changes on the part of all adult members of the *Eciton* colony. At dusk or slightly before workers stop carrying food into the old bivouac and start carrying it, along with their own larvae, in an outward direction to some new bivouac site along the pheromone-impregnated trails (Figure 16–6). Eventually, usually after most of the larvae have been transported to the site, the queen herself makes the journey. This event usually transpires between 2000 and 2200 hours, well after nightfall. Just before the queen emerges from the bivouac, the workers on the trail nearby become distinctly more excited, and the column of running workers thickens beyond its usual width of 2–3 centimeters, soon widening to as much as 15 centimeters. Suddenly the queen appears in the thickest part. As she runs along she is crowded in by the "retinue," a shifting mob consisting of an unusual number of soldiers and darkly colored, unladen smaller workers. The members of the retinue jostle her, press in underfoot, climb on her back, and at times literally envelop her body in a solid mass. Even with this encumbrance, the queen moves easily to the new bivouac site. She is guided by the odor trail and can follow it by herself even if the surrounding workers are taken away. After passage the emigration tapers off, and it is usually finished by midnight. If the column is disturbed near the queen, she stops and is swiftly covered by a blanket of protecting workers. All New World army ants employ retinues during emigrations ready to react this way. The largest are formed by *Eciton burchelli* and other species that travel aboveground and hence are most exposed to predators (Rettenmeyer et al., 1978).

The activity cycle of *Eciton* colonies is truly endogenous. It is not linked to any known astronomical rhythm or weather event. It continues at an even tempo month after month, in both wet and dry seasons throughout the entire year. Propelled by the daily emigrations of the migratory phase, the colony drifts perpetually back and forth over the forest floor (Figure 16–7). The results of experiments performed by Schneirla (1971) indicate that the phases of the activity cycle are determined by the stages of development of the brood and their effect on worker behavior. When he deprived *Eciton* colonies in the early migratory phase of their callow workers, they lapsed into the relatively lethargic state characteristic of the statary phase, and emigrations ceased. Migratory behavior was not resumed until the larvae present at the start of the experiments had grown much larger and more active. In order to test further the role of larvae in the activation of the workers, Schneirla divided colony fragments into two parts of equal size, one part with larvae and the other without. Those workers left with larvae showed much greater continuous activity.

These results, while provocative, are not decisive and at best solve only half the problem. For if the activity cycle is controlled by the reproductive cycle, what controls the reproductive cycle? The logical place to look would seem to be the queen. By her astonishing capacity to lay all of her eggs in one brief burst, she creates the synchronization of brood development, which is the essential feature for the colonial control of the activity cycle. At first Schneirla (1944)

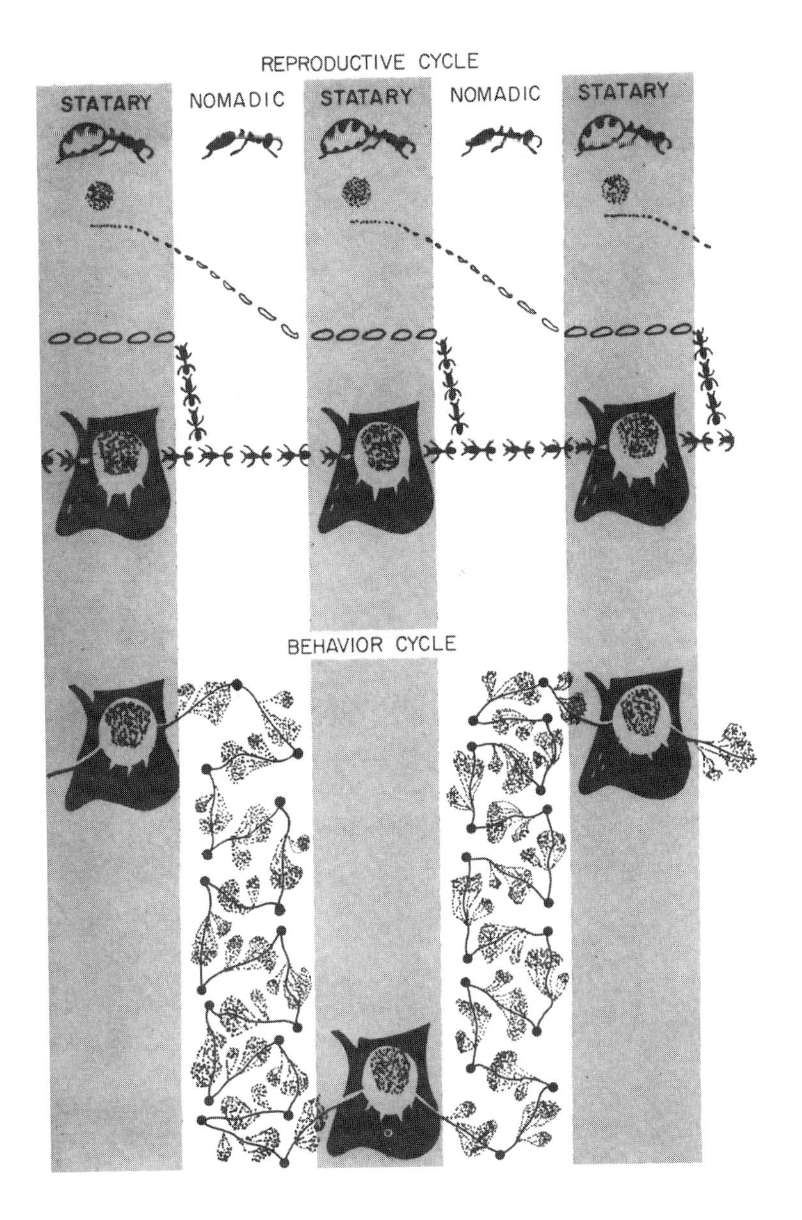


**FIGURE 16–4** The alternation of the statary and migratory (nomadic) phases of the *Eciton burchelli* colony, which consists of distinct but tightly synchronized reproductive and behavior cycles. During the statary phase the queen, shown at the top, generates and lays a large batch of eggs, all in a brief span of time; the eggs hatch into larvae; the pupae derived from the previous batch of eggs develop into adults; and, as indicated in the lower diagram, the colony remains in one bivouac site. During the migratory phase the larvae complete their development; the new workers emerge from their cocoons; and, as indicated in the lower diagram, the ants change their nest sites after the completion of each day's swarm raid. (From Wilson, 1971, redrawn from Schneirla and Piel, 1948.)

concluded that this reproductive effort by the queen is the "pacemaker," thus implying that the queen herself is the seat of an endogenous rhythm. Later, however, Schneirla (1949b, 1956b) modified his hypothesis by viewing the queen and her colony as reciprocally donating elements in an oscillating system.

> When each successive brood approaches larval maturity, the social-stimulative effect upon workers nears its peak. The workers thus energize and carry out some of the greatest raids in the nomadic phase, with their by-product larger and larger quantities of booty in the bivouac. But our histological studies show that, at the same time, more

**FIGURE 16–5** Queens of the army ant *Eciton hamatum*. The individual above rests with a major worker. Her deflated abdomen indicates that her colony is in the nomadic phase. The queen below is physogastric and laying eggs, typical of the statary phase. (The worker on her abdomen carries a parasitic *Planodiscus* mite on its middle leg.) (Photographs courtesy of C. W. Rettenmeyer.)

and more of the larvae (the largest first of all) soon reduce their feeding to zero as they begin to spin their cocoons. Thus in the last few days of each nomadic phase a food surplus inevitably arises. At this time the queen apparently begins to feed voraciously. It is probable that the queen does not overfeed automatically in the presence of plenty, but that she is started and maintained in the process by an augmented stimulation from the greatly enlivened worker population. Within the last few days of each nomadic phase, the queen's gaster begins to swell increasingly, first of all from a recrudescence of the fat bodies, then from an accelerating maturation of eggs. The overfeeding evidently continues into the statary phase, when, with colony food consumption greatly reduced after enclosure of the brood, smaller raids evidently bring in sufficient food to support the processes until the queen becomes maximally physogastric. These occurrences, which are regular and precise events in every *Eciton* colony, are adequate to prepare the queen for the massive egg-laying operation which begins about one week after the nomadic phase has ended. (1956b:401)

While this interpretation makes a pretty story, it is constructed with fragments of very circumstantial evidence. The crucial question is unanswered: whether the queen really is stimulated to feed in excess by the greater abundance of food or at least by the higher intensity of worker activity associated with the food, as Schneirla posited, or whether her increased feeding is timed by some other,

**FIGURE 16–6** During emigrations army ants sometimes create living bridges of their own bodies. In this photograph the workers of an *Eciton burchelli* colony can be seen linking their legs and, along the top of the bridge, hooking their tarsal claws together to form irregular systems of chains. A symbiotic silverfish, *Trichatelura manni,* is seen crossing the bridge in the center. The bridge was started above the top of the picture where the liana on the right joins the buttress on the left. The ants gradually moved the bridge downward while extending its length. The first step in such a bridge is only one ant in length. (From Wilson, 1971, photograph by C. W. Rettenmeyer.)

undetermined physiological event. Since work on *Eciton* physiology is still virtually nonexistent, and experimental evidence of any kind very sparse, one can do no more than reflect on these possibilities as competing hypotheses.

Another question of considerable interest, added to the inducement of queen oogenesis, is the stimulus that triggers the onset of the migratory phase. According to Schneirla's theory (actually a hypothesis) of brood stimulation, the migratory phase is initiated when workers become excited by the near-simultaneous eclosion of new, callow workers from the pupae. Migration is sustained thereafter by stimulation from the growing larvae. As illustrated in Figure 16–4, the mass eclosion at the start of the migratory phase coincides with the hatching of the egg mass. We may then ask which event, adult eclosion from the pupa or egg hatching, triggers the migratory phase? In an ingenious experiment conducted on *Neivamyrmex nigrescens,* Topoff et al. (1980) removed the larval brood of an early migratory colony and replaced it with the pupal brood of an early statary colony. The pupae eclosed well before the next batch of eggs laid by the host colony hatched, with the result that for several days the host colony was occupied by newly eclosed, callow workers but no larvae. It commenced its next migratory phase nonetheless, demonstrating that newly emerged adults are sufficient by themselves to drive this segment of the army ant cycle.

Very little is known concerning the actual communicative stimuli that mediate the activity cycles. In his voluminous theoretical writings on the subject, Schneirla often spoke of "trophallaxis" as the driving force of the cycles of army ants, but it is clear that he meant this term to be virtually synonymous with "communication" in the broadest sense. Apparently he had no clear ideas about the nature of the signals utilized. In earlier articles he attributed much of the stimulative effect of the larvae to their "squirming"; later he stressed the probable existence of pheromones as well. These speculations were based almost entirely on observations of the more obvious outward signs of communication, however, a level of study usually inadequate to distinguish even the sensory modalities employed in communication with insect colonies and unlikely to identify the signals employed. Lappano (1958) discovered that the labial glands of *burchelli* larvae become fully functional on the eighth or ninth migratory day, about the time raiding activity reaches its peak. She concluded that the labial glands are "probably" producing a pheromone that excites the worker. But again, the only evidence available is the stated coincidence in time of the two events. Our lack of knowledge of the semiotic basis of the *Eciton* cycle is due simply to the lack of any serious attempt to obtain it. This interesting subject does not seem likely to resist sustained experimental study; any such effort in the future is likely to yield exceptionally interesting results.

In his classic writings Schneirla was inadvertently misled by his failure to distinguish consistently between the ultimate causation and the proximate causation of the army ant cycle. It is possible and even likely that the adaptive value, hence the ultimate causation, is

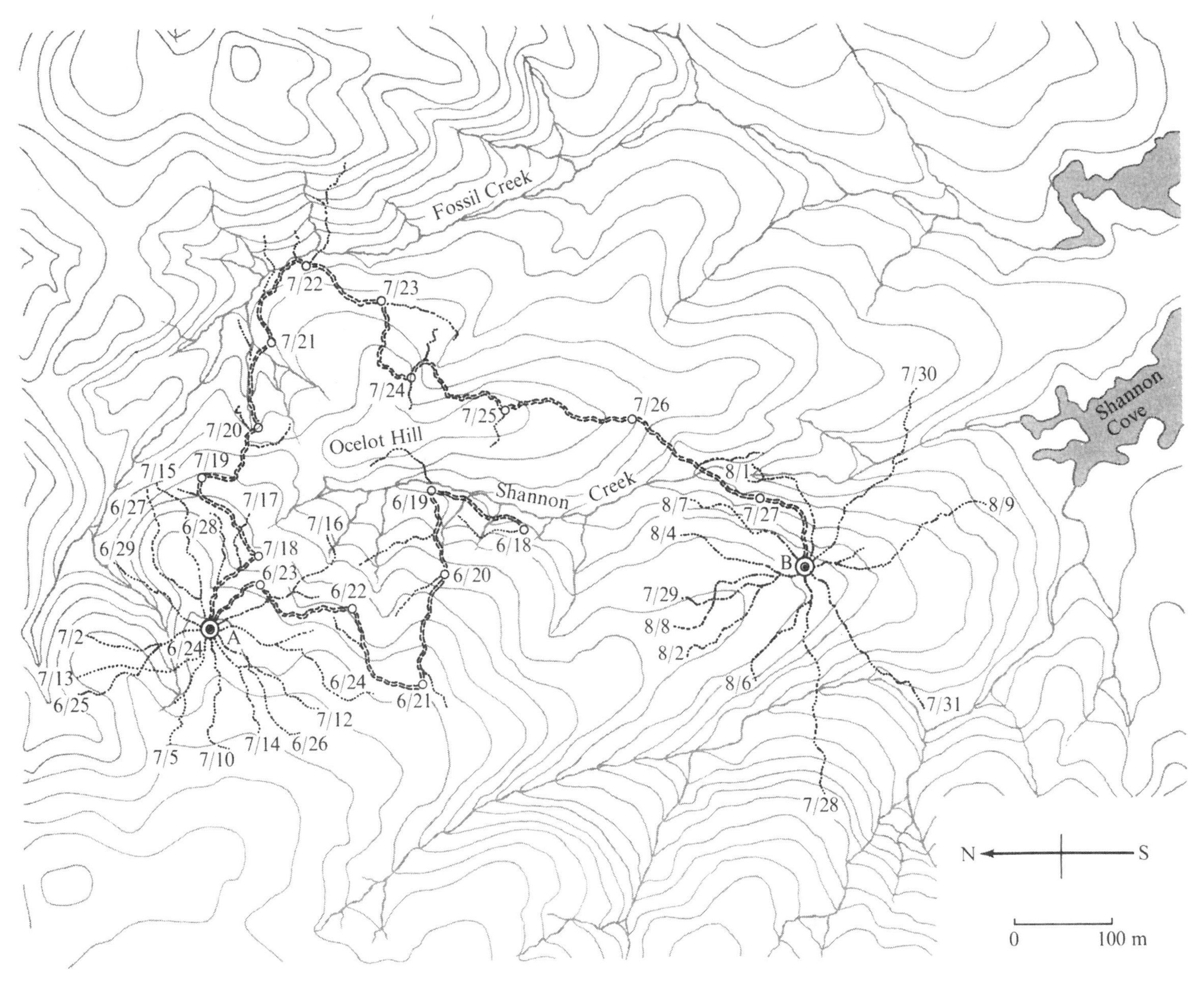


**FIGURE 16–7** The path used by a colony of the swarm raider *Eciton burchelli* on Barro Colorado Island, Panama, during two reproductive cycles (egg to adult). *A* and *B* are the persistent bivouac sites used by the ants throughout each of the two statary phases. Radiating from each are the routes taken on successive days by raiding swarms, which returned to the same bivouac sites at the end of each raid. The thick double line traces the route taken by the colony during its migratory phase, when it raided and relocated at a new bivouac site each day. (From Schneirla, 1971.)

the additional food made available to the colony when it emigrates frequently. However, biological systems often evolve so as to rely on endogenous rhythms to make the needed changes, rather than on a close reading of the environment from day to day. In other words the flush of callow workers becomes the token signal to the workers to initiate daily emigrations. They are the proximate cause of the emigrations, but the ultimate cause—the advantage emigrations give to emigration-prone genotypes in the ant population—remains the improved food supply. Although Schneirla occasionally mentioned that food availability might have been an important factor in the evolution of emigrations (1944, 1957b), the idea played no important role in his theoretical interpretation. After he had demonstrated the endogenous nature of the cycle, and its control by synchronous brood development, he dismissed the role of food depletion. The emigrations, he repeatedly asserted, are caused by the appearance of callow workers and the older larvae; they are not caused by food shortage. He meant only that food shortages are not the proximate cause.

**TABLE 16–1** Army ants in the narrow sense: biological characteristics of the best-known species of legionary ants.

| Species | Number of workers per colony | Nesting habits | Diet | Raiding pattern | Cycle | Authority |
|---|---|---|---|---|---|---|
| **SUBFAMILY PONERINAE** | | | | | | |
| *Leptogenys* sp. (close to *L. mutabilis*) | >30,000 | Subterranean bivouac sites or bivouac in leaf litter | Wide variety of arthropods | Fan-shaped raids, front 4–9 m across, tapering to trunk columns | Frequent emigrations but pattern irregular and not correlated with brood development | Maschwitz et al. (1989) |
| **SUBFAMILY DORYLINAE** | | | | | | |
| *Aenictus laeviceps* | 60,000–110,000 | Clustering in sheltered places on ground surface during migratory phase, movement underground during statary phase | Mostly other ants, social wasps, and termites; a few other arthropods | Weak dendritic columns, mostly over surface of ground; raids any time of day or night | *Eciton* pattern followed except that up to 5 or more emigrations occur daily when larvae are most active during migratory phase; considered a primitive condition; migratory phase lasts about 14 days, statary phase 28 days | Schneirla (1965, 1971), Gotwald (1978) |
| *Dorylus* (= *Anomma*) *wilverthi* | 2 million (Vosseler), 15–20 million (Raignier and van Boven) | Subterranean clustering with much excavation of soil | Many kinds of small animals, mostly arthropods; some vegetable matter | Swarms about 12 m wide in front tapering to trunk columns in rear; raids any time, usually at night | Emigrations irregular, average one every 20–25 days; continue day and night for 2–3 days for average of 223 m; occur when brood consists mostly of worker pupae | Vosseler (1905), Raignier and van Boven (1955), Gotwald (1978) |

*continued*

In his culminating synthesis Schneirla (1971) provided a more balanced view of evolution and physiological mechanisms. Toward the end of his life he also learned that food shortages are in fact among the proximate controls, at least to a minor degree in some army ant species. In his last field study, on the small Asian army ant *Aenictus*, he discovered that short-term variation in colony activity depends on the "alimentary condition prevalent in the brood" (Schneirla and Reyes, 1969). Specifically, the ants appear to emigrate more often when their food supply runs low. Topoff and Mirenda (1980a,b) later tested the impact of food supply on emigrations in *Neivamyrmex nigrescens* by a series of experiments. This was made possible by their feat of culturing army ant colonies in the laboratory long enough to follow the brood cycle under controlled conditions. Two colonies overfed with prey emigrated on only 28 percent of the days during the migratory phase of the cycle, while two underfed colonies emigrated on 62 percent of the migratory days. It is now evident that although *Neivamyrmex* colonies follow an endogenous cycle of roughly the *Eciton* type (see Table 16–1), the tempo of their migra-

**TABLE 16–1** *(continued)*

| Species | Number of workers per colony | Nesting habits | Diet | Raiding pattern | Cycle | Authority |
|---|---|---|---|---|---|---|
| **SUBFAMILY ECITONINAE** | | | | | | |
| *Eciton burchelli* | 150,000–700,000 | Exposed clustering above surface, often arboreal; no modification of bivouac site | Wide variety of arthropods, including immature stages of social wasps and other ants | Swarm 10–15 m in front, tapering to trunk column in rear; raid begins at daybreak, ends at dusk | Migratory phase (11–17 days) alternates with statary phase (19–22 days); emigrations daily during migratory phase, which begins when adults eclose from pupae | Schneirla (1957a, 1971), Rettenmeyer et al. (1983) |
| *E. hamatum* | 50,000–250,000 | Exposed clustering above surface, seldom arboreal; no modification of bivouac site | Immature stages of social wasps and other ants | Several dendritic columns lead away from bivouac site; raid begins at daybreak, ends at dusk | As in *E. burchelli*, except migratory phase lasts 16–18 days and statary phase 17–22 days | Schneirla (1957a, 1971), Rettenmeyer (1963a), Rettenmeyer et al. (1983) |
| *E. vagans* | ? | Mostly in preformed subterranean cavities such as abandoned ant nests, also in and under rotting logs; some excavation practiced | Primarily immature stages of other ants; a few other arthropods | As in *E. hamatum*, except that columns are partly subterranean and partly nocturnal | Emigrations with larvae and callow adults | Fiebrig (1907), Schneirla et al. (1954), Rettenmeyer (1963a) |
| *Labidus praedator* | "Probably" over 1 million | As in *E. vagans* | Wide variety of arthropods, including immature forms of other ants; some nuts and other vegetable matter | Swarm variable in size and shape; front usually less than 4 m across, tapering to trunk columns; much raiding underground, occurs day and night | Emigrations occur, but pattern apparently irregular and not closely correlated with brood development | Rettenmeyer (1963a), Lenko (1969), Rettenmeyer et al. (1983) |
| *Neivamyrmex nigrescens* | 10,000–140,000 | Mostly subterranean bivouac sites, often under rocks and logs; occasionally in rotting logs | Primarily other ants; also a few beetles and other small arthropods | Weak dendritic columns, mostly over surface of ground but partly subterranean; raids at night or on overcast days | *Eciton* pattern followed but less regularly, with emigrations less than daily during migratory phase, which lasts about 14 days and statary phase 19–21 days | Schneirla (1958, 1961, 1963), Rettenmeyer (1963a), Topoff and Mirenda (1978), Mirenda and Topoff (1980), Mirenda et al. (1980, 1982) |
| *Nomamyrmex esenbecki* | ? | Subterranean bivouac sites of unknown nature | Mostly brood of other ants; some other arthropods | Weak dendritic columns, mostly under objects on ground or underground | Probably irregular *Eciton* pattern, but evidence incomplete | Rettenmeyer (1963a) |

tory phase is strongly affected by the amount of food they harvest. The complex interactions mediating raiding and emigrations in ecitonine army ants have been reviewed by Topoff (1984).

To summarize very briefly to this point, we note that the colonies of *Eciton* and at least some other army ant species follow an endogenous cycle as Schneirla described it. The migratory phase is triggered at least in part by the emergence of new, callow adult workers. It is quite possible that the cycle was put in place genetically and is kept there through natural selection by the advantage it gives in overcoming food shortages. The frequency of raids is also fine-tuned in at least one species (*Neivamyrmex nigrescens*) by the day-to-day availability of food.

Colony multiplication in *Eciton*, first elucidated by Schneirla and Brown (1950, 1952), is a highly specialized and ponderous operation. Through most of the year the mother queen is the paramount center of attraction for the workers, even when she is in competition with the mature worker larvae toward the close of each migratory phase. By serving as the focal point of the aggregating workers, she literally holds the colony together. The situation changes drastically, however, when the annual sexual brood is produced early in the dry season. This kind of brood contains no workers, but, in *E. hamatum* at least, it consists of about 1,500 males and 6 new queens. Even when the sexual larvae are still very young, a large fraction of the worker force becomes affiliated with the brood as opposed to the mother queen. By the time the larvae are nearly mature, the bivouac can be found to consist of two approximately equal zones: a brood-free zone containing the queen and her affiliated workers, and a zone in which the rest of the workers hold the sexual brood. The colony has not yet split in any overt manner, but major behavioral differences between the two sections do exist. For example, if the queen is removed for a few hours at a time, she is readily accepted back into the brood-free zone from which she originated, but she is rejected by workers belonging to the other zone. Also, there is evidence that workers from the queen zone cannibalize brood from the other zone when they contact them.

The young queens are the first members of the sexual brood to emerge from the cocoons. The workers cluster excitedly over them, paying closest attention to the first one or two to appear (see Figure 16–8). Several days later the new adult males emerge from their cocoons. This event energizes the colony, sets off a maximum raid followed by emigration, and at last splits the bivouac. Raids are conducted along two radial trails from the old bivouac site. As they intensify during the day, the young queens and their nuclei of workers move out along one of the trails, while the old queen with her nucleus moves out along the other.

When the derivative swarm begins to cluster at the new bivouac site, only one of the virgin queens is able to make the journey to it. The others are held back by the clinging and clustering of small groups of workers. In Schneirla's expression, they are "sealed off" from the rest of the daughter colony. Eventually they are abandoned and left to die. Now there exist two colonies: one containing the old queen; the other containing the successful virgin, daughter queen. In a minority of cases the old queen is also superseded. That is, the old queen herself falls victim to the sealing-off operation, leaving both of the daughter colonies with new virgin queens. This seems to happen most often when the health and attractive power of the old queen begin to fail prior to colony fission. The maximum age of

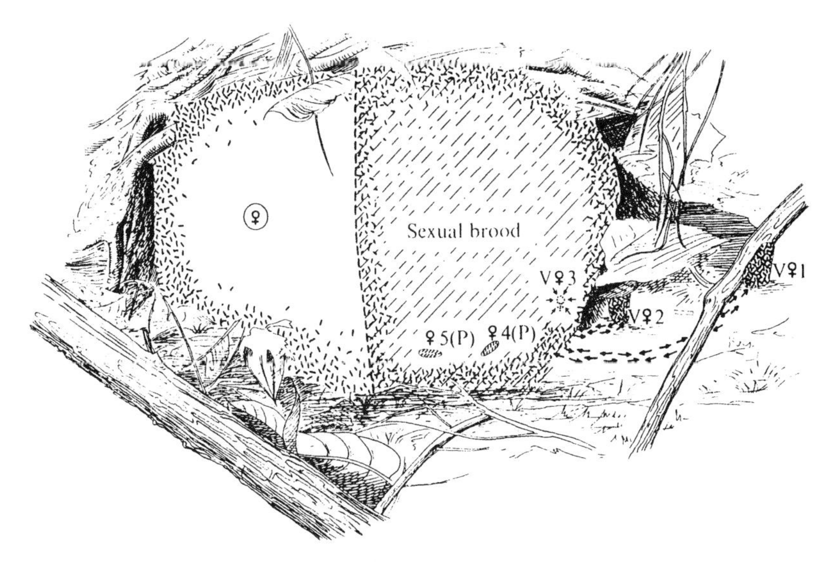


**FIGURE 16–8** An *Eciton hamatum* bivouac just prior to colony division. The left portion of the bivouac contains the mother queen but no brood, while the right portion contains the sexual brood. Two of the virgin queens (*V 1* and *V 2*) have emerged from their cocoons and moved to one side of the bivouac, to be attended by clusters of workers who still run back and forth to the bivouac along odor trails. A third virgin queen (*V 3*) has emerged more recently and is still confined within a knot of workers in the bivouac wall, while two others (*P*) remain in the pupal stage within cocoons. The males are also still in the pupal stage. (From Schneirla, 1956a.)

the *Eciton* queen is not known, but is believed to be relatively great for an insect; a marked queen of *E. burchelli*, for example, was recovered by Rettenmeyer after a period of four and a half years. The males, in contrast, enjoy only one to three weeks of adult existence. Shortly after their emergence they depart on flights away from the home bivouac in search of other colonies. Their bodies are heavily laden with exocrine glands resembling those of the queens. The new queens are fecundated within a few days of their emergence, and almost all of the males disappear within three weeks of that event. Rettenmeyer (1963a) has described an actual mating, and he has presented evidence that a queen sometimes mates more than once in her lifetime and may even mate annually. Two other matings, lasting two and ten hours, were observed by Schneirla (1971) after the ants had been removed for laboratory observation.

In 1987 Franks and Hölldobler proposed that in army ants an unusual form of sexual selection exists, in which the workers play a major role in selecting their sister's mate. Outbreeding appears to be the rule; Schneirla (1971) has suggested that *Eciton* males must fly before they can mate. Thus males have to enter alien colonies and break the worker barrier to get to the females. Under these circumstances workers are in a position to choose both the mother and father of their future nestmates. Franks and Hölldobler argued that the reason why workers should be involved in choosing mates for their queen is that they will later invest in the progeny of these males. Therefore the principle of sexual selection and female choice should apply to their preferences. To maximize their own inclusive fitness, the workers should actively choose the males that are most fertile. Sexual selection theory suggests that males may demonstrate that they will be donors of highly viable gene combinations, for example by being themselves large, robust, and vigorous. It is

**FIGURE 16–9** Mating in *Eciton* army ants. A male *E. burchelli,* who in this case has shed his wings, mates with a queen one to three years in age. (Photograph by C. W. Rettenmeyer.)

possible that worker involvement in sexual selection has favored males that are superficially similar in size and shape to their conspecific queen. Such is in fact the case in the polyphyletic lineages of army ants.

The unusual robustness of army ant males is well known. In Africa they are called sausage flies, and elsewhere they are commonly mistaken for wasps. It is possible that during evolutionary history, the gaster of the male became elongated and enlarged to house the massive sperm vesicles (Gotwald and Burdette, 1981) together with the ever-increasing glandular equipment needed to impress the workers. Hölldobler and Engel-Siegel (1982) discovered that ecitonine males are unusually well endowed with abdominal glands, especially between the tergites of the gaster. Such glands, which do not occur in workers, closely resemble those of the queen (Whelden, 1963). Fierce sexual competition is made likely by the highly skewed numerical sex ratio in army ants. Where only a few males succeed, the resulting selection has probably led to the evolution of the rich glandular equipment in males, because only the most attractive males were successful. In fact males have retinues of workers like the entourages of queens.

There is one further feature of the army ant syndrome that can be explained in terms of this unusual form of sexual selection and sexual competition. In many army ant species males lose their wings upon entering an alien colony (Figure 16–9). Dealate males have been found in the Old World army ants, including *Dorylus rubellus* (Savage, 1847, cited in Gotwald, 1982), as well as in the New World *Labidus praedator* (Rettenmeyer, 1963a) and species of *Eciton* (Schneirla, 1971).

## PHYLOGENY OF THE DORYLINE AND ECITONINE ARMY ANTS

The genus *Eciton* represents one of the furthest extensions of an evolutionary trend that began independently within each of many different groups of ants. According to Gotwald (1982), the "true" army ants of the subfamilies Dorylinae and Ecitoninae are triphyletic, or triple origined. Specifically, the combination of army ant behavior and the dichthadiigyne queen has arisen separately in the genera *Aenictus* and *Dorylus,* which constitute the subfamily Dorylinae, and a third time in *Eciton* and its relatives (*Cheliomyrmex, Labidus, Neivamyrmex,* and *Nomamyrmex*), which make up the subfamily Ecitoninae. The division between the two subfamilies is supported by a number of strong morphological differences, including even the Dufour's gland epithelium (Billen, 1985a).

On the basis of a phenetic study of *Dorylus,* Barr et al. (1985) recognized three groupings within that genus: *Dorylus* (including the "subgenera" *Anomma, Dichthadia,* and *Typhlopone*), *Alaopone,* and *Rhogmus.* These authors (and Gotwald, 1985) took the conservative position of placing all of the species under the single genus *Dorylus,* noting that "whether to recognize these groupings at the generic or subgeneric level must await the broader consideration which will be accorded the group in a complete revision." The higher taxonomic status of *Aenictus* is in similar limbo. If the genus is truly independent, having arisen out of a stock entirely separate from those of *Dorylus* and the ecitonines, it should logically either be placed in the same subfamily as the ancestral stock or accorded a subfamily of its own. But again, this decision awaits a full analysis of the Dorylinae and candidate ancestral groups, particularly the Ponerinae. Meanwhile Gotwald has provided sound biogeographic data to support his belief (1979) that *Dorylus* arose in Africa and spread into tropical Asia, while *Aenictus* arose in Asia and spread to Africa and Australia.

In the New World *Cheliomyrmex* is the phylogenetically most interesting genus. It is distinguished by a single waist segment, as opposed to the two waist segments possessed by all the other New World army ants. (Among Old World forms, *Dorylus* also has a single segment.) *Cheliomyrmex* is otherwise anatomically close to the remainder of the New World army ants (Gotwald, 1971). Wheeler, and later Gotwald and Kupiec (1975) and Gotwald (1978), considered the single-jointed waist primitive, and thus *Cheliomyrmex* itself primitive. *Cheliomyrmex* workers also have a median tooth on the tarsal claws and a simple pygidium (as opposed to a longitudinally impressed pygidium with terminal spines). Both of these traits are probably primitive (see Figure 16–10). *Cheliomyrmex* remains a rarely seen and enigmatic genus, however, and needs to be examined in much more detail before its phylogenetic position can be ascertained with confidence. Its biology and life cycle may also shed light on its position within the Ecitoninae.

The phylogenetic origins of the three army ant groups (Aenictini, Dorylini, Ecitonini plus Cheliomyrmecini) are more generally very poorly understood. A favorite view, originating with Emery (1895) and entertained more recently by Brown (1975) and Gotwald (1982), is that the army ants arose from the tribe Cerapachyini of the subfamily Ponerinae. One of the cerapachyine traits favorable to this hypothesis is the widespread habit of raiding other species of ants.

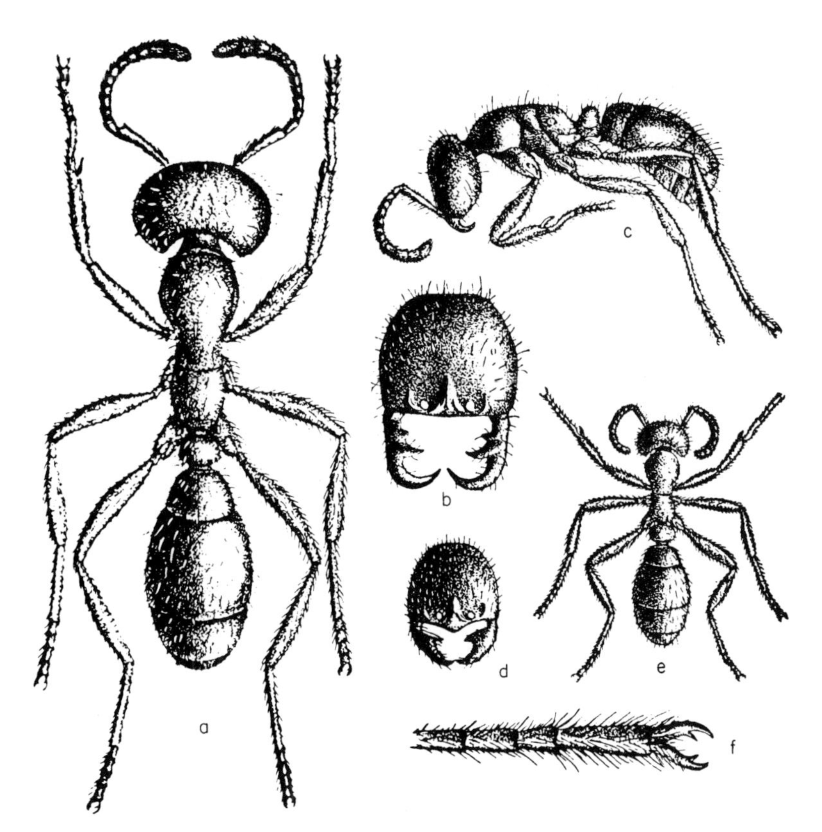


**FIGURE 16–10** The army ant *Cheliomyrmex morosus* (= *C. nortoni*) of Central America: (*a*) soldier; (*b*) head of soldier seen from above; (*c*) media; (*d*) head of media; (*e*) minor worker; (*f*) tarsus showing toothed claws. The genus is believed to be the most primitive of the New World true army ants, which constitute the subfamily Ecitoninae. (From Wheeler, 1910a.)

Again, however, both the anatomy and the behavior of the cerapachyines could be merely convergent to those of the doryline and ecitonine army ants.

The time when the three doryline and ecitonine lines originated is also unknown. A single species, *Neivamyrmex ectopus*, has been described from the Dominican Republic amber, which is late Oligocene or early Miocene in age (Wilson, 1985d). The two type workers are quite modern in aspect and closer in overall anatomy to species now living in Mexico and the southern United States than to the much richer Central and South American faunas. It is of special interest that army ants are today completely absent from the Dominican Republic and the remainder of the Greater Antilles. In other words they have retreated from this part of the world since no later than Oligocene times.

## WHAT IS AN ARMY ANT?

Behavior patterns characteristic of the doryline and ecitonine army ants also occur, at least to a limited extent, in the subfamilies Leptanillinae, Ponerinae, and Myrmicinae. In fact the best definition of the term "army ant" may well be a functional one, of the sort offered in the second edition of *Webster's New International Dictionary:* "Any species of ant that goes out in search of food in companies, particularly the driver and legionary ants." It should be added here, in order to clarify the vernacular nomenclature, that there has been a tendency by modern authors to use the terms "army ant" and "legionary ant" interchangeably. For these investigators, "legionary ant" refers first to the New World army ants of the genus *Eciton,* and second to the "driver ants" of the genus *Dorylus.* Most English-speaking authors of the late twentieth century have designated *Eciton* and its relatives as legionary ants, while employing "army ant" to cover all of the dorylines and ecitonines. We prefer to use the two terms synonymously in the broader functional sense (see Wilson, 1971).

Actually, the definition just quoted is incomplete. Upon closer examination of the subject one finds that there are really two discrete features that can be considered fundamental in army ant (legionary) behavior. These diagnostic features have been distinguished under the concepts of *migration* and *group predation* (Wilson, 1958e). Migration, or nomadism as Schneirla and Wilson called it, is defined as relatively frequent colony emigration. Most, if not all, ant species shift their nest site if the environment of the nest area becomes unfavorable, and some, for example the dolichoderine *Tapinoma sessile,* are exceptionally restless and may emigrate many times during the course of a single season. Yet none has been found that undertakes emigration so frequently or accomplishes it in such an orderly fashion to cover so much new territory as do the species of *Eciton* and the other, better-known dorylines.

Group predation includes group raiding and, usually, group transport in the process of hunting living prey. These two processes must be carefully differentiated since they involve quite different innate behavior patterns and are not invariably linked. Many ant species engage in the group transport of prey, meaning that two or more workers carry a single prey item back to the nest in a cooperative and efficient manner. Members of at least 27 genera representing most of the subfamilies use group transport (see Chapter 10). A few ponerines that specialize on termites also group-raid, including a few *Leptogenys,* as well as *Onychomyrmex,* a highly modified member of the ponerine tribe Amblyoponini found in tropical Australian forests, which group-raids to capture large arthropod prey (Figure 16–11). Other group-raiders include the tiny ants of the genus *Leptanilla* (subfamily Leptanillinae) and species of *Pheidologeton* (subfamily Myrmicinae) that swarm out to harvest a wide range of prey. Some species of the ponerine tribe Cerapachyini, which feed primarily or exclusively on other ants, may group-raid (Wilson, 1958e). Recent laboratory studies of *Cerapachys* indicate, however, that at least some of these ants employ conventional solitary scouting and group recruitment to organize their raids (Hölldobler, 1982b). The same is true of *Megaponera foetans* (Longhurst et al., 1979b), *Pachycondyla* (= *Termitopone*) *laevigata* (Hölldobler and Traniello, 1980b), and at least some *Leptogenys* species (Maschwitz and Mühlenberg, 1975; Maschwitz and Schönegge, 1983). The doryline and ecitonine "true" army ants display an extreme form of group foraging and raiding in which individual workers move exclusively in groups, venturing to run only short distances beyond the advancing edge before they turn back in. Their timidity is so strong that when masses of workers are dumped onto a clean flat surface or are cut off from the rest of the colony by rain, they commence "circular milling." In this bizarre formation workers go forward and inward with the crowd but not outward in a centrifugal direction, so that

**FIGURE 16–11** Raiding workers of *Onychomyrmex*, a genus of Australian amblyoponine army ants, have seized a large prey.

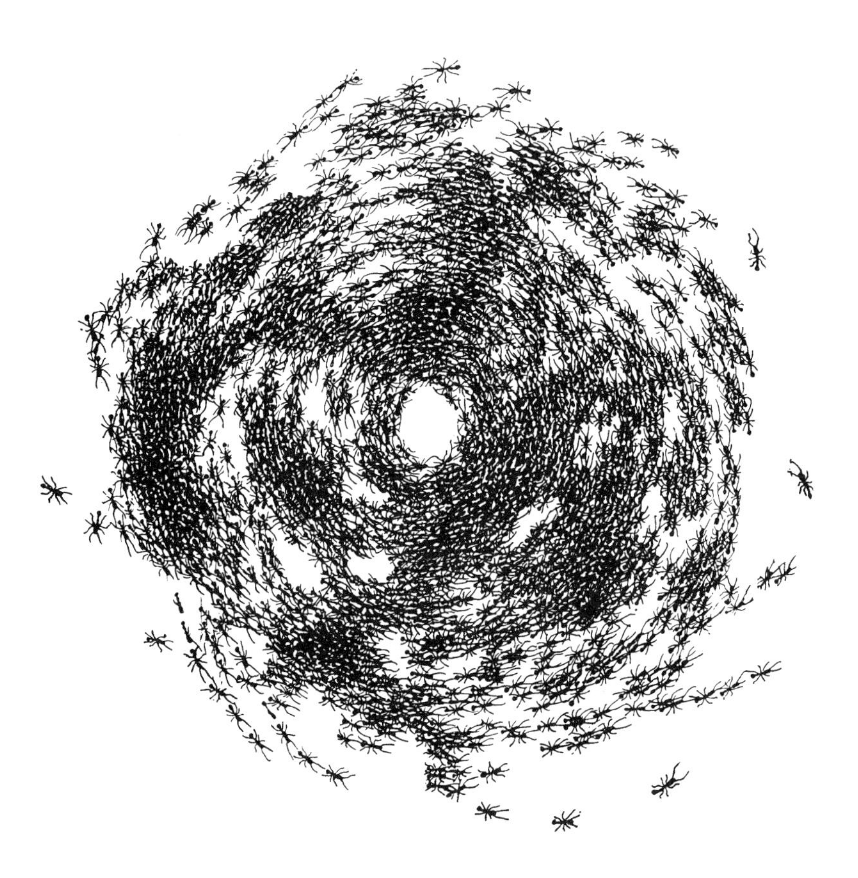

**FIGURE 16–12** A circular mill of the army ant *Labidus praedator*. This group was cut off from the rest of their colony by rain. The workers were so strongly attracted to each other that none developed enough centrifugal direction to lead the others out of the mill. After a day and a half, all were dead. (From Schneirla, 1971.)

the whole mass continues to circle round and round until all the ants are dead (Figure 16–12).

How much are migratory behavior and group predation associated in ants? We are still handicapped by a scarcity of information on emigration of any sort in most ant groups. Yet it can at least be established that the association of frequent emigration and group raiding, constituting the most general characteristics of army ant behavior, does exist in groups other than the Dorylinae and Ecitoninae. We have listed the known cases in Table 16–2. An additional characteristic of most of these species is the independent evolutionary development of the peculiar queen form known as the dichthadiigyne. As exemplified in Figure 16–13, the dichthadiigyne is a permanently wingless form with greatly reduced eyes, massive pedicel and abdomen, and strong legs. Her aberrant morphology contributes to two of her adaptations to migratory life: her capacity to deliver large quantities of eggs during a short span of time, and her ability to run under her own power from one bivouac site to another.

**TABLE 16-2** Army ants in the broad sense: genera and higher taxa of ants whose species show legionary behavior.

| Genus | Distribution | Described species (approx. no.) | Authority |
|---|---|---|---|
| **SUBFAMILY PONERINAE** | | | |
| **Tribe Amblyoponini** | | | |
| *Onychomyrmex* | Australia | 3 | Wheeler (1916a), Wilson (1958e), R. W. Taylor in Schneirla (1971), Hölldobler et al. (1982) |
| **Tribe Ponerini** | | | |
| *Leptogenys* (*processionalis* group) | Tropical Asia to Queensland | 10 | U. Maschwitz (personal communication) |
| *Simopelta* | Central and South America | 8 | Wilson (1958e), Gotwald and Brown (1966) |
| **SUBFAMILY DORYLINAE** | | | |
| **Tribe Aenictini** | | | |
| *Aenictus* | Africa to tropical Asia and Queensland | 50 | Wilson (1964) |
| **Tribe Dorylini** | | | |
| *Dorylus* (Subgenera: *Alaopone, Anomma, Dichthadia, Dorylus, Rhogmus, Typhlopone*) | Africa to tropical Asia | 54 | Emery (1910), Wheeler (1922), Raignier and van Boven (1955), Wilson (1964), Gotwald (1982) |
| **SUBFAMILY ECITONINAE** | | | |
| **Tribe Cheliomyrmecini** | | | |
| *Cheliomyrmex* | South America to southern Mexico | 5 | Wheeler (1921b), Gotwald (1971, 1982) |
| **Tribe Ecitonini** | | | |
| *Eciton* | South America to southern Mexico | 12 | Borgmeier (1955) |
| *Labidus* | South America to Texas | 8 | Creighton (1950), Borgmeier (1955) |
| *Neivamyrmex* | South America to Iowa and Virginia | 117 | Smith (1942b), Creighton (1950), Borgmeier (1955), Watkins (1976) |
| *Nomamyrmex* | South America to Texas | 2 | Creighton (1950), Borgmeier (1955) |
| **SUBFAMILY LEPTANILLINAE** | | | |
| *Leptanilla* | Africa, tropical Asia, Australia, South America | 10 | Wheeler and Wheeler (1965), Petersen (1968), Baroni Urbani (1977), Masuko (1987) |
| **SUBFAMILY MYRMICINAE** | | | |
| *Pheidologeton* | Tropical Asia | 20 | Ettershank (1966), Moffett (1984, 1986f, 1987a–c, 1988a,b, and personal communication) |

## THE DRIVER ANTS

The African driver ants of the genus *Dorylus* differ from the ecitonines in several significant details of the activity cycle, which are apparently caused by peculiarities in the queen caste (Raignier and van Boven, 1955). The queens are the largest of all ants (Figure 16–13). Those of *Dorylus* vary from 39 to 50 millimeters or more in total length and possess as many as 15,000 ovarioles capable of delivering 1 million to 2 million eggs in a month. The abdomen is in a permanent state of moderate physogastry. The queen lays eggs more or less continuously. Most are produced in bursts, however, that come in approximately three-week intervals and last five or six days. The ensuing development of the larval brood appears to have little effect on the inducement of emigration. In fact, larvae are usually outnumbered by pupae in the brood of emigrating colonies. No clocklike alternation between statary and migratory phases of the *Eciton* type is displayed by the colonies of *Dorylus*. The emigrations, which take several days to complete, are separated by statary periods that vary in duration from six days to two or three months. The bivouacs of *Dorylus* are also much more stable than those of *Eciton* and most other army ants. The colonies settle deeply into the soil at the end of the emigration, excavating labyrinthine systems of galleries and chambers to a depth of 1–4 meters. The colonies each contain millions of workers, as many as 22 million in the case of *Dorylus wilverthi*, according to Raignier and van Boven. The colonies are surprisingly numerous in parts of Africa, considering their huge individual size. Leroux (1977) calculated that *Dorylus nigricans* colonies occur at a density of about one colony per 10 hectares in the Guinea savanna of the Ivory Coast and three colonies per 10 hectares in the nearby forest.

From these secure nests the *Dorylus* send forth almost daily raids. The swarm pattern, illustrated in Figure 16–14, unfolds like a great pseudopodium. It engulfs all of the ground and low vegetation in its path, and then, after a few hours, drains back to the bivouac site. The advance is leaderless. The excited workers rush back and forth at an average speed of 4 centimeters per second. Those in the van press forward for a short distance and then retreat into the mass to give way to new advance runners. The columns resemble thick black ropes lying along the ground. A close examination shows them to be dozens or hundreds of workers wide. The ants are so dense that they pile on top of one another and run along on one another's backs, while some spill away from the column and form scattered crowds to either side, their antennae and mandibles pointed upward in threatening postures (Figure 16–15). The frontal swarm, which contains up to several millions of workers, advances at a rate of about 20 meters per hour. In his field notes Gotwald (1984–85) described a march in Gabon as follows: "The advancing swarm of worker ants moved with the effortlessness of a rain-swollen river. It flowed across the forest floor with singleness of purpose, altered in its course only by the most obtrusive of natural barriers . . . As the swarm progressed over the forest litter, small stationary groups of workers formed, giving the moving mass of foragers the appearance of an island-choked delta."

The ants sweep almost all forms of animal life before them, killing insects and larger creatures too sluggish to get out of the way. Thomas Savage's famous account of 1847 expresses the drama of the hunt:

> They will soon kill the largest animal if confined. They attack lizards, guanas, snakes, etc. with complete success. We have lost several animals by them,—monkeys, pigs, fowls, etc. The severity of their bite, increased to great intensity by vast numbers, it is impossible to conceive. We may easily believe that it would prove fatal to almost any animal in confinement. They have been known to destroy the *Python natalensis*, our largest serpent. When gorged with prey it lies powerless for days; then, monster as it is, it easily becomes their victim . . . Their entrance into a house is soon known by the simultaneous and universal movement of rats, mice, lizards, *Blapsidae, Blattidae* and of the numerous vermin that infest our dwellings. Not being agreed, they cannot dwell together, which modifies in a good measure the severity of the Driver's habits, and renders their visits sometimes (though very seldom in my view) desirable. Their ascent into our beds we sometimes prevent by placing the feet of the bedsteads into a basin of vinegar, or some other uncongenial fluid; this will generally be successful if the rooms are ceiled, or the floors overhead tight, otherwise they will drop down upon us, bringing along with them their noxious prey in the very act of contending for victory. They move over the house with a good degree of order unless disturbed, occasionally spreading abroad, ransacking one point after another, till, either having found something desirable, they collect upon it, when they can be destroyed "en masse" by hot water; or, disappointed, they abandon the premises as a barren spot, and seek some other more promising for exploration. When they are fairly in we give up the house, and try to wait with patience their pleasure, thankful, indeed, if permitted to remain within the narrow limits of our beds or chairs. They are decidedly carnivorous in their propensities. Fresh meat of all kinds is their favourite food; fresh oils they also love, especially that of the *Elais guiniensis*, either in the fruit or expressed. Under my observation they pass by milk, sugar, and pastry of all kinds, also salt meat; the latter, when boiled, they have eaten, but not with the zest of fresh. It is an incorrect statement, often made, that "they devour everything eatable" by us in our houses; there are many articles which form an exception. If a heap of rubbish comes within their route, they invariably explore it when larvae and insects of all orders may be seen borne off in triumph,—especially the former.

The dominance of the driver ants in sub-Saharan Africa has earned them special names in different African cultures: *siafu, ensanafu, kelelalu, bashikouay,* and *nkran,* among others. Their offensive power lies not in their stings, which are rarely if ever inserted, but in the powerful bite and shearing action of their mandibles. *Dorylus* was given its vernacular name by Savage, who wrote in 1847 that the ant "drives everything before it capable of muscular motion, so formidable is it from its numbers and bite . . . and, in distinction from other species of this country, may well take for its vulgar name that of Driver." According to the nineteenth-century explorer Paul Du Chaillu, cited by Gotwald (1984–85), criminals were exposed to *Dorylus* swarms as a cruel form of execution. While traveling in Ghana, Gotwald was told of an incident in which driver ants killed a baby left beneath a tree while its mother tilled the family garden.

Yet driver ants are not really the terror of the jungle as popularly conceived. Although the colony is an "animal" weighing in excess of 20 kilograms and possessing on the order of 20 million mouths, its raiders move over the ground at the rate of only a meter every 3 minutes. It is possible to watch the whole process at close range while seated comfortably in a camp chair (which of course must be periodically moved—carefully, and in the right direction!).

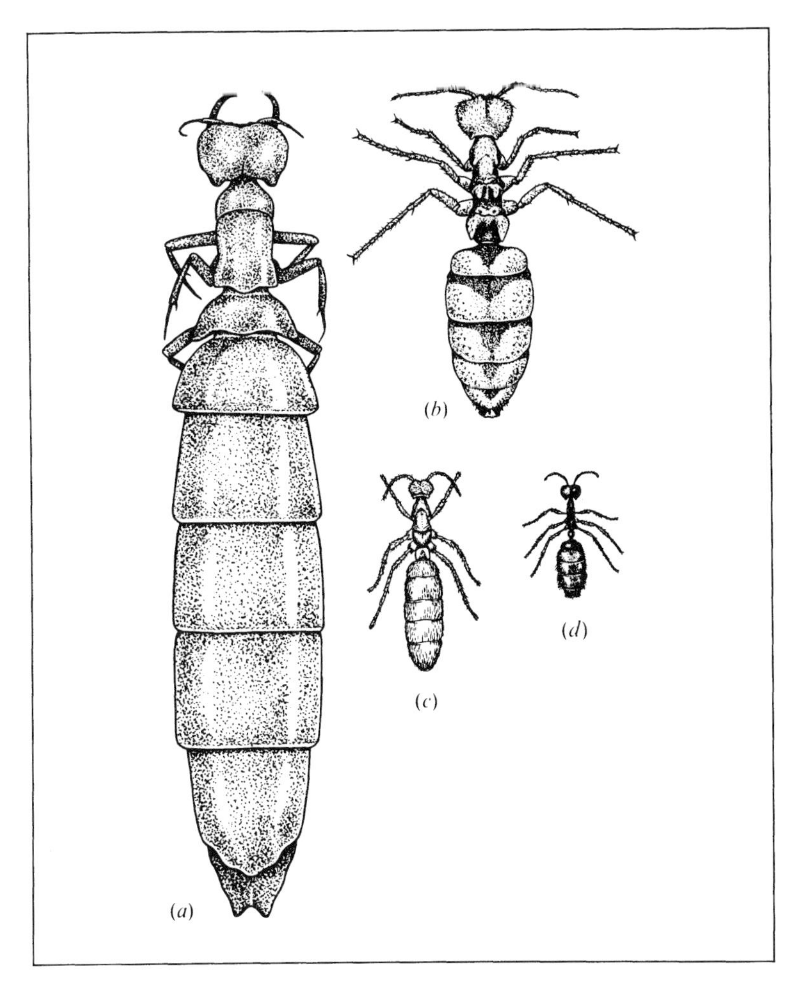


**FIGURE 16–13** The queens of four species of army ants. The larger the colony, the more eggs the queen must lay in each brood cycle to replenish dying workers, and hence the larger must be the queen's body. The body lengths are as follows: (*a*) *Dorylus wilverthi*, 52 millimeters; (*b*) *Eciton burchelli*, 21 millimeters; (*c*) *Neivamyrmex nigrescens*, 12.5 millimeters; and (*d*) *Aenictus gracilis*, 8 millimeters. Estimates of the number of eggs laid in one batch by each are 1–2 million; 225,000; 50,000; and 30,000 respectively. (From Schneirla, 1971.)

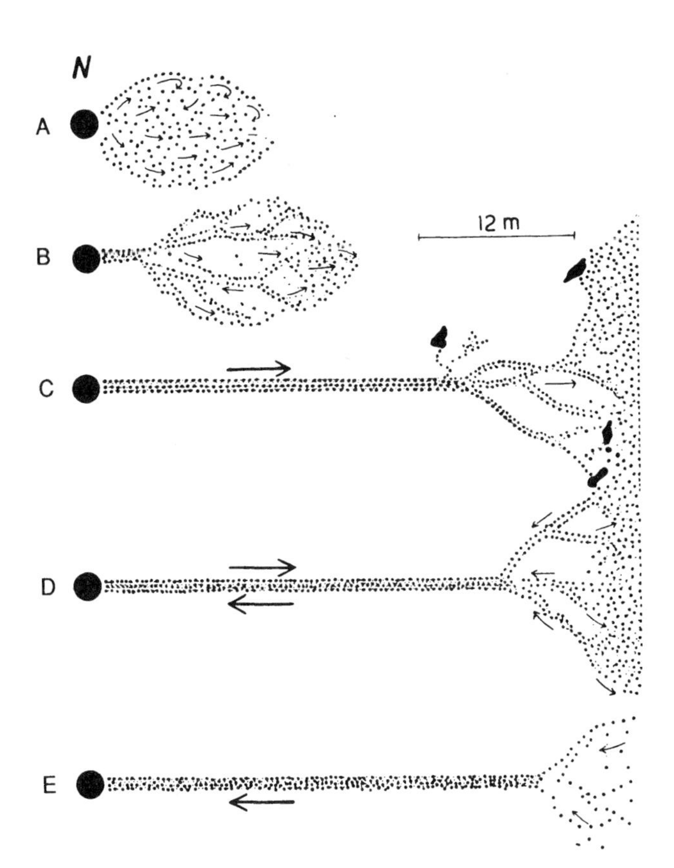


**FIGURE 16–14** The general pattern of advance (*A–C*) and retreat (*D, E*) of a swarm raid by a *Dorylus* colony from its bivouac site. (From Raignier and van Boven, 1955.)

**FIGURE 16–15** African driver ants (*Dorylus*) on the march. *Left:* A soldier stations herself near the foraging column and assumes a defensive posture. *Right:* A group of workers cooperate in transporting a large prey object. (From Gotwald, 1984–85; photographs courtesy of W. Gotwald.)

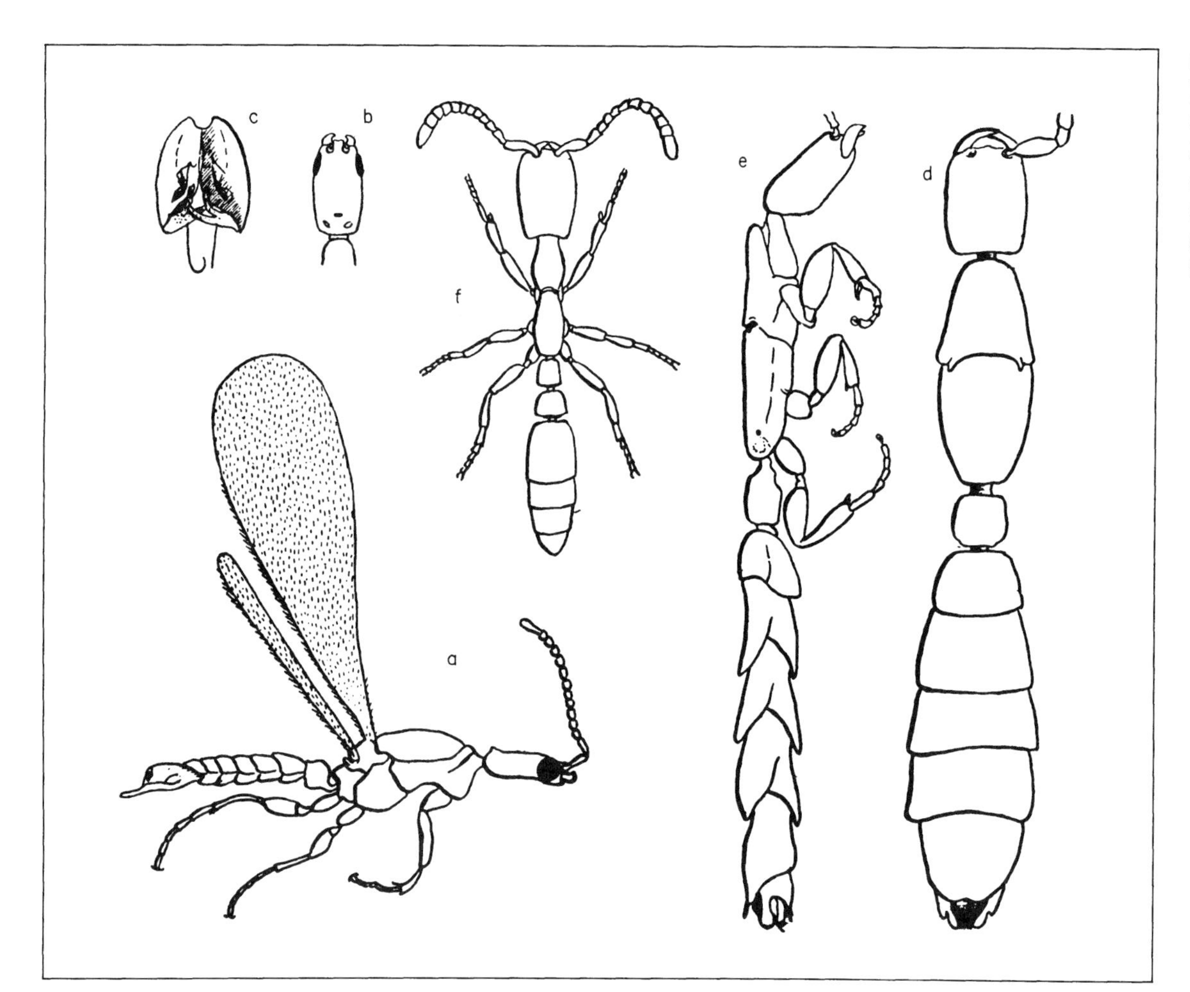


**FIGURE 16–16** Representatives of the Leptanillinae, the little-known group of subterranean ants with legionary habits. *Leptanilla minuscula:* (*a*) male; (*b, c*) head and genitalia of male. *L. revelieri:* (*d, e*) dichthadiiform queen; (*f*) worker. (From Wheeler, 1910a; after Emery and Santschi.)

Although most *Dorylus* species have very broad dietary habits, at least a few are specialized predators. Two colonies of *D. gerstaeckeri* studied by Gotwald (1974) fed only on earthworms. According to Leroux (1979), *Dorylus* of the "subgenus *Typhlopone,*" specifically either *D. badius* or *D. fulvus,* regularly raid colonies of *D. nigricans.*

*Dorylus* and the other Dorylinae and Ecitoninae have triumphed as legionary ants over all their competitors. An idea of their diversity can be gained from the synopsis presented in Table 16–1. They not only outnumber other kinds of legionary ants in both species and colonies, but they tend to exclude them altogether. Cerapachyines, for example, are relatively scarce throughout the continental tropics wherever dorylines and ecitonines abound, but they are much more common in remote places not yet reached by these advanced army ants—for example, Madagascar, Fiji, New Caledonia, and most of Australia.

## ARMY ANTS OUTSIDE THE DORYLINAE AND ECITONINAE

Because the dorylines and ecitonines are all fairly specialized in anatomy and social behavior, the best strategy for tracing the origins of migratory huntresses is the study of other ant taxa in early and intermediate stages of army ant evolution. Many of the species are less accessible by virtue of being rare or exotic, but important information has nevertheless accumulated steadily since 1960.

For example, members of the little subfamily Leptanillinae have long been considered army ants. But they have remained an enigma, because they are very rare and local. Most of the 33 species are known only from one or a few workers or males (Baroni Urbani, 1977). Neither of us has ever seen one alive, in spite of years of fieldwork in all parts of the tropics. W. L. Brown, probably the most widely traveled and productive ant collector of all time, has found only one colony during more than forty years.

The belief that leptanillines have legionary habits has come chiefly from their anatomy. The workers are tiny simulacra of dorylines or ecitonines in overall aspect, while the queens are typical dichthadiigynes remarkably convergent to those of the Dorylinae and Ecitoninae (see Figure 16–16). Their gasters, like those of ecitonine army ant queens, are richly endowed with intersegmental exocrine glands, believed to produce attractants for the workers (Hölldobler et al., 1989). In 1987, a breakthrough was achieved by Masuko, who succeeded in collecting eleven colonies of *Leptanilla japonica* in the broad-leafed forest at Cape Manazuru, Japan. The colonies each contained about a hundred workers and were strictly subterranean—a feature that perhaps explains the scarcity of leptanillines in collections. The ants are evidently specialized predators of geophilomorph centipedes, each of which is many times the size of a worker. The foragers follow trunk trails out from the nest, but it is not clear whether they move singly or in groups in the typical army ant manner. The workers possess a large median sternal gland at the seventh abdominal sternite. Although no experimental proof

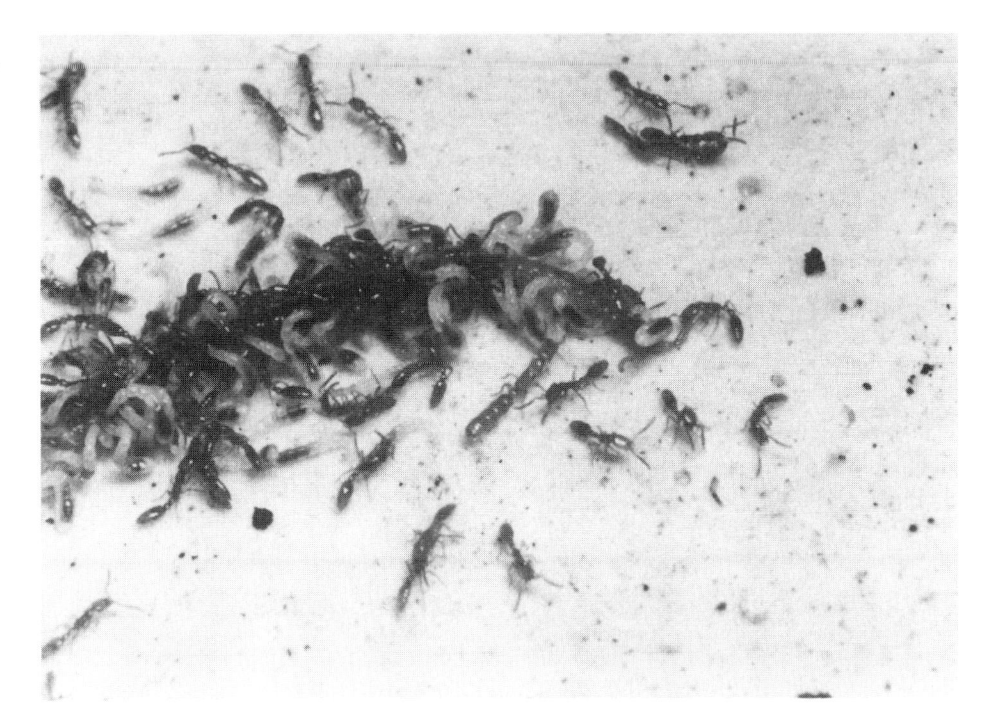

**FIGURE 16–17** *Above: Leptanilla japonica* feeding on a large geophilomorph centipede. The workers have carried the larvae to the large prey object. In the forefront the queen can be seen antennating a larva. *Below:* The physogastric queen of *Leptanilla japonica* sits in the middle of a group of last-instar larvae, holding one of the larvae in her mandibles. (Photographs courtesy of K. Masuko.)

has yet been attempted, it seems probable that the gland is the source of the trail pheromone (Hölldobler et al., 1989). The brood cycle is tightly synchronized in a long annual cycle: eggs, young larvae, and pupae are present in the summer; mature larvae appear during the winter and spring (the larvae pupate by summer).

In the laboratory *Leptanilla japonica* colonies promptly emigrate whenever they are disturbed. The swiftness of their response suggests that changes in nest site often occur in nature, perhaps approaching the army ant pattern. The transport of larvae is highly evolved and stylized. The workers do not use their mandibles to hold the larvae in the manner of other ants. Instead they use their lower mouthparts to grip a peculiar appendage projecting from the larval prothorax. *Leptanilla* larvae also possess a unique exudatory organ on each side of the third abdominal segment that provides hemolymph as a nutrient to the adults, especially the queens.

The *Leptanilla japonica* annual cycle has been interpreted by Masuko in the following way. While larvae are present, the workers intensively hunt centipedes (Figure 16–17a). The larvae consume these victims and grow quickly. During this period the queen's abdomen remains constricted. When the larvae mature, the colony passes into its oviposition phase. The queen feeds heavily on larval hemolymph from the abdominal exudatory organ and becomes physogastric (Figure 16–17b). Prey consumption ends, while the larvae pupate all at once and the queen lays the eggs of the next worker generation in a single batch.

The leptanilline ants are an especially promising group for the

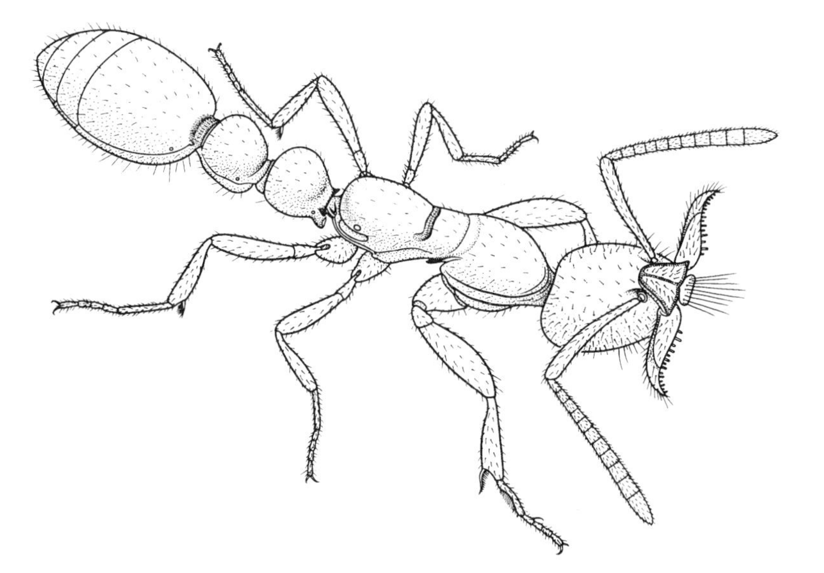

**FIGURE 16–18** The aberrant leptanilline ant *Protanilla wallacei* of Sabah defends previously captured prey by opening its mandibles 180 degrees and facing the intruder. When the sensitive labral hairs are touched, the mandibles snap shut. (Drawing by S. P. Kim from field notes and sketches by R. W. Taylor, reproduced with permission.)

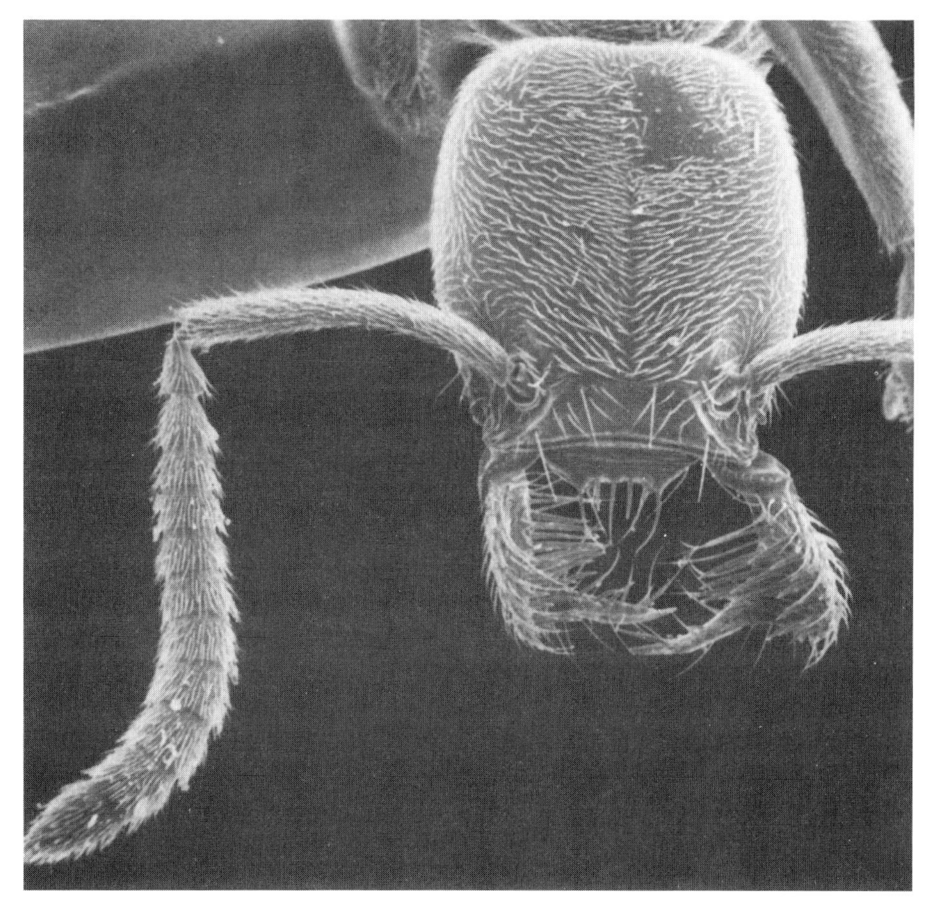

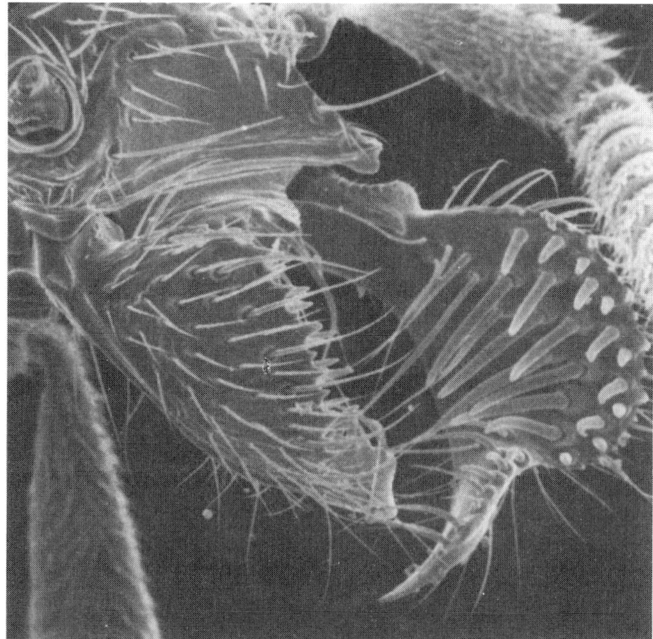

**FIGURE 16–19** Mandibles of the leptanilline ant *Anomalomyrma kubotai* of Japan are among the most specialized known in the ants. The blades are scoop-shaped and lined with pegs, the function of which is still unknown. Scanning electron microscope pictures of the whole head (*left*) and the mandibles (*right*). (R. W. Taylor, previously unpublished, used with permission.)

study of adaptive radiation of miniature group-predatory ants. Two new Asian genera have been identified by R. W. Taylor (personal communication). *Protanilla wallacei* has saber-shaped mandibles capable of being opened 180 degrees and a row of labral hairs that appear to serve as prey-seeking guides (Figure 16–18). *Anomalomyrma kubotai* has one of the most bizarre mandibles known in the ants: the blades are scoop-shaped and lined on the inner surface with thick, inward-directed pegs (Figure 16–19). The function of the latter armament is unknown.

Findings of equal significance have been made by U. Maschwitz and his collaborators in the ponerine genus *Leptogenys* (Maschwitz et al., 1989). They discovered several Malayan species that exhibit the complete army ant syndrome, including swarm raiding, temporary bivouacs and regular migrations to new hunting grounds. The brood cycle of the *Leptogenys* colonies is not synchronized, however. All brood stages were always present in the colony, and the single ergatoid queen did not exhibit physogastric cycles. One species close to *Leptogenys mutabilis* (in the well-known *processionalis* group) was studied in considerable detail. Each colony consists of more than 30,000 workers. During the day they stay in bivouacs, which are located in the soil or in the dense leaf litter. The foraging raids, often involving more than 20,000 ants, begin at dusk. They are not initiated by recruiting scouts or led by leader ants. A major raid fans out in one direction, progressing at approximately 5 meters per hour. The average distance the fan front moves away from the bivouac site is 22 meters (maximum 56 m). During one night a colony covers an area of approximately 300 square meters. The swarm raiders hunt a broad spectrum of prey, including a variety of insects, spiders, chilopods, diplopods, earthworms, and flatworms. The colonies move to new hunting grounds on the average of every 1.5 days (range to 10 days) and travel an average distance of 28 meters (maximum 50 m). Within the colony a host of myrmecophiles is found, including collembolans, flies, staphylinids, and isopods. These largely unstudied guests accompany the ants during colony emigrations (Figure 16–20).

Of equal novelty and interest are findings by Moffett (1984, 1986f, 1987a–c, 1988a,b) on the abundant Asian myrmicine *Pheidologeton diversus.* Colonies conduct swarm raids remarkably similar in some respects to those of *Eciton burchelli* and the *Dorylus* driver ants. The colonies are huge, with a single oversized queen and hundreds of thousands of workers. The foragers travel on one or two stable trunk trails that extend from 5 to over 100 meters from the nest. The trails sometimes last for weeks, with great numbers of ants traveling back and forth along them day and night. In this respect the *Pheidologeton* resemble many other kinds of ants that follow trunk trails to persistent food sources, such as the *Messor* and *Pogonomyrmex* that exploit seedfalls and the *Formica* and *Lasius* that visit clusters of honeydew-excreting aphids. Where the workers of these more conventional species wander away from the trails in large numbers to forage on their own, however, the *Pheidologeton* use the trails as departure points for column raids and occasionally swarm raids. As in other army ants, bouts of solitary hunting are restricted to the front of the raid, and even these brief forward sallies can be properly interpreted as part of a coordinated group effort. Workers otherwise almost never stray more than 5 centimeters away from the moving columns of ants. In short, they forage very much as do doryline and ecitonine army ants.

A *Pheidologeton* raid begins when ants move away as a group from the trunk trail. They form a narrow column that grows outward, like the pseudopod of an ameba, at the rate of 10–20 centimeters a minute. The explorers pay little attention to trails laid during previous raids, often traveling widely over previously unvisited terrain. After the column stretches between half a meter and 3 meters in length, some of the ants in the terminus spread out and progress slows down. In a minority of the cases, however, the terminus blossoms still further into a large, fan-shaped raid (see Figure 16–21). Behind the seething frontal edge of a full-blown raid, most of the ants run back and forth in a tapered network of feeder columns. These columns in turn funnel back into a single basal column, which lengthens as the swarm progresses.

Moffett (1987b) found that the large raiding swarms each contain tens of thousands of individuals and reach as far as 6 meters from the trunk trails. They closely resemble the formations of the swarm-raiding *Dorylus* and *Eciton* but travel outward at most only one-fifth as rapidly (1.5–2 m per hour as opposed to 10–20 m per hour). The *Pheidologeton,* like *Dorylus* and *Eciton,* are able to conquer exceptionally large and formidable prey, up to and including frogs, by overwhelming them with the sheer force of numbers. They also carry large objects rapidly back to the nest with well-coordinated group

**FIGURE 16–20** A worker of *Leptogenys* sp., a ponerine army ant, during colony emigration in Malaya. The ant carries a pupa, on which is riding a myrmecophilous isopod, *Exalloniscus maschwitzii.* (Photograph courtesy of U. Maschwitz.)

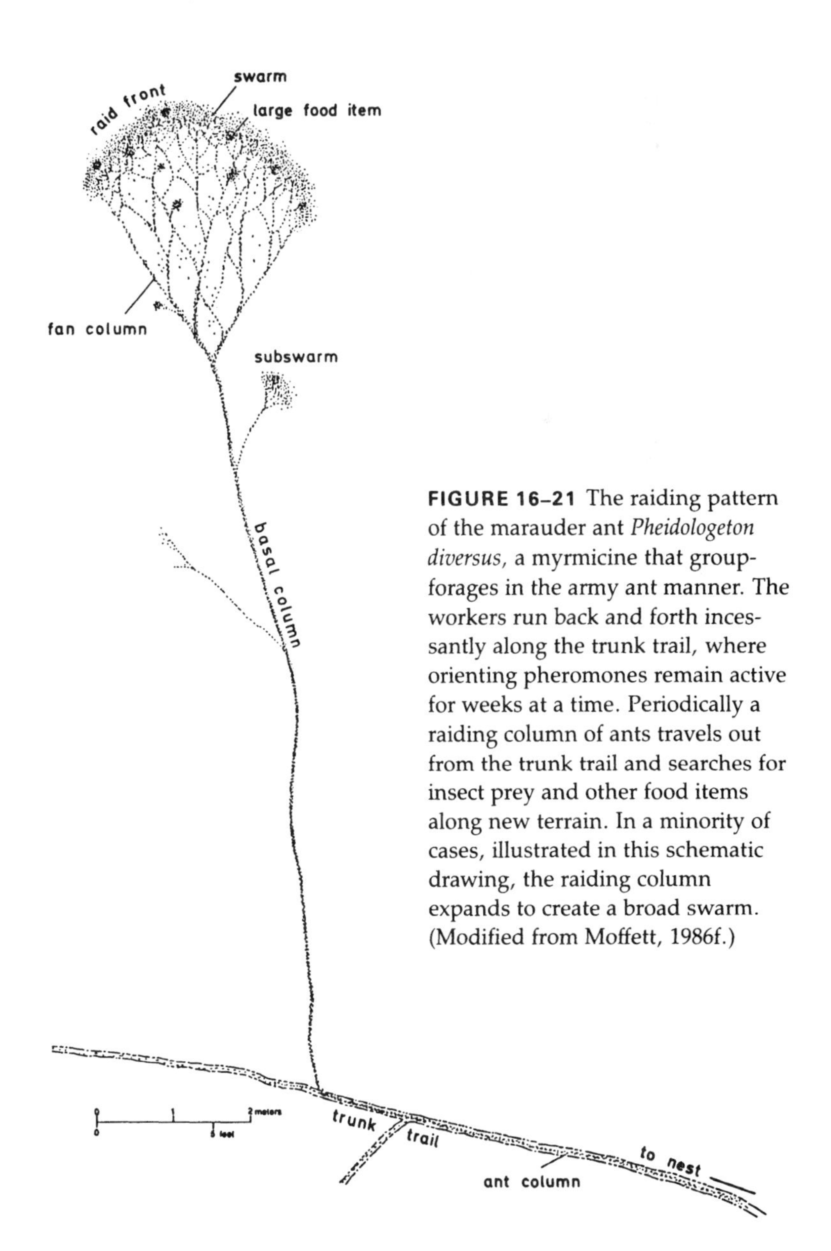


**FIGURE 16–21** The raiding pattern of the marauder ant *Pheidologeton diversus,* a myrmicine that group-forages in the army ant manner. The workers run back and forth incessantly along the trunk trail, where orienting pheromones remain active for weeks at a time. Periodically a raiding column of ants travels out from the trunk trail and searches for insect prey and other food items along new terrain. In a minority of cases, illustrated in this schematic drawing, the raiding column expands to create a broad swarm. (Modified from Moffett, 1986f.)

**FIGURE 16–22** Predation by the group-foraging Asian myrmicine *Pheidologeton diversus. Above:* Minor workers attempt to pin down a large termite soldier at the raid front; the termite has managed to kill two of its attackers. *Below:* a major arrives and crushes the termite's head. (From Moffett, 1987c.)

transport (see Figure 10–6). Because of these capabilities and the enormous size variation that occurs within the worker force of each colony (a 500-fold dry-weight ratio of largest to smallest workers), the ants are able to collect an impressive array of food (Figure 16–22 and Plate 21). While gangs are attacking earthworms, cockroaches, and other larger animals, smaller individuals are ferreting out tiny collembolans and flies. In addition the *Pheidologeton* gather seeds and fruits of many kinds. In fact usually about half of the material they carry to the nest is of vegetable origin. This complex foraging strategy works very well. *P. diversus* and other, similarly behaving members of the genus are among the most abundant and ecologically dominant ants over a large part of tropical Asia.

## THE ORIGIN OF LEGIONARY BEHAVIOR

Wilson (1958e) argued that the key to understanding the origin of legionary behavior lies in the adaptive significance of group raiding. Earlier writers had stated repeatedly that compact armies of ants are more efficient at flushing and capturing prey than are assemblages of foragers acting independently. This observation is certainly correct, but it is not the whole story. There is another, primary function of group raiding that becomes clear only when the prey preferences of the group-raiding ants are compared with those of predatory ants that forage in solitary fashion. Most nonlegionary ponerine species for which the food habits are known take living prey of approximately the same size as their worker caste or smaller. As a rule they must depend on relatively small animals that can be captured and retrieved by lone foraging workers. Group-raiding ants, on the other hand, feed on large arthropods or the brood of other social insects, prey not normally accessible to ants foraging solitarily. Thus, the species of *Leptanilla, Onychomyrmex,* and the *Leptogenys diminuta* group specialize on large arthropods; those of *Eciton* and *Dorylus* prey on a wide variety of arthropods that include social wasps and other ants; species of *Simopelta* raid other ants; and *Megaponera foetans* raids termites. The doughty little ecitonine *Neivamyrmex harrisi* evidently specializes on fire ants of the genus *Solenopsis* (Mirenda et al., 1980).

With this generalization in mind, we can fairly easily reconstruct the steps in evolution leading to the full-blown legionary behavior of the Dorylinae and Ecitoninae. The following scheme is modified from Wilson (1958e):

1. Group-recruitment raiding is developed to allow specialized feeding on other social insects. In this case group raids are initiated by successful scouts, who lead a raiding party to the discovered prey. This form of raiding is represented today in some *Leptogenys* species (Maschwitz and Mühlenberg, 1975; Maschwitz and Schönegge, 1983), *Pachycondyla (= Termitopone) laevigata* (Hölldobler and Traniello, 1980b), at least one *Cerapachys* species (Hölldobler, 1982b), and also *Megaponera foetens* (Longhurst et al., 1979b).

2. Group raids are initiated more autonomously, without the stimulus of recruiting leaders. They are usually more massive than the raiding parties organized by leader ants. This more advanced form of group raiding evidently developed to allow predation on large arthropods and other social insects and to cover a larger hunting area. Group raiding without frequent migrations occurs in *Pheidologeton* and possibly some of the Ponerinae. From his studies on *Pheidologeton,* Moffett (1988b) has suggested that the use of trunk trails in conjunction with recruitment was a forerunner of autonomous group raids.

3. Migratory behavior is either developed concurrently with group-raiding behavior or it is added shortly afterward. Large arthropods and social insects are more widely dispersed than other types of prey, and the group-predatory colony must constantly shift its trophophoric field to tap new food sources. With the acquisition of both group-raiding and migratory behavior, the species is now "legionary" in the full functional sense, at least roughly equivalent to the Dorylinae and Ecitoninae. Some of the group-raiding ponerines, including the Malayan *Leptogenys* just described, have evidently reached this adaptive level. Colony size in these species averages larger than in related, nonlegionary species, but it does not approach that attained by *Eciton* and *Dorylus.*

4. As group raiding becomes more efficient, large colony size becomes possible. This stage has been attained by many of the Dorylinae and Ecitoninae, including the species of *Aenictus* and *Neivamyrmex* and at least some of the column-raiding *Eciton (dulcius, hamatum, mexicanum, vagans).*

5. The diet may be expanded secondarily to include other smaller

and nonsocial arthropods and even small vertebrates and vegetable matter; concurrently, the colony size becomes extremely large. This is the stage reached by the driver ants of Africa and tropical Asia (*Dorylus*), the species of *Labidus*, and *Eciton burchelli*, most or all of which also utilize the technique of swarm raiding as opposed to column raiding. It is also partially approached by species of the myrmicine genus *Pheidologeton*, particularly *P. silenus*.

Two new variations have been added to the scheme of evolutionary grades proposed by Wilson. The discovery of swarm raiding in *Pheidologeton diversus* was a complete surprise. Besides being the first such case in the Myrmicinae, it shows that species can evolve to swarm raiding of the *Eciton burchelli* type without migratory behavior. The *Pheidologeton* evidently maintain themselves in stable foraging domains and nest sites by an extraordinarily broad diet, made possible by a heavy reliance on seeds and fruits and the ability to capture animal prey of almost all kinds and sizes.

A second complication is the correlation proposed by Gotwald (1978, 1982) between the soil zone in which legionary ants live and the degree of their prey specialization. In general, species that hunt underground or beneath the surface of rotting stumps and logs (hypogaeic foragers) are more specialized than those that hunt on the surface of the ground and vegetation (epigaeic foragers). Most legionary ant species are hypogaeic and hence relatively specialized. The majority of *Aenictus* for which data are available are predators of the immature stages of other ants. Hypogaeic *Dorylus* feed mostly on termites and other ants, and at least one or two species are predators of earthworms; and the species of *Neivamyrmex*, many of which are subterranean, specialize on ants. The known epigaeic *Aenictus* and *Dorylus* are generalized feeders, as are *Pheidologeton diversus*. If the correlation suggested by Gotwald holds, it is superimposed on the evolutionary grades defined by colony size, group raiding, and nomadism to account for part of the variance in degrees of prey specialization among species.

The data are too fragmentary to be certain, however. We have little or no information on the vast majority of species of dorylines and ecitonines, especially the less accessible hypogaeic species. The relation between diet, microhabitat, and social organization of the legionary ants is still a relatively unexplored and very promising subject. Rettenmeyer et al. (1983) have noted that as many as twenty ecitonine species are found together in some New World tropical forests. How, we may ask, can so many "Huns and Tartars" coexist?

# CHAPTER 17

# The Fungus Growers

Members of the myrmicine tribe Attini share with macrotermitine termites and certain wood-boring beetles the sophisticated habit of culturing and eating fungi. The Attini are a morphologically distinctive group limited to the New World, and most of the 12 genera and 190 species occur in the tropical portions of Mexico and Central and South America. Besides their unique behavior and the many peculiar behavioral and physiological changes associated with it, the Attini are distinguished from other ants by an unusual combination of anatomical traits, including the shape of the antennal segments; a less-than-absolute tendency toward hard, spinose, or tuberculate bodies; and a proportionately large, casement-like first gastral segment.

It is conceivable that fungus growing originated only once in a single ancestral attine living in South America during that continent's long period of geological isolation from late Mesozoic times to approximately 4 million years ago. Exactly when the event occurred is open to conjecture, but it was almost certainly prior to the Miocene Epoch. Extinct but modern-looking species of *Trachymyrmex* (Baroni Urbani, 1980) and *Cyphomyrmex* (Wilson, 1985h) have been found in Dominican amber, which is believed to date from either late Oligocene or early Miocene.

In Africa, southern Asia, and other parts of the Old World tropics, the Attini are replaced by fungus-growing termites (Macrotermitinae), which in their turn do not occur in the New World. No one can be sure whether this complementary global pattern is due to a mutual preemption involving competitive exclusion of one group by another or whether it is simply one more accidental outcome reflecting the extreme rarity of the evolutionary origin of fungus gardening. The latter possibility is more likely to be the case, which means that if attines were to be introduced today into the range of the macrotermitines, or vice versa, the two kinds of insects could coexist with little interference. This is possible because attines utilize insect excrement and fresh plant material for the most part, whereas the macrotermitines use dead plant material. Also, fungus-growing ants forage above ground, often even in trees; fungus-growing termites are primarily subterranean.

The Attini, where they exist, are an enormously successful group. One species, *Trachymyrmex septentrionalis,* ranges north to the pine barrens of New Jersey, while in the opposite direction several species of *Acromyrmex* penetrate to the cold temperate deserts of central Argentina. In the vast subtropical and tropical zones in between, attines are among the dominant ants. Many of the species gather pieces of fresh leaves and flowers to nourish the fungus gardens, and *Atta* and *Acromyrmex* rely on this source exclusively. Since they attack most kinds of vegetation, including crop plants, they are serious economic pests. The species of *Atta* in particular are among the scourges of tropical agriculture. They are familiar to local inhabitants as the *wiwi* in Nicaragua and Belize, the *bibijagua* in Cuba, the *hormiga arriera* in Mexico, the *bachac* in Trinidad, the *bachaco* in Venezuela, the *saúva* in Brazil, the *cushi* in Guyana, the *coqui* in Peru, and the leafcutting or parasol ant in most English-speaking countries, the last name alluding to the fact that an *Atta* worker holding a leaf fragment over her head gives the impression that she is carrying a parasol. The problems of agriculture in *Atta* country have been humorously epitomized in the following anecdote by V. Wolfgang von Hagen (1939), in connection with his attempt to grow a vegetable garden in Belize:

> My Indian servants, dusky, kinky-haired Miskito men, lamented all this work. It was useless, quoth a toothless elder, to plant anything but bananas or manioc, as the *Wiwis* were sure to cut off all the leaves. Without the slightest encouragement the Miskito Indians would launch forth on the tales of the ravages of the *Wiwi Laca,* but unswayed by the illustrations, like Pangloss I could only remark that all this was very well but let us cultivate our garden. In two weeks the carrots, the cabbages, the turnips were doing well. The carrots had unfurled their fernlike tops, the cabbage grew as if by magic. From our small palm-thatched house my wife and I cast admiring eyes over our jungle garden. Our minds called forth dishes of steaming vegetables to replace dehydrated greens and the inevitable beans and yucca. Even the toothless Miskito elder came by and admitted that white man's energy had overcome the lethargy of the Indian. Then the catastrophe fell upon us. We arose one morning and found our garden defoliated: every cabbage leaf was stripped, the naked stem was the only thing above the ground. Of the carrots nothing was seen. In the center of the garden, rising a foot in height, was a conical peak of earth, and about it were dry bits of earth, freshly excavated. Into a hole in the mound, ants, moving in quickened step, were carrying bits of our cabbage, tops of the carrots, the beans—in fact our entire garden was going down that hole. I could see the grinning face of the toothless Miskito Indian. *The Wiwis had come.*

The leafcutting ants of the genera *Atta* and *Acromyrmex* were preadapted for their role as agricultural pests by their ability to use many plant species with the aid of their symbiotic fungi, which serve as a sort of ancillary digestive system. The ants also build up high population densities, such as 5 colonies per hectare in *Atta vollenweideri* and 28 per hectare in *Atta capiguara,* with each colony containing a million or more workers (Fowler et al., 1986a,b).

Leafcutters are the dominant herbivores of the Neotropics, consuming far more vegetation than any other group of animals of comparable taxonomic diversity, including mammals, homopterans, and lepidopterans. The amount of vegetation cut from tropical forests by *Atta* alone has been calculated on the basis of 12 studies to lie between 12 and 17 percent of leaf production (Cherrett, 1986). Grass-cutting species of *Atta,* which are distinguished from other members of the genus by their short, massive mandibles, are equally voracious. Each colony of *A. capiguara* uses about 30–150 kilograms of dry matter each year; the figure of *A. vollenweideri* is 90–250 kilograms per year. *A. capiguara* reduces the commercial carrying capacity of pastureland, measured by the number of head of sustainable cattle, by as much as 10 percent (Fowler et al., 1986a).

Because of the catholicity of their diets, or rather the diets of their fungus, leafcutters have an extraordinarily diverse impact on agriculture. It includes the direct destruction of most kinds of crops, loss of land surface to the large nests (30–600 $m^2$ per nest, when soil erosion is included), accidents caused to animals and agricultural machinery, and highway and other right-of-way damage from excavation of the huge nests. Because of the variation in damage from one country to the next, the total loss caused by the ants is impossible to calculate, but it is probably in the billions of dollars. Yet research on leafcutters remains relatively neglected. According to Cherrett (1986), by the early 1980s only 1,250 articles had been written on *Atta* and *Acromyrmex,* as opposed to 10,000 on locusts.

Leafcutting ants have been important to the economy of Latin America throughout historical times. The early Portuguese colonists, who dubbed Brazil the kingdom of the ants, left behind such testaments to the *saúva* as the following: "If there is not much wine in this land it is because of ants which strip the leaves and fruit" (1587), "In a word, it is the worst scourge that farmers have" (1788), and "Either Brazil kills the *saúva* or the *saúva* will kill Brazil" (1822; cited by Mariconi, 1970). Deep within their huge nests, able to multiply themselves many times each year, the leafcutters are nearly invulnerable to anything but massive poisoning.

Because so many species thrive in cleared land and secondary forests, leafcutters as a whole have benefited by the advent of European civilization. The ubiquitous *Atta cephalotes,* for example, is specialized to live in forest gaps, and as a consequence it is able to invade subsistence farms and plantations from Mexico to Brazil (Cherrett and Peregrine, 1976). Prior to 1954 *Atta capiguara* around São Paulo was limited to a small savanna south of the city and had little or no economic impact in the area. When nearby forests were cleared for conversion to coffee plantations and then pastureland, the species spread rapidly and reached pest proportions. After *Acromyrmex octospinosus* was accidentally introduced into the West Indian island of Guadeloupe, shortly before 1954, it spread rapidly to become an important agricultural pest (Therrien et al., 1986). If any leafcutter ants, especially *Atta,* were to be established in sub-Saharan Africa or some other part of the Old World tropics, the result might be an ecological catastrophe. The terrestrial ecosystems of these continents are unprepared for a herbivore with the resiliency and proficiency of these highly organized insects.

In spite of the problems leafcutters cause, it would be a mistake to think of them as the uncompromising enemy of humankind. During millions of years of coevolution with their natural environment, they have become an integral part of the ecosystems of the New World tropics and warm temperate zones. They supplant to a large extent the populations of herbivorous mammals, which are relatively sparse through most of the New World tropics. They prune the vegetation, stimulate new plant growth, break down vegetable material rapidly, and turn and enrich the soil. In the tropical moist forests *Atta* are major deep excavators of soil and stimulators of root growth (Haines, 1978). If leafcutters were to be extirpated, a profound readjustment of the structure of forests and grasslands would result, including the extinction of at least a few species of plants and animals. Such considerations have led H. G. Fowler and his coworkers in Brazil (personal communication) to call for the protection of *Atta robusta,* a local forest-dwelling species in São Paulo State now endangered by rapid deforestation within its range.

Leafcutting ants are among the most advanced of all the social insects. During the past ten thousand years, a mere eyeblink in geological time, these insects have encountered the most advanced product of mammalian evolution from the Old World, *Homo sapiens.* Certain difficulties have arisen from this contact, with the great bulk of the losses occurring on the human side. In order to redress the balance, we need to learn a great deal more about the biology of our adversaries, paying particular attention to the weak points that undoubtedly occur in their complicated social systems. The goal, however, should be intelligent management of their populations and never their complete eradication. Our advantage—and responsibility—lies in the fact that we can think about these matters and they cannot.

## FUNGUS CULTURING

What happens to the vegetation after the *Atta* workers have carried it down their holes is a fascinating story that has been worked out through many decades of research. Bates, in *The Naturalist on the River Amazons* (1863), suggested that the ants use the leaves "to thatch the domes which cover the entrances to their subterranean dwellings, thereby protecting from the deluging rains the young broods in the nests beneath." Other early observers believed that the leaves are eaten or used to maintain a constant nest temperature by heat of fermentation. Belt was the first to surmise the far stranger truth. In *The Naturalist in Nicaragua* (1874) he described the garden chambers deep within the *Atta* nests as being

> always about three parts filled with a speckled brown, flocculent, spongy-looking mass of a light and loosely connected substance. Throughout these masses were numerous ants belonging to the smallest division of the workers, and which do not engage in leaf-carrying. Along with them were pupae and larvae, not gathered together, but dispersed, apparently irregularly, throughout the flocculent mass. This mass, which I have called the ant-food, proved, on examination, to be composed of minutely subdivided pieces of leaves, withered to a brown colour, and overgrown and lightly connected together by a minute white fungus that ramified in every direction throughout it . . . That they do not eat the leaves themselves I convinced myself; for I found near the tenanted chambers deserted ones filled with the refuse particles of leaves that had been exhausted as manure for the fungus, and were now left, and served as food for larvae of *Staphylinidae* and other beetles.

It was left to Alfred Möller (1893) to observe for the first time the actual eating of the fungi. He found that the tips of the hyphae produce peculiar spherical or ellipsoidal swellings (Figure 17–1) which are plucked and eaten. Möller called these objects "heads of kohlrabi" because of their fancied resemblance to the vegetable. Later Wheeler relabeled them gongylidia, and this name has stuck. A group of gongylidia, to complete the terminology, is sometimes referred to as a staphyla, while a piece of the peculiar morel-like fungus of *Cyphomyrmex rimosus* is called a bromatium. The gongylidial clusters of *Atta* and *Acromyrmex*, averaging about half a millimeter in diameter, were later observed to be eaten both by adult workers and larvae. The structures are rich in glycogen, in a form readily assimilated by the ants (Quinlan and Cherrett, 1979; Febvay and Kermarrec, 1983). Kermarrec et al. (1986) have described the gongylidium of the *Acromyrmex octospinosus* fungus as "a goat-skin bottle which has a thick wall covered with mucilage and is filled with a finely granulated mictoplasm that maintains its turgidity." It is a tank "filled with glycogen, hydrolases, and viral particles." About 56 percent of the dry weight of the mycelium as a whole, or interconnected mass of hyphae, of the *Atta colombica* fungus is available in the form of soluble nutrients, which include 27 percent carbohydrates, 4.7 percent free amino acids, 13 percent protein-bound amino acids, and 0.2 percent ergosterol and other lipids. The carbohydrates include trehalose, mannitol, arabinitol, and glucose, but no detectable polysaccharides (Martin et al., 1969a). Why the *Acromyrmex* fungus has abundant glycogen while the *Atta* fungus lacks it, if this reported difference actually exists, is not known.

As fresh leaves and other plant cuttings are brought into the nest, they are subjected to a process of degradation before being inserted into the garden substratum. First the ants lick and cut them into pieces 1–2 millimeters in diameter. They chew the fragments along the edges until the pieces become wet and pulpy, sometimes adding a droplet of clear anal liquid to the surface. Then, using side-to-side movements of the fore tarsi, they carefully insert the fragments into the substratum. Finally, the ants pluck tufts of mycelia from other parts of the garden and plant them on newly formed portions of the substratum. A newly inserted single leaf section 1 millimeter in diameter receives up to ten such tufts in five minutes. The transplanted mycelia grow rapidly, as much as 13 microns in length per hour. Within 24 hours they cover most of the substratal surface.

Michael Martin and his co-workers discovered that *Atta* workers contribute digestive enzymes in the fecal droplets they deposit on the fungus, including a chitinase, an α-amylase, and three proteinases (Martin, 1970; Martin et al., 1973). Subsequently, Boyd and Martin (1975) showed that the proteinases originate in the fungus and pass unaltered through the digestive tract of the ants back to the fungus. The ants avoid digesting fungal enzymes by the simple expedient of not secreting any digestive enzymes of their own. *Acromyrmex octospinosus* also lacks proteinases, but these smaller leafcutters produce their own chitinases in the labial glands (Febvay and Kermarrec, 1986). The metabolic capabilities of the attine ants and their symbiotic fungi have yet to be worked out in detail, but it is at least evident that the ants have lost some key enzymes. They depend heavily on their symbionts for many of their nutrients, while the fungi in turn depend on the ants for care and the recycling of some of the enzymes.

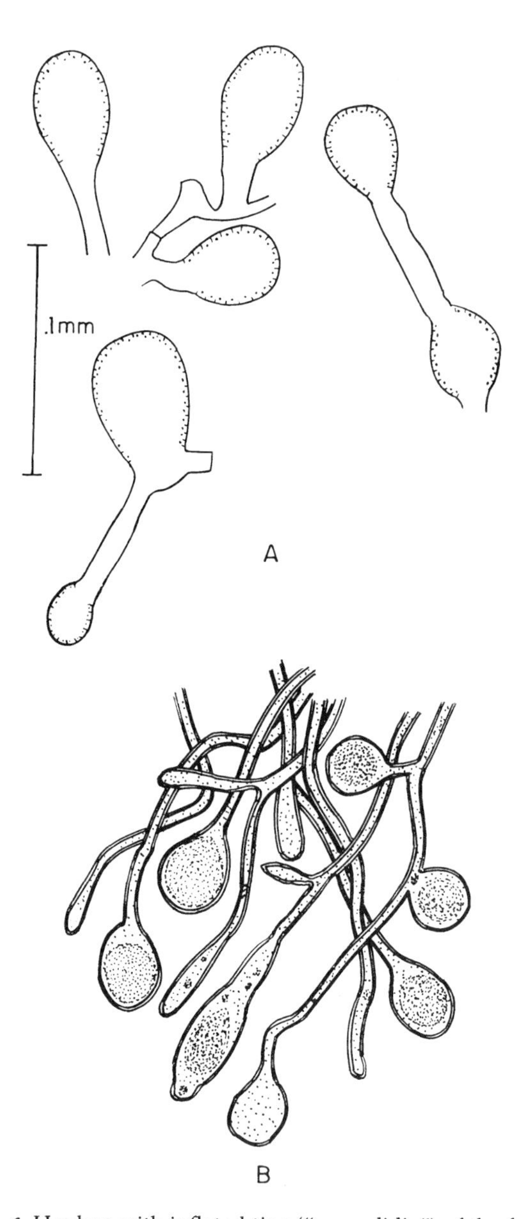


**FIGURE 17–1** Hyphae with inflated tips ("gongylidia") of the fungus eaten by (*A*) *Atta colombica* and (*B*) *Trachymyrmex jamaicensis*. The gongylidia are each 30–50 micrometers in diameter. (Redrawn from Weber, 1966.)

Still another chapter in leafcutter biology began with the revelation, by Barrer and Cherrett, that *Atta* and *Acromyrmex* workers feed directly on plant sap. As much as a third of the radioactivity in experimentally labeled leaves is absorbed directly by the ants as a result (Barrer and Cherrett, 1972; Littledyke and Cherrett, 1976). It may seem possible at first that the ants merely contribute the liquid to the fungus as added nutrient rather than assimilate it themselves. The sap must be crucial to the workers, however, because as Quinlan and Cherrett (1979) found, only 5 percent of their energy requirements are met by ingestion of juice of the fungal staphylae. In contrast, the larvae are able to subsist and grow entirely on the staphylae. These findings suggest that adult workers use only the juice of the staphylae, whereas the larvae use every part of the staphylae. The queen, to complete the story, is known to obtain at least a substantial part of her food from trophic eggs laid by workers and fed to her at frequent intervals.

In summary, the main properties of the leafcutter-fungus symbiosis can be stated as follows. Adult ants are fundamentally nectar feeders, predators, and scavengers. Their entire digestive system, from their peculiar infrabuccal and proventricular filters to the delicate midgut and limited spectrum of digestive enzymes, is geared to this dietary commitment. They are ill suited to be herbivores. The fungus, in exchange for protection and cultivation, digests the cellulose and other plant products normally inaccessible to leafcutters and shares part of the assimilable metabolic products with them. In the case of the "lower" attines, which do not cut leaves but use insect remains and excrement, the fungus converts the chitin and other products otherwise less available to ants.

Curiously, the single outstanding problem of attine biology, the identity and biological qualities of the symbiotic fungus, remains wrapped in mystery. The principal difficulty has been the reluctance—indeed, the near inability—of the fungus to form sporophores, the elaborate fruiting structures required for taxonomic diagnosis. Evidently the ants do not permit the fungi to form the mushrooms or other spore-bearing bodies under natural conditions. Instead the ants feed exclusively on the special gongylidial tips of the elementary mycelial clusters, a preference that appears to have resulted in loss of the ability of the fungus to produce sporophores. Reciprocally, the fungi utilize the ants for transport and do not have to depend on windborne spores to transfer themselves from nest to nest. Although Möller did not clarify this problem in *Atta*, he was lucky enough to discover sporophores growing from abandoned *Acromyrmex* nests on four separate occasions. These proved to be agaracine mushrooms, wine-red in color, which Möller formally named *Rozites gongylophora*. Mycologists have since confirmed their placement in the basidiomycete family Agaricaceae, but have transferred the species *gongylophora* to the genus *Leucocoprinus* (Heim, 1957; Kermarrec et al., 1986). Subsequent attempts by entomologists to locate sporophores in abandoned attine nests and to culture them in the laboratory from the gardens of various attine genera have rarely succeeded. The most notable advance was Weber's (1957b) use of a medium of sterile oats to rear sporophores of an apparent *Leucocoprinus* (= *Lepiota*) from mycelia originating from a *Cyphomyrmex costatus* garden. If future mycologists ever succeed in isolating a plant hormone or nutrient combination that enhances sporophore formation in fungi, dramatic further progress can be expected in this field.

The current evidence overall seems to support Roger Heim's opinion that the symbiotic fungus cultivated by all the attine ants is *Leucocoprinus gongylophorus*. The identification of this species or at least a set of closely similar forms placed variously in the basidiomycete genera *Leucocoprinus*, *Lepiota*, and *Rozites* has been confirmed by the rearing of sporophores from the garden mycelia in the attine genera *Atta*, *Cyphomyrmex*, and *Myrmicocrypta*. These ants represent almost the entire phylogenetic spread of the tribe Attini. According to Heim, it is unlikely that different attines picked up various leucocoprines here and there in the course of their evolution. In the absence of opposing strong evidence, this parsimonious hypothesis seems preferable to that of Weber (1979), who placed the fungi of the lower attines (such as *Cyphomyrmex* and *Myrmicocrypta*) in a separate genus, *Lepiota*. Lehmann (1975, 1976) offered a third, truly radical opinion, that the attine fungus is not a basidiomycete at all, but an ascomycete in the genus *Aspergillus* close to the symbiont of the fungus-growing beetles and Old World fungus-growing termites. This conclusion is probably too parsimonious. It is based on several tenuous morphological comparisons, including the supposedly primitive ascomycete appearance of the gongylidial swellings. It also flies in the face of contrary evidence based on sporophores cultured from attine mycelia. Weber (1979) has in fact identified an unusual *Aspergillus* in abnormal gardens of *Atta* and *Acromyrmex*, but it was strongly avoided by the ants and appeared to be a contaminant of unusually wet gardens. The only definite exception to the strict conformity of attines to *Leucocoprinus* or a closely related group of leucocoprine genera is the cultivation of a yeast by *Cyphomyrmex rimosus* (Wheeler, 1907b; Weber, 1979).

This last example may prove to be the tip of an iceberg in the study of the secondary microflora of the attine gardens. Although it is true that the ant cultures are dominated by a single fungus species, microorganisms also exist and may even participate in the symbiosis. Research to this end, reviewed by Kermarrec et al. (1986), has revealed that both yeasts and bacteria are in fact present in substantial numbers. The metabolic activity of these microorganisms remains uncertain, however. It is possible that bacteria assist in the lysis of cellulose into products that are more readily utilized by the fungi. At least six species of *Bacillus* have been identified inside nests of *Atta laevigata*, and the ants appear to inoculate fresh vegetation with these microorganisms during preparation of the substrate. It is equally possible that the bacteria parasitize the ant-fungus symbiosis, draining away some of the energy that would otherwise flow directly through the fungus to the ants. The parasitism hypothesis gains credence from a finding by Kermarrec and his co-workers that the symbiotic fungi of *Atta* and *Acromyrmex* secrete substances antagonistic to bacteria and other kinds of fungi.

Upon reflection it is impressive how nearly pure the ants keep the fungal growth in their nest chambers. They build this monoculture by a variety of techniques: the plucking out of alien fungi, the frequent inoculation of the *Leucocoprinus* mycelia onto fresh substrate, the manuring of the substrate with enzymes and nutrients to which the *Leucocoprinus* are especially adapted, the production of antibiotics to depress competing fungi and microorganisms, and the production of growth hormones. The last two methods entail an instinctive form of chemical engineering on the part of the ants. Maschwitz et al. (1970) and Schildknecht and Koob (1970) identified phenylacetic acid, D–3–hydroxydecanoic acid ("myrmicacin"), and indoleacetic acid in the secretions of the metapleural glands of *Atta sexdens* workers. They suggested that these compounds play different roles in the purification of the symbiotic fungus culture: phenylacetic acid suppresses bacterial growth, D–3-hydroxydecanoic acid inhibits the germination of spores of alien fungi, and indoleacetic acid, a plant hormone, stimulates mycelial growth. As Weber (1982) pointed out, this interpretation can be confirmed only by a demonstration that the components of the metapleural gland are actually present in the fungus gardens at bacteriostatic and fungistatic levels.

## THE LIFE CYCLE OF LEAFCUTTER ANTS

Leafcutting ants constitute 24 known species of *Acromyrmex* (Table 17–1) and 15 of *Atta* (Table 17–2). Because the *Atta* workers are so large and spectacular in their behavior, many entomologists have set out to study their life cycle and biology. These investigators include Möller, the pioneer in the subject, Forel, Goeldi, Huber, von Ihering, and Wheeler, all of whose publications are exhaustively reviewed in the classic 1907 study of the North American Attini by Wheeler. More recent researchers have included Autuori, Bitancourt, Bonetto, Borgmeier, Eidmann, Fowler, Geijskes, Gonçalves, Jacoby, Kerr, Moser, Stahel, Weber, and others; their work is carefully reviewed in Weber (1972, 1982) and in the symposium volume *Fire Ants and Leaf-cutting Ants* edited by Lofgren and Vander Meer (1986).

All of the *Atta* species appear to have basically the same colony life cycle. The nuptial flights of some species, such as the infamous *sexdens* of South America, take place in the afternoon, while *texana* of the southern United States and a few others hold their flights at night (Autuori, 1956; Moser, 1967a). Because the ponderous females work their way high into the air before the males approach them, actual matings have not been observed. Nevertheless, Kerr (1962), by counting sperm from the spermathecae of four newly mated *sexdens* queens with the aid of a hemocytometer, was able to show that each individual is inseminated by at least three to eight males. The actual estimated numbers of sperm varied among the queens he examined from 206 million to 320 million, seemingly more than enough to last an individual the ten or more years speculated to be the normal life span of an *Atta* queen.

During the nuptial flight and immediately afterward, as the queens attempt to start new colonies, mortality is extremely high. Out of 13,300 *Atta capiguara* founding colonies in Brazil, only 12 were alive three months later (Fowler et al., 1986b). From a start of 3,558 incipient *A. sexdens rubropilosa* colonies, only 90 or 2.5 percent were alive after three months (Autuori, 1950a). The survivorship of *rubropilosa* during the same time interval was 6.6 percent (Jacoby, 1944), while figures of 10 percent were obtained for *A. cephalotes* and zero percent for *A. capiguara* in Central America and Brazil respectively (Fowler et al., 1986b).

In 1898 von Ihering discovered the important mechanism by which the fungus is transferred from nest to nest. Before departing on the nuptial flight the *Atta sexdens* queen packs a small wad of mycelia into her infrabuccal chamber, a cavity located (in all ants including *Atta*) beneath the opening of the esophagus just to the rear of the base of the labium. Following the nuptial flight, which in Brazil may occur anytime from the end of October to the middle of December, the queen casts off her wings and quickly excavates a little nest in the soil. When finished, the nest consists of a narrow entrance gallery, 12–15 millimeters in diameter, which descends 20–30 centimeters to a single room 6 centimeters long and somewhat less in height. Onto the floor of this room, according to Jakob Huber (1905) and Autuori (1956), the queen now spits out the mycelial wad. By the third day fresh mycelia have begun to grow rapidly in all directions, and the queen has laid the first three to six eggs.

In the beginning the eggs and the little fungus garden are kept apart, but by the end of the second week, when more than 20 eggs are present and the fungal mass is ten times its original size, the two

**TABLE 17–1** Leafcutting ants of the genus *Acromyrmex*: distribution of the species.

| Species | Distribution |
|---|---|
| "Subgenus *Acromyrmex*" | |
| *Acromyrmex ambiguus* | Argentina, Brazil |
| *A. aspersus* | Argentina, Brazil, Peru, Colombia |
| *A. coronatus* | Bolivia and Brazil to Costa Rica |
| *A. crassispinus* | Argentina, Brazil, Paraguay |
| *A. diasi* | Brazil |
| *A. disciger* | Brazil |
| *A. gallardoi* | Argentina |
| *A. hispidus* | Argentina, Bolivia, Brazil |
| *A. hystrix* | Guianas, Brazil, Peru |
| *A. laticeps* | Bolivia, Uruguay, Brazil |
| *A. lobicornis* | Argentina, Bolivia, Brazil |
| *A. lundi* | Argentina, Bolivia, Brazil |
| *A. niger* | Brazil |
| *A. nobilis* | Brazil |
| *A. octospinosus* | Mexico to northern South America, Guadeloupe, Cuba |
| *A. rugosus* | Colombia to Argentina |
| *A. subterraneus* | Brazil and Peru to Argentina |
| "Subgenus *Moellerius*" | |
| *Acromyrmex heyeri* | Argentina, Brazil, Paraguay, Uruguay |
| *A. landolti* | Northern South America to Argentina |
| *A. mesopotamicus* | Argentina |
| *A. pulvereus* | Argentina |
| *A. silvestrii* | Argentina, Uruguay |
| *A. striatus* | Argentina, Bolivia, Brazil |
| *A. versicolor* | Arizona, Texas (United States), northern Mexico |

**TABLE 17–2** Leafcutting ants of the genus *Atta*: distribution of the species.

| Species | Distribution |
|---|---|
| *Atta bisphaerica* | Brazil |
| *A. capiguara* | Brazil, Paraguay |
| *A. cephalotes* | Southernmost Mexico to Ecuador and Brazil; Lesser Antilles as far north as Barbados |
| *A. colombica* | Guatemala to Colombia |
| *A. goiana* | Brazil |
| *A. insularis* | Cuba |
| *A. laevigata* | Colombia to Guianas to Paraguay |
| *A. mexicana* | Arizona (United States) to El Salvador |
| *A. opaciceps* | Brazil |
| *A. robusta* | Brazil |
| *A. saltensis* | Argentina, Bolivia, Paraguay |
| *A. sexdens* | Costa Rica to Argentina and Paraguay |
| *A. silvai* | Brazil |
| *A. texana* | Louisiana, Texas (United States) |
| *A. vollenweideri* | Argentina, Brazil, Bolivia |

are brought together. At the end of the first month the brood, now consisting of eggs, larvae, and possibly pupae as well, is embedded in the center of a mat of proliferating fungi. The first adult workers emerge sometime after 40 to 60 days. During all this time the queen cultivates the fungus garden herself. At intervals of an hour or so she tears out a small fragment of the garden, bends her abdomen forward between her legs, touches the fragment to the tip of the abdomen, and deposits onto it a clear yellowish or brownish droplet of fecal liquid (Figure 17–2). Then she carefully places the mycelial fragment back into the garden. Although the *Atta sexdens* queen does not sacrifice her own eggs as a culture medium, she does consume 90 percent of the eggs herself, and, when the larvae first hatch, they are fed with eggs thrust directly into their mouths. The queen apparently never consumes any of the growing fungus during the rearing of the first brood. Instead, she subsists entirely on her own catabolizing fat body and wing muscles. Soon after the first workers appear, they begin to feed themselves on the gongylidia. They also manure the fungal garden with their fecal emissions and feed their sister larvae with eggs laid by the mother queen. The eggs given to the larvae are larger than those permitted to hatch; a histological study by Bazire-Benazet (1957) has shown that they are in fact "omelets" formed in the oviducts by the fusion of two or more distinct but ill-formed eggs. After about a week the new workers dig their way up through the clogged entrance canal and start foraging on the ground in the immediate vicinity of the nest. Bits of leaves are brought in, chewed into pulp, and kneaded into the fungus garden. At about this time the queen ceases attending both brood and garden. She turns into a virtual egg-laying machine, in which state she remains for the rest of her life. Now for the first time the workers begin to collect gongylidia from the fungal mass and to feed them directly to the larvae.

The growth of the colony is at first very slow. During the second and third years it accelerates quickly and then tapers off as the colony starts to produce winged males and queens. Using data provided by Autuori, Bitancourt (1941) demonstrated that the growth of an *Atta sexdens* colony, if measured as the increasing number of nest entrances, closely fits the classic formula of logistic growth. This means that the rate of growth can be expressed as an elementary function of the population size times the difference between the population size at the given moment and the size finally reached by the colony. The essential qualities of colony growth, as they are understood at the present time, are illustrated in Figure 17–3.

The ultimate size reached by the *Atta* nests is enormous. Autuori's nest contained slightly more than 1,000 entrance holes at the end of the third year. Another 3-year-old nest excavated by Autuori (cited in Weber, 1966) contained 1,027 chambers, of which 390 were occupied by fungus gardens and ants. In only its first year of production of sexual forms, the colony had generated no fewer than 38,481 males and 5,339 virgin queens. Still another *A. sexdens* nest, 77 months old, contained 1,920 chambers of which 248 were occupied by fungus gardens and ants. The loose soil that had been brought out and piled on the ground by the ants during the excavation of their nest was shoveled off and measured. It occupied 22.72 cubic meters and weighed approximately 40,000 kilograms. Autuori also estimated that during the short life of the colony the workers had gathered no less than 5,892 kilograms of leaves to cultivate their fungus gardens! The garden substrate of an average *A. vollenweideri*

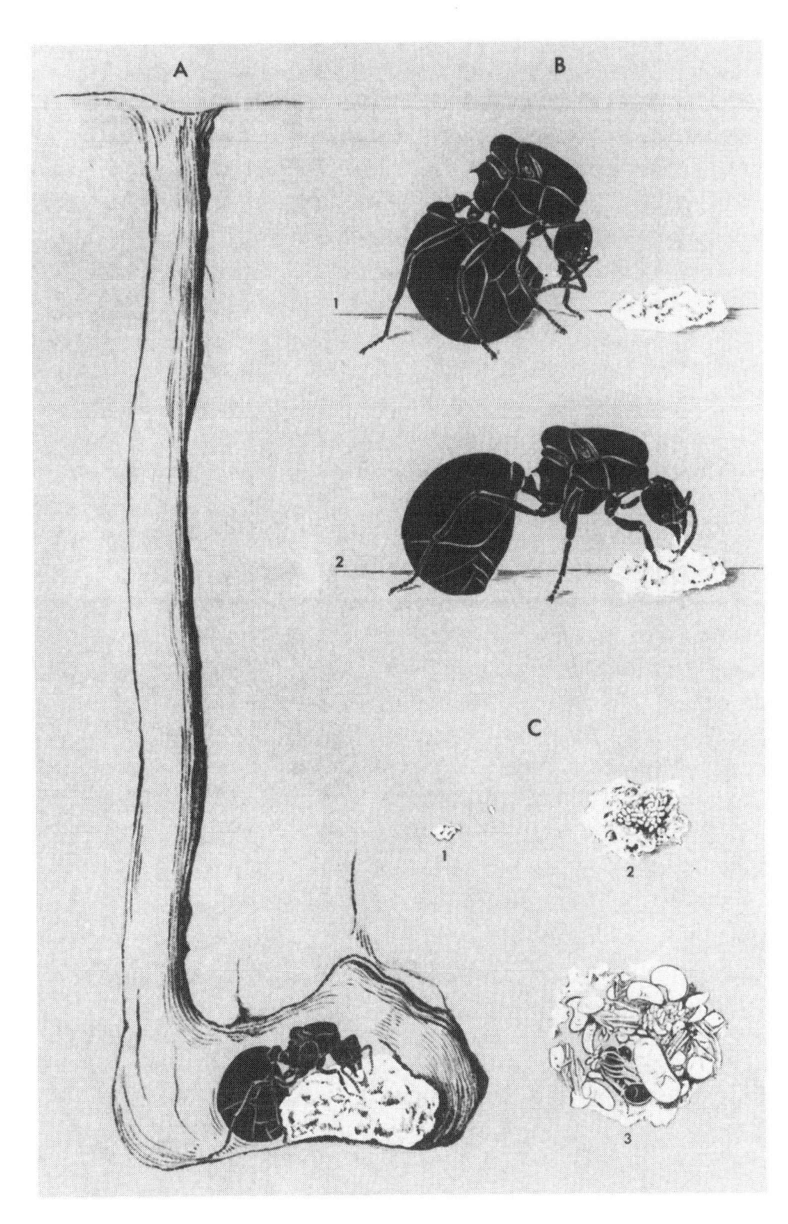


**FIGURE 17–2** Colony founding in *Atta:* (*A*) a queen in her first chamber with the beginning fungus garden; (*B*) the queen manures the garden by freeing a hyphal clump and applying an anal droplet to it; (*C*) three stages in the concurrent development of the fungus garden and first brood. (From Wilson, 1971, based on Huber, 1905, and Autuori, 1956; drawing by T. Hölldobler-Forsyth.)

nest, according to Jonkman (1980b), is built up of 182 million pieces of grass.

As these numbers suggest, the populations of old colonies of *Atta* are metropolitan in size. In publications over many years summarized by Fowler et al. (1986b), the numbers of workers in single colonies have been estimated as 1 million to 2.5 million in *Atta colombica,* 3.5 million in *A. laevigata,* 5 million to 8 million in *A. sexdens rubropilosa,* and 4 million to 7 million in *A. vollenweideri.*

The nests of mature colonies are also structures of extraordinary expanse and complexity, as documented in the studies of Eidmann (1935), Jacoby (1937, 1944), Stahel and Geijskes (1939), Moser (1963), and Jonkman (1980b). In well-drained soil the deepest galleries usually penetrate to more than 3 meters below the surface, and in some cases they descend to more than 6 meters. Their excavation for

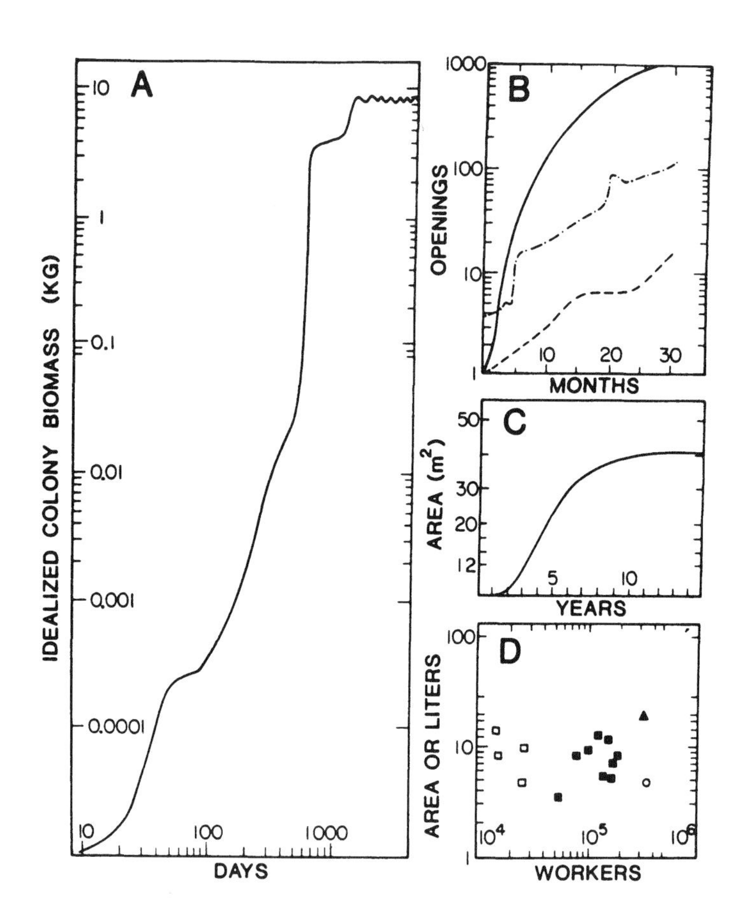


**FIGURE 17–3** The idealized properties of growth in colonies of leafcutter ants belonging to the genera *Atta* and *Acromyrmex*. (*A*) Growth in biomass during the first three years in species of *Atta*. (*B*) The relation between colony age and the number of nest openings, in two species of *Atta* (*top* and *center*) and one of *Acromyrmex* (*bottom*). (*C*) The relation between colony age and nest surface area in *Atta vollenweideri*. (*D*) Worker populations versus nest volume or surface area (the three symbols refer to different species of *Atta* and *Acromyrmex*). (Drawn from multiple sources by Fowler et al., 1986b.)

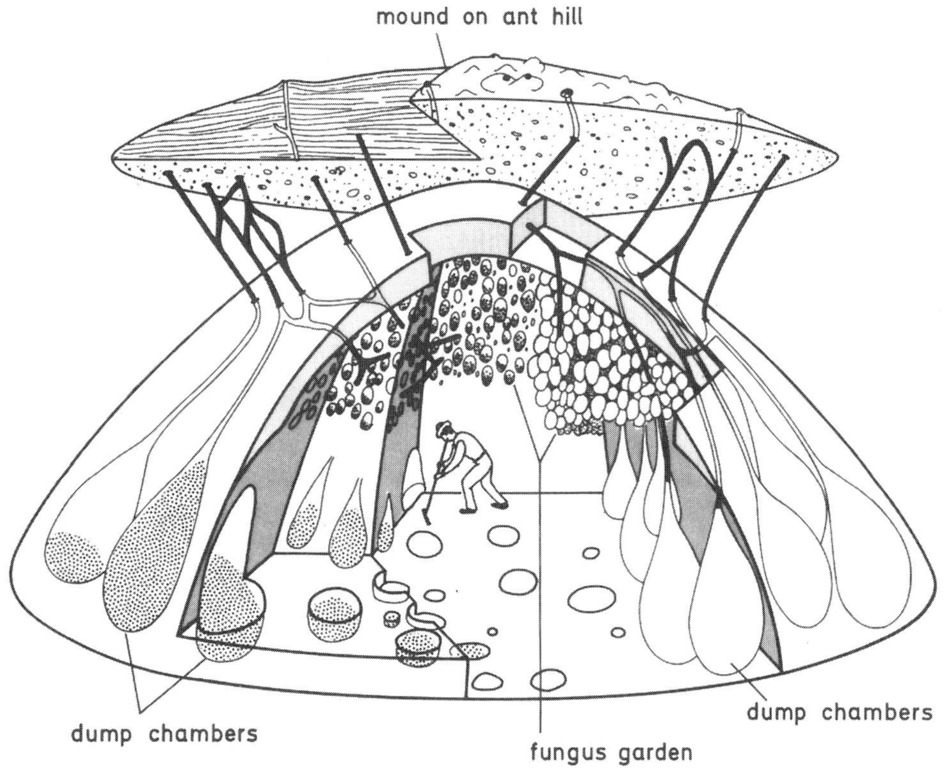


**FIGURE 17–4** The plan of a mature nest of the leafcutter ant *Atta vollenweideri*, based on actual excavations. The upper mound of soil was brought to the surface by the ants during the digging of the nest. The dump chambers contain exhausted substrate. The fungus is cultured in the fungus garden chambers. (Modified from J. C. M. Jonkman, in Weber, 1979.)

scientific study requires teams of laborers or, as in Moser's work on *A. texana* in Louisiana, the use of a bulldozer. Stahel and Geijskes systematically observed the movement of small puffs of smoke released over various of the nest entrances; they were thereby able to demonstrate the existence of a primitive ventilation system in the intact nests. Air, it was found, tends to pour into those nest openings located near the nest perimeter and to pass up out of the openings located closer to the nest center. The intake and exhaust openings are about equally numerous. A third kind of opening, through which no movement of air can be detected, is even more common.

There is a simple enough explanation for this pattern. Air is heated by metabolism more rapidly in the central zone of the nest, where the fungus gardens and ants are concentrated, and it therefore tends to rise through the central galleries. The movement in turn draws air from the remaining galleries, which are located in the peripheral zones. The "neutral" nest openings probably lead to blocked galleries or gallery systems with relatively few ants and fungus gardens. Thus the construction of large numbers of nest entrances in the *Atta* nests—a feature shared with only a few other kinds of ants—appears to be an adaptation to facilitate ventilation through the exceptionally large biomasses of the leafcutter colonies (see Figure 17–4).

## HOST SELECTION

The first impression one obtains in the field and laboratory is that *Atta* and *Acromyrmex* are indiscriminate in their choice of vegetation to serve as the fungal substrate. Under various circumstances many species accept fresh leaves, flowers, fruits, tubers, and stems of plants, as well as the endosperm of seeds. Colonies can be kept in laboratories indefinitely with a mix of processed cereals, such as rolled oats, and fresh leaves. Given a reasonable choice, however, the ants are moderately to strongly selective. Some are specialists on grasses. Examples include *Atta capiguara*, *Atta vollenweideri*, and *Acromyrmex landolti*. Others, including *Atta mexicana* and *Acromyrmex rugosus*, are specialists on dicots, while a few others, including *Atta laevigata* and *Acromyrmex lobicornis*, take both grasses and dicots (Fowler et al., 1986a). Of dicot harvesters and generalist species in Paraguay Schade (1973) wrote:

> I have seen large orange trees, full of semi-ripened fruit, completely denuded in one night. The fruit, which the ants did not touch, soon fell to the ground after having been sunburned for lack of shade. The ants seem to prefer some cultivated plants more than others: citrus trees, roses, violets, medlar trees, onions, carrots, strawberries, alfalfa, and peanuts. Somewhat less popular are avocado trees, mandioca (*Manihot utilissima*), peach trees, guava trees, mulberry leaves and fruits, privet leaves, flamboyant trees (*Delonix regia*), and many other trees, shrubs and plants of agricultural or ornamental value. Corn and beans are sampled somewhat less frequently. Members of the Compositae, Solanaceae, and Euphorbiaceae, especially the tallowtree (*Sapium* spp.), are frequently attacked. The castor-oil plant, *Ricinus communis*, a euphorb, is not touched.

In the Florencia Norte Forest of Costa Rica, Blanton and Ewel (1985) found that the dominant leafcutter, *Atta cephalotes*, attacked only 17 of 332 available plant species. They cut proportionately more woody than herbaceous species, more introduced species than natives, and a higher proportion of species with below-average water content. Comparable data concerning the selection of host plants were obtained for *Atta cephalotes* in Guyana by Cherrett (1968) and *Atta colombica* in Costa Rica by Rockwood (1976). Within the favored plant species, leafcutters prefer freshly sprouted shoots, leaves, and flowers. When preferred species decline in abundance, the colonies switch to less favored ones. For example, *Acromyrmex versicolor* foragers in the deserts of Arizona utilize freshly growing stems and leaves of dicots when they are available in the wet season but switch almost exclusively to grasses during the dry season (Gamboa, 1975). A similar seasonality in the choice of dicots is evident in detailed studies of *Atta texana* by Waller (1986).

To reach the plants of choice the workers follow trunk trails that commonly stretch for more than 100 meters from the nest. The record, cited for a colony of *Atta cephalotes* by Lewis et al. (1974a), is 250 meters. Successful scouts of *Atta* recruit nestmates with a powerful pheromone, 4-methylpyrrole-2-carboxylate, which they expel from the poison gland through the sting. Other still unidentified substances provide long-term orientation along the odor trails (see Chapter 7). Many of the *Atta* species clear broad highways along which their dense columns can travel unhindered. These "attian ways" are among the most conspicuous sights in the New World tropics.

What is the basis of host selection by the ants? In studies of the feeding behavior of *Atta* and *Acromyrmex*, Cherrett and Seaforth (1968) detected a wide range of plant phagostimulants, consisting principally of unidentified sapids and lipids. Such substances are unlikely to provide the sole basis of discrimination, however, because they occur widely and are generally efficacious as foodstuffs. Equally important, Howard (1987) has shown that *Atta cephalotes* is little influenced by energy content, moisture, or amount of nitrogen. The more likely basis of selectivity is the occurrence of greater concentrations of repellent substances in some plant species than in others. Hubbell and Wiemer (1983) and Howard and Wiemer (1986) have begun the important task of screening and identifying these compounds in the plants rejected by leafcutters. Their technique is to allow workers to forage through a random checkerboard array of rye flakes treated with extracts and synthetic compounds. Virtually all of the repellent substances discovered by this means have turned out to be terpenoids, including a great diversity of monoterpenoids, sesquiterpenoids, diterpernoids, and triterpenoids. A key question is, are the ants avoiding these substances because they are toxic to the foragers and substrate processors who drink the sap, or because they are poisonous to the symbiotic fungus? The latter effect may prove to be crucial, because many terpenoids have strong fungicidal activity. Howard et al. (1988) provided the first experimental evidence that three out of four terpenoid substances tested exhibit deleterious effects either on adult *Atta cephalotes* workers or their fungus. This study also indicated some correlation between deterrent ability and toxicity. A potential goal of future research is to establish the mode of action and relative effectiveness of these substances and their distribution through space and time in the thousands of plant species with which leafcutters regularly interact.

Another question of broad ecological interest is whether the leafcutter ants husband their resources by directing their attacks so as not to kill off too many plants close to home. Foragers have often been observed to shift their attentions from one tree to another without denuding any one of them. Columns frequently travel past intact food plants close to the nest to attack others far away. These facts led Cherrett (1968) to speculate that the ants sacrifice energetic efficiency in order to protect the host plants and thereby to gain a longer sustained yield. This intriguing and perfectly logical idea has been cast into some doubt, however, by the findings on repellent substances. The worker ants may simply be "shopping" among plants in order to locate those with the least toxic vegetation. The trunk trails, as Shepherd (1982) pointed out, tend to be laid to the temporarily most productive sites and to be changed around from time to time in a way that provides a high yield throughout the life span of the colony.

## THE ORIGIN OF THE ATTINI

The phylogenetic origin of the Attini remains a source of bafflement in spite of a century of speculation on the subject. One authoritative opinion was offered by Emery (1895), who on morphological evidence placed the Attini near *Ochetomyrmex* and *Wasmannia*. These taxa, together with the aberrant genus *Blepharidatta*, make up the tribe Ochetomyrmecini (Brown, 1953b). The ochetomyrmecines are exclusively Neotropical, which is at least consistent with the hypothesis of some kind of evolutionary link to the Attini. The overall morphological resemblance between the two tribes is not at all close, however, and in fact the Attini stand well apart from almost all other ants in their morphology. Forel (1902) offered the contrary opinion that the Attini stemmed from the Dacetini, which in the old, broad sense included the tribes Basicerotini and Stegomyrmecini. The larvae differ in morphology, however, and the most primitive known dacetines are the genera *Daceton* and *Orectognathus*, which forage aboveground and on vegetation in a way that distinguishes them from the primitive attines, which are soil-dwelling.

A much more likely candidate for an ancestral or cognate taxon among living ants is *Proatta butteli* of tropical Asia (Figure 17–5). The adults closely resemble some of the small attines, especially *Mycocepurus*, in their overall body form and the distinctive spines and tubercles that cover most of the head and body surfaces. When Forel (1912) originally described the genus, he erected a new tribe, the Proattini, to receive it. Emery (1921–22), in the authoritative *Genera Insectorum*, transferred *Proatta* to the Attini, reducing it to the rank of subtribe and implicitly recognizing its close relation to the fungus-growers. Most myrmecologists did not accept this placement, however. They believed that *Proatta* is not a true attine, attributing the outward traits it shares with the fungus-growers to convergent evolution. A new twist was added when Wheeler and Wheeler (1985b) were able to study *Proatta* larvae, which had been collected for the first time by Mark Moffett in Singapore and Malaysia. These authors concluded that "the larva of *Proatta* is definitely attine. We have a prejudice against attaching a small monotypic genus found locally in the Oriental Realm to a large widespread tribe in the Neotropical Realm; hence we had hoped that the larva would be either strongly attine or strongly non-attine. It is neither, but it is as good an attine as *Myrmicocrypta*. It lacks the coarse pinules on the mandibles, which is an attine character, but so does *Apterostigma*, which is otherwise like the higher attines." Moffett (1986d) found that *Proatta butteli* workers hunt small prey and scavenge for arthropod corpses. They also capture arthropods larger than themselves by a combination of rapid recruitment and group retrieval. Most important, however, they neither grow nor feed on fungi. Perhaps it is a disappointment to learn that *Proatta* is not a fungus-grower, but, as Wheeler and Wheeler remarked, "is it really necessary that the ancestral attine already have that habit?"

The peculiar position of *Proatta* leads us to the question of the evolutionary beginnings of fungus gardening. There are three competing hypotheses. The first, that of von Ihering (1898), proposes that attines originated from harvesting ants with slovenly habits:

> We know quite a number of ants, like the species of *Pheidole, Pogonomyrmex* and furthermore species of *Aphaenogaster* and even of *Lasius*, which carry in grains and seeds to be stored as food. Such grain carried in while still unripe, would necessarily mould and the ants feeding upon it would eat portions of the fungus. In doing this they might easily come to prefer the fungi to the seeds. If *Atta lundi* still garners grass seeds and in even greater than the natural proportion to the grass blades, this can only be regarded as a custom which has survived from a previous cultural stage.

In opposition, Forel (1902) suggested that the ancestral attines lived in rotting wood and gradually acquired the habit of eating the fungi they chanced to find growing on insect excrement left behind by wood-boring insects. A slight variant of this idea was offered by Weber (1956), who believed that the ants might have begun feeding on fungi which grew from their own feces. The third hypothesis, proposed by Garling (1979), is that the attine fungus arose from the fungi that live in mycorrhizal symbiosis with plant roots. She noted that most ectomycorrhizal fungi belong to the same group as the ant fungus, the Agaricales. The ant symbiosis could have arisen by repeated encounters that must have occurred between soil-dwelling ants and the ectomycorrhizal fungi living on the roots around their nests.

Something like von Ihering's slovenly-ant hypothesis finds support in one of the findings by Moffett on *Proatta butteli*. The colonies he studied accumulated substantial amounts of prey remains and other inedible refuse within their nest chambers, and this material formed substrate on which wild fungus grew profusely. If the ancestor of the Attini had a similar tendency to keep refuse in the nest—an unusual but not unique habit in ants generally—fungus gardening may have arisen when the ants began to feed on the mycelia taking root there. As shown in Table 17–3, most of the genera of small attines, which are thought to be relatively primitive among living species, use insect remains and other detritus as fungus substrates. But some also use insect excrement, a circumstance consistent with Forel's hypothesis.

The genera are listed in an order that reflects the idea, held by most students of the Attini since the time of Emery and Forel, that *Cyphomyrmex* is primitive, *Atta* is advanced, and the remaining genera occupy positions of varying degrees of intermediacy. Of course such a vertical array is bound to be an oversimplification, because the evolution of the Attini, like that of almost all other large animal groups whose histories are better known from the fossil record, almost certainly unfolded in a more complex, dendritic pattern. But the principal evolutionary trends do seem clear enough when considered separately, and they are at least loosely interconsistent. There is a gradual increase in body size and, in a few of the largest species, the appearance of well-marked worker polymorphism. The body develops certain unusual anatomical features such as tuberculation of the body surface, unusual hair structure, and cordate head shape. The mature colony size increases from small (that is, a few tens or hundreds of individuals) through medium (hundreds or thousands) to large (tens of thousands to millions), with a corresponding growth in the size and complexity of the nest structure.

Now if these trends do reflect a true evolutionary history, it is reasonable to suppose that feeding behavior also evolved in roughly the same direction, namely from *Cyphomyrmex* to *Atta* and the other, "higher" attine genera. And if that much is accepted, we can regard the use of nest refuse, including discarded arthropod remains, and insect feces as the culturing medium to be the primitive trait and the

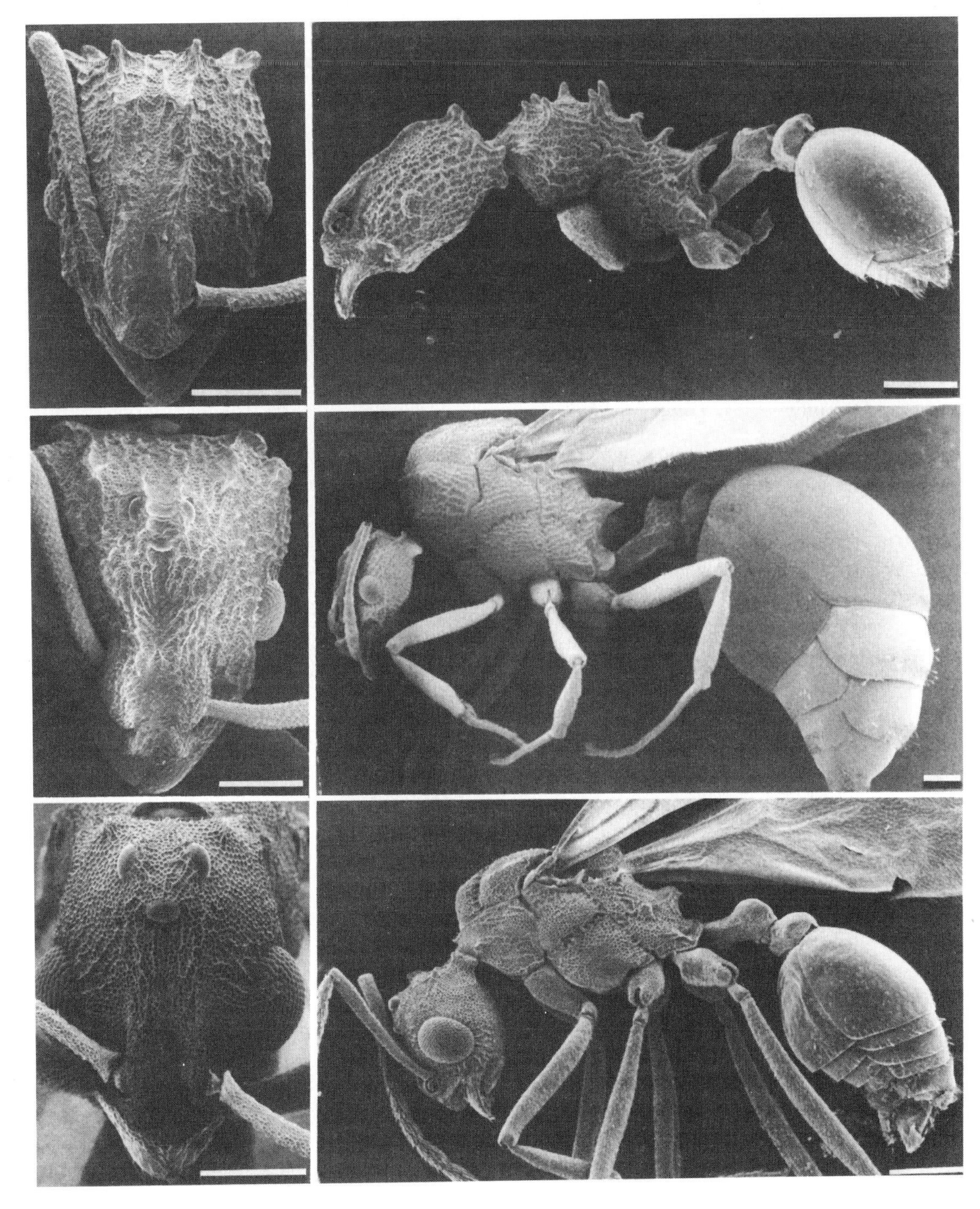

**FIGURE 17–5** Scanning electron micrographs of the tropical Asian ant *Proatta butteli*, an apparent relative of the fungus-growing ants. Frontal views of head and side views of entire body of worker (*top*), queen (*center*), and male (*bottom*). Scale bar = 0.25 mm. (From Moffett, 1986d.)

use of fresh vegetation to be the derived trait. It is therefore likely that a closer examination of the biology of *Cyphomyrmex*, along with that of the other presumably primitive attines and perhaps also of *Proatta*, will shed new light on the origin of the Attini and the fungus-culturing habit.

Whatever the beginnings of fungus growing, the event of greatest importance in the history of the Attini was the efficient utilization of all forms of fresh vegetation by the true leafcutter genera *Acromyrmex* and *Atta*. The achievement is unique within the entire animal kingdom. A close examination of the processing of the vegetation by *Atta cephalotes* and *Atta sexdens* reveals at least some of the reasons that fungus culturing of this particular kind is so rare (Wilson 1980a,b, 1983a,b). First of all, the ants must be relatively large. Workers of the two *Atta* species with head widths below 1.4 millimeters have difficulty cutting even the softest leaves and petals (the energetically most efficient size is 2.2 mm). Second, the workers have to be polymorphic. Individuals larger than about 1.2 millimeters are evidently unable to care for the minute fungi within the nest. Consequently the cultivation of the fungus entails a remarkable assembly-line operation, as follows. The medias (head width mode 2.2 mm) cut and retrieve the vegetation, smaller medias (1.6 mm) slice it into smaller pieces, still smaller ones (1.4 mm) degrade the pieces into small lumps, and then successively smaller minor workers (1.2–0.8 mm) place the lumps in the substrate, implant strands of fungus on fresh substrate, and care for the fungus as it proliferates (see Figures 8–28 to 8–30). In addition the ants use spe-

**TABLE 17–3** Characteristics of attine genera believed to exhibit consistent evolutionary trends within the group. The genera are listed in order of their presumed approximate phylogenetic position, with the first genus, *Cyphomyrmex*, being the most primitive. (Based mostly on Wheeler, 1907b; Weber, 1941b, 1946a, 1972, 1982).

| Genus | Morphology | Geographic range | Nest structure | Mature colony size | Garden substrate |
|---|---|---|---|---|---|
| *Cyphomyrmex rimosus* | Monomorphic; squamiform, appressed hairs; large, widely spaced frontal lobes; smooth body surface; small size | Mexico to Argentina, West Indies | Irregular cavity in soil or rotting wood | Small to medium | Insect feces |
| *Cyphomyrmex* , other species (30 spp.) | Monomorphic; hairs simple and sparse; large, widely spaced frontal lobes; smooth to tuberculate body; small size | Southern United States to Argentina | One symmetrical cell; usually in soil | Small | Insect feces, insect corpses, pieces of fruit |
| *Mycetophylax* (6 spp.) | Monomorphic; hairs sparse; smooth body surface; frontal lobes of medium size and spacing; small size | West Indies, South America (mostly Argentina) | One or two symmetrical cells in soil | Small | Dead grass |
| *Mycocepurus* (4 spp.) | Monomorphic; hairs sparse; spinose; frontal lobes approximated and small; small size | West Indies, Central and South America | One symmetrical cell in soil | Small | Insect feces |
| *Myrmicocrypta* (24 spp.) | Monomorphic; tuberculate thorax bearing squamiform hairs; frontal lobes approximated and small; small size | Mexico to Argentina | One large, symmetrical cell in soil or rotting wood | Medium | Vegetable matter, insect corpses |
| *Apterostigma* (27 spp.) | Monomorphic; abundant flexuous hairs; smooth body surface; small to medium size | Central and South America | One to several gardens surrounded (in some species) by very thin, mycelial shroud, built in the open under logs or in cavities under logs, loose bark, or stones | Small | Insect feces and dead, woody matter |
| *Sericomyrmex* (20 spp.) | Monomorphic; abundant flexuous hairs; body surface tuberculate; head cordate; medium size | Mexico to Brazil | One to several symmetrical cells in soil | Medium | Fruit, possibly dead vegetable matter |
| *Mycetosoritis* (3 spp.) | Monomorphic; body surface tuberculate with moderately abundant, curved hairs; small size | Texas (United States), Brazil to Argentina | Several symmetrical cells arranged vertically in soil | Small | Dead vegetable matter |
| *Trachymyrmex* (34 spp.) | Monomorphic to slightly polymorphic; body surface tuberculate with stiff, hooked hairs; small to medium size | Northern United States to Argentina, West Indies | Several symmetrical cells usually arranged vertically in soil | Small to medium | Insect feces, flower parts, dead vegetable matter |

*continued*

**TABLE 17–3** *(continued)*

| Genus | Morphology | Geographic range | Nest structure | Mature colony size | Garden substrate |
|---|---|---|---|---|---|
| *Acromyrmex* (24 spp.) | Polymorphic; body surface tuberculate with stiff hairs; occipital lobes developed; large size; strong queen-worker size difference | Southwestern United States to Argentina | Complex earthen nests with one very large or many chambers | Large | Fresh leaves, stems, flowers |
| *Acromyrmex* (= *Pseudoatta*) (1 sp.) | (A parasitic workerless genus possibly derived from *Acromyrmex*) | Argentina | — | — | — |
| *Atta* (15 spp.) | Strongly polymorphic; body surface partly tuberculate; large size; strong queen-worker size difference | Louisiana (United States) to Argentina | Complex earthen nests with many chambers | Large | Fresh leaves, stems, flowers |

cial procedures such as the recycling of chitinases and proteinases from the fungi (Boyd and Martin, 1975). Overall *Acromyrmex* and *Atta* have traveled a long path in evolution by mastering the technique of gardening and then shifting to a substitute of fresh vegetation. Because they depend on a fungus to accomplish much of their initial digestion and by this means can bypass the formidable array of terpenoids, alkaloids, and other defensive chemicals that deter most insect herbivores, the leafcutters have been able to exploit a very wide range of food plants, including most of the crop species grown in tropical regions.

## ANT–FUNGUS SYMBIOSES OUTSIDE THE ATTINI

A unique mutualistic symbiosis occurs between the European formicine ant *Lasius fuliginosus*, the "shining black ant" of some English-language literature or "glänzend schwarze Holzameise" in some German writings, and the ascomycete fungus *Cladosporium myrmecophilum*. The fungus grows exclusively in the walls of the *Lasius* carton nests, reinforcing them structurally. The ants exhibit specialized behaviors apparently directed at the cultivation of the fungus, and they transport inocula from one nest to another.

Colonies of *Lasius fuliginosus* transform large cavities in the soil and tree trunks by filling them with carton nests, whose internal structure is partitioned and resembles a sponge (see Figure 17–6). Maschwitz and Hölldobler (1970) found that the carton consists of particles of wood, dry vegetable material, and soil glued together with sugary secretions collected by the ants from aphids and other homopteran insects. The fungal mycelium grows through the walls of the carton and reinforces them in the same way that steel mesh or rods reinforce the walls of buildings.

The ants build their distinctive nest with a remarkable division of labor. Four castes based on age are employed. Workers of the first group, who are evidently older, collect the solid particles and carry them into the nest cavity, where they deposit them on the edge of the carton structure. The second group, also older, simultaneously collect homopteran honeydew and carry the liquid into the nest in their crop. Inside the nest they regurgitate it to nestmates who spend most of their time in nest construction and brood care. This third group of workers collect the solid particles from the edge of the carton and carry them to the construction site. There they regurgitate the sugary material onto the particles. They place the soaked particles onto the edge of the carton wall. During this process the workers often knead the fresh material with their mandibles while continuously touching the carton with their antennae and forelegs. The construction workers usually gather at specific sites where they line up along the edge of the wall in rows 4–5 centimeters long. Other ants, evidently constituting a fourth labor group, remove old particles from the lower surface of the carton wall and plant them on the upper, growing edge. Through this action they appear to be transferring fungal mycelia onto the new sugar bed, but the behavior could be a fortuitous outcome of the more ordinary distribution of nest materials—a common behavior of ants generally. The ants also crop the mycelium continuously. When the colony is separated from the carton, the fungal "lawn" sprouts into a furry mass. According to Lagerheim (1900), the carton fungus in central Europe, *Cladosporium myrmecophilum*, is known only from the nests of *Lasius fuliginosus*. The symbiosis appears to be truly mutualistic. Unlike the attines, however, the *Lasius* do not consume the fungus as food.

A possible second case of fungus cultivation outside the Attini occurs in the harvester ant *Messor* (= *Veromessor*) *pergandei*. Went et al. (1972) speculated that the ants feed on fungi growing on refuse

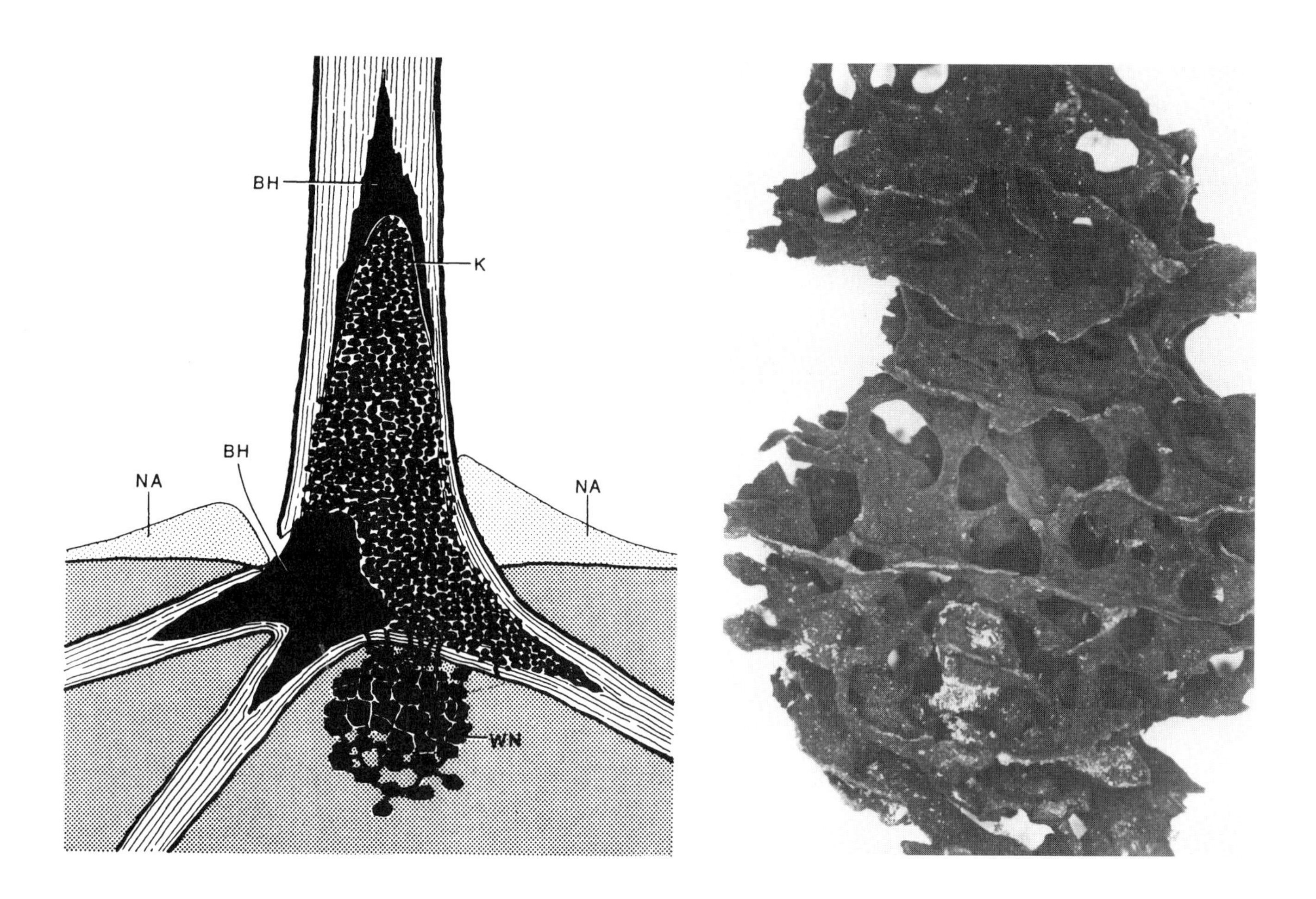


FIGURE 17–6 The European ant *Lasius fuliginosus* lives in carton nests infiltrated and strengthened by a symbiotic fungus. (*Left*) Schematic diagram of a typical nest. *BH*, tree cavity; *K*, carton nest; *NA*, nest refuse; *WN*, subterranean nest chambers, occupied primarily during the winter. (*Right*) Portion of the nest carton. (From Maschwitz and Hölldobler, 1970.)

within the nests. Their evidence, however, is circumstantial and tenuous. They noted that a very small percentage of items brought into the nests by foragers are arthropod exoskeletons and fragments of insect excrement. This material does not appear in the refuse piles surrounding the outside nest craters. Thus there exists a possibility, certainly worthy of further study, that the material is being used as a fungal substrate.

CHAPTER 18

# The Harvesting Ants

Harvesting ants are species that regularly use seeds as part of their diet. They constitute a broad assemblage representing many different evolutionary lines within the subfamilies Ponerinae, Myrmicinae, and Formicinae. They are distinguished from the even broader group of ant species that collect the myrmecochores, which are nutritive appendages fitted like caps or sheaths on the seeds. These latter foragers discard seeds as soon as they have detached the myrmecochores, anywhere between the plant and the nest; they are therefore major dispersers of myrmecochorous plants, as we showed in Chapter 14. Harvesting ants, in contrast, feed on the seeds themselves. Yet their effect on the plants they visit is not wholly negative. The "mistakes" they make, that is, the seeds they lose along the way or discard by accident at the nest, also disperse plants and compensate at least in part for the damage caused by seed predation. It is entirely possible that seed dispersal by harvesting ants preceded myrmecochore-aided seed dispersal during the coevolution of ants and plants (Rissing, 1986).

The harvesting of seeds by ants in deserts and grassland is bound to impress human beings who live by the same activity. The storing of seeds by ants in underground granaries has equal appeal. From the Book of Solomon to the writings of the ancient Greeks and Romans, ants were established very early in Western culture as the symbols of industriousness and prudence. The metaphorical view of these insects was shaped largely by the abundant Old World harvesters of the myrmicine genus *Messor.* One of us (Wilson, 1984a) paid a biologist's tribute to these ants and the tradition they engendered in the following way:

> Once on a tour of Old Jerusalem standing near the elevated site of Solomon's Throne, I looked down across the Jericho Road to the dark olive trees of Gethsemane and wondered which native Palestinian plants and animals might still be found in the shade underneath. Thinking of "Go to the ant, thou sluggard; consider her ways," I knelt on the cobblestones to watch harvester ants carry seeds down holes to their subterranean granaries, the same food-gathering activity that had impressed the Old Testament writer, and possibly the same species at the very same place. As I walked with my host back past the Temple Mount toward the Muslim Quarter, I made inner calculations of the number of ant species found within the city walls. There was a perfect logic to such eccentricity: the million-year history of Jerusalem is at least as compelling as its past three thousand years. (p. 6)

## HISTORY OF THE STUDY OF HARVESTING ANTS

The history of our understanding of these insects has consisted of two active periods with a conspicuous intervening hiatus. Ancient writers were well aware of harvesting by ants because they lived in the Mediterranean Region and the Middle East, where the phenomenon is prominent. The dominant species they encountered, as Wheeler (1910a) has pointed out, were undoubtedly *Messor barbarus,* which occurs throughout the Mediterranean littoral of Europe, Asia, and Africa southward to the Cape of Good Hope; *M. structor,* which is absent in Africa but ranges all the way from southern Europe to Java; and *M. arenarius,* which is abundant in the deserts of North Africa and the Middle East. These middle-sized, conspicuous ants are often serious grain pests, and it is to them that the writings of Solomon, Hesiod, Aesop, Plutarch, Horace, Virgil, Ovid, and Pliny almost certainly allude. The earliest work on ants published in the modern era, Wilde's *De Formica* (1615), merely repeats the accounts of these authors, in Latin. When Gould (1747), Latreille (1802), Huber (1810), and other early entomologists of the modern era began to study ants in the field, they saw no evidence of harvesting and consequently doubted or even denied the classical reports. This turn of events was due entirely to the fact that they lived in temperate Europe, where harvesting by ants is rare or nonexistent.

When Europeans began to report from warmer, drier climates, the phenomenon was quickly validated. At Poona, India, Sykes (1835) observed *Pheidole providens* bring rain-soaked grass seeds out of the nest and place them on the grass to dry. Jerdon (1854) confirmed the phenomenon in *P. providens, P. diffusa,* and *Solenopsis geminata,* and he saw the workers of these species collect seeds from different species of plants and store them in their nests. Moggridge (1873), during a sojourn in southern France, worked out the procedure of seed harvesting by *Messor barbarus* and *M. structor* in some detail. He found that the ants harvest the seeds of at least 18 families, and he confirmed reports of Plutarch and other ancient authors that the workers bite off the radicle to prevent germination, then store the deactivated seeds in granary chambers in the nests. He observed seed drying by the same method as that described by Sykes. In a remarkably modernistic twist, Moggridge also established that harvesters play an important role in dispersing plants by accidentally abandoning viable seeds in the nest vicinity or failing to deactivate them before they sprout. All of these key observations have been repeated many times by later observers, including Forel, André, Emery, Lameere, and, nowadays, an entire generation of younger researchers working on the ecology of harvesters.

Simultaneously, several early American entomologists addressed the subject of ant harvesting, including Buckley (1861), Lincecum (1862, 1866), McCook in his classic *Natural History of the Agricultural Ant of Texas* (1879), and Wheeler in his important synthesis of the subject in *Ants: Their Structure, Development and Behavior* (1910a). Each made closely similar observations on harvesting ants in the deserts of the southwestern United States. The favored subject was the ubiquitous genus *Pogonomyrmex,* but *Aphaenogaster* (= *Novomessor*), *Messor* (= *Veromessor*), and *Pheidole* were also included. One famous misconception of Lincecum's was that the Texas harvester *Pogonomyrmex molefaciens* deliberately sows the seeds of grasses of the genus *Aristida* around the periphery of its mound or crater nests and cultivates the crop in addition to collecting and storing the seeds in its granivores. Wheeler wrote:

> This notion, which even the Texan schoolboy has come to regard as a joke, has been widely cited, largely because Darwin stood sponsor for its publication in the Journal of the Linnean Society . . . Four years of nearly continuous observations of *molefaciens* and its nests enable me to suggest the probable source of Lincecum's misconception. If the nests of this ant can be studied during the cool winter months—and this is the only time to study them leisurely, as the cold subdues the fiery stings of their inhabitants—the seeds, which the ants have garnered in many of their chambers will often be found to have sprouted. Sometimes, in fact, the chambers are literally stuffed with dense wads of seedling grasses and other plants. On sunny days the ants may often be seen removing these seeds when they have sprouted too far to be fit for food and carrying them to the refuse heap, which is always at the periphery of the crater or cleared earthen disk. Here the seeds, thus rejected as inedible, often take root and in the spring form an arc or a complete circle of growing plants around the nest. (p. 286)

This interpretation of the "crop" as an adventitious growth of harvested seeds is consistent with the observations of Moggridge (1873) on *Messor.* It also fits studies of the relationships of *Pogonomyrmex* and other harvesters established by more recent researchers.

After the first fruitful period in the natural history of harvesters, extending roughly between 1860 and 1910, there was a lull in the study of these ants. An intense revival began in the 1970s when a new generation of ecologists recognized the convenience of *Messor, Monomorium* (= *Chelaner*), *Pogonomyrmex* and other harvesters for field and experimental studies in foraging and competition. This new work, which we discussed in Chapters 10 and 11, has grown into an important chapter of modern general ecology. We will now review more general aspects of the natural history and environmental importance of harvesting ants.

## THE DISTRIBUTION OF HARVESTING

The known harvesting ants are listed in Table 18–1, where it can be seen that the life habit is disproportionately concentrated in the Myrmicinae. Within that subfamily a great many genera phylogenetically remote from one another are represented, including such physically disparate forms as *Messor, Oxyopomyrmex, Meranoplus,* and *Pheidole.* The degree of commitment of harvesters to a seed diet varies across species to the greatest imaginable extent, from occasional and optional in the African *Atopomyrmex mocquerysi* to total or nearly so in *Monomorium* (= *Chelaner*) *whitei* of Australia and *Messor* (= *Veromessor*) *pergandei* of North America. Much of this range of variation occurs among species belonging to single genera, including *Rhytidoponera, Monomorium,* and *Pheidole.* The workers vary equally in temperament. At one extreme is *Pogonomyrmex,* viciously combative and with the most toxic venom of any known insect poison, at least with respect to mammals (Schmidt and Blum, 1978). *Pogonomyrmex* workers are so aggressive that they fight members of alien conspecific colonies 80 percent of the time when they encounter them while foraging, and fatalities are commonplace (Hölldobler, 1976a; De Vita, 1979; see Figure 10–13). At the opposite extreme are *Goniomma* and *Oxyopomyrmex,* whose workers are so timid and few in number that nests are difficult to find (Felix Santschi in Forel, 1904).

Harvesting ants are dominant elements in the deserts and drier grasslands in warm temperate and tropical regions around the world, especially in North America, Australia, the Sahara, and South Africa (Wehner, 1987). Seed-harvesting species compose more than half of all ant colonies in some Australian localities (Briese and Macauley, 1981). In the Namib Desert they make up more than 95 percent of the total forager biomass.

## AN EXTREME GRANIVORE

*Messor* (= *Veromessor*) *pergandei,* which has been studied intensively by Diane Davidson, Steven Rissing, and others over the past twenty years, is one of the most specialized granivores found among ants anywhere in the world. It flourishes in the deserts of southwestern Arizona, southern California, and Baja California. In Death Valley, one of the driest and hottest places in North America, the *Messor pergandei* are the most abundant ants; they have a biomass approximately equal to that of the total rodent population in the same area (Went et al., 1972). The toughness of the species in the face of harsh conditions is legendary among entomologists. In the Coachella Valley of California, colonies survived even after twelve successive years of severe drought (Tevis, 1958). The key to this success is the tendency of the ants to store large quantities of seeds underground, and their evident ability to subsist entirely on this food without the supplementation of arthropod prey or nectar.

The population of a mature colony is very large, ranging into the tens of thousands. A full census has not been taken, because no one has succeeded in excavating a complete nest. Tevis (1958) apparently has come the closest. He was able to follow one gallery for 4 meters before losing track of the nest system. Of their own effort Wheeler and Rissing (1975a) wrote, "In Deep Canyon and Death Valley we tried slicing off the top or digging in from the side toward the center of the nest; even with the enthusiastic assistance of several students, we were never able to dig quickly enough to find any large concentrations of workers." They succeeded in making a partial cast of a nest by pouring a casting resin into an entrance hole; the upper structure of the nest they revealed is shown in Figure 18–1. To accomplish more would require the planning and energy of an archaeological dig. The typical mature *Messor pergandei* nest, from what can be seen of it, has two or three active entrance holes 2

**TABLE 18–1** Harvester ants worldwide. These species are to be distinguished from ants that gather seeds to feed on elaiosomes (see Chapter 14). + + + indicates primary or exclusive reliance on seeds; + +, a substantial reliance but less than dependence; +, an occasional use of seeds.

| Species | Distribution | Degree of reliance | Comments | Source |
|---|---|---|---|---|
| **SUBFAMILY PONERINAE** | | | | |
| *Brachyponera senaarensis* | Central Africa and Asia, savanna | + + | 75 percent of food items are seeds, rest are arthropods | Lévieux and Diomande (1978b) |
| *Platythyrea* sp. | Tropical Africa, savanna | + | Documentation uncertain | Lévieux (1983b) |
| *Rhytidoponera tasmaniensis, R. victoriae* | Southeastern Australia, sclerophyllous forest | + + | Among major seed predators | Andersen (1985), Andersen and Ashton (1985) |
| *R. violacea* | Southwestern Australia, dry woodland | + + | | Majer (1982) |
| **SUBFAMILY MYRMICINAE** | | | | |
| *Acanthomyrmex ferox, A. notabilis* | Tropical Asia, rain forest | + + (?) | Colonies take seeds, apparently milled by large-headed majors; evidence still circumstantial | Moffett (1985b) |
| *Aphaenogaster* (= *Novomessor*) *albisetosus* | Southwestern United States | + | Accept seeds along with arthropod prey and corpses | B. Hölldobler (unpublished observations) |
| *Aphaenogaster* (= *Novomessor*) *cockerelli* | Southwestern United States, xeric habitats | + | Accept seeds along with arthropod prey and corpses | Wheeler (1910a), Davidson (1977b), Hölldobler et al. (1978), Whitford et al. (1980) |
| *Atopomyrmex mocquerysi* | West Africa, gallery forest | + | Very broad diet includes plant sap, seeds, and arthropod predators | Lévieux (1976c) |
| *Atta cephalotes* | Central America, lowland dry forest | + | Collect fig seeds, probably to use as substrate for symbiotic fungus | Roberts and Heithaus (1986) |
| *Goniomma* spp. | Mediterranean Region, xeric habitats | + + (?) | Relatively scarce | Emery (1921–22), Bernard (1968) |
| *Meranoplus* spp. | Australia, various habitats including mallee (semiarid scrub *Eucalyptus*) | + | Seeds a minor part of diet | Greenslade (1979), Andersen (1982), Briese (1982a), Campbell (1982) |
| *Messor* spp. | Africa, southern Europe, and Asia; xeric environments | + + | Among major seed predators of Old World xeric habitats | Délye (1971), Lévieux and Diomande (1978a), Lévieux (1979), Onoyama and Abe (1982), Hahn and Maschwitz (1985) |
| *Messor* (= *Veromessor*) *pergandei* | Southwestern United States, deserts | + + + | Obligatory granivores; take up to 27 percent of non-seed plant material in times of seed scarcity; workers occasionally collect insect remains, possibly not for food | Went et al. (1972), Bernstein (1975), Rissing and Wheeler (1976), Davidson (1978), Rissing (1981a) |
| *Monomorium chobauti, M. lameerei* | North Africa, deserts | + + | | Wheeler (1910a) |

*continued*

**TABLE 18–1** *(continued)*

| Species | Distribution | Degree of reliance | Comments | Source |
|---|---|---|---|---|
| *Monomorium* spp. | Australia, various habitats including mallee | + | Occasional to frequent seed predators | Andersen (1982), Campbell (1982) |
| *Monomorium* (= *Chelaner*) sp. nr. *flavigaster* | Southeastern Australia, sclerophyllous forest | + | | Andersen (1985) |
| *Monomorium* (= *Chelaner*) sp. nr. *kiliani* | Southeastern Australia, sclerophyllous forest | + + | Among principal seed predators | Andersen (1985), Andersen and Ashton (1985) |
| *Monomorium* (= *Chelaner*) *rothsteini*, and other spp. | Australia, especially in arid habitats | + + | Among principal seed predators | Greenslade (1979), Davison (1982) |
| *Monomorium* (= *Chelaner*) *whitei* | Australia, semiarid habitats | + + + | | Davison (1982) |
| *Oxyopomyrmex* spp. | Mediterranean Region, xeric habitats | + + (?) | Relatively scarce | Wheeler (1910a), Bernard (1968) |
| *Pheidole fallax*, *P. radoszkowskii* | Costa Rica, dry tropical forest | + | | Roberts and Heithaus (1986) |
| *P. latigena* | Southwestern Australia, dry woodland | + + | | Majer (1982) |
| *P. megacephala*, *P. pallidula* | Southeastern France | + | | Moggridge (1873) |
| *P. militicida* | Southwestern United States, xeric habitats | + + + | Among important seed predators | Creighton and Creighton (1959), Whitford (1978b), Hölldobler and Möglich (1980) |
| *Pheidole* 2 spp. | Southeastern Australia, sclerophyllous forest and mallee | + + | Among important seed predators | Andersen (1982, 1985), Andersen and Ashton (1985) |
| *Pheidole rhea*, *P. ridicula*, *P. xerophila*, and other spp. | Southwestern United States, xeric habitats | + + | Among important seed predators | Wheeler (1910a), Creighton (1966), Brown et al. (1979a,b), Davidson et al. (1985) |

*continued*

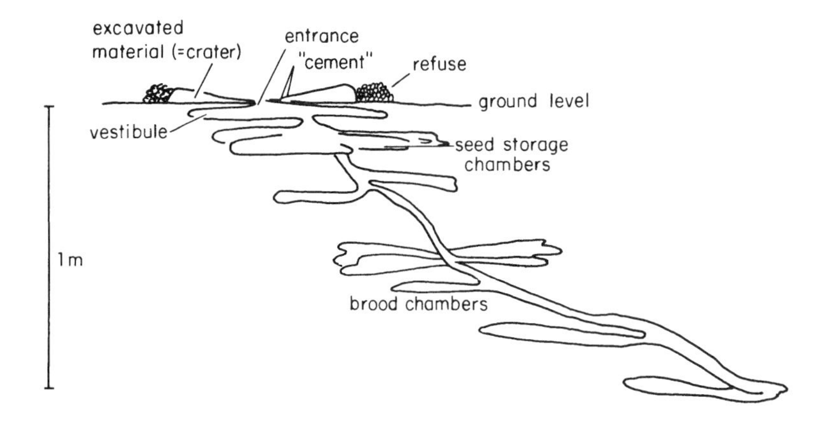


**FIGURE 18–1** The upper nest structure of a colony of the extreme desert granivore *Messor* (= *Veromessor*) *pergandei*. (Modified from Wheeler and Rissing, 1975a.)

by 4 centimeters across. Each entrance is surrounded by a crater of sand and fine gravel. The material nearest the entrance is held together by a yellowish cement. This material also contains some substances that the workers recognize as belonging to their nest. A neat pile of chaff comprising the husks of seeds forms a semicircle on the northern perimeter. Just below each entrance is a large chamber, or "vestibule," partly filled with chaff and a few seeds. Farther down are seed-storage chambers and finally rooms containing mixtures of larvae and pupae.

*Messor pergandei* workers harvest seeds from a wide spectrum of plants. Among those recorded to date are 14 genera of plants in the Coachella Valley (Tevis, 1958) and 24 genera comprising 29 species in Death Valley (Rissing and Wheeler, 1976). The ants are far from indiscriminate, however. They probe through piles of seeds before selecting one to take home, and they consistently choose larger than average grass seeds of each species offered (Rissing, 1981a). Species of *Chorizanthe*, *Franseria*, and *Lygodesmia* are less favored, perhaps because of their exceptionally hard seed coats. Workers also tend to "major," that is, to persist in harvesting one species of seed even when more desirable seeds are present in the same pile. In the nest the workers break the endosperm into fragments, which are then placed directly on the larvae. Went et al. (1972) suggested that the larvae metabolize the seed materials and regurgitate carbohydrate-

**TABLE 18–1** *(continued)*

| Species | Distribution | Degree of reliance | Comments | Source |
|---|---|---|---|---|
| *Pheidole* spp. | Central and South America, rain forest | + | | W. L. Brown and E. O. Wilson (unpublished notes) |
| *Pheidole* sp. nr. *sexspinosa* and another *Pheidole* sp. | New Guinea, lowland rain forest | + | Store abundant seeds in their nests | Wilson (1959c) |
| *Pheidologeton diversus* | Southeast Asia | + | Seeds an occasional part of diet | Moffett (1987b) |
| *Pogonomyrmex* spp. (about 60 spp.) | North America, Central and South America, and Haiti; xeric habitats | + + | Among most important seed predators of xeric environments in New World | Cole (1968), Hölldobler (1974, 1976a), Davidson (1977a,b), Hansen (1978), Whitford (1978a), MacKay (1981), Kugler (1984), Rissing (1986) |
| *Solenopsis geminata* | Throughout warmer parts of the New World, open woodland and scrub vegetation | + + | Regular seed predators, but depend more on insect prey and honeydew | Jerdon (1854), Wilson (1978), Risch and Carroll (1986) |
| *Tetramorium* sp. | Southern Australia, saltbush and grassland | + | | Briese (1982a) |
| **SUBFAMILY FORMICINAE** | | | | |
| *Melophorus* spp. | Australia, especially in arid habitats | + + | Some species are harvesters, others predators and scavengers | Greenslade (1979), Briese (1982a), Majer (1982) |
| *Prolasius* spp. | Southern Australia, temperate moist *Eucalyptus* forest | + | Three species eliminate over 60 percent of *Eucalyptus regnans* production | Ashton (1979) |

rich secretions back to the workers, in the manner later described in the Australian granivore *Monomorium* (= *Chelaner*) *whitei* by Davison (1982).

*Messor pergandei* workers observe two foraging periods in each day (Rissing and Wheeler, 1976). One begins in the early morning before there is any indication of daylight to the human observer and ends during the heat of late morning. The second period starts in the afternoon after the temperature has fallen from the midday high and continues until dusk, or even after dark on warm evenings. The workers emerge quickly in columns that follow preexisting trunk trails. On one occasion recorded by Wheeler and Rissing, a column containing about 17,000 ants extended 40 meters from the nest.

The success of *Messor pergandei* in the harshest American deserts is not due entirely to its heavy reliance on stored seeds, but in addition to three other traits that impart flexibility in the foraging strategies. First, the workers rely increasingly on individual searching when seed supplies are short, and on columns when seeds are encountered in patches. Second, the ants rotate their columns around the nest entrance in a way that brings them repeatedly to new patches and increases their yield over long stretches of time. Third, when the supply of desirable seeds is low, the workers turn to less desirable seeds and non-seed plant materials such as flower parts, leaves, and stems (Rissing and Wheeler, 1976). Davidson (1978) ascribed the considerable size variability of *M. pergandei* workers to an adaptive polymorphism in which small workers tend to collect small seeds and large workers favor large seeds. If true, this property of the worker caste would allow the ants to harvest seeds from multiple plant species more quickly and with greater energetic efficiency. In a later study, however, Rissing (1987) found that worker size accounts for less than 4 percent of the variance in the size of harvested seeds. He attributed the worker polymorphism to an annual cycle in the food made available to developing larvae. During the winter "triple crunch" (reduced seed availability, shorter foraging time, new queens and males being produced), worker-destined larvae evidently receive less food and end up smaller in size. If this interpretation is correct, the worker polymorphism of *Messor pergandei* should be more parsimoniously interpreted as a nonadaptive epiphenomenon rather than as an additional mechanism enhancing the flexibility of foraging.

## SEED SELECTION

All species of harvesting ants studied so far accept a wide array of plant seeds under natural conditions. The natural history of *Messor pergandei* shows why such latitude favors species most dependent on seeds for their livelihood. In the physically demanding and irregular environments in which harvesting ants live, few plant species can be depended upon to produce a profitable crop of seeds during any given year. This interpretation of the ecological significance of broad seed choice was nicely supported in a field study by Hahn and Maschwitz (1985) on *Messor rufitarsis*, an unusual harvester that occurs widely through central Europe. At the northern limit of its range, in the German state of Hessen, *M. rufitarsis* exists in scattered populations occupying open habitats in which exceptional numbers of plant species grow. From May to October, two or more of these species produce significant quantities of seeds, and the ants pass from one set to the next across the growing season, rather like a person traversing stepping stones across a pond.

In spite of the advantages of broad dietary choice, harvester ants do discriminate among seeds to some degree. All other things being equal, there is a general tendency to gather seeds of the kind that are most abundant (Davidson et al., 1980). Among species of *Pheidole* and *Pogonomyrmex* in the southwestern United States, a strong correlation exists between the size of the worker caste and the size of the seeds they prefer (Hölldobler, 1976a; Hansen, 1978; Chew and De Vita, 1980). In semi-arid Australia, *Monomorium* (= *Chelaner*) *rothsteini* collects a higher proportion of small seeds than does its competitor *Monomorium* (= *Chelaner*) *whitei* (Davison, 1982). In the savannas of northern Ivory Coast, *Messor galla* forages at night in the dry season while collecting seeds from the short and medium grasses *Monocymbium seresiiforme* and *Pennisetum hordeoides*, whereas *Messor regalis* is primarily nocturnal and harvests from the tall grass *Andropogon gayanus*. Preferences for the seeds of some plant species over those of others have also been reported in *Monomorium* (= *Chelaner*) (Davison, 1982), *Pheidole* (Mott and McKeon, 1977), *Pogonomyrmex* (Nickle and Neal, 1972; Whitford, 1978a), and *Solenopsis* (Risch and Carroll, 1986). In several cases the investigators observed a shift to other, less desirable seeds when those usually favored became less available. The chemical basis of the selectivity is unknown, although Buckley (1982b) noted the likely wide occurrence of repellents such as tannin (imposed on top of nutritive attractants), and Ashton (1979) reported the existence of sweet substances attractive to ants in the seeds of *Eucalyptus regnans* in Australia. On the other hand seeds may be protected by purely physical traits. Those of *Datura discolor* are among the largest and most energy rich of any ephemeral species of the southwestern American deserts, but they are evidently protected from harvester ants by their thick, heavily sculptured coats (O'Dowd and Hay, 1980).

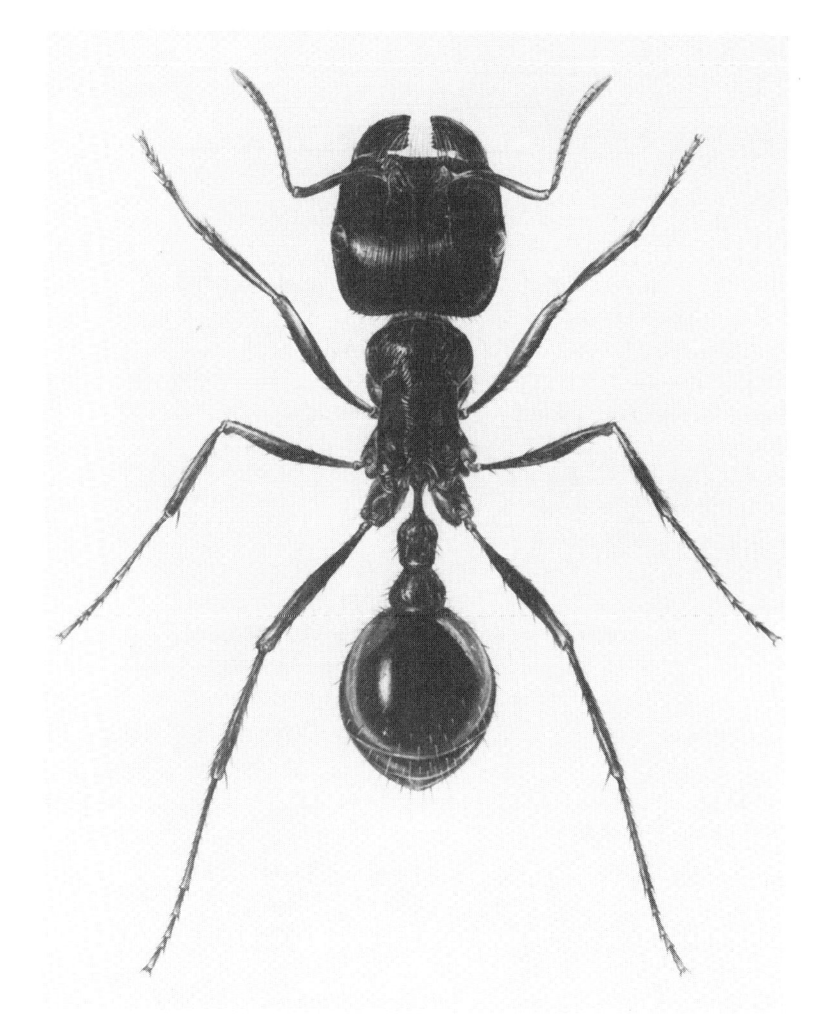

**FIGURE 18–2** Worker of the harvester ant *Pogonomyrmex rugosus*. (Painting by T. Hölldobler-Forsyth.)

## FORAGING PATTERNS

The commonest foraging strategy of harvesting ant species is a mix of individual foraging and column retrieval, adjusted according to need from one day to the next. This pattern is conspicuously displayed by the species of *Messor* and *Pogonomyrmex* (Figure 18–2). Scouts go forth to explore the terrain. They are guided variously by visual landmarks, by their compass direction relative to the sun, and by odor marks deposited in the vicinity of the nest exits (Hölldobler, 1971b). If a worker finds a solitary seed, she carries it back to the nest. If the ant encounters a patch of seeds, for example a seedfall beneath a grass clump, she carries one seed homeward while depositing an odor trail from the tip of her abdomen. Nestmates travel out to the seed patch along the trail, and while returning with burdens of their own they often add to the trail pheromone. In time, if the seedfall persists, the chemical deposits accumulate in sufficient strength to constitute a trunk trail along which large numbers of the ants travel back and forth. Even without reinforcement the trunk trail can remain active for days or weeks (see Figure 18–3). In *P. badius* and other *Pogonomyrmex* species, the original recruitment substance is emitted from the poison gland and the longer-lasting orientation substance from the Dufour's gland (Hölldobler and Wilson, 1970; Hölldobler, 1976a; see Figure 18–4). In *Messor rufitarsis*, which belongs to a wholly different stock of myrmicine ants, both pheromones come from the Dufour's gland (Hahn and Maschwitz, 1985). As food grows richer and more clumped in distribution, colonies shift more from individual orientation to foraging along trunk trails. Even in the best of times a few scouts wander away from the main routes. As a result they occasionally discover new seed

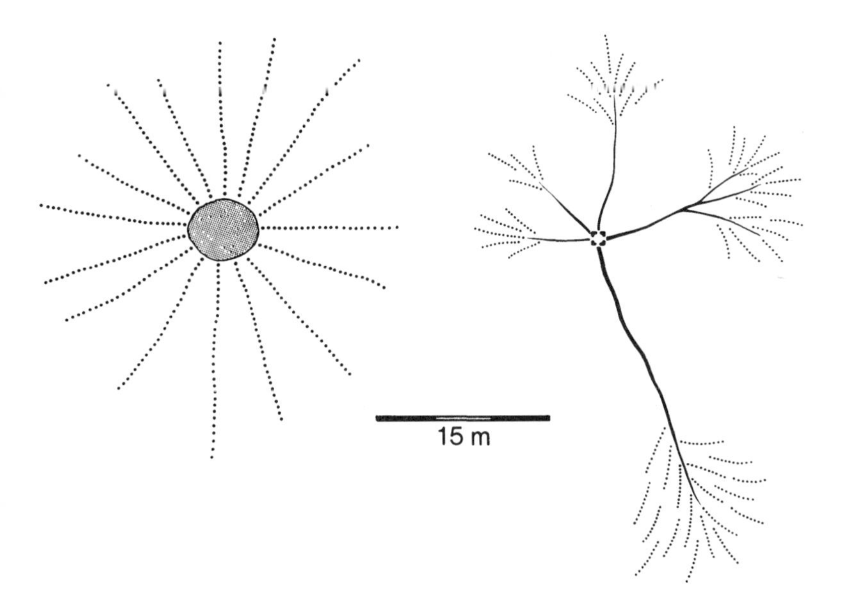


**FIGURE 18–3** The two major foraging patterns used by species of the harvesting ant genus *Pogonomyrmex*. The figure at the left depicts the pattern when foragers leave individually and disperse in all directions around the nest. This mode occurs in all *Pogonomyrmex* species studied, but it is the main one used by *P. maricopa* and *P. californicus*. The figure on the right depicts the trunk-route foraging pattern. Foragers leave the nest along relatively stable trunk routes and disperse on individual foraging excursions at the end of the route. They also return to the nest along the same central routes. This mode has been most frequently observed in *P. barbatus* and *P. rugosus*.

**FIGURE 18–4** A forager of *Pogonomyrmex badius* returns to a seed site while laying a chemical recruitment trail with her extruded sting. (From Hölldobler and Wilson, 1970.)

patches and together provide the colony with the flexibility of response needed to exploit all of the surrounding environment efficiently. Overall the density of foragers falls away steeply and exponentially with distance from the nest exit. In one census of *P. californicus* made by De Vita (1979), the modal density was 1.6 meters from the exit, and few if any foragers ranged beyond 13 meters.

Harvester ants using foraging columns shift them to avoid competition at food patches with colonies of the same and closely related species. The typical foraging pattern of these insects is an overdispersion of nest sites: the sites are spread out more uniformly than they probably would be as a result of chance alone, and the trunk trails of different colonies never overlap. Sometimes the trails come so close that their final branches interdigitate, but they still do not cross or terminate at the same seed patch (see Figure 10–14). This pattern has been documented in *Pogonomyrmex* (Hölldobler, 1976a) and the European *Messor* (Lévieux, 1979), and it is implied in descriptions of foraging in the American *M. pergandei* by Rissing and Wheeler (1976).

The commonest diel pattern of foraging in hot, dry climates is bimodal, with activity peaking in the cool of the morning and then again in the afternoon or early evening. But this trait is evidently very dependent on temperature or humidity. With the onset of the rainy (and cooler) season of Colombia, *Pogonomyrmex mayri* changes its pattern from bimodal to unimodal, and the peak shifts to midday (Kugler, 1984). During the rainy season in the Ivory Coast savannas, *Messor galla* and *M. regalis* change from mainly nocturnal foraging with a secondary peak in the morning to principally diurnal foraging (Lévieux, 1979). Harvesters living in cooler, moister environments, such as the Japanese population of *M. aciculatus*, reach their peak around the middle of the day (Onoyama, 1982).

A striking feature of harvester foraging is *Ortstreue*, or site tenacity, the persistent return of individual foragers to the same restricted area trip after trip and even day after day. When the colony harvests seeds along multiple trunk trails, workers showing *Ortstreue* behavior repeatedly choose one trail over others. The phenomenon has been documented in *Pogonomyrmex* (Hölldobler, 1976a), *Pheidole* (Hölldobler and Möglich, 1980), *Messor* (Onoyama, 1982; Onoyama and Abe, 1982), and *Monomorium* (= *Chelaner*) (Davison, 1982). A closely related behavior is majoring, in which individual workers persistently choose one kind of seed out of two or more available in the same foraging area. This kind of behavior is known to occur at least in *Monomorium* (= *Chelaner*) (Davison, 1982) and *Messor* (= *Veromessor*) (Rissing, 1981a). Both types of individual specialization, *Ortstreue* and majoring, certainly have the potential of increasing both individual efficiency and colony-wide efficiency, since they bypass the time-consuming procedures of exploration and prey choice. For this enhancement to occur, however, it is necessary for the species to add differentiation of worker choice among harvesting sites and seed type, as well as the capacity to shift rapidly to new sites and food items when old seedfalls are depleted. All of these properties have been documented in *Messor* and *Pogonomyrmex*, and they probably occur widely in other harvesting ants.

## EFFECTS OF HARVESTERS ON VEGETATION

There is general agreement among students of ant ecology that harvesters strongly alter the abundance and local distribution of flowering plants, especially in deserts, grasslands, and other xeric habitats where the ants are most abundant. They tip the balance in competition among some plant species and promote equilibria in others. They also rearrange the local distributions of the surviving species.

Under many circumstances, seed predation by ants reduces seed density and the subsequent vegetative mass of the plants. When ants were removed from experimental plots in Arizona by Brown et al. (1979b), annual plants were 50 percent denser after two seasons than in nearby control plots with their ant populations still intact. In eucalyptus woodland in southeastern Australia, seedling densities of *Eucalyptus baxteri* increased 15–fold after the ants were eliminated (Andersen, 1987).

On the other hand, ants often aid the exploited species by dispersing seeds more widely. *Pogonomyrmex rugosus* and *Messor* (= *Veromessor*) *pergandei* collect seeds of *Plantago insularis* and *Schismus arabicus* in Arizona deserts. Many seeds survive long enough to take root in the refuse piles around the ant nests. There the growing plants are at least five times denser on the average than in nearby sites away from the nests (Rissing, 1986). Thus the plants and the harvesters can be said to exist in a state of mutualism. The plants feed the ants a certain fraction of their seeds in return for which the ants transport another fraction to sites (the nest perimeter) that are relatively rich in nutrients and free of competitors.

Having established the countervailing forces of predation and dispersal as a general phenomenon in ant-plant interactions, ecologists have begun to define a far more complicated array of second-order effects. In the dry tropical forest of northwestern Costa Rica, species of *Pheidole* and *Atta* are secondary dispersers of fig seeds (*Ficus hondurensis*). The seeds are first scattered by birds, coatis, monkeys, and lizards as these animals feed on the fruit. The ants then rearrange the "seed shadow" from this primary dispersal by picking the seeds from the feces of the vertebrates and uneaten fruit fragments left behind. Some of the seeds transported by the ants germinate (Roberts and Heithaus, 1986). Thus there may be a sort of mutualism among plants, vertebrates, and ants, although it is doubtful that the ants coevolved to respond to seeds made available by the vertebrates.

At the opposite end of effectiveness, ants can exert powerful effects on competition and extinction. They can even serve as "keystone species," affecting plant community composition to an extent disproportionate to their numbers or biomass. An important example is the fire ant *Solenopsis geminata* in the annual cropping systems of the wet tropics in Mexico and Central America (Risch and Carroll, 1986). The seed abundance and plant biomass of weeds, especially grasses, are lowest in sites where the ants are present. In plots of corn and squash studied in Mexico, the ants reduced the number of arthropod individuals tenfold and the number of their species threefold. Such manifold effects have only begun to be explored. Risch and Carroll (1986) examined the impact of the fire ant on four pairwise combinations of weedy grasses and found that the ants generally preferred the seeds of one over the seeds of the other. The ultimate effects on plant biomass proved remarkably diverse. For one of the plant pairs the ants reversed the usual course of competition by differentially preying on the seeds of the dominant plant, allowing it to be excluded in the end by the subordinate. In two other cases the ants preferred seeds of the usually subordinate species, causing it to disappear more quickly. In the fourth combination the ants created a stable equilibrium by holding down the dominant just enough to allow the subordinate to survive.

Although the full cascade of effects of harvester ants on vegetation is obviously of major importance, it is still very poorly understood. Speaking of the Australian flora, Andersen and Ashton (1985) summarized the problems—and our general ignorance—of the ant-plant interactions in the following way:

> Although it is clear that ants can potentially destroy large numbers of seeds, there are many issues to resolve before their actual effect on seedling recruitment becomes known. For example, many sclerophyllous plants (such as species of *Leptospermum, Melaleuca* and *Kunzea*) have tiny seeds which might escape predation by falling amongst litter or by rapidly becoming incorporated into the soil. Second, seeds may avoid predation by falling during periods of low ant activity (*e.g.* winter) or by falling into areas where the activity of seed-eating ants is low. Investigations into patchiness of seed removal by ants indicate that even at sites where overall rates of removal are high, there are many places where removal is consistently low. Third, since most seed-eating ants are omnivorous, fluctuations in the availability of alternative food sources, such as insect prey, might have an important influence on removal rates. Fourth, stochastic events such as extensive seedfall in summer followed by an extended period of unseasonably cool and wet weather, might enable seeds to avoid predation, since rain reduces ant activity and promotes germination. Fifth, it has been assumed that all seeds removed by ants are ultimately destroyed, which obviously requires verification. Sixth, the importance of seed losses to ants depends on patterns of seedling recruitment; for example if there is high density-dependent seedling mortality, then seed predation may not be important. Finally, fire plays a key role in the reproductive biology of many sclerophyllous plants, and massive, fire-induced seedfall, resulting in predator satiation, might play an important role in successful seedling recruitment. It is also possible that ants are satiated in the absence of fire by plants which release their annual seed crop in a massive seasonal pulse.

To these complexities must be added competition of the harvester ants with other kinds of animals that exploit the seed crop. Exclusion experiments performed by J. H. Brown, D. W. Davidson, and their co-workers in the deserts of southern Arizona have documented these higher-order phenomena. At the Silver Bell alluvial plain 60 kilometers northwest of Tucson, these investigators removed ants from one series of fenced 0.1–hectare plots and rodents from another. Within a short time the number of ant colonies rose 71 percent in the rodent-free plots, while the rodents increased 20 percent in numbers of individuals (and 29 percent in biomass) in the absence of ants. These effects were evidently due to the greater number of seeds made available when one or the other of the two taxa was removed. Two years after the beginning of the experiment, however, the situation changed unexpectedly. The ants began to *decline* in the plots from which the rodents had been removed. The cause turned out to reside in the plants rather than the animals. With the rodents gone, the large-seeded plants normally favored by rodents began to increase, replacing the small-seeded plants favored by ants and thus reducing their food supply (Davidson et al., 1984). The same authors conducted similar experiments 250 kilometers to the east, on the Cave Creek alluvial plain near Portal, Arizona. In this area ants competed with rodents only for seeds produced during the less productive winter peak. When rodents were eliminated, *Pheidole xerophila* increased in numbers but *Pogonomyrmex desertorum* declined. Rodents did not change appreciably when relieved of ants. The reasons for the differences in the interactions between the ant species in each locality and, more generally, among all of the organisms across the two Arizona localities appear to entail: (1) seasonality in the production of seed resources and in their use by the two taxa; (2) specialization by ants and rodents on different density distributions of seeds; (3) "diffuse compensation" or compensation spread over many species populations; and (4) indirect interaction pathways, mediated through competing resource classes.

Finally, the higher-order phenomena of the kind disclosed in the Arizona experiments are likely to change in kaleidoscopic fashion in passing from one biome or continent to another. In Australia, for example, many species of ants and birds but few mammals specialize on seeds, whereas all three are prominent in North America. Whether or not the Australian harvester ant fauna is correspondingly more diverse and abundant, as Brown et al. (1979a) suppose, cannot be ascertained with existing field data. In South America, despite the absence of specialized seed-eating rodents, granivorous desert ants and birds are not conspicuously more diverse or abundant than in North America (Mares and Rosenzweig, 1978). A point of comparison is the prominent harvester ant genus *Pogonomyrmex*, represented by 22 species in the arid environments of North America (Cole, 1968) and 22 different species in comparably arid environments of the southern half of South America (Kempf, 1972b). The South American species are characterized by much smaller colonies, and they evidently occur in less dense populations.

# CHAPTER 19

# Weaver Ants

Among the thousands of social insects a few deserve to be called classic, because certain remarkable features in their behavior have prompted unusually careful and thorough studies. The honey bees, the bumblebees, the driver ants, the army ants, the leafcutter ants, the slavemaker ants, and the fungus-growing termites are all examples of classic social insects. The latest members of this select group are the weaver ants of the genus *Oecophylla*. These ants are relatively large, with bodies ranging up to 8 millimeters in length, and exclusively arboreal. The workers create natural enclosures for their nests by first pulling leaves together (see Figure 19–2) and then binding them into place with thousands of strands of larval silk woven into sheets. For this unusual procedure to succeed, the larvae must cooperate by surrendering their silk on cue, instead of saving it for the construction of their own cocoons. The workers bring nearly mature larvae to the building sites and employ them as living shuttles, moving them back and forth as they expel threads of silk from their labial glands.

Perhaps the first description of the biology and remarkable nest construction of *Oecophylla* was made by Joseph Banks, who accompanied Captain Cook on the voyage of the H.M.S. *Endeavour* in 1768 to Australia. In his *Journal*, Banks (cited in Musgrave, 1932) described his first encounters with the green tree ant in that part of New Holland now called New South Wales:

> Of insects there were but few sorts, and among them only the ants were troublesome to us. Mosquitoes, indeed, were in some places tolerably plentiful, but it was our good fortune never to stay any time in such places. The ants, however, made ample amends for the want of the mosquitoes; two sorts in particular, one green as a leaf, and living upon trees, where it built a nest, in size between that of a man's head and his fist, by bending the leaves together, and glueing them with whitish paperish substances which held them firmly together. In doing this their management was most curious: they bend down four leaves broader than a man's hand, and place them in such a direction as they choose. This requires a much larger force than these animals seem capable of; many thousands indeed are employed in the joint work. I have seen as many as could stand by one another, holding down such a leaf, each drawing down with all his might, while others within were employed to fasten the glue. How they had bent it down I had not the opportunity of seeing, but it was held down by main strength, I easily proved by disturbing a part of them, on which the leaf bursting from the rest, returned to its natural situation, and I had an opportunity of trying with my finger the strength that these little animals must have used to get it down.
>
> But industrious as they are, their courage, if possible excels their industry; if we accidentally shook the branches on which such nest was hung, thousands would immediately throw themselves down, many of which falling upon us made us sensible of their stings and revengeful disposition, especially if, as was often the case, they got possession of our necks and hair, their stings were by some esteemed not much less painful than those of a bee; the pain, however, lasted only a few seconds.

*Oecophylla* belongs to the subfamily Formicinae, the species of which lack a functional sting. The painful "sting" Banks described is actually the bite of the sharp and powerful mandibles, perhaps intensified by irritating secretions from the mandibular glands. The whitish glue that firmly holds the leaves together was described 137 years later as the salivary gland secretions of the *Oecophylla* larvae (Doflein, 1905).

The construction of communal silk nests has clearly contributed to the success of the *Oecophylla* weaver ants. It permits colonies to attain immense populations, in spite of the large size of the workers, because the ants are freed from the spatial limitations imposed on species that must live in beetles' burrows, leaf axils (the area between the stems of leaves and the parent branch), and other preformed vegetative cavities. This advance, along with the complex recruitment system that permits each colony to dominate up to several trees at the same time, has helped the weaver ants to become among the most abundant and successful social insects of the Old World tropics (Hölldobler and Wilson, 1977a,c,d, 1978; Hölldobler, 1979, 1983). A single species, *O. longinoda*, occurs across most of the forested portions of tropical Africa, while a second, closely related species, *O. smaragdina*, ranges from India to Queensland, Australia, and the Solomon Islands (Lokkers, 1986). The genus is ancient even by venerable insect standards: two species are known from Baltic amber of Oligocene age, dating back about 30 million years (Wheeler, 1914b). *O. leakeyi*, described from a fossil colony of Miocene age (approximately 15 million years old) found in Kenya, possessed a physical caste system very similar to that of the two living forms (Wilson and Taylor, 1964). In particular, the allometry by which the minor and major workers are differentiated is closely similar across the three species. Also, both the living and extinct species have a bimodal size-frequency distribution, with the major workers more common than the minor workers, and media workers present but scarce.

With their huge colonies and their ability to construct nests almost anywhere, the *Oecophylla* weaver ants have achieved a close control of their environment. The empire of the African species can be visualized by the following census made by Vanderplank (1960) of only a part of a colony on Zanzibar: in 192 leaf nests there were 62,694 workers; 18,498 worker larvae and pupae; and 657 queen lar-

**FIGURE 19–1** The caste system of the African weaver ant consists of three forms of adult female: a single queen, major workers that forage for food and perform a variety of other tasks, and minor workers that care for the eggs and younger larvae. (From Hölldobler and Wilson, 1977c; painting by T. Hölldobler-Forsyth.)

vae and pupae. The total mature and immature population was estimated to fall between 115,000 and 164,000. Yet this was by no means a large colony. Populations of a half million or more often occur, with nests extending through the crowns of up to three or more good-sized trees. Many nests contain scale insects that the weaver ants keep and protect for the sugary excrement they produce. Some of the peripheral nests are "barracks," containing mostly aging workers that sally out to attack alien ants and other intruders as soon as they cross the territorial borders (Hölldobler, 1983). The workers use no fewer than five recruitment systems, comprising differing chemical and tactile signals, to organize territorial defense, foraging, and the exploration of new terrain. Some of the pheromones come from rectal and sternal glands of a type unique to *Oecophylla* (see Chapter 7).

So far as we know, only one queen lives in each colony, and she is extremely attractive to major workers (Figure 19–1). Stimuli from her head, evidently chemical in nature, induce the workers to regurgitate and to present trophic eggs at frequent intervals. The queen suppresses the laying of viable eggs but not the laying of trophic eggs by the major workers; the effect is mediated by pheromones and persists in the corpse of the queen for as long as six months. When the live queen or her corpse is removed from the nest, some of the workers start to lay viable but unfertilized eggs. The head of the queen has exceptionally large propharyngeal and postpharyngeal glands as well as fully developed maxillary glands and mandibular glands. Her thorax and abdomen are lined with clusters of glandular cells drained by long ducts that exit from the exoskeleton. Since the mother queen does virtually nothing but eat and lay eggs, it seems likely that this unusual armamentarium of exocrine tissue is the source of her pheromones (Hölldobler and Wilson, 1983a).

The environmental control exercised by the weaver ants can be put to practical use. Records from southern China show that weaver ant nests have been gathered, sold, and placed in selected citrus trees to combat insect pests for approximately 1,700 years. The same basic techniques were repeatedly noted in the classical Chinese literature between A.D. 304 and 1795, and the practice is continued today as an alternative to chemical control in the provinces of Guangdong and Fujian (Huang and Yang, 1987). The weaver ant used for this purpose is the Asian species *Oecophylla smaragdina*. This utilization of weaver ants is the oldest known instance of the biological control of insects in the history of agriculture. D. Leston (personal communication) recommended employing the African species of weaver ant to control pests of tree crops such as cacao. Studies in Ghana have shown that the presence of weaver ants reduces the incidence of two of the most serious diseases of cacao, one caused by a virus and the other by a fungus. In both cases the pathogen is transmitted by mirid leaf bugs. The weaver ants evidently combat the diseases by attacking the bugs. The *Oecophylla* workers are also particularly effective in hunting insects that feed on the tissue and sap of trees.

## COMMUNAL NEST WEAVING

Our studies in the 1980s, building on those of other authors, have revealed an unexpectedly precise and stereotyped relation between the adult workers and the larvae. The larvae contribute all their silk to meet the colony's needs instead of their own. They produce large quantities of the material from enlarged silk glands at the beginning of the final instar rather than at the end, thus differing from cocoon-spinning ant species, and they never attempt to construct cocoons of their own (Wilson and Hölldobler, 1980; Hölldobler and Wilson, 1983b). The workers have taken over almost all the spinning movements from the larvae, turning them into passive dispensers of silk.

It would seem that close attention to the exceptional properties of *Oecophylla* nest weaving could shed new light on how cooperation and altruism operate in ant colonies, and especially on how larvae can function as an auxiliary caste. In addition, a second, equally interesting question is presented by the *Oecophylla* case: how could such extreme behavior have evolved in the first place? As is the case with the insect wing, the vertebrate eye, and other biological prodigies, it is hard to conceive how something so complicated and efficient in performance might be built from preexisting structures and processes. Fortunately, other phyletic lines of ants have evolved communal nest weaving independently and to variably lesser degrees than *Oecophylla,* raising the prospect of reconstructing the intermediate steps that lead to the extreme behavior of weaver ants. These lines are all within the Formicinae, the subfamily to which *Oecophylla* belongs. They include all the members of the small Neotropical genus *Dendromyrmex;* the two Neotropical species *Camponotus (Myrmobrachys) senex* and *C. (M.) formiciformis,* which are aberrant members of a large cosmopolitan genus; *C. (Karavaievia) gombaki, C. (K.) texens,* and probably other members of the tropical Asian subgenus *Karavaievia;* and various members of the large and diverse Old World tropical genus *Polyrhachis.*

Two additional but doubtful cases have been reported outside the Formicinae. According to Baroni Urbani (1978c), silk is used in the earthen nests of some Cuban species of *Leptothorax,* a genus of the subfamily Myrmicinae. He was uncertain, however, whether the material is obtained from larvae or from an extraneous source such as spiderwebs. Since no other myrmicine is known to produce silk under any circumstances, the latter alternative seems the more probable. Similarly, the use of silk to build nests was postulated for the Javan ant *Technomyrmex bicolor textor,* a member of the subfamily Dolichoderinae, in an early paper by Jacobson and Forel (1909). Again, however, the evidence is from casual field observations only, and the conclusion is rendered unlikely by the fact that no other dolichoderines are known to produce silk.

True silk weaving is easily confused with the construction of carton nests and shelters, which is a widespread practice in *Ectatomma, Crematogaster, Pheidole, Solenopsis, Monacis, Azteca, Tapinoma, Lasius,* and many other genera. These ants use bits of plant fibers and other detritus, often combined with soil, to construct galleries and chambers on the surface of vegetation. Although the structures are often quite elaborate, they do not contain silk.

We have studied the behavior of both living species of *Oecophylla* in much greater detail than did earlier entomologists, and have extended our investigations to the other, lesser-known nest-weaving genera, *Camponotus, Dendromyrmex,* and *Polyrhachis.* This research has made possible a preliminary characterization of the stages through which the separate evolving lines appear to have passed.

In piecing together the data, we utilized a now standard concept in organismic and evolutionary biology, the phylogenetic grade (Hölldobler and Wilson, 1983b). The four genera of formicine ants we considered are sufficiently distinct from one another on anatomical evidence to make it almost certain that the communal nest weaving displayed was in each case independently evolved. Thus it is proper to speak of the varying degrees of cooperative behavior and larval involvement not as the actual steps that led to the behavior of *Oecophylla* but as grades, or successively more advanced combinations of traits, through which autonomous evolving lines are likely to pass. Other combinations are possible, even though not now found in living species, and they may be the ones that were actually traversed by extreme forms such as *Oecophylla.* By examining the behavior of as many species and phyletic lines as possible, however, biologists are sometimes able to expose consistent trends and patterns that lend convincing weight to particular evolutionary reconstructions. This technique is especially promising in the case of insects, with several million living species to sample. Within this vast array there are 8,800 known species of ants, most of which have never been studied, making patterns of ant behavior exceptionally susceptible to the kind of analysis pursued in the case of nest weaving.

## THE HIGHEST GRADE OF COOPERATION

The studies conducted on *Oecophylla* prior to our own were reviewed by Wilson (1971) and Hemmingsen (1973). In essence, nest weaving with larval silk was discovered in *O. smaragdina* independently by H. N. Ridley in India and W. Saville-Kent in Australia, and was subsequently described at greater length in a famous paper by Doflein (1905). Increasingly detailed accounts of the behavior of *O. longinoda,* which is essentially similar to that of *O. smaragdina,* were provided by Ledoux (1950), Chauvin (1952), Sudd (1963), and Hölldobler and Wilson (1977c).

The sequence of behaviors by which the nests are constructed can be summarized as follows. Individual workers explore promising sites within the colony's territory, pulling at the edges and tips of leaves. When a worker succeeds in turning a portion of a leaf back on itself, or in drawing one leaf edge toward another, other workers in the vicinity join the effort. They line up in a row and pull together, or, in cases where a gap longer than an ant's body remains to be closed, they form a living chain by seizing one another's petiole (or "waist") and pulling as a single unit (Figure 19–2). Often rows of chains are aligned so as to exert a powerful combined force (Plates 22–23). The formation of such chains of ants to move objects requires intricate maneuvering and a high degree of coordination. So far as we know, the process is unique to *Oecophylla* among the social insects.

When the leaves have been maneuvered into a tent-like configuration, workers form rows and hold the leaves together (Figure 19–3). Another group of workers carries larvae from the interior of the existing nests and uses them as sources of silk to bind the leaves

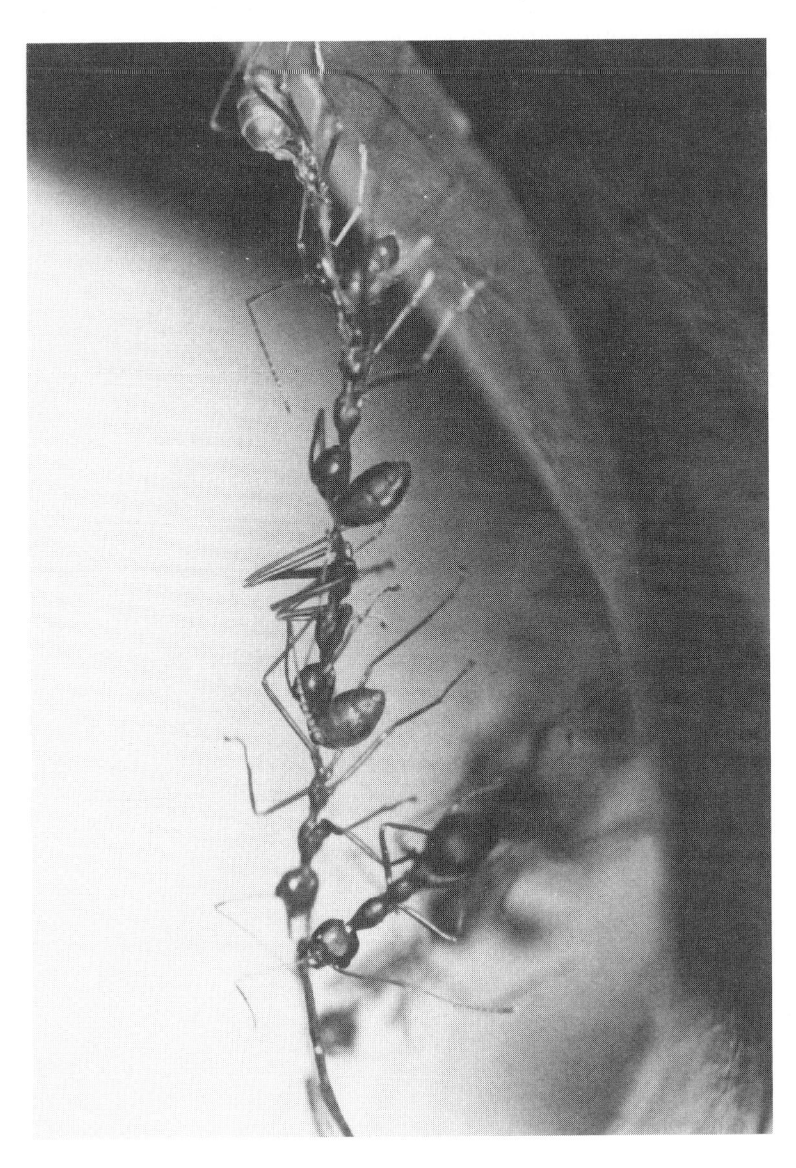

**FIGURE 19–2** To make a nest out of leaves and larval silk, workers of *Oecophylla longinoda*, the African weaver ant, first choose a pliable leaf. They then form a living chain by seizing one another's petiole and pulling on the leaves as a single unit. (From Hölldobler and Wilson, 1977c.)

together (Plate 24). Previous studies (Wilson and Hölldobler, 1980) showed that the *O. longinoda* larvae recruited for this purpose are all in the final of at least three instars, and have heads in excess of 0.5 millimeter wide. However, their bodies (exclusive of the rigid head capsule) are smaller than those of the larvae at the very end of the final instar, which are almost ready to turn into prepupae and commence adult development. Thus the larvae used in nest weaving are well along in development and possess large silk glands, but they have not yet reached full size and hence are more easily carried and manipulated by the workers.

In *O. longinoda* all the workers we observed with spinning larvae were majors, the larger adults that possess heads between 1.3 and 1.8 millimeters in width. Hemmingsen (1973) reported that majors of *O. smaragdina* perform the weaving toward the exterior, whereas minor workers—those with heads 1–1.2 millimeters wide—weave on the inner surfaces of the leaf cavities. We observed only major workers performing the task in *O. longinoda*, but admittedly our studies of interior activity were limited. Hemmingsen also recorded that exterior weaving is rare during the daytime but increases sharply at night, at least in the case of *O. smaragdina* working outdoors in Thailand. We saw frequent exterior weaving by *O. longinoda* during the day in a well-lit laboratory, as well as by *O. smaragdina* outdoors in Queensland.

In order to work out the details of the spinning process, we followed the entire sequence through a frame-by-frame analysis of 16-millimeter motion pictures taken at 25 frames per second. The most distinctive feature of the larval behavior, other than the release of the silk itself, is the rigidity with which the larva holds its body. There is no sign of the elaborate bending and stretching of the body or of the upward thrusting and side-to-side movements of the head that characterize cocoon spinning in other formicine ant larvae, particularly in *Formica* (Wallis, 1960; Schmidt and Gürsch, 1971). The larva keeps its body stiff, forming a straight line when viewed from above but a slightly curved, S-shaped line when seen from the side, with its head pointing obliquely downward as shown in Plate 24. Occasionally the larva extends its head for a very short distance when it is brought near the leaf surface, giving the impression that it is orienting itself more precisely at the instant before it releases the silk. The worker holds the larva in her mandibles between one-fourth and one-third of the way down the larva's body from the head, so that the head projects well out in front of the worker's mandibles.

The antennae of the adult workers are of an unusual conformation that facilitates tactile orientation along the edges of leaves and other vegetational surfaces. The last four segments are shorter relative to the eight segments closest to the body than in other ants we have examined, including even communal silk-spinning formicines such as *Camponotus senex* and *Polyrhachis acuta*. They are also unusually flexible and can be actively moved in various directions in a fashion seen in many solitary wasps.

As the worker approaches the edge of a leaf with a larva in her mandibles, the tips of her antennae are brought down to converge on the surface in front of her. For 0.2 ± 0.1 second ($x \pm$ SD, $n = 26$, involving a total of 4 workers), the antennae play along the surface, much in the manner of a blindfolded person feeling the edge of a table with his hands. Then the larva's head is touched to the surface and held in contact with the leaf for 1 second (0.9 ± 0.2 sec). During this interval the tips of the worker's antennae are vibrated around the larva's head, stroking the leaf surface and touching the larva's head about 10 times (9.2 ± 3.6). At some point the larva releases a minute quantity of silk, which attaches to the leaf surface.

About 0.2 second before the larva is lifted again, the worker spreads and raises her antennae. Then she carries the larva directly to the edge of the other leaf, causing the silk to be drawn out as a thread. While moving between leaves, the worker holds her antennae well away from the head of the larva. When she reaches the other leaf, she repeats the entire procedure exactly, except that the larva's head is held to the surface for only half a second (0.4 ± 0.01 sec); during this phase the worker's antennae touch the larva about 5 times (5.2 ± 2.4). In other words, the workers alternate between a longer time spent at one leaf surface and a shorter time spent at the opposing surface.

Thus, the weaving behavior of the *Oecophylla* worker is even more complicated, precise, and distinctive than earlier investigators real-

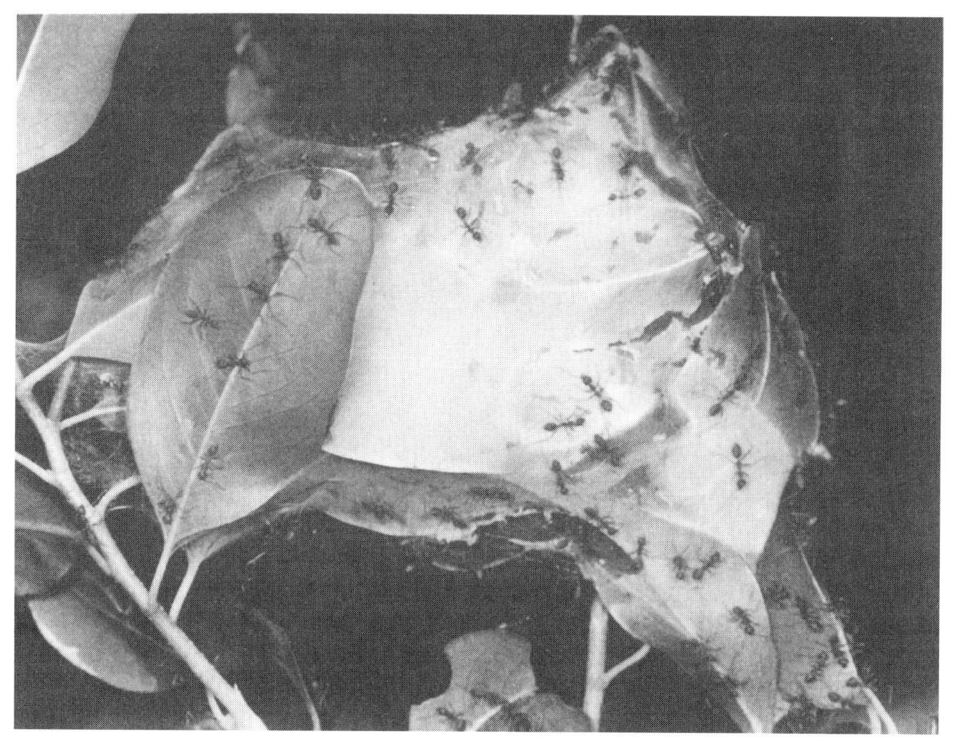

**FIGURE 19–3** Cooperative nest building in *Oecophylla* weaver ants. (*Above*) After maneuvering the leaves into a tent-like configuration, the workers hold the edges of the leaves together in preparation for silk weaving. (*Below*) The nest of the African weaver ant *O. longinoda* is formed basically of living leaves and stems bound together with larval silk. Some of the walls and galleries are constructed entirely of silk. (From Hölldobler and Wilson, 1983b.)

ized. The movements are rigidly stereotyped in form and sequence. The antennal tips are used for exact tactile orientation, a "topotaxis" somewhat similar to that employed by honey bee workers to assess the thickness of the waxen walls of the cells in the comb (Lindauer and Martin, 1969). The worker ant also appears to use her flexible antennal tips to communicate with the larva, presumably to induce it to release the silk at the right moment. Although we have no direct experimental proof of this effect, we can report an incidental observation consistent with it. One worker we filmed held the larva upside down, so that the front of the larva's head and its silk-gland openings could not touch the surface or be stroked by the antennal tips. The worker went through the entire sequence correctly, but the larva did not release any silk.

For its part, the larva has evolved distinctive traits and behaviors that serve communal weaving. It releases some signal, probably chemical, that identifies it as being in the correct phase of the final instar. When a worker picks it up, the larva assumes an unusual S-shaped posture. And when it is held against the surface of a leaf and touched by a worker's antennae, it releases silk, in a context and under circumstances quite out of the ordinary for most immature insects.

## INTERMEDIATE STEPS

The existence of communal nest weaving in *Polyrhachis* was discovered in the Asiatic species *P. (Myrmhopla) dives* by Jacobson (Jacobson and Wasmann, 1905). Few details of the behavior of these ants were available, however, until a study by Hölldobler (in Hölldobler and Wilson, 1983b).

A species of *Polyrhachis (Cyrtomyrma)*, tentatively classified near *doddi*, was observed in the vicinity of Port Douglas, Queensland, where its colonies are relatively abundant. The ants construct nests among the leaves and twigs of a wide variety of bushes and trees (Figure 19–4). Most of the units are built between two opposing leaves, but often only one leaf serves as a base or else the unit is entirely constructed of silk and is well apart from the nearest leaves.

*Polyrhachis* ants have never been observed to make chains of their own bodies or to line up in rows in the manner routine for *Oecophylla*. Occasionally a single *Polyrhachis* worker pulls and slightly bends the tip or edge of a leaf, but ordinarily the leaves are left in their natural position and walls of silk and debris are built between them.

The weaving of *Polyrhachis* also differs markedly from that of *Oecophylla*. The spinning larvae are considerably larger and appear to be at or near the end of the terminal instar (see Figure 19–4). The workers hold them gently from above, somewhere along the forward half of their body, and allow the larvae to perform all of the spinning movements. In laying silk on the nest wall, the larvae use a version of the cocoon-spinning movements previously observed in the larvae of *Formica* and other formicine ants. Like these more "typical" species, which do not engage in communal nest building, *Polyrhachis* larvae begin by protruding and retracting the head relative to the body segments while bending the forward part of the body downward. Approximately this much movement is also seen in *Oecophylla* larvae before they are touched to the surface of a leaf.

The *Polyrhachis* larvae are much more active, however, executing most of the spinning cycle in a sequence very similar to that displayed by cocoon-spinning formicines. Each larva begins with a period of bending and stretching, then returns to its original position through a series of arcs directed alternately to the left and to the right; in sum, its head traces a rough figure eight. Because the larva is held by the worker, the movements of its body are restricted. It cannot complete the "looping-the-loop" and axial rotary movements described by Wallis (1960), by which larvae of other formicine ants move around inside the cocoon to complete its construction. In fact, the *Polyrhachis* larvae do not build cocoons. They pupate in the naked state, having contributed all their expelled silk to the communal nest. In this regard they fall closer to the advanced *Oecophylla* grade than to the primitive *Dendromyrmex* one to be described shortly.

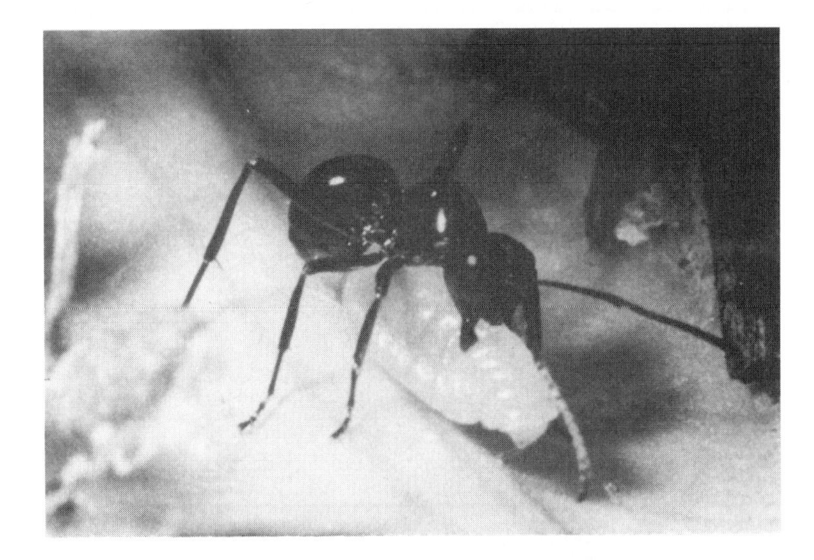

**FIGURE 19–4** A simple form of weaving is practiced by an Australian species of the formicine genus *Polyrhachis*. The worker holds a larva above the surface, allowing it to perform most of the weaving movements (*top*). The nest of the Australian *Polyrhachis* species consists of sheets of silk woven between leaves and twigs and reinforced by soil and dead vegetative particles (*center*). The interior of this type of nest has layers of silk tightly molded to the supporting leaf surface (*bottom*). (From Hölldobler and Wilson, 1983b.)

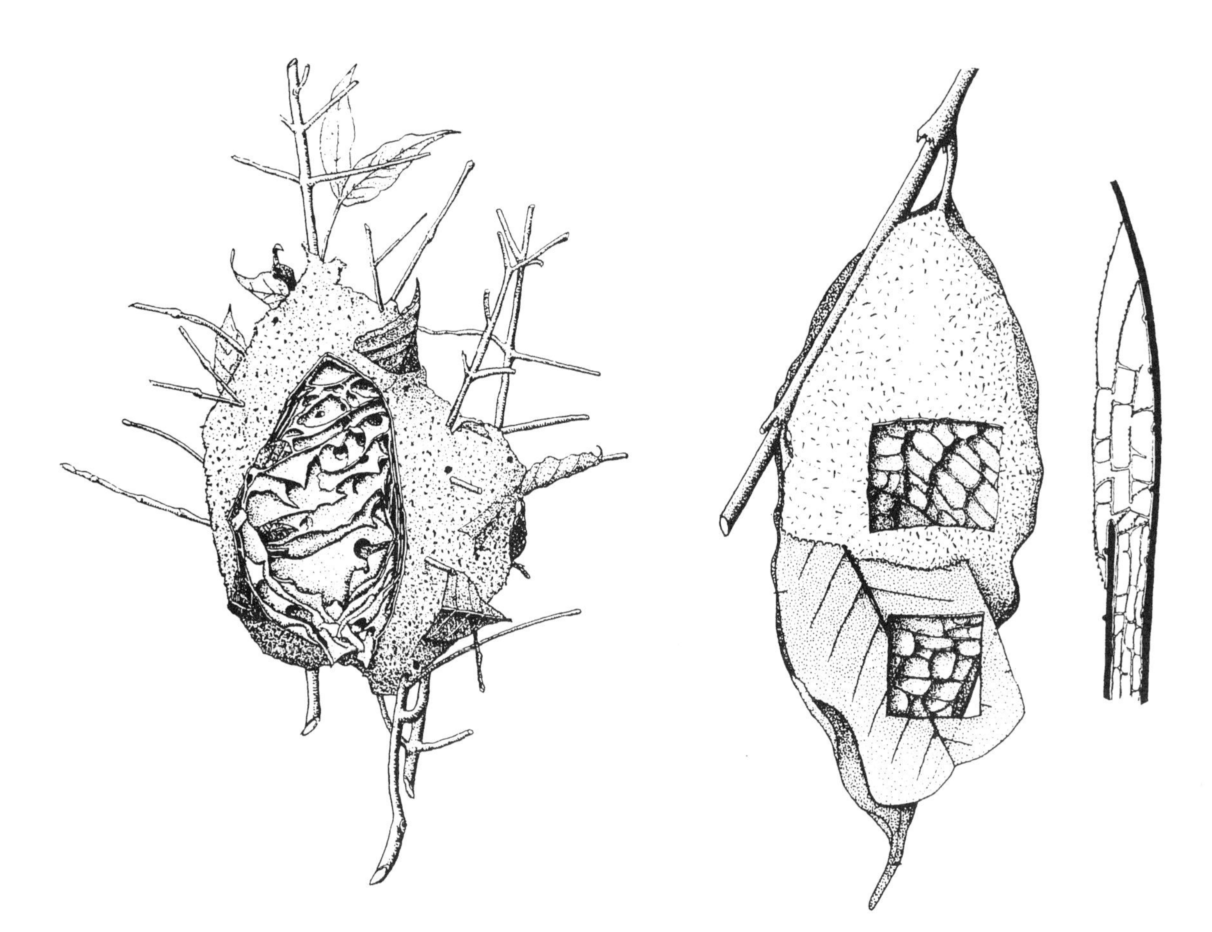

**FIGURE 19–5** The silk nest of *Camponotus senex*. (*Left*) Large nest with a cutaway portion of the multilayered silk wall, showing the internal chambers. (*Right*) The nest of a young colony built between two leaves of a coffee tree. Two cutaway portions of the outer silk cover (*above*) and the leaf surface (*below*) reveal the multiple chambers inside. At the far right is a section through the nest chambers, which occur in three layers. (From Schremmer, 1979a,b.)

*Polyrhachis* ants are also intermediate between *Oecophylla* and *Dendromyrmex* in another significant respect. The *Polyrhachis* workers do not move the larvae constantly, like living shuttles, as in *Oecophylla*, nor do they hold the larvae in one position for long periods of time or leave them to spin on their own as in *Dendromyrmex*. Rather, each spinning larva is held by a worker in one spot or moved slowly forward or to the side for a variable period of time (range 1–26 sec, mean 8 sec, SD 7.1 sec, $n$ = 29). After each such brief episode the larva is lifted up and carried to another spot inside the nest, where it is permitted to repeat the stereotyped spinning movements. While the larva is engaged in spinning, the worker touches the substrate, the silk, and the front half of the larva's body with her antennae. These antennal movements are less stereotyped than in *Oecophylla*, however.

The product of this coordinated activity is an irregular, wide-meshed network of silk extending throughout the nest. The construction usually begins with the attachment of the silk to the edge of a leaf or stem. As the spinning proceeds, some workers bring up small particles of soil and bark, wood chips, or dried leaf material that the ants have gathered on the ground below. They attach the detritus to the silk, often pushing particles into place with the front of their head, and then make the larvae spin additional silk around the particles to secure them more tightly to the wall of the nest. In this way a sturdy outside shell is built, consisting ultimately of several layers of silk reinforced by solid particles sealed into the fabric. The ants also weave an inside layer of pure silk, which covers the inner face of the outer wall and the surfaces of the supporting leaves and twigs. Reminiscent of wallpaper, this sheath is thin, very finely meshed, and tightly applied so as to follow the contours of the supporting surface closely. When viewed from inside, the nest of the *Polyrhachis* ant resembles a large communal cocoon (Figure 19–4).

A very brief description of the weaving behavior of *Polyrhachis (Myrmhopla) simplex* by Ofer (1970) suggests that this Israeli species constructs nests in a manner similar to that observed in the Queensland species. The genus *Polyrhachis* is very diverse and widespread, ranging from Africa to tropical Asia and the Solomon Islands. Many of the species spin communal nests, apparently of differing degrees of complexity. Recently, Yamauchi et al. (1987) demonstrated that colonies of *Polyrhachis dives* on Okinawa construct multiple nests out of leaves on the trees or grass on the ground and larval silk. They form truly multicolonial systems, with most nests further containing multiple queens. A steady traffic of workers, brood, and queens flows between the nests. Further study of the behavior of *Polyrhachis* weaver ants should prove highly rewarding.

A second intermediate grade is represented by *Camponotus (Myrmobrachys) senex*, which occurs in moist forested areas of South and Central America. With the closely related *C. (Myrmobrachys) formiciformis*, it is one of no more than six representatives of the very large and cosmopolitan genus *Camponotus* known to incorporate larval silk in nest construction (although admittedly very little information is available about most species of this genus), and in this respect must be regarded as an evolutionarily advanced form (Figure 19–5). The most complete account of the biology of *C. senex* to date is that of Schremmer (1972, 1979a,b).

Unlike the other weaver ants, *Camponotus senex* constructs its nest almost entirely of larval silk. The interior of the nest is a complex three-dimensional maze of many small chambers and connecting passageways. Leaves are often covered by the silken sheets, but they then die and shrivel, and thereafter serve as no more than internal supports. Like the Australian *Polyrhachis, C. senex* workers add small fragments of dead wood and dried leaves to the sheets of silk along the outer surface. The detritus is especially thick on the roof, where it serves to protect the nest from direct sunlight and rain.

As Schremmer stressed, chain formation by worker ants and other cooperative maneuvers among workers of the kind that characterize *Oecophylla* do not occur in *Camponotus senex*. The larvae employed in spinning are relatively large and most likely are near the end of the final instar. Although they contribute substantial amounts of silk collectively, they still spin individual cocoons—in contrast to both *Oecophylla* and the Australian *Polyrhachis*. Workers carrying spinning larvae can be seen most readily on the lower surfaces of the nest, where walls are thin and nest building is unusually active. During Schremmer's observations they were limited to the interior surface of the wall and consequently could be viewed only through the nascent sheets of silk. Although numerous workers were deployed on the outer surface of the same area at the same time, and were more or less evenly distributed and walked slowly about, they did not carry larvae and had no visible effect on the workers inside. Their function remains a mystery. They could be serving simply as guards.

Although Schremmer himself chose not to analyze the weaving behavior of *Camponotus senex* in any depth, we were able to make out some important details from a frame-by-frame analysis of his excellent film (Schremmer, 1972). In essence, *C. senex* appear to be very similar to the Australian *Polyrhachis* in this aspect of their behavior. Workers carry the larvae about slowly, pausing to hold them at strategic spots for extended periods. They do not contribute much to the contact between the heads of the larvae and the surface of the nest. Instead, again as in *Polyrhachis,* the larvae perform strong stretching and bending movements, with some lateral turning as well. When held over a promising bit of substrate, larvae appear to bring the head down repeatedly while expelling silk. We saw one larva perform six figure-eight movements in succession, each time touching its head to the same spot in what appeared to be typical weaving movements. The duration of the contact between its head and the substrate was measured in five of these cycles; the range was 0.4–1.5 seconds and averaged 0.8 second. During the spinning movements the workers play their antennae widely over the front part of the body of the larva and the adjacent substrate.

The nest weaving of *Camponotus senex,* then, is basically the same as that of the Australian *Polyrhachis.* The only relevant difference is that *C. senex* larvae construct individual cocoons and *Polyrhachis* larvae do not.

A wholly new example of communal weaving in the intermediate evolution grade has been discovered by Maschwitz et al. (1985a) in the tropical Asian subgenus *Karavaievia* of *Camponotus*. The behavior has been observed in only two rare and previously undescribed Malaysian species, *C. (K.) gombaki* and *C. (K.) texens,* but judging from an earlier ambiguous note of Viehmeyer (1915) on *C. (K.) dolichoderoides,* we may suppose that it is likely to be a general trait of this small but morphologically distinct group of species. The level of behavior in *Karavaievia* is approximately that of *C. senex.* Both of the Malaysian species weave their nests with the aid of fully grown last-instar larvae, which they hold anterior to the midlength of the body. When stimulated by antennal strokings of the workers, the larvae swing their heads back and forth while expelling silk threads. Like the workers of *C. senex* and species of *Polyrhachis,* those of *Karavaievia* do not conduct weaving movements of their own; they rely entirely on the initiative of the larvae held in their mandibles. As the weaving proceeds, a second group of workers gather sand particles, bits of detritus, and plant fragments and insert them into the fresh silk sheets. A third group bite into the loose silk, especially at the points of contact with the leaf surface, tightening and smoothing the sheet as a whole. A fourth group (in *texens* at least, which could be closely observed) transport honeydew-producing scale insects into the new pavilion. Thus a division of labor exists that is comparable to that displayed by the *Oecophylla* weaver ants during nest building.

Like *Polyrhachis* and *Camponotus senex* and unlike *Oecophylla,* the *Karavaievia* do not form chains of their bodies or pull leaves together as part of nest building. The silken pavilions are constructed directly on broad leaf surfaces or (in the case of *gombaki*) between leaves that are left otherwise unmodified. Both species build multiple single-chambered pavilions in the foliage of one or more trees. *C. (K.) texens* resembles *Oecophylla* in its territorial domination of the nest trees. It excludes other ants and destroys or appropriates any scale insects found outside the pavilions.

## THE SIMPLEST TYPE OF WEAVING

The seven species of *Dendromyrmex* weaver ants are concentrated in Brazil, but at least two species (*chartifex* and *fabricii*) range into Central America. The small colonies of these ants build oblong carton nests on the leaves of a variety of tree species in the rain forest (Weber, 1944b).

Wilson's 1981 study of the tree ants *Dendromyrmex chartifex* and *D. fabricii* has revealed a form of communal silk weaving that is the most elementary conceivable. The structure of the nests is reinforced with continuous sheets of larval silk (Figure 19–6). When the nest's walls are deliberately torn to test their strength, it can be seen that the silk helps to hold the carton together securely. Unlike *Oecophylla* larvae, those of *Dendromyrmex* contribute silk only at the end of the final instar, when they are fully grown and ready to pupate. Moreover, only part of the silk is used to make the nest. Although a few larvae become naked pupae, most enclose their own bodies with cocoons of variable thickness. Workers holding spinning larvae remain still while the larvae perform the weaving movements; in *Oecophylla,* the larvae are still and the workers move. Often the larvae add silk to the nest when lying on the surface unattended by workers. Overall, their nest-building movements differ from those of cocoon spinning only by a relatively small change in orientation. Not surprisingly, this facultative communal spinning results in a smaller contribution to the structure of the nest than is the case in *Oecophylla* and other advanced weaver ants.

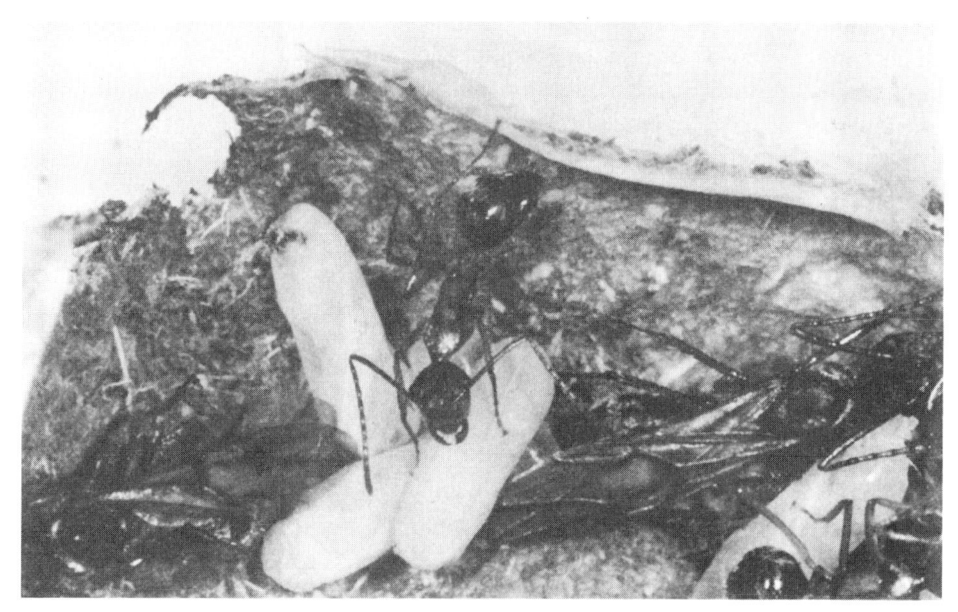

**FIGURE 19–6** The formicine genus *Dendromyrmex*, represented here by a Central American species, makes the simplest type of woven nest, a carton-like structure of chewed vegetative fibers reinforced with larval silk (*above*). The inside of the nest is lined with a relatively smooth silk cover (*below*). As can be seen in the photograph, the larvae also spin cocoons.

## ANATOMICAL CHANGES

The behavior of communally spinning ant larvae is clearly cooperative and altruistic. If general notions about the process of evolution are correct, we should expect to find some anatomical changes correlated with the behavioral modifications that produce this cooperation. Also, the degrees of change in the two kinds of traits should correspond to some extent. Finally, the alterations should be most marked in the labial glands, which produce the silk, and in the external spinning apparatus of the larva.

These predictions have generally been confirmed. *Oecophylla*, which has the most advanced cooperative behavior, also has the most modified external spinning apparatus. The labial glands of the spinning larvae of *Oecophylla* and *Polyrhachis* are much larger in proportion to the size of the larva's body than is the case in other formicine ant species whose larvae spin only individual cocoons (Karawajew, 1929; Wilson and Hölldobler, 1980). On the other hand, *Camponotus senex* larvae do not have larger labial glands than those of other *Camponotus* larvae. Schremmer (1979a) tried to fit this surprising result into the expected pattern by suggesting that the *C. senex* larvae produce silk for longer periods of time than other species that weave nests communally, and therefore do not need larger glands. This hypothesis has not yet been tested.

Until the 1980s little was known about the basic structure of the spinning apparatus of formicines. Using conventional histological sectioning of larvae in the ant genera *Formica* and *Lasius*, Schmidt and Gürsch (1970) concluded that the silk glands open to the outside by three tube-like projections, or nozzles. They were indeed able to pull three separate silken strands away from the heads of larvae with forceps. Our studies, however, which combine histology and use of the scanning electron microscope, have led us to draw a somewhat different picture. In general we found that the labial gland opens to the outside through a small slit with one nozzle at each end, as shown in Figure 19–7. This is the structure found in the Australian *Nothomyrmecia macrops*, which is considered to be the living species closest to the ancestors of the Formicinae, as well as in a diversity of formicines themselves. Among the formicines examined, including those engaged in communal nest weaving, only *Oecophylla* has a distinctly different external spinning apparatus. The labial-gland slit of these extremely advanced weaver ants is enlarged into a single nozzle, incorporating and largely obliterating the lateral nozzles. As a result it appears that each *Oecophylla* larva is capable of expelling a broad thread of silk—the kind of thread needed to create the powerful webs binding an arboreal nest together.

**FIGURE 19–7** Scanning electron micrographs reveal adaptations in the spinning apparatus of ant larvae in *Oecophylla* as opposed to less advanced kinds of ants. At the left, the head of an *O. longinoda* larva is shown from the side (*A*) and front (*B*); the arrows indicate the slit-shaped opening of the silk glands, which is modified substantially from that of the more primitive forms at the right. The reduced lateral nozzles in *O. longinoda* and the larger central nozzle are clearly visible in *C*. In *Nothomyrmecia macrops*, a living Australian ant thought to be similar to the earliest formicines, there is no central nozzle and the lateral nozzles are much more prominent; in *D*, the arrow points to the area enlarged in *E*. The silk-gland opening of the Australian weaver ant, a species of *Polyrhachis* (*F*), is similar in structure to that in *Nothomyrmecia*. (From Hölldobler and Wilson, 1983b.)

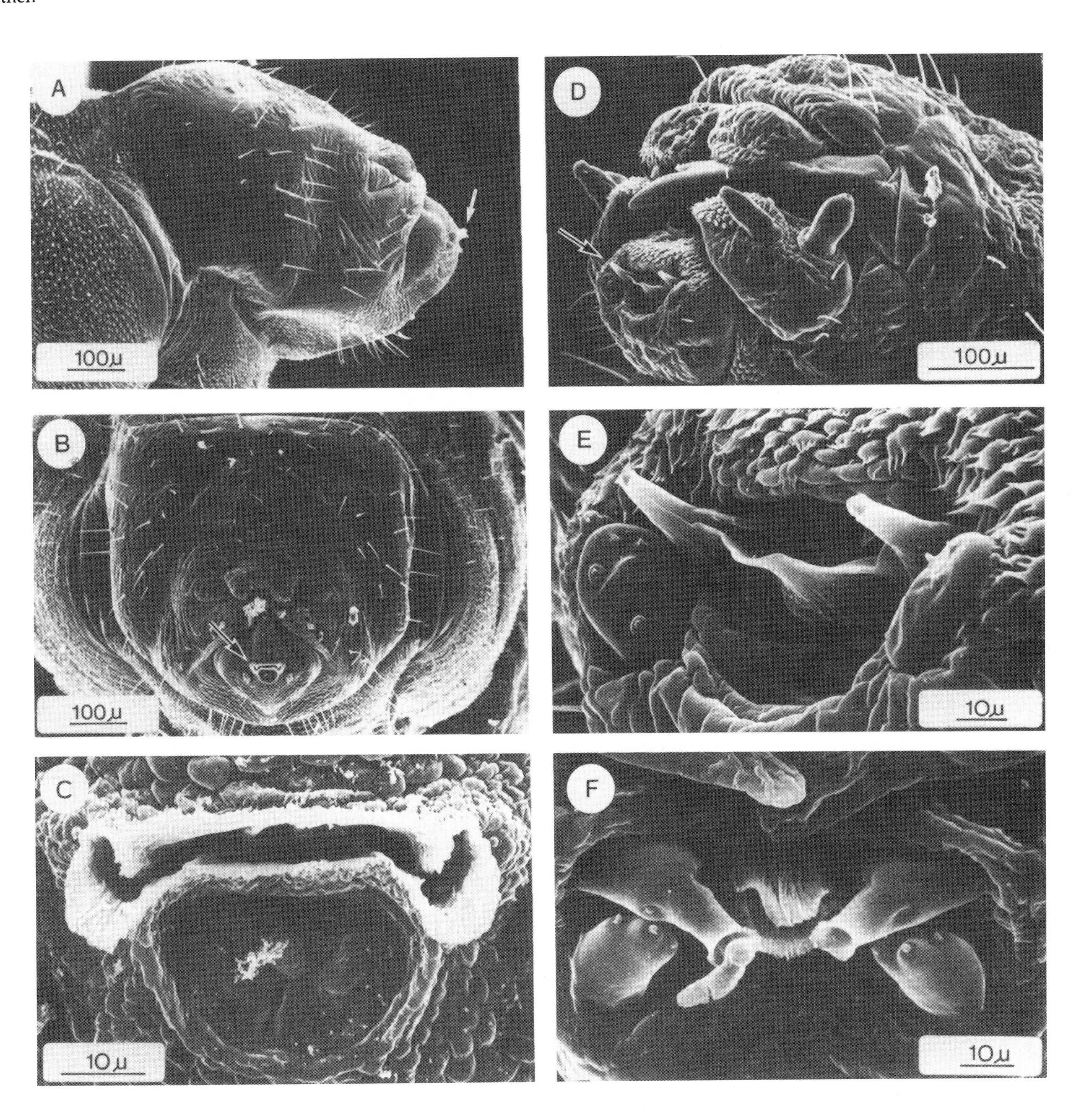

**TABLE 19–1** Grades of communal nest-weaving. (Modified from Hölldobler and Wilson, 1983b.)

| Species | Larvae contribute silk to nest | Workers always hold spinning larvae | Larvae no longer make individual cocoons | Workers repeatedly move larvae | Workers cooperate in adjusting substrate | Workers perform most spinning movements | Silk produced before end of final instar |
|---|---|---|---|---|---|---|---|
| Grade 1 | | | | | | | |
| *Dendromyrmex* spp. | + | – | – | – | – | – | – |
| Grade 2 | | | | | | | |
| *Polyrhachis ?doddi* | + | + | + | + | – | – | – |
| *Camponotus senex* | + | + | – | + | – | – | – |
| *C. texens* | + | + | – | + | – | – | – |
| Grade 3 | | | | | | | |
| *Oecophylla* spp. | + | + | + | + | + | + | + |

## THE UNCERTAIN CLIMB TOWARD COOPERATION

To summarize existing information on the evolution of communal spinning, we have defined the grades in Table 19–1 according to the presence or absence of particular traits associated with communal nest weaving. We believe that it is both realistic and useful to recognize three such stages. It is also realistic to suppose that the most advanced weaver ants, those of the genus *Oecophylla*, are derived from lines that passed through lower grades similar to, if not identical with, those exemplified by *Dendromyrmex, Polyrhachis*, and *Camponotus*.

On the other hand, we find it surprising that communal nest weaving has arisen only four times or so during the hundred million years of ant evolution. Even if new cases of this behavior are discovered in the future, the percentage of ant species that weave their nests communally will remain very small. It is equally puzzling that the most advanced grade was attained only once. The separate traits of *Oecophylla* nest weaving provide seemingly distinct advantages that should encourage their evolution in arboricole ants. The remarkable cooperative maneuvers of the workers allow the colony to arrange the substrate in the best positions for the addition of the silk bonds and sheets. By taking over control of the spinning movements from the larvae, the workers enormously increase the speed and efficiency with which the silk can be applied to critical sites. For their part the larvae have benefited the colony by moving the time when they produce silk forward in the final instar, thus surrendering once and for all the ability to construct personal cocoons while simultaneously allowing workers to carry and maneuver them more effectively because of their smaller size.

The case of *Dendromyrmex* is especially helpful in envisioning the first steps of the evolution in behavior that culminated in the communal nest weaving of *Oecophylla*. Although the contribution of the larvae to the structure of the nest is quite substantial, the only apparent change in their behavior is a relatively slight addition to their normal spinning cycle, so that the larva releases some silk onto the floor of the nest while weaving its individual cocoon. It is easy to imagine such a change occurring with the alteration of a single gene affecting the weaving program. Thus starting the evolution of a population toward communal weaving does not require a giant or otherwise improbable step.

There is another line of evidence indicating the general advantage of communal nest weaving and hence a relative ease of progression. We discovered that both male and female larvae contribute silk to the nest in the cases of *Oecophylla* (Wilson and Hölldobler, 1980) and *Dendromyrmex* (Wilson, 1981); male contribution has not yet been investigated in *Polyrhachis* and *Camponotus*. Because cooperation and altruism on the part of male ants is rare, males are always worthy of close examination. As Bartz (1982) showed, natural selection in the social Hymenoptera will favor the evolution of either male workers or female workers but not both, and the restrictive conditions imposed by the haplodiploid mode of sex determination—used by all Hymenoptera—favor all-female worker castes. In fact the sterile workers of hymenopterous societies are always female (Oster and Wilson, 1978). In boreal carpenter ants of the genus *Camponotus*, where the males do contribute some labor to the colony, it is in the form of food sharing, an apparent adaptation to the lengthy developmental cycle of *Camponotus*. The males are kept in the colonies from late summer or fall to the following spring, and it benefits both the colony and the individual males to exchange liquid food (Hölldobler, 1966).

The contribution of silk by male weaver-ant larvae is a comparable case. When the queens of *Oecophylla* and *Dendromyrmex* die, some of the workers lay eggs, which produce males exclusively (Hölldobler and Wilson, 1983a). Such queenless colonies can last for many months, until the last of the workers has died. During this period it is clearly advantageous for male larvae to add silk to the nest, for their own survival as well as that of the colony as a whole.

In summary, then, weaver ants exemplify very well an important problem of evolutionary theory: why so many intermediate species possess what appear to be "imperfect" or at least mechanically less efficient adaptations. Two hypotheses can be posed to explain the phenomena that are fully consistent with the manifest operation of natural selection in such cases. The first is that some species remain in the lower grades because countervailing pressures of selection come to balance the pressures that favor the further evolution of the trait. In particular, the tendency for larvae to collaborate in the construction of nests could be halted or even reversed in evolution if surrendering the ability to make cocoons reduces the larva's chance of survival. In other words, the lower grade might represent the optimum compromise between different pressures.

The second, quite different hypothesis is that the communal weavers are continuing to evolve—and will eventually attain or even surpass the level of *Oecophylla*—but species become extinct at a sufficiently high rate that most such evolutionary trends are curtailed before they can be consummated. Even a moderate frequency of extinction can result in a constant number of species dispersed across the various evolutionary grades.

At present we see no means of choosing between these two hypotheses or of originating still other, less conventional evolutionary explanations. The primary importance of phenomena such as communal nest weaving may lie in the prospects they offer for a deeper understanding of arrested evolution, the reasons why not all social creatures have attained what from our peculiar human viewpoint we have chosen to regard as the pinnacle of altruistic cooperation.

# CHAPTER 20

# Collecting, Culturing, Observing

This chapter provides a primer of simple techniques for studying ants for students and for a wide variety of field researchers who need to handle material quickly and efficiently. Our exposition is far from an exhaustive account. In the case of the culture of live colonies especially, specialized methods suited to the needs of particular species are often developed as part of research programs, and they can be found in the Materials and Methods section of the respective technical articles. What is offered here is a set of general procedures that we have found to work well over many years of experience, across almost all major groups of ants.

## HOW TO COLLECT ANTS

Collecting ants is simple and straightforward and can be conducted immediately by anyone. We routinely place specimens in 80 percent ethanol or isopropyl alcohol; the latter substance is especially useful because it can be obtained as rubbing alcohol without a prescription in many parts of the world. (An unusual but workable approach was taken by the late astronomer and amateur myrmecologist Harlow Shapley, who used to preserve ants in the strongest spirits of the country he visited. A worker of *Lasius niger* that he placed in vodka while dining with Stalin in the Kremlin is now in the Museum of Comparative Zoology at Harvard.) The vials we favor are small and slender, 55 millimeters long and 8 millimeters wide, dimensions that allow many to be kept in a small storage space or carried in a pocket or field pack. They are closed with neoprene stoppers, which permits preservation of "wet" material for many years. A few wider bottles, 55 millimeters long and 24 millimeters wide, are carried to accommodate the largest ants or exceptionally large series.

Workers should be collected whenever possible. They can be mixed, as to both colonies and species, if the ants are found foraging singly (this fact should be noted on the label). If the colony is discovered, however, a sample of at least 20 workers should be put together in a vial, along with up to 20 queens, 20 males, and 20 larvae if these can be captured. In emergencies, when the supply of vials is running low, you can place several nest series (that is, members of several colonies) in the same vial and separate them from one another by tight plugs of cotton. Up to four nest series can thus be accommodated in a typical, 55 x 8 millimeter vial. In small but clear letters, write a label of the following kind with a sharp pencil or indelible ink:

FLORIDA: Andytown, Broward Co.
VII–16–87. E.O.Wilson. Scrub hammock,
nesting in rotting palm trunk.

To pick up the ants, use stiff, narrow forceps with pointed (but not needle-sharp) tips. A pair of very sharp watchmaker's forceps, for example Dumont No. 5, can be carried for use with exceptionally small ants. A rapid, efficient method is to moisten the tip of the forceps with alcohol from the vial and touch it to the ant; this procedure fastens the specimen to the forceps long enough to transfer it to the liquid in the vial. Fine, flexible forceps can also be carried for the collection of live specimens, if these are needed for behavioral observation.

To conduct a general survey of a particular locality, continue collecting until no new species have been encountered for a period of several days. Work primarily during the day, but search through the same area at night with a flashlight or headlamp to pick up exclusively nocturnal foragers. A good collector can obtain a virtually complete list of the fauna in a site of 1 hectare within one to three days. Habitats with dense, complex vegetation, however, such as those in tropical rain forests, are likely to take much longer and require special techniques such as arboreal fogging with insecticides.

For ordinary arboreal collecting, rake branches and leaves back and forth with a strong sweep net. Then break open hollow dead twigs on bushes and trees. This will reveal colonies of species, especially those with nocturnal habits, not readily discovered in any other way. Often it is possible to make rapid, clean collections by snapping the inhabited twigs into short segments (3–6 mm long) and blowing the contents into the vial. An aspirator can also be used to suck up ants rapidly, particularly when the nest has just been broken open and the inhabitants are scattering. Care should be used in this technique, because many ants (especially dolichoderines and formicines) produce large quantities of formic acid, terpenoids, and other volatile toxic substances. The unwary collector is in danger of contracting formicosis, a painful but not fatal irritation of the throat, bronchial passage, and lungs.

For terrestrial species, collect workers foraging on the ground both daytimes and nighttimes. It is necessary to look closely for certain species that are small and slow-moving and hence difficult to see. A favorite technique of ours in sampling forest faunas is to lie prone, clear the loose leaves from a square meter of ground to expose the soil and humus, then simply watch for up to a half-hour for the most inconspicuous ants.

In open terrain look for crater nests and other excavations, and with a gardener's trowel dig down into them in search of colonies. Turn rocks and pieces of rotting wood on the ground to seek the species specialized for nesting in such protected sites. Tear open rotting logs and stumps, looking with special care beneath the bark for the small, inconspicuous species that abound in this microhabitat. Spread a ground cloth (a sheet of white cloth or plastic, 1 or 2 m on the side), and scatter leaf litter, humus, and topsoil over it. Break up rotting twigs and small tree branches buried within the litter. Where the humus and litter are relatively thick and moist, they are often the locus for a large part of the ant fauna and contain many inconspicuous and still poorly studied species.

The following technique has proved effective for collecting whole colonies that nest in small rotting logs and branches lying on the ground. Pick up a fragment of the decaying wood (say about 50 cm long), hold it above a photographer's developing pan or similar shallow-walled container, and strike the fragment with a trowel several times to shake out portions of the colony. Small pieces of the wood will also fall into the pan, but it is still much easier to locate and collect the ants, including entire colonies, than by ordinary excavation.

Slower but more thorough collecting of terrestrial ants can be accomplished with the aid of Berlese-Tullgren funnels, named after the Italian entomologist, A. Berlese, who invented them and the Swede, A. Tullgren, who modified and improved them. In simplest form the apparatus consists of a funnel topped by a wire-mesh screen onto which soil and litter are placed. As the material dries out, possibly aided by a light bulb or some other heat source placed above, the ants and other arthropods fall or slide down the smooth funnel sides into a collecting bottle partly filled with alcohol and suspended tightly under the lower spout of the funnel. Excellent descriptions of the Berlese-Tullgren methods and other extraction techniques are provided by Southwood (1966) and Mühlenberg (1976).

## PREPARATION FOR MUSEUM WORK

Ants can be stored indefinitely in alcohol, but it is best to prepare part of the nest series as pinned, dried specimens for convenient museum work. This step is especially important if the ants are to be given to a taxonomist for identification. It is also the best way to store them in museums as voucher specimens, to serve as references for field or laboratory research (all such studies should be taxonomically verifiable with voucher material). The standard method for preparing dried specimens is to glue each ant on the tip of a thin triangle of white, stiff paper. The tip should approach the right side of the ant and touch her ventral body surface beneath the coxae of the middle and hind legs. The droplet of glue should be small enough and placed so as not to obscure any other part of the body except a portion of the coxae and ventral alitrunk surface—which have relatively few features of taxonomic importance. Prior to this "pointing" procedure, an insect pin should be inserted through the broad ends of two or three of the paper triangles, so that two or three ants from the same colony can be mounted one to a triangle on each pin. A rectangular label with the locality data goes beneath the mounted ants, so that when you read the label, the triangles point to the left and the ants point away from you. An effort should be made to get a maximum diversity of castes on each pin: for example, queen, worker, and male, or major worker, media worker, and minor worker. In the case of large ants, it may be possible to mount only one or two ants to a pin; and in the case of *very* large ants, it is sometimes best simply to pass the insect pin directly through the center of the thorax.

## CULTURING ANTS

The culture and study of ants in the laboratory is a relatively simple operation. For many years we have used an economical arrangement that serves for both mass culturing and behavioral observation of a majority of species. The newly collected queenright colony is brought into the laboratory (preferably with some of the original nest material) and placed in plastic tubs chosen for size according to the size of the ants and the number of workers in the colony. For example, fire ant colonies (*Solenopsis* spp.) with populations of up to 20,000 are readily maintained in tubs about 50 centimeters long, 25 centimeters wide, and 15 centimeters deep. In order to prevent the ants from escaping, we use various means depending on the humidity of the room in which the ants are to be kept. The sides of the tub are coated with petroleum jelly, heavy mineral oil, talcum power, or, preferably, Fluon® (Northern Products Inc., Woonsocket, Rhode Island), a water-based material that is both effective (providing a silky smooth surface) and long-lasting (but unsatisfactory under humid conditions). The colony is allowed to settle into test tubes (15 cm long with inner diameters of 2.2 cm) into which water has been poured and then trapped at the bottom with tight cotton plugs, leaving about 10 centimeters of free air space from the plug to the mouth of the tube. The 10-centimeter segment is surrounded by aluminum foil to darken the air space and encourage the ants to move in (most do so promptly). It can be removed later to allow behavioral studies; most ant species adapt well to light at ordinary room intensities, carrying on brood care, food exchange, and other social activities in an apparently normal manner. The tubes are stacked at one end of the tub prior to placement of the colony, leaving most of the bottom surface of the tub bare to serve as a foraging arena.

The nest tubes can also be placed in closed plastic boxes, making it easier to keep the ambient air of the foraging arena moist and hence better suited for forest-dwelling species. The following dimensions are roughly correct for ant species with different-sized workers:

*Small.* 11 x 8.5 centimeters on the side and 6.2 centimeters deep. Very small ants such as *Adelomrymex, Cardiocondyla, Leptothorax,* small *Pheidole,* and *Strumigenys.* These species can also be cultured readily in small, round petri dishes (10 cm diameter, 1.5 cm depth).

*Medium.* 17 x 12 centimeters on the side and 6.2 centimeters deep. For example, *Aphaenogaster, Conomyrmex,* and *Formica.* Smaller colonies of *Camponotus, Messor,* and *Pogonomyrmex.*

*Large.* 45 x 22 centimeters on the side and 10 centimeters deep. For example, larger colonies of *Pheidole, Pogonomyrmex,* and *Solenopsis.*

Variations on the elementary test-tube arrangement can be adapted to ant species with unusual nesting habits. Colonies of arboreal stem-dwelling ants such as *Pseudomyrmex* and *Zacryptocerus* can be induced to move into glass tubes 10 centimeters long with diameters of 2–4 millimeters, the latter dimensions varied according to the size of the workers. The tubes are closed at one end with cotton plugs. The plugs can be kept moist, but in many cases this is not necessary because stem-dwelling ants are often adapted to dry nest interiors, and a small dish of water placed nearby is an adequate source. Each set of tubes containing a colony is then placed in a tub of the kind just described. Or the tubes can be bound horizontally in rows on a rack or potted plant, to simulate the natural environment.

Colonies of small fungus-growing ants such as *Apterostigma* and *Trachymyrmex* can be maintained easily in moistened tubes in tubs. Large fungus-growers, in other words leafcutter ants of the genera *Acromyrmex* and *Atta,* are better kept via a technique developed by Weber (1972). Newly inseminated queens or incipient colonies are collected in the field and transferred to a series of closed, clear plastic chambers each about 20 x 15 centimeters and 10 centimeters deep (ordinary refrigerator food receptacles with transparent walls serve very well). The chambers are connected by glass or plastic tubes 2.5 centimeters in diameter, allowing the ants to move readily from one chamber to the other. Foraging workers are permitted to collect fresh vegetation (possibly supplemented by dry cereal) either from empty chambers or from an open tub lined with Fluon or surrounded by a moat containing water or mineral oil. As the colony expands in size, the ants fill one chamber after another with the characteristic sponge-like masses of processed substratum, through which the symbiotic fungus grows luxuriantly. Except in the driest laboratory environments, no special water supply is needed, because the ants obtain all of the moisture required from the vegetation. Leaves from a wide range of plant species are accepted by the ants. In the northeastern United States we have most frequently used basswood (linden), oak, maple, and lilac; the last two are especially attractive to the foragers. The colonies deposit the exhausted substrate in some of the chambers, which can be removed and cleaned from time to time.

For close behavioral studies, more elaborate artificial nests are often required. One that works well for the great majority of ant species can be prepared as follows. Pour plaster of Paris to a depth of 2 centimeters in the bottom of a tub whose size suits the worker size and population of the colony under study (thus for minute ants such as thief ants or *Brachymyrmex* the container may be only 10 x 15 cm and 10 cm deep). As the plaster of Paris sets, carve into its surface 10 to 20 chambers that are roughly similar in size and proportion to the natural nest chambers of the colony to be cultured. In the case of some medium-sized *Pheidole* living in pieces of rotting wood, the chambers are typically ovoid or circular in shape and 1–4 centimeters across; hence chambers should be excavated that are about 2 x 3 centimeters and 1 centimeter deep. The artificial nest chambers are connected by galleries 5 millimeters wide and deep and covered tightly by a rectangular glass plate. Two to four exit galleries are cut from the outermost chambers to the remainder of the plaster of Paris surface, which serves as the foraging arena. Fragments of decaying wood and leaves from the vicinity of the original nest can be scattered over the surface to add to the "naturalness" of the microenvironment.

Additional culturing techniques adaptable to almost every known ant species are to be found in the works by Wheeler (1910a), Skaife (1961), and Gösswald (1985). Freeland (1958) invented an excellent vertical observation nest for *Myrmecia* and other very large ants. Wilson (1962b) designed a plastic nest that requires minimal maintenance and serves for the simultaneous observation of large colonies, for example those of fire ants, inside and outside the nest during foraging activity.

A completely defined synthetic diet for ants has been invented by Ettershank (1967). Diets and several mass-culturing techniques for various ant species have been reviewed by Carney (1970). We employ the Bhatkar diet (Bhatkar and Whitcomb, 1970), which is prepared as follows:

1 egg
62 ml honey
1 gm vitamins
1 gm minerals and salts
5 gm agar
500 ml water
Dissolve the agar in 250 ml boiling water. Let cool. With egg beater mix 250 ml water, honey, vitamins, minerals and egg until smooth. Add to this mixture, stirring constantly, the agar solution. Pour into petri dishes to set (0.5–1 cm deep). Store in refrigerator. The recipe fills four 15-cm diameter petri dishes, and is jelly-like in consistency.

Most insectivorous ant species thrive on this diet when fed three times weekly along with fragments of freshly killed insects, such as mealworms (*Tenebrio*), cockroaches (*Nauphoeta*), and crickets offered in small quantities. If the ants are also predators, they do especially well when allowed access to bottles containing *Drosophila* cultures, preferably flightless mutants. Alternatively, the *Drosophila* adults can be frozen and sprinkled onto the foraging arenas for the ants to discover.

Very often the food habits of ants newly brought to the laboratory are unknown. In the case of predatory species we have successfully employed the "cafeteria" method (e.g., Wilson, 1953a, 1955b; Hölldobler and Wilson, 1986a). Large numbers of arthropods and other potential prey are collected from the original nest vicinity by aspiration or Berlese funneling into moistened traps. These are released onto the foraging area of the captive colony amid some soil and leaf litter, where they can find shelter. A record is then kept of the proportions of various species captured by the ants and brought into the brood chamber. The method has proved very successful in detecting prey specialization. For example, *Belonopelta* and *Prionopelta* were discovered to be specialists on campodeid diplurans, and many species of *Smithistruma* and other small dacetines were found to be specialists on entomobryomorph collembolans.

## TRANSPORTING COLONIES

Colonies can be kept for days or weeks at a time in bottles or other tight containers, provided that certain elementary procedures are followed. The first, absolute rule is that the ants must be given a moist area into which to retreat—not soaking wet, with films or drops of water that might entrap the ants, but containing a zone with moist surface and saturated ambient air. The ideal retreat is part of the nest material itself, placed directly in the container preferably with a portion of the colony in it. A large piece of moistened (but not soaking) cotton wool or paper toweling should be added as backup. The rest of the container can be filled with nest material or loose-fitting paper toweling or other neutral material to prevent the colony from being knocked around excessively.

The colony should be uncrowded, in no case occupying more than 1 percent of the container volume. The lid of the container should be tightly fitted. Unless the colony is unusually active or aggressive, it is not necessary to punch holes in the lid to aerate the interior; in fact, this procedure risks excessive drying. Once or twice a day the lid can be removed and the container waved gently back and forth to freshen the air. The colony can be provided drops of sugar water and fragments of insects or other foods if the duration of the journey is more than several days. If ants appear dead after remaining in a closed container too long, they may be just narcotized by carbon dioxide. Expose them to the open air for a few hours to see if they can recover.

Because many countries have restrictions on the importation of live insects, it is prudent to check with the appropriate government agencies before collecting live colonies abroad. In the United States, for example, the U.S. Department of Agriculture (Animal and Plant Health Inspection Service, Plant Protection and Quarantine, Plant Importation and Technical Support) must furnish a permit, which has first been approved by the appropriate state officials. The entire procedure usually requires six to eight weeks. The permit is then presented to the appropriate customs officer upon reentry into the United States.

An increasing number of countries restrict the export of preserved and living specimens, including insects, and a special export permit may be required. Local regulations should always be consulted and respected.

## BREEDING NEW COLONIES

Reproductive forms can be easily reared in the laboratory, but those of most species cannot be induced to mate under the culturing conditions ordinarily employed. The reason is that the virgin queens and males must engage in extensive nuptial flights under an exacting regimen of temperature and humidity before they will copulate. The rule is not absolute, however. A few polygynous species mate in or close to the nest, so that laboratory colonies can be maintained and multiplied indefinitely. Examples include Pharaoh's ant (*Monomorium pharaonis;* see Peacock and Baxter, 1950, and Berndt and Eichler, 1987), a species of *Solenopsis (Diplorhoptrum)* from Ecuador we have had in culture for more than 12 years, *Xenomyrmex floridanus* (Hölldobler, 1971d), several species of *Cardiocondyla* (R. J. Stuart, personal communication), the slave-making myrmicine *Harpagoxenus sublaevis* and many other parasitic ants (Buschinger, 1972b, 1976a), the Argentine ant *Iridomyrmex humilis* (Smith, 1936a), and *Paratrechina longicornis.* Within these species it is possible to make careful studies of reproductive behavior, including bioassays of sex pheromones. In fact Karl Gösswald and his co-workers exploited this form of mating behavior to mass-produce queens in *Formica polyctena.* Colonies were then started in forests, where the workers protect the trees against pest insects (see Plate 4).

Buschinger (1975a) has gone so far as to conduct a genetic analysis by breeding experiments in the case of the ergatogynic locus of *Harpagoxenus sublaevis,* but in general such studies are handicapped by the relatively long time (sometimes several years) it takes to rear a mature colony from a newly inseminated queen. It is likely that future genetic studies will depend strongly on electrophoretic separation of enzymes and amino acid and nucleotide sequencing.

The nuptial flight could conceivably be short-circuited and fertile queens produced if queens were readily inseminated, a state that has been achieved by Cupp et al. (1973) in the case of the fire ant *Solenopsis invicta.* These authors kept reproductive forms in a warm (32° C), humid environment with an 18-hour photoperiod. They decapitated the male (an action that disinhibits copulatory movements), pinned it through the thorax, and set it in a parafilm mount with the abdomen pointing upward. The female was anesthetized with carbon dioxide. Her wings were grasped with jeweler's forceps and her abdomen stroked against the body in a front-to-rear motion until copulation occurred. Genetic analysis by breeding experiments could be greatly accelerated if artificial insemination were combined with precocious production of sexual forms through the treatment of young colonies with appropriate levels of disinhibitors or juvenile hormone. This technology, however, remains to be developed.

# Glossary

Included in this listing are some terms used generally in entomology, as well as most terms restricted to the study of social insects. The glossary was designed to make it possible to read this book with no more than an elementary background in biology and, specifically, to eliminate the need to refer to entomology textbooks. For illustrations of anatomy, and explanation of more technical terms used in description and taxonomy, see Chapters 2 and 7.

**Abdomen** The third, posteriormost major division of the body.

**Aciculate** Finely striate, as if scratched by a needle.

**Acidopore** In formicine ants, the circular exit of the poison gland formed by the margin of the terminal gastral sternite.

**Active space** The space within which the concentration of a pheromone (or any other behaviorally active substance) is at or above threshold concentration. The active space of a pheromone is, in fact, the signal itself.

**Aculeate** Pertaining to the Aculeata, or stinging Hymenoptera, a group including the bees, ants, and many of the wasps.

**Acuminate** Tapering to a fine point.

**Adaptive demography** Programmed schedules of individual birth, growth, and death resulting in frequency distributions of age and size in the worker caste that promote survival and reproduction of the colony as a whole.

**Adaptive radiation** The process of evolution by which species multiply, diverge into different niches (for example, species that are predators on different kinds of prey), and come to occupy the same or at least overlapping ranges.

**Adoption substance** A secretion presented by a social parasite that induces the host insects to accept the parasite as a member of their colony.

**Adult transport** The carrying or dragging of one adult social insect by a nestmate, usually during colony emigrations. In ants, adult transport is a very frequent and stereotyped form of behavior.

**Age polyethism** The regular changing of labor roles by colony members as they age.

**Aggregation** A group of individuals, comprising more than just a mated pair or a family, that has gathered in the same place but does not construct nests or rear offspring in a cooperative manner (as opposed to *Colony*).

**Aggression** A physical act or threat of action by one individual or colony that reduces the freedom or genetic fitness of another.

**Alarm-defense system** Defensive behavior which also functions as an alarm signaling device within the colony. Examples include the use by certain ant species of chemical defensive secretions that double as alarm pheromones.

**Alarm pheromone** A chemical substance exchanged among members of the same species that induces a state of alertness or alarm in the face of a common threat.

**Alarm-recruitment system** A communication system that rallies nestmates to some particular place to aid in the defense of the colony. An example is the odor trail system of lower termites, which is used to recruit colony members to the vicinity of intruders and breaks in the nest wall.

**Alate** Winged.

**Alitrunk** The mesosoma of the higher Hymenoptera, including the true thorax and (fused anteriorly to the thorax) the first abdominal segment.

**Allelic recognition** See *Recognition alleles.*

**Alloethism** The regular and disproportionate change in a particular category of behavior as a function of worker size.

**Allogrooming** Grooming directed at another individual, as opposed to self-grooming.

**Allometric growth** See *Allometry.*

**Allometric space** The set of joint measurements in two or more bodily dimensions. Each point in the space represents a particular anatomical type; certain points represent the anatomical types best able to meet particular contingencies in the environment.

**Allometry** Any size relation between two body parts that can be expressed by $y = bx^a$, where $a$ and $b$ are fitted constants. In the special case of isometry, $a = 1$ and the relative proportions of the body parts therefore remain constant with change in total body size. In all other cases ($a \neq 1$), the relative proportions change as total body size is varied (see Chapter 8).

**Allomone** A chemical substance or blend of substances used in communication among individuals belonging to different species. It evokes a response that is adaptively favorable to the emitter but not to the receiver (for example, a lure used by a predator in attracting its prey).

**Altruism** Self-destructive behavior performed for the benefit of others.

**Antbirds** Tropical birds of the family Formicariidae, which follow raiding swarms of army ants and feed on the prey stirred up by the ants.

**Antbutterflies** Butterflies of the family Nymphalidae, subfamily Ithomiinae, that follow the raiding swarms of army ants and feed on the droppings of antbirds.

**Antennal condyle** The narrowed, neck-like portion of the first antennal segment that connects to the head surface.

**Antennal fossa** The cavity or depression of the head into which the antenna is articulated.

**Antennation** Touching with the antennae. The movement can serve as a sensory probe or as a tactile signal to another insect.

**Ant garden** A cluster of epiphytic plants inhabited by ant colonies, which benefit from the association.

**Ant plant** Also known as a myrmecophyte; a species of plant with domatia, or specialized structures for housing ant colonies.

**Apodeme** An ingrowth or other rigid process of the exoskeleton, typically serving for muscle attachment.

**Appeasement substance** A secretion presented by a social parasite that reduces aggression in the host insects and aids the parasite's acceptance by the host colony.

**Appressed** Referring to a hair that runs parallel, or nearly parallel, to the body surface.

**Apterous** Wingless.

**Aril** See *Elaiosome.*

**Army ant** A member of an ant species that shows both nomadic and group-predatory behavior. In other words the nest site is changed at relatively frequent intervals, in some cases daily, and the workers forage in groups (also called legionary ant).

**Arolium** A cushion-like pad located between the tarsal claws and constituting part of the pretarsus.

**Arrhenotoky** The production of males from unfertilized eggs.

**Assembly** The calling together of colony members for any communal activity.

**Atrium** A chamber at the entrance of a body opening.

**Basad** Located at or near the base.

**Basitarsus** The proximal or basal segment of the tarsus.

**Bead gland** See *Pearl body.*

**Beccarian bodies** The pearl bodies produced by the stipules or young leaves of *Macaranga* and consumed by resident ants.

**Behavioral repertory** A list of the behavioral acts performed by an individual, caste, or species, sometimes with a specification of relative frequencies of the acts.

**Beltian bodies** The food bodies found on the tips of pinnules or rachises of *Acacia* and consumed by the resident *Pseudomyrmex* ants.

**Bidentate** Bearing two teeth, as on the anterior clypeal border.

**Bivouac** The mass of army ant workers within which the queen and brood find refuge.

**Blastogenesis** The origin of different caste traits from variation in either the ovarian environment of the egg or the nongenetic contents of the egg (opposed to genetic control of caste and *Trophogenesis).*

**Brachypterous** With proportionally reduced wings, incapable of full flight.

**Brood** The immature members of a colony collectively, including eggs, nymphs, larvae, and pupae. In the strict sense eggs and pupae are not members of the society, but they are nevertheless referred to as part of the brood.

**Budding** Colony multiplication by the departure of a relatively small force of workers from the main nest, accompanied by one or more queens. See also *Fission* and *Swarming.*

**Bulla** A blister-like structure, as for example the thin, convex roof of the metapleural gland cavity.

**Callow workers** Newly eclosed adult workers whose exoskeleton is still relatively soft and lightly pigmented.

**Carina** Elevated ridge on the body surface.

**Carinate** Possessing carinae, especially in parallel rows.

**Caste** Broadly defined, as in ergonomic theory (Chapter 8), any set of individuals of a particular morphological type or age group, or both, that performs specialized labor in the colony. More narrowly defined, any set of individuals in a given colony that is both morphologically distinct and specialized in behavior.

**Chain transport** The relaying of food from one worker to another in the course of transporting it back to the nest.

**Character** In taxonomy and a few other fields of biology, any kind of trait used for identification. A particular trait (such as two spines versus four) possessed by one individual or species as opposed to another is called a character state.

**Cladogram** A diagram showing nothing more than the sequence in which groups of organisms are interpreted to have originated and diverged in the course of evolution.

**Claustral colony founding** The procedure during which queens (or royal pairs in the case of termites) seal themselves off in cells and rear the first generation of workers on nutrients obtained mostly or entirely from their own storage tissues, including fat bodies and histolysed wing muscles.

**Clavate** Thickened, especially toward the tip.

**Cleptobiosis** The relation in which one species robs the food stores or scavenges in the refuse piles of another species, but does not nest in close association with it.

**Coefficient of relationship** Also called coefficient of relatedness or degree of relatedness. The probability that a gene possessed by one individual is also possessed by another individual through common descent in the previous few generations.

**Colony** A group of individuals, other than a single mated pair, which constructs nests or rears offspring in a cooperative manner (as opposed to *Aggregation*).

**Colony fission** The multiplication of colonies by the departure of one or more reproductive forms, accompanied by groups of workers, from the parental nest, leaving behind comparable units to perpetuate the "parental" colony. This mode is referred to occasionally as hesmosis in ant literature and sociotomy in termite literature. Swarming in honey bees can be regarded as a special form of colony fission.

**Colony odor** The odor found on the bodies of social insects which is peculiar to a given colony. By smelling the colony odor of another member of the same species, an insect is able to determine whether it is a nestmate (see *Nest odor* and *Species odor*).

**Column raid** A raid conducted by army ants in branching columns, the termini of which are headed by a relatively small group of workers laying chemical trails and capturing prey.

**Commensalism** Symbiosis in which members of one species are benefited while those of the other species are neither benefited nor harmed.

**Communication** Action on the part of one organism (or cell) that alters in an adaptive fashion the probability pattern of behavior in another organism (or cell).

**Compound nest** A nest containing colonies of two or more species of social insects, up to the point where the galleries of the nests anastomose and the adults sometimes intermingle but the broods of the species are still kept separate (see *Mixed nest*).

**Condyle** A structure that articulates an appendage to the body surface.

**Cordate** Heart-shaped, as the outline of the head.

**Corpora allata** Paired endocrine organs located just behind the brain; the source of juvenile hormone.

**Corrugated** Referring to a body surface that is wrinkled in appearance.

**Costa** An elevated ridge rounded at the crest.

**Costate** Bearing costae.

**Coxa** The basalmost segment of the leg.

**Crop** See *Social stomach*.

**Dealate** Referring to an individual who has shed her wings, as an inseminated female; also the condition of having shed the wings.

**Dealation** The removal of the wings by the queens (and also by males in the termites) during or immediately following the nuptial flight and prior to colony foundation.

**Declivity** A downward-sloping surface, as the posterior face of the propodeum.

**Decumbent** Referring to a hair that stands 10 to 40 degrees from the surface.

**Degree of relatedness** See *Coefficient of relationship*.

**Demography** The rate of growth and the age structure of populations, and the processes that determine these properties.

**Density dependence** The increase or decrease of the influence of a physiological or environmental factor on population growth as the density of the population increases.

**Dentate** Toothed, as the dentate inner borders of the mandibles.

**Denticulate** Furnished with minute teeth or tooth-like structures.

**Developmental cycle** The period from the birth of the eggs to the eclosion of the adult insect.

**Dichthadiiform ergatogyne (queen)** A member of an aberrant reproductive caste, limited to army ants, which is characterized by the possession of a wingless alitrunk, a huge gaster, and an expanded postpetiole.

**Dimorphism** In caste systems the existence in the same colony of two different forms, including two size classes, not connected by intermediates.

**Diphasic allometry** Polymorphism in which the allometric regression line, when plotted on a double logarithmic scale, "breaks" and consists of two segments of different slopes whose ends meet at an intermediate point.

**Disc** See *Imaginal disc*.

**Discriminators** Also called recognition labels. The genetically determined cues that permit individuals to be classified as kin or non-kin.

**Distad** Located toward the distal or farthest end.

**Distal** The farthest part away from the body, as the tip of the antenna.

**Division of labor** See *Polyethism*.

**Domatia** Also called myrmecodomatia; specialized structures, such as inflated stems, used by ant plants for the housing of ant colonies.

**Dominance hierarchy** Also known as dominance order. The physical domination of some members of a group by other members, in relatively orderly and long-lasting patterns. Except for the highest- and lowest-ranking individuals, a given member will dominate one or more of its companions and be dominated in turn by one or more of the others. The hierarchy is initiated and sustained by hostile behavior, albeit sometimes of a subtle and indirect nature.

**Dorsal** Pertaining to the dorsum or upper surface, versus ventral.

**Dorsum** The upper surface.

**Dorylophile** An obligatory guest of one of the army ants belonging to the tribe Dorylini.

**Driver ants** African legionary ants belonging to the genus *Dorylus*.

**Dufour's gland** An exocrine gland that empties at the base of the sting, also known as the accessory gland to the poison gland.

**Dulosis** The relation in which workers of a parasitic (dulotic) ant species raid the nests of another species, capture brood (usually in the form of pupae), and rear them as enslaved nestmates.

**Ecitophile** An obligatory guest of one of the army ants belonging to the tribe Ecitonini, especially of the genus *Eciton* itself.

**Eclosion** Emergence of the adult (imago) from the pupa; less commonly, the hatching of an egg.

**Ectosymbiont** A symbiont that associates with the host colonies during at least part of its life cycle in some relationship other than internal parasitism.

**Egg-microlarva pile** A clump of eggs and first-instar larvae (older larvae are characteristically separated into piles of their own).

**Elaiosome** Also called aril; specialized organ on seed that attracts ants.

**Elite** Referring to a colony member that displays greater than average initiative and activity.

**Emarginate** Notched, with a piece of any shape seemingly cut from the margin.

**Emery's rule** The rule that species of social parasites are very similar to their host species and therefore presumably closely related to them phylogenetically.

**Emigration** The movement of a colony from one nest site to another.

**Enemy specification** The exaggerated alarm response to species of ants and other arthropods that pose an unusually severe threat to the colony.

**Entire** In anatomy, referring to a smoothly unbroken margin, as opposed to *Emarginate*.

**Epigaeic** Living, or at least foraging, primarily above ground (as opposed to *Hypogaeic*).

**Epinotum** See *Propodeum*.

**Equals (=)** In taxonomy an = sign means that two species or other entities are the same, even though they were originally considered different. The name in parentheses following the = sign is the junior synonym, in other words, the more recently introduced; the first name, the senior synonym, is the one to use.

**Erect** Referring to a hair that stands straight up, or nearly so, from the body surface.

**Ergatogyne** Any form morphologically intermediate between the worker and the queen.

**Ergatoid male** See *Ergatomorphic male.*

**Ergatoid reproductive** A supplementary reproductive termite without a trace of wing buds, usually larval in external form, and with a distinctively rounded head (same as third-form reproductive, tertiary reproductive, and apterous neoteinic).

**Ergatomorphic male** An individual with normal male genitalia and a worker-like body.

**Ergonomics** The quantitative study of work, performance, and efficiency.

**Ergonomic stage** The stage of relatively rapid colony growth in which only workers are produced. It is preceded by the founding stage and followed by the reproductive stage (in which some males or virgin queens are produced).

**Ethocline** A series of different behaviors observed among related species and interpreted to represent stages in a single evolutionary trend.

**Ethogram** A complete description of the behavioral repertory of an individual, caste, or species with a specification of frequency of acts and the transition probabilities between them. See also *Behavioral repertory* and *Sociogram.*

**Eusocial** Applied to the condition, or to the group possessing it, in which individuals display all of the following three traits: cooperation in caring for the young; reproductive division of labor, with more or less sterile individuals working on behalf of individuals engaged in reproduction; and overlap of at least two generations of life stages capable of contributing to colony labor. This is the formal equivalent of the expressions "truly social" or "higher social," which are commonly used with less exact meaning.

**Excised** With a deep cut or notch, as on the margin of a segment.

**Exploratory trail** An odor trail laid more or less continuously by the advance workers of a foraging group. This kind of communication is used regularly by army ants (as opposed to *Recruitment trail*).

**Exudatoria** Finger-like appendages found on the larvae of certain ant species and on a variety of termitophiles, thought to be attractive to the adults.

**Facet** The ommatidium, one of the basic units of the compound eye.

**Facilitation** The same as social facilitation; see under *Group effect.*

**Falcate** Sickle-shaped or saber-shaped.

**Family** A higher taxonomic category, made up of a set of related genera.

**Female calling** The release of sex attractants by a reproductive female who stands in one place and "calls" males to her.

**Femur** The "thigh," or third segment of the leg away from the body.

**Fission** The multiplication of colonies by splitting of the worker force into two or more groups each of which depart with their own fertile queens. See also *Budding* and *Swarming.*

**Flood evacuation response** The mobilization and evacuation of the colony by alarm and recruitment when water encroaches on the nest.

**Formicarium** Same as *Formicary.*

**Formicary** A nest of ants. The term is also commonly applied to an ant mound or an artificial nest used in the laboratory to house ants.

**Fossa** A relatively large and deep pit on the body surface.

**Founding stage** The earliest period of the colony life cycle, in which the newly fecundated queen raises the first brood of workers. Succeeded by the *Ergonomic stage.*

**Fovea** A pit in the body surface.

**Foveate** Referring to a body surface bearing foveae.

**Frons** The area above the clypeus, approximately in the center of the front of the head; it often includes the frontal triangle, which is roughly triangular in form and demarcated by grooves.

**Frontal triangle** See *Frons.*

**Functional monogyny** The condition in which several mated queens coexist, but only one queen produces reproductive brood. The dominant queen suppresses the fertility of the other queens by behavioral or chemical dominance signals.

**Funiculus** All of the antenna except the first segment, or scape.

**Gamergate** A mated, egg-laying worker.

**Gaster** A special term occasionally applied to the metasoma, or terminal major body part, of ants.

**Gastral** Pertaining to the gaster; a common variant is gastric.

**Gena** The "cheek" of the head, the area between one of the compound eyes and the nearest antennal insertion.

**Genus** A set of similar species of relatively recent common ancestry.

**Gestalt model** The hypothetical system in kin recognition in which a common odor is created by pooling the recognition pheromones of some or all of the individuals belonging to a group. Others are classified as kin or non-kin according to the degree to which they possess the odor.

**Glabrous** Referring to a body surface that is smooth and shining.

**Gongylidium** A swollen hyphal tip of the symbiotic fungi cultured by attine ants. The ants pick the gongylidia as food. See also *Staphyla.*

**Grooming** Cleaning the bodies of nestmates (*Allogrooming*) or the individual's own body by licking and, in the case of self-grooming, wiping with the legs.

**Group effect** An alteration in behavior or physiology within a species brought about by signals that are directed in neither space nor time. A simple example is social facilitation, in which an activity increases merely from the sight or sound (or other form of stimulation) coming from other individuals engaged in the same activity.

**Group predation** The hunting and retrieving of living prey by groups of cooperating animals. A behavior pattern best developed in army ants.

**Group transport** The coordinated transport of a food item by two or more workers.

**Guest** A social symbiont.

**Gula** The central part of the lower surface of the head.

**Gynandromorph** An individual containing patches of both male and female tissue.

**Gyne** A female that is a member of the reproductive caste, whether functioning as a reproductive at the moment or not; hence a queen in the broad sense.

**Gynecoid** Queen-like; specifically a worker with some typically queen features such as an enlarged abdomen.

**Gynergate** A female containing patches of tissue of both the queen and worker castes.

**Haplodiploidy** The mode of sex determination in which males are derived from haploid eggs and females from diploid eggs.

**Haplometrosis** The founding of a colony by a single queen (as opposed to *Pleometrosis*).

**Harvesting ants** Ant species that feed substantially on seeds and store them in their nests. Many taxonomic groups have developed this habit independently in evolution.

**Hemimetabolous** Undergoing development which is gradual and lacks a sharp separation into larval, pupal, and adult stages. Termites, for example, are hemimetabolous (as opposed to *Holometabolous*).

**Hesmosis** Same as colony swarming, that is, the reproduction of colonies by the separation of colony fragments accompanied by fertile queens.

**Heterogony** See *Allometry.*

**Histolysis** The metabolic dissolution and absorption of a tissue, such as the wing muscles of the young queen during colony founding.

**Holometabolous** Undergoing a complete metamorphosis during development, with distinct larval, pupal, and adult stages. The Hymenoptera, for example, are holometabolous (as opposed to *Hemimetabolous*).

**Homeostasis** The maintenance of a steady state, especially a physiological or social steady state, by means of self-regulation through internal feedback responses.

**Homopteran** A member of, or pertaining to, the insect order Homoptera, which includes the aphids, jumping plant lice, treehoppers, spittlebugs, whiteflies, and related groups.

**Honeydew** A sugar-rich fluid derived from the phloem sap of plants and passed as excrement through the guts of sap-feeding aphids and other insects. Honeydew is the principal food of many kinds of ants.

**Honey pot (honeypot)** Referring to members of the specialized physiological caste of *Myrmecocystus* that store large quantities of sugary liquid in their hugely expanded crops.

**Humerus** The "shoulder," an anterior corner of the pronotum, or the anteriormost segment of the thorax.

**Hymenopteran** Pertaining to the insect order Hymenoptera; also, a member of the order, such as a wasp, bee, or ant.

**Hymenopterous** Pertaining to an individual or traits of individuals belonging to the insect order Hymenoptera, which comprises wasps, bees, and ants.

**Hypogaeic** Living primarily underground (subterranean) or at least beneath cover such as leaf litter, stones, and dead bark (cryptobiotic).

**Imaginal disc** A relatively undifferentiated tissue mass occurring in the body of a larva which is destined to develop later into an adult organ.

**Imago** The adult insect. In termites, the term is usually applied only to adult primary reproductives.

**Inclusive fitness** The sum of an individual's own fitness plus all the individual's influence on the fitness of relatives other than direct descendants, hence the total effect of individual selection and kin selection.

**Incrassate** Conspicuously swollen, especially near the tip, as an antenna.

**Individualistic model** The hypothetical system in which individuals judge others to be kin or non-kin according to whether they possess certain alleles that encode a particular recognition label. The alleles may be recognition alleles in the strict sense, controlling both the production of the pheromone and its perception, or they may prescribe a less direct phenotype matching system.

**Infrabuccal pocket** A cavity on the floor of the buccal chamber in which indigestible material accumulates and is compacted for later disposal.

**Ingluvial** Pertaining to the crop, the distensible middle portion of the foregut in which (in many species) liquid food is stored.

**Inquilinism** The relation in which a socially parasitic species spends the entire life cycle in the nests of its host species. Workers are either lacking or, if present, scarce and degenerate in behavior. This condition is sometimes referred to loosely as permanent parasitism.

**Insect society** In the strict sense, a colony of eusocial insects (ants, termites, eusocial wasps, or bees). In the broad sense adopted in this book, any group of presocial or eusocial insects.

**Insect sociobiology** The systematic study of all aspects of the biological basis of social behavior in insects.

**Instar** Any period between molts during the course of development.

**Internidal** Between nests; involving more than one nest whether of the same colony or different colonies.

**Intranidal** Within a single nest.

**Intrinsic rate of increase** Symbolized by $r$, the fraction by which a population is growing in each instant of time.

**Invitation signal** A signal used to induce another member of the same species to join in performing an act, such as following an odor trail or cooperating in adult transport.

**Isometry** The condition in which the sizes of two body parts remain constant relative to each other as total body size increases (isometry is a special case of *Allometry*).

**Kairomone** Any substance or blend of substances emitted by an organism that elicits a response adaptively favorable to the receiver but not the emitter.

**Kinopsis** Alarm communication or recruitment mediated by the sight of movement in nestmates alone.

**Kin recognition** Recognition of and discrimination toward various categories of kin.

**Kin selection** The selection of genes as a result of one or more individuals' favoring or disfavoring the survival and reproduction of relatives who possess the same genes by common descent. Usually excludes offspring.

**Labial palps** The segmented appendages arising from the labium, or lower "lip" of the head.

**Labium** The lower "lip," or lowermost mouthpart-bearing segment of insects, located just below the mandibles and the maxillae.

**Lamella** A thin, plate-like process.

**Lanuginous** Referring to hair that is woolly or down-like.

**Larva** An immature stage which is radically different in form from the adult; characteristic of the holometabolous insects, including the Hymenoptera. In the termites, the term is used in a special sense to designate an immature individual without any external trace of wing buds or soldier characteristics.

**Legionary ant** See *Army ant.*

**Lek** An aggregation of males at a site, as in *Pogonomyrmex* harvester ants, in which mating takes place year after year or at least repeatedly within the same year, and where males compete for access to females.

**Lestobiosis** The relation in which colonies of a small species nest in the walls of the nests of a larger species and enter the chambers of the larger species to prey on brood or to rob the food stores.

**Macraner** In species with two sizes of males, the larger of the two forms (contrast with *Micraner*).

**Macrogyne** In species with two sizes of queens, the larger of the two forms (contrast with *Microgyne*).

**Major worker** A member of the subcaste of largest workers, especially in ants. In ants the subcaste is usually specialized for defense, so that an adult belonging to it is often also referred to as a *Soldier* (q.v.; see also *Media worker* and *Minor worker*).

**Male aggregation syndrome** The mating pattern in which males from different nests gather in a group and queens join them to be inseminated.

**Mass communication** The transfer of information among groups of individuals of a kind that cannot be transmitted from a single individual to another. Examples include the spatial organization of army ant raids, the regulation of numbers of worker ants on odor trails, and certain aspects of the thermoregulation of nests.

**Maxilla** The second pair of jaws, usually kept folded beneath the principal pair of jaws, or mandibles.

**Maxillary palps** The pair of jointed appendages originating from the maxillae.

**Media worker** In polymorphic ant series involving three or more worker subcastes, an individual belonging to the medium-sized subcaste(s). See *Major worker* and *Minor worker*.

**Mesosoma (alitrunk)** The middle of the three major divisions of the insect body. In most insects it is the strict equivalent of the thorax, but in the higher Hymenoptera it includes the propodeum.

**Metapleural gland** A gland peculiar to the ants (Formicidae) found at the posteroventral angle of the metapleuron; it produces antibiotic substances.

**Metasoma** The hindmost of the three principal divisions of the insect body. In most insect groups it is the strict equivalent of the abdomen. In the higher Hymenoptera it is composed only of some of the abdominal segments, since the first segment (the propodeum) is fused with the thorax and has therefore become part of the mesosoma.

**Metrosis** The number of queens that starts a new colony; see *Haplometrosis* (one queen) and *Pleometrosis* (multiple queens).

**Micraner** In species with two sizes of males, the smaller of the two forms (contrast with *Macraner*).

**Microgyne** In species with two sizes of queens, the smaller of the two forms (contrast with *Macrogyne*).

**Minima** In ants a minor worker, and especially the smallest worker of the kind typically seen in founding colonies or at the low end of size variation of strongly and continuously polymorphic species.

**Minor worker** A member of the smallest worker subcaste, especially in ants. Same as *Minima* (see *Major worker* and *Media worker*).

**Mixed nest** A nest containing colonies of two or more species of social insects, in which mixing of both the adults and the brood occurs (see *Compound nest*).

**Modulatory communication** Communication that influences the behavior of receivers, not by forcing them into narrowly defined behavioral channels but by slightly shifting the probabilities of the performances of other behavioral acts.

**Molt (moult)** The casting off of the outgrown skin or exoskeleton in the process of growth. Also the cast-off skin itself. The word is further used as an intransitive verb to designate the performance of the behavior.

**Monogyny** The existence of only one functional queen in the nest (as opposed to *Polygyny*). Primary monogyny: monogyny through the founding of the colony by a single queen. Secondary monogyny: monogyny through the elimination of multiple founding queens until only one is left.

**Monomorphism** The existence within a species or colony of only a single worker subcaste.

**Monophasic allometry** Polymorphism in which the allometric regression line has a single slope; in ants use of the term also implies that the relation of some of the body parts measured is nonisometric (see Chapter 8).

**Mound nest** A nest at least part of which is constructed of a mound of soil or carton material that projects above the ground surface. The architecture of the mound is often elaborate, specific in plan to the species, and evidently adapted to contribute to microclimate control within the nest.

**Müllerian bodies** The food bodies produced on the bases of the petioles of *Cecropia* trees and consumed by the resident *Azteca* ants.

**Multicolonial** Pertaining to a population of social insects which is divided into colonies that recognize nest boundaries (as opposed to *Unicolonial*).

**Mutualism** Symbiosis that benefits the members of both of the participating species.

**Myrmecioid complex** One of two major taxonomic groups of ants sometimes recognized in classification; the name is based on the subfamily Myrmeciinae, one of the constituent taxa. It should not be confused with the subfamily Myrmicinae, which belongs to the poneroid complex (see Chapter 2).

**Myrmecochory** The dispersal of seeds by ants attracted by the elaiosomes.

**Myrmecodomatia** See *Domatia*.

**Myrmecology** The scientific study of ants.

**Myrmecophile** An organism that must spend at least part of its life cycle with ant colonies.

**Myrmecophytes** Higher plants that live in obligatory, mutualistic relationships with ants.

**Myrmecotrophy** The stimulation of growth in ant plants by nutrients carried to the plants by their guest ants.

**Nanitic worker** The dwarf workers produced from either the first ant broods or late ant broods that have been subjected to starvation. Nanitic workers occur in both monomorphic and polymorphic species.

**Nasus** The snout-like organ possessed by soldiers of some species in the Nasutitermitinae. The nasus is used to eject poisonous or sticky fluid at intruders.

**Nasute soldier** A soldier termite possessing a nasus.

**Necrophoresis** Transport of dead members of the colony away from the nest.

**Nest odor** The distinctive odor of a nest, by which its inhabitants can distinguish their own nest from those belonging to other colonies or at least from the surrounding environment. Certain insects, for example honey bees and some ants, can orient toward the nest by means of the odor. It is possible that the nest odor is the same as the *Colony odor* in some cases. The nest odor of honey bees is often referred to as the hive aura or hive odor.

**Nest robbing** Same as *Cleptobiosis*.

**Niche** The combined range of environmental variables, such as temperature, humidity, and food items, within which a species can exist and reproduce.

**Node** A rounded, knob-like structure, such as the petiolar node (the upper rounded portion of the petiole).

**Nomadic phase** The period (as opposed to the *Statary phase*) in the activity cycle of an army ant colony during which the colony forages more actively for food and moves frequently from one bivouac site to another. At this time the queen does not lay eggs, and the bulk of the brood is in the larval stage.

**Nomadism** The relatively frequent movement by an entire colony from one nest site to another.

**Nuptial flight** The mating flight of the winged queens and males.

**Nymph** In general entomology, the young stage of any insect species with hemimetabolous development. In termites, used in a slightly more restricted sense to designate immature individuals who possess external wing buds and enlarged gonads and who are capable of developing into functional reproductives by further molting.

**Occipital lobes** The rear corners of the head.

**Ocellus** One of the three simple eyes of adult insects, located on or near the center line of the dorsal surface of the head. The ocelli should be distinguished from the laterally placed compound eyes.

**Odor trail** A chemical trace laid down by one insect and followed by another. The odorous material is referred to either as the *Trail pheromone* or the *Trail substance*.

**Oligogyny** The occurrence in a single colony of two to several functional queens. Oligogyny is characterized by worker tolerance toward more than one queen, and antagonism among queens, so that multiple queens cannot coexist in the same immediate vicinity and must spread out.

**Ommatidium.** One of the basic visual units, or facets, of the insect compound eye. The ommatidia are bounded externally by the facets that together make up the glassy, rounded outer surface of the eye.

**Oophagy** Egg cannibalism; the eating by a colony member of her own eggs or those laid by a nestmate.

**Ortstreue** The tendency to return repeatedly to the same site during foraging or guard duty.

**Ovariole** One of the egg tubes which, together, form the ovary in female insects.

**Palpation** Touching with the labial or maxillary palps. The movement can serve as a sensory probe or as a tactile signal to another insect.

**Palp formula** The number of segments in the maxillary and labial palps respectively. Thus 6,4 would mean 6 segments in the maxillary palp and 4 in the labial palp.

**Parabiosis** The utilization of the same nest and sometimes even the same odor trails by colonies of different species, which nevertheless keep their brood separate.

**Parasitism** Symbiosis in which members of one species exist at the expense of members of another species, usually without going so far as to cause their deaths.

**Parasitoid** A parasite that slowly kills the victim, this event occurring near the end of the parasite's larval development. The term is also used as an adjective.

**Partially claustral colony founding** The procedure during which the queen founds the colony by isolating herself in a chamber but occasionally leaves to forage for part of her food supply.

**Patrolling** The act of investigating the nest interior and outer nest surface.

**Pearl body** One of a heterogeneous group of food bodies with a pearl-like luster and high concentration of lipids, apparently used by plants to attract and support ants. (Also called bead glands.)

**Pectinate** Comb-like or bearing a comb, as the tarsal spurs.

**Pedicel** The waist of the ant, made up of either one segment (the petiole) or two segments (the petiole plus the postpetiole), or the second segment of the antenna from the base outward.

**Pedunculate** Stalk-like, or set on a stalk or peduncle, as the waist of many ant species.

**Permanent social parasitism** See *Inquilinism*.

**Petiole** The first segment of the waist of aculeate Hymenoptera; it is in fact the second abdominal segment, since the first abdominal segment (propodeum) is fused to the thorax.

**Phenotype matching** The process by which an individual learns cues (such as recognition pheromones) from either itself or its kin and then matches them with cues provided by other individuals in order to classify them as kin or non-kin.

**Pheromone** A chemical substance or a blend of substances, usually a glandular secretion, which is used in communication within a species. One individual releases the material as a signal and another responds after tasting or smelling it. Primer pheromones alter the physiology of individuals and prepare them for new behavioral repertories. Releaser pheromones evoke responses directly.

**Phragmosis** The condition in which the head or tip of the abdomen is truncated and is used as a living plug for the nest entrance. Occurs in ants and termites, usually in the soldier caste.

**Physogastry** The swelling of the abdomen to an unusual degree due to the hypertrophy of fat bodies, ovaries, or both.

**Pilosity** The longer, stouter hairs, or setae, which stand out above the shorter, usually finer hairs that constitute the pubescence.

**Pleometrosis** The founding of a colony by multiple queens.

**Plesiobiosis** The close proximity of two or more nests, accompanied by little or no direct communication between the colonies inhabiting them.

**Plumose** Referring to hairs that are multiply branched and hence feather-like in appearance.

**Polycalic** Same as *Polydomous*.

**Polydomous** Pertaining to single colonies that occupy more than one nest.

**Polyethism** Division of labor among members of a colony. In social insects a distinction can be made between caste polyethism, in which morphological castes are specialized to serve different functions, and age polyethism, in which the same individual passes through different forms of specialization as she grows older.

**Polygyny** The coexistence in the same colony of two or more egg-laying queens (as opposed to *Monogyny*). When multiple queens found a colony together, the condition is referred to as primary polygyny. When supplementary queens are added after colony foundation, the condition is referred to as secondary polygyny. The coexistence of only two or several queens is sometimes called *Oligogyny.*

**Polymorphism** In social insects, the coexistence of two or more functionally different castes within the same sex. In ants it is possible to define polymorphism somewhat more precisely as the occurrence of nonisometric relative growth occurring over a sufficient range of size variation within a normal mature colony to produce individuals of distinctly different proportions at the extremes of the size range.

**Poneroid complex** One of two major taxonomic groups of ants sometimes recognized in classification; the name derives from the subfamily Ponerinae, one of the constituent taxa (see Chapter 2).

**Postpetiole** In certain ants, the second segment of the waist. This is in fact the third abdominal segment, since the first abdominal segment (propodeum) is fused to the thorax.

**Preadaptation** Any previously existing anatomical structure, physiological process, or behavior pattern that makes new forms of evolutionary adaptation more likely.

**Presocial** Applied to the condition, or to the group possessing it, in which individuals display some degree of social behavior short of eusociality. Presocial species are either subsocial, that is, the parents care for their own nymphs and larvae, or else parasocial, that is, one or two of the following three traits is shown: cooperation in care of young, reproductive division of labor, overlap of generations of life stages that contribute to colony labor.

**Pretarsus** The terminal segment of the leg, consisting usually of a pair of lateral claws and the arolium.

**Proctodeum** The hindgut and Malpighian tubules of an insect.

**Propodeum** In the higher Hymenoptera, the first abdominal segment when it is fused with the alitrunk. Also called epinotum.

**Proventriculus** The gizzard, a muscular tubular organ just posterior to the crop and anterior to the midgut (ventriculus).

**Proximal** Closest with reference to the body.

**Pruinose** Having the appearance of being frosted or lightly dusted in the manner of a plum.

**Pseudopolygyny** The condition in which several dealated virgin queens coexist with one egg-laying mated queen.

**Pterothorax** The wing-bearing portion of the thorax of winged insects, that is, the mesothorax and the metathorax.

**Pubescence** Short, fine hairs, typically forming a second layer beneath the pilosity, or longer, coarser hairs.

**Punctate** Referring to a surface bearing fine punctures like pinpricks.

**Pupa** The inactive instar of the holometabolous insects (including the Hymenoptera) during which development into the final adult form is completed.

**Pygidium** The last complete tergite (upper segmental plate) of the abdomen, regardless of its numerical designation.

**Queen** A member of the reproductive caste in semi-social or eusocial species. The existence of a queen caste presupposes the existence also of a worker caste at some stage of the colony life cycle. Queens may or may not be morphologically different from workers.

**Queen control** The inhibitory influence of the queen on the reproductive activities of the workers and other queens.

**Queenright** Referring to a colony that contains a functional queen.

**Queen substances** Originally, the set of pheromones by which the queen continuously attracts and controls the reproductive activities of the workers.

**Recognition alleles** Alleles hypothesized to encode the production of a recognition cue and simultaneously the ability to recognize the cue in others, leading to the discrimination of kin from non-kin.

**Recruitment** A special form of assembly in which members of a society are directed to some point in space where work is required.

**Recruitment trail** An odor trail laid by scout workers and used to recruit nestmates to a food find, a desirable new nest site, a breach in the nest wall, or some other place where the assistance of many workers is needed (as opposed to *Exploratory trail*).

**Recumbent** Referring to a hair lying on the body surface.

**Relative growth** The relative increase of one body part with respect to another as total body size is varied. *Allometry* is a special form of relative growth.

**Replete** An individual ant whose crop is greatly distended with liquid food, to the extent that the abdominal segments are pulled apart and the intersegmental membranes are stretched tight. Repletes usually serve as living reservoirs, regurgitating food on demand to their nestmates. Also used as an adjective, e.g., replete worker.

**Reproductives** Males and fertile females, including queens and laying workers.

**Reproductive stage** The stage in colony growth during which males and virgin queens are produced.

**Reticulate** Covered with a network of carinae, striae, or rugae.

**Retinue** A group of workers, not necessarily permanent or even long lasting in composition, who closely attend the queen.

**Ritualization** The evolutionary modification of a behavior pattern that turns it into a signal used in communication or at least improves its efficiency as a signal.

**Role** A pattern of behavior displayed by certain members of a society that has an effect on other members in a way that divides labor.

**Ruga** A wrinkle on the body surface.

**Rugoreticulate** With rugae (wrinkles) forming a network or grid.

**Rugose** Referring to a surface bearing multiple wrinkles, running approximately in parallel.

**Sclerite** A portion of the body wall bounded by sutures.

**Scout** A worker searching outside the nest for food or, in the case of slave-making species, a host colony suitable for raiding.

**Scrobe** A groove or other cavity into which an appendage can be protectively folded; ordinarily refers to the antennal scrobe found in some ant species.

**Self-grooming** See *Grooming*.

**Semiochemicals** Chemicals used in communication within or between species.

**Sensillum** In insects, a simple sense organ or one of the structural units of a compound sense organ.

**Serrate** With teeth along the edge; saw-like.

**Shagreened** Referring to a surface that is covered with a fine but close-set and irregular roughness.

**Slavery** Same as *Dulosis.*

**Social bucket** Refers to the process by which liquid food is carried between the mandibles, thence to be shared with nestmates by mouth-to-mouth contact.

**Social facilitation** See *Group effect.*

**Social homeostasis** The maintenance of steady states at the level of the society either by control of the nest microclimate or by regulation of the population density, behavior, and physiology of the group members as a whole.

**Social insect** In the strict sense, a "true social insect" is one that belongs to a eusocial species; in other words, it is an ant, a termite, or one of the eusocial wasps or bees. In the broad sense, a social insect is one that belongs to either a presocial or eusocial species.

**Social parasitism** The coexistence in the same nest of two species of social insects, one of which is parasitically dependent on the other. The term can also be applied loosely to the relation between symphiles and their social insect hosts.

**Social stomach** The first segment of the gastral gut, also called crop, where liquid food can be stored and from which it can be passed to nestmates by regurgitation.

**Society** A group of individuals belonging to the same species and organized in a cooperative manner. Some amount of reciprocal communication among the members is implied.

**Sociobiology** The systematic study of the biological basis of all aspects of social behavior.

**Sociogenesis** The collective processes and patterns that lead to the development of a colony throughout its life cycle.

**Sociogenetics** Pertaining to the hereditary basis of social behavior and caste systems, and to the study of that basis.

**Sociogram** The complete repertories of all castes of a colony, with a specification of the frequencies of the acts and of the interactions among the castes.

**Sociotomy** Same as *Colony fission.*

**Soldier** A member of a worker subcaste specialized for colony defense.

**Species odor** The odor found on the bodies of social insects which is peculiar to a given species. It is possible that the species odor is merely the less distinctive component of a larger mixture comprising the *Colony odor.*

**Spur** A spine-like appendage, often paired and/or pectinate, at the end of the tibia.

**Squamate** Scale-shaped, used commonly to describe a form of hair.

**Staphyla** A group of gongylidia, the swollen hyphal tips produced by fungi that live in symbiosis with attine ants.

**Statary phase** The period (as opposed to the *Nomadic phase*) in the activity cycle of an army ant colony during which the colony is relatively quiescent and does not move from site to site. At this time the queen lays the eggs and the bulk of the brood is in the egg and pupal stages.

**Sternal** Pertaining to the sternum or lower portion of the body or body part.

**Sternite** A ventral sclerite; in other words, a portion of the body wall bounded by sutures and located in a ventral position (as opposed to *Tergite*).

**Stigmergy** The guidance of work performed by individual colony members by the evidences of work previously accomplished rather than by direct signals from nestmates.

**Stochastic theory of mass behavior** The theory that transition probabilities in the behavior of individual social insects are programmed to produce optimal mass responses of the colony, that the probabilities have been determined by selection at the colony level, and that they represent a sensitive adaptation to the particular environmental conditions in which the species has existed during recent evolutionary time.

**Stomodeal trophallaxis** The exchange of liquid food mouth-to-mouth. The food is either regurgitated from the crop (the first segment of the gastral gut) or released from glands in the head or thorax.

**Stomodeum** The foregut of an insect.

**Stria** A fine impressed line on the body surface.

**Striate** Referring to a surface bearing multiple striae, or impressed lines.

**Stridulation** The production of sound by rubbing one part of the body surface against another.

**Strigilation** The licking of secretions from the body of another animal.

**Subapterous** Wings reduced in size and less than fully functional; see also *Brachypterous.*

**Suberect** Referring to a hair that stands at an angle of about 45 degrees from the body surface.

**Subgenus** One or more distinctive species of common phylogenetic origin within a genus.

**Subsocial** Applied to the condition, or to the group showing it, in which the adults care for their nymphs or larvae for some period of time (see also *Presocial*).

**Sulcus** A deep furrow or groove.

**Supercolony** A unicolonial population, in which workers move freely from one nest to another, so that the entire population is a single colony.

**Superorganism** Any society, such as the colony of a eusocial insect species, possessing features of organization analogous to the physiological properties of a single organism. The insect colony, for example, is divided into reproductive castes (analogous to gonads) and worker castes (analogous to somatic tissue); it may exchange nutrients by trophallaxis (analogous to the circulatory system); and so forth.

**Surface pheromone** A pheromone with an active space restricted so close to the body of the sending organism that direct contact, or something approaching it, must be made with the body in order to perceive the pheromone. Examples include the colony odors of many species.

**Swarming** Colony reproduction in which one or more queens and a number of workers separate to found a new colony. When the queen or queens are accompanied by a small number of workers and they leave the main parental nest, the process is called *Budding.* When major portions of the colony separate, each with one or more queens (as in army ants), the process is called *Fission.* Swarming also applies to the mass exodus from the nests of reproductive forms at the beginning of the nuptial flight.

**Symbiont** An organism that lives in symbiosis with another species.

**Symbiosis** The intimate, relatively protracted, dependent relationship of members of one species with those of another. The three principal kinds of symbiosis are *Commensalism, Mutualism,* and *Parasitism.*

**Symphile** A symbiont, in particular a solitary insect or other kind of arthropod, which is accepted to some extent by an insect colony and communicates with it amicably. Most symphiles are licked, fed, or transported to the host brood chambers, or treated to a combination of all three.

**Synechthran** A symbiont, usually a scavenger, a parasite, or a predator, that is treated with hostility by the host colony.

**Synoekete** A symbiont that is treated with indifference by the host colony.

**Tandem calling** The release of a pheromone by a leader ant that recruits a nestmate for tandem running (see below).

**Tandem running** A form of communication, used by the workers of certain ant species during exploration or recruitment, in which one individual follows closely behind another, frequently contacting the abdomen of the leader with her antennae.

**Tarsation** Touching with the tarsi, especially the touching of another insect as a tactile signal.

**Tarsus** The foot of an insect; the one- to five-segmented appendage attached to the tibia, or lower leg segment.

**Taxon** Any taxonomic entity, such as a particular species or genus.

**Temporal polyethism** Same as *Age polyethism.*

**Temporary social parasitism** Parasitism in which a queen of one species enters an alien nest, usually belonging to another species, kills or renders infertile the resident queen, and takes her place. The population of the colony then becomes increasingly dominated by the offspring of the parasite queen as the host workers die from natural causes.

**Tergal** Pertaining to the dorsal (upper) surface.

**Tergite** A dorsal sclerite; in other words, a portion of the body wall bounded by sutures and located in a dorsal position (as opposed to *Sternite*).

**Termitarium** A termite nest. Also, an artificial nest used in the laboratory to house termites.

**Termitology** The scientific study of termites.

**Termitophile** An organism that must spend at least part of its life cycle with termite colonies.

**Territorial pheromone** A substance deposited on or around the nest that is colony specific or species specific and aids in the exclusion of alien colonies.

**Territory** An area occupied more or less exclusively by an animal or group of animals (such as an ant colony) by means of repulsion through overt defense or aggressive advertisement.

**Thelytoky** The production of females from unfertilized eggs.

**Tibia** The fourth division of the leg, between the femur ("thigh") and tarsus ("foot").

**Totipotent** Capable of performing all essential tasks; the founding queens of some species are totipotent or nearly so.

**Trail parasitism** See *Trophic parasitism.*

**Trail pheromone** A substance laid down in the form of a trail by one animal and followed by another member of the same species.

**Trail substance** Same as *Trail pheromone.*

**Tribe** The taxonomic category between genus and subfamily, hence a set of similar genera of common phylogenetic origin.

**Trichome** A tuft of long, often yellow or golden hairs associated with glandular areas on the body surfaces of many myrmecophilous beetles. The hairs are believed to aid in the dissemination of attractants.

**Triphasic allometry** Polymorphism in which the allometric regression line, when plotted on a double logarithmic scale, breaks at two points and consists of three segments. In ants, the two terminal segments usually have slight to moderately high slopes and the middle segment has a very high slope (see Chapter 8).

**Trochanter** The short second division of the leg (away from the body), between the coxa and the femur.

**Trophallactic appeasement** The use of liquid food offerings (trophallaxis) to appease other, potentially hostile workers.

**Trophallaxis** The exchange of alimentary liquid among colony members and guest organisms, either mutually or unilaterally. In stomodeal trophallaxis the material originates from the mouth; in proctodeal trophallaxis it originates from the anus.

**Trophic egg** An egg, usually degenerate in form and inviable, which is fed to other members of the colony.

**Trophic parasitism** The intrusion of one species into the social system of another (as, for example, by utilization of the trail system) in order to steal food.

**Trophobiosis** The relationship in which ants receive honeydew from aphids and other homopterans, or the caterpillars of certain lycaenid and riodinid butterflies, and in return provide these insects with protection. The insects supplying the honeydew are referred to as trophobionts.

**Trophogenesis** The origin of different caste traits from differential feeding of the immature stages (as opposed to genetic control of castes and *Blastogenesis*).

**Tuberculate** Covered with tubercles (small thick spines or pimple-like structures).

**Unicolonial** Pertaining to a population of social insects in which there are no behavioral colony boundaries (as opposed to *Multicolonial).*

**Venter** The lower surface.

**Vertex** The upper surface of the head between the eyes, frons, and occiput.

**War** Overt aggression between groups of workers from different colonies that results in the appropriation of territorial space or nest sites.

**Weaver ants** Species of ants, such as members of the genus *Oecophylla,* that use larval silk in construction of the nest.

**Worker** A member of the non-reproductive laboring caste in semi-social and eusocial species. The existence of a worker caste presupposes the existence also of royal (reproductive) castes. In termites, the term is used in a more restricted sense to designate individuals in the family Termitidae that completely lack wings and have reduced pterothorax, eyes, and genital apparatus.

**Xenobiosis** The relation in which colonies of one species live in the nests of another species and move freely among the hosts, obtaining food from them by regurgitation or other means but still keeping their brood separate.

# Bibliography


Abbott, A. 1978. Nutrient dynamics of ants. *In* M. V. Brian, ed., *Production ecology of ants and termites* (International Biological Programme, no. 13), pp. 233–244. Cambridge University Press, New York.

Abe, T. 1971. On the food sharing among four species of ants in a sandy grassland, I: Food and foraging behavior. *Japanese Journal of Ecology*, 20(6): 219–230.

——— 1982. Ecological role of termites in a tropical rain forest. *In* M. D. Breed, C. D. Michener, and H. E. Evans, eds., *The biology of social insects* (Proceedings of the Ninth Congress of the International Union for the Study of Social Insects, Boulder, Colorado, 1982), pp. 71–75. Westview Press, Boulder.

Abraham, M. 1980. Comportement individuel lors de déménagements successifs chez *Myrmica rubra* L. *In* D. Cherix, ed., *Ecologie des insectes sociaux* (L'Union Internationale pour l'Etude des Insectes Sociaux, Section française, Compte Rendu Colloque Annuel, Lausanne, 1979), pp. 17–19. Cherix et Filanosa S.A., Nyon, Switzerland.

Abraham, M., and J. M. Pasteels. 1980. Social behaviour during nest-moving in the ant *Myrmica rubra* L. (Hym. Form.). *Insectes Sociaux*, 27(2): 127–147.

Abraham, M., J.-L. Deneubourg, and J. M. Pasteels. 1984. Idiosyncrasie lors du déménagement de *Myrmica rubra* L. (Hymenoptera Formicidae). *Actes des Colloques Insectes Sociaux* (L'Union Internationale pour l'Etude des Insectes Sociaux, Section française, Compte Rendu Colloque Annuel, Les Eyzies, 1983), 1: 19–25.

Adams, E. S., and J. F. A. Traniello. 1981. Chemical interference competition by *Monomorium minimum* (Hymenoptera: Formicidae). *Oecologia*, 51(2): 265–270.

Addicott, J. F. 1979. A multispecies aphid-ant association: density dependence and species-specific effects. *Canadian Journal of Zoology*, 57(3): 558–569.

Adlerz, G. 1884. Myrmecologiska studier, I: *Formicoxenus nitidulus* Nyl. *Öfversigt af Kongliga Vetenskaps-Akademiens Förhandlingar*, 41(8): 43–64.

——— 1896. Myrmecologiska studier, III: *Tomognathus sublaevis* Mayr. *Bihang Till Kongliga Svenska Vetenskaps-Akademiens Handlingar*, 21(Afdelning 4)(4): 1–76.

Adlung, K. G. 1966. A critical evaluation of the European research on use of red wood ants (*Formica rufa* group) for the protection of forests against harmful insects. *Zeitschrift für Angewandte Entomologie*, 57(2): 167–189.

Agbogba, C. 1984. Observations sur le comportement de marche en tandem chez deux espèces de fourmis ponérines: *Mesoponera caffraria* (Smith) et *Hypoponera* sp. (Hym. Formicidae). *Insectes Sociaux*, 31(3): 264–276.

Agosti, D., and C. A. Collingwood. 1987a. A provisional list of the Balkan ants (Hym., Formicidae) and a key to the worker caste, I: Synonymic list. *Mitteilungen der Schweizerischen Entomologischen Gesellschaft*, 60: 51–62.

——— 1987b. A provisional list of the Balkan ants (Hym. Formicidae) with a key to the worker caste, II: Key to the worker caste, including the European species without the Iberian. *Mitteilungen der Schweizerischen Entomologischen Gesellschaft*, 60: 261–293.

Agosti, D., and E. Hauschteck-Jungen. 1987. Polymorphism of males in *Formica exsecta* Nyl. (Hym.: Formicidae). *Insectes Sociaux*, 34(4): 280–290.

Akre, R. D. 1968. The behavior of *Euxenister* and *Pulvinister*, histerid beetles associated with army ants. *Pan-Pacific Entomologist*, 44(2): 87–101.

Akre, R. D., and W. B. Hill. 1973. Behavior of *Adranes taylori*, a myrmecophilous beetle associated with *Lasius sitkaensis* in the Pacific Northwest (Coleoptera: Pselaphidae; Hymenoptera: Formicidae). *Journal of the Kansas Entomological Society*, 46(4): 526–536.

Akre, R. D., and C. W. Rettenmeyer. 1966. Behavior of Staphylinidae associated with army ants (Formicidae: Ecitonini). *Journal of the Kansas Entomological Society*, 39(4): 745–782.

——— 1968. Trail-following by guests of army ants (Hymenoptera: Formicidae: Ecitonini). *Journal of the Kansas Entomological Society*, 41(2): 165–174.

Akre, R. D., and R. L. Torgerson. 1968. The behavior of *Diploeciton nevermanni*, a staphylinid beetle associated with army ants. *Psyche*, 75(3): 211–215.

Akre, R. D., G. Alpert, and T. Alpert. 1973. Life cycle and behavior of *Microdon cothurnatus* in Washington (Diptera: Syrphidae). *Journal of the Kansas Entomological Society*, 46(3): 327–338.

Aktaç, N. 1976. Studies on the myrmecofauna of Turkey, I: Ants of Siirt, Bodrum and Trabzon. *Istanbul Üniversitesi Fen Fakültesi Mecmuasi*, ser. B, 41(1–4): 115–135.

Alayo, D. P. 1974. Introducción al estudio de los himenópteros de Cuba. Superfamilia Formicóidea. *Serie Biológica, Academia de Ciencias de Cuba, Instituto de Zoología* (Havana), no. 53, 58 pp.

Alexander, R. D. 1974. The evolution of social behavior. *Annual Review of Ecology and Systematics*, 5: 325–383.

Alexander, R. D., and P. W. Sherman. 1977. Local mate competition and parental investment in social insects. *Science*, 196: 494–500.

Allee, W. C. 1931. *Animal aggregations: a study in general sociology.* University of Chicago Press, Chicago. ix + 431 pp.

——— 1938. *The social life of animals.* W. W. Norton, New York. 293 pp.

Allies, A. B. 1984. The cuckoo ant *Leptothorax kutteri:* the behaviour and reproductive strategy of a workerless parasite. B.Sc. thesis, University of Bath, Eng.

Allies, A. B., A. F. G. Bourke, and N. R. Franks. 1986. Propaganda substances in the cuckoo ant *Leptothorax kutteri* and the slave-maker *Harpagoxenus sublaevis*. *Journal of Chemical Ecology*, 12(6): 1285–93.

Alloway, T. M. 1979. Raiding behaviour of two species of slave-making ants, *Harpagoxenus americanus* (Emery) and *Leptothorax duloticus* (Wesson). *Animal Behaviour*, 27(2): 202–210.

——— 1980. The origins of slavery in leptothoracine ants (Hymenoptera: Formicidae). *American Naturalist*, 115(2): 247–261.

——— 1982. How the slave-making ant *Harpagoxenus americanus* (Emery) affects the pupa-acceptance behavior of its slaves. *In* M. D. Breed, C. D. Michener, and H. E. Evans, eds., *The biology of social insects* (Proceedings of the Ninth Congress of the International Union for the Study of Social Insects, Boulder, Colorado, 1982), pp. 261–265. Westview Press, Boulder.

Alloway, T. M., and M. G. Del Rio Pesado. 1983. Behavior of the slave-making ant, *Harpagoxenus americanus* (Emery), and its host species under "seminatural" laboratory conditions (Hymenoptera: Formicidae). *Psyche*, 90(4): 425–436.

Alloway, T. M., A. Buschinger, M. Talbot, R. Stuart, and C. Thomas. 1982. Polygyny and polydomy in three North American species of the ant genus *Leptothorax* Mayr (Hymenoptera: Formicidae). *Psyche*, 89(3–4): 249–274.

Allport, F. H. 1924. *Social psychology.* Houghton Mifflin, Boston. xiv + 453 pp.

Alpert, G. D. 1981. A comparative study of the symbiotic relationships between scarab beetles of the genus *Cremastocheilus* and their host ants, 2 vols. Ph.D. diss., Harvard University. 588 pp.

Alpert, G. D., and R. D. Akre. 1973. Distribution, abundance, and behavior of the inquiline ant *Leptothorax diversipilosus. Annals of the Entomological Society of America*, 66(4): 753–760.

Alpert, G. D., and P. O. Ritcher. 1975. Notes on the life cycle and myrmecophilous adaptations of *Cremastocheilus armatus* (Coleoptera: Scarabaeidae). *Psyche*, 82(3–4): 283–291.

Andersen, A. N. 1982. Seed removal by ants in the mallee of northwestern Victoria. *In* R. C. Buckley, ed., *Ant-plant interactions in Australia*, pp. 31–43. Dr. W. Junk, The Hague.

——— 1985. Seed predation by insects in sclerophyllous vegetation at Wilson's Promontory, Victoria, with particular reference to ants. Ph.D. diss., School of Botany, University of Melbourne. 241 pp.

——— 1986a. Diversity, seasonality and community organization of ants at adjacent heath and woodland sites in south-eastern Australia. *Australian Journal of Zoology*, 34(1): 53–64.

——— 1986b. Patterns of ant community organization in mesic southeastern Australia. *Australian Journal of Ecology*, 11(1): 87–97.

——— 1987. Effects of seed predation by ants on seedling densities at a woodland site in SE Australia. *Oikos*, 48(2): 171–174.

Andersen, A. N., and D. H. Ashton. 1985. Rates of seed removal by ants at heath and woodland sites in southeastern Australia. *Australian Journal of Ecology*, 10(4): 381–390.

Andersson, H. 1974. Studies on the myrmecophilous fly, *Glabellula arctica* (Zett.) (Dipt. Bombyliidae). *Entomologica Scandinavica*, 5(1): 29–38.

Andersson, M. 1984. The evolution of eusociality. *Annual Review of Ecology and Systematics*, 15: 165–189.

Andrasfalvy, A. 1961. Mitteilungen über Daten des Hochzeitsfluges verschiedener Ameisenarten in Ungarn und Ergebnisse von Versuchen der Koloniegründung im Formicar bei diesen Arten. *Insectes Sociaux*, 8(4): 299–310.

Andrews, C. C. 1983. *Melinaea lilis imitata* (Melineas, army ant butterfly). *In* D. H. Janzen, ed., *Costa Rican natural history*, pp. 736–738. University of Chicago Press, Chicago.

Andrews, E. A. 1926. Sequential distribution of *Formica exsectoides* Forel. *Psyche*, 33(6): 127–150.

——— 1927. Ant-mounds as to temperature and sunshine. *Journal of Morphology and Physiology*, 44(1): 1–20.

——— 1930. Honeydew reflexes. *Physiological Zoölogy*, 3(4): 467–484.

Aoki, S. 1977. *Colophina clematis* (Homoptera, Pemphigidae), an aphid species with "soldiers." *Kontyû* (Tokyo), 45(2): 276–282.

——— 1982. Soldiers and altruistic dispersal in aphids. *In* M. D. Breed, C. D. Michener, and H. E. Evans, eds., *The biology of social insects* (Proceedings of the Ninth Congress of the International Union for the Study of Social Insects, Boulder, Colorado, 1982), pp. 154–158. Westview Press, Boulder.

——— 1987. Evolution of sterile soldiers in aphids. *In* Y. Itō, J. L. Brown, and J. Kikkawa, eds., *Animal societies: theories and facts*, pp. 53–65. Japanese Scientific Societies Press, Tokyo.

Argano, R., and G. L. Pesce. 1974. Un Elumino mirmecofilo di Sardegna: *Typhloschizidium cottarellii* n. sp. (Isopoda, Oniscoidea, Armadillidiidae). *Fragmenta Entomologica*, 9(4): 283–291.

Armstrong, J. A. 1979. Biotic pollination mechanisms in the Australian flora: a review. *New Zealand Journal of Botany*, 17(4): 467–508.

Arnett, R. H., Jr. 1985. *American insects: a handbook of the insects of America north of Mexico.* Van Nostrand Reinhold, New York. xiv + 850 pp.

Arnold, G. 1915–1926. A monograph of the Formicidae of South Africa [6 parts, plus an appendix]. *Annals of the South African Museum*, 14(1–6): 1–766; 23(2): 191–295.

Arnoldi, K. V. 1930. Studien über die Systematik der Ameisen, VI: Eine neue parasitische Ameise, mit Bezugnahme auf die Frage nach der Entstehung der Gattungsmerkmale bei den parasitären Ameisen. *Zoologischer Anzeiger*, 91(9–12): 267–283.

——— 1932. Biologische Beobachtungen an der neuen paläarktischen Sklavenhalterameisen *Rossomyrmex proformicarum* K. Arn., nebst einigen Bemerkungen über die Beförderungsweise der Ameisen. *Zeitschrift für Morphologie und Ökologie der Tiere*, 24(2): 319–326.

——— 1968. Important additions to the myrmecofauna (Hymenoptera, Formicidae) of the USSR, with some new descriptions. *Zoologicheskii Zhurnal*, 47(12): 1800–22. [In Russian.]

Ashton, D. H. 1979. Seed harvesting by ants in forests of *Eucalyptus regnans* F. Muell. in central Victoria. *Australian Journal of Ecology*, 4(3): 265–277.

Athias-Henriot, C. 1947. Recherches sur les larves de quelques fourmis d'Algérie. *Bulletin Biologique de la France et de la Belgique*, 81(3–4): 247–272.

Atsatt, P. R. 1981a. Ant-dependent food plant selection of the mistletoe butterfly *Ogyris amaryllis* (Lycaenidae). *Oecologia*, 48(1): 60–63.

——— 1981b. Lycaenid butterflies and ants: selection for enemy-free space. *American Naturalist*, 118(5): 638–654.

Attygalle, A. B., and E. D. Morgan. 1983. Trail pheromone of the ant *Tetramorium caespitum* L. *Naturwissenschaften*, 70(7): 364–365.

——— 1985. Ant trail pheromones. *Advances in Insect Physiology*, 18: 1–30.

Attygalle, A. B., R. P. Evershed, E. D. Morgan, and M.-C. Cammaerts. 1983. Dufour gland secretions of workers of the ants *Myrmica sulcinodis* and *Myrmica lobicornis*, and comparison with six other species of *Myrmica. Insect Biochemistry*, 13(5): 507–512.

Attygalle, A. B., J. P. J. Billen, and E. D. Morgan. 1985. The postpharyngeal gland of workers of *Solenopsis geminata* (Hymenoptera: Formicidae). *Actes des Colloques Insectes Sociaux* (L'Union Internationale pour l'Etude des Insectes Sociaux, Section française, Compte Rendu Colloque Annuel, Diepenbeek, 1984), 2: 79–86.

Attygalle, A. B., M.-C. Cammaerts, R. Cammaerts, E. D. Morgan, and D. G. Ollett. 1986. Chemical and ethological studies of the trail pheromones of the ant *Manica rubida* (Hymenoptera: Formicidae). *Physiological Entomology*, 11(2): 125–132.

Attygalle, A. B., O. Vostrowsky, H. J. Bestmann, S. Steghaus-Kovac, and U. Maschwitz. 1988. (*3R,4S*)-4-Methyl-3-heptanol, the trail pheromone of the ant *Leptogenys diminuta. Naturwissenschaften*, 75(6): 315–317.

Attygalle, A. B., B. Siegel, O. Vostrowsky, H. J. Bestmann, and U. Maschwitz. 1989. Chemical composition and function of the metapleural gland secretion of the ant, *Crematogaster deformis* Smith (Hymenoptera: Myrmicinae). *Journal of Chemical Ecology*, 15(1): 317–329.

Auclair, J. L. 1963. Aphid feeding and nutrition. *Annual Review of Entomology*, 8: 439–490.

Autuori, M. 1950a. Contribuição para o conhecimento da saúva (*Atta* spp.—Hymenoptera-Formicidae), V: Número de formas aladas e redução dos sauveiros iniciais. *Archivos do Instituto Biológico, São Paulo*, 19(22): 325–331.

——— 1950b. Longevididade de uma colônia de saúva (*Atta sexdens rubropilosa* Forel, 1908) em condições de laboratório. *Ciência e Cultura*, 2(4): 285–286.

——— 1956. La fondation des sociétés chez les fourmis champignonnistes du genre "*Atta*" (Hym. Formicidae). *In* M. Autuori et al., eds., *L'instinct dans le comportement des animaux et de l'homme*, pp. 77–104. Masson et Cie, Paris.

——— 1974. Der Staat der Blattschneiderameisen. *In* G. H. Schmidt, ed., *Sozialpolymorphismus bei Insekten*, pp. 631–656. Wissenschaftliche Verlagsgesellschaft MBH, Stuttgart.

Ayre, G. L. 1958. Notes on insects found in or near nests of *Formica subnitens* Creighton (Hymenoptera: Formicidae) in British Columbia. *Insectes Sociaux*, 5(1): 1–7.

——— 1963a. Feeding behaviour and digestion in *Camponotus herculeanus* (L.) (Hymenoptera: Formicidae). *Entomologia Experimentalis et Applicata*, 6(3): 165–170.

——— 1963b. Response to movement by *Formica polyctena* Först. *Nature*, 199: 405–406.

——— 1969. Comparative studies on the behavior of three species of ants (Hymenoptera: Formicidae), II: Trail formation and group foraging. *Canadian Entomologist*, 101(2): 118–128.

Ayre, G. L., and M. S. Blum. 1971. Attraction and alarm of ants (*Camponotus* spp.—Hymenoptera: Formicidae) by pheromones. *Physiological Zoölogy*, 44(2): 77–83.

Ayyar, P. N. K. 1937. A new carton-building species of ant in South India, *Crematogaster dohrni artifex*, Mayr. *Journal of the Bombay Natural History Society*, 39(2): 291–308.

Baldridge, R. S., C. W. Rettenmeyer, and J. F. Watkins II. 1980. Seasonal, nocturnal and diurnal flight periodicities of Nearctic army ant males (Hymenoptera: Formicidae). *Journal of the Kansas Entomological Society*, 53(1): 189–204.

Bałazy, S., A. Lenoir, and J. Wiśniewski. 1986. *Aegeritella roussillonensis* n. sp. (Hyphomycetales, Blastosporae) une espèce nouvelle de champignon epizoique sur les fourmis *Cataglyphis cursor* (Fonscolombe) (Hymenoptera, Formicidae) en France. *Cryptogamie, Mycologie*, 7(1): 37–45.

Banck, L. J. 1927. *Contributions to myrmecophily, I: An anatomical-histological and experimental-biological study of* Thorictus foreli *Wasm.* Saint-Paul, Fribourg. 83 pp. [Cited by Plath, 1935.]

Banks, C. J., and H. L. Nixon. 1958. Effects of the ant, *Lasius niger* L., on the feeding and excretion of the bean aphid, *Aphis fabae* Scop. *Journal of Experimental Biology*, 35(4): 703–711.

Barbier, M., and B. Delage. 1967. Le contenu des glandes pharyngiennes de la fourmi *Messor capitatus* Latr. (Insecte, Hyménoptère Formicidé). *Comptes Rendus de l'Académie des Sciences, Paris*, ser. D, 264(11): 1520–22.

Barker, J. F. 1979. Endocrine basis of wing casting and flight muscle histolysis in the fire ant *Solenopsis invicta*. *Experientia*, 35(4): 552–554.

Barlin, M. R., M. S. Blum, and J. M. Brand. 1976. Species-specificity studies on the trail pheromone of the carpenter ant, *Camponotus pennsylvanicus* (Hymenoptera: Formicidae). *Journal of the Georgia Entomological Society*, 11(2): 162–164.

Baroni Urbani, C. 1968a. Domination et monogynie fonctionnelle dans une société digynique de *Myrmecina graminicola* Latr. *Insectes Sociaux*, 15(4): 407–412.

——— 1968b. Studi sulla mirmecofauna d'Italia, IV: La fauna mirmecologica delle Isole Maltesi ed il suo significato ecologico e biogeografico. *Annali del Museo Civico di Storia Naturale "Giacomo Doria" di Genova*, 77: 408–559.

——— 1969. Gli *Strongylognathus* del gruppo *huberi* nell'Europa occidentale: saggio di una revisione basata sulla casta operaia (Hymenoptera Formicidae). *Bolletino della Società Entomologica Italiana*, 99–101(7–8): 132–168.

——— 1970. Ueber die funktionelle Monogynie der Ameise *Myrmecina graminicola* (Latr.). *Insectes Sociaux*, 18(3): 219–222.

——— 1971a. Catalogo delle specie di Formicidae d'Italia. *Memorie della Società Entomologica Italiana*, 50: 1–287.

——— 1971b. Einige Homonymien in der Familie Formicidae (Hymenoptera). *Mitteilungen der Schweizerischen Entomologischen Gesellschaft*, 44(3–4): 360–362.

——— 1971c. Studien zur Ameisenfauna Italiens, XI: Die Ameisen des Toskanischen Archipels. Betrachtungen zur Herkunft der Inselfaunen. *Revue Suisse de Zoologie*, 78(4): 1037–67.

——— 1973. Die Gattung *Xenometra*, ein objektives Synonym (Hymenoptera, Formicidae). *Mitteilungen der Schweizerischen Entomologischen Gesellschaft*, 46(3–4): 199–201.

——— 1974. Polymorphismus in der Ameisengattung *Camponotus* aus morphologischer Sicht. *In* G. H. Schmidt, ed., *Sozialpolymorphismus bei Insekten*, pp. 543–564. Wissenschaftliche Verlagsgesellschaft MBH, Stuttgart.

——— 1975a. Contributo alla conoscenza dei generi *Belonopelta* Mayr e *Leiopelta* gen. n. (Hymenoptera: Formicidae). *Mitteilungen der Schweizerischen Entomologischen Gesellschaft*, 48(3–4): 295–310.

——— 1975b. Primi reperti del genere *Calyptomyrmex* Emery nel subcontinente Indiano. *Entomologica Basiliensia*, 1: 395–411.

——— 1976a. Le formiche dell'arcipelago della Galita (Tunisia). *Redia*, 59: 207–223.

——— 1976b. Réinterprétation du polymorphisme de la caste ouvrière chez les fourmis à l'aide de la régression polynomiale. *Revue Suisse de Zoologie*, 83(1): 105–110.

——— 1977. Materiali per una revisione della sottofamiglia Leptanillinae Emery (Hymenoptera: Formicidae). *Entomologica Basiliensia*, 2: 427–488.

——— 1978a. Adult populations in ant colonies. *In* M. V. Brian, ed., *Production ecology of ants and termites*, pp. 334–335. Cambridge University Press, New York.

——— 1978b. Contributo alla conoscenza del genere *Amblyopone* Erichson (Hymenoptera: Formicidae). *Mitteilungen der Schweizerischen Entomologischen Gesellschaft*, 51(1): 39–51.

——— 1978c. Materiali per una revisione dei *Leptothorax* neotropicali appartenenti al sottogenere *Macromischa* Roger, n. comb. (Hymenoptera: Formicidae). *Entomologica Basiliensia*, 3: 395–618.

——— 1980. First description of fossil gardening ants. *Stuttgarter Beiträge zur Naturkunde*, ser. B, 54: 1–13.

Baroni Urbani, C., and N. Aktaç. 1981. The competition for food and circadian succession in the ant fauna of a representative Anatolian semisteppic environment. *Mitteilungen der Schweizerischen Entomologischen Gesellschaft*, 54(1): 33–56.

Baroni Urbani, C., and C. A. Collingwood. 1977. The zoogeography of ants (Hymenoptera Formicidae) in northern Europe. *Acta Zoologica Fennica*, 152: 1–34.

Baroni Urbani, C., and P. B. Kannowski. 1974. Patterns in the red imported fire ant settlement of a Louisiana pasture: some demographic parameters, interspecific competition and food sharing. *Environmental Entomology*, 3(5): 755–760.

Baroni Urbani, C., and E. O. Wilson. 1987. The fossil members of the ant tribe Leptomyrmecini (Hymenoptera: Formicidae). *Psyche*, 94(1–2): 1–8.

Barr, D., and W. H. Gotwald. 1982. Phenetic affinities of males of the army ant genus *Dorylus* (Hymenoptera: Formicidae: Dorylinae). *Canadian Journal of Zoology*, 60(11): 2652–58.

Barr, D., J. van Boven, and W. H. Gotwald. 1985. Phenetic studies of African army ant queens of the genus *Dorylus* (Hymenoptera: Formicidae). *Systematic Entomology*, 10(1): 1–10.

Barrer, P. M., and J. M. Cherrett. 1972. Some factors affecting the site and pattern of leaf-cutting activity in the ant *Atta cephalotes* L. *Journal of Entomology*, 47(1): 15–27.

Barrett, K. E. J. 1977. Provisional distribution maps of ants in the British Isles. Appendix to M. V. Brian, *Ants*, pp. 203–216. Collins, London.

——— 1979. *Hymenoptera Formicidae*. *In* J. Heath, ed., *Provisional atlas of the insects of the British Isles*, part 5. Biological Records Centre, Institute of Terrestrial Ecology, Monks Wood, Huntingdon, Eng. [Cited by Elmes and Wardlaw, 1982.]

Bartholomew, G. A., J. R. B. Lighton, and D. H. Feener, Jr. 1988. Energetics of trail-running, load carriage, and emigration in the column-raiding army ant *Eciton hamatum*. *Physiological Zoölogy*, 61(1): 57–68.

Bartz, S. H. 1982. On the evolution of male workers in the Hymenoptera. *Behavioral Ecology and Sociobiology*, 11(3): 223–228.

——— 1983. Kin selection and relatedness in social insects. Ph.D. diss., Harvard University. 182 pp.

Bartz, S. H., and B. Hölldobler. 1982. Colony founding in *Myrmecocystus mimicus* Wheeler (Hymenoptera: Formicidae) and the evolution of foundress associations. *Behavioral Ecology and Sociobiology*, 10(2): 137–147.

Basilewsky, P. 1952. Les Cossyphodidae de l'Afrique Noire (Coleoptera, Heteromeroidea, Tenebrionaria). *Publicações Culturais da Companhia de Diamantes de Angola, Museu do Dundo*, no. 14, pp. 7–16.

Bates, H. W. 1863. *The naturalist on the River Amazons*, 2 vols. John Murray, London. ix + 351 pp; vi + 423 pp.

Bausenwein, F. 1960. Untersuchungen über sekretorische Drüsen des Kopf- und Brustabschnittes in der *Formica rufa*-Gruppe. *Časopis Československé Společnosti Entomologické*, 57(1): 31–57.

Bazire-Benazet, M. 1957. Sur la formation de l'oeuf alimentaire chez *Atta* sexdens rubropilosa, Forel, 1908 (Hym. Formicidae). *Comptes Rendus de l'Académie des Sciences, Paris,* 244(9): 1277–80.

Beattie, A. J. 1985. *The evolutionary ecology of ant-plant mutualisms.* Cambridge University Press, New York. x + 182 pp.

Beattie, A. J., and D. C. Culver. 1977. Effects of the mound nests of the ant, *Formica obscuripes,* on the surrounding vegetation. *American Midland Naturalist,* 97(2): 390–399.

——— 1981. The guild of myrmecochores in the herbaceous flora of West Virginia forests. *Ecology,* 62(1): 107–115.

Beattie, A. J., C. L. Turnbull, R. B. Knox, and E. G. Williams. 1984. Ant inhibition of pollen function: a possible reason why ant pollination is rare. *American Journal of Botany,* 71(3): 421–426.

Beattie, A. J., C. L. Turnbull, T. Hough, S. Jobson, and R. B. Knox. 1985. The vulnerability of pollen and fungal spores to ant secretions: evidence and some evolutionary implications. *American Journal of Botany,* 72(4): 606–614.

Beattie, A. J., C. L. Turnbull, T. Hough, and R. B. Knox. 1986. Antibiotic production: a possible function for the metapleural glands of ants (Hymenoptera: Formicidae). *Annals of the Entomological Society of America,* 79(3): 448–450.

Beccari, O. 1884. Piante ospitatrici, ossia piante formicarie della Malesia e della Papuasia. *Malesia* (Genoa), 2: 1–340.

Beck, H. 1972. Vergleichende histologische Untersuchungen an *Polyergus rufescens* Latr. und *Raptiformica sanguinea* Latr. *Insectes Sociaux,* 19(4): 301–342.

Beck, L. 1971. Bodenzoologische Gliederung und Charakterisierung des amazonischen Regenswaldes. *Amazoniana,* 3(1): 69–132.

Beier, M. 1970. Myrmecophile Pseudoskorpione aus Brasilien. *Annalen des Naturhistorischen Museums in Wien,* 74: 51–56.

Bellas, T., and B. Hölldobler. 1985. Constituents of mandibular and Dufour's glands of an Australian *Polyrhachis* weaver ant. *Journal of Chemical Ecology,* 11(4): 525–538.

Belt, T. 1874. *The naturalist in Nicaragua.* John Murray, London. xvi + 403 pp.

Ben-Dov, Y. 1978. *Andaspis formicarum* n. sp. (Homoptera, Diaspididae) associated with a species of *Melissotarsus* (Hymenoptera, Formicidae) in South Africa. *Insectes Sociaux,* 25(4): 315–321.

Benjamin, R. K. 1971. Introduction and supplement to Roland Thaxter's contribution towards a monograph of the Laboulbeniaceae. *Bibliotheca Mycologia,* 30: 1–155.

Bennett, B., and M. D. Breed. 1985. On the association between *Pentaclethra macroloba* (Mimosaceae) and *Paraponera clavata* (Hymenoptera: Formicidae) colonies. *Biotropica,* 17(3): 253–255.

Benson, W. W. 1985. Amazon ant-plants. *In* G. Prance and T. Lovejoy, eds., *Amazonia,* pp. 239–266. Pergamon Press, Elmsford, N.Y.

Benthuysen, J. L., and M. S. Blum. 1974. Quantitative sensitivity of the ant *Pogonomyrmex barbatus* to the enantiomers of its alarm pheromone. *Journal of the Georgia Entomological Society,* 9(4): 235–238.

Bentley, B. L. 1977. Extrafloral nectaries and protection by pugnacious bodyguards. *Annual Review of Ecology and Systematics,* 8: 407–427.

Bequaert, J. 1922a. The predaceous enemies of ants. *Bulletin of the American Museum of Natural History,* 45:271–331.

——— 1922b. Ants in their diverse relations to the plant world. *Bulletin of the American Museum of Natural History,* 45: 333–584.

Beraldo, M. J. A. H., and E. G. Mendes. 1982. The influence of temperature on oxygen consumption rates of workers of two leaf cutting ants, *Atta laevigata* (F. Smith, 1858) and *Atta sexdens rubropilosa* (Forel, 1908). *Comparative Biochemistry and Physiology,* ser. A, 71(3): 419–424.

Berg, R. Y. 1972. Dispersal ecology of *Vancouveria* (Berberidaceae). *American Journal of Botany,* 59(2): 109–122.

——— 1975. Myrmecochorous plants in Australia and their dispersal by ants. *Australian Journal of Botany,* 23(3): 475–508.

——— 1979. Legume, seed, and myrmecochorous dispersal in *Kennedia* and *Hardenbergia* (Fabaceae), with a remark on the Durian theory. *Norwegian Journal of Botany,* 26(4): 229–254.

Bergström, G., and J. Löfqvist. 1970. Chemical basis for odour communication in four species of *Lasius* ants. *Journal of Insect Physiology,* 16(12): 2353–75.

——— 1972. Similarities between the Dufour gland secretions of the ants *Camponotus ligniperda* (Latr.) and *Camponotus herculeanus* (L.) (Hym.). *Entomologica Scandinavica,* 3(2): 225–238.

——— 1973. Chemical congruence of the complex odoriferous secretions from Dufour's gland in three species of ants of the genus *Formica. Journal of Insect Physiology,* 19(4): 877–907.

Berkelheimer, R. C. 1984. An electrophoretic analysis of queen number in three species of dolichoderine ants. *Insectes Sociaux,* 31(2): 132–141.

Bernard, F. 1948. Les insectes sociaux du Fezzân: comportement et biogéographie. *Institut de Recherches Sahariennes de l'Université d'Alger, Mission Scientifique du Fezzân (1944–1945),* part V (Zoologie), pp. 85–201.

——— 1951. Adaptations au milieu chez les fourmis sahariennes. *Bulletin de la Société d'Histoire Naturelle de Toulouse,* 86(1–2): 88–96.

——— 1953. Les fourmis du Tassili des Ajjer (Sahara Central). *Institut de Recherches Sahariennes de l'Université d'Alger, Mission Scientifique au Tassili des Ajjer (1949),* part I (Recherches zoologiques et médicales), pp. 1–132.

——— 1956. Révision des fourmis paléarctiques du genre *Cardiocondyla. Bulletin de la Société d'Histoire Naturelle de l'Afrique du Nord,* 47(7–8): 299–306.

——— 1968. *Les fourmis (Hymenoptera Formicidae) d'Europe occidentale et septentrionale* (Faune de l'Europe et du Bassin Méditerranéen, no. 3). Masson et Cie, Paris. 411 pp.

——— 1981. Revision of the genus *Messor* (harvesting ants) on a biometrical basis. *In* P. E. Howse and J.-L. Clément, eds., *Biosystematics of social insects,* pp. 141–145. Academic Press, New York.

Berndt, K. P., and W. Eichler. 1987. Die Pharaoameise, *Monomorium pharaonis* (L.) (Hym., Myrmicidae). *Mitteilungen aus den Zoologischen Museum in Berlin,* 63(1): 1–188.

Berndt, K. P., and J. Nitschmann. 1979. The physiology of reproduction in the pharaoh's ant (*Monomorium pharaonis* L.), 2: The unmated queens. *Insectes Sociaux,* 26(2): 137–145.

Bernstein, R. A. 1974. Seasonal food abundance and foraging activity in some desert ants. *American Naturalist,* 108: 490–498.

——— 1975. Foraging strategies of ants in response to variable food density. *Ecology,* 56(1): 213–219.

——— 1976. The adaptive value of polymorphism in an alpine ant, *Formica neorufibarbis gelida* Wheeler. *Psyche,* 83(2): 180–184.

——— 1978. Slavery in the subfamily Dolichoderinae (F. Formicidae) and its ecological consequences. *Experientia,* 34: 1281–82.

——— 1979a. Evolution of niche breadth in populations of ants. *American Naturalist,* 114(4): 533–544.

——— 1979b. Schedules of foraging activity in species of ants. *Journal of Animal Ecology,* 48(3): 921–930.

Bernstein, S., and R. A. Bernstein. 1969. Relationships between foraging efficiency and the size of the head and component brain and sensory structures in the red wood ant. *Brain Research,* 16(1): 85–104.

Berton, F., and A. Lenoir. 1986. Fermeture des sociétés parthenogénétiques de *Cataglyphis cursor* (Hymenoptera, Formicidae). *Actes des Colloques Insectes Sociaux* (L'Union Internationale pour l'Etude des Insectes Sociaux, Section française, Compte Rendu Colloque Annuel, Vaison la Romaine, 1985), 3: 197–209.

Besuchet, C. 1972. Les Coléoptères Aculagnathides. *Revue Suisse de Zoologie,* 79(1): 99–145.

Bethe, A. 1898. Dürfen wir den Ameisen und Bienen psychische Qualitäten zuschreiben? *Pflügers Archiv für die Gesamte Physiologie,* 70: 15–100.

Betrem, J. G. 1960. Ueber die Systematik der *Formica rufa*-Gruppe. *Tijdschrift voor Entomologie,* 103(1–2): 51–81.

Bhatkar, A. P. 1979a. Evidence of intercolonial food exchange in fire ants and other Myrmicinae, using radioactive phosphorus. *Experientia,* 35: 1172–73.

——— 1979b. Trophallactic appeasement in ants from distant colonies. *Folia Entomológica Mexicana,* 41: 135–143.

——— 1983. Interspecific trophallaxis in ants, its ecological and evolutionary significance. *In* P. Jaisson, ed., *Social insects in the tropics*, vol. 2, pp. 105–123. Université Paris-Nord, Paris.

Bhatkar, A. P., and W. J. Kloft. 1977. Evidence, using radioactive phosphorus, of interspecific food exchange in ants. *Nature*, 265: 140–142.

Bhatkar, A. P., and W. H. Whitcomb. 1970. Artificial diet for rearing various species of ants. *Florida Entomologist*, 53(4): 229–232.

Bier, K. 1952. Zur scheinbaren Thelytokie der Ameisengattung *Lasius*. *Naturwissenschaften*, 39(18): 433.

——— 1953. Beziehungen zwischen Nährzellkerngrösse und Ausbildung ribonukleinsäurehaltiger Strukturen in den Oocyten von *Formica rufa rufopratensis minor* Gösswald. *In* W. Herre, ed., *Verhandlungen der Deutschen Zoologischen Gesellschaft* (Freiburg, 1952) (*Zoologischer Anzeiger* Supplement, vol. 17), pp. 369–374. Akademische Verlagsgesellschaft, Geest & Portig K.-G., Leipzig.

——— 1954a. Über den Saisondimorphismus der Oogenese von *Formica rufa rufo-pratensis minor* Gössw. und dessen Bedeutung für die Kastendetermination. *Biologisches Zentralblatt*, 73(3–4): 170–190.

——— 1954b. Über den Einfluss der Königin auf die Arbeiterinnenfertilität im Ameisenstaat. *Insectes Sociaux*, 1(1): 7–19.

——— 1956. Arbeiterinnenfertilität und Aufzucht von Geschlechtstieren als Regulationsleistung des Ameisenstaates. *Insectes Sociaux*, 3(1): 177–184.

——— 1958a. Die Bedeutung der Jungarbeiterinnen für die Geschlechtstieraufzucht im Ameisenstaat. *Biologisches Zentralblatt*, 77(3): 257–265.

——— 1958b. Die Regulation der Sexualität in den Insektenstaaten. *Ergebnisse der Biologie*, 20: 97–126.

Bigley, W. S., and S. B. Vinson. 1975. Characterization of a brood pheromone isolated from the sexual brood of the imported fire ant, *Solenopsis invicta*. *Annals of the Entomological Society of America*, 68(2): 301–304.

Billen, J. P. J. 1982a. Ovariole development in workers of *Formica sanguinea* Latr. (Hymenoptera: Formicidae). *Insectes Sociaux*, 29(1): 86–94.

——— 1982b. The Dufour gland closing apparatus in *Formica sanguinea* Latreille (Hymenoptera, Formicidae). *Zoomorphology*, 99(3): 235–244.

——— 1984a. Morphology of the tibial gland in the ant *Crematogaster scutellaris*. *Naturwissenschaften*, 71(6): 324–325.

——— 1984b. Stratification in the nest of the slave-making ant *Formica sanguinea* Latreille, 1798 (Hymenoptera, Formicidae). *Annales de la Société Royale Zoologique de Belgique*, 114(2): 215–225.

——— 1985a. Comparative ultrastructure of the poison and Dufour glands in Old and New World army ants (Hymenoptera, Formicidae). *Actes des Colloques Insectes Sociaux* (L'Union Internationale pour l'Etude des Insectes Sociaux, Section française, Compte Rendu Colloque Annuel, Diepenbeek, 1984), 2: 17–26.

——— 1985b. Ultrastructure de la glande de Pavan chez *Dolichoderus quadripunctatus* (L.) (Hymenoptera: Formicidae). *Actes des Colloques Insectes Sociaux* (L'Union Internationale pour l'Etude des Insectes Sociaux, Section française, Compte Rendu Colloque Annuel, Diepenbeek, 1984), 2: 87–95.

——— 1986a. Comparative morphology and ultrastructure of the Dufour gland in ants (Hymenoptera: Formicidae). *Entomologia Generalis*, 11(3–4): 165–181.

——— 1986b. Morphology and ultrastructure of the Dufour's and venom gland in the ant, *Myrmica rubra* (L.) (Hymenoptera: Formicidae). *International Journal of Insect Morphology & Embryology*, 15(1–2): 13–25.

——— 1986c. Morphology and ultrastructure of the abdominal glands in dolichoderine ants (Hymenoptera, Formicidae). *Insectes Sociaux*, 33(3): 278–295.

Billen, J. P. J., R. P. Evershed, A. B. Attygalle, E. D. Morgan, and D. G. Ollett. 1986. Contents of Dufour glands of workers of three species of *Tetramorium* (Hymenoptera: Formicidae). *Journal of Chemical Ecology*, 12(3): 669–685.

Bingham, C. T. 1903. *Hymenoptera*, vol. 2: *Ants and cuckoo-wasps* (The fauna of British India, including Ceylon and Burma). Taylor and Francis, London. xix + 506 pp.

——— 1907. *Butterflies*, vol. 2 (The fauna of British India, including Ceylon and Burma). Taylor and Francis, London. viii + 480 pp.

Bitancourt, A. A. 1941. Expressão matematica do crescimento de formigueiros de "*Atta sexdens rubropilosa*" representado pelo aumento do numero de olheiros. *Archivos do Instituto Biológico, São Paulo*, 12(16): 229–236.

Blair, K. G. 1929. Some new species of myrmecophilous Tenebrionidae (Col.). *Zoologischer Anzeiger*, 82: 238–247.

Blanton, C. M., and J. J. Ewel. 1985. Leaf-cutting ant herbivory in successional and agricultural tropical ecosystems. *Ecology*, 66(3): 861–869.

Blum, M. S. 1966. The source and specificity of trail pheromones in *Termitopone*, *Monomorium* and *Huberia*, and their relation to those of some other ants. *Proceedings of the Royal Entomological Society of London*, ser. A, 41(10–12): 155–160.

——— 1969. Alarm pheromones. *Annual Review of Entomology*, 14: 57–80.

——— 1970. The chemical basis of insect sociality. *In* M. Beroza, ed., *Chemicals controlling insect behavior*, pp. 61–94. Academic Press, New York.

——— 1974a. Myrmicine trail pheromones: specificity, source and significance. *Journal of the New York Entomological Society*, 82(2): 141–147.

——— 1974b. Pheromonal sociality in the Hymenoptera. *In* M. C. Birch, ed., *Pheromones*, pp. 222–249. North-Holland, Amsterdam.

——— 1981a. *Chemical defenses of arthropods*. Academic Press, New York. xii + 562 pp.

——— 1981b. Sex pheromones in social insects: chemotaxonomic potential. *In* P. E. Howse and J.-L. Clément, eds., *Biosystematics of social insects* (Systematics Association special volume no. 19), pp. 163–174. Academic Press, New York.

——— 1982. Pheromonal bases of insect sociality: communications, conundrums and caveats. *Les Colloques de l'Institut National de la Recherche Agronomique* (INRA) (Les Médiateurs Chimiques Agissant Comportement Insectes, Versailles, 1981), no. 7, pp. 149–162.

——— 1985. Alkaloidal ant venoms: chemistry and biological activities. *In* P. A. Hedin, ed., *Bioregulators for pest control* (American Chemical Society Symposium Series, no. 276), pp. 393–408. American Chemical Society, Washington, D.C.

——— 1987. The basis and evolutionary significance of recognitive olfactory acuity in insect societies. *In* J. M. Pasteels and J.-L. Deneubourg, eds., *From individual to collective behavior in social insects* (*Experientia* Supplement, vol. 54), pp. 277–293. Birkhäuser Verlag, Basel.

Blum, M. S., and H. R. Hermann. 1978a. Venoms and venom apparatuses of the Formicidae: Myrmeciinae, Ponerinae, Dorylinae, Pseudomyrmecinae, Myrmicinae, and Formicinae. *In* G. V. R. Born, O. Eichler, A. Farah, H. Herken, and A. D. Welch, eds., *Handbook of experimental pharmacology*, pp. 801–869. Springer-Verlag, New York.

——— 1978b. Venoms and venom apparatuses of the Formicidae: Dolichoderinae and Aneuretinae. *In* G. V. R. Born, O. Eichler, A. Farah, H. Herken, and A. D. Welch, eds., *Handbook of experimental pharmacology*, pp. 871–894. Springer-Verlag, New York.

Blum, M. S., and C. A. Portocarrero. 1964. Chemical releasers of social behavior, IV: The hindgut as the source of the odor trail pheromone in the Neotropical army ant genus *Eciton*. *Annals of the Entomological Society of America*, 57(6): 793–794.

——— 1966. Chemical releasers of social behavior, X: An attine trail substance in the venom of a non-trail laying myrmicine, *Daceton armigerum* (Latreille). *Psyche*, 73(2): 150–155.

Blum, M. S., and G. N. Ross. 1965. Chemical releasers of social behaviour, V: Source, specificity, and properties of the odour trail pheromone of *Tetramorium guineense* (F.) (Formicidae: Myrmicinae). *Journal of Insect Physiology*, 11(7): 857–868.

Blum, M. S., and S. L. Warter. 1966. Chemical releasers of social behavior, VII: The isolation of 2-heptanone from *Conomyrma pyramica* (Hymenoptera: Formicidae: Dolichoderinae) and its modus operandi as a releaser of alarm and digging behavior. *Annals of the Entomological Society of America*, 59(4): 774–779.

Blum, M. S., and E. O. Wilson. 1964. The anatomical source of trail substances in formicine ants. *Psyche*, 71(1): 28–31.

Blum, M. S., S. L. Warter, R. S. Monroe, and J. C. Chidester. 1963. Chemical releasers of social behaviour, I: Methyl-*n*-amyl ketone in *Iridomyrmex pruinosus* (Roger) (Formicidae: Dolichoderinae). *Journal of Insect Physiology*, 9(6): 881–885.

Blum, M. S., J. C. Moser, and A. D. Cordero. 1964. Chemical releasers of social behavior, II: Source and specificity of the odor trail substances in four attine genera (Hymenoptera: Formicidae). *Psyche*, 71(1): 1–7.

Blum, M. S., S. L. Warter, and J. G. Traynham. 1966. Chemical releasers of social behaviour, VI: The relation of structure to activity of ketones as releasers of alarm for *Iridomyrmex pruinosus* (Roger). *Journal of Insect Physiology*, 12(4): 419–427.

Blum, M. S., F. Padovani, and E. Amante. 1968a. Alkanones and terpenes in the mandibular glands of *Atta* species (Hymenoptera: Formicidae). *Comparative Biochemistry and Physiology*, 26(1): 291–299.

Blum, M. S., F. Padovani, H. R. Hermann, and P. B. Kannowski. 1968b. Chemical releasers of social behavior, XI: Terpenes in the mandibular glands of *Lasius umbratus*. *Annals of the Entomological Society of America*, 61(6): 1354–59.

Blum, M. S., R. M. Crewe, J. H. Sudd, and A. W. Garrison. 1969. 2-Hexenal: isolation and function in a *Crematogaster* (*Atopogyne*) sp. *Journal of the Georgia Entomological Society*, 4(4): 145–148.

Blum, M. S., R. M. Crewe, and J. M. Pasteels. 1971. Defensive secretion of *Lomechusa strumosa*, a myrmecophilous beetle. *Annals of the Entomological Society of America*, 64(4): 975–976.

Blum, M. S., T. H. Jones, B. Hölldobler, H. M. Fales, and T. Jaouni. 1980. Alkaloidal venom mace: offensive use by a thief ant. *Naturwissenschaften*, 67(3): 144–145.

Boch, R., and R. A. Morse. 1974. Discrimination of familiar and foreign queens by honey bee swarms. *Annals of the Entomological Society of America*, 67(4): 709–711.

——— 1979. Individual recognition of queens by honey bee swarms. *Annals of the Entomological Society of America*, 72(1): 51–53.

——— 1981. Effects of artificial odors and pheromones on queen discrimination by honey bees (*Apis mellifera* L.). *Annals of the Entomological Society of America*, 74(1): 66–67.

——— 1982. Genetic factor in queen recognition odors of honey bees. *Annals of the Entomological Society of America*, 75(6): 654–656.

Bolivar, I. 1905. Les blattes myrmécophiles. *Mitteilungen der Schweizerischen Entomologischen Gesellschaft*, 11(3): 134–141.

Bolton, B. 1971. Two new species of the ant genus *Epitritus* from Ghana, with a key to the world species (Hym. Formicidae). *Entomologist's Monthly Magazine*, 107: 205–208.

——— 1973a. The ant genera of West Africa: a synonymic synopsis with keys (Hymenoptera: Formicidae). *Bulletin of the British Museum (Natural History), Entomology*, 27(6): 319–368.

——— 1973b. The ant genus *Polyrhachis* F. Smith in the Ethiopian Region (Hymenoptera: Formicidae). *Bulletin of the British Museum (Natural History), Entomology*, 28(5): 285–369.

——— 1973c. A remarkable new arboreal ant genus (Hym., Formicidae) from West Africa. *Entomologist's Monthly Magazine*, 108: 234–237.

——— 1974a. A revision of the Palaeotropical arboreal ant genus *Cataulacus* F. Smith (Hymenoptera: Formicidae). *Bulletin of the British Museum (Natural History), Entomology*, 30(1): 1–105.

——— 1974b. A revision of the ponerine ant genus *Plectroctena* F. Smith (Hymenoptera: Formicidae). *Bulletin of the British Museum (Natural History), Entomology*, 30(6): 311–338.

——— 1975a. A revision of the ant genus *Leptogenys* Roger (Hymenoptera: Formicidae) in the Ethiopian Region with a review of the Malagasy species. *Bulletin of the British Museum (Natural History), Entomology*, 31(7): 237–305.

——— 1975b. A revision of the African ponerine ant genus *Psalidomyrmex* André (Hymenoptera: Formicidae). *Bulletin of the British Museum (Natural History), Entomology*, 32(1): 3–16.

——— 1975c. The *sexspinosa*-group of the ant genus *Polyrhachis* F. Smith (Hym. Formicidae). *Journal of Entomology*, ser. B, 44(1): 1–14.

——— 1976. The ant tribe Tetramoriini (Hymenoptera: Formicidae): constituent genera, review of smaller genera and revision of *Triglyphothrix* Forel. *Bulletin of the British Museum (Natural History), Entomology*, 34(5): 281–378.

——— 1977. The ant tribe Tetramoriini (Hymenoptera: Formicidae): the genus *Tetramorium* Mayr in the Oriental and Indo-Australian regions, and in Australia. *Bulletin of the British Museum (Natural History), Entomology*, 36(2): 67–151.

——— 1979. The ant tribe Tetramoriini (Hymenoptera: Formicidae): the genus *Tetramorium* Mayr in the Malagasy Region and in the New World. *Bulletin of the British Museum (Natural History), Entomology*, 38(4): 129–181.

——— 1980. The ant tribe Tetramoriini (Hymenoptera: Formicidae): the genus *Tetramorium* Mayr in the Ethiopian zoogeographical region. *Bulletin of the British Museum (Natural History), Entomology*, 40(3): 193–384.

——— 1981a. A revision of the ant genera *Meranoplus* F. Smith, *Dicroaspis* Emery and *Calyptomyrmex* Emery (Hymenoptera: Formicidae) in the Ethiopian zoogeographical region. *Bulletin of the British Museum (Natural History), Entomology*, 42(2): 43–81.

——— 1981b. A revision of six minor genera of Myrmicinae (Hymenoptera: Formicidae) in the Ethiopian zoogeographical region. *Bulletin of the British Museum (Natural History), Entomology*, 43(4): 245–307.

——— 1982. Afrotropical species of the myrmicine ant genera *Cardiocondyla*, *Leptothorax*, *Melissotarsus*, *Messor* and *Cataulacus* (Formicidae). *Bulletin of the British Museum (Natural History), Entomology*, 45(4): 307–370.

——— 1983. The Afrotropical dacetine ants (Formicidae). *Bulletin of the British Museum (Natural History), Entomology*, 46(4): 267–416.

——— 1984. Diagnosis and relationships of the myrmicine ant genus *Ishakidris* gen. n. (Hymenoptera: Formicidae). *Systematic Entomology*, 9(4): 373–382.

——— 1986a. A taxonomic review of the tetramoriine ant genus *Rhoptromyrmex* (Hymenoptera: Formicidae). *Systematic Entomology*, 11(1): 1–17.

——— 1986b. Apterous females and shift of dispersal strategy in the *Monomorium salomonis*-group (Hymenoptera: Formicidae). *Journal of Natural History*, 20(2): 267–272.

——— 1987. A review of the *Solenopsis* genus-group and revision of Afrotropical *Monomorium* Mayr (Hymenoptera: Formicidae). *Bulletin of the British Museum (Natural History), Entomology*, 54(3): 263–452.

——— 1988a. A new socially parasitic *Myrmica*, with a reassessment of the genus. *Systematic Entomology*, 13(1): 1–11.

——— 1988b. A review of *Paratopula* Wheeler, a forgotten genus of myrmicine ants. *Entomologist's Monthly Magazine*, 124: 127–143.

——— 1988c. *Secostruma*, a new subterranean tetramoriine ant genus (Hymenoptera: Formicidae). *Systematic Entomology*, 13(3): 263–270.

Bolton, B., W. H. Gotwald, and J. M. Leroux. 1976. A new West African ant of the genus *Plectroctena* with ecological notes (Hymenoptera: Formicidae). *Annales de l'Université d'Abidjan*, ser. E, 9: 371–381.

Bonavita-Cougourdan, A. 1983. Activité antennaire et flux trophallactique chez la fourmi *Camponotus vagus* Scop. (Hymenoptera, Formicidae). *Insectes Sociaux*, 30(4): 423–442.

Bonavita-Cougourdan, A., and L. Morel. 1984. Les activités antennaires au cours des contacts trophallactiques chez la fourmi *Camponotus vagus* Scop. Ont-elles valeur de signal? *Insectes Sociaux*, 31(2): 113–131.

——— 1985. Polyethism in social interactions in ants. *Behavioural Processes*, 11(4): 425–433.

——— 1988. Interindividual variability and idiosyncrasy in social behaviours in the ant *Camponotus vagus* Scop. *Ethology*, 77(1): 58–66.

Bonavita-Cougourdan, A., J.-L. Clément, and C. Lange. 1987a. Nestmate recognition: the role of cuticular hydrocarbons in the ant *Camponotus vagus* Scop. *Journal of Entomological Science*, 22(1): 1–10.

——— 1987b. Subcaste discrimination in the ant *Camponotus vagus* Scop. *In* J. Eder and H. Rembold, eds., *Chemistry and biology of social insects* (Proceedings of the Tenth International Congress of the International Union for the Study of Social Insects, Munich, 1986), p. 475. Verlag J. Peperny, Munich.

——— 1988. Reconnaissance des larves chez la Fourmi *Camponotus vagus* Scop. phénotypes larvaires des spectres d'hydrocarbures cuticulaires. *Comptes Rendus de l'Académie des Sciences, Paris*, ser. 3, 306(9): 299–305.

Bond, W. J., and G. J. Breytenbach. 1985. Ants, rodents and seed predation in Proteaceae. *South African Journal of Zoology*, 20(3): 150–154.

Bond, W. J., and P. Slingsby. 1983. Seed dispersal by ants in shrublands of

the Cape Province and its evolutionary implications. *South African Journal of Science*, 79(6) (June): 231–233.
——— 1984. Collapse of an ant-plant mutualism: the Argentine ant (*Iridomyrmex humilis*) and myrmecochorous Proteaceae. *Ecology*, 65(4): 1031–37.
Boomsma, J. J., and J. A. Isaaks. 1985. Energy investment and respiration in queens and males of *Lasius niger* (Hymenoptera: Formicidae). *Behavioral Ecology and Sociobiology*, 18(1): 19–27.
Boomsma, J. J., and A. Leusink. 1981. Weather conditions during nuptial flights of four European ant species. *Oecologia*, 50(2): 236–241.
Boomsma, J. J., and A. J. van Loon. 1982. Structure and diversity of ant communities in successive coastal dune valleys. *Journal of Animal Ecology*, 51(3): 957–974.
Boomsma, J. J., A. A. Mabelis, M. G. M. Verbeek, and E. C. Los. 1987. Insular biogeography and distribution ecology of ants on the Frisian Islands. *Journal of Biogeography*, 14(1): 21–37.
Borgmeier, T. 1925. Novos subsidios para o conhecimento da familia Phoridae. *Archivos do Museu Nacional, Rio de Janeiro*, 25: 85–281.
——— 1929. Über attophile Phoriden. *Zoologischer Anzeiger*, 82: 493–517.
——— 1931. Sobre alguns coleopteros ecitophilos do Brasil (Staphylinidae). *Revista de Entomologia*, 1(3): 355–367.
——— 1939. Sobre alguns Diapriideos myrmecophilos, principalmente do Brazil (Hym. Diapriidae). *Revista de Entomologia*, 10(3): 530–545.
——— 1948. Zur Kenntnis der bei *Eciton* lebenden myrmekophilen Histeriden (Col.). *Revista de Entomologia*, 19(3): 377–400.
——— 1949. Formigas novas ou pouco conhecidas de Costa Rica e da Argentina (Hymenoptera, Formicidae). *Revista Brasileira de Biologia*, 9(2): 201–210.
——— 1954. Two interesting dacetine ants from Brazil. *Revista Brasileira de Biologia*, 14(3): 279–284.
——— 1955. Die Wanderameisen der Neotropischen Region (Hym. Formicidae). *Studia Entomologica*, 3: 1–716.
——— 1959. Revision der Gattung *Atta* Fabricius (Hymenoptera, Formicidae). *Studia Entomologica*, n.s., 2(1–4): 321–390.
——— 1963. New or little known *Coniceromyia*, and some other Neotropical or Paleotropical Phoridae (Dipt.). *Studia Entomologica*, n.s., 6(1–4): 449–480.
——— 1968. A catalogue of the Phoridae of the world (Diptera, Phoridae). *Studia Entomologica*, n.s., 11(1–4): 1–367.
——— 1971. Supplement to a catalog of the Phoridae of the world (Diptera). *Studia Entomologica*, n.s., 14(1–4): 177–224.
Borgmeier, T., and A. P. Prado. 1975. New or little known Neotropical phorid flies, with description of eight new genera (Diptera: Phoridae). *Studia Entomologica*, n.s., 18(1): 3–90.
Bossert, W. H., and E. O. Wilson. 1963. The analysis of olfactory communication among animals. *Journal of Theoretical Biology*, 5(3): 443–469.
Bourke, A. F. G. 1988a. Dominance orders, worker reproduction and queen-worker conflict in the slave-making ant *Harpagoxenus sublaevis*. *Behavioral Ecology and Sociobiology*, 23(5): 323–333.
——— 1988b. Worker reproduction in the higher eusocial Hymenoptera. *Quarterly Review of Biology*, 63(3): 291–311.
Bourke, A. F. G., T. M. van der Have, and N. R. Franks. 1988. Sex ratio determination and worker reproduction in the slave-making ant *Harpagoxenus sublaevis*. *Behavioral Ecology and Sociobiology*, 23(4): 233–245.
Bourne, R. A. 1973. A taxonomic study of the ant genus *Lasius* Fabricius in the British Isles (Hymenoptera: Formicidae). *Journal of Entomology*, ser. B, 42(1): 17–27.
Boven, J. K. A. van. 1970. *Myrmica faniensis*, une nouvelle espèce parasite (Hymenoptera, Formicidae). *Bulletin & Annales de la Société Royale d'Entomologie de Belgique*, 106(1–3): 127–132.
Boyd, N. D., and M. M. Martin. 1975. Faecal proteinases of the fungus-growing ant, *Atta texana:* their fungal origin and ecological significance. *Journal of Insect Physiology*, 21(11): 1815–20.
Bradbury, J. W. 1985. Contrasts between insects and vertebrates in the evolution of male display, female choice, and lek mating. *In* B. Hölldobler and M. Lindauer, eds., *Experimental behavioral ecology and sociobiology* (Fortschritte der Zoologie, no. 31), pp. 273–289. Sinauer Associates, Sunderland, Mass.
Bradley, G. A. 1972. Transplanting *Formica obscuripes* and *Dolichoderus taschenbergi* (Hymenoptera: Formicidae) colonies in Jack Pine stands of southeastern Manitoba. *Canadian Entomologist*, 104(2): 245–249.
——— 1973. Interference between nest populations of *Formica obscuripes* and *Dolichoderus taschenbergi* (Hymenoptera: Formicidae). *Canadian Entomologist*, 105(2): 1525–28.
Bradshaw, J. W. S. 1981. The physicochemical transmission of two components of a multiple chemical signal in the African weaver ant (*Oecophylla longinoda*). *Animal Behaviour*, 29(2): 581–585.
Bradshaw, J. W. S., and P. E. Howse. 1984. Sociochemicals of ants. *In* W. J. Bell and R. T. Cardé, eds., *Chemical ecology of insects*, pp. 429–473. Chapman and Hall, London.
Bradshaw, J. W. S., R. Baker, and P. E. Howse. 1975. Multicomponent alarm pheromones of the weaver ant. *Nature*, 258: 230–231.
——— 1979. Multicomponent alarm pheromones in the mandibular glands of major workers of the African weaver ant, *Oecophylla longinoda*. *Physiological Entomology*, 4(1): 15–25.
Bradshaw, J. W. S., P. E. Howse, and R. Baker. 1986. A novel autostimulatory pheromone regulating transport of leaves in *Atta cephalotes*. *Animal Behaviour*, 34(1): 234–240.
Brand, J. M., M. S. Blum, H. M. Fales, and J. M. Pasteels. 1973a. The chemistry of the defensive secretion of the beetle *Drusilla canaliculata*. *Journal of Insect Physiology*, 19(2): 369–382.
Brand, J. M., R. M. Duffield, J. G. MacConnell, M. S. Blum, and H. M. Fales. 1973b. Caste-specific compounds in male carpenter ants. *Science*, 179: 388–389.
Brand, J. M., H. M. Fales, E. A. Sokoloski, J. G. MacConnell, M. S. Blum, and R. M. Duffield. 1973c. Identification of mellein in the mandibular gland secretions of carpenter ants. *Life Sciences*, 13(3): 201–211.
Brand, J. M., M. S. Blum, H. A. Lloyd, and D. J. C. Fletcher. 1974. Monoterpene hydrocarbons in the poison gland secretion of the ant *Myrmicaria natalensis* (Hymenoptera: Formicidae). *Annals of the Entomological Society of America*, 67(3): 525–526.
Brandão, C. R. F. 1978. Division of labor within the worker caste of *Formica perpilosa* Wheeler (Hymenoptera: Formicidae). *Psyche*, 85(2–3): 229–237.
——— 1983. Sequential ethograms along colony development of *Odontomachus affinis* Guérin (Hymenoptera, Formicidae, Ponerinae). *Insectes Sociaux*, 30(2): 193–203.
——— 1987. Queenlessness in *Megalomyrmex* (Formicidae: Myrmicinae), with a discussion on the effects of the loss of true queens in ants. *In* J. Eder and H. Rembold, eds., *Chemistry and biology of social insects* (Proceedings of the Tenth International Congress of the International Union for the Study of Social Insects, Munich, 1986), pp. 111–112. Verlag J. Peperny, Munich.
Brantjes, N. B. M. 1981. Ant, bee and fly pollination in *Epipactis palustris* (L.) Crantz (Orchidaceae). *Acta Botanica Neerlandica*, 30(1–2): 59–68.
Brauns, H., and H. Bickhardt. 1914. Descriptions of some new species of myrmecophilous beetles from southern Rhodesia together with a description of a new species of *Acritus*. *Proceedings of the Rhodesia Scientific Association*, 13(3): 32–42.
Breed, M. D. 1981. Individual recognition and learning of queen odors by worker honeybees. *Proceedings of the National Academy of Sciences of the United States of America*, 78(4): 2635–37.
——— 1983. Nestmate recognition in honey bees. *Animal Behaviour*, 31(1): 86–91.
——— 1987. Multiple inputs in the nestmate discrimination system of the honey bee. *In* J. Eder and H. Rembold, eds., *Chemistry and biology of social insects* (Proceedings of the Tenth International Congress of the International Union for the Study of Social Insects, Munich, 1986), pp. 461–462. Verlag J. Peperny, Munich.
Breed, M. D., and B. Bennett. 1987. Kin recognition in highly eusocial insects. *In* D. J. C. Fletcher and C. D. Michener, eds., *Kin recognition in animals*, pp. 243–285. John Wiley, New York.
Breed, M. D., and J. M. Harrison. 1988. Worker size, ovary development

and division of labor in the giant tropical ant, *Paraponera clavata* (Hymenoptera: Formicidae). *Journal of the Kansas Entomological Society*, 61(3): 285–291.

Breed, M. D., H. H. W. Velthuis, and G. E. Robinson. 1984. Do worker honey bees discriminate among unrelated and related larval phenotypes? *Annals of the Entomological Society of America*, 77(6): 737–739.

Breed, M. D., L. Butler, and T. M. Stiller. 1985. Kin discrimination by worker honey bees in genetically mixed groups. *Proceedings of the National Academy of Sciences of the United States of America*, 82(9): 3058–61.

Breed, M. D., J. H. Fewell, A. J. Moore, and K. R. Williams. 1987. Graded recruitment in a ponerine ant. *Behavioral Ecology and Sociobiology*, 20(6): 407–411.

Brian, M. V. 1950. The stable winter population structure in species of *Myrmica*. *Journal of Animal Ecology*, 19(2): 119–123.

——— 1951a. Caste determination in a myrmicine ant. *Experientia*, 7(5): 182–183.

——— 1951b. Summer population changes in colonies of the ant *Myrmica*. *Physiologia Comparata et Oecologia*, 2(3): 248–262.

——— 1952a. Interaction between ant colonies at an artificial nest-site. *Entomologist's Monthly Magazine*, 88: 84–88.

——— 1952b. The structure of a dense natural ant population. *Journal of Animal Ecology*, 21(1): 12–24.

——— 1953. Oviposition by workers of the ant *Myrmica*. *Physiologia Comparata et Oecologia*, 3(1): 25–36.

——— 1955. Studies of caste in *Myrmica rubra* L., 2: The growth of workers and intercastes. *Insectes Sociaux*, 2(1): 1–34.

——— 1956a. Segregation of species of the ant genus *Myrmica*. *Journal of Animal Ecology*, 25(2): 319–337.

——— 1956b. Studies of caste differentiation in *Myrmica rubra* L., 4: Controlled larval nutrition. *Insectes Sociaux*, 3(3): 369–394.

——— 1956c. The natural density of *Myrmica rubra* and associated ants in West Scotland. *Insectes Sociaux*, 3(4): 473–487.

——— 1957a. Caste determination in social insects. *Annual Review of Entomology*, 2: 107–120.

——— 1957b. The growth and development of colonies of the ant *Myrmica*. *Insectes Sociaux*, 4(3): 177–190.

——— 1963. Studies of caste differentiation in *Myrmica rubra* L., 6: Factors influencing the course of female development in the early third instar. *Insectes Sociaux*, 10(2): 91–102.

——— 1965a. Caste differentiation in social insects. *Symposium of the Zoological Society of London*, 14: 13–38.

——— 1965b. *Social insect populations*. Academic Press, London. vii + 135 pp.

——— 1968. Regulation of sexual production in an ant society. *Colloques Internationaux du Centre National de la Recherche Scientifique* (Paris, 1967), no. 173, pp. 61–76.

——— 1969. Male production in the ant *Myrmica rubra* L. *Insectes Sociaux*, 16(4): 249–268.

——— 1972. Population turnover in wild colonies. *Ekologia Polska*, 20(5): 43–53.

——— 1973. Temperature choice and its relevance to brood survival and caste determination in the ant *Myrmica rubra* L. *Physiological Zoölogy*, 46(4): 245–252.

——— 1975. Larval recognition by workers of the ant *Myrmica*. *Animal Behaviour*, 23(4): 745–756.

——— 1979a. Caste differentiation and division of labor. *In* H. R. Hermann, ed., *Social insects*, vol. 1, pp. 121–222. Academic Press, New York.

——— 1979b. Habitat differences in sexual production by two co-existent ants. *Journal of Animal Ecology*, 48(3): 943–953.

——— 1983. *Social insects: ecology and behavioural biology*. Chapman and Hall, New York. x + 377 pp.

——— 1985. Comparative aspects of caste differentiation in social insects. *In* J. A. L. Watson, B. M. Okot-Kotber, and C. Noirot, eds., *Caste differentiation in social insects*, pp. 385–398. Pergamon Press, New York.

——— 1986a. Bonding between workers and queens in the ant genus *Myrmica*. *Animal Behaviour*, 34(4): 1135–45.

——— 1986b. The importance of daylength and queens for larval care by young workers of the ant *Myrmica rubra* L. *Physiological Entomology*, 11(3): 239–249.

Brian, M. V., and A. Abbott. 1977. The control of food flow in a society of the ant *Myrmica rubra* L. *Animal Behaviour*, 25(4): 1047–55.

Brian, M. V., and M. S. Blum. 1969. The influence of *Myrmica* queen head extracts on larval growth. *Journal of Insect Physiology*, 15(12): 2213–23.

Brian, M. V., and A. D. Brian. 1951. Insolation and ant populations in the west of Scotland. *Transactions of the Royal Entomological Society of London*, 102(6): 303–330.

——— 1955. On the two forms macrogyna and microgyna of the ant *Myrmica rubra* L. *Evolution*, 9(3): 280–290.

Brian, M. V., and G. W. Elmes. 1974. Production by the ant *Tetramorium caespitum* in a southern English heath. *Journal of Animal Ecology*, 43(3): 889–903.

Brian, M. V., and E. J. M. Evesham. 1982. The role of young workers in *Myrmica* colony development. *In* M. D. Breed, C. D. Michener, and H. E. Evans, eds., *The biology of social insects* (Proceedings of the Ninth Congress of the International Union for the Study of Social Insects, Boulder, Colorado, 1982), pp. 228–232. Westview Press, Boulder.

Brian, M. V., and J. Hibble. 1963. Larval size and the influence of the queen on growth in *Myrmica*. *Insectes Sociaux*, 10(1): 71–81.

——— 1964. Studies of caste differentiation in *Myrmica rubra* L., 7: Caste bias, queen age and influence. *Insectes Sociaux*, 11(3): 223–238.

Brian, M. V., and C. Rigby. 1978. The trophic eggs of *Myrmica rubra* L. *Insectes Sociaux*, 25(1): 89–110.

Brian, M. V., J. Hibble, and D. J. Stradling. 1965. Ant pattern and density in a southern English heath. *Journal of Animal Ecology*, 34(3): 545–555.

Brian, M. V., J. Hibble, and A. F. Kelly. 1966. The dispersion of ant species in a southern English heath. *Journal of Animal Ecology*, 35(2): 281–290.

Brian, M. V., G. Elmes, and A. F. Kelly. 1967. Populations of the ant *Tetramorium caespitum* Latreille. *Journal of Animal Ecology*, 36(2): 337–342.

Bridwell, J. C. 1920. Some notes on Hawaiian and other Bethylidae (Hymenoptera) with the description of a new genus and species. *Proceedings of the Hawaiian Entomological Society*, 4(2): 291–314.

Briese, D. T. 1982a. Relationship between the seed-harvesting ants and the plant community in a semi-arid environment. *In* R. C. Buckley, ed., *Ant-plant interactions in Australia*, pp. 11–24. Dr. W. Junk, The Hague.

——— 1982b. The effect of ants on the soil of a semi-arid saltbush habitat. *Insectes Sociaux*, 29(2 bis): 375–382.

——— 1983. Different modes of reproductive behaviour (including a description of colony fission) in a species of *Chelaner* (Hymenoptera: Formicidae). *Insectes Sociaux*, 30(3): 308–316.

Briese, D. T., and B. J. Macauley. 1980. Temporal structure of an ant community in semi-arid Australia. *Australian Journal of Ecology*, 5(2): 121–134.

——— 1981. Food collection within an ant community in semi-arid Australia, with special reference to seed harvesters. *Australian Journal of Ecology*, 6(1): 1–19.

Bristow, C. M. 1983. Treehoppers transfer parental care to ants: a new benefit of mutualism. *Science*, 220: 532–533.

——— 1984. Differential benefits from ant attendance to two species of Homoptera on New York ironweed. *Journal of Animal Ecology*, 53(3): 715–726.

Brossut, R. 1976. Etude morphologique de la blatte myrmécophile *Attaphila fungicola* Wheeler. *Insectes Sociaux*, 23(2): 167–174.

Brothers, D. J. 1975. Phylogeny and classification of the aculeate Hymenoptera, with special reference to Mutillidae. *University of Kansas Science Bulletin*, 50(11): 483–648.

Brough, E. J. 1978. The multifunctional role of the mandibular gland secretion of an Australian desert ant, *Calomyrmex* (Hymenoptera: Formicidae). *Zeitschrift für Tierpsychologie*, 46(3): 279–297.

Broughton, W. B. 1963. Method in bio-acoustic terminology. *In* R.-G. Busnel, ed., *Acoustic behaviour of animals*, pp. 3–24. Elsevier, Amsterdam.

Brown, E. S. 1959. Immature nutfall of coconuts in the Solomon Islands, II: Changes in ant populations and their relation to vegetation. *Bulletin of Entomological Research*, 50(3): 523–558.

Brown, J. H., D. W. Davidson, and O. J. Reichman. 1979a. An experimental study of competition between seed-eating desert rodents and ants. *American Zoologist*, 19(4): 1129–43.
Brown, J. H., O. J. Reichman, and D. W. Davidson. 1979b. Granivory in desert ecosystems. *Annual Review of Ecology and Systematics*, 10: 201–227.
Brown, J. L. 1964. The evolution of diversity in avian territorial systems. *Wilson Bulletin*, 76(2): 160–169.
Brown, J. L., and G. H. Orians. 1970. Spacing patterns in mobile animals. *Annual Review of Ecology and Systematics*, 1: 239–262.
Brown, R. L. 1976. Behavioral observations on *Aethalion reticulatum* (Hem., Aethalionidae) and associated ants. *Insectes Sociaux*, 23(2): 99–108.
Brown, W. L. 1945. An unusual behavior pattern observed in a Szechuanese ant. *Journal of the West China Border Research Society*, ser. B, 15: 185–186.
——— 1948. A preliminary generic revision of the higher Dacetini (Hymenoptera: Formicidae). *Transactions of the American Entomological Society*, 74(2): 101–129.
——— 1949. Revision of the ant tribe Dacetini, I: Fauna of Japan, China and Taiwan. *Mushi*, 20(1): 1–25.
——— 1950. Revision of the ant tribe Dacetini, II: *Glamyromyrmex* Wheeler and closely related small genera. *Transactions of the American Entomological Society*, 76(1): 27–36.
——— 1952a. Revision of the ant genus *Serrastruma*. *Bulletin of the Museum of Comparative Zoology, Harvard*, 107(2): 67–86.
——— 1952b. The dacetine ant genus *Mesostruma* Brown. *Transactions of the Royal Society of South Australia*, 75: 9–13.
——— 1953a. A revision of the dacetine ant genus *Orectognathus*. *Memoirs of the Queensland Museum*, 13(1): 84–104.
——— 1953b. Characters and synonymies among the genera of ants, Pt. II. *Breviora*, no. 18, 8 pp.
——— 1953c. Revisionary notes on the ant genus *Myrmecia* of Australia. *Bulletin of the Museum of Comparative Zoology, Harvard*, 111(1): 1–35.
——— 1953d. Revisionary studies in the ant tribe Dacetini. *American Midland Naturalist*, 50(1): 1–137.
——— 1954a. Remarks on the internal phylogeny and subfamily classification of the family Formicidae. *Insectes Sociaux*, 1(1): 21–31.
——— 1954b. The ant genus *Strumigenys* Fred. Smith in the Ethiopian and Malagasy Regions. *Bulletin of the Museum of Comparative Zoology, Harvard*, 112(1): 1–34.
——— 1955a. A revision of the Australian ant genus *Notoncus* Emery, with notes on the other genera of Melophorini. *Bulletin of the Museum of Comparative Zoology, Harvard*, 113(6): 471–494.
——— 1955b. Ant taxonomy. *In* E. L. Kessel, ed., *A century of progress in the natural sciences, 1853–1953*, pp. 569–572. California Academy of Sciences, San Francisco.
——— 1955c. The first social parasite in the ant tribe Dacetini. *Insectes Sociaux*, 2(3): 181–186.
——— 1957. Predation of arthropod eggs by the ant genera *Proceratium* and *Discothyrea*. *Psyche*, 64(3): 115.
——— 1958a. A review of the ants of New Zealand. *Acta Hymenopterologica, Tokyo*, 1(1): 1–50.
——— 1958b. Contributions toward a reclassification of the Formicidae, II: Tribe Ectatommini (Hymenoptera). *Bulletin of the Museum of Comparative Zoology, Harvard*, 118(5): 173–362.
——— 1959a. A revision of the dacetine ant genus *Neostruma*. *Breviora*, no. 107, 13 pp.
——— 1959b. The Neotropical species of the ant genus *Strumigenys* Fr. Smith: group of *gundlachi* (Roger). *Psyche*, 66(3): 37–52.
——— 1960a. Ants, acacias and browsing mammals. *Ecology*, 41(3): 587–592.
——— 1960b. Contributions toward a reclassification of the Formicidae, III: Tribe Amblyoponini (Hymenoptera). *Bulletin of the Museum of Comparative Zoology, Harvard*, 122(4): 144–230.
——— 1962. The Neotropical species of the ant genus *Strumigenys* Fr. Smith: synopsis and key to species. *Psyche*, 69(4): 238–267.
——— 1964a. Revision of *Rhoptromyrmex*. *Pilot Register of Zoology* (Cornell University), cards nos. 11–19.
——— 1964b. The ant genus *Smithistruma:* a first supplement to the world revision (Hymenoptera: Formicidae). *Transactions of the American Entomological Society*, 89(3–4): 183–200.
——— 1965. Contributions to a reclassification of the Formicidae, IV: Tribe Typhlomyrmecini (Hymenoptera). *Psyche*, 72(1): 65–78.
——— 1967. A new *Pheidole* with reversed phragmosis (Hymenoptera: Formicidae). *Psyche*, 74(4): 331–339.
——— 1968. An hypothesis concerning the function of the metapleural glands in ants. *American Naturalist*, 102: 188–191.
——— 1969. *Strumigenys lopotyle* species nov. *Pilot Register of Zoology* (Cornell University), card no. 27.
——— 1972. *Asketogenys acubecca*, a new genus and species of dacetine ants from Malaya (Hymenoptera: Formicidae). *Psyche*, 79(1–2): 23–26.
——— 1973a. A comparison of the Hylean and Congo–West African rain forest ant faunas. *In* B. J. Meggers, E. S. Ayensu, and W. D. Duckworth, eds., *Tropical forest ecosystems in Africa and South America: a comparative review*, pp. 161–185. Smithsonian Institution Press, Washington, D.C.
——— 1973b. A new species of *Miccostruma* (Hymenoptera: Formicidae) from West Africa, with notes on the genus. *Journal of the Kansas Entomological Society*, 46(1): 32–35.
——— 1974a. A remarkable new island isolate in the ant genus *Proceratium* (Hymenoptera: Formicidae). *Psyche*, 81(1): 70–83.
——— 1974b. A supplement to the revision of the ant genus *Basiceros* (Hymenoptera: Formicidae). *Journal of the New York Entomological Society*, 82(2): 131–140.
——— 1974c. *Concoctio* genus nov., *Concoctio concenta* species nov. *Pilot Register of Zoology* (Cornell University), cards nos. 29 and 30.
——— 1974d. *Dolioponera fustigera* species nov. Insecta: Hymenoptera: Formicidae. *Pilot Register of Zoology* (Cornell University), card no. 32.
——— 1974e. *Novomessor manni* a synonym of *Aphaenogaster ensifera* (Hymenoptera: Formicidae). *Entomological News*, 85(2): 45–47.
——— 1975. Contributions toward a reclassification of the Formicidae, V: Ponerinae, tribes Platythyreini, Cerapachyini, Cylindromyrmecini, Acanthostichini, and Aenictogitini. *Search—Agriculture, Entomology* (Ithaca, N.Y.), 5(1): 1–116.
——— 1976a. *Cladarogenys* genus nov., *Cladarogenys lasia* species nov. *Pilot Register of Zoology* (Cornell University), cards nos. 33 and 34.
——— 1976b. Contributions toward a reclassification of the Formicidae, Part VI: Ponerinae, tribe Ponerini, subtribe Odontomachiti; Section A: Introduction, subtribal characters, genus *Odontomachus*. *Studia Entomologica*, n.s., 19(1–4): 67–171.
——— 1977a. An aberrant new genus of myrmicine ant from Madagascar. *Psyche*, 84(3–4): 218–224.
——— 1977b. A supplement to the world revision of *Odontomachus* (Hymenoptera: Formicidae). *Psyche*, 84(3–4): 281–285.
——— 1978. Contributions toward a reclassification of the Formicidae, Part VI: Ponerinae, tribe Ponerini, subtribe Odontomachiti; Section B: Genus *Anochetus* and bibliography. *Studia Entomologica*, n.s., 20(1–4): 549–638.
——— 1979. A remarkable new species of *Proceratium*, with dietary and other notes on the genus (Hymenoptera: Formicidae). *Psyche*, 86(4): 337–346.
——— 1980. *Protalaridris* genus nov. *Pilot Register of Zoology* (Cornell University), card no. 36, 2 pp.
——— 1985. *Indomyrma daspyx*, new genus and species, a myrmicine ant from peninsular India (Hymenoptera: Formicidae). *Israel Journal of Entomology*, 19(1): 37–49.
——— 1987. *Neoclystopsenella* (Bethylidae), a synonym of *Tapinoma* (Formicidae). *Psyche*, 94(3–4): 337.
Brown, W. L., and R. G. Boisvert. 1978. The dacetine ant genus *Pentastruma* (Hymenoptera: Formicidae). *Psyche*, 85(2–3): 201–207.
Brown, W. L., and F. M. Carpenter. 1978. A restudy of two ants from the Sicilian amber. *Psyche*, 85(4): 417–423.
Brown, W. L., and W. W. Kempf. 1960. A world revision of the ant tribe Basicerotini (Hym. Formicidae). *Studia Entomologica*, n.s., 3(1–4): 161–250.
——— 1967. *Tatuidris*, a remarkable new genus of Formicidae (Hymenoptera). *Psyche*, 74(3): 183–190.

——— 1969. A revision of the Neotropical dacetine ant genus *Acanthognathus* (Hymenoptera: Formicidae). *Psyche,* 76(2): 87–109.
Brown, W. L., and W. L. Nutting. 1949. Wing venation and the phylogeny of the Formicidae. *Transactions of the American Entomological Society,* 75(3–4): 113–134.
Brown, W. L., and R. W. Taylor. 1970. Superfamily Formicoidea. *In* D. F. Waterhouse, ed., *The Insects of Australia: a textbook for students and research workers,* pp. 951–959. Melbourne University Press, Carlton, Victoria.
Brown, W. L., and E. O. Wilson. 1956. Character displacement. *Systematic Zoology,* 5(2): 49–64.
——— 1957. *Dacetinops,* a new ant genus from New Guinea. *Breviora,* no. 77, 7 pp.
——— 1959. The evolution of the dacetine ants. *Quarterly Review of Biology,* 34(4): 278–294.
Brown, W. L., T. Eisner, and R. H. Whittaker. 1970a. Allomones and kairomones: transpecific chemical messengers. *BioScience,* 20(1): 21–22.
Brown, W. L., W. H. Gotwald, and J. Lévieux. 1970b. A new genus of ponerine ants from West Africa (Hymenoptera: Formicidae) with ecological notes. *Psyche,* 77(3): 259–275.
Brown, W. V., and B. P. Moore. 1979. Volatile secretory products of an Australian formicine ant of the genus *Calomyrmex* (Hymenoptera: Formicidae). *Insect Biochemistry,* 9(5): 451–460.
Bruch, C. 1916. Contribución al estudio de las hormigas de la Provincia de San Luis. *Revista del Museo de La Plata,* 23: 291–357.
——— 1917. Nuevas capturas de insectos mirmecófilos. *Physis* (Buenos Aires), 3(15): 458–465.
——— 1926. Nuevos histéridos ecitófilos (Col.). *Revista del Museo de La Plata,* 29: 17–33.
——— 1929. Neue myrmekophile Histeriden und Verzeichnis der aus Argentinien bekannten Ameisengäste. *Zoologischer Anzeiger,* 82: 421–437.
——— 1930. Histéridos huéspedes de *Pheidole. Revista de la Sociedad Entomologica Argentina,* 3(1): 1–12.
——— 1932a. Notas biológicas y sistematicas acerca de "Bruchomyrma acutidens" Santschi. *Revista del Museo de La Plata,* 33: 31–55.
——— 1932b. Descripción de un género y especie neuva de una hormiga parásita (Formicidae). *Revista del Museo de La Plata,* 33: 271–275.
——— 1932c. Un género nuevo de histérido mirmecófilo (Coleoptera). *Revista del Museo de La Plata,* 33: 277–281.
Brues, C. T. 1902. New and little-known guests of the Texan legionary ants. *American Naturalist,* 36: 365–378.
Brun, R. 1914. *Die Raumorientierung der Ameisen und das Orientierungsproblem im allgemeinen.* Gustav Fischer, Jena. viii + 234 pp.
——— 1952. Das zentralnervensystem von *Teleutomyrmex schneideri* Kutt. (Hym. Formicid.). *Mitteilungen der Schweizerischen Entomologischen Gesellschaft,* 25(2): 73–86.
Bruniquel, S. 1972. La ponte de la fourmi *Aphaenogaster subterranea* (Latr.): oeufs reproducteurs–oeufs alimentaires. *Comptes Rendus de l'Académie des Sciences, Paris,* ser. D, 275(3): 397–399.
Bruyn, G. J. de. 1978. Food territories in *Formica polyctena* (Först.). *Netherland Journal of Zoology,* 28(1): 55–61.
Bruyn, G. J. de, and A. A. Mabelis. 1972. Predation and aggression as possible regulatory mechanisms in *Formica. Ekologia Polska,* 20(10): 93–101.
Buchner, P. 1965. *Endosymbiosis of animals with plant microorganisms.* Interscience Publishers, John Wiley, New York. xvii + 909 pp.
Buckingham, E. N. 1911. Division of labor among ants. *Proceedings of the American Academy of Arts and Sciences,* 46(18): 425–507.
Buckle, G. R., and L. Greenberg. 1981. Nestmate recognition in sweat bees (*Lasioglossum zephyrum*): does an individual recognize its own odour or only odours of its nestmates? *Animal Behaviour,* 29(3): 802–809.
Buckley, R. C., ed. 1982a. *Ant-plant interactions in Australia.* Dr. W. Junk, The Hague. x + 162 pp.
Buckley, R. C. 1982b. Ant-plant interactions: a world review. *In* R. C. Buckley, ed., *Ant-plant interactions in Australia,* pp. 111–141. Dr. W. Junk, The Hague.
——— 1982c. A world bibliography of ant-plant interactions. *In* R. C. Buckley, ed., *Ant-plant interactions in Australia,* pp. 143–162. Dr. W. Junk, The Hague.
——— 1983. Interaction between ants and membracid bugs decreases growth and seed set of host plant bearing extrafloral nectaries. *Oecologia,* 58(1): 132–136.
Buckley, S. B. 1861. *Myrmica (Atta) molefaciens,* "stinging ant" or "mound-making ant," of Texas. *Proceedings of the Academy of Natural Sciences of Philadelphia* (1860), pp. 445–447.
Buddenbrock, W. von. 1917. Die Lichtkompassbewegung bei den Insekten, inbesondere den Schmetterlingsraupen. *Sitzungsberichte der Heidelberger Akademie der Wissenschaften,* ser. B, 1: 3–26.
Bünzli, G. H. 1935. Untersuchungen über coccidophile Ameisen aus den Kaffeefeldern von Surinam. *Mitteilungen der Schweizerischen Entomologischen Gesellschaft,* 16(6–7): 453–593.
Buren, W. F. 1944. A list of Iowa ants. *Iowa State College Journal of Science,* 18(3): 277–312.
——— 1958. A review of the species of *Crematogaster, sensu stricto,* in North America (Hymenoptera: Formicidae), Part I. *Journal of the New York Entomological Society,* 66(3–4): 119–134.
——— 1968a. Some fundamental taxonomic problems in *Formica* (Hymenoptera: Formicidae). *Journal of the Georgia Entomological Society,* 3(2): 25–40.
——— 1968b. A review of the species of *Crematogaster, sensu stricto,* in North America (Hymenoptera, Formicidae), Part II: Descriptions of new species. *Journal of the Georgia Entomological Society,* 3(3): 91–121.
——— 1972. Revisionary studies on the taxonomy of the imported fire ants. *Journal of the Georgia Entomological Society,* 7(1): 1–26.
Buren, W. F., H. R. Hermann, and M. S. Blum. 1970. The widespread occurrence of mandibular grooves in aculeate Hymenoptera. *Journal of the Georgia Entomological Society,* 5(4): 185–196.
Buren, W. F., J. C. Nickerson, and C. R. Thompson. 1975. Mixed nests of *Conomyrma insana* and *C. flavopecta*—evidence of parasitism (Hymenoptera: Formicidae). *Psyche,* 82(3–4): 306–314.
Buren, W. F., M. A. Naves, and T. C. Carlysle. 1977. False phragmosis and apparent specialization for subterranean warfare in *Pheidole lamia* Wheeler (Hymenoptera: Formicidae). *Journal of the Georgia Entomological Society,* 12(2): 100–108.
Burges, R. J. 1979. A rare fly and its parasitic behavior toward an ant (Diptera: Phoridae, Hymenoptera: Formicidae). *Florida Entomologist,* 62(4): 413–414.
Burkhardt, J. F. 1983. Foraging strategies in the ant, *Pheidole dentata:* the influence of colony size on the organization of an ant colony. M.A. thesis, Department of Zoology, Duke University. v + 56 pp.
Buschinger, A. 1966a. Untersuchungen an *Harpagoxenus sublaevis* Nyl. (Hym., Formicidae), I: Freilandbeobachtungen zu Verbreitung und Lebensweise. *Insectes Sociaux,* 13(1): 5–16.
——— 1966b. Untersuchungen an *Harpagoxenus sublaevis* Nyl. (Hym. Formicidae), II: Haltung und Brutaufzucht. *Insectes Sociaux,* 13(4): 311–322.
——— 1967. Verbreitung und Auswirkungen von Mono- und Polygynie bei Arten der Gattung *Leptothorax* Mayr (Hymenoptera, Formicidae). Doctoral diss., Bayerische Julius-Maximilians-Universität, Würzburg. iv + 114 pp.
——— 1968a. "Locksterzeln" begattungsbereiter ergatoider Weibchen von *Harpagoxenus sublaevis* Nyl. (Hymenoptera: Formicidae). *Experientia,* 24: 297.
——— 1968b. Untersuchungen an *Harpagoxenus sublaevis* Nyl. (Hymenoptera, Formicidae), III: Kopula, Koloniegründung, Raubzüge. *Insectes Sociaux,* 15(1): 89–104.
——— 1968c. Mono- und Polygynie bei Arten der Gattung *Leptothorax* Mayr (Hymenoptera Formicidae). *Insectes Sociaux,* 15(3): 217–226.
——— 1970a. Neue Vorstellungen zur Evolution des Sozialparasitismus und der Dulosis bei Ameisen (Hym., Formicidae). *Biologisches Zentralblatt,* 89(3): 273–299.
——— 1970b. Zur Frage der Monogynie oder Polygynie bei *Myrmecina graminicola* (Latr.) (Hym., Form.). *Insectes Sociaux,* 17(3): 177–181.

——— 1971a. "Locksterzeln" und Kopula der sozialparasitischen Ameise *Leptothorax kutteri* Buschinger (Hym. Form.). *Zoologischer Anzeiger*, 186(3–4): 242–248.
——— 1971b. Weitere Untersuchungen zum Begattungsverhalten sozialparasitischer Ameisen (*Harpagoxenus sublaevis* Nyl. und *Doronomyrmex pacis* Kutter, Hym. Formicidae). *Zoologischer Anzeiger*, 187(3–4): 184–198.
——— 1972a. Giftdrüsensekret als Sexualpheromon bei der Ameise *Harpagoxenus sublaevis*. *Naturwissenschaften*, 59(7): 313–314.
——— 1972b. Kreuzung zweier sozialparasitischer Ameisenarten, *Doronomyrmex pacis* Kutter und *Leptothorax kutteri* Buschinger (Hym. Formicidae). *Zoologischer Anzeiger*, 189(3–4): 169–179.
——— 1973a. Transport und Ansetzen von Larven an Beutestücke bei der Ameise *Aphaenogaster subterranea* (Latr.) (Hym., Formicidae). *Zoologischer Anzeiger*, 190(1–2): 63–66.
——— 1973b. Ameisen des Tribus Leptothoracini (Hym., Formicidae) als Zwischenwirte von Cestoden. *Zoologischer Anzeiger*, 191(5–6): 369–380.
——— 1974a. Monogynie und Polygynie in Insektensozietäten. *In* G. H. Schmidt, ed., *Sozialpolymorphismus bei Insekten*, pp. 862–896. Wissenschaftliche Verlagsgesellschaft MBH, Stuttgart.
——— 1974b. Polymorphismus und Polyethismus sozialparasitischer Hymenopteren. *In* G. H. Schmidt, ed., *Sozialpolymorphismus bei Insekten*, pp. 897–934. Wissenschaftliche Verlagsgesellschaft MBH, Stuttgart.
——— 1974c. Zur Biologie der sozialparasitischen Ameise *Leptothorax goesswaldi* Kutter (Hym., Formicidae). *Insectes Sociaux*, 21(2): 133–144.
——— 1974d. Experimente und Beobachtungen zur Gründung und Entwicklung neuer Sozietäten der sklavenhaltenden Ameise *Harpagoxenus sublaevis* (Nyl.). *Insectes Sociaux*, 21(4): 381–406.
——— 1975a. Eine genetische Komponente im Polymorphismus der dulotischen Ameise *Harpagoxenus sublaevis*. *Naturwissenschaften*, 62(5): 239.
——— 1975b. Sexual pheromones in ants. *In* C. Noirot, P. E. Howse, and G. Le Masne, eds., *Pheromones and defensive secretions in social insects* (Proceedings of a Symposium of the International Union for the Study of Social Insects, Dijon, 1975), pp. 225–233. Imprimerie de l'Université, Dijon.
——— 1976a. Eine Methode zur Zucht der Gastameise *Formicoxenus nitidulus* (Nyl.) mit *Leptothorax acervorum* (Fabr.) als "Wirtsameise" (Hym., Form.). *Insectes Sociaux*, 23(3): 205–214.
——— 1976b. Giftdrüsensekret als Sexualpheromon bei der Gastameise *Formicoxenus nitidulus* (Nyl.) (Hym., Form.). *Insectes Sociaux*, 23(3): 215–226.
——— 1978. Genetisch bedingte Entstehung geflügelter Weibchen bei der sklavenhaltenden Ameise *Harpagoxenus sublaevis* (Nyl.) (Hym. Form.). *Insectes Sociaux*, 25(2): 163–172.
——— 1979a. Functional monogyny in the American guest ant *Formicoxenus hirticornis* (Emery) ( = *Leptothorax hirticornis*), (Hym., Form.). *Insectes Sociaux*, 26(1): 61–68.
——— 1979b. *Doronomyrmex pocahontas* n. sp., a parasitic ant from Alberta, Canada (Hym. Formicidae). *Insectes Sociaux*, 26(3): 216–222.
——— 1981. Biological and systematic relationships of social parasitic Leptothoracini from Europe and North America. *In* P. E. Howse and J.-L. Clément, eds., *Biosystematics of social insects* (Systematics Association special volume no. 19), pp. 211–222. Academic Press, New York.
——— 1982a. *Epimyrma goesswaldi* Menozzi 1931—*Epimyrma ravouxi* (André, 1896)—morphologischer und biologischer Nachweis der Synonymie (Hym., Formicidae). *Zoologischer Anzeiger*, 208(5–6): 352–358.
——— 1982b. *Leptothorax faberi* n. sp., an apparently parasitic ant from Jasper National Park, Canada (Hymenoptera: Formicidae). *Psyche*, 89(3–4): 197–209.
——— 1983. Sexual behavior and slave raiding of the dulotic ant, *Harpagoxenus sublaevis* (Nyl.) under field conditions (Hym., Formicidae). *Insectes Sociaux*, 30(3): 235–240.
——— 1985. The *Epimyrma* species of Corsica (Hymenoptera, Formicidae). *Spixiana*, 8(3): 277–280.
——— 1986. Evolution of social parasitism in ants. *Trends in Ecology and Evolution*, 1(6): 155–160.
——— 1987a. Biological arguments for a systematic rearrangement of the ant tribe Leptothoracini. *In* J. Eder and H. Rembold, eds., *Chemistry and biology of social insects* (Proceedings of the Tenth International Congress of the International Union for the Study of Social Insects, Munich, 1986), p. 43. Verlag J. Peperny, Munich.
——— 1987b. Polymorphism and reproductive division of labor in advanced ants. *In* J. Eder and H. Rembold, eds., *Chemistry and biology of social insects* (Proceedings of the Tenth International Congress of the International Union for the Study of Social Insects, Munich, 1986), pp. 257–258. Verlag J. Peperny, Munich.
——— 1987c. *Teleutomyrmex schneideri* Kutter 1950 and other parasitic ants found in the Pyrenees. *Spixiana*, 10(1): 81–83.
Buschinger, A., and T. M. Alloway. 1977. Population structure and polymorphism in the slave-making ant *Harpagoxenus americanus* (Emery) (Hymenoptera: Formicidae). *Psyche*, 84(3–4): 233–242.
——— 1978. Caste polymorphism in *Harpagoxenus canadensis* M. R. Smith (Hym., Formicidae). *Insectes Sociaux*, 25(4): 339–350.
——— 1979. Sexual behaviour in the slave-making ant, *Harpagoxenus canadensis* M. R. Smith, and sexual pheromone experiments with *H. canadensis*, *H. americanus* (Emery), and *H. sublaevis* (Nylander) (Hymenoptera; Formicidae). *Zeitschrift für Tierpsychologie*, 49(2): 113–119.
Buschinger, A., and A. Francoeur. 1983. The guest ant, *Symmyrmica chamberlini*, rediscovered near Salt Lake City, Utah (Hymenoptera, Formicidae). *Psyche*, 90(3): 297–305.
Buschinger, A., and U. Maschwitz. 1984. Defensive behavior and defensive mechanisms in ants. *In* H. R. Hermann, ed., *Defensive mechanisms in social insects*, pp. 95–150. Praeger, New York.
Buschinger, A., and E. Pfeifer. 1988. Effects of nutrition on brood production and slavery in ants (Hymenoptera, Formicidae). *Insectes Sociaux*, 35(1): 61–69.
Buschinger, A., and H. Stoewesand. 1971. Teratologische Untersuchungen an Ameisen. *Beiträge zur Entomologie*, 21(1–2): 211–241.
Buschinger, A., and U. Winter. 1975. Der Polymorphismus der sklavenhaltende Ameise *Harpagoxenus sublaevis* (Nyl.). *Insectes Sociaux*, 22(4): 333–362.
——— 1976. Funktionelle Monogynie bei der Gastameise *Formicoxenus nitidulus* (Nyl.) (Hym., Form.). *Insectes Sociaux*, 23(4): 549–558.
——— 1977. Rekrutierung von Nestgenossen mittels Tandemlaufen bei Sklavenraubzügen der dulotischen Ameise *Harpagoxenus sublaevis* (Nyl.). *Insectes Sociaux*, 24(2): 183–190.
——— 1978. Echte Arbeiterinnen, fertile Arbeiterinnen und sterile Wirtsweibchen in Völkern der dulotischen Ameise *Harpagoxenus sublaevis* (Nyl.) (Hym., Form.). *Insectes Sociaux*, 25(1): 63–78.
——— 1983a. *Myrmicinosporidium durum* Hölldobler 1933, Parasit bei Ameisen (Hym. Formicidae), in Frankreich, der Schweiz und Jugoslawien wieder gefunden. *Zoologischer Anzeiger*, 210(5–6): 393–398.
——— 1983b. Population studies of the dulotic ant, *Epimyrma ravouxi*, and the degenerate slavemaker, *E. kraussei* (Hymenoptera: Formicidae). *Entomologia Generalis*, 8(4): 251–266.
——— 1985. Life history and male morphology of the workerless parasitic ant *Epimyrma corsica* (Hymenoptera: Formicidae). *Entomologia Generalis*, 10(2): 65–75.
Buschinger, A., G. Frenz, and M. Wunderlich. 1975. Untersuchungen zur Geschlechtstierproduktion der dulotischen Ameise *Harpagoxenus sublaevis* (Nyl.) (Hym. Formicidae). *Insectes Sociaux*, 22(2): 169–182.
Buschinger, A., W. Ehrhardt, and U. Winter. 1980a. The organization of slave raids in dulotic ants—a comparative study (Hymenoptera; Formicidae). *Zeitschrift für Tierpsychologie*, 53(2): 245–264.
Buschinger, A., A. Francoeur, and K. Fischer. 1980b. Functional monogyny, sexual behavior, and karyotype of the guest ant, *Leptothorax provancheri* Emery (Hymenoptera, Formicidae). *Psyche*, 87(1–2): 1–12.
Buschinger, A., W. Ehrhardt, and K. Fischer. 1981. *Doronomyrmex pacis*, *Epimyrma stumperi* und *E. goesswaldi* (Hym., Formicidae) neu für Frankreich. *Insectes Sociaux*, 28(1): 67–70.
Buschinger, A., U. Winter, and W. Faber. 1983. The biology of *Myrmoxenus gordiagini* Ruzsky, a slave-making ant (Hymenoptera, Formicidae). *Psyche*, 90(4): 335–342.

Buschinger, A., K. Fischer, H.-P. Guthy, K. Jessen, and U. Winter. 1986. Biosystematic revision of *Epimyrma kraussei, E. vandeli,* and *E. foreli* (Hymenoptera: Formicidae). *Psyche,* 93(3–4): 253–276.

Buschinger, A., J. Heinze, K. Jessen, P. Douwes, and U. Winter. 1987. First European record of a queen ant carrying a mealybug during her mating flight. *Naturwissenschaften,* 74(3): 139–140.

Buschinger, A., W. Ehrhardt, K. Fischer, and J. Ofer. 1988. The slave-making ant genus *Chalepoxenus* (Hymenoptera, Formicidae), I: Review of literature, range, slave species. *Zoologische Jahrbücher, Abteilung für Systematik Ökologie und Geographie der Tiere,* 115(3): 383–401.

Buser, M. W., C. Baroni Urbani, and E. Schillinger. 1987. Quantitative aspects of recruitment to new food by a seed-harvesting ant (*Messor capitatus* Latreille). *In* J. M. Pasteels and J.-L. Deneubourg, eds., *From individual to collective behavior in social insects* (*Experientia* Supplement, vol. 54), pp. 139–154. Birkhäuser Verlag, Basel.

Büsgen, M. 1891. Der Honigtau: Biologische Studien an Pflanzen und Pflanzenläusen. *Jenaische Zeitschrift für Naturwissenschaft,* 25: 339–428.

Busher, C. E., P. Calabi, and J. F. A. Traniello. 1985. Polymorphism and division of labor in the Neotropical ant *Camponotus sericeiventris* Guerin (Hymenoptera: Formicidae). *Annals of the Entomological Society of America,* 78(2): 221–228.

Butenandt, A., B. Linzen, and M. Lindauer. 1959. Über einen Duftstoff aus der Mandibeldrüse der Blattschneiderameise *Atta sexdens rubropilosa* Forel. *Archives d'Anatomie Microscopique et de Morphologie Expérimentale,* 48: 13–19.

Byron, P. A., E. R. Byron, and R. A. Bernstein. 1980. Evidence of competition between two species of desert ants. *Insectes Sociaux,* 27(4): 351–360.

Caetano, F. H., and C. da Cruz-Landim. 1985. Presence of microorganisms in the alimentary canal of ants of the tribe Cephalotini (Myrmicinae): location and relationship with intestinal structures. *Naturalia* (São Paulo), 10: 37–47.

Cagniant, H. 1968. Description d'*Epimyrma algeriana* (nov. sp.) (Hyménoptères Formicidae, Myrmicinae), fourmi parasite: représentation des trois castes: quelques observations biologiques, écologiques et éthologiques. *Insectes Sociaux,* 15(2): 157–170.

——— 1970. Une nouvelle fourmi parasite d'Algérie: *Sifolinia kabylica* (nov. sp.). *Insectes Sociaux,* 17(1): 39–48.

——— 1979. La parthénogénèse thélytoque et arrhénotoque chez la fourmi *Cataglyphis cursor* Fonsc. (Hym. Form.): cycle biologique en élevage des colonies avec reine et des colonies sans reine. *Insectes Sociaux,* 26(1): 51–60.

——— 1982. La parthénogénèse thélytoque et arrhénotoque chez la fourmi *Cataglyphis cursor* Fonscolombe (Hymenoptera, Formicidae): étude des oeufs pondus par les reines et les ouvrières: morphologie, devenir, influence sur le déterminisme de la caste reine. *Insectes Sociaux,* 29(2): 175–188.

——— 1983. Contribution à la connaissance des fourmis marocaines *Chalepoxenus tramieri,* nov. sp. *Nouvelle Revue d'Entomologie,* 13(3): 319–322.

Calabi, P. 1986. Division of labor in the ant *Pheidole dentata:* the role of colony demography and behavioral flexibility. Ph.D. diss., Boston University. xiii + 144 pp.

Calabi, P., J. F. A. Traniello, and M. H. Werner. 1983. Age polyethism: its occurrence in the ant *Pheidole hortensis,* and some general considerations. *Psyche,* 90(4): 395–412.

Calderone, N. W., and R. E. Page, Jr. 1988. Genotypic variability in age polyethism and task specialization in honey bee, *Apis mellifera* (Hymenoptera: Apidae). *Behavioral Ecology and Sociobiology,* 22(1): 17–25.

Callaghan, C. J. 1977. Studies on Restinga butterflies, I: Life cycle and immature biology of *Menander felsina* (Riodinidae), a myrmecophilous metalmark. *Journal of the Lepidopterists' Society,* 31(3): 173–182.

——— 1979. A new genus and a new subspecies of Riodinidae from southern Brasil. *Bulletin of the Allyn Museum,* no. 53, 7 pp.

——— 1981. Notes on the immature biology of two myrmecophilous Lycaenidae: *Juditha molpe* (Riodininae) and *Panthiades bitias* (Lycaeninae). *Journal of Research on the Lepidoptera,* 20(1): 36–42.

Cammaerts, M.-C. 1977. Recruitment to food in *Myrmica rubra* L. *Proceedings of the Eighth International Congress of the International Union for the Study of Social Insects* (Wageningen, 1977), p. 294.

——— 1982. Une source inédite de phéromone chez *Myrmica rubra* L. *Insectes Sociaux,* 29(4): 524–534.

Cammaerts, M.-C., and R. Cammaerts. 1980. Food recruitment strategies of the ants *Myrmica sabuleti* and *Myrmica ruginodis. Behavioural Processes,* 5(3): 251–270.

——— 1981. Food-gathering method of the ant *Myrmica rugulosa* including an original recruitment system. *Biology of Behaviour,* 6(3): 239–254.

Cammaerts, M.-C., E. D. Morgan, and R. Tyler. 1977. Territorial marking in the ant *Myrmica rubra* L. (Formicidae). *Biology of Behaviour,* 2(3): 263–272.

Cammaerts, M.-C., R. P. Evershed, and E. D. Morgan. 1981. Comparative study of the mandibular gland secretion of four species of *Myrmica* ants. *Journal of Insect Physiology,* 27(4): 225–231.

——— 1982. Mandibular gland secretions of workers of *Myrmica rugulosa* and *M. schencki:* comparison with four other *Myrmica* species. *Physiological Entomology,* 7(2): 119–125.

——— 1983. The volatile components of the mandibular gland secretion of workers of the ants *Myrmica lobicornis* and *Myrmica sulcinodis. Journal of Insect Physiology,* 29(8): 659–664.

Cammaerts, R. 1974. Le système glandulaire tégumentaire du coléoptère myrmécophile *Claviger testaceus* Preyssler, 1790 (Pselaphidae). *Zeitschrift für Morphologie der Tiere,* 77(3): 187–219.

——— 1977. Secretions of a beetle inducing regurgitation in its host ant. *Proceedings of the Eighth International Congress of the International Union for the Study of Social Insects* (Wageningen, 1977), p. 295.

Cammaerts-Tricot, M.-C. 1973. Phéromones agrégreant les ouvrières de *Myrmica rubra. Journal of Insect Physiology,* 19(6): 1299–1315.

——— 1974a. Piste et phéromone attractive chez la fourmi *Myrmica rubra. Journal of Comparative Physiology,* 88(4): 373–382.

——— 1974b. Production and perception of attractive pheromones by differently aged workers of *Myrmica rubra* (Hymenoptera Formicidae). *Insectes Sociaux,* 21(3): 235–247.

——— 1974c. Recrutement d'ouvrières, chez *Myrmica rubra,* par les phéromones de l'appareil à venin. *Behaviour,* 50(1–2): 111–122.

——— 1975. Ontogenesis of the defence reactions in the workers of *Myrmica rubra* L. (Hymenoptera: Formicidae). *Animal Behaviour,* 23(1): 124–130.

Cammaerts-Tricot, M.-C., and J.-C. Verhaeghe. 1974. Ontogenesis of trail pheromone production and trail following behaviour in the workers of *Myrmica rubra* L. (Formicidae). *Insectes Sociaux,* 21(3): 275–282.

Campbell, J. M. 1966. A revision of the genus *Lobopoda* (Coleoptera: Alleculidae) in North America and the West Indies. *Illinois Biological Monographs,* 37: 1–203.

Campbell, M. H. 1982. Restricting losses of aerially sown seed due to seed-harvesting ants. *In* R. C. Buckley, ed., *Ant-plant interactions in Australia,* pp. 25–30. Dr. W. Junk, The Hague.

Carlin, N. F. 1981. Polymorphism and division of labor in the dacetine ant *Orectognathus versicolor* (Hymenoptera: Formicidae). *Psyche,* 88(3–4): 231–244.

——— 1988. Species, kin and other forms of recognition in the brood discrimination behavior of ants. *In* J. C. Trager, ed., *Advances in myrmecology,* pp. 267–295. E. J. Brill, Leiden.

Carlin, N. F., and B. Hölldobler. 1983. Nestmate and kin recognition in interspecific mixed colonies of ants. *Science,* 222: 1027–29.

——— 1986. The kin recognition system of carpenter ants (*Camponotus* spp.), I: Hierarchical cues in small colonies. *Behavioral Ecology and Sociobiology,* 19(2): 123–134.

——— 1987. The kin recognition system of carpenter ants (*Camponotus* spp.), II: Larger colonies. *Behavioral Ecology and Sociobiology,* 20(3): 209–217.

——— 1988. Influence of virgin queens on kin recognition in the carpenter ant *Camponotus floridanus* (Hymenoptera: Formicidae). *Insectes Sociaux,* 35(2): 191–197.

Carlin, N. F., and A. B. Johnston. 1984. Learned enemy specification in the defense recruitment system of an ant. *Naturwissenschaften,* 71(3): 156–157.

Carlin, N. F., and P. H. Schwartz. 1989. Pre-imaginal experience and nestmate brood recognition in the carpenter ant, *Camponotus floridanus*. *Animal Behaviour*, 38(2): 89–95.

Carlin, N. F., R. Halpern, B. Hölldobler, and P. Schwartz. 1987a. Early learning and the recognition of conspecific cocoons by carpenter ants (*Camponotus* spp.). *Ethology*, 75(4): 306–316.

Carlin, N. F., B. Hölldobler, and D. S. Gladstein. 1987b. The kin recognition system of carpenter ants (*Camponotus* spp.), III: Within-colony discrimination. *Behavioral Ecology and Sociobiology*, 20(3): 219–227.

Carlson, D. M., and J. B. Gentry. 1973. Effects of shading on the migratory behavior of the Florida harvester ant, *Pogonomyrmex badius*. *Ecology*, 54(2): 452–453.

Carney, W. P. 1970. Laboratory maintenance of carpenter ants. *Annals of the Entomological Society of America*, 63(1): 332–334.

Carpenter, F. M. 1930. The fossil ants of North America. *Bulletin of the Museum of Comparative Zoology, Harvard*, 70(1): 1–66.

——— 1989. *Treatise on invertebrate paleontology, Part R, Arthropoda IV, Hexapoda: classes Collembola, Protura, Diplura, Insecta*. University of Kansas and Geological Society of America, Lawrence. Forthcoming.

Carpenter, J. M. 1989. Testing scenarios: wasp social behavior. *Cladistics*, 5(2): 131–144.

Carr, C. A. H. 1962. Further studies on the influence of the queen in ants of the genus *Myrmica*. *Insectes Sociaux*, 9(3): 197–211.

Carroll, C. R. 1979. A comparative study of two ant faunas: the stem-nesting ant communities of Liberia, West Africa and Costa Rica, Central America. *American Naturalist*, 113(4): 552–561.

Carroll, C. R., and D. H. Janzen. 1973. Ecology of foraging by ants. *Annual Review of Ecology and Systematics*, 4: 231–257.

Carthy, J. D. 1950. Odour trails of *Acanthomyops fuliginosus*. *Nature*, 166: 154.

——— 1951a. The orientation of two allied species of British ant, I: Visual direction finding in *Acanthomyops (Lasius) niger*. *Behaviour*, 3(4): 275–303.

——— 1951b. The orientation of two allied species of British ant, II: Odour trail laying and following in *Acanthomyops (Lasius) fuliginosus*. *Behaviour*, 3(4): 304–318.

Casevitz-Weulersse, J. 1984. Les larves à expansions laterales de *Crematogaster (Acrocoelia) scutellaris* (Olivier) (Hym. Formicidae). *Actes des Colloques Insectes Sociaux* (L'Union Internationale pour l'Etude des Insectes Sociaux, Section française, Compte Rendu Colloque Annuel, Les Eyzies, 1983), 1: 131–138.

Casnati, G., A. Ricca, and M. Pavan. 1967. Sulla secrezione difensiva delle glandole mandibolari di *Paltothyreus tarsatus* (Fabr.) (Hymenoptera: Formicidae). *Chimica e l'Industria* (Milan), 49: 57–58.

Cavill, G. W. K., and H. Hinterberger. 1960a. Dolichoderine ant extractives. *In* M. Pavan and T. Eisner, eds., *Verhandlungen*, vol. 3, *Symposium 3: Chemie der Insekten und Symposium 4: Chemische verteidigungsmechanismen bei arthropoden*, pp. 53–59. XI Internationaler Kongress für Entomologie, Wien 1960, c/o Istituto di Entomologia Agraria dell'Università di Pavia, Italy.

——— 1960b. The chemistry of ants, IV: Terpenoid constituents of some *Dolichoderus* and *Iridomyrmex* species. *Australian Journal of Chemistry*, 13(4): 514–519.

Cavill, G. W. K., D. L. Ford, and H. D. Locksley. 1956. The chemistry of ants, I: Terpenoid constituents of some Australian *Iridomyrmex* species. *Australian Journal of Chemistry*, 9(2): 288–293.

Cavill, G. W. K., P. L. Robertson, and N. W. Davies. 1979. An Argentine ant aggregation factor. *Experientia*, 35: 889–890.

Cavill, G. W. K., N. W. Davies, and F. J. McDonald. 1980. Characterization of aggregation factors and associated compounds from the Argentine ant, *Iridomyrmex humilis*. *Journal of Chemical Ecology*, 6(2): 371–384.

Cazier, M. A., and M. A. Mortenson. 1965. Bionomical observations on myrmecophilous beetles of the genus *Cremastocheilus* (Coleoptera: Scarabaeidae). *Journal of the Kansas Entomological Society*, 38(1): 19–44.

Cazier, M. A., and M. Statham. 1962. The behavior and habits of the myrmecophilous scarab *Cremastocheilus stathamae* Cazier with notes on other species (Coleoptera: Scarabaeidae). *Journal of the New York Entomological Society*, 70(3): 125–149.

Chadab, R., and C. W. Rettenmeyer. 1975. Mass recruitment by army ants. *Science*, 188: 1124–25.

Chadha, M. S., T. Eisner, A. Monro, and J. Meinwald. 1962. Defence mechanisms of arthropods, VII: Citronellal and citral in the mandibular gland secretion of the ant *Acanthomyops claviger* (Roger). *Journal of Insect Physiology*, 8(2): 175–179.

Champagne, P., J.-L. Deneubourg, J.-C. Verhaeghe, and J. M. Pasteels. 1984. Techniques d'étude des sequences comportementales appliquées à l'analyse du recrutement alimentaire chez les fourmis. *Actes des Colloques Insectes Sociaux* (L'Union Internationale pour l'Etude des Insectes Sociaux, Section française, Compte Rendu Colloque Annuel, Les Eyzies, 1983), 1: 31–37.

Chapman, J. W. 1964. Studies on the ecology of the army ants of the Philippines genus *Aenictus* Schuckard (Hymenoptera: Formicidae). *Philippine Journal of Science*, 93(4): 551–595.

Chapman, J. W., and S. R. Capco. 1951. Check list of the ants (Hymenoptera: Formicidae) of Asia. *Monographs of the Institute of Science and Technology, Manila*, no. 1, 327 pp.

Chapman, T. A. 1920. Contributions to the life history of *Lycaena euphemus* Hb. *Transactions of the Entomological Society of London*, 1919(3–4): 450–465.

Charlesworth, B. 1978. Some models of the evolution of altruistic behaviour between siblings. *Journal of Theoretical Biology*, 72(2): 297–319.

Charnov, E. L. 1978. Evolution of eusocial behavior: offspring choice or parental parasitism? *Journal of Theoretical Biology*, 75(4): 451–465.

Chauvin, R. 1952. Sur la reconstruction du nid chez les fourmis Oecophylles (*Oecophylla longinoda* L.). *Behaviour*, 4(3): 190–201.

——— 1959. Contribution à l'étude de la construction du dôme chez *Formica rufa*, II. *Insectes Sociaux*, 6(1): 1–11.

——— 1964. Expériences sur l'"apprentissage par équipe" du labyrinthe chez *Formica polyctena*. *Insectes Sociaux*, 11(1): 1–20.

Chen, S. C. 1937a. Social modification of the activity of ants in nest-building. *Physiological Zoölogy*, 10(4): 420–436.

——— 1937b. The leaders and followers among the ants in nest-building. *Physiological Zoölogy*, 10(4): 437–455.

Cherix, D. 1980. Note preliminaire sur la structure, la phenologie et le régime alimentaire d'une super-colonie de *Formica lugubris* Zett. *Insectes Sociaux*, 27(3): 226–236.

——— 1983. Intraspecific variations of alarm pheromones between two populations of the red wood ant *Formica lugubris* Zett. (Hymenoptera, Formicidae). *Mitteilungen der Schweizerischen Entomologischen Gesellschaft*, 56(1–2): 57–65.

——— 1986. *Les fourmis des bois ou fourmis rousses*. Atlas Visuels Payot, Lausanne. 64 pp.

Cherix, D., and J. D. Bourne. 1980. A field study on a super-colony of the red wood ant *Formica lugubris* Zett. in relation to other predatory arthropods (spiders, harvestmen and ants). *Revue Suisse de Zoologie*, 87(4): 955–973.

Cherix, D., and G. Gris. 1978. Relations et aggressivité chez *Formica lugubris* Zett. dans le Jura (Hymenoptera, Formicidae). *In* B. Pisarski and E. Kryżanowska, eds., *Competition in social insects* (Proceedings of the Eighth Meeting), pp. 7–12. Social Insects Section of the Polish Entomological Society, Puławy.

Cherrett, J. M. 1968. The foraging behaviour of *Atta cephalotes* L. (Hymenoptera: Formicidae), I: Foraging pattern and plant species attacked in tropical rain forest. *Journal of Animal Ecology*, 37(2): 387–403.

——— 1982. The economic importance of leaf-cutting ants. *In* M. D. Breed, C. D. Michener, and H. E. Evans, eds., *The biology of social insects* (Proceedings of the Ninth Congress of the International Union for the Study of Social Insects, Boulder, Colorado, 1982), pp. 114–118. Westview Press, Boulder.

——— 1986. History of the leaf-cutting ant problem. *In* C. S. Lofgren and R. K. Vander Meer, eds., *Fire ants and leaf-cutting ants: biology and management*, pp. 10–17. Westview Press, Boulder.

Cherrett, J. M., and D. J. Peregrine. 1976. A review of the status of leaf-cutting ants and their control. *Annals of Applied Biology*, 84: 124–133.

Cherrett, J. M., and C. E. Seaforth. 1968. Phytochemical arrestants for the leaf-cutting ants, *Atta cephalotes* (L.) and *Acromyrmex octospinosus* (Reich), with some notes on the ants' response. *Bulletin of Entomological Research,* 59: 615–625.

Chew, A. E., and R. M. Chew. 1980. Body size as a determinant of small-scale distributions of ants in evergreen woodland, southeastern Arizona. *Insectes Sociaux,* 27(3): 189–202.

Chew, R. M. 1979. Mammalian predation on honey ants, *Myrmecocystus* (Formicidae). *Southwestern Naturalist,* 24(4): 677–682.

——— 1987. Population dynamics of colonies of three species of ants in desertified grassland, southeastern Arizona, 1958–1981. *American Midland Naturalist,* 118(1): 177–188.

Chew, R. M., and J. De Vita. 1980. Foraging characteristics of a desert ant assemblage: functional morphology and species separation. *Journal of Arid Environments,* 3(1): 75–83.

Chillcott, J. G. 1965. New species and stages of Nearctic *Fannia* R. D. (Diptera: Muscidae) associated with nests of Hymenoptera. *Canadian Entomologist,* 97(6): 640–647.

China, W. E. 1928. A remarkable bug which lures ants to their destruction. *Natural History Magazine,* 1(6): 209–213.

Choe, J. C. 1988. Worker reproduction and social evolution in ants (Hymenoptera: Formicidae). *In* J. C. Trager, ed., *Advances in myrmecology,* pp. 163–187. E. J. Brill, Leiden.

Claassens, A. J. M., and C. G. C. Dickson. 1977. A study of the myrmecophilous behaviour of the immature stages of *Aloeides thyra* (L.) (Lep.: Lycaenidae) with special reference to the function of the retractile tubercles and with additional notes on the general biology of the species. *Entomologist's Record and Journal of Variation,* 89(9): 225–231.

Clark, D. B., C. Guayasamín, O. Pazmiño, C. Donoso, and Y. Páez de Villacís. 1982. The tramp ant *Wasmannia auropunctata:* autecology and effects on ant diversity and distribution on Santa Cruz Island, Galápagos. *Biotropica,* 14(3): 196–207.

Clark, G. C., and C. G. C. Dickson. 1971. *Life histories of the South African lycaenid butterflies.* Purnell and Sons, Capetown. xvi + 272 pp.

Clark, J. 1921. Notes on Western Australian ant-nest beetles. *Journal and Proceedings of the Royal Society of Western Australia,* 6(2): 97–104.

——— 1936. A revision of Australian species of *Rhytidoponera* Mayr (Formicidae). *Memoirs of the National Museum of Victoria,* 9: 14–89.

——— 1951. *The Formicidae of Australia,* vol. 1: *Subfamily Myrmeciinae.* Commonwealth Scientific and Industrial Research Organization, Melbourne. 230 pp.

Clark, W. H., and D. C. Prusso. 1986. *Desmidiospora myrmecophila* found infesting the ant *Camponotus semitestaceus. Mycologia,* 78(5): 865–866.

Clément, J.-L., A. Bonavita-Cougourdan, and C. Lange. 1987. Nestmate recognition and cuticular hydrocarbons in *Camponotus vagus* Scop. *In* J. Eder and H. Rembold, eds., *Chemistry and Biology of Social Insects* (Proceedings of the Tenth International Congress of the International Union for the Study of Social Insects, Munich, 1986), pp. 473–474. Verlag J. Peperny, Munich.

Coenen-Stass, D., B. Schaarschmidt, and I. Lamprecht. 1980. Temperature distribution and calorimetric determination of heat production in the nest of the wood ant, *Formica polyctena* (Hymenoptera, Formicidae). *Ecology,* 61(2): 238–244.

Cole, A. C. 1932. The rebuilding of mounds of the ant, *Pogonomyrmex occidentalis,* Cress. *Ohio Journal of Science,* 32(3): 245–246.

——— 1940. A guide to the ants of the Great Smoky Mountains National Park, Tennessee. *American Midland Naturalist,* 24(1): 1–88.

——— 1956. Studies of Nevada ants, II: A new species of *Lasius (Chthonolasius)* (Hymenoptera: Formicidae). *Journal of the Tennessee Academy of Science,* 31(1): 26–27.

——— 1957. *Paramyrmica,* a new North American genus of ants allied to *Myrmica* Latreille (Hymenoptera: Formicidae). *Journal of the Tennessee Academy of Science,* 32(1): 37–42.

——— 1968. *Pogonomyrmex harvester ants: a study of the genus in North America.* University of Tennessee Press, Knoxville. x + 222 pp.

Cole, A. C., and J. W. Jones. 1948. A study of the weaver ant, *Oecophylla smaragdina* (Fab.). *American Midland Naturalist,* 39(3): 641–651.

Cole, B. J. 1980. Repertoire convergence in two mangrove ants, *Zacryptocerus varians* and *Camponotus (Colobopsis)* sp. *Insectes Sociaux,* 27(3): 265–275.

——— 1981. Dominance hierarchies in *Leptothorax* ants. *Science,* 212: 83–84.

——— 1983a. Assembly of mangrove ant communities: patterns of geographical distribution. *Journal of Animal Ecology,* 52(2): 339–347.

——— 1983b. Multiple mating and the evolution of social behavior in the Hymenoptera. *Behavioral Ecology and Sociobiology,* 12(3): 191–201.

——— 1985. Size and behavior in ants: constraints on complexity. *Proceedings of the National Academy of Sciences of the United States of America,* 82(24): 8548–51.

——— 1986. The social behavior of *Leptothorax allardycei* (Hymenoptera, Formicidae): time budgets and the evolution of worker reproduction. *Behavioral Ecology and Sociobiology,* 18(3): 165–173.

Collingwood, C. A. 1956. A rare parasitic ant (Hym., Formicidae) in France. *Entomologist's Monthly Magazine,* 92: 197.

——— 1958a. A key to the species of ants (Hymenoptera, Formicidae) found in Britain. *Transactions of the Society for British Entomology,* 13(5): 69–96.

——— 1958b. The ants of the genus *Myrmica* in Britain. *Proceedings of the Royal Entomological Society of London,* ser. A, 33(4–6): 65–75.

——— 1958c. A survey of Irish Formicidae. *Proceedings of the Royal Irish Academy,* ser. B, 59(11): 213–219.

——— 1961. Ants in the Scottish Highlands. *Scottish Naturalist,* 70(1): 12–21.

——— 1964. The identification and distribution of British ants (Hym. Formicidae), 1: A revised key to the species found in Britain. *Transactions of the Society for British Entomology,* 16(3): 93–114.

——— 1979. *The Formicidae (Hymenoptera) of Fennoscandia and Denmark* (Fauna Entomologica Scandinavica, no. 8). Scandinavian Science Press, Klampenborg, Denmark. 174 pp.

——— 1985. Hymenoptera: fam. Formicidae of Saudi Arabia. *In* W. Büttiker and F. Krupp, eds., *Fauna of Saudi Arabia,* vol. 7, pp. 230–302. Pro Entomologia, Natural History Museum, Basel.

Collingwood, C. A., and K. E. J. Barrett. 1964. The identification and distribution of British ants (Hym. Formicidae), 2: The vice-county distribution of indigenous ants in the British Isles. *Transactions of the Society for British Entomology,* 16(3): 114–121.

Collingwood, C. A., and I. H. H. Yarrow. 1969. A survey of Iberian Formicidae (Hymenoptera). *EOS* (Revista Española de Entomología), 44(1968): 53–101.

Collins, H. L., and G. P. Markin. 1971. Inquilines and other arthropods collected from nests of the imported fire ant, *Solenopsis saevissima richteri. Annals of the Entomological Society of America,* 64(6): 1376–80.

Colombel, P. 1972. Recherches sur la biologie et l'ethologie d'*Odontomachus haematodes* L. (Hym. Formicoidea, Poneridae): biologie des ouvrières. *Insectes Sociaux,* 19(3): 171–194.

——— 1974. L'élevage artificiel du couvain d'*Odontomachus haematodes* L. (Hym. Form. Ponerinae) et la différenciation trophogénique des ouvrières et des reines. *Comptes Rendus de l'Académie des Sciences, Paris,* ser. D, 279(5): 489–491.

——— 1978. Biologie d'*Odontomachus haematodes* L. (Hym. Form.): determinisme de la caste femelle. *Insectes Sociaux,* 25(2): 141–151.

Colwell, R. K., and E. R. Fuentes. 1975. Experimental studies of the niche. *Annual Review of Ecology and Systematics,* 6: 281–310.

Common, I. F. B., and D. F. Waterhouse. 1972. *Butterflies of Australia.* Angus and Robertson, Sydney. 498 pp.

Connell, J. H. 1975. Some mechanisms producing structure in natural communities: a model and evidence from field experiments. *In* M. L. Cody and J. M. Diamond, eds., *Ecology and evolution of communities,* pp. 460–490. Harvard University Press, Cambridge, Mass.

Conway, J. R. 1977. Analysis of clear and dark amber repletes of the honey ant, *Myrmecocystus mexicanus hortideorum. Annals of the Entomological Society of America,* 70(3): 367–369.

Corbara, B., D. Fresneau, J.-P. Lachaud, Y. Leclerc, and G. Goodall. 1986. An automated photographic technique for behavioural investigations of social insects. *Behavioural Processes,* 13(3): 237–249.

Cordero, A. D. 1963. An unusual behavior of the leafcutting ant queen *Acromyrmex octospinosa* (Reich). *Revista de Biologia Tropicale,* 11(2): 221–222.

Corn, M. L. 1976. The ecology and behavior of *Cephalotes atratus*, a Neotropical ant (Hymenoptera: Formicidae). Ph.D. diss., Harvard University. 200 pp.

——— 1980. Polymorphism and polyethism in the Neotropical ant *Cephalotes atratus* (L.). *Insectes Sociaux*, 27(1): 29–42.

Corner, E. J. H. 1976. The climbing species of *Ficus:* derivation and evolution. *Philosophical Transactions of the Royal Society of London*, ser. B, 273(925): 359–386.

Cory, E. N., and E. E. Haviland. 1938. Population studies of *Formica exsectoides* Forel. *Annals of the Entomological Society of America*, 31(1): 50–56.

Costa Lima, A. da. 1962. Micro-coleóptero representante da nova subfamília Plaumanniolinae (Col., Ptinidae). *Revista Brasileira de Biologia*, 22(4): 413–418.

Cottrell, C. B. 1984. Aphytophagy in butterflies: its relationship to myrmecophily. *Zoological Journal of the Linnean Society*, 80(1): 1–57.

Craig, R. 1979. Parental manipulation, kin selection, and the evolution of altruism. *Evolution*, 33(1): 319–334.

——— 1980. Sex ratio changes and the evolution of eusociality in the Hymenoptera: simulation and games theory studies. *Journal of Theoretical Biology*, 87(1): 55–70.

Craig, R., and R. H. Crozier. 1979. Relatedness in the polygynous ant *Myrmecia pilosula*. *Evolution*, 33(1): 335–341.

Crawford, D. L., and S. W. Rissing. 1983. Regulation of recruitment by individual scouts in *Formica oreas* Wheeler (Hymenoptera, Formicidae). *Insectes Sociaux*, 30(2): 177–183.

Crawley, W. C. 1909. Queens of *Lasius umbratus*, Nyl., accepted by colonies of *Lasius niger*, L. *Entomologist's Monthly Magazine*, 45: 94–99.

Creighton, W. S. 1929. Further notes on the habits of *Harpagoxenus americanus*. *Psyche*, 36(1): 48–50.

——— 1930. The New World species of the genus *Solenopsis* (Hymenop. Formicidae). *Proceedings of the American Academy of Arts and Sciences*, 66(2): 37–182.

——— 1934. Descriptions of three new North American ants with certain ecological observations on previously described forms. *Psyche*, 41(4): 185–200.

——— 1950. The ants of North America. *Bulletin of the Museum of Comparative Zoology, Harvard*, 104: 1–585.

——— 1955. Observations on *Pseudomyrmex elongata* Mayr (Hymenoptera: Formicidae). *Journal of the New York Entomological Society*, 63(4): 17–20.

——— 1963. Further studies on the habits of *Cryptocerus texanus* Santschi (Hymenoptera: Formicidae). *Psyche*, 70(3): 133–143.

——— 1966. The habits of *Pheidole ridicula* Wheeler with remarks on habit patterns in the genus *Pheidole* (Hymenoptera: Formicidae). *Psyche*, 73(1): 1–7.

——— 1967. Studies on free colonies of *Cryptocerus texanus* Santschi (Hymenoptera: Formicidae). *Psyche*, 74(1): 34–41.

Creighton, W. S., and R. H. Crandall. 1954. New data on the habits of *Myrmecocystus melliger* Forel. *The Biological Review, City College of New York*, 16(1): 2–6.

Creighton, W. S., and M. P. Creighton. 1959. The habits of *Pheidole militicida* Wheeler (Hymenoptera: Formicidae). *Psyche*, 66(1–2): 1–12.

Creighton, W. S., and R. E. Gregg. 1954. Studies on the habits and distribution of *Cryptocerus texanus* Santschi (Hymenoptera: Formicidae). *Psyche*, 61(2): 41–57.

——— 1955. New and little-known species of *Pheidole* (Hymenoptera: Formicidae) from the southwestern United States and northern Mexico. *Colorado University Studies, Biological Series*, 3: 1–46.

Creighton, W. S., and W. L. Nutting. 1965. The habits and distribution of *Cryptocerus rohweri* Wheeler (Hymenoptera: Formicidae). *Psyche*, 72(1): 59–64.

Crewe, R. M., and M. S. Blum. 1971. 6-Methyl-5-hepten-2-one: chemotaxonomic significance in an *Iridomyrmex* sp. (Hymenoptera: Formicidae). *Annals of the Entomological Society of America*, 64(5): 1007–10.

——— 1972. Alarm pheromones of the Attini: their phylogenetic significance. *Journal of Insect Physiology*, 18(1): 31–42.

Crewe, R. M., J. M. Brand, and D. J. C. Fletcher. 1969. Identification of an alarm pheromone in the ant *Crematogaster peringueyi*. *Annals of the Entomological Society of America*, 62(5): 1212.

Crewe, R. M., J. M. Brand, D. J. C. Fletcher, and S. H. Eggers. 1970. The mandibular gland chemistry of some South African species of *Crematogaster* (Hymenoptera: Formicidae). *Journal of the Georgia Entomological Society*, 5(1): 42–47.

Crosland, M. W. J. 1988. Effect of a gregarine parasite on the color of *Myrmecia pilosula* (Hymenoptera: Formicidae). *Annals of the Entomological Society of America*, 81(3): 481–484.

Crosland, M. W. J., and R. H. Crozier. 1986. *Myrmecia pilosula*, an ant with only one pair of chromosomes. *Science*, 231: 1278.

Cross, J. H., R. C. Byler, U. Ravid, R. M. Silverstein, S. W. Robinson, P. M. Baker, J. S. De Oliveira, A. R. Jutsum, and J. M. Cherrett. 1979. The major component of the trail pheromone of the leaf-cutting ant, *Atta sexdens rubropilosa* Forel: 3-ethyl-2,5-dimethylpyrazine. *Journal of Chemical Ecology*, 5(2): 187–203.

Cross, J. H., J. R. West, R. M. Silverstein, A. R. Jutsum, and J. M. Cherrett. 1982. Trail pheromone of the leaf-cutting ant *Acromyrmex octospinosus* (Reich), (Formicidae: Myrmicinae). *Journal of Chemical Ecology*, 8(8): 1119–24.

Crowell, K. L. 1968. Rates of competitive exclusion by the Argentine ant in Bermuda. *Ecology*, 49(3): 551–555.

Crozier, R. H. 1970. Karyotypes of twenty-one ant species (Hymenoptera; Formicidae), with reviews of the known ant karyotypes. *Canadian Journal of Genetics and Cytology*, 12(1): 109–128.

——— 1971. Heterozygosity and sex determination in haplo-diploidy. *American Naturalist*, 105: 399–412.

——— 1973. Apparent differential selection at an isozyme locus between queens and workers of the ant *Aphaenogaster rudis*. *Genetics*, 73(2): 313–318.

——— 1974. Allozyme analysis of reproductive strategy in the ant *Aphaenogaster rudis*. *Isozyme Bulletin*, 7: 18.

——— 1975. *Hymenoptera*. *In* J. Bernard, ed., *Animal cytogenetics*, vol. 3: *Insecta 7*. Gebrüder Borntraeger, Berlin. vi + 95 pp.

——— 1977. Evolutionary genetics of the Hymenoptera. *Annual Review of Entomology*, 22: 263–288.

——— 1979. Genetics of sociality. *In* H. R. Hermann, ed., *Social insects*, vol. 1, pp. 223–286. Academic Press, New York.

——— 1981. Genetic aspects of ant evolution. *In* W. R. Atchley and D. Woodruff, eds., *Evolution and speciation: essays in honor of M. J. D. White*, pp. 356–370. Cambridge University Press, New York.

——— 1982. On insects and insects: twists and turns in our understanding of the evolution of eusociality. *In* M. D. Breed, C. D. Michener, and H. E. Evans, eds., *The biology of social insects* (Proceedings of the Ninth Congress of the International Union for the Study of Social Insects, Boulder, Colorado, 1982), pp. 4–9. Westview Press, Boulder.

——— 1987a. Genetic aspects of kin recognition: concepts, models, and synthesis. *In* D. J. C. Fletcher and C. D. Michener, eds., *Kin recognition in animals*, pp. 55–73. John Wiley, New York.

——— 1987b. Towards a sociogenetics of social insects. *In* J. Eder and H. Rembold, eds., *Chemistry and biology of social insects* (Proceedings of the Tenth International Congress of the International Union for the Study of Social Insects, Munich, 1986), pp. 325–328. Verlag J. Peperny, Munich.

Crozier, R. H., and D. Brückner. 1981. Sperm clumping and the population genetics of Hymenoptera. *American Naturalist*, 117(4): 561–563.

Crozier, R. H., and M. W. Dix. 1979. Analysis of two genetic models for the innate components of colony odor in social Hymenoptera. *Behavioral Ecology and Sociobiology*, 4(3): 217–224.

Crozier, R. H., and R. E. Page. 1985. On being the right size: male contributions and multiple mating in social Hymenoptera. *Behavioral Ecology and Sociobiology*, 18(2): 105–115.

Crozier, R. H., P. Pamilo, and Y. C. Crozier. 1984. Relatedness and microgeographic genetic variation in *Rhytidoponera mayri*, an Australian arid-zone ant. *Behavioral Ecology and Sociobiology*, 15(2): 143–150.

Crozier, R. H., P. Pamilo, R. W. Taylor, and Y. C. Crozier. 1986. Evolutionary patterns in some putative Australian species in the ant genus *Rhytidoponera. Australian Journal of Zoology,* 34(4): 535–560.

Culver, D. C., and A. J. Beattie. 1978. Myrmecochory in *Viola:* dynamics of seed-ant interactions in some West Virginia species. *Journal of Ecology,* 66(1): 53–72.

——— 1980. The fate of *Viola* seeds dispersed by ants. *American Journal of Botany,* 67(5): 710–714.

Cupp, E. W., J. O'Neal, G. Kearney, and G. P. Markin. 1973. Forced copulation of imported fire ant reproductives. *Annals of the Entomological Society of America,* 66(4): 743–745.

Cutler, B. 1980. Ant predation by *Habrocestum pulex* (Hentz) (Araneae: Salticidae). *Zoologischer Anzeiger,* 204(1–2): 97–101.

Czechowski, W. 1979. Competition between *Lasius niger* (L.) and *Myrmica rugulosa* Nyl. (Hymenoptera, Formicidae). *Annales Zoologici, Warszawa,* 34(14): 437–451.

Dajoz, R. 1977. *Coléoptères: Colydiidae et Anommatidae paléarctiques* (Faune de l'Europe et du Bassin Méditerranéen, no. 8). Masson, Paris. 280 pp.

Darlington, P. J. 1950. Paussid beetles. *Transactions of the American Entomological Society,* 76(2): 47–142.

——— 1971. The carabid beetles of New Guinea, Part IV: General considerations; analysis and history of fauna; taxonomic supplement. *Bulletin of the Museum of Comparative Zoology, Harvard,* 142(2): 129–337.

Dartigues, D., and L. Passera. 1979a. La ponte des ouvrières chez la fourmi *Camponotus aethiops* Latreille [Hym. Formicidae]. *Annales de la Société Entomologique de France,* n.s., 15(1): 109–116.

——— 1979b. Polymorphisme larvaire et chronologie de l'apparition des castes femelles chez *Camponotus aethiops* Latreille (Hymenoptera, Formicidae). *Bulletin de la Société Zoologique de France,* 104(2): 197–207.

Darwin, C. 1859. *On the origin of species.* (Facsimile of 1st edition, 1964). Harvard University Press, Cambridge, Mass. xxvii + 502 pp.

Das, G. M. 1959. Observations on the association of ants with coccids of tea. *Bulletin of Entomological Research,* 50(3): 437–448.

Davidson, D. W. 1977a. Species diversity and community organization in desert seed-eating ants. *Ecology,* 58(4): 711–724.

——— 1977b. Foraging ecology and community organization in desert seed-eating ants. *Ecology,* 58(4): 725–737.

——— 1978. Size variability in the worker caste of a social insect (*Veromessor pergandei* Mayr) as a function of the competitive environment. *American Naturalist,* 112: 523–532.

——— 1979. Experimental tests of the optimal diet in two social insects. *Behavioral Ecology and Sociobiology,* 4(1): 35–41.

——— 1982. Sexual selection in harvester ants (Hymenoptera: Formicidae: *Pogonomyrmex*). *Behavioral Ecology and Sociobiology,* 10(4): 245–250.

——— 1985. An experimental study of diffuse competition in harvester ants. *American Naturalist,* 125(4): 500–506.

——— 1988. Ecological studies of Neotropical ant gardens. *Ecology,* 69(4): 1138–52.

Davidson, D. W., and W. W. Epstein. 1989. Epiphytic associations with ants. *In* U. Lüttge, ed., *Vascular plants as epiphytes,* pp. 200–233. Springer-Verlag, New York.

Davidson, D. W., and S. R. Morton. 1981. Myrmecochory in some plants (*F. chenopodiaceae*) of the Australian arid zone. *Oecologia,* 50(3): 357–366.

Davidson, D. W., J. H. Brown, and R. S. Inouye. 1980. Competition and the structure of granivore communities. *BioScience,* 30(4): 233–238.

Davidson, D. W., R. S. Inouye, and J. H. Brown. 1984. Granivory in a desert ecosystem: experimental evidence for indirect facilitation of ants by rodents. *Ecology,* 65(6): 1780–86.

Davidson, D. W., D. A. Samson, and R. S. Inouye. 1985. Granivory in the Chihuahuan Desert: interactions within and between trophic levels. *Ecology,* 66(2): 486–502.

Davidson, D. W., J. T. Longino, and R. R. Snelling. 1988. Pruning of host plant neighbors by ants: an experimental approach. *Ecology,* 69(3): 801–808.

Davidson, D. W., R. R. Snelling, and J. T. Longino. 1989. Competition among ants for myrmecophytes and the significance of plant trichomes. *Biotropica,* 21(1): 64–73.

Davies, N. B., and A. I. Houston. 1984. Territory economics. *In* J. R. Krebs and N. B. Davies, eds., *Behavioural ecology: an evolutionary approach,* 2nd ed., pp. 148–169. Sinauer Associates, Sunderland, Mass.

Davison, E. A. 1982. Seed utilization by harvester ants. *In* R. C. Buckley, ed., *Ant-plant interactions in Australia,* pp. 1–6. Dr. W. Junk, The Hague.

Dawkins, R. 1976. *The selfish gene.* Oxford University Press, New York. xvi + 224 pp.

Degen, A. A., M. Gersani, Y. Avivi, and N. Weisbrot. 1986. Honeydew intake of the weaver ant *Polyrhachis simplex* (Hymenoptera: Formicidae) attending the aphid *Chaitophorous populialbae* (Homoptera: Aphididae). *Insectes Sociaux,* 33(2): 211–215.

De Haro, A. 1983. Valor adaptativo del transporte mútuo en sociedades de *Cataglyphis iberica* (Hym. Formicidae). *I Congreso Ibérico de Entomología, León, Spain,* pp. 339–348.

Dejean, A. 1980a. Le comportement de prédation de *Serrastruma serrula* Santschi (Formicidae, Myrmicinae), I: Capacité de détection des ouvrières, analyse des phases comportementales. *Annales des Sciences Naturelles, Zoologie,* ser. 13, 2(3): 131–143.

——— 1980b. Le comportement de prédation de *Serrastruma serrula* Santschi (Formicidae, Myrmicinae), II: Analyse séquentielle. *Annales des Sciences Naturelles, Zoologie,* ser. 13, 2(3): 145–150.

——— 1982. Quelques aspects de la prédation chez des fourmis de la tribu des Dacetini (Formicidae—Myrmicinae). Ph.D. diss., Université Paul Sabatier, Toulouse. 263 pp.

——— 1985a. Etude eco-éthologique de la prédation chez les fourmis du genre *Smithistruma* (Formicidae-Myrmicinae-Dacetini), II: Attraction des proies principales (Collemboles). *Insectes Sociaux,* 32(2): 158–172.

——— 1985b. Microévolution du comportement de capture des proies chez les dacetines de la sous-tribu des Strumigeniti (Hymenoptera, Formicidae, Myrmicinae). *Actes des Colloques Insectes Sociaux* (L'Union Internationale pour l'Etude des Insectes Sociaux, Section française, Compte Rendu Colloque Annuel, Diepenbeek, 1984), 2: 239–247.

Dejean, A., and L. Passera. 1974. Ponte des ouvrières et inhibition royale chez la fourmi *Temnothorax recedens* (Nyl.) (Formicidae, Myrmicinae). *Insectes Sociaux,* 21(4): 343–356.

Dejean, A., D. Masens, K. Kanika, M. Nsudi, and M. Buka. 1984. Première approche des modalites du retour au nid chez les ouvrières, chasseresses d'*Odontomachus troglodytes* Santschi (Formicidae, Ponerinae). *Actes des Colloques Insectes Sociaux* (L'Union Internationale pour l'Etude des Insectes Sociaux, Section française, Compte Rendu Colloque Annuel, Les Eyzies, 1983). 1: 39–47.

Dejean, A., D. Masens, K. Kanika, M. Nsudi, and R. Gunumina. 1986. Les termites et les fourmis, animaux dominants de la faune du sol de plusieurs formations forestières et herbeuses du Zaïre. *Actes des Colloques Insectes Sociaux* (L'Union Internationale pour l'Etude des Insectes Sociaux, Section française, Compte Rendu Colloque Annuel, Vaison la Romaine, 1985), 3: 273–283.

Delage, B. 1968. Recherches sur les fourmis moissoneuses du Bassin Aquitain: éthologie, physiologie de l'alimentation. *Annales des Sciences Naturelles, Zoologie,* ser. 12, 10(2): 197–265.

Delage-Darchen, B. 1971. Contribution à l'étude écologique d'une savane de Côte d'Ivoire (Lamto): les fourmis des strates herbacée et arborée. *Biologia Gabonica,* 7(4): 461–496.

——— 1972a. Une fourmi de Côte-d'Ivoire: *Melissotarsus titubans* Del., n. sp. *Insectes Sociaux,* 19(3): 213–226.

——— 1972b. Le polymorphisme larvaire chez les fourmis *Nematocrema* d'Afrique. *Insectes Sociaux,* 19(3): 259–278.

——— 1974a. Ecologie et biologie de *Crematogaster impressa* Emery, fourmi savanicole d'Afrique. *Insectes Sociaux,* 21(1): 13–34.

——— 1974b. Polymorphismus in der Ameisengattung *Messor* und ein Vergleich mit *Pheidole. In* G. H. Schmidt, ed., *Sozialpolymorphismus bei Insekten,* pp. 590–603. Wissenschaftliche Verlagsgesellschaft MBH, Stuttgart.

——— 1976. Les glandes post-pharyngiennes des fourmis: connaissances actuelles sur leur structure, leur fonctionnement, leur rôle. *L'Année Biologique*, 15(1–2): 63–76.

Delamare Deboutteville, C. 1948. Recherches sur les Collemboles termitophiles et myrmécophiles (Ecologie, Ethologie, Systématique). *Archives de Zoologie Expérimentale et Générale*, 85(5): 261–425.

Deligne, J., A. Quennedey, and M. S. Blum. 1981. The enemies and defense mechanisms of termites. *In* H. R. Hermann, ed., *Social insects*, vol. 2, pp. 1–76. Academic Press, New York.

Del Rio Pesado, M. G., and T. M. Alloway. 1983. Polydomy in the slave-making ant, *Harpagoxenus americanus* (Emery) (Hymenoptera: Formicidae). *Psyche*, 90(1–2): 151–162.

Délye, G. 1957. Observations sur la fourmi saharienne *Cataglyphis bombycina* Rog. *Insectes Sociaux*, 4(2): 77–82.

——— 1968. Recherches sur l'écologie, la physiologie et l'éthologie des fourmis du Sahara. Doctoral diss., Université d'Aix-Marseille. 155 pp.

——— 1971. Observations sur le nid et le comportement constructeur de *Messor arenarius* (Hyménoptères Formicidae). *Insectes Sociaux*, 18(1): 15–20.

Deneubourg, J.-L., J. M. Pasteels, and J.-C. Verhaeghe. 1983. Probabilistic behaviour in ants: a strategy of errors? *Journal of Theoretical Biology*, 105(2): 259–271.

Deneubourg, J.-L., S. Goss, J. M. Pasteels, D. Fresneau, and J.-P. Lachaud. 1987. Self-organization mechanisms in ant societies, II: Learning in foraging and division of labor. *In* J. M. Pasteels and J.-L. Deneubourg, eds., *From individual to collective behavior in social insects* (*Experientia* Supplement, vol. 54), pp. 177–196. Birkhäuser Verlag, Basel.

De Vita, J. 1979. Mechanisms of interference and foraging among colonies of the harvesting ant *Pogonomyrmex californicus* in the Mojave Desert. *Ecology* 60(4): 729–737.

De Vries, P. J. 1984. Of crazy-ants and Curetinae: are *Curetis* butterflies tended by ants? *Zoological Journal of the Linnean Society*, 80(1): 59–66.

De Vries, P. J., D. J. Harvey, and I. J. Kitching. 1986. The ant associated epidermal organs on the larva of the lycaenid butterfly *Curetis regula* Evans. *Journal of Natural History*, 20(3): 621–633.

De Vroey, C. 1979. Aggression and Gause's law in ants. *Physiological Entomology*, 4(3): 217–222.

Deyrup, M. A., N. Carlin, J. Trager, and G. Umphrey. 1988. A review of the ants of the Florida Keys. *Florida Entomologist*, 71(2): 163–176.

Diamond, J., and T. J. Case, eds. 1986. *Community ecology*. Harper & Row, New York. xxii + 665 pp.

Dingle, H. 1972. Aggressive behavior in stomatopods and the use of information theory in the analysis of animal communication. *In* H. E. Winn and B. L. Olla, eds., *Behavior of marine animals: current perspectives in research*, vol. 1, pp. 126–156. Plenum Press, New York.

Dinitz, J. L. M., and C. R. F. Brandão. 1989. Feeding behavior of *Thaumatomyrmex*. *Notes from Underground* (Cambridge, Mass.), no. 2, p. 13.

Dixon, A. F. G. 1985. *Aphid ecology*. Chapman and Hall, New York. ix + 157 pp.

Dlussky, G. M. 1965. Protected territory of ants (Hymenoptera: Formicidae). *Zhurnal Obshchei Biologii*, 26(4): 479–489. [In Russian with English summary.]

——— 1967. *Ants of the genus Formica*. Izdatelstvo "Nauka," Moscow. 237 pp. [In Russian.]

——— 1975. Formicoidea, Formicidae, Sphecomyrminae. *In* A. P. Rasnitsyn, ed., *The higher Hymenoptera of the Mesozoic* (Transactions of the Paleontological Institute, no. 147), pp. 114–122. Academy of Sciences of the USSR, Moscow. [In Russian.]

——— 1981. New Miocene ants from USSR. *Trudy Paleontologischeskogo Instituta Akademii Nauk SSSR*, 183: 64–83. [In Russian.]

——— 1983. A new family of Upper Cretaceous Hymenoptera: an "intermediate link" between the ants and the scolioids. *Paleontologischeskii Zhurnal*, 3(3): 65–78. [In Russian with English summary.]

Dlussky, G. M., and N. B. Chernyshova. 1975. Alarm pheromone of the ant *Formica aquilonia* Yarrow (Hymenoptera: Formicidae). *Vestnik Moskovskogo Gosudarstvennogo Universiteta*, ser. 6, 30: 41–46. [In Russian.]

Dlussky, G. M., and B. Pisarski. 1971. Rewizja polskich gatunków mrówek (Hymenoptera: Formicidae) z rodzaju *Formica* L. *Fragmenta Faunistica, Warzsawa*, 16(12): 145–224.

Dobrzańska, J. 1959. Studies on the division of labour in ants genus *Formica*. *Acta Biologiae Experimentalis*, 19: 57–81.

——— 1966. The control of the territory by *Lasius fuliginosus* Latr. *Acta Biologiae Experimentalis*, 26(2): 193–213.

——— 1973. Ethological studies on polycalic colonies of the ants *Formica exsecta* Nyl. *Acta Neurobiologiae Experimentalis*, 33(3): 597–622.

Dobrzański, J. 1956. Badania nad zmysłem czasu u mrówek [Investigations on time sense in ants]. *Folia Biologica*, 4(3–4): 385–397.

——— 1961. Sur l'éthologie guerrière de *Formica sanguinea* Latr. (Hyménoptère, Formicidae). *Acta Biologiae Experimentalis*, 21: 53–73.

——— 1965. Genesis of social parasitism among ants. *Acta Biologiae Experimentalis*, 25(1): 59–71.

——— 1966. Contribution to the ethology of *Leptothorax acervorum* (Hymenoptera: Formicidae). *Acta Biologiae Experimentalis*, 26(1): 71–78.

Dobrzański, J., and J. Dobrzańska. 1975. Ethological studies in the ant *Tetramorium caespitum* Mayr, II: Interspecific relationships. *Acta Neurobiologiae Experimentalis*, 35: 311–317.

Dodd, F. P. 1902. Contribution to the life history of *Liphyra brassolis* Westw. *Entomologist*, 35: 153–156, 184–188.

——— 1912. Some remarkable ant-friend Lepidoptera of Queensland. *Transactions of the Entomological Society of London*, 1911(3–4): 577–590.

Doflein, F. 1905. Beobachtungen an den Weberameisen (*Oecophylla smaragdina*). *Biologisches Centralblatt*, 25(15): 497–507.

Dondale, C. D., and J. H. Redner. 1972. A synonym proposed in *Perimones*, a synonym rejected in *Walckenaera*, and a new species described in *Cochlembolus* (Araneida: Erigonidae). *Canadian Entomologist*, 104(10): 1643–47.

Donisthorpe, H. St. J. K. 1915. *British ants, their life-history and classification*. William Brendon and Son, Plymouth, Eng. xvi + 379 pp.

——— 1920. British Oligocene ants. *Annals and Magazine of Natural History*, ser. 9, 6: 81–94.

——— 1927. *The guests of British ants, their habits and life histories*. George Routledge & Sons, London. xxiii + 244 pp.

——— 1936. The oldest insect on record. *Entomologist's Record and Journal of Variation*, 48(1): 1–2.

——— 1946. *Ireneopone gibber* (Hym., Formicidae), a new genus and species of myrmicine ant from Mauritius. *Entomologist's Monthly Magazine*, 82: 242–243.

Douglas, A., and W. L. Brown. 1959. *Myrmecia inquilina* new species: the first parasite among the lower ants. *Insectes Sociaux*, 6(1): 13–19.

Douwes, P. 1977. *Sifolinia karavajevi*, en för Sverige ny myra (Hym., Formicidae). *Entomologisk Tidskrift*, 98(4): 147–148.

Downey, J. C. 1962. Myrmecophily in *Plebejus (Icaricia) icarioides* (Lepid.: Lycaenidae). *Entomological News*, 73(3): 57–66.

——— 1966. Sound production in pupae of Lycaenidae. *Journal of the Lepidopterists' Society*, 20(3): 129–155.

Downey, J. C., and A. C. Allyn. 1973. Butterfly ultrastructure, 1: Sound production and associated abdominal structures in pupae of Lycaenidae and Riodinidae. *Bulletin of the Allyn Museum*, no. 14, 47 pp.

——— 1979. Morphology and biology of the immature stages of *Leptotes cassius theonus* (Lucas) (Lepid.: Lycaenidae). *Bulletin of the Allyn Museum*, no. 55, 27 pp.

Drake, C. J., and N. T. Davis. 1959. The morphology, phylogeny, and higher classification of the family Tingidae, including the description of a new genus and species of the subfamily Vianaidinae (Hemiptera: Heteroptera). *Entomologica Americana*, n.s., 39: 1–100.

Driessen, G. J. J., A. T. Van Raalte, and G. J. De Bruyn. 1984. Cannibalism in the red wood ant, *Formica polyctena* (Hymenoptera: Formicidae). *Oecologia*, 63(1): 13–22.

Droual, R. 1983. The organization of nest evacuation in *Pheidole desertorum* Wheeler and *P. hyatti* Emery (Hymenoptera: Formicidae). *Behavioral Ecology and Sociobiology*, 12(3): 203–208.

——— 1984. Anti-predator behaviour in the ant *Pheidole desertorum:* the importance of multiple nests. *Animal Behaviour,* 32(4): 1054–58.

Droual, R., and H. Topoff. 1981. The emigration behavior of two species of the genus *Pheidole* (Formicidae: Myrmicinae). *Psyche,* 88(1–2): 135–150.

Drummond, B. A., III. 1976. Butterflies associated with an army ant swarm raid in Honduras. *Journal of the Lepidopterists' Society,* 30(3): 237–238.

DuBois, M. B. 1981. Two new species of inquilinous *Monomorium* from North America (Hymenoptera: Formicidae). *University of Kansas Science Bulletin,* 52(3): 31–37.

——— 1986. A revision of the native New World species of the ant genus *Monomorium* (*minimum* group) (Hymenoptera: Formicidae). *University of Kansas Science Bulletin,* 53(2): 65–119.

Dubuc, C., and M. Meudec. 1984. Etude experimentale de relations entre l'activité d'affouragement et l'émigration chez la fourmi *Tapinoma erraticum. Actes des Colloques Insectes Sociaux* (L'Union Internationale pour l'Etude des Insectes Sociaux, Section française, Compte Rendu Colloque Annuel, Les Eyzies, 1983), 1: 57–66.

Duelli, P. 1973. Astrotaktisches Heimfindevermögen tragender und getragener Ameisen (*Cataglyphis bicolor* Fabr., Hymenoptera, Formicidae). *Revue Suisse de Zoologie,* 80(3): 712–719.

——— 1977. Das soziale Trageverhalten bei neotropischen Ameisen der Gattung *Pseudomyrmex* (Hym., Formicidae): eine Verhaltensnorm als Hinweis für Phylogenie und Taxonomie? *Insectes Sociaux,* 24(4): 359–365.

Duffield, R. M. 1981. Biology of *Microdon fuscipennis* (Diptera: Syrphidae) with interpretations of the reproductive strategies of *Microdon* species found north of Mexico. *Proceedings of the Entomological Society of Washington,* 83(4): 716–724.

Duffield, R. M., and M. S. Blum. 1973. 4-Methyl-3-heptanone: identification and function in *Neoponera villosa* (Hymenoptera: Formicidae). *Annals of the Entomological Society of America,* 66(6): 1357.

——— 1975. Methyl 6-methyl salicylate: identification and function in a ponerine ant (*Gnamptogenys pleurodon*). *Experientia,* 31(4): 466.

Duffield, R. M., J. W. Wheeler, and M. S. Blum. 1980. Methyl anthranilate: identification and possible function in *Aphaenogaster fulva* and *Xenomyrmex floridanus. Florida Entomologist,* 63(2): 203–206.

Du Merle, P. 1982. Fréquentation des strates arbustive et arborescente par les fourmis en montagne méditerranéenne française. *Insectes Sociaux,* 29(3): 422–444.

Dumpert, K. 1972. Alarmstoffrezeptoren auf der Antenne von *Lasius fuliginosus* (Latr.) (Hymenoptera, Formicidae). *Zeitschrift für Vergleichende Physiologie,* 76(4): 403–425.

——— 1978. *Das Sozialleben der Ameisen.* Verlag Paul Parey, Berlin. 253 pp. (1978. *The social biology of ants,* trans. C. Johnson. Pitman, London. vi + 298 pp.)

——— 1985. *Camponotus (Karavaievia) texens* sp. n. and *C. (K.) gombaki* sp. n. from Malaysia in comparison with other *Karavaievia* species (Hymenoptera: Formicinae). *Psyche,* 92(4): 557–573.

Duviard, D., and P. Segeren. 1974. La colonisation d'un myrmécophyte, le parasolier, par *Crematogaster* spp. (Myrmicinae) en Côte d'Ivoire forestière. *Insectes Sociaux,* 21(2): 191–212.

Dybas, H. S. 1962. Myrmecophiles of the beetle family Limulodidae. *Proceedings of the North Central Branch of the Entomological Society of America,* 17: 15–16.

Eberhard, W. G. 1978. Mating swarms of a South American *Acropyga* (Hymenoptera: Formicidae). *Entomological News,* 89(1–2): 14–16.

Echols, H. W. 1966. Assimilation and transfer of Mirex in colonies of Texas leaf-cutting ants. *Journal of Economic Entomology,* 59(6): 1336–38.

Eckloff, W. 1978. Wechselbeziehungen zwischen Pflanzenläusen und Ameisen. *Biologie in Unserer Zeit,* 8: 48–53.

Edwards, J. P., and J. Chambers. 1984. Identification and source of a queen-specific chemical in the Pharaoh's ant, *Monomorium pharaonis* (L.). *Journal of Chemical Ecology,* 10(12): 1731–47.

Ehrhardt, H.-J. 1962. Ablage übergrosser Eier durch Arbeiterinnen von *Formica polyctena* Förster (Ins., Hym.) in Gegenwart von Königinnen. *Naturwissenschaften,* 49(22): 524–525.

——— 1970. Die Bedeutung von Königinnen mit steter arrhenotoker Parthenogenese für die Männchenerzeugung in den Staaten von *Formica polyctena* Foerster (Hymenoptera, Formicidae). Doctoral diss., Bayerische Julius-Maximilians-Universität, Würzburg. 106 pp.

Ehrhardt, P. 1962. Untersuchungen zur Stoffwechselphysiologie von *Megoura viciae* Buckt., einer phloemsaugenden Aphide. *Zeitschrift für Vergleichende Physiologie,* 46(2): 169–211.

Ehrhardt, S. 1931. Über Arbeitsteilung bei *Myrmica-* and *Messor*-Arten. *Zeitschrift für Morphologie und Ökologie der Tiere,* 20(4): 755–812.

Ehrhardt, W. 1982. Untersuchungen zum Raubzugverhalten der sozialparasitischen Ameise *Chalepoxenus muellerianus* (Finzi) (Hym., Formicidae). *Zoologischer Anzeiger,* 208(3–4): 145–160.

Ehrlich, P. R. 1958. The comparative morphology, phylogeny, and higher classification of the butterflies (Lepidoptera: Papilionoidea). *University of Kansas Science Bulletin,* 39(8): 305–370.

Eibl-Eibesfeldt, I., and E. Eibl-Eibesfeldt. 1967. Das Parasitenabwehren der Minima-Arbeiterinnen der Blattschneider-Ameise (*Atta cephalotes*). *Zeitschrift für Tierpsychologie,* 24(3): 278–281.

Eickwort, K. R. 1969. Differential variation of males and females in *Polistes exclamans. Evolution,* 23(3): 391–405.

Eidmann, H. 1928. Weitere Beobachtungen über die Koloniegründung einheimischer Ameisen. *Zeitschrift für Vergleichende Physiologie,* 7(1): 39–55.

——— 1935. Zur Kenntnis der Blattschneiderameise *Atta sexdens* L., insbesondere ihre Ökologie. *Zeitschrift für Angewandte Entomologie,* 22(2–3): 185–241.

——— 1937. Die Gäste und Gastverhältnisse der Blattschneiderameise *Atta sexdens* L. *Zeitschrift für Morphologie und Ökologie der Tiere,* 32(3): 391–462.

——— 1938. Zur Kenntnis der Lebensweise der Blattschneiderameise *Acromyrmex subterraneus* For. var *eidmanni* Santschi und ihrer Gäste. *Revista de Entomologia,* 8(3–4): 291–314.

Eisner, T. 1957. A comparative morphological study of the proventriculus of ants (Hymenoptera: Formicidae). *Bulletin of the Museum of Comparative Zoology, Harvard,* 116(8): 439–490.

Eisner, T., and W. L. Brown. 1958. The evolution and social significance of the ant proventriculus. *Proceedings of the Tenth International Congress of Entomology* (Montreal, 1956), 2: 503–508.

Eisner, T., and G. M. Happ. 1962. The infrabuccal pocket of a formicine ant: a social filtration device. *Psyche,* 69(3): 107–116.

Eisner, T., K. Hicks, M. Eisner, and D. S. Robson. 1978. "Wolf-in-sheep's-clothing" strategy of a predaceous insect larva. *Science,* 199: 790–794.

Elgar, M. A., and N. E. Pierce. 1988. Mating success and fecundity in an ant-tended lycaenid butterfly. *In* T. H. Clutton-Brock, ed., *Reproductive success: studies of individual variation in contrasting breeding systems,* pp. 59–75. University of Chicago Press, Chicago.

Elgert, B., and R. Rosengren. 1977. The guest ant *Formicoxenus nitidulus* follows the scent trail of its wood ant host (Hymenoptera, Formicidae). *Memoranda Societatis pro Fauna et Flora Fennica,* 53(1): 35–38.

Elias, T. S. 1983. Extrafloral nectaries: their structure and distribution. *In* B. L. Bentley and T. S. Elias, eds., *The biology of nectaries,* pp. 174–203. Columbia University Press, New York.

Eliot, J. N. 1973. The higher classification of the Lycaenidae (Lepidoptera): a tentative arrangement. *Bulletin of the British Museum (Natural History), Entomology,* 28(6): 373–505.

Ellis, W. N. 1967. Studies on Neotropical Collembola, I: Some Collembola from Guatemala. *Beaufortia,* 14: 93–107.

Elmes, G. W. 1973. Observations on the density of queens in natural colonies of *Myrmica rubra* L. (Hymenoptera: Formicidae). *Journal of Animal Ecology,* 42(3): 761–771.

——— 1983. Some experimental observations on the parasitic *Myrmica hirsuta* Elmes. *Insectes Sociaux,* 30(2): 221–234.

——— 1987a. Temporal variation in colony populations of the ant *Myrmica sulcinodis,* I: Changes in queen number, worker number and spring production. *Journal of Animal Ecology,* 56(2): 559–571.

——— 1987b. Temporal variation in colony populations of the ant *Myrmica sulcinodis,* II: Sexual production and sex ratios. *Journal of Animal Ecology,* 56(2): 573–583.

Elmes, G. W., and J. C. Wardlaw. 1982. A population study of the ants *Myrmica sabuleti* and *Myrmica scabrinodis*, living at two sites in the south of England, I: A comparison of colony populations. *Journal of Animal Ecology*, 51(2): 651–664.

——— 1983. A comparison of the effect of a queen upon the development of large hibernated larvae of six species of the genus *Myrmica* (Hym. Formicidae). *Insectes Sociaux*, 30(2): 134–148.

Elton, C. S. 1927. *Animal ecology.* Sidgwick and Jackson, London. xx + 207 pp.

——— 1932. Territory among wood ants (*Formica rufa* L.) at Picket Hill. *Journal of Animal Ecology*, 1(1): 69–76.

——— 1933. *The ecology of animals.* Methuen, London. 97 pp.

El-Ziady, S. 1960. Further effects of *Lasius niger* L. on *Aphis fabae* Scopoli. *Proceedings of the Royal Entomological Society of London*, ser. A, 35(1–3): 30–38.

El-Ziady, S., and J. S. Kennedy. 1956. Beneficial effects of the common garden ant, *Lasius niger* L., on the black bean aphid, *Aphis fabae* Scopoli. *Proceedings of the Royal Entomological Society of London*, ser. A, 31(4–6): 61–65.

Elzinga, R. J. 1978. Holdfast mechanisms in certain uropodine mites (Acarina: Uropodina). *Annals of the Entomological Society of America*, 71(6): 896–900.

Elzinga, R. J., and C. W. Rettenmeyer. 1975. Seven new species of *Circocylliba* (Acarina: Uropodina) found on army ants. *Acarologia*, 16(4): 595–611.

Emden, F. van. 1936. Eine interessante zwischen Carabidae und Paussidae vermittelnde Käferlarve. *Arbeiten über Physiologische und Angewandte Entomologie aus Berlin-Dahlem*, 3: 250–256.

Emerton, J. H. 1911. New spiders from New England. *Transactions of the Connecticut Academy of Arts and Sciences*, 16: 383–407.

Emery, C. 1891. Le formiche dell'ambra Siciliana nel Museo Mineralogico dell'Università di Bologna. *Memorie della Reale Accademia delle Scienze dell'Istituto di Bologna*, ser. 5, 1: 1–26.

——— 1893. Studio monografico sul genere *Azteca* Forel. *Memorie della Reale Accademia della Scienze dell'Istituto di Bologna*, ser. 5, 3: 319–352.

——— 1895. Die Gattung *Dorylus* Fab. und die systematische Eintheilung der Formiciden. *Zoologische Jahrbücher, Abteilung für Systematik, Geographie und Biologie der Thiere*, 8(5): 685–778.

——— 1897. Formiche raccolte nella Nuova Guinea dal Dott. Lamberto Loria. *Annali del Museo Civico di Storia Naturale di Genova*, 38: 546–594.

——— 1898. Beiträge zur Kenntnis der palaearktischen Ameisen. *Öfversigt af Finska Vetenskaps-societetens Förhandlingar*, 40: 124–151.

——— 1899. Glanures myrmécologiques [Hymén.]. *Bulletin de la Société Entomologique de France*, no. 2, pp. 17–20.

——— 1909. Über den Ursprung der dulotischen, parasitischen und myrmekophilen Ameisen. *Biologisches Centralblatt*, 29(11): 352–362.

——— 1910. *Hymenoptera, fam. Formicidae, subfam. Dorylinae. In* P. Wytsman, ed., *Genera Insectorum*, no. 102. P. Wytsman, Zoologiste, 43 rue Saint-Aphonse, Brussels. 34 pp.

——— 1911a. Beobachtungen und Versuche an *Polyergus rufescens. Biologisches Centralblatt*, 31(20): 625–642.

——— 1911b. *Hymenoptera, fam. Formicidae, subfam. Ponerinae. In* P. Wytsman, ed., *Genera Insectorum*, no. 118. V. Verteneuil & L. Desmet, Brussels. 125 pp.

——— 1912. *Hymenoptera, fam. Formicidae, subfam. Dolichoderinae. In* P. Wytsman, ed., *Genera Insectorum*, no. 137. V. Verteneuil & L. Desmet, Brussels. 50 pp.

——— 1914. Les fourmis de la Nouvelle-Calédonie et des îles Loyalty. *In* F. Sarasin and J. Roux, eds., *Nova Caledonia: recherches scientifiques en Nouvelle Calédonie et aux îles Loyalty*, ser. A, *Zoologie*, vol. I, book 4, pp. 393–435. C. W. Kreidels Verlag, Wiesbaden.

——— 1921. Quels sont les facteurs du polymorphisme du sexe féminin chez les fourmis? *Revue Générale des Sciences Pures et Appliquées*, 32(24): 737–741.

——— 1921–22. *Hymenoptera, fam. Formicidae, subfam. Myrmicinae. In* P. Wytsman, ed., *Genera Insectorum*, no. 174. Louis Desmet-Verteneuil, Brussels. 397 pp.

——— 1925a. *Hymenoptera, fam. Formicidae, subfam. Formicinae. In* P. Wytsman, ed., *Genera Insectorum*, no. 183. Louis Desmet-Verteneuil, Brussels. 302 pp.

——— 1925b. Les espèces européennes et orientales du genre *Bothriomyrmex. Bulletin de la Société Vaudoise des Sciences Naturelles*, 56: 5–22.

Emlen, J. M. 1966. The role of time and energy in food preference. *American Naturalist*, 100: 611–617.

Emmert, W. 1968. Die Postembryonalentwicklung sekretorischer Kopfdrüsen von *Formica pratensis* Retz. und *Apis mellifica* L. (Ins., Hym.). *Zeitschrift für Morphologie der Tiere*, 63(1): 1–62.

——— 1969. Entwicklungsleistungen fragmentierter Labialdrüsen-Imaginalanlagen von *Formica pratensis* Retz. (Hymenoptera). *Wilhem Roux' Archiv*, 162(2): 97–113.

Erber, J. 1976. Retrograde amnesia in honeybees (*Apis mellifera carnica*). *Journal of Comparative and Physiological Psychology*, 90(1): 41–46.

Erickson, J. M. 1971. The displacement of native ant species by the introduced Argentine ant *Iridomyrmex humilis* Mayr. *Psyche*, 78(4): 257–266.

——— 1972. Mark-recapture techniques for population estimates of *Pogonomyrmex* ant colonies: an evaluation of the $^{32}P$ technique. *Annals of the Entomological Society of America*, 65(1): 57–61.

Errard, C. 1984. Evolution, en fonction de l'âge, des relations sociales dans les colonies mixtes hétérospécifiques chez les fourmis des genres *Camponotus* et *Pseudomyrmex. Insectes Sociaux*, 31(2): 185–198.

Errard, C., and P. Jaisson. 1984. Etude des relations sociales dans les colonies mixtes hétérospécifiques chez les fourmis (Hymenoptera, Formicidae). *Folia Entomológica Mexicana*, 61: 135–146.

Errard, C., and J. M. Jallon. 1987. An investigation of the development of the chemical factors in ants intra-society recognition. *In* J. Eder and H. Rembold, eds., *Chemistry and biology of social insects* (Proceedings of the Tenth International Congress of the International Union for the Study of Social Insects, Munich, 1986), p. 478. Verlag J. Peperny, Munich.

Erwin, T. L. 1981. A synopsis of the immature stages of Pseudomorphini (Coleoptera: Carabidae) with notes on tribal affinities and behavior in relation to life with ants. *Coleopterists Bulletin*, 35(1): 53–68.

Escherich, K. 1905. Das System der Lepismatiden. *Zoologica, Stuttgart*, 18(43): 1–164.

——— 1907. Neue Beobachtungen über *Paussus* in *Erythrea. Zeitschrift für Wissenschaftliche Insektenbiologie*, 3(1): 1–8.

——— 1917. *Die Ameise: Schilderung ihrer Lebensweise*, 2nd ed. Verlag von Friedr. Vieweg & Sohn, Braunschweig. xvi + 348 pp.

Escherich, K., and C. Emery. 1897. Zur Kenntnis der Myrmekophilen Kleinasiens, I: Coleopteren (mit einem Verzeichniss der in Kleinasien gesammelten Ameisen und einer Neubeschreibung). *Wiener Entomologische Zeitung*, 16(9): 229–239.

Espadaler, X. 1981. *Sifolinia lemasnei* (Bernard, 1968) en España (Hymenoptera, Formicidae). *Boletín de la Asociación Española de Entomología* (1980), 4: 121–124.

——— 1982a. *Epimyrma bernardi* n.sp., a new parasitic ant. *Spixiana*, 5(1): 1–6.

——— 1982b. *Myrmicinosporidium* sp., parasite interne des fourmis: étude au meb de la structure externe. *In* A. de Haro and X. Espadaler, eds., *La Communication chez les sociétés d'insectes* (Colloque Internationale de l'Union Internationale pour l'Etude des Insectes Sociaux, Section française, Barcelona, 1982), pp. 239–241. Universidad Autónoma de Barcelona, Bellaterra.

Espadaler, X., and J. L. Nieves. 1983. Hormigas ("Hymenoptera, Formicidae") pobladoras de agallas abandonadas de cinípidos ("Hymenoptera, Cynipidae") sobre "Quercus" sp. en la península Ibérica. *Boletín de la Estación Central de Ecología*, 12(23): 89–93.

Espadaler, X., and J. Wiśniewski. 1987. *Aegeritella superficialis* Bał. et Wiś. and *A. tuberculata* Bał. et Wiś. (Deuteromycetes), epizoic fungi on two *Formica* (Hymenoptera: Formicidae) species in the Iberian Peninsula. *Butlletí de la Institució Catalana d'Història Natural*, 54(sec. bot., 6): 31–35.

Ettershank, G. 1966. A generic revision of the world Myrmicinae related to *Solenopsis* and *Pheidologeton* (Hymenoptera: Formicidae). *Australian Journal of Zoology*, 14(1): 73–171.

——— 1967. A completely defined synthetic diet for ants (Hym., Formicidae). *Entomologist's Monthly Magazine,* 103: 66–67.

——— 1968. The three-dimensional gallery structure of the nest of the meat ant *Iridomyrmex purpureus* (Sm.) (Hymenoptera: Formicidae). *Australian Journal of Zoology,* 16(4): 715–723.

——— 1971. Some aspects of the ecology and nest microclimatology of the meat ant, *Iridomyrmex purpureus* (Sm.). *Proceedings of the Royal Society of Victoria,* 84(1): 137–152.

Ettershank, G., and J. A. Ettershank. 1982. Ritualised fighting in the meat ant *Iridomyrmex purpureus* (Smith) (Hymenoptera: Formicidae). *Journal of the Australian Entomological Society,* 21(2): 97–102.

Evans, H. C., and D. Leston. 1971. A ponerine ant (Hym., Formicidae) associated with Homoptera on cocoa in Ghana. *Bulletin of Entomological Research,* 61(2): 357–362.

Evans, H. C., and R. A. Samson. 1982. *Cordyceps* species and their anamorphs pathogenic on ants (Formicidae) in tropical forest ecosystems, I: The *cephalotes* (Myrmicinae) complex. *Transactions of the British Mycological Society,* 79(3): 431–453.

——— 1984. *Cordyceps* species and their anamorphs pathogenic on ants (Formicidae) in tropical forest ecosystems, II: The *Camponotus* (Formicinae) complex. *Transactions of the British Mycological Society,* 82(1): 127–150.

Evans, H. E. 1958. The evolution of social life in wasps. *Proceedings of the Tenth International Congress of Entomology* (Montreal, 1956), 2: 449–457.

——— 1962. A review of nesting behavior of digger wasps of the genus *Aphilanthops,* with special attention to the mechanics of prey carriage. *Behaviour,* 19(3): 239–260.

——— 1964. A synopsis of the American Bethylidae (Hymenoptera, Aculeata). *Bulletin of the Museum of Comparative Zoology, Harvard,* 132(1): 1–222.

——— 1970. A new genus of ant-mimicking spider wasps from Australia (Hymenoptera, Pompilidae). *Psyche,* 77(3): 303–307.

——— 1973. Studies on Neotropical Pompilidae (Hymenoptera), IX: The genera of Auplopodini. *Pysche,* 80(3): 212–226.

——— 1977a. Extrinsic versus intrinsic factors in the evolution of insect sociality. *BioScience,* 27(9): 613–617.

——— 1977b. Prey specificity in *Clypeadon* (Hymenoptera: Sphecidae). *Pan-Pacific Entomologist,* 53(2): 144.

Evans, H. E., and M. J. West-Eberhard. 1970. *The wasps.* University of Michigan Press, Ann Arbor. vi + 265 pp.

Evers, C. A., and T. D. Seeley. 1986. Kin discrimination and aggression in honey bee colonies with laying workers. *Animal Behaviour,* 34(3): 924–925.

Evershed, R. P., and E. D. Morgan. 1983. The amounts of trail pheromone substances in the venom of workers of four species of attine ants. *Insect Biochemistry,* 13(5): 469–474.

Evershed, R. P., E. D. Morgan, and M.-C. Cammaerts. 1982. 3-Ethyl-2,5-dimethylpyrazine, the trail pheromone from the venom gland of eight species of *Myrmica* ants. *Insect Biochemistry,* 12(4): 383–391.

Evesham, E. J. M. 1984a. Queen distribution movements and interactions in a semi-natural nest of the ant *Myrmica rubra* L. *Insectes Sociaux,* 31(1): 5–19.

——— 1984b. The sensitization of young workers to queens in the ant *Myrmica rubra* L. *Animal Behaviour,* 32(3): 782–789.

Faber, W. 1967. Beiträge zur Kenntnis sozialparasitischer Ameisen, 1: *Lasius* (*Austrolasius* n. sg.) *reginae* n. sp., eine neue temporär sozialparasitische Erdameise aus Österreich (Hym. Formicidae). *Pflanzenschutz-Berichte,* 36(5–7): 73–107.

——— 1969. Beiträge zur Kenntnis sozialparasitischer Ameisen, 2: *Aporomyrmex ampeloni* nov. gen., nov. spec. (Hym. Formicidae), ein neuer permanenter Sozialparasit bei *Plagiolepis vindobonensis* Lomnicki aus Österreich. *Pflanzenschutz-Berichte,* 39(3–6): 39–100.

Faegri, K., and L. van der Pijl. 1979. *The principles of pollination ecology,* 3rd rev. ed. Pergamon Press, Oxford. xii + 244 pp.

Fage, L. 1938. Quelques Arachnides provenant de fourmilières ou de termitières du Costa Rica. *Bulletin du Muséum National d'Histoire Naturelle, Paris,* ser. 2, 10(4): 369–376.

Fagen, R. M. 1981. *Animal play behavior.* Oxford University Press, New York. xviii + 684 pp.

Fagen, R. M., and R. N. Goldman. 1977. Behavioural catalogue analysis methods. *Animal Behaviour,* 25(2): 261–274.

Fales, H. M., M. S. Blum, R. M. Crewe, and J. M. Brand. 1972. Alarm pheromones in the genus *Manica* derived from the mandibular gland. *Journal of Insect Physiology,* 18(6): 1077–88.

Fall, H. C. 1912. Four new myrmecophilous Coleoptera. *Psyche,* 19(1): 9–12.

——— 1937. The North American species of *Nemadus* Thom., with descriptions of new species (Coleoptera, Silphidae). *Journal of the New York Entomological Society,* 45(3–4): 335–340.

Fanfani, A., and M. Dazzini Valcurone. 1984. Nuovi dati relativi alla "glandola di Pavan" in *Iridomyrmex humilis* Mayr (Formicidae Dolichoderinae). *Pubblicazioni dell'Istituto di Entomologia dell'Università di Pavia,* 28: 1–9.

——— 1986. Glandole delle ponerine e ricerche sulle glandole del gastro di *Megaponera foetens* (Fabr.) (Hymenoptera: Formicidae). *Accademia Nazionale dei Lincei Richerche Biologiche in Sierra Leone,* 260(4): 115–132.

Farish, D. J. 1972. The evolutionary implications of qualitative variation in the grooming behaviour of the Hymenoptera (Insecta). *Animal Behaviour,* 20(4): 662–676.

Farquharson, C. O. 1918. *Harpagomyia* and other Diptera fed by *Cremastogaster* ants in S. Nigeria. *Proceedings of the Entomological Society of London* (1918), pp. xxix–xxxix.

Febvay, G., and A. Kermarrec. 1981. Morphologie et fonctionnement du filtre infrabuccal chez une attine *Acromyrmex octospinosus* (Reich) (Hymenoptera: Formicidae): rôle de la poche infrabuccale. *International Journal of Insect Morphology & Embryology,* 10(5–6): 441–449.

——— 1983. Enzymes digestives de la fourmi attine *Acromyrmex octospinosus* (Reich): caractérisation des amylases, maltase et tréhalase des glandes labiales et de l'intestin moyen. *Comptes Rendus de l'Académie des Sciences, Paris,* ser. 3, 296(9): 453–456.

——— 1986. Digestive physiology of leaf-cutting ants. *In* C. S. Lofgren and R. K. Vander Meer, eds., *Fire ants and leaf-cutting ants: biology and management,* pp. 274–288. Westview Press, Boulder.

Febvay, G., F. Mallet, and A. Kermarrec. 1984. Attractivité du couvain et comportement des ouvrières de la fourmi attine *Acromyrmex octospinosus* (Reich) (Hym. Formicidae). *Actes des Colloques Insectes Sociaux* (L'Union Internationale pour l'Etude des Insectes Sociaux, Section française, Compte Rendu Colloque Annuel, Les Eyzies, 1983), 1: 79–86.

Feener, D. H. 1981. Competition between ant species: outcome controlled by parasitic flies. *Science,* 214: 815–817.

——— 1986. Alarm-recruitment behaviour in *Pheidole militicida* (Hymenoptera: Formicidae). *Ecological Entomology,* 11(1): 67–74.

——— 1987a. Response of *Pheidole morrisi* to two species of enemy ants, and a general model of defense behavior in *Pheidole* (Hymenoptera: Formicidae). *Journal of the Kansas Entomological Society,* 60(4): 569–575.

——— 1987b. Size-selective oviposition in *Pseudacteon crawfordi* (Diptera: Phoridae), a parasite of fire ants. *Annals of the Entomological Society of America,* 80(2): 148–151.

——— 1988. Effects of parasites on foraging and defense behavior of a termitophagous ant, *Pheidole titanus* Wheeler (Hymenoptera: Formicidae). *Behavioral Ecology and Sociobiology,* 22(6): 421–427.

Fellers, J. H. 1987. Interference and exploitation in a guild of woodland ants. *Ecology,* 68(5): 1466–78.

Fellers, J. H., and G. M. Fellers. 1976. Tool use in a social insect and its implications for competitive interactions. *Science,* 192: 70–72.

——— 1982. Scavenging rates of invertebrates in an eastern dediduous forest. *American Midland Naturalist,* 107(2): 389–392.

Fent, K., and R. Wehner. 1985. Ocelli: a celestial compass in the desert ant *Cataglyphis. Science,* 228: 192–194.

Ferrara, F., U. Maschwitz, S. Steghaus-Kovač, and S. Taiti. 1987. The genus *Exalloniscus* Stebbing, 1911 (Crustacea, Oniscidea) and its relationship with social insects. *Pubblicazioni dell'Istituto di Entomologia dell'Università di Pavia,* 36: 43–46.

Fiebrig, K. 1907. Nachtrag zu: Eine Wespen zerstörende Ameise aus Paraguay. *Zeitschrift für Wissenschaftliche Insektenbiologie*, 3(5–6): 154–156.

Fiedler, K. 1987. Quantitative Untersuchungen zur Myrmekophilie der Präimaginalstadien zweirer Bläulingsarten (Lepidoptera: Lycaenidae). Diplom-Arbeit, Fachbereich Biology, Universität Frankfurt. 91 pp.

Fiedler, K., and U. Maschwitz. 1987. Functional analysis of the myrmecophilous relationships between ants (Hymenoptera: Formicidae) and lycaenids (Lepidoptera: Lycaenidae), III: New aspects of the function of the retractile tentacular organs of lycaenid larvae. *Zoologische Beiträge*, n.s., 31(3): 409–416.

——— 1988. Functional analysis of the myrmecophilous relationships between ants (Hymenoptera: Formicidae) and lycaenids (Lepidoptera: Lycaenidae), II: Lycaenid larvae as trophobiotic partners of ants—a quantitative approach. *Oecologia*, 75(2): 204–206.

——— 1989. Functional analysis of the myrmecophilous relationships between ants (Hymenoptera: Formicidae) and lycaenids (Lepidoptera: Lycaenidae), I: Release of food recruitment in ants by lycaenid larvae and pupae. *Ethology*, 80(1–4): 71–80.

Fielde, A. M. 1903. Artificial mixed nests of ants. *Biological Bulletin of the Marine Biological Laboratory, Woods Hole*, 5(6): 320–325.

——— 1904a. Power of recognition among ants. *Biological Bulletin of the Marine Biological Laboratory, Woods Hole*, 7(5): 227–250.

——— 1904b. Tenacity of life in ants. *Biological Bulletin of the Marine Biological Laboratory, Woods Hole*, 7(6): 300–309.

——— 1905a. Observations on the progeny of virgin ants. *Biological Bulletin of the Marine Biological Laboratory, Woods Hole*, 9(6): 355–360.

——— 1905b. The progressive odor of ants. *Biological Bulletin of the Marine Biological Laboratory, Woods Hole*, 10(1): 1–16.

Fielde, A. M., and G. H. Parker. 1904. The reactions of ants to material vibrations. *Proceedings of the Academy of Natural Sciences of Philadelphia*, 56(2): 642–650.

Finnegan, R. J. 1975. Introduction of a predacious red wood ant, *Formica lugubris* (Hymenoptera: Formicidae), from Italy to Eastern Canada. *Canadian Entomologist*, 107(12): 1271–74.

——— 1977. Establishment of a predacious red wood ant, *Formica obscuripes* (Hymenoptera: Formicidae), from Manitoba to Eastern Canada. *Canadian Entomologist*, 109(8): 1145–48.

Fittkau, E. J., and H. Klinge. 1973. On biomass and trophic structure of the central Amazonian rain forest ecosystem. *Biotropica*, 5(1): 2–14.

Fletcher, D. J. C. 1971. The glandular source and social functions of trail pheromones in two species of ants (*Leptogenys*). *Journal of Entomology*, ser. A, 46(1): 27–37.

——— 1973. "Army ant" behaviour in the Ponerinae: a re-assessment. *Proceedings of the Seventh Congress of the Internationaal Union for the Study of Social Insects* (London, 1973), pp. 116–121.

Fletcher, D. J. C., and M. S. Blum. 1981. Pheromonal control of dealation and oogenesis in virgin queen fire ants. *Science*, 212: 73–75.

——— 1983a. Regulation of queen number by workers in colonies of social insects. *Science*, 219: 312–314.

——— 1983b. The inhibitory pheromone of queen fire ants: effects of disinhibition on dealation and oviposition by virgin queens. *Journal of Comparative Physiology*, 153(4): 467–475.

Fletcher, D. J. C., and J. M. Brand. 1968. Source of the trail pheromone and method of trail laying in the ant *Crematogaster peringueyi*. *Journal of Insect Physiology*, 14(6): 783–788.

Fletcher, D. J. C., and C. D. Michener, eds. 1987. *Kin recognition in animals*. John Wiley, New York. x + 465 pp.

Fletcher, D. J. C., and K. G. Ross. 1985. Regulation of reproduction in eusocial Hymenoptera. *Annual Review of Entomology*, 30: 319–343.

Fletcher, D. J. C., M. S. Blum, T. V. Whitt, and N. Temple. 1980. Monogyny and polygyny in the fire ant, *Solenopsis invicta*. *Annals of the Entomological Society of America*, 73(6): 658–661.

Fletcher, D. J. C., D. Cherix, and M. S. Blum. 1983. Some factors influencing dealation by virgin queen fire ants. *Insectes Sociaux*, 30(4): 443–454.

Fluker, S. S., and J. W. Beardsley. 1970. Sympatric associations of three ants: *Iridomyrmex humilis*, *Pheidole megacephala*, and *Anoplolepis longipes* in Hawaii. *Annals of the Entomological Society of America*, 63(5): 1290–96.

Folsom, J. W. 1923. Termitophilous Apterygota from British Guiana. *Zoologica, New York*, 3(19): 383–402.

Forbes, S. A. 1906. The corn root-aphis and its attendant ant (*Aphis maidiradicis* Forbes and *Lasius niger* L., var. *americanus* Emery). *Bulletin, U.S. Department of Agriculture, Division of Entomology*, 60: 29–39.

Forel, A. 1869. Observations sur les moeurs du *Solenopsis fugax*. *Mitteilungen der Schweizerischen Entomologischen Gesellschaft*, 3(3): 105–128.

——— 1874. *Les fourmis de la Suisse*. Société Helvétique des Sciences Naturelles, Zurich. iv + 452 pp. [Revised and corrected, Imprimerie Coopérative, La Chaux-de-Fonds (1920). xvi + 333 pp.]

——— 1878. Etudes myrmécologiques en 1878 (première partie) avec l'anatomie du gésier des fourmis. *Bulletin de la Société Vaudoise des Sciences Naturelles*, 15(80): 337–392.

——— 1886. Etudes myrmécologiques en 1886. *Annales de la Société Entomologique de Belgique*, 30: 131–215.

——— 1891. *Les Formicides*. *In* A. Grandidier, ed., *Histoire, physique, naturelle et politique de Madagascar*, vol. 20: *Histoire naturelle des hyménoptères*, part 2. Imprimerie Nationale, Paris. vi + 280 pp.

——— 1898. La parabiose chez les fourmis. *Bulletin de la Société Vaudoise des Sciences Naturelles*, 34(130): 380–384.

——— 1901. Fourmis termitophages, lestobiose, *Atta tardigrada*, sous-genres d'*Euponera*. *Annales de la Société Entomologique de Belgique*, 45: 389–398.

——— 1902. Beispiele phylogenetischer Wirkungen und Rückwirkungen bei den Instinkten und dem Körpenbau der Ameisen als Belege für die Evolutionslehre und die psychophysiologische Identitätslehre. *Journal für Psychologie und Neurologie*, 1(3): 99–110.

——— 1904. Miscellanea myrmécologiques. *Revue Suisse de Zoologie*, 12(1): 1–52.

——— 1906. Moeurs des fourmis parasites des genres *Wheeleria* et *Bothriomyrmex*. *Revue Suisse de Zoologie*, 14(1): 51–69.

——— 1910. Glanures myrmécologiques. *Annales de la Société Entomologique de Belgique*, 54: 6–32.

——— 1912. Descriptions provisoires de genres, sous-genres et espèces de Formicides des Indes orientales. *Revue Suisse de Zoologie*, 20(15): 761–774.

——— 1928. *The social world of the ants compared with that of man*, 2 vols., trans. C. K. Ogden. G. P. Putnam's Sons, London. xlv + 551; xx + 445 pp.

Forrest, H. F. 1963. The production of audible sound by common ants and its possible uses in communication, with species reference to stridulation. *In* H. F. Forrest, Three problems in invertebrate behavior, pp. 221–350. Ph.D. diss., Rutgers University.

Forsyth, A. 1981. Sex ratio and parental investment in an ant population. *Evolution*, 35(6): 1252–53.

Fortelius, W., P. Pamilo, R. Rosengren, and L. Sundström. 1987. Male size dimorphism and alternative reproductive tactics in *Formica exsecta* ants (Hymenoptera, Formicidae). *Annales Zoologici Fennici*, 24(1): 45–54.

Fowler, H. G. 1981. On the emigration of leaf-cutting ant colonies. *Biotropica*, 13(4): 316.

Fowler, H. G., and R. B. Roberts. 1983. Anomalous social dominance among queens of *Camponotus ferrugineus* (Hymenoptera: Formicidae). *Journal of Natural History*, 17(2): 185–187.

Fowler, H. G., L. C. Forti, V. Pereira-da-Silva, and N. B. Saes. 1986a. Economics of grass-cutting ants. *In* C. S. Lofgren and R. K. Vander Meer, eds., *Fire ants and leaf-cutting ants: biology and management*, pp. 18–35. Westview Press, Boulder.

Fowler, H. G., V. Pereira-da-Silva, L. C. Forti, and N. B. Saes. 1986b. Population dynamics of leaf-cutting ants: a brief review. *In* C. S. Lofgren and R. K. Vander Meer, eds., *Fire ants and leaf-cutting ants: biology and management*, pp. 123–145. Westview Press, Boulder.

Fox, M. D., and B. J. Fox. 1982. Evidence for interspecific competition influencing ant species diversity in a regenerating heathland. *In* R. C. Buckley, ed., *Ant-plant interactions in Australia*, pp. 99–110. Dr. W. Junk, The Hague.

Francfort, R. 1945. Quelques phénomènes illustrant l'influence de la fourmilière sur les fourmis isolées. *Bulletin de la Société Entomologique de France,* 50(7): 95–96.

Francke, O. F., and J. C. Cokendolpher. 1986. Temperature tolerances of the red imported fire ant. *In* C. S. Lofgren and R. K. Vander Meer, eds., *Fire ants and leaf-cutting ants: biology and management,* pp. 104–113. Westview Press, Boulder.

Francke, W., M. Bühring, and K. Horstmann. 1980. Untersuchungen über Pheromone bei *Formica polyctena* (Förster). *Zeitschrift für Naturforschung,* ser. C, 35: 829–831.

Francoeur, A. 1965. Ecologie des populations de fourmis dans un bois de chênes rouges et d'érables rouges. *Naturaliste Canadien,* 92(10–11): 263–276.

——— 1968. Une nouvelle espèce du genre *Myrmica* au Québec (Formicidae, Hymenoptera). *Naturaliste Canadien,* 95(3): 727–730.

——— 1973. Révision taxonomique des espèces néarctiques du groupe *fusca,* genre *Formica* (Formicidae, Hymenoptera). *Mémoires de la Société Entomologique du Québec,* no. 3, 316 pp.

——— 1974. Notes for a revision of the ant genus *Formica,* 1: New identifications and synonymies for some Nearctic specimens from Emery, Forel and Mayr collections. *Entomological News,* 85(9–10): 257–264.

——— 1979. Les fourmis du Québec. *Annales de la Société Entomologique du Québec,* 24(1): 12–47.

——— 1981. Le groupe néarctique *Myrmica lampra* (Formicidae, Hymenoptera). *Canadian Entomologist,* 113(8): 755–759.

Francoeur, A., and R. Loiselle. 1988. Évolution du strigile chez les formicides (Hyménoptères). *Naturaliste Canadien,* 115(3–4): 333–353.

Francoeur, A., and R. R. Snelling. 1979. Notes for a revision of the ant genus *Formica,* 2: Reidentifications for some specimens from the T. W. Cook Collection and new distribution data (Hymenoptera: Formicidae). *Contributions in Science, Natural History Museum of Los Angeles County,* no. 309, 7 pp.

Francoeur, A., R. Loiselle, and A. Buschinger. 1985. Biosystématique de la tribu Leptothoracini (Formicidae, Hymenoptera), 1: Le genre *Formicoxenus* dans la région holarctique. *Naturaliste Canadien,* 112(3): 343–403.

Frank, S. A. 1987. Variable sex ratio among colonies of ants. *Behavioral Ecology and Sociobiology,* 20(3): 195–201.

Franks, N. R. 1985. Reproduction, foraging efficiency and worker polymorphism in army ants. *In* B. Hölldobler and M. Lindauer, eds., *Experimental behavioral ecology and sociobiology* (Fortschritte der Zoologie, no. 31), pp. 91–107. Sinauer Associates, Sunderland, Mass.

——— 1986. Teams in social insects: group retrieval of prey by army ants (*Eciton burchelli,* Hymenoptera: Formicidae). *Behavioral Ecology and Sociobiology,* 18(6): 425–429.

Franks, N. R., and W. H. Bossert. 1983. The influence of swarm raiding army ants on the patchiness and diversity of a tropical leaf litter ant community. *In* E. L. Sutton, T. C. Whitmore, and A. C. Chadwick, eds., *Tropical rain forest: ecology and management,* pp. 151–163. Blackwell, Oxford.

Franks, N. R., and S. Bryant. 1987. Rhythmical patterns of activity within the nests of ants. *In* J. Eder and H. Rembold, eds., *Chemistry and biology of social insects* (Proceedings of the Tenth International Congress of the International Union for the Study of Social Insects, Munich, 1986), pp. 122–123. Verlag J. Peperny, Munich.

Franks, N. R., and C. R. Fletcher. 1983. Spatial patterns in army ant foraging and migration: *Eciton burchelli* on Barro Colorado, Panama. *Behavioral Ecology and Sociobiology,* 12(4): 261–270.

Franks, N. R., and B. Hölldobler. 1987. Sexual competition during colony reproduction in army ants. *Biological Journal of the Linnean Society,* 30(3): 229–243.

Franks, N. R., and P. J. Norris. 1987. Constraints on the division of labour in ants: D'Arcy Thompson's cartesian transformations applied to worker polymorphism. *In* J. M. Pasteels and J.-L. Deneubourg, eds., *From individual to collective behavior in social insects* (*Experientia* Supplement, vol. 54), pp. 253–270. Birkhäuser Verlag, Basel.

Franks, N. R., and T. Scovell. 1983. Dominance and reproductive success among slave-making worker ants. *Nature,* 304: 724–725.

Franzl, S., M. Locke, and P. Huie. 1984. Lenticles: innervated secretory structures that are expressed at every other larval moult. *Tissue and Cell,* 16(2): 251–268.

Freeland, J. 1958. Biological and social patterns in the Australian bulldog ants of the genus *Myrmecia. Australian Journal of Zoology,* 6(1): 1–18.

Frehland, E., B. Kleutsch, and H. Markl. 1985. Modelling a two-dimensional random alarm process. *BioSystems,* 18(2): 197–208.

Fresneau, D. 1984. Etude du polyethisme et du développement des ovaires dans une colonie de *Pachycondyla obscuricornis* (Hym. Formicidae). *Actes des Colloques Insectes Sociaux* (L'Union Internationale pour l'Etude des Insectes Sociaux, Section française, Compte Rendu Colloque Annuel, Les Eyzies, 1983), 1: 87–92.

——— 1985. Individual foraging and path fidelity in a ponerine ant. *Insectes Sociaux,* 32(2): 109–116.

Fresneau, D., and J. P. Lachaud. 1985. La régulation sociale: données préliminaires sur les facteurs individuels controlant l'organisation des taches chez *Neoponera apicalis* (Hym. Formicidae, Ponerinae). *Actes des Colloques Insectes Sociaux* (L'Union Internationale pour l'Etude des Insectes Sociaux, Section française, Compte Rendu Colloque Annuel, Diepenbeek, 1984), 2: 185–193.

Fridman, S., and E. Avital. 1983. Foraging by queens of *Cataglyphis bicolor nigra* (Hymenoptera: Formicidae): an unusual phenomenon among the Formicinae. *Israel Journal of Zoology,* 32(4): 229–230.

Frisch, K. von. 1950. Die Sonne als Kompaß im Leben der Bienen. *Experientia,* 6(6): 210–221.

Fritz, R. S. 1982. An ant-treehopper mutualism: effects of *Formica subsericea* on the survival of *Vanduzea arquata. Ecological Entomology,* 7(3): 267–276.

Frohawk, F. W. 1916. Further observations on the last stage of the larva of *Lycaena arion. Transactions of the Entomological Society of London,* 1915(3–4): 313–316.

Frumhoff, P. C. 1987. The social consequences of polyandry in honey bees *Apis mellifera* L. Ph.D. diss., University of California, Davis. 95 pp.

Frumhoff, P. C., and J. Baker. 1988. A genetic component to division of labour within honey bee colonies. *Nature,* 333: 358–361.

Frumhoff, P. C., and S. Schneider. 1987. The social consequences of honey bee polyandry: the effects of kinship on worker interactions within colonies. *Animal Behaviour,* 35(2): 255–262.

Fuchs, S. 1976a. The response to vibrations of the substrate and reactions to the specific drumming in colonies of carpenter ants (*Camponotus,* Formicidae, Hymenoptera). *Behavioral Ecology and Sociobiology,* 1(2): 155–184.

——— 1976b. An informational analysis of the alarm communication by drumming behavior in nests of carpenter ants (*Camponotus,* Formicidae, Hymenoptera). *Behavioral Ecology and Sociobiology,* 1(3): 315–336.

Fukuda, H., K. Kubo, K. Takeshi, A. Takahashi, M. Takahashi, B. Tanaka, M. Wakabayashi, and T. Shirozu. 1978. *Insects' life in Japan,* vol. 3: *Butterflies.* Hoikusha, Osaka.

Fukumoto, Y. 1983. A new method for studying the successive change of colony composition of ants in the field. *Biological Magazine of Okinawa,* 21: 27–31. [In Japanese with English summary.]

Fukumoto, Y., and T. Abe. 1983. Social organization of colony movement in the tropical ponerine ant, *Diacamma rugosum* (Le Guillon). *Journal of Ethology,* 1(1–2): 101–108.

Futuyma, D. J. 1986. *Evolutionary biology,* 2nd ed. Sinauer Associates, Sunderland, Mass. xiv + 600 pp.

Gadagkar, R. 1985. Kin recognition in social insects and other animals—a review of recent findings and a consideration of their relevance for the theory of kin selection. *Proceedings of the Indian Academy of Sciences (Animal Sciences),* 94(6): 587–621.

Gadeceau, E. 1907. Les plantes myrmécophiles. *La Nature,* 35(1793): 295–298.

Gadgil, M., and W. H. Bossert. 1970. Life historical consequences of natural selection. *American Naturalist,* 104: 1–24.

Gallardo, A. 1916a. Las hormigas de la República Argentina: subfamilia Dolichoderinas. *Anales del Museo Nacional de Historia Natural de Buenos Aires,* 28: 1–130.

——— 1916b. Notes systématiques et éthologiques sur les fourmis attines de la République Argentine. *Anales del Museo Nacional de Historia Natural de Buenos Aires,* 28: 317–344.
——— 1918. Las hormigas de la República Argentina: subfamilia Ponerinas. *Anales del Museo Nacional de Historia Natural de Buenos Aires,* 30: 1–112.
——— 1920. Las hormigas de la República Argentina: subfamilia Dorilinas. *Anales del Museo Nacional de Historia Natural de Buenos Aires,* 30: 281–410.
——— 1932. Las hormigas de la República Argentina: subfamilia Mirmicinas. *Anales del Museo Nacional de Historia Natural de Buenos Aires,* 37: 37–170.
Gallé, L. 1975. Factors stabilizing the ant populations (Hymenoptera: Formicidae) in the grass associations of the Tisza Basin. *Tiscia* (Szeged, Hung.), 10: 61–66.
——— 1978. Respiration as one of the manifestations of the group effect in ants. *Acta Biologica* (Szeged, Hung.), n.s., 24(1–4): 111–114.
Gamboa, G. J. 1975. Foraging and leaf-cutting of the desert gardening ant *Acromyrmex versicolor versicolor* (Pergande) (Hymenoptera: Formicidae). *Oecologia,* 20(1): 103–110.
Gamboa, G. J., H. K. Reeve, and D. W. Pfennig. 1986. The evolution and ontogeny of nestmate recognition in social wasps. *Annual Review of Entomology,* 31: 431–454.
Garling, L. 1979. Origin of ant-fungus mutualism: a new hypothesis. *Biotropica,* 11(4): 284–291.
Garnett, W. B., R. D. Akre, and G. Sehlke. 1985. Cocoon mimicry and predation by myrmecophilous Diptera (Diptera: Syrphidae). *Florida Entomologist,* 68(4): 615–621.
Garrett, S. D. 1956. *Biology of root-infecting fungi.* Cambridge University Press, Cambridge. xi + 293 pp.
Gascuel, J., M.-H. Pham-Delègue, G. Arnold, and C. Masson. 1987. Evidence for a sensitive period during the development of the olfactory system in honey bees: anatomical, functional and behavioural data. *In* J. Eder and H. Rembold, eds., *Chemistry and biology of social insects* (Proceedings of the Tenth International Congress of the International Union for the Study of Social Insects, Munich, 1986), pp. 226–227. Verlag J. Peperny, Munich.
Gause, G. F., and A. A. Witt. 1935. Behavior of mixed populations and the problem of natural selection. *American Naturalist,* 69: 596–609.
Gentry, J. B. 1974. Response to predation by colonies of the Florida harvester ant, *Pogonomyrmex badius. Ecology,* 55(6): 1328–38.
Gentry, J. B., and K. L. Stiritz. 1972. The role of the Florida harvester ant, *Pogonomyrmex badius,* in old field mineral nutrient relationships. *Environmental Entomology,* 1(1): 39–41.
Getz, W. M. 1982. An analysis of learned kin recognition in Hymenoptera. *Journal of Theoretical Biology,* 99(3): 585–597.
Getz, W. M., and K. B. Smith. 1983. Genetic kin recognition: honey bees discriminate between full and half sisters. *Nature,* 302: 147–148.
——— 1986. Honey bee kin recognition: learning self and nestmate phenotypes. *Animal Behaviour,* 34(6): 1617–26.
Getz, W. M., D. Brückner, and T. R. Parisian. 1982. Kin structure and the swarming behavior of the honey bee *Apis mellifera. Behavioral Ecology and Sociobiology,* 10(4): 265–270.
Getz, W. M., D. Brückner, and K. B. Smith. 1986. Conditioning honeybees to discriminate between heritable odors from full and half sisters. *Journal of Comparative Physiology,* ser. A, 159(2): 251–256.
Ghent, R. L. 1961. Adaptive refinements in the chemical defensive mechanisms of certain Formicinae. Ph.D. diss., Cornell University. [Cited by Parry and Morgan, 1979.]
Ghorpade, K. D. 1975. A remarkable predacious cetoniid, *Spilophorus maculatus* (Gory & Percheron), from southern India (Coleoptera: Scarabaeidae). *Coleopterists Bulletin,* 29(4): 226–230.
Glancey, B. M. 1986. The queen recognition pheromone of *Solenopsis invicta. In* C. S. Lofgren and R. K. Vander Meer, eds., *Fire ants and leaf-cutting ants: biology and management,* pp. 223–230. Westview Press, Boulder.
Glancey, B. M., C. E. Stringer, C. H. Craig, P. M. Bishop, and B. B. Martin. 1973. Evidence of a replete caste in the fire ant *Solenopsis invicta. Annals of the Entomological Society of America,* 66(1): 233–234.
Glancey, B. M., C. S. Lofgren, J. R. Rocca, and J. H. Tumlinson. 1982. Behavior of disrupted colonies of *Solenopsis invicta* towards queens and pheromone-treated surrogate queens placed outside the nests. *Sociobiology,* 7(3): 283–288.
Glunn, F. J., D. F. Howard, and W. R. Tschinkel. 1981. Food preferences in colonies of the fire ant *Solenopsis invicta. Insectes Sociaux,* 28(2): 217–222.
Goetsch, W. 1938. Kolonie-Gründung und Kasten-Bildung im Ameisenstaat. *Forschungen und Fortschritte,* 14(19): 223–224.
——— 1953. *Vergleichende Biologie der Insekten-Staaten.* Geest & Portig K.-G., Leipzig. viii + 482 pp.
Goetsch, W., and H. Eisner. 1930. Beiträge zur Biologie Körnersammelnder Ameisen, II. *Zeitschrift für Morphologie und Ökologie der Tiere,* 16(3–4): 371–452.
Goetsch, W., and Br. Käthner. 1937. Die Koloniegründung der Formicinen und ihre experimentelle Beeinflussung. *Zeitschrift für Morphologie und Ökologie der Tiere,* 33(2): 201–260.
Goidanich, A. 1959. Le migrazioni coatte mirmecogene dello *Stomaphis quercus* Linnaeus, Afide olociclico monoico omotopo (Hemiptera Aphidoidea Lachnidae). *Bollettino dell'Istituto di Entomologia della Università degli Studi di Bologna,* 23: 93–131.
Goldstein, E. L. 1975. Island biogeography of ants. *Evolution,* 29(4): 750–762.
Golley, F. B., and J. B. Gentry. 1964. Bioenergetics of the southern harvester ant, *Pogonomyrmex badius. Ecology,* 45(2): 217–225.
Gonçalves, C. R. 1961. O gênero *Acromyrmex* no Brasil (Hym. Formicidae). *Studia Entomologica,* n.s., 4(1–4): 113–180.
Goodloe, L., and R. Sanwald. 1985. Host specificity in colony-founding by *Polyergus lucidus* queens (Hymenoptera: Formicidae). *Psyche,* 92(2–3): 297–302.
Goodloe, L., and H. Topoff. 1987. Pupa acceptance by slaves of the social-parasitic ant *Polyergus* (Hymenoptera: Formicidae). *Psyche,* 94(3–4): 293–302.
Goodloe, L., R. Sanwald, and H. Topoff. 1987. Host specificity in raiding behavior of the slave-making ant *Polyergus lucidus. Psyche,* 94(1–2): 39–44.
Gordon, D. M. 1983a. Daily rhythms in social activities of the harvester ant, *Pogonomyrmex badius. Psyche,* 90(4): 413–423.
——— 1983b. The relation of recruitment rate to activity rhythms in the harvester ant, *Pogonomyrmex barbatus* (F. Smith) (Hymenoptera: Formicidae). *Journal of the Kansas Entomological Society,* 56(3): 277–285.
——— 1984a. The harvester ant (*Pogonomyrmex badius*) midden: refuse or boundary? *Ecological Entomology,* 9(4): 403–412.
——— 1984b. The persistence of role in exterior workers of the harvester ant, *Pogonomyrmex badius. Psyche,* 91(3–4): 251–265.
——— 1986. The dynamics of the daily round of the harvester ant colony (*Pogonomyrmex barbatus*). *Animal Behaviour,* 34(5): 1402–19.
——— 1987. Group-level dynamics in harvester ants: young colonies and the role of patrolling. *Animal Behaviour,* 35(3): 833–843.
——— 1988a. Group-level exploration tactics in fire ants. *Behaviour,* 104(1–2): 162–175.
——— 1988b. Nest-plugging: interference competition in desert ants (*Novomessor cockerelli* and *Pogonomyrmex barbatus*). *Oecologia,* 75(1): 114–118.
Goss, S., J. L. Deneubourg, J. M. Pasteels, and G. Josens. 1989. A model of noncooperative foraging in social insects. *American Naturalist,* 134(2): 273–287.
Gösswald, K. 1932. Ökologische Studien über die Ameisenfauna des mittleren Maingebietes. *Zeitschrift für Wissenschaftliche Zoologie,* 142(1–2): 1–156.
——— 1933. Weitere Untersuchungen über die Biologie von *Epimyrma goesswaldi* Men. und Bemerkungen über andere parasitische Ameisen. *Zeitschrift für Wissenschaftliche Zoologie,* 144(2): 262–288.
——— 1934. Ueber Ameisengäste und -schmarotzer des mittleren Maingebietes. *Entomologische Zeitschrift,* 48(15): 119–120.
——— 1938a. Grundsätzliches über parasitische Ameisen unter besonderer Berücksichtigung der abhängigen Koloniegründung von *Lasius umbratus mixtus* Nyl. *Zeitschrift für Wissenschaftliche Zoologie,* 151(1): 101–148.
——— 1938b. Über bisher unbekannte, durch den Parasitismus der Mermi-

thiden (Nemat.) verursachte Formveränderungen bei Ameisen. *Zeitschrift für Parasitenkunde,* 10(1): 138–152.

——— 1938c. Über den Einfluss von verschiedener Temperatur und Luftfeuchtigkeit auf die Lebensäusserungen der Ameisen, I: Die Lebensdauer ökologisch verschiedener Ameisenarten unter dem Einfluss bestimmter Luftfeuchtigkeit und Temperatur. *Zeitschrift für Wissenschaftliche Zoologie,* 151(3): 337–381.

——— 1950. Pflege des Ameisenparasiten *Tamiclea globula* Meig. (Dipt.) durch den Wirt mit Bemerkungen über den Stoffwechsel in der parasitierten Ameise. *Verhandlungen der Deutschen Zoologen* (Mainz, 1949), pp. 256–264.

——— 1951a. *Die rote Waldameise im Dienste der Waldhygiene: Forstwirtschaftliche Bedeutung, Nutzung, Lebensweise, Zucht, Vermehrung und Schutz.* Metta Kinau Verlag, Luneburg. 160 pp.

——— 1951b. Zur Ameisenfauna des Mittleren Maingebietes mit Bemerkungen über Veränderungen seit 25 Jahren. *Zoologische Jahrbücher, Abteilung für Systematik, Ökologie und Geographie der Tiere,* 80(5–6): 507–532.

——— 1951c. Versuche zum Sozialparasitismus der Ameisen bei der Gattung *Formica* L. *Zoologische Jahrbücher, Abteilung für Systematik, Ökologie und Geographie der Tiere,* 80(5–6): 533–582.

——— 1953. Histologische Untersuchungen an der arbeiterlosen Ameise *Teleutomyrmex schneideri* Kutter (Hym. Formicidae). *Mitteilungen der Schweizerischen Entomologischen Gesellschaften,* 26(2): 81–128.

——— 1954. Über die Wirtschaftlichkeit des Masseneinsatzes der Roten Waldameise. *Zeitschrift für Angewandte Zoologie,* 41: 145–185.

——— 1957. Über die biologischen Grundlagen der Zucht und Anweiselung junger Königinnen der Kleinen Roten Waldameise nebst praktischen Erfahrungen. *Waldhygiene,* 2: 33–53.

——— 1964. Der Ameisenstaat. *Naturwissenschaft und Medizin,* 1(3): 18–30.

——— 1985. *Organisation und Leben der Ameisen.* Wissenschaftliche Verlagsgesellschaft MBH, Stuttgart. 355 pp.

Gösswald, K., and K. Bier. 1953. Untersuchungen zur Kastendetermination in der Gattung *Formica,* 2: Die Aufzucht von Geschlechtstieren bei *Formica rufa pratensis* (Retz.). *Zoologischer Anzeiger,* 151(7–8): 126–134.

——— 1955. Beeinflussung des Geschlechtsverhältnisses durch Temperatureinwirkung bei *Formica rufa* L. *Naturwissenschaften,* 42(5): 133–134.

——— 1957. Untersuchungen zur Kastendetermination in der Gattung *Formica,* 5: Der Einfluss der Temperatur auf die Eiablage und Geschlechtsbestimmung. *Insectes Sociaux,* 4(4): 335–348.

Gösswald, K., and K. Horstmann. 1966. Untersuchungen über den Einfluss der kleinen roten Waldameise (*Formica polyctena* Foerster) auf den Massenwechsel des grünen Eichenwicklers (*Tortrix viridana* L.). *Waldhygiene,* 6: 230–255.

Gösswald, K., and W. Kloft. 1960a. Untersuchungen mit radioaktiven Isotopen an Waldameisen. *Entomophaga,* 5(1): 33–41.

——— 1960b. Neuere Untersuchungen über die sozialen Wechselbeziehungen im Ameisenvolk, durchgeführt mit Radio-Isotopen. *Zoologische Beiträge,* n.s., 5(2–3): 519–556.

——— 1963. Tracer experiments on food exchange in ants and termites. *In Proceedings of a symposium on radiation and radioisotopes applied to insects of agricultural importance* (Athens), pp. 25–42. International Atomic Energy Agency, Vienna.

Gösswald, K., and G. H. Schmidt. 1960. Untersuchungen zum Flügelabwurf und Begattungsverhalten einiger *Formica*-Arten (Ins. Hym.) im Hinblick auf ihre systematische Differenzierung. *Insectes Sociaux,* 7(4): 297–321.

Gotwald, W. H. 1969. Comparative morphological studies of the ants, with particular reference to the mouthparts (Hymenoptera: Formicidae). *Memoirs of the Cornell University Agricultural Experiment Station,* no. 408, 150 pp.

——— 1971. Phylogenetic affinities of the ant genus *Cheliomyrmex* (Hymenoptera: Formicidae). *Journal of the New York Entomological Society,* 79(3): 161–173.

——— 1974. Predatory behavior and food preferences of driver ants in selected African habitats. *Annals of the Entomological Society of America,* 67(6): 877–886.

——— 1978. Trophic ecology and adaptation in tropical Old World ants of the subfamily Dorylinae (Hymenoptera: Formicidae). *Biotropica,* 10(3): 161–169.

——— 1979. Phylogenetic implications of army ant zoogeography. *Annals of the Entomological Society of America,* 72(4): 462–467.

——— 1982. Army ants. *In* H. R. Hermann, ed., *Social insects,* vol. 4, pp. 157–254. Academic Press, New York.

——— 1985. Reflections on the evolution of army ants (Hymenoptera, Formicidae). *Actes des Colloques Insectes Sociaux* (L'Union Internationale pour l'Etude des Insectes Sociaux, Section française, Compte Rendu Colloque Annuel, Diepenbeek, 1984), 2: 7–16.

——— 1984–85. Death on the march: army ants in action. *Rotunda* (Royal Ontario Museum, Toronto), 17(3): 37–41.

——— 1986. The beneficial economic role of ants. *In* S. B. Vinson, ed., *Economic impact and control of social insects,* pp. 290–313. Praeger, New York.

Gotwald, W. H., and W. L. Brown. 1966. The ant genus *Simopelta* (Hymenoptera: Formicidae). *Psyche,* 73(4): 261–277.

Gotwald, W. H., and A. W. Burdette. 1981. Morphology of the male internal reproductive system in army ants: phylogenetic implications (Hymenoptera: Formicidae). *Proceedings of the Entomological Society of Washington,* 83(1): 72–92.

Gotwald, W. H., and B. M. Kupiec. 1975. Taxonomic implications of doryline worker ant morphology: *Cheliomyrmex morosus* (Hymenoptera: Formicidae). *Annals of the Entomological Society of America,* 68(6): 961–971.

Gotwald, W. H., and J. Lévieux. 1972. Taxonomy and biology of a new West African ant belonging to the genus *Amblyopone* (Hymenoptera: Formicidae). *Annals of the Entomological Society of America,* 65(2): 383–396.

Gotwald, W. H., and R. F. Schaefer. 1982. Taxonomic implications of doryline worker ant morphology: *Dorylus* subgenus *Anomma* (Hymenoptera: Formicidae). *Sociobiology,* 7(2): 187–204.

Gould, S. J. 1966. Allometry and size in ontogeny and phylogeny. *Biological Reviews of the Cambridge Philosophical Society,* 41(4): 587–640.

Gould, W. 1747. *An account of English ants.* A. Millar, London. xv + 109 pp.

Grabensberger, W. 1933. Untersuchungen über das Zeitgedächtnis der Ameisen und Termiten. *Zeitschrift für Vergleichende Physiologie,* 20(1–2): 1–54.

Graedel, T. E., and T. Eisner. 1988. Atmospheric formic acid from formicine ants: a preliminary assessment. *Tellus,* ser. B, 40(5): 335–339.

Graf, I., and B. Hölldobler. 1964. Untersuchungen zur Frage der Holzverwertung als Nahrung bei holzzerstörenden Roßameisen (*Camponotus ligniperda* Latr. und *Camponotus herculeanus* L.) unter Berücksichtigung der Cellulase-Aktivität. *Zeitschrift für Angewandte Entomologie,* 55(1): 77–80.

Grant, P. R. 1986. *Ecology and evolution of Darwin's finches.* Princeton University Press, Princeton, N.J. xiv + 458 pp.

Grassé, P.-P. 1946. Sociétés animales et effet de groupe. *Experientia,* 2(3): 77–82.

——— 1986. *Termitologia: anatomie-physiologie-biologie-systématique des termites,* vol. 3. Masson, Paris. xii + 715 pp.

Graur, D. 1985. Gene diversity in Hymenoptera. *Evolution,* 39(1): 190–199.

Graves, R. F., W. H. Clark, and A. B. Gurney. 1976. First record of *Myrmecophila* (Orthoptera: Gryllidae) from Mexico with notes on the ant hosts *Veromessor julianus* and *Aphaenogaster mutica* (Hymenoptera: Formicidae). *Journal of the Idaho Academy of Science,* 12: 97–100.

Gray, B. 1971a. Notes on the biology of the ant species *Myrmecia dispar* (Clark) (Hymenoptera: Formicidae). *Insectes Sociaux,* 18(2): 71–80.

——— 1971b. Notes on the field behaviour of two ant species *Myrmecia desertorum* Wheeler and *Myrmecia dispar* (Clark) (Hymenoptera: Formicidae). *Insectes Sociaux,* 18(2): 81–94.

——— 1971c. A morphometric study of the ant species, *Myrmecia dispar* (Clark) (Hymenoptera: Formicidae). *Insectes Sociaux,* 18(2): 95–109.

——— 1974a. Nest structure and populations of *Myrmecia* (Hymenoptera: Formicidae), with observations on the capture of prey. *Insectes Sociaux,* 21(1): 107–120.

——— 1974b. Associated fauna found in nests of *Myrmecia* (Hymenoptera: Formicidae). *Insectes Sociaux,* 21(3): 289–300.

Gray, R. A. 1952. Composition of honeydew excreted by pineapple mealybugs. *Science*, 115: 129–133.

Greaves, T., and R. D. Hughes. 1974. The population biology of the meat ant. *Journal of the Australian Entomological Society*, 13(4): 329–351.

Green, E. E. 1900. Note on the attractive properties of certain larval Hemiptera. *Entomologist's Monthly Magazine*, 36: 185.

Greenaway, P. 1981. Temperature limits to trailing activity in the Australian arid-zone meat ant *Iridomyrmex purpureus* form *viridiaeneus*. *Australian Journal of Zoology*, 29(4): 621–630.

Greenberg, L. 1979. Genetic component of bee odor in kin recognition. *Science*, 206: 1095–97.

Greenberg, L., D. J. C. Fletcher, and S. B. Vinson. 1985. Differences in worker size and mound distribution in monogynous and polygynous colonies of the fire ant *Solenopsis invicta* Buren. *Journal of the Kansas Entomological Society*, 58(1): 9–18.

Greenslade, P. J. M. 1971. Interspecific competition and frequency changes among ants in Solomon Islands coconut plantations. *Journal of Applied Ecology*, 8(2): 323–352.

——— 1972. Comparative ecology of four tropical ant species. *Insectes Sociaux*, 19(3): 195–212.

——— 1975a. Dispersion and history of a population of the meat ant *Iridomyrmex purpureus* (Hymenoptera: Formicidae). *Australian Journal of Zoology*, 23(4): 495–510.

——— 1975b. Short-term change in a population of the meat ant *Iridomyrmex purpureus* (Hymenoptera: Formicidae). *Australian Journal of Zoology*, 23(4): 511–522.

——— 1976. The meat ant *Iridomyrmex purpureus* (Hymenoptera: Formicidae) as a dominant member of ant communities. *Journal of the Australian Entomological Society*, 15(2): 237–240.

——— 1979. *A guide to ants of South Australia*. South Australian Museum, Adelaide. x + 44 pp.

Greenslade, P. J. M., and R. B. Halliday. 1983. Colony dispersion and relationships of meat ants *Iridomyrmex purpureus* and allies in an arid locality in South Australia. *Insectes Sociaux*, 30(1): 82–99.

Gregg, R. E. 1942. The origin of castes in ants with special reference to *Pheidole morrisi* Forel. *Ecology*, 23(3): 295–308.

——— 1954. Geographical distribution of the genus *Myrmoteras*, including the description of a new species (Hymenoptera: Formicidae). *Psyche*, 61(1): 20–30.

——— 1958a. Key to the species of *Pheidole* (Hymenoptera: Formicidae) in the United States. *Journal of the New York Entomological Society*, 66(1): 7–48.

——— 1958b. Two new species of *Metapone* from Madagascar. *Proceedings of the Entomological Society of Washington*, 60(3): 111–121.

——— 1961. The status of certain myrmicine ants in western North America with a consideration of the genus *Paramyrmica* Cole (Hymenoptera: Formicidae). *Journal of the New York Entomological Society*, 69(4): 209–220.

——— 1963. *The ants of Colorado*. University of Colorado Press, Boulder. xvi + 792 pp.

Grimalski, V. I. 1960. On the role of the ant *Formica rufa* in forest biocoenoses in the eastern Polessye of the Ukraine. *Zoologicheskii Zhurnal*, 39: 398. [In Russian.]

Gryllenberg, G., and R. Rosengren. 1984. The oxygen consumption of submerged *Formica* queens (Hymenoptera, Formicidae) as related to habitat and hydrochoric transport. *Annales Entomologici Fennici*, 50(3): 76–80.

Haber, W. A., G. W. Frankie, H. G. Baker, I. Baker, and S. Koptur. 1981. Ants like flower nectar. *Biotropica*, 13(3): 211–214.

Haddow, A. J., I. H. H. Yarrow, G. A. Lancaster, and P. S. Corbet. 1966. Nocturnal flight cycle in the males of African doryline ants (Hymenoptera: Formicidae). *Proceedings of the Royal Entomological Society of London*, ser. A, 41(7–9): 103–106.

Hagan, H. R. 1954. The reproductive system of the army-ant queen, *Eciton* (*Eciton*), part 3: The oöcyte cycle. *American Museum Novitates*, no. 1665, 20 pp.

Hagen, V. W. von. 1939. The ant that carries a parasol. *Natural History*, 43(1): 27–32.

Hagmann, G. 1907. Beobachtungen über einen myrmekophilen Schmetterling am Amazonenstrom. *Biologisches Centralblatt*, 27(11): 337–341.

Hahn, M., and U. Maschwitz. 1985. Foraging strategies and recruitment behaviour in the European harvester ant *Messor rufitarsis* (F.). *Oecologia*, 68(1): 45–51.

Haines, B. L. 1978. Element and energy flows through colonies of the leaf-cutting ant, *Atta colombica*, in Panama. *Biotropica*, 10(4): 270–277.

Hairston, N. G., F. E. Smith, and L. B. Slobodkin. 1960. Community structure, population control, and competition. *American Naturalist*, 94: 421–425.

Haldane, J. B. S., and H. Spurway. 1954. A statistical analysis of communication in "Apis mellifera" and a comparison with communication in other animals. *Insectes Sociaux*, 1(3): 247–283.

Hamilton, W. D. 1964. The genetical evolution of social behaviour, I, II. *Journal of Theoretical Biology*, 7(1): 1–52.

——— 1967. Extraordinary sex ratios. *Science*, 156: 477–488.

——— 1979. Wingless and fighting males in fig wasps and other insects. *In* M. S. Blum and N. A. Blum, eds., *Sexual selection and reproductive competition in insects*, pp. 167–220. Academic Press, New York.

Handel, S. N. 1978. The competitive relationship of three woodland sedges and its bearing on the evolution of ant-dispersal of *Carex pedunculata*. *Evolution*, 32(1): 151–163.

Handel, S. N., S. B. Fisch, and G. E. Schatz. 1981. Ants disperse a majority of herbs in a mesic forest community in New York State. *Bulletin of the Torrey Botanical Club*, 108(4): 430–437.

Hangartner, W. 1967. Spezifität und Inaktivierung des Spurpheromons von *Lasius fuliginosus* Latr. und Orientierung der Arbeiterinnen im Duftfeld. *Zeitschrift für Vergleichende Physiologie*, 57(2): 103–136.

——— 1969a. Carbon dioxide, a releaser for digging behavior in *Solenopsis geminata* (Hymenoptera: Formicidae). *Psyche*, 76(1): 58–67.

——— 1969b. Structure and variability of the individual odor trail in *Solenopsis geminata* Fabr. (Hymenoptera, Formicidae). *Zeitschrift für Vergleichende Physiologie*, 62(1): 111–120.

——— 1969c. Trail laying in the subterranean ant, *Acanthomyops interjectus*. *Journal of Insect Physiology*, 15(1): 1–4.

Hangartner, W., and S. Bernstein. 1964. Über die Geruchsspur von *Lasius fuliginosus* zwischen Nest und Futterquelle. *Experientia*, 20(7): 392–393.

Hangartner, W., J. M. Reichson, and E. O. Wilson. 1970. Orientation to nest material by the ant, *Pogonomyrmex badius* (Latreille). *Animal Behaviour*, 18(2): 331–334.

Hanna, N. H. C. 1975. Contribution à l'étude de la biologie et de la polygynie de la fourmi *Tapinoma simrothi phoenicium* Emery. *Comptes Rendus de l'Académie des Sciences, Paris*, ser. D, 281(14): 1003–5.

Hansen, S. R. 1978. Resource utilization and coexistence of three species of *Pogonomyrmex* ants in an Upper Sonoran grassland community. *Oecologia*, 35(1): 109–117.

Hare, J. F., and T. M. Alloway. 1987. Early learning and brood discrimination in leptothoracine ants (Hymenoptera: Formicidae). *Animal Behaviour*, 35(6): 1720–24.

Harkness, R. D. 1977a. Further observations on the relation between an ant, *Cataglyphis bicolor* (F.) (Hym., Formicidae) and a spider, *Zodarium frenatum* (Simon) (Araneae, Zodariidae). *Entomologist's Monthly Magazine*, 112: 111–121.

——— 1977b. The carrying of ants (*Cataglyphis bicolor* Fab.) by others of the same nest. *Journal of Zoology*, 183(4): 419–430.

Harkness, R. D., and R. Wehner. 1977. *Cataglyphis*. *Endeavour*, n.s., 1(3–4): 115–121.

Harris, R. A. 1979. A glossary of surface sculpturing. *Occasional Papers in Entomology, Department of Food and Agriculture, State of California*, no. 28, 31 pp.

Harrison, J. M., and M. D. Breed. 1987. Temporal learning in the giant tropical ant, *Paraponera clavata*. *Physiological Entomology*, 12(3): 317–320.

Harrison, J. S., and J. B. Gentry. 1981. Foraging pattern, colony distribu-

tion, and foraging range of the Florida harvester ant, *Pogonomyrmex badius*. *Ecology*, 62(6): 1467–73.
Hashmi, A. A. 1973. A revision of the Neotropical ant subgenus *Myrmothrix* of genus *Camponotus* (Hymenoptera: Formicidae). *Studia Entomologica*, n.s., 16(1–4): 1–140.
Haskins, C. P. 1928. Notes on the behavior and habits of *Stigmatomma pallipes* Haldeman. *Journal of the New York Entomological Society*, 36(2): 179–184.
——— 1939. *Of ants and men*. Prentice-Hall, New York. vii + 244 pp.
——— 1941. Note on the method of colony foundation of the ponerine ant *Bothroponera soror* Emery. *Journal of the New York Entomological Society*, 49(2): 211–216.
——— 1960. Note on the natural longevity of fertile females of *Aphaenogaster picea*. *Journal of the New York Entomological Society*, 68(2): 66–67.
——— 1970. Researches in the biology and social behavior of primitive ants. *In* L. R. Aronson, E. Tobach, D. S. Lehrman, and J. S. Rosenblatt, eds., *Development and evolution of behavior*, pp. 355–388. W. H. Freeman, San Francisco.
——— 1978. Sexual calling behavior in highly primitive ants. *Psyche*, 85(4): 407–415.
——— 1984. The ant and her world. *National Geographic*, 165(June): 774–777.
Haskins, C. P., and E. V. Enzmann. 1938. Studies of certain sociological and physiological features in the Formicidae. *Annals of the New York Academy of Sciences*, 37(2): 97–162.
——— 1945. On the occurrence of impaternate females in the Formicidae. *Journal of the New York Entomological Society*, 53(4): 263–277.
Haskins, C. P., and E. F. Haskins. 1950a. Notes on the biology and social behavior of the archaic ponerine ants of the genera *Myrmecia* and *Promyrmecia*. *Annals of the Entomological Society of America*, 43(4): 461–491.
——— 1950b. Note on the method of colony foundation of the ponerine ant *Brachyponera (Euponera) lutea* Mayr. *Psyche*, 57(1): 1–9.
——— 1951. Notes on the method of colony foundation of the ponerine ant *Amblyopone australis* Erichson. *American Midland Naturalist*, 45(2): 432–445.
——— 1955. The pattern of colony foundation in the archaic ant *Myrmecia regularis*. *Insectes Sociaux*, 2(2): 115–126.
——— 1964. Notes on the biology and social behavior of *Myrmecia inquilina*: the only known myrmeciine social parasite. *Insectes Sociaux*, 11(3): 267–282.
——— 1965. *Pheidole megacephala* and *Iridomyrmex humilis* in Bermuda—equilibrium or slow replacement? *Ecology*, 46(5): 736–740.
——— 1979. Worker compatibilities within and between populations of *Rhytidoponera metallica*. *Psyche*, 86(4): 299–312.
——— 1980. Notes on female and worker survivorship in the archaic ant genus *Myrmecia*. *Insectes Sociaux*, 27(4): 345–350.
——— 1983. Situation and location-specific factors in the compatibility response in *Rhytidoponera metallica* (Hymenoptera: Formicidae: Ponerinae). *Psyche*, 90(1–2): 163–174.
Haskins, C. P., and R. M. Whelden. 1954. Note on the exchange of ingluvial food in the genus *Myrmecia*. *Insectes Sociaux*, 1(1): 33–37.
——— 1965. "Queenlessness," worker sibship, and colony versus population structure in the formicid genus *Rhytidoponera*. *Psyche*, 72(1): 87–112.
Haskins, C. P., and P. A. Zahl. 1971. The reproductive pattern of *Dinoponera grandis* Roger (Hymenoptera, Ponerinae) with notes on the ethology of the species. *Psyche*, 78(1–2): 1–11.
Hatch, M. H. 1933. Studies on the Leptodiridae (Catopidae) with descriptions of new species. *Journal of the New York Entomological Society*, 41(1–2): 187–239.
Hauschteck-Jungen, E., and H. Jungen. 1976. Ant chromosomes, I: The genus *Formica*. *Insectes Sociaux*, 23(4): 513–524.
——— 1983. Ant chromosomes, II: Karyotypes of western Palearctic species. *Insectes Sociaux*, 30(2): 149–164.
Have, T. M. van der, J. J. Boomsma, and S. B. J. Menken. 1988. Sex-investment ratios and relatedness in the monogynous ant *Lasius niger* (L.). *Evolution*, 42(1): 160–172.
Hayashi, N., and H. Komae. 1977. The trail and alarm pheromones of the ant, *Pristomyrmex pungens* Mayr. *Experientia*, 33(4): 424–425.
Headley, A. E. 1943. Population studies of two species of ants, *Leptothorax longispinosus* Roger and *Leptothorax curvispinosus* Mayr. *Annals of the Entomological Society of America*, 36(4): 743–753.
——— 1949. A population study of the ant *Aphaenogaster fulva* ssp. *aquia* Buckley (Hymenoptera, Formicidae). *Annals of the Entomological Society of America*, 42(3): 265–272.
Hecht, O. 1924. Embryonalentwicklung und Symbiose bei *Camponotus ligniperda*. *Zeitschrift für Wissenschaftliche Zoologie*, 122(2): 173–204.
Hefetz, A. 1985. Mandibular gland secretions as alarm pheromones in two species of the desert ant *Cataglyphis*. *Zeitschrift für Naturforschung*, ser. C, 40: 665–666.
Hefetz, A., and M. S. Blum. 1978. Biosynthesis and accumulation of formic acid in the poison gland of the carpenter ant *Camponotus pennsylvanicus*. *Science*, 201: 454–455.
Hefetz, A., and T. Orion. 1982. Pheromones of ants of Israel, I: The alarm-defense system of some larger Formicinae. *Israel Journal of Entomology*, 16: 87–97.
Heim, R. 1957. A propos du *Rozites gongylophora* A. Möller. *Revue de Mycologie*, 22(3): 293–299.
Heimann, M. 1963. Zum Wärmehaushalt der kleinen roten Waldameise (*Formica polyctena* Foerst.). *Waldhygiene*, 5: 1–21.
Heinrich, B. 1979. *Bumblebee economics*. Harvard University Press, Cambridge, Mass. x + 245 pp.
——— 1984. Learning in invertebrates. *In* P. Marler and H. S. Terrace, eds., *The biology of learning*, pp. 135–147. Dahlem Konferenzen, Berlin.
Heinrich, B., and M. J. E. Heinrich. 1984. The pit-trapping foraging strategy of the ant lion, *Myrmeleon immaculatus* DeGeer (Neuroptera: Myrmeleontidae). *Behavioral Ecology and Sociobiology*, 14(2): 151–160.
Helava, J. V. T., H. F. Howden, and A. J. Ritchie. 1985. A review of the New World genera of the myrmecophilous and termitophilous subfamily Hetaeriinae (Coleoptera: Histeridae). *Sociobiology*, 10(2): 127–386.
Hemmingsen, A. M. 1973. Nocturnal weaving on nest surface and division of labour in weaver ants (*Oecophylla smaragdina* Fabricius, 1775). *Videnskabelige Meddelelser fra Dansk Naturhistorisk Forening*, 136: 49–56.
Henderson, G., and R. D. Akre. 1986a. Biology of the myrmecophilous cricket, *Myrmecophila manni* (Orthoptera: Grillidae). *Journal of the Kansas Entomological Society*, 59(3): 454–467.
——— 1986b. Dominance hierarchies in *Myrmecophila manni* (Orthoptera: Gryllidae). *Pan-Pacific Entomologist*, 62(1): 24–28.
——— 1986c. Morphology of *Myrmecophila manni*, a myrmecophilous cricket (Orthoptera: Gryllidae). *Journal of the Entomological Society of British Columbia*, 83: 57–62.
Hennaut-Riche, B., G. Josens, and J. Pasteels. 1980. L'approvisionnement du nid chez *Lasius fuliginosus*: pistes, cycles d'activité et spécialisation territoriale des ouvrières. *In* D. Cherix, ed., *Ecologie des insectes sociaux* (L'Union Internationale pour l'Etude des Insectes Sociaux, Section française, Compte Rendu Colloque Annuel, Lausanne, 1979), pp. 71–78. Cherix et Filanosa S. A., Nyon, Switzerland.
Henning, S. F. 1983a. Biological groups within the Lycaenidae (Lepidoptera). *Journal of the Entomological Society of Southern Africa*, 46(1): 65–85.
——— 1983b. Chemical communication between lycaenid larvae (Lepidoptera: Lycaenidae) and ants (Hymenoptera: Formicidae). *Journal of the Entomological Society of Southern Africa*, 46(2): 341–366.
Herbers, J. M. 1977. Behavioral constancy in *Formica obscuripes* (Hymenoptera: Formicidae). *Annals of the Entomological Society of America*, 70(4): 485–486.
——— 1979. Caste-biased polyethism in a mound-building ant species. *American Midland Naturalist*, 101(1): 69–75.
——— 1980. On caste ratios in ant colonies: population responses to changing environments. *Evolution*, 34(3): 575–585.
——— 1981a. Reliability theory and foraging by ants. *Journal of Theoretical Biology*, 89(1): 175–189.
——— 1981b. Time resources and laziness in animals. *Oecologia*, 49(2): 252–262.
——— 1983. Social organization in *Leptothorax* ants: within- and between-species patterns. *Psyche*, 90(4): 361–386.

——— 1984. Queen-worker conflict and eusocial evolution in a polygynous ant species. *Evolution*, 38(3): 631–643.
——— 1985. Seasonal structuring of a north temperate ant community. *Insectes Sociaux*, 32(3): 224–240.
——— 1986a. Effects of ecological parameters on queen number in *Leptothorax longispinosus* (Hymenoptera: Formicidae). *Journal of the Kansas Entomological Society*, 59(4): 675–686.
——— 1986b. Nest site limitation and facultative polygyny in the ant *Leptothorax longispinosus*. *Behavioral Ecology and Sociobiology*, 19(2): 115–122.
Herbers, J. M., and M. Cunningham. 1983. Social organization in *Leptothorax longispinosus* Mayr. *Animal Behaviour*, 31(3): 759–771.
Herbers, J. M., S. C. Adamowicz, and S. D. Helms. 1985. Seasonal changes in social organization of *Aphaenogaster rudis* (Hymenoptera: Formicidae). *Sociobiology*, 10(1): 1–16.
Hermann, H. R. 1971. Sting autotomy, a defensive mechanism in certain social Hymenoptera. *Insectes Sociaux*, 18(2): 111–120.
——— 1975. Crepuscular and nocturnal activities of *Paraponera clavata* (Hymenoptera: Formicidae: Ponerinae). *Entomological News*, 86(5–6): 94–98.
——— 1984a. Defensive mechanisms: general considerations. *In* H. R. Hermann, ed., *Defensive mechanisms in social insects*, pp. 1–31. Praeger, New York.
——— 1984b. Elaboration and reduction of the venom apparatus in aculeate Hymenoptera. *In* H. R. Hermann, ed., *Defensive mechanisms in social insects*, pp. 201–249. Praeger, New York.
Hermann, H. R., and M. S. Blum. 1967a. The morphology and histology of the hymenopterous poison apparatus, II: *Pogonomyrmex badius* (Formicidae). *Annals of the Entomological Society of America*, 60(3): 661–668.
——— 1967b. The morphology and histology of the hymenopterous poison apparatus, III: *Eciton hamatum* (Formicidae). *Annals of the Entomological Society of America*, 60(6): 1282–91.
——— 1981. Defensive mechanisms in the social Hymenoptera. *In* H. R. Hermann, ed., *Social insects*, vol. 2, pp. 77–197. Academic Press, New York.
Hermann, H. R., A. N. Hunt, and W. F. Buren. 1971. Mandibular gland and mandibular groove in *Polistes annularis* (L.) and *Vespula maculata* (L.) (Hymenoptera: Vespidae). *International Journal of Insect Morphology & Embryology*, 1(1): 43–49.
Herzig, J. 1937. Ameisen und Blattläuse. *Zeitschrift für Angewandte Entomologie*, 24: 604–615.
Heyde, K. 1924. Die Entwicklung der psychischen Fähigkeiten bei Ameisen und ihr Verhalten bei abgeänderten biologischen Bedingungen. *Biologisches Zentralblatt*, 44(11): 623–654.
Hickman, J. C. 1974. Pollination by ants: a low-energy system. *Science*, 184: 1290–92.
Higashi, S. 1974. Worker polyethism related with body size in a polydomous red wood ant, *Formica (Formica) yessensis* Forel. *Journal of the Faculty of Science, Hokkaido University*, ser. 6, 19(3): 695–705.
——— 1983. Polygyny and nuptial flight of *Formica (Formica) yessensis* Forel at Ishikari Coast, Hokkaido, Japan. *Insectes Sociaux*, 30(3): 287–297.
Higashi, S., and K. Yamauchi. 1979. Influence of a supercolonial ant *Formica (Formica) yessensis* Forel on the distribution of other ants in Ishikari Coast. *Japanese Journal of Ecology*, 29(3): 257–264.
Hill, W. B., R. D. Akre, and J. D. Huber. 1976. Structure of some epidermal glands in the myrmecophilous beetle *Adranes taylori* (Coleoptera: Pselaphidae). *Journal of the Kansas Entomological Society*, 49(3): 367–384.
Hindwood, K. A. 1959. The nesting of birds in the nests of social insects. *Emu*, 59(1): 1–36.
Hingston, R. W. G. 1929. *Instinct and intelligence*. Macmillan, New York. xv + 296 pp.
Hinton, H. E. 1951. Myrmecophilous Lycaenidae and other Lepidoptera—a summary. *Proceedings and Transactions of the South London Entomological and Natural History Society* (1949–50), pp. 111–175.
——— 1977. Subsocial behaviour and biology of some Mexican membracid bugs. *Ecological Entomology*, 2(1): 61–79.
Hocking, B. 1970. Insect associations with the swollen thorn acacias. *Transactions of the Royal Entomological Society of London*, 122(7): 211–255.
Hodgson, E. S. 1955. An ecological study of the behavior of the leaf-cutting ant *Atta cephalotes*. *Ecology*, 36(2): 293–304.
Hofstadter, D. R. 1979. *Gödel, Escher, Bach: an eternal golden braid*. Basic Books, New York. xxii + 727 pp.
Hohorst, B. 1972. Entwicklung und Ausbildung der Ovarien bei Arbeiterinnen von *Formica (Serviformica) rufibarbis* Fabricius (Hymenoptera: Formicidae). *Insectes Sociaux*, 19(4): 389–402.
Hohorst, W., and G. Graefe. 1961. Ameisen—obligatorische Zwischenwirte des Lanzettegels (*Dicrocoelium dendriticum*). *Naturwissenschaften*, 48(7): 229–230.
Hohorst, W., and G. Lämmler. 1962. Experimentelle Dicrocoeliose-Studien. *Zeitschrift für Tropenmedizin und Parasitologie*, 13(4): 377–397.
Hölldobler, B. 1961. Temperaturunabhängige rhythmische Erscheinungen bei Rossameisenkolonien (*Camponotus ligniperda* Latr. und *Camponotus herculeanus* L.) (Hym. Form.). *Insectes Sociaux*, 8(1): 13–22.
——— 1962. Zur Frage der Oligogynie bei *Camponotus ligniperda* Latr. und *Camponotus herculeanus* L. (Hym. Formicidae). *Zeitschrift für Angewandte Entomologie*, 49(4): 337–352.
——— 1964. Untersuchungen zum Verhalten der Ameisenmännchen während der imaginalen Lebenszeit. *Experientia*, 20(6): 329.
——— 1965. Das soziale Verhalten der Ameisenmännchen und seine Bedeutung für die Organisation der Ameisenstaaten. Doctoral diss., Bayerische Julius-Maximilians-Universität, Würzburg. 123 pp.
——— 1966. Futterverteilung durch Männchen in Ameisenstaat. *Zeitschrift für Vergleichende Physiologie*, 52(4): 430–455.
——— 1967. Zur Physiologie der Gast-Wirt-Beziehungen (Myrmecophilie) bei Ameisen, I: Das Gastverhältnis der *Atemeles*- und *Lomechusa*-Larven (Col. Staphylinidae) zu *Formica* (Hym. Formicidae). *Zeitschrift für Vergleichenden Physiologie*, 56(1): 1–121.
——— 1968a. Der Glanzkäfer als "Wegelagerer" an Ameisenstrassen. *Naturwissenschaften*, 55(8): 397.
——— 1968b. Verhaltensphysiologische Untersuchungen zur Myrmecophilie einiger Staphylinidenlarven. *In* W. Herre, ed., *Verhandlungen der Deutschen Zoologischen Gesellschaft* (Heidelberg, 1967) (*Zoologischer Anzeiger* Supplement, vol. 31), pp. 428–434. Akademische Verlagsgesellschaft, Geest & Portig K.-G., Leipzig.
——— 1969. Host finding by odor in the myrmecophilic beetle *Atemeles pubicollis* Bris. (Staphylinidae). *Science*, 166: 757–758.
——— 1970a. Chemische Verständigung im Insektenstaat am Beispiel der Hautflügler (Hymenoptera). *Umschau*, 70(21): 663–669.
——— 1970b. Orientierungsmechanismen des Ameisengastes *Atemeles* (Coleoptera, Staphylinidae) bei der Wirtssuche. *In* W. Herre, ed., *Verhandlungen der Zoologischen Gesellschaft* (Würzburg, 1969) (*Zoologischer Anzeiger* Supplement, vol. 33), pp. 580–585. Akademische Verlagsgesellschaft, Geest & Portig K.-G., Leipzig.
——— 1970c. Zur Physiologie der Gast-Wirt-Beziehungen (Myrmecophilie) bei Ameisen, II: Das Gastverhältnis des imaginalen *Atemeles pubicollis* Bris. (Col. Staphylinidae) zu *Myrmica* und *Formica* (Hym. Formicidae). *Zeitschrift für Vergleichende Physiologie*, 66(2): 215–250.
——— 1971a. Communication between ants and their guests. *Scientific American*, 224(3)(March): 86–93.
——— 1971b. Homing in the harvester ant *Pogonomyrmex badius*. *Science*, 171: 1149–51.
——— 1971c. Recruitment behavior in *Camponotus socius* (Hym. Formicidae). *Zeitschrift für Vergleichende Physiologie*, 75(2): 123–142.
——— 1971d. Sex pheromone in the ant *Xenomyrmex floridanus*. *Journal of Insect Physiology*, 17(8): 1497–99.
——— 1971e. *Steatoda fulva* (Theridiidae), a spider that feeds on harvester ants. *Psyche*, 77(2): 202–208.
——— 1972. Verhaltensphysiologische Adaptationen an ökologische Nischen in Ameisennester. *In* W. Rathmayer, ed., *Verhandlungen der Deutschen Zoologischen Gesellschaft* (Helgoland, 1971), vol. 65, pp. 137–144. Gustav Fischer Verlag, Stuttgart.
——— 1973a. Chemische Strategie beim Nahrungserwerb der Diebsameise (*Solenopsis fugax* Latr.) und der Pharaoameise (*Monomorium pharaonis* L.). *Oecologia*, 11(4): 371–380.

——— 1973b. *Formica sanguinea* (Formicidae): Futterbetteln. *Encyclopaedia Cinematographica*, film no. E2013. Institut für den Wissenschaftlichen Film, Göttingen. 11 pp.

——— 1973c. Zur Ethologie der chemischen Verständigung bei Ameisen. *Nova Acta Leopoldina*, 37(2): 259–292.

——— 1974. Home range orientation and territoriality in harvesting ants. *Proceedings of the National Academy of Sciences of the United States of America*, 71(8): 3274–77.

——— 1976a. Recruitment behavior, home range orientation and territoriality in harvester ants, *Pogonomyrmex*. *Behavioral Ecology and Sociobiology*, 1(1): 3–44.

——— 1976b. The behavioral ecology of mating in harvester ants (Hymenoptera: Formicidae: *Pogonomyrmex*). *Behavioral Ecology and Sociobiology*, 1(4): 405–423.

——— 1976c. Tournaments and slavery in a desert ant. *Science*, 192: 912–914.

——— 1977. Communication in social Hymenoptera. *In* T. A. Sebeok, ed., *How animals communicate*, pp. 418–471. Indiana University Press, Bloomington.

——— 1978. Ethological aspects of chemical communication in ants. *Advances in the Study of Behavior*, 8: 75–115.

——— 1979. Territories of the African weaver ant (*Oecophylla longinoda* [Latreille]): a field study. *Zeitschrift für Tierpsychologie*, 51(2): 201–213.

——— 1980. Canopy orientation: a new kind of orientation in ants. *Science*, 210: 86–88.

——— 1981a. Foraging and spatiotemporal territories in the honey ant *Myrmecocystus mimicus* Wheeler (Hymenoptera: Formicidae). *Behavioral Ecology and Sociobiology*, 9(4): 301–314.

——— 1981b. Trail communication in the dacetine ant *Orectognathus versicolor* (Hymenoptera: Formicidae). *Psyche*, 88(3–4): 245–257.

——— 1981c. Zur Evolution von Rekrutierungssignalen bei Ameisen. *Nova Acta Leopoldina*, n.s., 245: 431–447.

——— 1982a. Chemical communication in ants: new exocrine glands and their behavioral function. *In* M. D. Breed, C. D. Michener, and H. E. Evans, eds., *The biology of social insects* (Proceedings of the Ninth Congress of the International Union for the Study of Social Insects, Boulder, Colorado, 1982), pp. 312–317. Westview Press, Boulder.

——— 1982b. Communication, raiding behavior and prey storage in *Cerapachys* (Hymenoptera: Formicidae). *Psyche*, 89(1–2): 3–23.

——— 1982c. Interference strategy of *Iridomyrmex pruinosum* (Hymenoptera: Formicidae) during foraging. *Oecologia*, 52(2): 208–213.

——— 1982d. The cloacal gland, a new pheromone gland in ants. *Naturwissenschaften*, 69(4): 186–187.

——— 1983. Territorial behavior in the green tree ant (*Oecophylla smaragdina*). *Biotropica*, 15(4): 241–250.

——— 1984a. A new exocrine gland in the slave raiding ant genus *Polyergus*. *Psyche*, 91(3–4): 225–235.

——— 1984b. Communication during foraging and nest-relocation in the African stink ant, *Paltothyreus tarsatus* Fabr. (Hymenoptera, Formicidae, Ponerinae). *Zeitschrift für Tierpsychologie*, 65(1): 40–52.

——— 1984c. Evolution of insect communication. *In* T. Lewis, ed., *Insect communication* (Symposium of the Royal Entomological Society of London, no. 12), pp. 349–377. Academic Press, London.

——— 1984d. The wonderfully diverse ways of the ant. *National Geographic*, 165(6)(June): 778–813.

——— 1985. Liquid food transmission and antennation signals in ponerine ants. *Israel Journal of Entomology*, 19: 89–99.

——— 1986a. Food robbing in ants, a form of interference competition. *Oecologia*, 69(1): 12–15.

——— 1986b. Konkurrenzverhalten und Territorialität in Ameisenpopulationen. *In* T. Eisner, B. Hölldobler, and M. Lindauer, eds., *Chemische Ökologie, Territorialität, Gegenseitige Verständigung* (Information processing in animals, vol. 3), pp. 25–70. Gustav Fischer Verlag, New York.

——— 1987. Communication and competition in ant communities. *In* S. Kawano, J. H. Connell, and T. Hidaka, eds., *Evolution and coadaptation in biotic communities* (Proceedings of the Second International Symposium held in conjunction with the International Prize for Biology, Tokyo, 1986), pp. 95–124. Tokyo University Press, Tokyo.

Hölldobler, B., and S. H. Bartz. 1985. Sociobiology of reproduction in ants. *In* B. Hölldobler and M. Lindauer, eds., *Experimental behavioral ecology and sociobiology* (Fortschritte der Zoologie, no. 31), pp. 237–257. Sinauer Associates, Sunderland, Mass.

Hölldobler, B., and N. F. Carlin. 1985. Colony founding, queen dominance and oligogyny in the Australian meat ant *Iridomyrmex purpureus*. *Behavioral Ecology and Sociobiology*, 18(1): 45–58.

——— 1987. Anonymity and specificity in the chemical communication signals of social insects. *Journal of Comparative Physiology*, ser. A, 161(4): 567–581.

Hölldobler, B., and H. Engel. 1978. Tergal and sternal glands in ants. *Psyche*, 85(4): 285–330.

Hölldobler, B., and H. Engel-Siegel. 1982. Tergal and sternal glands in male ants. *Psyche*, 89(1–2): 113–132.

——— 1984. On the metapleural gland of ants. *Psyche*, 91(3–4): 201–224.

Hölldobler, B., and C. P. Haskins. 1977. Sexual calling behavior in primitive ants. *Science*, 195: 793–794.

Hölldobler, B., and C. J. Lumsden. 1980. Territorial strategies in ants. *Science*, 210: 732–739.

Hölldobler, B., and U. Maschwitz. 1965. Der Hochzeitsschwarm der Rossameise *Camponotus herculeanus* L. (Hym. Formicidae). *Zeitschrift für Vergleichende Physiologie*, 50(5): 551–568.

Hölldobler, B., and C. D. Michener. 1980. Mechanisms of identification and discrimination in social Hymenoptera. *In* H. Markl, ed., *Evolution of social behavior: hypotheses and empirical tests*, pp. 35–58. Verlag Chemie, Weinheim.

Hölldobler, B., and M. Möglich. 1980. The foraging system of *Pheidole militicida* (Hymenoptera: Formicidae). *Insectes Sociaux*, 27(3): 237–264.

Hölldobler, B., and R. W. Taylor. 1983. A behavioral study of the primitive ant *Nothomyrmecia macrops* Clark. *Insectes Sociaux*, 30(4): 384–401.

Hölldobler, B., and J. F. A. Traniello. 1980a. Tandem running pheromone in ponerine ants. *Naturwissenschaften*, 67(7): 360.

——— 1980b. The pygidial gland and chemical recruitment communication in *Pachycondyla* (= *Termitopone*) *laevigata*. *Journal of Chemical Ecology*, 6(5): 883–893.

Hölldobler, B., and E. O. Wilson. 1970. Recruitment trails in the harvester ant *Pogonomyrex badius*. *Psyche*, 77(4): 385–399.

——— 1977a. Colony-specific territorial pheromone in the African weaver ant *Oecophylla longinoda* (Latreille). *Proceedings of the National Academy of Sciences of the United States of America*, 74(5): 2072–75.

——— 1977b. The number of queens: an important trait in ant evolution. *Naturwissenschaften*, 64(1): 8–15.

——— 1977c. Weaver ants. *Scientific American*, 237(6)(December): 146–154.

——— 1977d. Weaver ants: social establishment and maintenance of territory. *Science*, 195: 900–902.

——— 1978. The multiple recruitment systems of the African weaver ant *Oecophylla longinoda* (Latreille) (Hymenoptera: Formicidae). *Behavioral Ecology and Sociobiology*, 3(1): 19–60.

——— 1983a. Queen control in colonies of weaver ants (Hymenoptera: Formicidae). *Annals of the Entomological Society of America*, 76(2): 235–238.

——— 1983b. The evolution of communal nest-weaving in ants. *American Scientist*, 71(5): 490–499.

——— 1986a. Ecology and behavior of the primitive cryptobiotic ant *Prionopelta amabilis* (Hymenoptera: Formicidae). *Insectes Sociaux*, 33(1): 45–58.

——— 1986b. Nest area exploration and recognition in leafcutter ants (*Atta cephalotes*). *Journal of Insect Physiology*, 32(2): 143–150.

——— 1986c. Soil-binding pilosity and camouflage in ants of the tribes Basicerotini and Stegomyrmecini (Hymenoptera: Formicidae). *Zoomorphology*, 106(1): 12–20.

Hölldobler, B., and M. Wüst. 1973. Ein Sexualpheromon bei der Pharaoameise *Monomorium pharaonis* (L.). *Zeitschrift für Tierpsychologie*, 32(1): 1–9.

Hölldobler, B., M. Möglich, and U. Maschwitz. 1973. *Bothroponera tesserinoda* (Formicidae) Tandemlauf beim Nestumzug. *Encyclopaedia Cinematogra-*

*phica*, film no. E2040. Institut für den Wissenschaftlichen Film, Göttingen. 14 pp.

——— 1974. Communication by tandem running in the ant *Camponotus sericeus*. *Journal of Comparative Physiology*, 90(2): 105–127.

Hölldobler, B., R. Stanton, and H. Engel. 1976. A new exocrine gland in *Novomessor* (Hymenoptera: Formicidae) and its possible significance as a taxonomic character. *Psyche*, 83(1): 32–41.

Hölldobler, B., R. C. Stanton, and H. Markl. 1978. Recruitment and food-retrieving behavior in *Novomessor* (Formicidae, Hymenoptera), I: Chemical signals. *Behavioral Ecology and Sociobiology*, 4(2): 163–181.

Hölldobler, B., M. Möglich, and U. Maschwitz. 1981. Myrmecophilic relationship of *Pella* (Coleoptera: Staphylinidae) to *Lasius fuliginosus* (Hymenoptera: Formicidae). *Psyche*, 88(3–4): 347–374.

Hölldobler, B., H. Engel, and R. W. Taylor. 1982. A new sternal gland in ants and its function in chemical communication. *Naturwissenschaften*, 69(2): 90–91.

Hölldobler, B., J. M. Palmer, K. Masuko, and W. L. Brown. 1989. New exocrine glands in the legionary ants of the genus *Leptanilla* (Hymenoptera, Formicidae, Leptanillinae). *Zoomorphology*, 108(5): 255–261.

Hölldobler, K. 1928. Zur Biologie der diebischen Zwergameise (*Solenopsis fugax*) und ihrer Gäste. *Biologisches Zentralblatt*, 48(3): 129–142.

——— 1929a. Über die Entwicklung der Schwirrfliege *Xanthogramma citrofasciatum* im Neste von *Lasius alienus* und *niger*. *Zoologischer Anzeiger*, 82: 171–176.

——— 1929b. Über eine merkwürdige Parasitenerkrankung von *Solenopsis fugax*. *Zeitschrift für Parasitenkunde*, 2(1): 67–72.

——— 1933. Weitere Mitteilungen über Haplosporidien in Ameisen. *Zeitschrift für Parasitenkunde*, 6(1): 91–100.

——— 1938. Weitere Beiträge zur Koloniengründung der Ameisen. *Zoologischer Anzeiger*, 121(3–4): 66–72.

——— 1947. Studien über die Ameisengrille (*Myrmecophila acervorum* Panzer) im mittleren Maingebiet. *Mitteilungen der Schweizerischen Entomologischen Gesellschaft*, 20(7): 607–648.

——— 1948. Über ein parasitologisches Problem: Die Gastpflege der Ameisen und die Symphilieinstinkte. *Zeitschrift für Parasitenkunde*, 14(1–2): 3–26.

——— 1951. Parasitologische Probleme beim Studium der Ameisenstaaten. *Berichte der Physikalisch-Medizinischen Gesellschaft zu Würzburg*, n.s., 65: 112–116.

——— 1953. Beobachtungen über die Koloniengründung von *Lasius umbratus umbratus* Nyl. *Zeitschrift für Angewandte Entomologie*, 34(4): 598–606.

——— 1965. Springende Ameisen. *Mitteilungen der Schweizerischen Entomologischen Gesellschaft*, 38(1–2): 80–81.

Hollingsworth, M. J. 1960. Studies on the polymorphic workers of the army ant *Dorylus (Anomma) nigricans* Illiger. *Insectes Sociaux*, 7(1): 17–37.

Holmes, W. G., and P. W. Sherman. 1983. Kin recognition in animals. *American Scientist*, 71(1): 46–55.

Holmgren, N. 1908. Über einige myrmecophile Insekten aus Bolivia und Peru. *Zoologischer Anzeiger*, 33(11): 337–349.

Hong, Y. 1983. Fossil insects in the diatoms of Shanweng. *Bulletin of the Tianjin Institute* (Tianjin Institute of Geology and Mineral Resources), 8: 1–11.

Hong, Y.-C., T.-C. Yang, S.-T. Wang, S.-E. Wang, Y.-K. Li, M.-R. Sun, H.-C. Sun, and N.-C. Tu. 1974. Stratigraphy and paleontology of Fushun Coalfield, Liaoning Province. *Acta Geologica Sinica*, 1974(2): 113–149.

Horn, H. S. 1968. The adaptive significance of colonial nesting in the Brewer's blackbird (*Euphagus cyanocephalus*). *Ecology*, 49(4): 682–694.

Horstmann, K. 1970. Untersuchungen über den Nahrungserwerb der Waldameisen (*Formica polyctena* Foerster) im Eichenwald, I: Zusammensetzung der Nahrung, Abhängigkeit von Witterungsfaktoren und von der Tageszeit. *Oecologia*, 5(2): 138–157.

——— 1972. Untersuchungen über den Nahrungserwerb der Waldameisen (*Formica polyctena* Foerster) im Eichenwald, II: Abhängigkeit von Jahresverlauf und vom Nahrungsangebot. *Oecologia*, 8(4): 371–390.

——— 1973. Untersuchungen zur Arbeitsteilung unter den Außendienstarbeiterinnen der Waldameise *Formica polyctena* Foerster. *Zeitschrift für Tierpsychologie*, 32(5): 532–543.

——— 1974. Untersuchungen über den Nahrungserwerb der Waldameisen (*Formica polyctena* Foerster) im Eichenwald, III: Jahresbilanz. *Oecologia*, 15(2): 187–204.

——— 1982a. Die Energiebilanz der Waldameisen (*Formica polyctena* Foerster) in einem Eichenwald. *Insectes Sociaux*, 29(3): 402–421.

——— 1982b. Spurorientierung bei Waldameisen (*Formica polyctena* Förster). *Zeitschrift für Naturforschung*, ser. C, 37: 348–349.

Horstmann, K., and A. Bitter. 1984. Spur-Rekrutierung zu Beute bei Waldameisen (*Formica polyctena* Förster). *Waldhygiene*, 15: 225–232.

Horstmann, K., and H. Schmid. 1986. Temperature regulation in nests of the wood ant, *Formica polyctena* (Hymenoptera: Formicidae). *Entomologia Generalis*, 11(3–4): 229–236.

Horstmann, K., A. Bitter, and P. Ulsamer. 1982. Nahrungsalarm bei Waldameisen (*Formica polyctena* Förster). *Insectes Sociaux*, 29(1): 44–66.

Horvitz, C. C., and A. J. Beattie. 1980. Ant dispersal of *Calathea* (Marantaceae) seeds by carnivorous ponerines (Formicidae) in a tropical rain forest. *American Journal of Botany*, 67(3): 321–326.

Horvitz, C. C., and D. W. Schemske. 1984. Effects of ants and an ant-tended herbivore on seed production of a Neotropical herb. *Ecology*, 65(5): 1369–78.

——— 1986. Ant-nest soil and seedling growth in a Neotropical ant-dispersed herb. *Oecologia*, 70(2): 318–320.

Howard, D. F., and W. R. Tschinkel. 1976. Aspects of necrophoric behavior in the red imported fire ant, *Solenopsis invicta*. *Behaviour*, 56(1–2): 158–180.

——— 1981. The flow of food in colonies of the fire ant, *Solenopsis invicta*: a multifactorial study. *Physiological Entomology*, 6(3): 297–306.

Howard, J. J. 1987. Leafcutting ant diet selection: the role of nutrients, water, and secondary chemistry. *Ecology*, 68(3): 503–515.

Howard, J. J., and D. F. Wiemer. 1986. Chemical ecology of host plant selection by the leaf-cutting ant, *Atta cephalotes*. *In* C. S. Lofgren and R. K. Vander Meer, eds., *Fire ants and leaf-cutting ants: biology and management*, pp. 260–273. Westview Press, Boulder.

Howard, J. J., J. Cazin, and D. F. Wiemer. 1988. Toxicity of terpenoid deterrents to the leafcutting ant *Atta cephalotes* and its mutualistic fungus. *Journal of Chemical Ecology*, 14(1): 59–69.

Huang, H. T., and P. Yang. 1987. The ancient cultured citrus ant. *BioScience*, 37(9): 665–671.

Hubbell, S. P., and D. F. Wiemer. 1983. Host plant selection by an attine ant. *In* P. Jaisson, ed., *Social insects in the tropics*, vol. 2, pp. 133–154. Université Paris-Nord, Paris.

Huber, J. 1905. Über die Koloniengründung bei *Atta sexdens*. *Biologisches Centralblatt*, 25(18): 606–619; (19): 625–635.

Huber, P. 1810. *Recherches sur les moeurs des fourmis indigènes*. J. J. Paschoud, Paris. xiii + 328 pp.

Hung, A. C.-F. 1973. Reproductive biology in dulotic ants: preliminary report (Hymenoptera: Formicidae). *Entomological News*, 84(8): 253–259.

Hunt, J. H. 1974. Temporal activity patterns in two competing ant species (Hymenoptera: Formicidae). *Psyche*, 81(2): 237–242.

——— 1983. Foraging and morphology in ants: the role of vertebrate predators as agents of natural selection. *In* P. Jaisson, ed., *Social insects in the tropics*, vol. 2, pp. 83–104. Université Paris-Nord, Paris.

Hunt, J. H., and R. R. Snelling. 1975. A checklist of the ants of Arizona. *Arizona Academy of Science*, 10(1): 20–23.

Hunt, J. H., I. Baker, and H. G. Baker. 1982. Similarity of amino acids in nectar and larval saliva: the nutritional basis for trophallaxis in social wasps. *Evolution*, 36(6): 1318–22.

Hunter, P. E. 1964. Three new species of *Laelaspis* from North America (Acarina: Laelaptidae). *Journal of the Kansas Entomological Society*, 37(4): 293–301.

Hustache, A., and C. Bruch. 1936. Descripción y notas biológicas acerca de un curculiónido mirmecófilo (Col. Curcul.). *Revista de Entomologia*, 6(3–4): 332–338.

Huwyler, S., K. Grob, and M. Viscontini. 1975. The trail pheromone of the ant, *Lasius fuliginosus:* identification of six components. *Journal of Insect Physiology,* 21(2): 299–304.

Huxley, C. R. 1978. The ant-plants *Myrmecodia* and *Hydnophytum* (Rubiaceae), and the relationships between their morphology, ant occupants, physiology and ecology. *New Phytologist,* 80(1): 231–268.

——— 1980. Symbiosis between ants and epiphytes. *Biological Reviews of the Cambridge Philosophical Society,* 55(3): 321–340.

——— 1982. Ant-epiphytes of Australia. *In* R. C. Buckley, ed., *Ant-plant interactions in Australia,* pp. 63–73. Dr. W. Junk, The Hague.

——— 1986. Evolution of benevolent ant-plant relationships. *In* B. Juniper and T. R. E. Southwood, eds., *Insects and the plant surface,* pp. 257–282. Edward Arnold, London.

Huxley, J. S. 1932. *Problems of relative growth.* Dial Press, New York. xix + 276 pp.

Ihering, H. von. 1898. Die Anlage neuer Colonien und Pilzgärten bei *Atta sexdens. Zoologischer Anzeiger,* 21: 238–245.

Illingworth, J. F. 1917. Economic aspects of our predaceous ant (*Pheidole megacephala*). *Proceedings of the Hawaiian Entomological Society,* 3(4): 349–368.

Imai, H. T., R. H. Crozier, and R. W. Taylor. 1977. Karyotype evolution in Australian ants. *Chromosoma,* 59(4): 341–393.

Imai, H. T., C. Baroni Urbani, M. Kubota, G. P. Sharma, M. N. Narasimhanna, B. C. Das, A. K. Sharma, A. Sharma, G. B. Deodikar, V. G. Vaidya, and M. R. Rajasekarasetty. 1984. Karyological survey of Indian ants. *Japanese Journal of Genetics,* 59(1): 1–32.

Imamura, S. 1982. Social modifications of work efficiency in digging by the ant, *Formica (Formica) yessensis* Forel. *Journal of the Faculty of Science, Hokkaido University,* ser. 6, 23(1): 128–142.

Ishay, J., and R. Ikan. 1968. Gluconeogenesis in the Oriental hornet *Vespa orientalis* F. *Ecology,* 49(1): 169–171.

Isingrini, M. 1987. La reconnaissance coloniale des larves chez la fourmi *Cataglyphis cursor* (Hymenoptera, Formicidae). *Insectes Sociaux,* 34(1): 20–27.

Isingrini, M., A. Lenoir, and P. Jaisson. 1985. Preimaginal learning as a basis of colony-brood recognition in the ant *Cataglyphis cursor. Proceedings of the National Academy of Sciences of the United States of America,* 82(24): 8545–47.

Itow, T., K. Kobayashi, M. Kubota, K. Ogata, H. T. Imai, and R. H. Crozier. 1984. The reproductive cycle of the queenless ant *Pristomyrmex pungens. Insectes Sociaux,* 31(1): 87–102.

Iwanami, Y., and T. Iwadare. 1978. Inhibiting effects of myrmicacin on pollen growth and pollen tube mitosis. *Botanical Gazette,* 139(1): 42–45.

Jackson, D. A. 1984. Ant distribution patterns in a Cameroonian cocoa plantation: investigation of the ant mosaic hypothesis. *Oecologia,* 62(3): 318–324.

Jackson, W. B. 1957. Microclimatic patterns in the army ant bivouac. *Ecology,* 38(2): 276–285.

Jacobs-Jessen, U. F. 1959. Zur Orientierung der Hummeln und einiger anderer Hymenopteren. *Zeitschrift für Vergleichende Physiologie,* 41(6): 597–641.

Jacobson, E. 1909. Ein Moskito als Gast und diebischer Schmarotzer der *Cremastogaster difformis* Smith und eine andere schmarotzende Fliege. *Tijdschrift voor Entomologie,* 52: 158–164.

——— 1911. Biological notes on the hemipteron *Ptilocerus ochraceus. Tijdschrift voor Entomologie,* 54: 175–179.

Jacobson, E., and A. Forel. 1909. Ameisen aus Java und Krakatau beobachtet und gesammelt. *Notes from the Leyden Museum,* 31(3–4)(note 15): 221–253.

Jacobson, E., and E. Wasmann. 1905. Beobachtungen ueber *Polyrhachis dives* auf Java, die ihre Larven zum Spinnen der Nester benutzt. *Notes from the Leyden Museum,* 25(3)(note 9): 133–140.

Jacobson, H. R., D. H. Kistner, and F. A. Abdel-Galil. 1987. A redescription of the myrmecophilous genera *Probeyeria, Beyeria* and the description of a closely related new genus from Arizona (Coleoptera: Staphylinidae). *Sociobiology,* 13(3): 307–338.

Jacoby, M. 1937. Das räumliche Wachsen des *Atta*-Nestes vom 50. bis zum 90. Tage (Hym. Formicidae). *Revista de Entomologia,* 7(4): 416–425.

——— 1944. Observaçōs e experiências sôbre *Atta sexdens rubropilosa* Forel visando facilitar seu combate. *Boletim do Ministério da Agricultura, Industria e Comercia, Rio de Janeiro* (May 1943), pp. 1–55.

——— 1952. Die Erforschung des Nestes der Blattschneider-Ameise *Atta sexdens rubropilosa* Forel (mittels des Ausgußverfahrens in Zement), Teil I. *Zeitschrift für Angewandte Entomologie,* 34(2): 145–169.

Jaffe, K. 1984. Negentropy and the evolution of chemical recruitment in ants (Hymenoptera: Formicidae). *Journal of Theoretical Biology,* 106(4): 587–604.

——— 1987. Evolution of territoriality and nestmate recognition systems in ants. *In* J. M. Pasteels and J.-L. Deneubourg, eds., *From individual to collective behavior in social insects* (*Experientia* Supplement, vol. 54), pp. 295–311. Birkhäuser Verlag, Basel.

Jaffe, K., and P. E. Howse. 1979. The mass recruitment system of the leaf-cutting ant, *Atta cephalotes* (L.). *Animal Behaviour,* 27(3): 930–939.

Jaffe, K., and M. Marquez. 1987. On agonistic behaviour among workers of the ponerine ant *Ectatomma ruidum* (Hymenoptera: Formicidae). *Insectes Sociaux,* 34(2): 87–95.

Jaffe, K., and H. Puche. 1984. Colony-specific territorial marking with the metapleural gland secretion in the ant *Solenopsis geminata* (Fabr). *Journal of Insect Physiology,* 30(4): 265–270.

Jaffe, K., and C. Sanchez. 1984a. Comportamiento alimentario y sistema de recrutamiento en la hormiga *Camponotus rufipes* (Hymenoptera: Formicidae). *Acta Cientifica Venezolana,* 35: 270–277.

——— 1984b. On the nestmate-recognition system and territorial marking behaviour in the ant *Camponotus rufipes. Insectes Sociaux,* 31(3): 302–315.

Jaffe, K., and G. Villegas. 1985. On the communication systems of the fungus-growing ant *Trachymyrmex urichi. Insectes Sociaux,* 32(3): 257–274.

Jaffe, K., M. Bazire-Benazet, and P. E. Howse. 1979. An integumentary pheromone-secreting gland in *Atta* sp.: territorial marking with a colony-specific pheromone in *Atta cephalotes. Journal of Insect Physiology,* 25(10): 833–839.

Jaffe, K., M. E. Lopez, and W. Aragort. 1986. On the communication systems of the ants *Pseudomyrmex termitarius* and *P. triplarinus. Insectes Sociaux,* 33(2): 105–117.

Jaisson, P. 1972a. Nouvelles expériences sur l'agressivité chez les fourmis: existence probable d'une substance active inhibitrice de l'agressivité et attractive sécrétée par la jeune formicine. *Comptes Rendus de l'Académie des Sciences, Paris,* ser. D, 274(2): 302–305.

——— 1972b. Mise en évidence d'une phéromone d'attractivité produite par la jeune ouvrière *Formica* (Hymenoptera: Formicidae). *Comptes Rendus de l'Académie des Sciences, Paris,* ser. D, 274(3): 429–432.

——— 1972c. Sobre el determinismo del comportamiento en las hormigas del género *Atta. Folia Entomológica Mexicana,* 23–24: 108–110.

——— 1975. L'imprégnation dans l'ontogenèse des comportements de soins aux cocons chez la jeune fourmi rousse (*Formica polyctena* Först.). *Behaviour,* 52(1–2): 1–37.

——— 1980. Environmental preference induced experimentally in ants (Hymenoptera: Formicidae). *Nature,* 286: 388–389.

——— 1985. Social behavior. *In* G. A. Kerkut and L. I. Gilbert, eds., *Comprehensive insect physiology, biochemistry and pharmacology,* vol. 9, *Behaviour,* pp. 673–694. Pergamon Press, Oxford.

Jaisson, P., and D. Fresneau. 1978. The sensitivity and responsiveness of ants to their cocoons in relation to age and methods of measurement. *Animal Behaviour,* 26(4): 1064–71.

Jander, R. 1957. Die optische Richtungsorientierung der Roten Waldameise (*Formica rufa* L.). *Zeitschrift für Vergleichende Physiologie,* 40(2): 162–238.

Janet, C. 1896. Sur les rapports des Lépismides myrmécophiles avec les fourmis. *Comptes Rendus de l'Académie des Sciences, Paris,* 122(13): 799–802.

——— 1897a. *Rapports des animaux myrmécophiles avec les fourmis.* H. Ducourtieux, Limoges. 98 pp.

——— 1897b. Sur les rapports de l'*Antennophorus uhlmanni* Haller avec le *Lasius mixtus* Nyl. *Comptes Rendus de l'Académie des Sciences, Paris,* 124(11): 583–585.

——— 1898. *Système glandulaire tégumentaire de la* Myrmica rubra*: observations diverses sur les fourmis.* G. Carré et C. Naud. 28 pp.

——— 1904. *Observations sur les fourmis.* Ducourtieux et Gout, Limoges. 68 pp.

——— 1907. *Anatomie du corselet et histolyse des muscles vibrateurs, après le vol nuptial chez la reine de la fourmi* (Lasius niger). Ducourtieux et Gout, Limoges. 149 pp.
Janssens, E. 1949. Sur la massue antennaire de *Paussus* Linné et genres voisins. *Bulletin de l'Institut Royal des Sciences Naturelles de Belgique,* 25(22): 1–9.
Janzen, D. H. 1966. Coevolution of mutualism between ants and acacias in Central America. *Evolution,* 20(3): 249–275.
——— 1967. Interaction of the bull's-horn acacia (*Acacia cornigera* L.) with an ant inhabitant (*Pseudomyrmex ferruginea* F. Smith) in eastern Mexico. *University of Kansas Science Bulletin,* 47(6): 315–558.
——— 1969. Allelopathy by myrmecophytes: the ant *Azteca* as an allelopathic agent of *Cecropia. Ecology,* 50(1): 147–153.
——— 1972. Protection of *Barteria* (Passifloraceae) by *Pachysima* ants (Pseudomyrmecinae) in a Nigerian rain forest. *Ecology,* 53(5): 885–892.
——— 1973a. Dissolution of mutualism between *Cecropia* and its *Azteca* ants. *Biotropica,* 5(1): 15–28.
——— 1973b. Evolution of polygynous obligate acacia-ants in western Mexico. *Journal of Animal Ecology,* 42(3): 727–750.
——— 1974. Epiphytic myrmecophytes in Sarawak: mutualism through the feeding of plants by ants. *Biotropica,* 6(4): 237–259.
——— 1975. *Pseudomyrmex nigropilosa:* a parasite of a mutualism. *Science,* 188: 936–937.
——— 1977. Why don't ants visit flowers? *Biotropica,* 9(4): 252.
——— 1979. New horizons in the biology of plant defenses. *In* G. A. Rosenthal and D. H. Janzen, eds., *Herbivores: their interaction with secondary plant metabolites,* pp. 331–350. Academic Press, New York.
Janzen, D. H., and C. R. Carroll. 1983. *Paraponera clavata* (bala, giant tropical ant). *In* D. H. Janzen, ed., *Costa Rican natural history,* pp. 752–753. University of Chicago Press, Chicago.
Jarvis, J. U. M. 1981. Eusociality in a mammal: cooperative breeding in naked mole-rat colonies. *Science,* 212: 571–573.
Jayasuriya, A. K., and J. F. A. Traniello. 1985. The biology of the primitive ant *Aneuretus simoni* (Emery) (Formicidae: Aneuretinae), I: Distribution, abundance, colony structure, and foraging ecology. *Insectes Sociaux,* 32(4): 363–374.
Jeanne, R. L. 1972. Social biology of the Neotropical wasp *Mischocyttarus drewseni. Bulletin of the Museum of Comparative Zoology, Harvard,* 144(3): 63–150.
——— 1979. A latitudinal gradient in rates of ant predation. *Ecology,* 60(6): 1211–24.
Jeannel, R. 1936. Monographie des Catopidae (Insectes Coléoptères). *Mémoires du Muséum National d'Histoire Naturelle, Paris,* n.s., 1: 1–433.
Jell, P. A., and P. M. Duncan. 1986. Invertebrates, mainly insects, from the freshwater, Lower Cretaceous, Koonwarra fossil bed (Korumberra Group), South Gippsland, Victoria. *In* P. A. Jell and J. Roberts, eds., *Plants and invertebrates from the Lower Cretaceous Koonwarra fossil bed, South Gippsland, Victoria,* pp. 189–191. Association of Australasian Paleontologists, Sydney.
Jensen, T. F., and I. Holm-Jensen. 1980. Energetic cost of running in workers of three ant species, *Formica fusca* L., *Formica rufa* L., and *Camponotus herculeanus* L. (Hymenoptera, Formicidae). *Journal of Comparative Physiology,* 137(2): 151–156.
Jerdon, T. C. 1854. A catalogue of the species of ants found in Southern India. *Annals and Magazine of Natural History,* ser. 2, 13: 45–56, 100–110.
Jessen, K., and U. Maschwitz. 1983. Abdominaldrüsen bei *Pachycondyla tridentata* (Smith): Formicidae, Ponerinae. *Insectes Sociaux,* 30(2): 123–133.
——— 1985. Individual specific trails in the ant *Pachycondyla tesserinoda* (Formicidae, Ponerinae). *Naturwissenschaften,* 72(10): 549–550.
——— 1986. Orientation and recruitment behavior in the ponerine ant *Pachycondyla tesserinoda* (Emery): laying of individual-specific trails during tandem running. *Behavioral Ecology and Sociobiology,* 19(3): 151–155.
Jessen, K., U. Maschwitz, and M. Hahn. 1979. Neue Abdominaldrüsen bei Ameisen, I: Ponerini (Formicidae: Ponerinae). *Zoomorphologie,* 94(1): 49–66.
Johnson, B., and P. R. Birks. 1960. Studies on wing polymorphism in aphids, I: The developmental process involved in the production of the different forms. *Entomologia Experimentalis et Applicata,* 3(4): 327–339.
Johnson, R. A. 1954. The behavior of birds attending army ant raids on Barro Colorado Island, Panama Canal Zone. *Proceedings of the Linnaean Society of New York,* nos. 63–65, pp. 41–70.
Johnson, S. J., and P. S. Valentine. 1986. Observations on *Liphyra brassolis* Westwood (Lepidoptera: Lycaenidae) in North Queensland. *Australian Entomological Magazine,* 13(1–2): 22–26.
Johnston, A. B., and E. O. Wilson. 1985. Correlations of variation in the major/minor ratio of the ant, *Pheidole dentata* (Hymenoptera: Formicidae). *Annals of the Entomological Society of America,* 78(1): 8–11.
Johnston, N. C., J. H. Law, and N. Weaver. 1965. Metabolism of 9-ketodec-2-enoic acid by worker honeybees (*Apis mellifera* L.). *Biochemistry,* 4(8): 1615–21.
Jolivet, P. 1952. Quelques données sur la myrmécophilie des Clytrides (Col. Chrysomeloidea). *Bulletin de l'Institut Royal des Sciences Naturelles de Belgique,* 28(8): 1–12.
——— 1986. *Les fourmis et les plantes: un exemple de coévolution.* Boubée, Paris. 254 pp.
Jones, C. R. 1929. Studies on ants and their relation to aphids. *Bulletin, Colorado State University Agricultural Experiment Station,* no. 341, 96 pp.
Jones, R. J. 1979. Expansion of the nest of *Nasutitermes costalis. Insectes Sociaux,* 26(4): 322–342.
——— 1980. Gallery construction by *Nasutitermes costalis:* polyethism and the behavior of individuals. *Insectes Sociaux,* 27(1): 5–28.
Jones, T. H., M. S. Blum, and H. M. Fales. 1982a. Ant venom alkaloids from *Solenopsis* and *Monomorium* species: recent developments. *Tetrahedron,* 38(13): 1949–58.
Jones, T. H., M. S. Blum, R. W. Howard, C. A. McDaniel, H. M. Fales, M. B. DuBois, and J. Torres. 1982b. Venom chemistry of ants in the genus *Monomorium. Journal of Chemical Ecology,* 8(1): 285–300.
Jones, T. H., M. S. Blum, A. N. Andersen, H. M. Fales, and P. Escoubas. 1988. Novel 2-ethyl-5-alkylpyrrolidines in the venom of an Australian ant of the genus *Monomorium. Journal of Chemical Ecology,* 14(1): 35–45.
Jonkman, J. C. M. 1980a. Average vegetative requirement, colony size and estimated impact of *Atta vollenweideri* on cattle-raising in Paraguay. *Zeitschrift für Angewandte Entomologie,* 89(2): 135–143.
——— 1980b. The external and internal structure and growth of nests of the leaf-cutting ant *Atta vollenweideri* Forel, 1893 (Hym.: Formicidae), part I. *Zeitschrift für Angewandte Entomologie,* 89(2): 158–173.
Jordan, K. H. C. 1913. Zur Morphologie und Biologie der myrmecophilen Gattungen *Lomechusa* und *Atemeles* und einiger verwandter Formen. *Zeitschrift für Wissenschaftliche Zoologie,* 107(2): 346–386.
——— 1937. Zur Biologie von *Eremocoris abietis* einer myrmecophilen Heteroptere (Mit einer Übersicht über die bei Ameisen vorkommenden Wanzen). *Stettiner Entomologische Zeitung,* 98(1): 23–33.
Joseph, K. J., and S. B. Mathad. 1963. A new genus of termitophilous Atelurinae (Thysanura: Nicoletidae) from India. *Insectes Sociaux,* 10(4): 379–386.
Jouvenaz, D. P. 1986. Diseases of fire ants: problems and opportunities. *In* C. S. Lofgren and R. K. Vander Meer, eds., *Fire ants and leaf-cutting ants: biology and management,* pp. 327–338. Westview Press, Boulder.
Jouvenaz, D. P., W. A. Banks, and C. S. Lofgren. 1974. Fire ants: attraction of workers to queen secretions. *Annals of the Entomological Society of America,* 67(3): 442–444.
Jutsum, A. R. 1979. Interspecific aggression in leaf-cutting ants. *Animal Behaviour,* 27(3): 833–838.
Jutsum, A. R., T. S. Saunders, and J. M. Cherrett. 1979. Intraspecific aggression in the leaf-cutting ant *Acromyrmex octospinosus. Animal Behaviour,* 27(3): 839–844.

Kaczmarek, W. 1953. Badania nad zespolami mrówek lesnych. *Ekologia Polska,* 1: 69–86.
Kalmus, H., and C. R. Ribbands. 1952. The origin of the odours by which honeybees distinguish their companions. *Proceedings of the Royal Society,* ser. B, 140: 50–59.

Kannowski, P. B. 1959a. The flight activities and colony-founding behavior of bog ants in southeastern Michigan. *Insectes Sociaux*, 6(2): 115–162.

——— 1959b. The use of radioactive phosphorus in the study of colony distribution of the ant *Lasius minutus*. *Ecology*, 40(1): 162–165.

——— 1963. The flight activities of formicine ants. *Symposia Genetica et Biologica Italica*, 12: 74–102.

Karawajew, W. 1906. Weitere Beobachtungen über Arten der Gattung *Antennophorus*. *Mémoires de la Société des Naturalistes de Kiew*, 20(2): 209–230.

——— 1929. Die Spinndrüsen der Weberameisen (Hym. Formicid.). *Zoologischer Anzeiger*, 82: 247–256.

Karlson, P., and A. Butenandt. 1959. Pheromones (ectohormones) in insects. *Annual Review of Entomology*, 4: 39–58.

Karlson, P., and M. Lüscher. 1959. "Pheromones," a new term for a class of biologically active substances. *Nature*, 183: 55–56.

Kaszab, Z. 1973. Zwei neue myrmecophile Tenebrioniden-Arten (Coleoptera) aus Brasilen. *Studia Entomologica*, n.s., 16(1–4): 315–320.

Katō, M. 1939. The diurnal rhythm of temperature in the mound of an ant, *Formica truncorum truncorum* var. *yessenni* Forel, widely distributed at Mt. Hakkōda. *Science Reports of the Tōhoku Imperial University*, ser. 4 (biol.), 14(1): 53–64.

Kaudewitz, F. 1955. Zum Gastverhältnis zwischen *Cremastogaster scutellaris* Ol. mit *Camponotus lateralis bicolor* Ol. *Biologisches Zentralblatt*, 74(1–2): 69–87.

Keay, R. W. J. 1958. *Randia* and *Gardenia* in West Africa. *Bulletin du Jardin Botanique de l'Etat Bruxelles*, 28(1): 15–72.

Kemner, N. A. 1923. Hyphaenosymphilie, eine neue, merkwürdige Art von Myrmecophilie bei einem neuen myrmekophilen Schmetterling (*Wurthia aurivilli*, n.sp.) aus Java beobachtet. *Arkiv für Zoologie*, 15(15): 1–28.

Kempf, W. W. 1951. A taxonomic study on the ant tribe Cephalotini (Hymenoptera: Formicidae). *Revista de Entomologia*, 22(1–3): 1–244.

——— 1958a. New studies of the ant tribe Cephalotini (Hym. Formicidae). *Studia Entomologica*, n.s., 1(1–2): 1–176.

——— 1958b. Estudos sôbre *Pseudomyrmex*, II (Hymenoptera: Formicidae). *Studia Entomologica*, n.s., 1(3–4): 433–462.

——— 1958c. The ants of the tribe Dacetini in the State of São Paulo, Brazil, with the description of a new species of *Strumigenys* (Hymenoptera: Formicidae). *Studia Entomologica*, n.s., 1(3–4): 553–560.

——— 1958d. Sôbre algumas formigas neotrópicas do gênero *Leptothorax* Mayr (Hymenoptera: Formicidae). *Anais da Academia Brasileira de Ciências*, 30(1): 91–102.

——— 1959a. A revision of the Neotropical ant genus *Monacis* Roger (Hym., Formicidae). *Studia Entomologica*, n.s., 2(1–4): 225–270.

——— 1959b. A synopsis of the New World species belonging to the *Nesomyrmex*-group of the ant genus *Leptothorax* Mayr (Hymenoptera: Formicidae). *Studia Entomologica*, n.s., 2(1–4): 391–432.

——— 1960a. Estudo sôbre *Pseudomyrmex*, I (Hymenoptera: Formicidae). *Revista Brasileira de Entomologia*, 9: 5–32.

——— 1960b. "*Phalacromyrmex*," a new ant genus from southern Brazil (Hymenoptera, Formicidae). *Revista Brasileira de Biologia*, 20(1): 89–92.

——— 1960c. A review of the ant genus "*Mycetarotes*" Emery (Hymenoptera, Formicidae). *Revista Brasileira de Biologia*, 20(3): 277–283.

——— 1960d. Miscellaneous studies on Neotropical ants. *Studia Entomologica*, n.s., 3(1–4): 417–466.

——— 1961a. Estudos sôbre *Pseudomyrmex*, III (Hymenoptera: Formicidae). *Studia Entomologica*, n.s., 4(1–4): 369–408.

——— 1961b. A survey of the ants of the soil fauna in Surinam (Hymenoptera: Formicidae). *Studia Entomologica*, n.s., 4(1–4): 481–524.

——— 1962. Retoques à classificação das formigas neotropicais do gênero *Heteroponera* Mayr (Hym., Formicidae). *Papéis Avulsos do Departamento de Zoologia, São Paulo*, 15(4): 29–47.

——— 1963a. A review of the ant genus *Mycocepurus* Forel (Hym. Formicidae). *Studia Entomologica*, n.s., 6(1–4): 417–432.

——— 1963b. Additions to the Neotropical ant genus *Rogeria* Emery, with a key to the hitherto recorded South American species. *Revista Brasileira de Biologia*, 23(2): 189–196.

——— 1964a. A revision of the Neotropical ants of the genus *Cyphomyrmex* Mayr, part I: Group of *strigatus* Mayr (Hym. Formicidae). *Studia Entomologica*, n.s., 7(1–4): 1–44.

——— 1964b. Miscellaneous studies of Neotropical ants, III (Hym. Formicidae). *Studia Entomologica*, n.s., 7(1–4): 45–71.

——— 1964c. On the number of ant species in the Neotropical region. *Studia Entomologica*, n.s., 7(1–4): 481–482.

——— 1965. A revision of the Neotropical fungus-growing ants of the genus *Cyphomyrmex* Mayr, part II: Group of *rimosus* (Spinola) (Hym. Formicidae). *Studia Entomologica*, n.s., 8(1–4): 161–200.

——— 1967a. A new revisionary note on the genus *Paracryptocerus* Emery (Hym. Formicidae). *Studia Entomologica*, n.s., 10(1–4): 361–368.

——— 1967b. Estudos sôbre *Pseudomyrmex*, IV (Hymenoptera: Formicidae). *Revista Brasileira de Entomologia*, 12: 1–12.

——— 1968. Miscellaneous studies on Neotropical ants, IV (Hymenoptera, Formicidae). *Studia Entomologica*, n.s., 11(1–4): 369–415.

——— 1969. Miscellaneous studies on Neotropical ants, V (Hymenoptera, Formicidae). *Studia Entomologica*, n.s., 12(1–4): 273–296.

——— 1970a. Catálogo das formigas do Chile. *Papéis Avulsos Zoologia, São Paulo*, 23(3): 17–43.

——— 1970b. Taxonomic notes on ants of the genus *Megalomyrmex* Forel, with the description of new species (Hym. Formicidae). *Studia Entomologica*, n.s., 13(1–4): 353–364.

——— 1971. A preliminary review of the ponerine ant genus *Dinoponera* Roger (Hym. Formicidae). *Studia Entomologica*, n.s., 14(1–4): 369–394.

——— 1972a. A new species of the dolichoderine ant genus *Monacis* Roger, from the Amazon, with further remarks on the genus (Hymenoptera, Formicidae). *Revista Brasileira de Biologia*, 32(2): 251–254.

——— 1972b. Catálogo abreviado das formigas da Região Neotropical (Hym. Formicidae). *Studia Entomologica*, n.s., 15(1–4): 3–344.

——— 1972c. A study of some Neotropical ants of genus *Pheidole* Westwood, I (Hym. Formicidae). *Studia Entomologica*, n.s., 15(1–4): 449–464.

——— 1973. A revision of the Neotropical myrmicine ant genus *Hylomyrma* Forel (Hymenoptera, Formicidae). *Studia Entomologica*, n.s., 16(1–4): 225–260.

——— 1974. A review of the Neotropical ant genus *Oxyepoecus* Santschi (Hymenoptera: Formicidae). *Studia Entomologica*, n.s., 17(1–4): 471–512.

——— 1975. A revision of the Neotropical ponerine ant genus *Thaumatomyrmex* Mayr (Hymenoptera: Formicidae). *Studia Entomologica*, n.s., 18(1–4): 95–126.

Kempf, W. W., and W. W. Brown. 1970. Two new ants of tribe Ectatommini from Colombia (Hym. Formicidae). *Studia Entomologica*, n.s., 13(1–4): 311–320.

Kennedy, C. H., and M. Talbot. 1939. Notes on the hypogaeic ant, *Proceratium silaceum* Roger. *Proceedings of the Indiana Academy of Sciences*, 48: 202–210.

Kermarrec, A., M. Decharme, and G. Febvay. 1986. Leaf-cutting ant symbiotic fungi: a synthesis of recent research. *In* C. S. Lofgren and R. K. Vander Meer, eds., *Fire ants and leaf-cutting ants: biology and management*, pp. 231–246. Westview Press, Boulder.

Kerr, W. E. 1962. Tendências evolutivas na reproduçao dos himenópteros sociais. *Arquivos do Museu Nacional, Rio de Janeiro*, 52: 115–116.

——— 1967. Genetic structure of the populations of Hymenoptera. *Ciência e Cultura*, 19(1): 39–44.

Khalifman, I. A. 1961. *Use of ants for forest protection*. Lesnoe Khozyaistvo, Moscow, no. 2. [Cited by Beattie, 1985.]

Kiil, V. 1934. Untersuchungen über Arbeitsteilung bei Ameisen (*Formica rufa* L., *Camponotus herculeanus* L. und *C. ligniperda* Latr.). *Biologisches Zentralblatt*, 54(3–4): 114–146.

Kim, B.-J. 1986. A systematic study of ants in Is. Ullungdo of Korea on the basis of external fine features. *Journal of Natural Science, Won Kwang University, Iri, Korea*, 5(2): 84–94.

Kim, B.-J., and C.-W. Kim. 1983a. A review of myrmicine ants from Korea on the basis of external fine features (Hym., Formicidae). *Thesis of the National Academy of Sciences, Korea*, 22: 51–90.

——— 1983b. A systematic revision of the genus *Formica* in Korea on the basis of external fine features (Hym.: Formicidae). *Entomological Research Bulletin, Korea University,* 9: 57–67.
——— 1986. On the one new species, *Camponotus jejuensis* (n. sp.) from Korea (Hym., Formicidae). *Korean Journal of Entomology,* 16(2): 139–144.
Kim, C.-W., and B.-J. Kim. 1982. A taxonomical study of the subfamily Myrmicinae (Formicidae) from Korea. *Annual Report of Biological Research, Jeonbug National University, Korea,* 3: 95–110.
King, R. L., and R. M. Sallee. 1953. On the duration of nests of *Formica obscuripes* Forel. *Proceedings of the Iowa Academy of Science,* 60: 656–659.
——— 1956. On the half-life of nests of *Formica obscuripes* Forel. *Proceedings of the Iowa Academy of Science,* 63: 721–723.
——— 1957. Mixed colonies in ants: third report. *Proceedings of the Iowa Academy of Science,* 64: 667–669.
——— 1962. Further studies on mixed colonies in ants. *Proceedings of the Iowa Academy of Science,* 69: 531–539.
King, T. J. 1977. The plant ecology of ant-hills in calcareous grasslands, I: Patterns of species in relation to ant-hills in southern England. *Journal of Ecology,* 65(1): 235–256.
Kinomura, K., and K. Yamauchi. 1987. Fighting and mating behaviors of dimorphic males in the ant *Cardiocondyla wroughtoni. Journal of Ethology,* 5(1): 75–81.
Kipyatkov, V. S. 1979. Ecology of photoperiodicity in the ant *Myrmica rubra* L. (Hymenoptera, Formicidae), I: Seasonal changes in the photoperiodic responses. *Review of Entomology of the USSR,* 58(3): 490–499.
——— 1988. *Myrmica* prepares for winter: a new ant pheromone has been identified. *Science in the USSR,* 1:76–83.
Kirchner, W. 1964. Jahreszyklische Untersuchungen zur Reservestoffspeicherung und Überlebensfähigkeit adulter Waldameisenarbeiterinnen (Gen. *Formica,* Hym. Formicidae). *Zoologische Jahrbücher, Abteilung für Allgemeine Zoologie und Physiologie der Tiere,* 71(1): 1–72.
Kistner, D. H. 1958. The evolution of the Pygostenini (Coleoptera, Staphylinidae). *Annales du Musée Royal du Congo Belge,* ser. 8 (Zoology), 68: 5–198.
——— 1966a. A revision of the African species of the aleocharine tribe Dorylomimini (Coleoptera: Staphylinidae), II: The genera *Dorylomimus, Dorylonannus, Dorylogaster, Dorylobactrus,* and *Mimanomma,* with notes on their behavior. *Annals of the Entomological Society of America,* 59(2): 320–340.
——— 1966b. A revision of the myrmecophilous tribe Deremini (Coleoptera: Staphylinidae), part I: The *Dorylopora* complex and their behavior. *Annals of the Entomological Society of America,* 59(2): 341–358.
——— 1968. Revision of the myrmecophilous species of the tribe Myrmedoniini, part II: The genera *Aenictonia* and *Anommatochara*—their relationship and behavior. *Annals of the Entomological Society of America,* 61(4): 971–986.
——— 1972. Studies of Japanese myrmecophiles, part I: The genera *Pella* and *Falagria* (Coleoptera: Staphylinidae). *In* Z. Hidaka, ed., *Entomological essays to commemorate the retirement of Professor K. Yasumatsu (1971),* pp. 141–165. Hokuryukan, Tokyo.
——— 1979. Social and evolutionary significance of social insect symbionts. *In* H. R. Hermann, ed., *Social insects,* vol. 1, pp. 339–413. Academic Press, New York.
——— 1982. The social insects' bestiary. *In* H. R. Hermann, ed., *Social insects,* vol. 3, pp. 1–244. Academic Press, New York.
——— 1983. A new genus and twelve new species of ant mimics associated with *Pheidologeton* (Coleoptera, Staphylinidae; Hymenoptera, Formicidae). *Sociobiology,* 8(2): 155–198.
Kistner, D. H., and M. S. Blum. 1971. Alarm pheromone of *Lasius (Dendrolasius) spathepus* (Hymenoptera: Formicidae) and its possible mimicry by two species of *Pella* (Coleoptera: Staphylinidae). *Annals of the Entomological Society of America,* 64(3): 589–594.
Kistner, D. H., and H. R. Jacobson. 1975a. A review of the myrmecophilous Staphylinidae associated with *Aenictus* in Africa and the Orient (Coleoptera; Hymenoptera, Formicidae) with notes on their behavior and glands. *Sociobiology,* 1(1): 21–73.
——— 1975b. The natural history of the myrmecophilous tribe, Pygostenini (Coleoptera: Staphylinidae). *Sociobiology,* 1(3): 151–379.
——— 1979. Revision of the myrmecophilous tribe Deremini, III: The remainder of the genera with notes on behavior, ultrastructure, glands and phylogeny (Coleoptera, Staphylinidae). *Sociobiology,* 3(3): 141–391.
Kistner, D. H., and R. Zimmerman. 1986. A new genus of ant mimic from Sulawesi and its relationship to *Pheigetoxenus* (Coleoptera: Staphylinidae). *Sociobiology,* 11(3): 325–338.
Kitching, R. L. 1981. Egg clustering and the Southern Hemisphere lycaenids: comments on a paper by N. E. Stamp. *American Naturalist,* 118(3): 423–425.
——— 1983. Myrmecophilous organs of the larvae and pupae of the lycaenid butterfly *Jalmenus evagoras* (Donovan). *Journal of Natural History,* 17(3): 471–481.
——— 1987. Aspects of the natural history of the lycaenid butterfly *Allotinus major* in Sulawesi. *Journal of Natural History,* 21(3): 535–544.
Kitching, R. L., and B. Luke. 1985. The myrmecophilous organs of the larvae of some British Lycaenidae (Lepidoptera): a comparative study. *Journal of Natural History,* 19(2): 259–276.
Klahn, J. E. 1979. Philopatric and nonphilopatric foundress associations in the social wasp *Polistes fuscatus. Behavioral Ecology and Sociobiology,* 5(4): 417–424.
Kleine, R. 1924. Die Myrmekophilie der Brenthidae. *Zoologische Jahrbücher, Abteilung für Systematik, Geographie und Biologie der Tiere,* 49(3): 197–228.
Kleinfeldt, S. E. 1978. Ant-gardens: the interaction of *Codonanthe crassifolia* (Gesneriaceae) and *Crematogaster longispina* (Formicidae). *Ecology,* 59(3): 449–456.
——— 1986. Ant-gardens: mutual exploitation. *In* B. Juniper and T. R. E. Southwood, eds., *Insects and the plant surface,* pp. 283–294. Edward Arnold, London.
Kleinjan, J. E., and T. E. Mittler. 1975. A chemical influence of ants on wing development in aphids. *Entomologia Experimentalis et Applicata,* 18(3): 384–388.
Kloft, W. J. 1949. Über den Einfluss von Mermisparasitismus auf den Stoffwechsel und die Organbildung bei Ameisen. *Zeitschrift für Parasitenkunde,* 14(4): 390–422.
——— 1953. Die Bedeutung einiger Pflanzenläuse in der Lebensgemeinschaft des Waldes. *Mitteilungen der Biologischen Zentralanstalt für Land- und Forstwirtschaft Berlin-Dahlem,* 75: 136–140.
——— 1959a. Versuch einer Analyse der trophobiotischen Beziehungen von Ameisen und Aphiden. *Biologisches Zentralblatt,* 78(6): 863–870.
——— 1959b. Zur Nestbautätigkeit der Roten Waldameisen. *Waldhygiene,* 3–4: 94–98.
——— 1960a. Die Trophobiose zwischen Waldameisen und Pflanzenläusen mit Untersuchungen über die Wechselwirkungen zwischen Pflanzenläusen und Pflanzengeweben. *Entomophaga,* 5(1): 43–54.
——— 1960b. Wechselwirkungen zwischen pflanzensaugenden Insekten und den von ihnen besogenen Pflanzengeweben, Teil I. *Zeitschrift für Angewandte Entomologie,* 45(4): 337–381.
——— 1960c. Wechselwirkungen zwischen pflanzensaugenden Insekten und den von ihnen besogenen Pflanzengeweben, Teil II. *Zeitschrift für Angewandte Entomologie,* 46(1): 42–70.
——— 1983. Interspecific trophallactic relations between ants of different species, genera and subfamilies—an important strategy in population ecology. *Annals of Entomology (India),* 1: 85–86.
——— 1987. Trophallaxis zwischen benachbarten Nestern gleicher und verschiedener Arten als Mittel der Konfliktvermeidung bei Ameisen. *Waldhygiene,* 17: 43–48.
Kloft, W. J., R. W. Woodruff, and E. S. Kloft. 1979. *Formica integra,* IV: Exchange of food and trichome secretions between worker ants and the inquiline beetle, *Cremastocheilus castaneus. Tijdschrift voor Entomologie,* 122(3): 47–57.
Kloft, W. J., A. Maurizio, and W. Kaeser, eds. 1985. *Waldtracht und Waldhonig in der Imkerei.* Franz Ehrenwirth Verlag, Munich. 329 pp.
Klotz, J. H. 1984. Diel differences in foraging in two ant species (Hymenop-

tera: Formicidae). *Journal of the Kansas Entomological Society*, 57(1): 111–118.
——— 1986. Social facilitation among digging ants (*Formica subsericea*). *Journal of the Kansas Entomological Society*, 59(3): 537–541.
Kneitz, G. 1964a. Saisonales Trageverhalten bei *Formica polyctena* Foerst. (Formicidae, Gen. *Formica*). *Insectes Sociaux*, 11(2): 105–129.
——— 1964b. Untersuchungen zum Aufbau und zur Erhaltung des Nestwärmehaushaltes bei *Formica polyctena* Foerst. (Hymenoptera: Formicidae). Doctoral diss., Bayerische Julius-Maximilians-Universität, Würzburg. 156 pp.
——— 1970a. Jahreszeitliche Veränderungen der Ovariolenzustände in der Arbeiterinnenkaste des Waldameisenstaates von *Formica polyctena* Foerst. (Hymenoptera, Formicidae). *In* W. Herre, ed., *Verhandlungen der [Deutschen] Zoologischen Gesellschaft* (Würzburg, 1969) (*Zoologischer Anzeiger* Supplement, vol. 33), pp. 209–215. Akademische Verlagsgesellschaft, Geest & Portig K.-G., Leipzig.
——— 1970b. Saisonale Veränderungen des Nestwärmehaushaltes bei Waldameisen in Abhängigkeit von der Konstitution und dem Verhalten der Arbeiterinnen als Beispiel vorteilhafter Anpassung eines Insektenstaates an das Jahreszeitenklima. *In* W. Rathmayer, ed., *Verhandlungen der Deutschen Zoologischen Gesellschaft* (Köln, 1970), vol. 64, pp. 318–322. Gustav Fischer Verlag, Stuttgart.
Kolb, G. 1959. Untersuchungen über die Kernverhältnisse und morphologischen Eigenschaften symbiontischer Mikroorganismen bei verschiedenen Insekten. *Zeitschrift für Morphologie und Ökologie der Tiere*, 48(1): 1–71.
Kolbe, W. 1969. Käfer im Wirkungsbereich der Roten Waldameise. *Entomologische Zeitschrift*, 79(18): 269–278.
——— 1971. Untersuchungen über die Bindung von *Zyras humeralis* (Coleoptera, Staphylinidae) an Waldameisen. *Entomologische Blätter*, 67(3): 129–136.
Kolbe, W., and M. G. Proske. 1973. Iso-Valeriansäure im Abwehrsekret von *Zyras humeralis* Grav. (Coleoptera, Staphylinidae). *Entomologische Blätter*, 69(1): 57–60.
Koltermann, R. 1974. Periodicity in the activity and learning performance of the honeybee. *In* L. B. Browne, ed., *Experimental analysis of insect behavior*, pp. 218–227. Springer-Verlag, Berlin.
Kondoh, M. 1968. Bioeconomic studies on the colony of an ant species, *Formica japonica* Motschulsky, 1: Nest structure and seasonal change of the colony members. *Japanese Journal of Ecology*, 18(3): 124–133.
——— 1977. On the difference of vitality among worker ants under starvation. *Proceedings of the Eighth International Congress of the International Union for the Study of Social Insects* (Wageningen), pp. 69–70.
Koptur, S. 1984. Experimental evidence for defense of *Inga* (Mimosoideae) saplings by ants. *Ecology*, 65(6): 1787–93.
Kostermans, A. J. G. H. 1957. Lauraceae. *Reinwardtia*, 4(2): 193–256.
Krebs, J. R., and N. B. Davies. 1984. *Behavioural ecology: an evolutionary approach*, 2nd ed. Sinauer Associates, Sunderland, Mass. xi + 493 pp.
Krikken, J. 1972. Species of the South American genus *Lomanoxia* (Coleoptera: Aphodiidae). *Studies on the Fauna of Suriname and Other Guyanas*, 13: 68–83.
——— 1976. New genera of New World Cremastocheilini, with revisional notes (Coleoptera: Cetoniidae). *Zoologische Mededelingen* (Rijksmuseum van Natuurlijke Historie te Leiden), 49(25): 307–315.
Kristensen, N. P. 1976. Remarks on the family-level phylogeny of butterflies (Insecta, Lepidoptera, Rhopalocera). *Zeitschrift für Zoologische Systematik und Evolutionsforschung*, 14(1): 25–33.
Kugler, C. 1978a. A comparative study of the myrmicine sting apparatus (Hymenoptera, Formicidae). *Studia Entomologica*, n.s., 20(1–4): 413–548.
——— 1978b. Pygidial glands in the myrmicine ants (Hymenoptera, Formicidae). *Insectes Sociaux*, 25(3): 267–274.
——— 1979. Alarm and defense: a function for the pygidial gland of the myrmicine ant, *Pheidole biconstricta*. *Annals of the Entomological Society of America*, 72(4): 532–536.
——— 1984. Ecology of the ant *Pogonomyrmex mayri:* foraging and competition. *Biotropica*, 16(3): 227–234.
——— 1986. Stings of ants of the tribe Pheidologetini (Myrmicinae). *Insecta Mundi*, 1(4): 221–230.
Kugler, C., and W. L. Brown. 1982. Revisionary & other studies on the ant genus *Ectatomma*, including the descriptions of two new species. *Search—Agriculture* (Ithaca, N.Y.), no. 24, 7 pp.
Kugler, J. 1983. The males of *Cardiocondyla* Emery (Hymenoptera: Formicidae) with the description of the winged male of *Cardiocondyla wroughtoni* (Forel). *Israel Journal of Entomology*, 17: 1–21.
——— 1986. The Leptanillinae (Hymenoptera: Formicidae) of Israel and a description of a new species from India. *Israel Journal of Entomology*, 20: 45–57.
Kunkel, H. 1967. Systematische Übersicht über die Verteilung zweier Ernährungsformtypen bei den Sternorrhynchen (Rhynchota, Insecta). *Zeitschrift für Angewandte Zoologie*, 54(1): 37–74.
——— 1973. Die Kotabgabe der Aphiden (Aphidina, Hemiptera) unter Einfluss von Ameisen. *Bonner Zoologische Beiträge*, 24(1–2): 105–121.
Kunkel, H., and W. J. Kloft. 1977. Fortschritte auf dem Gebiet der Honigtau-Forschung. *Apidologie*, 8(4): 369–391.
——— 1985. Die Honigtau-Erzeuger des Waldes. *In* W. J. Kloft, A. Maurizio, and A. Kaeser, eds., *Waldtracht und Waldhonig in der Imkerei*, pp. 48–267. Franz Ehrenwirth Verlag, Munich.
Kusnezov, N. 1951a. "Dinergatogina" en *Oligomyrmex bruchi* Santschi (Hymenoptera Formicidae). *Revista de la Sociedad Entomológica Argentina*, 15: 177–181.
——— 1951b. El género *Pogonomyrmex* Mary. *Acta Zoológica Lilloana*, 11: 227–333.
——— 1951c. Un caso de evolucion eruptiva *Eriopheidole symbiotica* nov. gen. nov. sp. (Hymenoptera, Formicidae). *Memorias del Museo de Entre Rio, Paraná, Republica Argentina*, 29(zool.): 7–31.
——— 1952. Acerca de las hormigas simbióticas del género *Martia* Forel (Hymenoptera, Formicidae). *Acta Zoológica Lilloana*, 10: 717–722.
——— 1953. Lista de las hormigas de Tucumán con descripcion de dos nuevos generos (Hymenoptera, Formicidae). *Acta Zoológica Lilloana*, 13: 327–339.
——— 1954. Un género nuevo de hormigas (*Paranamyrma solenopsidis* nov. gen. nov. sp.) y los problemas relacionados (Hymenoptera, Formicidae). *Memorias del Museo de Entre Rios, Paraná, Republica Argentina*, 30(zool.): 7–21.
——— 1957a. Die Solenopsidinen-Gattungen von Südamerika (Hymenoptera, Formicidae). *Zoologischer Anzeiger*, 158(11–12): 266–280.
——— 1957b. Numbers of species of ants in faunae of different latitudes. *Evolution*, 11(3): 298–299.
——— 1962. El género *Acanthostichus* Mayr (Hymenoptera, Formicidae). *Acta Zoológica Lilloana*, 18: 121–138.
Kutter, H. 1913. Ein weiterer Beitrag zur Frage der sozialparasitischen Koloniegründung von *F. rufa* L. Zugleich ein Beitrag zur Biologie von *F. cinerea*. *Zeitschrift für Wissenschaftliche Insektenbiologie*, 9(6–7): 193–196.
——— 1923. Die Sklavenräuber *Strongylognathus huberi* For. ssp. *alpinus* Wheeler. *Revue Suisse de Zoologie*, 30(15): 387–424.
——— 1931. *Forelophilus*, eine neue Ameisengattung. *Mitteilungen der Schweizerischen Entomologischen Gesellschaft*, 15(5): 193–195.
——— 1945. Eine neue Ameisengattung. *Mitteilungen der Schweizerischen Entomologischen Gesellschaft*, 19(10): 485–487.
——— 1950a. Über eine neue, extrem parasitische Ameise, 1: Mitteilung. *Mitteilungen der Schweizerischen Entomologischen Gesellschaft*, 23(2): 81–94.
——— 1950b. Über zwei neue Ameisen: *Chalepoxenus insubricus* u. *Epimyrma stumperi*. *Mitteilungen der Schweizerischen Entomologischen Gesellschaft*, 23(3): 337–346.
——— 1951. *Epimyrma stumperi* Kutter (Hym. Formicid.), 2: Mitteilung. *Mitteilungen der Schweizerischen Entomologischen Gesellschaft*, 24(2): 153–174.
——— 1952. Über *Plagiolepis xene* Stärcke (Hym. Formicid.). *Mitteilungen der Schweizerischen Entomologischen Gesellschaft*, 25(2): 57–72.
——— 1956. Beiträge zur Biologie palaearktischer *Coptoformica* (Hym. Form.). *Mitteilungen der Schweizerischen Entomologischen Gesellschaft*, 29(1): 1–18.
——— 1957. Zur Kenntnis schweizerischer Coptoformicaarten (Hym. Form.), 2: Mitteilung. *Mitteilungen der Schweizerischen Entomologischen Gesellschaft*, 30(1): 1–24.

——— 1958. Über die Modifikationen bei Ameisenarbeiterinnen, welche durch den Parasitismus von Mermithiden (Nematod.) verursacht worden sind. *Mitteilungen der Schweizerischen Entomologischen Gesellschaft*, 31(3–4): 313–316.

——— 1963a. Miscellanea myrmecologica, I. *Mitteilungen der Schweizerischen Entomologischen Gesellschaft*, 36(1–2): 129–137.

——— 1963b. Miscellanea myrmecologica, II. *Mitteilungen der Schweizerischen Entomologischen Gesellschaft*, 36(4): 321–329.

——— 1964. Miscellanea myrmecologica, III. *Mitteilungen der Schweizerischen Entomologischen Gesellschaft*, 37(3): 127–137.

——— 1967. Beschreibung neuer Sozialparasiten von *Leptothorax acervorum* F. (Formicidae). *Mitteilungen der Schweizerischen Entomologischen Gesellschaft*, 40(1–2): 78–91.

——— 1969. Die sozialparasitischen Ameisen der Schweiz (Neujahrsblatt 1969, no. 171). *Vierteljahrsschrift der Naturforschenden Gesellschaft in Zürich*, 113[5]: 1–62.

——— 1971. Taxonomische Studien an Schweizer Ameisen (Hymenopt., Formicidae). *Mitteilungen der Schweizerischen Entomologischen Gesellschaft*, 43(3–4): 258–271.

——— 1972. Über *Xenhyboma mystes* Santschi. *Mitteilungen der Schweizerischen Entomologischen Gesellschaft*, 45(4): 321–324.

——— 1973a. Über die morphologischen Beziehungen der Gattung *Myrmica* zu ihren Satellitengenera *Sifolinia* Em., *Symbiomyrma* Arnoldi und *Sommimyrma* Menozzi (Hymenoptera, Formicidae). *Mitteilungen der Schweizerischen Entomologischen Gesellschaft*, 46(3–4): 253–268.

——— 1973b. Beitrag zur Lösung taxonomischer Probleme in der Gattung *Epimyrma* (Hymenoptera Formicidae). *Mitteilungen der Schweizerischen Entomologischen Gesellschaft*, 46(3–4): 281–289.

——— 1975. Über die Waldameisenfauna der Türkei. *Mitteilungen der Schweizerischen Entomologischen Gesellschaft*, 48(1–2): 159–163.

——— 1977. *Hymenoptera: Formicidae. In* W. Sauter, ed., *Insecta Helvetica: Fauna*, vol. 6. Schweizerische Entomologische Gesellschaft, Zürich. 298 pp.

——— 1978. *Hymenoptera: Formicidae. In* W. Sauter, ed., *Insecta Helvetica: Fauna*, vol. 6a, Ergänzungsband. Schweizerische Entomologische Gesellschaft, Zürich. 112 pp.

Kutter, H., and R. Stumper. 1969. Hermann Appel, ein leidgeadelter Entomologie (1892–1966). *Proceedings of the Sixth International Congress of the International Union for the Study of Social Insects* (Bern), pp. 275–279.

Kwait, E. C., and H. Topoff. 1983. Emigration raids by slave-making ants: a rapid-transit system for colony relocation (Hymenoptera: Formicidae). *Psyche*, 90(3): 307–312.

Lachaud, J.-P., and D. Fresneau. 1987. Social regulation in ponerine ants. *In* J. M. Pasteels and J.-L. Deneubourg, eds., *From individual to collective behavior in social insects* (*Experientia* Supplement, vol. 54), pp. 197–217. Birkhäuser Verlag, Basel.

Lacher, V. 1967. Verhaltensreaktionen der Bienenarbeiterin bei Dressur auf Kohlendioxid. *Zeitschrift für Vergleichende Physiologie*, 54(1): 74–84.

Lagerheim, G. 1900. Über *Lasius fuliginosus* und seine Pilzzucht. *Entomologisk Tidskrift*, 21: 17–29.

Laine, K. J., and P. Niemelä. 1980. The influence of ants on the survival of mountain birches during an *Oporinia autumnata* (Lep., Geometridae) outbreak. *Oecologia*, 47(1): 39–42.

LaMon, B., and H. Topoff. 1981. Avoiding predation by army ants: defensive behaviours of three ant species of the genus *Camponotus*. *Animal Behaviour*, 29(4): 1070–81.

——— 1984. Social facilitation of eclosion in the fire ant, *Solenopsis invicta*. *Developmental Psychobiology*, 18(5): 367–374.

Lane, A. P. 1977. Tandem running in *Leptothorax unifasciatus* (Formicidae, Myrmicinae): new data concerning recruitment and orientation in this species. *Proceedings of the Eighth International Congress of the International Union for the Study of Social Insects* (Wageningen), pp. 65–66.

Lange, R. 1958. Die deutsche Arten der *Formica rufa*-Gruppe. *Zoologischer Anzeiger*, 161(9–10): 238–243.

——— 1960. Über die Futterweitergabe zwischen Angehörigen verschiedener Waldameisen. *Zeitschrift für Tierpsychologie*, 17(4): 389–401.

——— 1967. Die Nahrungsverteilung unter den Arbeiterinnen des Waldameisenstaates. *Zeitschrift für Tierpsychologie*, 24(5): 513–545.

Lappano, E. R. 1958. A morphological study of larval development in polymorphic all-worker broods of the army ant *Eciton burchelli*. *Insectes Sociaux*, 5(1): 31–66.

Latreille, P. A. 1802. *Histoire naturelle des fourmis*. Théophile Barrois père, Libraire, Paris. xvi + 445 pp.

——— 1805. *Histoire naturelle, générale et particulière, des crustacés et des insectes*, vol. 13. F. Dufart, Paris. 432 pp.

Lattke, J. E. 1986. Notes on the ant genus *Hypoclinea* Mayr, with descriptions of three new species (Hymenoptera: Formicidae). *Revista de Biologia Tropical*, 34(2): 259–265.

Lavigne, R. J. 1969. Bionomics and nest structure of *Pogonomyrmex occidentalis* (Hymenoptera: Formicidae). *Annals of the Entomological Society of America*, 62(5): 1166–75.

Law, J. H., and F. E. Regnier. 1971. Pheromones. *Annual Review of Biochemistry*, 40: 533–548.

Law, J. H., E. O. Wilson, and J. A. McCloskey. 1965. Biochemical polymorphism in ants. *Science*, 149: 544–546.

Lawrence, J. F., and H. Reichardt. 1969. The myrmecophilous Ptinidae (Coleoptera), with a key to Australian species. *Bulletin of the Museum of Comparative Zoology, Harvard*, 138(1): 1–27.

Lawrence, J. F., and K. Stephan. 1975. The North American Cerylonidae (Coleoptera: Clavicornia). *Psyche*, 82(2): 131–166.

Lea, A. M. 1905. On *Nepharis* and other ants' nest beetles taken by Mr. J. C. Goudie at Birchip. *Proceedings of the Royal Society of Victoria*, n.s., 17(2): 371–385.

——— 1910. Australian and Tasmanian Coleoptera inhabiting or resorting to the nests of ants, bees, and termites. *Proceedings of the Royal Society of Victoria*, n.s., 23(1): 116–230.

——— 1912. Australian and Tasmanian Coleoptera inhabiting or resorting to the nests of ants, bees and termites. *Proceedings of the Royal Society of Victoria, Supplement*, n.s., 25(1): 31–78.

——— 1919. Notes on some miscellaneous Coleoptera, with descriptions of new species, pt. V. *Transactions and Proceedings of the Royal Society of South Australia*, 43: 166–261.

Ledoux, A. 1949. Le cycle évolutif de la fourmi fileuse (*Oecophylla longinoda* Latr.). *Comptes Rendus de l'Académie des Sciences, Paris*, 229(3): 246–248.

——— 1950. Recherche sur la biologie de la fourmi fileuse (*Oecophylla longinoda* Latr.). *Annales des Sciences Naturelles*, ser. 11, 12(3–4): 313–461.

——— 1967. Action de la température sur l'activité *d'Aphaenogaster senilis (testaceo-pilosa)* Mayr (Hym. Formicoidea). *Insectes Sociaux*, 14(2): 131–156.

Ledoux, A., and D. Dargagnon. 1973. Le formation des castes chez la fourmi *Aphaenogaster senilis* Mayr. *Comptes Rendus de l'Académie des Sciences, Paris*, ser. D, 276(4): 551–553.

Lee, J. 1938. Division of labor among the workers of the Asiatic carpenter ants (*Camponotus japonicus* var. *aterrimus*). *Peking Natural History Bulletin*, 13(2): 137–145.

Lehmann, J. 1975. Ansatz zu einer allgemeinen Lösung des "Ambrosiapilz"-Problems. *Waldhygiene*, 11(2): 41–47.

——— 1976. Neue Erkenntnisse über die Nahrungspilze von *Attini (Myrmicinae, Hymenoptera)* und *Macrotermitinae (Isoptera)*. *Waldhygiene*, 11(5): 133–152.

Le Masne, G. 1941. *Tubicera lichtwardti* Schmitz (Dipt. Phoridae), hôte de *Plagiolepis pygmaea* Latr. (Hym. Formicidae). *Bulletin de la Société Entomologique de France*, 46: 110–111.

——— 1952. Les échanges alimentaires entre adultes chez la fourmi *Ponera eduardi* Forel. *Comptes Rendus de l'Académie des Sciences, Paris*, 235(23): 1549–51.

——— 1953. Observations sur les relations entre le couvain et les adultes chez les fourmis. *Annales des Sciences Naturelles*, ser. 11, 15(1): 1–56.

——— 1956a. La signification des reproducteurs aptères chez la fourmi *Ponera eduardi* Forel. *Insectes Sociaux*, 3(2): 239–259.

——— 1956b. Recherches sur les fourmis parasites: *Plagiolepis grassei* et l'évolution des *Plagiolepis* parasites. *Comptes Rendus de l'Académie des Sciences, Paris,* 243(7): 673–675.

——— 1961a. Recherches sur la biologie des animaux myrmécophiles: l'adoption des *Paussus favieri* Fairm. par une nouvelle société de *Pheidole pallidula* Nyl. *Comptes Rendus de l'Académie des Sciences, Paris,* 253(15): 1621–23.

——— 1961b. Recherches sur la biologie des animaux myrmécophiles, IV: Observations sur le comportement de *Paussus favieri* Fairm., hôte de la fourmi *Pheidole pallidula* Nyl. *Annales de la Faculté des Sciences de Marseille,* 31: 111–130.

——— 1970a. Recherches sur la biologie des animaux myrmécophiles, IV: Le comportement de *Dichillus minutus* Sol. (Col. Tenebrionidae), hôte de la fourmi *Pheidole pallidula* Nyl.: un cas de myrmécophilie facultative. *Comptes Rendus de l'Académie des Sciences, Paris,* ser. D, 270(10): 1377–80.

——— 1970b. Recherches sur la biologie des fourmis parasites: les relations des ouvrières de *Chalepoxenus* avec leurs hôtes. *Comptes Rendus de l'Académie des Sciences, Paris,* ser. D, 271(10): 1038–41.

——— 1970c. Recherches sur la biologie des fourmis parasites: le comportement agressif des ouvrières de *Chalepoxenus*. *Comptes Rendus de l'Académie des Sciences, Paris,* ser. D, 271(13): 1119–21.

Le Masne, G., and A. Bonavita. 1967. Colony-founding according to archaic type (with repeated foraging) in the ant *Neomyrma rubida* Latr. *Proceedings of the Tenth International Ethological Conference* (Stockholm, 1967), mimeographed summary.

Le Masne, G., and A. Bonavita-Cougourdan. 1972. Premiers résultats d'une irradiation prolongée au césium sur les populations de fourmis en Haute-Provence. *Ekologia Polska,* 20(14): 129–144.

Le Masne, G., and C. Torossian. 1965. Observations sur le comportement du Coléoptère myrmécophile *Amorphocephalus coronatus* Germar (Brenthidae) hôte des *Camponotus*. *Insectes Sociaux,* 12(2): 185–194.

Le Moli, F. 1978. Social influence on the acquisition of behavioural patterns in the ant *Formica rufa* L. *Bolletino di Zoologia,* 45(4): 399–404.

——— 1980. On the origin of slaves in dulotic ant societies. *Bolletino di Zoologia,* 47(1–2): 207–212.

Le Moli, F., and A. Mori. 1982. Early learning and cocoon nursing behaviour in the red wood ant *Formica lugubris* Zett. (Hymenoptera: Formicidae). *Bolletino di Zoologia,* 49(1): 93–97.

——— 1986. The aggression test as a possible taxonomic tool in the *Formica rufa* group. *Aggressive Behavior,* 12: 93–102.

——— 1987. The problem of enslaved ant species: origin and behavior. *In* J. M. Pasteels and J.-L. Deneubourg, eds., *From individual to collective behavior in social insects* (*Experientia* Supplement, vol. 54), pp. 333–363. Birkhäuser Verlag, Basel.

Le Moli, F., and S. Parmigiani. 1982. Intraspecific combat in the red wood ant (*Formica lugubris,* Zett.). *Aggressive Behavior,* 8: 145–148.

Le Moli, F., and M. Passetti. 1977. The effect of early learning on recognition, acceptance and care of cocoons in the ant *Formica rufa* L. *Atti dellà Società Italiana di Scienze Naturali e del Museo Civile di Storia Naturale, Milano,* 118(1): 49–64.

——— 1978. Olfactory learning phenomena and cocoon nursing behaviour in the ant *Formica rufa* L. *Bolletino di Zoologia,* 45(4): 389–397.

Le Moli, F., A. Mori, and S. Parmigiani. 1982. Agonistic behavior of *Formica rufa* L. (Hymenoptera Formicidae). *Monitore Zoologico Italiano,* n.s., 16: 325–331.

Lenczewski, B. 1985. Natural history, colonization and survival in a northern fungus-gardening ant, *Trachymyrmex septentrionalis* (Attini). M.S. thesis, Florida State University, Tallahassee.

Lenko, K. 1969. An army ant attacking the "Guaiá" crab in Brazil. *Entomological News,* 80(1): 6.

Lenoir, A. 1979a. Early influence and division of labour in the ant *Lasius niger*. *Abstracts of the Sixteenth International Conference of Ethology* (Vancouver, B.C.).

——— 1979b. Feeding behaviour in young societies of the ant *Tapinoma erraticum* L.: trophallaxis and polyethism. *Insectes Sociaux,* 26(1): 19–37.

——— 1979c. Le comportement alimentaire et la division du travail chez la fourmi *Lasius niger*. *Bulletin Biologique de la France et de la Belgique,* 113(2–3): 79–314.

——— 1981. Brood retrieving in the ant, *Lasius niger* L. *Sociobiology,* 6(2): 153–178.

——— 1982. An informational analysis of antennal communication during trophallaxis in the ant *Myrmica rubra*. *Behavioural Processes,* 7(1): 27–35.

——— 1984. Brood-colony recognition in *Cataglyphis cursor* worker ants (Hymenoptera: Formicidae). *Animal Behaviour,* 32(3): 942–944.

Lenoir, A., and H. Ataya. 1983. Polyéthisme et répartition des niveaux d'activité chez la fourmi *Lasius niger* L. *Zeitschrift für Tierpsychologie,* 63(2–3): 213–232.

Lenoir, A., and H. Cagniant. 1986. Role of worker thelytoky in colonies of the ant *Cataglyphis cursor* (Hymenoptera: Formicidae). *Entomologia Generalis,* 11(3–4): 153–157.

Lenoir, A., and P. Jaisson. 1982. Evolution et rôle des communications antennaires chez les insectes sociaux. *In* P. Jaisson, ed., *Social insects in the tropics,* vol. 1, pp. 157–180. Université Paris-Nord, Paris.

Lenoir, A., and J.-C. Mardon. 1978. Note sur l'application de l'analyse des correspondances à la division du travail chez les fourmis. *Comptes Rendus de l'Académie des Sciences, Paris,* ser. D, 287(5): 555–558.

Lenoir, A., L. Querard, and F. Berton. 1987a. Colony founding and role of parthenogenesis in *Cataglyphis cursor* ants (Hymenoptera—Formicidae). *In* J. Eder and H. Rembold, eds., *Chemistry and biology of social insects* (Proceedings of the Tenth International Congress of the International Union for the Study of Social Insects, Munich, 1986), p. 260. Verlag J. Peperny, Munich.

Lenoir, A., M. Isingrini, and M. Nowbahari. 1987b. Colony recognition in the ant *Cataglyphis cursor* (Hymenoptera, Formicidae). *In* J. Eder and H. Rembold, eds., *Chemistry and biology of social insects* (Proceedings of the Tenth International Congress of the International Union for the Study of Social Insects, Munich, 1986), pp. 476–477. Verlag J. Peperny, Munich.

Le Roux, A. M., and G. Le Roux. 1979. Activité et agressivité chez des ouvrières de *Myrmica laevinodis* Nyl. (Hymenoptère, Formicides): modification en fonction du groupement et de l'expérience individuelle. *Insectes Sociaux,* 26(4): 354–363.

Leroux, J. M. 1977. Densité des colonies et observations sur les nids de dorylines *Anomma nigricans* Illiger (Hym. Formicidae) dans la région de Lamto (Côte d'Ivoire). *Bulletin de la Société Zoologique Française,* 102: 51–62.

——— 1979. Sur quelques modalités de disparition des colonies d'*Anomma nigricans* Illiger (Formicidae Dorylinae) dans la région de Lamto (Côte d'Ivoire). *Insectes Sociaux,* 26(2): 93–100.

Leston, D. 1973a. Ants and tropical tree crops. *Proceedings of the Royal Entomological Society of London,* ser. C, 38(6): 1.

——— 1973b. Ecological consequences of the tropical ant mosaic. *Proceedings of the Seventh International Congress of the International Union for the Study of Social Insects* (London, 1973), pp. 235–242.

——— 1973c. The ant mosaic-tropical tree crops and the limiting of pests and diseases. *Pest Articles and News Summaries* (Centre for Overseas Pest Research, London), 19: 311–341.

——— 1978. A Neotropical ant mosaic. *Annals of the Entomological Society of America,* 71(4): 649–653.

Letendre, M., and L. Huot. 1972. Considérations préliminaires en vue de la révision taxonomique des fourmis du groupe *microgyna,* genre *Formica* (Hymenoptera: Formicidae). *Annales de la Société Entomologique du Québec,* 17(3): 117–132.

Letourneau, D. K. 1983. Passive aggression: an alternative hypothesis for the *Piper-Pheidole* association. *Oecologia,* 60(1): 122–126.

Letourneau, D. K., and J. C. Choe. 1987. Homopteran attendance by wasps and ants: the stochastic nature of interactions. *Psyche,* 94(1–2): 81–91.

Leuthold, R. H. 1968a. A tibial gland scent-trail and trail-laying behavior in the ant *Crematogaster ashmeadi* Mayr. *Psyche,* 75(3): 233–250.

——— 1968b. Recruitment to food in the ant *Crematogaster ashmeadi*. *Psyche,* 75(4): 334–350.

Lévieux, J. 1966. Traits généraux du peuplement en fourmis terricoles d'une savane de Côte d'Ivoire. *Comptes Rendus de l'Académie des Sciences, Paris*, ser. D, 262(14): 1583–85.

——— 1972. Le rôle des fourmis dans les réseaux trophiques d'une savane préforestière de Côte d'Ivoire. *Annales de l'Université d'Abidjan*, ser. E, 5(1): 143–240.

——— 1973. Etude du peuplement en fourmis terricoles d'une savane préforestière de Côte d'Ivoire. *Revue d'Ecologie et de Biologie du Sol*, 10(3): 379–428.

——— 1976a. Deux aspects de l'action des fourmis (Hymenoptera, Formicidae) sur le sol d'une savane préforestière de Côte d'Ivoire. *Bulletin d'Ecologie*, 7(3): 283–295.

——— 1976b. Etude de la structure du nid de quelques espèces terricoles de fourmis tropicales. *Annales de l'Université d'Abidjan*, ser. C, 12: 23–33.

——— 1976c. La nutrition des fourmis tropicales, III: Cycle d'activité et régime alimentaire d'*Atopomyrmex mocquerysi* André (Hymenoptera Formicidae, Myrmicinae). *Annales de l'Université d'Abidjan*, ser. E, 9: 338–349.

——— 1976d. La nutrition des fourmis tropicales, IV: Cycle d'activité et régime alimentaire de *Platythyrea conradti* Emery (Hymenoptera Formicidae, Ponerinae). *Annales de l'Université d'Abidjan*, ser. E, 9: 352–365.

——— 1977. La nutrition des fourmis tropicales, V: Eléments de synthèse: les modes d'exploitation de la biocoenose. *Insectes Sociaux*, 24(3): 235–260.

——— 1979. La nutrition des fourmis granivores, IV: Cycle d'activité et régime alimentaire de *Messor galla* et de *Messor (= Cratomyrmex) regalis* en saison des pluies fluctuations annuelles: discussion. *Insectes Sociaux*, 26(4): 279–294.

——— 1982. A comparison of the ground dwelling ant populations between a Guinea savanna and an evergreen rain forest of the Ivory Coast. *In* M. D. Breed, C. D. Michener, and H. E. Evans, eds., *The biology of social insects* (Proceedings of the Ninth Congress of the International Union for the Study of Social Insects, Boulder, Colorado, 1982), pp. 48–53. Westview Press, Boulder.

——— 1983a. Mode d'exploitation des ressources alimentaires épigées de savanes africaines par la fourmi *Myrmicaria eumenoides* Gerstaecker. *Insectes Sociaux*, 30(2): 165–176.

——— 1983b. The soil fauna of tropical savannas, IV: The ants. *In* F. Bourlière, ed., *Tropical savannas*, pp 525–540. Elsevier, Amsterdam.

Lévieux, J., and T. Diomande. 1987a. La nutrition des fourmis granivores, I: Cycle d'activité et régime alimentaire de *Messor galla* et de *Messor (= Cratomyrmex) regalis* (Hymenoptera, Formicidae). *Insectes Sociaux*, 25(2): 127–139.

——— 1978b. La nutrition des fourmis granivores, II: Cycle d'activité et régime alimentaire de *Brachyponera senaarensis* (Mayr) (Hymenoptera Formicidae). *Insectes Sociaux*, 25(3): 187–196.

Levine, S. H. 1976. Competitive interactions in ecosystems. *American Naturalist*, 110: 903–910.

Levings, S. C. 1983. Seasonal, annual, and among-site variation in the ground ant community of a deciduous tropical forest: some causes of patchy species distributions. *Ecological Monographs*, 53(4): 435–455.

Levings, S. C., and N. R. Franks. 1982. Patterns of nest dispersion in a tropical ground ant community. *Ecology*, 63(2): 338–344.

Levings, S. C., and J. F. A. Traniello. 1981. Territoriality, nest dispersion, and community structure in ants. *Psyche*, 88(3–4): 265–319.

Levings, S. C., and D. M. Windsor. 1984. Litter moisture content as a determinant of litter arthropod distribution and abundance during the dry season on Barro Colorado Island, Panama. *Biotropica*, 16(2): 125–131.

Levins, R., M. L. Pressick, and H. Heatwole. 1973. Coexistence patterns in insular ants. *American Scientist*, 61(4): 463–472.

Lewis, T., G. V. Pollard, and G. C. Dibley. 1974a. Rhythmic foraging in the leaf-cutting ant *Atta cephalotes* (L.) (Formicidae: Attini). *Journal of Animal Ecology*, 43(1): 129–141.

——— 1974b. Micro-environmental factors affecting diel patterns of foraging in the leaf-cutting ant *Atta cephalotes* (L.) (Formicidae: Attini). *Journal of Animal Ecology*, 43(1): 143–153.

Li, C. C. 1955. *Population genetics*. University of Chicago Press, Chicago. xi + 366 pp.

Lieberburg, I., P. M. Kranz, and A. Seip. 1975. Bermudian ants revisited: the status and interaction of *Pheidole megacephala* and *Iridomyrmex humilis*. *Ecology*, 56(2): 473–478.

Lighton, J. R. B., G. A. Bartholomew, and D. H. Feener. 1987. Energetics of locomotion and load carriage and a model of the energy cost of foraging in the leaf-cutting ant *Atta colombica* Guér. *Physiological Zoölogy*, 60(5): 524–537.

Lilienstern, M. 1932. Beiträge zur Bakteriensymbiose der Ameisen. *Zeitschrift für Morphologie und Ökologie der Tiere*, 26(1–2): 110–134.

Lin, N., and C. D. Michener. 1972. Evolution of sociality in insects. *Quarterly Review of Biology*, 47(2): 131–159.

Lincecum, G. 1862. Notice on the habits of the "agricultural ant" of Texas ["stinging ant" or "mound-making ant," *Myrmica (Atta) molefaciens*, Buckley], communicated by Charles Darwin. *Journal of the Linnean Society, Zoology*, 6: 29–31.

——— 1866. On the agricultural ant of Texas (*Myrmica molefaciens*). *Proceedings of the Academy of Natural Sciences of Philadelphia* (1866), pp. 323–331.

Lindauer, M., and H. Martin. 1969. Special sensory performances in the orientation of the honey bee. *In* M. Marois, ed., *Theoretical physics and biology*, pp. 332–338. North-Holland, Amsterdam.

Linsenmair, K. E. 1985. Individual and family recognition in subsocial arthropods, in particular in the desert isopod *Hemilepistus reaumuri*. *In* B. Hölldobler and M. Lindauer, eds., *Experimental behavioral ecology and sociobiology* (Fortschritte der Zoologie, no. 31), pp. 411–436. Sinauer Associates, Sunderland, Mass.

——— 1987. Kin recognition in subsocial arthropods, in particular in the desert isopod *Hemilepistus reaumuri*. *In* D. J. C. Fletcher and C. D. Michener, eds., *Kin recognition in animals*, pp. 121–208. John Wiley, New York.

Littledyke, M., and J. M. Cherrett. 1976. Direct ingestion of plant sap from cut leaves by the leaf-cutting ants *Atta cephalotes* (L.) and *Acromyrmex octospinosus* (Reich) (Formicidae, Attini). *Bulletin of Entomological Research*, 66(2): 205–217.

Livingstone, D. 1857. *Missionary travels and researches in South Africa*. John Murray, London. x + 687 pp.

Lloyd, H. A., N. R. Schmuff, and A. Hefetz. 1984. Chemistry of the male mandibular gland secretion of the carpenter ant, *Camponotus thoracicus fellah* Emery. *Comparative Biochemistry and Physiology*, ser. B, 78(3): 687–689.

——— 1986. Chemistry of the anal glands of *Bothriomyrmex syrius* Forel: olfactory mimetism and temporary social parasitism. *Comparative Biochemistry and Physiology*, ser. B, 83(1): 71–73.

Loeschcke, V. 1985. Coevolution and invasion in competitive guilds. *American Naturalist*, 126(4): 505–520.

Lofgren, C. S., and R. K. Vander Meer, eds. 1986. *Fire ants and leaf-cutting ants: biology and management*. Westview Press, Boulder. xv + 435 pp.

Löfqvist, J. 1976. Formic acid and saturated hydrocarbons as alarm pheromones for the ant *Formica rufa*. *Journal of Insect Physiology*, 22(10): 1331–46.

Lokkers, C. 1986. The distribution of the weaver ant, *Oecophylla smaragdina* (Fabricius) (Hymenoptera: Formicidae) in Northern Australia. *Australian Journal of Zoology*, 34(5): 683–687.

Long, W. H. 1902. New species of *Ceratopogon*. *Biological Bulletin of the Marine Biological Laboratory, Woods Hole*, 3(1–2): 3–14.

Longhurst, C., and P. E. Howse. 1979a. Some aspects of the biology of the males of *Megaponera foetens* (Fab.) (Hymenoptera: Formicidae). *Insectes Sociaux*, 26(2): 85–91.

——— 1979b. Foraging, recruitment and emigration in *Megaponera foetans* (Fab.) (Hymenoptera: Formicidae) from the Nigerian Guinea savanna. *Insectes Sociaux*, 26(3): 204–215.

Longhurst, C., R. A. Johnson, and T. G. Wood. 1978. Predation by *Megaponera foetens* (Fabr.) (Hymenoptera: Formicidae) on termites in the Nigerian southern Guinea savanna. *Oecologia*, 32(1): 101–107.

——— 1979a. Foraging, recruitment and predation by *Decamorium uelense* (Santschi) (Formicidae: Myrmicinae) on termites in southern Guinea savanna, Nigeria. *Oecologia*, 38(1): 83–91.

Longhurst, C., R. Baker, and P. E. Howse. 1979b. Termite predation by *Me-*

*gaponera foetens* (Fab.) (Hymenoptera: Formicidae): coordination of raids by glandular secretions. *Journal of Chemical Ecology*, 5(5): 703–725.
——— 1980. A multicomponent mandibular gland secretion in the ponerine ant *Bothroponera soror* (Emery). *Journal of Insect Physiology*, 26(8): 551–555.
Longino, J. T. 1986. Ants provide substrate for epiphytes. *Selbyana*, 9: 100–103.
——— 1988. Notes on the taxonomy of the Neotropical ant genus *Thaumatomyrmex* Mayr (Hymenoptera: Formicidae). *In* J. C. Trager, ed., *Advances in myrmecology*, pp. 35–42. E. J. Brill, Leiden.
Longino, J. T., and J. Wheeler. 1987. Ants in live oak galls in Texas. *National Geographic Research*, 3(1) (Winter): 125–127.
Lopez, A., and J.-C. Bonaric. 1977. Notes sur une nymphe myrmécophile du genre *Microdon* [Diptera, Syrphidae]: éthologie et structure tégumentaire. *Annales de la Société Entomologique de France*, n.s., 13(1): 131–137.
Lowe, G. H. 1948. Some observations on the habits of a Malayan ant of the genus *Carebara*. *Proceedings of the Royal Entomological Society of London*, ser. A, 23(4–6): 51–53.
Lu, K. L., and M. R. Mesler. 1981. Ant dispersal of a Neotropical forest floor gesneriad. *Biotropica*, 13(2): 159–160.
Lubbock, J. 1894. *Ants, bees, and wasps: a record of observations on the habits of the social Hymenoptera*, rev. ed. D. Appleton, New York. xix + 448 pp.
Lumsden, C. J. 1982. The social regulation of physical caste: the superorganism revived. *Journal of Theoretical Biology*, 95(4): 749–781.
Lumsden, C. J., and B. Hölldobler. 1983. Ritualized combat and intercolony communication in ants. *Journal of Theoretical Biology*, 100(1): 81–98.
Lund, A. W. 1831. Lettre sur les habitudes de quelques fourmis du Brésil, addressée à M. Audouin. *Annales des Sciences Naturelles*, 23: 113–138.
Lüönd, B., and R. Lüönd. 1981. Insect dispersal of pollen and fruits in *Ajuga*. *Candollea*, 36(1): 167–179.
Lutz, H. 1986. Eine neue Unterfamilie der Formicidae (Insecta: Hymenoptera) aus dem mittel-eozänen Ölschiefer der "Grube Messel" bei Darmstadt (Deutschland, S-Hessen). *Senckenbergiana Lethaea*, 67(1–4): 177–218.
Lyford, W. H. 1963. Importance of ants to brown podzolic soil genesis in New England. *Harvard Forest Paper* (Petersham, Mass.), no. 7, 18 pp.
Lynch, J. F., E. C. Balinsky, and S. G. Vail. 1980. Foraging patterns in three sympatric forest ant species, *Prenolepis imparis*, *Paratrechina melanderi* and *Aphaenogaster rudis* (Hymenoptera: Formicidae). *Ecological Entomology*, 5(4): 353–371.

Mabelis, A. A. 1979a. Nest splitting by the red wood ant (*Formica polyctena* Foerster). *Netherlands Journal of Zoology*, 29(1): 109–125.
——— 1979b. Wood ant wars: the relationship between aggression and predation in the red wood ant (*Formica polyctena* Först.). *Netherlands Journal of Zoology*, 29(4): 451–620.
——— 1984. Interference between wood ants and other ant species (Hymenoptera, Formicidae). *Netherlands Journal of Zoology*, 34(1): 1–20.
——— 1986. Why do young queens fly? (Hymenoptera, Formicidae). *Proceedings of the Third European Congress of Entomology* (Amsterdam), pp. 24–29.
MacArthur, R. H., and E. R. Pianka. 1966. On optimal use of a patchy environment. *American Naturalist*, 100: 603–609.
MacArthur, R. H., and E. O. Wilson. 1967. *The theory of island biogeography.* Princeton Unversity Press, Princeton, N.J. xi + 203 pp.
Macedo, M., and G. T. Prance. 1978. Notes on the vegetation of Amazonia, II: The dispersal of plants in Amazonian white sand campinas: the campinas as functional islands. *Brittonia*, 30(2): 203–215.
Macevicz, S. 1979. Some consequences of Fisher's sex ratio principle for social Hymenoptera that reproduce by colony fission. *American Naturalist*, 113(3): 363–371.
Macgregor, E. C. 1948. Odour as a basis for orientated movement in ants. *Behaviour*, 1(3–4): 267–296.
MacKay, W. P. 1981. A comparison of the nest phenologies of three species of *Pogonomyrmex* harvester ants (Hymenoptera: Formicidae). *Psyche*, 88(1–2): 25–74.
——— 1982a. An altitudinal comparison of oxygen consumption rates in three species of *Pogonomyrmex* harvester ants (Hymenoptera: Formicidae). *Physiological Zoölogy*, 55(4): 367–377.
——— 1982b. The effect of predation of western widow spiders (Araneae: Theridiidae) on harvester ants (Hymenoptera: Formicidae). *Oecologia*, 53(3): 406–411.
——— 1983a. Beetles associated with the harvester ants, *Pogonomyrmex montanus*, *P. subnitidus* and *P. rugosus* (Hymenoptera: Formicidae). *Coleopterists Bulletin*, 37(3): 239–246.
——— 1983b. Stratification of workers in harvester ant nests (Hymenoptera: Formicidae). *Journal of the Kansas Entomological Society*, 56(4): 538–542.
MacKay, W. P., and E. MacKay. 1982. Coexistence and competitive displacement involving two native ant species (Hymenoptera: Formicidae). *Southwestern Naturalist*, 27(2): 135–142.
——— 1985. Temperature modifications of the nest of *Pogonomyrmex montanus* (Hymenoptera: Formicidae). *Southwestern Naturalist*, 30(2): 307–309.
MacKay, W. P., and S. Van Vactor. 1985. New host record for the social parasite *Pogonomyrmex anergismus* (Hymenoptera: Formicidae). *Proceedings of the Entomological Society of Washington*, 87(4): 863.
MacKay, W. P., F. Pérez-Domínguez, L. I. Valdez, and P. V. Orozco. 1984. La biologia de *Crematogaster larreae* Buren (Hymenoptera: Formicidae). *Folia Entomológia Mexicana*, 62: 75–80.
Madison, M. 1979. Additional observations on ant-gardens in Amazonas. *Selbyana*, 5(2): 107–115.
Maidl, F. 1934. *Die Lebensgewohnheiten und Instinkte der staatenbildenden Insekten.* Fritz Wagner, Vienna. x + 823 pp.
Majer, J. D. 1976a. The maintenance of the ant mosaic in Ghana cocoa farms. *Journal of Applied Ecology*, 13(1): 123–144.
——— 1976b. The ant mosaic in Ghana cocoa farms: further structural considerations. *Journal of Applied Ecology*, 13(1): 145–155.
——— 1976c. The influence of ants and ant manipulation on the cocoa farm fauna. *Journal of Applied Ecology*, 13(1): 157–175.
——— 1982. Ant-plant interactions in the Darling botanical district of Western Australia. *In* R. C. Buckley, ed., *Ant-plant interactions in Australia*, pp. 45–61. Dr. W. Junk, The Hague.
Malicky, H. 1969. Versuch einer Analyse der ökologischen Beziehungen zwischen Lycaeniden (Lepidoptera) und Formiciden (Hymenoptera). *Tijdschrift voor Entomologie*, 112(8): 213–298.
——— 1970a. New aspects on the association between lycaenid larvae (Lycaenidae) and ants (Formicidae, Hymenoptera). *Journal of the Lepidopterists' Society*, 24(3): 190–202.
——— 1970b. Unterschiede im Agriffsverhalten von *Formica*-Arten (Hymenoptera, Formicidae) gegenüber Lycaenidenraupen (Lepidoptera). *Insectes Sociaux*, 17(2): 121–124.
Malyshev, S. I. 1960. The history and the conditions of the origin of instincts in ants (Hymenoptera, Formicoidea). *Trudy Vseo͡iuznogo Entomologicheskoe Obshchestvo*, 47: 5–52. [In Russian.]
——— 1968. *Genesis of the Hymenoptera and the phases of their evolution*. Trans. from Russian by B. Haigh and ed. by O. W. Richards and B. Uvarov. Methuen, London. viii + 319 pp.
Mamsch, E., and K. Bier. 1966. Das Verhalten von Ameisenarbeiterinnen gegenüber der Königin nach vorangegangener Weisellosigkeit. *Insectes Sociaux*, 13(4): 277–284.
Maneval, H. 1940. Observations sur un Aphidiidae (Hym.) myrmécophile: description du genre et de l'espèce. *Bulletin Mensuel de la Société Linnéene de Lyon*, 9(1): 9–14.
Manfredi, P. 1949. Miriapodi mirmecofili. *Natura, Milano*, 40(3–4): 82–83.
Mann, W. M. 1912a. Parabiosis in Brazilian ants. *Psyche*, 19(1): 36–41.
——— 1912b. Note on a guest of *Eciton hamatum* Fabr. *Psyche*, 19(3): 98–100.
——— 1914. Some myrmecophilous insects from Mexico. *Psyche*, 21(6): 171–184.
——— 1915. Some myrmecophilous insects from Hayti. *Psyche*, 22(5): 161–166.
——— 1919. The ants of the British Solomon Islands. *Bulletin of the Museum of Comparative Zoology, Harvard*, 63(7): 273–391.
——— 1921. The ants of the Fiji Islands. *Bulletin of the Museum of Comparative Zoology, Harvard*, 64(5): 401–499.

——— 1923a. Two new ants from Bolivia. *Psyche*, 30(1): 13–18.
——— 1923b. Two serphoid guests of *Eciton* (Hym.). *Proceedings of the Entomological Society of Washington*, 25(9): 181–182.
——— 1926. Some new Neotropical ants. *Psyche*, 33(4–5): 97–107.
Mares, M. A., and M. L. Rosenzweig. 1978. Granivory in North and South American deserts: rodents, birds, and ants. *Ecology*, 59(2): 235–241.
Mariconi, F. A. M. 1970. *As sauvas*. Editora Agronomica "Ceres," São Paulo. 167 pp. [Cited by Cherrett, 1986.]
Marikovsky, P. I. 1962a. On some features of behavior of the ants *Formica rufa* L. infected with fungous disease. *Insectes Sociaux*, 9(2): 173–179.
——— 1962b. On intraspecific relations of *Formica rufa* L. (Hymenoptera, Formicidae). *Entomological Review*, 41: 47–51.
——— 1963. A new species of the ant *Polyergus nigerrimus* Marik., sp. n. (Hymenoptera, Formicidae) and some features of its biology. *Entomologicheskoe Obozrenie*, 42(1): 110–113.
——— 1974. The biology of the ant *Rossomyrmex proformicarum* K. W. Arnoldi (1928). *Insectes Sociaux*, 21(3): 301–308.
——— 1979. Ants of the Semireche desert. *Nauka, Alma Ata*, 1979: 1–263. [In Russian.]
Markin, G. P. 1970. Food distribution within laboratory colonies of the Argentine ant, *Iridomyrmex humilis* (Mayr). *Insectes Sociaux*, 17(2): 127–157.
Markin, G. P., and J. H. Dillier. 1971. The seasonal life cycle of the imported fire ant, *Solenopsis saevissima richteri*, on the Gulf Coast of Mississippi. *Annals of the Entomological Society of America*, 64(3): 562–565.
Markin, G. P., J. H. Dillier, S. O. Hill, M. S. Blum, and H. R. Hermann. 1971. Nuptial flight and flight ranges of the imported fire ant, *Solenopsis saevissima richteri* (Hymenoptera: Formicidae). *Journal of the Georgia Entomological Society*, 6(3): 145–156.
Markin, G. P., H. L. Collins, and J. H. Dillier. 1972. Colony founding by queens of the red imported fire ant, *Solenopsis invicta*. *Annals of the Entomological Society of America*, 65(5): 1053–58.
Markin, G. P., J. H. Dillier, and H. L. Collins. 1973. Growth and development of colonies of the red imported fire ant, *Solenopsis invicta*. *Annals of the Entomological Society of America*, 66(4): 803–808.
Markin, G. P., J. O'Neal, J. H. Dillier, and H. L. Collins. 1974. Regional variation in the seasonal activity of the imported fire ant, *Solenopsis saevissima richteri*. *Environmental Entomology*, 3(3): 446–452.
Markl, H. 1962. Borstenfelder an den Gelenken als Schweresinnesorgane bei Ameisen und anderen Hymenopteren. *Zeitschrift für Vergleichende Physiologie*, 45(5): 475–569.
——— 1965a. Stridulation in leaf-cutting ants. *Science*, 149: 1392–93.
——— 1965b. Wie orientieren sich Ameisen nach der Schwerkraft? *Umschau*, 65(6): 185–188.
——— 1967. Die Verständigung durch Stridulationssignale bei Blattschneiderameisen, I: Die biologische Bedeutung der Stridulation. *Zeitschrift für Vergleichende Physiologie*, 57(3): 299–330.
——— 1968. Die Verständigung durch Stridulationssignale bei Blattschneiderameisen, II: Erzeugung und Eigenschaften der Signale. *Zeitschrift für Vergleichende Physiologie*, 60(2): 103–150.
——— 1973. The evolution of stridulatory communication in ants. *Proceedings of the Seventh Congress of the International Union for the Study of Social Insects* (London, 1973), pp. 258–265.
——— 1983. Vibrational communication. *In* F. Huber and H. Markl, eds., *Neuroethology and behavioral physiology*, pp. 332–353. Springer-Verlag, Heidelberg.
——— 1985. Manipulation, modulation, information, cognition: some of the riddles of communication. *In* B. Hölldobler and M. Lindauer, eds., *Experimental behavioral ecology and sociobiology* (Fortschritte der Zoologie, no. 31), pp. 163–194. Sinauer Associates, Sunderland, Mass.
Markl, H., and S. Fuchs. 1972. Klopfsignale mit Alarmfunktion bei Roßameisen (*Camponotus*, Formicidae, Hymenoptera). *Zeitschrift für Vergleichende Physiologie*, 76(2) 204–225.
Markl, H., and B. Hölldobler. 1978. Recruitment and food-retrieving behavior in *Novomessor* (Formicidae, Hymenoptera), II: Vibration signals. *Behavioral Ecology and Sociobiology*, 4(2): 183–216.
Markl, H., B. Hölldobler, and T. Hölldobler. 1977. Mating behavior and sound production in harvester ants (*Pogonomyrmex*, Formicidae). *Insectes Sociaux*, 24(2): 191–212.
Marlin, J. C. 1968. Notes on a new method of colony formation employed by *Polyergus lucidus lucidus* Mayr (Hymenoptera: Formicidae). *Transactions of the Illinois State Academy of Science*, 61(2): 207–209.
Marsh, A. C. 1985a. Aspects of the ecology of Namib desert ants. Ph.D. diss., University of Cape Town. [Cited by Wehner, 1987.]
——— 1985b. Microclimatic factors influencing foraging patterns and success of the thermophilic desert ant, *Ocymyrmex barbiger*. *Insectes Sociaux*, 32(3): 286–296.
Martin, H. 1964. Zur Nahorientierung der Biene im Duftfeld: Zugleich ein Nachweis für die Osmotropotaxis bei Insekten. *Zeitschrift für Vergleichende Physiologie*, 48(5): 481–533.
Martin, M. M. 1970. The biochemical basis of the fungus–attine ant symbiosis. *Science*, 169: 16–20.
Martin, M. M., and J. S. Martin. 1971. The presence of protease activity in the rectal fluid of primitive attine ants. *Journal of Insect Physiology*, 17(10): 1897–1906.
Martin, M. M., and N. A. Weber. 1969. The cellulose-utilizing capability of the fungus cultured by the attine ant *Atta colombica tonsipes*. *Annals of the Entomological Society of America*, 62(6): 1386–87.
Martin, M. M., R. M. Carman, and J. G. MacConnell. 1969a. Nutrients derived from the fungus cultured by the fungus-growing ant *Atta colombica tonsipes*. *Annals of the Entomological Society of America*, 62(1): 11–13.
Martin, M. M., J. G. MacConnell, and G. R. Gale. 1969b. The chemical basis for the attine ant–fungus symbiosis: absence of antibiotics. *Annals of the Entomological Society of America*, 62(2): 386–388.
Martin, M. M., M. J. Gieselmann, and J. S. Martin. 1973. Rectal enzymes of attine ants: α-amylase and chitinase. *Journal of Insect Physiology*, 19(7): 1409–16.
Maschwitz, U. 1964. Gefahrenalarmstoffe und Gefahrenalarmierung bei sozialen Hymenopteren. *Zeitschrift für Vergleichende Physiologie*, 47(6): 596–655.
——— 1966. Das Speichelsekret der Wespenlarven und seine biologische Bedeutung. *Zeitschrift für Vergleichende Physiologie*, 53(3): 228–252.
——— 1974. Vergleichende Untersuchungen zur Funktion der Ameisenmetathorakaldrüse. *Oecologia*, 16(4): 303–310.
——— 1981. Predatory behavior and its correlation to recruitment behavior, morphology and nesting habits in three species of ponerine ants. *In* F. G. Barth, ed., *Neurobiology and strategies of adaptation* (Joint Symposium, Hebrew University of Jerusalem and Johann-Wolfgang-Goethe-Universität, Frankfurt), pp. 52–59.
Maschwitz, U., and K. Fiedler. 1988. Koexistenz, Symbiose, Parasitismus: Erfolgsstrategien der Bläulinge. *Spektrum der Wissenschaft*, 1988(May): 56–66.
Maschwitz, U., and H. Hänel. 1985. The migrating herdsman *Dolichoderus (Diabolus) cuspidatus*: an ant with a novel mode of life. *Behavioral Ecology and Sociobiology*, 17(2): 171–184.
Maschwitz, U., and B. Hölldobler. 1970. Der Kartonnestbau bei *Lasius fuliginosus* Latr. (Hym. Formicidae). *Zeitschrift für Vergleichende Physiologie*, 66(2): 176–189.
Maschwitz, U., and R. Klinger. 1974. Trophobiontische Beziehungen zwischen Wanzen und Ameisen. *Insectes Sociaux*, 21(2): 163–166.
Maschwitz, U., and E. Maschwitz. 1974. Platzende Arbeiterinnen: eine neue Art der Feindabwehr bei sozialen Hautflüglern. *Oecologia*, 14(3): 289–294.
Maschwitz, U., and M. Mühlenberg. 1973. *Camponotus rufoglaucus*, eine wegelagernde Ameise. *Zoologischer Anzeiger*, 19(5–6): 364–368.
——— 1975. Zur Jagdstrategie einiger orientalischer *Leptogenys*-Arten (Formicidae: Ponerinae). *Oecologia*, 20(1): 65–83.
Maschwitz, U., and P. Schönegge. 1977. Recruitment gland of *Leptogenys chinensis*: a new type of pheromone gland in ants. *Naturwissenschaften*, 64(11): 589–590.
——— 1980. Fliegen als Bente- und Bruträuber bei Ameisen. *Insectes Sociaux*, 27(1): 1–4.
——— 1983. Forage communication, nest moving recruitment, and prey

specialization in the oriental ponerine *Leptogenys chinensis. Oecologia*, 57(1–2): 175–182.

Maschwitz, U., K. Koob, and H. Schildknecht. 1970. Ein Beitrag zur Funktion der Metathoracaldrüse der Ameisen. *Journal of Insect Physiology*, 16(2): 387–404.

Maschwitz, U., B. Hölldobler, and M. Möglich. 1974. Tandemlaufen als Rekrutierungsverhalten bei *Bothroponera tesserinoda* Forel (Formicidae: Ponerinae). *Zeitschrift für Tierpsychologie*, 35(2): 113–123.

Maschwitz, U., M. Wüst, and K. Schurian. 1975. Bläulingsraupen als Zuckerlieferanten für Ameisen. *Oecologia*, 18(1): 17–21.

Maschwitz, U., M. Hahn, and P. Schönegge. 1979. Paralysis of prey in ponerine ants. *Naturwissenschaften*, 66(4): 213–214.

Maschwitz, U., M. Schroth, H. Hänel, and Y. P. Tho. 1984. Lycaenids parasitizing symbiotic plant-ant partnerships. *Oecologia*, 64(1): 78–80.

Maschwitz, U., K. Dumpert, and G. Schmidt. 1985a. Silk pavilions of two *Camponotus (Karavaievia)* species from Malaysia: description of a new nesting type in ants (Formicidae: Formicinae). *Zeitschrift für Tierpsychologie*, 69(3): 237–249.

Maschwitz, U., K. Dumpert, and P. Sebastian. 1985b. Morphological and behavioural adaptations of homopterophagous blues (Lepidoptera: Lycaenidae). *Entomologia Generalis*, 11(1–2): 85–90.

Maschwitz, U., M. Schroth, H. Hänel, and Y. P. Tho. 1985c. Aspects of the larval biology of myrmecophilous lycaenids from West Malaysia (Lepidoptera). *Nachrichten des Entomologischen Verains Apollo*, n.s., 6(4): 181–200.

Maschwitz, U., K. Dumpert, and K. R. Tuck. 1986a. Ants feeding on anal exudate from tortricid larvae: a new type of trophobiosis. *Journal of Natural History*, 20(5): 1041–50.

Maschwitz, U., S. Lenz, and A. Buschinger. 1986b. Individual specific trails in the ant *Leptothorax affinis* (Formicidae: Myrmicinae). *Experientia*, 42(10): 1173–74.

Maschwitz, U., K. Jessen, and S. Knecht. 1986c. Tandem recruitment and trail laying in the ponerine ant *Diacamma rugosum:* signal analysis. *Ethology*, 7(1): 30–41.

Maschwitz, U., B. Fiala, and W. R. Dolling. 1987. New trophobiotic symbioses of ants with South East Asian bugs. *Journal of Natural History*, 21(5): 1097–1107.

Maschwitz, U., S. Steghaus-Kovac, R. Gaube, and H. Hänel. 1989. A South East Asia ponerine ant of the genus *Leptogenys* (Hym. Form.) with army ant habits. *Behavioral Ecology and Sociobiology*, 24(5): 305–316.

Masner, L. 1959. A revision of ecitophilous Diapriid-genus *Mimopria* Holmgren (Hym., Proctotrupoidea). *Insectes Sociaux*, 6(4): 361–367.

——— 1976. Notes on the ecitophilous diapriid genus *Mimopria* Holmgren (Hymenoptera: Proctotrupoidea, Diapriidae). *Canadian Entomologist*, 108(2): 123–126.

——— 1977. A new genus of ecitophilous diapriid wasps from Arizona (Hymenoptera: Proctotrupoidea: Diapriidae). *Canadian Entomologist*, 109(1): 33–36.

Masson, C. 1974. Quelques données sur l'ultrastructure de récepteurs gustatifs de l'antenne de la fourmi *Camponotus vagus* Scop. (Hymenoptera, Formicidae). *Zeitschrift für Morphologie der Tiere*, 77(3): 235–243.

Masson, C., and G. Arnold. 1984. Ontogeny, maturation and plasticity of the olfactory system in the workerbee. *Journal of Insect Physiology*, 30(1): 7–14.

Masters, W. M., J. Tautz, N. H. Fletcher, and H. Markl. 1983. Body vibration and sound production in an insect (*Atta sexdens*) without specialized radiating structures. *Journal of Comparative Physiology*, ser. A, 150(2): 239–249.

Masuko, K. 1984. Studies on the predatory biology of Oriental dacetine ants (Hymenoptera: Formicidae), I: Some Japanese species of *Strumigenys*, *Pentastruma*, and *Epitritus*, and a Malaysian *Labidogenys*, with special reference to hunting tactics in short-mandibulate forms. *Insectes Sociaux*, 31(4): 429–451.

——— 1986. Larval hemolymph feeding: a nondestructive parental cannibalism in the primitive ant *Amblyopone silvestrii* Wheeler (Hymenoptera: Formicidae). *Behavioral Ecology and Sociobiology*, 19(4): 249–255.

——— 1987. *Leptanilla japonica:* the first bionomic information on the enigmatic ant subfamily Leptanillinae. *In* J. Eder and H. Rembold, eds., *Chemistry and biology of social insects* (Proceedings of the Tenth International Congress of the International Union for the Study of Social Insects, Munich, 1986), pp. 597–598. Verlag J. Peperny, Munich.

Mattson, W. J. 1980. Herbivory in relation to plant nitrogen content. *Annual Review of Ecology and Systematics*, 11: 119–161.

Maurizio, A. 1985. Honigtau—Honigtauhonig. *In* W. Kloft, A. Maurizio, and W. Kaeser, eds., *Waldtracht und Waldhonig in der Imkerei*, pp. 268–295. Franz Ehrenwirth Verlag, Munich.

Maynard Smith, J. 1982. *Evolution and the theory of games*. Cambridge University Press, New York. viii + 224 pp.

Mayr, E. 1974. Behavior programs and evolutionary strategies. *American Scientist*, 62(6): 650–659.

McCluskey, E. S. 1958. Daily rhythms in male harvester and Argentine ants. *Science*, 128: 536–537.

——— 1965. Circadian rhythms in male ants of five diverse species. *Science*, 150: 1037–39.

——— 1967. Circadian rhythms in female ants, and loss after mating flight. *Comparative Biochemistry and Physiology*, 23(2): 665–677.

——— 1974. Generic diversity in phase of rhythm in myrmicine ants. *Journal of the New York Entomological Society*, 82(2): 93–102.

McCluskey, E. S., and W. L. Brown. 1972. Rhythms and other biology of the giant tropical ant *Paraponera*. *Psyche*, 79(4): 335–347.

McCluskey, E. S., and C. E. Carter. 1969. Loss of rhythmic activity in female ants caused by mating. *Comparative Biochemistry and Physiology*, 31(2): 217–226.

McCluskey, E. S., and S.-M. A. Soong. 1979. Rhythm variables as taxonomic characters in ants. *Psyche*, 86(1): 91–102.

McCook, H. C. 1879. *The natural history of the agricultural ant of Texas: a monograph of the habits, architecture, and structure of* Pogonomyrmex barbatus. Academy of Natural Sciences, Philadelphia. 208 pp.

——— 1880. Combats and nidification of the pavement ant, *Tetramorium caespitum*. *Proceedings of the Academy of Natural Sciences of Philadelphia* (1879), pp. 156–161.

——— 1882. *The honey ants of the Garden of the Gods, and the occident ants of the American Plains*. J. B. Lippincott, Philadelphia. 188 pp.

McCubbin, C. 1971. *Australian butterflies*. Nelson, Melbourne. xxx + 206 pp.

McDonald, P., and H. Topoff. 1985. Social regulation and behavioral development in the ant, *Novomessor albisetosus* (Mayr). *Journal of Comparative Psychology*, 99(1): 3–14.

——— 1986. The development of defensive behavior against predation by army ants. *Developmental Psychobiology*, 19(4): 351–367.

McEvoy, P. B. 1979. Advantages and disadvantages to group living in treehoppers (Homoptera: Membracidae). *Miscellaneous Publications of the Entomological Society of America*, 11: 1–13.

McGurk, D. J., J. Frost, E. J. Eisenbraun, K. Vick, W. A. Drew, and J. Young. 1966. Volatile compounds in ants: identification of 4-methyl-3-heptanone from *Pogonomyrmex* ants. *Journal of Insect Physiology*, 12(11): 1435–41.

McGurk, D. J., J. Frost, G. R. Waller, E. J. Eisenbraun, K. Vick, W. A. Drew, and J. Young. 1968. Iridodial isomer variation in dolichoderine ants. *Journal of Insect Physiology*, 14(6): 841–845.

McIver, J. D. 1987. On the myrmecomorph *Coquillettia insignis* Uhler (Hemiptera: Miridae): arthropod predators as operators in an ant-mimetic system. *Zoological Journal of the Linnean Society*, 90(2): 133–144.

McKey, D. 1984. Interaction of the ant-plant *Leonardoxa africana*, (Caesalpiniaceae) with its obligate inhabitants in a rainforest in Cameroon. *Biotropica*, 16(2): 81–99.

——— 1988. Promising new directions in the study of ant-plant mutualisms. *In* W. Greuter and B. Zimmer, eds., *Proceedings of the XIV International Botanical Congress, Berlin, 1987*, pp. 335–355. Koeltz Scientific Books, Königstein.

——— 1989. Interactions between ants and leguminous plants. *In* C. H. Stirton and J. L. Zarucchi, eds., *Advances in legume biology* (Monographs in Systematic Botany, no. 29), pp. 673–718. Missouri Botanical Garden, St. Louis.

Mei, M. 1987. *Myrmica samnitica* n. sp.: una nuova formica parassita dell'Appennino Abruzzese (Hymenoptera, Formicidae). *Fragmenta Entomologica,* 19(2): 457–469.

Meinwald, J., D. F. Wiemer, and B. Hölldobler. 1983. Pygidial gland secretions of the ponerine ant *Rhytidoponera metallica*. *Naturwissenschaften,* 70(1): 46–47.

Menozzi, C. 1930. Formiche della Somalia Italiana meridionale. *Memorie della Società Entomologica Italiana,* 9: 76–130.

Menzel, R. 1979. Behavioural access to short-term memory in bees. *Nature,* 281: 368–369.

——— 1985. Learning in honey bees in an ecological and behavioral context. *In* B. Hölldobler and M. Lindauer, eds., *Experimental behavioral ecology and sociobiology* (Fortschritte der Zoologie, no. 31), pp. 55–74. Sinauer Associates, Sunderland, Mass.

Menzel, R., K. Moch, G. Wladarz, and M. Lindauer. 1969. Tagesperiodische Ablagerungen in der Endokutikula der Honigbiene. *Biologisches Zentralblatt,* 88(1): 61–67.

Mercier, B., L. Passera, and J.-P. Suzzoni. 1985a. Etude de la polygynie chez la fourmi *Plagiolepis pygmaea* Latr. (Hym. Formicidae), I: La fécondité des reines en condition expérimentale monogyne. *Insectes Sociaux,* 32(4): 335–348.

——— 1985b. Etude de la polygynie chez la fourmi *Plagiolepis pygmaea* Latr. (Hym. Formicidae), II: La fécondité des reines en condition expérimentale polygyne. *Insectes Sociaux,* 32(4): 349–362.

Merlin, P., J. C. Braekman, D. Daloze, and J. M. Pasteels. 1988. Tetraponerines, toxic alkaloids in the venom of the Neo-Guinean pseudomyrmecine ant *Tetraponera* sp. *Journal of Chemical Ecology,* 14(2): 517–527.

Messina, F. J. 1981. Plant protection as a consequence of an ant-membracid mutualism: interactions on goldenrod (*Solidago* sp.). *Ecology,* 62(6): 1433–40.

Meudec, M. 1973. Note sur les variations individuelles du comportement de transport du couvain chez les ouvrières de *Tapinoma erraticum* Latr. *Comptes Rendus de l'Académie des Sciences, Paris,* ser. D, 277(3): 357–360.

——— 1977. Le comportement de transport du couvain lors d'une perturbation du nid chez *Tapinoma erraticum* (Dolichoderinae): rôle de l'individu. *Insectes Sociaux,* 24(4): 345–352.

——— 1978. Response to and transport of brood by workers of *Tapinoma erraticum*. *Behavioral Processes,* 3(3): 199–209.

Meudec, M., and A. Lenoir. 1982. Social responses to variation in food supply and nest suitability in ants (*Tapinoma erraticum*). *Animal Behaviour,* 30(1): 284–292.

Meyer, J. 1966. Essai d'application de certains modèles cybernétiques à la coordination chez les insectes sociaux. *Insectes Sociaux,* 13(2): 127–138.

Michener, C. D. 1958. The evolution of social behavior in bees. *Proceedings of the Tenth International Congress of Entomology* (Montreal, 1956), 2: 441–447.

——— 1964. Reproductive efficiency in relation to colony size in hymenopterous societies. *Insectes Sociaux,* 11(4): 317–341.

——— 1969. Comparative social behavior of bees. *Annual Review of Entomology,* 14: 299–342.

——— 1974. *The social behavior of the bees: a comparative study.* Belknap Press of Harvard University Press, Cambridge, Mass. xii + 404 pp.

Michener, C. D., and D. J. Brothers. 1974. Were workers of eusocial Hymenoptera initially altruistic or oppressed? *Proceedings of the National Academy of Sciences of the United States of America,* 71(3): 671–674.

Michod, R. E. 1982. The theory of kin selection. *Annual Review of Ecology and Systematics,* 13: 23–55.

Michod, R. E., and W. D. Hamilton. 1980. Coefficients of relatedness in sociobiology. *Nature,* 288: 694–697.

Milewski, A. V., and W. J. Bond. 1982. Convergence of myrmecochory in mediterranean Australia and South Africa. *In* R. C. Buckley, ed., *Ant-plant interactions in Australia,* pp. 89–98. Dr. W. Junk, The Hague.

Minsky, M. 1986. *The society of mind.* Simon and Schuster, New York. 339 pp.

Mintzer, A. 1979a. Colony foundation and pleometrosis in *Camponotus* (Hymenoptera: Formicidae). *Pan-Pacific Entomologist,* 55(2): 81–89.

——— 1979b. Foraging activity of the Mexican leafcutting ant *Atta mexicana* (F. Smith), in a Sonoran Desert habitat (Hymenoptera, Formicidae). *Insectes Sociaux,* 26(4): 364–372.

——— 1980. Simultaneous use of a foraging trail by two leafcutter ant species in the Sonoran desert. *Journal of the New York Entomological Society,* 88(2): 102–105.

——— 1982a. Copulatory behavior and mate selection in the harvester ant, *Pogonomyrmex californicus* (Hymenoptera: Formicidae). *Annals of the Entomological Society of America,* 75(3): 323–326.

——— 1982b. Nestmate recognition and incompatibility between colonies of the acacia-ant *Pseudomyrmex ferruginea*. *Behavioral Ecology and Sociobiology,* 10(3): 165–168.

——— 1987. Primary polygyny in the ant *Atta texana:* number and weight of females and colony foundation success in the laboratory. *Insectes Sociaux,* 34(2): 108–117.

Mintzer, A., and S. B. Vinson. 1985a. Cooperative colony foundation by females of the leafcutting ant *Atta texana* in the laboratory. *Journal of the New York Entomological Society,* 93(3): 1047–51.

——— 1985b. Kinship and incompatibility between colonies of the acacia ant *Pseudomyrmex ferruginea*. *Behavioral Ecology and Sociobiology,* 17(1): 75–78.

Mirenda, J. T., and H. Topoff. 1980. Nomadic behavior of army ants in a desert-grassland habitat. *Behavioral Ecology and Sociobiology,* 7(2): 129–135.

Mirenda, J. T., and S. B. Vinson. 1981. Division of labour and specification of castes in the red imported fire ant *Solenopsis invicta* Buren. *Animal Behaviour,* 29(2): 410–420.

Mirenda, J. T., D. G. Eakins, K. Gravelle, and H. Topoff. 1980. Predatory behavior and prey selection by army ants in a desert-grassland habitat. *Behavioral Ecology and Sociobiology,* 7(2): 119–127.

Mirenda, J. T., D. G. Eakins, and H. Topoff. 1982. Relationship of raiding and emigration in the Nearctic army ant *Neivamyrmex nigrescens* Cresson. *Insectes Sociaux,* 29(2 bis): 308–331.

Mittler, T. E. 1957. Studies on the feeding and nutrition of *Tuberolachnus salignus* (Gmelin) (Homoptera, Aphididae), I: The uptake of phloem sap. *Journal of Experimental Biology,* 34(3): 334–341.

——— 1958. Studies on the feeding and nutrition of *Tuberolachnus salignus* (Gmelin) (Homoptera, Aphididae), III: The nitrogen economy. *Journal of Experimental Biology,* 35(3): 626–638.

Moffett, M. W. 1984. Swarm raiding in a myrmicine ant. *Naturwissenschaften,* 71(11): 588–590.

——— 1985a. An Indian ant's novel method for obtaining water. *National Geographic Research,* 1(1)(Winter): 146–149.

——— 1985b. Behavioral notes on the Asiatic harvesting ants *Acanthomyrmex notabilis* and *A. ferox*. *Psyche,* 92(2–3): 165–179.

——— 1985c. Revision of the genus *Myrmoteras* (Hymenoptera: Formicidae). *Bulletin of the Museum of Comparative Zoology, Harvard,* 151(1): 1–53.

——— 1986a. Mandibles that snap: notes on the ant *Mystrium camillae* Emery. *Biotropica,* 18(4): 361–362.

——— 1986b. Revision of the myrmicine genus *Acanthomyrmex* (Hymenoptera: Formicidae). *Bulletin of the Museum of Comparative Zoology, Harvard,* 151(2): 55–89.

——— 1986c. Trap-jaw predation and other observations on two species of *Myrmoteras* (Hymenoptera: Formicidae). *Insectes Sociaux,* 33(1): 85–99.

——— 1986d. Behavior of the group-predatory ant *Proatta butteli* (Hymenoptera: Formicidae): an Old World relative of the attine ants. *Insectes Sociaux,* 33(4): 444–457.

——— 1986e. Observations on *Lophomyrmex* ants from Kalimantan, Java and Malaysia. *Malayan Nature Journal,* 39(3): 207–211.

——— 1986f. Marauders of the jungle floor. *National Geographic,* 170(2)(February): 273–286.

——— 1986g. Notes on the behavior of the dimorphic ant *Oligomyrmex overbecki* (Hymenoptera: Formicidae). *Psyche,* 93(1–2): 107–116.

——— 1986h. Evidence of workers serving as queens in the genus *Diacamma* (Hymenoptera: Formicidae). *Psyche,* 93(1–2): 151–152.

——— 1987a. Ants that go with the flow: a new method of orientation by mass communication. *Naturwissenschaften,* 74(11): 551–553.

——— 1987b. Sociobiology of the ants of the genus *Pheidologeton*. Ph.D. diss., Harvard University. 284 pp.

——— 1987c. Division of labor and diet in the extremely polymorphic ant *Pheidologeton diversus*. *National Geographic Research*, 3(3)(Summer): 282–304.

——— 1988a. Foraging behavior in the Malayan swarm-raiding ant *Pheidologeton silenus* (Hymenoptera: Formicidae: Myrmicinae). *Annals of the Entomological Society of America*, 81(2): 356–361.

——— 1988b. Foraging dynamics in the group-hunting myrmicine ant, *Pheidologeton diversus*. *Journal of Insect Behavior*, 1(3): 309–331.

Moggridge, J. T. 1873. *Harvesting ants and trap-door spiders: notes and observations on their habits and dwellings*. L. Reeve, London. xi + 156 pp.

——— 1874. *Supplement to harvesting ants and trap-door spiders*. L. Reeve, London. ix + pp. 157–304.

Möglich, M. 1978. Social organization of nest emigration in *Leptothorax* (Hym., Form.). *Insectes Sociaux*, 25(3): 205–225.

——— 1979. Tandem calling pheromone in the genus *Leptothorax* (Hymenoptera: Formicidae): behavioral analysis of specificity. *Journal of Chemical Ecology*, 5(1): 35–52.

Möglich, M., and G. D. Alpert. 1979. Stone dropping by *Conomyrma bicolor* (Hymenoptera: Formicidae): a new technique of interference competition. *Behavioral Ecology and Sociobiology*, 6(2): 105–113.

Möglich, M., and B. Hölldobler. 1974. Social carrying behavior and division of labor during nest moving in ants. *Psyche*, 81(2): 219–236.

——— 1975. Communication and orientation during foraging and emigration in the ant *Formica fusca*. *Journal of Comparative Physiology*, ser. A, 101(4): 275–288.

Möglich, M., U. Maschwitz, and B. Hölldobler. 1974. Tandem calling: a new kind of signal in ant communication. *Science*, 186: 1046–47.

Möller, A. 1893. Die Pilzgärten einiger südamerikanischer Ameisen. *Botanische Mittheilungen aus den Tropen*, no. 6, vi + 127 pp.

Monastersky, R. 1987. Ants and the atmosphere: no picnic. *Science News*, 131: 345.

Monteith, G. B. 1986. Some curious insect-plant associations in Queensland. *Queensland Naturalist*, 26(5–6): 105–114.

Moore, B. P. 1974. The larval habits of two species of *Sphallomorpha* Westwood (Coleoptera: Carabidae, Pseudomorphinae). *Journal of the Australian Entomological Society*, 13(3): 179–183.

Mordwilko, A. 1907. Die Ameisen und Blattläuse in ihren gegenseitigen Beziehungen und das Zusammenleben von Lebewesen überhaupt. *Biologisches Centralblatt*, 27(7): 212–224.

Moreau, R. E. 1966. *The bird faunas of Africa and its islands*. Academic Press, New York. ix + 424 pp.

Morel, L. 1984. Comportement trophallactique de la jeune ouvrière de *Camponotus vagus* Scop. (Insecte Hyménoptère, Formicoidea): analyse quantitative de la transmission de substance. *Comptes Rendus de l'Académie des Sciences, Paris*, ser. 3, 299(7): 245–248.

Morel, L., and R. K. Vander Meer. 1987. Nestmate recognition in *Camponotus floridanus:* behavioral and chemical evidence for the role of age and social experience. *In* J. Eder and H. Rembold, eds., *Chemistry and biology of social insects* (Proceedings of the Tenth International Congress of the International Union for the Study of Social Insects, Munich, 1986), pp. 471–472. Verlag J. Peperny, Munich.

Morel, L., R. K. Vander Meer, and B. K. Lavine. 1988. Ontogeny of nestmate recognition cues in the red carpenter ant (*Camponotus floridanus*): behavioral and chemical evidence for the role of age and social experience. *Behavioral Ecology and Sociobiology*, 22(3): 175–183.

Morgan, E. D. 1984. Chemical words and phrases in the language of pheromones for foraging and recruitment. *In* T. Lewis, ed., *Insect communication* (Symposium of the Royal Entomological Society of London, no. 12), pp. 169–194. Academic Press, London.

Morgan, E. D., and D. G. Ollett. 1987. Methyl 6-methylsalicylate, trail pheromone of the ant *Tetramorium impurum*. *Naturwissenschaften*, 74(12): 596–597.

Morgan, E. D., M. R. Inwood, and M.-C. Cammaerts. 1978. The mandibular gland secretion of the ant *Myrmica scabrinodis*. *Physiological Entomology*, 3(2): 107–114.

Morrill, W. L. 1974a. Dispersal of red imported fire ants by water. *Florida Entomologist*, 57(1): 39–42.

——— 1974b. Production and flight of alate red imported fire ants. *Environmental Entomology*, 3(2): 265–271.

Moser, J. C. 1963. Contents and structure of *Atta texana* nests in summer. *Annals of the Entomological Society of America*, 56(3): 286–291.

——— 1964. Inquiline roach responds to trail-marking substance of leaf-cutting ants. *Science*, 143: 1048–49.

——— 1967a. Mating activities of *Atta texana* (Hymenoptera, Formicidae). *Insectes Sociaux*, 14(3): 295–312.

——— 1967b. Trails of the leafcutters. *Natural History*, 76: 32–35.

Moser, J. C., and M. S. Blum. 1963. Trail marking substance of the Texas leaf-cutting ant: source and potency. *Science*, 140: 1228.

Moser, J. C., and S. E. Neff. 1971. *Pholeomyia comans* (Diptera: Milichiidae) an associate of *Atta texana:* larval anatomy and notes on biology. *Zeitschrift für Angewandte Entomologie*, 69(4): 343–348.

Moser, J. C., R. C. Brownlee, and R. Silverstein. 1968. Alarm pheromones of the ant *Atta texana*. *Journal of Insect Physiology*, 14(4): 529–535.

Mott, J. J., and G. M. McKeon. 1977. A note on the selection of seed types by harvester ants in northern Australia. *Australian Journal of Ecology*, 2(2): 231–235.

Mou, Y. C. 1938. Morphologische und histologische Studien über Paussidendrüsen. *Zoologische Jahrbücher, Abteilung für Anatomie und Ontogenie der Tiere*, 64(2–3): 287–346.

Mühlenberg, M. 1976. *Freilandökologie*. Quelle und Meyer, Heidelberg. 214 pp.

Mühlenberg, M., and U. Maschwitz. 1976. *Acanthaspis bistillata* und *Acanthaspis concinnula* (Reduviidae): Maskierung (Freilandaufnahmen). *Encyclopaedia Cinematographica*, film no. E1937/1975. Institut für den Wissenschaftlichen Film, Göttingen. 12 pp.

Muir, D. A. 1954. Ants *Myrmica rubra* L. and *M. scabrinodis* Nylander as intermediate hosts of a cestode. *Nature*, 173: 688–689.

Müller, W. 1886. Beobachtungen an Wanderameisen (*Eciton hamatum* Fabr.). *Kosmos* (Stuttgart), 18: 81–93.

Murphy, D. H. 1973. Preliminary key to genera of Malaysian ants, based on workers. *Bukit Timah Survey Report* (Department of Zoology, University of Singapore), 2(suppl. 1), pp. 1–27.

Murray, B. G. 1971. The ecological consequences of interspecific territorial behavior in birds. *Ecology*, 52(3): 414–423.

Murrell, K. F. H. 1965. *Ergonomics: man in his working environment*. Chapman and Hall, London. xix + 496 pp.

Musgrave, A. 1932. *Bibliography of Australian entomology 1775–1930, with biographical notes on authors and collectors*. Royal Zoological Society of New South Wales, Sydney. viii + 380 pp.

Myers, J. G. 1929. The nesting-together of birds, wasps and ants. *Proceedings of the Entomological Society of London*, 4(2): 80–88.

——— 1935. Nesting associations of birds with social insects. *Transactions of the Royal Entomological Society of London*, 83(1): 11–22.

Naarmann, H. 1963. Untersuchungen über Bildung und Weitergabe von Drüsensekreten bei *Formica* (Hymenopt. Formicidae) mit Hilfe der Radioisotopenmethode. *Experientia*, 19(8): 412–413.

Najt, J. 1987. Le collembole fossile *Paleosminthurus juliae* est un hyménoptère. *Revue Française d'Entomologie*, n.s., 9(4): 152–154.

Nakamura, S., H. Miki-Hirosige, and Y. Iwanami. 1982. Ultrastructural study of *Camellia japonica* pollen treated with myrmicacin, an ant-origin inhibitor. *American Journal of Botany*, 69(4): 538–545.

Nault, L. R., M. E. Montgomery, and W. S. Bowers. 1976. Ant-aphid association: role of aphid alarm pheromone. *Science*, 192: 1349–51.

Naves, M. A. 1985. A monograph of the genus *Pheidole* in Florida (Hymenoptera: Formicidae). *Insecta Mundi*, 1(2): 53–89.

Nelson, D. R., C. L. Fatland, R. W. Howard, C. A. McDaniel, and G. J. Blomquist. 1980. Re-analysis of the cuticular methylalkanes of *Solenopsis invicta* and *S. richteri*. *Insect Biochemistry*, 10(4): 409–418.

Newcomer, E. J. 1912. Some observations on the relations of ants and lycaenid caterpillars, and a description of the relational organs of the latter. *Journal of the New York Entomological Society*, 20(1): 31–36.

Nickerson, J. C., H. L. Cromroy, W. H. Whitcomb, and J. A. Cornell. 1975.

Colony organization and queen numbers in two species of *Conomyrma*. *Annals of the Entomological Society of America*, 68(6): 1083–85.

Nickerson, J. C., C. A. Rolph Kay, L. L. Buschman, and W. H. Whitcomb. 1977. The presence of *Spissistilus festinus* as a factor affecting egg predation by ants in soybeans. *Florida Entomologist*, 60(3): 193–199.

Nickle, D. A., and T. M. Neal. 1972. Observations on the foraging behavior of the southern harvester ant *Pogonomyrmex badius*. *Florida Entomologist*, 55(1): 65–66.

Nielsen, M. G. 1986. Respiratory rates of ants from different climatic areas. *Journal of Insect Physiology*, 32(2): 125–131.

——— 1987. The ant fauna (Hymenoptera: Formicidae) in northern and interior Alaska: a survey along the Trans-Alaskan Pipeline and a few highways. *Entomological News*, 98(2): 74–88.

Nielsen, M. G., T. F. Jensen, and I. Holm-Jensen. 1982. Effect of load carriage on the respiratory metabolism of running worker ants of *Camponotus herculeanus* (Formicidae). *Oikos*, 39(2): 137–142.

Nielsen, M. G., N. Skyberg, and G. Peakin. 1985a. Respiration in the sexuals of the ant *Lasius flavus*. *Physiological Entomology*, 10(2): 199–204.

——— 1985b. Respiration of ant queens (Hymenoptera, Formicidae). *Actes des Colloques Insectes Sociaux* (L'Union Internationale pour l'Etude des Insectes Sociaux, Section française, Compte Rendu Colloque Annuel, Diepenbeek, 1984), 2: 133–139.

Nijhout, H. F., and D. E. Wheeler. 1982. Juvenile hormone and the physiological basis of insect polymorphisms. *Quarterly Review of Biology*, 57(2): 109–133.

Nixon, G. E. 1951. *The association of ants with aphids and coccids*. Commonwealth Institute of Entomology, London. 36 pp.

Nonacs, P. 1986a. Ant reproductive strategies and sex allocation theory. *Quarterly Review of Biology*, 61(1): 1–21.

——— 1986b. Sex-ratio determination within colonies of ants. *Evolution*, 40(1): 199–204.

——— 1988. Queen number in colonies of social Hymenoptera as a kin-selected adaptation. *Evolution*, 42(3): 566–580.

Noonan, G. R. 1982. Notes on interactions between the spider *Eilica puno* (Gnaphosidae) and the ant *Camponotus inca* in the Peruvian Andes. *Biotropica* 14(2): 145–148.

Noonan, K. C. 1986. Recognition of queen larvae by worker honey bees (*Apis mellifera*). *Ethology*, 73(4): 295–306.

Nordlund, D. A. 1981. Semiochemicals: a review of the terminology. *In* D. A. Nordlund, R. L. Jones, and W. J. Lewis, eds., *Semiochemicals*, pp. 13–27. John Wiley, New York.

Novak, V. 1948. On the question of the origin of pathological individuals (pseudogynes) in ants of the genus *Formica*. *Věstník Československé Zoologické Společnosti*, 12: 97–131. [In Czech with English summary.]

Obin, M. S. 1986. Nestmate recognition cues in laboratory and field colonies of *Solenopsis invicta* Buren (Hymenoptera: Formicidae): effect of environment and role of cuticular hydrocarbons. *Journal of Chemical Ecology*, 12(9): 1965–75.

Obin, M. S., and R. K. Vander Meer. 1985. Gaster flagging by fire ants (*Solenopsis* spp.): functional significance of venom dispersal behavior. *Journal of Chemical Ecology*, 11(12): 1757–68.

——— 1988. Sources of nestmate recognition cues in the imported fire ant *Solenopsis invicta* Buren (Hymenoptera: Formicidae). *Animal Behaviour*, 36(5): 1361–70.

——— 1989. Nestmate recognition in fire ants (*Solenopsis invicta* Buren): do queens label workers? *Ethology*, 80(1–4): 255–264.

O'Dowd, D. J., and M. E. Hay. 1980. Mutualism between harvester ants and a desert ephemeral: seed escape from rodents. *Ecology*, 61(3): 531–540.

Oeser, R. 1961. Vergleichend-morphologische Untersuchungen über den Ovipositor der Hymenopteren. *Mitteilungen aus dem Zoologischen Museum in Berlin*, 37(1): 1–119.

Ofer, J. 1970. *Polyrhachis simplex*, the weaver ant of Israel. *Insectes Sociaux*, 17(1): 49–82.

Ogata, K. 1987. A generic synopsis of the poneroid complex of the family Formicidae in Japan (Hymenoptera), Part I: Subfamilies Ponerinae and Cerapachyinae. *Esakia*, 25: 97–132.

Ohly-Wüst, M. 1977. Soziale Wechselbeziehungen zwischen Larven und Arbeiterinnen im Ameisenstaat, mit besonderer Beachtung der Trophallaxis. Doctoral diss., Johann-Wolfgang-Goethe-Universität, Frankfurt. 137 pp.

Oinonen, E. A. 1956. On the ants of the rocks and their contribution to the afforestation of rocks in southern Finland. *Acta Entomologica Fennica*, no. 12, 212 pp. [In Finnish with English summary.]

Oke, C. 1932. Aculagnathidae: a new family of Coleoptera. *Proceedings of the Royal Society of Victoria*, n.s., 44(1): 22–24.

Økland, F. 1930. Studien über die Arbeitsteilung und die Teilung des Arbeitsgebietes bei der Roten Waldameise (*Formica rufa* L.). *Zeitschrift für Morphologie und Ökologie der Tiere*, 20(1): 63–131.

——— 1934. Utvandring og overvintring hos den røde skogmaur (*Formica rufa* L.). *Norsk Entomologisk Tidsskrift*, 3(5): 316–327.

Oliveira, P. S. 1985. On the mimetic association between nymphs of *Hyalymenus* spp. (Hemiptera: Alydidae) and ants. *Zoological Journal of the Linnean Society*, 83(4): 371–384.

——— 1986. Ant-mimicry in some spiders from Brazil. *Bulletin de la Société Zoologique de France*, 111(3–4): 297–311.

——— 1988. Ant-mimicry in some Brazilian salticid and clubionid spiders (Araneae: Salticidae, Clubionidae). *Biological Journal of the Linnean Society*, 33(1): 1–15.

Oliveira, P. S., and H. F. Leitão-Filho. 1987. Extrafloral nectaries: their taxonomic distribution and abundance in the woody flora of cerrado vegetation in southeast Brazil. *Biotropica*, 19(2): 140–148.

Oliveira, P. S., and I. Sazima. 1984. The adaptive bases of ant-mimicry in a Neotropical aphantochilid spider (Araneae: Aphantochilidae). *Biological Journal of the Linnean Society*, 22(2): 145–155.

——— 1985. Ant-hunting behaviour in spiders with emphasis on *Strophius nigricans* (Thomisidae). *Bulletin of the British Arachnological Society*, 6(7): 309–312.

Oliveira, P. S., A. T. Oliveira-Filho, and R. Cintra. 1987a. Ant foraging on ant-inhabited *Triplaris* (Polygonaceae) in western Brazil: a field experiment using live termite-baits. *Journal of Tropical Ecology*, 3(3): 193–200.

Oliveira, P. S., A. F. da Silva, and A. B. Martins. 1987b. Ant foraging on extrafloral nectaries of *Qualea grandiflora* (Vochysiaceae) in cerrado vegetation: ants as potential antiherbivore agents. *Oecologia*, 74(2): 228–230.

Olubajo, O., R. M. Duffield, and J. W. Wheeler. 1980. 4-Heptanone in the mandibular gland secretion of the Nearctic ant, *Zacryptocerus varians* (Hymenoptera: Formicidae). *Annals of the Entomological Society of America*, 73(1): 93–94.

O'Neal, J., and J. P. Markin. 1973. Brood nutrition and parental relationships of the imported fire ant *Solenopsis invicta*. *Journal of the Georgia Entomological Society*, 8(4): 294–303.

O'Neill, M. C. A., and A. G. Robinson. 1977. Ant-aphid associations in the province of Manitoba. *Manitoba Entomologist*, 11: 74–88.

Onoyama, K. 1982. Foraging behavior of the harvester ant *Messor aciculatus*, with special reference to foraging sites and diel activity of individual ants. *Japanese Journal of Ecology*, 32(4): 453–461.

Onoyama, K., and T. Abe. 1982. Foraging behavior of the harvester ant *Messor aciculatus* in relation to the amount and distribution of food. *Japanese Journal of Ecology*, 32(3): 383–393.

Orians, G. H., and N. E. Pearson. 1979. On the theory of central place foraging. *In* D. J. Horn, G. R. Stairs, and R. D. Mitchell, eds., *Analyis of ecological systems*, pp. 155–177. Ohio State University Press, Columbus.

Orr, P. H. 1985. Form and function of the mound in *Formica glacialis*. M.S. thesis, Case Western Reserve University, Cleveland. x + 314 pp.

Oster, G. F., and E. O. Wilson. 1978. *Caste and ecology in the social insects* (Monographs in Population Biology, no. 12). Princeton University Press, Princeton, N.J. xv + 352 pp.

Otto, D. 1958. Über die Arbeitsteilung im Staate von *Formica rufa rufopratensis minor* Gössw. und ihre verhaltensphysiologischen Grundlagen: ein Beitrag zur Biologie der Roten Waldameise. *Wissenschaftliche Abhand-*

*lungen der Deutschen Akademie der Landwirtschaftswissenschaften zu Berlin,* 30: 1–169.
——— 1960. Zur Erscheinung der Arbeiterinnenfertilität und Parthenogenese bei der Kahlrückigen Roten Waldameise (*Formica polyctena* Först.), (Hym.). *Deutsche Entomologisches Zeitschrift,* n.s., 7(1–2): 1–9.
——— 1962. *Die roten Waldameisen.* A. Ziemsen Verlag, Wittenberg, Lutherstadt. 151 pp.
Overal, W. L. 1980. Observations on colony founding and migration of *Dinoponera gigantea. Journal of the Georgia Entomological Society,* 15(4): 466–469.
——— 1986. Recrutamento e divisão de trabalho em colônias naturais da formiga *Ectatomma quadridens* (Fabr.) (Hymenoptera: Formicidae: Ponerinae). *Boletim do Museu Paraense Emilio Goeldi Zoologia,* 2(2): 113–135.
Overal, W. L., and A. G. Bandeira. 1985. Nota sobre hábitos de *Cylindromyrmex striatus* Mayr, 1870, na Amazônia (Formicidae, Ponerinae). *Revista Brasileira de Entomologia,* 29(3–4): 521–522.

Page, R. E. 1986. Sperm utilization in social insects. *Annual Review of Entomology,* 31: 297–320.
Page, R. E., and M. D. Breed. 1987. Kin recognition in social bees. *Trends in Ecology and Evolution,* 2(9): 272–275.
Page, R. E., and E. H. Erickson. 1984. Selective rearing of queens by worker honey bees: kin or nestmate recognition? *Annals of the Entomological Society of America,* 77(5): 578–580.
Page, R. E., and R. A. Metcalf. 1982. Multiple mating, sperm utilization, and social evolution. *American Naturalist,* 119(2): 263–281.
Palma-Valli, G., and G. Délye. 1981. Contrôle neuro-endocrine de la ponte chez les reines de *Camponotus lateralis* Olivier (Hyménoptères Formicidae). *Insectes Sociaux,* 28(2): 167–181.
Pamilo, P. 1981. Genetic organization of *Formica sanguinea* populations. *Behavioral Ecology and Sociobiology,* 9(1): 45–50.
——— 1982a. Genetic evolution of sex ratios in eusocial Hymenoptera: allele frequency simulations. *American Naturalist,* 119(5): 638–656.
——— 1982b. Genetic population structure in polygynous *Formica* ants. *Heredity,* 48(1): 95–106.
——— 1982c. Multiple mating in *Formica* ants. *Hereditas,* 97(1): 37–45.
——— 1984a. Genetic relatedness and evolution of insect sociality. *Behavioral Ecology and Sociobiology,* 15(4): 241–248.
——— 1984b. Genotypic correlation and regression in social groups: multiple alleles, multiple loci and subdivided populations. *Genetics,* 107(2): 307–320.
Pamilo, P., and R. H. Crozier. 1982. Measuring genetic relatedness in natural populations: methodology. *Theoretical Population Biology,* 21(2): 171–193.
Pamilo, P., and R. Rosengren. 1983. Sex ratio strategies in *Formica* ants. *Oikos,* 40(1): 24–35.
——— 1984. Evolution of nesting strategies of ants: genetic evidence from different population types of *Formica* ants. *Biological Journal of the Linnean Society,* 21(3): 331–348.
Pamilo, P., and S.-L. Varvio-Aho. 1979. Genetic structure of nests in the ant *Formica sanguinea. Behavioral Ecology and Sociobiology,* 6(2): 91–98.
Pamilo, P., R. H. Crozier, and J. Fraser. 1985. Inter-nest interactions, nest autonomy, and reproductive specialization in an Australian arid-zone ant, *Rhytidoponera* sp. 12. *Psyche,* 92(2–3): 217–236.
Pardi, L. 1948. Dominance order in *Polistes* wasps. *Physiological Zoölogy,* 21(1): 1–13.
Park, O. 1929. Ecological observations upon the myrmecocoles of *Formica ulkei* Emery, especially *Leptinus testaceus* Mueller. *Psyche,* 36(3): 195–215.
——— 1932. The myrmecocoles of *Lasius umbratus mixtus aphidicola* Walsh. *Annals of the Entomological Society of America,* 25(1): 77–88.
——— 1933. Ecological study of the ptiliid myrmecocole, *Limulodes paradoxus* Matthews. *Annals of the Entomological Society of America,* 26(2): 255–261.
——— 1964. Observations upon the behavior of myrmecophilous pselaphid beetles. *Pedobiologia,* 4(3): 129–137.
Parker, G. A. 1970. Sperm competition and its evolutionary consequences in the insects. *Biological Reviews of the Cambridge Philosophical Society,* 45(4): 525–567.
Parry, K., and E. D. Morgan. 1979. Pheromones of ants: a review. *Physiological Entomology,* 4(2): 161–189.
Passera, L. 1964. Données biologiques sur la fourmi parasite *Plagiolepis xene* Stärcke. *Insectes Sociaux,* 11(1): 59–70.
——— 1966. Fécondité des femelles au sein de la myrmécobiose *Plagiolepis pygmaea* Latr.–*Plagiolepis xene* Star. (Hyménoptères, Formicidae). *Comptes Rendus de l'Académie des Sciences, Paris,* ser. D, 263(21): 1600–3.
——— 1968a. Les stades larvaires de la caste ouvrière chez la fourmi *Plagiolepis pygmaea* Latr. (Hyménoptère, Formicidae). *Bulletin de la Société Zoologique de France,* 93(3): 357–365.
——— 1968b. Observations biologiques sur la fourmi *Plagiolepis grassei* Le Masne Passera, parasite social de *Plagiolepis pygmaea* Latr. (Hym. Formicidae). *Insectes Sociaux,* 15(4): 327–336.
——— 1973. Origine des soldats dans les sociétés de *Pheidole pallidula* Nyl. (Formicidae, Myrmicinae). *Proceedings of the Seventh International Congress of the International Union for the Study of Social Insects* (London), pp. 305–309.
——— 1974a. Différenciation des soldats chez la fourmi *Pheidole pallidula* Nyl. (Formicidae Myrmicinae). *Insectes Sociaux,* 21(1): 71–86.
——— 1974b. Kastendetermination bei der Ameise *Plagiolepis pygmaea* Latr. *In* G. H. Schmidt, ed., *Sozialpolymorphismus bei Insekten,* pp. 513–532. Wissenschaftliche Verlagsgesellschaft MBH, Stuttgart.
——— 1978. Une nouvelle catégorie d'oeufs alimentaires: les oeufs alimentaires émis par les reines vierges de *Pheidole pallidula* (Nyl.) (Formicidae, Myrmicinae). *Insectes Sociaux,* 25(2): 117–126.
——— 1980a. La ponte d'oeufs préorientés chez la fourmi *Pheidole pallidula* (Nyl.) (Hymenoptera Formicidae). *Insectes Sociaux,* 27(1): 79–95.
——— 1980b. La fonction inhibitrice des reines de la fourmi *Plagiolepis pygmaea* Latr.: rôle des pheromones. *Insectes Sociaux,* 27(3): 212–225.
——— 1984. *L'organisation sociale des fourmis.* Privat, Toulouse. 360 pp.
——— 1985. Soldier determination in ants of the genus *Pheidole. In* J. A. L. Watson, B. M. Okot-Kotber, and C. Noirot, eds., *Caste differentiation in social insects,* pp. 331–346. Pergamon Press, New York.
Passera, L., J. Bitsch, and C. Bressac. 1968. Observations histologiques sur la formation des oeufs alimentaires et des oeufs reproducteurs chez les ouvrières de *Plagiolepis pygmaea* Latr. (Hymenoptera Formicidae). *Comptes Rendus de l'Académie des Sciences, Paris,* ser. D, 266(24): 2270–72.
Pasteels, J. M. 1968. Le système glandulaire tégumentaire des Aleocharinae (Coleoptera, Staphylinidae) et son évolution chez les espèces termitophiles du genre *Termitella. Archives de Biologie, Liège,* 79(3): 381–469.
Pasteels, J. M., J. C. Verhaeghe, J. C. Braekman, D. Daloze, and B. Tursch. 1980. Caste-dependent pheromones in the head of the ant *Tetramorium caespitum. Journal of Chemical Ecology,* 6(2): 467–472.
Pasteels, J. M., J. C. Verhaeghe, and J.-L. Deneubourg. 1982. The adaptive value of probabilistic behavior during food recruitment in ants: experimental and theoretical approaches. *In* M. D. Breed, C. D. Michener and H. E. Evans, eds., *The biology of social insects* (Proceedings of the Ninth Congress of the International Union for the Study of Social Insects, Boulder, Colorado, 1982), pp. 297–301. Westview Press, Boulder.
Pasteels, J. M., J.-L. Deneubourg, and S. Goss. 1987. Transmission and amplification of information in a changing environment: the case of insect societies. *In* I. Prigogine and M. Sanglier, eds., *Laws of nature and human conduct,* pp. 129–156. Groupe Opérationnel de Recherche de Documentation et d'Etude sur la Science (G.O.R.D.E.S.), Brussels.
Patrizi, S. 1948. Contribuzioni alla conoscenza delle formiche e dei mirmicofili dell'Africa Orientale, V: Note etologiche su *Myrmechusa* Wasmann (Coleoptera Staphylinidae). *Bollettino dell'Istituto di Entomologia della Università degli Studi di Bologna,* 17: 168–173.
Paulian, R. 1948. Observations sur les Coléoptères commensaux d'*Anomma nigricans* en Côte d'Ivoire. *Annales des Sciences Naturelles,* ser. 11, 10(1): 79–102.
Paulsen, R. 1969. Zur Funktion der Propharynx- Postpharynx- und Labil-

drüsen von *Formica polyctena* Foerst. (Hymenoptera, Formicidae). Doctoral diss., Bayerische Julius-Maximilians-Universität, Würzburg. 90 pp.

Pavan, M. 1955. Studi sui Formicidae, I: Contributo alla conoscenza degli organi gastrali nei Dolichoderinae. *Natura, Milano*, 46: 135–145.

——— 1956. Studi sui Formicidae, II: Sull'origine, significato biologico e isolamento della dendrolasina. *La Ricerca Scientifica*, 26(1): 144–150.

Pavan, M., and G. Ronchetti. 1955. Studi sulla morfologia esterna e anatomia interna dell'operaia di *Iridomyrmex humilis* Mayr e ricerche chimiche e biologiche sulla iridomirmecina. *Atti della Società Italiana di Scienze Naturali, Milano*, 94(3–4): 379–477.

Pavan, M., and R. Trave. 1958. Etudes sur les Formicidae, IV: Sur le venin du dolichodéride *Tapinoma nigerrimum* Nyl. *Insectes Sociaux*, 5(3): 299–308.

Pavlova, Z. F. 1977. Earth hummocks inhabited by ants, as the principal microstructures of lake-coastal biogeocenoses. *Ekologiia* (Sverdlovsk, Nauka), no. 5, pp. 62–71. [In Russian.]

Peacock, A. D., and A. T. Baxter. 1950. Studies in Pharaoh's ant, *Monomorium pharaonis* (L.), 3: Life history. *Entomologist's Monthly Magazine*, 86: 171–178.

Peacock, A. D., I. C. Smith, D. W. Hall, and A. T. Baxter. 1954. Studies in Pharaoh's ant, *Monomorium pharaonis* (L.), 8: Male production by parthenogenesis. *Entomologist's Monthly Magazine*, 90: 154–158.

Peakall, R., A. J. Beattie, and S. H. James. 1987. Pseudocopulation of an orchid by male ants: a test of two hypotheses accounting for the rarity of ant pollination. *Oecologia*, 73(4): 522–524.

Peakin, G. J., and G. Josens. 1978. Respiration and energy flow. *In* M. V. Brian, ed., *Production ecology of ants and termites* (International Biology Programme, no. 13), pp. 111–163. Cambridge University Press, Cambridge.

Peakin, G., M. G. Nielsen, N. Skyberg, and J. Pedersen. 1985. Respiration in the larvae of the ants *Myrmica scabrinodis* and *Lasius flavus*. *Physiological Entomology*, 10(2): 205–214.

Pearson, B. 1981. The electrophoretic determination of *Myrmica rubra* microgynes as a social parasite: possible significance in the evolution of ant social parasites. *In* P. E. Howse and J.-L. Clément, eds., *Biosystematics of social insects* (Systematics Association special volume no. 19), pp. 75–84. Academic Press, New York.

——— 1983. Intra-colonial relatedness amongst workers in a population of nests of the polygynous ant, *Myrmica rubra* Latreille. *Behavioral Ecology and Sociobiology*, 12(1): 1–4.

Peck, S. B. 1973. A systematic revision and the evolutionary biology of the *Ptomaphagus (Adelops)* beetles of North America (Coleoptera; Leiodidae: Catopinae), with emphasis on cave-inhabiting species. *Bulletin of the Museum of Comparative Zoology, Harvard*, 145(2): 29–162.

——— 1976. The myrmecophilous beetle genus *Echinocoleus* in the southwestern United States (Leiodidae: Catopinae). *Psyche*, 83(1): 51–62.

Peeters, C. P. 1987a. The diversity of reproductive systems in ponerine ants. *In* J. Eder and H. Rembold, eds., *Chemistry and biology of social insects* (Proceedings of the Tenth International Congress of the International Union for the Study of Social Insects, Munich, 1986), pp. 253–254. Verlag J. Peperny, Munich.

——— 1987b. The reproductive division of labour in the queenless ant *Rhytidoponera* sp. 12. *Insectes Sociaux*, 34(2): 75–86.

Peeters, C. P., and R. M. Crewe. 1984. Insemination controls the reproduction division of labour in a ponerine ant. *Naturwissenschaften*, 71(1): 50–51.

——— 1985. Worker reproduction in the ponerine ant *Ophthalmopone berthoudi:* an alternative form of eusocial organization. *Behavioral Ecology and Sociobiology*, 18(1): 29–37.

——— 1986a. Male biology in the queenless ponerine ant *Ophthalmopone berthoudi* (Hymenoptera: Formicidae). *Psyche*, 93(3–4): 277–284.

——— 1986b. Queenright and queenless breeding systems within the genus *Pachycondyla* (Hymenoptera: Formicidae). *Journal of the Entomological Society of Southern Africa*, 49(2): 251–255.

——— 1987. Foraging and recruitment in ponerine ants: solitary hunting in the queenless *Ophthalmopone berthoudi* (Hymenoptera: Formicidae). *Psyche*, 94(1–2): 201–214.

Percy, J. E., and J. Weatherston. 1974. Gland structure and pheromone production in insects. *In* M. C. Birch, ed., *Pheromones*, pp. 11–34. North-Holland, Amsterdam.

Pérez-Bautista, M., J.-P. Lachaud, and D. Fresneau. 1985. La division del trabajo en la hormiga primitiva *Neoponera villosa* (Hymenoptera: Formicidae). *Folia Entomológica Mexicana*, 65: 119–130.

Peru, L. 1984. Individus teratologiques chez les fourmis *Leptothorax*. *Actes des Colloques Insectes Sociaux* (L'Union Internationale pour l'Etude des Insectes Sociaux, Section française, Compte Rendu Colloque Annuel, Les Eyzies, 1983), 1: 141–145.

Pętal, J. 1978. The role of ants in ecosystems. *In* M. V. Brian, ed., *Production ecology of ants and termites* (International Biology Programme, no. 13), pp. 293–325. Cambridge University Press, New York.

——— 1980. Intraspecific competition as an adaptation to food resources in an ant population. *Insectes Sociaux*, 27(3): 279.

Petersen, B. 1968. Some novelties in presumed males of Leptanillinae (Hym., Formicidae). *Entomologiske Meddelelser*, 36(6): 577–598.

——— 1977. Pollination by ants in the alpine tundra of Colorado. *Transactions of the Illinois State Academy of Science*, 70(3–4): 349–355.

Petersen-Braun, M. 1975. Untersuchungen zur sozialen Organisation der Pharaoameise *Monomorium pharaonis* (L.) (Hymenoptera, Formicidae), I: Der Brutzyklus und seine Steuerung durch populationseigene Faktoren. *Insectes Sociaux*, 22(3): 269–292.

——— 1977. Untersuchungen zur sozialen Organisation der Pharaoameise *Monomorium pharaonis* L., II: Die Kastendeterminierung. *Insectes Sociaux*, 24(4): 303–318.

——— 1982. Intraspezifisches Aggressionsverhalten bei der Pharaoameise *Monomorium pharaonis* L. (Hymenoptera, Formicidae). *Insectes Sociaux*, 29(1): 25–33.

Petersen-Braun, M., and A. Buschinger. 1975. Enstehung und Funktion eines thorakalen Kropfes bei Formiciden-Königinnen. *Insectes Sociaux*, 22(1): 51–66.

Petralia, R. S., and S. B. Vinson. 1979a. Comparative anatomy of the ventral region of ant larvae, and its relation to feeding behavior. *Psyche*, 86(4): 375–394.

——— 1979b. Developmental morphology of larvae and eggs of the imported fire ant, *Solenopsis invicta*. *Annals of the Entomological Society of America*, 72(4): 472–484.

Petralia, R. S., A. A. Sorenson, and S. B. Vinson. 1980. The labial gland system of larvae of the imported fire ant, *Solenopsis invicta* Buren: ultrastructure and enzyme analysis. *Cell and Tissue Research*, 206(1): 145–156.

Phillips, S. A., and S. B. Vinson. 1980. Comparative morphology of glands associated with the head among castes of the red imported fire ant, *Solenopsis invicta* Buren. *Journal of the Georgia Entomological Society*, 15(2): 215–226.

Pickering, J. 1980. Sex ratio, social behavior and ecology in *Polistes* (Hymenoptera, Vespidae), *Pachysomoides* (Hymenoptera, Ichneumonidae), and *Plasmodium* (Protozoa, Haemosporidia). Ph.D. diss., Harvard University. 362 pp.

Picquet, N. 1958. Contribution à l'étude des larves de Formicidae de la Côte d'Or. *Travaux du Laboratoire de Zoologie et de la Station Aquicole Grimaldi de la Faculté des Sciences de Dijon*, no. 23, 48 pp.

Pierce, N. E. 1983. The ecology and evolution of symbioses between lycaenid butterflies and ants. Ph.D. diss., Harvard University. x + 274 pp.

——— 1984. Amplified species diversity: a case study of an Australian lycaenid butterfly and its attendant ants. *In* R. I. Vane-Wright and P. R. Ackery, eds., *The biology of butterflies* (Symposium of the Royal Entomological Society of London, no. 11), pp. 197–200. Academic Press, London.

——— 1985. Lycaenid butterflies and ants: selection for nitrogen-fixing and other protein-rich food plants. *American Naturalist*, 125(6): 888–895.

——— 1987. The evolution and biogeography of associations between lycaenid butterflies and ants. *In* P. H. Harvey and L. Partridge, eds., *Oxford surveys in evolutionary biology*, vol. 4, pp. 89–116. Oxford University Press, Oxford.

Pierce, N. E., and S. Easteal. 1986. The selective advantage of attendant ants

for the larvae of a lycaenid butterfly, *Glaucopsyche lygdamus*. *Journal of Animal Ecology*, 55(2): 451–462.
Pierce, N. E., and M. A. Elgar. 1985. The influence of ants on host plant selection by *Jalmenus evagoras*, a myrmecophilous lycaenid butterfly. *Behavioral Ecology and Sociobiology*, 16(3): 209–222.
Pierce, N. E., and P. S. Mead. 1981. Parasitoids as selective agents in the symbiosis between lycaenid butterfly larvae and ants. *Science*, 211: 1185–87.
Pierce, N. E., and W. R. Young. 1986. Lycaenid butterflies and ants: two-species stable equilibria in mutualisitic, commensal, and parasitic interactions. *American Naturalist*, 128(2): 216–227.
Pierce, N. E., R. L. Kitching, R. C. Buckley, M. F. J. Taylor, and K. F. Benbow. 1987. The costs and benefits of cooperation between the Australian lycaenid butterfly, *Jalmenus evagoras*, and its attendant ants. *Behavioral Ecology and Sociobiology*, 21(4): 237–248.
Pierce, N. E., P. J. Rogers, and P. Vowles. 1989. Amino acids secreted by lycaenid larvae as rewards for attendant ants. Unpublished manuscript, cited with permission.
Piéron, H. 1904. Du rôle du sens musculaire dans l'orientation de quelques espèces des fourmis. *Bulletin de l'Institut Général Psychologique*, 4(2): 168–186.
Pijl, L. van der. 1955. Some remarks on myrmecophytes. *Phytomorphology*, 5(2–3): 190–200.
Pisarski, B. 1963. Nouvelle espèce du genre *Harpagoxenus* Forel de la Mongolie (Hymenoptera, Formicidae). *Bulletin de l'Académie Polonaise des Sciences*, ser. 2 (zool.), 11(1): 39–41.
——— 1966. Etudes sur les fourmis du genre *Strongylognathus* Mayr (Hymenoptera, Formicidae). *Annales Zoologici, Warszawa*, 23(22): 509–523.
——— 1972. La structure des colonies polycaliques de *Formica (Coptoformica) exsecta* Nyl. *Ekologia Polska*, 20(12): 111–116.
——— 1973. *La structure sociale de la* Formica (Coptoformica) exsecta *Nyl. (Hymenoptera: Formicidae) et son influence sur la morphologie, l'ecologie et l'éthologie de l'espèce*. Institute of Zoology, Polish Academy of Sciences, Warsaw. 134 pp. [In Polish with French summary.]
Pisarski, B., ed. 1982. *Structure et organisation des sociétés de fourmis de l'espèce* Formica (Coptoformica) exsecta *Nyl. (Hymenoptera, Formicidae)* (Memorabilia Zoologica, vol. 38). Institute of Zoology, Polish Academy of Sciences, Warsaw. 281 pp.
Pisarski, B., and K. Vepsäläinen. 1981. Organization of ant communities in the Tvärminne Archipelago. *In Regulation of social insects populations* (Proceedings of the Tenth Symposium), 4 pp. Social Insects Section of the Polish Entomological Society and Institute of Zoology of the Polish Academy of Sciences, Skierniewice.
Pisarski, B., K. Vepsäläinen, E. Ranta, S. Ås, Y. Haila, and J. Tiainen. 1982. A comparison of two methods of sampling island ant communities. *Annales Entomologici Fennici*, 48(3): 75–80.
Plateaux, L. 1960a. Adoptions expérimentales de larves entre des fourmis de genres différents: *Leptothorax nylanderi* Förster et *Solenopsis fugax* Latreille. *Insectes Sociaux*, 7(2): 163–170.
——— 1960b. Adoptions expérimentales de larves entre des fourmis de genres différents, II: *Myrmica laevinodis* Nylander et *Anergates atratulus* Schenck. *Insectes Sociaux*, 7(3): 221–226.
——— 1960c. Adoptions expérimentales de larves entre des fourmis de genres différents, III: *Anergates atratulus* Schenck et *Solenopsis fugax* Latreille, IV: *Leptothorax nylanderi* Förster et *Tetramorium caespitum* L. *Insectes Sociaux* 7(4): 345–348.
——— 1970. Sur le polymorphisme social de la fourmi *Leptothorax nylanderi* (Förster), I: Morphologie et biologie comparées des castes. *Annales des Sciences Naturelles*, ser. 12, 12(4): 373–478.
——— 1971. Sur le polymorphisme social de la fourmi *Leptothorax nylanderi* (Förster), II: Activité des ouvrières et déterminisme des castes. *Annales des Sciences Naturelles*, ser. 12, 13(1): 1–90.
——— 1972. Sur les modifications produites chez une fourmi par la présence d'un parasite cestode. *Annales des Sciences Naturelles*, ser. 12, 14(3): 203–220.
——— 1978. L'essaimage de quelques fourmis *Leptothorax*: rôles de l'éclairement et de divers autres facteurs: effet sur l'isolement reproductif et la répartition géographique. *Annales des Sciences Naturelles*, ser. 12, 20(1): 129–164.
——— 1981. The *pallens* morph of the ant *Leptothorax nylanderi*: description, formal genetics, and study of populations. *In* P. E. Howse and J.-L. Clément, eds., *Biosystematics of social insects* (Systematics Association special volume no. 19), pp. 63–74. Academic Press, New York.
——— 1986. Comparaison des cycles saisonniers, des durées des sociétés et des productions des trois espèces de fourmis *Leptothorax (Myrafant)* du groupe *nylanderi*. *Actes des Colloques Insectes Sociaux* (L'Union Internationale pour l'Etude des Insectes Sociaux, Section française, Compte Rendu Colloque Annuel, Vaison la Romaine, 1985), 3: 221–234.
——— 1987. Reproductive isolation in ants of the genus *Leptothorax*, subgenus *Myrafant*. *In* J. Eder and H. Rembold, eds., *Chemistry and biology of social insects* (Proceedings of the Tenth International Congress of the International Union for the Study of Social Insects, Munich, 1986), pp. 33–34. Verlag J. Peperny, Munich.
Plath, O. E. 1935. Insect societies. *In* C. Murchison, ed., *A handbook of social psychology*, pp. 83–141. Clark University Press, Worcester, Mass.
Plsek, R. W., J. C. Kroll, and J. F. Watkins II. 1969. Observations of carabid beetles, *Helluomorphoides texanus*, in columns of army ants and laboratory experiments on their behavior. *Journal of the Kansas Entomological Society*, 42(4): 452–456.
Pohl, L. 1957. Vergleichende anatomisch-histologische Untersuchungen an *Lepisma saccharina* Linné und der myrmecophilen *Atelura formicaria* Heyden (Beitrag zur Myrmecophilie, ester Abschnitt). *Insectes Sociaux*, 4(4): 349–363.
Poldi, B. 1963. Alcune osservazioni sul *Proceratium melinum* Rog. e sulla fuzione della particolare struttura del gastro. *Atti dell'Accademia Nazionale Italiana di Entomologia Rendiconti*, 11: 221–229.
Pollock, G. P., and S. W. Rissing. 1985. Mating season and colony foundation of the seed-harvester ant, *Veromessor pergandei*. *Psyche*, 92(1): 125–134.
Pontin, A. J. 1960. Field experiments on colony foundation by *Lasius niger* (L.) and *L. flavus* (F.) (Hym., Formicidae). *Insectes Sociaux*, 7(3): 227–230.
——— 1961. Population stabilization and competition between the ants *Lasius flavus* (F.) and *L. niger* (L.). *Journal of Animal Ecology*, 30(1): 47–54.
——— 1963. Further considerations of competition and the ecology of the ants *Lasius flavus* (F.) and *L. niger* (L.). *Journal of Animal Ecology*, 32(3): 565–574.
Porter, S. D. 1985. *Masoncus* spider: a miniature predator of *Collembola* in harvester ant colonies. *Psyche*, 92(1): 145–150.
——— 1986. Revised respiration rates for the southern harvester ant, *Pogonomyrmex badius*. *Comparative Biochemistry and Physiology*, ser. A, 83(1): 197–198.
Porter, S. D., and D. A. Eastmond. 1982. *Euryopis coki* (Theridiidae), a spider that preys on *Pogonomyrmex* ants. *Journal of Arachnology*, 10(3): 275–277.
Porter, S. D., and C. D. Jorgensen. 1981. Foragers of the harvester ant, *Pogonomyrmex owyheei*: a disposable caste? *Behavioral Ecology and Sociobiology*, 9(4): 247–256.
——— 1988. Longevity of harvester ant colonies in southern Idaho. *Journal of Range Management*, 41(2): 104–107.
Porter, S. D., and W. R. Tschinkel. 1985. Fire ant polymorphism: the ergonomics of brood production. *Behavioral Ecology and Sociobiology*, 16(4): 323–336.
——— 1986. Adaptive value of nanitic workers in newly founded red imported fire ant colonies (Hymenoptera: Formicidae). *Annals of the Entomological Society of America*, 79(4): 723–726.
——— 1987. Foraging in *Solenopsis invicta* (Hymenoptera: Formicidae): effects of weather and season. *Environmental Entomology*, 16(3): 802–808.
Portier, P., and M. Duval. 1929. Recherches sur la teneur en gaz carbonique de l'atmosphère interne des fourmilières. *Comptes Rendus des séances de la Société de Biologie, Paris*, 102(35): 906–908.
Post, D. C., and R. L. Jeanne. 1982. Recognition of former nestmates during

colony founding by the social wasp *Polistes fuscatus* (Hymenoptera: Vespidae). *Behavioral Ecology and Sociobiology*, 11(4): 283–285.

Prescott, H. W. 1973. Longevity of *Lasius flavus* (F.) (Hym., Formicidae): a sequel. *Entomologist's Monthly Magazine*, 109: 124.

Pricer, J. L. 1908. The life history of the carpenter ant. *Biological Bulletin of the Marine Biological Laboratory, Woods Hole*, 14(3): 177–218.

Principi, M. M. 1946. Contributi allo studio dei Neurotteri italiani, IV: *Nothochrysa italica* Rossi. *Bollettino dell'Istituto di Entomologia della Università degli Studi di Bologna*, 15: 85–102.

Princis, K. 1960. Zur Systematik der Blattarien. *EOS* (Revista Española de Entomología), 36(4): 427–449.

Prins, A. J. 1983. A new ant genus from Southern Africa (Hymenoptera, Formicidae). *Annals of the South African Museum*, 94(1): 1–11.

Provost, E. 1979. Etude de la fermeture de la société de fourmis chez diverses espèces de *Leptothorax* et chez *Camponotus lateralis* (Hyménoptères: Formicidae). *Comptes Rendus de l'Académie des Sciences, Paris*, ser. D, 288(4): 429–432.

——— 1985. Etude de la fermeture de la société chez les fourmis, I: Analyse des interactions entre ouvrières de sociétés différentes, lors de rencontres expérimentales, chez des fourmis du genre *Leptothorax* et chez *Camponotus lateralis* Ol. *Insectes Sociaux*, 32(4): 445–462.

——— 1987. Role of the queen in the intra-colonial aggressivity and nestmate recognition in *Leptothorax lichtensteini* ants. *In* J. Eder and H. Rembold, eds., *Chemistry and biology of social insects* (Proceedings of the Tenth International Congress of the International Union for the Study of Social Insects, Munich, 1986), p. 479. Verlag J. Peperny, Munich.

Pudlo, R. J., A. J. Beattie, and D. C. Culver. 1980. Population consequences of changes in an ant-seed mutualism in *Sanguinaria canadensis*. *Oecologia*, 46(1): 32–37.

Pyke, G. H., H. R. Pulliam, and E. L. Charnov. 1977. Optimal foraging: a selective review of theory and tests. *Quarterly Review of Biology*, 52(2): 137–154.

Quennedey, A. 1975. La guerre chimique chez les termites. *La Recherche*, 6(54): 274–276.

——— 1984. Morphology and ultrastructure of termite defense glands. *In* H. R. Hermann, ed., *Defensive mechanisms in social insects*, pp. 151–200. Praeger, New York.

Quilico, A., F. Piozzi, and M. Pavan. 1956. Sulla dendrolasina. *La Ricerca Scientifica*, 26(1): 177–180.

——— 1957. The structure of dendrolasin. *Tetrahedron*, 1(3): 177–185.

Quilico, A., P. Grünanger, and M. Pavan. 1960. Sul componente odoroso del veleno del formicide *Myrmicaria natalensis* Fred. [Smith]. *In* M. Pavan and T. Eisner, eds., *Verhandlungen*, vol. 3, *Symposium 3: Chemie der Insekten und Symposium 4: Chemische verteidigungsmechanismen bei arthropoden*, pp. 66–68. XI Internationaler Kongress für Entomologie, Wien 1960, c/o Istituto di Entomologia Agraria dell'Università di Pavia, Italy.

Quinlan, R. J., and J. M. Cherrett. 1979. The role of fungus in the diet of the leaf-cutting ant *Atta cephalotes* (L.). *Ecological Entomology*, 4(2): 151–160.

Radchenko, A. G. 1985. Ants of the genus *Strongylognathus* (Hymenoptera, Formicidae) in the European part of the USSR. *Zoologicheskii Zhurnal*, 64(10): 1514–23. [In Russian with English summary.]

Raffy, A. 1929. L'atmosphère interne des fourmilières contient-elle de l'oxyde de carbone? *Comptes Rendus des séances de la Société Biologie, Paris*, 102(35): 908–909.

Raignier, A. 1948. L'économie thermique d'une colonie polycalique de la fourmi des bois *Formica rufa polyctena* Foerst. (Hyménoptères, Formicides). *La Cellule* (Recueil de cytologie et d'histologie), 51(3): 279–368.

——— 1972. Sur l'origine des nouvelles sociétés des fourmis voyageuses Africaines (Hyménoptères Formicidae, Dorylinae). *Insectes Sociaux*, 19(2): 153–170.

Raignier, A., and J. van Boven. 1955. Etude taxonomique, biologique et biométrique des *Dorylus* du sous-genre *Anomma* (Hymenoptera: Formicidae). *Annales du Musée Royal du Congo Belge*, n.s. 4° (sciences zoologiques), 2: 1–359.

Raignier, A., J. van Boven, and R. Ceusters. 1974. Der Polymorphismus der afrikanischen Wanderameisen unter biometrischen und biologischen Gesichtspunkten. *In* G. H. Schmidt, ed., *Sozialpolymorphismus bei Insekten*, pp. 668–693. Wissenschaftliche Verlagsgesellschaft MBH, Stuttgart.

Ranta, E., K. Vepsäläinen, S. Ås, Y. Haila, B. Pisarski, and J. Tiainen. 1983. Island biogeography of ants (Hymenoptera, Formicidae) in four Fennoscandian archipelagoes. *Acta Entomologica Fennica*, 42: 64.

Ratcliffe, B. C. 1976. Notes on the biology of *Euphoriaspis hirtipes* (Horn) and descriptions of the larva and pupa (Coleoptera: Scarabaeidae). *Coleopterists Bulletin*, 30(3): 217–225.

Rathmayer, W. 1978. Venoms of Sphecidae, Pompilidae, Mutilidae, and Bethylidae. *In* S. Bettini, ed., *Handbook of experimental pharmacology*, vol. 48: *Arthropod venoms*, pp. 661–690. Springer-Verlag, Heidelberg.

Ray, T. S., and C. C. Andrews. 1980. Antbutterflies: butterflies that follow army ants to feed on antbird droppings. *Science*, 210: 1147–48.

Rayner, A. D. M., and N. R. Franks. 1987. Evolutionary and ecological parallels between ants and fungi. *Trends in Ecology and Evolution*, 2(5): 127–133.

Réaumur, R. A. F. de. 1926. *The natural history of ants* (From an unpublished manuscript in the Archives of the Academy of Sciences of Paris, written sometime between October 1742 and January 1743; trans. W. M. Wheeler, with annotations). A. A. Knopf, New York. xvii + 280 pp.

Redford, K. H. 1987. Ants and termites as food: patterns of mammalian myrmecophagy. *In* H. H. Genoways, ed., *Current mammalogy*, vol. 1, pp. 349–399. Plenum Press, New York.

Regnier, F. E. 1971. Semiochemicals—structure and function. *Biology of Reproduction*, 4(3): 309–326.

Regnier, F. E., and E. O. Wilson. 1968. The alarm-defence system of the ant *Acanthomyops claviger*. *Journal of Insect Physiology*, 14(7): 955–970.

——— 1969. The alarm-defence system of the ant *Lasius alienus*. *Journal of Insect Physiology*, 15(5): 893–898.

——— 1971. Chemical communication and "propaganda" in slave-maker ants. *Science*, 172: 267–269.

Regnier, F. E., M. Nieh, and B. Hölldobler. 1973. The volatile Dufour's gland components of the harvester ants *Pogonomyrmex rugosus* and *P. barbatus*. *Journal of Insect Physiology*, 19(5): 981–992.

Reichenbach, H. 1902. Ueber Parthenogenese bei Ameisen und andere Beobachtungen an Ameisenkolonien in künstlichen Nestern. *Biologisches Zentralblatt*, 22(14–15): 461–465.

Reichensperger, A. 1915. Myrmekophilen und Termitophilen aus Natal und Zululand gesammelt von Dr. I. Trägårdh. *Meddelanden från Göteborgs Musei Zoologiska Afdelning*, 5: 1–19.

——— 1922. Neue afrikanische Paussiden und Termitophilen (Pauss., Staphyl., Endomych.) (Col.). *Entomologische Mitteilungen*, 11(1): 22–35; (2): 76–83.

——— 1924. Neue südamerikanische Histeriden als Gäste von Wanderameisen und Termiten, II. *Revue Suisse de Zoologie*, 31(4): 117–152.

——— 1939. Beiträge zur Kenntnis der Myrmecophilen- und Termitophilenfauna Brasiliens und Costa Ricas, VI (Col. Hist. Staph.). *Revista de Entomologia*, 10(1): 97–137.

——— 1954. Paussiden-Studien, II. *Zoologische Jahrbücher, Abteilung für Systematik, Ökologie und Geographie der Tiere*, 83(1–2): 111–128.

——— 1958. Coleoptera Paussidae. *South African Animal Life*, 5: 456–463.

Reichle, F. 1943. Untersuchen über Frequenzrhythmen bei Ameisen. *Zeitschrift für Vergleichende Physiologie*, 30: 227–251.

Reiskind, J. 1965. A revision of the ant tribe Cardiocondylini (Hymenoptera, Formicidae), I: The genus *Prosopidris* Wheeler. *Psyche*, 72(1): 79–86.

——— 1977. Ant-mimicry in Panamanian clubionid and salticid spiders (Araneae: Clubionidae, Salticidae). *Biotropica*, 9(1): 1–8.

Reitter, E. 1889. Zwei neue Coleopteren-Gattungen aus Transkaukasien. *Wiener Entomologische Zeitung*, 8(9): 289–292.

Rembold, H. 1964. Die Kastenentstehung bei der Honigbiene, *Apis mellifica* L. *Naturwissenschaften*, 51(3): 49–54.

Rettenmeyer, C. W. 1960. Behavior, abundance and host specificity of mites found on Neotropical army ants (Acarina; Formicidae: Dorylinae). *Proceedings of the Eleventh International Congress of Entomology* (Vienna, 1960), 1: 610–612.

——— 1961. Arthropods associated with Neotropical army ants with a review of the behavior of these ants (Arthropoda: Formicidae: Dorylinae). Ph.D. diss., University of Kansas, Lawrence. xv + 605 pp.

——— 1962a. Notes on host specificity and behavior of myrmecophilous macrochelid mites. *Journal of the Kansas Entomological Society,* 35(4): 358–360.

——— 1962b. The diversity of arthropods found with Neotropical army ants and observations on the behavior of representative species. *Proceedings of the North Central Branch of the Entomological Society of America,* 17: 14–15.

——— 1962c. The behavior of millipeds found with Neotropical army ants. *Journal of the Kansas Entomological Society,* 35(4): 377–384.

——— 1963a. Behavioral studies of army ants. *University of Kansas Science Bulletin,* 44(9): 281–465.

——— 1963b. The behavior of Thysanura found with army ants. *Annals of the Entomological Society of America,* 56(2): 170–174.

——— 1970. Insect mimicry. *Annual Review of Entomology,* 15: 43–74.

Rettenmeyer, C. W., and R. D. Akre. 1968. Ectosymbiosis between phorid flies and army ants. *Annals of the Entomological Society of America,* 61(5): 1317–26.

Rettenmeyer, C. W., H. Topoff, and J. Mirenda. 1978. Queen retinues of army ants. *Annals of the Entomological Society of America,* 71(4): 519–528.

Rettenmeyer, C. W., R. Chadab-Crepet, M. G. Naumann, and L. Morales. 1983. Comparative foraging by Neotropical army ants. *In* P. Jaisson, ed., *Social insects in the tropics,* vol. 2, pp. 59–73. Université Paris-Nord, Paris.

Reyne, A. 1954. *Hippeococcus* a new genus of Pseudococcidae from Java with peculiar habits. *Zoologische Mededelingen* (Rijksmuseum van Natuurlijke Historie te Leiden), 32(21): 233–257.

Reznikova, J. I. 1975. Non-antagonistic relationships of ants occupying similar ecological niches. *Zoologischesekii Zhurnal,* 54(7): 1020–30. [In Russian with English summary.]

——— 1982. Interspecific communication between ants. *Behaviour,* 80(1–2): 84–95.

Rhoades, W. C., and D. R. Davis. 1967. Effects of meteorological factors on the biology and control of the imported fire ant. *Journal of Economic Entomology,* 60(2): 554–558.

Rice, B., and M. Westoby. 1986. Evidence against the hypothesis that ant-dispersed seeds reach nutrient-enriched microsites. *Ecology,* 67(5): 1270–74.

Richards, O. W. 1968. Sphaerocerid flies associating with doryline ants, collected by Dr. D. H. Kistner. *Transactions of the Royal Entomological Society of London,* 120(7): 183–198.

Richards, O. W., and M. J. Richards. 1951. Observations on the social wasps of South America (Hymenoptera Vespidae). *Transactions of the Royal Entomological Society of London,* 102(1): 1–170.

Rickson, F. R. 1977. Progressive loss of ant-related traits of *Cecropia peltata* on selected Caribbean islands. *American Journal of Botany,* 64(5): 585–592.

——— 1979. Absorption of animal tissue breakdown products into a plant stem—the feeding of a plant by ants. *American Journal of Botany,* 66(1): 87–90.

Ridley, H. N. 1930. *The dispersal of plants throughout the world.* L. Reeve, Ashford, Kent. xx + 744 pp.

Riley, R. G., R. M. Silverstein, and J. C. Moser. 1974a. Biological responses of *Atta texana* to its alarm pheromone and the enantiomer of the pheromone. *Science,* 183: 760–762.

Riley, R. G., R. M. Silverstein, B. Carroll, and R. Carroll. 1974b. Methyl 4-methylpyrrole-2-carboxylate: a volatile trail pheromone from the leaf-cutting ant, *Atta cephalotes. Journal of Insect Physiology,* 20(4): 651–654.

Risch, S. J., and C. R. Carroll. 1986. Effects of seed predation by a tropical ant on competition among weeds. *Ecology,* 67(5): 1319–27.

Risch, S. J., and F. R. Rickson. 1981. Mutualism in which ants must be present before plants produce food bodies. *Nature,* 291: 149–150.

Risch, S., M. McClure, J. Vandermeer, and S. Waltz. 1977. Mutualism between three species of tropical *Piper* (Piperaceae) and their ant inhabitants. *American Midland Naturalist,* 98(2): 433–444.

Rissing, S. W. 1981a. Foraging specializations of individual seed-harvester ants. *Behavioral Ecology and Sociobiology,* 9(2): 149–152.

——— 1981b. Prey preferences in the desert horned lizard: influence of prey foraging method and aggressive behavior. *Ecology,* 62(4): 1031–40.

——— 1982. Foraging velocity of seed-harvester ants, *Veromessor pergandei* (Hymenoptera: Formicidae). *Environmental Entomology,* 11(4): 905–907.

——— 1983. Natural history of the workerless inquiline ant *Pogonomyrmex colei* (Hymenoptera: Formicidae). *Psyche,* 90(3): 321–332.

——— 1984. Replete caste production and allometry of workers in the honey ant, *Myrmecocystus mexicanus* Wesmael (Hymenoptera: Formicidae). *Journal of the Kansas Entomological Society,* 57(2): 347–350.

——— 1986. Indirect effects of granivory by harvester ants: plant species composition and reproductive increase near ant nests. *Oecologia,* 68(2): 231–234.

——— 1987. Annual cycles in worker size of the seed-harvester ant *Veromessor pergandei* (Hymenoptera: Formicidae). *Behavioral Ecology and Sociobiology,* 20(2): 117–124.

Rissing, S. W., and G. B. Pollock. 1984. Worker size variability and foraging efficiency in *Veromessor pergandei* (Hymenoptera: Formicidae). *Behavioral Ecology and Sociobiology,* 15(2): 121–126.

——— 1986. Social interaction among pleometrotic queens of *Veromessor pergandei* (Hymenoptera: Formicidae) during colony foundation. *Animal Behaviour,* 34(1): 226–233.

——— 1987. Queen aggression, pleometrotic advantage and brood raiding in the ant *Veromessor pergandei* (Hymenoptera: Formicidae). *Animal Behaviour,* 35(4): 975–981.

Rissing, S. W., and J. Wheeler. 1976. Foraging responses of *Veromessor pergandei* to changes in seed production (Hymenoptera: Formicidae). *Pan-Pacific Entomologist,* 52(1): 63–72.

Rissing, S. W., R. A. Johnson, and G. B. Pollock. 1986. Natal nest distribution and pleometrosis in the desert leaf-cutter ant *Acromyrmex versicolor* (Pergande) (Hymenoptera: Formicidae). *Psyche,* 93(3–4): 177–186.

Ritcher, P. O. 1958. Biology of Scarabaeidae. *Annual Review of Entomology,* 3: 311–334.

Ritter, F. J., I. E. M. Brüggemann-Rotgans, P. E. J. Verwiel, C. J. Persoons, and E. Talman. 1977a. Trail pheromone of the Pharaoh's ant, *Monomorium pharaonis:* isolation and identification of faranal, a terpenoid related to juvenile hormone II. *Tetrahedron Letters,* no. 30, pp. 2617–18.

Ritter, F. J., I. E. M. Brüggemann-Rotgans, C. J. Persoons, E. Talman, A. M. van Osten, and P. E. J. Verwiel. 1977b. Evaluation of social insect pheromones in pest control with special reference to subterranean termites and Pharaoh's ants. *In* N. R. McFarlane, ed., *Crop protection agents: their biological evaluation,* pp. 195–216. Academic Press, London. [Cited by Parry and Morgan, 1979.]

Robbins, R. K., and A. Aiello. 1982. Foodplant and oviposition records for Panamanian Lycaenidae and Riodinidae. *Journal of the Lepidopterists' Society,* 36(2): 65–75.

Roberts, J. T., and E. R. Heithaus. 1986. Ants rearrange the vertebrate-generated seed shadow of a Neotropical fig tree. *Ecology,* 67(4): 1046–51.

Robertson, P. L. 1971. Pheromones involved in aggressive behaviour in the ant, *Myrmecia gulosa. Journal of Insect Physiology,* 17(4): 691–715.

Robertson, P. L., M. L. Dudzinski, and C. J. Orton. 1980. Exocrine gland involvement in trailing behaviour in the Argentine ant (Formicidae: Dolichoderinae). *Animal Behaviour,* 28(4): 1255–73.

Robinson, G. E., and R. E. Page. 1988. Genetic determination of guarding and undertaking in honey-bee colonies. *Nature,* 333: 356–358.

Robinson, S. W., and J. M. Cherrett. 1974. Laboratory investigations to evaluate the possible use of brood pheromones of the leaf-cutting ant *Atta cephalotes* (L.) (Formicidae, Attini) as a component in an attractive bait. *Bulletin of Entomological Research,* 63(3): 519–529.

Rockwood, L. L. 1973. Distribution, density, and dispersion of two species of *Atta* (Hymenoptera: Formicidae) in Guanacaste province, Costa Rica. *Journal of Animal Ecology*, 42(3): 803–817.

——— 1976. Plant selection and foraging patterns in two species of leaf-cutting ants (*Atta*). *Ecology*, 57(1): 48–61.

Roepke, W. 1925. Eine neue myrmekophile Tineïde aus Java: *Hypophrictoides dolichoderella* n. g. n. sp. *Tijdschrift voor Entomologie*, 68: 175–194.

——— 1930. Ueber einen merkwürdigen Fall von "Myrmekophilie," bei einer Ameise (*Cladomyrma* sp.?) auf Sumatra beobachtet. *Miscellanea Zoologica Sumatrana*, 45: 1–3.

Roger, J. 1862. Beiträge zur Kenntnis der Ameisen-Fauna der Mittelmeerländer. *Berliner Entomologische Zeitschrift*, 6(1–2): 255–262.

Rogers, L. E. 1974. Foraging activity of the western harvester ant in the shortgrass plains ecosystem. *Environmental Entomology*, 3(3): 420–424.

Rogers, L. E., R. Lavigne, and J. L. Miller. 1972. Bioenergetics of the western harvester ant in the shortgrass plains ecosystem. *Environmental Entomology*, 1(6): 763–768.

Rogerson, C. T. 1970. The hypocrealean fungi (Ascomycetes, Hypocreales). *Mycologia*, 62(5): 865–910.

Room, P. M. 1971. The relative distributions of ant species in Ghana's cocoa farms. *Journal of Animal Ecology*, 40(3): 735–751.

Roonwal, M. L. 1954. On the structure and population of the nest of the common tree ant, *Crematogaster dohrni rogenhoferi* Mayr (Hymenoptera, Formicidae). *Journal of the Bombay Natural History Society*, 52(2–3): 354–364.

Röseler, P.-F. 1974. Grössenpolymorphismus, Geschlechtsregulation und Stabilisierung der Kasten im Hummelvolk. *In* G. H. Schmidt, ed., *Sozialpolymorphismus bei Insekten*, pp. 298–335. Wissenschaftliche Verlagsgesellschaft MBH, Stuttgart.

Röseler, P.-F., I. Röseler, A. Strambi, and R. Augier. 1984. Influence of insect hormones on the establishment of dominance hierarchies among foundresses of the paper wasp, *Polistes gallicus*. *Behavioral Ecology and Sociobiology*, 15(2): 133–142.

Rosengren, R. 1971. Route fidelity, visual memory and recruitment behaviour in foraging wood ants of the genus *Formica* (Hymenoptera, Formicidae). *Acta Zoologica Fennica*, no. 133, 106 pp.

——— 1977a. Foraging strategy of wood ants (*Formica rufa* group), I: Age polyethism and topographic traditions. *Acta Zoologica Fennica*, no. 149, 30 pp.

——— 1977b. Foraging strategy of wood ants (*Formica rufa* group), II: Nocturnal orientation and diel periodicity. *Acta Zoologica Fennica*, no. 150, 30 pp.

——— 1986. Competition and coexistence in an insular ant community—a manipulation experiment (Hymenoptera: Formicidae). *Annales Zoologici Fennici*, 23(3): 297–302.

——— 1987. Polyethic structure of the foraging/guarding system of red wood ants (*Formica* s. str.). *In* J. Eder and H. Rembold, eds., *Chemistry and biology of social insects* (Proceedings of the Tenth International Congress of the International Union for the Study of Social Insects, Munich, 1986), pp. 118–119. Verlag J. Peperny, Munich.

Rosengren, R., and D. Cherix. 1981. The pupa-carrying test as a taxonomic tool in the *Formica rufa* group. *In* P. Howse and J. Clément, eds., *Biosystematics of social insects* (Systematics Association special vol. 19), pp. 263–281. Academic Press, New York.

Rosengren, R., and W. Fortelius. 1986a. Light:dark-induced activity rhythms in *Formica* ants (Hymenoptera: Formicidae). *Entomologia Generalis*, 11(3–4): 221–228.

——— 1986b. Ortstreue in foraging ants of the *Formica rufa* group—hierarchy of orienting cues and long-term memory. *Insectes Sociaux*, 33(3): 306–337.

Rosengren, R., and P. Pamilo. 1983. The evolution of polygyny and polydomy in mound-building *Formica* ants. *Acta Entomologica Fennica*, 42: 65–77.

——— 1986. Sex ratio strategy as related to queen number, dispersal behaviour and habitat quality in Formica ants (Hymenoptera: Formicidae). *Entomologia Generalis*, 11(3–4): 139–151.

Rosengren, R., D. Cherix, and P. Pamilo. 1985. Insular ecology of the red wood ant *Formica truncorum* Fabr., I: Polydomous nesting, population size and foraging. *Mitteilungen der Schweizerischen Entomologischen Gesellschaft*, 58(1–2): 147–175.

——— 1986a. Insular ecology of the red wood ant *Formica truncorum* Fabr., II: Distribution, reproductive strategy and competition. *Mitteilungen der Schweizerischen Entomologischen Gesellschaft*, 59(1–2): 63–94.

Rosengren, R., W. Fortelius, K. Lindström, and A. Luther. 1986b. Phenology and causation of nest heating and thermoregulation in red wood ants of the *Formica rufa* group studied in coniferous forest habitats in southern Finland. *Annales Zoologici Fennici*, 24: 147–155.

Ross, G. N. 1966. Life-history studies on Mexican butterflies, IV: The ecology and ethology of *Anatole rossi*, a myrmecophilous metalmark (Lepidoptera: Riodinidae). *Annals of the Entomological Society of America*, 59(5): 985–1004.

Ross, K. G. 1988. Differential reproduction in multiple-queen colonies of the fire ant *Solenopsis invicta* (Hymenoptera: Formicidae). *Behavioral Ecology and Sociobiology*, 23(6): 341–355.

Ross, K. G., and D. J. C. Fletcher. 1985a. Comparative study of genetic and social structure in two forms of the fire ant *Solenopsis invicta* (Hymenoptera: Formicidae). *Behavioral Ecology and Sociobiology*, 17(4): 349–356.

——— 1985b. Genetic origin of male diploidy in the fire ant, *Solenopsis invicta* (Hymenoptera: Formicidae), and its evolutionary significance. *Evolution*, 39(4): 888–903.

Ross, N. M., and G. J. Gamboa. 1981. Nestmate discrimination in social wasps (*Polistes metricus*, Hymenoptera: Vespidae). *Behavioral Ecology and Sociobiology*, 9(3): 163–165.

Rossel, S., and R. Wehner. 1982. The bee's map of the e-vector pattern in the sky. *Proceedings of the National Academy of Sciences of the United States of America*, 79(14): 4451–55.

——— 1984. How bees analyse the polarization patterns in the sky: experiments and model. *Journal of Comparative Physiology*, ser. A, 154(5): 607–615.

Roth, L. M., and E. R. Willis. 1960. The biotic associations of cockroaches. *Smithsonian Miscellaneous Collections*, vol. 141 (whole volume), vi + 470 pp.

Ruzsky, M. 1902. Neue Ameisen aus Russland. *Zoologische Jahrbücher, Abteilung für Systematik, Geographie und Biologie der Thiere*, 17(3): 469–484.

Ryti, R. T., and T. J. Case. 1984. Spatial arrangement and diet overlap between colonies of desert ants. *Oecologia*, 62(3): 401–404.

——— 1986. Overdispersion of ant colonies: a test of hypotheses. *Oecologia*, 69(3): 446–453.

Sabrosky, C. W. 1959. A revision of the genus *Pholeomyia* in North America (Diptera, Milichiidae). *Annals of the Entomological Society of America*, 52(3): 316–331.

Sakagami, S. F., and K. Hayashida. 1962. Work efficiency in heterospecific ant groups composed of hosts and their labour parasites. *Animal Behaviour*, 10(1–2): 96–104.

Salt, G. 1929. A contribution to the ethology of the Meliponinae. *Transactions of the Entomological Society of London*, 77(2): 431–470.

Samsinak, K. 1960. Ueber einige myrmekophile Milben aus der Familie Acaridae. *Časopis Československé Společnosti Entomologické*, 57(2): 185–192.

Samson, P. R., and C. F. O'Brien. 1981. Predation on *Ogyris genoveva* (Lepidoptera: Lycaenidae) by meat ants. *Australian Entomological Magazine*, 8(2–3): 21.

Samson, R. A., H. C. Evans, and E. S. Hoekstra. 1982. Notes on entomogenous fungi from Ghana, VI: The genus *Cordyceps*. *Proceedings of the Koninklijke Nederlandse Akademie van Wetenschappen*, ser. C, 85(4): 589–605.

Sanders, C. J. 1964. The biology of carpenter ants in New Brunswick. *Canadian Entomologist*, 96(6): 894–909.

——— 1970. The distribution of carpenter ant colonies in the spruce-fir forests of northwestern Ontario. *Ecology*, 51(5): 865–873.

Santschi, F. 1906. A propos des moeurs parasitiques temporaires des fourmis du genre *Bothriomyrmex*. *Annales de la Société Entomologique de France*, 75(3): 363–392.

——— 1911. Observations et remarques critiques sur le mécanisme de l'orientation chez les fourmis. *Revue Suisse de Zoologie,* 19(13): 303–338.

——— 1914. Fourmis du Natal et du Zoulouland récoltées par le Dr. I. Trägårdh. *Meddelanden från Göteborgs Musei Zoologiska Afdelning,* 3: 1–47.

——— 1917. Fourmis nouvelles de la Colonie du Cap, du Natal et de Rhodesia. *Annales de la Société Entomologique de France* (1916), 85(3): 279–296.

——— 1920. Fourmis du genre "Bothriomyrmex" Emery (systématique et moeurs). *Revue Zoologique Africaine,* 7(3): 201–224.

——— 1923. L'orientation sidérale des fourmis, et quelques considérations sur leurs différentes possibilités d'orientation. *Mémoires de la Société Vaudoise des Sciences Naturelles,* 1(4): 137–176.

——— 1930. Un nouveau genre de fourmi parasite sans ouvrières de l'Argentine. *Revista de la Sociedad Entomológica Argentina,* 3(2): 81–83.

——— 1931. Fourmis de Cuba et de Panama. *Revista de Entomologia,* 1(3): 265–282.

——— 1932. Quelques fourmis inédités de l'Amérique centrale et Cuba. *Revista de Entomologia,* 2(4): 410–414.

——— 1935. Fourmis du Musée du Congo Belge. *Revue de Zoologie et de Botanique Africaines,* 27(2): 254–285.

——— 1936. Fourmis nouvelles ou intéressantes de la République Argentine. *Revista de Entomologia,* 6(3–4): 402–421.

Savage, J. M., ed. 1982. *Ecological aspects of development in the humid tropics.* National Academy Press, Washington, D.C. ix + 297 pp.

Savage, T. S. 1847. On the habits of the "drivers" or visiting ants of West Africa. *Transactions of the Entomological Society of London,* 5(1): 1–15.

Savis, P. 1819. Osservazioni sopra *Blatta acervorum* di Panzer, *Gryllus myrmecophilus* nobis. *Bibliotheka Italiana,* 25(44): 3217–28.

Savolainen, R., and K. Vepsäläinen. 1988. A competition hierarchy among boreal ants: impact on resource partitioning and community structure. *Oikos,* 51(2): 135–155.

Schade, F. H. 1973. The ecology and control of the leaf-cutting ants of Paraguay. *In* J. R. Gorham, ed., *Paraguay ecological essays,* pp. 77–95. Academy of the Arts and Sciences of the Americas, Miami.

Schedl, W. 1975. Zur Kenntnis der Eidonomie und Verbreitung von *Eremoxemus chan* Semenow, 1892 (Insecta: Coleoptera, Brenthidae). *Berichte des Naturwissenschaftlich-Medizinischen Verains Innsbruck,* 62: 83–88. [Cited in Kistner, 1982.]

Scherba, G. 1958. Reproduction, nest orientation and population structure of an aggregation of mound nests of *Formica ulkei* Emery ("Formicidae"). *Insectes Sociaux,* 5(2): 201–213.

——— 1959. Moisture regulation in mound nests of the ant, *Formica ulkei* Emery. *American Midland Naturalist,* 61(2): 499–508.

——— 1961. Nest structure and reproduction in the mound-building ant *Formica opaciventris* Emery in Wyoming. *Journal of the New York Entomological Society,* 69(2): 71–87.

——— 1962. Mound temperatures of the ant *Formica ulkei* Emery. *American Midland Naturalist,* 67(2): 373–385.

——— 1964. Species replacement as a factor affecting distribution of *Formica opaciventris* Emery (Hymenoptera: Formicidae). *Journal of the New York Entomological Society,* 72(4): 231–237.

——— 1965. Observations on *Microtus* nesting in ant mounds. *Psyche,* 72(2): 127–132.

Schildknecht, H., and K. Koob. 1970. Plant bioregulators in the metathoracic glands of myrmicine ants. *Angewandte Chemie* (International Edition), 9(2): 173.

——— 1971. Myrmicacin, the first insect herbicide. *Angewandte Chemie* (International Edition), 10(2): 124–125.

Schilliger, E., and C. Baroni Urbani. 1985. Morphologie de l'organe de stridulation et sonogrammes comparés chez les ouvrières de deux espèces de fourmis moissonneuses du genre *Messor* (Hymenoptera, Formicidae). *Bulletin de la Société Vaudoise des Sciences Naturelles,* 77(4): 377–383.

Schimmer, F. 1909. Beitrag zu einer Monographie der Gryllodeengattung *Myrmecophila* Latr. *Zeitschrift für Wissenschaftliche Zoologie,* 93(3): 409–534.

Schmid-Hempel, P. 1982. Foraging ecology and colony structure of two sympatric species of desert ants—*Cataglyphis bicolor* and *Cataglyphis albicans.* Doctoral diss., University of Zürich.

——— 1984. Individually different foraging methods in the desert ant *Cataglyphis bicolor* (Hymenoptera, Formicidae). *Behavioral Ecology and Sociobiology,* 14(4): 263–271.

——— 1987. Foraging characteristics of the desert ant *Cataglyphis. In* J. M. Pasteels and J.-L. Deneubourg, eds., *From individual to collective behavior in social insects* (*Experientia* Supplement, vol. 54), pp. 43–61. Birkhäuser Verlag, Basel.

——— 1990. Reproductive competition and the evolution of work load in social insects. *American Naturalist,* 135(4): 501–526.

Schmid-Hempel, P., and R. Schmid-Hempel. 1984. Life duration and turnover of foragers in the ant *Cataglyphis bicolor* (Hymenoptera, Formicidae). *Insectes Sociaux,* 31(4): 345–360.

Schmidt, G. H. 1964. Aktivitätsphasen bekannter Hormondrüsen während der Metamorphose von *Formica polyctena* Foerst. (Hym. Ins.). *Insectes Sociaux,* 11(1): 41–57.

——— 1974a. Steuerung der Kastenbildung und Geschlechtsregulation im Waldameisenstaat. *In* G. H. Schmidt, ed., *Sozialpolymorphismus bei Insekten,* pp. 404–512. Wissenschaftliche Verlagsgesellschaft MBH, Stuttgart.

Schmidt, G. H., ed. 1974b. *Sozialpolymorphismus bei Insekten: Probleme der Kastenbildung im Tierreich.* Wissenschaftliche Verlagsgesellschaft MBH, Stuttgart. xxiv + 974 pp.

Schmidt, G. H. 1982. Egg dimorphism and male production in *Formica polyctena* Foerster. *In* M. D. Breed, C. D. Michener, and H. E. Evans, eds., *The biology of social insects* (Proceedings of the Ninth Congress of the International Union for the Study of Social Insects, Boulder, Colorado, 1982), pp. 243–247. Westview Press, Boulder.

Schmidt, G. H., and E. Gürsch. 1970. Zur Struktur des Spinnorgans einiger Ameisenlarven (Hymenoptera, Formicidae). *Zeitschrift für Morphologie und Ökologie der Tiere,* 67(2): 172–182.

——— 1971. Analyse der Spinnbewegungen der Larve von *Formica pratensis* Retz. (Form. Hym. Ins.). *Zeitschrift für Tierpsychologie,* 28(1): 19–32.

Schmidt, G. H., and I. Winkler. 1984. Über die makromolekulare Basis der unterschiedlichen Prädisposition von Waldameiseneiern. *In* F. G. Barth, ed., *Verhandlungen der Deutschen Zoologischen Gesellschaft* (Gießen, 1984), vol. 77, p. 153. Gustav Fischer Verlag, Stuttgart.

Schmidt, J. O. 1982. Biochemistry of insect venoms. *Annual Review of Entomology,* 27: 339–368.

——— 1986. Chemistry, pharmacology, and chemical ecology of ant venoms. *In* T. Piek, ed., *Venoms of the Hymenoptera,* pp. 425–508. Academic Press, New York.

Schmidt, J. O., and M. S. Blum. 1978. A harvester ant venom: chemistry and pharmacology. *Science,* 200: 1064–66.

Schmitz, H. 1950. Myrmekophile und termitophile Phoriden (Diptera) von S. Patrizi und F. Meneghetii in Afrika gesammelt. *Bolletino dell'Istituto di Entomologia della Università degli Studi di Bologna,* 18: 128–166.

——— 1958. Acht neue und einige bekannte Phoriden aus Angola und dem Belgischen Kongo (Phoridae, Diptera). *Publicações Culturais da Companhia de Diamantes de Angola, Museu do Dundo,* no. 40, pp. 13–61.

Schneider, G., and W. Hohorst. 1971. Wanderung der Metacercarien des Lanzett-Egels in Ameisen. *Naturwissenschaften,* 58(6): 327–328.

Schneirla, T. C. 1933a. Some important features of ant learning. *Zeitschrift für Vergleichende Physiologie,* 19(3): 439–452.

——— 1933b. Studies on army ants in Panama. *Journal of Comparative Psychology,* 15(2): 267–299.

——— 1938. A theory of army-ant behavior based upon the analysis of activities in a representative species. *Journal of Comparative Psychology,* 25(1): 51–90.

——— 1940. Further studies on the army-ant behavior pattern: mass-organization in the swarm-raiders. *Journal of Comparative Psychology,* 29(3): 401–460.

——— 1941. Social organization in insects, as related to individual function. *Psychological Review,* 48(6): 465–486.

——— 1943. The nature of ant learning, II: The intermediate stage of segmental maze adjustment. *Journal of Comparative Psychology*, 35(2): 149–176.
——— 1944. The reproductive functions of the army-ant queen as pacemakers of the group behavior pattern. *Journal of the New York Entomological Society*, 52(2): 153–192.
——— 1946a. Ant learning as a problem in comparative psychology. *In* P. L. Harriman, ed., *Twentieth century psychology*, pp. 276–305. Philosophical Library, New York.
——— 1946b. Problems in the biopsychology of social organization. *Journal of Abnormal and Social Psychology*, 41(4): 385–402.
——— 1949a. Army-ant life and behavior under dry-season conditions, 3: The course of reproduction and colony behavior. *Bulletin of the American Museum of Natural History*, 94(1): 1–81.
——— 1949b. Problems in the environmental adaptation of some New-World species of doryline ants. *Anales del Instituto de Biología, Universidad de Mexico*, 20(1–2): 371–384.
——— 1952. Basic correlations and coordinations in insect societies with special reference to ants. *Colloques Internationaux du Centre National de la Recherche Scientifique* (Paris, 1950), no. 34, pp. 247–269.
——— 1953a. Modifiability in insect behavior. *In* K. D. Roeder, ed., *Insect physiology*, pp. 723–747. John Wiley & Sons, New York.
——— 1953b. The army-ant queen: keystone in a social system. *Bulletin de la Troisième Congrès de l'Union Internationale pour l'Etude des Insectes Sociaux*, 1: 29–41.
——— 1956a. A preliminary survey of colony division and related processes in two species of terrestrial army ants. *Insectes Sociaux*, 3(1): 49–69.
——— 1956b. The army ants. *Report of the Smithsonian Institution for 1955*, pp. 379–406.
——— 1957a. A comparison of species and genera in the ant subfamiy Dorylinae with respect to functional pattern. *Insectes Sociaux*, 4(3): 259–298.
——— 1957b. Theoretical consideration of cyclic processes in doryline ants. *Proceedings of the American Philosophical Society*, 101(1): 106–133.
——— 1958. The behavior and biology of certain Nearctic army ants: last part of the functional season, southeastern Arizona. *Insectes Sociaux*, 5(2): 215–255.
——— 1961. The behavior and biology of certain Nearctic doryline ants—sexual broods and colony division in *Neivamyrmex nigrescens*. *Zeitschrift für Tierpsychologie*, 18(1): 1–32.
——— 1963. The behaviour and biology of certain Nearctic army ants: springtime resurgence of cyclic function—southeastern Arizona. *Animal Behaviour*, 11(4): 583–595.
——— 1965. Cyclic functions in genera of legionary ants (subfamily Dorylinae). *Proceedings of the Twelfth International Congress of Entomology* (London, 1964), pp. 336–338.
——— 1971. *Army ants: a study in social organization*. Edited by H. R. Topoff. W. H. Freeman, San Francisco. xxii + 349 pp.
Schneirla, T. C., and R. Z. Brown. 1950. Army-ant life and behavior under dry-season conditions, 4: Further investigation of cyclic processes in behavioral and reproductive functions. *Bulletin of the American Museum of Natural History*, 95(5): 263–353.
——— 1952. Sexual broods and the production of young queens in two species of army ants. *Zoologica, New York*, 37(1): 5–32.
Schneirla, T. C., and G. Piel. 1948. The army ant. *Scientific American*, 178(6)(June): 16–23.
Schneirla, T. C., and A. Y. Reyes. 1969. Emigrations and related behaviour in two surface-adapted species of the old-world doryline ant, *Aenictus*. *Animal Behaviour*, 17(1): 87–103.
Schneirla, T. C., R. Z. Brown, and F. C. Brown. 1954. The bivouac or temporary nest as an adaptive factor in certain terrestrial species of army ants. *Ecological Monographs*, 24(3): 269–296.
Schneirla, T. C., R. R. Gianutsos, and B. S. Pasternack. 1968. Comparative allometry in the larval broods of three army-ant genera, and differential growth as related to colony behavior. *American Naturalist*, 102: 533–554.
Schoener, T. W. 1971. Theory of feeding strategies. *Annual Review of Ecology and Systematics*, 2: 369–404.
——— 1979. Generality of the size-distance relation in models of optimal feeding. *American Naturalist*, 114(6): 902–914.
——— 1983. Field experiments on interspecific competition. *American Naturalist*, 122(2): 240–285.
——— 1986. Overview: kinds of ecological communities—ecology becomes pluralist. *In* J. Diamond and T. J. Case, eds., *Community ecology*, pp. 467–479. Harper & Row, New York.
Schremmer, F. 1972. Die südamerikanische Weberameise *Camponotus senex* (Freilandaufnahmen). *Encyclopaedia Cinematographica*, film no. W1161. Institut für den Wissenschaftlichen Film, Göttingen.
——— 1978. Zur Bionomie und Morphologie der myrmecophilen Raupe und Puppe der neotropischen Tagfalter-Art *Hamearis erostratus* (Lepidoptera: Riodinidae). *Entomologica Germanica*, 4(2): 113–121.
——— 1979a. Das Nest der neotropischen Weberameise *Camponotus (Myrmobrachys) senex* Smith (Hymenoptera, Formicidae). *Zoologischer Anzeiger*, 203(5–6): 273–282.
——— 1979b. Die nahezu unbekannte neotropische Weberameise *Camponotus (Myrmobrachys) senex* (Hymenoptera: Formicidae). *Entomologia Generalis*, 5(4): 363–378.
——— 1984. Untersuchungen und Beobachtungen zur Ökoethologie der Pflanzenameise *Pseudomyrmex triplarinus*, welche die Ameisenbäume der Gattung *Triplaris* bewohnt. *Zoologische Jahrbücher, Abteilung für Systematik, Ökologie, und Geographie der Tiere*, 111(3): 385–410.
Schroth, M., and U. Maschwitz. 1984. Zur Larvalbiologie und Wirtsfindung von *Maculinea teleius* (Lepidoptera: Lycaenidae), eines Parasiten von *Myrmica laevinodis* (Hymenoptera: Formicidae). *Entomologia Generalis*, 9(4): 225–230.
Schuh, R. T. 1973. The Orthotylinae and Phylinae (Hemiptera: Miridae) of South Africa with a phylogenetic analysis of the ant-mimetic tribes of the two subfamilies for the world. *Entomologica Americana*, 47: 1–332.
Schumacher, A., and W. G. Whitford. 1974. The foraging ecology of two species of Chihuahuan desert ants: *Formica perpilosa* and *Trachymyrmex smithi neomexicanus* (Hymenoptera Formicidae). *Insectes Sociaux*, 21(3): 317–330.
Schupp, E. W. 1986. *Azteca* protection of *Cecropia:* ant occupation benefits juvenile trees. *Oecologia*, 70(3): 379–385.
Schwarz, E. A. 1890. Myrmecophilous Coleoptera found in temperate North America. *Proceedings of the Entomological Society of Washington*, 1(4): 237–247.
Seeley, T. D. 1982. Adaptive significance of the age polyethism schedule in honeybee colonies. *Behavioral Ecology and Sociobiology*, 11(4): 287–293.
——— 1985. *Honeybee ecology: a study of adaptation in social life*. Princeton University Press, Princeton, N.J. x + 201 pp.
Seevers, C. H. 1965. The systematics, evolution and zoogeography of staphylinid beetles associated with army ants (Coleoptera, Staphylinidae). *Fieldiana, Zoology*, 47(2): 137–351.
Seevers, C. H., and H. S. Dybas. 1943. A synopsis of the Limulodidae (Coleoptera): a new family proposed for myrmecophiles of the subfamilies Limulodinae (Ptiliidae) and Cephaloplectinae (Staphylinidae). *Annals of the Entomological Society of America*, 36(3): 546–586.
Seger, J. 1981. Kinship and covariance. *Journal of Theoretical Biology*, 91(1): 191–213.
Seidel, J. L. 1988. The monoterpenes of *Gutierrezia sarothrae:* chemical interactions between ants and plants in Neotropical ant-gardens. Ph.D. diss., University of Utah, Salt Lake City. [Cited by Davidson, 1988.]
Selman, B. J. 1962. Remarkable new chrysomeloids found in the nests of arboreal ants in Tanganyika (Coleoptera: Clytridae and Cryptocephalidae). *Annals and Magazine of Natural History*, ser. 13, 5(53): 295–299.
Sernander, R. 1906. Entwurf einer Monographie der europäischen Myrmekochoren. *Kungliga Svenska Vetenskapsakademiens Handlingar*, 41(7): 1–410.
Sheata, M. N., and A. H. Kaschef. 1971. Foraging activities of *Messor aegyptiacus* Emery (Hym., Formicidae). *Insectes Sociaux*, 18(4): 215–226.
Shepard, M., and F. Gibson. 1972. Spider-ant symbiosis: *Cotinusa* spp. (Araneida: Salticidae) and *Tapinoma melanocephalum* (Hymenoptera: Formicidae). *Canadian Entomologist*, 104(12): 1951–54.

Shepherd, J. D. 1982. Trunk trails and the searching strategy of a leaf-cutter ant, *Atta colombica. Behavioral Ecology and Sociobiology*, 11(2): 77–84.
Sherman, P. W. 1979. Insect chromosome numbers and eusociality. *American Naturalist*, 113(6): 925–935.
Sherman, P. W., and W. G. Holmes. 1985. Kin recognition: issues and evidence. *In* B. Hölldobler and M. Lindauer, eds., *Experimental behavioral ecology and sociobiology* (Fortschritte der Zoologie, no. 31), pp. 437–460. Sinauer Associates, Sunderland, Mass.
Sherman, P. W., T. D. Seeley, and H. K. Reeve. 1988. Parasites, pathogens, and polyandry in social Hymenoptera. *American Naturalist*, 131(4): 602–610.
Silva, M. M. T. G. da. 1972. Contribuição ao estudo da biologia de *Eciton burchelli* Westwood (Hymenoptera: Formicidae). Doctoral diss., University of São Paulo, Brazil.
Silvestri, F. 1903. Contribuzioni alla conoscenza dei Mirmecofili, I: Osservazioni su alcuni mirmecofili dei dintorni di Portici. *Annuario del Museo Zoologie della R. Università di Napoli*, n.s., 1(13): 1–5.
Simberloff, D. S. 1983. Competition theory, hypothesis-testing, and other community ecological buzzwords. *American Naturalist*, 122(5): 626–635.
——— 1984. The great god of competition. *The Sciences*, 24(4)(July/August): 16–22.
Simberloff, D. S., and E. O. Wilson. 1969. Experimental zoogeography of islands: the colonization of empty islands. *Ecology*, 50(2): 278–296.
Simon, H. A. 1981. *The sciences of the artificial*, 2nd ed. MIT Press, Cambridge, Mass. xiv + 247 pp.
Skaife, S. H. 1961. *The study of ants*. Longmans, Green, London. vi + 178 pp.
Skellam, J. G., M. V. Brian, and J. R. Proctor. 1959. The simultaneous growth of interacting systems. *Acta Biotheoretica*, 13(2–3): 131–144.
Skinner, G. J. 1980a. Territory, trail structure and activity patterns in the wood-ant, *Formica rufa* (Hymenoptera: Formicidae) in limestone woodland in north-west England. *Journal of Animal Ecology*, 49: 381–394.
——— 1980b. The feeding habits of the wood-ant, *Formica rufa* (Hymenoptera: Formicidae) in limestone woodland in north-west England. *Journal of Animal Ecology*, 49: 417–433.
Skinner, G. J., and J. B. Whittaker. 1981. An experimental investigation of inter-relationships between the wood-ant (*Formica rufa*) and some tree-canopy herbivores. *Journal of Animal Ecology*, 50(1): 313–326.
Skwarra, E. 1934. Ökologie der Lebensgemeinschaften mexikanischer Ameisenpflanzen. *Zeitschrift für Morphologie und Ökologie der Tiere*, 29(2): 306–373.
Smallwood, J. 1982. Nest relocations in ants. *Insectes Sociaux*, 29(2): 138–147.
Smeeton, L. 1981. The source of males in *Myrmica rubra* L. (Hym. Formicidae). *Insectes Sociaux*, 28(3): 263–278.
Smirnov, V. I. 1966. *Ants as a factor in forest protection against oak leaf roller*. Lesnoe Khozyaistvo, Moscow, no. 2. [Cited by Beattie, 1985.]
Smith, B. H., M. L. Ronsheim, and K. R. Swartz. 1986. Reproductive ecology of *Jeffersonia diphylla* Berberidaceae). *American Journal of Botany*, 73(10): 1416–26.
Smith, D. R. 1979. Superfamily Formicoidea. *In* K. V. Krombein, P. D. Hurd, D. R. Smith, and B. D. Burks, eds., *Catalog of Hymenoptera in America North of Mexico*, vol. 2, pp. 1323–1467. Smithsonian Institution Press, Washington, D.C.
Smith, F. 1941. A note on noctuid larvae found in ant's nests (Lepidoptera; Hymenoptera: Formicidae). *Entomological News*, 52(4): 109.
——— 1944. Nutritional requirements of *Camponotus* ants. *Annals of the Entomological Society of America*, 37(4): 401–408.
Smith, J. B. 1886. Ants' nests and their inhabitants. *American Naturalist*, 20(8): 679–687.
Smith, K. G. V. 1969. Further data on the oviposition by the genus *Stylogaster* Macquart (Diptera: Conopidae, Stylogasterinae) upon adult calyptrate Diptera associated with ants and animal dung. *Proceedings of the Royal Entomological Society, London*, ser. A, 44(1–3): 35–37.
Smith, M. R. 1931. A revision of the genus *Strumigenys* of America, north of Mexico, based on a study of the workers (Hymn.: Formicidae). *Annals of the Entomological Society of America*, 24(4): 686–710.
——— 1934. Ponerine ants of the genus *Euponera* in the United States. *Annals of the Entomological Society of America*, 27(4): 557–564.
——— 1936a. Distribution of the Argentine ant in the United States and suggestions for its control or eradication. *Circular, U. S. Department of Agriculture*, no. 387, 39 pp.
——— 1936b. The ants of Puerto Rico. *University of Puerto Rico, Journal of Agriculture*, 20(4): 819–875.
——— 1942a. A new apparently parasitic ant. *Proceedings of the Entomological Society of Washington*, 44(4): 59–61.
——— 1942b. The legionary ants of the United States belonging to *Eciton* subgenus *Neivamyrmex* Borgmeier. *American Midland Naturalist*, 27(3): 537–590.
——— 1943. A generic and subgeneric synopsis of the male ants of the United States. *American Midland Naturalist*, 30(2): 273–321.
——— 1947a. A generic and subgeneric synopsis of the United States ants, based on the workers (Hymenoptera: Formicidae). *American Midland Naturalist*, 37(3): 521–647.
——— 1947b. A new genus and species of ant from Guatemala (Hymenoptera, Formicidae). *Journal of the New York Entomological Society*, 55(4): 281–284.
——— 1948. A new genus and species of ant from India (Hymenoptera: Formicidae). *Journal of the New York Entomological Society*, 56(4): 205–208.
——— 1949. A new species of *Camponotus*, subg. *Colobopsis* from Mexico (Hymenoptera: Formicidae). *Journal of the New York Entomological Society*, 57(3): 177–181.
——— 1954. Ants of the Bimini Island group, Bahamas, British West Indies (Hymenoptera: Formicidae). *American Museum Novitates*, no. 1671, 16 pp.
——— 1957. Revision of the genus *Stenamma* Westwood in America north of Mexico (Hymenoptera, Formicidae). *American Midland Naturalist*, 57(1): 133–174.
——— 1965. House-infesting ants of the eastern United States. *Technical Bulletin, Agricultural Research Service, U. S. Department of Agriculture*, no. 1326, 105 pp.
Smith, M. R., and M. W. Wing. 1954. Redescription of *Discothyrea testacea* Roger, a little-known North American ant, with notes on the genus (Hymenoptera: Formicidae). *Journal of the New York Entomological Society*, 62(2): 105–112.
Smith, R. H., and M. R. Shaw. 1980. Haplodiploid sex ratios and the mutation rate. *Nature*, 287: 728–729.
Snelling, R. R. 1963. The United States species of fire ants of the genus *Solenopsis*, subgenus *Solenopsis* Westwood with synonymy of *Solenopsis aurea* Wheeler. *Occasional Papers of the Bureau of Entomology, California Department of Agriculture*, no. 3, 11 pp.
——— 1965. Studies on California ants, 1: *Leptothorax hirticornis* Emery, a new host and descriptions of the female and ergatoid male (Hymenoptera: Formicidae). *Bulletin of the Southern California Academy of Science*, 64(1): 16–21.
——— 1973a. Two ant genera new to the United States (Hymenoptera: Formicidae). *Contributions in Science, Natural History Museum of Los Angeles County*, no. 236, 8 pp.
——— 1973b. The ant genus *Conomyrma* in the United States (Hymenoptera: Formicidae). *Contributions in Science, Natural History Museum of Los Angeles County*, no. 238, 6 pp.
——— 1973c. Studies on California ants, 7: The genus *Stenamma* (Hymenoptera: Formicidae). *Contributions in Science, Natural History Museum of Los Angeles County*, no. 245, 38 pp.
——— 1975. Descriptions of new Chilean ant taxa (Hymenoptera: Formicidae). *Contributions in Science, Natural History Museum of Los Angeles County*, no. 274, 19 pp.
——— 1976. A revision of the honey ants, genus *Myrmecocystus* (Hymenoptera: Formicidae). *Science Bulletin, Natural History Museum of Los Angeles County*, no. 24, 163 pp.
——— 1979. *Aphomomyrmex* and a related new genus of arboreal African

ants (Hymenoptera: Formicidae). *Contributions in Science, Natural History Museum of Los Angeles County*, no. 316, 8 pp.

——— 1981. Systematics of social Hymenoptera. *In* H. R. Hermann, ed., *Social insects*, vol. 2, pp. 369–474. Academic Press, New York.

——— 1982. A revision of the honey ants, genus *Myrmecocystus*, first supplement (Hymenoptera: Formicidae). *Bulletin of the Southern California Academy of Sciences*, 81(2): 69–86.

——— 1988. Taxonomic notes on Nearctic species of *Camponotus*, subgenus *Myrmentoma* (Hymenoptera: Formicidae). *In* J. C. Trager, ed., *Advances in myrmecology*, pp. 55–78. E. J. Brill, Leiden.

Snelling, R. R., and J. H. Hunt. 1975. The ants of Chile (Hymenoptera: Formicidae). *Revista Chilena Entomología*, 9: 63–129.

Sokolowski, A., and J. Wiśniewski. 1975. Teratologische Untersuchungen an Ameisen-Arbeiterinnen aus der *Formica rufa*-Gruppe (Hym., Formicidae). *Insectes Sociaux*, 22(2): 117–134.

Solbrig, O. T., and P. D. Cantino. 1975. Reproductive adaptations in *Prosopis* (Leguminosae, Mimosoideae). *Journal of the Arnold Arboretum, Harvard*, 56(2): 185–210.

Sorenson, A. A., R. S. Kamas, and S. B. Vinson. 1983. The influence of oral secretions from larvae on levels of proteinases in colony members of *Solenopsis invicta* Buren (Hymenoptera: Formicidae). *Journal of Insect Physiology*, 29(2): 163–168.

Sorenson, A. A., T. M. Busch, and S. B. Vinson. 1985. Control of food influx by temporal subcastes in the fire ant, *Solenopsis invicta*. *Behavioral Ecology and Sociobiology*, 17(3): 191–198.

Sörensen, U., and G. H. Schmidt. 1987. Vergleichende Untersuchungen zum Beuteeintrag der Waldameisen (Genus: *Formica*, Hymenoptera) in der Bredstedter Geest (Schleswig-Holstein). *Zeitschrift für Angewandte Entomologie*, 103(2): 153–177.

Soulié, J. 1955. Facteurs du milieu agissant sur l'activité des colonies de récolte chez la fourmi *Cremastogaster scutellaris* Ol (Hymenoptera, Formicoidea). *Insectes Sociaux*, 2(2): 173–177.

——— 1960. Des considérations écologiques peuvent-elles apporter une contribution à la connaissance du cycle biologique des colonies de *Cremastogaster* (Hymenoptera—Formicoidea)? *Insectes Sociaux*, 7(3): 283–295.

Southwood, T. R. E. 1966. *Ecological methods: with particular reference to the study of insect populations*. Methuen, London. xviii + 391 pp.

Spangler, H. G. 1967. Ant stridulations and their synchronization with abdominal movement. *Science*, 155: 1687–89.

Spradbery, J. P. 1973. *Wasps: an account of the biology and natural history of solitary and social wasps*. Sidgwick and Jackson, London. xvi + 408 pp.

Stäger, R. 1925. Das Leben der Gastameise (*Formicoxenus nitidulus* Nyl.) in neuer Beleuchtung. *Zeitschrift für Morphologie und Ökologie der Tiere*, 3(2–3): 452–476.

——— 1931. Über das Mitteilungsvermögen der Waldameise beim Auffinden und Transport eines Beutestückes. *Zeitschrift für Wissenschaftliche Insektenbiologie*, 26(4–6): 125–137.

Stahel, G., and D. C. Geijskes. 1939. Ueber den Bau der Nester von *Atta cephalotes* L. und *Atta sexdens* L. (Hym. Formicidae). *Revista de Entomologia*, 10(1): 27–78.

Starr, C. K. 1979. Origin and evolution of insect sociality: a review of modern theory. *In* H. R. Hermann, ed., *Social insects*, vol. 1, pp. 35–79. Academic Press, New York.

——— 1981. Trail-sharing by two species of *Polyrhachis* (Hymenoptera: Formicidae). *Philippine Entomologist*, 5(1): 5–8.

——— 1985. Enabling mechanisms in the origin of sociality in the Hymenoptera—the sting's the thing. *Annals of the Entomological Society of America*, 78(6): 836–840.

Starý, P. 1970. *Biology of aphid parasites (Hymenoptera: Aphidiidae) with respect to integrated control* (Series Entomologica, vol. 6). Dr. W. Junk, The Hague. 643 pp.

Stebaev, I. V., and J. I. Reznikova. 1972. Two interaction types of ants living in steppe ecosystem in South Siberia, USSR. *Ekologia Polska*, 20(11): 103–109.

Steenis, C. G. G. J. van. 1967. Miscellaneous botanical notes XVIII. *Blumea*, 15(1): 144–155.

Steiner, A. 1926. Temperaturmessung in den Nestern der Waldameise (*Formica rufa* var. *rufo-pratensis* For.) und der Wegameise (*Lasius niger* L.) während des Winters. *Mitteilungen der Naturforschenden Gesellschaft in Bern* (1925), pp. 1–12.

——— 1929. Temperaturuntersuchungen in Ameisennestern mit Erdkuppeln, im Nest von *Formica exsecta* Nyl. und in Nestern unter Steinen. *Zeitschrift für Vergleichende Physiologie*, 9(1): 1–66.

Stephens, D. W., and J. R. Krebs. 1986. *Foraging theory*. Princeton University Press, Princeton, N.J. xiv + 247 pp.

Stephens, K. 1968. Notes on additional distribution and ecology of *Euxestus punctatus* LeC. (Coleoptera: Colydiidae). *Coleopterists' Bulletin*, 22(1): 19.

Stevens, P. F. 1975. Review of *Chisocheton* (Meliaceae) in Papuasia. *Contributions from Herbarium Australiense*, no. 11, 55 pp.

Steyn, J. J. 1954. The pugnacious ant (*Anoplolepis custodiens* Smith) and its relation to the control of citrus scales at Letaba. *Memoirs of the Entomological Society of Southern Africa*, no. 3, iii + 96 pp.

Stitz, H. 1939. *Hautflüger oder Hymenoptera, I: Ameisen oder Formicidae* (Die Tierwelt Deutschlands und der angrenzenden Meeresteile, no. 37). Verlag G. Fischer, Jena. 428 pp.

Stout, J. 1979. An association of an ant, a mealy bug, and an understory tree from a Costa Rican rain forest. *Biotropica*, 11(4): 309–311.

Strassmann, J. E. 1983. Nest fidelity and group size among foundresses of *Polistes annularis* (Hymenoptera: Vespidae). *Journal of the Kansas Entomological Society*, 56(4): 621–634.

Strickland, A. H. 1947. Coccids attacking cacao (*Theobroma cacao*, L.), in West Africa, with descriptions of five new species. *Bulletin of Entomological Research*, 38(3): 497–523.

Strokov, V. V. 1956. *Techniques of using fauna for forest protections*. Goslesbumizdat, Moscow. [In Russian; cited by Beattie, 1985.]

Stuart, R. J. 1981. Abdominal trophallaxis in the slave-making ant, *Harpagoxenus americanus* (Hymenoptera: Formicidae). *Psyche*, 88(3–4): 331–334.

——— 1982. Territoriality and the origin of slave raiding in leptothoracine ants. *Science*, 215: 1262–63.

——— 1984. Experiments on colony foundation in the slave-making ant *Harpagoxenus canadensis* M. R. Smith (Hymenoptera; Formicidae). *Canadian Journal of Zoology*, 62(10): 1995–2001.

——— 1985a. Nestmate recognition in leptothoracine ants: exploring the dynamics of a complex phenomenon. Ph.D. diss., University of Toronto. 302 pp.

——— 1985b. Spontaneous polydomy in laboratory colonies of the ant *Leptothorax curvispinosus* Mayr (Hymenoptera; Formicidae). *Psyche*, 92(1): 71–81.

——— 1986. An early record of tandem running in leptothoracine ants: Gottfrid Adlerz, 1896. *Psyche*, 93(1–2): 103–106.

——— 1987a. Individual workers produce colony-specific nestmate recognition cues in the ant, *Leptothorax curvispinosus*. *Animal Behaviour*, 35(4): 1062–69.

——— 1987b. Nestmate recognition in leptothoracine ants: testing Fielde's progressive odor hypothesis. *Ethology*, 76(2): 116–123.

——— 1987c. Transient nestmate recognition cues contribute to a multicolonial population structure in the ant, *Leptothorax curvispinosus*. *Behavioral Ecology and Sociobiology*, 21(4): 229–235.

——— 1988a. Collective cues as a basis for nestmate recognition in polygynous leptothoracine ants. *Proceedings of the National Academy of Sciences of the United States of America*, 85(12): 4572–75.

——— 1988b. Development and evolution in nestmate recognition systems of social insects. *In* G. Greenberg and E. Tobach, eds., *Evolution of social behavior and integrative levels*, pp. 177–195. Lawrence Erlbaum Associates, Hillsdale, N.J.

Stuart, R. J., and T. M. Alloway. 1982. Territoriality and the origin of slave raiding in leptothoracine ants. *Science*, 215: 1262–63.

——— 1983. The slave-making ant, *Harpagoxenus canadensis* M. R. Smith, and its host species, *Leptothorax muscorum* (Nylander): slave raiding and territoriality. *Behaviour,* 85(1–2): 58–90.

——— 1985. Behavioural evolution and domestic degeneration in obligatory slave-making ants (Hymenoptera: Formicidae: Leptothoracini). *Animal Behaviour,* 33(4): 1080–88.

Stuart, R. J., A. Francoeur, and R. Loiselle. 1987a. Fighting males in the ant genus *Cardiocondyla*. *In* J. Eder and H. Rembold, eds., *Chemistry and biology of social insects* (Proceedings of the Tenth International Congress of the International Union for the Study of Social Insects, Munich, 1986), pp. 551–552. Verlag J. Peperny, Munich.

——— 1987b. Lethal fighting among dimorphic males of the ant, *Cardiocondyla wroughtoni*. *Naturwissenschaften,* 74(11): 548–549.

Stumper, R. 1921. Etudes sur les fourmis, III: Recherches sur l'éthologie du *Formicoxenus nitidulus* Nyl. *Bulletin de la Société Entomologique de Belgique,* 3: 90–97.

——— 1949. Etudes myrmécologiques, IX: Nouvelles observations sur l'éthologie de *Formicoxenus nitidulus* Nyl. *Bulletin de la Société des Naturalistes Luxembourgeois,* n.s., 43: 242–248.

——— 1950. Les associations complexes des fourmis: commensalisme, symbiose et parasitisme. *Bulletin Biologique de la France et de la Belgique,* 84(4): 376–399.

——— 1961. Radiobiologische Untersuchungen über den sozialen Nahrungshaushalt der Honigameise *Proformica nasuta* (Nyl). *Naturwissenschaften,* 48(24): 735–736.

——— 1962. Sur un effet de groupe chez les femelles de *Camponotus vagus* (Scopoli). *Insectes Sociaux,* 9(4): 329–333.

Stumper, R., and H. Kutter. 1951. Sur l'éthologie du nouveau myrmécobionte *Epimyrma stumperi* (nov. spec. Kutter). *Comptes Rendus de l'Académie des Sciences, Paris,* 233(17): 983–985.

Sturdza, S. A. 1942. Beobachtungen über die stimulierende Wirkung lebhaft beweglicher Ameisen auf träge Ameisen. *Bulletin de la Section Scientifique de l'Académie Roumaine,* 24: 543–546.

Sturtevant, A. H. 1927. The social parasitism of the ant *Harpagoxenus americanus*. *Psyche,* 34(1): 1–9.

Sudd, J. H. 1957. Communication and recruitment in Pharaoh's ant, *Monomorium pharaonis* (L.). *Animal Behaviour,* 5(3): 104–109.

——— 1960. The transport of prey by an ant, *Pheidole crassinoda* Em. *Behaviour,* 16(3–4): 295–308.

——— 1963. How insects work in groups. *Discovery, London,* 24(6)(June): 15–19.

——— 1965. The transport of prey by ants. *Behaviour,* 25(3–4): 234–271.

——— 1972. The absence of social enhancement of digging in pairs of ants (*Formica lemani* Bondroit). *Animal Behaviour,* 20(4): 813–819.

——— 1983. The distribution of foraging wood-ants (*Formica lugubris* Zett.) in relation to the distribution of aphids. *Insectes Sociaux,* 30(3): 298–307.

Sudd, J. H., and N. R. Franks. 1987. *The behavioural ecology of ants.* Chapman and Hall, New York. x + 206 pp.

Sudd, J. H., J. M. Douglas, T. Gaynard, D. M. Murray, and J. M. Stockdale. 1977. The distribution of wood-ants (*Formica lugubris* Zetterstedt) in a northern English forest. *Ecological Entomology,* 2(4): 301–313.

Summerlin, J. W. 1978. Beetles of the genera *Myrmecaphodius, Rhyssemus,* and *Blapstinus* in Texas fire ant nests. *Southwestern Entomologist,* 3(1): 27–29.

Suzzoni, J.-P., and H. Cagniant. 1975. Etude histologique des voies génitales chez l'ouvrière et la reine de *Cataglyphis cursor* Fonsc. (Hyménoptère Formicidae, Formicinae): arguments en faveur d'une parthénogenèse thélytoque chez cette espèce. *Insectes Sociaux,* 22(1): 83–92.

Suzzoni, J. P., L. Passera, and A. Strambi. 1980. Ecdysteroid titre and caste determination in the ant, *Pheidole pallidula* (Nyl.) (Hymenoptera: Formicidae). *Experientia,* 36(10): 1228–29.

Swain, R. B. 1977. The natural history of *Monacis,* a genus of Neotropical ants (Hymenoptera: Formicidae). Ph.D. diss., Harvard University. 258 pp.

——— 1980. Trophic competition among parabiotic ants. *Insectes Sociaux,* 27(4): 377–390.

Sykes, W. H. 1835. Descriptions of new species of Indian ants. *Transactions of the Entomological Society of London,* 1(2): 99–107.

Szabó-Patay, J. 1928. A kapus-hangya. *Termeszettudományi Közlöny, Budapest,* pp. 215–219.

Szlep, R., and T. Jacobi. 1967. The mechanism of recruitment to mass foraging in colonies of *Monomorium venustum* Smith, *M. subopacum* ssp. *phoenicium* Em., *Tapinoma israelis* For. and *T. simrothi* v. *phoenicium* Em. *Insectes Sociaux,* 14(1): 25–40.

Szlep-Fessel, R. 1970. The regulatory mechanism in mass foraging and recruitment of soldiers of *Pheidole*. *Insectes Sociaux,* 17(4): 233–244.

Taber, S. W. 1986. Karyological and scanning electron microscopic studies of the ant genus *Pogonomyrmex*. M.S. thesis, Texas Technological University, Lubbock. 71 pp.

Taber, S. W., and O. F. Francke. 1986. A bilateral gynandromorph of the western harvester ant, *Pogonomyrmex occidentalis* (Hymenoptera: Formicidae). *Southwestern Naturalist,* 31(2): 274–276.

Tafuri, J. F. 1955. Growth and polymorphism in the larva of the army ant (*Eciton (E.) hamatum* Fabricius). *Journal of the New York Entomological Society,* 63: 21–41.

Takada, H., and Y. Hashimoto. 1985. Association of the root aphid parasitoids *Aclitus sappaphis* and *Paralipsis eikoae* (Hymenoptera: Aphidiidae) with the aphid-attending ants *Pheidole fervida* and *Lasius niger* (Hymenoptera, Formicidae). *Kontyû, Tokyo,* 53(1): 150–160.

Taki, A. 1976. Colony founding of *Messor aciculatum* (Fr. Smith) (Hymenoptera: Formicidae) by single and grouped queens. *Physiology and Ecology Japan,* 17(1–2): 503–512.

——— 1987. The trophic eggs of colony founding ant queens. *In* J. Eder and H. Rembold, eds., *Chemistry and biology of social insects* (Proceedings of the Tenth International Congress of the International Union for the Study of Social Insects, Munich, 1986), p. 268. Verlag J. Peperny, Munich.

Talbot, M. 1934. Distribution of ant species in the Chicago region with reference to ecological factors and physiological toleration. *Ecology,* 15(4): 416–439.

——— 1943a. Population studies of the ant, *Prenolepis imparis* Say. *Ecology,* 24(1): 31–44.

——— 1943b. Response of the ant *Prenolepis imparis* Say to temperature and humidity changes. *Ecology,* 24(3): 345–352.

——— 1945. Population studies of the ant *Myrmica schencki* ssp. *emeryana* Forel. *Annals of the Entomological Society of America,* 38(3): 365–372.

——— 1946. Daily fluctuations in aboveground activity of three species of ants. *Ecology,* 27(1): 65–70.

——— 1948. A comparison of two ants of the genus *Formica*. *Ecology,* 29(3): 316–325.

——— 1951. Populations and hibernating conditions of the ant *Aphaenogaster (Attomyrma) rudis* Emery (Hymenoptera: Formicidae). *Annals of the Entomological Society of America,* 44(3): 302–307.

——— 1954. Populations of the ant *Aphaenogaster (Attomyrma) treatae* Forel on abandoned fields on the Edwin S. George Reserve. *Contributions from the Laboratory of Vertebrate Biology of the University of Michigan,* 69: 1–9.

——— 1957. Population studies of the slave-making ant *Leptothorax duloticus* and its slave, *Leptothorax curvispinosus*. *Ecology,* 38(3): 449–456.

——— 1967. Slave-raids of the ant *Polyergus lucidus* Mayr. *Psyche,* 74(4): 299–313.

——— 1972. Flights and swarms of the ant *Formica obscuripes* Forel. *Journal of the Kansas Entomological Society,* 45(2): 254–258.

——— 1975. A list of the ants (Hymenoptera: Formicidae) of the Edwin S. George Reserve, Livingston County, Michigan. *Great Lakes Entomologist,* 8(4): 245–246.

——— 1976. The natural history of the workerless ant parasite *Formica talbotae*. *Psyche,* 83(3–4): 282–288.

Talbot, M., and C. H. Kennedy. 1940. The slave-making ant, *Formica sanguinea subintegra* Emery, its raids, nuptial flights and nest structure. *Annals of the Entomological Society of America,* 33(3): 560–577.

Tanner, J. E. 1892. *Oecodoma cephalotes,* second paper. *Journal of the Trinidad Field Naturalists' Club,* 1(5): 123–127.

Taylor, F. 1977. Foraging behavior of ants: experiments with two species of myrmicine ants. *Behavioral Ecology and Sociobiology,* 2(2): 147–167.

Taylor, P. D., and A. Sauer. 1980. The selective advantage of sex-ratio homeostasis. *American Naturalist,* 116(2): 305–310.

Taylor, R. W. 1962a. New Australian dacetine ants of the genera *Mesostruma* Brown and *Codiomyrmex* Wheeler (Hymenoptera: Formicidae). *Breviora,* no. 152, 10 pp.

——— 1962b. The ants of the Three Kings Islands. *Records of the Auckland Institute and Museum,* 5(5–6): 251–254.

——— 1965a. A monographic revision of the rare tropicopolitan ant genus *Probolomyrmex* Mayr (Hymenoptera: Formicidae). *Transactions of the Royal Entomological Society of London,* 117(12): 345–365.

——— 1965b. The Australian ants of the genus *Pristomyrmex,* with a case of apparent character displacement. *Psyche,* 72(1): 35–54.

——— 1965c. A second African species of the dacetine ant genus *Codiomyrmex. Psyche,* 72(3): 225–228.

——— 1965d. Notes on the Indo-Australian ants of genus *Simopone* Forel (Hymenoptera—Formicidae). *Psyche,* 72(4): 287–290.

——— 1965e. New Melanesian ants of the genera *Simopone* and *Amblyopone* (Hymenoptera—Formicidae) of zoogeographic significance. *Breviora,* no. 221, 11 pp.

——— 1967a. A monographic revision of the ant genus *Ponera* Latreille (Hymenoptera: Formicidae). *Pacific Insects Monograph,* no. 13, 112 pp.

——— 1967b. Entomological survey of the Cook Islands and Niue, 1: Hymenoptera—Formicidae. *New Zealand Journal of Science,* 10(4): 1092–95.

——— 1967c. The Australian workerless inquiline ant *Strumigenys xenos* Brown (Hymenoptera—Formicidae) recorded from New Zealand. *New Zealand Entomologist,* 4(1): 47–49.

——— 1968a. A new Malayan species of the ant genus *Epitritus,* and a related new genus from Singapore (Hymenoptera: Formicidae). *Journal of the Australian Entomological Society,* 7(2): 130–134.

——— 1968b. Notes on the Indo-Australian basicerotine ants (Hymenoptera: Formicidae). *Australian Journal of Zoology,* 16(2): 333–348.

——— 1969. The identity of *Dorylozelus mjobergi* Forel (Hymenoptera: Formicidae). *Journal of the Australian Entomological Society,* 8(2): 131–133.

——— 1970a. Notes on some Australian and Melanesian basicerotine ants (Hymenoptera: Formicidae). *Journal of the Australian Entomological Society,* 9(1): 49–52.

——— 1970b. Characterization of the Australian endemic ant genus *Peronomyrmex* Viehmeyer (Hymenoptera: Formicidae). *Journal of the Australian Entomological Society,* 9(3): 209–211.

——— 1971. The ants of the Kermadec Islands. *New Zealand Entomologist,* 5(1): 81–82.

——— 1973. Ants of the Australian genus *Mesostruma* Brown (Hymenoptera: Formicidae). *Journal of the Australian Entomological Society,* 12(1): 24–38.

——— 1976a. Superfam. Formicoidea: fam. Formicidae. *Annales du Musée Royal de l'Afrique Centrale,* Sciences Zoologiques, no. 215, pp. 192–199.

——— 1976b. The ants of Rennell and Bellona Islands. *Natural History of Rennell Island, British Solomon Islands,* 7: 73–90.

——— 1977. New ants of the Australasian genus *Orectognathus,* with a key to the known species (Hymenoptera: Formicidae). *Australian Journal of Zoology,* 25(3): 581–612.

——— 1978a. A taxonomic guide to the ant genus *Orectognathus* (Hymenoptera: Formicidae). *Commonwealth Scientific and Industrial Research Organisation, Division of Entomology Reports, Canberra, Australia,* no. 3, 11 pp.

——— 1978b. Melanesian ants of the genus *Amblyopone* (Hymenoptera: Formicidae). *Australian Journal of Zoology,* 26(4): 823–839.

——— 1978c. *Nothomyrmecia macrops:* a living-fossil ant rediscovered. *Science,* 201: 979–985.

——— 1979. Notes on the Russian endemic ant genus *Aulacopone* Arnoldi (Hymenoptera: Formicidae). *Psyche,* 86(4): 353–361.

——— 1980. Australian and Melanesian ants of the genus *Eurhopalothrix* Brown and Kempf—notes and new species (Hymenoptera: Formicidae). *Journal of the Australian Entomological Society,* 19(3): 229–239.

——— 1985. The ants of the Papuasian genus *Dacetinops* (Hymenoptera: Formicidae: Myrmicinae). *In* G. E. Ball, ed., *Phylogeny and zoogeography of beetles and ants,* pp. 41–67. Dr. W. Junk Publishing, Dordrecht, Netherlands.

——— 1987. A checklist of the ants of Australia, New Caledonia, and New Zealand. *Commonwealth Scientific and Industrial Research Organisation, Division of Entomology Reports, Canberra, Australia,* no. 41, 92 pp.

Taylor, R. W., and E. O. Wilson. 1961. Ants from three remote oceanic islands. *Psyche,* 68(4): 137–144.

Terron, G. 1967. Description des castes de *Tetraponera anthracina* Santschi (Hym., Formicidae, Promyrmicinae). *Insectes Sociaux,* 14(4): 339–348.

——— 1969. Mise en évidence du parasitisme temporaire de *Tetraponera anthracina* Santschi par *Tetraponera ledouxi* nov. spec. (Hym. Formicidae, Promyrmicinae). *Annales de la Faculté des Sciences du Cameroun,* 3: 113–115.

——— 1970. Recherches morphologiques et biologiques sur *Tetraponera anthracina* Santschi et sur son parasite social temporaire *Tetraponera ledouxi* Terron (Hym. Formicidae, Promyrmicinae). Ph.D. diss., Université Paul Sabatier, Toulouse. 313 pp. [Cited by Dejean and Passera, 1974.]

——— 1971. Description des castes de *Tetraponera nasuta* Bernard (Hym. Formicidae, Promyrmicinae). *Annales de la Faculté des Sciences du Cameroun,* 6: 73–84.

——— 1972a. La ponte des ouvrières fécondées chez une fourmi camerounaise du genre *Technomyrmex* Mayr: mise en évidence d'une descendance ouvrière. *Comptes Rendus de l'Académie des Sciences, Paris,* ser. D, 274(10): 1516–17.

——— 1972b. Observations sur les mâles ergatoïdes et des mâles ailés chez une fourmi du genre *Technomyrmex* Mayr (Hym., Formicidae Dolichoderinae). *Annales de la Faculté des Sciences du Cameroun,* 10: 107–120.

——— 1977. Evolution des colonies de *Tetraponera anthracina* Santschi (Formicidae Pseudomyrmecinae) avec reines. *Bulletin Biologique de la France et de la Belgique,* 111(2): 115–181.

Tevis, L. 1958. Interrelations between the harvester ant *Veromessor pergandei* (Mayr) and some desert ephemerals. *Ecology,* 39(4): 695–704.

Thaxter, R. 1888. The Entomophthoreae of the United States. *Memoirs read before the Boston Society of Natural History,* 4(6): 133–201.

——— 1908. Contribution toward a monograph of the Laboulbeniaceae, pt. II. *Memoirs of the American Academy of Arts and Sciences.* 13: 217–469.

Therrien, P., J. N. McNeil, W. G. Wellington, and G. Febvay. 1986. Ecological studies of the leaf-cutting ant, *Acromyrmex octospinosus,* in Guadeloupe. *In* C. S. Lofgren and R. K. Vander Meer, eds., *Fire ants and leaf-cutting ants: biology and management,* pp. 172–183. Westview Press, Boulder.

Thomas, D. W. 1988. The influence of aggressive ants on fruit removal in the tropical tree, *Ficus capensis* (Moraceae). *Biotropica,* 20(1): 49–53.

Thomas, J. 1980. Why did the Large Blue become extinct in Britain? *Oryx,* 15(3): 243–247.

Thompson, F. C. 1981. Revisionary notes on Nearctic *Microdon* flies (Diptera: Syrphidae). *Proceedings of the Entomological Society of Washington,* 83(4): 725–758.

Thompson, J. N. 1981. Reversed animal-plant interactions: the evolution of insectivorous and ant-fed plants. *Biological Journal of the Linnean Society,* 16(2): 147–155.

Thompson, M. J., B. M. Glancey, W. E. Robbins, C. S. Lofgren, S. R. Dutky, J. Kochansky, R. K. Vander Meer, and A. R. Glover. 1981. Major hydrocarbons of the post-pharyngeal glands of mated queens of the red imported fire ant *Solenopsis invicta. Lipids,* 16(7): 485–495.

Thorpe, W. H. 1942. Observations on *Stomoxys ochrosoma* Speiser (Diptera Muscidae) as an associate of army ants (Dorylinae) in East Africa. *Proceedings of the Royal Entomological Society, London,* ser. A, 17(4–6): 38–41.

——— 1963. *Learning and instinct in animals,* 2nd ed. Methuen, London. x + 558 pp.

Tilman, D. 1978. Cherries, ants and tent caterpillars: timing of nectar production in relation to susceptibility of caterpillars to ant predation. *Ecology,* 59(4): 686–692.

Tinaut Ranera, J. A. 1981. *Rossomyrmex minuchae* nov. sp. (Hym. Formicidae) encontrada en Sierra Nevada, España. *Boletin de la Asociación Española de Entomología,* 4: 195–203.

Tohmé, G. 1972. Ecologie, biologie de la reproduction et éthologie de *Messor ebeninus* (Forel) (Hymenoptera, Formicoidea, Myrmicidae). Ph.D. diss., Université Paul Sabatier, Toulouse. 336 pp.

Tohmé, G., and H. Tohmé. 1978. Accroissement de la société et longévité de la reine et des ouvrières chez *Messor semirufus* (André) (Hym. Formicoidea). *Comptes Rendus de l'Académie des Sciences, Paris,* ser. D, 286(12): 961–963.

Tohmé, H., and G. Tohmé. 1979. Le genre *Epixenus* Emery (Hymenoptera, Formicidae, Myrmicinae) et ses principaux représentants au Liban et en Syrie. *Bulletin du Muséum National d'Histoire Naturelle, Paris,* sect. A (Zoologie), ser. 4, 1(4): 1087–1108.

Tomalski, M. D., M. S. Blum, T. H. Jones, H. M. Fales, D. F. Howard, and L. Passera. 1987. Chemistry and functions of exocrine secretions of the ants *Tapinoma melanocephalum* and *T. erraticum. Journal of Chemical Ecology,* 13(2): 253–263.

Topoff, H. R. 1969. A unique predatory association between carabid beetles of the genus *Helluomorphoides* and colonies of the army ant *Neivamyrmex nigrescens. Psyche,* 76(4): 375–381.

——— 1971. Polymorphism in army ants related to division of labor and colony cyclic behavior. *American Naturalist,* 105: 529–548.

——— 1984. Social organization of raiding and emigrations in army ants. *Advances in the Study of Behavior,* 14: 81–126.

——— 1985. Effect of overfeeding on raiding behavior in the western slave-making ant *Polyergus breviceps. National Geographic Research,* 1(3)(Summer): 437–441.

Topoff, H. R., and J. Mirenda. 1978. Precocial behaviour of callow workers of the army ant *Neivamyrmex nigrescens:* importance of stimulation by adults during mass recruitment. *Animal Behaviour,* 26(3): 698–706.

——— 1980a. Army ants do not eat and run: influence of food supply on emigration behaviour in *Neivamyrmex nigrescens. Animal Behaviour,* 28(4): 1040–45.

——— 1980b. Army ants on the move: relation between food supply and emigration frequency. *Science,* 207: 1099–1100.

Topoff, H. R., K. Lawson, and P. Richards. 1972. Trail following and its development in the Neotropical army ant genus *Eciton* (Hymenoptera: Formicidae: Dorylinae). *Psyche,* 79(4): 357–364.

Topoff, H. R., J. Mirenda, R. Droual, and S. Herrick. 1980. Onset of the nomadic phase in the army ant *Neivamyrmex nigrescens* (Cresson) (Hym. Form.): distinguishing between callow and larval excitation by brood substitution. *Insectes Sociaux,* 27(2): 175–179.

Topoff, H. R., B. LaMon, L. Goodloe, and M. Goldstein. 1984. Social and orientation behavior of *Polyergus breviceps* during slave-making raids. *Behavioral Ecology and Sociobiology,* 15(4): 273–279.

——— 1985a. Ecology of raiding behavior in the western slave-making ant *Polyergus breviceps* (Formicidae). *Southwestern Naturalist,* 30(2): 259–267.

Topoff, H. R., M. Inez-Pagani, L. Mack, and M. Goldstein. 1985b. Behavioral ecology of the slave-making ant, *Polyergus breviceps,* in a desert habitat. *Southwestern Naturalist,* 30(2): 289–295.

Topoff, H. R., M. Pagani, M. Goldstein, and L. Mack. 1985c. Orientation behavior of the slave-making ant *Polyergus breviceps* in an oak-woodland habitat. *Journal of the New York Entomological Society,* 93(3): 1041–46.

Topoff, H. R., D. Bodoni, P. Sherman, and L. Goodloe. 1987. The role of scouting in slave raids by *Polyergus breviceps* (Hymenoptera: Formicidae). *Psyche,* 94(3–4): 261–270.

Topoff, H. R., S. Cover, L. Greenberg, L. Goodloe, and P. Sherman. 1988. Colony founding by queens of the obligatory slave-making ant, *Polyergus breviceps:* the role of the Dufour's gland. *Ethology,* 78(3): 209–218.

Torgerson, R. L., and R. D. Akre. 1969. Reproductive morphology and behavior of a thysanuran, *Trichatelura manni,* associated with army ants. *Annals of the Entomological Society of America,* 62(6): 1367–74.

——— 1970. The persistence of army ant chemical trails and their significance in the ecitonine-ecitophile association (Formicidae: Ecitonini). *Melanderia,* 5: 1–28.

Torossian, C. 1959. Les échanges trophallactiques proctodéaux chez la fourmi *Dolichoderus quadripunctatus* (Hyménoptère—Formicoidea). *Insectes Sociaux,* 6(4): 369–374.

——— 1961. Les échanges trophallactiques proctodéaux chez la fourmi d'Argentine: *Iridomyrmex humilis* (Hym. Form. Dolichoderidae). *Insectes Sociaux,* 8(2): 189–191.

——— 1965. [Comment.] *Compte Rendu de la Cinqième Congrès de l'Union Internationale pour l'Etude des Insectes Sociaux* (Toulouse, 1965), p. 301.

——— 1968. Recherches sur la biologie et l'éthologie de *Dolichoderus quadripunctatus* (Hym. Form. Dolichoderidae). *Insectes Sociaux,* 15(1): 51–72.

——— 1972. Etude biologique des fourmis forestières peuplant les galles de Cynipidae des chênes, III: Rôle et importance numérique des femelles fondatrices. *Insectes Sociaux,* 19(1): 25–38.

——— 1973. Etude des communications antennaires chez les Formicoidea: analyse du comportement trophallactique lors d'échanges alimentaires practiqués entre ouvrières de la fourmi *Dolichoderus quadripunctatus. Comptes Rendus de l'Académie des Sciences, Paris,* ser. D, 277(14): 1381–84.

——— 1974. Polymorphismus und Kastendifferenzierung bei Dolichoderiden. *In* G. H. Schmidt, ed., *Sozialpolymorphismus bei Insekten,* pp. 657–667. Wissenschaftliche Verlagsgesellschaft MBH, Stuttgart.

——— 1978. La ponte d'oeufs abortifs chez les ouvrières de la fourmi *Dolichoderus quadripunctatus. Bulletin de la Société d'Histoire Naturelle de Toulouse,* 114(1–2): 207–211.

Torres, J. A. 1984a. Niches and coexistence of ant communities in Puerto Rico: repeated patterns. *Biotropica,* 16(4): 284–295.

——— 1984b. Diversity and distribution of ant communities in Puerto Rico. *Biotropica,* 16(4): 296–303.

Trager, J. C. 1984. A revision of the genus *Paratrechina* (Hymenoptera: Formicidae) of the continental United States. *Sociobiology,* 9(2): 49–162.

——— 1988. A revision of *Conomyrma* (Hymenoptera: Formicidae) from the southeastern United States, especially Florida, with keys to the species. *Florida Entomologist,* 71(1): 11–29.

Traniello, J. F. A. 1977. Recruitment behavior, orientation, and the organization of foraging in the carpenter ant *Camponotus pennsylvanicus* DeGeer (Hymenoptera: Formicidae). *Behavioral Ecology and Sociobiology,* 2(1): 61–79.

——— 1978. Caste in a primitive ant: absence of age polyethism in *Amblyopone. Science,* 202: 770–772.

——— 1980. Colony specificity in the trail pheromone of an ant. *Naturwissenschaften,* 67(7): 361–362.

——— 1981. Enemy deterrence in the recruitment strategy of a termite: soldier-organized foraging in *Nasutitermes costalis. Proceedings of the National Academy of Sciences of the United States of America,* 78(3): 1976–79.

——— 1982. Population structure and social organization in the primitive ant *Amblyopone pallipes* (Hymenoptera: Formicidae). *Psyche,* 89(1–2): 65–80.

——— 1983. Social organization and foraging success in *Lasius neoniger* (Hymenoptera: Formicidae): behavioral and ecological aspects of recruitment communication. *Oecologia,* 59(1): 94–100.

——— 1987a. Comparative foraging ecology of North Temperate ants: the role of worker size and cooperative foraging in prey selection. *Insectes Sociaux,* 34(2): 118–130.

——— 1987b. Social and individual responses to environmental factors in ants. *In* J. M. Pasteels and J.-L. Deneubourg, eds., *From individual to collective behavior in social insects* (*Experientia* Supplement, vol. 54), pp. 63–80. Birkhäuser Verlag, Basel.

——— 1988. Variation in foraging behavior among workers of the ant *Formica schaufussi:* ecological correlates of search behavior and the modification of search pattern. *In* R. L. Jeanne, ed., *Interindividual behavioral variability in social insects,* pp. 91–112. Westview Press, Boulder.

Traniello, J. F. A., and S. N. Beshers. 1985. Species-specific alarm/recruitment responses in a Neotropical termite. *Naturwissenschaften*, 72(9): 491–492.

Traniello, J. F. A., and B. Hölldobler. 1984. Chemical communication during tandem running in *Pachycondyla obscuricornis* (Hymenoptera: Formicidae). *Journal of Chemical Ecology*, 10(5): 783–794.

Traniello, J. F. A., and A. K. Jayasuriya. 1981a. Chemical communication in the primitive ant *Aneuretus simoni:* the role of the sternal and pygidial glands. *Journal of Chemical Ecology*, 7(6): 1023–33.

——— 1981b. The sternal gland and recruitment communication in the primitive ant *Aneuretus simoni*. *Experientia*, 37: 46.

——— 1985. The biology of the primitive ant *Aneuretus simoni* (Emery) (Formicidae: Aneuretinae), II: The social ethogram and division of labor. *Insectes Sociaux*, 32(4): 375–388.

Traniello, J. F. A., and S. C. Levings. 1986. Intra- and intercolony patterns of nest dispersion in the ant *Lasius neoniger:* correlations with territoriality and foraging ecology. *Oecologia*, 69(3): 413–419.

Traniello, J. F. A., M. S. Fujita, and R. V. Bowen. 1984. Ant foraging behavior: ambient temperature influences prey selection. *Behavioral Ecology and Sociobiology*, 15(1): 65–68.

Trave, R., and M. Pavan. 1956. Veleni degli insetti: principi estratti dalla formica *Tapinoma nigerrimum* Nyl. *Chemica e l'Industria* (Milan), 38: 1015–19.

Tricot, M.-C., J. M. Pasteels, and B. Tursch. 1972. Phéromones stimulant et inhibant l'agressivité chez *Myrmica rubra*. *Journal of Insect Physiology*, 18(3): 499–509.

Trivers, R. L. 1971. The evolution of reciprocal altruism. *Quarterly Review of Biology*, 46(1): 35–57.

Trivers, R. L., and H. Hare. 1976. Haplodiploidy and the evolution of the social insects. *Science*, 191: 249–263.

Tschinkel, W. R. 1986. The ecological nature of the fire ant: some aspects of colony function and some unanswered questions. *In* C. S. Lofgren and R. K. Vander Meer, eds., *Fire ants and leaf-cutting ants: biology and management*, pp. 72–87. Westview Press, Boulder.

——— 1987a. Relationship between ovariole number and spermathecal sperm count in ant queens: a new allometry. *Annals of the Entomological Society of America*, 80(2): 208–211.

——— 1987b. Fire ant queen longevity and age: estimation by sperm depletion. *Annals of the Entomological Society of America*, 80(2): 263–266.

——— 1987c. Seasonal life history and nest architecture of a winter-active ant, *Prenolepis imparis*. *Insectes Sociaux*, 34(3): 143–164.

——— 1988a. Colony growth and the ontogeny of worker polymorphism in the fire ant, *Solenopsis invicta*. *Behavioral Ecology and Sociobiology*, 22(2): 103–115.

——— 1988b. Distribution of the fire ants *Solenopsis invicta* and *S. geminata* (Hymenoptera: Formicidae) in northern Florida in relation to habitat and disturbance. *Annals of the Entomological Society of America*, 81(1): 76–81.

——— 1988c. Social control of egg-laying rate in queens of the fire ant, *Solenopsis invicta*. *Physiological Entomology*, 13(3): 327–350.

Tschinkel, W. R., and D. F. Howard. 1978. Queen replacement in orphaned colonies of the fire ant, *Solenopsis invicta*. *Behavioral Ecology and Sociobiology*, 3(3): 297–310.

——— 1983. Colony founding by pleometrosis in the fire ant, *Solenopsis invicta*. *Behavioral Ecology and Sociobiology*, 12(2): 103–113.

Tschinkel, W. R., and S. D. Porter. 1988. Efficiency of sperm use in queens of the fire ant, *Solenopsis invicta* (Hymenoptera: Formicidae). *Annals of the Entomological Society of America*, 81(5): 777–781.

Tsuji, K., and Y. Itō. 1986. Territoriality in a queenless ant, *Pristomyrmex pungens* (Hymenoptera: Myrmicinae). *Applied Entomology and Zoology*, 21(3): 377–381.

Tulloch, G. S. 1935. Morphological studies of the thorax of the ant. *Entomologica Americana*, n.s., 15(3): 93–131.

Tumlinson, J. H., R. M. Silverstein, J. C. Moser, R. G. Brownlee, and J. M. Ruth. 1971. Identification of the trail pheromone of a leaf-cutting ant, *Atta texana*. *Nature*, 234: 348–349.

Tumlinson, J. H., J. C. Moser, R. M. Silverstein, R. G. Brownlee, and J. M. Ruth. 1972. A volatile trail pheromone of the leaf-cutting ant, *Atta texana*. *Journal of Insect Physiology*, 18(5): 809–814.

Turner, A. J. 1913. Studies in Australian Lepidoptera, Pyralidae. *Proceedings of the Royal Society of Queensland*, 24: 111–163.

Uezu, K. 1977. On the foraging activity of *Diacamma rugosus* (Le Guillon). *Biological Magazine of Okinawa*, 15: 5–17.

Ule, E. 1902. Ameisengärten im Amazonasgebiet. *Botanische Jahrbucher für Systematik, Pflanzengeschichte und Pflanzengeographien*, 30(2)(Beiblatt Nr. 68): 45–52.

——— 1905. Wechselbeziehungen zwischen Ameisen und Pflanzen: Nach einer Arbeit von A. Forel über die von mir im Amazonas-Gebiet gesammelten Ameisen. *Flora*, 94: 491–497.

——— 1906. Ameisenpflanzen. *Botanische Jahrbucher für Systematik, Pflanzengeschichte und Pflanzengeographien*, 37(1): 335–352.

Uyenoyama, M. K., and M. W. Feldman. 1981. On relatedness and adaptive topography in kin selection. *Theoretical Population Biology*, 19(1): 87–123.

Vandermeer, J. 1980. Indirect mutualism: variations on a theme by Stephen Levine. *American Naturalist*, 116(3): 441–448.

Vander Meer, R. K. 1983. Semiochemicals and the red imported fire ant (*Solenopsis invicta* Buren) (Hymenoptera: Formicidae). *Florida Entomologist*, 66(1): 139–161.

——— 1986a. The trail pheromone complex of *Solenopsis invicta* and *Solenopsis richteri*. *In* C. S. Lofgren and R. K. Vander Meer, eds., *Fire ants and leaf-cutting ants: biology and management*, pp. 201–210. Westview Press, Boulder.

——— 1986b. Chemical taxonomy as a tool for separating *Solenopsis* spp. *In* C. S. Lofgren and R. K. Vander Meer, eds., *Fire ants and leaf-cutting ants: biology and management*, pp. 316–326. Westview Press, Boulder.

Vander Meer, R. K., and L. Morel. 1988. Brood pheromones in ants. *In* J. C. Trager, ed., *Advances in myrmecology*, pp. 491–513. E. J. Brill, Leiden.

Vander Meer, R. K., and D. P. Wojcik. 1982. Chemical mimicry in the myrmecophilous beetle *Myrmecaphodius excavaticollis*. *Science*, 218: 806–808.

Vander Meer, R. K., B. M. Glancey, C. S. Lofgren, A. Glover, J. H. Tumlinson, and J. Rocca. 1980. The poison sac of red imported fire ant queens: source of a pheromone attractant. *Annals of the Entomological Society of America*, 73(5): 609–612.

Vander Meer, R. K., F. D. Williams, and C. S. Lofgren. 1981. Hydrocarbon components of the trail pheromone of the red imported fire ant, *Solenopsis invicta*. *Tetrahedron Letters*, 22(18): 1651–54.

Vander Meer, R. K., B. M. Glancey, and C. S. Lofgren. 1982. Biochemical changes in the crop, oesophagus and postpharyngeal gland of colony-founding red imported fire ant queens (*Solenopsis invicta*). *Insect Biochemistry*, 12(1): 123–127.

Vanderplank, F. L. 1960. The bionomics and ecology of the red tree ant *Oecophylla* sp., and its relationship to the coconut bug *Pseudotheraptus wayi* Brown (Coreidae). *Journal of Animal Ecology*, 29(1): 15–33.

Vane-Wright, R. I. 1978. Ecological and behavioural origins of diversity in butterflies. *In* L. A. Mound and N. Waloff, eds., *Diversity of insect faunas* (Symposium of the Royal Entomological Society of London, no. 9), pp. 56–70. Blackwell, London.

Van Pelt, A. F. 1953. Notes on the above-ground activity and a mating flight of *Pogonomyrmex badius* (Latr.). *Journal of the Tennessee Academy of Science*, 28(2): 164–168.

——— 1956. The ecology of the ants of the Welaka Reserve, Florida (Hymenoptera: Formicidae). *American Midland Naturalist*, 56(2): 358–387.

——— 1976. Nest relocation in the ant *Pogonomyrmex barbatus*. *Annals of the Entomological Society of America*, 69(3): 493.

Van Pelt, A. F., and S. A. Van Pelt. 1972. *Microdon* (Diptera: Syrphidae) in nests of *Monomorium* (Hymenoptera: Formicidae) in Texas. *Annals of the Entomological Society of America*, 65(4): 977–979.

Van Vorhis Key, S. E., and T. C. Baker. 1982a. Trail-following responses of the Argentine ant, *Iridomyrmex humilis* (Mayr), to a synthetic trail pheromone component and analogs. *Journal of Chemical Ecology*, 8(1): 3–14.

——— 1982b. Specificity of laboratory trail following by the Argentine ant, *Iridomyrmex humilis* (Mayr), to (Z)-9-hexadecenal, analogs, and gaster extract. *Journal of Chemical Ecology,* 8(7): 1057–63.

Van Vorhis Key, S. E., L. K. Gaston, and T. C. Baker. 1981. Effects of gaster extract trail concentration on the trail following behaviour of the Argentine ant, *Iridomyrmex humilis* (Mayr). *Journal of Insect Physiology,* 27(6): 363–370.

Vargo, E. L., and D. J. C. Fletcher. 1986a. Evidence of pheromonal queen control over the production of male and female sexuals in the fire ant, *Solenopsis invicta. Journal of Comparative Physiology,* ser. A, 159(6): 741–749.

——— 1986b. Queen number and the production of sexuals in the fire ant, *Solenopsis invicta* (Hymenoptera: Formicidae). *Behavioral Ecology and Sociobiology,* 19(1): 41–47.

——— 1987. Effect of queen number on the production of sexuals in natural populations of the fire ant, *Solenopsis invicta. Physiological Entomology,* 12(1): 109–116.

Vepsäläinen, K., and B. Pisarski. 1981. The taxonomy of the *Formica rufa* group: chaos before order. *In* P. E. Howse and J.-L. Clément, eds., *Biosystematics of social insects* (Systematics Association special volume no. 19), pp. 27–35. Academic Press, New York.

——— 1982. Assembly of island ant communities. *Annales Zoologici Fennici,* 19(4): 327–335.

Verhaeghe, J. C. 1982. Food recruitment in *Tetramorium impurum* (Hymenoptera: Formicidae). *Insectes Sociaux,* 29(1): 67–85.

Verhaeghe, J. C., and J.-L. Deneubourg. 1983. Experimental study and modelling of food recruitment in the ant *Tetramorium impurum* (Hym. Form.). *Insectes Sociaux,* 30(3): 347–360.

Via, S. E. 1977. Visually mediated snapping in the bulldog ant: a perceptual ambiguity between size and distance. *Journal of Comparative Physiology,* 121(1): 33–51.

Viana, M. J., and J. A. Haedo Rossi. 1957. Primer hallazgo en el hemisferio sur de Formicidae extinguidos y catalogo mundial de los Formicidae fosiles. *Ameghiniana* (Buenos Aires), 1(1–2): 108–113.

Vick, K. W., W. A. Drew, E. J. Eisenbraun, and D. J. McGurk. 1969. Comparative effectiveness of aliphatic ketones in eliciting alarm behavior in *Pogonomyrmex barbatus* and *P. comanche. Annals of the Entomological Society of America,* 62(2): 380–381.

Viehmeyer, H. 1908. Zur Koloniegründung der parasitischen Ameisen. *Biologisches Centralblatt,* 28(1): 18–32.

——— 1915. Ameisen von Singapore. *Archiv für Naturgeschichte,* ser. A, 81(8): 108–168.

——— 1921. Die mitteleuropäischen Beobachtungen von *Harpagoxenus sublevis* Mayr. *Biologisches Zentralblatt,* 41(6): 269–278.

Visscher, P. K. 1986. Kinship discrimination in queen rearing by honey bees (*Apis mellifera*). *Behavioral Ecology and Sociobiology,* 18(6): 453–460.

Vogel, P., and F. von Brockhusen-Holzer. 1984. Ants as prey of juvenile *Anolis lineatopus* (Rept., Iguanidae) in prey choice experiments. *Zeitschrift für Tierpsychologie,* 65(1): 66–76.

Voss, C. 1967. Über das Formensehen der roten Waldameise (*Formica rufa*—Gruppe). *Zeitschrift für Vergleichende Physiologie,* 55(3): 225–254.

Voss, S. H. 1981. Trophic egg production in virgin fire ant queens. *Journal of the Georgia Entomological Society,* 16(4): 437–440.

Vosseler, J. 1905. Die ostafrikanische Treiberameise (Siafu). *Der Pflanzer* (Dar-es-Salam), no. 19, pp. 289–302.

Vowles, D. M. 1950. Sensitivity of ants to polarized light. *Nature,* 165: 282–283.

——— 1954. The orientation of ants, II: Orientation to light, gravity, and polarized light. *Journal of Experimental Biology,* 31(3): 356–375.

Waldman, B. 1987. Mechanisms of kin recognition. *Journal of Theoretical Biology,* 128(2): 159–185.

Waldman, B., P. C. Frumhoff, and P. W. Sherman. 1988. Problems of kin recognition. *Trends in Ecology and Evolution,* 3(1): 8–13.

Walker, J., and J. Stamps. 1986. A test of optimal caste ratio theory using the ant *Camponotus (Colobopsis) impressus. Ecology,* 67(4): 1052–62.

Waller, D. A. 1986. The foraging ecology of *Atta texana* in Texas. *In* C. S. Lofgren and R. K. Vander Meer, eds., *Fire ants and leaf-cutting ants: biology and management,* pp. 146–158. Westview Press, Boulder.

Wallis, D. I. 1960. Spinning movements in the larvae of the ant, *Formica fusca. Insectes Sociaux,* 7(2): 187–199.

Waloff, N. 1957. The effect of the number of queens of the ant *Lasius flavus* (Fab.) (Hym., Formicidae) on their survival and on the rate of development of the first brood. *Insectes Sociaux,* 4(4): 391–408.

Waloff, N., and R. E. Blackith. 1962. The growth and distribution of the mounds of *Lasius flavus* (Fabricius) (Hym: Formicidae) in Silwood Park, Berkshire. *Journal of Animal Ecology,* 31(3): 421–437.

Walsh, J. P., and W. R. Tschinkel. 1974. Brood recognition by contact pheromone in the red imported fire ant, *Solenopsis invicta. Animal Behaviour,* 22(3): 695–704.

Walther, J. R. 1981a. Die Morphologie und Feinstruktur der Sinnesorgane auf den Antennengeisseln der Männchen, Weibchen und Arbeiterinnen der Roten Waldameisen *Formica rufa* Linné 1759 mit einem Vergleich der antennalen Sensillenmuster weiterer Formicoidea. Ph.D. diss., Freie Universität, Berlin. 309 pp.

——— 1981b. Cuticular sense organs as characters in phylogenetic research. *Mitteilungen der Deutschen Gesellschaft für Allgemeine und Angewandte Entomologie,* 3: 146–150.

Ward, P. S. 1980. A systematic revision of the *Rhytidoponera impressa* group (Hymenoptera: Formicidae) in Australia and New Guinea. *Australian Journal of Zoology,* 28(3): 475–498.

——— 1981a. Ecology and life history of the *Rhytidoponera impressa* group (Hymenoptera: Formicidae), I: Habitats, nest sites, and foraging behavior. *Psyche,* 88(1–2): 89–108.

——— 1981b. Ecology and life history of the *Rhytidoponera impressa* group (Hymenoptera: Formicidae), II: Colony origin, seasonal cycles, and reproduction. *Psyche,* 88(1–2): 109–126.

——— 1983a. Genetic relatedness and colony organization in a species complex of ponerine ants, I: Phenotypic and genotypic composition of colonies. *Behavioral Ecology and Sociobiology,* 12(4): 285–299.

——— 1983b. Genetic relatedness and colony organization in a species complex of ponerine ants, II: Patterns of sex ratio investment. *Behavioral Ecology and Sociobiology,* 12(4): 301–307.

——— 1984. A revision of the ant genus *Rhytidoponera* (Hymenoptera: Formicidae) in New Caledonia. *Australian Journal of Zoology,* 32(1): 131–175.

——— 1985. The Nearctic species of the genus *Pseudomyrmex* (Hymenoptera: Formicidae). *Quaestiones Entomologicae,* 21(2): 209–246.

——— 1986. Functional queens in the Australian greenhead ant, *Rhytidoponera metallica* (Hymenoptera: Formicidae). *Psyche,* 93(1–2): 1–12.

——— 1987. Distribution of the introduced Argentine ant (*Iridomyrmex humilis*) in natural habitats of the Lower Sacramento Valley and its effects on the indigenous ant fauna. *Hilgardia,* 55(2): 1–16.

——— 1988. Mesic elements in the western Nearctic ant fauna: taxonomic and biological notes on *Amblyopone, Proceratium,* and *Smithistruma* (Hymenoptera: Formicidae). *Journal of the Kansas Entomological Society,* 61(1): 102–124.

Ward, P. S., and R. W. Taylor. 1981. Allozyme variation, colony structure and genetic relatedness in the primitive ant *Nothomyrmecia macrops* Clark (Hymenoptera: Formicidae). *Journal of the Australian Entomological Society,* 20(3): 177–183.

Wasmann, E. 1886. Über die Lebensweise einiger Ameisengäste, I. *Deutsche Entomologische Zeitschrift,* 30(1): 49–66.

——— 1889. Nachträgliche Bemerkungen zu *Ecitochara* und *Ecitomorpha. Deutsche Entomologische Zeitschrift,* 83(2): 414.

——— 1890. Vergleichende Studien über Ameisengäste und Termitengäste. *Tijdschrift voor Entomologie,* 33: 27–96.

——— 1891. *Die zusammengesetzen Nester und gemischten Kolonien der Ameisen.* Aschendorffschen Buchdruckerei, Münster in Westphalien. vii + 262 pp.

——— 1892. Zur Biologie einiger Ameisengäste. *Deutsche Entomologische Zeitschrift,* 36(2): 347–351.

——— 1894a. Die europäischen *Dinarda*, mit Beschreibung einer neuen deutschen Art. *Deutsche Entomologische Zeitschrift*, 38(2): 275–280.
——— 1894b. *Kritisches Verzeichniss der myrmecophilen und termitophilen Arthropoden.* Felix Dames, Berlin. xi + 231 pp.
——— 1895. Zur Kenntnis einiger schwieriger *Thorictus*-Arten. *Deutsche Entomologische Zeitschrift*, 39(1): 41–44.
——— 1899a. Die psychischen Fähigkeiten der Ameisen. *Zoologica, Stuttgart*, 11(26): 1–133.
——— 1899b. Neue Termitophilen und Myrmecophilen aus Indien. *Deutsche Entomologische Zeitschrift*, no. 1, pp. 145–169.
——— 1901. Zur Lebensweise der Ameisengrillen (*Myrmecophila*). 115 Beitrag zur Kenntnis der Myrmekophilen und Termitophilen. *Natur und Offenbarung*, no. 47, 24 pp.
——— 1902. Zur Kenntnis der myrmecophilen *Antennophorus* und anderer auf Ameisen und Termiten reitender Acarinen. *Zoologischer Anzeiger*, 25: 66–76.
——— 1903. Zur näheren Kenntnis des echten Gastverhältnisses (Symphilie) bei den Ameisen- und Termitengästen. *Biologisches Zentralblatt*, 23: 63–72, 195–207, 232–248, 261–276, 298–310.
——— 1905. Zur Lebensweise einiger in- und ausländischer Ameisengäste (148. Beitrag zur Kenntnis der Myrmecophilen und Termitophilen). *Zeitschrift für Wissenschaftliche Insektenbiologie*, 10: 329–336, 384–390, 418–428.
——— 1908. Weitere Beiträge zum sozialen Parasitismus und der Sklaverei bei den Ameisen. *Biologisches Centralblatt*, 28(8): 257–271.
——— 1910a. Die Doppelwirtigkeit der *Atemeles*. *Deutsche Entomologische National-Bibliothek*, 1: 1–11.
——— 1910b. Nachträge zum sozialen Parasitismus und der Sklaverei bei den Ameisen. *Biologisches Centralblatt*, 30(13): 453–524.
——— 1915a. Anergatides Kohli, eine neue arbeiterlose Schmarotzerameise vom oberen Kongo (Hym., Form.). *Entomologische Mitteilungen*, 4(10–12): 279–288.
——— 1915b. Neue Beiträge zur Biologie von *Lomechusa* und *Atemeles*, mit kritischen Bemerkungen über das echte Gastverhältnis. *Zeitschrift für Wissenschaftliche Zoologie*, 114(2): 233–402.
——— 1917. Neue Anpassungstypen bei Dorylinengästen Afrikas (Col., Staphylinidae). *Zeitschrift für Wissenschaftliche Zoologie*, 117(2): 257–360.
——— 1920. *Die Gastpflege der Ameisen.* Gebrüder Borntraeger, Berlin. xvii + 176 pp.
——— 1925. Die Ameisenmimikry—Ein exakter Beitrag zum Mimikryproblem und zur Theorie der Anpassung. (250. Beitrag zur Kenntnis der Myrmecophilen). *Abhandlungen der Theoretischen Biologie*, 19: i-xii, 1–164.
——— 1930. Zur Biologie von *Myrmedonia* (*Zyras*). *Entomologische Berichte*, 8(176): 150–151.
Wasmann, E., and H. Brauns. 1925. New genera and species of South African myrmecophilous and termitophilous beetles. *South African Journal of Natural History*, 5(1): 101–118.
Watkins, J. F., II. 1964. Laboratory experiments on the trail following of army ants of the genus *Neivamyrmex* (Formicidae: Dorylinae). *Journal of the Kansas Entomological Society*, 37(1): 22–28.
——— 1976. *The identification and distribution of New World army ants (Dorylinae: Formicidae).* Markham Press Fund of Baylor University Press, Waco, Tex. x + 102 pp.
——— 1982. The army ants of Mexico (Hymenoptera: Formicidae: Ecitoninae). *Journal of the Kansas Entomological Society*, 55(2): 197–247.
Watkins, J. F., II, and T. W. Cole. 1966. The attraction of army ant workers to secretions of their queens. *Texas Journal of Science*, 18(3): 254–265.
Watt, J. C. 1967. The families Perimylopidae and Dacoderidae (Coleoptera, Heteromera). *Proceedings of the Royal Entomological Society of London*, ser. B, 36(7–8): 109–118.
Way, M. J. 1953. The relationship between certain ant species with particular reference to biological control of the coreid, *Theraptus* sp. *Bulletin of Entomological Research*, 44(4): 669–691.
——— 1954a. Studies of the life history and ecology of the ant *Oecophylla longinoda* Latreille. *Bulletin of Entomological Research*, 45(1): 93–112.
——— 1954b. Studies on the association of the ant *Oecophylla longinoda* (Latr.) (Formicidae) with the scale insect *Saissetia zanzibarensis* Williams (Coccidae). *Bulletin of Entomological Research*, 45(1): 113–134.
——— 1963. Mutualism between ants and honeydew-producing Homoptera. *Annual Review of Entomology*, 8: 307–344.
Weaver, N. 1957. Effects of larval age on dimorphic differentiation of the female honey bee. *Annals of the Entomological Society of America*, 50(3): 283–294.
——— 1966. Physiology of caste determination. *Annual Review of Entomology*, 11: 79–102.
Weber, N. A. 1937. The sting of an ant. *American Journal of Tropical Medicine*, 17(5): 765–768.
——— 1939. The sting of the ant, *Paraponera clavata*. *Science*, 89: 127–128.
——— 1940. The biology of the fungus-growing ants, Part VI: Key to *Cyphomyrmex*, new Attini and a new guest ant. *Revista de Entomologia*, 11(1–2): 406–427.
——— 1941a. Four new genera of Ethiopian and Neotropical Formicidae. *Annals of the Entomological Society of America*, 34(1): 183–194.
——— 1941b. The biology of the fungus-growing ants, Pt. VII: The Barro Colorado Island, Canal Zone, species. *Revista de Entomologia*, 12(1–2): 93–130.
——— 1942. A neuropterous myrmecophile, *Nadiva valida* Erichs. *Psyche*, 49(1–2): 1–3.
——— 1943. Parabiosis in Neotropical "ant gardens." *Ecology*, 24(3): 400–404.
——— 1944a. The Neotropical coccid-tending ants of the genus *Acropyga* Roger. *Annals of the Entomological Society of America*, 37(1): 89–122.
——— 1944b. The tree ants (*Dendromyrmex*) of South and Central America. *Ecology*, 25(1): 117–120.
——— 1946a. The biology of the fungus-growing ants, Pt. IX: The British Guiana species. *Revista de Entomologia*, 17(1–2): 114–172.
——— 1946b. Two common ponerine ants of possible economic significance, *Ectatomma tuberculatum* (Olivier) and *E. ruidum* Roger. *Proceedings of the Entomological Society of Washington*, 48(1): 1–16.
——— 1947a. A revision of the North American ants of the genus *Myrmica* Latreille with a synopsis of the Palearctic species, I. *Annals of the Entomological Society of America*, 40(3): 437–474.
——— 1947b. Lower Orinoco River fungus-growing ants (Hymenoptera: Formicidae, Attini). *Boletin de Entomologia Venezolana*, 6(2–4): 143–161.
——— 1948. A revision of the North American ants of the genus *Myrmica* Latreille with a synopsis of the Palearctic species, II. *Annals of the Entomological Society of America*, 41(2): 267–308.
——— 1949. New ponerine ants from equatorial Africa. *American Museum Novitates*, no. 1398, 9 pp.
——— 1955. Pure cultures of fungi produced by ants. *Science*, 121: 109.
——— 1956. Treatment of substrate by fungus-growing ants. *Anatomical Record*, 125(3): 604–605.
——— 1957a. Dry season adaptations of fungus-growing ants and their fungi. *Anatomical Record*, 128(3): 638.
——— 1957b. Fungus-growing ants and their fungi: *Cyphomyrmex costatus*. *Ecology*, 38(3): 480 494.
——— 1957c. Weeding as a factor in fungus culture by ants. *Anatomical Record*, 128(3): 638.
——— 1959. The stings of the harvesting ant, *Pogonomyrmex occidentalis* (Cresson), with a note on populations (Hymenoptera). *Entomological News*, 70(4): 85–90.
——— 1961. Use of poison by the ant, *Tapinoma nigerrimum* (Hymenoptera: Formicidae). *Proceedings of the Entomological Society of Washington*, 63(3): 217–218.
——— 1966. Fungus-growing ants. *Science*, 153: 587–604.
——— 1967. Growth of a colony of the fungus-growing ant *Sericomyrmex urichi* (Hymenoptera: Formicidae). *Annals of the Entomological Society of America*, 60(6): 1328–29.
——— 1972. *Gardening ants: the attines* (Memoirs of the American Philosophical Society, vol. 92). American Philosophical Society, Philadelphia. xx + 146 pp.

——— 1979. Fungus-culturing by ants. *In* L. R. Batra, ed., *Insect-fungus symbiosis: mutualism and commensalism*, pp. 77–116. Allanheld and Osmun, Montclair, N.J.
——— 1982. Fungus ants. *In* H. R. Hermann, ed., *Social insects*, vol. 4, pp. 255–363. Academic Press, New York.
Webster, R. P., and M. C. Nielsen. 1984. Myrmecophily in the Edward's hairstreak butterfly *Satyrium edwardsii* (Lycaenidae). *Journal of the Lepidopterists' Society*, 38(2): 124–133.
Wehner, R. 1981. Spatial vision in arthropods. *In* H. Autrum, ed., *Handbook of sensory physiology*, vol. 7/6C, pp. 287–616. Springer-Verlag, Heidelberg.
——— 1982. Himmelsnavigation bei Insekten: Neurophysiologie und Verhalten (Neujahrsblatt 1982, no. 184). *Vierteljahrsschrift der Naturforschenden Gesellschaft in Zürich*, 126(5): 1–132.
——— 1983a. Celestial and terrestrial navigation: human strategies—insect strategies. *In* F. Huber and H. Markl, eds., *Neuroethology and behavioral physiology*, pp. 366–381. Springer-Verlag, Heidelberg.
——— 1983b. Taxonomie, Funktionsmorphologie und Zoogeographie der saharischen Wüstenameise *Cataglyphis fortis* (Forel 1902) stat. nov. (Insecta: Hymenoptera: Formicidae). *Senckenbergiana Biologica*, 64(1–3): 89–132.
——— 1987. Spatial organization of foraging behavior in individually searching desert ants, *Cataglyphis* (Sahara Desert) and *Ocymyrmex* (Namib Desert). *In* J. M. Pasteels and J.-L. Deneubourg, eds., *From individual to collective behavior in social insects* (*Experientia* Supplement, vol. 54), pp. 15–42. Birkhäuser Verlag, Basel.
Wehner, R., and B. Lanfranconi. 1981. What do the ants know about the rotation of the sky? *Nature*, 293: 731–733.
Wehner, R., and M. Müller. 1985. Does interocular transfer occur in visual navigation by ants? *Nature*, 315: 228–229.
Wehner, R., and S. Rossel. 1985. The bee's celestial compass—a case study in behavioural neurobiology. *In* B. Hölldobler and M. Lindauer, eds., *Experimental behavioral ecology and sociobiology* (Fortschritte der Zoologie, no. 31), pp. 11–53. Sinauer Associates, Sunderland, Mass.
Wehner, R., R. D. Harkness, and P. Schmid-Hempel. 1983. *Foraging strategies in individual searching ants* Cataglyphis bicolor *(Hymenoptera: Formicidae)* (Information Processing in Animals, vol. 1). Gustav Fischer Verlag, New York. iv + 79 pp.
Weir, J. S. 1958a. Polyethism in workers of the ant *Myrmica*, I. *Insectes Sociaux*, 5(1): 97–128.
——— 1958b. Polyethism in workers of the ant *Myrmica*, II. *Insectes Sociaux*, 5(3): 315–339.
——— 1959a. Egg masses and early larval growth in *Myrmica*. *Insectes Sociaux*, 6(2): 187–201.
——— 1959b. The influence of worker age on trophogenic larval dormancy in the ant *Myrmica*. *Insectes Sociaux*, 6(3): 271–290.
Wellenstein, G. 1928. Beiträge zur Biologie der roten Waldameise (*Formica rufa* L.) mit besonderer Berücksichtigung klimatischer und forstlicher Verhaltnisse. *Zeitschrift für Angewandte Entomologie*, 14: 1–68.
——— 1952. Die Ernährungsbiologie der roten Waldameise (*Formica rufa* L.). *Zeitschrift für Pflanzenkrankheiten und Pflanzenschutz*, 59: 430–451.
——— 1977. Die Grundlagen der Waldtracht und Möglichkeiten ihrer bienenwirtschaftlichen Nutzung. *Zeitschrift für Angewandte Zoologie*, 64(3): 291–309.
Went, F. W., J. Wheeler, and G. C. Wheeler. 1972. Feeding and digestion in some ants (*Veromessor* and *Manica*). *BioScience*, 22(2): 82–88.
Wesson, L. G. 1936. Contributions toward the biology of *Strumigenys pergandei*: a new food relationship among ants. *Entomological News*, 47(7): 171–174.
——— 1937. A slave-making *Leptothorax* (Hymen.: Formicidae). *Entomological News*, 48(5): 125–129.
——— 1939. Contributions to the natural history of *Harpagoxenus americanus* Emery (Hymenoptera: Formicidae). *Transactions of the American Entomological Society*, 65(2): 97–122.
——— 1940. Observations on *Leptothorax duloticus*. *Bulletin of the Brooklyn Entomological Society*, 35(3): 73–83.
West, M. J. 1967. Foundress associations in polistine wasps: dominance hierarchies and the evolution of social behavior. *Science*, 157: 1584–85.
West-Eberhard, M. J. 1975. The evolution of social behavior by kin selection. *Quarterly Review of Biology*, 50(1): 1–33.
——— 1978. Polygyny and the evolution of social behavior in wasps. *Journal of the Kansas Entomological Society*, 51(4): 832–856.
——— 1979. Sexual selection, social competition, and evolution. *Proceedings of the American Philosophical Society*, 123(4): 222–234.
——— 1981. Intragroup selection and the evolution of insect societies. *In* R. D. Alexander and D. W. Tinkle, eds., *Natural selection and social behavior*, pp. 3–17. Chiron Press, Concord, Mass.
——— 1982. Communication in social wasps: predicted and observed patterns, with a note on the significance of behavioral and ontogenetic flexibility for theories of worker "altrusim." *In* A. de Haro and X. Espadaler, eds., *La communication chez les sociétés d'insectes*, pp. 13–36. Universidad Autónoma de Barcelona, Bellaterra.
——— 1987. Flexible strategy and social evolution. *In* Y. Itō, J. L. Brown, and J. Kikkawa, eds., *Animal societies: theories and facts*, pp. 35–51. Japan Scientific Societies Press, Tokyo.
Westoby, M., B. Rice, J. M. Shelley, D. Haig, and J. L. Kohen. 1982. Plants' use of ants for dispersal at West Head, New South Wales. *In* R. C. Buckley, ed., *Ant-plant interactions in Australia*, pp. 75–87. Dr. W. Junk, The Hague.
Westwood, J. O. 1838. *The entomologist's text book*. W. S. Orr, London. x + 432 pp.
Wettstein, R. von. 1889. Über die Compositen der österreichisch-ungarischen Flora mit zuckerabscheidenden Hüllschuppen. *Sitzungsberichte der Kaiserlichen Akademie der Wissenschaften, Mathematisch-Naturwissenschaftliche Classe*, ser. 1, 97(7): 570–589.
Weyer, F. 1929. Die Eiablage bei *Formica rufa*-Arbeiterinnen. *Zoologischer Anzeiger*, 84(9–10): 253–256.
Wheeler, D. E. 1984. Behavior of the ant, *Procryptocerus scabriusculus* (Hymenoptera: Formicidae), with comparisons to other cephalotines. *Psyche*, 91(3–4): 171–192.
——— 1986a. Developmental and physiological determinants of caste in social Hymenoptera: evolutionary implications. *American Naturalist*, 128(1): 13–34.
——— 1986b. *Ectatomma tuberculatum:* foraging biology and association with *Crematogaster* (Hymenoptera: Formicidae). *Annals of the Entomological Society of America*, 79(2): 300–303.
——— 1986c. Polymorphism and division of labor in *Azteca chartifex laticeps* (Hymenoptera: Formicidae). *Journal of the Kansas Entomological Society*, 59(3): 542–548.
Wheeler, D. E., and B. Hölldobler. 1985. Cryptic phragmosis: the structural modifications. *Psyche*, 92(4): 337–353.
Wheeler, D. E., and H. F. Nijhout. 1981. Soldier determination in ants: new role for juvenile hormone. *Science*, 213: 361–363.
——— 1983. Soldier determination in *Pheidole bicarinata:* effect of methoprene on caste and size within castes. *Journal of Insect Physiology*, 29(11): 847–854.
——— 1984. Soldier determination in *Pheidole bicarinata:* inhibition by adult soldiers. *Journal of Insect Physiology*, 30(2): 127–135.
Wheeler, G. C., and J. Wheeler. 1951. The ant larvae of the subfamily Dolichoderinae (Hymenoptera, Formicidae). *Proceedings of the Entomological Society of Washington*, 53(4): 169–210.
——— 1953a. The ant larvae of the subfamily Formicinae, Part I. *Annals of the Entomological Society of America*, 46(1): 126–171.
——— 1953b. The ant larvae of the subfamily Formicinae, Part II. *Annals of the Entomological Society of America*, 46(2): 175–217.
——— 1956. The ant larvae of the subfamily Pseudomyrmecinae (Hymenoptera: Formicidae). *Annals of the Entomological Society of America*, 49(4): 374–398.
——— 1960a. The ant larvae of the subfamily Myrmicinae. *Annals of the Entomological Society of America*, 53(1): 98–110.
——— 1960b. Techniques for the study of ant larvae. *Psyche*, 67(4): 87–94.

——— 1963. *The ants of North Dakota.* University of North Dakota Press, Grand Forks. viii + 326 pp.
——— 1964a. The ant larvae of the subfamily Dorylinae: supplement. *Proceedings of the Entomological Society of Washington,* 66(3): 129–137.
——— 1964b. The ant larvae of the subfamily Ponerinae: supplement. *Annals of the Entomological Society of America,* 57(4): 443–462.
——— 1965. The ant larvae of the subfamily Leptanillinae (Hymenoptera, Formicidae). *Psyche,* 72(1): 24–34.
——— 1968. The rediscovery of *Manica parasitica* (Hymenoptera: Formicidae). *Pan-Pacific Entomologist,* 44(1): 71–72.
——— 1971. Ant larvae of the subfamily Ponerinae: second supplement. *Annals of the Entomological Society of America,* 64(6): 1197–1217.
——— 1973. *Ants of Deep Canyon.* University of California Press, Berkeley. xiv + 162 pp.
——— 1976. *Ant larvae: review and synthesis* (Memoirs of the Entomological Society of Washington, no. 7). Entomological Society, Washington, D.C. vi + 108 pp.
——— 1979. Larvae of the social Hymenoptera. *In* H. R. Hermann, ed., *Social insects,* vol. 1, pp. 287–338. Academic Press, New York.
——— 1985a. A simplified conspectus of the Formicidae. *Transactions of the American Entomological Society,* 111(2): 255–264.
——— 1985b. The larva of *Proatta. Psyche,* 92(4): 447–450.
——— 1986a. *The ants of Nevada.* Los Angeles County Museum of Natural History, Los Angeles. 138 pp.
——— 1986b. Young larvae of *Eciton* (Hymenoptera: Formicidae: Dorylinae). *Psyche,* 93(3–4): 341–349.
——— 1987. A checklist of the ants of South Dakota. *Prairie Naturalist,* 19(3): 199–208.
Wheeler, J., and S. W. Rissing. 1975a. Natural history of *Veromessor pergandei,* I: The nest (Hymenoptera: Formicidae). *Pan-Pacific Entomologist,* 51(3): 205–216.
——— 1975b. Natural history of *Veromessor pergandei,* II: Behavior (Hymenoptera: Formicidae). *Pan-Pacific Entomologist,* 51(4): 303–314.
Wheeler, J. W., and M. S. Blum. 1973. Alkylpyrazine alarm pheromones in ponerine ants. *Science,* 182: 501–503.
Wheeler, J. W., S. L. Evans, M. S. Blum, and R. L. Torgerson. 1975. Cyclopentyl ketones: identification and function in *Azteca* ants. *Science,* 187: 254–255.
Wheeler, W. M. 1900a. A new myrmecophile from the mushroom gardens of the Texan leaf-cutting ant. *American Naturalist,* 34: 851–862.
——— 1900b. The habits of *Myrmecophila nebrascensis* Bruner. *Psyche,* 9: 111–115.
——— 1901a. The compound and mixed nests of American ants. *American Naturalist,* 35: 431–448, 513–539, 701–724, 791–818.
——— 1901b. An extraordinary ant-guest. *American Naturalist,* 35: 1007–16.
——— 1903a. Ethological observations on an American ant (*Leptothorax emersoni* Wheeler). *Journal für Psychologie und Neurologie,* 2(1): 31–47.
——— 1903b. The origin of female and worker ants from the eggs of parthenogenetic workers. *Science,* 18: 830–833.
——— 1904a. A crustacean-eating ant (*Leptogenys elongata* Buckley). *Biological Bulletin of the Marine Biological Laboratory, Woods Hole,* 6(6): 251–259.
——— 1904b. A new type of social parasitism among ants. *Bulletin of the American Museum of Natural History,* 20(30): 347–375.
——— 1905. Worker ants with vestiges of wings. *Bulletin of the American Museum of Natural History,* 21(24): 405–408.
——— 1906. On the founding of colonies by queen ants, with special reference to the parasitic and slave-making species. *Bulletin of the American Museum of Natural History,* 22(4): 33–105.
——— 1907a. The polymorphism of ants, with an account of some singular abnormalities due to parasitism. *Bulletin of the American Museum of Natural History,* 23(1): 1–93.
——— 1907b. The fungus-growing ants of North America. *Bulletin of the American Museum of Natural History,* 23(31): 669–807.
——— 1908a. Honey ants, with a revision of the American Myrmecocysti. *Bulletin of the American Museum of Natural History,* 24(20): 345–397.
——— 1908b. Studies on myrmecophiles, I: *Cremastochilus. Journal of the New York Entomological Society,* 16(2): 68–79.
——— 1908c. Studies on myrmecophiles, II: *Hetaerius. Journal of the New York Entomological Society,* 16(3): 135–143.
——— 1910a. *Ants: their structure, development and behavior.* Columbia University Press, New York. xxv + 663 pp.
——— 1910b. Two new myrmecophilous mites of the genus *Antennophorus. Psyche,* 17(1): 1–6.
——— 1910c. Colonies of ants (*Lasius neoniger* Emery) infested with *Laboulbenis formicarum* Thaxter. *Psyche,* 17(3): 83–86.
——— 1911. Notes on the myrmecophilous beetles of the genus *Xenodusa,* with a description of the larva of *X. cava* (Leconte). *Journal of the New York Entomological Society,* 19(3): 163–169.
——— 1913. The ants of Cuba. *Bulletin of the Museum of Comparative Zoology, Harvard,* 54(17): 475–505.
——— 1914a. Notes on the habits of *Liomyrmex. Psyche,* 21(2): 75–76.
——— 1914b. The ants of the Baltic amber. *Schriften der Physikalisch-Ökonomischen Gesellschaft zu Königsberg,* 55: 1–142.
——— 1915. Two new genera of myrmicine ants from Brazil. *Bulletin of the Museum of Comparative Zoology, Harvard,* 59(7): 483–491.
——— 1916a. The Australian ants of the genus *Onychomyrmex. Bulletin of the Museum of Comparative Zoology, Harvard,* 60(2): 45–54.
——— 1916b. Ants collected in Trinidad by Professor Roland Thaxter, Mr. F. W. Urich, and others. *Bulletin of the Museum of Comparative Zoology, Harvard,* 60(8): 323–330.
——— 1916c. The marriage-flight of a bull-dog ant (*Myrmecia sanguinea* F. Smith). *Journal of Animal Behavior,* 6(1): 70–73.
——— 1918a. A study of some ant larvae, with a consideration of the origin and meaning of the social habit among insects. *Proceedings of the American Philosophical Society,* 57(4): 293–343.
——— 1918b. The ants of the genus *Opisthopsis* Emery. *Bulletin of the Museum of Comparative Zoology, Harvard,* 62(7): 341–362.
——— 1918c. The Australian ants of the ponerine tribe Cerapachyini. *Proceedings of the American Academy of Arts and Sciences,* 53(3): 215–265.
——— 1919a. A new paper-making *Crematogaster* from the southeastern United States. *Psyche,* 26(4): 107–112.
——— 1919b. The ants of Borneo. *Bulletin of the Museum of Comparative Zoology, Harvard,* 63(3): 45–147.
——— 1919c. The parasitic Aculeata, a study in evolution. *Proceedings of the American Philosophical Society,* 58(1): 1–40.
——— 1921a. A new case of parabiosis and the "ant gardens" of British Guiana. *Ecology,* 2(2): 89–103.
——— 1921b. Observations on army ants in British Guiana. *Proceedings of the American Academy of Arts and Sciences,* 56(8): 291–328.
——— 1922. Ants of the American Museum Congo Expedition, a contribution to the myrmecology of Africa, I: On the distribution of the ants of the Ethiopian and Malagasy regions; II: The ants collected by the American Museum Congo Expedition; VII: Keys to the genera and subgenera of ants; VIII: A synonymic list of the ants of the Ethiopian region; IX: A synonymic list of the ants of the Malagasy region. *Bulletin of the American Museum of Natural History,* 45(1): 13–27, 39–269, 631–710, 711–1004, 1005–1055.
——— 1925. A new guest-ant and other new Formicidae from Barro Colorado Island, Panama. *Biological Bulletin of the Marine Biological Laboratory, Woods Hole,* 49(3): 150–181.
——— 1927. The physiognomy of insects. *Quarterly Review of Biology,* 2(1): 1–36.
——— 1928. *The social insects: their origin and evolution.* Kegan Paul, Treanch, Trubner and Co., London. xviii + 378 pp.
——— 1929a. The identity of the ant genera *Gesomyrmex* Mayr and *Dimorphomyrmex* Ernest André. *Psyche,* 36(1): 1–12.
——— 1929b. Three new genera of ants from the Dutch East Indies. *American Museum Novitates,* no. 349, 8 pp.
——— 1930. *Demons of the dust.* W. W. Norton, New York. viii + 378 pp.

——— 1930–31. A list of the known Chinese ants. *Peking Natural History Bulletin* 5(1): 53–81.
——— 1933a. A second parasitic *Crematogaster. Psyche*, 40(2): 83–86.
——— 1933b. *Colony-founding among ants, with an account of some primitive Australian species.* Harvard University Press, Cambridge, Mass. × + 179 pp.
——— 1934a. A second revision of the ants of the genus *Leptomyrmex* Mayr. *Bulletin of the Museum of Comparative Zoology, Harvard*, 77(3): 67–118.
——— 1934b. Formicidae of the Templeton Crocker Expedition, 1933. *Proceedings of the California Academy of Sciences*, 21(14): 173–181.
——— 1935. Two new genera of myrmicine ants from Papua and the Philippines. *Proceedings of the New England Zoological Club*, 15:1–9.
——— 1936a. A singular *Crematogaster* from Guatemala. *Psyche*, 43(2–3): 40–48.
——— 1936b. Ants from Hispaniola and Mona Island. *Bulletin of the Museum of Comparative Zoology, Harvard*, 80(2): 193–211.
——— 1936c. Ecological relations of ponerine and other ants to termites. *Proceedings of the American Academy of Arts and Sciences*, 71(3): 159–243.
——— 1937a. Ants mostly from the mountains of Cuba. *Bulletin of the Museum of Comparative Zoology, Harvard*, 81(3): 439–465.
——— 1937b. *Mosaics and other anomalies among ants.* Harvard University Press, Cambridge, Mass. 95 pp.
——— 1942. Studies of Neotropical ant-plants and their ants. *Bulletin of the Museum of Comparative Zoology, Harvard*, 90(1): 1–262.
Wheeler, W. M., and I. W. Bailey. 1920. The feeding habits of pseudomyrmine and other ants. *Transactions of the American Philosophical Society*, n.s., 22(4): 235–279.
Wheeler, W. M., and J. W. Chapman. 1922. The mating of *Diacamma. Psyche*, 29(5–6): 203–211.
Wheeler, W. M., and W. M. Mann. 1914. The ants of Haiti. *Bulletin of the American Museum of Natural History*, 33(1): 1–61.
Whelden, R. M. 1960. The anatomy of *Rhytidoponera metallica* F. Smith (Hymenoptera: Formicidae). *Annals of the Entomological Society of America*, 53(6): 793–808.
——— 1963. Anatomy of adult queen and workers of army ants *Eciton burchelli* Westw. and *E. hamatum* Fabr. (Hymenoptera: Formicidae). *Journal of the New York Entomological Society*, 71(1): 14–30; (2): 90–115; (3): 158–178; (4): 246–261.
Whitcomb, W. H., A. Bhatkar, and J. C. Nickerson. 1973. Predators of *Solenopsis invicta* queens prior to successful colony establishment. *Environmental Entomology*, 2(6): 1101–3.
Whitford, W. G. 1978a. Foraging in seed-harvester ants *Pogonomyrmex* spp. *Ecology*, 59(1): 185–189.
——— 1978b. Structure and seasonal activity of Chihuahua Desert ant communities. *Insectes Sociaux*, 25(1): 79–88.
Whitford, W. G., and G. Ettershank. 1975. Factors affecting foraging activity in Chihuahuan desert harvester ants. *Environmental Entomology*, 4(5): 689–696.
Whitford, W. G., P. Johnson, and J. Ramirez. 1976. Comparative ecology of the harvester ants *Pogonomyrmex barbatus* (F. Smith) and *Pogonomyrmex rugosus* (Emery). *Insectes Sociaux*, 23(2): 117–132.
Whitford, W. G., E. Depree, and P. Johnson. 1980. Foraging ecology of two Chihuahuan desert ant species: *Novomessor cockerelli* and *Novomessor albisetosus. Insectes Sociaux*, 27(2): 148–156.
Whitford, W. G., D. Schaeffer, and W. Wisdom. 1986. Soil movement by desert ants. *Southwestern Naturalist*, 31(2): 273–274.
Wilde, J. 1615. *De Formica, Liber Unus.* Amberg near Schönfeld. 108 pp.
Wilde, J. de, and J. Beetsma. 1982. The physiology of caste development in social insects. *Advances in Insect Physiology*, 16: 167–246.
Wildemuth, V. L., and E. G. Davis. 1931. The red harvester ant and how to subdue it. *Farmer's Bulletin, U. S. Department of Agriculture*, no. 1668, 12 pp.
Willey, R. B., and W. L. Brown. 1983. New species of the ant genus *Myopias* (Hymenoptera: Formicidae: Ponerinae). *Psyche*, 90(3): 249–285.
Williams, E. C. 1941. An ecological study of the floor fauna of the Panama rain forest. *Bulletin of the Chicago Academy of Sciences*, 6(4): 63–124.
Williams, G. C. 1966. *Adaptation and natural selection: a critique of some current evolutionary thought.* Princeton University Press, Princeton, N.J. x + 307 pp.
Williams, G. C., and D. C. Williams. 1957. Natural selection of individually harmful social adaptations among sibs with special reference to social insects. *Evolution*, 11(1): 32–39.
Williams, T., and N. R. Franks. 1988. Population size and growth rate, sex ratio and behaviour in the ant isopod, *Platyarthrus hoffmannseggi. Journal of Zoology, London*, 215(4): 703–717.
Willis, E. O. 1967. The behavior of bicolored antbirds. *University of California Publications in Zoology*, 79: 1–127.
Willis, E. O., and Y. Oniki. 1978. Birds and army ants. *Annual Review of Ecology and Systematics*, 9: 243–263.
Wilson, E. O. 1950. Notes on the food habits of *Strumigenys louisianae* Roger (Hymenoptera: Formicidae). *Bulletin of the Brooklyn Entomological Society*, 45(3): 85–86.
——— 1952. Notes on *Leptothorax bradleyi* Wheeler and *L. wheeleri* M. R. Smith (Hymenoptera: Formicidae). *Entomological News*, 63(3): 67–71.
——— 1953a. The ecology of some North American dacetine ants. *Annals of the Entomological Society of America*, 46(4): 479–495.
——— 1953b. The origin and evolution of polymorphism in ants. *Quarterly Review of Biology*, 28(2): 136–156.
——— 1955a. A monographic revision of the ant genus *Lasius. Bulletin of the Museum of Comparative Zoology, Harvard*, 113(1): 1–201.
——— 1955b. Ecology and behavior of the ant *Belonopelta deletrix* Mann (Hymenoptera: Formicidae). *Psyche*, 62(2): 82–87.
——— 1955c. Division of labor in a nest of the slave-making ant *Formica wheeleri* Creighton. *Psyche*, 62(3): 130–133.
——— 1956. Feeding behavior in the ant *Rhopalothrix biroi* Szabó. *Psyche*, 63(1): 21–23.
——— 1957a. The discovery of cerapachyine ants on New Caledonia, with the description of new species of *Phyracaces* and *Sphinctomyrmex. Breviora*, no. 74, 9 pp.
——— 1957b. The organization of a nuptial flight of the ant *Pheidole sitarches* Wheeler. *Psyche*, 64(2): 46–50.
——— 1958a. A chemical releaser of alarm and digging behavior in the ant *Pogonomyrmex badius* (Latreille). *Psyche*, 65(2–3): 41–51.
——— 1958b. Observations on the behavior of the cerapachyine ants. *Insectes Sociaux*, 5(1): 129–140.
——— 1958c. Studies on the ant fauna of Melanesia, I: The tribe Leptogenyini; II: The tribes Amblyoponini and Platythyreini. *Bulletin of the Museum of Comparative Zoology, Harvard*, 118(3): 98–153.
——— 1958d. Studies on the ant fauna of Melanesia, III: *Rhytidoponera* in western Melanesia and the Moluccas; IV: The tribe Ponerini. *Bulletin of the Museum of Comparative Zoology, Harvard*, 119(4): 300–371.
——— 1958e. The beginnings of nomadic and group-predatory behavior in the ponerine ants. *Evolution*, 12(1): 24–31.
——— 1959a. Adaptive shift and dispersal in a tropical ant fauna. *Evolution*, 13(1): 122–144.
——— 1959b. Communication by tandem running in the ant genus *Cardiocondyla. Psyche*, 66(3): 29–34.
——— 1959c. Some ecological characteristics of ants in New Guinea rain forests. *Ecology*, 40(3): 437–447.
——— 1959d. Source and possible nature of the odor trail of fire ants. *Science*, 129: 643–644.
——— 1959e. Studies on the ant fauna of Melanesia, V: The tribe Odontomachini. *Bulletin of the Museum of Comparative Zoology, Harvard*, 120(5): 483–510.
——— 1959f. Studies on the ant fauna of Melanesia, VI: The tribe Cerapachyini. *Pacific Insects*, 1(1): 39–57.
——— 1961. The nature of the taxon cycle in the Melanesian ant fauna. *American Naturalist*, 95: 169–193.

——— 1962a. Behavior of *Daceton armigerum* (Latreille), with a classification of self-grooming movements in ants. *Bulletin of the Museum of Comparative Zoology, Harvard*, 127(7): 401–422.
——— 1962b. Chemical communication among workers of the fire ant *Solenopsis saevissima* (Fr. Smith), 1: The organization of mass-foraging; 2: An information analysis of the odour trail; 3: The experimental induction of social responses. *Animal Behaviour*, 10(1–2): 134–147, 148–158, 159–164.
——— 1962c. The ants of Rennell and Bellona Islands. *Natural History of Rennell Island, British Solomon Islands*, 4: 13–23.
——— 1962d. The Trinidad cave ant *Erebomyrma* (= *Spelaeomyrmex*) *urichi* (Wheeler), with a comment on cavernicolous ants in general. *Psyche*, 69(2): 63–72.
——— 1963. Social modifications related to rareness in ant species. *Evolution*, 17(2): 249–253.
——— 1964. The true army ants of the Indo-Australian area (Hymenoptera: Formicidae: Dorylinae). *Pacific Insects*, 6(3): 427–483.
——— 1965a. Chemical communication in social insects. *Science*, 149: 1064–71.
——— 1965b. Trail sharing in ants. *Psyche*, 72(1): 2–7.
——— 1966. Behaviour of social insects. *In* P. T. Haskell, ed., *Insect behaviour* (Symposium of the Royal Entomological Society of London, no. 3), pp. 81–96. Royal Entomological Society, London.
——— 1968. The ergonomics of caste in the social insects. *American Naturalist*, 102: 41–66.
——— 1969. The species equilibrium. *Brookhaven Symposia in Biology*, 22: 38–47.
——— 1971. *The insect societies*. Belknap Press of Harvard University Press, Cambridge, Mass. x + 548 pp.
——— 1974a. Aversive behavior and competition within colonies of the ant *Leptothorax curvispinosus*. *Annals of the Entomological Society of America*, 67(5): 777–780.
——— 1974b. The population consequences of polygyny in the ant *Leptothorax curvispinosus*. *Annals of the Entomological Society of America*, 67(5): 781–786.
——— 1974c. The soldier of the ant, *Camponotus (Colobopsis) fraxinicola*, as a trophic caste. *Psyche*, 81(1): 182–188.
——— 1975a. *Leptothorax duloticus* and the beginnings of slavery in ants. *Evolution*, 29(1): 108–119.
——— 1975b. *Sociobiology: the new synthesis*. Belknap Press of Harvard University Press, Cambridge, Mass. x + 697 pp.
——— 1975c. Some central problems of sociobiology. *Social Science Information*, 14(6): 5–18. [Also, *in* R. M. May, ed., *Theoretical ecology*, 1st ed., pp. 205–217. W. B. Saunders, Philadelphia (1976).]
——— 1976a. A social ethogram of the Neotropical arboreal ant *Zacryptocerus varians* (Fr. Smith). *Animal Behaviour*, 24(2): 354–363.
——— 1976b. The organization of colony defense in the ant *Pheidole dentata* Mayr (Hymenoptera: Formicidae). *Behavioral Ecology and Sociobiology*, 1(1): 63–81.
——— 1976c. Behavioral discretization and the number of castes in an ant species. *Behavioral Ecology and Sociobiology*, 1(2): 141–154.
——— 1976d. The first workerless parasite in the ant genus *Formica* (Hymenoptera: Formicidae). *Psyche*, 83(3–4): 277–281.
——— 1976e. Which are the most prevalent ant genera? *Studia Entomologica*, n.s., 19(1–4): 187–200.
——— 1978. Division of labor in fire ants based on physical castes (Hymenoptera: Formicidae: *Solenopsis*). *Journal of the Kansas Entomological Society*, 51(4): 615–636.
——— 1980a. Caste and division of labor in leaf-cutter ants (Hymenoptera: Formicidae: *Atta*), I: The overall pattern in *A. sexdens*. *Behavioral Ecology and Sociobiology*, 7(2): 143–156.
——— 1980b. Caste and division of labor in leaf-cutter ants (Hymenoptera: Formicidae: *Atta*), II: The ergonomic optimization of leaf cutting. *Behavioral Ecology and Sociobiology*, 7(2): 157–165.
——— 1981. Communal silk-spinning by larvae of *Dendromyrmex* tree-ants (Hymenoptera: Formicidae). *Insectes Sociaux*, 28(2): 182–190.
——— 1983a. Caste and division of labor in leaf-cutter ants (Hymenoptera: Formicidae: *Atta*), III: Ergonomic resiliency in foraging by *A. cephalotes*. *Behavioral Ecology and Sociobiology*, 14(1): 47–54.
——— 1983b. Caste and division of labor in leaf-cutter ants (Hymenoptera: Formicidae: *Atta*), IV: Colony ontogeny of *A. cephalotes*. *Behavioral Ecology and Sociobiology*, 14(1): 55–60.
——— 1984a. *Biophilia*. Harvard University Press, Cambridge, Mass. 157 pp.
——— 1984b. The relation between caste ratios and division of labor in the ant genus *Pheidole* (Hymenoptera: Formicidae). *Behavioral Ecology and Sociobiology*, 16(2): 89–98.
——— 1984c. Tropical social parasites in the ant genus *Pheidole*, with an analysis of the anatomical parasitic syndrome (Hymenoptera: Formicidae). *Insectes Sociaux*, 31(3): 316–334.
——— 1985a. Between-caste aversion as a basis for division of labor in the ant *Pheidole pubiventris* (Hymenoptera: Formicidae). *Behavioral Ecology and Sociobiology*, 17(1): 35–37.
——— 1985b. Altruism and ants. *Discover*, 6(8)(August): 46–51.
——— 1985c. Ants of the Dominican amber (Hymenoptera: Formicidae), 1: Two new myrmicine genera and an aberrant *Pheidole*. *Psyche*, 92(1): 1–9.
——— 1985d. Ants of the Dominican amber (Hymenoptera: Formicidae), 2: The first fossil army ants. *Psyche*, 92(1): 11–16.
——— 1985e. Ants of the Dominican amber (Hymenoptera: Formicidae), 3: The subfamily Dolichoderinae. *Psyche*, 92(1): 17–37.
——— 1985f. Ants from the Cretaceous and Eocene amber of North America. *Psyche*, 92(2–3): 205–216.
——— 1985g. The sociogenesis of insect colonies. *Science*, 228: 1489–1495.
——— 1985h. Invasion and extinction in the West Indian ant fauna: evidence from the Dominican amber. *Science*, 229: 265–267.
——— 1986a. Caste and division of labor in *Erebomyrma*, a genus of dimorphic ants (Hymenoptera: Formicidae: Myrmicinae). *Insectes Sociaux*, 33(1): 59–69.
——— 1986b. The defining traits of fire ants and leaf-cutting ants. *In* C. S. Lofgren and R. K. Vander Meer, eds., *Fire ants and leaf-cutting ants: biology and management*, pp. 1–9. Westview Press, Boulder.
——— 1986c. The organization of flood evacuation in the ant genus *Pheidole* (Hymenoptera: Formicidae). *Insectes Sociaux*, 33(4): 458–469.
——— 1987a. The arboreal ant fauna of Peruvian Amazon forests: a first assessment. *Biotropica*, 19(3): 245–251.
——— 1987b. Causes of ecological success: the case of the ants (The Sixth Tansley Lecture). *Journal of Animal Ecology*, 56(1): 1–9.
——— 1987c. The earliest known ants: an analysis of the Cretaceous species and an inference concerning their social organization. *Paleobiology*, 13(1): 44–53.
——— 1988. The biogeography of the West Indian ants (Hymenoptera: Formicidae). *In* J. Liebherr, ed., *Zoogeography of Caribbean insects*, pp. 214–230. Cornell University Press, Ithaca, N.Y.
——— 1989. *Chimaeridris*, a new genus of hook-mandibled myrmicine ants from tropical Asia (Hymenoptera: Formicidae). *Insectes Sociaux*, 36(1): 62–69.
Wilson, E. O., and W. H. Bossert. 1963. Chemical communication among animals. *Recent Progress in Hormone Research*, 19: 673–716.
Wilson, E. O., and W. L. Brown. 1956. New parasitic ants of the genus *Kyidris*, with notes on ecology and behavior. *Insectes Sociaux*, 3(3): 439–454.
——— 1958a. Recent changes in the introduced population of the fire ant *Solenopsis saevissima* (Fr. Smith). *Evolution*, 12(2): 211–218.
——— 1958b. The worker caste of the parasitic ant *Monomorium metoecus* Brown and Wilson, with notes on behavior. *Entomological News*, 69(2): 33–38.
——— 1984. Behavior of the cryptobiotic predaceous ant *Eurhopalothrix heliscata*, n. sp. (Hymenoptera: Formicidae: Basicerotini). *Insectes Sociaux*, 31(4): 408–428.
Wilson, E. O., and J. H. Eads. 1949. A report on the imported fire ant *Solenopsis saevissima* var. *richteri* Forel in Alabama. A special report to the Alabama Department of Conservation, Montgomery, Alabama, July 16, 1949. vi + 53 pp.

Wilson, E. O., and T. Eisner. 1957. Quantitative studies of liquid food transmission in ants. *Insectes Sociaux*, 4(2): 157–166.
Wilson, E. O., and R. M. Fagen. 1974. On the estimation of total behavioral repertories in ants. *Journal of the New York Entomological Society*, 82(2): 106–112.
Wilson, E. O., and B. Hölldobler. 1980. Sex differences in cooperative silk-spinning by weaver ant larvae. *Proceedings of the National Academy of Sciences of the United States of America*, 77(4): 2343–47.
——— 1985. Caste-specific techniques of defense in the polymorphic ant *Pheidole embolopyx* (Hymenoptera: Formicidae). *Insectes Sociaux*, 32(1): 3–22.
——— 1986. Ecology and behavior of the Neotropical cryptobiotic ant *Basiceros manni* (Hymenoptera: Formicidae: Basicerotini). *Insectes Sociaux*, 33(1): 70–84.
——— 1988. Dense heterarchies and mass communication as the basis of organization in ant colonies. *Trends in Ecology and Evolution*, 3(3): 65–68.
Wilson, E. O., and G. L. Hunt. 1966. Habitat selection by queens of two field-dwelling species of ants. *Ecology*, 47(3): 485–487.
——— 1967. Ant fauna of Futuna and Wallis Islands, stepping stones to Polynesia. *Pacific Insects*, 9(4): 563–584.
Wilson, E. O., and M. Pavan. 1959. Glandular sources and specificity of some chemical releasers of social behavior in dolichoderine ants. *Psyche*, 66(4): 70–76.
Wilson, E. O., and F. E. Regnier. 1971. The evolution of the alarm-defense system in the formicine ants. *American Naturalist*, 105: 279–289.
Wilson, E. O., and R. W. Taylor. 1964. A fossil ant colony: new evidence of social antiquity. *Psyche*, 71(2): 93–103.
——— 1967a. An estimate of the potential evolutionary increase in species density in the Polynesian ant fauna. *Evolution*, 21(1): 1–10.
——— 1967b. The ants of Polynesia (Hymenoptera: Formicidae). *Pacific Insects Monograph*, no. 14, 109 pp.
Wilson, E. O., T. Eisner, and B. D. Valentine. 1954. The beetle genus *Paralimulodes* Bruch in North America, with notes on morphology and behavior (Coleoptera: Limulodidae). *Psyche*, 61(4): 154–161.
Wilson, E. O., T. Eisner, G. C. Wheeler, and J. Wheeler. 1956. *Aneuretus simoni* Emery, a major link in ant evolution. *Bulletin of the Museum of Comparative Zoology, Harvard*, 115(3): 81–99.
Wilson, E. O., N. I. Durlach, and L. M. Roth. 1958. Chemical releasers of necrophoric behavior in ants. *Psyche*, 65(4): 108–114.
Wilson, E. O., F. M. Carpenter, and W. L. Brown. 1967a. The first Mesozoic ants. *Science*, 157: 1038–40.
——— 1967b. The first Mesozoic ants, with the description of a new subfamily. *Psyche*, 74(1): 1–19.
Wilson, N. L., J. H. Dillier, and G. P. Markin. 1971. Foraging territories of imported fire ants. *Annals of the Entomological Society of America*, 64(3): 660–665.
Windsor, J. K. 1964. Three scarabaeid genera found in nests of *Formica obscuripes* Forel in Colorado. *Bulletin of the Southern California Academy of Sciences*, 63(4): 205–209.
Wing, M. W. 1949. A new *Formica* from northern Maine, with a discussion of its supposed type of social parasitism (Hymenoptera: Formicidae). *Canadian Entomologist*, 81(1): 13–17.
——— 1951. A new genus and species of myrmecophilous Diapriidae with taxonomic and biological notes on related forms. *Transactions of the Royal Entomological Society of London*, 102(3): 195–210.
——— 1968. Taxonomic revision of the Nearctic genus *Acanthomyops* (Hymenoptera: Formicidae). *Memoirs of the Cornell University Agricultural Experiment Station*, no. 405, 173 pp.
Winter, U. 1979a. Untersuchungen zum Raubzugverhalten der dulotischen Ameise *Harpagoxenus sublaevis* (Nyl.). *Insectes Sociaux*, 26(2): 123–125.
——— 1979b. *Epimyrma goesswaldi* Menozzi, eine sklavenhaltende Ameise. *Naturwissenschaften*, 66(11): 581–582.
Winter, U., and A. Buschinger. 1983. The reproductive biology of a slavemaker ant, *Epimyrma ravouxi*, and a degenerate slavemaker, *E. kraussei* (Hymenoptera: Formicidae). *Entomologia Generalis*, 9(1–2): 1–15.
——— 1986. Genetically mediated queen polymorphism and caste determination in the slave-making ant, *Harpagoxenus sublaevis* (Hymenoptera: Formicidae). *Entomologia Generalis*, 11(3–4): 125–137.
Wiśniewski, J. 1963. Analiza skladu gatunkowego chraszczy wystepujacych winrowiskach *Formica rufa* L. i *Formica polyctena* Först. (Hymenoptera: Formicidae). [An analysis of the beetle species occurring in ant colonies of *Formica rufa* L. and *Formica polyctena* Först. (Hymenoptera: Formicidae).] *Polskie Pismo Entomologiczne*, 33(1): 183–193. [In Polish with English summary.]
Wojcik, D. P., W. A. Banks, D. M. Hicks, and J. W. Summerlin. 1977. Fire ant myrmecophiles: new hosts and distribution of *Myrmecaphodius excavaticollis* (Blanchard) and *Euparia castanea* Serville (Coleoptera: Scarabaeidae). *Coleopterists Bulletin*, 31(4): 329–334.
Wojcik, D. P., W. A. Banks, and D. H. Habeck. 1978. Fire ant myrmecophiles: flight periods of *Myrmecaphodius excavaticollis* (Blanchard) and *Euparia castanea* Serville (Coleoptera: Scarabaeidae). *Coleopterists Bulletin*, 32(1): 59–64.
Wojcik, D. P., D. P. Jouvenaz, and C. S. Lofgren. 1987. First report of a parasitic fly (Diptera: Phoridae) from a red imported fire ant (*Solenopsis invicta*) alate female (Hymenoptera: Formicidae). *Florida Entomologist*, 70(1): 181–182.
Wood, L. A., and W. R. Tschinkel. 1981. Quantification and modification of worker size variation in the fire ant *Solenopsis invicta*. *Insectes Sociaux*, 28(2): 117–128.
Wood, T. K. 1977. Role of parent females and attendant ants in maturation of the treehopper, *Entylia bactriana* (Homoptera: Membracidae). *Sociobiology*, 2(4): 257–272.
——— 1982. Ant-attended nymphal aggregations in the *Enchenopa binotata* complex (Homoptera: Membracidae). *Annals of the Entomological Society of America*, 75(6): 649–653.
Woodruff, R. E., and O. L. Cartwright. 1967. A review of the genus *Euparixia* with description of a new species from nests of leaf-cutting ants in Louisiana (Coleoptera: Scarabaeidae). *Proceedings of the U.S. National Museum*, 123(3616): 1–21.
Woyciechowski, M. 1987. The phenology of nuptial flights ants (Hymenoptera, Formicidae). *Acta Zoologica Cracoviensia*, 30(10): 137–140.
Woyciechowski, M., and A. Łomnicki. 1987. Multiple mating of queens and the sterility of workers among eusocial Hymenoptera. *Journal of Theoretical Biology*, 128(3): 317–327.
Wright, D. M. 1983. Life history and morphology of the immature stages of the Bog Copper butterfly *Lycaena epixanthe* (Bsd and Le C.) (Lepidoptera: Lycaenidae). *Journal of Research on the Lepidoptera*, 22(1): 47–100.
Wroughton, R. C. 1892. Our ants. *Journal of the Bombay Natural History Society*, 7(1): 13–60.
Wüst, M. 1973. Stomodeale und proctodeale Sekrete von Ameisenlarven und ihre biologische Bedeutung. *Proceedings of the Seventh Congress of the International Union for the Study of Social Insects* (London), pp. 412–417.
Wyatt, R. 1981. Ant-pollination of the granite outcrop endemic *Diamorpha smallii* (Crassulaceae). *American Journal of Botany*, 68(9): 1212–17.
Wygodzinsky, P. 1961. A new genus of termitophilous Atelurinae from South Africa (Thysanura: Nicoletiidae). *Journal of the Entomological Society of Southern Africa*, 24(1): 104–109.
Wynne-Edwards, V. C. 1962. *Animal dispersion in relation to social behaviour*. Oliver and Boyd, Edinburgh. xi + 653 pp.
——— 1986. *Evolution through group selection*. Blackwell Scientific Publications, Oxford. xii + 386 pp.

Yamauchi, K., and K. Hayashida. 1968. Taxonomic studies on the genus *Lasius* in Hokkaido, with ethological and ecological notes (Formicidae, Hymenoptera), 1: The subgenus *Dendrolasius* or Jet Black Ants. *Journal of the Faculty of Science, Hokkaido University*, ser. 4 (zool.), 16(3): 396–412.
Yamauchi, K., K. Kinomura, and S. Miyake. 1981. Sociobiological studies of the polygynic ant *Lasius sakagamii*, I: General features of its polydomous system. *Insectes Sociaux*, 28(3): 279–296.

Yamauchi, K., Y. Itō, K. Kinomura, and H. Takamine. 1987. Polycalic colonies of the weaver ant *Polyrhachis dives. Kontyū, Tokyo,* 55(3): 410–420.

Yarrow, I. H. H. 1955. The British ants allied to *Formica rufa* L. (Hym., Formicidae). *Transactions of the Society for British Entomology,* 12(1): 1–48.

Yasumatsu, K. 1937. *Lasius fuliginosus* (Latreille) var. *spathepus* (Wheeler) and its synechtrans *Zyras comes* Sharp and *Zyras cognathus* Märkel var. *japonicus* Sharp. *Nippon no Kochu,* 1: 47–51. [In Japanese.]

Yasuno, M. 1963. The study of the ant population in the grassland at Mt. Hakkōda, I. *Ecological Review,* 16: 83–91.

——— 1964. The study of the ant population in the grassland at Mt. Hakkōda, II: The distribution pattern of ant nests at the Kayano grassland. *Science Reports of the Tōhoku University,* ser. 4 (biol.), 30(1): 43–55.

——— 1965a. The study of the ant population in the grassland at Mt. Hakkōda, V: The interspecific and intraspecific relation in the formation of the ant population, with special reference to the effect of the removal of *Formica truncorum yessensis. Science Reports of the Tōhoku University,* ser. 4 (biol.), 31(3): 181–194.

——— 1965b. Territory of ants in the Kayano grassland at Mt. Hakkōda. *Science Reports of the Tōhoku University,* ser. 4 (biol.), 31(3): 195–206.

Young, A. M. 1977. Notes on the foraging of the giant tropical ant *Paraponera clavata* (Formicidae: Ponerinae) on two plants in tropical wet forest. *Journal of the Georgia Entomological Society,* 12(1): 41–51.

Young, A. M., and H. R. Hermann. 1980. Notes on foraging of the giant tropical ant *Paraponera clavata* (Hymenoptera: Formicidae: Ponerinae). *Journal of the Kansas Entomological Society,* 53(1): 35–55.

Zahn, M. 1958. Temperatursinn, Wärmehaushalt und Bauweise der roten Waldameisen (*Formica rufa* L.). *Zoologische Beiträge,* n.s., 3(2): 127–194.

Zakharov, A. A. 1972. *Vnutrividovya Otnosheniya u Murav'ev.* Nauka, Moscow. 216 pp.

——— 1975. Dinamitseskaya plotnost'i povedenie murav'ev. [Dynamic population density and ant behaviour.] *Zhurnal Obshchei Biologii,* 36(2): 243–250. [In Russian with English summary.]

——— 1977. Adaptatsiya sem'i murav'ev k usloviyam obitaniya. *Adaptatsiya Pochvennykh Zhivotnykh k Usloviyam Sredy,* pp. 61–81. Nauka, Moscow.

Zakharov, A. A., T. A. Orlova, A. A. Suvorov, and A. V. Demchenko. 1983. The structure of federation in ants *Formica aquilonia. Zoological Journal,* 62(12): 1807–17. [In Russian with English summary.]

Ziegler, H., and S. Penth. 1977. Zur Kenntnis der Zusammensetzung des Honigtaues. *Apidologie,* 8(4): 419–426.

Zikán, J. F. 1942. Algo sobre a simbiose de *Mydas* com *Atta. Rodrignesia,* 6: 61–67.

Zoebelein, G. 1956. Der Honigtau als Nahrung der Insekten, Part II. *Zeitschrift für Angewandte Entomologie,* 39(2): 129–167.

Zwölfer, H. 1958. Zur Systematik, Biologie und Ökologie unterirdisch lebender Aphiden (Homoptera, Aphidoidea) (Anoeciinae, Tetraneurini, Pemphigini und Fordinae), Part IV: Ökologische und systematische Erörterungen. *Zeitschrift für Angewandte Entomologie,* 43(1): 1–52.

# Acknowledgments


We wrote this book with the intent of making the fullest possible use of the work of our colleagues, and we were rewarded during its preparation by their warm support and cooperation. All seemed to agree that there is pressing need for a systematic, readily accessible summary of the existing knowledge of myrmecology, rapidly emerging as a field on its own. The degree to which we succeeded in producing a synthesis owes a great deal to the help of others.

Technical help and advice of various kinds were obtained from Barry Bolton, William L. Brown, Alfred Buschinger, Norman Carlin, James M. Carpenter, Jonathan Coddington, Stefan Cover, Konrad Fiedler, André Francoeur, William H. Gotwald, Roger Kitching, Martin Lindauer, David Maddison, Hubert Markl, Ulrich Maschwitz, Doyle McKey, Mark W. Moffett, D. H. "Paddy" Murphy, Volker Neese, Jacqueline M. Palmer, Dan Perlman, Naomi E. Pierce, Sanford D. Porter, A. J. Prins, Carl W. Rettenmeyer, Steven O. Shattuck, Roy R. Snelling, Robert W. Taylor, F. C. Thompson, James C. Trager, James F. A. Traniello, Philip S. Ward, Rüdiger Wehner, and Norman E. Woodley. Various of the chapters were critically read by Murray S. Blum, Barry Bolton, William L. Brown, Alfred Buschinger, Norman Carlin, Jae C. Choe, Ross H. Crozier, Diane W. Davidson, Konrad Fiedler, Hubert Markl, Ulrich Maschwitz, Mark W. Moffett, Christian Peeters, Dan Perlman, Naomi E. Pierce, Sanford D. Porter, Peter F. Stevens, Robin J. Stuart, Walter Tschinkel, and Diana E. Wheeler.

In the end, after much agonizing and consultation on our own part, most of the keys were written by Barry Bolton, Stefan Cover, and Robert W. Taylor. This contribution is especially welcome because these entomologists are in the midst of wide-ranging reviews of the Old World, New World, and Australian faunas respectively. Other important elements in the keys were provided by Roy Snelling, as specified in the introductory captions. We are grateful to several additional colleagues who tested the Neotropical key during a 1987 Entomological Society of America workshop.

To the illustrations of genera in Chapter 2 Robert W. Taylor contributed a set of beautiful, previously unpublished drawings of Australian ants. The artists were R. J. Kohout and F. Nanninga, and the copyright belongs to the Commonwealth Scientific and Industrial Research Organization, Canberra, Australia. Alfonse Coleman, Hiltrud Engel-Siegel, Elfriede Kaiser, and Jacqueline M. Palmer provided timely and indispensable aid in preparing prints of the numerous illustrations herein.

Caryl P. and Edna F. Haskins made a generous gift toward the publication expenses of the book.

A great deal of credit goes to Kathleen M. Horton, who checked the bibliographic references, assisted in other forms of library research, and typed the manuscript through three complicated, difficult drafts, all with a critical eye and a high level of accuracy. The appearance of *The Ants* would have been substantially delayed without her invaluable experience and skill in the preparation of large, complicated volumes such as this one.

In addressing some topics we have utilized portions of our own earlier writings where no revision because of new information was necessary. With the permission of the publishers where required, excerpts were taken from the following articles and books: N. F. Carlin and B. Hölldobler, "The Kin Recognition System of Carpenter Ants (*Camponotus* spp.), I: Hierarchical Cues in Small Colonies," *Behavioral Ecology and Sociobiology*, 19(2): 123–134, 1986 (Springer-Verlag, Heidelberg); B. Hölldobler, "Ethological Aspects of Chemical Communication in Ants," *Advances in the Study of Behavior*, 8: 75–115, 1978 (Academic Press, New York); B. Hölldobler, "Evolution of Insect Communication" in T. Lewis, ed., *Insect Communication*, pp. 349–377 (Academic Press, New York, 1984); B. Hölldobler, "Liquid Food Transmission and Antennation Signals in Ponerine Ants," *Israel Journal of Entomology*, 19: 89–99, 1985 (Entomological Society of Israel); B. Hölldobler and S. Bartz, "Sociobiology of Reproduction in Ants," *Fortschritte der Zoologie*, 31: 237–257, 1985 (G. Fischer Verlag, Stuttgart); B. Hölldobler and N. F. Carlin, "Anonymity and Specificity in the Chemical Communication Signals of Social Insects," *Journal of Comparative Physiology*, ser. A, 161: 567–581, 1987 (Springer-Verlag, Heidelberg); B. Hölldobler and C. J. Lumsden, "Territorial Strategies in Ants," *Science*, 210: 732–739, 1980 (American Association for the Advancement of Science); B. Hölldobler and E. O. Wilson, "The Number of Queens: An Important Trait in Ant Evolution," *Naturwissenschaften*, 64: 8–15, 1977 (Springer-Verlag, Heidelberg); B. Hölldobler and E. O. Wilson, "The Multiple Recruitment Systems of the African Weaver Ant *Oecophylla longinoda* (Latreille) (Hymenoptera: Formicidae)," *Behavioral Ecology and Sociobiology*, 3: 19–60, 1978 (Springer-Verlag, Heidelberg); B. Hölldobler and E. O. Wilson, "The Evolution of Communal Nest Weaving in Ants," *American Scientist*, 71: 490–499, 1983 (Sigma Xi, New Haven, Conn.); B. Hölldobler, M. Möglich, and U. Maschwitz, "Myrmecophilic Relationship of *Pella* (Coleoptera: Staphyliniidae) to *Lasius fuliginosus* (Hymenoptera: Formicidae)," *Psyche*, 88: 347–374, 1981 (Cambridge Entomological Society, Cambridge, Mass.); R. H. MacArthur and E. O. Wilson, *The Theory of Island Biogeography*, 1967 (Princeton University Press, Princeton, N.J.); G. F. Oster and E. O. Wilson, *Caste and Ecology in the Social Insects*, 1978 (Princeton University Press, Princeton, N.J.); E. O. Wilson, "The Ecology of Some North American Dacetine Ants," *Annals of the Entomological Society of America*, 46: 479–495, 1953 (Entomological Society of America); E. O. Wilson, *The Insect Societies*, 1971 (Belknap Press of Harvard University Press, Cambridge, Mass.); E. O. Wilson, *Sociobiology: The New Synthesis*, 1975 (Belknap Press of Harvard University Press, Cambridge, Mass.); E. O. Wilson, "The Evolution of Caste Systems in Social Insects," *Proceedings of the American Philosophical Society*, 123: 204–210, 1979 (American Philosophical Society); E. O. Wilson, "Caste and Division of Labor in Leaf-cutter Ants (Hymenoptera: Formicidae: *Atta*), I: The Overall Pattern in *A. sexdens*," *Behavioral Ecology and Sociobiology*, 7: 143–156, 1980 (Springer-Verlag, Heidelberg); E. O. Wilson, "The Principles of Caste Evolution," *Fortschritte der Zoologie*, 31: 307–324, 1985 (G. Fischer Verlag, Stuttgart); E. O. Wilson, "Kin Recognition: An Introductory Synopsis," in D. J. C. Fletcher and C. D. Michener, eds., *Kin Recognition in Animals*, pp. 7–18, 1987 (John Wiley & Sons, New York); E. O. Wilson and W. L. Brown, "Behavior of the Cryptobiotic Predaceous Ant *Eurhopalothrix heliscata*, n. sp. (Hymenoptera: Formicidae: Basicerotini)," *Insectes Sociaux*, 31: 408–428, 1984 (L'Union Internationale pour l'Étude des Insectes Sociaux, and Masson, Paris); E. O. Wilson and R. M. Fagen, "On the Estimation of Total Behavioral Repertories in Ants," *Journal of the New York Entomological Society,* 88: 106–112, 1974 (New York Entomological Society).

Our own research reported in this book has been supported over the years by grants from the United States National Science Foundation, the German Science Foundation, the National Geographic Society, and the Alexander von Humboldt Foundation, as well as by fellowships from the John Simon Guggenheim Foundation. John de Cuevas assisted both financially and with collaboration during field studies. The National Geographic Society gave generous help in the production of the color illustrations, including especially those published earlier in the *National Geographic Magazine.*

For permission to reproduce or adapt figures and tables, we are grateful to the following publishers.

The illustrations of *Epitritus minimus* on page 106 and of *Ireneopone gibber* on page 109 are reprinted from the *Entomologist's Monthly Magazine* by permission of Gem Publishing Company, Brightwell, Wallingford, Oxfordshire, England.

*Figure 2–8:* From W. H. Gotwald and B. M. Kupiec, "Taxonomic Implications of Doryline Worker Ant Morphology: *Cheliomyrmex morosus* (Hymenoptera: Formicidae)," *Annals of the Entomological Society of America,* 68(6): 961–971, 1975. Adapted with permission from the Annals of the Entomological Society of America.

*Figure 3–5:* From B. Hölldobler and C. P. Haskins, "Sexual Calling Behavior in Primitive Ants," *Science,* 195: 793–794, 1977. Copyright 1977 by the AAAS.

*Table 3–1:* From R. E. Page, "Sperm Utilization in Social Insects," *Annual Review of Entomology,* 31: 297–320, 1986. Copyright 1986 by Annual Reviews, Inc. Adapted with permission.

*Figure 4–7:* From B. Hölldobler and E. O. Wilson, "Queen Control in Colonies of Weaver Ants (Hymenoptera: Formicidae)," *Annals of the Entomological Society of America,* 76(2): 235–238, 1983. Adapted with permission from the Annals of the Entomological Society of America.

*Figures 7–26, 7–28, 7–45:* From A. Buschinger and U. Maschwitz, "Defensive Behavior and Defensive Mechanisms in Ants," in H. R. Hermann, ed., *Defensive Mechanisms in Social Insects,* pp. 95–150, 1984 (Praeger, New York). Copyright 1984 by Praeger Publishers. Reprinted with permission.

*Figure 7–36:* From M. Möglich, U. Maschwitz, and B. Hölldobler, "Tandem Calling: A New Kind of Signal in Ant Communication," *Science,* 186: 1046–47, 1974. Copyright 1974 by the AAAS.

*Figure 8–20:* From E. O. Wilson, "The Sociogenesis of Insect Colonies," *Science,* 228: 1489–95, 1985. Copyright 1985 by the AAAS.

*Figure 9–13:* From R. D. Harkness and R. Wehner, "*Cataglyphis,*" *Endeavour,* n.s., 1(3–4): 115–121, 1977. Reprinted with permission from Pergamon Press.

*Figures 10–9, 10–10, 10–11, 10–12, 10–15:* From B. Hölldobler and C. J. Lumsden, "Territorial Strategies in Ants," *Science,* 210: 732–739, 1980. Copyright 1980 by the AAAS.

*Figure 10–21:* From N. F. Carlin and B. Hölldobler, "Nestmate and Kin Recognition in Interspecific Mixed Colonies of Ants," *Science,* 222: 1027–29, 1983. Copyright 1983 by the AAAS.

*Figures 13–17, 13–19, 13–28:* From B. Hölldobler, "Communication between Ants and Their Guests," *Scientific American,* 224(3)(March): 86–93, 1971. Copyright 1971 by Scientific American. Reprinted with permission.

*Figure 13–35:* From N. E. Pierce and P. S. Mead, "Parasitoids as Selective Agents in the Symbiosis between Lycaenid Butterfly Larvae and Ants," *Science,* 211: 1185–87, 1981. Copyright 1981 by the AAAS.

*Figures 14–4, 14–7:* From C. R. Huxley, "Symbiosis between Ants and Epiphytes," *Biological Reviews of the Cambridge Philosophical Society,* 55(3): 321–340, 1980. Reprinted with permission of Cambridge University Press.

*Table 14–2:* From A. J. Beattie, *The Evolutionary Ecology of Ant-Plant Mutualisms,* 1985 (Cambridge University Press, New York). Reprinted with permission of Cambridge University Press.

*Figures 16–7, 16–12, 16–13:* From T. C. Schneirla, *Army Ants: A Study in Social Organization,* ed. H. R. Topoff, 1971 (W. H. Freeman, San Francisco). Copyright 1971 W. H. Freeman. Reprinted with permission.

# Index

*Library of Congress Cataloging in Publication Data*

Hölldobler, Bert, 1936-
The ants / Bert Hölldobler and Edward O. Wilson
p. cm.
Bibliography: p.
Includes index.
ISBN 0-674-04075-9 (alk. paper)
1. Ants. I. Wilson, Edward Osborne, 1929– . II. Title.
QL568.F7H57 1990
595.79'6—dc19

89-30653
CIP